ALGEBRA AND TRIGONOMETRY

Quadratic equation: $\quad ax^2 + bx + c = 0 \qquad\qquad x = \dfrac{-b + \sqrt{b^2 - 4ac}}{2a}$

Right triangle:

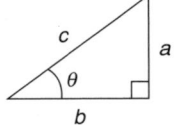

Pythagorean theorem: $\quad c^2 = a^2 + b^2$

$$\sin\theta = \frac{a}{c}$$

$$\cos\theta = \frac{b}{c}$$

$$\tan\theta = \frac{a}{b}$$

$$\sin(\alpha \pm \beta) = \sin\alpha\cos\beta \pm \cos\alpha\sin\beta$$

$$\cos(\alpha \pm \beta) = \cos\alpha\cos\beta \mp \sin\alpha\sin\beta$$

$$\cos\alpha + \cos\beta = 2\cos\left(\frac{\alpha + \beta}{2}\right)\cos\left(\frac{\alpha - \beta}{2}\right)$$

COMMONLY USED PREFIXES FOR POWERS OF 10

Power	Prefix	Abbreviation	Example	Power	Prefix	Abbreviation	Example
10^{-18}	atto-	a		10^{1}	deka-	da	(rarely used in physics)
10^{-15}	femto-	f		10^{2}	hecto-	h	(rarely used in physics)
10^{-12}	pico-	p	picofarad (pF)	10^{3}	kilo-	k	kilogram (kg)
10^{-9}	nano-	n	nanometer (nm)	10^{6}	mega-	M	megawatt (MW)
10^{-6}	micro-	μ	micrometer (μm)	10^{9}	giga-	G	gigajoule (GJ)
10^{-3}	milli-	m	millimeter (mm)	10^{12}	tera-	T	
10^{-2}	centi-	c	centimeter (cm)	10^{15}	peta-	P	
10^{-1}	deci-	d	decibel (dB)	10^{18}	exa-	E	

CONVERSION FACTORS OCCASIONALLY NEEDED

Length

1 inch = 2.54 cm

1 mile = 1.653 km

1 light-year = 9.461×10^{15} m

1 AU = 1.496×10^{11} m

Time

1 d = 8.6400×10^{4} s

1 y = 3.156×10^{7} s

Speed

1 km/h = 0.2778 m/s

Volume

1 m^3 = 1000 liters

Energy

1 eV = 1.602×10^{-19} J

Pressure

1 atm = 1.013×10^{5} Pa

University Physics

UNIVERSITY PHYSICS

RONALD LANE REESE
Washington and Lee University

Brooks/Cole Publishing Company

I(T)P® An International Thomson Publishing Company

Pacific Grove • Albany • Belmont • Boston • Cincinnati
Johannesburg • London • Madrid • Melbourne • Mexico City
New York • Scottsdale • Singapore • Tokyo • Toronto

Sponsoring Editors: *Keith Dodson and Beth Wilbur*
Editorial Assistant: *Nancy Conti*
Production Coordinator: *Tessa McGlasson Avila*
Production Management: *Electronic Publishing Services Inc., NYC*
Marketing: *Steve Catalano*
Interior Design: *Electronic Publishing Services Inc., NYC*

Cover Design: *Roy R. Neuhaus*
Cover Photo: *Ronald Lane Reese*
Interior Illustration: *Electronic Publishing Services Inc., NYC*
Photo Researcher: *Electronic Publishing Services Inc., NYC*
Typesetting: *Electronic Publishing Services Inc., NYC*
Printing and Binding: *Von Hoffman Printing Company*

For more information, contact:
BROOKS/COLE PUBLISHING COMPANY
511 Forest Lodge Road
Pacific Grove, CA 93950
USA

International Thomson Publishing Europe
Berkshire House 168-173
High Holborn
London WC1V 7AA
England

Thomas Nelson Australia
102 Dodds Street
South Melbourne, 3205
Victoria, Australia

Nelson Canada
1120 Birchmount Road
Scarborough, Ontario
Canada M1K 5G4

International Thomson Editores
Seneca 53
Col. Polanco
11560 México, D. F., México

International Thomson Publishing GmbH
Königswinterer Strasse 418
53227 Bonn
Germany

International Thomson Publishing Asia
60 Albert Street
#15-01 Albert Complex
Singapore 189969

International Thomson Publishing Japan
Hirakawacho Kyowa Building, 3F
2-2-1 Hirakawacho
Chiyoda-ku, Tokyo 102
Japan

Printed in the United States of America.

10 9 8 7 6 5 4 3 2

Library of Congress Cataloging-in-Publication Data

Reese, Ronald Lane
      University physics / Ronald Lane Reese.
         p.   cm.
      Includes index.
      ISBN 0-534-24655-9
      1. Physics.   I. Title.
   QC21.5.R435   1998
   530—dc21   98-41666
                                CIP

Magna opera Domini:
exquisita in omnes voluntates ejus

[*Psalmi* CXI, v 2]*

The past:
In loving memory of my mother
Edith Lemberg Reese
(1906–1984)

and mother-in-law
Bertha Marie Carlson
(1907–1981)

and in honor of my father
Harold Augustus Reese Sr.
(1906–)

The present:
With grateful thanks for the love
and devotion of my wife
Edith Joanne Carlson Reese

The future:
For the priceless blessings
of a wonderful son and daughter
Daniel Austin Reese
Anna-Loren Reese

* The Hexaplar Psalter, Samuel Bagster and Sons, London, 1843.

PREFACE

GOALS

In recent years much active discussion and debate has revolved around just what body of knowledge and skills science and engineering students should take from a university physics course. An obvious related issue is how best to achieve the desired learning goals. As the primary instructional resource for the student outside the classroom and the professor's office, the textbook naturally has been a focus of these discussions. Over the last 30 years or so a "standard model" of university physics textbook has evolved to the point of extensive refinement. Several generations of future scientists and engineers have been introduced to the powerful ideas of physics through these texts, and certainly we should acknowledge the many strengths of these books while considering *how we can improve on them as learning tools for today's physics students.*

When writing this text, I decided from the onset that the text should follow the twin educational commandments of Alfred North Whitehead: "Do not teach too many subjects, and what you teach, teach thoroughly." I believe that Whitehead's statement, at a basic level, reflects two of the most common themes emerging from what has come to be known as the physics reform movement. Thus, while this text was not written as a reform textbook, it nonetheless embraces the spirit of many reform goals, such as better integration of modern physics topics, a stronger emphasis on conceptual understanding, and an attention to different learning styles. Most importantly, however, this book is written for students, to allow them not only to learn the tools that physics provides but also to see why they work and the beauty of the ideas that underlie them.

TEXT OVERVIEW

A *Focused Perspective*

One of the great triumphs of physics is the amount of understanding that comes from a relatively small investment of fundamental ideas and principles. Students, however, often see the course as a random assortment of 25 to 30 topics deemed worthy of chapter status. Unifying concepts, such as conservation laws and field theory, can be lost amidst the mountainous amount of material. Students frequently fail to see just how little must be known to describe as much of nature as possible. Thus, a central goal of this text is to help students develop a thorough *understanding* of the *principles* of the basic areas of physics: kinematics, dynamics, waves, thermodynamics, electromagnetism, optics, relativity, and modern physics. It is better to build technical knowledge upon a firm foundation of fundamental principles than on a large collection of mere formulas.

Since most of us do not innately discern simplifying patterns and connections when faced with the seemingly complex, we become good and experienced students of physics through steady practice. This is a fundamental pedagogical issue, and one that this text addresses clearly through focusing on many of the difficulties encountered by students when studying physics, problems mentioned by

Arnold Arons in *A Guide to Introductory Physics Teaching*, and by others in the educational literature of physics. Thus, the book

- continually integrates the most significant material from previous chapters into new material, in keeping with Arons's admonition to "spiral back" frequently, for greater insight and retention.
- provides an accurate conceptual understanding of fundamental physical principles by placing great emphasis on these principles and how they arose.
- recognizes and points out the limits of applicability of the theories and equations of physics. It can be just as important, after all, to know what doesn't apply as what does.
- stresses connections between topics by incorporating many aspects of contemporary physics into a mix of traditional topics. This goal is carried through in all aspects of the text—exposition, examples, questions, problems, and investigative projects.

A *Thorough Development*

Some recent texts have jockeyed to outdo each other by reducing the number of overall pages. While brevity is often a laudatory goal, it can sometimes also work to defeat other, more important purposes. For true conceptual understanding to take place, a "fewer pages is more" approach can make the physics learning experience similar to trying to extrapolate the beauty and subtleties of a Shakespearean drama by reading a summary of the plot line. This text, while no longer than many other university physics texts, has been written with the primary philosophy that students need a text that lays a careful, detailed groundwork for strong conceptual understanding and the development of mature problem-solving skills. For example, much research has recently been done on the different learning styles that students apply when first studying new material, but for a text to try to implement pedagogical structure to these different learning styles and goals (such as multiple problem-solving approaches or collaborative learning techniques), it is inevitable that the lesser goal of brevity must be sacrificed. In a similar vein, students often complain that the examples in the text do not prepare them well for the more challenging homework problems, where more than one idea may be addressed. Page length can be kept down by focusing on just the most straightforward examples, but students also need to see how the principles can be applied to more involved scenarios. I have placed special emphasis on thoroughly preparing students for the homework sets through strong emphasis (and reemphasis) on problem-solving techniques, by frequent references to and explanations of common misunderstandings, and by providing a set of examples that address both single-concept problem solving and the application of fundamental principles to longer, multiconcept problems. The ability to question whether results are reasonable has been fostered throughout these examples.

The text contains an ample selection of sections from which individual instructors can design a course compatible with their academic institution and student audience. Numerous sections, typically at the end of chapters, are listed as optional (designated by a *) and may be omitted by instructors preferring a leaner course. Others may want to choose their own path, including some of the optional sections while omitting others.

Features

STYLE

Physics is a great story, and in this text I have attempted to tell that story in as lively, clear, and precise a manner as possible. Students sometimes fail to see how the topics connect to each other or to the world outside the classroom. Thus, I have placed great emphasis on introducing each new topic by describing how it relates to experiences and phenomena with which the student is already familiar or to topics previously discussed. There is also no reason that reading a physics text shouldn't be fun (or at least not a chore). My philosophy is that occasional lapses into whimsy are a small price to pay if the result is that students stay more engaged with the reading. Finally, by filling in the details that are sometimes left unstated, this text should help students better bridge the gaps where misconceptions can arise.

STRATEGIC EXAMPLES AND OTHER EXAMPLES

A strong emphasis has been placed on beginning almost all Examples with a few, fundamental principles and equations, rather than specialized equations of secondary importance. Strategic Examples address the application of fundamental principles to longer problems; they are discussed in great detail, which students find particularly helpful in developing their own problem-solving abilities. Moreover, many of the end-of-chapter problems mirror the methodological details of the Strategic Examples.

A unique feature of this text is that many of the Examples are solved in more than one way. All too often students suffer from the perception that they must be doing a problem incorrectly because a fellow student or even the professor has set it up differently. By working selected problems using different choices of signs, coordinate axes, or even overall approach, these Examples

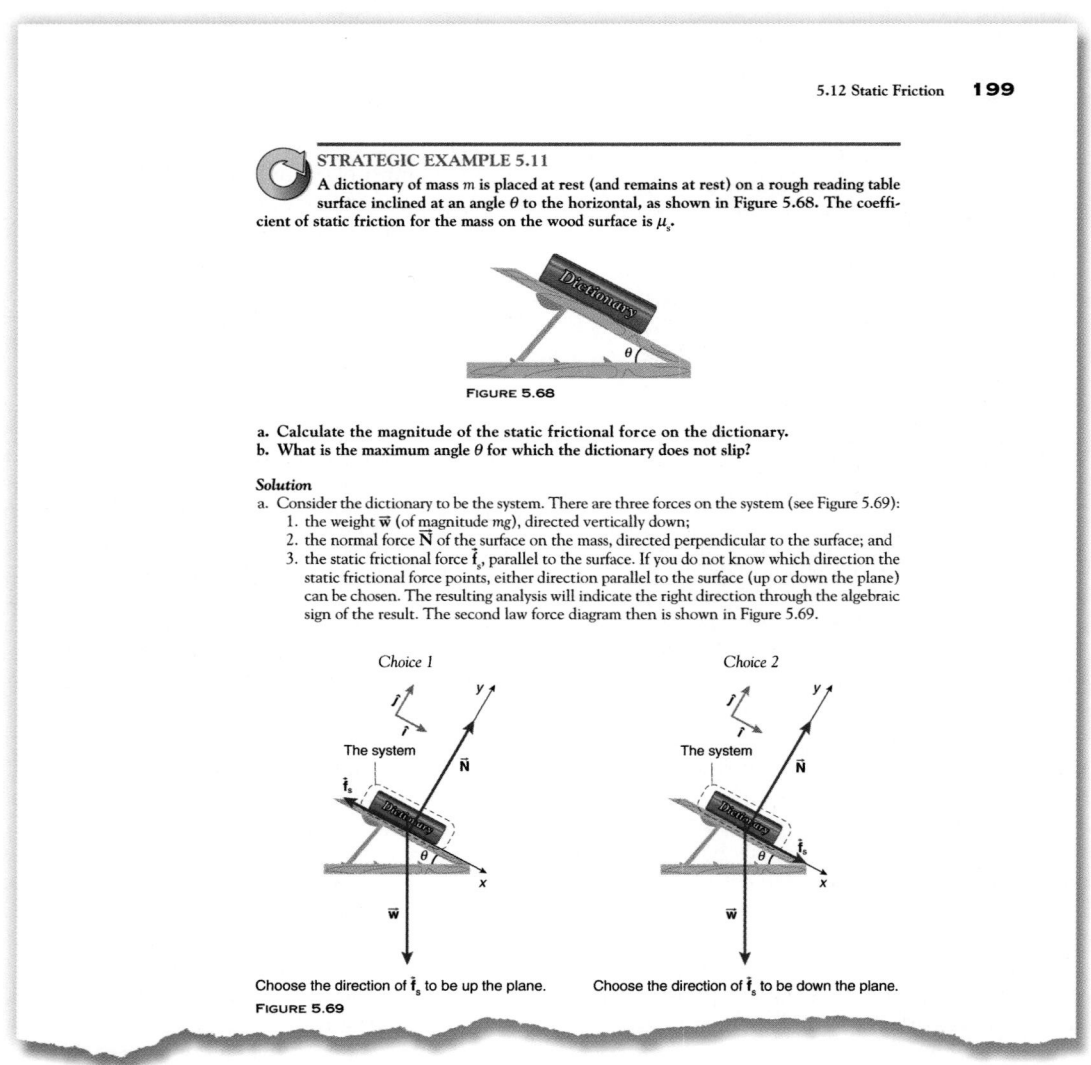

5.12 Static Friction   **199**

STRATEGIC EXAMPLE 5.11

A dictionary of mass m is placed at rest (and remains at rest) on a rough reading table surface inclined at an angle θ to the horizontal, as shown in Figure 5.68. The coefficient of static friction for the mass on the wood surface is μ_s.

FIGURE 5.68

a. Calculate the magnitude of the static frictional force on the dictionary.
b. What is the maximum angle θ for which the dictionary does not slip?

Solution

a. Consider the dictionary to be the system. There are three forces on the system (see Figure 5.69):
 1. the weight $\vec{w}$ (of magnitude mg), directed vertically down;
 2. the normal force $\vec{N}$ of the surface on the mass, directed perpendicular to the surface; and
 3. the static frictional force $\vec{f}_s$, parallel to the surface. If you do not know which direction the static frictional force points, either direction parallel to the surface (up or down the plane) can be chosen. The resulting analysis will indicate the right direction through the algebraic sign of the result. The second law force diagram then is shown in Figure 5.69.

Choice 1

The system

Choose the direction of $\vec{f}_s$ to be up the plane.

Choice 2

The system

Choose the direction of $\vec{f}_s$ to be down the plane.

FIGURE 5.69

also help students develop intuition on how certain choices can simplify later calculations compared with other alternatives.

CONCEPTUAL NOTES
Throughout the text, key points of each section have been highlighted with shading. Importantly, these Conceptual Notes are not just the most important equations—they focus on the principal ideas and concepts (and sometimes equations) that a student should take from each section. My students have found the Conceptual Notes very useful as a reviewing tool for tests and quizzes.

PROBLEM-SOLVING TACTICS
In addition to useful problem-solving hints, the Problem-Solving Tactics also provide warnings to students about common errors and how to avoid them. Often these important tips of the trade

are also integrated into the text discussion. For example, specific Problem-Solving Tactics are often cross-referenced in some of the Examples. At the end of each chapter a summary of that chapter's Problem-Solving Tactics is included, with a page reference to the related text discussion for each tactic.

QUESTIONS
A common student lament is that "I understand the material; I just can't do the problems." The questions within and at the end of each chapter test a student's understanding of concepts before the student is asked to apply these concepts to more complex or quantitative situations in the problems. Some of these questions entail a short qualitative explanation, whereas others may require a short back-of-the-envelope calculation or even a quick and dirty experiment to determine the approximate magnitude of a quantity.

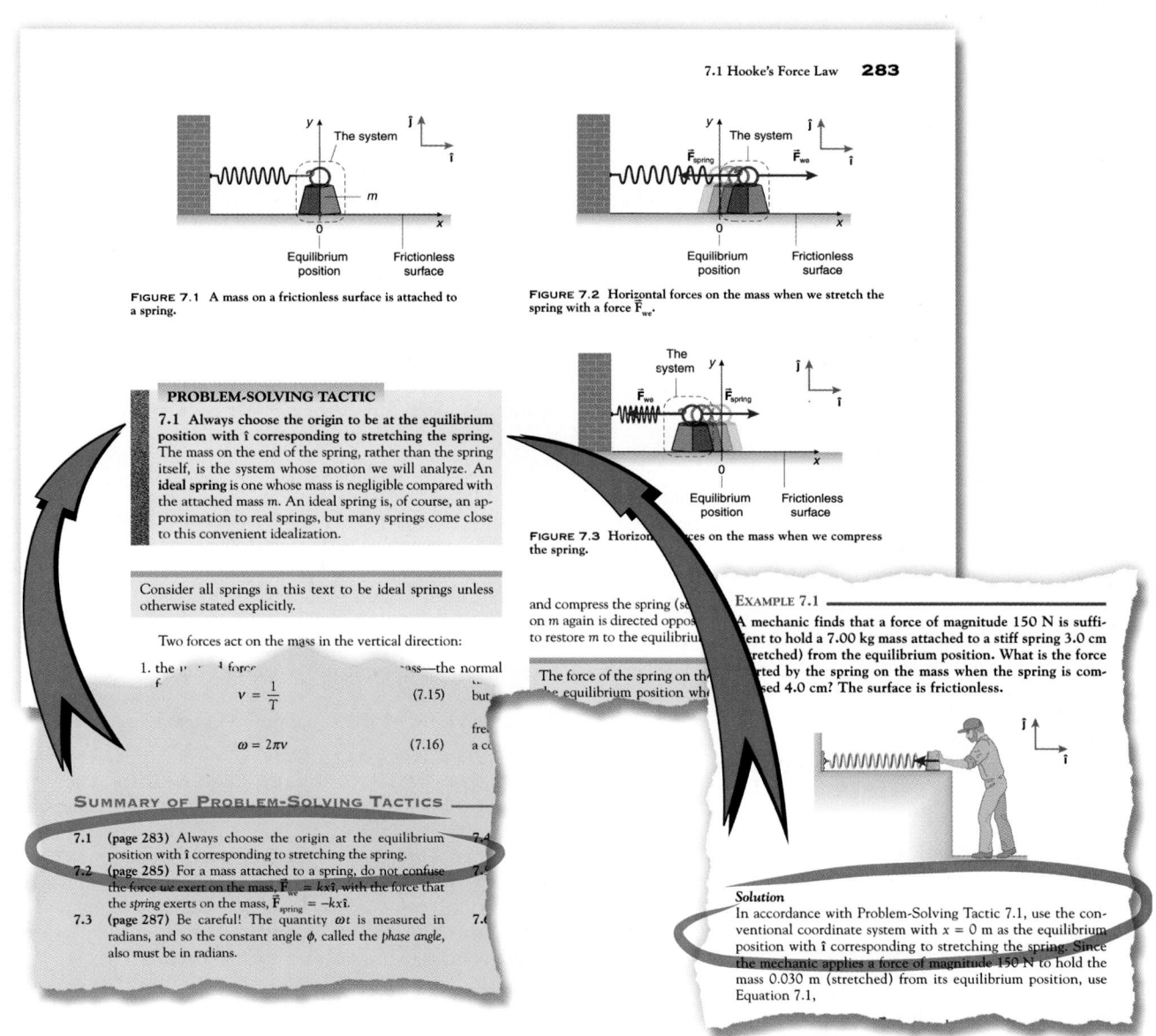

7.1 Hooke's Force Law **283**

FIGURE 7.1 A mass on a frictionless surface is attached to a spring.

FIGURE 7.2 Horizontal forces on the mass when we stretch the spring with a force $\vec{F}_{we}$.

FIGURE 7.3 Horizontal forces on the mass when we compress the spring.

PROBLEM-SOLVING TACTIC

7.1 Always choose the origin to be at the equilibrium position with $\hat{i}$ corresponding to stretching the spring. The mass on the end of the spring, rather than the spring itself, is the system whose motion we will analyze. An **ideal spring** is one whose mass is negligible compared with the attached mass m. An ideal spring is, of course, an approximation to real springs, but many springs come close to this convenient idealization.

Consider all springs in this text to be ideal springs unless otherwise stated explicitly.

Two forces act on the mass in the vertical direction:

1. the normal force

$$v = \frac{1}{T} \tag{7.15}$$

$$\omega = 2\pi v \tag{7.16}$$

SUMMARY OF PROBLEM-SOLVING TACTICS

7.1 (page 283) Always choose the origin at the equilibrium position with $\hat{i}$ corresponding to stretching the spring.
7.2 (page 285) For a mass attached to a spring, do not confuse the force *we* exert on the mass, $\vec{F}_{we} = kx\hat{i}$, with the force that the *spring* exerts on the mass, $\vec{F}_{spring} = -kx\hat{i}$.
7.3 (page 287) Be careful! The quantity ωt is measured in radians, and so the constant angle ϕ, called the *phase angle*, also must be in radians.

EXAMPLE 7.1

A mechanic finds that a force of magnitude 150 N is sufficient to hold a 7.00 kg mass attached to a stiff spring 3.0 cm (stretched) from the equilibrium position. What is the force exerted by the spring on the mass when the spring is compressed 4.0 cm? The surface is frictionless.

Solution
In accordance with Problem-Solving Tactic 7.1, use the conventional coordinate system with $x = 0$ m as the equilibrium position with $\hat{i}$ corresponding to stretching the spring. Since the mechanic applies a force of magnitude 150 N to hold the mass 0.030 m (stretched) from its equilibrium position, use Equation 7.1,

PROBLEMS

The development of successful problem-solving techniques is an essential goal for all introductory physics students. To help students hone these abilities, many of the problems involve a multistep approach in which students are guided through the problem with specific questions that enable them first to find all the pieces and then to put them together. Additionally, many of the problems mimic the approach of the Strategic Examples so that there is a strong correlation between the presentation of the material and the problems that a student is expected to be able to solve.

Since the real world is awash with information, the problems occasionally include irrelevant or superfluous information. This teaches a student to discriminate between what is needed and what is not. It also may be necessary to consult an appropriate table in the chapter or on the front and back inside covers to find numerical values of standard constants or parameters. In both the examples and problems, attention is paid to the consistent use of significant figures.

Three levels of difficulty are provided, with unbulleted, bulleted (•), and double-bulleted (❸) problems representing straightforward, moderate, and more difficult problems, respectively. Those problems with red numbers include answers at the back of the book and are solved in the Student's Solutions Manual. I have personally solved all of the problems, and the problem sets have been additionally fine-tuned by actual classroom usage over the course of several years.

INVESTIGATIVE PROJECTS

These projects are highly amenable to collaborative group work and are of the following types:

- *Expanded Horizons*—These projects are well suited to journal club research, discussion, and supplementary reading.
- *Lab and Field Work*—Doing physics is an important part of studying physics. In these projects, students are asked to design and carry out experiments either with other students or the professor.
- *Communicating Physics*—A key developmental goal for any student is the ability to write about and discuss technical topics. Practice on these written and oral communication skills is provided by these project topics, which are ideal for writing-intensive assignments, public speaking, and community service opportunities.

The projects are interesting to read, even if never performed or assigned, since they indicate the breadth and depth of applications of chapter material. As such, they can help stimulate inquiry, class discussion, and faculty–student interaction. Most of the projects are provided with references that serve as a guide (and entryway) to the appropriate literature.

SUMMARIES

Each chapter concludes with an extensive summary that, when combined with the Conceptual Notes and the Summary of Problem-Solving Tactics, provides an ideal in-text study guide for the student.

QUOTATIONS

I have used these frequently throughout the text to cast the subject matter in a different light, be it serious or whimsical. Great writing (communication!) from the past is central to a real understanding of any discipline, even physics.

MATHEMATICAL LEVEL

This text assumes a familiarity with calculus comparable to what a student would obtain from a high school calculus course (with or without advanced placement credit). Of course, additional calculus is useful when taken concurrently with this course. No prior knowledge of physics is presumed.

PRECISION

Effective, unambiguous communication in physics requires clear and consistent use of the technical vocabulary and a solid understanding of the meaning of the technical notation. This tenet has informed the presentation throughout the book.

Chapter Contents

CHAPTER 1 PRELUDES

An overview of physics is presented along with an introduction to measurement standards of the SI unit system, distinguishing them from common units of convenience. The various meanings of the equal sign are discussed as well as estimation and order of magnitude calculations. The distinction is made between precision and accuracy. The notion of significant figures is discussed in the context of common mathematical operations such as multiplication (and division) and addition (and subtraction). Having made these points, the text does not ignore their use and makes consistent use of significant figures throughout its examples and problems so that students realize their importance even outside a laboratory context.

CHAPTER 2 A MATHEMATICAL TOOLBOX: AN INTRODUCTION TO VECTOR ANALYSIS

The proper and consistent use of vectors is very important to success in physics. This chapter and the rest of the book distinguish clearly among a vector, its magnitude, and its components with respect to a chosen coordinate system. Vector addition and subtraction are designated by boldface **+** and **−** to distinguish the operations clearly from their scalar counterparts, a source of much student confusion in problem solving.

CHAPTER 3 KINEMATICS I: RECTILINEAR MOTION

The notion of a particle is addressed. A one-dimensional vector approach is used so its extension to two- and three-dimensional motion in Chapter 4 is seamless and painless. The choices that a student must make in establishing a coordinate system for a problem and the consequences of that choice in tailoring the (few) fundamental kinematic equations to a problem are stressed throughout. Consistent use of vectors and vector terminology takes the mystery out of the choice for the signs associated with the various terms in the equations of kinematics.

CHAPTER 4 KINEMATICS II: MOTION IN TWO AND THREE DIMENSIONS

The vector approach of Chapter 2 easily allows extension to motion in two and three dimensions. Relative velocity addition is examined. Uniform and nonuniform circular motion are approached by introducing both the angular velocity and angular acceleration vectors so students are ready for more advanced work with these vectors in upper-division mechanics or dynamics courses in physics and engineering.

CHAPTER 5 NEWTON'S LAWS OF MOTION

An overview of fundamental particles and forces is presented. The concept of force and its measurement are introduced stressing how important it is to define clearly the system under

INVESTIGATIVE PROJECTS

A. Expanded Horizons

1. Investigate the dynamics associated with throwing a bola, a traditional South American hunting weapon.
   D. L. Mathieson, "Wrap up rotational motion by throwing a bola," *The Physics Teacher, 30, #3*, pages 180–181 (March 1992).

2. Yo-yos are fascinating examples of spin and orbital motion. Investigate the physics of the yo-yo.
   William Boudreau, "Cheap and simple yo-yos," *The Physics Teacher, 28, #2*, page 92 (February 1990).
   Wolfgang Bürger, "The yo-yo: a toy flywheel," *American Scientist, 72, #2*, pages 137–142 (March–April 1984).
   Edward Zuckerman, "Quest for the perfect yo-yo," *Science Digest, 93, #7*, pages 54–55, 60 (July 1985).

3. Investigate the use of an ultracentrifuge to determine molecular masses of complex molecules.
   I. W. Richardson and Ejler B. Neergaard, *Physics for Biology and Medicine* (Wiley-Interscience, New York, 1972), pages 158–160.

4. Investigate the dynamics associated with boomerangs.
   Vernon Barger and Martin Olsson, *Classical Mechanics: A Modern Perspective* (2nd edition, McGraw-Hill, New York, 1995), pages 195–202.
   Allen L. King, "Project boomerang," *American Journal of Physics, 43, #9*, pages 770–773 (September 1975).
   Henk Vos, "Straight boomerang of balsa wood and its physics," *American Journal of Physics, 53, #6*, pages 524–527 (June 1985).
   Michael Hanson, "The flight of the boomerang," *The Physics Teacher, 28, #3*, pages 142–147 (March 1990).
   Jacques Thomas, "Why boomerangs boomerang," *New Scientist, 99, #1376*, pages 838–843

⋯ tigate the physi⋯
⋯ H. D. ⋯

9. Measurements of the moment of inertia are common in physics laboratories. On the other hand, it is quite another story to *measure* the moment of inertia of the Moon or a planet such as Venus or Mars; they are not easily tinkered with in the lab! Measurements of such moments of inertia enable astronomers and geologists to model the interiors of the moons and planets. How might the moments of inertia of moons and planets be measured?
   William B. Hubbard, *Planetary Interiors* (Van Nostrand Reinhold, New York, 1984).
   Ralph Snyder, "Two-density model of the Earth," *American Journal of Physics, 54, #6*, pages 511–513 (June 1986).

10. Investigate the physics of the rotating, orbiting space colony stations proposed by Gerald K. O'Neill.
    Gerald K. O'Neill, "The colonization of space," *Physics Today, 27, #9*, pages 32–40 (September 1974).

11. High angular speed flywheels are used as energy storage devices in hybrid race cars. Investigate the state of this technology.
    William B. Scott, "Satellite control concepts bolster civil, defense systems," *Aviation Week and Space Technology, 142*, pages 43, 46 (6 March 1995).

B. Lab and Field Work

12. Many science museums have a dramatic pendulum in their entrance foyers called a Foucault pendulum. The slow rotation of the plane of oscillation of the pendulum is experimental proof that the Earth rotates. Investigate the dynamics of a Foucault pendulum to explain why this is the case. Imagine such a

C. Communicating Physics

16. Take two quarters (or any two identical coins). Orient the two coins so they are both heads up with both profiles of George Washington upright as indicated in Figure I.16a. Roll one of the coins, say the one on the left, around *half* the circumference of the other coin as indicated. It is amazing to discover that the profile of the rolling coin is *not* upside down but right

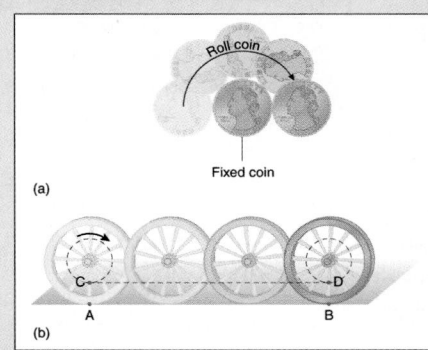

Roll coin

Fixed coin

(a)

C ← → D

A   B

(b)

FIGURE I.16

side up after this maneuver. Explain why. This phenomenon is related to another curious aspect of rolling motion. Suppose a wheel (see Figure I.16b) rolls through one complete turn, carrying it from point A to point B. The distance between A and B is, therefore, equal to the circumference of the wheel. Notice, however, that point C, along the radius from the center of the wheel *also* executes one revolution as the wheel moves from A to B. Since the distance AB is equal to the distance CD, "we are confronted with the evident absurdity that the circumference of the small circle is equal to the circumference of the large circle." Resolve this paradox.
   See James R. Newman, *The World of Mathematics* (Simon & Schuster, New York, 1956), volume 3, pages 1937–1939; the quotation is from this work, page 1938.

17. Call or visit a tire retailer and determine the manufacturer of a dynamic tire balancing machine. Communicate with engineers at the manufacturer to learn the physics associated with dynamic balancing of tires. Write a summary of your findings appropriate for an audience of your peers.
    Richard C. Smith, "Static vs spin balancing of automobile wheels," *American Journal of Physics, 40, #1*, pages 199–201 (January 1972).

18. The tippe top is a toy that has fascinated even great physicists like Niels Bohr and Wolfgang Pauli, perhaps even you. Demonstrate and explain its peculiar dynamics.
    Richard J. Cohen, "The tippe top revisited," *American Journal of Physics, 45, #1*, pages 12–17 (January 1977).
    Ivars Peterson, "Topsy-turvy top," *Science News, 146*, page 108 (13 August 1994).

consideration. The significance and importance of all three of Newton's laws of motion are addressed. Both second law and third law force diagrams are discussed. The intricacies of special forces such as weight, tension, and static and kinetic friction are explored. Inertial and noninertial reference frames are contrasted.

Chapter 6 The Gravitational Force and the Gravitational Field

Newton's law of universal gravitation is *not* presented as a *fait accompli* as if inscribed in stone; rather the process by which Newton deduced the law is explored. The gravitational shell theorems are discussed and applied (and proved in an appendix). Kepler's laws of planetary motion are discussed, along with a convenient simplification to the third law commonly used in astronomy (using customized units: years and astronomical units). Newton's form for Kepler's third law is derived. The concept of the gravitational field is introduced so that parallels with it may be exploited later when studying electricity and magnetism. Gauss's law for the gravitational field is proved, so that further parallels with electromagnetism can be made. Many of the problems consider contemporary astronomical applications.

CHAPTER 7 HOOKE'S FORCE LAW AND SIMPLE HARMONIC OSCILLATION

Hooke's law for springlike forces is introduced. A horizontal and a vertical spring (with the additional gravitational force) are compared. Simple harmonic oscillation and its relationship to uniform circular motion are discussed. The simple pendulum is introduced as well as the oscillatory gravitational motion through a uniform sphere. Damped simple harmonic oscillation and forced oscillation with resonance are explored.

CHAPTER 8 WORK, ENERGY, AND THE CWE THEOREM

Students typically think of work and energy as the same thing. The similarities and distinctions between work and energy are explored as well as the concept of power. The classical work–energy theorem (called the CWE theorem) and its limitations caused by the neglect of thermal effects are thoroughly examined to set the stage for a more general and encompassing conservation of energy theorem when we consider thermodynamics in Chapter 13. The importance of the choice made for the zero of a potential energy function is emphasized. The connection between the local form for the gravitational potential energy (mgy, with $\hat{\jmath}$ up) and the more general form ($-GMm/r^2$) is discussed. Applications to astrophysical problems such as the escape speed and black holes are explored. The concept of energy diagrams also is introduced at this early juncture to set the stage for their later use in modern physics.

CHAPTER 9 IMPULSE, MOMENTUM, AND COLLISIONS

The general principles are stressed rather than a plethora of specialized equations for collisions. The contemporary idea of force transmission by particle exchange is explored by means of a classical example for repulsive forces. The center of mass is introduced and the dynamics of a system of particles is explored.

CHAPTER 10 SPIN AND ORBITAL MOTION

The similarities and distinctions between spin and orbital motion are explored. The rotational dynamics of rigid bodies with at least one symmetry axis through the center of mass is examined, emphasizing the parallels to analogous equations in linear dynamics. The shape of the spinning Earth and the precession of tops and of the spinning Earth also are explored. Rolling motion and a model of a wheel also are examined to explain the difficult but common observation that less force is needed to roll rather than to drag a massive system.

CHAPTER 11 SOLIDS AND FLUIDS

The mechanical properties of solids and fluids are investigated. The variation of pressure with depth or height in a liquid is examined, leading to an equation giving students great freedom to approach a problem with many different coordinate choices. Archimedes' principle and the stability of floating systems are explored (why things such as submarines, ships, or poles will float in one orientation but are unstable in another orientation). Bernoulli's principle is derived from the CWE theorem. Capillary action, nonideal fluids, and viscous flow also are discussed.

CHAPTER 12 WAVES

General waves and their wavefunction, waveform, and oscillatory behavior are discussed. Nonsinusoidal periodic waves are discussed before sinusoidal periodic waves so the distinction between the oscillatory behavior of the wave at a fixed place can be clearly distinguished from its waveform at a fixed time. The classical wave equation is derived so that it can be contrasted with the Schrödinger equation in Chapter 27. Waves on strings are introduced. A unique section explores the nature of a sound wave and the relationship between the particle position and the pressure or density wave. The measurement of sound intensity and sound level is discussed so that students become aware of common sound levels that can damage their hearing. The acoustic Doppler effect treats motion of the source and/or observer as well as the effect of a wind. Superpositions of waves to form standing waves are applied to strings and both open and closed pipes. The superposition leading to wave beats is explored as well as the distinction between phase and group speeds. A simplified introduction to Fourier analysis leads to wave uncertainty relations that appear later in Chapter 27 as the Heisenberg uncertainty principle.

CHAPTER 13 TEMPERATURE, HEAT TRANSFER, AND THE FIRST LAW OF THERMODYNAMICS

The definition of a simple thermodynamic system is presented. The intuitive yet difficult concepts of temperature and heat transfer are introduced. Thermal effects in solids, liquids, and gases as well as mechanisms of heat transfer are examined. A general statement of energy conservation is developed that specializes to the CWE theorem of mechanics and to the first law of thermodynamics. Various thermodynamic processes for gases are explored.

CHAPTER 14 KINETIC THEORY

The kinetic theory of an ideal gas is presented as well as its limitations. The notion of degrees of freedom and the effect of quantum mechanics on the effective number of degrees of freedom are discussed. Adiabatic processes for ideal gases also are presented.

CHAPTER 15 THE SECOND LAW OF THERMODYNAMICS

The need for this great unifying principle is discussed in the context of why some things happen and others do not. Thermodynamic models for engines and refrigerators are presented and related to the second law. The nonintuitive concept of entropy is carefully developed as well as a classical model that explores the Boltzmann statistical interpretation of the meaning of entropy.

CHAPTER 16 ELECTRIC CHARGES, ELECTRIC FORCES, AND THE ELECTRIC FIELD

The chapter begins with an exploration of how electrical effects were distinguished from magnetic effects. The question of just what is meant by the term electric charge is confronted by a careful exploration of the experiments that led to the discovery of the two types of charge property and why Franklin's subsequent naming of them (as positive and negative charges) was particularly useful and convenient. Charge quantification is distinguished from charge quantization. The concept of the electric field is developed by exploring the similarities and differences between electricity and gravitation. Gauss's law for the electric field is developed from the parallel law for the gravitational field.

CHAPTER 17 ELECTRIC POTENTIAL ENERGY AND THE ELECTRIC POTENTIAL

The often confused and subtle distinction between these two concepts is thoroughly explored. The electron-volt energy unit is introduced and its convenience illustrated. Lightning rods also are discussed.

CHAPTER 18 CIRCUIT ELEMENTS, INDEPENDENT VOLTAGE SOURCES, AND CAPACITORS

CHAPTER 19 ELECTRIC CURRENT, RESISTANCE, AND DC CIRCUIT ANALYSIS

The terminology and methodology used for circuit analysis conforms to the standard conventions used in electrical engineering so the transition between physics and electronics can be made easily. The text clearly explains why positive charge flowing one way is equivalent to negative charge flowing in the opposite direction, a point of much mystification to students.

CHAPTER 20 MAGNETIC FORCES AND THE MAGNETIC FIELD

The need for a magnetic field is introduced by contrasting it with the electric field and its effects on electric charge. North and south magnetic poles are defined clearly rather than assumed to be obvious or innate. Applications include a velocity selector, mass spectrometer, and the Hall effect for determining the sign of the charge carriers of a current. Magnetic forces on currents lead to the torque on a current loop and the electric motor. The source of a magnetic field is introduced by exploring the parallels with both gravitational and electric fields. Gauss's law for the magnetic field, Ampere's law, the concept of a displacement current, and the Ampere–Maxwell law are discussed. The magnetic field of the Earth and how its reversals were discovered (via sea-floor sediments) also are explored to connect with another exciting discipline of the sciences that students see as remote from physics. *Nothing* is remote from physics!

CHAPTER 21 FARADAY'S LAW OF ELECTROMAGNETIC INDUCTION

The technological importance of Faraday's law is presented, leading to the development of an ac generator. The Maxwell equations are celebrated. The Maxwell equations in a vacuum are examined, leading to self-sustaining electromagnetic waves and the identification of such waves with light. Inductors and ideal transformers as standard circuit elements are explored using the standard engineering conventions.

CHAPTER 22 SINUSOIDAL AC CIRCUIT ANALYSIS

The typical approach to ac circuits in physics makes them seem impossibly complicated to students. In contrast, a brief introduction to complex variables permits the treatment of sinusoidal ac circuits via an extension of dc circuit analysis techniques, as is standard practice in electrical engineering. The use of current, potential difference, and voltage source phasors and the concept of impedance mean that sinusoidal ac circuit analysis then is reduced to the algebra of complex numbers.

CHAPTER 23 GEOMETRIC OPTICS

The simple Cartesian sign convention is used for mirrors, single surface refraction, and lenses, rather than a host of different, complex, and difficult to memorize mirror and lens conventions. Applications include the vertebrate eye, cameras, microscopes, and telescopes.

CHAPTER 24 PHYSICAL OPTICS

Interference via wavefront division (single, double, and multiple slit experiments as well as diffraction gratings) and amplitude division (thin-film interference) all are explored. Polarization and optical activity are discussed.

CHAPTER 25 THE SPECIAL THEORY OF RELATIVITY

Classical Galilean relativity is reviewed as well as the need for change. With the two postulates of special relativity, time dilation and length contraction are explored and used to derive the Lorentz transformation equations. The apparent relativistic paradox that *each* reference frame measures clocks in the *other* reference frame to run slow and lengths parallel to the motion to be shorter is confronted directly and resolved with a specific example. The existence of superluminal jets in astrophysics is found to be an optical illusion. The relativistic Doppler effect is explored, leading to the startling realization that for a source approaching with a nonzero impact parameter, the transition from a blue to a red shift occurs *before* the source is transverse to the line of sight. Questions of energy, momentum, the CWE theorem, and the relationship among mass, energy, and particles all are explored. The reason that the speed of light is an unreachable speed limit for material particles is discussed. The so-called mass–energy equivalence is clearly and properly addressed. Space–time diagrams are introduced and used to show why travel into the past (an idea with much student interest in view of contemporary culture) is forbidden in special relativity. The electromagnetic implications of relativity also are examined. The general theory of relativity and its classical tests are discussed using a qualitative approach.

CHAPTER 26 AN APERITIF: MODERN PHYSICS

The fortuitous discoveries of the electron, x-rays, and radioactivity are explored. The nuclear model of the atom is developed from the viewpoint that it was quite a radical proposal by Rutherford, rather than being simply obvious. The photoelectric effect and Compton scattering are used to justify the existence of the photon. The Bohr model and its limitations are explored. The biological effects of radiation and dosage units are discussed. The de Broglie hypothesis is introduced and questions raised about the meaning of a particle-wave.

CHAPTER 27 AN INTRODUCTION TO QUANTUM MECHANICS

The Heisenberg uncertainty principle is explored as well as the famous double slit experiment. The meaning of the wavefunction is assessed. Heuristic arguments lead to the Schrödinger equation.

A COMPLETE ANCILLARY PACKAGE

The following comprehensive teaching and learning package accompanies this book.

For the Student

MEDIA RESOURCES

Brooks/Cole Physics Resource Center is Brooks/Cole's website for physics, which contains a homepage for *University Physics*. All information is arranged according to the text's table of contents. Students can access flash cards for all glossary terms, supplementary practice and conceptual problems, practice quizzes for every chapter, and hyperlinks that relate to each chapter's contents.

InfoTrac® College Edition is an online library available FREE with each copy of each volume of *University Physics*. (Due to license restrictions, *InfoTrac College Edition* is only available to college students in North America upon the purchase of a new book.) It gives students access to full-length articles—not simply abstracts—from more than 700 scholarly and popular periodicals, updated daily and dating back as much as four years. Student subscribers receive a personalized account ID that gives them four months of unlimited Internet access—at any hour of the day—to readings from *Discover*, *Science World*, and *American Health* magazines.

OTHER STUDY AIDS
Student Solutions Manual in two volumes by Ronald Lane Reese, Mark D. Semon, and Robin B. S. Brooks includes answers and solutions to every other odd numbered end-of-chapter problem.

FOR THE INSTRUCTOR
Complete Solutions Manual in two volumes, by Ronald Lane Reese, Mark D. Semon, and Robin B. S. Brooks, contains answers and solutions to all end-of-chapter problems in the text.

ASSESSMENT TOOLS AND MATERIALS
Test Items for University Physics, by Frank Steckel, includes a copy of the test questions provided electronically in *Thomson World Class Learning™ Testing Tools*, Review Exercise Worksheets, and answers to the test item questions. The notation of the test items carefully follows that of the main text.

Thomson World Class Learning™ Testing Tools is a fully integrated suite of test creation, delivery, and classroom management tools. This invaluable set of tools includes World Class Test, Test Online, and World Class Manager software. World Class Test allows instructors to create dynamic questions that regenerate the values of variables and calculations between multiple versions of the same test. Tests, practice tests, and quizzes created in World Class Test can be delivered via paper, diskette or local hard drive, LAN (Local Area Network), or the Internet. All testing results can then be integrated into a complete classroom management tool with scoring, gradebook, and reporting capabilities.

With World Class Test, instructors can create a test from an existing bank of objective questions including multiple-choice, true/false, and matching questions or instructors can also easily edit existing questions and add their own questions and graphics. The online system can automatically score *objective* questions. *Subjective* essay and fill-in-the-blank questions that the instructor evaluates can also be added. Results can be scored, merged with final test results, and entered automatically into the gradebook.

Using *World Class Course*, you can quickly and easily create and update a web page specifically for a course or class. Post your own course information, office hours, lesson information, assignments, sample tests, and links to rich web content.

PRESENTATION TOOLS AND ONLINE RESOURCES
Transparencies in full color include more than 200 illustrations from the text, enlarged for use in the classroom and lecture halls.

CNN Physics Video, produced by Turner Learning, can stimulate and engage your students by launching a lecture, sparking a discussion, or demonstrating an application. Each physics-related segment from recent CNN broadcasts clearly demonstrates the relevancy of physics to everyday life.

With *Brooks/Cole's PhysicsLink*, a cross-platform CD-ROM, creating lectures has never been easier. Using multi-tiered indexing, search capabilities, and a comprehensive resource bank that includes glossary, graphs, tables, illustrations, photographs, and animations, instructors can conduct a quick search to incorporate these materials into presentations and tests. And, any *PhysicsLink* file can be posted to the web for easy student reference.

WebAssignOnline homework, a versatile, web-based homework delivery system, saves time grading and recording homework assignments and provides students with individual practice and instant feedback on their work. It delivers, collects, grades, and records customized homework assignments over the Internet. Assignments can be customized so each student can receive a unique question to solve. Access to *WebAssign* is secured by passwords and each student has access only to his or her record. *WebAssign* ©1998–99 by North Carolina State University.

ACKNOWLEDGMENTS

A wise man will hear, and will increase learning;
and a man of understanding shall attain unto wise counsels.

Proverbs 1:5

My interest in physics was sparked long ago by two gifted mentors at Middlebury College: Benjamin F. Wissler, whose wonderfully friendly Cheshire-cat-like grin I still can see today, and Chung-Ying Chih, whose appreciation of elegance was always apparent in his approach to physics. Their excitement for the subject was contagious and their rigor and demands upon their students legendary. Subsequent mentors included Herman Z. Cummins (then at The Johns Hopkins University, now at CUNY) and my colleagues at Washington and Lee University and elsewhere, gentlemen and gentlewomen whose gifts for teaching and research continue to be admirable role models, worthy of emulation.

The unrequited help of many persons involved with this project gives me great faith in the benevolence of humanity. My colleagues in the sciences, mathematics, and the libraries at Washington and Lee University withstood incessant questions about all manner of subjects. The conviviality, camaraderie, and good humor of the Department of Physics and Engineering are especially appreciated. All have fostered an academic environment where teaching and scholarship are emphasized in an atmosphere of mutual respect, dignity, and honor.

The reviewers of this manuscript are listed alphabetically below. Their thorough and insightful reading and frank and honest critiques were invaluable to the creation of this book.

Royal G. Albridge, Vanderbilt University
C. David Andereck, Ohio State University
Gordon Aubrecht, Ohio State University
Rene Bellwied, Wayne State University
Van Bluemel, Worcester Polytechnic Institute
Neal M. Cason, University of Notre Dame
Kenneth C. Clark, University of Washington
Richard M. Heinz, University of Indiana
Daniel G. Montague, Willamette University
Richard Muirhead, University of Washington
Richard Ditteon, Rose-Hulman Institute of Technology
Charles Scheer, University of Texas, Austin
Mark Semon, Bates College
William S. Smith, Boise State University
Karl Trappe, University of Texas, Austin
Ronald E. Zammit, California Polytechnic
    State University

Two reviewers deserve special accolades. Their dedication went well beyond the call.

Professor Kenneth C. Clark meticulously read and reread, critiqued and recritiqued *every* draft of the manuscript from its humblest beginnings, making innumerable suggestions for clean and clever ways to elucidate many phenomena. Special thanks also go to Professor Mark Semon, whose attention to detail in his reviews was similar: insightful, thoughtful, and brimming with constructive suggestions.

At Brooks/Cole, the dedication and diligence of the entire staff and the pre-production and production teams warrant praise. Their expertise exemplify the work of true professionals. Among them are Physics Editor Beth Wilbur, Senior Developmental Editor Keith Dodson, Production Editor Tessa McGlasson Avila, Assistant Editor Melissa Henderson, Editorial Assistants Georgia Jurickovich and Nancy Conti, and Cover Designer Roy Neuhaus. Also my sincere thanks to past editors Harvey Pantzis and Lisa Moller, Developmental Editors Maxine Effenson Chuck and Casey FitzSimons, Marketing Manager Steve Catalano, and Cartoonist Tom Wentzel.

The entire staff at Electronic Publishing Services deserves kudos for a wonderful job creating the book. They were a real pleasure to work with: Senior Production Editor Rob Anglin, Photo Researcher Francis Hogan, Copyeditor and Accuracy Checker Andrew Schwartz, Art Editor Michael Gutch, Creative Supervisor Stephanie McWilliams, Illustrator Matthew McAdams, their fine coterie of artists, Page Layout Specialists Linda Harms and Brent Burket, Operations Manager Patty O'Connell, Proofreader Cheryl Smith Robbins, and Indexer Lee Gable.

To my students, both recent and more venerable, I owe a *huge* and special debt. The many students here at Washington and Lee (as well as some at Bates College) who endured a photocopied manuscript and critiqued the early drafts deserve kudos and heartfelt thanks. Students Jennifer Strawbridge and William Kanner greatly assisted with the page proofs. I am humbly and profoundly grateful to all of you for your good humor, patience, enthusiasm, and encouragement. God bless you all.

During the decade this manuscript has been in preparation, the personal friendship of many people was a constant blessing. I particularly want to thank Charlotte and Chuck Gilmore, Susan and Doug Blevins, Karen Swan, Ned Wisnefski, Fran and Dick Hodges, Pastor Mark Graham and all other ministers at St. John Lutheran Church (Roanoke, Virginia), Bethany Arnold, Ed Reed, Lynn and Fred Genheimer, and the Chanthavongsa family. Finally, I treasure the warmth and love of my extended family: foremost, my dear and venerable father, as well as Harold Augustus Reese Jr., Christine Reese McCulloch, Elaine Hildebrand, Tom McCulloch, Robert Richards, Betty Fake, Liz Tisdale, Jackie and Mel Lockwood, Kristen Lockwood, Kim and Guy Boros, and Tory and George Bolten.

Clearly, any errors remaining within the text are my responsibility. I would be very grateful to readers who bring errors of any kind to my attention [reeser@wlu.edu]. I truly welcome all your comments, critiques, and suggestions.

Shalom aleichem

PREFACE TO STUDENTS

The supreme task of the physicist is to arrive at those universal elementary laws from which the cosmos can be built up by pure deduction. There is no logical path to these laws; only intuition, resting upon sympathetic understanding of experience, can reach them.

Albert Einstein (1879–1955)*

What a wonderful and exciting privilege it is to study and to teach physics! Physics is the bedrock of all the sciences and technology. Whether it be chemistry, geology, biology, medicine, engineering, or astronomy, our descriptions of nature involve understanding how particles move and interact individually and collectively. This understanding is the fundamental domain of physics. So, if you want to become a scientist, engineer, doctor, or a natural philosopher, or simply to understand nature at its most fundamental level as an intelligent citizen-scientist, you need to begin with a voyage through the foundations of physics.

The mission of this text is

- to present the subject in a logical, clear, and comprehensible style;
- to stress how *little* needs to be known in order to understand as much as possible;
- to recognize that many aspects of physics, while quantifiable, remain fundamentally abstract and mysterious—we do not have the answers to many profound questions; and
- to lighten the stress over technical gobbledygook with occasional humor.

I hope this book conveys the excitement of a fascinating search for a fundamental understanding of nature. The discipline of physics is, after all, the observation, explanation, and integration into a conceptual whole of as much as we can see, perceive, and infer during our all-too-short intellectual rendezvous with this amazing universe.

The beauty and coherence of physics may be obvious to professors, but may be less clear to you, our students, since physics likely seems less romantic than the beauty of the starlit sky. Understanding the celestial dances of the firmament, though, really is physics. So is understanding why the night sky is dark, why clouds are white, why the sky is blue, why bubbles appear in a bottle of beer or a glass of champagne, and what drives the wondrous biochemical processes that make life itself possible and gives us the opportunity to wonder at the beauty, not only of each other, but also of the natural world surrounding us.

We have an exciting time of study ahead. From a practical viewpoint, careful and thoughtful multiple readings of the text material are encouraged. Your first reading might omit the example problems in order to gain a conceptual overview of the material sufficient to address the Questions; a second reading then can include sufficient examples (particularly the Strategic Examples) for you to gain the necessary familiarity to approach the Problems.

In the conceptual development, examples, and problems, great emphasis is placed on the choices you have to make and the consequences of those choices. This is designed to gradually build confidence in your ability to tackle new and different situations. It will also teach you to be alert for the unexpected, or unanticipated result. Most significant discoveries in science begin with simply noticing something unexpected or peculiar. "What's that?" has led to many a significant "Aha!" in the history of science and technology.

---

*"Principles of Research," *Ideas and Opinions by Albert Einstein*, edited by Carl Seelig (Crown Publishers, New York, 1954), page 226.

ABOUT THE AUTHOR

Professor Reese teaches physics and astronomy at Washington and Lee University in Lexington, Virginia. He received his undergraduate degree in physics from Middlebury College and his Ph.D. in physics from The Johns Hopkins University. He has been teaching introductory physics, astronomy, and various advanced physics courses for almost 30 years. He also has performed consulting research on the interaction of visible and microwave electromagnetic radiation with matter at the Naval Research Laboratory in Washington, D.C., and has been a Visiting Fellow at University College, Oxford, during several sabbaticals.

Brief Contents

Contents

CHAPTER 4
Kinematics II 117
MOTION IN TWO AND THREE DIMENSIONS

CHAPTER 3
Kinematics I 73
RECTILINEAR MOTION

CHAPTER 5
Newton's Laws of Motion   169

CHAPTER 6
The Gravitational Force
and the Gravitational Field   231

CHAPTER 7
Hooke's Force Law and Simple Harmonic Oscillation   281

CHAPTER 8
Work, Energy, and the CWE Theorem   319

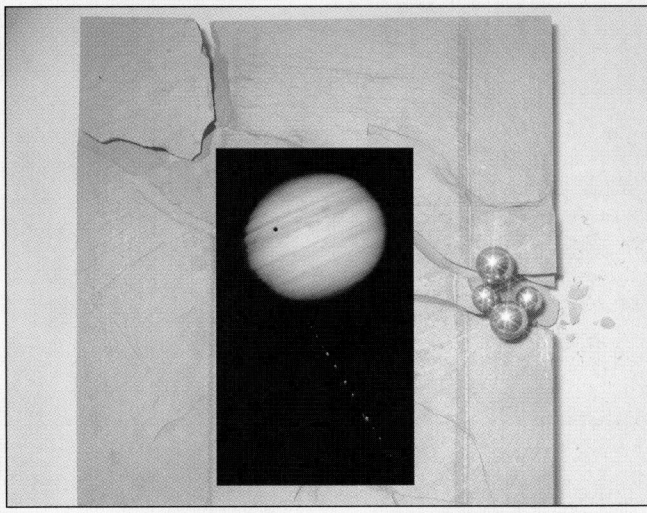

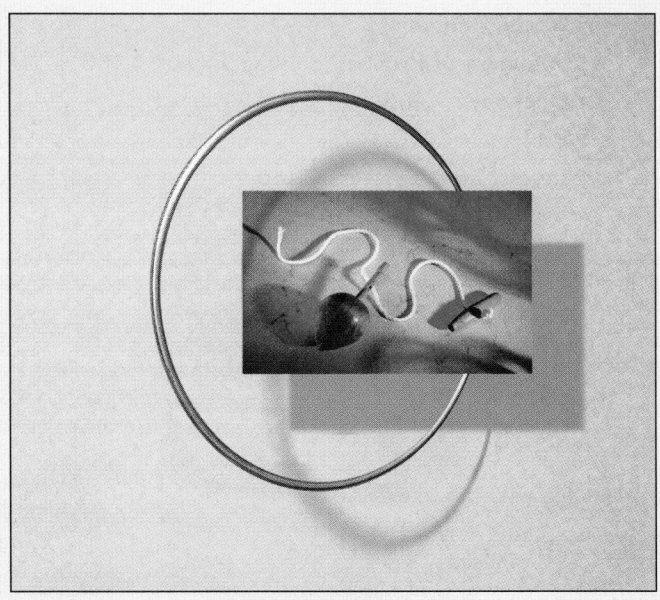

CHAPTER 9

Impulse, Momentum, and Collisions   381

CHAPTER 10

Spin and Orbital Motion   425

CHAPTER 11
Solids and Fluids   489

CHAPTER 12
Waves   531

CHAPTER 13

**Temperature, Heat Transfer, and the First
Law of Thermodynamics   587**

CHAPTER 14

Kinetic Theory   639

CHAPTER 15
The Second Law of Thermodynamics   667

CHAPTER 16
Electric Charges, Electrical Forces, and the Electric Field   705

CHAPTER 17
Electric Potential Energy and the Electric Potential   767

CHAPTER 18
Circuit Elements, Independent Voltage Sources, and Capacitors   805

CHAPTER 19
Electric Current, Resistance, and dc Circuit Analysis   835

CHAPTER 20
Magnetic Forces and the Magnetic Field   895

CHAPTER 21
Faraday's Law of
Electromagnetic Induction   **953**

CHAPTER 22
Sinusoidal ac Circuit Analysis   **1005**

CHAPTER 23
Geometric Optics   1041

CHAPTER 24
Physical Optics   1103

CHAPTER 25
The Special Theory of Relativity 1149

CHAPTER 26
An Aperitif: Modern Physics 1205

CHAPTER 27
An Introduction to
Quantum Mechanics   1249

EPILOGUE   1268

APPENDIX A
Proofs of the Gravitational Shell
Theorems   A.1

CHAPTER 1
PRELUDES

But thou shalt have a perfect and just weight, a perfect and just measure shalt thou have.

Deuteronomy 25:15

Art, religion, and the sciences are lenses through which we perceive the universe and attempt to come to terms with its existence—and our own. The word **science** stems from the Latin root *scientia,* meaning the state of knowing, or *knowledge.* In this context, art is the science of the way we see and depict beauty. Religion is the science of the mystical and supernatural. What we now mean by the word science, though, is the knowledge of the natural and physical world. One of the most incomprehensible things about the physical universe is that it *is* comprehensible, an observation made by Albert Einstein, among others.

1.1 NATURE AND MATHEMATICS: PHYSICS AS NATURAL PHILOSOPHY

Et harum scientarum porta et clavis est Mathematica.
[Of these sciences, the gate and key is mathematics.]
Roger Bacon (c. 1214–1294)*

The word **physics** comes from the Latin *physica* and the Greek *φυσικά,* meaning the knowledge or science of natural things. The study of physics is a tribute to the human intellect and its use of knowledge and reason to invent mathematical models of reality. These models mimic the actual happenings of nature, from the perceptible regimes of motion, forces, waves, and light to the less perceptible regimes of molecules, atoms, nuclei, elementary particles, the duality of particle-waves, and at the other extreme, the creation and evolution of the universe itself.

Physics is one of the liberal arts, combining elements of reason, philosophy, mathematics, language, and rhetoric. Historically, physics was called **natural philosophy**; we gave up much nuance when this phrase was abandoned during the 19th century. The term natural philosophy reflects the creative and dynamic interplay that exists in physics among experiment, theory, logic, insight, inspiration, symmetry, beauty, and language. It is in this spirit that we, as students and professors alike, study nature, constantly observing, changing, and modeling the world around us with all the real and theoretical tools that we, as an intelligent species, can invent. All of us are natural philosophers; we *all* are physicists.

Physics is timeless. The universe evolves and changes. Likewise, our *descriptions* of physics also change and improve as we develop better ways of observing and studying the universe with keener insight, more sophisticated equipment, more realistic approximations, and more encompassing theories. The underlying mechanisms of nature on the *macroscopic* level are immutable and exist with or without human scrutiny. On the other hand, at the *atomic* and *subatomic* levels, just the opposite may be true: the role of the observer and of measurement are inextricably interwoven into the quantum mechanical behavior of such systems, which is quite confounding!

Developing Theories

Physics and mathematics are the twin pillars on which we build modern science and its offspring, technology. While mathematics is essential to studying physics and engineering, mathematics

Opus Majus, Part 4, Chapter 1, translated by Robert Belle Burke (University of Pennsylvania Press, Philadelphia, 1928), Volume I, page 116.

itself is neither. In fact, we use the language of mathematics to describe and predict the behavior of the natural world, with nature itself as both guide and arbiter of our observations and physical theories. Most theories arise from and are nourished by observations and experiments[†]; all theories depend on experiments for confirmation or refutation. The goal of theory is to integrate as many observations of nature as possible under one conceptual umbrella—*to discover how little must be known to explain as much as possible.*

Most physical theories are *approximations* or models that accurately describe natural phenomena under certain limiting conditions and idealized circumstances. For example, classical dynamics is valid when the speeds involved are much less than the speed of light. Relativistic dynamics is appropriate for speeds approaching that of light and reduces to classical dynamics for small speeds. Physicists construct models whose primary purpose is to consistently account for and predict natural phenomena. A guiding principle of physics is expressed colloquially as the **KIS principle**: *Keep It Simple.* As long as it is consistent with observations in nature, simplicity is preferred over complexity. Observation is a necessary but not a sufficient condition for a good or a unique theory.

Theorists are like turtles: they must stick their necks out (i.e., make predictions) to get anywhere. We all do this; there is no gain without risk, no progress without change. Sometimes we get more physics out of a theory than we originally put into it; surprises are part of the game.

The characteristics of a good physical theory are shown in Figure 1.1.

Physics and Equations

If you feel that understanding physics amounts to memorizing myriad equations and formulas, and then searching for a "magic bullet" among them, you have missed the forest for the trees. Physics is only incidentally a process of learning what formula to use to solve a problem. After all, computers can solve equations easily, but the results are useless if the equations are the wrong

[†]It can be argued that Einstein's general theory of relativity arose more from aesthetic considerations and symmetry arguments than from experiments.

If a solution is too complex, it may not be the answer. A theory should account for observations with minimal complexity.

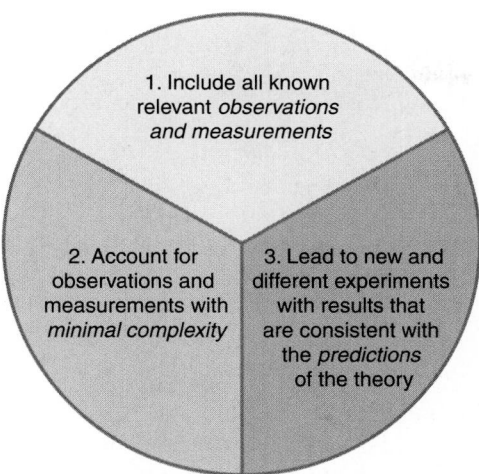

FIGURE 1.1 **Characteristics of a good physical theory.**

ones for the problem or are applied beyond their limitations. Computers can solve the wrong equations perfectly correctly. We must devise the theoretical equations to begin with and then tell the computers which equations to solve and under what conditions, and that is where physicists, scientists, and engineers come into play. "Doing physics" is the reasoning that leads from observations and experiments to a mathematical model or explanation of them. It is the reasoning process and the explanation that constitute the real physics. Mathematical results can be checked against experiments, but mathematical equations can sometimes camouflage what really is happening.

Language and Physics

Physics uses language as well as equations. It is through the written word that we communicate ideas and elucidate the natural world. If words were not necessary, this text would simply be a list of equations and formulas; surely such a mere list is not sufficient to understand nature.

For better or worse, physicists are like lawyers. Both use common words in special ways, with narrow and definite meanings that typically differ from their colloquial usage. For example, words such as velocity, force, momentum, work, and power have distinct meanings in physics, and these may vary from your intuitive understanding of them.

> *"When I use a word,"* Humpty Dumpty said in a rather scornful tone, *"it means just what I choose it to mean—neither more nor less."*
> Lewis Carroll (1832–1898)*

When *writing* physics, one must be as precise as when crafting a legal document. When *reading* physics or the law, you likewise must be careful, deliberate—and in a good mood too.

QUESTION 1

Mathematicians usually do not like to be called scientists. In what ways is mathematics similar to yet different from science or physics in particular?

Through the Looking Glass (Random House, New York, 1946), Chapter 6, page 94.

1.2 CONTEMPORARY PHYSICS: CLASSICAL AND MODERN

The history of physics encompasses two broad time periods in which **classical physics** and **modern physics** developed. Classical physics developed between 1600 and 1900. It embraces the general areas of physics known as mechanics, thermodynamics, and electromagnetism, culminating in relativity. Modern physics began developing between about 1890 and 1930, when it was realized that classical physics could not account for the newly discovered behavior of nature at the atomic and molecular level. Modern physics includes relativity[†] as well as quantum mechanics and most of the subsequent new physics discovered and developed during the 20th century.

Classical and modern physics together constitute **contemporary physics**. We use both classical and modern physics to account for physical phenomena of current interest. You should not think of classical physics as dead physics; we use classical physics today just as much as a hundred years ago. The flight of a baseball, the motion of a pendulum, the sound of a beautiful instrument, or a trip to the Moon are described using classical physics. On the other hand, we need modern physics to explain the structure of atoms and molecules, the process of light emission and absorption, and the origin of the universe itself.

The Emergence of Classical Physics

Classical mechanics has its roots in the work of Johannes Kepler, who, between 1600 and 1619, became the first person to describe quantitatively and accurately the elliptic paths of the planets around the Sun (see Figure 1.2).

The problem of the shape of the planetary orbits was not easy to solve, because all historical observations of the planets were made from the Earth, itself a moving planet. Although a crude *heliocentric* (Sun-centered) *model* of the solar system using circular orbits was proposed by the Greek astronomer Aristarchus in the third century B.C. and resurrected by Copernicus in 1542

[†]It can be argued that relativity is "the end" of classical physics, as well as "the beginning" of modern physics.

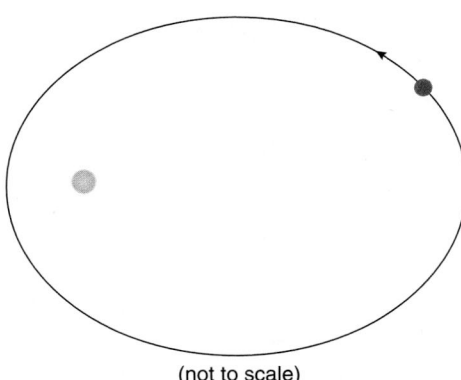

(not to scale)

FIGURE 1.2 **The path of a planet around the Sun is elliptical, with the Sun at one focus of the ellipse.**

(see Figure 1.3), the planetary positions predicted by Copernicus were *not* as accurate as those found using Ptolemy's complicated *geocentric* (Earth-centered) *model* of the second century A.D. (see Figure 1.4).

The pure Copernican model was inaccurate because Copernicus refused to abandon the flawed concept of early Greek thought that the natural path of a moving "perfect" celestial object, such as a planet, had to be in the shape of the "perfect" geometrical figure—a circle—with the planet moving at constant speed. Ironically, using the criteria for a good physical theory outlined in the previous section, a scientist living at the time of Copernicus would have been forced to reject the Copernican theory with its circular planetary orbits, since they did *not* account for the *observed* motions of the planets as accurately as the existing, but more complicated, geocentric theory of Ptolemy.

The change from the accurate but complex Ptolemaic geocentric model of the solar system to the accurate and simple heliocentric model of Kepler illustrates the characteristics of a good theory. With Kepler's work, accuracy *and* simplicity blossomed simultaneously from the heliocentric hypothesis. Kepler's model and mathematical methods began the development of physics as we know it. His work was a watershed of quantitative thought and a new worldview.

Observation of nature and experimental results typically precede the general theories that account for them. The extensive observations of the planets by Tycho Brahe just prior to 1600 provided Kepler with the data he needed to model the solar system accurately and precisely.*

---

*We distinguish between precision and accuracy in Section 1.8.

FIGURE 1.3 The Copernican heliocentric model of the solar system used circular orbits.

FIGURE 1.4 The earlier Ptolemaic geocentric model of the solar system also used circles.

In most instances, natural phenomena exist long before we observe them critically. For example, in the 17th century, Newton did not *discover* gravity; he *described* it quantitatively. Gravity was here from the beginning. Likewise, atoms existed long before we ever were able to describe them.

Kinematics and Dynamics

Kinematics, a quantitative description of motion, was invented by Galileo Galilei in the early 1600s after much observation and experimentation. We say *invented* rather than *discovered*, since physical theories do not lie lurking in nature, waiting to be discovered like an unknown planet or plant. Rather, *physical theories are inventions of the human intellect* that describe and account for observations of nature. Despite this distinction between the words invent and discover, the words will henceforth be used interchangeably in this text.

Galileo's kinematics enabled Isaac Newton to invent several *laws of motion* (a theory that explained the causes of motion, called *dynamics*) as well as a *law of universal gravitation* encompassing the ideas of both Galileo and Kepler. Newton's great synthesis was published in 1686 as the *Philosophiae Naturalis Principia Mathematica* (*Mathematical Principles of Natural Philosophy*), more commonly called simply the *Principia*. It was one of the most influential scientific books ever published, establishing classical kinematics and dynamics as the first comprehensive physical theory and illustrating the characteristics of a good theory shown in Figure 1.1. With its great predictive power, Newton's theory revolutionized our view of the natural world.

Electricity and Magnetism

Fundamental experiments in *electricity* and *magnetism* were performed by many investigators during the 18th and 19th centuries. Benjamin Franklin invented the names positive and negative for the property of matter we call electric charge. Charles Coulomb deduced the form of the electric force law between pointlike charges at rest. Christian Oersted studied magnetism. The experimental genius Michael Faraday discovered the *law of electromagnetic induction*, which made possible the electronic age. All this work culminated in 1864 with a seminal treatise by James Clerk Maxwell that integrated the heretofore separate theories of electrical and magnetic phenomena into a single *theory of electromagnetism*. Unexpectedly, his theory predicted the existence of electromagnetic waves, found to be identical to light. Electromagnetic theory was the second great physical theory of natural philosophy.

Thermodynamics

Thermodynamics arose during the 19th century from a need to understand the parameters associated with designing better steam engines during the Industrial Revolution. The fundamental research of Sadi Carnot into engine efficiency, of James Prescott Joule into the dissipation of mechanical energy and heat transfer, of Rudolf Clausius into entropy, of Ludwig Boltzmann into kinetic theory and the statistical meaning of entropy, and the culminating theoretical work of J. Willard Gibbs established the physical *laws of thermodynamics*.

Modern Physics

The dawn of the 20th century saw physicists heady with successes in mechanics, electromagnetism, and thermodynamics, almost smug with the successes of the seemingly all-encompassing theories of Newton, Maxwell, Boltzmann, and Gibbs. Some even predicted the end of physics was nigh. There seemed to be little left to discover.

Then, in 1905, Albert Einstein realized that Maxwell's electromagnetic theory implied that the Newtonian synthesis of motion and classical dynamics, the first triumph of classical physics and an unimpeachable theory for over two centuries, was only an *approximation* of a more general theory of space, time, and motion. This incredible discovery, now known as *relativity*, is a splendid example of getting more physics out of our equations than originally was put into them.

Several other unexpected and seemingly unrelated discoveries between 1895 and 1910 collectively shook the foundations of physics and our view of reality. Among these discoveries were the following:

a. The accidental discovery of a new penetrating radiation, called *x-rays*, in 1895 implied that there was more to learn about the emission of electromagnetic radiation and its interaction with matter. In 1901, Wilhelm Conrad Röntgen was awarded the first Nobel prize in physics for the discovery.*

b. The discovery of *radioactivity*, the spontaneous emission of particles by certain atoms, in 1896 clearly indicated that atoms, the heretofore "indivisible" elements of nature, had a divisible structure. Radioactivity also showed that microscopic processes were statistical in nature, similar to the statistical formulation of thermodynamics by Ludwig Boltzmann and J. Willard Gibbs. In 1903 Antoine Henri Becquerel was

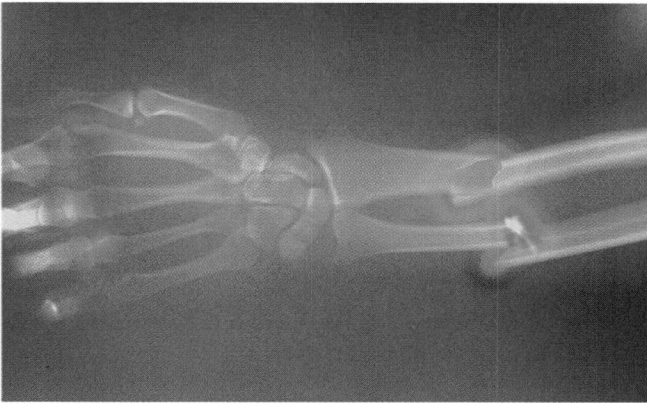

Medical applications of x-rays are legion.

awarded the Nobel prize in physics for the discovery of radioactivity. Very few individuals have won multiple Nobel prizes, but Marie Curie received *two* awards for her studies of radioactivity: one in physics in 1903 (along with her husband) and one in chemistry in 1911.[†‡]

c. The discovery of the negatively charged *electron* particle in 1896 indicated that atoms are composite structures made from more fundamental subunits, some of which must have negative and positive charge, since atoms themselves are

---

[†]John Bardeen won two awards in physics (in 1956 and again in 1972), and Linus Pauling won an award in chemistry (1954) and one in peace (1962).

[‡]One of Marie Curie's two daughters, Irène Joliet-Curie, also won a Nobel prize in chemistry, in 1935. Like mother, like daughter.

Radioactive carbon is used by archaeologists to date many historical remains such as this prehistoric hunter, discovered in a melting glacier in the Alps.

---

*The Nobel prizes are universally recognized as the most prestigious prizes in the fields of physics, chemistry, physiology and medicine, peace, and literature. The lucrative prizes were established by the will of Alfred Nobel (1833–1896), the inventor of dynamite. A Nobel prize in economics was established later.

electrically neutral. In 1906, J. J. Thomson received the Nobel prize for this discovery.

d. The shocking discovery of the *quantum of action* in 1901 resolved a troubling problem of an unexpectedly large amount of energy, known picturesquely as the "ultraviolet catastrophe," in the classical explanation of the electromagnetic radiation emitted by hot objects. The explanation introduced a hitherto unknown and mysterious fundamental physical constant into physics. The quantum of action now is called *Planck's constant, h*. In 1918, Max Planck became a Nobel laureate.

e. A new theory in 1905 of the *photoelectric effect*, the emission of electrons by metals when illuminated by appropriate light, led to a new and radically different view of light as a *particle-wave* called a *photon*. It was principally for this theoretical work that Einstein received the Nobel prize in 1921.

f. Between 1905 and 1912, the discovery of the *nucleus* of the atom, the invention of the atomic model of hydrogen, and the explanation of hydrogen's light emissions using the ideas of *quantization of angular momentum and energy* signaled that a fundamentally new type of physics was needed to explain the affairs of nature on the molecular, atomic, and nuclear scales. Ernest Rutherford received the Nobel award in chemistry in 1908 for the discovery of the nucleus, and Niels Bohr received the prize in physics in 1922 for his model of the hydrogen atom.*

The new physics that integrated all these and many other phenomena under one conceptual umbrella was *quantum mechanics*, developed during the 1920s by such giants of 20th-century physics as Erwin Schrödinger, Wolfgang Pauli, Werner Heisenberg, P. A. M. Dirac, Niels Bohr, and Enrico Fermi, all of whom eventually became Nobel laureates in physics.

Early in the 20th century, an idea emerged that gave many disparate areas of physics a unified leitmotif. You likely know that the three certainties of existence are death, taxes,

---

*One of Bohr's sons, Aage Bohr, also won the Nobel prize in physics (in 1975). Like father, like son.

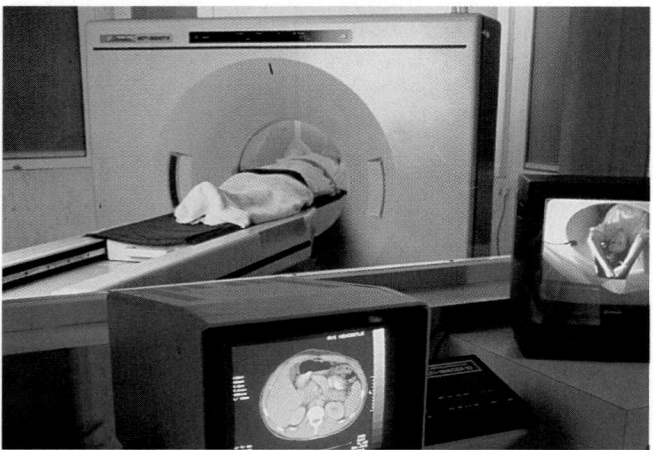

Medical imaging technologies are based on many discoveries in modern physics.

and change. When an isolated physical system changes in some way, certain properties of the physical system such as its total energy, charge, momentum, or angular momentum may be left unchanged. When such quantities are unchanged, we say they are *conserved* and there exists a *conservation law* for that property. The discovery of conservation laws for energy, charge, momentum, angular momentum, and other quantities was neither obvious nor easy. About 1920, an unsung heroine, Amalie Noether, realized that each conservation law is a manifestation of an abstract mathematical symmetry associated with the physical system. The conservation laws and their associated symmetry principles are common to both classical and modern physics and are unifying threads in the tapestry of physics.

Thus the era that saw the development of modern physics coincides with the first half of the 20th century. In the latter half of the 20th century, physicists also have made progress toward a theory to unify the *four fundamental forces of nature:* gravitation, electromagnetism, and two other forces, called the strong and weak nuclear forces, whose existence became apparent only as our knowledge of nuclear physics progressed. Although great strides have been made toward such a "theory of everything," it still is only a partially realized goal. The search awaits your participation.

1.3 STANDARDS FOR MEASUREMENT

There shall be one measure of wine throughout Our kingdom, and one of ale, and one measure of corn, to wit, the London quarter. . . . As with measures so shall it also be with weights.
Paragraph 35 of the *Magna Carta* (A.D. 1215)

The Congress shall have the Power To . . . fix the Standard of Weights and Measures
The Constitution of the United States of America, Article I, Section 8 (1789)

These quotations from some of our most cherished documents reflect the importance of **measurement standards**. These standards are significant not only to science but also to society—no one likes to be shortchanged at the grocery store or the petrol station.

Experiments are the key to the development of physical theories. Observations of experiments, of course, involve measurements of various physical quantities in terms of how many standard units of the quantity are present. It is quite remarkable to realize that centuries of the scientific enterprise indicate that *all* physical quantities ultimately can be described in terms of a surprisingly small number of primary physical parameters, including mass, length, time, electric current, and temperature. Physical concepts, such as area, volume, density, force, acceleration, pressure, and energy, that are expressed in terms of these primary parameters are known as **derived quantities**.

The primary parameters are not trivial to define or explain. Nonetheless, it is apparent that we need a set of standards for the measurement of them. The most widely used standards for measurement in physics are those of *Le Système International d'Unités* (the International System of Units, when translated from the French), known more simply as the **SI system**. The system is a

TABLE 1.1 The Primary Parameters of Mass, Length, and Time in the SI Unit System

Parameter	SI unit	Abbreviation
Mass	kilogram	kg
Length	meter	m
Time	second	s

refinement of the **metric system** that originated in France during the late 18th century amid the radical furor of the French Revolution.* The word *metric* comes from the Greek word μέτρον (*metron*), meaning measure.

The SI system is the standard system of units used in this text. The SI units used to quantify the primary parameters of mass, length, and time are indicated in Table 1.1.

The SI units for temperature and electric current (from which we get the unit for electric charge) will be defined later in the text, as the need arises. For now, the three parameters mass, length, and time are sufficient for our purposes.

Common prefixes associated with various SI units are indicated in Table 1.2.

Definitions are like belts. The shorter they are, the more elastic they need to be.

Stephen Toulmin (1922–)[†]

Mass

Newton was the first person to distinguish between mass and its weight; prior to Newton, the two terms had been considered synonymous.

Mass and weight are different physical concepts.

Mass is a measure of the resistance of an object to changes in its motion.[‡] It is still uncertain just what gives rise to this property of matter. The mass of an object has the same measure *anywhere* in

---

*The original standards of the metric system were supplanted as improvements in technology allowed more reproducible standards; the process is ongoing.
[†]*Foresight and Understanding* (Greenwood Press, Westport, Connecticut, 1981), Chapter 2, page 18.
[‡]An operational definition of mass is presented in Chapter 5.

TABLE 1.2 Common Prefixes Attached to SI Units

Factor	Prefix	Symbol
10^{12}	tera-	T
10^9	giga-	G
10^6	mega-	M
10^3	kilo-	k
10^1	deka-	dk**
10^{-1}	deci-	d
10^{-2}	centi-	c
10^{-3}	milli-	m
10^{-6}	micro-	μ
10^{-9}	nano-	n
10^{-12}	pico-	p
10^{-15}	femto-	f

**Rarely used in physics.

FIGURE 1.5 Your mass is independent of location.

the universe; your mass is the same whether you are on the Earth, the Moon, or the asteroid Hamburga, as indicated in Figure 1.5.[§]

On the other hand, **weight** is the measure of the gravitational force on a mass due to the simultaneous presence of *another* mass, be it the Earth, the Moon, the Sun, or an asteroid. A given mass has *different* weights on the surface of the Earth, far above or below the surface of the Earth, on the Moon, or near an asteroid (see Figure 1.6).

Weight depends on location in space; mass does not.

The standard SI unit for the measurement of mass is the **kilogram** (kg). See Figures 1.7 and 1.8. The kilogram is the mass of a single, special, custom-fabricated cylinder made of an alloy of platinum (90%) and iridium (10%).

[§]The asteroids are a collection of small, planet-like objects, most of which orbit the Sun between the orbits of Mars and Jupiter. More than 5000 asteroids are known. Most are numbered and named. Hamburga is number 449. An amusing astronomical parlor game is to create phrases from the asteroid names. Asteroids 904, 1603, 673, 991, and 449 spell out the phrase "Rockefellia Neva Edda McDonalda Hamburga."

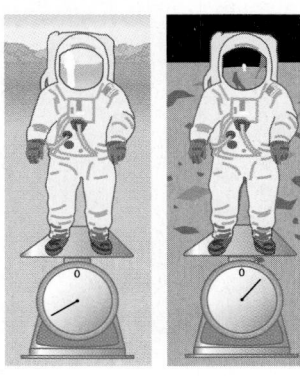

FIGURE 1.6 Your weight on the Earth is different from your weight on the Moon.

FIGURE 1.7 One kilogram of brass and one kilogram of cushion have the same mass.

The kilogram is the only SI unit still defined by a manufactured artifact rather than by a natural phenomenon or property. *The* cylinder defining the kilogram resides in a special environmental chamber at the International Bureau of Weights and Measures in Sèvres, France, near Paris.

Replicas of the standard kilogram are kept at the National Institute of Standards and Technology near Washington, D.C. (see Figure 1.8), but the standard that in fact defines the kilogram is the single cylinder at Sèvres. The height of the cylinder is equal to its diameter (both approximately 3.8948 centimeters at 0 degrees celsius) with slightly beveled edges. The platinum–iridium alloy was chosen because of its stability under variation in temperature and its resistance to chemical changes, such as oxidation (the chemical binding of a substance with oxygen).

FIGURE 1.8 A replica of the standard kilogram.

The size of the cylinder was chosen so that its mass was close to an older, obsolete definition of the kilogram, based on the amount of pure liquid water in a cube 10 cm on a side kept at special environmental conditions. Table 1.3 surveys a range of masses in kilograms.

Related to the mass standard is another SI unit known as the **mole** (mol). Twelve eggs are defined to be one dozen eggs, regardless of their mass, and 500 sheets of paper are defined to be a ream of paper, regardless of their size. A mole likewise is a fixed count of particles, atoms, or molecules, regardless of the particles being counted.

> One **mole** of any kind of particle, element, or molecule is Avogadro's number of the particles, atoms, or molecules. **Avogadro's number** is the number of atoms in exactly 12 grams of $^{12}_{6}C$, an isotope of carbon.

The subscript in the designation $^{12}_{6}C$ is the **atomic number** of the element: the number of protons in the nucleus of the atom. The number of protons in the nucleus identifies the particular chemical element; each carbon atom always has 6 protons in its nucleus but can have different numbers of neutrons. The superscript is the **atomic mass number** of the element and represents the sum of the number of protons and neutrons in the nucleus of the atom. The atomic mass number identifies a specific **isotope** of an element. The various isotopes of an element have the same number of protons in the nucleus but different numbers of neutrons. The common isotopes of carbon are $^{12}_{6}C$, with 6 protons and 6 neutrons in each nucleus, and $^{14}_{6}C$, with 6 protons and 8 neutrons in the nucleus of each atom.

Experiments indicate that Avogadro's number N_A is approximately

$$N_A = 6.022\ 137 \times 10^{23} \text{ particles/mole}$$

One mole of $^{12}_{6}C$ has a mass of exactly 12 grams and contains Avogadro's number of atoms. One mole of gold (Au) has Avogadro's number of gold atoms; one mole of water (H_2O) has Avogadro's number of water molecules. The mass in grams of

TABLE 1.3 A Range of Masses

The entire universe (very approximate)	$\sim 10^{53}$ kg
The Milky Way Galaxy	$\approx 3 \times 10^{41}$ kg
Sun	1.99×10^{30} kg
Earth	5.98×10^{24} kg
Moon of the Earth	7.36×10^{22} kg
Aircraft carrier	$\sim 3 \times 10^{8}$ kg
Boeing 747 jet (empty)	$\approx 1.6 \times 10^{5}$ kg
Physics student	≈ 60 kg
Grapefruit	~ 1 kg
Computer diskette	2.0×10^{-2} kg
Credit card	5×10^{-3} kg
Sand grain	$\sim 10^{-7}$ kg
Single living cell	$\sim 10^{-12}$ kg
HIV virus	$\sim 10^{-19}$ kg
Uranium atom	3.95×10^{-25} kg
Proton	1.67×10^{-27} kg
Electron	9.11×10^{-31} kg

one mole of a substance is numerically equal to the atomic mass number of the isotope or the **molecular mass** of the molecule.* The mass of one mole of any substance is called its **molar mass**. For example, one mole of gold $^{197}_{79}$Au has a mass of 197 g (= 0.197 kg) because the atomic mass number of this gold isotope is 197. Thus the molar mass of $^{197}_{79}$Au is 197 g and contains Avogadro's number of gold atoms. Water is composed of two hydrogen atoms, $^{1}_{1}$H, and one oxygen atom, $^{16}_{8}$O. The molecular mass is the sum of the atomic mass numbers of all the atoms in the molecule, here 1 + 1 + 16 = 18. Therefore, one mole of water has a mass of 18 g (= 1.8×10^{-2} kg) (see Figure 1.9). A mole of water contains Avogadro's number of water molecules.

QUESTION 2

From the standpoint of physics, what is wrong with these labels?

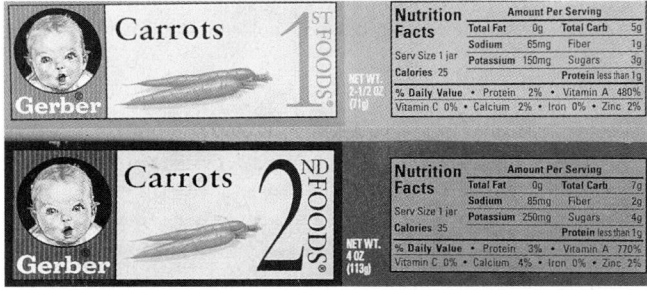

EXAMPLE 1.1

How many moles and how many gold atoms are in a gold brick of $^{197}_{79}$Au with a mass of 1.00 kg?

Solution
The mass of a mole of $^{197}_{79}$Au is 197 g = 0.197 kg. Therefore the 1.00 kg of gold has

$$\frac{1.00 \text{ kg}}{0.197 \text{ kg/mol}} = 5.08 \text{ mol}$$

Each mole has Avogadro's number of gold atoms, and so the number of gold atoms in the brick is

$$(5.08 \text{ mol})(6.02 \times 10^{23} \text{ atoms/mol}) = 3.06 \times 10^{24} \text{ atoms}$$

Time

For what is time? Who is able easily and briefly to explain that? Who is able so much as in thought to comprehend it, so as to express himself concerning it? And yet what in our usual discourse do we more familiarly and knowingly make mention of than time? And surely we understand it well enough, when we speak of it; we understand it also, when in speaking with another we hear it named. What is time then? If nobody asks me, I know; but if I were desirous to explain it to one that should ask me, plainly I know not.

St. Augustine (A.D. 354–430)[†]

FIGURE 1.9 One mole of water has a mass of 18 g.

It has been said, perhaps in jest, that time is nature's way of preventing everything from happening at once.

The standard SI unit for the measurement of time is the **second** (s). Originally, a second was defined to be 1/(24 × 60 × 60) = 1/86 400 of the average rotational period of the Earth with respect to the Sun (called a mean solar day). However, the rotational motion of the Earth is slightly variable and gradually, yet systematically, is slowing down.[‡] Therefore, the motion of the Earth is an unacceptable basis for a precise and invariant definition of a second. The development of atomic clocks led to the present determination of the second.

> A second is the time interval required for a specific kind of light emitted from the cesium atom $^{133}_{55}$Cs to undergo 9 192 931 770 oscillations. Do not memorize this number; it can be looked up if necessary.

While this definition may seem peculiar, it is based on the natural cesium atom, and provides a standard that is reproducible anywhere in the universe with suitable—if complex—experimental equipment.

The rotation of the Earth gradually gets out of step with the current definition of the second. Hence, an occasional *leap*

---

*The molar mass of an element listed in the periodic table of the elements accounts for the relative abundances of the various isotopes of the element (and other factors) and thus indicates molar masses to fractions of a gram per mole.

[†] *St. Augustine's Confessions*, translated by William Watts (Harvard University Press, Cambridge, Massachusetts, 1974), Book XI, #XIV, pages 237, 239.

[‡] The slowing down of the rotational motion of the Earth is caused by the tides; we discuss this briefly in Chapter 10, Problem 53.

second, similar to the extra day in February during leap years, is inserted into the calendar every few years, usually on the last day of June or December, so that noon according to the atomic clocks corresponds with noon astronomically.

Some time intervals of interest, expressed in seconds, are noted in Table 1.4.

Length

The SI standard unit for length is the **meter** (m). Originally, the meter was defined to be exactly one ten-millionth (1/10 000 000 ≡ 10^{-7}) of the distance between the north geographic pole of the Earth and the equator, measured along the curved path of the terrestrial meridian (a line of constant longitude) through Paris (specifically, the Paris Observatory). (See Figure 1.10.)

This definition proved to be unacceptable for two reasons: (1) the distance between the pole and equator is difficult to measure, and (2) the north and south geographic poles of the Earth wander slightly and erratically over the surface of the Earth, again increasing the difficulty of a precise measurement of the distance between the pole and equator.* The distance between two marks scratched on a straight platinum–iridium bar (made from the same alloy casting as the standard kilogram) then was used to define the meter, but this definition also was abandoned because of the difficulty of measuring accurately and precisely the distance between the positions of the centers of the two scratches on the bar. To avoid these difficulties, the meter now is determined using the defined speed of light and the definition of the second.

---

*The axis about which the Earth rotates defines the positions of the north and south geographic poles. (Of course, there is no real pole or axis sticking out of the Earth at the poles!) The area over which the position of the poles wanders is about the size of a tennis court. The reasons for the polar wanderings are quite complex but are related to seasonal changes in the distribution of mass on the surface of the Earth (relocations of snow and ice as they melt to liquid water) and complicated but unknown internal motions of materials deep within the Earth.

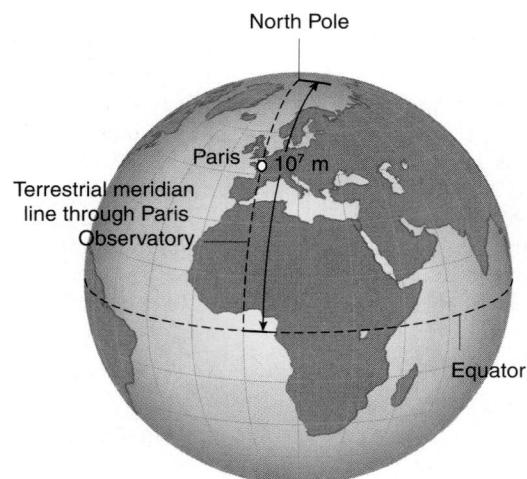

FIGURE 1.10 The old definition of the meter involved the distance between the north geographic pole and the equator.

A meter is the distance through which light travels in a vacuum during a time interval of exactly 1/299 792 458 second. You can look up this number if needed; do not memorize it.

The definition of the meter implies that the speed of light in vacuum is *exactly* 299 792 458 m/s.

A survey of various distances in meters is indicated in Table 1.5.

QUESTION 3

Describe the characteristics of a universe in which one of the primary physical parameters—mass, length, and time—did not exist.

TABLE 1.4 Time Intervals

Age of the universe	$\approx 6 \times 10^{17}$ s
Age of the solar system	$\approx 1.6 \times 10^{17}$ s
Age of a centenarian	3.2×10^{9} s
Age of an 18-year-old physics student	5.7×10^{8} s
One day	86 400 s
One 50-minute lecture	3.0×10^{3} s
Interval between heartbeats	≈ 1 s
One vibration corresponding to the musical note middle C	3.83×10^{-3} s
Time for light to travel one meter	3.33×10^{-9} s
Time for light to travel the approximate diameter of an atom	$\sim 10^{-18}$ s
Time for light to travel the approximate diameter of a proton (10^{-23} s is called a jiffy)	$\sim 10^{-23}$ s
Lifetime of the most unstable particles	$\sim 10^{-23}$ s
Planck time (the earliest time after the creation of the universe at which the laws of physics, as we currently know them, can be applied to the universe)	$\sim 10^{-43}$ s

TABLE 1.5 Lengths

Most distant object in the visible universe	$\sim 10^{26}$ m
Distance to the Andromeda Galaxy (a neighbor of the Milky Way Galaxy)	$\approx 2 \times 10^{22}$ m
Diameter of Milky Way Galaxy	$\sim 10^{21}$ m
Distance from Sun to nearest star	$\approx 4 \times 10^{16}$ m
Radius of the solar system	$\approx 6 \times 10^{12}$ m
Average distance between the Earth and the Sun	1.496×10^{11} m
Radius of the Sun	6.96×10^{8} m
Average distance between the Earth and Moon	3.84×10^{8} m
Average radius of the Earth	6.37×10^{6} m
Deepest ocean depth	$\approx 10^{4}$ m
Height of an average human	≈ 2 m
Diameter of a pencil lead	$\approx 10^{-3}$ m
Thickness of a sheet of paper	$\sim 10^{-4}$ m
Size of the bacterium *Escherichia coli* (*E. coli*)	$\approx 10^{-6}$ m
Diameter of an atom	$\sim 10^{-10}$ m
Diameter of the nucleus	$\sim 10^{-15}$ m

QUESTION 4 _____
How can we be certain that other primary physical parameters do or do not exist? Upon what factors should such a judgment be made?

1.4 UNITS OF CONVENIENCE AND UNIT CONVERSIONS

Occasionally, it is convenient to express physical quantities in **units of convenience** rather than in SI units. The common units such as the minute (min), hour (h), day (d), year (y), centimeter (cm), kilometer* (km), and gram (g) are units used for convenience, but they are not SI units. It is just easier to say you have a mass of one gram (1 g) rather than ten-to-the-minus-three kilograms (10^{-3} kg). Likewise, it is more convenient to say a trip lasts three hours (3.00 h) rather than 1.08×10^4 s. Common units of convenience are summarized in Table 1.6.

In this text we will write the units associated with each physical quantity in standard Roman type; for example, kilogram is represented as kg, meter as m, and second as s. In this way you can appropriately track the units through a calculation. Do not confuse *unit* abbreviations with algebraic *variables* representing various physical quantities that are denoted in italics. For example, m is used to represent the mass of an object, whereas m is the SI unit known as the meter. In later chapters we will see that V is the SI unit called the volt, whereas V is the algebraic symbol for both a volume and the electric potential. To avoid confusion between units and variables that use the same letter, you might choose to designate all units with parentheses [e.g., (m)], but that decision is up to you and your professor.

Unit conversions involve changing the measure of a quantity from one system of units to another, say from units of convenience to SI units or vice versa. You will need to learn to do unit conversions easily and with confidence, because this is an important skill in any science. Unit conversions are similar to expressing a collection of nickels, dimes, and quarters (units of convenience) in terms of dollars (the monetary standard).

When expressing unit conversions in words, the equations

$$1000 \text{ m} = 1 \text{ km}$$
$$1000 \text{ g} = 1 \text{ kg}$$
$$100 \text{ cm} = 1 \text{ m}$$

_____
*The proper pronunciation of this unit is kil′ə mēt′ər, *not* kə läm′ət ər.

are expressed as

one thousand meters *per* kilometer,
one thousand grams *per* kilogram, and
one hundred centimeters *per* meter.

The meaning of the word *per* is simply "in one," and so the expressions really mean, respectively,

one thousand meters *in one* kilometer,
one thousand grams *in one* kilogram, and
one hundred centimeters *in one* meter.

The word *per* also is used with certain derived units. For example:

> The **density** of an object is the mass contained in one unit of volume.

The *measure* of the density depends on the specific units employed for the measure of mass and volume. The density can be expressed in grams per cubic centimeter (g/cm³) or, in the SI unit system, the mass in kilograms contained in a volume of one cubic meter of the material, kilograms per cubic meter (kg/m³). The density of various substances is listed in Table 1.7.

When we say, for example, the density of mercury is 13.6×10^3 kg/m³, we usually say "thirteen thousand, six hundred kilograms *per* cubic meter." What we mean is there is a mass of 13.6×10^3 kg in a volume of one cubic meter of mercury.

> In this text, we will write the mathematical expressions of physical laws in forms appropriate for the use of SI units, with only rare exceptions.

Notice in the following examples that *units are treated as algebraic quantities* when making conversions and in the equations of physics.

> **PROBLEM-SOLVING TACTIC**
> **1.1 In solving most problems, be sure to convert units of convenience into standard SI units.**

TABLE 1.6 Units of Convenience and SI Units

	Abbreviation	SI units		Abbreviation	SI units		Abbreviation	SI units
Mass			*Length*			*Time*		
milligram	mg	$\equiv 10^{-6}$ kg	millimeter	mm	$\equiv 10^{-3}$ m	minute	min	$\equiv 60$ s
gram	g	$\equiv 10^{-3}$ kg	centimeter	cm	$\equiv 10^{-2}$ m	hour	h	$\equiv 3.6 \times 10^3$ s
kilogram	kg	SI standard unit	meter	m	SI standard unit	second	s	SI standard unit
metric ton	mt	$\equiv 10^3$ kg	kilometer	km	$\equiv 10^3$ m	day*	d	$\equiv 86\,400$ s
						year†	y	$\approx 3.16 \times 10^7$ s

*Occasionally, on those days when a leap second is introduced to keep the atomic definition of the second in step with the rotation of the Earth, the day has 86 401 s.

†The number is approximate, because the time it takes the Earth to orbit the Sun is approximately 365.256 37 d, called one sidereal year.

TABLE 1.7 Density of Various Substances

Substance	Density (kg/m³)
Air	1.3
Aluminum	2.70×10^3
Brass	8.4×10^3
Carbon	
Graphite	2.25×10^3
Diamond	3.5×10^3
Copper	8.9×10^3
Ethyl alcohol	0.79×10^3
Gasoline	0.67×10^3
Gold	19.3×10^3
Invar	8.00×10^3
Iron	7.8×10^3
Lead	11.0×10^3
Mercury	13.6×10^3
Uranium	18.7×10^3
Water	
Liquid	1.00×10^3
Ice	0.917×10^3
Sea water	1.02×10^3
Wood	
Red oak (*Quercus borealis*)	0.66×10^3
Sugar maple (*Acer saccharum*)	0.68×10^3
Eastern white pine (*Pinus strobus*)	0.37×10^3
Redwood (*Sequoia sempervirens*)	0.44×10^3

PROBLEM-SOLVING TACTICS

1.6 Realize that there may be more than one way to solve the problem. Of course, you may not realize this initially, but if a colleague does the same problem independently and comes up with the same answer using a different approach, it is likely you both are right. Such independent solutions are useful for checking calculations and typically provide additional insight.

1.7 Ask yourself if the answer you obtain is reasonable. When making any calculation, even when figuring your income taxes, it is a good idea to question whether the result seems reasonable. This may be difficult initially, but as you practice and study, your feeling for the approximate numerical values of various physical quantities develops.

QUESTION 5

Metric units are used in many manufacturing processes. This Bösendorfer grand piano is 2.25 m in length and 156 cm in width at its keyboard with 92 keys. The piano has a mass of 370 kg and a useful lifetime measured in gigaseconds, if properly cared for. Which of the specifications quoted here is not an SI unit?

PROBLEM-SOLVING TACTICS

1.2 Read and understand the problem. A slow and careful reading of any problem is critical to understand and distinguish between what information is given and what is to be determined. Be sure you understand any technical terms in the problem. A good way to test your understanding of a problem is to try to explain it orally in your own words to a roommate or friend. The process of explaining it to someone else usually indicates whether you know what you are talking about . . . or not! Try it frequently.

1.3 If appropriate, sketch a simple diagram of the actual arrangement of problem features. Label the relevant quantities with letter symbols. This stimulates concentration and planning.

1.4 Assess whether additional information is needed. It may be necessary to look up some information, such as conversion factors for various units, values of appropriate constants, or data from tables. The inside covers of the text have the numerical values of many useful constants in SI units.

1.5 Assess whether any given information is irrelevant. The real world is adrift in a sea of information, not knowledge. Some information presented within a problem may not be needed to answer the questions posed. Sorting relevant from irrelevant information is a critical skill to develop in any field, not only in the sciences.

STRATEGIC EXAMPLE 1.2

The funerary mask of the famous pharaoh Tutankhamen is made of almost pure gold. A reference indicates that the density of the most common gold isotope $^{197}_{79}\text{Au}$ is 19.3 grams per cubic centimeter. Convert the density to appropriate SI units.

Solution

Apply Problem-Solving Tactic 1.2: *Read and understand the problem.* We begin with 19.3 g/cm³ and are asked to find the density in SI units of kg/m³.

Keep in mind Problem-Solving Tactic 1.4: *Assess whether additional information is needed.* Since grams and centimeters are not SI units, you need the conversion factors for grams to kilograms:

$$10^3 \text{ g} = 1 \text{ kg}$$

and for centimeters to meters:

$$10^2 \text{ cm} = 1 \text{ m}$$

Now ponder Problem-Solving Tactic 1.5: *Assess whether any given information is irrelevant.* The information contained in the particular designation of the isotope of gold $^{197}_{79}$Au is not needed to solve the problem. The given density means there are 19.3 grams per cubic centimeter, or 19.3 grams in a volume of one cubic centimeter. Expand the third power of the centimeter unit:

$$19.3 \text{ g/cm}^3 = 19.3 \text{ g/(cm} \cdot \text{cm} \cdot \text{cm)}$$

Now convert each of the units of convenience into the appropriate SI unit:

$$19.3 \text{ g/cm}^3 = [19.3 \text{ g/(cm} \cdot \text{cm} \cdot \text{cm)}](10^{-3} \text{ kg/g})(10^2 \text{ cm/m})$$
$$\times (10^2 \text{ cm/m})(10^2 \text{ cm/m})$$
$$= 19.3 \times 10^3 \text{ kg/m}^3$$

You could also write this in scientific notation as

$$= 1.93 \times 10^4 \text{ kg/m}^3$$

Thus there are 1.93×10^4 kilograms per cubic meter—that is, 1.93×10^4 kilograms in a volume of one cubic meter of gold. The mass of one cubic meter of gold is 1.93×10^4 kilograms.

When finished with a calculation, try to bear in mind Problem-Solving Tactic 1.7: *Ask yourself if the answer is reasonable.* The given density represents the mass of gold in a volume of one cubic centimeter of material. The density in SI units is the mass in a volume of one cubic meter of material. Certainly, one cubic meter of gold has a mass much larger than one cubic centimeter of gold, and so the large numerical value of the density in SI units is reasonable.

EXAMPLE 1.3

Given the atomic mass number of gold (197), how many gold atoms are in a pure gold coin that has a mass 3.11×10^{-2} kg? This is one troy ounce of gold.

Solution
One mole of gold has a mass in grams numerically equal to its atomic mass number. Thus, one mole of gold has a mass of 197 g, or 1.97×10^{-1} kg using the proper SI unit, the kilogram; this is the molar mass of gold. The number of moles of gold in the coin is the mass ratio:

$$\frac{3.11 \times 10^{-2} \text{ kg}}{1.97 \times 10^{-1} \text{ kg/mol}} = 1.58 \times 10^{-1} \text{ mol}$$

A mole has Avogadro's number of particles. Therefore, the number N of gold atoms in the coin is the number of moles times the number of particles (here atoms) per mole (Avogadro's number):

$$N = (1.58 \times 10^{-1} \text{ mol})(6.02 \times 10^{23} \text{ atoms/mol})$$
$$= 9.51 \times 10^{22} \text{ atoms}$$

There are 9.51×10^{22} atoms in the gold coin.

EXAMPLE 1.4

How many gold atoms are in a pure gold nugget that has a volume of 2.10 cubic centimeters? The density of gold is 1.93×10^4 kg/m^3, and its atomic mass number is 197.

Solution
First, convert the 2.10 cubic centimeters to cubic meters:

$$2.10 \text{ cm}^3 = \frac{2.10 \text{ cm} \cdot \text{cm} \cdot \text{cm}}{(10^2 \text{ cm/m}) (10^2 \text{ cm/m}) (10^2 \text{ cm/m})}$$
$$= 2.10 \times 10^{-6} \text{ m}^3$$

The mass of the nugget is the product of its density times its volume:

$$(1.93 \times 10^4 \text{ kg/m}^3)(2.10 \times 10^{-6} \text{ m}^3) = 4.05 \times 10^{-2} \text{ kg}$$

Since the atomic mass number is 197, the mass of one mole of gold (its molar mass) is 197 g, or

$$(197 \text{ g/mol})(10^{-3} \text{ kg/g}) = 1.97 \times 10^{-1} \text{ kg/mol}$$

The number of moles of gold the nugget represents is the mass of the sample divided by the mass per mole:

$$\frac{4.05 \times 10^{-2} \text{ kg}}{1.97 \times 10^{-1} \text{ kg/mol}} = 2.06 \times 10^{-1} \text{ mol}$$

The number of atoms is the number of moles times the number of atoms per mole (Avogadro's number):

$(2.06 \times 10^{-1} \text{ mol})(6.02 \times 10^{23} \text{ atoms/mol})$

$\qquad = 1.24 \times 10^{23} \text{ atoms}$

Hence there are 1.24×10^{23} atoms in the gold nugget.

1.5 THE MEANING OF THE WORD DIMENSION

The word **dimension** has two different meanings in physics. One meaning is *geometrical*: a dimension is any of the least number of coordinates needed to specify the location of a point in space. Thus we live in a three-dimensional world, because we use three coordinates to locate a point in space. Certain physical concepts have the word dimension associated with them in this context: distances (not necessarily in a straight line) are one-dimensional, areas (not necessarily flat) are two-dimensional, and volumes are three-dimensional. (See Figure 1.11.) In Chapter 3, we will speak of motion in one, two, or three dimensions depending on the number of independent coordinates needed to specify the position of the particle under consideration.

The second meaning of the word dimension is quite different. We say the *dimensions* of the density of a material are *mass per unit volume*. Likewise, the dimensions of speed are those of distance divided by a time interval. The *measure* of density or speed depends on the unit system used. (See Table 1.8.) Density can be expressed in g/cm³ or in SI units of kg/m³. The speedometer in your car likely indicates speed in miles per hour (mi/h) or kilometers per hour (km/h). The measure of speed in the SI unit system is meters per second (m/s).

Equations in physics are such that

- both sides must have the same dimensions; and

- both sides must be expressed in the same units.

We will almost always use the SI system of units.

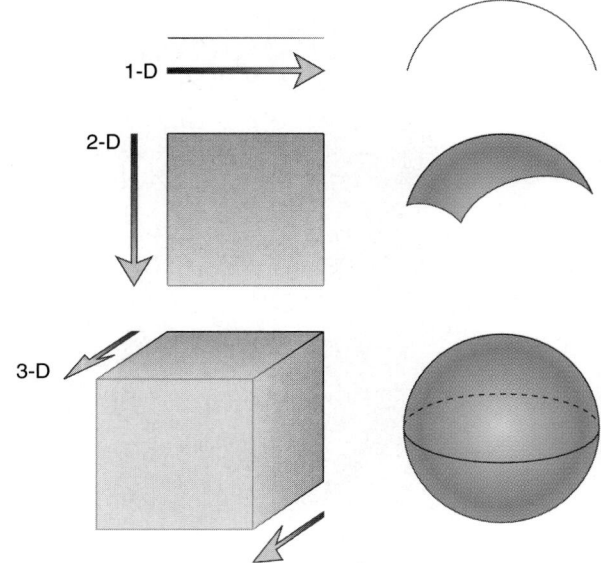

FIGURE 1.11 One, two, and three dimensions.

PROBLEM-SOLVING TACTIC

1.8 Check the dimensions or the units. The process of checking whether the dimensions or the units are the same on both sides of an equation or calculation is one aspect of **dimensional analysis**. If the dimensions or the units on both sides of an equation are *not* the same, you are guaranteed that something is incorrect. This is an extremely useful way of checking your work. Unfortunately, of course, Murphy's law* applies: if they *do* check out on both sides of an equation, there still is no guarantee that the result of your calculation is correct, since the analysis of the problem may have been incorrect. Sigh.

*There are many statements of Murphy's law. Among them: if something can go wrong, it will.

Some units, such as hectares and acres, are inherently two-dimensional.

Some units, such as liters or quarts, are inherently three-dimensional.

TABLE 1.8 Distinction Between Dimensions and Units

Dimensions	SI unit	Units of convenience
Mass	kg	g
Length	m	cm
Time	s	h, d, y
Mass per unit volume	kg/m^3	g/cm^3

Although both sides of every equation in physics must have the same units, we do *not* divide out the units to leave a purely numerical relationship. When we have an algebraic expression such as

$$3x^2 - xy = 5xz$$

there is an x in every term of the equation. Hence, algebraically, we divide by x to obtain

$$3x - y = 5z$$

On the other hand, when we have a physical expression such as

$$3 \text{ m} + 5 \text{ m} = 8 \text{ m}$$

we do *not* divide by the units, here m, on each side of the equation. To do so would leave a purely numerical equation,

$$3 + 5 = 8$$

where the information about the particular units has disappeared. To be meaningful, numerical relationships in physics must indicate the units.

> Conversions of units have identical dimensions on each side of the equal sign, but not the same units.

For example, when we have the conversion

$$100 \text{ cm} = 1 \text{ m}$$

both sides of the conversion have the same dimensions (length) but not the same units, since centimeters are on the left-hand side and meters are on the right-hand side of the equation. (See Figure 1.12.)

QUESTION 6

Artists such as M. C. Escher (1898–1972) occasionally play interesting tricks with geometrical dimensions. What makes this picture plausible yet impossible?

EXAMPLE 1.5

Your supervisor asks you for the formula for the surface area A of a right-circular cylinder. You and two colleagues suggest the following three versions of the formula, where r is the radius of the cylinder and ℓ is its length:

(1) $A = 2\pi r^2 + 2\pi r\ell$ (your suggestion)
(2) $A = r^2 + r\ell$ (colleague #2)
(3) $A = 2\pi r + 2\pi r\ell$ (colleague #3)

Which of the expressions are dimensionally correct?

Solution

Use Problem-Solving Tactic 1.8: *Check the dimensions and/or units*. Since the equations for the area of various geometric figures are independent of the specific units used, we check the dimensions of the equations. An area has the dimensions of a length squared. Notice that each term on the right-hand side of equations (1) and (2) has the product of two lengths. These two formulas are *both* dimensionally correct.

Notice that a dimensional analysis provides *no* information about the existence or correctness of purely numerical factors such as the 2 and π in all the equations. The first term of the third equation has only a single length, while the second term has the product of two lengths; it is the first term that makes the equation dimensionally inconsistent, because areas have dimensions of length squared. Hence the third equation for the area cannot be correct, and colleague #3 is fired by the supervisor for incompetence. Your suggestion, equation (1), actually is the correct expression for the surface area.

1.6 THE VARIOUS MEANINGS OF THE EQUAL SIGN

> *There was nothing there now except a single Commandment. It ran:*
>
> ALL ANIMALS ARE EQUAL
> BUT SOME ANIMALS ARE MORE EQUAL
> THAN OTHERS
>
> George Orwell (1903–1950)*

The equal sign is used in several contexts with distinctly different meanings. It is important to be aware of the subtle distinctions among the ways the equal sign is used in physics and other sciences.

The Equal Sign Is Used in Definitions, Occasionally Symbolized by ≡

In Chapter 3 we will introduce the concept of the average speed, v_{ave}, of a moving particle. When you drive home from campus, your average speed is the total distance d traveled (as

Animal Farm (Harcourt, Brace and Co., New York, 1946), page 112.

indicated by a trip odometer in your car) divided by the time interval Δt needed for the trip. Here we use the uppercase delta, Δ, to mean *difference in value of*. The quantity Δt thus means difference in the value of t, the elapsed time. The average speed is *defined* as

$$v_{\text{ave}} \equiv \frac{d}{\Delta t}$$

Rather than writing $d/\Delta t$, we use a new symbol v_{ave}. The special equal sign, $\equiv$, means that the quantity on the left-hand side is *defined* to be what is on the right-hand side of the symbol; the quantity $d/\Delta t$ is what we mean by the symbol v_{ave}.

The Equal Sign Is Used as an Equivalence in Conversions of Units

When we say that

$$1 \text{ m} = 100 \text{ cm}$$

we do not mean that $1 = 100$. A conversion means that both sides are equivalent measures of the same physical concept (here, a length or distance). As we noted in Section 1.5, *the two sides of a conversion have the same dimensions but not the same units*. The *dimensions* on both sides of the conversion $1 \text{ m} = 100 \text{ cm}$ are *lengths*; the *units* for measuring length are different (m on the left, cm on the right). Likewise, when we say that

$$1 \text{ kg} = 1000 \text{ g}$$

we do not mean that $1 = 1000$; rather, we mean one kilogram represents an equivalent measure of the same mass as one thousand grams. Both sides of the conversion have the same dimensions (mass), but not the same units.

Consider density. When we say $1 \text{ g/cm}^3 = 10^3 \text{ kg/m}^3$, we again have the same dimensions on each side (the mass divided by a length cubed), but different units. The mass of one cubic centimeter of material is not the same mass as that in one cubic meter of the material, but they each have the same density, no matter how we express it.

The Equal Sign Is Used in the Laws of Physics

Looking ahead, in Chapter 5 we will learn that one version of Newton's second law of motion is summarized by the algebraic equation

$$F_{\text{total}} = ma$$

where F_{total} is the magnitude of the total force on the mass m and a is the magnitude of its acceleration. The use of the equal sign in this equation is subtle. It means that if a total force of this magnitude acts on the mass, then m experiences an acceleration whose magnitude is calculated using this equation (namely, $a = F_{\text{total}}/m$). The expression ma is *not* a force, although it has the same dimensions as a force (because both sides of an equation in physics must have the same dimensions). The quantity ma is *equal* to the magnitude of the total force, but is *not* itself a force. The magnitude of the total force F_{total} on the mass (the left-hand side of the equation) produces a *response ma* (the right-hand side of the equation). What we mean is the magnitude of the total force F_{total} produces

Since all these masses of brass are made of the same material, they all have the same density.

a movement of m, which we find to be an acceleration of magnitude a. The two sides of the equation are, to use a clear but historically unfortunate phrase, *separate but equal*. Some of the fundamental laws of physics have this peculiar dichotomy on either side of the equal sign, because they are concerned with the relationships between different physical things.

> This text and more advanced texts typically use the standard equal sign (=) interchangeably for definitions, conversions, and the laws of physics.

Two other symbols are used to indicate, not equality in a strict sense, but *approximate equality*.

1. The symbol ~ means *on the order of*. For example, Table 1.3 indicates that the mass of the universe is on the order of 10^{53} kg, written $\sim\!10^{53}$ kg; it is known perhaps only to within a factor of 10. The symbol ~ indicates that the quantity is not precisely known. Thus we say the **order of magnitude** of the mass of the universe is about 10^{53} kg.

2. The symbol $\approx$ means *is approximately equal to*. For example, Table 1.3 indicates that the mass of the Milky Way Galaxy is approximately 3×10^{41} kg. This is written as $\approx\!3 \times 10^{41}$ kg.

Approximately equal, $\approx$, while not precise, indicates that the quantity is known with more precision than does ~.

1.7 ESTIMATION AND ORDER OF MAGNITUDE

> *. . . so easy it seemed*
> *Once found, which yet unfound most would have thought*
> *Impossible*
> John Milton (1608–1674)*

How many bricks were used to construct your science building? How many rice grains are in one kilogram of rice? These are not

---

**Paradise Lost, Book VI, lines 499–501.

easy numbers to come by unless one enjoys tedious counting. Nobody really cares about the rice grain problem, but someone had to estimate the number of bricks needed for the building in order to make a cost estimate and place the order for them.

In the sciences and other disciplines, it is often necessary to quantify something that at first glance may seem difficult to quantify. The great physicist Enrico Fermi was a master of the art of estimation; he even estimated the explosive yield of the first atomic bomb test on 16 July 1945 at Alamogordo, New Mexico, by observing the effect of the shockwave from the blast on bits of paper dropped from his hand (at some distance from ground zero!).

The importance of being able to make estimates transcends the discipline of physics and is useful in virtually any endeavor. You may want to estimate how much wallpaper to buy to paper your room. You may be asked to estimate a budget for a development project, or determine the popularity of a politician via polls. No matter what your intended career path, it is important for you to develop the ability to find an approximate value for many quantities.

Why are estimates useful?

1. It may be impossible or impractical to measure the quantity directly. For example, how many domestic cats are there in the United States?
2. The precise value of the quantity may not be needed. For example, what is the exterior wall area of your home so that you can purchase an appropriate amount of paint?
3. An estimate may be needed before a measurement is made. The estimate may determine the technique or instrument needed to perform a particular measurement. Should a meter stick or a long metric tape be used to measure the size of a swimming pool?
4. An estimate can serve as a check on a measurement or calculation. The estimate can be used to see if an instrument is working properly or a complex software package is yielding an appropriate answer to a sophisticated calculation. For instance, if an instrument for measuring the power output of a light source indicates 200 watts for a lighted match, your eyebrows should be raised about the calibration or performance of the instrument—200 watts is brighter than most desk lamps, hardly the case for a lighted match.

Quick estimates attempt to determine whether the value of the quantity to be estimated is on the order of 1, 10, 100, 10^{-1}, 10^{-2}, and so forth, to within perhaps a factor of 2 or 3. Such rough estimates, or order of magnitude calculations, also are colloquially referred to as back-of-the-envelope calculations; simply grab a piece of paper (the back of an envelope) and proceed. More precise estimates are needed for many applications in construction, design, and engineering.

EXAMPLE 1.6

The marketing manager of a pet food company needs an estimate of the number of domestic cats in the United States. Make such an estimate.

Solution

There are approximately three hundred million (3×10^8) people in the United States. If we assume there are an average of three people per household, this means there are about 1×10^8 family units. Some street smarts based on what is known about the local neighborhood tells us that perhaps one in five families has a cat, excluding multiple cats per family. Therefore, there are approximately

$$\text{number of cats} = \left(\frac{1}{5} \text{ cats/family}\right)(10^8 \text{ families})$$
$$= 2 \times 10^7 \text{ cats}$$

EXAMPLE 1.7

The American Association of Physics Teachers is planning a mailing to every physics professor in the United States. Estimate the number of physics professors teaching at colleges and universities in the U.S.

Solution

There are about three thousand (3×10^3) colleges and universities in the United States. Some of these likely have large physics departments with several dozen professors; some have small departments with perhaps 2 or 3 faculty. Since there are many more smaller colleges than larger colleges, perhaps 7 professors is a suitable average guesstimate for the number of physics faculty members per college or university. Thus the total is about

number of physics professors

$\approx$ (7 faculty/college)(3×10^3 colleges)

$\approx 2 \times 10^4$ faculty

or twenty thousand physics faculty nationwide.

EXAMPLE 1.8

The human population of the globe is about 6.0×10^9 people. If the entire human population were distributed uniformly over the dry land surface area of the Earth, approximately how far apart would the people be separated from one another?

Solution

Recall Problem-Solving Tactic 1.4: *Assess whether additional information is needed.* First, the surface area of the Earth needs to be calculated. To find the surface area, we need to know the radius of the Earth, which is not given in the statement of the problem, but can be found in the tables on the inside covers of the text. The average radius of the Earth is 6.37×10^6 m. The surface area of a sphere of radius r is $4\pi r^2$, and so the surface area of the Earth is

$$4\pi(6.37 \times 10^6 \text{ m})^2 = 5.10 \times 10^{14} \text{ m}^2$$

We also need to know the proportion of the area of the Earth that is dry land. This proportion is *not* given in the tables on the inside covers, but may be found by consulting an atlas; perhaps, though, you recall from your geography classes that about 3/4 of the surface of the Earth is covered with water, so that the dry land surface area is about 1/4 of the total surface area. Hence the dry land surface area of the Earth is about

$$(0.25)(5.10 \times 10^{14} \text{ m}^2) = 1.3 \times 10^{14} \text{ m}^2$$

The dry land surface area per person is then

$$\frac{1.3 \times 10^{14} \text{ m}^2}{6.0 \times 10^9 \text{ persons}} = 2.2 \times 10^4 \text{ m}^2/\text{person}$$

We can imagine each person at the center of a square of this area; the distance between each person is the same as the length of a side of such a square. The square root of 2.2×10^4 m^2 is about 1.5×10^2 m. The approximate distance between the people is about 1.5×10^2 m, or one hundred and fifty meters.

1.8 THE DISTINCTION BETWEEN PRECISION AND ACCURACY

Some quantities are known exactly, others less so. When we say we have one apple, there is no question that we have precisely one apple (as long as we are not eating it), since simple counting determines the number exactly. Although they can be counted, the number of apples in a cubic meter will vary depending on the size of the apples used to fill the volume. Other numbers, such as special irrational numbers in mathematics, have symbols to represent their precise value. Among them are the famous numbers

- the ratio between the circumference and diameter of a circle:

$$\pi = 3.141\ 592\ 6536\ldots$$

- the base of natural logarithms:

$$e = 2.718\ 281\ 8284\ldots$$

and other irrational numbers such as

- the ratio between the diagonal and side of a square:

$$\sqrt{2} = 1.414\ 2136\ldots$$

- other irrational square roots, such as

$$\sqrt{3} = 1.732\ 0508\ldots$$

Such numbers can be used in decimal form to whatever precision is desired.*

Physical quantities, however, involve measurements and they can *never* be made with infinite precision.

In physics we distinguish between the words **precision** and **accuracy**. To see the distinction, consider a rifle target as in Figure 1.13. Let the bull's-eye in the center of the target represent the actual or true value of the quantity to be measured, the value it really has if we could only measure it. The actual value may in fact be unknown. Individual mea-

surements of the quantity, under identical conditions, are represented schematically by individual shots on the target value. If the shots are closely clustered on the target but not near the bull's-eye (as in Figure 1.13), these results are described by saying the measurements are *precise*, because repetitive measurements yield substantially the same result. They are *not accurate*, however, since the value is not the bull's-eye—the actual value. In this case, there is a **systematic error** in the measurements. It is caused, perhaps, by a misaligned rifle sight, if one is talking about actual rifles and targets. In other cases systematic errors are caused by equipment that is not properly calibrated.

On the other hand, if the individual measurements (shots) are randomly scattered around the target (as in Figure 1.14), the measurements are *not precise* because they are scattered all about the target value (the bull's-eye). But a suitable average of these measurements yields a result that is *accurate*, since the values are centered on the bull's-eye and their average is close to the actual value of the quantity.

The hope of all experimentalists is shown in Figure 1.15. These results are both precise (closely clustered) and accurate (near the actual or true value). Physicists usually can agree on precision but may disagree about accuracy.

> The precision of a measurement is indicated by the number of digits written. The number of digits is called the number of **significant figures**.

For example, to determine the mass of an apple, a measurement is necessary. Thus, if the scale in Figure 1.16 is used to determine the mass, we estimate the mass as 0.2 kg.[†] There is one significant figure; the initial zero merely helps to locate the decimal point. The mass can be written in scientific notation as 2×10^{-1} kg, although 0.2 kg conveys the same information (one significant figure) in a more familiar form.

If the scale in Figure 1.17 is used, the mass is determined to be 0.25 kg = 2.5×10^{-1} kg; there are two significant figures. The more sensitive scale in Figure 1.18 might result in determining the mass to three significant figures: 0.250 kg = 2.50×10^{-1} kg.

[†]The scale actually measures the magnitude of the force of the Earth on the apple—its weight—which is proportional to its mass, as we will see later (in Chapters 5, 6, and 7). The dial on the scale is *calibrated* to indicate the mass directly.

---

*Of course, some square roots are exact: $\sqrt{49} = 7$.

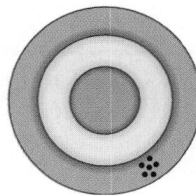

FIGURE 1.13 Precise but not accurate results.

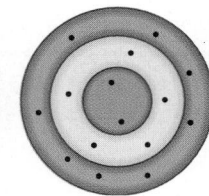

FIGURE 1.14 Not precise but accurate results.

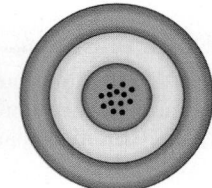

FIGURE 1.15 Precise and accurate results.

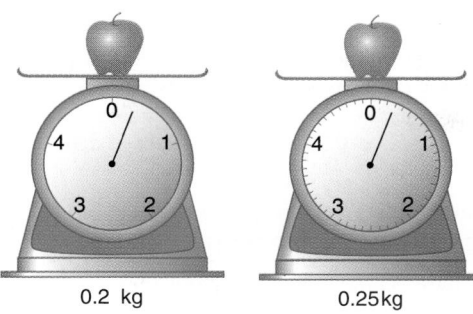

FIGURE 1.16 One significant figure.

FIGURE 1.17 Two significant figures.

FIGURE 1.19 Measure the volume.

The last zero here *is* significant, since it implies that the last digit is known (it is a zero and not some other number).

Now measure the *volume* of the same apple by dunking it in a graduated cylinder, as in Figure 1.19. Suppose the measurement of its volume yields 356 cm³ or, converting to cubic meters, 3.56×10^{-4} m³. Notice that the volume here has three significant figures. To determine the *average density* of the apple, we need to divide its mass by its volume; we will make this calculation in Example 1.9.

> The result of a calculation involving multiplication or division has no more significant digits than the quantity with the *least* number of significant digits.

When data with different numbers of significant figures are added or subtracted, we must be more careful. See Examples 1.10 and 1.11.

> For addition and subtraction, the result must have the same number of significant digits as the term with the smallest number of decimal places in the sum.

FIGURE 1.18 Three significant figures.

In other words, just because a calculator or computer indicates many digits, it does not follow that the result is more precise or has more significant figures than the data entered into the calculator.

> The data, not your calculator display, determine the precision of the result.

> To avoid confusion in this text, we assume that all data quoted have the indicated number of significant figures.

For example, in this text when we write that the mass of an object is 5.00 kg, three significant figures should be understood. If the mass is given as 5.0 kg, then only two significant figures are known.

> Purely numerical factors, such as 2, 4, 10, are written with no decimal point; they can be assumed to be exact.

Although the trigonometric functions such as sine, cosine, and tangent are mathematical in nature, in practice angles are determined from measurements. We will assume that all trigonometric functions have a number of significant figures equal to the number of significant figures in the angle itself.* See Example 1.12.

When making multiplication and division calculations, it is customary to round the answer to the appropriate number of significant figures at the conclusion of the calculation rather than at each step. With addition and subtraction, we must be more careful.

One gauge of the **uncertainty** of a measurement typically is half the place value of the last significant digit quoted. Thus, if you measure the length of your shoe and quote the result as 28 cm, the uncertainty of the measurement is half of one centimeter, the place value of the last significant digit in 28 cm (the 8 is in the ones place). The uncertainty of the measurement thus is ±0.5 cm. Uncertainties are quoted with ± values: 28 cm ±0.5 cm, meaning the actual value of the measurement is somewhere between 27.5 cm and 28.5 cm. If you make a more precise measurement and determine the length of the shoe to be 28.3 cm, the uncertainty of the measurement is half the value of 0.1 cm, or ±0.05 cm, since the last significant figure (3) is in the tenths place.

---

*This is not strictly true because the trigonometric functions are nonlinear functions of the angle. Nonetheless, the assumption is a reasonable first approximation that we will use throughout the text.

PROBLEM-SOLVING TACTICS

1.9 Do not confuse significant figures with decimal places. Use scientific notation as much as possible. The set of different numbers

$$4.56$$
$$0.456 = 4.56 \times 10^{-1}$$
$$0.0456 = 4.56 \times 10^{-2}$$
$$0.00456 = 4.56 \times 10^{-3}$$

all have three significant figures. The zeros are used merely to locate the decimal point and are *not* significant figures. The great advantage of scientific notation is that the number of significant figures is immediately apparent.

1.10 Quote your results using the appropriate number of significant figures. Some calculators can be set to display only a certain number of significant figures. When in this mode, the calculator actually is carrying along in its operations many more digits than are displayed. The display indicates the result automatically rounded to the preset number of significant figures. If you do not have or use this feature of your calculator, you must round any result to the appropriate number of significant figures. All examples and answers to problems in this text are quoted with the number of significant figures appropriate to the data given; you should try to do so as well.

QUESTION 7

Draw a bull's-eye target diagram with shots that are neither precise nor accurate.

QUESTION 8

How would you describe the results of this toddler trying to feed himself in terms of accuracy and precision?

EXAMPLE 1.9

Determine the average density of the apple shown in Figures 1.16–1.18 for the three cases where its mass is known to one, two, and three significant figures and the volume is known (in all cases) to be 3.56×10^{-4} m^3, that is, to three significant figures.

Solution

Remember Problem-Solving Tactic 1.10: *Use the appropriate number of significant figures to present your results.*

1. If the mass is known to one significant figure, then

$$\text{average density} = \frac{\text{mass}}{\text{volume}} = \frac{0.2 \text{ kg}}{3.56 \times 10^{-4} \text{ m}^3}$$

If you run this through your calculator, the result is

$$\text{average density} = 5.617\ 9775 \times 10^2 \text{ kg/m}^3$$

However, since the mass is known to only one significant figure while the volume is known to three, the average density must be quoted with only one significant figure. The other digits displayed on the calculator are meaningless. Hence, to one significant figure, the average density is, after rounding,

$$\text{average density} = 6 \times 10^2 \text{ kg/m}^3$$

One measure of the uncertainty of the result is half the place value of the last significant figure. So here the uncertainty of the result is $\pm 0.5 \times 10^2$ kg/m^3.

2. If the mass is known to two significant figures, the average density is

$$\text{average density} = \frac{0.25 \text{ kg}}{3.56 \times 10^{-4} \text{ m}^3}$$

Your calculator will yield

$$\text{average density} = 7.022\ 4719 \times 10^2 \text{ kg/m}^3$$

Since the mass is known to two significant figures, the average density can be quoted with only two significant figures:

$$\text{average density} = 7.0 \times 10^2 \text{ kg/m}^3$$

The uncertainty of the result is $\pm 0.05 \times 10^2$ kg/m^3, which is half the place value of the last significant figure.

3. If the mass is known to three significant figures, the average density can be quoted with three significant figures, because the volume also has three significant figures:

$$\text{average density} = \frac{0.250 \text{ kg}}{3.56 \times 10^{-4} \text{ m}^3}$$
$$= 7.02 \times 10^2 \text{ kg/m}^3$$

The uncertainty of the result is $\pm 0.005 \times 10^2$ kg/m^3, half the place value of the last significant digit.

EXAMPLE 1.10

Add the distance 2.30×10^{-2} m = 2.30 cm to 4×10^{-4} m = 0.04 cm.

Solution

The first distance has *three* significant figures (2, 3, 0). The last zero *is* significant because it implies the digit is known to be a zero. The second distance has *one* significant figure, 4 (the zeros in 0.04 cm merely locate the decimal point). Perform the addition by adding the columns:

$$2.30 \text{ cm}$$
$$+ \ 0.04 \text{ cm}$$
$$\overline{2.34 \text{ cm} = 2.34 \times 10^{-2} \text{ m}}$$

The result has three significant figures because both numbers were known with the *same precision* (to a tenth of a millimeter).

EXAMPLE 1.11

Add the distances 1.00 m and 6.78×10^{-4} m.

Solution
Each measurement has three significant figures. But in this case the first distance is not known with the same precision as the second one. The addition of the distances is

$$1.00 \text{ m}$$
$$+ \ 0.000678 \text{ m}$$
$$\overline{1.00 \text{ m}}$$

The result must be quoted as 1.00 m, *not* 1.000678 m, since the latter number (with *seven* significant figures) wrongly implies the distances were known with the same precision.

EXAMPLE 1.12

The angle between two stars is measured to be 49.0°.

a. How many significant figures are there in the measurement?
b. Find the sine, cosine, and tangent of the angle.

Solution
a. The angle has three significant figures. The zero is significant, since the angle was not measured to be either 48.9° or 49.1°.
b. Since the angle has three significant figures, you keep three significant figures in the trigonometric functions of the angle (see Figure 1.20):

$$\sin 49.0° = \frac{\text{opposite}}{\text{hypotenuse}} = 0.755 \quad (\text{not } 0.7547)$$

$$\cos 49.0° = \frac{\text{adjacent}}{\text{hypotenuse}} = 0.656 \quad (\text{not } 0.6561)$$

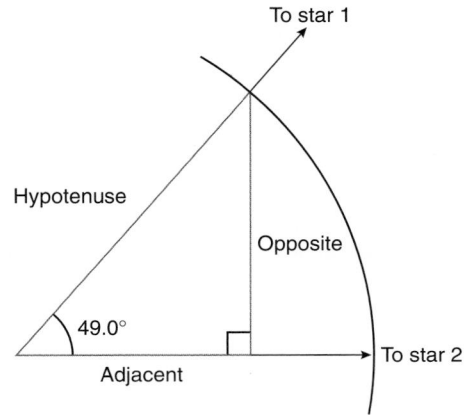

FIGURE 1.20

and

$$\tan 49.0° = \frac{\text{opposite}}{\text{adjacent}} = 1.15 \quad (\text{not } 1.150)$$

STRATEGIC EXAMPLE 1.13

A tabletop in a corporate boardroom is measured with a meter stick and found to have a length of 2.36 m and a width of 1.98 m (see Figure 1.21). The thickness is measured with a vernier caliper to be 3.02 cm (see Figure 1.22).

a. What is the area of the tabletop?
b. What is the volume of the tabletop?
c. What is the perimeter of the top?
d. What is the sum of the length and the thickness?

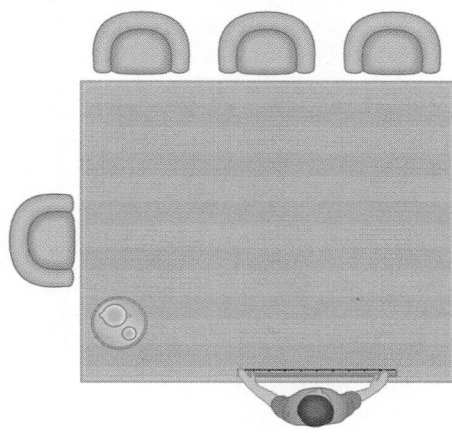

FIGURE 1.21

FIGURE 1.22

Solution
a.
$$\text{area} = (\text{length})(\text{width})$$
$$= (2.36 \text{ m})(1.98 \text{ m})$$
$$= 4.67 \text{ m}^2 \quad not \ 4.6728 \text{ m}^2$$

b.
$$\text{volume} = (\text{length})(\text{width})(\text{depth})$$
$$= (2.36 \text{ m})(1.98 \text{ m})(0.0302 \text{ m})$$
$$= 0.141 \text{ m}^3 \quad not \ 0.1411 \text{ m}^3$$

c.
$$\text{perimeter} = 2(\text{length}) + 2(\text{width})$$
$$= 4.72 \text{ m} + 3.96 \text{ m}$$
$$= 8.68 \text{ m}$$

d.
$$(\text{length}) + (\text{thickness}) = 2.36 \text{ m} + 0.0302 \text{ m}$$
$$= 2.39 \text{ m} \quad not \ 2.3902 \text{ m}$$

CHAPTER SUMMARY

The discipline of physics is divided into two broad areas:

- *classical physics*, which encompasses classical mechanics, electromagnetism, and thermodynamics; and
- *modern physics*, encompassing relativity and quantum mechanics.

Contemporary physics uses both classical and modern physics.

The standard system of units used for measurement in this text is *Le Système International d'Unités*, known as the *SI system*. Four of the fundamental units of this system are shown in Table 1.9.

Other units, such as the centimeter (cm), kilometer (km), gram (g), hour (h), day (d), and year (y), also are employed for convenience; it is important to know how to convert these *units of convenience* into the appropriate SI units.

Mass and *weight* are different physical concepts. Weight depends on location in space; mass does not.

There is a distinction between the word *dimension* (length, mass, time, speed, etc.) and the *units* used to specify the dimension (meter, kilogram, second, meters per second). Each side of an equation in physics must have the same dimensions and the same units. On the other hand, conversions, such as 1 m = 100 cm and 1 d = 86 400 s, have the same dimensions on each side, but not the same units.

The equal sign (=) is used for several purposes in physics:

a. in definitions (instead of the symbol ≡),
b. in conversions of units, and
c. in mathematical expressions representing physical laws.

The symbol (~) means *on the order of*, while the symbol (≈) means *approximately equal to*.

There is a distinction between *precision* and *accuracy* in physical measurements. *Precision* indicates the extent to which repetitive measurements of the same physical parameter, under identical conditions, yield the same value. The number of digits that have meaning in a measurement (*significant figures*) is crucial for correctly conveying the precision to which the quantity is known. *It is important to know and use the appropriate number of significant figures* associated with a measurement and subsequent calculations using such measurements.

Accuracy is the extent to which the measurements yield the true or actual value of the physical quantity; in many instances the actual value is unknown.

The ability to make *estimates* and *order of magnitude* calculations is important in physics—as well as in virtually any endeavor.

TABLE 1.9 Fundamental Units of the SI System

Physical quantity	SI unit	Definition
Mass	kilogram (kg)	One kilogram is the mass of a single, special cylinder of a platinum–iridium alloy.
Quantity	mole (mol)	One mole of a substance is Avogadro's number N_A of particles of the substance; Avogadro's number is 6.022×10^{23} particles/mol.
Time	second (s)	One second is the time interval for a specific kind of light emitted by the $^{133}_{55}$Cs atom to undergo 9 192 931 770 oscillations. Do not memorize this number; one can look it up if needed.
Length or distance	meter (m)	One meter is the distance through which light travels in exactly 1/299 792 458 seconds. Likewise, do not memorize this number; it can be looked up if needed.

SUMMARY OF PROBLEM-SOLVING TACTICS

1.1 **(page 11)** In solving most problems, be sure to convert units of convenience into standard SI units.
1.2 **(page 12)** Read and understand the problem.
1.3 **(page 12)** Sketch a diagram of the problem features.
1.4 **(page 12)** Assess whether additional information is needed.
1.5 **(page 12)** Assess whether any given information is irrelevant.
1.6 **(page 12)** Realize there may be more than one way to solve the problem.
1.7 **(page 12)** Ask yourself if the answer is reasonable.
1.8 **(page 14)** Check the dimensions or the units.
1.9 **(page 20)** Do not confuse significant figures with decimal places. Use scientific notation as much as possible.
1.10 **(page 20)** Quote your results using the appropriate number of significant figures.

QUESTIONS

1. (page 3); 2. (page 9); 3. (page 10); 4. (page 11); 5. (page 12); 6. (page 15); 7. (page 20); 8. (page 20)

9. Is the difference between physics and engineering the same as that between science and technology?

10. It has been said that science is always wrong. In what sense is this statement both true and false?

11. In what sense is the time signature in a musical score a time standard? (See Figure Q.11.) In what sense is it not?

12. How is the word meter used in a musical context? What is a metronome?

FIGURE Q.11

13. Is physics the only science with basic fundamental units of measurement? If not, are such basic units in other areas based on the SI system of units or other basic units? When

answering, consider several disciplines such as chemistry, biology, geology, astronomy, economics, psychology, mathematics, and various areas of engineering.

14. Discover the meaning of the peculiar English unit of length known as the *cubit*. What is the conversion from cubits to meters?

15. One of your elementary school teachers likely said that you cannot add apples to oranges. What, then, is the meaning of the following:

$$5 \text{ apples} + 3 \text{ oranges} = 8 \text{ fruit}$$

16. Astronomers use a continuous day count called the Julian Day number to calculate the number of days between two calendar dates. Each day has a unique Julian Day number associated with it. For example, 1 January A.D. 2000 has Julian Day number 2 451 545. The Julian Day number tells you how many days have elapsed from 1 January 4713 B.C. (which was Julian Day number 0) to the date in question. So, 2 451 545 days elapsed between 1 January 4713 B.C. and 1 January 2000. No kidding. The Julian Day number for any calendar date is found using tables in *The Astronomical Almanac* (published annually by the U.S. Naval Observatory and the Government Printing Office, Washington, D.C.); look in the index under "Julian Day Numbers." Use the *Astronomical Almanac* to determine the Julian Day number of your birth date and for today. A simple subtraction gives your age in days. From this number, calculate your age in seconds to an appropriate number of significant figures. Just for fun, tell your friends the results!

17. The use of occasional leap seconds means that the Julian Day count (Question 16) does not give the exact time interval between two time instants if there have been one or more intervening leap seconds. Explain why. Suggest an alternative continuous time count (analogous to the Julian Day count) that could give the precise time interval between two widely separated time instants.

18. Time is a physical quantity, since it can be measured. Refer to the quotation by St. Augustine in Section 1.3. If someone were to ask you what time really *is*, how would you respond? Where does time come from and where does it go? What is meant by the expression "at a point in time"?

19. The U.S. dollar is not an SI unit. Nonetheless, express the sticker price of your favorite sports car in kilodollars (kilobucks!). What is the cost per kilogram?

20. What is the approximate value of the national debt of the United States in teradollars as of this year?

21. Does the prefix in the term microwave (as in microwave oven) have the same meaning as the prefix micro- in the SI unit system? If not, what does it mean?

22. Compare and contrast the use of monetary standards in economics to the notion of physical standards for measurement in physics.

23. Just to be ornery, the height of horses is measured in a unit called *hands*; one hand is four inches. (See Figure Q.23.) What is the equivalent of a hand in meters?

24. When people say they have "to get the dimensions of the problem before attacking it," in what sense is the word dimension being used? How does it differ from the meanings of the word in physics?

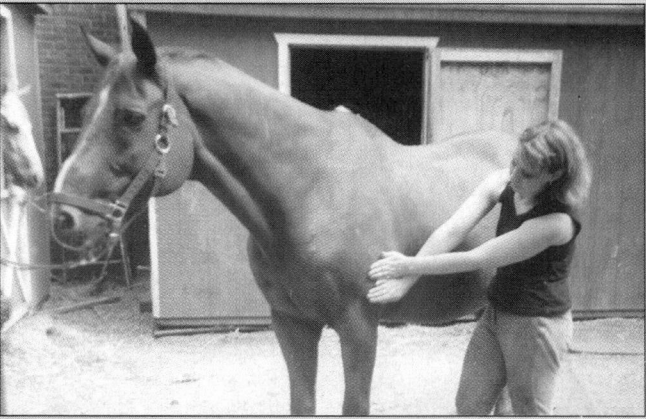
FIGURE Q.23

25. The kilogram is the only basic unit of the SI system defined in terms of a macroscopic artifact, the standard platinum–iridium cylinder at Sèvres, France. The *atomic mass unit* (u) is defined as 1/12 of the mass of a carbon-12 atom: $^{12}_{6}$C. Show that

$$1 \text{ u} = 1.6605 \times 10^{-27} \text{ kg}$$

Why do you think it is inconvenient to use atomic mass units for the masses of common objects such as tomatoes, students, and sports balls?

26. What factors have made the acceptance and common usage of the SI system of units so difficult in the United States?

27. Do you think it likely that ancient societies such as those in Africa, China, Egypt, Greece, Japan, Persia, and Rome had basic standards for measurement? If so, for what physical concepts?

28. Before wooden meter sticks are imprinted with the centimeter and millimeter markings indicating positions along the stick, it is essential that the wood be cured. Explain the consequences of not properly curing the wood beforehand.

29. Long steel measuring tapes were common in surveying years ago. What kind of systematic errors might be introduced into measurements made with such tapes between summer and winter? Surveying now is done with laser ranging systems.

30. The density of water is about 1.0×10^3 kg/m³, and mercury has a density of 13.6×10^3 kg/m³. On the same graph, qualitatively sketch how the mass m of a sample of each is a function of the volume V of the sample. Use m as the ordinate and V as the abscissa. Do you expect a linear relation between m and V or one that is curved? Be sure to indicate which line (curve?) corresponds to mercury and which to water. What is the meaning of the slope of each graph?

31. The prefixes kilo- and mega- in the words kilobyte and megabyte, used in connection with computers, do not have the same meaning as in the SI unit system. What do they mean? Why is there a difference in the meanings?

32. What are the drawbacks associated with the antiquated English system of units that make its use for scientific purposes inconvenient?

33. Suppose an extraterrestrial technical civilization were discovered. How could we communicate with them to compare our basic measurement standards for mass, length, and time with the different standards they likely have?

34. Is it possible to develop a complex social and technical civilization without the use of manipulative organs such as fingers? Can bee and ant colonies be considered to be technical civilizations? Give arguments for and against the proposition.

35. On what basis could it ever be said that the "end of physics" had arrived? (We do not mean the end of this course!)

36. How well can you count seconds? With a partner using a stopwatch or an accurate watch that you cannot see or hear, count out what you think is a sixty second interval and see how close you come.

37. How well can you tell mass by hefting? Search for a rock or dig an amount of dirt or sand that you think approximates a kilogram. Check to see how close you are.

38. Separate two pencils on your desk until they are what you think is a meter apart. Use a meter stick to see how close you are.

39. Approximately how many degrees of longitude correspond to a standard time zone? (See Figure Q.39.) When traveling eastward across the boundary between two time zones, do you set your watch ahead or behind? By how much? (There are locations where the time interval involved in crossing a time zone boundary is *half* the typical time interval. Newfoundland, central Australia, India, and Iran have such time zone boundaries.) Spacecraft circle the Earth in about 90 minutes, traveling eastward. Imagine yourself to be an astronaut in such a spacecraft, dutifully setting your watch accordingly as you traverse the time zone boundaries around the globe. After several orbits of the Earth, have you traveled into the past or future as the watch indicates? Explain. What demarcation is used on the globe so that such apparent time travel is not possible?

41. Suppose you and a friend are backpacking in the wilderness. Each of you has a watch, but the times indicated by the watches disagree. Using the two watches, is it possible to tell which watch indicates the correct time? If so, how? If not, why not? If a third person is along and also has a watch, does this change your conclusion? Explain.

42. If the standard kilogram at Sèvres, France, were destroyed by accident, design, warfare, or catastrophe, would this affect the SI system? If so, how? If not, why not?

43. If your watch does not keep accurate time, what units are typically used to express the rate at which it loses or gains time?

44. The arts are considered creative endeavors. Discuss how physics, engineering, and medicine also are creative enterprises. What about mathematics? Is the type of creativity in the arts, sciences, and mathematics the same? In what ways are they the same or different?

45. Time units such as the hours of the day, the days of the week, and the days of the month are counted using *modulo arithmetic*. For example, the hours of the day are counted from 1 to 12 (or 24) and then begin all over again; we say the hours are counted modulo 12 (or modulo 24). The days of the week are counted 1, 2, 3, 4, 5, 6, and 7, and then begin back at 1 again for the next week, and so are counted modulo 7. Cite three or four other common instances of modulo counting not associated with time units.

46. If one takes appropriate account of the number of significant digits at each step of a calculation, will the result differ from that obtained if the number of significant digits is accounted for only at the end of a calculation? Which procedure should be used?

47. When measuring the position of the edge of a book, a student notes what is seen in Figure Q.47. Is the position of the edge correctly given as 0.265 m, 0.266 m, or 0.2655 m? What is the basis of your judgment?

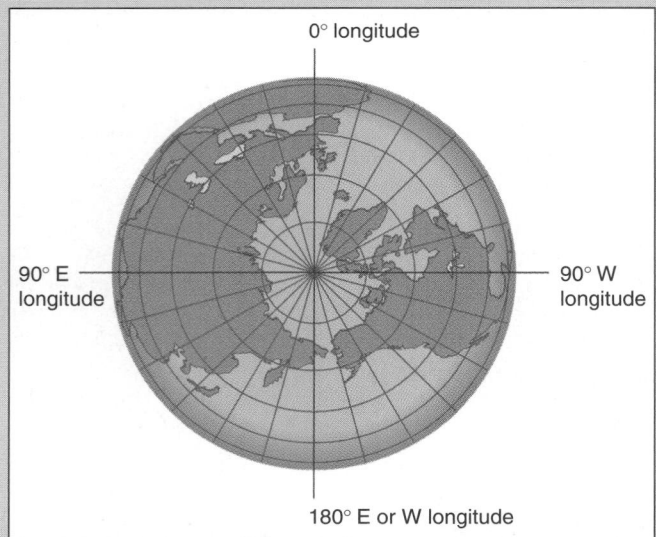

FIGURE Q.39

40. The determination of longitude used to be accomplished through astronomical observation of the position of the Sun in the local sky along with a clock set to keep Greenwich time (England). The meridian through Greenwich is 0° longitude. Describe how longitude can be determined approximately using such observations.

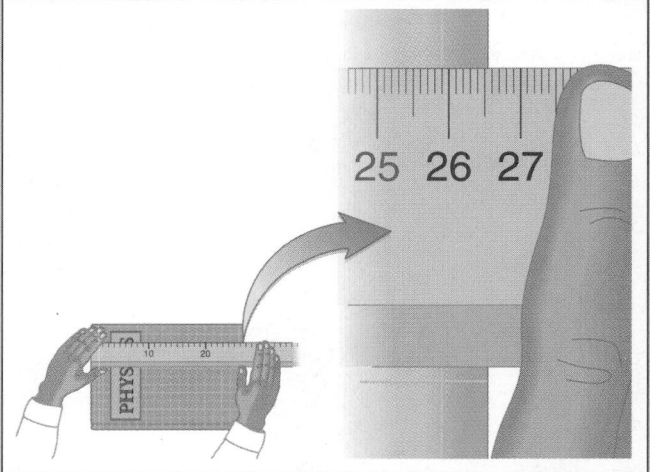

FIGURE Q.47

48. The typical shots hit by several golfers are described in the following way:
    - Golfer A hits shots that are precise, but not accurate.
    - Golfer B hits shots that are accurate, but not precise.
    - Golfer C hits shots that are neither precise nor accurate.
    - Golfer D hits shots that are both precise and accurate.

    Describe the characteristics of the shots made by each golfer.

PROBLEMS

Section 1.3 Standards for Measurement

1. What is the volume of the standard kilogram at Sèvres, France?

2. Determine the density of the platinum and iridium alloy used to fabricate the standard kilogram.

3. (a) With a suitable balance, determine the mass of a penny in kilograms. (b) What is the approximate monetary value of a kilogram of pennies? (c) Using the same scale, is it more precise to determine the mass of a single penny by using 1 or 100 pennies?

4. What time interval does it take light to travel 1.00 m in a vacuum?

5. One can have some fun with some of the SI prefixes. What are the following: 10^{-12} boos; 10^{-9} goats; 10^{-3} tary; 10^{-3} onaires; 10^6 phones; 10^{-6} phones; 10^6 bucks; 10^{-2} pedes; 10^{-6} film; 10^{12} pin; 10^6 lopolis; 10^1 cards; 10^{-6} fish; 10^{-6} waves; 10^{-1} mate; 10^{-6} economics; 2×10^3 mockingbird; 10^{12} bell; 10^{12} fermi; 10^6 lomaniacs; 10^{-3} nery; 10^{-6} scopes; 10^6 bytes
Many of these came from Robert L. Weber, *A Random Walk in Science* (Crane, Russak and Company, New York, 1973), page 61.

6. The hydrogen atom consists of a single proton in the nucleus and an electron. What percentage of the mass of a hydrogen atom is the electron? (Relevant data can be found on the inside covers of the text.)

7. Consider the Earth to be a sphere of radius 6.37×10^3 km. (a) What is the circumference of the Earth in kilometers? (b) What is the distance in meters between the pole and the equator measured along the surface of the Earth? (c) What is the straight-line distance through the Earth between a point on the equator and either pole?

8. Use the information on the inside covers to determine what percentage of the mass of the entire solar system resides in planets.

9. Compute approximate values for the following ratios to see which, if any, are of the same order of magnitude. Table 1.3 on page 8 has appropriate data.

    (a) $\dfrac{\text{mass of a proton}}{\text{mass of an electron}}$

    (b) $\dfrac{\text{mass of the Sun}}{\text{mass of the Earth}}$

    (c) $\dfrac{\text{mass of the Milky Way Galaxy}}{\text{mass of the Sun}}$

    (d) $\dfrac{\text{mass of the Universe}}{\text{mass of the Milky Way Galaxy}}$

10. Compute approximate values for the following ratios to see which, if any, are of the same order of magnitude. Table 1.5 on page 10 has appropriate data.

    (a) $\dfrac{\text{diameter of an atom}}{\text{diameter of the nucleus}}$

    (b) $\dfrac{\text{diameter of the planetary solar system}}{\text{diameter of the Earth}}$

    (c) $\dfrac{\text{diameter of the planetary solar system}}{\text{diameter of the Sun}}$

    (d) $\dfrac{\text{diameter of the Milky Way Galaxy}}{\text{diameter of the solar system}}$

    (e) $\dfrac{\text{diameter of the observable universe}}{\text{diameter of the Milky Way Galaxy}}$

11. Look at Table 1.3 on page 8. (a) Find the length of the edge of a cube of water whose mass is about a factor of 10^{40} smaller than the mass of the universe. The density of water is 1.0×10^3 kg/m³. (b) Find the length of the edge of a cube of water whose mass is about a factor of 10^{40} greater than the mass of an electron.

12. Look at Table 1.4 on page 10. What time interval in seconds is on the order of a factor of 10^{20} smaller than the age of the universe and on the order of 10^{20} greater than the lifetime of the most unstable particle?

13. Look at Table 1.5 on page 10. (a) What length is about a factor of 10^{20} smaller than the distance to the most distant objects in the universe? (b) What length is a factor of 10^{20} larger than the diameter of a typical nucleus?

‡14. Show that for a right-circular cylinder of fixed volume, its surface area is a minimum when the height of the cylinder is equal to the diameter of the circular cross section. This is the reason the standard kilogram cylinder was constructed with a diameter equal to its height: this minimizes the exposure of the material to the ambient atmosphere.

Sections 1.4 Units of Convenience and Unit Conversions
1.5 The Meaning of the Word Dimension
1.6 The Various Meanings of the Equal Sign

15. The cubical dice common to many games are approximately 1.0 cm on an edge. (See Figure P.15.) If the dice are stacked into a cubical array, what is the length of the edges of the cubical array that contains Avogadro's number of dice?

FIGURE P.15

16. Use a flexible tape measure to determine the circumference and diameter of the circular cross section of a soda can. What is the ratio between the circumference of the circle and its diameter? Does this ratio change if the size of the circle changes?

17. A peanut farmer has a rectangular field of length ℓ and width w and is planning next year's crop. If the length and width of the field are both tripled, (a) by what factor does the amount of fencing needed to enclose the field increase? (b) by what factor does the area under cultivation increase? (See Figure P.17.)

18. A truck tire has a radius three times as big as that of a tire for a compact car. (See Figure P.18.) Each tire rolls through one revolution. By what factor is the distance traveled by the truck tire greater than that traveled by the car tire?

19. A toy manufacturer is planning to make painted, cubical wooden blocks with sides of length ℓ and 3ℓ for a Montessori

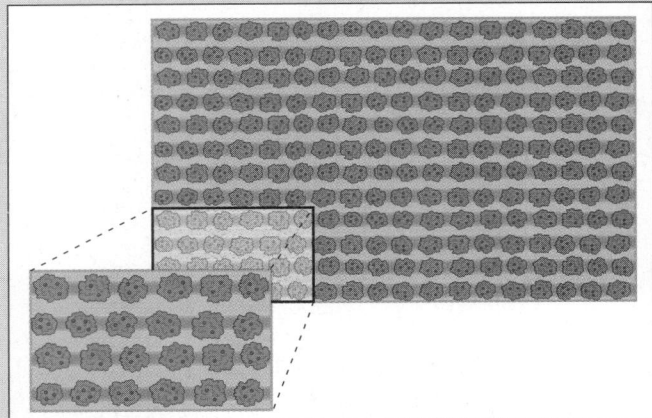

FIGURE P.17

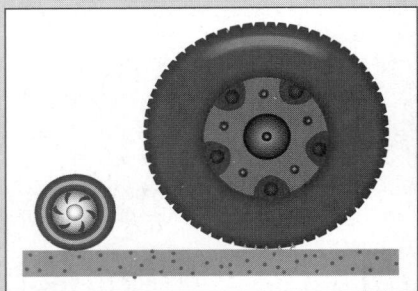

FIGURE P.18

preschool program. (See Figure P.19.) (a) What is the ratio of the amount of paint needed for a larger block to that required for a smaller block? (b) What is the ratio of the volume of wood needed for a larger block to that required for a smaller block?

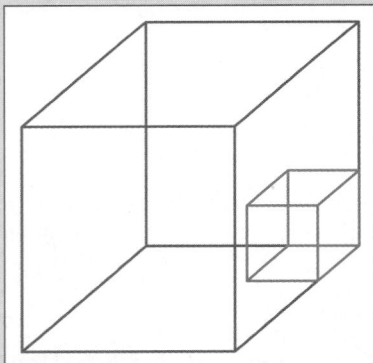

FIGURE P.19

20. Measure the circumference and the mass of a baseball and an inflated basketball. (a) What is the ratio of their radii? (b) What is the ratio of their surface areas? (c) What is the ratio of their volumes? (d) What is the ratio of their masses?

21. While inflating a spherical balloon, you note that the surface area triples with only two "blows." What lungs! (a) By what factor did the radius of the balloon increase? (b) By what factor did its volume increase?

22. Cepheid variable stars alternately increase and decrease their volume over time intervals ranging from days to weeks. If the volume of such a spherical star increases by 50% while in its expanding phase, (a) by what factor does its radius increase? (b) by what factor does its surface area increase?

23. As a community project, your dormitory is designing a square playground with a perimeter of 25.0 meters. The kids want it bigger, and so the perimeter of the square is increased to 36.0 meters. (See Figure P.23.) (a) What is the ratio of the larger perimeter to the smaller perimeter? (b) What is the ratio of the side of the larger square to the side of the smaller square? (c) What is the ratio of the area of the larger square to the area of the smaller square?

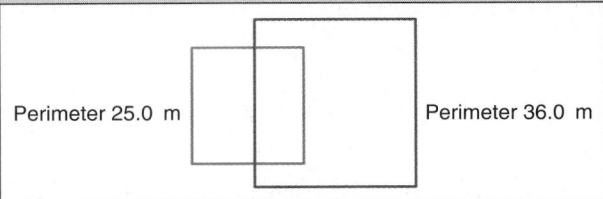

Perimeter 25.0 m   Perimeter 36.0 m

FIGURE P.23

24. (a) What is the ratio of the surface area of the Earth to the surface area of the Moon? Assume both are spherical. (See Figure P.24.) You will find appropriate data on the inside covers of the text. Does your result depend on the units used to measure the area? (b) Calculate the ratio of the volume of the Earth to the volume of the Moon.

FIGURE P.24

25. The speedometer on your slick new powder blue roadster indicates speed in miles per hour. (See Figure P.25.) What conversion factor allows you to change the speed of the car in miles per hour to meters per second (m/s)?

26. The posted speed limit on most rural interstate highways in the United States is 65 miles per hour. (See Figure P.26.) Convert this to kilometers per hour.

27. How many minutes are in 1.00 microcentury? Express the duration of your physics lecture in microcenturies. [This

FIGURE P.25

FIGURE P.26

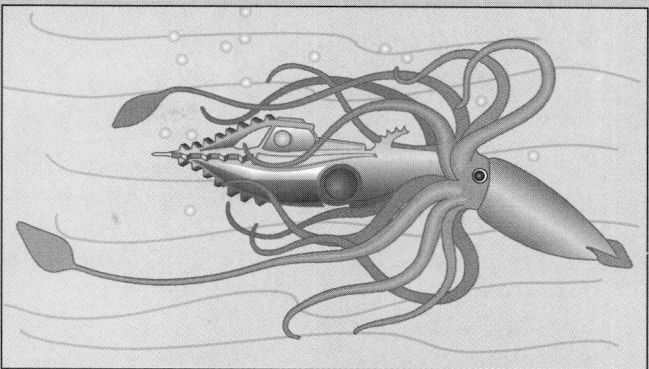

FIGURE P.28

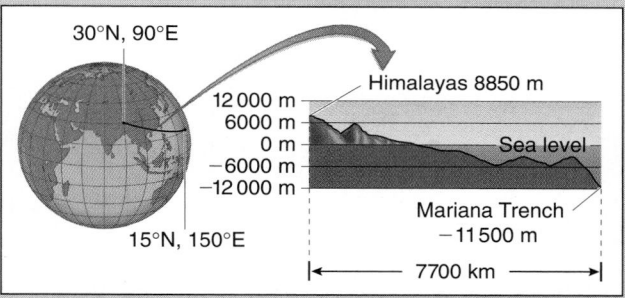

FIGURE P.30

comparison was first made by the Nobel prize–winning physicist Enrico Fermi (1901–1954).]

28. One league is three miles. Jules Verne wrote a book titled *Twenty Thousand Leagues Under the Sea.* How many kilometers is 20 000 leagues? (See Figure P.28.)

29. How many water molecules are there in 1.00 cm³ of water? The density of water is 1.0×10^3 kg/m³, and its molar mass is 18 g.

30. The Himalayas are the highest mountains on the Earth and rise up 8.85×10^3 m above sea level. The deepest ocean trench is the Mariana Trench, the bottom of which lies 1.15×10^4 m below sea level. (See Figure P.30.) The surface roughness of the Earth is measured by the sum of these distances. Express the surface roughness of the Earth as a percentage of the average radius of the Earth (6.37×10^6 m).

The smallness of this percentage means that the Earth is globally quite smooth.

31. How long does it take light to travel from the Sun to the Earth? Express your result in minutes.

32. Determine the number of seconds in one year. Is the result approximate or exact?

33. A *cord* is the standard unit of volume for stacked firewood in the United States. A cord of wood is a pile 4.00 feet high, 4.00 feet wide, and 8.00 feet long. (See Figure P.33.) Determine the volume of a cord in cubic meters.

34. Determine the number of meters in 1.00 mile.

35. If you moved one meter each day of your life, how far have you traveled during your lifetime to date?

36. After checking prices in the local supermarket, determine the cost per kilogram of filet mignon.

•37. A gold brick has a mass of 15.0 kg. (See Figure P.37.) The density of gold is 19.3×10^3 kg/m³. (a) What is the volume of the gold brick? (b) If an additional 3.0 kg of gold is added to the brick, what is the *increase* in the volume of the brick?

•38. At the end of their "life," stars collapse under gravitational forces and, under certain conditions, become spherical neutron stars or white dwarf stars. (a) A neutron star has a mass typically twice that of the Sun but a radius of only 10 km. Calculate the average density (in SI units) of such a neutron star. (b) A white dwarf star typically has a mass on the order of the mass of the Sun but with a radius on the order of the

FIGURE P.33

FIGURE P.37

radius of the Earth. Calculate the average density (in SI units) of a white dwarf with a mass equal to that of the Sun and a radius of 7.0×10^3 km.

•**39.** Consider the Earth to be a sphere of radius 6370 km. To mark the equator, a long red ribbon is wrapped around the Earth. (See Figure P.39.) Suppose an additional 1.000 km of ribbon is added to the length of the ribbon and the longer ribbon also is shaped into a circle centered on the Earth. What is the height of the new ribbon above the surface of the Earth?
The idea for this problem came from Arnold Arons, *A Guide to Introductory Physics Teaching* (John Wiley, New York, 1990), page 19; the problem originates with Kenneth Clark.

•**40.** The density of gold is 19.3×10^3 kg/m³, and a mole of gold has a mass of 0.197 kg. (a) Find the volume of a mole of gold. (b) If we somewhat unrealistically model the gold atoms as small cubes, what is the center-to-center distance between the atomic gold cubes?

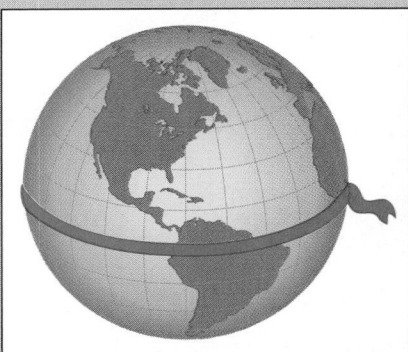

FIGURE P.39

•**41.** The population of the world is about 6.0 billion. Assume an average person can be modeled as a rectangular parallelepiped with edge sizes 0.200 m, 0.300 m, and 1.80 m. (See Figure P.41.) (a) What is the volume of the entire human population? (b) What are the dimensions of a cube that could accommodate the entire world population? (c) If everyone stood upright and elbow to elbow, what surface area would the entire human population cover? (d) What is the length of the side of a square whose area is equal to that calculated in part (c)?

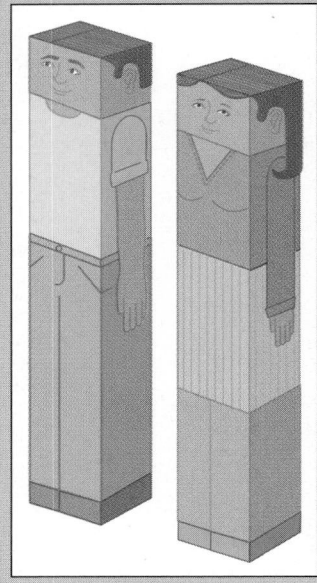

FIGURE P.41

•**42.** Assume the radius of the solar system is 50 times the distance between the Sun and the Earth. What is the radius of the solar system expressed in light-hours? A light-hour is the distance light travels during one hour.

•**43.** The distance between the Earth and Sun is called an astronomical unit (AU). Draw a line to represent the Earth–Sun distance, as in Figure P.43. Imagine a point P along the perpendicular bisector of this line. (a) At what distance along the perpendicular bisector is P if the Earth–Sun distance subtends an angle of 1.00 arc second measured at P? Express your result

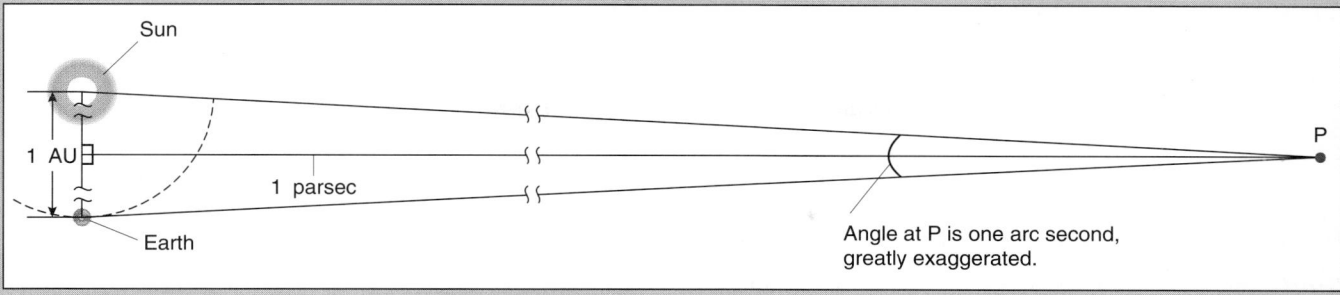

FIGURE P.43

in AU. There are exactly 3600 arc seconds in one degree. (b) In astronomy, the distance calculated in part (a) is defined to be one *parsec* (pc). How many astronomical units in one parsec? How many meters are in 1.00 pc? (c) A light-year (LY) is the distance light travels during one year. How many meters are in 1.00 LY? (d) How many light-years are in 1.00 pc?

•**44.** The brightest star in the night sky is the star α Canis Majoris, also known as Sirius (the Dog Star), and it lies 8.8 light-years away. If we represent the distance between the Earth and the Sun (1 AU) by a line 25 cm long, how far away is Sirius on this scale? (The results of the previous problems may be useful.)

•**45.** Calculate the speed of light in furlongs per fortnight. One furlong is exactly 1/8 of a mile, one mile is exactly 5280 feet, and a fortnight is exactly 2 weeks.

•**46.** A nautical mile is one arc minute of latitude. Given the radius of the Earth, determine the approximate number of meters in a nautical mile. (60 arc minutes = 1°.)

•**47.** A hectare is a metric unit of area equal to exactly 10^4 m². (a) What is the length of the side of a square field with an area of 1.0000 hectare? (b) An acre is an English unit of area equal to exactly 43 560 square feet. What is the length in feet of the side of a square field of area 1.0000 acre? (c) How many acres in 1.00 hectare? (d) How many hectares in 1.00 km²?

•**48.** (a) Use the data shown on the inside covers about the planets of the solar system to calculate the average densities of the planets. (b) The planets typically are grouped into two families: the large *Jovian planets* (Jupiter, Saturn, Uranus, and Neptune) and the *terrestrial planets* (Mercury, Venus, Earth, Mars, and Pluto). What conclusion can be drawn about the average densities of the Jovian planets versus those of the terrestrial planets? (c) How do the average densities of the Jovian planets compare with the average density of the Sun? What might this imply about the nature of the material comprising the Sun and that comprising the Jovian planets?

•**49.** Visit a local bank with a micrometer and a scale; as a courtesy, it may be advisable to phone ahead. With the permission of the bank manager, measure the thickness and the mass of a stack of new, crisp one-dollar bills. Measure the length and width of a single bill. (a) What is the mass of a single one-dollar bill? (b) What is the thickness of a single one-dollar bill? (c) What is the volume of a single one-dollar bill? (d) What is the mass of a stack of 1.00×10^9 one-dollar bills? (e) What is the height of a stack of 1.00×10^9 one-dollar bills? (f) What is the total area that could be covered by 1.00×10^9 one-dollar bills? (g) If 1.00×10^9 one-dollar bills were stored in and filled a cubical box, what is the length of an edge of the box?

•**50.** The national debt is on the order of $\$5.0 \times 10^{12}$. Calculate the height of a stack of one-dollar bills with a monetary value equal to the national debt. Compare your answer with the average distance between the Earth and Moon.

•**51.** You may have constructed a model of the atom when in elementary school. Suppose you model the nucleus with a marble that has a diameter of 1.0 cm. Use the data in Table 1.5 to determine the approximate diameter of the atom if your model is constructed to scale. What does this imply about most models of the atom that you see?

•**52.** You are commissioned by the local science museum to make a scale model of the solar system using a ball of diameter 1.00 m to represent the Sun. (a) Use the data on the inside covers to determine the sizes of the balls that are needed to represent each of the nine planets. (b) If the solar system is constructed to scale, determine the distances of the balls representing each planet from the ball representing the Sun. (c) In this scale model, what is the approximate distance to the star closest to the Sun in space, α Centauri, located 4.2 light-years from the Sun? Express your result in km.

•**53.** The average density of the universe is about 5×10^{-31} g/cm³. (a) Convert this density to SI units of kg/m³. (b) If a 60 kg physics professor is atomized and dispersed uniformly into a spherical volume large enough so the average density of the sphere is equal to the average density of the universe, what is the radius of such a sphere?

•**54.** Leap years occur every four years; coincidentally, leap years also are presidential election years in the United States. The first of January can occur on any day of the week. (a) If you collect calendars, how many different calendars exist? (b) If you collect calendars *in chronological order*, how many years elapse before the collection can be used again in the same order? This time period is called the *solar cycle* of the calendar.

•**55.** If the price of gold is $\$400$ for one troy ounce, what is the price per kilogram? One troy ounce has a mass of 3.11×10^{-2} kg.

Section 1.7 Estimation and Order of Magnitude

56. Estimate the order of magnitude of the mass in kilograms of (a) an adult elephant; (b) a large salmon; (c) a grape; (d) a grain of rice.

57. Estimate the time it takes (a) to swim 1 km; (b) to paddle a kayak around the world; (c) to get to sleep—zzzz . . . .

58. Estimate the distance from your location to (a) the nearest ice cream shop; (b) the sea; (c) the north pole.

59. (a) Estimate the number of heartbeats your heart has performed during your life to date. (b) Estimate the number of breaths you have taken during your life to date.

60. What is the current calendar year? Estimate the number of seconds that have elapsed since the year A.D. 1. Ironically, there is no year 0 in the calendar. The year 1 B.C. is followed immediately by the year A.D. 1. The current dating scheme originated in what we now call the year A.D. 532 by Dionysius Exiguus, although the method was not widely used until the Venerable Bede used it in his *Ecclesiastical History of the English People* several centuries later. Prior to Dionysius and Bede, Western dating was done on the Roman calendar, dating from the approximate founding of Rome by Romulus. The mathematical concept of zero is thought to have originated in India during the 4th century and only reached western Europe via the Arabs during the 12th century. Hence, when Dionysius Exiguus introduced the B.C. and A.D. dating scheme, zero was a number unknown to him. Our bookkeeping system of debits and credits also predated the ideas of zero and negative numbers.

61. Given the mass of the Sun (1.99×10^{30} kg) and the mass of a proton (1.67×10^{-27} kg), express the mass of the Sun in proton masses.

62. Estimate the number of bricks needed to build a wall 100 m in length and 1.0 m in height with bricks on both sides of the wall. (See Figure P.62.)

•63. (a) The Julian calendar (introduced by Julius Caesar in 46 B.C.) has three normal years, each with 365 d, followed by a leap year with 366 d. What is the average number of days per year on this calendar? (b) In 1582, Pope Gregory XIII introduced the Gregorian calendar so that the average duration of the calendar year would be closer to the actual duration of a year, 365.2422 d. He modified the Julian leap year rule in the following way. Century years not divisible by 400 (such as 1700, 1800, and 1900), which normally would be leap years in the Julian scheme since they are integrally divisible by 4 with no remainder, have the leap day omitted and are normal 365 d years. Century years that are divisible by 400 (such as 1600 and 2000) retain the leap year extra day. What is the average duration of the year with the Gregorian calendar reform?

•64. Enrico Fermi, one of the theoretical and experimental geniuses of 20th-century physics, was noted for his ability to make order of magnitude calculations. One of his more famous estimation problems concerned piano tuners. Estimate the number of piano tuners in the major metropolitan area nearest to you. What information source could you consult to see if your answer is about the right order of magnitude?

•65. You are employed as an estimator for a high-speed rail line to be constructed between New York and San Francisco. Such railroads use concrete ties rather than the wood ties of the early days of railroading. (a) Estimate the number of railroad ties needed for a parallel set of dual tracks, one for eastbound traffic, one for westbound traffic. (b) Estimate the volume of concrete needed to construct the ties.

•66. Estimate the number of soft drink containers (bottles and cans) used by the population of the United States during one year.

•67. A new tire of diameter 63.0 cm has a tread depth of 1.00 cm. (See Figure P.67.) The tire is designed to last 80×10^3 km before replacement, when the tread depth will be only 0.20 cm. (a) Approximately how much does the radius of the tire

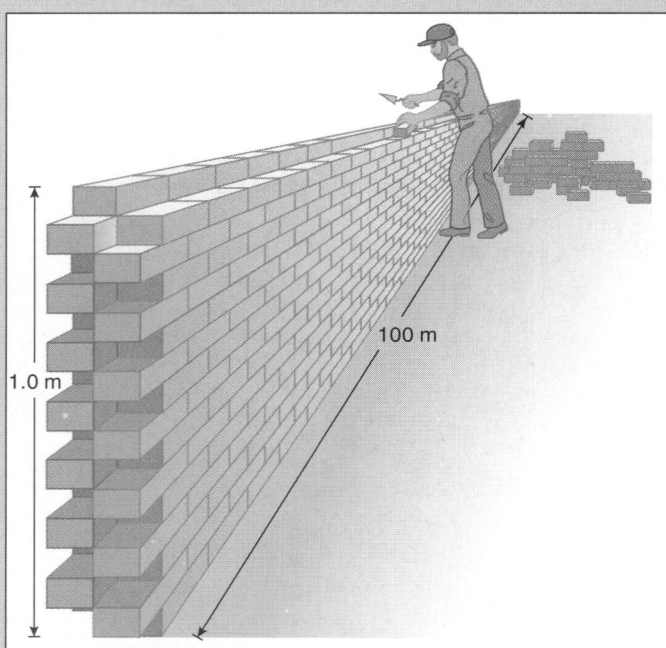

FIGURE P.62

FIGURE P.67

decrease with each revolution of the tire? (b) How many molecules are worn off the radius of the tire in each revolution of the tire? Assume the average diameter of the molecules making up the tire is about 10^{-10} m.

•**68.** Make and state some reasonable assumptions to estimate the volume of Mt. Everest, the summit of which is at an elevation of 8.85 km above sea level.

•69. Make and state some reasonable assumptions and estimate the volume of water in the oceans in cubic kilometers.

•**70.** Estimate the daily vehicular traffic count past the main intersection of your town or city each day.

•**71.** Use a popsicle stick or pencil, held in front of your eye at an appropriate distance, to barely but completely cover the Moon in the sky, as in Figure P.71. Make appropriate measurements and use your knowledge of geometry to measure the angular size of the Moon in degrees. For obvious reasons, ***do not do this for the Sun***.

FIGURE P.71

•**72.** Estimate the mass in kilograms of all the people on a fully loaded Boeing 747 jumbo jet.

•73. McDonalds claims to have served 100 billion burgers. Estimate the percentage of the population of the United States that visits McDonalds for a hamburger each day. Be sure to specify your assumptions.

‡**74.** While lost in a forest, you decide to while away the time while waiting for a rescue team to arrive. You notice that when you look horizontally through the forest, a tree trunk appears in every direction; some trees are nearby, of course, some farther away. The maximum distance that can be seen through the forest is called the *lookout limit*. (See Figure P.74.) Let $\langle A \rangle$ be the average land area occupied by a single tree including its canopy in the forest, and let $\langle w \rangle$ be the average width of a single tree trunk at eye-level height. (a) Show that an estimate of the lookout limit is

$$\frac{\langle A \rangle}{\langle w \rangle}$$

(b) Given the expression for the lookout limit in part (a), show that the number of trees that are within the lookout limit is

$$\frac{\pi \langle A \rangle}{\langle w \rangle^2}$$

(c) Evaluate the lookout limit and the total number of trees that can be seen if $\langle A \rangle = 100$ m^2 and $\langle w \rangle = 25$ cm. This problem is related to the cosmological problem known as the dark night sky paradox. The problem also is called *Olbers's paradox*, after Heinrich Olbers (1758–1840), although Johannes Kepler (1571–1630) may have been the first to state the problem in 1610 in his book *Conversation with the Starry Messenger*. The paradox goes like this: stars, of course, have a finite size; if there are enough stars, no matter what direction you look into space your line of sight will run into a star; thus the night sky should have stars in all directions and be as brilliant as the surface of the sun. So why is it dark at night? The resolution of the problem involves calculating the corresponding lookout limit for the universe: it turns out to be about 10^{23} light years. Since the age of the universe is estimated to be about 10 or 20 billion years, we can at best actually look out only 10 or 20 billion light-years. Then, too, many stars do not live for 20 billion years, so that all the lights are not on at the same time. Hence the night sky is dark.

This problem was inspired by Edward R. Harrison, *Cosmology: The Science of the Universe* (Cambridge University Press, Cambridge, England, 1981), Chapter 12, pages 249–265.

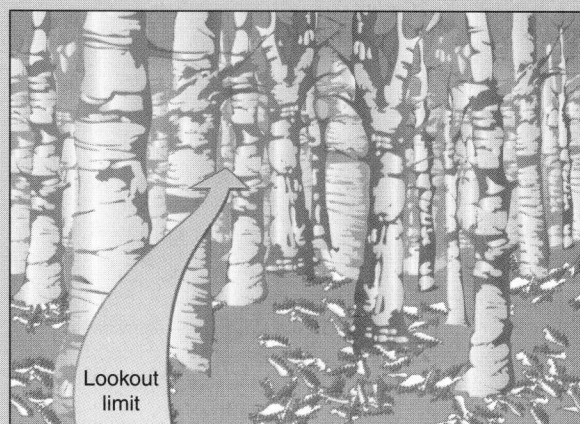

FIGURE P.74

Section 1.8 The Distinction Between Precision and Accuracy

75. The Census Bureau reports that there are 289 689 526 people in the United States as of some date. Then they say the count may be as much as 5 percent in error. Express the population count in scientific notation with an appropriate number of significant figures.

76. (a) What is the sum of the lengths 5.6 m, 0.22 m, 16.1 m, and 0.04 m? (b) What is the product of 3.46 m and 5.891 m? (c) What is π to six significant figures? (d) What is 7.83π? (e) How many significant figures are there in the product $\sqrt{2}\ \pi$? (f) What is $8.42 + \pi$?

77. Determine the sine, cosine, and tangent of each of the following angles with the appropriate number of significant figures:

$$78.0°$$
$$78.02°$$
$$78.024°$$

•**78.** Make a single graph with three curves plotting (i) the value of θ (in radians), (ii) sin θ, and (iii) tan θ, for values of θ (in degrees) from zero to thirty degrees along the abscissa. Note that for small angles, all three curves approach each other. (a) Do the curves ever cross each other over this range of angles? The *small angle approximation* states that for small angles θ, the sine of the angle, the tangent of the angle, and the angle itself (in radians) are approximately equal to each other. That is,

$$\theta \approx \sin \theta \approx \tan \theta \qquad \text{(for small } \theta \text{ in radians)}$$

(b) What is the largest angle θ such that these trigonometric functions are equal to each other to two significant figures? Express the angle in radians and degrees. (c) Why is the cosine of θ not part of the small angle approximation? That is, why is

$$\theta \approx \sin \theta \approx \tan \theta \neq \cos \theta \qquad \text{(for small } \theta \text{ in radians)?}$$

•**79.** Does the small angle approximation (see Problem 78) also imply that the cosecant of θ and the cotangent of θ are approximately equal for small angles θ in radians? In other words, is the following also true for small θ?

$$\frac{1}{\theta} \approx \csc \theta \approx \cot \theta \qquad \text{(for small } \theta \text{ in radians)}$$

On the same grid, plot (i) $1/\theta$, (ii) csc θ, and (iii) cot θ versus θ (as the abscissa) for angles between about 5° (say 0.09 rad) and 30° (0.52 rad) to confirm your conclusion.

•**80.** Measure the circumference of your neck, and express the result with an appropriate precision and uncertainty.

INVESTIGATIVE PROJECTS

A. Expanded Horizons

1. Investigate the historical origins of the metric and SI systems of units.
   Robert A. Nelson, "Foundations of the international system of units (SI)," *The Physics Teacher*, 19, #9, pages 596–613 (December 1981).
   J. L. Heilbron, "The politics of the meter stick," *American Journal of Physics*, 57, #11, pages 988–992 (November 1989).
   D. F. Bartlett, "Natural units: are they for everybody?" *American Journal of Physics*, 42, #2, pages 148–156 (February 1974).

2. *There is something fascinating about science. One gets such wholesale returns of conjecture out of such a trifling investment of fact.*

   Mark Twain (1835–1910)*

   Notice from Table 1.4 that the ratio between the approximate age of the universe divided by the lifetime of the most unstable particles is on the order of 10^{40}. Notice also from Table 1.5 that the ratio between the largest and smallest distances (the distance to the most distant object divided by the size of a typical nucleus) also is on the order of 10^{40}. Peculiar. Another funny coincidence: from Table 1.3 the mass of the universe divided by the mass of a proton is on the order of 10^{80}, or $(10^{40})^2$. We will see later that the ratio of the magnitude of the electrical force of attraction between an electron and a proton to the magnitude of the gravitational force of their attraction also is about 10^{40}. The approximate numerical coincidence of these diverse and apparently unrelated ratios is remarkable—but unrelated to the dreaded IRS 1040 form. Tables 1.3, 1.4, and 1.5 use the SI system to measure the physical quantities, but the ratios are the *same* whether we use fortnights, furlongs, slugs, or stones to measure the times, lengths, or masses. Some physicists and astronomers speculate that the approximate numerical coincidence of these ratios is more than accidental and may be a manifestation or hint of some yet unknown order in the universe. The reason for the coincidence may be the same as the answer to the following question: why is the universe 20 billion years old? The answer is because it took that long to find that out!

   Such speculation may be whimsical, but it should serve to make you wonder about what, if anything, might be implied by the numerical coincidences. This is the stuff that motivates physicists to search for unifying ideas and explanations. Check this out in the following reference.
   Edward R. Harrison, *Cosmology: The Science of the Universe* (Cambridge University Press, Cambridge, England, 1981), Chapter 17, pages 329–345.

3. Investigate the kinds of clocks used to indicate time with the greatest accuracy and precision and why such clocks are needed and useful.
   Wayne M. Itano and Norman F. Ramsey, "Accurate measurement of time," *Scientific American*, 269, #1, pages 56–65 (July 1993); other references are given in this article.

4. The construction of a calendar is not a trivial exercise. We use a calendar known as the Gregorian calendar, named after Pope Gregory XIII, who introduced it in 1582. Investigate the historical origins of the Gregorian calendar and the motivations for its peculiar leap year rules; see Problem 63.
   Gordon Moyer, "The Gregorian calendar," *Scientific American*, 246, #5, pages 144–152, 178 (May 1982).
   Alexander Philip, *The Calendar: Its History, Structure, and Improvement* (Cambridge University Press, Cambridge, England, 1921).
   Ronald Lane Reese and George Y. Chang, "The date and time of the vernal equinox: a graphical representation of the Gregorian calendar," *American Journal of Physics*, 55, #9, pages 848–849 (September 1987).
   W. C. Elmore, "Perpetual calendar," *American Journal of Physics*, 44, #5, pages 482–483 (May 1976).

5. Adult mammals of different species have a wide range of masses. Consult appropriate sources to determine the average mass and average life expectancy of a variety of mammals. By making an appropriate graph, investigate whether there is any correlation between the average mass of an adult mammal and its average life expectancy.

6. Investigate why clocks are needed to determine the longitude coordinate on the surface of the Earth. Research the work of John Harrison in developing the first accurate marine chronometers for determining longitude at sea, for which he won a state prize of £20 000 sterling in the latter part of the 18th century (but only after a lengthy legal battle).

Life on the Mississippi (J. R. Osgood, Boston, 1883), Chapter XVII.

Eric G. Forbes, "Who discovered longitude at sea?" *Sky and Telescope*, *41*, #1, pages 4–6 (January 1971).

Dava Sobel, *Longitude: The True Story of a Lone Genius Who Solved the Greatest Scientific Problem of His Time* (Walker, New York, 1995).

7. The gradual increase in the rotational period of the Earth (the time interval for one complete rotation) was discovered by noting discrepancies between ancient records of the locations where solar eclipses actually were observed and the locations calculated by extrapolating the motions of the Sun and Moon backward in time using a constant rotational period. Investigate the magnitude of the slowdown and its effect on eclipses.

Robert A. Nelson, "Foundations of the international system of units (SI)," *The Physics Teacher*, *19*, #9, pages 596–613 (December 1981).

Peter Borsche and Jurgen Sündermann, *Tidal Friction and the Earth's Rotation* (Springer-Verlag, New York, 1978).

F. Richard Stephenson, "Historical eclipses," *Scientific American*, *247*, #4, pages 170–183, 194 (October 1982).

8. You no doubt are familiar with the phenomenon of leap years, where an extra day is inserted into the month of February every four years because the orbital period of the Earth about the Sun is *approximately* one-fourth of a day more than an integral number of days (≈ 365.25 d). The extra one-fourth day accumulates to one day every four years. The accumulated day is inserted into February, since it is the shortest month. This process of inserting an extra day is called *intercalation*. The gradual slowdown in the rotation of the Earth (see Investigative Project 7) means that occasional *leap seconds* are introduced to keep time intervals based on atomic standards in step with those based on the astronomical spin and orbital motions of the Earth. Investigate more thoroughly why leap seconds are needed.

Robert A. Nelson, "Foundations of the international system of units (SI)," *The Physics Teacher*, *19*, #9, pages 596–613 (December 1981).

9. Compare and contrast the Gregorian calendar with the religious calendars of the Hebrew and Islamic faiths and those of other religions.

Frank Parise, *A Book of Calendars* (Facts on File, New York, 1982).

10. Astronomers use a peculiar time scale that is a continuous counting of days that ignores all aspects of the calendar. Each day has a unique *Julian Day number* associated with it that indicates how many days have elapsed since 1 January 4713 B.C. (which was Julian Day number 0). For example, 1 January 1998 is Julian Day 2 450 815, indicating that 2 450 815 days elapsed between 1 January 4713 B.C. and 1 January 1998. The Julian day count is based upon a 7980-year cycle known as the *Julian Period*. Investigate the peculiar 7980-year time interval and the origins of the Julian Period and the Julian Day count.

Ronald Lane Reese, Edwin D. Craun, and Steven M. Everett, "The origin of the Julian Period: an application of congruences and the Chinese Remainder Theorem," *American Journal of Physics*, *49*, #7, pages 658–661 (July 1981).

Ronald Lane Reese, Edwin D. Craun, and Charles W. Mason, "Twelfth-century origins of the Julian Period," *American Journal of Physics*, *51*, #1, page 73 (January 1983).

Ronald Lane Reese, Edwin D. Craun, and Michael Herrin, "New evidence concerning the origin of the Julian Period," *American Journal of Physics*, *59*, #11, page 1043 (November 1991).

11. Archaeologists have speculated about whether prehistoric cultures used a standard of length (a megalithic meter) in their construction activities. Based on his studies of ancient stone circles in England, the archaeoastronomer Alexander Thom has proposed they used a megalithic yard. Report on his investigations and the approximate length of a megalithic yard in meters.

Alexander Thom, "The megalithic unit of length," *Journal of the Royal Statistical Society* A, *125* Part 2, pages 243–251 (1962).

B. Lab and Field Work

12. You likely have marveled at the number of stars visible to the naked eye on a crisp, clear evening. Using a paper towel tube as an instrument, devise and describe a simple procedure to estimate the number of stars visible to the naked eye. (See Figure I.12.) Make the necessary observations on a clear, moonless evening, in a dark location, well away from local sources of light pollution. Report on your findings. You may be surprised by the result.

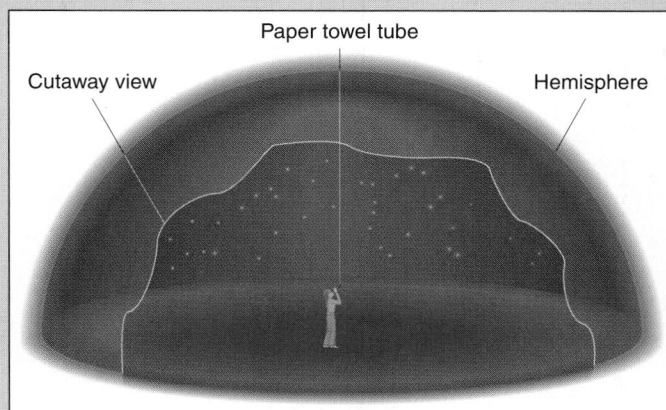

FIGURE I.12

13. You can estimate the diameter of a gasoline molecule in the following way. First, determine the density ρ of gasoline by appropriate measurements. Use a micropipette to make a droplet of known volume. Determine the mass m of the droplet of gasoline. Place the droplet of gasoline on a smooth puddle of water, and estimate the surface area A of the slick created by the gasoline on the water. (See Figure I.13.) Assume the layer of gasoline is one molecule thick and that the molecules are spherical in shape. (a) Show that an estimate of the radius r of a gasoline molecule is

$$r \approx \frac{3m}{4\rho A}$$

(b) Use the data secured to estimate the diameter of a gasoline molecule.

FIGURE I.13

14. Visit the office of a local surveyor and investigate the techniques and instruments used in the trade for measuring distances. To what precision can a 100 m distance be measured with such instruments?

15. The variety of automobiles available is quite staggering. Determine the sticker price and mass of, say, twenty different automobiles that span a variety of vehicle types and prices. Make two graphs, plotting the price per kilogram versus the mass and the sticker price of the automobiles. What conclusions can be drawn from your data?

16. With the information in Tables 1.3 and 1.5, plot the logarithm (base 10) of the mass, log *m*, versus the logarithm (base 10) of the size, log *d*. Approximate any missing masses or sizes, or find them in an appropriate reference. Do the points on the graph appear randomly over the diagram, or is there a pattern or trend? If there is a trend, can you find an empirical equation that roughly relates log *m* to log *d*?

C. Communicating Physics

17. Science is not value neutral. Ethics play a role in science as in politics or any field. Scientists have a collegial duty to each other and to society to be honest with the data and findings they report. Nonetheless, occasional instances of the deliberate falsification of data and results arise (fortunately, few) and are not just of recent vintage. What motivates such conduct, and what are the consequences of such unethical actions by scientists?

18. Science is not without controversy. Various issues highlight honest differences of opinion within the scientific community. Investigate one or more of the following issues, presenting an essay debating the issues involved: (a) the trial of Galileo; (b) the "discovery" of cold fusion; (c) the competition between funds for space research, particle accelerators, and those for social programs; (d) the ethics of radiation experiments on humans without informed consent; (e) whether it is moral to engage in weapons research and/or research for antiballistic missile defense; (f) alleged discrimination against women and minorities in research funding and academic advancement; (g) government versus commercial funding of basic research; (h) spectacular technological *faux pas* such as the flawed mirror of the Hubble space telescope (1990) and the loss of the Mars Observer (1993); (i) patent ownership by employees or the laboratories that employ them. Is discovery an individual or collective endeavor? (j) awarding and sharing credit for simultaneous discoveries; (k) should credit for discovery be assessed on the basis of the actual discovery date or upon the publication date of the results? (l) what research (if any) should be kept secret? (m) is scientific fraud a criminal or civil crime? (n) are computer virus hackers criminals? (o) should scientific results be first publicized to the press and then to the scientific community via research publications, or vice versa? (p) what are the responsibilities and obligations of a scientist to the profession and to the wider community? (q) what are the ethics of the planned obsolescence of products? (r) medical malpractice suits are increasingly common. Are such suits in science and engineering far behind? (s) should science be immune from issues of race and gender?

19. How is science as a creative process similar to yet different from the creative arts?

David Hawkins, "The creativity of science," *Science and the Creative Spirit* (University of Toronto Press, Toronto, 1958), pages 127–165.

20. Write an essay exploring the relationship between the sciences and the humanities.
C. P. Snow, *The Two Cultures and the Scientific Revolution* (Cambridge University Press, New York, 1959 and 1993).
Jacob Bronowski, *Science and Human Values* (Harper-Collins, New York, 1990).

21. Richard Feynman (1918–1988) a Nobel prize–winning physicist, once remarked that "science is the belief in the ignorance of experts." Do you agree or disagree with his opinion? Explain your answer in a well-reasoned essay.
Richard Feynman, "What is science?" *The Physics Teacher*, 7, #6, pages 313–320 (September 1969); quotation is on page 320.

22. When making measurements, explain how one can distinguish between the unavoidable uncertainty associated with every measurement and "human error." Is human error synonymous with a blunder?
John R. Taylor, *An Introduction to Error Analysis: The Study of Uncertainties in Physical Measurements* (2nd edition, University Science Books, Sausalito, California, 1997).

23. How has science been used in literature and literature in science?
F. E. L. Priestley, "Those scattered rays convergent : science and imagination in English literature," *Science and the Creative Spirit*, edited by Harcourt Brown (University of Toronto Press, Toronto, 1958), pages 53–88.
Alan Friedman, "Contemporary American physics fiction," *American Journal of Physics*, 47, #5, pages 392–395 (May 1979).
William H. Davenport, "Resource letter TLA-2: Technology, literature, and the arts," *American Journal of Physics*, 43, #1, pages 4–8 (January 1975).

"IT'S SAD... SHE DISCOVERED A NEW FUNDAMENTAL UNIT, BUT CAN'T FIND ANYTHING TO MEASURE USING IT."

CHAPTER 2

A MATHEMATICAL TOOLBOX
AN INTRODUCTION TO VECTOR ANALYSIS

**For the things of this world cannot be made known
without a knowledge of mathematics**
Roger Bacon (c. 1214–1294)*

While it is one of the languages of physics, mathematics is not what the enterprise of physics is about. Nonetheless, mathematics provides essential analytic tools used to describe our observations of nature and the results of experiments. In our study of motion and its causes, in our investigation of electric and magnetic phenomena, and in many other areas of physics, a working knowledge of vectors is necessary. It is one section of the mathematical toolbox that helps us describe a host of physical phenomena quantitatively.

This chapter inventories the vector portion of the mathematical toolbox. You will find this chapter essential for many succeeding chapters. It lays the foundation for almost every aspect of physics that we will examine. Consequently, this chapter warrants careful study. The time invested here will pay many dividends. When you master the various vector tools and techniques, you can use them with confidence, alacrity, and success in succeeding chapters as we apply them to a wide variety of physical situations. Should you already be familiar with these vector methods, proceed to Chapter 3 after a light skim through the toolbox.

2.1 SCALAR AND VECTOR QUANTITIES

What if angry vectors veer
Round your sleeping head, and form.
Robert Penn Warren (1905–1989)[†]

For thine arrows stick fast in me
Psalm 38:2

Our qualitative and quantitative ways of *describing* physical quantities depend on the phenomenon being considered. Certain physical concepts such as time, mass, and temperature can be defined using one numerical quantity; these are called **scalar quantities**. The particular numerical value of a scalar quantity depends on the specific physical concept represented, the standards used for measurement, and perhaps even the instant a measurement of the quantity is made.

> Physical concepts that require only one numerical quantity for their complete specification are known as scalar quantities.[‡]

Your mass may be 70 kilograms using SI units or, equivalently, 11 stone[§] in another unit system. Your mass typically increases with age and may vary slightly with time, depending on how many pizzas you just consumed. On the other hand, the mass of a fundamental particle, such as an electron or proton, is constant, independent of position and time; such a mass is a constant positive scalar quantity. Mass is intrinsically a positive scalar quantity (or zero); negative masses do not exist in nature.[#]

---

* (Chapter Opener) *Opus Majus*, translated by Robert Belle Burke (University of Pennsylvania Press, Philadelphia, 1928), Volume I, page 128.
[†] "Lullaby," lines 45–48, *Encounter*, May 1957, pages 13–14.
[‡] The word comes from the Latin *scalaris*, which means "resembling a staircase or ladder"; this is a metaphor for the real number line.
[§] One stone is a quaint English unit of mass that is approximately 6.4 kg.
[#] In Chapter 5 we shall see what observations lead us to conclude that mass is never negative.

The temperature outside your window in the dead of winter may be, say, −10 degrees celsius or, equivalently, 263 kelvin or 14 degrees fahrenheit. Chilly. The temperature at other times or locations, such as at Victoria Falls in Zimbabwe, may be different. The temperature at any given location is a positive or negative scalar quantity depending on the particular temperature scale used. We say the temperature is a scalar function of position and time.

Scalar physical quantities abound. The mathematics of such scalar quantities involves the familiar arithmetic, algebra, and calculus that you already know.

Many other physical concepts, such as those dealing with the motion of a system (which we discuss in detail in Chapters 3 and 4) and the pushes and pulls (forces) (Chapter 5) acting on it, have another aspect to them in conjunction with their simple numerical magnitude: *direction*. It makes a big difference *which direction* something is moving or pushed: a soccer ball headed toward your goal is quite different from one headed toward your opponent's goal, as shown in Figure 2.1.

The direction and speed of a ball and the direction and vigor of a push are examples of **vector quantities**.[¶]

> Vector quantities require for their complete specification a positive scalar quantity, called the **magnitude** of the vector, and a direction.

Vector quantities are represented visually by straight arrows, as in Figure 2.2. The arrow has a **head** or **tip** (i.e., the arrow point; the words head and tip can be used interchangeably) and a **tail** point (the opposite endpoint).

The length of the arrow is drawn proportional to the *magnitude*, or positive scalar numerical value, of the physical quantity (such as the number of meters per second a ball is moving). To make

---

[¶] The word *vector* comes from the Latin *vector*, meaning "a carrier." The word arises from the original description of the line from the Sun to a planet, a line that is "carried" around the Sun with the revolutionary motion of the planet.

FIGURE 2.1 Some physical quantities have directional attributes. The direction and speed of a ball have significant consequences in sports.

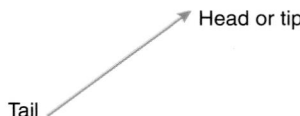

FIGURE 2.2 Visual representation of a vector: a straight arrow.

the drawing accurately, a scale must be introduced to indicate the relationship between, say, 1 cm of arrow length and the amount of the physical quantity represented by the arrow. The *direction* of the arrow representing the vector indicates the direction associated with the physical quantity. So we visualize a vector quantity as an arrow with a positive magnitude (or length) pointing in some direction. On the other hand, scalar physical quantities such as mass and temperature have no directional attributes and no arrow representations.

We must be careful here. All vectors have a magnitude and a direction, but *not* all concepts for which we can associate a magnitude and a direction are vectors. Later, in Section 2.5, we will examine a concept for which you can conceive a magnitude and an associated direction, but nonetheless is not a vector quantity. Hence, you have to be circumspect about which physical concepts constitute vector quantities. Strictly speaking, physical quantities are vectors if they have a magnitude and direction *and* obey the rules for vector operations, such as addition, that we establish in Section 2.4.

For our purposes in this text, *every physical quantity we consider is either a scalar quantity or a vector quantity* within the context that we will study them.* It is *crucial* that you know whether the physical quantity under consideration is a scalar or a vector quantity, because the arithmetic, algebra, and calculus of vectors are *different* from those of scalars, although the mathematical symbols used to represent the various operations look similar. Since the mathematics of vectors differs from that of scalars, you must learn the rules for the arithmetic, algebra, and calculus of vectors; that is the principal objective of this chapter. Fortunately, the rules for vector operations are not complicated.

An arrow can represent your position with respect to some origin or point of reference such as on a map, as in Figure 2.3. The direction to the top of the map is north while that to the right is east.† The length of the arrow representing your position is drawn proportional to the positive scalar distance you are from the origin (say, at the Washington Monument). The arrow direction indicates the direction *from* the origin *to* your location on the map. Such a vector is called a **position vector**, since it locates your position with respect to the chosen origin. We shall

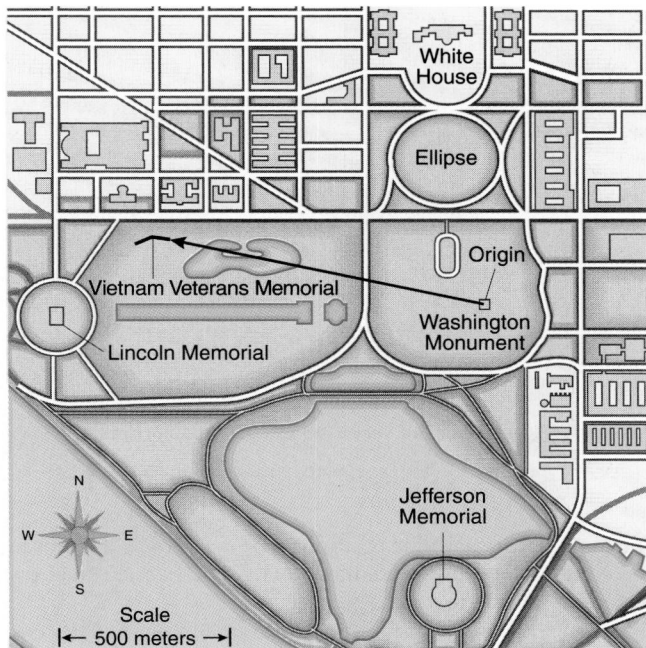

FIGURE 2.3 A position vector.

use such position vectors extensively to describe motion in Chapters 3 and 4.

A vector also can represent the direction and speed you are headed in your neat new car. We define the **velocity** as a vector whose magnitude is called the **speed** and whose direction is the direction of motion of the car. In this case the arrow length is drawn proportional to the positive scalar measure of how fast you are moving (your speed), say 20 m/s or 72 km/h. In Figure 2.4, the arrow is drawn with its tail at the center of the car; we consider the car to be represented by a point at this location. If the car is moving at 20 m/s east, we represent this information first by choosing a scale, say 1.0 cm of arrow length to represent 10 m/s. Therefore, the velocity vector, 20 m/s east, is represented by an arrow 2.0 cm long directed east. Note that the scale used to draw the physical size of the car can be different from that used to draw the velocity vector.

The velocity vector of a car moving twice as fast, 40 m/s, but going northeast, then is represented by an arrow twice as long, directed northeast, as in Figure 2.5.

Likewise, if you playfully push your roommate, an arrow is used to represent the direction of the push, with the arrow length

---

*Some physical concepts are neither scalars nor vectors; they have more involved mathematical descriptions and are called *tensors*. The word comes from the Latin *tensor*, meaning "to stretch," reflecting the use of tensor quantities in theories of the elasticity of solids. Certain concepts that we will treat as scalars, such as the index of refraction of an isotropic transparent material and the moment of inertia of a symmetric rigid body, actually are tensors for nonisotropic materials and asymmetric rigid bodies. We do not consider such materials and objects in this text. Scalar and vector quantities actually are special classes of tensor quantities.

†The convention for drawing maps with north at the top originated in the second century A.D. with the Alexandrian Greek astronomer Ptolemy. Recall from Chapter 1 that Ptolemy was the person who codified the geocentric theory of the solar system.

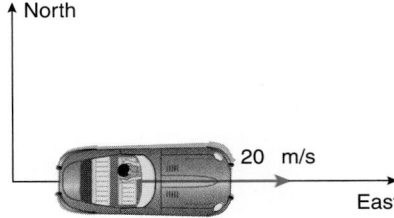

Scale 1.0 cm = 10 m/s for drawing this velocity vector

FIGURE 2.4 The velocity vector 20 m/s directed east.

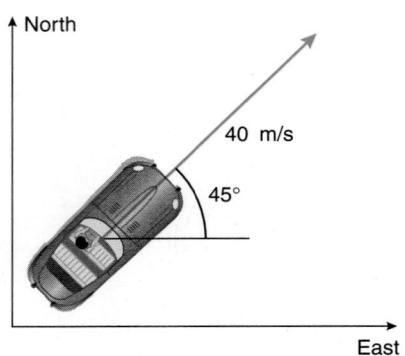

Scale 1.0 cm = 10 m/s
for drawing this velocity vector

FIGURE 2.5 The velocity vector 40 m/s directed northeast.

drawn proportional to the positive scalar magnitude of the push according to some predetermined standard push. Such an arrow represents a **force vector**.

Such visual representation drawings take considerable time and effort. It is convenient, for many purposes, to talk about the arrows representing vectors without actually drawing them to scale. A vector is named and indicated by a letter symbol in boldface with an arrow symbol over it. A position vector is designated as $\vec{r}$, a velocity vector $\vec{v}$, and a general vector $\vec{A}$. The small arrow symbol over the boldface symbolic letter *always* points to the right when symbolically representing a vector; it has no relationship to the actual direction of the vector in space. The phrase "vector $\vec{A}$" is redundant because the arrow over the boldface letter means it is a vector quantity. Nonetheless, we frequently say "vector $\vec{A}$" for emphasis.

The *positive scalar magnitude* of any vector is represented symbolically in two equivalent ways:

1. By the same letter symbol *without the arrow* and in *lightface italic* type. Thus the magnitude of vector $\vec{r}$ is written as r, the magnitude of vector $\vec{v}$ is written as v, and the magnitude of vector $\vec{A}$ is written A.
2. By surrounding the vector (boldface letter with its arrow) with absolute value signs, such as $|\vec{r}|$, $|\vec{v}|$, or $|\vec{A}|$.

Notationally, then, for *any* vector $\vec{A}$,

$$|\vec{A}| \quad \text{or} \quad A \tag{2.1}$$

are equivalent ways of saying the same thing: the magnitude of the vector symbolized by $\vec{A}$.

When $\vec{A}$ is written, we mean the *vector*, encompassing both its magnitude and direction. When A or $|\vec{A}|$ is written, only the *magnitude* of the vector is meant with no information about its direction. *These are important notational and semantic distinctions to remember.*

Vector	Magnitude of a vector
(magnitude *and* direction)	(positive scalar quantity)
$\vec{A}$	$|\vec{A}|$ or A

For example, when we say that the velocity of a car is $\vec{v}$, we mean the velocity vector with its associated magnitude and direction. When we write only v, we mean only the *magnitude* of the velocity vector, a positive scalar quantity known as the speed.

PROBLEM-SOLVING TACTICS

2.1 Scalars are not vectors, and vectors are not scalars. Lurking in your mind as you study should be the question, "Is this physical concept a vector quantity (i.e., with a magnitude and direction) or a scalar quantity?" If you keep this question in mind, you will avoid many of the pitfalls traditionally associated with understanding and applying physics.

2.2 Draw visual representations of vectors carefully. Decide on a scale to represent the vectors; draw them carefully to appropriate lengths with their proper orientations.

2.3 The magnitude of any vector always is a positive scalar quantity (or zero). Remember this!

QUESTION 1

How is the word *vector* used in a biological or medical context? (Consult a dictionary, if necessary.) In what way is its usage in biology and medicine similar to its usage in physics and mathematics? In what way(s) is it different?

2.2 MULTIPLICATION OF A VECTOR BY A SCALAR

If you pull on your car in a certain direction (whether it moves is irrelevant), we represent the pull by a vector $\vec{F}$ whose length is drawn proportional to the magnitude of the pull and directed in the direction of the pull, as in Figure 2.6.*

What happens if three of you, all of equal strength, pull on the car in the same direction? Of course, you get a pull three times what you alone can muster. We represent the total effect by an arrow three times as long as the original arrow of your pull alone, pointing in the same direction—that is, by $3\vec{F}$, as in Figure 2.7. This illustrates the need for the **multiplication of a vector by a scalar**.

---

*In Chapter 5, we will show that such pulls obey the rules of vector arithmetic, and so are vectors. For now, we assume such pulls are legitimate vectors for purposes of illustration.

Visual representation   Symbolic representation

$\vec{F}$

FIGURE 2.6 Vector $\vec{F}$ representing your pull on a car.

Visual representation   Symbolic representation

$$3\vec{F}$$

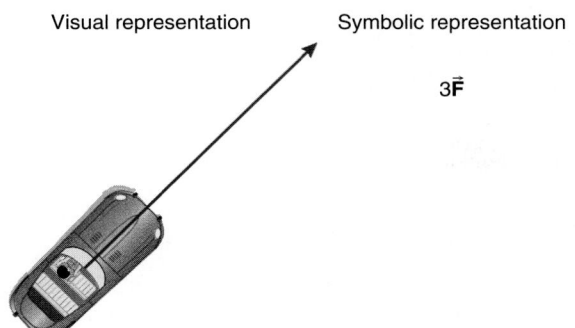

FIGURE 2.7 A vector that is three times your pull.

> If any vector $\vec{A}$, not only a push, is multiplied by a scalar α, the result is a vector $|\alpha|$ times as long as $\vec{A}$ itself; the new vector has a magnitude $|\alpha|$ times that of the original vector.

If the scalar multiplier is *positive*, the direction of the new vector is *unchanged*, as in Figure 2.8. However, if the scalar multiplier is *negative*, the direction of the vector is *reversed*, as shown in Figure 2.9.

> The **negative of any vector** $\vec{A}$, $(-1)\vec{A} \equiv -\vec{A}$, is a vector of the same magnitude as $\vec{A}$ but pointing in the opposite direction, as indicated in Figure 2.10.

Thus if $\vec{F}$ represents a push of some magnitude in a specific direction, $-\vec{F}$ is a push of the same magnitude but in the opposite direction. In Chapter 5 we shall find such push pairs (force pairs) are important in studying dynamics; they form the basis for Newton's third law of motion.

Two vectors that point in the same direction are called **parallel vectors** (see Figure 2.11). Two vectors that point in opposite directions are called **antiparallel vectors** (see Figure 2.12). Thus $-\vec{A}$ is antiparallel to $\vec{A}$, whatever the direction of $\vec{A}$.

The scalar zero times any vector is the **zero vector**. The zero vector has zero magnitude and no direction. We write the zero vector simply as a traditional zero, with no arrow over it, as the zero vector has no conceivable direction or imaginable visual representation:

$$0\vec{A} \equiv 0$$

We call the zero vector simply zero.

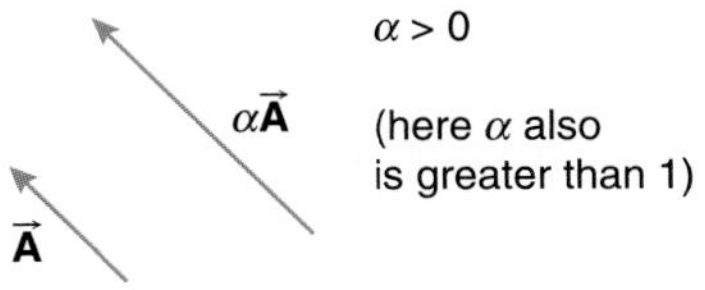

$\alpha > 0$
$\alpha\vec{A}$
(here α also is greater than 1)
$\vec{A}$

FIGURE 2.8 Multiplication of a vector by a *positive* scalar.

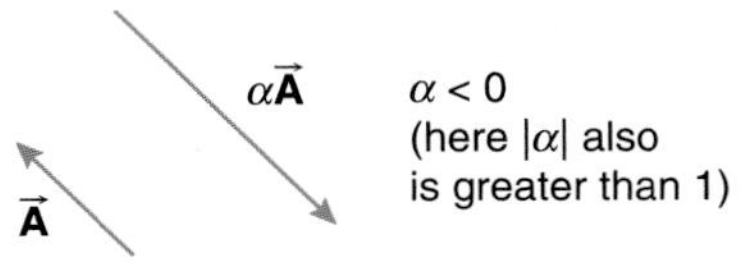

$\alpha\vec{A}$
$\alpha < 0$
(here $|\alpha|$ also is greater than 1)
$\vec{A}$

FIGURE 2.9 Multiplication of a vector by a *negative* scalar.

$-\vec{A}$
$\vec{A}$

FIGURE 2.10 Negative of a vector.

FIGURE 2.11
Parallel vectors.

FIGURE 2.12
Antiparallel vectors.

EXAMPLE 2.1

If $\vec{A}$ is the vector indicated in Figure 2.13, what is the vector $-3.0\vec{A}$?

$\vec{A}$

FIGURE 2.13

Solution

The vector $-3.0\vec{A}$ is a vector with a magnitude 3.0 times that of $\vec{A}$ itself (since magnitudes *always* are positive scalars), but $-3.0\vec{A}$ points in the direction opposite to $\vec{A}$ since the scalar multiplier is negative (see Figure 2.14).

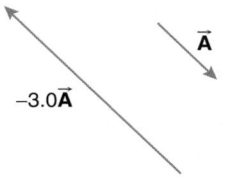

$\vec{A}$

$-3.0\vec{A}$

FIGURE 2.14

2.3 PARALLEL TRANSPORT OF VECTORS

For various mathematical purposes, such as addition or multiplication with other vectors,* you can imagine transporting a vector from one place to another, preserving both its direction in space and its magnitude, as shown in Figure 2.15. This procedure is called **parallel transport** of the vector.

The arrow is merely a visual *representation* of the vector physical quantity. A vector quantity has a meaning independent of the way you describe or draw it and is, in a sense, mathematically footloose.

---

*We consider these topics in the next few sections of the chapter.

FIGURE 2.15 Parallel transport of a vector preserves its length and direction.

2.4 VECTOR ADDITION BY GEOMETRIC METHODS: TAIL-TO-TIP METHOD

If you push on a refrigerator in one direction and a stronger buddy simultaneously pushes on it in another direction, what single push is equivalent to the two pushes at the same time (see Figure 2.16)? We schematically represent the pushes as acting on the center of the refrigerator, with the tails of both vectors located at this point, the same way we represent pulls. In this section we learn how to combine any two (or more) vectors of the same physical type.

Vectors with different magnitudes and directions that represent the same physical concept typically (but not always) are symbolized using the same boldface letter (either upper or lower case), adding subscripts or primes to distinguish them. Thus two different force vectors may be represented by $\vec{F}_1$ and $\vec{F}_2$, by $\vec{F}$ and $\vec{f}$, or by $\vec{F}$ and $\vec{F}'$. Two different velocity vectors may be designated by $\vec{v}_1$ and $\vec{v}_2$, or by $\vec{v}$ and $\vec{v}'$, two different position vectors by $\vec{r}_1$ and $\vec{r}_2$, and so on. You get the idea.

The rules for **vector addition** apply not just to pushes and pulls, but also to any vector quantities of the same physical type. Vectors of the same physical type always are represented in this text visually with arrows of the same color. When added together, two or more vectors representing the same physical concept, such as two or more position vectors or two or more forces, form a new vector called the **vector sum**, which is equivalent to all the others combined. We indicate the operation for forming the vector sum with the boldface addition symbol, **+** , to distinguish it from ordinary scalar addition (+).

Vectors (and scalars for that matter) that represent different physical quantities *cannot* be added together. You cannot add apples to oranges—this rule still applies. Thus do not attempt to add a force vector to a velocity vector yielding some strange interspecies combination. If you do attempt to add the wrong types of vectors together, the dimensions (and units) will be inconsistent, a "red flag" indicating that something is amiss. Such red flags are signals to stop and rethink what has been done; they keep us on the straight, narrow, and legitimate path of vector addition. So, for purposes of vector addition, vectors must be properly of the same physical species—that is, representing the same physical concept.

The simplest case of vector addition is to add a vector to itself: $\vec{F} + \vec{F}$. For example, this could represent two persons of equal strength pushing on a refrigerator in the same direction, as in Figure 2.17. The combined effect of this is found to be in the same direction as each push and twice the magnitude of either (see Figure 2.18). That is,

$$\vec{F} + \vec{F} = 2\vec{F}$$

where $2\vec{F}$ is a new vector, equivalent to the two vectors in the sum, with a magnitude twice that of $\vec{F}$. This special case is equivalent to the multiplication of the vector $\vec{F}$ by the scalar 2.

But suppose the two pushes are not of equal magnitude, nor in the same direction, say $\vec{F}_1$ and $\vec{F}_2$ in Figure 2.19. What is the effect of the pushes now? That is, what is the vector sum $\vec{F}_1 + \vec{F}_2$? The vector sum of the two vectors is defined to be a new vector found using the following geometric procedure, known as the **tail-to-tip method**:

1. Move the vector $\vec{F}_2$ by parallel transport—that is, preserving its magnitude and direction in space—so that its tail is next to the head or tip of $\vec{F}_1$ (i.e., *tail-to-tip*), as in Figure 2.20. This is an example of moving vectors using parallel transport for mathematical purposes.

2. Draw a new vector, $\vec{F}_{total}$, from the *tail* of the first vector $\vec{F}_1$ to the *tip* of the second vector $\vec{F}_2$, thus completing the triangle, as shown in Figure 2.21. The rule for drawing the new vector sum is from the tail of the first vector to the tip of the second vector: again, *tail-to-tip*. The new arrow, $\vec{F}_{total}$, is the vector that represents the vector sum of $\vec{F}_1$ and $\vec{F}_2$. Algebraically, the process of vector addition is written as

$$\vec{F}_{total} = \vec{F}_1 + \vec{F}_2 \qquad (2.2)$$

Note the use of the boldface plus sign (**+**) to represent vector addition to distinguish it from ordinary scalar addition.

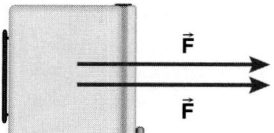

FIGURE 2.17 Two equal vectors.

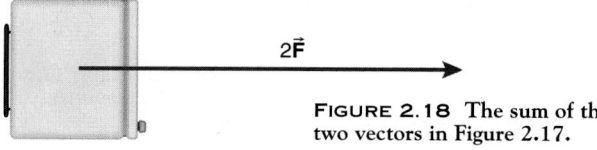

FIGURE 2.18 The sum of the two vectors in Figure 2.17.

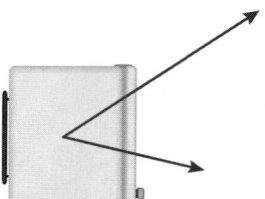

FIGURE 2.16 Two pushes of different magnitude and direction.

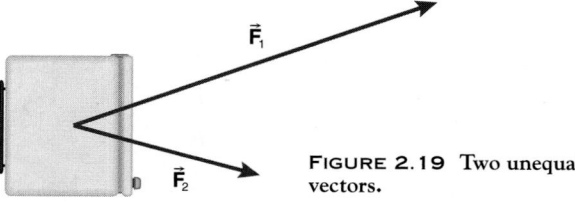

FIGURE 2.19 Two unequal vectors.

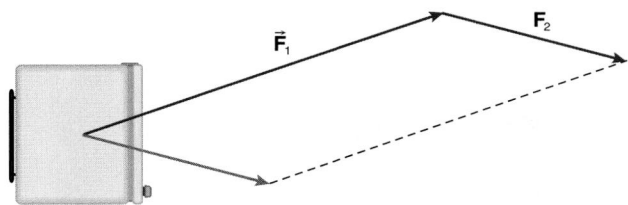

FIGURE 2.20 Parallel transport $\vec{\mathbf{F}}_2$ so its tail is next to the head of $\vec{\mathbf{F}}_1$.

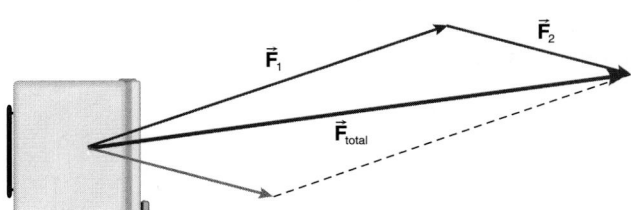

FIGURE 2.21 The vector sum $\vec{\mathbf{F}}_{\text{total}} = \vec{\mathbf{F}}_1 + \vec{\mathbf{F}}_2$.

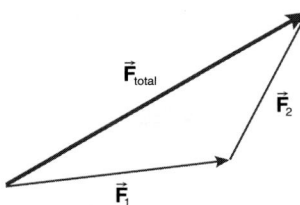

FIGURE 2.22 The magnitude (and direction) of the vector sum depends on the orientations of the vectors to be added.

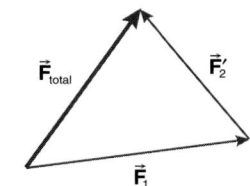

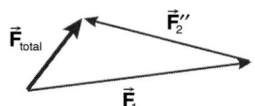

Vector arrow lengths are drawn proportional to the magnitudes of the vector quantities. Notice in the triangle of Figure 2.21 that the *magnitude* of the vector sum is *not* typically the sum of the magnitudes of the individual vectors included in the sum. That is, in most instances

$$F_{\text{total}} \neq F_1 + F_2 \qquad (2.3)$$

where the plus sign here indicates the *scalar addition* of the two scalar magnitudes. The magnitude of the vector sum *is* equal to the sum of the magnitudes *only* when the two vectors to be added are parallel.

The direction of the vector sum also typically is *not* in the direction of either vector in the sum. The magnitude (and direction) of a vector sum depends critically on the relative directions of the two vectors to be added. This is shown explicitly in Figure 2.22 for various orientations of two vectors of fixed magnitudes.

Vector addition is quite different from the addition of scalar quantities.

The order in which the vectors are added is not important. That is, you obtain the *same* vector whether you take $\vec{\mathbf{F}}_1 + \vec{\mathbf{F}}_2$ or $\vec{\mathbf{F}}_2 + \vec{\mathbf{F}}_1$; this is shown in Figure 2.23. In other words, the process of vector addition is commutative:

$$\vec{\mathbf{F}}_1 + \vec{\mathbf{F}}_2 = \vec{\mathbf{F}}_2 + \vec{\mathbf{F}}_1 \qquad (2.4)$$

If more than two vectors of the same type are to be added together, say

$$\vec{\mathbf{F}}_1 + \vec{\mathbf{F}}_2 + \vec{\mathbf{F}}_3 + \vec{\mathbf{F}}_4$$

they can be added sequentially in pairs in *any* order. Thus, the process of vector addition is associative.

PROBLEM-SOLVING TACTIC

2.5   There is a shortcut for the addition of many vectors. You can add several vectors, such as those in Figure 2.24, more easily by placing the tail of $\vec{\mathbf{F}}_2$ next to the tip of $\vec{\mathbf{F}}_1$ using parallel transport (see Figure 2.25), the tail of $\vec{\mathbf{F}}_3$ next to the tip of $\vec{\mathbf{F}}_2$, and the tail of $\vec{\mathbf{F}}_4$ next to the tip of $\vec{\mathbf{F}}_3$, and so forth (as in Figure 2.25). The vector sum of all the vectors is represented by an arrow from the tail of the first vector ($\vec{\mathbf{F}}_1$) to the tip of the last vector ($\vec{\mathbf{F}}_4$), as shown in Figure 2.26.

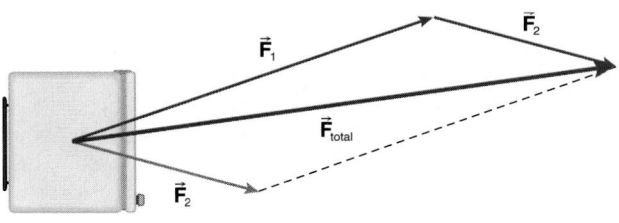

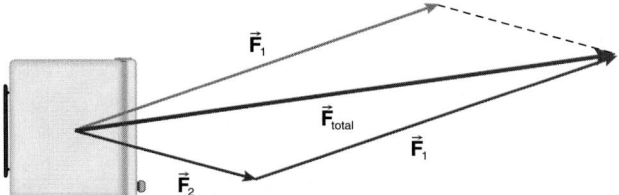

FIGURE 2.23 The order in which you add two vectors does not matter.

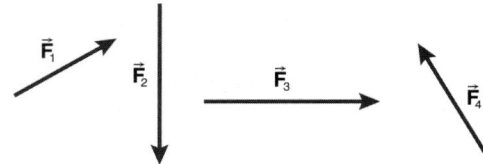

FIGURE 2.24 A collection of vectors to add.

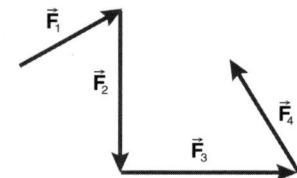

FIGURE 2.25 Parallel transport the vectors tail-to-tip.

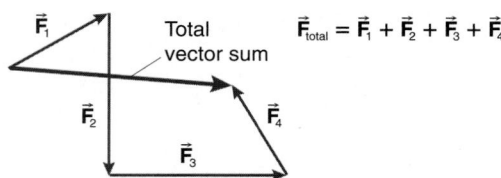

FIGURE 2.26 The vector sum $\vec{F}_{total}$ is drawn from the tail of the first vector to tip of the last.

QUESTION 2

Show how a vector of magnitude 1 can be added to another vector of magnitude 1 to equal a vector of magnitude 1. What geometrical figure do the three vector arrows form? You have learned how one plus one is one (for vectors)!

2.5 DETERMINING WHETHER A QUANTITY IS A VECTOR*

If you love pizzas, you likely think all hot pizzas are tasty; however, that does not mean all tasty things are hot pizzas. Likewise, although all vectors have both magnitude and direction, not all things with a magnitude and direction are vectors.

Simple arrows of any size can indicate direction without implying magnitude, as in Figure 2.27. If an arrow is used to represent a quantity that has a magnitude and an associated direction, the distinction between an arrow that is a vector and one that is not becomes increasingly subtle. Just because some physical concepts *can be* represented by an arrow does *not* ensure that the quantity is a vector.

For example, think of rotating an object about a specified axis, as indicated in Figure 2.28. The angle through which the

FIGURE 2.27 An arrow simply indicating direction is not a vector.

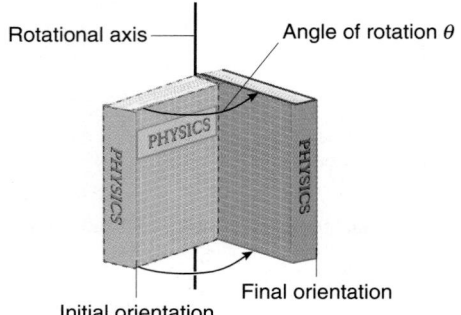

FIGURE 2.28 A rotation of an object about an axis.

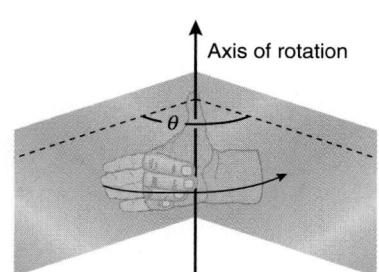

FIGURE 2.29 An arrow representing a rotation.

object is rotated is a measure of the magnitude of the rotation: the larger the angle, the larger the magnitude of the rotation. We cannot use the curved arrow in Figure 2.28 to represent the rotation, since vectors are represented visually only by *straight* arrows. A straight arrow will suffice if we direct it along the axis of rotation in the sense of the following *rotational right-hand rule*: Wrap the bunched fingers of your right hand around the axis in the sense the object is rotated about the axis, as shown in Figure 2.29. Then the extended right-hand thumb indicates the direction for the arrow representing the rotation.

We represent the rotation visually by a straight arrow along the axis of rotation whose length is proportional to the angle of rotation and whose direction is determined by the rotational right-hand rule. But even so, such rotation arrows are *not vectors*. Why? We will illustrate using a simple test.

Orient your textbook on a flat table with the binding facing you as in Figure 2.30. The Cartesian coordinate system in Figure 2.30 will assist you in visualizing and performing two rotations in succession:

1. Rotation 1: Rotate the book through an angle of 90° about the y-axis, as shown in Figure 2.31. The back cover of the text now faces you. Call this rotation Rot_1. The arrow representing this rotation is schematically indicated in Figure 2.31; the length of the arrow is proportional to 90°, and the direction of the arrow is toward increasing values of y according to the rotational right-hand rule.

2. Rotation 2: Rotate the book through 90° about the x-axis, as shown in Figure 2.32. The book now lies face up on the table. Call this rotation Rot_2. This rotation is represented by

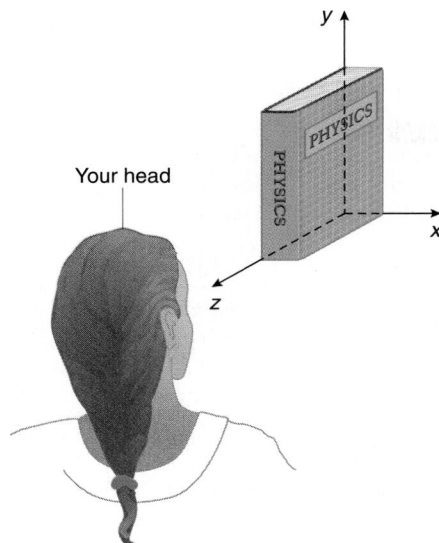

FIGURE 2.30 Your textbook with its binding facing you.

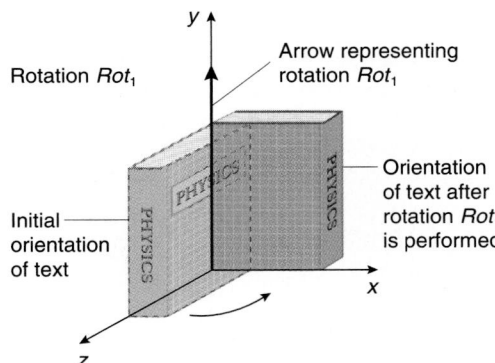

FIGURE 2.31 Rotate the book 90° about the *y*-axis.

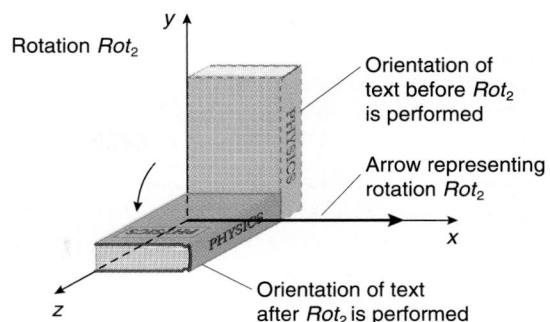

FIGURE 2.32 Rotate the book 90° about the *x*-axis.

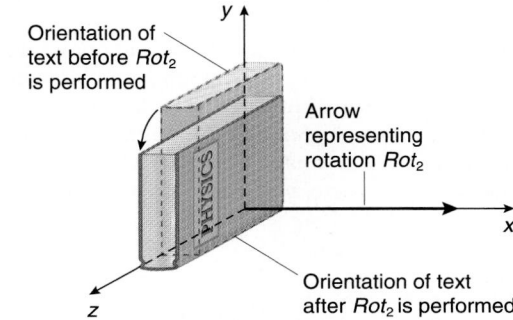

FIGURE 2.33 First rotate the book 90° about the *x*-axis.

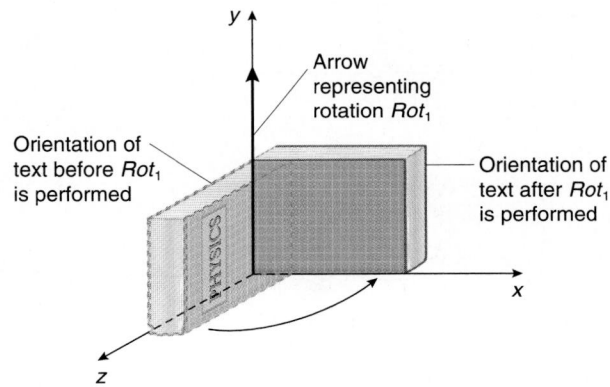

FIGURE 2.34 Next rotate the book 90° about the *y*-axis.

an arrow the same length as Rot_1 (since the two rotations each were of magnitude 90°) but with the Rot_2 arrow directed toward increasing values of *x* according to the rotational right-hand rule.

Thus the combination of rotation Rot_1 followed by rotation Rot_2, symbolized by $Rot_1 + Rot_2$, results in the book lying face up on the table as indicated in Figure 2.32. We still must determine whether the + sign here indicates what we mean by the vector addition, **+**.

We know that adding two *vectors* together in either order does not affect the outcome. Is the same true for these rotations? Let's see. Begin again with the text in the original orientation of Figure 2.30, but *change the order* in which the rotations are performed. That is, do rotation Rot_2 first, followed by Rot_1, symbolized by $Rot_2 + Rot_1$.

After rotation Rot_2 (90° about the *x*-axis), the text is oriented as shown in Figure 2.33. Then after the next rotation, Rot_1 (90° about the *y*-axis), the text is oriented as shown in Figure 2.34. The result of this new sequence of the same two rotations orients the textbook on its edge with the front cover

facing away from you. This is *different* from what we obtained with the same rotations performed in the original order!

We could blindly add the rotation arrows using the geometric vector addition rules, but *the result the arrows physically represent is not the same* when we take $Rot_1 + Rot_2$ instead of $Rot_2 + Rot_1$. In other words,

$$Rot_1 + Rot_2 \neq Rot_2 + Rot_1$$

The *order* or sequence of the rotations is critical to the result obtained; the rotations do not commute. The arrows of finite rotation thus do *not* obey the commutative vector addition rule (Equation 2.4). Hence these rotations are *not vector quantities* even though they can be represented visually by an arrow with a magnitude and a direction.

So, in conclusion, just because some physical concept *can* be represented visually by an arrow with a magnitude and a direction does *not* ensure that the physical concept is a vector quantity.

> To be a vector quantity, that quantity must also *obey the rules* we established for, among other things, vector arithmetic.

2.6 VECTOR DIFFERENCE BY GEOMETRIC METHODS

Finding the difference between two vectors is useful in many physical contexts. For example, if you are at the Vietnam Veterans Memorial in Washington, D.C., and are located by a position vector $\vec{r}_1$ with respect to the Washington Monument (as in Figure 2.35) and the White House is located by position vector $\vec{r}_2$ from the same origin, then what is the position vector $\vec{r}_3$ of the White House measured from your location? The three vectors are related by vector addition in the following way:

$$\vec{r}_1 + \vec{r}_3 = \vec{r}_2 \qquad (2.5)$$

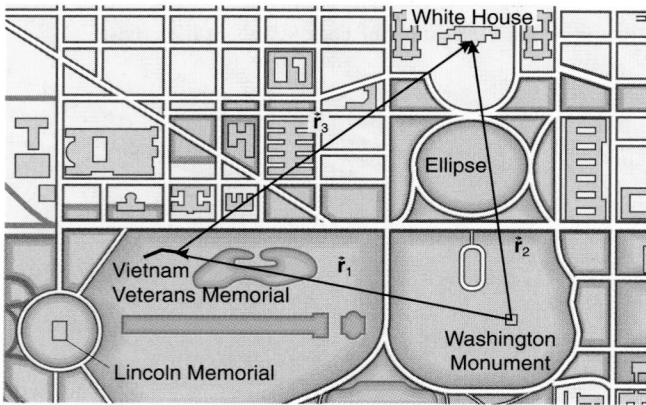

FIGURE 2.35 Three position vectors related by a vector sum.

Now transpose $\vec{r}_1$ in Equation 2.5 just as we would if this were ordinary algebra:

$$\vec{r}_3 = \vec{r}_2 - \vec{r}_1 \qquad (2.6)$$

We have to determine what we mean by the boldface minus sign ($-$), symbolizing a **vector difference**. We do this by noting in Figure 2.36 that both the magnitude and direction of vector $\vec{r}_3$ can be found by forming the *vector sum* of $\vec{r}_2$ with $-\vec{r}_1$:

$$\vec{r}_3 = \vec{r}_2 + (-\vec{r}_1) \qquad (2.7)$$

This means that the vector difference is executed geometrically by *adding* the vector $-\vec{r}_1$ to $\vec{r}_2$ using the *vector addition rules* in Section 2.4. That is,

$$\vec{r}_3 = \vec{r}_2 - \vec{r}_1$$
$$\equiv \vec{r}_2 + (-\vec{r}_1)$$

> The operation $+ (-\vec{r}_1)$ is identical in all respects to the operation implied by $- \vec{r}_1$.

We generalize to say that the process of finding the vector difference between *any* two vectors, $\vec{A}_2$ and $\vec{A}_1$ (see Figure 2.37), can be viewed in two steps.

1. To form $\vec{A} \equiv \vec{A}_2 - \vec{A}_1$, first form the vector $-\vec{A}_1$ by reversing the direction of the arrow of $\vec{A}_1$, preserving its length (magnitude) (see Step 1 in Figure 2.38).
2. Next *add* the vector $-\vec{A}_1$ to the vector $\vec{A}_2$ using the tail-to-tip method of vector addition, as in Step 2 in Figure 2.38.

The magnitude of the vector difference $\vec{A}$ is *not* typically the difference between the magnitudes of the two vectors $\vec{A}_2$ and $\vec{A}_1$. That is, in most instances

$$A \neq A_2 - A_1 \qquad (2.8)$$

Note that Equation 2.8 involves the *magnitudes* of the vectors, and the minus sign is ordinary scalar subtraction. The magnitude of the vector difference is equal to the absolute value of the difference between the individual magnitudes *only* when the two vectors to be subtracted are parallel or antiparallel.

The direction of the vector difference $\vec{A}$ also is *not* typically in the direction of either $\vec{A}_2$ or $\vec{A}_1$. The vector dif-

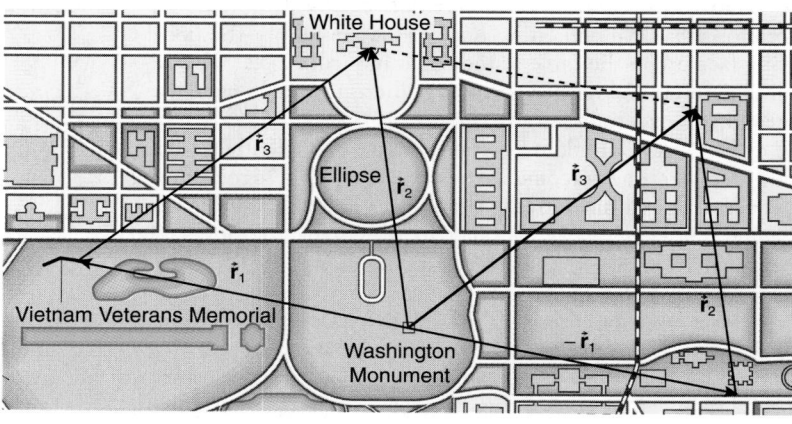

FIGURE 2.36 Another way to view the three vectors of Figure 2.35.

FIGURE 2.37 Two vectors.

Step 1

Step 2

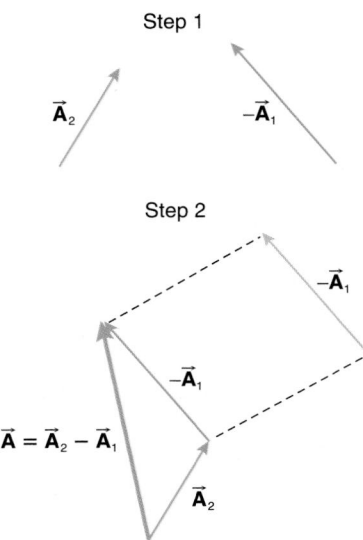

FIGURE 2.38 Reverse the direction of the vector to be subtracted and then add it to the other vector to form their difference.

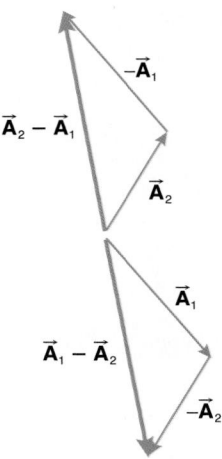

FIGURE 2.39 The vector $\vec{A}_2 - \vec{A}_1$ is the negative of the vector $\vec{A}_1 - \vec{A}_2$.

ference is in the direction of $\vec{A}_2$ only when the two vectors $\vec{A}_2$ and $\vec{A}_1$ are parallel *and* the magnitude of $\vec{A}_2$ is greater than the magnitude of $\vec{A}_1$; if they are parallel and $\vec{A}_1$ has the greater magnitude, then the difference points in the direction of $\vec{A}_1$.

The vector difference is different from the subtraction of scalar quantities.

Just as in scalar subtraction, the order of the terms in the vector difference affects the result (see Figure 2.39). That is,

$$\vec{A}_2 - \vec{A}_1 \neq \vec{A}_1 - \vec{A}_2$$

Notice from Figure 2.39 that $\vec{A}_2 - \vec{A}_1$ is the negative of $\vec{A}_1 - \vec{A}_2$:

$$\vec{A}_2 - \vec{A}_1 = -(\vec{A}_1 - \vec{A}_2) \tag{2.9}$$

If we subtract any vector from itself, we obtain the zero vector:

$$\vec{A}_1 - \vec{A}_1 = 0$$

PROBLEM-SOLVING TACTIC

2.6 The addition and subtraction of vector quantities is *different* from the addition and subtraction of scalars. This fact is *crucial* to remember!

QUESTION 3 _____

Show how a vector of magnitude 1 can be subtracted from another vector of magnitude 1 to equal a vector of magnitude 1.

2.7 THE SCALAR PRODUCT OF TWO VECTORS

Only vectors of the same physical type can be added to or subtracted from each other. But vectors of *different* types, representing different physical quantities, *can* be multiplied together, as can vectors of the same physical type. There are two distinct ways that two vectors are multiplied together; each way has its own rules, terminology, notation, and usefulness in physics. In this section we examine one of those ways, leaving the other for a subsequent section of the chapter.

We saw in Section 2.2 that a vector multiplied by a scalar α produces a vector with a magnitude $|\alpha|$ times the original vector. Multiplying a vector by a scalar produces another *vector*. Multiplying a vector by a scalar is *different* from what is meant by the **scalar product** in vector analysis. The scalar product is a way to multiply *two vectors* to yield a *scalar* result. The scalar product is useful for describing circular motion (as we will see in Chapter 4) and in many other physical contexts, but perhaps most importantly, in defining the physical concept of the work done by a force (as we shall see in Chapter 8).

The scalar product of any two vectors $\vec{A}$ and $\vec{B}$ is written as $\vec{A} \cdot \vec{B}$, with the product symbolized by the big fat dot $\cdot$. The scalar product also is commonly referred to as the *dot product* of two vectors, indicative of the symbol used to represent it. In this text we shall consistently use the term scalar product for this type of vector multiplication.

The scalar product of two vectors $\vec{A}$ and $\vec{B}$ is defined as

$$\boxed{\vec{A} \cdot \vec{B} \equiv AB \cos\theta} \qquad (2.10)$$

where A is the magnitude of $\vec{A}$, B is the magnitude of $\vec{B}$, and θ is the angle between the two vectors when they are drawn from a common point (i.e., with their tails together at the common point), as in Figure 2.40.

Parallel transport occasionally must be used (as illustrated in Figure 2.40) so that the two vectors have their tails together.

The angle between two vectors *always* is taken to be the *smaller* angle between the vectors when they are drawn from a common point (angle θ in Figure 2.40, not angle ϕ). With this convention, θ always is less than or equal to 180°.

The angle between two vectors always is considered a positive quantity and is always less than or equal to 180°.

The magnitudes of the two vectors on the right-hand side of Equation 2.10 are never negative. Since the *smaller* angle between the two vectors always will be somewhere between 0° and 180°, the cosine is either positive (for angles up to 90°) or negative (for angles greater than 90° but less than 180°).

Thus the scalar product can be either a positive or negative scalar quantity or zero, depending on the angle between the two vectors (see Figure 2.41).

PROBLEM-SOLVING TACTICS

2.7 The result of the scalar product of two vectors is a scalar quantity, not another vector quantity. Scalars and vectors are very different; do not confuse the two.

2.8 To see if two vectors are perpendicular to each other, take their scalar product. If the two vectors are perpendicular to each other, the angle between them is 90°. *Their scalar product is zero*, since cos 90° = 0. This is a wonderful test to see if two (nonzero) vectors are perpendicular.

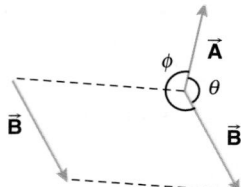

FIGURE 2.40 Parallel transport may be needed to determine the angle θ between two vectors.

The scalar product has a distributive property. That is,

$$\vec{A} \cdot (\vec{B}_1 + \vec{B}_2) = \vec{A} \cdot \vec{B}_1 + \vec{A} \cdot \vec{B}_2 \qquad (2.11)$$

This fact is comforting, for we can use our common algebraic knowledge to eliminate the parentheses. Notice, however, that the (+) sign on the *left*-hand side of Equation 2.11 represents the *vector addition* of $\vec{B}_1$ and $\vec{B}_2$, whereas the (+) sign on the *right*-hand side of this equation represents the addition of two *scalars*, $\vec{A} \cdot \vec{B}_1$ and $\vec{A} \cdot \vec{B}_2$.

There is a geometric way to interpret the scalar product. With $\vec{A}$ and $\vec{B}$ drawn with their tails together, if we drop a perpendicular from the tip of $\vec{B}$ to the line containing $\vec{A}$ (see Figure 2.42), the quantity $B \cos\theta$ is defined to be the **projection** or **component** of $\vec{B}$ on the line containing $\vec{A}$ (or on any line parallel to the line containing $\vec{A}$). Figure 2.42 illustrates why the word projection is used in this context. If we imagine sunlight or light from a distant slide projector shining from a direction perpendicular to $\vec{A}$, then the shadow of the vector $\vec{B}$ on the line containing $\vec{A}$ has length equal to the absolute value of the projection (or component) of $\vec{B}$ on the line containing $\vec{A}$.

If the angle θ between the vectors is acute ($\theta < 90°$), the projection is positive; if the angle is obtuse ($\theta > 90°$), the projection is negative since cos θ then is less than zero. The scalar product of the two vectors is the simple scalar multiplication of this projection ($B \cos\theta$) by the magnitude of $\vec{A}$:

$$\vec{A} \cdot \vec{B} = A(B \cos\theta)$$

We could also do the projection the other way around. Drop a perpendicular from the tip of $\vec{A}$ to the line containing $\vec{B}$, as in Figure 2.43. Then the quantity $A \cos\theta$ is the projection or component of $\vec{A}$ on the line containing $\vec{B}$ (or any line parallel to it). The scalar product then is the scalar multiplication of this projection ($A \cos\theta$) by the magnitude of $\vec{B}$:

$$\vec{A} \cdot \vec{B} = (A \cos\theta)B$$

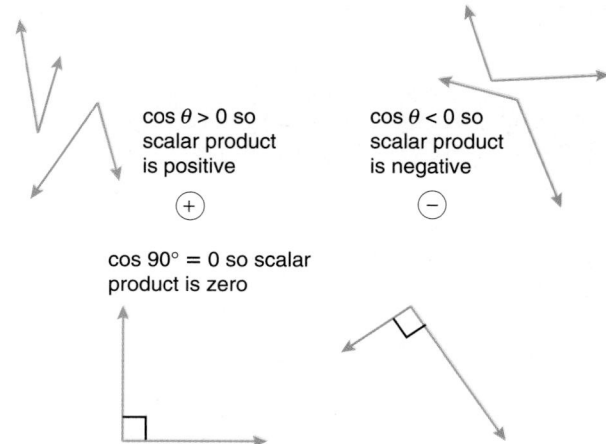

FIGURE 2.41 The scalar product can be positive, negative, or zero depending on the angle between the two vectors. Two vectors that are perpendicular to each other have a scalar product equal to zero.

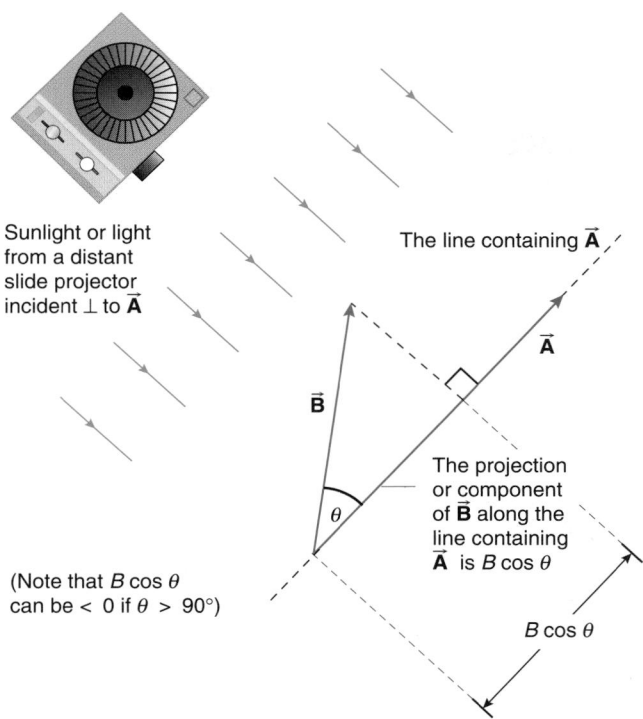

FIGURE 2.42 The projection or component of $\vec{B}$ on the line containing $\vec{A}$ is $B\cos\theta$.

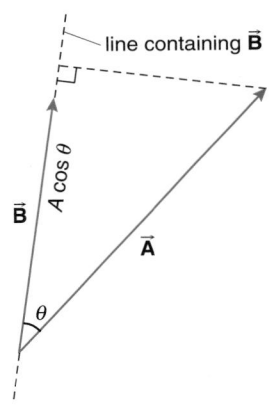

FIGURE 2.43 The projection or component of $\vec{A}$ on the line containing $\vec{B}$ (or any line parallel to it) is $A\cos\theta$.

QUESTION 4

Suppose you have a vector $\vec{A} \neq 0$. If $\vec{A} \cdot \vec{B} = 0$, does this necessarily mean that $\vec{B} = 0$? Explain.

EXAMPLE 2.2

Find the scalar product of the two vectors shown in Figure 2.44.

Solution

Parallel transport one of the vectors until their two tails are at a common point, as in Figure 2.45. The angle between the two vectors is 50.0°. The scalar product of the two vectors is found using Equation 2.10:

$$\vec{A} \cdot \vec{B} = (2.00)(5.00)\cos 50.0°$$

$$= 6.43$$

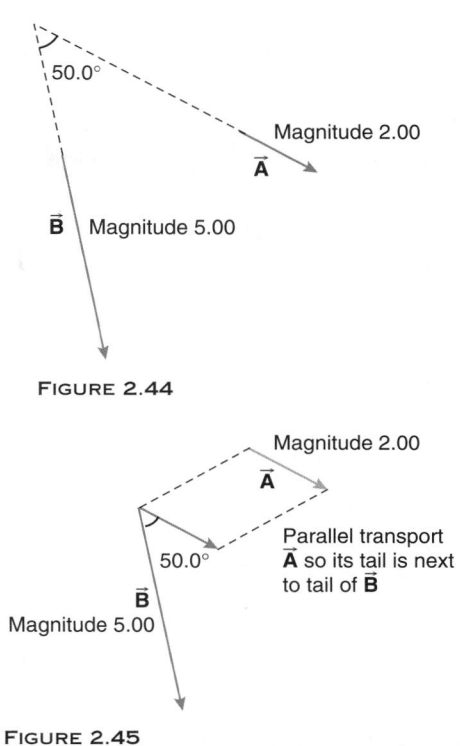

FIGURE 2.44

FIGURE 2.45

2.8 THE CARTESIAN COORDINATE SYSTEM AND THE CARTESIAN UNIT VECTORS

The magnitude of a vector is a positive scalar quantity. The direction of the vector is indicated by the direction of the arrow in its visual representation. One way that you can *specify* the direction of the vector is to choose a coordinate system.

In many situations in physics, we represent the three spatial dimensions in which we live with a **right-handed Cartesian coordinate system**. The three coordinate axes, known as the x-, y-, and z-axes, are perpendicular (i.e., orthogonal) to each other. To draw the three-dimensional coordinate system on a flat surface, at least one axis must be drawn skewed (in perspective) or at an angle, as in Figure 2.46. The axes are labeled so the coordinate values increase as you go from the origin to the x, y, or z label.

Positive values of x form what is called the $+x$-axis; negative values of x form the $-x$-axis. Similar conventions exist for the y- and z-axes. The Cartesian coordinate system shown in Figure 2.46 is called *right-handed* for the following reason: if you orient the index finger of your right hand along the direction that x increases in value and the middle finger along the direction y increases, then your extended thumb will indicate the direction that z increases. Be sure to use your right hand for this test; if you try this procedure in Figure 2.46 with your left hand, you will find that your left thumb does *not* point along the direction z increases; it will indicate the direction of decreasing values of z.

Equivalently, if in Figure 2.47 you bunch the four fingers of your right hand and point them along the direction that x increases in value, then turn them toward the direction y increases in value, then your extended right thumb will point in the direction that z increases in value.

We will always use right-handed coordinate systems.*

PROBLEM-SOLVING TACTIC

2.9 Always use right-handed coordinate systems. To use anything else is making life difficult for yourself and for others who must look over your work. For two-dimensional situations, it is *not* critical to check whether the coordinate system you choose is right-handed since the third, but irrelevant, coordinate dimension can be directed according to whatever choice you make for, say, the *x*- and *y*-axes. However, when introducing a three-dimensional Cartesian coordinate system, *always* check to be sure the coordinate system you introduce is right-handed.

Nothing is sacred about the specific spatial directions chosen for the *x*-, *y*-, and *z*-axes *as long as they form a right-handed system*. You need not always choose +*x* going to the right, +*y* going up, and +*z* coming toward you out of the page. A few legitimate and illegitimate coordinate choices are shown in Figure 2.48; you can test each with your right hand to verify that it is legitimate or illegitimate.

We want to be able to express any vector in terms of the directions you choose for the Cartesian coordinate axes. For this purpose, define three fundamental Cartesian vectors $\hat{\imath}$, $\hat{\jmath}$, and $\hat{k}$ as **basis vectors**, since we will see that any vector can be expressed in terms of them. The Cartesian basis vectors $\hat{\imath}$, $\hat{\jmath}$, and $\hat{k}$ also are called **Cartesian unit vectors** because each has a magnitude equal to the pure number 1.[†] The direction of each unit vector is as follows (see Figure 2.49).

> $\hat{\imath}$ indicates the direction that coordinate *x* increases in value. For $\hat{\imath}$, say "i-hat."
> $\hat{\jmath}$ indicates the direction that coordinate *y* increases in value. For $\hat{\jmath}$, say "j-hat."
> $\hat{k}$ indicates the direction that coordinate *z* increases in value. For $\hat{k}$, say "k-hat."

Notice in Figure 2.49 that parallel transport applies to the unit vectors; you need not consider their tails always to be at the origin. The vectors $-\hat{\imath}$, $-\hat{\jmath}$, and $-\hat{k}$ indicate the directions in which *x*, *y*, and *z*, respectively, decrease in value.

Unit vectors carry *no* dimensions and *no* units along with them; they are *dimensionless and unitless* vectors that serve merely to indicate directions. The unit vectors are designated differently from *physical* vectors such as position vectors and velocity

---

*If this seems like a bias in favor of right-handed persons, it is in a sense. However, left-handed persons have a distinct advantage: they can use their right hands to apply the right-hand rules without first setting down a pencil!

[†]In some fields of physics, notably solid state physics and crystallography, basis vectors are used that are neither unit vectors nor perpendicular (orthogonal) vectors. In such fields, it is more convenient to have basis vectors related to the crystal symmetry (see Problem 62).

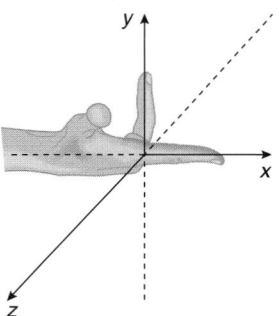

FIGURE 2.46 A right-handed Cartesian coordinate system.

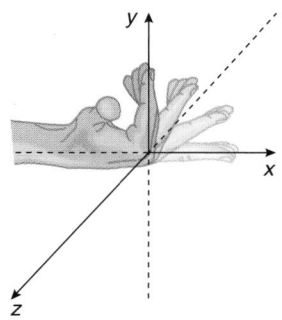

FIGURE 2.47 Another way to verify the right-handedness of a coordinate system.

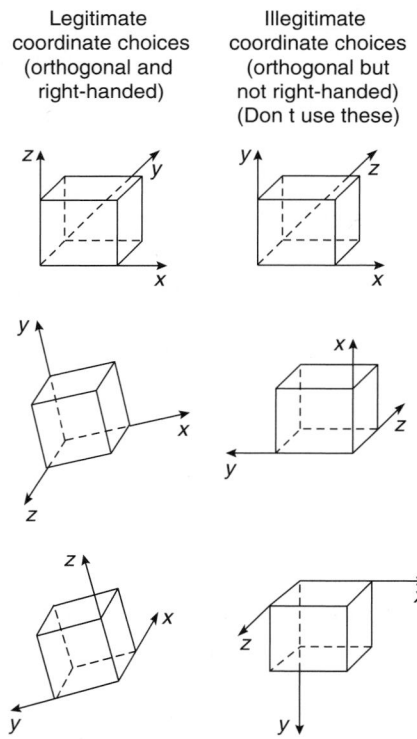

Legitimate coordinate choices (orthogonal and right-handed)	Illegitimate coordinate choices (orthogonal but not right-handed) (Don t use these)

FIGURE 2.48 Legitimate and illegitimate coordinate system choices.

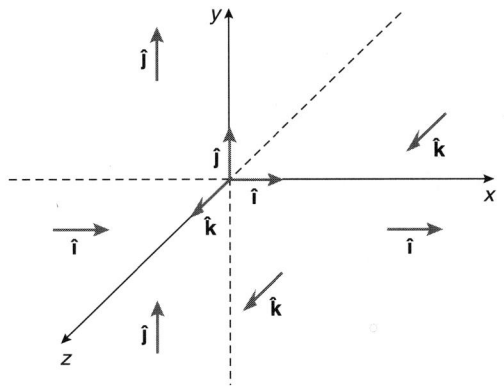

FIGURE 2.49 The Cartesian unit vectors $\hat{\imath}$, $\hat{\jmath}$, and $\hat{k}$.

vectors, which *do* have dimensions and units associated with them. Unit vectors do not have an arrow over them but rather a *hat* or circumflex.

Consider the scalar product of a Cartesian unit vector with itself. For example, the angle between $\hat{\imath}$ and $\hat{\imath}$ is 0°. Thus the scalar product of $\hat{\imath}$ with itself is, by definition (Equation 2.10), the product of the magnitudes of the two vectors times the cosine of the angle between them. Here the magnitude of each vector is 1 and the angle between them is 0°, so that

$$\hat{\imath} \cdot \hat{\imath} = (1)(1) \cos 0°$$
$$= 1$$

Now consider the scalar product of a Cartesian unit vector with a *different* Cartesian unit vector. For example, consider the scalar product of $\hat{\imath}$ with $\hat{\jmath}$, $\hat{\imath} \cdot \hat{\jmath}$. The angle between $\hat{\imath}$ and $\hat{\jmath}$ is 90°, since they are perpendicular. Thus their scalar product is

$$\hat{\imath} \cdot \hat{\jmath} = (1)(1) \cos 90°$$
$$= 0$$

All possible scalar products of the Cartesian unit vectors with each other are summarized in Table 2.1.

QUESTION 5

The Cartesian basis vectors $\hat{\imath}$, $\hat{\jmath}$, and $\hat{k}$ are called *unit vectors*, but they have no units associated with them. Explain what is meant by this apparent oxymoron.

2.9 THE CARTESIAN REPRESENTATION OF ANY VECTOR

Since the Cartesian coordinate system is used extensively in physics, we must know how to express any vector using the

Cartesian unit vectors $\hat{\imath}$, $\hat{\jmath}$, and $\hat{k}$ as a basis. In other words, we want to determine how to represent any vector $\vec{A}$ as an appropriate vector sum of the Cartesian unit vectors.

The Case of Two Dimensions

Consider a vector $\vec{A}$ confined to the *x–y* plane of a Cartesian coordinate system. If necessary, parallel transport the vector so its tail is at the origin, as in Figure 2.50. Take the scalar product of $\vec{A}$ with $\hat{\imath}$ for the fun of it; use Equation 2.10:

$$\vec{A} \cdot \hat{\imath} = (A)(1) \cos \theta$$
$$= A \cos \theta$$
$$\equiv A_x \qquad (2.12)$$

According to the geometric interpretation of the scalar product, $A \cos \theta$ ($\equiv A_x$) is the projection or **component** of $\vec{A}$ on the line containing the unit vector $\hat{\imath}$ (or any line parallel to it); see Figure 2.51. The specific projection or component A_x is known as the Cartesian *x-component* of $\vec{A}$.

Note that A_x is a *scalar* quantity. Directional sense is indicated by $\hat{\imath}$. Thus $A_x \hat{\imath}$ is the multiplication of the unit vector $\hat{\imath}$ by the scalar A_x. Likewise,

$$\vec{A} \cdot \hat{\jmath} = (A)(1) \cos(90° - \theta)$$
$$= A(\cos 90° \cos \theta + \sin 90° \sin \theta)$$
$$= A \sin \theta$$
$$\equiv A_y \qquad (2.13)$$

You can get from the first line of Equation 2.13 to the last line without using the double angle trigonometry formula by noting that in a right triangle, the cosine of one of the acute angles is equal to the sine of the other acute angle, as shown in Figure 2.52.

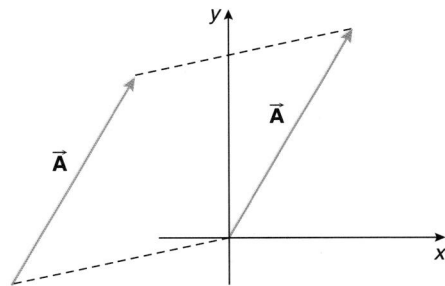

FIGURE 2.50 A vector in two dimensions, the *x–y* plane.

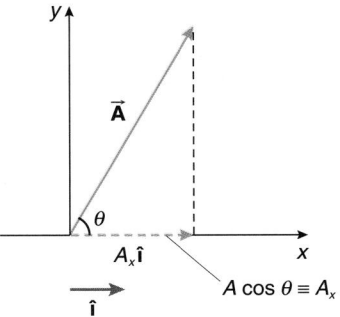

FIGURE 2.51 The *x*-component of $\vec{A}$.

TABLE 2.1 Scalar Product of the Cartesian Unit Vectors with Each Other

$\hat{\imath} \cdot \hat{\imath} = 1$	$\hat{\imath} \cdot \hat{\jmath} = 0$	$\hat{\imath} \cdot \hat{k} = 0$
$\hat{\jmath} \cdot \hat{\imath} = 0$	$\hat{\jmath} \cdot \hat{\jmath} = 1$	$\hat{\jmath} \cdot \hat{k} = 0$
$\hat{k} \cdot \hat{\imath} = 0$	$\hat{k} \cdot \hat{\jmath} = 0$	$\hat{k} \cdot \hat{k} = 1$

The scalar quantity A_y is the projection or component of $\vec{A}$ on the line containing $\hat{\jmath}$ (or any line parallel to it) and is called the Cartesian **y-component** of $\vec{A}$; see Figure 2.53.

The scalars A_x and A_y are known collectively as the **Cartesian components** of the vector $\vec{A}$ in two dimensions. Notice that since the scalar product is a scalar that is either positive, negative, or zero, *the components are scalar quantities that can be positive, negative, or zero.*

> The units and dimensions associated with the Cartesian components of $\vec{A}$ are those associated with vector $\vec{A}$ itself; the units are, say, meters only if $\vec{A}$ represents a directed distance between two points in space.

The *vector* $A_x\hat{\imath}$ is the scalar multiplication of the unit vector $\hat{\imath}$ by the scalar A_x; likewise the *vector* $A_y\hat{\jmath}$ is the scalar multiplication of the unit vector $\hat{\jmath}$ by the scalar A_y. In two dimensions, the vector $\vec{A}$ is written in terms of its components as the *vector*

sum of the mutually perpendicular vectors $A_x\hat{\imath}$ and $A_y\hat{\jmath}$, as shown in Figure 2.54. An alternative representation of this relationship is shown in Figure 2.55, with all three vectors with their tails at the origin.

PROBLEM-SOLVING TACTIC

2.10 Do not confuse a vector with its Cartesian components. The components are *scalar* quantities. The multiplication of the unit vectors by the Cartesian components leads to a vector sum that represents the vector.

$$\vec{A} = \underbrace{A_x}_{\text{components}}\hat{\imath} + \underbrace{A_y}_{}\hat{\jmath} \qquad (2.14)$$

vectors

Note several things from Figure 2.54 or 2.55:

a. The Pythagorean theorem means that we have

$$A^2 = A_x^2 + A_y^2 \qquad (2.15)$$

The magnitude of $\vec{A}$ thus is

$$A = (A_x^2 + A_y^2)^{1/2} \qquad \text{(use positive root only)} \qquad (2.16)$$

By convention, since the magnitude of a vector is never negative, *always* take the positive square root when using Equation 2.16.

b. Also note that the tangent of the angle θ that the two-dimensional vector makes with the direction of $\hat{\imath}$ is the ratio of the y- to the x-components:

$$\tan\theta = \frac{\text{opposite side}}{\text{adjacent side}}$$

$$= \frac{A_y}{A_x}$$

so

$$\theta = \tan^{-1}\left(\frac{A_y}{A_x}\right) \qquad (2.17)$$

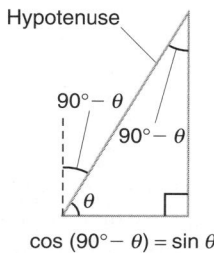

$$\cos(90° - \theta) = \sin\theta$$

FIGURE 2.52 In any right triangle, the cosine of one acute angle is equal to the sine of the other acute angle.

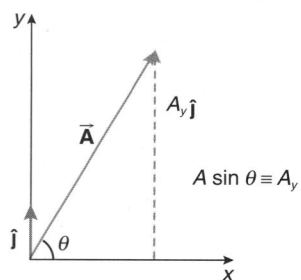

FIGURE 2.53 The y-component of $\vec{A}$.

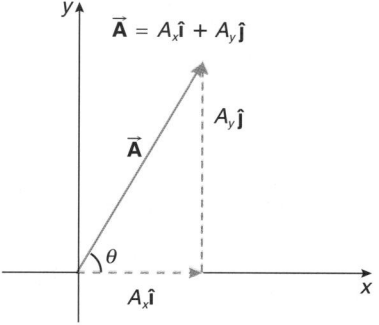

FIGURE 2.54 The two-dimensional vector $\vec{A}$ in Cartesian form.

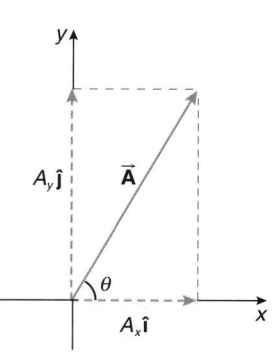

FIGURE 2.55 An alternative representation of $\vec{A}$.

2.11 The mathematical sign convention for angles on a two-dimensional Cartesian coordinate grid is used with Equation 2.17. If the angle θ is *positive*, the angle is measured *counterclockwise* from the positive *x*-axis. If the angle is negative, the angle is measured clockwise from the positive *x*-axis. Remember, however, that the angle between two vectors always is considered to be positive.

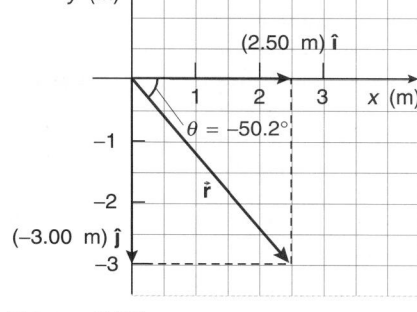

FIGURE 2.56

EXAMPLE 2.3

Given the two-dimensional position vector

$$\vec{r} = (2.50 \text{ m}) \, \hat{\imath} - (3.00 \text{ m}) \, \hat{\jmath}$$

a. Sketch the vectors $(2.50 \text{ m}) \, \hat{\imath}$, $(-3.00 \text{ m}) \, \hat{\jmath}$, and $\vec{r}$ on a two-dimensional Cartesian coordinate grid with the tail of each vector at the origin.
b. Find the magnitude of $\vec{r}$.
c. Find the angle that $\vec{r}$ makes with the positive *x*-direction.
d. If $\vec{r}$ is parallel transported to another location in the sketch, what happens to the answers to parts (b) and (c)?

Solution

a. The sketch of the vectors with their tails at the origin is given in Figure 2.56. By parallel transporting either $(2.50 \text{ m}) \hat{\imath}$ or $(-3.00 \text{ m}) \hat{\jmath}$, notice that $\vec{r}$ is the vector sum

$$\vec{r} = (2.50 \text{ m}) \, \hat{\imath} + (-3.00 \text{ m}) \, \hat{\jmath}$$

b. According to Equation 2.16, the magnitude of $\vec{r}$ is the square root of the sum of the squares of its components. From Section 2.6, $-(3.00 \text{ m}) \, \hat{\jmath}$ is equivalent to $+(-3.00 \text{ m}) \, \hat{\jmath}$. Hence the components of this vector are the scalars

$$r_x \equiv x = 2.50 \text{ m}$$
$$r_y \equiv y = -3.00 \text{ m}$$

The magnitude of the vector then is found from Equation 2.16:

$$r = [(2.50 \text{ m})^2 + (-3.00 \text{ m})^2]^{1/2}$$
$$= 3.91 \text{ m}$$

c. The angle θ that the vector makes with the positive *x*-direction (the direction of $\hat{\imath}$) is found from Equation 2.17:

$$\theta = \tan^{-1} \left(\frac{-3.00 \text{ m}}{2.50 \text{ m}} \right)$$
$$= -50.2°$$

or 50.2° measured clockwise from the positive *x*-axis according to Problem-Solving Tactic 2.11. The angle is indicated in Figure 2.56.
d. If the vector is parallel transported to another location in the sketch of part (a), its magnitude and direction are unchanged. The answers to parts (b) and (c) thus remain the same.

The Case of Three Dimensions

For the three-dimensional case, the three Cartesian components of any vector $\vec{A}$ are the scalar quantities:

$$\vec{A} \cdot \hat{\imath} \equiv A_x$$
$$\vec{A} \cdot \hat{\jmath} \equiv A_y \qquad (2.18)$$
$$\vec{A} \cdot \hat{k} \equiv A_z$$

They are the three scalar *projections or components* of $\vec{A}$ on the lines containing the three Cartesian coordinate axes (or lines parallel to them).

The vector $\vec{A}$ is the *vector sum* of three mutually perpendicular vectors:

components

$$\vec{A} = (A_x)\hat{\imath} + (A_y)\hat{\jmath} + (A_z)\hat{k} \qquad (2.19)$$

vectors

The vector sum indicated in Equation 2.19 is indicated schematically in Figure 2.57.

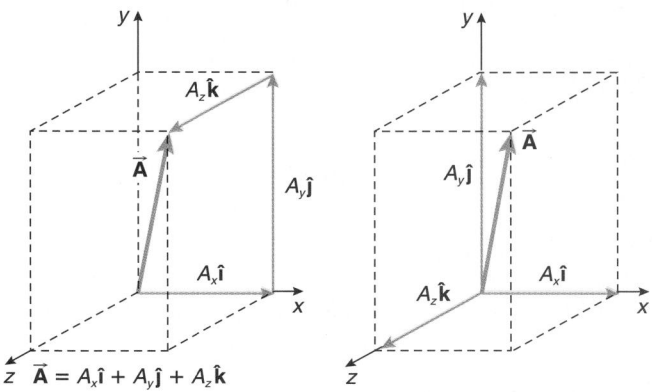

(a) The vector $\vec{A}$ is the vector sum of the vectors $A_x\hat{\imath}$, $A_y\hat{\jmath}$, and $A_z\hat{k}$.

(b) An equivalent view, parallel transporting $A_y\hat{\jmath}$ and $A_z\hat{k}$ to the axes themselves so that the tails of all the vectors are at a common point.

FIGURE 2.57 Two equivalent views of a vector $\vec{A}$ in three dimensions.

When $\vec{\mathbf{A}}$ is described as a sum of a set of mutually perpendicular vectors, we say that $\vec{\mathbf{A}}$ has been **resolved** into its *components*. The Cartesian components of $\vec{\mathbf{A}}$ are the *scalars* A_x, A_y, and A_z. The vector sum indicated in Equation 2.19 is equivalent to $\vec{\mathbf{A}}$ itself.

We can find the magnitude of any vector $\vec{\mathbf{A}}$ by taking the scalar product of $\vec{\mathbf{A}}$ with itself:

$$\vec{\mathbf{A}} \cdot \vec{\mathbf{A}} = (A_x\hat{\imath} + A_y\hat{\jmath} + A_z\hat{\mathbf{k}}) \cdot (A_x\hat{\imath} + A_y\hat{\jmath} + A_z\hat{\mathbf{k}})$$

Because of the distributive property of the scalar product operation, this results in nine scalar products of the unit vectors:

$$\begin{aligned}\vec{\mathbf{A}} \cdot \vec{\mathbf{A}} = &\; A_xA_x(\hat{\imath} \cdot \hat{\imath}) + A_xA_y(\hat{\imath} \cdot \hat{\jmath}) + A_xA_z(\hat{\imath} \cdot \hat{\mathbf{k}}) \\ &+ A_yA_x(\hat{\jmath} \cdot \hat{\imath}) + A_yA_y(\hat{\jmath} \cdot \hat{\jmath}) + A_yA_z(\hat{\jmath} \cdot \hat{\mathbf{k}}) \\ &+ A_zA_x(\hat{\mathbf{k}} \cdot \hat{\imath}) + A_zA_y(\hat{\mathbf{k}} \cdot \hat{\jmath}) + A_zA_z(\hat{\mathbf{k}} \cdot \hat{\mathbf{k}})\end{aligned}$$

Six of the scalar products of the unit vectors are zero; only three terms survive (refer back to Table 2.1). Something wonderfully simple emerges:

$$\vec{\mathbf{A}} \cdot \vec{\mathbf{A}} = A_x^2 + A_y^2 + A_z^2$$

But the Pythagorean theorem applied to Figure 2.57 indicates that

$$A^2 = A_x^2 + A_y^2 + A_z^2 \tag{2.20}$$

Thus we have proved that

$$\vec{\mathbf{A}} \cdot \vec{\mathbf{A}} = A^2 \tag{2.21}$$

After taking the positive square root, we have the magnitude of $\vec{\mathbf{A}}$:

$$\boxed{\begin{aligned}A &= (\vec{\mathbf{A}} \cdot \vec{\mathbf{A}})^{1/2} \\ &= (A_x^2 + A_y^2 + A_z^2)^{1/2} \quad \text{(use positive root only)}\end{aligned}} \tag{2.22}$$

Note from Equation 2.22 that a vector of magnitude zero (the zero vector) must have *all* its components equal to zero.

PROBLEM-SOLVING TACTIC

2.12 The magnitude of a vector $\vec{\mathbf{A}}$ that is expressed in terms of its Cartesian components is found by taking the square root of the scalar product of the vector with itself:

$$\begin{aligned}A &= (\vec{\mathbf{A}} \cdot \vec{\mathbf{A}})^{1/2} \\ &= (A_x^2 + A_y^2 + A_z^2)^{1/2} \quad \text{(use positive root only)}\end{aligned}$$

QUESTION 6

What are the three Cartesian components of the vector $\hat{\imath}$?

EXAMPLE 2.4

a. What are the Cartesian components of the velocity vector

$$\vec{\mathbf{v}} = (-2.20 \text{ m/s})\hat{\imath} - (3.30 \text{ m/s})\hat{\jmath} + (6.85 \text{ m/s})\hat{\mathbf{k}}$$

b. What is the magnitude of this velocity vector (the speed)?

Solution

a. The Cartesian components are the scalar coefficients of the Cartesian unit vectors. Remember that $-(3.30 \text{ m/s})\hat{\jmath}$ is equivalent to $+(-3.30 \text{ m/s})\hat{\jmath}$. Hence you have

$$v_x = -2.20 \text{ m/s}$$
$$v_y = -3.30 \text{ m/s}$$
$$v_z = 6.85 \text{ m/s}$$

b. The magnitude of the velocity vector is found by taking the square root of $\vec{\mathbf{v}} \cdot \vec{\mathbf{v}}$ (which is the same as using Equation 2.22):

$$\begin{aligned}v &= (\vec{\mathbf{v}} \cdot \vec{\mathbf{v}})^{1/2} \\ &= [(-2.20 \text{ m/s})^2 + (-3.30 \text{ m/s})^2 + (6.85 \text{ m/s})^2]^{1/2} \\ &= 7.92 \text{ m/s}\end{aligned}$$

2.10 MULTIPLICATION OF A VECTOR EXPRESSED IN CARTESIAN FORM BY A SCALAR

The multiplication of a vector $\vec{\mathbf{A}}$ by a scalar α changes all the Cartesian components by the same factor:

$$\begin{aligned}\alpha\vec{\mathbf{A}} &= \alpha(A_x\hat{\imath} + A_y\hat{\jmath} + A_z\hat{\mathbf{k}}) \\ &= \alpha A_x\hat{\imath} + \alpha A_y\hat{\jmath} + \alpha A_z\hat{\mathbf{k}}\end{aligned} \tag{2.23}$$

where α is any scalar (positive, negative, or zero).

EXAMPLE 2.5

If $\vec{\mathbf{B}} = -2.0\hat{\imath} + 4.0\hat{\jmath} - 3.0\hat{\mathbf{k}}$, what is $-2.2\vec{\mathbf{B}}$?

Solution

$$\begin{aligned}-2.2\vec{\mathbf{B}} &= -2.2(-2.0\hat{\imath} + 4.0\hat{\jmath} - 3.0\hat{\mathbf{k}}) \\ &= 4.4\hat{\imath} - 8.8\hat{\jmath} + 6.6\hat{\mathbf{k}}\end{aligned}$$

2.11 EXPRESSING VECTOR ADDITION AND SUBTRACTION IN CARTESIAN FORM

As we learned in Sections 2.4 and 2.6, vectors in Cartesian form may be added and subtracted provided the vectors are of the same type, representing the same kind of physical quantity, such as forces added to forces or velocities added to velocities. In particular, when the vectors are in Cartesian form we have

$$\vec{\mathbf{F}} + \vec{\mathbf{f}} = (F_x\hat{\imath} + F_y\hat{\jmath} + F_z\hat{\mathbf{k}}) + (f_x\hat{\imath} + f_y\hat{\jmath} + f_z\hat{\mathbf{k}})$$

Collect the various x-, y-, and z-components together:

$$\vec{\mathbf{F}} + \vec{\mathbf{f}} = (F_x + f_x)\hat{\imath} + (F_y + f_y)\hat{\jmath} + (F_z + f_z)\hat{\mathbf{k}} \tag{2.24}$$

Thus the components of the vector sum $\vec{\mathbf{F}} + \vec{\mathbf{f}}$ are the scalar sums of the respective components of the individual vectors.

The vector difference between $\vec{\mathbf{F}}$ and $\vec{\mathbf{f}}$ is similar:

$$\begin{aligned}\vec{\mathbf{F}} - \vec{\mathbf{f}} &= (F_x\hat{\imath} + F_y\hat{\jmath} + F_z\hat{\mathbf{k}}) - (f_x\hat{\imath} + f_y\hat{\jmath} + f_z\hat{\mathbf{k}}) \\ &= (F_x - f_x)\hat{\imath} + (F_y - f_y)\hat{\jmath} + (F_z - f_z)\hat{\mathbf{k}}\end{aligned} \tag{2.25}$$

Thus the components of the vector $\vec{F} - \vec{f}$ are the scalar differences of the respective scalar components of the individual vectors, which should be comforting to you from an algebraic viewpoint.

EXAMPLE 2.6

If

$$\vec{F} = 2\hat{\imath} - 3\hat{\jmath} + 4\hat{k}$$

and

$$\vec{f} = -2\hat{\imath} + 4\hat{\jmath} - 7\hat{k}$$

find (a) $\vec{F} + \vec{f}$ and (b) $\vec{F} - \vec{f}$.

Solution

a. Use Equation 2.24 and add the respective Cartesian components:

$$\vec{F} + \vec{f} = (F_x + f_x)\hat{\imath} + (F_y + f_y)\hat{\jmath} + (F_z + f_z)\hat{k}$$
$$= [2 + (-2)]\hat{\imath} + [-3 + 4]\hat{\jmath} + [4 + (-7)]\hat{k}$$
$$= 0\hat{\imath} + \hat{\jmath} - 3\hat{k}$$
$$= \hat{\jmath} - 3\hat{k}$$

b. Use Equation 2.25 and subtract the respective Cartesian components:

$$\vec{F} - \vec{f} = (F_x - f_x)\hat{\imath} + (F_y - f_y)\hat{\jmath} + (F_z - f_z)\hat{k}$$
$$= [2 - (-2)]\hat{\imath} + [-3 - 4]\hat{\jmath} + [4 - (-7)]\hat{k}$$
$$= 4\hat{\imath} - 7\hat{\jmath} + 11\hat{k}$$

2.12 THE SCALAR PRODUCT OF TWO VECTORS EXPRESSED IN CARTESIAN FORM

When two vectors $\vec{A}$ and $\vec{B}$ are expressed in Cartesian form, the scalar product becomes

$$\vec{A} \cdot \vec{B} = (A_x\hat{\imath} + A_y\hat{\jmath} + A_z\hat{k}) \cdot (B_x\hat{\imath} + B_y\hat{\jmath} + B_z\hat{k})$$

The scalar product simplifies considerably, since six of the nine scalar products of the unit vectors are zero (refer to Table 2.1), leaving only three surviving terms:

$$\boxed{\vec{A} \cdot \vec{B} = A_xB_x + A_yB_y + A_zB_z} \qquad (2.26)$$

Notice that the right-hand side of Equation 2.26 is an ordinary *scalar sum*; the result of a scalar product always is a scalar quantity.

> The scalar product of two vectors is the (scalar) sum of the products of their respective Cartesian components.

EXAMPLE 2.7

If

$$\vec{A} = 2\hat{\imath} - \hat{\jmath} + 4\hat{k}$$

and

$$\vec{B} = -3\hat{\imath} - 2\hat{\jmath} + 4\hat{k}$$

find $\vec{A} \cdot \vec{B}$.

Solution

Use Equation 2.26 and (scalar) sum the products of the respective Cartesian components:

$$\vec{A} \cdot \vec{B} = A_xB_x + A_yB_y + A_zB_z$$
$$= (2)(-3) + (-1)(-2) + (4)(4)$$
$$= -6 + 2 + 16$$
$$= 12$$

2.13 DETERMINING THE ANGLE BETWEEN TWO VECTORS EXPRESSED IN CARTESIAN FORM

We now have all the tools needed for some interesting geometric calculations. For example, consider two particular vectors $\vec{A}$ and $\vec{B}$ expressed in Cartesian form:

$$\vec{A} = 1.00\hat{\imath} - 2.00\hat{\jmath} + 4.00\hat{k}$$

and

$$\vec{B} = 2.00\hat{\imath} + 1.00\hat{\jmath} - 3.00\hat{k}$$

We want to find the angle θ between them. If we sketch the geometric situation, as in Figure 2.58, with the tails of the two vectors at the origin, the measure of the angle between the two vectors is not at all obvious. It is even difficult to estimate the angle from the three-dimensional sketch.

The angle θ between the two vectors is found by exploiting the scalar product, which involves the angle between them:

$$\vec{A} \cdot \vec{B} = AB \cos \theta$$

1. First, find the magnitude of the vector $\vec{A}$, using Equation 2.22:

$$A = (\vec{A} \cdot \vec{A})^{1/2} = (A_x^2 + A_y^2 + A_z^2)^{1/2}$$
$$= [(1.00)^2 + (-2.00)^2 + (4.00)^2]^{1/2}$$
$$= (1.00 + 4.00 + 16.0)^{1/2}$$
$$= (21.0)^{1/2}$$

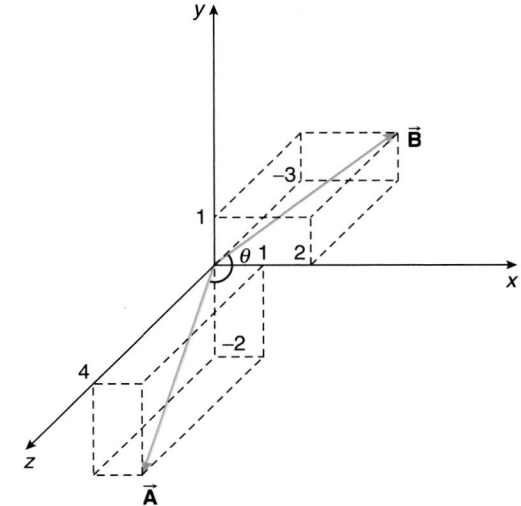

FIGURE 2.58 The vectors $\vec{A} = 1.00\hat{\imath} - 2.00\hat{\jmath} + 4.00\hat{k}$ and $\vec{B} = 2.00\hat{\imath} + 1.00\hat{\jmath} - 3.00\hat{k}$.

2. Then find the magnitude of $\vec{B}$ the same way:

$$B = (\vec{B} \cdot \vec{B})^{1/2} = (B_x^2 + B_y^2 + B_z^2)^{1/2}$$
$$= [(2.00)^2 + (1.00)^2 + (-3.00)^2]^{1/2}$$
$$= (4.00 + 1.00 + 9.00)^{1/2}$$
$$= (14.00)^{1/2}$$

3. Now find the scalar product of $\vec{A}$ and $\vec{B}$ using the Cartesian components of the vectors and Equation 2.26:

$$\vec{A} \cdot \vec{B} = A_x B_x + A_y B_y + A_z B_z$$
$$= (1.00)(2.00) + (-2.00)(1.00) + (4.00)(-3.00)$$
$$= 2.00 - 2.00 - 12.0$$
$$= -12.0$$

4. Now put the pieces together into the definition of the scalar product, Equation 2.10:

$$\vec{A} \cdot \vec{B} = AB \cos \theta$$

The left-hand side of the equation is -12.0 (from the Cartesian evaluation of the scalar product); on the right-hand side we know the magnitude of $\vec{A}$ is $(21.0)^{1/2}$ and $\vec{B}$ is $(14.00)^{1/2}$. Thus

$$-12.0 = (21.0)^{1/2}(14.00)^{1/2} \cos \theta$$

so that

$$\cos \theta = \frac{-12.0}{(21.0)^{1/2}(14.00)^{1/2}}$$
$$= -0.700$$

or

$$\theta = 134°$$

This technique is used to find the angle between *any* two vectors when they are expressed in Cartesian form.

STRATEGIC EXAMPLE 2.8

A carpenter has built a rectangular shipping container of dimensions 1.00 m by 1.00 m by 1.50 m, as shown in Figure 2.59. A triangular piece of wood is to be fitted diagonally across the box. What is the angle θ that the hypotenuse of the triangular piece of wood makes with the base of the container?

Solution
Use the coordinate choice indicated in Figure 2.59. The diagonal of the box (the hypotenuse of the triangle) has the Cartesian vector representation

$$\vec{D} = (1.00 \text{ m})\hat{\imath} + (1.50 \text{ m})\hat{\jmath} - (1.00 \text{ m})\hat{k}$$

The edge of the base of the triangular piece of wood has the Cartesian vector representation

$$\vec{d} = (1.00 \text{ m})\hat{\imath} - (1.00 \text{ m})\hat{k}$$

The angle θ is the angle between these two vectors.

1. The magnitude of $\vec{D}$ is

$$D = [(1.00 \text{ m})^2 + (1.50 \text{ m})^2 + (-1.00 \text{ m})^2]^{1/2}$$
$$= 2.06 \text{ m}$$

2. The magnitude of $\vec{d}$ is

$$d = [(1.00 \text{ m})^2 + (-1.00 \text{ m})^2]^{1/2}$$
$$= 1.41 \text{ m}$$

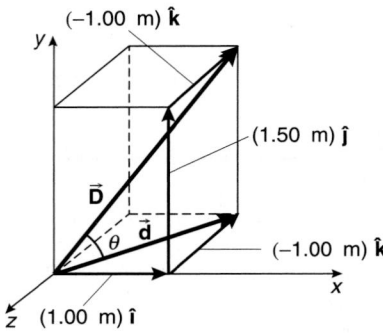

FIGURE 2.59

3. Use Equation 2.26 to find the scalar product $\vec{D} \cdot \vec{d}$:

$$\vec{D} \cdot \vec{d} = (1.00 \text{ m})(1.00 \text{ m}) + (1.50 \text{ m})(0 \text{ m})$$
$$+ (-1.00 \text{ m})(-1.00 \text{ m})$$
$$= 2.00 \text{ m}^2$$

4. Use the definition of the scalar product, Equation 2.10:

$$\vec{D} \cdot \vec{d} = Dd \cos \theta$$

This yields

$$2.00 \text{ m}^2 = (2.06 \text{ m})(1.41 \text{ m})\cos \theta$$

Solve for $\cos \theta$, and you find

$$\cos \theta = 0.689$$

or

$$\theta = 46.4°$$

So the angle between the diagonal and the base is $46.4°$.

2.14 EQUALITY OF TWO VECTORS

Note that if $\vec{F} - \vec{f} = 0$, then $\vec{F} = \vec{f}$. If two vectors are equal to each other, their respective components are equal. Thus the *single* vector equation

$$\vec{F} = \vec{f}$$

really is *three* different scalar equations, one for each of the three Cartesian components:

$$F_x = f_x$$
$$F_y = f_y$$
$$F_z = f_z$$

This is one of the beauties of vector notation; much information can be gleaned from a single vector equation.

QUESTION 7

Your roommate says that two vectors are "equal and opposite." Is this possible? Explain what your friend probably means.

2.15 VECTOR EQUATIONS

It is important to realize that an equation relating various vectors is independent of the particular coordinate system used to represent the vectors. For example, looking ahead to Chapter 5, we say that a vector $\vec{F}_{total}$ is equal to a scalar m times another vector $\vec{a}$:

$$\vec{F}_{total} = m\vec{a}$$

This mathematical relationship is valid independent of the coordinate system we choose to represent the various vectors algebraically. Once a coordinate system is chosen, we write *all* the vectors in the equation in terms of the *same* coordinate system.

Nature and natural phenomena on a macroscopic scale do not depend on the choice we scientists make of a coordinate system.*† Nature was "doing its thing" long before we physicists and engineers arrived on the scene with our mathematical toolboxes to describe physical phenomena. We will see that the most general *laws of physics* are equations relating scalars to other scalars (scalar equations) or vectors to other vectors (vector equations).

In view of what we mean by the equality of two vectors, a vector equation such as $\vec{F}_{total} = m\vec{a}$ actually represents *three scalar equations* relating the components of the vectors:

$$F_{x\,total} = ma_x$$
$$F_{y\,total} = ma_y$$
$$F_{z\,total} = ma_z$$

We will see in Chapter 5 that these equations are a statement of Newton's second law of motion.

2.16 THE VECTOR PRODUCT OF TWO VECTORS

The **vector product** of two vectors, as its name implies, creates a new vector from the two that form the product. We will use the vector product extensively to simplify the description of circular motion in Chapter 4, which is why we introduce it into the vector toolbox at this early juncture. The vector product has many other uses in physics as well, particularly in the study of orbital motion and spin motion (Chapter 10) and magnetism (Chapter 20). For now, we play with the vector product more abstractly, recognizing that its utility in physics will be apparent soon.

The vector product of two vectors $\vec{A}$ and $\vec{B}$ is written symbolically as

$$\vec{A} \times \vec{B}$$

where the boldfaced symbol $\times$ is used to distinguish the vector product from the scalar product of two vectors (symbolized by $\cdot$). The vector product commonly is called the *cross product*, reflecting the symbol used to designate this type of vector multiplication. We prefer the term vector product, to emphasize the vector nature of the result.

The new vector that results from the vector product has both a magnitude and a direction. The *magnitude* of the vector product is related to the magnitudes of $\vec{A}$ and $\vec{B}$ and the angle θ between them. As we learned in Section 2.7, the angle θ between the two vectors is the *smaller angle* measured when they are drawn from a common point (i.e., with their tails together), as shown in Figure 2.60. Parallel transport may be needed to get the tails of the two vectors together so θ can be determined. The angle θ is the *same* angle used in the definition of the scalar product.

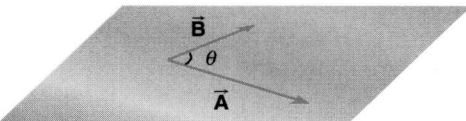

Plane of $\vec{A}$ and $\vec{B}$

FIGURE 2.60 Two vectors and the angle θ between them.

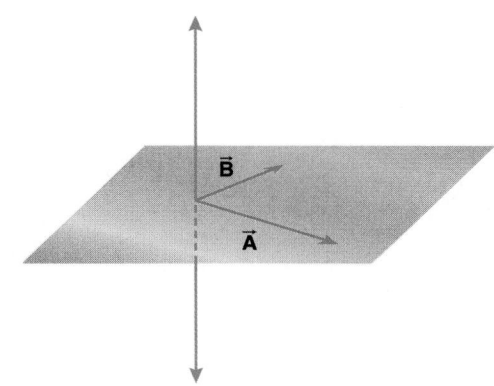

FIGURE 2.61 A perpendicular to the plane formed by $\vec{A}$ and $\vec{B}$ defines two directions.

The *magnitude* of the vector product is defined to be

$$|\vec{A} \times \vec{B}| \equiv AB \sin \theta \qquad (2.27)$$

Notice that the *sine* of the angle is used in the magnitude of the vector product; it was the cosine of the angle that determined the scalar product. The angle between the two vectors always is less than or equal to 180°. The sine is *never negative* over this range of angles. Since the magnitudes of $\vec{A}$ and $\vec{B}$ in the product also are positive scalars, the magnitude of the vector product, given by the right-hand side of Equation 2.27, is a scalar quantity greater than or equal to zero, as is the magnitude of *any* vector.‡

The *direction* of the vector product is defined to be perpendicular to *both* the vectors $\vec{A}$ and $\vec{B}$.

If the two vectors are not coplanar, use the parallel transport property to move either vector, preserving its length and orientation in space, until the tails of the two vectors emerge from a common point. The two vectors then define a plane, as shown in Figure 2.60. The direction of the vector product is perpendicular to this plane. But the perpendicular defines *two* directions (see Figure 2.61) and one must be singled out. The direction universally chosen is that determined by a **vector product right-hand rule**. There are two equivalent ways to envision this rule:

1. Orient the index finger of your right hand along the direction of the first vector of the product and the middle finger along the direction of the second vector, as in Figure 2.62. The

---

*Occasionally, a specific coordinate choice is made by convention.
†In quantum mechanics, nature can be more subtle!

‡On the other hand, the scalar product can be positive (when $\theta < 90°$), negative (when $\theta > 90°$), or zero (when $\theta = 90°$).

extended thumb then indicates the directional sense of the vector product $\vec{A} \times \vec{B}$.

2. Alternatively, take the four bunched fingers of your right hand and imagine swinging them from the direction of the first vector in the vector product to the direction of the second vector, as shown in Figure 2.63. The extended thumb of your right hand then indicates the appropriate direction of the vector product. Always choose the smaller of the two possible angles for swinging your fingers from the first vector to the second vector of the product.

These are the same sorts of right-hand rules that defined the right-handed Cartesian coordinate directions and the relative orientations of the Cartesian unit vectors $\hat{i}$, $\hat{j}$, and $\hat{k}$.

PROBLEM-SOLVING TACTIC

2.13 When taking the vector product of two vectors, remember that the result is a vector perpendicular to both. You must specify the magnitude and direction of the vector that results from the product, or express it in Cartesian form.

There are several geometric interpretations of the *magnitude* of the vector product.

1. The magnitude of the vector product is

$$AB \sin \theta$$

where θ is the angle between the two vectors when their tails are together at a common point. This magnitude is the area of the *parallelogram* whose sides are the two vectors, as shown in Figure 2.64. This area is measured in square meters *only if*

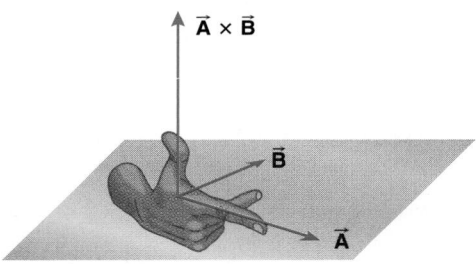

FIGURE 2.62 One way to determine the direction of the vector product $\vec{A} \times \vec{B}$.

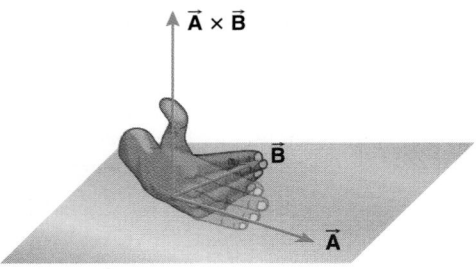

FIGURE 2.63 An equivalent way to find the direction of the vector product $\vec{A} \times \vec{B}$.

both vectors represent directed lengths expressed in meters; otherwise, the SI units attached to the area are those of vector $\vec{A}$ times those of vector $\vec{B}$.

2. Another way to interpret the magnitude of the vector product is as *twice* the area of the *triangle* formed by connecting the tips of the two vectors, as shown in Figure 2.65.

The two geometric interpretations of the vector product are not, of course, independent of one another since the area of the parallelogram in Figure 2.64 is twice the area of the triangle in Figure 2.65.

Since the angle between any vector $\vec{A}$ and *itself* is 0°, the magnitude of the vector product of $\vec{A}$ with itself is, from Equation 2.27,

$$|\vec{A} \times \vec{A}| = AA \sin 0°$$

Since $\sin 0° = 0$, the magnitude of $\vec{A} \times \vec{A}$ is zero. Since the magnitude is zero, the resulting vector is the zero vector; that is,

$$\vec{A} \times \vec{A} = 0 \qquad \text{for } any \text{ vector } \vec{A} \qquad (2.28)$$

The vector product of any vector with itself is zero.

If the vector product of two vectors is zero and *neither* vector is of zero magnitude, then the sine of the angle between the two vectors must be zero. That is, if

$$\vec{A} \times \vec{B} = 0$$

and $A \neq 0$ and $B \neq 0$, then

$$|\vec{A} \times \vec{B}| = AB \sin \theta = 0$$

implies

$$\sin \theta = 0$$

Thus θ is either 0° or 180°. The two vectors are either parallel (0° angle between them) or antiparallel (180° angle between them).

The order of the terms in a vector product is important.

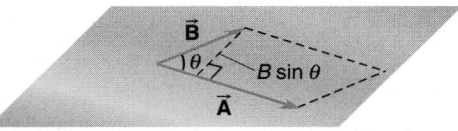

Area of parallelogram = $A (B \sin \theta) = |\vec{A} \times \vec{B}|$

FIGURE 2.64 The magnitude of the vector product is the area of the parallelogram formed by the two vectors.

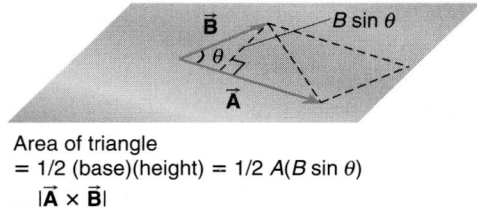

Area of triangle
= 1/2 (base)(height) = 1/2 $A(B \sin \theta)$
= $\dfrac{|\vec{A} \times \vec{B}|}{2}$

FIGURE 2.65 The magnitude of the vector product is twice the area of the triangle formed by connecting the tips of the vectors.

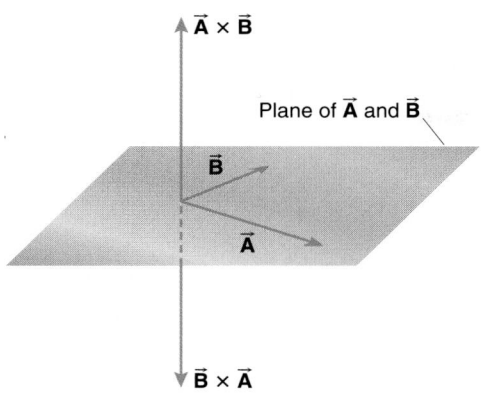

FIGURE 2.66 $\vec{A} \times \vec{B}$ and $\vec{B} \times \vec{A}$ have the same magnitude but point in opposite directions.

The terms of the vector product do *not* commute.

That is,

$$\vec{A} \times \vec{B} \neq \vec{B} \times \vec{A} \qquad (2.29)$$

Although $\vec{A} \times \vec{B}$ and $\vec{B} \times \vec{A}$ have the same magnitude, they point in opposite directions according to the vector product right-hand rule for determining the directions of a vector product. That is,

$$\vec{A} \times \vec{B} = -\vec{B} \times \vec{A} \qquad (2.30)$$

This is illustrated in Figure 2.66.

The order of the terms in a vector product is important, since $\vec{A} \times \vec{B} = -\vec{B} \times \vec{A}$.

The vector product obeys a distributive law as long as the *order* of the terms in the product is preserved. That is,

$$\vec{A} \times (\vec{B}_1 + \vec{B}_2) = (\vec{A} \times \vec{B}_1) + (\vec{A} \times \vec{B}_2)$$

The parentheses are immaterial on the right-hand side of the equation, since the vector products must be performed before the vector addition:

$$\vec{A} \times (\vec{B}_1 + \vec{B}_2) = \vec{A} \times \vec{B}_1 + \vec{A} \times \vec{B}_2 \qquad (2.31)$$

Notice that the **+** sign on both sides of this expression represents *vector* addition.

QUESTION 8

Suppose a vector $\vec{A}$ is not the zero vector. If $\vec{A} \times \vec{B} = 0$, does this necessarily mean that $\vec{B} = 0$? Explain.

EXAMPLE 2.9

For the two vectors indicated in Figure 2.67, find the magnitude of their vector product and indicate the direction of the vector product in a sketch.

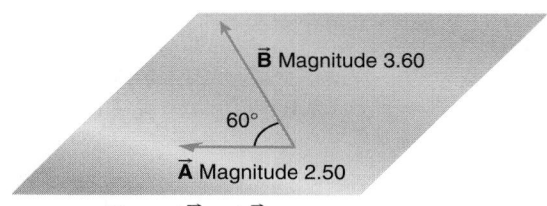

Plane of $\vec{A}$ and $\vec{B}$
FIGURE 2.67

Solution

The magnitude of their vector product is found from Equation 2.27:

$$
\begin{aligned}
|\vec{A} \times \vec{B}| &= AB \sin \theta \\
&= (2.50)(3.60) \sin 60° \\
&= 7.8
\end{aligned}
$$

The direction of the vector product is found using either vector product right-hand rule. Swing the fingers of your right hand from $\vec{A}$ to $\vec{B}$. Your extended right thumb indicates the direction of $\vec{A} \times \vec{B}$, shown in Figure 2.68.

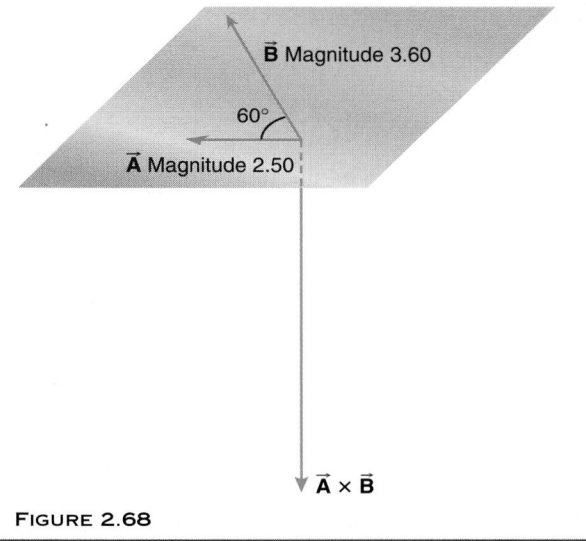

FIGURE 2.68

2.17 THE VECTOR PRODUCT OF TWO VECTORS EXPRESSED IN CARTESIAN FORM

The vector products of the Cartesian unit vectors are useful for the computation of vector products in Cartesian form. To evaluate the vector products of the Cartesian unit vectors, let's consider two specific examples.

1. To find $\hat{\imath} \times \hat{\imath}$, remember that the vector product of *any* vector with itself is zero. Hence $\hat{\imath} \times \hat{\imath} = 0$. Equivalently, since the angle between $\hat{\imath}$ and itself is 0°, the magnitude of $\hat{\imath} \times \hat{\imath}$ is

$$
\begin{aligned}
|\hat{\imath} \times \hat{\imath}| &= (1)(1) \sin 0° \\
&= 0
\end{aligned}
$$

2. To find $\hat{\imath} \times \hat{\jmath}$, note that the magnitude of the resulting vector is, by definition (Equation 2.27), the product of the magnitudes of the two vectors times the sine of the angle between them. Since both $\hat{\imath}$ and $\hat{\jmath}$ each have magnitude 1, and the angle between them is 90°, the magnitude of the vector product is

$$|\hat{\imath} \times \hat{\jmath}| = (1)(1) \sin 90°$$
$$= 1$$

The unit vectors are dimensionless, as is the sine, and so the magnitude of their vector product also is dimensionless. We use the vector product right-hand rule to find the direction of the vector product $\hat{\imath} \times \hat{\jmath}$; it is perpendicular to both $\hat{\imath}$ and $\hat{\jmath}$. We find that $\hat{\imath} \times \hat{\jmath}$ is directed toward increasing values of z, that is, parallel to $\hat{k}$. Thus

$$\hat{\imath} \times \hat{\jmath} = \hat{k}$$

since $\hat{k}$ is the vector of magnitude 1 that points toward increasing values of z.

The results of all possible vector products of the Cartesian unit vectors with each other are summarized in Table 2.2.

PROBLEM-SOLVING TACTIC

2.14 There is a mnemonic for remembering the vector products of the Cartesian unit vectors. Write the unit vectors $\hat{\imath}$, $\hat{\jmath}$, and $\hat{k}$ twice in succession:

$$\hat{\imath} \quad \hat{\jmath} \quad \hat{k} \quad \hat{\imath} \quad \hat{\jmath} \quad \hat{k}$$

Any time you form the vector product of one unit vector with the one to the right, the result is the next one to the right.

$$\hat{\imath} \times \hat{\jmath} = \hat{k}$$
$$\hat{\jmath} \times \hat{k} = \hat{\imath}$$
$$\hat{k} \times \hat{\imath} = \hat{\jmath}$$

This ordering of the terms is called **cyclic ordering**. If the vector product of one unit vector with another must be read from right to left in the mnemonic (backward or **anticyclic ordering**), the result is the *negative* of the next one to the left in the list. Thus

$$\hat{k} \times \hat{\jmath} = -\hat{\imath}$$
$$\hat{\jmath} \times \hat{\imath} = -\hat{k}$$
$$\hat{\imath} \times \hat{k} = -\hat{\jmath}$$

These relationships can be summarized symbolically by

$$\overset{\rightarrow \; + \; \rightarrow}{\hat{\imath} \quad \hat{\jmath} \quad \hat{k} \quad \hat{\imath} \quad \hat{\jmath} \quad \hat{k}}$$
$$\underset{\leftarrow \; - \; \leftarrow}{}$$

The vector product of two vectors that are expressed in Cartesian form gets complicated:

$$\vec{A} \times \vec{B} = (A_x\hat{\imath} + A_y\hat{\jmath} + A_z\hat{k}) \times (B_x\hat{\imath} + B_y\hat{\jmath} + B_z\hat{k})$$

The distributive property of the vector product is used to evaluate the nine vector products of the Cartesian unit vectors on the right-hand side of the equation. Remember that the *order* of the terms in the vector product is important! After the nine individual vector products of the Cartesian unit vectors are performed, three of which are zero, the result is, after collecting terms,

TABLE 2.2 Vector Products of the Cartesian Unit Vectors

$\hat{\imath} \times \hat{\imath} = 0$	$\hat{\imath} \times \hat{\jmath} = \hat{k}$	$\hat{\imath} \times \hat{k} = -\hat{\jmath}$
$\hat{\jmath} \times \hat{\imath} = -\hat{k}$	$\hat{\jmath} \times \hat{\jmath} = 0$	$\hat{\jmath} \times \hat{k} = \hat{\imath}$
$\hat{k} \times \hat{\imath} = \hat{\jmath}$	$\hat{k} \times \hat{\jmath} = -\hat{\imath}$	$\hat{k} \times \hat{k} = 0$

$$\vec{A} \times \vec{B} = (A_yB_z - A_zB_y)\hat{\imath} + (A_zB_x - A_xB_z)\hat{\jmath}$$
$$+ (A_xB_y - A_yB_x)\hat{k} \qquad (2.32)$$

The positive terms have the subscripts x, y, and z in cyclic order ($x\,y\,z\,x\,y\,z$, read left to right), whereas the negative terms have the subscripts in the reverse (anticyclic) order. The x-component of the vector product involves only the y- and z-components of the vectors involved in the product, the y-component of the product involves only the x- and z-components of the vectors in the product, and the z-component of the vector product involves only the x- and y-components.

If you know how to evaluate a 3×3 determinant, the vector product of two vectors in Cartesian form can be written elegantly as

$$\vec{A} \times \vec{B} = \begin{vmatrix} \hat{\imath} & \hat{\jmath} & \hat{k} \\ A_x & A_y & A_z \\ B_x & B_y & B_z \end{vmatrix} \qquad (2.33)$$

If a determinant is unfamiliar to you, just ignore it.

PROBLEM-SOLVING TACTICS

2.15 Do not memorize Equation 2.32. Evaluate the vector product of two vectors expressed in Cartesian form term-by-term using the mnemonic described in Problem-Solving Tactic 2.14. See Strategic Example 2.10.

2.16 A way to check the calculation of a vector product in Cartesian form is to recall that if two vectors are perpendicular to each other, their scalar product is zero. Since the vector product of two vectors is perpendicular to *both* vectors in the product, the *scalar product* of the vector $\vec{A} \times \vec{B}$ with either $\vec{A}$ or $\vec{B}$ must be zero; that is,

$$(\vec{A} \times \vec{B}) \cdot \vec{A} = 0$$

and

$$(\vec{A} \times \vec{B}) \cdot \vec{B} = 0$$

When calculating the vector product of two vectors, these scalar products serve as a useful check on whether the vector product was taken successfully (see Strategic Example 2.10).

 **STRATEGIC EXAMPLE 2.10**

If

$$\vec{A} = 2\hat{\imath} + 3\hat{\jmath} - 4\hat{k}$$

and

$$\vec{B} = -3\hat{\imath} + 2\hat{\jmath} - 3\hat{k}$$

(a) find $\vec{A} \times \vec{B}$ and (b) verify that $\vec{A} \times \vec{B}$ is perpendicular to both $\vec{A}$ and $\vec{B}$.

Solution

a. Use Problem-Solving Tactic 2.15; *do not memorize Equation 2.32*, just go ahead and take the vector product (although it is a bit tedious and you must be quite careful to watch out for all the signs):

$$\vec{A} \times \vec{B} = (2\hat{\imath} + 3\hat{\jmath} - 4\hat{k}) \times (-3\hat{\imath} + 2\hat{\jmath} - 3\hat{k})$$
$$= (2)(-3)(\hat{\imath} \times \hat{\imath}) + (2)(2)(\hat{\imath} \times \hat{\jmath}) + (2)(-3)(\hat{\imath} \times \hat{k})$$
$$+ (3)(-3)(\hat{\jmath} \times \hat{\imath}) + (3)(2)(\hat{\jmath} \times \hat{\jmath}) + (3)(-3)(\hat{\jmath} \times \hat{k})$$
$$+ (-4)(-3)(\hat{k} \times \hat{\imath}) + (-4)(2)(\hat{k} \times \hat{\jmath}) + (-4)(-3)(\hat{k} \times \hat{k})$$

But $\hat{\imath} \times \hat{\imath} = 0$, $\hat{\jmath} \times \hat{\jmath} = 0$, and $\hat{k} \times \hat{k} = 0$. Use the mnemonic mentioned in Problem-Solving Tactic 2.14 to evaluate the other vector products of the unit vectors; you find

$$\vec{A} \times \vec{B} = 4\hat{k} - 6(-\hat{\jmath}) - 9(-\hat{k}) - 9\hat{\imath} + 12\hat{\jmath} - 8(-\hat{\imath})$$
$$= -\hat{\imath} + 18\hat{\jmath} + 13\hat{k}$$

b. Use Problem-Solving Tactic 2.16 to check the calculation of a vector product in Cartesian form. Since $\vec{A} \times \vec{B}$ is perpendicular to $\vec{A}$, the scalar product of $\vec{A} \times \vec{B}$ with $\vec{A}$ should be zero:

$$(\vec{A} \times \vec{B}) \cdot \vec{A} = (-\hat{\imath} + 18\hat{\jmath} + 13\hat{k}) \cdot (2\hat{\imath} + 3\hat{\jmath} - 4\hat{k})$$
$$= (-1)(2) + (18)(3) + (13)(-4)$$
$$= -2 + 54 - 52$$
$$= 0$$

Hence $\vec{A} \times \vec{B}$ is perpendicular to $\vec{A}$. Since $\vec{A} \times \vec{B}$ also is perpendicular to $\vec{B}$, the scalar product of $\vec{A} \times \vec{B}$ with $\vec{B}$ should be zero:

$$(\vec{A} \times \vec{B}) \cdot \vec{B} = (-\hat{\imath} + 18\hat{\jmath} + 13\hat{k}) \cdot (-3\hat{\imath} + 2\hat{\jmath} - 3\hat{k})$$
$$= (-1)(-3) + (18)(2) + (13)(-3)$$
$$= 3 + 36 - 39$$
$$= 0$$

Hence $\vec{A} \times \vec{B}$ is perpendicular to $\vec{B}$.

2.18 VARIATION OF A VECTOR

But many that are first shall be last; and the last shall be first.
Matthew 19:30

Variations in vector quantities are ubiquitous and important in physics and engineering. If a vector quantity varies, its visual and Cartesian representations change. Vectors can change in several ways:

1. The *magnitude* of the vector may change, but its direction is preserved. The change may produce either an increase or decrease in magnitude. For example, in Figure 2.69 the speed of a car may increase or decrease (as indicated by its speedometer), causing a change in the magnitude of the velocity vector (which specifies the speed and direction of motion).

2. The *direction* of the vector changes, but its magnitude is preserved. For example, in Figure 2.70 a car moving at constant speed around a curve changes its direction of motion, as indicated by a compass attached to the dashboard.

3. The magnitude *and* direction of the vector *both* change. For example, in Figure 2.71 the car may brake or speed up as it executes a curve. The speedometer reading changes, indicating a change in the speed (the magnitude of the velocity), and the compass direction changes (indicating a change in the direction of the velocity vector).

The change in any vector always is defined to be the vector difference between the new (final) vector and the old (initial) vector.

$$\Delta \vec{A} \equiv \vec{A}_f - \vec{A}_i \qquad (2.34)$$

This definition applies whether the change is in time or space (hence the relevance of the quotation at the beginning of this section). The change in a vector $\vec{A}$ is symbolized by $\Delta \vec{A}$, where the use of the symbolic mathematical Δ indicates "change in" and not multiplication of $\vec{A}$ by some scalar factor Δ. So the expression $\Delta \vec{A}$ should be read as *change in the vector $\vec{A}$* and should be considered a single algebraic vector entity. Notice that by

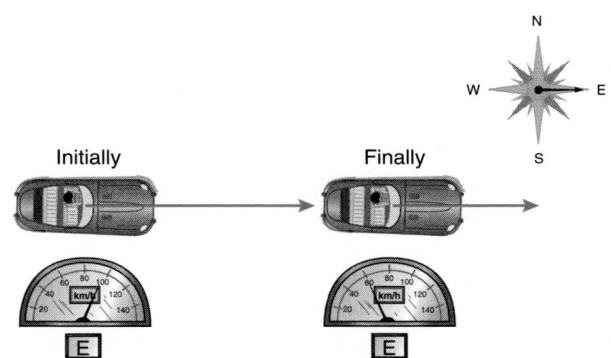

FIGURE 2.69 A car changing its speed but not its direction is changing its velocity vector.

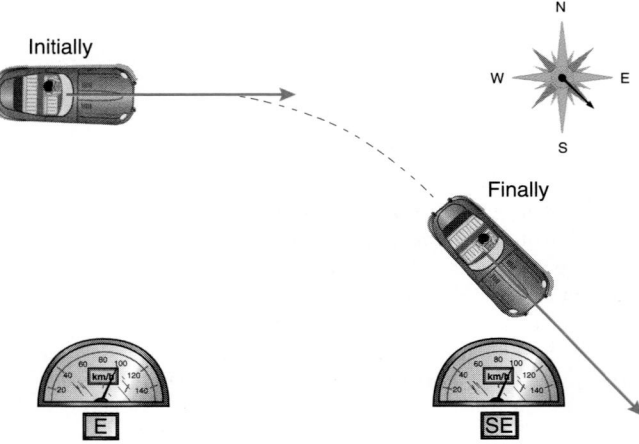

FIGURE 2.70 A car changing its direction but not its speed is changing its velocity vector.

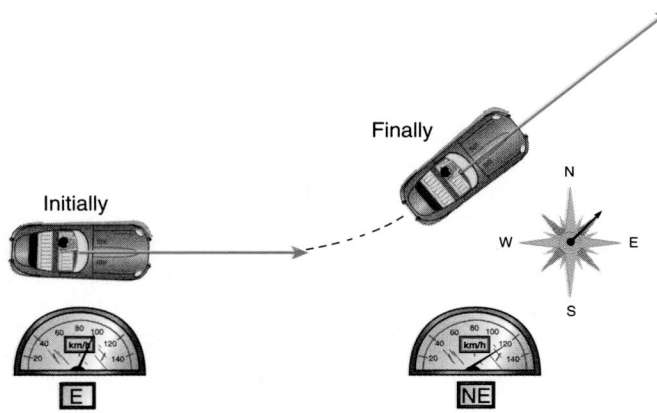

FIGURE 2.71 A car changing its speed and direction is changing its velocity vector.

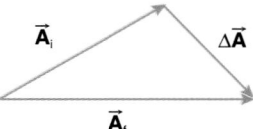

FIGURE 2.72 The change $\Delta\vec{A}$ in a vector $\vec{A}$.

transposing $\vec{A}_i$ in Equation 2.34, the change in the vector added *vectorially* to the initial vector produces the final vector, as shown geometrically in Figure 2.72:

$$\vec{A}_f = \vec{A}_i + \Delta\vec{A} \qquad (2.35)$$

> The magnitude and direction of the change in *any* vector quantity are independent of the coordinate system used to describe the change.

This important fact is apparent in Figure 2.72, which makes no reference to any specific choice of a coordinate system.

2.19 SOME ASPECTS OF VECTOR CALCULUS

In our study of motion in the next chapter, we will have frequent need to differentiate vectors that are functions of time. We will be interested in dynamic changes where the independent variable of interest is the time t. Or, for other purposes, the independent variable of interest may be a position coordinate such as the Cartesian coordinates x, y, or z or a radial coordinate r.

Whatever the independent variable, remember that the *same* rules of calculus apply for differentiating functions whether the operation is

$$\frac{d}{dt}, \frac{d}{dx}, \frac{d}{dy}, \frac{d}{dz}, \text{ or } \frac{d}{dr}$$

Differentiation of a vector expressed in Cartesian form proceeds as follows. For example, the time derivative of a vector $\vec{A}$ is

$$\frac{d\vec{A}}{dt} = \frac{d}{dt}(A_x\hat{\imath} + A_y\hat{\jmath} + A_z\hat{k})$$
$$= \frac{d}{dt}(A_x\hat{\imath}) + \frac{d}{dt}(A_y\hat{\jmath}) + \frac{d}{dt}(A_z\hat{k})$$

Each differentiation is performed in turn, according to the standard differentiation rules. The Cartesian unit vectors $\hat{\imath}$, $\hat{\jmath}$, and $\hat{k}$ each have *fixed* magnitude (= 1) and fixed direction; they are *constant, unchanging* vectors.* Thus the time derivative of $\vec{A}$ is

$$\frac{d\vec{A}}{dt} = \frac{dA_x}{dt}\hat{\imath} + \frac{dA_y}{dt}\hat{\jmath} + \frac{dA_z}{dt}\hat{k} \qquad (2.36)$$

> The derivative of a vector also is a vector, whose components are the derivatives of the respective scalar components of the original vector.

The derivative of *any constant vector* is *zero*. Recall that a constant vector has a constant magnitude *and* a fixed or unchanging direction. A vector that is of constant magnitude but changing direction is *not* a constant vector. A vector of fixed orientation but changing magnitude also is *not* a constant vector.

The distributive property applies to vector differentiation:

$$\frac{d}{dt}(\vec{A}_1 + \vec{A}_2) = \frac{d\vec{A}_1}{dt} + \frac{d\vec{A}_2}{dt} \qquad (2.37)$$

The differentiation of both the scalar and vector products proceeds according to the usual product rule in calculus. For example, differentiating a scalar product proceeds like

$$\frac{d}{dt}(\vec{A} \cdot \vec{B}) = \left(\frac{d\vec{A}}{dt}\right) \cdot \vec{B} + \vec{A} \cdot \left(\frac{d\vec{B}}{dt}\right) \qquad (2.38)$$

Note that the + sign is that for scalar addition.

For the vector product, one must be careful to preserve the order of the terms in the product since they do not commute. Otherwise the rule is the standard product rule:

$$\frac{d}{dt}(\vec{A} \times \vec{B}) = \left(\frac{d\vec{A}}{dt}\right) \times \vec{B} + \vec{A} \times \left(\frac{d\vec{B}}{dt}\right) \qquad (2.39)$$

Note that the + sign here is vector addition.

---

*This assumes that the *directions* of the coordinate axes in space do *not* change with time. A rotating coordinate system does not meet this condition; you will encounter rotating coordinate systems in more advanced courses in mechanics. We will always consider Cartesian coordinate systems with axes maintaining *fixed* directions in space, and so the time derivatives of the Cartesian unit vectors are zero.

QUESTION 9

Is the following expression legitimate? Explain why or why not.

$$\frac{d}{dt}(\vec{A} \cdot \vec{B}) = \vec{B} \cdot \left(\frac{d\vec{A}}{dt}\right) + \left(\frac{d\vec{B}}{dt}\right) \cdot \vec{A}$$

If the expression is not legitimate, correct it to make it a true statement.

EXAMPLE 2.11

Differentiate the following vector with respect to *t*:

$$\vec{A} = 5t^2\hat{\imath} + 6t\hat{\jmath} + 8\hat{k}$$

Solution

Differentiate each Cartesian component with respect to t using the standard differentiation rules; you obtain

$$\frac{d\vec{A}}{dt} = 5(2t)\hat{\imath} + 6(1)\hat{\jmath} + 0\hat{k}$$
$$= 10t\hat{\imath} + 6\hat{\jmath}$$

CHAPTER SUMMARY

A *scalar quantity* is completely specified by one numerical quantity that may be positive, negative, or zero.

A *vector quantity* is completely specified by its magnitude and direction in space; it also is specified by its three scalar components in an appropriate coordinate system (such as a Cartesian system).

The *magnitude of a vector* is a *positive scalar* quantity. A vector quantity can be visualized as a straight arrow whose length is drawn proportional to the magnitude of the vector. The magnitude of a vector $\vec{A}$ is symbolized by A or, equivalently, by $|\vec{A}|$.

Multiplying a vector $\vec{A}$ by a scalar quantity α produces a vector $\alpha\vec{A}$ whose magnitude is $|\alpha|$ times that of the original vector. The direction of $\alpha\vec{A}$ is the same as that of $\vec{A}$ if $\alpha > 0$. The direction of $\alpha\vec{A}$ is opposite to that of $\vec{A}$ if $\alpha < 0$.

Two vectors that point in the same direction are called *parallel* vectors. Two vectors that point in opposite directions are called *antiparallel* vectors.

A *unit vector* is a dimensionless vector of magnitude equal to 1. The Cartesian unit vectors $\hat{\imath}$, $\hat{\jmath}$, and $\hat{k}$ are directed toward increasing values of the coordinates x, y, and z, respectively.

A *component of a vector* is the projection of a vector on a specified coordinate axis or line. A component of a vector is a *scalar* quantity.

The Cartesian representation of a vector $\vec{A}$ has the form

$$\vec{A} = A_x\hat{\imath} + A_y\hat{\jmath} + A_z\hat{k} \qquad (2.19)$$

where A_x, A_y, and A_z are called the Cartesian components of $\vec{A}$. The Cartesian components are scalar quantities that may be positive, negative, or zero.

The rules for vector addition and subtraction are *not* the same as for scalar addition and subtraction. Arithmetic operations with vectors such as vector addition and vector subtraction of two vectors of the same physical type (two forces, two velocities, two accelerations, etc.) can be performed using either

i. geometric techniques via the *tail-to-tip method*; or
ii. analytical techniques when the vectors are expressed in Cartesian form:

$$\vec{F} \pm \vec{f} = (F_x \pm f_x)\hat{\imath} + (F_y \pm f_y)\hat{\jmath} + (F_z \pm f_z)\hat{k} \quad (2.24, 2.25)$$

The scalar product of any two vectors is a scalar quantity symbolized by $\vec{A} \cdot \vec{B}$. The scalar product is defined as

$$\vec{A} \cdot \vec{B} \equiv AB\cos\theta \qquad (2.10)$$

where θ is the angle ($\leq 180°$) between the two vectors when they are drawn from a common point. The angle θ between two vectors *always* is considered positive. Because of the cosine term, the scalar product is positive if the angle θ between the two vectors is acute ($\theta < 90°$), negative if obtuse ($\theta > 90°$), and zero if $\theta = 90°$.

If the vectors are expressed in Cartesian form, the scalar product is the scalar sum of the products of their respective components:

$$\vec{A} \cdot \vec{B} = A_xB_x + A_yB_y + A_zB_z \qquad (2.26)$$

The magnitude of a vector is the square root of the sum of the squares of its Cartesian components:

$$A = (A_x^2 + A_y^2 + A_z^2)^{1/2} \qquad (2.22)$$

If two vectors are perpendicular to each other, their scalar product is zero.

The vector product of any two vectors is a new vector symbolized by $\vec{A} \times \vec{B}$. The vector product is a vector of magnitude

$$|\vec{A} \times \vec{B}| \equiv AB\sin\theta \qquad (2.27)$$

whose direction is determined using the vector product right-hand rule. The vector product of any vector with itself is zero:

$$\vec{A} \times \vec{A} = 0 \qquad (2.28)$$

The vector product of two parallel or two antiparallel vectors is zero.

The vector product of two vectors in Cartesian form is

$$\vec{A} \times \vec{B} = (A_yB_z - A_zB_y)\hat{\imath} + (A_zB_x - A_xB_z)\hat{\jmath} + (A_xB_y - A_yB_x)\hat{k} \qquad (2.32)$$

but is more easily calculated and remembered using the mnemonic

$$\hat{\imath} \quad \hat{\jmath} \quad \hat{k} \quad \hat{\imath} \quad \hat{\jmath} \quad \hat{k}$$

for the Cartesian unit vectors.

A vector may vary in space and/or time in three ways:

- its magnitude may vary;
- its direction may vary; or
- both its magnitude and direction may vary.

The change in a vector quantity, $\Delta\vec{A}$, is defined as the vector difference

$$\Delta\vec{A} \equiv \vec{A}_f - \vec{A}_i \qquad (2.34)$$

Vectors that vary in space or time or both, as well as scalar products and vector products, can be differentiated using rules similar to the calculus of scalar functions.

With a thorough grounding in vector analysis, you now have a well-stocked toolbox for successful achievement in physics.

SUMMARY OF PROBLEM-SOLVING TACTICS

2.1 (page 38) Scalars are not vectors, and vectors are not scalars.

2.2 (page 38) Draw visual representations of vectors carefully.

2.3 (page 38) The magnitude of any vector quantity always is a positive scalar quantity (or zero).

2.4 (page 40) Parallel transport vectors carefully.

2.5 (page 41) There is a shortcut for the addition of many vectors.

2.6 (page 45) The addition and subtraction of vector quantities is different from the addition and subtraction of scalars.

2.7 (page 46) The result of the scalar product of two vectors is a scalar quantity, not another vector quantity.

2.8 (page 46) To see if two vectors are perpendicular to each other, take their scalar product and see if it is equal to zero.

2.9 (page 48) Always use right-handed coordinate systems.

2.10 (page 50) Do not confuse a vector with its Cartesian components (which are scalars).

2.11 (page 51) The mathematical sign convention for angles on a two-dimensional Cartesian coordinate grid is used for Equation 2.17. But remember that the angle between two vectors always is considered positive.

2.12 (page 52) The magnitude of any vector is found by taking the square root of the scalar product of the vector with itself.

2.13 (page 56) When taking the vector product of two vectors, remember that the result is a vector perpendicular to both.

2.14 (page 58) There is an easy way to remember the vector products of the Cartesian unit vectors:

$$\overset{\rightarrow \; + \; \rightarrow}{\hat{\imath} \quad \hat{\jmath} \quad \hat{k} \quad \hat{\imath} \quad \hat{\jmath} \quad \hat{k}}_{\leftarrow \; - \; \leftarrow}$$

2.15 (page 58) Do not memorize Equation 2.32; evaluate vector products in Cartesian form, term by term.

2.16 (page 58) A way to check the calculation of a vector product is to verify that

$$(\vec{A} \times \vec{B}) \cdot \vec{A} = 0$$
$$(\vec{A} \times \vec{B}) \cdot \vec{B} = 0$$

QUESTIONS

1. (page 38); 2. (page 42); 3. (page 45); 4. (page 47); 5. (page 49); 6. (page 52); 7. (page 54); 8. (page 57); 9. (page 61)

10. Explain the distinction between the operation implied by **+** and the operation implied by +. Likewise, distinguish between the operation implied by **−** and the operation implied by −.

11. Show how a vector of magnitude 2 can be added to a vector of magnitude 1 to yield a vector of magnitude 2. What geometrical figure do the three vector arrows form?

12. What superficially has a magnitude and a direction (other than the rotations discussed in Section 2.5), but is not a true vector because it violates one of the rules of vector arithmetic?

13. Explain why a vector equation has a triple meaning.

14. Distinguish between the meaning of zero as a scalar and the zero vector.

15. Two vectors $\vec{F}_1$ and $\vec{F}_2$ have magnitudes such that $F_1 \neq F_2$. Is it possible for

$$\vec{F}_1 + \vec{F}_2 = 0?$$

16. If two vectors sum to zero:

$$\vec{F} + \vec{f} = 0$$

what can be said (if anything) about the magnitudes and directions of the two vectors?

17. Suppose vector $\vec{F}$ is the sum of two vectors $\vec{F}_1$ and $\vec{F}_2$:

$$\vec{F} = \vec{F}_1 + \vec{F}_2$$

Under what special circumstances does this also mean $F = F_1 + F_2$? Why is it that F usually is *not* equal to $F_1 + F_2$?

18. Suppose vector $\vec{E}$ is the difference between two vectors:

$$\vec{E} = \vec{E}_1 - \vec{E}_2$$

Under what circumstances does this *also* mean $E = E_1 - E_2$? Under what circumstances is $E = E_2 - E_1$? Why is it that E usually is *not* equal to $E_1 - E_2$?

19. Show that if three vectors add to zero, that is, if

$$\vec{F}_1 + \vec{F}_2 + \vec{F}_3 = 0$$

then if the vectors are parallel transported to form their sum, the three vectors all lie in the same plane.

20. If $\vec{F}_1 + \vec{F}_2 + \vec{F}_3 = 0$, and $F_1 = F_2$, what is the maximum value of F_3? What is the minimum value of F_3?

21. Distinguish clearly between the multiplication of a vector by a scalar and the scalar product of two vectors.

22. Suppose you have a vector $\vec{A} \neq 0$. If $\vec{A} \cdot \vec{B} = 0$ *and* $\vec{A} \times \vec{B} = 0$, does this necessarily mean that $\vec{B} = 0$? Explain.

23. For any two arbitrary vectors $\vec{A}$ and $\vec{B}$, explain why the following two expressions are (a) both scalars and (b) both equal to zero:

$$\vec{A} \cdot (\vec{A} \times \vec{B})$$
$$\vec{B} \cdot (\vec{A} \times \vec{B})$$

No detailed calculations need be made.

24. Explain why the magnitude of a vector is always greater than or equal to the absolute value of its smallest Cartesian component.

25. What are the Cartesian components of the vector $\hat{\imath} - \hat{\jmath}$?

26. Is the following expression legitimate?

$$\vec{A} \times (\vec{F} + \hat{f}) = \vec{A} \times \hat{f} + \vec{A} \times \vec{F}$$

Explain why or why not.

27. Why is the following expression meaningless?

$$\vec{A} \times (\vec{B} \cdot \vec{C})$$

28. If the magnitude of $\vec{F}$ is 6 and that of $\hat{f}$ is 2, what restrictions (if any) can be placed on each of the following expressions: (a) $|\vec{F} + \hat{f}|$; (b) $|\vec{F} - \hat{f}|$; (c) $F + f$; (d) $F - f$; (e) $\vec{F} \cdot \hat{f}$; (f) $|\vec{F} \times \hat{f}|$.

29. Is it possible for a vector of zero magnitude to have a nonzero component? Explain.

30. If $\vec{A}$ is a vector and β is a scalar, explain why each of the following expressions is meaningless: (a) $\vec{A} \pm \beta$; (b) $\vec{A} \cdot \beta$; (c) $\vec{A} \times \beta$.

31. The Cartesian unit vectors $\hat{\imath}$, $\hat{\jmath}$, and $\hat{k}$ form a right-handed coordinate system. Do the vectors $-\hat{\imath}$, $-\hat{\jmath}$, and $-\hat{k}$ form a right-handed system?

32. A student is asked to take two given vectors and (a) form their sum; (b) find their components; (c) form their scalar product; and (d) form their vector product. For which of these vector operations is it absolutely necessary to specify a coordinate system?

33. In the context of our study of vectors, explain why the following product of two vectors has no meaning:

$$\vec{A}\vec{B}$$

34. One of the following expressions is correct and the other is incorrect. Explain.

$$\frac{d}{dt}(\vec{A} \times \vec{B}) = \vec{A} \times \left(\frac{d\vec{B}}{dt}\right) + \left(\frac{d\vec{A}}{dt}\right) \times \vec{B}$$

$$\frac{d}{dt}(\vec{A} \times \vec{B}) = \vec{B} \times \left(\frac{d\vec{A}}{dt}\right) + \vec{A} \times \left(\frac{d\vec{B}}{dt}\right)$$

By using appropriate minus signs, correct the incorrect expression.

35. Time intervals Δt have a magnitude

$$\Delta t = t_f - t_i$$

and a direction we can define as from the initial instant t_i to the final instant t_f. Why is a time interval not a vector?

36. Temperature can go up or down, yet it is not a vector. Explain why.

37. For two given vectors $\vec{A}$ and $\vec{B}$, show that there is more than one vector $\vec{C}$ that satisfies

$$A = \vec{B} \cdot \vec{C}$$

Also show that there is more than one vector $\vec{D}$ that satisfies

$$\vec{A} = \vec{B} \times \vec{D}$$

Explain why these results imply that we cannot define the vector division symbolized by $\vec{A}/\vec{B}$.

38. The previous question indicates why vector division is not well defined. On the other hand, for a given vector $\vec{A}$, explain why the following expressions are legitimate: (a) $1/A$; (b) $\vec{A}/4$.

PROBLEMS

Sections 2.1 Scalar and Vector Quantities
    2.2 Multiplication of a Vector by a Scalar
    2.3 Parallel Transport of Vectors
    2.4 Vector Addition by Geometric Methods:
       Tail-to-Tip Method
    2.5 Determining Whether a Quantity Is a Vector*
    2.6 Vector Difference by Geometric Methods

1. For the two vectors $\vec{F}_1$ and $\vec{F}_2$ indicated in Figure P.1, show geometrically that the arrow representing $\vec{F}_1 + \vec{F}_2$ has the same length and orientation as the arrow representing $\vec{F}_2 + \vec{F}_1$. Thus the order in which two vectors are added is immaterial.

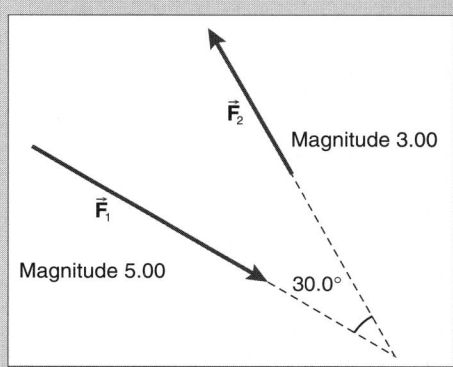

FIGURE P.1

2. For the two vectors indicated in Figure P.2, use geometric methods to find (a) $\vec{E}_1 + \vec{E}_2$; (b) $\vec{E}_1 - \vec{E}_2$; (c) $\vec{E}_2 - \vec{E}_1$. Also determine (d) $E_1 + E_2$; (e) $E_1 - E_2$; (f) $E_2 - E_1$; (g) $|\vec{E}_1 + \vec{E}_2|$; (h) $|\vec{E}_1 - \vec{E}_2|$; (i) $|\vec{E}_2 - \vec{E}_1|$.

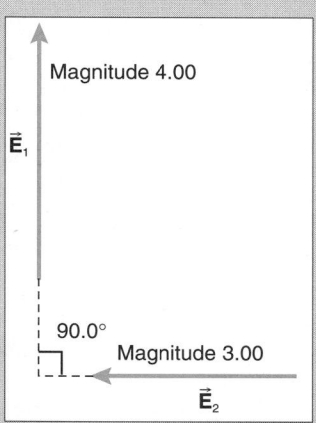

FIGURE P.2

3. For the vectors shown in Figure P.3, find (a) $\vec{F} + \hat{f}$; (b) $F + f$; (c) $|\vec{F} + \hat{f}|$; (d) $\vec{F} - \hat{f}$; (e) $F - f$; (f) $|\vec{F} - \hat{f}|$.

4. Draw the vectors shown in Figure P.4 to full scale. Then use the diagram to find (a) $F + f$; (b) $|\vec{F} + \hat{f}|$; (c) $F - f$; (d) $|\vec{F} - \hat{f}|$.

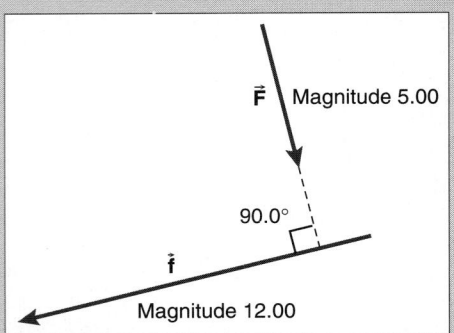

FIGURE P.3

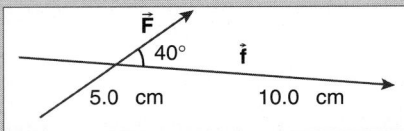

FIGURE P.4

- 5. On a level field, you begin at a wishing well and stroll east an unknown number of meters. Turning exactly northwest, you stroll again until you are 100 m due north of the wishing well. Sketch the situation. Determine the magnitudes of two vectors, one pointing east, one pointing exactly northwest that, when added, form a vector of magnitude 100 m pointing north.

- 6. For no good reason, you take 5.0 paces south, 3.0 paces west, and 6.0 paces exactly northeast. All the paces are of equal length. (a) Carefully sketch the situation. (b) At the end of the stroll, how many paces are you located south of the starting point? (c) How many paces are you east of the starting point? (d) How many paces are you from where you began? (e) Which of the answers to (b), (c), and (d), if any, depend on the *order* of the sequence of steps? (f) At the end of the small journey, in what direction are you relative to the starting point and the direction east? Specify an angle.

- 7. Let the vectors $\vec{r}_1$, $\vec{r}_2$, $\vec{r}_3$, $\vec{r}_4$, $\vec{r}_5$, and $\vec{r}_6$ be the position vectors from the center to the vertices of the regular hexagon as shown in Figure P.7. (a) Express each side (ab, bc, etc.) of the hexagon in terms of these vectors. (b) Show that

$$\vec{r}_1 + \vec{r}_2 + \vec{r}_3 + \vec{r}_4 + \vec{r}_5 + \vec{r}_6 = 0 \text{ m}$$

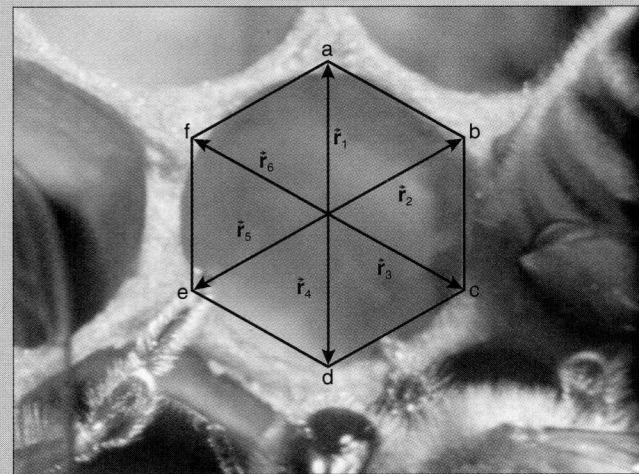

FIGURE P.7

Section 2.7 The Scalar Product of Two Vectors

8. Show that the expression

$$\frac{\vec{A} \cdot \vec{B}}{B}$$

represents the component of vector $\vec{A}$ on the line containing vector $\vec{B}$.

- 9. Suppose vector $\vec{R}$ has the same *magnitude* as vector $\vec{r}$ but they are *not* parallel or antiparallel to each other. Using an appropriate scalar product, show that the vector *sum* of the two vectors, $\vec{R} + \vec{r}$, is perpendicular to the vector *difference* of the two vectors, $\vec{R} - \vec{r}$. Does it matter which way the difference is taken? That is, will $\vec{R} + \vec{r}$ also be perpendicular to $\vec{r} - \vec{R}$?

Section 2.8 The Cartesian Coordinate System and the Cartesian Unit Vectors

10. Which of the coordinate systems in Figure P.10 are right-handed Cartesian coordinate systems and which are not?

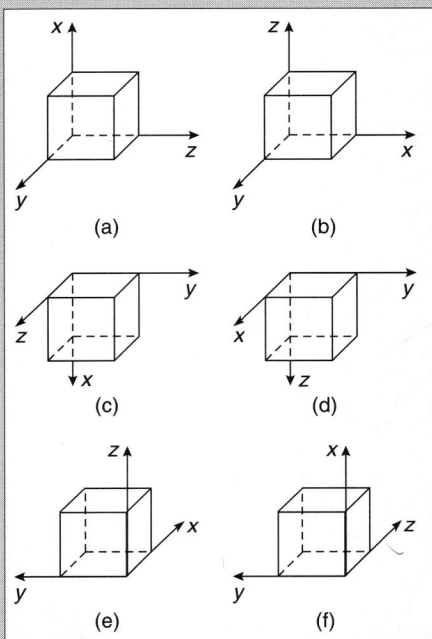

FIGURE P.10

Sections 2.9 The Cartesian Representation of Any Vector
2.10 Multiplication of a Vector Expressed in Cartesian Form by a Scalar
2.11 Expressing Vector Addition and Subtraction in Cartesian Form
2.12 The Scalar Product of Two Vectors Expressed in Cartesian Form
2.13 Determining the Angle Between Two Vectors Expressed in Cartesian Form

11. (a) What is the magnitude of the vector $\hat{i} + \hat{j} + \hat{k}$? (b) What is the magnitude of the vector $\hat{i} - \hat{j} + \hat{k}$? (c) Find a vector with only two Cartesian components that has the same magnitude. Is there only one solution?

12. A wizard creates the vector $\vec{A} = 2.0\hat{i} + 4.0\hat{j} - 3.0\hat{k}$. (a) Sketch a three-dimensional Cartesian coordinate system and indicate

the orientation of the vector, placing its tail at the origin. (b) What are the Cartesian components of the vector? (c) What is the magnitude of the vector? (d) Relabel the x-axis as the y-axis, the y-axis as the z-axis, and the z-axis as the x-axis. Is the new coordinate system right-handed? What is the new Cartesian form for the vector? What is the magnitude of the new Cartesian representation of the vector?

13. Use the definition of the scalar product to verify that the Cartesian unit vectors have the scalar products with each other that are given in Table 2.1.

14. If

$$\vec{A} = 2.0\hat{\imath} - 4.0\hat{\jmath} + 6.0\hat{k}$$

and

$$\vec{B} = -1.0\hat{\imath} + 3.0\hat{\jmath} + 3.0\hat{k}$$

(a) Find $\vec{A} \cdot \vec{B}$. (b) Find the magnitude of $\vec{A}$. (c) Find the magnitude of $\vec{B}$.

•**15.** An origin is on the ground and the x-axis is directed east, the y-axis is directed north, and the z-axis is directed vertically upward. The position of a turkey (*Meleagris gallopavo*) is determined to be

$$\vec{r}_1 = (3.0 \text{ m})\hat{\imath} + (4.0 \text{ m})\hat{\jmath} + (5.0 \text{ m})\hat{k}$$

A blue-footed booby (*Sula nebouxii*) is located at

$$\vec{r}_2 = (-4.0 \text{ m})\hat{\imath} - (2.0 \text{ m})\hat{\jmath} + (3.0 \text{ m})\hat{k}$$

(a) What is the difference in their heights above the ground? (b) How far is the turkey from the origin? (c) How far apart are the two birds?

•**16.** Imagine three position vectors:

$$\vec{r}_1 = (1.00 \text{ m})\hat{\imath} + (1.00 \text{ m})\hat{\jmath} + (1.00 \text{ m})\hat{k}$$
$$\vec{r}_2 = (2.00 \text{ m})\hat{\imath} - (1.00 \text{ m})\hat{\jmath} + (3.00 \text{ m})\hat{k}$$
$$\vec{r}_3 = (1.00 \text{ m})\hat{\imath} - (2.00 \text{ m})\hat{\jmath} - (2.00 \text{ m})\hat{k}$$

with their tails at the origin. The heads of the vectors locate the heads of three partridges (*Perdix perdix*) in a pear tree, P_1, P_2, and P_3, respectively. Schematically sketch the situation. (a) What are the coordinates of the points P_1, P_2, and P_3? (b) Find the vector $\vec{\ell}_1$ that points *from* P_1 *to* P_2. (c) Find the vector $\vec{\ell}_2$ that points *from* P_2 *to* P_3. (d) Find the vector $\vec{\ell}_3$ that points *from* P_3 *to* P_1. (e) Show explicitly that $\vec{\ell}_1 + \vec{\ell}_2 + \vec{\ell}_3 = 0$ m.

•**17.** Find a vector that points in the same direction as the vector

$$\hat{\imath} + \hat{\jmath} + \hat{k}$$

but is of magnitude 1. This problem illustrates how to construct a unit vector pointing along any arbitrary direction.

•**18.** Find a unit vector that points in the same direction as

$$\hat{\imath} + \hat{\jmath} + 2\hat{k}$$

•**19.** Show that a unit vector in the same direction as vector $\vec{A}$, where $\vec{A}$ is an unspecified vector, is $\vec{A}/A$.

•**20.** Consider the two vectors:

$$\vec{L} = 1\hat{\imath} + 2\hat{\jmath} + 3\hat{k}$$

and

$$\vec{\ell} = 4\hat{\imath} + 5\hat{\jmath} + 6\hat{k}$$

Find the value of the scalar α such that the vector

$$\vec{L} - \alpha\vec{\ell}$$

is *perpendicular* to $\vec{L}$.

•**21.** Let vector $\vec{r} = \vec{r}_1 + \vec{r}_2$. Sketch this relationship geometrically. Evaluate the expression

$$r^2 = (\vec{r}_1 + \vec{r}_2) \cdot (\vec{r}_1 + \vec{r}_2)$$

You may recognize the result as the law of cosines from trigonometry.

•**22.** Use the general Cartesian representation of the vectors to show that the scalar product of $\vec{A}$ with the vector sum of $\vec{F}$ and $\vec{f}$, that is,

$$\vec{A} \cdot (\vec{F} + \vec{f})$$

is the same as the scalar sum of $\vec{A} \cdot \vec{F}$ and $\vec{A} \cdot \vec{f}$.

•**23.** Find the angle between the two vectors

$$\vec{A} = 3.00\hat{\imath} + 4.00\hat{\jmath} + 5.00\hat{k}$$

and

$$\vec{B} = -1.00\hat{\imath} - 1.00\hat{\jmath} - 1.00\hat{k}$$

•**24.** Find the angle between the vectors

$$\vec{A} = 1.0\hat{\imath} + 2.0\hat{\jmath} - 1.0\hat{k}$$

and

$$\vec{B} = -1.0\hat{\imath} + 3.0\hat{\jmath} + 5.0\hat{k}$$

•**25.** (a) Find the angle that the vector $\vec{A} = 3.00\hat{\imath} + 4.00\hat{\jmath} + 5.00\hat{k}$ makes with the direction $\hat{\imath}$. (b) What angle does vector $\vec{A}$ make with $-\hat{\imath}$?

•**26.** Find the angle θ between the body diagonal of a cube and one of its edges (see Figure P.26) using the vector techniques of the scalar product.

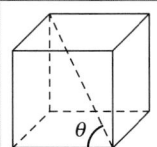

FIGURE P.26

•**27.** Find the angle ϕ between the body diagonal of a cube and a face diagonal of the cube using vector techniques. See Figure P.27.

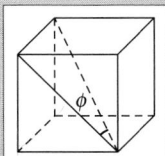

FIGURE P.27

•**28.** The principal body diagonal of a rectangular parallelopiped connects one corner with the opposite corner through the parallelopiped as shown in Figure P.28. Find the angle between the principal body diagonal of the rectangular parallelopiped and the side of length 8.0 cm.

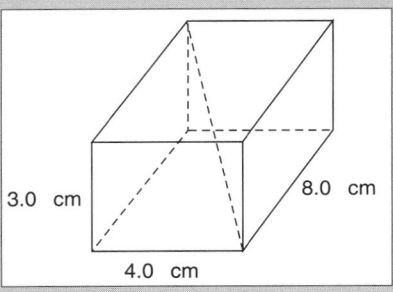

FIGURE P.28

•29. Consider the two vectors

$$\vec{A} = (\cos \alpha)\hat{\imath} + (\sin \alpha)\hat{\jmath}$$

and

$$\vec{B} = (\cos \beta)\hat{\imath} + (\sin \beta)\hat{\jmath}$$

where α and β are the respective angles that each vector makes with the +x-axis. (a) Show that each of these vectors has a magnitude equal to 1. (b) Make a schematic sketch of the vectors in the x–y plane. Make use of these vectors and the scalar product to derive the trigonometric identity

$$\cos(\alpha - \beta) = \cos \alpha \cos \beta + \sin \alpha \sin \beta$$

•30. Once upon a time, there was a vivid, vivacious vector $\vec{v}$ that had the Cartesian form

$$\vec{v} = 3.0\hat{\imath} + 5.0\hat{\jmath} - 4.0\hat{k}$$

and lived happily ever after. (a) What is the z-component of $\vec{v}$? (b) Calculate the magnitude of $\vec{v}$. (c) What angle does $\vec{v}$ make with the +y-axis? (d) Formulate a unit vector that points in the same direction as $\vec{v}$.

•31. Let the angle that a vector $\vec{B}$ makes with the +x-axis be α; the angle that $\vec{B}$ makes with the +y-axis be β; and the angle that $\vec{B}$ makes with the +z-axis be γ, as shown in Figure P.31. The *cosines* of these angles are known as the *direction cosines*. Show that

$$\cos^2\alpha + \cos^2\beta + \cos^2\gamma = 1$$

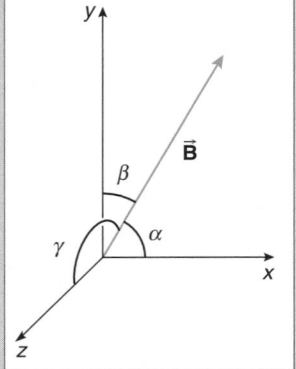

FIGURE P.31

‡32. Three vectors of magnitudes 5.00, 6.00, and 7.00 vector sum to zero. (a) Sketch the situation. (b) Find the angle between

• the vector of magnitude 5.00 and the vector of magnitude 6.00;
• the vector of magnitude 6.00 and the vector of magnitude 7.00;
• the vector of magnitude 7.00 and the vector of magnitude 5.00.

Remember that the angle between two vectors is the angle between them when they are drawn from a common point, with their tails tied together.

‡33. Two vectors of magnitude 12 and 16 have a vector sum of magnitude 20. (a) Find the angle the two vectors of magnitudes 12 and 16 make with each other. (b) Find the angle that each makes with their vector sum.

‡34. You can use the scalar product to determine the shortest distance between two locations on the Earth measured along the surface. Figure P.34 represents a section of the northern hemisphere of the Earth. A point such as A on the surface of the Earth is located (in spherical coordinates) by the three coordinates (R, θ, ϕ) where R is the radius of the Earth (here assumed to be spherical with $R = 6.37 \times 10^3$ km). (a) Show that with the indicated Cartesian coordinate system, the Cartesian position vector of point A is

$$\vec{R}_A = R[(\sin \theta \cos \phi)\hat{\imath} + (\sin \theta \sin \phi)\hat{\jmath} + (\cos \theta)\hat{k}]$$

where the angle θ is the complement of the latitude L of point A:

$$\theta = 90° - L$$

and ϕ is the longitude (with the convention that longitudes *east* are positive angles and longitudes *west* are negative angles. (b) Write the Cartesian position vector of New York (latitude 40.7° N, longitude 74.0° W). (c) Write the Cartesian position vector of Copenhagen, Denmark (latitude 55.7° N, longitude 12.5° E). (d) Use the scalar product technique to find the angle β between these two position vectors. (e) If β is measured in radians, show that the shortest distance between the cities measured on the surface of the Earth is $R\beta$. (f) Evaluate the distance between New York and Copenhagen in kilometers.

This problem was inspired by an article by B. Cameron Reed, "Dot products and great-circle distances," *The Physics Teacher*, 26, #5, pages 280–281 (May 1988).

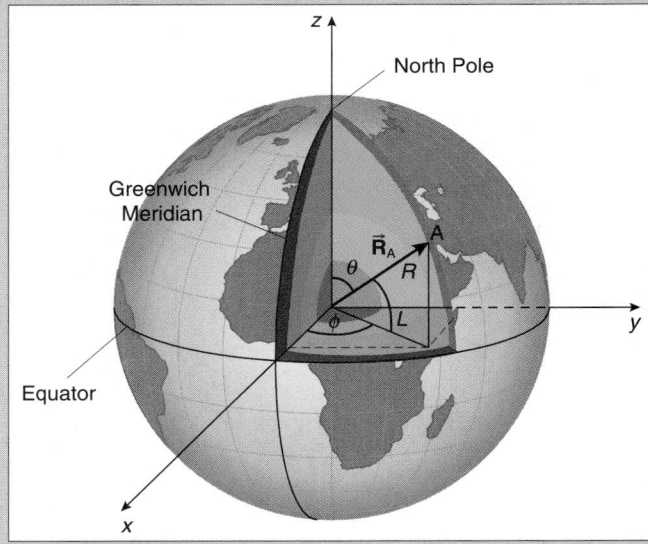

FIGURE P.34

‡35. Let the vectors $\vec{r}_1$, $\vec{r}_2$, $\vec{r}_3$, $\vec{r}_4$, and $\vec{r}_5$ be the position vectors from the center to the vertices of the Pentagon in Washington, D.C., as shown in Figure P.35. (a) Express each side (ab, bc,

etc.) of the Pentagon in terms of these vectors. (b) Show that

$$\vec{r}_1 + \vec{r}_2 + \vec{r}_3 + \vec{r}_4 + \vec{r}_5 = 0 \text{ m}$$

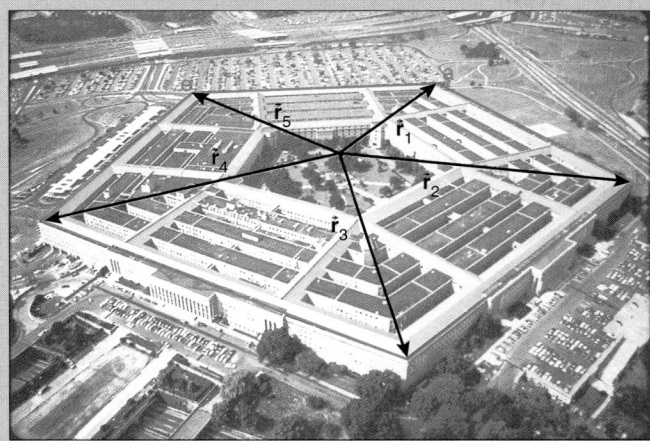

FIGURE P.35

Sections 2.14 Equality of Two Vectors
2.15 Vector Equations

36. If $\vec{F} = 3.00\vec{a}$ and $F_x = 4.00$, $F_y = -6.00$, and $F_z = 9.00$, find $\vec{a}$.

37. If

$$3.00\hat{i} + 4.00\hat{j} - 6.00\hat{k}$$
$$= (2.00 - \alpha)\hat{i} - (6.00 + \beta)\hat{j} + (3.00 - \gamma)\hat{k}$$

find the values of α, β, and γ.

Sections 2.16 The Vector Product of Two Vectors
2.17 The Vector Product of Two Vectors Expressed in Cartesian Form

38. Evaluate the following expressions: (a) $(\hat{i} + \hat{j}) \cdot (\hat{i} + \hat{k})$; (b) $(\hat{i} + \hat{j}) \times (\hat{i} + \hat{k})$.

39. Show that the magnitude of the vector defined by

$$\vec{A} \times (\vec{A} \times \vec{B})$$

is $A^2B \sin\theta$, where θ is the angle between $\vec{A}$ and $\vec{B}$.

40. Find the magnitude of the vector $(\hat{i} + \hat{j}) \times \hat{k}$.

41. Find a vector perpendicular to $\hat{i} + \hat{j}$. Is your solution unique?

42. Show that

$$\vec{A} \times \vec{B} = (A_x\hat{i} + A_y\hat{j} + A_z\hat{k}) \times (B_x\hat{i} + B_y\hat{j} + B_z\hat{k})$$

reduces to

$$\vec{A} \times \vec{B}$$
$$= (A_yB_z - A_zB_y)\hat{i} + (A_zB_x - A_xB_z)\hat{j} + (A_xB_y - A_yB_x)\hat{k}$$

43. Given the two vectors

$$\vec{A} = -\hat{i} + 2\hat{j} + 3\hat{k}$$

and

$$\vec{B} = 2\hat{i} + \hat{j} - \hat{k}$$

(a) Find $\vec{A} \times \vec{B}$. (b) Verify that $\vec{A} \times \vec{B}$ is indeed perpendicular to *both* $\vec{A}$ and $\vec{B}$ by showing

$$(\vec{A} \times \vec{B}) \cdot \vec{A} = 0$$

and

$$(\vec{A} \times \vec{B}) \cdot \vec{B} = 0$$

(c) Find the vector product $\vec{B} \times \vec{A}$. Think! You do *not* have to reevaluate the whole expression; the answer is almost right before your eyes if you evaluated $\vec{A} \times \vec{B}$ correctly.

44. Deep inside an ancient physics text, you discover two vectors:

$$\vec{A} = 2\hat{i} - \hat{j} + 4\hat{k}$$

and

$$\vec{B} = 5\hat{i} + 2\hat{j} - 2\hat{k}$$

Not content with these hoary relics, you are asked to find a new vector that is *perpendicular to both* $\vec{A}$ and $\vec{B}$. Is the solution unique?

45. You are given a vector

$$\vec{A} = -3.20\hat{i} + 2.40\hat{j}$$

and a vector $\vec{B}$ in the x–y plane of magnitude 5.20. Vector $\vec{B}$ makes an angle of 120° with the positive x-axis. (a) Find $\vec{A} \cdot \vec{B}$. (b) Find $\vec{A} \times \vec{B}$.

46. Find a vector that is perpendicular to both the vector

$$\hat{i} + \hat{j} + \hat{k}$$

and the vector

$$3\hat{i} - \hat{j} + 2\hat{k}$$

Are there other possible solutions?

•**47.** Find a vector that is perpendicular to the vector

$$3\hat{i} - 2\hat{j} + 4\hat{k}$$

Is the solution unique? Explain your answer with the assistance of a sketch.

•**48.** Create a vector of magnitude unity perpendicular to both $2.0\hat{i} + 1.5\hat{j}$ and $2.0\hat{j} + 3.2\hat{k}$. Is there only one solution?

•**49.** The vectors

$$\vec{A} = \hat{i} + 2\hat{j} - \hat{k}$$

and

$$\vec{C} = 3\hat{j} + 6\hat{k}$$

were created to appease the academic gods. (a) Make an offering by finding $\vec{A} \cdot \vec{C}$. (b) Find a vector $\vec{B}$ so that

$$\vec{A} \times \vec{B} = \vec{C}$$

Is the solution for $\vec{B}$ unique? (c) Using the result of part (b), what is $\vec{B} \cdot \vec{C}$?

•**50.** Show that

$$(\vec{A} \cdot \vec{B})^2 + |\vec{A} \times \vec{B}|^2 = A^2B^2$$

•**51.** For the three vectors indicated in Figure P.51, find (a) $\vec{B}_1 + \vec{B}_2 + \vec{B}_3$; (b) $\vec{B}_1 + \vec{B}_2 - 2\vec{B}_3$; (c) $\vec{B}_1 \cdot \vec{B}_2$; (d) $\vec{B}_1 \times \vec{B}_2$; (e) $\vec{B}_3 \times \vec{B}_1$; (f) $\vec{B}_3 \cdot \vec{B}_1$.

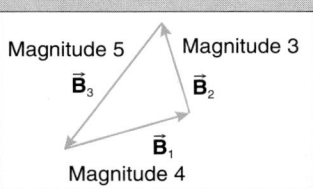

FIGURE P.51

•**52.** By evaluating the indicated vector and scalar products, show that

$$[\vec{A} \cdot (\vec{B} \times \hat{\imath})]\hat{\imath} + [\vec{A} \cdot (\vec{B} \times \hat{\jmath})]\hat{\jmath} + [\vec{A} \cdot (\vec{B} \times \hat{k})]\hat{k}$$
$$= (A_y B_z - A_z B_y)\hat{\imath} + (A_z B_x - A_x B_z)\hat{\jmath} + (A_x B_y - A_y B_x)\hat{k}$$

Thus, in view of Equation 2.32,

$$[\vec{A} \cdot (\vec{B} \times \hat{\imath})]\hat{\imath} + [\vec{A} \cdot (\vec{B} \times \hat{\jmath})]\hat{\jmath} + [\vec{A} \cdot (\vec{B} \times \hat{k})]\hat{k} = \vec{A} \times \vec{B}$$

•**53.** Use the Cartesian representation of each vector to show that $\vec{A} \times (\vec{B}_1 + \vec{B}_2)$ is equivalent (i.e., identical) to $\vec{A} \times \vec{B}_1 + \vec{A} \times \vec{B}_2$, so that the vector product obeys a distributive law:

$$\vec{A} \times (\vec{B}_1 + \vec{B}_2) = \vec{A} \times \vec{B}_1 + \vec{A} \times \vec{B}_2$$

•**54.** Three points on a large jungle gym have Cartesian coordinates (x, y, z) given by

$$P_1: \quad 1.00 \text{ m}, \ 2.00 \text{ m}, \ 3.00 \text{ m}$$
$$P_2: -2.00 \text{ m}, \ 1.00 \text{ m}, \ 4.00 \text{ m}$$
$$P_3: \quad 2.00 \text{ m}, -3.00 \text{ m}, \ 5.00 \text{ m}$$

(a) Let $\vec{\ell}_1$ be the vector from P_1 to P_2. Find the Cartesian representation of $\vec{\ell}_1$. (b) Let $\vec{\ell}_2$ be the vector from P_1 to P_3. Find the Cartesian representation of $\vec{\ell}_2$. (c) Relate the area of the triangle formed by the three points to the *magnitude* of the vector product of $\vec{\ell}_1$ and $\vec{\ell}_2$, and find the numerical value of the area.

•**55.** While cleaning out the lab, you discover two soiled vectors named

$$\vec{p}_1 = 2.1\hat{\imath} - 5.3\hat{\jmath} + 3.4\hat{k}$$

and

$$\vec{p}_2 = 3.6\hat{\imath} + 2.8\hat{\jmath} - 0.9\hat{k}$$

Find (a) $\vec{p}_1 + \vec{p}_2$; (b) $\vec{p}_1 - \vec{p}_2$; (c) the magnitude of $\vec{p}_1$; (d) the magnitude of $\vec{p}_2$; (e) the scalar product of $\vec{p}_1$ and $\vec{p}_2$; (f) the vector product $\vec{p}_1 \times \vec{p}_2$; (g) the angle between $\vec{p}_1$ and $\vec{p}_2$.

•**56.** Three vectors are expressed in Cartesian form:

$$\vec{A} = 4\hat{\jmath}$$
$$\vec{B} = \hat{\imath} - 3\hat{\jmath} + 2\hat{k}$$
$$\vec{C} = 2\hat{\imath} - \hat{\jmath} + 3\hat{k}$$

(a) For the pure thrill of it, evaluate

$$\vec{A} \times (\vec{B} \times \vec{C})$$

and show that it is equivalent to

$$(\vec{A} \cdot \vec{C})\vec{B} - (\vec{A} \cdot \vec{B})\vec{C}$$

This equivalence is true for *any* three vectors, not just the ones given. We state this without proof, but you should be able to verify the identity of the two expressions for three general vectors expressed in Cartesian form. (b) What is $\vec{A} \cdot [\vec{A} \times (\vec{B} \times \vec{C})]$? What does this result mean?

•**57.** Consider the vectors

$$\vec{A}_1 = 4.0\hat{\jmath}$$
$$\vec{A}_2 = 1.0\hat{\imath} - 3.0\hat{\jmath} + 2.0\hat{k}$$
$$\vec{A}_3 = 2.0\hat{\imath} - 1.0\hat{\jmath} + 3.0\hat{k}$$

(a) Find $\vec{A}_1 + \vec{A}_2 - 2\vec{A}_3$. (b) Find $\vec{A}_1 \cdot (\vec{A}_2 \times \vec{A}_3)$. (c) For no apparent reason, calculate the angle between $\vec{A}_2$ and $\vec{A}_3$.

‡**58.** Consider the following vector relationship:

$$\vec{F} = q(\vec{v} \times \vec{B})$$

where q is a scalar. In our study of magnetism (Chapter 20), we will see that this vector relationship gives the force $\vec{F}$ on a charge q moving at a velocity $\vec{v}$ in a magnetic field $\vec{B}$. Suppose you conduct some experiments and find the following:

If the vector $\vec{v} = \hat{\imath}$, then $\vec{F}/q = -4\hat{\jmath} - 2\hat{k}$.

If the vector $\vec{v} = \hat{\jmath}$, then $\vec{F}/q = 4\hat{\imath} - 3\hat{k}$.

If the vector $\vec{v} = \hat{k}$, then $\vec{F}/q = 2\hat{\imath} + 3\hat{\jmath}$.

Use this information to find the vector $\vec{B}$.

‡**59.** This problem illustrates how vector notation can be used to write some of the laws of physics. In geometric optics, if a ray of light (1) is incident on a boundary between two transparent media as in Figure P.59, some of the light is reflected, producing a reflected ray (2), and some of the light is transmitted, producing a refracted ray (3). Among other things, the law of reflection states that

$$\theta_1 = \theta_2$$

and the law of refraction states that

$$n_1 \sin \theta_1 = n_2 \sin \theta_3$$

where n_1 and n_2 are two pure numbers characteristic of the respective media (they are known as the indices of refraction). Let $\hat{r}_1$, $\hat{r}_2$, and $\hat{r}_3$ be dimensionless *unit vectors* pointing in the direction of the respective rays as indicated in Figure P.59. Furthermore, let $\hat{r}$ be a dimensionless *unit vector* perpendicular to the interface as shown. (a) Use these unit vectors to write the law of reflection as a scalar product. (b) Use these unit vectors, the indices of refraction, and the concept of a vector product to write a vector equation whose magnitude yields the law of refraction.

This problem was inspired by a problem in Mary L. Boas, *Mathematical Methods in the Physical Sciences* (2nd edition, Wiley, New York, 1983), page 243.

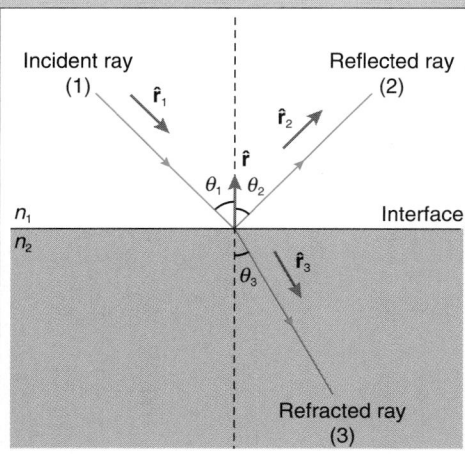

FIGURE P.59

‡**60.** The story in Figure P.60A appears in George Gamow's book *One Two Three . . . Infinity* (Bantam, New York, 1961), Chapter II.

There was a young and adventurous man who found among his great-grandfather's papers a piece of parchment that revealed the location of a hidden treasure. The instructions read:

"Sail to _____ North latitude and _____ West longitude[1] where thou wilt find a deserted island. There lieth a large meadow, not pent, on the north shore of the island where standeth a lonely oak and a lonely pine.[2] There thou wilt see also an old gallows on which we once were wont to hang traitors. Start thou from the gallows and walk to the oak counting thy steps. At the oak thou must turn *right* by a right angle and take the same number of steps. Put here a spike in the ground. Now must thou return to the gallows and walk to the pine counting thy steps. At the pine thou must turn *left* by a right angle and see that thou takest the same number of steps, and put another spike into the ground. Dig halfway between the spikes; the treasure is there."

The instructions were quite clear and explicit, so our young man chartered a ship and sailed to the South Seas. He found the island, the field, the oak and the pine, but to his great sorrow the gallows was gone. Too long a time had passed since the document had been written; rain and sun and wind had disintegrated the wood and returned it to the soil, leaving no trace even of the place where it once had stood.

Our adventurous young man fell into despair, then in an angry frenzy began to dig at random all over the field. But all his efforts were in vain: the island was too big! So he sailed back with empty hands. And the treasure is probably still there.

[1]The actual figures of longitude and latitude were given in the document but are omitted in this text, in order not to give away the secret.

[2]The names of the trees are also changed for the same reason as above. Obviously there would be other varieties of trees on a tropical treasure island.

FIGURE P.60A

The location of the treasure can be found by using vector analysis in the following way. Construct a Cartesian coordinate system as indicated in Figure P.60B with an origin halfway between the oak and pine trees with the x- and y-axes directed as indicated. The position vectors of the pine and oak trees then are

$$\vec{r}_{pine} = \ell\hat{i}$$

$$\vec{r}_{oak} = -\ell\hat{i}$$

Let the unknown location of the gallows be represented by the vector $\vec{R}$:

$$\vec{R} = R_x\hat{i} + R_y\hat{j}$$

(a) Show that the vector $\vec{L}_1$ *from* the gallows *to* the oak tree is given by

$$\vec{L}_1 = \vec{r}_{oak} - \vec{R}$$
$$= -(\ell + R_x)\hat{i} - R_y\hat{j}$$

Follow the map directions and use the concept of the vector product to show that the location of spike 1 is

$$\vec{R}_1 = \vec{L}_1 \times \hat{k}$$
$$= -R_y\hat{i} + (\ell + R_x)\hat{j}$$

with the tail of $\vec{R}_1$ located at the oak. (b) Show that the vector $\vec{L}_2$ *from* the gallows *to* the pine tree is given by

$$\vec{L}_2 = \vec{r}_{pine} - \vec{R}$$
$$= (\ell - R_x)\hat{i} - R_y\hat{j}$$

Follow the map directions and use the concept of the vector product to show that the location of spike 2 is

$$\vec{R}_2 = \hat{k} \times \vec{L}_2$$
$$= R_y\hat{i} + (\ell - R_x)\hat{j}$$

with the tail of $\vec{R}_2$ located at the pine. (c) Show that the position vector $\vec{r}$ of the treasure is

$$\vec{r} = \ell\hat{j}$$

or along the y-axis at a distance from the origin equal to half the distance between the trees.

This problem is adapted from R. Ramirez-Bon, "An interesting problem solved by vectors," *The Physics Teacher*, 28, #9, pages 594–595 (December 1990).

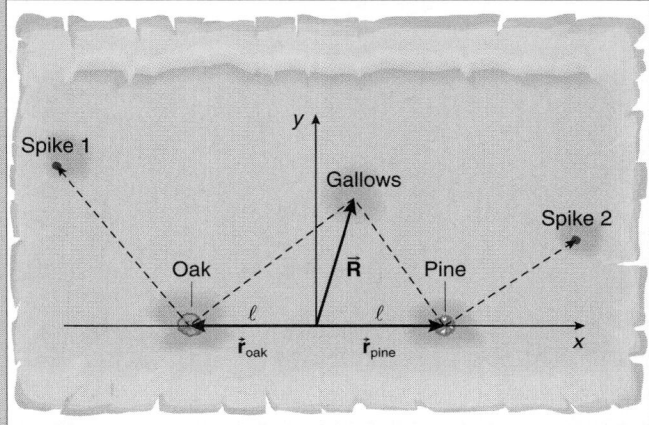

FIGURE P.60B

61. Three noncoplanar vectors with their tails at a common point can define a parallelopiped as shown in Figure P.61. (a) Show that $|\vec{B} \times \vec{C}|$ is the area of a parallelogram whose sides have lengths B and C. (b) Show that the expression $\vec{A} \cdot (\vec{B} \times \vec{C})$ represents the volume of the parallelopiped whose sides are formed from the vectors as in Figure P.61.

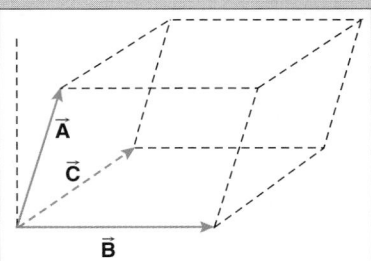

FIGURE P.61

62. Crystals are of interest to geologists, crystallographers, physicists, engineers, and jewelry makers. It is customary to locate the atoms defining the elementary or primitive structural arrangement of a crystal by means of what are called *primitive lattice vectors*. For example, many elements, such as sodium, chromium, and zirconium, have a crystalline structure known as the body-centered cubic lattice. For this structure, the primitive lattice vectors are

$$\vec{a} = \frac{\ell}{2}(\hat{i} + \hat{j} - \hat{k})$$

$$\vec{b} = \frac{\ell}{2}(-\hat{i} + \hat{j} + \hat{k})$$

$$\vec{c} = \frac{\ell}{2}(\hat{i} - \hat{j} + \hat{k})$$

where ℓ is the side of the elementary cubic structure and $\hat{i}$, $\hat{j}$, and $\hat{k}$ are the usual Cartesian unit vectors, here directed parallel to the edges of the elementary cubic structure. The locations of all the atoms in the crystal can be represented by a linear combination of these primitive lattice vectors with integral coefficients. The result of Problem 61 indicates that the volume of the primitive cell of the lattice, whose sides are the vectors $\vec{a}$, $\vec{b}$, and $\vec{c}$, is

$$\vec{a} \cdot (\vec{b} \times \vec{c})$$

This is a real, physical volume since the lattice vectors represent physical lengths. (a) Use the primitive lattice vectors defined for the body-centered cubic lattice to show that the volume of the primitive cell is

$$\vec{a} \cdot (\vec{b} \times \vec{c}) = \frac{\ell^3}{2}$$

Another useful set of vectors associated with the crystal are known as the *reciprocal lattice vectors*. They are defined in the following way:

$$\vec{a}' \equiv 2\pi \frac{\vec{b} \times \vec{c}}{\vec{a} \cdot (\vec{b} \times \vec{c})}$$

$$\vec{b}' \equiv 2\pi \frac{\vec{c} \times \vec{a}}{\vec{a} \cdot (\vec{b} \times \vec{c})}$$

$$\vec{c}' \equiv 2\pi \frac{\vec{a} \times \vec{b}}{\vec{a} \cdot (\vec{b} \times \vec{c})}$$

(For some purposes the factor of 2π is omitted from the definition, so that one needs to be careful about *what* definition is used for the reciprocal lattice vectors.) (b) Find the reciprocal lattice vectors for the body-centered cubic lattice. (c) Find the volume of the primitive cell formed by the reciprocal lattice vectors. That is, find

$$\vec{a}' \cdot (\vec{b}' \times \vec{c}')$$

(The result is not a real physical volume, since the dimensions of the reciprocal lattice vectors are not lengths but *inverse* lengths.) The primitive cell formed by the reciprocal lattice vectors is known in the trade as the *first Brillouin zone*.

Sections 2.18 Variation of a Vector
2.19 Some Aspects of Vector Calculus

63. Differentiate the vector $\vec{A}(t) = 3t^2\hat{i} + 2t\hat{j} + 8\hat{k}$ with respect to t.

•64. Differentiate the vector

$$\vec{A}(t) = [3 \sin (5t)]\hat{i} + [4 \cos (2t)]\hat{j} + 3e^{-4t}\hat{k}$$

with respect to t.

•65. Differentiate the vector

$$\vec{A}(t) = 5\sqrt{t}\,\hat{i} + 6t^{3/2}\hat{j}$$

with respect to t.

•66. Differentiate the vector

$$\vec{r}(t) = [A \cos (\omega t)]\hat{i} + [A \sin (\omega t)]\hat{j}$$

with respect to t, where ω is a constant.

•67. Use the Cartesian representation of each vector to show explicitly that

$$\frac{d}{dt}(\vec{F} + \vec{f})$$

the time derivative of the sum of $\vec{F}$ and $\vec{f}$, is the same as the vector sum of

$$\frac{d\vec{F}}{dt}$$

and

$$\frac{d\vec{f}}{dt}$$

In other words, first add $\vec{F}$ and $\vec{f}$ as vectors and then take the derivative of their sum. Then show that this result is identical to that found by taking the vector sum of the time derivative of $\vec{F}$ and the time derivative of $\vec{f}$.

•68. Use the Cartesian representation of the vectors to show that the time derivative of the scalar product of $\vec{A}$ and $\vec{B}$,

$$\frac{d}{dt}(\vec{A} \cdot \vec{B})$$

is the same as the scalar sum of

$$\frac{d\vec{A}}{dt} \cdot \vec{B}$$

and

$$\vec{A} \cdot \frac{d\vec{B}}{dt}$$

•69. Use the Cartesian representation of the vectors to show that the time derivative of the vector product of $\vec{A}$ and $\vec{B}$,

$$\frac{d}{dt}(\vec{A} \times \vec{B})$$

is the same as the vector sum of

$$\frac{d\vec{A}}{dt} \times \vec{B}$$

and

$$\vec{A} \times \frac{d\vec{B}}{dt}$$

Thus the derivative of a vector product obeys the usual product rule in calculus as long as we preserve the order of the terms.

INVESTIGATIVE PROJECTS

A. Expanded Horizons

1. Investigate the similarities between the addition and subtraction of two-dimensional vectors and the algebraic addition and subtraction of complex numbers.
   Ruel V. Churchill, *Complex Variables and Applications* (2nd edition, McGraw-Hill, New York, 1960), pages 4–5.
   W. E. Baylis, J. Huschilt, and Jiansu Wei, "Why *i*?" *American Journal of Physics*, 60, #9, pages 788–797 (September 1992).

2. Vector notation as we know it was invented only about a century ago. Should you be interested in researching its historical roots, a good place to begin is the following:
   A. P. Wills, *Vector Analysis with an Introduction to Tensor Analysis* (Dover, New York, 1958), Chapter 1.
   Michael J. Crowe, *A History of Vector Analysis* (Dover, New York, 1994).

3. Investigate the use of vectors in crystallography. In particular, discover the form for the basis vectors for various crystal symmetries and the corresponding reciprocal lattice vectors (see Problem 62).
   Charles Kittel, *Introduction to Solid State Physics* (7th edition, John Wiley, New York, 1996), Chapters 1 and 2.
   George R. Mitchell, "The reciprocal lattice—a demonstration," *American Journal of Physics*, 46, #5, pages 574–575 (May 1978).

4. Investigate the mathematical field of *Clifford algebra* and how it relates to the vector algebra of this chapter.
   Anthony Garrett, "An advertisement for Clifford algebra," *Physics World*, 5, #9, pages 13–14 (September 1992). This short article references tutorial articles about Clifford algebra.
   David Hestenes, *Clifford Algebra to Geometric Calculus: A Unified Language for Mathematics and Physics* (Kluwer Academic Press, Dordrecht, Netherlands 1987).

5. Investigate the analytical or algebraic extension of vector ideas in three dimensions to tensors using matrix techniques.
   Robert B. Leighton, *Principles of Modern Physics* (McGraw-Hill, New York, 1959), pages 15–22.

6. Study the general area of Cartesian tensors and the Einstein "dummy index" (repeated index) notational convention as a way of representing summations in vector and tensor analysis.
   Mary L. Boas, *Mathematical Methods in the Physical Sciences* (2nd edition, John Wiley, New York, 1983), pages 437–441.
   Robert B. Leighton, *Principles of Modern Physics* (McGraw-Hill, New York, 1959), pages 20–21.

7. Investigate the distinction between (a) scalars and *pseudoscalars* and (b) vectors and *pseudovectors* (also known as *axial vectors*).
   Llewelyn G. Chambers, *A Course in Vector Analysis* (Chapman and Hall, London, 1969), pages 58–60.
   Walter Hauser, "Vector products and pseudovectors," *American Journal of Physics*, 54, #2, pages 168–172 (February 1986).

8. Investigate the ideas of vector analysis in coordinate systems with axes that are *not* perpendicular to each other (as is common in crystallography). In particular, investigate the distinction between *covariant* and *contravariant* components of a vector.
   Mary L. Boas, *Mathematical Methods in the Physical Sciences* (2nd edition, John Wiley, New York, 1983), pages 447–449.
   George Arfken, *Mathematical Methods for Physicists* (Academic Press, New York, 1970), pages 184–187.

B. Communicating Physics

9. A local newspaper editor calls you as a technical expert and asks you to write a short but interesting feature for the science page that distinguishes between scalars and vectors for the edification of the nontechnical general public. Write such an article, keeping within the space limitations set by the editor: 300 words. Suggest an interesting title for your article.

CHAPTER 3

KINEMATICS I
RECTILINEAR MOTION

**Examine the beauty of mobility in a body—Numbers are turned
in time. Go within to the art whence these numbers proceed.
Look there for time and for place.**

St. Augustine (A.D. 354–430)*

If nothing moved, we might as well be rocks. Motion implies change, and it is change that makes life—and physics—viable and interesting. We move, soccer balls move (hopefully toward the right goal), the Earth moves, electrons move, and light *really* moves. The whole universe is a bewildering array of incessant motion. For this reason we begin our study of physics with a description of motion, called **kinematics**.[†]

Our study of motion begins with that of a single **particle**, which is the basic building block for the study of the motion of more complex systems. A particle is anything whose geometric size is ignored; it is represented by a point in space. Protons, electrons, atoms, molecules, soccer balls, cars, boats, *you,* the Earth, Moon, Sun, and stars all are particles if we represent them as pointlike objects whose geometric size is ignored.

The study of motion has roots deep in antiquity. Observations of the night sky with the moving Moon and wandering planets were some of the first inspirations for studying motion. The study was frustrating. Developing a consistent theory of motion was difficult because, as we mentioned in Chapter 1, physical theories do not exist in nature just waiting to be discovered by some lucky scientist, like a prospector sifting for precious gold nuggets. Physical theories are *creations* of the human intellect; they must be *invented* rather than discovered. In this sense, physics is like the arts and literature: it is a creative process. It builds on the best of the past to create new visions of nature.

The poetry of motion!

Kenneth Grahame (1859–1932)[‡]

When Johannes Kepler (1571–1630) first was able to decipher accurately and precisely the paths and positions of the wandering planets, including the Earth, during the early 1600s,[§]

---

*(Chapter Opener) *De Libero Arbitrio*, translated by Francis E. Tourscher (The Peter Reilly Company, Philadelphia, 1937), Book III, Chapter 16, page 205.

[†]The word comes from the Greek word $\kappa\acute{\iota}\nu\eta\mu\alpha$, meaning motion; the word *cinema* has the same Greek root.

[‡]*The Wind in the Willows* (Charles Scribner's, New York, 1908), Chapter II, "The Open Road," page 41.

[§]Kepler's three empirical laws of planetary motion will be examined in Chapter 6.

it seemed possible that the motions of the universe around us, even motions here on the Earth, were within the grasp of the human intellect. Once the Earth was seen by Nicholas Copernicus (1473–1543) and Johannes Kepler to be just another planet of the Sun, there was no longer any reason to think that mathematics could be applied only to the celestial realm and not to physical phenomena in our own backyard.

It was Galileo Galilei (1564–1642) who revolutionized the description of the motion of particles. The work of Galileo was a great turning point in the development of quantitative science, for it was one of the opening salvos of a scientific revolution that continues to this day. Galileo broke radically with the 2000-year-old—but incorrect—Aristotelian ideas that held that while mathematics (with the purity of geometry and numbers) could be applied to the "perfect," godlike, celestial realm of the stars (astronomy), it could not be applied to the mundane, changing phenomena of nature on the Earth.

Galileo's genius lay not only in realizing that mathematics was the key to the sciences, but also in bridging the intellectual chasm between observations (experiments) and mathematical models. He realized that *approximations* were needed to limit complexity. The natural world brims with effects that hinder generalizations, friction and the finite size of material objects being the foremost culprits impeding the development of an accurate, consistent theory of motion. Galileo was among the first to design experiments to test his ideas, experiments that enabled him to probe or isolate one factor from another in the description of motion. Galileo invented the description of motion using inspiration, experiment, and extrapolations from experiments in the real world to idealizations that approximated reality, such as situations without friction.

In this chapter and the next, we examine a quantitative theory of kinematics whose roots lie with Galileo. The underlying theory of the *causes of motion,* called **dynamics**, was developed after Galileo by Isaac Newton (1642–1727) and published in 1686 as the *Principia*. The Newtonian theory of dynamics will be examined in Chapter 5.

Johannes Kepler deciphered the structure of the solar system.

Galileo Galilei made many significant contributions to kinematics.

FIGURE 3.1 Examples of rectilinear motion.

3.1 RECTILINEAR MOTION

I saw before me a vast multitude of small Straight Lines . . . interspersed with other Beings still smaller . . . all moving to and fro in one and the same Straight Line.

Edwin Abbott Abbott (1838–1926)*

Roll a marble across the floor, drop a rock from a high bridge into a river below, toss a ball straight up and watch it ascend and descend, or shoot an electron down the straight beam tube of the Stanford Linear Accelerator (SLAC). All are examples of **rectilinear motion**, the motion of a particle along a straight line. Other examples of rectilinear motion are shown in Figure 3.1.

The generalization to two- and three-dimensional kinematics involves repetitions of the one-dimensional case two or three times, as we will see in Chapter 4. Although it is possible to study rectilinear motion without using vectors, the generalization to two and three dimensions is easier if we use vectors from the beginning, even for the one-dimensional situation. A vector approach also makes the algebraic signs of the numerical values of various terms in the few significant equations of kinematics a matter of the choice we make for the directions of the coordinate axes.

QUESTION 1

Is the universe as a whole a particle? Is it the "mother of all particles"? Is the universe *the* fundamental particle? Explain your response.

3.2 POSITION AND CHANGES IN POSITION

To meet a friend arriving by plane, you need to know two things: *when* and *where* it arrives. Incomplete information such as knowing the time of arrival but not the gate number, or vice versa, is frustrating and inconvenient. To describe the motion of a particle, we also begin by specifying where it is located at a specified time. To say that it is "here" at "such-and-such" a time defines an **event**. An event in physics is no more complicated than the information on an airport arrival marquee or an invitation to an exciting party: where and when need to be specified.

Imagine yourself in your car moving along a straight dragstrip, as in Figure 3.2. Let the car be represented as a particle, say, by a point at the center of the car. Let the x-axis of a Cartesian coordinate system be the one-dimensional straight path along which the car is moving, with $\hat{\mathbf{i}}$ directed to the right.

The origin may be chosen to be at any convenient point on the line, perhaps at the starting line of the dragstrip. The **position** of the car on the x-axis is indicated by the *coordinate x*

Flatland (Princeton University Press, Princeton, New Jersey, 1991), Part II, Section 13, pages 53–54.

of the particle (the center of the car) at any instant t. We use subscripts (or primes) to distinguish specific x-coordinates and specific instants. Imagine the particle to be at the coordinate x_i at the instant t_i, as in Figure 3.2; these need not be the origin or when the clock reads zero. Rather, x_i and t_i represent where the car is when we initialize, or begin our observations. The location of the car at this instant is specified by a one-dimensional position vector $\vec{\mathbf{r}}_i$ that locates the coordinate of the particle with respect to the (arbitrarily chosen) origin of the coordinate system as shown in Figure 3.3.

Specifying $\vec{\mathbf{r}}_i$ and t_i defines an initial event: where the car is with respect to the origin of the coordinate grid and when it is there. It is important to realize that the moving particle spends zero time at this position. The position vector is an *instantaneous* position vector that locates the particle at an instant in time.

> **Time instants** have no time duration and do not correspond to **time intervals**, even of very short duration.

Later, at the instant t_f, the car is found at the coordinate x_f; its new (instantaneous) position vector is $\vec{\mathbf{r}}_f = x_f\hat{\mathbf{i}}$, as shown in Figure 3.4. Specifying $\vec{\mathbf{r}}_f$ and t_f defines another event—another where and when.

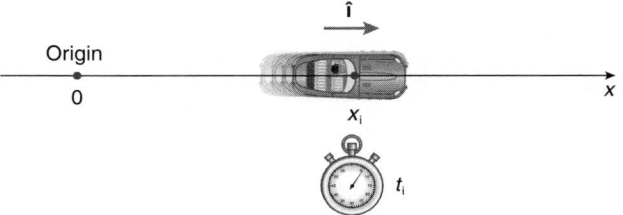

FIGURE 3.2 A car moving along a straight dragstrip.

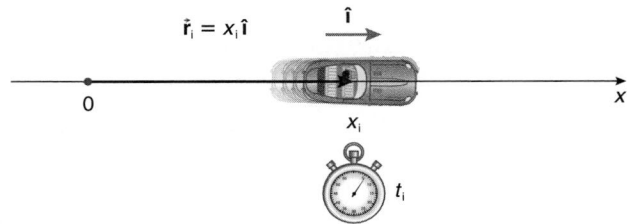

FIGURE 3.3 The initial position vector of the car is $\vec{\mathbf{r}}_i = x_i\hat{\mathbf{i}}$.

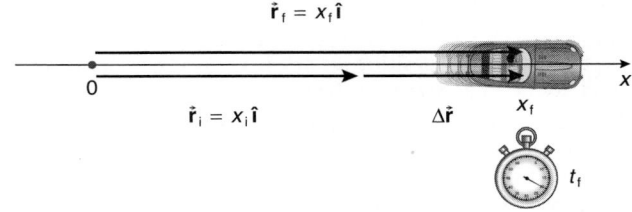

FIGURE 3.4 A subsequent position vector of the car is $\vec{\mathbf{r}}_f = x_f\hat{\mathbf{i}}$.

An event specifies where and when like this invitation to a Fancy Dress Ball.

A **time interval** Δt involves the subtraction of two different clock readings (time instants). The time interval between two events associated with the motion is

$$\Delta t \equiv t_f - t_i \qquad (3.1)$$

and is a *positive* scalar quantity since $t_f > t_i$. The change in the position vector $\Delta \vec{r}$ is the vector difference between the new (final) position vector and the old (initial) position vector:

$$\Delta \vec{r} \equiv \vec{r}_f - \vec{r}_i \qquad (3.2)$$

For rectilinear motion, this becomes

$$\begin{aligned} \Delta \vec{r} &= x_f \hat{i} - x_i \hat{i} \\ &= (x_f - x_i)\hat{i} \\ &= \Delta x\,\hat{i} \end{aligned} \qquad (3.3)$$

By transposing $\vec{r}_i$ in Equation 3.2, the three vectors are related by the vector sum:

$$\vec{r}_f = \vec{r}_i + \Delta \vec{r}$$

which is shown in Figure 3.4.

The magnitude of the change in the position vector is *not* necessarily equal to the total distance d traveled by the particle during the time interval Δt between the two events.

The change in the position vector is concerned *only* with the initial and final positions of the particle, not with the details of what happened to the particle during the time interval. Think of the total distance traveled as the reading of a trip odometer initially set to zero at the initial position of the particle. On the other hand, the magnitude of the change in the position vector is concerned only with where the particle ended up relative to where it began.

The distance traveled differs from the magnitude of the change in the position vector for rectilinear motion if the particle changes direction or moves back and forth along the axis while moving from $\vec{r}_i$ to $\vec{r}_f$.

Indeed, for a round trip, the initial and final position vectors of the particle are the same, and $|\Delta \vec{r}| = 0$ m, but the total distance d traveled by the car is not zero. Likewise, in another context, a beeline from your dorm to a pizza parlor and back during an interval Δt means $|\Delta \vec{r}| = 0$ m, but a nonzero trip distance d during the interval. Also, if you toss a marble vertically into the air and catch it at the position it was released, then $|\Delta \vec{r}| = 0$ m while $d \neq 0$ m for the interval Δt.

As we noted in Section 2.18, the magnitude and direction of the change in the position vector $\Delta \vec{r}$, indeed, of *any* vector, are independent of the coordinate system chosen to describe the change. However, the signs of the x-components of $\vec{r}$ and of $\Delta \vec{r}$ depend on the coordinate values of x_f and x_i; *these values depend on the choice you make for the direction of $\hat{i}$ and the location you choose for the origin.*

Since the magnitude and direction of the change in the position vector are independent of the coordinate choice, you can make things easier for yourself if you follow Problem-Solving Tactic 3.1.

PROBLEM-SOLVING TACTIC

3.1 For rectilinear motion, always choose a Cartesian coordinate axis to be coincident with the line followed by the particle.

For simplicity, we will always use the x-axis for rectilinear motion. The Cartesian origin can be chosen at any convenient position along the line of motion; it need not correspond to where the particle is when it begins its motion or to its position when we begin our observations. The direction corresponding to increasing values of the coordinate, that is, the direction of the unit vector $\hat{i}$, is ours to choose; it need not correspond to the initial direction of travel of the particle. With such coordinate conventions and choices, though, the description of rectilinear motion will be simpler than with other coordinate choices.

Notice that we have choices to make for

- the location of the origin and
- the direction for $\hat{i}$.

Such choices must be specified when solving problems.

The easiest way to specify such choices is with a sketch, as in Figure 3.5. Place 0 where you choose the origin to be located. Place $\hat{i}$ somewhere in the sketch or label x at the end of the coordinate line, or both. Bear in mind:

a. The position vector $\vec{r}$ locates the particle at an instant t;
b. The magnitude and direction of the *change* in the position vector $\Delta \vec{r}$ that occurs during a time interval Δt are independent of the specific coordinate system we choose; and

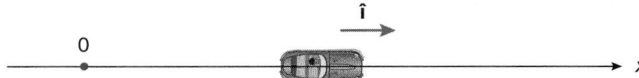

FIGURE 3.5 Specify the location of the origin and the direction of î and/or the x-axis.

c. The total distance d traveled by the particle during the interval Δt is a positive scalar quantity that, even for rectilinear motion, may differ from the magnitude of the change in the position vector if there is back-and-forth motion on the one-dimensional line during the interval Δt.

The position vector of a moving particle is a function of time, written as $\vec{r}(t)$. For one-dimensional (i.e., rectilinear) motion with the x-axis coincident with the line of motion of the particle,

$$\vec{r}(t) = x(t)\hat{\imath} \tag{3.4}$$

The scalar function $x(t)$ describes the time dependence of the co-ordinate x of the particle. At any specific instant, say t_1, the instantaneous position vector of the particle is $\vec{r}(t_1)$; we find this vector function by evaluating the scalar function $x(t)$ at the particular instant t_1:

$$\vec{r}(t_1) = x(t_1)\hat{\imath}$$

A graph of the x-component of the position vector of the particle as a function of time is called an x versus t graph. For your car moving along a dragstrip, an x versus t graph might appear as in Figure 3.6. The line on the graph representing $x(t)$ is called a **world-line**. An event is a point on the world-line.

PROBLEM-SOLVING TACTIC

3.2 Read and construct graphs carefully. Notice what is plotted versus what or, if you are making the graph, be sure to label what each axis represents. When we say to "plot x versus t," plot x vertically (as the ordinate) and t horizontally (as the abscissa). Pay attention to the units used along each axis of a graph or, if you are making the graph, indicate the appropriate units for the scale you use along each axis.

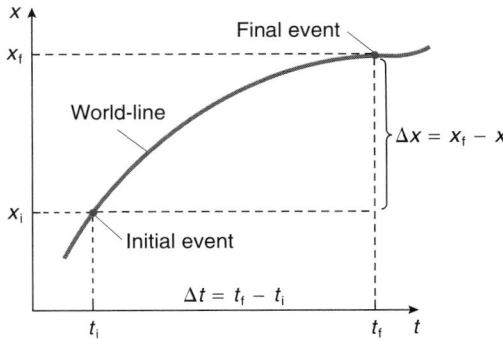

FIGURE 3.6 An x versus t graph.

QUESTION 2

Explain why the world-line of a particle on the x versus t graph in Figure Q.2 is meaningless.

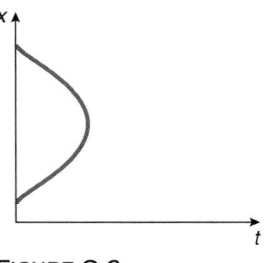

FIGURE Q.2

EXAMPLE 3.1

Your pet beagle is running happily along a wharf and you note her position vector at two instants, as in Figure 3.7.

a. Describe the two events and indicate them on an x versus t graph.
b. Specify the initial position vector.
c. Specify the final position vector.
d. Specify the time interval, the change in the position vector $\Delta\vec{r}$, and its magnitude and direction. Sketch $\Delta\vec{r}$.
e. What, if anything, can be said about the total distance traveled by the beagle-particle during the time interval between the two instants?

Solution

a. The dog initially had the x-coordinate 4.00 m when the stopwatch reading was 3.00 s. Later, when the stopwatch read 5.00 s, the x-coordinate of the beagle was −2.00 m. The two events are represented by points on the x versus t graph shown in Figure 3.8.
b. The initial position vector is $\vec{r}_i = (4.00\ \text{m})\hat{\imath}$.
c. The final position vector is $\vec{r}_f = (-2.00\ \text{m})\hat{\imath}$.

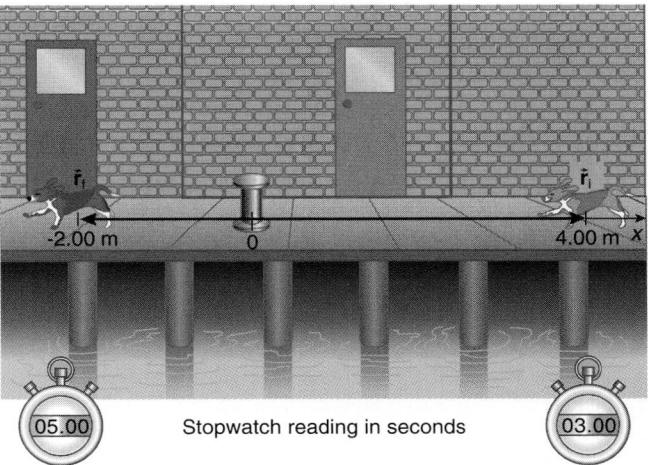

FIGURE 3.7 The position vectors of a beagle-particle at two instants indicated by a digital stopwatch.

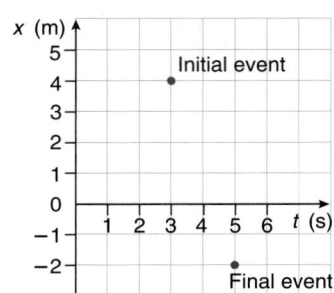

FIGURE 3.8

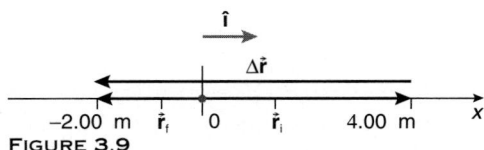

FIGURE 3.9

d. The time interval between the events is

$$\Delta t = t_f - t_i$$
$$= 5.00 \text{ s} - 3.00 \text{ s}$$
$$= 2.00 \text{ s}$$

The change in the position vector is

$$\Delta \vec{r} = \vec{r}_f - \vec{r}_i$$
$$= (-2.00 \text{ m})\hat{\imath} - (4.00 \text{ m})\hat{\imath}$$
$$= (-6.00 \text{ m})\hat{\imath}$$

The magnitude of this vector is 6.00 m, and it is directed to the left in Figure 3.9.

e. You know the positions of the beagle at two instants of time but are given no information about the detailed motion of the dog during the time interval between the two instants. If the dog proceeded from the initial position to the final position without any back-and-forth motion, then the total distance traveled is 6.00 m; if any back-and-forth motion or departures from the straight path occurred during the interval, the total distance traveled is greater than 6.00 m.

EXAMPLE 3.2

The position vector of a fleet-footed African cheetah (*Acinonyx jubatus*) is given by the following function of time:

$$\vec{r}(t) = x(t)\hat{\imath}$$
$$= [4.00 \text{ m} - (2.00 \text{ m/s})t - (3.00 \text{ m/s}^2)t^2]\hat{\imath}$$

where the time t is expressed in seconds on a stopwatch.

a. What is the position vector of the cheetah-particle at the instant the stopwatch commences, when $t = 0$ s?
b. What is the position vector of the cheetah when the stopwatch indicates $t = 3.00$ s?
c. What is the change in the position vector during the three second interval between these time instants?

d. Make a schematic sketch of the two position vectors and the change in the position vector.
e. Make an accurate sketch of the world-line of the cheetah on an x versus t graph between the two events described in parts (a) and (b).

Solution
a. To find the position vector when $t = 0$ s, substitute this instant into the expression for $\vec{r}(t)$:

$$\vec{r}(0 \text{ s}) = [4.00 \text{ m} - (2.00 \text{ m/s})(0 \text{ s})$$
$$- (3.00 \text{ m/s}^2)(0 \text{ s})^2]\hat{\imath}$$
$$= (4.00 \text{ m})\hat{\imath}$$

This is a vector of magnitude 4.00 m, directed in the same sense as $\hat{\imath}$, and it locates the position of the particle when $t = 0$ s.
b. Likewise, to find the position vector when $t = 3.00$ s, substitute the instant $t = 3.00$ s into the expression for $\vec{r}(t)$:

$$\vec{r}(3.00 \text{ s}) = [4.00 \text{ m} - (2.00 \text{ m/s})(3.00 \text{ s})$$
$$- (3.00 \text{ m/s}^2)(3.00 \text{ s})^2]\hat{\imath}$$
$$= (4.00 \text{ m} - 6.00 \text{ m} - 27.0 \text{ m})\hat{\imath}$$
$$= (-29.0 \text{ m})\hat{\imath}$$

This is a vector of magnitude 29.0 m, directed in the same sense as $-\hat{\imath}$, and represents the new position of the cheetah when $t = 3.00$ s.
c. The change in the position vector is

$$\Delta \vec{r} = \vec{r}_f - \vec{r}_i$$
$$= (-29.0 \text{ m})\hat{\imath} - (4.0 \text{ m})\hat{\imath}$$
$$= (-33.0 \text{ m})\hat{\imath}$$

This means the change in the position vector is a vector of magnitude 33.0 m, parallel to $-\hat{\imath}$ (to the left in Figure 3.10).
d. The three vectors are indicated schematically in Figure 3.10.
e. A graph of $x(t)$ versus t is constructed by substituting various values of t into the given equation and determining the corresponding values of x. This is summarized in the following table:

t (s)	x (m)
0	4.00
1.00	−1.00
2.00	−12.0
3.00	−29.0

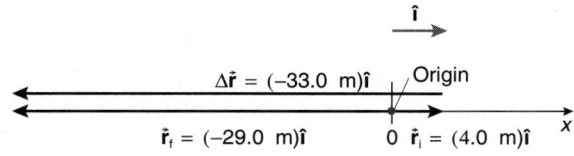

FIGURE 3.10

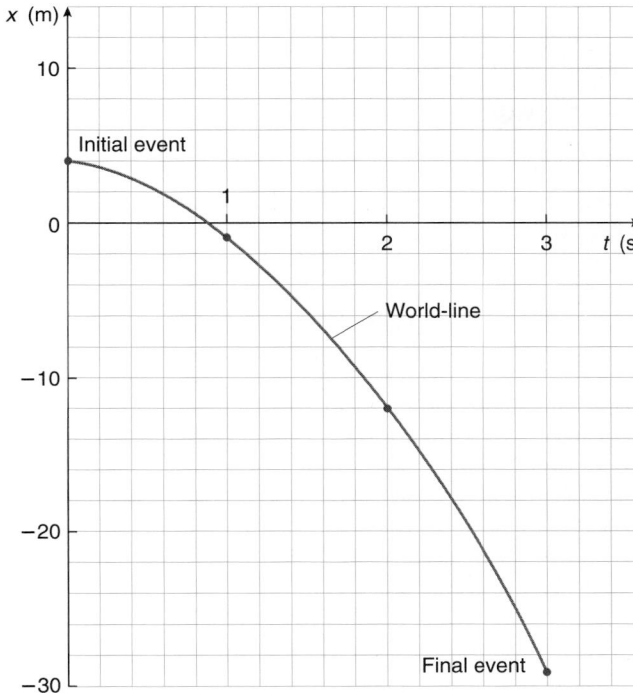

FIGURE 3.11

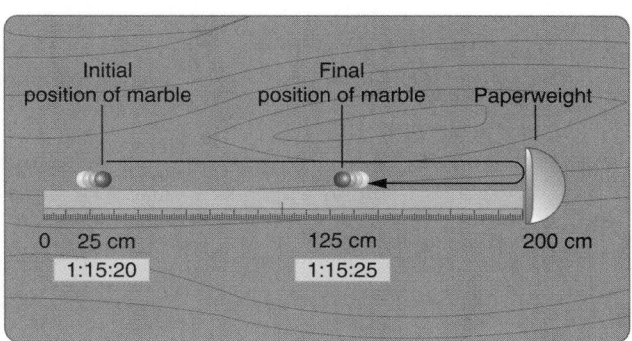

FIGURE 3.12

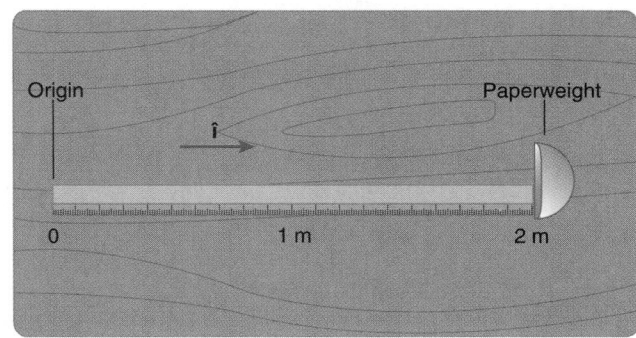

FIGURE 3.13

The curve is plotted in Figure 3.11. The two events described in parts (a) and (b) are indicated as the initial and final points on the world-line.

EXAMPLE 3.3

Lay a 2-meter stick on your desk and roll a marble slowly alongside it as illustrated in Figure 3.12. The marble is found at the 25 cm mark when a clock reads 1 h 15 min 20 s. The marble rebounds from a paperweight at the end of the 2-meter stick and is discovered at the 125 cm mark when the clock reads 1 h 15 min 25 s. In appropriate SI units, find the following:

a. the position vector of the marble at each of the two instants relative to the zero centimeter mark of the scale. Let the stick represent the x-axis;
b. the change in the position vector $\Delta\vec{r}$; and
c. the total distance traveled during the time interval between the two events.
d. Locate the two events on an appropriate x versus t graph. Sketch an appropriate world-line connecting the two events, describing the salient features of the world-line.

Solution

a. Use Problem-Solving Tactic 3.1 and *choose a coordinate axis for rectilinear motion to be coincident with the line followed by the particle.* Choose the origin of a Cartesian coordinate x to be coincident with the zero of the meter stick, since the problem seeks to find the position vectors with respect to this point (see Figure 3.13). You also must choose a direction for $\hat{\imath}$, that is, for increasing values of x. Here it is convenient to choose $\hat{\imath}$ to the

right, so that the meter stick positions correspond to increasing x-coordinates.

The position vector $\vec{r}_i$ at the instant t_i (clock reading 1 h 15 min 20 s) is

$$\vec{r}_i = (0.25 \text{ m})\hat{\imath}$$

using proper SI units. This is a vector of magnitude 0.25 m, directed in the same direction as $\hat{\imath}$. The position vector $\vec{r}_f$ at the instant t_f (clock reading 1 h 15 min 25 s) is

$$\vec{r}_f = (1.25 \text{ m})\hat{\imath}$$

This is a vector of magnitude 1.25 m, directed in the same direction as $\hat{\imath}$.

b. According to Equation 3.2, the change in the position vector is

$$\Delta\vec{r} \equiv \vec{r}_f - \vec{r}_i$$
$$= (1.25 \text{ m})\hat{\imath} - (0.25 \text{ m})\hat{\imath}$$
$$= (1.00 \text{ m})\hat{\imath}$$

The change in the position vector is a vector quantity. Here the magnitude of the change in the position vector is 1.00 m and its direction is to the right, parallel to the direction chosen for $\hat{\imath}$.

c. The total distance d traveled during the 5-second interval is the total distance from its initial position to the paperweight plus the distance from the paperweight to the final position. This is

$$d = 1.75 \text{ m} + 0.75 \text{ m}$$
$$= 2.50 \text{ m}$$

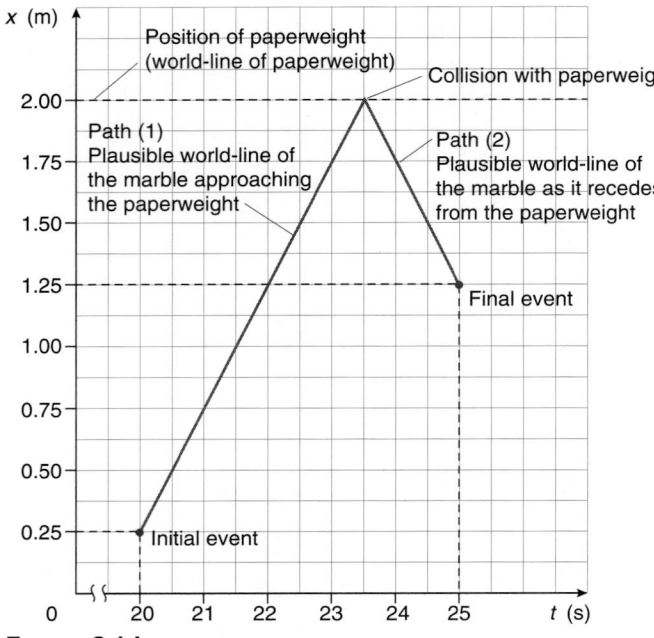

FIGURE 3.14

The total distance traveled always is a positive scalar quantity or zero; it is never negative. Notice that the total distance traveled (2.50 m) is not the same as the magnitude of the change in the position vector (1.00 m), since the particle executed back-and-forth motion along the line of its motion during the interval.

d. An *x* versus *t* graph is a graph of the world-line *x(t)*, with events represented as points on the world-line. Figure 3.14 shows the two events. Since the paperweight is located at the coordinate *x* = 2.00 m and the marble rebounds from the paperweight, a plausible world-line is shown. On path (1) of the world-line, the marble is approaching the paperweight; on path (2), the marble is receding from the paperweight, retracing its path. The sharp point on the world-line represents the collision event of the marble with the paperweight; here it is assumed the collision is instantaneous, taking zero time.

3.3 AVERAGE SPEED AND AVERAGE VELOCITY

Tempus est mensura motus rerum mobilium

[Time is the measure of movement]

Aristotle (384–322 B.C.)*

If your car is moving fast, it travels the length of a dragstrip during a short time interval Δ*t*; conversely, if the car is moving slowly, it takes a longer time interval to traverse the length of the strip.

Les Auctoritates Aristotlelis, un florilège médiéval étude historique et édition critique, editor Jacqueline Hamesse (Louvain, Paris, 1974), page 151; from Aristotle's *Physics.*

The **average speed** of a particle during a time interval Δ*t* is defined as the total distance *d* traveled by the particle divided by the time interval Δ*t* it takes to traverse the distance:

$$\text{average speed} = \frac{|d|}{|\Delta t|} = \frac{\text{total distance traveled}}{\text{time interval}} \quad (3.5)$$

The average speed is a positive scalar quantity, not a vector quantity.

Since both the total distance (the trip odometer reading) and the time interval are positive scalars, the absolute value signs in Equation 3.5 are superfluous; they are included only for emphasis. The dimensions of the average speed are distance divided by time; the SI units are m/s.

There is no reason that you could not define a measure of the motion to be the inverse of the ratio of Equation 3.5, $|\Delta t|/|d|$. You might call this ratio the average "sluggishness" of a particle. The average sluggishness is a measure of the time interval needed to traverse a given distance (say, 30 seconds per kilometer for your car), rather than the average speed, which is a measure of the distance traversed during a given time interval (120 kilometers per hour for your car). A large value of the average sluggishness means it takes a particle a long time interval to traverse a small distance. The contrast in usefulness between the average sluggishness and the average speed illustrates why we say that the theory of motion was *invented*, not discovered.

The **average velocity** $\vec{v}_{\text{ave}}$ of a particle during the time interval Δ*t* is defined as the ratio of the change in the position vector of the particle to the time interval during which the change occurred:

$$\vec{v}_{\text{ave}} = \frac{\Delta \vec{r}}{\Delta t} \quad (3.6)$$

The average velocity is a vector quantity.

The average velocity is a vector because it is a scalar multiple (1/Δ*t*) of the change in the position vector. The dimensions of the average velocity vector are the same as those of the average speed, distance divided by time—in SI units, m/s. However, do not confuse the average speed (which involves the total distance traveled, *d*) with the average velocity (which involves the change in the position vector, Δ$\vec{r}$); they are quite different concepts.

The magnitude of the average velocity is *not* necessarily equal to the average speed during the same time interval.

The inequality arises because the total distance *d* traveled by the particle during the time interval may be different from the magnitude of the change in the position vector that occurs during the same time interval. The quantity $|\Delta \vec{r}|$ always is less than or equal to *d*. For rectilinear motion, the inequality arises if the particle reverses direction or undergoes back-and-forth motion during the interval Δ*t*.

The position vector for one-dimensional motion has only a single component, because we always choose the coordinate axis to be coincident with the line followed by the particle.

The x-component of the average velocity, $v_{x\,ave}$, during an interval Δt, its only component for rectilinear motion,

$$\vec{\mathbf{v}}_{ave} = v_{x\,ave}\hat{\mathbf{i}}$$

is found from Equations 3.6 and 3.3 to be

$$v_{x\,ave} = \frac{x_f - x_i}{\Delta t}$$
$$= \frac{\Delta x}{\Delta t}$$

On an x versus t graph, such as Figure 3.15, the average velocity component $v_{x\,ave}$ is the slope of the straight line connecting the initial and final events. The average velocity component $v_{x\,ave}$ is found using *only* the initial and final events on the x versus t graph and *does not involve the detailed shape of the world-line connecting the events.*

The average velocity component $v_{x\,ave}$ is positive if $x_f > x_i$; then the average velocity is in the same sense as the direction chosen for $\hat{\mathbf{i}}$. If $x_f < x_i$, the average velocity component is negative, meaning the average velocity vector is in the same sense as $-\hat{\mathbf{i}}$ in the chosen coordinate system.

PROBLEM-SOLVING TACTIC

3.3 The sign of $v_{x\,ave}$ depends on the choice you make for the direction of $\hat{\mathbf{i}}$ and the values of x_f and x_i. Different coordinate choices change the way you *describe* the motion but not the underlying physics. The magnitude and direction in space of the *vector* $\vec{\mathbf{v}}_{ave}$ do not depend on the coordinate choice.

QUESTION 3

If your average car speed is 100 km/h, what is the corresponding measure of the average sluggishness of the car?

QUESTION 4

Your dog runs in a straight line east chasing a squirrel. Indicate a coordinate choice so that $v_{x\,ave}$ of your dog is positive. Indicate another coordinate choice so that $v_{x\,ave}$ is negative.

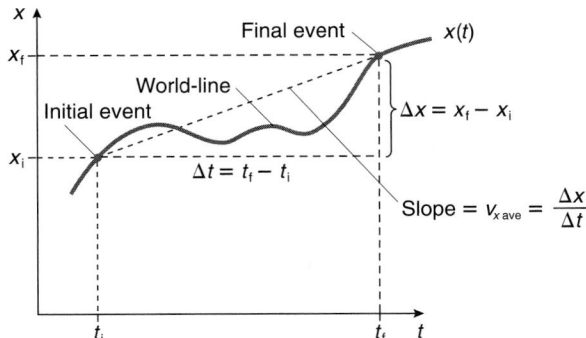

FIGURE 3.15 The average velocity component $v_{x\,ave}$ is the slope of the straight line connecting the initial and final events on an x versus t graph.

EXAMPLE 3.4

a. **Find the average speed of the marble in Example 3.3 during the 5 s interval.**
b. **Find the average velocity of the marble of Example 3.3 during the 5 s interval.**

Solution
a. The average speed is found by dividing the total distance the particle travels ($d = 2.50$ m) between the two events by the time interval between the events ($\Delta t = 5$ s):

$$\text{average speed} = \frac{2.50\ \text{m}}{5\ \text{s}}$$
$$= 0.5\ \text{m/s}$$

b. The average velocity is the ratio of the change in the position vector to the time interval during which the change occurred:

$$\vec{\mathbf{v}}_{ave} = \frac{\Delta\vec{\mathbf{r}}}{\Delta t}$$
$$= \frac{(1.00\ \text{m})\hat{\mathbf{i}}}{5\ \text{s}}$$
$$= (0.2\ \text{m/s})\hat{\mathbf{i}}$$

Notice that the average velocity is a vector, since the change in the position vector is a vector. The *magnitude* of the average velocity vector here is 0.2 m/s; this is *not the same as the average speed* [found to be 0.5 m/s in part (a)].

EXAMPLE 3.5

The Iditarod dogsled race from Anchorage to Nome, Alaska, covers a distance of 1864 km. The winning time is about 11 days.

a. **Find the average speed of the winner in SI units.**
b. **Qualitatively compare the average speed during the interval of the entire race with the likely average speed during the actual time during which the team was on the course and not resting or sleeping.**

Solution
a. The average speed is the total distance traveled divided by the time interval during which the distance was traversed:

$$\text{average speed} = \frac{1864\ \text{km}}{11\ \text{days}}$$
$$= \frac{(1864\ \text{km})(1000\ \text{m/km})}{(11\ \text{days})(86\,400\ \text{s/day})}$$
$$= 2.0\ \text{m/s}$$

This means that the winning team could complete the race by traveling at 2.0 m/s throughout the entire 11-day time interval.

b. Because the team had to stop, rest, and eat along the route, the time interval actually spent running the race was shorter than 11 days (perhaps about 7 days). Hence the average speed during the many intervals the team actually was on the race course was greater than the average speed for the duration of the entire trip.

STRATEGIC EXAMPLE 3.6

You drive your car 2.00 km down a dragstrip, then 2.50 km in the opposite direction, completing the excursion in 180 s.

a. **Choose an appropriate coordinate system to describe the motion.**
b. **Determine the initial and final position vectors and the change in the position vector during the time interval.**
c. **Find the average velocity.**
d. **Find the average speed.**
e. **Explain why the average speed of the car is not equal to the magnitude of the average velocity.**

Solution

a. Choose a coordinate system with the *x*-axis coincident with the dragstrip and an origin at the initial position of the car (Figure 3.16).

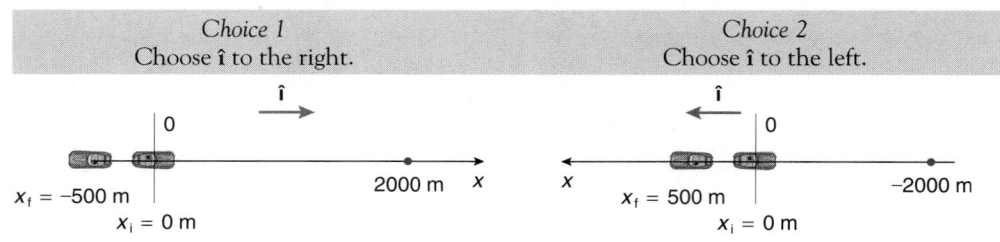

FIGURE 3.16

b. The initial position vector for the car is

$$\vec{r}_i = x_i \hat{i}$$

$\vec{r}_i = (0 \text{ m})\hat{i}$	$\vec{r}_i = (0 \text{ m})\hat{i}$

The final position vector of the car is

$$\vec{r}_f = x_f \hat{i}$$

$\vec{r}_f = (-500 \text{ m})\hat{i}$	$\vec{r}_f = (500 \text{ m})\hat{i}$

The change in the position vector is

$$\Delta\vec{r} = \vec{r}_f - \vec{r}_i$$

$\Delta\vec{r} = (-500 \text{ m})\hat{i} - (0 \text{ m})\hat{i}$ $= (-500 \text{ m})\hat{i}$	$\Delta\vec{r} = (500 \text{ m})\hat{i} - (0 \text{ m})\hat{i}$ $= (500 \text{ m})\hat{i}$

Notice that the magnitude of the change in the position vector is 500 m and is directed to the left for *both* coordinate choices. This illustrates how the magnitude and direction in space of the *change* in a vector are the same regardless of the coordinate system you choose.

c. The average velocity is the change in the position vector divided by the time interval during which the change occurred:

$$\vec{v}_{ave} = \frac{\Delta\vec{r}}{\Delta t}$$

$\vec{v}_{ave} = \dfrac{(-500 \text{ m})\hat{i}}{180 \text{ s}}$ $= (-2.78 \text{ m/s})\hat{i}$	$\vec{v}_{ave} = \dfrac{(500 \text{ m})\hat{i}}{180 \text{ s}}$ $= (2.78 \text{ m/s})\hat{i}$

Note that these results are consistent with Problem-Solving Tactic 3.3: *the sign of* $v_{x\,ave}$ *depends on the choice you make for the direction of* $\hat{i}$ *and the values of* x_f *and* x_i. However, the average velocity here has a magnitude of 2.78 m/s and is directed to the left, independent of whatever choice you made for the coordinate system.

d. The average speed is the total distance traveled divided by the time interval:

$$\text{average speed} = \frac{4.50 \times 10^3 \text{ m}}{180 \text{ s}}$$

$$= 25.0 \text{ m/s}$$

e. The magnitude of the average velocity, here 2.78 m/s, is not equal to the average speed, here 25.0 m/s, because the total distance traveled is not equal to the magnitude of the change in the position vector: the car went back and forth along the dragstrip during the interval Δt.

3.4 INSTANTANEOUS SPEED AND INSTANTANEOUS VELOCITY

In many instances we want to know more than averages; we want details about the dynamic motion of the particle on an instant-by-instant basis. Such considerations are the determining factors in "awarding" speeding tickets to motorists. Instantaneous information is gleaned by imagining the particle to be equipped with a speedometer like that in your car. In your car, a glance at the numerical reading of the speedometer at an *instant* t indicates the **instantaneous speed** of the car at that instant. The instantaneous speed of the particle is more commonly referred to as simply the **speed** v.

The number indicated by the speedometer indicates how far the car *would* travel during the next *one* unit interval of time *if* it continued moving *at the same speedometer reading* for that one unit interval of time. For example, if the speedometer in your car indicates the speed is 100 km/h, the car will move 100 km during the entire next 1 hour interval if the speed remains *constant* (on cruise control) during this interval. Traffic jams, traffic lights, or the unnerving sight of the flashing lights of the state police in the rearview mirror may intervene to prevent you from actually keeping the car at the same speedometer reading for the next one hour. But, *at the instant* the speedometer reading is noted, the speed of the car is 100 km/h. Converting the speedometer reading to meters per second, 100 km/h = 27.8 m/s, you see the car will move 27.8 meters during the next 1 second *if* the car continues uninterrupted at the same speed during the next one second interval.

Your car may have a compass that provides information about the instantaneous directional heading of the car.

The **instantaneous velocity** $\vec{\mathbf{v}}$ is a vector quantity whose magnitude v is equal to the instantaneous speed and whose instantaneous direction is in the same sense that the particle is moving at that instant.

The instantaneous velocity more commonly is called simply the **velocity** of the particle at the instant t. The velocity vector always is tangent to the trajectory or path followed by the particle, although this point is moot for rectilinear motion.

Speed and velocity are *not* the same physical concepts, even though they are both commonly measured with the same SI units, m/s. *Velocity is a vector quantity whose positive scalar magnitude is the speed.* Be sure you know and remember the distinction between the two terms.

Another way to think of the instantaneous velocity uses the notion of limits from calculus. Let $\vec{\mathbf{r}}(t) = x(t)\hat{\mathbf{i}}$ be the function of time that represents the one-dimensional position vector of the particle at any instant of time. Choose a particular instant $t_i = t$, so that

$$\vec{\mathbf{r}}_i = \vec{\mathbf{r}}(t)$$
$$= x(t)\hat{\mathbf{i}}$$

A short time interval Δt later, at the instant $t_f = t + \Delta t$, the position vector is

$$\vec{\mathbf{r}}_f = \vec{\mathbf{r}}(t + \Delta t)$$
$$= x(t + \Delta t)\hat{\mathbf{i}}$$

The *change in the position vector* during the time interval Δt then is

$$\Delta\vec{\mathbf{r}} = \vec{\mathbf{r}}(t + \Delta t) - \vec{\mathbf{r}}(t)$$
$$= [x(t + \Delta t) - x(t)]\hat{\mathbf{i}}$$

The average velocity $\vec{\mathbf{v}}_{\text{ave}}$ of the particle during the time interval Δt is, using Equation 3.6,

$$\vec{\mathbf{v}}_{\text{ave}} = \frac{\Delta\vec{\mathbf{r}}}{\Delta t}$$

$$= \frac{[x(t + \Delta t) - x(t)]\hat{\mathbf{i}}}{\Delta t}$$

Take the limit of this expression as the time interval Δt approaches zero; the average velocity thus is taken over shorter and shorter time interval deviations from the instant t:

$$\lim_{\Delta t \to 0 \text{ s}} \frac{\vec{\mathbf{r}}(t + \Delta t) - \vec{\mathbf{r}}(t)}{\Delta t} = \lim_{\Delta t \to 0 \text{ s}} \frac{[x(t + \Delta t) - x(t)]\hat{\mathbf{i}}}{\Delta t}$$

This process is shown graphically in the x versus t graph of Figure 3.17.

You may recognize this limit as one way of defining the derivative in calculus. This limiting process defines the instantaneous velocity $\vec{\mathbf{v}}(t)$ of the particle at the instant t. In other words,

$$\vec{\mathbf{v}}(t) \equiv \lim_{\Delta t \to 0 \text{ s}} \vec{\mathbf{v}}_{\text{ave}} = \frac{d\vec{\mathbf{r}}}{dt}$$

$$= \frac{dx}{dt}\hat{\mathbf{i}} \qquad \text{(for rectilinear motion)} \qquad (3.7)$$

As Δt approaches zero, the slope of the line between $x(t)$ and $x(t_i)$ approaches the slope of the tangent to that graph at t, which is $v_x(t)$.

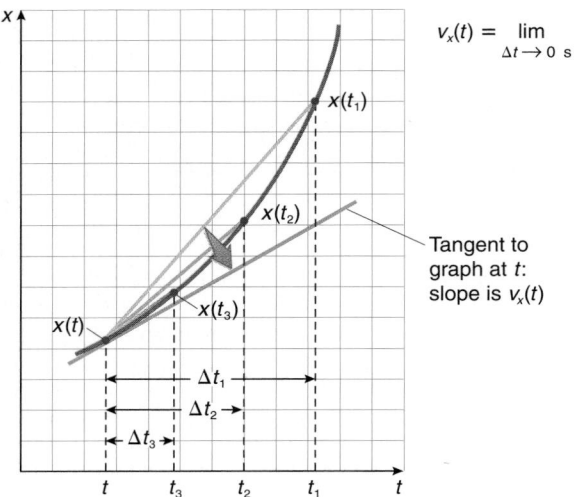

$$v_x(t) = \lim_{\Delta t \to 0 \text{ s}} v_{x\,\text{ave}}$$

FIGURE 3.17 The limiting process on an x versus t graph.

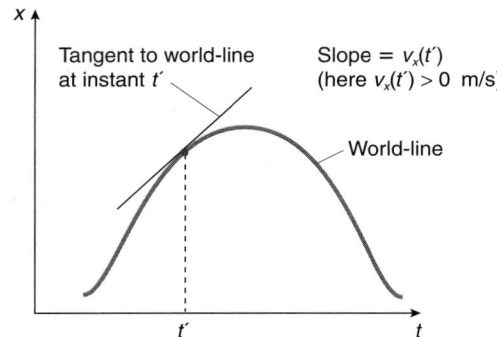

FIGURE 3.18 The instantaneous velocity component at an instant t' is equal to the slope of the tangent to the world-line at that instant. Note that the slope of the graph may change during a time interval, indicating that the velocity component is positive at some instants, negative at others, or even zero.

FIGURE 3.19

For rectilinear motion, the instantaneous velocity vector $\vec{v}(t) = v_x(t)\hat{\imath}$ has a single component $v_x(t)$, where, from Equation 3.7,

$$v_x(t) = \frac{dx(t)}{dt} \tag{3.8}$$

Since instants have no duration, the particle has the instantaneous velocity $v(t)$ for zero seconds.

In calculus you learned that the derivative of a function is the *slope* of a graph of the function. Exploiting this geometric interpretation of the derivative from calculus, Equation 3.8 means that the instantaneous velocity component $v_x(t')$ at any instant t' is the slope of the tangent to the world-line of the particle at that instant t'. See Figure 3.18. Note the distinction in Figure 3.17 between the average velocity *during* a time interval and the instantaneous velocity *at* an instant.

PROBLEM-SOLVING TACTIC

3.4 The choice you make for the coordinate system influences the sign of the velocity component v_x. As you are now aware, the choices you make for a coordinate system have consequences in the signs associated with the components of vectors used to describe the motion or the physics. This will be the case throughout our study of physics, not only in kinematics.

EXAMPLE 3.7 ─────────────

You and a friend both leave campus by car and head home in opposite directions along a straight stretch of inclined roadway, as shown in Figure 3.19. A short interval after getting under way, at time instant 2 h 26 min 36.0 s, both of you glance at the speedometers in your cars. Yours indicates 60 km/h while your friend's indicates 90 km/h.

a. Introduce an appropriate Cartesian coordinate system and write the instantaneous velocities of each car in terms of the coordinate system you choose. Use proper SI units.

b. Indicate on a sketch what is meant by the position vectors of the two cars. Is there sufficient information to determine the position vectors at the same instant?

Solution

a. The speedometers are read *at* a particular *instant* in time, the clock time 2 h 26 min 36.0 s, and so the speeds quoted are instantaneous speeds. Convert the speeds into proper SI units of m/s:

$$90 \text{ km/h} = \frac{(90 \text{ km/h}) (10^3 \text{ m/km})}{(3.600 \times 10^3 \text{ s/h})}$$

Notice that the factors of 10^3 in the numerator and denominator divide out, and so

$$90 \text{ km/h} = \frac{90}{3.600} \text{ m/s}$$
$$= 25 \text{ m/s}$$

In a similar way,

$$60 \text{ km/h} = 17 \text{ m/s}$$

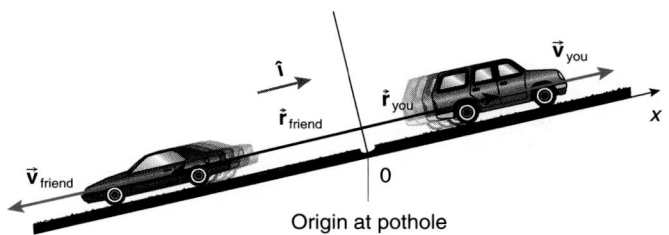
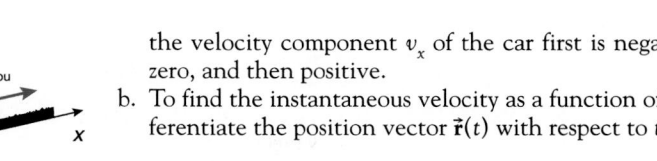

FIGURE 3.20

 Since the road is straight, choose a coordinate system with the direction of increasing x (the +x-axis) in the direction your car is moving, with the origin at some convenient pothole in the highway, as in Figure 3.20. Choose the x-axis to be along the road even though the road is not horizontal, because we want the axis to be along the line followed by the cars. Then the instantaneous velocities of the cars are

$$\text{your velocity:} \quad \vec{v}_{\text{you}} = (17 \text{ m/s})\hat{\imath}$$
$$\text{your friend's velocity:} \quad \vec{v}_{\text{friend}} = (25 \text{ m/s})(-\hat{\imath})$$
$$= (-25 \text{ m/s})\hat{\imath}$$

Draw the velocity vectors with their tails at the centers of the cars since you represent the cars as particles, imagining each car as a point.

b. The position vectors of the two cars are indicated in Figure 3.20. Since no information is given about where the two cars are located along the axis at the given instant, there is insufficient information to determine the Cartesian form for the two position vectors.

STRATEGIC EXAMPLE 3.8

The one-dimensional position vector of a sports car is given by

$$\vec{r}(t) = x(t)\hat{\imath}$$
$$= [4.00 \text{ m} - (2.00 \text{ m/s})t + (3.00 \text{ m/s}^2)t^2]\hat{\imath}$$

a. **Make an x versus t graph including at least the interval from $t = 0$ s to $t = 3.00$ s.**
b. **Find the velocity $\vec{v}(t)$ as a function of time.**
c. **Specify the Cartesian component $v_x(t)$ of the velocity at any instant t.**
d. **Make a graph of $v_x(t)$ versus t.**
e. **Find the velocity at the instant when $t = 2.00$ s.**
f. **Determine at what time the velocity is zero.**
g. **Describe the motion of the particle.**

Solution

a. An x versus t graph depicts $x(t)$ versus t. The graph is parabolic in shape since x is a quadratic function of t. See Figure 3.21.

The slope of the tangent to the curve at any instant t is the velocity component $v_x(t)$ at that instant. Notice that as t increases from zero, the slope of the world-line first is negative, then is zero at the instant represented by the lowest point on the graph, and then is positive. This indicates that

the velocity component v_x of the car first is negative, then zero, and then positive.

b. To find the instantaneous velocity as a function of time, differentiate the position vector $\vec{r}(t)$ with respect to t:

$$\vec{v}(t) = \frac{d\vec{r}(t)}{dt} = \frac{dx(t)}{dt}\hat{\imath}$$
$$= \frac{d}{dt}[4.00 \text{ m} - (2.00 \text{ m/s})t + (3.00 \text{ m/s}^2)t^2]\hat{\imath}$$
$$= [0 - 2.00 \text{ m/s} + (3.00 \text{ m/s}^2)2t]\hat{\imath}$$
$$= [-2.00 \text{ m/s} + (6.00 \text{ m/s}^2)t]\hat{\imath}$$

c. The x-component of the velocity $v_x(t)$ is the scalar function

$$v_x(t) = -2.00 \text{ m/s} + (6.00 \text{ m/s}^2)t$$

d. Notice that $v_x(t)$ is a linear function of t, since t appears only to the first power. The graph of $v_x(t)$, therefore, is a straight line with slope 6.00 m/s^2 and intercept −2.00 m/s. See Figure 3.22.
e. The velocity at the instant when $t = 2.00$ s is found by substituting this value for t into $\vec{v}(t)$:

$$\vec{v}(2.00 \text{ s}) = [-2.00 \text{ m/s} + (6.00 \text{ m/s}^2)(2.00 \text{ s})]\hat{\imath}$$
$$= (10.0 \text{ m/s})\hat{\imath}$$

The particle is moving with a speed of 10.0 m/s in the direction parallel to $\hat{\imath}$ (toward increasing values of x) at the instant when $t = 2.00$ s. The slope of the world-line is 10.0 m/s at the instant when $t = 2.00$ s (see Figure 3.21).

f. The instantaneous velocity is zero at those times when its component $v_x(t)$ is zero:

$$v_x(t) = 0 \text{ m/s} = -2.00 \text{ m/s} + (6.00 \text{ m/s}^2)t$$

and so

$$t = \frac{2.00 \text{ m/s}}{6.00 \text{ m/s}^2}$$
$$= 0.333 \text{ s}$$

So the particle is instantaneously at rest when $t = 0.333$ s. The slope of the tangent to the world-line (Figure 3.21) is zero at this instant, implying the velocity component is zero when $t = 0.333$ s.

g. Notice from the graph in Figure 3.22 that for times $t < 0.333$ s, the x-component v_x of the velocity vector is negative. This means that particle is moving in the same sense as $-\hat{\imath}$ for these times, or toward decreasing values of the coordinate x. For times $t > 0.333$ s, the x-component of the velocity is positive, which means the particle is moving in the same sense as $\hat{\imath}$, that is, in the direction of increasing values of x. Thus the particle reversed its direction at the instant $t = 0.333$ s. The car was, perhaps, initially backing up, then reversed direction when $t = 0.333$ s. The location of the particle at this instant is found from the position vector by substituting $t = 0.333$ s:

$$\vec{r}(t) = x(t)\hat{\imath}$$
$$= [4.00 \text{ m} - (2.00 \text{ m/s})t + (3.00 \text{ m/s}^2)t^2]\hat{\imath}$$

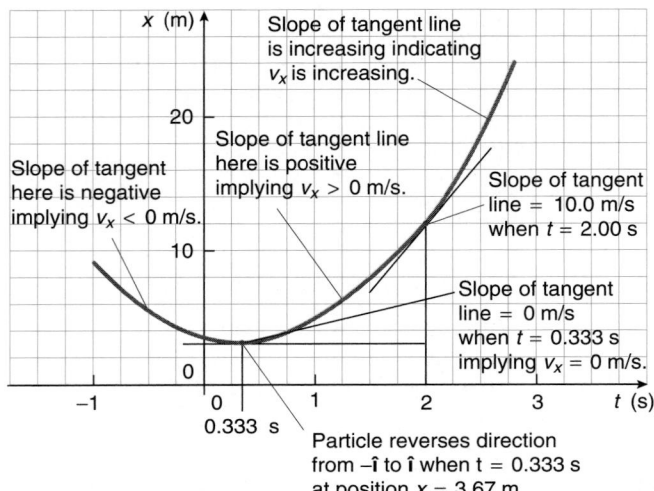

FIGURE 3.21 The *x* versus *t* graph.

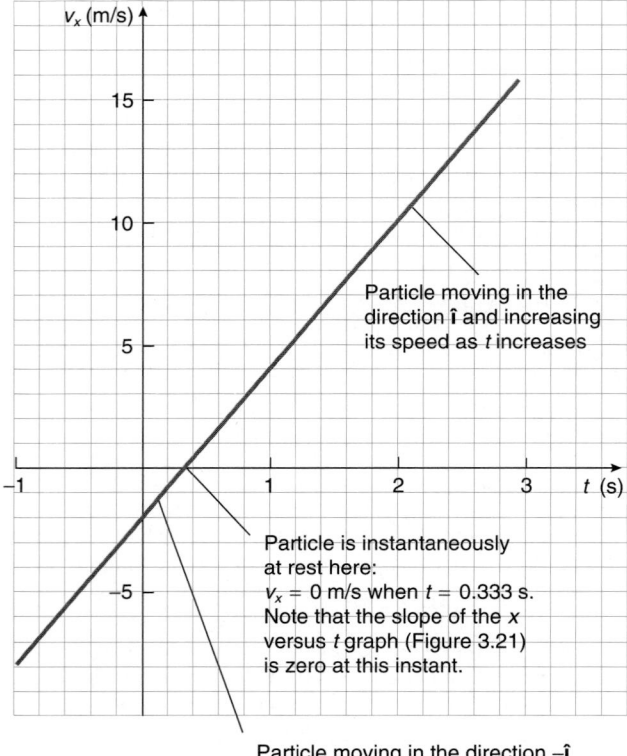

FIGURE 3.22 The v_x versus *t* graph.

Hence

$$\vec{\mathbf{r}}(0.333 \text{ s}) = (4.00 \text{ m} - 0.666 \text{ m} + 0.333 \text{ m})\hat{\mathbf{i}}$$
$$= (3.67 \text{ m})\hat{\mathbf{i}}$$

The particle is on the $+x$-axis 3.67 m from the origin when $t = 0.333$ s. This corresponds to the minimum on the *x* versus *t* graph shown in Figure 3.21. The velocity of the particle reverses direction from $-\hat{\mathbf{i}}$ to $\hat{\mathbf{i}}$ at the coordinate $x = 3.67$ m, where its velocity is instantaneously zero. Recall that *instants are not intervals of time*, not even very short ones.

3.5 AVERAGE ACCELERATION

You likely have felt the dramatic changes in the velocity of a jet when it is moving along a runway prior to takeoff or just after touchdown when landing. Likewise, when you toss a ball vertically upward, the velocity of the ball changes both in magnitude *and* direction over the ensuing seconds before impact. These examples illustrate that there is more to motion than just position and velocity, even for one-dimensional motion. *Changes in velocity are important too.* Any change in the velocity of a particle with time indicates the presence of an **acceleration**.*

Acceleration, though, is more than just the change in velocity. Certainly changing the velocity of a sports car from 0 km/h east to 120 km/h east during a 5 s interval on a dragstrip (see Figure 3.23) is quite different from executing the same change in velocity during a leisurely 90 s interval in a jalopy. Hence the time interval during which the change in velocity is accomplished also is relevant. We want to quantify this distinction.

Let the velocity of a particle change from $\vec{\mathbf{v}}_i$ at an instant t_i to $\vec{\mathbf{v}}_f$ at a later instant t_f.

> One way to quantify the effect of a change in the velocity during the interval is to form the ratio of the change in the velocity $\Delta\vec{\mathbf{v}} \equiv \vec{\mathbf{v}}_f - \vec{\mathbf{v}}_i$ to the interval $\Delta t = t_f - t_i$. This ratio is called the **average acceleration** $\vec{\mathbf{a}}_{\text{ave}}$ during the interval Δt:
>
> $$\vec{\mathbf{a}}_{\text{ave}} \equiv \frac{\Delta\vec{\mathbf{v}}}{\Delta t} \tag{3.9}$$

If a given change in velocity occurs during a short time interval, the average acceleration is larger than if it occurs during a longer time interval. The average acceleration is a vector quantity, since the change in velocity is a vector quantity.

---

*We will never use the word deceleration, meaning a decrease in the magnitude of the velocity during a time interval. *All* changes in velocity manifest accelerations.

FIGURE 3.23 A car changing its velocity on a dragstrip experiences an acceleration.

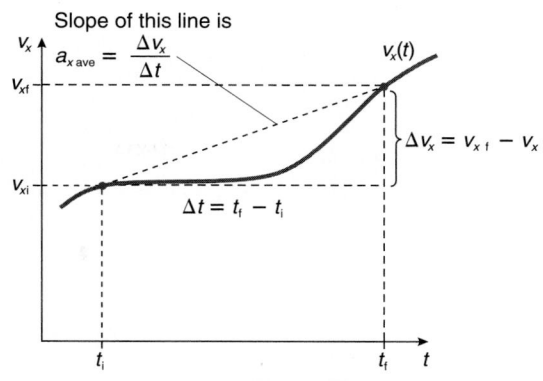

FIGURE 3.24 The average acceleration is the slope of the straight line connecting the initial and final points on a v_x versus t graph.

The units associated with the average acceleration in the SI unit system are those of the change in the velocity, m/s, divided by those for the time interval, s—that is, m/s^2.

For rectilinear motion, the velocity has a single component, $v_x(t)$, and so the change in the velocity also has a single component. This in turn implies that the average acceleration has a single component, $a_{x\,ave}$:

$$\vec{a}_{ave} = a_{x\,ave}\hat{i}$$

If a graph is made of $v_x(t)$ versus t, as in Figure 3.24, the average acceleration component during the interval Δt between t_i and t_f is the slope of the straight line connecting the two points with coordinates $(t_i, v_x(t_i))$ and $(t_f, v_x(t_f))$.

For notational simplicity, we write

$$v_x(t_i) \equiv v_{x\,i}$$

and

$$v_x(t_f) \equiv v_{x\,f}$$

so that

$$a_{x\,ave} \equiv \frac{v_{x\,f} - v_{x\,i}}{\Delta t} \qquad (3.10)$$

PROBLEM-SOLVING TACTIC

3.5 The sign of the average acceleration component $a_{x\,ave}$ depends on the choice you make for the direction of the unit vector $\hat{i}$ and the values of $v_{x\,f}$ and $v_{x\,i}$.

STRATEGIC EXAMPLE 3.9

A car moving up a long, straight hill has the speed 30.0 m/s at the instant 15.00 s. Later, at the instant 24.00 s, the speed of the car is observed to be 6.00 m/s in the opposite direction. See Figure 3.25.

a. Indicate an appropriate coordinate system in a sketch.
b. Find the average acceleration of the car during the time interval between the two instants.
c. Find the magnitude of the average acceleration of the car during the time interval.

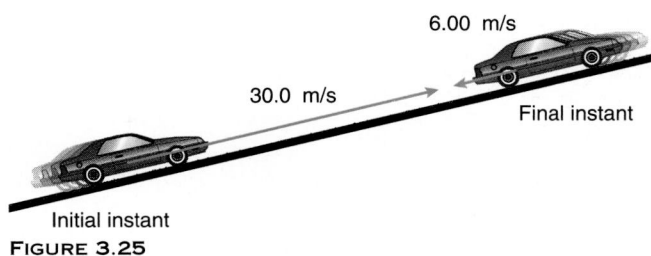

FIGURE 3.25

Solution

a. Model the car as a particle. Choose the *x*-axis to be the line along which the particle is moving. As shown in Figure 3.26, there are two possible choices for the direction of $\hat{\imath}$:

Choice 1	*Choice 2*
Choose $\hat{\imath}$ to be up the hill. Choose the origin to be anywhere along the slope.	Choose $\hat{\imath}$ to be down the hill. Choose the origin to be anywhere along the slope.

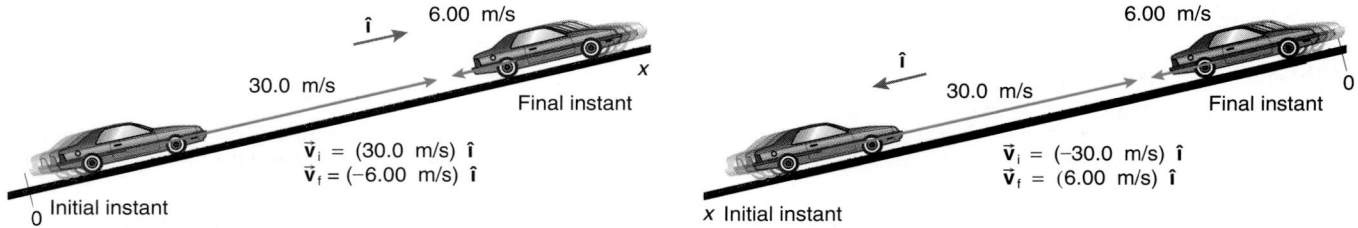

FIGURE 3.26

b. The average acceleration is the change in the velocity vector divided by the time interval during which the change occurs. The change in the velocity vector is

$$\Delta\vec{v} = \vec{v}_f - \vec{v}_i$$

$\Delta\vec{v} = (-6.00\ \text{m/s})\hat{\imath} - (30.0\ \text{m/s})\hat{\imath}$ $= (-36.0\ \text{m/s})\hat{\imath}$	$\Delta\vec{v} = (6.00\ \text{m/s})\hat{\imath} - (-30.0\ \text{m/s})\hat{\imath}$ $= (36.0\ \text{m/s})\hat{\imath}$

The time interval between the two instants is $\Delta t = 9.00$ s. The average acceleration during the time interval is

$$\vec{a}_{\text{ave}} \equiv \frac{\Delta\vec{v}}{\Delta t}$$

$= \dfrac{(-36.0\ \text{m/s})\hat{\imath}}{9.00\ \text{s}}$ $= (-4.00\ \text{m/s}^2)\hat{\imath}$ This vector is directed in the same sense as $-\hat{\imath}$, that is, toward the left.	$= \dfrac{(36.0\ \text{m/s})\hat{\imath}}{9.00\ \text{s}}$ $= (4.00\ \text{m/s}^2)\hat{\imath}$ This vector is directed in the same sense as $\hat{\imath}$, that is, toward the left.

This illustrates Problem-Solving Tactic 3.4: *the sign of the average acceleration component $a_{x\ \text{ave}}$ depends on the choice you make for the direction of the unit vector $\hat{\imath}$. Note that the average acceleration has the same magnitude and direction in space with either coordinate choice.*

c. Since there is only one component, the *magnitude* of the average acceleration is

$$a_{\text{ave}} = (a^2_{x\ \text{ave}})^{1/2}$$

$= [(-4.00\ \text{m/s}^2)^2]^{1/2}$ $= 4.00\ \text{m/s}^2$	$= [(4.00\ \text{m/s}^2)^2]^{1/2}$ $= 4.00\ \text{m/s}^2$

The magnitude of *any* vector is a scalar quantity greater than or equal to zero.

3.6 INSTANTANEOUS ACCELERATION

The thrill of riding in a sports car as it accelerates quickly or the horror of having to brake your car severely to avoid an accident is familiar to most of us. In the previous section we learned how to calculate the average acceleration. In this section we will discuss what is meant by an instantaneous acceleration.

In Section 3.4 we found it useful to compute the average velocity over shorter and shorter time intervals. In the limit as $\Delta t \to 0$ s, the average velocity became the instantaneous velocity at the instant t. We shall also find it convenient to use a similar limiting process with the average acceleration during a time interval Δt to find the **instantaneous acceleration**, commonly called simply the **acceleration**.

Let $\vec{\mathbf{v}}(t) = v_x(t)\hat{\mathbf{i}}$ be the vector that describes the instantaneous velocity of a particle as a function of time in one-dimensional motion. The scalar function $v_x(t)$ is the x-component of the velocity vector, its only component for rectilinear motion. The instantaneous velocity at any particular time instant t is found by substituting the particular value of t into $\vec{\mathbf{v}}(t) = v_x(t)\hat{\mathbf{i}}$. After a time interval Δt, the instant is $t + \Delta t$ and the instantaneous velocity is

$$\vec{\mathbf{v}}(t + \Delta t) = v_x(t + \Delta t)\hat{\mathbf{i}}$$

The *change* in the velocity vector during this interval is found from

$$\begin{aligned}
\Delta\vec{\mathbf{v}} &= \vec{\mathbf{v}}_f - \vec{\mathbf{v}}_i \\
&= \vec{\mathbf{v}}(t + \Delta t) - \vec{\mathbf{v}}(t) \\
&= [v_x(t + \Delta t) - v_x(t)]\hat{\mathbf{i}}
\end{aligned}$$

Use Equation 3.9 to find the average acceleration during the time interval Δt:

$$\begin{aligned}
\vec{\mathbf{a}}_{\text{ave}} &= \frac{\Delta\vec{\mathbf{v}}}{\Delta t} \\
&= \frac{[v_x(t + \Delta t) - v_x(t)]\hat{\mathbf{i}}}{\Delta t}
\end{aligned}$$

Now calculate the average acceleration over shorter and shorter intervals Δt from the instant t itself. This process is illustrated graphically in Figure 3.27.

Take the limit of the average acceleration as the time interval Δt approaches 0 s; we obtain the instantaneous acceleration at the instant t, $\vec{\mathbf{a}}(t)$:

$$\begin{aligned}
\vec{\mathbf{a}}(t) &\equiv \lim_{\Delta t \to 0 \text{ s}} \frac{\Delta\vec{\mathbf{v}}}{\Delta t} \\
&= \lim_{\Delta t \to 0 \text{ s}} \frac{[v_x(t + \Delta t) - v_x(t)]\hat{\mathbf{i}}}{\Delta t}
\end{aligned}$$

As Δt approaches zero, the slope of the line between $v(t)$ and $v(t_i)$ approaches the slope of the tangent to the graph at t, which is $a_x(t)$.

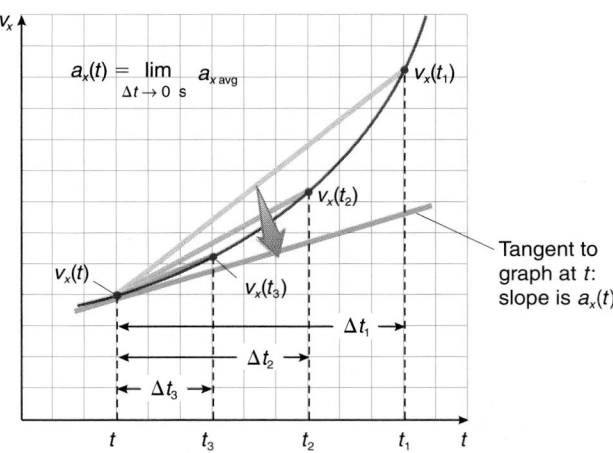

FIGURE 3.27 Graphical depiction of the limiting process.

This limit is the time derivative of the function $v_x(t)\hat{\mathbf{i}}$, that is, the time derivative of the instantaneous velocity vector.

> The instantaneous acceleration $\vec{\mathbf{a}}(t)$ of the particle is the time rate of change of the velocity vector:
>
> $$\vec{\mathbf{a}}(t) \equiv \frac{d\vec{\mathbf{v}}(t)}{dt} \qquad (3.11)$$
>
> The acceleration is a vector quantity.

Just as with the average acceleration, the SI units associated with the instantaneous acceleration are those of the velocity, m/s, divided by the time, s—or m/s^2.

For rectilinear motion, $\vec{\mathbf{v}}(t)$ has only one component $v_x(t)$, so that

$$\vec{\mathbf{a}}(t) = \frac{dv_x(t)}{dt}\hat{\mathbf{i}} \qquad (3.12)$$

But from Equation 3.8 we saw that

$$v_x(t) = \frac{dx(t)}{dt}$$

Make this substitution for $v_x(t)$ into Equation 3.12. You find that the instantaneous acceleration is the second derivative of the instantaneous position vector with respect to time:

$$\vec{\mathbf{a}}(t) = \frac{dv_x(t)}{dt}\hat{\mathbf{i}} = \frac{d^2x(t)}{dt^2}\hat{\mathbf{i}} = \frac{d^2\vec{\mathbf{r}}(t)}{dt^2} \qquad (3.13)$$

The acceleration vector for rectilinear motion has one component $a_x(t)$:

$$\vec{\mathbf{a}}(t) = a_x(t)\hat{\mathbf{i}}$$

where from Equation 3.13 we see that

$$a_x(t) = \frac{dv_x(t)}{dt} = \frac{d^2x(t)}{dt^2} \qquad (3.14)$$

Thus the x-component of the acceleration, $a_x(t)$, is the first time derivative of the x-component of the velocity, $v_x(t)$, or the second time derivative of the x-component of the position vector, $x(t)$.

The x-component of the acceleration, a_x, at any instant is the *slope* of the tangent line to the graph of v_x versus t at the instant in question, as indicated in Figure 3.28.

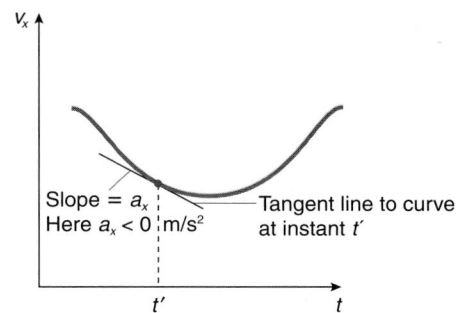

FIGURE 3.28 The acceleration component is the slope of the graph of the velocity component versus time.

FIGURE 3.29 A plumb bob.

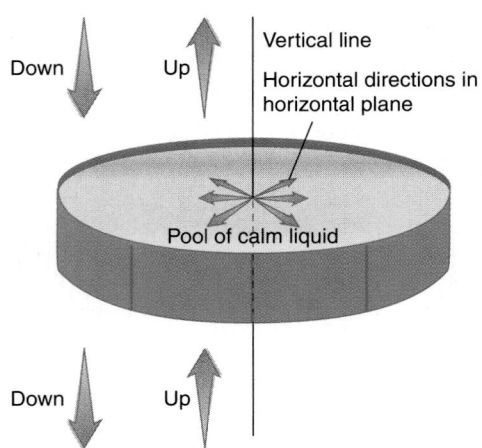

FIGURE 3.30 **Horizontal and vertical directions.**

To detect and measure the instantaneous acceleration, we need an **accelerometer**, a device analogous to a speedometer used to measure speed (the magnitude of the velocity). An accelerometer detects accelerations and, when suitably calibrated, indicates their magnitude. On the surface of the Earth, a crude but effective accelerometer consists of a simple mass suspended by a string, a device called a **plumb bob**, shown in Figure 3.29. This device detects accelerations directed *parallel* to the surface of the Earth— that is, in a **horizontal direction**. A horizontal direction is defined to be any direction parallel to the smooth surface of a liquid at rest on the Earth's surface (see Figure 3.30). Any plane parallel to the liquid surface is called a horizontal plane. A line perpendicular to a smooth liquid surface defines a **vertical direction**. The vertical directions up and down are defined using such a line.

Horizontal and vertical directions are purely local directions because of the curvature of the Earth itself. The vertically suspended plumb bob cannot be used to detect the existence of accelerations in the vertical direction.*

For example, if the plumb bob is attached to a particle either at rest or moving at constant velocity in a horizontal direction, the plumb bob hangs vertically. Try it yourself by tying a small mass to a piece of string and hanging it from the rearview mirror of your car. When at rest or traveling at constant speed in a fixed horizontal direction, the plumb bob hangs vertically (see Figure 3.31). In both situations, there is no acceleration in the horizontal plane since the velocity vector is not changing

---

*Other devices, such as a mass on a simple vertical spring, are used as accelerometers to detect accelerations in a vertical direction.

with time. On the other hand, when getting up to speed (see Figure 3.32) or stopping (Figure 3.33) when a traffic light changes, the plumb bob no longer hangs vertically.

If the velocity vector changes in magnitude, in direction, or in both magnitude and direction in a horizontal direction, the plumb bob no longer hangs in the vertical direction. If a snapshot of the plumb bob at an instant indicates it is not hanging in the vertical direction, there is a nonzero acceleration in the horizontal direction at that instant. Such effects also are apparent when observing a plumb bob in an airplane during the takeoff roll or braking along a horizontal runway.

The deviation of the plumb bob from the vertical is a measure of the magnitude of the acceleration; the direction of the deviation of the plumb bob from the vertical direction is *opposite* to the direction of the acceleration.[†] If the plumb bob

---

[†] In Chapter 5, we will see why the plumb bob is deflected in a direction opposite to that of the acceleration as well as how to calibrate the accelerometer so the acceleration in m/s² can be found from the angle of deviation of the plumb bob from the vertical.

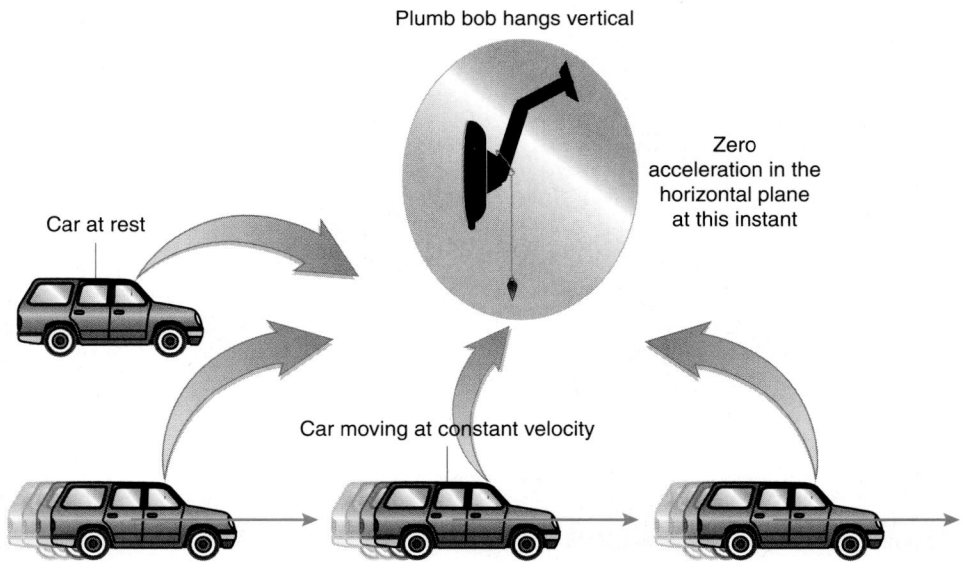

FIGURE 3.31 **If a plumb bob hangs vertical, there is zero acceleration in the horizontal direction.**

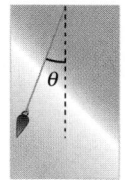

Plumb bob not vertical

Direction of deviation of
plumb bob from vertical is
opposite to the direction
of the acceleration of the
particle in the horizontal plane.

$\vec{a}$

FIGURE 3.32 Car speeding up in rectilinear motion.

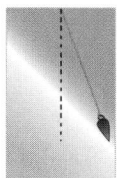

Plumb bob not vertical

Direction of deviation of
plumb bob from vertical is
opposite to the direction
of the acceleration of the
particle in the horizontal plane.

$\vec{a}$

 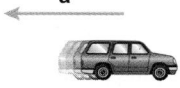 

FIGURE 3.33 Car slowing down in rectilinear motion.

is vertical at the instant you observe it, the instantaneous acceleration is zero in any direction in the horizontal plane.

STRATEGIC EXAMPLE 3.10

A traffic light turns red when a stopwatch indicates $t = 2.00$ s, and a car initially traveling at a constant velocity in a straight line toward the light at speed 50.0 km/h takes 5.00 s to come to a stop. Let $\hat{\imath}$ be in the direction the car is initially moving, as in Figure 3.34. Figure 3.35 is a graph modeling the x-component of the velocity of the car as a function of time.

a. What is the acceleration of the car at each instant between the times $t = 0$ s and $t = 2.00$ s on the stopwatch?
b. Estimate the instantaneous acceleration of the car at the instant $t = 3.00$ s on the stopwatch.
c. What is happening to the x-component of the acceleration of the car during the 5.00 s interval it takes the car to stop? Is the magnitude of the instantaneous acceleration of the car constant during the 5.00 s interval? Explain your answers.
d. Notice in Figure 3.35 that *two* tangent lines can be drawn to the curve at the instant $t = 2.00$ s. One of the tangent lines has zero slope, the other a negative slope. Does this mean that the particle has two different accelerations at that instant? Explain your answer.

Solution
a. On a plot of $v_x(t)$ versus t, the x-component of the acceleration is the slope of a line drawn tangent to the curve at the instant in question. The slope of the tangent lines at each

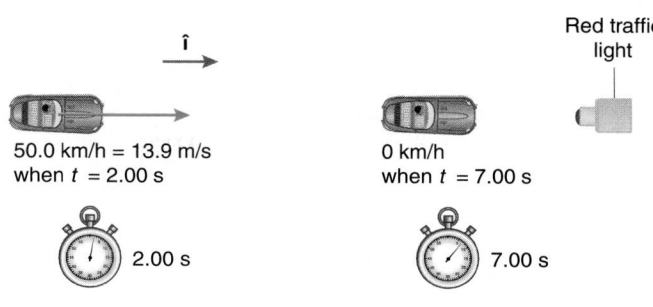

FIGURE 3.34

instant between $t = 0$ s and $t = 2.00$ s is zero. At each instant during this time interval, the instantaneous acceleration of the car is zero; the car has a constant x-component of the velocity.
b. Since you are to estimate the acceleration when $t = 3.00$ s, draw a tangent to the curve at this instant. See Figure 3.36. The tangent line has a negative slope that is approximately

$$a_x \approx \frac{-8.0 \ \text{m/s}}{2.00 \ \text{s}}$$
$$\approx -4.0 \ \text{m/s}^2$$

c. Figure 3.37 shows a series of tangent lines over the 5.00 s interval between 2.00 s and 7.00 s. Since the slopes of these lines change, the x-component of the acceleration of the car is not constant; it is becoming less negative. The magnitude of the acceleration (a positive scalar quantity, since the magnitude of a vector always is a positive scalar or zero) decreases over the interval.
d. The particle cannot have two different accelerations at the same instant. The graph *models* the behavior of the particle. Near the instant $t = 2.00$ s, a more realistic graph of the x-component of the velocity is shown in the blowup in

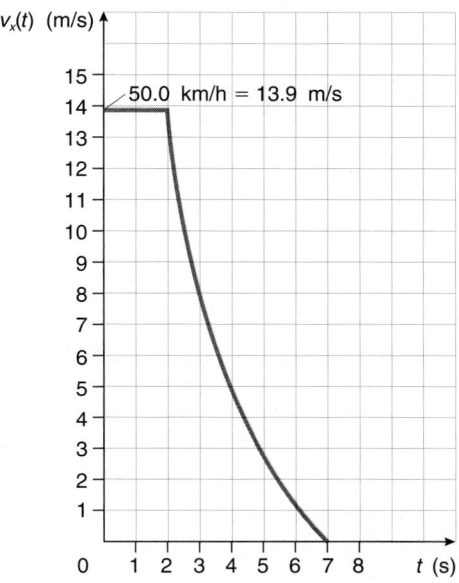

FIGURE 3.35

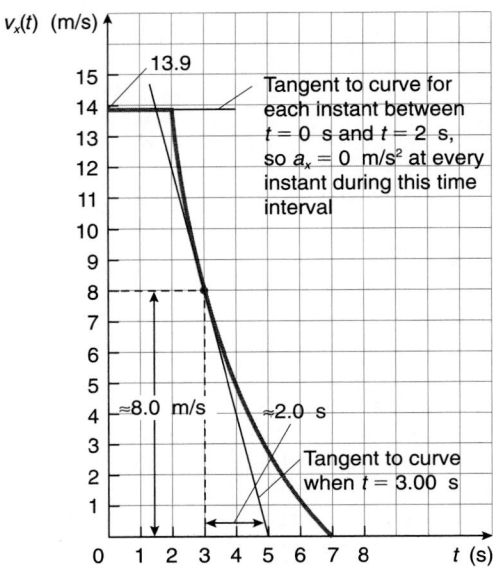

FIGURE 3.36

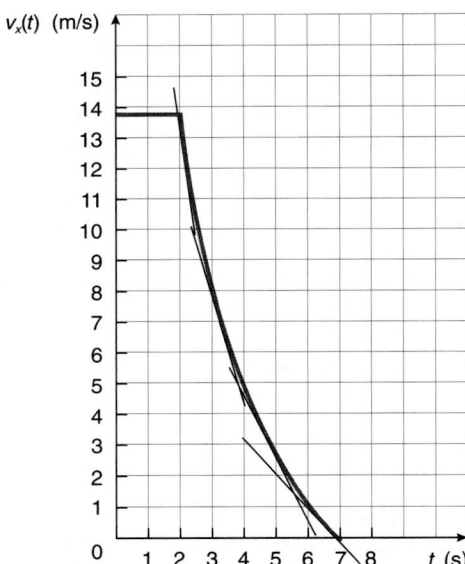

FIGURE 3.37

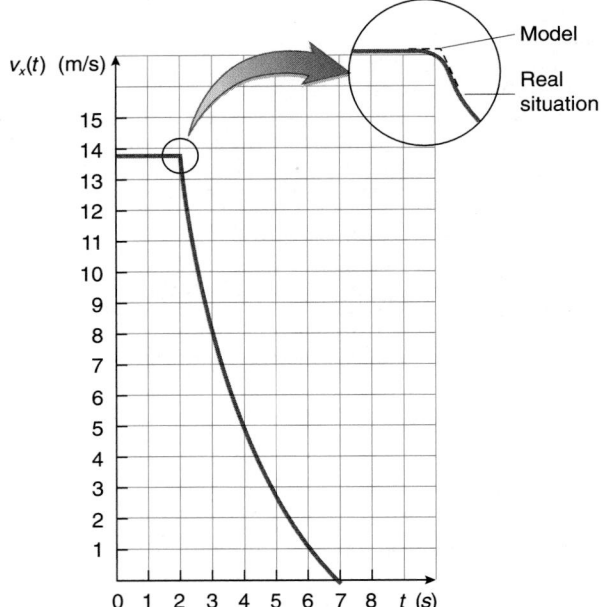

FIGURE 3.38

Figure 3.38. There is no discontinuity in the slope when $t = 2.00$ s; rather, there is a rounding of the cusp so that the slope changes continuously (and rapidly) rather than discontinuously.

3.7 RECTILINEAR MOTION WITH A CONSTANT ACCELERATION

The motion of an airplane accelerating along a runway for takeoff or after landing, the motion of a car braking for a red light or speeding up for a green light, and the **free-fall** motion of vertically moving objects near the surface of the Earth (when we ignore the effects of air resistance) are examples of the many kinematic situations that commonly are modeled to a first approximation using a constant acceleration.

> A constant acceleration means that *neither* the magnitude *nor* the direction of the acceleration vector changes with time or location.

We continue to choose the x-axis to be coincident with the line followed by the particle in rectilinear motion, whether the motion be horizontal (as when a car accelerates or brakes on a level highway), vertical (as when a ball is tossed upward or a rock is dropped off a bridge), or in another direction (as a car accelerating up or down a straight hill). As indicated in Figure 3.39, you have the freedom to choose:

1. The location of the origin of the coordinate system. The origin need not be at the position of the particle when we begin observations.
2. The direction taken for $\hat{\imath}$.

Write the *constant* acceleration vector as

$$\vec{\mathbf{a}} = a_x\hat{\imath} \qquad (3.15)$$

fixed in magnitude and direction in both space and time; that is, the acceleration component a_x *is a constant scalar quantity*, independent of both x and t. The numerical value of the acceleration component, a_x, may be positive or negative depending on the choice you make for the direction of $\hat{\imath}$ in a particular problem. We express the acceleration component a_x in SI units of m/s².

In many applications, the *initial* time t_i when we begin our observations is chosen to be *zero* ($t_i = 0$ s) and corresponds to starting a stopwatch for a problem.

> The particle is not necessarily at rest when $t = 0$ s.

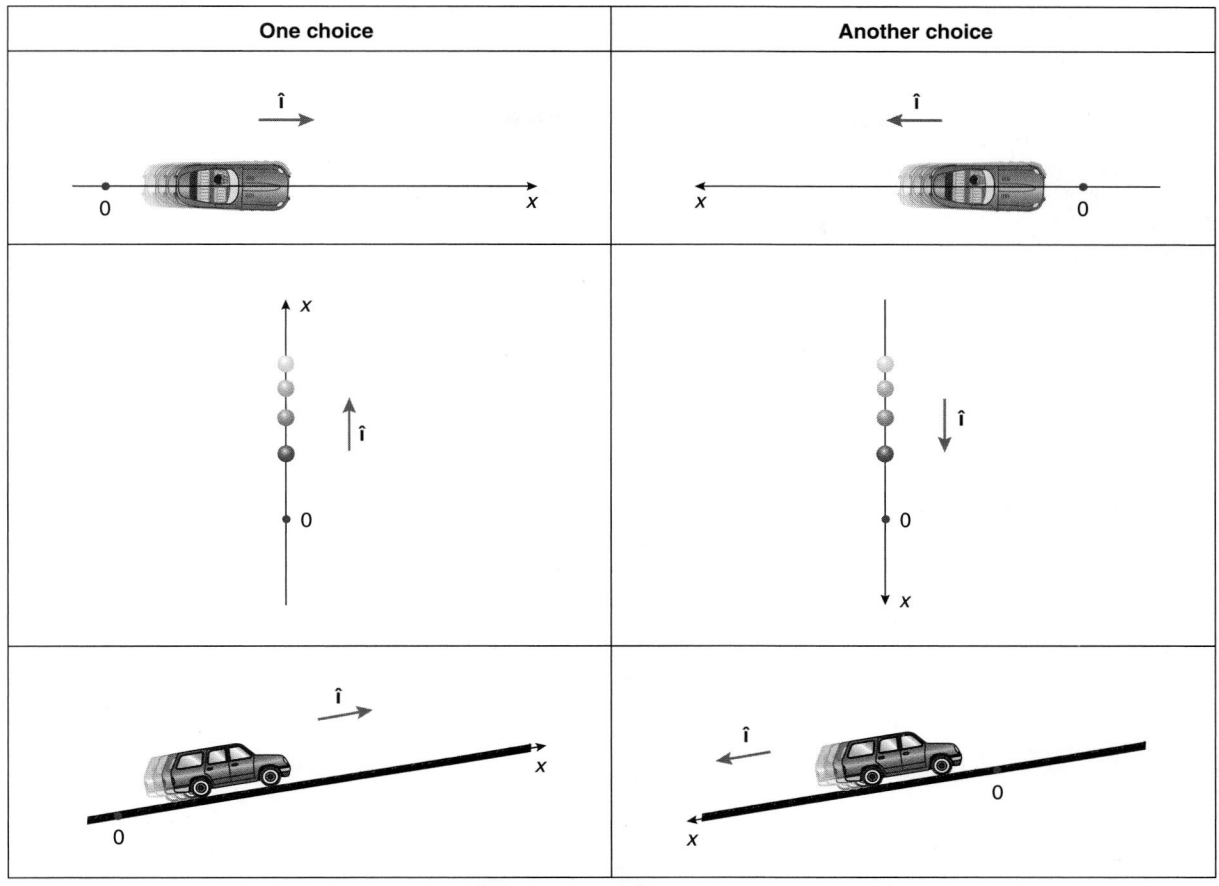

FIGURE 3.39 The *x*-axis is taken as the line of motion and the origin is chosen at any convenient point. The direction of î can be taken either way along the line of motion; these decisions are yours to make for each problem.

The instant $t = 0$ s is when we begin our observations of the particle. Likewise, we usually let the time instant t_f be a general time instant t and omit the subscript. With this convention, it is important to realize that t now has a *dual meaning*:

1. t means a specific time *instant* on a stopwatch—thus $x(t)$ and $v_x(t)$ are the position and velocity components at the instant t on the stopwatch; and
2. Since we chose the initial time instant to be the stopwatch reading 0 s, the value of t also is numerically equal to the *time interval* between the zero clock reading and the instant t:

$$\Delta t = t_f - t_i$$
$$= t - 0 \text{ s}$$
$$= t$$

As before, the *x*-component of the velocity vector at instant t is written $v_x(t)$, while the corresponding *x*-component of the position vector is $x(t)$.

We also introduce several notational abbreviations:

a. The quantity v_{x0} is the *x*-component of the velocity of the particle when we begin our observations when $t = 0$ s:

$$v_x(0 \text{ s}) \equiv v_{x0}$$

b. The quantity x_0 is the *x*-component of the position vector of the particle when we begin observations when $t = 0$ s:

$$x(0 \text{ s}) \equiv x_0$$

PROBLEM-SOLVING TACTIC

3.6 Know the meaning of the notation. To understand the meaning of technical notation is central to applying physics to a problem. Rather than just looking at the notation, envision its meaning in a short phrase or sentence. Say it aloud if possible (without disturbing others); if you can verbalize the meaning of a technical symbol, you probably know what it means. For example, when you see x_0, think "the *x*-coordinate of the particle when $t = 0$ s." When you see v_{x0}, think "the *x*-component of the velocity when $t = 0$ s." Study the kinematic notation just introduced; it will be used extensively.

Since the acceleration component a_x is constant, Equation 3.14,

$$\frac{dv_x(t)}{dt} = a_x$$

indicates that $v_x(t)$ must vary linearly with t:

$$v_x(t) = v_{x0} + a_x t \qquad (3.16)$$

Notice that when $t = 0$ s in Equation 3.16, the velocity component is v_{x0}, consistent with the new notation we have introduced. Differentiating Equation 3.16 with respect to t yields Equation 3.14.

A schematic graph of Equation 3.16 is linear (see Figure 3.40). The vertical intercept is the velocity component v_{x0} when $t = 0$ s. The constant slope of the linear graph is the constant acceleration component a_x. In Figure 3.40, $a_x > 0$ m/s². If a_x were negative, the line would have a negative slope and $v_x(t)$ would decrease as t increased.

Since the velocity component $v_x(t)$ varies linearly with t, the x-component of the *average velocity* during the interval $\Delta t = t - 0$ s $= t$ is*

$$v_{x\,\text{ave}} = \frac{v_{x0} + v(t)}{2} \qquad (3.17)$$

Adapted to our new notation, Equation 3.6 for the average velocity, written in terms of its (single) component, is

$$v_{x\,\text{ave}} = \frac{x(t) - x_0}{t - 0\ \text{s}} \qquad (3.18)$$

Equate Equations 3.17 and 3.18 for the average velocity component:

$$\frac{v_{x0} + v(t)}{2} = \frac{x(t) - x_0}{t}$$

Substituting for $v(t)$ using Equation 3.16, we find

$$\frac{v_{x0} + v_{x0} + a_x t}{2} = \frac{x(t) - x_0}{t}$$

Solve for $x(t)$:

$$x(t) = x_0 + v_{x0} t + \frac{1}{2} a_x t^2 \qquad (3.19)$$

---

*If the acceleration is not constant, Equation 3.17 cannot be used to find the average velocity component.

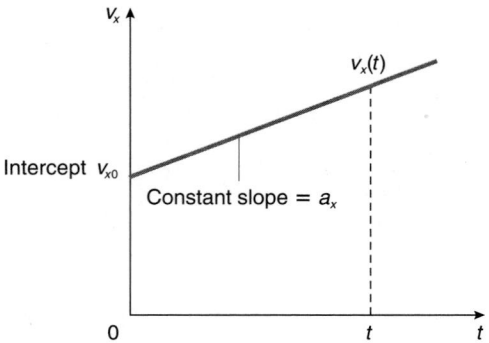

FIGURE 3.40 For a constant acceleration, a graph of v_x versus t is linear with intercept v_{x0} and slope a_x.

Therefore, one-dimensional motion with a constant acceleration is described by two fundamental equations that determine the components of the velocity, and position vectors:

One-dimensional motion with a constant acceleration

$$a_x = \text{constant} \qquad \text{where } \vec{\mathbf{a}} = a_x \hat{\mathbf{i}}$$

$$v_x(t) = v_{x0} + a_x t \qquad \text{where } \vec{\mathbf{v}}(t) = v_x(t)\hat{\mathbf{i}} \qquad (3.16)$$

$$x(t) = x_0 + v_{x0} t + \frac{1}{2} a_x t^2 \qquad \text{where } \vec{\mathbf{r}}(t) = x(t)\hat{\mathbf{i}} \qquad (3.19)$$

Remember:

x_0 is the x-coordinate of the particle when $t = 0$ s; it also is the x-component of the position vector of the particle when $t = 0$ s.

v_{x0} is the x-component of the velocity vector of the particle when $t = 0$ s

PROBLEM-SOLVING TACTICS

3.7 Know the limitations associated with using certain equations. Equations 3.16 and 3.19 are the important ones for motion with a *constant* acceleration (both in magnitude and direction). However, before using them, be sure the problem involves an acceleration that is constant, or can be modeled with a constant acceleration to a first approximation. If the acceleration is not constant, then these equations are useless and, if used, produce predictions that may not correspond with observations.

3.8 Avoid needless memorization of equations. Every problem involving one-dimensional motion with a constant acceleration can be solved using only Equations 3.16 and 3.19. No other equations are needed. You should begin any problem having a constant acceleration by considering only Equations 3.16 and 3.19.

3.9 Tailor the general equations to the specific choice you make for the coordinate system. This tailoring process will become apparent after studying several examples that follow. The flexibility involved in choosing the location and orientation of the Cartesian coordinate system is considered in Strategic Example 3.12. If you understand how to tailor the one-dimensional motion equations at constant acceleration to the specific choice of a Cartesian coordinate system for the problem at hand, no problem will stump you. Everything that can be asked about such motion is extracted from Equations 3.16 and 3.19. Remember, there is no sacred choice for the coordinate system. This allows you the freedom to tackle the situation however you want. That is part of the beauty of physics; it allows independent observers to choose how to describe the situation. The critical skill to develop, then, is the ability to determine and set up appropriate and correct equations.

The experiment performed in Example 3.11 (following) shows that near the surface of the Earth,[†] any object in vertical free-fall (ignoring air resistance effects) continually experiences a constant acceleration *directed vertically down*, regardless of whether the particle is moving vertically up or vertically down. We will see

---

[†]"Near" means any vertical distances traversed by the particle are negligible compared with the radius of the Earth (6370 km).

in Chapter 6 how this constant acceleration arises from the gravitational force of the Earth on the particle. The letter g designates the *magnitude* of this gravitational acceleration near the surface of the Earth; g is called the magnitude of the **local acceleration due to gravity**. In this text, we take the value of g to be*

$$g = 9.81 \text{ m/s}^2$$

PROBLEM-SOLVING TACTIC

3.10 Since g is the magnitude of the local gravitational acceleration, g always is a *positive* scalar quantity in this text.

If you choose $\hat{\imath}$ to be vertically up (see Figure 3.41), then

$$a_x = -g$$

If you choose $\hat{\imath}$ to be vertically down (see Figure 3.42), then

$$a_x = +g$$

QUESTION 5

Toss a ball straight up, and you will see it rise awhile and fall awhile. Is the acceleration vector always in the same direction, or is it directed one way for the flight up and another way for the flight down?

QUESTION 6

Some people might say that an acceleration component is negative and others might say, equally correctly, that it is positive. Explain the difference in viewpoints.

EXAMPLE 3.11

A student drops an egg from rest at an initial height of 2.00 m above the floor. An electronic timer indicates that the egg begins its impact with the floor 0.6385 s after release. Determine the magnitude of the acceleration of the egg during its fall, assuming it is constant. This is a reasonable assumption.

Solution

Use Problem-Solving Tactic 3.1, which states that *for rectilinear motion, always choose a Cartesian coordinate axis to be coincident with*

the line followed by the particle. Choose the x-axis to be coincident with the vertical path followed by the egg. Remember, there is nothing sacred about the orientation you choose for the x-axis in space; it can be horizontal, vertical, or whatever. You simply need to specify what orientation you choose. You can choose the origin to be anywhere along the vertical path. Suppose you choose the origin to be at the point of release of the bottom of the egg (or you can consider the egg as a particle of negligible geometric size). The choice for the direction of $\hat{\imath}$ can be either up or down; choose, say, the upward direction for $\hat{\imath}$, as in Figure 3.43.

Also use Problem-Solving Tactic 3.9 and *tailor the general equations to the specific choice made for the coordinate system*. The egg is released when $t = 0$ s with the bottom of the egg at the initial position $x_0 = 0$ m. The egg was released at rest, so that the initial x-component of the velocity is $v_{x0} = 0$ m/s. Hence Equation 3.19 for the x-component of the position vector at any instant t becomes

$$x(t) = x_0 + v_{x0}t + \frac{1}{2}a_x t^2$$
$$= 0 \text{ m} + (0 \text{ m/s})t + \frac{1}{2}a_x t^2$$
$$= \frac{1}{2}a_x t^2$$

At the instant $t = 0.6385$ s, the x-coordinate of the bottom of the egg is -2.00 m; this is the x-component of the position vector at that instant. Hence

$$-2.00 \text{ m} = \frac{1}{2}a_x(0.6385 \text{ s})^2$$

Solving for a_x, you obtain

$$a_x = -9.81 \text{ m/s}^2$$

The acceleration is

$$\vec{a} = a_x\hat{\imath}$$
$$= (-9.81 \text{ m/s}^2)\hat{\imath}$$

The acceleration is directed in the same sense as $-\hat{\imath}$. With the choice you made for the direction of $\hat{\imath}$ (up), the acceleration is directed vertically downward. The egg falls with an acceleration whose magnitude is 9.81 m/s².

---

*Since the Earth is only approximately spherical, the value of g depends slightly on your latitude on the surface of the Earth, varying from 9.78 m/s² at the equator to 9.83 m/s² at the poles. For simplicity, we ignore this slight variation of g in this text.

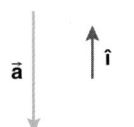

$$\vec{a} = a_x\hat{\imath} = -g\hat{\imath}$$

FIGURE 3.41 Negative gravitational acceleration component.

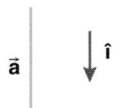

$$\vec{a} = a_x\hat{\imath} = g\hat{\imath}$$

FIGURE 3.42 Positive gravitational acceleration component.

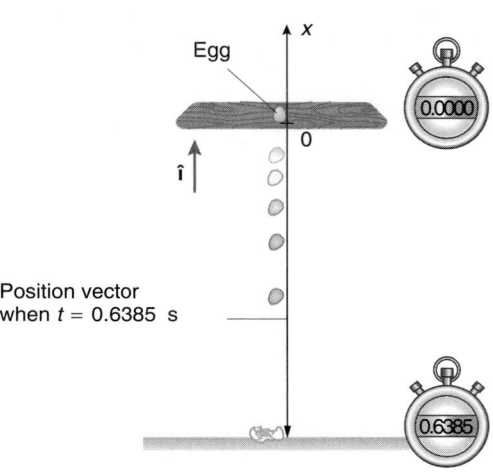

Egg

Position vector when $t = 0.6385$ s

FIGURE 3.43 The coordinate system choice.

STRATEGIC EXAMPLE 3.12

A geologist at the top of a 100-meter-deep crevasse cannot resist the temptation to hurl a gneiss rock down to the bottom. The rock has an initial downward speed of 10.0 m/s as it leaves the geologist's hand when $t = 0$ s. Find

a. the time interval during which the particle is in flight, called the time of flight; and
b. the speed of the particle at the instant just before impact.

Solution

The Cartesian coordinate system used to solve this problem can be chosen with the direction for increasing values of x directed either up or down. The origin also can be chosen to be anywhere along the vertical axis. To convince you of this freedom, we will solve the problem two ways using different choices for the coordinate system. Obviously, the time of flight and speed (a positive scalar) the instant just before impact are indifferent about the choice you make, and so the *same* answers will be obtained whatever choice is made for the coordinate system. However, the *description* of the motion is different in each case, since the description is tailored to the specific choice made for the coordinate system.

Choose a Cartesian coordinate system in either of two different ways as shown in Figures 3.44 and 3.45: The acceleration vector is directed downward and is of magnitude $g = 9.81$ m/s^2.

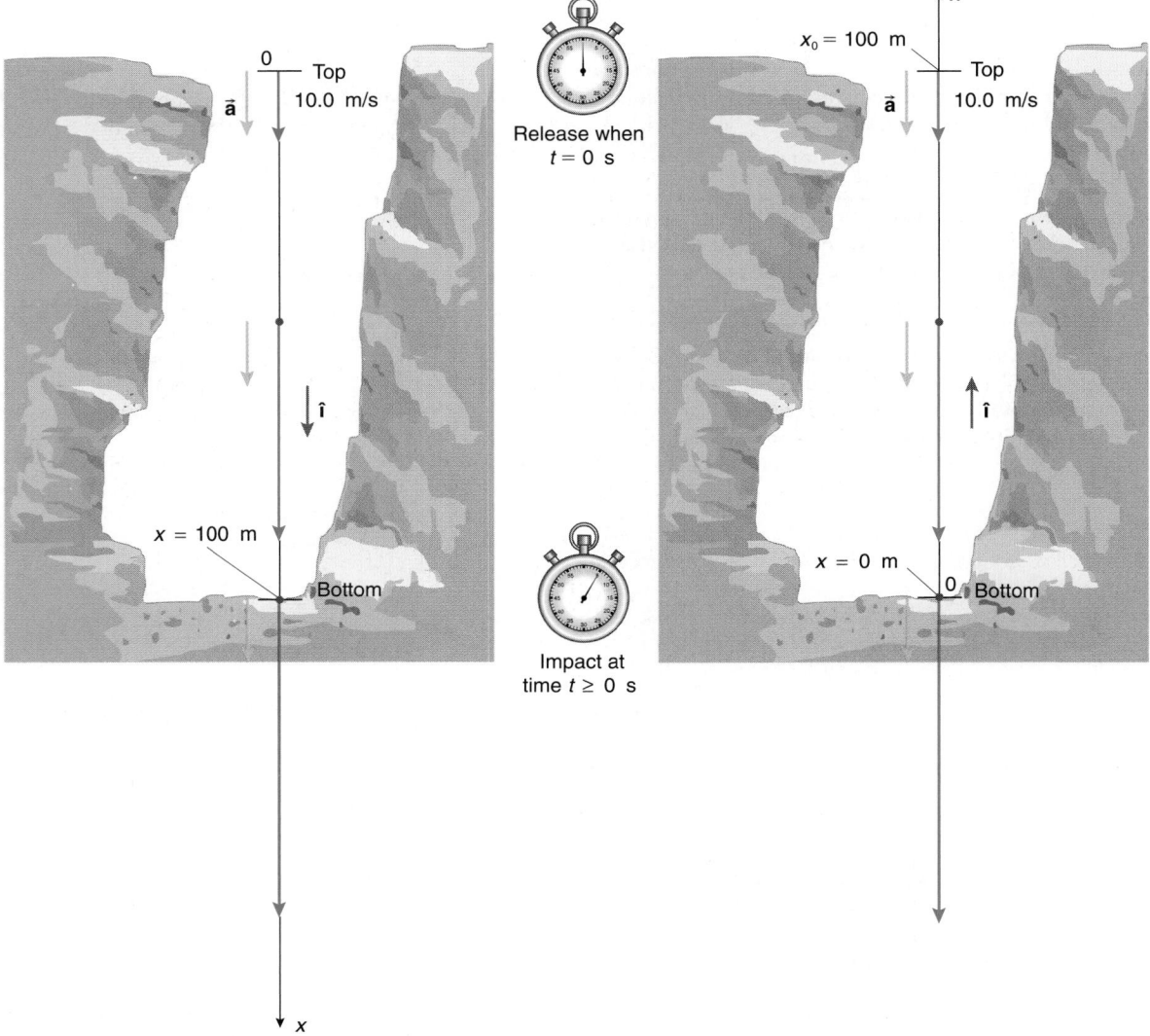

FIGURE 3.44 The origin is at the point of release and the direction of increasing values of x, $\hat{\imath}$, is vertically down.

FIGURE 3.45 The origin is at the bottom of the crevasse and $\hat{\imath}$ is vertically up.

Choice 1	*Choice 2*
The acceleration is toward increasing values of x, in the same direction as î. Hence the acceleration vector is	The acceleration is toward decreasing values of x, in the same direction as −î. Hence the acceleration is

$$\vec{a} = a_x \hat{i}$$
$$= (+9.81 \text{ m/s}^2)\hat{i}$$

$$\vec{a} = a_x \hat{i}$$
$$= (-9.81 \text{ m/s}^2)\hat{i}$$

The x-component of the acceleration is	The x-component of the acceleration is

$$a_x = +9.81 \text{ m/s}^2$$

$$a_x = -9.81 \text{ m/s}^2$$

Because the rock is not dropped from rest, but hurled down the crevasse, there is a nonzero initial velocity

directed toward increasing values of x, in the same direction as î:	directed toward decreasing values of x, in the same direction as −î:

$$v_{x0}\hat{i} = (10.0 \text{ m/s})\hat{i}$$

$$v_{x0}\hat{i} = (-10.0 \text{ m/s})\hat{i}$$

so the x-component of the initial velocity, v_{x0}, is

$$v_{x0} = +10.0 \text{ m/s}$$

$$v_{x0} = -10.0 \text{ m/s}$$

The origin of the coordinate system was chosen to be

at the point where the rock was released, so the initial x-component of the position vector, x_0, is zero:	at the bottom of the crevasse, so the initial x-component of the position vector, x_0, is 100 m:

$$x_0 = 0 \text{ m}$$

$$x_0 = 100 \text{ m}$$

The instant $t = 0$ s is when the rock is released. We made this choice to start the clock for any problem when we developed the two equations for one-dimensional motion with a constant acceleration, Equations 3.16 and 3.19.

At the instant t immediately before impact,

the rock is located at the coordinate	the rock is located at the coordinate

$$x(t) = +100 \text{ m}$$

$$x(t) = 0 \text{ m}$$

Notice that t plays a dual role: as an *interval* of time (from the $t = 0$ s initial clock reading, $\Delta t = t - 0 \text{ s} = t$) and as an *instant* when the particle has the position vector component $x(t)$ and velocity component $v_x(t)$.

The equation for the x-component of the position vector, Equation 3.19,

$$x(t) = x_0 + v_{x0}t + \frac{1}{2} a_x t^2$$

becomes (at the instant just before impact)

$100 \text{ m} = 0 \text{ m} + (10.0 \text{ m/s}) t$ $+ \frac{1}{2}(9.81 \text{ m/s}^2) t^2$	$0 \text{ m} = 100 \text{ m} + (-10.0 \text{ m/s}) t$ $+ \frac{1}{2}(-9.81 \text{ m/s}) t^2$

Notice that we use the *same* fundamental equations for both coordinate choices, but the values of the various substitutions are *different* because the coordinate systems differ.

With some arranging of terms, we have

$\frac{9.81 \text{ m/s}^2}{2} t^2 + (10.0 \text{ m/s}) t - 100 \text{ m} = 0 \text{ m}$	$\frac{9.81 \text{ m/s}^2}{2} t^2 + (10.0 \text{ m/s}) t - 100 \text{ m} = 0 \text{ m}$

which is quadratic in t. Notice that the equation for t is the same for *both* coordinate choices, illustrating once again that the physics is independent of the way you impose a coordinate system on a problem. The equation for t is solved using the quadratic formula. The two roots are

$$t = 3.61 \text{ s} \qquad \text{and} \qquad t = -5.65 \text{ s}$$

Since the impact time obviously is after the release (which was when $t = 0$ s), the positive solution is the one desired*:

$$t = 3.61 \text{ s}$$

The component of the velocity vector at the instant just before impact is found from Equation 3.16:

$$v_x = v_{x0} + a_x t$$

Substituting for v_{x0}, a_x, and t, you obtain

$v_x = 10.0$ m/s $+ (9.81$ m/s$^2)(3.61$ s$)$ $\quad = 45.4$ m/s	$v_x = -10.0$ m/s $+ (-9.81$ m/s$^2)(3.61$ s$)$ $\quad = -45.4$ m/s

The velocity vector at the instant of impact is $\vec{v} = v_x \hat{\imath}$:

$\vec{v} = (45.4$ m/s$)\hat{\imath}$	$\vec{v} = (-45.4$ m/s$)\hat{\imath}$
and is in the same sense as $\hat{\imath}$, vertically downward.	and is in the same sense as $-\hat{\imath}$, vertically downward.

The speed is the *magnitude* of the velocity vector, or 45.4 m/s. Notice that the velocity has the same magnitude and direction in space for *both* coordinate choices, but the way you *describe* the downward velocity vector depends on your choice of a Cartesian coordinate system.

EXAMPLE 3.13

How far has the rock in Example 3.12 traveled when its speed is 30.0 m/s?

Solution

Since there is no back-and-forth motion of the rock as it falls and the problem is one-dimensional, the magnitude of the change in the position vector is equivalent to the distance the rock travels. Specifically, you want the *change* in its position vector between the instant of release ($t = 0$ s) and the instant its speed is 30.0 m/s.

Choose the point of release as the origin of the coordinate system with the increasing values of x in the upward direction as shown in Figure 3.46. The coordinate choice differs from either of the choices used in Example 3.12. We illustrate this new choice to show the flexibility you have when making coordinate choices.

The initial position vector of the rock when $t = 0$ s then is $(0$ m$)\hat{\imath}$, so that $x_0 = 0$ m. Since $\hat{\imath}$ is up, the constant downward directed acceleration

$$\vec{a} = a_x \hat{\imath}$$

is

$$\vec{a} = (-9.81 \text{ m/s}^2)\hat{\imath}$$

and so the x-component of the acceleration is

$$a_x = -9.81 \text{ m/s}^2$$

From Equation 3.16, the velocity component $v_x(t)$ is

$$v_x(t) = v_{x0} + a_x t$$

The initial velocity component v_{x0} is -10.0 m/s, because the rock was hurled in the direction of decreasing values of x, in

the same direction as $-\hat{\imath}$. This problem is concerned with what is happening when the *speed* is 30.0 m/s. Since the rock is moving downward (toward decreasing values of x with your choice of the coordinate system), you must find the instant t when the velocity component $v_x(t)$ is -30.0 m/s. Use this information in Equation 3.16 to find

$$-30.0 \text{ m/s} = -10.0 \text{ m/s} + (-9.81 \text{ m/s}^2)t$$

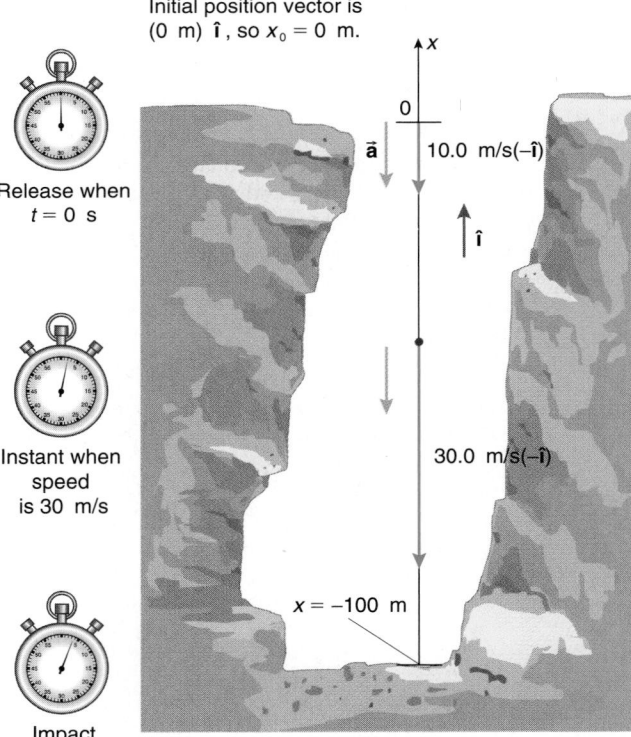

Initial position vector is $(0$ m$)$ $\hat{\imath}$, so $x_0 = 0$ m.

Release when $t = 0$ s

$\vec{a}$ 10.0 m/s$(-\hat{\imath})$

$\hat{\imath}$

Instant when speed is 30 m/s

30.0 m/s$(-\hat{\imath})$

Impact $x = -100$ m

FIGURE 3.46

---

*The negative root corresponds to when the particle was at $x = 100$ m (Choice 1) or $x = 0$ m (Choice 2) if it had been previously launched vertically upward from the bottom of the crevasse so that it would have a downward speed of 10 m/s at the top of the crevasse when $t = 0$ s.

Solving for t, you find

$$t = 2.04 \text{ s}$$

The position of the rock at any time is found from the equation for the x-component of the position vector, Equation 3.19:

$$x = x_0 + v_{x0}t + \frac{1}{2}a_xt^2$$

Since the rock was released from the origin, the initial position vector component x_0 is 0 m. Recall that $v_{x0} = -10.0$ m/s. Make these substitutions in Equation 3.19:

$$x = 0 \text{ m} + (-10.0 \text{ m/s})(2.04 \text{ s}) + \frac{1}{2}(-9.81 \text{ m/s}^2)(2.04 \text{ s})^2$$

or

$$x = -20.4 \text{ m} - 20.4 \text{ m}$$
$$= -40.8 \text{ m}$$

Thus the position vector of the rock when its *speed* is 30.0 m/s is

$$\vec{r} = x\hat{\imath}$$
$$= (-40.8 \text{ m})\hat{\imath}$$

The change in the position vector of the rock is then

$$\Delta\vec{r} = \vec{r}_f - \vec{r}_i$$
$$= (-40.8 \text{ m})\hat{\imath} - (0 \text{ m})\hat{\imath}$$
$$= (-40.8 \text{ m})\hat{\imath}$$

Hence the magnitude of the change in the position vector of the rock is 40.8 m when its *speed* is 30.0 m/s. Since the rock did not move back and forth along its path, 40.8 m is the distance through which the rock fell while achieving the speed of 30.0 m/s.

EXAMPLE 3.14

You are driving downhill at 110 km/h along a straight section of rural road when a fawn of a white-tailed deer (*Odocoileus virginianus*) springs in front of your car. Fortunately, you manage to stop the car 3.50 s after your foot first hits the brake pedal, and without hitting Bambi. The roadway makes an angle of 5.00° with the horizontal. Assume the acceleration of the car is constant during the 3.50 s time interval.

a. What is the acceleration?
b. What is the average acceleration during the 3.50 s interval?
c. Through what distance did the car move during the 3.50 s interval?

Solution

a. The 110 km/h initial speed is equivalent to

$$110 \text{ km/h} = (110 \text{ km/h})\frac{1000 \text{ m/km}}{3600 \text{ s/h}}$$
$$= 30.6 \text{ m/s}$$

Remember that for one-dimensional motion, you choose the x-axis to be coincident with the path followed by the particle, whatever its orientation. Set up a coordinate system with the origin at the initial position of the car when the braking began and $\hat{\imath}$ in the direction of motion of the car down the hill (see Figure 3.47). The initial velocity component v_{x0} of the car when $t = 0$ s is

$$v_{x0} = +30.6 \text{ m/s}$$

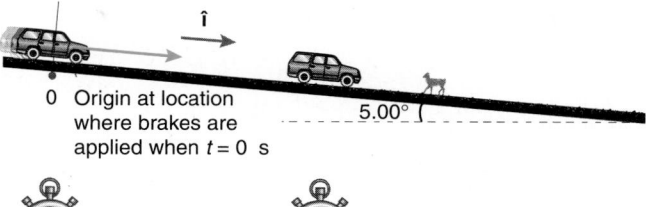

0 Origin at location where brakes are applied when $t = 0$ s   5.00°

$t = 0$ s $t = 3.50$ s

FIGURE 3.47

The velocity component of the car at time instant t is given by Equation 3.16:

$$v_x(t) = v_{x0} + a_xt$$

When $t = 3.50$ s, the velocity component v_x is zero, so that Equation 3.16 becomes

$$0 \text{ m/s} = 30.6 \text{ m/s} + a_x(3.50 \text{ s})$$

Solving for a_x, you find

$$a_x = -8.74 \text{ m/s}^2$$

Hence the constant acceleration is

$$\vec{a} = a_x\hat{\imath}$$
$$= (-8.74 \text{ m/s}^2)\hat{\imath}$$

The acceleration vector is directed opposite to the direction the car is moving during the time interval.

b. The average acceleration is given by Equation 3.9:

$$\vec{a}_{ave} = \frac{\Delta\vec{v}}{\Delta t}$$
$$= \frac{\vec{v}_f - \vec{v}_i}{\Delta t}$$
$$= \frac{(0 \text{ m/s})\hat{\imath} - (30.6 \text{ m/s})\hat{\imath}}{3.50 \text{ s}}$$
$$= (-8.74 \text{ m/s}^2)\hat{\imath}$$

Notice that the average acceleration during the 3.50 s interval is the same as the constant instantaneous acceleration at any instant during the interval. If the acceleration is constant, the average acceleration over any interval is equal to the constant acceleration during the interval.

c. Since there is no back-and-forth motion of the car during the time interval, the distance the car travels is equal to the magnitude of the change in its position vector. Note that the initial x-component of the position vector of the car, x_0, is 0 m, because the origin was chosen at this point. Use

$v_{x0} = 30.6$ m/s and $a_x = -8.74$ m/s² from part (a); then Equation 3.19 becomes

$$x = 0 \text{ m} + (30.6 \text{ m/s}) t + \frac{1}{2} (-8.74 \text{ m/s}^2) t^2$$

The position x of the car when it stops is found by substituting $t = 3.50$ s:

$$x = (30.6 \text{ m/s}) (3.50 \text{ s}) + \frac{1}{2} (-8.74 \text{ m/s}^2) (3.50 \text{ s})^2$$

$$= 107 \text{ m} - 53.5 \text{ m} = 53 \text{ m}$$

The change in the position vector of the car is

$$\Delta \vec{r} = \Delta x \, \hat{\imath}$$
$$= (53 \text{ m})\hat{\imath} - (0 \text{ m})\hat{\imath}$$
$$= (53 \text{ m})\hat{\imath}$$

The distance traveled by the car is the magnitude of this vector, 53 m, since the car did not move back and forth along the roadway during the 3.50 s interval it took to stop.

3.8 GEOMETRIC INTERPRETATIONS*

We found in Section 3.4 that the x-component of the velocity, $v_x(t)$, is related to the x-component of the position vector by

$$v_x(t) = \frac{dx(t)}{dt}$$

Hence the value of $v_x(t)$ at any instant t is the slope of a graph of $x(t)$ versus time at the instant t. Likewise, we saw in Section 3.6 that the x-component of the acceleration $a_x(t)$ is found from the velocity component via Equation 3.14:

$$a_x(t) = \frac{dv_x(t)}{dt}$$

Hence the value of a_x at any instant t is the slope of a graph of v_x versus time at the instant t. Thus a knowledge of the position vector component $x(t)$ enables us to find the velocity component $v_x(t)$ and then the acceleration component $a_x(t)$ by differentiation with respect to time.

We could equally well reverse the process and proceed from a knowledge of the acceleration component to the velocity component and then to the position vector component via integration. In other words, if the acceleration component $a_x(t)$ is known, then we integrate Equation 3.14 over time to find the velocity component $v_x(t)$:

$$\int_{v_x(t_i)}^{v_x(t_f)} dv_x = \int_{t_i}^{t_f} a_x(t) \, dt$$

where $v_x(t_i)$ is the x-component of the velocity at time t_i and $v_x(t_f)$ is the velocity component at time t_f. After integrating

the left-hand side according to the rules of calculus:

$$v_x(t_f) - v_x(t_i) = \int_{t_i}^{t_f} a_x(t) \, dt \qquad (3.20)$$

As we know from calculus, an integral is the geometric area under the curve of the function being integrated (the integrand). In other words, if we make a graph of the function $a_x(t)$ versus t, as shown in Figure 3.48, the *change* in the value of $v_x(t)$—that is, $\Delta v_x \equiv v_x(t_f) - v_x(t_i)$—is the area under the curve representing $a_x(t)$ between the times t_i and t_f.

Areas under the curve can be positive or negative. From calculus, if the curve is above the horizontal axis (see Figure 3.49), and we are integrating toward increasing values of the abscissa (here t), then the area between the curve and the axis is positive. If the curve dips below the axis, the area between the curve and the axis is negative.

Likewise, from Equation 3.8, the change in the position vector component is the time integral of the velocity component $v_x(t)$:

$$\int_{x(t_i)}^{x(t_f)} dx = \int_{t_i}^{t_f} v_x(t) \, dt$$

where $x(t_i)$ is the x-component of the position vector at time t_i and $x(t_f)$ is the x-component of the position vector at time t_f. After performing the integration on the left-hand side, we have

$$x(t_f) - x(t_i) = \int_{t_i}^{t_f} v_x(t) \, dt \qquad (3.21)$$

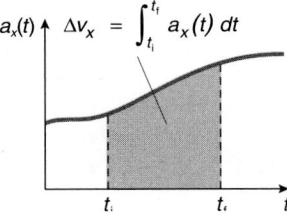

FIGURE 3.48 The change in the velocity component is the area under the curve of a_x versus t.

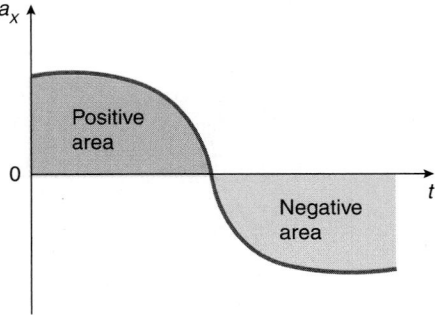

FIGURE 3.49 Areas under a curve can be positive or negative.

On a graph of $v_x(t)$ versus t, as in Figure 3.50, the *change* in the value of the x-component of the position vector—that is, $\Delta x = x(t_f) - x(t_i)$—is the area under the curve representing $v_x(t)$ between the times t_i and t_f.

If we let $t_i = 0$ s and $t_f = t$, then Equations 3.20 and 3.21 become

$$v_x(t) - v_{x0} = \int_{0\,s}^{t} a_x(t)\, dt \qquad (3.22)$$

and

$$x(t) - x_0 = \int_{0\,s}^{t} v_x(t)\, dt \qquad (3.23)$$

Such equations can be applied to situations in which the acceleration is not a constant vector.

If the acceleration is *constant*, we can use Equations 3.22 and 3.23 to get Equations 3.16 and 3.19 in a way different from that used in Section 3.7. According to Equation 3.22, to find the *change* in the x-component of the velocity, you integrate the constant acceleration component a_x:

$$v_x(t) - v_{x0} = \int_{0\,s}^{t} a_x\, dt = a_x \int_{0\,s}^{t} dt$$

where v_{x0} is the x-component of the velocity when $t = 0$ s, and $v_x(t)$ is the velocity component at instant t. After integrating, we have

$$v_x(t) - v_{x0} = a_x[t - 0\ \text{s}] \qquad (3.24)$$

$$\Delta v_x = a_x t$$

Geometrically, in Figure 3.51 Δv_x is the

$$\text{rectangular area} = (\text{width})(\text{length})$$
$$= a_x t$$

Note that if $a_x < 0$ m/s², the product $(\text{width})(\text{length}) = a_x t$ yields a negative area under the curve (between the curve and the axis).

Transposing v_{x0} in Equation 3.24, we obtain Equation 3.16 once again:

$$v_x(t) = v_{x0} + a_x t$$

To find the *change* in the x-component of the position vector, use Equation 3.23 and integrate the x-component of the velocity vector between 0 s and t:

$$x(t) - x_0 = \int_{0\,s}^{t} v_x(t)\, dt = \int_{0\,s}^{t} (v_{x0} + a_x t)\, dt$$

where both v_{x0} and a_x are constants. After integrating and transposing the x_0 term, we have Equation 3.19 once again:

$$x(t) = x_0 + v_{x0} t + \frac{1}{2} a_x t^2$$

The geometric interpretation of this equation on a v_x versus t graph is shown in Figure 3.52.

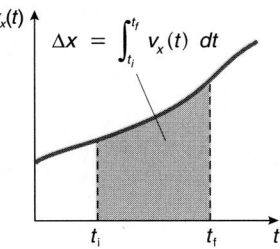

FIGURE 3.50 The change in the value of the position vector component is the area under the curve of v_x versus t.

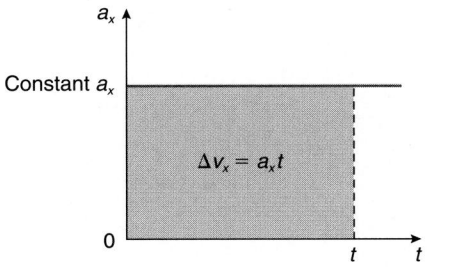

FIGURE 3.51 For a constant acceleration, the change in the velocity component is the rectangular area under the a_x versus t graph.

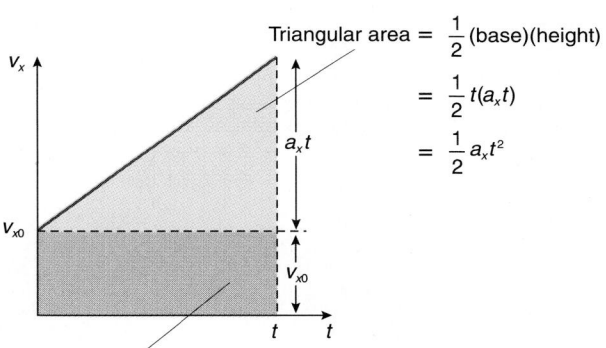

FIGURE 3.52 Geometric interpretation of Equation 3.19 on a v_x versus t graph. The change in the position vector component Δx is the area under the v_x versus t graph.

CHAPTER SUMMARY

Kinematics is the description of motion. *Dynamics* is the study of the causes of motion. *Rectilinear motion* is one-dimensional motion along a straight line.

A *particle* is anything whose geometrical size is ignored.

The location of a particle with respect to the origin of an arbitrarily chosen Cartesian coordinate system is specified by an *instantaneous position vector* $\vec{r}(t)$. For rectilinear motion, $\vec{r}(t) = x(t)\hat{\imath}$.

The *average speed* during an interval Δt is the total distance travelled divided by Δt. The *average velocity* of a particle during a time interval Δt is defined to be the ratio of the *change* in its position vector $\Delta\vec{r}$ during the time interval to the time interval itself:

$$\vec{v}_{\text{ave}} \equiv \frac{\Delta\vec{r}}{\Delta t} \qquad (3.6)$$

The *instantaneous velocity* of a particle, more commonly called the *velocity* $\vec{v}$, is the limit of its average velocity as the time interval Δt approaches zero; this is equivalent to the time derivative of the instantaneous position vector

$$\vec{v}(t) = \lim_{\Delta t \to 0 \text{ s}} \frac{\Delta\vec{r}}{\Delta t} = \frac{d\vec{r}(t)}{dt}$$

$$= \frac{dx(t)}{dt}\hat{\imath} \qquad \text{(for rectilinear motion)} \qquad (3.7)$$

The magnitude of the velocity is called the *speed*. The instantaneous velocity component at any time instant is the slope of the graph of the corresponding position vector component versus time at that instant. That is, v_x at instant t' is the slope of the graph of $x(t)$ versus t at the particular instant t'.

The *average acceleration* of a particle during a time interval Δt is the ratio of the change in its velocity during the time interval to the time interval itself:

$$\vec{a}_{\text{ave}} \equiv \frac{\Delta\vec{v}}{\Delta t} \qquad (3.9)$$

The *instantaneous acceleration*, commonly called the *acceleration* $\vec{a}$, is the limit of the average acceleration as the time interval Δt approaches zero; this is equivalent to the time derivative of the velocity, or the second time derivative of the position vector:

$$\vec{a}(t) = \lim_{\Delta t \to 0 \text{ s}} \frac{\Delta\vec{v}}{\Delta t} = \frac{d\vec{v}(t)}{dt} = \frac{d^2\vec{r}(t)}{dt^2} \qquad (3.13)$$

$$= \frac{dv_x(t)}{dt}\hat{\imath} = \frac{d^2x(t)}{dt^2}\hat{\imath} \qquad \text{(for rectilinear motion)}$$

The instantaneous acceleration component at any instant is the slope of the graph of the corresponding velocity component versus time at that instant. That is, a_x at instant t' is the slope of the graph of $v_x(t)$ versus t at the particular instant t'.

If the motion is with a *constant* acceleration and the initial time instant t_i is set equal to zero, then the velocity and position vector components of the particle at any instant t are

$$v_x(t) = v_{x0} + a_x t \qquad a_x \text{ constant} \quad (3.16)$$

$$x(t) = x_0 + v_{x0}t + \frac{1}{2}a_x t^2 \qquad a_x \text{ constant} \quad (3.19)$$

where a_x is the constant x-component of the acceleration, v_{x0} is the x-component of the velocity when $t = 0$ s, and x_0 is the x-component of the position vector of the particle when $t = 0$ s. The signs and numerical values of a_x, v_{x0}, and x_0 depend on the particular choices you make for the location of the origin for the x-axis of a Cartesian coordinate system, the direction chosen for $\hat{\imath}$, and the specifics of the problem at hand.

After particles are thrown vertically (up or down) or dropped vertically near the surface of the Earth, they experience *free-fall* and (neglecting air resistance) a *constant* acceleration, called the *local acceleration due to gravity*, of *magnitude* $g = 9.81$ m/s² directed vertically downward. The sign of the vertical acceleration component a_x depends on the choice you make for the direction of $\hat{\imath}$:

$$a_x = +g \qquad \text{if } \hat{\imath} \text{ is chosen vertically down}$$

$$a_x = -g \qquad \text{if } \hat{\imath} \text{ is chosen vertically up}$$

The free-fall acceleration due to gravity *always* is directed vertically down, whether the particle is rising, falling, or at the instant, between rising and falling, when it is at rest.

If a graph is made of $a_x(t)$ versus t, then the *change* in the velocity component between times t_i and t_f is the area under the graph of $a_x(t)$ between those times. That is,

$$v_x(t_f) - v_x(t_i) = \int_{t_i}^{t_f} a_x(t)\, dt \qquad (3.20)$$

If a graph is made of $v_x(t)$ versus t, then the *change* in the position vector component between times t_i and t_f is the area under the graph of $v_x(t)$ between those times. That is,

$$x(t_f) - x(t_i) = \int_{t_i}^{t_f} v_x(t)\, dt \qquad (3.21)$$

SUMMARY OF PROBLEM-SOLVING TACTICS

3.1 **(page 76)** Always choose a coordinate axis for rectilinear motion that is coincident with the line followed by the particle.

3.2 **(page 77)** Read and construct graphs carefully.

3.3 **(page 81)** The sign of $v_{x\text{ ave}}$ depends on the choice you make for the direction of $\hat{\imath}$.

3.4 **(page 84)** The choice you make for the coordinate system influences the sign of the velocity component v_x.

3.5 **(page 87)** The sign of the average acceleration component $a_{x\text{ ave}}$ depends on the direction you choose for $\hat{\imath}$.

3.6 **(page 93)** Know the meaning of the notation.

3.7 **(page 94)** Know the limitations associated with using certain equations.

3.8 **(page 94)** Avoid needless memorizing of equations.

3.9 **(page 94)** Tailor the general equations to the specific choice you make for the coordinate system.

3.10 **(page 95)** Since $g = 9.81$ m/s² is the magnitude of the local acceleration due to gravity, g always is a positive scalar quantity.

QUESTIONS

1. (page 75); 2. (page 77); 3. (page 81); 4. (page 81); 5. (page 95); 6. (page 95).

7. For rectilinear motion, if the change in the position vector $\Delta \vec{r}$ is in the same direction as $\hat{\imath}$, is it possible to conclude that the particle was moving toward increasing values of x during the *entire* time interval between t_i and t_f? Explain.

8. For rectilinear motion, if the change in the position vector $\Delta \vec{r}$ is in the same direction as $-\hat{\imath}$, is it possible to conclude that the particle was moving toward decreasing values of x during the *entire* time interval between t_i and t_f? Explain.

9. How would you sketch a situation such that the initial and final position vectors of a particle have the same magnitude, that is, $|\vec{r}_i| = |\vec{r}_f|$, but with $\Delta \vec{r} \neq 0$ m, so that $|\Delta \vec{r}| \neq 0$ m?

10. Explain why the magnitude of the change in the position vector $\Delta \vec{r}$ never can be greater than the total distance traveled. What implications does this result have for the relationship between the average speed and the magnitude of the average velocity?

11. For rectilinear motion, must all particles have *continuous* $x(t)$, $v_x(t)$, and $a_x(t)$ graphs?

12. Sketch a graph that has as the ordinate the x-coordinate of a particle and as abscissa the clock time t. Indicate schematically but clearly on the graph what is meant by (a) an event; (b) a time interval Δt; (c) no change in position during the time interval; (d) a positive change in the position vector component during the interval, where *both* the initial and final values of the position vector component are positive; (e) a positive change in the position vector component during the interval, where *both* the initial and final values of the position vector component are negative; (f) a positive change in the position component during the interval, where the initial position vector component is negative and the final position vector component is positive.

13. If the world-lines of two particles are drawn on the same graph and the two world-lines intersect, what does this mean physically? Can such intersections of world-lines happen to physical objects of finite size as opposed to their representations as particles?

14. A car is traveling on a straight road up a hill of constant slope. (a) If the needle of a common circularly shaped speedometer in the car maintains a fixed orientation pointing toward the numerical value 100 km/h on the scale, what is the car doing physically? Is its acceleration necessarily zero? (b) If the speedometer needle is moving clockwise, what is the car doing physically? (c) If the speedometer needle is moving counterclockwise, what is the car doing physically?

15. What data would you need to estimate the average speed at which a cornstalk grows? Make such an estimate. If you are more familiar with lawn grass, estimate the speed at which it grows.

16. The position coordinate x of a particle as a function of time is indicated in the graphs of Figure Q.16. Describe the motion of the particle in words. Try to mimic the motion of the particle with your hand.
Diagrams from Arnold B. Arons, *A Guide to Introductory Physics Teaching* (John Wiley, New York, 1990), pages 43–44.

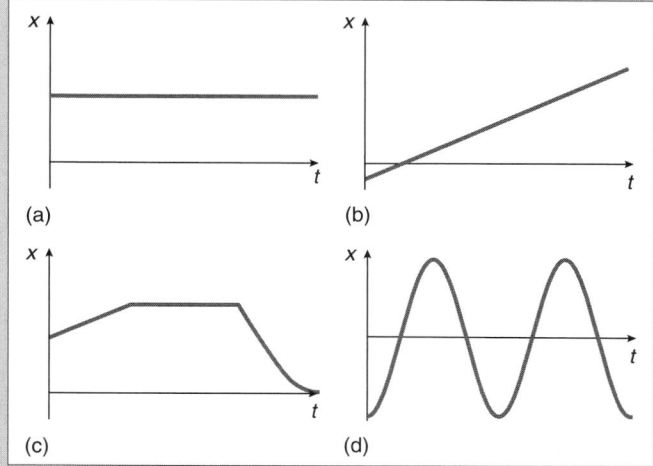

FIGURE Q.16

17. Sketch schematic x versus t graphs for each of the following rectilinear motion situations: (a) Your car is parked for ten minutes. (b) A father pushes a baby stroller across the street. (c) You walk in a straight line up to a postal window, purchase a stamp, and walk away along the same line. (d) A tethered dog runs back and forth along the length of a 15 m wire.

18. Sketch a graph that has the velocity component v_x of a particle with respect to a Cartesian origin as the ordinate and the time t as abscissa. Indicate schematically but clearly on the graph what is meant by (a) a time interval Δt; (b) no change in the velocity component during the time interval; (c) a positive change in the velocity component during the interval, where *both* the initial and final values of the velocity components are positive; (d) a positive change in the velocity component during the interval, where *both* the initial and final values of the velocity components are negative; (e) a positive change in the velocity component during the interval, where the initial velocity component is negative and the final velocity component is positive; (f) a negative change in the velocity component during the interval, where *both* the initial and final values are positive; (g) a negative change in the velocity component during the interval, where *both* the initial and final values of the velocity components are negative; (h) a negative change in the velocity component during the interval, where the initial velocity component is positive and the final velocity component is negative. (i) Is it possible to have a negative change in the velocity component during the interval, with the initial velocity component negative and the final velocity component positive? Explain.

19. In Section 3.3, we called the ratio

$$\frac{\Delta t}{d}$$

a measure of the average sluggishness of a particle, where d is the total distance traveled by a particle during the interval Δt. If this ratio is small, what is the particle doing physically? If the ratio is large, what is the particle doing physically? Use the data provided in Example 3.3 to find the marble's measure of this ratio in SI units.
Inspired by Arnold B. Arons, A *Guide to Introductory Physics Teaching* (John Wiley, New York, 1990), page 23.

20. Examine the ratio

$$\frac{\Delta t}{\Delta \vec{\mathbf{r}}}$$

In view of our study of vectors in Chapter 2, why is this ratio mathematically meaningless in contrast to the ratio examined in the previous question? Does the ratio

$$\frac{\Delta t}{|\Delta \vec{\mathbf{r}}|}$$

make mathematical sense? Why or why not? What does this ratio measure? How does this ratio differ from the ratio defining the average sluggishness? Why does the ratio $\Delta t / |\Delta \vec{\mathbf{r}}|$ provide less information than the ratio $\Delta \vec{\mathbf{r}}/\Delta t$ that was used to define the average velocity?

21. Consider the ratio

$$\frac{\Delta t}{|\Delta \vec{\mathbf{v}}|}$$

where $|\Delta \vec{\mathbf{v}}|$ is the magnitude of the change in the velocity of a particle during a time interval Δt. (a) Under what physical circumstances is this ratio large? small? (b) Invent an appropriate term to describe what this quantity represents physically. (c) Why does this ratio contain less information than the ratio

$$\frac{\Delta \vec{\mathbf{v}}}{\Delta t}$$

used to define the average acceleration?

22. Use the plumb bob accelerometer mentioned in Section 3.6 under the following circumstances. Describe the movement of the plumb bob with respect to the local vertical direction as viewed by a observer at rest with respect to the plumb bob. (a) A traffic light turns green and a car accelerates along a level, straight highway. (b) An elevator begins moving upward from rest to a speed of 2.0 m/s. (c) An automobile moves with a constant speed along a straight, level stretch of interstate highway.

23. Figure Q.23 is a graph of the x-component of the position vector of a particle as a function of time. (a) Does the particle have a nonzero velocity? If so, at what time instants? (b) Is the x-component of the velocity of the particle constant? Explain why or why not. (c) What is the velocity component v_x of the

particle at the instant $t = 0$ s? (d) What is the velocity component v_x of the particle at the instant $t = 3.00$ s? (e) At what instant is the position vector component x equal to zero? What is the velocity component v_x of the particle at this instant? (f) What is the x-component of the acceleration of the particle?

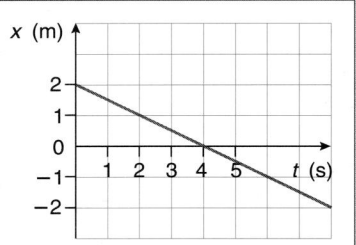

FIGURE Q.23

24. Figure Q.24 is a graph of the x-component of the velocity of a particle as a function of time. (a) Does the particle have a nonzero x-component of the acceleration? (b) Is the x-component of the acceleration of the particle *constant*? Explain why or why not. (c) What is the velocity component v_x of the particle at the instant $t = 0$ s? (d) What is the acceleration component a_x of the particle at the instant $t = 0$ s? (e) At what instant is the velocity component v_x zero? (f) What is the acceleration component a_x of the particle at the instant $t = 2.00$ s?

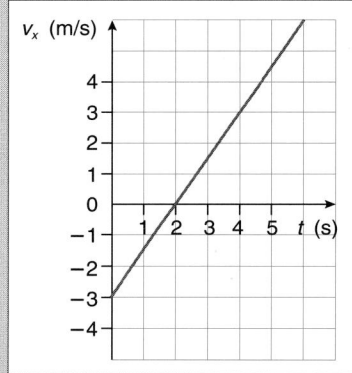

FIGURE Q.24

25. Toss an eraser vertically upward. At the instant when the eraser reaches its greatest elevation: (a) is its velocity equal to zero? Explain. (b) is its acceleration equal to zero? Explain.

26. Sketch a series of graphs of the velocity component v_x of a particle versus time that schematically indicate: (a) a situation where the particle has a zero velocity component at some instant but a nonzero *positive* acceleration component. Indicate this instant on the sketch; (b) a situation where the particle has a zero velocity component at some instant but a nonzero *negative* acceleration component. Indicate this instant on the sketch; (c) a situation where the particle has a velocity component equal to zero and an acceleration component equal to zero at some instant.

27. A car moves left to right along a level, straight highway. Photographs of the car are taken, separated by equal intervals of time Δt. Which of the series of photographs in Figure Q.27, if any, depicts the car (a) moving at constant velocity? (b) moving with a nonzero acceleration? (c) with a positive x-component of the velocity if $\hat{\imath}$ is to the right? (d) with a positive x-component of the velocity if $\hat{\imath}$ is to the left? (e) with a negative x-component of the velocity if $\hat{\imath}$ is to the right? (f) with a negative x-component of the velocity if $\hat{\imath}$ is to the left? (g) with a positive x-component of the acceleration if $\hat{\imath}$ is to the right? (h) with a positive x-component of the acceleration if $\hat{\imath}$ is to the left? (i) with a negative x-component of the acceleration if $\hat{\imath}$ is to the right? (j) with a negative x-component of the acceleration if $\hat{\imath}$ is to the left?

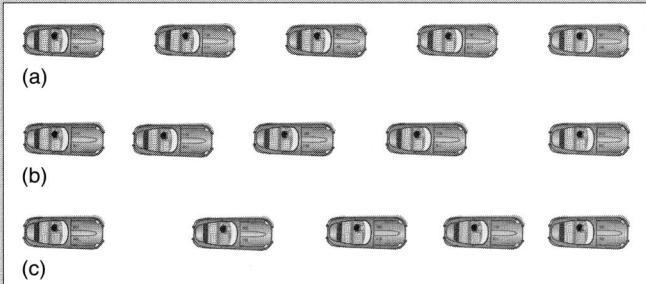

(a)

(b)

(c)

FIGURE Q.27

28. The velocity components v_x of two particles are plotted schematically on the same graph in Figure Q.28. (a) At what times, if any, do the objects have the same velocity component? Indicate such times with the designation t_A. (b) At what times, if any, do the particles have the same acceleration components a_x? Indicate such times with the designation t_B. (c) With just the information given, is it possible to tell whether the two particles simultaneously ever have the same position vector component? Explain your answer.

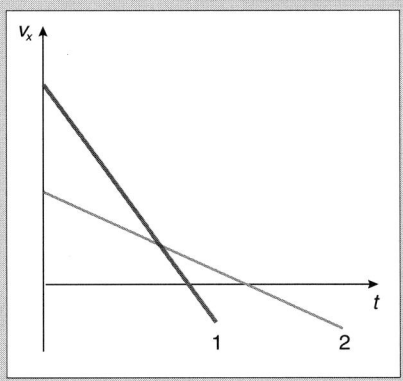

FIGURE Q.28

29. If a particle has the position vector $\vec{r}_i$ at time t_i and at time $t_f > t_i$ the particle has the *same* position vector, does this necessarily imply that the particle did not move during the interval Δt between t_i and t_f? Explain your answer.

30. Someone drops an egg from rest off a cliff while someone at the bottom of the same cliff throws a rock vertically upward. What can be said (if anything) about the accelerations of the two objects during their flights?

31. A particle is found to have an average velocity equal to zero during some interval Δt. What can be said (if anything) about the position vectors of the particle at the beginning and end of the time interval? What can be said (if anything) about changes in the instantaneous position vector *during* the time interval? Explain your answers.

32. A hammer is hurled vertically down from the roof while another hammer is simultaneously hurled vertically upward from the same initial location with the *same* initial speed. (a) If $\hat{\imath}$ is chosen to be directed vertically *upward*, roughly sketch a graph of the x-component of the velocity v_x of each hammer as a function of time. (No numbers need be placed on the graph.) (b) Make a similar graph if $\hat{\imath}$ is chosen to be vertically *downward*. (No numbers need be placed on the graph.)

33. A particle undergoes rectilinear motion. The velocity component v_x of the particle as a function of time is indicated in the graphs shown in Figure Q.33. Describe the motion of the particle in words and try to mimic the motion of the particle with your hand.
    Inspired by Arnold B. Arons, *A Guide to Introductory Physics Teaching* (John Wiley, New York, 1990), page 44.

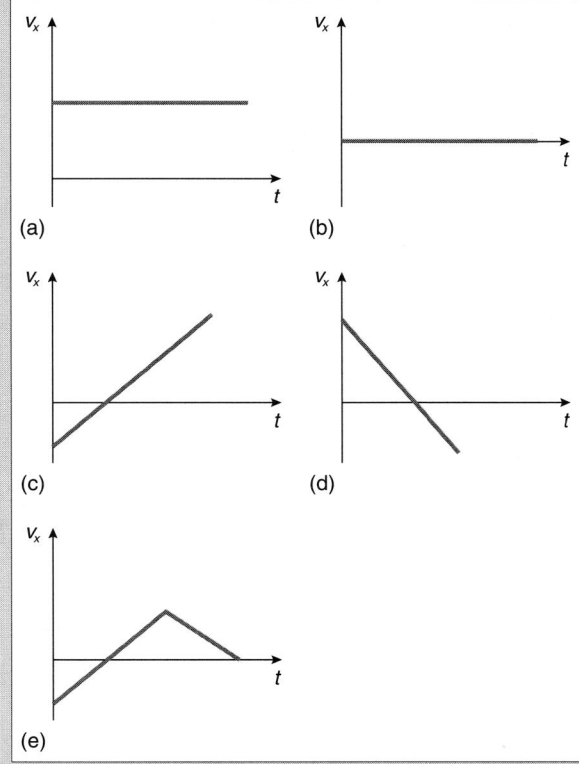

FIGURE Q.33

34. Figure Q.34 is a graph of the rectilinear positions of two box turtles (*Terrapene carolina*) as a function of time during a race. (a) Indicate on the sketch the times that the two turtles have

the same position. Label such times t_A. (b) Indicate on the sketch the times that the two turtles have the same velocity. Label such times t_B. (c) If either turtle is (instantaneously) at rest during the time interval shown, indicate which one and indicate the times this occurs on the sketch by t_C. (d) Is either turtle undergoing zero acceleration at some instant? If so, which one? Also indicate the corresponding instants. (e) Is either turtle undergoing accelerated motion? If so, which one? Is it possible to tell from the information given whether this acceleration is constant or not?

Inspired by Arnold Arons, *A Guide to Introductory Physics Teaching* (John Wiley, New York, 1990), page 46.

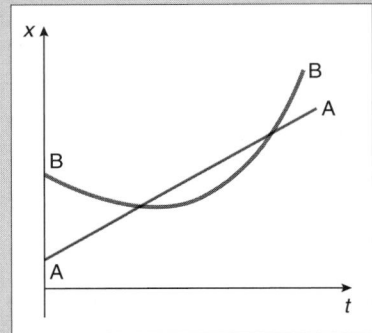

FIGURE Q.34

PROBLEMS

Ignore the effects of air resistance in all of the problems in this chapter.

Sections 3.1 Rectilinear Motion
3.2 Position and Changes in Position
3.3 Average Speed and Average Velocity

1. You throw a stick for your dog Sirius to fetch. Sirius runs from right next to you straight out to the stick, lying 25 m away, and brings it directly back to you like a good dog. The trip takes the dog 12 s and you reward Sirius with a pat on the head and a tasty dog biscuit. (a) What is the change in the position vector of the dog? (b) What is the total distance traveled by the dog? (c) What is the average speed of the dog? (d) What is the average velocity of the dog?

2. A pilot flies a Boeing 737 from Seattle, Washington, to Portland, Oregon, taking 60 min. See Figure P.2. After a 45 min layover in Portland, she flies on to San Francisco, south of Portland, with a flight time of 2.0 h. After a 1.0 h layover in San Francisco, the plane flies nonstop back to Seattle in 2.5 h. (a) Set up a coordinate system to analyze the motion, indicating the locations of all three cities. Assume all three cities lie on the same straight line. (b) After determining the distance between the cities from Figure P.2 and the accompanying scale, specify the position vectors of each city with respect to your chosen origin. (c) Determine the average speed in km/h of each flight. (d) Determine the average speed in km/h during the entire time interval from Seattle back to Seattle. (e) Determine the average velocity in km/h of each flight. (f) Determine the average velocity in km/h during the entire time interval from Seattle back to Seattle.

3. The Andromeda and Milky Way galaxies are approximately 2×10^6 light-years apart. (One light-year is the distance traveled by light during 1 year.) How long does it take for a cosmic ray proton, traveling at a constant speed equal to 99.99% the speed of light, to travel between these galaxies?

4. Many years ago, the Apollo lunar astronauts left a collection of corner cube reflectors on the lunar surface that reflect incident light back in the same direction from which it came. See Figure P.4. By measuring the time it takes a light pulse to travel from an Earth-based laser to the corner cube array on the Moon and back, we can measure the distance between the laser and the corner cube with great precision (with an uncertainty of only about a centimeter). Such measurements have

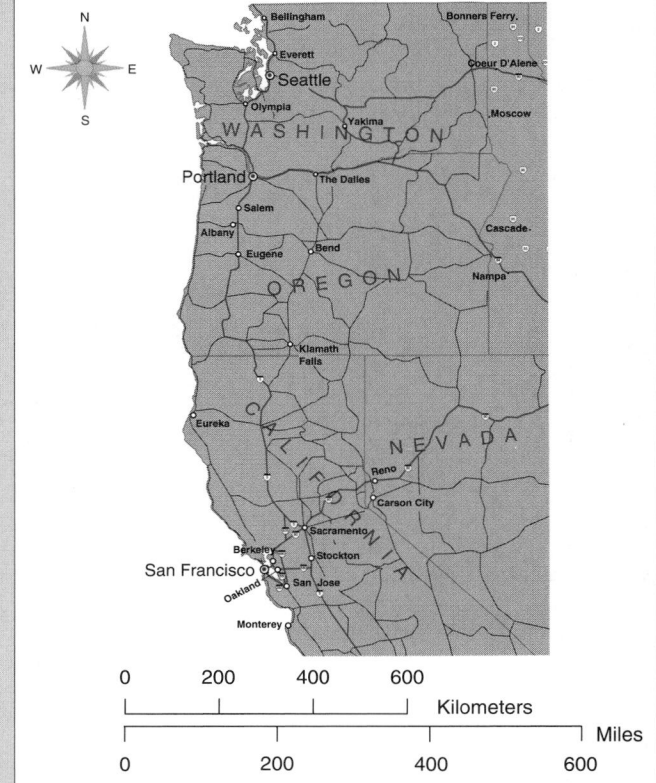

FIGURE P.2

confirmed that the Moon is slowly but systematically moving away from the Earth at a rate of several centimeters per year. (a) Approximately how long does it take light to make the round trip if the distance to the Moon is approximately 3.80×10^5 km? The speed of light is approximately 3.00×10^8 m/s. (b) For a round trip, what is the change in the position vector of the light? (c) What is the average velocity of the light for the round trip?

5. A salesperson hawking kitchen widgets drives about 4.0×10^4 km each year. Assuming an average speed of 60 km/h while in the car, how many weeks a year does the salesperson spend traveling in the car?

FIGURE P.4

FIGURE P.11

6. The slow geologic motion of the continental plates occurs at speeds on the order of one-tenth the rate at which your fingernails grow. Estimate the time in centuries it takes for the continental plates to move 1 meter.

• 7. A base runner is 3.0 m from first base, awaiting the next pitch to his teammate. The pitcher glances over his shoulder and fires the ball to first base at a speed of 120 km/h. Assume the ball travels in a straight line. The distance from the pitcher to the first baseman is 19 m. What is the minimum average speed needed for the base runner to successfully make it back to first base before the ball?

• 8. Your flight leaves in 65 min and you are 40 km from the airport—and fuming. You need 5.0 min to park the car and run to the gate. Unfortunately, heavy traffic on the way keeps your average speed to 25 km/h for the first 20 min of the drive. What must be your average speed for the rest of the trip to make the plane? Is this a minimum or a maximum average speed to make the plane?

• 9. When you see a deer run unexpectedly in front of your car, your reaction time (the time interval it takes you to step on the brake after first seeing danger) is about 0.30 s and your body height is 2.0 m. If your reaction time is caused by the time it takes electrical nerve impulses to be sent from your brain to your feet, what is the minimum speed of such signals? The actual speed of such nerve impulses is on the order of 10^2 m/s. About how long do they actually take to propagate from your head to your feet? What other factors might account for the relatively large reaction time?

• 10. Your average running speed is 8.0 m/s and your father's is 7.0 m/s. You both are located 75 m from the family minivan and begin running toward it at the same time. What is your separation distance when you win the race to the car?

Sections 3.4 Instantaneous Speed and Instantaneous Velocity
      3.5 Average Acceleration
      3.6 Instantaneous Acceleration
      3.7 Rectilinear Motion with a Constant Acceleration

11. A reckless, unroped mountain climber (see Figure P.11) falls off a vertical cliff at an elevation of 50 m from the rocks below. How long does the climber have to contemplate meeting his maker? What is the impact speed? As an old saying goes, it's not the fall that hurts, it's the sudden stop (that is, the large acceleration during the short interval of the stop).

12. Calculate the magnitude of the acceleration of a sports car that can drag race from rest to 100 km/h in 5.00 s. Assume the acceleration is constant, although typically this is not a good assumption for automobiles. Compute the ratio of the magnitude of this acceleration to the magnitude of the local acceleration due to gravity ($g = 9.81$ m/s²). Make a graph of the speed of the car versus time.

13. A large supertanker, initially moving at 30.0 km/h, is brought to a halt in 15.00 min after a lookout spies an icefield ahead. (a) Set up an appropriate coordinate system and determine the acceleration. (b) Through what distance did the tanker move while stopping?

14. A rubber chicken at a circus is tossed vertically upward with an initial speed of 10.0 m/s from an initial elevation of 2.00 m above the ground. (a) Sketch the situation. (b) Choose a coordinate system clearly indicating the origin and the direction chosen for an appropriate unit vector. (c) Tailor the one-dimensional, constant acceleration equations (Equations 3.16 and 3.19) to the particular choice you made for the coordinate system. (d) Find the time for the chicken to reach its maximum height. (e) What is the maximum height of the bird?

15. Solve the problem of Example 3.12 with the origin of the coordinate system midway down the crevasse and with $\hat{\imath}$ vertically upward (see Figure P.15). (a) For this coordinate choice, what are x_0, v_{x0}, and a_x? (b) What is the x-coordinate of the rock when it reaches the bottom of the crevasse? (c) Solve for the time it takes the rock to fall, the velocity (a vector), and the speed (a scalar) at the instant just before impact. The results for the time and speed should be the same as those obtained in Example 3.12.

16. The high-speed train in France known as the TGV (*Train à Grande Vitesse*, or Train of Great Speed) accelerates from rest to 200 km/h in 120 s along a straight section of track (see Figure P.16). (a) What is the magnitude of its acceleration, assuming it is constant? (b) Through what distance did it move while accelerating to this speed?

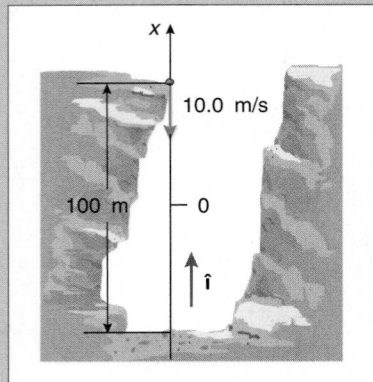

FIGURE P.15

FIGURE P.16

17. When $t = 0$ s, a flower pot on a window sill drops from rest to the sidewalk some 30.0 m below and shatters into smithereens, luckily missing all pedestrians. (a) How long was the pot in flight? (b) What is its maximum speed? (c) How fast is the pot moving when it is at a height of 10.0 m above the ground? (d) How far above the ground is the flower pot when $t = 1.00$ s?

18. After a grueling history test, your flight on U.S. Snarelines is ready for takeoff from Jack Daniels International Airport on its way to Tahiti. Sun and fun! The interval of the takeoff is 40.0 s and the plane uses 1.70 km of the runway. You find yourself with an uncontrollable urge to make a calculation. Find (a) the magnitude of the acceleration (assuming it is constant) in m/s² and (b) the takeoff speed in both m/s and km/h.

19. A Martian drops your shoe into the Olympus Mons volcanic caldera, (see Figure P. 19) and it falls 30.0 m in 4.00 s. (a) Find the magnitude of the (constant) acceleration due to gravity on the surface of the planet. (b) By forming a suitable ratio, compare this result with the magnitude of the local acceleration due to gravity on the surface of the Earth (9.81 m/s²).

20. A thief is cruising along a straight stretch of level highway in your stolen sports car at 120 km/h. See Figure P.20. A flashing light in the rearview mirror triggers a constant acceleration to 180 km/h in 4.00 s. Go get 'im! (a) Set up an appropriate coordinate system to analyze this problem. (b) Determine the appropriate component of the acceleration consistent with your coordinate choice. (c) Determine the distance over which the car moved while accelerating from 120 to 180 km/h.

21. The x-component of the position vector of a particle is shown in the graph in Figure P.21 as a function of time. (a) Estimate the velocity component v_x at the instant 3.0 s. (b) Is the velocity

FIGURE P.19

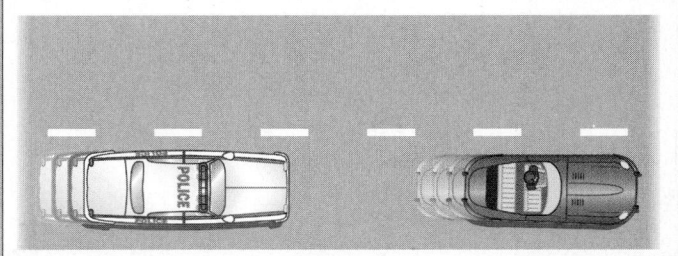

FIGURE P.20

component zero at any time? If so, estimate the time. If not, explain why not. (c) Is the particle always moving in the same direction along the x-axis? If so, explain what leads you to this conclusion. If not, determine the positions at which the particle changes direction.

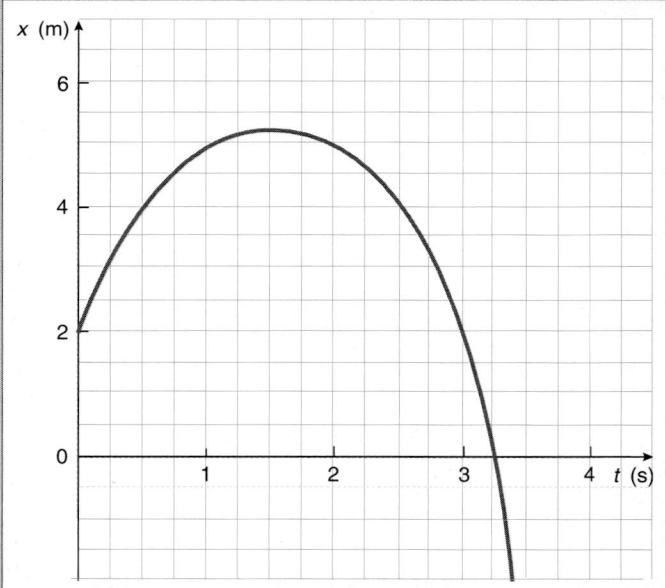

FIGURE P.21

22. The x-component of the velocity of a particle is graphed in Figure P.22 as a function of time. (a) Is the acceleration of the particle constant? Explain your answer. (b) Estimate the acceleration of the particle at the instant 1.0 s. (c) At what instants (if any) is the particle at rest? If such instants exist, is the acceleration of the particle also zero at these instants? Explain your answer.

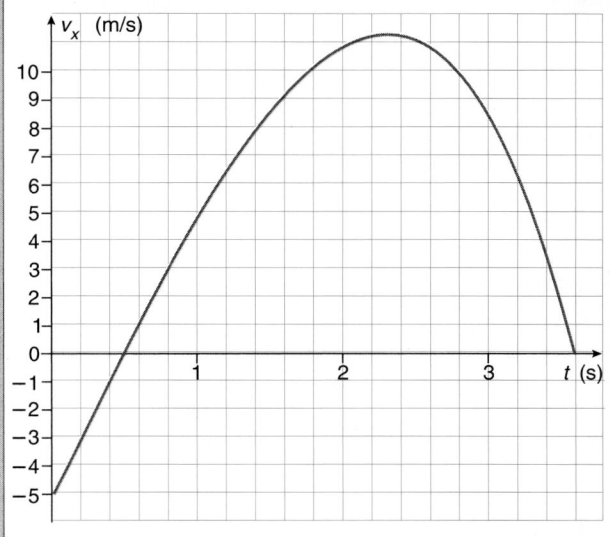

FIGURE P.22

23. When a radio wave impinges on the antenna of your car, electrons in the antenna move back and forth along the antenna with a velocity component v_x as shown schematically in Figure P.23. Roughly sketch the same graph and indicate the time instants when (a) the velocity component is zero; (b) the acceleration component a_x is zero; (c) the acceleration has its maximum magnitude.

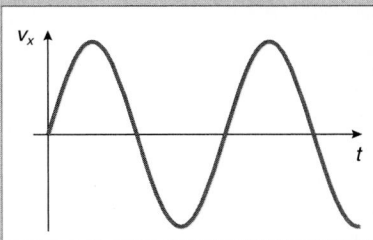

FIGURE P.23

24. A bullfrog (*Rana catesbeiana*) zaps a nearby tasty mosquito (*Culex pungens*) for dinner. The tongue of the frog flicks out 8.0 cm and is back with the mosquito inside the frog's mouth in 0.10 s total time. (a) What is the average velocity of the tongue during this interval? (b) What is the average speed of the tongue during the interval?

25. X-rays are created by slamming high-speed electrons into materials such as tungsten. The rapid acceleration of the electrons produces the x-rays. If an electron initially moving at a speed 3×10^6 m/s is brought to rest in 1×10^{-18} s, what is the magnitude of the average acceleration of the electron?

•**26.** A rock is thrown vertically upward at a speed v_0. It barely misses the edge of a cliff (see Figure P.26) and strikes the fluid in a foul-smelling river at a speed $2v_0$. How high is the cliff above the surface of the water? Express your answer in terms of v_0 and g.

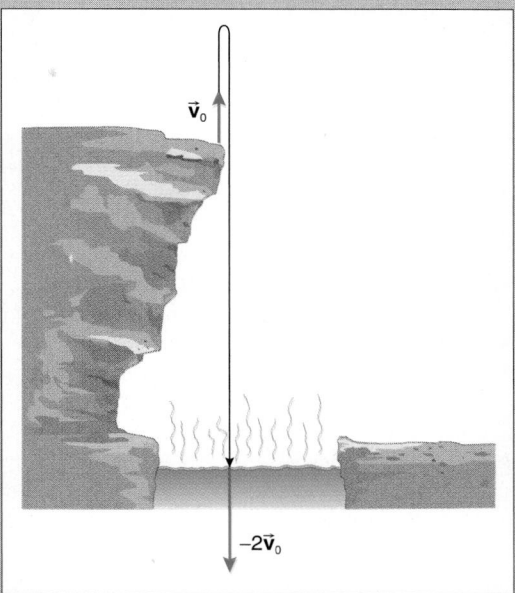

FIGURE P.26

•**27.** A ladybug (*Anatis quindecimpunctatum*) is observed crawling along a meter stick. The position of the ladybug at various times is recorded by an astute entomologist, yielding the following data:

Time (s)	Position (cm)
0.0	5.4
2.0	7.6
4.0	9.2
5.0	10.0
7.0	9.2
10.0	8.6
12.0	8.2
14.0	7.8

(a) Graph the position versus time. For (b)–(e), quote your results in appropriate SI units. Take the x-axis to be the meter stick with its origin at 0 m and with $\hat{\imath}$ in the direction the scale increases. Use the data or the graph to calculate the requested information. (b) What is the average velocity of the ladybug during the interval between 10.0 s and 14.0 s? (c) What is the velocity of the ladybug when $t = 12.0$ s? (d) At about what times is the velocity of the ladybug approximately 0 m/s? (e) What is the average acceleration of the ladybug during the interval from 2.0 to 12.0 s?

•**28.** A water balloon is thrown straight up and reaches its maximum height after 2.00 s. Sketch the situation and introduce an appropriate coordinate system to answer the following questions. (a) What is the acceleration of the balloon after it leaves the hand and is rising? (b) What is the acceleration of the balloon as it is falling? (c) What is the acceleration of the balloon at the instant it is motionless at its maximum height? (d) What was the initial velocity component of the balloon? (e) How high did the balloon rise above the point of its release?

•29. You are tooling along a straight, level road in a new sports car at a leisurely 30.0 m/s. Suddenly you realize that there is a barrier blocking a bridge ahead. You hit the brakes when $t = 0$ s with the car only 160 m from the barrier across the road. When $t = 5.0$ s, the car is still moving toward the bridge barrier but at only 15.0 m/s. You break into a sweat. Assume the acceleration of the car is constant. (a) At what time does the car come to rest? (b) Will the car collide with the bridge barrier? If not, how far is the car from the bridge barrier when the car stops?

•30. You are trailing the leader in a long cross-country ski race by only 3.0 m with both of you traveling at 20 km/h with a flat, straight 1000 m distance between the leader and the finish line. The leader is unable to increase her speed, but you feel a new surge of adrenalin and manage to win by a whisker. Assuming a constant acceleration, what was the magnitude of your acceleration during the last part of the race?

•31. You accidentally drop a peanut from a tower when $t = 0$ s. Oops. One second (1.00 s) later you decide that another peanut deserves the same treatment, and it too falls from the tower with zero initial speed. (a) Sketch on a single graph the speeds of the peanuts as functions of time. (b) While the peanuts are in flight, do they stay the same *distance* apart? Explain why or why not. (c) What is the distance between the peanuts when $t = 3.00$ s?

•32. Charles Walters* suggests slowing his motorcycle in an unusual way. He is traveling across a straight, level stretch of interstate in Kansas. Charles travels at 60.0 km/h for 1 minute, then quickly brakes briefly to achieve a speed of 59.0 km/h for the next minute, braking briefly to a speed of 58.0 km/h for the next minute, and so on until zero speed is achieved. (a) What total distance is covered in this process? (b) Introduce an appropriate coordinate system and determine the average acceleration of the motorcycle during this process.

•33. The x-component of the velocity of a garbage barge is indicated in the simplified graph shown in Figure P.33. (a) What is the average acceleration of the barge during the time interval between 0 min and 1.0 min on the clock? What is the instantaneous acceleration of the barge when $t = 0.5$ min? (b) What is the average acceleration of the barge during the time interval between 1.0 and 2.0 min on the clock? What is the instantaneous acceleration of the barge when $t = 1.5$ min? (c) What is the average acceleration of the barge during the time interval between 2.0 and 3.0 min on the clock? What is the instantaneous acceleration at the instant $t = 2.5$ min? (d) Explain what is happening to the barge in descriptive English. (e) Sketch a graph of the position x of the barge as a function of time consistent with and as complete as the information provided allows. Assume the barge initially is at the Cartesian origin.

•34. You throw your class ring vertically upward in a room with a high cathedral ceiling. It leaves your hand 1.2 m above the floor and reaches a maximum height of 3.0 m before falling to the floor. (a) How long a time interval did the ring rise? (b) What was the speed of the ring when it left your hand? (c) What is the duration of the entire time of flight? (d) What is the speed of the ring the instant before impact?

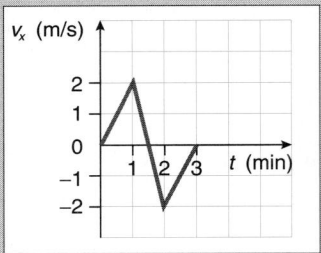

FIGURE P.33

•35. A cheetah (*Acinonyx jubatus*) is chasing an impala (*Aepyceros melampus*) for lunch. (See Figure P.35.) The cheetah is found at the positions along the $+x$-axis at 100, 200, 280, 340, and 380 m when a stopwatch reads $t = 20.0, 30.0, 40.0, 50.0,$ and 60.0 s, respectively. (a) Calculate the x-component of the average velocity during each 10.0 s interval. (b) Make a histogram plot of the x-component of the average velocity versus t. (c) Is it reasonable to conclude that the instantaneous acceleration is constant during the interval from $t = 20.0$ s to 60.0 s? (d) If you assume the instantaneous acceleration is constant, find a_x. (e) Assume that a_x also is constant for times less than 20.0 s and greater than 60.0 s. Find v_{x0} and x_0. (f) Find $x(t)$. (g) At what time will the exhausted cheetah come to rest if the data trend continues? (h) Where does the cheetah come to rest?

FIGURE P.35

•36. Figure P.36 is a graph of the instantaneous velocity component v_x of a chicken fleeing the axe of a farmer. The chicken is at $x = 0.0$ m when $t = 0.0$ s. (a) When is the chicken at rest? (b) What is the acceleration of the chicken when $t = 3.0$ s? (c) Make a graph of the acceleration component a_x as a function of time over the same time interval as in Figure P.36. (d) Where is the chicken when $t = 5.0$ s?

•37. The x-component of the velocity of an African kudu (*Tragelaphus imberbis*) is graphed in Figure P.37 as a function of time. (a) When is the animal at rest? (b) When is its acceleration equal to zero (if it ever is)? (c) What is its acceleration when $t = 4.5$ s? (d) How far does the animal travel during the first 6.0 s?

•38. A baseball is projected vertically upward with a speed equal to 29.4 m/s. Choose the direction of increasing values of x to be vertically upward. A stopwatch is started the instant the ball is released. (a) What is the velocity component v_x of the ball at each of the following instants: $t = 0.00$ s; $t = 1.00$ s; $t = 2.00$ s;

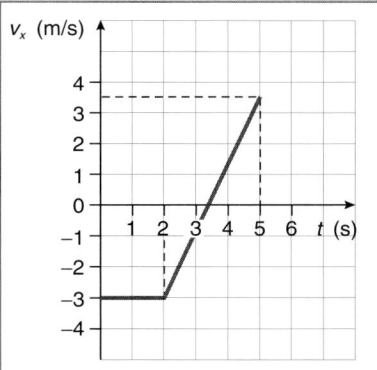

FIGURE P.36

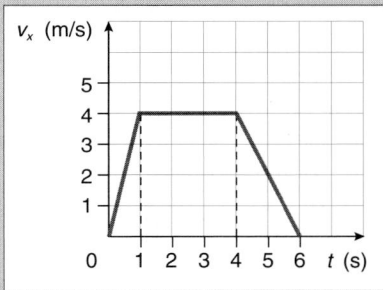

FIGURE P.37

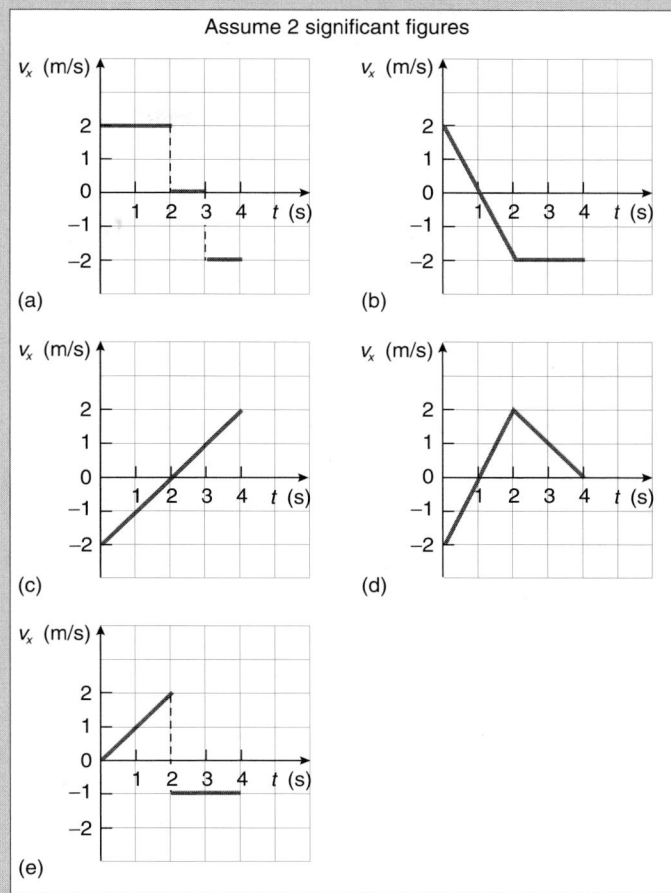

Assume 2 significant figures

(a)

(b)

(c)

(d)

(e)

FIGURE P.40

$t = 3.00$ s; $t = 4.00$ s? (b) Make a graph of v_x versus time from $t = 0.00$ s to 4.00 s. (c) What is the acceleration component a_x of the ball at each of the instants above?

• **39.** A rifle fires a bullet with a muzzle speed of 350 m/s directly into a fixed block of hardwood. The bullet is brought to a stop embedded 6.0 cm into the wood. Assume the acceleration of the bullet is constant. Calculate the magnitude of the acceleration and the time it takes the block to stop the bullet.

• **40.** The graphs in Figure P.40 depict the velocity component v_x of a rat (*Rattus rattus*) in a one-dimensional maze as a function of time. Your task is to make graphs of the corresponding (i) acceleration component a_x versus time and (ii) the *x*-component of the position vector versus time. In all cases assume $x = 0$ m when $t = 0$ s.

• **41.** The draw on Robin Hood's archery bow is 0.70 m. (See Figure P.41.) (a) Calculate the magnitude of the acceleration of an arrow that leaves the bow with a speed of 30 m/s; assume the arrow experiences a constant acceleration over the 0.70 m distance before it leaves the bow. (b) If the arrow is shot vertically upward and leaves the bow at a distance 1.80 m from the ground, find the maximum height to which the arrow rises, and the total time of flight until the arrow hits the ground (assume the archer moves out of the way!).

• **42.** *If* it were possible to maintain a uniform acceleration with a magnitude equal to 9.81 m/s², how long would it take to accelerate an object from rest to a speed approximately equal to the speed of light (3.00×10^8 m/s)? Express your result in years. Ignore any effects caused by the theory of relativity.

• **43.** A partner in the law firm of Filch, Robb & Steele notices an ambulance go by at a constant speed of 50 m/s along a level, straight roadway, headed for an accident. As the ambulance

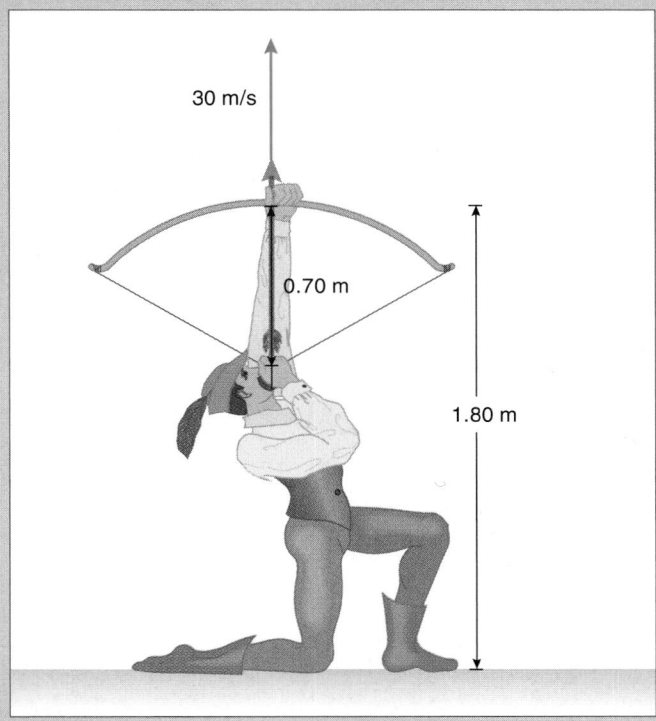

30 m/s

0.70 m

1.80 m

FIGURE P.41

passes, the lawyer sets out in hot pursuit of a new client, with zero initial speed and stopwatch initially set to zero, maintaining a constant acceleration of magnitude 1.0 m/s². (a) Set up an appropriate coordinate system to analyze the situation. (b) Set up an equation that describes the position of the ambulance as a function of time on the stopwatch. (c) Set up an equation that describes the position of the lawyer as a function of time on the stopwatch. (d) How much time elapses before the lawyer catches the ambulance? (e) What distance did the ambulance and lawyer travel during this time? (f) Draw a graph showing the x-component of the position vectors of the ambulance and lawyer as a function of time. Indicate the event that represents the lawyer catching the ambulance.

•44. Have your roommate hold a meter stick by one end as indicated in Figure P.44. Initially, the top of your hand is at, say, the 10.0 cm mark, ready to catch the stick when it is dropped. Without warning, your compatriot drops the stick. Without reaching downward, you catch it as quickly as possible and find it has fallen a distance d. Show that a measure of your reaction time t_{reaction}, the time between when you see the stick begin to fall and the time you catch it, is (in seconds)

$$t_{\text{reaction}} = (0.452 \text{ s/m}^{1/2})\sqrt{d}$$

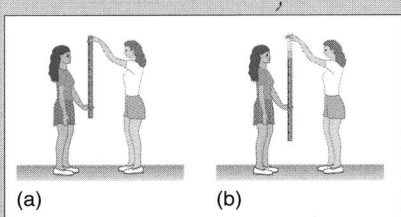

(a)   (b)

FIGURE P.44

•45. Alas, having dallied too long at Attitude Adjustment Hour (AAH!), two commuters arrive at a commuter rail platform just in time to see their train start from rest with an acceleration of magnitude 0.250 m/s². They are 30.0 m from the end of the last car, straight along the track, and both run at the furious (and constant) speed of 4.00 m/s to catch the train. A stopwatch monitors the time it takes them to catch the train. (a) Introduce an appropriate coordinate system. (b) At what *two* times will the commuters and the end of the rear car have the same position coordinate? (c) At each of the times calculated in part (b), describe what is passing what.

•46. For one-dimensional motion with a constant acceleration, show that the fundamental equations

$$v_x = v_{x0} + a_x t \qquad\qquad (3.16)$$

$$x = x_0 + v_{x0}t + \frac{1}{2}a_x t^2 \qquad\qquad (3.19)$$

imply that

$$v_x^2 = v_{x0}^2 + 2a_x(x - x_0)$$

•47. You toss a quarter into the air, releasing it at a distance of 2.0 m above the ground and simultaneously starting a stopwatch. The quarter is in the air for 3.0 s before striking the ground. (a) Sketch the situation. (b) Introduce an appropriate coordinate system, clearly indicating the origin and the direction

of an appropriate unit vector. (c) Tailor Equations 3.16 and 3.19 to your coordinate choice. (d) What was the initial velocity component of the quarter? (e) How high above the ground did the quarter rise? (f) At what speed did the quarter strike the ground?

•48. A dazed physics student is found to have the following positions at the indicated times:

t (s)	x (m)
0.00	0.00
1.00	1.00
2.00	3.00

(a) Is it possible for these data to be consistent with a *constant* acceleration? Explain the reasoning you use to come to your conclusion. (b) If a constant acceleration is possible, find the constant acceleration component a_x and v_{x0}.

•49. The Standard Linear Accelerator (SLAC) at Stanford University in California is 3.0 km long (see Figure P.49). Electrons are accelerated along the length of the straight beam tube. Assume the acceleration of the electrons is constant along the length of the accelerator. If the electrons begin at rest and emerge with a speed of 6.0×10^6 m/s, find (a) the time it takes each electron to traverse the tube and (b) the magnitude of the acceleration. The local acceleration due to gravity may be ignored; why is this a reasonable assumption?

FIGURE P.49

•50. You are asked to move a large basket of eggs initially at rest at the origin to a position at rest 16 m away along the x-axis. The eggs are speeded up with an acceleration of magnitude 1.0 m/s² for the first half of the trip and slowed down with an acceleration of equal magnitude for the latter half of the trip. (a) How long does the trip take? (b) Make a graph of the velocity component v_x of the eggs as a function of time. (c) What is the average velocity for the first half of the trip? for the second half?

•**51.** When a stopwatch is started, a tennis ball is tossed vertically up with an initial speed of 20.0 m/s. When the ball reaches its maximum height, a second ball is tossed straight up with the same initial speed of 20.0 m/s so that they collide in midair. Both balls are released at an elevation 1.50 m above the ground. (a) Choose $\hat{\imath}$ to be directed vertically up, with the origin at the ground. Sketch on the same graph the v_x components of the velocities of the two balls as functions of time. (b) Show that the speed with which the two balls approach each other before impact is constant and equal to 20.0 m/s. (c) At what elevation will the two tennis balls collide?

•**52.** A pitcher throws a searing 150 km/h fastball to the catcher, located 19.0 m away. Assume the ball follows a straight line from the pitcher to the catcher. (Ignore any vertical motion of the ball on the way to the plate.) (a) How long does it take the ball to reach the catcher? (b) As the ball leaves the pitcher's hand, a runner on first base attempts to steal second base, covering the 27.4 m distance in 4.00 s. After catching the ball from the pitcher, the catcher hurls the ball, after a 0.8 s delay, at 130 km/h to second base, 38.8 m away from him. (Again assume the ball follows a straight path.) Assuming the second baseman catches the ball and makes the tag, is the runner out or safe?

•**53.** A ball is tossed vertically upward with a speed of 25.0 m/s. (a) Indicate an appropriate coordinate system for analyzing its motion. (b) What is its maximum altitude above the point from which it was thrown? (c) How long does it take to return to the point of release? (d) Make a graph of the acceleration component a_x of the particle versus time, indicating appropriate numerical values along the axes. (e) Make a graph of the velocity component v_x of the particle versus time, indicating appropriate numerical values along the axes. (f) Make a graph of the position vector component x as a function of time, indicating appropriate numerical values on the axes.

•**54.** If speed is measured in body lengths per second, it may come as no surprise that the fastest insect in the world is the common American cockroach (*Periplaneta americana*). It can achieve speeds of about 50 body lengths per second. Assume a body length of 4.0 cm—jumbo size! (a) What is its fastest speed in m/s? (b) If the roach achieves this speed in 0.10 s, starting from rest, what is the magnitude of its acceleration assuming it is constant? Compare the magnitude of this acceleration with *g*. (c) If a 2.0 m tall human could run with a speed of 50 body lengths per second, what would be the approximate time for a 100 m sprint? The world record for a 100 m sprint is about 9.85 s.

•**55.** A basketball is dropped from rest 3.0 m above the court. After bouncing from the floor, the ball reaches a height of 1.5 m. (a) What is the speed of the ball the instant before hitting the floor? (b) What is the speed of the ball just after leaving the floor? (c) The ratio of speed after the bounce to the speed before the bounce is called the *coefficient of restitution* ε of the ball. Find ε for the basketball. (d) If the ball is in contact with the floor for 0.025 s, what is the magnitude of the average acceleration of the ball during this time interval?

•**56.** A traffic light turns green, and you zip off from rest in your car along a straight highway with an acceleration of magnitude 1.50 m/s² for 10.0 s, maintain a constant velocity for the next 30.0 s, and then slow to a stop with an acceleration of magnitude 2.00 m/s². How far did you travel?

•**57.** A pine cone drops from the mouth of an eastern gray squirrel (*Sciurus carolinensis*) while it is scampering at a speed 3.0 m/s up the vertical trunk of a tall pine tree. You notice that it takes the cone 1.5 s to reach the ground. How high was the squirrel when it dropped the cone while running?

•**58.** A naval gun projects an artillery shell from rest to a speed of 2.00 km/s over the 4.00 m length of a gun barrel. Assume the acceleration of the shell is constant over the length of the barrel. Determine the magnitude of the acceleration of the shell and the time interval the shell is in the barrel after firing.

•**59.** A kid drops from a tree limb 3.0 m above a swimming hole. (a) With what speed does the kid hit the water? (b) If the kid is brought to rest 1.5 m below the surface of the water, what is the magnitude of the acceleration of the child after contact with the water? Assume the acceleration is constant.

•**60.** You are standing on the high bridge over the New River gorge in southern West Virginia. The rail of the bridge is 269 m above the surface of the river below. Your buddy drops a rock from the bridge rail. You throw a rock down 1.00 s later from the same location and both rocks strike the water simultaneously. (a) How long is the first rock in flight? (b) With what speed did you throw the second rock?

•**61.** The worst nightmare of a skydiver is a parachute failure. There are then two options: scream all the way down or enjoy the view for a few all-too-fleeting moments. On rare occasions, people actually survive such falls, not typically by the intercession of divine providence so much as by the intercession of deep snow, steep slopes, thick brush, or the like. Assume the human body can withstand, without fatal consequences, brief accelerations with magnitudes of about 75 times the local acceleration due to gravity (75g). If a skydiver hits at 50 m/s, over what minimum distance is the skydiver brought to a stop under the action of such a large acceleration?

•**62.** A runner accelerates from the starting blocks to a speed of 10.0 m/s over the first 6.0 m of track in a 100 m sprint. (a) During what time interval does she cover the first 6.0 m if her acceleration is constant? (b) What is the magnitude of her acceleration during this time?

‡**63.** Hang a mass on a vertical spring; imagine an origin at the place where the mass is at rest with $\hat{\imath}$ directed *down* as shown in Figure P.63. Now set the mass into oscillation in the vertical direction by lowering the mass a distance A and letting it go. The subsequent motion is called simple harmonic oscillation and will be investigated in some detail in Chapter 7. We shall see that the position vector of a particle executing such one-dimensional oscillatory motion is given by the expression

$$\vec{r}(t) = [A \cos(\omega t)]\hat{\imath}$$

where A is expressed in meters, and ω is expressed in radians per second; both are constants. (a) Find the velocity vector and the acceleration vector as functions of time. (b) Is the acceleration constant? (c) What are the *greatest magnitudes* of the velocity and acceleration vectors? (d) What is the earliest (nonnegative) time that the position vector attains maximum magnitude? When the position vector has its greatest magnitude, what is the magnitude of the velocity vector? What is the magnitude of the acceleration vector at the same time? (e) At what time ($t \geq 0$ s) does the position vector first attain a magnitude of 0 m? At this time, what are the magnitudes of the velocity and acceleration vectors?

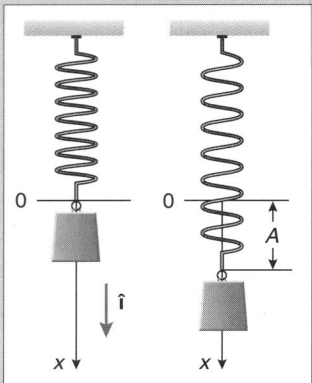

FIGURE P.63

64. When a ball bounces from the floor, the ratio of the speed at the instant just after the bounce to the speed at the instant just before the bounce is called the coefficient of restitution ε of the ball. (a) Show that if a ball, initially at rest, is dropped from a height h_0 and rebounds to a height h_1, the coefficient of restitution is

$$\varepsilon = \left(\frac{h_1}{h_0} \right)^{1/2}$$

(b) The height of rebound is difficult to measure precisely. An alternative way of determining the coefficient of restitution is to drop a ball from height h_0 and measure the total time T from its release to the instant of its *third* impact with the floor. Show that the coefficient of restitution is related to T by

$$T = \sqrt{\frac{2h_0}{g}} \left(1 + 2\varepsilon + 2\varepsilon^2 \right)$$

This problem was inspired by an experiment "Measuring gravity with a superball" by Mark St. John, *Thinking Like a Physicist* (Group in Science and Mathematics, University of California, Berkeley, 1978).

65. It is always fun to drop rocks down into a well. An enterprising physics student realizes that it is possible to determine the distance to the water surface of the well with a stopwatch. From a vantage point at the top of the well, the student drops a rock (with zero initial speed) into a deep wishing well and discovers that the total time from release to reception of the sound from the ensuing splash is 3.00 s. Allowing for the finite speed of sound in air, 343 m/s at 20 °C, calculate the distance to the water surface.

66. A stone is projected vertically upward at speed v_0 from the top of a cliff of height h. The stone just misses the edge of the cliff on the way down. A second stone is dropped from rest from the top of the cliff. How long should you wait before releasing the second stone so that they both arrive at the bottom of the cliff simultaneously?

67. Make and state some reasonable assumptions to estimate the magnitude of the acceleration experienced by the occupants of

a car traveling at 120 km/h if it collides with a massive bridge abutment. Explain how air bags are used to decrease the magnitudes of such accelerations.

Section 3.8 Geometric Interpretations*

•68. A particle initially at the origin and moving with an initial velocity (2.00 m/s)$\hat{\imath}$ experiences an acceleration given by

$$\vec{a} = (3.00 \ \text{m/s}^2)\hat{\imath}$$

(a) Make a graph of a_x versus t over the range of times from $t = 0$ s to 5.00 s. What is the area under the curve of a_x? What is the significance of the area under the graph from $t = 0$ s to 5.00 s? (b) Find the change in the velocity component over the same time interval and make an accurate graph of v_x versus t. What is the area under the curve between 0 s and 5.00 s? What is the significance of the area under the graph of the velocity component between $t = 0$ s and 5.00 s? (c) Find the change in the position vector component over the same time interval and make a graph of x versus t. What is the slope of this graph when $t = 0$ s?

•69. The x-component of the velocity of a particle in free-fall motion near the surface of the Earth is shown in Figure P.69. (a) Was $\hat{\imath}$ chosen to be vertically up or down? (b) Was the particle initially moving up or down? (c) At what instant was the particle at its greatest elevation? (d) What is the area under the curve between 0 s and 5.0 s? (e) What is the change in the x-component of the position vector of the particle during the time interval between 0 s and 5.0 s?

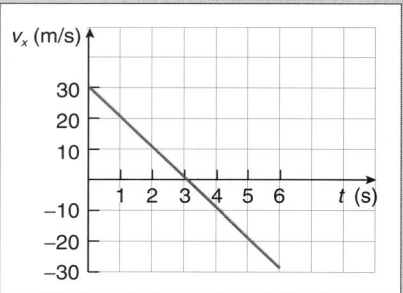

FIGURE P.69

70. A particle initially traveling with a velocity $v_{x0}\hat{\imath}$ is subjected (beginning when $t = 0$ s) to an acceleration

$$\vec{a} = a_0 e^{-\alpha t}\hat{\imath}$$

where α and a_0 are positive scalar constants, independent of time. (a) What are the SI units of the constant α? of a_0? (b) Find $v_x(t)$. (c) After a long time (i.e., as $t \to \infty$ s), what is the speed of the particle?

71. One model for the motion of a particle moving in a resistive medium suggests that the speed decreases exponentially according to the expression

$$v(t) = v_0 e^{-\beta t}$$

where v_0 is the speed of the particle when $t = 0$ s and β is a positive constant. (a) How long will it take the particle to reach half its initial speed? (b) What distance does the particle traverse during the time interval calculated in part (a)? (c) Through what distance does the particle move before it is brought to rest?

72. A theorist in your research group comes up with the following equation for the *x*-component of the position vector of a particle:

$$x(t) = (20.0 \text{ m/s})te^{-(2.00\text{s}^{-1})t}$$

You are assigned the following tasks: (a) Make a graph of *x* versus *t* for the interval from *t* = 0 s to 1.00 s. (b) Determine the (finite) position at which the particle is instantaneously at rest and see if the acceleration of the particle also is zero at this location. (c) Ascertain if the particle is initially moving parallel to $\hat{\imath}$ or in the opposite direction. (d) Find the acceleration component of the particle when *t* = 0 s.

INVESTIGATIVE PROJECTS

A. Expanded Horizons

1. The Greek philosopher Zeno of Elea (c. fifth century B.C.) posed a paradox now known as *Zeno's paradox*. The Greek hero Achilles starts at point A and attempts to catch a slowly moving tortoise initially at point B. When Achilles reaches point B, the tortoise has moved to point C; when Achilles reaches C, the tortoise is at yet another point D, and so on *ad infinitum*. How does Achilles ever manage to catch the tortoise? Investigate Zeno's paradox and its resolution.
   William I. McLaughlin, "Resolving Zeno's paradoxes," *Scientific American*, 271, #5, pages 84–89 (November 1994).
   James R. Newman, *The World of Mathematics* (Simon & Schuster, New York, 1956), Volume 1, pages 41, 94.

2. Investigate the physiological effects of accelerations on the human body. In particular, determine the approximate magnitude of accelerations that are lethal to humans.
   James G. Parker and Vita R. West (editors), *Bioastronautics Data Book* (2nd edition, NASA, Washington, D.C., 1973).
   Bernard Hoop, "Resource letter PPPP-1: Physical principles of physiological phenomena," *American Journal of Physics*, 55, #3, pages 204–210 (March 1987).

3. Investigate the idea of the time rate of change of the acceleration, known as the *jerk* $\vec{\jmath}$ (no kidding!). That is, the instantaneous jerk at instant *t* is:

   $$\vec{\jmath}(t) \equiv \frac{d\vec{a}(t)}{dt}$$

   where $\vec{a}$ is the acceleration. Does the jerk have any physical manifestations or consequences that warrant its use in kinematics? Is there any value in continuing the process of further differentiation, such as

   $$\frac{d\vec{\jmath}}{dt}$$

   On what basis should such a judgment be made?
   Steven H. Schot, "Jerk: the time rate of change of acceleration," *American Journal of Physics*, 46, #11, pages 1090–1094 (November, 1978).
   T. R. Sandin, "The jerk," *The Physics Teacher*, 28, #1, pages 36–40 (January, 1990).

4. Consult appropriate sources such as *The Guiness Book of Records* to determine the approximate fastest speeds for each of the following life forms: human, flying insect, crawling insect (see Problem 54), fish, marine mammal, land mammal, bird. If the fastest speeds are divided by the approximate length or size of the creature, do the resulting speeds per body size have about the same order of magnitude?

B. Lab and Field Work

5. If there is a safe, high place from which to drop a rock so that it falls vertically, design and perform an experiment to determine the magnitude of the local acceleration due to gravity g near the surface of the Earth. What is the precision of your result and what is the uncertainty associated with your measurement? If your physics laboratory has a photogate timer, compare the result you obtain using it with that obtained from measurements of a falling rock.

6. Design and perform an experiment to estimate the magnitude of the acceleration of a sprinter leaving the starting blocks at the beginning of a race. If a safe, public dragstrip is nearby, you also might try to determine the maximum magnitude of the acceleration of your car as its speed changes from 0 km/h to 60 km/h.

7. There likely is a subset of your class that keeps fit by jogging. Using as large a group of such students as possible, time a 100 m sprint by each student. See if there is a relationship between their times and body mass. Is there a difference between any such relationship for males and females?

8. Devise a method for forming water drops (raindrops) of constant size and known mass, perhaps using a graduated micropipette, available in most chemistry and biology departments. Design and perform an experiment to determine if such raindrops reach a limiting (constant) speed when falling through a known distance such as the height of your laboratory. Do the results depend on the mass of the raindrops? You might also see what happens if the water drops fall through a graduated cylinder of cooking or olive oil.
   Howard E. Evans II, "Raindrops keep falling on my head," *The Physics Teacher*, 29, #2, pages 120–121 (February 1991).

9. The kinematics of traffic flow are interesting to examine. Imagine a straight section of interstate with a speed limit of 105 km/h (≈ 65 miles per hour). Assume cars on the road travel at the speed limit if traffic conditions permit. If traffic is light, the average number of cars past any given kilometer post marker during 1 min is equal to the average number of cars per kilometer. Make a qualitative graph of the average number of cars per minute past a kilometer marker as a function of the average number of cars per kilometer. Take into account the average size of a car (say 5 m) as well as the fact that if the traffic becomes too heavy, gridlock ensues. Your state highway department may have data available to support your plot; or secure such data yourself with a well-crafted experiment.

This project was inspired by a letter by Clyde E. Holvenstot to *American Scientist*, 82, #6, pages 504–505 (November–December 1994).

10. Design and perform an experiment to measure the magnitude of the acceleration of an arrow as it is released from rest in an archery bow. Is the magnitude constant?

    W. C. Marlow, "Bow and arrow dynamics," *American Journal of Physics*, 49, #4, pages 320–333 (April 1981).

C. Communicating Physics

11. A student observes the x-component of the position vector of a particle undergoing rectilinear motion at constant acceleration. The data are graphed as a function of time with the results shown in Figure I.11. Discuss how the student should treat the data point that appears inconsistent with the other data. Can the student justify throwing out the "bad" data point? If so, how? If not, explain why. Should the bad data point simply be ignored, and not discussed?

    John R. Taylor, "Rejection of data: Chauvenet's criterion," Chapter 6 in *An Introduction to Error Analysis: The Study of Uncertainties in Physical Measurements* (2nd edition, University Science Books, Sausalito, California, 1997).

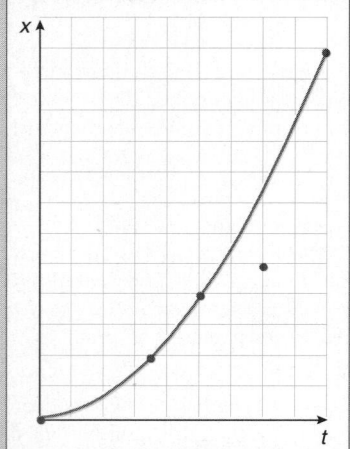

FIGURE I.11

12. The effect of air resistance on free-falling objects causes them not to fall with a constant acceleration of magnitude *g*. Write a research proposal to your physics department for an experiment to measure the effects of air resistance on free-falling objects. Indicate in your proposal (a) how you would attempt to measure the effect; (b) requests for appropriate equipment, indicating why the equipment is needed for the experiment;

(c) what objects will be used in the free-fall; (d) how your experimental results can distinguish between free-fall at constant acceleration and cases where the effects of air resistance are apparent; and (e) an estimate of the time needed to conduct the experiment and prepare a report of your findings. Perhaps if your proposal is judged well, your professor will excuse you from the regular laboratory to perform the experiment and report your findings.

13. Imagine a technical civilization that develops a theory of kinematics based not on speed measured in meters per second, but on the sluggishness alluded to in Section 3.3, measured in seconds per meter. Explain in some detail how such a civilization might formulate a concept analogous to the magnitude of the acceleration. Invent an appropriate word for the concept. How might the directional attributes of these new concepts be treated? Would such concepts obey the rules for vector arithmetic?

RECTILINEAR NONMOTION

CHAPTER 4

KINEMATICS II
MOTION IN TWO AND THREE DIMENSIONS

Such a life, with all vision limited to a Point and all motion to a
Straight Line, seemed to me inexpressibly dreary; . . .

Edwin Abbott Abbott (1838–1926)*

Life is pretty dull confined to one dimension: we would all be particles living and moving in the same rut. Sports fields and amusement park rides would be one-dimensional lines much like the gutter of a bowling alley. Boring. In this chapter we exploit our familiarity with vectors to extend our knowledge of rectilinear motion to the richness of motion in two- and three-dimensional space. We will see that the two- or three-dimensional motions of a particle are natural extensions of the one-dimensional kinematics of Chapter 3. We must learn to describe the kinematics of the multidimensional motion of a particle with dexterity, since such motion frequently is encountered in both our natural and technological environments.

The circular motion of a particle is two-dimensional motion in a plane. We see examples of circular motion in everything from kitchen blenders to merry-go-rounds, from tires and wheels to CD players and gears, from computer disk drives to centrifuges, from pinwheels to windmills, from cyclotron particle accelerators in physics to the rotation of planets, stars, and galaxies in the universe. Perhaps surprisingly, we shall see in this chapter that while the circular motion of a particle occurs in a two-dimensional plane in space, the description of the motion involves only one time-dependent coordinate.

4.1 THE POSITION, VELOCITY, AND ACCELERATION VECTORS IN TWO DIMENSIONS

Of the nature of flatland

Edwin Abbott Abbott[†]

When you toss a crumpled paper from your desk to a wastebasket or shoot a few hoops (see Figure 4.1), the motion of the

---

*(Chapter Opener) *Flatland* (Princeton University Press, Princeton, New Jersey, 1991), Part I, Section 1, page 3.
[†]*Flatland*, Part II, Section 13, page 56.

FIGURE 4.1 A basketball shot is motion in two dimensions.

particle is confined to a two-dimensional vertical plane. When you drive north and then east on a level plane, your motion also is confined to a plane in two dimensions. But since this travel consists of two successive rectilinear parts, it hardly qualifies as fully two dimensional. To the extent that a motion cannot be reduced to some one chosen dimension, we can consider it necessarily two or three dimensional.

When you *come about* while sailing, which means to reverse direction on the sea, or if you turn a level corner in your car, as illustrated in Figure 4.2, the motion is confined to a two-dimensional horizontal plane needing both *x*- and *y*-coordinates. Let the plane of the two-dimensional motion be the *x–y* plane of a Cartesian coordinate system. You have the freedom to choose orientations of the two coordinate axes in any convenient directions as long as they are mutually perpendicular, as shown in Figure 4.3.

> ### PROBLEM-SOLVING TACTIC
>
> **4.1 The specific coordinate choice is yours to make, but it needs to be explicitly indicated in a sketch and used consistently throughout a given problem.** You learned this in the previous chapter; it bears repeating, though. For two-dimensional motion, it is *not* necessary to check whether the coordinate system you choose is right-handed, since the third, but irrelevant, direction (*z*) always can be chosen to make the system right-handed. For example, $\hat{\mathbf{k}}$ is directed perpendicularly out of the page in Figures 4.3a and b, but into the page in Figure 4.3c. However, the *z*-axis is not needed to describe the two-dimensional motion, and so it is superfluous to specify its direction explicitly.

For two-dimensional motion, the two coordinates *x* and *y* locate the position of the particle whose motion is to be described. Since the particle is moving, both coordinates are functions of time: $x(t)$ and $y(t)$. The position vector of the particle locates the particle with respect to the origin you choose and is the vector sum:

$$\vec{\mathbf{r}}(t) = x(t)\hat{\mathbf{i}} + y(t)\hat{\mathbf{j}} \qquad (4.1)$$

Later, at the instant $t + \Delta t$, the particle has the coordinates $x(t + \Delta t)$ and $y(t + \Delta t)$. The new position vector is

$$\vec{\mathbf{r}}(t + \Delta t) = x(t + \Delta t)\hat{\mathbf{i}} + y(t + \Delta t)\hat{\mathbf{j}} \qquad (4.2)$$

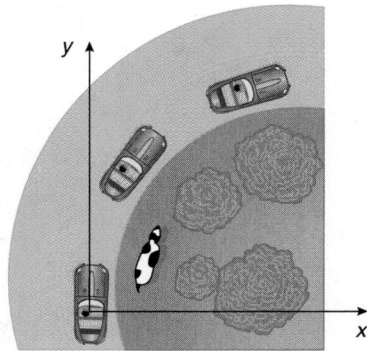

FIGURE 4.2 When you turn a level corner in your car, the motion is in two dimensions.

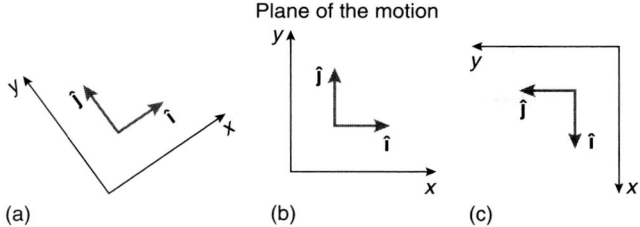

FIGURE 4.3 Some possible coordinate choices for motion in a plane.

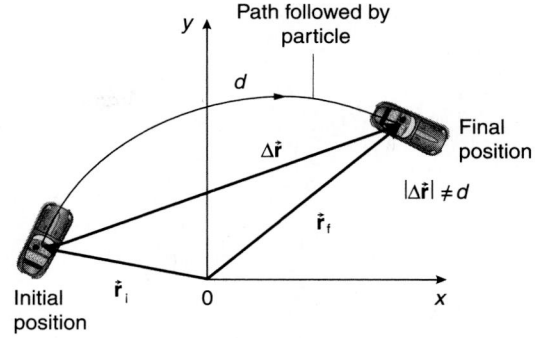

FIGURE 4.4 The total distance traveled, d, typically is not equal to the magnitude of the change in the position vector, $|\Delta\vec{r}|$.

The change in the position vector $\Delta\vec{r}$ of the particle is the vector difference

$$\Delta\vec{r} \equiv \vec{r}_f - \vec{r}_i$$
$$= \vec{r}(t + \Delta t) - \vec{r}(t)$$
$$= [x(t + \Delta t) - x(t)]\hat{i} + [y(t + \Delta t) - y(t)]\hat{j}$$

As with any vector, the magnitude and direction of the change in the position vector of a particle are independent of the choice you make for the coordinate system.

Just as in one-dimensional motion, the average speed involves the total distance d traveled by the particle during the interval Δt:

$$\text{average speed} \equiv \frac{d}{\Delta t}$$

The average velocity of the particle during the interval Δt involves the change in the position vector and is, from Equation 3.6,

$$\vec{v}_{ave} \equiv \frac{\Delta\vec{r}}{\Delta t}$$
$$= \frac{\vec{r}(t + \Delta t) - \vec{r}(t)}{\Delta t} \quad (4.3)$$

As in one dimension, the magnitude of the change in the position vector typically is not equal to the total distance traveled by the particle during the interval Δt, as shown in Figure 4.4. Consequently, the average speed typically is not equal to the magnitude of the average velocity.*

If we take the limit of average velocity for smaller and smaller time intervals Δt, we approach the instantaneous velocity $\vec{v}(t)$ as $\Delta t \rightarrow 0$ s:

$$\vec{v}(t) = \lim_{\Delta t \rightarrow 0 \text{ s}} \frac{\vec{r}(t + \Delta t) - \vec{r}(t)}{\Delta t} = \frac{d\vec{r}(t)}{dt} \quad (4.4)$$

Just as in rectilinear motion, the instantaneous velocity of the particle is the time derivative of the instantaneous position vector. As with rectilinear motion, the magnitude of the velocity

vector is the speed of the particle. The velocity vector always is tangent to the path followed by the particle in space. According to Equation 4.4, the velocity is

$$\vec{v}(t) = \frac{d\vec{r}(t)}{dt}$$
$$= \frac{dx(t)}{dt}\hat{i} + \frac{dy(t)}{dt}\hat{j}$$
$$\equiv v_x(t)\hat{i} + v_y(t)\hat{j} \quad (4.5)$$

where

$$v_x(t) \equiv \frac{dx(t)}{dt}$$
$$v_y(t) \equiv \frac{dy(t)}{dt} \quad (4.6)$$

The velocity components are the time derivatives of the corresponding components of the position vector.

Just as in rectilinear motion, the average acceleration during an interval Δt is related to the change in the velocity by

$$\vec{a}_{ave} = \frac{\Delta\vec{v}}{\Delta t} \quad (4.7)$$

The instantaneous acceleration is the limit of the average acceleration as the time interval approaches zero:

$$\vec{a}(t) = \lim_{\Delta t \rightarrow 0 \text{ s}} \vec{a}_{ave} = \lim_{\Delta t \rightarrow 0 \text{ s}} \frac{\Delta\vec{v}}{\Delta t}$$

This in turn is the time derivative of the velocity vector:

$$\vec{a}(t) = \frac{d\vec{v}(t)}{dt} = \frac{dv_x(t)}{dt}\hat{i} + \frac{dv_y(t)}{dt}\hat{j}$$
$$\equiv a_x(t)\hat{i} + a_y(t)\hat{j} \quad (4.8)$$

where

$$a_x(t) \equiv \frac{dv_x(t)}{dt}$$
$$a_y(t) \equiv \frac{dv_y(t)}{dt} \quad (4.9)$$

---

*Recall that for rectilinear motion, if the particle did not undergo back-and-forth motion during the interval Δt, the magnitude of the position vector was equal to the total distance traveled by the particle.

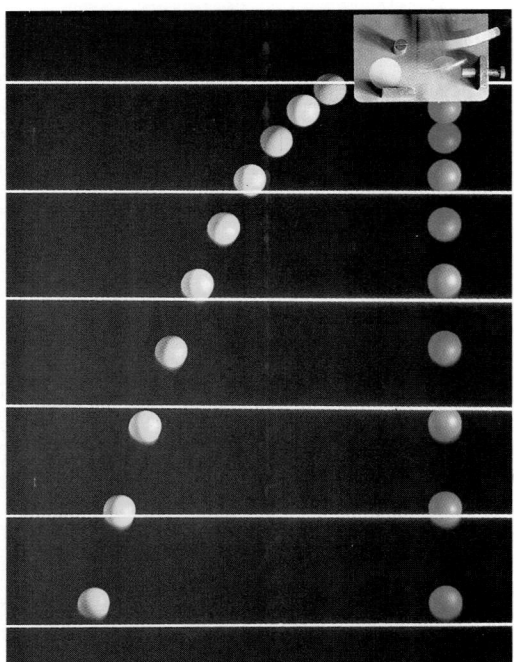

FIGURE 4.5 A marble is dropped from rest while another marble is simultaneously projected horizontally. The vertical motion is independent of the horizontal motion.

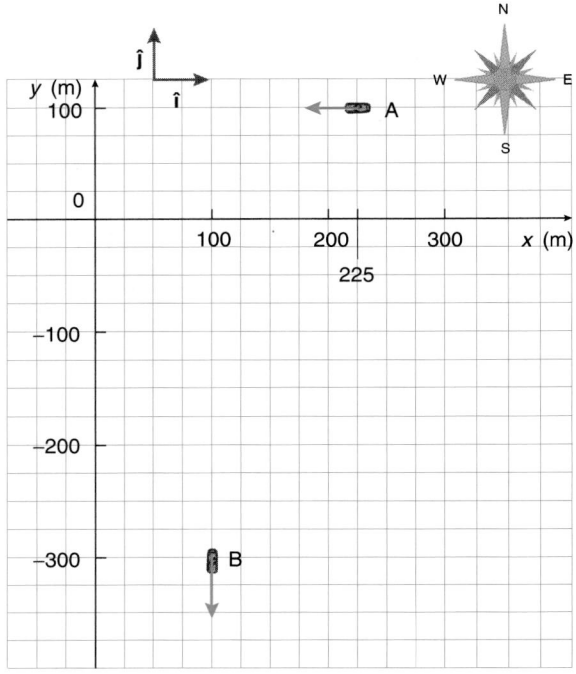

FIGURE 4.6

So the relationships between the position vector, the velocity, and the acceleration are the same as those found for the rectilinear case. The only difference is that we now have two dimensions, which means that the various vectors have two components.

With motion in two or more dimensions, a new question arises. Do the instantaneous values of a_y, v_y, and y depend on the values of a_x, v_x, or x at the same instant or other instants? In other words, are the motions along *perpendicular* coordinate axes *independent* of each other? To answer this question, we turn, as we must, to experiments, the final arbiters for questions in physics. Imagine what happens if a marble is dropped from rest while an identical marble is simultaneously projected horizontally (see Figure 4.5). The vertical motions of the two marbles are identical: they strike the floor simultaneously. The vertical motion is unaffected by the horizontal motion.

Such a kinematic experiment and others like it imply an important fact:

> Perpendicular motions are independent of each other.

The implication of this independence means we can analyze the motion along each coordinate axis *separately*, as we will see in the next section.

QUESTION 1

A particle moves at constant speed. Specify the additional circumstance(s) that lets you conclude its acceleration is zero.

EXAMPLE 4.1

Let $\hat{\imath}$ be directed east and $\hat{\jmath}$ be directed north, as in Figure 4.6. When a stopwatch indicates 2.00 s, your car is located at position A and moving west at a speed of 20.0 m/s. Later,

when the stopwatch indicates 17.00 s, the car is located at position B and moving south at 25.0 m/s.

a. Find the two position vectors of the car. Indicate them accurately on a sketch.
b. What is the change in the position vector? What is the magnitude of the change in the position vector? Indicate $\Delta\vec{r}$ on your sketch.
c. Determine the average velocity of the car during the time interval. What is the magnitude of the average velocity of the car?
d. What is the change in the velocity of the car?
e. What is the average acceleration of the car?

Solution

a. In Figure 4.6, the x- and y-coordinates of the car are initially $x = 225$ m and $y = 100$ m. The initial position vector thus is

$$\vec{r}_i = (225 \text{ m})\hat{\imath} + (100 \text{ m})\hat{\jmath}$$

Later the coordinates are $x = 100$ m and $y = -300$ m, and so the final position vector is

$$\vec{r}_f = (100 \text{ m})\hat{\imath} - (300 \text{ m})\hat{\jmath}$$

These position vectors are indicated in Figure 4.7.

b. The change in the position vector is

$$\begin{aligned}
\Delta\vec{r} &= \vec{r}_f - \vec{r}_i \\
&= [(100 \text{ m})\hat{\imath} - (300 \text{ m})\hat{\jmath}] - [(225 \text{ m})\hat{\imath} + (100 \text{ m})\hat{\jmath}] \\
&= (-125 \text{ m})\hat{\imath} - (400 \text{ m})\hat{\jmath}
\end{aligned}$$

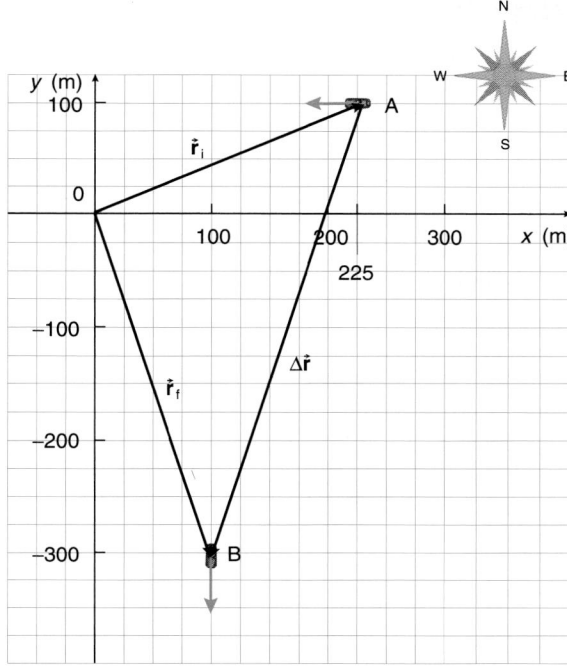

FIGURE 4.7

This vector also is indicated in Figure 4.7. The magnitude of this vector is

$$|\Delta\vec{\mathbf{r}}| = [(-125 \text{ m})^2 + (-400 \text{ m})^2]^{1/2}$$
$$= 419 \text{ m}$$

c. The average velocity is found from Equation 4.3:

$$\vec{\mathbf{v}}_{ave} = \frac{\Delta\vec{\mathbf{r}}}{\Delta t}$$

$$= \frac{(-125 \text{ m}) \,\hat{\mathbf{i}} - (400 \text{ m}) \,\hat{\mathbf{j}}}{17.00 \text{ s} - 2.00 \text{ s}}$$

$$= (-8.33 \text{ m/s}) \,\hat{\mathbf{i}} - (26.7 \text{ m/s}) \,\hat{\mathbf{j}}$$

The magnitude of the average velocity vector is

$$v_{ave} = [(-8.33 \text{ m/s})^2 + (-26.7 \text{ m/s})^2]^{1/2}$$
$$= 28.0 \text{ m/s}$$

d. The change in the velocity of the car is

$$\Delta\vec{\mathbf{v}} = \vec{\mathbf{v}}_f - \vec{\mathbf{v}}_i$$
$$= (-25.0 \text{ m/s})\hat{\mathbf{j}} - (-20.0 \text{ m/s})\hat{\mathbf{i}}$$
$$= (20.0 \text{ m/s})\hat{\mathbf{i}} - (25.0 \text{ m/s})\hat{\mathbf{j}}$$

e. The average acceleration of the car is, from Equation 4.7,

$$\vec{\mathbf{a}}_{ave} = \frac{\Delta\vec{\mathbf{v}}}{\Delta t}$$

$$= \frac{(20.0 \text{ m/s}) \,\hat{\mathbf{i}} - (25.0 \text{ m/s}) \,\hat{\mathbf{j}}}{17.00 \text{ s} - 2.00 \text{ s}}$$

$$= (1.33 \text{ m/s}^2) \,\hat{\mathbf{i}} - (1.67 \text{ m/s}^2) \,\hat{\mathbf{j}}$$

4.2 TWO-DIMENSIONAL MOTION WITH A CONSTANT ACCELERATION

Throw a rock into a pond or toss a ball to a friend. Shoot a basketball from the foul line or pitch a baseball. Slide a puck across an inclined airtable or watch a pinball move between stations in the maze. All are examples of **projectile motion**. If frictional effects are very small, all such motions have a constant acceleration. (Remember that a constant acceleration vector has a fixed magnitude and direction at all times and locations during the motion.)

> **PROBLEM-SOLVING TACTIC**
>
> **4.2 If the acceleration vector is constant, choose one of the coordinate axes to be the line containing the constant acceleration or a line parallel to it.** With this choice the acceleration has only a *single* component. The description of the motion then is simpler than otherwise would be the case.

> We will choose the y-axis to be the line containing the constant acceleration vector. Thus the acceleration vector is
>
> $$\vec{\mathbf{a}} = a_y \hat{\mathbf{j}}$$
>
> where the single component a_y is a constant scalar, and $a_x = 0 \text{ m/s}^2$.

> **PROBLEM-SOLVING TACTICS**
>
> **4.3 The sign of the acceleration component a_y depends on the direction you choose for $\hat{\mathbf{j}}$ along the line containing the acceleration vector.** See Figure 4.8.
>
> **4.4 The origin may be taken at any convenient location along the line chosen for the y-axis.**

Since motions along perpendicular directions are independent of each other, the equations we developed in the last chapter for studying rectilinear motion with a constant acceleration (Equations 3.16 and 3.19) can be used for *each* of the perpendicular coordinate directions with appropriate changes made for labeling the different coordinate axes.

If a particle has an acceleration component a_x equal to zero, the x-component of its velocity is, from Equation 3.16,

$$v_x(t) = v_{x0} + a_x t$$
$$= v_{x0} + (0 \text{ m/s}^2)t$$
$$= v_{x0} \qquad \text{(i.e., is constant)} \qquad (4.10)$$

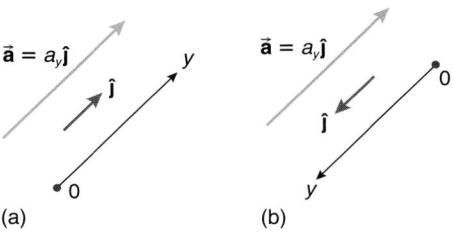

FIGURE 4.8 (a) If you choose $\hat{\mathbf{j}}$ to be parallel to $\vec{\mathbf{a}}$, then $a_y > 0 \text{ m/s}^2$. (b) If you choose $\hat{\mathbf{j}}$ to be antiparallel to $\vec{\mathbf{a}}$, then $a_y < 0 \text{ m/s}^2$.

> The x-component of the velocity is constant throughout the motion and equal to the initial x-component of the velocity v_{x0}, because $a_x = 0$ m/s^2.

A graph of the x-component of the velocity versus time (see Figure 4.9) is a horizontal line with zero slope, again since $a_x = 0$ m/s^2.

Equation 3.19 for the x-coordinate of the particle—with $a_x = 0$ m/s^2—indicates that

$$x(t) = x_0 + v_{x0}t \qquad (4.11)$$

where, as always, v_{x0} is the initial x-component of the velocity of the particle when $t = 0$ s and x_0 is the corresponding initial x-component of the position vector. The x-coordinate changes linearly with time, and so the x versus t graph is a straight line with slope v_{x0} and intercept x_0, as shown in Figure 4.10.

QUESTION 2

What would the x versus t graph (with $a_x = 0$ m/s^2) look like if $x_0 > 0$ m and $v_{x0} < 0$ m/s?

The equations for $y(t)$ and $v_y(t)$ have the identical form as the rectilinear motion equations for a constant acceleration (Equations 3.16 and 3.19), with merely a notational change to accommodate the different coordinate:

$$v_y(t) = v_{y0} + a_y t \qquad (4.12)$$

$$y(t) = y_0 + v_{y0}t + \frac{1}{2}a_y t^2 \qquad (4.13)$$

where v_{y0} is the initial y-component of the velocity when $t = 0$ s and y_0 is the initial y-component of the position vector. A graph of $v_y(t)$ versus t is a straight *inclined* line since $v_y(t)$ depends linearly on t (see Figure 4.11). The line has intercept v_{y0} and slope a_y.

QUESTION 3

What would the $v_y(t)$ versus t graph look like if $v_{y0} > 0$ m/s and $a_y < 0$ m/s^2?

A graph of $y(t)$ versus t is parabolic, since $y(t)$ is quadratic in t (see Figure 4.12). The slope at any instant t_1 is the y-component of the velocity of the particle, $v_y(t_1)$. The intercept is y_0.

We begin our observations when $t = 0$ s. The initial velocity vector has the Cartesian form

$$\vec{v}_0 = v_{x0}\hat{i} + v_{y0}\hat{j} \qquad (4.14)$$

and the initial position vector of the particle (when $t = 0$ s) has the Cartesian form

$$\vec{r}_0 = x_0\hat{i} + y_0\hat{j} \qquad (4.15)$$

> ### PROBLEM-SOLVING TACTIC
>
> **4.5 Notice that the equations for $x(t)$ and $v_x(t)$ have the same form as those for $y(t)$ and $v_y(t)$ but with an acceleration component a_x equal to zero.** By setting $a_x = 0$ m/s^2, we see that Equations 3.16 and 3.19 for $x(t)$ and $v_x(t)$ emerge from the general equations for rectilinear motion with a constant acceleration. There is no need to remember two equations for the x-axis motion and two more for the y-axis motion. Just remember Equations 3.16 and 3.19 and apply them to the x- and y-motions for two-dimensional motion.

The equation for the trajectory $y(x)$ of the particle in space (i.e., the path of the particle in space) is found by solving Equation 4.11 for t:

$$t = \frac{x - x_0}{v_{x0}}$$

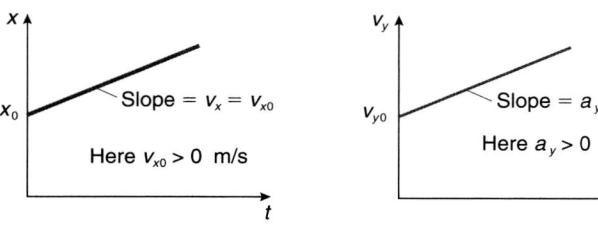

FIGURE 4.10 A graph of x versus t when $a_x = 0$ m/s^2.

FIGURE 4.11 A graph of $v_y(t)$ versus t.

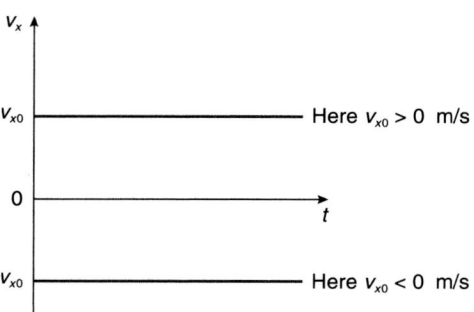

FIGURE 4.9 The x-component of the velocity is constant.

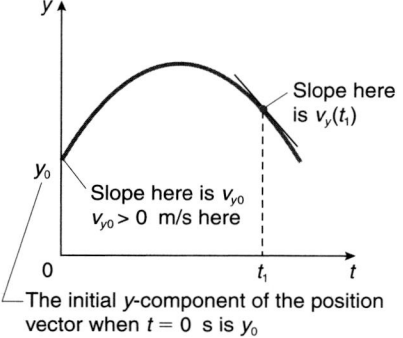

FIGURE 4.12 A graph of $y(t)$ versus t when $a_y < 0$ m/s^2.

and substituting for t in Equation 4.13. We obtain

$$y(x) = y_0 + v_{y0}\frac{x - x_0}{v_{x0}} + \frac{1}{2}a_y\left(\frac{x - x_0}{v_{x0}}\right)^2$$

> The equation for the trajectory $y(x)$ is quadratic in x, indicating that the path in space is *parabolic* in shape.* Two-dimensional motion with a constant acceleration produces parabolic trajectories (see Figure 4.13).

In Figure 4.13, the y-axis is a line containing the acceleration $\vec{a}$, in accordance with Problem-Solving Tactic 4.2. Also notice that the velocity vector at any instant is tangent to the trajectory (path) of the particle in space. The acceleration vector is constant at all points in space (and time). The x-component of the velocity is constant throughout the motion since the acceleration is purely in the y-direction.

Particles thrown or dropped near the surface of the Earth[†] are common cases of projectile motion. While in flight, we say they experience free-fall motion. For simplicity, we neglect the effects of air resistance. The simplification is an approximation to the real flight of such projectiles and means that they experience the constant, vertically downward local acceleration due to gravity, of magnitude $g = 9.81$ m/s^2.

For such projectiles, invoking Problem-Solving Tactic 4.2, choose the y-axis to be a vertical line, since it is along this line that the constant acceleration is directed. You may choose the specific left or right location of the y-axis, as well as the elevation of the origin for y. The x-axis is horizontal; its elevation is at the location of the origin chosen for y. The x-component of the acceleration is zero: $a_x = 0$ m/s^2. Problem-Solving Tactic 4.3 indicates that the algebraic sign of the vertical acceleration component a_y depends on the direction chosen for $\hat{\jmath}$ (see Figure 4.14).

The procedure for applying these ideas to the flight of projectiles, neglecting any resistive effects, is best illustrated by careful study of Examples 4.2 and 4.3 following.

---

*Do not confuse the parabolic shape of the trajectory in space, $y(x)$, with the parabolic shape of the $y(t)$ versus t graph: they represent different things!

[†] As we mentioned in Chapter 3, near the surface of the Earth means the total vertical extent of the motion is negligible compared with the radius of the Earth, 6370 km.

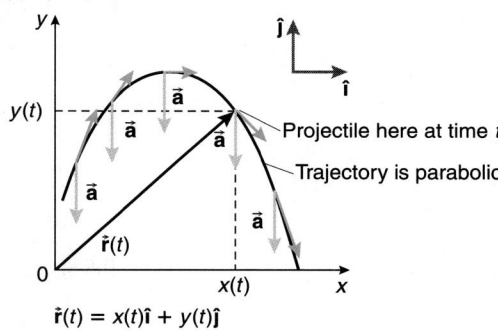

$$\vec{r}(t) = x(t)\hat{\imath} + y(t)\hat{\jmath}$$

FIGURE 4.13 Two-dimensional motion with a constant acceleration produces a parabolic trajectory. Here $a_y < 0$ m/s^2.

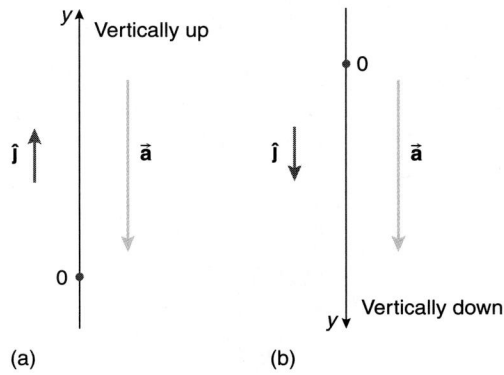

FIGURE 4.14 Free-fall motion. (a) If $\hat{\jmath}$ is up, then a_y is negative: $\vec{a} = a_y\hat{\jmath} = -g\hat{\jmath}$, and so $a_y = -g = -9.81$ m/s^2. (b) If $\hat{\jmath}$ is down, then a_y is positive: $\vec{a} = a_y\hat{\jmath} = g\hat{\jmath}$, and so $a_y = +g = 9.81$ m/s^2.

QUESTION 4

A juggler throws a juggling pin across a high room and simultaneously races across the room and barely manages to catch it. A neat trick if you can do it. (a) Which object (the pin or the juggler) had the greater average velocity during the interval or is it the same for both? Explain your answer. (b) Which object had the greater average speed during the interval or is it the same for both? Explain your answer.

STRATEGIC EXAMPLE 4.2

You kick a soccer ball at a speed of 10.0 m/s off a 100 m high cliff. The initial velocity vector is 30.0° above the horizontal direction as the ball leaves your foot, as shown in Figure 4.15.

a. Choose an appropriate coordinate system and write the equations for $v_x(t)$, $x(t)$, $v_y(t)$, and $y(t)$ consistent with your choice.
b. Determine the time of flight.
c. Find the speed of the ball the instant just before impact.
d. Find the horizontal distance of the impact point from the point the ball was kicked.
e. Determine the greatest height of the projectile above the plain on which the ball strikes.

Solution

a. We solve this problem with two different choices for a Cartesian coordinate system to show you the freedom you have in solving problems. Use Problem-Solving Tactic 4.2 and *choose the y-axis*

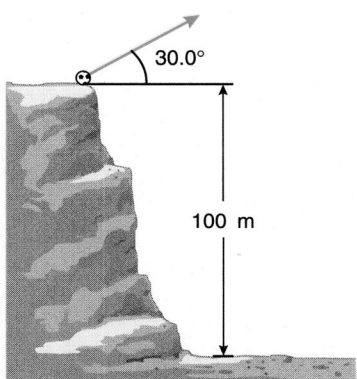

FIGURE 4.15

to be a line containing the constant acceleration vector. Likewise, using Problem-Solving Tactic 4.4, *the origin may be taken at any convenient location along the line chosen for the y-axis.*

Introduce a Cartesian coordinate system as in Figure 4.16.

Choice 1

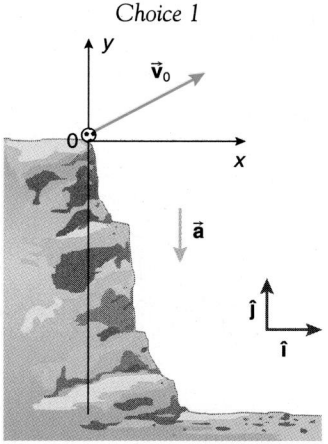

Let the origin be at the point the projectile leaves your foot, with $\hat{\jmath}$ vertically upward, and $\hat{\imath}$ horizontal and to the right.

FIGURE 4.16

Choice 2

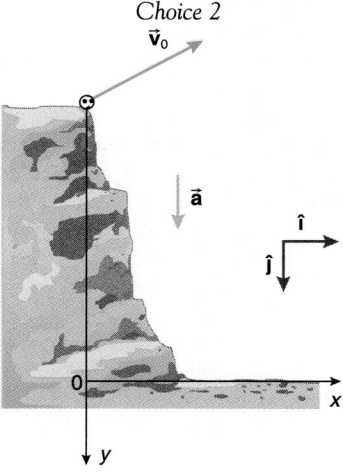

Let the origin be at the bottom of the cliff directly below the point the ball was kicked, with $\hat{\jmath}$ vertically down and $\hat{\imath}$ horizontal and to the right.

Recall from Problem-Solving Tactic 4.3 that *the sign of the acceleration component a_y depends on the direction you choose for $\hat{\jmath}$.* With the coordinate choice determined, the downward acceleration vector is

$$\vec{a} = (-9.81 \text{ m/s}^2)\hat{\jmath} \qquad\qquad\qquad \vec{a} = (+9.81 \text{ m/s}^2)\hat{\jmath}$$

The acceleration has a single nonzero component:

$$a_x = 0 \text{ m/s}^2 \qquad\qquad\qquad\qquad a_x = 0 \text{ m/s}^2$$
$$a_y = -9.81 \text{ m/s}^2 \qquad\qquad\qquad a_y = 9.81 \text{ m/s}^2$$

The initial velocity vector $\vec{v}_0$ of the ball when $t = 0$ s is expressed in Cartesian form as

$$\vec{v}_0 = [(10.0 \text{ m/s}) \cos 30°]\hat{\imath} \qquad\qquad \vec{v}_0 = [(10.0 \text{ m/s}) \cos 30°]\hat{\imath}$$
$$+ [(10.0 \text{ m/s}) \sin 30°]\hat{\jmath} \qquad\qquad\quad - [(10.0 \text{ m/s}) \sin 30°]\hat{\jmath}$$

Hence the two components of the initial velocity are

$$v_{x0} = (10.0 \text{ m/s}) \cos 30.0°$$
$$= 8.66 \text{ m/s}$$
$$v_{y0} = (10.0 \text{ m/s}) \sin 30.0°$$
$$= 5.00 \text{ m/s}$$

$$v_{x0} = (10.0 \text{ m/s}) \cos 30.0°$$
$$= 8.66 \text{ m/s}$$
$$v_{y0} = (-10.0 \text{ m/s}) \sin 30.0°$$
$$= -5.00 \text{ m/s}$$

The initial position vector of the ball

is zero since it is at the origin when $t = 0$ s.

is $(-100 \text{ m})\hat{\jmath}$ since it is at the top of the cliff when $t = 0$ s.

Therefore, the components of the initial position vector

are both zero:

$$x_0 = 0 \text{ m}$$
$$y_0 = 0 \text{ m}$$

are

$$x_0 = 0 \text{ m}$$
$$y_0 = -100 \text{ m}$$

Since the motions along the coordinate axes are independent of each other, you examine each direction separately. Look at the motion along the x-axis, where $a_x = 0 \text{ m/s}^2$. From Equation 4.10 with

$$v_{x0} = 8.66 \text{ m/s}$$

you obtain

$$v_x(t) = 8.66 \text{ m/s}$$

$$v_{x0} = 8.66 \text{ m/s}$$

you obtain

$$v_x(t) = 8.66 \text{ m/s}$$

This result is independent of time. The horizontal component of the velocity is constant throughout the entire motion because the acceleration component a_x is zero. The acceleration of the ball is in the vertical direction, y. The x-component of the position vector at any time is found using Equation 4.11, or, after substituting for x_0, v_{x0}, and a_x (= 0 m/s²):

$$x(t) = 0 + (8.66 \text{ m/s})t$$
$$= (8.66 \text{ m/s})t \qquad (1a)$$

$$x(t) = 0 + (8.66 \text{ m/s})t$$
$$= (8.66 \text{ m/s})t \qquad (1b)$$

The x-component of the position vector changes linearly with time since the velocity component v_x is constant.

Now examine the y-axis motion. The y-component of the velocity at any time is given by Equation 4.12:

$$v_y(t) = v_{y0} + a_y t$$

Substitute for v_{y0} and a_y from the preceding:

$$v_y(t) = 5.00 \text{ m/s} + (-9.81 \text{ m/s}^2)t$$
$$= 5.00 \text{ m/s} - (9.81 \text{ m/s}^2)t \quad (2a)$$

$$v_y(t) = -5.00 \text{ m/s} + (9.81 \text{ m/s}^2)t \quad (2b)$$

The y-component of the position vector is given by Equation 4.13:

$$y(t) = y_0 + v_{y0}t + \frac{1}{2} a_y t^2$$

which, after substituting, becomes

$$y(t) = 0 \text{ m} + (5.00 \text{ m/s})t$$
$$+ \frac{1}{2}(-9.81 \text{ m/s}^2)t^2$$
$$= (5.00 \text{ m/s})t - \frac{9.81 \text{ m/s}^2}{2}t^2 \quad (3a)$$

$$y(t) = -100 \text{ m} - (5.00 \text{ m/s})t$$
$$+ \frac{9.81 \text{ m/s}^2}{2}t^2 \qquad (3b)$$

This completes the tailoring of the kinematic equations for motion with a constant acceleration to the specific choice for the Cartesian coordinate system used to solve the problem. Now you bring in other relevant information.

b. At the impact point, the y-coordinate of the projectile is

$$y = -100 \text{ m} \qquad\qquad\qquad y = 0 \text{ m}$$

Substitute this value for y into equation (3) to determine the instant t of impact:

$$-100 \text{ m} = (5.00 \text{ m/s}) \, t - \frac{9.81 \text{ m/s}^2}{2} t^2 \qquad 0 \text{ m} = -100 \text{ m} - (5.00 \text{ m/s}) \, t + \frac{9.81 \text{ m/s}^2}{2} t^2$$

or

$$\frac{9.81 \text{ m/s}^2}{2} t^2 - (5.00 \text{ m/s}) \, t - 100 \text{ m} = 0 \text{ m} \qquad\qquad \frac{9.81 \text{ m/s}^2}{2} t^2 - (5.00 \text{ m/s}) \, t - 100 \text{ m} = 0 \text{ m}$$

The equation can be solved for t using the quadratic formula. Notice that you get the *same* equation for t with *both* coordinate choices; this is to be expected since the instant of impact will be the same, regardless of your coordinate choice. The numerical value of t also is equal to the time interval of the flight since, by convention, the clock was set to $t = 0$ s at launch. The two solutions for t are

$$5.05 \text{ s} \qquad \text{and} \qquad -4.03 \text{ s}$$

The positive solution must be chosen because the impact occurs *after* the instant of the kick (which was $t = 0$ s). Thus the time of flight is

$$t = 5.05 \text{ s}$$

The negative root corresponds to the other time the ball has the y-coordinate

$$y = -100 \text{ m} \qquad\qquad\qquad y = 0 \text{ m}$$

on the parabolic trajectory (see Figure 4.17). You can discard the negative root for t, because you began your observations when $t = 0$ s and the impact certainly comes after the kick.

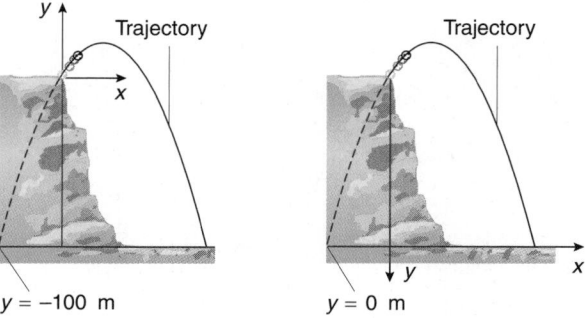

FIGURE 4.17 The meaning of the negative solution for t.

c. The x-component of the velocity is constant throughout the flight and is

$$v_x = 8.66 \text{ m/s} \qquad\qquad\qquad v_x = 8.66 \text{ m/s}$$

The y-component of the velocity at the instant of impact (when $t = 5.05$ s) is

$$v_y = 5.00 \text{ m/s} - (9.81 \text{ m/s}^2)t$$
$$= 5.00 \text{ m/s} - (9.81 \text{ m/s}^2)(5.05 \text{ s})$$
$$= -44.5 \text{ m/s}$$

$$v_y = -5.00 \text{ m/s} + (9.81 \text{ m/s}^2)t$$
$$= -5.00 \text{ m/s} + (9.81 \text{ m/s}^2)(5.05 \text{ s})$$
$$= +44.5 \text{ m/s}$$

So the velocity vector at the instant just before impact is

$$\vec{v} = (8.66 \text{ m/s})\hat{\imath} - (44.5 \text{ m/s})\hat{\jmath}$$

$$\vec{v} = (8.66 \text{ m/s})\hat{\imath} + (44.5 \text{ m/s})\hat{\jmath}$$

The *speed* at the instant just before impact is the magnitude of the velocity vector:

$$v = (v_x^2 + v_y^2)^{1/2}$$

$$v = [(8.66 \text{ m/s})^2 + (-44.5 \text{ m/s})^2]^{1/2}$$
$$= 45.3 \text{ m/s}$$

$$v = [(8.66 \text{ m/s})^2 + (44.5 \text{ m/s})^2]^{1/2}$$
$$= 45.3 \text{ m/s}$$

Notice that the speed at impact is the same with both coordinate choices.

d. The horizontal distance of the impact point from the point the ball was kicked is the x-component of the position vector of the impact point, found by substituting $t = 5.05$ s into equation (1) for the x-component of the position vector:

$$x_{\text{impact}} = 0 \text{ m} + (8.66 \text{ m/s})(5.05 \text{ s})$$
$$= 43.7 \text{ m}$$

$$x_{\text{impact}} = 0 \text{ m} + (8.66 \text{ m/s})(5.05 \text{ s})$$
$$= 43.7 \text{ m}$$

Since the x-axis origin in both coordinate systems is at the same horizontal position, you expect the x-coordinates of the impact point to be the same for both coordinate choices.

e. To determine the greatest height of the projectile above the plain on which the ball strikes, you need to know the instant the projectile reaches this point in space. The only thing you know about the projectile when it is at its greatest elevation is that its vertical velocity component v_y is zero. Use equation (2) for v_y at any instant t:

$$v_y(t) = 5.00 \text{ m/s} - (9.81 \text{ m/s}^2)t$$

$$v_y(t) = -5.00 \text{ m/s} + (9.81 \text{ m/s}^2)t$$

Solve for t when $v_y(t) = 0$ m/s:

$$0 \text{ m/s} = 5.00 \text{ m/s} - (9.81 \text{ m/s}^2)t$$

$$0 \text{ m/s} = 5.00 \text{ m/s} - (9.81 \text{ m/s}^2)t$$

Solving for t, for both coordinate choices you obtain

$$t = 0.510 \text{ s}$$

Since you are interested in the vertical height at this time, we find the y-component of the position vector when $t = 0.510$ s from equation (3):

$$y(t) = (5.00 \text{ m/s})t - \frac{9.81 \text{ m/s}^2}{2}t^2$$

$$y(t) = -100 \text{ m} - (5.00 \text{ m/s})t + \frac{9.81 \text{ m/s}^2}{2}t^2$$

Substitute $t = 0.510$ s to find the y-coordinate of the position vector at the peak of the trajectory:

$$y_{\text{max}} = 1.27 \text{ m}$$

$$y_{\text{max}} = -101 \text{ m}$$

Since the origin for y was at the point of release, located 100 m above the plain, the elevation of the projectile above the plain below is 100 m + 1.27 m = 101 m, since you can keep only three significant figures in the addition.

Since the origin for y was at the plain, the elevation of the projectile is the absolute value of this component of the position vector, or 101 m.

You get the same physical results regardless of the coordinate system used, but your description of the motion depends on the choices you make for the coordinate system.

EXAMPLE 4.3 ━━━━━━━━

You throw a baseball with an initial speed v_0 at an elevation angle θ across level terrain, as indicated in Figure 4.18. Your roommate catches the ball at the same height above the level ground that you released it. Find expressions for the time of flight t and for the horizontal range R_{horiz} in terms of v_0, θ, and g.

Solution

Use the Cartesian coordinate system given in Figure 4.18. The general equations, 4.10–4.13, for motion at constant acceleration in two dimensions are

$$v_x(t) = v_{x0}$$
$$x(t) = x_0 + v_{x0}t$$
$$v_y(t) = v_{y0} + a_y t$$
$$y(t) = y_0 + v_{y0}t + \frac{1}{2}a_y t^2$$

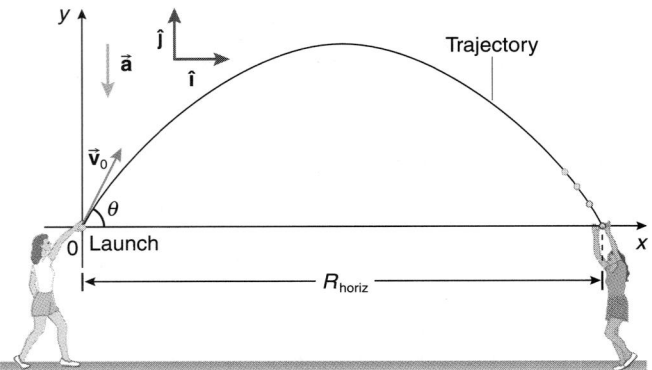

FIGURE 4.18

With the coordinate choice and with the acceleration (of magnitude $g = 9.81$ m/s^2) vertically downward, parallel to $-\hat{\jmath}$, you make the following identifications:

for the components of the initial position vector when $t = 0$ s:

$$x_0 = 0 \text{ m} \qquad y_0 = 0 \text{ m}$$

for the components of the initial velocity vector when $t = 0$ s:

$$v_{x0} = v_0 \cos\theta \qquad v_{y0} = v_0 \sin\theta$$

and for the constant acceleration component:

$$a_y = -g$$

Equations 4.10–4.13 describing the x- and y-motion then become

$$v_x(t) = v_0 \cos\theta$$
$$x(t) = (v_0 \cos\theta)t$$
$$v_y(t) = (v_0 \sin\theta) - gt$$
$$y(t) = (v_0 \sin\theta)t - \frac{1}{2}gt^2$$

At the instant just before the catch, the x-coordinate of the particle is R_{horiz} and its y-coordinate is 0 m. Hence at the instant t just before the catch the four preceding equations are

$$v_x(t) = v_0 \cos\theta \tag{1}$$
$$R_{\text{horiz}} = (v_0 \cos\theta)t \tag{2}$$

$$v_y(t) = (v_0 \sin\theta) - gt \tag{3}$$
$$0 \text{ m} = (v_0 \sin\theta)t - \frac{1}{2}gt^2 \tag{4}$$

Since $t = 0$ s corresponds to launch, t also is numerically equal to the time of flight in these equations.

Because you are interested in finding the time of flight t and the horizontal range R_{horiz}, you need to solve equations (2) and (4) simultaneously for R_{horiz} and t. To do this, first factor the equation (4) into the form

$$0 \text{ m} = t\left(v_0 \sin\theta - \frac{g}{2}t\right)$$

This equation has two roots that tell you when the y-coordinate of the projectile is zero:

$$t = 0 \text{ s}$$

which corresponds to launch, and

$$t = \frac{2v_0 \sin\theta}{g}$$

which corresponds to the catch when $x = R_{\text{horiz}}$. The latter root is the one desired.

The horizontal range is found by putting the time of flight into equation (2) for the horizontal range:

$$\begin{aligned}
R_{\text{horiz}} &= (v_0 \cos\theta)t \\
&= (v_0 \cos\theta)\frac{2v_0 \sin\theta}{g} \\
&= \frac{2v_0^2 \sin\theta \cos\theta}{g}
\end{aligned}$$

The double angle trigonometric identity

$$2\sin\theta \cos\theta = \sin 2\theta$$

enables you to put the expression for the horizontal range into the form

$$R_{\text{horiz}} = \frac{v_0^2 \sin 2\theta}{g} \tag{5}$$

Equation (5) is the expression for the range of a projectile *only when it is launched from and strikes on the same horizontal plane.* Many projectiles hit before or after the impact time for the horizontal range given by equation (5), as shown in Figure 4.19. In

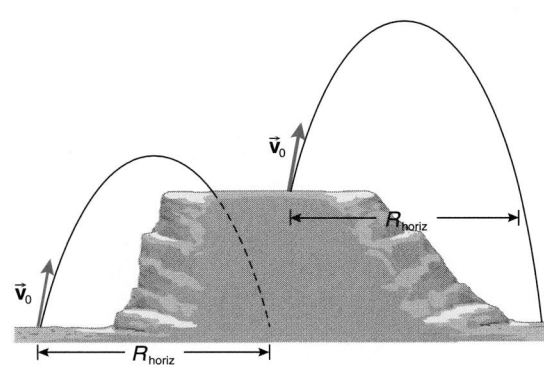

FIGURE 4.19 **Projectiles that strike before or after they would on a horizontal range.**

such situations, equation (5) for the horizontal range is useless and the horizontal distance covered by the projectile before impact must be determined from the specifics of the given situation.

4.3 MOTION IN THREE DIMENSIONS

> Stranger. *Exactly: you see you do not even know what Space is. You think it is of Two Dimensions only; but I have come to announce to you a Third . . . .*
>
> Edwin Abbott Abbott*

When navigating a difficult rapid with a kayak, you are moving in three dimensions. Likewise, when charged particles emitted by the Sun[†] encounter the magnetic field in the vicinity of the Earth, their motion resembles a curved spiral and they corkscrew into the atmosphere of the Earth, producing beautiful auroras from their collisions with molecules in the air (see Figure 4.20).

To extend the concepts of the position vector, velocity, and acceleration to motion in three dimensions, we introduce the third Cartesian coordinate axis, z. The position vector then is

$$\vec{r}(t) = x(t)\hat{\imath} + y(t)\hat{\jmath} + z(t)\hat{k} \qquad (4.16)$$

The definitions of the average speed,

$$\text{average speed} = \frac{d}{\Delta t}$$

the average velocity,

$$\vec{v}_{ave} = \frac{\Delta \vec{r}}{\Delta t}$$

and average acceleration,

$$\vec{a}_{ave} = \frac{\Delta \vec{v}}{\Delta t}$$

are the same as in one- or two-dimensional motion. Likewise, the velocity,

$$\vec{v}(t) = \frac{d\vec{r}(t)}{dt}$$

and acceleration,

$$\vec{a}(t) = \frac{d\vec{v}(t)}{dt}$$

also have the same form as in one or two dimensions, showing again the convenience of using vector notation. For three-dimensional motion, each vector has three possible Cartesian components.

4.4 RELATIVE VELOCITY ADDITION AND ACCELERATIONS

Kids find it fun to run up the down escalator or down the up escalator; the stairs are moving one way while they are moving the other way, each with different speeds. Likewise, if you ever have been in a raft, canoe, or kayak on a swiftly flowing stream, you know that your velocity with respect to the banks can differ greatly from your paddling velocity with respect to the water, sometimes with frightening consequences. Pilots in airplanes encounter similar effects; they continually must account for the speed and direction of the wind to determine their headings. Here we see how such effects are neatly accounted for using a vector description of motion.

Coordinate systems (such as the Cartesian coordinate system) and clocks are the "hangers" on which we drape a problem. A coordinate system with a set of synchronized clocks, all ticking at the same rate, is called a **reference frame**. Many choices are possible for a coordinate system. Thus, although the coordinate system is central to setting up the problem, the particular choice for the direction of, say, the Cartesian x-, y-, and z-axes is immaterial to the underlying physics, as we have seen. Different choices for the coordinate system lead to different *descriptions* of the motion, which may be more or less complex than for other coordinate choices, but the underlying *physics* is nonetheless the same.

Until now, the various coordinate system choices we made were all fixed or at rest with respect to each other. Here we relax that condition. Suppose you employ a fixed coordinate system with an origin at 0, perhaps on the ground, while your friends in a hot-air balloon, moving along with the wind, use a coordinate system with an origin at 0′ at the gondola, as in Figure 4.21. An airplane, represented as a particle, is flying at point P. We describe the position of the airplane by means of position vectors either with respect to 0 or with respect to 0′, as illustrated in Figure 4.21. Let $\vec{r}$ be the position vector of the airplane at P with respect to 0,

---

Flatland, Part II, Section 16, page 69.
[†]Such particle flows constitute what is called the *solar wind*.

FIGURE 4.20 An aurora.

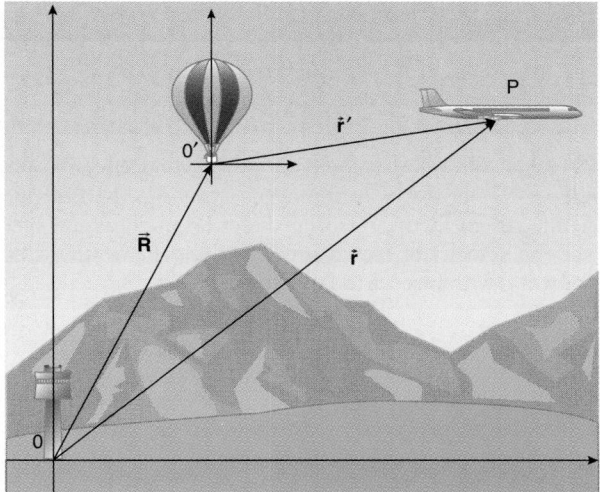

FIGURE 4.21 A plane located with respect to two different coordinate systems.

and let $\vec{\mathbf{r}}'$ be its position vector with respect to 0′. Vector $\vec{\mathbf{R}}$ locates the gondola origin 0′ with respect to the origin 0 on the ground.

From Figure 4.21 we can see that the three vectors are related by the vector sum

$$\vec{\mathbf{r}} = \vec{\mathbf{R}} + \vec{\mathbf{r}}' \qquad (4.17)$$

If *both* the airplane at point P and the gondola origin 0′ are moving, then all three position vectors are functions of time. In particular, we can take the time derivative of Equation 4.17 to obtain

$$\frac{d\vec{\mathbf{r}}}{dt} = \frac{d\vec{\mathbf{R}}}{dt} + \frac{d\vec{\mathbf{r}}'}{dt} \qquad (4.18)$$

The quantity

$$\frac{d\vec{\mathbf{r}}}{dt}$$

is the velocity of the particle at P *with respect to* 0. Call this velocity $\vec{\mathbf{v}}_{\mathrm{P0}}$, to be read as "the velocity of P with respect to 0." The quantity

$$\frac{d\vec{\mathbf{r}}'}{dt}$$

is the velocity of the particle *with respect to* 0′. Call this velocity $\vec{\mathbf{v}}_{\mathrm{P0'}}$, read as "the velocity of P with respect to 0′." And finally, the quantity

$$\frac{d\vec{\mathbf{R}}}{dt}$$

is the velocity of origin 0′ *with respect to* 0. Call this velocity $\vec{\mathbf{v}}_{0'0}$, read as "the velocity of 0′ with respect to 0." With these definitions, Equation 4.18 becomes

$$\vec{\mathbf{v}}_{\mathrm{P0}} = \vec{\mathbf{v}}_{0'0} + \vec{\mathbf{v}}_{\mathrm{P0'}} \qquad (4.19)$$

Because the order of the terms in a vector sum is immaterial, Equation 4.19 can be rearranged slightly into the form

$$\boxed{\vec{\mathbf{v}}_{\mathrm{P0}} = \vec{\mathbf{v}}_{\mathrm{P0'}} + \vec{\mathbf{v}}_{0'0}} \qquad (4.20)$$

Equation 4.20 is called the **relative velocity addition equation.**

There is an easy way to remember the relative velocity addition equation. On the right-hand side of the equation, the subscript in the middle (here 0′) is imagined to cancel out, leaving on the left-hand side a velocity of the first subscript (P) with respect to the last subscript (0):

$$\vec{\mathbf{v}}_{\mathrm{P0}} = \vec{\mathbf{v}}_{\mathrm{P0'}} + \vec{\mathbf{v}}_{0'0}$$

As you watch a car approach a bridge on which you are standing, the occupants of the car see you and the bridge approaching them in the opposite direction at the same speed. Hence the velocity of P with respect to 0 is the *negative* of the velocity of 0 with respect to P. That is

$$\vec{\mathbf{v}}_{\mathrm{P0}} = -\vec{\mathbf{v}}_{0\mathrm{P}}$$

So, if we interchange the two *subscripts*, a minus sign is introduced.

If a coordinate system 0′ is moving at *constant velocity* with respect to 0, then $\vec{\mathbf{v}}_{0'0}$ is a constant vector. When $\vec{\mathbf{v}}_{0'0}$ is a constant vector, its time derivative is zero:

$$\frac{d\vec{\mathbf{v}}_{0'0}}{dt} = 0 \ \mathrm{m/s}^2$$

Then if we differentiate Equation 4.20, we find that

$$\frac{d\vec{\mathbf{v}}_{\mathrm{P0}}}{dt} = \frac{d\vec{\mathbf{v}}_{\mathrm{P0'}}}{dt} + 0 \ \mathrm{m/s}^2$$

$$= \frac{d\vec{\mathbf{v}}_{\mathrm{P0'}}}{dt} \qquad (4.21)$$

The term on the left-hand side of Equation 4.21 is the acceleration $\vec{\mathbf{a}}_{\mathrm{P0}}$ of the particle at P with respect to the coordinate system with origin at 0. The right-hand side of Equation 4.21 is the acceleration $\vec{\mathbf{a}}_{\mathrm{P0'}}$ of the particle with respect to the coordinate system with origin at 0′.

The accelerations of a particle with respect to two coordinate systems that are moving at constant velocity with respect to each other are the same:

$$\vec{\mathbf{a}}_{\mathrm{P0}} = \vec{\mathbf{a}}_{\mathrm{P0'}} \qquad \text{(if } \vec{\mathbf{v}}_{0'0} \text{ is constant)} \qquad (4.22)$$

EXAMPLE 4.4

The water in a scenic river flows at a constant velocity $\vec{\mathbf{v}}_{\text{water ground}}$. A canoeist paddles at velocity $\vec{\mathbf{v}}_{\text{canoeist water}}$ with respect to the water. In what direction θ relative to the current should the canoeist paddle so that the resulting velocity of the canoe with respect to the ground is perpendicular to the river bank? The canoeist will then go straight across the river. See Figure 4.22. In canoeing parlance, the angle θ is called the angle of attack. Discuss the implications if the speed of the current, $v_{\text{water ground}}$, is greater than the paddling speed $v_{\text{canoeist water}}$ of the canoeist.

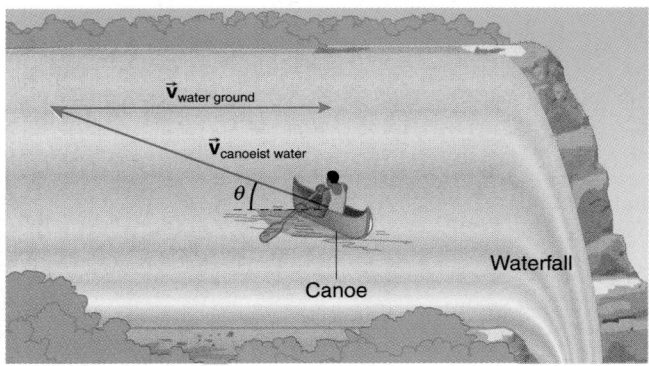

FIGURE 4.22

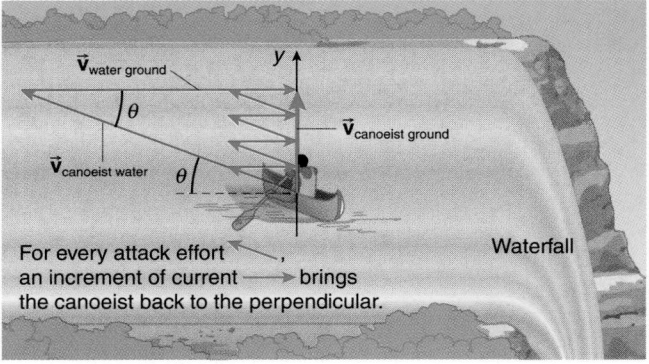

FIGURE 4.23

Whitewater ferrying

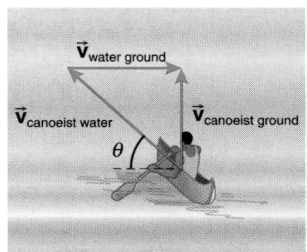 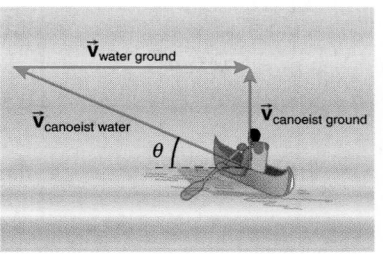 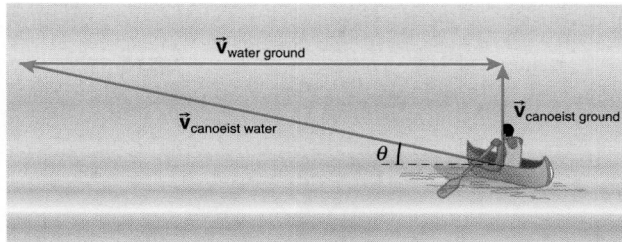

FIGURE 4.24 The faster the current, the smaller the angle of attack.

Solution
According to the relative velocity addition (Equation 4.20), the velocities are related by

$$\vec{v}_{\text{canoeist ground}} = \vec{v}_{\text{canoeist water}} + \vec{v}_{\text{water ground}}$$

To have the canoeist go straight across the river, the velocity vectors must form the right triangle shown in Figure 4.23. Use the right triangle in Figure 4.23 to find the cosine of the angle of attack in terms of the speeds:

$$\cos\theta = \frac{v_{\text{water ground}}}{v_{\text{canoeist water}}}$$

The faster the current, the smaller the angle θ of attack (see Figure 4.24). This is familiar to any whitewater canoeist (the maneuver is known as a ferry).

Notice that if $v_{\text{water ground}} > v_{\text{canoeist water}}$, then the preceding equation for the cosine indicates $\cos\theta > 1$, which is mathematically impossible for real angles, because a side of a right triangle cannot exceed the length of the hypotenuse. (Try to take the inverse cosine of a number greater than 1; your calculator will indicate an error.) Physically this means that if $v_{\text{water ground}} > v_{\text{canoeist water}}$, then no θ can be found. The canoeist cannot paddle fast enough for the boat to go straight across the river; in trying to cross it, the paddler then is inevitably swept downstream and perhaps over a waterfall.

STRATEGIC EXAMPLE 4.5
An airplane is in level flight at a speed of 600 km/h with respect to the surrounding air. A wind is blowing toward the northeast direction at a speed of 100 km/h, as shown in Figure 4.25.

a. What is the vector relationship among the velocity of the plane with respect to the ground, the velocity of the plane with respect to the air, and the wind velocity?
b. In what direction must the pilot aim the plane so that the velocity direction of the plane with respect to the ground is due north?

Solution
a. Let the velocity of the airplane (P) with respect to the surrounding air (O′) be

$$\vec{v}_{\text{PO}'} \equiv \vec{v}_{\text{plane air}}$$

Let the velocity of the air (O′) with respect to the ground (O) be written as

$$\vec{v}_{\text{O}'\text{O}} \equiv \vec{v}_{\text{air ground}}$$

Call the velocity of the plane (P) with respect to the ground (O)

$$\vec{v}_{\text{PO}} \equiv \vec{v}_{\text{plane ground}}$$

Then according to Equation 4.20, the various velocities are related by the vector sum

$$\vec{v}_{\text{plane ground}} = \vec{v}_{\text{plane air}} + \vec{v}_{\text{air ground}}$$

In accordance with the conceptual note following Equation 4.20, notice that on the right-hand side of this equation the subscript in the middle is imagined to cancel out, leaving the velocity of the first subscript with respect to the last subscript on the left-hand side of the equation.

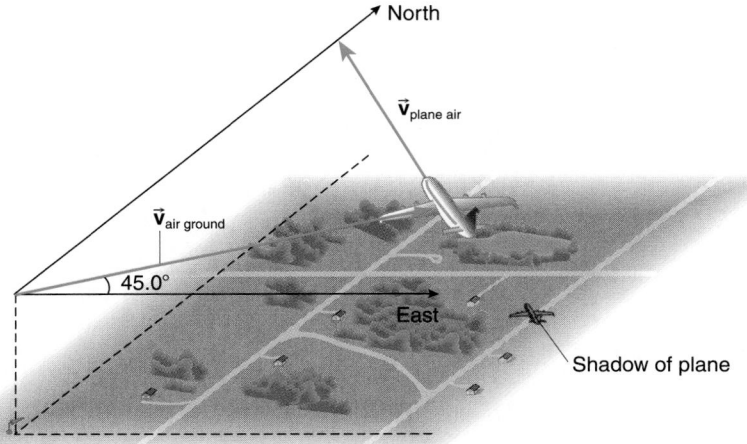

FIGURE 4.25

b. There are two ways to solve the problem.

Method 1

This method is geometric. The three vectors are related according to the sketch in Figure 4.26. Divide the triangle formed by the vectors in Figure 4.26 into two right triangles as shown. The common side of the two right triangles has a magnitude of

$$\frac{100}{\sqrt{2}} \text{ km/h}$$

or 70.7 km/h. You can use the Pythagorean theorem to find the remaining side of the upper right triangle in Figure 4.26, since the hypotenuse (600 km/h) and one side (70.7 km/h) are known; the remaining side is of magnitude 596 km/h. To this add the component of wind vector along the y-axis (70.7 km/h), so that the magnitude of the vector lying along the y-axis is

$$596 \text{ km/h} + 70.7 \text{ km/h} = 667 \text{ km/h}$$

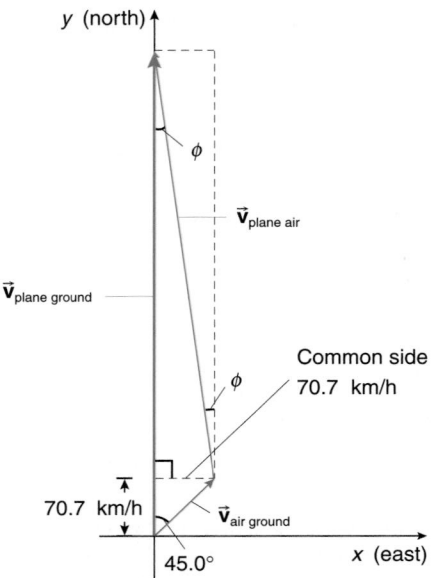

FIGURE 4.26

The angle that the velocity vector $\vec{\mathbf{v}}_{\text{plane air}}$ makes with the north direction then is found using trigonometry:

$$\sin \phi = \frac{70.7 \text{ km/h}}{600 \text{ km/h}}$$

$$\phi = 6.77°$$

The pilot should aim about 6.8° west of north.

Method 2

Establish the coordinate system as indicated in Figure 4.27 with the +x-axis directed east and the +y-axis directed north. Consider what you know about the three vectors individually. You know both the magnitude of the wind (100 km/h) and its direction (northeast). Thus you can write $\vec{\mathbf{v}}_{\text{air ground}}$ as its magnitude times a unit vector pointing in the northeast

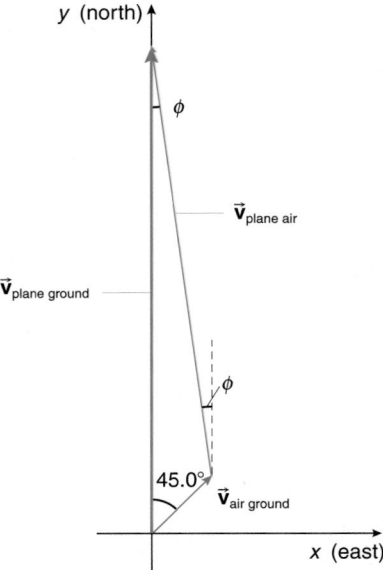

FIGURE 4.27

direction. The vector $\hat{\mathbf{i}} + \hat{\mathbf{j}}$ points in the northeast direction, but is not 1 unit long (its magnitude is $\sqrt{2}$). Thus the vector

$$\frac{\hat{\mathbf{i}} + \hat{\mathbf{j}}}{\sqrt{2}}$$

is a unit vector pointing in the northeast direction. Hence the wind is represented in Cartesian form as

$$\vec{\mathbf{v}}_{\text{air ground}} = (100 \text{ km/h}) \frac{\hat{\mathbf{i}} + \hat{\mathbf{j}}}{\sqrt{2}}$$
$$= (70.7 \text{ km/h}) \hat{\mathbf{i}} + (70.7 \text{ km/h}) \hat{\mathbf{j}}$$

Another way to find the vector representing the wind is to realize that the velocity of the wind makes an angle of 45° with both the x- and y-axes, because the wind is directed to the northeast. Since the wind speed is 100 km/h, the vector representing the wind velocity is

$$\vec{\mathbf{v}}_{\text{air ground}} = [(100 \text{ km/h}) \cos 45.0°]\hat{\mathbf{i}}$$
$$+ [(100 \text{ km/h}) \sin 45.0°]\hat{\mathbf{j}}$$
$$= (70.7 \text{ km/h})\hat{\mathbf{i}} + (70.7 \text{ km/h})\hat{\mathbf{j}}$$

You want the resulting velocity of the plane with respect to the ground to be in the northern direction (in the same direction as $+\hat{\mathbf{j}}$), but you do not know the magnitude of this vector. Call its magnitude v. Thus you write (see Figure 4.27)

$$\vec{\mathbf{v}}_{\text{plane ground}} = v\hat{\mathbf{j}}$$

You know the magnitude of the velocity of the plane with respect to the air (600 km/h), but not its direction. The vector relationship among the velocities is, from Equation 4.20,

$$v\hat{\mathbf{j}} = \vec{\mathbf{v}}_{\text{plane air}} + (70.7 \text{ km/h})\hat{\mathbf{i}} + (70.7 \text{ km/h})\hat{\mathbf{j}} \quad (1)$$

Rearrange equation (1) to isolate $\vec{\mathbf{v}}_{\text{plane air}}$:

$$\vec{\mathbf{v}}_{\text{plane air}} = (-70.7 \text{ km/h})\hat{\mathbf{i}} + (v - 70.7 \text{ km/h})\hat{\mathbf{j}} \quad (2)$$

The square of the magnitude of $\vec{v}_{\text{plane air}}$ is $(600 \text{ km/h})^2$. Use equation (2) to find the squared magnitude of $\vec{v}_{\text{plane air}}$:

$$(600 \text{ km/h})^2 = (-70.7 \text{ km/h})^2 + (v - 70.7 \text{ km/h})^2$$

which is quadratic in the unknown v. Use the quadratic formula and take the positive root to obtain a positive result for the speed v; you will find that*

$$v = 666 \text{ km/h}$$

This result is the speed of the plane with respect to the ground. You find $\vec{v}_{\text{plane air}}$ by using this value for v in equation (2) for $\vec{v}_{\text{plane air}}$:

$$\begin{aligned} \vec{v}_{\text{plane air}} &= (-70.7 \text{ km/h})\hat{\imath} + (666 \text{ km/h} - 70.7 \text{ km/h})\hat{\jmath} \\ &= (-70.7 \text{ km/h})\hat{\imath} + (595 \text{ km/h})\hat{\jmath} \end{aligned}$$

The angle ϕ that this vector makes with the northern direction (see Figure 4.27) is

$$\tan \phi = \frac{70.7 \text{ km/h}}{595 \text{ km/h}}$$

or

$$\phi = 6.78°$$

The pilot should aim in a direction about 6.8° west of north.

4.5 UNIFORM CIRCULAR MOTION: A FIRST LOOK

Round and round the circle

T. S. Eliot (1888–1965)[†]

Circular motion at constant speed is called **uniform circular motion**. Imagine yourself standing at rest on the outer rim of a constantly spinning merry-go-round. Your orbital path is a circle in the two-dimensional horizontal plane. In physics **spin** describes the motion of an extended object about an axis through itself.[‡] Thus we say the merry-go-round spins about an axis through its center (see Figure 4.28).

The motion of a particle about a point that is not coincident with the particle is called **orbital motion**. An example is you

---

*The discrepancy with Method 1 is due to rounding.
[†]*The Family Reunion*, Part II, Scene III, in T. S. Eliot, *The Complete Poems and Plays 1909–1950* (Harcourt, Brace and World, New York, 1971), page 293.
[‡]We define spin more precisely in Chapter 10.

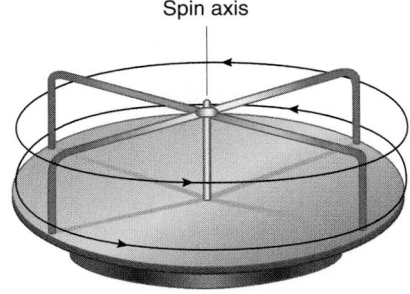

FIGURE 4.28 A merry-go-round spins about an axis through its center.

(modeled as a particle) traveling in a circular path about the center of the merry-go-round (see Figure 4.29).

Thus, in another context, we say the Earth *spins* on its axis each day, but *orbits* the Sun on a path that is approximately circular (see Figure 4.30).[§] In this example the spin axis and orbital axis are not the same.

We wish to quantify the rate of the spin of the merry-go-round or the rate at which a particle orbits the center of a circular trajectory. Every point of the merry-go-round turns through the same angle $\Delta\theta$ during an interval Δt. Hence the angular rate of the orbital motion of a particle at rest on the merry-go-round equals the angular rate of spin of the merry-go-round. The angular rate of spin or of orbital motion is called the **angular speed**, designated by ω (lower case Greek omega) and is specified in either of the following ways:

1. For convenience, we use the number of complete turns per unit of time. When a particle completes one circular trajectory, we say it has completed one **revolution** of its motion. Hence the angular speed may be specified in revolutions per minute (rev/min), which is not necessarily an integer. You may have a tachometer in your car that indicates rpm, the number of revolutions per minute that the crankshaft within the engine is spinning.

2. For most calculations in the sciences, it is preferable to use the number of radians per unit of time, for example, radians per second (rad/s), since the second is the SI unit of time and radians are the mathematically natural units for specifying angles. The angular speed in radians per second specifies how many radians of angle the particle would sweep out during one second if the particle moved at the same angular speed throughout the one second interval.

[§]We treat the orbital path of the Earth about the Sun approximately as a circle, though it is more accurately an ellipse. In Chapter 6 we analyze elliptical orbital motion.

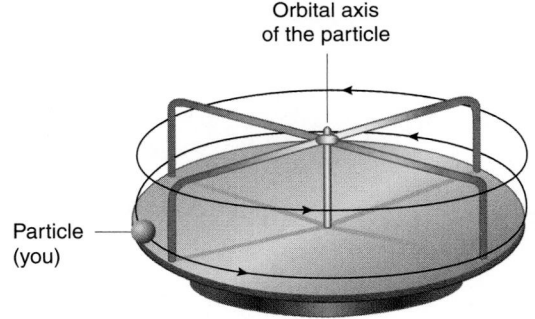

FIGURE 4.29 You orbit the same axis about which the merry-go-round spins.

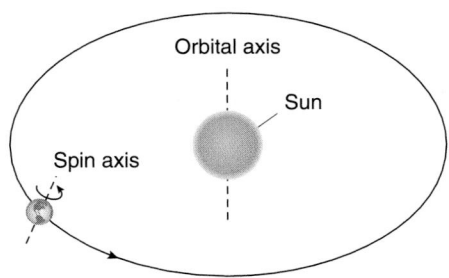

FIGURE 4.30 The Earth spins on its axis and orbits the Sun about a different axis.

Assume the angular speed ω of the merry-go-round is constant. Your velocity vector while riding around the circle is tangent to the circular path and is continually changing its direction, though its magnitude (the speed) is constant. Your speed in circular motion also is known as the **tangential speed** v, since your velocity vector always is tangent to the circular path. But a changing velocity implies a nonzero acceleration. What is the direction of your acceleration? To find out, use the plumb bob accelerometer introduced in Section 3.6. Recall that the deflection of the plumb bob and the acceleration are in *opposite* directions, and the greater the deflection angle, the greater the magnitude of the acceleration. Note that the plumb bob deflection is always away from the center of the merry-go-round (see Figure 4.31), and so the acceleration must be *toward the center* of the circular path.

Such an acceleration toward the center of a circle is called a **centripetal acceleration** $\vec{a}_c$.* We need to find how the magnitude of the centripetal acceleration depends on the tangential speed v and the distance r from the center of the circular path being followed by the particle.

Consider a particle moving in a circle of radius r at constant speed v. Let the particle move a short arc length $v\,\Delta t$ along the circle during a time interval Δt, as shown in Figure 4.32. Viewed from the center of the circle, the angle $\Delta\theta$ subtended by the arc length is, in radians,

$$\Delta\theta = \frac{v\,\Delta t}{r} \tag{4.23}$$

---

*The word centripetal comes from the Latin *centripetus* and means "center seeking."

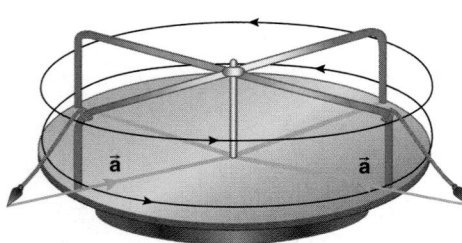

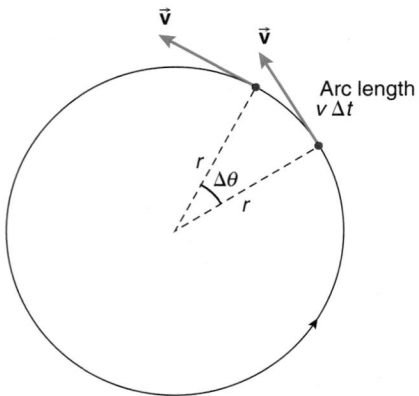

FIGURE 4.31 The plumb bob accelerometer indicates that the direction of the acceleration is toward the center of the circular motion.

FIGURE 4.32 A particle moves through a short arc length during a short interval Δt.

The angle through which the velocity vector turns during the interval Δt also is $\Delta\theta$, since the velocity always is tangent to the trajectory of the particle. Parallel transport the velocity vector at the beginning of the interval to the velocity vector at the end of the interval to form a velocity vector triangle, as in Figure 4.33.

The magnitude of the change in the velocity vector is the base of the isosceles triangle whose sides are v and whose apex angle is $\Delta\theta$. If the time interval is short, we can use the isosceles triangle formed by the velocity vectors to find another expression for $\Delta\theta$:

$$\Delta\theta \approx \frac{\Delta v}{v} \tag{4.24}$$

This approximation becomes better as the time interval Δt and the velocity change $\Delta\vec{v}$ become smaller. The magnitude of the average acceleration a_{ave} during the time interval is

$$a_{\text{ave}} = \frac{\Delta v}{\Delta t}$$

Using Equation 4.24, we get

$$a_{\text{ave}} \approx \frac{v\,\Delta\theta}{\Delta t}$$

Substituting for $\Delta\theta$ using Equation 4.23, we have

$$a_{\text{ave}} \approx \frac{v}{\Delta t}\frac{v\,\Delta t}{r}$$

$$\approx \frac{v^2}{r}$$

The direction of the average acceleration is the direction of $\Delta\vec{v}$, which (from Figures 4.32 and 4.33) is directed toward the center of the circle as Δt approaches zero. If the time interval is short, the average acceleration approaches the instantaneous acceleration and is equal to it in the limit as $\Delta t \to 0$ s.

The centripetal acceleration of the particle is of magnitude v^2/r and is directed toward the center of the circle:

$$a_c = \frac{v^2}{r} \tag{4.25}$$

The magnitude of the centripetal acceleration varies directly as the square of the speed and inversely with the radius.

QUESTION 5

You move to a larger radius on a merry-go-round and test for the magnitude of the centripetal acceleration with the plumb bob accelerometer. Based on such an experiment, does the magnitude of the centripetal acceleration increase, decrease, or remain the same?

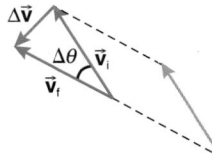

FIGURE 4.33 Parallel transport to form a velocity triangle.

EXAMPLE 4.6

Your car is executing a circular turn of radius 200 m at the constant speed 60.0 km/h. What is the magnitude of the centripetal acceleration of the car in appropriate SI units?

Solution

Convert the speed of the car to SI units:

$$60.0 \text{ km/h} = \frac{(60.0 \text{ km/h})(10^3 \text{ m/km})}{3600 \text{ s/h}}$$

$$= 16.7 \text{ m/s}$$

The magnitude of the centripetal acceleration is found from Equation 4.25:

$$a_c = \frac{v^2}{r}$$

$$= \frac{(16.7 \text{ m/s})^2}{200 \text{ m}}$$

$$= 1.39 \text{ m/s}^2$$

EXAMPLE 4.7

You are located at the outer rim of a merry-go-round of diameter 20.0 m and are orbiting at 6.00 rev/min.

a. What is your angular speed in rad/s?
b. What is your tangential speed in m/s?
c. What is the magnitude of your centripetal acceleration?

Solution

a. To find the angular speed ω, convert 6.00 rev/min to rad/s:

$$\omega = \frac{(6.00 \text{ rev/min})(2\pi \text{ rad/rev})}{60 \text{ s/min}}$$

$$= 0.628 \text{ rad/s}$$

b. The radius of the merry-go-round is 10.0 m. Its circumference is

$$2\pi r = 62.8 \text{ m}$$

The angular speed of 6.00 rev/min means you travel a distance of 6.00 times the circumference during one minute. Your speed v is

$$v = \frac{6.00 \ (62.8 \text{ m})}{60 \text{ s}}$$

$$= 6.28 \text{ m/s}$$

c. The magnitude of the centripetal acceleration is found from Equation 4.25:

$$a_c = \frac{v^2}{r}$$

$$= \frac{(6.28 \text{ m/s})^2}{10.0 \text{ m}}$$

$$= 3.94 \text{ m/s}^2$$

EXAMPLE 4.8

Imagine yourself standing at rest at a latitude angle β on the spinning Earth of radius R, as shown in Figure 4.34.

a. What is the radius of your circular orbital motion?

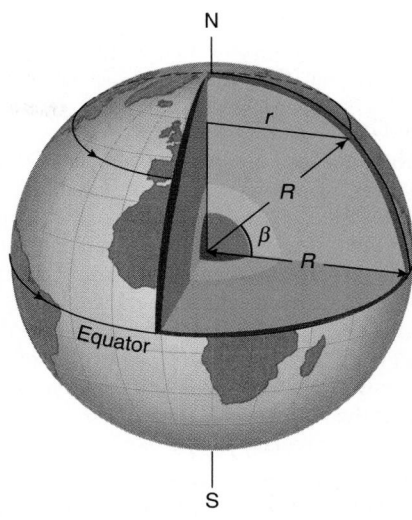

FIGURE 4.34

b. If the Earth rotates on its axis during a time T, what is your orbital speed in terms of R, β, and T?
c. What is your centripetal acceleration?
d. At what latitudes would your centripetal acceleration be zero?

Solution

a. From the geometry of Figure 4.34, the circular path along which you are moving has a radius $r = R \cos \beta$.
b. During the time interval T, you travel the circumference of a circle of radius $R \cos \beta$. Hence your speed v is

$$v = \frac{2\pi(R \cos \beta)}{T}$$

c. The centripetal acceleration is of magnitude

$$a_c = \frac{v^2}{r}$$

$$= \frac{4\pi^2 R \cos \beta}{T^2}$$

d. The centripetal acceleration has zero magnitude when $\cos \beta = 0$, and so $\beta = \pm 90°$. These latitudes correspond to the north and south geographic poles.

4.6 THE ANGULAR VELOCITY VECTOR

The angular speed of a particle in circular motion is the measure of the number of radians per second through which the particle moves. It is convenient to represent the angular speed ω as the magnitude of a vector called the **angular velocity $\vec{\omega}$** of the particle. The angular velocity vector will provide a convenient visual representation of the circular motion of a particle.

In Section 2.5 we saw that an arrow representing the *angle* of a rotation is *not* a vector, since the arrow did not obey the rules for vector arithmetic when rotations were taken about different axes in succession. However, if rotations all are about the same *fixed* axis in space, the angle of rotation *can* be modeled visually as a vector arrow as we attempted in Section 2.5, since then the arrows are confined to the single line in space representing the axis. It may come as a surprise to learn that the angular velocity that we define here *is* a vector quantity for more general rotations in space. Proving this assertion is not easy and is beyond the level and scope of this text.

Since the direction of motion of the particle continually is changing in the plane of its circular motion, no unique direction in this plane is a logical choice for the direction of the angular velocity vector. But at least the axis direction remains the same. Thus we choose a direction perpendicular to the plane of the circle, using a **circular motion right-hand rule**: wrap the four bunched fingers of your right hand around the circle in the sense that the particle is moving (see Figure 4.35). Then the extended thumb of your right hand indicates the appropriate perpendicular direction to take as the direction of the vector $\vec{\omega}$. We draw the angular velocity vector at the center of the circular path of the particle.

The direction of the angular velocity vector always is determined according to this circular motion right-hand rule. As we have discussed many times now, how you *describe* this direction in space depends on the choice you make for a coordinate system. We will *always* choose the z-axis to be the line containing $\vec{\omega}$.

PROBLEM-SOLVING TACTIC

4.6 Choose the z-axis to be the line containing $\vec{\omega}$. The magnitude of the angular velocity vector is the angular speed. Hence as a particle begins from rest and speeds up along a circular path (see Figure 4.36), the visual representation of the angular velocity vector $\vec{\omega}(t)$ gets longer, reflecting the increased angular speed of the particle, but the vector maintains the same direction in space.

4.7 THE GEOMETRY AND COORDINATES FOR DESCRIBING CIRCULAR MOTION

In Section 4.5 we learned that circular motion at constant speed is nonetheless accelerated motion. In this section we examine general circular motion, encompassing situations in which the particle is moving in a circle, but not necessarily at constant speed, such as your motion on a merry-go-round as it starts spinning from rest. We will see how to exploit vectors to describe circular motion of all types in an elegant, useful way.

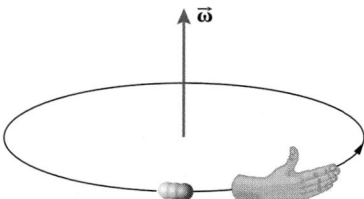

FIGURE 4.35 The circular motion right-hand rule for determining the direction of the angular velocity vector $\vec{\omega}$.

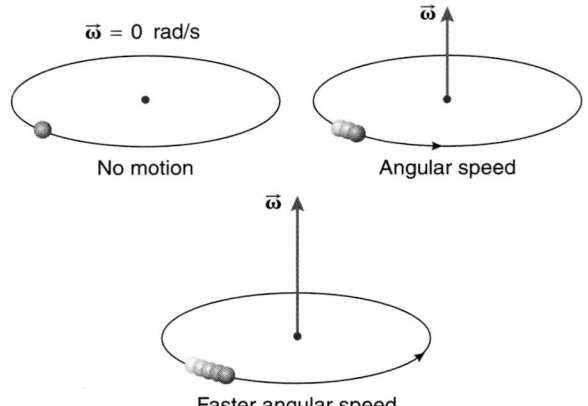

FIGURE 4.36 Increasing angular speed.

For a particle moving in a circle of fixed radius, the choice to make for the origin of an appropriate coordinate system is the center of the circle. You know that the choice of a coordinate system and its origin is arbitrary, at least in principle. However, some coordinate choices make the description of the problem far easier than others; this certainly is the case here in choosing the origin at the center of the circle. Any other choice makes the description of the circular motion of the particle a mathematical quagmire. So we relinquish freedom here, remembering that, all else being equal, simplicity is preferred over complexity.

It is customary to let the plane of the circular motion be the x–y plane of a Cartesian coordinate system; the z-axis of the coordinate system then is *perpendicular* to the plane of the motion and along the line containing the angular velocity vector $\vec{\omega}$. The angular velocity vector $\vec{\omega}(t)$ thus has a single component $\omega_z(t)$:

$$\vec{\omega}(t) = \omega_z(t)\hat{k} \qquad (4.26)$$

PROBLEM-SOLVING TACTIC

4.7 The direction you choose for $\hat{k}$ determines the sign of ω_z. See Figure 4.37.

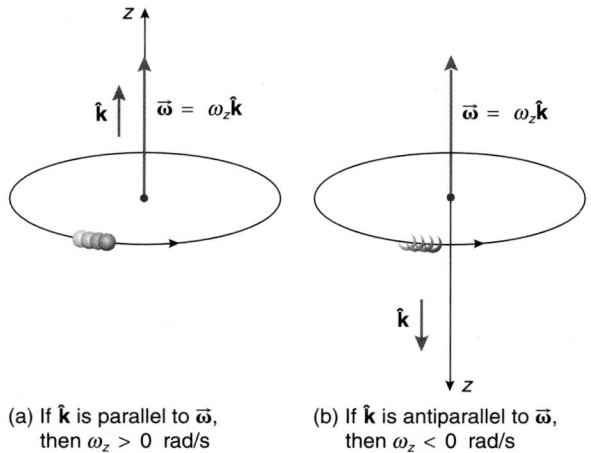

(a) If $\hat{k}$ is parallel to $\vec{\omega}$, then $\omega_z > 0$ rad/s

(b) If $\hat{k}$ is antiparallel to $\vec{\omega}$, then $\omega_z < 0$ rad/s

FIGURE 4.37 The relationship between $\vec{\omega}$ and the choice you make for $\hat{k}$.

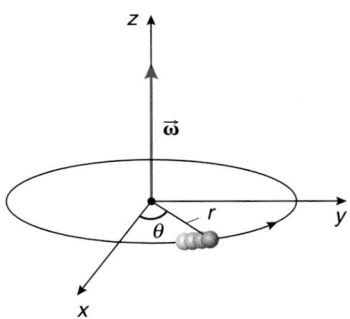

FIGURE 4.38 The angle θ is measured with respect to the *x*-axis.

The motion of a particle around a circle of fixed radius r is described by specifying an angular coordinate θ with respect to a fixed reference direction, taken to be the *x*-axis of a Cartesian coordinate system whose origin is at the center of the circle (see Figure 4.38). In this way, r and θ are the polar coordinates of the particle. Since the radius r of the fixed circle is constant, there is only one coordinate that changes with time, $\theta(t)$. Circular motion is *one-variable motion* since the angle θ is the only coordinate, other than the fixed radius of the circle, needed to locate the position of the particle. We will systematically exploit the constancy of r and the one-variable nature of circular motion for the kinematic description of circular motion.

4.8 THE POSITION VECTOR FOR CIRCULAR MOTION

With the x, y, and z coordinate conventions of the previous section, the position vector of the particle is confined to the x–y plane and is a two-dimensional vector, as shown in Figure 4.39:

$$\vec{r}(t) = x(t)\hat{\imath} + y(t)\hat{\jmath} \qquad (4.27)$$

The fact that the path is circular is not clear from $x(t)$ and $y(t)$. However, using the polar coordinates r and $\theta(t)$ we see that the Cartesian components are

$$x(t) = r \cos \theta(t)$$

and

$$y(t) = r \sin \theta(t)$$

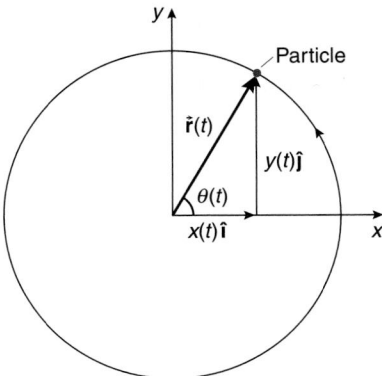

FIGURE 4.39 The position vector is confined to the *x*–*y* plane.

Hence the position vector is

$$\vec{r}(t) = [r \cos \theta(t)]\hat{\imath} + [r \sin \theta(t)]\hat{\jmath} \qquad (4.28)$$

Using Equation 4.28 to find the magnitude of $\vec{r}(t)$, we get

$$|\vec{r}(t)| = \{[r \cos \theta(t)]^2 + [r \sin \theta(t)]^2\}^{1/2}$$

$$= \{r^2[\cos^2 \theta(t) + \sin^2 \theta(t)]\}^{1/2}$$

$$= r$$

(The last step follows from the trigonometric identity $\cos^2 \theta + \sin^2 \theta = 1$.) Hence the magnitude of the position vector $\vec{r}(t)$ is the fixed radius r of the circle and does *not* depend on t.

> For circular motion the *magnitude* r of the position vector $\vec{r}(t)$ does *not* change with time. *Only* the angular coordinate $\theta(t)$ is a function of time.

4.9 THE VELOCITY AND ANGULAR VELOCITY IN CIRCULAR MOTION

We know that the velocity of a particle is the time rate of change (the time derivative) of its position vector. With Equation 4.28 for the position vector, the velocity of the particle is

$$\vec{v}(t) = \frac{d\vec{r}(t)}{dt}$$

$$= \frac{d}{dt}\left\{[r \cos \theta(t)]\,\hat{\imath} + [r \sin \theta(t)]\,\hat{\jmath}\right\}$$

Remember that the polar coordinate r is constant for circular motion. Hence the velocity is

$$\vec{v}(t) = r \frac{d}{dt}[\cos \theta(t)]\,\hat{\imath} + r \frac{d}{dt}[\sin \theta(t)]\,\hat{\jmath}$$

$$= r[-\sin \theta(t)]\frac{d\theta(t)}{dt}\,\hat{\imath} + r[\cos \theta(t)]\frac{d\theta(t)}{dt}\,\hat{\jmath} \quad (4.29)$$

The velocity vector is tangent to the circular trajectory and, because the particle is moving in a circle, $\vec{v}(t)$ is perpendicular to the position vector $\vec{r}(t)$ at all times. We can verify this by taking the *scalar product* of $\vec{v}(t)$ with $\vec{r}(t)$ using Equations 4.28 and 4.29; after some algebra, we see that the scalar product $\vec{r}(t) \cdot \vec{v}(t)$ indeed is zero, independent of t, which means that the two vectors are perpendicular at all times.

The speed of the particle is the magnitude of the velocity vector. The square of the magnitude of the velocity vector is the sum of the squares of its Cartesian components:

$$v^2 = \vec{v} \cdot \vec{v} = v_x^2 + v_y^2$$

Using Equation 4.29 for $\vec{v}(t)$, we have

$$[v(t)]^2 = \left[-r \sin \theta(t) \frac{d\theta(t)}{dt}\right]^2 + \left[r \cos \theta(t) \frac{d\theta(t)}{dt}\right]^2$$

$$= r^2\left[\frac{d\theta(t)}{dt}\right]^2 [\sin^2 \theta(t) + \cos^2 \theta(t)]$$

$$= r^2\left[\frac{d\theta(t)}{dt}\right]^2$$

The speed $v(t)$ of the particle therefore is

$$v(t) = r \left| \frac{d\theta(t)}{dt} \right| \tag{4.30}$$

The quantity

$$\left| \frac{d\theta(t)}{dt} \right|$$

is the absolute value of the time rate of change of the angular coordinate θ of the particle; this is the angular speed $\omega(t)$, the magnitude of the angular velocity vector. Hence Equation 4.30 can be written

$$v(t) = r\omega(t) \tag{4.31}$$

Take the magnitude of Equation 4.26:

$$\omega(t) = |\omega_z(t)|$$

Compare this with Equations 4.30 and 4.31; we make the identification

$$\omega_z(t) \equiv \frac{d\theta(t)}{dt} \tag{4.32}$$

with $\omega_z(t)$ specified in rad/s.

In Figure 4.40, note that the velocity vector $\vec{v}(t)$ is, at any instant t, perpendicular to *both* the angular velocity $\vec{\omega}(t)$ and the position vector $\vec{r}(t)$. This is reminiscent of the vector product of two vectors. Consider the vector product

$$\vec{\omega}(t) \times \vec{r}(t)$$

Since the angle between $\vec{\omega}(t)$ and $\vec{r}(t)$ always is 90°, the magnitude of their vector product is

$$\begin{aligned} |\vec{\omega}(t) \times \vec{r}(t)| &= \omega(t)r \sin 90° \\ &= \omega(t)r \\ &= |\omega_z(t)|r \end{aligned} \tag{4.33}$$

Remember that r, the magnitude of $\vec{r}(t)$, is *constant* and has no time dependence for circular motion. Compare Equation 4.33 with Equation 4.31; we conclude that the magnitude of the vector product $\vec{\omega}(t) \times \vec{r}(t)$ is just the speed $v(t)$. Use the vector product right-hand rule and notice that the vector product $\vec{\omega}(t) \times \vec{r}(t)$ is in the *same* direction as $\vec{v}(t)$ at any instant. Since $\vec{\omega}(t) \times \vec{r}(t)$ has the same magnitude as $\vec{v}(t)$ and points in the same direction as $\vec{v}(t)$ at any instant, $\vec{\omega}(t) \times \vec{r}(t)$ must be the same as the vector $\vec{v}(t)$.

For circular motion, the three vectors $\vec{v}(t)$, $\vec{\omega}(t)$, and $\vec{r}(t)$ *always* are related to each other via the vector product

$$\boxed{\vec{v}(t) = \vec{\omega}(t) \times \vec{r}(t)} \tag{4.34}$$

Since the three vectors are mutually perpendicular, taking magnitudes in Equation 4.34 gives

$$\boxed{v(t) = r\omega(t)} \tag{4.35}$$

Equations 4.34 and 4.35 are appropriate for

1. circular motion at constant speed v or, equivalently, constant angular velocity $\vec{\omega}$—that is, uniform circular motion; and

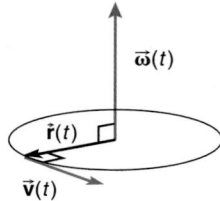

FIGURE 4.40 The velocity always is perpendicular to both the angular velocity and the position vector.

2. circular motion with a changing speed $v(t)$ and changing angular velocity vector $\vec{\omega}(t)$, called **nonuniform circular motion**.

4.10 UNIFORM CIRCULAR MOTION REVISITED

Whether it be a centrifuge whipping around at 50 000 rev/min, the rotation of the Earth once every 24 h, or the planet Venus orbiting the Sun* every 224.7 d, the circular motion of a particle at a constant speed has manifestations everywhere. Although the particle is moving along the circle at a constant speed v, the velocity vector $\vec{v}(t)$ is *not* constant, because the direction of motion of the particle is changing with time.

Since the speed v is constant and the radius of the circle r is fixed, Equation 4.30 implies that

$$\frac{d\theta}{dt}$$

is constant. Since

$$\omega_z = \frac{d\theta}{dt}$$

the z-component (the only component) of the angular velocity vector is constant.

For uniform circular motion, the angular velocity vector is a constant vector.

We find the acceleration by taking the time derivative of the velocity vector:

$$\vec{a}(t) = \frac{d\vec{v}(t)}{dt}$$

Using Equation 4.29 for $\vec{v}(t)$, we get

$$\vec{a}(t) = \frac{d}{dt} \left\{ \left[-r \frac{d\theta}{dt} \sin \theta(t) \right] \hat{\imath} + \left[r \frac{d\theta}{dt} \cos \theta(t) \right] \hat{\jmath} \right\}$$

Since the particle is moving in a circle, the radius r is constant. As we know from Section 4.5, if the particle is undergoing *uniform* circular motion,

$$\frac{d\theta}{dt} = \omega_z$$

---

*The shape of the orbit of a planet about the Sun actually is an ellipse, as we will see in Chapter 6. As a first approximation, the planetary orbits typically are considered to be circular, since their eccentricities are small (except for Pluto, Mercury, and Mars); the orbit of Venus is the most nearly circular of all the planets.

also is a constant. It is only $\theta(t)$ itself that varies with time. With both r and

$$\frac{d\theta}{dt}$$

constants, the differentiation affects only the trigonometric functions:

$$\vec{a}(t) = -\left\{r\frac{d\theta}{dt}[\cos\theta(t)]\frac{d\theta}{dt}\right\}\hat{\imath} + \left\{r\frac{d\theta}{dt}[-\sin\theta(t)]\frac{d\theta}{dt}\right\}\hat{\jmath}$$

$$= -\left(\frac{d\theta}{dt}\right)^2\left\{\left[r\cos\theta(t)\right]\hat{\imath} + \left[r\sin\theta(t)\right]\hat{\jmath}\right\}$$

Notice that the term in curly braces is just the position vector $\vec{r}(t)$ of the particle (Equation 4.28). Hence

$$\vec{a}(t) = -\left(\frac{d\theta}{dt}\right)^2\vec{r}(t)$$

Since

$$\frac{d\theta}{dt} = \omega_z$$

the constant angular velocity component, the expression for the acceleration becomes

$$\vec{a}(t) = -\omega_z^2\vec{r}(t) \qquad (4.36)$$

> The minus sign in Equation 4.36 indicates that the acceleration is antiparallel to the position vector $\vec{r}(t)$, just as our plumb bob accelerometer indicated. That is, $\vec{a}(t)$ is directed toward the center of the circle. This is the centripetal acceleration.

> The *magnitude* of the centripetal acceleration is constant and is written in terms of the (constant) angular velocity component ω_z as
>
> $$\boxed{a_c = \omega_z^2 r} \qquad (4.37)$$
>
> or, in terms of the speed v (by substituting for ω_z using $v = |\omega_z|r$ from Equation 4.31),
>
> $$a_c = \left(\frac{v}{r}\right)^2 r$$
>
> $$= \frac{v^2}{r} \qquad (4.38)$$

For a given circle size, the greater the value of the speed v of the particle, the greater the magnitude of the centripetal acceleration. For a given speed v, the centripetal acceleration is larger in magnitude for smaller circles (smaller r).

An equivalent expression for the centripetal acceleration can be found using Equation 4.34 and differentiating the vector product:

$$\vec{a}(t) = \frac{d\vec{v}(t)}{dt}$$

$$= \frac{d}{dt}[\vec{\omega} \times \vec{r}(t)]$$

Remember that the angular velocity $\vec{\omega}$ is a constant vector for *uniform* circular motion. Thus

$$\vec{a}(t) = \vec{\omega} \times \frac{d\vec{r}(t)}{dt}$$

$$= \vec{\omega} \times \vec{v}(t)$$

Use the vector product right-hand rule and the geometry of Figure 4.40 to note that the direction of $\vec{\omega} \times \vec{v}(t)$ is *opposite* to that of $\vec{r}(t)$. Hence the acceleration is directed toward the center of the circle—the centripetal acceleration:

$$\vec{a}_c(t) = \vec{\omega} \times \vec{v}(t) \qquad (4.39)$$

Because we will be using these results for uniform circular motion on many occasions, it is best to keep them fresh.

> Summarizing, the following are equivalent expressions for the centripetal acceleration in circular motion:
>
> $$\vec{a}_c(t) = \vec{\omega} \times \vec{v}(t) = -\omega_z^2\vec{r}(t) \qquad (4.39, 4.36)$$
>
> and for its constant magnitude:
>
> $$a_c = \frac{v^2}{r} = \omega_z^2 r \qquad (4.38, 4.37)$$

PROBLEM-SOLVING TACTIC

4.8 The seemingly different expressions for the centripetal acceleration are all variations of the same equation.

QUESTION 6

Explain carefully why the centripetal acceleration for uniform circular motion has a constant magnitude but is not a constant vector.

STRATEGIC EXAMPLE 4.9

A particle is moving at the constant speed of 15.0 m/s in a circle of radius 4.00 m.

a. Sketch the situation and indicate the direction of the angular velocity vector.
b. Find the angular speed.
c. Find the angular velocity vector $\vec{\omega}$ and its z-component ω_z.
d. Find the magnitude of the centripetal acceleration.

Solution

a. The circular motion right-hand rule means that the direction of the angular velocity vector is as shown in Figure 4.41. (Note that this is one of two possible sketches.)

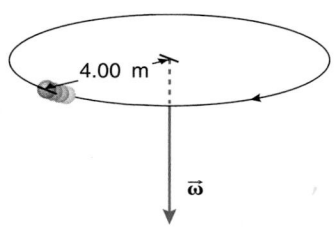

FIGURE 4.41

b. There are two ways to find the angular speed:

Method 1	Method 2
According to Equations 4.30 and 4.32, $$v = r|\omega_z|$$ Since $v = 15.0$ m/s and $r = 4.00$ m, we find that $$|\omega_z| = \frac{15.0 \text{ m/s}}{4.00 \text{ m}}$$ $$= 3.75 \text{ rad/s}$$ since ω_z is expressed in rad/s.	The angular speed indicates how many radians the particle sweeps out during one second at constant speed. Since the speed of the particle is constant and equal to 15.0 m/s, the particle traverses 15.0 m of circumference during one second. So the angular speed in radians per second is $$\omega = \frac{\text{arc length in one second}}{\text{radius}}$$ $$= \frac{15.0 \text{ m/s}}{4.00 \text{ m}}$$ $$= 3.75 \text{ rad/s}$$

c. To write the angular velocity vector, choose the direction of the z-axis to be along the line of the angular velocity (see Figure 4.42).

Choice 1	Choice 2

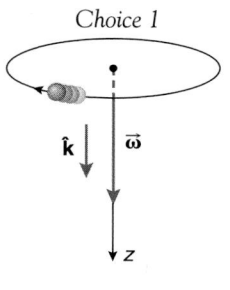

Take $\hat{\mathbf{k}}$ parallel to $\vec{\omega}$ so the z-component of the angular velocity is *positive*.

Take $\hat{\mathbf{k}}$ antiparallel to $\vec{\omega}$ so the z-component of the angular velocity is *negative*.

FIGURE 4.42

$\omega_z = 3.75$ rad/s $\vec{\omega} = (3.75 \text{ rad/s})\hat{\mathbf{k}}$	$\omega_z = -3.75$ rad/s $\vec{\omega} = (-3.75 \text{ rad/s})\hat{\mathbf{k}}$

d. The magnitude of the centripetal acceleration can be found in two ways:

Method 1	Method 2
$$a_c = \frac{v^2}{r}$$ $$= \frac{(15.0 \text{ m/s})^2}{4.00 \text{ m}}$$ $$= 56.3 \text{ m/s}^2$$	$$a_c = \omega_z^2 r$$ $$= (\pm 3.75 \text{ rad/s})^2(4.00 \text{ m})$$ The $\pm$ sign choice depends on which direction you chose for $\hat{\mathbf{k}}$ in part (c). For either choice: $$a_c = 56.3 \text{ m/s}^2$$

4.11 NONUNIFORM CIRCULAR MOTION AND THE ANGULAR ACCELERATION

It takes some time for a merry-go-round to get up to speed or to come to a stop. Likewise, when a biologist turns on a centrifuge, it may take several minutes for it to get up to speed or to coast to a stop. Similar things happen when you start a washing machine and dryer or monitor the engine rpm on your car as you speed up or slow down in traffic. There are many examples of nonuniform circular motion in the world around us.

If the speed of a particle moving in a fixed circle is not constant, then the velocity vector of the particle is changing in both direction *and* magnitude. For motion confined to a fixed plane in a circle of fixed radius, the *angular velocity vector* changes its magnitude but not its direction.

Here we can exploit our knowledge of vectors to great advantage. The acceleration of the particle is the time rate of change of the velocity vector:

$$\vec{a}(t) = \frac{d\vec{v}(t)}{dt}$$

But from Equation 4.34,

$$\vec{v}(t) = \vec{\omega}(t) \times \vec{r}(t)$$

and so the acceleration is

$$\vec{a}(t) = \frac{d}{dt}[\vec{\omega}(t) \times \vec{r}(t)]$$

Both the position vector and the angular velocity vector change with time in nonuniform circular motion. Thus there are two terms for the time derivative of the vector product. Remember to keep the terms in the vector product in the appropriate order when differentiating:

$$\vec{a}(t) = \frac{d\vec{\omega}(t)}{dt} \times \vec{r}(t) + \vec{\omega}(t) \times \frac{d\vec{r}(t)}{dt} \quad (4.40)$$

Since

$$\frac{d\vec{r}(t)}{dt} = \vec{v}(t)$$

the second term on the right-hand side of the acceleration equation is

$$\vec{a}_c = \vec{\omega}(t) \times \vec{v}(t) \quad (4.41)$$

which is the *centripetal acceleration* we encountered in the last section. Since the speed of the particle in the circle is changing with time, the magnitude of the centripetal acceleration is not constant, unlike the case for uniform circular motion. Since $\vec{\omega}(t)$ and $\vec{v}(t)$ are perpendicular, the magnitude of their vector product is

$$a_c(t) = \omega(t)v(t) \quad (4.42)$$

In view of Equation 4.35, this becomes

$$a_c = \frac{v(t)^2}{r} = |\omega_z(t)|^2 r \quad (4.43)$$

The centripetal acceleration has the same form as for uniform circular motion, but here its magnitude changes with time.

The first term on the right-hand side of Equation 4.40 is new. Let's see what is involved with this term. The time rate of change of the angular velocity vector is defined to be the **angular acceleration** vector, designated $\vec{\alpha}(t)$. That is,

$$\vec{\alpha}(t) \equiv \frac{d\vec{\omega}(t)}{dt} \quad (4.44)$$

Since $\vec{\omega}(t)$ is measured in rad/s, the angular acceleration is measured in rad/s².

The particle either is speeding up in its circular motion (thus increasing the length of the angular velocity arrow) or slowing down (decreasing the length of the angular velocity vector). The line along which the angular velocity vector lies does not change its orientation, since the motion is in a fixed plane. The angular acceleration vector $\vec{\alpha}(t)$ points in the direction of the *change* in the angular velocity vector $\vec{\omega}(t)$. Thus the angular acceleration vector $\vec{\alpha}(t)$ is directed either *parallel* to $\vec{\omega}(t)$ when the particle is speeding up along the circle, or *antiparallel* to $\vec{\omega}(t)$ when the particle is slowing down along the circle (see Figure 4.43).

Since the angular velocity vector is either parallel or antiparallel to $\hat{k}$ (according to the standard coordinate choice), the angular acceleration vector likewise has only a z-component:

$$\vec{\alpha}(t) = \alpha_z(t)\hat{k}$$

The acceleration term $\vec{\alpha}(t) \times \vec{r}(t)$ in Equation 4.40 has a direction determined using the vector product right-hand rule. Since $\vec{\alpha}(t)$ is perpendicular to the plane of the circle and $\vec{r}(t)$ is in the plane, the direction of the vector product $\vec{\alpha}(t) \times \vec{r}(t)$ always is *tangent* to the circular path and parallel to $\vec{v}(t)$ if the particle is speeding up (see Figure 4.44), or antiparallel to $\vec{v}(t)$ if the particle is slowing down in its circular motion (Figure 4.45).

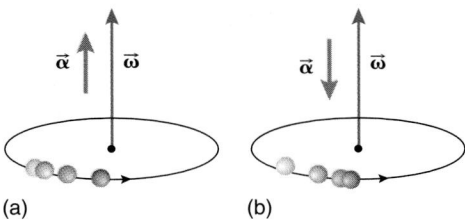

(a)   (b)

FIGURE 4.43 (a) Particle speeding up along the circle means $\vec{\alpha}$ is parallel to $\vec{\omega}$. (b) Particle slowing down along the circle means $\vec{\alpha}$ is antiparallel to $\vec{\omega}$.

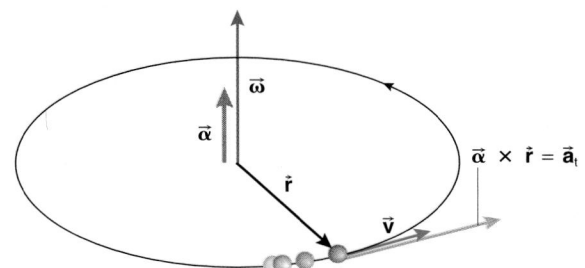

FIGURE 4.44 For a particle speeding up along the circle, $\vec{\alpha}(t) \times \vec{r}(t)$ is parallel to $\vec{v}(t)$.

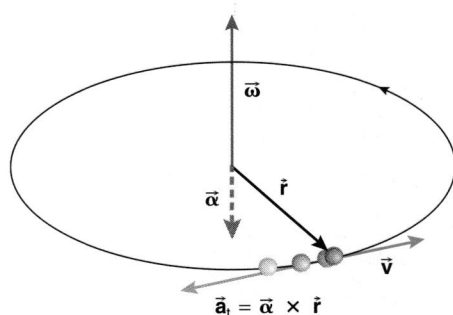

FIGURE 4.45 For a particle slowing down along the circle, $\vec{\alpha}(t) \times \vec{r}(t)$ is antiparallel to $\vec{v}(t)$.

Since $\vec{\alpha}(t) \times \vec{r}(t)$ is tangent to the circle, it is known as the **tangential acceleration** $\vec{a}_t$:

$$\vec{a}_t(t) \equiv \vec{\alpha}(t) \times \vec{r}(t) \qquad (4.45)$$

The tangential acceleration always is perpendicular to the position vector $\vec{r}(t)$, and so the magnitude of the vector product of the two vectors on the right-hand side of Equation 4.45 can be evaluated easily. Thus the magnitude of the tangential acceleration is

$$\begin{aligned} a_t(t) &= \alpha(t)r \sin 90° \\ &= \alpha(t)r \\ &= |\alpha_z(t)|r \qquad (4.46) \end{aligned}$$

Remember that the magnitude of the position vector is constant, since the particle is moving in a circle of fixed size.

Equation 4.40 for the acceleration is rewritten as

$$\vec{a}(t) = \vec{\alpha}(t) \times \vec{r}(t) + \vec{\omega}(t) \times \vec{v}(t) \qquad (4.47)$$

> The total acceleration $\vec{a}(t)$ is the vector sum of the two, mutually perpendicular, tangential and centripetal accelerations:
>
> $$\vec{a}(t) = \vec{a}_t(t) + \vec{a}_c(t) \qquad (4.48)$$

Since both the tangential and centripetal accelerations are in the plane of the circular motion, the total acceleration also is in this plane.

The direction of the total acceleration always is toward the inside of the circle, since the centripetal acceleration *always* is toward the center of the circle (see Figures 4.46 and 4.47).

QUESTION 7

If a particle travels around a circle, can the total acceleration ever be zero at any point along the trajectory? Explain your answer.

EXAMPLE 4.10

A particle moving in a circle of radius 2.00 m has a total acceleration of magnitude $a = 20.0$ m/s² at some instant directed as shown in Figure 4.48.

a. Find the magnitude of the centripetal acceleration at this instant.
b. Find the speed of the particle at this instant.
c. Find the magnitude of the angular velocity of the particle at this instant.

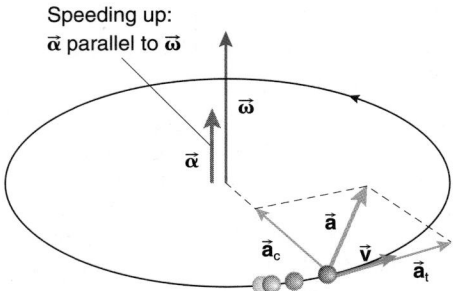

FIGURE 4.46 The bevy of vectors for a particle speeding up in circular motion.

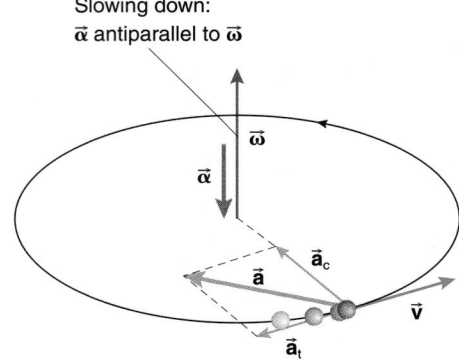

FIGURE 4.47 The bevy of vectors for a particle slowing down in circular motion.

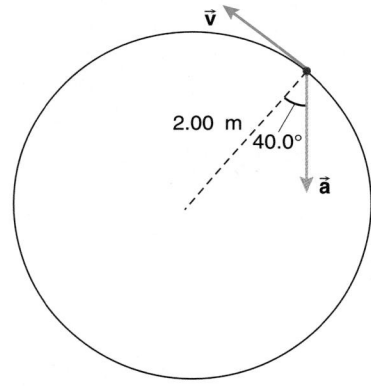

FIGURE 4.48

d. Find the magnitude of the tangential acceleration of the particle at this instant.
e. Find the magnitude of the angular acceleration of the particle at this instant.
f. Sketch the directions of the angular velocity vector and the angular acceleration vector at this instant.

Solution

a. The total acceleration is the vector sum of the mutually perpendicular centripetal and tangential accelerations, as indicated in Figure 4.49. The magnitude of the centripetal

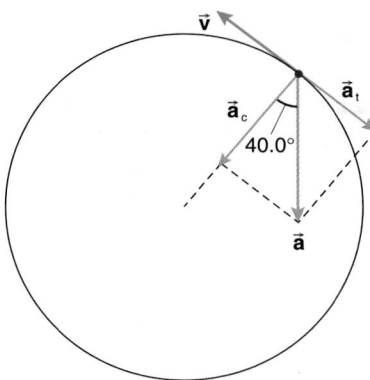

FIGURE 4.49

acceleration is the absolute value of the component of the total acceleration in the radial direction. Hence

$$a_c = a \cos 40.0°$$
$$= (20.0 \text{ m/s}^2) \cos 40.0°$$
$$= 15.3 \text{ m/s}^2$$

b. The magnitude of the centripetal acceleration is given by Equation 4.43:

$$a_c = \frac{v^2}{r}$$

so

$$v^2 = ra_c$$

Substitute for r and a_c [found in part (a)]:

$$v^2 = (2.00 \text{ m})(15.3 \text{ m/s}^2)$$
$$v = 5.53 \text{ m/s}$$

c. There are two ways to find the angular speed:

Method 1

The magnitude of the centripetal acceleration is, from Equation 4.37 (realizing that $|\omega_z| = \omega$),

$$a_c = \omega^2 r$$

Substitute for a_c and r:

$$15.3 \text{ m/s}^2 = \omega^2(2.00 \text{ m})$$

Solving for ω, you find

$$\omega = 2.77 \text{ rad/s}$$

Method 2

Use Equation 4.35 realizing that $|\omega_z| = \omega$:

$$v = \omega r$$

Substitute for v and r:

$$5.53 \text{ m/s} = \omega(2.00 \text{ m})$$

Solving for ω, you find

$$\omega = 2.77 \text{ rad/s}$$

d. From Figure 4.49, the tangential acceleration component of the total acceleration is of magnitude

$$a_t = a \sin 40.0°$$
$$= (20.0 \text{ m/s}^2) \sin 40.0°$$
$$= 12.9 \text{ m/s}^2$$

e. Equation 4.46 indicates that

$$a_t = \alpha r$$

Substitute for a_t and r:

$$12.9 \text{ m/s}^2 = \alpha(2.00 \text{ m})$$

Solve for α:

$$\alpha = 6.45 \text{ rad/s}^2$$

f. The direction of $\vec{\omega}$ is found using the circular motion right-hand rule. The tangential acceleration here is directed *opposite* to the velocity vector, so that the particle is slowing down in its rotational motion. Thus the angular acceleration vector $\vec{\alpha}$ is directed *opposite* to the direction of $\vec{\omega}$. The orientation of the two vectors is shown in Figure 4.50.

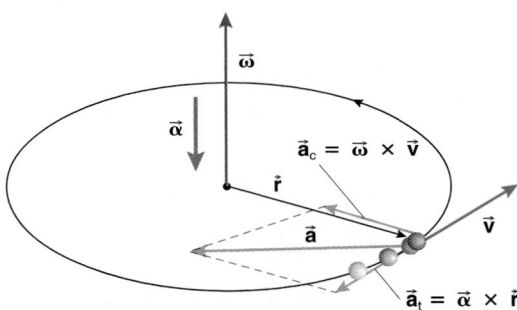

FIGURE 4.50

Notice that the vector product right-hand rule verifies the relative directions of the vectors in the vector relationships

$$\vec{a}_c = \vec{\omega} \times \vec{v}$$

and

$$\vec{a}_t = \vec{\alpha} \times \hat{r}$$

4.12 NONUNIFORM CIRCULAR MOTION WITH A CONSTANT ANGULAR ACCELERATION

Just as in rectilinear motion, many circular motions are modeled with a constant angular acceleration to a first approximation. The kinematics of such circular motion are similar to rectilinear motion, since they *both* are motions with one variable coordinate—in the one case θ, in the other case x. This means we can use the rectilinear motion equations with a constant acceleration as a guide to writing the circular motion equations for a constant angular acceleration:

For rectilinear motion with constant acceleration component a_x:

$$a_x = \frac{dv_x}{dt}$$

we found that

$$v_x(t) = v_{x0} + a_x t$$

Likewise

$$x(t) = x_0 + v_{x0}t + \frac{1}{2} a_x t^2$$

For circular motion with constant angular acceleration component α_z:

$$\alpha_z = \frac{d\omega_z}{dt}$$

Hence, analogously to rectilinear motion,

$$\boxed{\omega_z(t) = \omega_{z0} + \alpha_z t} \tag{4.49}$$

and

$$\boxed{\theta(t) = \theta_0 + \omega_{z0}t + \frac{1}{2} \alpha_z t^2} \tag{4.50}$$

4.9 Notice the formal similarity between the equations for rectilinear motion with a constant acceleration and nonuniform circular motion with a constant angular acceleration. The similarity arises because both motions are functions of a single coordinate. We replace the linear acceleration, velocity, and position with the corresponding angular counterparts:

- the constant acceleration component a_x with the constant *angular* acceleration component α_z;
- the initial velocity component v_{x0} with the initial *angular* velocity component ω_{z0}; and
- the initial position component x_0 with the initial *angular* position coordinate θ_0. For convenience, the angle θ_0, the value of θ when $t = 0$ s, conventionally is set equal to 0 rad.

QUESTION 8

A particle accelerates from rest beginning when $t = 0$ s around a circular track at a constant angular acceleration. (a) What is the ratio

$$\frac{a_c}{a_t}$$

when $t = 0$ s? (b) What does this imply about the initial direction of the *total* acceleration vector of the particle relative to the radial direction? (c) As the particle picks up speed around the circle, how does the direction of the *total* acceleration vector qualitatively change relative to the radial direction?

EXAMPLE 4.11

A centrifuge begins at rest and achieves an angular speed of 5.00×10^4 rev/min after 2.00 min (see Figure 4.51).

a. Sketch the direction of the angular velocity vector.
b. Specify an appropriate direction for the z-axis.
c. Find the z-component (the only component) of the angular acceleration, assuming it is constant.
d. Find the magnitude of the angular acceleration.

Solution

a. The direction of the angular velocity vector is determined from the circular motion right-hand rule (see Figure 4.51); if the spin is drawn in the opposite sense, the angular velocity vector $\vec{\omega}$ also is in the opposite direction.

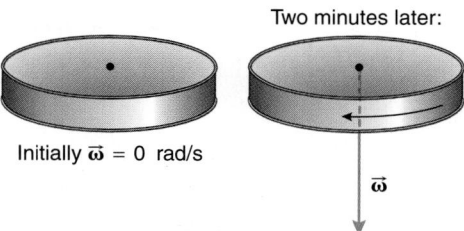

Two minutes later:

Initially $\vec{\omega} = 0$ rad/s

$\vec{\omega}$

FIGURE 4.51

b. The z-axis is taken along the line containing $\vec{\omega}$ (see Figure 4.52).

Choice 1

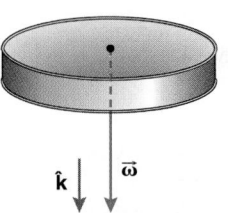

Choose $\hat{\mathbf{k}}$ to be in the direction of $\vec{\omega}$.

FIGURE 4.52

Choice 2

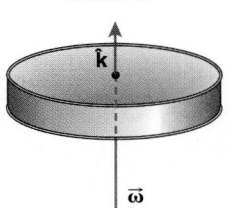

Choose $\hat{\mathbf{k}}$ to be in the opposite direction of $\vec{\omega}$.

c. The initial angular velocity is zero when $t = 0$ s, so

$\omega_{z0} = 0$ rad/s	$\omega_{z0} = 0$ rad/s

The final angular velocity component is

$\omega_z = \dfrac{(5.00 \times 10^4 \ \text{rev/min})(2\pi \ \text{rad/rev})}{60 \ \text{s/min}}$	$\omega_z = \dfrac{(-5.00 \times 10^4 \ \text{rev/min})(2\pi \ \text{rad/rev})}{60 \ \text{s/min}}$
$= 5.24 \times 10^3$ rad/s	$= -5.24 \times 10^3$ rad/s

Use Equation 4.49,

$$\omega_z = \omega_{z0} + \alpha_z t$$

and make the appropriate substitutions:

5.24×10^3 rad/s $= 0$ rad/s $+ \alpha_z(120 \ \text{s})$	-5.24×10^3 rad/s $= 0$ rad/s $+ \alpha_z(120 \ \text{s})$

Solving for α_z, you find

$\alpha_z = 43.7$ rad/s^2	$\alpha_z = -43.7$ rad/s^2

d. The magnitude of the angular acceleration is

$$|\alpha_z| = 43.7 \ \text{rad/s}^2$$

STRATEGIC EXAMPLE 4.12

Your parents' antique stereo turntable of radius 13.7 cm, initially spinning at 33.0 revolutions per minute, is shut off. The turntable coasts to a stop after 120 s. Assume a constant angular acceleration. Calculate the angular acceleration of the turntable and the number of revolutions through which it spins as it stops.

Solution

Stereo turntables rotate clockwise when viewed from the top. Use the circular motion right-hand rule to determine the direction of the angular velocity vector (see Figure 4.53). You find it points down.

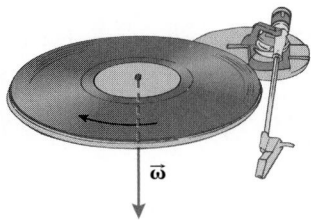

FIGURE 4.53

Choose a direction for $\hat{k}$ (see Figure 4.54):

Choice 1

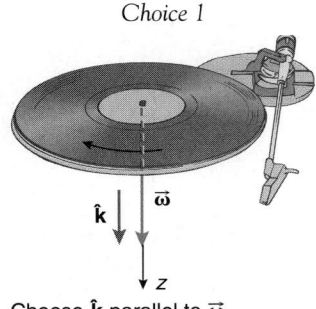

Choose $\hat{k}$ parallel to $\vec{\omega}$.

Choice 2

Choose $\hat{k}$ antiparallel to $\vec{\omega}$.

FIGURE 4.54

Since the turntable is slowing down, the magnitude of the angular velocity vector is decreasing, and so the angular velocity vector arrow is getting smaller during the interval. The angular acceleration vector points in the direction of the *change* in the angular velocity vector. Thus the angular acceleration vector is pointed opposite to the direction of the angular velocity vector itself, as shown in Figure 4.55.

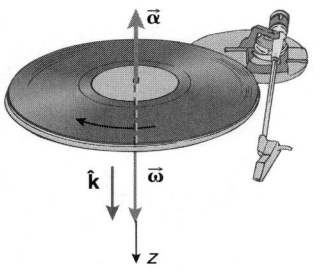

The angular acceleration $\vec{\alpha}$ points toward decreasing values of z, and so you anticipate that α_z will be negative.

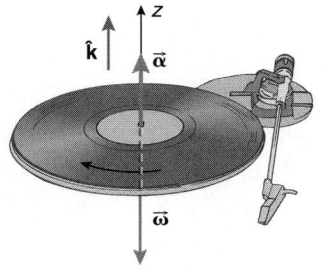

The angular acceleration $\vec{\alpha}$ points toward increasing values of z, and so you anticipate that α_z will be positive.

FIGURE 4.55

Convert the initial angular speed of 33.0 rev/min to radians per second (one revolution of the turntable is equivalent to 2π rad):

$$33.0 \text{ rev/min} = \frac{(33.0 \text{ rev/min})(2\pi \text{ rad/rev})}{60 \text{ s/min}}$$

$$= 3.46 \text{ rad/s}$$

Thus the initial angular velocity component is

$\omega_{z0} = 3.46$ rad/s	$\omega_{z0} = -3.46$ rad/s

After 120 s, the final angular velocity is zero. Hence Equation 4.49 for the angular velocity component is, when $t = 120$ s,

0 rad/s = 3.46 rad/s + α_z(120 s)	0 rad/s = -3.46 rad/s + α_z(120 s)

Solve for the angular acceleration component α_z:

$\alpha_z = -0.0288$ rad/s^2	$\alpha_z = 0.0288$ rad/s^2

The angular acceleration vector is

$$\vec{\alpha} = \alpha_z \hat{k}$$

$\vec{\alpha} = (-0.0288 \text{ rad/s}^2)\hat{k}$	$\vec{\alpha} = (0.0288 \text{ rad/s}^2)\hat{k}$
and so $\vec{\alpha}$ is in the same direction as $-\hat{k}$.	and so $\vec{\alpha}$ is in the same direction as $\hat{k}$.

In both cases the angular acceleration is *antiparallel* to $\vec{\omega}$, indicating the turntable is slowing down in its circular motion.

The total angle through which the turntable turns as it slows down is found from Equation 4.50 for the angular position:

$$\theta = \theta_0 + \omega_{z0}t + \frac{1}{2}\alpha_z t^2$$

Take the initial angular position of a point on the turntable as zero; hence $\theta_0 = 0$ rad. The angular position θ of the same point on the turntable when $t = 120$ s then is found by substituting for ω_{z0}, α_z, and t:

$\theta = 0$ rad $+ (3.46 \text{ rad/s})(120 \text{ s})$	$\theta = 0$ rad $- (3.46 \text{ rad/s})(120 \text{ s})$
$+ \frac{1}{2}(-0.0288 \text{ rad/s}^2)(120 \text{ s})^2$	$+ \frac{1}{2}(0.0288 \text{ rad/s}^2)(120 \text{ s})^2$
$= 208$ rad	$= -208$ rad

We indicate x- and y- axes in Figure 4.56, consistent with the choice for $\hat{k}$.

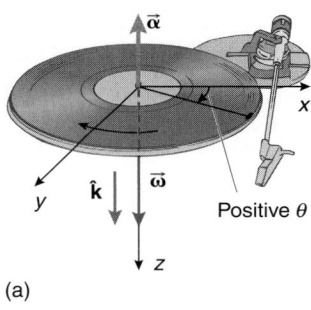

(a)

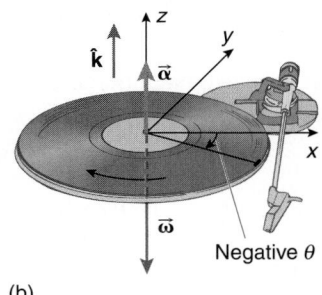

(b)

FIGURE 4.56

In polar coordinates, the angle θ is positive when measured counterclockwise from the x-axis, when viewed looking down the $+z$-axis, parallel to $-\hat{k}$.

The positive result for θ indicates that the polar angle θ is measured counterclockwise from the x-axis when sighting along $-\hat{k}$.	The negative result for θ indicates that the polar angle θ is measured clockwise from the x-axis when sighting along $-\hat{k}$.

The absolute value of the angle through which the point on the turntable turns is 208 rad. The number of revolutions is obtained by dividing the angle by 2π rad/rev. The total number of revolutions of the turntable is*

$$\text{number of revolutions} = \frac{\left| +208 \text{ rad} \right|}{2\pi \text{ rad/rev}}$$

$$= 33.1 \text{ rev}$$

---

*Note that the answer is actually exactly 33.0 rev: a record spinning at 33.0 rev/min slowing down to rest in 2 min (120 s) with constant acceleration has average angular speed 16.5 rev/min; since this average angular speed is during 2 min, the record travels 16.5 rev/min × 2 min = 33.0 rev. The discrepancy in the answer is due to rounding.

EXAMPLE 4.13

a. Find the magnitude of the tangential acceleration of a point on the rim of the spinning turntable of Example 4.12 as it slows down. Indicate the direction of the tangential acceleration vector on a sketch that also includes the direction of the velocity vector of the same point on the rim.
b. Determine the magnitude of the centripetal acceleration of a point on the rim the instant the turntable begins to slow down. Sketch the direction of the centripetal acceleration for the same point on the rim as the point considered in part (a).

Solution

a. The magnitude of the tangential acceleration is given by Equation 4.46:

$$a_t = \alpha r = |\alpha_z| r$$

We found the angular acceleration in Example 4.12 to be of magnitude

$$\alpha = |\alpha_z| = 0.0288 \text{ rad/s}^2$$
$$= 2.88 \times 10^{-2} \text{ rad/s}^2$$

The radius of the turntable is 13.7 cm = 0.137 m. Make these substitutions into Equation 4.46:

$$a_t = (2.88 \times 10^{-2} \text{ rad/s}^2)(0.137 \text{ m})$$
$$= 3.95 \times 10^{-3} \text{ m/s}^2$$

There are two ways we can determine the direction of the tangential acceleration for a point on the rim.

Method 1

Since the turntable is slowing down, the tangential acceleration must be directed opposite to the velocity vector of the point on the rim. See Figure 4.57.

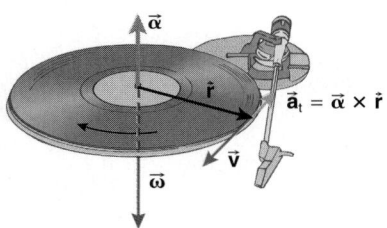

FIGURE 4.57

Method 2

Since the turntable is slowing down, $\vec{\alpha}$ must be directed opposite to $\vec{\omega}$. See Figure 4.57. The position vector $\vec{r}$ and velocity of a point on the rim are indicated. Since $\vec{a}_t = \vec{\alpha} \times \vec{r}$ the vector product right-hand rule indicates $\vec{\alpha} \times \vec{r}$ is opposite to $\vec{v}$, as shown in Figure 4.57.

b. The magnitude of the centripetal acceleration of a point on the rim is found using Equation 4.43:

$$a_c = |\omega_z(t)|^2 r$$

The angular speed at the instant the turntable begins to slow down is $|\omega_{z0}|$, which was found in Example 4.12 to be

$$|\omega_{z0}| = 3.46 \text{ rad/s}$$

We know the radius of the turntable is 0.137 m. Make these numerical substitutions:

$$a_c = (3.46 \text{ rad/s})^2(0.137 \text{ m})$$
$$= 1.64 \text{ m/s}^2$$

The orientation of the centripetal and tangential acceleration vectors for a point on the rim of the turntable is sketched in Figure 4.58. The total acceleration is the vector sum of the tangential and centripetal accelerations.

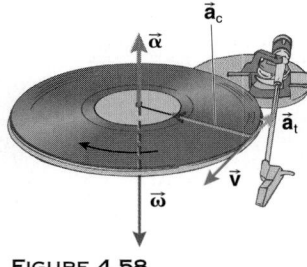

FIGURE 4.58

STRATEGIC EXAMPLE 4.14

A sports car moves around a horizontal circular curve of radius 200 m, as shown in Figure 4.59. While traversing the curve, the car changes speed at a constant rate from 36.0 km/h (= 10.0 m/s) to 90.0 km/h (= 25.0 m/s) during 20.0 s.

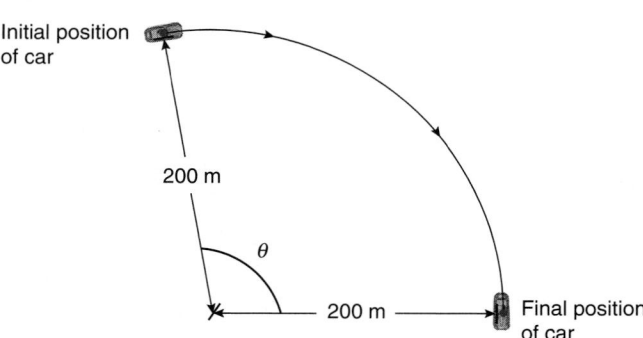

FIGURE 4.59

a. Find the magnitudes of the centripetal acceleration, the tangential acceleration, and the total acceleration of the car (i) the instant when the car has the speed 36.0 km/h, and (ii) the instant the car has the speed 90.0 km/h.
b. Find the angle ϕ between the direction of the total acceleration and the radial direction when the speed of the car is 36.0 km/h, and again when the speed is 90.0 km/h.
c. Find the angle θ through which the car moves during the 20.0 s interval and the length of highway traversed along the curve.

Solution

a. The total acceleration has two mutually perpendicular components at each instant: (1) the centripetal acceleration directed toward the center of the circle; and (2) the tangential acceleration directed tangent to the circle. Both are illustrated in Figure 4.60 for the two given instants. Equation 4.43 indicates the centripetal acceleration has a magnitude equal to

$$a_c = \frac{v^2}{r}$$

where v is the instantaneous speed of the car and r is the radius of the circular turn. At the instant the speed of the car is 36.0 km/h, or 10.0 m/s in SI units, the magnitude of the centripetal acceleration is

$$a_c = \frac{(10.0 \text{ m/s})^2}{200 \text{ m}}$$

$$= 0.500 \text{ m/s}^2$$

At the instant when the car reaches the speed of 90.0 km/h (= 25.0 m/s), the magnitude of the centripetal acceleration also is found by substituting the speed and radius into Equation 4.43:

$$a_c = \frac{(25.0 \text{ m/s})^2}{200 \text{ m}}$$

$$= 3.13 \text{ m/s}^2$$

We see that as the speed of the car increases around the curve, the centripetal acceleration *increases in magnitude*; the direction of the centripetal acceleration vector also changes continuously, but always points toward the center of the circle. Now we find the magnitude of the tangential acceleration.

Method 1

The magnitude of the tangential acceleration is *constant* as the car traverses the curve, since the speed of the car changes at a constant rate from 36.0 km/h to 90.0 km/h. Since the magnitude of the (instantaneous) tangential acceleration is constant, it is equal to the magnitude of the *average* tangential acceleration during the 20.0 s interval. The latter is found by dividing the change in the tangential speed of the car by the time elapsed. Hence the magnitude of the tangential acceleration here is

$$a_t = \frac{25.0 \text{ m/s} - 10.0 \text{ m/s}}{20.0 \text{ s}}$$

$$= 0.750 \text{ m/s}^2$$

and is the same at *every* instant during the 20.0 s time interval. The direction of the tangential acceleration changes continuously as the car traverses the curve, since, as its name implies, this acceleration is directed tangent to the circle.

Method 2

The tangential acceleration also may be found using the vector approach used in describing circular motion, although this way is more involved. Since the car speeds up at a constant rate around the curve, this is a problem of *nonuniform* circular motion at *constant* angular acceleration. From the circular motion right-hand rule, the angular velocity vector of the car as it traverses the curve is directed perpendicularly *into* the page. Let this also be the direction of the $+z$-axis, as shown in Figure 4.61. When the car has a speed of 36.0 km/h (= 10.0 m/s), the magnitude of the angular velocity vector is found from Equation 4.35:

$$v = r\omega$$

Thus

$$\omega = \frac{v}{r} = \frac{10.0 \text{ m/s}}{200 \text{ m}} = 0.0500 \text{ rad/s}$$

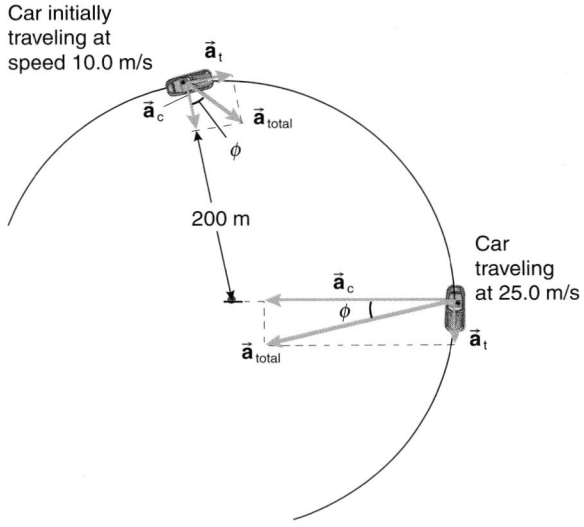

Car initially traveling at speed 10.0 m/s

Car traveling at 25.0 m/s

200 m

FIGURE 4.60

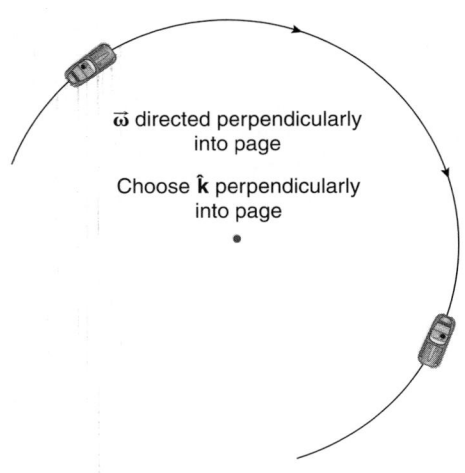

$\vec{\omega}$ directed perpendicularly into page

Choose $\hat{\mathbf{k}}$ perpendicularly into page

FIGURE 4.61

The angular velocity vector $\vec{\omega}$ of the sports car initially (when $t = 0$ s) is

$$\vec{\omega} = \omega_{z0}\hat{k}$$

so its z-component is

$$\omega_{z0} = 0.0500 \ \text{rad/s}$$

When $t = 20.0$ s, the speed of the car is 90.0 km/h (= 25.0 m/s) and the angular velocity component has increased to

$$\omega_z = \frac{v}{r} = \frac{25.0 \ \text{m/s}}{200 \ \text{m}} = 0.125 \ \text{rad/s}$$

The angular acceleration component α_z is found from the nonuniform circular motion equation (for constant angular accelerations), Equation 4.49:

$$\omega_z = \omega_{z0} + \alpha_z t$$

Solve for the angular acceleration component by substituting the values for the angular velocity components and the time:

$$0.125 \ \text{rad/s} = 0.0500 \ \text{rad/s} + \alpha_z(20.0 \ \text{s})$$

So

$$\alpha_z = 0.003 \ 75 \ \text{rad/s}^2$$

Since α_z is positive, the angular acceleration vector $\vec{\alpha}$ is directed along the $+z$ axis. Since $\vec{\alpha}$ is in the same direction as $\vec{\omega}$, the car is speeding up in its circular motion. The magnitude of the tangential acceleration is found from Equation 4.46:

$$a_t = \alpha r$$

Substitute for α and r:

$$a_t = (0.003 \ 75 \ \text{rad/s}^2)(200 \ \text{m})$$
$$= 0.750 \ \text{m/s}^2$$

b. The centripetal and tangential accelerations are shown schematically in Figure 4.60 at the beginning and end of the 20.0 s time interval. The total acceleration of the car at each instant is the vector sum of the tangential and centripetal accelerations at each instant. The angle ϕ that the total acceleration vector initially makes with the radial direction is (from Figure 4.60)

$$\tan \phi = \frac{a_t}{a_c}$$
$$= \frac{0.750 \ \text{m/s}^2}{0.500 \ \text{m/s}^2}$$
$$= 1.50$$

or

$$\phi = 56.3°$$

Notice that since the magnitude of the tangential acceleration initially is greater than that of the centripetal acceleration, we would expect the total acceleration to be inclined by more than 45° to the radial direction.

At the final instant, when the speed of car is 90.0 km/h, the angle ϕ that the total acceleration vector makes with the radial direction is (from Figure 4.60)

$$\tan \phi = \frac{a_t}{a_c}$$
$$= \frac{0.750 \ \text{m/s}^2}{3.13 \ \text{m/s}^2}$$
$$= 0.240$$

or

$$\phi = 13.5°$$

The angle now is less than 45°, which indicates that the magnitude of the centripetal acceleration is greater than that of the tangential acceleration at this instant.

c. The angle through which the car moves during the 20.0 s interval is found using the nonuniform circular motion equation for constant angular accelerations, Equation 4.50:

$$\theta = \theta_0 + \omega_{z0}t + \frac{1}{2}\alpha_z t^2$$

where θ_0 is the initial angular position of the car when $t = 0$ s; we set this initial angular position equal to zero: $\theta_0 = 0$ rad. Substituting

$$\omega_{z0} = 0.0500 \ \text{rad/s}$$
$$\alpha_z = 0.003 \ 75 \ \text{rad/s}^2$$

the angular position of the car when $t = 20.0$ s is found to be

$$\theta = 0 \ \text{rad} + (0.050 \ \text{rad/s})(20.0 \ \text{s})$$
$$+ \frac{(0.003 \ 75 \ \text{rad/s}^2)(20.0 \ \text{s})^2}{2}$$
$$= 1.75 \ \text{rad}$$

which is equivalent to 100°. With the definition of an angle in radians,

$$\theta = \frac{\text{arc length}}{\text{radius}} = \frac{s}{r}$$

the arc length s of the curve traversed by the car during the 20.0 s interval is

$$s = r\theta$$
$$= (200 \ \text{m})(1.75 \ \text{rad})$$
$$= 350 \ \text{m}$$

CHAPTER SUMMARY

The general form for the position vector of a moving particle is

$$\vec{r}(t) = x(t)\hat{i} + y(t)\hat{j} + z(t)\hat{k} \qquad (4.16)$$

As in rectilinear motion, the *average velocity* of a moving particle during a time interval Δt is the ratio of the change in the position vector $\Delta \vec{r}$ during the interval to the interval itself:

$$\vec{v}_{ave} \equiv \frac{\Delta \vec{r}}{\Delta t} \qquad (4.3)$$

Likewise, the *velocity* of the particle is the limit of the average velocity as the time interval Δt approaches zero; this is equivalent to the time derivative of the position vector:

$$\vec{v}(t) = \lim_{\Delta t \to 0 \text{ s}} \frac{\Delta \vec{r}}{\Delta t} = \frac{d\vec{r}(t)}{dt} \qquad (4.4)$$

The *average acceleration* of the particle during an interval Δt is the ratio of the change in the velocity during the time interval to the interval itself:

$$\vec{a}_{ave} \equiv \frac{\Delta \vec{v}}{\Delta t} \qquad (4.7)$$

The *acceleration* is the limit of the average acceleration as the time interval Δt approaches zero; this is equivalent to the time derivative of the velocity, or the second time derivative of the position vector:

$$\vec{a}(t) = \lim_{\Delta t \to 0 \text{ s}} \frac{\Delta \vec{v}}{\Delta t} = \frac{d\vec{v}(t)}{dt} = \frac{d^2\vec{r}(t)}{dt^2} \qquad (4.8)$$

These equations have the same form as those for rectilinear motion, but the vectors now have two or three components depending on whether the particle is moving in two or three dimensions.

Motions along perpendicular directions are independent of each other and can be analyzed separately.

For two-dimensional motion with a constant acceleration, choose the y-axis to be the line containing the acceleration vector. The acceleration then has only a single, constant component a_y, and the x- and y-components of the velocity and position vectors are

$$v_x(t) = v_{x0} \qquad \text{and is constant} \qquad (4.10)$$
$$x(t) = x_0 + v_{x0}t \qquad (4.11)$$
$$v_y(t) = v_{y0} + a_y t \qquad (4.12)$$
$$y(t) = y_0 + v_{y0}t + \frac{1}{2}a_y t^2 \qquad (4.13)$$

The signs of the various components depend on the directions you choose for the unit vectors $\hat{i}$ and $\hat{j}$.

If the motion of a particle P is referred to two coordinate systems with origins at 0 and 0', one of which (0') is moving, then the velocity $\vec{v}_{P0'}$ of the particle with respect to 0', and the velocity $\vec{v}_{P0}$ of the particle with respect to 0, are related by the equation

$$\vec{v}_{P0} = \vec{v}_{P0'} + \vec{v}_{0'0} \qquad (4.20)$$

where $\vec{v}_{0'0}$ is the velocity of 0' with respect to 0.

The *acceleration* of a particle is the *same* with respect to any reference frames moving at *constant velocity* with respect to each other.

The circular motion of a particle in the x–y plane is described by the position vector

$$\vec{r}(t) = [r\cos\theta(t)]\hat{i} + [r\sin\theta(t)]\hat{j} \qquad (4.28)$$

where r, the constant radius of the circle, and $\theta(t)$ are the polar coordinates of the particle. Circular motion is accelerated motion, since the velocity vector changes with time.

The direction of the *angular velocity vector* $\vec{\omega}$ always is chosen according to the *circular motion right-hand rule*. The direction conventionally chosen for the $+z$-axis is either parallel *or* antiparallel to $\vec{\omega}$; the specific choice is yours to make. Whichever you choose, the angular velocity then has a single component:

$$\vec{\omega} = \omega_z \hat{k}$$

The angular velocity is specified in rad/s. The time rate of change of the polar coordinate $\theta(t)$ is the z-component, the only component, of the angular velocity vector:

$$\omega_z(t) = \frac{d\theta(t)}{dt} \qquad (4.32)$$

The position, velocity, and angular velocity vectors in circular motion are related by the vector product:

$$\vec{v}(t) = \vec{\omega}(t) \times \vec{r}(t) \qquad (4.34)$$

All three vectors are mutually perpendicular at all times.

Uniform circular motion is circular motion at constant speed v. The angular velocity vector $\vec{\omega}$ is constant for such motion. The acceleration is directed toward the center of the circle and is called the *centripetal acceleration*:

$$\vec{a}_c(t) = \vec{\omega} \times \vec{v}(t) = -\omega_z^2 \vec{r}(t) \qquad (4.39, \ 4.36)$$

The magnitude of the centripetal acceleration is constant for uniform circular motion:

$$a_c = \frac{v^2}{r} = \omega_z^2 r \qquad (4.38, \ 4.37)$$

Nonuniform circular motion is circular motion with a changing speed. The acceleration then has two perpendicular components:

$$\vec{a}(t) = \vec{a}_c(t) + \vec{a}_t(t) \qquad (4.48)$$

The centripetal acceleration is directed toward the center of the circle and is

$$\vec{a}_c(t) \equiv \vec{\omega}(t) \times \vec{v}(t) \qquad (4.41)$$

but is not of constant magnitude, since the speed of the particle is a function of time. The magnitude of the centripetal acceleration is

$$a_c(t) = \frac{v(t)^2}{r} = \omega_z(t)^2 r \qquad (4.43)$$

The *tangential acceleration* is

$$\vec{a}_t(t) \equiv \vec{\alpha}(t) \times \vec{r}(t) \qquad (4.45)$$

where $\vec{\alpha}(t)$ is the *angular acceleration*—that is, the time rate of change of the angular velocity:

$$\vec{\alpha}(t) \equiv \frac{d\vec{\omega}(t)}{dt} \qquad (4.44)$$

The tangential acceleration is either parallel to $\vec{v}$ if the particle is speeding up, or antiparallel to $\vec{v}$ if the particle is slowing down.

Both $\vec{\omega}$ and $\vec{\alpha}$ always are perpendicular to the plane of the circular motion. Since we choose the z-axis to be along this line, $\vec{\omega}$ and $\vec{\alpha}$ both have only a z-component. The angular acceleration is specified in rad/s^2. The magnitude of the tangential acceleration is

$$a_t(t) = \alpha(t)r = |\alpha_z(t)|r \qquad (4.46)$$

The angular acceleration $\vec{\alpha}(t)$ is

1. parallel to $\vec{\omega}(t)$, when a particle is speeding up in its circular motion; or

2. antiparallel to $\vec{\omega}(t)$, when a particle is slowing down in its circular motion.

For circular motion at *constant angular acceleration* $\vec{\alpha} = \alpha_z \hat{k}$:

a. the z-component (and only component) of the angular velocity at any instant is

$$\omega_z(t) = \omega_{z0} + \alpha_z t \qquad (4.49)$$

where ω_{z0} is the z-component of the angular velocity when $t = 0$ s;

b. the angular coordinate of the particle $\theta(t)$ at any instant is

$$\theta(t) = \theta_0 + \omega_{z0}t + \frac{1}{2}\alpha_z t^2 \qquad (4.50)$$

where θ_0 is the angular position of the particle when $t = 0$ s; θ_0 usually is taken to be zero (i.e., $\theta_0 = 0$ rad).

The signs of α_z, ω_{z0}, and $\omega_z(t)$ depend on the direction you choose for $\hat{k}$.

SUMMARY OF PROBLEM-SOLVING TACTICS

4.1   (page 118) The specific coordinate choice is yours to make, but it needs to be explicitly indicated in a sketch and used consistently throughout a given problem.

4.2   (page 121) If the acceleration vector is constant, choose one of the coordinate axes to be the line containing the constant acceleration or a line parallel to it; we choose the line containing the constant acceleration to be the y-axis.

4.3   (page 121) The sign of the acceleration component a_y depends on the direction you choose for $\hat{j}$ along the line containing the acceleration vector.

4.4   (page 121) The origin may be taken at any convenient location along the line chosen for the y-axis.

4.5   (page 122) Notice that the equations for $x(t)$ and $v_x(t)$ have the same form as those for $y(t)$ and $v_y(t)$ but with an acceleration component a_x equal to zero.

4.6   (page 136) Choose the z-axis to be the line containing $\vec{\omega}$.

4.7   (page 136) The direction you choose for $\hat{k}$ determines the sign of ω_z.

4.8   (page 139) The seemingly different expressions for the centripetal acceleration are all variations of the same equation.

4.9   (page 144) Notice the formal similarity between the equations for rectilinear motion with a constant acceleration and nonuniform circular motion with a constant angular acceleration.

QUESTIONS

1. (page 120); 2. (page 122); 3. (page 122); 4. (page 123); 5. (page 134); 6. (page 139); 7. (page 142); 8. (page 144)

9. Why is the instantaneous speed the magnitude of the instantaneous velocity, whereas the average speed is *not* the magnitude of the average velocity?

10. Devise an example to show that the average velocity of a particle during a time interval Δt is *not* necessarily equal to

$$\frac{\vec{v}_f + \vec{v}_i}{2}$$

11. (a) Sketch the trajectory of a projectile that has zero speed at the top of its trajectory. (b) Sketch the trajectory of a projectile that has nonzero speed at the top of its trajectory.

12. Describe a situation where the average speed of a particle during a time interval Δt is *not* zero, but its average velocity *is* zero during the same time interval.

13. Is it possible to have the average speed equal to zero but the average velocity nonzero? If so, explain how. If not, explain why not.

14. Describe a physical situation in which (a) the velocity vector of a particle is perpendicular to its position vector; (b) the acceleration vector of the particle is perpendicular to its velocity vector; (c) the acceleration is perpendicular to the position vector; (d) the velocity is in the same direction as the acceleration; (e) the velocity and acceleration are in opposite directions.

15. Three fleas (*Ctenocephalides canis*) have a jumping contest and discover to their amazement that they each jump to the same elevation off the floor (see Figure Q.15) but land in different positions horizontally. Explain your reasoning in answering the following questions about the contest: (a) Which flea had the greatest initial speed? (b) Which flea was in the air for the longest time? (c) Which flea had the greatest magnitude of the acceleration while in flight? (d) Which flea had the horizontal velocity component of greatest absolute value? (e) Which flea had the vertical velocity component of greatest absolute value?

16. At what point along the trajectory of a projectile is its speed a minimum?

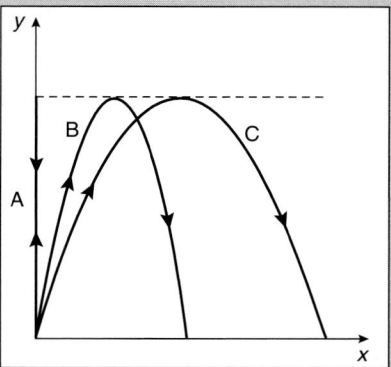

FIGURE Q.15

17. At what point along the trajectory of a projectile is the velocity perpendicular to the acceleration?

18. Can the velocity ever be parallel to the acceleration for some type of projectile motion? Explain your answer.

19. What factors influence the range of an athlete competing in the long jump versus one competing in the high jump? See Figure Q.19.

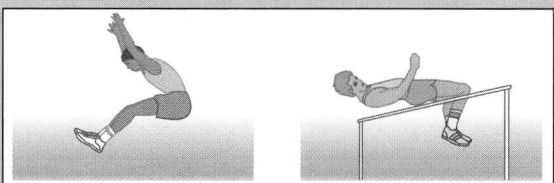

FIGURE Q.19

20. A vertically falling raindrop strikes loose dirt on a slope. Make a diagram that shows why dirt splashed down the hill travels farther along the slope than dirt splashed up the slope if the initial velocity vectors of the dirt make equal angles with the vertical direction. This process contributes to the slow erosion of mountains.

21. Two identical twins of equal strength are using identical canoes to race across a river of constant width; the current speed is less than the paddling speed of the twins. One twin keeps his canoe aimed directly across the river and is swept downstream as the crossing is made. The other twin ferries across by aiming the canoe so that the actual path followed by the canoe is directly across the river perpendicular to its banks. Who wins the race or is it a tie? Explain your reasoning. See Figure Q.21.

22. Sailors use strips of sailcloth attached to a mast or a stay to deduce wind directions. For a sailboat underway at sea, does the direction of the strip give a true indication of the direction to which the wind is blowing? Explain. See Figure Q.22.

23. While out for a stroll in the rain, why do you tilt the umbrella in the direction you are moving to keep drier?

24. You make one complete orbit riding on a merry-go-round during a time interval Δt. What are your average velocity and average acceleration?

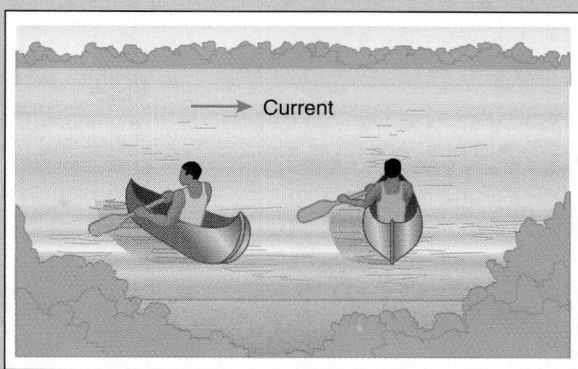

FIGURE Q.21

FIGURE Q.22

25. Make a schematic diagram of the Earth with its north and south geographic poles. Indicate on the diagram the direction of the angular velocity vector of a particle at rest on the surface of the Earth.

26. Is the centripetal acceleration of a particle located on the surface of the Earth directed toward the center of the Earth? If so, explain why; if not, explain why not.

27. Where on the surface of the spinning Earth does the centripetal acceleration of a particle at rest on its surface have the greatest magnitude? The least magnitude?

28. Does the angular speed of a particle at rest on the surface of the spinning Earth depend on either the latitude or longitude of the particle? Explain.

29. At what two geographic locations on the surface of the spinning Earth is the centripetal acceleration of a particle at rest on the surface equal to zero?

30. Is the magnitude of the centripetal acceleration of a particle on the surface of the spinning Earth a function of either the latitude or longitude of the particle?

31. As a resident of the spinning Earth, you are carried along with it, orbiting the line connecting the north and south poles. At what two geographic locations on the Earth is your motion spin rather than orbital motion?

32. Do the sweep second, minute, and hour hands of a clock turn at constant angular speed?

33. The angular velocity vector of a particle in circular motion is indicated in Figure Q.33. Indicate the direction the particle is moving around the circle.

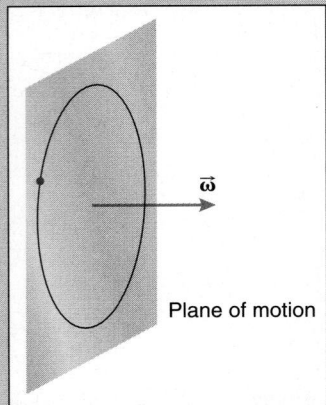

FIGURE Q.33

34. Is the angular velocity vector representing the daily rotational motion of the Earth in the same direction as the angular velocity vector associated with the orbital motion of the Earth around the Sun?

35. What is the ratio of the angular speed of the spinning Earth to the angular speed associated with the orbital motion of the Earth about the Sun?

36. Is the angular speed of a cassette tape the same throughout its playing time? Explain.

37. For nonuniform circular motion, can the *total* acceleration ever be instantaneously directed as shown in Figure Q.37? Explain your answer.

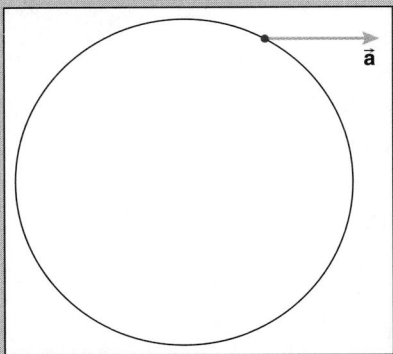

FIGURE Q.37

38. For circular motion, if the tangential acceleration is zero, the total acceleration is just the centripetal acceleration and the motion is *uniform* circular motion. Is it possible in circular motion to have the centripetal acceleration equal to zero but not the tangential acceleration? Explain.

39. Is it possible to have the angular velocity $\vec{\omega}$ equal to zero at some instant but a nonzero angular acceleration $\vec{\alpha}$? If so, how? If not, why not?

40. Is it possible to have the angular acceleration $\vec{\alpha}$ equal to zero at some instant but not the angular velocity $\vec{\omega}$? If so, how? If not, why not?

41. How would you *qualitatively describe* what is happening physically to an object in circular motion with a *time-dependent* angular acceleration $\vec{\alpha}(t)$? What does such an angular acceleration imply about the appearance of a graph of the angular velocity component versus time?

42. Does a particle at rest on the surface of the Earth experience a *tangential* acceleration because of the rotation of the Earth? If so, in what direction? If not, explain why not.

43. Is the angular speed of the film reel on a motion picture projector constant throughout the interval of the movie? Explain.

PROBLEMS

Neglect any effects caused by air resistance in these problems.

Section 4.1   The Position, Velocity, and Acceleration Vectors in Two Dimensions

1. A gypsy moth caterpillar (*Porthetria dispar*) inches along a crooked branch to a tasty oak leaf, wriggling 15 cm horizontally and then 30 cm along a section of the branch inclined at 30° to the horizontal as shown in Figure P.1. (a) Write the initial and final position vectors of the caterpillar in SI units using the crook in the branch as the origin for a set of horizontal and vertical coordinate axes. (b) If the caterpillar traverses the distance during 1.0 min,

• what is its average speed?
• what is its average velocity?
• what is the magnitude of its average velocity?

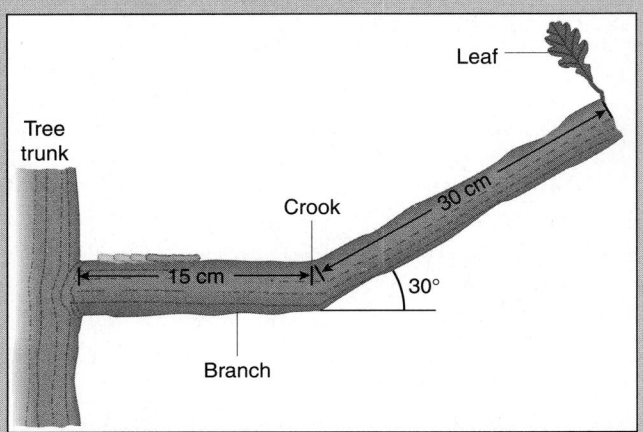

FIGURE P.1

2. A zoologist notices a monkey located at position $(20.0 \text{ m})\hat{j}$ intent on finding a banana. The monkey wanders to a tree situated at $(30.0 \text{ m})\hat{i} + (40.0 \text{ m})\hat{j}$ and then proceeds to a nearby vertical vine whose lower end is located at $(40.0 \text{ m})\hat{i} + (30.0 \text{ m})\hat{j}$. The x–y plane is the horizontal plane. (a) Sketch the situation on a Cartesian coordinate grid. (b) What is the change in the position vector of the monkey in moving from the tree to the vine? (c) The monkey climbs the vine to a height of 10.0 m and finally spies a banana at the origin. What is the distance between the monkey on the vine and the banana? (d) The monkey leaps from the vine and lands in a swampy pool at $(25.0 \text{ m})\hat{i} + (25.0 \text{ m})\hat{j}$. What is the distance between the tree and where the monkey landed in the pool?

3. Superman flies from his well-known telephone booth located at $(5.0 \text{ km})\hat{i} + (8.0 \text{ km})\hat{j}$ to the offices of *The Daily Planet* at $(-4.0 \text{ km})\hat{i} + (2.0 \text{ km})\hat{j}$ in 2.0 min. (a) Draw the two position vectors on a labeled coordinate grid. (b) What is the change in Superman's position vector in SI units? (c) What is his average velocity?

•**4.** As a race car driver, you complete one lap at an average speed $v_{1 \text{ ave}}$ and the second lap at the average speed $v_{2 \text{ ave}}$. (a) Show that the average speed v_{ave} over the two laps is

$$v_{\text{ave}} = \frac{2v_{1 \text{ ave}} v_{2 \text{ ave}}}{v_{1 \text{ ave}} + v_{2 \text{ ave}}}$$

The average speed v for the two laps is the *harmonic mean* of the average speeds $v_{1 \text{ ave}}$ and $v_{2 \text{ ave}}$ for the two individual laps. The harmonic mean n of any two numbers n_1 and n_2 is defined as the reciprocal of the arithmetic mean of the reciprocals of n_1 and n_2. That is

$$\frac{1}{n} = \frac{\frac{1}{n_1} + \frac{1}{n_2}}{2}$$

This can be rewritten as

$$\frac{2}{n} = \frac{1}{n_1} + \frac{1}{n_2}$$

or

$$n = \frac{2n_1 n_2}{n_1 + n_2}$$

(b) Find the average speed over two laps if the average speed on the first lap is 120 km/h while that on the second lap is 80 km/h. This problem was inspired by an article by P. Glaister, "Natural occurrences of the harmonic mean," *Physics Education*, 27, #4, page 181 (July 1992).

5. Consider two galaxies traveling radially outward with different speeds as seen from another galaxy at the origin as shown in Figure P.5. The two position vectors

$$\vec{r}_1 = v_1 t \hat{r}_1$$

and

$$\vec{r}_2 = v_2 t \hat{r}_2$$

locate the moving galaxies with respect to the one at the origin 0, where $\hat{r}_1$ and $\hat{r}_2$ are unit vectors in the radially outward directions and v_1 and v_2 are the respective speeds of the galaxies. Note that the speed of each galaxy is proportional to its distance from the origin. The position vectors make an angle θ with respect to each other. Let $\vec{r}$ be the vector from galaxy A to galaxy B. (a) Show that the magnitude of $\vec{r}$ is

$$r = (v_1^2 + v_2^2 - 2v_1 v_2 \cos \theta)^{1/2} t$$

This represents the distance between galaxies A and B. (b) Show that the speed with which galaxies A and B recede from each other (the magnitude of the velocity vector)

$$\vec{v} \equiv \frac{d\vec{r}}{dt}$$

is

$$v = (v_1^2 + v_2^2 - 2v_1 v_2 \cos \theta)^{1/2}$$
$$= \frac{r}{t}$$

Thus observers on galaxy A see the other galaxy B receding from them with a speed that is proportional to the separation of the galaxies. This shows that there is no privileged galaxy at the center of the universe. *Every* galaxy is a center and other galaxies move away from it at speeds proportional to their distances, much like raisins in a loaf of rising raisin bread. This exercise is the essence of the discovery made by Edwin Hubble in 1929 known as the *expansion of the universe*. A subtlety to the expansion, however, is not apparent in this exercise. The universe is not expanding *into* an empty space that already exists out there, since this would imply an edge to the universe with nothing beyond. The *space* between the galaxies is expanding, carrying the galaxies along with it, much like stretching a rubber sheet with buttons (galaxies) sewn onto it.

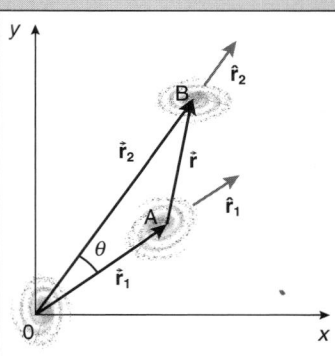

FIGURE P.5

Section 4.2 Two-Dimensional Motion with a Constant Acceleration

•**6.** An erupting volcano violently projects a boulder from a vent at an angle of 40° to the horizontal direction. The rock lands 6.0 km away as indicated in Figure P.6. (a) Introduce an appropriate coordinate system. (b) Tailor Equations 4.10–4.13 to the given information. (c) Determine the initial speed of the rock and the time of flight.

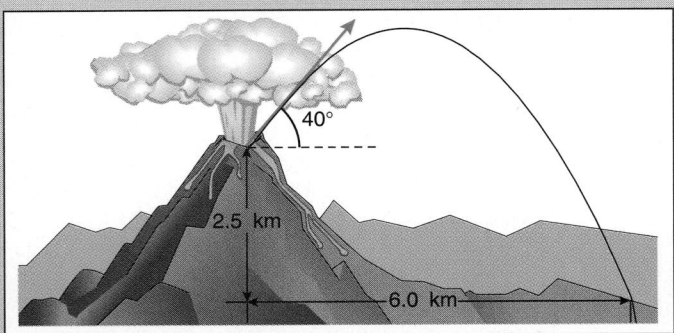

FIGURE P.6

•7. Slim Pickens hurls a baseball at a speed of 30.0 m/s at an angle of 60.0° to the horizontal as indicated in Figure P.7. The ball leaves Slim's hand at a point 2.50 m above the ground and 20.0 m from the base of a barrier 30.0 m high. (a) Indicate in a sketch an appropriate Cartesian coordinate system for analyzing the problem. (b) Tailor Equations 4.10–4.13 to the given information. (c) Does the ball hit or clear the vertical barrier? (d) Where does the ball first strike? (e) How long is the ball in the air before its first impact?

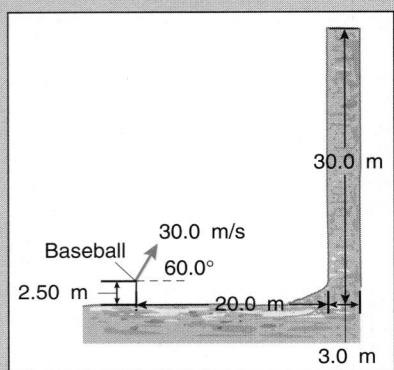

FIGURE P.7

•8. A rock is tossed off a cliff as indicated in Figure P.8. (a) Introduce a coordinate system to analyze the problem and tailor Equations 4.10–4.13 to the given information. (b) Find the time of flight. (c) Determine the impact speed of the rock.

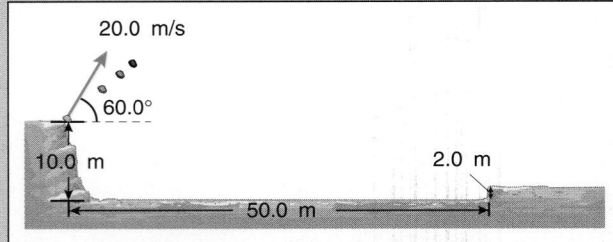

FIGURE P.8

•9. A soccer ball is kicked off a cliff at a speed of 20.0 m/s as indicated in Figure P.9. (a) In a sketch, indicate a choice for a coordinate system to analyze the problem. (b) Tailor the general Equations 4.10–4.13 to incorporate the specific infor-

mation given about the ball. (c) Determine the time of flight. (d) Determine where the ball hits the ground the first time.

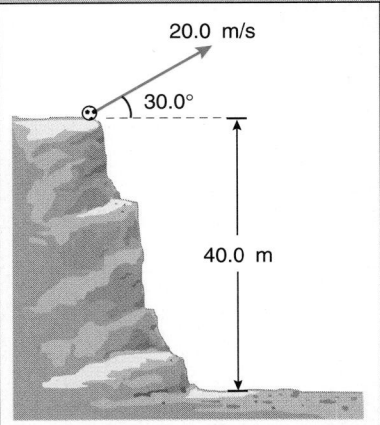

FIGURE P.9

•10. A young basketball player is attempting to make a shot. The ball leaves the hands of the player at an angle of 60° to the horizontal at an elevation of 2.0 m above the floor. The skillful player makes the shot with the ball traveling precisely through the center of the hoop as indicated in Figure P.10. To loud cheers, calculate the speed at which the ball left the hands of the player.

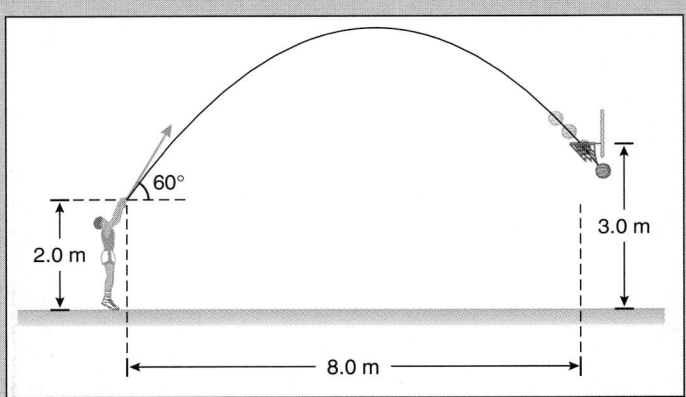

FIGURE P.10

•11. A projectile is fired across a level surface. The initial velocity $\vec{v}_0$ makes an angle θ with the horizontal (see Figure P.11). (a) Show that the maximum height $h_{\max}$ of the projectile is

$$h_{\max} = \frac{v_0^2 \sin^2 \theta}{2g}$$

(b) The horizontal range of the projectile was found in Example 4.3 to be

$$R_{\text{horiz}} = \frac{v_0^2 \sin 2\theta}{2g}$$

What angle θ makes the range of the projectile a maximum? What is the maximum range?

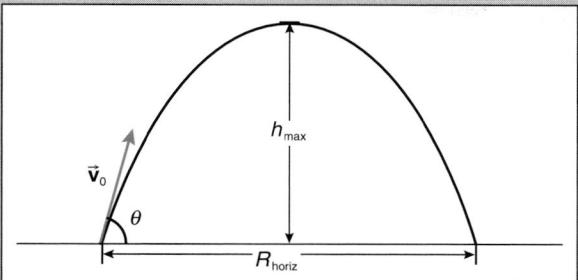

FIGURE P.11

•**12.** A beam of electrons, each with a speed of 5.00×10^6 m/s, is shot horizontally into a region where each electron experiences a vertically upward acceleration of magnitude 5.00×10^{13} m/s². The magnitude of this acceleration is so large that the magnitude of the local acceleration due to gravity may be ignored by comparison. Each electron experiences the large acceleration only while it traverses the region delineated in Figure P.12. (a) Set up an appropriate coordinate system for analyzing the motion of each electron within the region, and tailor Equations 4.10–4.13 to the given information. (b) For how long a time is each electron in the region? (c) What is the vertical deflection of each electron from its initial horizontal path? Such deflections of electron beams are the principle underlying CRT (cathode ray tube) displays such as computer monitors, TVs, and oscilloscopes.

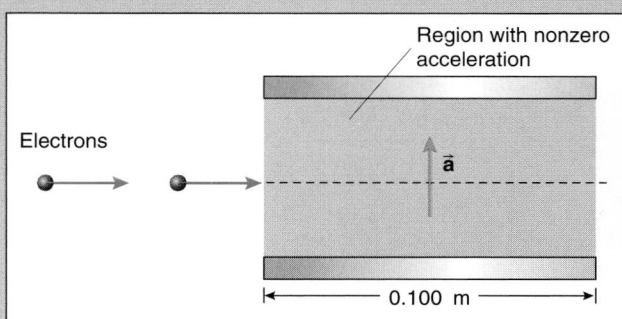

FIGURE P.12

•13. A famous soccer player, Hedley Foote, kicks the ball at a speed 15.0 m/s when $t = 0$ s; the initial velocity vector of the ball makes an angle of 60.0° with the horizontal direction. Out of curiosity, calculate the instants when the velocity vector of the ball makes an angle of 45.0° with the horizontal direction.

•**14.** The Chickens are playing the Turkeys in football and the score is tied at $\sqrt{43}$ apiece. Don't ask how. A placekicker is sent out for the Chickens with instructions from Coach Gung-ho to kick a field goal from 45.0 m out. The top of the crossbar on the goalpost is 3.05 m above the level playing field. The moment of truth arrives and the ball leaves the ground at an angle of 30.0° to the horizontal. (a) What is the minimum speed that the ball must have to make the field goal? Assume appropriate aiming. (b) How long does it take the ball to reach the crossbar?

•**15.** A pitcher throws a fastball. The ball is released horizontally from the pitcher's mound when the hand of the pitcher is at an elevation of 1.7 m above the level field. The ball is caught by the catcher 19.5 m away at an elevation of 0.90 m above the ground. See Figure P.15. Calculate the initial speed of the ball as it left the pitcher's hand.

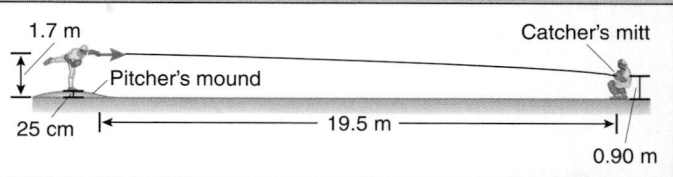

FIGURE P.15

•**16.** A 10-year-old launches a water balloon horizontally with an initial speed of 10.0 m/s from near the top of a tall tree 15.0 m above the ground. (a) What is the time of flight? (b) How far from the base of the tree does the water balloon hit? (c) What is the speed of the water balloon at the instant just before impact?

•**17.** A golfer launches a divot which hits the side of a golf cart as indicated in Figure P.17. (a) Indicate an appropriate coordinate system to analyze the flight of the divot. (b) Determine the initial speed of the divot. (c) Determine the time of flight of the divot.

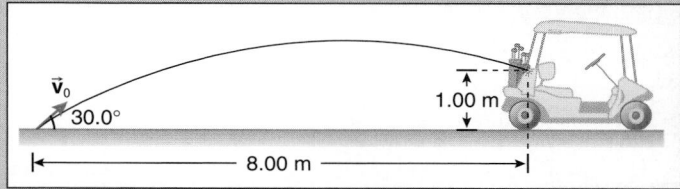

FIGURE P.17

•**18.** A dirt biker races up a 15.0° incline at a constant speed of 120 km/h. The end of the ramp is 3.00 m off the ground as indicated in Figure P.18. A 4.00 m high obstacle is located 20.0 m from the base of the ramp. (a) Indicate an appropriate coordinate system to attack the problem. (b) Will the daredevil clear the obstacle? If so, where will she land?

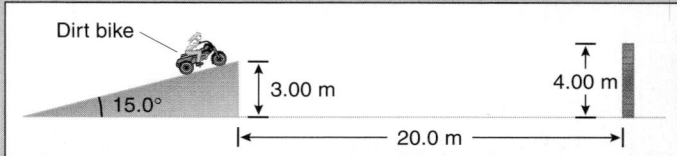

FIGURE P.18

•**19.** The Apollo 14 lunar astronaut Alan Shepard was the first golfer on the Moon. He screwed the head of a 6-iron to a geological sampling tool and, using one hand, whacked a golf ball across the lunar surface. The flight time of the ball was 40 s. There is no atmosphere on the Moon, so that the ball did not hook or slice, a golfer's paradise. The local acceleration due to gravity on the surface of the Moon is 0.166 ($\approx$ 1/6) that on the surface of the Earth. What was the greatest possible horizontal range of the lunar golf ball?

•**20.** Two identical pea shooters fire peas with an initial speed of 10.0 m/s. The pea shooters are located a distance d, one above the other as shown in Figure P.20. The lower pea emerges from the shooter at an angle of 20.0° to the horizontal, while the upper pea emerges horizontally. The peas both hit at the same point. (a) Express the horizontal range of the upper pea in terms of the height d. (b) What is the horizontal range of the lower pea? (c) Determine d.

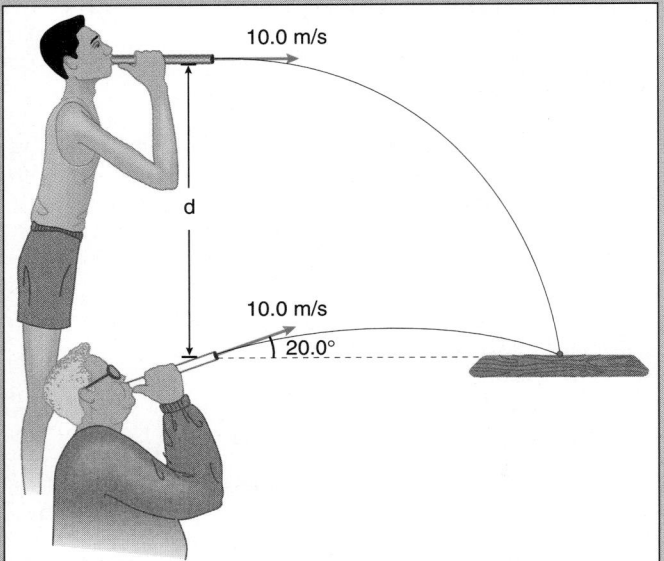

FIGURE P.20

•**21.** An Englishman named Hew Kennedy has constructed a full-scale reproduction of a medieval siege machine known as a *trebuchet*. A trebuchet originally was used to hurl large masses, such as putrid dead horses, over the walls of medieval castles. However, Kennedy uses his reconstruction to hurl pianos, dead sows, small cars, even flaming gasoline-filled toilets across a pasture on his estate near Shropshire, a feat described by a local paper as "Those Magnificent Men and Their Flaming Latrines." The record hurl for an upright piano is about 140 m; grand pianos average only about 90 m.* (a) Assuming an upright piano is launched at an altitude of 15 m and at an angle of 45° to the horizontal, what is the initial speed of the piano as it leaves the trebuchet? (b) The object to be launched is accelerated for about 1.5 s. Calculate the magnitude of the average acceleration to reach the speed calculated in part (a).

•**22.** A projectile is to be fired at angle θ to the horizontal. Place a Cartesian origin at the point of departure with the y-axis vertically up and the x-axis horizontal as shown in Figure P.22. The projectile must reach a peak height y_{peak}. Show that the initial speed v_0 of the projectile is

$$v_0 = \left(\frac{2gy_{\text{peak}}}{\sin^2 \theta} \right)^{1/2}$$

---

*Glynn Mapes, "A Scud it's not, but the trebuchet hurls a mean piano," *The Wall Street Journal*, 30 July 1991 (Volume CCXVIII, #21), pages A1 and A5.

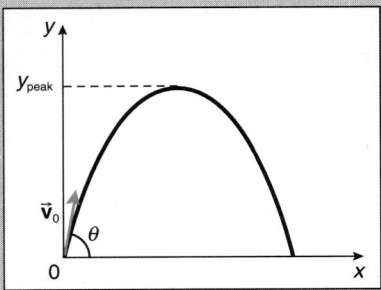

FIGURE P.22

(b) If the peak height of the projectile is to be 10.0 m and $\theta = 60.0°$, show that $v_0 = 16.2$ m/s.

•**23.** A projectile is fired with an initial speed of 16.2 m/s at 60.0° to the horizontal, as in Figure P.23. Use the coordinate system indicated. (a) Show that the peak height of the projectile is 10.0 m. (b) Show that if you pick any height y between 0 m and 10.0 m, say 5.0 m, there are two times later than 0 s when the projectile achieves this height. (c) Show that there is one time later than 0 s and one time before 0 s when the y-coordinate is less than 0 m, say y = −5.0 m. (d) Show that if you attempt to find the times the projectile is at a y-coordinate greater than the peak height of 10.0 m, say 15.0 m, the times are complex numbers. Interpret this result.

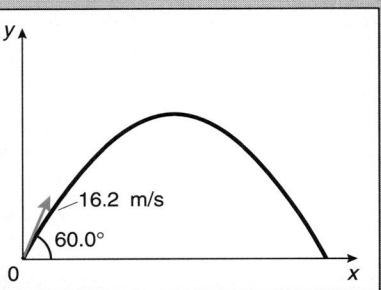

FIGURE P.23

•**24.** A projectile is launched as shown in Figure P.24 with initial velocity components v_{x0} and v_{y0}. Show that the x-coordinate of the peak of the trajectory, x_{peak}, is

$$x_{\text{peak}} = \frac{v_{x0} \, v_{y0}}{g}$$

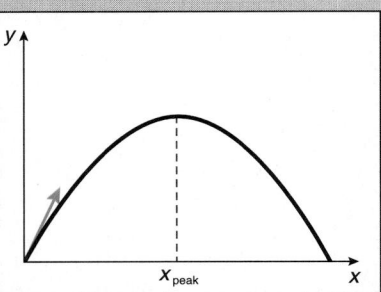

FIGURE P.24

25. A projectile is launched at speed v_0 at an angle θ (with the horizontal) from the bottom of a hill of constant slope β as shown in Figure P.25. Show that the range of the projectile up the slope is

$$\frac{2v_0^2 \cos\theta \, \sin(\theta - \beta)}{g \cos^2\beta}$$

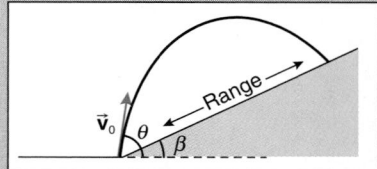

FIGURE P.25

26. The 18th green lies 200 m from a tee as indicated in Figure P.26. Golf pro Sandy Trappe clubs the ball heroically, and it leaves the tee making an angle of 40.0° with the horizontal. Regrettably the ball falls prey to a water hazard 180 m from the tee. (a) What was the speed of the errant ball as it left the tee? (b) If the ball left the tee at the same speed, but at an angle of 45.0° to the horizontal, could the ball have landed on the green? (c) Does the 45.0° launch angle ensure that the ball will travel the maximum horizontal distance? Explain your answer.

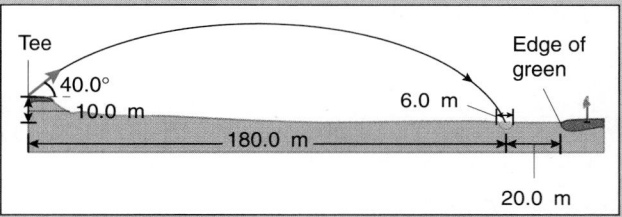

FIGURE P.26

27. A flea (*Ctenophalides canis*) jumps from level ground with an initial speed v_0 toward a nearby dog, but fails to reach the dog in one bound. At what angle θ should the flea jump so that its horizontal range is equal to the maximum height of the jump (see Figure P.27)?

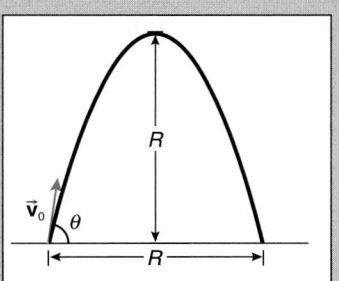

FIGURE P.27

28. Homer Ball, a designated hitter for the Physics Majors, connects with a baseball. The ball is caught 90.0 m from home plate 3.00 s later. The ball is hit and caught at an elevation of 1.00 m above the ground. (a) Find the angle that the ball makes with the horizontal direction as it leaves the bat.

(b) Find the initial speed of the ball. (c) Find the maximum height of the ball above the playing field.

29. A naturalist observes the South African clawed frog (*Xenopus laevis*) leap vertically to a height h. (a) With what speed did the frog leave the ground? Express your answer in terms of h and g. (b) If the frog used the same speed to leap horizontally for maximum range, what distance could it cover? (c) To what height does the frog ascend in making the leap for maximum horizontal range?

30. A football is kicked as pictured in Figure P.30. The initial velocity of the ball makes an angle θ with the horizontal direction. The angle ϕ locates the point of maximum height. Show that

$$\tan\phi = \frac{\tan\theta}{2}$$

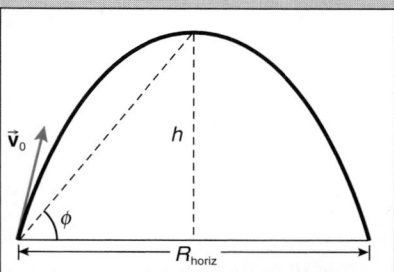

FIGURE P.30

31. A projectile is launched at speed v_0 toward a target located a distance R_{horiz} away over level terrain as indicated in Figure P.31. (a) Show that there are *two* possible launch angles, hence there is a high trajectory and a low trajectory, *except* when the launch angle is 45° for maximum horizontal range, in which case the two launch angles are identical to each other. The two launch angles are symmetrical about the 45° angle. (b) If $R_{\text{horiz}} = 100$ m, and $v_0 = 40.0$ m/s, find the two possible angles for launch. (c) Which launch angle is a quarterback on a football team likely to use?

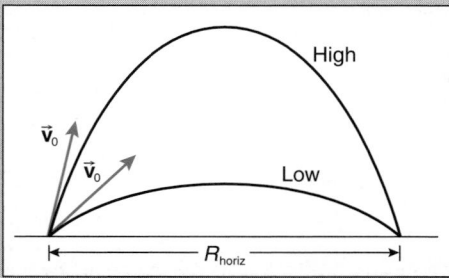

FIGURE P.31

32. (a) A projectile is launched over level terrain as indicated in Figure P.32. Let t_f be the time of flight of the projectile. (a) Show that the maximum height h_{max} to which the projectile rises is

$$h_{\text{max}} = \frac{gt_f^2}{8}$$

Thus if the time of flight of the projectile is known, the

maximum height can be calculated *without* knowing either the initial speed or the initial angle of the launch. (b) This has interesting implications in, say, baseball (if we ignore the effects of air resistance). Suppose a baseball is hit and caught at the same height above the ground, say 1.00 m. If the baseball is in the air for 4.00 s, what is the maximum height of the baseball above the field on its trajectory? Try this out on your friends at the next game you attend.

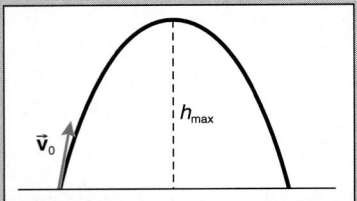

FIGURE P.32

‡33. One of the classic demonstrations in physics is known as the monkey–blowgun problem. A blowgun fires a small pellet at an object (a toy monkey or a tin can) suspended from the ceiling of a lab. See Figure P.33. The instant the pellet leaves the blowgun, the toy drops from its perch and falls vertically. Show that the pellet invariably hits the toy as long as the vertical distance of the fall is sufficiently large.

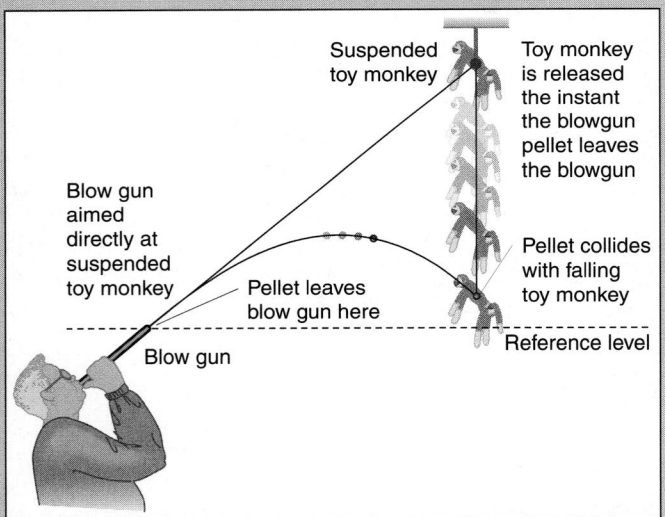

FIGURE P.33

‡34. An artillery mortar accidentally is shot approximately straight up, its initial direction being within 1.0° of the vertical direction. The initial speed of the shell is 400 m/s. Talk about a blunder! The gun crew stands still calculating for 1.00 s and then decides to make tracks in a straight line before the shell lands. (a) Assuming a nonexplosive impact of the shell, how far must they scatter horizontally from the launch point to ensure their safety? (b) With what average speed must they vamoose to get out of potential danger?

‡35. A projectile is launched as indicated in Figure P.35. (a) Show that the *y*- and *x*-coordinates of the projectile are related

by the equation

$$y = (\tan \theta)x - \frac{1}{2}\frac{g}{v_0^2 \cos^2 \theta}x^2$$

Since $y(x)$ is quadratic in x, the path of the projectile is a parabola. (b) In analytic geometry, the equation of a parabola passing through the origin, whose directrix is the line $y = c$ and with a focus at coordinates (x_f, y_f), is

$$2(c - y_f)y = (2x_f - x)x$$

Compare this equation with the result of part (a) to show that

$$x_f = \frac{v_0^2}{2g}\sin 2\theta$$

The mathematical definition of a parabola states that any point on the parabola is equidistant from both the directrix and the focus. Since the origin lies on the parabola, apply the definition mentioned in the previous sentence to the origin to show that

$$c^2 = x_f^2 + y_f^2$$

Use this relation in turn to show that

$$y_f = -\frac{v_0^2}{2g}\cos 2\theta$$

Use the equations for x_f and y_f to show that

$$c = \frac{v_0^2}{2g}$$

Recall that c is the distance of the directrix from the origin; notice that this result for the location of the directrix is the *same* as the height to which the projectile would rise if it were launched vertically. (c) Show that the polar coordinates (r_f, θ_f) of the focus are

$$r_f = \frac{v_0^2}{2g}$$

and

$$\theta_f = 2\theta - 90°$$

Since this last result can be rewritten as

$$90° - \theta = \theta - \theta_f$$

the angle that the position vector of the focus makes with the

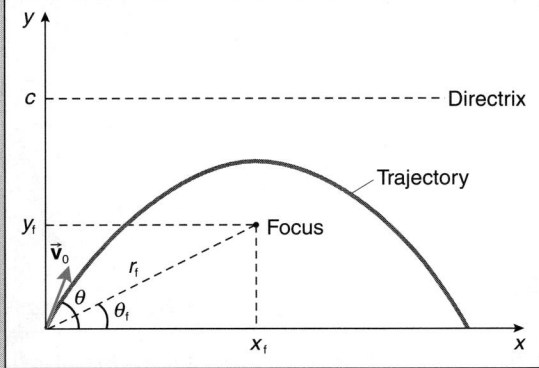

FIGURE P.35

initial angle of flight is equal to the angle between the vertical direction and the initial line of flight. (d) Show that the perpendicular distance of the projectile from the directrix (or the distance of the projectile from the focus) is, at any time,

$$\frac{v^2}{2g}$$

where v is the speed of the projectile.

This problem was inspired by an article by Larry D. Johnson, "The path of a projectile," *The Physics Teacher*, 30, #2, pages 104–105 (February 1992).

⬗36. A projectile is fired at initial speed v_0 from an elevation h above a level plain with the initial velocity vector making an angle θ with the horizontal as shown in Figure P.36a. (a) Show that the range R of the projectile satisfies the following equation:

$$0 \text{ m} = h + R \tan\theta - \frac{gR^2}{2v_0^2}\sec^2\theta$$

(b) By taking the ratio of the velocity components at the impact point, show that the impact angle ϕ satisfies

$$\tan\phi = \frac{(v_0^2\sin^2\theta + 2gh)^{1/2}}{v_0\cos\theta}$$

(c) For a fixed initial speed v_0, show that the angle θ that maximizes the range R is

$$\theta = \tan^{-1}\frac{v_0}{(v_0^2 + 2gh)^{1/2}}$$

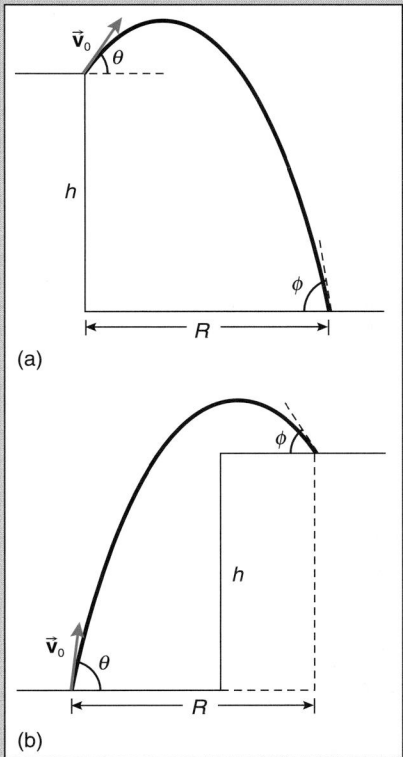

(a)

(b)

FIGURE P.36

(Hint: Find

$$\frac{dR}{d\theta}$$

using the equation in part (a).) Notice the numerator of the fraction in the equation for θ is the initial speed and the denominator is the impact speed of the projectile. (d) Show that the maximum range R_{max} then is

$$R_{max} = \frac{v_0}{g}(v_0^2 + 2gh)^{1/2}$$

(e) Show that when the projectile is launched at the angle for maximum range, the launch angle θ and the impact angle ϕ are related as follows:

$$\tan\phi = \cot\theta$$

This means that the launch and impact angles are complements of each other when the projectile has maximum range. (f) If the projectile is launched to a level plain at height h above the launch point as in Figure P.36b, will the arguments and results of parts (a) through (e) also apply?

This problem was inspired by an article by Carey S. Inouye and Eric W. T. Chong, "Maximum range of a projectile," *The Physics Teacher*, 30, #3, pages 168–169 (March 1992).

Sections 4.3 Motion in Three Dimensions
4.4 Relative Velocity Addition and Accelerations

37. If a swimmer swims at speed 5.00 km/h with respect to the water and a uniform current in a river is flowing at 4.00 km/h, in what direction must the swimmer aim so that the resulting path of the swimmer is directly across a river of width 100 m? How long will the crossing take?

38. A kayaker attempts to cross a channelized river in which the current has a uniform speed of 5.0 km/h across the 50.0 m width of the river. The kayaker is paddling at the frenetic speed of 10.0 km/h with respect to the water and wants to travel straight across the river to avoid a huge waterfall just downstream. (a) What is the resulting speed of the kayak with respect to the ground? (b) At what angle to the stream flow should the kayaker paddle so that the transit is made without moving upstream or downstream? (c) How long does the trip take?

•39. A pilot flies a plane at a speed of 700 km/h with respect to the surrounding air. A wind is blowing to the northeast at a speed of 100 km/h. In what direction must the plane be aimed so that the resulting direction of the plane is due north? What is the ground speed of the aircraft?

•40. Sergeant Pepper is on a bombing run in a helicopter aimed horizontally northward above a road. The speed of the helicopter is 40.0 m/s relative to the air. A wind is blowing to the west at a constant speed of 8.6 m/s relative to the ground. (a) What is the speed of the helicopter relative to the ground? (b) A water bomb is dropped from an altitude of 150 m in the direction of the Dean of Deans but misses its target by a whisker. How long will it take the bomb to hit the ground? (c) Where, relative to the helicopter, will the bomb hit? (d) What is the speed of the bomb the instant before impact?

•41. After an argument with a friend, you stomp off east at 1.50 m/s while he nonchalantly walks north at 1.00 m/s. (a) Introduce an appropriate coordinate system and calculate the velocity at which he recedes from you. (b) Does the answer to (a) depend on where you both are located?

•42. You head directly across a straight river of width 100 m, swimming at a speed of 1.5 m/s relative to the water. You arrive at the opposite bank having been swept downstream a distance of 50 m. (a) How long does the journey take? (b) What is the speed of the current? (c) What is your speed relative to the ground? (d) At what angle with respect to the upstream direction should you have headed to go straight across the stream without slipping downstream?

•43. A pilot of Fly-by-Night Airlines wishes to fly due southeast. A wind is blowing *to* the west at a speed of 120 km/h. The airspeed of the plane is 700 km/h. Determine (a) the speed of the plane with respect to the ground and (b) the angle the plane should be aimed with respect to the direction south.

•44. You are flying an airplane in a strong wind. An air traffic controller reports your velocity to be 700 km/h 30.0° east of north. The wind is reported to be 120 km/h due east. (a) Sketch the situation and introduce an appropriate coordinate system. (b) What is the velocity of the air with respect to the plane? (c) What is the speed of the air with respect to the plane?

•45. As an air traffic controller, you observe Flight 001 of U.S. Snarelines traveling due east at 600 km/h and Flight 002 of Continental Drift Airlines moving exactly northwest at 700 km/h. (a) What is the velocity of Flight 002 as seen by the pilot of Flight 001? (b) What is the speed of Flight 002 as seen by the pilot of Flight 001?

•46. An airplane flies at constant speed v relative to the air on a straight line from Washington, D.C., to Seattle and back, a round-trip distance of 2ℓ. Neglect layover time in Seattle. (a) If there is no wind, what is the round-trip flight time? (b) If there is a headwind of speed v_w going west and a tailwind of the same speed going east, what is the round-trip flight time? (c) Show that the round-trip flight time with a wind is greater than the round-trip flight time with no wind.

‡47. Two identical swimmers have stroke frequencies of one complete stroke per second (a right and left stroke is one complete stroke). They swim at a speed of 3.00 m/s in still water. They start swimming simultaneously from a fixed rock in a uniformly flowing river whose current speed is v. One swimmer goes 100 m downstream and returns to the starting point. The other swimmer moves directly across the stream (along a path perpendicular to the current) to a point 100 m across the river and also returns via the same route. Both swimmers execute perfect turns, taking zero time. The swimmer who went downstream and back returns to the starting point 5.00 s later than the cross-stream swimmer. (a) By how many stroke periods (the time for one right plus left stroke) does the cross-stream swimmer win? (b) Find expressions for the times the swimmers take to complete their courses in terms of the current speed v. (c) Calculate the speed of the current v from the given data.

‡48. Imagine snow falling vertically downward at constant speed v_s. A car is driving horizontally in a straight line through the snow at constant speed v. (a) Show that according to the driver, the snow appears to fall along a direction that makes an

angle θ with the vertical direction where

$$\tan \theta = \frac{v}{v_s}$$

If you tilt an umbrella at this angle while walking at speed v in snow (or rain) falling vertically at speed v_s, you keep driest. (b) This problem also has implications in astronomy. Imagine the snow to be starlight moving at constant speed c toward the solar system (see Figure P.48). The Earth, in its orbital motion, travels at speed v. Observers on the Earth see the light coming from a direction θ that is not in the direction of the true position of the star. The angle is found from

$$\tan \theta = \frac{v}{c}$$

Thus a telescope must be tilted in the direction of the orbital velocity of the Earth by an angle θ to see the star. Over the course of a year, the star executes a small circle of angular radius θ about its true position perpendicular to the plane of the orbit of the Earth. This effect is known as *stellar aberration*. The orbital speed of the Earth is about 30 km/s. Evaluate the angle θ. If the star is in the plane of the orbit, the aberrational path of the star is a line centered on the true position of the star. If the star is between the plane of the orbit and the perpendicular to the orbit, the aberrational path of the star is an ellipse.

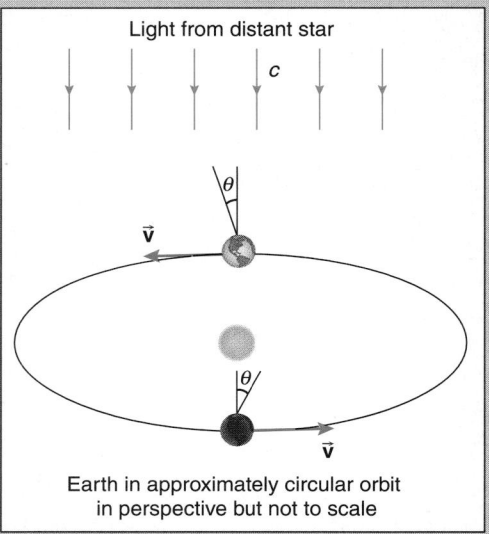

FIGURE P.48

Section 4.5 Uniform Circular Motion: A First Look

49. The tachometer on the dashboard of an automobile indicates the motor is turning at 3000 rev/min. What is the angular speed of the motor in rad/s?

50. Determine the angular speed in rad/s of a particle at rest on the surface of the rotating Earth.

51. A dentist has a new dental drill that spins at 5.00×10^4 rev/min. (a) Calculate the speed of a point on the rim of a drill bit of radius 0.500 mm. (b) Calculate the magnitude of the centripetal acceleration of a point on the rim of the drill bit and compare it with the magnitude of the local acceleration due to gravity, $g = 9.81$ m/s^2.

52. An object is executing circular motion at a constant speed of 1.00 km/s at an angular speed of 2.00π rad/s. What is the radius of the circular path of the object?

53. A fast sprinter can attain speeds of about 10 m/s. If the sprinter is running around a circular track at this (constant) speed, what radius circle results in a centripetal acceleration with a magnitude equal to $g = 9.81$ m/s²?

54. The orbit of the Earth around the Sun is approximately circular and of radius 1.496×10^8 km. It takes the Earth one year (about 365.25 d) to travel around the Sun. (a) Determine the speed of the Earth in its orbit in km/s. (b) Calculate the angular speed of the Earth in rad/s. (c) What is the magnitude of the centripetal acceleration (in m/s²) of the Earth in its orbit about the Sun?

55. The average distance between the Earth and the Moon is 3.84×10^5 km, about sixty times the radius of the Earth. Assume the orbit of the Moon is circular with the Moon completing one revolution in 27.322 d. (a) Calculate the magnitude of the centripetal acceleration of the Moon in m/s². (b) Calculate the ratio between the magnitudes of the centripetal acceleration of the Moon and the local acceleration due to gravity on the surface of the Earth:

$$\frac{a_c}{g}$$

The numerical value of this ratio is close to $1/3600 = 1/(60)^2$ and had great significance for Isaac Newton when he developed his theory of universal gravitation (as we will see in Chapter 6).

56. In the picturesque (but flawed) Bohr model of the hydrogen atom, the electron is imagined as orbiting the nucleus at a radius of 5.28×10^{-11} m with a constant speed of 2.18×10^6 m/s. (a) Calculate the magnitude of the centripetal acceleration of the electron. (b) How many times greater than the magnitude of our local acceleration due to gravity ($g = 9.81$ m/s²) is the magnitude of this centripetal acceleration? (c) Calculate the number of revolutions per second made by the electron in this orbit.

•57. Neutron stars are rather spectacular. They have radii on the order of only 10 km yet have masses on the order of *twice* the mass of our Sun. Neutron stars also spin quite fast; one such neutron star in the Crab Nebula (in the constellation Taurus) rotates 33 times each second. Calculate the centripetal acceleration of a particle on the equator of such a star.

•58. A physics student is located at latitude 38° North. (a) Calculate the centripetal acceleration of the student due to the rotation of the Earth. (b) At what places on the surface of the Earth is the centripetal acceleration due to the rotation of the Earth a maximum? (c) At what places is it a minimum?

•59. A particle is moving at constant speed 10.0 m/s in circles of various radii. (a) Make a graph of the magnitude of the centripetal acceleration of the particle versus the radius of the circle for radii between $r = 0$ m and 100 m. (b) At what radius is the magnitude of the centripetal acceleration of the particle equal to the magnitude of the local acceleration due to gravity ($g = 9.81$ m/s²)? What is the angular speed ω of the particle with speed 10.0 m/s at this radius?

•60. A particle is moving in a circle of radius 5.00 m. (a) Make a graph of the magnitude of the centripetal acceleration versus the speed of the particle for speeds ranging between 0 m/s and 10.0 m/s. What is the slope of the graph when $v = 0$ m/s? (b) Make a corresponding graph of the magnitude of the centripetal acceleration versus the *angular speed* ω of the particle for the same speeds indicated in part (a). What is the slope of this graph when $\omega = 0$ rad/s?

•61. The TGV (*Train à Grande Vitesse* or Train of Great Speed) high-speed trains between Paris and Lyon achieve speeds exceeding 300 km/h. If the magnitude of the centripetal acceleration of the train is not to exceed 4.0% of the magnitude of the local acceleration due to gravity ($g = 9.81$ m/s²), what is the minimum radius for a circular curve in the track? Explain why this is a *minimum* radius.

•62. The minute hand on Big Ben in London is 4.30 m long. (a) What is the speed of the tip of the minute hand? (b) What is the magnitude of the centripetal acceleration of a point at the tip of the minute hand? (c) Through what distance does the tip of the minute hand move during a year of 365 d?

•63. The Moon orbits the Earth in about 27.3 days, moving in a counterclockwise sense when viewed from above the plane of the orbit (*way* over the north pole of the Earth). Seen from the Earth, the Moon appears projected against a different star background each night. The Earth orbits the Sun in 365.25 days, also moving in a counterclockwise fashion when viewed from above the plane of the orbit. Because of the revolution of the Earth about the Sun, we see the Sun projected against a slightly different star background each day. The Sun and Moon both appear as circles in the sky with an angular size of about 0.5° when seen from the Earth. Assume the Moon and Earth move in circular orbits at constant speed. (a) About how many degrees per day does the Moon move in its orbital motion? (b) About how many degrees per day does the Sun appear to move against the distant star background as seen from the Earth? (c) Assume the Moon and Sun follow the same path when they are viewed against the distant star background (their paths actually are inclined to each other by about 5°). Both move in the same directional sense along this path. About how many days elapse between successive lunar passings of the Sun in the sky? This is the time interval of the phases of the Moon.

⁑64. Consider two planets in circular orbits, both planets moving at constant (but different) speeds about the Sun in a counterclockwise sense as indicated in Figure P.64a. The time it takes a planet to complete one circuit of its orbit is called the *sidereal period* of the planet. Let the sidereal period of the *inner* planet, called the inferior planet, be P_{inf} and that of the *outer* planet, called the superior planet, be P_{sup}. Let the planets have the initial configuration as indicated in Figure P.64a with both along a common Sun–planet line. The planets move at different speeds around their different orbits, with the inner planet moving faster. Their motion is analogous to a very unfair race between two mismatched runners on the inside and outside of a circular track with the faster runner on the inside and the slower runner on the outside. (a) Show that the time S it takes for the inner (and faster) planet to gain a full lap on the slower outer planet is

$$\frac{1}{S} = \frac{1}{P_{inf}} - \frac{1}{P_{sup}}$$

The time period S is known as the *synodic period* of the two planets. (b) Let the inner planet be the Earth ($P_{inf} = 1$ y) and the outer planet be Mars ($P_{sup} = 1.88$ y). Evaluate S. This is the time interval between successive *oppositions* of Mars (an *opposition* occurs when the planet and the Sun are in opposite directions as seen from the Earth). (c) Let the outer planet now be the Earth and the inner planet be Venus ($P_{inf} = 0.62$ y). Evaluate S. This is the time interval between successive *inferior conjunctions* of Venus (an *inferior conjunction* is when a planet closer to the Sun than the Earth is between the Sun and Earth). (d) If the planets begin as in Figure P.64b but move in *opposite* directions,* show that the time S when they next align along a common Sun–planet line is

$$\frac{1}{S} = \frac{1}{P_{inf}} + \frac{1}{P_{sup}}$$

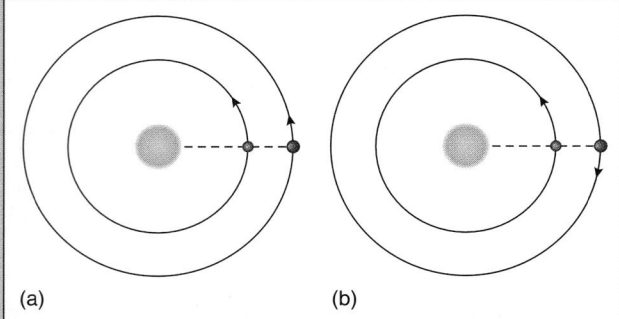

(a)   (b)

FIGURE P.64

Sections 4.6   The Angular Velocity Vector
4.7   The Geometry and Coordinates for Describing Circular Motion
4.8   The Position Vector for Circular Motion
4.9   The Velocity and Angular Velocity in Circular Motion
4.10   Uniform Circular Motion Revisited

65. (a) Calculate the angular speeds (in rad/s) of the second, minute, and hour hands of a clock; assume they rotate smoothly. (b) Do your answers for part (a) depend on the radius of the clock hands? (c) Do the angular velocity vectors of all the hands point in the same direction? Describe the direction of the angular velocity vector relative to the plane of the face of the clock. (d) What is the ratio of the angular speed of the second hand to the angular speed of the minute hand? (e) What is the ratio of the angular speed of the minute hand to the angular speed of the hour hand?

•66. The sweep second hand of a large clock is 10.0 cm long. Choose a coordinate system with its origin at the center of the clock and the y-axis toward the numeral 12 on the clock face and the x-axis toward the numeral 3, as shown in Figure P.66. (a) What is the velocity $\vec{v}$ of the tip of the second hand at the instant it points toward the number 3? (b) What is the change in the position vector of the tip of the second hand during the 15.0 s it takes to sweep from the numeral 12 to the numeral 3?

FIGURE P.66

(c) What is the acceleration of the tip of the second hand when it is at the numeral 3?

•67. Use Equation 4.28 for $\vec{r}(t)$:

$$\vec{r}(t) = [r \cos \theta(t)]\hat{\imath} + [r \sin \theta(t)]\hat{\jmath}$$

and Equation 4.26 for $\vec{\omega}$:

$$\vec{\omega}(t) = \omega_z(t)\hat{k}$$

to evaluate the expression

$$\vec{r}(t) \cdot [\vec{\omega}(t) \times \vec{r}(t)]$$

Explain why this result is another way to see that $\vec{v}(t)$ is always perpendicular to $\vec{r}(t)$ for circular motion.

Sections 4.11   Nonuniform Circular Motion and the Angular Acceleration
4.12   Nonuniform Circular Motion with a Constant Angular Acceleration

68. What is the angular acceleration $\vec{\alpha}$ of a particle on the surface of the rotating Earth?

69. An automotive engine shaft revs up from 3.00×10^3 rev/min to 4.50×10^3 rev/min during a 1.20 s interval. (a) Calculate the magnitude of the angular acceleration of the shaft assuming it is constant during the interval. (b) Determine the number of revolutions of the shaft during the interval.

•70. A particle moving clockwise in a circle with a radius of 3.00 m has a total acceleration at some instant of magnitude 15.0 m/s² directed as indicated in Figure P.70. (a) Find the magnitude of the centripetal acceleration at this instant. (b) Find the speed of the particle at this instant. (c) Find the angular speed of the particle at this instant. (d) Determine the magnitude of the tangential acceleration at this instant. (e) Determine the magnitude of the angular acceleration at this instant. (f) Make a sketch indicating the directions of $\vec{\omega}$, $\vec{\alpha}$, $\vec{r}$, and $\vec{v}$. (g) Is the particle speeding up or slowing down in its circular motion?

•71. A round object initially spinning clockwise on a fixed axis with an angular speed of 20.0 rad/s is subjected (when $t = 0$ s) to an angular acceleration of magnitude 0.500 rad/s² that slows it down in its circular motion. The angular acceleration ceases after 1.00 min. (a) Indicate in a schematic diagram the

---

*All the planets of the solar system move around the Sun in the same sense, not in opposite senses.

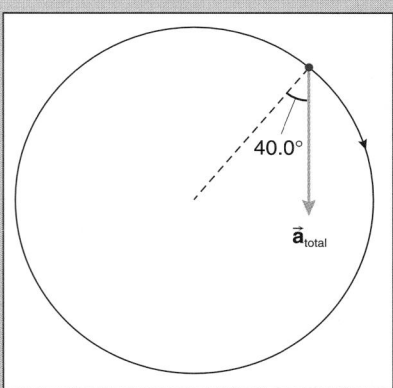

FIGURE P.70

rotational sense of the object, and the directions of the initial angular velocity vector and the angular acceleration vector. (b) Choose an appropriate direction for the +z-axis and indicate this direction on the sketch of part (a). (c) Tailor Equation 4.49,

$$\omega_z = \omega_{z0} + \alpha_z t$$

to the particulars of this problem. (d) Make a graph of ω_z versus t over the range of times from $t = 0$ s to 90.0 s. (e) At what time (if any) is the object instantaneously at rest? (f) *Describe* in words what happens to the object over the ninety second interval of the graph.

•**72.** An ice skater initially is spinning about a vertical axis at 150 rev/min. During 10.0 s, the rotational speed of the skater decreases at a constant rate to 100 rev/min. (a) Choose $\hat{\mathbf{k}}$ to be opposite to the direction of the angular velocity vector. Indicate this choice in an appropriate sketch. (b) What is the initial angular velocity component ω_{z0}? (c) What is the final angular velocity component? (d) Determine the angular acceleration component α_z. (e) Through how many revolutions did the skater turn during the 10.0 s interval?

•**73.** An ultracentrifuge accelerates from rest to 1.00×10^5 rev/min during 7.00 min. (a) Calculate the magnitude of the angular acceleration (assuming it is constant). (b) Through how many revolutions does the rotor turn in reaching this incredible angular speed? (c) The rotor has an effective radius of 8.00 cm. Calculate the magnitude of the centripetal acceleration of a point on the rim of the rotor when spinning at full speed. Compare this with the magnitude of the local acceleration due to gravity. (d) If it takes the rotor 4.00 min to brake to a stop from this angular speed, find the magnitude of the angular acceleration (assuming it is constant) and indicate on a sketch the orientation of the angular acceleration vector relative to the orientation of the angular velocity vector.

•**74.** A ceiling fan with blades 0.50 m long is rotating at 180 rev/min. The fan is shut off and coasts to a stop after 2.5 min. Let $\hat{\mathbf{k}}$ be in the direction of the angular velocity vector. (a) Determine the initial angular velocity vector of the fan in rad/s. (b) Assuming the angular acceleration is constant, determine the angular acceleration vector in rad/s². (c) Through how many revolutions does the fan turn in coming to rest? (d) When the fan was rotating at 180 rev/min, what was the magnitude of the centripetal acceleration of the points on the tips of the blades?

•**75.** A particle moving along a circle of radius 4.00 m has an instantaneous total acceleration of magnitude 40.0 m/s² directed as indicated in Figure P.75. (a) Indicate an appropriate direction for the z-axis. (b) What is the magnitude of the centripetal acceleration? (c) What is the magnitude of the tangential acceleration? (d) What is the angular velocity vector at the instant indicated? (e) What is the angular acceleration vector at the instant indicated? (f) Is the particle speeding up or slowing down in its circular motion? (g) Assume the angular acceleration is constant. Is the tangential acceleration then of constant magnitude? Is the centripetal acceleration of constant magnitude? (h) How long will it take the particle to come to rest?

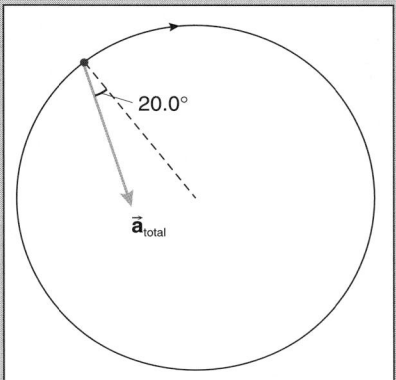

FIGURE P.75

•**76.** A particle moving in a circle of radius 2.00 m begins at rest when time $t = 0$ s and attains a speed of 10.0 m/s at the end of one revolution with a constant angular acceleration. (a) Determine the angular speed after one revolution. (b) Determine the magnitude of the angular acceleration and the time it takes to complete the first revolution. (c) At the instant the speed is 10.0 m/s, find

- the magnitude of the centripetal acceleration,
- the magnitude of the tangential acceleration,
- the magnitude of the *total* acceleration, and
- the angle that the total acceleration vector makes with the radial direction.

•**77.** A particle begins at rest on a circular track and is subjected to a constant angular acceleration of magnitude α beginning when $t = 0$ s. (a) Show that the magnitudes of the tangential and centripetal accelerations of the particle are *equal* when

$$t = \left(\frac{1}{\alpha}\right)^{1/2}$$

independent of the radius of the circular track. (b) What is the angle that the total acceleration vector makes with the radial direction at this time?

•**78.** Show that the ratio of the magnitudes of the tangential and centripetal accelerations in circular motion can be written in the form

$$\frac{a_t}{a_c} = \frac{\alpha}{\omega^2}$$

Notice that the ratio is equal to 1 when the angular speed ω has the value $\sqrt{\alpha}$.

•79. For the car considered in Example 4.14, find the instant during the 20 s interval when the tangential and centripetal accelerations have the same magnitude, so that the total acceleration vector makes an angle of 45° with the radial direction.

•80. A NASA engineer is designing a sophisticated merry-go-round apparatus capable of subjecting astronauts to centripetal accelerations with magnitudes up to five times the magnitude of the local acceleration due to gravity on the surface of the Earth—five *g*s, using the lingo of the trade. The radius of the machine is to be 20.0 m and the astronauts will be located on the rim of the device. (a) What is the speed of the chamber on the rim when the magnitude of the centripetal acceleration is 5.00*g*? (b) What is the angular speed of the device when the magnitude of the centripetal acceleration is 5.00*g*? (c) Determine the number of revolutions per minute executed by the device when the magnitude of the centripetal acceleration at the rim is 5.00*g*. (d) If the device is to get up to speed during one minute, determine the magnitude of a constant angular acceleration needed to accomplish the task.

•81. A turntable is moving at a constant angular speed of 0.500 rad/s. When $t = 0$ s, two centipedes (*Scutigera forceps*) at rest on opposite sides of the turntable (but not on the turntable) both spy a tasty morsel on the rim of the rotating turntable. See Figure P.81. The morsel is right next to the centipede named Millifoot when $t = 0$ s. Millifoot realizes, of course, that the morsel soon will be carried around by the turntable to the other centipede, named Multiplex, who is salivating eagerly in anticipation of the oncoming meal. Multiplex remains at rest while Millifoot sets off in hot pursuit of the morsel around the circle at a constant angular acceleration and manages to catch the morsel just as it arrives at the position of Multiplex. (a) Set up an appropriate direction for the +*z*-axis in a sketch. (b) At what time is the morsel next to Multiplex? (c) Determine the angular acceleration component α_z of Millifoot. (d) Make a graph of the angular velocity component of the morsel as a function of time. On the same graph, plot the angular velocity component of Millifoot. Indicate the point on the graph that corresponds to Millifoot catching the morsel.

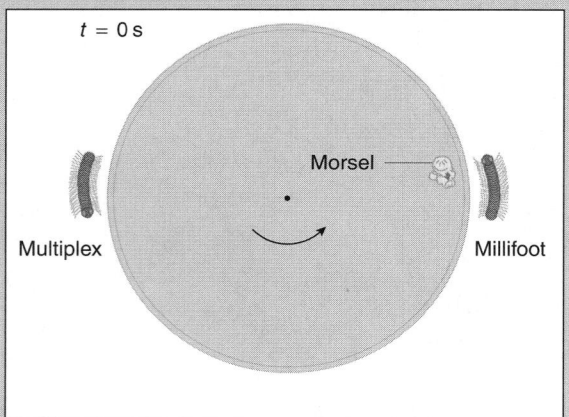

FIGURE P.81

•82. A particle with an initial angular velocity of
$$\omega_{z0}\hat{k} = (20.0 \text{ rad/s})\hat{k}$$
has a constant angular acceleration given by
$$\vec{\alpha} = (-2.00 \text{ rad/s}^2)\hat{k}$$
Consider the interval from $t = 0$ s to 10.0 s in what follows. (a) Make a schematic sketch of the situation indicating the rotational sense of the particle, the angular velocity vector, and the angular acceleration vector. Is the particle speeding up or slowing down in its rotational motion? (b) Make a graph of α_z versus t. (c) Find ω_z as a function of time. (d) What is the physical significance of the area under the α_z versus t curve between $t = 0$ s and 10.0 s? (e) Make a graph of ω_z versus t. (f) What is the physical significance of the area under the graph of ω_z versus t between $t = 0$ s and 10.0 s? (g) Through what angle does the particle turn during the interval?

•83. A car is traveling around a banked, circular curve of radius 150 m on a test track. At the instant when $t = 0$ s, the car is moving north, and its speed is 30.0 m/s but decreasing uniformly, so that after 5.0 s its *angular speed* will be 3/4 that it was when $t = 0$ s. (a) What is the angular speed of the car when $t = 0$ s? (b) What is the angular speed 5.0 s later? (c) Find the magnitude of the centripetal acceleration of the car when $t = 0$ s. (d) Find the magnitude of the centripetal acceleration of the car when $t = 5.00$ s. (e) Find the magnitude of the angular acceleration. (f) Find the magnitude of the tangential acceleration.

‡84. The position vector $\vec{r}(t) = x(t)\hat{i} + y(t)\hat{j}$ of a particle moving in the x–y plane is described by
$$\vec{r}(t) = [A \cos(\omega t)]\hat{i} + [B \sin(\omega t)]\hat{j}$$
where ω is a constant angular speed (in rad/s) and A and B are two distances (in m). (a) Make a sketch to indicate the location of the particle when $t = 0$ s. (b) On the same sketch indicate the location of the particle when $t = \pi/(2\omega)$. Notice that the particle is *not* moving in a circular path if $A \neq B$. (c) Find an expression for the velocity of the particle as a function of time. (d) Is the velocity vector $\vec{v}(t)$ perpendicular to $\vec{r}(t)$ at *all* times? Is the velocity vector perpendicular to $\vec{r}(t)$ at *some* times? If so, what are these times and where is the particle at these times? (e) Find an expression for the acceleration $\vec{a}(t)$ of the particle. (f) Is the acceleration antiparallel to $\vec{r}(t)$ at *all* times? (g) Is the acceleration *perpendicular* to $\vec{v}(t)$ at *all* times? Is the acceleration vector perpendicular to $\vec{v}(t)$ at *some* times? If so, what are these times and where is the particle at these times? (h) Make use of the trigonometric relation
$$\sin^2(\omega t) + \cos^2(\omega t) = 1$$
to find an expression for the orbital path of the particle in terms of the Cartesian coordinates x and y of the position vector, and the constants A and B but *not* the time t. What is the name for the *shape* of this path?

‡85. A particle is moving in a circle at a speed that is *not* constant. Consider Equation 4.28:
$$\vec{r}(t) = [r \cos \theta(t)]\hat{i} + [r \sin \theta(t)]\hat{j}$$

(a) Show that the acceleration vector can be written as

$$\vec{a}(t) = \left\{ [-r \sin \theta(t)] \frac{d^2\theta(t)}{dt^2} - [r \cos \theta(t)] \left[\frac{d\theta(t)}{dt} \right]^2 \right\} \hat{i}$$

$$+ \left\{ [r \cos \theta(t)] \frac{d^2\theta(t)}{dt^2} - [r \sin \theta(t)] \left[\frac{d\theta(t)}{dt} \right]^2 \right\} \hat{j}$$

$$\equiv a_x \hat{i} + a_y \hat{j}$$

(b) Introduce a new coordinate system with one axis directed radially outward (with unit vector $\hat{r}$) and the other tangent to the circle in the direction of increasing θ (with unit vector $\hat{\theta}$) as indicated in Figure P.85. Show that the component of the acceleration along the radial direction is

$$a_r = a_x \cos \theta + a_y \sin \theta$$

and the component of the acceleration in the tangential direction of increasing θ is

$$a_\theta = -a_x \sin \theta + a_y \cos \theta$$

(c) Let s be the arc length traveled by the particle along the circle, so that

$$s = r\theta$$

Show that the tangential acceleration component is the same as

$$\frac{d^2 s}{dt^2}$$

Show that the radial acceleration component is

$$-\frac{\left(\dfrac{ds}{dt} \right)^2}{r}$$

Notice that this is the centripetal acceleration component and is equal to $-v^2/r$.

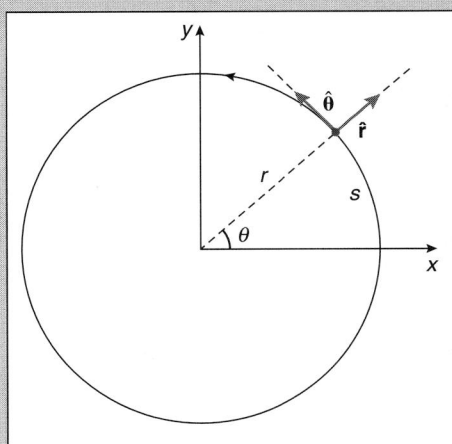

FIGURE P.85

86. You likely have noticed that as a cassette tape is played on a boom box or other noise generator, the radius of the tape left on the supply spool decreases slowly with time as the tape is drawn through the machine at a constant speed v_0. Choose $\hat{k}$ to be parallel to $\vec{\omega}$. Since the radius of the tape remaining changes slowly compared with the speed of the tape through the machine, to a good approximation we can write

$$v_0 = r \frac{d\theta}{dt}$$

where

$$\frac{d\theta}{dt} = \omega_z > 0 \text{ rad/s}$$

is the *angular speed* of the tape reel. Notice that the tape speed v_0 is constant, but r changes slowly with time; thus the *angular speed* of the spool also changes slowly with time. Assume that r represents the average radius of each coiled layer (since the tape actually is wound in a *spiral*). Let r_0 be the initial radius of the tape when $t = 0$ s and $\theta_0 = 0$ rad, where θ is the total angle of tape unwound from the spool. Represent the thickness of the tape by b. When the tape has unwound through a total angle θ, the radius remaining is

$$r = r_0 - b \frac{\theta}{2\pi} \qquad (1)$$

(a) Show that

$$\frac{dr}{dt} = -\frac{b}{2\pi} \frac{d\theta}{dt}$$

$$= -\frac{b}{2\pi} \frac{v_0}{r}$$

(b) Integrate the result of part (a) to show that

$$\pi(r_0^2 - r^2) = b v_0 t$$

What is the geometrical interpretation of the left-hand side of this equation? What is the physical interpretation of the right-hand side of this equation? (c) Solve the result of part (b) for r and substitute this for r in equation (1). Integrate the resulting expression to show that the total angle θ through which the tape has turned at time t is

$$\theta = \frac{2\pi}{b} \left\{ r_0 - \left[r_0^2 - \frac{b v_0 t}{\pi} \right]^{1/2} \right\}$$

This problem was inspired by a paper by R. Jordinson, "The angular motion of the supply spool on a tape recorder," *Physics Education*, 15, #5, pages 240–241 (July 1980).

87. The position vector of an electron moving in a magnetic field is described by the expression

$$\vec{r}(t) = [r \cos(\omega t)]\hat{i} + [r \sin(\omega t)]\hat{j} + v_{z0} t \hat{k}$$

where r, ω, and v_{z0} are positive constants. Make a schematic three-dimensional plot of the trajectory of the electron in space. Describe the motion in a concise sentence. When charged particles emanating from the Sun encounter the magnetic field of the Earth, they move in trajectories similar to this one. When the particles collide with molecules in the atmosphere of the Earth, auroras are produced.

INVESTIGATIVE PROJECTS

A. Expanded Horizons

1. Aspects of kinematics and projectile motion are involved in the sport of juggling. Investigate the physics of this skilled sport.
   Bengt Magnusson and Bruce Tiemann, "The physics of juggling," *The Physics Teacher*, 27, #8, pages 584–588 (November 1989).
   Eric Kincanon, "Juggling and the theorist," *The Physics Teacher*, 28, #4, pages 221–223 (April 1990).

2. Experienced sharpshooters know that if a rifle sight is adjusted for hitting the bull's-eye on a horizontal range, the bullet invariably will hit *above* the bull's-eye when aimed at the bull's-eye when firing at *either* an uphill *or* a downhill target placed the same straight-line distance away. Investigate why this is the case.
   Ole Anton Haugland, "A puzzle in elementary ballistics," *The Physics Teacher*, 21, #4, pages 246–248 (April 1983).
   H. A. Buckmaster, "Ideal ballistic trajectories revisited," *The American Journal of Physics*, 53, #7, pages 638–641 (July 1985).

3. The kinematics of living creatures is interesting. Land-based living things have different numbers of legs for locomotion. Snakes have no legs, humans have two, horses have four, insects have six, arachnids have eight, while centipedes and millipedes have many. How do snakes manage to move? What is the difference between millipedes and centipedes? Do more legs mean a greater speed per body length of the creature? Investigate the work of Robert J. Full on the kinematics of locomotion at the Poly-pedal Laboratory at Stanford University.
   Arthur Herschman, "Animal locomotion as evidence for the universal constancy of muscle tension," *American Journal of Physics*, 42, #9, pages 778–779 (September 1974).

B. Lab and Field Work

4. Most physics laboratories have a device called a ballistic pendulum. Part of the apparatus is a spring gun that can project a metal ball across the room. Devise a method for accurately measuring the range of the ball and use the data to determine the speed of the ball when projected from the spring gun.

5. For high jumps or long jumps, make a few measurements to determine the approximate speed of the athletes as well as the angle they commence flight.
   Herbert H. Lin, "Newtonian mechanics and the human body: some estimates of performance," *American Journal of Physics*, 46, #1, pages 15–18 (January 1978).
   Cliff Frohlich, "Effect of wind and altitude on record performance in foot races, pole vault, and long jump," *American Journal of Physics*, 53, #8, pages 726–730 (August 1985).

6. Devise a method for measuring the angular speed of a fan blade when at the low, medium, and high speed settings. Measure how long it takes the fan to reach these angular speeds from rest, as well as from low to medium, and medium to high settings. Determine the magnitude of the corresponding angular accelerations, assuming each is constant. Do you get the same magnitude angular acceleration for each change of angular speed? Determine the centripetal accelerations of a point on the rim at each of these settings.

7. Secure an automobile or marine tachometer from an automotive or marine graveyard and dissect it to determine how it is able to indicate the number of revolutions per minute.

C. Communicating Physics

8. On the basis of your intuitive ideas about resistive effects and with the knowledge that the density of the atmosphere decreases with altitude above the surface of the Earth, formulate a hypothesis about whether the actual horizontal range of a long-range artillery shell will be greater or less than that predicted by the analysis of projectiles in Example 4.3. Be sure to explain the reasoning that leads to your hypothesis.
   Michael A. Day and Martin H. Walker, "Experimenting with the national guard: field artillery gunnery," *The Physics Teacher*, 31, #3, pages 136–143 (March 1993).

9. Astronomers use the word *rotation* for what a physicist calls spin and the words *revolutionary motion* for what a physicist calls orbital motion. In a cogent paragraph, describe the meaning of the terms in language appropriate for a nontechnical audience such as a junior high school class, illustrating the distinctions with the motion of some everyday object. Use a thesaurus to see if other words convey the same meaning.

"YOU'RE LUCKY! YOU GET REPEAT BUSINESS!"

See Problem 21.

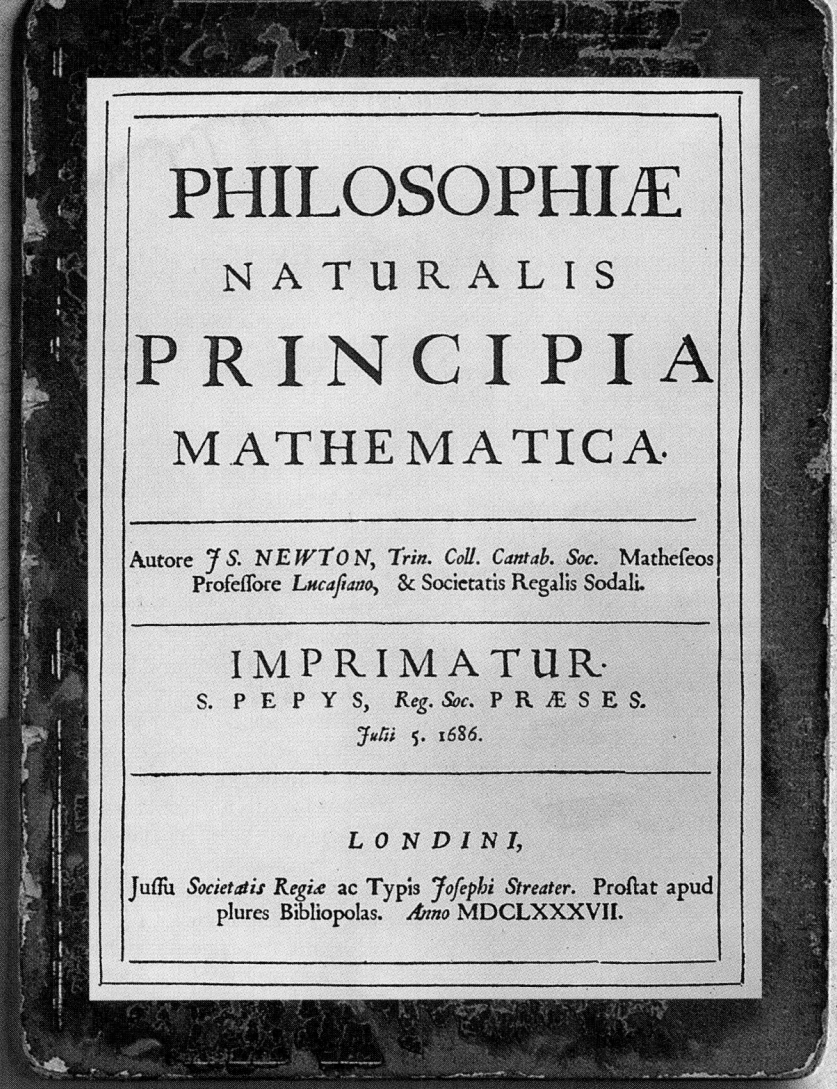

CHAPTER 5

NEWTON'S LAWS OF MOTION

Nature, and Nature's Laws lay hid in Night.
God said: *Let Newton be*! and All was *Light*.
*Alexander Pope (1688–1744)**

The description of motion, or kinematics, is half of the picture we wish to paint. Its complement is an understanding of the underlying *causes of changes in motion:* dynamics. What causes an acceleration? For that matter, what causes a constant velocity? You know, from common experience, that if you stop peddling your bicycle on level ground, you soon come to rest. Why? These are profound questions whose answers we address in this chapter.

Historically, dynamics was a sticky problem. When a pitcher accelerates a baseball from rest to its release, an arm is the cause of the acceleration. Likewise, when a car slams into a tree, the immediate cause of the brief, perhaps catastrophic acceleration of the car is the tree. Such causes are descriptive but are not sufficiently general to constitute a general theory for the causes of accelerations. Now think about dropping a pencil from your hand. The pencil falls to the floor. What causes the downward acceleration of the pencil? In this case, when you *release* the pencil, it accelerates, seemingly of its own accord; *nothing obvious* is in physical contact with the pencil during its fall to cause its acceleration![†] Likewise, it is not clear what causes the centripetal acceleration of the Earth in its approximately circular orbit around the Sun, since nothing physically connects the Earth to the Sun. Apparently, whatever causes accelerations can act either with or without direct physical contact.

The preceding examples illustrate why the search for possible causes for changes in motion was so difficult. As with kinematics, our scientific forebears were faced again with *inventing* new, unifying physical concepts to encompass a wide range of seemingly disparate physical phenomena whose only common thread was the kinematical description.

It was Isaac Newton (1642–1727) who, in 1686, first published a consistent theory of the motion of particles, based on the notion of **forces**. His theory is called **Newtonian mechanics**, or classical dynamics. While extraordinarily encompassing, it does have limitations (discussed in Section 5.7), as do all physical theories. Some of these limitations were discovered rather late in the game, during the early 20th century.

---

*(Chapter Opener) *Epitaph Intended for Sir Isaac Newton.*

[†]To say that *gravity* causes the rock to fall does not really provide an answer; this is essentially saying "it falls because it falls."

Isaac Newton also made significant contributions to our understanding of optics, as this stamp commemorates.

In this chapter, we first take a brief, modern look at fundamental particles and the fundamental forces of nature. We then elaborate on the concept of what we mean by a force through its action on a standard mass, the standard kilogram. Subsequently, we examine **Newton's laws of motion**, which form the basis for the study of classical dynamics.

Finally, later in the chapter, we also consider the peculiar aspects of several forces commonly encountered in many practical applications, realizing that these common forces are convenient simplifications of several of the fundamental forces of nature.

5.1 FUNDAMENTAL PARTICLES*

So, Nat'ralists observe, a Flea
Hath smaller Fleas that on him prey,
And these have smaller Fleas to bite 'em,
And so proceed ad infinitum

Jonathan Swift (1667–1745)[‡]

Although it is not obvious, there is an underlying lumpiness or graininess to the masses of everyday macroscopic objects. Even the ancient Greeks thought that there were tiny particles, called atoms,[§] that could not be subdivided into smaller particles. In other words, there are **fundamental particles** that constitute everything from atoms and molecules, to textbooks, soda cans, physics students, fleas, and other systems of interest, use, or annoyance.

Curiously, we now know that some of the fundamental particles in physics have *zero mass*, such as **photons** (the particles of light) and (perhaps) peculiar particles known as **neutrinos** (both of which we will encounter in Chapter 26).[#] Such zero-mass particles *always* travel at the speed of light (although the reasons for this are not apparent at this juncture in our study of physics) and have special dynamical properties that we study when we examine relativity (in Chapter 25).

With the discovery of the electron in 1896, we realized that the elemental atoms of the Greeks had structure. By the early 1930s, the fundamental particles of nature were thought to be electrons, protons (discovered in 1921), neutrons (discovered in 1932), and the massless photons of light (proposed in 1905). However, hosts of new fundamental particles were discovered during the mid-20th century, creating what has been described as an elementary particle zoo because of the variety and number of them (there are hundreds). We group the many particles into three families called **leptons**, **quarks**, and **mediating particles** (see Table 5.1). The familiar electron is a member of the lepton family and is thought to be the particle with the smallest nonzero mass. **Hadrons** are composite particles consisting of various combinations of members of the quark family. For example, the proton and neutron, each composed of three quarks, are hadrons. The proton is composed of two up quarks and one down quark,

---

[‡]*On Poetry: A Rapsody*, lines 337–340. *The Poems of Jonathan Swift*, edited by Harold Williams (Oxford at the Clarendon Press, Oxford, England, 1937), volume II, page 651.

[§]The word comes from the Greek ἄτομος, meaning "indivisible."

[#]It is still not known for certain whether neutrinos have a mass exactly equal to zero. Evidence is increasing for a small, nonzero mass.

TABLE 5.1 Names and Symbolic Representations of What Are Thought to Be the Fundamental Particles of Nature

	Leptons		Quarks		Mediating particles
e	electron	u	up	γ	photon
ν_e	electron neutrino	d	down	g_i	gluons (8 exist)
μ	muon	c	charm	W^-	
ν_μ	muon neutrino	s	strange	W^+	vector bosons
τ	tau	t	top	Z^0	
ν_τ	tau neutrino	b	bottom		graviton*

* Not yet found experimentally.

or $(u\ u\ d)$; the neutron is composed of one up and two down quarks, or $(u\ d\ d)$. The quark names are playfully picturesque rather than significant. The photon is an example of a member of the mediating particle family. Other exotic particles in the families need not concern us for now. Frankly, we do not know if this grouping of particles is the whole story or whether there exists yet another underlying arrangement of still more fundamental particles or families.

For each fundamental particle there is a corresponding **antiparticle**. Each particle and antiparticle have identical mass, but other properties differ, such as opposite electric charge. For example, the antielectron, called a **positron**, has the same mass as an ordinary electron, but has a positive charge. An **antiproton** has the same mass as a common proton, but has a negative charge. Some neutral particles are their own antiparticle. Particle and antiparticle pairs are created when there is sufficient energy available to "condense" and form their mass (see Chapter 25). For now, do not be overly concerned with these fundamental particles and their antiparticles for we must first master Newton's laws of motion for common macroscopic particles, such as baseballs and hockey pucks!

5.2 THE FUNDAMENTAL FORCES OF NATURE*

Forces of this kind . . . produce every action in this world.
Roger Bacon (c. 1214–1294)[†]

The search for simplicity and unity in physics is well illustrated by the attempts to enumerate the fundamental forces of nature. The problem is not an easy one.

The **gravitational force** was the first fundamental force to be recognized and described analytically (by Isaac Newton during the 17th century). We consider one aspect of the gravitational force in Section 5.10; Chapter 6 studies the gravitational force in some detail. The gravitational force is an example of a **long-range force**, for its effects are felt over immense distances. Gravitation is the force that dominates our immediate world, the structure of the solar system, and even the dynamics of the universe as a whole. Gravitation truly is a universal force, acting however weakly or strongly on every particle in the universe.

As the nature of electric and magnetic phenomena were investigated during the Renaissance, distinct electric and magnetic

forces were introduced. In the mid-19th century, James Clerk Maxwell integrated both forces into a single theory of electromagnetism. As the 20th century dawned, just two fundamental forces were known to exist in nature: the **electromagnetic force** and the gravitational force. The electromagnetic force, like gravitation, is a long-range force: electrical and magnetic effects can be felt over large distances. The long-range magnetic effects of the Earth are apparent on a simple compass needle, and the long range of electrical forces is spectacularly revealed in thunderstorms. We study the electric force in Chapter 16 and the magnetic force in Chapter 20; their connection is considered in Chapter 25.

The development of the nuclear model of the atom during the early 20th century led to the discovery of two additional fundamental forces, both **short-range forces**. The effects of these new forces are felt only over very small distances, comparable in size to the nuclei of atoms. The newly discovered forces were suggestively (but unimaginatively) named (1) the **strong force** (also known as the **nuclear force**), which manifests itself in interactions within the nucleus of the atom and among quarks; and (2) the **weak force**, which manifests itself in certain types of radioactive decay processes and among the leptons. The strong and weak forces we mention only for completeness; we will not consider them further.

How are short- and long-range forces transmitted between particles? For example, how is the gravitational force of the Earth on the Moon transmitted to the Moon? Likewise, how is the Earth aware of the gravitational presence of the Sun? There is no direct physical contact between these celestial bodies; the space between them is a vacuum, devoid of material particles. The same problem arises with the other fundamental forces. Newton assumed that the gravitational force was felt instantaneously and that changes or rearrangements of the mass of the Earth or Sun were instantaneously apparent at any distance. Such instantaneous response was called **action-at-a-distance**, and this concept troubled physicists for many years because of its implicit assumption that the effects of forces propagate at infinite speeds.

In the 19th century, Michael Faraday invented another useful mechanism for force transmission. He proposed that a massive particle, an electric charge, or a magnet establishes a peculiar condition of space, called a **field**, in the space surrounding itself. *Other* massive particles, charges, or magnets placed *in* the fields feel their effect at the place where these other particles are located in space. Each fundamental force has its own field; thus, we have gravitational fields that act on masses, electric fields that act on electric charges, and magnetic fields that act on moving charges and magnets. We will see the usefulness of the field concept when we study gravitation, electricity, and magnetism in greater detail. Changes are propagated via the field at finite speeds.

Another view, born of the age of quantum mechanics, imagines force transmission via the exchange of surrealistic elementary particles. The exchanged particles, the carriers of the force, are called mediating particles (refer back to Table 5.1) to disguise the difficulty in imagining how nature actually manages to pull this off. We return to this question again in Chapter 9 when we investigate collisions between particles. The mediating particles for each of the four fundamental forces are listed in Table 5.2.

[†]*Opus Majus*, translated by Robert Belle Burke (University of Pennsylvania Press, Philadelphia, 1928), volume I, page 130.

TABLE 5.2 Fundamental Forces and Their Mediating Particles

Fundamental force	Mediating particle(s)
1. Gravitational force	graviton (not yet been detected experimentally)
2. Electromagnetic force	photon (the particle of light)
3. Strong force	8 gluons (collectively designated g_i)
4. Weak force	3 vector bosons (called W^+, W^-, and Z^0)

The gravitational and electromagnetic forces are long-range because the mediating particles for these forces have zero mass. The strong and weak forces are short-range because the mediating particles have nonzero mass. In other words, the *greater* the mass of the mediating particle, the *shorter* the range of the force. The connection between the range of the force and the mass of the mediating particle cannot be understood without quantum mechanics, so do not worry if the connection is cryptic to you; it certainly is *not* obvious.

The holy grail of theoretical physics is to show how these four forces naturally arise as special cases of a single all-encompassing theoretical structure or theory. Progress has been made toward such a **Theory of Everything**, but the goal has not yet

been achieved. The progress toward unification is depicted schematically in Figure 5.1. The electromagnetic and weak forces were unified into a single, so-called **electroweak force** in the 1960s. This marriage of the forces is similar to, but more mathematically complicated than, the unification of the electric and magnetic forces by Maxwell during the 19th century. The strong force now is well on the way to being integrated into the structure. So far the gravitational force is the tough one, and it has eluded all attempts at integration into a grand scheme. We all can appreciate the common **rule of difficulty**: if you know how to do something, it is easy; if you do *not* know how to do something, it is hard. The unification problem is hard; nobody as yet knows how to do it all. The search continues. Keep your mind open and your imagination honed despite the orthodoxies thrust on you by current physical dogma. Nobel awards are lurking, but are yet unclaimed. Perhaps you will be the one to do it!

But hold on now. We are talking as if the notion of force and its measurement are obvious; they certainly are not. How do we go about defining what is meant by a force and how can it be measured? These topics we now address in Sections 5.3 and 5.4 by returning to the kinematics of a particle and the cause of changes in its motion.

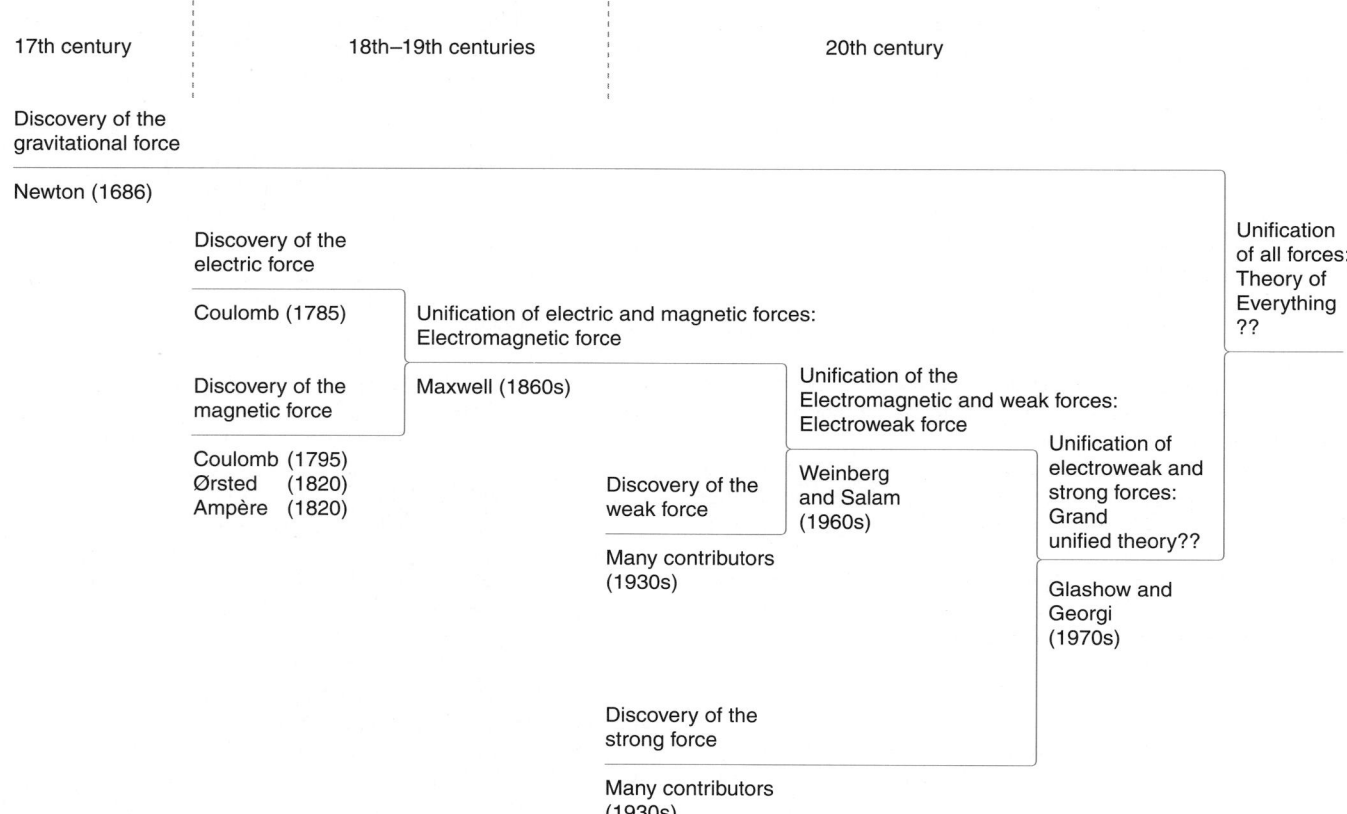

FIGURE 5.1 The quest for unification of the fundamental forces of physics. (Of course, the forces themselves have existed since very near "the beginning," but we only became aware of them during the indicated time frames.)

5.3 NEWTON'S FIRST LAW OF MOTION AND A QUALITATIVE CONCEPTION OF FORCE

Recall from Chapter 3 that a particle is any object whose geometric size is ignored; we think of a particle as represented by a point. The particle or particles whose motion is to be studied define a **system** and can be just about anything we want.

> **PROBLEM-SOLVING TACTIC**
>
> **5.1 It is essential to specify clearly the system whose motion is to be examined.** Think of drawing an imaginary surface around what you want to investigate; what is inside the imaginary surface is the system under consideration.

> In this chapter we consider only systems of constant mass.

> A *system that has an acceleration equal to zero is said to be in a* state of **mechanical equilibrium**.

The phrase mechanical equilibrium usually is contracted to simply the word equilibrium in classical mechanics.* The acceleration is the time rate of change of the velocity:

$$\vec{a} \equiv \frac{d\vec{v}}{dt}$$

Hence zero acceleration implies the system has a constant velocity vector, whose magnitude *and* direction do not change with time.

> A system with a constant velocity is in equilibrium.

A system that remains at rest has a constant velocity equal to zero, and so is a special case of a system in equilibrium.

Slide a hockey puck across the horizontal surface of your desk, as in Figure 5.2. Since we are interested in the motion of the hockey puck, define it to be the *system*. After release, the puck slows down and eventually comes to a stop. While doing so, it is not in equilibrium since its velocity changes as it slows; the puck experiences a nonzero acceleration. Now slide the puck across a smooth, ice hockey rink. The velocity of the puck still changes, albeit more slowly than before (Figure 5.3). The puck again is not in equilibrium, but its acceleration is less than when it was slid across your desk.

---

*Later, when we study thermodynamics, we introduce another distinct concept known as *thermodynamic equilibrium*.

Galileo Galilei (1564–1643) first extrapolated such situations to an idealized *gedanken* (thought) experiment on a horizontal surface with *no* **friction**.† On such a so-called **frictionless surface**, the motion of the puck continues indefinitely at constant velocity (zero acceleration). The puck then is in equilibrium.

> A system in equilibrium defines what we mean by **zero total force** on the system:
>
> | system in equilibrium | ⇔ | zero total force on the system |
> |---|---|---|
> | $\vec{a} = 0$ m/s^2 | | $\vec{F}_{total} = 0$ |
>
> The term *total force* and the term *net force* mean the same thing.

The expressions total force "*on* the system," "*acting on* the system," and "*exerted on* the system" are synonymous and are used interchangeably.

> **Newton's first law of motion** defines a state that we call zero total force on the system: *any system in mechanical equilibrium remains in mechanical equilibrium unless compelled to change that state by a* **nonzero** *total force acting on the system.*
>
> | system *not* in mechanical equilibrium | ⇔ | *nonzero* total force on the system |
> |---|---|---|
> | $\vec{a} \neq 0$ m/s^2 | | $\vec{F}_{total} \neq 0$ |

It is difficult to appreciate the revolutionary nature of Newton's first law of motion, since we now have the benefit of three centuries of hindsight. But prior to Galileo and Newton, it was thought that a nonzero total force was needed simply to maintain a constant velocity. Hands-on experience with everything from wine barrels to oxcarts indicated that to keep an object moving at constant velocity on a horizontal surface, a continual force (a push or pull) on the system is needed.‡ We

---

†We define what we mean by friction more carefully later in the chapter. The word friction stems from the Latin *fricatus*, meaning "rubbing."

‡This generalization was made despite the existence of things such as bows and arrows that called the premise into question: the arrow obviously continues to move once it has left the bow.

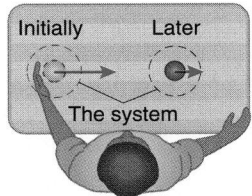

FIGURE 5.2 Slide a hockey puck across your desk.

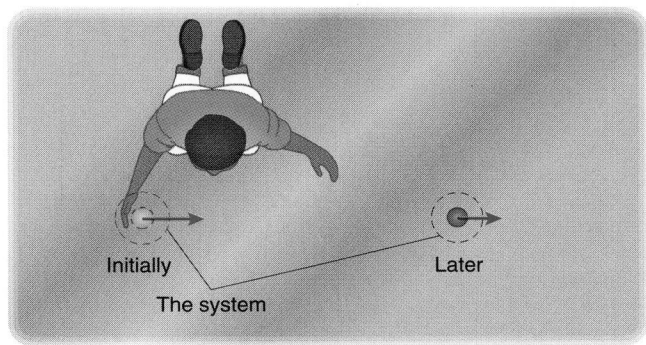

FIGURE 5.3 Slide a hockey puck across an ice rink.

now understand that our push or pull is not the only force or the total force in such situations. The only reason we need to continually push to maintain a constant velocity is to cancel the ever-present effects of frictional forces. The first law of motion states that *zero total force* exists when the system has a *constant velocity*; however, a *nonzero* total force is required to *change* the velocity. In other words, a nonzero total force on the system causes its acceleration.

Recall from Section 4.4 that a coordinate system, together with a set of synchronized clocks, all ticking at the same rate, constitute a reference frame. Those reference frames in which Newton's first law is found to be valid are called **inertial reference frames**; frames that are not inertial are called **noninertial reference frames**. These and other reference frames are examined in greater detail later in this chapter (Sections 5.7, 5.8, and 5.16).*

Thus Newton's first law gives a qualitative and quantitative definition of what we mean by zero total force; the law serves to define an inertial reference frame. Next we discuss a way to measure and quantify forces.

5.4 THE CONCEPT OF FORCE AND ITS MEASUREMENT

yea, every force entangles
Itself with strength

William Shakespeare (1564–1616)[†]

The mass of a system is a characteristic scalar property that indicates how sluggishly the system moves in response to forces (think of them as pushes or pulls) of various magnitudes; this property of the system (how slightly it responds to a force) is called **inertia**.[‡] The quantitative measure of inertia in the SI system is the kilogram. The mass of a system is the same whether it is here on the Earth, on the Moon, on Superman's fictional planet Krypton, or near the bright star Arcturus. The concept of mass really was invented to describe how a system responds to forces. Other perfectly legitimate ways of defining this characteristic inertial property of a system are possible, but they represent roads not taken in the history of physics (see Question 1 after reading this section).

Quantifying Force

Take the standard platinum–iridium mass cylinder at Sèvres, France, that defined the kilogram and place it on a horizontal frictionless surface; such a surface can be approximated by using

---

*Not all reference frames are inertial reference frames. We will see in Section 5.16 that if a reference frame is accelerating (such as a coordinate system attached to an airplane accelerating along a runway), then observations and experiments made with respect to this reference frame do *not* obey Newton's first law. A coordinate system attached to and rotating with a spinning turntable or merry-go-round is another example of a noninertial reference frame.

[†]*Antony and Cleopatra*, Act IV, Scene 14, lines 48–49.

[‡]The word comes from the Latin word *inertia*, meaning "inactivity."

an airtable in many physics laboratories. The standard mass is the system whose motion we want to examine. Pull horizontally on the standard mass with a spring of negligible mass whose stretch can be measured, as shown in Figure 5.4. The stretch of the spring helps us quantify the concept of force.

A stronger pull by you on the spring produces a greater stretch of the spring. If you vary the strength of the pull (as indicated by the stretch of the spring), you find that different accelerations are imparted to the standard kilogram. You can tell if the strength of the pull is constant by checking to see if the extension of the spring stays the same while the system is pulled. The strength of the pull together with its direction indicates that the pull has the salient characteristics of a vector quantity: a magnitude and a direction. Indeed, experiments show that (a) if you pull in a certain horizontal direction, the mass is accelerated in the *same* horizontal direction as the pull; and (b) if you repeat the experiment using the same mass, with the same strength of the pull, and in the same direction, the *same* acceleration always is produced.

> The strength of the pull needed to impart to the standard kilogram an acceleration of magnitude exactly 1 m/s^2 is defined to be a force of magnitude 1 **newton**, abbreviated N. See Figure 5.5.

The stretch of the spring corresponding to an acceleration of magnitude 1 m/s^2 of the standard kilogram can be marked on a scale adjacent to the spring. Thus an extension of the spring by

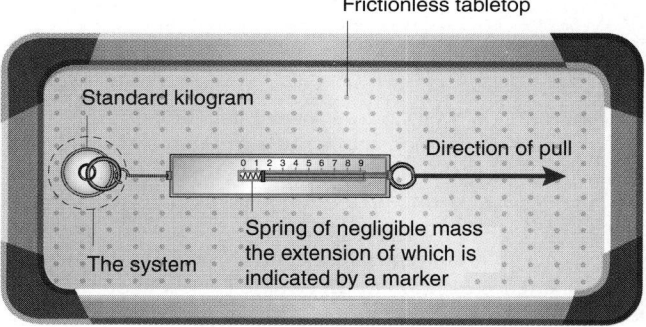

FIGURE 5.4 Pull on the standard kilogram with a spring.

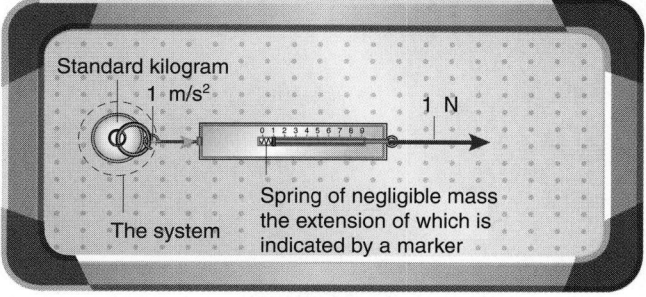

FIGURE 5.5 A total force of magnitude 1 N produces an acceleration of exactly 1 m/s^2 of the standard kilogram.

this amount corresponds to a force or pull of magnitude 1 N on *whatever* is attached to the spring.

Now perform another experiment: pull on the spring so the acceleration of the standard kilogram is precisely 2 m/s^2; note and mark the extension or stretch of the spring. This extension of the spring defines a force with a magnitude of precisely 2 N (see Figure 5.6). In this way the spring can be calibrated into an instrument capable of indicating the magnitude of a force in newtons. Further experiments are conducted with the calibrated spring on different systems. If the system consists of *two* replicas of the standard kilogram, the system has a mass of 2 kg. If the calibrated spring is used to pull and accelerate the 2 kg system, experiments indicate that

1. a force of magnitude 1 N imparts an acceleration of precisely 0.5 m/s^2 to the 2 kg mass; and
2. a force of magnitude 2 N imparts an acceleration of magnitude 1 m/s^2 to the 2 kg mass.

Indeed, the magnitude of the acceleration produced by a given magnitude force is *inversely* proportional to the mass of the system; the direction of the acceleration always is in the direction of the force or pull. For a system of constant mass, a graph of the magnitude of the force in N used in these experiments versus the magnitude of the resulting acceleration in m/s^2 (see Figure 5.7) is *linear* with zero intercept, and the *constant slope* of the graph is found to be numerically equal to the mass of the system in kg.

The slope of *every* such graph is positive; no system ever has been found with a negative slope for such a graph. Hence mass is always positive.

Force Is a Vector Quantity

Further experiments with *two* calibrated springs indicate that if a system on a horizontal frictionless surface is subjected to two pulls of the same magnitude in opposite horizontal directions, no acceleration results (see Figure 5.8). As Newton's first law states, with zero acceleration there must be zero total force on the system. The two applied forces, therefore, sum to zero like vectors, and so the total force on the system is zero.

Likewise, if the two pulls of the same magnitude are applied in the same horizontal direction (see Figure 5.9), the acceleration of the system is twice that with one pull of the same magnitude; the individual forces again add like vectors. An experiment with two forces of different magnitudes and directions (see Figure 5.10) produces an acceleration consistent with a total force that is the vector sum of the two.

> These experiments indicate that *forces are vectors* and the acceleration of the system depends on the vector sum of all forces on the system. The vector sum of all the forces on a system is the *total force* $\vec{\mathbf{F}}_{total}$ on the system.

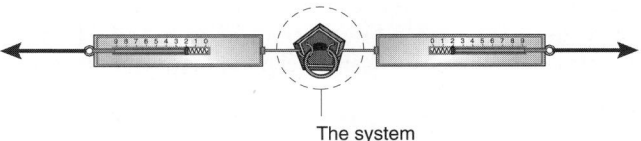

FIGURE 5.8 Two forces of equal magnitude and opposite direction produce zero acceleration of the system.

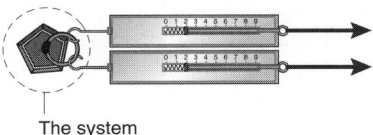

FIGURE 5.9 Two forces of equal magnitude and the same direction produce an acceleration twice that produced by either force acting alone on the system.

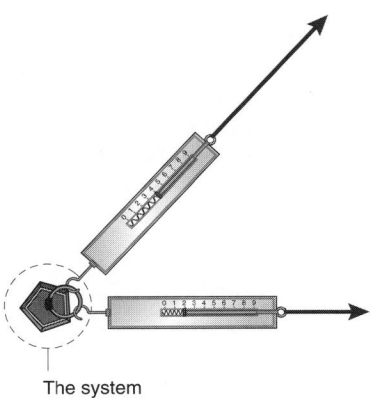

FIGURE 5.10 Two forces with different magnitudes and directions produce an acceleration consistent with a total force that is the vector sum of the two.

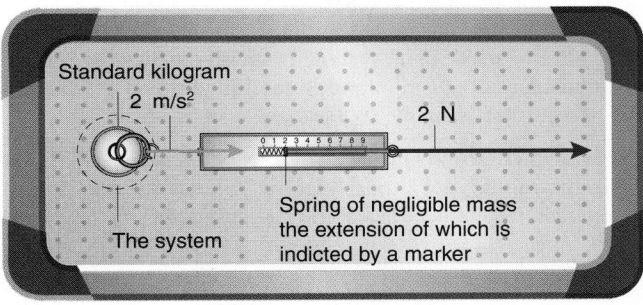

FIGURE 5.6 If the acceleration of the standard kilogram is exactly 2 m/s^2, the magnitude of the total force on it is exactly 2 N.

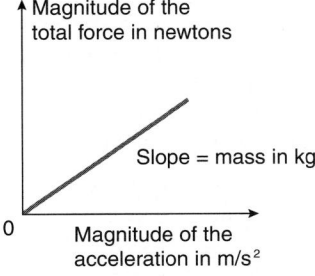

FIGURE 5.7 A graph of the magnitude of the total force on a system versus the magnitude of its acceleration is linear with zero intercept and a slope equal to the mass of the system.

5.5 NEWTON'S SECOND LAW OF MOTION

. . . it will be found that everything depends on the composition of the forces with which the particles of matter act upon one another; and from these very forces, as a matter of fact, all phenomena of Nature take their origin.

Roger Joseph Boscovich (1711–1787)*

The results of the experiments discussed in Section 5.4 and the graph in Figure 5.7 are summarized by **Newton's second law of motion**, one of the great and unifying laws of physics:

A system of constant mass m, subjected to a totality of individual forces whose vector sum is $\vec{F}_{total}$, experiences an acceleration $\vec{a}$ that is inversely proportional to the mass m of the system and is in the same direction as $\vec{F}_{total}$:

$$\vec{a} = \frac{\vec{F}_{total}}{m}$$

This is usually written as

$$\boxed{\vec{F}_{total} = m\vec{a}} \tag{5.1}$$

In terms of the base units of the SI system, one newton is equivalent to one kilogram meter per second squared ($kg \cdot m/s^2$), since (from Equation 5.1) a force of magnitude 1 N on the standard kilogram produces an acceleration of magnitude 1 m/s^2. That is,

$$1 \text{ N} = (1 \text{ kg})(1 \text{ m/s}^2)$$
$$= 1 \text{ kg} \cdot m/s^2$$

The characteristic scalar property of the system that appears in Newton's second law of motion is called the **inertial mass** of the system or, more colloquially, simply the **mass**.

Keep in mind that mass is an *abstract concept, defined by experiment*. We could just as easily have called the inertial property of matter by other names: "stuff" or "inertial charge"; but we choose to use the word *mass*.† The mass of a system relates the response of the system (its acceleration) to the total force *on* the system. Ultimately, we do not know what mass really is, or what gives rise to the property of inertia. Whatever this inherent inertial characteristic of matter really is, we call it mass. We will encounter other such abstract ideas in physics, invented to describe the results of various experiments and observations.‡

Newton's second law of motion, Equation 5.1, is a *vector equation*. We saw in Chapter 2 that a vector equation is equivalent to *three scalar equations*, one for each of the components of the vectors (with respect to some coordinate system). In a Cartesian coordinate system, the x-component of the left-hand side of the equation is equal to the x-component of the right-hand side of the equation, and likewise for the other two

Cartesian directions. That is, Equation 5.1 represents the following three scalar equations for the Cartesian components:

$$\begin{aligned} F_{x\,total} &= ma_x \\ F_{y\,total} &= ma_y \\ F_{z\,total} &= ma_z \end{aligned} \tag{5.2}$$

QUESTION 1

Suppose Newton did things differently and found that a particle has a scalar property χ associated with it such that the scalar multiplication of this characteristic property with the total force on the particle is equal to the acceleration of the particle; that is,

$$\chi\vec{F}_{total} = \vec{a}$$

(a) For a given total force $\vec{F}_{total}$ on the particle, large values of χ produce what kinds of accelerations? (A qualitative answer is OK.) What about small values of χ? (b) Invent a descriptive name that might describe this property χ of the particle. (c) If the magnitude of $\vec{F}_{total}$ is plotted versus the magnitude of $\vec{a}$, what does χ represent on the graph? (d) Suppose two particles, one with this property of value χ_1 and one with value χ_2, are fastened together to make a single system. What is the effective χ value of the combination? (e) What are the SI units associated with this property χ? (f) How is the property χ related to the scalar property known as the mass m of the particle, or is it related at all? (g) Which property, χ or m, is more *convenient* to use, or is this simply a question of which concept (χ or m) we are more familiar with? These questions should convince you that what we call the mass of a system is an inventive theoretical construct of our mind, not something inherently natural, waiting for someone to discover.

Newton's second law of motion, Equation 5.1, is laden with meaning and subtlety that belie its innocuous appearance. Before tackling applications of the law, we emphasize some of its subtle aspects.

1. The left-hand side of the equation is the *vector sum of all forces acting on the system* of mass m. The preposition "on" is key as well. *Only* the vector sum of the forces *on the system*§ is important for determining the motion of the system. We shall see later that forces *by* the system *on* other things are not involved in the motion of the system itself; they determine the motion of the other things.

2. Since Newton's second law is a vector equation, the direction of the total force vector and the direction of the acceleration vector are the same.

3. If the vector sum of all the forces *on* the system is zero, then the left-hand side of the second law of motion is zero:

$$0 \text{ N} = m\vec{a}$$

Theoria Philosophiae Naturalis (Venice, 1763), Part I, Section 5. [*A Theory of Natural Philosophy* (MIT Press, Cambridge, Massachusetts, 1966), page 20.]
†The word is believed to come from an ancient Greek word for "barley-cake."
‡Among them is the familiar concept of electric charge. We also do not know what gives rise to the property of matter we call electric charge.

---

§Remember that the phrases "on," "acting on," and "exerted on" a system are synonymous.

which implies that the acceleration vector $\vec{a}$ also is zero, and we get back Newton's first law of motion. You might be tempted to say that the first law is a special case of the second; do not be led into such temptation! As we mentioned in Section 5.2, Newton's first law *defines* what we mean by zero total force and an inertial reference frame, and so the first law has independent significance. As a matter of historical fact, the first law was discovered by Galileo.

4. Having $\vec{F}_{total}$ (the left-hand side of Newton's second law) equal to zero does *not* necessarily mean there are *no* forces on the system; it *does* mean that the vector sum of all forces on the system is zero.* Conversely, if the acceleration $\vec{a}$ of the system is zero, then the total force $\vec{F}_{total}$ on the system is zero and the system is in equilibrium.

5. The two sides of the equation are equal to each other, but the equality is of a peculiar sort. Imagine a wall of separation between the two sides of the equation: on the left are the physical forces acting on the system; on the right is simply the mass times the acceleration of the system, which we can think of as the *response* of the system to the forces on it. Think of the forces involved in $\vec{F}_{total}$ as distinct from the effects they cause. In other words, the total force is *not the same thing* as mass times an acceleration. The product $m\vec{a}$ is not what we mean by force; $m\vec{a}$ is the physical result of the action of forces on the system. The left-hand side represents a kind of action: the forces *on* the system. The right-hand side represents the response of the system to those forces. This wall of separation is a subtle point, but an important one; it will help you to avoid many pitfalls in applications of the second law of motion to physical situations.

> ### PROBLEM-SOLVING TACTIC
>
> **5.2 The two sides of Newton's second law represent different things.** The *left*-hand side of the law involves the vector sum of the *forces on the system*. The *right*-hand side is the product of the mass (a characteristic scalar property of the system) and the acceleration (a kinematic aspect of the system), which is the *response* of the system to the forces acting on it.

One of the best examples of the distinction between the left- and right-hand sides of the second law of motion arises in uniform circular motion. In this setting, the product of the mass times the magnitude of the centripetal acceleration

$$m\,\frac{v^2}{r}$$

has the units of a force ($\text{kg} \cdot \text{m/s}^2$ or N); but the product is *not* a force. Many texts and, regrettably, many physicists and engineers, call the product of the mass and the centripetal acceleration a "centripetal force." This is, in fact,

The result of an appropriate, short-lived, contact force on a baseball: smiles and a home run!

poor terminology that should be avoided. We will not use the term "centripetal force," since it is simply *not a force*. The left-hand side of the second law, then, involves forces such as the pulls measured by our springs or, as we shall see later, forces caused by tensions in cords or ropes, gravitational forces, pushes or pulls, and so on. We will examine the characteristics of these individual forces later in the chapter. The quantity mv^2/r is simply the product of the mass times the magnitude of the centripetal acceleration; it *is* the magnitude of the *right*-hand side of the second law equation for uniform circular motion.

6. There are two types of forces. For a ball hit by a bat, it is clear that the force of the bat on the ball exists only during the brief time interval during which they are in physical contact with each other; this is an example of a **contact force**, as we mentioned in the introduction to this chapter. Likewise, when you push a baby stroller with your hands, this force on the stroller exists only as long as your hands are in physical contact with it. On the other hand, the force of the Earth on a falling rock is an example of a force where physical contact between the two objects is not required; gravitation is a **noncontact force**. The other fundamental forces of nature (see Section 5.2) also are noncontact forces; contact forces are useful, macroscopic manifestations and simplifications of one or more of the fundamental forces (see Section 5.15).

EXAMPLE 5.1

A total force on a 10.0 kg mass produces an acceleration of magnitude 3.00 m/s² in the direction indicated in Figure 5.11. Find the magnitude and direction of the total force on the mass.

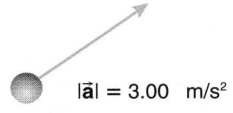

$|\vec{a}| = 3.00 \ \text{m/s}^2$

FIGURE 5.11

---

*Remember our experiment in Figure 5.8 with two springs pulling in opposite directions with equal magnitude forces; there are forces on the system, but their vector sum—that is, the total force—is zero.

Solution

Choose a coordinate system with the +x-axis parallel to the acceleration as indicated in Figure 5.12. The system is the 10.0 kg mass. The total force acting on the system is parallel to the acceleration. Apply Newton's second law of motion (Equation 5.1) to the system to find the total force:

$$\vec{F}_{total} = m\vec{a}$$
$$= (10.0 \text{ kg})(3.00 \text{ m/s}^2)\hat{i}$$
$$= (30.0 \text{ kg·m/s}^2)\hat{i}$$
$$= (30.0 \text{ N})\hat{i}$$

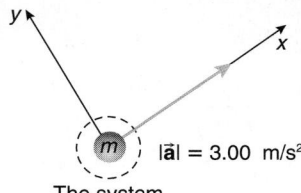

The system

FIGURE 5.12

STRATEGIC EXAMPLE 5.2

The only two forces acting on a 5.00 kg mass are indicated in Figure 5.13.

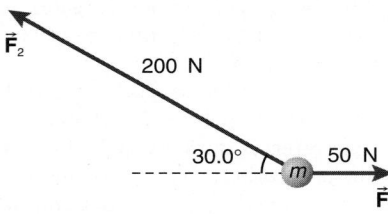

FIGURE 5.13

a. Determine the total force $\vec{F}_{total}$ acting on the mass.
b. Find the magnitude of the total force.
c. Find the acceleration of the mass.
d. Determine the magnitude of the acceleration.

Solution

Let the 5.00 kg mass be the system. Introduce the coordinate system indicated in Figure 5.14, with $\hat{i}$ parallel to $\vec{F}_1$.

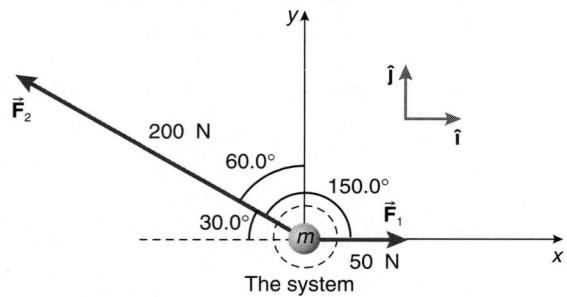

The system

FIGURE 5.14

a. Write each of the forces in Cartesian form. Since $\vec{F}_1$ is parallel to $\hat{i}$,

$$\vec{F}_1 = (50 \text{ N})\hat{i}$$

The Cartesian representation of $\vec{F}_2$ may be found in two ways.

Method 1
The Cartesian representation may be found from

$$\vec{F}_2 = (\vec{F}_2 \cdot \hat{i})\hat{i} + (\vec{F}_2 \cdot \hat{j})\hat{j}$$

The angle between $\vec{F}_2$ and $\hat{i}$ is 150.0° and the angle between $\vec{F}_2$ and $\hat{j}$ is 60.0°, and so you can evaluate the scalar products:

$$\vec{F}_2 = (200 \cos 150.0°)\hat{i} + (200 \cos 60.0°)\hat{j}$$

Method 2
Equivalently, you may find the Cartesian representation more geometrically from Figure 5.14:

$$\vec{F}_2 = (200 \cos 30.0°)(-\hat{i}) + (200 \sin 30.0°)\hat{j}$$

Using either method, you find

$$\vec{F}_2 = (-173 \text{ N})\hat{i} + (100 \text{ N})\hat{j}$$

The total force on the system is the vector sum of the two forces:

$$\vec{F}_{total} = \vec{F}_1 + \vec{F}_2$$
$$= (50 \text{ N})\hat{i} + [(-173 \text{ N})\hat{i} + (100 \text{ N})\hat{j}]$$
$$= (-123 \text{ N})\hat{i} + (100 \text{ N})\hat{j}$$

b. The magnitude of the total force is the magnitude of this vector:

$$F_{total} = [(-123 \text{ N})^2 + (100 \text{ N})^2]^{1/2}$$
$$= 159 \text{ N}$$

c. Newton's second law of motion enables you to find the acceleration.

Method 1
Use Equation 5.1,

$$\vec{F}_{total} = m\vec{a}$$

and solve for $\vec{a}$:

$$\vec{a} = \frac{(-123 \text{ N})\hat{i} + (100 \text{ N})\hat{j}}{5.00 \text{ kg}}$$
$$= (-24.6 \text{ m/s}^2)\hat{i} + (20.0 \text{ m/s}^2)\hat{j}$$

Method 2
An equivalent way of finding the acceleration is to write the second law for each coordinate direction separately using Equation 5.2. That is, along the x-direction we have a total force component $F_{x\,total} = -123$ N. The x-component of the acceleration, a_x, then is found using Equation 5.2:

$$F_{x\,total} = ma_x$$
$$-123 \text{ N} = (5.00 \text{ kg})a_x$$

Solving for a_x,

$$a_x = -24.6 \text{ m/s}^2$$

Correspondingly, along the y-direction the total force component is $F_{y \text{ total}} = 100$ N. The y-component of the acceleration, a_y, then is found using Equation 5.2:

$$F_{y \text{ total}} = ma_y$$
$$100 \text{ N} = (5.00 \text{ kg})a_y$$

Solving for a_y,

$$a_y = 20.0 \text{ m/s}^2$$

The acceleration vector is

$$\vec{\mathbf{a}} = a_x \hat{\mathbf{i}} + a_y \hat{\mathbf{j}}$$
$$= (-24.6 \text{ m/s}^2)\hat{\mathbf{i}} + (20.0 \text{ m/s}^2)\hat{\mathbf{j}}$$

d. The magnitude of the acceleration is the magnitude of this vector:

$$a = [(-24.6 \text{ m/s}^2)^2 + (20.0 \text{ m/s}^2)^2]^{1/2}$$
$$= 31.7 \text{ m/s}^2$$

5.6 NEWTON'S THIRD LAW OF MOTION

Then there's a pair of us!

Emily Dickinson (1830–1886)*

Part of Newton's genius lay in realizing that a complete theory of dynamics needed one additional tenet. An experiment will show why. When the spring pulls on the standard kilogram in Figures 5.4 and 5.5, the standard kilogram pulls on the spring (thus stretching the spring) with a force of equal magnitude but in the opposite direction. Try another experiment: bang your hand on the desk, thus exerting a force *on* the desk. The desk reciprocates and exerts a force *on* your hand, which is the force you feel when you do this exercise. Bang downward *on* the desk harder and the desk responds in kind with a correspondingly increased upward force *on* your hand. Continue banging even harder and eventually the force of the desk on your hand is so great that your hand hurts (ouch!); the force you exert on the desk only "hurts" the desk if you have a black belt in karate.

Newton's third law of motion is an insightful generalization of the common observations just made:

If a system A exerts a force on another system B, then B exerts a force of the same magnitude on A but in the opposite direction. That is,

$$\vec{\mathbf{F}}_{A \text{ on } B} = -\vec{\mathbf{F}}_{B \text{ on } A} \qquad (5.3)$$

In other words, *forces always occur in pairs*, called **third law force pairs**.

---

*Poem 288, line 3. *The Poems of Emily Dickinson*, edited by Thomas H. Johnson (The Belknap Press of Harvard University Press, Cambridge, Massachusetts, 1955), volume I, page 206.

This is nature's way of doing tit-for-tat. The law is colloquially called the action–reaction law, although this phrase conveys little information about the content of the third law and should be banished from the literature.[†]

There are several important aspects of the third law of motion that warrant emphasis:

1. Forces occur in *pairs* (like socks).
2. Each member of a given force pair has the *same magnitude* (the socks of a given pair are the same size).
3. The two members of a given force pair point in *opposite directions* (socks are *not* worn this way!).
4. The paired forces *act on different systems* (socks are worn on different feet).

Once you grasp these aspects of the third law of motion, you will be well on your way to being a master classical mechanic.

PROBLEM-SOLVING TACTICS

5.3 Watch the little words. The words *of*, *on*, and *by* have important ramifications in the Newtonian theory of motion, as we alluded to before. Frequently the preposition *of* is used synonymously with *by*, so that we say the force *of* A on B or, equivalently, the force *by* A on B.

5.4 Remember that the two forces in a third law force pair act on different systems. This is a critical point. If the two members of a third law force pair were to act on the same system, there would never be a nonzero total force on a system and it would always be in equilibrium, which is clearly not the case for all systems.

5.5 Decide what the systems are before you sort out third law force pairs. Consider only the forces acting on a chosen system when applying the second law of motion to it. Only *one* of every third law force pair is involved in applying the *second law of motion* to a particular system. If you are interested in the motion of system A, then the force *of* B *on* A is relevant; the other member of this third law force pair, the force *of* A *on* B, is irrelevant to system A, since this force does not act *on* system A.

When you bang your hand on the desk, the two systems are your hand and the desk. According to Newton's third law, the force you exert *on* the desk in the downward direction has a third law counterpart: the upward force of the desk *on* your hand (see Figure 5.15). The two members of this third law force pair have exactly the same magnitude, point in opposite directions, but *act on different systems*.

When you slam a home run, the bat exerts a huge force on the ball, but the ball exerts a force of *identical magnitude* on the bat in the opposite direction. The two forces represent a third law force pair: they are of equal magnitude, are directed in opposite directions, but act on different systems. Likewise, when a little car crashes into a truck, the force of the truck on the little

---

[†]Action–reaction is mentioned here only so you can avoid using the term!

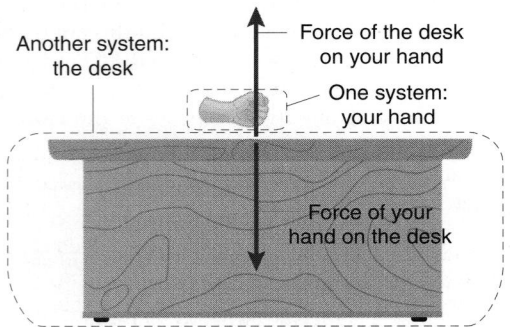

FIGURE 5.15 A third law force pair when you bang on a desk.

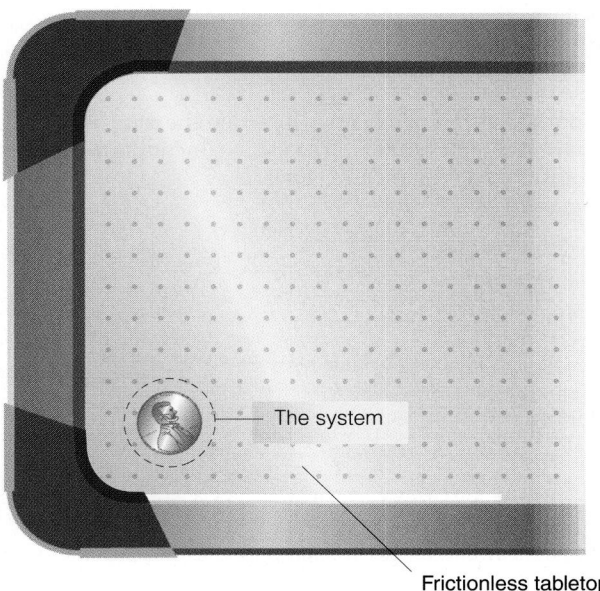

FIGURE 5.16 There is zero total force acting on a coin at rest on a table.

car and the force of the little car on the truck are a Newton's third law force pair; *the two forces have the same magnitude*! The *effect* of the force on each system is another matter.

QUESTION 2

A seer inquires: "According to Newton's third law of motion, forces always occur in pairs that are equal in magnitude and opposite in direction. This means every third law pair of forces vector sums to zero. How, then, can anything ever accelerate?" What is the flaw in this line of reasoning?

5.7 LIMITATIONS TO APPLYING NEWTON'S LAWS OF MOTION

There are several limitations associated with applying Newton's laws of motion.

Reference Frames

Newton's first law defined an inertial reference frame: a reference frame in which a system in equilibrium experiences zero total force. Newton's *second law* only can be used in such inertial reference frames. Let's look at an example of what *is* and what *is not* an inertial reference frame. Place a coin at rest on a level frictionless surface in the laboratory, as in Figure 5.16. The coin remains at rest. Its acceleration is equal to zero and there is zero total force acting on the coin. There *are* forces acting on the coin,* but the vector sum of these forces is zero.

So the laboratory is an inertial reference frame: Newton's first and second laws of motion are obeyed in this reference frame. Any coordinate grid attached to the laboratory serves as an appropriate reference frame for analyzing dynamical problems using Newton's laws.

Now imagine yourself in another laboratory that is spinning around a vertical axis in the first laboratory, much like a merry-go-round. Suppose a horizontal frictionless surface also exists in the spinning laboratory and physicist A. S. Tute in the *nonspinning* lab slowly places a coin (at rest) on the frictionless surface

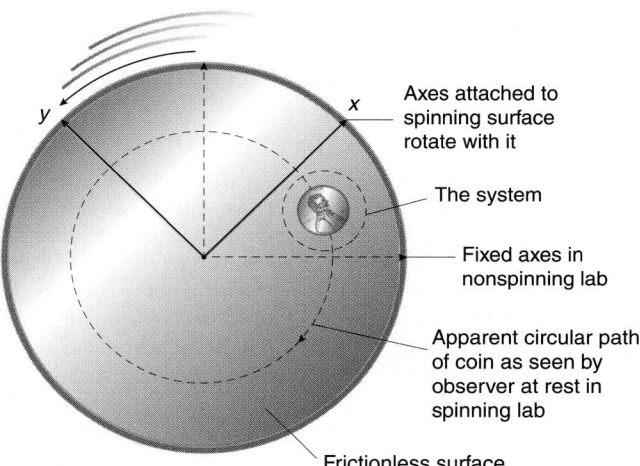

FIGURE 5.17 A coin placed on a frictionless, spinning surface appears to move in a circle with respect to axes attached to the spinning surface.

that is spinning (see Figure 5.17). In the nonspinning lab, the coin remains at rest where it was placed, in accordance with Newton's first and second laws. Yet, lo and behold, the coin is observed by physicist D. Z. Dean in the spinning lab to move in a circle with respect to local coordinate axes attached to the spinning lab!

In the case of the spinning lab, no horizontal force exists on the coin, yet the spinning observer says there indeed is a centripetal acceleration with respect to the coordinate grid in the spinning lab, because the coin is seen to be moving in a circle. Newton's second law of motion is *not* obeyed relative to the reference frame of the spinning lab; hence this is a noninertial reference frame.

---

*We shall see later that these forces are the downward gravitational force of the Earth on the coin (the weight of the coin) and the upward force of the surface on the coin.

> Noninertial reference frames are accelerated reference frames.

> Newton's laws of motion only can be applied in reference frames that are themselves not being accelerated, that is, only in inertial reference frames.

Speed Limits

Newton's second law of motion provides a way of determining the acceleration of a system from the total force acting on it. Newtonian dynamics implicitly assumes that *any* speed, however large, can be attained by the system under the influence of a rectilinear acceleration, given enough time. However, there actually is a speed limit in the universe: the speed of light, always called c (exactly 299 792 458 m/s). The existence of this speed limit implies that Newtonian dynamics must be modified for very large speeds. Such modifications for speeds that are significantly large fractions of the speed of light are a new dynamical theory, called *relativistic dynamics* (which we study in Chapter 25).

> Newtonian mechanics, therefore, is a good *approximation* as long as the speed of the system is much less than the speed of light.

Most common objects of macroscopic size have speeds significantly less than the speed of light, and so Newtonian mechanics is an excellent approximation.

Quantum Mechanics

Experiments originally performed during the early 20th century also indicated that Newtonian mechanics cannot describe or account for many phenomena on the atomic and nuclear scale. This problem instigated the development of a radically new mechanics, called *quantum mechanics*. The relationship between the two types of mechanics was first elucidated by Niels Bohr and others, but the details still are an active area of research today. We briefly consider some of the ideas of quantum mechanics in Chapter 27.

Force Propagation

The third law of motion assumes that forces propagate instantaneously through space and matter at infinite speed during zero time intervals. Change the mass of the Sun, and the third law implies that the effect is felt throughout the universe instantaneously. *This is not strictly true.* If you attach a mass to one end of a long slinky and suddenly pull on the other end, the mass does not accelerate at the instant of the pull, but at sometime later when the force finishes its propagation down the length of the slinky and "arrives" at the mass, as pictured in Figure 5.18.

FIGURE 5.18 Pull on one end of a slinky and a mass at the other end does not respond to the force until a short time interval later.

Likewise, gravitational forces or other action-at-a-distance forces (such as electrical and magnetic forces, discussed later in the text) are not felt everywhere instantaneously. In other words, forces somehow take time to propagate from one place to another. Newton's third law does not account for such propagation delays and, therefore, is not rigorously true over finite distances at all time instants. It is true on a point-by-point basis as the influence or force propagates through space or matter.

> In this text, we assume that the third law of motion is rigorously true; however, modifications must be made at more advanced levels of study to account for the finite propagation speeds.

QUESTION 3

Is the surface of the rotating Earth an inertial reference frame or only *approximately* such a reference frame? Explain.

5.8 INERTIAL REFERENCE FRAMES: DO THEY REALLY EXIST?*

We like to think that the laws of physics (whatever they may be) have the same form in any reference frame. Observers in different reference frames then can approach physical problems using the same mathematical forms for the fundamental laws of nature. Everyone, in any reference frame, can use the same physics book; the descriptions of phenomena may differ depending on the reference frame,[†] but the underlying laws of physics are the same. While a requirement that the laws of physics be the same in *every* reference frame is elegant, aesthetic, and philosophically satisfying, there are untoward mathematical side effects from such a demand. Unfortunately, to make the laws of physics have the same form in *all* reference frames forces us into mathematical quagmires that are inappropriate for an introductory course. Thus we restrict ourselves to a subset of reference frames in which the laws of physics are just a lot easier to write and apply.

As we have seen, there are two distinct types of reference frames:

1. Inertial reference frames are those in which Newton's first law of motion is obeyed. Reference frames either at rest or moving at constant velocity with respect to an inertial reference frame also are inertial frames.
2. Noninertial reference frames are those in which Newton's first law of motion is not obeyed. We mentioned one noninertial reference frame in Section 5.7: a spinning reference frame is an accelerated—and thus noninertial—reference frame. Coordinate systems that are undergoing any acceleration also are noninertial reference frames. Accelerated reference frames (noninertial frames) are considered briefly in Section 5.16; in such accelerated frames Newton's second law of motion must be modified to describe the motion of a system.

[†]For example, the description of the motion of a car depends on whether you are observing the auto from inside the car or outside along the road.

We saw in Section 4.4 that observers in two *different* inertial reference frames measure the *same* acceleration for a given system. Since both inertial reference frames measure the same acceleration for the system, they must agree on the vector sum of the forces $\vec{\mathbf{F}}_{total}$ on it. Newton's second law has the same form and describes the motion in *either* inertial reference frame. Since the total force on the system is the same in each inertial frame, Newton's third law of motion also applies within each inertial reference frame.

Do inertial reference frames actually exist? We can only approximately confirm it. The Earth spins on its axis, and so laboratories on the Earth actually are subjected to (small) centripetal accelerations toward the axis of rotation of the Earth. Therefore laboratories on the Earth are, strictly speaking, noninertial reference frames.

To describe precisely the motion of particles moving on or near the surface of the Earth, using coordinate axes attached to the surface, requires slight modifications to Newton's second law. The flavor of such modifications is considered in Section 5.16 for a rectilinear accelerated reference frame.

Likewise, the Earth itself revolves about the Sun, and so the Earth as a whole experiences a (small) centripetal acceleration toward the center of the Sun; thus, the Earth as a whole is a noninertial reference frame. Correspondingly, the Sun is in orbital motion about the center of the Milky Way Galaxy; our Galaxy also has complex motions as part of the so-called Local Group or cluster of galaxies; and so on. Reference frames attached to each of these objects thus are accelerated reference frames and, therefore, are noninertial reference frames in a strict sense. However, the magnitudes of these accelerations are quite small compared with that of the local acceleration due to gravity, $g = 9.81$ m/s^2.

> We assume the surface of the Earth is a rigorous inertial reference frame, but realize this is an approximation.

5.9 SECOND LAW AND THIRD LAW FORCE DIAGRAMS

. . . the contemplation falls into unfathomable ecstasies in view of all these decompositions of forces resulting in unity.
Victor Hugo (1802–1885)*

To apply the Newtonian laws of motion to a chosen system, we must *identify all forces acting on the system*. In many situations, the best choice for the system itself is dictated by the problem we want to solve; in other cases, it is less obvious and we have to choose an appropriate system with some thought.

Once the system itself is clearly defined, draw a schematic **second law force diagram**[†] that

1. identifies the system; and
2. illustrates all forces acting on the system with their directions indicated explicitly in the diagram.

Once all forces acting on the system are identified and appropriately directed in a second law force diagram, *choose a coordinate system*. Then write the forces as vectors with components in

---

Les Misérables (1862) (A. L. Burt, New York), Saint Denis, Book III, Chapter 3.
[†]Such second law force diagrams also are known as *free-body diagrams*.

terms of the chosen coordinate system *consistent* with their directions on the force diagram. Appropriate vector sums of all the forces are then made and the motion is described in analytical detail using Newton's second law of motion.

> **PROBLEM-SOLVING TACTICS**
>
> **5.6 Identify the system and indicate the forces acting on it in a sketch (a second law force diagram).**
>
> **5.7 If the direction of a particular force is unknown, presume a direction.** The analysis will indicate whether the presumed direction is the actual direction or not by the algebraic sign of the answer.

Occasionally, we also draw **third law force diagrams** to indicate schematically the pairs of forces involved in Newton's *third* law. Each such diagram always has only two forces (the third law force pair), which always are the same magnitude, are in opposite directions, and act on two different systems. See Examples 5.3 and 5.4, after reading the next section.

5.10 WEIGHT AND THE NORMAL FORCE OF A SURFACE

But thou hast ordained all things in measure, number, and weight.
Apocrypha, Wisdom of Solomon, 11:20

Remember from Chapter 1 that the mass of a system and its weight are very different. The mass is a scalar measure of the inertia property of the system, and its value is independent of its location in space. On the other hand, the weight $\vec{\mathbf{w}}$ is the gravitational force vector on the system and depends on its location in space.

We examine the nature of the gravitational force in some detail in Chapter 6. However, since the gravitational force is ubiquitous on the Earth, we need to develop a preliminary understanding of it at this juncture.

Recall from Chapter 3 that any freely falling object near the surface of the Earth experiences an acceleration of magnitude $g \approx 9.81$ m/s^2, directed vertically downward. This acceleration occurs because the gravitational force of the Earth on the falling object is the total force on it. Apply Newton's second law of motion, Equation 5.1, to the freely falling system. On the *left*-hand side of the equation, the only force acting on the falling mass m (neglecting air resistance) is its weight, so that here $\vec{\mathbf{F}}_{total} \equiv \vec{\mathbf{w}}$. On the *right*-hand side of the second law, the acceleration is of magnitude g, directed down. Using the *magnitudes* of the vectors, Equation 5.1, $\vec{\mathbf{F}}_{total} = m\vec{\mathbf{a}}$, becomes

$$w = mg \qquad (5.4)$$

The second law force diagram for the freely falling mass m has the single force $\vec{\mathbf{w}}$, indicated in Figure 5.19. Thus the gravitational

The system

FIGURE 5.19 A freely falling system accelerates because the gravitational force is the total force on it.

Your weight causes your acceleration when you skydive.

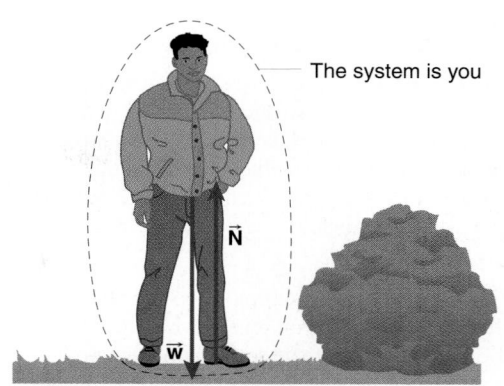

FIGURE 5.20 When you stand on the ground, the total force on you is zero.

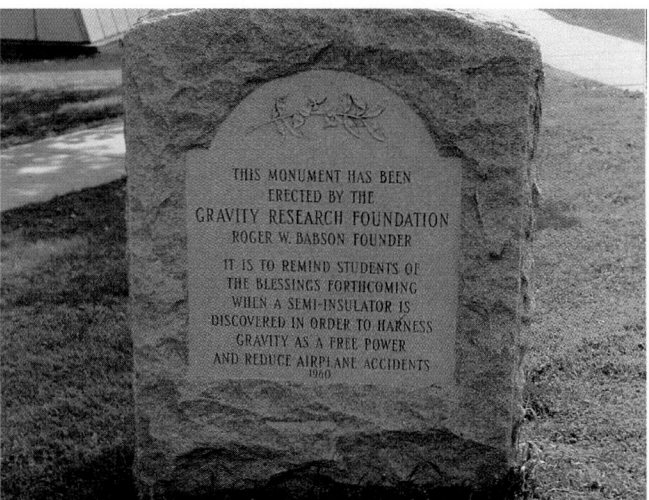

A monument on the campus of Middlebury College, Middlebury, Vermont. A similar monument exists at Colby College, Waterville, Maine.

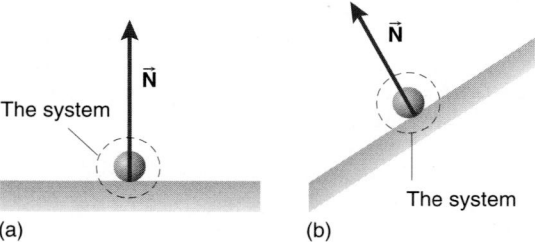

FIGURE 5.21 (a) If the surface is horizontal, then the normal force of the surface on a system is vertical (perpendicular to the surface). (b) If the surface is not horizontal, then the normal force of the surface on a system is *not* vertical, but it still is perpendicular to the surface.

force $\vec{w}$ of the Earth on a system of mass m has a *magnitude w* equal to mg and is directed vertically down.

> The gravitational force of the Earth acts on *every* object on or near the Earth, *whether falling or not*, and is of magnitude mg. Therefore every object near the surface of the Earth has at least one force acting on it, its weight $\vec{w}$.

It is interesting that we do not directly sense our weight. We tend to ignore the most familiar. When you jump off a rock wall or skydive from an airplane (before the parachute opens), you sense no pull in the downward direction as you fall. But the gravitational force of the Earth on you is there nonetheless, whether you are falling or simply standing on the surface of the Earth. You are not aware of your weight because the force never ceases to act on you; we simply take it for granted because it is impossible to eliminate it or insulate ourselves from it.

When you stand at rest on the surface of level ground (see Figure 5.20), you are not accelerating ($\vec{a} = 0$ m/s^2). Hence $\vec{F}_{\text{total}} = 0$ N from Newton's second law of motion.

Call the force that the surface exerts on you $\vec{N}$.* Then since $\vec{a} = 0$ m/s^2, the vector sum of $\vec{N}$ and the ever-present weight $\vec{w}$ must be zero. Equation 5.1, $\vec{F}_{\text{total}} = m\vec{a}$, becomes

$$\vec{N} + \vec{w} = m(0 \text{ m/s}^2)$$
$$= 0 \text{ N}$$

Transposing,

$$\vec{N} = -\vec{w}$$

The minus sign indicates that $\vec{w}$ and $\vec{N}$ are in opposite directions. Thus the surface of the level ground exerts a force on you equal in magnitude to your weight, but directed vertically upward since $\vec{w}$ is directed vertically down. The force you *do* feel is the force $\vec{N}$ of the level ground *on* you.

> The force that a surface exerts perpendicular to the surface is called a **normal force**, designated by $\vec{N}$. The word **normal** means perpendicular to the surface. Normal forces always are directed perpendicular to the surface, whatever its orientation (see Figure 5.21).

---

*Do not confuse the normal force $\vec{N}$, nor its magnitude N, with the SI unit for force, the newton, which we abbreviate as N. As mentioned in Chapter 1, some people write all units in parentheses to minimize the chances of falling for such ambiguities.

PROBLEM-SOLVING TACTIC

5.8 The normal force on a system always is directed perpendicular to the surface in contact with the system, but the magnitude of the normal force is not always equal to magnitude of the weight of the system. The magnitude of the normal force *must* be determined on a case-by-case basis. See Example 5.5.

When you jump off the top of a rock wall and briefly experience free-fall, the force you sensed when standing on the rock wall (the normal force of the surface on you) no longer acts on you, since it was a contact force. However, the gravitational force of the Earth on you (your weight) nonetheless still exists, even though you do not sense its pull. It is your weight that causes your acceleration when falling.

While undergoing free-fall, it is colloquially said that you are "weightless," but this terminology is a *misnomer* and dangerous for physics students. The Earth *never ceases* to exert a gravitational force on you in free-fall (or in other circumstances on or near the Earth); since the gravitational force of the Earth on you is your weight, you are *never* weightless.

Imagine yourself standing on a bathroom scale in free-fall (see Figure 5.22). During the free-fall the upward force that the surface of the scale exerts on you is *zero*, because the scale also is in free-fall with you. The scale exerts no upward force on you, and you exert no downward force on the scale. The two forces are a Newton's third law pair of forces, with each force of zero magnitude during free-fall. The scale reads zero during free-fall, and people commonly refer to this observation as weightlessness. The *sensation* of weightlessness, then, is what happens in free-fall. But in such free-fall, you still have weight; you are really normal-force-less, not weightless.

In terms of Newton's third law of motion, you are one system and the Earth is another system. Since the Earth exerts a gravitational force on you called your weight $\vec{w}$, *you* exert a gravitational force of equal magnitude *on the Earth* in the opposite direction, according to Newton's third law of motion, as shown in Figure 5.23. A third law force diagram includes the force of the Earth on you and the force of you on the Earth, since they are a Newton's third law force pair. This pair of forces exists whether you are in free-fall or not. You and the Earth are parts of another, larger system, a total system.

Likewise, when you stand on level ground, as in Figure 5.24, the ground exerts a normal force on you directed perpendicular to the surface of the ground, in this case vertically upward, and you exert a force of the same magnitude in the opposite direction *on the ground*; these two forces also are a Newton's third law force pair. In this case there also is another third law force pair: the gravitational force of the Earth on you (your weight) and the gravitational force of you on the Earth (this third law pair is identical to Figure 5.23).

QUESTION 4

Does your weight vary depending on whether you stand on a level or inclined surface? Explain your answer carefully.

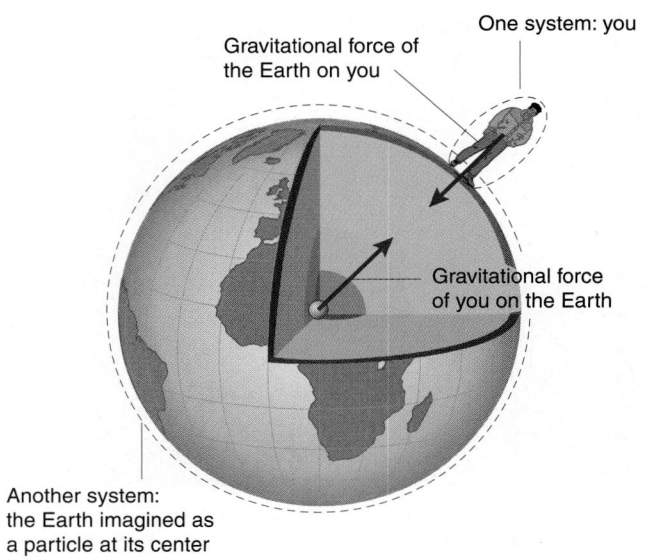

FIGURE 5.23 A third law force diagram of you and the Earth.

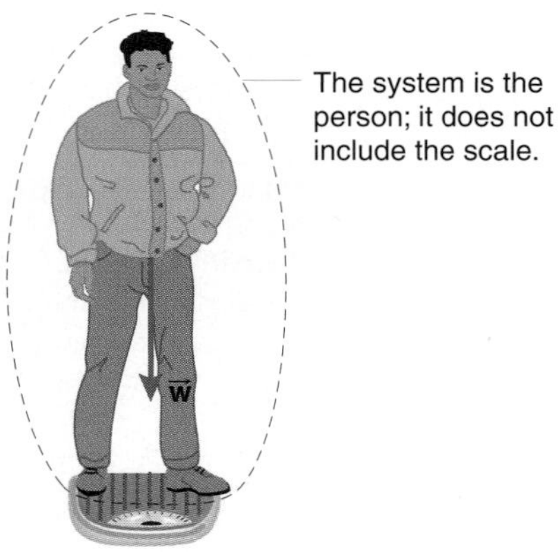

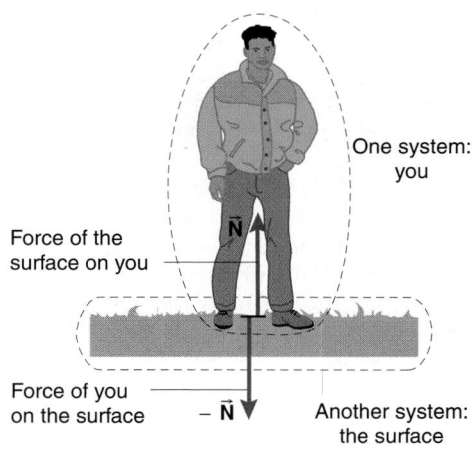

FIGURE 5.22 Free-fall with a bathroom scale contacting your feet.

FIGURE 5.24 A third law force pair if you are on the ground.

QUESTION 5

If you stand on a bathroom scale on a level surface and then when it is on an inclined surface, will the reading of the scale be the same in both cases? Why or why not? Explain your answers carefully. Does the scale really measure your weight? If not, what does it really measure?

EXAMPLE 5.3

Your mass is 70.0 kg, and you are standing at rest on level ground.

a. Make a second law force diagram indicating all forces acting on you.
b. What is your weight?
c. What is the normal force of the ground on you?
d. For *each* force on you, make a third law force diagram, identifying the forces in each third law pair.

Solution

a. You are the system and the two forces acting on you are (see Figure 5.25)

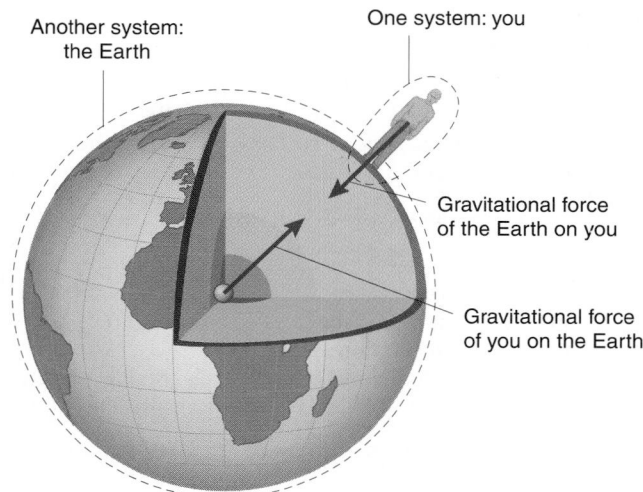

FIGURE 5.26

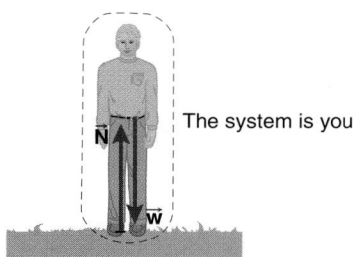

FIGURE 5.25

1. the gravitational force of the Earth on you (your weight), directed vertically down, and
2. the normal force of the surface on you, directed perpendicular to the surface, here vertically up.

b. From Equation 5.4, your weight has a magnitude w equal to

$$w = mg$$
$$= (70.0 \text{ kg})(9.81 \text{ m/s}^2)$$
$$= 687 \text{ N}$$

and is directed vertically down.

c. Since you are not accelerating in the vertical direction, $\vec{a} = 0$ m/s^2 and $\vec{F}_{total} = 0$ N according to Equation 5.1. Since here $\vec{F}_{total} = \vec{N} + \vec{w}$, you have $\vec{N} = -\vec{w}$. Hence the normal force of the ground on you is equal in magnitude to your weight (687 N) and is directed vertically upward.

d. One third law pair of forces is the gravitational force of the Earth on you (your weight) and the gravitational force of you on the Earth. These forces are of equal magnitude, point in opposite directions, and act on different systems, as shown in Figure 5.26. The other third law pair of forces is the normal force of the level surface on you and the force of you on the level surface. These forces are of equal magnitude, point in opposite directions, and act on different systems, as shown in Figure 5.27.

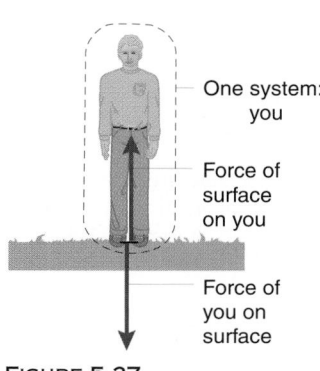

FIGURE 5.27

EXAMPLE 5.4

A tired dog named Canis Major, with a weight of magnitude 110 N, comes along and lies down for a nap on a 50.0 kg log lying on level ground, as indicated in Figure 5.28.

a. Draw a second law force diagram schematically indicating all forces acting on the log.
b. Find the magnitudes of these forces.
c. Specify the paired forces that are involved in Newton's third law and draw corresponding third law force diagrams.

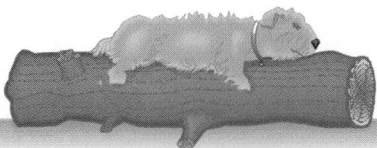

FIGURE 5.28

Solution

Let the log be the system.

a. There are three forces acting on the log. The second law force diagram includes
1. the gravitational force of the Earth on the log (the weight $\vec{w}$ of the log), of magnitude

$$mg = (50.0 \text{ kg})(9.81 \text{ m/s}^2)$$
$$= 491 \text{ N}$$

   directed vertically downward;
2. the force of the dog on the log, $\vec{F}_{dog}$—of magnitude 110 N, directed vertically downward; and
3. the normal force $\vec{N}$ of the surface on the log, directed perpendicular to the surface, here vertically upward.

These forces are shown in the second law force diagram of Figure 5.29. Choose $\hat{j}$ going vertically down, as indicated in Figure 5.29. Newton's second law of motion for the vertical direction states that

$$F_{y\,total}\hat{j} = ma_y\hat{j}$$

Since the log is not accelerating, $a_y = 0$ m/s², and so

$$(491 \text{ N})\hat{j} + (110 \text{ N})\hat{j} - N\hat{j} = m(0 \text{ m/s}^2)$$

Solving for N, you find

$$N = 601 \text{ N}$$

Notice that here the magnitude of the normal force of the surface on the log is not equal to the magnitude of the weight of the log, even though the surface is horizontal.

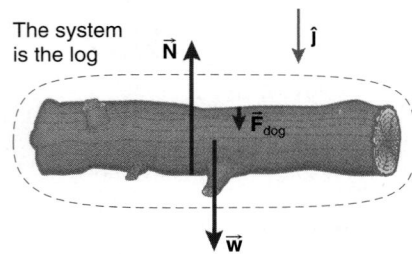

FIGURE 5.29

c. The paired forces in Newton's third law are the following:

Pair 1 The gravitational force *of* the Earth *on* the log and the gravitational force *of* the log *on* the Earth, indicated in the third law force diagram of Figure 5.30.

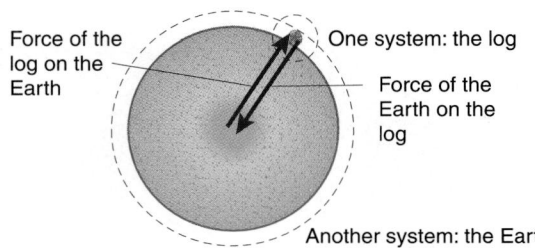

FIGURE 5.30

Pair 2 The force *of* the dog *on* the log and the force *of* the log *on* the dog, shown in Figure 5.31.

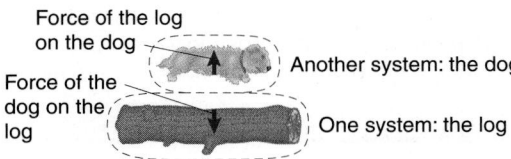

FIGURE 5.31

Pair 3 The normal force *of* the surface *on* the log and the force *of* the log *on* the surface, shown in Figure 5.32.

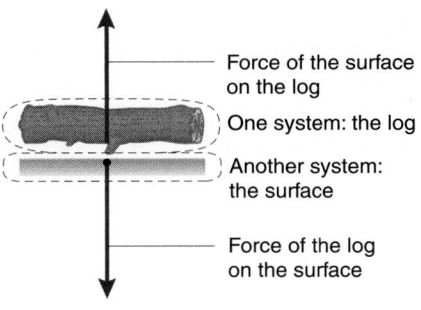

FIGURE 5.32

Notice that the forces in each third law force pair have the same magnitude, are in opposite directions, and act *on* different systems. Only one member of each pair is involved in analyzing the motion of the log.

STRATEGIC EXAMPLE 5.5

A cute pika (*Ochotona princeps*) of mass m slips at rest at the top of an icy frictionless plane inclined at an angle θ to the horizontal direction, as in Figure 5.33.

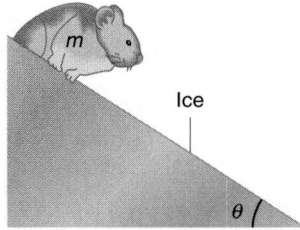

FIGURE 5.33

a. Make a second law force diagram indicating all forces acting on the pika.
b. What is the magnitude of the normal force on the pika?
c. What is the magnitude of the acceleration of the pika?
d. If the inclination angle increases to 90°, what do you anticipate the magnitude of the acceleration of the pika will be? Is this consistent with part (c) when $\theta = 90°$?

Solution

a. Use Problem-Solving Tactic 5.1 and *specify the system*. Let the system be the pika of mass m. Use Problem-Solving Tactic 5.6

and *identify the forces acting on the system and draw a second law force diagram*. There are two forces acting on the pika (see Figure 5.34):

1. the gravitational force of the Earth on the system (the weight $\vec{w}$ of the pika), of magnitude mg, directed vertically down; and
2. the normal force $\vec{N}$ of the surface on the pika, of magnitude N, directed perpendicular to the inclined surface.

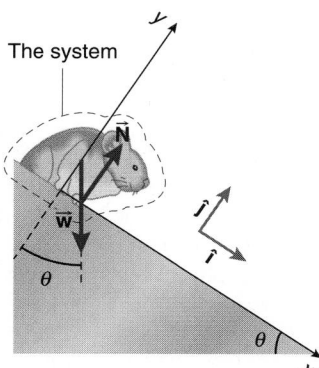

FIGURE 5.34

b. Introduce a coordinate system with axes parallel and perpendicular to the plane, since the motion is parallel to it (see Figure 5.34). The weight $\vec{w}$ makes an angle with a perpendicular to the inclined surface that is equal to the angle θ of the incline. This is shown in Figure 5.34 and makes use of the geometric fact that if two angles θ and ϕ have mutually perpendicular sides (see Figure 5.35), the angles are equal: $\theta = \phi$.

Hence $\vec{w}$ is expressed in Cartesian form as

$$\vec{w} = (mg \sin \theta)\hat{\imath} + (mg \cos \theta)(-\hat{\jmath})$$
$$= (mg \sin \theta)\hat{\imath} - (mg \cos \theta)\hat{\jmath}$$

The normal force is expressed in Cartesian form as

$$\vec{N} = N\hat{\jmath}$$

As the mass slides down the plane, its y-coordinate does not change; thus the velocity *and* acceleration have y-components equal to zero. With the y-component of the acceleration equal

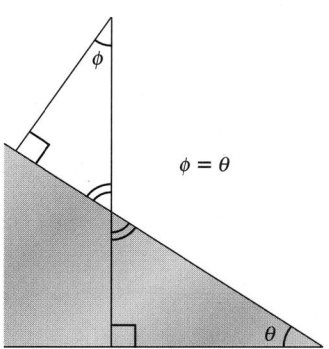

FIGURE 5.35

to zero, apply Newton's second law of motion, Equation 5.2, to the y-direction:

$$F_{y\,\text{total}} = ma_y$$
$$N - mg \cos \theta = m(0 \text{ m/s}^2)$$

Hence

$$N = mg \cos \theta$$

c. Apply Newton's second law of motion, Equation 5.2, to the x-direction:

$$F_{x\,\text{total}} = ma_x$$

Along the x-direction, there is only one force component ($mg \sin \theta$), and so the second law becomes

$$mg \sin \theta = ma_x$$

Solving for a_x, you obtain

$$a_x = g \sin \theta$$

Since $a_y = 0$ m/s^2, the magnitude of the total acceleration is

$$a = (a_x^2 + a_y^2)^{1/2}$$
$$= [(g \sin \theta)^2 + (0 \text{ m/s}^2)^2]^{1/2}$$
$$= g \sin \theta$$

Therefore the magnitude of the acceleration is $g \sin \theta$.

d. Notice that if the inclination angle is 0°, the magnitude of the acceleration of the pika is zero, since $\sin 0° = 0$. As the inclination angle increases to 90°, the acceleration increases in magnitude. When $\theta = 90°$, the magnitude of the acceleration should equal g, since the pika then is in free-fall. Indeed, when $\theta = 90°$,

$$a_x = g \sin 90° = g$$

A frictionless inclined plane thus provides a way of securing accelerations ranging in magnitude from 0 m/s^2 to 9.81 m/s^2. Galileo exploited this aspect of inclined planes when he began his studies of kinematics. By using small inclination angles, he could make the accelerations small enough for him to use the crude timing devices of the early 17th century.

5.11 TENSIONS IN ROPES, STRINGS, AND CABLES

Mountaineers use ropes for safety and excitement. Fishing would not be very successful without appropriate line. Steel cables are used to support signs and suspension bridges and for ski lifts. The famous cable cars of San Francisco and the gondolas that take you to the top of Table Mountain in Capetown also use cables. Since strings, ropes, and cables have frequent technological applications, it is useful to know how such devices transmit contact forces.

An Ideal String

Real ropes, strings, and cables are modeled, in a first approximation, by an **ideal string**.

An ideal string

a. has negligible mass;
b. does not stretch when pulled;
c. can pull but not push; and
d. pulls at any point only in a direction along the line of the string.

Imagine such an idealized string tied to a wall. A friend at the other end exerts a pull of magnitude T newtons horizontally (see Figure 5.36).

Examine the situation in the horizontal direction and consider the string to be the system under study. The string is not accelerating in the horizontal direction; hence Newton's second law of motion implies the total force $\vec{F}_{total}$ on the string in the horizontal direction must be zero. This means that the wall must exert a force on the string, and this force must be equal in magnitude to the force your friend exerts, but must point in the opposite direction; in this way, the two horizontal forces vector sum to zero. A second law force diagram of the forces on the string includes the force your friend exerts on the string and the force of the wall on the string, as shown in Figure 5.37.

Since the acceleration of the string is zero, the two forces are of equal magnitude and opposite direction, because their vector sum must be zero in Newton's second law. Even though the force your friend exerts on the string and the force the wall exerts on the string are equal in magnitude and are opposite in direction, they are *not* a Newton's third law pair of forces since they both act on the same system (the string).

Now *imagine* cutting the string at some point along its length so that it consists of two fictitious pieces. One way to visualize the forces at the cut is to imagine yourself holding the two pieces of the string as they are cut, as indicated in Figure 5.38. Looking at the figure, you must pull to the right on the piece of string on the left, and you must pull to the left on the piece of string on the right. Consider each piece to be a separate system and apply Newton's second law of motion to each. Each piece has zero acceleration, and so the vector sum of the forces in the horizontal direction on *each piece* must be zero. This means that for the piece on the right (see Figure 5.39), *at the point of the cut* there must be a force directed *to the left* on the segment, equal in magnitude to the force your friend exerts on the end of the string. As a result, the total force on this segment is zero in the horizontal direction. Likewise, for the segment on the left (see Figure 5.40), there must be a force *at the cut* that is directed to the right, equal in magnitude to the force of the wall on the string.

Notice that the original pull of magnitude T that your friend exerts *to* the right, *on* the right end of the right segment of the string (Figure 5.39), is equal in magnitude to the force (in Figure 5.40) that the right segment exerts on the left segment *at the cut* and *points in the same direction* as the force your friend exerts on the end of the string. [This is legalese in physics (quite a mouthful of a sentence); read it again slowly and carefully.] Therefore you can think of the force your friend exerts at the *end* of the string as *also* being exerted *at the imaginary cut* on the segment of string on the left side of Figure 5.38. The string, therefore, transmits the pull on its end,

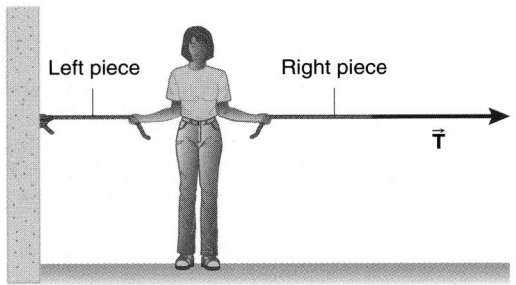

FIGURE 5.38 You hold both ends of the cut string.

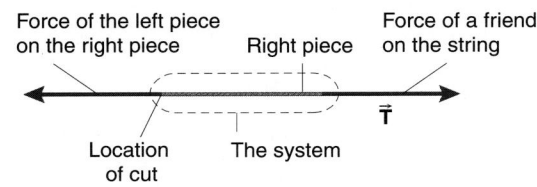

FIGURE 5.39 Second law force diagram for the piece of string on the right.

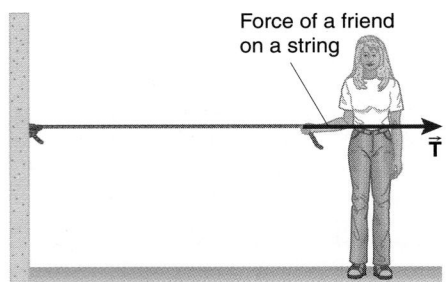

FIGURE 5.36 A friend pulls horizontally on a string tied to a wall.

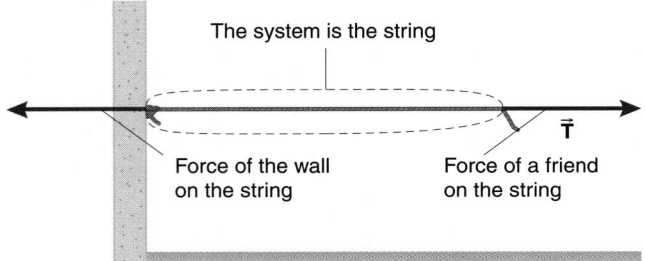

FIGURE 5.37 The horizontal forces on an ideal string tied to a wall and pulled at one end by your friend.

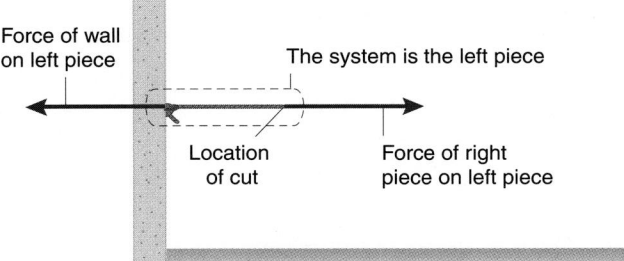

FIGURE 5.40 Second law force diagram for the piece of string on the left.

point by point along its length, since we could imagine the cut to be anywhere along the length of the string.

> The force exerted by any elemental length of the string on its neighbors at any point along the length of a string is called the **tension** *in the string.*

Thus at any point along the string we imagine a cut such that *at the cut*, the tension in the string acts *to the left* on the segment of the string on the right, and *to the right* on the segment of the string on the left, as indicated in Figure 5.41. The forces *at the imagined cut* are of equal magnitude, point in opposite directions, and act on different segments (or systems); thus, they form a third force law pair.

The ideal string transmits a pull on its end, undiminished in magnitude, along its length to the wall. When your friend pulls on the string, the *effect of the string* is as follows:

a. The string exerts a force on your friend in a direction opposite to the force she exerts on the string (see Figure 5.42), since these forces form a third law pair.
b. The string also exerts a force on the wall in a direction opposite to the force exerted by the wall on the string (see Figure 5.43), since these forces form a third law pair.

> **PROBLEM-SOLVING TACTIC**
>
> **5.9 An ideal string exerts forces on the things at its ends that are directed toward and along the string; these forces are equal in magnitude.** See Figure 5.44.

The forces that ropes, strings, and cables exert on objects at their ends also are called *tensions*, since they arise from the tension forces in such ropes, strings, or cables.

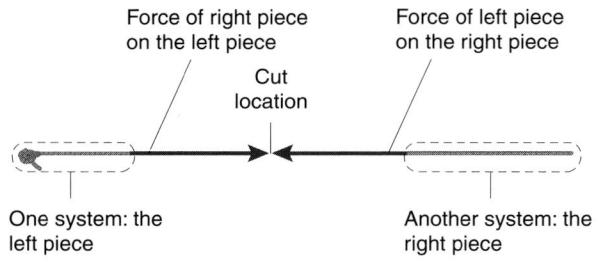

FIGURE 5.41 Third law forces at an imagined cut in an ideal string.

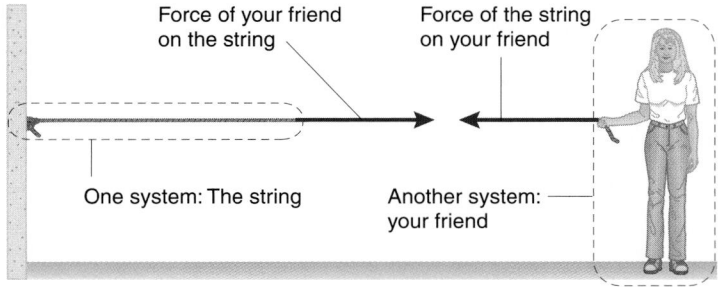

FIGURE 5.42 Third law force diagram at the end of the string pulled by your friend.

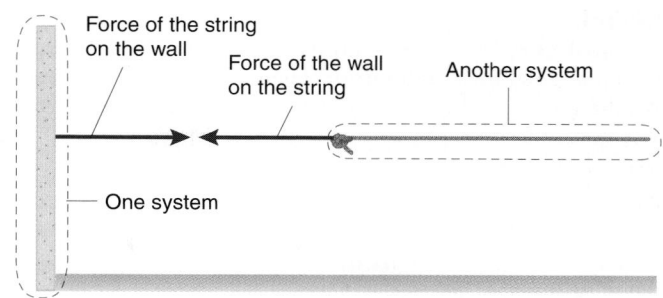

FIGURE 5.43 Third law force diagram at the point the string is attached to the wall.

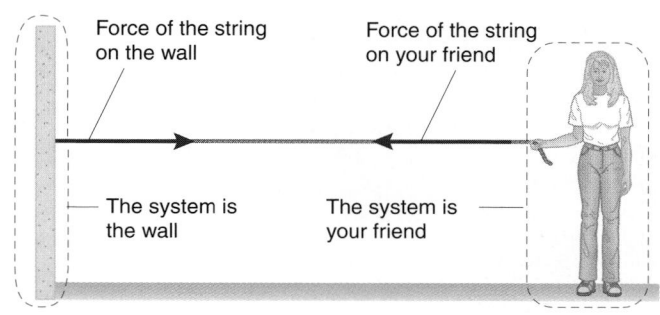

FIGURE 5.44 Force *by* a string *on* things at its ends.

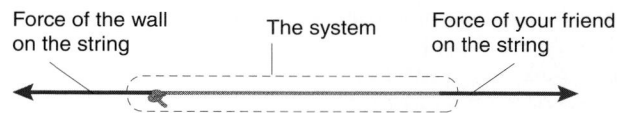

FIGURE 5.45 Forces *on* the string *by* things on its ends.

The *things on the ends of the string* exert forces *on the string,* directed along the string but outward from its ends; these forces are equal in magnitude (see Figure 5.45). The forces in Figure 5.45 are only of interest if you consider the string to be the system, which is not commonly the case (although we do this in the following subsection for a massive string). Typically, the objects on the ends of the string are of concern, not the string itself.

EXAMPLE 5.6 ━━━━━━━━━━━━━━━━━

Your dog Sirius does not want to budge from her favorite rug. You pull on her leash with a force of magnitude 150 N as shown in Figure 5.46. Consider the leash to be an ideal string.

a. What is the force of the leash on her?
b. What is the force of the leash on you?

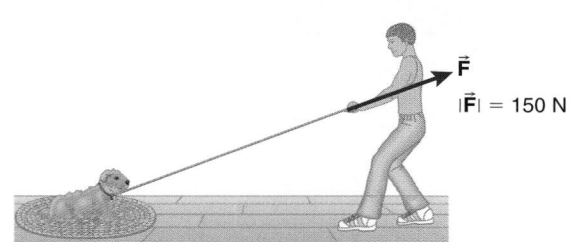

FIGURE 5.46

Solution

An ideal string exerts forces on the objects at its ends that are parallel to and toward the string. Hence the force of the leash on the dog is as shown in Figure 5.47, and the force of the leash on you is as in Figure 5.48.

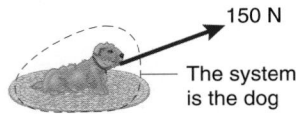

FIGURE 5.47 Force of the leash on the dog.

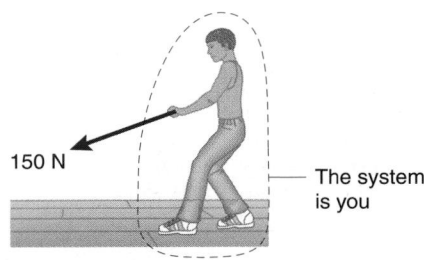

FIGURE 5.48 Force of the leash on you.

A String with Mass

What is the effect of a string whose mass is not negligible but otherwise exhibits the properties of an ideal string? Imagine a rope of mass m_{rope} attached to a block of mass m_{block} on a frictionless horizontal surface, as in Figure 5.49. Pull on the end of the rope with a horizontal force of magnitude T_1 newtons.

Examine the horizontal forces along the axis of the rope. Imagine cutting the rope at its juncture with the block; call T_2 the magnitude of the tension in the rope at the cut. We want to find the relationship between T_1 and T_2. We examine two different systems: (a) the rope and (b) the block.

a. The forces acting on the rope in the horizontal direction are shown in Figure 5.50. The pull of magnitude T_1 that we exert on the rope to the right and a force of magnitude T_2 on the rope

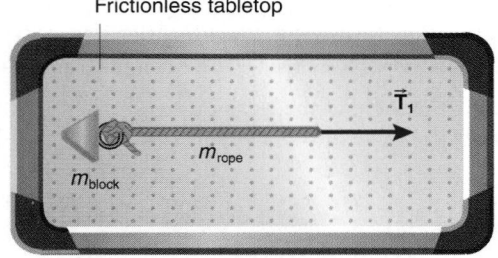

FIGURE 5.49 Pull on the end of massive string attached to a block.

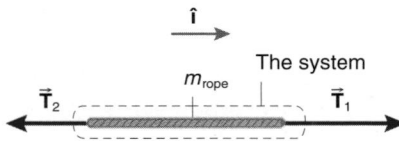

FIGURE 5.50 The horizontal forces on a massive rope.

by the block is directed to the left. Apply Newton's second law of motion to the whole rope as a system. Choose $\hat{\mathbf{i}}$ to be directed as indicated in Figure 5.50. Then Newton's second law becomes

$$F_{x\text{ total on rope}}\hat{\mathbf{i}} = m_{rope}a_x\hat{\mathbf{i}}$$

or

$$T_1\hat{\mathbf{i}} + T_2(-\hat{\mathbf{i}}) = m_{rope}a_x\hat{\mathbf{i}}$$

Simplifying yields

$$T_1 - T_2 = m_{rope}a_x \tag{5.5}$$

b. Now examine the block as a separate system. The only force acting on the block in the horizontal direction is the force of magnitude T_2 exerted on the block by the rope. Since the block is at the end of the rope, the force of the rope on the block is directed toward and along the rope, as shown in Figure 5.51. Applying Newton's second law of motion to the block, you have

$$T_2\hat{\mathbf{i}} = m_{block}a_x\hat{\mathbf{i}}$$

Hence

$$T_2 = m_{block}a_x$$

Solving for a_x,

$$a_x = \frac{T_2}{m_{block}}$$

If the rope does not stretch (one characteristic of an ideal string), the block and the rope have the same acceleration component. So substitute the expression for a_x into Equation 5.5 for the *rope*, yielding

$$T_1 - T_2 = m_{rope}\frac{T_2}{m_{block}}$$

Dividing both sides by T_2 and rearranging the terms slightly,

$$\frac{T_1}{T_2} = 1 + \frac{m_{rope}}{m_{block}}$$

The magnitudes of the forces acting on each end of the massive rope are *not* equal. However, if the mass of the rope is small compared with mass of the block, that is, if

$$\frac{m_{rope}}{m_{block}} \ll 1$$

the rope approaches the characteristics of an *idealized, massless string* and the magnitude of the tensions at each end of the ideal rope are equal, even if the rope and block are accelerating.

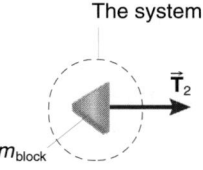

FIGURE 5.51 Horizontal forces on the block.

FIGURE 5.52 An ideal pulley changes the direction of the tension but not its magnitude.

Ideal, massless ropes, strings, wires, rods, or beams have the same magnitude of tension at all points along their lengths whether the system is in equilibrium or accelerating.

The effects of an ideal, massless, frictionless **pulley** on an ideal rope are the following:

a. The pulley *changes the direction* of the rope and, therefore, the direction *of the tension* force in the rope, as shown in Figure 5.52.

b. The ideal pulley *does not change the magnitude of the tension* in the rope. That is, the tension in the ideal rope is the *same* on both sides of the pulley.

c. An idealized rope does not stretch; therefore, if the rope is moving over the pulley, the speed at which rope comes onto the pulley is equal to the speed at which it leaves.

QUESTION 6

A mass is pulled vertically by a string (see Figure Q.6) and has an acceleration of magnitude $a > g$. Explain why the tension in the string is greater for an upward acceleration than for a downward acceleration.

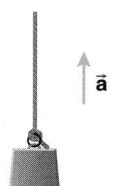

FIGURE Q.6

EXAMPLE 5.7

Two mountaineers of masses m_1 and m_2 strangely happen to be hanging at rest on one ideal rope in equilibrium over a cliff, as shown in Figure 5.53. Find the magnitudes of the tensions in the two sections of rope.

Solution

Let the person on the bottom be the m_1-system. The second law force diagram for this system includes (see Figure 5.54)

1. the weight $\vec{\mathbf{w}}_1$ of the person, of magnitude $m_1 g$, acting in the downward direction (this is the gravitational force of the Earth on the person); and

2. the force $\vec{\mathbf{T}}_1$ of the lower rope on the m_1-system, which has a magnitude equal to the magnitude of the tension in the lower rope.

Since the m_1-system is in equilibrium, the *right*-hand side of Newton's second law (Equation 5.1) is zero:

$$\vec{\mathbf{F}}_{1\,\text{total}} = m_1 \vec{\mathbf{a}}_1$$
$$m_1 g(-\hat{\mathbf{j}}) + T_1 \hat{\mathbf{j}} = m_1(0 \text{ m/s}^2)$$
$$= 0 \text{ N}$$

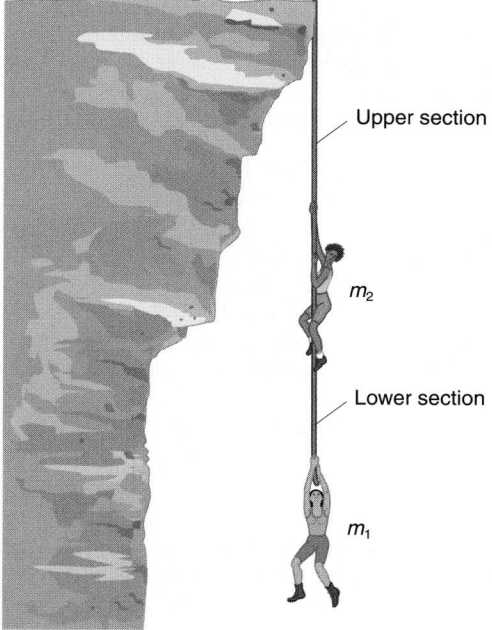

FIGURE 5.53

Thus the vector sum of the forces acting *on* the m_1-system (the *left*-hand side of the second law) must be zero. After transposing,

$$T_1 = m_1 g$$

The tension in the lower rope is of magnitude $m_1 g$.

There are two ways to find the tension in the upper rope.

Method 1

Call the upper person the m_2-system. The second law force diagram for this system includes the forces shown in Figure 5.55:

1. the weight $\vec{\mathbf{w}}_2$ of the person, of magnitude $m_2 g$, directed vertically down;

2. the force $\vec{\mathbf{T}}_1$ of the lower rope on the m_2-system; this force is equal in magnitude to the tension in the lower rope; and

3. the force $\vec{\mathbf{T}}_2$ of the upper rope on the m_2-system, which has a magnitude equal to the magnitude of the tension in the upper rope.

Since the mass of each rope is negligible, the tension in each rope has a constant magnitude along its respective length. Thus the magnitude of the force $\vec{\mathbf{T}}_1$ of the lower rope on m_2 is

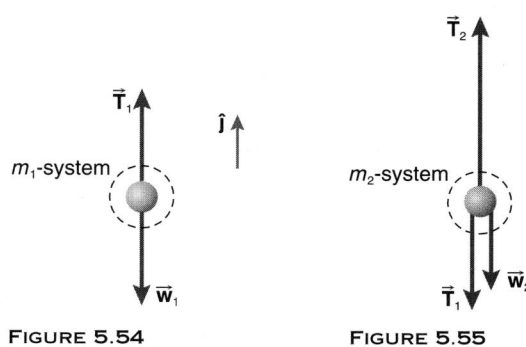

FIGURE 5.54   **FIGURE 5.55**

equal to the magnitude of the force of the lower rope on m_1, and both are equal to the magnitude of the tension in the lower rope. Since the m_2-system has zero acceleration, the right-hand side of the second law is zero:

$$\vec{F}_{2\,\text{total}} = m_2\vec{a}_2$$
$$T_2\hat{j} + m_2g(-\hat{j}) + T_1(-\hat{j}) = m_2(0 \text{ m/s}^2)$$
$$= 0 \text{ N}$$

Solving for T_2,

$$T_2 = m_2g + T_1$$

But T_1 is the magnitude of the tension in the lower rope, which is equal to m_1g (from our analysis of the m_1-system); thus

$$T_2 = m_2g + m_1g$$
$$= (m_1 + m_2)g$$

Method 2
Let both m_1 and m_2 be considered a *single* system called the $(m_1 + m_2)$-system. In this case the second law force diagram for the system includes the following forces (see Figure 5.56):

1. the weight $\vec{w}_{12}$ of the $(m_1 + m_2)$-system, which is of magnitude to $(m_1 + m_2)g$, directed vertically down; and
2. the force $\vec{T}_2$ of the upper rope on the $(m_1 + m_2)$-system, which is equal in magnitude to the tension in the rope leading to the top of the cliff.

Notice that the forces of the lower rope on m_1 and m_2 are *not* forces acting *on* the $(m_1 + m_2)$-system; these forces are *within* (or *internal* to) the system. Since the rope is of negligible mass, the tension in the rope between m_1 and m_2 is the same along its length; so the force of this rope acting on m_1 and the force of this rope acting on m_2 are of equal magnitude, point in opposite directions, and vector sum to zero. The internal forces can be considered as Newton's third law pairs within the larger $(m_1 + m_2)$-system.

The $(m_1 + m_2)$-system has zero acceleration, and so the right-hand side of Newton's second law, Equation 5.1, is zero:

$$\vec{F}_{\text{total}} = (m_1 + m_2)\vec{a}$$
$$(m_1 + m_2)g(-\hat{j}) + T_2\hat{j} = (m_1 + m_2)(0 \text{ m/s}^2)$$
$$= 0 \text{ N}$$

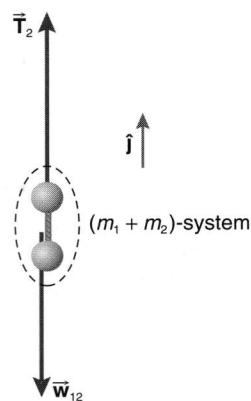

FIGURE 5.56

Note that the mass on the right-hand side of the second law is the mass of the *system*, here $m_1 + m_2$. Solving for T_2,

$$T_2 = (m_1 + m_2)g$$

You obtain the same result by either method. It *has* to come out that way: the physical world does not know, or care, which system you use to analyze the situation. But notice that with the $(m_1 + m_2)$-system, you gain *no* information about the tension in the lower rope *between* the two masses, since the forces this rope exerts on m_1 and m_2 are *internal* to the combined system.

STRATEGIC EXAMPLE 5.8

Two climbers of mass m_1 and m_2 are connected by an ideal rope and held in place by a force $\vec{F}$ directed up and parallel to an icy, frictionless slope that makes an angle θ with the horizontal (see Figure 5.57).

a. Consider each climber as a separate system and make second law force diagrams schematically indicating all forces acting on each system.
b. Determine the magnitude of the normal force of the surface on each climber, the magnitude of the tension in the rope between them, and the magnitude of the force $\vec{F}$ needed to hold the climbers in place at rest in terms of m_1, m_2, g, and θ.
c. Will the results of any of your calculations be different if the climbers are pulled up the incline at a constant velocity?

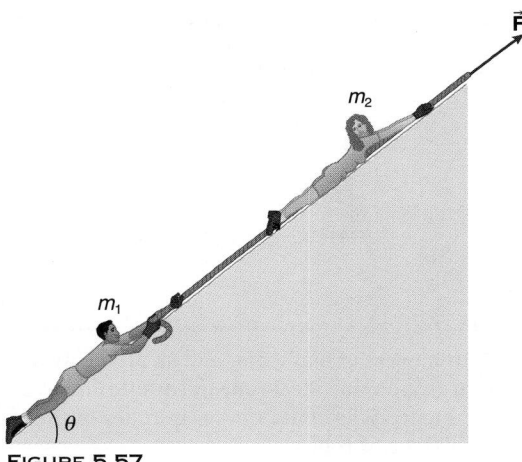

FIGURE 5.57

Solution
a. Let the climber of mass m_1 be the system. There are three forces acting on him (see Figure 5.58):
   1. the gravitational force of the Earth on him—his weight $\vec{w}_1$, of magnitude m_1g, directed vertically down;
   2. the normal force $\vec{N}_1$ of the surface on him, directed perpendicular to the surface; and
   3. the force of the rope on him, $\vec{T}_1$, equal in magnitude to the tension in the rope between them and directed parallel to and up the slope.

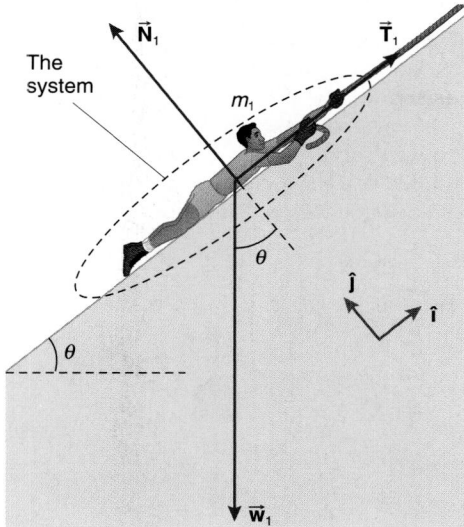

FIGURE 5.58

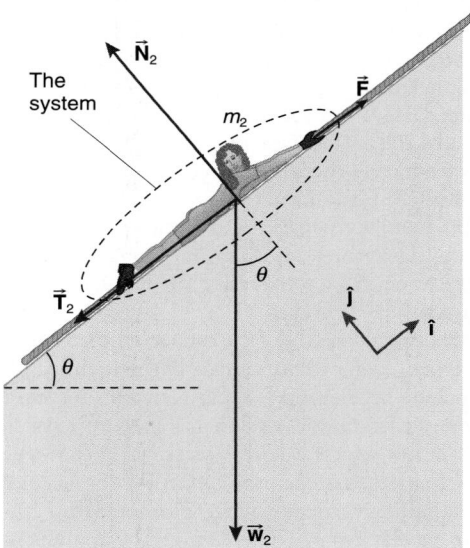

FIGURE 5.59

Let the climber of mass m_2 be another system. There are four forces acting on her (see Figure 5.59):

1. the gravitational force of the Earth on her—her weight $\vec{w}_2$, of magnitude to $m_2 g$, directed vertically down;
2. the normal force $\vec{N}_2$ of the surface on her, directed perpendicular to the surface;
3. the force of the lower rope on her, $\vec{T}_2$, equal in magnitude to the tension in the rope (and to the magnitude of $\vec{T}_1$, since the rope is ideal), directed parallel to and down the slope; and
4. the force $\vec{F}$ on m_2, directed parallel to and up the inclined plane, needed to hold the system at rest on the inclined plane.

b. The forces of the connecting rope on the two masses have the same magnitude since the rope is ideal; that is, $T_1 = T_2 = T$, the tension at any point along the length of the ideal rope between them. Introduce a coordinate system with axes parallel to and perpendicular to the plane, as shown in Figures 5.58 and 5.59. This coordinate choice is most convenient since any possible motion of the masses is parallel to the incline. The accelerations of both masses are zero since they are at rest, but this does not detract from the convenience of this choice for the coordinate system. With this coordinate choice, write the weight of each climber in Cartesian form as

$$\vec{w}_1 = (m_1 g \sin\theta)(-\hat{\imath}) + (m_1 g \cos\theta)(-\hat{\jmath})$$
$$= (-m_1 g \sin\theta)\hat{\imath} - (m_1 g \cos\theta)\hat{\jmath}$$

and

$$\vec{w}_2 = (m_2 g \sin\theta)(-\hat{\imath}) + (m_2 g \cos\theta)(-\hat{\jmath})$$
$$= (-m_2 g \sin\theta)\hat{\imath} - (m_2 g \cos\theta)\hat{\jmath}$$

The two normal forces are, in Cartesian form,

$$\vec{N}_1 = N_1 \hat{\jmath}$$
$$\vec{N}_2 = N_2 \hat{\jmath}$$

and the two forces of the connecting rope on each mass are

$$\vec{T}_1 = T\hat{\imath}$$
$$\vec{T}_2 = T(-\hat{\imath})$$

The force directed parallel to and up the inclined plane is

$$\vec{F} = F\hat{\imath}$$

Consider the mass m_1 as a system. The m_1-system is in equilibrium (zero acceleration). Since $a_x = 0$ m/s^2, Newton's second law of motion, Equation 5.2, becomes

$$F_{x\,\text{total}} = m_1 a_x$$
$$-m_1 g \sin\theta + T = m_1(0 \text{ m/s}^2)$$
$$= 0 \text{ N}$$

After transposing, the magnitude of the tension in the rope is found to be

$$T = m_1 g \sin\theta$$

Likewise, since the m_1-system is not accelerating along the y-direction, $a_y = 0$ m/s^2 and the sum of the force components along the y-direction must also be zero. The second law of motion, Equation 5.2, becomes

$$F_{y\,\text{total}} = m_1 a_y$$
$$N_1 - m_1 g \cos\theta = m_1(0 \text{ m/s}^2)$$
$$= 0 \text{ N}$$

After transposing, the normal force of the surface on mass m_1 is found to be

$$N_1 = m_1 g \cos\theta$$

Now apply Newton's second law of motion to the m_2-system. The m_2-system also is in equilibrium (zero acceleration); the sum of the force components *on* m_2 along the x-direction is zero since $a_x = 0$ m/s². Thus

$$F_{x\,total} = m_2 a_x$$
$$F - (m_2 g \sin \theta) - T = m_2(0 \text{ m/s}^2)$$
$$= 0 \text{ N}$$

Transposing,

$$F = m_2 g \sin \theta + T$$

Substituting for T, the magnitude of the force F is found to be

$$F = m_2 g \sin \theta + m_1 g \sin \theta$$
$$= (m_1 + m_2)g \sin \theta$$

Since m_2 also is not accelerating along the y-direction, $a_y = 0$ m/s² and the sum of the force components along the y-direction also is zero for mass m_2. The second law of motion, Equation 5.2, becomes

$$F_{y\,total} = m_2 a_y$$
$$N_2 - m_2 g \cos \theta = m_2(0 \text{ m/s}^2)$$
$$= 0 \text{ N}$$

After transposing, the normal force of the surface on the mass m_2 is found to be

$$N_2 = m_2 g \cos \theta$$

c. If the masses are pulled up the inclined plane at constant velocity, the system of two masses still is in equilibrium. The results we obtained, therefore, also apply to such a situation.

STRATEGIC EXAMPLE 5.9

Two masses m_1 and m_2 are connected by an ideal string passing over an ideal pulley, as shown in Figure 5.60. Mass m_2 is on an inclined frictionless surface.

a. Treat each mass as an individual system and make a second law force diagram indicating all forces acting on each mass.
b. Determine the acceleration of the system.
c. Find the magnitude of the tension in the ideal string connecting the two masses.

Solution

a. Choose m_1 to be a system. The second law force diagram for it has the two forces shown in Figure 5.61:
   1. the gravitational force of the Earth on m_1 (the weight $\vec{w}_1$ of m_1), of magnitude $m_1 g$, directed vertically downward; and
   2. the force $\vec{T}_1$ of the cord on m_1, equal in magnitude to the tension in the cord and directed vertically upward.

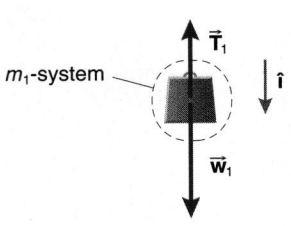

FIGURE 5.61

Choose m_2 to be another system. The second law force diagram for it includes the three forces shown in Figure 5.62:
   1. the gravitational force of the Earth on m_2 (the weight $\vec{w}_2$ of m_2), of magnitude $m_2 g$, directed vertically downward;
   2. the force $\vec{T}_2$ of the cord on m_2, which has a magnitude equal to that of the tension in the cord. Recall that the tension has the same magnitude everywhere along an ideal cord; the ideal pulley does not alter this situation but merely changes the direction in which the tension acts. Thus the force $\vec{T}_2$ on m_2 has the same *magnitude* as the force $\vec{T}_1$ of the cord on m_1; call their common magnitude T; and
   3. the normal force $\vec{N}$ of the surface on the mass m_2, directed perpendicular to the surface of the inclined plane.

b. To find the acceleration, apply Newton's second law to the motion of each mass considered as a separate system. But you do not know which way the system of masses will move! In such instances, *assume* a direction for the acceleration, say with m_1 falling vertically down. The actual direction that the system accelerates will emerge from the analysis.

Choose a coordinate system with $\hat{i}$ vertically down as indicated in Figure 5.61. Newton's second law, Equation 5.2, applied to the m_1-system implies that the vector sum of the forces in the

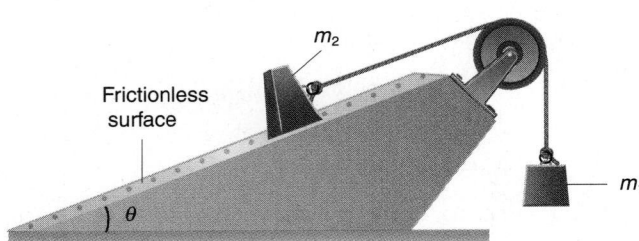

FIGURE 5.60

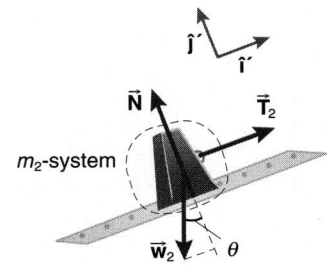

FIGURE 5.62

x-direction is equal to the mass of the system times its acceleration $a_x\hat{\imath}$:

$$m_1 g\hat{\imath} - T\hat{\imath} = m_1 a_x\hat{\imath}$$

Thus

$$m_1 g - T = m_1 a_x \qquad (1)$$

Notice that this equation has two unknowns, T and a_x. You need to find another equation relating T and a_x in order to find the solution. The remaining equation is found by applying Newton's second law to the m_2-system.

Choose a coordinate system with axes parallel and perpendicular to the incline, indicated in Figure 5.62, so the movement of m_2 in the positive x'-direction corresponds to m_1 moving in the positive x-direction. Write the weight $\vec{w}_2$ in terms of its Cartesian components parallel and perpendicular to the plane:

$$\vec{w}_2 = (m_2 g \sin\theta)(-\hat{\imath}') + (m_2 g \cos\theta)(-\hat{\jmath}')$$
$$= (-m_2 g \sin\theta)\hat{\imath}' - (m_2 g \cos\theta)\hat{\jmath}'$$

The acceleration of m_2 perpendicular to the plane is zero $(a_{y'} = 0 \text{ m/s}^2)$, and so the vector sum of the forces on m_2 in the direction perpendicular to the plane, according to Newton's second law, must vanish:

$$N\hat{\jmath}' - (m_2 g \cos\theta)\hat{\jmath}' = m_2 a_{y'}\hat{\jmath}'$$
$$= 0 \text{ N}$$

Thus you have

$$N = m_2 g \cos\theta$$

With the assumption made about the direction of the acceleration of m_1, the mass m_2 accelerates *up* the plane. The string is ideal and does not stretch, and so the acceleration of m_2 has the same *magnitude* as the acceleration of m_1. Look at the motion of m_2 parallel to the incline. The force $\vec{T}_2$ is directed up the plane while a component of the weight $(m_2 g \sin\theta)$ is directed down the plane. Apply Newton's second law to the m_2-system along the x'-direction:

$$T\hat{\imath}' - (m_2 g \sin\theta)\hat{\imath}' = m_2 a_{x'}\hat{\imath}'$$
$$T - m_2 g \sin\theta = m_2 a_{x'}$$

The acceleration components a_x and $a_{x'}$ have the same value since the string does not stretch; thus

$$T - m_2 g \sin\theta = m_2 a_x \qquad (2)$$

Now you can solve equations (1) and (2) simultaneously for T and a_x. First solve equation (1) for T:

$$T = m_1(g - a_x)$$

Then substitute for T into equation (2):

$$m_1(g - a_x) - m_2 g \sin\theta = m_2 a_x$$

Solving for a_x, you find

$$a_x = \frac{m_1 - m_2 \sin\theta}{m_1 + m_2} g \qquad (3)$$

Notice in equation (3) that the acceleration component a_x is *positive* if $m_1 > m_2 \sin\theta$. In other words, the system will move in the direction you assumed ($\hat{\imath}$ for m_1 and $\hat{\imath}'$ for m_2) if this inequality is satisfied. On the other hand if $m_1 < m_2 \sin\theta$, then the acceleration component a_x is *negative*. So the system accelerates in a direction *opposite* to the one you presumed to begin the calculation. Thus the *actual* direction of the acceleration emerges naturally from the analysis.

c. With the acceleration component in hand, you can determine the magnitude of the tension in the cord from either equation (1) or (2). Using equation (1), you find

$$T = m_1(g - a_x)$$

Substituting for a_x with equation (3),

$$T = m_1 g - m_1 \frac{m_1 - m_2 \sin\theta}{m_1 + m_2} g$$

After some algebraic simplification, you find that the tension in the rope is of magnitude

$$T = \frac{m_1 m_2(1 + \sin\theta)}{m_1 + m_2} g$$

Notice how you accounted for the forces by analyzing their components *consistent* with the directions of the forces indicated on the force diagram and the coordinate system(s) chosen.

EXAMPLE 5.10

Mexican Indians have an apparatus that, for spiritual purification, swings a person around a circle at constant angular speed ω, as shown in Figure 5.63. The length of each cable is ℓ and the radius of the circular hub is R.

a. Show that the angle θ that the cable makes with the vertical direction is independent of the mass of the patron and harness and is given by

$$\tan\theta = \frac{(R + \ell \sin\theta)\omega^2}{g}$$

b. If $R = 2.0$ m, $\ell = 3.0$ m, and the apparatus is spinning at 12.0 rev/min, find the angle θ.

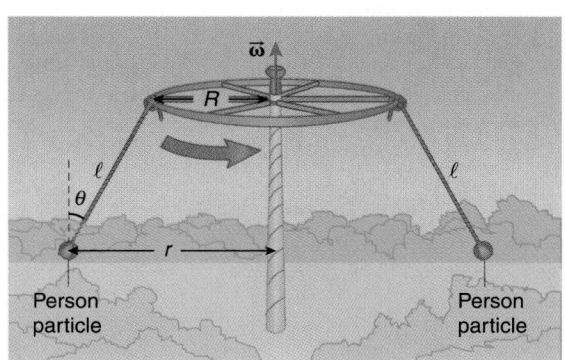

FIGURE 5.63

Solution

a. Consider a person and harness to be a single system of mass m. A second law force diagram on the system on the left includes the two forces shown in Figure 5.64:

1. the weight $\vec{w}$ of the system, of magnitude mg, directed vertically down; and
2. the force $\vec{T}$ of the cable on the system, equal in magnitude to the tension in the cable, directed along the cable making an angle θ with the vertical direction.

There is no centripetal force, since there is no such force. However, there *is* a centripetal acceleration.

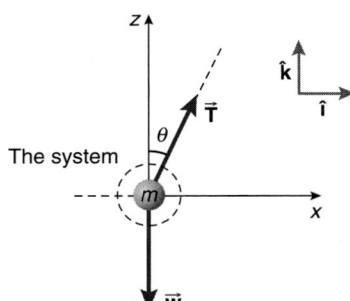

FIGURE 5.64

The coordinate system is fixed on the ground and does not rotate with the apparatus. We analyze the system when it is instantaneously on the *x*-axis as indicated in Figure 5.64.

The system is not accelerating in the vertical direction, so that $a_z = 0$ m/s^2 and the application of Newton's second law to this axis yields

$$F_{z\,\text{total}}\hat{k} = ma_z\hat{k}$$

$$(T\cos\theta)\hat{k} + mg(-\hat{k}) = m(0 \text{ m/s}^2)\hat{k}$$

$$= 0 \text{ N}$$

$$T\cos\theta = mg \qquad (1)$$

Along the *x*-direction, there is a force component $T\sin\theta$ directed toward $\hat{i}$ at the instant depicted in Figure 5.64. Since the system is in uniform circular motion, there is a centripetal acceleration of magnitude v^2/r, directed toward the center of the circle, toward $\hat{i}$ at this instant, where v is the speed of the system. Thus Newton's second law applied to the *x*-direction gives

$$F_{x\,\text{total}}\hat{i} = ma_x\hat{i}$$

$$(T\sin\theta)\,\hat{i} = m\frac{v^2}{r}\,\hat{i}$$

$$T\sin\theta = m\frac{v^2}{r} \qquad (2)$$

Divide equation (2) by equation (1); the magnitude of the tension is eliminated:

$$\frac{T\sin\theta}{T\cos\theta} = \frac{m\frac{v^2}{r}}{mg}$$

$$\tan\theta = \frac{v^2}{rg}$$

Note that the angle is independent of the mass of the system, so that all persons swing at the same angle with the vertical.

For circular motion, the speed v and the angular speed ω are related by Equation 4.34,

$$v = r\omega$$

Making this substitution into the expression for $\tan\theta$, you find

$$\tan\theta = \frac{r^2\omega^2}{rg}$$

$$= \frac{r\omega^2}{g}$$

From the geometry of Figure 5.63, $r = R + \ell\sin\theta$; thus

$$\tan\theta = \frac{(R + \ell\sin\theta)\omega^2}{g} \qquad (3)$$

b. The angular speed of the apparatus is 12.0 rev/min or

$$\omega = \frac{(12.0 \text{ rev/min})(2\pi \text{ rad/rev})}{60 \text{ s/min}}$$

$$= 1.26 \text{ rad/s}$$

Since $R = 2.0$ m, $\ell = 3.0$ m, and $g = 9.81$ m/s^2, equation (3) becomes

$$\tan\theta = \frac{[2.0 \text{ m} + (3.0 \text{ m})\sin\theta](1.26 \text{ rad/s})^2}{9.81 \text{ m/s}^2}$$

$$= 0.32 + 0.49\sin\theta$$

This is a transcendental equation for θ, not an uncommon outcome. There are two ways to find θ.

Method 1

A solution can be found graphically by making two graphs on the same sheet, one of $\tan\theta$ versus θ, the other of $0.32 + 0.49\sin\theta$ versus θ, as in Figure 5.65. The value of θ where the two graphs intersect is the value of θ that makes the sides of the equation equal.

Method 2

The angle θ also can be found by successive approximations, beginning with a value of θ that may be *reasonable*, such as 20°, then trying both higher and lower values of θ to see if they are closer to satisfying the equation, and walking-in on the solution this way.

Whichever method is chosen, you find

$$\theta = 29°$$

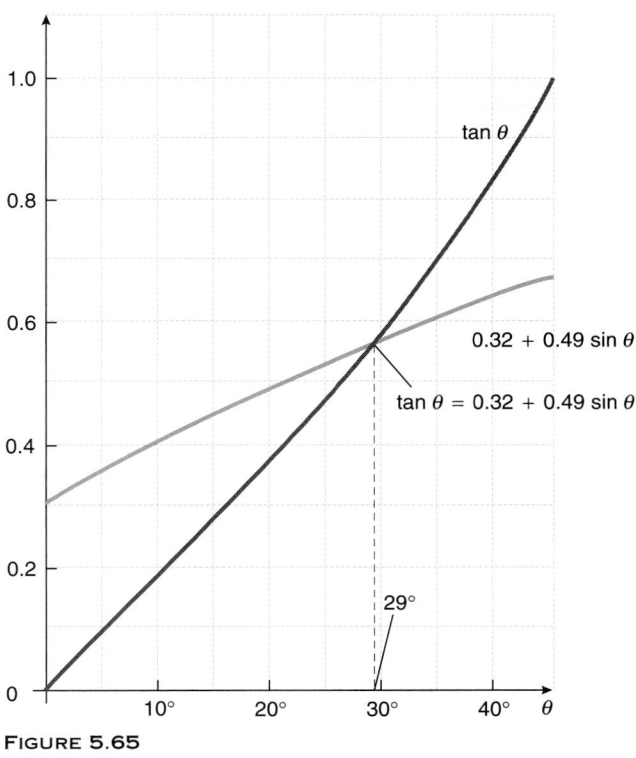

FIGURE 5.65

5.12 STATIC FRICTION

The obedience of matter is limited by friction.

Victor Hugo*

Frictional effects permeate the macroscopic world around us, and so we need to have a good understanding of them. As we noted in the introduction to this chapter, the existence of friction likely impeded the development of a consistent theory of motion. You should not think that frictional forces are always bad. Without friction, your car could never move, and neither could you, as you may know from attempting to accelerate from rest on an icy pavement. Friction is the proverbial two-edged sword: a blessing or a curse depending on the circumstances.

There are two general types of frictional forces: (1) a **static frictional force**, involving no motion (no sliding) at the places where a system stays in contact with a surface, such as the bottom of your hiking boots in contact with the trail, or the bottom of a tire in contact with the surface of a road; and (2) a **kinetic frictional force** that arises when one surface slides along a second surface, such as when your car wheels lock up and skid along the road, or when a moving automotive engine piston rubs against the cylinder containing it. Kinetic frictional forces also arise when a system moves through a fluid (such as air or water), as when you ride your bike at great speed.

A series of simple experiments will help you to see what is involved with static frictional effects.

Experiment 1: Imagine your calculus text lying at rest on your desk. The contact surface is *not* frictionless. The book is the

---

Les Misérables (1862) (A. L. Burt, New York), volume II, Jean Valjean, Book VI, The White Knight, IV, page 673.

system we want to analyze. We know of two forces on the text (see Figure 5.66):

1. the gravitational force of the Earth on the text, its weight $\vec{w}$, of magnitude mg, directed vertically downward; and
2. the normal force $\vec{N}$ of the surface on the text, directed perpendicular to the surface.

The second law of motion, Equation 5.1, becomes

$$\vec{F}_{total} = m\vec{a}$$
$$\vec{w} + \vec{N} = m(0 \text{ m/s}^2)$$
$$= 0 \text{ N}$$

Thus

$$\vec{N} = -\vec{w}$$

Since the book is not accelerating in the horizontal direction, the total force on it in the horizontal direction is zero; indeed, there are no forces at all in the horizontal direction.

Experiment 2: Now push slightly on the text in the horizontal direction with your finger. Since the motion in the vertical direction is independent of the motion in the horizontal direction, nothing changes in our previous analysis of the vertical motion (Experiment 1). You find the text *remains* in equilibrium despite the existence of the slight horizontal force you apply. The acceleration of the text along the horizontal direction is zero, since its velocity (equal to zero) is not changing with time. Newton's second law of motion implies that the vector sum of the forces *parallel* to the surface must be zero because the text is in equilibrium. One such horizontal force is the slight push $\vec{F}$ by your finger (see Figure 5.67). Another force *must* exist parallel to the surface so that the vector sum of the two forces is zero. This other force is the **static frictional force** $\vec{f}_s$.

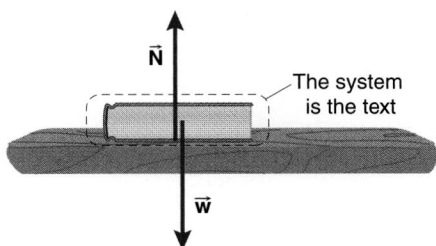

FIGURE 5.66 Forces on your text when it is at rest on your desk.

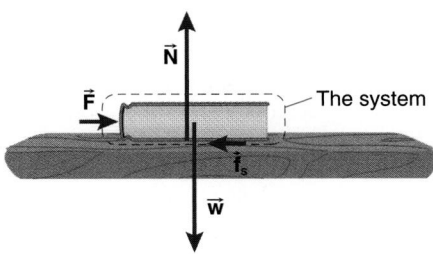

FIGURE 5.67 Forces on your text when you push it slightly in the horizontal direction.

The *total force* in the horizontal direction is zero, in keeping with our observation that the text is in equilibrium (zero acceleration). Therefore the static frictional force must be of the same magnitude as that of the force you exert on the text, but it must point in the opposite direction.

The static frictional force always acts in a direction parallel to the surface regardless of the orientation of the surface with respect to the horizontal direction. The static fictional force is directed to prevent the system from slipping.

Note that the static frictional force of the surface on the text and the force you exert on the text are equal in magnitude and opposite in direction; but they are *not* a Newton's third law force pair, since they act on the *same* system (the text). The two forces are of equal magnitude and opposite direction because of zero acceleration and the *second* law of motion.

Experiment 3: Now push a bit harder with your finger. The text will still be in equilibrium. The static frictional force matches your push newton-for-newton. The vector sum of the two forces remains zero because the acceleration of the text, once again, is zero. Push a bit harder and the static frictional force grows to match it. The static frictional force is seemingly quite "intelligent"; it knows exactly how hard you are pushing and matches the push appropriately! The vector sum of the forces on the book acting parallel to the surface stays zero because the acceleration is zero.

Eventually, however, if you push hard enough, the text reaches a state where it is on the verge of slipping across the table. When the text is on the verge of slipping, and *only* then, the ratio between the magnitude of this greatest static frictional force and the magnitude of the normal force is defined to be a pure number called the **coefficient of static friction** μ_s. This definition is applied only when the system is on the verge of slipping:

$$\frac{f_{s\,max}}{N} \equiv \mu_s \qquad \text{(system ready to slip)}$$

This is usually written as

$$f_{s\,max} = \mu_s N \qquad \text{(system ready to slip)} \qquad (5.6)$$

Equation 5.6 for $f_{s\,max}$ is *not* a vector equation. It is a *scalar* relationship between the *magnitudes* of the maximum static frictional force and the normal force; the two forces point in different directions. The normal force is directed perpendicular to the surface, while the static frictional force is parallel to the surface; the angle between the two forces $\vec{f}_{s\,max}$ and $\vec{N}$ is always 90°.*

The numerical value of the (dimensionless) coefficient of static friction μ_s depends on the specific materials that compose the two surfaces in contact, but to a first approximation it is independent of the area in contact. In other words, the coefficient

TABLE 5.3 Approximate Values of the Coefficient of Static Friction

Surfaces in contact	Coefficient of static friction μ_s when system is ready to slip
Glass on glass	0.9
Rubber on dry concrete	~1
Rubber on wet concrete	0.4
Steel on steel	0.6
Steel on steel (lubricated)	0.1
Wood on wood	0.3
Waxed ski on snow	0.1

of static friction for the text on the table has one value; the coefficient of static friction for the text on, say, a shag rug has quite another numerical value. Approximate numerical values of the coefficient of static friction for various surfaces in contact with each other are indicated in Table 5.3. Precise values must always be determined from experiments with the actual materials used. Example 5.11 illustrates a common experimental procedure for determining μ_s. The numerical value of the coefficient of static friction μ_s is the same *regardless* of the orientation of the surfaces with respect to the horizontal or vertical directions.

Recall that the static frictional force matched your pushes newton-for-newton up to the point where slipping was about to occur. The relationship given by Equation 5.6 is the *maximum* magnitude of the static frictional force, as indicated by the subscript.

The static frictional force always opposes *slippage* of surfaces in contact but may, under certain circumstances, actually be the force responsible for the *acceleration* of a system. Example 5.14 illustrates this perhaps counterintuitive idea.

Equation 5.6 for $f_{s\,max}$ is *not* a fundamental physical law on a par with, say, Newton's laws of motion. It is an empirical definition of the coefficient of static friction, used to model a complex molecular surface interaction in a straightforward way.

PROBLEM-SOLVING TACTIC

5.10 Equation 5.6 is useful only when the system is ready to slip across the surface. If the system is not on the verge of slipping, the magnitude f_s of the actual static frictional force is *less than* $f_{s\,max}$ and *cannot* be found using Equation 5.6.

QUESTION 7

In Experiment 2, if a graph is made of the magnitude of the static frictional force $\vec{f}_s$ versus the magnitude of your horizontal push $\vec{F}$ on the text, what is the slope of the graph? If the experiment were performed on an inclined surface, would the graph be the same?

QUESTION 8

For a given system in contact with a given surface, what would a graph of $f_{s\,max}$ versus N look like?

---

*It is possible to relate a vector in one direction to a vector pointing in another direction; we did this once via vector products, but that is a special case. The most general way to formulate this relationship is using matrices.

STRATEGIC EXAMPLE 5.11

A dictionary of mass m is placed at rest (and remains at rest) on a rough reading table surface inclined at an angle θ to the horizontal, as shown in Figure 5.68. The coefficient of static friction for the mass on the wood surface is μ_s.

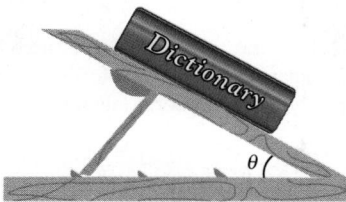

FIGURE 5.68

a. Calculate the magnitude of the static frictional force on the dictionary.
b. What is the maximum angle θ for which the dictionary does not slip?

Solution
a. Consider the dictionary to be the system. There are three forces on the system (see Figure 5.69):
    1. the weight $\vec{\mathbf{w}}$ (of magnitude mg), directed vertically down;
    2. the normal force $\vec{\mathbf{N}}$ of the surface on the mass, directed perpendicular to the surface; and
    3. the static frictional force $\vec{\mathbf{f}}_s$, parallel to the surface. If you do not know which direction the static frictional force points, either direction parallel to the surface (up or down the plane) can be chosen. The resulting analysis will indicate the right direction through the algebraic sign of the result. The second law force diagram then is shown in Figure 5.69.

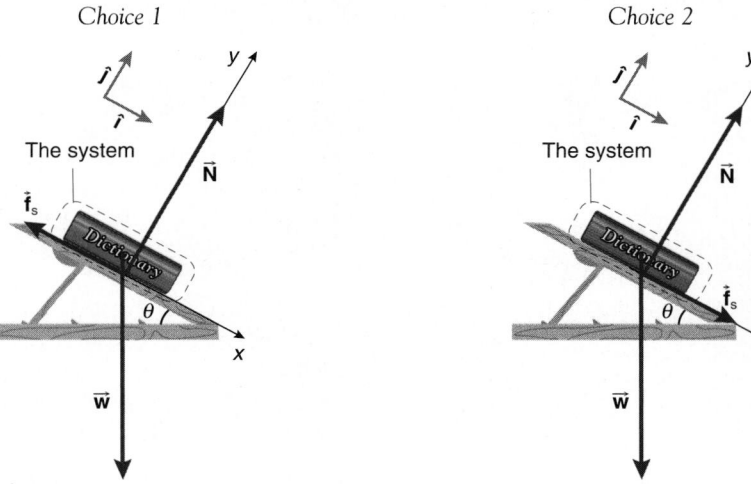

Choose the direction of $\vec{\mathbf{f}}_s$ to be up the plane.   Choose the direction of $\vec{\mathbf{f}}_s$ to be down the plane.
FIGURE 5.69

Choose the coordinate system indicated. There is nothing sacred about this choice, as we have seen on many occasions; but for inclined planes, it is usually convenient to choose coordinate axes parallel and perpendicular to the incline when the motion or potential motion of the system is up or down the inclined surface.

Write the forces in Cartesian form, *consistent* with the directions of the forces as indicated in the force diagram:
    1. The weight has components along both the x- and y-directions:

$$\vec{\mathbf{w}} = (mg \sin \theta)\hat{\mathbf{i}} + (mg \cos \theta)(-\hat{\mathbf{j}})$$

2. The normal force is

$$\vec{N} = N\hat{j}$$

3. The static frictional force is

$\vec{f}_s = f_s(-\hat{i})$	$\vec{f}_s = f_s\hat{i}$

If the dictionary remains at rest on the slope, its acceleration is zero. Newton's second law of motion then indicates that the vector sum of these forces must be zero. Sum the force vectors along the y-direction and apply Newton's second law with $a_y = 0$ m/s²:

$$F_{y\,total}\hat{j} = ma_y\hat{j}$$
$$N\hat{j} - (mg\cos\theta)\hat{j} = m(0 \text{ m/s}^2)\hat{j}$$
$$= 0 \text{ N}$$

Solving for N,

$$N = mg\cos\theta$$

Note that here the magnitude of the normal force N is less than mg.

The vector sum of the forces in the x-direction also must be zero according to Newton's second law of motion, because $a_x = 0$ m/s²:

$$F_{x\,total}\hat{i} = ma_x\hat{i}$$
$$= m(0 \text{ m/s}^2)\hat{i}$$
$$= 0 \text{ N}$$

$(mg\sin\theta)\hat{i} - f_s\hat{i} = 0$ N	$(mg\sin\theta)\hat{i} + f_s\hat{i} = 0$ N

or after transposing,

$f_s = mg\sin\theta$	$f_s = -mg\sin\theta$

This means that the static frictional force $\vec{f}_s$, which we wrote as $\vec{f}_s = f_s\hat{i}$, consistent with the second law force diagram, is

$$\vec{f}_s = (-mg\sin\theta)\hat{i}$$

$$= (mg\sin\theta)(-\hat{i})$$

which indicates that $\vec{f}_s$ actually is directed parallel to $(-\hat{i})$, or *up* the plane. Notice that the *magnitude* of the frictional force still is $mg\sin\theta$.

b. The expression for the magnitude of the static frictional force is valid for *any* angle θ as long as the mass is at rest. Notice that as θ increases, the magnitude of the static frictional force also increases, since $\sin\theta$ increases with increasing θ. Eventually, when the inclination angle θ is large enough, the system is on the verge of slipping. At this critical angle θ_{verge}, the static frictional force reaches its maximum magnitude:

$$f_{s\,max} = mg\sin\theta_{verge}$$

As the angle increases to θ_{verge}, the magnitude of the normal force *decreases* since $N = mg\cos\theta$. When the system is ready to slip, the corresponding magnitude of the normal force is $mg\cos\theta_{verge}$.

Divide the magnitude of the maximum static frictional force by the magnitude of the corresponding normal force when the system is ready to slip; you find

$$\frac{f_{s\,max}}{N} = \frac{mg\sin\theta_{verge}}{mg\cos\theta_{verge}}$$

$$= \tan\theta_{verge}$$

When the system is on the verge of slipping, the ratio of these force magnitudes is (by definition) the coefficient of static friction (Equation 5.6):

$$\frac{f_{s\,max}}{N} \equiv \mu_s$$

Equating the expressions for the ratio $f_{s\,max}/N$, you find that the coefficient of static friction is

$$\mu_s = \tan\theta_{verge} \qquad (1)$$

In other words, when the system is ready to slip, the tangent of the inclination angle is equal to the coefficient of static friction. Notice that if the angle θ_{verge} is greater than 45°, then the coefficient of static friction is greater than 1, which is entirely possible for some surfaces. Equation (1) is the basis for the experimental determination of the coefficient of static friction for two surfaces in contact.

STRATEGIC EXAMPLE 5.12

Take a dictionary of mass 5.00 kg and push it hard against a vertical wall with a horizontal force $\vec{F}$ of magnitude 200 N, so that the tome does not slip down the wall, as shown in Figure 5.70. The coefficient of static friction for the book in contact with the wall is $\mu_s = 0.300$.

a. Determine the magnitude of the normal force of the wall on the book.
b. Find the magnitude of the static frictional force on the book.
c. If the magnitude of the force $\vec{F}$ decreases until the book is ready to slip, what is the magnitude of $\vec{F}$ when slipping is imminent?

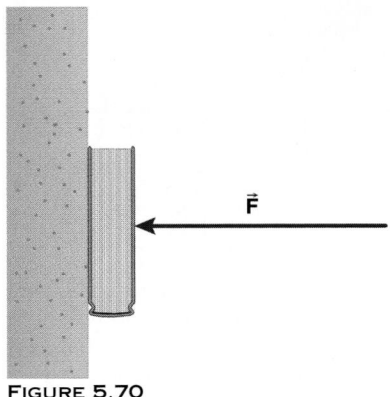

FIGURE 5.70

Solution
Consider the dictionary to be the system. There are four forces acting on the dictionary, shown in Figure 5.71:

1. the weight $\vec{w}$ of the book, of magnitude

$$mg = (5.00\ kg)(9.81\ m/s^2)$$
$$= 49.1\ N$$

directed vertically down;
2. the horizontal force $\vec{F}$ that you exert on the dictionary;
3. the normal force $\vec{N}$ of the surface on the book, directed perpendicular to the surface; and

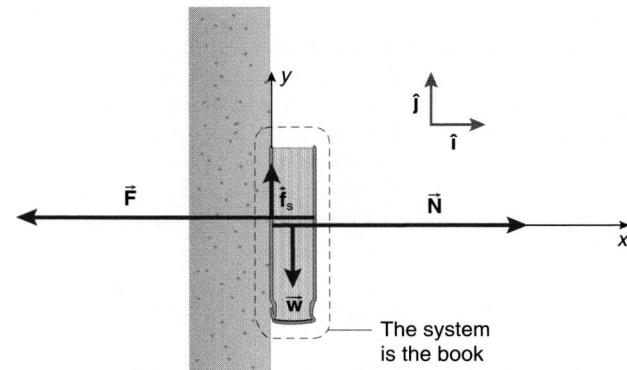

FIGURE 5.71

4. the static frictional force $\vec{f}_s$ on the book arising from its contact with the vertical wall. This force acts parallel to the surface to prevent the book from slipping down the wall; thus the static frictional force here acts vertically upward.

Introduce a coordinate system with horizontal and vertical axes, as indicated in Figure 5.71.

a. The book is in equilibrium and so the acceleration components a_x and a_y both are zero. Apply Newton's second law of motion, Equation 5.2, to the horizontal direction:

$$F_{x\,total}\hat{\imath} = ma_x\hat{\imath}$$
$$N\hat{\imath} + F(-\hat{\imath}) = m(0\ m/s^2)\hat{\imath}$$

Thus

$$N = F$$

The magnitude of the normal force is equal to the magnitude of the force you exert on the book, here 200 N. Notice that here the normal force of the surface on the book has *no* relationship whatever to the weight of the book, illustrating that the magnitude of the normal force must be determined on a case-by-case basis (Problem-Solving Tactic 5.8).

b. Apply Newton's second law of motion to the vertical direction:

$$F_{y\,total}\hat{\jmath} = ma_y\hat{\jmath}$$
$$mg(-\hat{\jmath}) + f_s\hat{\jmath} = m(0\ m/s^2)\hat{\jmath}$$

Thus

$$f_s = mg$$
$$= (5.00 \text{ kg})(9.81 \text{ m/s}^2)$$
$$= 49.1 \text{ N}$$

The magnitude of the static frictional force is equal to the magnitude of the weight. Notice that the magnitude of the static frictional force here is not found from Equation 5.6, because the book is not on the verge of slipping. Here Equation 5.6 would yield

$$f_{s\,max} = \mu_s N$$
$$= (0.300)(200 \text{ N})$$
$$= 60.0 \text{ N}$$

well above the *actual* magnitude of the static frictional force of 49.1 N.

c. Now ease your push $\vec{F}$ on the book until the book *is* on the verge of slipping down the wall. What magnitude of force F_{min} is needed so the book is on the verge of slipping? As you ease your push, *the magnitude of the static frictional force acting on the book does not change* and still is equal in magnitude to the weight of the book (49.1 N). The book will be on the verge of slipping when the normal force has decreased to a magnitude N', where

$$f_{s\,max} = \mu_s N'$$

The value of $f_{s\,max}$ is 49.1 N, *not* 60.0 N—be sure you understand why! Substitute for $f_{s\,max}$ and μ_s:

$$49.1 \text{ N} = 0.300 N'$$

Solve for N':

$$N' = 164 \text{ N}$$

In this example the normal force of the surface on the book is, at all times, equal in magnitude to that of the force you apply. The book is on the verge of slipping when the force you apply has a magnitude of 164 N.

EXAMPLE 5.13

The roadbeds underneath highways and railroad tracks typically are banked around high-speed curves, as shown in Figure 5.72. There is less wear on the tires, wheels, roadway, and rails if there is no static frictional force down the incline,

FIGURE 5.72 Curves on racetracks and highways are banked for a reason.

parallel to the surface. The ride inside the vehicle also is more comfortable, for there then is no tendency to slide across the seat. Aircraft also bank when turning for similar reasons.

a. Let the radius of the arc of a circular curve be r; your car is traveling at a speed v around the curve. What angle should the roadbed make with the horizontal to assure the least amount of wear? The angle is called the *banking angle* of the turn.
b. Calculate the banking angle for a car traveling at 100 km/h around a circular turn of radius 500 m.

Solution

a. Let the system be the mass of the wheel along with its share of the mass of the car and its contents; call this entire mass m. The forces acting on the system are the normal force of the surface and the weight, indicated in the second law force diagram (Figure 5.73). Remember that each perpendicular direction can be analyzed independently of the others. Here you are concerned only with the instantaneous motion in the plane of Figure 5.73. You can ignore any horizontal forces perpendicular to the plane of Figure 5.73.

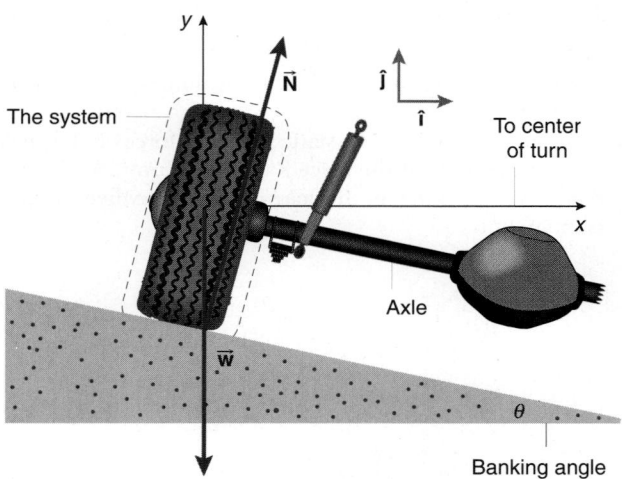

FIGURE 5.73

It is convenient to choose one of the coordinate axes to be in the same direction as the acceleration of the system. When you examined inclined planes, such as in Example 5.11, you chose the axes to be parallel and perpendicular to the incline, as in Figure 5.69, because any accelerations for that problem would be parallel to the incline. Here, however, the centripetal acceleration is toward the center of the curve; so choose coordinate axes that are horizontal and vertical. In this way, one coordinate axis is along the line of the acceleration. This is simpler than choosing the directions of the coordinate axes to be parallel and perpendicular to the slope. The preferred choice of axes is that which makes the problem simplest. Part of the skill of attacking a problem is making an appropriate choice for the axis directions, a skill you will develop with practice.

Since the car is moving around a curve, the vector sum of the forces in the x–y plane must provide the centripetal acceleration toward the center of the curve.

Express the normal force $\vec{N}$ in terms of its two Cartesian components along the *x*- and *y*-axes. The normal force makes an angle with the vertical direction equal to the banking angle θ (since two angles whose sides are mutually perpendicular are equal). Hence

$$\vec{N} = (N \sin \theta)\hat{\imath} + (N \cos \theta)\hat{\jmath}$$

The weight is

$$\vec{w} = -mg\hat{\jmath}$$

Analyze the motion along the *x*- and *y*-axes independently:

The *y*-coordinate of the wheel remains constant around the curve. Therefore there is no *y*-component to the acceleration ($a_y = 0 \text{ m/s}^2$). Newton's second law of motion in this direction yields

$$F_{y\ \text{total}} = ma_y$$
$$N \cos \theta - mg = m(0 \text{ m/s}^2)$$
$$= 0 \text{ N}$$

Solving for *N*,

$$N = \frac{mg}{\cos \theta}$$

Along the *x*-direction, the acceleration is the centripetal acceleration toward the center of the turn, of magnitude v^2/r. Applying Newton's second law of motion to the *x*-direction gives

$$F_{x\ \text{total}} = ma_x$$

$$N \sin \theta = m \frac{v^2}{r}$$

Substituting for *N*,

$$\frac{mg}{\cos \theta} \sin \theta = m \frac{v^2}{r}$$

Divide by the mass and recall that

$$\frac{\sin \theta}{\cos \theta} = \tan \theta$$

So

$$g \tan \theta = \frac{v^2}{r}$$

Thus

$$\tan \theta = \frac{v^2}{rg}$$

Notice that the proper banking angle is *independent* of the mass.
b. Convert 100 km/h into m/s:

$$100 \text{ km/h} = \frac{100 \text{ km/h} \times 10^3 \text{ m/km}}{3600 \text{ s/h}}$$

$$= 27.8 \text{ m/s}$$

Substituting this into the equation for the tangent of the banking angle gives

$$\tan \theta = \frac{(27.8 \text{ m/s})^2}{(500 \text{ m})(9.81 \text{ m/s}^2)}$$

$$= 0.158$$

Thus

$$\theta = 8.98°$$

The same principle is used for banking turns on bobsled runs. Should you wish to go around a curve where there is *no* static friction, the curve must be banked so the component of the normal force directed toward the center of the turn is able to cause the centripetal acceleration. If a bobsled does not have the appropriate angle when executing a turn, there is trouble, which is why much practice is needed to be a good bobsled driver.

STRATEGIC EXAMPLE 5.14

A Reese's* Peanut Butter Cup® candy bar of mass m_{PBC} rests on a tray of mass m_{tray} that is on a horizontal frictionless surface (see Figure 5.74). The coefficient of static friction for the candy bar on the tray surface is μ_s. A constant horizontal force $\vec{F}$ on the tray accelerates the tray and candy bar to the right.

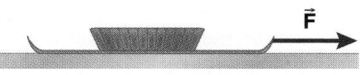

FIGURE 5.74

a. Draw a second law force diagram indicating the forces acting on the tray and candy bar, considering them as a single system.
b. Determine the magnitude of the acceleration in terms of the applied horizontal force and appropriate masses.
c. Sketch another second law force diagram indicating the forces acting on the candy bar, considering it as a separate system. What force accelerates the candy bar?
d. What is the maximum magnitude of the acceleration if the candy bar does not slip on the tray?
e. Considering the tray as a separate system, draw yet another second law force diagram indicating the forces acting on the tray. Use this force diagram to determine the magnitude of the acceleration of the tray, and show that the acceleration is the same as the result obtained in part (b).

Solution
a. The forces acting on the tray and candy bar system are (see Figure 5.75)
   1. the gravitational force of the Earth on the tray and candy bar system; this is the combined weight $\vec{w}$ of the tray and candy bar, of magnitude $(m_{\text{PBC}} + m_{\text{tray}})g$, directed vertically down;
   2. the normal force $\vec{N}$ of the surface on the tray and candy bar system, directed vertically upward; and

---
*No known relation to the author. Sigh.

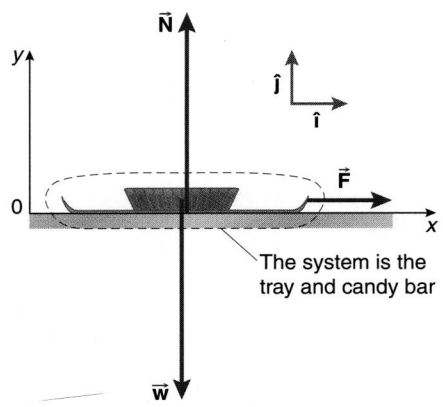

FIGURE 5.75

3. the horizontal force $\vec{F}$ applied to the tray and candy bar system.

There is no frictional force on the tray and candy bar system since it is on a frictionless surface.

b. Use the coordinate system indicated in Figure 5.75. The acceleration of the system is found by applying Newton's second law of motion to the tray and candy bar system in the horizontal direction:

$$F\hat{i} = (m_{\text{PBC}} + m_{\text{tray}})a_x\hat{i}$$

The mass on the right-hand side of Newton's second law is the mass of the system, here the tray and candy bar. Solve for the acceleration component a_x:

$$a_x = \frac{F}{m_{\text{PBC}} + m_{\text{tray}}} \tag{1}$$

Note that in the vertical direction, there is no acceleration $(a_y = 0 \text{ m/s}^2)$; thus Newton's second law applied to this direction yields

$$N\hat{j} + (m_{\text{PBC}} + m_{\text{tray}})g(-\hat{j}) = (m_{\text{PBC}} + m_{\text{tray}})(0 \text{ m/s}^2)\hat{j}$$
$$= 0 \text{ N}$$

Solving for N,

$$N = (m_{\text{PBC}} + m_{\text{tray}})g$$

c. The forces on the candy bar, considering it as a separate system, are the following (see Figure 5.76):
1. the gravitational force of the Earth on the candy bar (its weight $\vec{w}_{\text{PBC}}$) of magnitude $m_{\text{PBC}}g$, directed vertically down;
2. the normal force $\vec{N}'$ of the surface of the tray on the candy bar, directed vertically up; and
3. the static frictional force $\vec{f}_s$ of the surface of the tray on the candy bar.

You may be uncertain which way along the horizontal direction the static frictional force of the tray on the candy bar is directed. Is it parallel or antiparallel to the horizontal acceleration? The static frictional force opposes slippage. The candy bar (if it slips) will slip to the left if the applied force $\vec{F}$

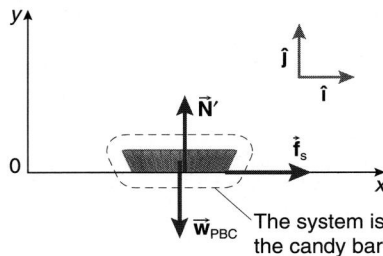

FIGURE 5.76

to the right on the tray is great enough; therefore, the direction of $\vec{f}_s$ is to the right.*

Now apply Newton's second law of motion to the candy bar system. In the vertical direction, there is no acceleration, so that $a_y = 0 \text{ m/s}^2$ and the second law of motion becomes

$$N'\hat{j} + m_{\text{PBC}}g(-\hat{j}) = m_{\text{PBC}}(0 \text{ m/s}^2)\hat{j}$$
$$= 0 \text{ N}$$

Thus

$$N' = m_{\text{PBC}}g \tag{2}$$

In the x-direction, Newton's second law implies

$$f_s\hat{i} = m_{\text{PBC}}a_x\hat{i}$$

Hence

$$f_s = m_{\text{PBC}}a_x \tag{3}$$

If the candy bar does not slip, its acceleration is identical to the acceleration of the tray, so that a_x is given by equation (1). Notice that the *static* frictional force on the candy bar causes its acceleration!

d. The maximum acceleration permitted is attained when the candy bar is ready to slip. The static frictional force then attains its maximum value, given by Equation 5.6:

$$f_{s\text{ max}} = \mu_s N'$$

Using equation (2) for N',

$$f_{s\text{ max}} = \mu_s m_{\text{PBC}}g$$

Using Newton's second law to determine the maximum acceleration,

$$f_{s\text{ max}}\hat{i} = m_{\text{PBC}}a_{x\text{ max}}\hat{i}$$

Substituting for $f_{s\text{ max}}$, you find

$$\mu_s m_{\text{PBC}}g = m_{\text{PBC}}a_{x\text{ max}}$$

---

*If you choose the reverse direction for $\vec{f}_s$ in the sketch, the result for the component of $\vec{f}_s$ will be negative, indicating that the direction of $\vec{f}_s$ is in fact opposite to the direction you chose in the second law force diagram.

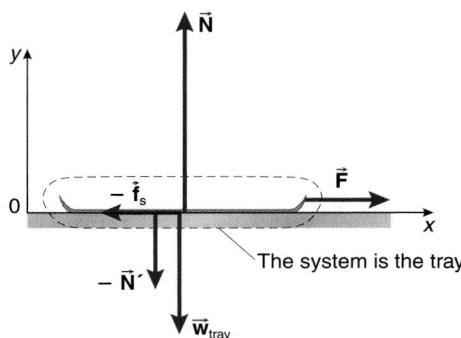

FIGURE 5.77

Thus

$$a_{x\ max} = \mu_s g$$

e. The forces acting on the tray, considering it as an individual system, are (see Figure 5.77)

1. the gravitational force of the Earth on the tray (the weight of the tray $\vec{w}_{tray}$), of magnitude $m_{tray}g$, directed vertically down;

2. the normal force $\vec{N}$ of the surface on the tray, directed vertically up;

3. the force of the candy bar on the tray, $-\vec{N}'$, of magnitude $m_{PBC}g$, directed vertically down; this force and the force $\vec{N}'$ of the tray (surface) on the candy bar [equation (2)] form a Newton's third law pair (notice that $\vec{N}'$ acts on the candy bar, *not* on the tray);

4. the horizontal force $\vec{F}$ exerted on the tray; and

5. the static frictional force $-\vec{f}_s$ of the candy bar on the top of the tray. This force is equal in magnitude, but opposite in direction to the static frictional force of the tray on the candy bar [equation (3)]; they form a Newton's third law pair. Thus the static frictional force of the candy bar on the tray is of magnitude $f_s = m_{PBC}a_x$ and is directed parallel to $-\hat{i}$.

Now apply Newton's second law of motion to the tray, considering it as an independent system. In the vertical direction we have, since $a_y = 0$ m/s²,

$$F_{y\ total}\hat{j} = m_{tray}a_y\hat{j}$$
$$m_{tray}g(-\hat{j}) + m_{PBC}g(-\hat{j}) + N\hat{j} = m_{tray}(0\ m/s^2)\hat{j}$$
$$= 0\ N$$

Solving for N,

$$N = (m_{PBC} + m_{tray})g$$

as we obtained before in part (a).

In the horizontal direction, Newton's second law implies

$$F\hat{i} + f_s(-\hat{i}) = m_{tray}a_x\hat{i}$$

But since $f_s = m_{PBC}a_x$, this becomes

$$F\hat{i} + m_{PBC}a_x(-\hat{i}) = m_{tray}a_x\hat{i}$$

Solving for a_x, you obtain

$$a_x = \frac{F}{m_{PBC} + m_{tray}}$$

the same as we found in part (a).

5.13 KINETIC FRICTION AT LOW SPEEDS

Solid surfaces in contact stick more effectively when at rest than when there is sliding motion parallel to the surfaces. Therefore, once a system begins sliding across a surface, the coefficient of friction changes from the static value μ_s to a smaller kinetic (motional) value μ_k. The ratio between the magnitude of the **kinetic frictional force** $\vec{f}_k$ and the magnitude of the normal force $\vec{N}$ defines the **coefficient of kinetic friction** μ_k:

$$\mu_k \equiv \frac{f_k}{N} \qquad \text{(system sliding)}$$

This is usually written as

$$f_k = \mu_k N \qquad \text{(system sliding)} \qquad (5.7)$$

As with the coefficient of static friction, numerical values for the coefficient of kinetic friction μ_k depend on the specific materials and are, to a first approximation, independent of the surface area in contact. When the system is sliding at low speeds,* the coefficient of kinetic friction is approximately constant, independent of speed. However, as the speed increases, the kinetic frictional force depends on the speed in complex ways that are typically determined empirically on a case-by-case basis. Approximate values for μ_k are tabulated in Table 5.4 for various surfaces.

> **PROBLEM-SOLVING TACTIC**
>
> **5.11** The kinetic frictional force on a system by a surface always is directed opposite to the velocity of the system; the static frictional force on a system is directed opposite to the attempted slippage of the system.

---

*The quantitative meaning of a low speed depends on the specific surfaces in contact.

TABLE 5.4 Approximate Values of the Coefficient of Kinetic Friction[†]

Surfaces in contact	Coefficient of kinetic friction μ_k when system is moving at low speed
Glass on glass	0.4
Rubber on dry concrete	0.8
Rubber on wet concrete	0.3
Steel on steel	0.5
Steel on steel (lubricated)	0.05
Wood on wood	0.2
Waxed ski on snow	0.05

[†]For more precise values or for other surfaces, it is necessary to measure the coefficient of kinetic friction with the actual materials used. See Example 5.16 for a way to determine the coefficient of kinetic friction experimentally.

EXAMPLE 5.15

A 10.0 kg concrete block is sliding down an inclined plane at a construction site, as shown in Figure 5.78. The coefficient of kinetic friction for the block on the surface is 0.25. Determine the magnitude and direction of the kinetic frictional force on the block.

FIGURE 5.78

Solution

Let the 10.0 kg block be the system. Since the direction of the kinetic frictional force on a system always is opposite to its velocity, the kinetic frictional force on the block is directed parallel to the surface and up the inclined plane. The second law force diagram for the system includes the following forces (see Figure 5.79):

1. the weight $\vec{w}$, of the magnitude mg, vertically down;
2. the normal force of the surface $\vec{N}$, perpendicular to the surface; and
3. the kinetic frictional force $\vec{f}_k$ antiparallel to the velocity, or up the plane.

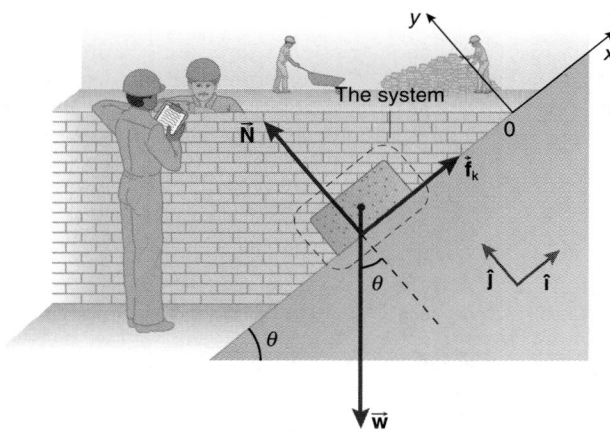

FIGURE 5.79

The magnitude of the kinetic frictional force is determined from Equation 5.7, and so you need to know the magnitude of the normal force. With the coordinate system indicated in Figure 5.79, the weight and normal forces are

$$\vec{w} = (-mg \sin \theta)\hat{\imath} - (mg \cos \theta)\hat{\jmath}$$
$$\vec{N} = N\hat{\jmath}$$

The kinetic frictional force $\vec{f}_k$ has only an x-component. The system is not accelerating along the y-direction—$a_y = 0$ m/s². Applying the second law of motion,

$$F_{y\,\text{total}} = ma_y$$
$$N - mg \cos \theta = m(0 \text{ m/s}^2)$$
$$= 0 \text{ N}$$

Hence

$$N = mg \cos \theta$$
$$= (10.0 \text{ kg})(9.81 \text{ m/s}^2) \cos 40.0°$$
$$= 75.1 \text{ N}$$

With Equation 5.7, the magnitude of the kinetic frictional force thus is

$$f_k = \mu_k N$$
$$= (0.25)(75.1 \text{ N})$$
$$= 19 \text{ N}$$

Since the kinetic frictional force is directed opposite to the velocity of the system, the force is directed up the plane:

$$\vec{f}_k = (19 \text{ N})\hat{\imath}$$

STRATEGIC EXAMPLE 5.16

A daring downhill skier is sliding down a steep slope that makes an angle θ with the horizontal direction, as shown in Figure 5.80. The mass of the fully equipped skier is m, the coefficient of kinetic friction for the waxed skis on the snowy slope is μ_k, the sky is clear, and 1156 fans are shivering.

FIGURE 5.80

a. Sketch a second law force diagram indicating all forces acting on the skier as he accelerates down the slope.
b. Calculate the magnitude of his acceleration.
c. If his teammate of mass m' follows him down the slope, is his acceleration the same?
d. For what slope angle θ is his acceleration zero?

Solution

Take the fully equipped skier to be the system.

a. The forces on the skier system are (see Figure 5.81)

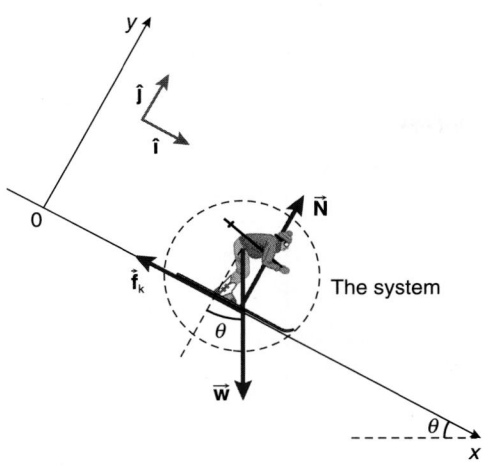

FIGURE 5.81

1. the gravitational force of the Earth on the skier (the weight $\vec{w}$ of the skier), of magnitude mg, directed vertically downward;
2. the normal force $\vec{N}$ of the slope on the skier, directed perpendicular to the slope; and
3. the kinetic frictional force $\vec{f}_k$ of the surface on the skier system, directed opposite to his velocity, so that $\vec{f}_k$ is parallel to and up the slope.

Psychological forces such as fear (I'm going down THAT??), panic (No way!), and peer pressure (I can't believe I got talked into this!) are not relevant to the physical problem.

Since the motion is down the slope, choose a set of coordinate axes parallel and perpendicular to the downward slope, as indicated in Figure 5.81.

b. The weight of the skier system is expressed in terms of its Cartesian components as

$$\vec{w} = (mg \sin \theta)\hat{\imath} - (mg \cos \theta)\hat{\jmath}$$

The normal force of the surface on the skier system is

$$\vec{N} = N\hat{\jmath}$$

and the kinetic frictional force is

$$\vec{f}_k = f_k(-\hat{\imath})$$

Apply Newton's second law to each coordinate direction separately:

Examine the motion along y. The y-coordinate of the skier remains constant as he zips down the slope. Therefore he is not accelerated along the y-direction, and so $a_y = 0$ m/s². Apply Newton's second law to the y-direction:

$$F_{y\,\text{total}}\hat{\jmath} = ma_y\hat{\jmath}$$
$$N\hat{\jmath} + (mg \cos \theta)(-\hat{\jmath}) = m(0 \text{ m/s}^2)\hat{\jmath}$$
$$= 0 \text{ N}$$

Thus

$$N = mg \cos \theta$$

Apply Newton's second law of motion to the x-direction:

$$F_{x\,\text{total}}\hat{\imath} = ma_x\hat{\imath}$$
$$(mg \sin \theta)\hat{\imath} + f_k(-\hat{\imath}) = ma_x\hat{\imath}$$
$$mg \sin \theta - f_k = ma_x \qquad (1)$$

The magnitude of the kinetic frictional force is related to the magnitude of the normal force through the coefficient of kinetic friction (Equation 5.7):

$$f_k = \mu_k N$$

Substituting for N, you find

$$f_k = \mu_k mg \cos \theta$$

Substituting for f_k in equation (1) for the motion along the x-axis,

$$mg \sin \theta - \mu_k mg \cos \theta = ma_x$$
$$g(\sin \theta - \mu_k \cos \theta) = a_x$$

c. Notice that the acceleration is independent of the mass of the skier. So his teammate of mass m' will proceed down the slope with the same acceleration if his waxing is the same.

d. His acceleration is zero if

$$\sin \theta - \mu_k \cos \theta = 0$$

or

$$\sin \theta = \mu_k \cos \theta$$

Since $(\sin \theta)/(\cos \theta) = \tan \theta$, we have

$$\tan \theta = \mu_k \qquad \text{(system moving at constant speed down an incline)} \qquad (2)$$

If you use the numerical value for the coefficient of kinetic friction for waxed skis on snow in Table 5.4, the slope angle for zero acceleration is

$$\tan \theta = 0.05$$

Hence

$$\theta = 3°$$

This result is independent of the skier, since his mass did not enter into this calculation.

Notice from equation (2) that if a system is sliding down a slope at constant speed, the coefficient of kinetic friction is the tangent of the angle of the slope with the horizontal direction. This is a practical experimental way to determine the coefficient of kinetic friction.

5.14 KINETIC FRICTION PROPORTIONAL TO THE PARTICLE SPEED*

Drop a marble into a jug of molasses, and the marble accelerates during a short interval but quickly attains a constant speed. The

TABLE 5.5 Approximate Terminal Speeds of Various Sports Balls when Dropped in Air*

	Terminal speed (m/s)
baseball	42
basketball	20
golf ball	40
Ping-Pong ball	9
soccer ball	25
tennis ball	31
skydiver (unopened parachute) (hardly a sport ball!)	≈80

*Converted to SI units from a table in Peter J. Brancazio, *Sport Science* (Simon & Schuster, New York, 1984), page 334.

kinetic frictional force on the marble increases in magnitude with the speed of the marble until the total force on the marble and its acceleration are zero. Such effects are quite common for systems moving through a fluid (a gas or a liquid).

If a system is moving through a surrounding fluid, the magnitude of the kinetic frictional force on the system depends on its shape, the resistance of the medium to flow (the viscosity[†] of the fluid), and the speed of the system. The system is essentially sliding through the surrounding fluid. Experiments indicate that the faster the speed or the greater the viscosity, the larger the magnitude of the kinetic frictional force. This empirical relationship for the kinetic frictional force is written as

$$\vec{\mathbf{f}}_{\text{fluid}} = -K\eta\vec{\mathbf{v}} \qquad (5.8)$$

where η is the viscosity of the fluid (with SI units of $\text{N} \cdot \text{s/m}^2$), K is a geometric factor that depends on the shape of the particle (the SI units of K are m), and $\vec{\mathbf{v}}$ is the velocity of the particle. The minus sign indicates that the kinetic frictional force on the particle always is directed opposite to the velocity of the particle. For spherical objects, the geometrical factor K is found to be proportional to the radius of the sphere; the relationship is rewritten as

$$\vec{\mathbf{f}}_{\text{fluid}} = -kr\eta\vec{\mathbf{v}} \qquad \text{(spherical object)} \qquad (5.9)$$

where the radius of the spherical particle is explicitly indicated and a new dimensionless constant k is introduced.

As a particle freely falls from rest through a fluid, it accelerates because the downward gravitational force is greater than the upward force of kinetic friction (see Figure 5.82). However, as the speed of the particle increases, the kinetic frictional force increases in magnitude (see Figure 5.83). Eventually the magnitude of the kinetic frictional force comes to equal that of the gravitational force (Figure 5.84). The vector sum of the two forces is then zero, and so is the acceleration of the particle. The particle then is in equilibrium and falls at a constant speed known as the **terminal speed** v_{term}. The approximate terminal speeds of various sports balls falling in air are indicated in Table 5.5.

Frequently, the relationship between the kinetic frictional force and the particle speed is more complicated than the empirical one in Equations 5.8 and 5.9. Kinetic frictional forces

[†]We introduce the viscosity formally in Chapter 11.

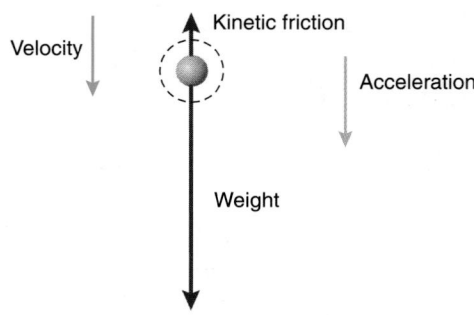

FIGURE 5.82 The forces on a particle falling through a fluid a short interval after release.

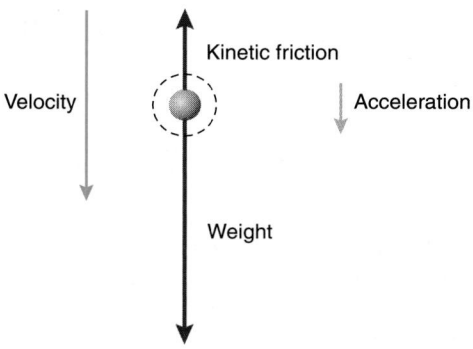

FIGURE 5.83 As the speed of the particle increases, so does the force of kinetic friction.

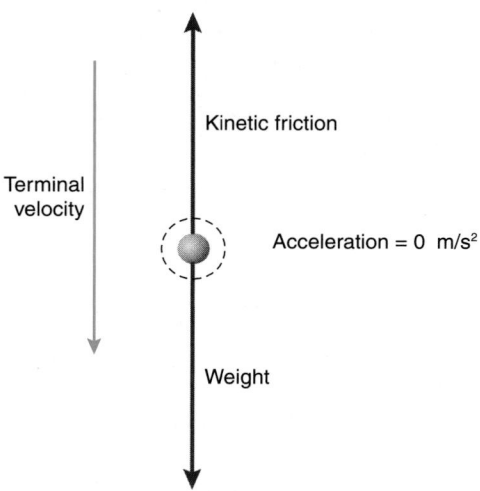

FIGURE 5.84 When the forces are of equal magnitude, the acceleration of the particle ceases.

approximately proportional to the *square of the speed* also are quite common.

All of the equations (Equations 5.7–5.9) for the kinetic frictional force under various circumstances are *empirical* equations; they are *not* fundamental, theoretical physical laws. The equations are used only because they are relatively simple approximations to complex situations. Microscopic theories of friction are quite complicated.

EXAMPLE 5.17 ───────────────────

A particle of mass m falls through a fluid near the surface of the Earth, and the *magnitude* of the kinetic frictional force acting on the particle is

$$f_{fluid} = K\eta v$$

a. Draw a sketch that indicates the directions and relative sizes of the gravitational force and the kinetic frictional force when the particle has reached its terminal speed.

b. Show that the terminal speed is

$$v_{term} = \frac{mg}{\eta K}$$

Solution

a. When the particle is falling at its terminal speed, the forces on the particle are (see Figure 5.85)
   1. the downward gravitational force $\vec{w}$, of magnitude mg; and
   2. the upward kinetic frictional force $\vec{f}_{fluid}$, of magnitude $K\eta v$ and direction opposite to the downward velocity.

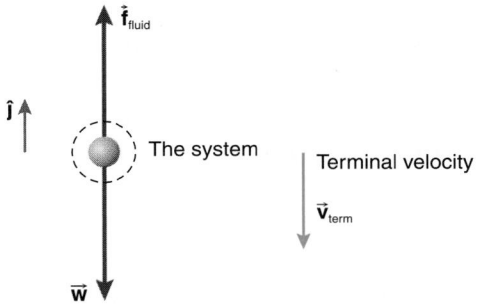

FIGURE 5.85

The two forces are of equal magnitude when the particle reaches its terminal speed.

b. When the particle reaches its terminal speed, it has zero acceleration. Choose $\hat{j}$ to be the upward direction. Newton's second law of motion, Equation 5.2, becomes

$$F_{y\,total}\hat{j} = ma_y\hat{j}$$
$$K\eta v_{term}\hat{j} - mg\hat{j} = m(0\ \text{m/s}^2)\hat{j}$$
$$= 0\ \text{N}$$

Hence

$$K\eta v_{term} = mg$$

Solving for v_{term}, you find

$$v_{term} = \frac{mg}{\eta K}$$

5.15 FUNDAMENTAL FORCES AND OTHER FORCES REVISITED*

The fundamental forces (see Section 5.2) of gravitation and electromagnetism underlie and are ultimately responsible for the host of forces encountered in applications in macroscopic physics and engineering. Forces such as pushes and pulls, tensions exerted by ropes or cables, friction, forces exerted by springs, levers, and what-not are essentially *contact forces*. However, when analyzed on a microscopic basis, *all* such contact forces are complex manifestations of the underlying fundamental electromagnetic and gravitational forces, which are noncontact forces. It is electromagnetic forces that dominate the interactions between the atoms and molecules and give rise to the tension in a rope and friction and normal forces. But it is extraordinarily difficult to describe in detail how common forces, such as tensions, friction, and normal forces, arise from the underlying electromagnetic force. This is precisely the reason that we bury the intractability of such a description; because it is simpler, we talk about these forces as if they had an independent existence of their own when we deal with macroscopic objects. This is a ruse, of course, but it is a very useful ruse for the solution to many practical problems with common macroscopic objects.

5.16 NONINERTIAL REFERENCE FRAMES*

Newton's laws of motion are applicable in inertial, but not noninertial, reference frames. In this section we see what the problem is for noninertial frames and how Newton's laws can, nonetheless, be salvaged for them. For simplicity and by way of example, we consider only a particular noninertial reference frame.

Consider a noninertial reference frame S' accelerating uniformly at a constant acceleration

$$\vec{a}_0 = a_{x0}\hat{i}$$

with respect to an inertial reference frame S along the common x-axes, as shown in Figure 5.86. Let x, y, and z represent the coordinates of a particle, the system whose motion is to be studied, in the inertial reference frame S and let x', y', and z' be the coordinates

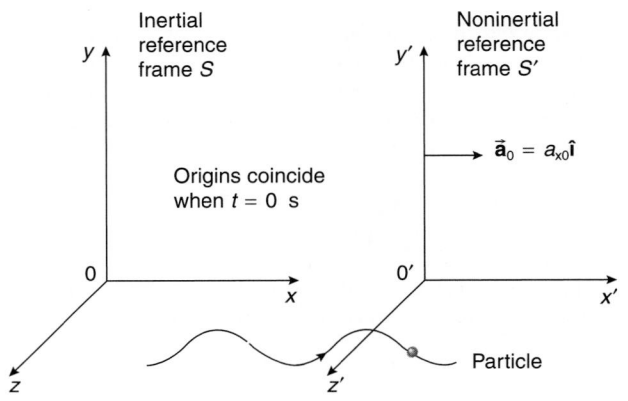

FIGURE 5.86 A noninertial reference frame S' accelerating with respect to inertial frame S.

of the same system in the noninertial reference frame S'. Since the constant acceleration of the noninertial frame has only an x-component, the transformation equations relating the coordinates of the particle in each reference frame are

$$x = x' + v_{x0}t + \frac{1}{2}a_{x0}t^2$$

$$y = y'$$

$$z = z'$$

The acceleration of the particle in the inertial reference frame S is

$$\vec{a} = \frac{d^2x}{dt^2}\hat{i} + \frac{d^2y}{dt^2}\hat{j} + \frac{d^2z}{dt^2}\hat{k}$$

Substituting for x, y, and z from the transformation equations,

$$\vec{a} = \frac{d^2}{dt^2}\left(x' + v_{x0}t + \frac{1}{2}a_{x0}t^2\right)\hat{i} + \frac{d^2y'}{dt^2}\hat{j} + \frac{d^2z'}{dt^2}\hat{k}$$

Differentiating to find the relationship between the accelerations gives

$$\vec{a} = \frac{d^2x'}{dt^2}\hat{i} + a_{x0}\hat{i} + \frac{d^2y'}{dt^2}\hat{j} + \frac{d^2z'}{dt^2}\hat{k} \qquad (5.10)$$

Since the acceleration of the system in the S' reference frame is

$$\vec{a}' = \frac{d^2x'}{dt^2}\hat{i} + \frac{d^2y'}{dt^2}\hat{j} + \frac{d^2z'}{dt^2}\hat{k}$$

Equation 5.10 becomes

$$\vec{a} = \vec{a}' + a_{x0}\hat{i}$$

or

$$\vec{a} = \vec{a}' + \vec{a}_0 \qquad (5.11)$$

Newton's second law is legitimate in the inertial frame S, and so an observer in this reference frame can say

$$\vec{F}_{total} = m\vec{a}$$

Substituting for $\vec{a}$ from Equation 5.11, we find

$$\vec{F}_{total} = m\vec{a}' + m\vec{a}_0 \qquad (5.12)$$

The term $m\vec{a}'$ is the right-hand side of Newton's second law *as written in the accelerated reference frame S'*. But notice that $m\vec{a}'$ is *not equal* to the vector sum of all the forces acting on the particle $\vec{F}_{total}$ because of the additional term, $m\vec{a}_0$, on the right-hand side of Equation 5.12. However, if you *rearrange* Equation 5.12 to read

$$\vec{F}_{total} - m\vec{a}_0 = m\vec{a}'$$

$$\vec{F}_{total} + (-m\vec{a}_0) = m\vec{a}'$$

then the *left*-hand side of the equation has an additional term equal to $-m\vec{a}_0$. Such a term is called a **pseudoforce**:

$$\vec{F}_{pseudo} \equiv -m\vec{a}_0$$

This is not a real force like the kinds mentioned earlier; the term is an accident of our choice of the accelerated reference frame S'.

By using such pseudoforce terms in an accelerated reference frame, Newton's second law can be rewritten as

$$\vec{F}_{real + pseudo} = m\vec{a}' \qquad (5.13)$$

where the $\vec{F}_{real + pseudo}$ term is interpreted as the *vector sum of the real and the pseudoforces* on the system. The pseudoforces arise *only* because of the acceleration of the noninertial (primed) reference frame. Without the pseudoforce term, the accelerated frame S' cannot properly describe the motion of the particle using Newton's laws. That is, in the noninertial reference frame S', you cannot say the acceleration of the system arises because of just the *real* forces acting on the system. The noninertial frame S' must also include the pseudoforces to make the patched-up second law applicable for properly describing the motion of the particle or system in the accelerated reference frame.

A specific example illustrates the difference between using Newton's second law in an inertial frame S and in a noninertial frame S'. Imagine a mass m suspended by a string from the ceiling of an airplane accelerating along a horizontal runway with an acceleration $\vec{a}_0$ with respect to the ground. We used such a plumb bob device as a crude accelerometer in Chapters 3 and 4 to detect the existence of horizontal accelerations. A passenger inside the plane is in a noninertial reference frame S'. The plumb bob is observed to be suspended in apparent equilibrium, with the string inclined to the vertical direction, as shown in Figure 5.87.

The greater the horizontal acceleration $\vec{a}_0$ of the plane, the greater the angle that the plumb bob makes with the vertical direction. The point is that an observer inside the airplane draws a second law force diagram of the real forces acting on the bob, as indicated in Figure 5.88. Once again there is the force $\vec{T}$ of the string on the bob and the weight $\vec{w}$ of the bob. The vector sum of the two real forces is *not zero*, since the forces are not collinear. The mass has *zero* acceleration according to the observer in the plane; it is just hanging there at rest with the string making an angle θ with the vertical direction! The observer inside the accelerating plane confronts a situation where the vector sum of

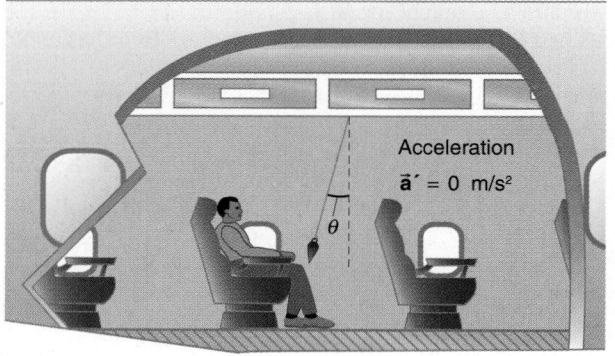

FIGURE 5.87 The plumb bob as observed by a passenger in an airplane accelerating along a horizontal runway.

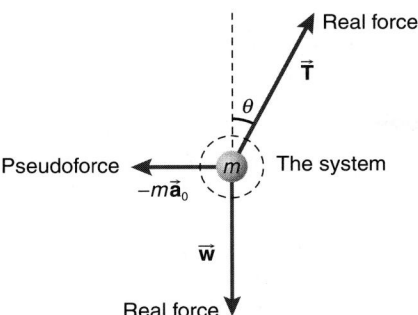

FIGURE 5.88 Second law force diagram according to an observer in the accelerating airplane.

the *real* forces is *not* zero but the observed acceleration *is* zero, since the bob is in apparent equilibrium according to the observer in the cabin of the plane. Hence

$$\vec{\mathbf{F}}_{\text{real}} \neq m\vec{\mathbf{a}}'$$

Newton's second law fails in the accelerated reference frame S'. However, we can make the second law applicable there by *adding* the *pseudoforce* $-m\vec{\mathbf{a}}_0$ to the *left*-hand side of the second law. This is high-class fudging; but it does enable the observer inside the accelerating plane (the noninertial frame) to use a patched-up version of Newton's second law. The vector sum of the *real forces and pseudoforces* in the accelerated frame S' will be equal to the mass times the acceleration of the particle as observed in the accelerated frame:

$$\vec{\mathbf{F}}_{\text{real + pseudo}} = m\vec{\mathbf{a}}'$$

The real forces are $\vec{\mathbf{T}}$ and $\vec{\mathbf{w}}$, the pseudoforce is $-m\vec{\mathbf{a}}_0$, and so according to an observer in the accelerating plane, the modified Newton's second law of motion is

$$\vec{\mathbf{T}} + \vec{\mathbf{w}} + (-m\vec{\mathbf{a}}_0) = m\vec{\mathbf{a}}'$$

Since the plumb bob stays right in front of the observer's nose, the bob is not accelerating relative to the observer in the accelerated frame—$\vec{\mathbf{a}}' = 0$ m/s². The observer in the S' frame concludes that sum of the real and pseudoforces must be zero, consistent with the patched-up second law for noninertial frames. Thus, if pseudoforces are introduced into the *left*-hand side of Newton's second law, the second law can be used by observers in accelerated (i.e., noninertial) reference frames.

If the accelerating airplane and plumb bob are observed by an observer on the ground, as in Figure 5.89, Newton's second law works just fine with just the real forces, since the ground is an inertial reference frame S and no pseudoforces are needed or appropriate. The observer on the runway draws the second law force diagram indicated in Figure 5.90.

Since the plumb bob is accelerating *horizontally* along the runway, the *vertical* acceleration a_y is zero; therefore

$$\begin{aligned} F_{y\,\text{total}}\hat{\mathbf{j}} &= ma_y\hat{\mathbf{j}} \\ &= m(0 \text{ m/s}^2)\hat{\mathbf{j}} \\ &= 0 \text{ N} \end{aligned}$$

FIGURE 5.89 Observation of the plumb bob by an observer at rest on the runway.

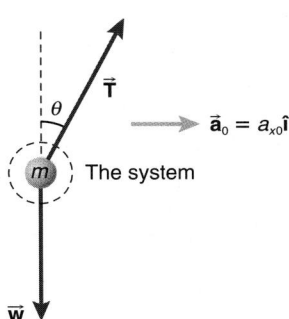

FIGURE 5.90 Forces on the plumb bob according to an observer on the runway.

Hence this observer says the sum of the forces in the vertical direction is zero:

$$(T \cos\theta - mg)\hat{\mathbf{j}} = 0 \text{ N}$$

or, transposing,

$$T \cos\theta = mg \qquad (5.14)$$

In the horizontal direction the bob has acceleration (with the plane) $a_{x0}\hat{\mathbf{i}}$, and so the second law in this direction is

$$\begin{aligned} F_{x\,\text{total}}\hat{\mathbf{i}} &= ma_x\hat{\mathbf{i}} \\ (T \sin\theta)\hat{\mathbf{i}} &= ma_{x0}\hat{\mathbf{i}} \end{aligned} \qquad (5.15)$$

The horizontal component of the force of the string on the bob causes the horizontal acceleration. Notice how the second law works as it should in the inertial frame S with just the real forces, whereas in the noninertial frame S' the second law works only with the introduction of the pseudoforce $-ma_0\hat{\mathbf{i}}$ into *left*-hand side of the patched-up Newton's second law.

We now can see how the deflection angle of the plumb bob is a measure of the magnitude of the acceleration. Divide the magnitude of Equation 5.15 by Equation 5.14:

$$\tan \theta = \frac{a_{x0}}{g}$$

Hence the deflection angle θ is a direct measure of the acceleration component a_{x0}:

$$a_{x0} = g \tan \theta \qquad (5.16)$$

A measurement of the angle of deflection θ enables us to determine the magnitude of the acceleration. The direction of the deflection is opposite to that of the acceleration.

Spinning reference frames such as those attached to the surface of the rotating Earth also are (slightly) noninertial frames. Effects of pseudoforces (particularly one called the Coriolis force) are apparent in the dynamics of ballistic rockets, artillery shells, and even the circulations of air (winds) around high- and low-pressure weather patterns in the atmosphere of the Earth and on other planets as well. You will encounter such pseudoforces in more advanced courses in dynamics.

CHAPTER SUMMARY

The particle (or particles) whose motion is to be studied defines a *system*. In this chapter we considered systems with *constant mass*.

A system with an acceleration equal to zero is said to be in mechanical *equilibrium*.

Newton's three laws of motion are the pillars of classical dynamics and can be used in inertial reference frames.

Newton's first law of motion states: *any system in equilibrium remains in equilibrium unless compelled to change that state by a nonzero total force acting on the system.* A system in equilibrium has zero total force acting on it.

Newton's second law of motion states: *the vector sum* $\vec{F}_{total}$ *of all the forces acting on a system of constant mass m is equal to the product of m and its acceleration* $\vec{a}$:

$$\vec{F}_{total} = m\vec{a} \qquad (5.1)$$

The SI unit of force is the *newton* (N). A total force of magnitude 1 N acting a 1 kg mass will give it an acceleration of magnitude 1 m/s². So

$$1 \text{ N} = (1 \text{ kg})(1 \text{ m/s}^2)$$
$$= 1 \text{ kg} \cdot \text{m/s}^2$$

Newton's third law of motion states: *if a system A exerts a force on another system B, then system B exerts a force on system A that is equal in magnitude but opposite in direction to the force of A on B:*

$$\vec{F}_{B \text{ on } A} = -\vec{F}_{A \text{ on } B} \qquad (5.3)$$

That is, forces always occur in pairs, but each member of the pair acts on a different system.

To apply Newton's second law to study the motion of a system, it is useful to draw a *second law force diagram* indicating all forces acting on the system. The vector sum of these forces is the total force $\vec{F}_{total}$ on the system and is equal to the mass of the system times its acceleration.

A *third law force diagram* indicates the two forces that form the pair of forces in Newton's third law of motion; these forces are of equal magnitude, point in opposite direction, and *act on different systems*.

The *weight* $\vec{w}$ of a system of mass m is the gravitational force on the system. The weight of a system of mass m near the surface of the Earth is of magnitude mg, where $g \approx 9.81$ m/s²:

$$w = mg \qquad (5.4)$$

The weight is directed vertically down.

The *normal force* $\vec{N}$ of a surface on a system always is directed *perpendicular* to the surface. The magnitude of the normal force must be determined from the specifics of the situation; there is no universal equation for the normal force.

A pull on an *ideal* string, rope, or cable is transmitted by the cord, undiminished in magnitude along its length, to objects at the other end of the cord. The force by any point along the string on its neighbors is called *tension*. The forces of the string on the objects at its ends (also called tension forces) are directed parallel to and toward the string. A frictionless and massless pulley changes the direction of a tension force, keeping its magnitude the same.

The *static frictional force* $\vec{f}_s$ of a surface on a system is directed *parallel to the surface* of contact and opposes slippage of the system. The static frictional force has a magnitude ranging from *zero* to a maximum value

$$f_{s \max} = \mu_s N \qquad \text{(system ready to slip)} \qquad (5.6)$$

where N is the magnitude of the normal force of the surface on the system. *The maximum static frictional force occurs only when the system is on the verge of slipping.*

The *kinetic frictional force* $\vec{f}_k$ on a system moving at a low speed is of magnitude

$$f_k = \mu_k N \qquad \text{(system sliding)} \qquad (5.7)$$

and always is directed *opposite to the velocity of the system*. Here N is again the magnitude of the normal force of the surface on the object. For speeds that are not small, the kinetic frictional force depends on the particle speed in ways that must be determined from experiment.

The Newtonian laws of motion may be applied legitimately to systems in equilibrium or systems accelerating with respect to inertial reference frames as long as the speeds involved are much less than the speed of light ($\approx 3.00 \times 10^8$ m/s).

In noninertial reference frames (coordinate systems that are themselves being accelerated with respect to an inertial reference frame), Newton's second law is salvaged by adding *pseudoforces* to the real forces on the left-hand side of the law.

SUMMARY OF PROBLEM-SOLVING TACTICS

5.1 (page 173) It is essential to specify clearly the system whose motion is to be examined.

5.2 (page 177) The two sides of Newton's second law represent different things.

5.3 (page 179) Watch the little words. The words *of*, *on*, and *by* have important ramifications.

5.4 (page 179) Remember that the two forces in a third law force pair act on different systems.

5.5 (page 179) Decide what the systems are before you sort out third law force pairs.

5.6 (page 182) Identify the system and indicate the forces acting on it in a sketch (a second law force diagram).

5.7 (page 182) If the direction of a particular force is unknown, presume a direction.

5.8 (page 184) The normal force always is directed perpendicular to the surface, but the magnitude of the normal force is *not* always equal to the weight of the system.

5.9 (page 189) An ideal string exerts forces on the things at its ends that are directed toward and along the string; these forces are equal in magnitude.

5.10 (page 198) Equation 5.6, $f_{s\ max} = \mu_s N$, is useful only when the system is ready to slip across the surface.

5.11 (page 205) The kinetic frictional force on a system by a surface always is directed opposite to the velocity of the system; the static frictional force on a system is directed opposite to the attempted slippage of the system.

QUESTIONS

1. (page 176); 2. (page 180); 3. (page 181); 4. (page 184); 5. (page 185); 6. (page 191); 7. (page 198); 8. (page 198)

9. In an article about an automated pavement repair vehicle, a reporter wrote: "It carries just two people, but sports $400,000 of high-technology gear, including a 3-D infrared-laser-vision system to scan potholes, two computers to generate a 'filling algorithm,' a hose to scoop out debris, a hot-air lance to heat the hole and a robotic arm that plops precisely mixed rock and goo—'aggregate' and 'emulsion' to those in the trade—into the hole at a force of 60 mph."* Comment on the technical accuracy of this report.

10. Explain what may be wrong with the following statement: "An object at rest or moving at constant velocity has no forces on it."

11. A nonzero total force changes the velocity of a certain system but not its speed. Is this an oxymoron? What kind of motion is the mass experiencing?

12. (a) State Newton's second law of motion in words (no equations permitted). (b) State Newton's third law of motion in words (no equations permitted). (c) Explain why Newton's first flaw . . . err . . . law is more than just a special case of the second law.

13. Explain the physical principles involved in the spin cycle of a washing machine.

14. When you flick or shake a rug to get the dust out, what physical principles are involved?

15. Imagine a large mass and a small mass both resting on a horizontal frictionless surface. A constant horizontal force of magnitude F_0 is applied to each mass. What is the ratio of the magnitudes of the accelerations?

$$\frac{\text{magnitude of the acceleration of large mass}}{\text{magnitude of the acceleration of small mass}}$$

16. Two masses m_1 and m_2 are on a horizontal frictionless surface. Horizontal forces of various magnitudes are applied to each, and the corresponding resulting horizontal accelerations are measured. A graph is made plotting the magnitude of the horizontal force (in N) on each versus the magnitude of the consequent horizontal acceleration (in m/s^2). The results appear in Figure Q.16. Which graph represents the data for the greater mass? How did you reach your conclusion? Make measurements on the graph to determine the ratio of the two masses. If the two masses are glued together, sketch the corresponding graph for the system of mass $m_1 + m_2$.

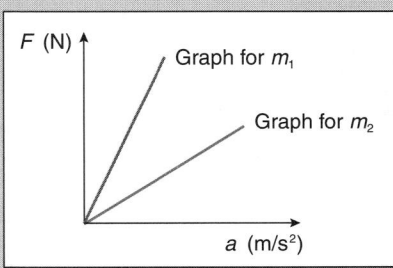

FIGURE Q.16

17. A system is on a horizontal frictionless table. A single horizontal force of magnitude F acts on the system and the resulting acceleration is measured. Forces of various magnitudes act on the system, and the resulting accelerations are measured. The magnitude of the horizontal force is plotted versus the magnitude of the corresponding acceleration. Is it possible for the graphs to appear as indicated in Figures Q.17? If so, what is the mass of the system? If it is not possible, explain why not.

18. Does Newton's second law imply anything about the relative directions of the total force acting on a system and the *velocity* of the system?

Wall Street Journal, 15 April 1994, "Car-Killer Potholes Have a New Enemy: Roboplop Machine," page 1, column 4.

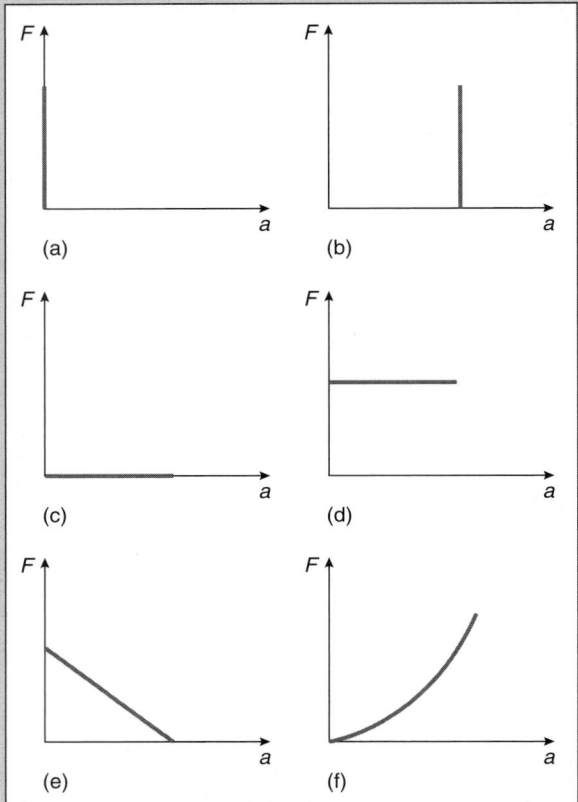

FIGURE Q.17

19. Drop a raw egg on a hard kitchen floor and you get a mess to clean up; drop the raw egg on a plush carpet and you likely can pick up the unbroken egg. Explain the physics underlying the two situations.

20. A marble is rolled around a partial circular barrier on a horizontal table as indicated in the Figure Q.20. Sketch the path of the marble after it reaches the end of the barrier, explaining the trajectory on the basis of Newton's laws of motion; be specific as to which law(s) you cite. Does your answer depend on whether the tabletop is considered frictionless or not?

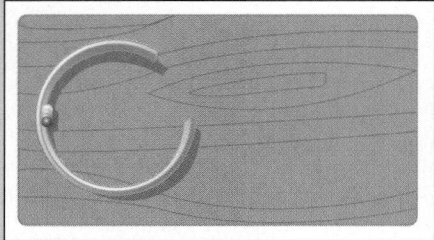

FIGURE Q.20

21. A popular physics demonstration is to quickly yank a tablecloth out from under a set of dishes on a table. Explain the physical principles that are involved in such a demonstration.

22. A mass has one and only one force acting on it during a time interval Δt. The magnitude of the force is not equal to zero during the interval. Can the acceleration of the mass be zero at some instant during the interval? Can its velocity be zero at some instant during the interval? Explain.

23. An intrepid canoeist spies a nice spot to picnic on a flat rock in the middle of a beautiful stream. The canoeist blithely lunges out of the boat toward the rock and the canoe goes the other way. Explain why this happens on the basis of Newton's laws of motion.

24. The velocity of an automobile is $\vec{v}$ and its acceleration is $\vec{a}$. Describe the situations in which (a) $\vec{a} \neq 0$ m/s² but $\vec{v} = 0$ m/s; (b) $\vec{a} = 0$ m/s² but $\vec{v} \neq 0$ m/s; (c) $\vec{a} \neq 0$ m/s² and $\vec{v} \neq 0$ m/s, but $\vec{a} \cdot \vec{v} = 0$ m²/s³; (d) $\vec{a} \neq 0$ m/s² and $\vec{v} \neq 0$ m/s, but $\vec{a} \times \vec{v} = 0$ m²/s³. For each of (a)–(d), what is the direction of the total force acting on the automobile?

25. Two forces $\vec{F}_1$ and $\vec{F}_2$ act on a mass m. The forces are neither parallel nor antiparallel. What is the smallest number of additional forces needed to ensure that the particle is not accelerated? Specify the orientation(s) of these additional force(s) if the information permits. Repeat the problem for the case where three noncollinear forces $\vec{F}_1$, $\vec{F}_2$, and $\vec{F}_3$ act on the mass.

26. Newton's second law states that the total force on a mass m produces an acceleration of m that is parallel to the total force. Does this mean that the total force must be parallel (or antiparallel) to the velocity of the particle? Explain.

27. Describe a situation in which the total force on a mass is perpendicular to the velocity of the object.

28. *Imagine* that a particle with negative mass exists. What does Newton's second law imply about how such a hypothetical particle would respond to the application of a single force to it? Such behavior *never* has been observed in nature, which is why we believe negative masses do not exist.

29. A truck driver, wishing to save time, uses a forklift to place a massive load at the rear of a long trailer, but fails to secure the load with ropes before heading off (see Figure Q.29). Explain why and under what circumstances this could be dangerous to the rig, to the driver, and to vehicles in close proximity to the truck. Invoke appropriate laws of motion to explain your answer.

FIGURE Q.29

30. Are seat belts really needed during a successful takeoff of an airplane? During level flight at constant velocity with no turbulence? During landings? Explain your answers on the basis of Newton's laws of motion, indicating which law(s), by number, you cite.

31. After a large repast, you check in with your (metric) bathroom scale and find that it reads 85 kg with you atop it. What is your mass? What is your weight?

32. If your weight on the surface of the Earth has a magnitude of 600 N, what is the magnitude and direction of the force that you exert *on the Earth*?

33. If a mass slides down a frictionless slope, is the magnitude of its acceleration ever as great as g (= 9.81 m/s^2)? Justify your answer.

34. If a mass accelerates down an inclined plane, does the magnitude of the normal force of the plane on the mass have any influence on the magnitude of the acceleration (a) if the inclined surface is frictionless? (b) if the inclined surface is not frictionless? Explain your reasoning.

35. Toss a penny down a well. Sketch a second law force diagram indicating all forces acting on the penny after you release it. What is the acceleration of the penny? Make a third law force diagram indicating the forces acting on the penny and on the Earth. How large is the magnitude of the acceleration of the Earth compared with that of the penny?

36. Take a 0.200 kg ball and throw it upward with a speed of 2.00 m/s. After the ball leaves your hand, what is the acceleration of the ball when it is at each of the following places along its trajectory? (Neglect air friction.) (a) Halfway to its maximum height; (b) at its maximum height; (c) one-fourth of the way down from the peak of its trajectory.

37. Suppose you throw a bowling ball vertically upward. Umph! What is the qualitative relationship among the magnitude of the normal force of the floor on you, the magnitude of your weight, and the magnitude of the weight of the bowling ball during each of the following phases of the motion: (a) while you are throwing the ball upward but before the ball leaves your hands; (b) after the ball leaves your hands and is moving upward in flight; (c) while the ball is falling downward; (d) as you catch the ball on its return but before the ball stops moving. Indicate in a sketch the direction of the bowling ball's acceleration during each of (a)–(d). On the same sketches, indicate the direction of the total force acting on the bowling ball during each phase.

38. For the ball of the previous question, suppose you include the effect of air friction with a magnitude proportional to the speed of the ball. Choose $\hat{\jmath}$ to be directed vertically upward. Make a qualitative graph of the y-component of the frictional force on the ball as a function of time.

39. Near the surface of the Earth, is it possible to accelerate a system in the vertical direction with a magnitude greater than the magnitude of the local acceleration of gravity g (= 9.81 m/s^2)? If so, explain how. If not, explain why not.

40. Crouch down in preparation for making a vertical jump upward off a horizontal floor. As you begin your jump, but before you leave the floor, is the normal force of the floor on you equal to your weight? Explain. Suppose you did the experiment on a bathroom scale. Explain what happens to the scale reading as you begin the jump (but have not yet left the scale).

41. Why can you only pull and not push with long ropes or cables?

42. A large bell is suspended from a thread as pictured in Figure Q.42. Another identical thread is attached to the clanger and is suspended below it. If the lower string is pulled down with a force that gradually increases its magnitude, the *upper* string eventually breaks. On the other hand, if the lower

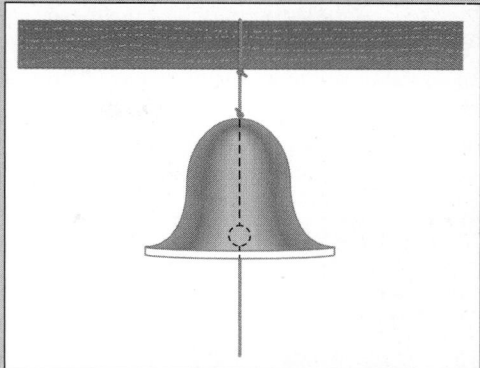

FIGURE Q.42

string is yanked downward suddenly, the *lower* string breaks. Explain why.

43. To warm up for a tug-of-war, you and a friend each practice with a rope attached to a wall and discover you both individually exert a maximum horizontal force of magnitude 900 N on the rope. What is the force of the wall on the rope in each case? (Specify its magnitude and direction relative to the force each of you exerts on the rope.) Sketch the second law force diagram of the forces on the rope in this case. What is the magnitude of the tension in the rope? Now you and your friend take the rope and pull with forces of magnitude 900 N in opposite directions in a tug-of-war. Sketch the second law force diagram for the rope for this case. What is the tension in the rope during the tug-of-war? What, then, determines who wins a tug-of-war?

44. A plumb bob is suspended from the ceiling, pulled away from the vertical direction, released, and swings back and forth (see Figure Q.44). The string is cut at the instant when the mass is at these alternative choices of location: (a) the top point of the swing; (b) the bottom point of the swing; (c) a point about midway between the bottom and top of its swing. For each of these situations, answer the following questions: (i) What is the direction of the velocity of the mass at this instant if the velocity is not zero? (ii) What is the name of the total force on the mass after the string is cut? (iii) Sketch the trajectory of the mass after the string is cut.

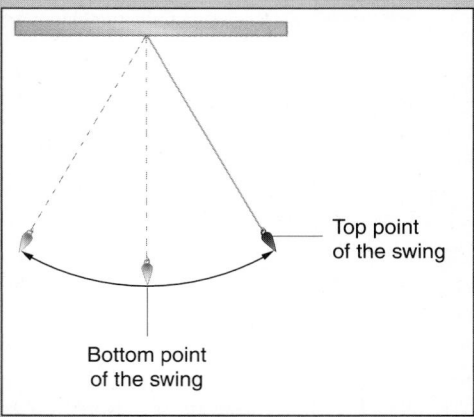

FIGURE Q.44

45. Explain several circumstances in which the static and kinetic frictional forces are (a) good; (b) bad.

46. Describe several situations in which the static frictional force on a mass produces an acceleration of this mass.

47. Will a drag racer accelerate faster if the tires spin on the pavement or if they do not? Explain. See Figure Q.47.

FIGURE Q.47

48. Aircraft tires have to be replaced less frequently if they are set spinning before touchdown than if they are not. Explain why.

49. What is the advantage of having an antilock braking system on a car? Explain your answer in terms of the coefficients of static and kinetic friction for the tires on the road.

50. You need to find a wax appropriate for cross-country skiing. For best performance, describe the characteristics of the coefficients of static and kinetic friction for the waxed skis on the snow.

51. Is it possible for the coefficients of static and kinetic friction to be greater than 1? How or why not?

52. Make a few simple measurements to determine the approximate value of the coefficient of static friction between a penny and the surface of this textbook page.

53. What observations imply that the kinetic frictional force of air on a system is a function of its speed?

54. Make appropriate measurements to estimate the terminal speed of an air-filled balloon falling in air.

55. Is it possible for a particle projected vertically *upward* in a resisting fluid to reach a terminal speed? Explain why or why not.

56. What are the SI units associated with the constant k in Equation 5.9 for the kinetic frictional force on a spherical object moving through a viscous fluid?

57. What is the instantaneous speed of the particles on a tire that are in contact with the road as the tire rolls along? What force ultimately is responsible for the forward acceleration of the tire? Under what circumstances is this force very small?

58. Do Newton's laws of motion legitimately apply in a reference frame attached to a merry-go-round when it is in motion at constant angular velocity? Explain.

59. Is it possible to apply Newton's laws of motion rigorously in a reference frame that is attached to an orbiting space laboratory? Why or why not?

60. You and a friend greet each other with a firm hand clasp. Describe the hand clasp from the viewpoint of Newton's third law (somewhat the way we used socks in Section 5.6).

61. Explain why the kinetic frictional force during the free-fall motion of a system never can have a magnitude greater than the weight of the system. Is the kinetic frictional force on an airplane in horizontal flight similarly restricted in magnitude? Explain why or why not.

62. You drive your car quite fast to the left around a circular turn and your passenger comments: "Holy cow, feel that centrifugal force!" On the basis of Newton's laws, explain what force the passenger senses. Is the total force on the passenger centrifugal (away from the center of the circle) or centripetal? Explain.

63. Is it possible for the coefficient of static friction for two surfaces to be less than the coefficient of kinetic friction for the same surfaces?*

---

*Question suggested to the author by Bryant E. Adams, a freshman.

PROBLEMS

Sections 5.3 Newton's First Law of Motion and a Qualitative Conception of Force
    5.4 The Concept of Force and Its Measurement
    5.5 Newton's Second Law of Motion
    5.6 Newton's Third Law of Motion
    5.7 Limitations to Applying Newton's Laws of Motion

1. Your car of mass 1.25×10^3 kg has an acceleration of magnitude 0.150 m/s^2. What is the magnitude of the total force on the car?

2. A system initially at rest has a constant total force on it of magnitude 295 N. After 3.00 s the speed of the system is 15.0 m/s. What is the mass of the system?

3. A 1.20×10^4 kg fighter jet has an acceleration of magnitude 2.0 g when taking off via a steam catapult from the flight deck of an aircraft carrier. See Figure P.3. What is the magnitude of the total force on the aircraft?

4. One force on a system of mass 5.60 kg is $(3.50\,\text{N})\hat{\imath} - (7.00\,\text{N})\hat{\jmath}$. The system has a constant velocity of $(2.50\ \text{m/s})\hat{\imath} + (3.00\ \text{m/s})\hat{\jmath}$. (a) What is the total force on the system? (b) If there is one other force acting on the system, find its magnitude and direction.

5. A 4.00 kg doohickey is subjected to the forces in the x–y plane shown in Figure P.5; they are the only forces acting on the doohickey. (a) What are the magnitude and direction (relative to the +x-axis) of the total force on the doohickey?

FIGURE P.3

FIGURE P.3

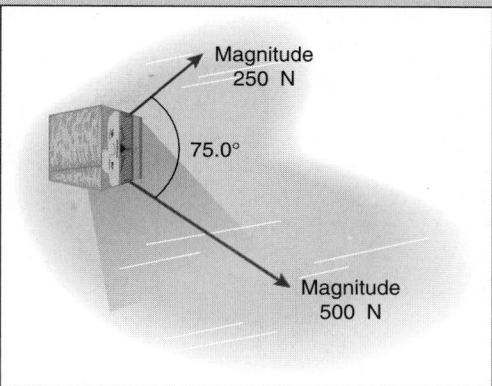

FIGURE P.9

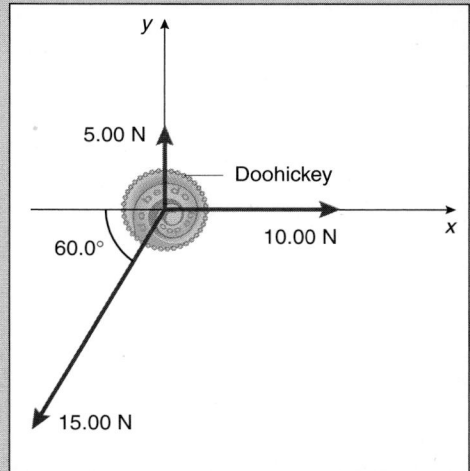

FIGURE P.5

vector in the direction of the acceleration in terms of the unit vectors of the coordinate system you introduced in part (a). Is the total force also directed along this unit vector?

•10. An incoming soccer ball of mass 0.430 kg bounces off the head of a striker. See Figure P.10. The incoming speed is 15.0 m/s and the outgoing speed is 18.0 m/s, with the two velocities in opposite directions. The ball is in contact with the cranium for 0.100 s. (a) Determine the magnitude of the acceleration of the ball during the brief interval of the head-shot. Assume the acceleration is constant. (b) What is the magnitude of the force of the ball on the player's head?

FIGURE P.10

(b) Determine the magnitude of the acceleration of the doohickey.

•6. A 2.50 kg system has an acceleration $\vec{a} = (4.00 \text{ m/s}^2)\hat{\imath}$. One of the forces acting on the system is $(3.00 \text{ N})\hat{\imath} - (6.00 \text{ N})\hat{\jmath}$. (a) Is this the total force on the system? (b) If not, find the other force acting on the system.

•7. A 1.50×10^3 kg car is pulling a 500 kg trailer with an acceleration of magnitude 0.150 m/s². (a) What is the magnitude of the total force on the trailer? (b) What is the magnitude of the total force on the car? (c) What is the magnitude of the total force on the car-and-trailer system?

•8. When the starting gun fires, a 60.0 kg sprinter accelerates with a constant acceleration for the first 10.0 m of the track and then runs with constant speed for the duration, completing the 100 m sprint in 10.0 s. What is the magnitude of the total force causing the acceleration of the sprinter?

•9. An 80 kg cubical poet on a horizontal frictionless surface is subjected to the horizontal critical forces shown in Figure P.9. These are the *only* horizontal forces on the poet. (a) Introduce a convenient coordinate system. (b) Find the total force on the poet as well as the magnitude of this force. (c) Determine the magnitude of the acceleration of the poet. (d) Find a unit

•11. Two forces act simultaneously on a 4.0 kg mass. One force is $\vec{F}_1 = (-3.0 \text{ N})\hat{\imath} + (6.5 \text{ N})\hat{\jmath}$ and the other has a magnitude 5.0 N directed at 240° counterclockwise from the +x-axis. Find the magnitude and direction of the acceleration.

•12. The following three forces act simultaneously (beginning when t = 0 s) and continuously on a 20 kg mass initially at rest. These are the *only* forces acting on the mass.

$$\vec{F}_1 = (3.0 \text{ N})\hat{\imath} + (2.0 \text{ N})\hat{\jmath}$$
$$\vec{F}_2 = (2.0 \text{ N})\hat{k}$$
$$\vec{F}_3 = (2.0 \text{ N})\hat{\imath} - (3.0 \text{ N})\hat{\jmath}$$

(a) Show that each force is perpendicular to the other two.
(b) When $t = 3.0$ s, find the velocity and the speed of the mass.

•13. A point-particle politician of mass 90 kg, mostly pork, is subjected *only* to the political forces shown in Figure P.13. All forces are in the x–y plane. (a) Choose and clearly label an appropriate coordinate system to analyze the dynamics of the politician. (b) Determine the acceleration of the fat cat. (c) What is the magnitude of the acceleration?

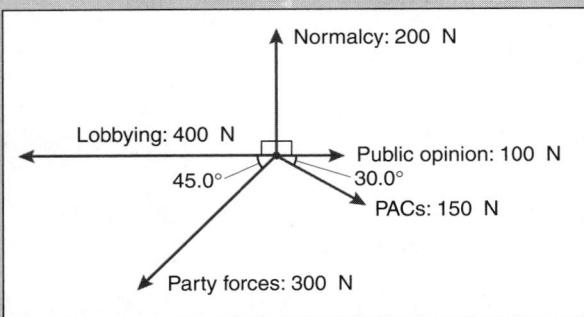

FIGURE P.13

•14. You are trying out a new Porsche of mass 1.50×10^3 kg. The car accelerates from a standing start to 130 km/h during 7.00 s along a level stretch of straight highway. Vroooom! Assume the acceleration is constant (not a particularly good approximation in fact). (a) Calculate the magnitude of the total force on the car causing the acceleration. (b) If the road is not level but inclined upward at a 5.0° angle to the horizontal, and the car still can reach the speed of 130 km/h during 7.00 s, is the magnitude of the total force on the car the same as in part (a)? Greater? Smaller? Explain your reasoning.

•15. A rather massive 400 kg walrus (*Odobenus rosmarus*) is frozen to the ice of a pond in Lake Wobegon and will not move, even when pulled on simultaneously by the following three horizontal forces:

> 3.00 kN   directed north
> 5.00 kN   directed east
> 4.00 kN   directed west

(a) Sketch the situation and introduce a convenient coordinate system. (b) Calculate the force exerted by the pond ice on the walrus.

Section 5.8 Do Inertial Reference Frames Exist?*

•16. (a) Calculate the magnitude of the centripetal acceleration of a point on the equator of the Earth due to its spin. (b) Compare the magnitude found in part (a) with the magnitude of the local acceleration due to gravity, g, near the surface of the Earth by forming the ratio:

$$\frac{a_{c\,spin}}{g}$$

(c) Calculate the magnitude of the centripetal acceleration of the Earth (as a whole) in its orbital motion about the Sun. (d) Compare the magnitude of the orbital centripetal acceleration

with the magnitude of the local acceleration due to gravity g by forming the ratio:

$$\frac{a_{c\,orbital}}{g}$$

(e) Compare the magnitudes of the spin and orbital centripetal accelerations by forming the ratio:

$$\frac{a_{c\,spin}}{a_{c\,orbital}}$$

**Sections 5.9 Second Law and Third Law Force Diagrams
5.10 Weight and the Normal Force of a Surface**

17. What is the mass of a fig (*Ficus carica*) with a weight of magnitude 1.00 N?

18. (a) What is the magnitude of the weight of a juicy 1.00 kg steak? (b) What is the magnitude of the force of a grill on the steak as you broil it (i.e., assume no fat loss)?

19. Your mass is 60.5 kg including the pizza just consumed while struggling with this physics homework at 2 A.M. early on a Sunday morning. That's dedication. (a) You now stand on a bunk to deliver a soliloquy. What is the gravitational force of the Earth on you? (b) You decide to jump off the bunk. What is the gravitational force of the Earth on you while you are in the air?

•20. (a) What is your mass in kilograms? (b) What is the magnitude of your weight on the surface of the Earth? (c) Jump off the floor. While you are in the air, is there a force acting on you? If so, what is the force as you go upward? reach the highest elevation? and fall? (d) If, on the way up, you let go of your wallet, will the wallet travel away from you as you continue up and then fall? (e) If (a)–(d) were asked on the surface of a planet where the local acceleration due to gravity was 4.50 m/s², which answers (if any) are different and what are the new answers?

•21. A 1.50 kg rock is in the middle of the bottom of an open bucket of mass 2.00 kg. The bucket and rock rotate in a *vertical* circle of radius 1.20 m at a constant speed of 4.00 m/s. (a) Make a second law force diagram indicating the forces on the rock when it is at the lowest point in its motion. (b) What is the magnitude of the total force on the rock when the bucket is at the lowest point of its circular path? (c) What is the force of the bucket on the rock at this lowest point? (d) At the highest point on the circular path, will the rock fall from the bottom of the bucket? Indicate your reasoning with an appropriate calculation and explanation.

•22. Muscles. On the ground you are just barely able to lift 100 kg. Inside an elevator, you discover that you can lift 110 kg (barely)! (a) What is the direction of the acceleration of the elevator relative to the ground? Sketch a second law force diagram showing the forces on the mass you lift in the elevator. (b) Calculate the magnitude of the acceleration of the elevator.

•23. You are sitting in the rear seat of an automobile that accelerates uniformly from rest to a speed of 120 km/h during 6.00 s along a straight stretch of level highway. The seat has a horizontal surface and a vertical seat back. Your mass is 75.0 kg. (a) Sketch a second law force diagram indicating all forces on you in the horizontal and vertical directions from the viewpoint of an observer at rest outside the car on the side of the highway. (b) Determine the magnitude of your acceleration. (c) What is the magnitude of the total force on you? In which direction does the total force act? (d) Identify the forces that are the Newton's third law pairs to the forces identified in (a) and sketch third law force diagrams indicating these forces and the systems on which they act.

•24. A car traveling at constant speed reaches the crest of a huge frost-heave (hill) that is a cylindrical cross section of radius 70 m as shown in Figure P.24. (a) At the instant when the car is at the crest of the hill, make a second law force diagram indicating schematically all forces acting on the car. (b) What is the maximum speed the car can travel without becoming airborne at the crest of the hill? Express your result in km/h.

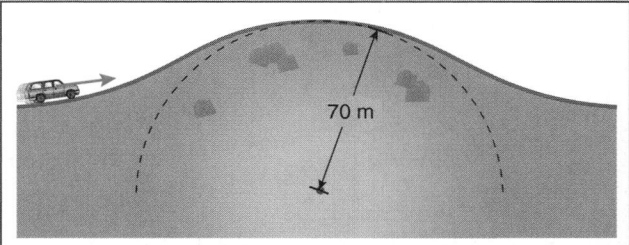

70 m

FIGURE P.24

•25. (a) What inclination angle is needed for a frictionless inclined plane so that masses released on the plane experience an acceleration of magnitude precisely 1.00 m/s²? (b) If the mass is given an initial speed of 2.00 m/s down the plane, is its magnitude of acceleration the same as in part (a) after it is released? (c) If a 1.00 kg mass is placed at rest on this incline and released, what is the magnitude of the total force on the mass?

•26. A 51.0 kg crate is pushed up a frictionless 20.0° incline with a constant horizontal force (see Figure P.26). The magnitude of the acceleration of the crate is 0.100 m/s². (a) What are the magnitude and direction of the total force on the crate? (b) What is the magnitude of the weight of the crate? (c) What is the magnitude of the horizontal force applied? (d) What is the magnitude of the normal force of the surface on the crate?

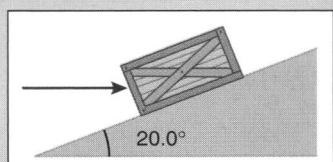

20.0°

FIGURE P.26

‡27. A chain of total mass m has n_0 identical links. The chain is suspended vertically from the top link and is at rest with no links touching the floor. (a) What is the mass of a single link? What is the magnitude of the weight of a single link? (b) What is the magnitude of the upward force on the link at the bottom of the chain (call this link number 1) by the next link up?

(c) What is the magnitude of the upward force on the link at the top of the chain (link number n_0)? (d) What is the magnitude of the upward force acting on link number n, counting from the bottom of the chain? (e) If the chain has an upward acceleration of magnitude a, recalculate your answers to parts (a)–(d).

‡28. A large Ferris wheel of radius 15.0 m is rotating wildly out of control at 4.00 rev/min. See Figure P.28. A student of mass 80.0 kg is becoming increasingly uncomfortable with the ride. On the other hand, you (of mass 50.0 kg) are rather enjoying the wild ride. To calm your companion's panic, you decide to explain soothingly the physics of the effects you both are experiencing on the Ferris wheel. You first point out that during the entire ride, there are only two forces acting on each of you: (1) your weight and (2) the normal force of the bench seat on your derriere. (a) When the chair is at the top of the circle, which of the two forces has the greater magnitude? What is the magnitude of the difference between your weight and the normal force of the seat on your derriere? (b) When the chair is at the bottom of the circle, which of the two forces is greater in magnitude? What is the difference in the magnitudes of the forces on you at this location? (c) When the chair is not at the top or bottom of the circle, the two forces are no longer antiparallel to each other, since the chair swivels. At what two places in the motion is the deviation of the direction of the normal force from the local vertical a maximum? At these locations, what is the angle θ between the direction of this force and the local vertical direction? (d) At what angular speed must the wheel rotate so you both are apparently "weightless" at the top of the spin? Quote the result in rev/min. Are you truly weightless at the top of the spin? Explain.

TICKETS TICKETS

FIGURE P.28

Section 5.11 Tensions in Ropes, Strings, and Cables
Unless otherwise stated, assume all ropes, strings, and cables are ideal.

29. A fishing line will break if the magnitude of the tension in it exceeds 100 N. What is the greatest mass that can be hung vertically from the string?

30. A towing firm owned by Ima Wheeler is attempting to pull your car horizontally and frictionlessly out of an icy culvert. You and the car have a combined mass of 1355 kg. The cable attached to the car will break if its tension exceeds 4.00 kN.

What is the magnitude of the maximum acceleration that can be given to you and the car?

•**31.** Two masses m_1 and m_2 are connected by a string and are pulled across a horizontal frictionless surface by a horizontal force of magnitude F_0, as shown in Figure P.31. Find an expression for the magnitude of the tension in the cord connecting the masses.

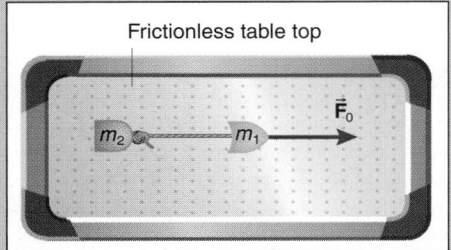

FIGURE P.31

•**32.** A string will break if its tension exceeds 500 N. (a) Is it possible to hang a 35.0 kg stone on the string? (b) If part (a) indicates that the mass can be hung from the string, what is the *maximum* magnitude of acceleration vertically upward that can be given to the mass by the attached string without its breaking? (c) What maximum magnitude of acceleration vertically down can be imparted to the stone without breaking the string? (d) If the rock lies on a horizontal frictionless surface, what is the maximum horizontal acceleration that a horizontal pull with the string can give to the stone?

•**33.** A rope is hanging down from the ceiling of a gym. While standing on the floor, you grab the rope and pull down on the rope with a force of magnitude 100 N. Assume your mass is 70.0 kg. (a) What are the magnitude and direction of the force that the rope exerts on you? (b) Draw a second law force diagram indicating all forces acting on you. (c) For every force acting on you in part (b), state in words the corresponding force of a Newton's third law pair. (d) What is the magnitude of the gravitational force of the Earth on you? (e) What is the magnitude of the normal force that the floor exerts on you? (f) If you did this experiment while standing on a bathroom scale (calibrated in kilograms), the magnitude of which force can be determined from the scale reading? What is the scale reading?

•**34.** An intrepid mountain climber is resting in the awkward situation depicted in Figure P.34. (a) Draw a second law force diagram indicating schematically all forces acting on the climber. (b) Find the magnitude of the tensions in the two supporting ropes.

•**35.** A 200 kg meteorite is suspended in a science museum by two cables as shown in Figure P.35. (a) Sketch a second law force diagram considering the meteorite as the system. (b) Determine the magnitude of the forces the cables exert on the meteorite.

•**36.** A 1.20×10^4 kg helicopter is vertically lifting a 3.00×10^3 kg load (see Figure P.36). The upward acceleration is of magnitude 0.500 m/s². (a) Define an appropriate system and determine the magnitude of the tension in the vertical cable. (b) Define an appropriate system and determine the magnitude of the upward force of the air on the helicopter blades.

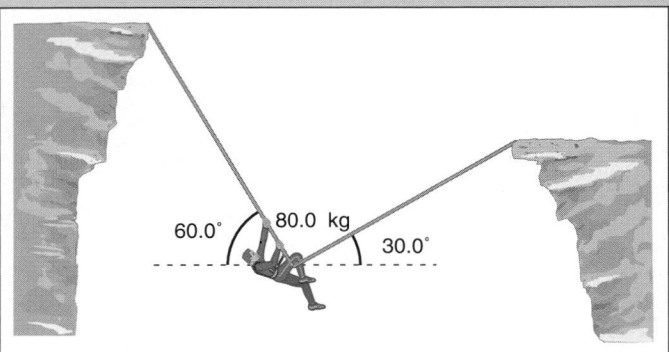

FIGURE P.34

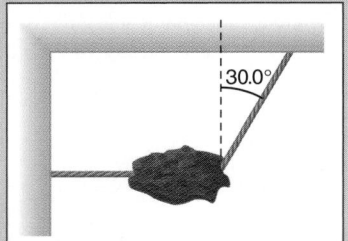

FIGURE P.35

FIGURE P.36

•**37.** Two masses, m_1 and m_2, hang over an ideal pulley and the system is free to move (see Figure P.37). Such an arrangement is called an *Atwood's machine*. (a) Show that the acceleration $\vec{a}$ of the system of two masses has a magnitude

$$a = \frac{|m_2 - m_1|g}{m_2 + m_1}$$

(b) Show that the tension in the cord is of magnitude

$$\frac{2m_1m_2g}{m_1 + m_2}$$

•**38.** Two unfortunate frozen, cubical mountain climbers connected by a rope find themselves sliding over an icy (frictionless) precipice as indicated in Figure P.38. Rather than help them,

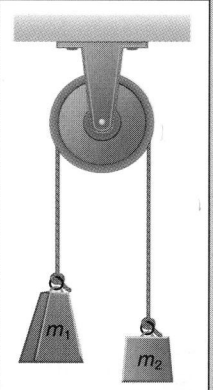

FIGURE P.37

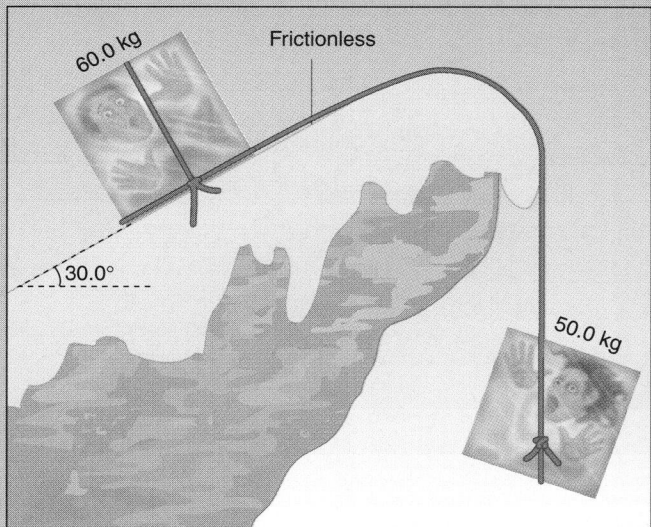

FIGURE P.38

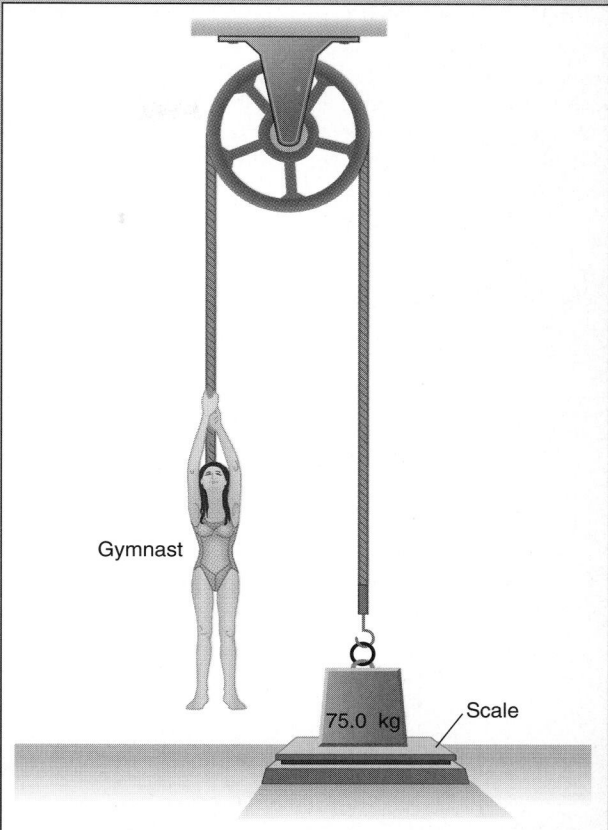

FIGURE P.39

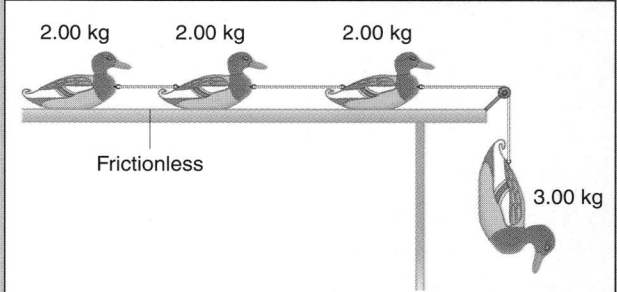

FIGURE P.40

solve the following problems: (a) Consider the system to be the pair of climbers. Sketch a second law force diagram indicating the forces acting on this system. Write Newton's second law for this system to determine the magnitude and direction of the acceleration of the system. (b) Now sketch second law force diagrams indicating the forces acting on each climber considered as an individual system. (c) By considering the motion of either climber individually, find the magnitude of the tension in the rope.

•39. A 50.0 kg gymnast hangs on one side of a rope over an ideal pulley (see Figure P.39). The other end of the rope is attached to a 75.0 kg mass at rest on a scale on the floor. What is the magnitude of the minimum upward acceleration of the gymnast sufficient to have the scale indicate zero?

•40. Three 2.00 kg decoys on a frictionless table are connected in series by strings as indicated in Figure P.40. The final string passes over an ideal pulley at the edge of the table and suspends the mother of all decoys with a mass of 3.00 kg. The ducks are initially all at rest. (a) Which string has the least tension and why? (b) Which string has the greatest tension? Show that the greatest tension has a magnitude less than 29.4 N. (c) Do you

expect the magnitude of the acceleration of the ducky system to be less than, greater than, or equal to g? (d) Calculate the magnitude of the acceleration of the system of ducks.

•41. A string made from braided dental floss has a maximum permissible tension of 500 N and is used by a 60.0 kg inmate to escape prison (this actually happened!). The inmate starts at rest and slides vertically down the string for 8.0 m. (a) Make a second law force diagram indicating schematically all forces acting on the inmate in the descent. (b) Determine the magnitude of the acceleration of the inmate so the string barely does not break. Is this a minimum value or a maximum value? Explain. (c) What is the minimum speed of the former inmate at the end of the 8.0 m descent if the string does not break?

•**42.** A mass m_1 lies on a fixed, smooth (frictionless) cylinder. An ideal cord attached to m_1 passes over the cylinder to mass m_2 as pictured in Figure P.42. (a) Make a second law force diagram indicating schematically all forces acting on m_2. (b) Make a second law force diagram indicating schematically all forces acting on m_1. (c) Show that the system is in equilibrium if

$$\sin \theta = \frac{m_2}{m_1}$$

(d) Evaluate θ if $m_1 = 40$ kg and $m_2 = 10$ kg.

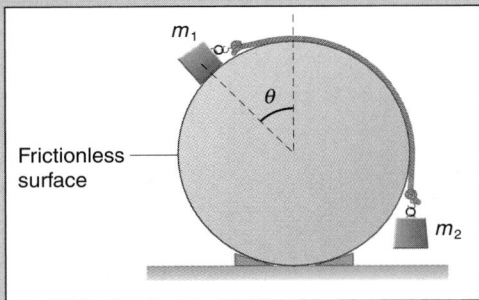

FIGURE P.42

•**43.** You are standing in an elevator that is accelerating vertically upward. (a) Sketch a second law force diagram indicating all forces acting on you. (b) Which force has the greatest magnitude? (c) Specify with a verbal description the forces that are the Newton's third law pairs to the forces delineated in part (a). Sketch corresponding third law force diagrams indicating the systems on which these forces act. (d) Repeat parts (a) and (b) for a situation in which the elevator is accelerating downward.

•**44.** A 500 kg elevator car is moving up in the world with an acceleration of 0.50 m/s² by means of a cable. (a) Make a second law force diagram indicating schematically all forces acting on the elevator car. (b) Determine the magnitude of the tension in the cable.

•**45.** A 100 kg oaf is in a 1.00×10^3 kg elevator car accelerating upward at 1.50 m/s². (a) Determine the magnitude of the tension in the cable supporting the elevator car. (b) What is the force of the floor of the elevator car on the oaf? Is this force equal in magnitude to that of the weight of the oaf? Explain why or why not.

•**46.** Two gargoyles, each of mass 2.00 kg, are hanging from the ceiling of an elevator by means of strings as indicated in Figure P.46. The elevator is moving downward with an acceleration of magnitude 2.0 m/s². (a) Choose an appropriate system, sketch an appropriate second law force diagram, and determine the magnitude of the tension in the lower string. (b) Choose an appropriate system, sketch an appropriate second law force diagram, and determine the magnitude of the tension in the upper string.

•**47.** Wahoo! You are swinging a mass m at speed v around on a string in a circle of radius r whose plane is 1.00 m above the ground (see Figure P.47). The string makes an angle θ with the vertical direction. (a) Make a second law force diagram in-

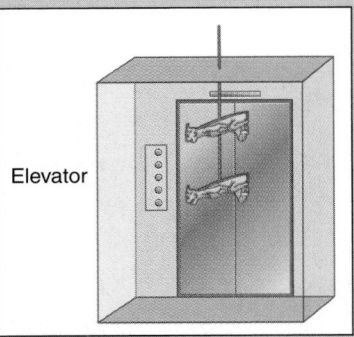

Elevator

FIGURE P.46

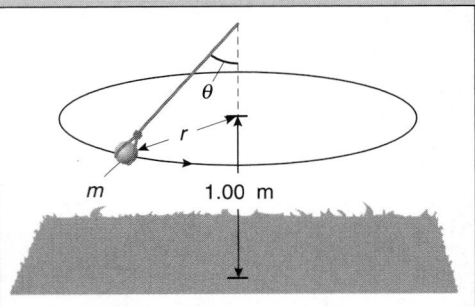

FIGURE P.47

dicating all forces acting on the mass at a particular instant in its motion. Indicate the direction to the center of its circular path. (b) What is the *direction* of the acceleration of the mass? (c) Apply Newton's second law to the horizontal and vertical directions to show that

$$\tan \theta = \frac{v^2}{rg}$$

(d) If the angle $\theta = 47.4°$ and the radius of the circle is 1.50 m, find the speed of the mass. (e) If the mass is 1.50 kg, what is the magnitude of the tension in the string? (f) The string breaks unexpectedly when the mass is moving exactly eastward. Where does the mass hit the ground?

•**48.** A 3.00 kg mass attached to a string is swung around, at constant speed, in a circle of radius 0.500 m by means of a massless string of length 2.00 m as indicated in Figure P.48. (a) Draw a second law force diagram schematically indicating the forces on the mass at some instant. (b) Is the total force on

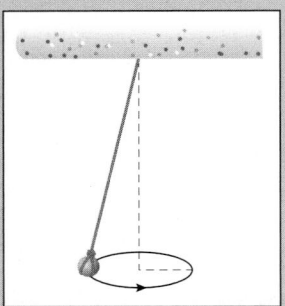

FIGURE P.48

the mass zero? Explain. (c) Find the magnitude of the tension in the string. (d) What is the speed of the mass? (e) What is the time needed for the mass to complete one circular trajectory?

•49. The Dean of Deans (whose mass is 80.0 kg) steps into an express elevator car headed for a meeting of the Committee on Committees, the mother of all committees. The elevator initially is at rest. (a) Sketch a second law force diagram indicating schematically all forces on the Dean. (b) What is the magnitude of the gravitational force of the Earth on the Dean? (c) Find the magnitude of the force that the floor exerts on the Dean while the elevator car is at rest. In (d)–(h), the elevator car now gets under way with an upward acceleration of 1.00 m/s^2. (d) Sketch a second law force diagram indicating schematically all forces on the Dean. (e) What is the magnitude of the weight of the Dean during this acceleration? (f) Find the magnitude of the force of the floor on the Dean during this upward acceleration. (g) If the Dean were standing on a scale, what would the scale read during this acceleration? Express your result in newtons. How do you reconcile the scale reading with the answer to part (e)? In (h)–(k), as the appropriate floor is approached, the elevator car slows down with an acceleration of magnitude of 1.00 m/s^2. (h) Sketch a second law force diagram indicating schematically all forces on the Dean. (i) What is the magnitude of the gravitational force of the Earth on the Dean during this phase of the motion? (j) Calculate the magnitude of the force of the floor on the Dean during this phase of the motion. (k) If the Dean were standing on a scale, what would the scale read during this phase of the motion? Express your result in newtons.

•50. An igneous rock of mass 3.0 kg is moving in uniform circular motion in a horizontal plane 2.0 m above the ground as indicated in Figure P.50. The radius of the orbit is 0.80 m. (a) What is the magnitude of the force of the cord on the mass? (b) What is the magnitude of the acceleration of the rock? (c) What is the angular speed ω of the rock? (d) How many revolutions per second does the rock make? (e) If the cord breaks, how long will it take the rock to hit the ground?

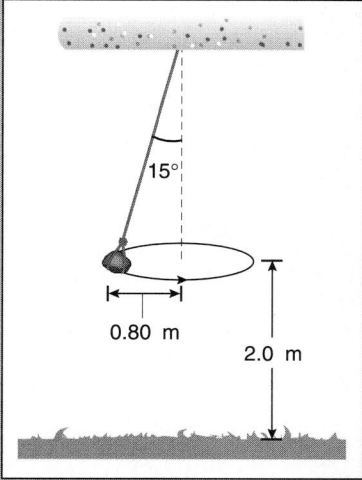

FIGURE P.50

•51. You are headed up to the offices of the famous law firm, Dewey, Cheatam, and Howe, and take an elevator. Strangely, there is a scale on the floor of the elevator car and you stand on the scale. With the elevator car accelerating upward with an

acceleration of magnitude 0.981 m/s^2, the scale reads 600 N. (a) Sketch a second law force diagram indicating the forces acting on you. (b) Determine your mass. (c) If the total mass of the elevator car and its contents is 2.00×10^3 kg, what is the magnitude of the tension in the cable on which it hangs while the system is accelerating upward? (d) A loose 3.00 g screw falls out of the ceiling (while the elevator is accelerating upward) and falls to the floor 1.80 m below. Calculate the total force on the screw while it falls through the air.

•52. A *massive* cable is attached to a 30.0 kg mass on a frictionless surface and a horizontal force of magnitude F accelerates the arrangement as shown in Figure P.52. If the force of the cable on the mass has a magnitude 0.900F, what is the mass of the cable?

FIGURE P.52

Sections 5.12 Static Friction
5.13 Kinetic Friction at Low Speeds
5.14 Kinetic Friction Proportional to the Particle Speed*

53. If a horizontal force of magnitude 350 N is exerted on a 50 kg mushroom anchor, the anchor is on the verge of slipping across the surface of a dock. What is the coefficient of static friction for the anchor on the dock surface?

54. A 10.0 m long chain of mass 25 kg is dragged at constant velocity across a flat beach by means of a horizontal force of magnitude 270 N. What is the coefficient of kinetic friction for the chain on the beach surface?

55. You find that a horizontal force of magnitude 250 N is able to slide a 30 kg mass across a horizontal surface with an acceleration of magnitude 0.500 m/s^2. What is the coefficient of kinetic friction for the mass on the surface?

56. The coefficients of friction for a file cabinet on the floor are $\mu_s = 0.45$ and $\mu_k = 0.30$. You push horizontally with a force of magnitude 75 N on a 100 kg file cabinet and are unable to move it. What is the force of static friction by the floor on the cabinet?

57. A skier can coast directly down a 5° straight slope at constant velocity. What is the coefficient of friction for the skis and the snow surfaces? Is this the coefficient of static or kinetic friction?

58. An exit ramp with a circular curve of radius 200 m, banked at an angle of 20°, is covered with ice. At what speed (in km/h) could a car execute the turn successfully without slipping down or up the curve and hitting the guard rails?

•59. While executing a level, unbanked circular curve to the right of radius 200 m at constant speed 100 km/h, a passenger of mass 55 kg in your car discovers that she is ready to slip across the horizontal seat. What is the coefficient of static friction for the passenger on the seat?

•**60.** Your 2.50 kg physics text lies on a reading surface that is inclined to the horizontal at an angle of 35° as shown in Figure P.60. While enjoying a reverie about a fantastic physics lecture, you determine that a force of magnitude 20 N directed parallel to and up the plane is barely sufficient to slide the text up the incline at constant speed. Amazing stuff. (a) Draw a second law force diagram indicating schematically all forces acting on the text. (b) Determine the coefficient of friction for the text on the surface. (c) Is this the coefficient of static or kinetic friction?

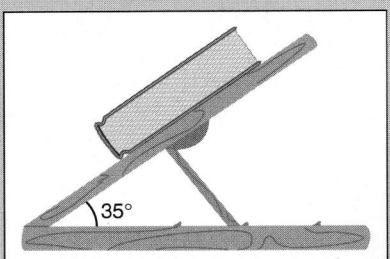

35°

FIGURE P.60

•**61.** The static and kinetic coefficients of friction for a 50 kg mass on a table surface are $\mu_s = 0.20$ and $\mu_k = 0.15$. See Figure P.61. The pulley is ideal. The system is at rest in equilibrium. (a) For each mass, sketch a second law force diagram indicating schematically all forces that are acting on each mass. (b) How large can the mass m be and not move the system? What is the magnitude of the frictional force on the 50 kg mass when this situation exists? (c) If m is only half the maximum mass calculated in part (b), what is the magnitude of the frictional force on the 50 kg block? Indicate whether this is a force of static friction or a force of kinetic friction.

50 kg

m

FIGURE P.61

•**62.** You try to stand on a 30° slope on which the coefficients of static and kinetic friction for your shoes on the slope are 0.55 and 0.45 respectively. Your mass is 65.0 kg. (a) Can you stand motionless on one foot and/or two feet, or will you slip? (b) If you do not slip and slide down the plane, what is the magnitude of the static frictional force? If you do slip, find the magnitude of your acceleration. Will you continue to slide all the way down the slope?

•**63.** (a) A circular turn of radius 3000 m is being designed for high-speed trains traveling at speeds of 200 km/h. Calculate the appropriate banking angle of the turn for minimum wear on the track. (b) If trains execute the turn at speeds *slower* than the design speed, which rail experiences greater wear (the rail on the inside of the turn or the rail on the outside of the turn)? Explain your reasoning.

•**64.** Your car of mass 1250 kg makes a circular turn of radius 150 m while traveling at a constant speed of 35 km/h. The curve is not banked but is horizontal. (a) What is the magnitude of the centripetal acceleration of the car? (b) What is the magnitude of the tangential acceleration of the car? (c) What magnitude force is required to keep the car going around the curve? What provides this force? (d) What is the minimum value of the coefficient of static friction required for the car to make the turn without skidding?

•**65.** You are on the indicated inclined plane shown in Figure P.65 and are connected to mass m by means of an ideal rope passing over an ideal pulley. Your mass is 70 kg and the coefficients of friction for you on the plane are $\mu_s = 0.40$ and $\mu_k = 0.35$. (a) If $m = 0$ kg, draw a second law force diagram indicating all forces acting on you. Will you slide down the plane? Explain your reasoning. (b) If $m \neq 0$ kg, draw a second law force diagram indicating all forces acting on you. Draw a similar second law force diagram for mass m. What mass m enables you to accelerate *up* the plane with an acceleration of magnitude 1.50 m/s²? (c) Under the conditions of part (b), determine the tension in the rope.

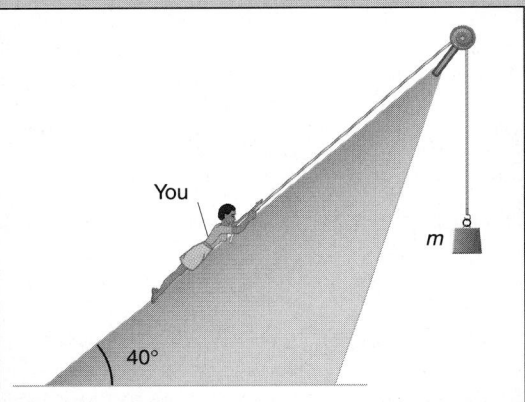

You

m

40°

FIGURE P.65

•**66.** Talus slopes are formed at the base of large cliffs because of the accumulation of rocks that break off the cliff and fall into a debris pile at the base. See Figure P.66. Sand piles also keep getting steeper until slippage of the particles occurs. If the coefficient of static friction between the particles is μ_s, find the angle θ that the talus slope makes with horizontal direction.

FIGURE P.66

•**67.** An old stereo turntable of 15.0 cm radius turns at 33.0 rev/min while mounted on a 30.0° incline as indicated in Figure P.67. (a) If a mass m can be placed anywhere on the turntable without slipping, where is the most critical place on the disk where slipping might occur? Why? (b) Calculate the least possible coefficient of friction that must exist if no slipping occurs. Is your answer independent of m? Is this a static or kinetic coefficient of friction?

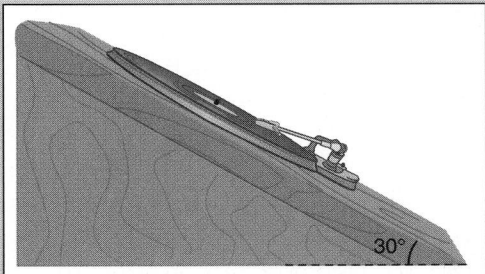

30°

FIGURE P.67

•**68.** A mass m is placed on a rough cylinder as shown in Figure P.68. The coefficient of static friction for the mass on the cylinder is 0.15. (a) Draw a second law force diagram indicating all forces acting on the mass. (b) Find the angle θ at which the mass is just ready to slip off the cylinder. (c) Is the angle different if the rough surface is a sphere?

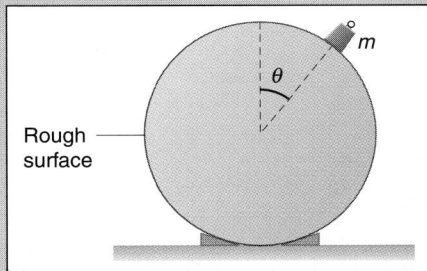

θ m

Rough surface

FIGURE P.68

•**69.** A 500 kg crate of apples is in the middle of the flat, horizontal bed of a pickup truck, but is not tied down. The coefficients of friction for the crate on the bed of the truck are $\mu_s = 0.35$ and $\mu_k = 0.25$. What is the magnitude of the maximum horizontal acceleration of the truck so that the crate does not slide across the bed?

•**70.** A 100 kg container is on a loading ramp inclined to the horizontal by 20.0° as shown in Figure P.70. The coefficients of friction for the crate on the slope are $\mu_s = 0.300$ and $\mu_k = 0.250$. (a) Draw a second law force diagram indicating the forces on the crate. (b) Will the mass slip down the ramp of its own accord? If so, what minimum magnitude force up the plane must be used to keep the crate in place? If not, what magnitude of force is needed down the plane to get the crate to slide?

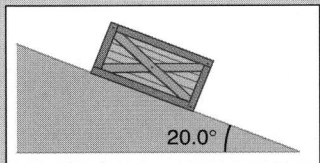

20.0°

FIGURE P.70

•**71.** A frustrated parent of mass 70 kg is dragging a child of mass 20 kg having a temper tantrum across level but rough ground. The parent exerts a force of magnitude 80 N on the child; the force is upward and forward at an angle of 45° to the horizontal (see Figure P.71). The parent and child are moving at constant speed of 1.5 m/s in a straight line homeward from a video arcade. (a) Draw a second law force diagram indicating all forces on the child. (b) Calculate the frictional force of the ground on the child. (c) Find the magnitude of the normal force of the surface on the child. Is the normal force equal in magnitude to that of the weight of the child? Explain why or why not. (d) Find the coefficient of friction for the child on the ground. Is this the coefficient of static or kinetic friction?

FIGURE P.71

•**72.** Two spherical students of identical 60 kg mass are dragged by Professor Mediocratus to a final exam in classics as indicated in Figure P.72. The coefficient of kinetic friction for the students on the floor is 0.30. The tension in the rope between the students is 300 N. (a) Choose an appropriate system and calculate the initial acceleration of the students. (b) Calculate the magnitude of the tension in the rope between the leading student and professor.

Exams

FIGURE P.72

•**73.** A fast merry-go-round is turning at a constant angular speed of 1.00 rad/s. Wheeee! You are standing on the merry-go-round minding a young sibling. The coefficient of static friction for your shoes on the merry-go-round is 0.80 and the coefficient of kinetic friction is 0.50. (a) What is the direction of the total force acting on you and why is it in this direction? (b) If you can stand 2.0 m from the axis of rotation, with what speed are you moving? (c) How far from the axis can you stand without slipping? (d) If you threw a ball straight up and parallel to the axis of rotation, describe the path of the ball (i) from the viewpoint of an observer on the ground, and (ii) from your viewpoint on the merry-go-round.

•74. A nerd of mass 50 kg sits in a chair of mass 5.0 kg on a platform of mass 100 kg that is accelerated off a stage in a straight line with an acceleration of magnitude 1.0 m/s². (a) What coefficient of static friction will keep the chair from slipping on the platform? (b) The chair exerts a frictional force on the platform, equal in magnitude but opposite in direction to the frictional force of the platform on the chair. How, then, can there be any acceleration of either one? Explain. (c) Is the platform an inertial reference frame? Explain.

•75. A large 5.0 kg bone rests on a level surface and the coefficients of friction for the bone on the surface are $\mu_s = 0.40$ and $\mu_k = 0.20$. A dog is pulling horizontally with a force of magnitude 30 N on the bone as indicated in Figure P.75. (a) Draw a second law force diagram indicating all forces acting on the bone. (b) Choose an appropriate coordinate system and indicate it clearly in a sketch. (c) What is the normal force on the bone? (d) What is the *maximum* magnitude of the static frictional force on the bone? (e) Will the bone accelerate? If so, determine its acceleration. If not, explain why it does not accelerate. (f) Determine the magnitude of the frictional force on the bone. (g) How many newtons of force must the dog use to give the bone an acceleration of magnitude 1.0 m/s²?

FIGURE P.75

•76. A train consists of 50 identical hopper cars each with a mass of 2.00×10^4 kg. The train lacks a caboose; the hopper cars are numbered, with car number 1 being the first car behind the engines. A series of three locomotives at the head of the train provide a force on the first hopper car of magnitude 100 kN in the horizontal direction. The acceleration of the train has a magnitude 6.00×10^{-2} m/s² on a straight stretch of level track. Chug, chug, chug; I think I can, I think I can, I think I can do this problem. (a) What is the magnitude of the vertical force exerted by the track on *each* hopper car? (b) The frictional force exerted by the track on each car is the same; what is the magnitude of the frictional force on each car? (c) What is the magnitude of the force that the coupling at the rear of car 24 exerts on the coupling at the front of car 25? What is the magnitude of the force that the coupling at the front of car 25 exerts on the coupling at the rear of car number 24? Do these forces form a Newton's third law pair?

•77. A slapshot sends a hockey puck of mass m in a straight line with an initial speed v_0 across a vast expanse of ice. The coefficient of kinetic friction for the puck on the ice is μ_k. (a) Set up an appropriate coordinate system to analyze the problem. (b) Make a second law force diagram appropriate for the puck as it slides across the ice. (c) Find the velocity component as a function of time. (d) How long does it take the puck to come to rest? (e) Find an expression for the position of the puck as a function of time with respect to the origin chosen for your coordinate system. (f) How far does the puck travel before

it stops? (g) Evaluate the distance determined in part (f) if $v_0 = 25$ m/s and $\mu_k = 0.050$.

•78. An ideal physics text of mass 2.5 kg is placed on a rough inclined plane as indicated in Figure P.78. The coefficients of static and kinetic friction for the text on the plane are 0.30 and 0.20 respectively. An ideal string is attached to the text and passes over an ideal pulley to another mass m. (a) Sketch a second law force diagram indicating schematically all forces acting on the text. (b) Make a similar diagram for the mass m. (c) What is the maximum mass m that can be attached to the cord before the text starts to slip? (d) If m is 10.0 kg, determine the magnitude of the acceleration of the text.

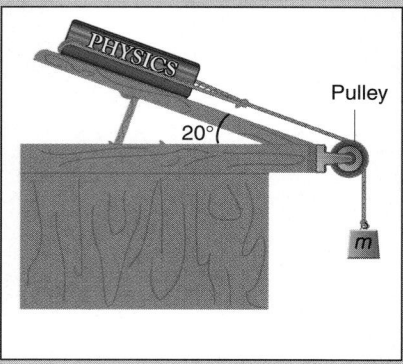

FIGURE P.78

•79. A 20 kg mass is attached to a 30 kg mass by means of an ideal cord as indicated in Figure P.79. Both masses are on a level surface, and the coefficients of friction for the masses on the surface are $\mu_s = 0.40$ and $\mu_k = 0.20$. A dog is pulling horizontally with a force of magnitude 30 N as indicated. (a) Draw a second law force diagram indicating all forces acting on the 20 kg mass. (b) Draw a second law force diagram indicating all forces acting on the 30 kg mass. (c) Define a system consisting of *both* masses. Draw a second law force diagram indicating all forces acting on the combined system. (d) Will the system accelerate? If so, determine the magnitude of the acceleration. (e) What magnitude of force is needed for the system of two masses to have an acceleration of magnitude 1.2 m/s²?

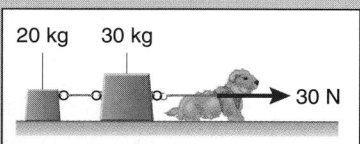

FIGURE P.79

•80. A chicken feather and a fist-sized rock are dropped at the same time from a height 50 m in the air. (a) Why does one object reach the ground before the other? (b) Sketch qualitatively accurate graphs of the speeds of the two objects as functions of time. (c) Sketch qualitatively accurate graphs of the magnitudes of the accelerations of the two objects as functions of time.

•81. An automobile tire of mass m is spinning at angular speed ω_0 while held at rest a slight distance above a horizontal road. The tire then is released onto the roadway and skids away from its initial position (see Figure P.81). (a) Sketch a second law force diagram indicating all forces on the tire while it is skidding. (b) While the tire is skidding, name the force that accelerates the tire in the horizontal direction. (c) Show that while the tire is skidding, the horizontal acceleration is of magnitude $\mu_k g$, where μ_k is the coefficient of kinetic friction for the tire on the road.

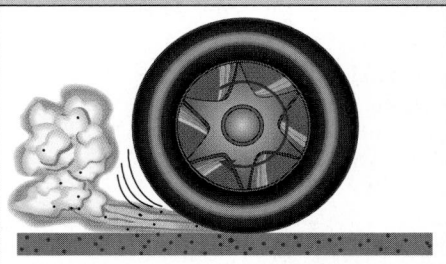

FIGURE P.81

•82. Before pulling the rip cord on a parachute that fails to open, a skydiving physics student of mass 50.0 kg, falling vertically, somehow observes that air friction has caused the magnitude of her acceleration at some instant to be only $0.250g$. (a) Draw a second law force diagram indicating all forces acting on the unfortunate lass. (b) Calculate the magnitude of the frictional force acting on the student. (c) Sometime later, with the parachute now deployed but not open, she reaches a terminal speed (in more ways than one) of 150 km/h. What is her acceleration when the terminal speed is achieved? If the magnitude of the total kinetic frictional force at this speed is assumed to be proportional to the speed:

$$f_k = Kv$$

find the constant K and indicate its units.

•83. Amusement parks occasionally have a large cylindrical tube mounted to spin about the vertical axis of the cylinder as shown in Figure P.83. Patrons stand against the wall of the cylinder and the apparatus is set into rotation. At a certain critical angular speed, the floor on which the patrons are standing can be lowered and the patrons are left hanging on the wall (but on the verge of slipping) due to the force of static friction on them. Let R be the radius of the apparatus, m the mass of a patron, and μ_s the coefficient of static friction for a patron on the wall. (a) Consider a patron to be the system whose motion is to be analyzed. Make a sketch of a second law force diagram indicating all forces acting on the patron. (b) Is there an acceleration in the vertical direction? What does this imply about the relationship between the static frictional force on the patron and the weight of the patron? Are these forces a third law force pair? (c) Is there an acceleration of the patron in the horizontal direction? How is this acceleration related to the speed of the patron? (d) How is the angular speed ω of the cylinder related to the tangential speed of the patron? If the patron is ready to slip, find the angular speed ω of the tube in terms of R, μ_s, and g. Express your result in rev/s.

‡84. Two masses, m_1 and m_2, are connected by an ideal cord as shown in Figure P.84. The cord passes over an ideal pulley. A student determines that when the surface is inclined at an angle θ to the horizontal, m_1 is just on the verge of slipping to the *right* up the incline. (a) Draw a second law force diagram indicating schematically all forces on mass m_2. (b) Draw a second law force diagram indicating schematically all forces on mass m_1. (c) Introduce appropriate coordinate system(s) and show that the coefficient of static friction for m_1 on the inclined plane is

$$\mu_s = \frac{|m_2 - m_1 \sin \theta|}{m_1 \cos \theta}$$

Just to be mean, the inclination angle of the inclined plane now is increased to angle ϕ so that the mass m_1 is just on the verge of slipping to the *left* down the incline. (d) Draw a second law force diagram indicating schematically all forces on m_2. (e) Draw a second law force diagram indicating schematically all forces on m_1. (f) Show that the coefficient of static friction for m_1 on the inclined plane is

$$\mu_s = \frac{|m_1 \sin \phi - m_2|}{m_1 \cos \phi}$$

Notice that if the suspended mass $m_2 = 0$ kg, then $\phi = \theta$ and this equation reduces to

$$\mu_s = \tan \phi = \tan \theta$$

in agreement with what was derived in Example 5.11. (g) Is this the *same* coefficient of friction as that determined in part (c)? Why or why not? (h) Show that the ratio of the two masses is given by the rather elegant expression

$$\frac{m_2}{m_1} = \frac{\tan \theta + \tan \phi}{\sec \theta + \sec \phi}$$

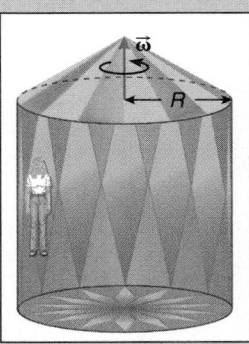

FIGURE P.83

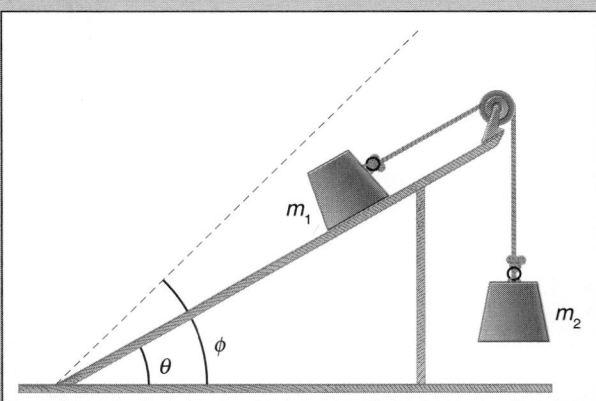

FIGURE P.84

‡85. For gung-ho students who have completed the previous problem, consider this: the mass ratio m_2/m_1 must be less than unity. Why? Likewise, the problem also indicates that the angle ϕ must be greater than the angle θ. For a *given* angle θ, then, the angle ϕ has a range from a minimum value of $\phi_{min} = \theta$ to a maximum value of $\phi_{max} = 90°$. (a) Show that when $\phi_{min} = \theta$, the mass ratio m_2/m_1 has a value equal to $\sin \theta$. (b) Show that when ϕ has its maximum value of 90°, the mass ratio is equal to unity. Thus, for a given θ, the *possible* values of the mass ratio are located between $\sin \theta$ and 1. (c) Make a graph indicating the possible values of the mass ratio as a function of θ.

‡86. Assume the kinetic frictional force on a falling mass m is proportional to its speed v, with a proportionality constant β. Choose $\hat{\jmath}$ to be vertically downward. (a) Show that Newton's second law of motion yields

$$m \frac{dv_y}{dt} = mg - \beta v_y$$

(b) When the mass reaches its terminal speed, what is

$$\frac{dv_y}{dt}$$

(c) Show that the terminal speed is

$$v_{term} = \frac{mg}{\beta}$$

(d) If the mass is dropped from rest, show that

$$v_y(t) = v_{term}(1 - e^{-\beta t/m})$$

Sections 5.15 Fundamental Forces and Other Forces Revisited*
5.16 Noninertial Reference Frames*

•87. Calibrate the scale of the plumb bob accelerometer mentioned in Section 3.6. That is, mark the circular arc of the

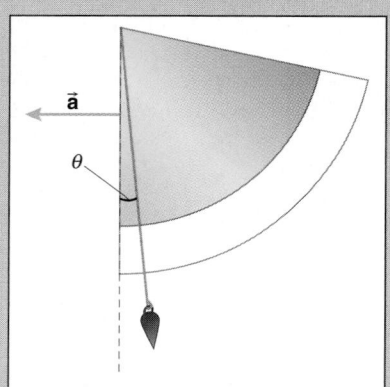

FIGURE P.87

accelerometer in Figure P.87 so that if the plumb bob is hung from the center of the circular arc, the scale on the circular arc indicates horizontal accelerations of magnitudes from 0 to 20.0 m/s². Is the scale linear with angle? Explain why or why not.

•88. For the plumb bob accelerometer referred to in Section 3.6 and calibrated in the previous problem: (a) What magnitude acceleration in the horizontal direction results in a deviation angle of the plumb bob from the vertical direction of 63.4°? (b) If an acceleration in the horizontal direction is of magnitude 98.1 m/s², what is the deviation angle θ of the plumb bob accelerometer?

•89. The equatorial radius of Jupiter is 7.19×10^4 km, more than ten times as great as that of the Earth. The local acceleration due to gravity near the surface* of Jupiter is of magnitude 24.5 m/s², two and a half times as great as that on the surface of the Earth. Jupiter rotates quite fast: a complete rotation in only 9 h 50 min. Calculate the ratio between the magnitudes of the centripetal acceleration on the Jovian equator and the local acceleration of gravity near the surface of Jupiter. Comment on whether the effects of pseudoforces can be neglected on Jupiter as much as they can be on the Earth in analyzing the motion of particles.

---

*Unlike the Earth, Jupiter does not have a well-defined surface.

INVESTIGATIVE PROJECTS

A. Expanded Horizons

1. Investigate historical ideas regarding motion and the causes of motion, in particular so-called *natural motion* proposed by the ancient Greek philosopher Aristotle, and how these ideas differed from those of Galileo and Newton.

   Arnold B. Arons, *Development of Concepts of Physics* (Addison-Wesley, Reading, Massachusetts, 1965), Chapter 2, pages 35–61.

   Giorgio De Santillana, *The Origins of Scientific Thought* (University of Chicago Press, Chicago, 1961).

   Herbert Butterfield, *The Origins of Modern Science 1300–1800* (MacMillan, New York, 1956).

   Allan Franklin, "Inertia in the Middle Ages," *The Physics Teacher, 16*, #4, pages 201–208 (April 1978).

   Lindsay Judson (editor), *Aristotle's Physics* (Oxford at the Clarendon Press, Oxford, 1991).

   Robert J. Whitaker, "Aristotle is not dead: student understanding of trajectory motion," *American Journal of Physics, 51*, #4, pages 352–357 (April 1983).

2. The remarkable forces involved in karate and other Asian martial arts are rather dramatic; they can exceed 3 kN in magnitude. Boards break and concrete blocks shatter because of the huge force of the human hand on them. Newton's third law implies the blocks and boards exert forces of the *same magnitude* on the hand! Investigate how the human hand can withstand these tremendous forces without injury.

   Jearl Walker, "The amateur scientist," *Scientific American, 243*, #1 (July 1980), pages 150–162; additional references are on page 162.

   Michael S. Field, Ronald E. McNair, and Stephen R. Wilk, "The physics of karate," *Scientific American, 240*, #4 (April 1979), pages 150–158; additional references are on page 190.

   Jearl D. Walker, "Karate strikes," *American Journal of Physics, 43*, #10, pages 845–849 (October 1975).

   Donn F. Draeger and Robert W. Smith, *Asian Fighting Arts* (Kodansha International, Tokyo, 1969).

3. Investigate the dynamics of particles moving at speeds *faster* than the so-called terminal speeds in a fluid medium such as air or water.

   Ernest Zebrowski Jr., "Superterminal velocities," *The Physics Teacher, 27*, #8, pages 618–619 (November 1989).

4. Plot the terminal speeds in Table 5.5 versus the cross-sectional area of the various sports balls. Formulate a hypothesis to account for the graph and any deviations of a ball from any trends apparent.

B. Lab and Field Work

5. Design and perform an experiment to determine the static coefficient of friction for two Velcro surfaces in contact. Also investigate whether or not μ_s is independent of the surface area in contact. Velcro was invented in 1948 by the Swiss engineer Georges deMestral.

   Judith Stone, "Velcro: the final frontier," *Discover, 9*, #5 (May 1988), pages 82–84.

6. Design an experiment to investigate the frictional characteristics of belts and capstans when the belt or rope is on the verge of slipping. Theory indicates that the frictional force is exponentially related to the total angle of contact between the belt and groove or between the rope and capstan; see if your experiments confirm the theory.

   George L. Hazelton, "A force amplifier: the capstan," *The Physics Teacher, 14*, #7, pages 432–433 (October 1976).

   William F. Riley and Leroy D. Sturges, *Engineering Mechanics: Statics* (John Wiley, New York, 1993), pages 400–402.

   Eugene Levin, "Friction experiments with a capstan," *American Journal of Physics, 59*, #1, pages 80–84 (January 1991).

   Clifford Bettis, "Capstan experiment," *American Journal of Physics, 49*, #11, pages 1080–1081 (November 1981).

7. Investigate the speed-dependent kinetic frictional effects of the air on the flight of baseballs, golf balls, tennis balls, and other balls used in various sports.

   Cliff Frohlich, "Aerodynamic drag crisis and its possible effect on the flight of baseballs," *American Journal of Physics, 52*, #4, pages 325–334 (April 1984).

   Peter J. Brancazio, "Looking into Chapman's homer: the physics of judging a fly ball," *American Journal of Physics, 53*, #9, pages 849–855 (September 1985).

   Joseph M. Zayas, "Experimental determination of the coefficient of drag of a tennis ball," *American Journal of Physics, 54*, #7, pages 622–625 (July 1986).

8. Design and perform an experiment to determine the coefficient of viscosity of air using spherical air-filled balloons falling at their terminal speeds.

9. Visit a local orthopedic medical practice and/or orthodontist and investigate the various ways in which traction is applied to various parts of the body and teeth. In particular, investigate the mechanics of the following traction devices: Bryant traction; Russell traction; and Sayre traction.

   Marian Williams and Herbert R. Lissner, *Biomechanics of Human Motion* (Saunders, Philadelphia, 1962).

C. Communicating Physics

10. Newton's laws had consequences that rippled well beyond the sciences, even into political and social arenas. Write a précis about the following article, discussing how Newton's mechanistic view of nature influenced the framing of the Constitution of the United States.

    John Patrick Diggins, "Science and the American experiment: how Newton's laws shaped the Constitution," *The Sciences, 27*, #6 (November–December 1987), pages 28–31.

11. Think about the following provocative statement: *science is always wrong*. Write an essay explaining why the statement is in some sense true and in other ways false. Most professional scientists submit their writing to colleagues for a critique before submission to a professional journal, after which the article typically is sent out for peer review to determine if it is acceptable for publication. If your instructor permits, let a colleague critique your essay before you submit it to your professor.

12. The conflict between science and religion arose around the time of Copernicus (16th century), and continued with the famous trial of Galileo (17th century) down to the present day. Science and religion clearly are very different intellectual endeavors. Nonetheless, in the spirit of good debate, think about and write an essay that explores how science and religion are *similar*. If your instructor permits, have a good discussion of the topic with your friends before writing the essay.

    Albert Einstein, *Ideas and Opinions* (Crown, New York, 1954), pages 36–54.

    Jacob Needleman, *A Sense of the Cosmos: The Encounter of Modern Science and Ancient Truth* (Open Court Publishers, Peru, Illinois, 1995).

CHAPTER 6
THE GRAVITATIONAL FORCE AND THE GRAVITATIONAL FIELD

> . . . the Newtonian principle of gravitation is now more firmly established, on the basis of reason, than it would be were the government to step in, and make it an article of necessary faith. Reason and experiment have been indulged, and error has fled before them.
>
> *Thomas Jefferson (1743–1826)* *

The gravitational force governs much of the great, large-scale natural drama of the universe. Gravitation determines many of the features of star formation, planetary systems, galaxies, and the universe as a whole. Gravitation was the first of the four fundamental and primordial forces (see Section 5.2) to be described quantitatively, and its characterization was another magnificent achievement for Newton. Newton's law of universal gravitation was the scientific culmination of the Renaissance, for it brought order and harmony to our understanding of the cosmos.

There are two likely reasons that the gravitational force was the first fundamental force discovered:

1. The force determines the structure of the solar system, a problem that intrigued scholars since the dawn of civilization.
2. The force is all-encompassing; it is apparent in virtually everything you do from simply dropping a pencil to lugging your books to the library for yet another long night of study. Gravity never ceases its action, but this very fact meant that for millennia the force was not recognized—it was all but impossible to imagine situations without it.

With ingenuity and inspiration, Newton realized that the gravitational influence of the Earth was weaker at the distance of the Moon but was nonetheless felt by it (see Section 6.1). Thus the gravitational force was not a purely local interaction but extended into space at least as far as the Moon. Occasionally we get more out of our thoughts and equations than we originally put into them: in a single stroke, Newton realized that the gravitational force likely explained the dynamics of the solar system and the universe itself. The gravitational force thus has a very long range: its effects are felt from microscopic to cosmological distances. Unlike the other fundamental forces of nature, gravitation is a truly *universal* force that acts, however weakly or strongly, on *every* particle in the universe as we mentioned in Section 5.2.

In this chapter we first see how Newton deduced the law of gravitation and will explore the nature and features of the gravitational force. We also examine some of its implications in the local environment of the Earth, within the solar system, and beyond. Lastly we investigate a convenient concept, the *gravitational field*, for describing how the gravitational force acts on masses that are not in direct physical contact with each other, such as astronomical masses separated by billions of kilometers or even light-years of empty space.

6.1 HOW DID NEWTON DEDUCE THE GRAVITATIONAL FORCE LAW?

After dinner, the weather being warm, we went into the garden and drank thea, under the shade of some appletrees, only he and myself. Amidst other discourse, he told me, he was just in the same situation, as when formerly, the notion of gravitation came into his mind. It was occasion'd by the fall of an apple, as he sat in a contemplative mood.
William Stukeley (1687–1765)[†]

*(Chapter Opener) *Notes on the State of Virginia*, Query 17, "Religion" (Prichard and Hall, Philadelphia, 1788).

[†]*Memoirs of Sir Isaac Newton's Life* (1752), pages 19–20. Recent edition published by Taylor and Francis, London (1936).

Sir Isaac Newton is pictured on this British £1 note.

Newton realized that both a falling apple and the distant orbiting Moon accelerate toward the Earth. The force causing these accelerations he called the **gravitational force**. The ratio of the magnitudes of the accelerations of the apple and the Moon gave a clue about how the gravitational force depended on distance. Consideration of Newton's second and third laws of motion, coupled with experiments on falling objects, provided information about the mass dependence of the force. Here we examine these arguments.

Dependence on Distance

For simplicity, assume the Moon moves in uniform circular motion about the Earth, as shown in Figure 6.1.[‡] The Moon experiences a centripetal acceleration of magnitude

$$a_c = \frac{v^2}{r}$$

where v is the speed of the Moon in its orbit and r is the orbital radius, measured from the center of the Earth to the center of the Moon.

The assumption that the Moon is in uniform circular motion means the instantaneous speed v of the Moon at any instant is equal to the average speed during any interval, found by dividing the distance traversed by the corresponding time interval. Since the Moon makes one circumference of its orbit ($2\pi r$) in a time equal to its orbital **period** T, the average speed of the Moon in its orbit is

$$v = \frac{2\pi r}{T}$$

[‡] A circle is a reasonable first approximation; a second approximation is an ellipse. The *actual* orbital path of the Moon is quite complicated because of the gravitational effects of the Sun and the major planets on the Moon. The effects of these other masses on the motion of the Moon was a problem that worried even Newton and "made his head ache and kept him awake so often that he would think of it no more" [attributed to Edmund Halley; see Roger R. Bate, Donald D. Mueller, and Jerry E. White, *Fundamentals of Astrophysics* (Dover, New York, 1971), page 322].

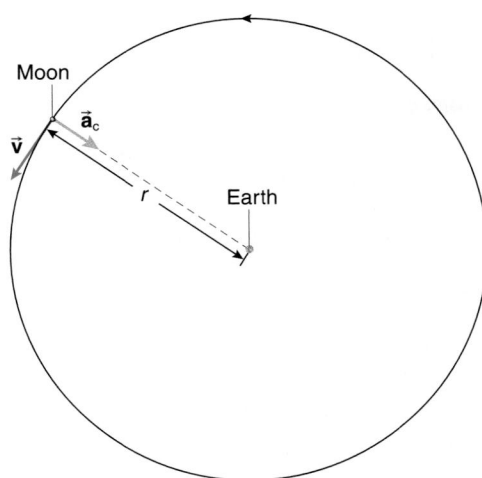

FIGURE 6.1 The Moon in a circular orbit experiences a centripetal acceleration.

Thus the magnitude of the centripetal acceleration of the Moon is

$$a_c = \frac{\left(\dfrac{2\pi r}{T}\right)^2}{r}$$

$$= \frac{4\pi^2 r}{T^2} \qquad (6.1)$$

Both r and T are determined from astronomical observations and long-distance surveying techniques. The radius of the lunar orbit is 3.84×10^8 m and the orbital period of revolution is 27.3 days (= 2.36×10^6 s). Upon substitution of these numerical values into Equation 6.1, the magnitude of the centripetal acceleration of the Moon is found to be

$$a_c = 2.72 \times 10^{-3} \text{ m/s}^2$$

or just less than 3 mm/s².

The magnitude of the acceleration due to gravity near the surface of the Earth is found, experimentally, to be about 9.81 m/s². The ratio of the magnitude of the acceleration due to gravity on the surface of the Earth to its magnitude at the distance of the lunar orbit is

$$\frac{a_{\text{near the surface of Earth}}}{a_{\text{at the distance of the Moon}}} = \frac{9.81 \text{ m/s}^2}{2.72 \times 10^{-3} \text{ m/s}^2}$$

$$\approx 3.60 \times 10^3$$

Hence the magnitude of the acceleration due to gravity at the distance of the Moon is about 1/3600 that on the surface of the Earth. Newton realized that the distance from the center of the Earth to the lunar orbit is about 60 times the radius of the Earth:

$$\frac{\text{radius of lunar orbit}}{\text{radius of the Earth}} = \frac{3.84 \times 10^8 \text{ m}}{6.37 \times 10^6 \text{ m}}$$

$$\approx 60$$

Note that $(60)^2 = 3600$.

Newton concluded that since the *acceleration* caused by the gravitational force, acting alone, evidently decreased in magnitude proportionally to the inverse square of the distance, the magnitude of the gravitational force must do likewise because, according to Newton's second law of motion, the force (acting alone) is proportional to the acceleration. That is,

$$F_{\text{grav}} \propto \frac{1}{r^2}$$

where r is the center-to-center distance between the Earth and the Moon.

Dependence on Mass

Since the only common physical property of the Earth and Moon is *mass*, it was reasonable to suppose that their masses also play a role in the magnitude of the gravitational force they exert on each other. Let M be the mass of the Earth and m that of the Moon. Newton's third law of motion states that if the Earth (as one system) exerts a force $\vec{F}_{\text{M on } m}$ on the Moon (another system), then the Moon exerts a force $\vec{F}_{m \text{ on M}}$ on the Earth of the *same magnitude*, but in the opposite direction:

$$\vec{F}_{\text{M on } m} = -\vec{F}_{m \text{ on M}} \qquad (6.2)$$

Thus if the magnitude of the gravitational force depends on mass, it must involve the masses of *both* the Earth and Moon in a *symmetric* way, so that if the masses are interchanged in the gravitational force law, the magnitude of the force on each is unchanged. The force could depend on *either* the *sum* of the masses (M + m), *or* the *product* (Mm), or perhaps some power n of the sum or product:

$$(M + m)^n \qquad \text{or} \qquad (Mm)^n$$

since these forms are the same if the masses are interchanged in their algebraic position in the expression [to $(m + M)^n$ or $(mM)^n$]. Therefore the possibilities for the mass dependence of the magnitude of the gravitational force are

$$F_{\text{grav}} \propto (M + m)^n \qquad \text{or} \qquad F_{\text{grav}} \propto (Mm)^n$$

By appealing to the *second law of motion* and experiment, Newton ruled out all possibilities except the first power of the product of the masses. Why is that? The argument is the following: From the experiments of Galileo, experiments easily repeated today, it is known that the *acceleration* of *any* mass m produced by the *sole* action of the gravitational force of another given mass M (say, the mass of the Earth) is *independent* of the mass m. Rocks, socks, and blocks fall with the same acceleration during free-fall motion (in a vacuum). We can use this observation to determine the mass dependence of the gravitational force.

Assume the magnitude of the gravitational force of each mass on the other depends on, say, the *sum* of the masses to some power n, with a proportionality constant K. Then, when the gravitational force is the only force on m, the second law of motion applied to system m,

$$\vec{F}_{grav} = m\vec{a}_{grav}$$

yields, using the magnitudes of the vectors,

$$K(M + m)^n = ma_{grav}$$

Hence the magnitude of the acceleration of m is

$$a_{grav} = \frac{K(M + m)^n}{m}$$

This magnitude of acceleration depends on m no matter what the exponent n happens to be,* which contradicts experiment. Therefore the magnitude of the gravitational force cannot depend on the *sum* of the masses or a power of the sum.

On the other hand, assume the magnitude of the gravitational force of M on m depends on the *product* of the masses raised to a power n, with a proportionality constant K'. When the gravitational force is the only force on m, the second law of motion applied to m gives (again, using magnitudes)

$$F_{grav} = ma_{grav}$$
$$K'(Mm)^n = ma_{grav}$$

Solving for the magnitude of the acceleration of m,

$$a_{grav} = K'\frac{(Mm)^n}{m}$$
$$= K'M^n m^{n-1}$$

This magnitude of acceleration is independent of m *if and only if* $n = 1$. *Only* the simple product Mm means that the gravitational acceleration of m caused by mass M is independent of m, confirming experimental observations.

> Therefore the gravitational force *must* depend on the simple product of the masses.

Now combine the distance and mass dependence into a single relation: the magnitude of the gravitational force must depend directly on the product of the masses and inversely on the square of their separation:

$$F_{grav} \propto \frac{Mm}{r^2}$$

Rewriting this relation with a constant of proportionality, G, yields

$$\boxed{F_{grav} = \frac{GMm}{r^2}} \tag{6.3}$$

---

*Of course if the exponent $n = 0$, then $(M + m)^n = 1$; but then the gravitational force would have no mass dependence at all.

The constant of proportionality G is the **universal constant of gravitation**. Equation 6.3 is called **Newton's law of universal gravitation**.

> The gravitational force law is called an **inverse square law** because the exponent of r in the denominator is 2.

The inverse square nature of the force has remarkable consequences (see Section 6.15). Nature was kind to create the gravitational force as an inverse square law; gravitational physics would be very different and difficult with *any* other exponent for the power of r.

> **PROBLEM-SOLVING TACTIC**
>
> **6.1** Watch your m & M's. Even though they have the same magnitude, be sure you understand which force is needed: the force of M on m, or the force of m on M. Use Equation 6.3 to find the magnitude of the gravitational force on each of the two pointlike systems of mass m and M. To write the forces as vectors, you will need to introduce a coordinate system.

6.2 NEWTON'S LAW OF UNIVERSAL GRAVITATION

Newton's law of universal gravitation describes the remarkable observation that any two pointlike particles in the universe exert a mutually attractive force on each other that is proportional to the product of their masses and inversely proportional to the square of the distance of their separation.

The *magnitude* of the gravitational force *on each particle* is given by Equation 6.3 as shown in Figure 6.2. The mutually attractive forces are directed along the line joining the particles and represent a Newton's third law force pair; Figure 6.2 is a third law force diagram.

The universal constant of gravitation G is a fundamental constant of nature whose value is determined in the laboratory (with difficulty) by directly measuring the slight force between two known masses separated by a known distance. The value of G, in SI units, is approximately

$$G \approx 6.67 \times 10^{-11} \ \text{N} \cdot \text{m}^2/\text{kg}^2$$

Newton mentions that even after he deduced Equation 6.3, he still did not know what gravity actually *is*; he could describe only what it *does*. Gravity itself was still just as much of a mystery after Newton as it was before him. *Why* does

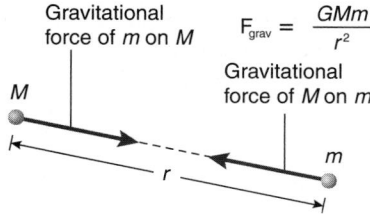

FIGURE 6.2 **The gravitational force of two pointlike masses on each other.**

gravity act this way and not some other way? *How* is the force transmitted between the masses across, say, the vacuum of space? Gravity itself (along with the other fundamental forces of nature) is an abstract, almost unfathomable concept that greatly intrigued Einstein and led him to develop his general theory of relativity.

Several aspects of the gravitational force need to be noted.

> Equation 6.3 yields the *magnitude* of the gravitational force of two *pointlike* masses on each other.

Newton worried about what happens if the objects have appreciable physical size; we will, too, in Sections 6.3 and 6.4. In fact, as long as the sizes of the particles are much less than their separation, Equation 6.3 describes the interaction quite well; nevertheless, as we will see, it is an approximation unless the masses are spherical in shape and have a spherically symmetric mass distribution.

> Equation 6.3 is not an independent law of motion (like Newton's three laws of motion); rather it is an analytical expression, based on reason and experiment, for the magnitude of a specific force: the gravitational force of each particle on the other.

> The gravitational force always appears in nature as an *attractive force*, directed along the line joining the two particles, as indicated in Figure 6.2.

While there are particles of *zero* mass, there are *no* particles with negative masses in nature.

QUESTION 1

If negative masses existed, what does Newton's second law of motion imply about their dynamics (the direction of the total force on a particle relative to the direction of its consequent acceleration)?

Looking ahead, the gravitational force contrasts strikingly with the electrical force (Chapter 16). Electrical forces are observed to be *either* attractive *or* repulsive. Therefore there must be two distinct types of electric charge, called positive charge and negative charge, names invented by Benjamin Franklin. The same kind of charges (like charges) *repel* each other; different kinds of charges (unlike charges) attract each other. Since there is only one kind of mass, the gravitational force must always be either attractive *or* repulsive; experiments indicate it is attractive, quite different from electricity. "Like" masses attract; there is no possibility for "unlike" masses in gravitation.

> The masses that appear in Equation 6.3, the law of universal gravitation, are called, formally, **gravitational masses**.

The mass m that appears on the right-hand side of Newton's second law of motion (Equation 5.1),

$$\vec{F}_{total} = m\vec{a}$$

is the *inertial mass* of a system, as we mentioned in Section 5.5. Newton was concerned about whether the gravitational mass and the inertial mass of a system were, in fact, the same or not. There is no *a priori* reason to think they are the same, since the second law of motion is a different type of statement than Equation 6.3 for the magnitude of the gravitational force. The second law indicates how a particle (via its inertial mass) responds when *any* total force whatsoever acts on the particle. The mass in the gravitational force law of Equation 6.3 provides a way to calculate the magnitude of a *particular* force: the gravitational force.

What do experiments indicate is the relationship between the inertial and gravitational masses? Newton's second law of motion states that the magnitude of the total force on a system and the magnitude of its acceleration are related by

$$F_{total} = m_{inertial}a$$

where we explicitly write a subscript indicating that the right-hand side of the second law involves the inertial mass of the system. If, say, the gravitational force of the Earth (with gravitational mass M_{grav}) is the only force acting on the system, then the left-hand side of the second law is the gravitational force, and the second law becomes

$$\frac{GM_{grav}\,m_{grav}}{r^2} = m_{inertial}\,a$$

where m_{grav} is the gravitational mass of the system whose inertial mass is $m_{inertial}$. So the acceleration of the system has a magnitude

$$a = \frac{GM_{grav}}{r^2}\,\frac{m_{grav}}{m_{inertial}} \tag{6.4}$$

The acceleration depends on the ratio of the gravitational and inertial masses. Experiments with the greatest precision indicate that *all* freely falling masses (in a vacuum) have the *same* acceleration, the so-called acceleration due to gravity, if the gravitational force is the only force acting on them. In a vacuum, feathers fall with the same gravitational acceleration as rocks. This observation implies that the ratio between the gravitational and inertial masses in Equation 6.4 must be equal to 1. The latest experiments indicate that the gravitational and inertial masses of the system are, in fact, equal to each other to better than 1 part in 10^{12}:

$$\frac{m_{inertial}}{m_{grav}} = 1.000\,000\,000\,000\,?$$

Equation 6.4 then indicates that the acceleration due to the gravitational force of the Earth alone is independent of the mass of the system (but depends on the mass of the Earth), in conformity with experiment. About 1915, Einstein elevated the experimental equality of the gravitational and inertial masses of a system to the status of a postulate called the **principle of equivalence**:

> The inertial and gravitational masses of a system are *exactly equal*:
> $$m_{grav} \equiv m_{inertial} \tag{6.5}$$

The universal gravitational constant G is the first of a number of fundamental physical constants of nature that we will encounter. Other such fundamental constants likely are familiar to you, among them the speed of light and the charge on the electron. The numerical value of G depends of course on the particular unit system used to measure the forces, the masses, and the distances in Equation 6.3. This is also true for other fundamental physical constants. For example, the speed of light is $2.997\ 924\ 58 \times 10^8$ m/s (exactly) in SI units, or about 186 000 miles per second in peculiarly American units, or more quaintly 1.80×10^{12} furlongs per fortnight; but the plethora of different unit systems is of our doing and does not make the fundamental constants any less fundamental.

Among the fundamental constants of physics, the gravitational constant G is known with the least precision (fewest significant figures) because of the weakness of the gravitational interaction. The smallness of G reflects the experimental observation that the magnitudes of the gravitational forces of common everyday objects on each other are extremely small; see Example 6.1. Therefore experiments to determine the value of G by directly measuring the gravitational force between two laboratory masses are notoriously difficult. There is precious little gravitational attraction between you and this text (perhaps in more ways than one), because the masses involved are so small; the mutual gravitational force of attraction on you and the book is of negligible magnitude compared with other forces on you and the text. Your *weight* (the gravitational force of the Earth on you) is substantial only because the mass of the Earth is huge.

Experiments to determine the numerical value of G by directly measuring the gravitational force of two laboratory masses on each other first were undertaken by Sir Henry Cavendish (1731–1810) in 1798, over a century after Newton. A modern-day version of the experimental apparatus, known as a Cavendish balance, is shown in Figure 6.3 along with a brief description of the procedure involved in the measurement.

There have been recent theoretical speculations about whether the fundamental constants of nature really are constant or change as the universe itself ages and evolves. The constant G has been of particular interest in these speculations, since the gravitational force dominates the large-scale dynamical aspects of the universe itself.* The fundamental philosophical question here is whether the laws of physics themselves might evolve or change along with inexorable changes in the universe as a whole. There is as yet *no* definitive experimental evidence to confirm these speculations about variations in G or in other fundamental constants. It seems that the physics "now" was the physics "then" and will be the physics in the future. Thank goodness. Just imagine how wacky the world and the universe would be if science changed with time (or were subject to change by Congress or your state legislature)! We like to seek out the unchanging statements and rules about the natural world. We are quite confident that the laws of physics, whatever they may be, are now and will be as they were in the beginning.

> Experiments indicate that the gravitational force follows a **principle of linear superposition**.

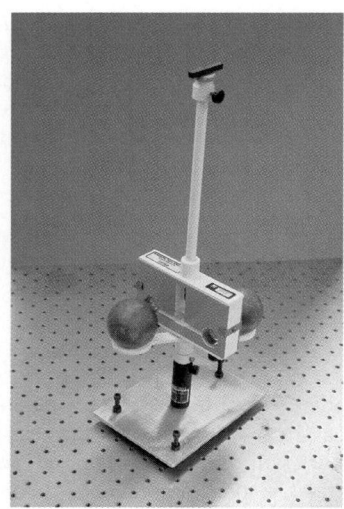

FIGURE 6.3 The Cavendish balance used for the experimental determination of the numerical value of the universal gravitational constant G. The apparatus consists of a dumbbell arrangement of two small lead spheres, each of mass *m*, suspended from its center by a thin fiber so the dumbbell is free to turn in the horizontal plane. The arrangement is allowed to hang until it is at rest. Two other larger lead spheres, each of mass M, then are mounted near the small spheres and attract them, causing the dumbbell to rotate. Careful measurements of the rotation provide a way of directly calculating the force in newtons of each sphere on the other. In Section 6.3 we will see that a uniform spherical mass acts, for purposes of calculating its gravitational force on other masses located outside the sphere itself, as if it were a point mass located at the center of the sphere. By measuring the force, the masses of the spheres, and the separation between their centers, the only unknown quantity in Equation 6.3 is the universal gravitational constant G. The numerical value of G may thus be determined. There is a more sophisticated version of the experiment that involves switching the locations of the large spheres and then measuring the oscillation period of the dumbbell and its new equilibrium position. Although the procedure is straightforward, the experiment is not easy to perform; small vibrations in the laboratory induced by people walking up nearby stairs, motor vehicles lumbering by, and other nearby disturbances affect the apparatus and make data collection challenging.

The gravitational forces of two particles on each other are independent of forces arising from the presence of other masses.[†] The principle means that if, say, three particles are considered, the total gravitational force on one of them is the vector sum of the individual gravitational forces on that particle by the others, *taken one at a time*. In other words, the total gravitational force on, say, a particle of mass *m* is the vector sum of the gravitational force of particle M on *m* and the gravitational force of particle M′ on *m*, as shown in Figure 6.4. That is,

$$\vec{F}_{\text{total gravitational force on } m} = \vec{F}_{\text{gravitational force of M on } m}$$
$$+ \vec{F}_{\text{gravitational force of M}' \text{ on } m}$$

Of course, the same applies to the total force on any one of the other particles: they are calculated by vector summing all the

---

*Edward R. Harrison, *Cosmology* (Cambridge University Press, Cambridge, England, 1981), pages 323–324, 336–337.

[†] The statement is true as long as the presence of other masses does not affect the shape of either of the given two masses.

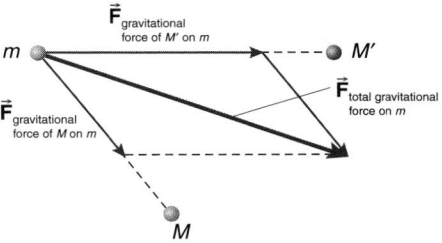

FIGURE 6.4 The principle of linear superposition.

individual gravitational forces on it arising from the *other* particles present. Example 6.2 illustrates the procedure.

QUESTION 2

In T. S. Eliot's *Old Possum's Book of Practical Cats*, there is the couplet:

Macavity, Macavity, there's no one like Macavity,
He's broken every human law, he breaks the law of gravity
"Macavity, The Mystery Cat," lines 5–6

What is the difference between a human law and a law of nature? Is the phenomenon of gravitation a human or natural law? Is Equation 6.3 a human law, since it was deduced by Newton? Explain why Macavity's last feat is science fiction.

EXAMPLE 6.1

A 6.00 kg mass is located 0.500 m from a 3.00 kg mass, as shown in Figure 6.5. Assume they are both pointlike masses.

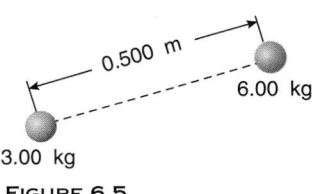

FIGURE 6.5

a. Find the gravitational force of the 6.00 kg mass on the 3.00 kg mass.
b. What is the gravitational force of the 3.00 kg mass on the 6.00 kg mass?

Solution
a. The *magnitude* of the gravitational force that each exerts on the other is found using Equation 6.3:

$$F = \frac{GMm}{r^2}$$

$$= \frac{(6.67 \times 10^{-11} \text{ N} \cdot \text{m}^2/\text{kg}^2)(6.00 \text{ kg})(3.00 \text{ kg})}{(0.500 \text{ m})^2}$$

$$= 4.80 \times 10^{-9} \text{ N}$$

Note the very small magnitude for the force: only a few nanonewtons.

The force of the 6.00 kg mass on the 3.00 kg mass is directed toward the 6.00 kg mass. Use the coordinate system indicated in Figure 6.6:

$$\vec{F}_{\text{on 3 kg mass}} = (4.80 \times 10^{-9} \text{ N})\hat{\imath}$$

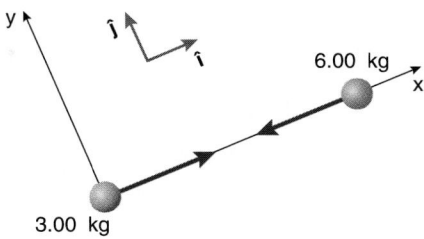

FIGURE 6.6

b. The gravitational force of the 3.00 kg mass on the 6.00 kg mass has the same magnitude but is directed toward the 3.00 kg mass, as shown in Figure 6.6. The two forces form a Newton's third law pair of forces: they have the same magnitude, are in opposite directions, and act on different systems.

$$\vec{F}_{\text{on 6 kg mass}} = (4.80 \times 10^{-9} \text{ N})(-\hat{\imath})$$
$$= (-4.80 \times 10^{-9} \text{ N})\hat{\imath}$$

 STRATEGIC EXAMPLE 6.2

Three masses are arranged as shown in Figure 6.7. Assume the masses are pointlike masses.

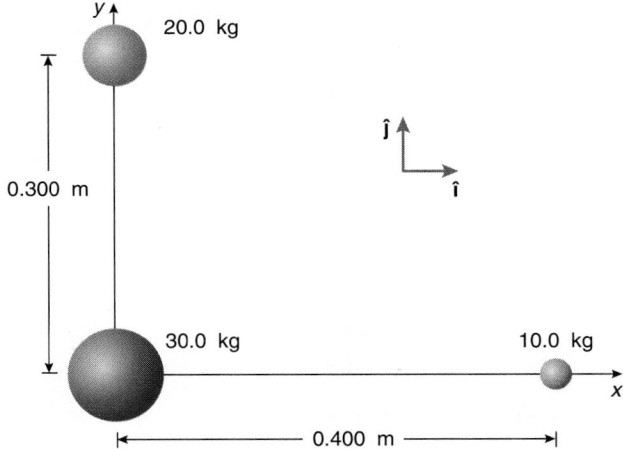

FIGURE 6.7

a. Find the gravitational force of the 30.0 kg mass on the 10.0 kg mass.
b. Find the gravitational force of the 20.0 kg mass on the 10.0 kg mass.
c. What is the *total* gravitational force on the 10.0 kg mass caused by the two other masses?
d. Find the magnitude of the total gravitational force on the 10.0 kg mass.
e. Find the angle that the total gravitational force on the 10.0 kg mass makes with î in Figure 6.7.

Solution

a. Let the 10.0 kg mass be the system. We consider the effect of the 30.0 kg mass on the 10.0 kg mass as if the 20.0 kg mass were not present, as shown in Figure 6.8. The *magnitude* of the gravitational force of the 30.0 kg mass on the 10.0 kg mass is found from Equation 6.3:

$$F_1 = \frac{GMm}{r^2}$$
$$= \frac{(6.67 \times 10^{-11} \ \text{N} \cdot \text{m}^2/\text{kg}^2) \,(30.0 \ \text{kg}) \,(10.0 \ \text{kg})}{(0.400 \ \text{m})^2}$$
$$= 1.25 \times 10^{-7} \ \text{N}$$

Use the coordinate system indicated in Figure 6.8. The gravitational force $\vec{\mathbf{F}}_1$ then is

$$\vec{\mathbf{F}}_1 = (-1.25 \times 10^{-7} \ \text{N})\hat{\imath}$$

b. Now consider the effect of the 20.0 kg mass on the 10.0 kg mass as if the 30.0 kg mass were not present, as in Figure 6.9. Use the Pythagorean theorem to find the distance between the 20.0 kg and 10.0 kg masses:

$$[(0.300 \ \text{m})^2 + (0.400 \ \text{m})^2]^{1/2} = 0.500 \ \text{m}$$

The *magnitude* of the gravitational force of the 20.0 kg mass on the 10.0 kg mass is found from Equation 6.3:

$$F_2 = \frac{(6.67 \times 10^{-11} \ \text{N} \cdot \text{m}^2/\text{kg}^2) \,(20.0 \ \text{kg}) \,(10.0 \ \text{kg})}{(0.500 \ \text{m})^2}$$
$$= 5.34 \times 10^{-8} \ \text{N}$$

Use the coordinate system indicated in Figure 6.9 to express this force in Cartesian form:

$$\vec{\mathbf{F}}_2 = (5.34 \times 10^{-8} \ \text{N})[(\cos \theta)(-\hat{\imath}) + (\sin \theta)\hat{\jmath}]$$

The sine and cosine of θ are found from the geometry of the arrangement in Figures 6.8 and 6.9:

$$\cos \theta = 0.800 \qquad \text{and} \qquad \sin \theta = 0.600$$

Thus

$$\vec{\mathbf{F}}_2 = (5.34 \times 10^{-8} \ \text{N})(-0.800\hat{\imath} + 0.600\hat{\jmath})$$
$$= (-4.27 \times 10^{-8} \ \text{N})\hat{\imath} + (3.20 \times 10^{-8} \ \text{N})\hat{\jmath}$$

c. The *total* gravitational force on the 10.0 kg mass caused by the other two masses together is the *vector sum* of $\vec{\mathbf{F}}_1$ and $\vec{\mathbf{F}}_2$ according to the principle of superposition (see Figure 6.10).

$$\vec{\mathbf{F}}_{\text{total}} = \vec{\mathbf{F}}_1 + \vec{\mathbf{F}}_2 = (-1.25 \times 10^{-7} \ \text{N})\hat{\imath}$$
$$+ [(-4.27 \times 10^{-8} \ \text{N})\hat{\imath} + (3.20 \times 10^{-8} \ \text{N})\hat{\jmath}]$$
$$= (-1.68 \times 10^{-7} \ \text{N})\hat{\imath} + (3.20 \times 10^{-8} \ \text{N})\hat{\jmath}$$

d. The magnitude of the total gravitational force on the 10.0 kg mass is the magnitude of $\vec{\mathbf{F}}_{\text{total}}$:

$$F_{\text{total}} = [(-1.68 \times 10^{-7} \ \text{N})^2 + (3.20 \times 10^{-8} \ \text{N})^2]^{1/2}$$
$$= 1.71 \times 10^{-7} \ \text{N}$$

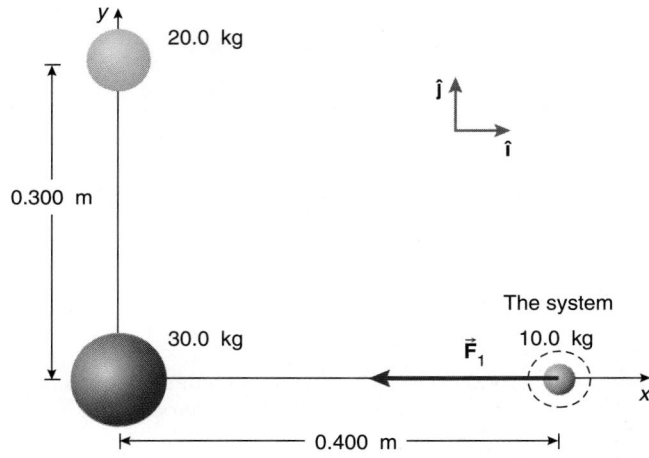

FIGURE 6.8

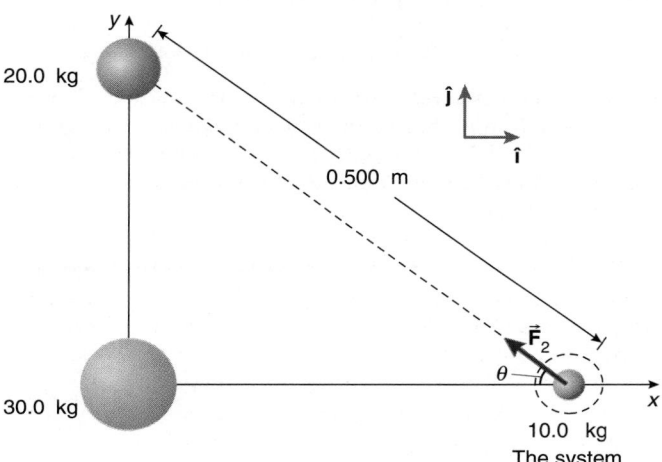

FIGURE 6.9

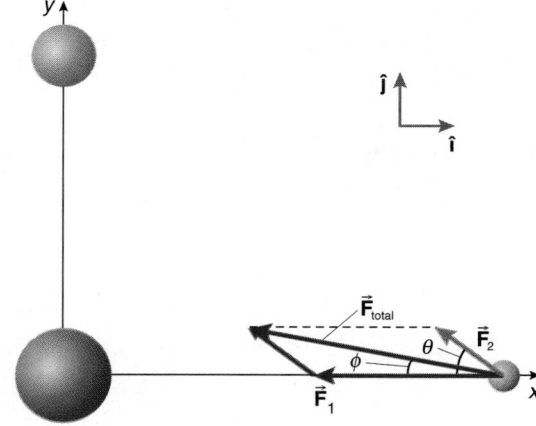

FIGURE 6.10

e. The angle ϕ in Figure 6.10 is found from the components of F_{total}:

$$\tan \phi = \left| \frac{F_{y\,total}}{F_{x\,total}} \right|$$

$$= \frac{3.20 \times 10^{-8} \text{ N}}{1.68 \times 10^{-7} \text{ N}}$$

$$= 0.190$$

So

$$\phi = 10.8°$$

The angle the total force makes with $\hat{\imath}$ is, therefore,

$$180° - \phi = 169.2°$$

6.3 GRAVITATIONAL FORCE OF A UNIFORM SPHERICAL SHELL ON A PARTICLE

Newton's law of universal gravitation describes the magnitude and direction of the force of two pointlike particles on each other. Newton was concerned about the gravitational effects of extended objects, because he was interested particularly in the gravitational influences of the Earth, Moon, Sun, and the planets, all objects of considerable size. Many of these celestial objects are roughly spherical in shape, and so it is of some interest to consider the gravitational effects of spherically symmetric systems. It seemed reasonable to think that a spherical mass of constant density, a so-called *uniform sphere*, could be treated as if it were a particle located at the center of the sphere. In fact, this is the case but it is not easy to show rigorously.

The case for a sphere is best considered by first examining the gravitational force of a uniform, thin, spherical shell of mass M on a pointlike particle of mass m. The results then can be extended to solid spheres.

Appendix A provides detailed proofs of the following two important gravitational shell theorems:

If a pointlike particle of mass m is located anywhere within a thin, uniform, hollow spherical shell of mass M, the total gravitational force of the shell on m is equal to zero (see Figure 6.11).

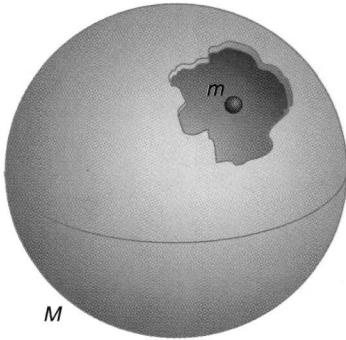

FIGURE 6.11 The gravitational force of a uniform spherical shell on a mass m inside it is zero.

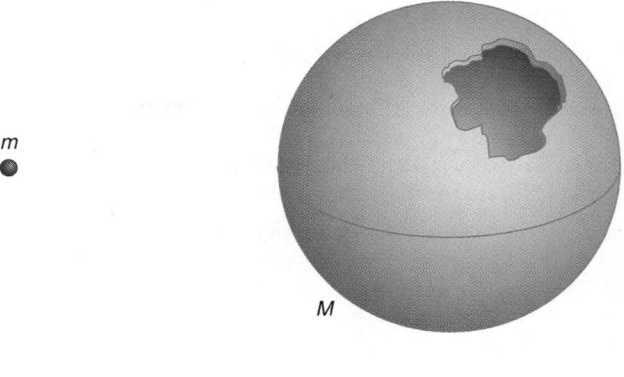

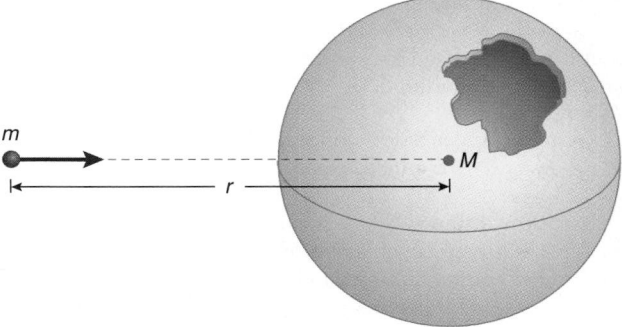

FIGURE 6.12 A mass m outside a uniform spherical shell experiences a gravitational force as if the mass of the shell were located at its center.

QUESTION 3
What is the gravitational force of m on the shell?

If a pointlike particle of mass m is located anywhere outside a thin, uniform, hollow spherical shell of mass M, the total gravitational force of the shell on m is the same as that of a pointlike mass M located at the center of the shell, as shown in Figure 6.12.

In order to prove the shell theorems, Newton had to invent the calculus. Ironically, he kept the results of his work to himself for years before Edmund Halley finally persuaded him to publish them.*

6.4 GRAVITATIONAL FORCE OF A UNIFORM SPHERE ON A PARTICLE

The gravitational shell theorems are used to determine the gravitational force of a uniform sphere of mass M on a pointlike particle of mass m. We consider two situations: (1) a point mass m located outside the sphere, and (2) a point mass m located within the volume enclosed by the sphere itself.

---

*Edmund Halley (1656–1743) was the first person to realize that comets orbited the Sun and periodically returned. He predicted that what we now call Comet Halley would return in 1758, but did not live to see his prediction vindicated. Comet Halley orbits the Sun in a highly elliptical orbit with a period of about 76 years and will next appear about the year A.D. 2061.

A Point Mass *m* Located Outside the Sphere

If the point mass *m* is located *outside* a uniform sphere of radius *R* and total mass M, divide the uniform sphere into a nested network of concentric, thin, uniform spherical shells, much like an onion, as shown in Figure 6.13. The magnitude of the gravitational force of any shell on *m* is calculated as if the mass of the shell were located at the center of the sphere.

Therefore, for purposes of calculating the magnitude of the gravitational force of the sphere on *m*, the *entire* mass of the sphere acts as if it were concentrated at the sphere's center, as shown in Figure 6.14.

This result is wonderful!

Thus the magnitude of the gravitational force of the sphere on *m* is just the magnitude of the force between two pointlike particles as in Equation 6.3:

$$F = \frac{GmM}{r^2} \quad \begin{array}{l}(m \text{ outside a uniform sphere } or \text{ a spherically} \\ \text{symmetric mass distribution})\end{array} \quad (6.6)$$

where *r* is the distance between *m* and the center of the sphere. The force of the sphere on *m* is directed toward the center of the sphere, since the gravitational force always is an attractive force.

You can use similar arguments for *nonuniform spheres* with a mass distribution that is *spherically symmetric*. The term spherically symmetric means that the density of the sphere depends *only* on the radial coordinate *r*, not on any angles such as the spherical coordinates θ and ϕ. We imagine breaking up the spherically symmetric mass distribution into thin shells, each of which is uniform; again the mass of each shell, whatever it is, acts as if its mass were concentrated at the center of the sphere. So the effect once again is just to have the entire mass of the sphere acting at the center. The Earth is *not* a uniform sphere; but the distribution of mass within the Earth is, to a good first approximation, spherically symmetric. Thus, to calculate the gravitational force of the Earth on particles outside the Earth, the entire mass of the Earth is imagined to be located at the center of the Earth. Other planets and stars also are spherical to a first approximation.*

---

*Jupiter and Saturn are the most oblate planets; the difference between their polar and equatorial diameters is about 10%. For the Earth, the difference is less than 1%. The Sun has no measurable difference between its polar and equatorial diameters.

Like the Earth, they also are not uniform spheres; but their mass distribution is approximately spherically symmetric.

Thus the entire mass of a spherically symmetric mass distribution is imagined to be located at the center of the sphere for purposes of calculating the gravitational influence of the sphere on exterior particles such as planets, moons, and spacecraft.

Of course, according to Newton's third law of motion, the mass *m* exerts a force of the same magnitude on the sphere in the direction opposite to that of the force of the sphere on *m*; the two forces are a third law pair of forces.

A Point Mass *m* Located Within the Volume Enclosed by the Sphere Itself

Let the point mass *m* lie at some distance *r* from the center but *inside* a sphere of total mass M with a spherically symmetric mass distribution, as shown in Figure 6.15. Imagine the portion of the mass of the sphere that lies more distant from its center than *m* to be composed of a nested network of concentric uniform spherical shells, as indicated in Figure 6.15. Each individual shell exerts *zero* total gravitational force on *m*, since *m* is interior to each of these overlying shells.

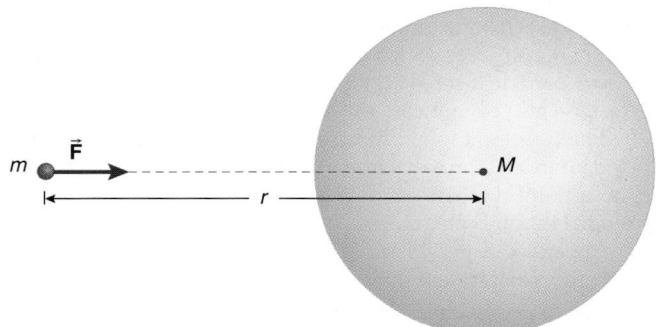

FIGURE 6.14 The entire sphere can be imagined as a pointlike particle of mass M located at the center of the sphere.

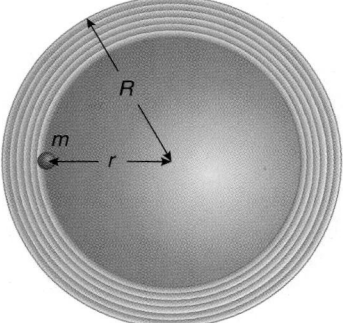

FIGURE 6.15 Mass *m* inside a solid sphere; the overlying mass has no gravitational effect on *m*.

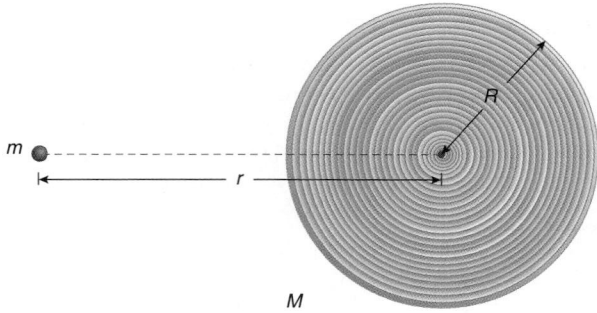

FIGURE 6.13 Divide the solid sphere into a concentric nest of shells.

Thus the overlying portion of the sphere contributes *zero* total gravitational force on *m*.

Imagine the portion of the mass of the sphere that lies closer to the center than *m* also as a nested network of uniform spherical shells, as shown in Figure 6.16. In this case, *m* is outside the shells, and so the mass of these shells can be considered to be located at the center of the sphere for purposes of calculating the gravitational force on *m*.

Thus the magnitude of the total gravitational force of the entire sphere on *m* is due *only* to that portion of the total mass of the sphere that lies closer to the center than *m* itself.

If the sphere is uniform, the fraction of M that lies within *r* of the center is the ratio of the volume of this smaller sphere to the volume of the entire sphere:

$$\frac{\frac{4}{3}\pi r^3}{\frac{4}{3}\pi R^3} = \frac{r^3}{R^3}$$

So the magnitude of the total gravitational force of the sphere on *m* is the gravitational attraction between *m* and a mass $(r^3/R^3)M$ located at the center of the sphere. From Equation 6.3, this is just

$$F = \frac{Gm\left(\dfrac{r^3}{R^3}M\right)}{r^2}$$

$$= \frac{GmM}{R^3}r \qquad (m \text{ within a uniform sphere } only) \qquad (6.7)$$

Thus the magnitude of the gravitational force on *m* is directly proportional to the distance of *m* from the center of the sphere.

The magnitude of the gravitational force exerted by a uniform spherical mass on a particle *m* is plotted as a function of the radial distance of *m* from the center of the sphere in Figure 6.17. Notice that the magnitude of the force on *m* increases linearly with the distance *r* until *m* is at the surface of the sphere, at which point the magnitude of the force reaches its maximum;

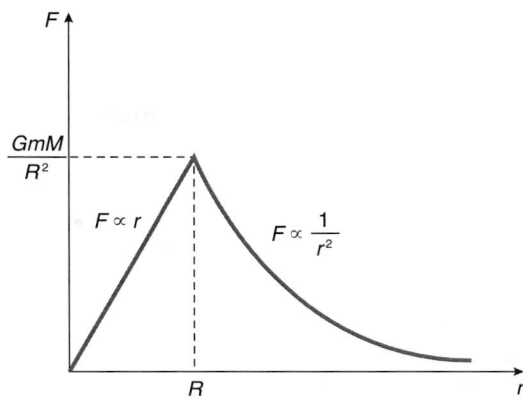

FIGURE 6.17 A schematic graph of the magnitude of the gravitational force on *m* versus its radial distance *r* from the center of a uniform sphere of radius R and mass M.

outside the sphere, the magnitude decreases as the inverse square of the distance: $1/r^2$.

If the spherical mass is *not* uniform, Equation 6.7 does not apply. However, if the mass distribution (i.e., the density) is *spherically symmetric*, one can still state that the gravitational force on *m* arises only from the material between *m* and the center of the sphere, but Equation 6.7 is inappropriate for such cases. As always, the equations of physics have strings attached to their use, and we must be aware of the limitations associated with every equation modeling physical phenomena.

QUESTION 4

Assume the Earth is a uniform sphere. Make qualitatively accurate graphs indicating how the mass and the magnitude of the weight of a rock vary as a function of its distance from the *center* of the Earth, from $r = 0$ m to $r \gg R_{\text{Earth}}$.

6.5 MEASURING THE MASS OF THE EARTH

It is hard to imagine how the mass of the Earth is determined, because the traditional technique for determining the mass of common objects by using a laboratory balance simply is inapplicable for the Earth itself. Here we see how the mass of the Earth and that of other astronomical bodies is found.

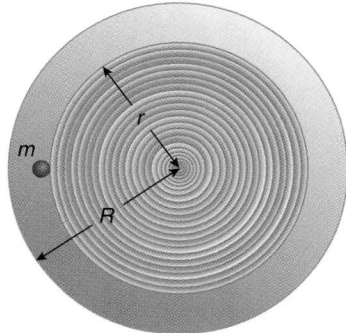

FIGURE 6.16 Only the mass of the sphere interior to *m* has a nonzero contribution to the gravitational force on *m*.

How to measure the mass of the Earth takes some thought.

The gravitational force on a particle of mass m was defined to be the weight of the particle. Model the Earth as a sphere, to a first approximation,* with a spherically symmetric mass distribution, so that we can use the results of the previous section to find the gravitational force of the Earth on m.

Let the mass m be outside the Earth but close to its surface. The distance r is the distance of m from the *center* of the Earth; r is essentially equal to the radius R of the Earth. Then the gravitational force of the spherical Earth on m is of *magnitude*

$$\frac{GmM}{R^2} \tag{6.8}$$

where M is the mass of the Earth. The force on m is directed, of course, toward the center of the Earth. Previously (in Chapter 5), when m is close to the surface of the Earth, we wrote the magnitude of the weight of a mass m as

$$w = mg \tag{6.9}$$

where g is the magnitude of the local acceleration due to gravity. Equations 6.8 and 6.9 are equivalent expressions for the *identical* force. The magnitudes must be the same:

$$mg = \frac{GmM}{R^2}$$

Hence

$$g = \frac{GM}{R^2} \tag{6.10}$$

The local acceleration due to gravity g is easily measured in the laboratory. A more precise experimental value of the universal gravitational constant is $G = 6.672 \times 10^{-11}$ N·m^2/kg^2 and the effective value of the radius of the Earth is 6.378×10^6 m. Therefore Equation 6.10 can be used to calculate the mass of the Earth. We find

$$M = \frac{gR^2}{G}$$
$$= \frac{(9.81 \text{ m/s}^2)\,(6.378 \times 10^6)^2}{6.672 \times 10^{-11} \text{ N·m}^2/\text{kg}^2}$$
$$= 5.98 \times 10^{24} \text{ kg}$$

This technique is quite versatile. If the magnitude a of the acceleration of a spacecraft is measured at a known distance r from the center of a planet or moon of mass M, then M is found using a generalization of the equation above:

$$M = \frac{ar^2}{G} \tag{6.11}$$

6.6 ARTICIAL SATELLITES OF THE EARTH

Stand on a high mountain and begin throwing rocks horizontally off the top. The rocks, of course, fall to the surface of the Earth.

*The shape is an oblate spheroid to a second approximation, since the polar and equatorial diameters of the Earth differ by 43 km. This distortion of its shape occurs because of the rotation of the Earth as we will see in Chapter 10.

Pitch the rocks with greater initial horizontal velocity and the rocks have a greater range before colliding with the surface of the Earth. Keep pitching the rocks with ever greater speed and the range keeps right on increasing. Neglect any frictional effects due to the air in the low atmosphere and any biological limitations on your throwing arm. Eventually, if the speed of the rock is sufficient, the projectile will circle the Earth, as shown in Figure 6.18.

The rock is falling but continuously *missing* the Earth! Figure 6.18 actually is from Newton's *Principia* and clearly indicates that Newton, in the late 17th century, anticipated the feasibility of artificial **satellites**[†] of the Earth.

[†]The word *satellite*, meaning a small secondary object (i.e., a "moon") in orbit about a larger body, first was used in this context by Johannes Kepler in 1611, referring to the four so-called "secondary planets" of Jupiter discovered by Galileo in 1610. Galileo called the Jovian moons the *sidera Medicaea*, or Medici stars, to curry favor with the Medici family that dominated Florence and Tuscany in Italy during the Renaissance. The four moons now are called the Galilean moons of Jupiter.

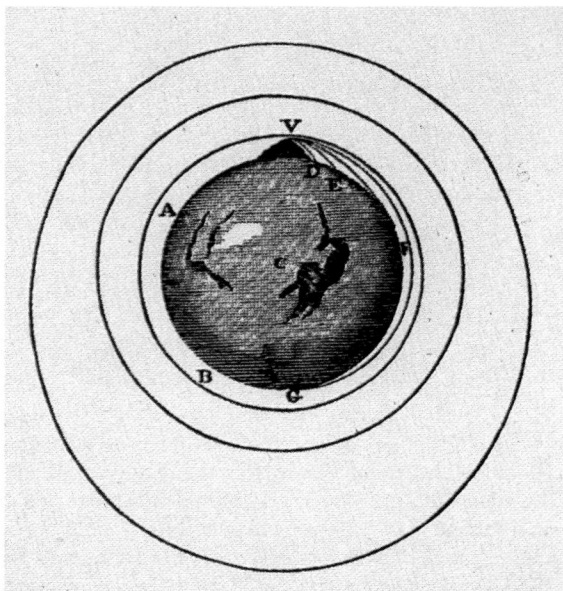

FIGURE 6.18 Launching a satellite of the Earth, as depicted by Newton in the *Principia*.

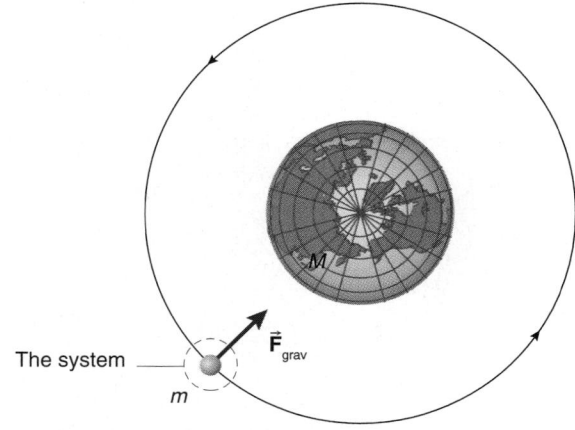

FIGURE 6.19 A satellite orbiting the Earth.

Consider a satellite in circular orbit about the Earth (as in Figure 6.19), well above the atmosphere, so that we can avoid the frictional effects of the air. The *only* force acting on the satellite system is the gravitational force of the Earth. The mass of the satellite typically is constant and the total force and acceleration are in the same direction (toward the center of the circular orbit). Newton's second law relates the total force to the acceleration; in terms of their magnitudes,

$$F_{\text{total}} = ma$$

Here F_{total} is the magnitude of the gravitational force of the Earth on the satellite, since this is the only force on it.

Since the satellite is in a circular orbit, the acceleration is the centripetal acceleration. The second law of motion thus becomes

$$\frac{GmM}{r^2} = m\frac{v^2}{r} \qquad (6.12)$$

The period T of the satellite is the time for one complete orbit. The average speed during any interval is equal to the constant instantaneous speed for uniform circular motion. So the speed of the satellite is the distance traveled, say the circumference of the circular orbit, divided by the time interval it takes to travel this distance, which is the period:

$$v = \frac{2\pi r}{T}$$

Substituting for v in Equation 6.12, we find

$$\frac{GmM}{r^2} = \frac{m\left(\dfrac{2\pi r}{T}\right)^2}{r}$$

Solving for T^2,

$$T^2 = \frac{4\pi^2}{GM}r^3 \qquad (6.13)$$

The square of the period of the satellite is proportional to the cube of the radius of its orbit. Equation 6.13 is a special simple case of **Kepler's third law of planetary motion**. Equation 6.13 is legitimate as long as the mass of the satellite is negligible compared with the mass of the planet, so that the acceleration of the mother planet is negligible when acted on by the third law counterpart to the gravitational force on the satellite. The approximation $m \ll M$ is excellent for any artificial satellite but is not necessarily a good approximation for natural satellites (the moons) of the planets. We will develop a more general statement of Kepler's third law of planetary motion in Section 6.11.

A few points to clarify:

1. One point is semantics. We defined the weight of a system of mass m to be the gravitational force on the system. A satellite certainly is subjected to the gravitational force of the Earth. Indeed, it is the gravitational force of the Earth on the satellite that gives the satellite its centripetal acceleration. The satellite is *not weightless*; indeed, it is the weight that keeps the satellite circling in its orbit! If the gravitational force somehow ceased its action on the satellite beginning at some instant, the satellite would proceed in a straight line tangent to the orbit. A satellite in orbit is in free-fall motion around the Earth, continually missing it. It is undergoing the same type of free-fall experienced when you jump off a low wall or a high cliff (though the latter is not recommended). If you do such a foolish thing while simultaneously standing on a scale (whose reading is a measure not of your weight but of the magnitude of the normal force of you on the scale), the scale will read zero because it too is in free-fall. You are normal-force-less, *not* weightless, when undergoing free-fall motion. Objects within a satellite (such as astronauts) experience normal-forcelessness because they are in free-fall motion around the Earth along with the satellite itself.

2. Any ordinary projectile, such as a baseball thrown from the surface of the Earth, actually is in orbit about the center of the Earth. The projectile is just *unable to complete the orbit* because the surface of the Earth gets in the way. You can have some fun with this idea: just wager to your friends that you can launch an Earth satellite with your bare hands. Guarantee them double if you lose the wager. Collect the bets. With the money safely in your pocket, simply toss a rock into the air. That's all there is to it; you have *launched* a satellite of the Earth. While the rock is in flight, it is a satellite of the Earth; there is no arguing with that fact. But... but... but, your friends may claim. But nothing. You never claimed that the satellite would *complete* an orbit; *that* is another matter! You simply claimed that you could *launch* it.

3. There is nothing sacred about the third rock from the Sun (the Earth!) in this argument about satellites. With the appropriate mass for M used in Equation 6.13, the arguments equally well apply to satellites or spacecraft in circular orbit about Jupiter, Mars, Venus, the Sun, or Superman's fictional home planet Krypton.

Astronauts in orbit are not weightless but normal-force-less.

What are *two* technical errors in the following journalistic description of a space shuttle mission in orbit about the Earth?

Their research, along with most of the nine other experiments on the shuttle, is designed to measure how various materials behave or are created in a weightless environment, outside Earth's gravitational pull.

Boston Globe (1 October 1988)

STRATEGIC EXAMPLE 6.3

a. What is the magnitude of the gravitational force of the Earth on an IRS surveillance satellite of mass 1040 kg that is in a grazing circular orbit 100 km above the *surface* of the Earth?
b. Find the magnitude of the acceleration of a satellite in such a grazing orbit. The order of magnitude of the result should not be unexpected. Explain why.
c. Determine the speed of a satellite in such an orbit.
d. Calculate the period of the satellite.

Solution
a. The magnitude of the gravitational force of the Earth on the satellite is found from Newton's law of universal gravitation, Equation 6.3:

$$F_{grav} = \frac{GMm}{r^2}$$

The distance r is the distance of the satellite from the *center* of the Earth, equal to 100 km *plus* the radius of the Earth. That is,

$$r = 0.100 \times 10^6 \text{ m} + 6.37 \times 10^6 \text{ m}$$
$$= 6.47 \times 10^6 \text{ m}$$

Let M be the mass of the Earth and m the mass of the satellite. Hence the magnitude of the force is

$$F_{grav} = \frac{(6.67 \times 10^{-11} \text{ N·m}^2/\text{kg}^2)(5.98 \times 10^{24} \text{ kg})(1040 \text{ kg})}{(6.47 \times 10^6 \text{ m})^2}$$

$$= 9.91 \times 10^3 \text{ N}$$

b. Determine the magnitude of the centripetal acceleration from Newton's second law; use the magnitudes of the vectors:

$$F_{total} = ma$$
$$9.91 \times 10^3 \text{ N} = (1040 \text{ kg})a$$

So

$$a = 9.53 \text{ m/s}^2$$

The satellite is in free-fall motion and the magnitude of the acceleration due to gravity 100 km above the surface of the Earth should be slightly less than the value at the surface, 9.81 m/s².
c. The speed of the satellite is found from the magnitude of its centripetal acceleration:

$$a_c = \frac{v^2}{r}$$

where r is the radius of its orbit.

$$9.53 \text{ m/s}^2 = \frac{v^2}{6.47 \times 10^6 \text{ m}}$$

Solving for v,

$$v = 7.85 \times 10^3 \text{ m/s}$$
$$= 7.85 \text{ km/s}$$

The satellite is really moving fast!
d. The period is the time for one complete orbit.

Method 1
The satellite traverses the circumference of its circular orbit ($2\pi r$) at the constant speed v, and so the time it takes is

$$T = \frac{2\pi r}{v}$$
$$= \frac{2\pi \times 6.47 \times 10^6 \text{ m}}{7.85 \times 10^3 \text{ m/s}}$$
$$= 5.18 \times 10^3 \text{ s}$$

or 86.3 minutes.

Method 2
Use Kepler's third law of planetary motion, Equation 6.13:

$$T^2 = \frac{4\pi^2}{GM} r^3$$
$$= \frac{4\pi^2 (6.47 \times 10^6 \text{ m})^3}{(6.67 \times 10^{-11} \text{ N·m}^2/\text{kg}^2)(5.98 \times 10^{24} \text{ kg})}$$
$$= 2.68 \times 10^7 \text{ s}^2$$

Taking the square root, you find

$$T = 5.18 \times 10^3 \text{ s}$$

6.7 KEPLER'S FIRST LAW OF PLANETARY MOTION AND THE GEOMETRY OF ELLIPSES

Almost a century before Newton, between about 1600 and 1620, Johannes Kepler (1571–1630) discovered three empirical laws of planetary motion that, for the first time, accurately described the heliocentric orbital motion of the planets in the solar system. Kepler made these discoveries by tediously analyzing the precise observational planetary position data collected painstakingly for 20 years during the late 16th century, before telescopes, by his mentor Tycho Brahe (1546–1601). All three of the planetary laws of Kepler were subsequently shown by Newton to be consequences of Newton's laws of motion and his law of universal gravitation. Kepler's laws, therefore, are not independent laws of motion. This illustrates how succeeding laws can build on and supplant earlier laws. We study Kepler's laws here to gain an appreciation for the beauty and structure of the solar system in which we reside. The laws also are applicable to any two-body gravitational interaction elsewhere in the cosmos, such as a

The planetary data secured by Tycho Brahe enabled Kepler to decipher the structure of the solar system.

moon or satellite in orbit around a mother planet, or the newly discovered planets orbiting stars other than the Sun.

> **Kepler's first law** of planetary motion states that the path of a planet orbiting the Sun is an ellipse with the Sun at one focus of the ellipse; the other focus is empty.

The empty focus perplexed and annoyed Kepler, but is not physically significant. Newton showed that the inverse square nature of the gravitational force law permits orbits that are *any* of the **conic sections** familiar to you from your study of analytic geometry: circles, ellipses, parabolas, and hyperbolas. More advanced courses in classical mechanics will show you how the conic section orbital shapes stem from the law of universal gravitation. Here we will look at only several characteristic features of the circular and elliptical orbits that are so common in orbital dynamics.

An **ellipse** is defined mathematically as the locus of points, the sum of whose distances from two fixed points, F_1 and F_2, known as the **foci**, is a constant.* Consider a planet in an elliptical orbit about the Sun, as in Figure 6.20. The Sun is at one focus of the

---

*Do not confuse the foci labels F_1 and F_2 with the magnitudes of forces.

ellipse, say F_1. Let the planet be at point P in its orbit. The distance of either focus from the geometric center of the ellipse is c.[†]

The definition of an ellipse implies that the sum of the distance r of the planet at P from the Sun and the distance s of the planet from the other (empty) focus is a constant. Call the constant $2a$, where the numerical factor of 2 is introduced for later convenience.[‡] So the definition of an ellipse implies that

$$r + s \equiv 2a \qquad (6.14)$$

Imagine the planet at the special point P_1 in Figure 6.20, located at one end of the **major axis** of the ellipse, closest to the Sun, called the **perihelion** point. Equation 6.14 then becomes

$$P_1F_1 + P_1F_2 = 2a \qquad (6.15)$$

Examine Figure 6.20; you can see that $P_1F_2 = P_1F_1 + 2c$, and so Equation 6.15 becomes

$$P_1F_1 + (P_1F_1 + 2c) = 2a \qquad (6.16)$$

But because of the symmetry of the ellipse, P_1F_1 is the same as P_2F_2, where P_2 is at the other end of the major axis; when at P_2, the planet is furthest from the Sun, at its **aphelion** point. Making this substitution into Equation 6.16, you get

$$P_1F_1 + (P_2F_2 + 2c) = 2a$$

The left-hand side of this equation is the sum of line segments that form the major axis of the ellipse (P_1P_2). Thus $2a$ is equal to the major axis of the ellipse, and the physical meaning of the constant a is *half the major axis*, known as the **semimajor axis** of the ellipse.

The **eccentricity** ε of the ellipse is defined to be the ratio of c to a:

$$\varepsilon \equiv \frac{c}{a} \qquad (6.17)$$

If the two foci of the ellipse are coincident, so $2c = 0$ m (or $c = 0$ m), then the eccentricity $\varepsilon = 0$ as well. The resulting geometrical figure is a circle. Thus a circle is an ellipse with zero eccentricity. At the other extreme, if c approaches a, then the ellipse becomes more and more elongated. When $c = a$, making $\varepsilon = 1$, the resulting geometrical figure is a parabola, which is not a closed curve.

The general equation of an ellipse in polar coordinates is found in the following way. Let $\vec{r}$ be the position vector of the planet in its orbit using the Sun as the origin, as shown in Figure 6.21. Let $\vec{c}$ be a vector from the Sun to the center of the ellipse. Let $\vec{s}$ be a vector from the second (empty) focus to the planet.

The rules of vector addition indicate that

$$\vec{r} = 2\vec{c} + \vec{s}$$

Thus

$$\vec{s} = \vec{r} - 2\vec{c}$$

[†] Do not confuse this distance c with the speed of light, also called c.

[‡] Do not confuse the constant a here with the magnitude of the acceleration. Here we will see that the constant a is the semimajor axis of the ellipse. It is usually clear from the context (if only from a dimensional analysis) whether a means the magnitude of the acceleration or a means the semimajor axis. Regrettably, there is a certain amount of unavoidable notational overlap in every technical field.

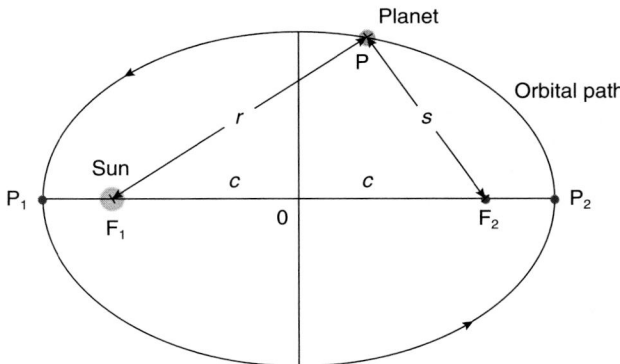

FIGURE 6.20 A planet in an elliptical orbit about the Sun.

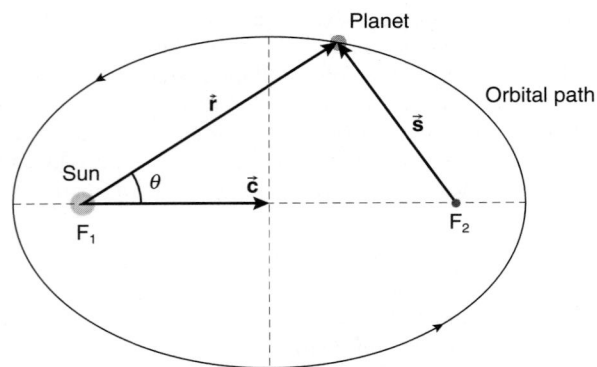

FIGURE 6.21 Position vector of a planet in an elliptical orbit.

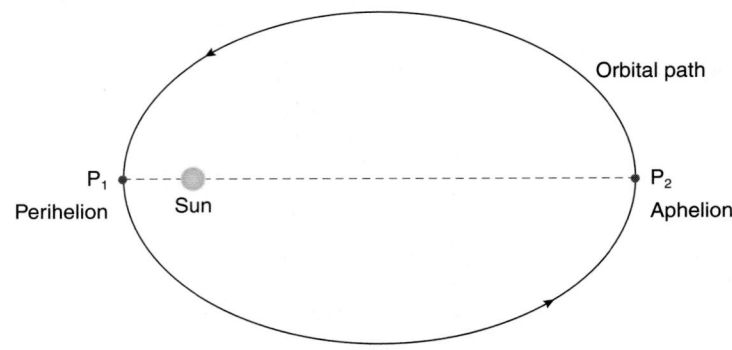

FIGURE 6.22

Take the scalar product of $\vec{s}$ with itself to find the square of its magnitude:

$$s^2 = \vec{s} \cdot \vec{s} = (\vec{r} - 2\vec{c}) \cdot (\vec{r} - 2\vec{c})$$
$$= r^2 + 4c^2 - 4\,\vec{r} \cdot \vec{c} \qquad (6.18)$$

But from the definition of an ellipse,

$$s + r = 2a$$

and so

$$s = 2a - r$$

From the definition of the eccentricity, Equation 6.17, $c = \varepsilon a$. Making these substitutions for s and c into Equation 6.18, we find

$$(2a - r)^2 = r^2 + 4(\varepsilon a)^2 - 4r(\varepsilon a)\cos\theta$$

where θ is the polar coordinate angle between $\vec{r}$ and $\vec{c}$, shown in Figure 6.21. After performing the indicated multiplications, we obtain

$$4a^2 - 4ra + r^2 = r^2 + 4\varepsilon^2 a^2 - 4r\varepsilon a \cos\theta$$

The quadratic terms in r subtract out. Solving this expression for r, we find

$$r = \frac{a(1 - \varepsilon^2)}{1 - \varepsilon \cos\theta} \qquad (6.19)$$

This is the equation for an ellipse in polar coordinates. The polar coordinate r is measured from the focus and θ is measured with respect to the major axis (see Figure 6.21).

EXAMPLE 6.4 ⎯⎯⎯⎯⎯⎯⎯⎯⎯⎯⎯

When a planet is at position P_1 in Figure 6.22, the planet is at perihelion, its closest distance to the Sun. When the planet is at the position P_2, the planet is at aphelion, its greatest distance from the Sun.

a. What is the angle θ in the polar equation for an ellipse, Equation 6.19, when the planet is at aphelion?
b. Use the polar equation for the ellipse to show that the aphelion distance is related to the semimajor axis and the eccentricity of the ellipse by

$$r_{\text{aph}} = a(1 + \varepsilon)$$

c. What is the angle θ when the planet is at perihelion?

d. Show that the perihelion distance is related to the semi-major axis a and the eccentricity by

$$r_{\text{peri}} = a(1 - \varepsilon)$$

Solution
a. When the planet is at aphelion, point P_2, the angle θ is equal to $0°$, as shown in Figure 6.23.

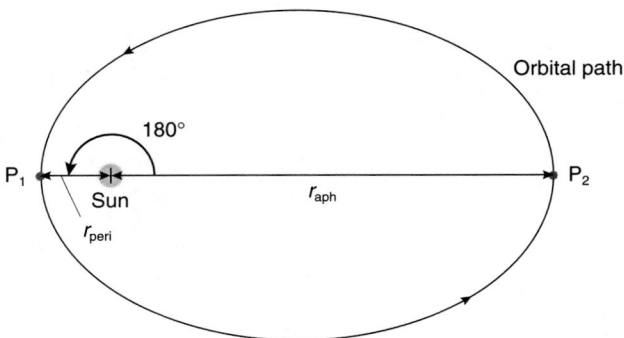

FIGURE 6.23

b. Substitute $\theta = 0°$ into the polar equation for an ellipse, Equation 6.19:

$$r_{\text{aph}} = \frac{a(1 - \varepsilon^2)}{1 - \varepsilon \cos 0°}$$
$$= \frac{a(1 - \varepsilon)(1 + \varepsilon)}{1 - \varepsilon}$$
$$= a(1 + \varepsilon)$$

c. When the planet is at perihelion, the angle $\theta = 180°$, as shown in Figure 6.23.
d. Substitute $\theta = 180°$ into the polar equation for an ellipse, Equation 6.19:

$$r_{\text{peri}} = \frac{a(1 - \varepsilon^2)}{1 - \varepsilon \cos 180°}$$
$$= \frac{a(1 - \varepsilon)(1 + \varepsilon)}{1 + \varepsilon}$$
$$= a(1 - \varepsilon)$$

Note that the *ratio* of the aphelion and perihelion distances depends only on the eccentricity:

$$\frac{r_{\text{aph}}}{r_{\text{peri}}} = \frac{a(1 + \varepsilon)}{a(1 - \varepsilon)}$$

$$= \frac{1 + \varepsilon}{1 - \varepsilon} \tag{1}$$

EXAMPLE 6.5

The eccentricity of the orbit of the Earth about the Sun is 0.0167 and the semimajor axis of the orbit is 1.496×10^8 km. Determine the aphelion and perihelion distances of the Earth from the Sun.

Solution
In Example 6.4, you showed that

$$r_{\text{aph}} = a(1 + \varepsilon)$$

and

$$r_{\text{peri}} = a(1 - \varepsilon)$$

Substituting numerical values for a and ε, you find

$$r_{\text{aph}} = (1.496 \times 10^8 \text{ km})(1 + 0.0167)$$
$$= 1.521 \times 10^8 \text{ km}$$

and

$$r_{\text{peri}} = (1.496 \times 10^8 \text{ km})(1 - 0.0167)$$
$$= 1.471 \times 10^8 \text{ km}$$

For cultural enrichment, in 1999 the Earth was at perihelion on 3 January and at aphelion on 6 July.

6.8 SPATIAL AVERAGE POSITION OF A PLANET IN AN ELLIPTICAL ORBIT*

The distance of a planet from the Sun varies between the extremes of the perihelion and aphelion distances. What is the average of all the possible spatial distances of the planet in its orbit?

Such a spatial average might be estimated by drawing n lines from the Sun to equally spaced positions of the planet along its orbital path, as in Figure 6.24. The n distances then are summed, and an estimate of the spatial average found by dividing the sum by n.[†] If you let the number of such lines increase indefinitely by letting the spacing Δs in Figure 6.24 approach zero, you obtain a true spatial average of all the possible positions of the planet in its orbit.

There is, however, an easier way to determine such a spatial average. The following imaginative procedure enables you to calculate the spatial average without undue effort.

Consider two possible positions of the planet on its orbit, points P and P′ in Figure 6.25, that are located *symmetrically* with respect to the **minor axis** of the ellipse. When the planet is at P its distance from the Sun is r, and when the planet is at

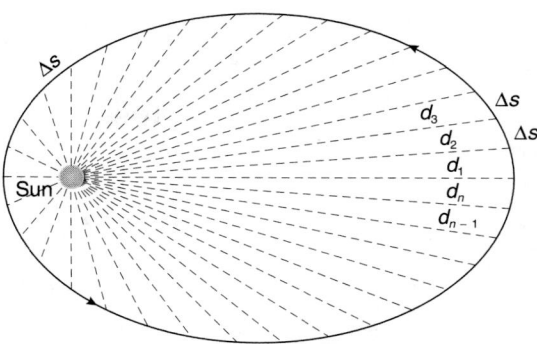

FIGURE 6.24 Lines from the Sun to points equally spaced along the orbital path.

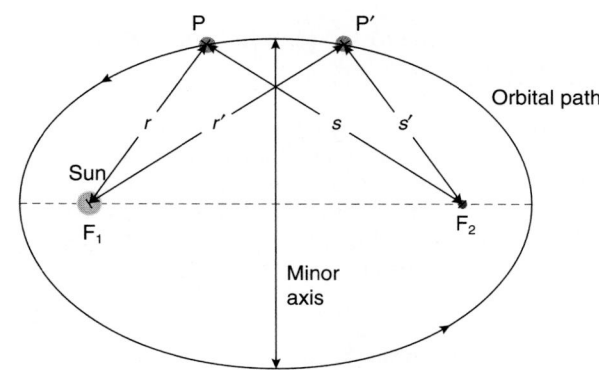

FIGURE 6.25 Locate two points P and P′ symmetrically with respect to the minor axis.

P′ its distance from the Sun is r'. The symmetrical placement of P and P′ means that $r' = s$. Thus the average of these two distances becomes

$$\frac{r + s}{2}$$

But the definition of the ellipse, Equation 6.14, indicates that $r + s = 2a$. Thus the average is $2a/2$ or just a, the semimajor axis. But the points P and P′ were *any* two points on the orbit symmetrically placed. So the average of *all* such paired points also is just the semimajor axis.

> Thus the spatial average of the distances of the planet from the Sun is equal to the semimajor axis of the ellipse.

6.9 KEPLER'S SECOND LAW OF PLANETARY MOTION

Kepler also discovered another aspect of the orbital motions of the planets, now known as **Kepler's second law** (although chronologically, it actually was the first one he discovered).

[†]In fact, this was the way Kepler found this average in the early 1600s.

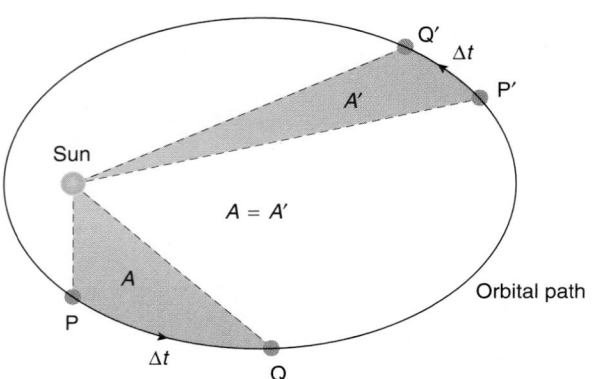

FIGURE 6.26 Kepler's second law of planetary motion.

A line from the Sun to a given planet sweeps out equal areas during equal time intervals.

If it takes the same time interval Δt for the planet to move between the positions at points P and Q and between P′ and Q′ in Figure 6.26, then the areas A and A′ also are equal.

This law implies that the speed of a planet when near its perihelion point (i.e., closest to the Sun) must be greater than when near its aphelion point (most distant from the Sun).

QUESTION 6

Imagine a line from the Sun to the Earth and another line from the Sun to Mars. One week elapses. Are the areas swept out by the lines equal? Explain your answer.

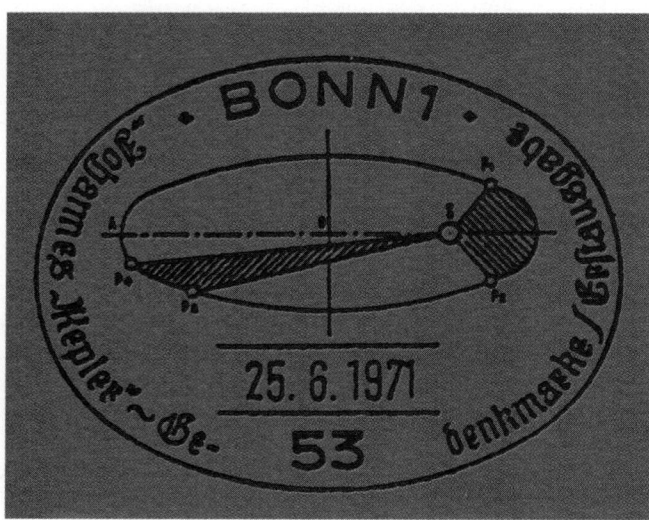

A postmark depicting Kepler's second law of planetary motion.

6.10 CENTRAL FORCES, ORBITAL ANGULAR MOMENTUM, AND KEPLER'S SECOND LAW*

A force that acts along the line between two particles is called a **central force**. The gravitational force $\vec{\mathbf{F}}$ of the Sun on a planet is antiparallel to the position vector $\vec{\mathbf{r}}$ locating the planet, using the Sun as an origin, as indicated in Figure 6.27. Thus the gravitational force is a central force. The magnitude of the *vector product* $\vec{\mathbf{r}} \times \vec{\mathbf{F}}$ must be zero, because the sine of the 180° angle between two antiparallel vectors is zero ($\sin 180° = 0$):

$$|\vec{\mathbf{r}} \times \vec{\mathbf{F}}| = rF \sin 180°$$
$$= 0 \text{ N·m} \qquad (6.20)$$

Notice that this result is true for *any central force*, regardless of the functional form of the force.

The central gravitational force is the only force on the planet. Newton's second law of motion indicates that

$$\vec{\mathbf{F}}_{\text{central}} = m\vec{\mathbf{a}}$$

where $\vec{\mathbf{a}}$ is the acceleration of the planet. Substituting $m\vec{\mathbf{a}}$ for the force in the vector product $\vec{\mathbf{r}} \times \vec{\mathbf{F}}$ and using Equation 6.20, you have

$$\vec{\mathbf{r}} \times (m\vec{\mathbf{a}}) = 0 \text{ N·m} \qquad (6.21)$$

Look for a moment at the time derivative of the vector product $\vec{\mathbf{r}} \times (m\vec{\mathbf{v}})$:

$$\frac{d}{dt}[\vec{\mathbf{r}} \times (m\vec{\mathbf{v}})]$$

Remember that when differentiating you must preserve the order of the terms in the vector product and that m is constant. The differentiation proceeds as follows:

$$\frac{d}{dt}[\vec{\mathbf{r}} \times (m\vec{\mathbf{v}})] = \frac{d\vec{\mathbf{r}}}{dt} \times (m\vec{\mathbf{v}}) + \vec{\mathbf{r}} \times \left(m\frac{d\vec{\mathbf{v}}}{dt}\right) \quad (6.22)$$

However, since

$$\frac{d\vec{\mathbf{r}}}{dt} = \vec{\mathbf{v}} \qquad \text{and} \qquad \frac{d\vec{\mathbf{v}}}{dt} = \vec{\mathbf{a}},$$

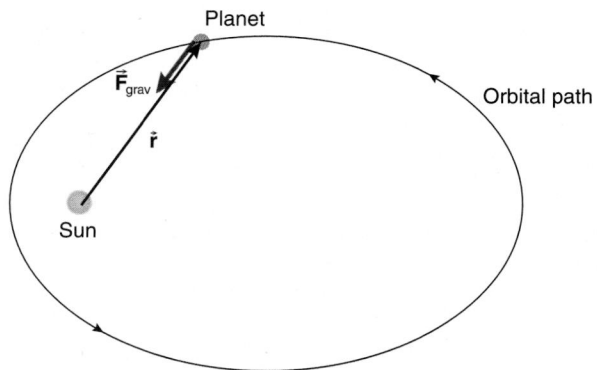

FIGURE 6.27 The gravitational force acts along the line connecting the two particles.

Equation 6.22 becomes

$$\frac{d}{dt}[\vec{r} \times (m\vec{v})] = \vec{v} \times (m\vec{v}) + \vec{r} \times (m\vec{a}) \qquad (6.23)$$

But the vector product of *any* vector with itself is zero, and so

$$\vec{v} \times (m\vec{v}) = m(\vec{v} \times \vec{v}) = 0 \ \text{kg·m}^2/\text{s}^2$$

and Equation 6.23 simplifies to

$$\frac{d}{dt}[\vec{r} \times (m\vec{v})] = \vec{r} \times (m\vec{a})$$

But $\vec{r} \times (m\vec{a}) = 0 \ \text{kg·m}^2/\text{s}^2$ from Equation 6.21, because $\vec{r}$ and $\vec{a}$ are antiparallel vectors for any central force. Thus we have

$$\frac{d}{dt}[\vec{r} \times (m\vec{v})] = 0 \ \text{kg·m}^2/\text{s}^2 \qquad (6.24)$$

> The product $m\vec{v}$ is defined to the **momentum $\vec{p}$** of the mass m:*
>
> $$\vec{p} \equiv m\vec{v} \qquad (6.25)$$

The momentum is a vector parallel to the velocity. Using this definition in Equation 6.24, we find

$$\frac{d}{dt}(\vec{r} \times \vec{p}) = 0 \ \text{kg·m}^2/\text{s}^2$$

If the time derivative of a vector is zero, the vector does not change with time and is a *constant vector*. Thus the vector product $\vec{r} \times \vec{p}$ does not change with time for *any central force*, acting alone, on m. If we think of a planet moving about the Sun in an elliptical orbit, the vector $\vec{r}$ changes with time (both in magnitude and direction), the vector $\vec{p}$ changes with time (since the velocity of the planet changes in both magnitude and direction in an elliptical orbit), but *the vector $\vec{r} \times \vec{p}$ does not change with time for a particle under the influence of a central force.* We say such a vector is **conserved.**

> The word *conserved* means the quantity is *constant at all times and locations.*

Thus $\vec{r} \times \vec{p}$ is quite a special vector; it occurs in many contexts in physics and is called the **orbital angular momentum $\vec{L}$:†**

> $$\vec{L} \equiv \vec{r} \times \vec{p} \qquad (6.26)$$

> The orbital angular momentum of a particle (or planet) under the influence of a central force is conserved; that is,
>
> $$\frac{d\vec{L}}{dt} = 0 \ \text{kg·m}^2/\text{s}^2$$

---

*For now, we take this definition on an ad hoc basis; we introduce the momentum of a particle in a more pedagogical fashion in Chapter 9.
† We explore the orbital angular momentum in more detail in Chapter 10.

Since $\vec{L}$ is a constant vector for central force motion, and $\vec{L}$ is perpendicular to *both* $\vec{r}$ and $\vec{p}$ (from the definition of the vector product), *the motion of the particle under the influence of a central force must be confined to a plane.*

An interesting geometrical consequence of the constancy of the orbital angular momentum vector is Kepler's law of areas. To see this connection, consider a planet moving in an elliptical orbit, as in Figure 6.28. Let $d\vec{r}$ be the *change* in the position vector of the planet during a time interval dt. The position vector of the planet sweeps out a small differential area dA during this time. The small area swept out is essentially triangular in shape.

The geometric formula for the area of a triangle is

$$dA = \frac{1}{2}(\text{base})(\text{height})$$

The base of the skinny triangle is $r\,d\theta$ and the height is r; thus, the differential area is

$$dA = \frac{1}{2}(r\,d\theta)\,r$$

This expression is the same as the *magnitude* of the vector

$$\frac{1}{2}\vec{r} \times d\vec{r}$$

because the magnitude of this vector product is

$$\frac{1}{2}r\,dr\sin\phi$$

where ϕ is the angle between $\vec{r}$ and $d\vec{r}$ as indicated in Figure 6.28. But $dr\sin\phi$ is the same as the base of the little triangle $(r\,d\theta)$. Thus we create a differential *area vector*

$$d\vec{A} = \frac{1}{2}\vec{r} \times d\vec{r}$$

whose magnitude is the differential area dA; its direction is *perpendicular* to the plane of the orbit from the vector product right-hand rule (since $\vec{r}$ and $d\vec{r}$ are in the plane of the orbit).

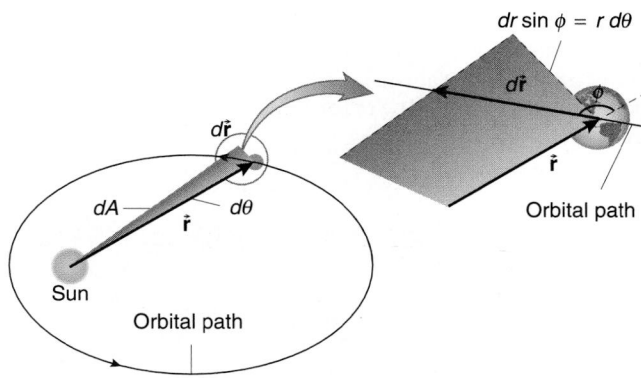

FIGURE 6.28 A line from the Sun to the planet sweeps out a small triangular area during a short interval dt.

If we divide the differential area vector by the differential time dt, we obtain

$$\frac{d\vec{A}}{dt} = \frac{1}{2}\frac{\vec{r} \times d\vec{r}}{dt}$$

$$= \frac{1}{2}\vec{r} \times \frac{d\vec{r}}{dt}$$

$$= \frac{1}{2}\vec{r} \times \vec{v} \qquad (6.27)$$

The orbital angular momentum vector is $\vec{L} = \vec{r} \times \vec{p} = \vec{r} \times m\vec{v}$; hence, $\vec{r} \times \vec{v} = \vec{L}/m$ and Equation 6.27 becomes

$$\frac{d\vec{A}}{dt} = \frac{\vec{L}}{2m} \qquad (6.28)$$

Since $\vec{L}$ is a constant vector and m is just the (constant) mass of the planet, the area vector has a constant time derivative. That is, A itself must vary linearly with the time t. This means that a line from the Sun to the planet sweeps out equal areas during equal time intervals, which is Kepler's law of areas.

To avoid any misconceptions, note that the constant orbital angular momentum vector and the constant rate at which the area is swept out by the position vector to the planet are characteristic of the *individual planet* whose motion is being considered. The amounts of angular momentum of the various planets differ from each other because they have different orbits and different masses.

EXAMPLE 6.6

What is the direction of the orbital angular momentum vector $\vec{L}$ for the planetary motion depicted in the Figure 6.29? Is the magnitude or direction of $\vec{L}$ different at another position of the planet in its orbit?

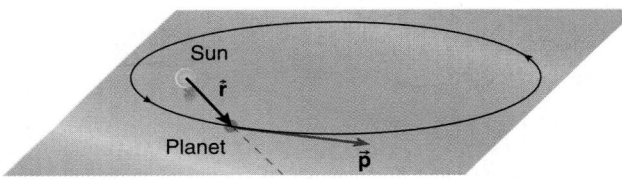

Plane of elliptical orbit
FIGURE 6.29

Solution
To find the direction of the angular momentum $\vec{L} = \vec{r} \times \vec{p}$, use the vector product right-hand rule. The vector $\vec{r} \times \vec{p}$ is directed perpendicular to the plane of the orbit of the planet as indicated in Figure 6.30.

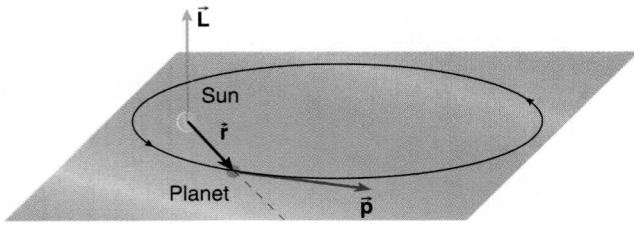

Plane of elliptical orbit
FIGURE 6.30

Since the angular momentum $\vec{L}$ is a constant vector, it has the same magnitude and direction for all positions of the planet in its orbit.

EXAMPLE 6.7

Consider a planet at its aphelion and perihelion points, as in Figure 6.31.

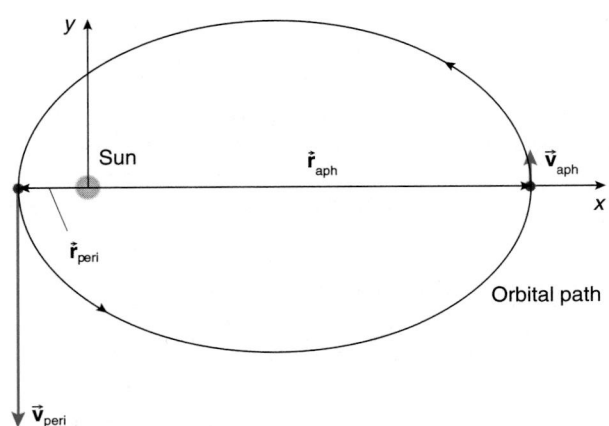

FIGURE 6.31

a. What is the angle between the position vectors and the respective velocity vectors of the planet at these locations?
b. By applying conservation of the orbital angular momentum of the planet at these two locations, show that the ratio of the *speeds* of the planet at the perihelion and aphelion locations is

$$\frac{v_{peri}}{v_{aph}} = \frac{1 + \varepsilon}{1 - \varepsilon}$$

Solution
a. The velocity vector always is tangent to the orbital path. At the aphelion and perihelion points in the orbit (but nowhere else in an elliptical orbit), the angle between $\vec{r}$ and $\vec{v}$ is 90°.
b. Since the orbital angular momentum vector is conserved, it must be the same at the perihelion and aphelion locations. Use the coordinate system indicated in Figure 6.31. The angular momentum of the planet of mass m when at aphelion is

$$\vec{L}_{aph} = \vec{r}_{aph} \times (m\vec{v}_{aph})$$

$$= mr_{aph}\hat{\imath} \times (v_{aph}\hat{\jmath})$$

$$= mr_{aph}v_{aph}\hat{k}$$

Likewise, at perihelion

$$\vec{L}_{peri} = \vec{r}_{peri} \times (m\vec{v}_{peri})$$

$$= mr_{peri}(-\hat{\imath}) \times [v_{peri}(-\hat{\jmath})]$$

$$= mr_{peri}v_{peri}\hat{k}$$

The angular momentum is a constant vector throughout the motion, and so

$$\vec{L}_{aph} = \vec{L}_{peri}$$

$$mr_{aph}v_{aph}\hat{k} = mr_{peri}v_{peri}\hat{k}$$

Divide by m and rearrange this to obtain the ratio you want:

$$\frac{v_{\text{peri}}}{v_{\text{aph}}} = \frac{r_{\text{aph}}}{r_{\text{peri}}} \qquad (1)$$

Equation (1) in Example 6.4 showed that the ratio of the perihelion and aphelion distances is

$$\frac{r_{\text{aph}}}{r_{\text{peri}}} = \frac{1 + \varepsilon}{1 - \varepsilon}$$

Equation (1) here thus becomes

$$\frac{v_{\text{peri}}}{v_{\text{aph}}} = \frac{1 + \varepsilon}{1 - \varepsilon}$$

The speed at perihelion thus is faster than at aphelion.

6.11 NEWTON'S FORM FOR KEPLER'S THIRD LAW OF PLANETARY MOTION

According to Newton's law of universal gravitation, two spherically symmetric masses m and M (such as the Moon and the Earth) exert gravitational forces on each other directed along the line between their centers (see Figure 6.32). The gravitational forces on each are equal in magnitude and opposite in direction, in accordance with Newton's third law. Assume the gravitational force is the only force acting on each mass.

The gravitational force of M on m results in an acceleration of m according to Newton's second law. The magnitudes of the force and acceleration are related by

$$F_{\text{on } m} = ma$$

But the magnitude of the total force on m is the magnitude of the gravitational force of M on m, so that

$$\frac{GmM}{r^2} = ma$$

Consider also the gravitational force of m on M and the resulting acceleration a' of M. Once again, the magnitudes are related by the second law of motion:

$$F_{\text{on } M} = Ma'$$

The magnitude of the total force on M is the magnitude of the gravitational force of m on M, and so

$$\frac{GmM}{r^2} = Ma'$$

where a' is the acceleration of M.

Notice that the *same magnitude of force acts on each mass* (because of Newton's third law of motion) but the *magnitude of the acceleration of each mass is different* (as deduced from Newton's second law of motion, applied to each as an individual system):

Magnitude of acceleration of m:

$$a = \frac{GM}{r^2}$$

Magnitude of acceleration of M:

$$a' = \frac{Gm}{r^2}$$

These expressions imply that the magnitude of the acceleration of each mass is proportional to the *other* mass. Take the ratio of the magnitudes of the accelerations of the two masses:

$$\frac{\text{magnitude of acceleration of } m}{\text{magnitude of acceleration of } M} = \frac{a}{a'}$$

$$= \frac{\dfrac{GM}{r^2}}{\dfrac{Gm}{r^2}}$$

$$= \frac{M}{m} \qquad (6.29)$$

Let m be a trivial mass compared with M, say the mass of a small rock compared with the mass of the Earth. The acceleration of the *rock* is proportional to the mass of the *Earth*. Since the mass of the Earth is so much larger than that of the small rock, the acceleration of the rock is significant. The rock exerts an *equal* gravitational *force* on the Earth; however, the acceleration of the *Earth* is proportional to the mass of the *rock*. Since the mass of the rock is so much smaller than that of the Earth, the acceleration of the Earth toward the rock is negligible. This is why we could neglect the gravitational effects *of* a satellite *on* the Earth in Section 6.6; the acceleration of the Earth is minuscule compared with that of the satellite. This argument also is precisely the reason we neglect any motion of the Earth when considering the motions of common objects subjected to the Earth's gravitational force.

However, if the two masses m and M are comparable in magnitude, the accelerations are comparable as well according to Equation 6.29. What happens then? Let the two masses be separated by a distance r, as in Figure 6.33. Imagine a point 0 on the line connecting the masses, and let r_1 be the distance of m from 0 and r_2 the distance of M from 0. Suppose 0 is the special point at which the products of the masses and their respective distances from 0 are equal; then 0 is called the **center of mass** point.

FIGURE 6.32 Two masses m and M exert gravitational forces on each other.

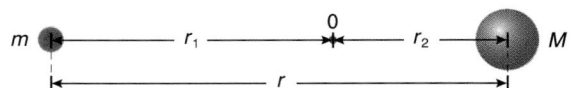

FIGURE 6.33 Two masses separated by a distance r.

FIGURE 6.34 Two masses released at rest.

The location of the center of mass point is defined by the equation*:

$$mr_1 = Mr_2 \qquad (6.30)$$

The center of mass point thus is located closer to the more massive object.[†]

If the two masses are released from rest, as in Figure 6.34, they accelerate toward each other along a straight line and eventually collide. Notice that if we differentiate Equation 6.30 twice with respect to time, we get the relationship between the magnitudes of the accelerations of the two objects (Equation 6.29, slightly rearranged): $ma = Ma'$. If we neglect the physical sizes of the two masses, they collide at the center of mass point of the two-body system.

So if the objects initially are at rest, and the only forces on them are their mutual gravitational forces, they will collide after some time interval. The collision occurs at the center of mass (if we neglect the different finite sizes of the two objects). When a little mass m is released near the Earth and falls, the center of mass of the system is effectively at the center of the Earth; but the collision occurs on the surface because of the size of the Earth.

A collision of the masses may be avoided if the objects initially are not at rest but are moving in a direction not coincident with the line between them. In such situations, Newton showed that both objects move in paths that

1. are similar in shape,
2. are one of the conic sections, and
3. describe motion about the center of mass point (which is fixed in space).

The special case of circular orbits about the center of mass point is the easiest to consider (see Figure 6.35), and so we examine it in some detail. Our objective is to find a relationship between the period T of the orbital motion and the separation r of the masses for such circular orbital motion of the masses about the center of mass point.

The magnitudes of the gravitational forces (1) of M on m and (2) of m on M are the *same* by Newton's third law. We assume both masses have appropriate spherical symmetry to use the point-mass gravitational force law. According to Newton's second law, the single force on each causes an acceleration in the

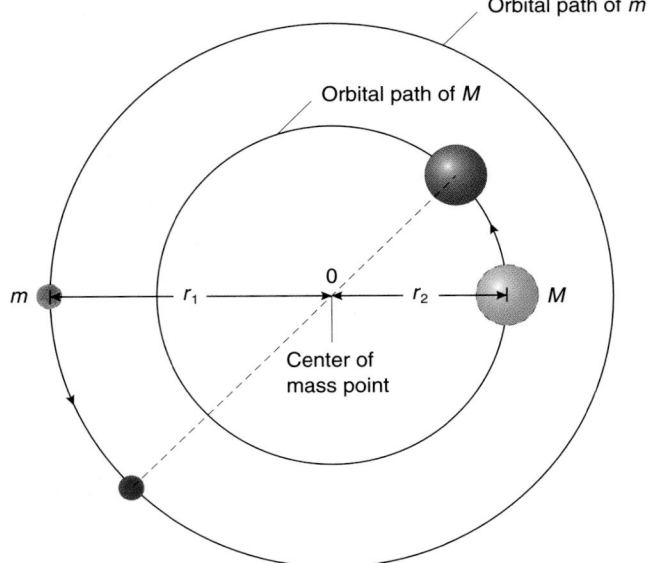

FIGURE 6.35 Circular orbits about the center of mass.

same direction as the force. Since the orbits are circular, the acceleration of each mass is a centripetal acceleration.

For the motion of m (use Newton's second law of motion with m as the system), the magnitudes of the force and acceleration are related via

$$\frac{GMm}{r^2} = ma$$

Substituting for a the magnitude of the centripetal acceleration of m, we obtain

$$\frac{GMm}{r^2} = m\frac{v_1^2}{r_1} \qquad (6.31)$$

where v_1 is the orbital speed of m in its orbit of radius r_1 about the center of mass.

Likewise, for the motion of M apply Newton's second law to M, with M as a separate system. The magnitudes of the force and acceleration are related by

$$\frac{GMm}{r^2} = Ma'$$

Again, substitute for a' the magnitude of the centripetal acceleration of M:

$$\frac{GMm}{r^2} = M\frac{v_2^2}{r_2} \qquad (6.32)$$

where v_2 is the speed of M in its orbit of radius r_2 about the center of mass.

Both masses have the same orbital period of revolution, since the center of mass must lie on the line between them.[‡] The time for

---

*The location of the center of mass for a collection of particles (not just two) will be considered in greater detail in Chapter 9.
[†] You can think of the center of mass point as the location of a "balance point" along the line between the two masses, just like the balance point of an imaginary (massless) see-saw with the two masses m and M on each end. To do this, one would need a third great mass equidistant from both of these masses to provide the gravitational forces to enable the balance to operate.

[‡] If the periods differed, eventually both masses could be on the same side of the center of mass, which is ridiculous. Why?

one orbital revolution is the period T. The constant speed of each mass is found from the distance traveled during one period (the circumference of the circular orbit) divided by the time it takes (one period); that is:

For the speed of m:

$$v_1 = \frac{2\pi r_1}{T}$$

For the speed of M:

$$v_2 = \frac{2\pi r_2}{T}$$

Substitute these expressions for the speeds in the second law equations of each mass. Equation 6.31 for m becomes

$$\frac{GMm}{r^2} = \frac{m}{r_1}\left(\frac{2\pi r_1}{T}\right)^2$$

or

$$\frac{GM}{r^2} = \frac{4\pi^2 r_1}{T^2} \qquad (6.33)$$

Equation 6.32 for M becomes

$$\frac{GMm}{r^2} = \frac{M}{r_2}\left(\frac{2\pi r_2}{T}\right)^2$$

or

$$\frac{Gm}{r^2} = \frac{4\pi^2 r_2}{T^2} \qquad (6.34)$$

Now solve Equations 6.33 and 6.34 for the radius of each orbit:

$$r_1 = \frac{GMT^2}{4\pi^2 r^2} \qquad \text{(orbital radius of } m) \qquad (6.35)$$

and

$$r_2 = \frac{GmT^2}{4\pi^2 r^2} \qquad \text{(orbital radius of M)} \qquad (6.36)$$

The orbital radius of each mass is proportional to the mass of the *other* object. The *smaller* mass m has the *larger* orbital radius about the center of mass point.

The separation distance r is the sum of the orbital radii:

$$r = r_1 + r_2$$

Use Equation 6.35 to substitute for r_1 and Equation 6.36 to substitute for r_2. The separation distance r thus is

$$r = \frac{GMT^2}{4\pi^2 r^2} + \frac{GmT^2}{4\pi^2 r^2}$$

After factoring, we have

$$r = \frac{G(M + m)T^2}{4\pi^2 r^2}$$

Solve for T^2:

$$T^2 = \frac{4\pi^2}{G(M + m)}r^3 \qquad \text{(circular orbits)} \qquad (6.37)$$

The square of the period is proportional to the cube of the center-to-center separation of the two masses. Equation 6.37 is **Newton's form for Kepler's third law** of planetary motion for *circular orbits* about the center of mass point. Notice that this law of Kepler is not an independent law; it is a natural consequence of the general Newtonian theory of motion and the law of universal gravitation. The other two empirical laws of planetary motion discovered by Kepler about 75 years before Newton also are not independent laws, but follow from Newton's general theory of motion and law of gravitation.

Rearrange Equation 6.30 to form the mass ratio:

$$\frac{m}{M} = \frac{r_2}{r_1}$$

As the mass ratio m/M becomes smaller and smaller, the ratio of the radii of the orbits of the two masses also decreases. In particular, as the mass ratio decreases, the orbital radius r_2 of M becomes smaller relative to the orbital radius r_1 of m, as shown in the sequence of sketches in Figure 6.36.

When m is *negligible* compared with M, the center of mass point is essentially located at the position of the center of the more massive object M. In this case m orbits M with an orbital radius $r_1 \approx r$, since $r_2 \approx 0$ m. The orbital motion of M is negligible because the center of mass point is essentially coincident with M itself.

When $m << M$, the constant of proportionality in Equation 6.37 becomes approximately

$$\frac{4\pi^2}{G(M + m)} \approx \frac{4\pi^2}{GM} \qquad \text{for } m << M$$

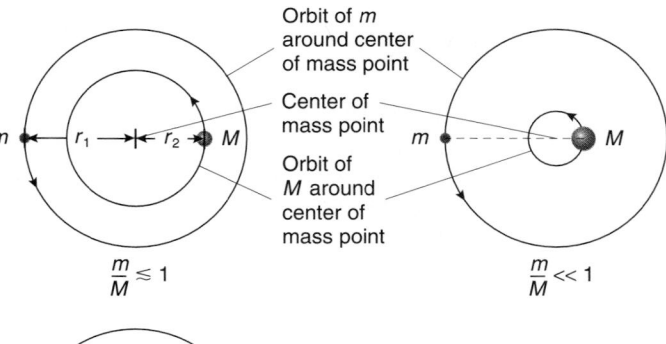

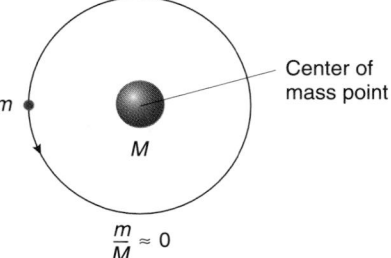

FIGURE 6.36 As the mass ratio m/M becomes smaller, the orbital radius of M decreases.

Then Newton's form of Kepler's third law, Equation 6.37, becomes

$$T^2 = \frac{4\pi^2}{GM} r^3 \qquad m \ll M, \text{ circular orbits} \qquad (6.38)$$

where r is the radius of the circular orbit of m about M. This is the result we derived in Section 6.6 when considering artificial satellites of the Earth, which satisfy the condition $m \ll M$.

Newton's form for Kepler's third law was derived here for the special case of *circular orbits*. When $m \ll M$ and the orbit of m about M is *elliptical*, Equation 6.38 has exactly the same form with r replaced by a, the semimajor axis (see Figure 6.37):*

$$T^2 = \frac{4\pi^2}{GM} a^3 \qquad m \ll M, \text{ elliptical orbit} \qquad (6.39)$$

Equation 6.39 is a more general expression for the Newtonian form of Kepler's third law of planetary motion.

Within the solar system, the mass of any planet m is considerably less than the mass M of the Sun,[†] so that the condition $m \ll M$ is well satisfied. The square of the period of a planet is proportional to the cube of the semimajor axis of its orbit, with a proportionality constant essentially independent of the planetary mass m. That is, for a mass m in an elliptical orbit of semimajor axis a about another much more massive object M, the square of the period T is proportional to the cube of the semimajor axis of the orbit (Equation 6.39). This equation certainly is applicable to *any* artificial satellite orbit about the Earth, or spacecraft or planetary orbits about the Sun, since the condition $m \ll M$ is well satisfied.

6.12 CUSTOMIZED UNITS

For the orbits of the planets or spacecraft about the Sun, the mass m of the planet or spacecraft is very small compared with the mass $M_\odot$ of the Sun. Equation 6.39 thus takes the form

$$T^2 = \frac{4\pi^2}{GM_\odot} a^3 \qquad m_{\text{orbiting object}} \ll M_\odot \qquad (6.40)$$

---

*We leave the proof of this statement to more advanced courses in dynamics.
[†]The most massive planet (Jupiter) has a mass that is somewhat less than 10^{-3} times the mass of the Sun, so that $m/M \approx 10^{-3}$. The mass of the Earth is on the order of 10^{-6} times the mass of the Sun.

where T is the period of the planet or spacecraft in its orbit about the Sun, a is the semimajor axis of the orbit, and $M_\odot$ is the mass of the Sun.

The semimajor axis of the orbit of the Earth about the Sun is defined to be one **astronomical unit** (AU):

$$1 \text{ AU} \approx 1.496 \times 10^{11} \text{ m}$$

The period of the Earth in its orbit is, of course, one year.[‡] If we measure the period, not in seconds, but in years, and measure the semimajor axis of the planetary orbit in astronomical units rather than meters, then Equation 6.40 has the following form when applied to the Earth:

$$(1 \text{ y})^2 = \frac{4\pi^2}{GM_\odot} (1 \text{ AU})^3$$

Thus

$$\frac{4\pi^2}{GM_\odot} = 1 \text{ y}^2/\text{AU}^3$$

Newton's form for Kepler's third law (Equation 6.40) then takes on the simple form:

$$T^2 = (1 \text{ y}^2/\text{AU}^3)a^3 \qquad (6.41)$$

T *must* be in y

a *must* be in AU

for *any* object in orbit about the Sun.

This legitimate trick of creating a simplified equation using *customized units* is employed in various branches of the sciences for equations that are frequently used. But once again, in using such equations (indeed, all equations), realize that you must be careful to know and consider any strings attached to their use.

PROBLEM-SOLVING TACTIC

6.2 Be sure to express the period T in years and the semimajor axis in astronomical units when using Equation 6.41. The proportionality constant on the right-hand side of Equation 6.41 is written explicitly as $1 \text{ y}^2/\text{AU}^3$ to remind you of the appropriate units to use. Equation 6.41 is *not valid* in SI units. Use Equation 6.39 when using SI units with T in s and a in m.

EXAMPLE 6.8

The orbital period of Jupiter is 11.87 y. Find the semimajor axis of the orbit of Jupiter.

[‡] The true orbital period of the Earth about the Sun is the *sidereal year*, equal to approximately 365.256 37 days. The *calendar year*, also known as the *tropical year* or the year of the seasons, is slightly shorter and is approximately 365.242 199 days. We neglect the small difference between the two years in this text except where explicitly mentioned.

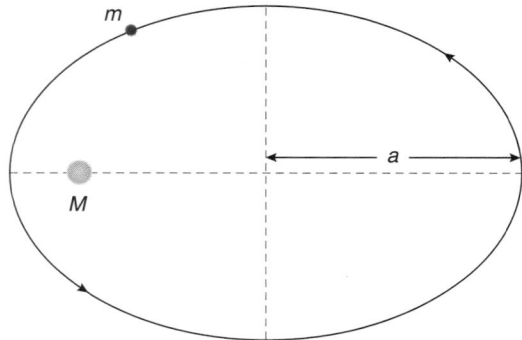

FIGURE 6.37 Elliptical orbit.

The largest planet in the solar system is Jupiter.

Solution

Method 1

Use the customized version of Kepler's third law, Equation 6.41:

$$T^2 = (1 \ y^2/AU^3)a^3$$

where T is in y and a is in AU. Substituting $T = 11.87$ y, you find a in AU:

$$(11.87 \ y)^2 = (1 \ y^2/AU^3)a^3$$

Solving for a,

$$a = 5.204 \ AU$$

If desired, this result can be converted to meters using the conversion factor between AU and meters

$$1 \ AU = 1.496 \times 10^{11} \ m$$

Hence

$$a = (5.204 \ AU) \times (1.496 \times 10^{11} \ m/AU)$$
$$= 7.785 \times 10^{11} \ m$$

Method 2

Use Equation 6.39:

$$T^2 = \frac{4\pi^2}{GM} a^3$$

which necessitates SI units: T in *seconds* and M equal to the mass of the Sun in kg:

$$T = 11.87 \ y$$
$$= 3.746 \times 10^8 \ s$$
$$G = 6.67 \times 10^{-11} \ N \cdot m^2/kg^2$$
$$M_\odot = 1.99 \times 10^{30} \ kg$$

The semimajor axis emerges in meters. Making the substitutions, you get

$$(3.746 \times 10^8 \ s)^2$$
$$= \frac{4\pi^2 a^3}{(6.67 \times 10^{-11} \ N \cdot m^2/kg^2)(1.99 \times 10^{30} \ kg)}$$

Solving for a,

$$a = 7.78 \times 10^{11} \ m$$

This is more awkward computationally, but equally legitimate. The result can be converted to AU, if desired, using the conversion factor

$$1 \ AU = 1.496 \times 10^{11} \ m$$

STRATEGIC EXAMPLE 6.9

A far-out spacecraft is in an elliptical orbit about the Sun with the perihelion point at the orbit of the Earth and the aphelion point at the orbit of Saturn, as shown in Figure 6.38. Such orbits are used by spacecraft making direct trips to the outer planets of the solar system. Similar orbits are used to visit the inner planets, but the aphelion point then is at the orbit of the Earth and the perihelion point is at the orbit of the inner planet. Once the spacecraft is inserted into its orbit (which necessitates a large rocket at launch), the spacecraft proceeds unpowered along its orbit of the Sun.

The planet Saturn and its spectacular ring system.

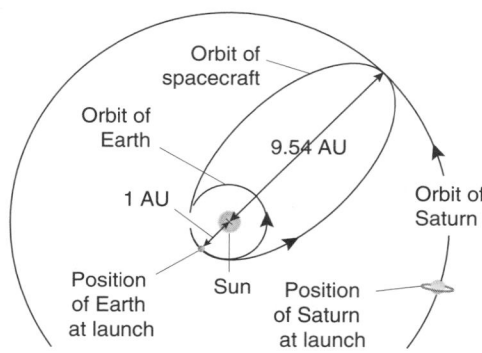

FIGURE 6.38

Consider the orbits of the Earth and Saturn to be circular. The radius of the orbit of Saturn is 9.54 AU.

a. Find the time it takes for the spacecraft to go from the orbit of the Earth to that of Saturn.
b. Determine the eccentricity of the orbit of the spacecraft.

Solution

a. The major axis of the orbit of the spacecraft is

$$9.54 \ AU + 1.00 \ AU = 10.54 \ AU$$

The semimajor axis a of the orbit of the spacecraft is then

$$a = \frac{10.54 \ AU}{2}$$
$$= 5.27 \ AU$$

Use Equation 6.41:

$$T^2 = (1 \ y^2/AU^3)a^3$$
$$= (1 \ y^2/AU^3)(5.27 \ AU)^3$$

Solving for T,

$$T = 12.1 \ y$$

To travel from the Earth to Saturn, the spacecraft only traverses *half* its orbit, from one end of its semimajor axis to the other end. Hence the time of flight to Saturn is *half* the period of the spacecraft in its complete orbit, or

$$t_{\text{flight}} = 6.05 \ y$$

This problem also illustrates the reason behind *launch windows*. When the spacecraft arrives at the orbit of Saturn, it would be nice if Saturn also was in the vicinity; it would be quite embarrassing if this were not the case and the planet was located elsewhere in its orbit! Therefore the spacecraft must be launched at such a time that when it arrives at the orbit of Saturn, the planet Saturn itself also is at that point in its orbit.

b. To determine the eccentricity ε of the orbit of the spacecraft, use the result of Example 6.4, knowing that the perihelion distance of the orbit of the spacecraft is 1.00 AU and its aphelion distance is 9.54 AU:

$$\frac{r_{\text{aph}}}{r_{\text{peri}}} = \frac{1 + \varepsilon}{1 - \varepsilon}$$

Solving for ε, you obtain

$$\varepsilon = 0.810$$

6.13 THE GRAVITATIONAL FIELD

It is inconceivable, that inanimate brute Matter should, without the Mediation of something else, which is not material, operate upon, and affect other Matter, without mutual Contact. . .

Isaac Newton (1642–1727)*

Newton clearly was troubled by the idea of the gravitational force magically acting over vast distances without any apparent physical mechanism for conveying the force between the masses. The force may as well have been transmitted by angels—or the devil. Such action-at-a-distance never really was resolved by him. The problem simply was ignored, for no conceptual framework existed to approach the problem adequately. Nonetheless, one could solve many problems of interest without worrying about such subtleties.

During the 19th century an idea emerged that attempted to explain action-at-a-distance via an intermediary known as a **field**. The concept of a field still is very useful for conveying the action of each of the fundamental forces (see Section 5.2). Thus it is important to consider the concept of a **gravitational field** in some detail; it serves as a paradigm for the treatment of the other fundamental forces, such as the electric and magnetic forces.

Consider the gravitational interaction between two point-like particles of mass m and M (or particles with appropriate spherical symmetry). Let the mass $m << M$.[†] Mass m experiences a gravitational force from the presence of mass M located a

---

*Letter to Richard Bentley, 25 February 1692(3), *Isaac Newton's Papers and Letters on Natural Philosophy*, edited by I. Bernard Cohen (Cambridge, Massachusetts, 1958), pages 300–309; quote is on page 302.

[†]This condition implies that the presence of m does *not* (via gravitational forces) affect the distribution or geometrical shape of the mass M, when M is not a pointlike mass.

distance r away, as shown in Figure 6.39. The gravitational force of M on m is of magnitude

$$\frac{GmM}{r^2}$$

and is directed, of course, toward M.

Suppose we simply rearrange the expression for the magnitude of the gravitational force on m as

$$m\left(\frac{GM}{r^2}\right)$$

The stuff in the parentheses can be attributed to just M itself and the space surrounding M. In other words, we imagine M influencing space itself, setting up a "condition of space" or field at each point. Now if m is placed at a point in space, the *field* of M *at the point where m is placed* acts on m to produce the force on m. The field that M establishes at all points in space, by virtue of its mass, is called the gravitational field $\vec{\mathbf{g}}$. It conveys the gravitational force. In other words, the very existence of M sets up or establishes the gravitational field at *all* points in space *whether the other mass m is present or not*. The gravitational field established by a point mass M at any point (say a distance r away from M) is a vector directed toward M and is of magnitude

$$\frac{GM}{r^2}$$

Let $\hat{\mathbf{r}}$ be a dimensionless unit vector, analogous to the familiar Cartesian $\hat{\mathbf{i}}$, $\hat{\mathbf{j}}$, and $\hat{\mathbf{k}}$ unit vectors, but pointing *radially away* from M, as shown in Figure 6.40. Then the gravitational field vector established in space by a point mass M is

$$\vec{\mathbf{g}}(r) = -\frac{GM}{r^2}\hat{\mathbf{r}} \tag{6.42}$$

> If mass m now is placed at a point in space where the gravitational field is $\vec{\mathbf{g}}$, m experiences a force given by the product of m times the value of the gravitational field at the location where m is placed. That is,
>
> $$\vec{\mathbf{F}}_{\text{grav force on } m} = m[\text{gravitational field wherever } m \text{ is placed}]$$
> $$= m\vec{\mathbf{g}} \tag{6.43}$$

> Equation 6.42 for the gravitational field of a point mass M is *not* the only kind of gravitational field. The specific analytical form for the gravitational field established by a mass M depends on how the mass M is shaped and distributed within its volume.

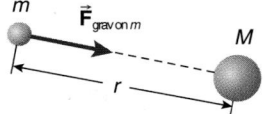

FIGURE 6.39 The gravitational force of M on m.

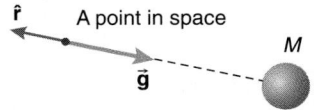

FIGURE 6.40 The unit vector $\hat{\mathbf{r}}$ points radially away from M, while the gravitational field of M points toward it.

The expression for $\vec{\mathbf{g}}$ in Equation 6.42 is just a specific example of a gravitational field: that established by a pointlike mass M. Other specific types of gravitational fields are listed in Table 6.1. *Whatever* the specific form of the gravitational field $\vec{\mathbf{g}}$ at a certain point in space, any mass m, placed at the point where the value of the gravitational field is $\vec{\mathbf{g}}$, experiences a gravitational force given by Equation 6.43. The gravitational field $\vec{\mathbf{g}}$ into which m is placed is established by *other* masses, not by m itself.

A few things to note:

1. Rearrange Equation 6.43 to find:

$$\vec{\mathbf{g}}_{\text{at the position of } m} = \frac{\vec{\mathbf{F}}_{\text{on } m}}{m}$$

Hence the gravitational field can be imagined as the *force per unit mass* (the force on one kilogram) at a given point in space.

Notice the similarity between Equation 6.43 and Newton's second law of motion when the gravitational force of M on m is the *only* force acting on m: $\vec{\mathbf{F}}_{\text{on } m} = m\vec{\mathbf{a}}$. The field has the same magnitude and direction as the acceleration caused by the total gravitational force on m when m is placed at the point in question.

Hence the gravitational field at a point in space is the same as the acceleration due to gravity at that location.

TABLE 6.1 Gravitational Fields Established in Space by Various Mass Distributions

Mass distribution	Location	Gravitational field
Point mass M	gravitational field a distance r away from M	$\vec{\mathbf{g}}(r) = \dfrac{-GM}{r^2}\,\hat{\mathbf{r}}$
Uniform spherical shell	gravitational field *inside* the shell	$\vec{\mathbf{g}}(r) = 0 \ (\text{m/s}^2)$
	gravitational field *outside* the shell a distance r away	$\vec{\mathbf{g}}(r) = \dfrac{-GM}{r^2}\,\hat{\mathbf{r}}$
Uniform sphere (either a uniform or a spherically symmetric distribution of mass)	gravitational field *outside* the sphere a distance r away	$\vec{\mathbf{g}}(r) = \dfrac{-GM}{r^2}\,\hat{\mathbf{r}}$
Uniform sphere	gravitational field *inside* the sphere a distance r from the center	$\vec{\mathbf{g}}(r) = \dfrac{-GM\,r}{R^3}\,\hat{\mathbf{r}}$
Gravitational field	near the surface of the Earth	(a constant vector) $g = 9.81 \ \text{m/s}^2$ directed vertically down
Gravitational field of a mass ring	a distance z from the plane of the ring along the symmetry axis of the ring	$\vec{\mathbf{g}}(z) = \dfrac{-GM\,z}{(R^2 + z^2)^{3/2}}\,\hat{\mathbf{k}}$

When near the surface of the Earth, the gravitational field has a magnitude 9.81 m/s^2 directed vertically down, but at other locations in space, the magnitude of the gravitational field of the Earth (or that due to other masses) is different.

PROBLEM-SOLVING TACTIC

6.3 Do not confuse the magnitude of the gravitational field $\vec{\mathbf{g}}(r)$ at any location in space with g, the magnitude of the local acceleration due to gravity near the surface of the Earth. The field $\vec{\mathbf{g}}(r)$ has a magnitude $g = 9.81$ m/s^2 only when near the surface of the Earth.

The gravitational field is a vector quantity since it is a force per unit mass (or, equivalently, the local acceleration due to gravity).

2. Since gravitational *forces* follow a principle of linear superposition, gravitational *fields* also must obey the same type of superposition rule because the field is a force per unit mass. Thus the total gravitational field established at some point in space by a number of masses is the *vector sum* of the individual gravitational fields at that point established by each mass acting individually, as shown in Figure 6.41.

The gravitational field established in space by an *extended mass* is found by dividing the extended mass into small pointlike differential masses dM (see Figure 6.42). The field $d\vec{\mathbf{g}}$ established at a point P in space by this differential mass dM is

$$d\vec{\mathbf{g}} = -\frac{G\,dM}{r^2}\,\hat{\mathbf{r}} \qquad (6.44)$$

where r is the distance between dM and P and $\hat{\mathbf{r}}$ is a unit vector pointing radially away from dM.

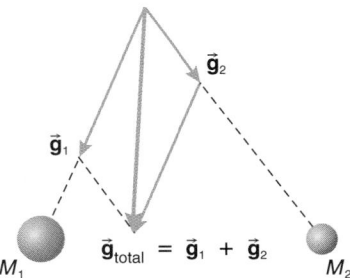

FIGURE 6.41 The principle of superposition applies to gravitational fields.

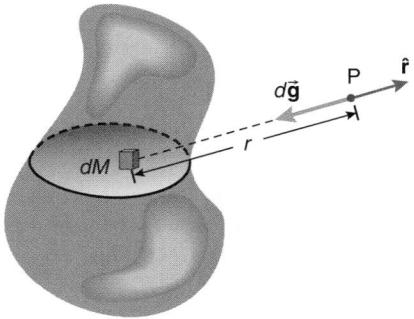

FIGURE 6.42 A differential mass establishes a differential gravitational field at a point in space.

The total gravitational field established by the extended mass at point P in space then is found by integrating Equation 6.44 over the physical extent of the mass distribution M:

$$\vec{g} = -G \int_{\text{mass distribution}} \frac{dM}{r^2} \hat{r} \qquad (6.45)$$

This integral is a vector integration and care must be taken when executing it. Example 6.12 illustrates how to perform such integrations. The gravitational field established by an extended mass is a sophisticated application of the principle of superposition. Since m cannot exert a gravitational force on itself, if m is placed *in* a gravitational field, the field affecting m is *not* the field established by m itself, but is the field established by all *other* masses.

3. The influence of a mass M *on* another mass *m* is thought of as being conveyed by the field. If M changes its location at some instant, the change in the field of M at a particular point in space occurs at a later time. Einstein's theory of general relativity predicts that such changes in the gravitational field propagate through space at the speed of light. Thus if, say, the Sun somehow disappeared at some instant, the gravitational field of the Sun at the position of the Earth would be unaffected for eight minutes, the light-travel-time from the Sun to the Earth, before ceasing. As mentioned earlier, we will not be concerned in this text with such propagation delays, but you should be aware that they exist.

4. If it looks as if we have merely taken one abstract idea for conveying the force, action-at-a-distance, and substituted the equally abstract idea of a field, you are right to some extent. The field solves the problem of instantaneous propagation, but it still does not answer the question of what gravity *is* or why it acts this way. But the field idea is a very useful abstraction.

One of the reasons for introducing the idea of a gravitational field in the space surrounding a mass M is that it enables us to talk about the gravitational influence of M itself without the bother of having another mass (say *m*) in the picture. In other words, with the gravitational field, we can begin to talk about the gravitational characteristics of M all by itself. A problem that initially had *two* masses (M and *m*) and *two* forces (of M *on m* and of *m on* M) now just has *one* mass (say M) and its very own gravitational field.*

QUESTION 7 _____

Is it possible to measure directly the gravitational field, or is it determined only from its action on a mass placed in the field?

EXAMPLE 6.10 ━━━━━━━━━━

a. Find the gravitational field established by the Earth at a point in space located a distance from the center of the Earth equal to twice the Earth's radius.

---

*Of course, *m* produces its own gravitational field too, which acts on the other mass M.

b. What is the acceleration due to gravity at this location?
c. What is the gravitational force on a **1.00 kg** mass at this location?
d. What is the gravitational force of the Earth on a **500 kg** space-cow chewing its cud at this unlikely location in the field?

Solution

a. The Earth is approximately spherical in shape, and so, from Table 6.1, the gravitational field established by the Earth in space is

$$\vec{g} = -\frac{GM}{r^2}\hat{r}$$

where M is the mass of the Earth (5.98×10^{24} kg) and r is the distance from the center of the Earth to the point where we want the field, here two Earth radii:

$$r = 2(6.37 \times 10^6 \text{ m})$$
$$= 12.7 \times 10^6 \text{ m}$$

Substituting into the expression for the field,

$$\vec{g} = -\frac{(6.67 \times 10^{-11} \text{ N}\cdot\text{m}^2/\text{kg}^2)(5.98 \times 10^{24} \text{ kg})}{(12.7 \times 10^6 \text{ m})^2}\hat{r}$$
$$= (-2.47 \text{ m/s}^2)\hat{r}$$

b. The gravitational field *is* the acceleration due to gravity at this location, and so the acceleration due to gravity at this location is

$$(-2.47 \text{ m/s}^2)\hat{r}$$

c. The gravitational force on a mass m at a location where the field is $\vec{g}$ is

$$\vec{F} = m\vec{g}$$
$$= (1.00 \text{ kg})(-2.47 \text{ m/s}^2)\hat{r}$$
$$= (-2.47 \text{ N})\hat{r}$$

and is directed toward the center of the Earth. Notice that the gravitational force *on one kilogram* is numerically equal to the value of the field; thus the field can be considered to be the gravitational force per kilogram at the point.

d. The gravitational force of the Earth on a 500 kg space-cow at this location is the product of the mass of the cow and the gravitational field of the Earth at this point in space:

$$\vec{F} = m\vec{g}$$
$$= (500 \text{ kg})(-2.47 \text{ m/s}^2)\hat{r}$$
$$= (-1.24 \times 10^3 \text{ N})\hat{r}$$

STRATEGIC EXAMPLE 6.11

a. What is the total gravitational field established by both the Earth and Moon at the point P in Figure 6.43?
b. What is the magnitude of the total gravitational field at this point?
c. What is the acceleration of a 1200 kg spacecraft at this location? What is the magnitude of this acceleration?

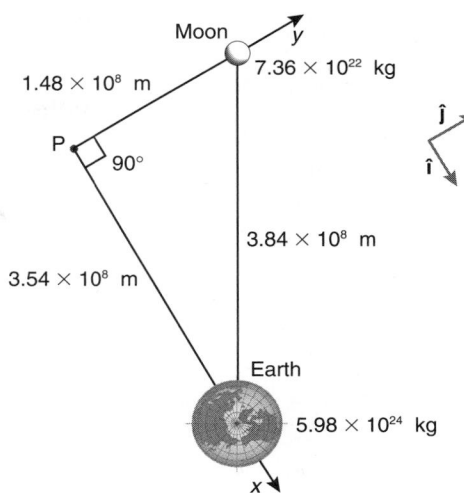

FIGURE 6.43

d. What is the total gravitational force on the spacecraft at this location? What is the magnitude of the total gravitational force on the spacecraft?

Solution

a. Consider both the Earth and Moon to be spherical masses. Use the principle of superposition to find the total field by first finding, individually, the field established by the Moon at this location (as if the Earth were not present) and the field established by the Earth at this location (as if the Moon were not present).

The gravitational field established by the Moon at this location is of magnitude

$$g_{Moon} = \frac{GM_{Moon}}{r^2}$$

where r is the distance between the center of the Moon and the point P, here 1.48×10^8 m. Substituting numerical values, you find

$$g_{Moon} = \frac{(6.67 \times 10^{-11} \text{ N} \cdot \text{m}^2/\text{kg}^2)(7.36 \times 10^{22} \text{ kg})}{(1.48 \times 10^8 \text{ m})^2}$$
$$= 2.24 \times 10^{-4} \text{ m/s}^2$$

This field is directed toward the center of the Moon. Use the coordinate system indicated in Figure 6.44. The gravitational field established by the Moon at the point P is

$$\vec{g}_{Moon} = (2.24 \times 10^{-4} \text{ m/s}^2)\,\hat{\jmath}$$

The magnitude of the gravitational field established by the Earth at the point P also is of the form

$$g_{Earth} = \frac{GM_{Earth}}{r^2}$$

where now r is the distance from the center of the Earth to P, here 3.54×10^8 m. Making the substitutions,

$$g_{Earth} = \frac{(6.67 \times 10^{-11} \text{ N} \cdot \text{m}^2/\text{kg}^2)(5.98 \times 10^{24} \text{ kg})}{(3.54 \times 10^8 \text{ m})^2}$$
$$= 3.18 \times 10^{-3} \text{ m/s}^2$$

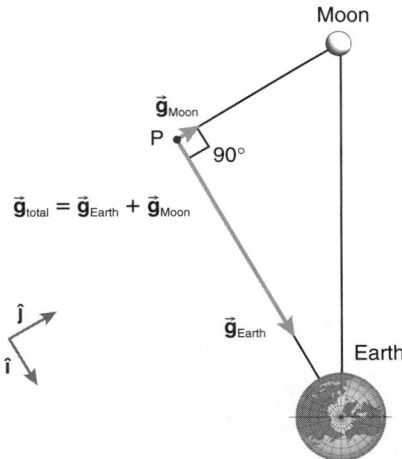

FIGURE 6.44

This field is directed toward the center of the Earth. Use the coordinate system in Figure 6.44; the gravitational field established by the Earth at point P is

$$\vec{g}_{Earth} = (3.18 \times 10^{-3} \text{ m/s}^2)\,\hat{\imath}$$

The *total* gravitational field at the point P is the *vector sum* of the two fields, as indicated in Figure 6.45:

$$\vec{g}_{total} = \vec{g}_{Earth} + \vec{g}_{Moon}$$
$$= (3.18 \times 10^{-3} \text{ m/s}^2)\,\hat{\imath} + (0.22 \times 10^{-3} \text{ m/s}^2)\,\hat{\jmath}$$

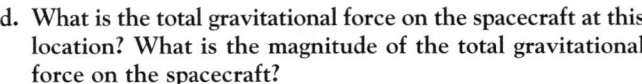

FIGURE 6.45

b. The magnitude of the total field at P is the magnitude of the vector $\vec{g}_{total}$:

$$g_{total} = [(3.18 \times 10^{-3} \text{ m/s}^2)^2 + (0.22 \times 10^{-3} \text{ m/s}^2)^2]^{1/2}$$
$$= 3.19 \times 10^{-3} \text{ m/s}^2$$

c. The total gravitational field at P *is* the acceleration due to gravity at this location:

$$\vec{g}_{total} = (3.18 \times 10^{-3} \text{ m/s}^2)\,\hat{\imath} + (0.22 \times 10^{-3} \text{ m/s}^2)\,\hat{\jmath}$$

The magnitude of the acceleration of the spacecraft is the magnitude of the field:

$$g_{\text{total}} = 3.19 \times 10^{-3} \text{ m/s}^2$$

The acceleration is independent of the mass of the spacecraft.

d. The gravitational force on the spacecraft when it is at this location is found using Equation 6.43:

$$\vec{F} = m\vec{g}_{\text{total}}$$
$$= (1200 \text{ kg})[(3.18 \times 10^{-3} \text{ m/s}^2)\hat{\imath} + (0.22 \times 10^{-3} \text{ m/s}^2)\hat{\jmath}]$$
$$= (3.82 \text{ N})\hat{\imath} + (0.26 \text{ N})\hat{\jmath}$$

The magnitude of this force is
$$F = [(3.82 \text{ N})^2 + (0.26 \text{ N})^2]^{1/2}$$
$$= 3.83 \text{ N}$$

The magnitude of the force also could be found using

$$F = mg_{\text{total}}$$
$$= (1200 \text{ kg})(3.19 \times 10^{-3} \text{ m/s}^2)$$
$$= 3.83 \text{ N}$$

STRATEGIC EXAMPLE 6.12

A mass M is distributed uniformly in the shape of a circular ring of radius R, as shown in Figure 6.46. Such mass rings are used to model the individual rings of the planets Jupiter, Saturn, Uranus, and Neptune.

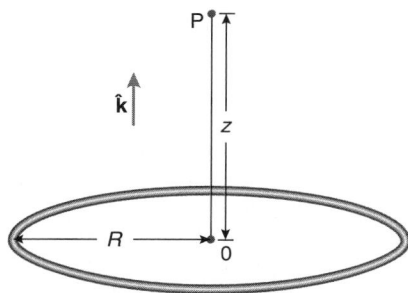

FIGURE 6.46

The concentric ring system of Saturn consists of a vast collection of orbiting rocks. The rings are not solid.

a. Find the gravitational field established by the ring along the symmetry axis of the ring at a distance z from the plane of the ring.

b. What is the gravitational field at the center of the ring?

Solution

a. The ring is an extended mass distribution; you need to use Equation 6.45 and perform a vector integration. On every meter of the circumference of the ring there is a mass λ, where

$$\lambda = \frac{M}{2\pi R}$$

The mass per meter λ is known as a linear mass density. A small differential length ds of the circumference of the ring, therefore, has a mass dM equal to the product of ds times the linear mass density λ:

$$dM = \lambda \, ds$$

This pointlike differential mass produces a gravitational field at the point P that is of magnitude

$$dg = G \frac{dM}{r^2}$$

and is directed toward the mass dM, as shown in Figure 6.47.

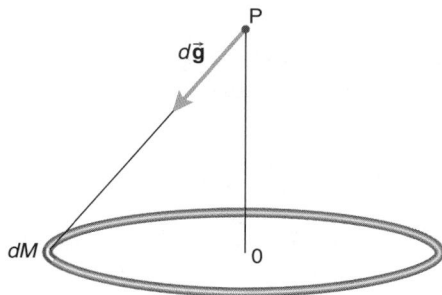

FIGURE 6.47

Another similar differential mass on the opposite side of the ring produces a similar field at P directed toward it (see Figure 6.48). Notice in Figure 6.48 that only the components of these fields parallel to $-\hat{k}$ survive a vector addition of the fields; the components parallel to the plane of the ring vector sum to zero. Hence the portion of the field produced by dM at P that survives the vector summation (integration) will be directed along $-\hat{k}$ and is

$$d\vec{g} = \left(G \frac{dM}{r^2} \cos \theta \right)(-\hat{k})$$

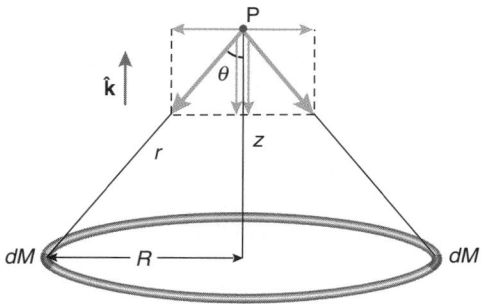

FIGURE 6.48

Substitute for dM and integrate over the extent of the ring. This is the total gravitational field of the ring at P:

$$\vec{g} = G \int_{ring} \left(\frac{\lambda \, ds}{r^2} \cos \theta \right) (-\hat{k})$$

When you integrate around the ring mass distribution, every mass element on the ring is the same distance r from the point P. Likewise $\cos \theta$ also is constant, since the angle θ is unchanged as you go around the ring. The linear mass density λ is constant too. Hence the integration becomes

$$\vec{g} = -\frac{G\lambda \cos \theta}{r^2} \int_{ring} ds \, \hat{k}$$

The integral of ds around the ring is the circumference of the ring, $2\pi R$. Hence the gravitational field of the ring at P is

$$\vec{g} = -\frac{G\lambda \cos \theta}{r^2} 2\pi R \hat{k}$$

Now substitute for $\lambda = M/(2\pi R)$. With the Pythagorean theorem you have

$$r^2 = R^2 + z^2$$

and geometrically (see Figure 6.48)

$$\cos \theta = \frac{z}{r} = \frac{z}{(R^2 + z^2)^{1/2}}$$

After these substitutions, the gravitational field established by the ring at P is found to be

$$\vec{g}(z) = -\frac{GMz}{(R^2 + z^2)^{3/2}} \hat{k} \tag{1}$$

Notice that for $z > 0$ m the field is directed toward the ring (parallel to $-\hat{k}$). For $z < 0$ m the field *also* is directed toward the ring (parallel to $+\hat{k}$).

b. There are two ways to find the gravitational field at the center of the ring.

Method 1

At the center of the ring where $z = 0$ m, equation (1) indicates the field is zero.

Method 2

Using symmetry considerations, you also can deduce that the field will be zero at the center of the ring since mass elements on opposite sides of the ring produce gravitational fields in opposing directions, as shown in Figure 6.49.

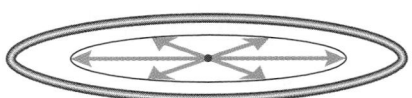

FIGURE 6.49

A mass placed at the center of the ring will experience zero gravitational force, since the field is zero at that location. Problem 67 asks you to show that the maximum magnitude of the field along the symmetry axis is at the point where $|z| = R/\sqrt{2}$.

6.14 THE FLUX OF A VECTOR*

The gravitational field has an interesting property associated with the masses that establish the field. To investigate this property of the gravitational field, it is necessary to introduce a new physical concept associated with a vector: its **flux**.

A differential surface area dS is a tiny area of differential size, indicated schematically in Figure 6.50. We use the letter S for the area here, meaning Surface. Since the differential area is arbitrarily small (the case with all differentials), the area is imagined to be *locally flat*, even if it is part of a larger curved surface.

We occasionally make a *vector* out of such a tiny differential area. Take the differential area as the *magnitude* of the differential area vector. In the plane of the differential area, there are infinitely many different directions that might be chosen and it is impossible to single out a unique direction for the area vector. But two unique directions are those lying along the perpendicular to the locally flat differential area, as shown in Figure 6.51. Either one of these perpendicular directions could serve as the direction of the differential area vector. However, if the differential area is part of a larger surface that also is a **closed surface**, then the convention is to choose the direction pointing to the *outside* of the enclosed volume, as in Figure 6.52.

What is a closed surface? Imagine submerging such a surface in water. A closed surface will not get wet on the inside; there are no holes. So, for example, an empty but sealed milk carton has a closed surface. An empty milk carton without its cap is not a closed surface. If the sealed carton has even a pin prick in it, the surface is not closed. A doughnut has a surface that is closed even though the doughnut is said to have a hole. We will encounter closed surfaces in a number of contexts.

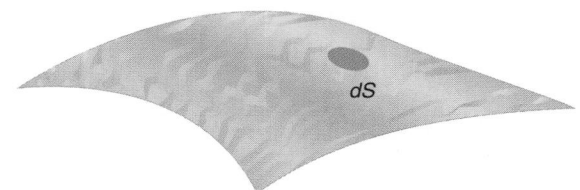

FIGURE 6.50 A differential area dS.

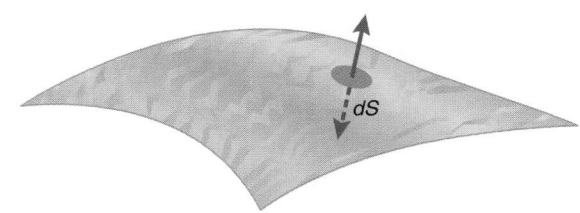

FIGURE 6.51 Two directions are associated with the perpendicular to the differential area.

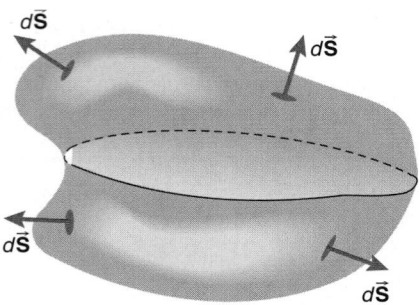

FIGURE 6.52 If the larger surface is closed, the differential area vector points to the outside of the enclosed volume.

Thus a differential area vector $d\vec{S}$ is locally flat, has a magnitude equal to the differential area in m², and is directed perpendicular to the surface in an outward direction if the larger surface is closed. If the larger surface is not closed, we are free to choose the direction of $d\vec{S}$ to be along either direction of the perpendicular to the differential surface area.

Now we tackle the idea of the flux of a vector, say, the flux of $\vec{g}$, the gravitational field vector, where $\vec{g}$ exists at the location of the differential area vector $d\vec{S}$. We first define what is meant by the flux in a mathematical sense and then explore its meaning to get a feeling for the terminology and a mental picture of what the flux represents.

The differential flux $d\Phi$ of the vector $\vec{g}$ through the differential area dS is defined to be the scalar product of $\vec{g}$ with $d\vec{S}$:

$$d\Phi \equiv \vec{g} \cdot d\vec{S} \qquad (6.46)$$

The value of $\vec{g}$ at the location of the differential area is used in this expression. We describe Equation 6.46 in the following way: the differential flux of the vector $\vec{g}$ *through* the differential area is $d\Phi$.

Note that the flux is a scalar quantity.

Use the definition of the scalar product of two vectors; the differential flux through the differential area is

$$d\Phi = g \, dS \cos \theta$$

where θ is the angle between the vector $\vec{g}$ and the differential area vector $d\vec{S}$, as shown in Figure 6.53. The differential flux $d\Phi$ may be positive, negative, or zero depending on the orientation of $\vec{g}$ relative to $d\vec{S}$.

Let's look at a few situations to get a feeling for the meaning of the flux. The word flux implies flow of something through the differential area dS.* While this has implications of fluids moving through the differential area dS rather than gravitational fields, the concept of flux finds wide use particularly with field vectors such as the gravitational field vector $\vec{g}$.†

If the angle between the vectors $\vec{g}$ and $d\vec{S}$ is 90°, as in Figure 6.54, then the cosine is zero and the differential flux of $\vec{g}$ through dS is zero:

$$d\Phi = g \, dS \cos 90° = 0 \ \ \text{m}^3/\text{s}^2$$

In this case the arrow representing the vector $\vec{g}$ does not pass through, or "thread," the little area if you imagine the differential area to be like the eye of a needle.

If the angle between the vectors $\vec{g}$ and $d\vec{S}$ is 0°, as in Figure 6.55, then the differential flux of the vector $\vec{g}$ through dS is

$$d\Phi = g \, dS \cos 0° = g \, dS$$

In this case, the arrow representing the vector $\vec{g}$ threads the little area dS "head-on." This is the geometry used when one really does try to thread a needle with a piece of thread.

If the angle between $\vec{g}$ and $d\vec{S}$ is somewhere between 0° and 90°, as in Figure 6.56, the differential flux has a value between its maximum value ($g \, dS$) and zero.

The largest negative value of the differential flux occurs when the angle between the two vectors is 180°, as in Figure 6.57. For this case,

$$d\Phi = \vec{g} \cdot d\vec{S} = g \, dS \cos 180°$$
$$= -g \, dS$$

---

*The word flux comes from the Lain *fluxus*, to flow. It originally was associated with the amount of fluid flowing per second through an area placed at various orientations to the direction of the fluid flow, but the concept has been generalized to encompass other vectors as well.

† We also will use the flux concept later with electric and magnetic fields.

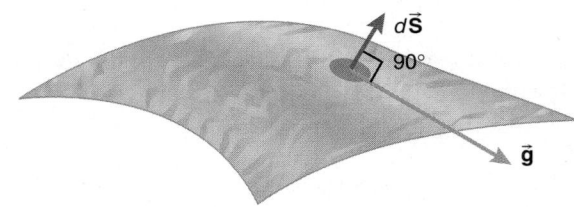

FIGURE 6.54 Zero differential flux of $\vec{g}$ through dS.

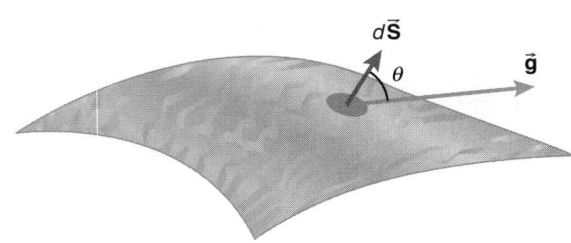

FIGURE 6.53 The differential flux of $\vec{g}$ through dS is the scalar product $\vec{g} \cdot d\vec{S}$.

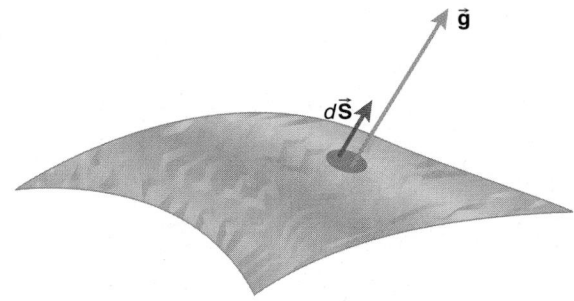

FIGURE 6.55 Maximum differential flux of $\vec{g}$ through dS.

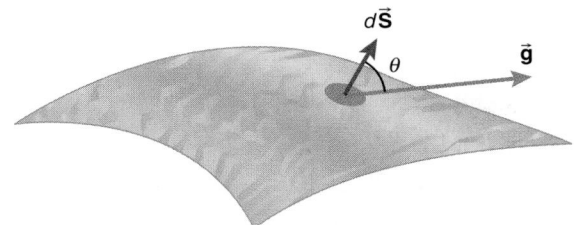

FIGURE 6.56 An intermediate case.

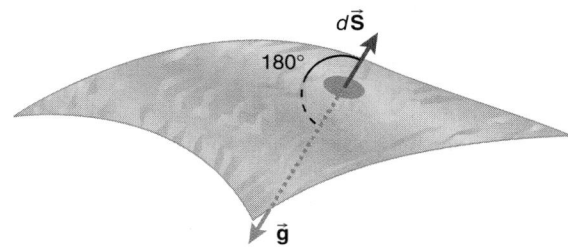

FIGURE 6.57 Largest negative differential flux of $\vec{\mathbf{g}}$ through dS.

In this case the vector $\vec{\mathbf{g}}$ threads the little area dS backward, or opposite to the direction of the differential area vector $d\vec{\mathbf{S}}$.

Hence the differential flux of $\vec{\mathbf{g}}$ through the differential area either can be positive, negative, or zero depending on the relative orientation of the vector $\vec{\mathbf{g}}$ to $d\vec{\mathbf{S}}$. The flux is a measure of the extent to which the vector $\vec{\mathbf{g}}$ threads or flows through the differential area.

To find the total flux Φ of the vector $\vec{\mathbf{g}}$ through a finite area S, we must integrate the differential flux over the entire area S:

$$\Phi = \int_S d\Phi = \int_S \vec{\mathbf{g}} \cdot d\vec{\mathbf{S}} = \int_S g \, dS \cos\theta \qquad (6.47)$$

This integration may be difficult or easy depending on the shape of the surface S and the analytical form of the field. As the differential area dS scans a large surface, the value of the vector $\vec{\mathbf{g}}$ may change and the angle $\vec{\mathbf{g}}$ makes with the local $d\vec{\mathbf{S}}$ also may change, so that the integration can be difficult. Certain types of fields may have a high degree of symmetry, dictating surfaces through which the procedure of flux integration is tractable. We often *choose* the surface over which to calculate the flux, and it is best to make the job as easy as possible by choosing the surfaces dictated by the symmetry of the problem. The choice is accomplished by trying to select larger areas over which the vector $\vec{\mathbf{g}}$ has constant magnitude and maintains a fixed orientation with respect to all the differential areas on the surface.

EXAMPLE 6.13

a. Find the flux of the local gravitational field near the surface of the Earth through each of the five surfaces of the inclined plane shown in Figure 6.58.
b. What is the total flux through the entire closed surface?

Solution

a. Take $\hat{\mathbf{j}}$ to be vertically upward. Then the constant gravitational field near the surface of the Earth is

$$\vec{\mathbf{g}} = -g\hat{\mathbf{j}}$$

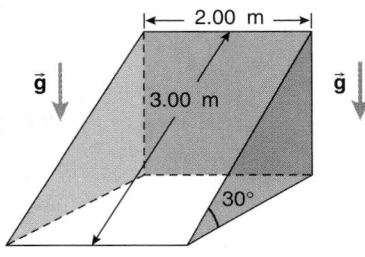

FIGURE 6.58

where g is 9.81 m/s² near the surface of the Earth. Since the larger surface is closed, all the differential area vectors must be chosen (by convention) to be directed to the outside of the closed surface.

Note that on each of the three vertical surfaces, every differential area vector $d\vec{\mathbf{S}}$ is perpendicular to the direction of the field, as shown in Figure 6.59. Hence, at every $d\vec{\mathbf{S}}$ on each vertical surface,

$$\vec{\mathbf{g}} \cdot d\vec{\mathbf{S}} = 0 \ \text{m}^3/\text{s}^2$$

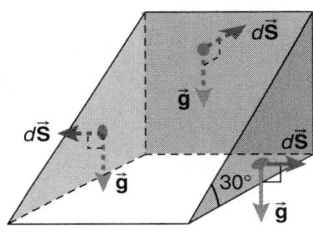

FIGURE 6.59

Hence there is no flux of $\vec{\mathbf{g}}$ through the vertical parts of the larger closed surface:

$$\Phi_{\text{vert}} = 0 \ \text{m}^3/\text{s}^2$$

Note also that the lines and arrows representing $\vec{\mathbf{g}}$ nowhere thread these vertical surfaces.

All the differential area vectors on the inclined surface are perpendicular to the inclined plane (see Figure 6.60). Since the larger area in question is closed, you have to choose the outward direction for all the $d\vec{\mathbf{S}}$ on this inclined surface. Use the coordinate system indicated in Figure 6.60 to write the differential area vector as

$$d\vec{\mathbf{S}} = dS \, [(-\sin\theta)\,\hat{\mathbf{i}} + (\cos\theta)\,\hat{\mathbf{j}}]$$

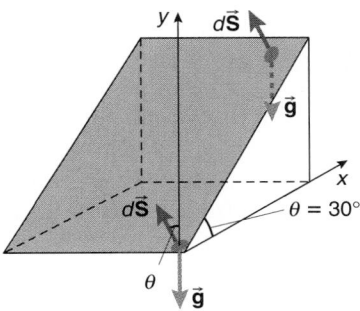

FIGURE 6.60

According to Equation 6.47, the flux of $\vec{g}$ through the area S is

$$\Phi_{incl} = \int_S \vec{g} \cdot d\vec{S}$$
$$= \int_S (-g\hat{j}) \cdot dS \, [(-\sin\theta)\,\hat{i} + (\cos\theta)\,\hat{j}]$$
$$= -\int_S g \, dS \cos\theta$$

Since g and $\cos\theta$ are constant over the entire area S, the flux becomes

$$\Phi_{incl} = -gS \cos\theta$$
$$= (-9.81 \text{ m/s}^2)(6.00 \text{ m}^2) \cos 30.0°$$
$$= -51.0 \text{ m}^3/\text{s}^2$$

The negative value of the flux indicates that the gravitational field vector here passes into the larger closed surface through the inclined surface.

On the bottom surface, all the differential area vectors are written as

$$d\vec{S} = -dS \, \hat{j}$$

since the outward direction from the closed surface must be chosen (see Figure 6.61). The flux of $\vec{g}$ through this area then is

$$\Phi_{btm} = \int_S \vec{g} \cdot d\vec{S}$$
$$= \int_S (-g\hat{j}) \cdot (-dS \, \hat{j})$$
$$= \int_S g \, dS$$

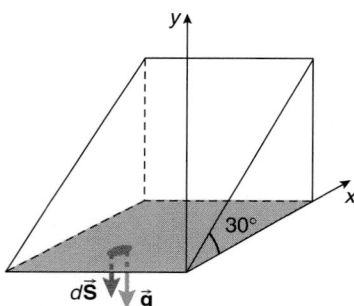

FIGURE 6.61

Since g is constant over the entire lower surface, the flux through this surface becomes

$$\Phi_{btm} = gS$$
$$= (9.81 \text{ m/s}^2)[(2.00 \text{ m})(3.00 \text{ m}) \cos 30.0°]$$
$$= 51.0 \text{ m}^3/\text{s}^2$$

The positive result indicates that the gravitational field threads out of the larger closed surface through the bottom surface.

b. The total flux of the gravitational field is the sum of the fluxes through each of the parts of the larger closed surface:

$$\Phi_{total} = \Phi_{incl} + \Phi_{vert} + \Phi_{btm}$$
$$= -51.0 \text{ m}^3/\text{s}^2 + 0 \text{ m}^3/\text{s}^2 + 51.0 \text{ m}^3/\text{s}^2$$
$$= 0 \text{ m}^3/\text{s}^2$$

Although the gravitational field vectors enter the closed surface through the inclined surface, they depart through the bottom, so that the total flux of the gravitational field of the Earth through the closed surface is zero.

With some reflection, you should be able to show that the flux of a constant vector through *any* closed surface is zero.

6.15 GAUSS'S LAW FOR THE GRAVITATIONAL FIELD*

Although one can calculate the flux of *any* vector through *any* surface, the flux of many vectors is not physically meaningful. For some vectors, however, the flux of the vector through a surface does have consequence; the gravitational field is one of those lucky vectors.

It will take some thoughtful study to find the total flux of *any* gravitational field through *any* closed surface; we will see, however, that the *result* is remarkably simple. The flux of the gravitational field through a closed surface involves calculating the integral over the surface of the scalar product of $\vec{g}$ with a differential area element $d\vec{S}$:

$$\Phi_{clsd \atop surface\ S} = \int_{clsd \atop surface\ S} \vec{g} \cdot d\vec{S}$$

The closed surface can be *any* closed surface we please to choose. It can be a real surface or a surface that we simply *imagine*.

To find the flux of the gravitational field through a closed surface, it first is convenient to extend the idea of angular measure to three dimensions. The radian measure of an angle is defined as the ratio between an arc length s of a circle of radius r and the radius itself (see Figure 6.62):

$$\theta \equiv \frac{s}{r}$$

where θ is measured at the center of the circle.

Since the circumference of a circle is $2\pi r$, there are

$$\frac{2\pi r}{r} = 2\pi \text{ rad}$$

subtended by a complete circle.

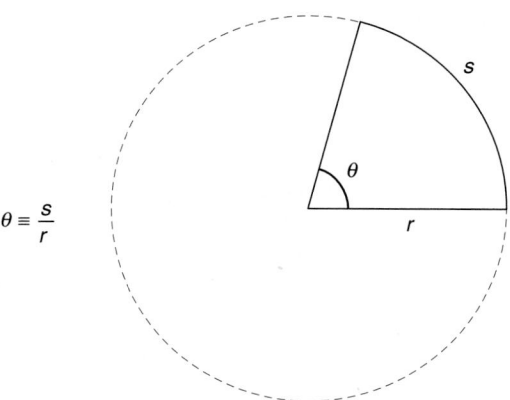

FIGURE 6.62 An angle measured in radians.

Instead of an arc on a circle, consider an area A on a sphere, as shown in Figure 6.63. The **solid angle** Ω measured at the center of the sphere, subtended by the area A is defined to be the ratio between the area A on the sphere and the square of the radius of the sphere:

$$\Omega \equiv \frac{A}{r^2} \qquad (6.48)$$

The units for measuring solid angles are **steradians** (sr). Steradians are dimensionless units, as are radians. Since the surface area of a sphere is $4\pi r^2$, the number of steradians subtended by the whole sphere is

$$\frac{4\pi r^2}{r^2} = 4\pi \ \text{sr}$$

A differential area dA on a sphere subtends a differential solid angle $d\Omega$ (see Figure 6.64), where

$$d\Omega = \frac{dA}{r^2}$$

If a differential area dS is *not* on a sphere of radius r, nor perpendicular to $\vec{r}$, the differential solid angle $d\Omega$ subtended by dS from a reference point 0 then is the quotient of the area of dS *projected* perpendicular to the radial direction (see Figure 6.65), divided by the square of the radial distance. The projected area dA involves the angle ϕ between the differential area vector $d\vec{S}$ and the radial direction. Indeed, $dA = d\vec{S} \cdot \hat{r}$.

Hence the solid angle $d\Omega$ subtended by dS is

$$d\Omega = \frac{dA}{r^2} = \frac{d\vec{S} \cdot \hat{r}}{r^2} = \frac{dS \cos \phi}{r^2} \qquad (6.49)$$

Now we have the tools to find the flux of any gravitational field through any closed surface.

Let a *closed surface* of *any arbitrary shape* surround (i.e., enclose) a particle of mass M, as in Figure 6.66. The mass M may be located anywhere within the closed surface. The flux of the gravitational field of M through this closed surface is

$$\Phi_{\substack{\text{clsd surface} \\ S \ \text{encl M}}} = \int_{\substack{\text{clsd} \\ \text{surface } S}} \vec{g} \cdot d\vec{S}$$

Look at the integrand:

$$\vec{g} \cdot d\vec{S} = g \, dS \cos \theta$$

where $\vec{g}$ is the gravitational field vector created by mass M at the location of the differential area $d\vec{S}$. Here θ is the angle between the direction of $\vec{g}$ and the direction of the differential area vector $d\vec{S}$ as shown in Figure 6.66. A point mass M creates a gravitational field given by Equation 6.42. The *magnitude* of the gravitational field established by M is

$$g = \frac{GM}{r^2}$$

The differential area vector is normal to the surface and points *outward*, since the surface is a closed surface. From Figure 6.66, note that

$$\cos \theta = \cos(\pi - \phi) = -\cos \phi$$

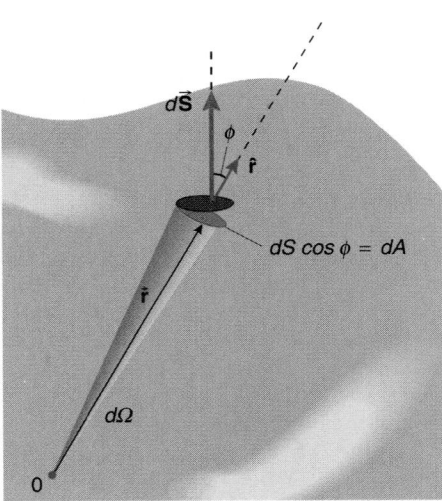

FIGURE 6.65 A differential area dS not part of a sphere also subtends a differential solid angle $d\Omega$ from a reference point 0.

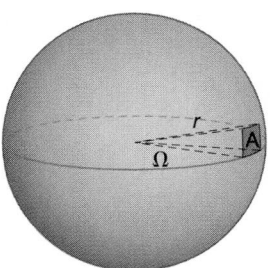

FIGURE 6.63 An area A on a sphere of radius R.

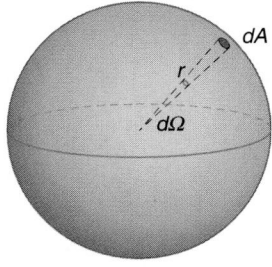

FIGURE 6.64 A differential area dA on a sphere subtends a solid angle $d\Omega$.

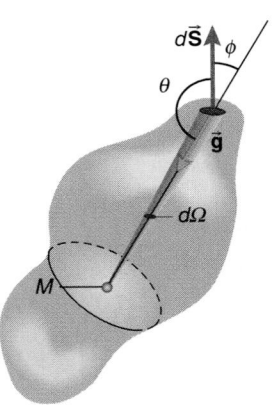

FIGURE 6.66 A closed surface surrounding a mass M.

Therefore the integrand in the flux integral is

$$g \, dS \cos \theta = \frac{GM}{r^2} dS \, (-\cos \phi)$$

$$= -GM \frac{dS \cos \phi}{r^2} \qquad (6.50)$$

But $(dS \cos \phi)/r^2$ is the differential solid angle $d\Omega$ subtended by the differential area dS as seen from M (Equation 6.49). Thus the integrand for the flux integral is

$$\vec{\mathbf{g}} \cdot d\vec{\mathbf{S}} = -GM \, d\Omega$$

Therefore the flux of the gravitational field of M through the entire closed surface is

$$\Phi_{\substack{\text{clsd surface} \\ S \text{ encl M}}} = \int_{\substack{\text{clsd} \\ \text{surface}}} \vec{\mathbf{g}} \cdot d\vec{\mathbf{S}} = -GM \int_{\substack{\text{clsd} \\ \text{surface}}} d\Omega$$

But the integral of the solid angle over the whole closed surface is the solid angle subtended by the whole surface as seen from M. Since the closed surface completely surrounds M, the solid angle subtended by the closed surface is the *same* as the solid angle subtended by a sphere enclosing M—namely, 4π sr. Thus the integral of the solid angle over the whole closed surface is simply 4π sr. Hence the flux of the gravitational field $\vec{\mathbf{g}}$ through an arbitrarily shaped closed surface surrounding M is

$$\Phi_{\text{clsd surface encl M}} = -4\pi GM \qquad (6.51)$$

where we omit the steradian (sr) unit associated with the 4π, since it is dimensionless. Equation 6.51 is a remarkable result because it is independent of the surface shape *or* of where the mass M is located within the surface. If you study this derivation carefully (now, how else would you?), you will see that the simplicity of the result arises only because of the *inverse square* nature of the gravitational field: the r^2 term in Equation 6.50 came from the gravitational field vector of the pointlike mass M. We see how kind nature has been to formulate gravitation as an inverse square law!

What does the flux of the gravitational field mean? It is a measure of the extent or total amount that the gravitational field vector threads through the surface. In the case of a closed surface surrounding M, the gravitational field vector $\vec{\mathbf{g}}$ of M comes into the closed surface without any corresponding exits. The minus sign in Equation 6.51 for the flux indicates that the field threads into the closed surface rather than threading outward (exiting).

If some other mass M′ is located within the closed surface instead of M, its gravitational field produces a flux through the closed surface that is equal to

$$-4\pi GM'$$

because the flux through the surface is independent of the location of the mass within the surface. Thanks to the principle of superposition, if *both* M and M′ are within the surface, the total flux of the total field through the closed surface is the sum of the individual fluxes:

$$\Phi_{\substack{\text{total} \\ \text{gravitational} \\ \text{field}}} \text{ (total through S)} = -4\pi GM - 4\pi GM'$$

So the total flux of the gravitational field through a closed surface is a measure of the total mass enclosed *anywhere* within the surface:

flux of the total
gravitational $= -4\pi G[\text{total mass enclosed by the surface}]$
field through a
closed surface

or, more mathematically,

$$\int_{\substack{\text{clsd} \\ \text{surface } S}} \vec{\mathbf{g}} \cdot d\vec{\mathbf{S}} = -4\pi GM_{\text{within } S} \qquad (6.52)$$

We now address the question of what happens to the flux of the gravitational field through a closed surface if it does *not* have M inside, as in Figure 6.67. Something quite remarkable happens. Let this total surface be composed of two arbitrary parts A and B, as shown in Figure 6.68. Note the directions in Figure 6.68 of the outward differential area vectors on surface B. Next construct an *extension* of the surface to enclose M, as in Figure 6.69. You can consider the extension to be a closed surface composed of parts B and C. Note that the outward differential area vectors of area B, when considered as part of the extended closed surface enclosing M, are opposite to those in Figure 6.68.

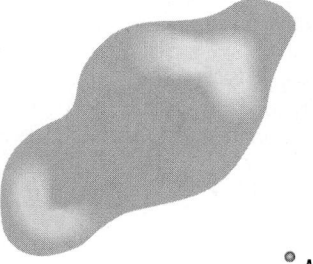

FIGURE 6.67 A mass M outside a closed surface.

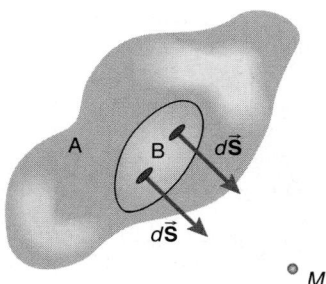

FIGURE 6.68 Divide the closed surface into two arbitrary parts, A and B.

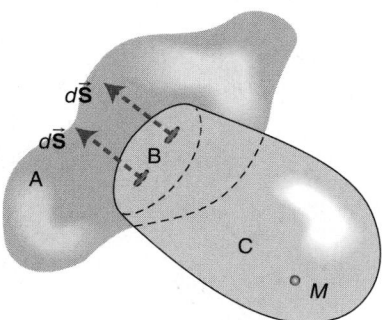

FIGURE 6.69 An extension surface C creates two closed surfaces: surface A with C and surface C with B.

Since we already know the value of the total flux threading a complete surface that encloses M, we have two cases here:

$$\Phi_A + \Phi_C = -4\pi GM \qquad (6.53)$$

$$\Phi_C + (-\Phi_B) = -4\pi GM \qquad (6.54)$$

The minus sign for Φ_B arises because the outward differential areas to surface B, when considered with surface C (forming a closed surface enclosing M), are opposite in direction to those when B is considered with surface A (forming the closed surface *not* enclosing M). Subtracting Equation 6.54 from Equation 6.53,

$$\Phi_A + \Phi_B = 0 \ \text{m}^3/\text{s}^2$$

Finally, since A plus B define our original empty closed surface of arbitrary shape, we have proved that the flux of the gravitational field through any arbitrarily shaped closed surface that excludes M vanishes. The gravitational field has no *net* threading of the closed surface if M is on the *outside* of the closed surface.

These are remarkable results. We summarize them here:

> If M is *inside* the closed surface (whatever the shape of the surface or the location of M inside), the flux of the gravitational field through the surface is
>
> $$\Phi_{\text{clsd surface, M inside}} = -4\pi GM$$

> If M is anywhere *outside* the closed surface, the flux of the gravitational field through the closed surface is zero:
>
> $$\Phi_{\text{clsd surface, M outside}} = 0 \ \text{m}^3/\text{s}^2$$

Both results can be summarized in the following way:

$$\Phi_{\text{clsd surface } S} = -4\pi GM_{\text{within } S} \qquad (6.55)$$

This is known as **Gauss's law for the gravitational field**. The law is named for its discoverer, Carl Friedrich Gauss (1777–1855), whose almost unbelievably imaginative and productive contributions illuminated 19th-century physics, mathematics, and astronomy.

PROBLEM-SOLVING TACTIC

6.4 If you know the amount of mass within a closed surface S, the value of the flux of the gravitational field through the surface is found from the right-hand side of Gauss's law, Equation 6.55. It is not necessary to perform a complicated integral for the flux.

Despite poor and humble roots, Gauss was famous for his intellectual achievements when only in his mid-twenties.

STRATEGIC EXAMPLE 6.14

Consider the Earth–Moon system. The total gravitational field at any point in space is the vector sum of the gravitational field of the Earth and the gravitational field of the Moon at the point. Imagine a closed surface of arbitrary shape surrounding just the Earth, as indicated in Figure 6.70. What is the flux of the *total* gravitational field through this closed surface?

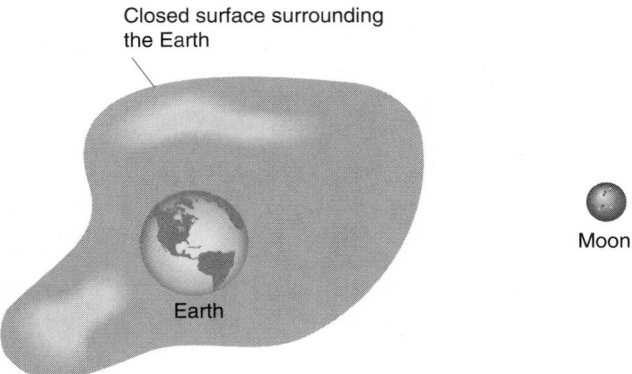

Closed surface surrounding the Earth

Earth

Moon

FIGURE 6.70

Solution

A direct calculation of the flux integral of the total gravitational field through this closed surface is formidable, because the total gravitational field of the Earth and Moon is a complicated function of position over the indicated surface. The angle the total gravitational field vector makes with differential area vectors on the imaginary closed surface also is a complicated function of position on the closed surface. So a brute force calculation of the flux of the *total* gravitational field is something to avoid like the plague. However, Gauss's law for the gravitational field, Equation 6.55, states that the numerical value of that complicated integral for the flux of the *total* gravitational field through this weird closed surface is very simply related to the mass enclosed within the closed surface. Here the only mass within the closed surface is the mass of the Earth, and so it is only the gravitational field established by the Earth that contributes to the flux of the *total* gravitational field through the indicated closed surface; the gravitational field established by the Moon contributes nothing to the flux because the Moon lies outside the indicated closed surface. So use Gauss's law for the gravitational field:

$$\Phi_{\text{clsd surface } S} = -4\pi GM_{\text{within } S}$$
$$= -4\pi GM_{\text{Earth}}$$

Substituting numerical values for G and M_{Earth}, you find

$$\Phi_{\text{clsd surface } S} = -4\pi(6.67 \times 10^{-11} \ \text{N} \cdot \text{m}^2/\text{kg}^2)(5.98 \times 10^{24} \ \text{kg})$$
$$= -5.01 \times 10^{15} \ \text{N} \cdot \text{m}^2/\text{kg}$$
$$= -5.01 \times 10^{15} \ \text{m}^3/\text{s}^2$$

CHAPTER SUMMARY

Newton's law of universal gravitation states that two pointlike masses m and M exert attractive forces on each other, called gravitational forces, of equal magnitude, given by

$$F = \frac{GMm}{r^2} \tag{6.3}$$

where r is the separation of the masses and G is the *universal gravitational constant*

$$G \approx 6.67 \times 10^{-11} \ \text{N} \cdot \text{m}^2/\text{kg}^2$$

The gravitational force acts along the line between the two masses and so is a *central force*.

The *principle of linear superposition* states that when two or more masses exert gravitational forces on another mass m, the total gravitational force on m is the vector sum of the individual gravitational forces on m.

The total gravitational force of a uniform spherical shell of mass M on a mass m located anywhere *inside* the shell is equal to *zero*. If the mass m is located *outside* the shell, the effect of the shell on m is identical to that of a point mass, equal to the mass M of the shell, located at the center of the shell.

The total gravitational force of a spherically symmetric mass M on a pointlike mass m located *outside* the sphere is equivalent to that obtained by considering the sphere as a pointlike mass M located at the center of the sphere. If a mass m is located *inside* the sphere of radius R at a distance $r < R$, only that part of the mass of the sphere that is *closer* to the center than m is effective in exerting a gravitational force on m.

The experimental observation that *all* masses near the surface of the Earth experience the *same* gravitational acceleration (when the gravitational force is the only force on the system) implies that the gravitational mass of a system (the mass appearing in Newton's law of universal gravitation) is identical to the inertial mass of the system (the mass appearing on the right-hand side of Newton's second law of motion).

Kepler's empirical laws of planetary motion describe several characteristics of the orbital motion of a system of mass m around another mass M, where $m \ll M$. The laws of Kepler describe the orbital motion of planets about the Sun, of spacecraft in orbit about the Sun, of artificial satellites of the Earth, and of planetary moons as long as the mass of the moon is much less than the mass of the mother planet. Applied to the planets of the Sun, the laws are as follows:

Kepler's first law: The shape of the orbit of a planet about the Sun is an ellipse with the Sun at one focus of the ellipse; the second focus is empty.

Kepler's second law: A line from the Sun to a given planet sweeps out equal areas during equal time intervals. This law arises because the gravitational force is a *central force*—that is, one directed along the line between two particles.

Kepler's third law: The square of the period of a planet is proportional to the cube of the semimajor axis of its orbit.

Kepler's laws are consequences of Newton's general theory of motion and Newton's law of universal gravitation.

Two objects of comparable mass (for example, two stars, or a planet with a massive moon) both execute orbits about the center of mass of the two-mass system. The center of mass is located closer to the more massive of the pair. The orbits of the two masses have similar shapes, but the orbit of the more massive object of the pair is smaller in size.

For *circular orbits*, the *Newtonian form* for Kepler's third law is

$$T^2 = \frac{4\pi^2}{G(M + m)} r^3 \tag{6.37}$$

where r is the distance between the two masses. The separation distance r is the sum of the radii of the circular orbits of the two masses about their *center of mass*.

For the planets of the Sun, for spacecraft in orbit about the Sun, or for spacecraft in orbit about the Earth or another very massive object (M), the orbital size and acceleration of the very massive object about the center of mass is negligible, since the center of mass is essentially coincident with the mass M itself. For these situations, the Newtonian form of Kepler's third law can be *simplified* (since m is *negligible* compared with M) and *generalized* to include elliptical orbits as well as circular orbits:

$$T^2 = \frac{4\pi^2}{GM} a^3 \qquad (m \ll M) \tag{6.39}$$

where a is the semimajor axis of the orbit of m about the (stationary) mass M. A circle is an ellipse of zero eccentricity; its semimajor axis is identical to the radius of the circle.

For masses in orbit about the Sun, Kepler's third law may be simplified further to

$$T^2 = (1 \ \text{y}^2/\text{AU}^3)a^3 \tag{6.41}$$

where the period T must be expressed in years and the semimajor axis a in astronomical units (AU).

By definition, 1 AU is the semimajor axis of the orbit of the Earth:

$$1 \ \text{AU} \approx 1.496 \times 10^{11} \ \text{m}$$

A mass M establishes a *gravitational field* $\vec{g}$ in space. The gravitational field depends on the *shape* of M and how mass is *distributed* within its shape. The gravitational fields produced by various mass shapes and distributions are tabulated in Table 6.1. *The gravitational field at any location is identical to the acceleration due to gravity at that location.* The field at any location also may be thought of as the gravitational force per kilogram at that location in space.

If another mass m is placed at a location where the gravitational field has the value $\vec{g}$, then m experiences a gravitational force $\vec{F}$ equal to

$$\vec{F} = m\vec{g} \tag{6.43}$$

The *flux* Φ of the gravitational field through any surface S is

$$\Phi = \int_S \vec{g} \cdot d\vec{S} \tag{6.47}$$

The flux of the gravitational field through S is a measure of the extent to which the gravitational field vector threads the surface.

Gauss's law for the gravitational field states that the flux Φ of the gravitational field through any *closed surface* S is proportional to the mass enclosed within S:

$$\Phi_{\substack{\text{clsd} \\ \text{surface } S}} = -4\pi G M_{\text{within } S} \tag{6.55}$$

SUMMARY OF PROBLEM-SOLVING TACTICS

6.1 **(page 234)** Watch your m & M's. Be sure you understand which force is needed: the force on m or the force on M.

6.2 **(page 254)** Be sure to express T in years and a in astronomical units when using Equation 6.41:

$$T^2 = (1 \ y^2/AU^3)a^3 \qquad (6.41)$$

6.3 **(page 257)** Do not confuse the magnitude of the gravitational field $\vec{g}(r)$ at *any* location in space with g, the magnitude of the local acceleration due to gravity near the surface of the Earth.

6.4 **(page 267)** If you know the amount of mass within a closed surface S, the value of the flux of the gravitational field through the surface is found from the right-hand side of Gauss's law, Equation 6.55:

$$\Phi_{\substack{\text{clsd} \\ \text{surface } S}} = -4\pi GM_{\text{within } S} \qquad (6.55)$$

It is then not necessary to perform the typically complicated integration for the flux.

QUESTIONS

1. (page 235); 2. (page 237); 3. (page 239); 4. (page 241); 5. (page 244); 6. (page 248); 7. (page 258)

8. Carefully explain why the gravitational force of the Earth on a mass m is *directly proportional* to m while the free-fall acceleration is *independent* of m.

9. The Ring Nebula in the constellation Lyra (see Figure Q.9) is a spherically symmetric expanding shell of gas of mass m ejected from a central star of mass M.* (a) What is the total gravitational force of the shell on the central star? (b) What is the total gravitational force of the central star on the shell? (c) Are these a Newton's third law pair of forces? (d) The shell is outside the star; why then is the force of the central star on the shell equal to zero?

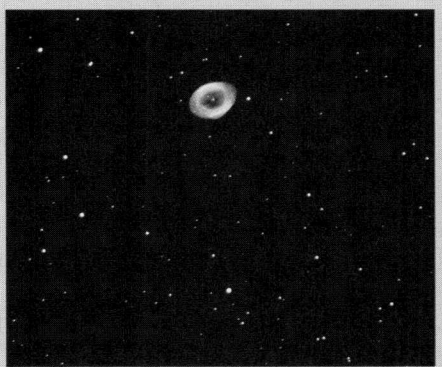

FIGURE Q.9

10. Imagine yourself located somewhere in the empty interior of a hypothetical hollow planet. Describe the trajectory of a pea shot from a peashooter.

11. A reporter for a magazine once wrote[†]:

> Ah, but the worse is yet to come. Also locked up in a French vault is a block of metal that weighs exactly one kilogram.

Explain why the first sentence accurately describes the second sentence.

---

*The shell has the appearance of a ring because there is more material along our line of sight around the edges of the shell than along the line of sight to the center of the shell.

[†] *Smithsonian*, 24, #12 (March 1994), page 21.

12. A mass m is inside a uniform spherical shell of mass M. Another mass m_0 lies outside the shell as shown in Figure Q.12. (a) What is the gravitational force of the shell on m? (b) Is the gravitational force of m_0 on m also zero, since m lies inside the shell? In other words, does the shell act as a gravitational insulator, effectively *shielding* m from the influence of m_0?

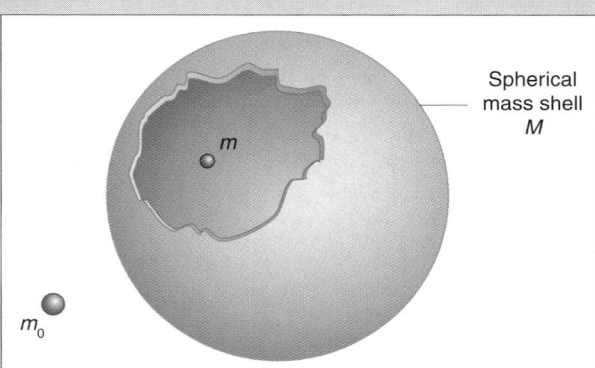

Spherical
mass shell
M

m_0

FIGURE Q.12

13. The Earth and the Moon exert gravitational forces of equal magnitude on each other. If a fraction of the mass of the Earth were transferred to the Moon, would the magnitude of their mutual gravitational force increase, decrease, or remain the same? Explain your reasoning. (See Problem 6.)

14. If the mass of the Earth remained constant but its radius doubled, would this affect the orbit of the Moon? Explain.

15. If the gravitational force on masses on the surface of the Earth were much stronger than it actually is (say, because the Earth had the same size but a larger mass density and total mass), discuss the influence this might have had on the evolution of human anatomy.

16. If gravity did not exist, could life be possible? Explain.

17. Despite massive backpacks and spacesuits, the Apollo astronauts were able to jump much higher on the Moon than on the Earth without all the accoutrements. Carefully explain why.

18. Consult an astronomy text to discover why the sidereal year is slightly longer than the tropical year as mentioned in the footnote in Section 6.12 on page 254.

19. A north–south line in the sky passing through your zenith is known as your *local celestial meridian*. The time between successive celestial meridian passages of the Sun is called an *apparent solar day*. Will the interval of an apparent solar day about the time the Earth is at perihelion (3 January) be longer, shorter, or the same as the apparent solar day about the time the Earth is at aphelion (6 July). Explain.

20. Comet Halley has an elliptical orbit of high eccentricity about the Sun. The comet is visible to the naked eye for only several months out of its 76 y period. Explain how Kepler's law of areas might account for this.

21. The Earth is closest to the Sun (at its perihelion distance) on 3 January and most distant from the Sun (at its aphelion distance) on 6 July. What, then, causes the seasons? Consult an astronomy text if necessary.

22. How does Kepler's law of areas account for the unequal length of the seasons? See the following table:

Northern hemisphere season	Southern hemisphere season	Approximate calendar dates	Length of the season
spring	fall	20 March–21 June	92.76 days
summer	winter	21 June–23 September	93.65 days
fall	spring	23 September–21 December	89.83 days
winter	summer	21 December–20 March	89.00 days
		Total	365.24 days

23. Which geometrical figure has the greater area: an ellipse with semimajor axis a and eccentricity $\varepsilon = 0$ (a circle!) or an ellipse of semimajor axis a and eccentricity $\varepsilon \neq 0$?

24. How does Kepler's law of areas explain why a planet has the greatest speed when it is at its perihelion distance?

25. The time it takes a planet to go from one end of the major axis of its orbit to the other end is half the orbital period of the planet. Explain why the time to go from one end of the *minor* axis to the other end is *not* half the period.

26. Two satellites are in circular orbit around the Earth, one at a radius r from the center of the Earth, the other at radius $2r$. What is the *ratio* of their periods? Speeds? Accelerations?

27. Earth satellites usually are launched to the east. Why is this more advantageous than launching them to the west?

28. Explain why the orbital planes of all the planets must pass through the center of the Sun.

29. Explain why an orbiting astronaut experiences what is colloquially but unfortunately called "weightlessness."

30. An orbiting astronaut is not, strictly speaking, weightless. But a geologist located in the empty interior of a hypothetical hollow planet, *is* truly weightless. Explain why.

31. Explain why it is not possible to place a satellite in an orbit that follows a path directly above either the Tropics of Cancer or Capricorn nor the Arctic or Antarctic Circles.

32. Explain why it is easier to determine the mass of a planet that has a moon than to find the mass of a planet with no moon. (Only Mercury and Venus lack any moons.)

33. If a planetary moon has appreciable mass compared with the mother planet, Newton's form for Kepler's third law (Equation 6.37) only enables you to determine the *sum* of the mass of the planet and the mass of the moon. What additional information is necessary to determine the masses of the individual objects?

34. For a planet in an elliptical orbit around the Sun, will the *temporal* average of the various positions of the planet from the Sun (the average of its distance from the Sun spaced at equal time intervals) be equal to the semimajor axis of the orbit? Explain why or why not. If not, will the temporal average likely be less than or greater than the semimajor axis?

35. Weather and communication satellites are placed in circular orbits such that they hover or stay above certain locations on the surface of the Earth. (a) Explain why the satellite does not fall down to the surface of the Earth. (b) What is the orbital angular speed of the satellite? (c) Could such a weather satellite be placed so that it hovered directly over the Big Apple (New York)? Why or why not?

36. The three laws of planetary motion discovered by Kepler concern the shape of an orbit, the variation of the speed around an orbit, and the period of the orbit. If the numerical value of the gravitational constant G were *different* from what it is, which of the Keplerian laws would be affected and which would be unaffected? Explain.

37. Newton showed that the possible shapes of the paths followed by objects subjected to (only) the gravitational force of the Sun are not only circles and ellipses, but also parabolas and hyperbolas. Kepler's *first* law is generalized to state that the Sun is located at either (i) the center of a circular path; (ii) the single focus of a parabolic path; (iii) one of the two foci for an elliptical path (the other focus being empty); or (iv) one focus of a hyperbolic path (the other focus and sheet of the hyperbola being empty). (a) Kepler's *second* law (the law of areas) was formulated by Kepler for elliptical (and circular) orbits. Does Kepler's second law *also* apply to parabolic and hyperbolic paths of objects about the Sun? Explain why or why not. (b) Kepler's *third* law concerns the orbital period. Is it meaningful to talk about a *period* associated with parabolic or hyperbolic paths? Explain.

38. Explain how a geologist might exploit *variations* in the gravitational field over the surface of the Earth to detect the presence of a large subterranean oil pool. (The density of oil is only a fraction of the density of rock.)

39. Show that the SI units for the flux Φ of the gravitational field are either $N \cdot m^2/kg$ or m^3/s^2.

40. If the flux of the gravitational field through a closed surface is zero, does this imply there is zero mass within the surface? Does such a zero flux through a closed surface imply the gravitational field *itself* is zero at every point on the closed surface? Explain.

41. *Imagine* that negative masses exist as well as positive masses. If a closed surface surrounded such a negative mass, what does Gauss's law for the gravitational field imply about the flux of the gravitational field through the closed surface? What does this mean in terms of the direction of the gravitational field of such a hypothetical negative mass? If a normal positive mass particle were placed in the gravitational field of such a hypothetical negative mass, what would be the direction of the gravitational force on the normal positive mass particle relative to the line between the two particles?

PROBLEMS

Sections 6.1 How Did Newton Deduce the Gravitational Force Law?
   6.2 Newton's Law of Universal Gravitation

1. Two 2.50 kg pointlike masses are separated by 1.50 m. What is the magnitude of the gravitational force that each exerts on the other?

2. What is the magnitude of the gravitational force of a 1.00 kg young goshawk (*Accipiter gentilis*) on the Earth? In what direction does this force point?

3. A classics student of mass 70.0 kg is 0.50 m from a physics student of mass 50.0 kg. Calculate the approximate magnitude of the gravitational force that each exerts on the other. Explain why your result is only approximately equal to the magnitude of the actual gravitational force of each on the other.

•4. When the giant planet Jupiter ($M_{\text{Jupiter}} = 1.90 \times 10^{27}$ kg) is closest to the Earth, it is about 5.90×10^8 km from us. (a) Calculate the magnitude of the gravitational force of Jupiter on a 70.0 kg astrologer when Jupiter is closest to the Earth. Assume both Jupiter and the astrologer are pointlike particles. So much for the influence of the planets on our daily lives. (b) What mass m placed 10.0 m from the astrologer would produce a gravitational force on the astrologer of the same magnitude as the force of Jupiter calculated in part (a)?

•5. The orbit of the Earth about the Sun is circular to a first approximation (the eccentricity of the elliptical orbit is only about $\varepsilon = 0.0167$). (a) Using Newton's second law of motion and the gravitational force law (among other things), determine the orbital speed of the Earth in terms of the mass of the Sun, the radius of the orbit, and other appropriate constants. Consider both the Earth and Sun to be pointlike particles and the Sun to be fixed in position. Evaluate the speed and express it in km/s. (b) If the Earth suddenly split gently into two halves of equal mass, what would happen to the orbital speed of each piece, assuming the radius of the orbit were unchanged?

‡6. Two masses m_1 and m_2 exert gravitational forces of equal magnitude on each other. Show that if the total mass $M = m_1 + m_2$ is fixed, the magnitude of the mutual gravitational force on each of the two masses is a maximum when they have equal mass.

‡7. A small mass m is on the surface of the Earth as indicated in Figure P.7. The mass is located along the line joining the centers of the Earth and Moon. (a) What is the magnitude of the gravitational force *of the Moon* on the mass m? (b) What is the magnitude of the gravitational force of the Moon on mass m if m were located at the center of the Earth? (c) The difference between the magnitude of the force of the Moon on m calculated in part (a) and that calculated in part (b) is known as the magnitude of the *tidal force* $\vec{F}_{\text{tidal}}$ on m due to the Moon; $\vec{F}_{\text{tidal}}$ is directed toward the Moon. Show that the magnitude of the tidal force on m at this location can be written approximately as

$$F_{\text{tidal}} \approx GMm\,\frac{2R}{d^3}$$

where M is the mass of the *Moon*. Assume $d \gg R$. The tidal force of the Moon on a mass m thus depends on (1) the size of the object on which m finds itself (in this case, the radius R of the Earth), and (2) the inverse *cube* of the distance d between the centers of the Earth and Moon. (d) Show that if a similar calculation is made for the tidal force of the Sun on m, the result is approximately 2.2 times smaller than the tidal force of the Moon on m. So both the Moon and the Sun cause tidal forces on the Earth. These tidal forces cause the familiar ocean tides.

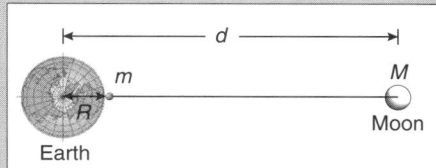

FIGURE P.7

‡8. In 1850 the French mathematician Édouard Roche showed that if a massive moon of substantial physical size ventured too close to a planet, the moon would be broken up by tidal forces. The moon must be of sufficient size and mass that its shape is determined by its self-gravitation rather than by other forces. Typically this means a generally spherical moon whose radius is several hundreds of kilometers. The distance of closest approach for such a moon is called the *Roche limit*. For a moon and planet of the *same density*, the Roche limit is equal to 2.455 planetary radii. A crude calculation can estimate the order of magnitude of the Roche limit using the approximations for the tidal forces in Problem 7 and crude assumptions about the shapes of the objects. Imagine a small mass m on the surface of a spherical moon of mass M' and radius R as indicated in Figure P.8. The moon is located a distance d from a planet of mass M and radius R_0. The magnitude of the tidal force on the mass m due to the *planet* is (using the results of Problem 7)

$$\frac{GMm\,2R}{d^3}$$

This force is directed toward the planet. The magnitude of the gravitational force of attraction of the moon for the mass m (the self-gravitational force) is

$$\frac{GM'm}{R^2}$$

by Newton's law of universal gravitation. This force is directed toward the moon. If the magnitudes of the two forces are set equal to one another, the moon is about to break up under the action of the tidal force. (a) Write the equation that gives this critical condition. (b) Show that the distance of separation d at which the tidal force on m is equal in magnitude to the force of the moon on m satisfies

$$d^3 = \frac{M\,2R^3}{M'}$$

(c) Show that if the spherical planet and moon have the *same density*, then

$$d^3 = 2R_0^3$$

or

$$d \approx 1.26R_0$$

where R_0 is the radius of the *planet*. This calculation means that if the moon ventures closer than 1.26 planetary radii to the planet of radius R_0, the moon will break up under the action of the tidal forces of the planet. This result disagrees with the exact calculation by Roche by about a factor of 2, which is not surprising because of the rather extreme assumptions made about the shapes of the objects and the tidal forces when the objects are so close together. While the calculation is crude, it does give an order of magnitude result. The rings of Saturn are within the Roche limit of the planet, whereas its first sizable moon (Mimas, with a radius of about 200 km) is outside the Roche limit.

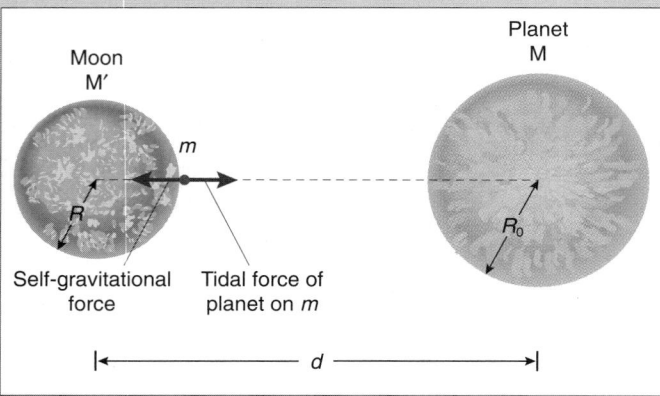

FIGURE P.8

Sections 6.3 Gravitational Force of a Uniform Spherical Shell on a Particle
6.4 Gravitational Force of a Uniform Sphere on a Particle

9. A pair of 10.0 cm diameter spheres of masses 4.00 kg and 0.400 kg are placed 10.1 cm apart, center-to-center. (a) What is the magnitude of the gravitational force of the 4.00 kg sphere on the 0.400 kg sphere? (b) What is the magnitude of the gravitational force of the 0.400 kg sphere on the 4.00 kg sphere? (c) If these forces are the *only* forces acting on the spheres, what is the magnitude of the acceleration of each sphere?

• 10. A Cavendish balance (see Figure 6.3), used to determine the value of G, uses a small lead sphere of radius 0.500 cm and a larger lead sphere of radius 3.00 cm. The spheres are separated by 4.00 cm center-to-center. The density of lead is 11.0×10^3 kg/m³. (a) What is the magnitude of the gravitational force of each sphere on the other? (b) What is the magnitude of the weight of the smaller sphere? (c) What is the ratio of the magnitude of the weight of the smaller sphere to the magnitude of the gravitational force of the larger sphere on it? This ratio indicates why the Cavendish experiment is not easy to do.

• 11. At what altitude above the surface of the Earth is the magnitude of the acceleration due to gravity 1.00% less than its value on the surface of the Earth?

• 12. A mass m is inside a uniform spherical shell of mass M' and a mass M is outside the shell as shown in Figure P.12. What is the magnitude of the total gravitational force on m?

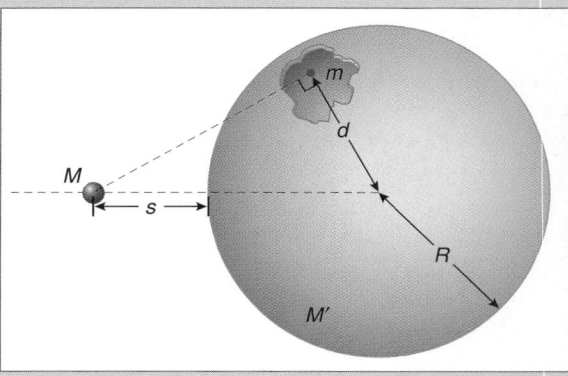

FIGURE P.12

• 13. (a) A spherical asteroid has a radius R and average density ρ. (a) Show that the magnitude of the acceleration due to gravity on the surface of the asteroid is $(4/3)\pi GR\rho$. (b) The asteroid 107 Camilla has a diameter of 252 km and an average density of 4.5×10^3 kg/m³. What is the magnitude of the local acceleration due to gravity on the surface of 107 Camilla? (c) If you can high jump 2.0 m on the surface of the Earth, how high a jump could you make on 107 Camilla? How long would you be off the surface of the asteroid in making the jump?

• 14. A certain neutron star has a radius of 10.0 km and a mass of 4.00×10^{30} kg, about twice the mass of the Sun. (a) Calculate the magnitude of the acceleration of an 80.0 kg student foolish enough to be 100 km from the center of the neutron star. (b) How many times g is the magnitude of this acceleration? (c) If the student is in a circular orbit of radius 100 km about the neutron star, use Newton's laws to find the orbital period.

• 15. Imagine a spherical planet of radius R with an average density of 5.00×10^3 kg/m³; this is approximately the average density of the Earth. An 80.0 kg astronaut is on the surface of the planet. The magnitude of the gravitational force of the planet on the astronaut is only 1.00 N. What is the radius of the planet?

• 16. (a) Given the mass and radius of the Earth, calculate the average density of the Earth. (b) If the mass of the Earth remained constant, but the radius of the Earth changed to increase the magnitude of the acceleration due to gravity on the surface of the Earth to twice its current value (i.e., to 19.6 m/s²), what would be the new radius of the Earth? Assume a uniform density.

‡ 17. Two identical, uniform spherical masses, each with mass m and radius R, are placed in contact with each other. (a) What must be the value of the density ρ so that the gravitational force of one mass on the other has a magnitude F? (b) If F = 1.00 N, do *any* of the natural elements have sufficient density for 1.00 cm radii spheres? Consult a reference that gives the densities of the natural elements. Surprisingly, the most dense of the 92 natural elements is not uranium ($\rho = 18.7 \times 10^3$ kg/m³) but osmium ($\rho = 22.5 \times 10^3$ kg/m³). (c) If the spheres are composed of osmium, what radius R is needed to produce a gravitational force of magnitude 1.00 N by one sphere on the other? What is the mass of each sphere in this case?

‡ 18. A mass m is at rest on the equator of a spinning, spherical planet. (a) Sketch a second law force diagram indicating all the forces acting on the mass m. (b) What happens, qualitatively, to the magnitude of the normal force of the surface on the mass as the spin angular speed of the planet increases?

(c) If the planet spins fast enough, the normal force of the surface on the mass vanishes and the centripetal acceleration is provided by the gravitational force alone. Under these circumstances, the gravitational force is barely able to keep the planet together. For a planet of average density ρ, show that the critical period of spin T_{crit} is

$$T_{crit} = \left(\frac{3\pi}{G\rho} \right)^{1/2}$$

Thus, to prevent a planet from breaking up,* the *rotational period* of the planet T must be greater than T_{crit}. Many moons within the solar system have average densities of about 3.0×10^3 kg/m³. Calculate a numerical value for T_{crit} for this average density. No objects within the solar system rotate with periods this short: they would have come apart. (d) The same arguments can be used for stars as well as planets. Show that for a star of radius R, spinning at *frequency* ν (the number of rotations per second), the *minimum* mass that the star must have to be stable against rotational breakup is

$$M_{min} = \frac{4\pi^2}{G} R^3 \nu^2$$

Explain why this is a minimum mass. (e) Exotic stars known as neutron stars have radii of about 10 km. The neutron star at the center of the Crab Nebula in the constellation Taurus is spinning 33 times each second. Using the results of part (d), calculate the *minimum* mass for this neutron star. Express your result in terms of solar mass units: 1 solar mass = 1.99×10^{30} kg. Real neutron stars have masses between about 1.4 and 3.5 solar masses, considerably greater than the minimum mass calculated here.

Sections 6.5 Measuring the Mass of the Earth
6.6 Artificial Satellites of the Earth

•19. Calculate the period of a satellite of the Earth in a circular orbit with a radius 200 km greater than the mean radius of the Earth.

•20. Two satellites are in circular orbit about the Earth at radii r_1 and r_2. (a) Find the *ratio* between their orbital periods. (b) Evaluate this ratio if $r_2 = 2r_1$.

•21. A spacecraft on the surface of Mars discovers that a freely falling screw has an acceleration of magnitude 3.776 m/s². The radius of Mars is known to be 3.37×10^6 m. Use these data to calculate the mass of Mars. Express your result both in kilograms and as a fraction of the mass of the Earth.

•22. (a) Calculate the radius of the orbit of a satellite in circular orbit about the Earth with a period of one day. The planes of the orbits of such satellites typically are made coincident with the equatorial plane of the Earth and the motion is in the same directional sense that the Earth rotates (i.e., eastward). In this way the satellite remains above a fixed geographical position above the equator. Many such geosynchronous satellites are communication satellites, since they assist in transmitting radio, television, and telephone signals between continents. Three such satellites spaced 120° apart in approximately the same orbit constitute a global grid capable of

relaying messages virtually instantaneously over most of the Earth. (b) Find the orbital speed of the satellite. (c) What is the magnitude of the acceleration of the satellite?

•23. Two satellites are launched into circular equatorial orbits at essentially equal radii r about the Earth. One satellite moves east and the other west. Sketch the situation. (a) Will the satellites have identical periods? (b) At how many locations in your sketch will they pass each other? (c) Will the places where the satellites pass each other remain over the same locations on the surface of the Earth? (d) Which satellite (if either) will have the shorter time interval between successive passages over the same place on the equator of the Earth?

‡24. On a clear day, you can see forever. . . . Well, not quite. Model the Earth as a sphere of radius R (= 6370 km). Seen from an altitude h above the surface of the Earth (see Figure P.24), the horizon lies a distance s away measured along the surface of the Earth. (a) Using the figure and your knowledge of geometry, show that

$$h = R(\sec \theta - 1)$$

where $\theta = s/R$. (b) For small angles θ,

$$\cos \theta = 1 - \frac{\theta^2}{2}$$

Also, by definition $\sec \theta = 1/\cos \theta$. Use these facts and the binomial expansion to show that, for small angles,

$$\sec \theta \approx 1 + \frac{\theta^2}{2}$$

(c) Use the results of parts (a) and (b) to show that an approximate expression for the distance s to the horizon is

$$s = (2Rh)^{1/2}$$

(d) Show that when at a height $h = 1.00$ km, the distance s to the horizon is approximately 113 km.

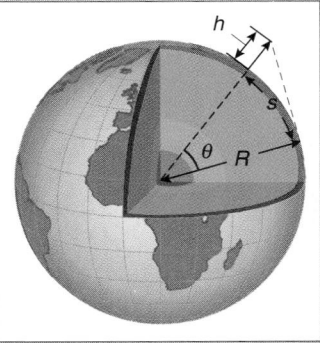

FIGURE P.24

Sections 6.7 Kepler's First Law of Planetary Motion and the Geometry of Ellipses
6.8 Spatial Average Position of a Planet in an Elliptical Orbit*

•25. The eccentricity of the orbit of the Earth about the Sun is $\varepsilon = 0.0167$. The semimajor axis of the orbit is 1.496×10^8 km.

---

*This is colloquially referred to as a breakup due to centrifugal pseudoforces.

(a) What is the separation of the foci of the orbit of the Earth?
(b) Is the center of the orbit inside or outside the Sun (the radius of the Sun is 6.96×10^5 km). (c) What are the perihelion and aphelion distances of the Earth from the Sun?
(d) In a scale model of the orbit of the Earth, let 10.0 cm represent the semimajor axis; what is the separation of the foci of the ellipse representing the orbit in this model?

•26. (a) On the same polar plot, make a series of accurate graphs of ellipses, all with a semimajor axis of 1.00 AU, appropriately scaled, with eccentricities ε between 0 and 0.900, in increments of 0.100. Include the orbits of Venus (0.723 AU), Mercury (0.387 AU), and Mars (1.52 AU), assuming they are circular. Locate the Sun on your graph. (b) Which elliptical orbits have perihelions closer to the Sun than Mercury? Do any of these orbits have aphelions beyond the orbit of Mars?

•27. Use the polar equation for an ellipse

$$r = \frac{a(1 - \varepsilon^2)}{1 - \varepsilon \cos \theta}$$

to describe the position r of a planet from the Sun (located at one of the foci of the ellipse). The value of r ranges between a minimum at the perihelion distance and a maximum at the aphelion distance. Show that when the planet is on the ends of the minor axis of the ellipse, the distance of the planet from the Sun is equal to the semimajor axis of the orbit, and that $\cos \theta = \pm \varepsilon$ at these positions.

•28. What is the eccentricity of the elliptical orbit of a planet whose aphelion distance is twice its perihelion distance?

•29. The eccentricity of the lunar orbit about the Earth is $\varepsilon = 0.055$. Calculate the ratio of the angle subtended by the diameter of the Moon at perigee (when it is closest to the Earth) to the angle subtended by the lunar diameter at apogee (when it is most distant from the Earth). Imagine yourself at the center of the Earth.

•30. A planet is in an elliptical orbit of semimajor axis a about the Sun. Use Equation 6.19 for the polar equation for the elliptical orbit:

$$r = \frac{a(1 - \varepsilon^2)}{1 - \varepsilon \cos \theta}$$

to prove that it is impossible for the planet to be located at a distance r from the Sun equal to half the aphelion distance unless $\varepsilon \geq 1/3$. Of course, ε also must be less than 1 for an elliptical orbit.

•31. The orbit of Pluto, with an eccentricity of 0.248, is the most eccentric of all the major planets. The semimajor axis of Pluto is 39.4 AU. (a) How far is the center of the elliptical orbit of Pluto from the center of the Sun? (b) Determine the perihelion and aphelion distances of Pluto. This should explain why Pluto occasionally is closer to the Sun than Neptune (most recently, between 1979 and 1999). The orbit of Neptune has a semimajor axis of 30.1 AU but an eccentricity of only 0.0086.

•32. The area A of an ellipse is

$$A = \pi a b$$

where a is the semimajor axis and b is the semiminor axis.
(a) Use the geometrical properties of an ellipse to show that the area also can be expressed in terms of the eccentricity as

$$A = \pi a^2 (1 - \varepsilon^2)^{1/2}$$

(b) For a fixed semimajor axis, say equal to 1.00 AU, make a graph of the area of the ellipse as a function of its eccentricity.

•33. (a) Show that for an ellipse of eccentricity ε, the ratio of the major to minor axes is

$$\frac{\text{major axis}}{\text{minor axis}} = \frac{1}{(1 - \varepsilon^2)^{1/2}}$$

(b) What is the eccentricity of an ellipse if the ratio of the major axis to the minor axis is 1.01? 1.10 ?

‡34. Prove that the area of an ellipse of semimajor axis a and semiminor axis b is equal to $\pi a b$.

‡35. Construct a Cartesian coordinate system with an origin at the geometrical *center* of an ellipse with the x- and y-axes oriented along the major and minor axes respectively. Show that the Cartesian equation for an ellipse

$$\frac{x^2}{a^2} + \frac{y^2}{b^2} = 1$$

(where a is the semimajor axis and b is the semiminor axis) is the same as the polar expression for an ellipse:

$$r = \frac{a(1 - \varepsilon^2)}{1 - \varepsilon \cos \theta}$$

where ε is the eccentricity of the ellipse and the origin for the polar coordinates r and θ is at one focus.

‡36. Prove that in the plane of any ellipse, the normal to the ellipse at any point P bisects the angle formed by lines from P to the two foci.

Sections 6.9 Kepler's Second Law of Planetary Motion
6.10 Central Forces, Orbital Angular Momentum, and Kepler's Second Law*

37. The eccentricity of the orbit of the Earth is $\varepsilon_{\text{Earth}} = 0.0167$. Find the ratio of the speed of the Earth when at perihelion to its speed at aphelion.

•38. Calculate the surface area of the spherical Earth. How long a time is necessary for a line from the Sun to the center of the Earth to sweep out an area equal to the surface area of the Earth? Assume the orbit of the Earth is circular.

•39. A planet is in an elliptical orbit about the Sun with an eccentricity ε as indicated in Figure P.39. Let the semimajor and semiminor axes of the ellipse be a and b respectively. (a) Let the speed of the planet at perihelion (point P) be u, its speed at aphelion (point A) be v, and its speed at point C be V. At point C the planet is at its greatest distance from the major axis. By applying conservation of the orbital angular momentum of the planet at the points P, A, and C, show that

$$Vb = ua(1 - \varepsilon) = va(1 + \varepsilon)$$

(b) Use the expressions [from part (a)]

$$V = \frac{ua(1 - \varepsilon)}{b}$$

and

$$V = \frac{va(1 + \varepsilon)}{b}$$

to form an expression for V^2 and show that

$$V^2 = uv$$

This problem was inspired by a paper by Ira M. Freeman, "An interesting property of the Kepler ellipse," *American Journal of Physics*, 45, #6, pages 585–586 (June 1977).

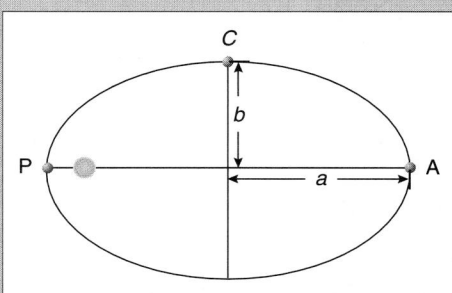

FIGURE P.39

•**40.** A planet of mass m is in an elliptical orbit about the Sun ($m \ll M_{Sun}$) with an orbital period T. Let L be the magnitude of the orbital angular momentum of the planet and let A be the area of the orbit. Show that

$$L = \frac{2mA}{T}$$

Sections 6.11 Newton's Form for Kepler's Third Law of Planetary Motion
6.12 Customized Units

41. Most of the asteroids have orbits between Mars and Jupiter and have semimajor axes of about 2.8 AU. Determine the periods of the asteroids in these orbits.

42. The orbit of Pluto has a semimajor axis of 39.4 AU. Determine the period of Pluto. Pluto was discovered in 1930. In what year A.D. will Pluto first complete one orbital trip since its discovery?

43. Comet Swift-Tuttle (first discovered in 1862) has a period of about 130 y. It has been predicted that this comet may collide with the Earth in A.D. 2126; so . . . if you have nothing else to worry about . . . don't lose sleep; the prediction subsequently was found to be erroneous. What is the semimajor axis of the orbit of the comet in AU?

44. (a) What is the semimajor axis of a hypothetical planet of the Sun with a period equal to 10.0 y? (b) What is the period of a hypothetical planet of the Sun with a semimajor axis equal to 10.0 AU?

•45. The famous Comet Halley has a perihelion distance of 0.57 AU and a period of 75.6 y. (a) Determine the semimajor axis of its orbit. (b) What is the eccentricity of the orbit? (c) Find

the aphelion distance of the comet. (d) Determine the ratio of the maximum speed to the minimum speed of the comet.

•**46.** If the mass of the Sun were doubled, at what distance from the Sun (in AU) would a planet have an orbital period of 1.00 y?

•**47.** If the semimajor axis of an orbit about the Sun increases by a factor of 2, by what factor does the period change? Will the factor be the same for orbits about the Earth?

•**48.** The mass of the Sun is 3.30×10^5 times the mass of the Earth. Find the location of the center of mass of the Earth–Sun system and compare this with the radius of the Sun, 6.96×10^5 km.

•49. The spatial average distance between the Earth and the Moon, center-to-center, is about 3.84×10^8 m. The mass of the Earth is about 80 times the mass of the Moon. Determine the location of the center of mass of the Earth–Moon system: (a) relative to the center of the Earth; (b) relative to the *surface* of the Earth.

•**50.** What would be the orbital period of a rather hot grand piano in circular orbit about the Sun with a radius equal to the radius of the Sun?

•**51.** A planet is in a circular orbit about the Sun. Show that the square of the period of the planet is proportional to the 3/2 power of the *area* of the orbit.

•**52.** The orbit of Venus, with an eccentricity of 0.007, is the most circular of all the planets. The semimajor axis of the orbit is 0.723 AU. (a) How far is the center of the elliptical orbit of Venus from the center of the Sun? (b) Determine the perihelion and aphelion distances of Venus from the Sun. (c) Determine the period of the planet in its orbit.

•53. A peculiar triple star system is discovered in the constellation Physics Major (see Figure P.53). Two stars, called Ψ and Ψ^*, each of mass $M/4$, are at opposite points of the same circular orbit about a central star of mass M. (a) Determine the magnitude of the gravitational force of star Ψ^* on star Ψ. (b) Determine the magnitude of the gravitational force of the central star on star Ψ. (c) What is the magnitude of the total gravitational force on star Ψ? (d) Find an expression for the speed of star Ψ in terms of the radius r of its orbit and the period T of its orbital motion. (e) Find an expression for the square of the orbital period of star Ψ in terms of the various masses, the radius of the orbit, and the universal constant of gravitation G.

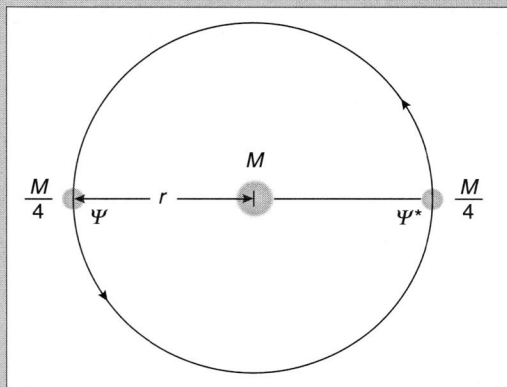

FIGURE P.53

•**54.** (a) Construct a graph of the square of the period (T^2) versus the cube of the semimajor axis (a^3) for the planets of the Sun.

Use the data located on the inside covers. (b) Is the graph linear? Why or why not? (c) What is the slope of the graph? (d) What does the graph represent or show?

•55. A spacecraft is placed in an elliptical orbit about the Sun. The perihelion distance is that of the Earth (1 AU) and its aphelion distance is the distance to the nearby star, Proxima Centauri, located approximately 2.7×10^5 AU from the Sun. (a) Sketch the situation. (b) Calculate the number of years it takes the spacecraft to make a one-way trip from its perihelion point to its aphelion point; neglect the gravitational influence of Proxima Centauri on the spacecraft. Anyone care to go along for the ride? (c) What is the approximate eccentricity of the orbit of the spacecraft?

•56. The time it takes for a mass m to fall from a distance d (expressed in AU) to, say, the Sun, can be approximated in the following way. Imagine m in a very eccentric elliptical orbit ($\varepsilon \approx 1$) about the Sun as indicated in Figure P.56. The semimajor axis a of the orbit then is approximately $a \approx d/2$. Use Kepler's third law with the customized units (of Section 6.12) to show that the time in years for *half* the orbit (aphelion to perihelion) to be completed is

$$t = \frac{T}{2} = \left(\frac{1}{2\sqrt{8}} \ \text{y}/\text{AU}^{3/2} \right) d^{3/2}$$

Estimate this time for an object located at the distance of the Earth from the Sun ($d = 1.00$ AU).
This problem was inspired by an article by John S. Dilsaver and Joseph R. Siler, "The solar swan dive," *The Physics Teacher*, 29, #1, pages 28–29 (January 1991).

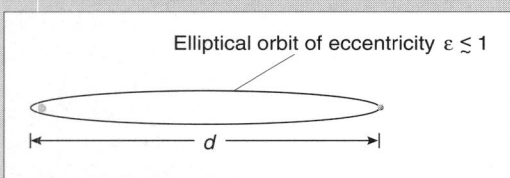

Elliptical orbit of eccentricity $\varepsilon \lesssim 1$

d

FIGURE P.56

•57. Plaskett's binary (also known by the more exotic but less romantic name HD 47129), in the constellation Monoceros, consists of two blue-giant stars. Assume they are of equal mass and in circular orbits about their center of mass located midway between them. The orbital speed of each star has been found to be 250 km/s and the period of their orbit is 14.4 d. Find the mass of each star and express it as a multiple of the mass of our Sun ($M_{Sun} = 1.99 \times 10^{30}$ kg). These are among the most massive stars known.

•58. The orbits of the Earth and Venus are essentially circular. A spacecraft is in an orbit about the Sun with the perihelion point at the orbit of Venus and the aphelion point at the orbit of the Earth. The radius of the orbit of Venus is 0.723 AU. (a) Calculate the flight time for the spacecraft from the Earth to Venus. (b) Determine the eccentricity of the orbit of the spacecraft.

•59. The eccentricity of the orbit of Pluto is the largest of all the major planets: $\varepsilon_{Pluto} = 0.248$. The semimajor axis of Pluto is 39.4 AU. (a) Determine the perihelion and aphelion distances of Pluto from the Sun. (b) The eccentricity of the orbit of the planet Neptune is 0.0100 and its semimajor axis is 30.1 AU. Is

it possible for the planet Pluto to be closer to the Sun than Neptune? (c) Calculate the flight time in years for a spacecraft to go from the Earth to Pluto if the orbit of the spacecraft has its perihelion point at the orbit of the Earth and its aphelion point at the orbit of Pluto when Pluto is at its perihelion point. Assume the orbit of the Earth is circular. (d) Find the eccentricity of the orbit of the spacecraft.

‡60. A planet is in an elliptical orbit about the Sun with a semimajor axis a and eccentricity ε. The mass m of the planet is negligible with respect to the mass M of the Sun. Show that the square of the period of the planet is equal to the following:

$$\frac{4\sqrt{\pi}}{GM(1 - \varepsilon^2)^{3/4}} \ A^{3/2}$$

where A is the area of the elliptical orbit (see Problem 32 for an expression for the area of an ellipse).

‡61. The solar system as a whole is in orbit about the center of the Milky Way Galaxy. The distance of the Sun from the center of the galaxy is about 2.5×10^4 light-years.* Assume for simplicity that the orbit of the solar system is circular. Astronomers also have been able to determine, independently, that the speed of the solar system in its galactic orbit is about 230 km/s. (a) How many years does it take the solar system to complete one galactic orbit? (b) The age of the solar system is about 5.0×10^9 y. How many orbits about the galactic center has the solar system completed in its lifetime? (c) Assume a spherical distribution of matter between us and the center of the galaxy.† How much total mass exists between the orbit of the solar system and the center of the galaxy? The mass of the Sun is 1.99×10^{30} kg. Express your result both in kilograms and in units of solar masses. By using sampling techniques, astronomers can estimate the number of the stars (luminous matter) between the Sun and the center of the galaxy. The results are significantly less than the number deduced by arguments such as those used in this problem. This has convinced many astronomers that, in addition to stars, there is a significant amount of invisible, nonluminous, so-called *dark matter* in the galaxy. (d) What effect does material in the galaxy located at distances *beyond* the galactic orbit of the solar system have on the orbital motion of the solar system about the center of the galaxy?

‡62. One of the ancient Greek models of the solar system hypothesized an identical planet (called the antichthon [ἀντίχθων], or counterearth), orbiting the Sun in the *same* orbit as the Earth but on the opposite side of the Sun as in Figure P.62. Consider the orbit to be circular in shape and of radius r. (a) Sketch a second law force diagram indicating all the forces acting on the Earth if such a counterearth existed. (b) Using Newton's second law of motion, find an expression for the orbital period of the Earth in terms of the mass M of the Sun, the mass m of the counterearth, the radius of the orbit of the Earth and counterearth, and appropriate constants. (c) Does the presence of a counterearth mean that the period of the Earth in its orbit about the Sun is less or greater than the period of the Earth without the existence of a counterearth? (d) With a circular orbit, is it possible to see the counterearth at any time

---

*One light-year is the distance travelled by light during one year.
† In fact this is *not* a good assumption, but making it serves to give us an order of magnitude answer.

from the Earth, say during a total eclipse of the Sun? The Sun has a diameter about 100 times that of the Earth. Explain your reasoning. (e) If the Earth and counterearth were in the same *elliptical* orbit, would it be possible for the Earth at some time to catch a direct glimpse of the counterearth during a total solar eclipse? Explain your reasoning.

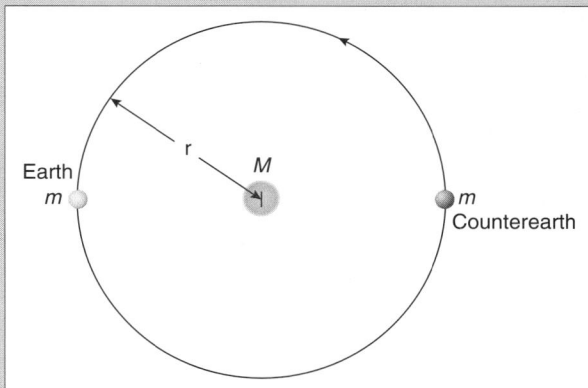

FIGURE P.62

‡63. The Sun exerts the dominant gravitational force on the Earth. The gravitational effects of the other planets on the motion of the Earth are small but measurable and are known as *perturbations*. Two distinct types of effects can take place: (1) permanent, long-term effects depending on the mere existence of another planet or the shape and location of the orbit of such a planet; these effects are known as *secular perturbations*; and (2) short-term effects depending on the detailed actual relative positions of the planets in their orbits; these effects are known as *periodic perturbations*. (a) Does the presence of the planets Mercury and Venus mean that the Earth orbits the Sun in less time or in more time than the time calculated assuming these planets did not exist? Is this effect a secular or a periodic perturbation? (b) Are the perturbations of *outer* planets on *inner* planets likely to be secular or periodic perturbations? Discrepancies between the observed and predicted motion of the planet Uranus, even after accounting for the effects of all the known planets, were used by John C. Adams (1819–1892) and Urbain J. J. LeVerrier (1811–1877) to predict the existence of a trans-Uranian planet (a planet beyond the orbit of Uranus). On the basis of their calculations and predictions, the planet Neptune was discovered on 23 September 1846 by Johann G. Galle (1812–1910) and Heinrich L. d'Arrest (1822–1875) on the first night they began the search! Ironically, discrepancies between the predicted and observed motion of the planet Mercury led LeVerrier to infer the existence of a planet closer to the Sun than Mercury, one that he even named: Vulcan. Despite numerous reported "sightings" of Vulcan, none has stood the test of time. Vulcan does not exist. The discrepancies in the motion of the planet Mercury were neatly and completely explained by Einstein as a crucial test of his general theory of relativity in 1914.

Section 6.13 The Gravitational Field

64. What are the magnitude and direction of the gravitational field of the Earth at a point $3R$ above the *surface* of the Earth, where R is the radius of the Earth?

•65. (a) At what point between the Earth and the Moon is their total gravitational field equal to zero? Express your answer in terms of the distance d between the centers of the Earth and Moon. This special point is known as the L_1 Lagrangian point. Hint:

$$\frac{\text{mass of Earth}}{\text{mass of Moon}} = 81.3$$

(b) If a mass at this location is moved slightly toward the Moon or toward the Earth, will the mass return to the L_1 point under the influence of only gravitational forces?

•66. (a) Find the magnitude of the total gravitational field at the point P in Figure P.66. (b) What is the magnitude of the acceleration experienced by a 4.00 kg salt lick at point P? (c) What is the magnitude of the total gravitational force on the salt lick if it is placed at P?

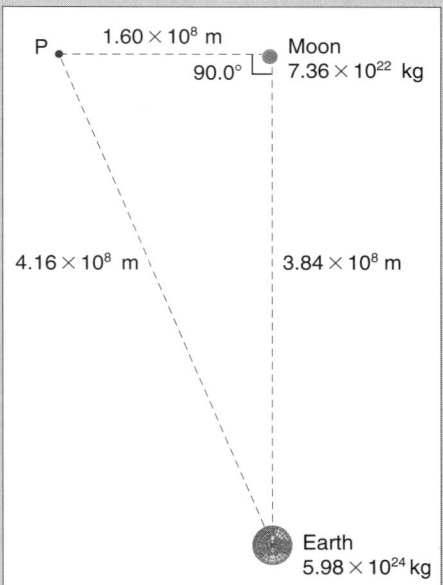

FIGURE P.66

•67. For the mass ring of Example 6.12, notice from Table 6.1 that when $z = 0$ m, the gravitational field is zero. Also at very large distances z from the ring, the field becomes vanishingly small. The magnitude of the gravitational field therefore has a maximum value somewhere along both the positive and negative z-axes. Show that the gravitational field has its maximum magnitude on the axis where $|z| = R/\sqrt{2}$.

•68. Let the mass of the Earth be $M_\oplus$ and let its radius be $R_\oplus$. (a) Show that the *ratio* between the magnitude of the gravitational field on the surface of the Earth and that on the surface of an object of radius R and mass M is

$$\frac{M}{M_\oplus}\left(\frac{R_\oplus}{R}\right)^2$$

This ratio is a pure number called the number of gs. (b) Evaluate the numerical value of this ratio for a white dwarf star that has a mass 1.50 times that of the Sun and a radius equal to the radius of the Earth. (c) Evaluate this ratio for a neutron star with a radius of 10.0 km and a mass 2.00 times that of the Sun.

‡69. Precise measurements of the gravitational field of the Earth indicate that it has a magnitude of 9.832 m/s² at the poles and 9.780 m/s² on the equator, indicating that the polar diameter of the Earth is slightly smaller than the equatorial diameter. Use these data, the value of G, the average radius of the Earth, and the mass of the Earth to estimate the difference between the equatorial and polar diameters of the Earth.

‡70. A mass M is in the shape of a thin uniform disk of radius R. Let the z-axis represent the symmetry axis of the disk as indicated in Figure P.70. The Milky Way Galaxy is modeled as such a mass disk to a first approximation. (a) Show that the gravitational field of the disk at a coordinate z along the symmetry axis of the disk is

$$\vec{g} = -\frac{2GM}{R^2}\left[1 - \frac{z}{(R^2 + z^2)^{1/2}}\right]\hat{k}$$

(b) Show that for $z \gg R$ the expression for $\vec{g}$ in part (a) becomes

$$\vec{g} \approx -\frac{GM}{z^2}\hat{k} \qquad (z \gg R)$$

That is, for distances very far from the disk along the symmetry axis, the gravitational field of the disk is essentially that of a point mass (since at great distances, the disk looks like a point). (c) Let σ be the surface mass density of the disk (the number of kilograms per square meter of the disk), so that

$$\sigma = \frac{M}{\pi R^2}$$

Show that as $z \to 0$ m along the positive z-axis, the gravitational field in part (a) approaches

$$\vec{g} \approx -2\pi G\sigma\hat{k}$$

This is a constant (uniform) field, directed downward. For very small z, the disk looks essentially like an infinite plane and resembles the situation on the surface of the Earth. On the surface of the Earth, the Earth appears to be a flat, infinite plane and we have a constant (uniform) gravitational field directed downward.

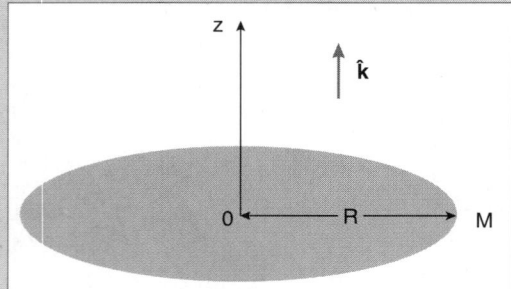

FIGURE P.70

‡71. The Earth is not, in fact, a sphere of uniform density. A high-density core is surrounded by a shell or mantle of lower-density material. Suppose we model a planet of radius R as indicated in Figure P.71. A core of density 2ρ and radius $3R/4$ is surrounded by a mantle of density ρ and thickness

$R/4$. Let M be the mass of the *core* and m be the mass of the *mantle*. (This is *not* an accurate model of the interior structure of the Earth but makes for an interesting and tractable problem.) (a) Show that the mass of the core can be expressed as

$$M = \frac{9}{8}\pi R^3\rho$$

(b) Show that the mass of the mantle can be expressed as

$$m = \frac{37}{48}\pi R^3\rho$$

(c) Show that the total mass of the planet (core + mantle) is

$$\frac{91}{48}\pi R^3\rho$$

(d) Show that the magnitude of the gravitational field at the surface of the planet is

$$g_{\text{surface}} = \frac{91}{48}\pi GR\rho$$
$$= 1.896\,\pi GR\rho$$

(e) Show that the magnitude of the gravitational field at the interface between the core and the mantle is

$$g_i = 2\pi GR\rho$$

Note that the magnitude of the field (the magnitude of the local acceleration of gravity) at the interface between the core and the mantle of the planet is greater than the surface field by about 5%.

This problem is based on the following two papers:
Donald G. Ivey, "Gravity in the real world," *The Physics Teacher*, 30, #4, pages 242–243 (April 1992).
Laurent Hodges, "Gravitational field strength inside the Earth," *American Journal of Physics*, 59, #10, pages 954–956 (October 1991).

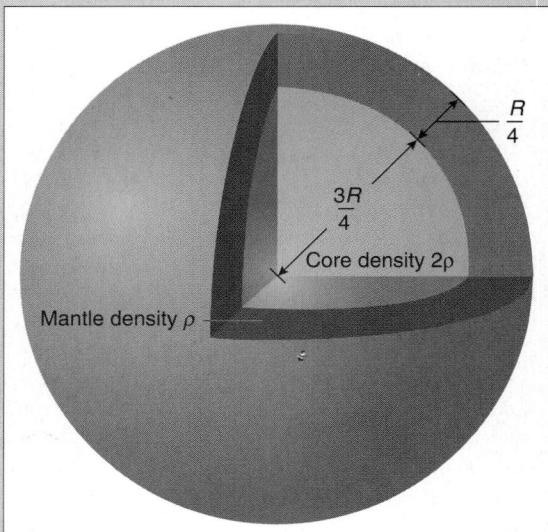

FIGURE P.71

Sections 6.14 The Flux of a Vector*
6.15 Gauss's Law for the Gravitational Field*

72. What is the flux of the gravitational field of mass m through the truncated cone shown in Figure P.72?

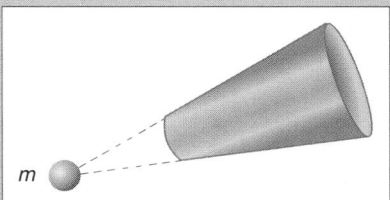

FIGURE P.72

•**73.** What is the flux of the total gravitational field of the Earth and Moon through a closed surface that encloses the Moon? If another surface encloses both the Earth and Moon, by how much is the flux of the gravitational field changed?

•**74.** (a) What is the flux of the total gravitational field caused by the Sun and all the planets through a closed surface that encloses only the Sun? (b) If the surface is expanded so that the entire solar system is enclosed by the surface, does the flux of the gravitational field change significantly?

•**75.** A thin uniform shell has a mass M and radius R as indicated in Figure P.75a. (a) Imagine a concentric spherical closed surface S of radius $r < R$. How much mass is within the surface S? What is the total flux of the gravitational field through this closed surface S? (b) Now imagine a closed surface of *any* shape, drawn *anywhere* inside the mass shell so long as it is totally within the mass shell as indicated in Figure P.75b. What is the mass within such a closed surface? What is the flux of $\vec{g}$ through any such surface? What does this imply about the value of the gravitational field on the closed surface within the shell? In this way, Gauss's law for the gravitational field may be used to prove one of the gravitational shell theorems.

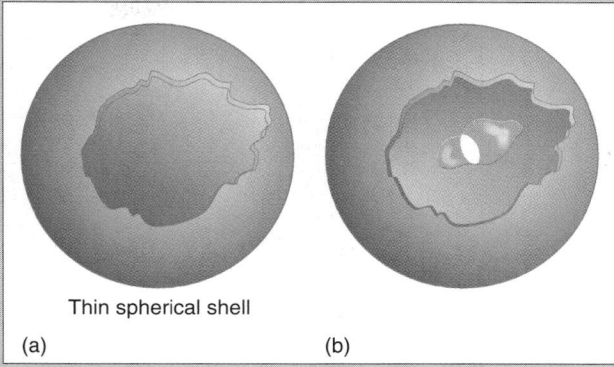

Thin spherical shell

(a)   (b)

FIGURE P.75

•**76.** Viewed from the center of a sphere, the sphere subtends an angle of 4π sr. If one marks out an area on the surface of a sphere that is one degree by one degree (see Figure P.76), the area subtends one *square degree*. Show that 4π sr is equivalent to approximately 41 253 square degrees.

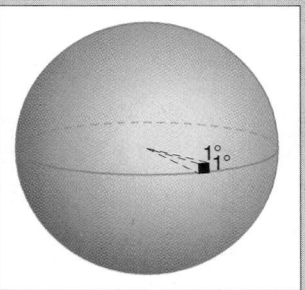

FIGURE P.76

INVESTIGATIVE PROJECTS

A. Expanded Horizons

1. Investigate the relationship between the *inertial mass* (the mass that appears on the right-hand side of Newton's second law of motion) and the *gravitational mass* (the mass that appears in Newton's law of universal gravitation). The apparent equality of these masses is known as the principle of equivalence.
   Wolfgang Rindler, *Essential Relativity* (Springer-Verlag, New York, 1995).
   P. W. Worden Jr. and C. W. F. Everitt, Resource letter GI-1: Gravity and inertia," *American Journal of Physics*, 50, #6, pages 494–500 (June 1982); contains several references to this topic.
   Hans C. Ohanian, "What is the principle of equivalence," *American Journal of Physics*, 45, #10, pages 903–909 (October 1977).
   Albert Einstein and Leopold Infeld, *The Evolution of Physics* (Simon & Schuster, New York, 1967).

2. Investigate *Mach's principle*, the idea that the inertial mass of a system is caused by the existence of the rest of the universe.
   Wolfgang Rindler, *Essential Relativity* (Springer-Verlag, New York, 1995).
   J. David Nightingale, "Specific consequences of Mach's principle," *American Journal of Physics*, 45, #4, pages 376–379 (April 1977).
   J. Barbour and H. Pfester (editors), *From Newton's Bucket to Quantum Gravity* (Birkhauser Boston, Cambridge, Massachusetts, 1996).

3. Investigate experimental attempts to verify the inverse square law nature of the gravitational force law. That is, how close to exactly 2 is the exponent in the denominator of the expression for the magnitude of the gravitational force? Also research the consequences of *not* having the exponent exactly equal to 2.
   P. W. Worden Jr. and C. W. F. Everitt, Resource letter GI-1: Gravity and inertia," *American Journal of Physics*, 50, #6, pages 494–500 (June 1982); contains several references to this topic.
   G. T. Gillies, "Resource letter MNG-1: Measurements of Newtonian gravitation," *American Journal of Physics*, 58, #6, pages 525–534 (June 1990); this is an extensive guide to the gravitational literature.

4. During the 1970s and 1980s, there was speculation about the existence of a *fifth* fundamental force that acted like a weak, *repulsive* gravity. Investigate the state of these speculations.
   Allan Franklin, *The Rise and Fall of the Fifth Force* (American Institute of Physics, New York, 1993).
   A. P. French, "Is there a fifth fundamental force?" *The Physics Teacher*, 24, #5, pages 270–273 (May 1986).
   Robert Pool, "Was Newton wrong?" *Science*, 241, #4867, pages 789–790 (12 August 1988).

Clifford M. Will, "Is it twilight time for the fifth force?" *Sky and Telescope*, 80, #5, pages 472–479 (November 1990).

B. Lab and Field Work

5. Investigate experimental techniques used to determine the value of G, the universal constant of gravitation. In particular, use a *Cavendish balance* to measure G. To how many significant figures can G be determined in your experiment?

   Leybold-Heraeus, Instruction sheet 332 10, "Gravitation torsion balance," (1978) and an earlier version: Leybold physics leaflets, "Mechanics, Gravitation" #DC 531.51, a and b (1959); see also E. M. Purcell, "Gravitational torsion balance," *American Journal of Physics*, 25, #6, pages 393–394 (September 1957).

   Michael S. Saulnier and David Frisch, "Measurements of the gravitational constant without torsion," *American Journal of Physics*, 57, #5, pages 417–420 (May 1989).

   B. E. Clotfelter, "The Cavendish experiment as Cavendish knew it," *American Journal of Physics*, 55, #3, pages 210–213 (March 1987).

   Eric M. Rogers, *Physics for the Inquiring Mind* (Princeton University Press, Princeton, New Jersey, 1973), pages 336–339.

   Kirtley F. Mather and Shirley L. Mason, *Source Book in Geology* (McGraw-Hill, New York, 1939), pages 103–107; this is a reprint of the original experimental results of Cavendish.

   G. T. Gillies, "Resource letter MNG-1: Measurements of Newtonian gravitation," *American Journal of Physics*, 58, #6, pages 525–534 (June 1990); this is an extensive guide to the gravitational literature.

   Jesse W. Beams, "Finding a better value for G," *Physics Today*, 24, #5, pages 34–40 (May 1971).

   John Maddox, "Continuing doubt on gravitation," *Nature*, 310, page 723 (30 August 1984).

   George T. Gillies and Alvin J. Sanders, "Getting the measure of gravity," *Sky and Telescope*, 85, #4, pages 28–32 (April 1993).

6. The true orbital period of the Earth about the Sun, called the *sidereal year*, is slightly longer than the *calendar year* (also known as the tropical year) (see the footnote in Section 6.12). Investigate the reason for this difference. Design an extended experiment to measure these time intervals.

   Eric Chaisson and Steve McMillan, *Astronomy Today* (2nd edition, Prentice Hall, Upper Saddle River, New Jersey, 1996), pages 11–12, 21–22.

7. Design an experiment to measure the time interval between successive meridian passages of the Sun and the same interval for stars. Are the two intervals the same? What is the difference in the time intervals?

   Eric Chaisson and Steve McMillan, *Astronomy Today* (2nd edition, Prentice Hall, Upper Saddle River, New Jersey, 1996), page 10.

C. Communicating Physics

8. The fine structure of the gravitational field of the Earth and other planets can be inferred from slight deviations in satellite motions about the planets or, in the case of the Earth, from direct, precise measurements of the local field at various locations on the surface. Such data yield information about the interior of the planets. As part of a large grant proposal to fund such research, your supervisor assigns you the task of writing a layperson's introduction describing the use of such detailed knowledge of the gravitational field in geophysics and planetary physics.

Paul A. Bender, "Resource letter SEG-1: Solid-earth geophysics," *American Journal of Physics*, 44, #10, pages 903–911 (October 1976); this contains many references to these topics.

9. You have been commissioned by the local science museum to design a scale model of the solar system. The Sun is to be modeled as a sphere with a diameter of 1.00 m. Use the data on the inside covers to specify the diameters of the planets and their distances from the solar sphere. Report your results in tabular form.

10. Occasionally, an important discovery eludes even the best scientists. The planets Mercury, Venus, Earth, Mars, Jupiter, and Saturn have been known since antiquity. The planet Uranus was discovered accidentally in 1781 by William Herschel; the planet Neptune was discovered by Johann G. Galle and Heinrich L. d'Arrest in 1846, based on calculations and predictions by John C. Adams and Urbain J. J. LeVerrier; but Galileo inadvertently observed it on 28 December 1612. How did Galileo miss out on the discovery of Neptune when he observed it? Pluto was discovered in 1930 by Clyde Tombaugh.

    Charles T. Kowal and Stillman Drake, "Galileo's observations of Neptune," *Nature*, 287, pages 311–313, (25 September 1980).

    Stillman Drake and Charles T. Kowal, "Galileo's sighting of Neptune," *Scientific American*, 243, #6, pages 74–81, 246, (December 1980).

    Sky and Telescope, 60, #5, page 363, (November 1980).

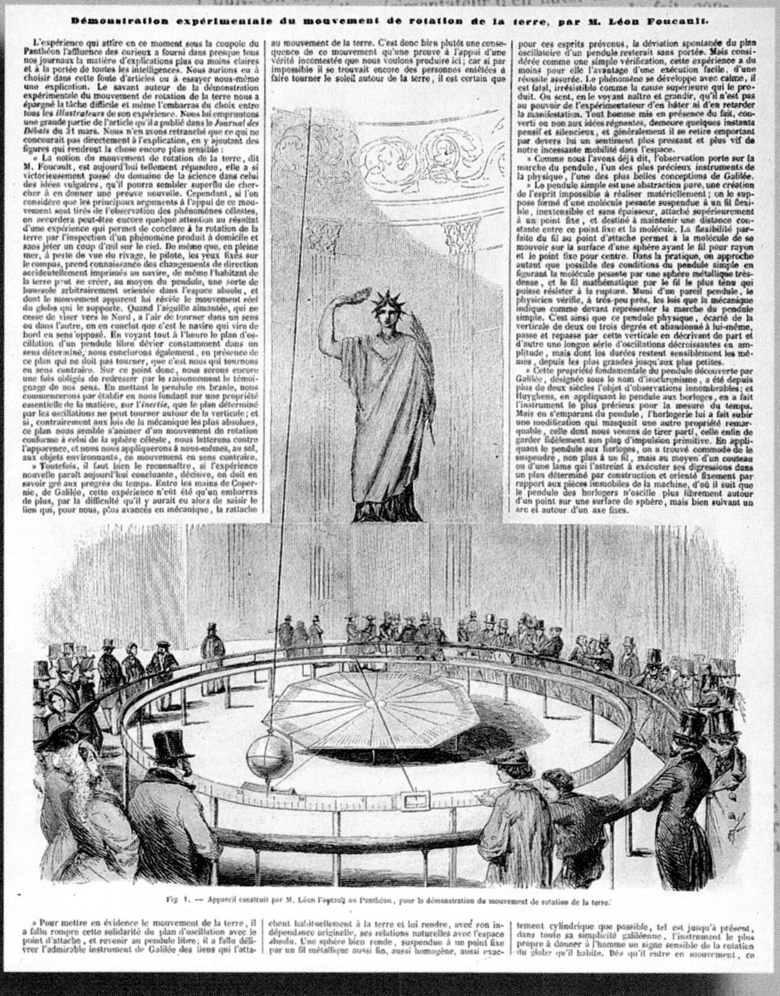

Démonstration expérimentale du mouvement de rotation de la terre, par M. Léon Foucault.

L'expérience qui attire en ce moment sous la coupole du Panthéon l'affluence des curieux a fourni dans presque tous nos journaux la matière d'explications plus ou moins claires et à la portée de toutes les intelligences. Nous aurions eu à choisir dans cette foule d'articles ou à essayer nous-même une explication. Le savant auteur de la démonstration expérimentale du mouvement de rotation de la terre nous a épargné la tâche difficile et même l'embarras du choix entre tous les *illustrateurs* de son expérience. Nous lui empruntons une grande partie de l'article qu'il a publié dans le *Journal des Débats* du 31 mars. Nous n'en avons retranché que ce qui ne concourait pas directement à l'explication, en y ajoutant des figures qui rendront la chose encore plus sensible :

« La notion du mouvement de rotation de la terre, dit M. Foucault, est aujourd'hui tellement répandue, elle a si victorieusement passé du domaine de la science dans celui des idées vulgaires, qu'il pourra sembler superflu de chercher à en donner une preuve nouvelle. Cependant, si l'on considère que les principaux arguments à l'appui de ce mouvement sont tirés de l'observation des phénomènes célestes, on accordera peut-être encore quelque attention au résultat d'une expérience qui permet de conclure à la rotation de la terre par l'inspection d'un phénomène produit à domicile et sans jeter un coup d'œil sur le ciel. De même que, en pleine mer, à perte de vue du rivage, le pilote, les yeux fixés sur le compas, prend connaissance des changements de direction accidentellement imprimés au navire, de même l'habitant de la terre peut se créer, au moyen du pendule, une sorte de boussole arbitrairement orientée dans l'espace absolu, et dont le mouvement apparent lui révèle le mouvement réel du globe qui le supporte. Quand l'aiguille aimantée, qui ne cesse de viser vers le Nord, a l'air de tourner dans un sens ou dans l'autre, on en conclut que c'est le navire qui vire de bord en sens opposé. En voyant tout à l'heure le plan d'oscillation d'un pendule libre dévier constamment dans un sens déterminé, nous conclurons également, en présence de ce plan qui ne doit pas tourner, que c'est nous qui tournons en sens contraire. Sur ce point donc, nous serons encore une fois obligés de redresser par le raisonnement le témoignage de nos sens. En mettant le pendule en branle, nous commencerons par établir en nous fondant sur une propriété essentielle de la matière, sur l'*inertie*, que le plan déterminé par les oscillations ne peut tourner autour de la verticale ; et si, contrairement aux lois de la mécanique les plus absolues, ce plan nous semble s'animer d'un mouvement de rotation conforme à celui de la sphère céleste, nous lutterons contre l'apparence, et nous nous appliquerons à nous-mêmes, au sol, aux objets environnants, ce mouvement en sens contraire.

» Toutefois, il faut bien le reconnaître, si l'expérience nouvelle paraît aujourd'hui concluante, décisive, on doit en savoir gré aux progrès du temps. Entre les mains de Copernic, de Galilée, cette expérience n'eût été qu'un embarras de plus, par la difficulté qu'il y aurait eu alors de saisir le lien qui, pour nous, plus avancés en mécanique, la rattache au mouvement de la terre. C'est donc bien plutôt une conséquence de ce mouvement qu'une preuve à l'appui d'une vérité incontestée que nous voulons produire ici ; car si par impossible il se trouvait encore des personnes entêtées à faire tourner le soleil autour de la terre, il est certain que pour ces esprits prévenus, la déviation spontanée du plan oscillatoire d'un pendule resterait sans portée. Mais considérée comme une simple vérification, cette expérience a du moins pour elle l'avantage d'une exécution facile, d'une réussite assurée. Le phénomène se développe avec calme, il est fatal, irrésistible comme la cause supérieure qui le produit. On sent, en le voyant naître et grandir, qu'il n'est pas au pouvoir de l'expérimentateur d'en hâter ni d'en retarder la manifestation. Tout homme mis en présence du fait, converti ou non aux idées régnantes, demeure quelques instants pensif et silencieux, et généralement il se retire emportant par devers lui un sentiment plus pressant et plus vif de notre incessante mobilité dans l'espace.

» Comme nous l'avons déjà dit, l'observation porte sur la marche du pendule, l'un des plus précieux instruments de la physique, l'une des plus belles conceptions de Galilée.

» Le pendule simple est une abstraction pure, une création de l'esprit impossible à réaliser matériellement ; on le suppose formé d'une molécule pesante suspendue à un fil flexible, inextensible et sans épaisseur, attaché supérieurement à un point fixe, et destiné à maintenir une distance constante entre ce point fixe et la molécule. La flexibilité parfaite du fil au point d'attache permet à la molécule de se mouvoir sur la surface d'une sphère ayant le fil pour rayon et le point fixe pour centre. Dans la pratique, on approche autant que possible des conditions du pendule simple en figurant la molécule pesante par une sphère métallique très-dense, et le fil mathématique par le fil le plus ténu qui puisse résister à la rupture. Muni d'un pareil pendule, le physicien vérifie, à très-peu près, les lois que la mécanique indique comme devant représenter la marche du pendule simple. C'est ainsi que ce pendule physique, écarté de la verticale de deux ou trois degrés et abandonné à lui-même, passe et repasse par cette verticale en décrivant de part et d'autre une longue série d'oscillations décroissantes en amplitude, mais dont les durées restent sensiblement les mêmes, depuis les plus grandes jusqu'aux plus petites.

» Cette propriété fondamentale du pendule découverte par Galilée, désignée sous le nom d'isochronisme, a été depuis plus de deux siècles l'objet d'observations innombrables ; et Huyghens, en appliquant le pendule aux horloges, en a fait l'instrument le plus précieux pour la mesure du temps. Mais en s'emparant du pendule, l'horlogerie lui a fait subir une modification qui masquait une autre propriété remarquable, celle dont nous venons de tirer parti, celle enfin de garder fidèlement son plan d'impulsion primitive. En appliquant le pendule aux horloges, on a trouvé commode de le suspendre, non plus à un fil, mais au moyen d'un couteau ou d'une lame qui l'astreint à exécuter ses digressions dans un plan déterminé par construction et orienté fixement par rapport aux pièces immobiles de la machine, d'où il suit que le pendule des horlogers n'oscille plus librement autour d'un point sur une surface de sphère, mais bien suivant un arc et autour d'un axe fixes.

Fig. 1. — Appareil construit par M. Léon Foucault au Panthéon, pour la démonstration du mouvement de rotation de la terre.

» Pour mettre en évidence le mouvement de la terre, il a fallu rompre cette solidarité du plan d'oscillation avec le point d'attache, et revenir au pendule libre ; il a fallu délivrer l'admirable instrument de Galilée des liens qui l'attachent habituellement à la terre et lui rendre, avec son indépendance originelle, ses relations naturelles avec l'espace absolu. Une sphère bien ronde, suspendue à un point fixe par un fil métallique aussi fin, aussi homogène, aussi exactement cylindrique que possible, tel est jusqu'à présent, dans toute sa simplicité galiléenne, l'instrument le plus propre à donner à l'homme un signe sensible de la rotation du globe qu'il habite. Dès qu'il entre en mouvement, ce

CHAPTER 7

Hooke's Force Law and Simple Harmonic Oscillation

An oscillation is a motion that cannot make up its mind which way it wants to go.

An anonymous fourth grader

In 1660 Robert Hooke (1635–1703) described in a quantitative fashion the force that arises when a common, helically coiled spring is stretched or compressed. His empirical discovery now is called Hooke's force law. Working at Oxford University, Hooke was a contemporary of Newton at rival Cambridge University. Unfortunately, as a result of controversies regarding the priority of several discoveries,* the two gentlemen had a relationship filled with personal animosity, disrespect, and thinly veiled insults. Even great physicists are human.

In this chapter we study the force of a helical spring on a mass attached to it. If our concern about the force exerted by a simple, mundane spring on a mass appears to be a bit arcane and antiquated, in some respects it is just that. But many physical systems exhibit similar behavior and are *modeled* by a Hooke's force law to a first approximation. More advanced courses in physics and engineering return repeatedly to Hooke's force law as a foundation for understanding many complex oscillatory motions on both macroscopic and microscopic scales. For example:

1. If atoms are displaced slightly from their equilibrium positions in molecules, they experience springlike elastic restoring forces proportional to their distance from equilibrium in the molecule.
2. If you attempt to pull a quark out of a proton or other hadron, the quark experiences a restoring force that increases with distance to a crude first approximation.
3. If a beam is loaded with mass, the weight causes the beam to sag a distance that is proportional to the load.
4. If a rod is twisted, the restoring force is proportional to the twist.
5. If a steel cable on a bridge is stretched under tension, there is a restoring force proportional to the stretch.

---

*The discovery of gravitation was among them.

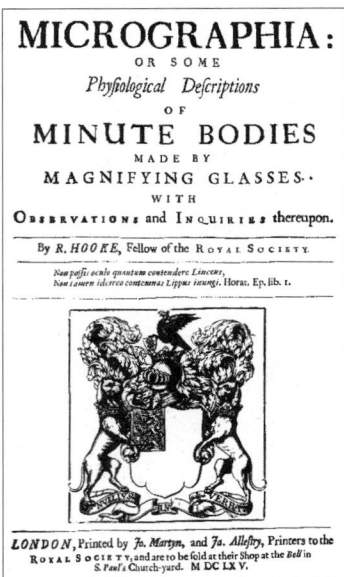

Robert Hooke wrote the first important treatise on observations with a microscope, coining the biological term *cell.*

Thus the relationship discovered by Hooke for simple springs has applications far beyond whatever he imagined during the 17th century. The force law is encountered in classical and quantum mechanics (for small excursions from stable equilibrium positions) as well as in many areas of engineering. As you pursue your career in the sciences, you will see Hooke's force law used to model many complex physical phenomena; so this is an important force to master even in an introductory course.

We begin by examining the force law discovered by Hooke and studying the one-dimensional oscillatory dynamics of a mass subjected to only such a force. We then investigate several different oscillatory phenomena that, at first glance, have little to do with masses attached to springs but, upon closer inspection, are camouflaged manifestations of Hooke's force law, revealing that it frequently is encountered in nature well beyond its original domain of a simple spring. Finally, conjuring up memories of your childhood experiences on a swing, we investigate realistic oscillations that either die out over time or continue because of the periodic application of additional forces.

7.1 HOOKE'S FORCE LAW

Springs have many applications, from ballpoint pen retractors and automobiles to the docking collar for the space shuttle when connecting with the Russian Mir spacecraft.

Consider a mass m attached to one end of a horizontally oriented spring, as shown in Figure 7.1. The other end of the spring is secured to a rigid, immovable wall. For simplicity, let the surface on which the mass m rests be frictionless. Choose $\hat{\jmath}$ to be vertically upward, as indicated in Figure 7.1. The position of the mass when the spring is neither stretched nor compressed is the **equilibrium position**, since there is zero acceleration and zero total force on m at this location.

Take the origin of a Cartesian coordinate system to be at the equilibrium position of the mass, with the +x-axis ($\hat{\imath}$) corresponding to *stretching* the spring, as in Figure 7.1. This choice for the direction of $\hat{\imath}$ is the conventional choice, which we will use consistently throughout our treatment in this chapter and elsewhere in the text when considering springs.

Hidden springs are used for the space shuttle docking collar on the Mir spacecraft; what look like springs in this photo actually are coiled data cables.

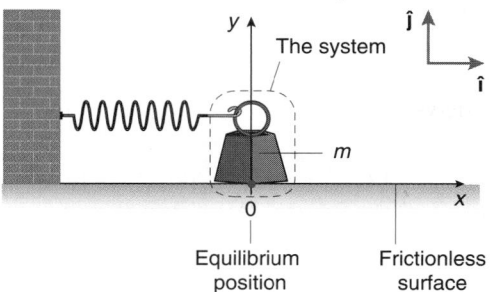

FIGURE 7.1 A mass on a frictionless surface is attached to a spring.

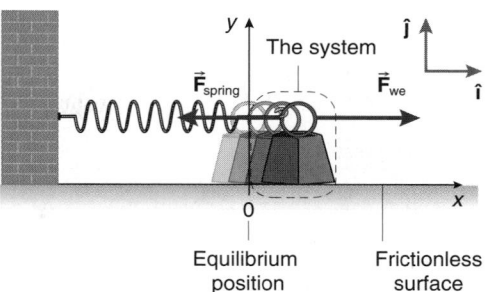

FIGURE 7.2 Horizontal forces on the mass when we stretch the spring with a force $\vec{F}_{we}$.

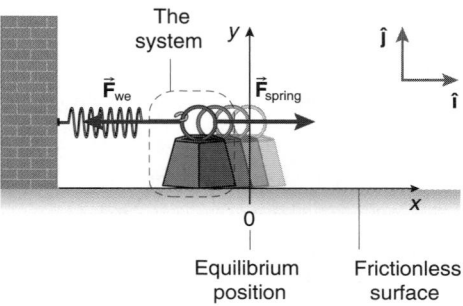

FIGURE 7.3 Horizontal forces on the mass when we compress the spring.

Consider all springs in this text to be ideal springs unless otherwise stated explicitly.

Two forces act on the mass in the vertical direction:

1. the upward force of the surface on the mass—the normal force; and

2. the downward gravitational force of the Earth on m—the weight of mass m.

These vertical forces, though, are irrelevant to motion in the horizontal direction since motions in perpendicular directions are independent. It is the motion parallel to the axis of the spring that we wish to examine.

We stretch or compress the spring horizontally with a force $\vec{F}_{we}$ and hold the mass at rest at some position coordinate x. Since the origin was chosen at the equilibrium position of the mass, x also represents the x-component of the position vector of the system and measures the extent to which the spring is either stretched ($x > 0$ m) or compressed ($x < 0$ m). In the horizontal direction there are two forces acting on the system, shown in Figure 2.2:

1. the force $\vec{F}_{we}$ that we exert on the mass to stretch the spring;

2. the force $\vec{F}_{spring}$ of the spring on the mass.

Since the surface is frictionless, there is no frictional force parallel to the surface.

If we pull on the mass to stretch the spring, the force of the spring on m is directed opposite to our pull. Notice that the force of the spring on m is directed toward the equilibrium position. If instead of pulling on m and stretching the spring, we push on m

and compress the spring (see Figure 7.3), the force of the spring on m again is directed opposite to ours and once again attempts to restore m to the equilibrium position.

The force of the spring on the mass always is directed toward the equilibrium position whether the spring is stretched or compressed.

For this reason the force of the spring on m occasionally is called a **restoring force**, since it acts to restore the mass to its equilibrium position.

In the late 17th century, Robert Hooke discovered an empirical relationship between the stretch (or compression) of a spring and the force $\vec{F}_{we}$ that we exert on m to hold it at rest at coordinate x. To stretch the spring, we clearly have to provide a force on m in the same direction as î; to compress the spring we need to push on m with a force directed along $-$î. Hooke discovered that a graph of the x-component of the force $\vec{F}_{we}$ that we need to exert on m to keep it at a given value of x is directly proportional to x. In other words, the force component depends *linearly* on x, as shown in Figure 7.4.

Let the slope of such a linear plot be k. Hooke also noted that if we gradually decrease the magnitude of the x-component of the force we exert on m, the force component retraces its path on the graph; and so when the force we provide is zero, m is once again at the original equilibrium position. The graph thus has an intercept equal to zero. Furthermore, he discovered that the force on m to hold it at a given x-coordinate is, remarkably, *independent* of m and depends *only* on the particular spring used in the experiment.

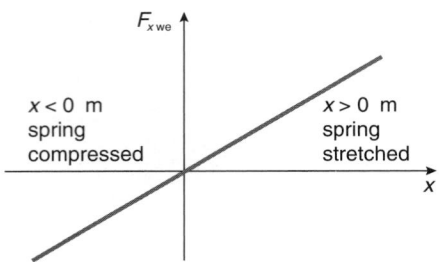

FIGURE 7.4 A graph of the *x*-component of $\vec{\mathbf{F}}_{we}$ (its only component) versus *x*.

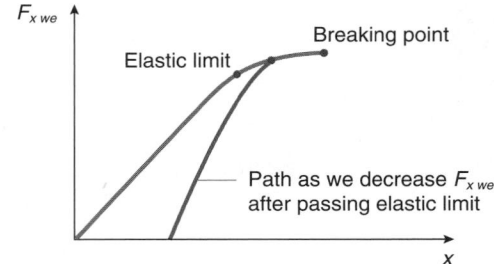

FIGURE 7.5 Stretch a spring too far and it is permanently deformed.

Since the equilibrium position of the mass on the spring is the origin of the Cartesian coordinate system chosen (Figure 7.1), we express Hooke's empirical discoveries by saying that the *x*-component of $\vec{\mathbf{F}}_{we}$ (its only component) is directly proportional to *x* itself; that is,*

$$\vec{\mathbf{F}}_{we} = kx\hat{\mathbf{i}} \qquad (7.1)$$

where the proportionality constant *k*, the slope of the graph in Figure 7.4, is characteristic of the particular spring used in the experiment and is independent of the mass attached to the spring.

> The scalar constant *k* is called the **spring constant** and has the SI units of N/m.† The spring constant *k* represents the number of newtons of force needed to stretch (or compress) the spring one meter.

All springs have a positive spring constant. The larger the value of *k*, the harder it is for us to hold the mass on the spring at a given coordinate *x*. A spring with a large spring constant *k* is therefore a **stiff spring**.

If you ever have played with a small spring, such as that inside many retractable ballpoint pens, you are aware that if you stretch it too far, beyond what is called its **elastic limit**, the spring is permanently deformed and will not relax to its original equilibrium position if you subsequently remove the force you exert on *m*. If the spring is stretched still further, eventually it breaks. A schematic graph of the force component exerted versus the stretch or elongation of the spring from its relaxed position is shown in Figure 7.5.

The linear portion of the graph (including the linear section corresponding to compressing the spring) is called the **Hooke's law region**. The spring is said to be **elastic** if the stretch is confined to the Hooke's law region of the graph. In fact the graphs of many strangely behaving springlike forces, if repeatable, approach the linear shape if the departure from equilibrium is within small enough limits. Springs in special shapes like coils provide quite extensive ranges of linear behavior, and they are

examples of our ideal springs. We will only consider springs confined to the Hooke's law region.

Imagine the mass *held* at position *x*, as in Figure 7.6. The mass is not accelerated in the horizontal direction, and so the total force on the mass in the horizontal direction must be zero according to Newton's second law of motion. Thus, when $a_x = 0$ m/s², the vector sum of *the force that we exert* on the mass and *the force of the spring* on the mass must be zero:

$$\vec{\mathbf{F}}_{we} + \vec{\mathbf{F}}_{spring} = 0 \text{ N}$$

If we restrict ourselves to the Hooke's law region of the graph, and substitute for $\vec{\mathbf{F}}_{we}$ from Equation 7.1, we have

$$kx\hat{\mathbf{i}} + \vec{\mathbf{F}}_{spring} = 0 \text{ N}$$

which states that the force of the spring on the mass in the Hooke's law region‡ is

$$\vec{\mathbf{F}}_{spring} = -kx\hat{\mathbf{i}} \qquad (7.2)$$

Notice that if the spring is stretched, *x* is positive, and the force of the spring on *m* is toward $-\hat{\mathbf{i}}$ (i.e., toward the equilibrium position). If the spring is compressed, *x* is negative, and the force of the spring on *m* is toward $+\hat{\mathbf{i}}$ (again, toward the equilibrium position). A graph of the *x*-component (the only component) of the force of the spring on *m* as a function of *x* is shown in Figure 7.7.

Notice that this graph has a negative slope equal to $-k$, the opposite of the slope of the graph of the component of $\vec{\mathbf{F}}_{we}$ versus *x* in Figure 7.4.

> Thus *the force of the spring* on *m*
>
> 1. is proportional to the amount of the stretch or compression of the spring and is always directed toward the equilibrium position; and
>
> 2. is independent of the mass *m* attached to the spring; the force depends only on the spring constant *k* and the stretch or compression of the spring *x*.

---

*Do not confuse the spring constant *k* (a scalar) with the Cartesian *unit vector* $\hat{\mathbf{k}}$.

†The spring constant also is known as the *force constant* of the spring. We will not use this terminology.

---

‡Problem 7 invites you to show that if the direction for $\hat{\mathbf{i}}$ is chosen to correspond to *compressing* the spring, the Hooke's law expression for the force of the spring on *m* *still* is $\vec{\mathbf{F}}_{spring} = -kx\hat{\mathbf{i}}$, unchanged in form.

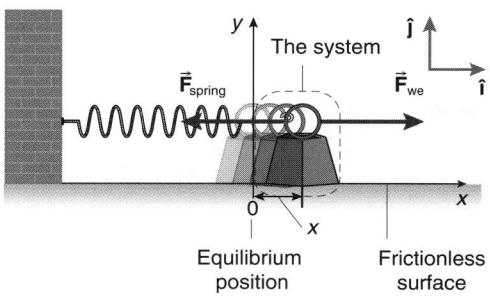

FIGURE 7.6 Stretch and hold the mass at position x.

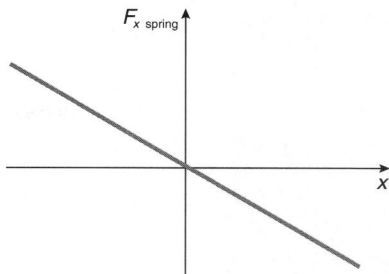

FIGURE 7.7 The x-component of $\vec{F}_{spring}$ (its only component) versus x.

PROBLEM-SOLVING TACTIC

7.2 For a mass attached to a spring, do not confuse the force we exert on the mass, $\vec{F}_{we} = kx\hat{i}$, with the force that the spring exerts on the mass, $\vec{F}_{spring} = -kx\hat{i}$. Be sure you know which force is appropriate in a problem.

QUESTION 1

When we pull on a mass attached to a horizontal spring on a frictionless surface and hold the mass at some position coordinate x, the force of the spring on m is equal in magnitude and opposite in direction to the force we exert on m. Are these forces a Newton's third law force pair? Explain why or why not.

EXAMPLE 7.1

A mechanic finds that a force of magnitude 150 N is sufficient to hold a 7.00 kg mass attached to a stiff spring 3.0 cm (stretched) from the equilibrium position. What is the force exerted by the spring on the mass when the spring is compressed 4.0 cm? The surface is frictionless.

Solution

In accordance with Problem-Solving Tactic 7.1, use the conventional coordinate system with $x = 0$ m as the equilibrium position with $\hat{i}$ corresponding to stretching the spring. Since the mechanic applies a force of magnitude 150 N to hold the mass 0.030 m (stretched) from its equilibrium position, use Equation 7.1,

$$\vec{F}_{mechanic} = kx\hat{i}$$

to find

$$(150 \text{ N})\hat{i} = k(0.030 \text{ m})\hat{i}$$

The spring constant k is

$$k = 5.0 \times 10^3 \text{ N/m}$$

Aware of Problem-Solving Tactic 7.2, do not confuse the force *the mechanic* exerts on the mass attached to the spring, $\vec{F}_{mechanic} = kx\hat{i}$, with the force that the *spring* exerts on the mass. The force of the spring is given by Equation 7.2:

$$\vec{F}_{spring} = -kx\hat{i}$$

When the spring is compressed 4.0 cm, $x = -0.040$ m, and so

$$\vec{F}_{spring} = (-5.0 \times 10^3 \text{ N/m})(-0.040 \text{ m})\hat{i}$$
$$= (2.0 \times 10^2 \text{ N})\hat{i}$$

The force of the spring always is directed toward the equilibrium position. Since the spring is compressed, the force of the spring is parallel to $\hat{i}$.

EXAMPLE 7.2

A force of magnitude 2.50×10^3 N is able to hold a 10.0 kg mass attached to one end of a spring a distance of 10 cm from its equilibrium position. The spring is stretched.

a. What is the spring constant k of the spring?
b. What is the force of the spring on m?
c. What is the force needed to hold the mass at a location with the spring compressed by 10 cm? What then is the force of the spring on the mass at this location?

Solution

a. Take the equilibrium position of the mass as the origin for the x-coordinate with $\hat{i}$ corresponding to stretching the spring. The force needed to hold the mass at the coordinate $x = 0.10$ m then is, from Equation 7.1,

$$\vec{F}_{we} = kx\hat{i}$$
$$(2.50 \times 10^3 \text{ N})\hat{i} = k(0.10 \text{ m})\hat{i}$$

So the spring constant k is

$$k = \frac{2.50 \times 10^3 \ \text{N}}{0.10 \ \text{m}}$$

$$= 2.5 \times 10^4 \ \text{N/m}$$

b. Since the mass is not accelerating, the force of the spring and the force we exert on the mass must vector sum to zero according to Newton's second law:

$$\vec{\mathbf{F}}_{\text{spring}} + \vec{\mathbf{F}}_{\text{we}} = 0 \ \text{N}$$

So $\vec{\mathbf{F}}_{\text{spring}}$ is opposite to the direction we pull and of the same magnitude as our pull. Since the force we applied was $(2.50 \times 10^3 \ \text{N})\hat{\imath}$, the force of the spring on m is

$$\vec{\mathbf{F}}_{\text{spring}} = (-2.5 \times 10^3 \ \text{N})\hat{\imath}$$

This is parallel to $-\hat{\imath}$, because the force of the spring on the mass always is directed toward the equilibrium position and the spring was stretched.

c. When the spring is compressed the same distance, we need to provide a force

$$\vec{\mathbf{F}}_{\text{we}} = (-2.5 \times 10^3 \ \text{N})\hat{\imath}$$

to hold the mass at this location. Since the mass is not accelerating, the second law of motion implies

$$\vec{\mathbf{F}}_{\text{spring}} + \vec{\mathbf{F}}_{\text{we}} = 0 \ \text{N}$$

The force of the spring on the mass thus is

$$\vec{\mathbf{F}}_{\text{spring}} = (2.5 \times 10^3 \ \text{N})\hat{\imath}$$

and, again, is directed toward the equilibrium position.

7.2 SIMPLE HARMONIC OSCILLATION

> *O damnéd vacillating state!*
>
> Alfred, Lord Tennyson (1809–1892)*

Many physical systems **oscillate**,[†] moving or swinging back and forth with a steady, repetitive rhythm, much like the tides. Masses attached to springs, playground swings, clock pendulums, tennis balls in a good rally, and professors pacing in front of a classroom all oscillate, or move back and forth. Here we examine a special kind of oscillation called **simple harmonic oscillation**. Not all oscillatory phenomena exhibit simple harmonic oscillation, but simple harmonic oscillation underlies the analy-

sis of almost every type of oscillation. It is important to know the characteristic features of simple harmonic oscillations and to be able to recognize whether an oscillatory motion is simple harmonic or not (which we study in Section 7.5). The oscillatory behavior of a mass on an ideal spring is the paradigm for simple harmonic oscillation.

Once again, we begin with a mass m attached to an ideal spring, lying on a horizontal frictionless surface. We pull the mass attached to the spring to the right a distance x_0 and then suddenly (instantaneously) release the mass at rest, as in Figure 7.8.

When we suddenly remove our force $\vec{\mathbf{F}}_{\text{we}}$ on the mass, the force of the spring $\vec{\mathbf{F}}_{\text{spring}}$ on m still exists, since the spring is stretched. From experience, we know what happens when our force is removed: the mass oscillates back and forth. Since the surface is frictionless and the spring is ideal, the oscillatory motion continues indefinitely. The oscillatory motion of m when the force of the spring is the *only* force on m in a given direction is simple harmonic oscillation, and the arrangement is called a **simple harmonic oscillator**. We describe this oscillatory motion of m in more detail using Newton's laws of motion. It typifies many cases of oscillation.

After the mass is released and located at coordinate x, the only force on the mass in the horizontal direction is the force of the spring on the mass, $\vec{\mathbf{F}}_{\text{spring}} = -kx\hat{\imath}$. Newton's second law of motion, applied to the horizontal motion of m, then is

$$-kx\hat{\imath} = ma_x\hat{\imath} \tag{7.3}$$

The position vector of the mass is

$$\vec{\mathbf{r}}(t) = x(t)\hat{\imath}$$

The x-component of the acceleration, a_x, is the second derivative of the x-component of the position vector with respect to time:

$$a_x = \frac{d^2 x}{dt^2}$$

Make this substitution for the acceleration component in Newton's second law of motion for m. Equation 7.3 then becomes

$$-kx = m \frac{d^2 x}{dt^2} \tag{7.4}$$

We want to discover how x varies with time. That is, we want to find the function $x(t)$ that satisfies this equation. Equation 7.4 is

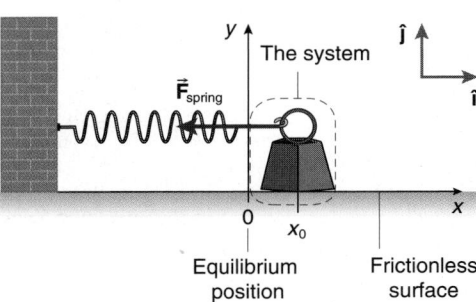

FIGURE 7.8 We pull a mass from the equilibrium position to coordinate x_0 and release the mass at rest.

Supposed Confessions of a Second-Rate Sensitive Mind Not in Unity with Itself, line 207. *The Poetical Works of Alfred Tennyson* (James R. Osgood and Company, Boston, 1877), page 285.

[†]The word comes from the Latin *oscillare*, meaning "to swing."

an example of a **differential equation**: an equation involving a function [in this case $x(t)$] and its derivatives (in this case just the second derivative). There are many types and kinds of differential equations; they are encountered frequently in physics and engineering. Many techniques exist for solving such equations, that is, for finding the function like our $x(t)$. You likely have not yet had a full-fledged course in these techniques for solving differential equations. But you will see that with your growing or existing knowledge of calculus, you can deduce a solution rather than use the formal techniques for solving such equations employed in higher-level mathematics, physics, and engineering courses.

The statement of Newton's second law for our problem, Equation 7.4, indicates that the second derivative of the (as yet) unknown function $x(t)$ is proportional to the function $x(t)$ itself. We need to find a function whose second derivative brings the *same* function back again, with perhaps a few constants thrown in for good measure. The mathematical functions that come to mind are the sines and cosines.* The first derivative of the sine is the cosine, but the second derivative of the sine brings back the sine again, apart from some purely numerical factors such as minus signs and other constants. Likewise the first derivative of the cosine is proportional to the sine, but the second derivative of the cosine brings back the cosine once again (apart from minus signs and constants).

Sine and cosine functions certainly oscillate as functions of their arguments, and so their use here may not be out of hand. Hence the solution we need for $x(t)$ likely is a sine or cosine function or some additive combination of them. We could try a solution for $x(t)$ such as

$$x(t) = a \cos \omega t + b \sin \omega t \qquad (7.5)$$

where ω, a, and b are some constants, independent of time, whose values are chosen to suit a given physical situation. Since t represents time, the constant ω must have the dimensions of inverse time, to make the argument of the trigonometric functions (the angle) dimensionless. Equation 7.5 will work just fine for these purposes. However, with no loss of generality, it is more convenient in the long run (although it is not apparent at this juncture) to use a single cosine function of the form

$$x(t) = A \cos(\omega t + \phi) \qquad (7.6)$$

where ω, A, and ϕ are three constants, independent of time, whose values must be determined by relating them to the oscil-

lator; ω and A are positive. Example 7.3 following shows that Equations 7.5 and 7.6 are equivalent expressions for $x(t)$.

PROBLEM-SOLVING TACTIC

7.3 Be careful! The quantity ωt is measured in radians, and so the constant angle ϕ, called the phase angle, also must be in radians. Thus radians are added to radians in the argument of the cosine.[†] Since the time t has the SI units of seconds and ωt is in radians, the units of ω are rad/s.

The quantity ω is called the **angular frequency** of the oscillation for reasons we will explore. It is customary *always* to indicate the SI units of the angular frequency ω explicitly as rad/s.

Recall that the maximum value of the cosine is 1; hence the maximum value of x in Equation 7.6 is A. The constant A in Equation 7.6 is called the **amplitude** of the oscillation, since it is the largest value that the x-coordinate of the oscillator can attain. Recall that x is the x-component of the position vector of the oscillator. Since the cosine can take on values only between (and including) -1 and $+1$, the range of x-coordinates visited at some time by the oscillator is twice the amplitude, or 2A.

QUESTION 2

A tennis ball is hit back and forth across the net by two excellent tennis players in fixed positions. Is the back-and-forth motion of the ball simple harmonic oscillation? Explain your reasoning.

EXAMPLE 7.3

Show that Equation 7.6 is equivalent to Equation 7.5.

Solution
Use the trigonometric formula for the cosine of the sum of two angles,

$$\cos(\alpha + \beta) = \cos \alpha \cos \beta - \sin \alpha \sin \beta$$

to rewrite Equation 7.6:

$$x(t) = A \cos(\omega t + \phi)$$
$$= A \cos \omega t \cos \phi - A \sin \omega t \sin \phi$$

This is identical to Equation 7.5 if

$$a = A \cos \phi \qquad (1)$$

and

$$b = -A \sin \phi \qquad (2)$$

---

*Exponential functions involving the base of natural logarithms e also have this property. However, thanks to the Euler identity,
$$e^{i\theta} = \cos \theta + i \sin \theta$$
and its complex conjugate,
$$e^{-i\theta} = \cos \theta - i \sin \theta$$
the sines and cosines are combinations of exponentials with complex exponents. If you add the two complex exponential expressions, you find that
$$\cos \theta = \frac{e^{i\theta} + e^{-i\theta}}{2}$$
If the expressions are subtracted, you find that
$$\sin \theta = \frac{e^{i\theta} - e^{-i\theta}}{2i}$$

[†]In some disciplines, the angle ϕ occasionally is given in *degrees* for convenience. This is done frequently by electrical engineers. You must be aware that to perform the angular addition of ωt with ϕ, *both* must be expressed in the same angular units.

Divide equation (2) by equation (1) to eliminate A; this yields

$$\tan \phi = -\frac{b}{a}$$

Thus

$$\phi = \tan^{-1}\left(-\frac{b}{a}\right) \quad (3)$$

Use the trigonometric identity

$$\sin^2 \phi + \cos^2 \phi = 1$$

and equations (1) and (2) for $\cos \phi$ and $\sin \phi$, to discover that

$$\frac{b^2}{A^2} + \frac{a^2}{A^2} = 1$$

Solve for A:

$$A = (a^2 + b^2)^{1/2} \quad (4)$$

This shows that we can use

$$x(t) = A \cos(\omega t + \phi)$$

rather than

$$x(t) = a \cos \omega t + b \sin \omega t$$

without any loss of generality. Both expressions have two adjustable constants—A and ϕ, or a and b; so each expression will be equally able to be adjusted to match the starting conditions.

EXAMPLE 7.4 _____

Find the choices that need to be made for A and ϕ so that

$$x(t) = b \sin \omega t$$

can be written in the form

$$x(t) = A \cos(\omega t + \phi)$$

Solution
We use the trigonometric formula for the cosine of the sum of two angles,

$$\cos(\alpha + \beta) = \cos \alpha \cos \beta - \sin \alpha \sin \beta$$

to write

$$\begin{aligned} x(t) &= A \cos(\omega t + \phi) \\ &= A(\cos \omega t \cos \phi - \sin \omega t \sin \phi) \\ &= (A \cos \phi) \cos \omega t - (A \sin \phi) \sin \omega t \end{aligned}$$

Since this must be identical to $x(t) = b \sin \omega t$, we must have

$$A \cos \phi = 0 \text{ m} \quad (1)$$

and

$$-A \sin \phi = b \quad (2)$$

Equation (1) means we could have $A = 0$ m; but this choice is not appropriate since it implies $x(t) \equiv 0$ m for all times t. Hence, since $A \neq 0$ m, to satisfy equation (1), the angle ϕ can be $\pm \pi/2$ rad because $\cos(\pm \pi/2 \text{ rad}) = 0$. We must choose $\phi = -\pi/2$ rad to make A positive; then equation (2) implies that

$$-A \sin\left(-\frac{\pi}{2} \text{ rad}\right) = b$$

That is,

$$A = b$$

Hence $x(t) = b \sin \omega t$ is equivalent to

$$x(t) = b \cos\left(\omega t - \frac{\pi}{2} \text{ rad}\right)$$

_____

We have to see how to determine the constants ω, A, and ϕ in Equation 7.6. We know two things about the motion:

1. Equation 7.4 governs the motion at all times, since the equation stems from Newton's second law; and
2. We know how the mass was released, in our case, at position x_0 with zero speed.

Rearrange Equation 7.4:

$$m\frac{d^2x}{dt^2} + kx = 0 \text{ N}$$

Now divide by m:

$$\boxed{\frac{d^2x}{dt^2} + \frac{k}{m}x = 0 \text{ m/s}^2} \quad (7.7)$$

This is the **equation for simple harmonic oscillation**. The equation must be satisfied by $x(t)$ at all times t, since it describes the dynamical behavior of the particle according to Newton's second law of motion. Take the function that we think is a solution, Equation 7.6,

$$x(t) = A \cos(\omega t + \phi)$$

and find its second derivative with respect to time (remember that A, ω, and ϕ are constants, independent of time). The first derivative of x with respect to time is the x-component (the only component) of the velocity:

$$\frac{dx}{dt} = -A\omega \sin(\omega t + \phi) \quad (7.8)$$

The second derivative is the x-component (the only component) of the acceleration:

$$\begin{aligned} \frac{d^2x}{dt^2} &= -A\omega^2 \cos(\omega t + \phi) \\ &= -\omega^2 A \cos(\omega t + \phi) \\ &= -\omega^2 x(t) \end{aligned} \quad (7.9)$$

Substitute the second derivative into Equation 7.7 for the simple harmonic oscillator:

$$-\omega^2 x(t) + \frac{k}{m} x(t) = 0 \ \mathrm{m/s^2}$$

After factoring, we have

$$\left(-\omega^2 + \frac{k}{m}\right) x(t) = 0 \ \mathrm{m/s^2}$$

This relation must be true at *any* time *t*. If $x(t)$ were *always* zero, we would not have much of a solution: there would be no oscillation. Indeed we know that $x(t)$ is not going to be zero at all times, since we stretched the spring and released it. Hence at all times *t* we must have

$$-\omega^2 + \frac{k}{m} = 0 \ \mathrm{rad^2/s^2}$$

or

$$\boxed{\omega = \left(\frac{k}{m}\right)^{1/2}} \qquad (7.10)$$

The constant angular frequency ω is related to the spring constant *k* and the mass *m*. Recall that the units of ω always are expressed as rad/s. Notice that the differential equation describing the oscillation, Equation 7.7, also may be written cleanly using the angular frequency ω as

$$\boxed{\frac{d^2x}{dt^2} + \omega^2 x = 0 \ \mathrm{m/s^2}} \qquad (7.11)$$

The coefficient of the term involving *x* in the differential equation is the *square* of the angular frequency; we will use this fact in later examples of oscillatory motion.

> Any system of mass *m* whose position coordinate *x* satisfies Equation 7.7, or equivalently Equation 7.11, is executing simple harmonic oscillation.

We will return to investigate the meaning of ω after we discover more about the other constants we have to find: A and ϕ. The two constants A and ϕ are determined from the **initial conditions**. Such conditions also are called **boundary conditions**. Boundary conditions need not be the conditions when $t = 0$ s (which are called initial conditions). Boundary conditions also can be certain features of the motion at another specified time. In our case, the way in which the oscillation was initiated enables us to determine A and ϕ. For instance, we initiated the oscillation by stretching the spring to coordinate x_0 and releasing it with zero initial speed. Thus when $t = 0$ s, *x* must equal x_0. That is, from Equation 7.6,

$$x(0 \ \mathrm{s}) = x_0 = A \cos[\omega(0 \ \mathrm{s}) + \phi]$$
$$x_0 = A \cos \phi \qquad (7.12)$$

Since $x(t) = A \cos(\omega t + \phi)$, the *x*-component of the velocity (its only component) is

$$v_x(t) = \frac{dx(t)}{dt}$$
$$= -A\omega \sin(\omega t + \phi)$$

The mass was released when $t = 0$ s, at which time the velocity component was zero; thus,

$$0 \ \mathrm{m/s} = -A\omega \sin[\omega(0 \ \mathrm{s}) + \phi]$$
$$= -A\omega \sin \phi$$

The constant ω is not zero (earlier we found it was equal to $(k/m)^{1/2}$). Since A itself is not zero (if it *were* zero, our solution would be zero at all times, which is not the case), we must have $\phi = 0$ rad since $\sin(0 \ \mathrm{rad}) = 0$.

Thus $\phi = 0$ rad, and, from Equation 7.12,

$$x_0 = A \cos \phi = A \cos(0 \ \mathrm{rad})$$
$$= A \qquad (7.13)$$

PROBLEM-SOLVING TACTIC

7.4 The way the oscillator is released provides the information needed to determine the amplitude A and phase angle ϕ. Recall that the angular frequency ω depends only on the particular spring used (through the spring constant *k*) and the mass on the end of the spring; ω is *independent* of the initial conditions. See Example 7.6.

We have found that $x(t) = A \cos \omega t$, where $A = x_0$, the initial position of the mass *m* when it was released at rest. The position vector of the mass is $\vec{r}(t) = x(t)\hat{\imath}$. A graph of $x(t)$ is shown in Figure 7.9. We see that the range of *x*-coordinates visited by the oscillator is *twice* the amplitude or 2A. Note that when time $t = 0$ s, $x = A = x_0$, which is consistent with how we released the oscillator. Note also that when $\omega t = 2\pi$ rad, the mass is back at the position where it was released.

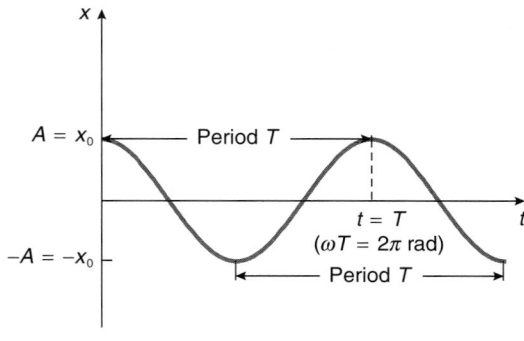

FIGURE 7.9 A graph of $x(t)$ versus t.

The time for the oscillator to make one complete cycle of its motion is the **period** T of the oscillation. Thus the period must satisfy the equation

$$\omega T = 2\pi \ \text{rad}$$

Hence

$$T = \frac{2\pi \ \text{rad}}{\omega} \tag{7.14}$$

The **frequency** ν of the motion* is the number of cycles of the motion completed during one second. The SI units for frequency are s^{-1}; but since the frequency is so commonly used in discussing oscillatory motion, the frequency unit s^{-1} is renamed a hertz (Hz), after Heinrich Hertz (1857–1894), discoverer of radio waves (among other things). Since the period is the time for one complete oscillation, the frequency and period are reciprocals of each other:

$$\boxed{\nu = \frac{1}{T}} \tag{7.15}$$

Notice also that since $T = (2\pi\,\text{rad})/\omega$, the frequency ν is related to the angular frequency ω by

$$\nu = \frac{\omega}{2\pi \ \text{rad}}$$

or equivalently,

$$\boxed{\omega = 2\pi\nu} \tag{7.16}$$

where we suppress the radian units of the factor 2π because they are dimensionless.

PROBLEM-SOLVING TACTIC

7.5 The SI units for the frequency ν are always given in hertz (Hz) ($= s^{-1}$); the SI units for the angular frequency ω are always given in rad/s ($= s^{-1}$). The frequency and angular frequency have the same dimensions (inverse time) but different units; we clearly distinguish between them by using Hz when we mean the frequency ν and rad/s when we mean the angular frequency ω.

The velocity component of the oscillator at any time is

$$v_x(t) = \frac{dx(t)}{dt}$$
$$= -A\omega \ \sin(\omega t)$$

A graph of the velocity component $v_x(t)$ is shown in Figure 7.10. Compare the graphs of $x(t)$ and $v_x(t)$. Notice that the *speed* reaches its maximum value ($A\omega$) when the oscillator is instantaneously at the *equilibrium* position $x = 0$ m.

The acceleration component $a_x(t)$ of the oscillator is found by differentiating the velocity component $v_x(t)$ with respect to time:

$$a_x(t) = \frac{dv_x(t)}{dt}$$

---

*The frequency of a system undergoing simple harmonic oscillation under the action of a Hooke's law force also is called the **natural oscillation frequency** of the oscillator.

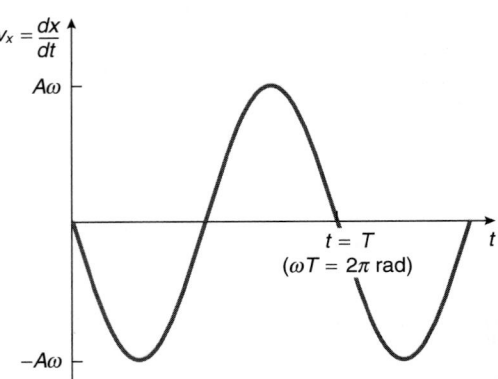

FIGURE 7.10 A graph of the velocity component $v_x(t)$ versus t.

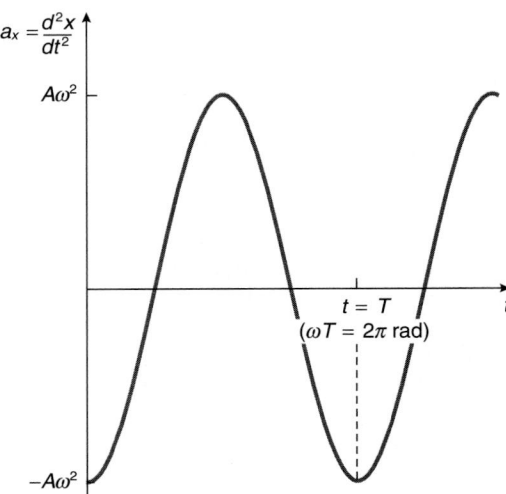

FIGURE 7.11 A graph of the acceleration component $a_x(t)$ versus t.

Since $v_x(t) = -A\omega \sin \omega t$,

$$a_x(t) = \frac{d}{dt}[-A\omega \sin(\omega t)]$$
$$= -A\omega^2 \cos(\omega t)$$

A graph of the acceleration component is shown in Figure 7.11. Notice that the acceleration component of a simple harmonic oscillator is *not constant*.

By comparing the graphs in Figures 7.9–7.11, we see that the magnitude of the maximum acceleration is $A\omega^2$. This occurs when the oscillator is instantaneously at its maximum distance from the equilibrium position (where $x = \pm A$) and its speed is zero. At these positions (where $x = \pm A$), the force of the spring on the mass is greatest, so that one expects the acceleration to be greatest at these positions too.

PROBLEM-SOLVING TACTIC

7.6 If you know $x(t)$, then $v_x(t)$ is its first time derivative and $a_x(t)$ is its second time derivative. The maximum speed is $A\omega$. The magnitude of the maximum acceleration is $A\omega^2$.

STRATEGIC EXAMPLE 7.5

A 0.500 kg passive, ruffed grouse (*Bonasa umbellus*) on a frictionless horizontal surface is attached to an ideal spring and is found to complete one oscillation every 2.00 s. The range of the oscillation is measured to be 0.40 m. The bird has zero speed when $t = 0$ s.

a. Determine the period T, the frequency v, and angular frequency ω of the oscillation.
b. Find the spring constant k of the spring.
c. Determine the amplitude A and phase angle ϕ of the oscillation.
d. Find $x(t)$, where $x = 0$ m is the equilibrium position of the mass on the spring.
e. Determine $v_x(t)$ and the maximum speed.
f. Determine $a_x(t)$ and the magnitude of the maximum acceleration.

Solution

a. The bird mass executes one complete oscillation during 2.00 s. The period T of the oscillation is the time interval for one complete oscillation, and so

$$T = 2.00 \text{ s}$$

The frequency v is the number of oscillations during one second, or the inverse of the period:

$$v = \frac{1}{T}$$

$$= \frac{1}{2.00 \text{ s}}$$

$$= 0.500 \text{ Hz}$$

The frequency v *always* is expressed in hertz (Hz) in SI units.

The angular frequency ω is related to v by Equation 7.16:

$$\omega = 2\pi v$$

$$= (2\pi \text{ rad})(0.500 \text{ Hz})$$

$$= 3.14 \text{ rad/s}$$

The angular frequency ω *always* is expressed in rad/s in SI units.

b. The angular frequency ω is related to the spring constant k and the mass m on the spring via Equation 7.10:

$$\omega = \left(\frac{k}{m}\right)^{1/2}$$

Solving for k, you find

$$k = m\omega^2$$

$$= (0.500 \text{ kg})(3.14 \text{ rad/s})^2$$

$$= 4.93 \text{ N/m}$$

The radian unit has no dimensions, and so k can be expressed in terms of N/m. The SI units for the spring constant *always* are expressed as N/m rather than kg/s^2. You should convince yourself that kg/s^2 are equivalent to N/m.

c. The range of the oscillation is 0.40 m. This is *twice* the amplitude A. Thus the amplitude A of the oscillation is

$$A = 0.20 \text{ m}$$

The mass has zero speed when $t = 0$ s; this information is used to determine the angle ϕ in $x(t) = A \cos(\omega t + \phi)$.

The velocity component is

$$v_x(t) = \frac{dx(t)}{dt}$$

$$= -A\omega \sin(\omega t + \phi)$$

When $t = 0$ s, $v_x(t) = 0$ m/s, because the oscillator was released at rest; hence

$$0 \text{ m/s} = -A\omega \sin(0 \text{ rad} + \phi)$$

This implies that

$$\phi = 0 \text{ rad}$$

d. With A and ϕ now determined, the position at any time is

$$x(t) = (0.20 \text{ m}) \cos[(3.14 \text{ rad/s}) t]$$

e. The velocity component is

$$v_x(t) = \frac{dx(t)}{dt}$$

Use the expression for $x(t)$ in (d) to find

$$v_x(t) = (0.20 \text{ m})(-1) \sin[(3.14 \text{ rad/s}) t](3.14 \text{ rad/s})$$

$$= (-0.63 \text{ m/s}) \sin[(3.14 \text{ rad/s}) t]$$

Since the maximum value of the sine is 1, the maximum speed is 0.63 m/s; this is $A\omega$, according to Equation 7.8 and Figure 7.10.

f. The acceleration component $a_x(t)$ is the time derivative of $v_x(t)$:

$$a_x(t) = \frac{dv_x(t)}{dt}$$

$$= (-0.63 \text{ m/s}) \cos[(3.14 \text{ rad/s}) t] (3.14 \text{ rad/s})$$

$$= (-2.0 \text{ m/s}^2) \cos[(3.14 \text{ rad/s}) t]$$

Since the maximum value of the cosine is 1, the magnitude of the maximum acceleration is 2.0 m/s^2; this is $A\omega^2$ (see Equation 7.9 and Figure 7.11).

STRATEGIC EXAMPLE 7.6

A 1.50 kg grapefruit is on a horizontal frictionless surface and attached to a spring with spring constant 75.0 N/m. With $x = 0$ m as the equilibrium position of the mass and $\hat{\imath}$ in the direction of stretching the spring, the moving mass is released when $t = 0$ s at $x = 0.200$ m with an initial velocity $(-0.500 \text{ m/s})\hat{\imath}$.

a. **Find the constants ω, A, and ϕ in Equation 7.6.**
b. **Make a graph of $x(t)$ versus t during two periods of the motion.**
c. **Find and plot $v_x(t)$ over the same domain for t.**
d. **Find and plot $a_x(t)$ over the same domain for t.**

Solution

a. The angular frequency depends only on m and k, not on how the mass is released. In particular, from Equation 7.10 you have

$$\omega = \left(\frac{k}{m}\right)^{1/2}$$

$$= \left(\frac{75.0 \text{ N/m}}{1.50 \text{ kg}}\right)^{1/2}$$

$$= 7.07 \text{ rad/s}$$

Since the oscillator is released when $t = 0$ s at the position $x = 0.200$ m, Equation 7.6 becomes

$$0.200 \text{ m} = A \cos \phi \qquad (1)$$

Differentiate Equation 7.6 to find the velocity component at any time:

$$v_x(t) = -A\omega \sin(\omega t + \phi)$$

According to the given information, when $t = 0$ s, the velocity component is −0.500 m/s, and so

$$-0.500 \text{ m/s} = -A\omega \sin \phi$$
$$0.500 \text{ m/s} = A\omega \sin \phi \qquad (2)$$

Divide equation (2) by equation (1) to eliminate A; you get

$$\frac{0.500 \text{ m/s}}{0.200 \text{ m}} = \frac{A\omega \sin \phi}{A \cos \phi}$$

$$2.50 \text{ s}^{-1} = \omega \tan \phi$$

Since $\omega = 7.07$ rad/s,

$$\tan \phi = 0.354$$

Thus

$$\phi = 0.340 \text{ rad}$$

Knowing both ω and ϕ, you find A using either equation (1) or (2). The result is

$$A = 0.212 \text{ m}$$

The position at any time, Equation 7.6, then is

$$x(t) = (0.212 \text{ m}) \cos[(7.07 \text{ rad/s})t + 0.340 \text{ rad}] \qquad (3)$$

b. The frequency v of the oscillation is related to ω by Equation 7.16; so

$$v = \frac{\omega}{2\pi}$$

$$= \frac{7.07 \text{ rad/s}}{2\pi \text{ rad}}$$

$$= 1.13 \text{ Hz}$$

The period T is the inverse of the frequency:

$$T = \frac{1}{v} = \frac{2\pi \text{ rad}}{7.07 \text{ rad/s}}$$

$$= 0.889 \text{ s}$$

A plot of $x(t)$, given by equation (3), encompassing two periods, from $t = 0$ s to 1.78 s is shown in Figure 7.12.

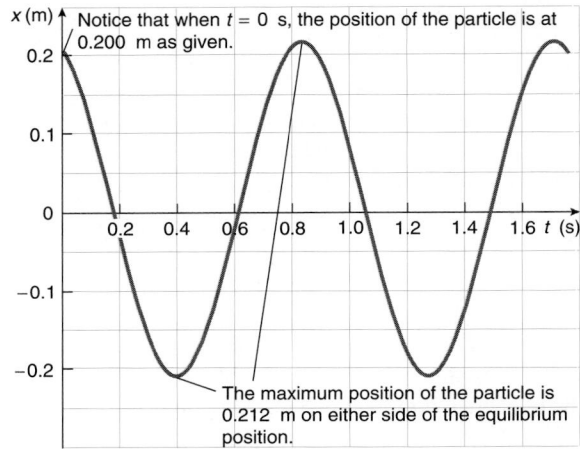

FIGURE 7.12

c. The velocity component at any time is found by differentiating Equation 7.6 with respect to time:

$$v_x(t) = \frac{dx(t)}{dt}$$

$$= -A\omega \sin(\omega t + \phi)$$

$$= (-1.50 \text{ m/s}) \sin[(7.07 \text{ rad/s})t + 0.340 \text{ rad}] \qquad (4)$$

One could also differentiate equation (3) with respect to time and obtain the same result. A graph of this over two periods is shown in Figure 7.13.

d. The acceleration component at any time is the second derivative of Equation 7.6 with respect to time:

$$a_x(t) = -A\omega^2 \cos(\omega t + \phi)$$

$$= (-10.6 \text{ m/s}^2) \cos[(7.07 \text{ rad/s})t + 0.340 \text{ rad}]$$

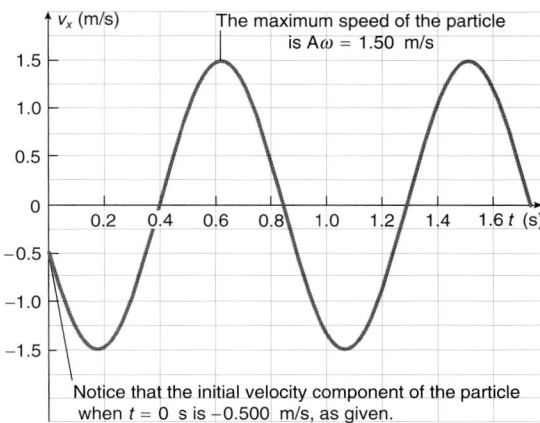

Notice that the initial velocity component of the particle when $t = 0$ s is −0.500 m/s, as given.

FIGURE 7.13

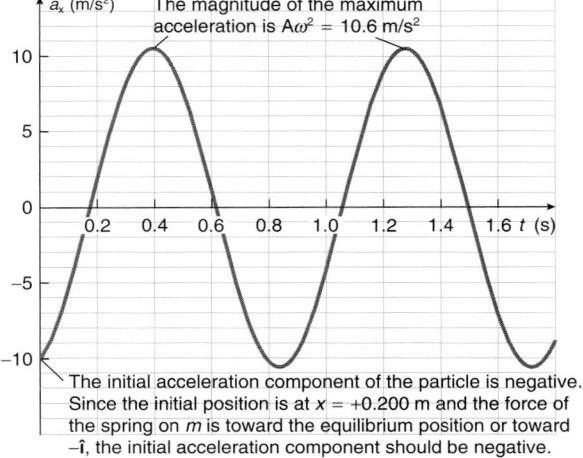

The initial acceleration component of the particle is negative. Since the initial position is at $x = +0.200$ m and the force of the spring on m is toward the equilibrium position or toward $-\hat{\imath}$, the initial acceleration component should be negative.

FIGURE 7.14

One also can obtain the same result by differentiating equation (4) once with respect to time. A graph of the acceleration component is shown in Figure 7.14.

7.3 A VERTICALLY ORIENTED SPRING

In the previous section we found that a mass m displaced from its equilibrium position on a spring oriented *horizontally*, on a *frictionless plane*, executes simple harmonic oscillation about the equilibrium position of the spring.

If the same spring now hangs by one end in the *vertical* direction, will a mass suspended from it still execute simple harmonic motion? When the spring is vertical, there are *two* vertical forces acting on the mass after it is released: the force

of the spring on m and the *constant* gravitational force of the Earth on m (its weight). How does the additional constant force affect the motion?

Let $x' = 0$ m correspond to the position of the bottom of the vertically oriented spring when *no* mass m is hung from it, as shown in Figure 7.15. The origin of the x'-axis is at this position $0'$. By convention, the positive x'-direction is taken to be in the direction of stretching the spring, and so $\hat{\imath}$ in Figure 7.15 is directed vertically *down*, as indicated.

A mass m now is carefully hung on the end of the spring. The spring stretches so that the mass is in equilibrium at coordinate x'_e, shown in Figure 7.16. There are two forces acting on the mass:

1. The gravitational force of the Earth on m, the weight $\vec{w}$ of the mass. In terms of the coordinate system chosen,

$$\vec{w} = mg\hat{\imath}$$

2. The force of the spring on the mass is

$$\vec{F}_{spring} = -kx'_e\hat{\imath}$$

since the end of the spring was stretched to coordinate x'_e when the mass was hung on it.

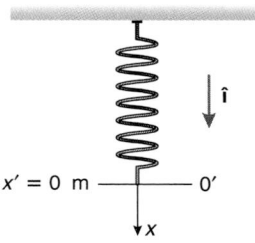

FIGURE 7.15 A vertical spring.

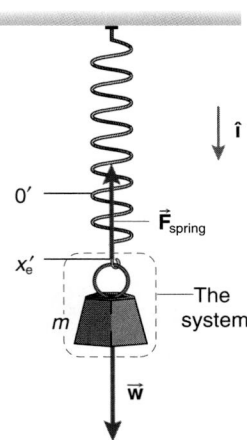

FIGURE 7.16 A mass on the vertical spring stretches it and is in equilibrium at coordinate x'_e.

The mass m is in equilibrium (zero acceleration, $a_{x'} = 0$ m/s^2). Hence Newton's second law of motion becomes

$$F_{x'\,\text{total}} = ma_{x'}$$
$$-kx_e' + mg = m(0 \text{ m/s}^2)$$

Transposing, we find

$$kx_e' = mg \qquad (7.17)$$

Now introduce a new x-axis with its origin 0 at the new equilibrium position x_e'. Pull the mass m down an additional distance x, *measured from the new equilibrium position*. Release the mass m with zero initial speed suddenly at this location. After release, there again are *two* forces acting on m, as shown in Figure 7.17.

1. The force of the spring on m. Since the total stretch of the spring is now $x_e' + x$, the (upward) force of the spring on m is

$$-k(x_e' + x)\hat{\imath}$$

2. The constant gravitational force of the Earth on m, its weight $\vec{\mathbf{w}}$. The weight of the mass m is unchanged, and equal to

$$mg\hat{\imath}$$

Thus the *total force* acting on m in the vertical direction is

$$-k(x_e' + x)\hat{\imath} + mg\hat{\imath}$$

This is the left-hand side of Newton's second law. Newton's second law becomes

$$(-kx_e' - kx + mg)\hat{\imath} = ma_x\hat{\imath}$$

Using Equation 7.17 to substitute for kx_e', we have

$$-mg - kx + mg = ma_x$$

and so

$$-kx = ma_x$$

Since

$$a_x = \frac{d^2x}{dt^2}$$

we have, after transposing,

$$m\frac{d^2x}{dt^2} + kx = 0 \text{ N}$$

Dividing by m, you may be surprised to find

$$\frac{d^2x}{dt^2} + \frac{k}{m}x = 0 \text{ m/s}^2$$

This is *identical* to the equation we found for the spring oscillating in the horizontal direction on a frictionless surface (Equation 7.7). The angular frequency for oscillations in the horizontal direction on a frictionless plane was (Equation 7.10)

$$\omega = \left(\frac{k}{m}\right)^{1/2}$$

and the angular frequency of the oscillation in the vertical direction is the *same*.

> The only change when the spring is hung vertically is that the simple harmonic oscillation of m in the vertical direction is about the new equilibrium position of the mass when it is hung on the vertical spring. The constant force of the Earth on m has no effect on the angular frequency of the oscillation; it merely shifts the equilibrium position of the oscillation.

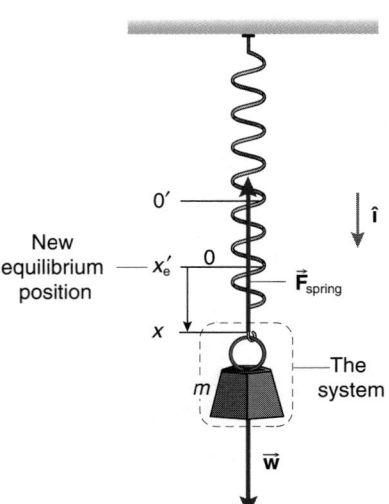

FIGURE 7.17 The mass is pulled down an additional distance x from the new equilibrium position and released at rest.

The frequency ν of the oscillation and the period T of the oscillation are simply related to the angular frequency ω by Equations 7.15 and 7.16, as before:

$$\omega = 2\pi\nu = \frac{2\pi}{T}$$

Hence we can conclude:

> The frequency and period are unchanged by hanging the spring in the vertical direction.

QUESTION 3

A mass m undergoes vertical simple harmonic oscillation with frequency ν on a spring located on the surface of the Earth. The system then is put into orbit about the Earth. Will the mass execute simple harmonic oscillation with the same frequency ν while in orbit? Explain your reasoning.

EXAMPLE 7.7

An ideal spring with spring constant 150 N/m is hung vertically and has an unstretched length of 0.250 m. A 2.00 kg mass is hung from the spring. Find the equilibrium position of the mass with respect to the suspension point of the spring. If the mass is set into vertical oscillation, determine the angular frequency ω and the frequency ν of oscillation about this equilibrium position.

Solution
In equilibrium, the spring has been slowly stretched until the upward force of the spring on the mass is equal in magnitude to the downward force of the Earth on the mass (its weight). The mass is unaccelerated in equilibrium.

With the $+x'$-axis downward (the direction that the spring will stretch when the mass is attached), let $x' = 0$ m correspond to the position of the bottom of the spring when no mass is attached, as in Figure 7.18a. Let x_e' be the new coordinate of the end of the spring when the mass is attached and in equilibrium, as in Figure 7.18b.

Newton's second law becomes

$$-kx_e'\hat{\imath} + mg\hat{\imath} = m(0 \text{ m/s}^2)\hat{\imath}$$

Hence

$$kx_e' = mg$$

and so

$$x_e' = \frac{mg}{k}$$

Substituting for m, g, and k, we have

$$x_e' = \frac{(2.00 \text{ kg})(9.81 \text{ m/s}^2)}{150 \text{ N/m}}$$

$$= 0.131 \text{ m}$$

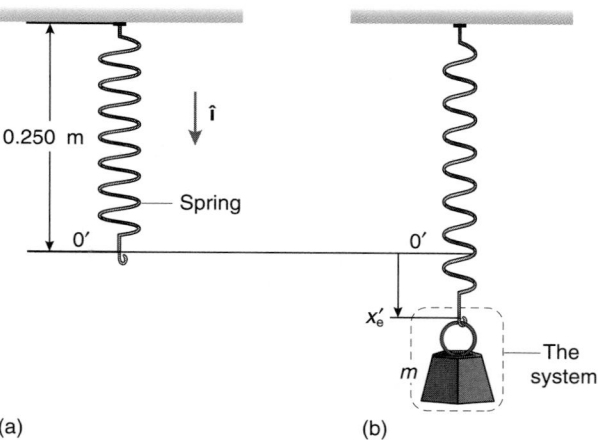

(a)   (b)

FIGURE 7.18

The mass is in equilibrium $0.131 \text{ m} + 0.250 \text{ m} = 0.381 \text{ m}$ from the point of support of the spring. It is about this equilibrium position that the mass m will execute simple harmonic oscillation if the mass is, say, pulled down further and released.

The angular frequency of the oscillation ω is, from Equation 7.10,

$$\omega = \left(\frac{k}{m}\right)^{1/2}$$

$$= \left(\frac{150 \text{ N/m}}{2.00 \text{ kg}}\right)^{1/2}$$

$$= 8.66 \text{ rad/s}$$

The frequency of oscillation ν is found using Equation 7.16:

$$\nu = \frac{\omega}{2\pi \text{ rad}}$$

$$= 1.38 \text{ Hz}$$

EXAMPLE 7.8

A mass m is on the end of an ideal spring (with a spring constant k) and is released at the equilibrium position ($x = 0$ m) when $t = 0$ s with an initial velocity $v_0\hat{\imath}$.

a. Determine the constants A, ω, and ϕ in the general solution for $x(t)$:

$$x(t) = A\cos(\omega t + \phi)$$

b. Make a graph of $x(t)$ versus t if

$$m = 0.200 \text{ kg}$$
$$k = 10.0 \text{ N/m}$$
$$v_0 = 0.500 \text{ m/s}$$

over the domain $0 \text{ s} \leq t \leq 2.0 \text{ s}$.

c. What is the period of the oscillation?

Solution

a. Since the particle is located at $x = 0$ m when $t = 0$ s, substitute these values into Equation 7.6, $x(t) = A\cos(\omega t + \phi)$. This yields

$$0 \text{ m} = A\cos\phi$$

Since $A \neq 0$ m, we could have $\phi = \pm\pi/2$ rad; choose $-\pi/2$ rad so A will be positive (see below).

We also are given the velocity of the mass when $t = 0$ s. The velocity component $v_x(t)$ at any time is

$$\begin{aligned} v_x(t) &= \frac{dx(t)}{dt} \\ &= -A\omega\sin(\omega t + \phi) \end{aligned}$$

Since $\phi = -\pi/2$ rad and $v_x = v_0$ when $t = 0$ s, make these substitutions to obtain

$$v_0 = -A\omega\sin\left(-\frac{\pi}{2}\text{ rad}\right)$$

Hence

$$A = \frac{v_0}{\omega}$$

The solution for $x(t)$ thus is

$$x(t) = \frac{v_0}{\omega}\cos\left(\omega t - \frac{\pi}{2}\text{ rad}\right)$$

where $\omega = (k/m)^{1/2}$.

b. With the data provided,

$$\begin{aligned} \omega &= \left(\frac{k}{m}\right)^{1/2} \\ &= \left(\frac{10.0 \text{ N/m}}{0.200 \text{ kg}}\right)^{1/2} \\ &= 7.07 \text{ rad/s} \end{aligned}$$

Likewise

$$\begin{aligned} A &= \frac{v_0}{\omega} \\ &= \frac{0.500 \text{ m/s}}{7.07 \text{ rad/s}} \\ &= 0.0707 \text{ m} \end{aligned}$$

Therefore

$$x(t) = (0.0707 \text{ m})\cos\left[(7.07 \text{ rad/s})\,t - \frac{\pi}{2}\text{ rad}\right]$$

This is graphed in Figure 7.19.

c. The period of the oscillation is

$$T = \frac{2\pi \text{ rad}}{\omega}$$

where $\omega = 7.07$ rad/s, and so

$$T = 0.889 \text{ s}$$

7.4 Connection between Simple Harmonic Oscillation and Uniform Circular Motion

There is an interesting connection between simple harmonic oscillation and uniform circular motion (Sections 4.5 and 4.10). The position vector of a particle executing uniform circular motion at constant angular velocity $\vec{\omega}$ in a circle of radius r is Equation 4.28 with $\theta(t) = \omega t$, where ω is the magnitude of the angular velocity vector (the angular speed):

$$\vec{r}(t) = (r\cos\omega t)\hat{\imath} + (r\sin\omega t)\hat{\jmath}$$

Let the radius r of the circle be A, so that the position vector is (see Figure 7.20)

$$\vec{r}(t) = (A\cos\omega t)\hat{\imath} + (A\sin\omega t)\hat{\jmath}$$

FIGURE 7.19

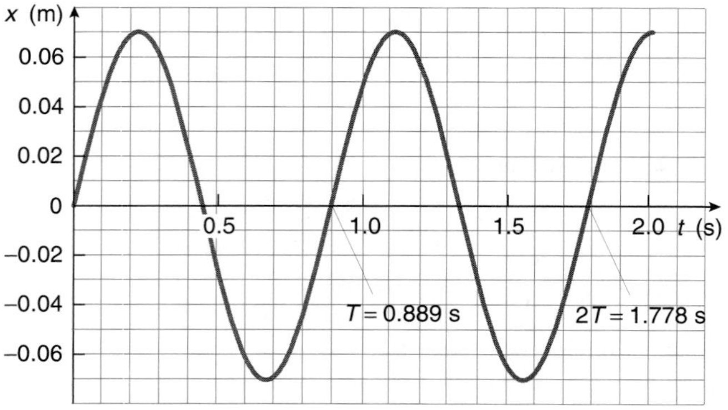

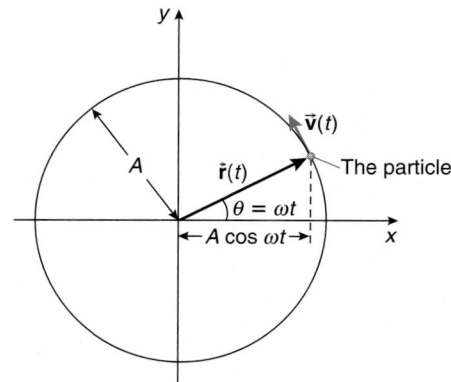

FIGURE 7.20 The position vector for uniform circular motion.

The x-component of the position vector is

$$x(t) = A \cos \omega t$$

This is the same description we found for the motion of the mass on the end of a spring executing *simple harmonic oscillation* at angular frequency ω and amplitude A with a phase angle $\phi = 0$ rad.

> We always can describe simple harmonic oscillation along a line as a component of a uniform circular motion position vector in a common plane of the motion.

If the light from a distant spotlight is directed vertically down on the particle executing uniform circular motion in a vertical plane, the *shadow* of the particle on a horizontal surface executes simple harmonic oscillation. The shadow represents the projection of the uniform circular motion of the particle onto a horizontal line; see Figure 7.21. It is because of this similarity that the constant

$$\omega = \left(\frac{k}{m} \right)^{1/2}$$

is called the angular frequency of the simple harmonic oscillation.

> • When referring to *simple harmonic oscillation*, ω is the *angular frequency* of the oscillation.
> • When referring to *uniform circular motion*, ω is the *angular speed* (the magnitude of the angular velocity).

QUESTION 4

A particle is executing uniform circular motion at constant speed in a vertical circle. Is the projection of the motion along a line in the plane of the circle but inclined at 45° to the horizontal simple harmonic oscillation?

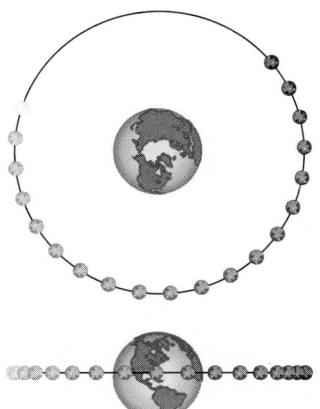

FIGURE 7.21 A satellite in a circular orbit at constant speed. Viewed in the plane of the orbit, the motion is simple harmonic oscillation.

7.5 HOW TO DETERMINE IF AN OSCILLATORY MOTION IS SIMPLE HARMONIC OSCILLATION

Not every type of oscillatory motion is simple harmonic oscillation.

> Simple harmonic oscillation is a very particular kind of oscillation.

Oscillatory motion *is* simple harmonic oscillation *only* if either of two equivalent conditions is met:

1. If there is a total force on a system that is proportional to the position coordinate of the system measured from its equilibrium position with an effective spring constant k,

$$\vec{F}_{\text{total}} = -kx\hat{\imath}$$

   whether the force is from an actual spring or not, and if this force, or this force and a *constant* force along the same axis,* are the *only* forces acting in the x-direction.
2. If, in applying Newton's second law to a system, we find a differential equation *identical in form* to the differential equation for a simple harmonic oscillator (Equation 7.11),

$$\frac{d^2x}{dt^2} + \omega^2 x = 0 \ \text{m/s}^2$$

   then the system will execute simple harmonic oscillation. The coefficient of the linear term x in the differential equation is the square of the angular frequency ω of the simple harmonic oscillation.

If the motion is found to be simple harmonic oscillation by virtue of one these conditions, then the angular frequency of the oscillation ω and the effective spring constant k are related by Equation 7.10:

$$\omega = \left(\frac{k}{m} \right)^{1/2}$$

and the period T, frequency ν, and angular frequency ω are related by Equations 7.15 and 7.16:

$$T = \frac{1}{\nu}$$

$$\omega = 2\pi\nu$$

We also can simply write down the expression for $x(t)$ (Equation 7.6),

$$x(t) = A \cos(\omega t + \phi)$$

without going through all the formalities again. The velocity component $v_x(t)$ and the acceleration component $a_x(t)$ then can be found by differentiating $x(t)$ once [for $v_x(t)$] and twice [for $a_x(t)$].

---

*Such as the constant gravitational force with a spring oriented in the vertical direction; see Section 7.3.

7.6 THE SIMPLE PENDULUM

A **simple pendulum** consists of a mass particle suspended on an ideal string, as shown in Figure 7.22. The pendulum swinging methodically in a grandfather's clock is a simple pendulum to a first approximation. A playground swing is another example. Is such oscillatory motion simple harmonic oscillation? Can you and your date "swing together" on side-by-side swings of the same length? How does the period of oscillation depend on the mass of the person in the swing and its length? Here we examine the physics of a simple pendulum.

If the mass is brought slightly to the side and released, the pendulum swings back and forth much like a swing. Let's look at this motion in more detail. The mass on the end of the string is the system. Draw a second law force diagram indicating all the forces acting on this system; there are two such forces, as shown in Figure 7.23:

1. the force $\vec{T}$ of the string on m (equal in magnitude to the tension in the string); and
2. the gravitational force of the Earth on m, its weight $\vec{w}$.

A coordinate system is introduced in Figure 7.23 with the origin at the equilibrium position of the pendulum at the bottom of its path. The force of the string on m is written in terms of its Cartesian components as*

$$\vec{T} = (-T \sin \theta)\hat{\imath} + (T \cos \theta)\hat{\jmath}$$

The weight is

$$\vec{w} = -mg\hat{\jmath}$$

Along the y-axis, Newton's second law is

$$F_{y\,total}\hat{\jmath} = ma_y\hat{\jmath}$$

where $F_{y\,total}\hat{\jmath}$ is the vector sum of the forces along the y-direction and a_y is the corresponding acceleration component. The vector sum of the forces along the y-direction is

$$(-mg + T \cos \theta)\hat{\jmath}$$

So for the y-direction, the Newton's second law of motion is

$$-mg + T \cos \theta = m \frac{d^2y}{dt^2} \qquad (7.18)$$

Geometrically, from Figure 7.23,

$$\cos \theta = \frac{\ell - y}{\ell}$$

---

*Do not confuse the notation for the tension force of the string $\vec{T}$ (a force), or its scalar magnitude T, with the *period* T of simple harmonic oscillation (a scalar time in seconds). These are the conventional symbols for each. It should be clear, if only from a dimensional analysis, whether T means a force magnitude or a time.

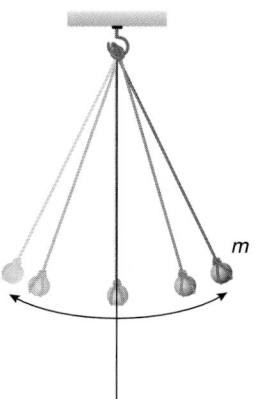

FIGURE 7.22 A simple pendulum.

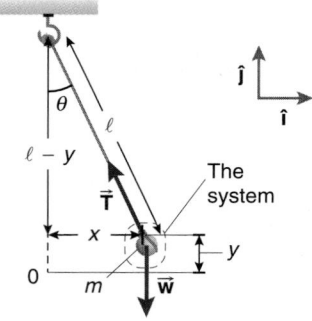

FIGURE 7.23 Forces on the system.

The small-amplitude oscillation of a swing is simple harmonic oscillation.

Thus Equation 7.18 for the y-motion is

$$-mg + T\frac{\ell - y}{\ell} = m\frac{d^2y}{dt^2}$$

In the x-direction Newton's second law is

$$F_{x\,total}\hat{\imath} = ma_x\hat{\imath}$$

where $F_{x\,total}\hat{\imath}$ is the vector sum of the forces along the x-direction and a_x is the corresponding acceleration component. There is only one force along the x-direction:

$$(-T\sin\theta)\hat{\imath}$$

Thus Newton's second law applied to the x-direction is

$$-T\sin\theta = m\frac{d^2x}{dt^2} \qquad (7.19)$$

Geometrically, from Figure 7.23, $\sin\theta = x/\ell$. Hence Equation 7.19 becomes

$$-T\frac{x}{\ell} = m\frac{d^2x}{dt^2} \qquad (7.20)$$

If the deviation of the pendulum from the equilibrium position is small (meaning that the angle θ made by the cord with the vertical direction is small), then the vertical motion is negligible compared with the horizontal motion along x. With this small-angle approximation, then, we neglect the acceleration in the vertical direction:

$$\frac{d^2y}{dt^2} \approx 0 \text{ m/s}^2$$

Also, for small angles θ, $y \ll \ell$, and so the expression for the cosine of θ becomes

$$\cos\theta = \frac{\ell - y}{\ell} \approx 1$$

The equation for the vertical (y) motion (Equation 7.18) then becomes

$$-mg + T \approx 0 \text{ N}$$

Hence

$$T \approx mg$$

for the magnitude of the force of the string on the mass.

Under these approximations, the equation in the horizontal direction, Equation 7.20, becomes (after substituting for T)

$$-mg\frac{x}{\ell} = m\frac{d^2x}{dt^2}$$

Notice that with this approximation, the force acting to restore the pendulum to its equilibrium position [$= (-mgx/\ell)\hat{\imath}$] is proportional to the horizontal position of the mass from the equilibrium position. This is a Hooke's law type of force where the effective spring constant k is mg/ℓ.

The equation governing the motion of the pendulum in the horizontal direction becomes, after transposing,

$$\frac{d^2x}{dt^2} + \frac{g}{\ell}x = 0 \text{ m/s}^2 \qquad (7.21)$$

which has the same form as the equation for the motion of a mass on the end of a spring (Equation 7.11). A pendulum swinging through small angles with respect to the vertical thus is essentially a simple harmonic oscillator. The solution for $x(t)$ is

$$x(t) = A\cos(\omega t + \phi)$$

The coefficient of the term involving x in Equation 7.21 is the square of the angular frequency of the oscillation. That is, comparing Equation 7.21 with Equation 7.11, we see that

$$\omega^2 = \frac{g}{\ell}$$

or

$$\omega = \left(\frac{g}{\ell}\right)^{1/2}$$

The period of the pendulum is, using Equations 7.15 and 7.16,

$$T = \frac{2\pi}{\omega}$$
$$= 2\pi\left(\frac{\ell}{g}\right)^{1/2} \qquad (7.22)$$

The period of the pendulum is independent of the mass on its end and independent of the amplitude of the oscillation (as long as the amplitude is small compared with ℓ).

The greater the length ℓ of the pendulum, the longer the period T; the period is proportional to the square root of the length. The qualitative dependence of the period on the length is easily remembered if you think about the dynamical behavior of your childhood backyard swing: those of long length take more time to go back-and-forth than those of short length.

A simple pendulum is a simple harmonic oscillator *only* to the extent you can neglect the vertical motion—that is, when the angle the pendulum makes with the vertical is small.

Hooke's law is a valid approximation for the simple pendulum as long as we stay within certain limits—here, small angles θ with the vertical. If θ becomes appreciable, the approximations we made are no longer valid. The motion of the pendulum, while still oscillatory, *is no longer simple harmonic oscillation* because the equations for x and y (Equations 7.18 and 7.19) then become more complicated. For large angles, the resulting force along x is not proportional to x, and so is not a Hooke's law force. We also cannot ignore the motion along the vertical direction y for large angles θ. For large angles θ, the solutions for the motion of the pendulum, $x(t)$ and $y(t)$, are complicated and beyond the scope of this course. Recall, also, that a spring can be approximated by a Hooke's force law *only* if we stay within the elastic limit of the spring. Our simple descriptions of nature have limits and are valid only to the extent we stay within them. Recognizing the existence of limits, as well as what the limits actually are, is a good lesson to learn in life as well as science.

QUESTION 5

A pendulum of length 1.00 m is hung inside a spacecraft orbiting the Earth every 100 min. What is the period of oscillation of the pendulum? Explain your answer.

EXAMPLE 7.9

Calculate the length of a swing whose period is 5.0 s for small oscillation amplitudes.

Solution

The expression for the period of a simple pendulum is Equation 7.22:

$$T = 2\pi \left(\frac{\ell}{g} \right)^{1/2}$$

Square this expression to find

$$T^2 = 4\pi^2 \frac{\ell}{g}$$

Solving for ℓ,

$$\ell = \frac{T^2 g}{4\pi^2}$$

Substituting for g and the period $T = 5.0$ s,

$$\ell = \frac{(5.0 \text{ s})^2 (9.81 \text{ m/s}^2)}{4\pi^2}$$

$$= 6.2 \text{ m}$$

Perhaps you were lucky enough to have a swing this long in your backyard!

EXAMPLE 7.10

You can measure g with a simple pendulum. Set up a simple pendulum of length 1.76 m in your room. With a stopwatch, you find that it executes 25 complete small oscillations during 66.4 s.

a. What value of g do these data imply?
b. If the oscillation amplitude were reduced to half its value, what would be the period?

Solution

a. Since 25 oscillations took 66.4 s, the time for one oscillation, the period, is

$$T = \frac{66.4 \text{ s}}{25}$$

$$= 2.66 \text{ s}$$

Since the period of the pendulum is related to its length and the value of g via Equation 7.22, solve that equation for g:

$$g = 4\pi^2 \frac{\ell}{T^2}$$

Substituting for ℓ and T, you have

$$g = 4\pi^2 \frac{1.76 \text{ m}}{(2.66 \text{ s})^2}$$

$$= 9.82 \text{ m/s}^2$$

b. The period is unchanged, because the period is independent of the amplitude for small oscillations.

EXAMPLE 7.11

What is the period of a simple pendulum whose length is 1.00 m?

Solution

The period T of a simple pendulum of length ℓ is, from Equation 7.22,

$$T = 2\pi \left(\frac{\ell}{g} \right)^{1/2}$$

Substituting numerical values for ℓ and g,

$$T = 2\pi \left(\frac{1.00 \text{ m}}{9.81 \text{ m/s}^2} \right)^{1/2}$$

$$= 2.01 \text{ s}$$

It is quite coincidental that the period of a one meter pendulum is very close to two seconds, making each swing from one side to the other take about one second. Thomas Jefferson proposed a standard unit of length based on a pendulum whose period was precisely 2 s, so that each swing had a duration of one second. Jefferson's standard was not adopted (see Problem 51).

EXAMPLE 7.12

a. Show that if the length of a simple pendulum is increased by a small amount $\Delta\ell$, the period of the pendulum increases by a small amount ΔT, where

$$\Delta T \approx \frac{\pi}{(g\ell)^{1/2}} \Delta\ell$$

b. If length of a pendulum 1.00 m long is increased by 1.0 cm, what is the change in the period?

Solution

a. The period and length of a pendulum are related by Equation 7.22:

$$T = 2\pi \left(\frac{\ell}{g} \right)^{1/2}$$

Take the derivative of T with respect to ℓ:

$$\frac{dT}{d\ell} = \frac{\pi}{(g\ell)^{1/2}}$$

Approximate the differentials with the small changes; that is, let $dT \approx \Delta T$ and $d\ell \approx \Delta \ell$, so that

$$\Delta T \approx \frac{\pi}{(g\ell)^{1/2}} \Delta \ell$$

b. For a 1.0 cm increase in the length of a 1.00 m pendulum, the increase in the period is

$$\Delta T = \frac{\pi \ (0.010 \ \text{m})}{[(9.81 \ \text{m/s}^2) \ (1.00 \ \text{m})]^{1/2}}$$

$$= 0.010 \ \text{s}$$

7.7 THROUGH A FICTIONAL EARTH IN 42 MINUTES

I'll put a girdle round the earth
In forty minutes.

William Shakespeare (1564–1616)*

In Chapter 6 we saw that as long as the mass distribution within the Earth is radially symmetric, the gravitational force of the Earth on particles on its surface or at distances from the center that are greater than the radius of the Earth is found by considering the Earth as a point particle in the gravitational force law. However, for a particle *within the Earth*, only that part of the mass of the Earth *closer* to the center than the particle produces a nonzero gravitational force on the particle.

The Earth is *not* a uniform sphere, although its mass distribution is approximately radially symmetric. The Earth has a dense nickel–iron core surrounded by material (the mantle) of lesser density. This means that the results we obtained for the magnitude of the gravitational force on a particle within a uniform sphere (Equation 6.7) cannot in fact be used for particles within the Earth. But what the heck; assume that the Earth *is* a uniform sphere, knowing full well that the assumption is only a crude first approximation. If we make this gross simplification, some rather interesting physics emerges for particles within this fictional model of the Earth.

Imagine you somehow fulfilled your childhood fantasy of digging a hole right through the center of the fictional Earth model (see Figure 7.24). A rock of mass m then is dropped into the hole. The rock falls down the hole until it reaches the center of the Earth, then goes up the hole on the opposite side.

Set up a coordinate system with its origin at the center of the fictional model Earth with an axis aligned with the hole, as in Figure 7.24. Neglect all air resistance (another unreasonable assumption) and assume the walls of the hole are frictionless.

The magnitude of the gravitational force on m is given by Equation 6.7, because the rock of mass m is within the uniform Earth. In accordance with our assumption about neglecting frictional effects, the gravitational force is the only force acting on the particle along the x-direction. With this choice of coordinates, the gravitational force of the Earth on m is

$$-\frac{GmM}{R^3} x\hat{\imath} \tag{7.23}$$

Notice that this force is proportional to the distance the mass m is from the origin and is directed toward the origin (at the center of the Earth) when x is positive *and* when x is negative. That is, the force is a Hooke's force law,

$$-kx\hat{\imath} \tag{7.24}$$

with an effective spring constant found by comparing Equation 7.23 with Equation 7.24:

$$k \equiv \frac{GmM}{R^3}$$

Hence the rock executes simple harmonic oscillation about the equilibrium position of m at the center of the fictional model Earth! Since we analyzed such motion in some detail, we immediately can write down the angular frequency of the oscillation, since it is related to the effective spring constant and the mass m on the *gravitational* spring. With Equation 7.10, we have

$$\omega = \left(\frac{k}{m}\right)^{1/2}$$

$$= \left(\frac{GM}{R^3}\right)^{1/2}$$

The period of the oscillation is found from Equations 7.15 and 7.16:

$$T = \frac{2\pi}{\omega}$$

$$= 2\pi \left(\frac{R^3}{GM}\right)^{1/2} \tag{7.25}$$

which is independent of the mass of the rock dropped down the hole.

Another way to reach the same conclusion is to look at Newton's second law for the rock. Since the gravitational force is the only force acting on the rock, Newton's second law becomes

$$F_{x \ \text{total}} = ma_x$$

$$-\frac{GmM}{R^3} x = m \frac{d^2x}{dt^2}$$

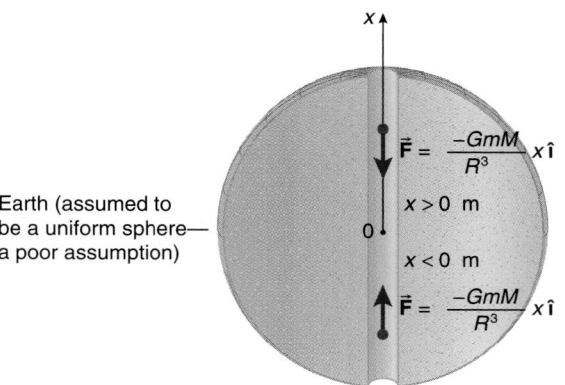

Earth (assumed to be a uniform sphere—a poor assumption)

$\vec{F} = \dfrac{-GmM}{R^3} x\hat{\imath}$

$x > 0$ m

$x < 0$ m

$\vec{F} = \dfrac{-GmM}{R^3} x\hat{\imath}$

FIGURE 7.24 A frictionless hole through the center of the Earth.

---

*A Midsummer Night's Dream, Act II, Scene 1, lines 175–176.

The equation of motion for the rock is thus

$$\frac{d^2x}{dt^2} + \frac{GM}{R^3} x = 0 \text{ m/s}^2$$

which is identical in form to the simple harmonic oscillator equation (Equation 7.11). When the differential equation is in this form, the coefficient of x in the differential equation is ω^2. So once again we find for the angular frequency of the oscillation:

$$\omega = \left(\frac{GM}{R^3}\right)^{1/2}$$

If we substitute the numerical values for G, the radius R of the Earth, and the mass M of the Earth into Equation 7.25 for the period of the oscillation, we find

$$T = 2\pi \left(\frac{R^3}{GM}\right)^{1/2}$$
$$= 2\pi \left[\frac{(6.37 \times 10^6 \text{ m})^3}{(6.67 \times 10^{-11} \text{ N·m}^2/\text{kg}^2)(5.98 \times 10^{24} \text{ kg})}\right]^{1/2}$$
$$= 5.06 \times 10^3 \text{ s}$$

or about 84.3 min. The particle can go *one-way* through the Earth in half the period, or just a tad over 42 min. This is the ultimate in express mail service: the mail delivers itself with no effort on our part other than dropping it down the letter slot. Of course there are a few minor engineering difficulties associated with constructing the slot through the center of the Earth, but . . . .

What happens if the frictionless tunnel dug through the fictional model Earth does *not* pass through the center of the Earth, as in Figure 7.25? The frictionless walls ensure that the rock slides freely in the hole. There are now two forces acting on the rock within the tunnel:

1. the gravitational force of the Earth on the rock; and
2. the normal force of the tunnel wall on the rock.

Set up the coordinate system indicated in Figure 7.25. The weight of the rock is written in terms of its Cartesian components as

$$-\frac{GmM}{R^3} r[(\cos \theta)\,\hat{\imath} + (\sin \theta)\,\hat{\jmath}]$$

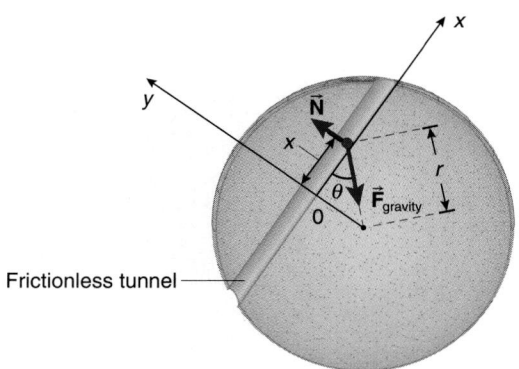

FIGURE 7.25 A frictionless tunnel between any two points on the surface of the Earth.

The rock is not accelerating in the y-direction, and so the vector sum of the two forces along y is zero according to Newton's second law:

$$\left[N - \frac{GmM}{R^3} r(\sin \theta)\right]\hat{\jmath} = 0 \text{ N}$$

There is only one force along the x-direction:

$$-\frac{GmM}{R^3} r(\cos \theta)\,\hat{\imath}$$

But from the geometry of Figure 7.25, $\cos \theta = x/r$ and the force in the x-direction is

$$-\frac{GmM}{R^3} r \frac{x}{r}\,\hat{\imath}$$

or

$$-\frac{GmM}{R^3} x\hat{\imath}$$

Notice that the force on m is proportional to its distance from the midpoint of the tunnel and is directed toward the origin at the center of the tunnel. Once again we have a Hooke's force law, and this is the only force on the mass in the x-direction. Therefore the mass executes simple harmonic oscillation about the midpoint of the tunnel. Indeed, the effective spring constant k for the problem is

$$k = \frac{GmM}{R^3}$$

just as with the tunnel through the center of the Earth. Therefore the period T of the oscillation is the same as for a tunnel through the center of the Earth, Equation 7.25. Hence the period of the motion through the tunnel is *independent of the path of the tunnel*. The period again is 84.3 min. A one-way trip through such a tunnel to *anywhere* on the Earth takes a little over 42 min. Such a rapid transit system need have only one timetable: 42 min to any- and everywhere. What is more, since the tunnel is frictionless, the trip uses no fuel.

7.8 DAMPED OSCILLATIONS*

The oscillations of the simple harmonic oscillator, described by Equation 7.6, continue indefinitely and do not decrease in amplitude with time. If you hang a mass on a spring, set it into oscillation, and plot the position of the mass as a function of time (see Figure 7.26), it is quite apparent that the oscillations do not continue indefinitely; they gradually decrease in amplitude and die out.

Even if we treat the spring as ideal, the decrease in amplitude arises because of frictional effects in the medium in which the mass is moving while oscillating. Such an oscillator is called a **damped harmonic oscillator**. Such oscillations are very common in the natural world and can be observed in everything from pendulums to diving boards.

Now imagine placing such an oscillator in thick, gooey molasses, pulling the mass from its equilibrium position, and then releasing the mass, as shown in Figure 7.27. The mass slowly returns to its equilibrium position without overshooting it, so

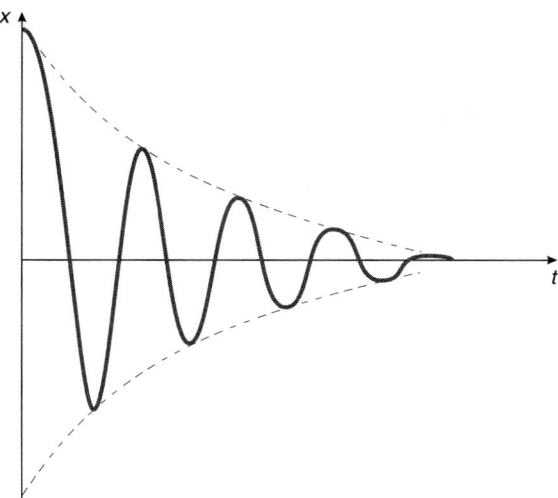

FIGURE 7.26 An oscillation that gradually dies out.

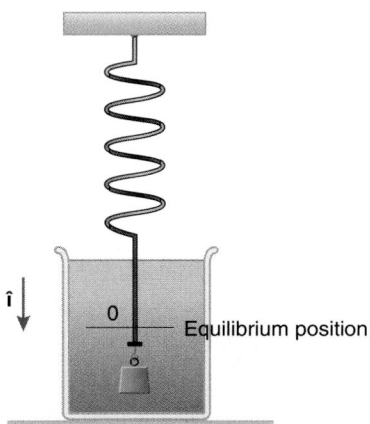

FIGURE 7.27 An oscillator in thick molasses.

that the mass is never really oscillating; this behavior is called **overdamped oscillation**. A graph of the position of the mass as a function of time appears as in Figure 7.28.

Suppose you mix water with the molasses to thin it, and repeat the experiment. When sufficient water is added to thin the molasses, the mass eventually reaches a state such that if more water is mixed, the mass, when released, will overshoot its equilibrium position before eventually coming back to rest at the equilibrium position. At this critical point of barely no overshoot, the oscillator is called a **critically damped oscillator**. If the molasses is thinned still more, the system behaves like the damped oscillator we described initially.

Here we look at such realistic oscillations in more detail. There are two forces on the mass:

1. the force of the spring,

$$\vec{\mathbf{F}}_{spring} = -kx\hat{\mathbf{i}}$$

where we use the standard coordinate system with positive x corresponding to stretching the spring; and
2. a kinetic frictional force.

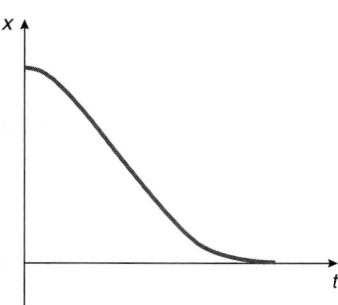

FIGURE 7.28 A plot of x versus t for an overdamped oscillator.

In a first and realistic approximation, the frictional effects are modeled by a frictional force proportional to but opposite in direction to the velocity of the mass:

$$\vec{\mathbf{f}}_k = -\beta v_x \hat{\mathbf{i}}$$

Newton's second law then becomes

$$-kx\hat{\mathbf{i}} - \beta v_x \hat{\mathbf{i}} = ma_x \hat{\mathbf{i}} \qquad (7.26)$$

Since

$$v_x = \frac{dx}{dt}$$

and

$$a_x = \frac{d^2x}{dt^2}$$

Equation 7.26 becomes, after minor rearranging,

$$m\frac{d^2x}{dt^2} + \beta\frac{dx}{dt} + kx = 0 \text{ N} \qquad (7.27)$$

This is a more complicated differential equation than the simple harmonic oscillator equation. The mathematical technique for solving such an equation likely is unfamiliar to you, and so we will have to resort to some smoke and mirrors to find the function $x(t)$ that satisfies Equation 7.27. If you examine Figure 7.26, the damped oscillator has a sinusoidal-like oscillation, $\cos(\omega t + \phi)$, reminiscent of the simple harmonic oscillator, but whose amplitude decreases (decays) with time, perhaps as an exponential: $e^{-\alpha t}$ as an educated guess. One might, then, concoct a solution to Equation 7.27 of the form

$$x(t) = Ae^{-\alpha t}\cos(\omega t + \phi) \qquad (7.28)$$

In fact, this is just the thing to do. Differentiate Equation 7.28 to find

$$\frac{dx}{dt} \qquad \text{and} \qquad \frac{d^2x}{dt^2}$$

which will be rather messy looking expressions. Put these derivatives into Equation 7.27, and isolate the sine and cosine terms. After some tedious algebra, you should secure the following ungainly expression:

$$(m\alpha^2 - m\omega^2 - \beta\alpha + k)Ae^{-\alpha t}\cos(\omega t + \phi)$$
$$+ (2m\alpha\omega - \beta\omega)Ae^{-\alpha t}\sin(\omega t + \phi) = 0 \text{ N} \qquad (7.29)$$

Recall that this equation stemmed from Newton's second law, so it must be true at *any* and *every* instant *t*. The only way Equation 7.29 can be satisfied at all instants *t* is if *each* expression in parentheses is zero separately. That is,

$$2m\alpha\omega - \beta\omega = 0 \text{ kg/s}^2 \qquad (7.30)$$

and

$$m\alpha^2 - m\omega^2 - \beta\alpha + k = 0 \text{ kg/s}^2 \qquad (7.31)$$

Solve Equation 7.30 for α:

$$\alpha = \frac{\beta}{2m}$$

Use this in Equation 7.31, and then solve it for ω, obtaining

$$\omega = \left[\frac{k}{m} - \left(\frac{\beta}{2m}\right)^2\right]^{1/2} \qquad (7.32)$$

If there is no frictional force, $\beta = 0$ kg/s and the oscillation is simple harmonic with angular frequency $\omega_0 = (k/m)^{1/2}$. So we rewrite Equation 7.32 for the angular frequency of the damped oscillator as

$$\omega = \left[\omega_0^2 - \left(\frac{\beta}{2m}\right)^2\right]^{1/2} \qquad (7.33)$$

If the magnitude of the frictional force is small compared with that of the force of the spring, then β is small and the system oscillates with an angular frequency close to that of the undamped simple harmonic oscillator but with an amplitude that decreases with time. When the value of β is such that

$$\frac{\beta}{2m} = \omega_0$$

then the angular frequency ω in Equation 7.33 is zero and the oscillator is critically damped; the cosine term in Equation 7.28 then is constant, and so $x(t)$ decreases exponentially with time. If the resistive force is so great that $\beta/(2m) > \omega$, then the system is overdamped and $x(t)$ again decreases exponentially with time. For both the overdamped and critically damped conditions, the mass simply returns to and then remains at the equilibrium position without overshooting it.

7.9 FORCED OSCILLATIONS AND RESONANCE*

Think of a pendulum as a playground swing. Your little sister is on the swing and wants to be pushed. If you give her a small push and then wait for a minute or so before giving her another push, the amplitude of the motion of the swing will not be great and soon complaints will begin to emanate. For this situation, the frequency of your pushes (number of pushes per second) is small and the resulting amplitude of the swing is small. So you change strategies: now you execute a series of very fast pushes, say 10 pushes per second. Once again, the amplitude of the oscillation of the swing is not very large and your sister again starts to complain: "Come on, do it right!"

What is "right"? You push the swing every time it comes back to you. Under these circumstances the amplitude of the oscillation can grow to be quite large. What you have done is to apply a force to the oscillator at a frequency that matches the **natural oscillation frequency** of the oscillator. The natural oscillation frequency is the frequency of the swing when allowed to oscillate under the influence of *only* the Hooke's force law. The *additional* force that you exert on the oscillator (here by your occasional pushing) produces what is called **forced oscillation**. If the frequency of the applied force matches (is equal to) the natural oscillation frequency of the system, then a large-amplitude oscillation of the system results. Forced oscillation with matched frequencies produces **resonance**, large-amplitude oscillations. Resonance is a very general phenomenon with oscillators.

As another illustration of the idea of forced oscillation and resonance, fill a bathtub partially with water. Now use a paddle or a board to slosh the water back and forth in the tub, creating a "water oscillation." If the sloshing is done at too large or too small a frequency, only small-amplitude oscillations of the water are generated in the tub. But if the sloshing is done at just the right frequency (equal to the natural oscillation frequency of the water in the tub), then large-amplitude oscillations of the water are generated.

Any small child has the innate ability to find the right sloshing (forcing) frequency, with attendant parental consternation over the resulting resonance and flooded bathroom! The natural oscillation frequency of the water in the tub depends on the geometric shape of the tub, the depth of the water, and the presence of obstacles (people or rocks) in the tub itself. Since the frequency of the sloshing needed to create large water oscillations in the tub must match the natural oscillation frequency of the water in the tub, the effective frequency is different for different tubs and depths of water. In the Bay of Fundy,* a natural tub, the natural oscillation frequency of the water is close to the sloshing frequency of the *tides*; a large tidal rise and fall therefore exists in the Bay—a large tidal oscillation. The tidal rise and fall approaches 15 *meters*! Other bays have natural oscillation frequencies that are quite different from the tidal frequency, so that tidal oscillations of the water there are not as dramatic.

---

*The Bay of Fundy is located between New Brunswick and Nova Scotia.

The tidal frequency closely matches the natural oscillation frequency of the Bay of Fundy, resulting in large-amplitude tidal oscillations.

Complex structures such as bridges and buildings have many natural frequencies of oscillation. Resonance effects occur if a periodic (repetitive) force happens to coincide with one of the natural frequencies. Occasionally, disastrous effects result. Engineers must design structures to ensure that forces caused by winds, earthquakes, and the like do not induce or excite resonance effects in the structure. Tall buildings generally have lower natural frequencies of oscillation than do short buildings. Take two vertical meter sticks with one clamped at its end and the other at the midpoint; tweak them both. The one with the longer free length oscillates at a smaller natural frequency than does the one of shorter length. Earthquake motions are often closer to the natural frequencies of shorter buildings, and so these structures can be affected more severely than taller structures.

Forced oscillations and resonance effects are present in many aspects of the natural world and are involved electrically in everything from radios and TVs to lasers and medical imaging technologies. Detailed quantitative treatments of forced oscillations in a wide variety of systems (in mechanics, electromagnetism, optics, and quantum mechanics) will be encountered in more advanced courses in physics and engineering. Here we examine the behavior of such a forced oscillator in a semiquantitative way.

Assume there are three forces acting on a forced oscillator:

1. the applied force $F_{x\,\text{applied}}\hat{\imath}$;
2. frictional damping opposite in direction to the velocity; and
3. the Hooke's force law restoring force.

If, as before, we assume frictional damping proportional to the velocity, Newton's second law becomes

$$F_{x\,\text{applied}} - \beta v_x - kx = ma_x$$

or with a slight rearrangement,

$$m\frac{d^2x}{dt^2} + \beta\frac{dx}{dt} + kx = F_{x\,\text{applied}} \qquad (7.34)$$

In many situations, the applied force has an x-component that is sinusoidal:

$$F_{x\,\text{applied}} = F_0 \cos \omega t$$

Forced oscillations induced by earthquakes cause even more damage if they match the natural oscillation frequency of a building.

The properly timed push you give your sister on a swing is *not* quite like this, since your push is applied only briefly near the peak amplitudes of the oscillation. Eventually, after a sinusoidal force is applied to the oscillator, the steady-state solution to Equation 7.34 has the form

$$x(t) = A(\omega) \cos(\omega t + \phi) \qquad (7.35)$$

where the amplitude is a function of the applied angular frequency. Using a procedure similar to that which we used for damped oscillations (substitution of the solution into the differential equation), and after a significant bit of differentiation and algebra, you will find that the amplitude depends on ω according to

$$A(\omega) = \frac{\dfrac{F_0}{m}}{\left[(\omega^2 - \omega_0^2)^2 + \left(\dfrac{\beta\omega}{m}\right)^2\right]^{1/2}} \qquad (7.36)$$

A qualitative graph of $A(\omega)$ versus ω (see Figure 7.29) confirms that the peak amplitude, as we know from our qualitative discussion, occurs when the angular frequency ω of the applied force is equal to the natural angular frequency ω_0 of the oscillator with no damping, which defines the *resonance* condition.

If there is no damping, $\beta = 0$ kg/s and the amplitude at resonance is infinite according to Equation 7.36. Since all oscillators in nature exhibit some damping, however small, an infinite amplitude never actually is encountered. As the damping increases, the amplitude of the oscillation at resonance decreases and the amplitude function becomes less sharply peaked.

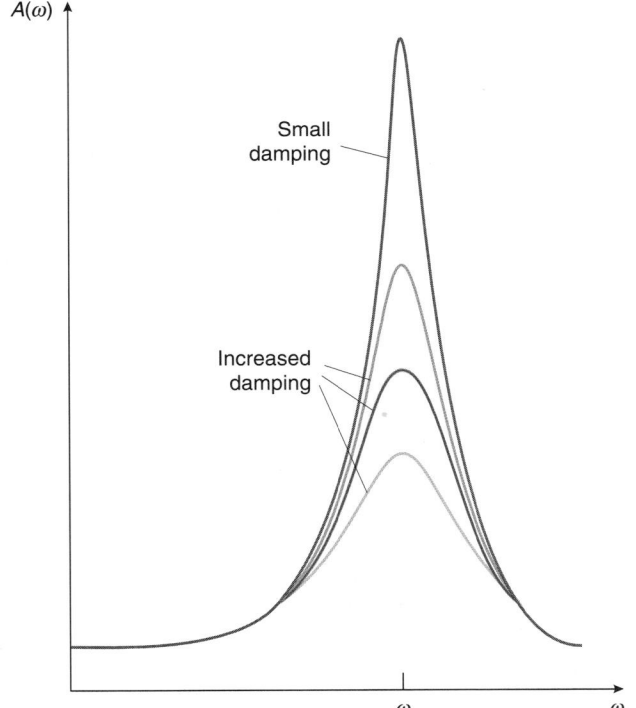

FIGURE 7.29 A graph of $A(\omega)$ versus ω.

CHAPTER SUMMARY

The force of a spring on an attached mass m is proportional to the position x of the mass m relative to its equilibrium position on the spring, taken to be $x = 0$ m:

$$\vec{F}_{\text{spring}} = -kx\hat{i} \qquad (7.2)$$

where k is a characteristic of the specific spring and is called the *spring constant*, expressed in N/m. By convention, $\hat{i}$ is chosen to be in the direction of stretching the spring. The force of the spring on m is independent of m and depends only on the stretch or compression of the spring and the spring constant. This empirical force law is known as *Hooke's force law* and is legitimate as long as the spring does not exceed its *elastic limit*.

If the mass is set into oscillation about its equilibrium position on the spring (whether the spring is horizontal or vertical), the position of the mass at any time $x(t)$ satisfies a differential equation of the form

$$\frac{d^2x}{dt^2} + \omega^2 x = 0 \text{ m/s}^2 \qquad (7.11)$$

which is found from applying Newton's second law of motion to the mass m. Equation 7.11 is called the *simple harmonic oscillator equation*. Systems that obey Equation 7.11 are called *simple harmonic oscillators*. The solution of this differential equation is

$$x(t) = A\cos(\omega t + \phi) \qquad (7.6)$$

where A is the *amplitude* of the oscillation; its value is determined from the initial conditions of the oscillation; ω is the *angular frequency* of the oscillation in rad/s, given by

$$\omega = \left(\frac{k}{m}\right)^{1/2} \qquad (7.10)$$

and ϕ is a *phase angle* whose value is determined from the initial conditions of the oscillation.

Both the product ωt and ϕ are expressed in radians. The *period* of the oscillation T, its *frequency* ν, and the angular frequency ω are related by

$$\nu = \frac{1}{T} \qquad (7.15)$$

$$\omega = 2\pi\nu \qquad (7.16)$$

The velocity component $v_x(t)$ of the oscillator at any time is found by differentiating $x(t)$ with respect to time:

$$v_x(t) = \frac{dx(t)}{dt}$$
$$= -A\omega\sin(\omega t + \phi)$$

The acceleration component at any time is found by differentiating the velocity component with respect to time:

$$a_x(t) = \frac{dv_x(t)}{dt} = \frac{d^2x(t)}{dt^2} = -\omega^2 x(t)$$
$$= -A\omega^2\cos(\omega t + \phi)$$

Hooke's law is important because such elastic forces are used to model many interactions in physics, engineering, and chemistry. Any situation in which you can recognize (a) that the *total* force on a system is proportional to the departure of the system from its equilibrium position—that is,

$$F_{x\text{ total}} = -kx$$

or equivalently, (b) that in applying Newton's second law to the system you find a differential equation that has the form of Equation 7.11,

$$\frac{d^2x}{dt^2} + \omega^2 x = 0 \text{ m/s}^2 \qquad (7.11)$$

is a system that you can conclude will execute simple harmonic oscillation. The angular frequency ω, the frequency ν, the period T, the position $x(t)$, the velocity component $v_x(t)$, and the acceleration component $a_x(t)$ are found using the equations just presented for these important parameters of the motion.

A simple pendulum is approximately a simple harmonic oscillator for small-angle oscillations with respect to the vertical direction. The period of a simple pendulum is independent of its mass and is proportional to the square root of its length ℓ:

$$T = 2\pi\left(\frac{\ell}{g}\right)^{1/2} \qquad (7.22)$$

An oscillator subjected to frictional forces is called a *damped harmonic oscillator* and the amplitude of its oscillation decreases with time. A *critically damped* or *overdamped* oscillator does not truly oscillate but merely returns to the equilibrium position without overshooting.

Large amplitudes result if a force is applied to a system at a frequency that matches a *natural oscillation frequency* of the system, a condition known as *resonance* under *forced oscillation*.

SUMMARY OF PROBLEM-SOLVING TACTICS

7.1 **(page 283)** Always choose the origin at the equilibrium position with $\hat{i}$ corresponding to stretching the spring.

7.2 **(page 285)** For a mass attached to a spring, do not confuse the force *we* exert on the mass, $\vec{F}_{\text{we}} = kx\hat{i}$, with the force that the *spring* exerts on the mass, $\vec{F}_{\text{spring}} = -kx\hat{i}$.

7.3 **(page 287)** Be careful! The quantity ωt is measured in radians, and so the constant angle ϕ, called the *phase angle*, also must be in radians.

7.4 **(page 289)** The way the oscillator is released provides the information to determine the amplitude A and phase angle ϕ.

7.5 **(page 290)** The SI units for the frequency ν are always given in hertz (Hz); the SI units for the angular frequency ω are always given in rad/s.

7.6 **(page 290)** If you know $x(t)$, then $v_x(t)$ is its first time derivative and $a_x(t)$ is its second time derivative. The maximum speed is $A\omega$. The magnitude of the maximum acceleration is $A\omega^2$.

QUESTIONS

1. (page 285); 2. (page 287); 3. (page 295); 4. (page 297); 5. (page 300)

6. A mass m is attached in turn to two different springs. Measurements are made of the x-component $F_{x\text{ we}}$ of the force we need to exert on m to hold the mass at various positions along a meter stick scale beside each spring. Plots are made of the x-component of the force versus the position on the meter stick; see Figure Q.6. For each case, answer the following: (a) Does the spring obey Hooke's force law? Why or why not? (b) Draw the corresponding graph of the x-component of the *force of the spring* on m versus position on the meter stick. (c) If the spring does obey Hooke's force law, what needs to be done *to the data* to make the graph of the force component we exert on m look like Figure 7.4?

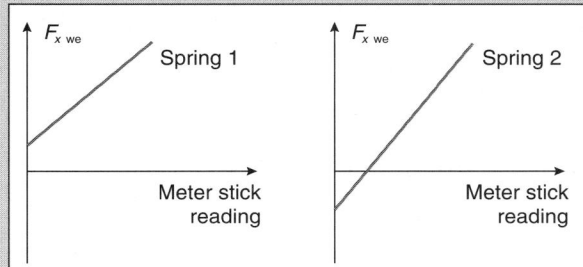

FIGURE Q.6

7. A mass m on a frictionless horizontal surface is attached to a horizontal spring. When we pull and hold the mass at some position coordinate x, where $x = 0$ m is the equilibrium position of the mass, our force on m and the force of the spring on m are of equal magnitude. If we gradually increase our pull on the mass, the mass gradually moves to a larger x-coordinate and the spring force on m gradually increases in magnitude, matching our force newton-for-newton. If we gradually decrease the magnitude of the force exerted on m, the spring force also decreases. Recall that the magnitude of the static frictional force also matches forces newton-for-newton (see Section 5.12); the magnitude of the static frictional force on m when it is on a rough surface also increases as we push harder, and decreases as we push less. Compare and contrast the force of a spring on m when it is on a frictionless surface to the static frictional force on m when it is on a rough surface. Explain why the static frictional force *cannot* produce simple harmonic oscillation of a mass.

8. A mass is attached to one end of a spring on a frictionless surface inclined at an angle θ to the horizontal (see Figure Q.8). When displaced from its equilibrium position, will the mass oscillate at the same angular frequency as it would if the spring were either horizontal (on a frictionless surface) or vertical? Justify your answer.

9. A mass is executing simple harmonic oscillation with amplitude A. What is the total distance traveled by the mass during one period of the oscillation? Does your answer depend

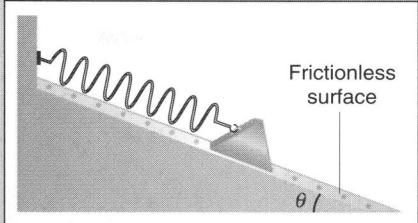

FIGURE Q.8

on the time instant from which the period of the oscillation is measured?

10. A mass is executing simple harmonic oscillation according to the equation

$$x(t) = A \cos \omega t$$

The period of the oscillation is T. Answer the following questions for each of the following time intervals: (i) 0 s $< t < T/4$; (ii) $T/4 < t < T/2$; (iii) $T/2 < t < 3T/4$; (iv) $3T/4 < t < T$. (a) Is the position vector parallel or antiparallel to the velocity? (b) Is the position vector parallel or antiparallel to the acceleration? (c) Is the velocity parallel or antiparallel to the acceleration?

11. A particle is executing simple harmonic motion according to the equation $x(t) = A \cos \omega t$. (a) What is the maximum speed of the particle? (b) What is the magnitude of the maximum acceleration of the particle? (c) At what positions is the *speed* of the particle *half* the maximum speed? (d) At what positions is the magnitude of the acceleration of the particle *half* the magnitude of the maximum acceleration?

12. One astronaut on the Moon has a spring with a known spring constant k while another astronaut has simply a string of known length ℓ. An interesting rock is found. Describe how the astronaut with the spring can determine the mass of the rock from observations and data secured with the spring without knowing the magnitude of the acceleration due to gravity on the Moon. Is it possible for the astronaut with the string to check the result? Explain how or why not.

13. You have a variety of masses available but only one spring. What can be done to double the maximum speed for simple harmonic oscillation at a fixed amplitude A?

14. The amplitude of a simple harmonic oscillation is tripled. Describe how this affects (a) the frequency of the oscillation; (b) the maximum speed of the oscillation; and (c) the magnitude of the maximum acceleration.

15. Make a list of several everyday examples of simple harmonic oscillation. Make a similar list of oscillating objects that do *not* exhibit simple harmonic oscillation.

16. A superball is dropped vertically and bounces up and down on the floor. Is the motion oscillatory? Is the motion damped simple harmonic oscillation? Explain your answers.

17. The x-component of the force needed to stretch or compress a mass m (attached to a peculiar spring) to a coordinate x (and hold it there) is plotted versus x with the results indicated in

Figure Q.17. The coordinate $x = 0$ m corresponds to the equilibrium position of the mass with $\hat{\imath}$ corresponding to stretching the spring. (a) *Describe* how this peculiar spring differs from an ordinary spring that obeys Hooke's law. (b) Think of a physical system that might behave this way.

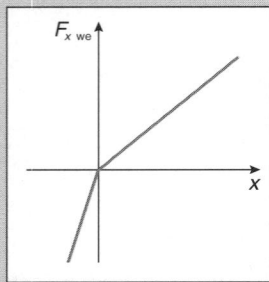

FIGURE Q.17

18. Two springs are oriented vertically with the position of the bottom of each spring designated as $x = 0$ m and $\hat{\imath}$ directed vertically downward. Various masses are suspended from each spring in turn and the position of the bottom of the spring is noted when the mass is in equilibrium. The magnitude of the weight of each mass is plotted versus the corresponding position of the bottom of the spring from its massless equilibrium position. The resulting graphs appear in Figure Q.18. (a) Which spring has the greater spring constant k? (b) When the most mass is hung from each spring and the mass is set into vertical simple harmonic oscillation, which spring will cause the mass to oscillate with the largest frequency? With the greatest angular frequency? With the longest period? (c) With equal masses on each spring, does the information given provide a way to determine which spring gives the mass the greatest speed? The greatest acceleration? Explain your answer.

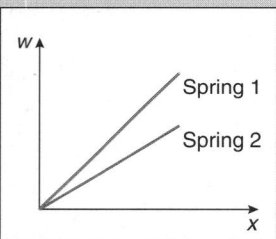

FIGURE Q.18

19. The mass of your car typically appears on its registration form. Push the car down quickly on one corner and estimate the period of oscillation. Knowing the mass and period, estimate the spring constant of each of the four springs, assuming they are equal.

20. If the length of a simple pendulum is tripled, by what factor does the period change?

21. A simple pendulum is hung from the ceiling inside an elevator at rest. Its period of oscillation is T_0. Describe qualitatively what happens to the period of the pendulum (if anything) when the elevator is: (a) moving upward at constant velocity; (b) moving downward at constant velocity; (c) moving upward with a constant acceleration; (d) moving downward with a constant acceleration.

22. The mass of a simple pendulum is doubled but the length of the pendulum is unchanged. By what factor does the period of the pendulum change?

23. The determination of longitude at sea necessitates an accurate clock. The first accurate marine chronometers were invented by John Harrison in the latter part of the 18th century. Several of his clocks are on display at the former Royal Observatory in Greenwich, England. Explain why a pendulum clock is rather useless as an accurate timekeeping device on a ship.

24. If the length of a simple pendulum increases by 1%, by what factor is the period of the pendulum changed? Does the period increase or decrease?

25. The period of a mass m oscillating on a spring is

$$2\pi \left(\frac{m}{k} \right)^{1/2}$$

and the period of a mass m on a simple pendulum executing small-angle oscillations is

$$2\pi \left(\frac{\ell}{g} \right)^{1/2}$$

Since both motions are simple harmonic, why does one period depend on the mass while the other does not?

26. A family moves its heirloom grandfather pendulum clock from Miami (at sea level) to Denver (elevation about 1500 m) without making adjustments in the length of the pendulum. If the temperature is the same in Miami and in Denver, will the clock run slow or fast in Denver? Should the length of the pendulum be slightly increased or decreased for it to keep accurate time?

27. We have studied several relationships involving force components or magnitudes of forces that have similar algebraic forms; among them are

$$F_{x\,total} = ma_x$$
$$f_{s\,max} = \mu_s N$$

$$f_k = \mu_k N$$
$$F_{x\,sprg} = -kx$$

$$w = mg$$
$$\vec{F}_{grav} = m\vec{g}$$

(a) Carefully explain what is meant by each of these expressions. (b) Each expression involves a particular constant (m, μ_s, μ_k, k); what are the SI units of each constant and what property of what objects does the constant quantify? (c) What are the *limitations* associated with *each* expression? (d) Is it possible to rank the expressions in the order of their importance or priority in physics? Justify your answer.

28. After a diver jumps off a diving board, the board oscillates at its natural frequency of oscillation. Qualitatively, how does the natural oscillation frequency of the diving board depend on the length of the board? (You can easily simulate the diving board by clamping a meter stick or ruler down to a table with part of its length hanging off the table and tweaking the free end of the stick. Vary the length of the free end of the stick and observe the effect on the frequency.)

29. Qualitatively, how would the natural frequencies of oscillation of a tall structure compare with those of a short structure? Earthquakes exert oscillating forces on all structures. To avoid significant structural damage, what should be the relationship between the shaking frequencies associated with earthquakes and the natural frequencies of oscillations of buildings?

30. Explain why automotive suspension systems typically are designed for critical damping of oscillations induced by road bumps.

31. The frequency of damped oscillatory motion is less than that of undamped oscillatory motion. Explain why this is reasonable in view of the forces acting on the system.

32. Why is it inadvisable to have large numbers of military troops marching in step across bridges?

33. A mass is attached to a peculiar spring. Every 10.0 s the x-component of the force we need to pull and hold m at a given coordinate x is suddenly changed; its value is quickly determined and then plotted versus x. The original position of the mass is at the coordinate $x = 0$ m when $t = 0$ s. The *chronological* sequence of data taken in the experiment appears in the table on the right.

Position coordinate x in cm of mass m	x-component of the force in N needed to act on m to keep m at the specified x-coordinate
0	0
2	4
5	16
10	28
15	34
20	40
15	38
10	37
5	34
0	30
−5	22
−10	0
−15	−30
−20	−40
−15	−38
−10	−37
−5	−34
0	−30
5	−22
10	0
15	30
20	40

(a) Make a graph of $F_{x\,we}$ versus x (in *meters*), indicating the path the mass follows *chronologically* on the graph. (b) Using the graph as a guide, *describe* what is happening during the experiment. (c) Does this spring obey Hooke's force law? Explain your answer. Such a peculiar spring exhibits what are called *hysteresis* effects. Can you think of a physical system that might behave in a way similar to this?

PROBLEMS

Section 7.1 Hooke's Force Law

1. A force of magnitude 5.00 N stretches a spring 10 cm. What is the spring constant of the spring?

2. You have a spring with a spring constant of 225 N/m. What magnitude force is needed to stretch the spring 5.0 cm?

•3. An inquisitive student hangs various masses from a stretchy cord and records the position of the end of the cord from the original position of its end, located at the 2.0 cm mark on a meter stick. The following data are secured:

Mass hung on the cord in kg	Position of the end of the cord in cm
0.100	7.7
0.200	10.1
0.300	11.9
0.400	13.4
0.500	14.8

(a) Make a graph of the magnitude of the force acting on the end on the cord versus the meter stick record of the *position* of the end of the cord. (b) Does the string obey Hooke's force law? Explain why or why not.

•4. An experiment is performed in which we measure the force needed to hold a 0.250 kg mass (attached to a spring) away from its equilibrium position on the spring. The equilibrium position of the mass on the spring is at $x = 0$ m and $x > 0$ m corresponds to stretching the spring. The mass is held at rest at various x-coordinates over the range -0.20 m $< x <$ $+0.20$ m. The spring constant is 10 N/m. On the same graph, accurately plot: (a) the x-component of the force of the spring on m; (b) the x-component of the force we exert on the mass; and (c) the x-component of the vector sum of the two forces on the mass versus x over the domain -0.20 m $< x <$ $+0.20$ m. Label each coordinate axis appropriately and be sure to distinguish among the graphs (a), (b), and (c).

•5. A horizontal spring is attached to a 15.5 kg mass on a rough horizontal surface with coefficients of friction $\mu_s = 0.25$ and $\mu_k = 0.20$. We pull the spring manually and find that when the spring is stretched 0.150 m, the mass remains at rest; but when the stretch is 0.151 m, the mass slips. What is the spring constant?

•6. A cage holds a 150 g mass, which is attached to one end of a spring. The other end of the spring is attached to the opposite end of the cage. The spring is stretched 4.0 cm as the mass is whirled at 25.0 rev/s in a circle of radius 6.0 cm about the center of the cage on a horizontal frictionless surface as indicated in Figure P.6. What is the spring constant?

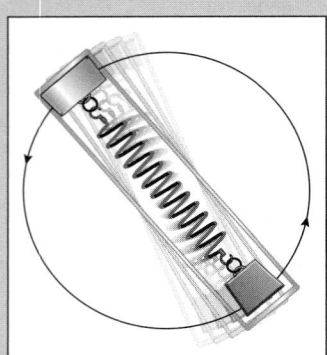

FIGURE P.6

•7. Suppose, as is conventional, we take $x = 0$ m as the equilibrium position of a mass m attached to a spring. If $\hat{\imath}$ is chosen to be in the direction corresponding to *compressing* the spring, show that the force of the spring on m has the form

$$\vec{F}_{spring} = -kx\hat{\imath}$$

This expression for $\vec{F}_{spring}$ is the *same* form as when $\hat{\imath}$ corresponded to *stretching* the spring (Equation 7.2). Thus the direction chosen for $\hat{\imath}$ does *not* affect the results we obtained in this chapter.

•8. Show that if two identical springs, each with spring constant k, are connected in parallel as indicated in Figure P.8, the effective spring constant of the pair is $2k$.

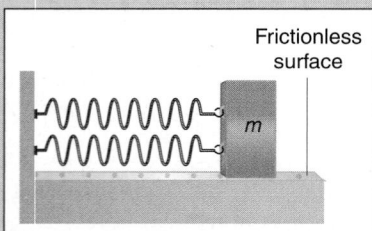

FIGURE P.8

•9. Two springs with spring constants k_1 and k_2 are connected as shown in Figure P.9 with a mass m attached on a frictionless surface. We pull on m with a force $\vec{F}_{we}$ and hold m at rest. The mass m moves a distance x and point A moves a distance x' from their equilibrium positions. (a) By examining the forces on point A, show that

$$x' = \frac{k_2}{k_1 + k_2} x$$

(b) By examining the forces on m, show that if the system is modeled with a single spring stretching m a distance x, the effective spring constant of the single spring is

$$k_{eff} = \frac{k_1 k_2}{k_1 + k_2}$$

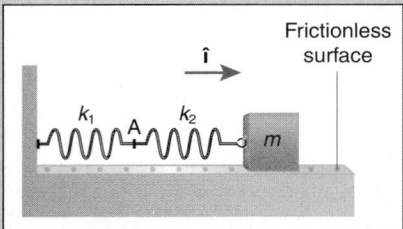

FIGURE P.9

•10. A spring with a spring constant k is cut transverse to its axis into two equal parts and then patched together again, so that the original spring can be modeled as two half-springs in series with each other (see the previous problem). Show that the spring constant of each half-spring is $2k$. Thus one way to create a stiffer spring (one with a larger spring constant) is to cut and remount the spring, so that it has fewer coils.

•11. A spring gun is used to project a 100 g mass horizontally across a frictionless surface. The spring is compressed 5.0 cm and has a spring constant of 30 N/m. What is the initial magnitude of the acceleration of the mass the instant the spring gun fires? Is the acceleration of the mass constant while it still is in contact with the spring?

Section 7.2 Simple Harmonic Oscillation

•12. A fleet-footed politician of mass 50.0 kg is observed oscillating back and forth on a slippery, frictionless issue according to the expression

$$x(t) = (6.0 \text{ m}) \cos\left[(8.00 \text{ rad/s}) t + \frac{\pi}{4} \text{ rad}\right]$$

Physics can describe anything! (a) Find the amplitude A of the oscillation. (b) Find the angular frequency ω of the oscillation. (c) Find the frequency ν of the oscillation. (d) Find the period T of the oscillation. (e) Find the effective spring constant k associated with the oscillation. (f) Find the position x of the politician when $t = 2.00$ s. (g) Determine $v_x(t)$ and find the velocity component of the hack when $t = 2.00$ s. (h) Determine $a_x(t)$ and find the acceleration component of the pol when $t = 2.00$ s. (i) Make graphs of $x(t)$, $v_x(t)$, and $a_x(t)$ versus t during at least two periods of the motion, beginning when $t = 0$ s. Verify your answers to parts (f)–(h) from these graphs.

•13. A 2.00 kg mass executes simple harmonic oscillation with an amplitude of 5.00 cm. The maximum speed of the mass is 2.00 m/s. (a) Determine the angular frequency ω of the oscillation. (b) Find the frequency ν of the oscillation. (c) If the mass is at $x = 0$ m when $t = 0$ s, find $x(t)$. (d) What is the position of the mass when $t = 1.00$ s?

•14. The position of a mass undergoing simple harmonic oscillation is described by

$$x(t) = A \cos \omega t$$

(a) Qualitatively sketch the graph of $x(t)$ versus t during two periods of the motion. (b) Qualitatively sketch a graph of the velocity component $v_x(t)$ versus t during two periods of the motion. (c) Qualitatively sketch a graph of the acceleration component $a_x(t)$ versus t during two periods of the motion. On each graph indicate the value of the maximum position, speed, and magnitude of acceleration.

•15. A mass m rests on a block of mass M that is attached to a horizontal spring as indicated in Figure P.15. The block M rests on a frictionless surface. The coefficient of static friction for the contact surfaces of the two masses is μ_s and the coefficient of kinetic friction is μ_k. The system is set into simple harmonic oscillation with amplitude A. What spring constant ensures that m is on the verge of slipping? If m is then increased, will the mass slip or not? Explain.

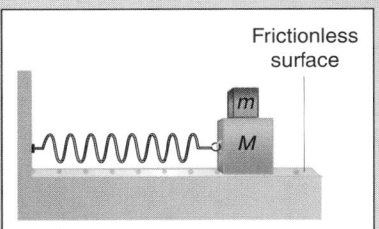

Frictionless surface

FIGURE P.15

•16. The position of a mass m undergoing simple harmonic oscillation is described by $x(t) = A \cos \omega t$. Make a graph of the *distance* traveled by the mass as a function of time during one period of its motion.

•17. A 1.50 kg mass on a horizontal frictionless surface is attached to a horizontal spring with spring constant $k = 200$ N/m. The mass is in equilibrium at $x = 0$ m. The mass is released when $t = 0$ s at coordinate $x = 0.100$ m with a velocity (2.00 m/s)$\hat{\imath}$. (a) Find the constants A, ω, and ϕ in

$$x(t) = A \cos(\omega t + \phi)$$

(b) Find the period T of the oscillation. (c) Determine the maximum speed of the oscillation and the magnitude of the maximum acceleration. (d) Plot $x(t)$ during the time interval $t = 0$ s to $t = 2T$.

•18. A manufacturer claims that a device is capable of withstanding accelerations of magnitude at least equal to 50 times the magnitude of the local acceleration due to gravity g (= 9.81 m/s²) without destruction. As an engineer in a testing laboratory, you decide to test the claim by attaching the device to a spring and subjecting it to simple harmonic oscillation of amplitude 0.10 m. What oscillation frequency will subject the device to an acceleration of magnitude $50g$ at some times during the oscillation?

•19. A 500 g mass is undergoing simple harmonic consternation . . . oscillation (!) . . . that is described by the following equation for its position $x(t)$ from equilibrium:

$$x(t) = (0.50 \text{ m}) \cos\left[(6.0\pi \text{ rad/s})t + \frac{\pi \text{ rad}}{6.0}\right]$$

(a) What is the amplitude A of the oscillation? (b) What is the angular frequency ω of the oscillation? (c) What is the frequency ν of the oscillation? (d) What is the period T of the oscillation? (e) What is the fall constant. . . rather, the spring constant k? (f) What is the position of the oscillator when $t = 0$ s? (g) At what time $t > 0$ s is the oscillator *first* at maximum distance from equilibrium? (h) What is the maximum speed of the oscillator? (i) What is the magnitude of the maximum acceleration of the oscillator?

•20. A mass is executing simple harmonic oscillation with a frequency of 9.0 Hz and amplitude 0.20 m about an equilibrium position at $x = 0$ m. When $t = 0$ s, the mass is located at the x-coordinate 0.10 m. (a) Find $x(t)$ and plot it versus t during a time interval of two periods beginning when $t = 0$ s. (b) Find $v_x(t)$ and plot it versus t during a time interval of two periods from $t = 0$ s. (c) Find $a_x(t)$ and plot it versus t during a time interval of two periods from $t = 0$ s.

•21. In Example 7.8, determine specific expressions for $v_x(t)$ and $a_x(t)$ and plot them versus t for the domain 0 s $\leq t \leq 2.0$ s as Figure 7.19 does for $x(t)$.

•22. A can of cola of mass $m = 0.250$ kg on a horizontal frictionless surface is attached to the end of a horizontal spring (with spring constant $k = 15.0$ N/m) and is released at the equilibrium position $x = 0$ m when $t = 0$ s with an initial velocity $v_0\hat{\imath} = (0.500 \text{ m/s})\hat{\imath}$. (a) Determine the constants A, ω, and ϕ in the general equation for $x(t)$:

$$x(t) = A \cos(\omega t + \phi)$$

(b) Determine the period T of the oscillation. (c) Find and plot $x(t)$, $v_x(t)$, and $a_x(t)$ versus t for the domain 0 s $\leq t \leq 2T$.

‡23. Real springs have mass. Thus when a mass m is attached to a spring of mass m_s and executes simple harmonic oscillation, part of the mass of the spring also is in oscillation. To account for the mass of the spring, let $M = m + \gamma m_s$, where γ represents the *fraction* of the mass of the spring that participates in the oscillation. The period of the oscillation T then is

$$T = 2\pi \left(\frac{M}{k}\right)^{1/2}$$

Square and substitute for M; you will find that

$$T^2 = \frac{4\pi^2}{k} m + \frac{4\pi^2}{k} \gamma m_s$$

A student uses a spring of mass 0.180 kg and measures the period of oscillation T when various masses m are suspended from the spring. The following data are secured:

Mass m in kg attached to spring	Measured period of oscillation in s
0.050	0.75
0.150	1.00
0.250	1.20
0.350	1.35
0.450	1.51
0.550	1.65
0.650	1.77
0.750	1.88
0.850	2.00
0.950	2.10

(a) Plot the *square* of the period (T^2) versus the mass m attached to the spring. In view of the foregoing expression for T^2, this graph should be linear. Find the best-fit straight line through the data. What is the slope of the graph? What is the intercept along the T^2-axis? (b) Use the slope and intercept to determine (i) the spring constant k of the spring and (ii) the fraction γ of the mass of the spring that participates in the os-cillation.

‡24. The force of a spring on a mass m when it is a distance x from the equilibrium position at $x = 0$ m is

$$\vec{F}_{spring} = -kx\hat{\imath}$$

Suppose, instead, that the only force acting on mass m was a force given by

$$\vec{F} = kx\hat{\imath}$$

(a) Qualitatively describe the force $\vec{F}$ (its magnitude and direction) as the position of the mass moves toward increasing positive values of x. (b) Qualitatively describe the force $\vec{F}$ (its magnitude and direction) as the position of the mass moves toward increasingly negative values of x. (c) Use Newton's second law to show that the differential equation that describes the motion of the mass when $\vec{F}$ is the only force acting on it is

$$\frac{d^2x}{dt^2} - \frac{k}{m}x = 0 \text{ m/s}^2$$

(d) Does $x(t) = A\cos(\omega t + \phi)$ satisfy this differential equation? If so, what is the value of ω that emerges from this? (e) Does the expression

$$x(t) = Ae^{\omega t}$$

satisfy the differential equation? If so, what is ω? Qualitatively describe what happens to the particle for times $t > 0$ s if $x(t) = Ae^{\omega t}$. (f) Do you think that such a force actually exists in nature? Explain.

‡25. Two masses are undergoing simple harmonic oscillation accor-ding to the equations

$$x_1(t) = (0.150 \text{ m}) \cos\left[(0.250 \text{ rad/s}) t + \frac{\pi}{6} \text{ rad}\right]$$

$$x_2(t) = (0.200 \text{ m}) \cos\left[(0.350 \text{ rad/s}) t + \frac{\pi}{4} \text{ rad}\right]$$

(a) Find the period of each oscillation. (b) On the same graph, plot $x_1(t)$ and $x_2(t)$ versus t during $t = 0$ s to 60 s and determine the first six instants at which they have the same position. Is the interval between these instants the same? (c) By changing one of the amplitudes, see if the instants when the oscillators have the same position depend on the amplitudes.

‡26. Use the two oscillations in Problem 25 to make a graph of their superposition $x(t) = x_1(t) + x_2(t)$ during $t = 0$ s to 200 s. The result illustrates how more complex oscillations are imagined as composed of superpositions (sums and differences) of various simple harmonic oscillations with different amplitudes, fre-quencies, and phase angles. The process of determining the simple harmonic oscillations that constitute a complex oscil-lation is called *Fourier analysis*.

Section 7.3 A Vertically Oriented Spring

27. A 300 g mass is suspended from a vertical spring and the spring extends a distance of 0.15 m. (a) Determine the spring constant k. (b) Find the angular frequency ω and the period T at which the mass will oscillate on the spring.

•28. A medieval damsel in distress is on top of a tower of height h. An elastic bungee cord that obeys Hooke's law, with $k = 40.0$ N/m, hangs from the top of the tower and reaches halfway to the ground. The 60.0 kg damsel climbs down the cord and when she reaches the end amazingly finds that she is safely on the ground. What is the height of the tower from which she descended?

•29. A 2.00 kg mass is suspended from a vertical spring of spring constant 100 N/m and unstretched length 0.300 m. Find the position, relative to the fixed end of the spring, of the point about which the mass will make simple harmonic oscillations.

•30. A vertical spring is attached to the ceiling. An 8.00 kg frozen turkey is attached to the bottom of the spring and causes the spring to stretch an additional 10 cm. The turkey then is pulled down 4.0 cm beyond the equilibrium position and released at rest. Take the positive x-direction to be downward. (a) What is the spring constant? (b) If the oscillation is described by $x(t) = A \cos(\omega t + \phi)$, find A, ω, and ϕ. (c) What is the maximum speed of the turkey? (d) What is the mag-nitude of its maximum acceleration?

•31. A leprechaun of mass 70.0 kg sits on the end of a long diving board of negligible mass and the end is depressed 25 cm. (a) What is the effective spring constant k of this system if we model it as a simple spring? (b) At what frequency could this system oscillate? (c) What oscillation amplitude would allow the leprechaun to reach an upward acceleration equal in magnitude to that of the local acceleration due to gravity g?

•32. A 10.0 kg mass attached to a spring is dragged at constant speed up a rough surface ($\mu_s = 0.30$, $\mu_k = 0.25$) inclined at 10° to the horizontal as in Figure P.32. The spring is stretched

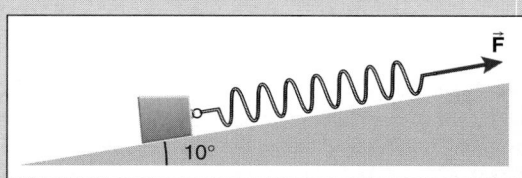

FIGURE P.32

8.0 cm. (a) What is the spring constant of the spring? (b) If the spring drags the mass at constant speed down the slope, what is the stretch of the spring?

•33. A 0.250 kg mass is suspended by a string from a 0.500 kg mass as shown in Figure P.33. The 0.500 kg mass is attached to a spring with a spring constant of 50.0 N/m. (a) What is the frequency of simple harmonic oscillation? (b) At what point in the oscillation is the tension in the string a minimum? (c) At what point in the oscillation is the tension in the string a maximum? (d) What amplitude oscillation ensures the tension in the string is zero at some instant during the motion?

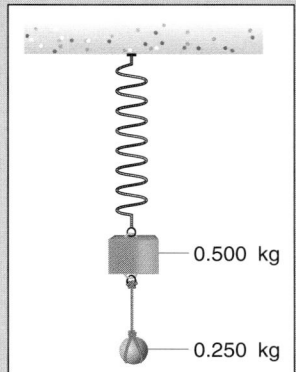

0.500 kg

0.250 kg

FIGURE P.33

•34. A troll of mass m is placed on top of a piston executing simple harmonic motion in a vertical direction with an amplitude A. (a) At what position in the simple harmonic motion of the piston is the troll likely to leave the surface of the piston? (b) If the frequency of the simple harmonic oscillation of the piston is small, the troll on top of the piston is able to follow the oscillation. On the other hand, if the frequency of the simple harmonic oscillation is too large, the troll will bounce atop the piston. What is the maximum frequency of simple harmonic oscillation so that the troll barely remains continuously in contact with the piston at all times during the motion?

•35. A 1.50 kg mass is attached to a vertical spring. A student gradually pushes slowly up on the mass until the upward force of the student on the mass is equal to the weight of the mass. With the mass at rest, the student then suddenly removes the supporting hand and notes that the mass executes simple harmonic oscillation over a vertical range of 1.20 m. Determine the spring constant k and the frequency of oscillation v.

•36. Two identical springs, each with spring constant k, are attached to two identical masses m with the lower mass suspended by its spring from the upper mass. The top of the upper spring is pulled upward in the vertical direction with a force sufficient to cause an acceleration of magnitude a. The unstretched length of each spring is ℓ_0. (a) What is the change in the length of the lower spring? (b) What is the change in the length of the upper spring?

•37. A 1.5 kg mass oscillates with an amplitude of 0.10 m on a vertical spring whose spring constant is 350 N/m. (a) What is the maximum speed of the mass? (b) What is the magnitude of the maximum acceleration of the mass?

•38. A 0.250 kg sugar plum fairy is attached to a vertically suspended spring whose spring constant is 2.00 N/m. The sugar plum fairy is pulled down 10 cm and released at rest. (a) Find $x(t)$. (b) At what position is the fairy when $t = 3.5$ s? (c) What is the velocity component of the fairy when $t = 3.5$ s? (d) What is the acceleration component of the fairy when $t = 3.5$ s?

$38. A *massive spring* with mass m and spring constant k is horizontal and neither stretched nor compressed. Its length is measured to be ℓ_0. Show that the increase in the length of the spring, because of its weight, if it is hung in the vertical direction is

$$\frac{mg}{2k}$$

Hint: Generalize Problems 9 and/or 10 and consider the given spring to be a collection of n springs in series, each with spring constant nk, where n is an integer much greater than 1.

Section 7.4 Connection between Simple Harmonic Oscillation and Uniform Circular Motion

•40. A phonograph record with a diameter of 30.0 cm has a young cricket (*Acheta domestica*) sitting on its rim. The record and cricket are turning at 33.3 rev/min. The setting Sun casts a shadow of the cricket on a wall directly facing the Sun. (a) Does the shadow move in simple harmonic motion? (b) What is the amplitude of the motion of the shadow? (c) What is its period? (d) What is the maximum speed of the shadow? (e) What is the magnitude of the maximum acceleration of the shadow? (f) When the acceleration of the shadow is zero, is the velocity of the shadow also zero? (g) What are the forces acting on the cricket? Sketch an appropriate second law force diagram.

•41. A mass is executing uniform circular motion counterclockwise in a vertical circle of radius R at angular speed ω. Place the origin at the center of the circle, with the x-axis horizontal and in the plane of the circle with $\hat{\imath}$ to the right; choose the y-axis vertical with $\hat{\jmath}$ up. When $t = 0$ s, the particle is on the $+x$-axis. The motion is projected onto a line that passes through the center of the circle in the plane of the circle and is inclined at 30.0° to $\hat{\imath}$; see Figure P.41. Write the simple harmonic motion along the line in the standard form

$$A \cos(\omega t + \phi)$$

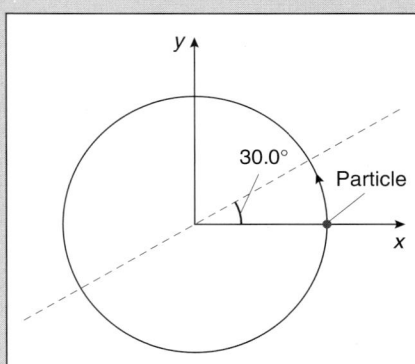

FIGURE P.41

•42. A 0.150 kg mass is attached to one end of a horizontal spring as indicated in Figure P.42. The mass lies on a frictionless surface. The unstretched length of the spring is 5.00 cm. The spring and mass then are whirled in a horizontal circle at an angular speed of 50.0 rad/s about a vertical axis through the other end of the spring. It is found that the mass extends a distance of 3.00 cm from its position when at rest. (a) Find the spring constant k of the spring. (b) A spotlight directed horizontally projects a shadow of m onto a wall. Is the motion of the shadow simple harmonic oscillation? If so, what is the angular frequency ω of the oscillation of the shadow? What is the effective spring constant of the simple harmonic oscillation of the shadow?

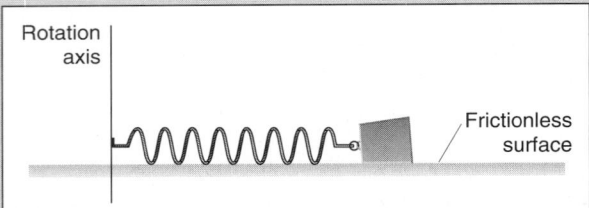

FIGURE P.42

•43. A line from the Sun to the star Regulus in the constellation Leo lies essentially in the plane of the orbit of the Earth about the Sun. Assume the orbit of the Earth is circular. Let the Earth be on this line when $t = 0$ s (about 21 February and again on 23 August). Write the projection of the motion of the Earth on this line in the form $x(t) = A \cos(\omega t + \phi)$ using SI units.

‡44. You have an unknown mass m, a spring whose spring constant is unknown, and a 1.000 kg standard mass. The mass m and the standard mass are attached to opposite ends of the spring. The spring is compressed on a frictionless horizontal surface and both masses released at rest when $t = 0$ s. You simultaneously determine the magnitude of the initial acceleration of m to be 9.00 m/s² while that of the standard 1.000 kg mass is 4.50 m/s² in the opposite direction. What is the unknown mass m?

Section 7.5 How to Determine if an Oscillatory Motion Is Simple Harmonic Oscillation

45. A beagle runs back and forth between two children, making one round trip every 10 s on a regular basis. (a) What is the period of the dog? (b) What is the frequency of the oscillation of the dog? (c) Is the motion of the dog simple harmonic oscillation or not? Explain.

•46. A mythical hero of mass m is strung between two springs as shown in Figure P.46. The spring constants are k_1 and k_2 as indicated. At the position $x = 0$ m, the hero is subjected to no force from either spring. The gods move the hero to position x. (a) What is the force of spring 1 on the hero? (b) What is the force of spring 2 on the hero? (c) What is the total force on the hero in the horizontal direction due to the two springs? (d) The horizontal surface is frictionless. When the gods release the hero at rest, apply Newton's second law to the hero and cast the result into the standard form for the differential equation for simple harmonic oscillation (a form like Equation 7.11).

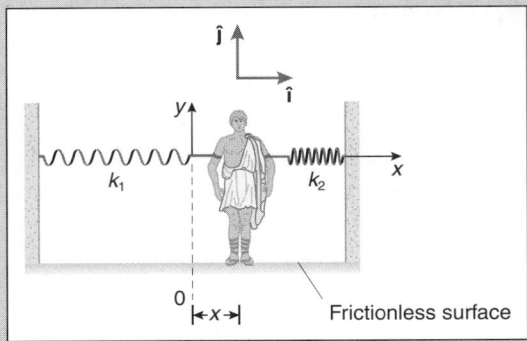

FIGURE P.46

(e) Determine the period of oscillation of the hero from the equation obtained in part (d).

•47. A particle in a frictionless hemispherical bowl of radius R is moved slightly away from the bottom and released at rest. Show that particle executes simple harmonic oscillation about the bottom of the bowl and find the period of the oscillation.

‡48. A plumb bob of length ℓ is swung so that the mass on the end of the string moves in a horizontal circle at constant angular speed with the string making a constant, small angle θ with the vertical direction as shown in Figure P.48. Show that if the motion is viewed sideways in the plane of the horizontal circle, the motion is simple harmonic and of period

$$T = 2\pi \left(\frac{\ell \cos \theta}{g} \right)^{1/2}$$

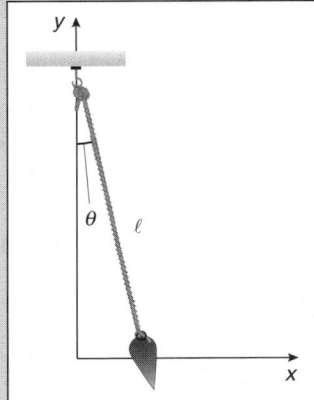

FIGURE P.48

‡49. A piston moves inside a cylindrical tube and is connected by a tie-rod to the rim of a circular shaft that is turning at constant angular speed ω (see Figure P.49). Show that the oscillatory motion of the piston is *not* simple harmonic oscillation.

‡50. A spring with spring constant k is attached to a mass m that is confined to move along a frictionless rail oriented perpendicular to the axis of the spring as indicated in Figure P.50. The spring is initially unstretched and of length ℓ_0 when the mass is at the position $x = 0$ m in the indicated coordinate system.

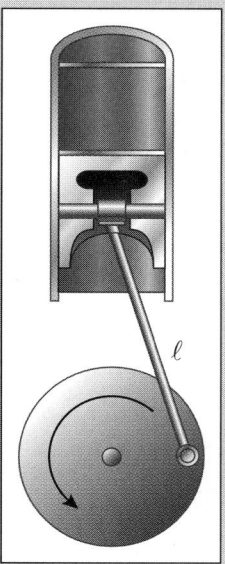

FIGURE P.49

(a) If the mass m is pulled out a distance x along the rail, show that the new total length of the spring is

$$(\ell_0^2 + x^2)^{1/2}$$

(b) Show that the x-component of the force that the spring exerts on the mass is

$$F_x = -k[(\ell_0^2 + x^2)^{1/2} - \ell_0]\cos\theta$$

(c) Show that for *small* coordinates x along the rail, the horizontal force component on the mass m is approximately

$$F_x \approx -\frac{k}{2\ell_0^2}x^3$$

If the mass is released from the point x, oscillations occur but the oscillations are *not* simple harmonic oscillations. Indeed, the period of this oscillator is *inversely* proportional to the amplitude of the oscillator, not independent of the amplitude as is the case for simple harmonic oscillation. This is an example of a *nonlinear* or *anharmonic* oscillator.
This problem was inspired by a paper by Alan Cromer, "The x^3 oscillator," *The Physics Teacher*, 30, #4, pages 249–250 (April 1992).

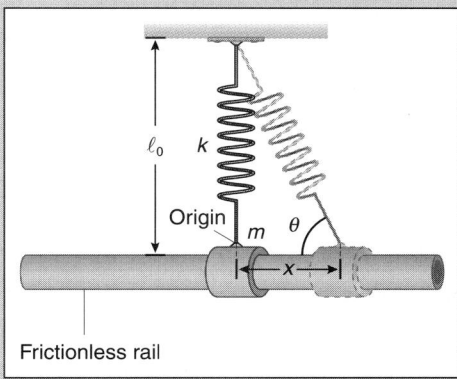

FIGURE P.50

Section 7.6 The Simple Pendulum

51. Thomas Jefferson, as Secretary of State under George Washington, was asked to devise a decimal system of weights and measures. Jefferson proposed a standard of length based on a simple pendulum whose *half period* was *one second*, so that $T = 2.00$ s. A clock based on such a pendulum would tick once with each swing. If we take the magnitude of local acceleration due to gravity to be precisely 9.81 m/s², what length pendulum has a period of precisely 2.00 s? The length is very close to but not the same as a familiar standard of length, developed in France during the French Revolution.

52. The numerical value of the magnitude of the acceleration due to gravity g varies slightly with geographic location and altitude on the Earth. What value of g will give a simple pendulum with a length of precisely 1.000 m a period of 2.000 s? The largest values of g on the surface of the Earth are at the geographic poles, where $g \approx 9.83$ m/s², and the smallest values are on the equator, where $g \approx 9.78$ m/s².

53. In 1851, Jean Bernard Léon Foucault (1819–1868) suspended a very long pendulum (67.5 m) from the inside of the dome of the Panthéon in Paris to demonstrate the rotation of the Earth. (See chapter opening photo.) Such Foucault pendulums are a prominent feature of many science museums. (See Figure P.53.) What was the oscillation period of his pendulum?

FIGURE P.53

•**54.** The horizontal motion of a simple pendulum is given by the equation

$$x(t) = (0.050 \text{ m})\cos\left[\left(\frac{\pi}{2.00}\text{ rad/s}\right)t + \frac{\pi}{6.00}\text{ rad}\right]$$

(a) What is the frequency ν of the oscillation? (b) What is the length of the pendulum?

•**55.** Before the advent of accurate clocks in the early 17th century, a crude instrument for measuring the pulse rate consisted of a pendulum whose length was adjusted until *half* its period matched the pulse rate of a patient. The length of the pendulum was recorded as a measure of the pulse rate. Construct a simple pendulum with a piece of string and a convenient mass. With the assistance of a buddy, adjust the length of the pendulum until its half period matches your pulse rate. From this length, determine your pulse rate in beats per minute, the now standard medical unit for expressing pulse rate.

●56. For a simple pendulum, at what angle from the vertical is the vertical rise 0.100 of the horizontal distance? See Figure P.56.

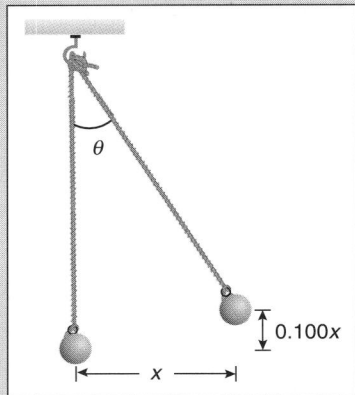

FIGURE P.56

●57. Two playground swings with lengths $\ell_1 > \ell_2$ have periods T_1 and T_2 respectively. They both are drawn aside together and released at the same instant to swing in small oscillations. After what time interval will they again be in step with each other?

●58. A simple pendulum that would have a period of 1.00 s on the surface of the Earth is carried to the Moon and set into oscillation on its surface. (a) Look up the radius and mass of the Moon and calculate the magnitude of the local acceleration due to gravity on the surface of the Moon. (b) Determine the period of oscillation of the pendulum on the lunar surface.

●59. If a pendulum 1.00 m long were somehow set into oscillation on the surface of a white dwarf star with a mass equal to that of the Sun and a radius equal to that of the Earth, determine what the frequency of oscillation would be.

‡60. Show that if a pendulum of length ℓ and period T is moved to a location where the magnitude of the local acceleration due to gravity differs by a small amount Δg, the change ΔT in the period of the pendulum is approximately

$$\Delta T \approx -\frac{T}{2g}\Delta g$$

‡61. The pendulum inside a grandfather clock has a *half* period of 1.0000 s at a location where the magnitude of the local acceleration of gravity is 9.800 m/s². The clock carefully is moved to another location at the same temperature and is found to run slow by 89.0 s per day. What is the magnitude of the local acceleration due to gravity at the new location?

‡62. A pendulum of length ℓ is released from rest when its string makes a small angle θ_0 with the vertical direction, as shown in Figure P.62. Find an expression for the angular velocity component of the pendulum as a function of time.

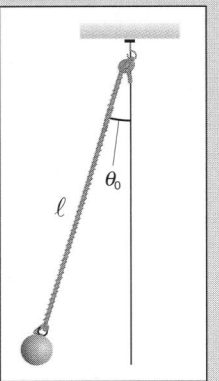

FIGURE P.62

Section 7.7 Through a Fictional Earth in 42 Minutes

63. Let your latitude be θ north and your longitude be ϕ west on the surface of the Earth. If you dig a vertical hole down through and beyond the center of the Earth, at what latitude and longitude does the hole emerge?

64. Determine the time for a mass m to make a one-way trip through a frictionless hole tunneled through a nonrotating neutron star with a mass twice that of the Sun and a radius of 10.0 km. Assume the star is a uniform sphere.

65. If you keep the mass of the Earth constant, what radius would permit a mass to make a one-way trip through a tunnel through the Earth in 1.00 min?

Section 7.8 Damped Oscillations*

66. What are the SI units of the damping constant β in Equation 7.26?

●67. A 0.950 kg mass hangs vertically from a spring that has a spring constant of 8.50 N/m. The mass is set into vertical oscillation and after 600 s you find that the amplitude of the oscillation is 1/10 that of the initial amplitude. What is the damping constant β associated with the motion?

●68. Show that if the position of a damped oscillator is given by

$$x(t) = Ae^{-\alpha t}\cos \omega t$$

where $\alpha = \beta/(2m)$, the time interval for the amplitude to decrease to half its initial value when $t = 0$ s is

$$\frac{2m}{\beta}\ln 2$$

●69. Hang a known mass from a string of known length to make a simple pendulum. Set the pendulum swinging in small-angle oscillations and time how long it takes for the amplitude to decrease to half its initial value. Use this measurement to determine the damping constant β for the pendulum.

Section 7.9 Forced Oscillations and Resonance*

70. What is the resonant frequency of a 0.500 kg mass attached to a spring with a spring constant $k = 10.0$ N/m?

71. What is the resonant frequency of a swing whose length is 3.50 m?

•**72.** The suspension system of your 1.20×10^3 kg car has a natural oscillation frequency of 2.0 Hz without functioning shock absorbers. The shock absorbers usually provide critical damping to the system. Along a worn section of concrete highway, you notice that the concrete joints, spaced 10.0 m apart, set up a violent oscillation of the car making it almost uncontrollable. At what speed are you driving?

•**73.** Country dirt roads frequently become rippled and when you drive over them at a certain speed, the bumps set up resonance in the car's suspension system, making control of the car problematic. If the ripples are 0.50 m apart and you find resonance occurs when you are driving at 25 km/h over them, what is the natural oscillation frequency of the suspension system on your car?

•**74.** At resonance, find the amplitude and maximum speed of a forced harmonic oscillator whose position coordinate $x(t)$ is given by Equation 7.35.

INVESTIGATIVE PROJECTS

A. Expanded Horizons

1. The superposition of two simple harmonic oscillations at right angles to each other produces some interesting patterns known as *Lissajous figures*, named for Jules Antione Lissajous (1822–1880) who first described them in 1855. Investigate this topic to determine how the various patterns are formed.
Thomas B. Greenslade Jr., "All about Lissajous figures," *The Physics Teacher*, 31, #6, pages 364–370 (September 1993).
William Fogg Osgood, *Mechanics* (Macmillan, New York, 1937), pages 182–190.
Forest K. Harris, *Electrical Measurements*, (John Wiley, New York, 1952), pages 644–647.
Gaylord P. Harnwell, *Principles of Electricity and Electromagnetism* (2nd edition, McGraw-Hill, New York, 1949), pages 256–258.
Brian A. Smith, "Lissajous figures," *Physics Education*, 16, #1, pages 38–46 (January 1981).
Mu-Shiang Wu and W. H. Tsai, "Corrections for Lissajous figures in books," *American Journal of Physics*, 52, #7, pages 657–658 (July 1984).
P. Jasselette and J. Vandermeulen, "More on Lissajous figures," *American Journal of Physics*, 54, #2, pages 182–183 (February 1986).
A. D. Crowell, "Motion of the Earth as viewed from the Moon and the Y-suspended pendulum," *American Journal of Physics*, 49, #5, pages 452–454 (May 1981).
Robert H. Romer, "A double pendulum 'art machine,'" *American Journal of Physics*, 38, #9, pages 1116–1121 (September 1970).

2. The superposition of two or more simple harmonic oscillations in the same plane with different amplitudes and frequencies results in complex oscillations (see Problem 26). Such superpositions form the basis for the *Fourier analysis* of complex, but periodic oscillations. Investigate this important area of more advanced analysis of oscillations.
Mary L. Boas, *Mathematical Methods in the Physical Sciences* (2nd edition, John Wiley, New York, 1983), Chapter 7, pages 297–335.

B. Lab and Field Work

3. The Lissajous patterns in Project 1 can be generated with an oscilloscope and a function generator or using a suitable software graphics package. Generate a set of such patterns with either or both of these methods.

4. The period of a simple pendulum of length ℓ for small oscillations is

$$T = 2\pi \left(\frac{\ell}{g}\right)^{1/2}$$

If the angle that a simple pendulum initially makes with the vertical is *not* small, the period of the pendulum no longer is independent of the amplitude and the equation for T becomes, approximately,

$$T \approx 2\pi \left(\frac{\ell}{g}\right)^{1/2} \left(1 + \frac{1}{16}\theta_0^2\right)$$

where θ_0 is the angle the pendulum initially makes with the vertical direction when released (at rest). When $\theta_0 = \pi/2$ rad, how much does the period differ from the small-angle result? Design and perform an experiment to see if you can detect the variation of T with θ.
Grant R. Fowles and George L. Cassiday, *Analytical Mechanics* (5th edition, Saunders, Fort Worth, 1993), page 107.
Keith R. Symon, *Mechanics* (3rd edition, Addison-Wesley, Reading, Massachusetts, 1971), pages 212–215.

5. Investigate the Foucault pendulum experiment (see Problem 53) used to detect the rotation of the Earth. If a suitable deep stairwell exists, set up a long Foucault pendulum using a large mass on its end. The wire supporting the pendulum must be freely gimballed with a device similar to the swivel used on fishing line. The pendulum should be released with as little perturbing force as possible by drawing the ball aside with a lightweight cloth sling tied with a strong but thin thread to a nearby wall. Break the string with a lighted match to set the pendulum into oscillation. Observe and measure the rate at which the plane of oscillation appears to rotate due to the rotation of the Earth. Compare your results with the predicted time period for the rotation of the plane of the pendulum:

$$\frac{24 \text{ h}}{\sin \theta}$$

where θ is your latitude.
Grant R. Fowles and George L. Cassiday, *Analytical Mechanics* (5th edition, Saunders, Forth Worth, 1993), pages 177–178.

6. Borrow an archery bow from a local sporting goods store or a friend, or use your own if you have one. Design and perform an experiment to determine if the draw of the bow obeys Hooke's force law.
W. C. Marlow, "Bow and arrow dynamics," *American Journal of Physics*, 49, #4, pages 320–333 (April 1981).

7. Design and perform an experiment to measure whether the vertical deflection of a diving board obeys Hooke's law over the common range of its deflection.

R. L. Page, "The mechanics of swimming and diving," *The Physics Teacher*, 14, #2, pages 72–80 (February 1976).

C. Communicating Physics

8. The personal differences between Hooke and Newton illustrate that scientists are human, with all the complexities associated with the human species. Should you wish to investigate the sociological aspects of the relationship between these two great scientists on matters of the priority of discoveries, see the following references. Comment on the likelihood of whether it is possible to have such strong disagreements without them becoming personal in nature.

Gale E. Christianson, *In the Presence of the Creator* (The Free Press, New York, 1984); see the listing under Hooke in the index for extensive pages references.

A. Rupert Hall, *Isaac Newton: Adventurer in Thought* (Blackwell, Oxford, 1992); see the index under Hooke for page references.

Robert K. Merton, *On the Shoulders of Giants* (University of Chicago Press, Chicago, 1993).

Richard S. Westfall, *Never at Rest: A Biography of Isaac Newton* (Cambridge University Press, Cambridge, England, 1980).

CHAPTER 8

WORK, ENERGY, AND THE CWE THEOREM

Our fantastic civilization . . . [is one] obsessed with power, which explains its whole world in terms of energy . . . .

Henry Beston (1888–1968) *

Whave analyzed the dynamics of a system by considering the forces on it and using Newton's second law of motion. As you have learned, forces are vectors; therefore, you must carefully employ the rules for manipulating such force vectors to determine the resulting acceleration of the system. In this chapter, we will examine another valuable approach to dynamics, one that uses the *scalar* physical concepts of work and energy.

Our intuitive notions are a hindrance to understanding the concepts of work and energy. This is not surprising, for if you look up the word *work* in the *Oxford English Dictionary*, you will find more than eleven pages (in *very* small print) of various meanings of the word; almost a page is devoted to various meanings of the word *energy*. As we have seen in other contexts, we cannot let our everyday understanding of words interfere with their precise meaning in physics. Common ideas of what constitutes work in many cases bear little resemblance to the concept of work in the physical sciences. For example, when you say that you have "much work to do" to prepare for an examination, or when you say that a new computer "works fine," you are using the word in ways quite differently from the way you would use the term in physics. We will see that a system *cannot* be said to *have* work. *Work is not a property of a system* in the same sense that mass is a property; rather work is done *on* or *to* the system.

Our intuitive understanding of energy also differs from its meaning in physics. Colloquially, energy is associated with vigor. In physics, *energy is a property of a system*. There are many different types (denominations) of energy as we will see: kinetic energy, potential energy, mechanical energy, internal energy, and others, each having a precise meaning (much like one, five, ten, and twenty dollar bills all are denominations of money). A system *has* energy, much like you have an amount of money (and care little about the denominations of the currency). In this chapter we explore the important, close, but distinct relationship between work and energy.

8.1 MOTIVATION FOR INTRODUCING THE CONCEPTS OF WORK AND ENERGY

Aren't Newton's laws sufficient for every purpose? Why bother to introduce the intertwined concepts of work and energy? Their *raison d'être* is twofold:

1. Work and energy are defined as *scalar* quantities. On a practical level, scalars are easier to use than vectors in solving problems. We use the concepts of work and energy to reformulate Newton's second law of motion into a new, encompassing scalar equation quite different from the vector equation $\vec{F}_{\text{total}} = m\vec{a}$. We call the new scalar equivalent of Newton's second law the *CWE theorem*. The initials stem from its traditional name: the Classical Work–Energy

theorem.[†] The CWE theorem provides a powerful additional tool to understand the natural world.
2. The introduction of the scalar concepts of work and energy leads to a *totally new physical principle*, distinct from Newton's second law of motion: conservation of energy. There is no such corresponding conservation law for forces. The CWE theorem is a special case of energy conservation; a general statement of energy conservation must await our study of thermodynamics in Chapter 13.

We make the following assumptions:

We restrict ourselves to systems that are represented as single particles with ignorable characteristics of size, internal structure, internal motions, and deformations; we also ignore thermal effects.

If you recall, these restrictions are synonymous with what we mean by the word particle.[‡][§] We also ignore internal complications such as engines or biochemical processes that necessarily lead to chemical and thermodynamic considerations involving heat transfer.

The forces on a physical system act both in *time* and *space*. The forces on a system may move with the system over *spatial distances* as the system travels from an initial to a final position. We explore one spatial aspect of forces in this chapter, leading us on a journey to the CWE theorem.[#] Forces also act during various *time intervals*, be they long or short; the temporal aspect of forces will be explored in Chapter 9.

Forces underlie changes in the velocity of a system, and so it is not surprising that the idea of the physical *work* done on a system is related to the concept of force and to changes in the position vector of the system. Since both are vector quantities, however, we invent a new scalar physical quantity called the *work done on a system by a force* using the mathematical concept of the scalar product.

In this chapter we explore how to calculate the work done on a system by any force and how the work done by various forces is related to changes in two kinds of energy: *potential energy* and *kinetic energy*. We then formulate the CWE theorem and use it to solve many problems that are much more difficult to attack using only Newton's second law of motion. Later in the chapter we explore the temporal aspects of the work done on a system, which gives rise to the concepts of *average power* and *instantaneous power*.

[†]In keeping with new trends in pedagogy, we prefer not to use the traditional term, because the theorem is not, strictly speaking, a universal statement about the concepts of work and energy. For our purposes, however, the name *CWE theorem* bridges the semantic gap between the traditional phrase and a necessary new term for this concept.

[‡]If you roll a cylinder down an inclined plane, the system (the cylinder) rotates. A car whose tires are rolling along a road also is a system with internal motions (the rotation of the tires and the innards of the engine). We do not consider such systems in this chapter except to the extent that such internal motions can be neglected and the system treated as a simple single particle.

[§]If you jump off the floor or push yourself off from a wall, you are a system that is changing its shape (i.e., is deformed). Analysis of such a system is not amenable to the techniques outlined in this chapter.

[#]Another spatial aspect of forces, a *vector* quantity known as torque, will be examined in Chapter 10.

(Chapter Opener) The Outermost House (Doubleday, Doran, New York, 1931), page 168.

8.2 THE WORK DONE BY ANY FORCE

Who first invented work . . .?

Charles Lamb (1775–1834)*

Moving day. You are loading a van with all your worldly possessions; there is a lot of real, physical work to be done. With a rope, pull a crate a distance *s* up a rough ramp, as in Figure 8.1. Let the crate be the system. There are several forces acting on it:

1. its weight $\vec{\mathbf{w}}$;
2. the normal force $\vec{\mathbf{N}}$ of the surface;
3. the force $\vec{\mathbf{T}}$ of the rope on the crate; and
4. the kinetic frictional force $\vec{\mathbf{f}}_k$.

> The differential **work** dW done by *any* force $\vec{\mathbf{F}}$ on a system is defined to be the scalar product of $\vec{\mathbf{F}}$ with the differential change in the position vector $d\vec{\mathbf{r}}$ of the point of application of the force on the system:
>
> $$dW \equiv \vec{\mathbf{F}} \cdot d\vec{\mathbf{r}} \qquad (8.1)$$
>
> Since the force $\vec{\mathbf{F}}$ acts *on* the system, we likewise say that the work done by $\vec{\mathbf{F}}$ is work done *on* the system.

Since the system is represented as a particle, the change in the position vector of the point of application of the force is equivalent to the change in the position vector of the system itself.[†]

Notice that the definition of the differential work says *nothing* about the time interval dt during which the position vector of the system (i.e., point of application of the force) changes by $d\vec{\mathbf{r}}$. Therefore the work of $\vec{\mathbf{F}}$ on the system involves the force in purely a *spatial* context.

As in our initial example, many forces may act on the system. Hence you can calculate

- the work done by *each individual force* on the system; or
- the work done by *all the forces* on the system—that is, by the *total force* $\vec{\mathbf{F}}_{\text{total}}$ on the system.

---

*Quoted from the *Examiner* in a letter by Charles Lamb to Bernard Barton, dated 11 September 1822; *The Letters of Charles Lamb*, edited by Alfred Ainger (Macmillan, London, 1900), volume 3, letter CCXV, page 64.

[†]This is not necessarily the case if the system undergoes rotation or is deformed; it is for this reason that we exclude such cases from consideration in this introduction to the concepts of work and energy.

First we see how to calculate the work done on the system by any individual force; then we can find the work done on the system by the total force $\vec{\mathbf{F}}_{\text{total}}$ and investigate how the total work done on the system affects the system itself.

As the system moves along a path from an initial position to a final position (see Figure 8.2), the work done by a particular force $\vec{\mathbf{F}}$ is the integral of its differential work along the specific path followed by the system. That is, the work done on the system by $\vec{\mathbf{F}}$ as the system moves in space along a path from an initial position to a final position is

$$W = \int_i^f \vec{\mathbf{F}} \cdot d\vec{\mathbf{r}} \qquad (8.2)$$

Notice in Equation 8.2 that the lower limit of the definite integral is the initial position and the upper limit is the final position. We say W is the work done on the system by the force $\vec{\mathbf{F}}$ over (or along) the path of travel from the initial to final position. The phrases *over the path* and *along the path* mean the same thing. The integral in Equation 8.2 is known as a **line integral**, although the path followed by the system need not be straight. The scalar product uses the value of $\vec{\mathbf{F}}$ at the location of each differential $d\vec{\mathbf{r}}$ along the path followed by the system as it moves from the initial to final positions. Since a force $\vec{\mathbf{F}}$ may change its magnitude and/or direction at various points along the system's path, the integration in Equation 8.2 must be performed carefully. In Figure 8.1, for example, the kinetic frictional force, the normal force, and the weight are constant forces on the crate along the path on the ramp, but we might change the direction and magnitude of our pull as it moves along the path.

> **PROBLEM-SOLVING TACTIC**
>
> **8.1** The differential change $d\vec{\mathbf{r}}$ in the position vector of the system is written as
>
> $$d\vec{\mathbf{r}} = dx\,\hat{\mathbf{i}} + dy\,\hat{\mathbf{j}} + dz\,\hat{\mathbf{k}} \qquad (8.3)$$
>
> **whether the particle is moving toward increasing or decreasing values of the coordinates.** The limits of integration account for the actual direction the particle moves in going from its initial position (the lower limit) to its final position (the upper limit). Examples 8.1 and 8.2 illustrate this point. If the motion is purely along the direction you call *x*, then $d\vec{\mathbf{r}} = dx\,\hat{\mathbf{i}}$,[‡] whether the particle is moving toward increasing or decreasing values of *x*.

[‡]This is because $dy = dz = 0$ m when the motion is only along the *x*-axis.

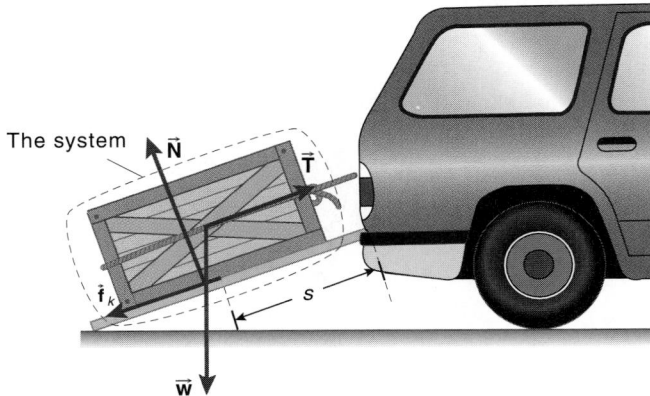

FIGURE 8.1 Forces on a crate being pulled up a ramp.

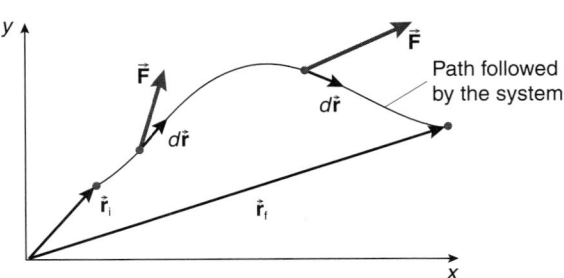

FIGURE 8.2 The work done by a force $\vec{\mathbf{F}}$ is the integral of $\vec{\mathbf{F}} \cdot d\vec{\mathbf{r}}$ along the path followed by the system.

There are several important points to note about the definition of work:

> *Do not* think of work in the sense of simply a force times a distance. The orientation of the force vector $\vec{F}$ relative to the differential change in the position vector $d\vec{r}$ is critical.

For example, the work done by the normal force of the surface on the crate in Figure 8.1 is *not* Ns. Think of the differential work as the *scalar product* of the force and the differential change in the position vector. In Figure 8.1 the normal force $\vec{N}$ is perpendicular to $d\vec{r}$ at every point along the entire path from the initial to final positions of the crate along the ramp; thus, the scalar product $\vec{N} \cdot d\vec{r}$ is zero at each and every point along the entire path and *zero* work is done by this force.

> Work is a *scalar* quantity.

This is one reason for introducing the concept, though not the most important one. It is convenient to have a scalar quantity, so that we can use it without all the baggage that usually accompanies the manipulation of vectors.

The dimensions for work are those of force multiplied by distance or N·m in the SI unit system.

> Since the concept of work is used prolifically, the SI units, N·m, *when associated with work or energy,** are called *joules* (J).

The unit is named after James Prescott Joule[†] (1818–1889), a 19th-century wealthy English brewer-turned-scientist who spent much time formulating the ideas of work and energy in physics, while enjoying the financial and liquid fruits of his trade.

Because of the scalar product in the definition, the differential work dW done by a force $\vec{F}$ on a system as it changes its position vector by $d\vec{r}$ may be

1. positive, if the angle between $\vec{F}$ and $d\vec{r}$ is acute (less than 90°);

---

*In Chapter 10 we introduce a new vector concept called torque that also has the SI units of N·m but *never* is expressed in joules.

[†]Although Joule pronounced his name to rhyme with owl, the joule energy unit is now almost universally pronounced to rhyme with tool. We won't object if you prefer it Joule's way.

James Prescott Joule

2. negative, if the angle between $\vec{F}$ and $d\vec{r}$ is obtuse (greater than 90° but less than or equal to 180°); or
3. zero (if the angle between $\vec{F}$ and $d\vec{r}$ is equal to 90°).

The work done by a force over a finite path also may be positive, negative, or zero.

QUESTION 1

> If you push and hold your textbook very hard against the rigid wall of your room for ten minutes, is any work done by the force you exert on the book? Explain your answer.

EXAMPLE 8.1

You drag your textbook 0.75 m across your level desk with a constant force of magnitude 6.0 N, oriented at 30° to the horizontal direction.

a. What is the work done by the force you exert on the book?
b. What is the work done by the gravitational force of the Earth on the book?
c. What is the work of the normal force of the surface on the book?

Solution

The book is the system. Introduce a coordinate system and draw a second law force diagram indicating the forces on the book, as in Figure 8.3.

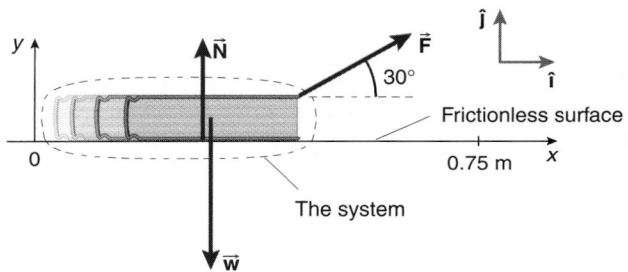

FIGURE 8.3

a. The constant force you exert on the book is

$$\vec{F} = (6.0 \text{ N})(\cos 30°)\hat{\imath} + (6.0 \text{ N})(\sin 30°)\hat{\jmath}$$
$$= (5.2 \text{ N})\hat{\imath} + (3.0 \text{ N})\hat{\jmath}$$

Since the book moves only along the *x*-axis, the differential change in the position vector of the book is written as

$$d\vec{r} = dx\,\hat{\imath}$$

The work done by this force then is

$$W = \int_{0 \text{ m}}^{0.75 \text{ m}} [(5.2 \text{ N})\,\hat{\imath} + (3.0 \text{ N})\,\hat{\jmath}] \cdot dx\,\hat{\imath}$$
$$= (5.2 \text{ N}) \int_{0 \text{ m}}^{0.75 \text{ m}} dx$$
$$= (5.2 \text{ N})(0.75 \text{ m} - 0 \text{ m})$$
$$= 3.9 \text{ J}$$

Notice that the component of the force perpendicular to the path followed by the book did no work. Only the component of the force along the direction of the path does work.

b. The weight $\vec{\mathbf{w}} = -mg\hat{\jmath}$ is perpendicular to $d\vec{\mathbf{r}} = dx\,\hat{\imath}$ at each and every point along the path. Hence

$$\vec{\mathbf{w}} \cdot d\vec{\mathbf{r}} = -mg\hat{\jmath} \cdot dx\,\hat{\imath}$$
$$= 0\ \text{J}$$

at each and every point along the path. The weight does zero work on the book along this path.

c. The normal force $\vec{\mathbf{N}} = N\hat{\jmath}$ also is perpendicular to $d\vec{\mathbf{r}} = dx\,\hat{\imath}$ at each and every point along the path. Therefore

$$\vec{\mathbf{N}} \cdot d\vec{\mathbf{r}} = N\hat{\jmath} \cdot dx\,\hat{\imath}$$
$$= 0\ \text{J}$$

at each and every point along the path. The normal force does zero work on the book along this path.

EXAMPLE 8.2

A 0.100 kg trinket is attached to a spring with spring constant 40.0 N/m and undergoes simple harmonic oscillation with an amplitude of 0.150 m before the glazed eyes of a dazed tourist. As usual, the equilibrium position is taken to be $x = 0$ m with $\hat{\imath}$ in the direction of stretching the spring. What is the work done by the force of the spring on the trinket as it moves from the equilibrium position to $x = -0.150$ m?

Solution

The trinket is the system. The work done is the integral of $\vec{\mathbf{F}}_{\text{spring}} \cdot d\vec{\mathbf{r}}$ from the initial to final positions. The force of the spring on the trinket is given by Equation 7.2:

$$\vec{\mathbf{F}}_{\text{spring}} = -kx\hat{\imath}$$

According to Problem-Solving Tactic 8.1, since the motion is purely along x, take $d\vec{\mathbf{r}} = dx\,\hat{\imath}$ and let the limits of integration take care of the direction the system is moving along the x-axis:

$$W = \int_{i}^{f} (-kx\hat{\imath}) \cdot dx\,\hat{\imath}$$
$$= -k \int_{0\ \text{m}}^{-0.150\ \text{m}} x\ dx$$

After integrating, you have

$$W = -k\frac{x^2}{2}\Bigg|_{0\ \text{m}}^{-0.150\ \text{m}}$$
$$= -\frac{40.0\ \text{N/m}}{2}[(-0.150\ \text{m})^2 - (0\ \text{m})^2]$$
$$= -0.450\ \text{J}$$

Notice that the force of the spring and the differential change in the position vector were in opposite directions over the entire path; this accounts for the negative result for this work.

8.3 THE WORK DONE BY A CONSTANT FORCE

The weight of a system of constant mass is a constant vector no matter how the system moves about, as long as it remains close to

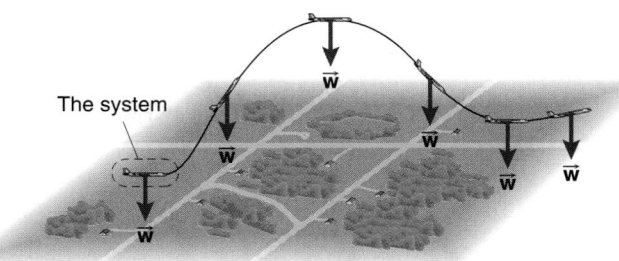

FIGURE 8.4 The weight is a constant vector as a system moves along any path near the surface of the Earth.

the surface of the Earth (see Figure 8.4). This is one of many situations that involve constant force vectors, and so it is useful to consider the special case of the work done on a system by a constant force as the system moves along a path that is *not* necessarily straight.

> A constant force vector has both a constant magnitude *and* a constant direction.

For example, as you move a crate along a curved path, as in Figure 8.5, the weight of the crate *is* a constant force along the curved path, because its magnitude and direction are the same at all points along the path. However, the normal force and the kinetic frictional force are *not* constant forces because their magnitudes and directions change along this path.

To calculate the work done on the system by a constant force, begin with the general definition of the work, Equation 8.2:

$$W = \int_{i}^{f} \vec{\mathbf{F}} \cdot d\vec{\mathbf{r}}$$

It is useful to write the constant force $\vec{\mathbf{F}}$ in terms of its constant Cartesian components:

$$\vec{\mathbf{F}} = F_x\hat{\imath} + F_y\hat{\jmath} + F_z\hat{k}$$

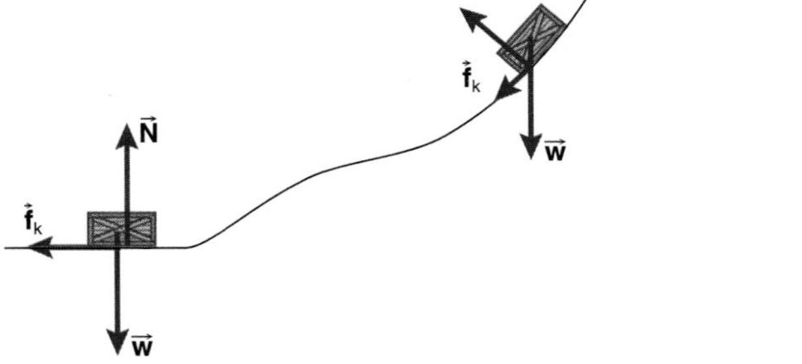

$$\vec{\mathbf{N}} = 0\ \text{N}$$
$$\vec{\mathbf{f}}_k = 0\ \text{N}$$

FIGURE 8.5 The weight of the crate is a constant force along this surface and path, but the normal force and the kinetic frictional force are not.

We also express the change in the position vector $d\vec{\mathbf{r}}$ in its most general Cartesian form as (Equation 8.3)

$$d\vec{\mathbf{r}} = dx\,\hat{\mathbf{i}} + dy\,\hat{\mathbf{j}} + dz\,\hat{\mathbf{k}}$$

Making these substitutions into Equation 8.2, we get

$$W = \int_i^f [F_x\hat{\mathbf{i}} + F_y\hat{\mathbf{j}} + F_z\hat{\mathbf{k}}] \cdot [dx\,\hat{\mathbf{i}} + dy\,\hat{\mathbf{j}} + dz\,\hat{\mathbf{k}}]$$

$$= \int_i^f (F_x\,dx + F_y\,dy + F_z\,dz) \qquad (8.4)$$

Expressed in Cartesian form, the initial and final position vectors of the system are

$$\vec{\mathbf{r}}_i = x_i\hat{\mathbf{i}} + y_i\hat{\mathbf{j}} + z_i\hat{\mathbf{k}}$$
$$\vec{\mathbf{r}}_f = x_f\hat{\mathbf{i}} + y_f\hat{\mathbf{j}} + z_f\hat{\mathbf{k}}$$

Substitute the appropriate limits for the integrations into Equation 8.4. The work done is

$$W = \int_{x_i}^{x_f} F_x\,dx + \int_{y_i}^{y_f} F_y\,dy + \int_{z_i}^{z_f} F_z\,dz$$

Since the force components of a constant force are themselves constant, we have

$$W = F_x\int_{x_i}^{x_f} dx + F_y\int_{y_i}^{y_f} dy + F_z\int_{z_i}^{z_f} dz$$

$$= F_x(x_f - x_i) + F_y(y_f - y_i) + F_z(z_f - z_i) \qquad (8.5)$$

Let $\Delta\vec{\mathbf{r}}$ be the *change* in the position vector of the system:

$$\Delta\vec{\mathbf{r}} \equiv \vec{\mathbf{r}}_f - \vec{\mathbf{r}}_i$$
$$= (x_f - x_i)\hat{\mathbf{i}} + (y_f - y_i)\hat{\mathbf{j}} + (z_f - z_i)\hat{\mathbf{k}}$$

in Cartesian form.

> The work done on the system by the constant force, Equation 8.5, is expressed succinctly as
>
> $$\boxed{W = \vec{\mathbf{F}} \cdot \Delta\vec{\mathbf{r}} \qquad \text{(constant force } \vec{\mathbf{F}})} \qquad (8.6)$$

PROBLEM-SOLVING TACTIC

8.2 The work done by a constant force, Equation 8.6, can be used even if the path between the initial and final positions is not straight. However, take care: the force must be a *constant force*, with both a fixed magnitude and direction over the *entire path* followed by the system between its initial and final positions.

QUESTION 2

You push Victor Hugo's massive novel *Les Misérables* at constant speed along a path on your level desk. Describe a path along which the kinetic frictional force on the book is a constant force. Describe a path along which the kinetic frictional force is not constant.

EXAMPLE 8.3

A force $\vec{\mathbf{F}} = (-30.0\,\text{N})\hat{\mathbf{i}} + (20.0\,\text{N})\hat{\mathbf{j}}$ acts on a 10.0 kg golf bag as it moves from $x = 10.00$ m, $y = 0$ m to $x = 5.00$ m, $y = 2.00$ m

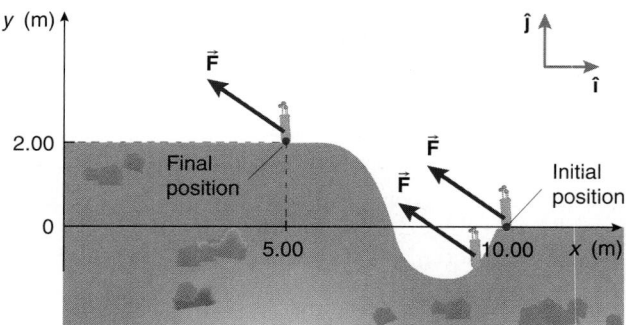

FIGURE 8.6

along the path indicated in Figure 8.6. Find the work done on the golf bag system by this force over this path.

Solution

Method 1

Notice that the force is a *constant vector*; it has the same magnitude and direction along the entire path followed by the system between the initial and final positions. Hence you can calculate the work by using Equation 8.6:

$$W = \vec{\mathbf{F}} \cdot \Delta\vec{\mathbf{r}} \qquad \text{(constant force } \vec{\mathbf{F}})$$

The initial and final position vectors of the system are

$$\vec{\mathbf{r}}_i = (10.00\,\text{m})\hat{\mathbf{i}}$$
$$\vec{\mathbf{r}}_f = (5.00\,\text{m})\hat{\mathbf{i}} + (2.00\,\text{m})\hat{\mathbf{j}}$$

Therefore the change in the position vector is

$$\Delta\vec{\mathbf{r}} = \vec{\mathbf{r}}_f - \vec{\mathbf{r}}_i$$
$$= [(5.00\,\text{m})\hat{\mathbf{i}} + (2.00\,\text{m})\hat{\mathbf{j}}] - (10.00\,\text{m})\hat{\mathbf{i}}$$
$$= (-5.00\,\text{m})\hat{\mathbf{i}} + (2.00\,\text{m})\hat{\mathbf{j}}$$

Use Equation 8.6 to find the work:

$$W = \vec{\mathbf{F}} \cdot \Delta\vec{\mathbf{r}}$$
$$= [(-30.0\,\text{N})\hat{\mathbf{i}} + (20.0\,\text{N})\hat{\mathbf{j}}] \cdot [(-5.00\,\text{m})\hat{\mathbf{i}} + (2.00\,\text{m})\hat{\mathbf{j}}]$$

Evaluating the various scalar products, you find

$$W = 150\,\text{J} + 40.0\,\text{J}$$
$$= 190\,\text{J}$$

Method 2

The definition of the work done on the system by any force over a path is given by Equation 8.2:

$$W = \int_i^f \vec{\mathbf{F}} \cdot d\vec{\mathbf{r}}$$

Notice that the force has an x-component directed toward decreasing values of x and the system is moving toward decreasing values of x. Thus the x-component of the force is directed parallel to the direction the system moves on the x-axis while traveling from the initial to the final position; you can anticipate that the work done by the x-component of the force will be positive. Since the y-coordinates of the system first decrease and then increase, it is perhaps not as clear what is the sign of the work done by the y-component of the force.

Since the system is moving in two dimensions, the differential change in the position vector $d\vec{r}$ is written as

$$d\vec{r} = dx\,\hat{\imath} + dy\,\hat{\jmath}$$

even though the particle is moving toward decreasing values of x and the value of y first decreases and then increases. The limits of integration account for all the changes in position. The work done by this force over this path is

$$W = \int_{i}^{f}[(-30.0\ \text{N})\,\hat{\imath} + (20.0\ \text{N})\,\hat{\jmath}] \cdot (dx\,\hat{\imath} + dy\,\hat{\jmath})$$

Evaluating the scalar products in the integrand, you obtain

$$W = (-30.0\ \text{N})\int_{10.00\ \text{m}}^{5.00\ \text{m}} dx + (20.0\ \text{N})\int_{0\ \text{m}}^{2.00\ \text{m}} dy$$

$$= (-30.0\ \text{N})\,x\,\Big|_{10.00\ \text{m}}^{5.00\ \text{m}} + (20.0\ \text{N})\,y\,\Big|_{0\ \text{m}}^{2.00\ \text{m}}$$

$$= (-30.0\ \text{N})(5.00\ \text{m} - 10.00\ \text{m})$$
$$+ (20.0\ \text{N})(2.00\ \text{m} - 0\ \text{m})$$

$$= 150\ \text{J} + 40.0\ \text{J}$$

$$= 190\ \text{J}$$

EXAMPLE 8.4

A satellite is in a circular orbit of radius r about the Earth. Calculate the work done by the gravitational force of the Earth on the satellite around one complete orbit.

Solution
The work done by the gravitational force around one complete orbital path is the integral of the scalar product of the gravitational force and the differential change in the position vector $d\vec{r}$ along the orbital path:

$$W = \int_{\substack{\text{one} \\ \text{orbital path}}} \vec{F}_{\text{grav}} \cdot d\vec{r}$$

Look at the integrand:

$$\vec{F}_{\text{grav}} \cdot d\vec{r}$$

As the satellite moves along its circular path, the gravitational force of the Earth on the satellite (which is directed radially toward the Earth) is perpendicular to $d\vec{r}$ at every point along the path, as shown in Figure 8.7. Thus the scalar product in the integrand is

$$\vec{F}_{\text{grav}} \cdot d\vec{r} = F_{\text{grav}}\,dr\cos 90°$$
$$= 0\ \text{J}$$

Thus the integrand is zero at every point along the path of the orbit. Therefore the work done by the gravitational force around the circular orbit is zero:

$$W = \int_{\substack{\text{one} \\ \text{orbital path}}} \vec{F}_{\text{grav}} \cdot d\vec{r}$$
$$= 0\ \text{J}$$

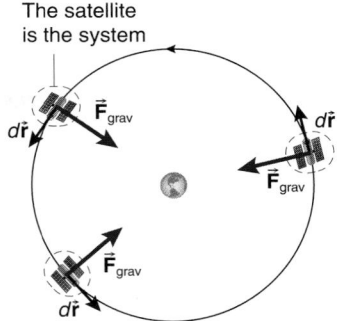

The satellite is the system

FIGURE 8.7

You might be tempted to use Equation 8.6 to solve this problem:

$$W = \vec{F} \cdot \Delta\vec{r} \qquad (\text{constant force } \vec{F})$$

However, while the *magnitude* of the force here is constant over the entire path, this force is *not a constant vector* since *its direction changes* along the path. Hence the criterion for using Equation 8.6 is *not* satisfied and it is incorrect to use it to calculate the work done on the system by $\vec{F}_{\text{grav}}$. Ironically, since we wanted to calculate the work done by $\vec{F}_{\text{grav}}$ around one complete orbital path, the change in the position vector $\Delta\vec{r}$ of the system here is *zero* because the initial and final position vectors are identical. With $\Delta\vec{r} = 0$ m, Equation 8.6 gives the correct answer, but the equation *cannot* be used legitimately, since $\vec{F}_{\text{grav}}$ is not a constant vector. Blindly using Equation 8.6 here for the work gives the right answer, but for the wrong reasons! This occasionally happens and illustrates that simply securing the right answer is not necessarily a good test of understanding.

QUESTION 3
Note that the work done by the gravitational force in Example 8.4 is *not* the magnitude of the force times the distance, $F_{\text{grav}}(2\pi r)$. Why?

8.4 THE WORK DONE BY THE TOTAL FORCE

Many systems have more than one force acting on them. The total force on a system is the vector sum of all the individual forces on it:

$$\vec{F}_{\text{total}} = \vec{F}_1 + \vec{F}_2 + \vec{F}_3 + \cdots$$

where $\vec{F}_1, \vec{F}_2, \vec{F}_3, \ldots$ are the individual forces on the system. The total work done on the system along some path is the work done by the total force:

$$W_{\text{total}} = \int_{i}^{f} \vec{F}_{\text{total}} \cdot d\vec{r}$$

Substitute the vector sum of the individual forces for $\vec{F}_{\text{total}}$:

$$W_{\text{total}} = \int_{i}^{f} (\vec{F}_1 + \vec{F}_2 + \vec{F}_3 + \cdots) \cdot d\vec{r}$$

$$= \int_{i}^{f} \vec{F}_1 \cdot d\vec{r} + \int_{i}^{f} \vec{F}_2 \cdot d\vec{r} + \int_{i}^{f} \vec{F}_3 \cdot d\vec{r} + \cdots$$

The integrals on the right-hand side of this expression are just the work done by the individual forces themselves. Thus

$$W_{\text{total}} = W_1 + W_2 + W_3 + \cdots$$

> The work done by the total force acting on the system is the algebraic, scalar sum of the work done by the individual forces.

The work done by an individual force may be positive, negative, or zero, depending on the specific situation, and so W_{total} may be positive, negative, or zero.

STRATEGIC EXAMPLE 8.5

A crate of mass m containing a new lab instrument is dragged by enthusiastic physics students a distance s along a straight, level corridor from an elevator to a physics lab. The students maintain a constant pull of magnitude T newtons applied to the crate by means of a rope inclined to the horizontal at angle θ. The velocity of the crate along the straight path is *not* constant. The coefficient of kinetic friction for the crate on the floor is μ_k. Calculate the work done by each force on the crate and the work done by the total force on the crate.

Solution

Consider the crate to be the system and represent it schematically as a particle. A second law force diagram (see Figure 8.8) includes

1. the gravitational force of the Earth on the crate—the weight $\vec{w}$ of the crate;
2. the normal force $\vec{N}$ of the surface on the crate;
3. the force of the rope $\vec{T}$ on the crate; and
4. the kinetic frictional force $\vec{f}_k$.

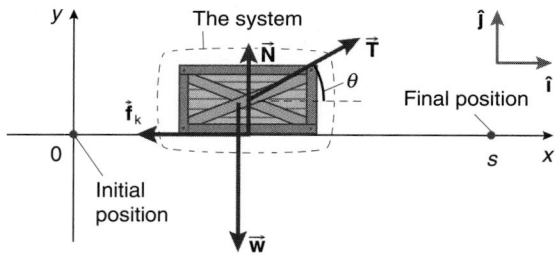

FIGURE 8.8

For convenience introduce a coordinate system with the origin at the initial position of the crate, so that $\vec{r}_i = 0$ m. The final position of the crate is $\vec{r}_f = s\hat{i}$. The change in the position vector of the crate thus is

$$\Delta\vec{r} = s\hat{i} - 0 \text{ m}$$
$$= s\hat{i}$$

The various forces on the crate are written in terms of their Cartesian components as

1. the weight: $\vec{w} = -mg\hat{j}$;
2. the normal force: $\vec{N} = N\hat{j}$; and
3. the force of the rope: $\vec{T} = (T \cos \theta)\hat{i} + (T \sin \theta)\hat{j}$.

Since the kinetic frictional force always is directed opposite to the velocity, this force is toward decreasing values of x:

4. the kinetic frictional force: $\vec{f}_k = -f_k\hat{i}$.

Each force is constant along the entire path, and so in each case you can use Equation 8.6 for the work done by a constant force:

$$W = \vec{F} \cdot \Delta\vec{r} \qquad \text{(constant force } \vec{F})$$

The work done by the normal force is

$$W_{\text{normal}} = N\hat{j} \cdot s\hat{i}$$
$$= 0 \text{ J}$$

The work done by the weight is

$$W_{\text{weight}} = -mg\hat{j} \cdot s\hat{i}$$
$$= 0 \text{ J}$$

The work done by the normal force $\vec{N}$ and the work done by the weight $\vec{w}$ are both zero because the horizontal path followed by the system is perpendicular to both of these forces at each and every point along the path.

The magnitude of the kinetic frictional force $\vec{f}_k$ is related to the magnitude of the normal force by the empirical equation

$$f_k = \mu_k N$$

The magnitude of the normal force is found by applying Newton's second law to the vertical direction, realizing that the system is not accelerating in this direction:

$$F_{y\,\text{total}} = ma_y$$
$$-mg + N + T \sin \theta = m(0 \text{ m/s}^2)$$

So

$$N = mg - T \sin \theta$$

Thus the kinetic frictional force is of magnitude

$$f_k = \mu_k(mg - T \sin \theta)$$

The work done by the kinetic frictional force is

$$W_{\text{frictn}} = (-f_k\hat{i}) \cdot s\hat{i}$$
$$= -f_k s$$
$$= -\mu_k(mg - T \sin \theta)s$$

Notice that the work done by the kinetic frictional force is *negative*, indicating that the force and the change in the position vector were in opposite directions.

The work done by the force of the rope on the system is

$$W_{\text{rope}} = [(T \cos \theta)\hat{i} + (T \sin \theta)\hat{j}] \cdot s\hat{i}$$
$$= Ts \cos \theta$$

Since the velocity of the crate is not constant, the total force on it is not zero. There are two ways you can find the work done by the total force.

Method 1

The work done by the *total* force is the *algebraic, scalar sum* of the work done by all the individual forces. Thus

$$W_{\text{total}} = W_{\text{weight}} + W_{\text{normal}} + W_{\text{frictn}} + W_{\text{rope}}$$

or

$$W_{total} = 0 \text{ J} + 0 \text{ J} + [-\mu_k(mg - T \sin \theta)s] + Ts \cos \theta$$
$$= Ts(\cos \theta + \mu_k \sin \theta) - \mu_k mgs$$

Method 2

The work done by the *total* force is

$$W_{total} = \int_i^f \vec{F}_{total} \cdot d\vec{r}$$

The total force is the vector sum of all the forces on the crate:

$$\vec{F}_{total} = \vec{w} + \vec{N} + \vec{T} + \vec{f}_k$$
$$= -mg\hat{j} + N\hat{j} + [(T \cos \theta)\hat{i} + (T \sin \theta)\hat{j}] + (-f_k\hat{i})$$
$$= (T \cos \theta - f_k)\hat{i} + (N + T \sin \theta - mg)\hat{j}$$

The total force is a constant vector along this path, and so you can use Equation 8.6 to find its work:

$$W_{total} = \vec{F}_{total} \cdot \Delta \vec{r}$$
$$= [(T \cos \theta - f_k)\hat{i} + (N + T \sin \theta - mg)\hat{j}] \cdot s\hat{i}$$
$$= (T \cos \theta - f_k)s$$

Since $f_k = \mu_k(mg - T \sin \theta)$, the total work is

$$W_{total} = [T \cos \theta - \mu_k(mg - T \sin \theta)]s$$
$$= Ts(\cos \theta + \mu_k \sin \theta) - \mu_k mgs$$

Notice that the y-component of the total force is everywhere perpendicular to the path and, thus, does no work; only the component of the total force along the path does nonzero work.

EXAMPLE 8.6 _____

A physics professor moves a mass m attached to a spring, stretching the spring very slowly along the entire path as m moves through a distance s along a frictionless surface from its equilibrium position (at $x = 0$ m), as shown in Figure 8.9. At any position the acceleration of the mass is essentially zero. Calculate the work done on the system of mass m:

a. by the force of the professor on m; and
b. by the force of the spring on m.
c. What is the work done by all the forces on the system?

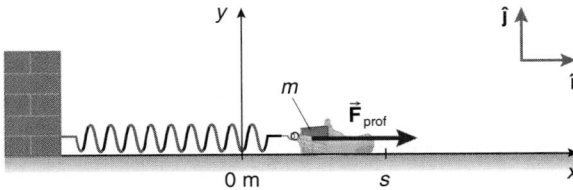

FIGURE 8.9

Solution

Consider the mass m to be the system. Although the spring is deformed (stretched), the spring is not the system. Use the

coordinate system indicated in Figures 8.9 and 8.10. The forces on the system are as follows:

1. The force of the spring on m is, according to Equation 7.2,

$$\vec{F}_{spring} = -kx\hat{i}$$

2. Likewise, from Equation 7.1, the professor must exert a force

$$\vec{F}_{prof} = kx\hat{i}$$

on the system at any given position x.
3. The weight of the system, given by

$$\vec{w} = -mg\hat{j}$$

4. The normal force of the surface on the system, given by

$$\vec{N} = N\hat{j}$$

The force of the professor and that of the spring on the system are *not* constant forces over the path; you cannot use Equation 8.6 for a constant force but must calculate their work from the general expression for the work of any force, Equation 8.2.

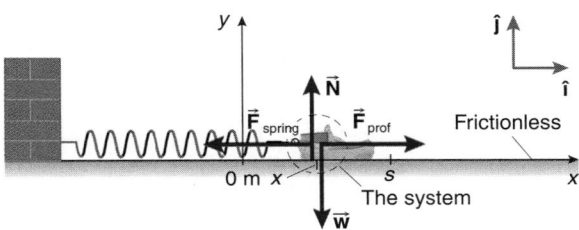

FIGURE 8.10

Since the motion is one-dimensional, the differential change in the position vector $d\vec{r}$ is, from Equation 8.3,

$$d\vec{r} = dx\,\hat{i}$$

a. Assume that the spring remains within its elastic limit. The work done by the force of the professor on the mass is then

$$W_{prof} = \int_i^f \vec{F}_{prof} \cdot d\vec{r}$$
$$= \int_{0\text{ m}}^s kx\,\hat{i} \cdot dx\,\hat{i}$$
$$= \int_{0\text{ m}}^s kx\,dx$$
$$= \left. \frac{kx^2}{2} \right|_{0\text{ m}}^s$$
$$= \frac{ks^2}{2}$$

b. The force of the spring on m also is not a constant force. Hence the work done by this force on the system must be found using Equation 8.2, the general expression for the work of any force. The work is

$$W_{spring} = \int_i^f \vec{F}_{spring} \cdot d\vec{r}$$

$$= \int_{0\ m}^s (-kx\hat{\imath}) \cdot dx\,\hat{\imath}$$

$$= -k \int_{0\ m}^s x\ dx$$

$$= -k\frac{x^2}{2} \Big|_{0\ m}^s$$

$$= -\frac{ks^2}{2}$$

There are two other forces acting on the mass: the normal force $\vec{N}$ of the surface and the weight $\vec{w}$. These are *constant forces*. The work done by both of these forces is *zero* because both forces are perpendicular to the path followed by the system at each and every point along the path.

c. There are two ways to find the total work done by all the forces on the system:

Method 1

The work done by the total force is the *algebraic, scalar sum* of the work done on the system by the individual forces. The total work done by all the forces on the system is

$$W_{total} = W_{spring} + W_{prof} + W_{weight} + W_{normal}$$

$$= -\frac{ks^2}{2} + \frac{ks^2}{2} + 0\ J + 0\ J$$

$$= 0\ J$$

Method 2

Use Equation 8.2,

$$W_{total} = \int_i^f \vec{F}_{total} \cdot d\vec{r}$$

and note that the *total force* on the system is zero at all points along the path (because the system is not accelerating). Therefore the total work must be zero.

Does $W_{total} = 0\ J$ mean that no work is done by the professor pulling on the mass? No, that work is equal to the work done by the force of the professor on the system, $W_{prof} = ks^2/2$.

8.5 GEOMETRIC INTERPRETATION OF THE WORK DONE BY A FORCE

There is a geometric way to visualize the work done by a force on a system, which ties in well with your growing knowledge of calculus, where integrals can be viewed as areas under a curve. If a system moves from x_i to x_f along the x-coordinate axis and

$$\vec{F} = F_x\hat{\imath} + F_y\hat{\jmath} + F_z\hat{k}$$

is one of the forces on the system, the work done on the system by $\vec{F}$ is, according to the definition of work, Equation 8.2,

$$W = \int_i^f \vec{F} \cdot d\vec{r}$$

$$= \int_i^f (F_x\hat{\imath} + F_y\hat{\jmath} + F_z\hat{k}) \cdot dx\,\hat{\imath}$$

After evaluating the scalar products, we find

$$W = \int_{x_i}^{x_f} F_x\ dx$$

Figure 8.11 is a schematic graph of the x-component of the force as a function of x.

The work done on the system by the force component F_x as the system moves from x_i to x_f is the area under the curve between x_i and x_f.

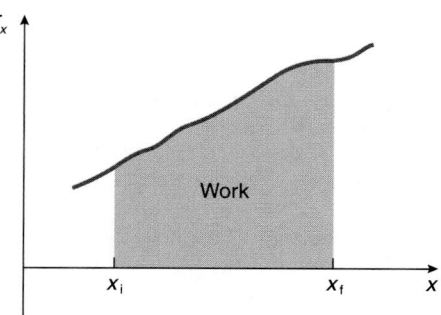

FIGURE 8.11   The work done by F_x is the area under the curve of F_x versus x between the initial and final x-coordinate positions.

Such areas may be positive or negative. Whether an area is positive or negative depends on two things: the direction the system is moving along the coordinate axis and whether the force component is positive or negative, as shown in Figure 8.12.

If the force

$$\vec{F} = F_x\hat{\imath} + F_y\hat{\jmath} + F_z\hat{k}$$

acts on a system while the system moves from position vector

$$\vec{r}_i = x_i\hat{\imath} + y_i\hat{\jmath} + z_i\hat{k}$$

to position vector

$$\vec{r}_f = x_f\hat{\imath} + y_f\hat{\jmath} + z_f\hat{k}$$

the work done by this force is

$$W = \int_i^f (F_x\hat{\imath} + F_y\hat{\jmath} + F_z\hat{k}) \cdot (dx\,\hat{\imath} + dy\,\hat{\jmath} + dz\,\hat{k})$$

$$= \int_{x_i}^{x_f} F_x\ dx + \int_{y_i}^{y_f} F_y\ dy + \int_{z_i}^{z_f} F_z\ dz$$

Each term is the area under the curve of the graph of that force component versus the corresponding coordinate.

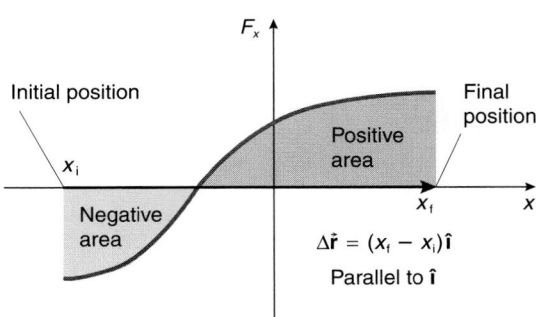 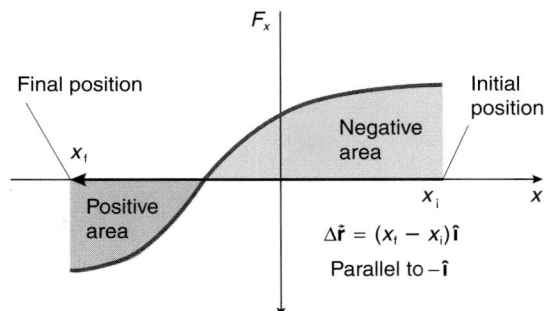

FIGURE 8.12 Positive or negative work (area) depends on both the sign of the force component and the direction of $\Delta \vec{r}$.

EXAMPLE 8.7

A force $\vec{F} = (3.00 \text{ N})\hat{\imath}$ acts on a system that moves along the x-axis from position $x_i = 3.00$ m to $x_f = 5.00$ m.

a. Graph F_x versus x.
b. Use the graph to find the work done by this force as the system moves between the two coordinates.

Solution

a. The x-component of the force is $F_x = 3.00$ N and is constant, and so a plot of F_x versus x is a horizontal line; see Figure 8.13.
b. The work done by F_x is the area under the curve between the two coordinates. This is the shaded rectangular area in Figure 8.13, which is

$$W = (3.00 \text{ N})(2.00 \text{ m})$$
$$= 6.00 \text{ J}$$

The area under the curve is positive since the system is moving parallel to $+\hat{\imath}$ and the force component is positive. We could anticipate a positive result since the force is in the same direction as the change in the position vector of the system.

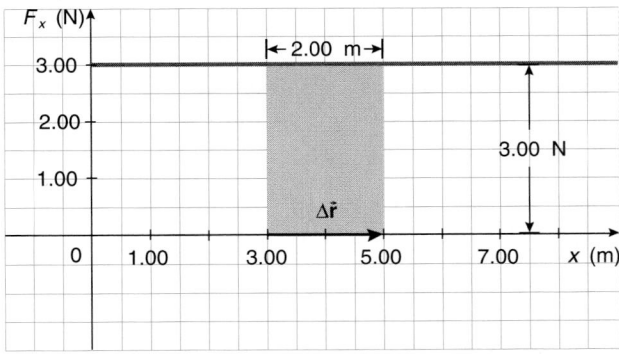

FIGURE 8.13

EXAMPLE 8.8

A force whose only component is F_x acts on a system; a graph of F_x versus x is shown in Figure 8.14. Use the graph to find the work done on the system by the force as the system moves along the x-axis from $x = -2.0$ m to $x = +10.0$ m.

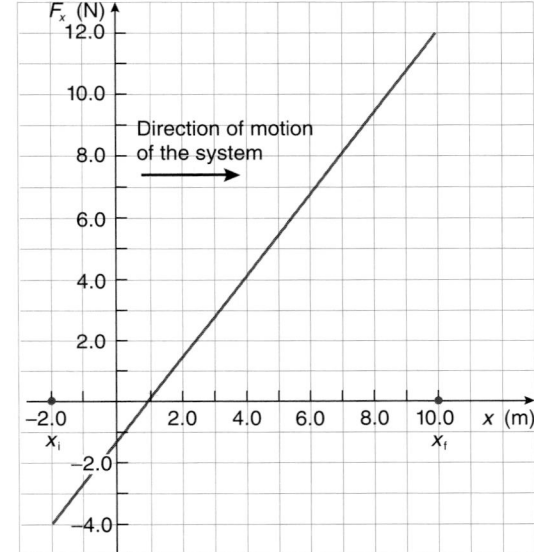

FIGURE 8.14

Solution

The work done is the area under the graph between $x_i = -2.0$ m and $x_f = +10.0$ m. The area under the curve between -2.0 m and 1.0 m is *negative*, because the system is moving parallel to $\hat{\imath}$ and the force component for these values of x is negative. Since the region is triangularly shaped, the area is

$$\text{area} = \frac{1}{2} (\text{base})(\text{height})$$

$$= \frac{1}{2} (3.0 \text{ m})(-4.0 \text{ N})$$

$$= -6.0 \text{ J}$$

The area between $+1.0$ m and $+10.0$ m is positive, because the system is moving parallel to $\hat{\imath}$ and the force component is positive for these values of x. This region also is triangularly shaped, and so the area is

$$\text{area} = \frac{1}{2} (9.0 \text{ m}) (12.0 \text{ N})$$

$$= 54 \text{ J}$$

The total area between $x_i = -2.0$ m and $x_f = +10.0$ m is the algebraic sum of these areas. Therefore the work W done by this force is

$$W = -6.0 \text{ J} + 54 \text{ J}$$
$$= 48 \text{ J}$$

EXAMPLE 8.9

A force has only an *x*-component, and a graph of F_x versus x is shown in Figure 8.15; the curves are arcs of circles. Use the graph to find the work done by this force on a system as it moves along the *x*-axis from $x_i = 10.00$ m to $x_f = -5.00$ m.

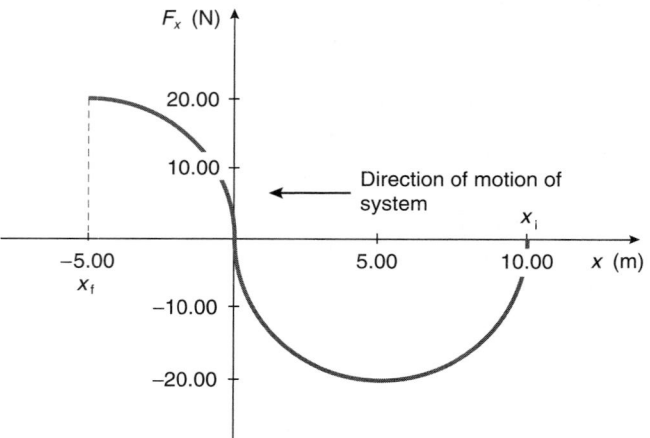

FIGURE 8.15

Solution

The work done by this force is the area under the curve between $x_i = 10.00$ m and $x_f = -5.00$ m. The semicircular area under the curve between $x = 10.00$ m and $x = 0$ m is positive, because the system is moving parallel to $-\hat{\imath}$ and the force component also is negative for these values of *x*. This semicircular area is equal to

$$\text{area} = \frac{1}{2}\pi(-20.0 \text{ N})(-5.00 \text{ m})$$
$$= 157 \text{ J}$$

The area under the curve between $x = 0$ m and $x = -5.00$ m is negative, because the system is moving parallel to $-\hat{\imath}$ but the force component is positive for these values of *x*. This area is equal to

$$\text{area} = \frac{1}{4}\pi(20.0 \text{ N})(-5.00 \text{ m})$$
$$= -78.5 \text{ J}$$

The total work is the algebraic sum of the areas:

$$W = -78.5 \text{ J} + 157 \text{ J}$$
$$= 78 \text{ J}$$

8.6 CONSERVATIVE, NONCONSERVATIVE, AND ZERO-WORK FORCES

Forces come in three work-related "flavors": conservative forces, nonconservative forces, and zero-work forces.

> A force is called a **conservative force**
> - if the work done by the force on any system as it moves between two points is *independent of the path* followed by the system between the two points; or equivalently,
> - if the work done by the force on any system *around any closed path is zero*. A closed path (round-trip path) is one that begins and ends at the same point in space.

The work done by a force around a *closed path* is written mathematically as

$$\oint \vec{F} \cdot d\vec{r}$$

where the elegant little circle through the integral sign is an indication that the integration is done around a *closed path*. If

$$\oint \vec{F} \cdot d\vec{r} = 0 \text{ J} \qquad (8.7)$$

for every closed path, then the force $\vec{F}$ is conservative.

You will see the equivalence of these definitions as we consider several examples of conservative forces in the following section. We shall see that the work done by any conservative force on a system as it moves between two points depends only on the locations of the initial and final positions of the system.

> If the work done by a force around a closed path is *not zero*, or equivalently, if the work done by a force as a system moves between two points depends on the particular path taken between the two points, then the force is called a **nonconservative force.**[*]

> Forces that *always do zero work* are called **zero-work forces.**

QUESTION 4

A simple pendulum swings back and forth once during its oscillation period. During one period, what work is done on the mass system attached to the string by the force of the string on the mass? Can you conclude that this force is conservative? Describe a physical situation in which the force of a string on a mass does nonzero work around a closed path.

---

[*]There are no corresponding "liberal forces" in the physics lexicon!

8.7 EXAMPLES OF CONSERVATIVE, NONCONSERVATIVE, AND ZERO-WORK FORCES

In this section we determine whether some of the common forces we considered in Chapters 5–7 are conservative, nonconservative, or zero-work forces.

The Local Gravitational Force on a Mass near the Surface of the Earth

The local gravitational force of the Earth on a system of mass m is its weight, of magnitude mg, where g is the magnitude of the gravitational field at the surface of the Earth (the magnitude of the acceleration due to gravity near the surface: $g = 9.81$ m/s^2). This force is a constant vector. To see if this force is conservative, calculate the work done by the force on a system as it moves along an arbitrary path between two points A and B, as shown in Figure 8.16.

Set up a Cartesian coordinate system with $\hat{\jmath}$ directed up, as shown in Figure 8.16. The gravitational force on m is

$$\vec{F} = -mg\hat{\jmath}$$

Other forces must also be involved in moving the particle. However, we focus on the gravitational force and the work it alone does on the system as it moves between the two points. Since the gravitational force is a *constant force* near the surface of the Earth, you can calculate the work done by this force using Equation 8.6 for the work done by a constant force:

$$W = \vec{F} \cdot \Delta\vec{r} \qquad \text{(constant force } \vec{F})$$

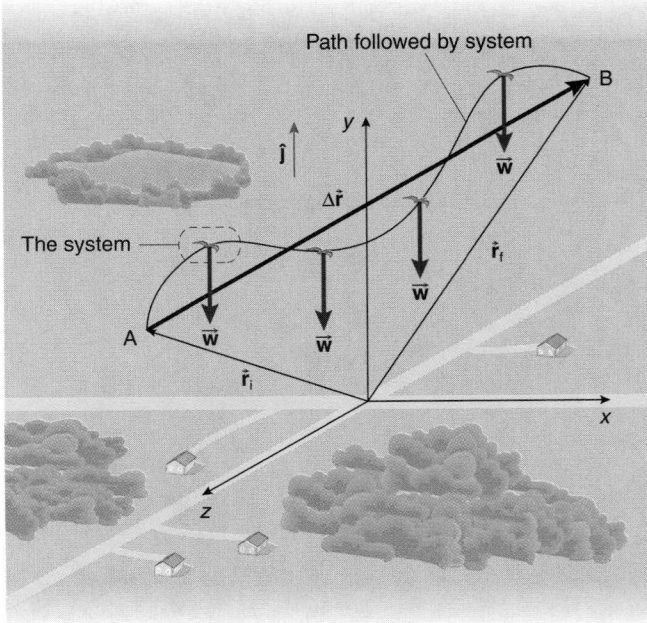

FIGURE 8.16 The weight is a constant vector along any path near the surface of the Earth.

where

$$\begin{aligned}
\Delta\vec{r} &= \vec{r}_f - \vec{r}_i \\
&= (x_f\hat{\imath} + y_f\hat{\jmath} + z_f\hat{k}) - (x_i\hat{\imath} + y_i\hat{\jmath} + z_i\hat{k}) \\
&= (x_f - x_i)\hat{\imath} + (y_f - y_i)\hat{\jmath} + (z_f - z_i)\hat{k}
\end{aligned}$$

Thus the work done by this force is

$$\begin{aligned}
W_{local} &= (-mg\hat{\jmath}) \cdot [(x_f - x_i)\hat{\imath} + (y_f - y_i)\hat{\jmath} + (z_f - z_i)\hat{k}] \\
&= -mg(y_f - y_i) \qquad (\hat{\jmath} \text{ directed up}) \qquad (8.8)
\end{aligned}$$

Moreover, since the gravitational force is a constant force along *any* path between the initial and final positions, the result for W is independent of the path.

> Hence the local gravitational force is a conservative force.

There are several things to note about Equation 8.8:

a. The work is *negative* if the system *increases its elevation*—that is, if $y_f > y_i$. The negative answer can be anticipated since the gravitational force is *antiparallel* to the vertical change in the position vector over any path along which the system increases its elevation. The work is *positive* if the system *decreases its elevation* along a path—that is, if $y_f < y_i$; in this case, the gravitational force is *parallel* to the vertical change in the position vector. For any calculation of work, try to anticipate whether the result will be positive or negative on the basis of the relative directions of the force and the change in the position vector.

b. The work done by the local gravitational force depends *only* on the difference between the final and initial vertical positions of the particle; this will be important later (in Sections 8.8 and 8.9). The work done by the local gravitational force does *not* depend on differences in the horizontal coordinates of the system. The work done is *independent of the path* between the two points.

c. Notice that if the system is taken on a round trip around *any closed path*, the final and initial positions of the system are the same. In particular, $y_f = y_i$ and the work done by the local gravitational force is *zero*. Hence the work done by the local gravitational force around *any closed path* is zero. Thus we see explicitly that the two ways we defined a conservative force at the beginning of Section 8.6 here are equivalent.

Our results here can be generalized.

> Any force that is a constant vector over an arbitrary *closed path* is a conservative force.

The General Form for the Gravitational Force

Unlike the constant gravitational force near the surface of the Earth, the general form for the gravitational force is *not a constant force*. Thus we have to see if this force is conservative by calculating the work done as a system of mass m moves between two points using Equation 8.2 rather than Equation 8.6. If the work is independent of the path, the force is conservative.

Let's calculate the work done by the gravitational force on a system of mass m when it changes its position from point A (a distance r_i from mass M) to point B (a distance r_f from mass M), as shown in Figure 8.17. With an origin located at mass M, the gravitational force of M on m is, from Equation 6.6,

$$\vec{F}_{grav} = -\frac{GMm}{r^2}\,\hat{r}$$

where $\hat{r}$ is a unit vector pointing radially away from M, as shown in Figure 8.17.

How shall we get the system from A to B? Take the funky crooked path shown in Figure 8.18. Each segment of the path either is the arc of a circle centered on M or is along the radial direction itself.

The work done along each circular arc segment is zero, because the gravitational force is perpendicular to all the circular arc path segments $d\vec{r}$ at every point along them. So the only work done on the system by the gravitational force occurs when the system moves along the radial segments. We can accumulate all the contributions along the radial segments into a single integral since

$$\int_{r_i}^{r_1}\vec{F}\cdot d\vec{r} + \int_{r_1}^{r_2}\vec{F}\cdot d\vec{r} + \int_{r_2}^{r_3}\vec{F}\cdot d\vec{r} + \cdots + \int_{r_n}^{r_f}\vec{F}\cdot d\vec{r}$$

$$= \int_{r_i}^{r_f}\vec{F}\cdot d\vec{r}$$

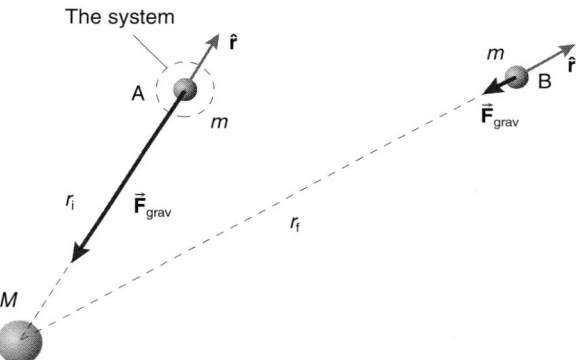

FIGURE 8.17 **The gravitational force varies in magnitude and direction as you take m from A to B.**

Thus the work done by the gravitational force in moving the system from A to B is

$$W = \int_{r_i}^{r_f}\left(-\frac{GmM}{r^2}\,\hat{r}\right)\cdot dr\,\hat{r}$$

In accordance with Problem-Solving Tactic 8.1, notice that we write $d\vec{r}$, the change in the position vector, as $dr\,\hat{r}$, because the limits of integration account for whether we are moving inward or outward in the radial direction.

Since the scalar product of a unit vector with itself is 1 ($\hat{r}\cdot\hat{r} = 1$), the work done is

$$W = -GmM\int_{r_i}^{r_f}\frac{dr}{r^2}$$

$$= -GmM\left(-\frac{1}{r}\right)\Big|_{r_i}^{r_f}$$

$$= -GmM\left(-\frac{1}{r_f} + \frac{1}{r_i}\right) \tag{8.9}$$

The work done by the gravitational force depends only on the initial and final radial coordinate positions of the system of mass m relative to M. Note also that we could have chosen *any* other similarly segmented path to get from A to B with the same result.

A smooth path between the initial and final positions (see Figure 8.19) may be broken up into billions and billions of small radial and circular arc segments. Thus the argument made for the segmented path also applies to such smooth paths.* Therefore the work done by the gravitational force on the system as it moves from A to B is *independent of the path*.

> We can conclude that the general form for the gravitational force is a conservative force.

The work of the gravitational force depends only on the relative scalar radial position coordinates of A and B (and the constants

---

*This procedure is the standard limiting process in calculus.

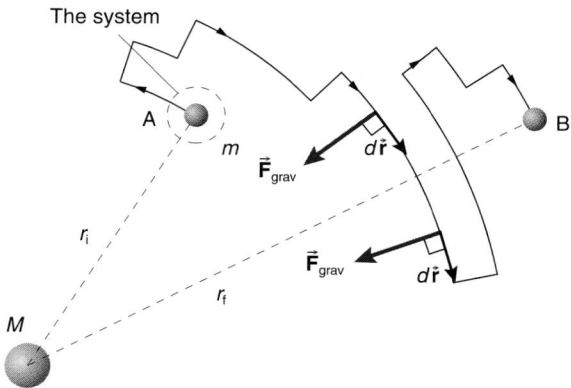

FIGURE 8.18 **A path from A to B consisting of radial and circular segments.**

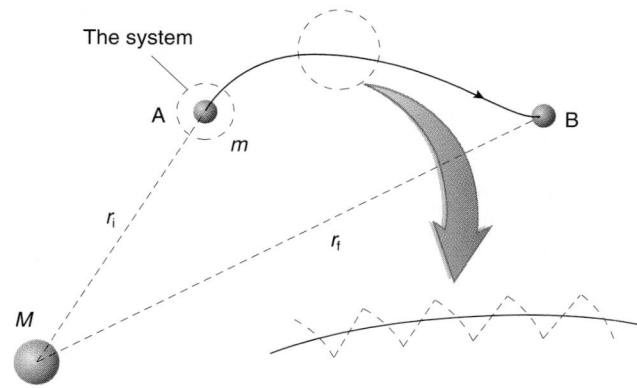

FIGURE 8.19 **A smooth path can be broken up into many small radial and circular segments.**

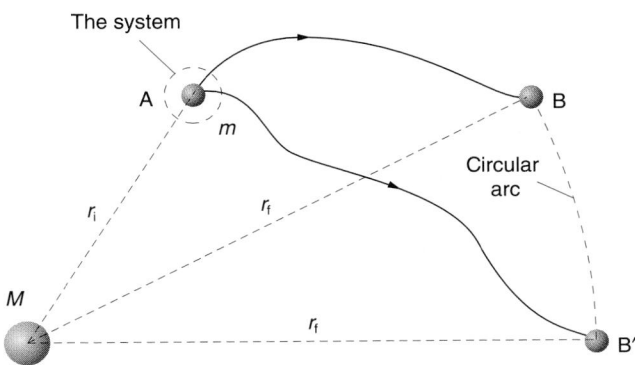

FIGURE 8.20 The work done by the gravitational force from A to B is the same as from A to B′ if B′ is the same distance from M as B.

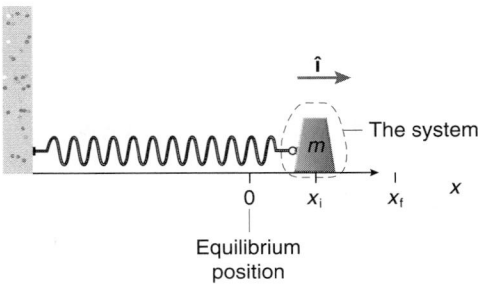

FIGURE 8.21 A spring is stretched.

G, M, and m). The work done by the gravitational force on m in going from A to B′ in Figure 8.20 is the same as the work done in getting m from A to B because B′ has the same scalar radial coordinate as B.

If $r_f > r_i$, the work done by the gravitational force is *negative*. In this case the gravitational force is in the direction opposite to the outward radial change in the position vector of the system. On the other hand, if $r_f < r_i$, then the gravitational work is *positive*. In this case the gravitational force is in the same direction as the inward radial change in the position vector of the system.

A Central Force

Recall from Section 6.10 that a central force is a force that always is directed along the line connecting two particles. With some thought, the arguments used in the preceding subsection to show that the general form for the gravitational force is conservative can be used for *any central force*. Nonzero work is done by a central force only when the system changes its radial distance with respect to the source of the force; no work is done in moving perpendicular to the radial direction. Likewise, the work done by any central force depends only on the radial position coordinates of the initial and final points of the path.

> Thus any central force is a conservative force.

Hooke's Force Law

We show here that Hooke's force law (see Chapter 7), important in simple harmonic oscillation, is conservative. In Figure 8.21, a system of mass m, attached to a spring characterized by spring constant k, is stretched from position x_i to position x_f, with $x = 0$ m corresponding to the equilibrium position of m on the spring (as usual).

We want to calculate the work done on the system by the force of the spring. In fact we already made such a calculation in Example 8.2. With the usual coordinate system for Hooke's law, the force of the spring on m is (from Equation 7.2)

$$\vec{F}_{spring} = -kx\hat{\imath}$$

This is *not* a constant force. Therefore to find the work we cannot use Equation 8.6 for the work of a constant force; we must use the general expression for the work of any force given by Equation 8.2:

$$W = \int_i^f \vec{F} \cdot d\vec{r}$$

Since the path is one-dimensional and along the x-axis, $d\vec{r}$ is

$$d\vec{r} = dx\,\hat{\imath}$$

because, as always, the limits of integration account for the specific direction the system moves along the axis. Thus the work done by the force of the spring on m is

$$
\begin{aligned}
W &= \int_{x_i}^{x_f}(-kx\hat{\imath}) \cdot (dx\,\hat{\imath}) \\
&= -k \int_{x_i}^{x_f} x\,dx \\
&= -k\frac{x^2}{2}\bigg|_{x_i}^{x_f} \\
&= -k\frac{x_f^2}{2} + k\frac{x_i^2}{2}
\end{aligned}
\tag{8.10}
$$

Notice that the result depends only on the initial and final positions of the system attached to the spring. In particular, the spring does negative work on the system if $|x_f| > |x_i|$—that is, when the system is moving away from the equilibrium position, for then the force of the spring on the system and the change in its position vector are antiparallel. On the other hand, the work done by the spring on the system is positive if $|x_f| < |x_i|$—that is, when the force of the spring on the system is parallel to its change in position vector.

We also can calculate the work done by the force of the spring on the mass around a closed path by setting $x_f = x_i$ in Equation 8.10. The result is that the work done by the force of the spring on m around a closed path is zero.

> Hence Hooke's force law is a conservative force.

The Kinetic Frictional Force

Though many forces are conservative, not all fit that criterion. A good example of a nonconservative force is the kinetic frictional force. Let's examine why this is the case.

To calculate the work done by the kinetic frictional force on a system of mass m around a closed path, move m along a straight-line distance s from point A to point B and then back to point A, as shown in Figure 8.22.

While other forces are involved in the motion, we are interested only in the work done by the kinetic frictional force. We use the specific form for the kinetic frictional force that is independent of the speed of the system; however, the nonconservative nature of the force does not depend on this choice for the specific form for the force.

Since the kinetic frictional force $\vec{f}_k$ is constant along the path from A to B, the work done by it as the system moves from A to B is found by using Equation 8.6 for the work of a constant force:

$$W = \vec{f}_k \cdot \Delta\vec{r}$$

As the system moves from A to B, we anticipate that the kinetic frictional work will be *negative*, because the force is directed opposite to the change in the position vector of the system along the path from A to B. Use the Cartesian coordinate system indicated in Figure 8.22; the kinetic frictional force along the path from A to B then is

$$\vec{f}_k = -f_k\hat{\imath}$$

where f_k is the magnitude of the kinetic frictional force. The change in the position vector of the system is

$$\Delta\vec{r} = s\hat{\imath}$$

The work done by the kinetic frictional force in going from A to B is

$$W = (-f_k\hat{\imath}) \cdot s\hat{\imath}$$
$$= -f_k s$$

Now calculate the work W' done by the kinetic frictional force as the mass is moved from B back to A, thus completing a closed path. For this case, we *also* anticipate a negative result for the frictional work, because the kinetic frictional force once again is directed opposite to the change in the position vector of the system along the path from B back to A. In moving from B back to A, the kinetic frictional force is constant along this path and is

$$\vec{f}_k = f_k\hat{\imath}$$

since it is directed parallel to $\hat{\imath}$ as we move from B back to A. The change in the position vector of the system now is

$$\Delta\vec{r} = -s\hat{\imath}$$

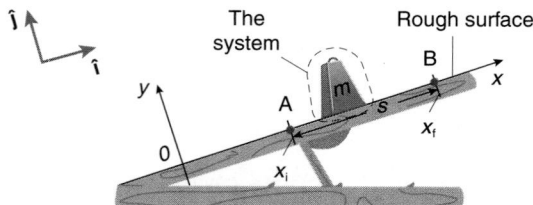

FIGURE 8.22 Move a mass from A to B and back to A.

So the work done by the kinetic frictional force from B back to A along this path is

$$W' = (f_k\hat{\imath}) \cdot (-s\hat{\imath})$$
$$= -f_k s$$

The *total work* done by the kinetic frictional force around the *closed path* from A to B and back to A again then is the sum of the work along each segment:

$$W_{\text{clsd path}} = W + W'$$
$$= -f_k s - f_k s$$
$$= -2f_k s$$

The work done around the closed path is *not zero*.

> Although we specifically considered only the speed-independent kinetic frictional force, *all* kinetic frictional forces are nonconservative.

If you are tempted to calculate the work done by the kinetic frictional force around a closed path using Equation 8.6 for the work done by a constant force, keep the following in mind. While the frictional force has a constant *magnitude* over the entire closed path, the force does *not* have a constant direction, and so it is *not* a constant vector over the *entire* closed path. This is quite unlike the constant local gravitational force, which is a constant vector over any closed path near the surface of the Earth and so is conservative.

QUESTION 5

Can you say that the work done by the kinetic frictional force *always* is negative?

Zero-Work Forces

Certain forces always do zero work and hence are called zero-work forces. Among them are the following:

a. The *static frictional force*. There is a subtle aspect to the work done by the static frictional force on a system. The static frictional force, by its very definition, involves no motion at the point(s) in contact; thus, the work done by static frictional forces is zero.

b. The *normal force* of a rigid surface. The normal force that a surface exerts on a system is perpendicular to the surface. If the system moves parallel to the surface, the normal force is perpendicular to the change in the position vector and so always does zero work. On the other hand, if the system moves perpendicular to the surface, thus leaving the surface (such as in jumping), the force exists only as long as the system is in contact with the surface (there is no motion at the point of contact); and so the normal force still does zero work.

8.8 THE CONCEPT OF POTENTIAL ENERGY

In Section 8.7, we showed that several common forces are conservative forces:

a. the *gravitational force* (in all its forms, whether the local gravitational force or the more general gravitational force);

b. *Hooke's force law* associated with springs and simple harmonic oscillation;
c. *any central force*; and
d. any force that is *constant over any and every closed path.*

We associate the word *work* with an *action* that is applied to the system externally. The action is applied by forces acting on the system as it moves from one place to another. This action is *transferred* to the system across our imaginary boundary that defines the system. It is useful and convenient to associate this action with the *change* in some *property* associated with the system itself. This property of the system is called its **energy**.

To exploit the position dependence of the work done by *conservative forces* on a system, a position-dependent property of the system called its **potential energy** is used. The work done by each conservative force has associated with it a change in a particular potential energy function.

> We define the work done on a system by a specific conservative force as the *negative* of the change in a potential energy function associated with the force. *Changes* always are taken as final minus initial values:
>
> $$W_{consrv} \equiv -(PE_f - PE_i)$$
> $$\equiv -\Delta PE \qquad (8.11)$$

The reason for the seemingly pesky negative sign in the definition will become apparent in Section 8.13. For now, just swallow hard and take the minus sign as part of the definition.

Several things to note immediately:

> *Each* conservative force has its own associated potential energy function. *Only* conservative forces have potential energy functions associated with them. Nonconservative forces do *not* have potential energy functions associated with them.

> Since the work done by a conservative force depends only on the initial and final positions of the particle, the change in the potential energy function also depends only on these positions. *Therefore potential energy functions are functions of the position of the system in space.*

> The work done by a conservative force on the particle *is* the negative of the change in the potential energy of the particle; the work done by the conservative force and the change in the associated potential energy function are not different things, they are the *same* thing.

> Potential energies (and all other energies) have the same dimensions as work (the product of a force and a distance); the SI units are joules (J).

Let's resurrect some of the results of the previous section regarding the work done by specific conservative forces and see what potential energy function is associated with each conservative force on a system.

8.9 THE GRAVITATIONAL POTENTIAL ENERGY OF A SYSTEM NEAR THE SURFACE OF THE EARTH

We discovered in Section 8.7 (Equation 8.8) that the work done by the gravitational force near the surface of the Earth in moving a system of mass m from $\vec{r}_i$ to $\vec{r}_f$ depends only on the difference in the elevation of the system along the y-coordinate axis (with $\hat{j}$ directed vertically up):

$$W_{local} = -mg(y_f - y_i)$$
$$= -(mgy_f - mgy_i) \qquad (\hat{j} \text{ directed up}) \qquad (8.12)$$

We can choose the origin for the vertical y-axis at *any* convenient elevation; it does not have to be on the ground (see Figure 8.23).

According to Equation 8.11, the work done by a conservative force is defined to be the negative of the change in the associated potential energy function of the system:

$$W_{local} \equiv -(PE_f - PE_i)$$

Comparing Equation 8.12 with Equation 8.11, we identify the final potential energy of the system with the term mgy_f and its initial potential energy with mgy_i.

> Since these points can be located anywhere near the surface of the Earth, a system of mass m located at the vertical coordinate y has a gravitational potential energy of mgy:
>
> $$PE_{local} = mgy \qquad (\hat{j} \text{ directed up}) \qquad (8.13)$$

We say the system *has* potential energy by virtue of its *position in space* at vertical coordinate y near the surface of the Earth.

Note several things about this local gravitational potential energy:

a. The local gravitational potential energy of the system is a *scalar* property of the system that is a function of its *position* in space.
b. The system has zero local gravitational potential energy at the elevation where we choose the y-coordinate to be zero; the potential energy is zero on the ground *only* if we choose the origin for the y-coordinate axis to be on the ground.

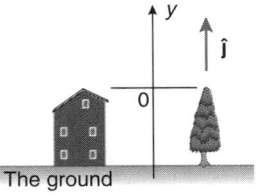

FIGURE 8.23 Choose $\hat{j}$ to be vertically up; you can choose the origin to be at any convenient elevation.

c. The local gravitational potential energy of a system is positive, negative, or zero depending on whether the y-coordinate of the system is positive, negative, or zero respectively.

d. *Changes* in the potential energy are equal to the negative of the work done on the system by the local gravitational force.

e. *If* the gravitational force is the *only* force acting on the system *and* the system is released *at rest*, then the system will move initially in a direction to lower its gravitational potential energy. That is, if m is released at rest some distance above the ground, so that the weight is the only force acting on m, then the local gravitational force moves m (doing positive work on m) in such a direction that it decreases the gravitational potential energy of the system. Since $PE = mgy$, with $\hat{j}$ up, the system decreases its potential energy by decreasing its y-coordinate; cutting through the technical jargon, this argument is a physicist's way of saying *m falls*.

QUESTION 6

Three students are having a disagreement about the gravitational potential energy of a cola bottle at rest on a desk one meter above the floor. One says the potential energy of the bottle is zero, another argues it is positive, and the third says it is negative. Explain the details about how they each may be right within the context of the choices they make for a coordinate system.

STRATEGIC EXAMPLE 8.10

A 6.00 kg medicine ball is located 2.50 m above the ground.

a. **Find the change in the gravitational potential energy of the ball if it is moved to a point located 0.50 m above the ground.**

b. **What is the work done by the gravitational force on the ball over this change in position?**

c. **Does the work done by the gravitational force depend on the specific path used to get from one point to the other?**

Solution

a. Since the ball is near the surface of the Earth, use the local form for the gravitational potential energy, Equation 8.13:

$$PE = mgy \qquad (\hat{j} \text{ directed up})$$

You need to choose an elevation for the origin for the y-coordinate; this choice must be clearly indicated in a sketch. We show here the consequences of two choices for the y-coordinate origin to convince you that the *same* change in the potential energy emerges regardless of where the origin for y is chosen.

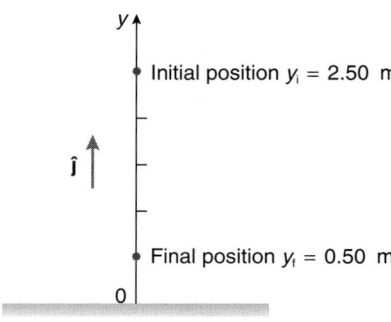

FIGURE 8.24

Method 1

Choose the elevation of the origin to be at ground level, as indicated in Figure 8.24. The initial y-coordinate of the mass then is $y_i = 2.50$ m and the initial potential energy of the mass is

$$\begin{aligned} PE_i &= mgy_i \\ &= (6.00 \text{ kg})(9.81 \text{ m/s}^2)(2.50 \text{ m}) \\ &= 147 \text{ J} \end{aligned}$$

The final y-coordinate of the mass now is $y_f = 0.50$ m, and so the final potential energy of the mass is

$$\begin{aligned} PE_f &= mgy_f \\ &= (6.00 \text{ kg})(9.81 \text{ m/s}^2)(0.50 \text{ m}) \\ &= 29 \text{ J} \end{aligned}$$

The *change* in the gravitational potential energy then is

$$\begin{aligned} \Delta PE &\equiv PE_f - PE_i \\ &= 29 \text{ J} - 147 \text{ J} \\ &= -118 \text{ J} \end{aligned}$$

Method 2

Choose the elevation of the origin to be at the 2.50 m elevation above the ground, as in Figure 8.25. Then the initial y-coordinate of the mass is $y_i = 0$ m, and the initial potential energy of the mass is

$$\begin{aligned} PE_i &= mgy_i \\ &= (6.00 \text{ kg})(9.81 \text{ m/s}^2)(0 \text{ m}) \\ &= 0 \text{ J} \end{aligned}$$

When the mass is finally at the elevation 0.50 m above the ground, the y-coordinate of the mass is $y_f = -2.00$ m and its potential energy is

$$\begin{aligned} PE_f &= mgy_f \\ &= (6.00 \text{ kg})(9.81 \text{ m/s}^2)(-2.00 \text{ m}) \\ &= -118 \text{ J} \end{aligned}$$

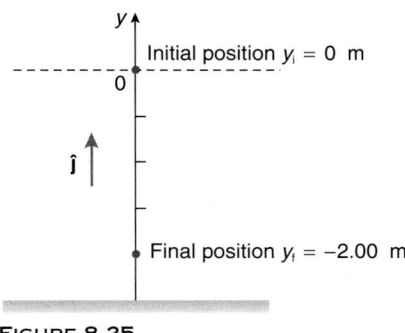

FIGURE 8.25

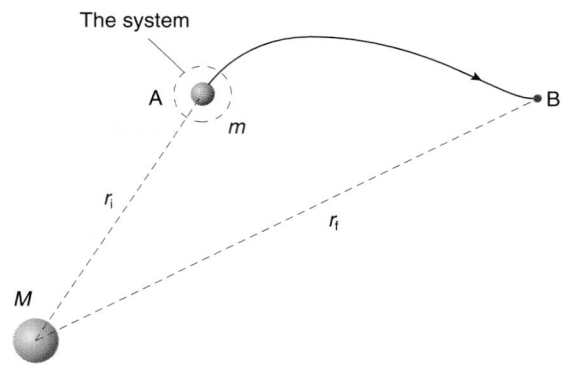

FIGURE 8.26 A mass m is moved from r_i to r_f in the gravitational field of another mass M.

The *change* in the potential energy of the mass then is

$$\Delta PE = PE_f - PE_i$$
$$= -118 \text{ J} - 0 \text{ J}$$
$$= -118 \text{ J}$$

Notice that the change in the potential energy of the system is the *same* for both coordinate choices. It is only *changes* in the potential energy that are physically significant.

b. By definition, the work done by the gravitational force on m is the *negative* of the change in the gravitational potential energy of m:

$$W_{grav} \equiv -\Delta PE$$
$$= -(-118 \text{ J})$$
$$= 118 \text{ J}$$

You could anticipate a positive result for the work done by the gravitational force, since the force is in the same direction as the vertical change in the position vector of the mass.

c. Since the gravitational force is a conservative force, the work done by the gravitational force is *independent* of the specific path used to get the system between the two points.

8.10 THE GENERAL FORM FOR THE GRAVITATIONAL POTENTIAL ENERGY

In Section 8.7, we found (Equation 8.9) that the work done by the gravitational force in moving a system of mass m from a position with radial coordinate r_i to a position with radial coordinate r_f (both measured with an origin at another mass M; see Figure 8.26) is

$$W_{grav} = -GmM\left(-\frac{1}{r_f} + \frac{1}{r_i}\right)$$

This can be rewritten even more awkwardly as

$$W_{grav} = -\left[\left(-\frac{GmM}{r_f}\right) - \left(-\frac{GmM}{r_i}\right)\right] \quad (8.14)$$

According to Equation 8.11, the work done by a conservative force is the negative of the change in the potential energy of the system:

$$W_{grav} \equiv -(PE_f - PE_i)$$

Compare Equation 8.14 with Equation 8.11 and identify

$$PE_f \text{ with } -\frac{GmM}{r_f}$$

and

$$PE_i \text{ with } -\frac{GmM}{r_i}$$

Therefore a system of mass m that finds itself at some position r from a mass M can be described in several ways:

- the system is at some position r in the gravitational field of M;

- there is a gravitational force of M on the system m.

We also can say that the system of mass m, because of its position relative to M, has a gravitational potential energy given by

$$PE_{grav} = -\frac{GmM}{r} \quad (8.15)$$

Note that:

a. The potential energy is a *scalar* property of the system that is a function of its position in space with respect to M.

b. The position coordinate r is raised to the *first* power in the denominator; r is raised to the *second* power in the expression for the magnitude of the gravitational *force* on m (and M), and in the expression for gravitational *field* of M.

PROBLEM-SOLVING TACTIC

8.4 When using Equation 8.15 for the gravitational potential energy, choose the origin of the coordinate system to be at M; no other choice for the origin should be made. This is the conventional choice for such two-particle gravitational interactions.

Note that with this choice of coordinates, the gravitational potential energy of *m* is *always negative*. The place where the gravitational potential energy is *zero* is at infinity ($r = \infty$ m), *not* the origin for *r*. The gravitational potential energy of *m* increases as its distance (r) from M increases: the potential energy becomes less negative as *r* increases.

If *m* is released *at rest* from some point and the gravitational force is the *only* force acting on *m*, then it moves in response to the force in a direction that *decreases* its gravitational potential energy. To do this, *m* falls toward M, decreasing *r* and making the potential energy of *m* more negative.

Any central force will have a potential energy function that depends only on the spatial coordinate *r*:

$$PE = PE(r) \qquad \text{(any central force)} \qquad (8.16)$$

EXAMPLE 8.11

A 500 kg satellite is initially in a circular orbit with a radius equal to twice the radius of the Earth. The satellite is moved to a circular orbit with a radius equal to three times the radius of the Earth.

a. Find the gravitational potential energy of the satellite in each orbit.
b. Find the change in the gravitational potential energy of the satellite.
c. Find the work done by the gravitational force as the satellite is moved between the two locations. Does this work depend on the path used to take the satellite between the two orbits?

Solution

a. Since the satellite is *not* located near the surface of the Earth, you *cannot* use the local expression for the gravitational potential energy, Equation 8.13. Rather you must use the more general expression of Equation 8.15:

$$PE_{grav} = -\frac{GmM}{r}$$

The radius of the Earth is $R_\oplus = 6.37 \times 10^6$ m. (The subscript $\oplus$ is the astronomical symbol for the Earth.) The satellite initially is located where

$$r_i = 2R_\oplus = 1.27 \times 10^7 \text{ m}$$

The mass of the Earth is $M_\oplus = 5.98 \times 10^{24}$ kg. Make these substitutions into the expression for the potential energy:

$$
\begin{aligned}
PE_i &= -\frac{GmM}{r_i} \\
&= -\frac{(6.67 \times 10^{-11} \text{ N} \cdot \text{m}^2/\text{kg}^2)(500 \text{ kg})(5.98 \times 10^{24} \text{ kg})}{1.27 \times 10^7 \text{ m}} \\
&= -1.57 \times 10^{10} \text{ J}
\end{aligned}
$$

Since the orbit of the satellite is circular, the gravitational potential energy of the satellite is the same at all positions in its orbit. When the satellite is moved to the circular orbit at $r = 3R_\oplus$, or

$$r_f = 1.91 \times 10^7 \text{ m}$$

the gravitational potential energy is

$$
\begin{aligned}
PE_f &= -\frac{GmM}{r_f} \\
&= -\frac{(6.67 \times 10^{-11} \text{ N} \cdot \text{m}^2/\text{kg}^2)(500 \text{ kg})(5.98 \times 10^{24} \text{ kg})}{1.91 \times 10^7 \text{ m}} \\
&= -1.04 \times 10^{10} \text{ J}
\end{aligned}
$$

Once again, the gravitational potential energy is constant at all points in the new circular orbit.

b. The *change* in the gravitational potential energy is

$$
\begin{aligned}
\Delta PE &= PE_f - PE_i \\
&= -1.04 \times 10^{10} \text{ J} - (-1.57 \times 10^{10} \text{ J}) \\
&= 5.3 \times 10^9 \text{ J}
\end{aligned}
$$

c. The work done by the gravitational force on the mass is the *negative* of the change in its potential energy:

$$
\begin{aligned}
W_{grav} &\equiv -\Delta PE \\
&= -5.3 \times 10^9 \text{ J}
\end{aligned}
$$

You could have anticipated the negative result, since here the gravitational force on the mass is directed *opposite* to the radial direction that the mass moved.

The work done by the gravitational force is *independent* of the path used, since the gravitational force is a conservative force.

8.11 THE RELATIONSHIP BETWEEN THE LOCAL FORM FOR THE GRAVITATIONAL POTENTIAL ENERGY AND THE MORE GENERAL FORM*

The most general form for the gravitational potential energy of a system of mass *m* located outside the Earth at a distance *r* from its center is given by Equation 8.15:

$$PE_{grav} = -\frac{GmM}{r}$$

where M is the mass of the Earth. Near the surface of the Earth, we use another expression for the gravitational potential energy, Equation 8.13:

$$PE_{local} = mgy \qquad (\hat{j} \text{ directed up})$$

where *g* is the magnitude of the local acceleration due to gravity near the surface of the Earth, 9.81 m/s². The two forms for the potential energy must be related to each other, since the force involved in both situations is the gravitational force of the Earth on the system. Here we show the connection between the two expressions for the gravitational potential energy.

For a system of mass m near the surface of the Earth, we write the distance r of the system from the *center* of the Earth as

$$r = R + y$$

where R is the distance from the center of the Earth to the origin of the Cartesian coordinate system we choose. As you can see in Figure 8.27, the distance R is equal to the radius of the Earth $R_\oplus$ plus the distance h that the origin of our coordinate system is above the surface of the Earth:

$$R = R_\oplus + h$$

Near the surface of the Earth, $h \ll R_\oplus$ and so

$$R \approx R_\oplus$$

In fact R is fixed once the origin for the y-coordinate axis is chosen. Substitute $r = R + y$ into the general expression for the gravitational potential energy of m (Equation 8.15):

$$\text{PE}_{\text{grav}} = -\frac{GMm}{R + y}$$

We can show how this is related to PE_{local}. To accomplish this, factor R out of the denominator:

$$\text{PE}_{\text{grav}} = -\frac{GMm}{R\left(1 + \dfrac{y}{R}\right)}$$

and rewrite this as

$$\text{PE}_{\text{grav}} = -\frac{GMm}{R}\left(1 + \frac{y}{R}\right)^{-1} \tag{8.17}$$

Now we use the **binomial theorem** to expand the quantity in brackets. This theorem states that

$$(1 + x)^n = 1 + nx + \frac{n(n-1)}{2!}x^2 + \frac{n(n-1)(n-2)}{3!}x^3 + \cdots \tag{8.18}$$

The binomial expansion implies that

$$\left(1 + \frac{y}{R}\right)^{-1} = 1 - \frac{y}{R} + \left(\frac{y}{R}\right)^2 - \cdots$$

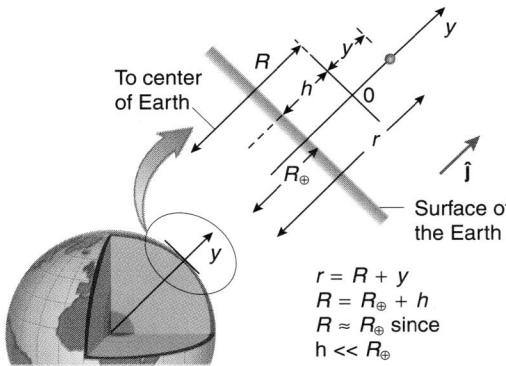

FIGURE 8.27 A coordinate system near the surface of the Earth.

However, since y is much less than R, we can neglect powers of (y/R) higher than the first power and keep just the first two terms of the expansion. Thus

$$\left(1 + \frac{y}{R}\right)^{-1} \approx 1 - \frac{y}{R}$$

Substituting this into Equation 8.17 for the potential energy, we get

$$\begin{aligned}\text{PE}_{\text{grav}} &= -\frac{GMm}{R}\left(1 - \frac{y}{R}\right) \\ &= -\frac{GMm}{R} + \frac{GM}{R^2}my\end{aligned}$$

The first term on the right-hand side is a *constant* (since R is fixed). This constant represents the potential energy of m at the location chosen for the y-axis origin when using Equation 8.15. In the second term on the right-hand side, the quantity

$$\frac{GM}{R^2} \approx \frac{GM}{R_\oplus^2}$$

is g, the magnitude of the local acceleration due to gravity near the surface of the Earth (as we found in Section 6.5). Thus the gravitational potential energy is

$$\begin{aligned}\text{PE}_{\text{grav}} &= -\frac{GMm}{R} + mgy \\ &= -\frac{GMm}{R} + \text{PE}_{\text{local}}\end{aligned}$$

Remember that only *changes* in the potential energy are physically significant; so the constant term,

$$-\frac{GMm}{R}$$

subtracts out when *differences* in potential energy are taken. Thus, for regions near the surface of the Earth, as long as the y-coordinate of the mass does not become large with respect to the radius of the Earth, you can use mgy as the gravitational potential energy with $\hat{\jmath}$ chosen in the vertically upward direction.

8.12 THE POTENTIAL ENERGY FUNCTION ASSOCIATED WITH HOOKE'S FORCE LAW

We discovered in Section 8.7 (Equation 8.10) that the work done by the force of a spring on a system of mass m as the system moves from x_i to x_f is

$$W_{\text{spring}} = -k\frac{x_f^2}{2} + k\frac{x_i^2}{2}$$

or with a slight rearrangement,

$$W_{\text{spring}} = -\left(k\frac{x_f^2}{2} - k\frac{x_i^2}{2}\right) \tag{8.19}$$

The work done by a conservative force is the negative of the change in the potential energy (Equation 8.11):

$$W_{spring} \equiv -(PE_f - PE_i)$$

Compare Equation 8.19 with Equation 8.11 and identify the various potential energy terms.

> The potential energy function associated with a system of mass m attached to a spring when the system is located at coordinate x (where $x = 0$ m is the equilibrium position of m) is
>
> $$PE_{spring} = \frac{1}{2}kx^2 \qquad (8.20)$$

A few things to note about this potential energy function:

a. Once again we see that the potential energy is a scalar property of the system that is a function of its position in space.
b. The potential energy of this system *is independent of its mass*; the potential energy depends only on the spring constant k and the location of the system relative to the equilibrium position.
c. The potential energy is zero at the equilibrium position of the system ($x = 0$ m), where we chose the origin of our coordinate system.
d. Since the spring constant k is always positive, and $x^2 \geq 0$ m^2, this potential energy function is never negative.
e. The work done by the force of the spring on the system is the negative of the *change* in the potential energy of the system.
f. If the mass m is released *at rest* at some coordinate x and the spring force is the only force in the x-direction acting on m, then m moves initially in a direction to decrease its potential energy. In doing this, m approaches the equilibrium position.

Table 8.1 summarizes the potential energy function associated with each conservative force.

TABLE 8.1   Summary of Potential Energy Functions

(1) Gravitational potential energy when the system of mass m is always near the surface of the Earth:

$$PE = mgy \quad (\hat{j} \text{ up}) \qquad (8.13)$$

(you choose the elevation of the y-coordinate origin)

(2) General form for the gravitational potential energy of a system of mass m located a distance r from another pointlike mass M or an appropriate sphere:

$$PE = -\frac{GmM}{r} \qquad (8.15)$$

(origin for r must be chosen at M)

(3) The potential energy of a system of mass m attached to a spring with spring constant k when m is at coordinate x:

$$PE = \frac{1}{2}kx^2 \qquad (8.20)$$

(where $x = 0$ m is the equilibrium position of m)

EXAMPLE 8.12 ━━━━━━━━━━

A 2.00 kg mass on a frictionless horizontal surface is attached to a spring with a spring constant $k = 75.0$ N/m. The mass is initially at its equilibrium position ($x = 0$ m) and is moved to the coordinate $x = -0.30$ m.

a. Find the potential energy of the mass at each location.
b. Find the change in the potential energy of the mass.
c. Find the work done by the force of the spring on m when it is moved between these two locations.

Solution

a. The potential energy function for a system of mass m on a spring is given by Equation 8.20:

$$PE = \frac{1}{2}kx^2$$

When the mass is initially at the coordinate $x_i = 0$ m, the potential energy is

$$PE_i = \frac{1}{2}(75.0 \text{ N/m})(0 \text{ m})^2$$
$$= 0 \text{ J}$$

When the mass is at the coordinate $x_f = -0.30$ m, the potential energy is

$$PE_f = \frac{1}{2}(75.0 \text{ N/m})(-0.30 \text{ m})^2$$
$$= 3.4 \text{ J}$$

b. The *change* in the potential energy is

$$\Delta PE = PE_f - PE_i$$
$$= 3.4 \text{ J} - 0 \text{ J}$$
$$= 3.4 \text{ J}$$

c. The work done by the force of the spring on m is equal to the *negative* of the change in the potential energy of m:

$$W_{spring} \equiv -\Delta PE$$
$$= -3.4 \text{ J}$$

You could anticipate the negative result, because the force of the spring on m is opposite to the direction that m was moved from $x = 0$ m to $x = -0.30$ m.

8.13 THE CWE THEOREM

Energy is Eternal Delight

William Blake (1757–1827)*

The CWE theorem is a special case of an important physical principle that is the *raison d'être* for introducing energy concepts: conservation of energy. This central theorem of classical mechanics stems from Newton's second law of motion.

Recall from Section 8.1 that we consider pointlike, nondeformable, single-particle systems of constant mass with no

---

The Marriage of Heaven and Hell, Plate 4, The voice of the Devil. In *William Blake's Writings*, edited by Gerald E. Bentley Jr., Volume 1, Engraved and Etched Writings (Oxford at the Clarendon Press, Oxford, England, 1978), page 78.

internal motions, and ignore thermal effects. If a system has a nonzero total force acting on it, the system experiences a nonzero acceleration according to Newton's second law of motion. Accelerations imply changes in the velocity of the system.

The dynamics of the system are governed by Newton's second law of motion:

$$\vec{\mathbf{F}}_{\text{total}} = m\vec{\mathbf{a}}$$
$$= m\frac{d\vec{\mathbf{v}}}{dt}$$

where $\vec{\mathbf{F}}_{\text{total}}$ is the vector sum of all the forces acting on the system and $\vec{\mathbf{v}}$ is its velocity. Let's calculate the total work W_{total} done by the total force $\vec{\mathbf{F}}_{\text{total}}$ on the system as it moves from some initial position to some final position. According to the general definition of work, Equation 8.2, W_{total} is

$$W_{\text{total}} = \int_{i}^{f} \vec{\mathbf{F}}_{\text{total}} \cdot d\vec{\mathbf{r}}$$

Use Newton's second law for $\vec{\mathbf{F}}_{\text{total}}$:

$$W_{\text{total}} = m\int_{i}^{f} \frac{d\vec{\mathbf{v}}}{dt} \cdot d\vec{\mathbf{r}} \qquad (8.21)$$

The velocity $\vec{\mathbf{v}}$ is the time rate of change of the position vector:

$$\vec{\mathbf{v}} = \frac{d\vec{\mathbf{r}}}{dt}$$

So the differential change $d\vec{\mathbf{r}}$ in the position vector can be written as

$$d\vec{\mathbf{r}} = \vec{\mathbf{v}}\, dt$$

Then substituting $\vec{\mathbf{v}}\, dt$ for $d\vec{\mathbf{r}}$ in Equation 8.21 changes the integration from one over position to one over time and then to one over the *speed*:

$$W_{\text{total}} = m\int_{i\text{ time}}^{f\text{ time}} \frac{d\vec{\mathbf{v}}}{dt} \cdot \vec{\mathbf{v}}\, dt$$
$$= m\int_{i\text{ speed}}^{f\text{ speed}} d\vec{\mathbf{v}} \cdot \vec{\mathbf{v}} \qquad (8.22)$$

To perform the integration on the right-hand side of Equation 8.22, look at the following derivative just for the fun of it:

$$\frac{d}{dt}(\vec{\mathbf{v}} \cdot \vec{\mathbf{v}})$$

Use the product rule to differentiate the scalar product:

$$\frac{d}{dt}(\vec{\mathbf{v}} \cdot \vec{\mathbf{v}}) = \frac{d\vec{\mathbf{v}}}{dt} \cdot \vec{\mathbf{v}} + \vec{\mathbf{v}} \cdot \frac{d\vec{\mathbf{v}}}{dt}$$
$$= 2\frac{d\vec{\mathbf{v}}}{dt} \cdot \vec{\mathbf{v}}$$

Hence

$$d\vec{\mathbf{v}} \cdot \vec{\mathbf{v}} = \frac{1}{2}d(\vec{\mathbf{v}} \cdot \vec{\mathbf{v}}) \qquad (8.23)$$

The left-hand side of Equation 8.23 is the integrand in Equation 8.22. After this substitution, Equation 8.22 becomes

$$W_{\text{total}} = m\int_{i\text{ speed}}^{f\text{ speed}} \frac{d(\vec{\mathbf{v}} \cdot \vec{\mathbf{v}})}{2}$$

Notice that $\vec{\mathbf{v}} \cdot \vec{\mathbf{v}} = v^2$, the square of the speed of the particle. Hence the integrand is a perfect differential:

$$W_{\text{total}} = \frac{1}{2}m\int_{i\text{ speed}}^{f\text{ speed}} d(v^2)$$
$$= \frac{1}{2}mv^2 \Big|_{i\text{ speed}}^{f\text{ speed}}$$
$$= \frac{1}{2}mv_{\text{f}}^2 - \frac{1}{2}mv_{\text{i}}^2 \qquad (8.24)$$

The right-hand side of this expression is defined to be the change in the classical translational **kinetic energy** of the system of mass m:

$$\Delta KE \equiv KE_{\text{f}} - KE_{\text{i}}$$
$$= \frac{1}{2}mv_{\text{f}}^2 - \frac{1}{2}mv_{\text{i}}^2$$

The kinetic energy is a scalar property of a system that is proportional to the square of its speed v:*

$$KE \equiv \frac{1}{2}mv^2 \qquad (8.25)$$

The classical translational kinetic energy of the system typically is called, more simply, the kinetic energy of the system.

A few things should be noted about the kinetic energy of a system of mass m:

1. The kinetic energy is a *scalar* property of the system.
2. Since $m \geq 0$ kg and $v^2 \geq 0$ m²/s², the kinetic energy of a system is *never negative*.
3. Equation 8.25 for the kinetic energy of the system can be used as long as the speed v of the system is *small compared with the speed of light* (although this is not a fact that arises from our analysis). When the speed of the system becomes an appreciable fraction of the speed of light (universally called c), the expression for the kinetic energy is more elaborate than Equation 8.25, but reduces to it for speeds $v \ll c$.[†]
4. Since the SI units for work are joules, the SI units for the kinetic energy also are joules (J).

---

*The special significance of the quantity mv^2 first was recognized by Christian Huygens (1629–1695) in 1668, before Newton published his *Principia*. Wilhelm von Leibnitz (1646–1716), co-inventor of calculus, also recognized the significance of this quantity and gave it the name *vis viva* (meaning "living force").

[†]The reason a modification of the kinetic energy is needed for so-called relativistic speeds will have to await our treatment of relativistic dynamics in Chapter 25.

We rewrite Equation 8.24 as

$$\boxed{W_{\text{total}} = \Delta KE} \qquad (8.26)$$

This statement is the **CWE theorem**. It states that the effect of the work done by all the forces acting on a point-particle system is to change the value of the kinetic energy property of the system.

Think of the total work done on a system as a way of transferring energy to the system via the action of forces on the system. In other words, *work is a form of energy transfer*. A system does not "*have*" an amount of work by virtue of its position in space or time. We *cannot* say "the work *of* the system"; we *do* say the "work done *on* a system" by forces as it moves from one location to another.

If the system has a speed v, its kinetic energy is $(1/2)mv^2$. Thus a system *can* be said to *have* kinetic energy if $v \neq 0$ m/s. As a result of the energy transfer by means of the work done on the system by all the forces on the system, the CWE theorem states that the system changes its kinetic energy.

If nonzero total work is done *on* a system, the kinetic energy of the system changes; energy is transferred *to* the system. Such energy transfer may be positive, negative, or zero. If the total work W_{total} done *on* the system is positive (> 0 J), the energy transfer is positive and the system *increases* its kinetic energy (since $\Delta KE > 0$ J). If the total work W_{total} done on the system is negative, the energy transfer is negative and the kinetic energy change is negative—the system *loses* kinetic energy.

Several aspects of the CWE theorem should be noted:

1. It is critical to realize that the left-hand side of Equation 8.26, W_{total}, is the work done by *all the forces* acting on the system, not just some of the forces. Knowing what the forces are on the system and drawing a second law force diagram to indicate these forces still are essential.
2. Since the CWE theorem was derived from Newton's second law of motion (a vector equation), the CWE theorem is, in a sense, a scalar equivalent of the second law of motion.

Some forces are conservative forces. It is useful to separate the work done by *all* the forces acting on the system, W_{total}, into two categories:

1. the total work done by all *conservative* forces on the system: W_{consrv}; and
2. the total work done by all *other* forces on the system: W_{other}.

Then the CWE theorem, Equation 8.26, is rewritten as

$$W_{\text{other}} + W_{\text{consrv}} = \Delta KE$$

The work done by a conservative force was defined to be the *negative* of the change in the potential energy function associated with that conservative force (see Equation 8.11):

$$W_{\text{consrv}} \equiv -\Delta PE$$

If more than one conservative force acts on the system, the work done by *each* conservative force is the negative of the change in *each* corresponding potential energy function. Thus W_{consrv} gathers together all the (negative) changes in the potential

energy functions. Substitute for the work done by all the conservative forces acting on the system into the CWE theorem:

$$W_{\text{other}} + (-\Delta PE) = \Delta KE$$

Now you can finally see why the negative sign was introduced into the definition of the work done by conservative forces ($= -\Delta PE$). Transposing the changes in the potential energy functions to the right-hand side of the CWE theorem, we find

$$\boxed{W_{\text{other}} = \Delta KE + \Delta PE} \qquad (8.27)$$

or more explicitly,

$$W_{\text{other}} = (KE_f - KE_i) + (PE_f - PE_i)$$

Now collect the terms associated with the initial and final positions of the system:

$$W_{\text{other}} = (KE_f + PE_f) - (KE_i + PE_i)$$

The CWE theorem then becomes

$$\boxed{W_{\text{other}} = \Delta(KE + PE)} \qquad (8.28)$$

In other words, the work done by the "other" forces on the system (that is, by forces that we do not classify as conservative forces) is equal to the change in the sum of the kinetic and potential energies of the system between the initial and final positions.

The sum of the kinetic and potential energies of a system ($KE + PE$) at any point in space is called the **total mechanical energy** E of the system. So the work done by the "other" forces is equal to the change in the total mechanical energy of the system:

$$\boxed{W_{\text{other}} = \Delta E} \qquad (8.29)$$

PROBLEM-SOLVING TACTIC

8.5 Equations 8.26, 8.27, 8.28, and 8.29 are all equivalent expressions for the CWE theorem. The theorem is an extraordinarily useful result. Remember it. More importantly, use it. In terms of importance, the CWE theorem ranks right up there with Newton's second law of motion. It is one of the reasons that the physical concept of work is such a useful invention.

A frequent, special case of the general CWE theorem arises if either of the following situations arise:

a. if the forces acting on the system are all conservative forces; or
b. if the work done by all "other" forces acting on the system is zero—that is, $W_{\text{other}} = 0$ J.

For *either* case, the work done by nonconservative and zero-work forces (i.e., "other" forces) is zero, so that the left-hand side of the CWE theorem is zero in Equations 8.27, 8.28, and 8.29.

Rewriting Equation 8.28 with the left-hand side equal to zero puts the CWE theorem into a more restricted form:

$$0 \text{ J} = \Delta(\text{KE} + \text{PE}) \qquad (W_{\text{other}} = 0 \text{ J}) \qquad (8.30)$$
$$= \Delta E$$

In such situations [specified in (a) and (b)], the total mechanical energy of the system is conserved or unchanged throughout the motion; this *restricted version* of the CWE theorem is called **conservation of mechanical energy**.

QUESTION 7

If the total work done on a system is zero, what (if anything) can be said about the speed of the system?

EXAMPLE 8.13

A meteoroid of mass 2.09×10^4 kg is moving at a speed of 1.20×10^4 km/h; this meteoroid is a rock only about a meter in diameter. Meteoroids of about this mass occasionally collide with the atmosphere of the Earth, creating spectacular fireballs.

a. Find the kinetic energy of the meteoroid.
b. What mass m, moving at a speed of 10.0 m/s ($\approx$ 36 km/h), has the same kinetic energy?

Solution
a. First convert the speed from km/h to m/s to use proper SI units for the speed:

$$1.20 \times 10^4 \text{ km/h} = \frac{(1.20 \times 10^4 \text{ km/h})(10^3 \text{ m/km})}{3.600 \times 10^3 \text{ s/h}}$$
$$= 3.33 \times 10^3 \text{ m/s}$$

The kinetic energy of the meteoroid then is

$$\text{KE} = \frac{1}{2} mv^2$$
$$= \frac{1}{2}(2.09 \times 10^4 \text{ kg})(3.33 \times 10^3 \text{ m/s})^2$$
$$= 1.16 \times 10^{11} \text{ J}$$

This is a rather substantial amount of kinetic energy! The explosion of one metric ton (1 metric ton = 10^3 kg) of TNT*

*Trinitrotoluene ($C_7H_5N_3O_6$).

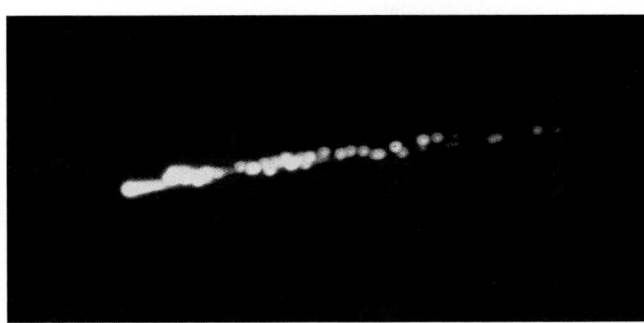

A meteoroid entering the atmosphere of the Earth.

produces about 4.9×10^9 J, and so the kinetic energy of this meteoroid has an energy equivalent to the explosion of about 24 metric tons of TNT.

b. A mass m moving at a speed of 10.0 m/s has kinetic energy

$$\text{KE} = \frac{1}{2} mv^2$$

Substituting 1.16×10^{11} J on the left-hand side and $v = 10.0$ m/s on the right-hand side, you have

$$1.16 \times 10^{11} \text{ J} = \frac{1}{2} m \,(100 \text{ m}^2/\text{s}^2)$$

Solving for m,

$$m = 2.32 \times 10^9 \text{ kg}$$

This is 2.32 *million* metric tons, about the mass of twenty fully loaded oil supertankers.

EXAMPLE 8.14

An American bald eagle (*Haliaeetus leucocephalus*) of mass 4.00 kg has a velocity $\vec{v} = (5.00 \text{ m/s})\hat{\imath} - (12.00 \text{ m/s})\hat{\jmath}$. What is the kinetic energy of the bird?

Solution
To determine the kinetic energy, we need the *speed* of the bird, that is, the magnitude of the velocity vector. Evaluating $v^2 = \vec{v} \cdot \vec{v}$, you find

$$v^2 = [(5.00 \text{ m/s})\hat{\imath} - (12.00 \text{ m/s})\hat{\jmath}] \cdot [(5.00 \text{ m/s})\hat{\imath} - (12.00 \text{ m/s})\hat{\jmath}]$$
$$= (5.00 \text{ m/s})^2 + (-12.00 \text{ m/s})^2$$
$$= 169 \text{ m}^2/\text{s}^2$$

Hence the kinetic energy is

$$\text{KE} = \frac{1}{2} mv^2$$
$$= \frac{1}{2}(4.00 \text{ kg})(169 \text{ m}^2/\text{s}^2)$$
$$= 338 \text{ J}$$

EXAMPLE 8.15

A car is moving at speed v_i. By what factor must its speed increase to *double* its kinetic energy?

Solution
Let the mass of the system be m and let its initial speed be v_i. Its initial kinetic energy then is

$$\text{KE}_i = \frac{1}{2} mv_i^2$$

When its speed is increased to v_f, its kinetic energy is

$$\text{KE}_f = \frac{1}{2} mv_f^2$$

and this must equal $2\,\mathrm{KE_i}$. That is,

$$\frac{1}{2}\,mv_{\mathrm{f}}^2 = 2\left(\frac{1}{2}\,mv_{\mathrm{i}}^2\right)$$

Hence

$$v_{\mathrm{f}}^2 = 2v_{\mathrm{i}}^2$$

or

$$v_{\mathrm{f}} = \sqrt{2}\,v_{\mathrm{i}}$$

Therefore, to double the kinetic energy, the speed of the car must be increased by the factor $\sqrt{2} \approx 1.41$. For example, increasing the speed from 60 km/h to 85 km/h *doubles* the kinetic energy of the car, which is why car crashes rapidly become more dangerous as the speed of the car increases.

STRATEGIC EXAMPLE 8.16

Beginning at rest, a stale bagel of mass 0.250 kg slides down a frictionless slope of straight-line length 2.00 m inclined at an angle of 30.0° to the horizontal before entering the awaiting mouth of a domestic pig (*Sus scrofa*); see Figure 8.28. The slope has a frictionless bump partway down.

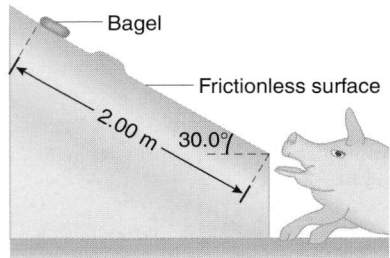

FIGURE 8.28

a. Sketch a second law force diagram indicating all the forces acting on the bagel during the motion.
b. Identify which forces are conservative (if any).
c. What work is done by "other" (i.e., nonconservative and zero-work) forces?
d. Apply the CWE theorem to determine the speed of the bagel at the bottom of the slope.

Solution

a. The forces acting on the bagel during the slippery slide are shown in Figure 8.29:

  1. the gravitational force of the Earth on the bagel—that is, the weight $\vec{\mathbf{w}}$ of the bagel; and
  2. the normal force $\vec{\mathbf{N}}$ of the surface on the bagel.

  The surface is frictionless so there is no frictional force.

b. The gravitational force is a conservative force. The work done by the gravitational force on the bagel is accounted for by a gravitational potential energy term in the CWE theorem. The work done by the normal force must be accounted for in the W_{other} term on the left-hand side of the CWE theorem.
c. The work done by the normal force is the work done by "other" forces on the system, since it is a zero-work force:

$$W_{\mathrm{normal}} = \int_{\mathrm{i}}^{\mathrm{f}} \vec{\mathbf{N}} \cdot d\vec{\mathbf{r}}$$

The normal force is perpendicular to the surface at every point along the path of the bagel, even over the bump. Hence the integrand is zero at every point along the path and the normal force does *zero work*; it is a zero-work force.

d. The presence of the bump makes it difficult, if not impossible, to apply Newton's second law, since the shape of the bump is not known. Use the CWE theorem instead. The appropriate potential energy function to use is the one associated with the gravitational force on a mass near the surface of the Earth:

$$PE_{\mathrm{local}} = mgy \qquad (\hat{\jmath}\ \text{directed up})$$

You can choose the elevation for the y-axis origin to be anywhere you like for this potential energy function. Choose the origin to be at the bottom of the incline, as shown in Figure 8.30. The horizontal placement of the vertical y-axis is not crucial. At the initial position of the bagel at the top of the incline, the potential energy of the bagel is

$$PE_{\mathrm{i}} = mgy_{\mathrm{i}}$$

Geometrically,

$$y_{\mathrm{i}} = (2.00\ \text{m}) \sin 30.0°$$
$$= 1.00\ \text{m}$$

Hence

$$PE_{\mathrm{i}} = (0.250\ \text{kg})(9.81\ \text{m/s}^2)(1.00\ \text{m})$$
$$= 2.45\ \text{J}$$

The initial kinetic energy of the bagel is zero since the initial speed is zero:

$$KE_{\mathrm{i}} = \frac{1}{2}\,mv_{\mathrm{i}}^2$$
$$= 0\ \text{J}$$

At the bottom of the incline the y-coordinate of the bagel is zero, and so the final potential energy of the bagel is

$$PE_{\mathrm{f}} = mgy_{\mathrm{f}}$$
$$= (0.250\ \text{kg})(9.81\ \text{m/s}^2)(0\ \text{m})$$
$$= 0\ \text{J}$$

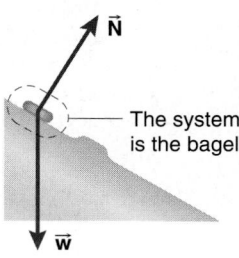

The system is the bagel

FIGURE 8.29

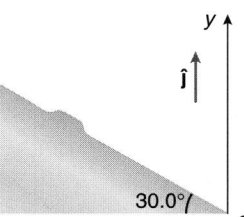

FIGURE 8.30

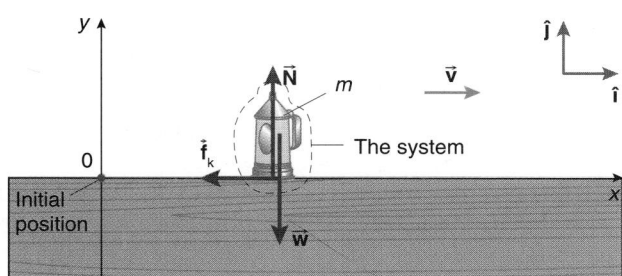

FIGURE 8.31

Let v be the unknown speed of the bagel at the bottom of the incline. The kinetic energy of the bagel at the bottom of the incline is

$$KE_f = \frac{1}{2} mv^2$$

$$= \frac{1}{2} (0.250 \text{ kg}) v^2$$

$$= (0.125 \text{ kg}) v^2$$

Making the appropriate substitutions into the CWE theorem,

$$W_{other} = \Delta(KE + PE)$$
$$= (KE_f + PE_f) - (KE_i + PE_i)$$

you get

$$0 \text{ J} = [(0.125 \text{ kg})v^2 + 0 \text{ J}] - (0 \text{ J} + 2.45 \text{ J})$$

Solving for v^2,

$$v^2 = \frac{2.45 \text{ J}}{0.125 \text{ kg}}$$

$$= 19.6 \text{ m}^2/\text{s}^2$$

The speed v thus is

$$v = 4.43 \text{ m/s}$$

STRATEGIC EXAMPLE 8.17

At the Longbranch saloon, a full beer stein of mass 2.00 kg is slid horizontally (and without rotation) across the bar with an initial speed of 3.00 m/s. If the stein travels a distance 1.53 m before stopping without losing a drop of amber liquid, find the coefficient of kinetic friction μ_k for the stein on the bar surface.

Solution

Our strategy is to apply the CWE theorem to the problem using the beer stein as the system. We find the changes in the kinetic and potential energies of the stein and set this equal to the work done by the "other" forces in the CWE theorem. The forces acting on the stein as it slides across the bar are (see Figure 8.31)

1. the gravitational force of the Earth on the stein, its weight $\vec{w}$ of magnitude mg, directed vertically down;

2. the normal force $\vec{N}$ of the surface of the bar on the stein, directed vertically up; and
3. the kinetic frictional force $\vec{f}_k$ on the stein, directed horizontally in the direction opposite to its velocity.

The normal force and the weight are perpendicular to the path of the stein at every point on its route. Thus both of these forces do zero work on the stein. Since the gravitational force is conservative, the work done by the gravitational force is equal to the negative of the change in the gravitational potential energy of the stein. The appropriate gravitational potential energy is mgy (with $\hat{j}$ directed up). The change in the gravitational potential energy of the stein is zero because its vertical coordinate is the same at both the initial and final positions. (The vertical coordinate y of the stein is in fact constant everywhere along the path.) Thus

$$\Delta PE = 0 \text{ J}$$

The change in the kinetic energy is

$$\Delta KE = KE_f - KE_i$$

The final speed of the stein is zero, so that

$$KE_f = \frac{1}{2} mv_f^2 = 0 \text{ J}$$

The initial kinetic energy is

$$KE_i = \frac{1}{2} mv_i^2$$

$$= \frac{1}{2} (2.00 \text{ kg})(3.00 \text{ m/s})^2$$

$$= 9.00 \text{ J}$$

Hence the change in the kinetic energy of the stein is

$$\Delta KE = KE_f - KE_i$$
$$= 0 \text{ J} - 9.00 \text{ J}$$
$$= -9.00 \text{ J}$$

Thus the CWE theorem becomes

$$W_{other} = \Delta KE + \Delta PE$$
$$= -9.00 \text{ J} + 0 \text{ J}$$
$$= -9.00 \text{ J} \qquad (1)$$

The nonconservative force here is the kinetic frictional force on the stein. The magnitude of the kinetic frictional force $\vec{f}_k$ is related to the magnitude of the normal force $\vec{N}$ by

$$f_k = \mu_k N$$

Since there is no acceleration of the stein in the vertical direction, $a_y = 0$ m/s². Applying Newton's second law of motion to the y-direction,

$$F_{y\ total} = ma_y$$
$$N - mg = m(0 \ \text{m/s}^2)$$

and so

$$N = mg$$

Thus the kinetic frictional force on the stein is of magnitude

$$f_k = \mu_k N$$
$$= \mu_k mg$$

Use the coordinate system indicated in Figure 8.31; the kinetic frictional force is

$$\vec{f}_k = -\mu_k mg\hat{\imath}$$

Since the kinetic frictional force is a *constant vector* over the *entire* one-way straight line path of the stein, you can use Equation 8.6 to find the work of the kinetic frictional force:

$$\begin{aligned} W &= \vec{F} \cdot \Delta\vec{r} \qquad (\vec{F} \text{ constant}) \\ &= \vec{f}_k \cdot (\vec{r}_f - \vec{r}_i) \\ &= -\mu_k mg\hat{\imath} \cdot [(1.53 \ \text{m})\hat{\imath} - (0 \ \text{m})\hat{\imath}] \\ &= (-1.53 \ \text{m})\mu_k mg \end{aligned}$$

Substituting for m and g, you find

$$W_{\text{frictn}} = -30.0\mu_k \ \text{J}$$

The work done by the kinetic frictional force is negative since the force is antiparallel to the change in the position vector of the system.

The work done by the kinetic frictional force is the left-hand side of the CWE theorem since this is the work done by nonconservative forces, the "other" forces, acting on the system. Equation (1) then becomes

$$(-30.0 \ \text{J})\mu_k = -9.00 \ \text{J}$$

Solving for μ_k,

$$\mu_k = 0.300$$

STRATEGIC EXAMPLE 8.18

A particle of mass m is at a height h_0 above the ground at the top of a smooth, frictionless slide and plane, as in Figure 8.32. The mass is released at rest, slides down the incline, and encounters a spring (with spring constant k). Determine the maximum extent to which the spring is compressed before it hurls the mass back in the direction whence it came.

Solution

Choose a coordinate system with its origin at the uncompressed end of the spring, as shown in Figure 8.33. Let mass m be the system. You could break the problem into two parts:

a. the descent down and then along the smooth slide, and
b. the encounter with the spring;

and use the CWE theorem to examine each part separately; you might try this as an exercise. However, it is easier to solve the problem with a *single* application of the CWE theorem. Take the initial position of the system of mass m to be when it is at the top of the incline and the final position when it is (instantaneously) at rest with the spring at maximum compression.

The forces acting on the system are as follows:

1. The gravitational force of the Earth on the system (the weight $\vec{w}$ of the system). This is a conservative force, and so the work done by this force is accounted for in the CWE theorem (Equation 8.28) by the change in the gravitational potential energy term.
2. The normal force $\vec{N}$ of the incline and horizontal surface on the system. This force changes its magnitude and direction as the system goes down the incline and onto the horizontal surface. Nonetheless, the normal force at all points along the path of the system does *zero work*, since it is at all points perpendicular to the path of the system; as always, it is a zero-work force.
3. The force of the spring on the system when it encounters the spring. This is a conservative force, and the work done on the system by this force is accounted for by the change in a spring potential energy term in the CWE theorem.

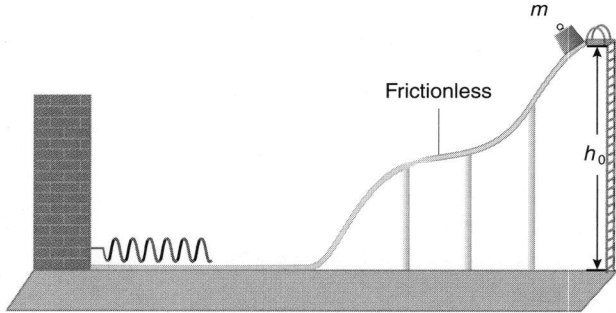

FIGURE 8.32

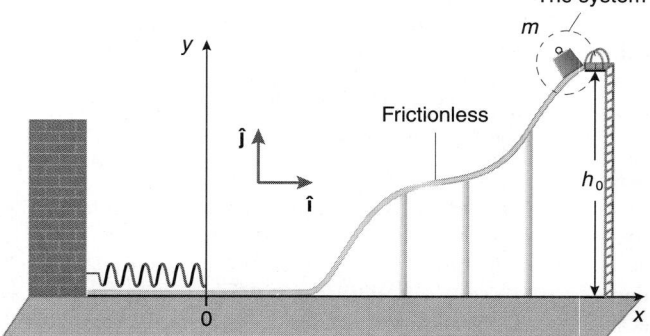

FIGURE 8.33

In applying the CWE theorem here, there are two potential energy terms, one associated with the gravitational force and one with the force of the spring on the system. Hence the theorem is

$$W_{\text{other}} = (\text{KE} + \text{PE}_{\text{grav}} + \text{PE}_{\text{spring}})_f - (\text{KE} + \text{PE}_{\text{grav}} + \text{PE}_{\text{spring}})_i$$

The term W_{other} is zero, since no work is done by the normal force and this is the only force on the system "other" than conservative forces. Since the system was released at rest, the initial kinetic energy is zero:

$$\text{KE}_i = 0 \text{ J}$$

The initial gravitational potential energy is mgy_i. With the coordinate system indicated in Figure 8.33, $y_i = h_0$, and you have

$$\text{PE}_{\text{grav i}} = mgh_0$$

The initial potential energy of the system associated with the spring is zero since the spring is not (yet) compressed:

$$\text{PE}_{\text{sprg i}} = 0 \text{ J}$$

At the final position of the system, with the spring under maximum compression, the system is instantaneously at rest, and so

$$\text{KE}_f = 0 \text{ J}$$

The final gravitational potential energy of the system, mgy_f, also is zero since $y_f = 0$ m:

$$\text{PE}_{\text{grav f}} = 0 \text{ J}$$

The spring is compressed to coordinate $x < 0$ m, and so the potential energy of the system because of the spring is

$$\text{PE}_{\text{sprg f}} = \frac{1}{2} kx^2$$

Substituting into the CWE theorem,

$$W_{\text{other}} = (\text{KE} + \text{PE}_{\text{grav}} + \text{PE}_{\text{spring}})_f - (\text{KE} + \text{PE}_{\text{grav}} + \text{PE}_{\text{spring}})_i$$

$$0 \text{ J} = \left(0 \text{ J} + 0 \text{ J} + \frac{1}{2} kx^2\right) - (0 \text{ J} + mgh_0 + 0 \text{ J})$$

Solve for x. The negative root must be chosen because the spring is compressed:

$$x = -\left(\frac{2mgh_0}{k}\right)^{1/2}$$

8.14 THE ESCAPE SPEED

The law of gravity says no fair jumping up without coming back down.

Anonymous fourth grader

Throw a ball vertically upward and it comes back down again. If the ball is thrown upward with a greater speed, it travels higher and spends a longer time in flight, but eventually it comes back

to haunt us again. However, if thrown vertically upward with sufficient speed, equal to or greater than what is called the **escape speed**, the mass will never return. So if you *really* want to get rid of something, just project it vertically upward with a speed exceeding the escape speed and your troubles are over.

> The escape speed is the minimum speed needed to project a mass m upward from the surface of another mass M (such as a planet, asteroid, or star) so that m never returns.

The purpose of the huge rocket engines for interplanetary space probes is to provide the "kick" (the initial kinetic energy) needed to propel the spacecraft to the escape speed, so that it may escape from the Earth. A few meteorites from Mars, found in Antarctica, escaped from the red planet during an ancient collision of a small asteroid with Mars, propelling them into space, eventually to land on the Earth.

The CWE theorem can determine the escape speed. Throw a system of mass m radially outward from the surface of a spherically symmetric planet or star of mass M and radius R, as shown in Figure 8.34. For simplicity, disregard any kinetic frictional forces from the atmosphere as the mass m escapes, although this

A rocket must propel the payload of an interplanetary space probe to a speed at least equal to the escape speed. This is the Mars Global Surveyor Mission launched 11 November 1996.

This meteorite reached the Earth by being blasted from Mars by an impact that gave the rock a speed greater than the escape speed from Mars.

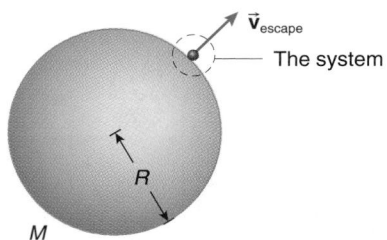

FIGURE 8.34 A mass m is propelled upward from a spherical mass M.

assumption is a bit unrealistic on planets with an atmosphere. While in its unpowered flight, the only force on the system is the gravitational force of the central mass M; there is no work done by nonconservative forces and the *left*-hand side of the CWE theorem, W_{other}, is zero:

$$W_{other} = 0 \text{ J}$$

We want to find the minimum possible launch speed necessary for mass m to escape from M; this is the escape speed v_{escape}. We launch mass m with an initial kinetic energy

$$KE_i = \frac{1}{2} mv_{escape}^2$$

Since m is under the gravitational influence of mass M, the mass m initially also has gravitational potential energy:

$$PE_i = -\frac{GMm}{R}$$

where we have to use the general expression for the gravitational potential energy, Equation 8.15, since m is not going to be confined to the local vicinity of the surface of M. Indeed, that's the point: we want to project m from M to great distances.

If m just barely escapes from M, the final speed of m at an infinite distance away is zero; thus the final kinetic energy of m is zero:

$$KE_f = 0 \text{ J}$$

Since m can go (effectively) infinitely far away from M after it escapes, $r_f = \infty$ m, and the final gravitational potential energy of m,

$$PE_f = -\frac{GMm}{r_f}$$

approaches zero as $r_f \to \infty$ m:

$$PE_f = 0 \text{ J}$$

Apply the CWE theorem:

$$
\begin{aligned}
W_{other} &= \Delta(KE + PE) \\
&= (KE + PE)_f - (KE + PE)_i
\end{aligned}
$$

$$0 \text{ J} = (0 \text{ J} + 0 \text{ J}) - \left[\frac{1}{2} mv_{escape}^2 + \left(-\frac{GMm}{R} \right) \right]$$

Hence

$$\frac{1}{2} mv_{escape}^2 = \frac{GMm}{R} \tag{8.31}$$

Solve for the escape speed v_{escape}:

$$v_{escape} = \left(\frac{2GM}{R} \right)^{1/2} \tag{8.32}$$

Notice that the escape speed does *not* depend on the mass m of the system propelled upward: it is the *same* speed for all masses escaping from M.

Systems (such as you, me, baseballs, and footballs) with speeds less than the escape speed of the Earth have kinetic energies that are too small to escape—fortunately!

Examine Equation 8.31. In order to escape, the initial kinetic energy of m must at least be equal to the absolute value of its initial gravitational potential energy. If not, the mass m is "bound" to M and will return to the surface when thrown upward. With this in mind, the absolute value of the gravitational potential energy of m, the quantity

$$\left| -\frac{GMm}{R} \right|$$

is called the **binding energy** of m when it is at rest on the surface of M. More generally, the binding energy of a system at any location is the additional energy the system needs to barely escape (with final speed zero) to an infinite distance away.

The escape speed from other planets or stars can be found in terms of the escape speed from the Earth by taking the ratio of the two escape speeds. The escape speed from a planet or star of mass M and radius R is

$$v_{escape} = \left(\frac{2GM}{R} \right)^{1/2}$$

The escape speed from the Earth (symbolized with the subscript $\oplus$) is

$$v_\oplus = \left(\frac{2GM_\oplus}{R_\oplus} \right)^{1/2}$$

Take the *ratio* of the two expressions; this eliminates G and numerical constants:

$$\frac{v_{escape}}{v_\oplus} = \left(\frac{\dfrac{M}{M_\oplus}}{\dfrac{R}{R_\oplus}} \right)^{1/2}$$

$$= \left(\frac{\text{mass ratio}}{\text{radius ratio}} \right)^{1/2} \qquad (8.33)$$

Thus the relative magnitudes of the escape speeds can be found if the mass and radius ratios are known.

EXAMPLE 8.19

Calculate the escape speed from the surface of the Earth.

Solution
The escape speed from the surface of the Earth is given by Equation 8.32:

$$v_{escape} = \left(\frac{2GM_\oplus}{R_\oplus} \right)^{1/2}$$

where $M_\oplus$ is the mass of the Earth and $R_\oplus$ is its radius. Substituting numerical values for these quantities and for G, you get

$$v_{escape} = \left[\frac{2 \times (6.67 \times 10^{-11} \ \text{N} \cdot \text{m}^2/\text{kg}^2)(5.98 \times 10^{24} \ \text{kg})}{6.37 \times 10^6 \ \text{m}} \right]^{1/2}$$

$$= 1.12 \times 10^4 \ \text{m/s}$$

or 11.2 km/s. Thus to send a spacecraft off the Earth to, say, Mars, a minimum speed of 11.2 km/s is needed.

EXAMPLE 8.20

Jupiter is 318 times as massive as the Earth; the radius of the giant planet is 10.9 times that of the Earth. Find the escape speed from Jupiter in terms of the escape speed from the Earth.

Solution
The mass ratio is

$$\frac{M_{Jupiter}}{M_\oplus} = 318$$

The radius ratio is

$$\frac{R_{Jupiter}}{R_\oplus} = 10.9$$

Use Equation 8.33; the ratio of the escape speeds is

$$\frac{v_{escape}}{v_\oplus} = \left(\frac{318}{10.9} \right)^{1/2}$$

$$= 5.40$$

Hence the escape speed from Jupiter is 5.40 times that from the Earth. Since the escape speed from the Earth is 11.2 km/s (see Example 8.19), it takes a minimum speed of 5.40 × 11.2 km/s =

60.5 km/s to escape from Jupiter. Don't get too close or you'll never come back! Escape speeds for other planets and even stars can be calculated using similar ratio techniques.

EXAMPLE 8.21

a. Calculate the binding energy of a 1.00 kg mass on the surface of the Earth.
b. How much kinetic energy is needed for this mass to escape from the Earth?

Solution
a. The binding energy is the absolute value of the gravitational potential energy of the mass on the surface of the Earth. Here you *cannot* use the local expression for the gravitational potential energy (mgy); rather you must use the more general expression for the gravitational potential energy, given by Equation 8.15:

$$PE = -\frac{GM_\oplus m}{R_\oplus}$$

where $M_\oplus$ is the mass of the Earth, $R_\oplus$ is the radius of the Earth, and m is the 1.00 kg mass whose binding energy we are asked to find. Making these substitutions, you find

$$|PE| = \frac{(6.67 \times 10^{-11} \ \text{N} \cdot \text{m}^2/\text{kg}^2)(5.98 \times 10^{24} \ \text{kg})(1.00 \ \text{kg})}{6.37 \times 10^6 \ \text{m}}$$

$$= 6.26 \times 10^7 \ \text{J}$$

b. The binding energy is equal to the amount of kinetic energy needed to escape (barely) from the Earth; therefore, the 1.00 kg mass needs a kinetic energy of 6.26×10^7 J to escape (barely) from the surface of the Earth. Since this example involved a 1.00 kg mass, the binding energy of a mass m is

$$\text{binding energy} = m(6.26 \times 10^7 \ \text{J/kg})$$

This large amount of energy is why NASA puts a premium on minimizing the mass of payloads: it costs real money for the energy. It pays to be an astronaut of small mass!

8.15 BLACK HOLES*

Confinement to the Black Hole . . . to be reserved for cases of Drunkenness, Riot, Violence, or Insolence to Superiors
   1844 Regulations and Orders of the Army 121; from the *Oxford English Dictionary.*

. . . — black it stood as Night
Fierce as ten Furies, terrible as Hell

John Milton (1608–1674)[†]

During the 18th century, John Michell (1724?–1793) realized that the escape speed from a sufficiently massive and compact star might exceed the speed of light (only measured as finite during the early part of that century), producing what we now call a **black hole.**

[†] *Paradise Lost*, Book II, lines 670–671.

The following argument for the possible existence of black holes is based on Newtonian mechanics. The expression for the escape speed from a spherically symmetric mass M of radius R is given by Equation 8.32:

$$v_{escape} = \left(\frac{2GM}{R} \right)^{1/2} \tag{8.32}$$

If the mass M somehow is compressed into a smaller spherical volume, decreasing its radius R as in Figure 8.35, the escape speed increases.

> If the mass M is compressed to such a small size that the escape speed *exceeds* the speed of light, the mass M forms a black hole.

Not even light can escape from a black hole, since $v_{escape} > c$. The critical radius to which a mass M must be compressed for the escape speed to *equal* the speed of light is known as the **Schwarzschild radius** R_s of the mass, named for Karl Schwarzschild (1873–1916), who first deduced the expression from Einstein's then new theory of general relativity. The Schwarzschild radius of a mass M is found from the equation

$$v_{escape} = \text{speed of light} \equiv c$$

Substitute for v_{escape} from Equation 8.32:

$$\left(\frac{2GM}{R_s} \right)^{1/2} = c$$

Solving for the Schwarzschild radius R_s, we find

$$R_s = \frac{2GM}{c^2} \tag{8.34}$$

In principle, anything can become a black hole if the radius R of the mass M is less than R_s. So if *you* want to turn into a black hole, just curl yourself up into a size smaller than your Schwarzschild radius R_s. Needless to say, thanks to the large value of the speed of light (in the denominator of the expression for R_s), the Schwarzschild radius for ordinary masses is *very* small (see Problem 25). However, for sufficiently massive stars at the end of their lifetimes, when the star exhausts its nuclear fusion fuel supplies, no known force can prevent a complete gravitational collapse of the core of the star and a black hole forms. If you use

the result of Example 8.22 (below), a black hole with a mass of three times that of the Sun has a Schwarzschild radius of about 9 km. Venture nearer to a black hole than the Schwarzschild radius and you can kiss the rest of the universe goodbye*; you never can get out, since the escape speed from the black hole is greater than the speed of light.[†‡] *Supermassive* black holes, with masses millions of times greater than the mass of the Sun, are thought to be located at the center of many galaxies, including our own Milky Way Galaxy.

A word of caution. We used classical nonrelativistic dynamics to "derive" the expression for the Schwarzschild radius, Equation 8.34. This derivation is hocus-pocus. We have noted on several occasions that Newtonian dynamics is restricted to speeds much smaller than the speed of light. Ironically, the complicated, full-blown relativistic calculation for the Schwarzschild radius yields precisely the same result as Equation 8.34. This is quite fortuitous for our purposes here, since such coincidences between nonrelativistic and relativistic equations are rare. The classical derivation is an illustration of getting the right answer, but with incorrect physics; the end rarely justifies the means, but here we could not resist!

EXAMPLE 8.22

Calculate the Schwarzschild radius of a hypothetical black hole with the mass equal to that of the Sun.

Solution
The Schwarzschild radius of a black hole of mass M is given by Equation 8.34:

$$R_s = \frac{2GM}{c^2}$$

Substituting for the mass of the Sun, the speed of light, and the universal gravitational constant, you get

$$R_s = \frac{2 \times (6.67 \times 10^{-11} \ \text{N·m}^2/\text{kg}^2) \, (1.99 \times 10^{30} \ \text{kg})}{(3.00 \times 10^8 \ \text{m/s})^2}$$
$$= 2.95 \times 10^3 \ \text{m}$$

or about 3 km. For reasons that are not readily explainable at this level, *real* black holes must have masses *greater* than about three times the mass of the Sun. The Sun, therefore, will *not* become a black hole when it exhausts the nuclear fuel in its core about A.D. 5×10^9. Real black holes that result from the evolution of stars have Schwarzschild radii of at least about 3×3 km = 9 km.

---

*Actually, by the time you reach the Schwarzschild radius you will be dissociated into your constituent particles by the enormous gravitational tidal forces in the vicinity of the hole; you reach the Schwarzschild radius as a hailstorm of little particles.

[†] There is an upper limit to the speed of any mass m: the speed of light. The reason the speed of light is an upper bound will become apparent when we study relativity in Chapter 25.

[‡] In the process of entering a black hole, there are frictional forces that cause material to lose kinetic and potential energy, with the result that the infalling material is captured by the hole. Once inside the black hole, there is no way to escape; it is impossible for matter to attain the escape speed, since it is greater than the speed of light.

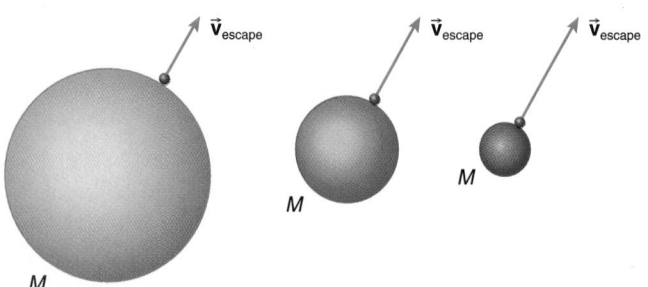

FIGURE 8.35 Decrease the radius of a mass M and the escape speed increases.

8.16 LIMITATIONS OF THE CWE THEOREM: TWO PARADOXICAL EXAMPLES*

We have restricted ourselves to systems that can be represented as simple pointlike particles with no internal motions, deformations, or thermal effects. Here we consider two examples of systems that do not meet these criteria to show that the CWE theorem, useful as it is, is not the last word on energy considerations.[†]

Case 1: A Car Accelerates from Rest

Take your car out for a spin. The car, initially at rest, accelerates to speed v_0 along a horizontal section of straight highway. What does the CWE theorem have to say about this phenomenon? The car is a complex system with internal moving parts (the engine and tires) and thermal effects (the combustion of the fuel). The entire car moves through some distance, but the part of the tires in contact with the road surface is always at rest along the path followed by the car (unless skidding occurs).

Choose a coordinate system with $\hat{\jmath}$ vertically upward and $\hat{\imath}$ horizontal and in the direction the car is moving, as shown in Figure 8.36. The origin is at the point where the car begins to accelerate. The forces acting on the car are as follows (see Figure 8.36):

1. The gravitational force of the Earth on the car, its weight $\vec{w}$, directed vertically downward:

$$\vec{w} = -mg\hat{\jmath}$$

The work done by the conservative gravitational force here is zero, since $\vec{w}$ is everywhere perpendicular to the horizontal path followed by the car.

2. The normal force $\vec{N}$ of the road surface on each tire of the car, directed perpendicular to the surface, so that

$$\vec{N} = N\hat{\jmath}$$

These normal forces, as always, also do zero work because they are perpendicular to the path followed by the car at every point along its path.

3. The frictional force of the road surface on the drive wheels of the car is the static frictional force because the points of contact of the tires with the road are *at rest*:

$$\vec{f}_s = f_s\hat{\imath}$$

[†] The full story is revealed in Chapter 13.

It is this force that causes the car to accelerate. However, since there is no motion at the point of application of the static frictional force of the road on the tires, the static frictional force does zero work.

Each of the forces on the car does *zero work*. The change in the kinetic energy of the car is

$$\Delta KE \equiv KE_f - KE_i$$
$$= \frac{1}{2}mv_0^2 - 0 \text{ J}$$
$$= \frac{1}{2}mv_0^2$$

Apply the CWE theorem to the car:

$$W_{total} = \Delta KE$$
$$0 \text{ J} = \frac{1}{2}mv_0^2$$

The left-hand side of the theorem apparently is zero but the right-hand side is not! Certainly there is something wrong with applying the CWE theorem to the motion, which should be unsettling to you. Applying Newton's second law of motion to the system *is* appropriate, since it is the static frictional force that accelerates the car. Without the static frictional force on the car, it simply would not change its velocity or accelerate (on an icy, frictionless road, this is indeed what happens).

Case 2: Push Off from a Wall

Another example of a problem arising from the CWE theorem is the following. Imagine yourself on a frictionless horizontal surface (such as an ice rink) next to a vertical wall, as shown in Figure 8.37. Push off from the wall with your arms, attaining a speed v in the horizontal direction. What does the CWE theorem have to say about this situation?

This is an example of a system (you) that *deforms* as your arms compress and extend during the time they are in contact with the wall; the system is not a simple pointlike particle. There are three forces on you during the push-off, shown in Figure 8.37:

1. your weight $\vec{w}$, directed vertically down;
2. the normal force $\vec{N}_{rink}$ of the surface of the ice rink on you, directed vertically upward; and
3. the normal force $\vec{N}_{wall}$ of the wall on you, directed horizontally.

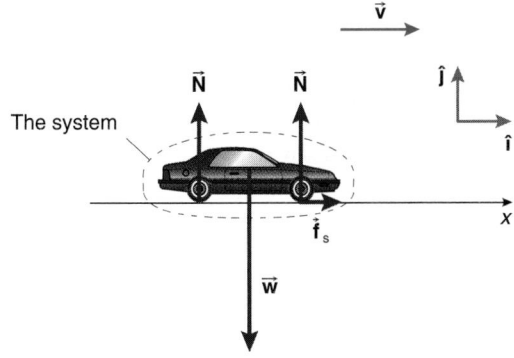

FIGURE 8.36 Forces on the car.

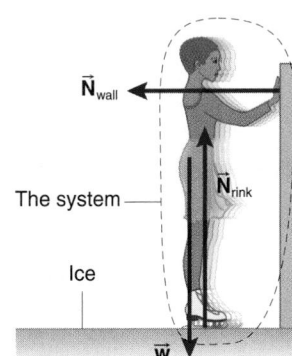

FIGURE 8.37 Forces on you when you push off from a wall.

The weight and the normal force of the rink each do zero work, because they are perpendicular to the change in the position vector of the system at every point along the horizontal path during the push-off. The normal force of the wall also does zero work, since the force exists only at the point of contact between you and the wall and there is no motion at the point of application of this normal force during the push-off.

Apply the CWE theorem:

$$W_{\text{total}} = \Delta KE$$
$$= KE_f - KE_i$$
$$0 \text{ J} = \frac{1}{2} mv^2 - 0 \text{ J}$$

Once again, the CWE theorem yields a ridiculous result. Just as in Case 1, there is something fishy going on here with the CWE theorem.

There is *nothing wrong with applying Newton's second law* to these cases. But yet there clearly *is* something incomplete with the CWE theorem. The clue to both examples is that something, somewhere is quite literally getting warm. Thermodynamic effects cannot be ignored.

One way to resolve or patch up* the CWE theorem for these cases is to imagine or invent a new kind of fictional work, called the **pseudowork** done by the static frictional force in Case 1 and the normal force of the wall in Case 2. In other words, separate the left-hand side of the CWE theorem, W_{total}, into two contributions—the *real work done* by forces and *pseudowork* done by forces that do no real work:

$$W_{\text{total}} \equiv W_{\text{real}} + W_{\text{pseudo}}$$

In this way, the CWE theorem becomes

$$W_{\text{real}} + W_{\text{pseudo}} = \Delta KE$$

The pseudowork of the forces that do no real work is found by integrating the force over the path followed by the system as if it were a pointlike particle. For example, the pseudowork of the static frictional force in Case 1 is

$$W_{\text{pseudo}} = \int_i^f \vec{\mathbf{f}}_s \cdot d\vec{\mathbf{r}}$$

For Case 1, the pseudowork of the frictional force is equal to the change in the kinetic energy of the car. For Case 2, the pseudowork of the normal force of the wall is equal to the change in the kinetic energy of the skater.

The source of the apparent pseudowork of these forces lies *within the system* itself. For the car, the energy that changes its kinetic energy ultimately comes from chemical reactions inside the engine. For the skater, the energy comes from biochemical reactions that drive the muscles.

> The paradoxes arising from the CWE theorem are resolved by treating the system as a more complex system than the particle-systems we typically have considered. A more general statement about energy conservation than the CWE theorem is needed; we study this more general statement in Chapter 13.

---

*Another way to describe this is high-class fudging.

8.17 THE SIMPLE HARMONIC OSCILLATOR REVISITED

A system of mass m attached to an ideal spring on a horizontal frictionless surface undergoes simple harmonic oscillation. The only force on the system that does nonzero work is the force of the spring, the Hooke's law conservative force.

The work done by the force of the spring is accounted for by the change in a potential energy term in the CWE theorem. For a simple harmonic oscillator system, the *left*-hand side of the CWE theorem, W_{other} in Equation 8.28, is zero and the theorem takes the form

$$0 \text{ J} = \Delta(KE + PE)$$

The total mechanical energy E is the sum of the kinetic and potential energies. Thus the CWE theorem indicates that the total mechanical energy E of a simple harmonic oscillator is *constant* or *conserved*. That is, $0 \text{ J} = \Delta E$, and so E is constant. Since the simple harmonic oscillator is such an important system in physics, we want to calculate the total mechanical energy of *any* simple harmonic oscillator, be it a mass m attached to a spring or an atom executing simple harmonic oscillation about its equilibrium position in a molecule.

Recall that for *any* simple harmonic oscillator (Chapter 7) the position x of the oscillator as a function of time is given by Equation 7.6,

$$x(t) = A \cos(\omega t + \phi)$$

where ω is the angular frequency $[= (k/m)^{1/2}$ for a mass m attached to a spring with spring constant $k]$; A is the amplitude of the oscillation and ϕ is a phase angle. Both A and ϕ depend on how the oscillator is initially released. The position $x = 0$ m corresponds to the equilibrium position of the oscillator.

The potential energy of the oscillator at any position x is (according to Equation 8.20)

$$PE(t) = \frac{1}{2} kx(t)^2$$
$$= \frac{1}{2} kA^2 \cos^2(\omega t + \phi)$$

The potential energy varies with time.

To find the kinetic energy, we need to know the speed of the oscillator. There is only one velocity component, v_x, and this is

$$v_x(t) = \frac{dx(t)}{dt}$$
$$= -A\omega \sin(\omega t + \phi)$$

Hence the kinetic energy of the system is

$$KE(t) = \frac{1}{2} mv_x(t)^2$$
$$= \frac{1}{2} mA^2 \omega^2 \sin^2(\omega t + \phi)$$

But since $\omega = (k/m)^{1/2}$, the kinetic energy is

$$KE(t) = \frac{1}{2} kA^2 \sin^2(\omega t + \phi)$$

The kinetic energy also varies with time.

The total mechanical energy E of the oscillator is the sum of the kinetic and potential energies:

$$E = KE(t) + PE(t)$$

Substitute for the kinetic and potential energy of the harmonic oscillator; after factoring, we find

$$E = \frac{1}{2}kA^2[\cos^2(\omega t + \phi) + \sin^2(\omega t + \phi)]$$

But the sum of the squares of the sine and cosine of the same angle always is equal to 1. Thus the total mechanical energy of the oscillator is

$$E = \frac{1}{2}kA^2 \qquad (8.35)$$

Since the time t does not appear in this expression, the total mechanical energy E of a simple harmonic oscillator is independent of time, confirming what the CWE theorem has to say about the system: $\Delta E = 0$ J, and so E is *constant*.

> The total mechanical energy of a simple harmonic oscillation is conserved; only the mix or proportion of the energy that is kinetic or potential energy changes with time.

If the oscillator is released at $x = A$ when $t = 0$ s, then the phase angle $\phi = 0$ rad and potential and kinetic energy of the oscillator are

$$PE(t) = \frac{1}{2}kA^2\cos^2(\omega t)$$

$$KE(t) = \frac{1}{2}kA^2\sin^2(\omega t)$$

with their sum equal to $E = (1/2)kA^2$.

The potential energy, kinetic energy, and total mechanical energy of this oscillator are plotted versus time in Figure 8.38. If the phase angle $\phi \neq 0$ rad, the potential and kinetic energy graphs are shifted from the previous case, but their sum is still the constant $E = (1/2)kA^2$. This is illustrated in Figure 8.39 for the case when $\phi = \pi/6$ rad.

Since

$$\omega^2 = \frac{k}{m}$$

the total mechanical energy of the oscillator also can be expressed in terms of the angular frequency ω:

$$E = \frac{1}{2}m\omega^2A^2$$

If we convert from the angular frequency ω (in rad/s) to the frequency ν (in Hz) using

$$\omega = 2\pi\nu$$

the total mechanical energy of the oscillator also can be written as

$$E = 2\pi^2m\nu^2A^2 \qquad (8.36)$$

8.17 The Simple Harmonic Oscillator Revisited

> The total mechanical energy of a simple harmonic oscillator is proportional to the product of the square of the frequency and the square of the amplitude.

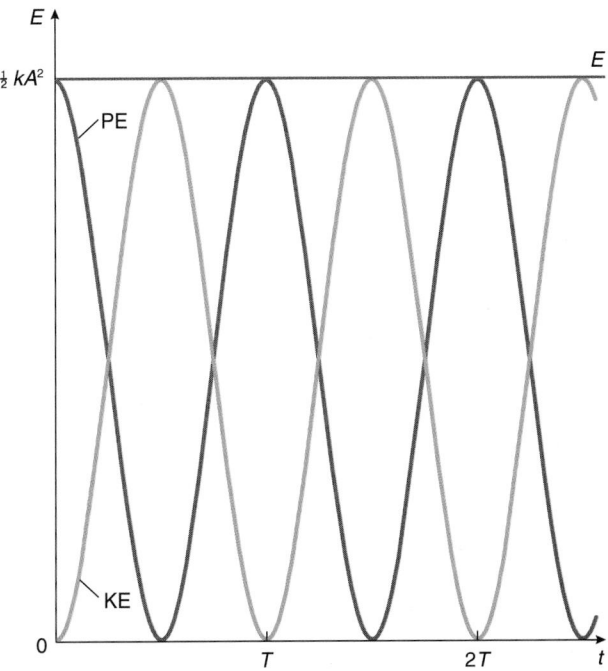

FIGURE 8.38 The potential and kinetic energies of a simple harmonic oscillator as functions of time if $\phi = 0$ rad; note that their sum, the total mechanical energy, is constant.

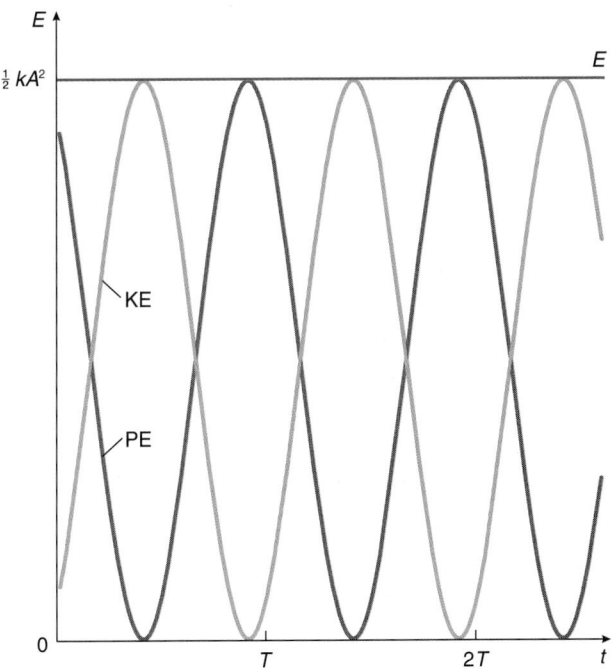

FIGURE 8.39 The potential and kinetic energies of a simple harmonic oscillator as functions of time if $\phi = \pi/6$ rad; note that their sum, the total mechanical energy, is constant.

Since the total mechanical energy of the oscillator is fixed, we can find the speed of the oscillator as a function of position x by writing

$$E = KE + PE$$

Substituting, we get

$$\frac{1}{2} kA^2 = \frac{1}{2} mv^2 + \frac{1}{2} kx^2$$

Solve for v (take the positive square root since the speed is a positive scalar):

$$v(x) = \left[\frac{k}{m} \left(A^2 - x^2 \right) \right]^{1/2} \qquad (8.37)$$

Remember the foregoing results apply to *any* simple harmonic oscillation with angular frequency ω and amplitude A.

EXAMPLE 8.23

A 3.00 kg mass is attached to a spring with spring constant $k = 300$ N/m and is executing simple harmonic motion with an amplitude of 0.40 m.

a. Find the total mechanical energy of the oscillator.
b. Make a graph of the *speed* v of the oscillator as a function of its position x.

Solution

a. The total mechanical energy of the oscillator is given by Equation 8.35:

$$E = \frac{1}{2} kA^2$$

The amplitude $A = 0.40$ m and $k = 300$ N/m, and so

$$E = \frac{1}{2} (300 \text{ N/m})(0.40 \text{ m})^2$$
$$= 24 \text{ J}$$

b. The speed v as a function of position x is found from Equation 8.37:

$$v(x) = \left[\frac{k}{m} \left(A^2 - x^2 \right) \right]^{1/2}$$
$$= \left\{ \frac{300 \text{ N/m}}{3.00 \text{ kg}} \left[(0.40 \text{ m})^2 - x^2 \right] \right\}^{1/2}$$
$$= (10.0 \text{ s}^{-1})(0.16 \text{ m}^2 - x^2)^{1/2}$$

The domain of x is -0.40 m $\leq x \leq 0.40$ m. The function $v(x)$ is parabolic and is plotted in Figure 8.40. On the other hand, the velocity component $v_x(t)$ (plotted in Chapter 7) is sinusoidal. Notice in Figure 8.40 that the maximum speed of the simple harmonic oscillator occurs at the equilibrium position $x = 0$ m, and zero speed occurs at maximum amplitude.

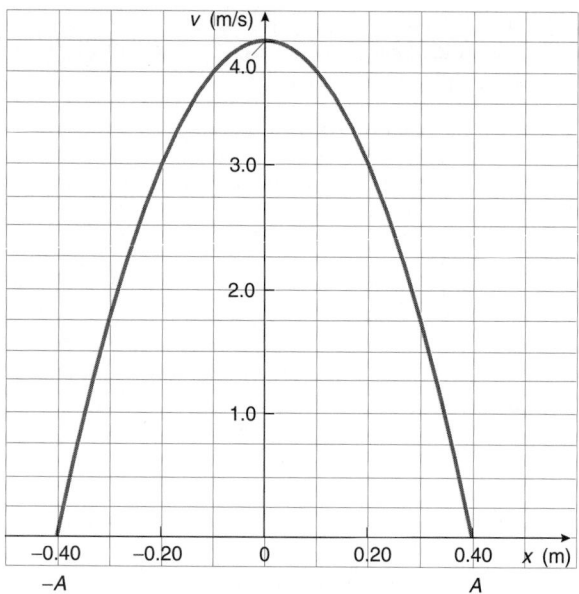

FIGURE 8.40 The speed of a simple harmonic oscillator as a function of position.

8.18 THE AVERAGE AND INSTANTANEOUS POWER OF A FORCE

I sell here, Sir, what all the world desires to have—POWER.
Matthew Boulton (1728–1809)*

As part of a work-study project, you have been hired to lift a number of bricks from the ground up to a height where a mason is laying bricks for a new science building at your university. You could pick up the bricks a few at a time and lift them, or you could take a nearby forklift and lift the whole collection of bricks at one time and get the job done more quickly. The forces that you exert or the forklift exerts on the bricks do the same amount of work on the collection of bricks, but there clearly is a difference between the two situations.

The difference involves the time interval required to do the work on the brick pile. The time rate at which work is done is called **power**. Since work is a way in which energy is transferred from external agents to the system, power represents the time rate at which energy is transferred to the system. Let's take a look at this definition from a quantitative point of view.

> The **average power** P_{ave} transferred by a force to a system is defined to be the ratio of the work done by the force on the system to the time interval Δt during which the work is accomplished:
>
> $$P_{ave} = \frac{\text{work done on system by } \vec{F} \text{ during time interval } \Delta t}{\Delta t}$$
>
> $$(8.38)$$

*Quoted when speaking to Boswell of his engineering works (22 March 1776). In James Boswell, *Life of Samuel Johnson*, edited by George Birkbeck Hill (revised and enlarged edition by Lawrence F. Powell, Oxford at the Clarendon Press, Oxford, England, 1979), volume 2, page 459.

For the brick problem, the average power in the two situations is clearly different. The forklift accomplished the same work on the bricks during a shorter time than you could; it did the job using more average power. The forklift can do more work during a given time interval than you can.

The dimensions of power are those of work divided by time, or J/s using SI units. Since the concept of power is so important in many contexts, the SI unit of power, J/s, is called a **watt** (W), after James Watt (1736–1819), the inventor of the steam engine. Although you probably are more familiar with the power unit through its use in electrical contexts (e.g., 100 W light bulbs), the unit is a general one in the SI system.*

> The time rate at which a force $\vec{F}$ does work is called its **instantaneous power**.

The instantaneous power (typically called simply the **power**) transferred to a system by a force doing work on the system is found in the following way. The differential work dW done by a force $\vec{F}$ when the system changes its position vector by $d\vec{r}$ is, by definition (Equation 8.1),

$$dW = \vec{F} \cdot d\vec{r}$$

The differential change in the position vector $d\vec{r}$ is related to the instantaneous velocity $\vec{v}$:

$$\vec{v} = \frac{d\vec{r}}{dt}$$

Rearranging,

$$d\vec{r} = \vec{v} \, dt$$

---

*Another unit of power that may be familiar is the *horsepower* (hp). The horsepower is an antiquated English unit (with all the quaint peculiarities of that abominable system of weights and measures) introduced by James Watt to measure the efficiency of his newly invented steam engines in the early 19th century in terms of a "standard horse." For those interested, the conversion factor is 1 hp = 746 W ≈ 0.75 kW. We shall not use the horsepower unit in this text.

A forklift is more powerful than you because it can do more work per unit time.

Substituting for $d\vec{r}$ in the expression for the differential work done by $\vec{F}$, we find

$$dW = \vec{F} \cdot \vec{v} \, dt$$

The power P is

$$P \equiv \frac{dW}{dt} = \vec{F} \cdot \vec{v} \qquad (8.39)$$

Power is a *scalar* quantity. Think of the power provided by a force as a measure of the time rate of doing work by that force on the system. Notice that if the force is perpendicular to the velocity, the scalar product vanishes and the instantaneous power of the force is zero.

Take the definition of the instantaneous power of a force:

$$P = \frac{dW}{dt}$$

Rearrange:

$$dW = P \, dt$$

Finally, integrate between time t_i and time t_f:

$$W = \int_{t_i}^{t_f} P \, dt \qquad (8.40)$$

This can be interpreted geometrically in the following way. If a graph is made of the instantaneous power $P(t)$ versus t, as in Figure 8.41, the work done by the force during the interval from t_i to t_f is the area under the graph between the two times.

This complements the geometric interpretation of work, discussed in Section 8.5, where we saw that the work done by a force component F_x between *positions* x_i and x_f can be interpreted as the area under a graph of the force component F_x versus x between the two positions.

It is important to mention that the word power (as well as the words work, force, and energy) is used in many contexts *outside* the sciences that convey meanings unrelated to the concept in physics and engineering. Colloquial usages typically convey an *incorrect* impression of what is meant by power in physics (e.g., political power, power lunches, power play). As a scientist or engineer, you will recognize these uses differ from the technical meaning of power.

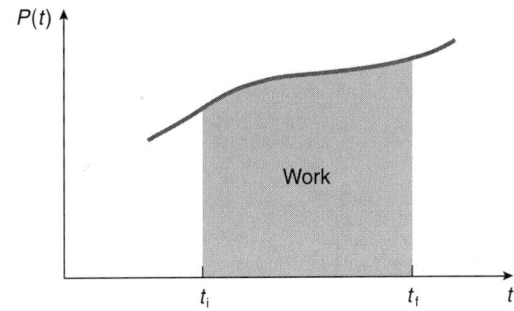

FIGURE 8.41 The work is the area under the graph of $P(t)$ versus t.

QUESTION 8

A system of mass m executes a closed path during a time interval Δt; one of the forces acting on the system is a conservative force. What is the average power of the conservative force during the time interval?

EXAMPLE 8.24

One hundred 2.00 kg bricks are to be lifted up 1.50 m from a pallet. A mason accomplishes the task during 10.0 min and a forklift does the same task in 3.00 s. Calculate the average power of the mason and that of the forklift.

Solution

The system is the 100 bricks. The forces on the bricks are

1. the gravitational force of the Earth on the bricks; and
2. the force of the mason on the bricks (for the case of the mason lifting the bricks), or (for the case of the forklift) the force of the forklift on the bricks.

The work done by the gravitational force in the process is accounted for by the change in the potential energy term in the CWE theorem. The work done by the mason or forklift must be accounted for on the *left*-hand side of the CWE theorem, since these forces have no potential energy function associated with them. Assume that the kinetic energy of the bricks is unchanged; they are picked up at rest and delivered at rest:

$$\Delta KE = 0 \text{ J}$$

Choose the coordinate system indicated in Figure 8.42 with $\hat{j}$ up and the origin at the bricks on the pallet. Since $y_i = 0$ m and $y_f = 1.50$ m, the change in the potential energy of the bricks is

$$\Delta PE = PE_f - PE_i$$
$$= mgy_f - mgy_i$$
$$= (100 \text{ bricks})(2.00 \text{ kg/brick})(9.81 \text{ m/s}^2)(1.50 \text{ m}) - 0 \text{ J}$$
$$= 2.94 \times 10^3 \text{ J}$$

Apply the CWE theorem:

$$W_{\text{mason or forklift}} = \Delta KE + \Delta PE$$
$$= 0 \text{ J} + 2.94 \times 10^3 \text{ J}$$
$$= 2.94 \times 10^3 \text{ J}$$

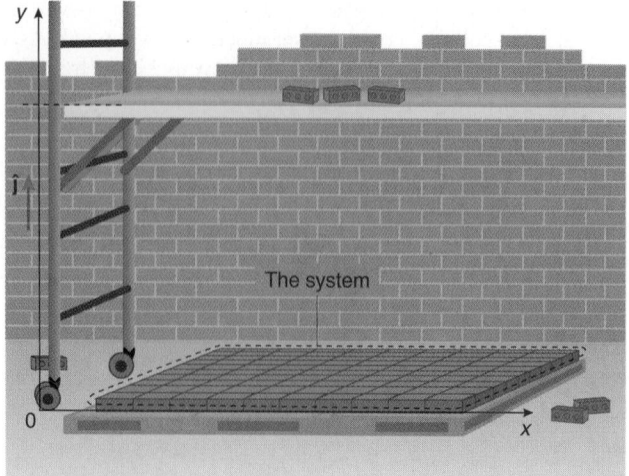

FIGURE 8.42

The mason accomplished the work on the bricks during 10.0 min or $\Delta t_{\text{mason}} = 600$ s, and so the average power of the mason is

$$P_{\text{ave mason}} = \frac{W_{\text{mason}}}{\Delta t_{\text{mason}}}$$
$$= \frac{2.94 \times 10^3 \text{ J}}{600 \text{ s}}$$
$$= 4.90 \text{ W}$$

The forklift also did the same work (actually more since it raised the pallet too), but during a time interval of 3.00 s, and so the average power of the forklift is

$$P_{\text{ave forklift}} = \frac{W_{\text{forklift}}}{\Delta t_{\text{forklift}}}$$
$$= \frac{2.94 \times 10^3 \text{ J}}{3.00 \text{ s}}$$
$$= 980 \text{ W}$$

EXAMPLE 8.25

What is the (instantaneous) power provided by the gravitational force on a particle of mass m in *circular* orbit about the Sun (of mass M); see Figure 8.43.

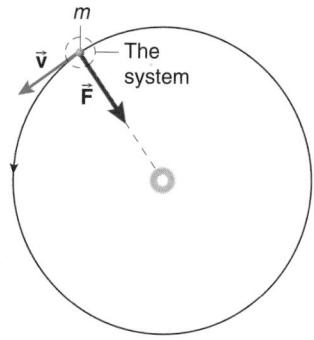

FIGURE 8.43

Solution

The instantaneous power is given by Equation 8.39:

$$P = \vec{F} \cdot \vec{v}$$

Note that the gravitational force is perpendicular to the velocity of m at every point in the circular orbit. Hence, at any instant, $\vec{F} \cdot \vec{v} = 0$ W, and the instantaneous power at all times and places around the circular orbit is equal to zero.

EXAMPLE 8.26

Hydroelectric power plants use falling water as a source of energy to turn generators to produce electricity. Consider a mass m of water released at rest that falls through a distance h.

a. Find the average power of the gravitational force during the interval of the fall.
b. Find the instantaneous power of the gravitational force the instant just before impact.

Solution

a. Take m to be the system. To find the average power, you need to know the work done on the system by the gravitational force and the time interval Δt during which this work was done. The time interval of the fall is found from the one-dimensional motion equations at constant acceleration (Chapter 3). Choose $\hat{\jmath}$ up and with the y-origin at bottom of the fall, as shown in Figure 8.44. Then the one-dimensional motion equation for constant acceleration,

$$y = y_0 + v_{y0}t + \frac{1}{2}a_y t^2$$

has

$$y_0 = h$$
$$v_{y0} = 0 \text{ m/s}$$
$$a_y = -g$$

At impact $y = 0$ m, so that

$$0 \text{ m} = h + \frac{1}{2}(-g)t^2$$

Solving for t (and taking the positive root, since the impact occurs after the mass is released),

$$t = \left(\frac{2h}{g}\right)^{1/2}$$

The work done by the gravitational force during the fall is the negative of the change in the gravitational potential energy. Use the coordinate system indicated in Figure 8.44:

$$\begin{aligned} W_{grav} &= -\Delta PE \\ &= -(PE_f - PE_i) \\ &= -(0 \text{ J} - mgh) \\ &= mgh \end{aligned}$$

The average power of the gravitational force during the interval of the fall then is

$$\begin{aligned} P_{ave} &= \frac{W_{grav}}{t} \\ &= \frac{mgh}{\left(\dfrac{2h}{g}\right)^{1/2}} \\ &= \frac{m}{\sqrt{2}}\sqrt{h}\,g^{3/2} \end{aligned}$$

b. Use the CWE theorem to calculate the speed of the mass just before impact. For variety, choose another coordinate system, the one shown in Figure 8.45.

During the fall the only force acting on the mass is the gravitational force. The work done by the gravitational force is accounted for by the potential energy term in the CWE theorem. The left-hand side of the theorem is zero, because there are no other forces acting on the mass during the fall. The initial kinetic energy is zero, since the mass was released from rest. The initial gravitational potential energy also is zero, because the mass was released at a point where its y-coordinate is zero. The final kinetic energy of the particle at the instant of impact is

$$KE_f = \frac{1}{2}mv^2$$

The final gravitational potential energy is equal to mgy_f where $y_f = -h$:

$$\begin{aligned} PE_f &= mg(-h) \\ &= -mgh \end{aligned}$$

The outflow from a hydroelectric power facility.

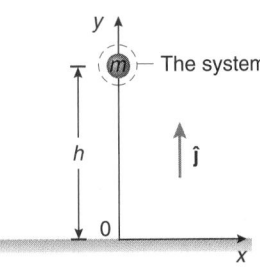

FIGURE 8.44

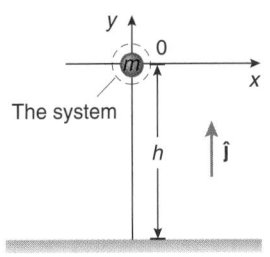

FIGURE 8.45

Therefore the CWE theorem assumes the form

$$W_{\text{other}} = \Delta(\text{KE} + \text{PE})$$

$$0 \text{ J} = (\text{KE} + \text{PE})_f - (\text{KE} + \text{PE})_i$$

$$= \left[\frac{1}{2} mv^2 + (-mgh) \right] - (0 \text{ J} + 0 \text{ J})$$

Hence

$$v = (2gh)^{1/2}$$

To calculate the power you need vector expressions for both the force and the velocity. With the coordinate system in Figure 8.45, the gravitational force on m is

$$\vec{\mathbf{F}}_{\text{grav}} = -mg\hat{\mathbf{j}}$$

and the velocity vector just before impact is

$$\vec{\mathbf{v}} = (2gh)^{1/2}(-\hat{\mathbf{j}})$$

Thus the instantaneous power P provided by the gravitational force is given by Equation 8.39:

$$P = \vec{\mathbf{F}}_{\text{grav}} \cdot \vec{\mathbf{v}}$$
$$= -mg\hat{\mathbf{j}} \cdot [-(2gh)^{1/2}\hat{\mathbf{j}}]$$
$$= \sqrt{2h} \, mg^{3/2}$$

Substitute representative numerical values to get a feeling for the quantity of this power: say $m = 2.00$ kg and $h = 30.0$ m. Then

$$P = [2(30.0 \text{ m})]^{1/2}(2.00 \text{ kg})(9.81 \text{ m/s}^2)^{3/2}$$
$$= 476 \text{ W}$$

8.19 THE POWER OF THE TOTAL FORCE ACTING ON A SYSTEM

It is the total force that causes the acceleration of a system. In this section we will see that the power of the total force dictates the time rate of change of the kinetic energy of the system.

The kinetic energy of a system of mass m when it has a speed v is given by Equation 8.25:

$$\text{KE} = \frac{1}{2} mv^2$$

Since $v^2 = \vec{\mathbf{v}} \cdot \vec{\mathbf{v}}$, the kinetic energy may be rewritten as

$$\text{KE} = \frac{1}{2} m \, \vec{\mathbf{v}} \cdot \vec{\mathbf{v}}$$

The time rate at which the kinetic energy of the system changes is the time derivative:

$$\frac{d}{dt}(\text{KE}) = \frac{d}{dt}\left(\frac{1}{2} m \, \vec{\mathbf{v}} \cdot \vec{\mathbf{v}} \right)$$

Consider a system of constant mass (we have considered only such systems). Differentiate the scalar product on the right-hand side with respect to time, obtaining

$$\frac{d}{dt}(\text{KE}) = \frac{1}{2} m \left(\frac{d\vec{\mathbf{v}}}{dt} \cdot \vec{\mathbf{v}} + \vec{\mathbf{v}} \cdot \frac{d\vec{\mathbf{v}}}{dt} \right)$$

The two terms on the right-hand side are equal. Thus

$$\frac{d}{dt}(\text{KE}) = m \frac{d\vec{\mathbf{v}}}{dt} \cdot \vec{\mathbf{v}} \qquad (8.41)$$

The *total force* acting on the system is, by Newton's second law,

$$\vec{\mathbf{F}}_{\text{total}} = m \frac{d\vec{\mathbf{v}}}{dt}$$

Substitute this in Equation 8.41 to obtain the time rate of change of the kinetic energy:

$$\frac{d}{dt}(\text{KE}) = \vec{\mathbf{F}}_{\text{total}} \cdot \vec{\mathbf{v}}$$

The right-hand side of this equation is the (instantaneous) power of the *total force*, i.e., the instantaneous total power.

> The instantaneous *total power* is equal to the time rate of change of the kinetic energy of the system:
>
> $$\frac{d}{dt}[\text{KE}(t)] = P_{\text{total}}(t) \qquad (8.42)$$

QUESTION 9

A graph is made of the kinetic energy of a system versus t. What is the physical significance of the slope of the graph?

EXAMPLE 8.27

A spaceship of essentially constant mass m, initially moving at velocity $v_{x0}\hat{\mathbf{i}}$ is briefly subjected to a constant acceleration $a_x\hat{\mathbf{i}}$ by its rocket motors. Find the power of the total force on the spacecraft during its acceleration.

Solution
The spacecraft is the system. The velocity component of the system as a function of time is found from the one-dimensional kinematic equations for a constant acceleration (Chapter 3):

$$v_x(t) = v_{x0} + a_x t$$

Method 1
The power is the scalar product of the total force and the velocity:

$$P = \vec{\mathbf{F}}_{\text{total}} \cdot \vec{\mathbf{v}}$$

The total force is found using Newton's second law:

$$\vec{\mathbf{F}}_{\text{total}} = ma_x\hat{\mathbf{i}}$$

Hence the power is

$$P(t) = ma_x\hat{\mathbf{i}} \cdot (v_{x0} + a_x t)\hat{\mathbf{i}}$$
$$= ma_x(v_{x0} + a_x t)$$

Method 2

The power of the total force is the time rate of change of the kinetic energy of the system. The kinetic energy of the system is

$$KE(t) = \frac{1}{2} m \left[v_x(t) \right]^2$$

$$= \frac{1}{2} m (v_{x0} + a_x t)^2$$

$$= \frac{1}{2} m (v_{x0}^2 + 2 v_{x0} a_x t + a_x^2 t^2)$$

Take the time derivative of the kinetic energy to find the power of the total force:

$$P_{\text{total}}(t) = \frac{dKE(t)}{dt}$$

$$= m(v_{x0} a_x + a_x^2 t)$$

$$= m a_x (v_{x0} + a_x t)$$

8.20 MOTION UNDER THE INFLUENCE OF CONSERVATIVE FORCES ONLY: ENERGY DIAGRAMS*

It is convenient to conceptualize the motion of a system subjected only to conservative forces with an **energy diagram**. Such diagrams are used extensively in contemporary physics, particularly in quantum mechanics. It is useful to explore them first in a classical context.

Consider a system that satisfies *either* of the following conditions:

a. the *only* forces acting on a particle are conservative forces; or
b. if there are "other" forces acting, the work done by all these other forces sums to zero—so $W_{\text{other}} = 0$ J.

Under *either* of these conditions, the CWE theorem of Equation 8.28 reduces to Equation 8.30:

$$0 \text{ J} = \Delta(KE + PE)$$

The total mechanical energy of the system is conserved, or unchanged during the motion. If the potential energy of the particle changes, the kinetic energy compensates, so that the sum of the two energies remains constant.

Let's see what this means from both an algebraic and graphical perspective. Substitute the kinetic energy, $KE = mv^2/2$, into the definition of the total mechanical energy:

$$E = \frac{mv^2}{2} + PE$$

Now solve for the speed:

$$v = \left[\frac{2}{m} \left(E - PE \right) \right]^{1/2} \qquad (8.43)$$

For the speed of the particle to be a *real number*, we must have

$$E \geq PE$$

at all times and places during the motion. If $E < PE$, the speed is an *imaginary number*; while this is mathematically legitimate, it is not physically permissible because any physical measurement of the speed yields a real number.

> Thus the total mechanical energy of the system must at all times and places be greater than or equal to its potential energy.

Look at several specific examples to see what this condition on E means.

Simple Harmonic Oscillator

We saw in Section 8.17 that the total mechanical energy E of a simple harmonic oscillator is constant. The potential energy of a simple harmonic oscillator was found in Section 8.12 to be (Equation 8.20)

$$PE = \frac{1}{2} kx^2$$

where $x = 0$ m represents the equilibrium position of the oscillator.

Make a graph of the potential energy as a function of x (see Figure 8.46). The graph is parabolic in shape because of the quadratic dependence on x. Plot the specific value of the total mechanical energy E of the given simple harmonic oscillator on the same diagram. Since E is fixed, the graph of E is a horizontal line at the appropriate number of joules on the vertical axis. The line representing E intercepts this potential energy curve at two places, whose x-coordinates are called **turning points**.

The possible x-coordinates of the oscillator must lie between the turning points, since it is only for these values of x that $E \geq PE$, leading to real values of the speed according to Equation 8.43. Thus the x-coordinate of the oscillator is bounded (confined) by the turning points. For values of x outside these bounds, E is less than the PE, and at those positions the speed of the oscillator becomes imaginary according to the speed equation (Equation 8.43). Thus, for a given E in Figure 8.46, portions of the x-axis are prohibited. We say the particle is **bound**, because it is confined to a finite region of the x-axis between the turning points.

The maximum value of x is the amplitude of the motion; the turning points are at $x = \pm A$. At any permitted position coordinate x, the potential and kinetic energies sum to the total mechanical energy as indicated in Figure 8.46. When the x-coordinate of the oscillator is at either turning point, the potential energy has increased so that it is equal to the (fixed) value of the total energy E; the kinetic energy of the oscillator there is zero and the system is instantaneously at rest. At the two turning points, the oscillator has its maximum amplitude and is reversing direction.

Suppose the oscillator is initially at rest at the equilibrium position, $x = 0$ m. The potential energy is zero, since $x = 0$ m; the kinetic energy is zero, since $v = 0$ m/s; and the total mechanical energy is zero, since $E = KE + PE = 0$ J + 0 J. The oscillator will not move, because the two turning points are both at $x = 0$ m in Figure 8.46. We mentioned previously that if a conservative force is the only force on a system and the system is released at rest, the system will tend to move *initially* in a direction to lower its potential energy. With the particle initially at $x = 0$ m and released at rest, the potential energy already is as

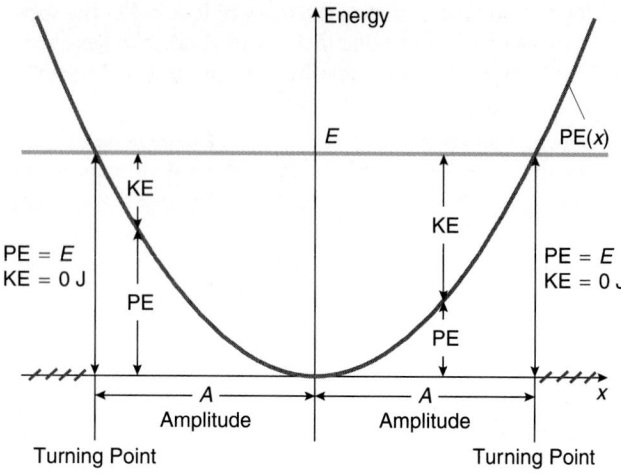

FIGURE 8.46 A graph of $PE(x)$ versus x for a simple harmonic oscillator.

low as it can be, and so no motion will ensue. We also know that if $x = 0$ m, the spring exerts no force on the particle and, therefore, the particle will not change its velocity.

On the other hand, when the oscillator is pulled to $x = A$ and then released at rest, the total mechanical energy of the oscillator is

$$E = \frac{1}{2} kA^2$$

This corresponds to releasing the oscillator at one of the turning points. The oscillator moves initially in a direction to lower its potential energy—that is, toward decreasing values of $|x|$. The motion is bounded by the two turning points at $x = \pm A$.

In this one-dimensional situation, the differential work done by the force of the spring on the system is

$$dW = \vec{\mathbf{F}}_{spring} \cdot d\vec{\mathbf{r}}$$
$$= F_x \hat{\mathbf{i}} \cdot dx\, \hat{\mathbf{i}}$$
$$= F_x\, dx \qquad (8.44)$$

The differential work done by a conservative force on the system is the *negative* of the differential change in the potential energy of the system (from the definition of the potential energy):

$$dW = -d(PE)$$

After substituting for dW, Equation 8.44 becomes

$$-d(PE) = F_x\, dx$$

and the x-component of the force is

$$F_x = -\frac{d}{dx}(PE) \qquad (8.45)$$

The component of the force of the spring is the negative of the slope of the potential energy curve.

The potential energy function for Hooke's force law is

$$PE = \frac{1}{2} kx^2$$

We can find the force component by taking the negative spatial derivative of the potential energy function:

$$F_x = -\frac{d}{dx}\left(\frac{1}{2} kx^2\right)$$
$$= -kx$$

That is,

$$\vec{\mathbf{F}}_{spring} = F_x \hat{\mathbf{i}}$$
$$= -kx\hat{\mathbf{i}}$$

as before.

At the bottom of the curve, the slope is zero and so the force component is zero. When $x > 0$ m, the slope is positive, so the expression for the force component F_x is negative, indicating that the force is directed toward $-\hat{\mathbf{i}}$. When $x < 0$ m, the slope is negative, and the force component becomes positive or directed along $+\hat{\mathbf{i}}$.

QUESTION 10 _____
A graph of the acceleration component a_x of a simple harmonic oscillator is made as a function of x. What is the shape of the curve?

The Gravitational Force

The gravitational force is a conservative force. If the gravitational force is the only force on a system, the CWE theorem becomes

$$W_{other} = \Delta(KE + PE)$$
$$0\ J = \Delta(KE + PE)$$

The total mechanical energy $E = KE + PE$ of the system is conserved.

The most general form for the potential energy function of a system of mass m located a distance r from another pointlike or spherical mass M is, from Equation 8.15,*

$$PE = -\frac{GmM}{r}$$

_____
*The masses m and M must be either point masses or spheres with spherically symmetric mass distributions. The center-to-center position of m and M must be greater than the sum of the radii of the spheres to use this expression for the potential energy. A different form for the potential energy (see Problem 20) is used for values of r less than the radius of the sphere.

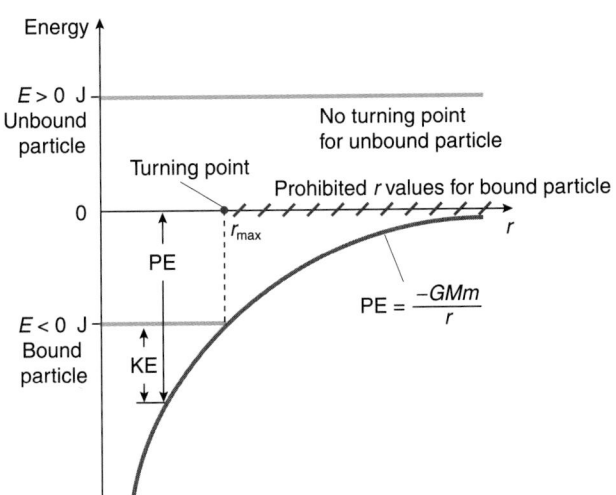

FIGURE 8.47 A graph of the gravitational potential energy as a function of r.

A graph of this potential energy as a function of r is shown in Figure 8.47. Recall that the radial coordinate r is always greater than or equal to zero, and so only positive values of r are plotted. According to Equation 8.43, the total mechanical energy E must be greater than or equal to the potential energy in order to have a real number value for the speed for the system.

For the value of the total mechanical energy $E < 0$ J indicated in Figure 8.47, the indicated region along the r-axis is prohibited, since for these r-coordinates the total mechanical energy E is less than the potential energy and the speed is imaginary. There is a turning point where the line representing E intersects the potential energy curve. What this means physically is that if E is fixed, there are many possible orbits of m about M (ellipses with various eccentricities), but all with an upper bound on r at the turning point.

At any given permitted r, the values of the kinetic and potential energies are schematically indicated in Figure 8.47. The fixed energy E is the sum of the kinetic and potential energies. Notice also that if the total mechanical energy E is greater than zero, there are no turning points and no bound on the motion of the particle: any r value from zero to infinity is permitted. Particles with $E \geq 0$ J are called **unbound** particles; their orbits are straight lines and the conic sections known as parabolas (for $E = 0$ J) and hyperbolas (for $E > 0$ J) for inverse-square-law forces. For negative total energies ($E < 0$ J), the motion is bounded (there is a maximum value that r can attain). Particles with negative total energies are bound particles, because the radial coordinate has an upper bound. The orbits of particles with negative total energies are circles (with M at the center), ellipses (with M at one focus), and straight lines (with M on the line) for attractive inverse-square-law forces.

Once again the differential work done by the force is

$$dW = \vec{\mathbf{F}} \cdot d\vec{\mathbf{r}}$$

Write the central gravitational force of M on m as

$$\vec{\mathbf{F}} = F_r \hat{\mathbf{r}}$$

(where $F_r < 0$ N, since gravitation is an attractive force). The gravitational force (like any central force) has *only* a radial component.

The only work done by the gravitational force on m occurs when m moves in the radial direction—that is, when $d\vec{\mathbf{r}} = dr\,\hat{\mathbf{r}}$:

$$\begin{aligned} dW &= F_r \hat{\mathbf{r}} \cdot dr\,\hat{\mathbf{r}} \\ &= F_r\,dr \end{aligned} \tag{8.46}$$

But the differential work done by a conservative force is the negative of the differential change in the potential energy:

$$dW = -d(\text{PE})$$

This substitution for dW in Equation 8.46 gives

$$-d(\text{PE}) = F_r\,dr$$

and so the radial component (and only component) of the force is

$$F_r = -\frac{d(\text{PE})}{dr} \tag{8.47}$$

Here again the force component is the negative of the slope of the potential energy function. Use the potential energy function,

$$\text{PE} = -\frac{GmM}{r}$$

and take the negative of its derivative with respect to r to obtain the radial force component F_r:

$$F_r = -\frac{d}{dr}\left(-\frac{GmM}{r} \right)$$

We find

$$F_r = -\frac{GmM}{r^2}$$

The force on m then is

$$\begin{aligned} \vec{\mathbf{F}} &= F_r \hat{\mathbf{r}} \\ &= -\frac{GmM}{r^2}\,\hat{\mathbf{r}} \end{aligned}$$

Voilà. Figure 8.47 has a positive slope throughout its extent, so F_r is negative, indicating that the gravitational force always is an attractive force.

Central Forces

The gravitational force is a central force. Any central force has only a radial force component:

$$\vec{\mathbf{F}} = F_r \hat{\mathbf{r}}$$

Likewise, the potential energy function associated with a central force depends only on r:

$$PE = PE(r)$$

> The force component F_r is the negative of the slope of the graph of the potential energy versus r.

Nothing in the derivation of Equation 8.46,

$$F_r = -\frac{d(PE)}{dr}$$

depends on anything other than having the potential energy be a function of r alone: $PE = PE(r)$.

The force is repulsive ($F_r > 0$ N) where the slope of the potential energy curve is negative; the force is attractive ($F_r < 0$ N) where the slope is positive. Again the motion is restricted to regions where $E \geq PE$ to keep the speed of the system real rather than imaginary.

EXAMPLE 8.28

The gravitational potential energy of a mass m near the surface of the Earth is (Equation 8.13)

$$PE = mgy \qquad (\hat{\jmath} \text{ directed up})$$

a. Make a schematic graph of the potential energy as a function of y for both positive and negative values of y.
b. What is the value of the y-intercept of the graph?
c. What is the slope of this graph?
d. How is the slope related to the gravitational force on m?
e. Show that the gravitational force

$$\vec{F} = F_y \hat{\jmath}$$

has a component F_y that is

$$F_y = -\frac{d(PE)}{dy}$$

Solution

a. The graph of the potential energy as a function of y is shown in Figure 8.48.
b. The y-intercept of the graph is zero.
c. The slope of the graph is the coefficient of y in the equation $PE = mgy$. Thus the slope is mg.
d. The *negative* of the slope is the gravitational force component F_y on m.
e. Since

$$F_y = -\frac{d(PE)}{dy}$$

substitute for the potential energy:

$$F_y = -\frac{d(mgy)}{dy}$$
$$= -mg$$

Hence the force is

$$\vec{F} = F_y \hat{\jmath}$$
$$= -mg\hat{\jmath}$$

Since $\hat{\jmath}$ is vertically upward, the gravitational force on m is directed vertically downward.

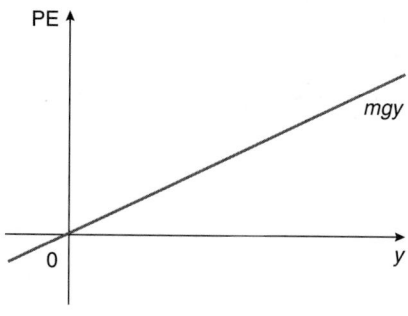

FIGURE 8.48

STRATEGIC EXAMPLE 8.29

A stone of mass 0.500 kg is thrown vertically upward with an initial speed of 6.00 m/s from a point at an elevation of 2.00 m above the ground. Choose the $\hat{\jmath}$ to be up with an origin at the ground.

a. Find the total mechanical energy of the stone at launch.
b. Does the total mechanical energy of the stone change during its flight?
c. Make a detailed graph of the potential energy of the stone as a function of y.
d. Plot the total mechanical energy of the stone on the potential energy graph.
e. Locate the turning point on the graph and explain its physical significance.
f. Locate the initial y-coordinate of the stone on the graph and indicate on the graph what represents the initial kinetic and potential energies of the stone.

Solution

a. The total mechanical energy of the stone at launch is the sum of its initial kinetic and potential energies:

$$E = KE + PE$$
$$= \frac{1}{2}mv_i^2 + mgy_i$$

Since the y-axis origin is on the ground,

$$y_i = 2.00 \text{ m}$$

Also you are given $v_i = 6.00$ m/s. Hence

$$E = \frac{1}{2}(0.500 \text{ kg})(6.00 \text{ m/s})^2$$
$$+ (0.500 \text{ kg})(9.81 \text{ m/s}^2)(2.00 \text{ m})$$
$$= 9.00 \text{ J} + 9.81 \text{ J}$$
$$= 18.81 \text{ J}$$

b. Since only the gravitational force acts on the stone during its flight, $W_{\text{other}} = 0$ J in the CWE theorem, and so the theorem becomes

$$0 \text{ J} = \Delta(\text{KE} + \text{PE})$$

The total mechanical energy of the stone *does not change* throughout the flight.

c. The potential energy of the stone as a function of y is

$$\text{PE} = mgy$$
$$= (0.500 \text{ kg})(9.81 \text{ m/s}^2)y$$
$$= (4.91 \text{ J/m})y$$

This potential energy function is graphed in Figure 8.49.

d. The total mechanical energy E is a horizontal line on the graph with a y-axis intercept of 18.81 J.

e. From the graph, the turning point is the y-coordinate where the line representing E intersects the potential energy curve. This intersection occurs at about $y_{\text{turn pt}} \approx 3.8$ m. At the turning point, $E = \text{PE}$, so that the KE of the stone there is zero. The stone is instantaneously at rest at this coordinate, and so this is the y-coordinate of the maximum height of the stone. Since we chose the y-axis origin at the ground, the maximum height of the stone is about 3.8 m.

The stone cannot have y-coordinates greater than the turning point, and so $y > 3.8$ m is a prohibited region.

f. The initial y-coordinate of the stone was $y_i = 2.00$ m. The initial potential energy is represented by the vertical distance

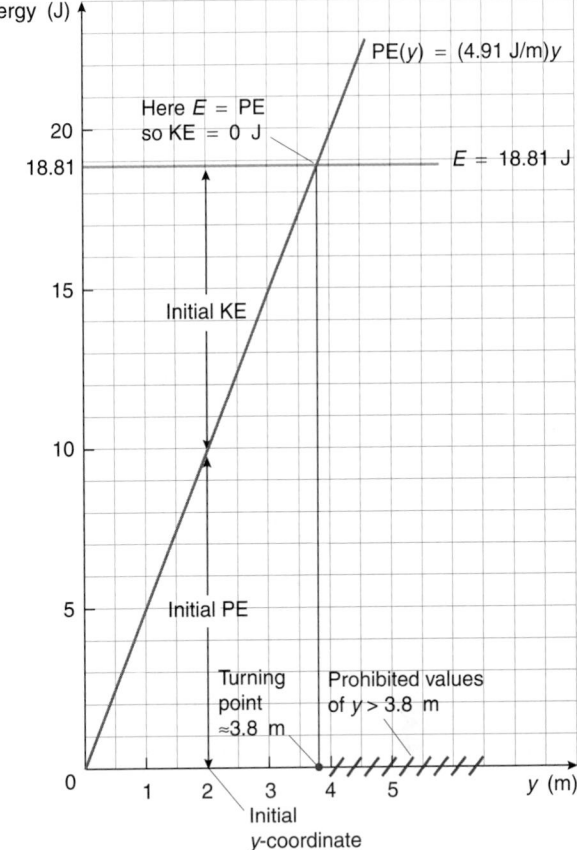

FIGURE 8.49

on the graph between the horizontal axis (y) at $y = 2.00$ m and the line representing the potential energy. Since $E = \text{PE} + \text{KE}$, the initial kinetic energy is the vertical distance on the graph between the potential energy (at $y = 2.00$ m) and the line representing E.

CHAPTER SUMMARY

The *work* W done by any force $\vec{\mathbf{F}}$ on a system of mass m is

$$W = \int_i^f \vec{\mathbf{F}} \cdot d\vec{\mathbf{r}} \qquad (8.2)$$

where the integral is taken along the path followed by the system between its initial and final positions. The SI units of work are N·m, called *joules* (J).

If a force is a *constant force* (i.e., a constant vector), along a path (whether straight or curved), then the work done by such a force on a system is

$$W = \vec{\mathbf{F}} \cdot \Delta\vec{\mathbf{r}} \qquad (\text{constant force } \vec{\mathbf{F}}) \qquad (8.6)$$

where $\Delta\vec{\mathbf{r}}$ is the *change* in the position vector of the system:

$$\Delta\vec{\mathbf{r}} = \vec{\mathbf{r}}_f - \vec{\mathbf{r}}_i$$

A system *cannot* be said to *have* work, so that *work is not a property of a system*. On the other hand, *energy is a property of a system*. Changes in the energy of a system are related to the work done on the system by the forces acting on it.

Forces are classified as *conservative*, *nonconservative*, or *zero-work* forces. The work done by a *conservative force* on a system as it moves between two positions is *independent of the path* followed by the system between the two points. Equivalently, the work done by a conservative force around any closed path is zero. If the work done by a force around an arbitrarily chosen closed path is not equal to zero, the force is a *nonconservative force*. Certain forces always do zero work and are known as *zero-work forces*. The normal force of a rigid surface on a system and the static frictional force are zero-work forces.

The work done by a conservative force is defined to be equal to the negative of the change in a potential energy function associated with the conservative force:

$$W_{consrv} \equiv -\Delta PE \qquad (8.11)$$

The potential energy of a system depends on its *location* in space with respect to a reference position where PE = 0 J.

Important potential energy functions to know are those associated with the following conservative forces (on a system of mass m):

a. For the *gravitational force* on the mass m when it is always near the surface of the Earth, the potential energy of m is

$$PE_{local} = mgy \qquad (\hat{\jmath} \text{ directed up})$$

where $y = 0$ m may be placed at any convenient elevation.

b. For the more general gravitational force of M on m, the potential energy of m when it is located a distance r from M is

$$PE_{grav} = -\frac{GmM}{r} \qquad (8.15)$$

where the origin for r is at M and the zero of potential energy always is where $r = \infty$ m.

c. For Hooke's force law associated with a mass m on a spring, the potential energy of m is

$$PE_{spring} = \frac{1}{2} kx^2 \qquad (8.20)$$

where $x = 0$ m is the equilibrium position of m and k is the spring constant.

The CWE *theorem* states that the work, W_{total}, done by *the total force* on a system of mass m is equal to the change in the kinetic energy of the system:

$$W_{total} = \Delta KE \qquad (8.26)$$

Since the forces acting on a system are either conservative forces, nonconservative ("other") forces, or zero-work forces, the total work done by the total force is broken up into two contributions:

$$W_{other} + W_{consrv} = \Delta KE$$

Since $W_{consrv} \equiv -\Delta PE$, the CWE theorem also can be put into the form

$$W_{other} = \Delta KE + \Delta PE$$
$$= \Delta (KE + PE) \qquad (8.28)$$

The sum of the kinetic and potential energies of a system is called the *total mechanical energy E* of the system, and so the CWE theorem has yet another variation:

$$W_{other} = \Delta E \qquad (8.29)$$

If $W_{other} = 0$ J, then the total mechanical energy E of the system is *conserved*, or constant. The condition $W_{other} = 0$ J is met if either

(1) *all* the forces on the system are conservative or zero-work forces, or (2) the total work done by "other" (e.g., nonconservative) forces on the system is zero.

The *escape speed* from a spherical mass M of radius R is

$$v_{escape} = \left(\frac{2GM}{R} \right)^{1/2} \qquad (8.32)$$

A simple harmonic oscillator has a total mechanical energy E that is constant and equal to

$$E = \frac{1}{2} kA^2 \qquad (8.35)$$

or

$$E = 2\pi^2 m v^2 A^2 \qquad (8.36)$$

The average power P_{ave} of a force $\vec{F}$ on a system is defined to be

$$P_{ave} \equiv \frac{\text{work done by } \vec{F} \text{ during a time interval } \Delta t}{\Delta t} \qquad (8.38)$$

The instantaneous power P of a force $\vec{F}$ on a system is

$$P = \vec{F} \cdot \vec{v} \qquad (8.39)$$

where $\vec{v}$ is the velocity of the system. The SI unit for power is the J/s or a *watt* (W).

The (instantaneous) *total power* P_{total} of the total force on a system is the time rate of change of the kinetic energy of the system:

$$P_{total} = \frac{d(KE)}{dt} \qquad (8.42)$$

Interesting information about a system for which $W_{other} = 0$ J can be obtained from an *energy diagram* in which the potential energy of the system is graphed as a function of its position. The CWE theorem implies that the speed of the system is

$$v = \left[\frac{2}{m} \left(E - PE \right) \right]^{1/2} \qquad (8.43)$$

To have a real (rather than imaginary) value for v, the (constant) total mechanical energy E must satisfy

$$E \geq PE$$

The coordinates on such energy diagrams where $E = PE$ are called *turning points*. Coordinates where $E < PE$ are prohibited.

At any point on such a potential energy graph, the *negative of the slope* is the force component associated with that potential energy function:

$$F_x = -\frac{d(PE)}{dx} \qquad (8.45)$$

SUMMARY OF PROBLEM-SOLVING TACTICS

8.1 **(page 321)** The differential change $d\vec{r}$ in the position vector of the system is written as

$$d\vec{r} = dx\,\hat{\imath} + dy\,\hat{\jmath} + dz\,\hat{k} \qquad (8.3)$$

whether the particle is moving toward increasing or decreasing values of the coordinates. The limits of integration account for the actual direction the particle moves in going

from its initial position (the lower limit) to its final position (the upper limit).

8.2 **(page 324)** The work done by a constant force, Equation 8.6, can be used even if the path between the initial and final positions is not straight.

8.3 **(page 336)** Equation 8.13 for the local gravitational potential energy must have $\hat{\jmath}$ directed up, since that is the choice we made in deriving the equation; you can choose y = 0 m to be at any elevation you want.

8.4 **(page 337)** When using Equation 8.15 for the gravitational potential energy, choose the origin of the coordinate system to be at M; no other choice for the origin should be made.

8.5 **(page 342)** Equations 8.26, 8.27, 8.28, and 8.29 are all equivalent expressions for the CWE theorem.

QUESTIONS

1. (page 322); 2. (page 324); 3. (page 325); 4. (page 330); 5. (page 334); 6. (page 336); 7. (page 343); 8. (page 356); 9. (page 358); 10. (page 360)

11. Can the normal force of a surface on a system ever do nonzero work?

12. Is the kinetic frictional force a constant force if the path followed by a system is not a straight-line path? Explain.

13. A raindrop is falling through the air at a terminal speed. (Refer back to Section 5.14.) What forces act on the raindrop? What is the relationship between the amounts of work done on the raindrop by these forces?

14. The normal force of a surface and the static frictional force both are zero-work forces, but typically for different reasons. Explain why each is a zero-work force.

15. A student slowly lifts a book from the floor up to a library shelf without changing the kinetic energy of the book. Newton's second law implies that the force of the Earth on the book (its weight) and the force of the student on the book are of equal magnitude and opposite direction. Are these a Newton's third law pair of forces? Explain. The total force on the book is zero, indicating zero total work was done. Yet certainly *some* work was done, since the book is now on the shelf. Resolve this dilemma.

16. A system of mass m lies on a rough table. The system is dragged at a constant low speed between two points A and B on the table. The kinetic frictional force on the system is independent of its low speed. The path is *not* a straight line. Carefully present arguments to show that the work done by the kinetic frictional force of magnitude f_k is equal to

$$W_{\text{A to B}} = -f_k s$$

where s is the total length of the path taken between the two points. Thus the work done by the kinetic frictional force depends on the specific path taken between the two points, and so it is not a conservative force.

17. The orbital path of a satellite is an ellipse (see Figure Q.17). The work done by the gravitational force around one orbit is

$$W = \int_{\substack{\text{one} \\ \text{orbital path}}} \vec{F}_{\text{grav}} \cdot d\vec{r}$$

(a) At what points along the orbit is the integrand zero? (b) Is the *integrand* zero at *every* point along the orbital path? (c) Is the work done by the gravitational force over one complete orbital path zero? Explain why or why not.

18. A mass m is located somewhere inside a thin, uniform, hollow spherical shell. Give a convincing argument to show that the gravitational potential energy of m is constant, independent of its position within the shell.

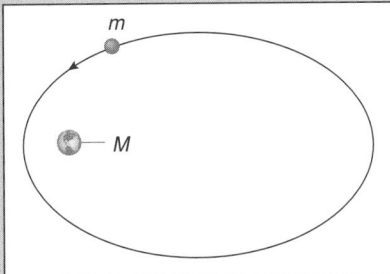

FIGURE Q.17

19. A student applies the CWE theorem and surprisingly discovers from a calculation that the kinetic energy of the system is negative at some location. What does a negative kinetic energy imply about the speed of the system? It is *certain* that the student made an error in applying the CWE theorem. Why?

20. By what factor must the speed of a system of mass m be increased to *triple* its kinetic energy?

21. A legislator argues that a statewide speed limit should not be increased from 100 km/h to 140 km/h because "the force of a collision (with a stationary object) doubles." Does it? Explain your reasoning.

22. A rock is dropped from rest. An identical rock is thrown downward with an initial speed of 5.00 m/s. Is the work done by the total force on each rock over the first meter of its flight (after their release) the same? Find the change in the speed of each rock after they have traveled a distance of one meter from the point of their release. The gravitational force did the same work on each; why are the changes in the speeds different? Are the changes in the kinetic energy of both rocks over the one meter path the same? Explain.

23. A ball at rest is dropped from your hand and falls to the floor. Use the CWE theorem to explain why the ball never can bounce up to a height greater than that from which it was released. How can the ball be released from your hand so it *does* bounce to a higher elevation than the point of release?

24. Let $\Delta\vec{v} = \vec{v}_f - \vec{v}_i$. Explain why the change in the kinetic energy is *not*

$$\frac{1}{2} m(\Delta v)^2$$

25. Is the force $\vec{F} = (1.00 \text{ N})\hat{\imath} - (1.00 \text{ N})\hat{\jmath}$ a conservative force? Explain why or why not.

26. A physics student throws this (!) textbook off a cliff in one of the ways shown in Figure Q.26. He wants the impact speed to be as large as possible. Arg! If the *initial* speed of the text is the same along each trajectory, along which is the impact speed the greatest or does it matter? Justify your answer.

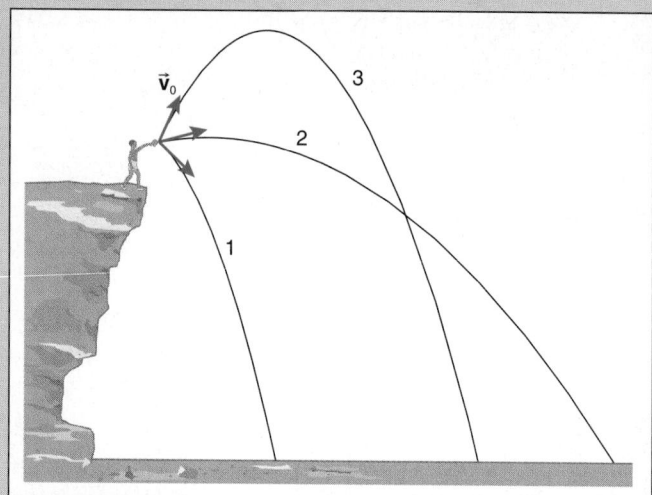

FIGURE Q.26

Q.31. (a) Make a corresponding rough sketch of the velocity component v_x as a function of time. (b) Make a corresponding rough sketch of the acceleration component a_x as a function of time. (c) Make a corresponding rough sketch of the force component F_x as a function of time. (d) Make a corresponding rough sketch of the kinetic energy as a function of time. (e) Make a corresponding rough sketch of the potential energy as a function of time.

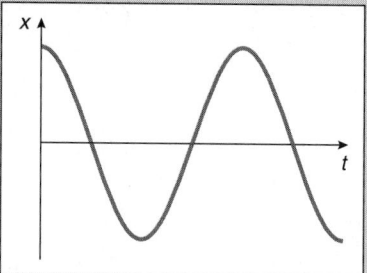

FIGURE Q.31

27. Consider an inertial reference frame S' moving at constant velocity $v\hat{\imath}$ with respect to another inertial reference frame S, with their respective coordinate axes parallel to each other, and their x-axes coincident as shown in Figure Q.27. A system of mass m is moving in reference frame S' with an initial velocity $v_0\hat{\imath}'$. (a) What is the kinetic energy of the system as measured in S'? (b) What is the kinetic energy of the system as measured in reference frame S? (c) We showed in Chapter 4 that the acceleration of a system is the *same* in both inertial frames, so that both frames will measure the *same* total force acting on the system. If the total force is not zero, will the two reference frames agree on the total work done by the total force acting on the system during a one second interval? Why or why not?

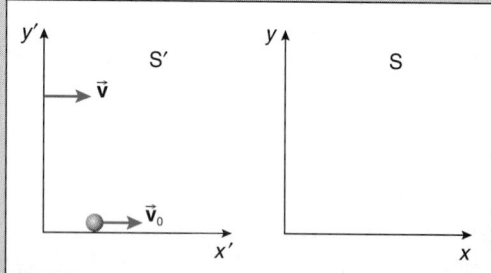

FIGURE Q.27

28. Can the CWE theorem be applied in noninertial reference frames? Explain.

29. Say the work done by all the nonconservative forces on a system is negative as it moves between two points. What happens (qualitatively) to the total mechanical energy of the system? Describe a physical situation that illustrates this scenario.

30. The nonconservative forces on a system are found to do positive work on the system. What happens (qualitatively) to the total mechanical energy of the system? Describe a physical situation that illustrates this scenario.

31. A rough sketch of the position x of a mass executing simple harmonic oscillation as a function of time is shown in Figure

32. For an elliptical orbit of a planet about the Sun, name several physical quantities that are *constant* at all locations in the orbit and several physical quantities that depend on where the planet is located in its orbit.

33. A baseball of mass m is tossed vertically up into the air and caught at the same place on the way down. Ignoring the effects of air friction, what is the work done by the gravitational force on the ball over the path followed by the ball? If you consider air friction effects, does this change the answer to the previous question? If air friction *cannot* be neglected, what can be said (if anything) about the speed of the ball as it is caught compared with the launch speed? Justify your answer via the CWE theorem.

34. For a particle of mass m in an elliptical orbit about the Sun, is the instantaneous power of the gravitational force on m equal to zero at *every* point along the orbit? If not, are there points where it *is* equal to zero? If so, what points? Is the average power of the gravitational force over one period of the orbit equal to zero? Over what portions of the orbit is the work done by the gravitational force positive? Negative?

35. A log of mass 5.00 kg is dropped from rest off a cliff of height 30.0 m. What is the *initial* power of the gravitational force on the log?

36. A graph is made of the instantaneous power $P(t)$ versus time t. What is the physical significance of the area under the graph between $t = 0$ s and some other time $t > 0$ s?

37. Are the colloquial uses of the words work, energy, and power synonymous? What is meant by each term in physics? Think of several other scientific words that have different meanings in a colloquial context.

38. A graph of the potential energy of an α-particle (the nucleus of a helium atom) in the immediate vicinity of a massive nucleus has the schematic form indicated in Figure Q.38. If the total energy E of the α-particle is that indicated in the

sketch, qualitatively describe the motion of the α-particle (i) if it is located in the region to the left of the hump on the diagram, and (ii) if it is located to the right of the hump. Explain why it is (classically) impossible for the α-particle to be in the region between r_1 and r_2. One of the strange features of quantum mechanics is that it is possible for the α-particle to "tunnel" through the region where it is (classically) forbidden to be; among other things, this phenomenon describes radioactivity.

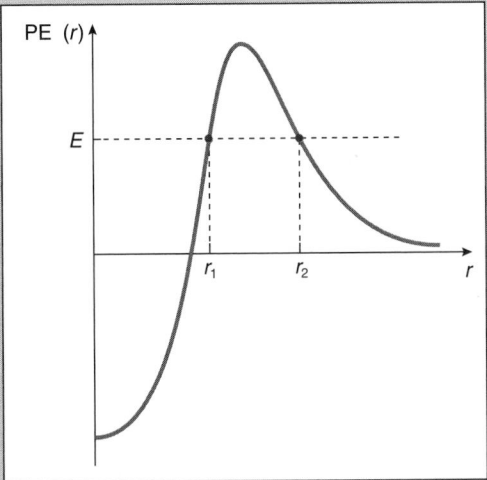

FIGURE Q.38

39. A satellite is in *circular* orbit about the Earth. How is the satellite represented on an energy diagram like Figure 8.47? How is a satellite in an *elliptical* orbit represented on such a diagram?

40. Only conservative forces act on a certain system. The total potential energy of the system is plotted as a function of a coordinate x; Figure Q.40 is obtained. The total mechanical energy of the system also is indicated. What can be said about the motion of the system?

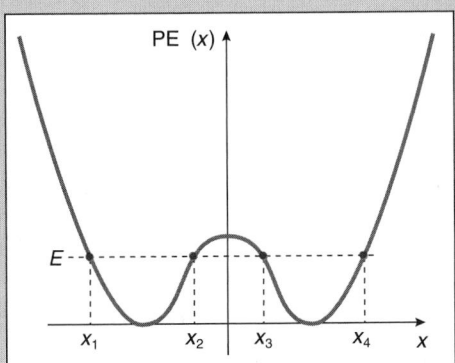

FIGURE Q.40

41. The strong force, also known as the nuclear force, is one of the fundamental forces of nature, as mentioned in Section 5.2. The strong force is conservative, which means that there is a potential energy associated with it. A nuclear particle such as a neutron has a potential energy that is a function of its position relative to the center of the nucleus. Write the strong force as

$$\vec{F}_{strong} = F_r \hat{r}$$

The potential energy of a neutron at a distance r from the center of a nucleus has the general schematic dependence on r indicated in Figure Q.41; in the figure R is the radius of the nucleus (about 10^{-15} m). (a) For distances $r > R + \Delta r$, what is the magnitude and direction of the strong force? (b) For distances $R < r < R + \Delta r$, what can be said about the magnitude and direction of the strong force? (c) For distances $R - \Delta r < r < R$, what can be said about the magnitude and direction of the strong force? (d) For distances $r < R - \Delta r$, what can be said about the magnitude and direction of the strong force? (e) Describe how the strong force on the neutron is quite *different* from the gravitational force of the nucleus on the neutron.

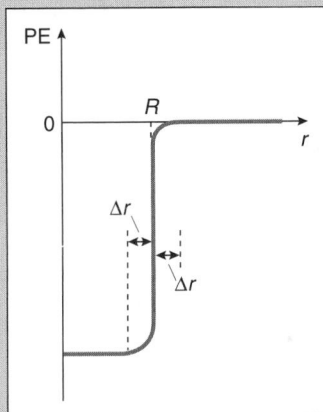

FIGURE Q.41

42. For forced oscillations, give a convincing argument to explain why the power transferred to the oscillator by a sinusoidal applied force is a maximum if the applied force component has the same frequency and phase angle as the velocity component.

43. Rewrite the following passage using proper physical terminology:

Master Andry raised his eyes, seemed for a moment to be measuring the height of the column, the weight of the rascal, mentally multiplying that weight by the square of the velocity, and was silent.

Victor Hugo, *Notre-dame de Paris*, Volume I, Book I, Chapter I, "The Great Hall" (Estes and Lauriat, Boston, 1892), page 11.

PROBLEMS

Sections 8.1 Motivation for Introducing the Concepts of Work and Energy
8.2 The Work Done by Any Force
8.3 The Work Done by a Constant Force
8.4 The Work Done by the Total Force
8.5 Geometric Interpretation of the Work Done by a Force*

1. A force $\vec{F} = (15.0\ \text{N})\hat{i}$ is one of the forces on a pumpkin (*Cucurbita pepo*) of mass 10.0 kg. For no apparent reason, calculate the work done by this force on the pumpkin for each of the changes in its position vector in Figure P.1.

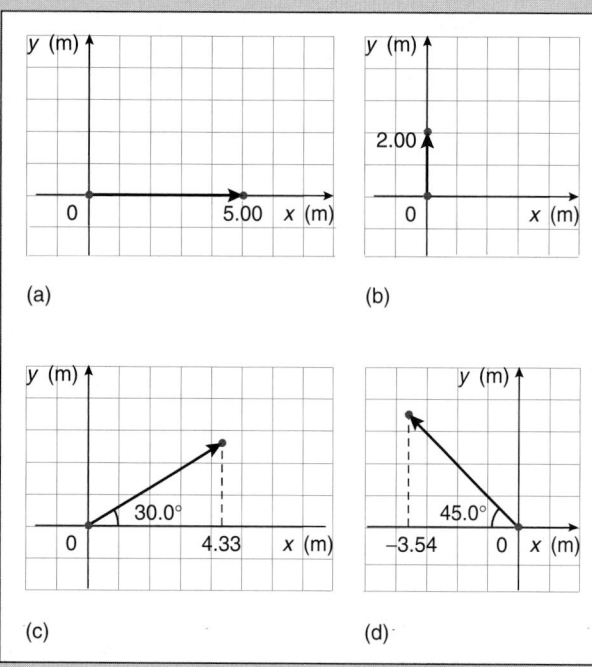

(a)

(b)

(c)

(d)

FIGURE P.1

2. A constant force $\vec{F} = (20.0\ \text{N})\hat{i}$ pulls a 5.00 kg monster zucchini along the x-coordinate axis from $x = 2.00$ m to $x = 5.00$ m. Find the work done by $\vec{F}$ on the squash over this path.

3. A force $\vec{F} = (5.00\ \text{N})\hat{j}$ acts on a 2.0 kg bundle of homework problems while it moves along the y-axis from the coordinate $y_1 = 4.00$ m to $y_2 = -2.00$ m. (a) Graph F_y versus y. (b) Find the work done by this force on the bundle between the two points.

4. A force whose only component is F_x acts on 1.50 kg system; a graph of F_x versus x appears in Figure P.4. Find the work done on the system by this force while the system moves along the x-axis from $x = -2.00$ m to $x = +10.00$ m.

5. A force $\vec{F} = -(6.00\ \text{N/m})x\hat{i}$ acts on a 3.0 kg boxful of nuts and bolts as it moves along the z-axis between the initial position $z_1 = 9.00$ m and final position $z_2 = 2.00$ m. (a) Make a graph of F_x versus x. (b) Find the work done by this force as the system moves between the two positions.

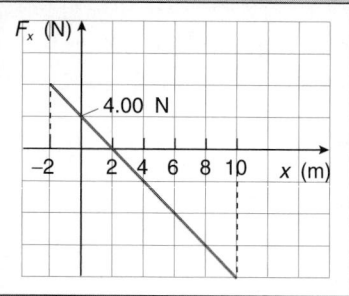

FIGURE P.4

6. A 5.00 kg watermelon (*Citrullus vulgaris*) moves from $x = 8.00$ m, $y = 2.00$ m, $z = 0$ m to $x = -4.00$ m, $y = 1.00$ m, $z = 0$ m by moving along the *curved* path shown in Figure P.6. One of the forces acting on the watermelon system is $\vec{F} = (60.0\ \text{N})\hat{i} - (40.0\ \text{N})\hat{j} + (25.0\ \text{N})\hat{k}$. Calculate the work done on the system by this force as the system moves from its initial position to final position.

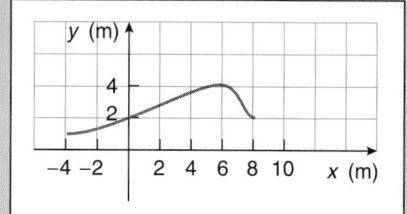

FIGURE P.6

7. A 0.450 kg quahog clam (*Venus mercenaria*) moves from $x = -2.00$ m, $y = 1.00$ m to $x = 3.00$ m, $y = 1.00$ m along the path indicated in Figure P.7. One of the forces acting on the clam system is $\vec{F} = (20.0\ \text{N})\hat{i} + (10.0\ \text{N})\hat{j}$. What is the work done on the system by this force over the path followed by the clam?

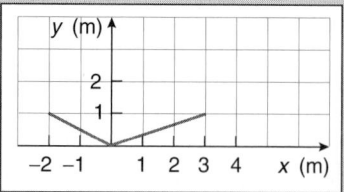

FIGURE P.7

8. A force $\vec{F} = (-30.0\ \text{N})\hat{i} + (20.0\ \text{N})\hat{j}$ acts on a 10.0 kg log as it moves from $x = 10.0$ m to $x = 5.0$ m along the x-coordinate axis. Find the work done on the log by this force over this path.

9. A force on a system has only an x-component F_x; a graph of this component as a function of position x is shown in Figure P.9. Calculate the work done by this force if the system moves along the x-axis from $x = 0$ m to $x = 10.0$ m.

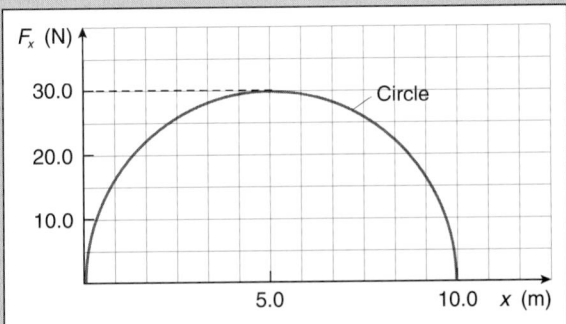

FIGURE P.9

10. Your professor is having her office cleaned and asks you to slide a 175 kg filing cabinet 6.5 m across a rough floor in a straight line. You push with a horizontal force of magnitude 280 N and do the job for brownie points. How much work do you do?

•11. The force $\vec{\mathbf{F}} = (1.00\,\text{N})\hat{\mathbf{i}} - (1.00\,\text{N})\hat{\mathbf{j}}$ is one of the forces on a 4.00 kg system located at the origin. (a) Sketch the direction of the force on a suitable coordinate grid. (b) Indicate a path that the system may follow for 2.00 m so that $\vec{\mathbf{F}}$ does *zero* work on the system along the path. If the system returns to the origin along this path, is the work done by this force zero along the return trip? Explain your reasoning. (c) Indicate a 2.00 m long path that the system may follow so that $\vec{\mathbf{F}}$ does the maximum amount of *positive work* on the system along the path. What is the work done by the force along this path? If the system returns to the origin along this path, does the force do positive work along the return trip? Explain. (d) Indicate a 2.00 m long path that the system may follow so that $\vec{\mathbf{F}}$ does the maximum amount of *negative work* on the system along the path. What is the work done by this force along this path? If the system returns to the origin along this path, does the force do negative work along the return trip? Explain.

•12. A professor moves a 3.00 kg mass attached to a spring (with $k = 10.0$ N/m) very slowly along a straight path from $x = -0.10$ m to $x = +0.15$ m; see Figure P.12. At any position, the acceleration of the mass is essentially zero. Calculate the work done on the mass m: (a) by the force of the professor on m; and (b) by the force of the spring on m. (c) What is the work done by the sum of the two forces on the system?

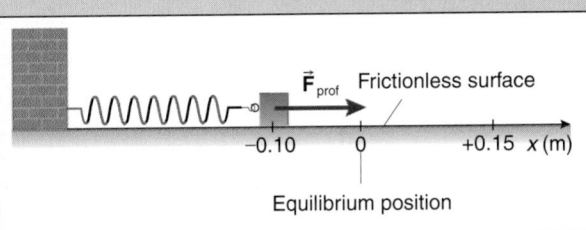

FIGURE P.12

•13. Each graph in Figure P.13 describes the one-dimensional motion of a 5.00 kg system during a 4.00 s interval. For each case, what is the work done on the system by the total force on the system during the interval?

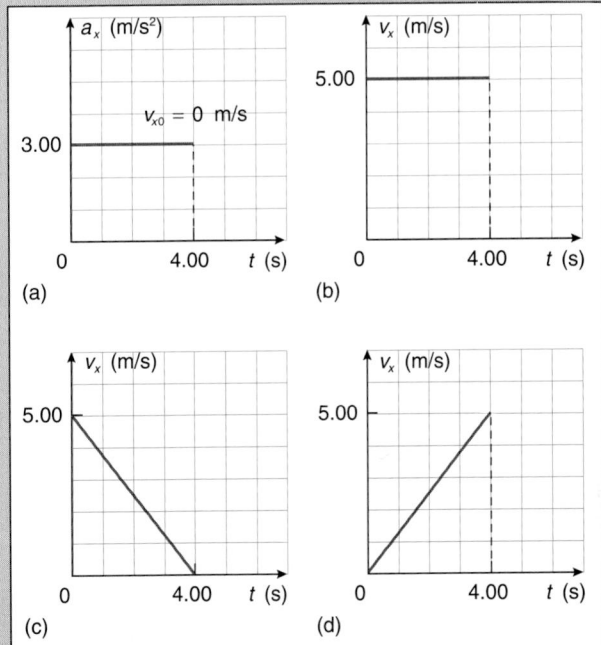

FIGURE P.13

•14. Show that the work done by the gravitational force in moving a system of mass m at a distance r_i from the center of a uniform sphere (of mass M) to a distance r_f is

$$W = -\frac{GmM}{R^3}\left(\frac{r_f^2}{2} - \frac{r_i^2}{2}\right)$$

where both r_i and r_f are less than the radius R of the sphere (see Figure P.14). What does this imply about the work done by this form of the gravitational force around a closed path?

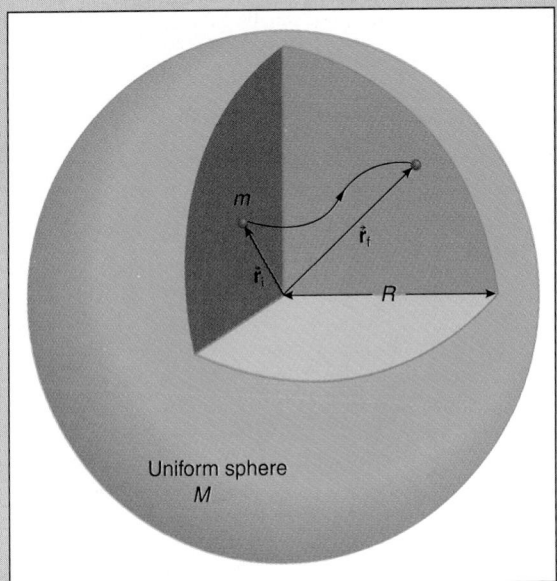

FIGURE P.14

Sections 8.6 Conservative, Nonconservative, and Zero-Work Forces

8.7 Examples of Conservative, Nonconservative, and Zero-Work Forces

8.8 The Concept of Potential Energy

8.9 The Gravitational Potential Energy of a System near the Surface of the Earth

8.10 The General Form for the Gravitational Potential Energy

8.11 The Relationship between the Local Form for the Gravitational Potential Energy and the More General Form*

8.12 The Potential Energy Function Associated with Hooke's Force Law

•15. A force $\vec{F} = (2.00\text{ N})\hat{i} + (1.00\text{ N})\hat{j}$ is one force on a system as it executes the rectangular path shown in Figure P.15. (a) Find the work done on the system by this force if the path is traversed in the clockwise sense. (b) Find the work done by this force if the loop is traversed in the counterclockwise sense. (c) Is this force conservative?

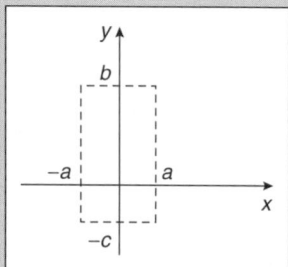

FIGURE P.15

•16. A force $\vec{F} = -kx\hat{i} - ky\hat{j}$ is one of the forces on a particle of mass m. The quantity k is constant. This force represents a two-dimensional Hooke's force law. (a) Is this force conservative? (b) What is an appropriate potential energy function for this force?

•17. Show that if the distance of a particle of mass m from another mass M increases by a small amount Δr, the change in the gravitational potential energy, ΔPE, of m is approximately

$$\Delta\text{PE} \approx \frac{GMm}{r^2}\,\Delta r$$

•18. How high above the surface of the Earth must a mass be located so its gravitational potential energy is only 1.00% of its value on the surface?

•19. A 3.00 kg mass is on a rough horizontal table. The coefficient of kinetic friction for the mass on the table surface is $\mu_k = 0.25$. The mass is pushed slowly between points A and B via the two different routes shown in Figure P.19. Find the work done by the kinetic frictional force along each path. The work is not the same along each path; hence the work done by the kinetic frictional force depends on the path. Thus the kinetic frictional force is not conservative.

•20. (a) Show that the gravitational potential energy of a system of mass m located inside a uniform sphere of radius R and mass M can be taken to be of the form

$$\text{PE} = \frac{1}{2}\frac{GmM}{R^3}\,r^2$$

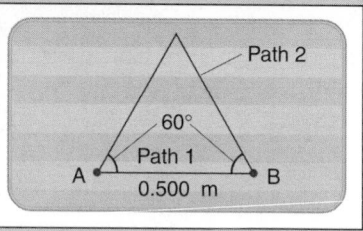

FIGURE P.19

You can make the calculation easier by thinking of the result of the harmonic oscillator problem; why? (b) Where is this potential energy function equal to zero?

•21. If the position of a mass m attached to a spring changes by a small amount Δx, show that the change in the potential energy ΔPE of the mass is approximately

$$\Delta\text{PE} \approx kx\,\Delta x$$

•22. A system of mass m has a gravitational potential energy mgy, where $y > 0$ m and $\hat{j}$ is directed vertically up. If the mass then is attached to a horizontal spring of spring constant k, by how much must the spring be extended or compressed so that the potential energy of the mass on the spring is equal to its gravitational potential energy?

Sections 8.13 The CWE Theorem

8.14 The Escape Speed

8.15 Black Holes*

8.16 Limitations of the CWE Theorem: Two Paradoxical Examples*

23. The velocity of a system of mass 3.00 kg is measured to be

$$(3.00\text{ m/s})\hat{i} - (2.00\text{ m/s})\hat{j} + (5.00\text{ m/s})\hat{k}$$

when $t = 3.00$ s. Later, when $t = 6.00$ s, the velocity is found to be

$$(-4.00\text{ m/s})\hat{i} + (6.00\text{ m/s})\hat{j} - (8.00\text{ m/s})\hat{k}$$

(a) What is the change in the velocity of the system? (b) What is the change in the kinetic energy of the system? (c) Why is the change in the kinetic energy *not* equal to

$$\frac{1}{2}m(\Delta v)^2$$

(d) What is the work done by all the forces on the system?

24. A spacecraft with a mass of 1.20×10^3 kg is to be sent to Mars from the surface of the Earth. What minimum kinetic energy is required to send the spacecraft on its way? (Consider only the gravitational influence of the Earth on the spacecraft.)

25. Show that the Schwarzschild radius for a 75 kg physics student is about $R_s \approx 10^{-25}$ m. This is *ten* orders of magnitude smaller than the size of a typical atomic nucleus. Teeny tiny for sure. So it is *quite* impossible to implode yourself to form a black hole, thank goodness.

•26. Starting from rest at point (1) in Figure P.26, a 4.00 kg tooth fairy slides down the indicated frictionless surface. The zero of gravitational potential energy is chosen to be at the elevation indicated. Take the fairy to be the system. At each of the points labeled (1), (2), (3), and (4), specify the values of:

(a) the total mechanical energy of the system; (b) the potential energy of the system; (c) the kinetic energy of the system; and (d) the speed of the system.

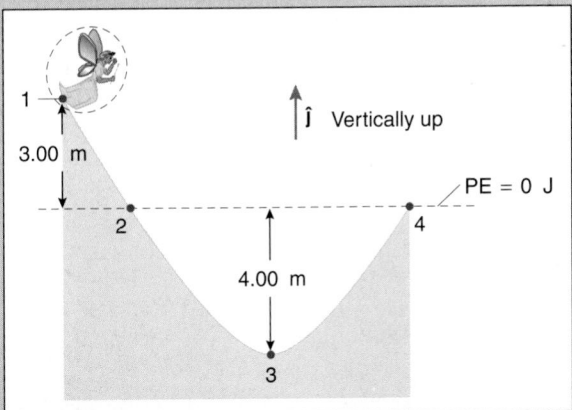

FIGURE P.26

•27. A 0.75 kg upside down bicycle helmet slides 8.0 m down the rough incline indicated in Figure P.27 at a constant small speed of 0.005 m/s. Use the CWE theorem to determine the coefficient of kinetic friction for the helmet on the surface.

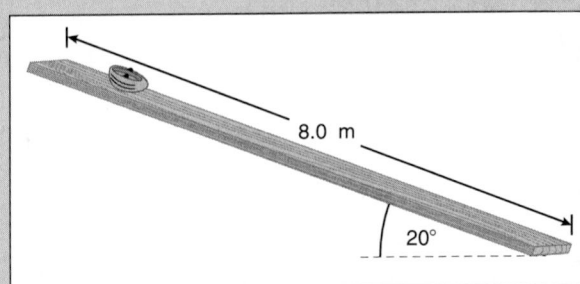

FIGURE P.27

•28. A 100 kg warthog (*Phacochoerus aethiopics*) squeals with delight as it slides from rest down the greased (frictionless) slope and plane in Figure P.28a. The hog encounters a spring that compresses 150 cm before bringing the hog to rest for an instant. (a) What is the spring constant of the spring? (b) The hog is sent back in the opposite direction by the spring. To what height will the hog rebound up the slope? (c) Suppose the hog begins anew at the 5.00 m height but now encounters a 2.00 m stretch of level rough ground on the way to the spring and back (see Figure P.28b). The coefficient of kinetic friction for the hog on the rough ground is 0.30. How much will the spring now compress? To what height will the wild bacon rise on the rebound?

•29. Latter-day infidels have succeeded in constructing a modern Tower of Babel with a height equal to the radius of the Earth. To celebrate the accomplishment, the barbarians are celebrating with a sacrifice to appease the gravitational gods. The peter-principled Dean of Deans (*Maximus incompetus*) is dropped from the tower with zero initial speed and plummets toward the Earth. Neglect all histrionics and atmospheric effects. Determine the speed at impact with the surface of the Earth.

•30. It is thought that *supermassive* black holes exist at the center of many galaxies. Such black holes have masses of 10^6 times that of the our Sun, or more. Calculate the Schwarzschild radius of a black hole with a mass of 1.0×10^6 solar masses and compare the radius with the average distance of the Earth from the Sun (1 AU = 1.496×10^{11} m).

•31. A system of two identical springs (each with spring constant k) is used to accelerate a mass m from rest to as great a speed as possible on a frictionless surface. Various arrangements of the springs are possible: (i) use just one spring; (ii) use two springs in series; (iii) use two springs in parallel. See Figure P.31. Regardless of which design is used, the spring system can be set (cocked) the *same* distance x_0 from the equilibrium position of m. (a) Find the maximum potential energy of m in each spring system. (b) Find an expression for the maximum speed of m as it leaves each system. (c) To give m the greatest possible speed, which arrangement should be used (or does it make any difference)?

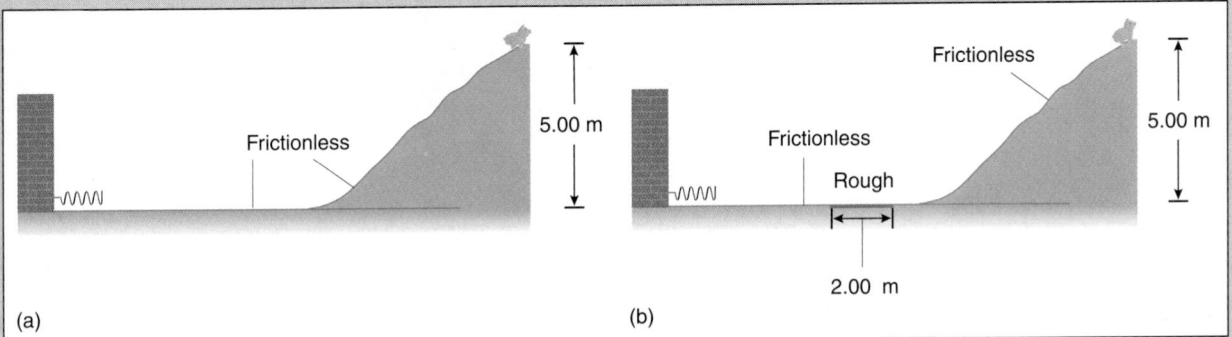

FIGURE P.28

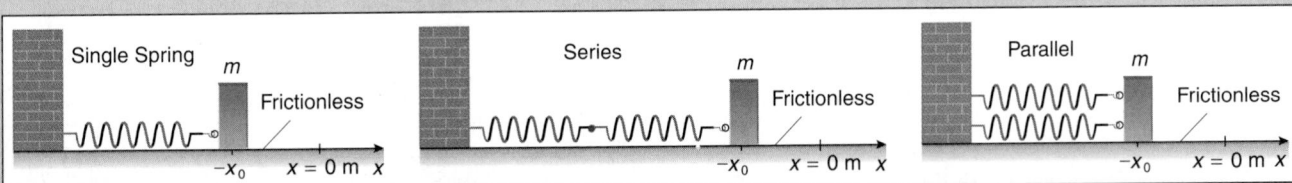

FIGURE P.31

•32. The ancient Greek god Zeus decides to drop in for an unannounced inspection. Let the unknown mass of Zeus be $m > 0$ kg. He begins his fall at zero speed from an infinite distance and free-falls, under the influence of the gravitational force of the Earth alone, all the way down to Athens (latitude 38°N, longitude 24°E) on the surface of the Earth. At what speed does Zeus arrive? How does this speed compare with the escape speed from the Earth?

•33. (a) Calculate the orbital speed of a hypothetical, frictionless satellite in circular orbit several meters above the surface of a perfectly spherical Earth (a so-called "grazing" orbit). (b) What is the *ratio* of the orbital speed of the satellite to its escape speed from the surface?

•34. (a) What is the binding energy of a 250 kg satellite of the Earth in a circular orbit with a radius equal to twice the radius of the Earth? (b) What additional energy must be given to the satellite to have it escape from the gravitational influence of the Earth?

•35. From the surface of the Earth, Superman throws a baseball vertically upward with a speed equal to *half* the escape speed. Neglecting air frictional effects, what is the maximum height of the ball? Express your result in terms of multiples of the radius of the Earth.

•36. An 80.0 kg sphere is suspended by a wire of length 25.0 m from the ceiling of a science museum as indicated in Figure P.36. A *horizontal* force of magnitude F is applied to the ball, moving it very slowly at constant speed until the wire makes an angle with the vertical direction equal to 35°. (a) Sketch a second law force diagram indicating all the forces on the ball at any point along the path. (b) Is the force needed to accomplish the task constant in magnitude along the path followed by the ball? (c) Use the CWE theorem to find the work done by the force $\vec{F}$.

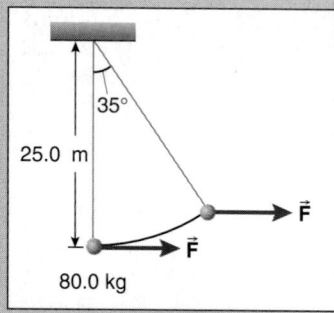

FIGURE P.36

•37. A very fast sprinter can run with a speed of about 10 m/s. A spherical asteroid has a density equal to the density of the Earth, about 5.3×10^3 kg/m³. What is the maximum radius for the asteroid so the escape speed from it does not exceed 10 m/s?

•38. Uh oh. You carelessly forget to take the car keys from the ignition, and your car (of mass 1.50×10^3 kg) is in neutral with the emergency brake off. As you rush to another exciting physics lecture, your car starts from rest and rolls away in the sloping parking lot; gravity never rests. You turn around and notice to your horror that the car, after descending 2.00 m in vertical height and 50.0 m along the slope of the lot, is moving

at 4.00 m/s just before it crashes into the very expensive car of your professor. Consider the car to be a particle; consider yourself in deep trouble. (a) What is the change in the kinetic energy of the car? (b) What is the change in the gravitational potential energy of the car? (c) What is the work done by forces other than the gravitational force in the process?

•39. A 60.0 kg engine block head is held against a spring ($k = 4.00 \times 10^3$ N/m) while it is compressed horizontally through a distance of 1.50 m. The block head is released at rest when $t = 0$ s and slides horizontally over a smooth surface. Along the way, the block head encounters a 2.00 m rough stretch of ground characterized by $\mu_k = 0.200$ and $\mu_s = 0.300$, as shown in Figure P.39. After passing over the rough ground, another smooth (frictionless) section is encountered as well as another spring ($k' = 3.00 \times 10^3$ N/m). This spring, of course, reverses the direction of the block head. (a) What is the initial total mechanical energy of the block head when $t = 0$ s? (b) How much work is done by the kinetic frictional force on the block head during *each* passage over the rough ground? (c) What is the maximum compression of the spring on the right when the block head encounters it for the first time? (d) How many times will the block head pass *completely* over the rough stretch of ground before coming to a halt?

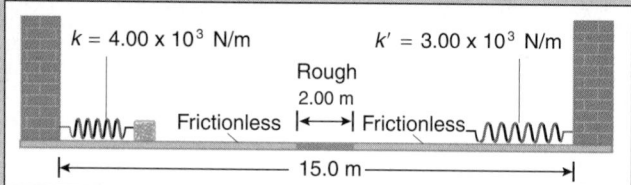

FIGURE P.39

•40. A hiker pushes a 100 kg boulder off a cliff as pictured in Figure P.40. (This is not a particularly cool thing to do!) The rock falls vertically for 30.0 m and then slides down a talus slope. At the bottom of the slope and at a speed of 2.00 m/s, the rock passes some astonished park rangers who subsequently set off in hot pursuit of the dastardly hiker. (a) How much work was done by the kinetic frictional force as the rock slid down the slope to the point where it passed the rangers? (b) Assuming the kinetic frictional force is of constant magnitude, find the coefficient of kinetic friction for the boulder on the slope.

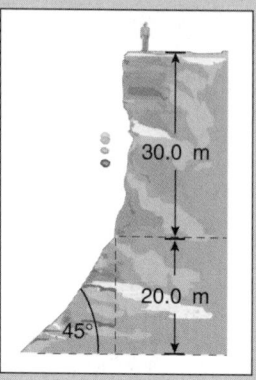

FIGURE P.40

•41. Consider the orbit of the Earth (mass *m*) about the Sun (mass M) to be circular and of radius *r*. In terms of G, *m*, M, and *r*, find (for the sheer joy of it): (a) the potential energy of the Earth; (b) the orbital speed of the Earth; and (c) the orbital kinetic energy of the Earth. (d) What is the *ratio* of the kinetic energy to the potential energy of the Earth? The result you obtain for this ratio is a consequence of a general theorem in (advanced) mechanics known as the *virial theorem*.

•42. A 1.00×10^3 kg surveillance satellite is placed in a circular orbit with an altitude of 910 km above the *surface* of the Earth. (a) Calculate the magnitude of the gravitational field of the Earth at the orbital radius of the satellite. (b) Calculate the magnitude of the gravitational force of the Earth on the satellite in its orbit. (c) Determine the speed of the satellite in its orbit in m/s. (d) What is the period of the satellite in hours? (e) Approximately how many revolutions around the Earth does the satellite make per day? (f) What is the kinetic energy of the satellite? (g) What is its gravitational potential energy? (h) What is the *ratio* of the kinetic energy of the satellite to its gravitational potential energy?

•43. Hi ho, hi ho, it's off to work we go. . . . A student flings a 2.00 kg physics tome down a rough inclined plane with an initial speed of 2.00 m/s as indicated in Figure P.43. Amazingly, as luck would have it, the book stops right at the bottom of the incline, just before the precipice. (a) What is the work done by the gravitational force on the book? (b) What is the work done by the normal force of the plane on the book? (c) What is the change in the kinetic energy of the book? (d) What is the change in the gravitational potential energy of the book? (e) What is the value of the coefficient of kinetic friction for the book on the incline?

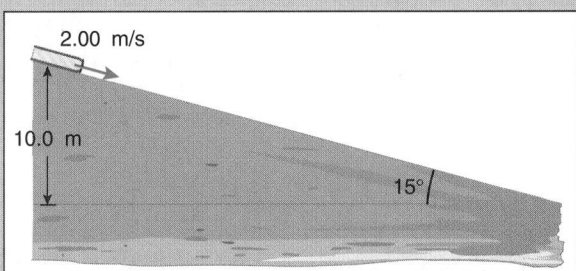

FIGURE P.43

•44. Consider the orbit of the Earth about the Sun to be circular. Let *m* be the mass of the Earth, M the mass of the Sun, and *r* the radius of the orbit of the Earth. (a) Derive an expression for the orbital speed *v* of the Earth in terms of M, *r*, and appropriate physical constants. (b) The Greek Titan, Atlas, screeches the Earth to a halt in its orbit and then kicks it with a speed *just* sufficient to expel the Earth entirely from the solar system. Neglecting the influence of the other planets, derive an expression for the minimum speed v_{escape} the Earth needs to escape to infinity from its former orbit at distance *r*. (c) What is the *ratio* of this escape speed to the orbital speed?

•45. You have been assigned the exciting task of designing a new roller coaster. The track is to include a loop-the-loop of radius *R* as shown in Figure P.45. To a first approximation, ignore all frictional effects. The approach to the loop is made by starting the roller coaster vehicle (of mass *m*) at zero speed down an incline of height *h* as indicated. (a) What is the speed of the vehicle at the beginning of the loop (point A)? (b) What is the speed of the

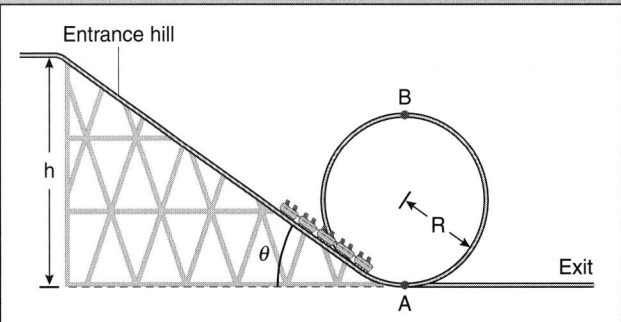

FIGURE P.45

vehicle at point B if the force of the track on the vehicle at this point is zero? (c) Find the ratio *h/R* so that the force of the track on the vehicle at point B just becomes zero. (d) What is the speed of the vehicle as it exits the loop? (e) What is the total work done by the gravitational force on the vehicle from the top of the incline to the exit from the loop? (f) Which of your answers to parts (a)–(e) depend on the mass *m* of the vehicle?

•46. A huge 10.0 kg otter (*Lutra canadensis*) is released at rest at the top of a frictionless inclined plane as shown in Figure P.46. The otter slides onto a horizontal surface with a coefficient of kinetic friction $\mu_k = 0.20$ for the otter tummy on the surface. (a) What is the speed of the otter as it reaches the bottom of the incline? (b) How far will it slide along the horizontal surface before coming to rest?

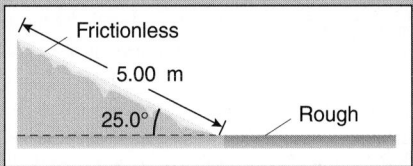

FIGURE P.46

•47. A popular recreation for the fearless is jumping from tall structures while attached to an elastic bungee cord to ensure that less than catastrophic consequences result from the fall. Model the bungee cord as an ideal spring. A student of mass *m* drops off a tower of height *h*, tethered to the bungee cord. The

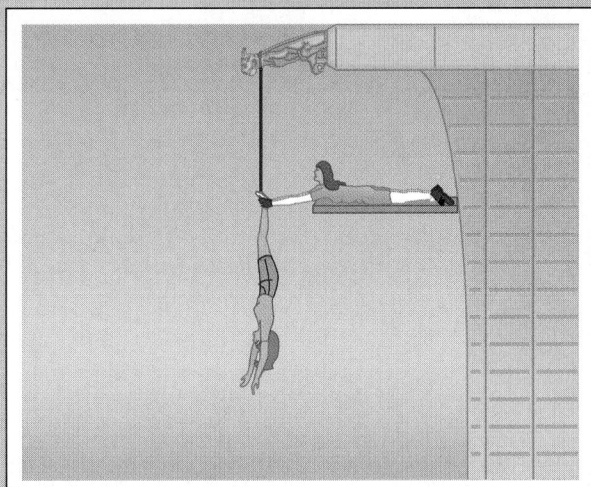

FIGURE P.47

elasticity of the bungee cord is chosen so that its effective spring constant is $k = 2mg/h$. The jump is begun at zero speed at altitude h with the bungee cord fastened vertically above her, barely taut (with no slack), but under no initial tension. See Figure P.47. (a) Choose a convenient coordinate system to analyze the problem. (b) What is the initial total mechanical energy of the student? (c) Show that the student will arrive at the ground with zero kinetic energy. Whew! (d) At what height above the ground is the speed of the student a maximum? (e) Will the *same* bungee cord be safe to use for a jump from the same tower (of height h) for a person with mass $m' < m$? With mass $m' > m$? Explain your reasoning.

•**48.** A spring gun is used in the lab to project a 0.070 kg mass from rest to a horizontal speed of 7.0 m/s. The spring is compressed a distance of 5.0 cm before release. Assume frictional forces are zero. (a) What is the kinetic energy of the projectile as it emerges from the gun? (b) What is the potential energy of the mass on the fully compressed spring? (c) What is the spring constant of the spring?

•49. A projectile is launched at initial speed v_0 at an angle θ with the horizontal direction. Neglect air friction and assume the trajectory is confined to a region near the surface of the Earth. Use the CWE theorem to show that the maximum vertical height of the projectile at the top of its flight path is

$$\frac{v_0^2 \sin^2 \theta}{2g}$$

•**50.** A train-forming railyard is designed so that boxcars of mass 3.00×10^4 kg, moving at an initial speed of 1.50 m/s, roll down an incline of height 2.00 m and across a level section of track before coupling at a speed of 0.500 m/s with other boxcars. See Figure P.50. Consider the boxcars to be particles, neglecting the rolling motion of the wheels. (a) What is the work done by the gravitational force on each boxcar? (b) What is work done by the normal force of the track on each boxcar? (c) What is the work done by other forces on each boxcar in the process?

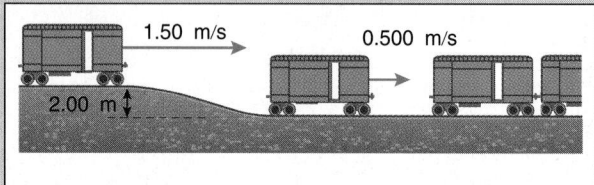

•**51.** A spacecraft of mass m is in a circular orbit about the Earth at a distance r from the center of the Earth (of mass M). (a) Use Newton's second law of motion to show that the speed v of the satellite satisfies

$$v^2 = \frac{GM}{r}$$

(b) Show that the total mechanical energy E of the spacecraft is

$$E = -\frac{1}{2} \frac{GMm}{r}$$

NASA wants to *change* the orbit of the spacecraft to one with an orbital radius of $2r$. (c) What is the total mechanical energy of the spacecraft in the new orbit? (d) How much energy must be given to the spacecraft to change its orbit from r to $2r$?

•**52.** (a) Use Equation 8.32 for the escape speed from a uniform spherical object of radius R and mass M to show that the escape speed from the surface of the Sun is about 1/500 the speed of light. (b) Assume M to be a uniform sphere. Show that the expression for the escape speed can be written in terms of the density ρ of the sphere in the following way:

$$v_{escape} = R \left(\frac{8\pi}{3} G\rho \right)^{1/2}$$

In 1784 John Michell realized that if the radius of the Sun were increased by a factor of 500 while keeping its density constant, that "all light emitted from such a body would be made to return to it, by its power of gravity,"* thus anticipating the idea of a black hole.

•53. A satellite of mass m is in an elliptical orbit about a planet of mass M ($m \ll M$), as shown in Figure P.53. The only force acting on the satellite is the gravitational force of M. The speed of the satellite at A is v_1 and its speed at B is v_2. (a) What is the potential energy of the satellite at point A? At point B? (b) What is the magnitude of the orbital angular momentum of the satellite at point A? At point B? (If necessary, see Section 6.10 for the meaning of orbital angular momentum.) Recall that one consequence of Kepler's second law of planetary motion, and indeed of all central force motion, is that the orbital angular momentum of the satellite is conserved: the magnitude (and direction) of the orbital angular momentum vector is the same at both A and B. (c) Use the CWE theorem and conservation of orbital angular momentum to show that the total mechanical energy E of the satellite can be written as

$$E = -\frac{GMm}{2a}$$

where $a = (r_1 + r_2)/2$ is the semimajor axis of the elliptical orbit. This result indicates that the total mechanical energy of

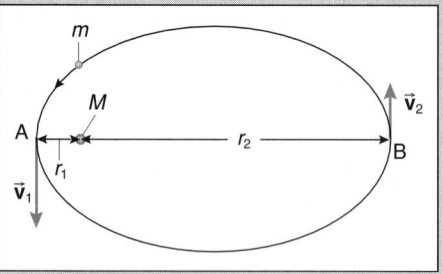

---

*See Simon Schaffer, "John Michell and black holes," *Journal for the History of Astronomy, 10*, Part 1, #27, pages 42–43 (February 1979). Notice how the word "power" is *not* used by Michell in the way we now define the power of a force.

the orbit of a satellite (or planet about the Sun) depends only on the semimajor axis of the orbit and not other factors such as the eccentricity ε.

•54. A spring with $k = 500$ N/m is suspended vertically as indicated in Figure P.54. A pan of negligible mass is on the end of the spring. Let the position of the pan in this configuration be taken as the origin of a y-coordinate axis with $\hat{\jmath}$ directed up. (a) A 2.00 kg mass is dropped (with zero initial speed) into the pan from a position 0.500 m above the pan. What is the initial total mechanical energy of the mass? (b) After the mass is in the pan, show that the mass is instantaneously at rest where $y = -0.241$ m and $y = +0.163$ m. (c) If the 2.00 kg mass is simply placed (rather than dropped) into the pan, show that the mass is in equilibrium in the pan where $y = -0.039$ m. Notice that when the mass is dropped into the pan, the oscillation is symmetric about the position of the pan determined in part (c) and of amplitude 0.202 m. Thus, if a mass is hung vertically from a vertical spring, the mass undergoes simple harmonic oscillation about the new equilibrium position of the mass on the spring, as you learned in Section 7.3.

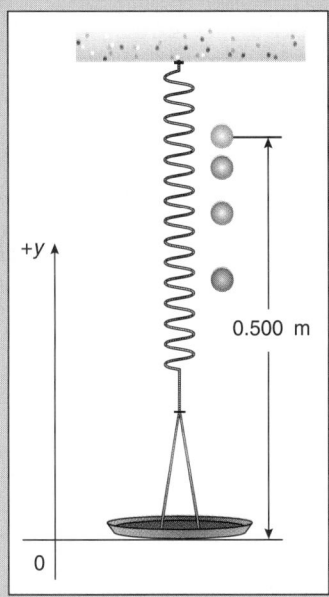

+y

0.500 m

0

FIGURE P.54

•55. The many fragments of Comet Shoemaker–Levy 9 plunged into the atmosphere of Jupiter in July 1994 with huge amounts of kinetic energy. Consider a fragment modeled as a spherical ice mass 1.00 km in diameter; the density of ice is 917 kg/m³. The incident speed of the fragment is 50 km/s. (a) What is the mass of the fragment? (b) What is its kinetic energy? (c) One metric ton (1000 kg) of the explosive TNT* has an energy yield of about 4.9×10^9 J. What is the incident kinetic energy of the comet fragment expressed in equivalent metric tons of TNT? (d) If such a comet fragment were approaching the Earth, comment on whether nuclear missiles could appreciably change the kinetic energy of the fragment.

---

*Trinitrotoluene ($C_7H_5N_3O_6$).

•56. Jupiter has a mass of 1.90×10^{27} kg and rotates once every 9.84 h. A 1200 kg spacecraft is placed into an equatorial circular orbit about Jupiter with the same rotational period, so that the satellite remains above a fixed point on the Jovian equator. (The giant planet has no solid surface.) (a) Use Newton's second law to determine the radius of the orbit of the satellite. (b) What is the orbital speed of the satellite? (c) What is the total mechanical energy of the satellite? (d) To what speed would the satellite need to be propelled to escape from this orbit to a great distance from Jupiter?

‡57. A system of mass m is traveling at a large initial speed v_0. A small amount of work ΔW_{total} is done by all the forces acting on the system. Begin with the CWE theorem in the form

$$W_{total} = \Delta KE$$

and show that the change in the speed Δv of the system is approximately given by

$$\Delta v \approx \frac{\Delta W_{total}}{mv_0}$$

Use this equation to estimate the change in the speed of a 10.0 kg mass initially traveling at 30.0 m/s when 1.00 J of work is done on it by the total force on the system.

‡58. A system of mass m is undergoing vertical motion near the surface of the Earth with the gravitational force of the Earth the *only* force on the system. For such vertical motion near the surface of the Earth, the total mechanical energy of a system of mass m is

$$E = KE + PE$$
$$= \frac{1}{2} m \left(\frac{dy}{dt} \right)^2 + mgy$$

where $\hat{\jmath}$ is vertically up. The system has a constant vertical (downward) acceleration and the position of the system at any time is

$$y = y_0 + v_{0y}t - \frac{1}{2} gt^2$$

Use this in the foregoing expression for E to show that E is independent of time (that is, the total mechanical energy is conserved) and to find E in terms of y_0, v_{y0}, m, and g. The result should not be unexpected.

‡59. We all likely have enjoyed the childhood experience of dropping a marble down a set of hardwood steps. The tap, tap, tap of the marble as it works its way down the steps is a delight to almost any child—or adult. Imagine a marble dropped down a set of steps of step-height h. The rebounding marble bounces to a height equal to the height h of the previous step as indicated in Figure P.59. (Use this figure to help solve part (b) only.) The *coefficient of restitution* ε of the bouncing marble is defined to be

$$\varepsilon = \frac{\text{speed immediately after the bounce}}{\text{speed immediately before the bounce}}$$

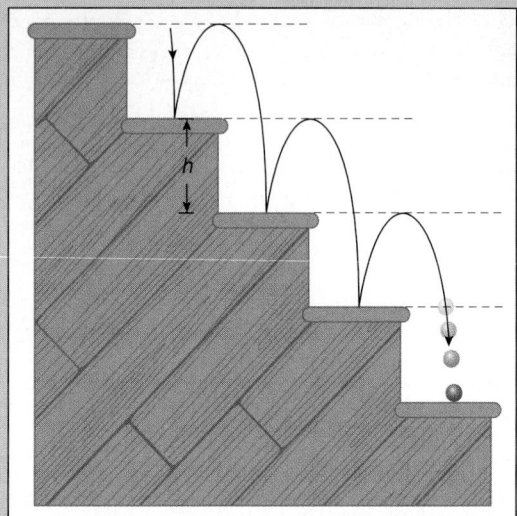

FIGURE P.59

(a) If the coefficient of restitution is zero, what is happening? (b) Neglect the horizontal speed of the marble. Show that the coefficient of restitution for the situation indicated in Figure P.59 is

$$\varepsilon = \left(\frac{1}{2}\right)^{1/2}$$

(c) Sketch what happens to the marble if its coefficient of restitution is equal to 1.

The inspiration for this problem came from Robert F. Kingsbury, *Elements of Physics* (Van Nostrand, Princeton, New Jersey, 1965), page 172.

•**60.** The rules governing the construction and composition of tennis balls are quite stringent. When dropped from rest vertically from a height of 2.54 m, the ball must, after bouncing from a concrete floor, rebound to a height of at least 1.35 m, but no more than 1.47 m. What is the range of permissible coefficients of restitution of tennis balls? (See Problem 59 for a definition of the coefficient of restitution.)

⁃**61.** A mass executing simple harmonic oscillation is described by

$$x(t) = A \cos \omega t$$

(a) Find the velocity component v_x of the oscillator as a function of time. (b) The potential energy of the oscillator,

$$PE = \frac{1}{2} kx^2$$

is time dependent. The average value of a time-dependent function $f(t)$ during an interval of time T is found using the following recipe:

$$f_{ave} = \frac{1}{T} \int_0^T f(t) \, dt$$

Find the time average value of the potential energy of the oscillator during one period of the oscillation. (c) The kinetic energy of the oscillator is

$$KE = \frac{1}{2} mv^2 = \frac{1}{2} mv_x^2$$

This also is time dependent. Find the time average value of the kinetic energy of the oscillator during a period of the oscillation. (d) Show that the time average value of the potential energy of the oscillator is equal to the time average value of the kinetic energy of the oscillator. That is, show that

$$PE_{ave} = KE_{ave}$$

This result is an application of a general theorem in (advanced) mechanics known as the *virial theorem*.

⁃**62.** A particle of mass m is dropped (from rest) from a height h in the uniform gravitational field near the surface of the Earth as shown in Figure P.62. For convenience, the indicated coordinate system has been chosen to analyze this problem. (a) For review, show that the y-component of the velocity is

$$v_y = -gt$$

and the y-coordinate of the particle is

$$y(t) = h - \frac{1}{2} gt^2$$

where g is the magnitude of the local acceleration due to gravity (here $g = 9.81$ m/s²). Show also that the time of flight of the particle is

$$T = \left(\frac{2h}{g}\right)^{1/2}$$

(b) The potential energy of the particle is time dependent. The following equation gives the average value of a time-dependent function $f(t)$:

$$f_{ave} = \frac{1}{T} \int_0^T f(t) \, dt$$

Use this equation to show that the time average value of the potential energy, during the time of flight T, is

$$PE_{ave} = \frac{2}{3} mgh$$

(c) The kinetic energy of the particle depends on the time. Show that the time average value of the kinetic energy of the particle, averaged over the time of flight, is

$$KE_{ave} = \frac{1}{3} mgh$$

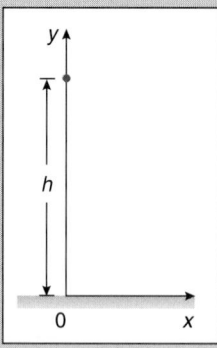

FIGURE P.62

The results of parts (b) and (c) indicate that

$$KE_{ave} = \frac{1}{2} PE_{ave}$$

which is a consequence of a theorem in (advanced) mechanics known as the *virial theorem*.

Section 8.17 The Simple Harmonic Oscillator Revisited

63. (a) What is the total mechanical energy of a 2.00 kg mass executing simple harmonic oscillation of amplitude 0.30 m on a spring of spring constant 40 N/m? (b) What is the period of the oscillation?

•64. For the simple harmonic oscillator of Problem 63, what are the kinetic and potential energies of the mass system when it is 0.10 m from its equilibrium position?

•65. A 4.00 kg mass is attached to a spring and executes simple harmonic oscillation with a period of 1.50 s. The total mechanical energy of the system is 12.0 J. (a) What is the spring constant? (b) Determine the amplitude of the oscillation.

•66. A simple harmonic oscillator has an amplitude A. At what coordinate(s) from the equilibrium position (as origin) is the kinetic energy equal to the potential energy?

•67. When a mass executing simple harmonic oscillation has a position equal to half the amplitude, what fraction of the total mechanical energy of the oscillator is kinetic?

•68. One end of a spring is placed on the floor with the axis of the spring vertical. The length of the spring is 60 cm. A 2.00 kg physics text is placed on the spring at rest and the spring compresses to three-fourths of its original length. The text then is raised slightly and dropped (from rest), compressing the spring now to one-fourth of its original length. Find the height above the floor from which the text was released.

Sections 8.18 The Average and Instantaneous Power of a Force
8.19 The Power of the Total Force Acting on a System

•69. A 2.00 kg mass falls from rest a distance of 100 m near the surface of the Earth. (a) What is the average power of the gravitational force during the time of the fall? (b) What is the instantaneous power of the gravitational force at the instant just before impact?

•70. You are playing with your little sister on a swing. At first, you pull the swing back and release her (without pushing—i.e., at rest) at an elevation that is a distance d above the bottom of the path of the swing. She is not content. So you then release her (again without pushing) a distance $2d$ above the low point of the motion. (a) By what factor will the maximum speed of the swing increase? (b) What is the power provided by the force of the ropes on the swing when the swing has its maximum speed? (c) What can be said (if anything) about the power provided by the force of the ropes on the swing at other points along the path of the motion?

•71. A mass m is dropped at rest from a height h when $t = 0$ s. Show that the instantaneous power of the gravitational force on m at any instant t is

$$P(t) = mg^2t$$

•72. When $t = 0$ s, a mass m is projected at speed v_0 vertically upward. (a) Show that the instantaneous power of the gravitational force at any instant t is

$$P(t) = mg^2t - mgv_0$$

(b) Sketch a graph of $P(t)$ versus t, labeling appropriate points. (c) At what point in the motion is the power zero? (d) What prevents the power from increasing indefinitely? (e) What is the physical significance of the area under the $P(t)$ versus t graph?

•73. An 80.0 kg premedical student decides to liven things up by sliding (from a standing start) down a slippery (i.e., frictionless) 25° mud slope as indicated in Figure P.73. At the end of the slope, the student encounters a horizontal stretch of terrain where the coefficient of kinetic friction for the student's derriere on the turf is $\mu_k = 0.20$. (a) Set up an appropriate coordinate system clearly indicating the origin and the direction of the coordinates axes. (b) Write an expression for the total mechanical energy of the premed at the top of the slide. (c) What is the speed of the premed as she leaves the incline? (d) How far will she travel along the rough horizontal plane before coming to a stop? (e) What is the total work done by the normal force of the surface on the student (i) during the descent and (ii) along the horizontal segment of the path? (f) What is the total work done by the kinetic frictional force on the student along the horizontal section? (g) What is the power of the kinetic frictional force as the student commences the horizontal segment? Is this power constant over the entire horizontal segment? Explain your reasoning.

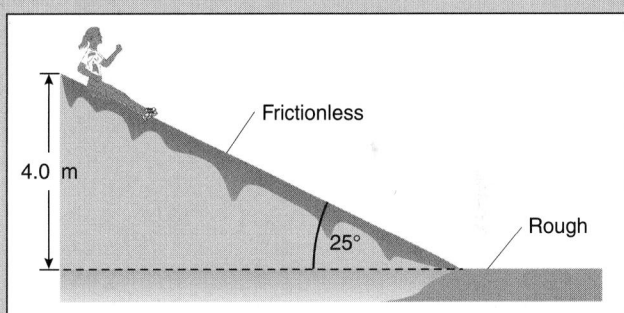

FIGURE P.73

•74. The position coordinate of a 2.00 kg mass executing simple harmonic oscillation is described by the equation

$$x(t) = (0.50 \text{ m}) \cos[(7.00 \text{ rad/s})t]$$

where $x = 0$ m is the equilibrium position. (a) What is the spring constant? (b) When the mass is at the maximum distance from the equilibrium position, what is the power of the spring force? (c) When the mass goes through its equilibrium position, what is the power of the spring force? (d) Break up the oscillation into four successive parts:

Part 1: from $x = A$ to $x = 0$ m
Part 2: from $x = 0$ m to $x = -A$
Part 3: from $x = -A$ to $x = 0$ m
Part 4: from $x = 0$ m to $x = A$

During what parts of the motion is the power of the spring force positive? Explain why. (Do not evaluate the power specifically.)

During what parts of the motion is the power of the spring force negative? Explain why. (Do not evaluate the power specifically.)

•75. For each case shown in Figure P.13 (Problem 13), answer the following questions. (a) Find an expression for the kinetic energy as a function of time that is valid during the four second interval. (b) What is the average power of the total force on the system? (c) Find $P(t)$ at any instant during the four second interval. Is the instantaneous power of the total force constant during the four second interval? Explain why or why not.

•76. A passenger elevator car with a total mass of 1.20×10^3 kg is zipping upward at a constant speed of 4.00 m/s ferrying health inspectors to the famous catering firm of Cook, Wells, Dunn, and Burns. (a) What is the kinetic energy of the elevator car? (b) What is the force of the cable on the elevator car? (c) What is the power of the force of the cable? (d) What is the power of the gravitational force on the elevator car? (e) What is the power of the total force on the elevator car?

•77. A 1.00×10^3 kg sports car on a level road accelerates from rest to a speed of 150 km/h during 8.00 s. Consider the car to be a particle. What is the average power of the total force on the car during this time interval?

•78. A mass m is attached to a spring of spring constant k and is oscillating horizontally at angular frequency ω on a frictionless surface. The mass is released at rest when $t = 0$ s at the position $x = A$. (a) What is the position x of the oscillator as a function of time? (b) What is the velocity component

$$v_x = \frac{dx}{dt}$$

as a function of time? (c) What is the instantaneous power of the spring force when $t = 0$ s? Explain your answer. (d) What is the instantaneous power of the spring force when the oscillator is at position $x = 0$ m? Explain your answer. The results of parts (c) and (d) imply that the oscillator reaches a peak power sometime during the first quarter of a complete oscillation. (e) Show that the instantaneous power $P(t)$ of the oscillator force is

$$P(t) = \frac{k\omega A^2}{2} \sin(2\omega t)$$

(f) Show that the power is at a *maximum* for the first time when t is equal to one-eighth the period of oscillation. (g) What is the average power of the oscillator force during one period of the oscillation? (h) Make graphs of $F_{x\,\text{sprg}}(t)$, $P(t)$ [the result of part (e)], and $v_x(t)$ versus time during one period of the oscillation if the mass is 1.00 kg, the spring constant is 49.0 N/m, and the amplitude of the oscillation of 0.100 m. What conclusions can be drawn from the graphs?

•79. A particle of mass m, initially moving with a velocity $v_{x0}\hat{\mathbf{i}}$, is subjected when $t \geq 0$ s to a constant acceleration $a_x\hat{\mathbf{i}}$. (a) Find the kinetic energy as a function of time, $\text{KE}(t)$. (b) Determine the power of the total force by evaluating $\vec{\mathbf{F}}_{\text{total}} \cdot \vec{\mathbf{v}}$ and show that this is the same as the time rate of change of the kinetic energy, confirming Equation 8.42.

Section 8.20 Motion Under the Influence of Conservative Forces Only: Energy Diagrams*

•80. A 3.00 kg brick is hurled vertically upward at an initial speed of 5.00 m/s from a height of 20.0 m. Choose $\hat{\mathbf{j}}$ up and place the y-coordinate origin at ground level. (a) Find the total mechanical energy E of the brick. (b) Make a graph of the potential energy of the brick as a function of y. (c) Plot the total mechanical energy E on the graph. (d) For a general y-coordinate, indicate on the graph what represents the kinetic and potential energies of the brick. (e) Is there a turning point? (f) What y-coordinates are prohibited for the brick?

•81. The potential energy function of a conservative force acting on a system moving in one dimension along the x-axis is described by

$$PE = C|x|$$

where C is a positive constant. The conservative force is the only force on the system. (a) What are the SI units of C? (b) What is the x-component of the force on the system? (c) Make a schematic graph of the potential energy as a function of x. (d) Let the total mechanical energy E of the system be greater than zero. Sketch a representative value for $E > 0$ J on your graph of the potential energy function. Do turning points exist? If so, how many? Indicate prohibited values of x on the graph. (e) Describe the motion of the system. Is the motion simple harmonic oscillation? In what ways is the motion similar to simple harmonic oscillation? In what ways is it different?

•82. A pendulum of mass m and length ℓ makes an angle θ with the vertical direction. (a) Find an expression for the gravitational potential energy of m as a function of θ so that the potential energy is zero when $\theta = 0$ rad. (b) Make a sketch of this potential energy function in units of $mg\ell$ as a function of θ over the range

$$-\frac{\pi}{2} \text{ rad} \leq \theta \leq \frac{\pi}{2} \text{ rad}$$

(c) Let PE_{max} be the maximum potential energy of m over this range of θ. Let the total mechanical energy E of m be constant and have a value of $E = \text{PE}_{\text{max}}/2$. What are the turning points of a pendulum with such a value for the total mechanical energy?

‡83. The total force on a system of mass m varies with position according to

$$\vec{\mathbf{F}} = F_0 \cos\left(\frac{2\pi x}{\ell}\right)\hat{\mathbf{i}}$$

(a) Calculate the work done on the system by this force as the system moves from $x = 0$ m to $x = \ell/2$. (b) Calculate the work done on the system by this force as the system moves from $x = \ell/2$ to $x = 0$ m. (c) Is the force conservative? If so, proceed with the following questions. If not, you have the rest of the afternoon off. (d) Show that the following potential energy function is appropriate for this force:

$$PE(x) = -\frac{F_0\ell}{2\pi} \sin\left(\frac{2\pi x}{\ell}\right)$$

(e) Sketch the potential energy function over the domain $-\ell \leq x \leq \ell$. (f) The system is released at $x = 0$ m with zero speed. What is the total mechanical energy E of the system? Plot this value of E on the energy diagram of part (c). (g) Describe the motion of the system, calling particular attention to whether the system is bounded in its motion by turning points (and if so at what x values), and whether the system is oscillatory (and if so, whether the oscillation is simple harmonic or not). (h) If the total mechanical energy of the system somehow is decreased slightly to $E' < E$, what happens to the motion of the system? (i) For what values of E is the motion unbounded?

INVESTIGATIVE PROJECTS

A. Expanded Horizons

1. Investigate the relationship between *physical work* as we have defined it in physics and *isometric* work done by muscles when, for example, your arms hold a set of barbells at rest over your head or you push hard against an unyielding wall.
Bruce Abernethy, Vaughan Kippers, Laurel Traeger Mackinnon, Robert J. Neal, and Stephanie Hanrahan, *The Biophysical Foundations of Human Movement* (Human Kinetics Publishers, Champaign, Illinois, 1997).
Colin J. Pennycuick, *Newton Rules Biology* (Oxford University Press, Oxford, England, 1992).
Theodore C. Ruch and Harry D. Patton, *Physiology and Biophysics* (W. B. Saunders, Philadelphia, 1965).

2. Investigate the notion of *scaling* and *mechanical similarity* and how these ideas can be applied to (a) gravitational potential energy and (b) the potential energy of a simple harmonic oscillator to deduce many of the features of motion under these forces. Also investigate notions of scaling in biology.
Lev D. Landau and Evgenii M. Lifshitz, *Mechanics* (Addison-Wesley, Reading, Massachusetts, 1960), pages 22–24.
Herbert Lin, "Fundamentals of zoological scaling," *American Journal of Physics*, 50, #1, pages 72–81 (January 1982).
William H. Press, "Man's size in terms of fundamental constants," *American Journal of Physics*, 48, #8, pages 597–598 (August 1980).

3. Several problems in this chapter (Problems 41, 61, and 62) referred to a general theorem in mechanics known as the *virial theorem*. Find out what this theorem says in a general context and connect it to the results of these problems.
Lev D. Landau and Evgenii M. Lifshitz, *Mechanics* (Addison-Wesley, Reading, Massachusetts, 1960), pages 23–24.
Jerry B. Marion and Stephen T. Thornton, *Classical Dynamics of Particles and Systems* (Saunders, Philadelphia, 1995).
Peter Kleban, "Virial theorem and scale transformations," *American Journal of Physics*, 47, #10, pages 883–888 (October 1979).

4. Investigate the possibility that the universe itself is a black hole. We might be enjoying life inside one!
William J. Kaufmann III, *Universe* (3rd edition, W. H. Freeman, New York, 1990), page 546.

5. During the 19th century, Hermann von Helmholtz developed a theory that the Sun derived its energy from a gradual self-gravitational collapse now known as Helmholtz contraction. Investigate this theory and estimate the lifetime of the Sun if its energy were derived from this source. The energy source of the Sun actually is hydrogen fusion. Yet the mechanism of Helmholtz contraction is important in the *formation* of stars from the collapse of huge gaseous nebulae as well as in the late stages of stellar evolution.
Elske v. P. Smith and Kenneth C. Jacobs, *Introductory Astronomy and Astrophysics* (W. B. Saunders, Philadelphia, 1973), pages 366–367.

B. Lab and Field Work

6. Determine the spring constant of a spring that is available in your physics laboratory (or the departmental attic!). Use the spring to design two harmonic oscillators, one with 1.0 J and the other with 10.0 J of total mechanical energy.

7. With appropriate equipment, design and perform an experiment to measure the average power of the force a weightlifter exerts on a set of barbells when lifting them from the floor to overhead.

C. Communicating Physics

8. Look up the words *work* and *energy* in the *Oxford English Dictionary* and read through the plethora of their meanings in the English language. If you are multilingual, consider whether these words are as laden with meaning in other languages. Write an essay comparing, contrasting, and discussing the various meanings and how they differ from their meanings in physics, either in English or another language.

9. As you read the newspaper or news magazines, begin a collection of technical bloopers consisting of misuses of words like force, work, energy, and power. See how many you can find during, say, the next month.

CHAPTER 9
IMPULSE, MOMENTUM,
AND COLLISIONS

**Or by collision of two bodies grinde
The Air attrite to Fire**

John Milton (1608–1674)*

ollisions occur all around us. Whether marbles or pool, bowling or golf, baseball or football, tennis or paddle-ball, collisions are involved in some of our most cherished recreational sports. Cars occasionally collide with each other, sometimes with disastrous consequences on the occupants, but almost always on their wallets (though not the wallets of their lawyers). On a grand astronomical scale, the spectacular collisions of the pieces of Comet Shoemaker–Levy 9 with Jupiter during the summer of 1994 were extraordinarily violent.

It is likely that a stupendous collision of a large asteroid (about the size of Mars) with the Earth early in its history gave rise to the Moon. The collision of a smaller asteroid (about 10 km in diameter) with the Earth about 65 million years ago likely caused the mass extinctions that wiped out the dinosaurs.[†] Collisions of asteroids with the Earth are not only a thing of the past: they inevitably will occur again. Even whole galaxies, each with hundreds of billions of stars, collide with each other.[‡]

At the opposite extreme, collisions are ubiquitous in nuclear, atomic, and molecular physics. Atoms and molecules collide incessantly in gases. Physicists even use collisions of particles to fathom the ultimate nature of matter and to glean the characteristics of the fundamental forces of nature. There simply is no avoiding collisions, and so it is important to understand the physics of collision phenomena. We shall see that the concept of *momentum* is particularly useful for this purpose.[§]

In Chapter 5 we saw how Newton's second law relates the total force on a system to its acceleration. In this chapter we see that more generally, Newton's second law of motion relates the total force to the time rate of change of the system's momentum. In Chapter 8 we examined how the work done by the total force

acting during motion over a *distance* changes the *kinetic energy* of the system. Here we see that if we change our viewpoint and examine how the total force acts during a *time interval*, the *momentum* of the system changes.

Some forces on a system always are present, whereas others act only for limited time intervals. For example, in the vicinity of the Earth, the gravitational force of the Earth on a system of mass m never ceases its attraction, even for an instant. Although this force is ever present, it may not be the only force on the system. Other forces on the system may come and go as we or circumstances dictate. For example, you push your car only rarely, one hopes. The normal force of a surface on a system exists only as long as the system is in contact with the surface. Likewise, in the collision of a baseball with a bat, the force of the bat on the ball is present only during the very short interval they are in physical contact. The *impulse–momentum theorem* relates the change in the momentum of a system to the total force on it and the time interval during which this total force acts.

Until now our attention has focused on systems of just one or two distinct mass particles. We treated the system as a single particle of negligible size (were it a grand piano, a massive dictionary, or a subatomic particle) and examined its dynamical motion as a whole. We ignored the distribution of mass within the system itself. Toward the end of this chapter, we begin to consider systems either of *many* particles or of masses of *finite* size composed of an essentially continuous distribution of atoms and/or molecules. We begin to ask to what extent the physical size and composition of the system affect its motion. The dynamics of complex systems of many particles is simplified considerably by inventing the concept of the *center of mass*, a special point in space whose motion typifies a more complex system.[#] The concept of the center of mass will prove useful in the next chapter as well, where we investigate rotational motion in detail.

---

*(Chapter Opener) *Paradise Lost*, Book X, lines 1072–1073.

[†] The Chicxulub impact crater from this collision spans land and sea on the Yucatan Peninsula in Mexico.

[‡] A dwarf galaxy presently is colliding with our own Milky Way Galaxy, but on the opposite side of the galactic center.

[§] We introduced the concept of momentum briefly in Section 6.10. Here we introduce momentum in a more formal way.

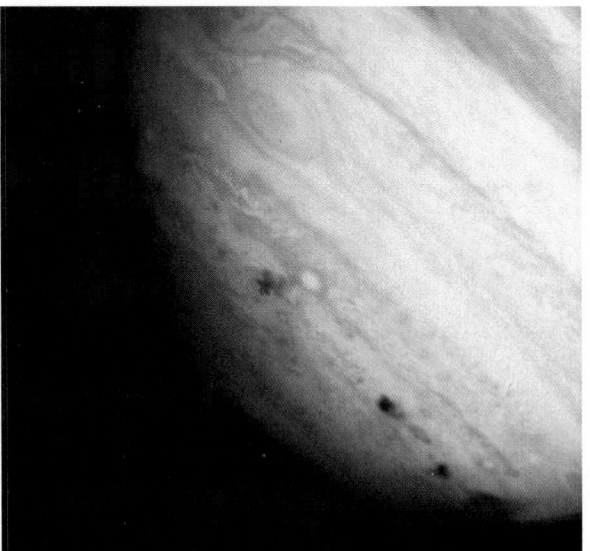

The collisions of the pieces of Comet Shoemaker–Levy 9 with Jupiter produced this series of huge dark blotches in the atmosphere of the giant planet.

9.1 MOMENTUM AND NEWTON'S SECOND LAW OF MOTION

There is something qualitatively different about getting a 10 kg baby stroller moving from rest to a speed of 2 m/s and getting a 15 000 kg truck moving from rest to the same speed. It is doubtless easier for you to change the velocity of the baby stroller than that of the truck. The distinction between the situations has to do with their **momentum**.

> The momentum $\vec{p}$ of a mass m moving at a velocity $\vec{v}$ is[¶]
>
> $$\boxed{\vec{p} \equiv m\vec{v}} \qquad (9.1)$$
>
> The momentum is thus a vector quantity and its direction is always parallel to the velocity vector. The SI units of momentum are kg·m/s.

[#] We introduced the idea of the center of mass of a two-particle system in Chapter 6 when studying orbital motion. Here we will generalize to systems of many particles.

[¶] Equation 9.1 defines the *classical momentum* of a system, valid as long as the speed of the system is much less than the speed of light—that is, $v \ll c$. For systems having speeds that are significant fractions of the speed of light, the definition of the momentum must be modified. The reasons for the modification are examined in Chapter 25.

Systems with the same velocity but different masses have different momenta, as with the baby stroller and truck. Also we can agree from experience that a larger force is needed to *change* the momentum of the truck than the momentum of the baby stroller during the same time interval (e.g., bring them both from rest to velocity $\vec{v}$ during the same time interval). So forces are involved in changing the momentum of a system.

We introduced Newton's second law of motion in Equation 5.1 as

$$\vec{F}_{total} = m\vec{a}$$

where $\vec{F}_{total}$ is the vector sum of all the forces acting on a system of mass m and $\vec{a}$ is the system's acceleration. Actually Newton formulated the second law of motion using the *momentum*.* The momentum of a single-particle system changes at a rate equal to the total force on it:

$$\vec{F}_{total} = \frac{d\vec{p}}{dt}$$

Now think of the total mass of a system being accelerated as composed of lots of different masses, perhaps fastened together or perhaps not, and each with its own distinct momentum. The vector sum of all the individual momenta is the total momentum $\vec{p}_{total}$ of the entire system of masses. Each individual momentum changes at a rate equal to the total force on that piece. Hence the grand total momentum $\vec{p}_{total}$ of the whole system changes at a rate equal to the vector sum $\vec{F}_{total}$ of all the forces acting on all the pieces. This result,

$$\boxed{\vec{F}_{total} = \frac{d\vec{p}_{total}}{dt}} \qquad (9.2)$$

holds true for any system, be it a single point-particle mass, a huge boulder, or a wild collection of pieces flying helter-skelter in all sorts of directions.

Equation 9.2 is the most general statement of Newton's second law of motion.

For systems of *constant* mass, Equations 5.1 and 9.2 for the second law are equivalent. To confirm this, substitute the definition of the momentum $\vec{p}$ from Equation 9.1 into the right-hand side of Equation 9.2; we obtain Equation 5.1:

$$\begin{aligned}
\vec{F}_{total} &= \frac{d\vec{p}_{total}}{dt} \\
&= \frac{d}{dt}(m\vec{v}) \\
&= m\frac{d\vec{v}}{dt} \qquad (\text{constant } m) \\
&= m\vec{a} \qquad (\text{constant } m)
\end{aligned}$$

The effects of the total force acting during a time interval on even systems of varying mass are investigated using Equation 9.2, since it is the most general statement of Newton's second law of motion.

_____
*Newton called it the *state of motion*, a term no longer used.

We focus our attention here not on the individual forces on the system, but rather on the total force $\vec{F}_{total}$, because it is the total force that determines the rate at which the total momentum of the system changes. For example, if a baby stroller and a truck both accelerate from rest to velocity $\vec{v}$ during the same time interval, their momenta both change, but by very different amounts. Clearly, during the same time interval, it will take a total force of greater magnitude to change the momentum of the truck than that of the baby stroller because, according to Equation 9.2, the total force is equal to the time rate of change of the momentum.

QUESTION 1

Two masses have equal kinetic energies. Under what circumstances (if any) are the momenta of the particles the same?

EXAMPLE 9.1 _____

An 18.2 kg stone is used in the Scottish game of curling. If the stone has a speed of 5.00 m/s and is moving toward increasing values of a coordinate *x*, what is the momentum of the stone?

Players coax the stone used in the game of curling.

Solution

The velocity of the stone is $(5.00 \text{ m/s})\hat{\imath}$; so its momentum is (from Equation 9.1)

$$\begin{aligned}
\vec{p} &= m\vec{v} \\
&= (18.2 \text{ kg})(5.00 \text{ m/s})\hat{\imath} \\
&= (91.0 \text{ kg}\cdot\text{m/s})\hat{\imath}
\end{aligned}$$

EXAMPLE 9.2 _____

A 600 kg African black rhinoceros (*Diceros bicornis*) is charging a stalled safari vehicle filled with terrified tourists. The rhino has a velocity $\vec{v} = (5.00 \text{ m/s})\hat{\imath} + (6.00 \text{ m/s})\hat{\jmath}$ with respect to a convenient Cartesian coordinate system with $\hat{\imath}$ north and $\hat{\jmath}$ west, laid out by a wildlife biologist.

a. Find the momentum of the rhino and the magnitude of its momentum.
b. If a baseball of mass 0.145 kg has momentum of the same magnitude, what is its speed?

Solution

a. Use Equation 9.1. The momentum is

$$\vec{p} = m\vec{v}$$
$$= (600\ \text{kg})[(5.00\ \text{m/s})\hat{\imath} + (6.00\ \text{m/s})\hat{\jmath}]$$
$$= (3.00 \times 10^3\ \text{kg·m/s})\hat{\imath} + (3.60 \times 10^3\ \text{kg·m/s})\hat{\jmath}$$

The magnitude of this vector is

$$p = [(3.00 \times 10^3\ \text{kg·m/s})^2 + (3.60 \times 10^3\ \text{kg·m/s})^2]^{1/2}$$
$$= 4.69 \times 10^3\ \text{kg·m/s}$$

b. The speed of the baseball is found from the magnitude of its momentum:

$$p = mv$$

So

$$4.69 \times 10^3\ \text{kg·m/s} = (0.145\ \text{kg})v$$

Solving for v, you find

$$v = 3.23 \times 10^4\ \text{m/s}$$
$$= 32.3\ \text{km/s}$$

This is almost three times its escape speed! A *real* fastball.

9.2 IMPULSE–MOMENTUM THEOREM

The longer the total force acts on a system, the longer the system is subjected to an acceleration and the greater the change in its velocity. Changes in velocity imply changes in momentum, according to the definition of the momentum, Equation 9.1. Here we find the relationship between the total force, the interval during which it acts, and the resulting change in the momentum of the system.

Let $\vec{F}_{\text{total}}$ be the total force on a system during a time interval Δt. Of course there is no guarantee that the total force is constant during the interval; the total force may well be time dependent. For example, the force of a bat on a baseball is not of constant magnitude during the brief interval they are in contact. Likewise the gravitational force on a planet in an elliptical orbit changes its magnitude and direction as the planet assumes different positions in space at various times, as shown in Figure 9.1. So we must account for situations where the total force is time dependent, $\vec{F}_{\text{total}}(t)$.

Rearrange the general form of Newton's second law, Equation 9.2, into the differential form

$$\vec{F}_{\text{total}}(t)\ dt = d\vec{p}$$

Integrate over the time interval between t_i and t_f. Due to the action of the total force, the total momentum of the system changes from $\vec{p}_i$ to $\vec{p}_f$ during the interval:

$$\int_{t_i}^{t_f} \vec{F}_{\text{total}}(t)\ dt = \int_{\vec{p}_i}^{\vec{p}_f} d\vec{p}$$

The integration on the left-hand side cannot be performed without full knowledge of how $\vec{F}_{\text{total}}(t)$ varies with time. On the

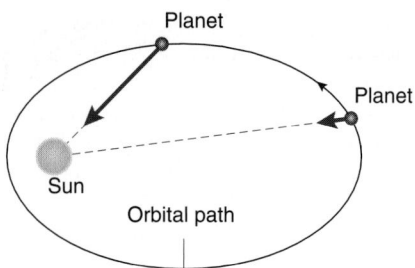

FIGURE 9.1 The gravitational force of the Sun on a planet in an elliptical orbit changes its magnitude and direction as the planet assumes different positions in its orbit.

other hand, the integration on the right-hand side *can* be performed easily: the integrand is a perfect differential. Hence we have

$$\int_{t_i}^{t_f} \vec{F}_{\text{total}}(t)\ dt = \vec{p}_f - \vec{p}_i \qquad (9.3)$$

The left-hand side of this relationship, the integral of the total force over the interval Δt, is called the **impulse** $\vec{I}$ of the total force.* Since force is a vector quantity, impulse is also a vector quantity[†]:

$$\boxed{\vec{I} \equiv \int_{t_i}^{t_f} \vec{F}_{\text{total}}(t)\ dt} \qquad (9.4)$$

This definition is a vector equation and so represents three equations, one for each of the Cartesian components of the impulse. That is,

$$I_x = \int_{t_i}^{t_f} F_{x\,\text{total}}(t)\ dt$$

with similar equations for the y- and z-components. The SI units of impulse are N·s, which are equivalent to kg·m/s, the units for momentum.

The right-hand side of Equation 9.3 is the change in the momentum $\Delta\vec{p}$ of the system.

Hence Equation 9.3 grandly states that the impulse of the total force acting on the system is equal to the change in the total momentum of the system:

$$\boxed{\vec{I} = \Delta\vec{p}} \qquad (9.5)$$

The total impulse vector $\vec{I}$ has the *same* direction as the *change* in the momentum vector $\Delta\vec{p}$ and they both have identical magnitudes. Equation 9.5 is known as the **impulse–momentum theorem**.

As we have seen, to evaluate the integral for the impulse, defined by Equation 9.4, the time dependence of the force must

---

*The name impulse arises from the Latin *impulsus*, the past participle of *impellere*, to impel (here, to change the momentum of a system).

[†]We also could calculate the impulse provided by each individual force acting on the system using a similar integration. The total impulse then is the vector sum of the impulses of all the individual forces.

be known; this integration is thus often difficult to perform. This difficulty certainly is the case for a bat hitting a baseball, for example. However, if the total force is constant, it can be factored out of the integral and the integration for the impulse can be done explicitly (as we will see in Example 9.3).

If the way a component of the total force on the system varies with time is known and a graph is made of that component versus t, as in Figure 9.2, the corresponding component of the impulse of the total force between t_i and t_f is the area under the force component curve between the two times. Such areas may be positive or negative, of course, and the corresponding impulse component is likewise positive or negative.

Think of the collision of a bat with a ball, and take the ball to be the system. The force of the bat on the ball during the collision typically is much greater than the weight of the ball (see Example 9.4). Thus, to a good approximation, the total force on the ball during the collision is just the force of the bat on the ball. This force increases and decreases dramatically over the brief interval of the collision as indicated schematically in Figure 9.3. Here, for convenience, we represent the total force as if it has only one component. Forces that exist only for short intervals are called **impulsive forces**, in keeping with the colloquial meaning of the word impulse.

Often it is useful to calculate a time-averaged force component, say $F_{x\,ave}$; it provides an estimate of the order of magnitude of the actual force component during the interval Δt. By definition the average force component $F_{x\,ave}$, during the

same time interval Δt, has the same area under its curve as does the actual total force component $F_{x\,total}$:

$$F_{x\,ave}\,\Delta t \equiv \int_{t_i}^{t_f} F_{x\,total}\,dt \equiv I_x \qquad (9.6)$$

The impulse–momentum theorem (Equation 9.5) indicates the x-component of the total impulse is equal to the x-component of the change in the momentum of the particle. Therefore the expression for the x-component of the average total force becomes

$$F_{x\,ave} = \frac{\Delta p_x}{\Delta t} \qquad (9.7)$$

This equation enables us to calculate the average total force *if* the interval during which it exists can be estimated in some way.

QUESTION 2

How can a small force on a system produce a large impulse? Conversely, how can a large force on a system produce a small impulse?

STRATEGIC EXAMPLE 9.3

A pickled egg of mass 0.250 kg of the extinct dodo bird (*Raphus cucullatus*) falls from rest off a museum shelf and plunges 2.00 m to the floor. Humpty Dumpty!

a. Use Equation 9.4 to calculate the impulse of the gravitational force of the Earth on the egg during the interval of the fall.
b. Determine the speed of the egg at the instant just before impact, and from this calculate the change in the momentum of the egg during its fall.
c. The egg is brought to a halt by the floor during 0.010 s. Find the impulse of the total force on the egg during the crash. Find the time-averaged total force on the egg during the crash and compare the magnitude of this force with the magnitude of the weight of the egg.

Solution

a. The gravitational force of the Earth on the egg (its weight) is a constant force near the surface of the Earth. Use the coordinate system indicated in Figure 9.4. The gravitational force is $-mg\hat{\jmath}$. The time of fall is found from the one-dimensional motion equation for a constant acceleration:

$$y = y_0 + v_{y0}t + \frac{1}{2}a_y t^2$$

Here

$$y_0 = 2.00 \text{ m}$$
$$v_{y0} = 0 \text{ m/s}$$
$$a_y = -9.81 \text{ m/s}^2$$

The impact occurs where $y = 0$ m at time t. Making these substitutions,

$$0 \text{ m} = 2.00 \text{ m} + (0 \text{ m/s})\,t + \frac{1}{2}(-9.81 \text{ m/s}^2)\,t^2$$

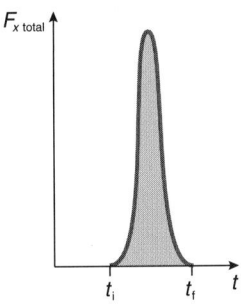

FIGURE 9.2 The x-component of the impulse during Δt is the area under the graph of $F_{x\,total}$ versus t between t_i and t_f.

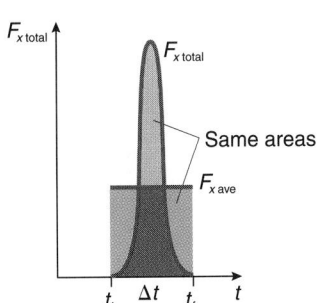

FIGURE 9.3 The area (component of the impulse) under the graph of the actual force component versus t is equal to the area under the average force component.

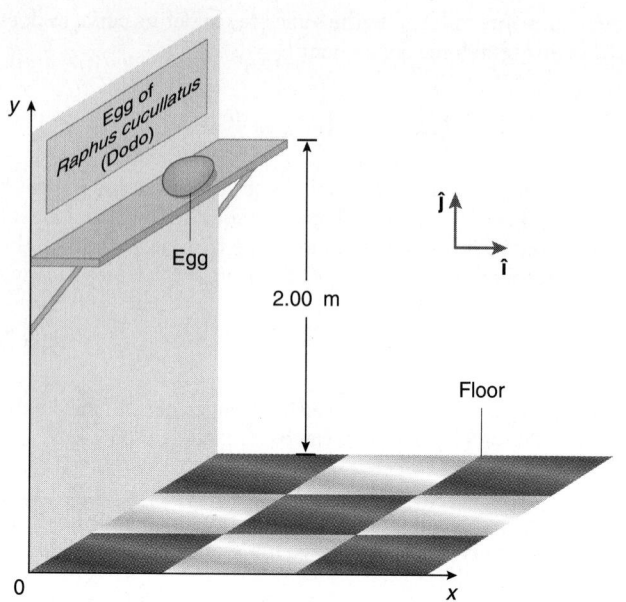

FIGURE 9.4

Solving for t, taking the positive root, you have

$$t = 0.639 \text{ s}$$

The integral for the impulse, Equation 9.4, then is

$$\vec{\mathbf{I}} = \int_{0 \text{ s}}^{t} (-mg\hat{\jmath}) \, dt$$

Since the force is constant, you find

$$\vec{\mathbf{I}} = -mg(t - 0 \text{ s})\hat{\jmath}$$

Substituting for m, g, and $t = 0.639$ s for the time of fall, you find the impulse to be

$$\vec{\mathbf{I}} = (-0.250 \text{ kg})(9.81 \text{ m/s}^2)(0.639 \text{ s})\hat{\jmath}$$
$$= (-1.57 \text{ kg·m/s})\hat{\jmath}$$

b. The velocity component at impact is found using

$$v_y = v_{y0} + a_y t$$
$$= 0 \text{ m/s} + (-9.81 \text{ m/s}^2)(0.639 \text{ s})$$
$$= -6.27 \text{ m/s}$$

The final momentum of the egg is thus

$$\vec{\mathbf{p}}_f = m\vec{\mathbf{v}}_f$$
$$= mv_y\hat{\jmath}$$
$$= (0.250 \text{ kg})(-6.27 \text{ m/s})\hat{\jmath}$$
$$= (-1.57 \text{ kg·m/s})\hat{\jmath}$$

The initial momentum of the egg is zero, because it began its fall with zero speed:

$$\vec{\mathbf{p}}_i = (0 \text{ kg·m/s})\hat{\jmath}$$

The change in the momentum of the egg is

$$\Delta\vec{\mathbf{p}} = \vec{\mathbf{p}}_f - \vec{\mathbf{p}}_i$$
$$= (-1.57 \text{ kg·m/s})\hat{\jmath} - (0 \text{ kg·m/s})\hat{\jmath}$$
$$= (-1.57 \text{ kg·m/s})\hat{\jmath}$$

Notice that the change in the momentum of the egg is equal to the impulse, consistent with Equation 9.5, the impulse–momentum theorem.

c. During the brief interval of the crash, take the initial time to be when the egg first makes contact with the floor and the final time to be when the egg stops. This time interval is given as $\Delta t = 0.010$ s. The impulse of the total force cannot be found by integrating $\vec{\mathbf{F}}_{\text{total}}(t)$ over the time interval, because the details of how the total force on the egg varies with time during the crash interval are unknown. However, the impulse–momentum theorem states the total impulse $\vec{\mathbf{I}}$ is equal to the change in the momentum of the system, and this we can calculate.

For this phase of the problem, the initial momentum of the egg is the momentum at the instant of contact:

$$\vec{\mathbf{p}}_i = m\vec{\mathbf{v}}_{\text{impact}}$$
$$= (0.250 \text{ kg})(-6.27 \text{ m/s})\hat{\jmath}$$
$$= (-1.57 \text{ kg·m/s})\hat{\jmath}$$

The final momentum is zero, since the egg is then at rest:

$$\vec{\mathbf{p}}_f = (0 \text{ kg·m/s})\hat{\jmath}$$

The change in the momentum of the egg is

$$\Delta\vec{\mathbf{p}} = \vec{\mathbf{p}}_f - \vec{\mathbf{p}}_i$$
$$= (0 \text{ kg·m/s})\hat{\jmath} - (-1.57 \text{ kg·m/s})\hat{\jmath}$$
$$= (1.57 \text{ kg·m/s})\hat{\jmath}$$

Since $\vec{\mathbf{I}} = \Delta\vec{\mathbf{p}}$, the impulse of the total force during the crash is

$$\vec{\mathbf{I}} = (1.57 \text{ kg·m/s})\hat{\jmath}$$

The average force during the crash is found using the vector version of Equation 9.6:

$$\vec{\mathbf{F}}_{\text{ave}} = \frac{\vec{\mathbf{I}}}{\Delta t}$$
$$= \frac{(1.57 \text{ kg·m/s})\hat{\jmath}}{0.010 \text{ s}}$$
$$= (1.6 \times 10^2 \text{ N})\hat{\jmath}$$

This average total force is the vector sum of the force of the floor on the egg (the normal force) and the gravitational force of the Earth on the egg (its weight). The magnitude of the average total force is 1.6×10^2 N and the magnitude of the weight of the egg is mg, here 2.45 N. The ratio of the magnitude of the average total force during the crash to the magnitude of the weight is thus

$$\frac{1.6 \times 10^2 \text{ N}}{2.45 \text{ N}} = 65$$

Notice that the magnitude of the average total force on the egg during the crash is much greater than the magnitude of its weight.

EXAMPLE 9.4

A baseball of mass 0.145 kg arrives at home plate with a horizontal speed 20.0 m/s. The batter hits a line drive. Thwack! Immediately after the ball leaves the bat, the speed of the ball is 50.0 m/s in the opposite horizontal direction.

a. Calculate the impulse imparted to the ball by the total force on the ball.
b. Estimate the magnitude of the average total force on the ball if the collision of the ball with the bat lasts for 1.00×10^{-3} s.
c. Compare the magnitude of the average total force with the magnitude of the weight of the ball.

Solution
a. Diagram the situation and introduce the coordinate system shown in Figure 9.5. The ball is the system. The change in the momentum of the ball is

$$\Delta\vec{p} = \vec{p}_f - \vec{p}_i$$
$$= m\vec{v}_f - m\vec{v}_i$$
$$= (0.145 \text{ kg})(50.0 \text{ m/s})(-\hat{i}) - (0.145 \text{ kg})(20.0 \text{ m/s})\hat{i}$$
$$= (-10.2 \text{ kg·m/s})\hat{i}$$

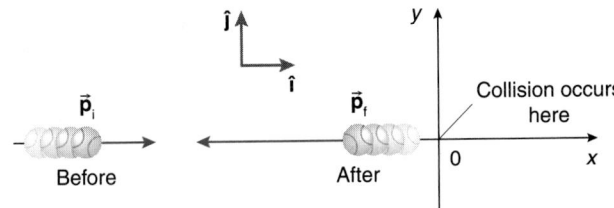

FIGURE 9.5

According to the impulse–momentum theorem, Equation 9.5, the impulse of the total force is equal to the change in the momentum, and so

$$\vec{I} = (-10.2 \text{ kg·m/s})\hat{i}$$

b. Use the vector version of Equation 9.6 to find the average total force:

$$\vec{F}_{ave} = \frac{\vec{I}}{\Delta t}$$
$$= \frac{(-10.2 \text{ kg·m/s}) \hat{i}}{1.00 \times 10^{-3} \text{ s}}$$
$$= (-1.02 \times 10^4 \text{ N}) \hat{i}$$

c. Take the ratio of the magnitude of the average total force to the magnitude of the weight of the ball:

$$\frac{F_{ave}}{mg} = \frac{1.02 \times 10^4 \text{ N}}{(0.145 \text{ kg})(9.81 \text{ m/s}^2)}$$
$$= 7.17 \times 10^3$$

Would you agree that the magnitude of the average total force on the ball during the collision is *much* greater than that of the weight of the ball?

EXAMPLE 9.5

An incoming horizontal pitch at speed v is hit. The baseball of mass m leaves the bat in the vertical direction at speed $2v$.

a. What is the impulse of the total force on the ball?
b. What is the magnitude of the impulse?

Solution
a. The ball is the system. Use the coordinate system indicated in Figure 9.6. The initial momentum of the ball is

$$\vec{p}_i = mv\hat{i}$$

and its final momentum is

$$\vec{p}_f = m(2v)\hat{j}$$

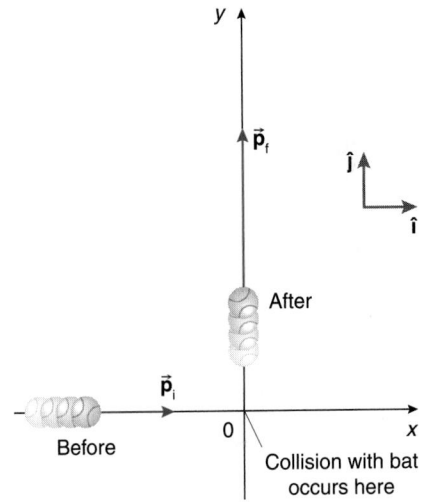

FIGURE 9.6

The change in momentum of the ball is then

$$\Delta\vec{p} = \vec{p}_f - \vec{p}_i$$
$$= 2mv\hat{j} - mv\hat{i}$$
$$= mv(-\hat{i} + 2\hat{j})$$

According to the impulse–momentum theorem, Equation 9.5, the impulse of the total force on a system is equal to the change in the momentum of the system, and so

$$\vec{I} = mv(-\hat{i} + 2\hat{j})$$

Notice that the direction of the impulse is *not* necessarily along the direction of either momentum vector; the impulse is in the direction of the *change* in the momentum.

b. The magnitude of the impulse is

$$I = mv[(-1)^2 + 2^2]^{1/2}$$
$$= \sqrt{5}\,mv$$

9.3 THE ROCKET: A SYSTEM WITH VARIABLE MASS*

An infamous editorial blunder:

> . . . after the rocket quits our air and really starts on its longer journey [to the Moon] its flight would neither be accelerated nor maintained by the explosion of the charges it then might have left. To claim that it would be is to deny a fundamental law of dynamics. . . . That Professor Goddard [inventor of the liquid fueled rocket] . . . does not know the relation of action to reaction, and of the need to have something better than a vacuum against which to react—to say that would be absurd. Of course he only seems to lack the knowledge ladled out daily in high schools.
>
> *New York Times* (1920)*

The liftoff of a rocket propelling a payload into orbit is both a spectacular sight and a technological triumph. Most of the mass of the rocket on the launch pad is fuel. During the powered phase of the launch, the mass of the rocket system changes dramatically as exhaust gases are expelled furiously at great speed. A rocket can propel itself in the vacuum of outer space by expelling gases at great speed in a direction opposite to that of the rocket itself. Everything needed for the combustion of its fuel is carried within the rocket. Jet airplanes use the same physical principle of expelling gases at high speed but are restricted to lower altitudes to use atmospheric oxygen for the combustion of the fuel. Jets are more complex systems than rockets because they accumulate mass with their intakes and expel even more mass (combustion by-products) at great speed. Here we examine the physics of rockets, our first encounter with systems of variable mass.

For simplicity, consider a single-stage rocket moving in one dimension, say, along the *x*-axis parallel to $\hat{\mathbf{i}}$; see Figure 9.7. Let the system be the rocket with any remaining fuel. Call the mass of the system $m(t)$, where we explicitly indicate that the mass varies with time. Its corresponding velocity relative to a fixed (inertial) coordinate system is $v_x(t)\hat{\mathbf{i}}$.

Assume that the exhaust gases are ejected at constant speed v_{exh} *relative to the rocket*. The velocity of the rocket and the velocity of the exhaust gases are in opposite directions. Hence the velocity of the exhaust gases relative to the fixed inertial coordinate system is

$$v_x(t)\hat{\mathbf{i}} - v_{\text{exh}}\hat{\mathbf{i}} = [v_x(t) - v_{\text{exh}}]\hat{\mathbf{i}}$$

At an instant *t* the momentum of the rocket (with the remainder of its unburned fuel) is

$$m(t)v_x(t)\hat{\mathbf{i}} \qquad (9.8)$$

During a short differential time interval *dt*, the mass of the rocket *changes* by *dm*. The mass change *dm* is *negative* since the rocket loses

---

*Editorial, "A Severe Strain on Credulity," 13 January 1920, page 12, column 5.

Rockets are systems that dramatically change their mass with time.

mass; this does *not* imply that mass is negative: *dm* is the *change* in the mass of the rocket. The mass of the exhausted gases is −*dm* (which is positive since *dm* itself is negative). At the end of the interval *dt*, the mass of the rocket is $m(t) + dm$ (remember that *dm* is a negative quantity) and its velocity is $[v_x(t) + dv_x]\hat{\mathbf{i}}$. Hence the momentum of the rocket and residual fuel in its tanks is (see Figure 9.8)

$$[m(t) + dm][v_x(t) + dv_x]\hat{\mathbf{i}}$$

The momentum of the fuel expelled during *dt* is

$$(-dm)[v_x(t) - v_{\text{exh}}]\hat{\mathbf{i}}$$

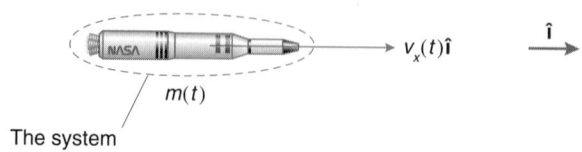

FIGURE 9.7 The rocket system at instant *t*.

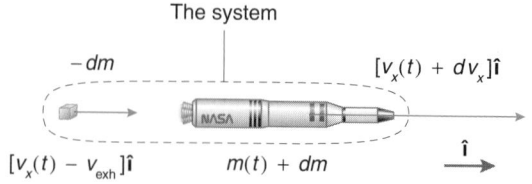

FIGURE 9.8 The rocket system and expelled fuel at instant $t + dt$.

At the end of the interval dt the momentum of the total system is the vector sum

$$[m(t) + dm][v_x(t) + dv_x]\hat{\imath} + (-dm)[v_x(t) - v_{\text{exh}}]\hat{\imath}$$
$$= [m(t)v_x(t) + m(t)\,dv_x + dm\,v_x(t) - dm\,v_x(t) + v_{\text{exh}}\,dm]\hat{\imath}$$
$$= [m(t)v_x(t) + m(t)\,dv_x + v_{\text{exh}}\,dm]\hat{\imath} \qquad (9.9)$$

where we neglect the term $dm\,dv_x\,\hat{\imath}$ involving the product of two differentials, since as differentials approach zero, their product vanishes even faster. The differential change $dp_x\,\hat{\imath}$ in the momentum of the system is, therefore, the difference between Equations 9.9 and 9.8:

$$dp_x\,\hat{\imath} = [m(t)\,dv_x + v_{\text{exh}}\,dm]\hat{\imath}$$

The impulse–momentum theorem (Equation 9.5) states that the differential change in the x-component of the momentum must equal the x-component of the differential impulse of the total force on the system during the interval dt:

$$[m(t)\,dv_x + v_{\text{exh}}\,dm]\hat{\imath} = F_{x\,\text{total}}\,dt\,\hat{\imath}$$

Hence

$$F_{x\,\text{total}} = m(t)\frac{dv_x}{dt} + v_{\text{exh}}\frac{dm}{dt} \qquad (9.10)$$

The x-component of the acceleration of the rocket is

$$\frac{dv_x}{dt}$$

Equation 9.10 can be rewritten as

$$F_{x\,\text{total}} - v_{\text{exh}}\frac{dm}{dt} = m(t)\frac{dv_x}{dt} \qquad (9.11)$$

Equation 9.11 is known as the **rocket equation**. Since the rocket is losing mass,

$$\frac{dm}{dt} < 0 \text{ kg/s}$$

and the term

$$-v_{\text{exh}}\frac{dm}{dt}$$

actually is a force component in the same direction as the acceleration component. The term

$$v_{\text{exh}}\left|\frac{dm}{dt}\right| \qquad (9.12)$$

is called the **thrust** of the rocket and arises because the mass of the rocket is not constant. The thrust can be increased by increasing the speed v_{exh} of the exhaust gases relative to the rocket or by increasing the rate at which the rocket loses mass.

For a launch near the surface of the Earth, $F_{x\,\text{total}}$ must include the weight of the rocket (which varies with time during the flight) and speed-dependent kinetic frictional forces caused by air resistance. At launch the thrust must be greater than the weight of the rocket for it to leave the launch pad.

If we restrict ourselves to an isolated rocket, far away from any other large masses like the Earth, or to a direction perpendicular to gravitational forces, $F_{x\,\text{total}} = 0$ N and the rocket equation, 9.11, becomes

$$-v_{\text{exh}}\frac{dm}{dt} = m(t)\frac{dv_x}{dt}$$

Rearranging, we find

$$-\frac{dm}{m} = \frac{dv_x}{v_{\text{exh}}} \qquad (9.13)$$

Let the initial speed of the rocket be zero when its mass is m_0. Integrate Equation 9.13:

$$-\int_{m_0}^{m}\frac{dm}{m} = \int_{0\text{ m/s}}^{v_x}\frac{dv_x}{v_{\text{exh}}}$$

Since the speed of the exhaust gases is constant, we have

$$\ln m_0 - \ln m = \frac{v_x}{v_{\text{exh}}}$$

Hence

$$v_x = v_{\text{exh}}\,\ln\left(\frac{m_0}{m}\right) \qquad (9.14)$$

Separate the initial mass of the rocket into payload and fuel:

$$m_0 = m_{\text{payload}} + m_{\text{fuel}}$$

When all fuel is exhausted, the mass of the rocket is just m_{payload}. Equation 9.14 then indicates that the maximum speed of the rocket is

$$v_x = v_{\text{exh}}\,\ln\left(\frac{m_{\text{fuel}} + m_{\text{payload}}}{m_{\text{payload}}}\right)$$
$$= v_{\text{exh}}\,\ln\left(\frac{m_{\text{fuel}}}{m_{\text{payload}}} + 1\right) \qquad (9.15)$$

From this result we can see that to attain a large final speed, the fuel to payload mass ratio must be large. Since in deriving Equation 9.15 we ignored the presence of forces such as the weight of the rocket and kinetic friction, you *cannot* use simply Equation 9.15 for satellite launches from the surface of the Earth.

EXAMPLE 9.6

A rocket motor expels exhaust gases at a rate of 1.5×10^3 kg/s with a speed of 3.0 km/s. What is the thrust of the motor?

Solution

The thrust is the product of the speed of the exhaust gases relative to the rocket and the absolute value of the rate of mass loss:

$$v_{\text{exh}}\left|\frac{dm}{dt}\right| = (3.0 \times 10^3 \text{ m/s})(1.5 \times 10^3 \text{ kg/s})$$
$$= 4.5 \times 10^6 \text{ N}$$

EXAMPLE 9.7

Well away from any other masses, a rocket with a fuel to payload mass ratio of $m_{fuel}/m_{payload} = 4.0$ is fired. The exhaust gases leave the rocket at a speed of 2.0 km/s relative to the rocket. If it is initially at rest with respect to an inertial reference frame, what is the speed of the rocket in this frame when the fuel is exhausted?

Solution

The problem meets the conditions attached to the use of Equation 9.15. Thus

$$v_x = v_{exh} \ln \left(\frac{m_{fuel}}{m_{payload}} + 1 \right)$$
$$= (2.0 \times 10^3 \text{ m/s})(\ln 5.0)$$
$$= 3.2 \times 10^3 \text{ m/s}$$

The speed is 3.2 km/s.

9.4 CONSERVATION OF MOMENTUM

If the total force acting on a system is zero, the system is said to be mechanically **isolated**. For such isolated systems, Equation 9.2 becomes

$$0 \text{ N} = \frac{d\vec{p}_{total}}{dt}$$

The zero time derivative implies that $\vec{p}_{total}$ must be a constant vector.

> When $\vec{p}_{total}$ is constant, we say the system exhibits **conservation of momentum**.*

Nonzero forces of one particle within the system on other particles within it may exist; these are called **internal forces**. Such internal forces may change the momentum of the individual particles within the system, but the *total* momentum of the system remains fixed; it is just reapportioned among these particles in different ways. Since the internal forces of one particle on another within the system always are paired according to Newton's third law of motion, they are of equal magnitude and opposite direction; so the internal forces vector sum to zero and cannot affect the total momentum. Only external forces from outside the system (i.e., a nonzero total force) can change its total momentum.

We shall see that conservation of momentum, like the CWE theorem, is a very important physical theorem. In particular, in the next section we see that the total momentum is conserved in collisions if the system you define is large enough to include all the particles participating in the collision.

*The expression really should be called conservation of the *total* momentum $\vec{p}_{total}$ of the system, since the *individual* particle momenta may change; it is the *total* momentum of the system (the vector sum of the individual particle momenta that make up the system) that is unvarying.

9.5 COLLISIONS

We already have considered a few examples of collisions as examples of impulsive forces. We now present a systematic, general approach to collision phenomena.

If a system has zero total force on it ($\vec{F}_{total} = 0$ N), then the time rate of change of the total momentum of the system also is zero according to Newton's second law of motion, expressed in Equation 9.2:

$$0 \text{ N} = \frac{d\vec{p}_{total}}{dt}$$

This implies that the total momentum of the system must be a constant vector; that is, the total momentum of the system is conserved.

Take *all* the particles involved in a collision as a single system, as shown in Figure 9.9. The total momentum of the system just before the collision is the vector sum of the individual momenta of the particles; for a two-particle collision, this is

$$\vec{p}_{total \ bfr} \equiv \vec{p}_{1 \ bfr} + \vec{p}_{2 \ bfr}$$

Likewise the total momentum of the system after the collision is the vector sum of the individual momenta just after the collision; for a two-particle collision, this is

$$\vec{p}_{total \ aft} \equiv \vec{p}_{1 \ aft} + \vec{p}_{2 \ aft}$$

If the total force on the system is zero, conservation of momentum means that the total momentum of the system the instant before the collision is equal to the total momentum of the system the instant after the collision; that is,

$$\vec{p}_{total \ bfr} = \vec{p}_{total \ aft} \qquad (9.16)$$

> In a collision, the *individual momenta* of the particles *do change*, but the *total momentum of the system of colliding particles does not.*

Why do the individual particle momenta change? Consider the collision between a baseball and a bat. Take the ball as a single-particle system. During the collision, there are two forces on the ball: (1) the large force of the bat on the ball and (2) the comparatively small weight of the ball (the gravitational force of the Earth on the ball). During the brief time interval of the collision, the force of the bat on the ball causes a much greater impulse on the ball than the impulse of the weight. Hence, to an excellent approximation, the change in the momentum of the ball is determined by the impulse of the bat; we can essentially

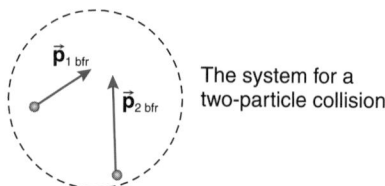

The system for a two-particle collision

FIGURE 9.9 Take the collection of all the particles in a collision as the system.

neglect the impulse of the gravitational force. In any case, the total force on the ball (considered as a single-particle system) during the collision is not zero; so according to Newton's second law of motion (applied to the single-particle ball system), the momentum of this system (the ball) is not conserved.

It is only by considering *all* the particles involved in a collision as a single system that Newton's second law leads to conservation of the total momentum of the system.

> Conservation of the total momentum is exact as long as the vector sum of all the forces on the system is zero.

Conservation of the total momentum can be applied to two-particle collisions or two-million-particle collisions as long as the system is defined to include all the particles participating in the collision.

Occasionally there are nonzero total forces on a system of all particles involved in a collision. For example, when two balls collide in midair, the gravitational force of the Earth on the two balls (their weight) is, of course, ever present. However, during the typically brief time interval of the collision, the gravitational forces usually are negligible compared with the forces of the particles on each other arising from the collision. Hence, during the collision, the dynamics of each particle is dominated by the forces involved in the collision. These internal collision forces are paired according to Newton's third law and vector sum to zero when you consider the system to be all the particles involved in the collision. Conservation of the total momentum thus is an excellent approximation for analyzing the dynamics of the system of colliding particles during the brief time interval between the instant before and the instant after the collision. Once the brief collision is over, however, the dynamics of each particle is determined by the gravitational force on each ball, since the forces on each associated with the collision no longer exist.

It is kinetic energy considerations that enable us to classify collisions into two broad regimes.

Elastic Collisions

> If the total kinetic energy of the particles is conserved, then the collision is called an **elastic collision**.

In other words, for elastic collisions there is no change in the total kinetic energy of the system of particles:

$$\Delta KE_{total} = 0 \text{ J}$$

Thus

$$KE_{total\ aft} = KE_{total\ bfr}$$

where $KE_{total\ aft}$ is the (scalar) sum of the scalar kinetic energies of the particles just after the collision and $KE_{total\ bfr}$ is the sum of the kinetic energies of the particles just before the collision.

If the kinetic energy is conserved (unchanged) before and after the collision, then the CWE theorem, Equation 8.26, becomes

$$W_{total} = \Delta KE_{total}$$
$$= 0 \text{ J}$$

Thus the work done by the total force acting during the collision is zero according to the CWE theorem. The forces involved in the collision therefore must be either conservative forces (such as Hooke's-law elastic forces) or forces doing zero work.

During the brief time interval of the collision, the kinetic energy is *not* conserved; some or all of the initial total kinetic energy of the particles is stored during the brief collision as the potential energy of various conservative forces (such as Hooke's law elastic forces, or electrical potential energy on a microscopic scale). But as the time interval of collision ends, this potential energy all is dumped back into kinetic energy. No loss of kinetic energy arises if we look at the kinetic energy the instant just before the collision and the kinetic energy the instant just after the collision. There are no thermal effects, and no work is done by nonconservative forces during an elastic collision. Collisions between macroscopic particles are rarely, if ever, elastic but can come close to that ideal in certain cases.

Inelastic Collisions

> If the total kinetic energy of the particles involved in a collision is *not* conserved, the collision is called an **inelastic collision**. Collisions of macroscopic particles typically are inelastic. *If the particles stick together after the collision*, the collision is called a **completely inelastic collision**.

The famous and important CWE theorem of Chapter 8 *cannot* account for this lost kinetic energy of the system. Why? The forces involved in the collision are paired according to Newton's third law (they are internal to the system of particles involved in the collision). The total force on the system of particles during the collision thus is zero. If the total force is zero, the work done by the total force is zero. Hence the left-hand side of the CWE theorem,

$$W_{total} = \Delta KE_{total}$$

is zero but the right-hand side is not! The loss of kinetic energy typically manifests itself as thermal energy (what physicists call the thermodynamic internal energy, as we will see in Chapter 13).* The CWE theorem of classical mechanics *neglects* such thermal effects. We return to this question once again in Chapter 13 to see how the problem finally is resolved.

> The things to remember about collisions are the following:
>
> a. The total momentum of the system is conserved in *both* elastic and inelastic collisions as long as
>   • the system is defined to include all the particles participating in the collision, and
>   • the total external force on the system is either zero or can be neglected compared with forces present during the collision itself.
> b. The total kinetic energy is conserved *only* for elastic collisions.
> c. A collision that is not elastic is called inelastic.
> d. If the particles stick together after the collision, it is called a completely inelastic collision.

---

*Some energy also may appear initially as sound waves.

PROBLEM-SOLVING TACTIC

9.1 None of the many specialized equations for various types of specific collisions enumerated in the examples is worth remembering; just remember the general points we have highlighted. Such specialized equations come with so many strings attached to them, restricting their usage, that by the time you pick your way through the strings to find out if a particular collision problem fits the special circumstances appropriate for the equation, you likely could derive the equation from the general principles just outlined. The memorization of equations is not what physics is all about.*

We look at several examples to illustrate the approach used to solve collision problems. *The results we obtain are not as important as the method used to set up the problems*, since each individual collision problem must be analyzed from scratch.

QUESTION 3

Two masses, one initially at rest, collide. Is it possible for *both* masses to be at rest after the collision? Explain.

EXAMPLE 9.8

Two particles of masses m and $2m$ approach each other with equal speeds along a line on a frictionless surface and make a completely inelastic collision in one dimension, as shown in Figure 9.10. After the collision the composite particle moves along the line of their approach. Find the velocity of the composite particle.

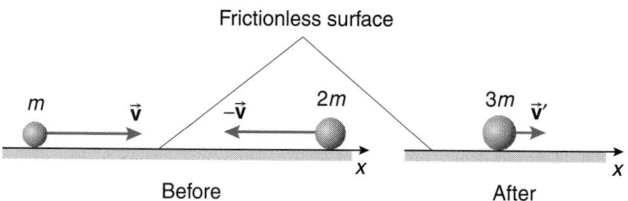

FIGURE 9.10

Solution

Define the system to be the two particles undergoing the collision. Introduce the coordinate system shown in Figure 9.11.

There are external forces acting on the system, namely the respective weights of the particles and the normal forces of the surface on them. However, the vector sum of these external forces acting on the system is zero by Newton's second law, because there is no acceleration along the vertical direction. The forces that each particle exerts on the other during the collision are *internal forces*; they vector sum to zero according to Newton's third law. Since the total force on the system is zero, the total momentum of the system is conserved: the total vector momentum of the system of particles is constant.

The individual momenta of the particles before the collision are

$$\vec{p}_{1 \text{ bfr}} = mv\hat{i}$$

*In an introductory physics course, quite frankly, it may *seem* like you are awash in equations; but there are in fact only a few fundamental physical principles. A good understanding of the latter makes it unnecessary to memorize many specialized equations because you can figure them out! This is part of what makes physics fun! (?)

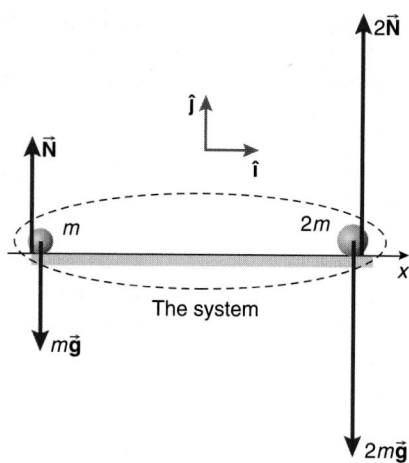

FIGURE 9.11

and

$$\vec{p}_{2 \text{ bfr}} = 2mv(-\hat{i})$$
$$= -2mv\hat{i}$$

The total momentum of the system before the collision is thus

$$\vec{p}_{\text{total bfr}} = \vec{p}_{1 \text{ bfr}} + \vec{p}_{2 \text{ bfr}}$$
$$= mv\hat{i} - 2mv\hat{i}$$
$$= -mv\hat{i}$$

Since the total momentum of this system is conserved, the momentum after the collision must be the same as the momentum before the collision. Let $\vec{v}'$ be the velocity of the composite particle system after the collision. Since the mass of the system is $m + 2m = 3m$, the total momentum of the system after the collision is, by the definition of momentum,

$$\vec{p}_{\text{total aft}} = 3m\vec{v}'$$

Applying momentum conservation,

$$\vec{p}_{\text{total aft}} = \vec{p}_{\text{total bfr}}$$
$$3m\vec{v}' = -mv\hat{i}$$

Solving for $\vec{v}'$,

$$\vec{v}' = -\frac{v}{3}\hat{i}$$

The minus sign indicates the velocity of the composite particle actually is opposite to that pictured in Figure 9.10. Notice that if we were to define the system to be only, say, the particle on the left (of mass m) (see Figure 9.11), then the momentum of *that* system is *not* conserved. This particle does suffer a change in momentum during the collision; the change arises from the force of the other particle (of mass $2m$) on it, a force that is now *external* to this single-particle system.

STRATEGIC EXAMPLE 9.9

Two particles make a one-dimensional, head-on elastic collision on a frictionless surface, as shown in Figure 9.12. One particle, of mass m_2, initially is at rest at the origin of the coordinate system introduced to analyze the problem.

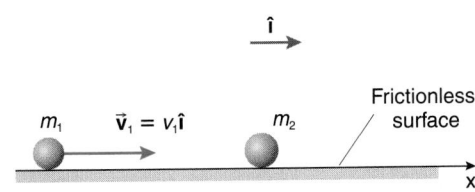

FIGURE 9.12

a. **Find the velocity components v_1' and v_2' of the particles after the collision.**
b. **Discuss the results if $m_2 \gg m_1$.**
c. **Discuss the results if $m_1 \gg m_2$.**

Solution
a. The system is the two particles involved in the collision. The total vector momentum before collision is

$$\vec{P}_{\text{total bfr}} = m_1 v_1 \hat{\imath} + m_2(0 \text{ m/s})\hat{\imath}$$
$$= m_1 v_1 \hat{\imath}$$

The total vector momentum after the collision is

$$\vec{P}_{\text{total aft}} = m_1 v_1' \hat{\imath} + m_2 v_2' \hat{\imath}$$

Since the total force on the system of two particles is zero, the total momentum is conserved:

$$\vec{P}_{\text{total bfr}} = \vec{P}_{\text{total aft}}$$

That is,

$$m_1 v_1 \hat{\imath} = m_1 v_1' \hat{\imath} + m_2 v_2' \hat{\imath}$$
$$m_1 v_1 = m_1 v_1' + m_2 v_2' \quad (1)$$

Since the collision is elastic, the total kinetic energy also is conserved. That is, the total kinetic energy before the collision is equal to the total kinetic energy after the collision:

$$KE_{\text{total bfr}} = KE_{\text{total aft}}$$

or

$$\frac{1}{2} m_1 v_1^2 = \frac{1}{2} m_1 v_1'^2 + \frac{1}{2} m_2 v_2'^2$$

Simplifying slightly,

$$m_1 v_1^2 = m_1 v_1'^2 + m_2 v_2'^2 \quad (2)$$

That's it for the physics: we conserved momentum and also kinetic energy, since *this* collision is an *elastic* collision. The problem now is mathematical. This is typically the case in collision problems: the mathematics is more tedious than the physics. We tune up our algebraic prowess. Kramer's rule for solving sets of simultaneous *linear* equations is of no use, because the kinetic energy equation is nonlinear in the

velocity components. We resort to brute force algebraic techniques or algebraic software.

The algebraic solution proceeds as follows: Take equation (2) stemming from the conservation of kinetic energy and transpose the term in particle 1 from the right to the left side, yielding

$$m_1(v_1^2 - v_1'^2) = m_2 v_2'^2$$

Now factor the difference of squares:

$$m_1(v_1 + v_1')(v_1 - v_1') = m_2 v_2'^2 \quad (3)$$

Make a similar transposition in the momentum component equation (1):

$$m_1(v_1 - v_1') = m_2 v_2' \quad (4)$$

Now divide equation (3) by equation (4), assuming $v_1 - v_1' \neq 0$ m/s. (If $v_1 = v_1'$, the problem then is quite uninteresting, since it means that particle m_1 has the same velocity after the collision. In such a case, no collision actually takes place: m_1 misses m_2.) The division yields

$$v_1 + v_1' = v_2' \quad (5)$$

Now substitute for v_2' in the right-hand side of equation (4):

$$m_1(v_1 - v_1') = m_2 v_2'$$
$$= m_2(v_1 + v_1')$$

Solve for v_1':

$$v_1' = \frac{m_1 - m_2}{m_1 + m_2} v_1$$

Then use equation (5) to find v_2':

$$v_2' = v_1 + v_1'$$
$$= v_1 + \frac{m_1 - m_2}{m_1 + m_2} v_1$$

Place things over a common denominator, and you finally get

$$v_2' = \frac{2m_1}{m_1 + m_2} v_1$$

Whew.
b. Suppose $m_2 \gg m_1$. This is like a BB pellet or a Ping-Pong ball colliding head-on with a stationary bowling ball. Such elastic collisions of particles with immobile objects will be important when we examine the kinetic theory of gases in Chapter 14. The velocity components after the collision are then

$$v_1' = \frac{m_1 - m_2}{m_1 + m_2} v_1$$
$$\approx -v_1 \quad \text{since } m_2 \gg m_1$$

and

$$v_2' = \frac{2m_1}{m_1 + m_2}$$

$$\approx 0 \text{ m/s} \quad \text{since } m_2 \gg m_1$$

The incident particle m_1 merely reverses direction and m_2 is unmoved by the whole affair. No surprises here.

c. On the other hand suppose $m_1 \gg m_2$, so that the mass of the incident particle is huge with respect to that of the stationary target particle sitting placidly at the origin. This is like a cannonball colliding head-on with a stationary BB pellet. In this case the expressions just derived indicate that the velocity components of the respective particles after the collision are

$$v_1' = \frac{m_1 - m_2}{m_1 + m_2} v_1$$

$$\approx v_1 \quad \text{since } m_1 \gg m_2$$

and

$$v_2' = \frac{2m_1}{m_1 + m_2} v_1$$

$$\approx 2v_1 \quad \text{since } m_1 \gg m_2$$

The huge incident mass essentially is unaffected by the encounter and the tiny target mass m_2 zips off at *twice* the speed of the incident behemoth! What does this case tell you about the speed of a baseball relative to the speed of the bat when the ball is hit from its position on a "T-ball" stand?

STRATEGIC EXAMPLE 9.10

Two particles, each with speed v, make a completely inelastic collision. The collision is not head-on, as shown in Figure 9.13.

a. Find the velocity of the composite particle after the collision, and the angle ϕ that this velocity makes with the x-axis in Figure 9.13.

b. What is the change in the total kinetic energy of the particles?

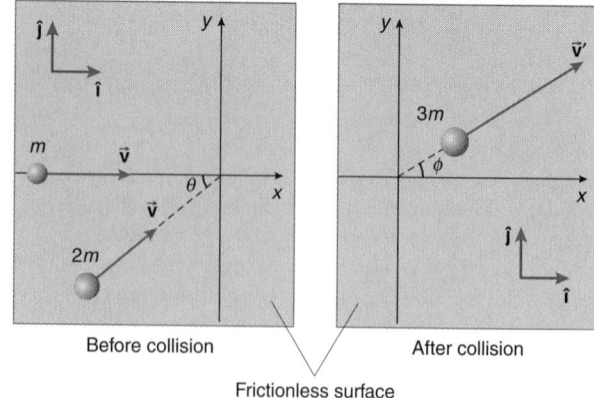

Before collision   After collision

Frictionless surface

FIGURE 9.13

Solution

Consider the system to be the two particles involved in the collision.

a. Momentum is conserved in collisions, so that the *total* momentum before the collision is equal to the *total* momentum after the collision. Remember that momentum is a *vector*, so that

$$\vec{P}_{\text{total bfr}} = \vec{P}_{\text{total aft}}$$

Use the coordinate system indicated in Figure 9.13. The *individual* momenta of the particles before the collision are

$$\vec{P}_{1 \text{ bfr}} = mv\hat{\imath}$$
$$\vec{P}_{2 \text{ bfr}} = (2mv \cos \theta)\hat{\imath} + (2mv \sin \theta)\hat{\jmath}$$

After the collision the composite particle has mass $3m$; let its unknown velocity be $\vec{v}'$. The total momentum of the system after the collision is the momentum of this single composite particle:

$$\vec{P}_{\text{total aft}} = 3m\vec{v}'$$

Conserve the total momentum:

$$\vec{P}_{\text{total aft}} = \vec{P}_{\text{total bfr}}$$
$$3m\vec{v}' = mv\hat{\imath} + [(2mv \cos \theta)\hat{\imath} + (2mv \sin \theta)\hat{\jmath}]$$

The mass m is common to every term and may be divided out. Solving for $\vec{v}'$, you find

$$\vec{v}' = \frac{1}{3} v \left[\left(1 + 2 \cos \theta \right) \hat{\imath} + \left(2 \sin \theta \right) \hat{\jmath} \right]$$

The angle ϕ that the velocity of the composite particle makes with the x-axis is found from the ratio of the components of the velocity $\vec{v}'$ indicated in Figure 9.13:

$$\tan \phi = \frac{v_y'}{v_x'}$$

$$= \frac{2 \sin \theta}{1 + 2 \cos \theta}$$

$$\phi = \tan^{-1}\left(\frac{2 \sin \theta}{1 + 2 \cos \theta} \right)$$

b. The total kinetic energy after the collision is the kinetic energy of the single composite particle of mass $3m$:

$$KE_{\text{total aft}} = \frac{1}{2} (3m) v'^2$$

Substituting for v', you have

$$KE_{\text{total aft}} = \frac{1}{2} (3m) \frac{v^2}{9} \left[\left(1 + 2 \cos \theta \right)^2 + \left(2 \sin \theta \right)^2 \right]$$

After a bit of algebraic simplification, you find

$$KE_{\text{total aft}} = \frac{mv^2}{6} (1 + 4 \cos \theta + 4 \cos^2 \theta + 4 \sin^2 \theta)$$

Since $\sin^2 \theta + \cos^2 \theta = 1$, the expression in the bracket can be simplified, yielding

$$KE_{\text{total aft}} = \frac{mv^2}{6}(5 + 4\cos\theta)$$

The kinetic energy before the collision is the sum of the kinetic energies of the two individual particles involved in the collision:

$$\begin{aligned} KE_{\text{total bfr}} &= \frac{mv^2}{2} + \frac{(2m)v^2}{2} \\ &= \frac{3mv^2}{2} \end{aligned}$$

Thus the *change* in the kinetic energy is

$$\begin{aligned} KE_{\text{total aft}} - KE_{\text{total bfr}} &= \frac{mv^2}{6}(5 + 4\cos\theta) - \frac{3mv^2}{2} \\ &= \frac{mv^2}{6}(5 + 4\cos\theta - 9) \\ &= -\frac{mv^2}{6}4(1 - \cos\theta) \end{aligned}$$

The change of kinetic energy is negative. Kinetic energy was lost in the collision since $KE_{\text{total aft}} < KE_{\text{total bfr}}$. In Chapter 13 we will see what happens to this lost kinetic energy.

EXAMPLE 9.11

Two small, hard spheres of *equal mass* collide elastically, as shown in Figure 9.14. One sphere is at rest before the collision. If the spheres do not collide head-on, after the collision they emerge making angles θ and ϕ with the incident direction as indicated. Find the angle β between the paths followed by the masses after the collision. That is, find the *sum* of the angles θ and ϕ.

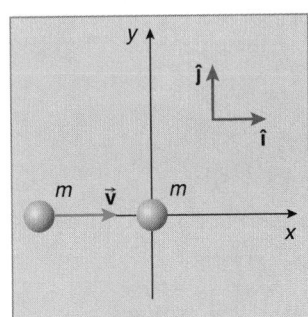
Before collision

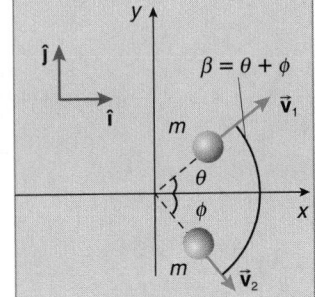
After collision

FIGURE 9.14

Solution
As with many things, there is an easy way and a hard way to solve this problem.

Method 1
Let us see what is involved if we tackle the problem in a straight-forward fashion by conserving momentum and, since the collision

is elastic, kinetic energy as well. With the choice of coordinates indicated in Figure 9.14, the total momentum of the particles before the collision is

$$\begin{aligned} \vec{\mathbf{p}}_{\text{total bfr}} &= mv\hat{\imath} + m(0 \text{ m/s})\hat{\imath} \\ &= mv\hat{\imath} \end{aligned}$$

After the collision the total momentum is

$$\begin{aligned} \vec{\mathbf{p}}_{\text{total aft}} &= \vec{\mathbf{p}}_{1\,\text{aft}} + \vec{\mathbf{p}}_{2\,\text{aft}} \\ &= (mv_1 \cos\theta)\hat{\imath} + (mv_1 \sin\theta)\hat{\jmath} \\ &\quad + (mv_2 \cos\phi)\hat{\imath} - (mv_2 \sin\phi)\hat{\jmath} \end{aligned}$$

where v_1 and v_2 are the speeds of the particles after the collision. Momentum conservation means that

$$\vec{\mathbf{p}}_{\text{total bfr}} = \vec{\mathbf{p}}_{\text{total aft}}$$

Two vectors are equal to each other if and only if all of their respective components are each equal to each other. Equate the x-components of the total momenta:

$$\begin{aligned} mv &= mv_1 \cos\theta + mv_2 \cos\phi \\ v &= v_1 \cos\theta + v_2 \cos\phi \end{aligned} \qquad (1)$$

Equate the y-components:

$$0 \text{ kg·m/s} = mv_1 \sin\theta - mv_2 \sin\phi$$

so

$$v_1 \sin\theta = v_2 \sin\phi \qquad (2)$$

Since the collision is elastic, the kinetic energy also is conserved. That is,

$$\frac{mv^2}{2} = \frac{mv_1^2}{2} + \frac{mv_2^2}{2}$$

which reduces to

$$v^2 = v_1^2 + v_2^2 \qquad (3)$$

This essentially ends the physics of the problem. We now have a mathematical problem, and a rather messy one at that. You have to eliminate the speeds and find the sum of the two angles: $\beta = \theta + \phi$. You have three equations to work with: equations (1) and (2) from conservation of momentum, and equation (3) from conservation of kinetic energy. If you like algebraic morasses, this is a good one to curl up with and wade through when you are in a good mood. This approach is cumbersome but legitimate, what a physicist likely might call correct, but inelegant.

Method 2
An easier way to do the problem is, amazingly, not to get so involved. In particular write momentum conservation in vector form *without* reference to a particular coordinate choice:

$$\begin{aligned} m\vec{\mathbf{v}} &= m\vec{\mathbf{v}}_1 + m\vec{\mathbf{v}}_2 \\ \vec{\mathbf{v}} &= \vec{\mathbf{v}}_1 + \vec{\mathbf{v}}_2 \end{aligned}$$

Take the scalar product of this equation with itself:

$$\vec{\mathbf{v}} \cdot \vec{\mathbf{v}} = (\vec{\mathbf{v}}_1 + \vec{\mathbf{v}}_2) \cdot (\vec{\mathbf{v}}_1 + \vec{\mathbf{v}}_2)$$

So

$$v^2 = v_1^2 + v_2^2 + 2\ \vec{v}_1 \cdot \vec{v}_2 \qquad (4)$$

But the conservation of kinetic energy equation [equation (3)] indicates that

$$v^2 = v_1^2 + v_2^2$$

Comparing equation (4) with this equation indicates that the scalar product $\vec{v}_1 \cdot \vec{v}_2$ must be zero:

$$\vec{v}_1 \cdot \vec{v}_2 = 0\ \text{m}^2/\text{s}^2$$

Hence

$$v_1 v_2 \cos \beta = 0\ \text{m}^2/\text{s}^2$$

Since neither v_1 nor v_2 is zero, $\cos \beta$ must be zero, which means $\beta = 90°$. The two velocity vectors are perpendicular to each other! After the collision the particles emerge with an angle $\beta = 90°$ between their paths. Wow. That was fast and the result is quite remarkable!

This example illustrates that occasionally the straightforward way of solving a problem may not necessarily be the easiest; some thought and inspiration occasionally yield imaginative solutions. Physicists (and mathematicians too) would call the vector shortcut quite elegant, even beautiful, because of its compactness and clarity compared with the algebraic drudgery of the more straightforward but mathematically cumbersome procedure of Method 1.

9.6 DISINTEGRATIONS AND EXPLOSIONS

The mental image we have of explosions and disintegrations is of many particles flying off in all directions. *BOOM!* The investigations into the tragic explosion and disintegration of TWA Flight 800 in the summer of 1996 and the bombing of Pan Am Flight 103 in 1988 attempted to discover what happened by extrapolating backward from the distribution of the recovered pieces.

For explosions and disintegrations the original system, considered as a single particle, suddenly breaks up into many particles zipping off in all directions. Disintegrations, such as those that occur in the radioactive decay of certain nuclei, are completely inelastic collisions in reverse; a video of the events, run backward, would look like a completely inelastic collision. Explosions also can be considered as inelastic collisions in reverse, to the extent that the mass of the explosive itself can be ignored.*

Define the system to be all particles involved in the disintegration or explosion. The *internal* forces that exist by and on the particles that compose the system during the short time interval of the disintegration or explosion typically are *much* greater than *external* forces on the particles. External forces, such

as gravitational forces of the Earth on the particles (their weight), typically are negligible by comparison. Since the internal forces are paired according to Newton's third law and vector sum to zero, the total force on the system of particles is zero (or negligible) during the disintegration or explosion. As a consequence, the total momentum of the system is conserved, to an excellent approximation. In other words, the total momentum of the system at the instant before the disintegration or explosion is equal to the total momentum of all the particles at the instant immediately after the event:

$$\vec{P}_{\text{total bfr}} = \vec{P}_{\text{total aft}}$$

The total kinetic energy of the system is *not conserved*, just as for any inelastic collision. The source of the resulting kinetic energy of the particles comes from chemical, mechanical, or nuclear sources associated with the disintegration or the explosive material. Energy considerations for explosions are accounted for by a very general statement of energy conservation to be introduced later, in Chapter 13.

EXAMPLE 9.12

A 15.0 kg object moving vertically upward at speed 10.0 m/s suddenly explodes into three pieces of mass 2.0 kg, 3.0 kg, and 10.0 kg. Immediately after the breakup, the 2.0 kg mass is found by astute, hawk-eyed physics students to be moving vertically upward at a speed of 20.0 m/s and the 3.0 kg chunk is discovered to be moving horizontally at a speed of 5.0 m/s. Find the velocity and speed of the 10.0 kg piece.

Solution

Conserve the momentum of the system of particles before and after the disintegration. Use the coordinate system indicated in Figure 9.15. The total momentum of the system before the disintegration is the momentum of the single particle of mass 15.0 kg:

$$\vec{P}_{\text{total bfr}} = (15.0\ \text{kg})(10.0\ \text{m/s})\hat{j}$$
$$= (150\ \text{kg·m/s})\hat{j}$$

The total momentum of the system after the explosion is the vector sum of the momenta of the three particles. The momentum of the 2.0 kg mass is

$$\vec{P}_{1\ \text{aft}} = (2.0\ \text{kg})(20.0\ \text{m/s})\hat{j}$$
$$= (40\ \text{kg·m/s})\hat{j}$$

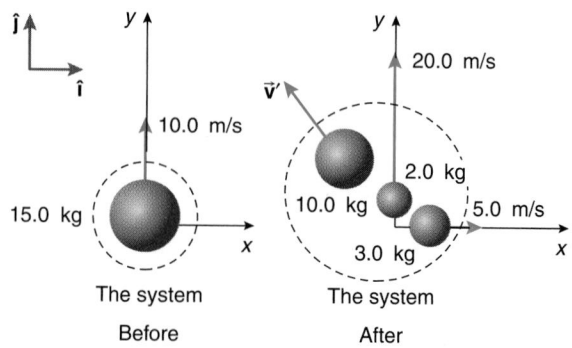

The system
Before

The system
After

FIGURE 9.15

---

*This assumption is *not*, in many cases, a good approximation for many explosions. Certainly in much ordnance the mass of the explosive is much greater than the mass of the casing.

The momentum of the 3.0 kg chunk is

$$\vec{\mathbf{p}}_{2\text{ aft}} = (3.0 \text{ kg})(5.0 \text{ m/s})\hat{\mathbf{i}}$$
$$= (15 \text{ kg·m/s})\hat{\mathbf{i}}$$

Let $\vec{\mathbf{v}}'$ be the velocity of the 10.0 kg piece. Its momentum after the disintegration then is

$$\vec{\mathbf{p}}_{3\text{ aft}} = (10.0 \text{ kg})\vec{\mathbf{v}}'$$

The total momentum after the disintegration thus is

$$\vec{\mathbf{p}}_{\text{total aft}} = (15 \text{ kg·m/s})\hat{\mathbf{i}} + (40 \text{ kg·m/s})\hat{\mathbf{j}} + (10 \text{ kg})\vec{\mathbf{v}}'$$

Conserve momentum:

$$\vec{\mathbf{p}}_{\text{total bfr}} = \vec{\mathbf{p}}_{\text{total aft}}$$
$$(150 \text{ kg·m/s})\hat{\mathbf{j}} = (15 \text{ kg·m/s})\hat{\mathbf{i}} + (40 \text{ kg·m/s})\hat{\mathbf{j}} + (10 \text{ kg})\vec{\mathbf{v}}'$$

Solving for $\vec{\mathbf{v}}'$, you obtain

$$\vec{\mathbf{v}}' = (-1.5 \text{ m/s})\hat{\mathbf{i}} + (11.0 \text{ m/s})\hat{\mathbf{j}}$$

The speed of the 10.0 kg is the magnitude of its velocity vector, or

$$v' = [(-1.5 \text{ m/s})^2 + (11.0 \text{ m/s})^2]^{1/2}$$
$$= 11.1 \text{ m/s}$$

9.7 THE CENTRIPETAL ACCELERATION REVISITED*

There is an interesting way to think of the centripetal acceleration of a particle in circular motion that stems from our study of collisions and the impulse–momentum theorem. Imagine a particle of mass m moving at speed v on a frictionless surface inside a fixed horizontal circular hoop of radius r, as shown in Figure 9.16.

Assume the collisions of the particle with the circular hoop are elastic. Since the kinetic energy is conserved in elastic collisions and the hoop is fixed, the particle maintains a constant speed before and after each such collision. The magnitude of the momentum of the particle is therefore constant but, clearly, its direction is changed with each collision with the hoop. Thus *the momentum of the particle is not conserved in its collision with the hoop.* (See Question 33.) Viewed from the center of the circle, successive collisions are separated by an angle θ, as indicated in Figure 9.16.

Examine the collision at point A; use the coordinate system indicated in Figure 9.16 to write the momentum of the particle before and after the collision at point A:

$$\vec{\mathbf{p}}_{\text{bfr}} = \left[mv\cos\left(\frac{\theta}{2}\right) \right]\hat{\mathbf{i}} + \left[mv\sin\left(\frac{\theta}{2}\right) \right]\hat{\mathbf{j}}$$

$$\vec{\mathbf{p}}_{\text{aft}} = \left[mv\cos\left(\frac{\theta}{2}\right) \right]\hat{\mathbf{i}} - \left[mv\sin\left(\frac{\theta}{2}\right) \right]\hat{\mathbf{j}}$$

The change in the momentum $\Delta\vec{\mathbf{p}}$ of the particle is

$$\Delta\vec{\mathbf{p}} \equiv \vec{\mathbf{p}}_{\text{aft}} - \vec{\mathbf{p}}_{\text{bfr}}$$
$$= \left[-2mv\sin\left(\frac{\theta}{2}\right) \right]\hat{\mathbf{j}}$$

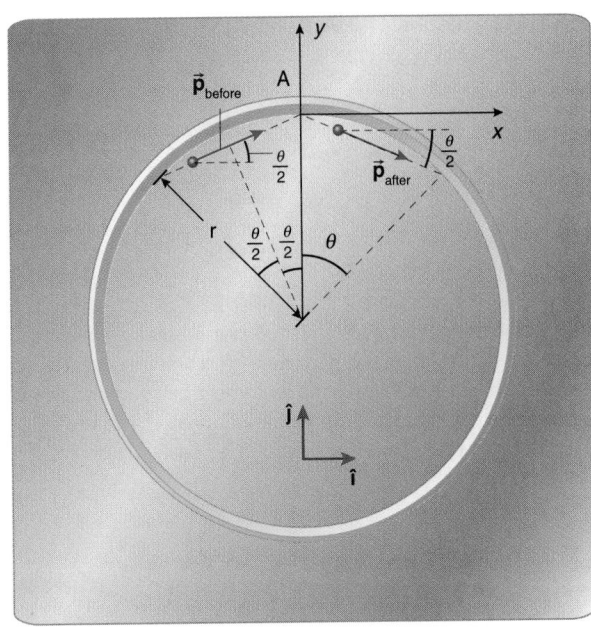

FIGURE 9.16 Geometry for a collision of the particle with the hoop.

Notice that the change in the momentum of the particle is directed toward the center of the circle. The impulse–momentum theorem (Equation 9.5) implies that the impulse of the force of the hoop on the particle also must be directed toward the center of the circle.

The *magnitude* of the change in the momentum of the particle is

$$\Delta p = 2mv\sin\left(\frac{\theta}{2}\right)$$

The hoop exerts an impulse on the particle that, according to the impulse–momentum theorem, is equal to the change in the momentum of the particle. Use Equation 9.6, which expresses the impulse as the product of the magnitude of the average force $\vec{\mathbf{F}}_{\text{ave}}$ and the time interval of the collision Δt, in the impulse–momentum theorem:

$$F_{\text{ave}}\,\Delta t = \Delta p$$
$$= 2mv\sin\left(\frac{\theta}{2}\right) \qquad (9.17)$$

The time interval between successive collisions, $\Delta t'$, is the distance between successive collisions divided by the speed of the particle. The distance between successive collisions is, using the geometry of Figure 9.16, equal to $2r\sin(\theta/2)$. Hence the time interval between successive collisions is

$$\Delta t' = \frac{2r}{v}\sin\left(\frac{\theta}{2}\right) \qquad (9.18)$$

Now imagine the particle moving so the time interval between successive collisions, $\Delta t'$, approaches the time interval associated with the collision of the particle with the hoop, Δt; the particle then is essentially in continuous contact with the hoop and in circular motion. Use Equation 9.18 for $\Delta t'$ to substitute for Δt in Equation 9.17; we get

$$F_{ave} \frac{2r}{v} \sin\left(\frac{\theta}{2}\right) = 2mv \sin\left(\frac{\theta}{2}\right)$$

or

$$F_{ave} = m\frac{v^2}{r}$$

This we recognize as Newton's second law applied to the particle: the force of the hoop on the particle is the total force on the particle in the horizontal plane, and so it must be equal to the mass m times the acceleration. Thus we can identify v^2/r as the magnitude of the centripetal acceleration of the particle. Newton made this argument in his *Principia*.[†]

9.8 AN ALTERNATIVE WAY TO LOOK AT FORCE TRANSMISSION*

When we studied gravitation in Chapter 6, the gravitational field was introduced as a way of understanding how the force mysteriously is transmitted even over long, empty distances between masses not in physical contact with each other. The idea of a field also is very useful for electromagnetism, as we will see in Chapters 16 and 20.

Our study of collisions can give us an inkling of an alternative way to visualize force transmission: via the exchange of **mediating particles**. The mediating particles associated with the four fundamental forces were listed in Table 5.2. The example we examine next illustrates how a repulsive force can be conveyed by particle exchange.

Consider a series of collision experiments involving a 100.0 kg astronaut and a 1000 kg spacecraft both initially floating freely at rest in outer space, well away from any other masses such as the Earth, other planets, and the Sun. In this way the astronaut and spacecraft are isolated in space. We neglect the small gravitational attraction of the astronaut and spacecraft on each other.

Experiment 1: One-Particle Exchange

Imagine the 100.0 kg astronaut to be 2.00 m from the spacecraft. The astronaut initially has zero velocity relative to the spacecraft, as shown in Figure 9.17. A launcher on the spacecraft sends out a 1.00 kg canister to the astronaut at a speed of, say, 2.00 m/s.

The astronaut catches the canister 1.00 s after launch. The catch can be viewed as a completely inelastic collision of the astronaut and the canister. Define the system to be the canister and the astronaut; momentum is conserved in the collision. The momentum of the astronaut before the collision is zero, because the astronaut had zero velocity:

$$\vec{P}_{ast\ bfr} = (0\ \text{kg·m/s})\hat{\imath}$$

[†] Newton's argument is in Book I, Proposition IV. The author is indebted to Arnold Arons, *A Guide to Introductory Physics Teaching* (Wiley, New York, 1990), pages 138–140 for bringing this argument to his attention.

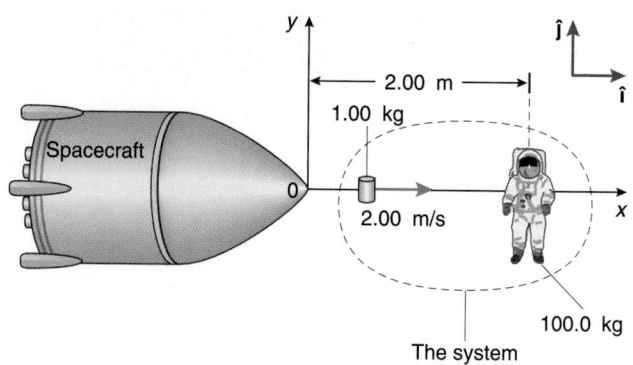

FIGURE 9.17 A canister is sent from the spacecraft to the astronaut.

The momentum of the canister before the collision is (with the choice of coordinate axes in Figure 9.17)

$$\vec{P}_{can\ bfr} = (1.00\ \text{kg})(2.00\ \text{m/s})\hat{\imath}$$

Therefore the total initial momentum of the astronaut–canister system before the collision is

$$\vec{P}_{total\ bfr} = \vec{P}_{can\ bfr} + \vec{P}_{ast\ bfr}$$
$$= (2.00\ \text{kg·m/s})\hat{\imath} + (0\ \text{kg·m/s})\hat{\imath}$$

Since the astronaut caught and held the canister, the total momentum of the single, composite particle after the collision is

$$\vec{P}_{total\ aft} = (100.0\ \text{kg} + 1.00\ \text{kg})\vec{v}$$
$$= (101.0\ \text{kg})\vec{v}$$

where $\vec{v}$ is the common velocity of the astronaut and the canister after the collision.

Momentum conservation in the collision means that the total momentum before the collision is equal to total momentum after the collision:

$$\vec{P}_{total\ bfr} = \vec{P}_{total\ aft}$$
$$(2.00\ \text{kg·m/s})\hat{\imath} = (101.0\ \text{kg})\vec{v}$$

So the velocity $\vec{v}$ of the astronaut and canister after the collision is

$$\vec{v} = \frac{2.00\ \text{kg·m/s}}{101.0\ \text{kg}}\hat{\imath}$$
$$= (0.0198\ \text{m/s})\hat{\imath}$$

A graph of the velocity component of the astronaut as a function of time is shown in Figure 9.18. The velocity component of the astronaut is zero until the canister touches the hand of the astronaut. After the short interval of the collision, the canister and astronaut have the same velocity component.

Experiment 2: Two-Particle Exchange

Instead of the previous collision experiment, suppose the spacecraft successively threw *two* canisters each of mass 0.500 kg, half that of the single canister in Experiment 1, at the same speed of 2.00 m/s. Each of the canisters now has *half* of the momentum of the original larger 1.00 kg canister. The astronaut catches each smaller canister in turn. The total change in the momentum of

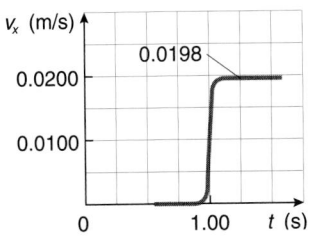

FIGURE 9.18 Velocity component of the astronaut as a function of time for the one-canister experiment.

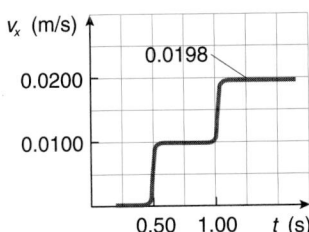

FIGURE 9.19 Velocity component of the astronaut as a function of time for the two-canister experiment.

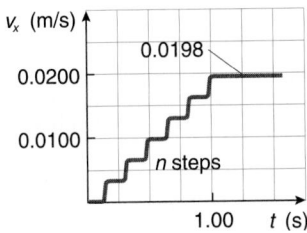

FIGURE 9.20 Velocity component of the astronaut as a function of time for the n-canister experiment.

the astronaut is the *same* as in the first experiment; the momentum just changes via two smaller impulses rather than in one larger impulse. So the final velocity of the astronaut is the same as in the first (single-canister) experiment. You should be able to verify this statement by considering the two collisions of the canisters with the astronaut.

Suppose the two canisters were thrown so they were caught successively at times 0.500 s and 1.00 s by the astronaut. A plot of the velocity component of the astronaut as a function of time is shown in Figure 9.19.

Experiment 3: *n*-Particle Exchange

Now divide the original 1.00 kg canister into n smaller canisters, each of mass $(1.00 \text{ kg})/n$. Each is thrown at the astronaut successively, each at speed 2.00 m/s, and all caught within one second. The astronaut experiences n successive inelastic collisions and the velocity and momentum of the astronaut change in n small steps (see Figure 9.20).

Recall that the acceleration component is the slope of the corresponding velocity component versus t graph. If n is very large, then the acceleration is almost continuous (and constant)

during the one second interval. The astronaut experiences something more like a steady push rather than a series of jerky impulses from the collisions. The astronaut likely would say a *force* was pushing on her.

According to the most general statement of Newton's second law, Equation 9.2, the total force on a system is equal to the time rate of change of the total momentum of the system:

$$\vec{\mathbf{F}}_{\text{total}} = \frac{d\vec{\mathbf{p}}_{\text{total}}}{dt}$$

Let's calculate the average force on the astronaut (considered as a single-particle system) during the one second time interval. The momentum change of the astronaut during the one second interval was

$$\Delta\vec{\mathbf{p}} = (2.00 \text{ kg·m/s})\hat{\mathbf{i}}$$

The impulse on the astronaut is, from the vector version of Equation 9.6,

$$\vec{\mathbf{I}} = \vec{\mathbf{F}}_{\text{ave}} \Delta t$$

According to the impulse–momentum theorem, the impulse is equal to the change in the momentum:

$$\vec{\mathbf{F}}_{\text{ave}}(1.00 \text{ s}) = (2.00 \text{ kg·m/s})\hat{\mathbf{i}}$$
$$\vec{\mathbf{F}}_{\text{ave}} = (2.00 \text{ N})\hat{\mathbf{i}}$$

Experiment 4: Effect on the Spacecraft

Now let's look at the other side of the picture: the spacecraft that was projecting the large canister or smaller canisters toward the astronaut. Examine the first of the experiments we performed. The spacecraft initially is at rest before it emits the canister. The emission of the canister also can be viewed as a collision, since if we play a video of the event backward in time, the time-reversed event has all the character of a completely inelastic collision.

Define the system to be the spacecraft and the emitted canister. The spacecraft throws out a 1.00 kg canister at a speed of 2.00 m/s. Initially the spacecraft and canister each have zero momentum since both are not moving. The total momentum of the spacecraft–canister system before the collision is zero:

$$\vec{\mathbf{P}}_{\text{total bfr}} = (0 \text{ kg·m/s})\hat{\mathbf{i}}$$

After the emission, however, the canister has a momentum of

$$(2.00 \text{ kg·m/s})\hat{\mathbf{i}}$$

and the spacecraft recoils with a velocity $\vec{\mathbf{v}}'$ and momentum

$$(999 \text{ kg})\vec{\mathbf{v}}'$$

The total momentum after the emission is

$$\vec{\mathbf{P}}_{\text{total aft}} = (2.00 \text{ kg·m/s})\hat{\mathbf{i}} + (999 \text{ kg})\vec{\mathbf{v}}'$$

Conservation of momentum says the total momentum before the interaction must be equal to the total momentum after the interaction. So

$$(0 \text{ kg·m/s})\hat{\mathbf{i}} = (2.00 \text{ kg·m/s})\hat{\mathbf{i}} + (999 \text{ kg})\vec{\mathbf{v}}'$$

and the velocity of the recoiling spacecraft is

$$\vec{\mathbf{v}}' = \frac{-2.00 \ \text{kg} \cdot \text{m/s}}{999 \ \text{kg}} \ \hat{\mathbf{i}}$$

$$= (-0.002 \ 00 \ \text{m/s}) \ \hat{\mathbf{i}}$$

The momentum change of the recoiling spacecraft has exactly the same magnitude as the momentum change of the astronaut, but the two vectors point in opposite directions.

We could analyze the other experiments (which transferred momenta by using many particle canisters) for their effects on the spacecraft as well but you likely can see that the game is essentially the same. For the *n*-particle exchange, we attribute the change in the momentum of the astronaut to the impulse of the canisters on the astronaut (thanks to the reception or *absorption* of the many little particles); the momentum change of the spacecraft also can be attributed to an impulse on the spacecraft (thanks to the *emission* of many little particles). The two impulses and the two average forces are of the same magnitude but are oppositely directed. This equality is equivalent to Newton's third law of motion.

These experiments have several important features that touch on many major issues in contemporary physics. Since the exchanged particles travel at finite speeds, forces take time to propagate from one object to another. In other words, when the spacecraft emits the exchanged particles (the canisters), the astronaut does not feel the effects until they are received some finite time later. Such propagation delays are important in many areas of physics. These effects bothered Newton himself, but he was unable to come to grips with the problem appropriately.

The astronaut and spacecraft move away from each other as a result of the particle exchange, much like a *repulsive force*. This was easy to see. However, using classical physics, it is quite difficult to imagine how an attractive force is conveyed by means of particle exchange. Somehow nature manages to do it, but we lack the conceptual framework to describe such interactions with simple classical models. Quantum mechanics is another game entirely. So here we are in an *introductory* physics course and find ourselves face-to-face with a difficult question we cannot answer easily. Maybe you can figure it out. No one else has been able to do this as yet.* Think about it. This problem should convince you that there is a difficult underlying subtlety about forces and their effects.

We will not pursue further this view of force transmission via particle exchange, but this example should serve to make you aware that there are various ways to model the way forces are transmitted. In this text, we will be content only to examine the effects of the forces themselves on the motion of particles and the field mechanism of force transmission; that is trouble enough for now!

9.9 THE CENTER OF MASS

How do we deal neatly with a collection of masses at different locations as a single system? The most representative location of the system is called the **center of mass** point. The center of

---

*There is a simplistic (but unsatisfactory) way of imagining an attractive force arising from particle exchange: two individuals on ice skates are each holding a basketball (the mediating particle). If the two individuals snatch or grab the basketballs from each other, the individuals likely will be drawn toward each other on the ice. But this crude model necessitates almost immediate contact between the particles to be attracted, whereas attractive forces in physics can be exerted on particles over vast distances.

mass of an extended system is the point whose dynamics typifies the system as a whole when it is treated as a particle. We will find that this concept leads to useful simplifications of the dynamics of an extended system of many particles. For example, when a high-board diver or gymnast executes a complicated maneuver, their center of mass follows the trajectory you would calculate when treating the system as a simple particle (see Figure 9.21).

How do you locate the center of mass point of an extended system? Think of this analogy. The center of population of the United States is not just the average of the positions of all the towns, because some towns have more people than others; Dixville Notch, New Hampshire, is not Los Angeles! In the averaging, a town of 10 000 persons should count ten times as much as a town of 1000. And so it is with the concept of the center of mass of a collection of particles. Each particle carries an importance proportional to its mass. The sum of the products of the mass of each particle and its position vector gives the grand total mass m_{total} times the position vector of the center of mass (see Figure 9.22):

$$m_1\vec{\mathbf{r}}_1 + m_2\vec{\mathbf{r}}_2 + m_3\vec{\mathbf{r}}_3 + \cdots \equiv m_{\text{total}}\vec{\mathbf{r}}_{\text{CM}}$$

We write this equation more compactly as

$$\sum_i m_i\vec{\mathbf{r}}_i \equiv m_{\text{total}}\vec{\mathbf{r}}_{\text{CM}} \tag{9.19}$$

where $\vec{\mathbf{r}}_i$ is the position vector of the *i*th particle (of mass m_i) with respect to the origin of the chosen coordinate system.

> Rearranging Equation 9.19 slightly to isolate $\vec{\mathbf{r}}_{\text{CM}}$, we see that the position vector of the center of mass is
>
> $$\vec{\mathbf{r}}_{\text{CM}} \equiv \frac{1}{m_{\text{total}}} \sum_i m_i\vec{\mathbf{r}}_i \tag{9.20}$$

Since two vectors are equal only if all their respective components are equal, the equation defining the location of the center

FIGURE 9.21 When executing this complicated maneuver, the center of mass of this gymnast follows a parabolic path after leaving the apparatus.

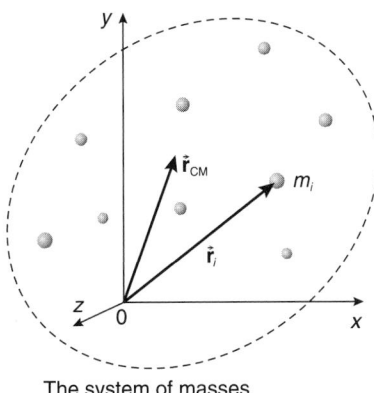

The system of masses

FIGURE 9.22 The vector $\vec{r}_{CM}$ locates the position of the center of mass.

of mass is really three scalar equations, one for each (Cartesian) component of the center of mass:

$$x_{CM} \equiv \frac{1}{m_{\text{total}}} \sum_i m_i x_i$$

$$y_{CM} \equiv \frac{1}{m_{\text{total}}} \sum_i m_i y_i \qquad (9.21)$$

$$z_{CM} \equiv \frac{1}{m_{\text{total}}} \sum_i m_i z_i$$

The center of mass is a *point in space*; there may or may not be any mass actually located at the center of mass.

It is useful to consider common objects as essentially continuous distributions of matter; this is an excellent approximation for macroscopic objects, because atoms are so small and close together. For such essentially continuous distributions of matter, the discrete summations (Equations 9.21) defining the components of the position vector of the center of mass become integrals over the physical extent of the object:

$$x_{CM} \equiv \frac{1}{m_{\text{total}}} \int x \, dm$$

$$y_{CM} \equiv \frac{1}{m_{\text{total}}} \int y \, dm \qquad (9.22)$$

$$z_{CM} \equiv \frac{1}{m_{\text{total}}} \int z \, dm$$

where dm is the differential amount of mass in a differential volume dV located at the position vector whose components are x, y, and z. Let ρ be the mass density at the location of the volume element dV; then

$$dm = \rho \, dV$$

If the system is an extended mass of *uniform composition* (i.e., of constant density ρ), then the geometric *symmetry* of the system (if it has any!) often can be exploited to locate the center of mass quickly—without the agony of performing the summations or integrations indicated in Equations 9.21 or 9.22. For example:

1. A uniform rectangular plate has its center of mass at the geometric center of the plate (see Figure 9.23). We can see this as follows. If we choose to locate the origin of a coordinate system at the geometric center of the plate, then any bit of mass dm

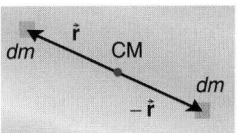

FIGURE 9.23 The center of mass of a uniform rectangular plate is at its geometric center.

with position vector $\vec{r}$ has a symmetrically placed counterpart located at $-\vec{r}$. The two terms sum to zero: $dm \, \vec{r} + dm \, (-\vec{r}) = 0 \text{ kg} \cdot \text{m}$. The mass of the entire plate can be accounted for this way, and so the summations in Equations 9.21 or, really, the integrals in Equations 9.22 are zero. The position vector of the center of mass is zero; hence the origin chosen at the geometric center of the plate is coincident with the center of mass point, since its position vector is zero.

2. Similar symmetry arguments are used to show that the center of mass of

   • a uniform spherical shell,
   • a uniform sphere, or
   • a sphere with a spherically symmetric mass distribution

   each is located at the geometric center of the sphere.

Occasionally the center of mass can be found using several legitimate tricks of the trade:

Trick 1: Segmented mass method

a. Segment the extended mass into regular geometric pieces and locate the center of mass of each piece using symmetry considerations.
b. Find the fraction of the total mass represented by each segment.
c. Then treat the collection of mass segments as a collection of pointlike particles each located at the center of mass of the respective segment. The center of mass of the entire system then is found using Equation 9.20 for a collection of particles. See Example 9.14 (Method 1) for an illustration of this procedure.

Trick 2: Strategic subtraction method

Another way is to create voids or holes by judicious subtraction of regularly shaped pieces. See Example 9.14 (Method 2) for an illustration of this technique.

Trick 3: An experimental method

If the system is irregular in shape or not of uniform composition, the analytic techniques we have outlined become intractable in a practical sense. In such situations the location of the center of mass can be determined *experimentally* by suspending the object freely from two arbitrary points. For example, consider an irregular two-dimensional (planar) object. When the object is freely suspended, the mass will hang with its center of mass on the vertical line through the pivot, and below it, as shown in Figure 9.24.

This position minimizes the gravitational potential energy of the system. A vertical line is drawn on the object indicating that the center of mass is located somewhere along the line. The object then is suspended freely from any other point and another such line drawn, as in Figure 9.25. The intersection of the two lines locates the center of mass in the plane of the object. If the mass is not planar, extrapolate the lines drawn outside the object to their point of convergence within the mass.

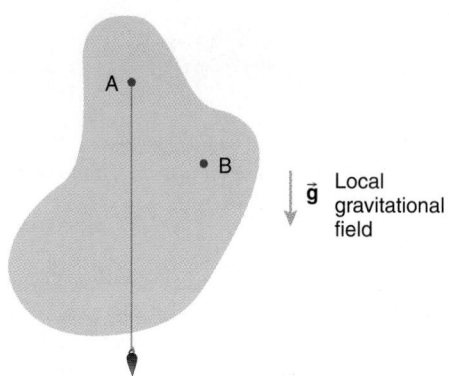

FIGURE 9.24 The center of mass lies on the vertical line through the pivot and below it.

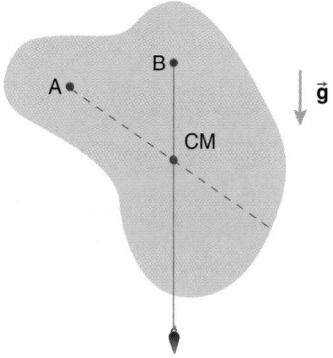

FIGURE 9.25 Suspend the system from another point.

QUESTION 4

Where is the location of the center of mass of the ring system surrounding the planet Saturn? Where is the center of mass of a hula hoop?

EXAMPLE 9.13

Find the location of the center of mass of the system of three particles shown in Figure 9.26.

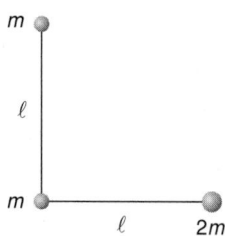

FIGURE 9.26

Solution

Introduce a coordinate system as in Figure 9.27. The position vectors of the three masses are then

$$\vec{r}_1 = \ell\,\hat{i}$$
$$\vec{r}_2 = \ell\,\hat{j}$$
$$\vec{r}_3 = (0 \text{ m})\,\hat{j}$$

The position vector of the center of mass is found using Equation 9.20. Since there are three masses, there are three terms in the summation:

$$\vec{r}_{CM} = \frac{1}{m_{total}} (m_1\vec{r}_1 + m_2\vec{r}_2 + m_3\vec{r}_3)$$

where m_{total} is the total mass of the system, $4m$. Substituting the masses and their respective position vectors,

$$\vec{r}_{CM} = \frac{1}{4m}\left[2m\ell\,\hat{i} + m\ell\,\hat{j} + m(0 \text{ m})\,\hat{j}\right]$$
$$= \frac{\ell}{4}(2\hat{i} + \hat{j})$$

This position is indicated in Figure 9.27. For this system, there is no mass located at the center of mass.

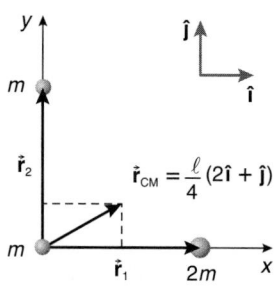

FIGURE 9.27

STRATEGIC EXAMPLE 9.14

Find the location of the center of mass of the uniform plate of mass m, shown in Figure 9.28.

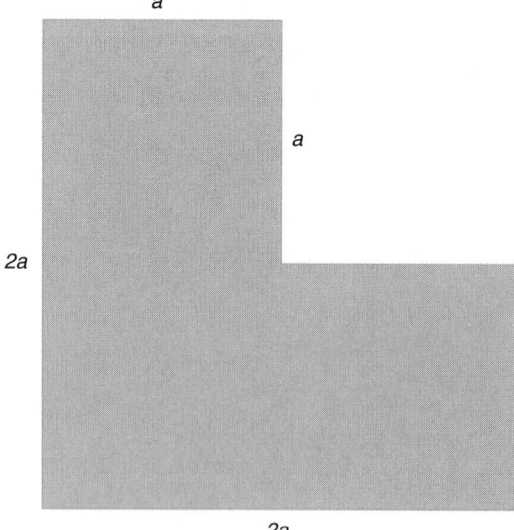

FIGURE 9.28

Solution

Method 1: The segmented mass method

Consider the system to be composed of three uniform square plates each with sides a, as shown in Figure 9.29. By symmetry, the center of mass of each segment is at the geometric center of each square plate. Since the plates all are of equal area, the mass of each plate segment is $m/3$.

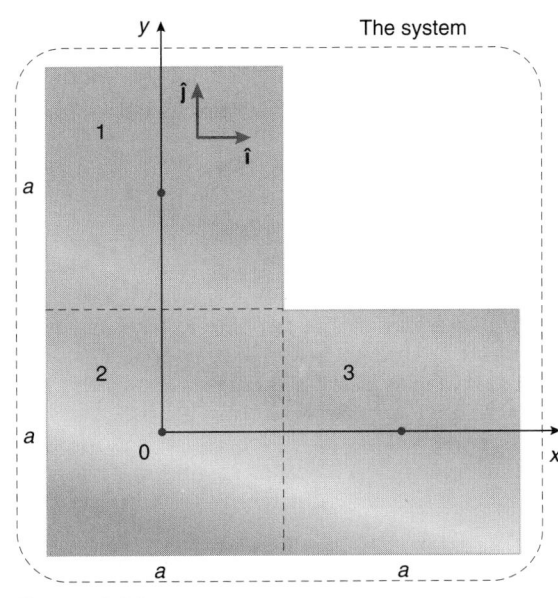

FIGURE 9.29

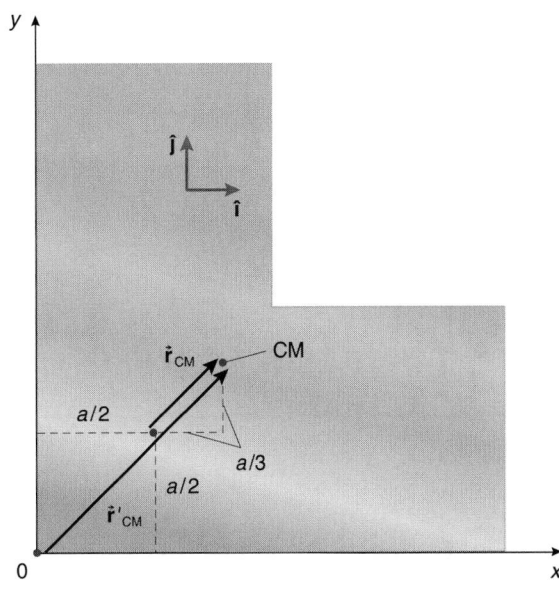

FIGURE 9.30

Choose the origin of the coordinate system to be at the center of the square on the lower left, as indicated in Figure 9.29. The position vector of the center of mass of each of the segments then is

$$\vec{\mathbf{r}}_1 = a\hat{\mathbf{j}}$$
$$\vec{\mathbf{r}}_2 = (0 \text{ m})\hat{\mathbf{i}}$$
$$\vec{\mathbf{r}}_3 = a\hat{\mathbf{i}}$$

The position vector of the center of mass of the three-particle system then is, from Equation 9.20,

$$
\begin{aligned}
\vec{\mathbf{r}}_{\text{CM}} &= \frac{1}{m_{\text{total}}} \sum_i m_i \vec{\mathbf{r}}_i \\
&= \frac{1}{m_{\text{total}}} (m_1\vec{\mathbf{r}}_1 + m_2\vec{\mathbf{r}}_2 + m_3\vec{\mathbf{r}}_3) \\
&= \frac{1}{m} \left[\frac{m}{3} a\hat{\mathbf{j}} + \frac{m}{3}(0 \text{ m})\hat{\mathbf{i}} + \frac{m}{3} a\hat{\mathbf{i}} \right] \\
&= \frac{a}{3}(\hat{\mathbf{i}} + \hat{\mathbf{j}})
\end{aligned}
$$

The position of the center of mass with respect to the lower left corner means $a/2$ must be added to each component in the last line, as shown in Figure 9.30. Hence, measured from the corner, the position vector of the center of mass is

$$
\begin{aligned}
\vec{\mathbf{r}}'_{\text{CM}} &= \left(\frac{a}{2} + \frac{a}{3} \right)\hat{\mathbf{i}} + \left(\frac{a}{2} + \frac{a}{3} \right)\hat{\mathbf{j}} \\
&= \frac{5}{6} a\hat{\mathbf{i}} + \frac{5}{6} a\hat{\mathbf{j}}
\end{aligned}
$$

Method 2: The strategic subtraction method

Consider the original system to be what is left when a square plate, of sides $2a$, has a square with sides a cut from it, as shown in Figure 9.31. The mass of the large square is $(4/3)m$, while that

of the piece to be cut away is $m/3$. The location of the center of mass of each square plate is at its respective geometric center. Choose the origin of the coordinate system to be at the lower left corner, as shown in Figure 9.31.

The position vector of the center of mass of the larger plate thus is

$$\vec{\mathbf{r}}_1 = a\hat{\mathbf{i}} + a\hat{\mathbf{j}}$$

while that of the smaller plate is

$$\vec{\mathbf{r}}_2 = \frac{3}{2} a\hat{\mathbf{i}} + \frac{3}{2} a\hat{\mathbf{j}}$$

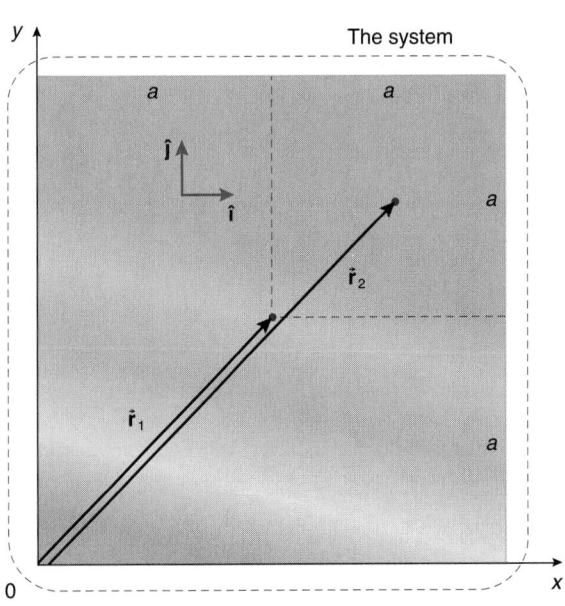

FIGURE 9.31

The problem has been reduced to a two-mass system, except that we *subtract* the contribution of the smaller square from that of the larger square:

$$\vec{\mathbf{r}}'_{CM} = \frac{1}{m_{total}}(m_1\vec{\mathbf{r}}_1 - m_2\vec{\mathbf{r}}_2)$$

$$= \frac{1}{\left(\dfrac{4}{3}m - \dfrac{m}{3}\right)}\left[\frac{4}{3}m(a\hat{\mathbf{i}} + a\hat{\mathbf{j}}) - \frac{m}{3}\left(\frac{3}{2}a\hat{\mathbf{i}} + \frac{3}{2}a\hat{\mathbf{j}}\right)\right]$$

$$= \frac{5}{6}a\hat{\mathbf{i}} + \frac{5}{6}a\hat{\mathbf{j}}$$

9.10 DYNAMICS OF A SYSTEM OF PARTICLES

With the concept of the center of mass, there is a great simplification in store for us. Here we want to see what happens to a system of many particles when forces act on them. It is important, of course, to decide first what particles constitute our chosen system.

The location of the center of mass of a system of particles is defined by Equation 9.20:

$$\vec{\mathbf{r}}_{CM} = \frac{1}{m_{total}}\sum_i m_i\vec{\mathbf{r}}_i$$

where m_{total} is the total mass of the collection of particles. Multiplying this equation by m_{total}, we have

$$\sum_i m_i\vec{\mathbf{r}}_i = m_{total}\vec{\mathbf{r}}_{CM}$$

To see what happens to the position vectors as functions of time, differentiate with respect to time (assume that the mass of each particle is constant):

$$\sum_i m_i\frac{d\vec{\mathbf{r}}_i}{dt} = m_{total}\frac{d\vec{\mathbf{r}}_{CM}}{dt}$$

The time derivatives of the position vectors are the respective velocities of the particles, and so we have

$$\sum_i m_i\vec{\mathbf{v}}_i = m_{total}\vec{\mathbf{v}}_{CM} \qquad (9.23)$$

The momentum of each individual particle is $\vec{\mathbf{p}}_i = m_i\vec{\mathbf{v}}_i$, so that the summation on the left-hand side of Equation 9.23 is the vector sum of the individual particle momenta, which is the total momentum $\vec{\mathbf{p}}_{total}$ of the system of particles:

$$\vec{\mathbf{p}}_{total} = m_{total}\vec{\mathbf{v}}_{CM} \qquad (9.24)$$

Equation 9.24 means that the total momentum of the system of particles is the same as the product of the mass of the entire system, m_{total}, and the velocity of the center of mass point, $\vec{\mathbf{v}}_{CM}$.

What a nice simplification!

Notice that if the center of mass of the system is not moving—that is, if $\vec{\mathbf{v}}_{CM} = 0$ m/s—then the *vector sum* of the individual particle momenta also is zero, even if the individual momenta themselves are not zero. For example, consider the case of an air-filled balloon at rest: the individual atoms each are moving and, therefore, have individual momenta $\vec{\mathbf{p}}_i$; yet the vector sum of all the $\vec{\mathbf{p}}_i$ is zero because the center of mass of the air in the stationary balloon is at rest.

How do we handle all the forces acting on a many-particle system? If Equation 9.23 is differentiated with respect to time, we find

$$\frac{d}{dt}\left(\sum_i m_i\vec{\mathbf{v}}_i\right) = \frac{d}{dt}\left(m_{total}\vec{\mathbf{v}}_{CM}\right) \qquad (9.25)$$

Look at a typical particle within the system, say the *i*th particle. According to Newton's second law of motion, the product

$$m_i\frac{d\vec{\mathbf{v}}_i}{dt}$$

must be equal to the vector sum of *all* the forces acting on *this* particle, $\vec{\mathbf{F}}_{i\,total}$. Let's include what is happening to all the particles at once. Equation 9.25 then becomes

$$\sum_i \vec{\mathbf{F}}_{i\,total} = m_{total}\frac{d\vec{\mathbf{v}}_{CM}}{dt} \qquad (9.26)$$

The forces on the *i*th particle arise from two distinct sources:

1. Forces on the *i*th particle by the *other* particles of the system. Call the vector sum of these forces on the *i*th particle $\vec{\mathbf{F}}_{i\,total\,int}$ since they arise from other particles *within* (i.e., *internal* to) the system of particles under consideration.
2. Forces on the *i*th particle arising from sources *outside* the system itself. Call the vector sum of these forces on the *i*th particle originating from outside (i.e., *external* to) the system $\vec{\mathbf{F}}_{i\,total\,ext}$.

Classifying the forces in this way in Equation 9.26, we have

$$\sum_i \left(\vec{\mathbf{F}}_{i\,total\,int} + \vec{\mathbf{F}}_{i\,total\,ext}\right) = m_{total}\frac{d\vec{\mathbf{v}}_{CM}}{dt} \qquad (9.27)$$

The first summation on the left-hand side of this equation,

$$\sum_i \vec{\mathbf{F}}_{i\,total\,int}$$

gives the grand sum of all the forces by each particle on each of the other particles of the system. Fortunately, as Newton's third

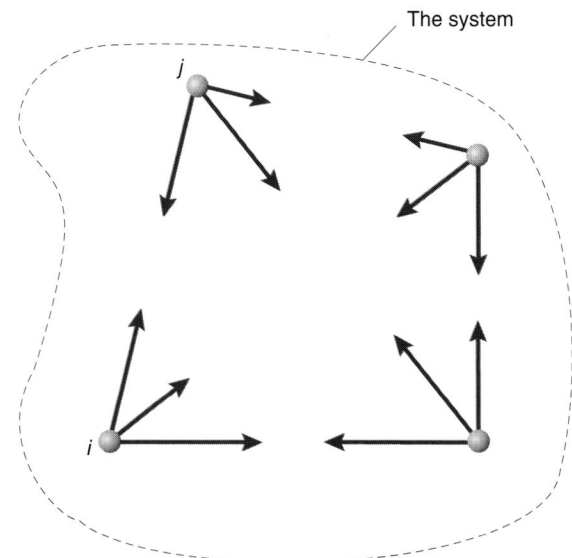

FIGURE 9.32 Internal forces occur in third law force pairs.

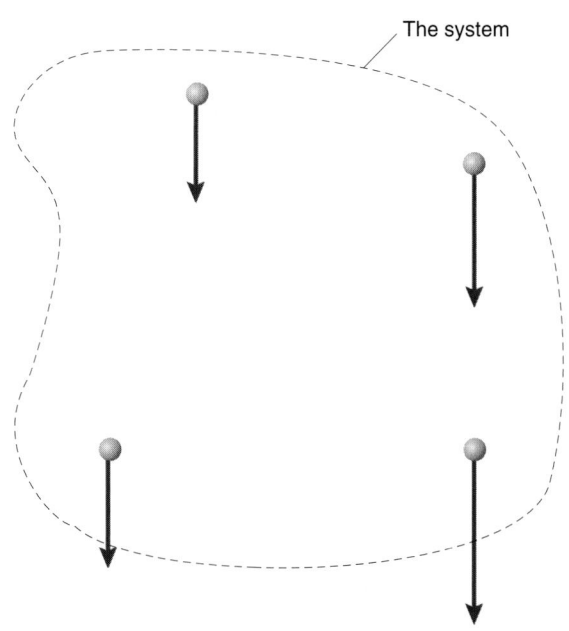

FIGURE 9.33 External forces on the system may not sum to zero.

law of motion states, all forces occur in pairs of equal magnitude and opposite direction (though the pair of forces act on different objects). Hence the force of particle j on particle i is equal in magnitude and opposite in direction to the force of particle i on particle j, as shown in Figure 9.32.

Since *both* these forces are in the sum of the *internal* forces, they will vector sum to zero. *All* of the internal forces are so paired; thus, the vector sum of all the internal forces is zero:

$$\sum_i \vec{F}_{i\text{ total int}} = 0 \text{ N} \qquad \text{(because of Newton's third law)}$$

This certainly helps simplify things!

The corresponding vector sum of the applied *external* forces on the particles of the system need *not* be zero, since only *one* member of each external force third law force pair acts on the *i*th particle. For example, the force of the Earth on the *i*th particle (its weight) is present in the sum, but the force of the *i*th particle *on* the external Earth is *not* in the sum if the Earth is not part of the system you chose. See Figure 9.33. There is no Newton's third law pairwise summation of the external forces since only *one* member of the external force pairs is a force on a particle of the system.

Incorporating these results into Equation 9.27, we have

$$\sum_i \vec{F}_{i\text{ total ext}} = m_{\text{total}} \frac{d\vec{v}_{\text{CM}}}{dt}$$

Let $\vec{F}_{\text{total}}$ be the vector sum of all the *external* forces on *all the particles of the system*; the overall motion of the system has a simple description:

$$\boxed{\vec{F}_{\text{total}} = m_{\text{total}} \frac{d\vec{v}_{\text{CM}}}{dt}} \qquad (9.28)$$

The vector sum of the total external forces on the system produces an acceleration of the center of mass of the system inversely proportional to the total mass of the system. This is why the center of mass concept is so useful: the motion of the entire system is determined by imagining the system as a *single particle* of mass m_{total} located at the center of mass point.

Since $m_{\text{total}}\vec{v}_{\text{CM}}$ is the total momentum $\vec{p}_{\text{total}}$ of the system, we can write Equation 9.28 as

$$\vec{F}_{\text{total}} = \frac{d\vec{p}_{\text{total}}}{dt} \qquad (9.29)$$

which is just Newton's second law for the system of particles treated as a *single particle* located at the center of mass point.

QUESTION 5
A handful of marbles is pitched into the air simultaneously. What is the acceleration of the center of mass of the system of marbles while they are in the air?

9.11 KINETIC ENERGY OF A SYSTEM OF PARTICLES

The center of mass also provides a useful simplification when finding the total kinetic energy of a system of many moving particles. Choose an origin 0 at some convenient point in space and locate a typical particle in the system by means of a position vector $\vec{r}_i$ originating at 0 and another position vector $\vec{r}_i'$

originating at the center of mass, as shown in Figure 9.34. The two vectors are related by

$$\vec{r}_i = \vec{r}_{CM} + \vec{r}'_i \qquad (9.30)$$

where $\vec{r}_{CM}$ locates the center of mass with respect to the origin 0.

Differentiate the position vectors with respect to time to find the relationship between the velocities:

$$\vec{v}_i = \vec{v}_{CM} + \vec{v}'_i \qquad (9.31)$$

The kinetic energy of the *i*th particle as measured in the reference frame with its origin at 0 is

$$\frac{1}{2}m_i v_i^2$$

The total kinetic energy of the whole system is the scalar sum of the kinetic energies of each of the particles:

$$KE_{total} = \sum_i \frac{1}{2}m_i v_i^2$$

$$= \sum_i \frac{1}{2}m_i \vec{v}_i \cdot \vec{v}_i$$

Substituting for $\vec{v}_i$ using Equation 9.31, we obtain

$$KE_{total} = \sum_i \frac{1}{2}m_i (\vec{v}_{CM} + \vec{v}'_i) \cdot (\vec{v}_{CM} + \vec{v}'_i) \qquad (9.32)$$

But

$$(\vec{v}_{CM} + \vec{v}'_i) \cdot (\vec{v}_{CM} + \vec{v}'_i) = v_{CM}^2 + v_i'^2 + 2\,\vec{v}_{CM} \cdot \vec{v}'_i$$

The total kinetic energy thus is

$$KE_{total} = \sum_i \frac{1}{2}m_i\left(v_{CM}^2 + v_i'^2 + 2\vec{v}_{CM} \cdot \vec{v}'_i\right)$$

$$= \frac{1}{2}\sum_i m_i v_{CM}^2 + \sum_i \frac{1}{2}m_i v_i'^2 + \vec{v}_{CM} \cdot \sum_i m_i\vec{v}'_i \qquad (9.33)$$

In the first summation, the sum of all the masses is the total mass m_{total} of the system of particles and there is only one v_{CM}. The second summation is the kinetic energy of the system of particles with respect to the center of mass. For the third sum, consider the following. Since each $\vec{r}'_i$ is measured with respect to the center of mass, Equation 9.20 implies

$$\sum_i m_i\vec{r}'_i = 0 \ \text{kg·m}$$

If we differentiate this with respect to time, we find

$$\sum_i m_i\vec{v}'_i = 0 \ \text{kg·m/s}$$

and so the third summation in Equation 9.33 is zero. The total kinetic energy of the system of particle is then

$$KE_{total} = \frac{1}{2}m_{total}v_{CM}^2 + \sum_i \frac{1}{2}m_i v_i'^2 \qquad (9.34)$$

Therefore the total kinetic energy of the system is the sum of

a. the kinetic energy of the center of mass, as if the total mass of the system were located all at that point; and
b. the kinetic energy of the particles with respect to (wrt) a reference frame with its origin at the center of mass.

That is,

$$KE_{total} = KE_{of\,CM} + KE_{wrt\,CM} \qquad (9.35)$$

EXAMPLE 9.15

An 8.00 kg runaway snowboard is approaching you at a speed of 12.00 m/s while a 4.00 kg small sled is coming at you from the opposite direction at a speed of 3.00 m/s on a frozen lake at the base of a ski area; see Figure 9.35.

a. What is the total kinetic energy of the two masses?
b. What is the velocity of their center of mass with respect to you?
c. What is the velocity of each mass with respect to the center of mass?

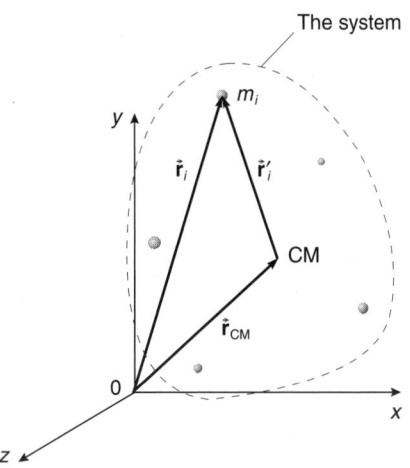

FIGURE 9.34 Locate a given particle with two position vectors.

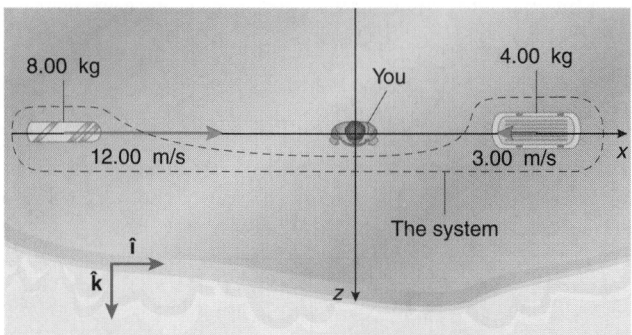

FIGURE 9.35

d. What is the kinetic energy of the two masses when viewed from the center of mass point?

e. Show that the total kinetic energy you found in part (a) is equal to the sum of the kinetic energy of the center of mass point and the kinetic energy with respect to the center of mass point, as predicted by Equation 9.35.

Solution

a. You determine the total kinetic energy by taking the sum of the kinetic energy of each mass individually:

$$KE_{total} = \frac{1}{2}(8.00 \text{ kg})(12.00 \text{ m/s})^2 + \frac{1}{2}(4.00 \text{ kg})(3.00 \text{ m/s})^2$$
$$= 594 \text{ J}$$

b. The position vector of the center of mass of the two particles is (from Equation 9.20)

$$\vec{r}_{CM} = \frac{1}{m_1 + m_2}(m_1\vec{r}_1 + m_2\vec{r}_2)$$

If you differentiate this equation with respect to time, you find an expression for the velocity of the center of mass in terms of the velocities of the particles:

$$\vec{v}_{CM} = \frac{1}{m_1 + m_2}(m_1\vec{v}_1 + m_2\vec{v}_2)$$

Using the coordinate system indicated in Figure 9.35,

$$\vec{v}_{CM} = \frac{(8.00 \text{ kg})(12.00 \text{ m/s})\,\hat{\imath} + (4.00 \text{ kg})(-3.00 \text{ m/s})\,\hat{\imath}}{8.00 \text{ kg} + 4.00 \text{ kg}}$$
$$= (7.00 \text{ m/s})\,\hat{\imath}$$

c. Since the 8.00 kg mass is moving parallel to $\hat{\imath}$ at speed 12.00 m/s and the center of mass is also moving parallel to $\hat{\imath}$ but at speed 7.00 m/s, the velocity of the 8.00 kg mass with respect to the center of mass is $(5.00 \text{ m/s})\hat{\imath}$. Since the 4.00 kg is moving parallel to $-\hat{\imath}$ at speed 3.00 m/s while the center of mass is approaching it at speed 7.00 m/s, the velocity of the 4.00 kg mass with respect to the center of mass is $(-10.00 \text{ m/s})\hat{\imath}$.

d. The kinetic energy of the two particles with respect to the center of mass is calculated using their speeds with respect to the center of mass, found in part (c):

$$\frac{1}{2}(8.00 \text{ kg})(5.00 \text{ m/s})^2 + \frac{1}{2}(4.00 \text{ kg})(10.00 \text{ m/s})^2 = 300 \text{ J}$$

e. You determine the kinetic energy of the center of mass point by imagining the mass of the whole system at that point, moving with speed v_{CM}:

$$KE_{CM} = \frac{1}{2}(8.00 \text{ kg} + 4.00 \text{ kg})(7.00 \text{ m/s})^2$$
$$= 294 \text{ J}$$

Equation 9.35 computes the total kinetic energy of the two masses as the sum of the kinetic energy of the center of mass

point (which you just found) and the kinetic energy of the particles with respect to the center of mass (found in part (d)):

$$KE_{total} = 294 \text{ J} + 300 \text{ J}$$
$$= 594 \text{ J}$$

This is the same as the kinetic energy you found directly in part (a); hence Equation 9.35 correctly determines the total kinetic energy.

9.12 THE VELOCITY OF THE CENTER OF MASS FOR COLLISIONS*

When various masses collide in a complicated manner, does the velocity of the center of mass before and after the collision change? Perhaps surprisingly, the velocity of the center of mass is conserved (i.e., unchanged) if the external forces on the system of colliding particles during the collision are negligible compared with the internal collision forces. Let's now show why this is true.

Consider a system of two particles of masses m_1 and m_2 moving with velocities $\vec{v}_{1 \text{ bfr}}$ and $\vec{v}_{2 \text{ bfr}}$, respectively, and subjected to zero (or negligible) total external force. Use the coordinate system indicated in Figure 9.36.

The position vector of the center of mass of the two-particle system is, using Equation 9.20,

$$\vec{r}_{CM} = \frac{1}{m_1 + m_2}(m_1\vec{r}_1 + m_2\vec{r}_2)$$

The velocity of the center of mass is found by differentiating the equation for $\vec{r}_{CM}$ with respect to time:

$$\vec{v}_{CM} = \frac{d\vec{r}_{CM}}{dt}$$
$$= \frac{1}{m_1 + m_2}(m_1\vec{v}_{1 \text{ bfr}} + m_2\vec{v}_{2 \text{ bfr}})$$

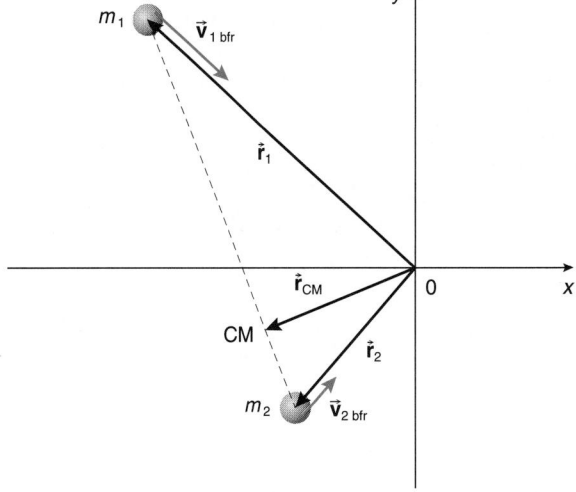

FIGURE 9.36 Two particles approaching a collision.

Rearrange this slightly to find

$$(m_1 + m_2)\vec{v}_{CM} = m_1\vec{v}_{1\,bfr} + m_2\vec{v}_{2\,bfr} \qquad (9.36)$$

The terms on the right-hand side of Equation 9.36 are the momenta of the individual particles; the term on the left-hand side of the equation is the *momentum of the center of mass*, if you imagine the entire mass of the system located at that point.

If the two particles undergo a collision and emerge with velocity vectors $\vec{v}_{1\,aft}$ and $\vec{v}_{2\,aft}$, the total momentum after the collision is

$$m_1\vec{v}_{1\,aft} + m_2\vec{v}_{2\,aft}$$

But we know that momentum is conserved in all collisions where external forces are negligible compared with the internal forces arising from the collision. Therefore

$$m_1\vec{v}_{1\,bfr} + m_2\vec{v}_{2\,bfr} = m_1\vec{v}_{1\,aft} + m_2\vec{v}_{2\,aft}$$

Using this equality in Equation 9.36, we find

$$(m_1 + m_2)\vec{v}_{CM} = m_1\vec{v}_{1\,bfr} + m_2\vec{v}_{2\,bfr}$$
$$= m_1\vec{v}_{1\,aft} + m_2\vec{v}_{2\,aft}$$

> This means that the velocity of the center of mass is the same before and after the collision.

Although we derived this result considering just a two-particle collision, the result is true for any number of colliding particles because conservation of momentum, the key step, is true for all collisions (with negligible external forces).

QUESTION 6

> If the velocity of the center of mass is *unchanged* by a collision, why, then, do two colliding cars invariably come to rest after a collision even if their brakes are not used?

9.13 THE CENTER OF MASS REFERENCE FRAME*

There is great flexibility in choosing the origin for a coordinate system. A reference frame with its origin at the center of mass point of a system is called the **center of mass reference frame**. If the motion of a system is described from the vantage point of the center of mass reference frame, we find that the total momentum of the system is *zero*.

To see that this is the case, recall from Equation 9.20 that the position vector of the center of mass is

$$\vec{r}_{CM} = \frac{1}{m_{total}}\sum_i m_i\vec{r}_i$$

where m_{total} is the mass of the entire system of particles. If we choose the origin for the coordinate system to be at the center of mass itself, then the position vector of the center of mass point is zero; that is, $\vec{r}_{CM} = 0$ m, or

$$0\text{ m} = \frac{1}{m_{total}}\sum_i m_i\vec{r}_i'$$

where the $\vec{r}_i'$ locate each particle with respect to the center of mass as the origin. Since $m_{total} \neq 0$ kg, the summation must be zero:

$$0\text{ kg}\cdot\text{m} = \sum_i m_i\vec{r}_i'$$

Differentiating this equation with respect to time, we find

$$0\text{ kg}\cdot\text{m/s} = \sum_i m_i\vec{v}_i'$$

where $\vec{v}_i'$ is the velocity of the *i*th particle relative to the center of mass frame. Each term in the sum is the momentum $\vec{p}_i'$ of the respective particle in the center of mass frame, and so the summation is the total momentum $\vec{p}_{total\,wrt\,CM}$ of the system when measured with respect to the center of mass; that is,

$$0\text{ kg}\cdot\text{m/s} = \sum_i \vec{p}_i'$$
$$= \vec{p}_{total\,wrt\,CM}$$

> Thus the total momentum of the system is zero when measured with respect to the center of mass. For this reason, the center of mass reference frame, with its origin at the center of mass, also is known as a **zero momentum reference frame**.

Whether such a reference frame is an inertial (nonaccelerated) reference frame depends on the absence of any appreciable external force.

QUESTION 7

> Since the total momentum of a system is zero in its center of mass reference frame, why is the total kinetic energy *not* zero (unless the particles all are at rest in this reference frame)?

STRATEGIC EXAMPLE 9.16

Two particles are sliding on a frictionless ice rink toward an impending collision, as shown in Figure 9.37. Thanks to some Velcro, the particles make a one-dimensional, completely inelastic collision, as indicated in Figure 9.38.

a. Find the velocity of the center of mass point according to an observer at rest on the ice rink.
b. Find the momentum of each particle with respect to the center of mass reference frame.
c. What is the total momentum of the two particles in the center of mass reference frame?
d. Find the velocity of the composite particle in the laboratory reference frame.

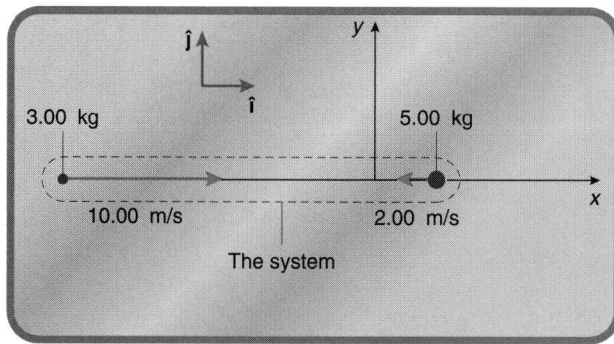

FIGURE 9.37

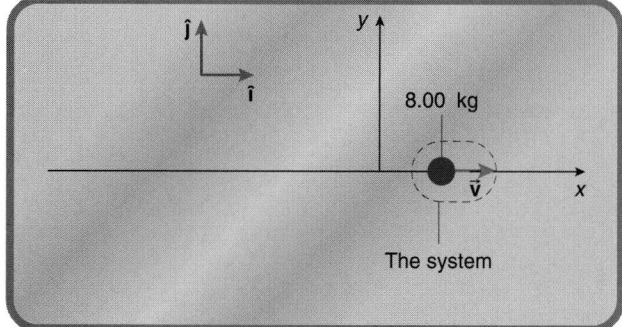

FIGURE 9.38

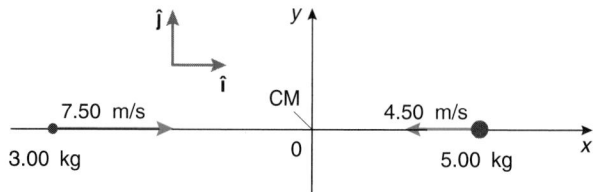

FIGURE 9.39

velocity of the 3.00 kg particle with respect to the center of mass point is $(7.50 \text{ m/s})\hat{\imath}$.

The momentum of the 3.00 kg particle in the center of mass reference frame then is

$$\vec{\mathbf{p}}_1' = (3.00 \text{ kg})(7.50 \text{ m/s})\hat{\imath}$$
$$= (22.5 \text{ kg·m/s})\hat{\imath}$$

The 5.00 kg particle is moving to the left at a speed of 2.00 m/s while the center of mass point is moving to the right at a speed of 2.50 m/s. Hence the speed of the 5.00 kg particle with respect to the center of mass is the sum, or 4.50 m/s. Viewed from the center of mass, the 5.00 kg particle is moving to the left—that is, toward $-\hat{\imath}$. Hence the velocity of the 5.00 kg particle in the center of mass reference frame is $(-4.50 \text{ m/s})\hat{\imath}$, as shown in Figure 9.39. The momentum of the 5.00 kg particle in the center of mass reference frame is

$$\vec{\mathbf{p}}_2' = (5.00 \text{ kg})(-4.50 \text{ m/s})\hat{\imath}$$
$$= (-22.5 \text{ kg·m/s})\hat{\imath}$$

c. The total momentum of the system with respect to the center of mass point is

$$\vec{\mathbf{P}}_{\text{total wrt CM}} = \vec{\mathbf{p}}_1' + \vec{\mathbf{p}}_2'$$
$$= (22.5 \text{ kg·m/s})\hat{\imath} + (-22.5 \text{ kg·m/s})\hat{\imath}$$
$$= 0 \text{ kg·m/s}$$

This calculation shows explicitly that the total momentum of the system in its center of mass reference frame is zero.

d. The velocity $\vec{\mathbf{v}}$ of the composite particle in the laboratory is found by conserving the total momentum of the system in the ice rink reference frame:

$$\vec{\mathbf{P}}_{\text{total bfr}} = \vec{\mathbf{P}}_{\text{total aft}}$$
$$(3.00 \text{ kg})(10.00 \text{ m/s})\hat{\imath} + (5.00 \text{ kg})(2.00 \text{ m/s})(-\hat{\imath})$$
$$= (3.00 \text{ kg} + 5.00 \text{ kg})\vec{\mathbf{v}}$$

Solving for $\vec{\mathbf{v}}$, you find

$$\vec{\mathbf{v}} = (2.50 \text{ m/s})\hat{\imath}$$

This is the same as the velocity of the center of mass, calculated in part (a). Why? Since the collision is completely inelastic, the center of mass point is coincident with the composite particle itself. Therefore, after the collision, the velocity of the particle and the velocity of the center of mass point must be the same: they are the same point!

Solution

a. The position vector of the center of mass point in the lab is found from Equation 9.20:

$$\vec{\mathbf{r}}_{\text{CM}} = \frac{1}{m_{\text{total}}} \sum_i m_i \vec{\mathbf{r}}_i$$

The velocity of the center of mass point is found by differentiating this with respect to time:

$$\vec{\mathbf{v}}_{\text{CM}} = \frac{1}{m_{\text{total}}} \sum_i m_i \vec{\mathbf{v}}_i$$

Using the coordinate system indicated in Figure 9.37, this becomes

$$\vec{\mathbf{v}}_{\text{CM}} = \frac{(3.00 \text{ kg})(10.00 \text{ m/s})\hat{\imath} + (5.00 \text{ kg})(2.00 \text{ m/s})(-\hat{\imath})}{3.00 \text{ kg} + 5.00 \text{ kg}}$$
$$= (2.50 \text{ m/s})\hat{\imath}$$

The center of mass point is moving parallel to $\hat{\imath}$ at a constant speed of 2.50 m/s as seen by the observer at rest on the ice rink.

b. Since the 3.00 kg particle is moving to the right at a speed of 10.00 m/s and the center of mass point is moving to the right at a lower speed of 2.50 m/s, the speed of the 3.00 kg particle with respect to the center of mass point is the difference, or 7.50 m/s, also to the right, as shown in Figure 9.39. Thus the

CHAPTER SUMMARY

The classical *momentum* $\vec{p}$ of a particle of mass m is defined to be

$$\vec{p} \equiv m\vec{v} \tag{9.1}$$

where $\vec{v}$ is its velocity. This relationship is valid as long as the speed of the particle is much less than the speed of light.

The most general statement of Newton's second law of motion is

$$\vec{F}_{total} = \frac{d\vec{p}_{total}}{dt} \tag{9.2}$$

where $\vec{p}_{total}$ is the total momentum of the system.

The *impulse* $\vec{I}$ imparted by a force $\vec{F}(t)$ during a time interval Δt is the integral of the force over the time interval:

$$\vec{I} \equiv \int_{t_i}^{t_f} \vec{F}(t)\, dt \tag{9.4}$$

$$= \vec{F}_{ave}\, \Delta t \tag{9.6}$$

where $\Delta t \equiv t_f - t_i$.

The *impulse–momentum theorem* states that impulse of the *total force* on a system is equal to the change in the total momentum of the system:

$$\vec{I} = \Delta\vec{p} \tag{9.5}$$

If the total force $\vec{F}_{total}$ on a system of one or more particles is equal to zero, then the total momentum of the system is a *constant vector*; that is, the total momentum is *conserved*. This is known as *conservation of momentum*. The individual momenta of the particles of the system may change because of the existence of forces internal to the system, but the total momentum of the system does not change.

The total momentum of a system of particles is conserved in collisions, disintegrations, and explosions; that is,

$$\vec{p}_{total\ bfr} = \vec{p}_{total\ aft} \tag{9.16}$$

Collisions are classified in the following way:

1. *Elastic collisions*: The total kinetic energy of the system of particles is conserved; that is,

$$KE_{total\ bfr} = KE_{total\ aft}$$

2. *Inelastic collisions*: The total kinetic energy of the system of particles is *not* conserved; that is,

$$KE_{total\ bfr} \neq KE_{total\ aft}$$

A collision is *completely inelastic* if all the particles stick together after the collision.

The position vector $\vec{r}_{CM}$ of the *center of mass* point of a system of particles is determined from

$$\vec{r}_{CM} = \frac{1}{m_{total}} \sum_i m_i \vec{r}_i \tag{9.20}$$

where m_{total} is the total mass of the system,

$$m_{total} = \sum_i m_i$$

and $\vec{r}_i$ is the position vector of the ith particle (of mass m_i) with respect to the origin of a coordinate system.

For a system of one or more particles, the total force on the system changes the velocity of the center of mass:

$$\vec{F}_{total} = m_{total} \frac{d\vec{v}_{CM}}{dt} \tag{9.28}$$

The total kinetic energy of a system of particles is the sum of the kinetic energy of the center of mass (imagining the total mass of the system to be located at that point) and the kinetic energy of the particles with respect to the center of mass:

$$KE_{total} = KE_{of\ CM} + KE_{wrt\ CM} \tag{9.35}$$

$$= \frac{1}{2} m_{total} v_{CM}^2 + \sum_i \frac{1}{2} m_i v_i'^2 \tag{9.34}$$

If a coordinate system has its origin at the center of mass point, the reference frame is called the *center of mass reference frame*. The total momentum of a system of particles in its center of mass reference frame is zero, and so such a reference frame also is known as a *zero momentum reference frame*. The center of mass reference frame is an inertial frame if the total (external) force on the system is zero.

In collisions, disintegrations, and explosions, the velocity of the center of mass point is the same immediately before and after the collision, disintegration, or explosion.

SUMMARY OF PROBLEM-SOLVING TACTICS

9.1 **(page 392)** None of the many specialized equations for various types of specific collisions is worth remembering; just remember the general points highlighted on pages 390–391.

QUESTIONS

1. (page 383); 2. (page 385); 3. (page 392); 4. (page 402); 5. (page 405); 6. (page 408); 7. (page 408)

8. Is it ever possible for the momentum to be *antiparallel* to the velocity vector? Explain why or why not.

9. A large mass M and a small mass m have equal kinetic energies. Which has the greater momentum?

10. Compare and contrast the colloquial meanings of the words *momentum* and *impulse* with their meanings in physics.

11. What condition must be satisfied for you to say the total momentum of a system is conserved?

12. The forces involved in automobile collisions can be lethal if too great. Explain how air bags are able to decrease the magnitudes of average forces in collisions and thereby save lives.

13. What is wrong with the physics in the editorial statement quoted at the beginning of Section 9.3?

14. When a rock is dropped, its momentum is not conserved. Explain why.

15. Is it possible for a system of two moving masses to collide, leaving both at rest? Explain.

16. If the total kinetic energy of a system is conserved, what can be said about the work done by the total force on the system?

17. It *is* possible to conserve the total momentum of a system but not its total kinetic energy (e.g., in inelastic collisions). Is it possible to conserve the total kinetic energy of a system but *not* the total momentum? Explain.

18. In an elastic collision, is the potential energy of the system conserved? Is the potential energy of the system the same at all instants during the time interval of the collision?

19. The total momentum of a system is conserved if the total force on the system is zero. If the total force on the system is zero and remains zero, we also can say that there is a "total force conservation law." Why is this "total force conservation law" *not* as useful as momentum conservation in dynamical situations?

20. A physicist observes an unstable particle at rest. Suddenly the particle breaks up (we say *decays*) into at least two fragments, perhaps more. The paths and velocity vectors of two of the resulting fragments are noted and sketched in Figure Q.20. In which cases (if any) can the physicist be certain of the existence of at least one other additional, unobserved particle associated with the decay? Explain your reasoning. For each case, what can be said (if anything) about the relative masses of the fragments?

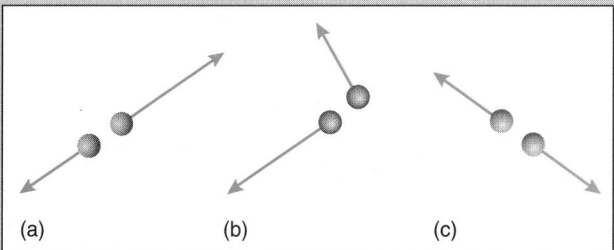

(a)   (b)   (c)

FIGURE Q.20

21. A piece of bubble gum is thrown at and sticks to a wall. Is the momentum of the gum conserved? What happens to the momentum of the gum?

22. In a collision, the occupants of a massive car are more likely to survive than the occupants of a compact car of small mass. Give a critique of why this is so. What can be said about the magnitude of the force that each car exerts on the other at each instant during the collision?

23. Explain why a single particle at rest cannot disintegrate (or decay) into *two* particles, one of which is at rest. Is a decay into three particles, with one at rest, possible?

24. If a single particle at rest disintegrates (or decays) into two particles, explain why the two particles *must* go off in opposite directions.

25. A small compact car collides head-on with a huge trailer truck moving in the opposite direction. What is the relationship between the force of the car on the truck and the force of the truck on the car during the collision? Explain your answer.

26. How could you design a rifle to minimize its recoil when it is fired?

27. Explain the physics associated with the executive toy known as Newton's cradle, pictured in Figure Q.27.

FIGURE Q.27

28. A mass m moving at speed v collides elastically with an identical mass at rest. Compare the initial kinetic energy associated with this collision with one in which the two masses approach each other, both initially with speed v.

29. Is it possible for a mass to have nonzero momentum yet zero total mechanical energy? Is it possible for a mass to have nonzero total mechanical energy yet zero momentum? Explain your answers.

30. A radioactive nucleus at rest emits a β^- particle (an electron). The direction of the emitted electron and that of the recoiling so-called daughter nucleus are found experimentally to be *not* in opposite directions. Explain why this observation means that an *additional particle* must have been emitted in the decay process. The third particle is known as an antineutrino.

31. Explain the underlying physical principles behind jet engines and rocket propulsion.

32. Is it possible for high jumpers to pass *over* the bar while their centers of mass pass *under* the bar? Explain your answer, perhaps with a series of sketches.

33. In the argument of Section 9.7, since the particle makes elastic collisions with the circular hoop, why is the momentum of the particle *not conserved* in each collision?

34. While meditating. . . . A small impulse applied to a roll of toilet paper can unwind the entire roll. A large impulse rips off a single sheet. Explain.

35. Must there be mass at the center of mass?

36. What configuration of planets in the solar system places the center of mass of the solar system at the greatest distance from the Sun? (The planets of the solar system are more or less confined to within a few degrees of the orbital plane of the Earth, known as the ecliptic plane.)

37. Sketch the shape of several objects whose center of mass lies outside the object itself.

38. An imaginary sphere of radius R is barely circumscribed around a massive object of arbitrary shape. No part of the mass lies outside the circumscribed sphere. Is the center of mass of the object *guaranteed* to be within the sphere? Explain why or why not.

39. Stand up straight with both feet slightly spread apart. If you raise one foot off the floor, what must happen to the rest of your torso to prevent yourself from falling?

40. Explain why it is not possible to stand next to a wall with one foot in front of the other *and* with the sides of both feet touching the wall.

41. A traditional time-keeping hourglass is placed on a sensitive scale to determine its mass (see Figure Q.41a). The hourglass now is inverted (see Figure Q.41b). Has the mass of the hourglass changed? Will the scale indicate the same value as in Figure Q.41a? Explain.

42. If the motion of a system is described using the center of mass reference frame, is the total momentum of the system zero even if the center of mass point is undergoing a nonzero acceleration as seen in the laboratory? Justify your answer.

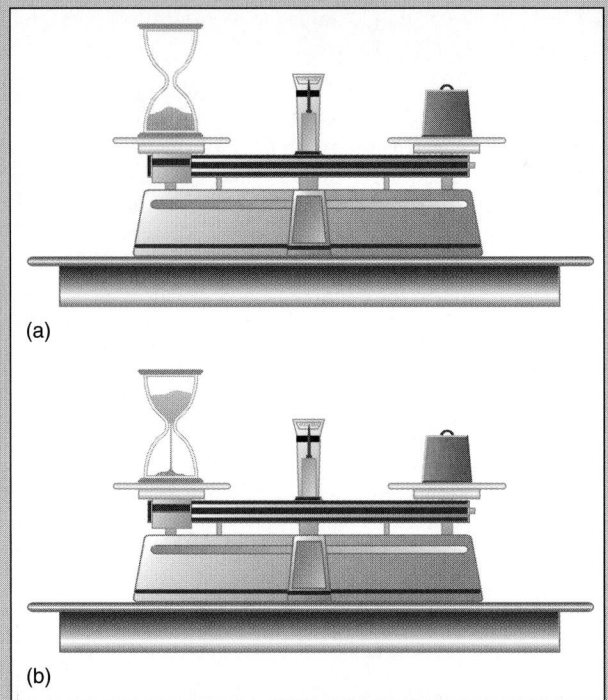

(a)

(b)

FIGURE Q.41

PROBLEMS

Section 9.1 Momentum and Newton's Second Law of Motion

1. A rock-filled dump truck with a mass of 2.0×10^4 kg is creeping along at a speed of only 1.0 km/h. How fast would you have to run to have the same magnitude of momentum as that of the truck? Assume your mass is 60 kg. Express your speed in km/h.

2. A 100 g bullet has a speed of 400 m/s. How fast would you have to run to have the same magnitude of momentum? Assume your mass is 60 kg.

3. (a) The α-particle (a helium nucleus with mass 6.68×10^{-27} kg) emitted in the radioactive decay of $^{238}_{92}$U has a speed of 1.4×10^7 m/s. Consider this speed to be small compared with the speed of light (it is only about 5% of the speed of light). What is the magnitude of the momentum of the α-particle? (b) What is the speed of a 1.00 g mass that has the same magnitude of momentum? (c) How long would it take the 1.00 g mass to traverse 1.00 m at this speed? Express your result in years.

4. (a) Show that the kinetic energy of a system of mass m can be written in terms of the magnitude of its momentum as

$$KE = \frac{p^2}{2m}$$

(b) Show that the kinetic energy of the system also can be written as

$$KE = \frac{pv}{2}$$

where v is the speed of the system. The first of these expressions is used quite frequently; the second expression almost never is seen because p is a function of v as well.

5. How fast must a 60.0 kg physics student run so she has the same magnitude of momentum as a 1.00×10^3 kg automobile moving at 1.00 km/h?

6. Your car of mass 1.20×10^3 kg has a horizontal speed of 100 km/h in a northwest direction. A Cartesian coordinate grid has $\hat{\mathbf{i}}$ east and $\hat{\mathbf{j}}$ north, as in Figure P.6. (a) What is the momentum of the car? (b) What is the magnitude of the momentum?

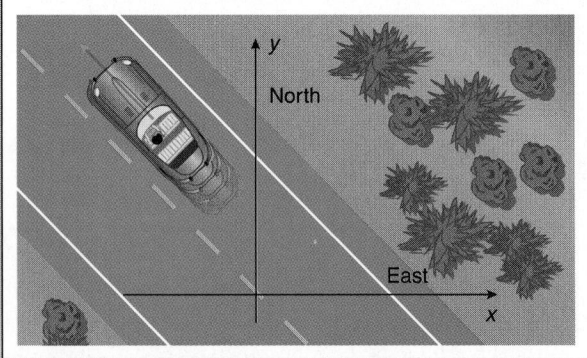

FIGURE P.6

7. A 50 caliber machine gun* fires bullets, each with a mass of about 45 g, at a speed of about 8.0×10^2 m/s. (a) What is the magnitude of the momentum of each bullet? (b) How fast would a 70 kg soldier have to run to have the same magnitude of momentum?

---

*The caliber number gives the diameter of the bullet in hundredths of an inch. A 50 caliber bullet has a diameter of 0.50 inch.

•8. A 2.73 kg bag of bread flour accidentally is dropped from rest from a height 2.00 m above the floor of a supermarket. What is the magnitude of the momentum of the bag the instant before its impact with the floor?

•9. For breakfast, a herring gull (*Larus argentatus*) flying horizontally at speed 15.0 m/s drops a tasty cockleshell (*Dinocardium robustum*) of mass 0.200 kg from a height of 20.0 m above a flat rocky beach as in Figure P.9. What is the magnitude of the momentum of the bivalve the instant before it strikes the beach?

FIGURE P.9

Section 9.2 Impulse–Momentum Theorem

10. Show that the SI units for impulse, N·s, are the same as those for momentum, kg·m/s, and therefore that the impulse–momentum theorem has the same SI units on each side of the equation.

11. A fire hose discharges water at a rate of 15.0 kg/s with a speed of 20 m/s as in Figure P.11. What magnitude of force is needed to hold the nozzle stationary?

FIGURE P.11

•12. Golf pro Guy Boros whacks a golf ball of mass 45.0 g. The ball leaves the tee at an angle of 45.0° to the horizontal and with a speed of 60.0 m/s. (a) Choose an appropriate coordinate system. (b) What impulse was given to the ball? (c) If the club is in contact with the ball for 1.00 ms, what is the magnitude of the average force of the club on the ball? See Figure P.12.

FIGURE P.12

•13. An ostrich (*Struthio camelus*) egg of mass 1.00 kg is dropped from rest from a height of 10.0 m. (a) Choose a convenient coordinate system and indicate it clearly in a diagram. (b) What is the sum of the kinetic and potential energies of the egg at a height of 5.0 m? (c) Does the momentum of the egg change at a constant rate with time as it falls, or not? Explain. (d) Find the impulse imparted to the egg by the gravitational force during the fall to the instant before impact.

•14. A supertanker of mass 1.0×10^5 metric tons traveling at 25 km/h runs aground on a reef and is brought to a complete stop in 8.0 s. Crunch. Call out the environmental lawyers. What is the magnitude of the average horizontal force of the reef on the tanker? (1 metric ton = 1000 kg)

•15. A golf club hits a golf ball of mass 0.045 kg and gives it a speed of 50 m/s. Estimate the magnitude of the average force of the club on the ball if the collision lasts for 1.0 ms.

•16. A battleship at rest fires a 400 mm diameter cylindrically symmetric shell with a mass of 1.00×10^3 kg. BOOM! The barrel is aimed off the stern and elevated at an angle of 30.0° to the horizontal. The shell emerges from the gun with a speed of 600 m/s. (a) What impulse is given initially to the shell in the firing? (b) What impulse is given to the battleship in the firing? (c) If the ship has a mass of 5.0×10^4 metric tons, what is the recoil speed of the battleship? Assume the ship can move both horizontally and vertically.

•17. The time dependence of a force component $F_x(t)$ on a system is indicated in Figure P.17. (a) What is the impulse of the force during the first 2.00 s? (b) What is the average force during the first 2.00 s? (c) What is the impulse of the force during the first 4.00 s? (d) What is the average force during the first 4.00 s?

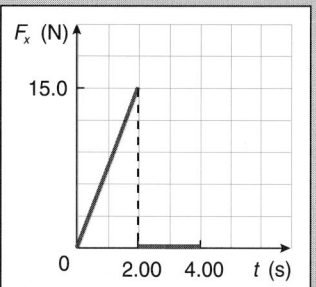

FIGURE P.17

•**18.** (a) During the time interval from $t = 0$ s to 4.0 s, a total force whose only component is $F_x(t)$ acts on a moonshine jug system; see Figure P.18. Find the impulse of the total force. (b) What is the change in the momentum of the system during the same time interval?

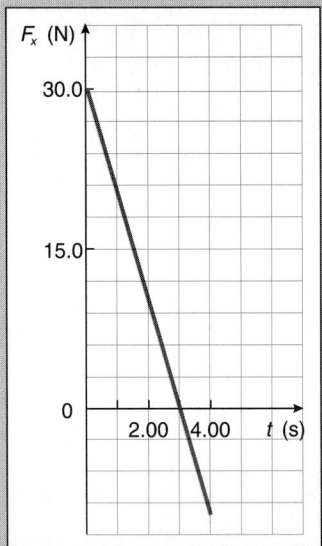

FIGURE P.18

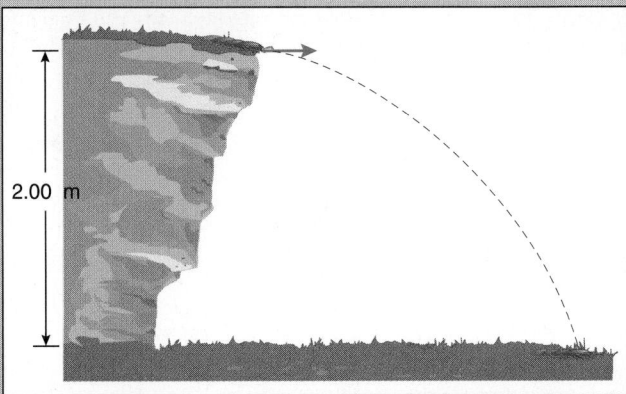

FIGURE P.20

•**19.** A 15 g cricket (*Acheta domestica*) leaps through a 2.20 m horizontal distance. The initial velocity of the cricket makes a 30° angle with the horizontal direction. If it takes the cricket 0.010 s to leave the ground, what is the magnitude of the impulse of the total force on the cricket that gives it the speed needed for the jump?

•**20.** A cockroach (*Periplaneta americana*) of mass 0.0100 kg runs off a precipice at speed 1.50 m/s and falls to the ground 2.00 m below as shown in Figure P.20. (a) Indicate a convenient coordinate system. (b) What is the initial momentum of the roach? (c) What is the momentum of the roach at the instant just before impact? (d) What is the *change* in the momentum of the roach? (e) What force provides the impulse to change the momentum of the roach? (f) Calculate the impulse of this force using

$$\vec{I} = \int_{t_i}^{t_f} \vec{F} \, dt$$

and show that the result is the same as the change in the momentum calculated in part (d).

•**21.** There is an old adage that goes: "It's not the fall that hurts, it's the sudden stop." Estimate the magnitude of the acceleration experienced by a 60 kg painter during the sudden stop associated with a fall off a scaffold at an elevation of 8.0 m. Assume the painter is brought to rest in 0.15 s.

•**22.** Marbles of mass m are thrown at speed v at a rigid wall as shown in Figure P.22. Assume the collision of each marble

with the wall is elastic. (a) What is the momentum change of each marble? (b) What is the impulse of the force of the wall on each marble? (c) What is the impulse of the force of a marble on the wall? (d) If the marbles are thrown at a rate of n marbles each second, what is the average force of the marbles on the wall? (e) If the marbles hit the wall in a randomly distributed manner over an area A, what is the average force per square meter of the marbles on the wall? The magnitude of the force per unit area is called the *pressure*.

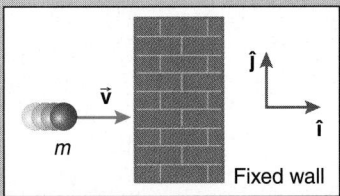

FIGURE P.22

•**23.** Repeat Problem 22 but consider the marbles and wall to be covered with contact cement, so that the collisions of the marbles with the wall are completely inelastic collisions.

•**24.** Repeat Problem 22 but assume that the marbles encounter and rebound from the wall at an angle θ with respect to a normal line as shown in Figure P.24. Assume the collisions with the wall are elastic.

Section 9.3 The Rocket: A System with Variable Mass*

25. What fuel to payload ratio is need for a rocket to achieve a final speed equal to that of its exhaust gases if the rocket is launched at rest well away from any other large masses?

26. The exhaust gases leave a certain rocket with a speed of 2.5 km/s with respect to the rocket. What fuel to payload ratio is needed for the rocket to achieve a final speed equal to the escape speed from the Earth if the rocket is initially at rest well away from any other masses (including the Earth)? Will the fuel to

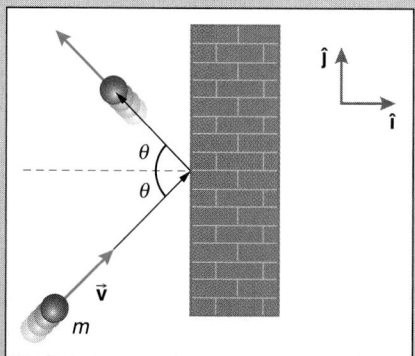

FIGURE P.24

payload ratio have to be greater or the same if the rocket actually is launched from the Earth to achieve the escape speed? Explain your answer.

• **27.** The Boeing 747-400 has the maximum, fully loaded takeoff mass of any civilian aircraft: a whopping 3.956×10^5 kg. A newly developed jet engine for this plane can produce 333 kN of thrust. Using four such engines, what is the magnitude of the maximum acceleration of the fully loaded aircraft? Neglect effects of air resistance on the plane, the mass of air needed by each engine per second for combustion, as well as the (comparatively small) mass loss during takeoff.

‡**28.** A fire hose fixed to a support delivers a horizontal stream of water at a speed v_0 and a mass flow rate of β kilograms per second. (a) The water stream is directed against a wall along a perpendicular to the wall and is brought to rest. What is the magnitude of the force of the water on the wall? (b) What is the momentum per unit length of the stream? (c) The stream now is directed into a hole in the back of a tank car (see Figure P.28) that is free to roll from rest (with no frictional work done). The tank car has an initial mass m_0. Let the total mass of the tank car be m (tank car plus accumulated water). Let v be the x-component of the velocity of the tank car at time t. Show that

$$m \, dv = (v_0 - v) \, dm$$

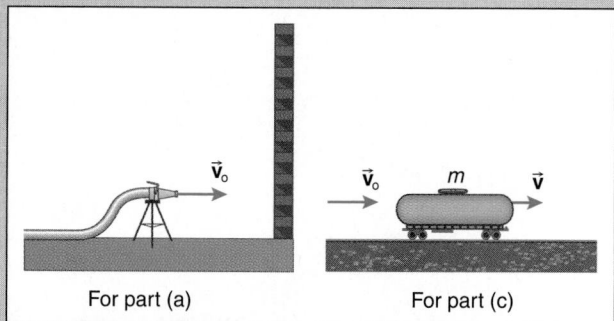

FIGURE P.28

(d) Integrate this equation to show that

$$v = \left(1 - \frac{m_0}{m}\right)v_0$$

‡**29.** A massive anchor chain of length ℓ is held vertical from its top link so that its lowest link is barely in contact with the horizontal ground. The chain is dropped. When the top link has fallen a distance y (see Figure P.29), show that the magnitude of the normal force of the ground on the chain is $3mgy/\ell$.

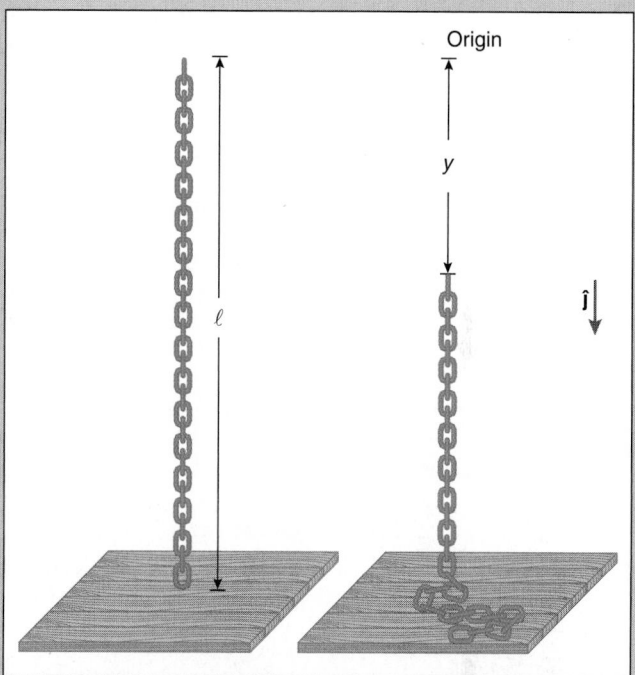

FIGURE P.29

Sections 9.4 Conservation of Momentum
9.5 Collisions
9.6 Disintegrations and Explosions
9.7 The Centripetal Acceleration Revisited*
9.8 An Alternative Way to Look at Force Transmission*

30. You are running at speed v_0 to another dynamic physics lecture and unfortunately collide in a completely inelastic, one-dimensional manner with a buddy of equal mass at rest. What fraction of the initial kinetic energy is lost in the collision?

• **31.** Two particles, of masses m and $2m$, are incident as indicated in Figure P.31. The particles collide and, thanks to the wonders of contact cement, stick together after the collision. (a) What type of collision is this? (b) Introduce a Cartesian coordinate system to analyze the problem; use the impact point as the origin and clearly indicate the directions of the positive coordinate axes. (c) What is the total momentum of the system of two particles before the collision? (d) What is the velocity

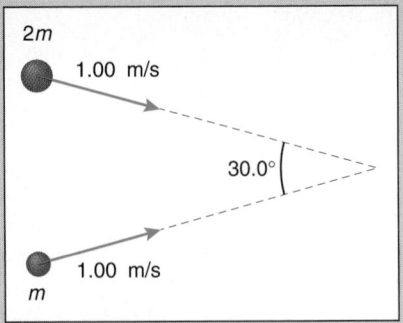

<figure>**FIGURE P.31**</figure>

vector of the composite particle after the collision? (e) What is the speed of the composite particle after the collision? (f) What is the angle that the velocity vector of the composite particle makes with the original direction of m after the impact? (g) What is the change in the kinetic energy of the system?

•32. Trouble has come to River City. Pool. A hustler takes a cue and sends the cue ball at speed v_0 slamming into the 6 ball initially at rest (see Figure P.32). The collision is elastic and the 6 ball clunks into a pocket of the table. The two balls have equal mass. Treat the balls as particles, neglecting any rotation. Show that the speed of the rebounding cue ball is $0.500v_0$ and that the speed of the 6 ball after the collision is $0.865v_0$.

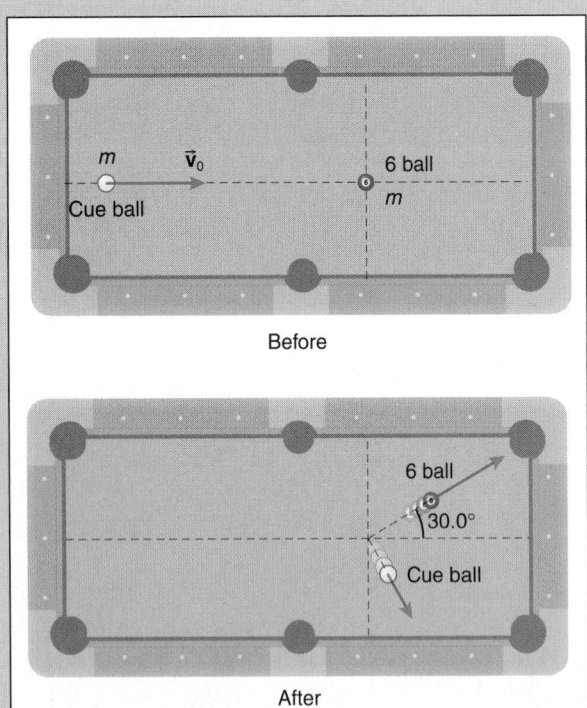

<figure>**FIGURE P.32**</figure>

•33. A 1.00×10^3 kg car is moving toward increasing values of x along the x-axis and strikes a stationary 4.00×10^3 kg truck. CRASH! Immediately after the collision, the car is motionless and the truck is moving with the velocity of $(2.00 \text{ m/s})\hat{\imath}$.

Assume only collision forces are relevant. (a) What was the original velocity of the car? (b) Was the collision elastic? Support your answer with a calculation. (c) Was the momentum of the *car* conserved? Explain.

•34. Two pucks slide on a frictionless, horizontal airtable. One puck has a mass of 0.200 kg and a velocity of $(3.00 \text{ m/s})\hat{\imath} - (4.50 \text{ m/s})\hat{\jmath}$. The second puck has mass 0.250 kg and velocity $(-7.10 \text{ m/s})\hat{\imath}$. The pucks collide, leaving the first puck at rest. (a) Find the velocity of the second puck after the collision. (b) What is the change in the kinetic energy of the system of two pucks?

•35. A quarterback of mass 100 kg, running at speed 10.0 m/s toward increasing values of x, runs (stupidly) directly into the arms of a 150 kg tackler moving in the opposite direction at 5.0 m/s. The impact occurs in one dimension in midair and lasts for 0.20 s, after which the players are locked together. (a) Is the momentum of *each individual* conserved in the collision? Explain. (b) What is the velocity of the composite mass after the collision? (c) What is the impulse imparted to the quarterback by the tackler? (d) What is the magnitude of the average force exerted by the tackler on the quarterback? Compare this with the magnitude of the weight of the quarterback.

•36. A cart with a mass of 400 kg is sliding across a frictionless ice pond with a speed of 10.0 m/s. A horrendous downpour dumps a mass m of rain (falling vertically) into the cart, slowing it to 8.00 m/s. Define an appropriate system. (a) Is the horizontal component of the momentum of your system conserved in this process? Explain. (b) Find the mass of rain that fell into the cart.

•37. Two cars of equal mass $m = 1.00 \times 10^3$ kg are approaching an intersection at speeds of 60.0 km/h and 40.0 km/h as indicated in Figure P.37. Neither driver yields the right of way and the two jerks collide. The cars stick together on impact. (a) What kind of a collision is this? (b) What is the total momentum of

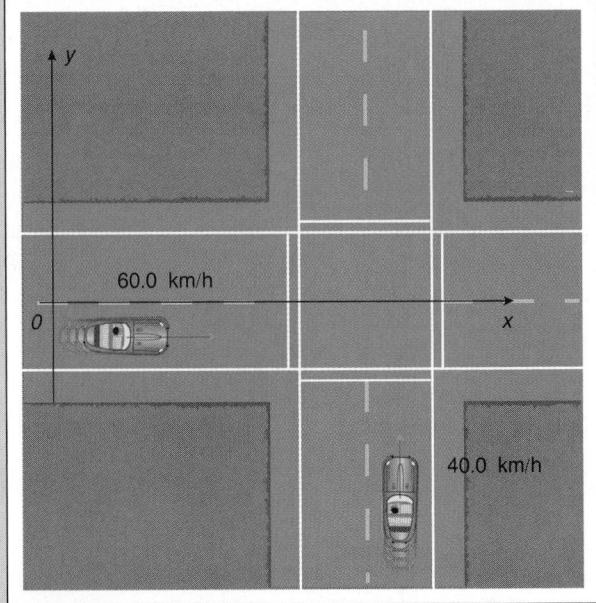

<figure>**FIGURE P.37**</figure>

the system of two cars before the collision? (c) What is the total momentum after the collision? (d) What is the speed of the crumpled composite car immediately after the collision? (e) What was the change in the kinetic energy of the system of two cars in the collision?

•38. A 10.0 g bullet is fired into and remains in a 3.00 kg block of wood located 1.00 m from the edge of a frictionless table as indicated in Figure P.38. The block slides along the table surface, falls off the table, and lands 2.50 m from the base of the table (of height 75.0 cm). (a) Does the table move? Explain why or why not. (b) Determine the speed of the bullet the instant before it hits the block.

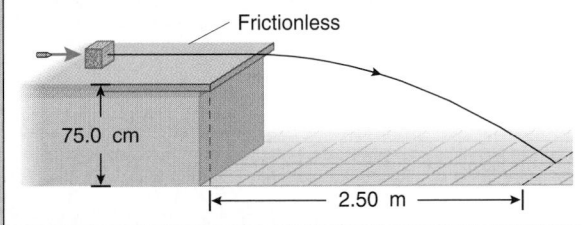

Frictionless

75.0 cm

2.50 m

FIGURE P.38

•39. Five equal masses are free to slide along a horizontal frictionless surface as indicated in Figure P.39. Initially four of the masses are at rest and the fifth is approaching the others at a speed v. At each collision the particles stick together. (a) What is the total momentum of the system of five masses before the first collision? After the last collision? (b) Calculate the speed of the composite particle after each collision in turn. (c) Find the change in the kinetic energy of the system of five masses, taking the final kinetic energy as that after the last collision and the initial KE as that before the first collision.

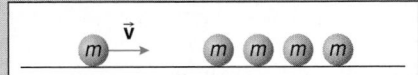

m $\vec{v}$ m m m m

FIGURE P.39

•40. Thanks to front and side air bags, a speeding sports car of mass 1.50×10^3 kg makes a nonfatal, completely inelastic collision with a truck of mass 3.00×10^3 kg [which includes its load of turkeys (*Meleagris gallopavo*)] traveling at speed 25.0 m/s. After the collision the vehicle pair is moving (with uninjured drivers) at an angle of 70.0° with respect to the initial direction of the car (as indicated in Figure P.40) along with some *very* agitated gobblers. (a) Establish a coordinate system to analyze the problem, clearly labeling the positive direction of the axes. (b) Determine the *speed* of the smashed car–truck system immediately after the collision. (c) Determine the initial speed of the car.

•41. Your 1050 kg car is skidding across a frictionless ice pond in northern Minnesota with a velocity 22 m/s east. Unfortunately another car of mass 900 kg is sliding at speed 18 m/s directed 30° south of east and collides with and sticks to your car. Crunch! Now cry. (a) After the collision, what is the velocity of the vehicles? (b) What is the change in the kinetic energy of the two-car system as a result of the collision?

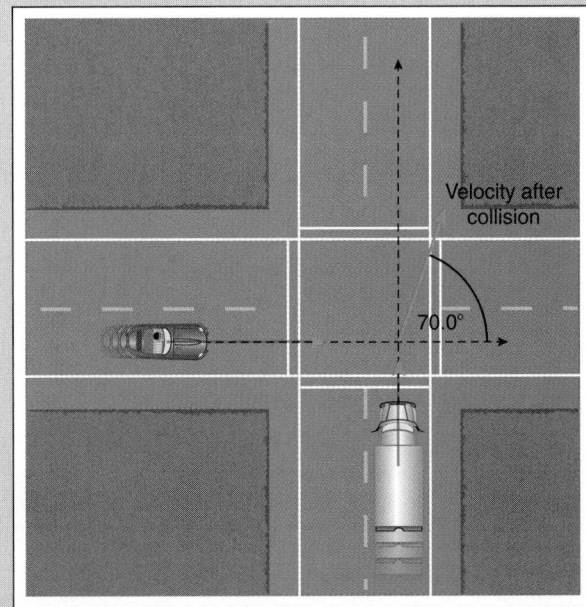

Velocity after collision

70.0°

FIGURE P.40

•42. The muzzle speed of a bullet can be determined using a device called a ballistic pendulum, shown in Figure P.42. A bullet of mass m moving at speed v encounters a large mass M hanging vertically as a pendulum at rest. The mass M absorbs the bullet. The hanging mass (now consisting of M + m) then swings to some height h above the initial position of the pendulum as shown. (a) Show that the initial speed v' of the pendulum (with the embedded bullet) after impact is

$$v' = \frac{mv}{M + m}$$

(b) Show that the muzzle speed v of the bullet is

$$v = \frac{M + m}{m}(2gh)^{1/2}$$

(c) If h = 10.0 cm, M = 2.50 kg, and m = 10.0 g, find v.

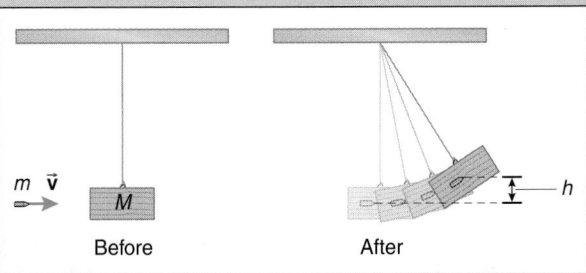

m $\vec{v}$

M

h

Before   After

FIGURE P.42

•43. There has been increasing concern about the inevitability of a collision between the Earth and an incident asteroid. (a) If a vertically incoming asteroid has a speed of 50 km/s and makes a completely inelastic collision with the Earth, estimate the mass of an asteroid that will give the Earth a recoil speed of 1.0 mm/s. (b) If the density of the asteroid is 5.0×10^3 kg/m^3 and it is spherical in shape, what is its diameter?

•**44.** A collection of identical pool balls (each of mass 0.165 kg) is at rest on a pool table. A ball of unknown mass is incident upon the collection at speed 1.50 m/s. As a result of the elastic collision, three of the identical pool balls (initially at rest) move off as shown in Figure P.44. All the other balls, including the incident ball, are at rest. Treat the balls as particles. (a) What is the speed of the ball labeled A? (b) What is the mass of the unknown ball?

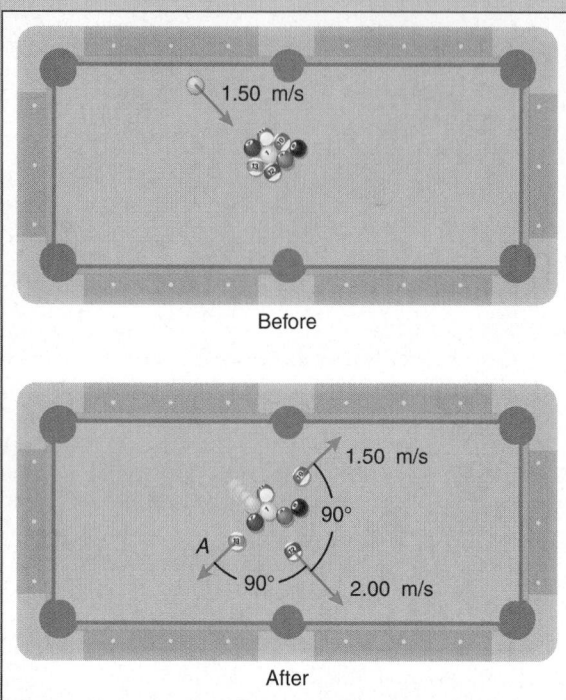

FIGURE P.44

•**45.** You are tooling along in a 1.00×10^3 kg sports car in a dense fog. Uh oh. You suddenly collide into the rear end of an 1.80×10^3 kg behemoth that was moving at 15.0 m/s in the same direction. Your cars stick together and move at 20.0 m/s immediately after the collision. Consider the motion of all objects in this problem to be in a straight line with only collision forces relevant. (a) Were you exceeding the legal speed limit of 30.0 m/s (roughly 110 km/h) before the crash? (b) An instant after the foregoing collision, the pair of cars strikes yet another car (of mass 1.50×10^3 kg) head-on that had been traveling at 40.0 m/s in the opposite direction. All the cars stick together. What is the speed of the system of three mangled cars just after this collision?

•**46.** A Cadillac of mass 2250 kg going east collides with a Geo of mass 1200 kg headed north at 25.0 km/h on a level highway. The cars remain tangled and slide to rest after traveling 25.0 m in a straight line at an angle of 20.0° north of east. The coefficient of kinetic friction for the tires on the road is 0.20. How fast was the Cadillac traveling before the unfortunate collision?

•**47.** Two particles collide at the origin of a frictionless horizontal *x–y* plane. Prior to the collision, the mass $m_1 = 3.50$ kg was moving along the *x*-axis toward increasing values of *x* with speed 4.00 m/s, while mass $m_2 = 5.00$ kg was moving along the *y*-axis toward increasing values of *y* at a speed 1.50 m/s, as shown in Figure P.47. After the collision, m_1 is at rest at the

origin. (a) Find the velocity of m_2 after the collision. (b) Is the collision elastic or inelastic?

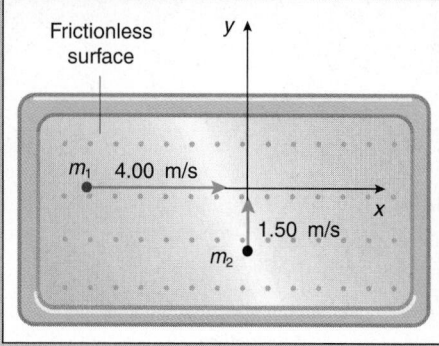

FIGURE P.47

•**48.** A coin of mass *m* is released from rest on a frictionless plane inclined to the horizontal at angle θ. After sliding a distance *d* to the bottom of the incline, the coin encounters a rough horizontal surface. The coefficient of kinetic friction for the coin on the surface is μ_k. The coin slides a short distance *s* and then collides completely inelastically with a coin of mass M that is initially at rest. Find an expression for each of the following in terms of the given quantities. Indicate explicitly what physical principles (such as the CWE theorem, conservation of momentum, etc.) you use for each part. (a) What is the speed of the coin of mass *m* at the bottom of the incline? (b) What is the speed of the coin of mass *m* just before the collision? (c) What is the speed of the coins immediately after the collision?

•**49.** A block of mass *m* is released at rest on a frictionless inclined plane as indicated in Figure P.49. The horizontal floor is rough (the coefficients of friction for *m* on the floor surface are $\mu_s = 0.40$ and $\mu_k = 0.25$). The mass *m* slides 0.20 m and makes a completely inelastic collision with another mass M that is initially at rest. The composite particle has a speed 0.25 m/s immediately after the collision. Find the ratio M/m of the two masses.

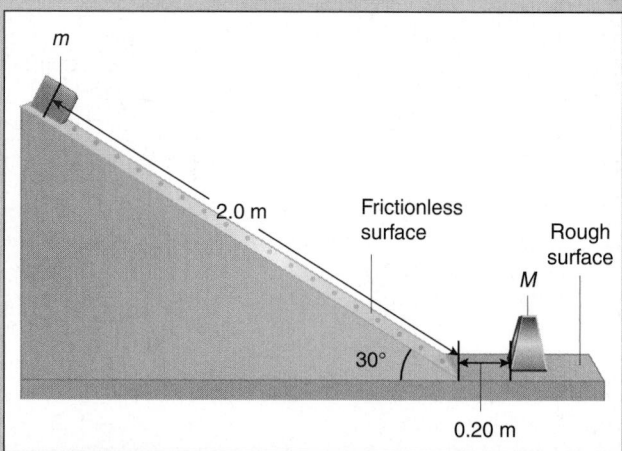

FIGURE P.49

‡**50.** A system of two small spheres of identical mass *m* collide obliquely as shown in Figure P.50. The collision is inelastic. Initially one of the masses is at rest. The momenta of the par-

ticles immediately *after* the collision are $\vec{p}'_1$ and $\vec{p}'_2$. The angle between the paths of the masses after the collision is β. Show that the change in the kinetic energy of the system is

$$\Delta KE = -\frac{p'_1 p'_2}{m} \cos \beta$$

Notice that for an *elastic* collision, $\Delta KE = 0$ J and $\beta = 90°$ as we found in Example 9.11. (Hint: Write the momentum conservation equation in vector form and the kinetic energies in terms of the magnitudes of the momenta. Use a technique similar to the short and elegant Method 2 in Example 9.11.)

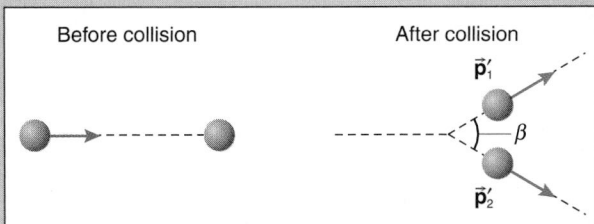

Before collision | After collision

$\vec{p}'_1$

β

$\vec{p}'_2$

FIGURE P.50

51. The curved inclines of two identical wedge-shaped masses M smoothly merge with a horizontal, frictionless surface as indicated in Figure P.51. The curved surfaces of the inclines also are frictionless. Another mass m is released at rest from a height h on one of the curved, smooth surfaces of the wedges. (a) Consider first the motion of m from its initial height h down to the horizontal plane. Is the total mechanical energy of m and M conserved in the descent? If so, write an equation representing this. Is the total momentum of the system of m and M conserved in the descent? If so, write an appropriate equation expressing this conservation when m reaches the horizontal surface. (b) After m reaches the horizontal plane, it then encounters the right-hand wedge and ascends it to some height h'. Is the total mechanical energy of the system of m and the right-hand wedge conserved in the ascent? If so, write an equation expressing this relationship. Is the total momentum of m and the right-hand wedge M conserved in the ascent? If so, write an equation expressing this relationship. (c) Show that the height h' to which m ascends the right-hand wedge is

$$h' = \frac{M^2}{(M + m)^2} h$$

(d) What mass ratio M/m results in $h' = h/4$?
This problem was inspired by a problem in *Quantum*, 3, #2 (November–December 1992), pages 27, 55–56.

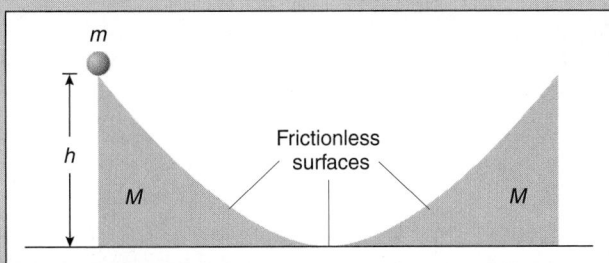

m

h

Frictionless surfaces

M

M

FIGURE P.51

52. A particle of mass M at rest disintegrates into *three* particles. Two of the particles (of masses m and M') move off with speeds v and v' respectively and directions indicated in Figure P.52. Show that the angle ϕ, locating the trajectory of the third particle, is given by the expression

$$\tan \phi = \frac{mv}{M'v'} \sec \theta - \tan \theta$$

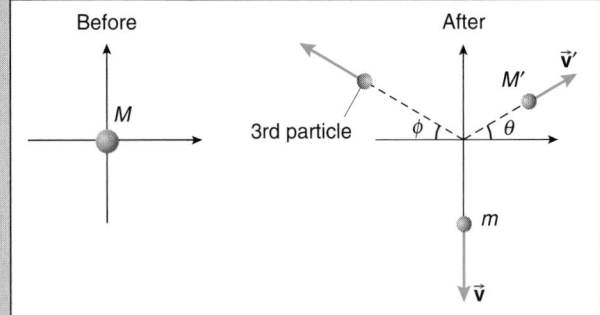

Before | After

M

$\vec{v}'$

M'

3rd particle

ϕ θ

m

$\vec{v}$

FIGURE P.52

53. For elastic collisions, show that the relative speed of approach before the collision (the speed of one particle with respect to the other particle) is equal to the relative speed of recession after the collision.

54. A ball of large mass M and a ball of small mass m are dropped from the same height h above the floor as shown in Figure P.54a. Neglect the (small) radii of the balls in this problem. (a) Show that if no mechanical energy is lost in the bounce of the large ball from the floor, then the speed with which the large ball rebounds is

$$v = (2gh)^{1/2}$$

(b) The instant after the large mass rebounds from the floor, the small mass [which is still momentarily traveling downward at the speed v given in the answer to part (a)] makes an elastic collision with the large ball rising at the same speed v. Imagine yourself on the large ball just after it rebounds from the floor (see Figure P.54b). What is the speed of approach of the small ball when viewed by an observer on the large ball? (c) Relative to an observer on the large ball, what is the speed at which the small ball moves *after* its elastic collision with the large ball? (d) Show that with respect to the *ground*, the small ball rebounds from this collision with a speed $3v$. (e) Show that the height to which the small ball rebounds is $9h$. This problem can be used to explain why interplanetary space probes occasionally make encounters with planets in maneuvers known colloquially as slingshot effects or gravitational pumping. Imagine a spacecraft traveling at speed v and at a distance r from a planet traveling at speed v' in the opposite direction, as in Figure P.54c. The spacecraft makes an "elastic collision" with the planet by being whipped around the planet by the gravitational interaction. The relative speed of approach of the spacecraft (viewed from the *planet*) is $v' + v$. The speed at which the spacecraft recedes from the encounter, when at the same distance as in Figure P.54d, also is $v' + v$ with respect to the *planet*. But the speed of the spacecraft relative to the reference frame of Figure P.54e is $v' + v$ plus the speed v' of the planet itself—that is,

$$(v' + v) + v'$$

or

$$2v' + v$$

The change in the speed of receding spacecraft is twice the speed of the planet! This example also qualitatively illustrates the extraordinary speeds of matter ejected from supernova explosions of massive stars.

This problem was inspired by G. Stroink, "Superball problem," *The Physics Teacher*, 21, #7, page 466 (October 1983).

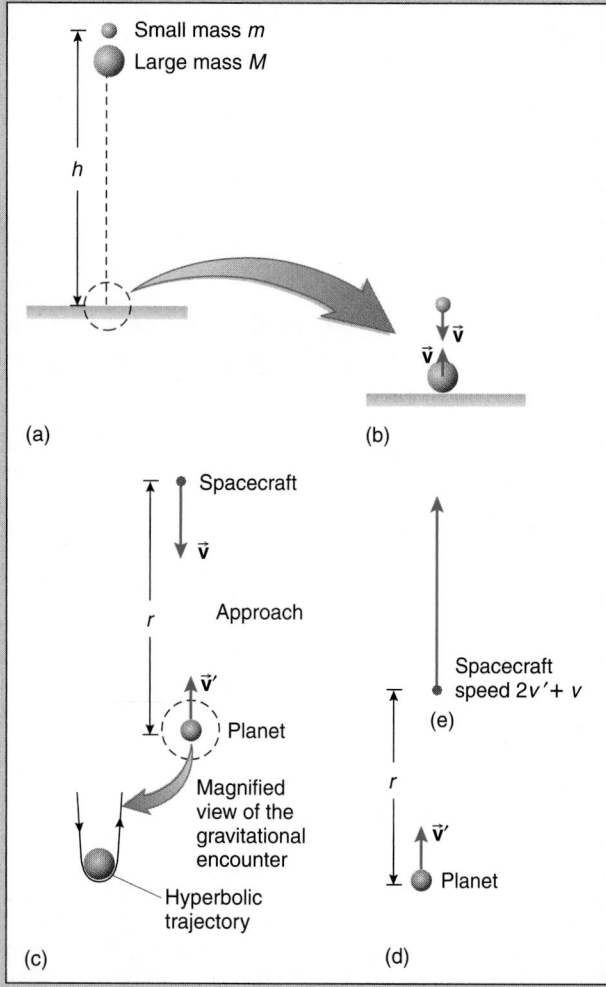

FIGURE P.54

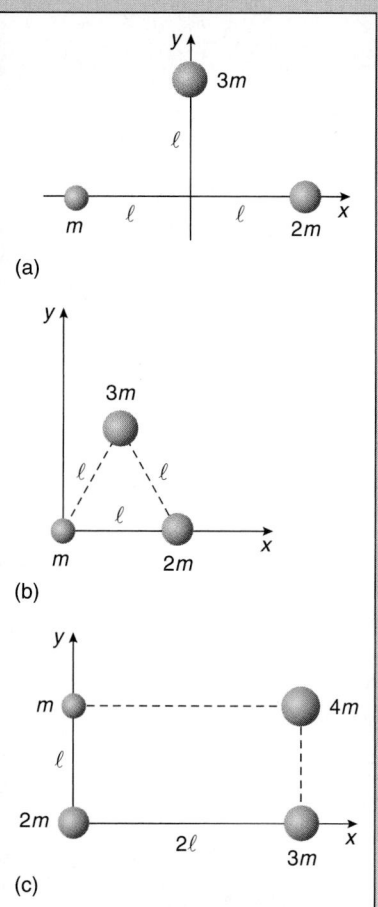

FIGURE P.56

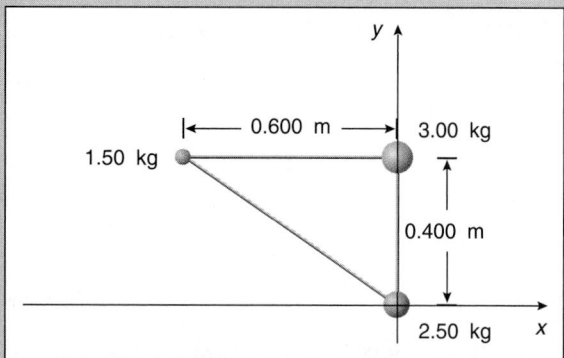

FIGURE P.57

Section 9.9 The Center of Mass

55. A 3.0 kg sack of sugar is located at coordinates $x = 3.0$ m, $y = -4.0$ m. A 4.0 kg sack of industrial grade coffee is located at $x = -2.0$ m, $y = 2.2$ m. Determine the position vector of the center of mass of the two-particle system.

56. For amusement (?), find the location of the center of mass of the collections of particles shown in Figure P.56.

• **57.** Three particles are held in a rigid triangle by thin rods of negligible mass, as shown in Figure P.57. Find the coordinates of the center of mass of the system of three particles.

• **58.** The doohickey shown in Figure P.58 is constructed from a uniform plate of steel and is to be used in a sophisticated ornamental doodad. Find the center of mass of the doohickey in *two different ways* to be sure that your calculations are correct.

• **59.** Find the center of mass of a system of identical particles distributed at the vertices of a regular tetrahedron with sides of length ℓ.

• **60.** A uniform circular plate of material of radius R has an off-center hole cut out as indicated in Figure P.60. The total mass of the plate with the hole is m. The radius of the hole is $R/2$. Find the location of the center of mass of the plate.

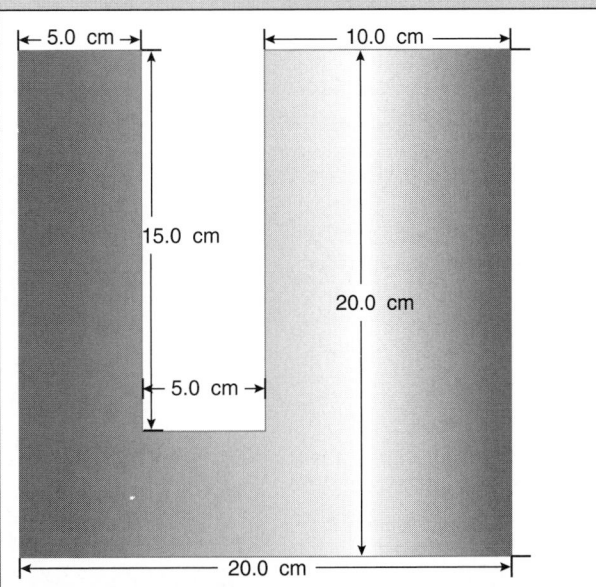

FIGURE P.58

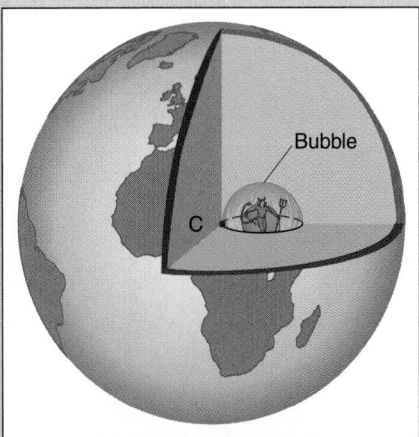

FIGURE P.61

•63. A collection of six uniform planks each of mass m and length ℓ are stacked according to the scheme indicated in Figure P.63. (a) Show that the center of mass of the *top five planks* is located directly over the right edge of the bottom plank. The system will not topple over, even though the top plank is lying totally beyond the edge of the table! (Note that only *five* planks really are needed to accomplish this feat.) You might like to try this stacking experimentally with, say, six meter sticks. It is a good idea to keep them separated by slightly less than the indicated offset distances as a safety precaution in building the stack. (b) The stable stacking can be continued with more planks. If a seventh plank were added at the *bottom* of the stack, what would be its offset relative to the plank directly above it? The general rule governing the offsets is that the right edge of the bottom plank is directly under the center of mass of the collection of planks on top of it.

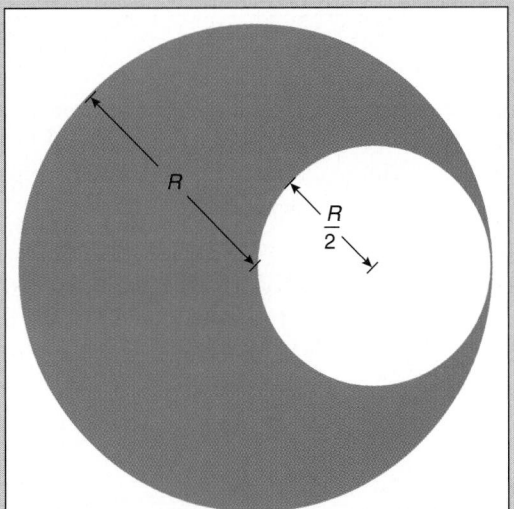

FIGURE P.60

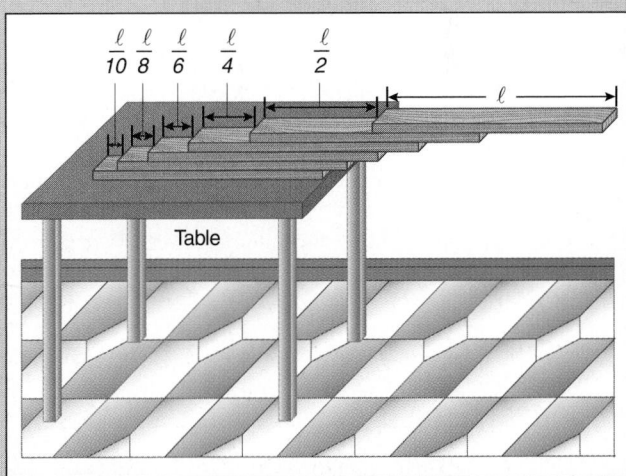

FIGURE P.63

Section 9.10 Dynamics of a System of Particles

•64. An 80.0 kg ΦΒΚ student stands inside a 20.0 kg canoe that can slide freely on a frictionless water surface; see Figure P.64. Both are initially at rest. The student then runs like crazy at a speed of 3.0 m/s with respect to the canoe along its length (ignore the thwarts of the canoe). (a) Describe and explain what

•61. For millennia Hell was thought to be located inside the Earth. (Many students now think it is located at their university during finals.) Suppose we model the quaint Hell-inside-the-Earth cosmology with a uniform sphere of radius R that has an off-center, hellish, spherical bubble of radius $R/4$ as indicated in Figure P.61. The Devil orders you to find the position vector of the center of mass of the system using the center of the large sphere C as the origin of your coordinate axis; choose the x-axis to be directed from the origin to the center of the bubble representing Hell.

•62. An open-topped, thin-walled, cubical box with sides of length ℓ is constructed from sheet metal. Determine the location of the center of mass of the box.

happens to the center of mass of the student–canoe system. (b) How fast is the student moving relative to the water? How fast is the canoe moving relative to the water?

FIGURE P.64

•65. A father (of mass 70 kg) faces his young daughter (of mass 20 kg) on a level roller rink. Dad pushes his daughter and she rolls backward at a speed of 2.5 m/s. Determine the speed of her Dad and the direction of his motion relative to hers.

FIGURE P.65

•66. A 2.00 kg mass and a 3.00 kg mass are used to compress opposite ends of an ideal spring with spring constant 1.50×10^3 N/m on a frictionless table; see Figure P.66. The spring is compressed 40.0 cm from its unstretched length and released with both masses initially at rest. Define the system to be the two masses and the ideal spring. (a) Explain why the momentum of the system is conserved. (b) Explain why the total mechanical energy of the system is conserved. (c) Determine the speeds of the masses as they leave the spring.

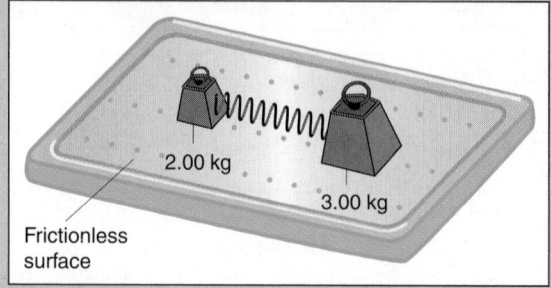

FIGURE P.66

Section 9.11 Kinetic Energy of a System of Particles

•67. A 5.00 kg dingbat is approaching a stationary 3.00 kg Cheshire cat at a speed of 10.0 m/s as shown in Figure P.67. (a) What is the total kinetic energy of the system of two particles? (b) What is the velocity of the center of mass? (c) What is the kinetic energy of the center of mass? (d) Find the velocities of the two particles with respect to the center of mass. (e) Calculate the kinetic energy of the two particles with respect to the center of mass point. (f) Use your calculations to check Equation 9.35.

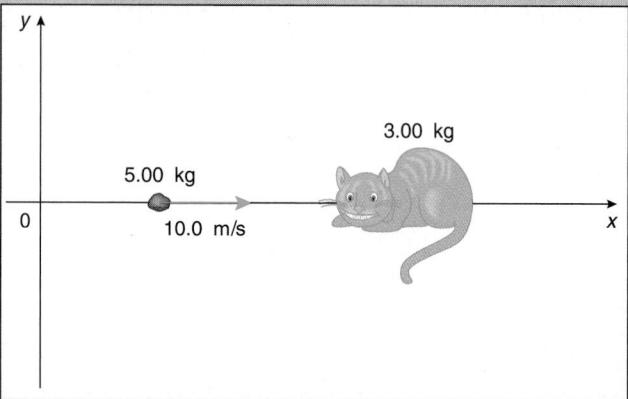

FIGURE P.67

⁞68. Two barred owls (*Syrnium varium*) are dive-bombing a small opossum (*Didelphis virginiana*) at rest, as shown in Figure P.68. (a) What is the total kinetic energy of the system of two owls? (b) What is the velocity of the center of mass of the owls? (c) What is the kinetic energy of the center of mass of the owls? (d) Find the velocities of the two owls with respect to their center of mass. (e) Calculate the kinetic energy of the two owls with respect to their center of mass. (f) Use your calculations to check Equation 9.35.

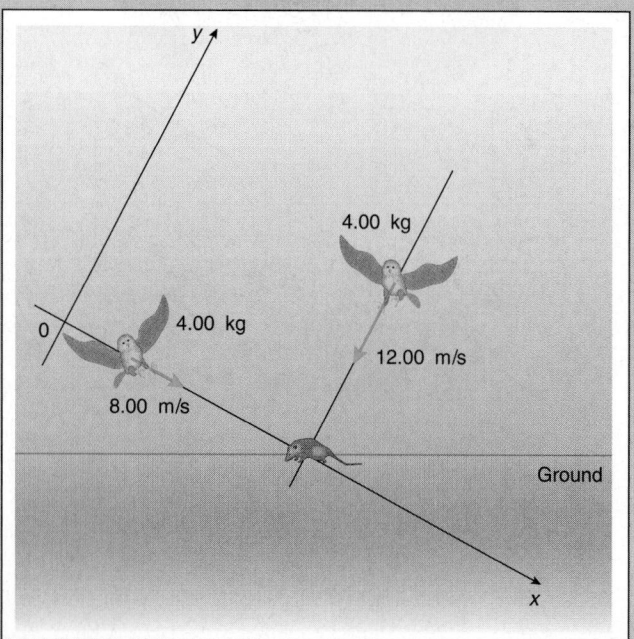

FIGURE P.68

Sections 9.12 The Velocity of the Center of Mass for Collisions*
9.13 The Center of Mass Reference Frame*

•69. A loaded coal car with a mass of 120 metric tons traveling at 2.0 m/s collides and couples with an empty, stationary coal car with a mass of 10 metric tons. (a) Determine the final speed of the two cars. (b) What is the change in the kinetic energy of the system of two cars in the collision? (c) What is the speed of the center of mass of the system of two cars throughout the process? (1 metric ton = 10^3 kg)

•70. A 20 kg projectile is launched over level ground near the surface of the Earth with a speed of 50 m/s and at an angle of 60° with the horizontal. At the top of its trajectory, the projectile suddenly breaks into two pieces, each of mass 10 kg. One of the 10 kg pieces is instantaneously at rest after the explosion (before falling). (a) Introduce an appropriate coordinate system and find the velocity and speed of the remaining 10 kg piece. (b) Sketch the trajectory of the center of mass point from launch to the instant of the breakup, and from the instant of the breakup to the instant when the center of mass point strikes the ground. Place appropriate horizontal distances on your sketch. (c) How far apart are the two 10 kg pieces when they hit the ground?

•71. A mass m is at rest. It disintegrates into *two* particles of mass m_1 and m_2 such that $m = m_1 + m_2$. (a) Is kinetic energy conserved in this process? (b) Show that the two particles m_1 and m_2 *must* move off in *opposite* directions in order to conserve momentum. This problem has implications in radioactivity and other types of particle decay. If the initial particle is at rest and decays into two particles, the two decay particles move off in opposite directions. In β-decay of a radioactive nucleus at rest, the recoiling (so-called daughter) nucleus and the emitted electron do *not* move off in opposite directions. This is a strong clue that β-decay produces not two particles but three (or perhaps more). The third particle is called a neutrino. Neutrinos finally were detected in the 1950s. Neutrinos are very difficult to detect because they interact extraordinarily weakly with matter.

INVESTIGATIVE PROJECTS

A. Expanded Horizons

1. Investigate methods used to reduce injuries from automobile accidents. Include aspects of automotive design, the use of seat belts, air bags, energy absorbing bumpers, collapsible front ends and steering columns, and so forth.
   M. A. Paulo, "An occupant restraint primer," *Forensic Accident Investigation: Motor Vehicles*, edited by Thomas L. Bohan and Arthur C. Damask (Michie Butterworth, Charlottesville, Virginia, 1995).
   James S. Baker and Lynn B. Fricke, *Traffic Accident Investigation Manual*, volume 1 (Northwestern University Traffic Institute, Evanston, Illinois, 1986).
   Lynn B. Fricke, *Traffic Accident Reconstruction*; volume 2 of the *Traffic Accident Investigation Manual* (Northwestern University Traffic Institute, Evanston, Illinois, 1990).

2. A mass m makes an elastic collision with a rigid wall. Will the collision also be elastic if the wall is moving at speed v_{wall} along the direction of its normal line? Will the angle the ball makes with the normal line as it rebounds be the same, less than, or greater than the angle at which it approached the moving wall?

B. Lab and Field Work

3. Design and perform an experiment to estimate the order of magnitude of the impulse delivered to a nail when hit by a hammer as it is driven into a 2 × 4 or other board.

4. Design and perform an experiment to estimate the magnitude of the impulse and average force used by a karate or tae kwon do practitioner when they shatter a board with their hand or foot.
   George A. Amann and Floyd T. Holt, "Karate demonstration," *The Physics Teacher*, 23, #1, page 40 (January 1985).
   S. R. Wilk, R. E. McNair, and M. S. Feld, "The physics of karate," *American Journal of Physics*, 51, #9, pages 783–790 (September 1983).
   Jearl D. Walker, "Karate strikes," *American Journal of Physics*, 43, #10, pages 845–849 (October 1975).

5. Your physics department may have an airtable and force transducers that electronically measure the magnitude of forces.
   Use this equipment to demonstrate that in a two-particle collision between two different masses, the magnitude of the force that each exerts on the other during the collision is the same. Compare this magnitude of force with that of the weight of each mass.

6. If a ball in a game of billiards hits the edge of the table at an angle θ with respect to a horizontal normal line to the edge, design and perform an experiment to see if the ball rebounds at the same angle with the normal line.
   R. D. Edge, "The errant pool balls," *The Physics Teacher*, 20, #1, pages 50–52 (January 1982).

7. Design a system that can stop a raw egg without breaking it when the egg is dropped from a height of about 5 m. Estimate the impulse and total force on the egg during the portion of its motion when its velocity is decreasing with time.

8. See how many meter sticks you can stack according to the scheme in Problem 63 to determine how far out from the table the top meter stick can project from the table.

9. Physicists occasionally are called as expert witnesses in accident investigations and other aspects of forensic science. Ask the head of your physics department if any local faculty members have been involved in such activities; you also might acquire names of local physicists from neighboring legal firms. Write a report about an actual investigation.
   Arthur L. Robinson, "Mastering the physics of disaster," *Physics Today*, 47, #12, pages 47–48 (December 1994).
   P. K. Tao, *Experiments, Devices, and Techniques: The Physics of Traffic Accident Investigation* (Oxford University Press, Hong Kong, 1987).
   Tim Folger, "Road scholar," *Discover*, 12, #8, pages 62–65 (August 1991).
   Vincent Lytle, "After the crash," *Discover*, 17, #5, pages 98–101 (May 1996).
   See the last two references for Project 1 as well.

10. A favorite recreational pastime at physics picnics is a water balloon toss. Two people toss a water-filled balloon back and

forth, gradually moving apart until one of them either misses or the water balloon breaks while being caught. (a) Secure a balloon from an appropriate source, fill it partially with water, and tie it shut. Determine its mass. By placing various weights on the filled balloon on a suitably protected flat surface, estimate the magnitude of the force needed to break the balloon. (b) Making reasonable assumptions, estimate the maximum separation between the balloon tossers when the water balloon breaks while being caught.

C. Communicating Physics

11. Interview one or more members of the physics department at your college or university to learn more about force transmission via the exchange of mediating particles. You likely will want to seek out theoretical physicists. See if they can explain to you how an attractive force might be transmitted by such particle exchange. Write a report about your findings appropriate for the audience of your campus newspaper. Perhaps you can become their science reporter!

12. Suppose an asteroid with a diameter of 1.0 km and an average density of 5.0×10^3 kg/m^3 is detected approaching the Earth at a relative approach speed of 30 km/s. The estimated impact time is only one week away. As an intern for the scientific advisor to the President of the United States, you are asked to evaluate whether firing one, several, or even the entire nuclear arsenal of the United States (~10,000 warheads with an average energy yield of 100 metric kilotons of TNT) might be effective in deflecting the asteroid sufficiently for it to miss the Earth. Organize a technical answer to the question, stating clearly any assumptions you make in coming to a conclusion whether the scheme is hopeless. The energy yield of TNT is about 4.9×10^9 J per metric ton (see Example 8.13).

ELASTIC COLLISION

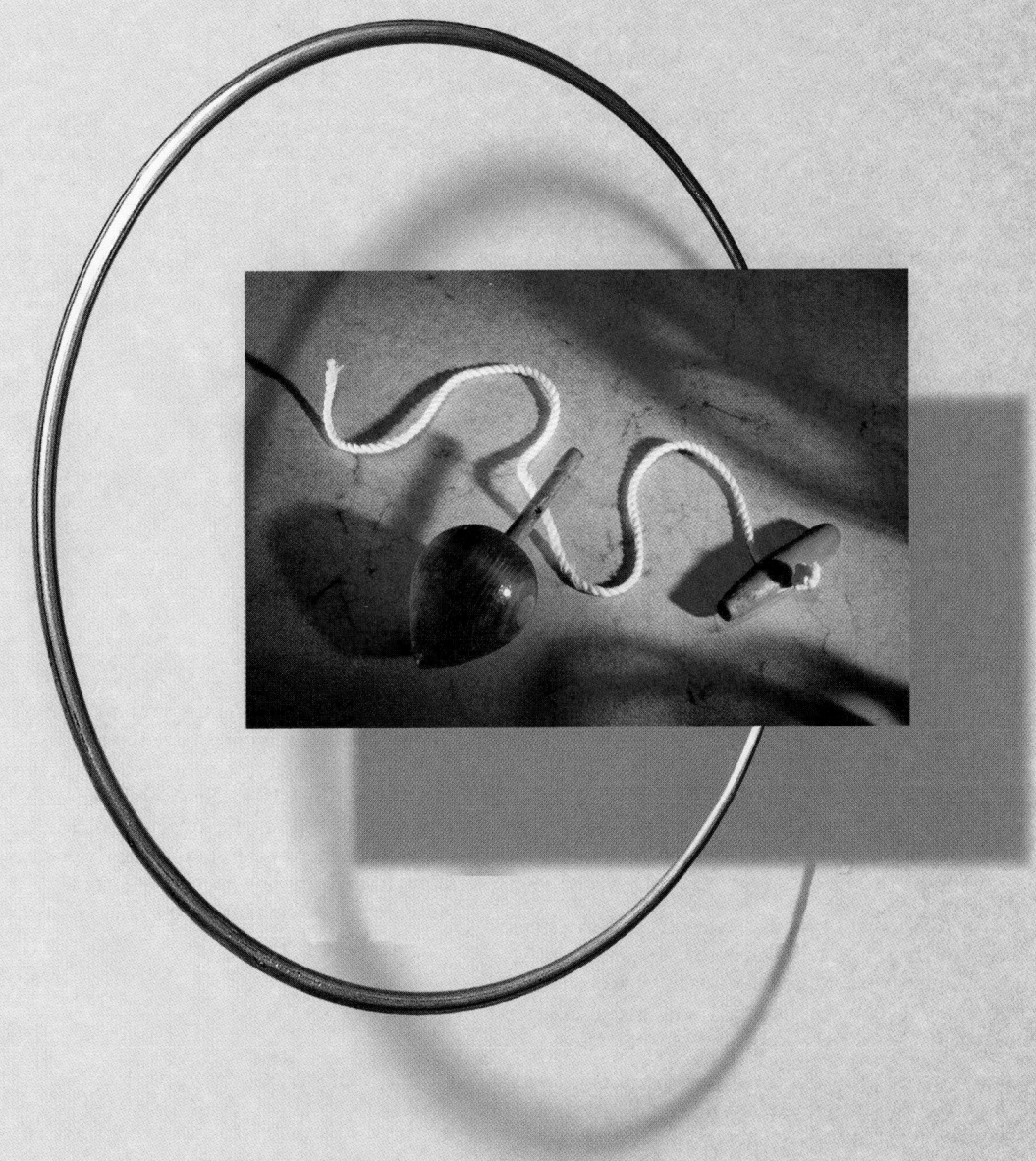

CHAPTER 10

SPIN AND ORBITAL MOTION

Surely the devil of rotations is afoot.
James Watt (1736–1819) *

Rotational motion is apparent everywhere. Wheels and gears turn. Ice skaters and ballet dancers impress us with their pirouettes. Gymnasts and hot-dog skiers perform amazing flips in flight. Helicopter rotors, lawn mower and automobile engines, turbines, and jet engines all exhibit rotational motion. The planets spin while simultaneously orbiting the Sun. Electrons in atoms also "spin" and "orbit" the nucleus.

In Chapter 4 we studied the *kinematics* of circular rotational motion and introduced the angular velocity and angular acceleration vectors. In this chapter we study the *dynamics* of spin and orbital motion; we will see that it has many parallels with rectilinear dynamics. Kinematic similarities were apparent in Chapter 4, where the description of circular motion was seen to be a kind of generalized one-dimensional motion involving angular coordinates rather than rectilinear coordinates. The kinematic equations used to describe rotation were similar to those used to describe motion in one dimension. In this chapter, we will see further parallels and find a relationship that is the rotational analog of Newton's second law of motion.

We begin by examining the orbital motion of a single pointlike particle. We use this as a paradigm for a more complex spinning system of many particles: a symmetric rigid body. Finally, we look into simultaneous spin and orbital motion, along the way investigating several special examples of rotational motion. These include the precession of spinning tops and the Earth, synchronous rotation, rolling motion, and wheels. We conclude by noting two conditions necessary for the equilibrium of rigid bodies.

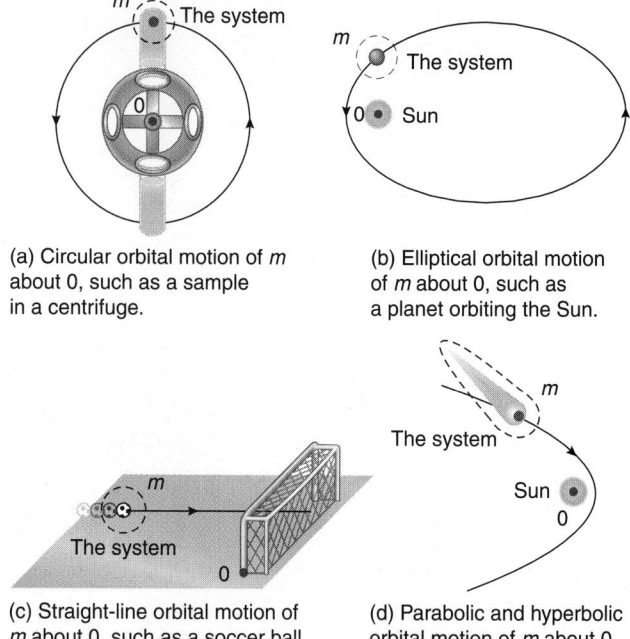

(a) Circular orbital motion of *m* about 0, such as a sample in a centrifuge.

(b) Elliptical orbital motion of *m* about 0, such as a planet orbiting the Sun.

(c) Straight-line orbital motion of *m* about 0, such as a soccer ball passing a goalpost.

(d) Parabolic and hyperbolic orbital motion of *m* about 0, such as a nonperiodic comet about the Sun.

FIGURE 10.1   Orbital motion of *m* about the point 0.

10.1   THE DISTINCTION BETWEEN SPIN AND ORBITAL MOTION

> We use the word **spin** to describe rotational motion of a system about an axis through its center of mass.

Thus we say the Earth spins on its axis, or a CD spins in a stereo system. A **rigid body** is a system composed of many pointlike particles that maintain fixed distances from each other at all times. Each pointlike particle of a spinning rigid body system executes circular motion about an axis through the center of mass.

> If the center of mass of the system *also* is moving in space from the perspective of a particular reference frame, we say the center of mass is executing **orbital** motion with respect to the origin of the coordinate system.

The motion of the center of mass need not be circular to be called orbital motion, as shown in Figure 10.1. Orbital motion describes *motion of the center of mass*, even though its path need *not* be a closed path in many cases.

A traditional example of orbital motion is the annual motion of the Earth about the Sun; this orbit is elliptical in shape. In many cases we can consider the orbit of the Earth to be

circular because the eccentricity of the orbit is small ($\varepsilon = 0.016$). The Earth also spins once a day about an axis through its center of mass. The actual spin period with respect to the distant stars is 23 h 56 min 4.10 s. Thus the Earth exhibits both spin and orbital motion, as shown in Figure 10.2.

For the Earth, the spin angular velocity $\vec{\boldsymbol{\omega}}_{\text{spin}}$ is quite different from the orbital angular velocity $\vec{\boldsymbol{\omega}}_{\text{orbit}}$ in both magnitude and direction. The Earth spins once in about 24 h, and so its angular speed of spin is

$$\omega_{\text{spin}} = \frac{2\pi \text{ rad}}{(24 \text{ h})(3600 \text{ s/h})}$$
$$= 7.3 \times 10^{-5} \text{ rad/s}$$

It takes the Earth 365.25 days to complete one orbit of the Sun, and so its average orbital angular speed is

$$\omega_{\text{orbit}} = \frac{2\pi \text{ rad}}{(365.25 \text{ d})(86\,400 \text{ s/d})}$$
$$= 1.9910 \times 10^{-7} \text{ rad/s}$$

The plane of the equator of the Earth is inclined to the plane of its orbit about the Sun by about 23.5°; hence, the two angular velocity vectors also have an angle between them of 23.5°.

Rolling wheels also exhibit both spin and orbital motion. Skidding, locked wheels exhibit no spin, but do have orbital motion with respect to the ground because the center of mass of each wheel is moving.

In this chapter, we begin our study of rotational motion by examining the orbital motion of a single pointlike particle; later,

---

*(Chapter Opener) Quoted in Henry W. Dickinson and Rhys Jenkins, *James Watt and the Steam Engine* (Oxford at the Clarendon Press, Oxford, England, 1927), page 159.

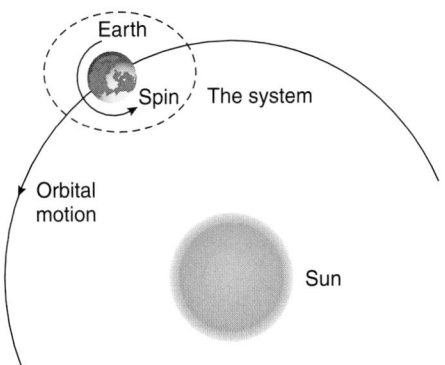

FIGURE 10.2 The Earth spins on an axis through its center of mass and simultaneously orbits the Sun.

we shall see that the single particle also may represent the center of mass point of a more complex system. We then examine the spinning motion of a system of particles about an axis coincident with a symmetry axis through the center of mass, when the center of mass itself is at rest (so there is no orbital motion). Finally, we examine what happens when a system undergoes simultaneous spin and orbital motion.

QUESTION 1

Consider two stars of equal mass in circular orbit about their center of mass point as in Figure Q.1. Consider the stars to be particles and ignore any spin associated with each. (a) If you consider the system to be *one* of the stars, is the system undergoing orbital or spin motion? (b) If you consider the system to be *both* stars, is the system undergoing orbital or spin motion?

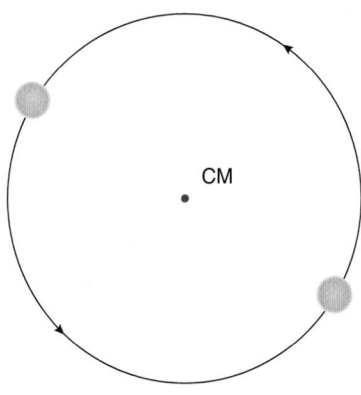

FIGURE Q.1

10.2 THE ORBITAL ANGULAR MOMENTUM OF A PARTICLE

One of the principle parameters associated with rotational motion (in all its hideous variations) is **angular momentum**. We briefly encountered the orbital angular momentum of a particle in Section 6.10 where we found the orbital angular momentum of each planet was constant, a manifestation of Kepler's empirical second law of

planetary motion.* In this section we develop the concept of angular momentum more leisurely; in the next few sections we discover why it is so important for spin and orbital dynamics.

Recall from Chapter 9 that a pointlike particle of mass m moving with velocity $\vec{v}$ has a (classical) momentum $\vec{p}$ given by (Equation 9.1)

$$\vec{p} \equiv m\vec{v}$$

The velocity of the particle is measured with respect to a reference frame with an origin 0 at some convenient point of your choosing, as in Figure 10.3.

The momentum of a particle is important because, as we saw in the last chapter, the most general statement of Newton's second law of motion relates the total force on the particle to the time rate of change of its momentum; angular momentum has a similar importance.

Whatever the path or trajectory of the particle, be it a straight line, a curved path, or a closed orbital path, the **orbital angular momentum $\vec{L}$** of the particle at any position with respect to a reference point (such as the origin of your coordinate choice) is

$$\boxed{\vec{L} \equiv \vec{r} \times \vec{p}} \qquad (10.1)$$

Since the position vector $\vec{r}$ is measured with respect to the origin at 0, we say the orbital angular momentum of the particle is taken *about* the point 0. The dimensions of the orbital angular momentum are those of distance times those of the momentum; using SI units, m(kg·m/s) = kg·m²/s.

We shall see in this chapter that the total angular momentum of a system is closely connected to a rotational analog of Newton's second law of motion. Bear with us (i.e., hang in there) as we develop this unifying and useful idea.

QUESTION 2

A certain particle has nonzero momentum $\vec{p}$ and nonzero orbital angular momentum $\vec{L}$. The angle between $\vec{r}$ and $\vec{p}$ is neither 0° nor 180°. What is the angle between the vectors $\vec{p}$ and $\vec{L}$?

---

*A line from the Sun to a given planet sweeps out equal areas during equal times. [Section 6.10 was optional; if you skipped it, there is *no* need to go back and read it now (unless you want to!).]

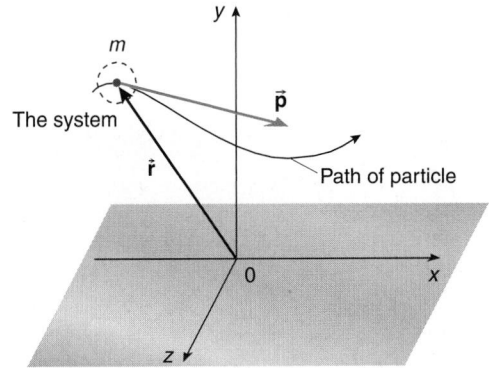

FIGURE 10.3 The momentum of a particle and its position vector with respect to the origin at 0.

EXAMPLE 10.1

A small 8.00 kg Australian wild dog called a dingo (*Canis dingo*) is running with a velocity $(-6.0 \text{ m/s})\hat{\jmath}$ along the y-axis as indicated in Figure 10.4. Find the momentum and the orbital angular momentum of the dingo with respect to a terrified rabbit (*Lepus timidus*) located at the origin in Figure 10.4.

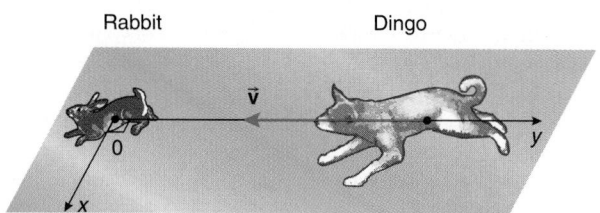

Rabbit   Dingo

FIGURE 10.4

Solution

The dingo is the system. The momentum of the dingo is

$$\vec{p} = m\vec{v}$$
$$= (8.00 \text{ kg})(-6.0 \text{ m/s})\hat{\jmath}$$
$$= (-48 \text{ kg·m/s})\hat{\jmath}$$

The orbital angular momentum of the dingo is found from Equation 10.1:

$$\vec{L} = \vec{r} \times \vec{p}$$

The position vector of the dingo when it is at coordinate y on the y-axis is

$$\vec{r} = y\hat{\jmath}$$

The orbital angular momentum of the dingo with respect to the rabbit at the origin thus is

$$\vec{L} = y\hat{\jmath} \times (-48 \text{ kg·m/s})\hat{\jmath}$$
$$= 0 \text{ kg·m}^2/\text{s}$$

regardless of the y-coordinate of the dingo.

EXAMPLE 10.2

A 2.00 kg partridge (*Perdix perdix*) is flying at a constant velocity $\vec{v} = (5.00 \text{ m/s})\hat{\imath}$ along a straight line past a pear tree at the origin,

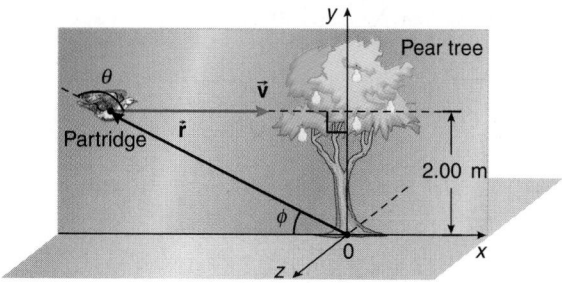

FIGURE 10.5

as shown in Figure 10.5. Find the orbital angular momentum of the partridge with respect to the base of the pear tree as an origin.

Solution

Method 1

The momentum $\vec{p}$ of the partridge is given by Equation 9.1:

$$\vec{p} = m\vec{v}$$
$$= mv\hat{\imath}$$

When the partridge is a distance r from the pear tree, the position vector of the bird is

$$\vec{r} = (-r \cos \phi)\hat{\imath} + (r \sin \phi)\hat{\jmath}$$

Hence the orbital angular momentum is

$$\vec{L} = \vec{r} \times \vec{p}$$
$$= [(-r \cos \phi)\hat{\imath} + (r \sin \phi)\hat{\jmath}] \times mv\hat{\imath}$$
$$= -(r \sin \phi)mv\hat{k}$$

Geometrically, from Figure 10.5, $r \sin \phi = 2.00$ m; this is the perpendicular distance between the line of motion and the origin. Hence the orbital angular momentum is

$$\vec{L} = -(2.00 \text{ m})(2.00 \text{ kg})(5.00 \text{ m/s})\hat{k}$$
$$= (-20.0 \text{ kg·m}^2/\text{s})\hat{k}$$

Notice (perhaps with some surprise!) that the orbital angular momentum here is *constant* and independent of the location of the partridge along the straight-line path.

Method 2

You also could determine the orbital angular momentum by first finding its magnitude and then its direction. The magnitude of the orbital angular momentum is

$$L = |\vec{r} \times \vec{p}|$$
$$= rp \sin \theta$$

where θ is the angle between the vectors $\vec{r}$ and $\vec{p}$ (with their tails at a common point). Geometrically, from Figure 10.5, $\theta = \pi - \phi$, where we measure the angles in radians, so that

$$L = rp \sin(\pi - \phi)$$
$$= rp(\sin \pi \cos \phi - \cos \pi \sin \phi)$$
$$= rp \sin \phi$$
$$= (r \sin \phi)(mv)$$

From the geometry of Figure 10.5, $r \sin \phi = 2.00$ m. So the orbital angular momentum has a magnitude

$$L = (2.00 \text{ m})(2.00 \text{ kg})(5.00 \text{ m/s})$$
$$= 20.0 \text{ kg·m}^2/\text{s}$$

The vector product right-hand rule indicates that $\vec{L}$ must be directed as shown in Figure 10.6: parallel to $-\hat{k}$. Hence

$$\vec{L} = (-20.0 \text{ kg·m}^2/\text{s})\hat{k}$$

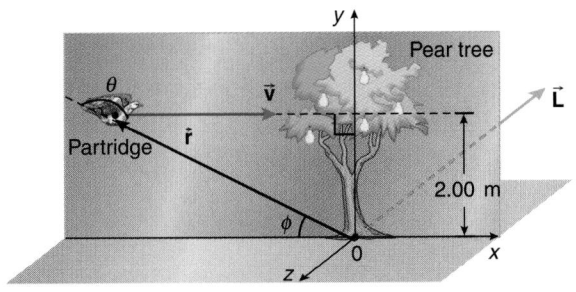

FIGURE 10.6

10.3 THE CIRCULAR ORBITAL MOTION OF A SINGLE PARTICLE

Here we consider both uniform and nonuniform circular motion of a single particle, since this will form the basis for the spinning rotational motion of more complex, many-particle systems such as rigid bodies. The *circular* orbital motion of a planet about the Sun is an example of circular motion that is uniform (i.e., with constant angular speed). The circular orbital motion of a sample in a centrifuge is another example of circular orbital motion, although *not* necessarily at constant angular speed.

Let r be the radius of the circular motion of the single particle, as shown in Figure 10.7. Choose the center of the circle as the origin. The direction of the angular velocity vector is found from the circular motion right-hand rule given in Chapter 4.

The orbital angular momentum of the particle is, from Equation 10.1,

$$\vec{L} = \vec{r} \times \vec{p}$$

where $\vec{r}$ is the position vector of the particle and $\vec{p}$ is its momentum. We say the particle has angular momentum $\vec{L}$ about the point chosen as origin, here the center of the circle. Thanks to the vector product right-hand rule, the orbital angular momentum vector of the particle is in the same direction as the orbital angular velocity vector of the particle. Both $\vec{L}$ and $\vec{\omega}$ are perpendicular to the plane of the circular motion, as shown in Figure 10.7.

Since the orbital angular momentum and orbital angular velocity vectors are parallel, perhaps they are related to each other in some other way. Substitute for the momentum of the particle into the orbital angular momentum:

$$\vec{L} = \vec{r} \times m\vec{v}$$
$$= m\vec{r} \times \vec{v}$$

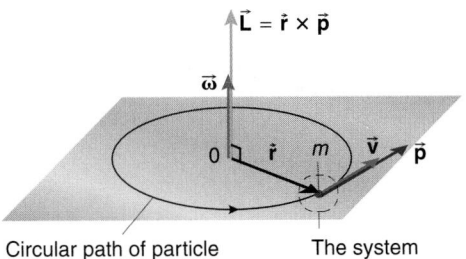

FIGURE 10.7 Circular motion of a particle and the vectors associated with it.

For circular motion $\vec{r}$ is perpendicular to $\vec{v}$; hence, the magnitude of the angular momentum is

$$L = mrv \sin 90°$$
$$= mvr \qquad (10.2)$$

The velocity $\vec{v}$ and angular velocity $\vec{\omega}$ in circular motion are related according to Equation 4.34:

$$\vec{v} = \vec{\omega} \times \vec{r}$$

For circular motion $\vec{\omega}$ is perpendicular to $\vec{r}$, and so the speed is

$$v = \omega r \sin 90°$$
$$= \omega r \qquad (10.3)$$

We use Equation 10.3 to substitute for v in Equation 10.2:

$$L = m(\omega r)r$$
$$= mr^2 \omega$$

But we know $\vec{L}$ is parallel to $\vec{\omega}$, so that we can write

$$\vec{L} = mr^2 \vec{\omega} \qquad (10.4)$$

The orbital angular momentum of the particle is proportional to the orbital angular velocity; the proportionality constant is called the **moment of inertia** I of the particle.* The moment of inertia of the particle about the axis through the center of the circle is, therefore,

$$I \equiv mr^2 \quad \text{(orbital motion of a particle only)} \quad (10.5)$$

The SI units of the moment of inertia are kg·m².

Notice that the moment of inertia of the particle depends on *both* its mass and its location with respect to the axis of rotation. The moment of inertia does *not* depend on the angular velocity vector.

Equation 10.4 for the orbital angular momentum of the particle about the origin can be rewritten using the moment of inertia as

$$\vec{L} = I\vec{\omega} \qquad (10.6)$$

For *uniform* circular motion, the angular velocity is constant; thus, according to Equation 10.6, the angular momentum is constant as well. For *nonuniform* circular motion, where there exists a nonzero angular acceleration $\vec{\alpha}$, the angular velocity vector is not constant; therefore, the angular momentum is not constant either.

We want to see what causes angular accelerations and, consequently, changes in the angular momentum. Forces cause changes in the momentum via Newton's second law of motion. Forces also are involved in causing changes in the angular momentum, but in a more subtle way.

If the total force on the particle is directed toward the center of the circle, the angular acceleration $\vec{\alpha}$ is zero. This is uniform

---

*Do not confuse the (scalar) moment of inertia I with the magnitude of the vector impulse $\vec{I}$ of a force, discussed in Chapter 9.

circular motion with a purely centripetal acceleration. Note that the vector product of $\vec{r}$ and $\vec{F}_{total}$ is zero in this case, because the two vectors are antiparallel (see Figure 10.8).

We saw in Chapter 4 that nonzero angular acceleration means the particle experiences a *total acceleration* that is *not* directed toward the center of the circle; the total acceleration then has both centripetal and tangential components. Thus, from Newton's second law of motion, the total force is *not* directed toward the center of the circle for nonuniform circular motion, as shown in Figure 10.9. The particle in circular motion can either speed up or slow down in its circular motion if the total force on the particle is not directed toward the center of the circle. This condition also implies that the vector product of $\vec{r}$ with $\vec{F}_{total}$ is *not* zero.

Hence the vector product of $\vec{r}$ and $\vec{F}_{total}$ is an indicator of whether $\vec{F}_{total}$ produces an angular acceleration or not. To account for the effect of a force in causing angular accelerations, we introduce a new physical concept: the **torque** of a force.* One can calculate the torque of an individual force or of the total force.

Whatever the force $\vec{F}$, its torque is defined to be

$$\vec{\tau} \equiv \vec{r} \times \vec{F} \qquad (10.7)$$

where $\vec{r}$ is the *position vector of the point of application of the force with respect to the chosen origin* for the coordinate system (see Figure 10.10). If the motion is circular, the origin is always taken to be the center of the circle.

The SI units of torque are N·m. While N·m are equivalent to joules (J), we *always* quote the SI units of torque as N·m, reserving joules specifically for work or energy (scalar quantities).

---

*The word comes from the Latin *torquis*, a twisted neck chain worn by ancient Britons, Gauls, and Germans.

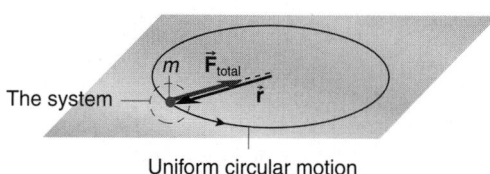

FIGURE 10.8 In uniform circular motion, the position vector of a particle is antiparallel to the total force on it.

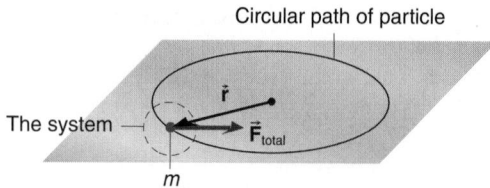

FIGURE 10.9 In nonuniform circular motion, the position vector of a particle and the total force on it are not antiparallel.

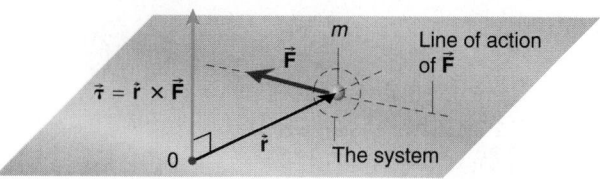

FIGURE 10.10 The torque $\vec{\tau}$ is perpendicular to both the position vector $\vec{r}$ and the force $\vec{F}$.

The torque is a vector quantity and, because of its definition as a vector product, its direction always is perpendicular to both $\vec{r}$ and $\vec{F}$; see Figure 10.10. The line along which the force lies is called the **line of action** of the force.

The torque is a measure of how effective the force is in causing an angular acceleration of the system about the point 0; a nonzero torque means there will be a nonzero angular acceleration *if the force $\vec{F}$ acts alone.*

The torque is zero if $\vec{r}$ and $\vec{F}$ are either parallel or antiparallel, as in Figure 10.11. In such cases the force $\vec{F}$ has no ability to cause an angular acceleration of the system about the point 0.

PROBLEM-SOLVING TACTIC

10.1 If the line of action of a force goes through a point 0, the force has zero torque about that point. The torque is a maximum when the two vectors $\vec{r}$ and $\vec{F}$ are perpendicular to each other, as in Figure 10.12. In this case the force $\vec{F}$ is most effective in causing angular acceleration of the system about 0.

The magnitude of the torque is

$$\tau = rF \sin \theta$$

where θ is the angle between the vectors $\vec{r}$ and $\vec{F}$ (when their tails are at a common point). The quantity $r \sin \theta$ is the perpendicular distance from the point 0 to the line of action of $\vec{F}$. This

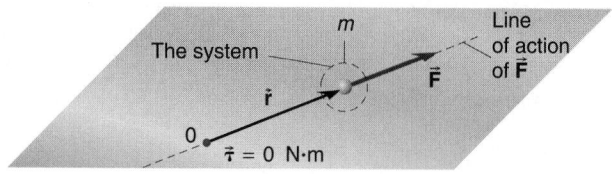

FIGURE 10.11 A force that has zero torque about the point 0.

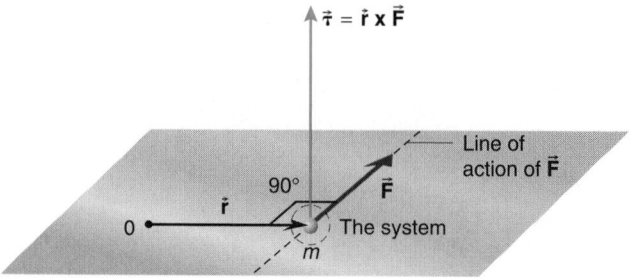

FIGURE 10.12 The torque of $\vec{F}$ about 0 is a maximum when $\vec{F}$ is perpendicular to $\vec{r}$.

perpendicular distance is called the **moment arm** of the force, as shown in Figure 10.13.*

We shall see that the torque of the total force on the particle is closely related to changes in the angular momentum of the particle, in much the way the total force is related to changes in the momentum of the particle via Newton's second law of motion. To show this we first must find the torque of the total force $\vec{\mathbf{F}}_{\text{total}}$ on the particle. Calculate the torque of the total force about the origin of your coordinate system:

$$\vec{\boldsymbol{\tau}}_{\text{total}} = \vec{\mathbf{r}} \times \vec{\mathbf{F}}_{\text{total}}$$

Since we say the total force acts on the particle, we also say that the total torque acts on the particle. Substitute

$$\frac{d\vec{\mathbf{p}}}{dt}$$

for the total force, according to Newton's second law of motion:

$$\vec{\boldsymbol{\tau}}_{\text{total}} = \vec{\mathbf{r}} \times \frac{d\vec{\mathbf{p}}}{dt} \quad (10.9)$$

As an aside, note that

$$\frac{d}{dt}(\vec{\mathbf{r}} \times \vec{\mathbf{p}}) = \frac{d\vec{\mathbf{r}}}{dt} \times \vec{\mathbf{p}} + \vec{\mathbf{r}} \times \frac{d\vec{\mathbf{p}}}{dt} \quad (10.10)$$

The first term on the right-hand side of Equation 10.10 is zero because the vector product of any vector with itself is zero:

$$\frac{d\vec{\mathbf{r}}}{dt} \times \vec{\mathbf{p}} = \vec{\mathbf{v}} \times m\vec{\mathbf{v}}$$
$$= m\vec{\mathbf{v}} \times \vec{\mathbf{v}}$$
$$= 0 \ \text{kg} \cdot \text{m}^2/\text{s}^2$$

Hence Equation 10.10 becomes

$$\frac{d}{dt}(\vec{\mathbf{r}} \times \vec{\mathbf{p}}) = \vec{\mathbf{r}} \times \frac{d\vec{\mathbf{p}}}{dt} \quad (10.11)$$

We use Equation 10.11 in Equation 10.9 for the total torque and obtain

$$\vec{\boldsymbol{\tau}}_{\text{total}} = \frac{d}{dt}(\vec{\mathbf{r}} \times \vec{\mathbf{p}})$$

*The term *lever arm* also is used synonymously with the term moment arm.

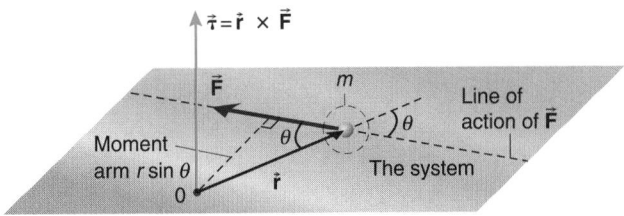

FIGURE 10.13 The moment arm of a force.

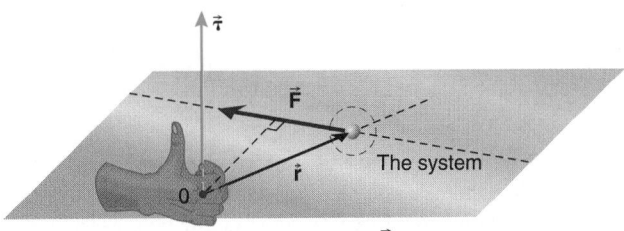

FIGURE 10.14 If you place your right-hand thumb along the direction of the torque, the right-hand fingers will indicate the sense that the force will tend to accelerate (or rotate) the system about 0.

But $\vec{\mathbf{r}} \times \vec{\mathbf{p}}$ is the orbital angular momentum $\vec{\mathbf{L}}$ of the particle about the origin. Making this substitution, we discover the following:

The total torque on the particle is equal to the time rate of change of its angular momentum:

$$\boxed{\vec{\boldsymbol{\tau}}_{\text{total}} = \frac{d\vec{\mathbf{L}}}{dt}} \quad (10.12)$$

This equation is the rotational counterpart of Newton's second law of motion

$$\vec{\mathbf{F}}_{\text{total}} = \frac{d\vec{\mathbf{p}}}{dt}$$

where the *torque* is the counterpart of the force and the *angular momentum* plays the part of the momentum.

Since the angular momentum is proportional to the angular velocity via Equation 10.6, Equation 10.12 also can be written as

$$\vec{\boldsymbol{\tau}}_{\text{total}} = \frac{d}{dt}(I\vec{\boldsymbol{\omega}})$$

The moment of inertia I is constant as long as the particle is of fixed mass and is in circular orbital motion. Hence

$$\vec{\boldsymbol{\tau}}_{\text{total}} = I\frac{d\vec{\boldsymbol{\omega}}}{dt}$$

But

$$\frac{d\vec{\boldsymbol{\omega}}}{dt}$$

is the angular acceleration $\vec{\boldsymbol{\alpha}}$.

Hence we find that

$$\vec{\tau}_{\text{total}} = I\vec{\alpha} \qquad (\text{constant } I) \qquad (10.13)$$

Equation 10.13 is the rotational counterpart of $\vec{F}_{\text{total}} = m\vec{a}$. Equation 10.13 can be used only when the moment of inertia I of the particle is constant, just as $\vec{F}_{\text{total}} = m\vec{a}$ only can be used if m is constant.

QUESTION 3

If $\vec{\omega} = 0$ rad/s at some instant, does this imply $\vec{\tau}_{\text{total}} = 0\,\text{N·m}$ at that instant? If $\vec{\alpha} = 0$ rad/s² at some instant, does this imply that $\vec{\tau}_{\text{total}} = 0\,\text{N·m}$ at that instant?

EXAMPLE 10.3

A mechanic tightens a bolt on a V-8 engine head with a wrench. He applies a tangential force 20 cm from the bolt, as shown in Figure 10.15. The bolt will break if the applied torque about the axis of the bolt exceeds 80 N·m in magnitude. What is the maximum magnitude of force that the mechanic can apply without breaking the bolt?

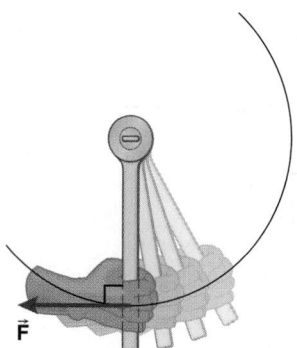

FIGURE 10.15

Solution

Method 1
The magnitude of the torque is

$$\tau = |\vec{r} \times \vec{F}|$$
$$= rF\sin\theta$$

where $\vec{r}$ is the vector from the point about which the torque is taken (here the axis of the bolt) to the point of application of the force $\vec{F}$ (here 20 cm from the bolt); θ is the angle between $\vec{r}$ and $\vec{F}$—here $\theta = 90°$. Since the magnitude of the torque cannot exceed 80 N·m, you have

$$80\ \text{N·m} = (0.20\ \text{m})F\sin 90°$$

Solving for F,

$$F = 4.0 \times 10^2\ \text{N}$$

Method 2
The moment arm of the force is the perpendicular distance from the origin (here at the center of the circle) to the line of action

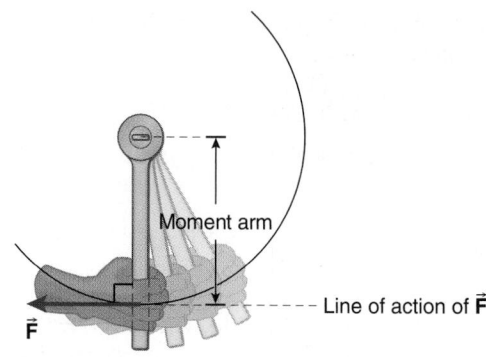

FIGURE 10.16

of the force, as shown in Figure 10.16. Here the moment arm is the radius of the circle. The magnitude of the torque is the product of the magnitude of the force and its moment arm:

$$\tau = F[\text{moment arm}]$$
$$80\ \text{N·m} = F(0.20\ \text{m})$$

Solving for F,

$$F = 4.0 \times 10^2\ \text{N}$$

For safety many pieces of machinery, such as snow blowers and jet engine mounts have bolts, called shear pins, that are designed to break (or shear off) if they experience forces that exceed some specified maximum magnitude.

EXAMPLE 10.4

A huge centrifuge is used to subject astronauts to large accelerations. It consists of two 1500 kg chambers on opposite ends of a rigid boom of length 20 m, much like a dumbbell. The chamber and boom system is brought from rest to an angular speed of 12 rev/min during 1.5 min by means of a force of constant magnitude applied 2.0 m from the center of mass and perpendicular to the boom, as shown in Figure 10.17. Consider the chambers to be pointlike masses at the end of the boom; neglect the mass of the boom.

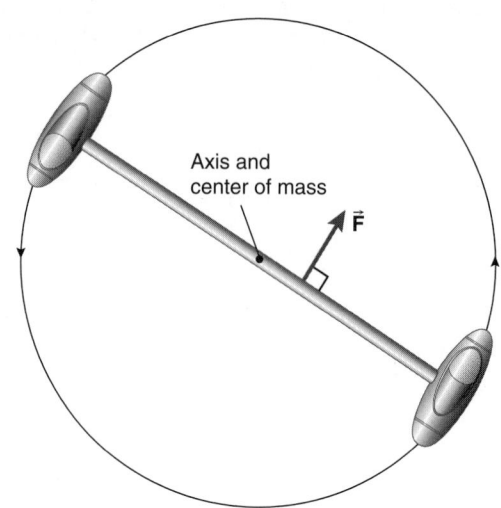

FIGURE 10.17

a. Find the torque of the force.
b. Find the magnitude of the force.

Solution
Each chamber is a particle in circular motion. Since the moment of inertia is a scalar and both particles have the same angular speed, the total moment of inertia of the two-particle system is

$$I = mr^2 + mr^2$$
$$= 2mr^2$$
$$= 2(1500 \text{ kg})(10 \text{ m})^2$$
$$= 3.0 \times 10^5 \text{ kg·m}^2$$

Choose the z-axis to be parallel to the angular velocity vector, whose direction is found from the circular motion right-hand rule; $\hat{k}$ is perpendicularly out of the page in Figure 10.17. The final angular velocity component of the system is

$$\omega_z = \frac{(12 \text{ rev/min})(2\pi \text{ rad/rev})}{60 \text{ s/min}}$$
$$= 1.3 \text{ rad/s}$$

Use Equation 4.49 to find the constant angular acceleration component; the initial angular velocity component is 0 rad/s:

$$\omega_z = \omega_{z0} + \alpha_z t$$
$$1.3 \text{ rad/s} = 0 \text{ rad/s} + \alpha_z(90 \text{ s})$$

Solving for α_z,

$$\alpha_z = 1.4 \times 10^{-2} \text{ rad/s}^2$$

The constant angular acceleration is

$$\vec{\alpha} = \alpha_z \hat{k}$$
$$= (1.4 \times 10^{-2} \text{ rad/s}^2)\hat{k}$$

a. The torque is in the same direction as the angular acceleration, according to Equation 10.13:

$$\vec{\tau} = I\vec{\alpha}$$
$$= I\alpha_z \hat{k}$$
$$= (3.0 \times 10^5 \text{ kg·m}^2)(1.4 \times 10^{-2} \text{ rad/s}^2)\hat{k}$$
$$= (4.2 \times 10^3 \text{ N·m})\hat{k}$$

b. The force is applied perpendicular to the boom and 2.0 m from the axis, so that

$$\tau = |\vec{r} \times \vec{F}|$$
$$= rF \sin 90°$$
$$= rF$$

where $\vec{r}$ locates the point of application of $\vec{F}$ with the axis as an origin; here $r = 2.0$ m. Solving for the magnitude of the force,

$$F = \frac{\tau}{r}$$
$$= \frac{4.2 \times 10^3 \text{ N·m}}{2.0 \text{ m}}$$
$$= 2.1 \times 10^3 \text{ N}$$

EXAMPLE 10.5

A hammer-throw athlete whirls a 7.30 kg mass in a circle of radius 1.20 m at a constant speed of 8.00 m/s, barely above the ground, as shown in Figure 10.18.

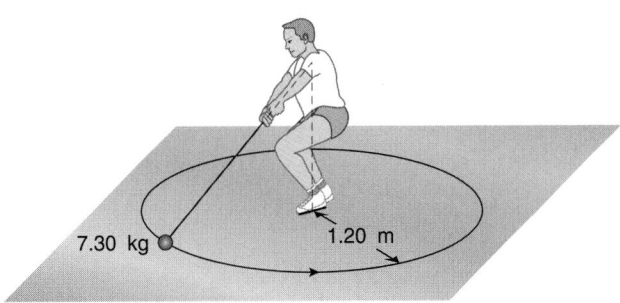

7.30 kg   1.20 m

FIGURE 10.18

a. Does the mass have spin and/or orbital angular momentum?
b. Determine the orbital angular momentum of the particle with respect to the center of the circle.

Solution
a. The mass has orbital angular momentum because the center of mass of the single particle is moving. Since the particle also rotates once during each revolution, it also has some spin angular momentum.
b. There are several ways to calculate the orbital angular momentum $\vec{L} = \vec{r} \times \vec{p}$. Here $\vec{r}$ locates the particle with respect to the center of the circle, since you want the angular momentum about this point.

Method 1
Find the magnitude of angular momentum using the magnitude of the vector product $\vec{L} = \vec{r} \times \vec{p}$:

$$L = rp \sin \theta$$
$$= rmv \sin \theta$$

In this case the angle between $\vec{r}$ and $\vec{p}$ is always 90°. Substituting for r, m, v, and θ, you obtain

$$L = (1.20 \text{ m})(7.30 \text{ kg})(8.00 \text{ m/s}) \sin 90°$$
$$= 70.1 \text{ kg·m}^2/\text{s}$$

The direction of $\vec{L}$ is found by applying the vector product right-hand rule to $\vec{r} \times \vec{p}$; see Figure 10.19.

Method 2
Use Equation 10.6, $\vec{L} = I\vec{\omega}$, to find the angular momentum. The moment of inertia of a point particle is

$$I = mr^2$$

The angular velocity $\vec{\omega}$ is found from the linear velocity using Equation 4.34:

$$\vec{v} = \vec{\omega} \times \vec{r}$$

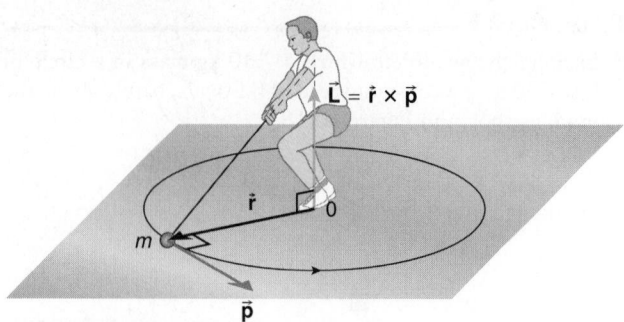

FIGURE 10.19

Since $\vec{\omega}$ is perpendicular to $\vec{r}$ in circular motion, the magnitudes of these vectors are related by

$$v = \omega r$$

or

$$\omega = \frac{v}{r}$$

Notice that the angular speed is the magnitude of the component of the velocity perpendicular to $\vec{r}$ (which is just v itself for circular motion) divided by r. Thus the magnitude of the orbital angular momentum $\vec{L}$ is

$$L = I\omega$$

$$= mr^2 \frac{v}{r}$$

$$= mrv$$

$$= (7.30 \text{ kg})(1.20 \text{ m})(8.00 \text{ m/s})$$

$$= 70.1 \text{ kg}\cdot\text{m}^2/\text{s}$$

The direction of $\vec{L}$ is the same as the direction of $\vec{\omega}$, since $\vec{L} = I\vec{\omega}$. The direction of $\vec{\omega}$ is found using the circular motion right-hand rule: wrap the fingers of your right hand around the center of the circle in the sense the particle is moving; the extended right-hand thumb indicates the direction of $\vec{\omega}$. Thus $\vec{\omega}$ is perpendicular to the plane of the circle and so is $\vec{L}$, as shown in Figure 10.20.

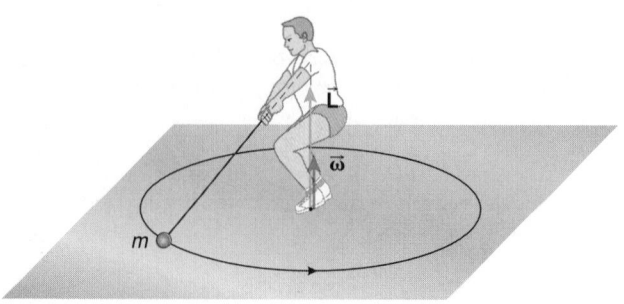

FIGURE 10.20

STRATEGIC EXAMPLE 10.6

A 3.00 kg mass on a frictionless horizontal surface is attached to a string of length 50.0 cm that is fixed at one end but can slide around the axis. The mass initially is at rest and the string is taut. For times $t \geq 0$ s, a force of magnitude 5.00 N is applied tangentially as indicated in Figure 10.21.

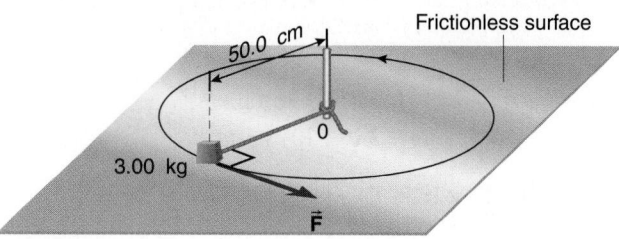

FIGURE 10.21

a. What are the forces acting on the mass?
b. What is the torque of the total force on the mass?
c. What is the moment of inertia of the mass about the fixed end of the string?
d. What is the angular acceleration of the mass?
e. Is the angular momentum of the mass constant? Explain your answer.
f. What is the angular momentum of the mass when $t = 3.00$ s?

Solution
Consider the 3.00 kg mass as a particle that is the system.

a. The forces on the system are (see Figure 10.22)
   1. the gravitational force of the Earth on the system (its weight $\vec{w}$), directed vertically down;
   2. the normal force $\vec{N}$ of the surface on the system, directed perpendicular to the surface, here vertically upward;
   3. the applied tangential force $\vec{F}$ of constant magnitude 5.00 N; and
   4. the force $\vec{T}$ of the string on the system, directed toward the center of the circle.

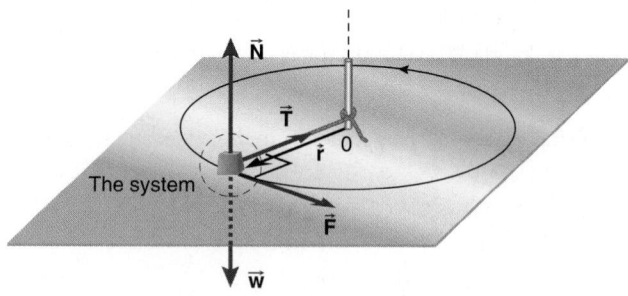

FIGURE 10.22

b. There is no acceleration in the vertical direction, and so the vector sum of the weight and normal force is zero. The vector sum of the weight and the normal force produces zero torque since the vector sum of these forces is zero and they are colinear. Since the tension is directed toward the center of the circle, antiparallel to $\vec{r}$, the torque of the tension also is zero:

$$\vec{r} \times \vec{T} = 0 \text{ N}\cdot\text{m}$$

The total torque on the mass about the axis of its rotation thus is equal to the torque of the tangential force. Examine Figure 10.23; the torque of this force is given by

$$\vec{\tau}_{total} = \vec{r} \times \vec{F}$$

where $\vec{r}$ is the vector from the origin to the point where $\vec{F}$ is applied. Using the vector product right-hand rule, the direction of the torque is perpendicular (upward in Figure 10.23) to the plane of the motion.

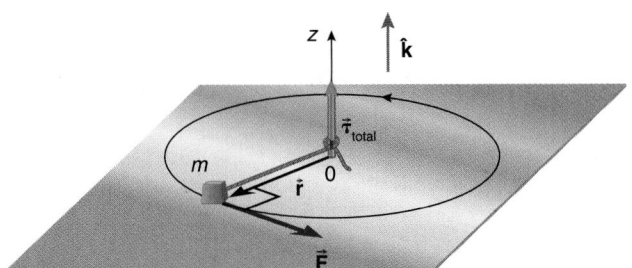

FIGURE 10.23

The magnitude of the torque is

$$\tau_{total} = rF \sin 90°$$
$$= (0.500 \text{ m})(5.00 \text{ N})(1)$$
$$= 2.50 \text{ N·m}$$

With the coordinate system in Figure 10.23, the total torque is

$$\vec{\tau}_{total} = (2.50 \text{ N·m})\hat{k}$$

c. The moment of inertia of the 3.00 kg particle is found using Equation 10.5 for a pointlike mass in orbital motion:

$$I = mr^2$$
$$= (3.00 \text{ kg})(0.500 \text{ m})^2$$
$$= 0.750 \text{ kg·m}^2$$

d. The angular acceleration of the particle is found from Equation 10.13,

$$\vec{\tau}_{total} = I\vec{\alpha}$$
$$(2.50 \text{ N·m})\hat{k} = (0.750 \text{ kg·m}^2)\vec{\alpha}$$

Solving for $\vec{\alpha}$ and simplifying the units,

$$\vec{\alpha} = (3.33 \text{ rad/s}^2)\hat{k}$$

e. Equation 10.12 implies that the orbital angular momentum of the mass is *not constant* since the total torque is not zero.

f. To determine the angular momentum when $t = 3.00$ s, use Equation 10.6,

$$\vec{L} = I\vec{\omega}$$

You have to find the angular velocity when $t = 3.00$ s. The angular acceleration of the mass is constant, so that you can use Equation 4.49 for circular motion with a constant angular acceleration:

$$\omega_z(t) = \omega_{z0} + \alpha_z t$$

When $t = 0$ s, the mass had no angular speed, so that $\omega_{z0} = 0$ rad/s. From part (d), you see that $\alpha_z = 3.33$ rad/s^2. Hence, when $t = 3.00$ s, the angular velocity component is

$$\omega_z = 0 \text{ rad/s} + (3.33 \text{ rad/s}^2)(3.00 \text{ s})$$
$$= 9.99 \text{ rad/s}$$

The angular velocity then is

$$\vec{\omega} = \omega_z \hat{k}$$
$$= (9.99 \text{ rad/s})\hat{k}$$

The orbital angular momentum then is

$$\vec{L} = I\vec{\omega}$$
$$= (0.750 \text{ kg·m}^2)(9.99 \text{ rad/s})\hat{k}$$
$$= (7.49 \text{ kg·m}^2/\text{s})\hat{k}$$

10.4 NONCIRCULAR ORBITAL MOTION

Equation 10.12,

$$\vec{\tau}_{total} = \frac{d\vec{L}}{dt}$$

also can be applied to noncircular orbital motion, such as a particle moving in a straight line or the familiar case of the elliptical orbital motion of a planet about the Sun. For the latter case, take the Sun as the origin of a coordinate system and consider the planet as a particle, as in Figure 10.24. Recall from Chapter 6 that the gravitational force is a *central force*—that is, it acts along the line connecting the two particles. The total torque on the planet is zero: $\vec{r}$ is always along the same line as the gravitational force $\vec{F}$, so that $\vec{r} \times \vec{F} = 0$ N·m.

Thus Equation 10.12 becomes

$$0 \text{ N·m} = \frac{d\vec{L}}{dt}$$

That is, the orbital angular momentum of the planet is constant, which is equivalent to Kepler's second law of planetary motion as we demonstrated in Section 6.10.

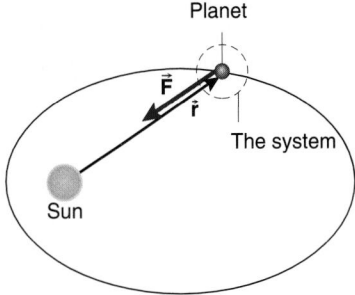

FIGURE 10.24 The gravitational force on a planet is a central force with zero torque.

If the planet is in a circular orbit (Venus very nearly fits this case), then the moment of inertia of the planet about the Sun ($I = mr^2$) is constant and the angular velocity is too. On the other hand, if the planet or comet is moving in an orbit that is elliptical, parabolic, or hyperbolic, the moment of inertia of the particle changes (since r changes). However, the magnitude of the angular velocity changes to compensate, and so the angular momentum of the particle, given by Equation 10.6,

$$\vec{\mathbf{L}} = I\vec{\boldsymbol{\omega}}$$

remains constant (as it must because the total torque on the planet is zero). Since the angular velocity of a planet in an elliptical orbit changes, the angular acceleration $\vec{\boldsymbol{\alpha}}$ of the planet is *not* zero. However, you *cannot* use Equation 10.13,

$$\vec{\boldsymbol{\tau}}_{\text{total}} = I\vec{\boldsymbol{\alpha}}$$

for such planets. The total torque on the planet *is* zero, because the gravitational force is a central force. The left-hand side of Equation 10.13 is zero, but the right-hand side is not! Clearly this result is ridiculous. You cannot use Equation 10.13 for such noncircular orbital motion because the equation assumes that the moment of inertia of the particle is constant. For noncircular motion I is *not* constant, because the distance of the particle from the origin changes with time. Just as $\vec{\mathbf{F}}_{\text{total}} = m\vec{\mathbf{a}}$ *cannot* be used for a system with varying mass (the more general equation

$$\vec{\mathbf{F}}_{\text{total}} = \frac{d\vec{\mathbf{p}}}{dt}$$

must be used instead), $\vec{\boldsymbol{\tau}}_{\text{total}} = I\vec{\boldsymbol{\alpha}}$ *cannot* be used for a system with a time-varying moment of inertia (the more general equation

$$\vec{\boldsymbol{\tau}}_{\text{total}} = \frac{d\vec{\mathbf{L}}}{dt}$$

must be used instead). Note the satisfying parallelism.

10.5 RIGID BODIES AND SYMMETRY AXES

You boil it in sawdust: You salt it in glue:
You condense it with locusts and tape:
Still keeping one principal object in view —
To preserve its symmetrical shape

Lewis Carroll (1832–1898)*

We now know how to calculate the angular momentum and the torque of the total force on a single-particle system as well as its consequent angular acceleration. In this section we consider more complex systems containing *many* particles.

A rigid body system is not deformable but maintains its shape. With a few exceptions,[†] we restrict ourselves to multi-

---

The Hunting of the Snark, Fit the Fifth, The Beaver's Lesson, lines 93–96 (Macmillan, New York, 1927), page 38.

[†] The exceptions we consider are either (a) single-particle systems, (b) swarms of unconnected single particles, or (c) systems that deform under the action of *internal forces* in such a way that the system at all times still is symmetric about the axis of symmetry. For example, a figure skater who gradually brings in both arms or a star that gradually expands or contracts are deformable systems but, at any instant, are symmetric about a symmetry axis.

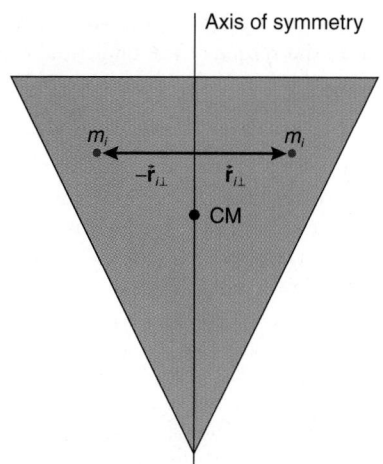

FIGURE 10.25 A symmetric rigid body has particles located symmetrically with respect to an axis of symmetry.

particle systems that are rigid bodies. A perfectly rigid body is an idealization, since all objects deform to a greater or lesser degree under the action of various forces, as we will see in Chapter 11. Nonetheless, the idea of a rigid body is extremely useful for simplifying the dynamics of rotational motion. For simplicity we also restrict ourselves to rigid bodies that possess at least one **axis of symmetry**.

An axis of symmetry can be imagined as a line in space such that for each pointlike particle m_i making up part of the rigid body, located at position vector $\vec{\mathbf{r}}_{i\perp}$, there is a corresponding pointlike particle of equal mass located at position vector $-\vec{\mathbf{r}}_{i\perp}$; see Figure 10.25.

Both position vectors are measured *perpendicular* to the axis of symmetry (hence the subscript ⊥) and have the *axis* as origin. Such a symmetry axis always passes through the center of mass of the system.

We shall see in the next section that the assumption of a rigid body with at least one axis of symmetry greatly simplifies the dynamical description of rotational motion. A *general* description of the rotational motion of *un*symmetric rigid bodies and nonrigid bodies is beyond the scope of an introductory physics course, which may be a relief to you!

QUESTION 4

Sketch one of the symmetry axes through the center of mass of a uniform, square plate.

10.6 SPIN ANGULAR MOMENTUM OF A RIGID BODY

Systems rotating about an axis through the center of mass exhibit spinning motion or, more simply, *spin*. The rotational motion of a CD, a windmill pinwheel, an ultracentrifuge rotor, or even a common (empty) clothes dryer all are examples of symmetric rigid bodies undergoing spin. Each pointlike particle of the rigid body system undergoes circular orbital motion with the same angular velocity.

We consider the spinning motion of a system of many pointlike particles of a *rigid body* with at least one axis of symmetry, as in Figure 10.25. The symmetry axis passes through the center of mass point of the system. For the time being we restrict ourselves to situations in which two conditions hold:

> 1. *The center of mass of the system is at rest.*
>
> We make this restriction because spin is the simplest kind of rigid body rotational motion and the kinematics of spin is an extension of the kinematic description of circular motion that we encountered in Chapter 4.
>
> 2. Also, for simplicity, we assume *the rotational axis of the system coincides with the symmetry axis of the system.*

Later we relax both of these restrictions, when we have developed the conceptual tools to do so.

All the pointlike particles that make up the rigid body have the *same* angular velocity $\vec{\boldsymbol{\omega}}$. The direction of the angular velocity vector is found using the familiar circular motion right-hand rule of Chapter 4. We shall see that the total angular momentum of such a system is proportional to the angular velocity vector, just as for single-particle orbital motion (Equation 10.6).

Take the fixed center of mass point to be the origin of a coordinate system (see Figure 10.26). A pointlike particle of mass m_i (that is part of the rigid body system) has an angular momentum $\vec{\mathbf{L}}_i$ that is the vector product of the position vector $\vec{\mathbf{r}}_i$ of the particle with its momentum $\vec{\mathbf{p}}_i$:

$$\vec{\mathbf{L}}_i \equiv \vec{\mathbf{r}}_i \times \vec{\mathbf{p}}_i$$
$$= \vec{\mathbf{r}}_i \times m_i\vec{\mathbf{v}}_i$$

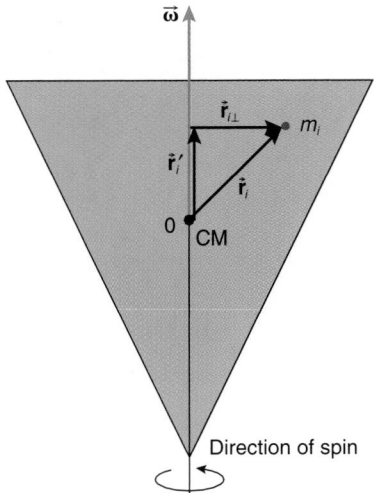

FIGURE 10.26 The relationship between the position vectors $\vec{\mathbf{r}}_i$, $\vec{\mathbf{r}}_{i\perp}$, and $\vec{\mathbf{r}}_i'$.

The *total* angular momentum $\vec{\mathbf{L}}$ of the spinning system is the vector sum of the orbital angular momenta of all the pointlike particles that make up the system:

$$\vec{\mathbf{L}} = \sum_i \vec{\mathbf{L}}_i$$
$$= \sum_i m_i\vec{\mathbf{r}}_i \times \vec{\mathbf{v}}_i \qquad (10.14)$$

Each particle of the rigid body executes *circular* motion about the rotational axis. Let $r_{i\perp}$ be the radius of the circle for the *i*th particle, as shown in Figure 10.26. The position vector $\vec{\mathbf{r}}_{i\perp}$ locates the particle *with respect to the axis of rotation* rather than the center of mass point.

The velocity $\vec{\mathbf{v}}_i$ of the *i*th particle is related to the angular velocity $\vec{\boldsymbol{\omega}}$ and $\vec{\mathbf{r}}_{i\perp}$ according to Equation 4.34 (modified slightly for the new notation here):

$$\vec{\mathbf{v}}_i = \vec{\boldsymbol{\omega}} \times \vec{\mathbf{r}}_{i\perp}$$

Also note in Figure 10.26 that we can write $\vec{\mathbf{r}}_i$ as the vector sum of $\vec{\mathbf{r}}_i'$ and $\vec{\mathbf{r}}_{i\perp}$, where $\vec{\mathbf{r}}_i'$ locates the tail of $\vec{\mathbf{r}}_{i\perp}$ with respect to the center of mass point:

$$\vec{\mathbf{r}}_i = \vec{\mathbf{r}}_i' + \vec{\mathbf{r}}_{i\perp}$$

Making these substitutions for $\vec{\mathbf{v}}_i$ and $\vec{\mathbf{r}}_i$ into Equation 10.14 for the spin angular momentum, we have quite an algebraic mouthful:

$$\vec{\mathbf{L}} = \sum_i m_i(\vec{\mathbf{r}}_i' + \vec{\mathbf{r}}_{i\perp}) \times (\vec{\boldsymbol{\omega}} \times \vec{\mathbf{r}}_{i\perp})$$

Expand this into two summations:

$$\vec{\mathbf{L}} = \sum_i m_i\vec{\mathbf{r}}_i' \times (\vec{\boldsymbol{\omega}} \times \vec{\mathbf{r}}_{i\perp}) + \sum_i m_i\vec{\mathbf{r}}_{i\perp} \times (\vec{\boldsymbol{\omega}} \times \vec{\mathbf{r}}_{i\perp}) \quad (10.15)$$

Now stare carefully at Figure 10.26. The vector product $\vec{\boldsymbol{\omega}} \times \vec{\mathbf{r}}_{i\perp}$ is directed perpendicularly into the page. The vector product of $\vec{\mathbf{r}}_i'$ with $\vec{\boldsymbol{\omega}} \times \vec{\mathbf{r}}_{i\perp}$ thus is parallel to $-\vec{\mathbf{r}}_{i\perp}$ and, since the vectors are all mutually perpendicular, is of magnitude $r_i'\omega r_{i\perp}$. Therefore the first summation on the right-hand side of Equation 10.15 can be rewritten as

$$-\sum_i m_i r_i'\omega\vec{\mathbf{r}}_{i\perp}$$

Likewise, by using the vector product right-hand rule, the term $\vec{\mathbf{r}}_{i\perp} \times (\vec{\boldsymbol{\omega}} \times \vec{\mathbf{r}}_{i\perp})$ in the second summation of Equation 10.15 is directed parallel to $\vec{\boldsymbol{\omega}}$ and is of magnitude $r_{i\perp}^2\omega$, since the vectors are mutually perpendicular. Therefore the second summation in Equation 10.15 can be written as

$$\sum_i m_i r_{i\perp}^2\vec{\boldsymbol{\omega}}$$

In this way Equation 10.15 can be rewritten as

$$\vec{\mathbf{L}} = -\sum_i m_i r_i'\omega\vec{\mathbf{r}}_{i\perp} + \sum_i m_i r_{i\perp}^2\vec{\boldsymbol{\omega}} \qquad (10.16)$$

This expression looks rather ghastly, and might mirror your feelings if your eyes have glazed over. Because of the presence of the first summation in Equation 10.16, the spin angular momentum vector is *not* necessarily parallel to the angular velocity vector! This is bad news for simplicity. The vector $\vec{\mathbf{L}}$ is an involved vector summation of the angular velocity $\vec{\boldsymbol{\omega}}$ and a host of other vectors $\vec{\mathbf{r}}_{i\perp}$. Herein lies the complexity of rotational motion for extended objects of *arbitrary* shape. The rotation of an oddly shaped object about *any* axis of rotation is beyond the scope of this text. By restricting ourselves to *symmetric* rigid bodies, we can spare ourselves much complication and still produce quite useful results.

For symmetric rigid bodies with the rotational axis coincident with a symmetry axis, the first summation in Equation 10.16 is *zero*. Why? Since the object is symmetric about the rotational axis, for every particle located at position vector $\vec{\mathbf{r}}_{i\perp}$ with respect to the axis, there is an identical particle located at $-\vec{\mathbf{r}}_{i\perp}$ as in Figure 10.25. Hence, when you sum over all the particles, the terms of the first summation pairwise add to zero. Indeed this is the very reason that we choose to consider only the rotation of rigid bodies that have an axis of symmetry! For unsymmetrical systems, the first summation is *not* zero and the dynamical description of their rotation is much more complicated, because the angular momentum vector then is not parallel to the angular velocity vector.*

Therefore, for symmetric rigid bodies spinning about an axis coincident with the symmetry axis, the total spin angular momentum of the system, Equation 10.16, reduces to

$$\vec{\mathbf{L}} = \sum_i m_i r_{i\perp}^2 \, \vec{\boldsymbol{\omega}}$$

Since the angular velocity is a common factor of every term in the summation, we have

$$\vec{\mathbf{L}} = \left(\sum_i m_i r_{i\perp}^2 \right) \vec{\boldsymbol{\omega}}$$

The summation in parentheses is the *moment of inertia* I_{CM} of the rigid body about a rotational axis through the center of mass and parallel to the symmetry axis of the rigid body. That is, by definition

$$I_{CM} = \sum_i m_i r_{i\perp}^2 \qquad (10.17)$$

The moment of inertia of a rigid body thus depends on not only its mass but also how this mass is distributed with respect to the axis of rotation. The greater the distance of a mass point from the axis of rotation, the greater its contribution to the moment of inertia. In Section 10.8 we will see how to calculate the moment of inertia of several common symmetric rigid bodies; some values will be tabulated there in Table 10.1. The moment of inertia of a symmetric rigid body about a given symmetry axis is a characteristic of the particular rigid body shape, its mass, and that symmetry axis. It is quite possible for a rigid body to have more than one moment of inertia I_{CM}, but each one then is appropriate for a *different* symmetry axis.

For symmetric rigid bodies rotating about one of their symmetry axes, the spin angular momentum vector of the system is *parallel* to the spin angular velocity vector; the scalar moment of inertia I_{CM} relates the two:

$$\vec{\mathbf{L}}_{spin} = I_{CM} \vec{\boldsymbol{\omega}}_{spin} \qquad (10.18)$$

This equation is identical in form to Equation 10.6 for a *single* particle undergoing orbital motion, with an appropriate value for the moment of inertia.

10.7 THE TIME RATE OF CHANGE OF THE SPIN ANGULAR MOMENTUM

We want to see if the time rate of change of the spin angular momentum is equal to the total torque on the system, which is the case for a single particle. The total spin angular momentum of a system is the vector sum of the (circular orbital) angular momenta of all the individual particles that make up the system. As in the previous section, take the origin of a coordinate system to be at the stationary center of mass point of the system, as shown in Figure 10.27.

The spin angular momentum is

$$\vec{\mathbf{L}} = \sum_i \vec{\mathbf{L}}_i$$
$$= \sum_i \vec{\mathbf{r}}_i \times \vec{\mathbf{p}}_i$$

where $\vec{\mathbf{r}}_i$ is the position vector of the particle with momentum $\vec{\mathbf{p}}_i$. Substituting $m_i \vec{\mathbf{v}}_i$ for the momentum, we have

$$\vec{\mathbf{L}} = \sum_i m_i \vec{\mathbf{r}}_i \times \vec{\mathbf{v}}_i$$

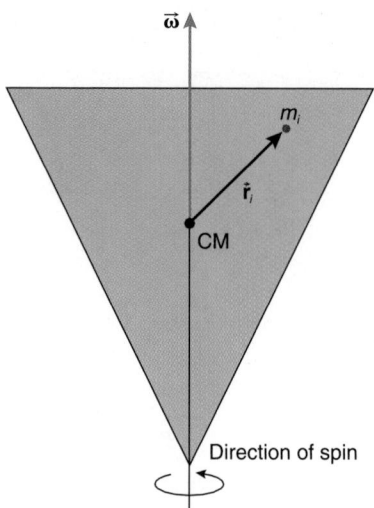

FIGURE 10.27 Use an origin at the center of mass.

---

*For unsymmetric rigid bodies, the moment of inertia is not a scalar but a second rank tensor. Such tensors are studied in more advanced courses in dynamics.

The time rate of change of the spin angular momentum is

$$\frac{d\vec{L}}{dt} = \sum_i \frac{d}{dt}\left(m_i\vec{r}_i \times \vec{v}_i\right)$$

Differentiate the vector product, remembering to preserve the order of the terms. We find

$$\frac{d\vec{L}}{dt} = \sum_i m_i\left(\frac{d\vec{r}_i}{dt} \times \vec{v}_i + \vec{r}_i \times \frac{d\vec{v}_i}{dt}\right)$$

Since

$$\frac{d\vec{r}_i}{dt} = \vec{v}_i$$

the first term in the bracket is $\vec{v}_i \times \vec{v}_i$, which is zero; you might explain why. This leaves you with

$$\frac{d\vec{L}}{dt} = \sum_i m_i\left(\vec{r}_i \times \frac{d\vec{v}_i}{dt}\right)$$

$$= \sum_i \vec{r}_i \times \frac{d\vec{p}_i}{dt}$$

According to Newton's second law of motion, the time rate of change of the momentum of the ith particle is equal to the total force $\vec{F}_{i\,\text{total}}$ on the ith particle. Hence

$$\frac{d\vec{L}}{dt} = \sum_i \vec{r}_i \times \vec{F}_{i\,\text{total}}$$

Each term in the sum is the total torque on each particle of the system.

The total force acting on any particle of the system can be written as the sum of two types of forces: forces originating from outside the system itself (*external* forces such as externally applied pushes or pulls) and forces originating from other particles within the system itself (*internal* forces). Let's look first at the torques produced by the internal forces between any two pointlike particles; the results can then be generalized easily to any number of such particles.

Let $\vec{F}_{1\,\text{int}}$ be the force on particle 1 due to particle 2 and let $\vec{F}_{2\,\text{int}}$ be the force on particle 2 due to particle 1. According to Newton's third law of motion, these forces have the same magnitude but point in opposite directions. However, there is nothing in the third law that indicates the specific direction along which these mutual forces act. *Central* internal forces act along the line between the particles,* as shown in Figure 10.28. For central internal forces between the particles, the total internal torque produced by the force of each particle on the other is

$$\vec{r}_1 \times \vec{F}_{1\,\text{int}} + \vec{r}_2 \times \vec{F}_{2\,\text{int}}$$

where the vectors $\vec{r}_1$ and $\vec{r}_2$ locate the points of application of the forces with respect to the center of mass point. Since $\vec{F}_{2\,\text{int}} = -\vec{F}_{1\,\text{int}}$

Noncentral internal forces do *not* act along the line between the particles. Although noncentral forces exist, they are more appropriately considered in advanced and specialized areas of physics. We will not consider them.

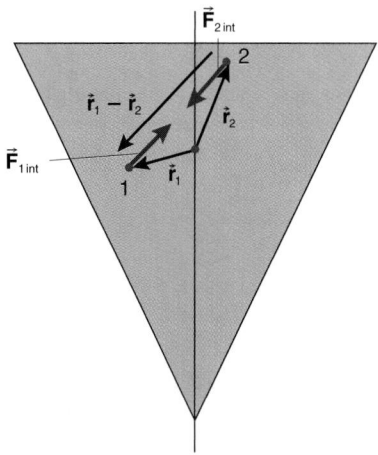

FIGURE 10.28 Central internal forces act along the line connecting the particles.

according to Newton's third law, we can eliminate $\vec{F}_{2\,\text{int}}$; so the sum of the internal torques of these forces is

$$\vec{r}_1 \times \vec{F}_{1\,\text{int}} + \vec{r}_2 \times (-\vec{F}_{1\,\text{int}})$$

or

$$(\vec{r}_1 - \vec{r}_2) \times \vec{F}_{1\,\text{int}}$$

Notice, in Figure 10.28, that the vector $\vec{r}_1 - \vec{r}_2$ points along the line between the two particles. However, the force $\vec{F}_{1\,\text{int}}$ also is directed along this line (as is $\vec{F}_{2\,\text{int}}$), since it is a central force. Thus the vector product is zero:

$$(\vec{r}_1 - \vec{r}_2) \times \vec{F}_{1\,\text{int}} = 0 \text{ N·m}$$

Therefore the *internal* torques between *every* pair of particles in the system *vanish*. Consequently the sum of the torques of the internal forces vanishes if the internal forces are central forces.

> The absence of internal torques means that the total torque on a system is caused only by forces external to the system itself.

> Thus the time rate of change of the spin angular momentum of the system is equal to sum of the torques of the external forces on the system (the total torque on the system due to external forces):
>
> $$\vec{\tau}_{\text{total ext}} = I_{\text{CM}}\frac{d\vec{\omega}}{dt}$$
>
> If the moment of inertia I_{CM} is constant, use Equation 10.18, $\vec{L} = I_{\text{CM}}\vec{\omega}$, and we have
>
> $$\vec{\tau}_{\text{total ext}} = \frac{d\vec{L}}{dt} \qquad (10.19)$$
>
> or in terms of the angular acceleration,
>
> $$\vec{\tau}_{\text{total ext}} = I_{\text{CM}}\vec{\alpha} \qquad (10.20)$$

After all this, you can rejoice because Equations 10.19 and 10.20 are similar to the earlier Equations 10.12 and 10.13 for a single particle executing orbital motion about a fixed point.

10.8 THE MOMENT OF INERTIA OF VARIOUS RIGID BODIES

Since the moment of inertia is so important to the dynamics of rotational motion, we must know how to calculate or find it for many common rigid bodies. In this section we see how this is accomplished.

Point Particle

The moment of inertia of a particle of mass m located a distance r from the axis of rotation (see Figure 10.29) we found to be given by Equation 10.5:

$$I = mr^2 \quad \text{(point particle)}$$

Collection of Point Particles

Since the moment of inertia is a scalar quantity, the moment of inertia of a collection of particles all with the same angular velocity (see Figure 10.30) is the sum of the moments of inertia of the individual particles about their common axis of rotation:

$$I = \sum_i m_i r_{i\perp}^2 \quad (10.21)$$

where $r_{i\perp}$ is the *perpendicular* distance of each mass m_i from the axis of rotation.

A Rigid Body

The moment of inertia of a rigid body is found by

a. breaking up the rigid body into a collection of differential masses dm, as in Figure 10.31;

b. finding the moment of inertia of each differential mass (a pointlike particle) about the axis of rotation:

$$r_\perp^2 \, dm$$

where $r_\perp$ is the distance of dm from the axis of rotation; and then

c. integrating this expression over the geometrical extent of the rigid body:

$$I = \int_{\text{object}} r_\perp^2 \, dm \quad (10.22)$$

If the mass is distributed along essentially one dimension (thin rods), this integration is a one-dimensional integral. Example 10.7 illustrates this procedure. The integral is a two-

TABLE 10.1 Moment of Inertia I_{CM} of Some Common Symmetric Rigid Bodies About an Axis Coincident with the Symmetry Axis Through the Center of Mass

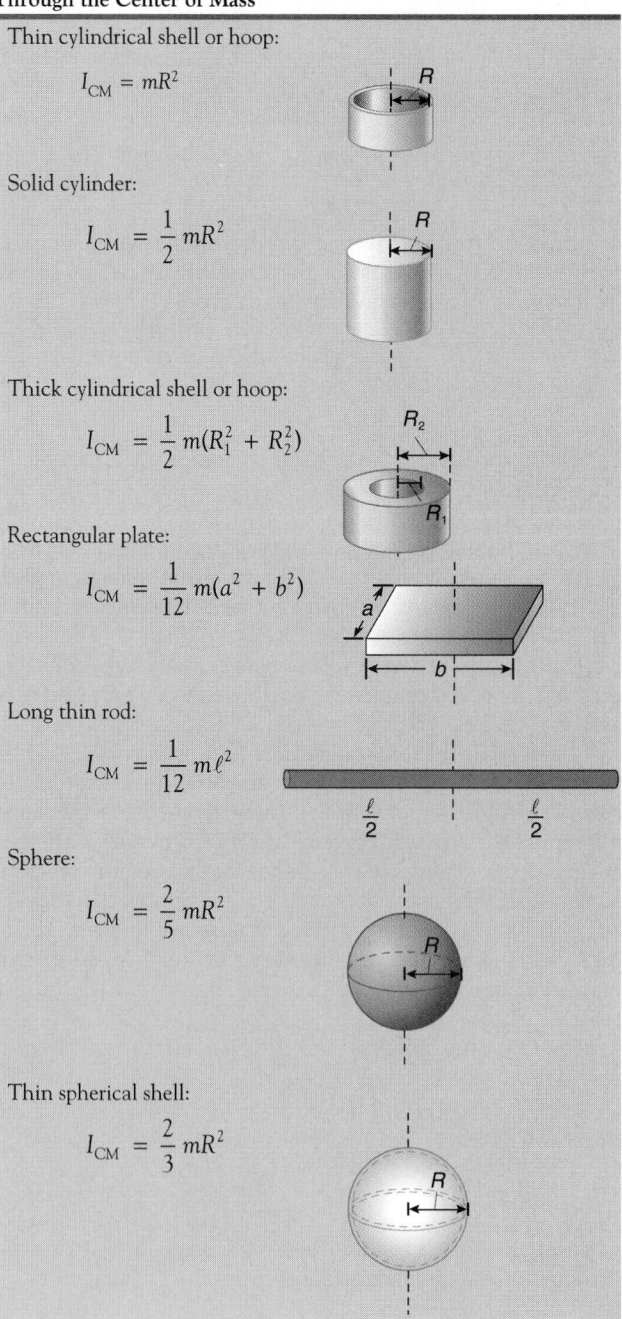

Thin cylindrical shell or hoop:
$$I_{CM} = mR^2$$

Solid cylinder:
$$I_{CM} = \frac{1}{2} mR^2$$

Thick cylindrical shell or hoop:
$$I_{CM} = \frac{1}{2} m(R_1^2 + R_2^2)$$

Rectangular plate:
$$I_{CM} = \frac{1}{12} m(a^2 + b^2)$$

Long thin rod:
$$I_{CM} = \frac{1}{12} m\ell^2$$

Sphere:
$$I_{CM} = \frac{2}{5} mR^2$$

Thin spherical shell:
$$I_{CM} = \frac{2}{3} mR^2$$

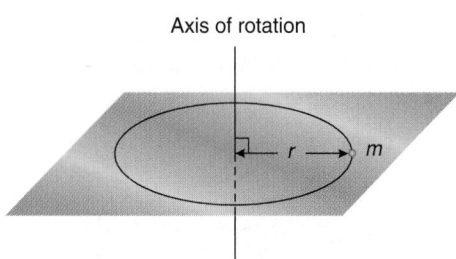

FIGURE 10.29 A pointlike particle located a distance r from an axis of rotation.

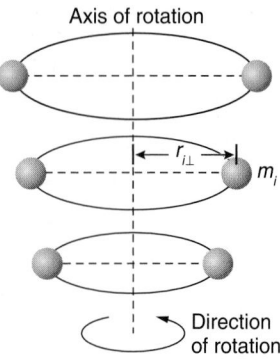

FIGURE 10.30 A collection of pointlike particles rotating about a common axis.

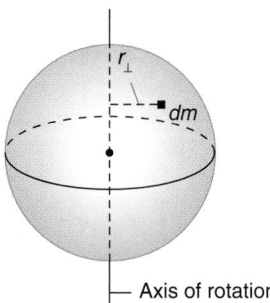

FIGURE 10.31 A solid, rigid body is considered as a collection of small differential mass elements dm.

dimensional integral if the object is planar. The integral is a three-dimensional volume integral if the object is three-dimensional (spheres, cubes, and so forth).

The moment of inertia is an intrinsically *positive quantity* that is the product of a purely numerical factor times the mass and the square of some characteristic dimension of the object. The moments of inertia of some common symmetrical rigid bodies are tabulated in Table 10.1.

The moments of inertia of the rigid bodies in Table 10.1 about axes *parallel* to, but not coincident with, the indicated axis will be discussed in Section 10.14.

> **PROBLEM-SOLVING TACTIC**
>
> **10.3** If the rigid body can be segmented into several symmetrically shaped pieces, such as in Figure 10.32, the moment of inertia of the object as a whole is the sum of the moments of inertia of all the segmented pieces provided the moments are about the same axis.

QUESTION 5

Explain why it is possible for some rigid bodies to have more than one moment of inertia I_{CM}. Give an example of such a rigid body.

EXAMPLE 10.7

Calculate the moment of inertia of a thin rod of mass m and length ℓ about a symmetry axis through the center of mass and perpendicular to the length of the rod, as shown in Figure 10.33.

Solution

The moment of inertia of an extended, continuous rigid body is found by integrating the moment of inertia of a differential mass dm over the object:

$$\int r_\perp^2 \, dm$$

where $r_\perp$ is the distance of dm from the axis of rotation (see Figure 10.34); here $r_\perp = x$.

The thin rod has a mass per unit length, λ, that is given by

$$\lambda \equiv \frac{m}{\ell}$$

Thus the differential amount of mass dm on a differential length dx of the rod is

$$dm = \lambda \, dx$$

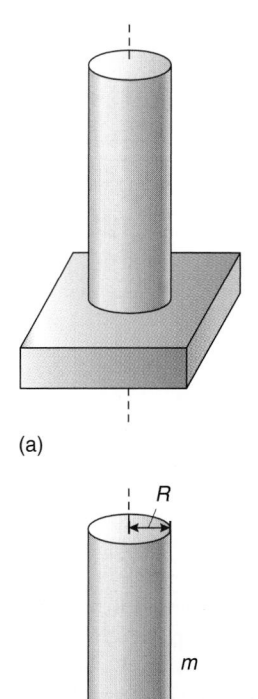

(a)

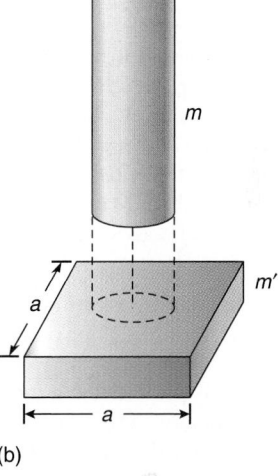

(b)

FIGURE 10.32 Segment a more complicated rigid body into smaller, symmetrical shapes whose moments of inertia are known. (a) Rigid body consisting of two symmetrically shaped pieces, a cylinder and a square plate. (b) Determine the moment of inertia of each symmetrical piece:

$$I_{cyldr} = \frac{1}{2} mR^2$$

$$I_{plate} = \frac{m'a^2}{6}$$

The total moment of inertia is the sum $I_{cyldr} + I_{plate}$.

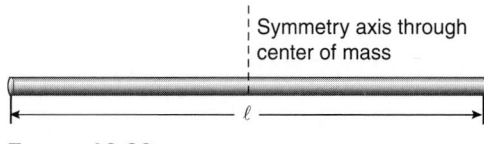

FIGURE 10.33

FIGURE 10.34

The differential bit of moment of inertia of this pointlike mass about the axis of rotation is

$$dI = x^2\, dm$$
$$= x^2 \lambda\, dx$$

Integrate this over the extent of the rod, from $x = -\ell/2$ to $x = \ell/2$, to find the moment of inertia of the rod I:

$$I = \lambda \int_{-\ell/2}^{\ell/2} x^2\, dx$$

$$= \lambda \left. \frac{x^3}{3} \right|_{-\ell/2}^{\ell/2}$$

$$= \frac{m\ell^2}{12}$$

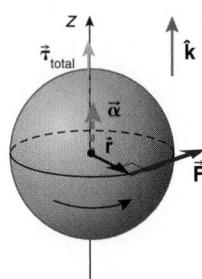

STRATEGIC EXAMPLE 10.8

A satellite is placed in a high-altitude orbit for precisely measuring large distances across the globe. The satellite consists of a uniform, solid sphere of Invar of mass 905 kg and radius 0.300 m. For stability it is to be set spinning about an axis through its center of mass. The final angular speed of the sphere is to be 300 rev/min. The sphere begins at rest and reaches its final angular speed in 30.0 s with a constant angular acceleration.

a. Determine the angular acceleration of the sphere.
b. Determine the total torque acting on the sphere that causes the angular acceleration.
c. If the total torque is provided by a force applied tangential to and along the equator of the sphere by small thrusters of negligible mass, what is the magnitude of the force required?

Solution
Introduce the coordinate system shown in Figure 10.35.

FIGURE 10.35

a. The final angular speed is

$$\frac{(300 \text{ rev/min})(2\pi \text{ rad/rev})}{60 \text{ s/min}} = 31.4 \text{ rad/s}$$

For a constant angular acceleration we have, from Equation 4.49,

$$\omega_z(t) = \omega_{z0} + \alpha_z t$$

Here ω_{z0} is zero, since the sphere begins at rest; also $\omega_z = 31.4$ rad/s when $t = 30.0$ s. Making these substitutions, you have

$$31.4 \text{ rad/s} = 0 \text{ rad/s} + \alpha_z(30.0 \text{ s})$$

Solving for α_z,

$$\alpha_z = 1.05 \text{ rad/s}^2$$

Hence the angular acceleration is

$$\vec{\alpha} = (1.05 \text{ rad/s}^2)\hat{k}$$

The angular acceleration vector is sketched in Figure 10.35.

b. The moment of inertia of a solid sphere about its symmetry axis is found in Table 10.1:

$$I_{\text{CM}} = \frac{2}{5} mR^2$$
$$= (0.400)(905 \text{ kg})(0.300 \text{ m})^2$$
$$= 32.6 \text{ kg} \cdot \text{m}^2$$

The total torque on the sphere is found from Equation 10.20:

$$\vec{\tau}_{\text{total}} = I_{\text{CM}}\vec{\alpha}$$
$$= (32.6 \text{ kg} \cdot \text{m}^2)(1.05 \text{ rad/s}^2)\hat{k}$$
$$= (34.2 \text{ N} \cdot \text{m})\hat{k}$$

The total torque vector also is sketched in Figure 10.35.

c. The total torque is

$$\vec{\tau}_{\text{total}} = \vec{r} \times \vec{F}$$

Since the force is applied tangentially to and along the equator of the sphere, $\vec{r}$ is perpendicular to $\vec{F}$ and

$$\vec{\tau}_{\text{total}} = rF\hat{k}$$
$$(34.2 \text{ N} \cdot \text{m})\hat{k} = (0.300 \text{ m})F\hat{k}$$

The magnitude of the total force thus is

$$F = 114 \text{ N}$$

10.9   THE KINETIC ENERGY OF A SPINNING SYSTEM

A spinning rigid body system has many pointlike particles in motion. The kinetic energy associated with the spinning rigid body is the sum of the kinetic energies of all its constituent particles. At first glance, this may appear to be a complicated summation, but in this section we find that the result is a rather elegant expression for the kinetic energy associated with the spin, known as the **kinetic energy of rotation**.

Figure 10.36 depicts a rigid body system spinning about an axis coincident with a symmetry axis through the center of mass point of the system. The total kinetic energy of the system is the sum of the kinetic energy of each pointlike particle:

$$KE_{\text{rot}} = \sum_i \frac{1}{2} m_i v_i^2 \tag{10.23}$$

Each particle is moving in a circle of radius $r_{i\perp}$ with angular velocity $\vec{\omega}$; the angular velocity is the *same* for all the particles because the system is a rigid body. For such circular spinning motion, the velocity of the ith particle is found from Equation 4.34, as we have seen before:

$$\vec{v}_i = \vec{\omega} \times \vec{r}_{i\perp}$$

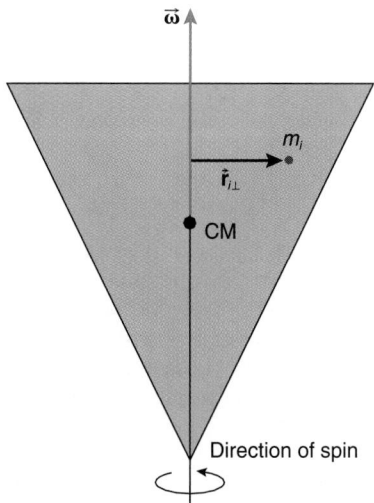

FIGURE 10.36 A rigid body spinning about a symmetry axis through its center of mass point.

where $\vec{\mathbf{r}}_{i\perp}$ is the perpendicular position vector of the ith particle from the axis of rotation. Since $\vec{\boldsymbol{\omega}}$ is perpendicular to $\vec{\mathbf{r}}_{i\perp}$, the speed of the ith particle is

$$v_i = \omega r_{i\perp}$$

Substituting for v_i in Equation 10.23 for the kinetic energy, we find

$$KE_{rot} = \sum_i \frac{1}{2} m_i \omega^2 r_{i\perp}^2$$

$$= \sum_i \frac{1}{2} m_i r_{i\perp}^2 \omega^2$$

Since the angular speed ω is the same for all the particles, ω^2 can be factored out of each term in the summation:

$$KE_{rot} = \frac{1}{2}\left(\sum_i m_i r_{i\perp}^2\right)\omega^2$$

Perhaps you recognize the expression in parentheses as the moment of inertia I_{CM} of the system relative to the symmetry axis passing through the center of mass (Equation 10.17).

> The rotational kinetic energy about the center of mass is
>
> $$KE_{rot} = \frac{1}{2} I_{CM} \omega^2 \qquad (10.24)$$

This has the same basic form as the kinetic energy

$$\frac{1}{2} m v^2$$

of a single particle of mass m, with the moment of inertia taking the place of the mass and the angular speed replacing the speed.

QUESTION 6

A sphere, a cylinder, and a thin-walled cylindrical pipe all have the same mass and radius and are spinning at the same angular speed about a symmetry axis through their respective centers of mass (for each cylinder, the symmetry axis coincident with the axis of the cylinder). Which has the greatest rotational kinetic energy? Which has the least?

EXAMPLE 10.9

Massive flywheels are used to store significant quantities of energy. Consider a flywheel disk of mass 150.0 kg and radius 25.0 cm spinning at 1000 rev/min about an axis perpendicular to the disk and passing through its center of mass, as shown in Figure 10.37.

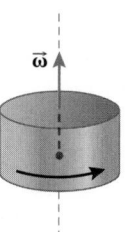

FIGURE 10.37

a. Find the rotational kinetic energy of the flywheel system.
b. At what speed would a 1.00 kg particle have the same kinetic energy?

Solution

a. The moment of inertia of a disk is, from Table 10.1,

$$I_{CM} = \frac{1}{2} m R^2$$

$$= \frac{1}{2}(150.0 \text{ kg})(0.250 \text{ m})^2$$

$$= 4.69 \text{ kg} \cdot \text{m}^2$$

The angular speed of the disk is

$$\omega = \frac{1000 \text{ rev/min}}{60 \text{ s/min}}(2\pi \text{ rad/rev})$$

$$= 104.7 \text{ rad/s}$$

The kinetic energy associated with the spin then is found from Equation 10.24:

$$KE_{rot} = \frac{1}{2} I_{CM} \omega^2$$

$$= \frac{1}{2}(4.69 \text{ kg} \cdot \text{m}^2)(104.7 \text{ rad/s})^2$$

$$= 2.57 \times 10^4 \text{ J}$$

b. A particle of mass m moving with speed v has kinetic energy

$$KE = \frac{1}{2} m v^2$$

If a 1.00 kg particle has 2.57×10^4 J of kinetic energy, its speed is determined from

$$2.57 \times 10^4 \text{ J} = \frac{1}{2}(1.00 \text{ kg}) v^2$$

Solving for v,

$$v = 227 \text{ m/s}$$

10.10 SPIN DISTORTS THE SHAPE OF THE EARTH*

To a first approximation, the Earth has a spherical shape. A completely fluid, *nonspinning* Earth would assume a spherical shape under its own self-gravitation. Careful measurements, however, indicate that the equatorial diameter of the Earth is about 43 km greater than its polar diameter. This means that the Earth is oblate and has an equatorial bulge or equatorial deviation from spherical shape. The Earth is more nearly shaped like an **oblate ellipsoid** rather than a sphere. An oblate ellipsoid is a solid figure obtained by rotating an ellipse about its minor axis; see Figure 10.38. (A **prolate ellipsoid** is the figure obtained by rotating an ellipse about its major axis, as shown in Figure 10.39.)

Newton was the first to realize that a spinning, fluid Earth assumes the shape of an oblate ellipsoid. In this section, we show why the spinning Earth cannot be spherical in shape but must become oblate.

Consider a system of mass m lying at rest on the surface of a stationary, *nonspinning*, spherical Earth. The mass experiences two forces (see Figure 10.40):

1. the gravitational force of the Earth on the system, its weight $\vec{\mathbf{w}}$; and
2. the normal force $\vec{\mathbf{N}}$ that the surface exerts on the system, directed perpendicular to the surface.

Since the mass is not accelerated, the two forces must be equal in magnitude and opposite in direction. Their vector sum is zero in accordance with Newton's second law of motion.

On the other hand, since the Earth really is spinning, it cannot remain spherical because the mass m on the surface is moving around a circular path and so must experience a centripetal

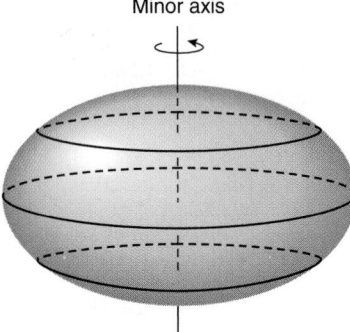

FIGURE 10.38 An oblate ellipsoid.

Minor axis

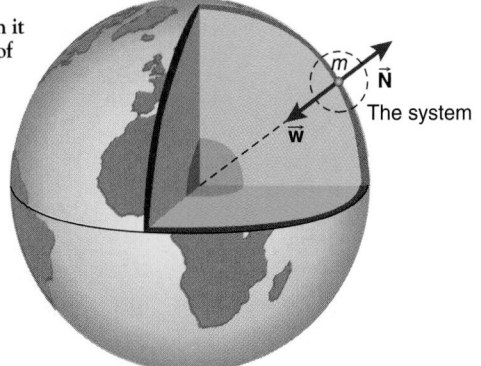

FIGURE 10.39 A prolate ellipsoid.

Major axis

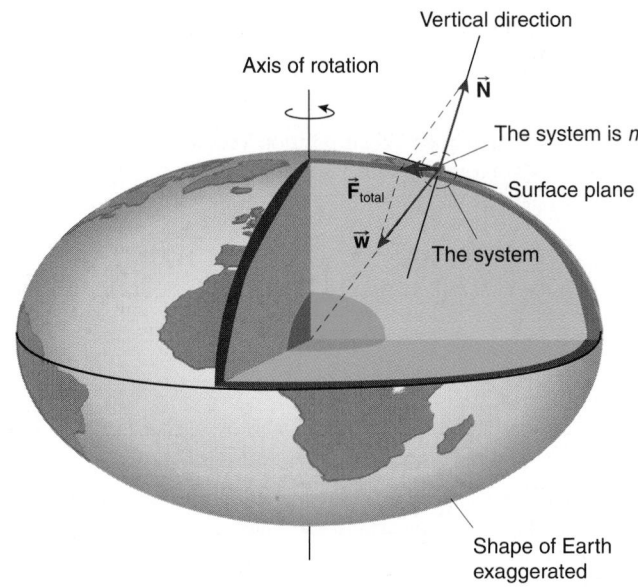

Vertical direction

Axis of rotation

$\vec{\mathbf{N}}$

The system is m

Surface plane

$\vec{\mathbf{F}}_{total}$

$\vec{\mathbf{w}}$

The system

Shape of Earth exaggerated

FIGURE 10.41 Forces on a mass m on the surface of the spinning Earth.

FIGURE 10.40 Forces on m when it is on the surface of a spherical, nonspinning Earth.

m

$\vec{\mathbf{N}}$

$\vec{\mathbf{w}}$

The system

The giant planet Saturn spins so fast that it is noticeably oblate.

acceleration toward the axis of rotation. A force is needed to provide this acceleration. The shape of the Earth distorts so that the vertical direction (a direction perpendicular to the surface of a liquid on the surface of the Earth) no longer passes through the center of the Earth (except for locations at the poles and the equator), as shown in Figure 10.41.

> The weight $\vec{\mathbf{w}}$ is no longer directed opposite to the normal force on a spinning Earth.

The shape of the Earth distorts so the vector sum of the two forces produces a total force directed toward the axis of rotation, which then provides the centripetal acceleration associated with the rotational motion of m about the axis. The faster the spin, the greater the distortion needed to provide the necessary force for the greater centripetal acceleration of the mass m. The giant planets Jupiter and Saturn spin very fast (Jupiter once in 9.84 h and Saturn once in 10.2 h) and have noticeable oblateness in telescopic images; the differences between the equatorial and polar diameters of these planets is almost 10%.

The reason the Earth distorts into an oblate ellipsoid rather than a prolate ellipsoid can be seen in Figure 10.42. The prolate ellipsoid shape would provide a vector sum of $\vec{\mathbf{w}}$ and $\vec{\mathbf{N}}$ that is directed *away* from the axis of rotation. As you know, however, for circular motion, the total force must be directed *toward* the center of the circle; a prolate ellipsoid shape cannot provide such a force.

The oblate shape of the Earth implies that the surface distance corresponding to a degree of latitude[†] is smaller near the

---

[†]Latitude is the angle between the local vertical direction (a direction perpendicular to the surface of liquid on the Earth's surface) and the perpendicular direction to the axis of spin. Equivalently, it is the angle between the local zenith and the place where the celestial equator (the plane of the Earth's equator projected onto the sky or celestial sphere) intersects the local celestial meridian (a north–south line through the zenith).

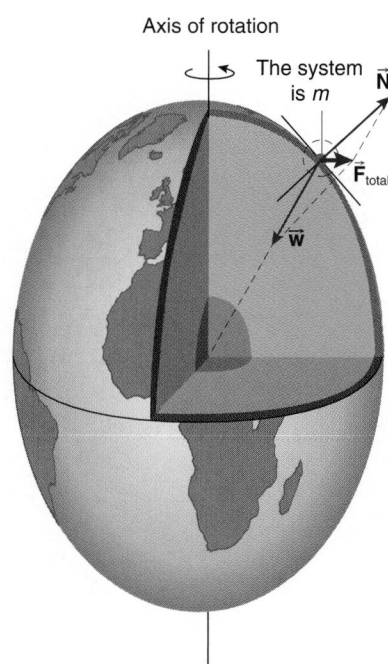

FIGURE 10.42 A mass m on a hypothetical prolate-ellipsoid-shaped planet.

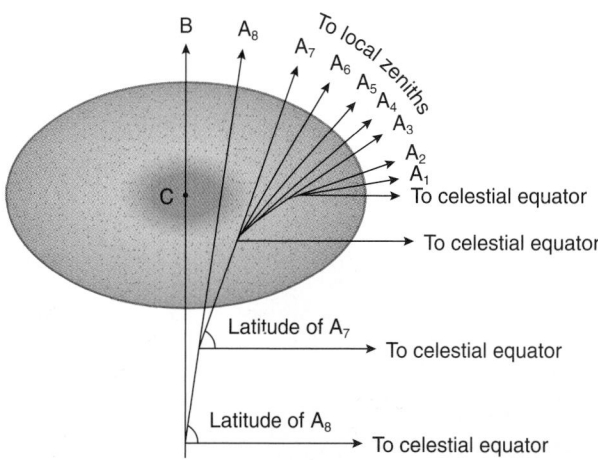

FIGURE 10.43 The distance corresponding to a degree of latitude is smaller near the equator than near the poles.

equator than near the poles, as shown in Figure 10.43. The reverse is true on a prolate ellipsoid. On the Earth, a degree of latitude represents 111.70 km near the pole and 110.57 km near the equator. The first convincing evidence for this difference came with the careful measurements made during several French expeditions to various parts of the Earth in the 1730s. The measurements were performed by Pierre Bouguer (1698–1758), Charles Marie de la Condamine (1701–1744), Louis Godin (1704–1760), and Pierre Louis Moreau de Maupertius (1698–1759).

10.11 THE PRECESSION OF A RAPIDLY SPINNING TOP*

One of the more remarkable physical systems in classical dynamics is a spinning top, familiar to us from the halcyon days of our childhood. A spinning top defies our preconceived notions of what should happen. It is wonderful to see how the Newtonian formulation of mechanics is able to account for the motions of such a marvelous toy. The same principles are used in the design of gyroscopic systems for navigating airplanes, spacecraft, and ships.

A simple top has an axis of symmetry passing through the center of mass. Several forces act on the top:

1. the weight $\vec{\mathbf{w}}$ (of magnitude mg), which can be thought of as acting on the top at its center of mass (see Figure 10.44); and
2. the force of the surface on the top at the fixed contact point between the top and the surface (the vector sum of the usual normal force of the surface and the static frictional force of the surface); this force is not shown in Figure 10.44.

The normal force of the surface and the static frictional force at the contact point produce zero torque about the contact point taken as an origin because their lines of action pass through the origin. Only the weight produces a torque about the point of contact:

$$\vec{\boldsymbol{\tau}}_{\text{total}} = \vec{\mathbf{r}} \times \vec{\mathbf{w}}$$

The vector $\vec{\mathbf{r}}$ is the vector from the origin (the point of contact) to the point of application of the weight (the center of mass of the top).

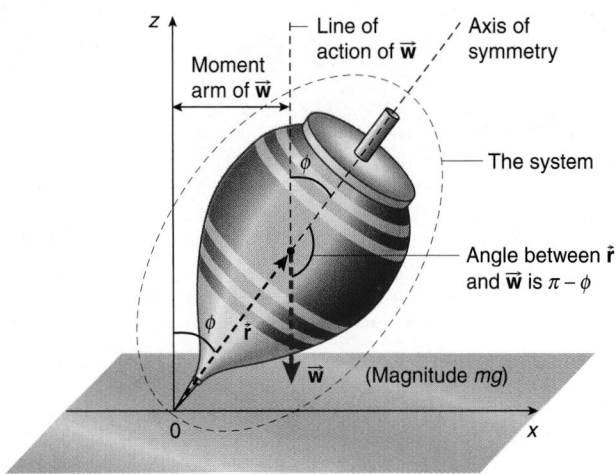

FIGURE 10.44 Only the weight has a nonzero torque about the point of contact of the top with the surface.

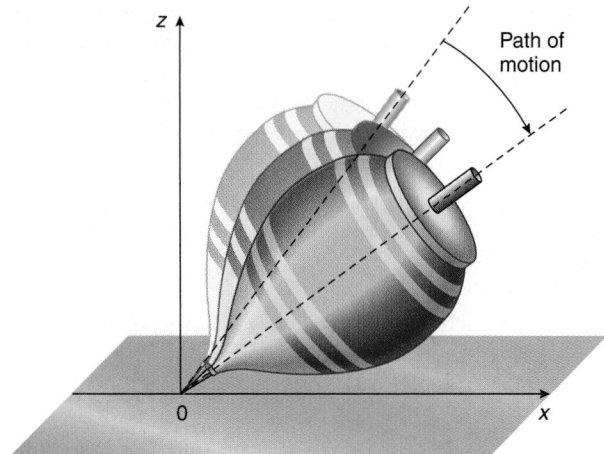

FIGURE 10.45 A nonspinning top falls over.

The magnitude of the total torque on the top is, from Figure 10.44,

$$\tau_{\text{total}} = r(mg) \sin(\pi - \phi)$$
$$= mgr \sin \phi$$

where ϕ is the angle between the symmetry axis of the top and the vertical direction.

The vector product always is perpendicular to the vectors forming the product, and so the direction of the torque is perpendicular to both $\vec{\mathbf{r}}$ and $\vec{\mathbf{w}}$. The direction of the torque is perpendicular to the plane formed by the symmetry axis of the top and the local vertical direction. *The torque on the top always is perpendicular to this plane regardless of the orientation of the top with respect to the particular coordinate system* of Figure 10.44.

According to Equation 10.19, the torque is equal to the time rate of change of the angular momentum:

$$\vec{\boldsymbol{\tau}}_{\text{total}} = \frac{d\vec{\mathbf{L}}}{dt}$$

Hence the *differential change* $d\vec{\mathbf{L}}$ in the angular momentum vector is in the same direction as the torque—that is, in a direction perpendicular to the plane of the symmetry axis and the local vertical direction.

We consider two cases.

A Nonspinning Top

In this case the top initially has zero angular velocity and zero angular momentum, since $\vec{\mathbf{L}} = I\vec{\boldsymbol{\omega}}$. The *change* in the angular momentum is in the direction of the torque according to Equation 10.19. The torque is directed along $\hat{\mathbf{j}}$ when the top is in the position indicated in Figures 10.44 and 10.45.

Because of the torque, the *change* in the angular momentum (initially zero) also is along $\hat{\mathbf{j}}$, perpendicular to the plane of the symmetry axis and the local vertical direction. This is precisely what happens when the top simply falls over and hits the table. In Figure 10.45 the top rotates in a clockwise sense as it falls over, and so the top develops a growing angular velocity vector

oriented along $\hat{\mathbf{j}}$ according to the circular motion right-hand rule. The angular momentum vector grows from zero and is parallel to the growing angular velocity vector, because $\vec{\mathbf{L}} = I\vec{\boldsymbol{\omega}}$. Both are in the same direction as the torque (perpendicular to the plane of the symmetry axis and the local vertical). Nothing surprising here: the top falls over because of the torque caused by its weight.

A Rapidly Spinning Top

In this case the top has a *nonzero* initial angular momentum vector directed along the symmetry axis of the top; see Figure 10.46. The angular momentum is related to the angular velocity through the moment of inertia I_{CM} of the top:

$$\vec{\mathbf{L}}_{\text{spin}} = I_{\text{CM}}\vec{\boldsymbol{\omega}}_{\text{spin}}$$

This relationship, though, is peripheral to understanding what happens to the top as a result of the torque. All we need to know is that there is an angular momentum vector directed along the symmetry axis. The torque is unchanged regardless of whether the top is spinning.

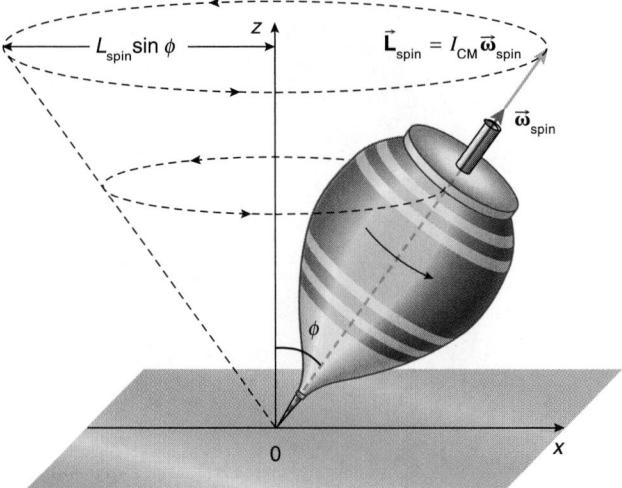

FIGURE 10.46 A rapidly spinning top.

But the *effect* of the torque now is quite different. The *change* in the angular momentum vector is (as always) in the *same* direction as the torque. The torque forces the *existing* angular momentum vector to move in a direction perpendicular to the plane of the symmetry axis and the local vertical direction! Since the change in the angular momentum vector is perpendicular to the angular momentum vector itself, the effect of the torque is to change the *direction* of the *existing* angular momentum vector but not its magnitude! The angular momentum vector is tied to the symmetry axis of the top, and so the top does *not* fall down: it moves in a direction perpendicular to the plane of the symmetry axis and the local vertical. As the top moves, the direction of the torque vector changes, because it is always perpendicular to the plane of the symmetry axis and the local vertical. The torque and the change in the angular momentum vector remain perpendicular to $\vec{L}_{spin}$ itself, and so the axis of the top moves (we say *precesses*) counterclockwise about the z-axis of Figure 10.46. The angular momentum vector $\vec{L}_{spin}$ and the top trace out a precessional cone.

We can calculate the angular speed of the precession. Notice the spin angular momentum vector $\vec{L}_{spin}$ and the change $d\vec{L}$ in this vector in Figure 10.47. From the geometry, the magnitude of the change is

$$dL = L_{spin} \sin \phi \, d\theta$$

Solving for $d\theta$,

$$d\theta = \frac{dL}{L_{spin} \sin \phi}$$

The angular speed of the precession is

$$\omega_{precs} = \frac{d\theta}{dt}$$

or upon substituting for $d\theta$,

$$\omega_{precs} = \frac{1}{L_{spin} \sin \phi} \frac{dL}{dt}$$

But

$$\frac{dL}{dt}$$

is equal to the magnitude of the torque (= $mgr \sin \phi$). With this substitution, the precessional angular speed is

$$\omega_{precs} = \frac{mgr \sin \phi}{L_{spin} \sin \phi}$$

$$= \frac{mgr}{L_{spin}}$$

The faster the spin given to the top, the greater L_{spin} and *the slower* or smaller the precessional angular speed. The precessional angular velocity is parallel to $\hat{k}$ in Figures 10.46 and 10.47.

Since the top has a precessional angular velocity, the center of mass of the top is in motion, and so the top also has a small amount of *orbital* angular momentum (directed along $\hat{k}$ in Figures 10.46 and 10.47). Thus the *total* angular momentum of the top,

$$\vec{L}_{total} = \vec{L}_{orbit} + \vec{L}_{spin}$$

is not, in a strict sense, coincident with the symmetry axis of the top; $\vec{L}_{spin}$ is directed along the symmetry axis while $\vec{L}_{orbit}$ is along $\hat{k}$. The torque is equal to the time rate of change of the *total* angular momentum. However, if the spin of the top is large, the precessional angular speed is small and, therefore, the magnitude of the orbital angular momentum of the center of mass also will be small compared with the magnitude of the spin angular momentum about the center of mass:

$$L_{orbit} \ll L_{spin}$$

Therefore the total angular momentum is essentially $\vec{L}_{spin}$ and our analysis of the dynamics is a good approximation. An exact analysis, particularly for slowly spinning tops, is more involved.

QUESTION 7

What happens to the direction of precession of a top if the spin of the top is in the direction opposite to that indicated in Figure 10.46?

10.12 THE PRECESSION OF THE SPINNING EARTH*

The Earth spins on its axis about once a day. The rotational axis of the Earth defines the geographic north and south poles. The spin angular momentum vector of the Earth is directed along the symmetry axis, as shown in Figure 10.48. You can think of the Earth as a huge spinning top, like the one we examined in the previous section.

The Earth is not precisely a sphere; it has a middle-aged spread (an equatorial bulge) caused by its rotation, as was discussed in Section 10.10. The equatorial diameter of the Earth is about 43 km larger than the polar diameter, and the Earth is roughly shaped like an oblate ellipsoid.

The orbital plane of the Earth in its motion about the Sun, known as the plane of the **ecliptic**, is *different* from the equatorial

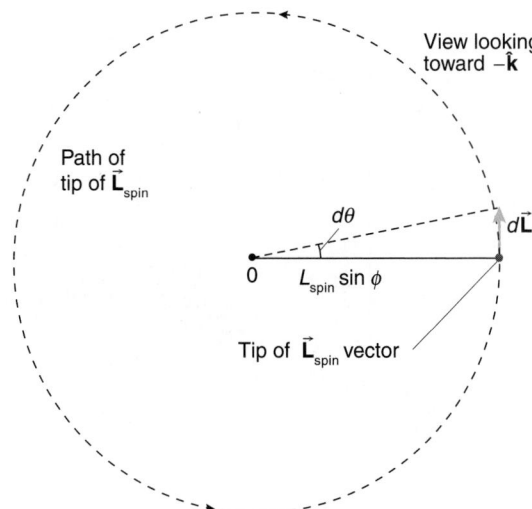

FIGURE 10.47 Top-down view of the precession.

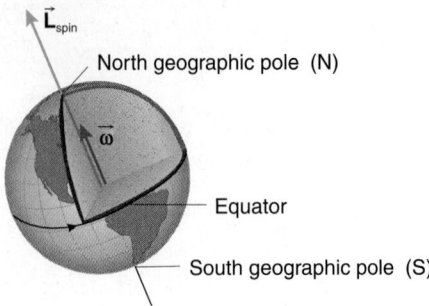

FIGURE 10.48 The spinning Earth.

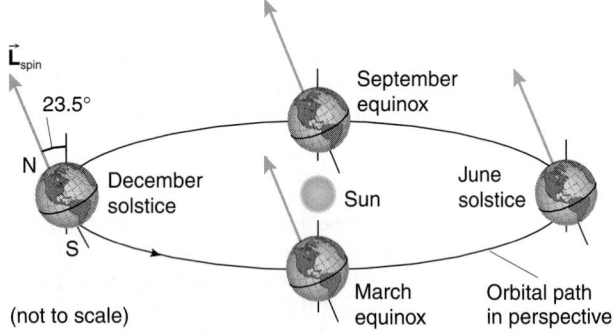

FIGURE 10.49 The orbital plane of the Earth (the ecliptic) is different from the equatorial plane.

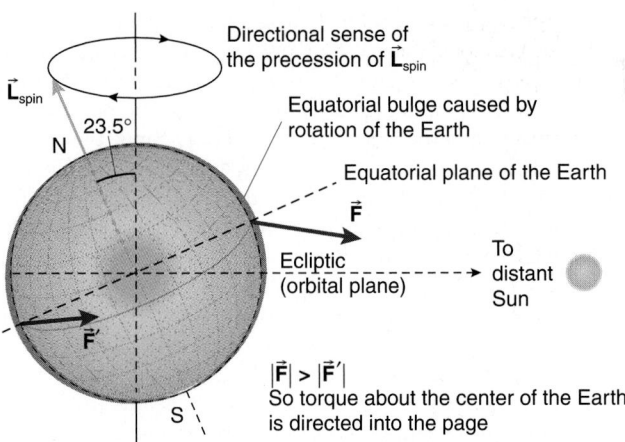

FIGURE 10.50 The force on the parts of the bulge closer to the Sun is of slightly greater magnitude than that on parts further from the Sun.

plane of the Earth, as shown in Figure 10.49. The two planes are inclined to each other by about 23.5°, an angle known as the **obliquity of the ecliptic**.

Consider the effect of the Sun on the equatorial bulge of the Earth. The gravitational attraction of the Sun on the part of the equatorial bulge in the hemisphere facing the Sun is slightly greater than that on the part of the bulge in the hemisphere opposite the Sun because of their slightly different distances from the Sun, stemming from the finite size of the Earth. The force on the parts of the bulge closer to the Sun and that on the parts further from the Sun also point in slightly different directions, as shown in Figure 10.50.

The forces of the Sun on the equatorial bulge produce torques about the center of the Earth. The resulting *total* torque is directed perpendicularly into the page of Figure 10.50.

Notice that if the Earth were spherical or if the angle between the equatorial plane and the ecliptic plane were zero, there would be no torque. Examining Figure 10.49, you can see that the torque is greatest at the times of the solstices (in June and December) and least at the times of the equinoxes (in March and September), so that it has a semiannual periodicity associated with it. The torque tries to force the equatorial bulge (and the Earth with it) into the plane of the ecliptic and align the polar axis of the Earth perpendicular to this plane. This is precisely what would occur if the Earth were not spinning.

The Earth *is* spinning, however, and so the *change* in the spin angular momentum vector (which is in the *same* direction

as the *torque*) is *perpendicular* to the spin angular momentum vector of the Earth, because the torque is perpendicular to $\vec{L}_{spin}$. As a result the rotational axis of the Earth and the spin angular momentum vector maintain a *fixed* angle with the ecliptic plane and slowly change their orientation in space, tracing out a precessional cone. The spinning Earth precesses. The speed of the precessional motion is quite small, 50.262 arc seconds (= 0.013 962°) per year, completing the precessional circle in 25 785 y.

The motion is similar to that of a spinning top, with one important difference. Set a top spinning counterclockwise (as seen from above) with its axis inclined to the vertical, as shown in Figure 10.51. The gravitational force of the Earth on the top does not cause the top to fall over (which is what would occur if the top were not spinning). The top precesses, the axis of the top tracing out a cone in space about the vertical direction. In the

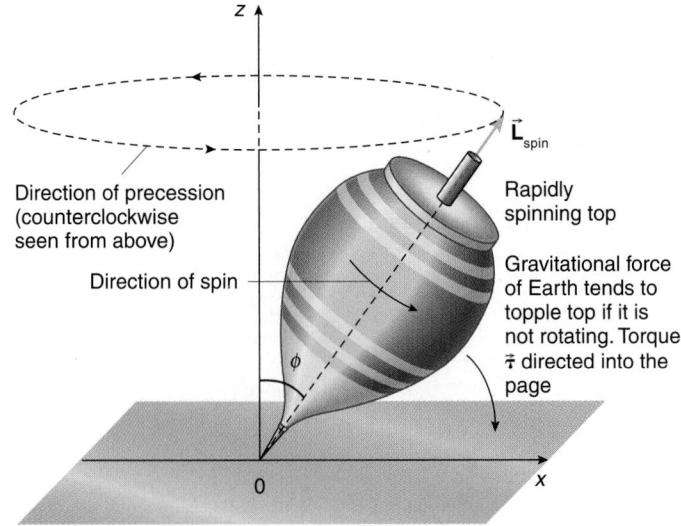

FIGURE 10.51 A counterclockwise spinning top precesses in the counterclockwise direction.

case depicted here, the precessional motion of the top also is *counterclockwise* about the vertical direction (when viewed from above), since the force is tending to topple the top, not to make it vertical or to "right" the top.

The precession of the spinning Earth is similar except that the precessional motion of the axis is in the *opposite direction*—

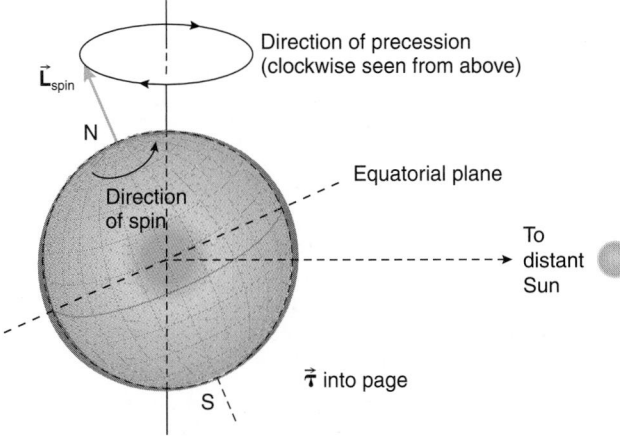

FIGURE 10.52 The spinning Earth precesses in the clockwise sense when viewed from above the Northern Hemisphere.

that is, *clockwise* (when seen from above) about a line perpendicular to the plane of the ecliptic; see Figure 10.52. In this case the torque tends to "right" the Earth, not topple it or tip it over as was the case for the top.

The precession of the rotational axis of the Earth has interesting observational consequences for astronomers on Earth. The point toward which the rotational axis of the Earth points in the heavens in the Northern Hemisphere is known as the **North Celestial Pole**, or NCP. The brightest star nearest the position of the NCP customarily is called the **pole star** (the north star for Northern Hemisphere observers). Precession means that the identity of the pole star changes with time. The north star is now Polaris (α Ursae Minoris). About 5000 years ago, when the Great Pyramid of Khufu was built in Egypt, the north star was Thuban (α Draconis).* The bright star Vega (α Lyrae) will be the north star for the class of A.D. 14 000 at your college or university. The path traced out by the North Celestial Pole in the sky over the 25 785-year precessional period is indicated in the star map of the northern sky shown in Figure 10.53.

---

*The sloping shaft into the interior of the Great Pyramid of Khufu (c. 2900 B.C.) at Giza (near modern Cairo) was oriented toward the pole star and so is parallel to the spin axis of the Earth. The shaft still points toward the pole star, but now it is Polaris, not Thuban.

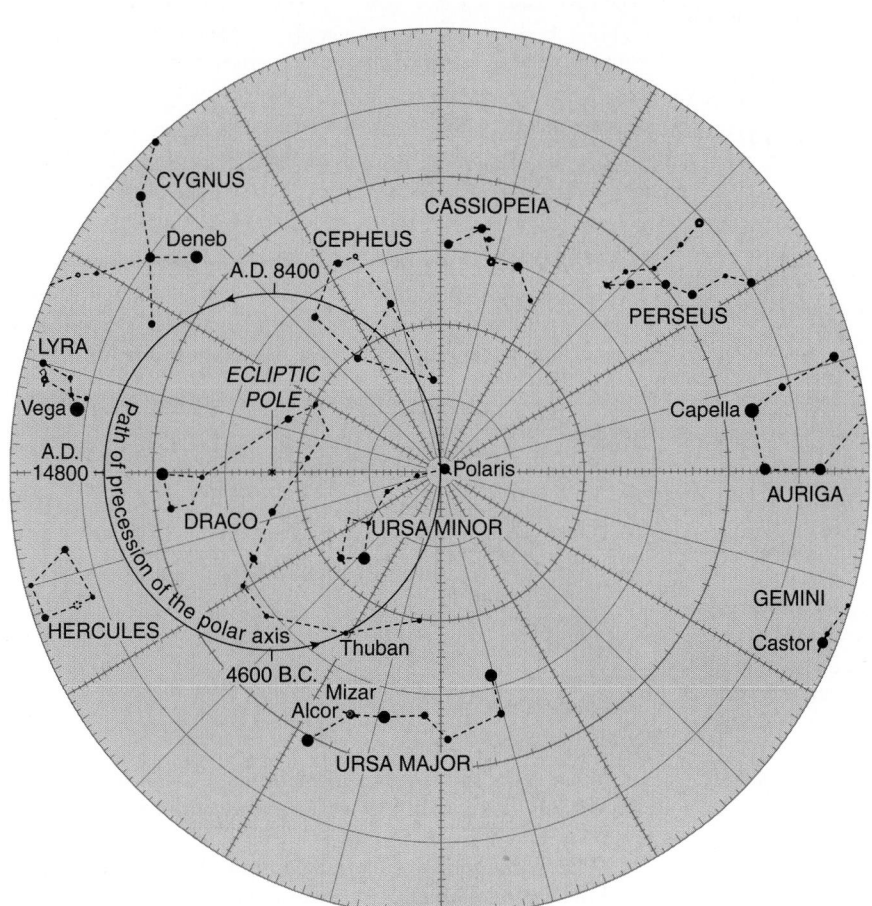

FIGURE 10.53 The precessional path of the polar axis of the Earth during the 25 785-year precessional period.

The ancient Greek astronomer Hipparchus (c. 125 B.C.) discovered the precession of the rotational axis of the Earth. The NCP moves only about 1° around the precessional circle during an average human lifetime of 70 years (much longer than the average lifetime at the time of Hipparchus!). The discovery gives you a feeling for how diligently ancient cultures observed the sky!

QUESTION 8

Imagine yourself at the north geographic pole of the Earth. As you look vertically up into the sky, does the slow precessional path of the axis of the spinning Earth trace out a clockwise or counterclockwise path during 25 785 y? Is your answer the same if you stand at the south geographic pole?

10.13 SIMULTANEOUS SPIN AND ORBITAL MOTION

There are many examples of systems that undergo both spin and orbital motion. The system spins about a symmetry axis through the center of mass point and, simultaneously, the center of mass is in motion. The tires of your car spin while their center of mass moves (orbital motion). A beautiful football pass has both spin and orbital motion. Each planet spins about its axis and orbits the Sun. The moons of the planets simultaneously spin on their axes while orbiting the mother planet.

A spinning system has rotational kinetic energy as we discovered in Section 10.9. The kinetic energy associated with the spin is (Equation 10.24 with appropriate subscripts added)

$$KE_{rot} = \frac{1}{2} I_{CM} \omega_{spin}^2$$

But there is additional kinetic energy associated with the motion of the center of mass. The kinetic energy associated with the motion of the center of mass is

$$KE = \frac{1}{2} m v_{CM}^2$$

or equivalently,

$$KE = \frac{1}{2} (m r_\perp^2) \omega_{orbit}^2$$

We showed in Section 9.11 that the total kinetic energy is the sum of the rotational kinetic energy about the center of mass and the kinetic energy associated with the motion of the center of mass itself:

$$KE_{total} = KE_{rot} + \frac{1}{2} m v_{CM}^2 \qquad (10.25)$$

Likewise, for a system with spin and orbital motion, the total angular momentum has two contributions. The spin angular momentum $\vec{L}_{spin}$ of the system is proportional to the spin angular velocity (Equation 10.18):

$$\vec{L}_{spin} = I_{CM} \vec{\omega}_{spin}$$

where I_{CM} is the moment of inertia of the system about the appropriate symmetry axis through the center of mass point, and $\vec{\omega}_{spin}$ is the spin angular velocity of the system about this axis. For orbital motion, the system is imagined to be a point particle, with a mass equal to the total mass m of the entire system, located at the center of mass point. The orbital angular momentum of the system is

$$\vec{L}_{orbit} = \vec{r} \times \vec{p}_{CM}$$

where $\vec{p}_{CM}$ is the momentum of the center of mass,

$$\vec{p}_{CM} = m \vec{v}_{CM}$$

(here $\vec{v}_{CM}$ is the velocity of the center of mass). Equivalently the orbital angular momentum can be found from Equation 10.4 (with appropriate subscripts added):

$$\vec{L}_{orbit} = m r_\perp^2 \vec{\omega}_{orbit}$$

where $\vec{\omega}_{orbit}$ is the angular velocity of the center of mass about an axis and $r_\perp$ is the perpendicular distance of the center of mass point from the axis of orbital motion.

The total angular momentum of the system is the vector sum of the spin and orbital angular momenta:

$$\vec{L}_{total} = \vec{L}_{spin} + \vec{L}_{orbit}$$
$$= I_{CM} \vec{\omega}_{spin} + m r_\perp^2 \vec{\omega}_{orbit} \qquad (10.26)$$

EXAMPLE 10.10

A 3.00 kg disk of radius 0.100 m is spinning at 15.0 rev/s while simultaneously zipping across a horizontal (frictionless) ice surface at a speed of 5.00 m/s, as shown in Figure 10.54.

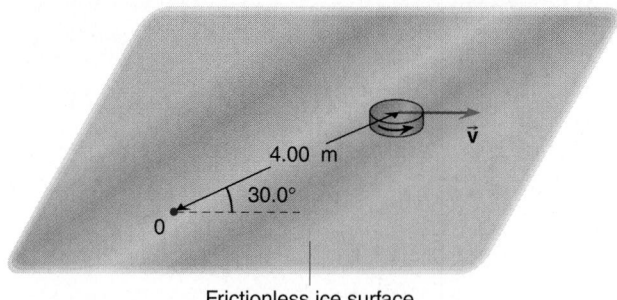

4.00 m
30.0°
0
v

Frictionless ice surface

FIGURE 10.54

a. What is the spin angular momentum of the disk?
b. What is the orbital angular momentum of the disk about the point 0?
c. What is the total angular momentum of the disk?
d. Find the total kinetic energy of the disk.

Solution

a. The moment of inertia of a disk about the symmetry axis perpendicular to the disk and through the center of mass is found from Table 10.1 to be

$$I_{CM} = \frac{1}{2}mR^2$$
$$= \frac{1}{2}(3.00 \text{ kg})(0.100 \text{ m})^2$$
$$= 1.50 \times 10^{-2} \text{ kg} \cdot \text{m}^2$$

The spin angular momentum is, from Equation 10.18,

$$\vec{L}_{spin} = I_{CM}\vec{\omega}_{spin}$$

Use the circular motion right-hand rule and the coordinate system indicated in Figure 10.55 to discover that the direction of the spin angular velocity vector is along $+\hat{k}$. Hence

$$\vec{L}_{spin} = I_{CM}\omega_{spin}\hat{k}$$

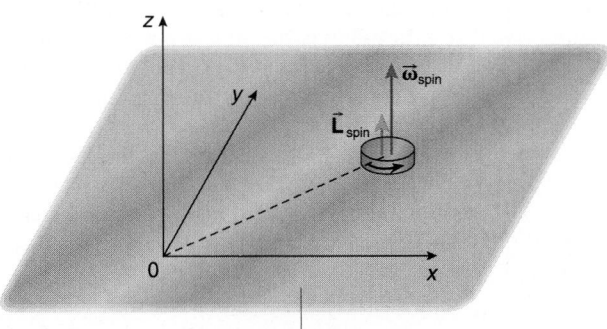

FIGURE 10.55

The spin angular speed is

$$\omega_{spin} = (15.0 \text{ rev/s})(2\pi \text{ rad/rev})$$
$$= 94.2 \text{ rad/s}$$

Hence the spin angular momentum is

$$\vec{L}_{spin} = I_{CM}\omega_{spin}\hat{k}$$
$$= (1.50 \times 10^{-2} \text{ kg} \cdot \text{m}^2)(94.2 \text{ rad/s})\hat{k}$$
$$= (1.41 \text{ kg} \cdot \text{m}^2/\text{s})\hat{k}$$

b. The orbital angular momentum of the disk about the point 0 is

$$\vec{L}_{orbit} = \vec{r} \times \vec{p}$$

Its magnitude is

$$L_{orbit} = rp \sin\theta$$
$$= (4.00 \text{ m})(3.00 \text{ kg})(5.00 \text{ m/s}) \sin 30.0°$$
$$= 30.0 \text{ kg} \cdot \text{m}^2/\text{s}$$

The direction of the orbital angular momentum is the direction of $\vec{r} \times \vec{p}$. From the vector product right-hand rule, this is toward $-\hat{k}$. Hence

$$\vec{L}_{orbit} = (-30.0 \text{ kg} \cdot \text{m}^2/\text{s})\hat{k}$$

c. The total angular momentum is the vector sum of the spin and orbital angular momenta (see Figure 10.56):

$$\vec{L}_{total} = \vec{L}_{spin} + \vec{L}_{orbit}$$
$$= (1.41 \text{ kg} \cdot \text{m}^2/\text{s})\hat{k} + (-30.0 \text{ kg} \cdot \text{m}^2/\text{s})\hat{k}$$
$$= (-28.6 \text{ kg} \cdot \text{m}^2/\text{s})\hat{k}$$

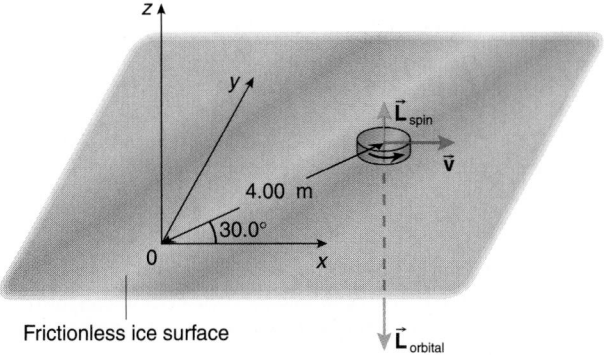

FIGURE 10.56

d. The total kinetic energy is the sum of the kinetic energy of rotation plus the kinetic energy associated with the translation of the center of mass:

$$KE_{total} = \frac{1}{2}I_{CM}\omega_{spin}^2 + \frac{1}{2}mv_{CM}^2$$
$$= \frac{(1.50 \times 10^{-2} \text{ kg} \cdot \text{m}^2)(94.2 \text{ rad/s})^2}{2}$$
$$+ \frac{(3.00 \text{ kg})(5.00 \text{ m/s})^2}{2}$$
$$= 66.6 \text{ J} + 37.5 \text{ J}$$
$$= 104.1 \text{ J}$$

10.14 SYNCHRONOUS ROTATION AND THE PARALLEL AXIS THEOREM

In many cases the orbital angular velocity $\vec{\omega}_{orbit}$ is different from the spin angular velocity $\vec{\omega}_{spin}$. For example, the spin angular velocity of the Earth $\vec{\omega}_{spin}$ is significantly greater than its orbital angular velocity $\vec{\omega}_{orbit}$ about the Sun; the two vectors also point in different directions. Other planets and some moons are similar in these respects.

Occasionally, however, a system has *equal* spin and orbital angular velocities:

$$\vec{\omega}_{spin} = \vec{\omega}_{orbit}$$

In this case:

1. the two vectors are parallel; and

2. the spin angular speed is equal to the orbital angular speed.

If these conditions are satisfied, the system is said to undergo **synchronous rotation**.

A good example of synchronous rotation is the motion of the Moon about the Earth*: as the Moon orbits the Earth, the Moon always keeps the *same* hemisphere facing the Earth.[†]

In synchronous rotation, as the system completes a revolution with respect to the orbital axis, the system *also* completes a spin revolution about an axis through its center of mass, as shown in Figure 10.57. If $\vec{\omega}_{spin}$ is equal to $\vec{\omega}_{orbit}$, call the common angular velocity $\vec{\omega}$. Then the total angular momentum of the system is, according to Equation 10.26,

$$\begin{aligned}\vec{L}_{total} &= I_{CM}\vec{\omega} + md^2\vec{\omega} \\ &= (I_{CM} + md^2)\vec{\omega}\end{aligned} \qquad (10.27)$$

Write the total angular momentum as

$$\vec{L}_{total} = I\vec{\omega} \qquad (10.28)$$

where I is the moment of inertia of the system about an axis parallel to the symmetry axis through the center of mass point, but not coincident with this symmetry axis. Compare Equations 10.27 and 10.28.

We see that the moment of inertia I of the system about an axis parallel to the symmetry axis of the system and separated from it by a perpendicular distance d is

$$I = I_{CM} + md^2 \qquad (10.29)$$

Equation 10.29 is known as the **parallel axis theorem**.

---

*Tidal gravitational forces of the Earth on the Moon caused the motion of the Moon to have this characteristic. For similar reasons, many (but not all) of the moons of the planets also exhibit synchronous rotation.

[†] The spin axis and the orbital axis are not quite parallel (they have an angle of about 6° between them), but for simplicity we assume they are parallel here. The 6° angle and the elliptical orbit of the Moon enable observers on the Earth to see approximately 59% of the lunar real estate as the Moon traverses its orbital path.

This theorem applies to any rigid body in synchronous rotation about an axis that is parallel to the symmetry axis through the center of mass. The parallel axis theorem can be applied to a system *only* when the spin and orbital angular velocities of the system are the same—that is, for synchronous rotation.

10.15 ROLLING MOTION WITHOUT SLIPPING

A rolling stone gathers no moss

Publius Syrus (c. 1st century B.C.)[‡]

The rolling motion of wheels is such a common feature of our technological society that we take it for granted. No one, least of all a college student, wants to be without wheels. In this section we consider rolling motion in some detail, for it is not as trivial as it may seem.

When a symmetric rigid body system undergoes rolling motion, it is simultaneously spinning about its center of mass while the center of mass is undergoing translation. Notice in Figure 10.57 that if the radius of the orbit of point A is zero, the system is *rolling without slipping* about the orbital rotational axis. No slipping means there is no sliding at the point of contact of the system with the orbital rotational axis. The wheels of a car that is moving forward while its tires are spinning madly, leaving behind black stripes on the road, is an example of rolling motion *with slipping*; we commonly call it skidding. Likewise, when drag racers spin the tires of a car, the tires are rolling with slipping at the point of contact between the tires and the road.

Rolling motion (without slipping) is a special case of synchronous rotation.

The spin and orbital axes are parallel and separated by the radius of the circularly shaped system undergoing rolling motion. The

---

[‡] *The Moral Sayings of Publius Syrus, a Roman Slave*, translated by Darius Lyman (Andrew J. Graham, New York, 1862), page 48 (saying #524).

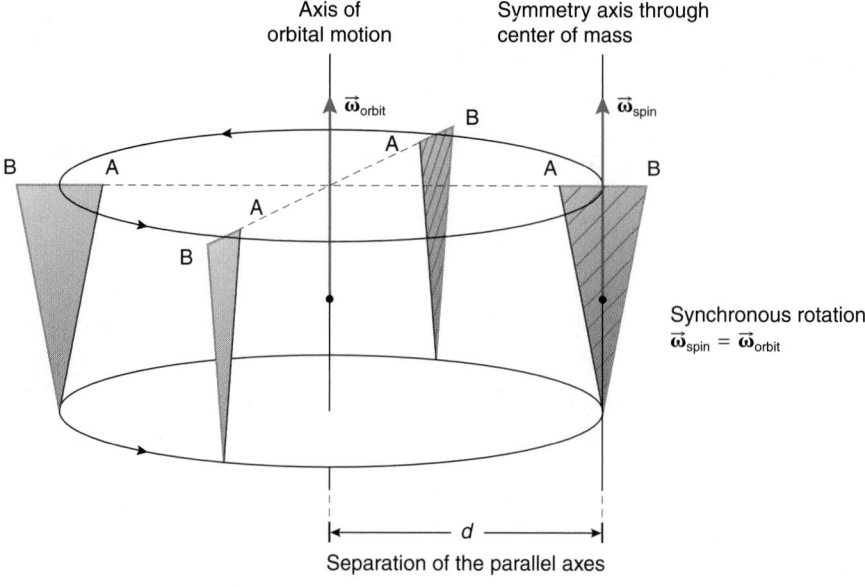

FIGURE 10.57 In synchronous rotation, a system keeps the same side facing the orbital axis.

moment of inertia I, found from the parallel axis theorem, then represents the moment of inertia of the system about the point of rolling contact (e.g., the point of contact between a tire and the road).

It is preferable for tires to roll without slipping on pavement, for then the appropriate coefficient of friction is the static value, rather than the smaller coefficient of kinetic friction. Tires rolling *with* slipping are skidding (such as on icy pavements), which can be quite dangerous for the occupants of the vehicle, not to say for the vehicle itself. Antilock braking systems are designed to ensure that tires roll without slipping during braking.

If a symmetric rigid body system is rolling without slipping, there are relationships between

1. the distance the center of mass travels and the corresponding angle through which the system rotates about the symmetry axis through the center of mass;
2. the speed of the center of mass and the angular speed of rotation; and
3. the magnitude of the acceleration of the center of mass and the magnitude of the angular acceleration of the system.

We want to determine what these relationships are.

Consider a round rigid body system such as a wheel, sphere, or other system with a circular cross section, as shown in Figure 10.58. After one complete rotation, the center of the system moves through a distance equal to the circumference of its circular cross section. In other words the center moves through a distance s, where

$$s = 2\pi R$$

If the system rotates through an angle θ (in radians), as in Figure 10.59, then the center moves through a distance

$$s = R\theta \tag{10.30}$$

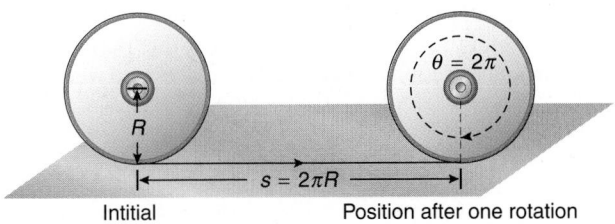

FIGURE 10.58 One complete revolution of a rolling system translates the center of mass by the circumference of the circular cross section.

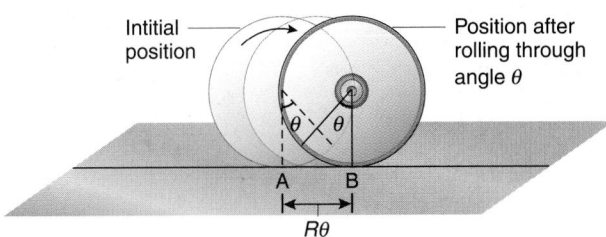

FIGURE 10.59 Rolling through an angle θ translates the center of mass by a distance $R\theta$.

Differentiating Equation 10.30 with respect to time, we find

$$\left| \frac{ds}{dt} \right| = R \left| \frac{d\theta}{dt} \right|$$

But

$$\left| \frac{ds}{dt} \right| = v_{center}$$

the speed of the center of the system. Likewise

$$\left| \frac{d\theta}{dt} \right| = \omega$$

the angular speed of the system about the rotational axis through the center of the circular cross section.

Thus the rolling condition is equivalent to the following relation between the translational speed of the center and the angular speed:

$$v_{center} = R\omega \tag{10.31}$$

Differentiation of Equation 10.31 relates the magnitude of the linear acceleration of the center of the wheel, $\vec{a}_{center}$, to the magnitude of the angular acceleration $\vec{\alpha}$ about the axis through the center of the wheel:

$$a_{center} = R\alpha \tag{10.32}$$

Equations 10.30, 10.31, or 10.32 imply that the object is rolling without slipping; therefore they are known as **rolling constraints**.

Consider an origin at 0 in Figure 10.60. The position vector $\vec{r}$ of any point P within or on the circular cross section is the sum of the vector $\vec{r}_{center}$ locating the center of the circle and the position vector $\vec{r}'$ locating the point P with respect to the center of the circle:

$$\vec{r} = \vec{r}_{center} + \vec{r}'$$

Differentiating this equation with respect to time, we find

$$\vec{v} = \vec{v}_{center} + \vec{v}' \tag{10.33}$$

where $\vec{v}_{center}$ is the velocity of the center of the wheel and $\vec{v}'$ is the velocity of the point P with respect to the center of the wheel. *For points on the circumference of the wheel, the magnitude of $\vec{v}'$ is the same as the magnitude of $\vec{v}_{center}$ because of the rolling*

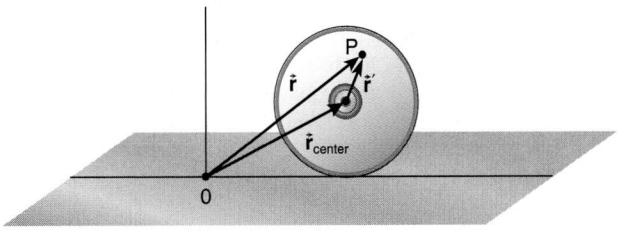

FIGURE 10.60 Geometry for describing rolling motion.

constraints. The velocity $\vec{v}$ obtained by performing the vector addition in Equation 10.33 for various points on the wheel is shown in Figure 10.61. Notice that the point of the wheel in contact with the surface is at rest while the point at the top of the wheel is moving with a velocity $2\vec{v}_{center}$.

Assume the center is also the center of mass. As we learned in Section 9.11, the *kinetic energy* of a rolling system has two contributions:

1. the kinetic energy associated with the motion of the center of mass:

$$KE_{trans} = \frac{1}{2}mv_{CM}^2$$

where v_{CM} is the speed of the center of mass; and

2. the kinetic energy associated with the spin about the center of mass:

$$KE_{rot} = \frac{1}{2}I_{CM}\omega^2$$

where I_{CM} is the moment of inertia of the object about the center of mass and ω is the angular speed of the rotation (spin) about the center of mass.

Thus the total kinetic energy is

$$KE_{total} = KE_{trans} + KE_{rot}$$
$$= \frac{1}{2}mv_{CM}^2 + \frac{1}{2}I_{CM}\omega^2$$

Use the rolling constraint, Equation 10.31, for v_{CM}:

$$KE_{total} = \frac{1}{2}mR^2\omega^2 + \frac{1}{2}I_{CM}\omega^2$$
$$= \frac{1}{2}(mR^2 + I_{CM})\omega^2$$

The parallel axis theorem, Equation 10.29, enables us to write the total kinetic energy as

$$KE_{total} = \frac{1}{2}I\omega^2 \qquad (10.34)$$

PROBLEM-SOLVING TACTIC

10.4 If you use the parallel axis theorem to determine the moment of inertia I of the system, the total kinetic energy of the system due to *both* its spin and orbital motion is given by Equation 10.34:

$$KE_{total} = \frac{1}{2}I\omega^2 \qquad (10.34)$$

QUESTION 9

An object of mass m, circular cross section (of radius R), and moment of inertia I_{CM} is *rolling* without slipping along a horizontal surface. (a) What is the *ratio* of the kinetic energy of translation to the kinetic energy of rotation? That is, find

$$\frac{KE_{trans}}{KE_{rot}}$$

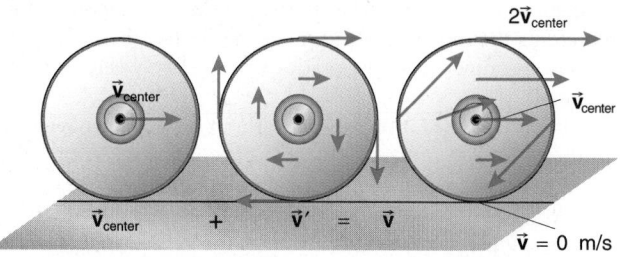

FIGURE 10.61 The velocity of the center, the velocity of points relative to the center, and the velocity with respect to the ground.

(b) Does the ratio depend on the speed of the object?
(c) Which of the objects listed in Table 10.1 with circular cross sections has the *greatest* value for this ratio? For which (if any) is the ratio equal to 1?

STRATEGIC EXAMPLE 10.11

Release at rest a spherical marble of mass m and radius R and let it roll down an inclined plane of vertical height h on your desk. Determine the speed of the marble at the base of the incline.

Solution

Consider the marble to be the system. The forces acting on the sphere are the following:

1. Its weight $\vec{w}$ of magnitude mg. The work done by the gravitational force is accounted for in the CWE theorem through the change in the gravitational potential energy of the center of mass of the sphere.
2. The normal force $\vec{N}$ of the surface on the sphere. As always, this force does zero work, because the normal force is directed perpendicular to the change in the position vector of the sphere at all points along its path.
3. The frictional force on the sphere by the surface. For rolling motion without slipping, the point of contact between the sphere and the surface is at rest, so that the frictional force is the *static* frictional force. This force does zero work.

The CWE theorem here takes the form

$$0 \text{ J} = \Delta(KE + PE)$$
$$= (KE + PE)_f - (KE + PE)_i$$

At the top of the incline the KE is initially zero:

$$KE_i = 0 \text{ J}$$

Take the origin of a coordinate system to be at ground level, as shown in Figure 10.62. The initial gravitational PE of the center of mass of the sphere is

$$PE_i = mg(h + R)$$

At the bottom of the incline, the gravitational PE of the center of mass of the sphere is

$$PE_f = mgR$$

The final KE may be found in two (equivalent) ways.

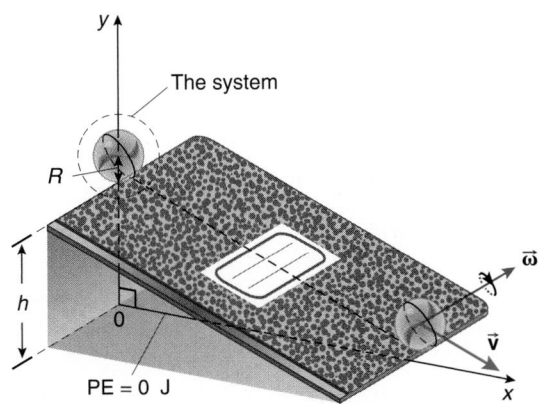

FIGURE 10.62

Method 1

Think of the KE as having two contributions: the motion of the center of mass and the spin about the center of mass. The KE of the center of mass is

$$\frac{1}{2} m v^2$$

and the rotational KE about the center of mass is

$$\frac{1}{2} I_{CM} \omega^2$$

where I_{CM} is the moment of inertia of a sphere about an axis passing through its center, found in Table 10.1:

$$I_{CM} = \frac{2}{5} m R^2$$

That is,

$$KE_f = \frac{1}{2} m v^2 + \frac{1}{2} \left(\frac{2}{5} m R^2 \right) \omega^2$$

Use the rolling constraint of Equation 10.31, for ω, and you have

$$KE_f = \frac{1}{2} m v^2 + \frac{1}{2} \left(\frac{2}{5} m R^2 \right) \left(\frac{v}{R} \right)^2$$

$$= \frac{7}{10} m v^2$$

Method 2

Realize that the sphere is instantaneously rotating about the point of contact between the sphere and the plane. With this view, the total KE then is found from Equation 10.34 where I is the moment of inertia of the sphere about an axis at the point of contact. This moment of inertia is obtained from the parallel axis theorem of Equation 10.29:

$$I = \frac{2}{5} m R^2 + m R^2$$

$$= \frac{7}{5} m R^2$$

Hence

$$KE_f = \frac{1}{2} \left(\frac{7}{5} m R^2 \right) \omega^2$$

Use the rolling constraint of Equation 10.31 for ω, and you obtain

$$KE_f = \frac{7}{10} m v^2$$

Having found an expression for the final KE, apply the CWE theorem:

$$\begin{aligned}
0 \text{ J} &= (KE + PE)_f - (KE + PE)_i \\
&= \left(\frac{7}{10} m v^2 + mgR \right) - \left(0 \text{ J} + mg\,[h + R] \right) \\
&= \frac{7}{10} m v^2 - mgh
\end{aligned}$$

Solving for v,

$$v = \left(\frac{10}{7} gh \right)^{1/2}$$

10.16 WHEELS*

The wheel is come full circle

William Shakespeare (1564–1616)[†]

It is difficult to imagine life without wheels, since they form the basis of many machines. Trains, trucks, cars, carts, Roman chariots, roller skates, roller blades, or roller bearings: wheels are a symbol of intelligent, technological life. The invention of the wheel certainly was one of the paramount discoveries in the history of technology. Here we form a crude model of an ideal wheel to see how the great advantage arises in using wheels to assist us in doing physical work.

A Stationary Wheel

First consider a wheel of mass M and radius R with a smaller hub of radius r, as shown in Figure 10.63. The wheel is free to turn about its center, which is fixed in position.

A mass m hangs from an ideal string that is wound around the central hub. Another ideal string is wrapped around the perimeter on which we pull with a force $\vec{F}$. For simplicity we neglect any friction in the axle bearing.

If the wheel has zero angular acceleration, the total torque on the wheel must also be zero. Choose the center of the wheel as the point about which to calculate the total torque. The force $\vec{F}'$ of the axle on the wheel and the weight $\vec{w}'$ of the wheel itself (of magnitude Mg) produce zero torque about this point, because their lines of action pass through the point about which we are

---
[†] *King Lear*, Act V, Scene 3, line 176.

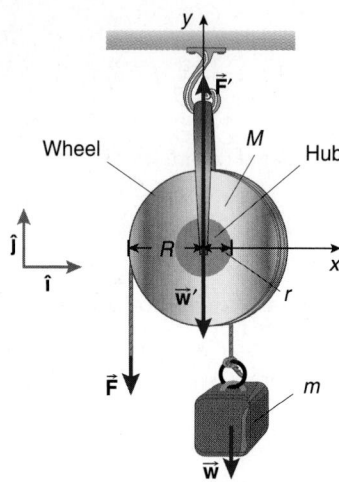

FIGURE 10.63 A wheel that can turn about its fixed center.

calculating the torque. The torques supplied by the hanging mass and the force $\vec{\mathbf{F}}$ are

$$\text{torque of } \vec{\mathbf{F}}: \quad -R\hat{\mathbf{i}} \times (-F\hat{\mathbf{j}}) = RF\hat{\mathbf{k}}$$
$$\text{torque of the weight of } m: \quad r\hat{\mathbf{i}} \times (-mg\hat{\mathbf{j}}) = -rmg\hat{\mathbf{k}}$$

The total torque must sum to zero, since there is zero angular acceleration:

$$(RF - rmg)\hat{\mathbf{k}} = 0 \text{ N·m}$$

Thus

$$RF = rmg$$

Rearranging this slightly, we find

$$F = \frac{r}{R}\, mg \qquad (10.35)$$

The magnitude of the force $\vec{\mathbf{F}}$ is *less* than the magnitude of the weight hanging from the hub by the factor r/R. The quantity r/R is known as the **mechanical advantage** of the wheel.*

Now let the force $\vec{\mathbf{F}}$ slowly raise the weight through one turn of the wheel, so that the kinetic energy of the system is unchanged throughout the process. Since the force $\vec{\mathbf{F}}$ is in the same direction as the change in the position vector of its point of application (the end of the rope), the work done by this constant force is

$$F(2\pi R)$$

The work done by the gravitational force is accounted for in the change in the potential energy of the mass m. Since the mass m is raised a distance $2\pi r$ when the wheel turns through one revolution, the *change* in the gravitational potential energy of the mass is

$$mg(2\pi r)$$

The CWE theorem states that

$$W_{\text{other}} = \Delta\text{KE} + \Delta\text{PE}$$

The work done by "other" forces is the work done by $\vec{\mathbf{F}}$. The change in the kinetic energy is zero (because the mass was raised at constant speed). Thus we have

$$F(2\pi R) = 0 \text{ J} + mg(2\pi r) \qquad (10.36)$$

The work we do is the same as if we simply raised the mass m a distance $2\pi r$ without the advantage of the wheel.

The real advantage of the wheel, then, is *not* in the work done on m, but rather that the force $\vec{\mathbf{F}}$ is smaller than that force used when the mass is raised without the assistance of the wheel.

The price we pay for using a smaller force $\vec{\mathbf{F}}$ is that it must move through a greater distance $(2\pi R)$ to accomplish the same amount of work as raising the mass directly through a distance $2\pi r$.

Notice that the result of the CWE theorem (Equation 10.36) also indicates that

$$F = \frac{r}{R}\, mg$$

as we found from considering the torques (Equation 10.35).

A Rolling Wheel

You likely know from personal experience that it is far easier to pull a mass when it is on wheels than to drag it. Here we consider a simplified model to explain why this is the case.

Consider a rather idealized case of a steel wheel (of radius R) moving on a steel rail, as on a railroad track. Let the steel wheel have an axle of radius r, around which the wheel can slip, as shown in Figure 10.64. Let the track be straight and horizontal.

Imagine the wheel to be locked, so that it cannot roll or spin. A horizontal force $\vec{\mathbf{F}}_{\text{drag}}$ is needed to drag the wheel along at constant speed. The kinetic frictional force on the locked wheel depends on the coefficient of kinetic friction μ_{k} (here for steel sliding on steel) and the normal force of the track on the wheel. The magnitude of the normal force of the track on the wheel here is equal to the magnitude of the weight of the wheel and its fraction of the load of the rail car; say this is of magnitude mg. If we drag the locked wheel through a distance

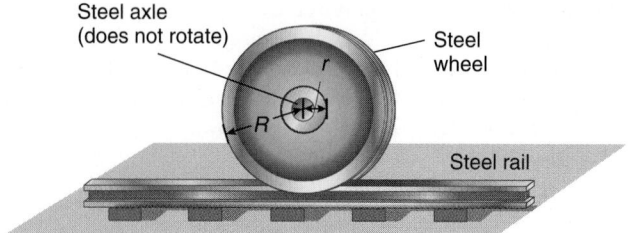

FIGURE 10.64 A rolling wheel.

---

*The term mechanical advantage also is used for other simple machines such as the lever, inclined plane, and screw.

equal to the circumference of the wheel, $2\pi R$, the work done by the force $\vec{\mathbf{F}}_{drag}$ is then

$$W_{drag} = F_{drag}(2\pi R)$$

The work done by the kinetic frictional force is *negative* because the change in the position vector is opposite to the direction of the force:

$$W_{frictn} = -(\mu_k mg)2\pi R$$

Now we apply the CWE theorem,

$$W_{other} = \Delta KE + \Delta PE$$

There is no change in either the kinetic energy or the gravitational potential energy, and so we have

$$W_{drag} + W_{frictn} = 0 \text{ J} + 0 \text{ J}$$

or

$$F_{drag}(2\pi R) - \mu_k mg(2\pi R) = 0 \text{ J}$$

Solving for F_{drag},

$$F_{drag} = \mu_k mg \qquad (10.37)$$

The force $\vec{\mathbf{F}}_{drag}$ is equal in magnitude to that of the *kinetic* frictional force on the steel wheel by the steel track.

Now allow the wheel to *roll* without slipping and pull the wheel through one revolution. The frictional force on the wheel at the rail now is the *static* frictional force, which does zero work because there is no motion at the point of contact. There only is steel-on-steel *kinetic* friction at the place where the wheel slides around the axle.* The kinetic frictional force certainly is no bigger than it was before:

$$\mu_k mg$$

since the axle must bear the weight of the load of mass m (the force likely is less since the axle does not need to bear the weight of half the wheel). The work done by the kinetic frictional force as the wheel turns through one revolution now is effectively[†]

$$W_{frictn} = -\mu_k mg(2\pi r)$$

where r is the radius of the *axle*. This work is negative because the kinetic frictional force opposes the motion. The work done by the applied horizontal force $\vec{\mathbf{F}}_{roll}$ after the wheel moves through one revolution is

$$W_{roll} = F_{roll}(2\pi R)$$

where R is the radius of the *wheel*. The CWE theorem states that

$$W_{other} = \Delta KE + \Delta PE$$

Since there is no change in either the kinetic or potential energies, the CWE theorem becomes

$$W_{roll} + W_{frictn} = 0 \text{ J} + 0 \text{ J}$$
$$F_{roll}(2\pi R) - \mu_k mg(2\pi r) = 0 \text{ J}$$

Solving for the magnitude of the rolling force,

$$F_{roll} = \frac{r}{R}\mu_k mg$$

We use Equation 10.37 to cast this in the form

$$F_{roll} = \frac{r}{R}F_{drag} \qquad (10.38)$$

The magnitude of the force needed to roll the wheel is less than that needed to drag it by the factor r/R, the mechanical advantage of the wheel.

It should be emphasized that Equation 10.38 is *not* a general relationship, but was derived for this simplified model of a horizontally rolling wheel. Equation 10.38 implies that the magnitude of the rolling force approaches zero for an axle with zero radius; this limit, of course, cannot be reached in practice.

Although this argument is somewhat oversimplified, it does explain, at least semiquantitatively, why it is easier to roll rather than drag things. Many other factors come into play in modeling the behavior of real wheels. Among these are (1) the use of roller bearings with no sliding contacts, which make F_{roll} even less than what we calculated here; (2) real wheels do not make point contact with the surface but deform and create a small area of contact, called the footprint of the wheel, easily visible with your car tires after a rain; and (3) kinetic energy is converted into thermodynamic internal energy by deformation forces, which is one reason that freely rolling wheels (those with no force $\vec{\mathbf{F}}_{roll}$ pulling them) eventually come to a stop. These factors serve to complicate the analysis of real wheels, and make for fertile fodder for thoughtful analysis of engineering applications in more advanced courses.

10.17 TOTAL ANGULAR MOMENTUM AND TORQUE

Newton's second law of motion states that the total force acting on a system is equal to the time rate of change of its momentum. We found there is a corresponding relation between the total torque and the time rate of change of the *spin* angular momentum (Equation 10.19), as well as for a single particle in *orbital* motion (Equation 10.12). In this section we will see that the *same* relationship exists when *both* spin and orbital motion exist, provided certain conditions are met.

A system of particles is shown in Figure 10.65. The origin at 0 is located in an inertial reference frame. Another reference point is at the point P, which may or may not be the center of mass of the system of particles. Point P is located with respect to 0 by means of the position vector $\vec{\mathbf{r}}_P$. Each particle of mass m_i in the system is located either via the position vector $\vec{\mathbf{r}}_i$ (with respect to origin 0) or the position vector $\vec{\mathbf{r}}_i'$ (with respect to the reference point P).

These position vectors are related to each other by the vector sum

$$\vec{\mathbf{r}}_i = \vec{\mathbf{r}}_P + \vec{\mathbf{r}}_i'$$

Rewrite this as

$$\vec{\mathbf{r}}_i' = \vec{\mathbf{r}}_i - \vec{\mathbf{r}}_P \qquad (10.39)$$

---

*For simplicity we assume no roller bearings are used.

[†] This is an approximation because the frictional force in this case is *distributed* around the entire circumference of the axle, which makes the actual calculation of the work more complicated than this simplified argument.

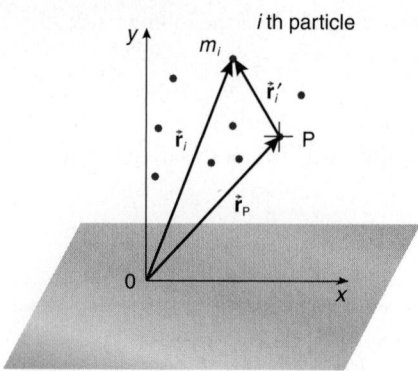

FIGURE 10.65 A particle is located via position vectors with respect to 0 and P.

The total angular momentum $\vec{\mathbf{L}}_P$ of the system about the point P is the vector sum of the individual angular momenta of the particles about the point P. That is,

$$\vec{\mathbf{L}}_P = \sum_i m_i \vec{\mathbf{r}}_i' \times \vec{\mathbf{v}}_i' \qquad (10.40)$$

where $\vec{\mathbf{v}}_i'$ is the velocity of m_i with respect to the reference point P. Now differentiate Equation 10.39 with respect to time:

$$\frac{d\vec{\mathbf{r}}_i'}{dt} = \frac{d\vec{\mathbf{r}}_i}{dt} - \frac{d\vec{\mathbf{r}}_P}{dt}$$
$$\vec{\mathbf{v}}_i' = \vec{\mathbf{v}}_i - \vec{\mathbf{v}}_P \qquad (10.41)$$

where $\vec{\mathbf{v}}_i$ is the velocity of m_i with respect to the origin 0 and $\vec{\mathbf{v}}_P$ is the velocity of the reference point P with respect to origin 0. Substituting for $\vec{\mathbf{r}}_i'$ using Equation 10.39 and for $\vec{\mathbf{v}}_i'$ using Equation 10.41, the expression for the angular momentum about the point P (Equation 10.40) becomes

$$\vec{\mathbf{L}}_P = \sum_i m_i (\vec{\mathbf{r}}_i - \vec{\mathbf{r}}_P) \times (\vec{\mathbf{v}}_i - \vec{\mathbf{v}}_P) \qquad (10.42)$$

Taking a cue from Newton's second law of motion, suppose we find the time rate of change of the angular momentum about the point P; that is, differentiate Equation 10.42 with respect to t. You have to be careful to preserve the order of the terms in the vector product:

$$\frac{d\vec{\mathbf{L}}_P}{dt} = \sum_i m_i \left[\frac{d}{dt} (\vec{\mathbf{r}}_i - \vec{\mathbf{r}}_P) \right] \times (\vec{\mathbf{v}}_i - \vec{\mathbf{v}}_P)$$
$$+ \sum_i m_i (\vec{\mathbf{r}}_i - \vec{\mathbf{r}}_P) \times \frac{d}{dt} (\vec{\mathbf{v}}_i - \vec{\mathbf{v}}_P)$$

The derivatives of the various position vectors are the various velocities. We find

$$\frac{d\vec{\mathbf{L}}_P}{dt} = \sum_i m_i (\vec{\mathbf{v}}_i - \vec{\mathbf{v}}_P) \times (\vec{\mathbf{v}}_i - \vec{\mathbf{v}}_P)$$
$$+ \sum_i (\vec{\mathbf{r}}_i - \vec{\mathbf{r}}_P) \times m_i \frac{d\vec{\mathbf{v}}_i}{dt}$$
$$- \sum_i (\vec{\mathbf{r}}_i - \vec{\mathbf{r}}_P) \times m_i \frac{d\vec{\mathbf{v}}_P}{dt} \qquad (10.43)$$

Since the vector product of any vector (here $\vec{\mathbf{v}}_i - \vec{\mathbf{v}}_P$) with itself is zero, the first summation vanishes, which is partial relief from the complexity.

In the second summation, the term

$$m_i \frac{d\vec{\mathbf{v}}_i}{dt}$$

can be written as

$$\frac{d\vec{\mathbf{p}}_i}{dt}$$

where $\vec{\mathbf{p}}_i$ is the momentum of particle m_i. In the third summation, the term

$$\frac{d\vec{\mathbf{v}}_P}{dt}$$

is the acceleration $\vec{\mathbf{a}}_P$ of the reference point P. Make these changes in Equation 10.43, which tidies it up a bit:

$$\frac{d\vec{\mathbf{L}}_P}{dt} = \sum_i (\vec{\mathbf{r}}_i - \vec{\mathbf{r}}_P) \times \frac{d\vec{\mathbf{p}}_i}{dt} - \sum_i (\vec{\mathbf{r}}_i - \vec{\mathbf{r}}_P) \times m_i \vec{\mathbf{a}}_P \qquad (10.44)$$

From Newton's second law, the time rate of change of the momentum of a particle,

$$\frac{d\vec{\mathbf{p}}_i}{dt}$$

is equal to the total force $\vec{\mathbf{F}}_{i\,\text{total}}$ acting on it. Thus Equation 10.44 becomes

$$\frac{d\vec{\mathbf{L}}_P}{dt} = \sum_i (\vec{\mathbf{r}}_i - \vec{\mathbf{r}}_P) \times \vec{\mathbf{F}}_{i\,\text{total}} - \sum_i (\vec{\mathbf{r}}_i - \vec{\mathbf{r}}_P) \times m_i \vec{\mathbf{a}}_P \qquad (10.45)$$

Using Equation 10.39 for the position vectors, Equation 10.45 becomes

$$\frac{d\vec{\mathbf{L}}_P}{dt} = \sum_i \vec{\mathbf{r}}_i' \times \vec{\mathbf{F}}_{i\,\text{total}} - \sum_i \vec{\mathbf{r}}_i' \times m_i \vec{\mathbf{a}}_P$$

The first summation is the *total torque* $\vec{\boldsymbol{\tau}}_P$ of the system of particles about the reference point P. Hence

$$\frac{d\vec{\mathbf{L}}_P}{dt} = \vec{\boldsymbol{\tau}}_P - \sum_i \vec{\mathbf{r}}_i' \times m_i \vec{\mathbf{a}}_P$$

Rearrange the terms in the summation slightly to obtain

$$\frac{d\vec{\mathbf{L}}_P}{dt} = \vec{\boldsymbol{\tau}}_P - \sum_i m_i \vec{\mathbf{r}}_i' \times \vec{\mathbf{a}}_P$$

Since the acceleration $\vec{\mathbf{a}}_P$ of the reference point P is independent of the particle locations and masses (P was just a *reference point*), the summation can be written as

$$\frac{d\vec{\mathbf{L}}_P}{dt} = \vec{\boldsymbol{\tau}}_P - \left(\sum_i m_i \vec{\mathbf{r}}_i' \right) \times \vec{\mathbf{a}}_P$$

The second term on the right-hand side is zero if any one of the following three conditions on the reference point P is met:

1. *The acceleration $\vec{\mathbf{a}}_P$ of the reference point P is zero.* If this is the case, the reference point P is in an inertial reference frame (just like the origin 0).

2. *The reference point P is the center of mass of the system of particles.* If this is the case, the summation vanishes because of the definition of the location of the center of mass (Equation 9.20):

$$\vec{\mathbf{r}}'_{CM} = \frac{1}{m_{total}} \sum_i m_i \vec{\mathbf{r}}'_i$$

That is, if P is the center of mass then $\vec{\mathbf{r}}'_{CM} = 0$ m, and so we have

$$\sum_i m_i \vec{\mathbf{r}}'_i = 0 \ \text{kg} \cdot \text{m}$$

3. *The acceleration $\vec{\mathbf{a}}_P$ of the point P is in the same direction as the vector*

$$\sum_i m_i \vec{\mathbf{r}}'_i$$

This is equivalent to saying that the acceleration $\vec{\mathbf{a}}_P$ is parallel or antiparallel to the vector (from P) locating the center of mass of the system.

In summary, if the reference point P

a. is in an inertial reference frame, or
b. is the center of mass of the system, or
c. has an acceleration parallel (or antiparallel) to a vector (from P) locating the center of mass,

then the time rate of change of the angular momentum of the system (about P) is equal to the total torque (about P):

$$\frac{d\vec{\mathbf{L}}_P}{dt} = \vec{\boldsymbol{\tau}}_P \qquad (10.46)$$

Now that is a welcome simplification in dealing with a horde of masses.

For a *rigid body*, the total torque is related to the angular acceleration via Equation 10.20:

$$\vec{\boldsymbol{\tau}}_{total} = I\vec{\boldsymbol{\alpha}}$$

Thus Equation 10.46 also can be written as

$$\frac{d\vec{\mathbf{L}}_P}{dt} = I\vec{\boldsymbol{\alpha}} \qquad (10.47)$$

STRATEGIC EXAMPLE 10.12

Wind a string around a rigid body with a circular cross section of radius R and mass m, as shown in Figure 10.66. Fix the string at one end and release the mass at rest. Determine the magnitude of the downward acceleration of the mass as it falls. Let I_{CM} be the moment of inertia about the symmetry axis through the center of mass.

Solution

Let the rigid body be the system. The system falls under the influence of two forces, shown in Figure 10.67:

1. its weight $\vec{\mathbf{w}}$ of magnitude mg; and
2. the force $\vec{\mathbf{T}}$ of the string on the system.

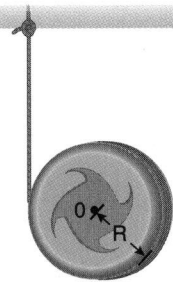

FIGURE 10.66

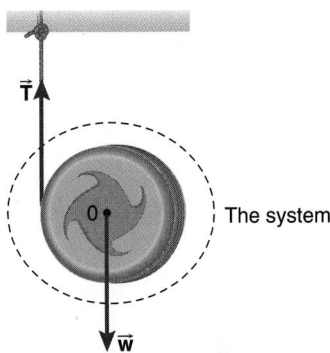

The system

FIGURE 10.67

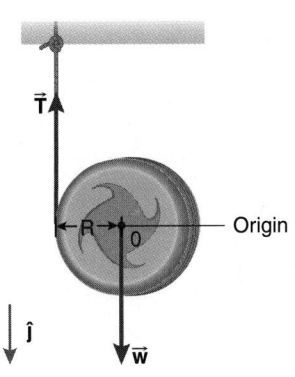

Origin

FIGURE 10.68

Method 1

Use the center of mass of the rigid body as the origin or the point about which to calculate the torque of the forces. With this choice (see Figure 10.68), the weight has zero torque since its line of action passes through the chosen origin. The torque of the force of the string is of magnitude TR, since the moment arm of the tension is the radius of the rigid body.

Thus the total torque on the system about the point at its center is of magnitude TR. The magnitude of the torque about the center of the disk is equal to the moment of inertia of the disk *about this point* times the magnitude of the angular acceleration $\vec{\boldsymbol{\alpha}}$ of the disk about this point according to Equations 10.46 and 10.47:

$$\tau = I\alpha$$

The moment of inertia of the disk to use here is the moment of inertia I_{CM} of the system about the point where we calculated the torque. This is because I_{CM} is moment of inertia of the object about its symmetry axis through the center of mass point. Since the mass rolls off the string, the magnitude of the angular acceleration $\vec{\alpha}$ and the magnitude of the downward acceleration $\vec{a}$ are related by the rolling constraint of Equation 10.32:

$$\alpha = \frac{a}{R}$$

Hence Equation 10.47 becomes

$$TR = I_{CM}\frac{a}{R} \qquad (1)$$

To find T, look at the translational motion of the system as a whole and apply Newton's second law. Taking $\hat{j}$ down, the total force component in the downward direction is

$$mg - T$$

and this must equal the mass of the object times the acceleration component a in this direction:

$$mg - T = ma$$

Thus

$$T = m(g - a)$$

Substituting this expression for T into equation (1), you obtain

$$m(g - a)R = I_{CM}\frac{a}{R}$$

Solving for a, you obtain

$$a = \frac{m}{m + \dfrac{I_{CM}}{R^2}}g$$

or

$$a = \frac{mR^2}{mR^2 + I_{CM}}g$$

Notice that the greater the moment of inertia, the smaller the downward acceleration.

Method 2

Choose the point on the rim of the rigid body where the string last is in contact with the object as the origin about which to calculate the torque, as shown in Figure 10.69. With this choice, the torque of the force $\vec{T}$ of the string is zero since its line of action passes through the chosen origin. Thus the total torque on

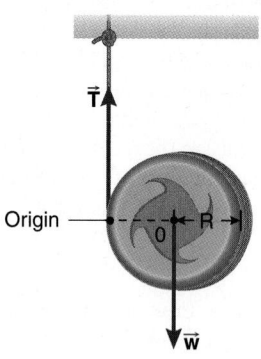

FIGURE 10.69

the object is the torque of the weight. The magnitude of the torque provided by the weight is mg times its moment arm R, or

$$mgR$$

The magnitude of the total torque about the origin is equal to the moment of inertia of the object *about this point* times the magnitude of the angular acceleration of the disk about this point, or

$$\tau = I\alpha \qquad (2)$$

In this case the moment of inertia I must be obtained from the parallel axis theorem because the rotational axis through the point about which we calculated the torque is parallel to but not coincident with the symmetry axis through the center of mass. So, by Equation 10.29,

$$I = I_{CM} + mR^2$$

Since the system is rolling off the string, the magnitude of the angular acceleration $\vec{\alpha}$ here is found using the rolling constraint, Equation 10.32:

$$\alpha = \frac{a}{R}$$

where a is the magnitude of the downward acceleration of the system. Substituting for the magnitude of the torque, the moment of inertia, and magnitude of the angular acceleration into equation (2), you find

$$mgR = \left(mR^2 + I_{CM}\right)\frac{a}{R}$$

Solving for a,

$$a = \frac{mR^2}{mR^2 + I_{CM}}g$$

10.18 CONSERVATION OF ANGULAR MOMENTUM

Conservation laws are important in physics not only because they indicate that some features of a system are unchanged as other factors associated with it do change, but also because such conservation laws provide a convenient tool for solving practical problems. Newton's second law of motion,

$$\vec{F}_{total} = \frac{d\vec{P}_{total}}{dt}$$

states that if the total force on a system is zero, then

$$0\text{ N} = \frac{d\vec{P}_{total}}{dt}$$

and the total momentum $\vec{p}_{total}$ of the system is a constant vector. This statement of *conservation of momentum* is one of the great conservation laws of physics; we used it extensively in analyzing collision phenomena in Chapter 9. Likewise the rotational counterpart of Newton's second law is, as we have seen,

$$\vec{\tau}_{total} = \frac{d\vec{L}_{total}}{dt}$$

If the total torque on a system is zero, then

$$0 \ \text{N·m} = \frac{d\vec{L}_{\text{total}}}{dt}$$

and then the total angular momentum $\vec{L}_{\text{total}}$ of the system must be a *constant vector*. This is **conservation of angular momentum**.

This conservation law also is of great importance and utility in physics.

EXAMPLE 10.13

An ice skater with a moment of inertia of 0.600 kg·m² about a vertical axis is spinning with an angular speed of 12.6 rad/s. The skater brings her arms in closer to the axis, decreasing her moment of inertia to 0.240 kg·m². Find her new angular speed.

Solution

Take the skater to be the system. Although the system is neither rigid nor frigid, the forces involved in the deformation of the system are internal forces. The two external forces on the skater, her weight and the normal force of the ice surface on her, are of equal magnitude and in opposite directions (but are *not* Newton's third law forces; why not?). Take the center of mass of the skater as the point about which to calculate the torques. The two external forces produce zero torque on the skater because their lines of action pass through the point about which we are taking the torque, the center of mass. Hence the total torque on the system is zero and the total angular momentum of the system is conserved:

$$\vec{L}_i = \vec{L}_f$$

The initial angular momentum of the skater is

$$\vec{L}_i = I_i \vec{\omega}_i$$

and, by the circular motion right-hand rule, is directed upward along the vertical direction, as shown in Figure 10.70.

The final angular momentum of the skater is

$$\vec{L}_f = I_f \vec{\omega}_f$$

Figure skaters change their angular speed of spin by changing their moment of inertia to exploit conservation of angular momentum.

Conservation of angular momentum for the skater means that

$$I_i \vec{\omega}_i = I_f \vec{\omega}_f$$
$$(0.600 \ \text{kg·m}^2)(12.6 \ \text{rad/s})\hat{k} = (0.240 \ \text{kg·m}^2)\vec{\omega}_f$$

Solving for $\vec{\omega}_f$,

$$\vec{\omega}_f = (31.5 \ \text{rad/s})\hat{k}$$

Her new angular speed is 31.5 rad/s.

STRATEGIC EXAMPLE 10.14

You (mass m) are incident at velocity $\vec{v}_0$ along a straight line tangent to a disk-shaped merry-go-round of mass M and radius R that is spinning with an angular speed ω_0 about a fixed axle; see Figure 10.71. You jump on and hold on to the rim of the disk.

a. Find the impulse provided by the axle to the system consisting of you and the merry-go-round.
b. Find the final angular speed of the merry-go-round after you jump on.
c. What incident speed v_0 is sufficient for you to bring the merry-go-round to a halt?

Solution

Let the system be you and the merry-go-round.

a. The collision of you with the merry-go-round is completely inelastic, and so we do *not* expect the kinetic energy of the

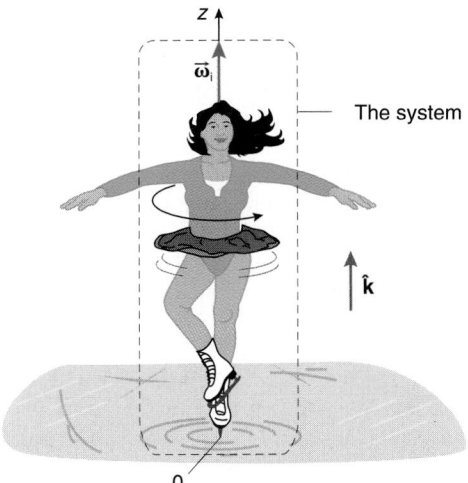

FIGURE 10.70

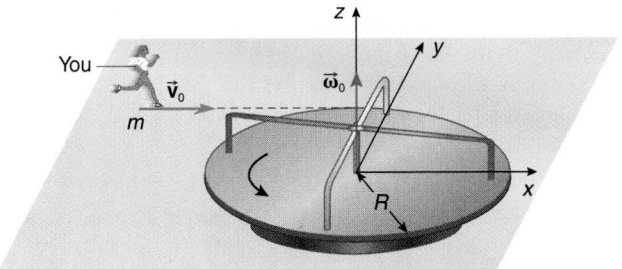

FIGURE 10.71

system to be conserved. If the axle of the disk were not fixed, the momentum of the system would be conserved. But because the axle is fixed, the system has no horizontal velocity after the collision. The momentum of the system thus is not conserved. An external force changes the momentum of the system. The force of you on the merry-go-round and the force of the merry-go-round on you are internal forces to the system; they form a Newton's third law pair (and sum to zero). However, the axle provides a horizontal external force on the system during the collision. The impulse provided by the force of the axle of the system equals the change in its momentum. The momentum of the system initially is due only to you; the merry-go-round has zero momentum because its center of mass is at rest. Use the coordinate system indicated in Figure 10.71. The initial momentum of the system is

$$mv_0\hat{\mathbf{i}}$$

The final momentum of the system is zero. The change in the momentum of the system is equal to the impulse in accordance with the impulse–momentum theorem of Chapter 9. Thus the impulse provided by the axle on the disk is

$$\begin{aligned} \vec{\mathbf{I}} &= \Delta\vec{\mathbf{p}} \\ &= \vec{\mathbf{p}}_f - \vec{\mathbf{p}}_i \\ &= (0 \text{ kg·m/s})\hat{\mathbf{i}} - mv_0\hat{\mathbf{i}} \\ &= -mv_0\hat{\mathbf{i}} \end{aligned}$$

b. Let's see if you can conserve the *angular momentum* of the system consisting of you and the merry-go-round. An external force (the force of the axle) acts on the system in the horizontal direction; however, if you use the center of the disk as an origin, the force of the axle on the system produces *zero* torque about this point, because the line of action of the force passes through the point about which you take the torque. Thus, if you measure the angular momentum of the system about the center of the disk, there are no external torques on the system. Thus Equation 10.46,

$$\vec{\boldsymbol{\tau}}_{\text{total}} = \frac{d\vec{\mathbf{L}}_{\text{total}}}{dt} \tag{10.46}$$

becomes

$$0 \text{ N·m} = \frac{d\vec{\mathbf{L}}_{\text{total}}}{dt}$$

which means $\vec{\mathbf{L}}_{\text{total}}$ is a constant vector. The total angular momentum of the system about the center of the disk *is* conserved:

$$\vec{\mathbf{L}}_{\text{total f}} = \vec{\mathbf{L}}_{\text{total i}}$$

The total angular momentum of the system initially consists of the vector sum of the orbital angular momentum of you and the spin angular momentum of the merry-go-round. The initial orbital angular momentum of you about the center of the disk can be found in two ways.

Method 1
Use the definition of the angular momentum of a point particle, Equation 10.1:

$$\vec{\mathbf{L}}_{\text{you}} = \vec{\mathbf{r}} \times \vec{\mathbf{p}}$$

where $\vec{\mathbf{p}}$ is your momentum: $\vec{\mathbf{p}} = mv_0\hat{\mathbf{i}}$ with the coordinate system indicated in Figure 10.72. The magnitude of $\vec{\mathbf{L}}_{\text{you}}$ is equal to the magnitude of $\vec{\mathbf{r}}$ times the magnitude of $\vec{\mathbf{p}}$ times the sine of the angle θ between the two vectors:

$$L_{\text{you}} = r(mv_0) \sin \theta$$

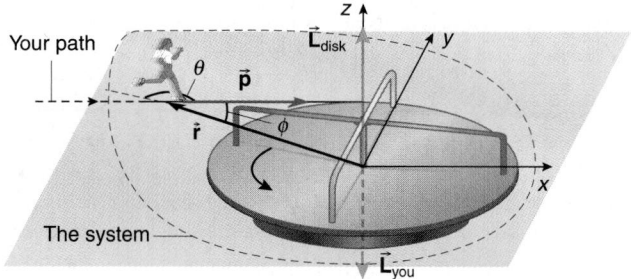

FIGURE 10.72

Geometrically,

$$\begin{aligned} r \sin \theta &= r \sin(180° - \phi) \\ &= r \sin \phi \\ &= R \end{aligned}$$

which is the radius of the disk. So

$$L_{\text{you}} = mv_0R$$

The direction of your angular momentum is found using the usual vector product right-hand rule: this vector product is directed along $-\hat{\mathbf{k}}$. Thus

$$\vec{\mathbf{L}}_{\text{you}} = -mv_0R\hat{\mathbf{k}}$$

Method 2
One might think of using the relationship in Equation 10.6,

$$\vec{\mathbf{L}} = I\vec{\boldsymbol{\omega}}$$

where I is your moment of inertia about the axis:

$$I = mr^2$$

The only trouble is that your distance from the axis is *changing*, so that your moment of inertia about this axis *also* is changing. The distance to the axis is decreasing, so that your moment of inertia is decreasing. Your angular velocity about the axis *also* is changing: when you are very far from the axis your angular velocity is small, and as you approach the merry-go-round your angular velocity about the axis increases. The two effects compensate for each other, but it is more involved to calculate the angular momentum using $\vec{\mathbf{L}} = I\vec{\boldsymbol{\omega}}$ at a general position for the incoming particle (you). However, the instant before you collide with the disk, you can use Equation 10.6 relatively easily. At this location your instantaneous moment of inertia (of the you-particle) about the axis is mR^2, since you are then a distance R from the axis. Your instantaneous angular speed about the axis is equal to

v_0/R. The instantaneous angular velocity then is (using the coordinate system indicated in Figure 10.71 or 10.72)

$$-\frac{v_0}{R}\,\hat{\mathbf{k}}$$

In this way your orbital angular momentum about the axis is found to be

$$\begin{aligned}
\vec{\mathbf{L}}_{\text{you}} &= I\vec{\boldsymbol{\omega}} \\
&= mR^2\left(-\frac{v_0}{R}\right)\hat{\mathbf{k}} \\
&= -mv_0R\,\hat{\mathbf{k}}
\end{aligned}$$

Having found your initial angular momentum, the initial spin angular momentum of the merry-go-round is found from Equation 10.18:

$$\vec{\mathbf{L}} = I_{\text{CM}}\vec{\boldsymbol{\omega}}$$

Use the circular motion right-hand rule to determine the direction of $\vec{\boldsymbol{\omega}}$. The angular velocity of the disk is directed along $+\hat{\mathbf{k}}$. So the angular velocity of the disk is

$$\vec{\boldsymbol{\omega}} = \omega_0\hat{\mathbf{k}}$$

The moment of inertia of the disk-shaped merry-go-round is found in Table 10.1; we neglect the rails:

$$I_{\text{CM}} = \frac{1}{2}MR^2$$

Thus the initial angular momentum of the disk is

$$\vec{\mathbf{L}}_{\text{disk}} = \frac{1}{2}MR^2\omega_0\hat{\mathbf{k}}$$

Hence the total angular momentum of the system initially is

$$\begin{aligned}
\vec{\mathbf{L}}_{\text{i}} &= \vec{\mathbf{L}}_{\text{disk}} + \vec{\mathbf{L}}_{\text{you}} \\
&= \left(\frac{1}{2}MR^2\omega_0 - mv_0R\right)\hat{\mathbf{k}}
\end{aligned}$$

Now you have to find the *final* angular momentum of the system. You can do this two ways.

Method 1

The merry-go-round now is spinning at an unknown angular velocity $\omega\hat{\mathbf{k}}$. Thus the final spin angular momentum of the merry-go-round is found using Equation 10.18 once again:

$$\frac{1}{2}MR^2\omega\hat{\mathbf{k}}$$

You are now on the rim of the merry-go-round, also rotating at the same angular velocity $\omega\hat{\mathbf{k}}$. Your orbital angular momentum can be found from Equation 10.6:

$$\vec{\mathbf{L}}_{\text{you}} = I_{\text{you}}\vec{\boldsymbol{\omega}}$$

Since I_{you} is just mR^2, your orbital angular momentum is

$$\vec{\mathbf{L}}_{\text{you}} = mR^2\omega\hat{\mathbf{k}}$$

The final total angular momentum of the system is the vector sum of your angular momentum and that of the merry-go-round:

$$\vec{\mathbf{L}}_{\text{f}} = \left(\frac{1}{2}MR^2 + mR^2\right)\omega\hat{\mathbf{k}}$$

Method 2

You also can find the final total angular momentum of the system by considering the system to be a composite of a disk, of moment of inertia

$$\frac{1}{2}MR^2$$

and a point particle (you!) on its rim (of moment of inertia mR^2). Since both objects are rotating about the same axis with the same angular velocity, the total moment of inertia of the system is the sum of the moments of inertia of its constituent pieces:

$$I_{\text{total}} = \frac{1}{2}MR^2 + mR^2$$

The final total angular momentum is

$$\begin{aligned}
\vec{\mathbf{L}}_{\text{f}} &= I_{\text{total}}\vec{\boldsymbol{\omega}} \\
&= \left(\frac{1}{2}MR^2 + mR^2\right)\omega\hat{\mathbf{k}}
\end{aligned}$$

which is what we obtained in Method 1.

The total angular momentum of the system is conserved:

$$\vec{\mathbf{L}}_{\text{f}} = \vec{\mathbf{L}}_{\text{i}}$$

so

$$\left(\frac{M}{2} + m\right)R^2\omega\hat{\mathbf{k}} = \left(\frac{1}{2}MR^2\omega_0 - mv_0R\right)\hat{\mathbf{k}}$$

From this you can find the final angular velocity component ω:

$$\omega = \frac{\dfrac{1}{2}MR^2\omega_0 - mv_0R}{\left(\dfrac{M}{2} + m\right)R^2} \tag{1}$$

c. If your speed v_0 is sufficiently large, the collision can bring the disk to rest. The final angular velocity component of the disk then is zero; therefore the numerator in equation (1) must be zero. This means that

$$\frac{1}{2}MR^2\omega_0 = mv_0R$$

Thus your initial speed v_0 that is sufficient to halt the rotation of the merry-go-round is

$$v_0 = \frac{MR\omega_0}{2m}$$

10.19 CONDITIONS FOR STATIC EQUILIBRIUM

Buildings, towers, bridges, and jungle gyms must be stable structures for safety, as well as to avoid lawsuits from lean and hungry packs of trial attorneys. What are some of the principles involved in a safe design? Certainly such structures should neither accelerate nor begin to rotate because of forces acting on them. Here we find there are but two conditions for static equilibrium that form the basis for the engineering analysis of all static structures.

Forces cause changes in momentum; torques cause changes in angular momentum. The two fundamental equations that govern the dynamics of a physical system are

Newton's second law of motion

$$\vec{F}_{\text{total}} = \frac{d\vec{P}_{\text{total}}}{dt} \quad \text{(most general form)}$$

$$= m\vec{a} \quad \text{(for systems of constant mass)}$$

and its rotational counterpart

$$\vec{\tau}_{\text{total}} = \frac{d\vec{L}_{\text{total}}}{dt} \quad \text{(most general form)}$$

$$= I\vec{\alpha} \quad \text{(for systems with constant moment of inertia } I\text{)}$$

We now can extend the concept of single-particle equilibrium, first introduced in Chapter 5, to that of many-particle systems. If the acceleration of the center of mass of the system is zero, and the angular acceleration of the system about a symmetry axis through its center of mass also is zero, the system is in **static equilibrium**.

Hence the **two conditions for static equilibrium** of a system are

1. the total force on the system must be zero: $\vec{F}_{\text{total}} = 0$ N; and

2. the total torque on the system must be zero: $\vec{\tau}_{\text{total}} = 0$ N·m.

Therefore the center of mass of a system in static equilibrium has a constant velocity: $\vec{v}_{\text{CM}}$ is constant. The angular velocity about a symmetry axis through the center of mass point also is constant: $\vec{\omega}$ is constant. The velocity of the center of mass and the angular velocity about the center of mass need not be zero; the important thing for static equilibrium is that both are *constant* vectors.

The acceleration $\vec{a}$ and angular acceleration $\vec{\alpha}$ both must be zero for a system to be in static equilibrium.

As we have seen, to calculate the torque on a system, we typically introduce a coordinate system and calculate the torque of the various forces with respect to the origin of that coordinate system. The motion of the system, however, could care less about where we choose the origin of our coordinate system. Thus we can choose *any point* as the origin about which to calculate the torques of the various forces. For static equilibrium, the vector sum of the torques of the forces must vanish *regardless* of where we choose to place the origin for the purpose of calculating the torques. This gives us great freedom when analyzing a problem.

PROBLEM-SOLVING TACTIC

10.5 **If the lines of action of several forces pass through a particular point, that is the point about which to calculate torques for static equilibrium problems.** The torques of the forces whose lines of action pass through the point are zero, because their moment arms are zero. This choice of a point about which to calculate torques is not obligatory, of course, but it does make the calculations easier to do.

QUESTION 10

Stand up straight with both feet spread about 0.5 m apart. If you raise one foot off the floor, explain what must happen to your center of mass to prevent yourself from falling over.

EXAMPLE 10.15

Two scales are positioned beneath the ends of a uniform board of mass M and length ℓ_0. A person of mass m is standing partway between the scales, a distance ℓ from one end, as shown in Figure 10.73. The board and person are at rest. What magnitude of force does each scale exert on the board?

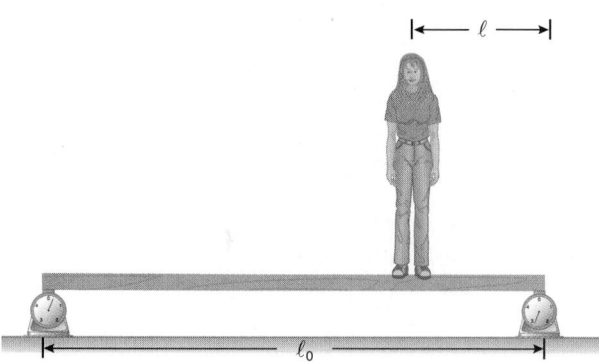

FIGURE 10.73

Solution

Let the board be the system. The forces on the board are (see Figure 10.74)

1. the force of the person acting down on the surface of the board, equal in magnitude and direction to the weight $\vec{w}$ of the person.

2. the weight $\vec{w}'$ of the board, of magnitude Mg; and

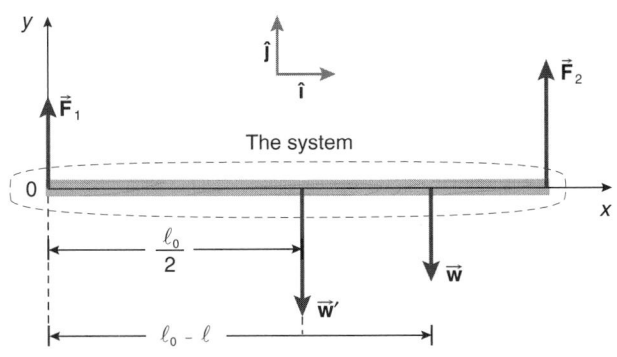

FIGURE 10.74

point about which to take the torques. The torque of the force $\vec{F}_1$ then is zero, since its line of action passes through the point about which we are taking the torque. The torque of the force of the person on the board is

$$\vec{\tau}_{\text{person}} = (\ell_0 - \ell)\hat{i} \times (-mg)\hat{j}$$
$$= -(\ell_0 - \ell)mg\hat{k}$$

The torque of the weight of the board is

$$\vec{\tau}_{\text{board}} = \frac{\ell_0}{2}\hat{i} \times (-Mg)\hat{j}$$
$$= -\frac{\ell_0}{2}Mg\hat{k}$$

The torque of the force $\vec{F}_2$ is

$$\vec{\tau}_{\text{of }F_2} = \ell_0\hat{i} \times F_2\hat{j}$$
$$= \ell_0 F_2\hat{k}$$

The vector sum of the torques must be zero since the system obviously has zero angular acceleration. The vector sum is

$$\ell_0 F_2\hat{k} - (\ell_0 - \ell)mg\hat{k} - \frac{\ell_0}{2}Mg\hat{k} = 0 \text{ N·m}$$

Solving for F_2, you find

$$F_2 = \left(1 - \frac{\ell}{\ell_0}\right)mg + \frac{Mg}{2}$$

Using equation (1) to find F_1, you find that

$$F_1 = \frac{\ell}{\ell_0}mg + \frac{Mg}{2}$$

3. the forces of the two scales on the board. Call these forces $\vec{F}_1$ and $\vec{F}_2$, directed as shown in Figure 10.74. Introduce the coordinate system indicated in that figure.

The first condition of equilibrium states that the vector sum of the forces must be zero:

$$F_1\hat{j} + F_2\hat{j} + mg(-\hat{j}) + Mg(-\hat{j}) = 0 \text{ N}$$

or

$$F_1 + F_2 = (m + M)g \quad (1)$$

Equation (1) alone is not sufficient to answer the question, since we have two unknowns: the force components F_1 and F_2. The second condition of equilibrium enables us to write another equation. The second condition of equilibrium states that the total torque about any point must vanish. Choose the origin of the coordinate system, located at the left end of the board, as the

CHAPTER SUMMARY

The rotational motion of a system about an axis through its center of mass is called *spin*. If the center of mass is moving with respect to a reference frame, the center of mass is executing *orbital motion* with respect to the origin of the coordinate system. Orbital motion need not be circular orbital motion.

The *orbital angular momentum* $\vec{L}$ of a particle about a point 0 is

$$\vec{L} \equiv \vec{r} \times \vec{p} \quad (10.1)$$

where $\vec{r}$ is the position vector of the particle with respect to 0 and $\vec{p}$ is the momentum of the particle.

The orbital angular momentum of a particle of mass m executing circular orbital motion is related to its angular velocity by

$$\vec{L} = I\vec{\omega} \quad (10.6)$$

where I is the *moment of inertia* of the point particle about the center of the circle:

$$I = mr^2 \quad \text{(single particle only)} \quad (10.5)$$

For a symmetric rigid body undergoing spin about a symmetry axis through the center of mass, we also have

$$\vec{L}_{\text{spin}} = I_{\text{CM}}\vec{\omega}_{\text{spin}} \quad (10.18)$$

where I_{CM} is the moment of inertia of the rigid body about the symmetry axis through its center of mass point. The moment of inertia is defined to be

$$I_{\text{CM}} = \sum_i m_i r_{i\perp}^2 \quad (10.17)$$

where $r_{i\perp}$ is the perpendicular distance of the point mass m_i from the axis of rotation. The moments of inertia I_{CM} for various symmetric rigid bodies are listed in Table 10.1 on page 440.

For a rigid body undergoing simultaneous spin and orbital motion, the total angular momentum of the system is the vector sum of the spin and orbital angular momenta. The orbital angular momentum is that of a point particle with a mass equal to the total mass of the system located at the center of mass point.

For a system undergoing *synchronous rotation*, the total angular momentum also is proportional to its angular velocity:

$$\vec{L}_{\text{total}} = I\vec{\omega} \quad (10.28)$$

where I then is found from the *parallel axis theorem*.

The *torque* $\vec{\tau}$ of any force $\vec{F}$ is defined to be

$$\vec{\tau} \equiv \vec{r} \times \vec{F} \quad (10.7)$$

where $\vec{r}$ is the position vector that locates the point of application of the force with respect to a chosen origin.

The total torque of a system about some point is equal to the time rate of change of the total angular momentum of the system about the same point:

$$\vec{\tau}_{\text{total}} = \frac{d\vec{L}_{\text{total}}}{dt} \qquad (10.12)$$

Equation 10.12 is the rotational analog of Newton's second law:

$$\vec{F}_{\text{total}} = \frac{d\vec{p}_{\text{total}}}{dt}$$

If the moment of inertia I is constant, the rotational analog of Newton's second law also may be written as

$$\vec{\tau}_{\text{total}} = I\,\vec{\alpha} \qquad \text{(constant } I) \qquad (10.13)$$

where $\vec{\alpha}$ is the angular acceleration. This is analogous to

$$\vec{F}_{\text{total}} = m\vec{a} \qquad \text{(constant } m)$$

The kinetic energy associated with the spin of a rigid body is

$$KE_{\text{rot}} = \frac{1}{2} I_{\text{CM}} \omega^2 \qquad (10.24)$$

The total kinetic energy is the sum of the kinetic energy of the center of mass point and the rotational kinetic energy of the system associated with the spin about the symmetry axis through the center of mass:

$$KE_{\text{total}} = KE_{\text{rot}} + \frac{1}{2} m v_{\text{CM}}^2 \qquad (10.25)$$

If the system is either rolling without slipping or undergoing synchronous rotation, its moment of inertia about an axis parallel to the symmetry axis through the center of mass point can be found from the *parallel axis theorem*:

$$I = I_{\text{CM}} + md^2 \qquad (10.29)$$

where d is the perpendicular distance between the symmetry axis and the parallel rotation axis. The total kinetic energy of the system then is

$$KE_{\text{total}} = \frac{1}{2} I\omega^2 \qquad (10.34)$$

For a system with circular cross section of radius R that is rolling without slipping, the *rolling constraints* are

$$s = R\theta \qquad (10.30)$$
$$v = R\omega \qquad (10.31)$$
$$a = R\alpha \qquad (10.32)$$

where s is the distance through which the center of mass of the system moves, θ is the angle through which the system turns while rolling, v is the speed of the center of mass point, ω is the angular speed about the symmetry axis through the center of mass point, a is the magnitude of the acceleration of the center of mass point, and α is the magnitude of the angular acceleration of the system about the symmetry axis through the center of mass point.

If the total torque on a system about some point is zero, the total angular momentum of the system about the same point is conserved. This is *conservation of angular momentum*.

A system is in *static equilibrium* if its acceleration $\vec{a}$ and angular acceleration $\vec{\alpha}$ both are zero. Newton's second law and its rotational counterpart mean that the *two conditions for the static equilibrium* of a rigid body are

$$\vec{F}_{\text{total}} = 0 \text{ N}$$

and

$$\vec{\tau}_{\text{total}} = 0 \text{ N·m}$$

SUMMARY OF PROBLEM-SOLVING TACTICS

10.1 (page 430) If the line of action of a force goes through a point 0, the force has zero torque about that point.

10.2 (page 431) The magnitude of the torque is the product of the magnitude of the force and its moment arm; the direction of the torque is determined from the vector product right-hand rule.

10.3 (page 441) If a rigid body can be segmented into several symmetrically shaped pieces, the moment of inertia of the object as a whole is the sum of the moments of inertia of all the segmented pieces provided the moments are about the same axis.

10.4 (page 454) If you use the parallel axis theorem to determine the moment of inertia I of the system, the total kinetic energy of the system due to *both* its spin and orbital motion is given by Equation 10.34:

$$KE_{\text{total}} = \frac{1}{2} I\omega^2 \qquad (10.34)$$

10.5 (page 464) If the lines of action of several forces pass through a particular point, that is the point about which to calculate torques for static equilibrium problems.

QUESTIONS

1. (page 427); 2. (page 427); 3. (page 432); 4. (page 436); 5. (page 441); 6. (page 443); 7. (page 447); 8. (page 450); 9. (page 454); 10. (page 464)

11. A particle is moving in a straight line. Sketch the situation. Indicate a point about which the particle has zero orbital angular momentum. Indicate a point about which the particle has nonzero orbital angular momentum.

12. Show that the SI units of angular momentum kg·m²/s are equivalent to J·s. Later in the course (in Chapter 26), we will see that the SI units of an important fundamental constant called Planck's constant, always symbolized as h, also are J·s: $h = 6.626 \times 10^{-34}$ J·s. This suggests that Planck's constant is in some way connected to angular momentum; we also will see that on the level of elementary

particles, the angular momentum is quantized in integral or half-integral units of $h/(2\pi)$. Those particles with spin angular momentum that are half-integral units of $h/(2\pi)$ are called *fermions*; those with integral units are called *bosons*. The combination $h/(2\pi)$ occurs so frequently in quantum physics that it is given its own special symbol, $\hbar$, pronounced "h-bar": $\hbar \equiv h/(2\pi)$.

13. Give a convincing argument to show that the center of mass point must lie on the symmetric axis (if one exists) of a rigid body with uniform density.

14. Sketch the shape of an object that has two distinct symmetry axes with different moments of inertia about each axis.

15. How many symmetry axes does a solid sphere have?

16. Sketch and indicate the number of symmetry axes through the center of mass point for each rigid body depicted in Figure Q.16. The objects are planar and of uniform composition.

FIGURE Q.16

17. The rotation of the Sun was discovered by Galileo in the early 17th century by observing sunspots. Suppose you notice several sunspots aligned as indicated in Figure Q.17a. Several weeks later, after the Sun has executed one rotation, the sunspots are observed as in Figure Q.17b. Can you conclude that the Sun is a rigid body? What assumption must be made to reach your conclusion?

18. A disk is mounted so it can rotate about a horizontal axis as indicated in Figure Q.18. (a) Indicate how *two* forces, of *equal magnitude*, can act on the rim of the disk so that the total force on it is *zero* but the total torque on it is *not* zero. (b) Indicate how *two* forces, of *equal magnitude*, can act on the rim of the disk so that the total force on it is *not* zero but the total torque on it *is* zero.

19. If a system is rotating, can you conclude that there is a nonzero total torque on the system? Explain your answer.

20. With respect to a fixed reference frame attached to the Sun (but not rotating with the Sun) (see Figure Q.20), is your speed on the Earth greater at noon or at midnight?

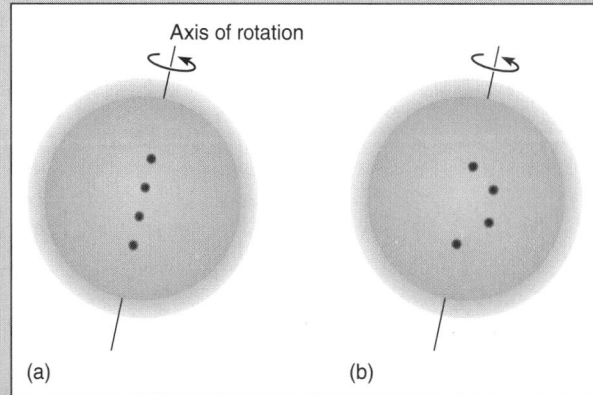

FIGURE Q.17

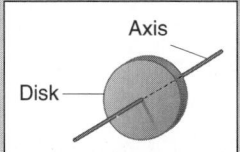

FIGURE Q.18

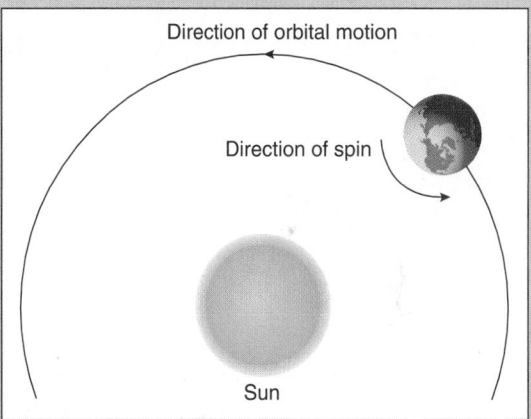

FIGURE Q.20

21. Two solid cylinders of the same radius and length are made of materials of different densities. The cylinders are released at rest at the top of an inclined plane. Figure out whether both cylinders reach the bottom of the incline simultaneously. If not, does the cylinder with the greater density reach the bottom first or second? Explain your reasoning.

22. The Mississippi River ("Big Muddy") carries large amounts of sediment south into the Gulf of Mexico. What does this do (if anything) to the moment of inertia of the Earth? Will this affect the length of the day? Explain your answers.

23. After making appropriate measurements, determine the approximate length of the record groove on an old-fashioned LP record.

24. Explain why the moment of inertia of an object is a minimum if the rotational axis passes through the center of mass of the object.

25. Consider a thin uniform disk of mass m and radius R and two possible axes of rotation: (i) an axis perpendicular to the disk through its center; (ii) an axis coincident with a diameter of the disk. Without making a detailed calculation, determine about which axis the moment of inertia of the disk is greater.

26. A rigid body rotates about a symmetry axis through its center of mass. Two points on the object lie at different radii from the center such that $r_1 > r_2$. (a) Which point (if either) has the greater angular speed? (b) Which point (if either) has the greater linear speed? (c) Which point (if either) has the greater centripetal acceleration?

27. Two interlocking gears of radii r_1 and r_2 have n_1 and n_2 teeth respectively. What is the ratio of the angular speeds of the two gears?

28. Two disks have the same thickness and radii, but are manufactured from materials of different densities. Which, if either, has the greater moment of inertia? Explain your reasoning.

29. The planar solids in Figure Q.29 have equal mass and the same thickness and height. Which has the greatest moment of inertia? Which has the smallest moment of inertia?

FIGURE Q.29

30. Explain how it is experimentally possible to distinguish between a raw egg and a hard-boiled egg without breaking the shell. Both eggs are at the same temperature.

31. When next in the weight room, take a barbell and spin it about the two axes indicated in Figure Q.31. Explain why it is easier for you to spin the barbell about the axis in Figure Q.31a than about the axis in Figure Q.31b.

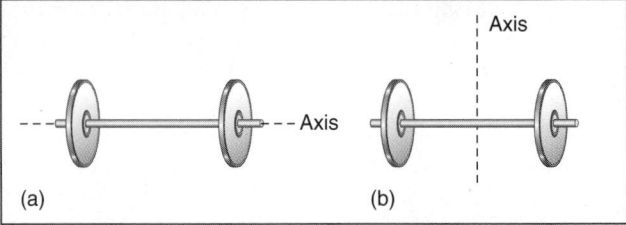

FIGURE Q.31

32. Why do divers tuck themselves into a ball-like shape when executing several turns in a dive? As the diver straightens out before entering the water (see Figure Q.32), does the diver actually stop rotating?

FIGURE Q.32

33. If the tread of a retreaded tire fails and breaks away from the tire itself, explain why it is likely that the failure occurs when the weakest part of the tread is at the top of the rolling motion of the tire.

34. A thin pillar on a level frictionless surface initially is upright but then falls. Sketch the trajectory of the center of mass of the pillar.

35. A flywheel is mounted in a special suitcase and set spinning at a large angular speed about a horizontal axis. A skycap picks up the suitcase and begins walking with it. Explain what happens when the skycap attempts to turn a corner with the suitcase.

36. A bicycle racer clearly wants to use a bicycle with a small mass. Explain why for a given frame it is beneficial to use tires with small moments of inertia.

37. Explain why tightrope walkers carry a long pole. (See Figure Q.37.) Why is the pole carried horizontally and not vertically?

FIGURE Q.37

38. A sphere, cylinder, and hoop all have the same radius and are made from the same material. They all are set rolling from rest from the same position down an inclined plane. Is there sufficient information to determine which wins the race to the bottom? Explain your reasoning. Does your answer change if the three objects also have the same mass?

39. A right-handed quarterback jumps vertically upward to make a critical pass. Describe and explain the motions of his lower body as he makes the pass.

40. What is the *ratio* of the (average) angular speeds of the second and minute hands of a clock to that of the hour hand? What is the direction of each of their angular velocity vectors?

41. Why do helicopters have at least two rotors, usually set at right angles to each other? (See Figure Q.41.)

FIGURE Q.41

42. In designing a car for a soapbox derby, is it advisable to design the wheels with large or small moments of inertia? Explain your answer.

43. Is it possible to have $\vec{F}_{total} = 0$ N but $\vec{\tau}_{total} \neq 0$ N·m? If so, describe such a situation.

44. Is it possible to have $\vec{F}_{total} \neq 0$ N but $\vec{\tau}_{total} = 0$ N·m? If so, describe such a situation.

45. When you buy a new tire for your car, you should be offered the option of having the tire balanced. (a) Why is this a good idea? (b) Tire stores offer the option of a static balance or a (more expensive) dynamic balance. Why is a dynamic balance better?
See Richard C. Smith, "Static vs. spin balancing of automobile wheels, *American Journal of Physics*, 40, #1, pages 199–201 (January 1972).

46. There is essentially zero torque on a spinning ice skater; hence according to

$$\vec{\tau}_{total} = \frac{d\vec{L}}{dt}$$

her angular momentum is constant or conserved. If the skater brings in her extended arms closer to her body, the angular speed increases dramatically, indicating the presence of a nonzero angular acceleration. The left-hand side of

$$\vec{\tau}_{total} = I_{CM}\vec{\alpha}$$

is zero, but the right-hand side is not. Resolve this dilemma with a clear explanation.

47. After studying hard, you and a friend decide to go cycling using identical bicycles. Your friend has a greater mass than you. If you both begin together from rest and coast down the same hill, which of you will reach the bottom of the incline first? Or will it be a tie? Justify your answer.

48. A twin engine airplane has propellers or jet engines on each side or wing. Why is it advantageous to have the engines spinning in opposite directions?

49. Since wheels are so advantageous, why did they not evolve in biological evolution?

50. Name a 20th-century technological discovery that might rival the invention of the wheel for its importance in technology.

51. The drive wheels of steam locomotives (see Figure Q.51) do *not* have a symmetric distribution of mass about the axle, unlike modern railroad locomotives. Explain why.

52. What elements of physics are involved in the design of *mobiles*? (See Figure Q.52.) Formulate a rule indicating where to place a vertical wire to hold all masses located below the wire in their desired locations in the mobile.

53. Is a particle in uniform circular motion in static equilibrium? Explain your answer.

FIGURE Q.51

FIGURE Q.52

PROBLEMS

Sections 10.1 The Distinction between Spin and Orbital Motion
10.2 The Orbital Angular Momentum of a Particle
10.3 The Circular Orbital Motion of a Single Particle
10.4 Noncircular Orbital Motion

1. What is the angular momentum of the fully loaded brown pelican (*Pelecanus occidentalis*) in Figure P.1 with respect to the flying fish (*Cypselurus heterurus*) at point Q? Is the angular momentum orbital or spin angular momentum?

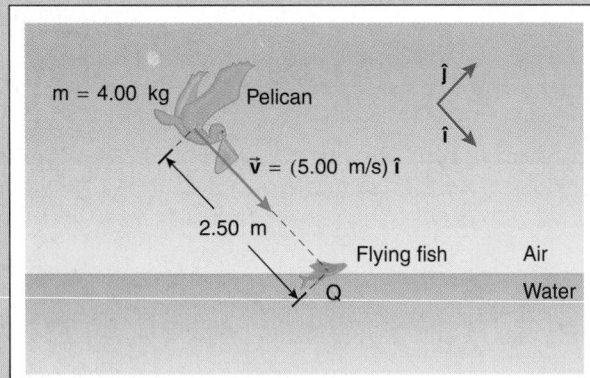

FIGURE P.1

2. Calculate the orbital angular momentum of a mute swan (*Cygnus olor*) of mass m about you (located at the point P), when the bird is at positions (a) and (b) in Figure P.2.

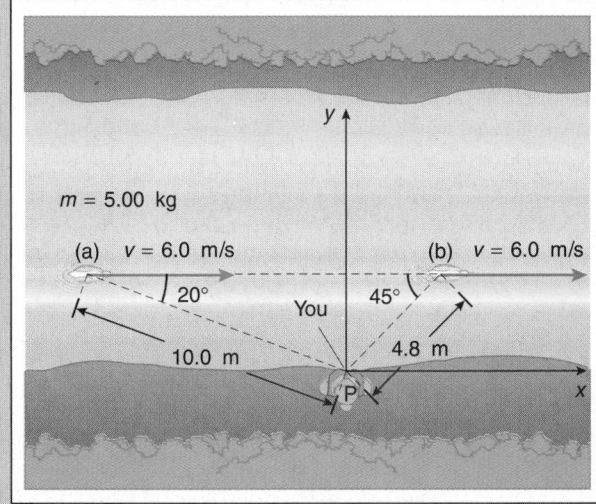

FIGURE P.2

•3. What is the magnitude of the orbital angular momentum about the center of the Earth of a 1000 kg satellite in a circular orbit 100 km above the *surface* of the Earth?

•4. A fantastic slapshot gives a hockey puck (of mass 0.170 kg) a speed of 40.0 m/s in a straight line toward a petrified goalie as indicated in Figure P.4. Consider the ice to be frictionless. (a) Calculate the momentum of the puck. (b) Is the momentum of the puck conserved along its trajectory? Why or why not? (c) What is the orbital angular momentum of the puck about the goalie when the puck is a distance of 2.0 m from the goalie? (d) Calculate the orbital angular momentum of the puck about the goalie when the puck is closest to him.

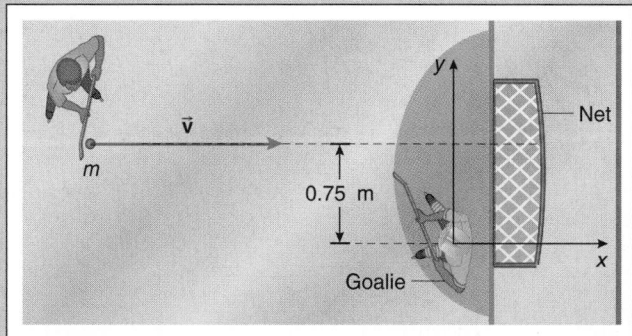

FIGURE P.4

a horizontal trajectory and is placed into circular orbit about the Earth (see Figure P.8). Clearly, we neglect air frictional effects! (a) What is the angular speed of the Ferris wheel if it is located at the north pole of the Earth? (b) If the Ferris wheel is located on the equator of the Earth and oriented so the particle is launched traveling *east*, is your answer to the question different? Why or why not? What is the new angular speed of the Ferris wheel if it is different from that calculated in part (a)? (c) If the Ferris wheel is located on the equator but the particle is launched traveling *west*, is your answer different? If so, find the new angular speed of the Ferris wheel.

•5. A planet is in an *elliptical* orbit about the Sun. The orbit has a semimajor axis a and an eccentricity ε. (a) Show that the ratio of the aphelion and perihelion distances is

$$\frac{r_{\text{aph}}}{r_{\text{peri}}} = \frac{1 + \varepsilon}{1 - \varepsilon}$$

(b) Show that the ratio of the moment of inertia of the planet (with respect to the Sun) at aphelion to that at perihelion is

$$\frac{I_{\text{aph}}}{I_{\text{peri}}} = \left(\frac{1 + \varepsilon}{1 - \varepsilon}\right)^2$$

•6. A planet of mass m is in circular motion about the Sun of mass M (where $M \gg m$). Since the mass of the Sun is so much larger than that of the planet, consider the Sun to be the origin and essentially fixed in position. Let r be the radius of the orbit of the planet and let T be its period. (a) Show that the magnitude of the orbital angular momentum $\vec{L}$ of the planet can be written as

$$L = \frac{2\pi mr^2}{T}$$

(b) In what direction is the orbital angular momentum vector? (c) Use the result of part (a) and the Newtonian form for Kepler's third law (Equation 6.37) to show that the magnitude of the orbital angular momentum also can be written in the form

$$L = m[G(M + m)r]^{1/2}$$
$$\approx m(GMr)^{1/2}$$

Thus the magnitude of the orbital angular momentum of the planet is proportional to the square root of the radius of the orbit.

•7. Two wheels of diameter 0.76 m are attached to opposite ends of an axle of length 2.0 m. The wheels roll around a circular track of inside radius 10 m. (a) Through what angle around the circular track must the axle assembly move so that the outer wheel makes one revolution more than the inner wheel? (b) What is the ratio of the angular speeds of spin of the two wheels? The differing angular speeds of the wheels is the reason for the differential in the drivetrain of a car or a truck.

⁑8. Imagine a huge Ferris wheel of radius 40.0 m spinning at angular speed ω. A small mass m emerges from the wheel with

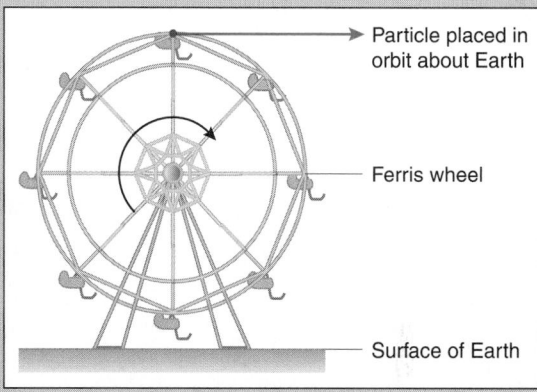

FIGURE P.8

Sections 10.5 Rigid Bodies and Symmetry Axes
   10.6 Spin Angular Momentum of a Rigid Body
   10.7 The Time Rate of Change of the Spin
    Angular Momentum
   10.8 The Moment of Inertia of Various
    Rigid Bodies
   10.9 The Kinetic Energy of a Spinning System

9. Imagine a rigid disk of radius r spinning at 1.00 rev/s. At what distance from the center of the disk is the tangential speed of a particle equal to the speed of light $c = 3.00 \times 10^8$ m/s? When we study relativity in Chapter 25, we shall see that it is not possible for material particles to travel at the speed of light; hence the disk cannot remain rigid.

•10. Calculate the total angular momentum of the system of particles pictured in Figure P.10 about the origin at 0.

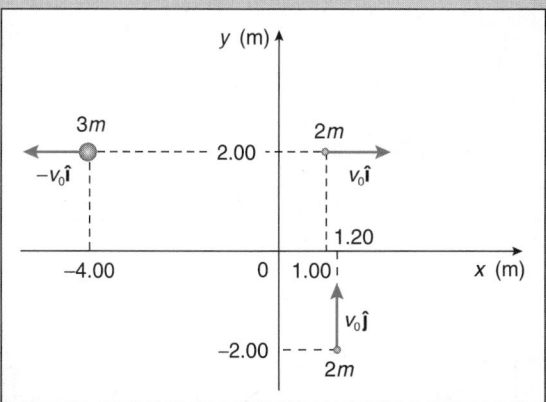

FIGURE P.10

•11. Calculate the moment of inertia of each particle system in Figure P.11 about the indicated axis of rotation.

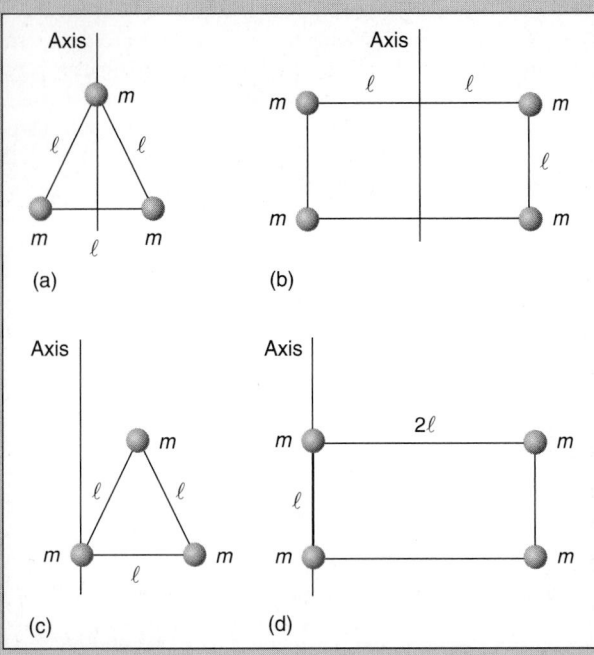

FIGURE P.11

•12. Find the center of mass of the collection of mass points in Figure P.12 and then find the moment of inertia of the system about an axis through the center of mass and parallel to the y-axis. Is the system of particles symmetric about this axis through the center of mass?

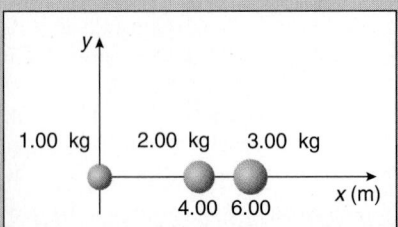

FIGURE P.12

•13. A uniform disk of mass m and radius R has an additional rim of mass m as well as four symmetrically placed masses, each of mass $m/4$, fastened at positions $R/2$ from the center as shown in Figure P.13. What is the total moment of inertia of the disk about an axis perpendicular to the disk through its center?

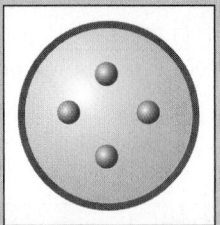

FIGURE P.13

•14. The total mass of the bolt shown in Figure P.14 is 200 g. (a) What is the mass of the rectangular plate that constitutes the head of the bolt? (b) What is the moment of inertia of the head of the bolt about the indicated symmetry axis? (c) What is the mass of the body of the bolt? (d) What is the moment of inertia of the body of the bolt about the indicated symmetry axis? (e) What is the total moment of inertia of the bolt about the indicated symmetry axis?

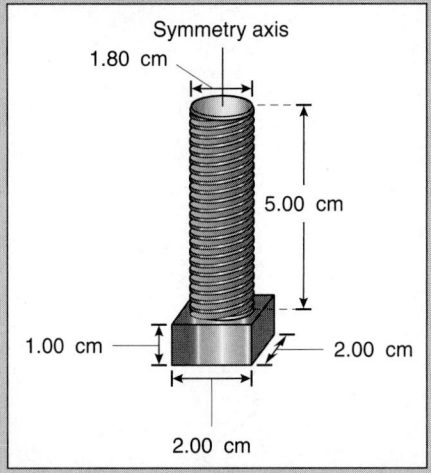

FIGURE P.14

•15. The moment of inertia of a uniform sphere of mass m and radius R about an axis through the center of mass is (see Table 10.1)

$$I_{CM \text{ uniform sphere}} = \frac{2}{5} mR^2 = 0.400 mR^2$$

The moment of inertia of a spherical shell of the same mass and radius is

$$I_{CM \text{ spherical shell}} = \frac{2}{3} mR^2 = 0.667 mR^2$$

Spherical planets and moons are neither uniform spheres nor spherical shells. Write the moment of inertia of such spherical masses about an axis through their center of mass as

$$I_{CM} = \beta mR^2$$

(a) What is the maximum value for β? Is there a minimum value for β? (b) As β decreases from the value 0.400 (the value for a uniform sphere), what does this imply about the mass distribution within the planet? Is it likely that a spherical planet or moon has value of β much larger than 0.400? What would this imply about the structure of the planet or moon? (c) Models of the mass distribution of the Earth, Moon, and Mars (inferred from detailed tracking of orbiting spacecraft) indicate that the numerical value of β is

for the Moon	0.391
for Mars	0.377
for the Earth	0.331

What conclusions can be drawn from this data about the comparative interior structure of these celestial objects?

•**16.** Consider a two-dimensional *planar* object as indicated in Figure P.16. Let the x- and y-axes be two perpendicular axes in the plane of the figure as indicated and let I_x and I_y be the moments of inertia of the system about each of these axes respectively.* Show that the moment of inertia I_z of the system about the z-axis, perpendicular to the plane of the object, is

$$I_z = I_x + I_y$$

This result is known as the *perpendicular axis theorem* and applies only to *planar* masses. Explain why this restriction is necessary.

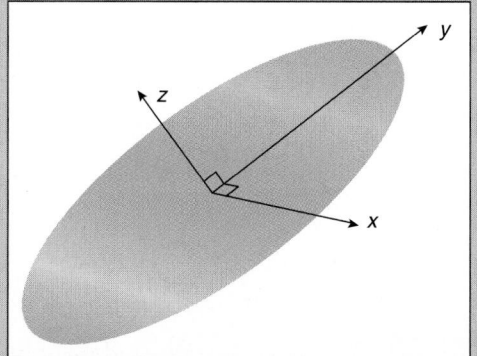

FIGURE P.16

•17. Use the perpendicular axis theorem of the previous problem to show that the moment of inertia of a circular hoop of mass m and radius R about the axis indicated in Figure P.17 is

$$\frac{1}{2} mR^2$$

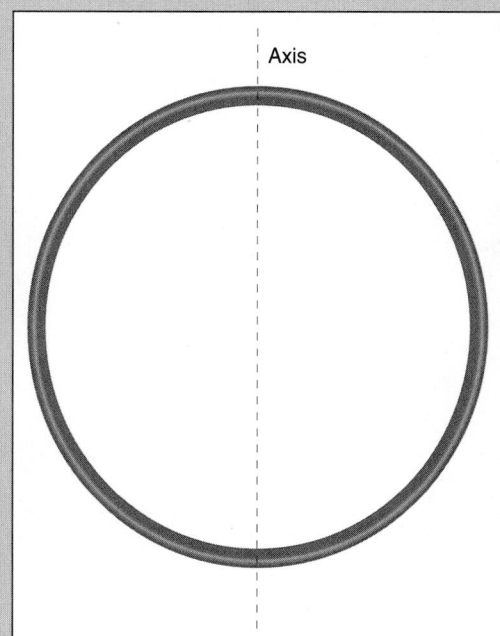

FIGURE P.17

•**18.** Professorial dreams. Two students are fighting over a physics text and exerting forces of equal magnitude F on it as indicated in Figure P.18. The mass of the text is 2.50 kg and its dimensions are 20 cm by 30 cm. Assume it can be modeled as a uniform and unusually dense rectangular plate. (a) What is the magnitude of the total torque about the center of mass of the text? (b) What is the magnitude of the initial angular acceleration of the text about its center of mass?

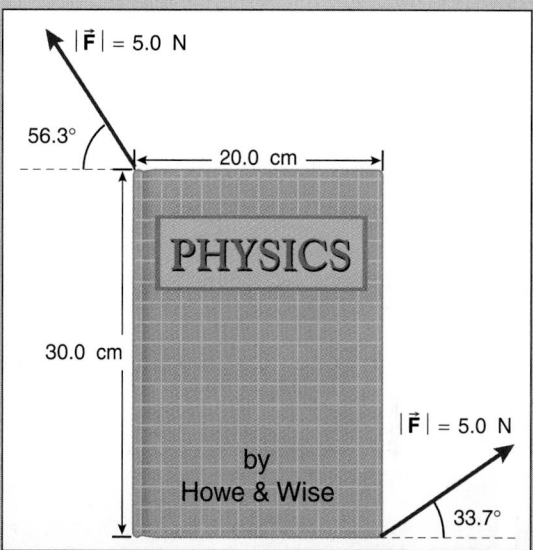

FIGURE P.18

•**19.** Assume the Sun is a uniform, rigid sphere (it is spherically shaped but actually is not of uniform density nor a rigid body). The Sun rotates once every 27 days.[†] (a) Calculate the rotational kinetic energy of the Sun. (b) The luminosity (power output) of the Sun is 3.826×10^{26} W, a rather powerful light bulb! For how many years could the Sun shine at its present luminosity if it were radiating away its rotational kinetic energy? The energy source of the Sun is not the rotational kinetic energy but rather the nuclear fusion of hydrogen nuclei, creating helium.

•**20.** Massive spinning flywheels (disks) can be used for storing energy. Consider a flywheel with a diameter of 1.00 m and a mass of 500 kg. A constant force of magnitude F is applied tangentially to the rim of the flywheel to accelerate it from rest to 4.00×10^3 rev/min during 3.00 min. (a) What is the magnitude of the angular acceleration of the flywheel during this interval? (b) What magnitude of torque is necessary to cause this angular acceleration? (c) What magnitude F of force is needed? (d) What is the rotational KE of the flywheel at the end of this acceleration? (e) What average power did the accelerating mechanism provide to the flywheel? (f) From what height h would you have to drop a mass of 500 kg for it to achieve a KE just before impact that is equal to the rotational KE of the flywheel?

•21. The total mass of your new sports car is 1.500×10^3 kg. The four wheels each have a mass of 35.0 kg and a radius of 25.0 cm.

---

*The quantities I_x, I_y, and I_z are *not* the components of a vector since the moment of inertia is a scalar, not a vector.

[†] The Sun actually exhibits *differential rotation*: the rotational period depends on solar *latitude* or the distance from the solar equator. The rotational period is about 25 days on the solar equator and about 35 days near the poles.

To a first approximation, consider the wheels as uniform disks. (a) When the car is zipping along a straight section of highway at 150 km/h attempting to elude the flashing blue lights visible in the rearview mirror, what is the total kinetic energy of the car? (b) What fraction of the total kinetic energy is the rotational kinetic energy of the wheels?

•22. Space stations have been proposed to accommodate the surplus population of the Earth. The initial design is for a hollow, uniform, cylindrical space station of diameter 3.00 km, length 10.00 km, and total mass of 1.20×10^{10} metric tons. (See Figure P.22.) The space station is to be spun about the symmetry axis coincident with the axis of the cylindrical shape. (a) What angular speed of rotation ω is needed to simulate the magnitude of the local acceleration due to gravity (9.81 m/s²) for objects on the perimeter of the space station? Express your result in rev/min. (b) What is the rotational kinetic energy of the space station? (c) Small, thrusting rockets mounted tangential to the circular cross section are to set the space station in rotational motion. Starting from rest, the spacecraft reaches the angular speed calculated in part (a) after 1.00 y. Each rocket is capable of exerting a force of magnitude 1000 N continuously over the year. How many thrusting rockets are needed?

FIGURE P.22

•23. A Maytag repairman models the washer drum of a washing machine as a thin cylindrical shell of mass 30.0 kg, radius 25.0 cm, and height 50.0 cm together with a uniform disk (the bottom of the drum) of mass 10.0 kg of the same radius, as indicated in Figure P.23. The washer is 3/4 filled with water (of density $\rho = 1.00 \times 10^3$ kg/m³) as it begins its spin cycle. The drain fails to open and the machine accelerates the load from rest to 10.0 rev/s during 60.0 s. Assume that the mass of water retains a cylindrical shape during the spin cycle (the top

surface actually assumes the shape of a paraboloid of revolution). (a) Find the moment of inertia of the tub. (b) Find the moment of inertia of the water. (c) Find the total moment of inertia of the tub and water system. (d) Calculate the magnitude of the angular acceleration. (e) What magnitude of torque is provided by the motor? (f) What is the kinetic energy of the system when rotating at 10.0 rev/s?

Direction of spin

FIGURE P.23

‡24. Show that the moment of inertia of a thick cylindrical shell of mass m, whose inner and outer radii are R_1 and R_2 respectively (see Figure P.24), about its symmetry axis is

$$\frac{1}{2} m \left(R_1^2 + R_2^2 \right)$$

Explain why this moment of inertia cannot be found by taking the moment of inertia of a disk of radius R_2 and subtracting that of a disk of radius R_1.

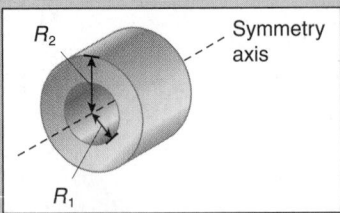

R_2 Symmetry axis

R_1

FIGURE P.24

‡25. At the center of the famous Crab Nebula in the constellation Taurus lies a compact (radius $R = 10$ km) but massive neutron star that spins with a period T of only 0.0333 s. Astronomers

have discovered that the period of the neutron star is gradually increasing at the rate of 3.65×10^{-8} s/d. Assume the star has a mass m of twice that of the Sun ($M_{Sun} = 1.99 \times 10^{30}$ kg). (a) Show that the rotational kinetic energy of the system is

$$KE_{rot} = \frac{4\pi^2}{5} \frac{mR^2}{T^2}$$

(b) Evaluate the rotational kinetic energy. (c) Use the result of part (a) to show that the rate of kinetic energy loss can be expressed as

$$\frac{d(KE)}{dt} = -\frac{2\,KE}{T}\frac{dT}{dt}$$

(d) Evaluate this rate of kinetic energy loss (power!) and compare it with the power output of the Sun (3.83×10^{26} W).

26. A mass m is attached to a rigid rod of negligible mass as in Figure P.26. The system is pivoted at point 0 and rotates about the indicated z-axis with angular velocity $\vec{\omega}$, maintaining a fixed angle θ with the axis. (a) Show that at the instant pictured the angular momentum of m about the pivot 0 is

$$\vec{L} = (-mr^2\omega \sin\theta \cos\theta)\hat{i} + (mr^2\omega \sin^2\theta)\hat{k}$$

(b) Note that $\vec{L}$ is *not* parallel to $\vec{\omega}$. Explain why. Is $\vec{L}$ parallel to $\vec{\omega}$ at any instant during the motion?

This problem was inspired by the article by Herman Erlichson, "Angular momentum and angular velocity," *The Physics Teacher, 32,* #5, pages 274–275 (May 1994).

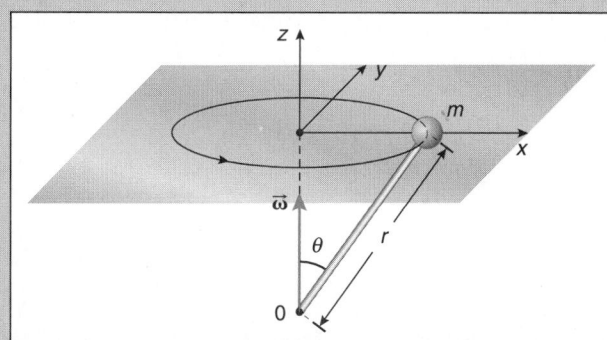

FIGURE P.26

Sections 10.11 The Precession of a Rapidly Spinning Top*
10.12 The Precession of the Spinning Earth*

27. A thin disk of mass m has a moment of inertia I about a symmetry axis perpendicular to the plane of the disk. The disk is spinning with a large angular speed ω. The disk is suspended from a swivel that permits the symmetry axis of the disk (a rod of negligible mass) to turn freely while keeping the position of an end of the axis fixed; see Figure P.27. The disk slowly precesses in a horizontal circle. (a) In a sketch, indicate the direction of the angular momentum vector of the disk. (b) What force provides a torque on the disk about the point of support? What is the direction of the total torque on the

disk about the point of support? (c) When viewed from above, does the disk precess in the clockwise or counterclockwise sense? (d) Show that the angular speed of the precession is

$$\frac{mgb}{I\omega}$$

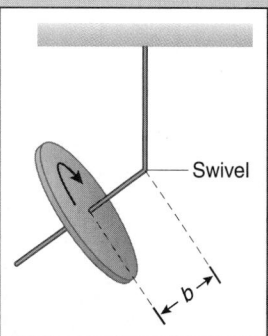

FIGURE P.27

28. The planet Venus spins in the sense opposite to the spin of the Earth and has a spin period of 243 days; see Figure P.28. The Earth has a spin period of one day. The radius of Venus is 6.05×10^3 km, close to that of the Earth (6.37×10^3 km). The rotational axis of Venus makes an angle of about 3.0° with a perpendicular to the plane of its orbit about the Sun (for the Earth, the corresponding angle is about 23.5°). (a) Do you expect the oblateness of Venus to be greater than, less than, or about the same as that of the Earth? Explain your answer. (b) Will the axis of Venus precess? If so, is the directional sense of the precession the same as that of the Earth's axis or in the opposite sense? Explain your answer. (c) Will the precessional period likely be greater or less than that of the Earth (for the Earth it is 25 785 years)? Explain your answer.

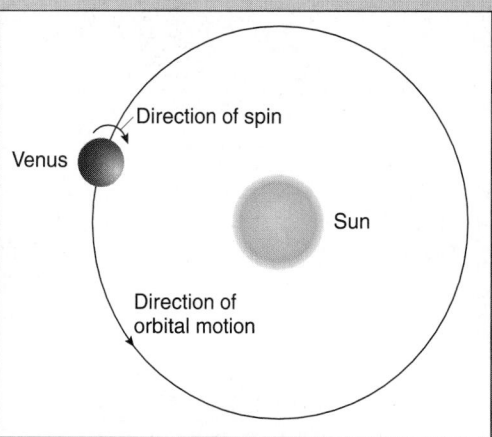

FIGURE P.28

Sections 10.13 Simultaneous Spin and Orbital Motion
10.14 Synchronous Rotation and the Parallel
Axis Theorem
10.15 Rolling Motion without Slipping
10.17 Total Angular Momentum and Torque
10.18 Conservation of Angular Momentum

29. Calculate the moment of inertia of each mass in Figure P.29 about the indicated axis.

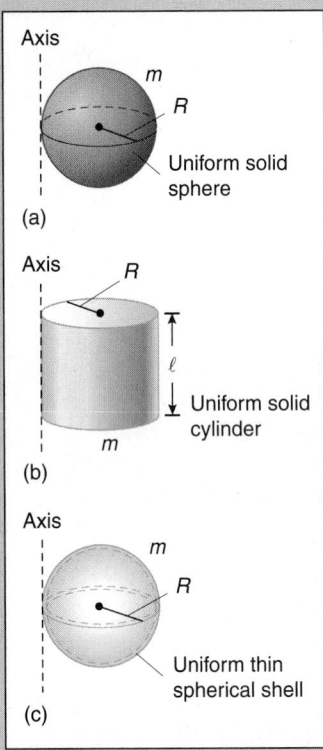

FIGURE P.29

•30. A solid spherical bowling ball has a mass of 8.5 kg and a radius of 10 cm. How much work is it necessary to do on the ball to get it to a speed of 6.0 m/s? Assume the sphere starts at rest and eventually is rolling without slipping on a horizontal surface.

•31. A sphere has a diameter equal to the length of an edge of a cube of the same mass. Both are released at rest at the top of an inclined plane with a smooth and rough path down the incline. The sphere rolls without slipping down the rough surface and the cube slides without friction down the smooth surface. (a) Which (if either) has the greater kinetic energy at the bottom of the incline? Justify your answer with a quantitative argument. (b) Which (if either) has the greater speed at the bottom of the incline? Justify your answer with a quantitative argument.

•32. A rod of length ℓ and mass m is free to rotate in a vertical plane about an axis at one end. The rod is released at rest in the horizontal orientation as shown in Figure P.32. (a) What is the moment of inertia of the rod about this axis?

(b) What is the initial torque of the gravitational force on the rod about this axis? (c) Specify any other forces that act on the rod as it rotates. Calculate their torques about the axis. (d) Specify the initial angular acceleration of the rod. (e) What is the torque of the gravitational force when the rod is vertically oriented? What is the angular acceleration when the rod has this orientation? (f) Use the CWE theorem to find the angular velocity of the rod when it has a vertical orientation. (g) Find the angular momentum of the rod when it has a vertical orientation.

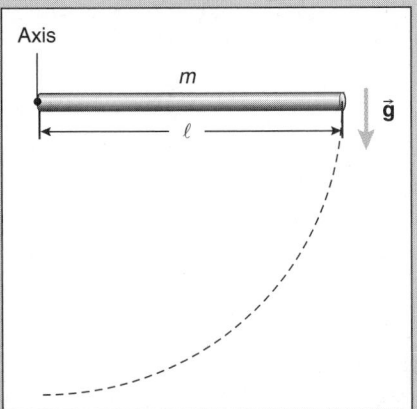

FIGURE P.32

•33. As a first approximation, consider the Earth to be spherical in shape with radius 6.37×10^3 km. (a) Calculate the angular speed of the spin of the Earth. (b) Calculate the magnitude of the spin angular momentum of the Earth. (c) Rivers such as the Mississippi carry large quantities of sediment from northern latitudes to more southern latitudes. Over long periods of time, what *qualitative effect* [if any] will such sediment transport toward the equator have on the length of a day? (No calculation is intended; simply explain what change takes places in the length of a day (if any) and why a change does or does not occur.) (d) Calculate the orbital angular speed of the Earth in its revolutionary motion about the Sun. Assume the orbit is circular. (e) Calculate the magnitude of the orbital angular momentum of the Earth about the Sun. (f) What is the ratio of the magnitudes of the orbital and spin angular momenta of the Earth?

•34. A frictionless human-powered merry-go-round has a radius of 4.0 m and a moment of inertia of 4.00×10^4 kg·m². Your kid sister (mass 30 kg) is located 3.0 m from the axis and asks you to give her a ride. You oblige by pushing tangentially along the rim of the merry-go-round with a force of magnitude 250 N. (a) What is the total moment of inertia of the system composed of the merry-go-round and your sister? (b) How long will it take you to get the system moving from rest until you reach a speed of 1.00 m/s? (c) Show the *ratio* of the kinetic energy of the system to the magnitude of its angular momentum is proportional to t, and so increases with time.

●**35.** A uniform plank of mass 100 kg is mounted so it can rotate about a horizontal axis as indicated in Figure P.35. Two physics students with masses 60.0 kg and 80.0 kg are attached to the ends of the plank as indicated. Consider the students as particles. (a) Calculate the moment of inertia of the system of the plank and two students about the indicated axis. (b) What is the magnitude of the total torque acting on the system about the indicated axis arising from gravitational forces? (c) Determine the magnitude of the initial angular acceleration of the system, assuming it is released initially at rest. (d) Is the angular acceleration constant as the plank turns? (e) Determine the angular speed of rotation of the system when the plank is vertical. (Assume unrealistically the students do not fall off!) (f) What is the magnitude of the angular acceleration of the system when the plank is vertical?

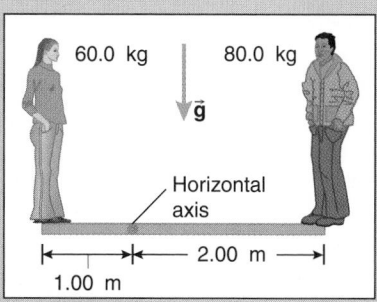

FIGURE P.35

●**36.** A disk of mass m and radius R is free to turn about a fixed, horizontal axle. The disk has an ideal string wrapped around its periphery from which another mass m (equal to the mass of the disk) is suspended, as indicated in Figure P.36. (a) What is the magnitude of the acceleration of the falling mass? (b) What is the magnitude of the angular acceleration of the disk?

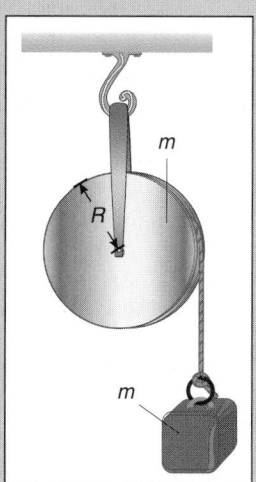

FIGURE P.36

●**37.** A thin, hollow sphere of mass m is filled with a liquid of mass $m/2$. The filled sphere is spinning freely with no friction at angular speed ω_0 about an axis through the center of mass point of the sphere. The hollow sphere and all parts of the fluid have the same angular speed ω_0. A leak develops along the axis of the sphere (see Figure P.37) and all the fluid leaks out. What is the angular speed of the now hollow sphere?

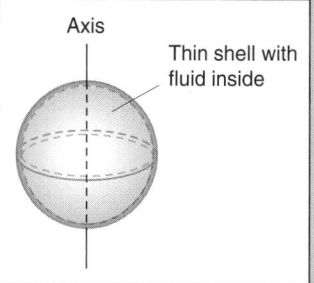

FIGURE P.37

●**38.** Two uniform spheres of identical mass and radius are placed on inclined planes at the same height h and inclination angle θ. One plane is rough and causes one sphere to roll down the plane; the other is frictionless, and so the sphere on it slides down the incline. (a) Find the ratio of the kinetic energies of the two spheres at the bottom of the inclines:

$$\frac{KE_{slide}}{KE_{roll}}$$

(b) Which sphere takes the greater time to reach the bottom of the incline or do they take the same time? (c) Find the ratio of the times it takes the spheres to reach the bottom of the inclines:

$$\frac{t_{slide}}{t_{roll}}$$

(d) Find the ratio of their speeds at the base of the inclines:

$$\frac{v_{slide}}{v_{roll}}$$

●**39.** Oops. You just lost your marbles. One marble in particular has a mass m and radius R and rolls without slipping down a 30° incline. (a) Show that the magnitude of the acceleration of the marble is independent of its mass and equals 0.36g. (b) If the marble began at rest, how far does it go during the 3.0 s it takes you to recoup all your other marbles?

●**40.** A cylinder is rolled without slipping from rest by a constant force of magnitude F_0 in the horizontal direction applied by a string wrapped around the cylinder as indicated in Figure P.40.

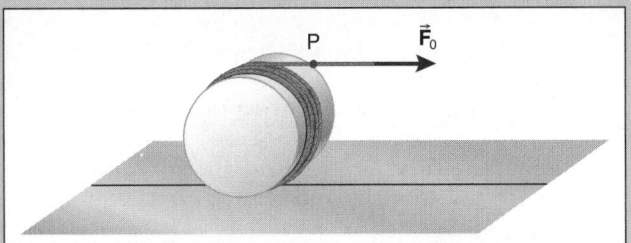

FIGURE P.40

(a) If a point P on the horizontal section of the string moves a distance x, how far does the center of the cylinder move? (b) If P moves a distance x, how much work is done by the force? (c) After P moves the distance x, what is the speed v of the center of mass of the cylinder?

•41. A solid sphere, a spherical shell, a disk, and a thin cylindrical shell all have the same radius R and the same mass m. They all are situated at rest at the top of an inclined plane of height h as shown in Figure P.41. When $t = 0$ s, they all begin to roll, without slipping, down the plane. (a) Write the moment of inertia of each object as

$$I = \beta m R^2$$

with the appropriate value of β for each as found in Table 10.1. Show that the speed of the objects at the bottom of the incline can be written as

$$v = \left(\frac{2gh}{1 + \beta} \right)^{1/2}$$

(b) At the bottom of the incline the kinetic energy of each object has two contributions: (1) the kinetic energy of the center of mass and (2) rotational kinetic energy. Rank the objects in the order of largest to smallest values of the rotational and translational kinetic energy that they have at the bottom of the incline. (c) Indicate the order in which the objects arrive at the base of the incline.

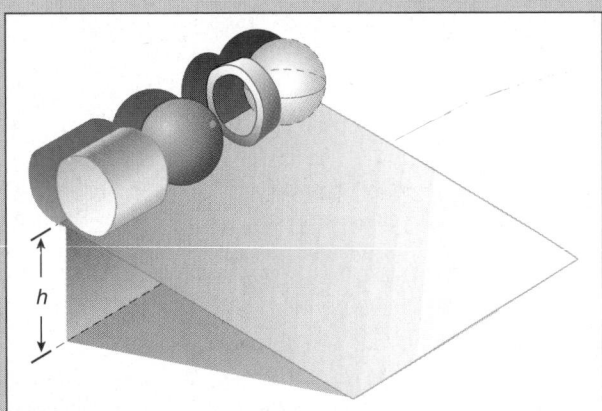

FIGURE P.41

•42. Two hoops of radii R_1 and R_2 have identical masses. The centers of the hoops are traveling at the same speed v parallel to a level surface along which the hoops are rolling. (a) What is the ratio of the magnitudes of the momenta of the two hoops? (b) What is the ratio of the magnitudes of the angular momenta of the hoops taken about their respective centers of mass? (c) What is the ratio of the kinetic energies of the hoops? (d) If the hoops roll without slipping to a stop by going up an incline, what is the ratio of the change in the heights attained by the center of mass point of each hoop?

•43. You are driving a car of mass m at a constant speed v around a horizontal circular track of radius r. On this track the magnitude of the frictional force of the road on the tires of the car can be as great as half the magnitude of the weight of the car before slipping occurs. (a) Specify the forces that act on the car. (b) What is the direction of the total force that acts on the car? (c) Is the momentum of the car conserved? Explain. (d) What is the maximum possible speed of the car such that slipping does not occur?

•44. During World War II, the British mounted an unorthodox and successful night air attack on the Mohne Dam in Nazi Germany. Large, almost cylindrical, horizontal bombs were dropped at a speed of 360 km/h, at low altitude (18 m), and about 550 m from the dam itself. Much like flat stones, the bombs skipped across the lake behind the dam (about eight skips!), came to rest just behind the dam itself, sank, and exploded; the resulting shock wave destroyed the dam. The mass of each bomb was about 4.0×10^3 kg and its dimensions were approximately 0.70 m in radius with a length of 2.1 m. When released, each bomb had a spin about its horizontal cylindrical symmetry axis of 500 rev/min. (a) Find the moment of inertia of each bomb, assuming it is of uniform density. (b) Calculate the work needed to get each bomb from an initial angular speed of zero to the given angular speed. (c) Comment on the likely reason for spinning the bombs before they were dropped.

A fascinating account, complete with other aspects of the physics underlying the mission, can be found in an article by Arthur Stinner, "Physics and the dambusters," *Physics Education*, 24, #5, pages 260–267 (September 1989). The mission was recounted in a book by Paul Brickhill, *The Dambusters* (Ballantine, New York, 1951) and made into a movie in 1954.

•45. A 200 kg uniform beam hangs from a hinge and is supported by a cable as shown in Figure P.45. Time inexorably takes its toll and the cable snaps. (a) What is the magnitude of the total torque about the hinge the instant after the cable breaks? (b) What is the magnitude of the initial angular acceleration of the beam? (c) Is the angular acceleration of the beam constant as it swings down? Explain why or why not. (d) What is the angular speed of the beam just before it crashes into the vertical wall?

•46. An ice skater spinning with an angular speed of 1.50 rev/s is holding two 3.00 kg masses out at a radius of 0.80 m from her axis of symmetry using her two arms (of negligible mass). The *total* moment of inertia of the skater and the two 3.00 kg masses is 6.00 kg·m². (a) Find the moment of inertia of the skater herself. (b) If the two 3.00 kg masses are brought in to

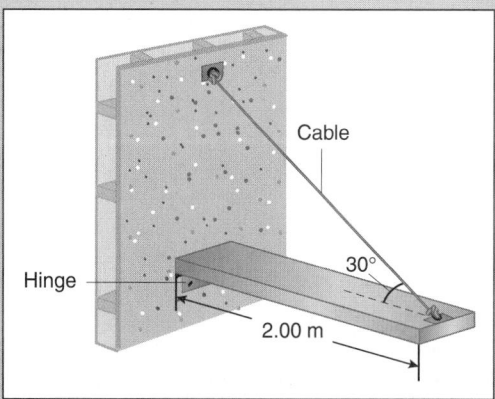

FIGURE P.45

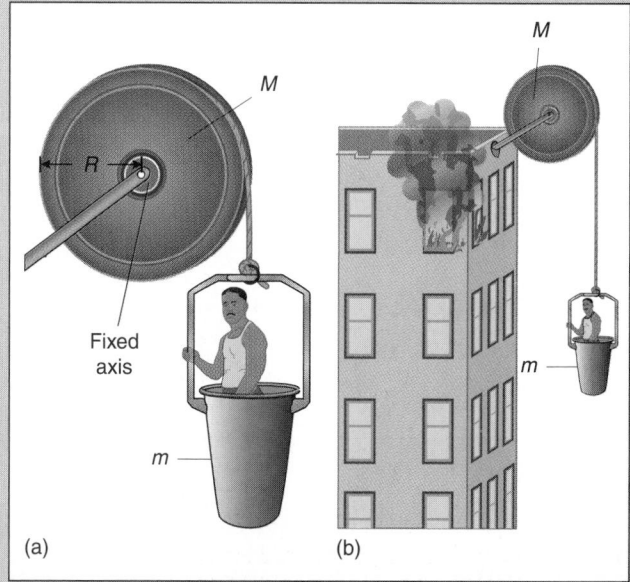

(a)

(b)

FIGURE P.48

a radius of 0.25 m, what is the new angular speed of the skater in rev/s? (c) If the two masses were dropped with her arms left extended, instead of being pulled in, would her spin rate change? If so, to what value in rev/s?

•**47.** An empty flat-bottomed cylindrical tub with vertical sides is freely spinning about its symmetry axis at 180 rev/min. The bottom of the tub has a mass of 5.0 kg and a radius of 25 cm, while the cylindrical sides have a mass of 6.0 kg. Sand gradually and gently is shoveled uniformly into the tub. Neglect any initial kinetic energy of the sand. As the tub fills, the sand assumes the same angular speed as the tub. Assume the sand in the tub has a cylindrical volume. (a) What is the moment of inertia of the tub about its symmetry axis? (b) What is the initial kinetic energy of the tub? (c) What mass of sand must be added to the tub to reduce the angular speed to 60 rev/min? (d) What is the kinetic energy of the rotating tub–sand system when its angular speed is 60 rev/min? (e) What fraction of the initial kinetic energy of the tub–sand system is part (d)?

•**48.** A mass m is hung from an ideal cord wrapped around the periphery of a disk of mass M and radius R as indicated in Figure P.48a. The disk is free to turn on a fixed horizontal axle with negligible friction. (a) Sketch a second law force diagram indicating all the forces acting on m. (b) What is the magnitude of the total torque on the disk about its axis? (c) Apply Newton's second law of motion to m and the rotational counterpart of Newton's second law to the spin of the disk to show that the downward acceleration of the mass m is of magnitude

$$a = \frac{1}{1 + \dfrac{M}{2m}} g$$

(d) What must be the ratio of M to m so the magnitude of the downward acceleration of m is only 0.010g? (e) If such a device is used as a fire escape on a building (see Figure P.48b) for a person of mass 80 kg, what must be the mass of the disk to satisfy the conditions of part (d)? Neglect the mass of the bucket compared with that of the person. If the person is to hit

the ground with a speed no greater than 2.0 m/s, what is the maximum height of a building that can employ such a system?

•**49.** Use conservation of angular momentum to find the ratio between the angular speeds of a planet at perihelion and aphelion in terms of the eccentricity of its orbit. That is, find

$$\frac{\omega_{\text{aph}}}{\omega_{\text{peri}}}$$

•**50.** After getting a flat tire fixed, you are hurriedly rolling the tire back to your car, simmering because you now are at least an hour late for a lavish party and your dress is, well, a sight to be seen. Model the tire (no pun intended) as a uniform disk of mass m and radius R; we better not model you. Beginning at rest at the base of an incline, you roll the tire up the incline to a vertical height h so the tire has a speed v when you reach your car (see Figure P.50). Way to go! What work is done by your constant force on the tire to accomplish this? And now you are sweaty too.

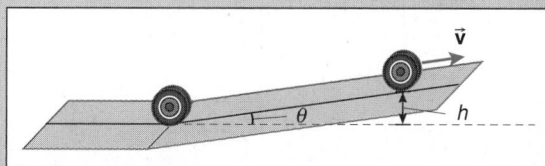

FIGURE P.50

•**51.** A skater is rotating at an angular speed of 1.00 rad/s on smooth ice with both arms and legs outstretched (as indicated in Figure P.51). When the arms and legs are returned to their normal positions, the angular speed of the skater increases to

FIGURE P.51

10.0 rad/s. (a) Is the angular momentum of the skater conserved in this change? Explain. (b) By what multiplying factor does the magnitude of the angular momentum of the skater change? (c) By what factor does the moment of inertia of the skater change? (d) Is kinetic energy of the skater conserved in this change? (e) By what multiplying factor does the kinetic energy of the skater change?

•**52.** A log rolling politician is (to first order) in the shape of a cylinder of mass 60.0 kg and radius 20.0 cm. An ideal cord is wrapped around the circumference of the good-ol'boy with one end attached to the ceiling of the dome of the Capitol as in Figure P.52. The politician is released at rest and becomes totally unraveled due to the lack of a spin doctor. (a) Sketch a second law force diagram indicating the forces on the politician. (b) What is the magnitude of the downward acceleration of the chap? (c) What is the magnitude of the angular acceleration? (d) What is the angular speed 1.50 s after release? (e) What is the speed of the center of mass 1.50 s after release?

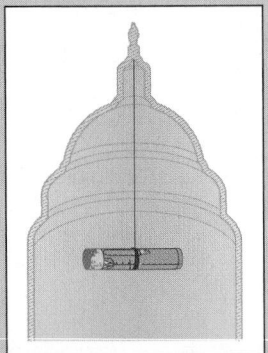

FIGURE P.52

•53. Consider the Earth–Moon system. Both objects spin about axes through their respective centers of mass; both also orbit the center of mass of the Earth–Moon system. Assume that both the Earth and Moon are spherical in shape and that their orbits about the center of mass are circular in shape. The orbital period of the Earth and Moon about the center of mass is known as the sidereal period of the Moon and is 27.322 d. The Moon also is in synchronous rotation; that is, its period of rotation is the same as the period of its orbital motion about the center of mass of the Earth–Moon system. (This equality is not coincidental but came about because of tidal forces of the Earth acting on the Moon over the long history of the Earth–Moon system.) The rotational period of the Earth is 23.933 h. (a) Find the location of the center of mass of the Earth–Moon system. (b) Calculate the magnitude of the orbital angular momentum of the Moon about the center of mass of the Earth–Moon system. (c) Calculate the magnitude of the orbital angular momentum of the Earth about the center of mass of the Earth–Moon system. (d) Calculate the magnitude of the spin angular momentum of the Moon. (e) Calculate the magnitude of the spin angular momentum of the Earth. Notice that most of the total angular momentum of the Earth is *spin* angular momentum while most of total angular momentum of the Moon is *orbital* angular momentum. For the Earth–Moon system, the gravitational tidal forces of the Earth on the Moon and of the Moon on the Earth are *internal forces*. To a first approximation, there are no external torques on the Earth–Moon system, and so the total angular momentum of the Earth–Moon system is conserved. The tidal force of the Moon on the Earth produces an internal torque on the Earth whose effect is to decrease the angular speed of the spin of the Earth, thus lengthening the day.* This, of course, decreases the spin angular momentum of the Earth, the dominant contribution to the angular momentum of the Earth. To keep the *total* angular momentum of the Earth–Moon system constant, the angular momentum of the Moon must increase. Since most of the angular momentum of the Moon is orbital angular momentum, the orbital angular momentum of the Moon increases. The effect of this increase in the orbital angular momentum of the Moon is to increase the distance between the Earth and the Moon by about 3.8 centimeters per year. [See Problem 6, part (c).] Tidal forces thus gradually slow the rotation of the Earth (lengthening the day by several thousandths of a second per century) and increase the distance between the Earth and the Moon (by several meters per century). Both effects have been measured directly! Eventually the duration of an Earth day will be equal to the orbital period of the Moon, and both will be about 53 (current) days long [see Arthur Eisenkraft and Larry D. Kirkpatrick, "When days are months," *Quantum*, 1, #2, page 45 (November–December 1990), for a way to calculate this time period].

•**54.** A kid of mass 30 kg is at the center of a playground merry-go-round that is rotating (on frictionless bearings) at 20 rev/min. The mass of the merry-go-round is 150 kg and its radius is

---

*This is because the fast rotational speed of the Earth carries the tidal bulges of the Earth ahead of the Earth–Moon line. The lagging Moon exerts gravitational forces on these tidal bulges that tend to slow or brake the rotation of the Earth.

2.00 m. Model the merry-go-round as a disk and the kid as a point particle. The kid slowly crawls to the rim. Find the new angular speed of the merry-go-round in revolutions per minute.

•55. A uniform beam of length ℓ is in a vertical position with its lower end on a rough surface that prevents this end from slipping. The beam topples. Show that at the instant before impact with the floor, the angular speed of the beam about its fixed end is

$$\omega = \left(\frac{3g}{\ell}\right)^{1/2}$$

•56. (a) Model a star as a rigid uniform spherical mass of radius R_0 spinning with a period T_0. (In fact, such a model of a star like the Sun as a rigid body is *not* very realistic, but we will go ahead with it anyway.) If the star collapses under gravitational forces to a radius R and loses no mass nor angular momentum, what is the new period of rotation of the star? (b) Consider a star with twice the mass of the Sun, yet with an initial radius that is equal to the radius of the Sun and with a initial period of rotation of 25.0 d. If this star collapses and becomes a neutron star with a radius of 10.0 km, what is the new rotational period of the star? Assume (unrealistically) the star satisfies the conditions of the model in part (a) of this problem. Such small neutron stars are created by the gravitational collapse of massive stars when they exhaust their options for generating energy via nuclear fusion reactions in their cores. In fact, in the formation of neutron stars via supernova explosions, a significant percentage of the mass of the star is lost to the system, perhaps as much as 95%. This means, of course, that the initial star had to have a mass considerably greater than what we assumed in part (b). As a result of the mass loss and the concomitant loss of the angular momentum associated with the expelled material, neutron stars do not spin as rapidly as the calculation in part (b) would indicate. Nonetheless, neutron stars (which always have masses between about 1.5 and 3.5 times the mass of the Sun, and radii of only about 10 km) *do* rotate quite quickly: the neutron star in the middle of the Crab Nebula in the constellation Taurus (the bull) rotates 30 times each second!

•57. A solid sphere and a hollow spherical shell have equal masses and radii. (a) Without consulting a table or making a detailed calculation, which object has the larger moment of inertia? Explain your reasoning. (b) How could you distinguish one from the other by rolling them (without slipping) down an inclined plane? Justify your answer with an appropriate technical argument.

•58. A 20 kg pointlike mass and a 30 kg pointlike mass are separated by 0.80 m, connected by a string under tension, and spinning clockwise at 8.0 rad/s on a frictionless frozen pond. (a) Determine the location of the center of mass of the system. (b) What is the total angular momentum of the system about the center of mass point? (c) What is the total kinetic energy of rotation about the center of mass point? (d) If the connecting string slips and becomes 1.6 m long, what is the new angular speed of the particles about the center of mass point? (e) If the string breaks, what happens to the kinetic energy and to the angular momentum?

•59. Model a Ferris wheel by means of point masses $m = 100$ kg connected together by thin rigid rods of negligible total mass as shown in Figure P.59. (a) Determine the moment of inertia of the Ferris wheel about its axis. The Ferris wheel initially is not moving. Another mass $M = 200$ kg now is dropped from a height $h = 50.0$ m above the ground, lands in the indicated bucket, and sticks there. (b) What is the magnitude of the orbital angular momentum of the falling mass M about the axis of the Ferris wheel the instant before M hits the bucket? (c) Determine the angular speed of the system the instant after the mass M hits the bucket.

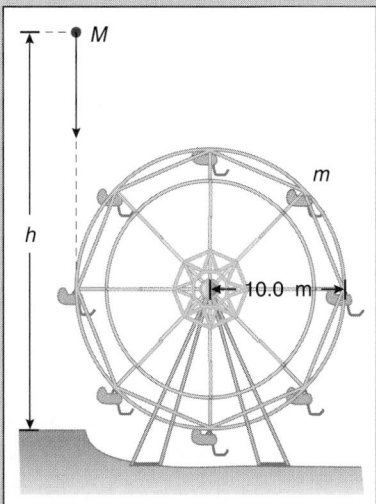

FIGURE P.59

•60. A girl of mass 30.0 kg, initially running tangentially at a speed of 5.00 m/s, jumps onto a disk-shaped playground merry-go-round initially at rest as shown in Figure P.60. The mass of the merry-go-round is 100 kg and its radius is 1.50 m. Find the angular speed of the merry-go-round after the girl jumps on.

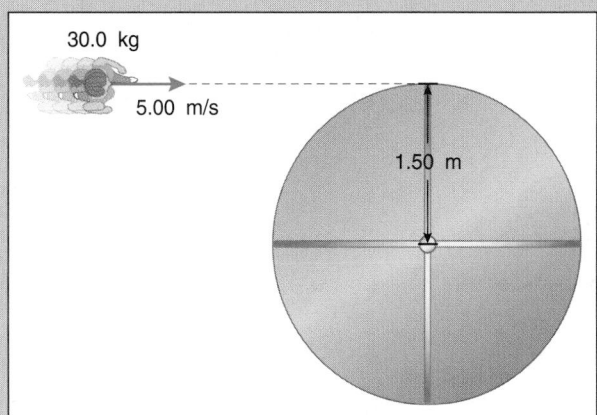

FIGURE P.60

•61. You spy a common house mouse (*Mus musculus*) of mass *m* at rest on the rim of a lazy Susan (also at rest). The mass of the lazy Susan is *M* and its radius is *R*. The fleet-footed rodent is frightened into running clockwise around the rim of the lazy Susan at speed v_0 relative to the surface of the lazy Susan. In what rotational sense does the lazy Susan rotate? What is the angular speed of the lazy Susan as measured by an observant and terrified cook in the kitchen? Assume frictionless bearings.

•62. A varsity-sized tarantula (*Eurypelma californica*) of mass 100 g is hanging above a stereo turntable. The platter is of mass 2.00 kg and radius 15.0 cm and is spinning with no friction at an antiquated 33.0 rev/min with the motor off. The spider decides to make the turntable into a merry-go-round, drops from a height of 2.00 mm, and sticks to the rim of the turntable. (a) Calculate the magnitude of the initial spin angular momentum of the platter of the turntable. (b) What is the total moment of inertia of the platter and spider system when the spider is riding the merry-go-round? (c) Determine the new angular speed of the platter in rev/min.

•63. A neutron star with a mass of twice that of the Sun and a radius of 10 km rotates 30 times each second. Neutron stars realistically can be modeled as uniform, solid stars. (a) Calculate the rotational kinetic energy of the star. The result you obtain is greater than the entire energy output of the Sun during its 10 billion year lifetime! (b) Calculate the magnitude of the spin angular momentum of the neutron star. (c) Suppose the star shrinks 1.0 m in radius. What is the *ratio* of the new period of rotation to the initial period of rotation? Radio astronomers can detect changes in the rotational period of such stars corresponding to a mere 1 mm change in radius. Such changes occur occasionally in the rotational periods of neutron stars and are thought to be caused by "starquakes" that slightly change the radius of the star.

•64. The planet Venus has the following characteristics:

Rotational (spin) period	243 d
Radius	6051 km
Mass	4.87×10^{24} kg
Orbital period	225 d
Orbital radius	1.08×10^8 km

The spin is in the sense opposite to the orbital motion, as shown in Figure P.64. Consider the planet to be spherical in shape and, for simplicity, of uniform composition. The orbit of Venus is essentially a circle (the eccentricity of the orbit is only 0.007). The axis of rotation is essentially perpendicular to the plane of its orbit (the inclination of the axis to a perpendicular to its orbital plane is only 3.39°; we neglect this inclination in the present problem). (a) Find the magnitude of the spin angular momentum of the planet. (b) Find the magnitude of the orbital angular momentum of the planet. (c) Find the magnitude of the total angular momentum of the planet. Is most of the angular momentum orbital or spin angular momentum? (d) Find the total kinetic energy of the planet. Is most of the kinetic energy associated with the spin or with the orbital motion?

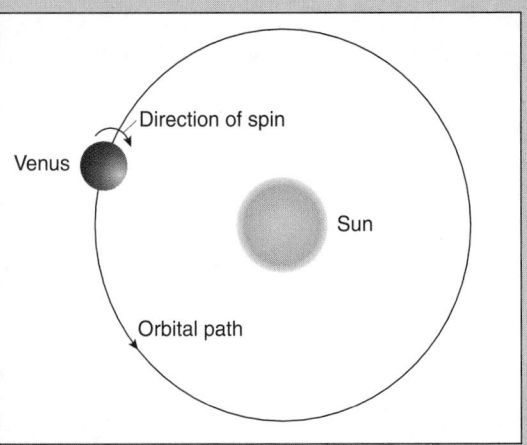

FIGURE P.64

•65. The Sun rotates on its axis about once every 27 days. (a) Calculate the magnitude of the spin angular momentum of the Sun. (b) The spin angular momentum of the Sun represents only about 2% of the *total* angular momentum of the solar system. The planets have 98% of the total angular momentum of the solar system (in both orbital angular momentum associated with their orbital motion and the orbital motion of their moons and spin angular momentum associated with the spin of the various planets and moons). Using the result of part (a), what is the magnitude of the total angular momentum of the solar system? (c) Assume that we could somehow transfer the angular momentum of the planets and their moons to the Sun. If the Sun contained 100% of the total angular momentum of the solar system, in other words, if we increased the magnitude of the spin angular momentum of the Sun by a factor of 50, what would be the rotational period of the Sun? When the solar system was formed about five billion years ago from the gravitational collapse of a giant nebula, conservation of angular momentum indicates that the Sun should have been spinning much faster than the result you obtained in part (c). This means that angular momentum was lost in the collapse of the primitive solar nebula. The missing angular momentum was carried off by mass that was lost to the system during the early epochs of the formation of the solar system.

•66. A disk with moment of inertia I_1 is rotating with initial angular speed ω_0; a second disk with moment of inertia I_2 initially is not rotating (see Figure P.66). The arrangement is much like a LP record ready to drop onto an unpowered, freely spinning turntable. The second disk drops onto the first and friction between them brings them to a common angular speed ω. Show that

$$\omega = \frac{I_1}{I_1 + I_2} \omega_0$$

‡67. A hula hoop of mass *m* and radius *R* is hung over a peg as indicated in Figure P.67. (a) What is the moment of inertia of the

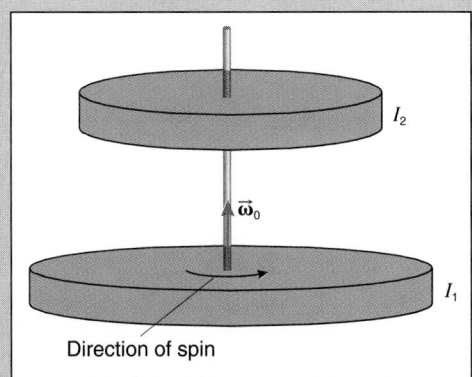

FIGURE P.66

hula hoop about the peg? (b) If the hoop is moved so that a line from the peg to the center of the hoop makes an angle θ with the vertical direction, what is the torque on the hoop about the peg? (c) For small angles θ, will the system execute simple harmonic oscillation? If so, what is the period of the oscillation?

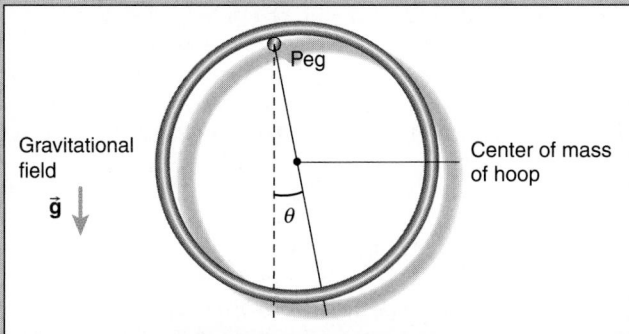

FIGURE P.67

68. A wheel of radius R is rolling at constant speed v_0 along a level road. (a) Find an expression for the speed v of a particle on the rim of the wheel as a function of the angle θ in Figure P.68. Test the expression to show that it gives the appropriate values of v when $\theta = 0°$ and when $\theta = 180°$. (b) Make a graph of v versus θ.

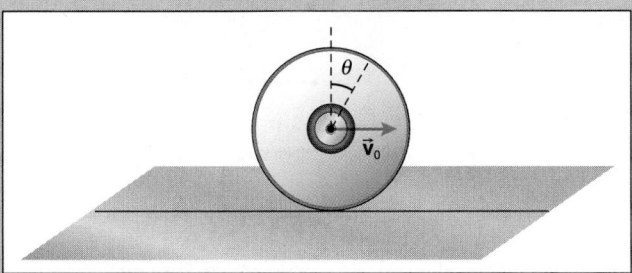

FIGURE P.68

69. One way to determine the moment of inertia of a disk of radius R and mass M in the laboratory is to suspend a mass m from it to give the disk an angular acceleration. The mass m is attached to a string that is wound around a small hub of radius r, as shown in Figure P.69. The mass and moment of inertia of the hub can be neglected. The mass m is released and takes a time t to fall a distance h to the floor. The total torque on the disk is due to the torque of the tension in the string and the small but unknown frictional torque $\vec{\tau}_{frictn}$ of the axle on the disk. Once the mass m hits the floor, the disk slowly stops spinning during an additional time t' because of the sole influence of the small frictional torque on the disk. Show that the moment of inertia of the disk is given by

$$I = \frac{mr^2 \left(\dfrac{gt^2}{2h} - 1 \right)}{1 + \dfrac{t}{t'}}$$

where m is the falling mass and r is the radius of the hub. In an experiment the value of I measured can be compared with the theoretical value of the moment of inertia of a disk of mass M and radius R:

$$\frac{1}{2} MR^2$$

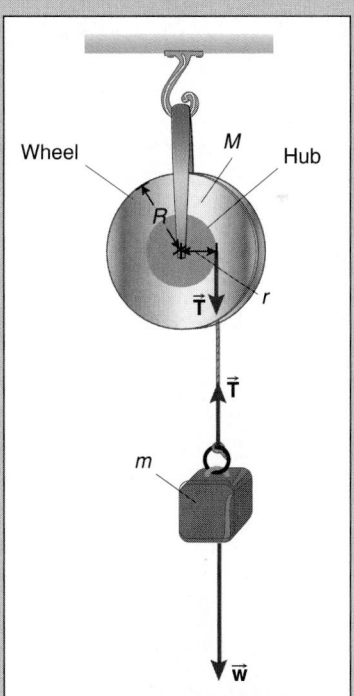

FIGURE P.69

70. A marble of radius r rolls without slipping off a fixed globe of radius R, beginning from rest at the top of the globe (see

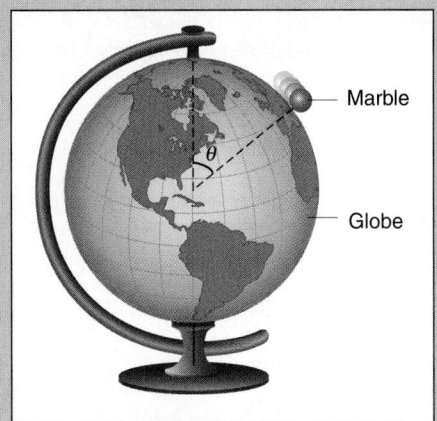

FIGURE P.70

Figure P.70). Show that the angle θ at which the marble leaves the globe is given by

$$\theta = \cos^{-1}\left(\frac{10}{17}\right)$$

:71. The mass of the polar ice caps on the Earth is on the order of $m = 6 \times 10^{18}$ kg. The Antarctic ice cap has a mass of $0.9m$; the Arctic ice cap has a mass of $0.1m$. Model the polar ice caps as two point masses coincident with the rotational axis of the Earth as in Figure P.71a. Should sufficient global warming occur, the ice caps will melt and the resulting liquid water will spread over the ocean areas of the Earth, raising sea levels around the globe. Assume the Earth is a uniform sphere. (a) Make the naive assumption that the water from the ice caps is spread *uniformly* over the surface of the Earth (i.e., ignore continental land forms), thus increasing its effective radius (see Figure P.71b). If the ice caps mass were spread uniformly over the surface of the Earth, by how much would the present radius of the Earth (6.37×10^3 km) increase? Express your result in meters. What does this bode for owners of ocean waterfront property? (b) The redistributed mass changes the moment of inertia of the Earth. If all the polar ice melts according to the assumptions in part (a), what is the increase in the rotational period of the Earth? Express your result in seconds.

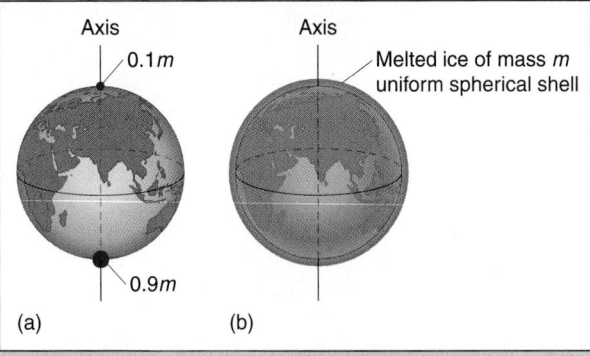

FIGURE P.71

Section 10.19 Conditions for Static Equilibrium

72. A uniform beam of mass m and length ℓ is hung from two cables, one at the end of the beam and the other $5\ell/8$ of the way to the other end as shown in Figure P.72. Determine the magnitudes of the forces the cables exert on the beam.

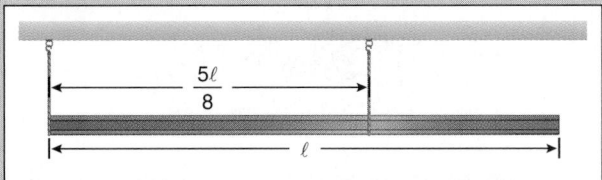

FIGURE P.72

•73. A uniform ladder of mass 20 kg and length 5.0 m is leaning against a smooth (frictionless) wall. The coefficient of static friction for the ladder on the ground is $\mu_s = 0.30$. What is the minimum angle θ that the ladder can make with the ground without slipping?

•74. A massive 250 kg door is hung from two hinges in the portico of a cathedral as indicated in Figure P.74. Each hinge exerts both a horizontal and a vertical force on the door to keep it in equilibrium. The door is of uniform composition. (a) Sketch a second law force diagram indicating the forces acting on the door. (b) Find the horizontal and vertical components of the forces that each hinge exerts on the door.

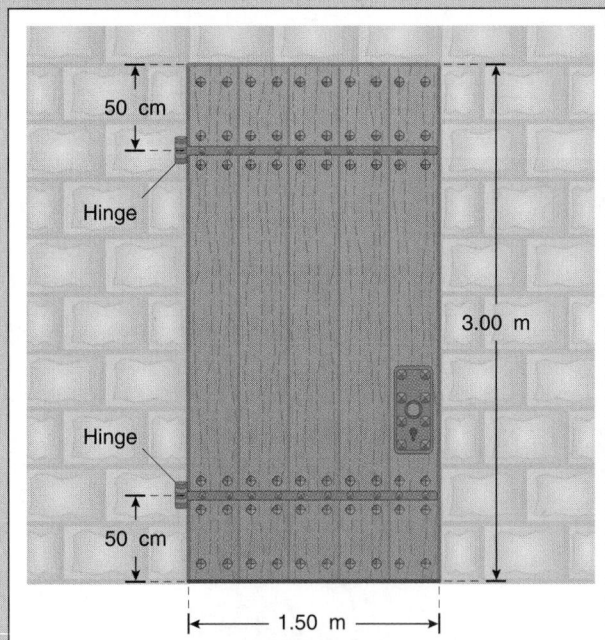

FIGURE P.74

•75. An unsharpened pencil of length ℓ and mass m is leaning against a smooth (i.e., frictionless) vertical surface. The pencil is approximately uniform in composition. The static coefficient of friction for the pencil on the horizontal surface is μ_s. (a) Draw a second law force diagram indicating all the forces acting on the pencil. (b) What is the smallest angle θ that the pencil can make with the horizontal direction without slipping? (c) You might try this situation experimentally on your desk using a protractor to measure the minimum angle θ as a way to find the coefficient of static friction. To satisfy the conditions of a frictionless vertical surface, is it better to have the eraser end of the pencil on the horizontal or on the vertical surface?

•**76.** An extended object of mass 5.00 kg is subjected *only* to the forces shown in Figure P.76. (a) What is the total force acting on the mass? Will the object accelerate? (b) Calculate the torque of each force about the point P in Figure P.76. (c) What is the total torque acting on the object? (d) Is the object in equilibrium?

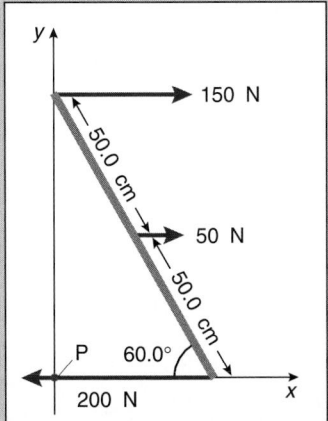

FIGURE P.76

•77. Uh oh. You (mass 70.0 kg) decide to tempt fate. A massive uniform plank of length 4.00 m and mass 100 kg lies flat but extends 1.50 m beyond the edge of the physics building as shown in Figure P.77. What is the maximum distance from the edge of the building that you can walk along the plank without disaster?

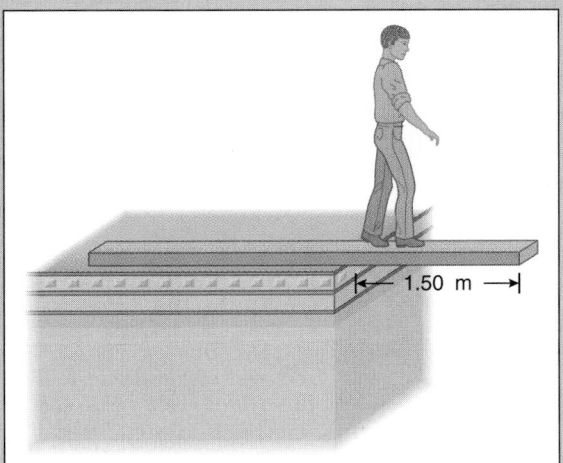

FIGURE P.77

•**78.** While hiking, you notice a dead tree trunk 4.00 m long resting on level ground. Always prepared to do some interesting physics, you discover experimentally that a force of 900 N is barely sufficient to lift one end while a force of 450 N barely is sufficient to lift the other end. What is the mass of the tree trunk?

•**79.** A crude model of an arm and its biceps muscle is shown in Figure P.79. Let the mass of the arm and biceps muscle be negligible (the classic 40.9 kg weakling). (a) What magnitude of force must the biceps muscle provide for the arm to hold a 10 kg mass in the orientation shown? (b) The mechanical advantage of a lever such as this is defined as the ratio between the magnitude of the force of the load (the weight of the 10 kg mass) and the magnitude of the force exerted by the muscle. Determine the mechanical advantage of this lever system. The arm is a very inefficient lever, since its mechanical advantage is so small. The great advantage of the system arises in the ability of the arm to move freely and throw objects at high speed. If the biceps were attached to the lower arm close to the hand, the mechanical advantage of the muscle would increase (we could hold greater masses in the given orientation) but our throwing abilities (controlled by the triceps on the back of the upper arm) would be severely compromised; no fastballs!

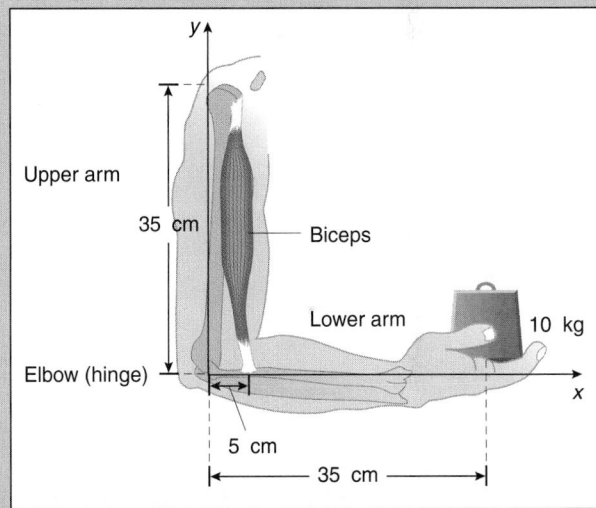

FIGURE P.79

INVESTIGATIVE PROJECTS

A. Expanded Horizons

1. Investigate the dynamics associated with throwing a bola, a traditional South American hunting weapon.
   D. L. Mathieson, "Wrap up rotational motion by throwing a bola," *The Physics Teacher*, 30, #3, pages 180–181 (March 1992).

2. Yo-yos are fascinating examples of spin and orbital motion. Investigate the physics of the yo-yo.
   William Boudreau, "Cheap and simple yo-yos," *The Physics Teacher*, 28, #2, page 92 (February 1990).
   Wolfgang Bürger, "The yo-yo: a toy flywheel," *American Scientist*, 72, #2, pages 137–142 (March–April 1984).
   Edward Zuckerman, "Quest for the perfect yo-yo," *Science Digest*, 93, #7, pages 54–55, 60 (July 1985).

3. Investigate the use of an ultracentrifuge to determine molecular masses of complex molecules.
   I. W. Richardson and Ejler B. Neergaard, *Physics for Biology and Medicine* (Wiley-Interscience, New York, 1972), pages 158–160.

4. Investigate the dynamics associated with boomerangs.
   Vernon Barger and Martin Olsson, *Classical Mechanics: A Modern Perspective* (2nd edition, McGraw-Hill, New York, 1995), pages 195–202.
   Allen L. King, "Project boomerang," *American Journal of Physics*, 43, #9, pages 770–773 (September 1975).
   Henk Vos, "Straight boomerang of balsa wood and its physics," *American Journal of Physics*, 53, #6, pages 524–527 (June 1985).
   Michael Hanson, "The flight of the boomerang," *The Physics Teacher*, 28, #3, pages 142–147 (March 1990).
   Jacques Thomas, "Why boomerangs boomerang," *New Scientist*, 99, #1376, pages 838–843 (22 September 1983).

5. Investigate the physics associated with riding a bicycle.
   J. Lowell and H. D. McKell, "The stability of bicycles," *American Journal of Physics*, 50, #12, pages 1106–1112 (December 1982).
   David E. H. Jones, "The stability of the bicycle," *Physics Today*, 23, #4, pages 34–40 (April 1970).
   Thomas B. Greenslade Jr., "More bicycle physics," *The Physics Teacher*, 21, #6, pages 360–363 (September 1983).
   Daniel Kirshner, "Some nonexplanations of bicycle stability," *American Journal of Physics*, 48, #1, pages 36–38 (January 1980).

6. Falling chimneys tend to break at characteristic locations along their length before they hit the ground. Investigate the causes of this phenomenon.
   Ernest L. Madsen, "Theory of the chimney breaking while falling," *American Journal of Physics*, 45, #2, pages 182–184 (February 1977).
   Arthur Taber Jones, "The falling chimney," *American Journal of Physics*, 14, #4, page 275 (July–August 1946).
   Albert A. Bartlett, "More on the falling chimney," *The Physics Teacher*, 14, #6, pages 351–353 (September 1976); this reference has a bibliography on the subject.

7. Investigate the static equilibrium and dynamics of tightrope walking.
   Robert B. Prigo, "Classroom tightrope walking," *American Journal of Physics*, 50, #5, pages 471–473 (May 1982).

8. Investigate the physics associated with "pumping" a swing.
   William B. Case and Mark A. Swanson, "The pumping of a swing from the seated position," *American Journal of Physics*, 58, #5, pages 463–467 (May 1990).
   Stephen M. Curry, "How children swing," *American Journal of Physics*, 44, #10, pages 924–926 (October 1976).

9. Measurements of the moment of inertia are common in physics laboratories. On the other hand, it is quite another story to *measure* the moment of inertia of the Moon or a planet such as Venus or Mars; they are not easily tinkered with in the lab! Measurements of such moments of inertia enable astronomers and geologists to model the interiors of the moons and planets. How might the moments of inertia of moons and planets be measured?
   William B. Hubbard, *Planetary Interiors* (Van Nostrand Reinhold, New York, 1984).
   Ralph Snyder, "Two-density model of the Earth," *American Journal of Physics*, 54, #6, pages 511–513 (June 1986).

10. Investigate the physics of the rotating, orbiting space colony stations proposed by Gerald K. O'Neill.
    Gerald K. O'Neill, "The colonization of space," *Physics Today*, 27, #9, pages 32–40 (September 1974).

11. High angular speed flywheels are used as energy storage devices in hybrid race cars. Investigate the state of this technology.
    William B. Scott, "Satellite control concepts bolster civil, defense systems," *Aviation Week and Space Technology*, 142, pages 43, 46 (6 March 1995).

B. Lab and Field Work

12. Many science museums have a dramatic pendulum in their entrance foyers called a Foucault pendulum. The slow rotation of the plane of oscillation of the pendulum is experimental proof that the Earth rotates. Investigate the dynamics of a Foucault pendulum to explain why this is the case. Imagine such a pendulum at the north pole, for simplicity. Construct a Foucault pendulum in a large open stairwell and measure the period associated with the apparent rotation of its plane of oscillation. See also Project 5 in Chapter 7.
    Charles Kittel, Walter D. Knight, and Malvin S. Ruderman, *Mechanics*, Berkeley Physics Course, volume 1 (McGraw-Hill, New York, 1965), pages 77–79.
    H. Richard Crane, "The Foucault pendulum as a murder weapon and a physicist's delight," *The Physics Teacher*, 28, #5, pages 264–269 (May 1990).
    Anthony P. French, "The Foucault pendulum," *The Physics Teacher*, 16, #1, pages 61–62 (January 1978).
    H. Richard Crane, "Foucault pendulum wall clock," *American Journal of Physics*, 63, #1, pages 33–39 (January 1995).

13. Several basic aspects of rotational motion can be demonstrated easily with common materials. Create and explain several of the demonstrations mentioned in the article listed here.
    George Lehmberg, "Investigating rotational inertia using coffee cans," *Science Teacher*, 60, #3, pages 68–69 (March 1993).

14. Design and perform an experiment to measure the moment of inertia of an automobile tire when mounted on its hub. Problem 69 may be of use in this regard. Specify and discuss the assumptions you make about the tire.
    Richard C. Smith, "General physics and the automobile tire," *American Journal of Physics*, 46, #8, pages 858–859 (August 1978).

15. Plan and make suitable astronomical observations to measure the rotational period of the Earth with respect to the stars and with respect to the Sun. Explain the difference in the values you obtain for these two time periods.

C. Communicating Physics

16. Take two quarters (or any two identical coins). Orient the two coins so they are both heads up with both profiles of George Washington upright as indicated in Figure I.16a. Roll one of the coins, say the one on the left, around *half* the circumference of the other coin as indicated. It is amazing to discover that the profile of the rolling coin is *not* upside down but right

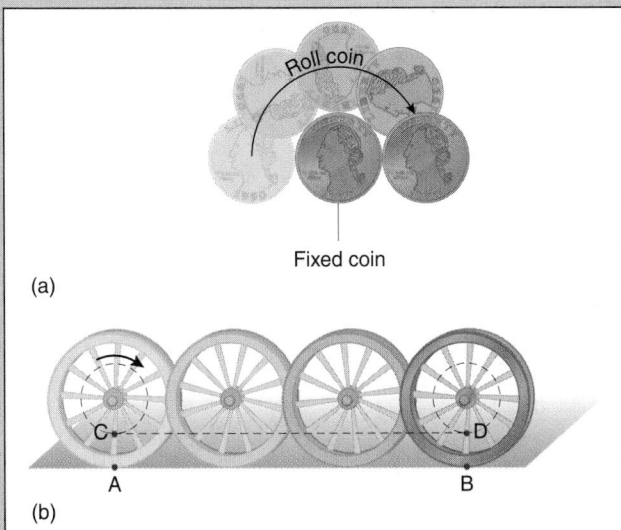

(a)

(b)

FIGURE I.16

side up after this maneuver. Explain why. This phenomenon is related to another curious aspect of rolling motion. Suppose a wheel (see Figure I.16b) rolls through one complete turn, carrying it from point A to point B. The distance between A and B is, therefore, equal to the circumference of the wheel. Notice, however, that point C, along the radius from the center of the wheel *also* executes one revolution as the wheel moves from A to B. Since the distance AB is equal to the distance CD, "we are confronted with the evident absurdity that the circumference of the small circle is equal to the circumference of the large circle." Resolve this paradox.

See James R. Newman, *The World of Mathematics* (Simon & Schuster, New York, 1956), volume 3, pages 1937–1939; the quotation is from this work, page 1938.

17. Call or visit a tire retailer and determine the manufacturer of a dynamic tire balancing machine. Communicate with engineers at the manufacturer to learn the physics associated with dynamic balancing of tires. Write a summary of your findings appropriate for an audience of your peers.

Richard C. Smith, "Static vs spin balancing of automobile wheels," *American Journal of Physics*, 40, #1, pages 199–201 (January 1972).

18. The tippe top is a toy that has fascinated even great physicists like Niels Bohr and Wolfgang Pauli, perhaps even you. Demonstrate and explain its peculiar dynamics.

Richard J. Cohen, "The tippe top revisited," *American Journal of Physics*, 45, #1, pages 12–17 (January 1977).

Ivars Peterson, "Topsy-turvy top," *Science News*, 146, page 108 (13 August 1994).

CHAPTER 11

SOLIDS AND FLUIDS

Archimedes' lost principle: When a body is immersed in water, the phone will ring.

Don Simanek*

Who has not marveled at the wonders of the sea, the dynamics of hurricanes and tornados, the thrill of a whitewater river adventure, or our ability to build structures of great size and beauty such as the Golden Gate Bridge, the grand arch in St. Louis, the World Trade Center, a giant jumbo jet, or the space shuttle?

Solids and fluids (liquids and gases) are the tangible part of the natural world in which we live. The study of the mechanical and dynamical properties of materials is essential for the advancement of our technological prowess as well as for an appreciation of the raw elements of nature and our environment. In this chapter we investigate some of the basic classical mechanical properties of solids and fluids.

11.1 STATES OF MATTER

We typically classify macroscopic matter (large collections of atoms or molecules) into one of four **states of matter**: solids, liquids, gases, and plasmas. These states are difficult to define precisely, since some materials, even common ones, defy such elementary characterizations and exhibit characteristics of two or more of them. Physicists and engineers have developed branches of their disciplines devoted to studying the static and dynamical behavior of each of the various states of matter. Thus you hear of disciplines such as fluid mechanics, solid mechanics, condensed matter physics, plasma physics, and materials science. The essential physics underlying them all is the same: forces, the response of the system to those forces, and energy considerations.

A **solid** usually is thought of as a semirigid collection of atoms or molecules that maintains a definite shape and volume. Bricks, baseball bats, and boulders are examples of typical solids.

---

*(Chapter Opener) *The Physics Teacher*, 27, #7, page 536 (October 1989).

A rock is a solid. Duh! But when heated to a high temperature, rocks can exhibit the flowlike properties of a liquid or even vaporize.

A **liquid** typically is imagined as a collection of atoms or molecules that has a density comparable to that of solids, maintains a definite volume, but also flows and assumes the shape of any container, though the liquid may not necessarily fill it. The milk in a partially filled container illustrates the properties of a typical liquid.

A **gas** typically is imagined as a loose collection of atoms or molecules that also exhibits flow characteristics but completely fills a container of any shape or volume in which it is placed. Gases are more easily compressed or expanded than solids and liquids.

Another unusual state of matter is a **plasma**. Plasmas exist typically at temperatures high enough for some electrons to be liberated from their parent atoms. A plasma is a gas- or liquid-like collection of relatively independent electrons and positive ions that is, overall, electrically neutral. Matter is found in the plasma

The World Trade Center in New York City with a spectacular 19th-century structural gift from the French in the foreground.

Milk is a liquid, but on another level, it is a colloidal suspension of small, hydrophobic particles.

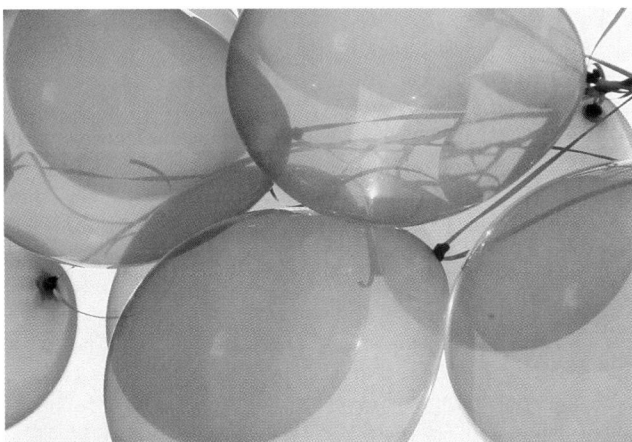

These balloons are filled with helium gas. The gas liquefies at the chilly temperature of a few degrees above the absolute zero of temperature.

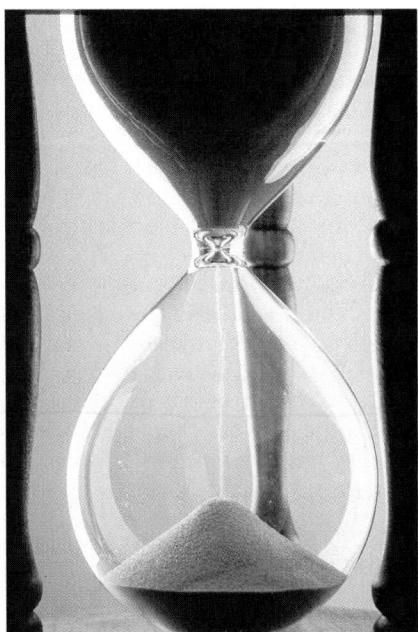

Particulate materials like sand and grains exhibit properties of both solids and liquids.

Plasmas exist inside neon lights.

Even solids like this glacier exhibit properties of liquids, such as flow.

state inside the Sun and other stars (but not all stars). Plasmas also exist inside neon lights and laser tubes that employ gases such as neon, carbon dioxide, xenon, argon, or gas mixtures such as helium with neon.

Such simple characterizations of materials are not always valid. Water, for example, can exist as a solid, liquid, gas, or combinations of them depending on the temperature and pressure; it is the same with many other materials. Materials such as talcum powder, dry sand, cement, or harvested grains are composed of small macroscopic solid particles that are able to flow and can take the shape of their container, similar to liquids (think of an hourglass). Likewise ice, typically thought of as a solid, also can flow, albeit slowly, as it does in large glacial formations.

Asphalt, typically thought of as a solid, also can flow slowly, particularly on hot days, creating annoying ripples in the pavement. Materials that flow (typically gases and liquids) usually are called **fluids**. So most matter is not as easily categorized as you might have thought.

In all these states of matter, *internal* forces exist among the particles of the material. The magnitude or strength of these internal forces of interaction among the particles is greatest in solids, less in liquids, and smallest in gases. To a large extent,

the response of the system to applied *external* forces determines whether a substance is properly treated as a solid, liquid, or gas.

11.2 STRESS, STRAIN, AND YOUNG'S MODULUS FOR SOLIDS

Stretch a cord too tightly, and it will be likely to break.
Publius Syrus (c. 1st century B.C.)*

The stress and strain of the academic year likely are by now taking their toll on you. In this section we see yet another example of how physics takes common words from our everyday discourse and uses them in a different, technical context.

The Moral Sayings of Publius Syrus, a Roman Slave, translated by Darius Lyman (Andrew J. Graham, New York 1862), page 52 (saying #570).

Perhaps even as a child you noticed that when you sit on a rope swing, the ropes stretch slightly. Likewise, if a bulldozer pulls on a wire rope attached to a massive tree stump, the wire rope stretches. What factors influence the amount of the stretch?

Rigid bodies in fact deform when subjected to forces. The rigid bodies that we presumed in Chapter 10 when studying spin and orbital motion are idealizations that, strictly speaking, actually do not exist. We now seek to characterize the deformations of materials under forces.

Consider a long, thin, essentially one-dimensional solid such as a rope or wire, as in Figure 11.1. The rope is hung vertically from a fixed support. We neglect the small weight of the rope. A force of magnitude F is applied to stretch the rope along its length. An applied force that attempts to stretch the rope is called a **tensile force**.* Experimental observations indicate that a tensile force of magnitude F stretches the rope by an amount $\Delta \ell$. The original experiments investigating such stretching, made by Robert Hooke during the 17th century, discovered that the elongation $\Delta \ell$ of a wire or rope is proportional to the magnitude of the applied force $\vec{F}$. That is,

$$F = k \, \Delta \ell \qquad (11.1)$$

where k is a proportionality constant that depends on the specific material composition of the rope and on geometric factors such as its length and cross-sectional area. The constant k has the SI units N/m and is effectively the spring constant of the material, because of the similarity of Equation 11.1 to the force we exert to stretch a Hooke's force law spring (Equation 7.1). So perhaps it is no surprise that a solid is a large collection of atoms or molecules whose interactions are modeled (as a first approximation) by imaginary Hooke's law springs.

As in most situations, there are limits. If the applied tensile force becomes too large, the rope or wire is permanently deformed (stretched) and will not return to its original length when the force is removed, much like a spring under similar circumstances. The rope or wire then is said to be inelastic. A schematic graph of the magnitude of the applied tensile force versus the elongation $\Delta \ell$ is shown in Figure 11.2.

The linear region of the graph is called the **elastic region** and it is in this region that Equation 11.1 is valid. The constant k

---

*A force in the opposite direction is called a compressional force; but it is hardly possible to apply such a force along a long thin rope.

in Equation 11.1 is the slope of the linear region. The curved region is the **inelastic region**. Eventually, if the applied tensile force becomes too great, the rope breaks; this is very similar to the behavior of springs we investigated in Chapter 7. Just as with springs, a large value of k (or spring constant) means a very stiff rope or wire, one that is difficult to stretch. Steel wires can be subjected to large forces before reaching the elastic limit; fishing line reaches its elastic limit with the application of relatively small forces.

The elongation depends not only on the magnitude of the force applied, but also on the composition of the rope or wire and its geometric dimensions. Wires of the same material but of different lengths ℓ and different cross-sectional areas A have different elongations when subjected to the same tensile force $\vec{F}$. Experiments indicate that for a tensile force of a given magnitude, the longer the rope (of fixed cross section), the greater the elongation $\Delta \ell$; in contrast, for a given length ℓ, the greater the cross-sectional area, the less the elongation $\Delta \ell$. These experimental observations are summarized by saying that $\Delta \ell$ is proportional to F and ℓ, but inversely proportional to the cross-sectional area A. That is,

$$\Delta \ell = \frac{1}{E} \frac{F \ell}{A} \qquad (11.2)$$

This is the same as Equation 11.1 with

$$k = \frac{EA}{\ell}$$

For reasons of convenience, the constant of proportionality in Equation 11.2 is written as $1/E$. The larger the value of E, the stiffer the wire or rope, meaning the smaller the deformation $\Delta \ell$ for a given force of magnitude F. By incorporating these geometric considerations, the constant E is left dependent *only* on the material from which the wire is formed. The constant E is called **Young's modulus** of the material.

The magnitude of the applied tensile force $\vec{F}$ divided by the cross-sectional area A of the rope or wire is called the **tensile stress** on it:

$$\text{tensile stress} \equiv \frac{F}{A}$$

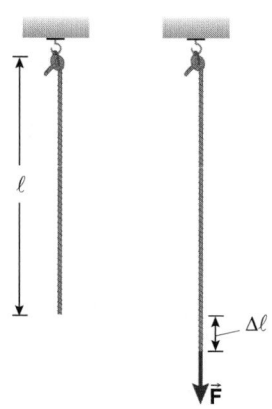

FIGURE 11.1 A force of magnitude F stretches a rope or wire.

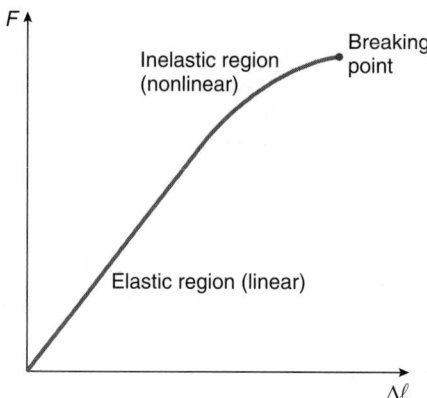

FIGURE 11.2 A schematic graph of the magnitude of the tensile force versus the elongation it produces.

The change in length of the rope $\Delta\ell$ divided by the length of the rope ℓ is called the **tensile strain**:

$$\text{tensile strain} \equiv \frac{\Delta\ell}{\ell}$$

Take the ratio of the tensile stress to the strain:

$$\frac{\text{tensile stress}}{\text{tensile strain}} = \frac{F/A}{\Delta\ell/\ell}$$

Using Equation 11.2, we find that the ratio of the tensile stress to tensile strain is Young's modulus E:

$$E \equiv \frac{\text{tensile stress}}{\text{tensile strain}}$$

$$= \frac{F/A}{\Delta\ell/\ell} \qquad (11.3)$$

Values of Young's modulus for various materials are found in Table 11.1. Note that the SI units for Young's modulus are the same as for tensile stress, N/m^2, since the strain is dimensionless.

In three dimensions, there are other ways in which forces can be applied to materials to cause deformation. For example, consider a solid rectangular parallelopiped, as in Figure 11.3. If a pair of forces, each of magnitude F, is applied to the material over and parallel to opposite surfaces of areas A and separation ℓ, as in Figure 11.4, the material is subjected to what is called a **shear stress.**

The applied forces typically are distributed over the area A rather than at a single point. The shear stress then is F/A. A shear stress deforms a solid initially in the shape of a rectangular parallelopiped into a *non*rectangular parallelopiped. The deformation $\Delta\ell$ divided by the separation ℓ of the two surfaces is known as the **shear strain**.

The ratio of the shear stress to the shear strain is the **shear modulus** G of the material:

$$G \equiv \frac{\text{shear stress}}{\text{shear strain}} = \frac{F/A}{\Delta\ell/\ell} \qquad (11.4)$$

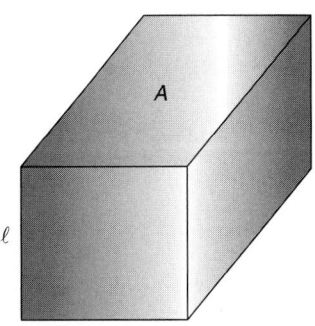

FIGURE 11.3 A solid rectangular parallelopiped.

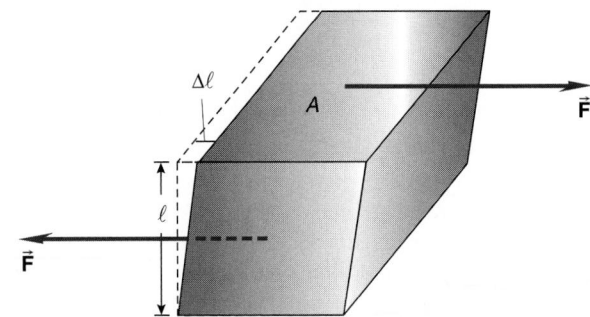

FIGURE 11.4 A shear stress applied to a solid material.

Values of the shear modulus depend on the material and are tabulated in Table 11.1. The SI units for the shear modulus are N/m^2, since the strain is dimensionless.

Although solids and liquids typically are thought of as incompressible, when subjected to compressive forces per unit area, they deform (or are squeezed) into smaller volumes (even if slightly). Let such a material be subjected to uniform squeezing from all directions, as in Figure 11.5. The magnitude of the applied compressive force per unit area is the **pressure** (it could also be considered a volume stress). Let the volume of the material originally be V_i and its change ΔV. The change in the volume is

$$\Delta V = V_f - V_i$$

TABLE 11.1 Moduli of Various Materials

Material	Young's modulus (N/m^2 $\equiv$ Pa)	Shear modulus (N/m^2 $\equiv$ Pa)	Bulk modulus (N/m^2 $\equiv$ Pa)
Aluminum	7.0×10^{10}	2.4×10^{10}	7.0×10^{10}
Boron carbide	35×10^{10}		
Brass	9.0×10^{10}	3.5×10^{10}	6.1×10^{10}
Copper	11×10^{10}	4.2×10^{10}	14×10^{10}
Gold	7.9×10^{10}		
Kevlar	$6.2\text{–}12.4 \times 10^{10}$		
Lead	1.5×10^{10}	0.5×10^{10}	
Mercury	—	—	2.7×10^{10}
Nylon	2×10^{9}		
Platinum	17×10^{10}	6.4×10^{10}	
Silicon carbide	41×10^{10}		
Silver	7.8×10^{10}	2.6×10^{10}	
Steel	20×10^{10}	8.1×10^{10}	16×10^{10}
Tungsten	35×10^{10}	15×10^{10}	20×10^{10}
Water	—	—	0.20×10^{10}

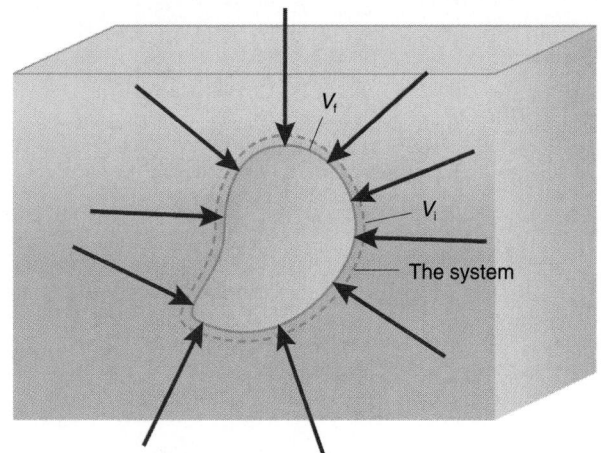

FIGURE 11.5 Squeezing a solid or liquid changes its volume slightly.

and, since the material is compressed into a smaller volume, the change in the volume is *negative*. The ratio of $\Delta V / V_i$ is called the **volume strain**. The ratio of the pressure (volume stress) to the volume strain is the **bulk modulus** B of the material:

$$B \equiv -\frac{\text{pressure}}{\text{volume strain}} \qquad (11.5)$$

The minus sign is introduced so the bulk modulus B is positive. The bulk modulus for various materials also is tabulated in Table 11.1. The SI units for the bulk modulus are the same as those for pressure: N/m^2. The inverse of the bulk modulus is called the **compressibility** β:

$$\beta \equiv \frac{1}{B} \qquad (11.6)$$

Notice that all three moduli are defined as the ratio of an applied stress to the resulting strain in the material, and all three moduli have the SI units of N/m^2.

If the applied stress of this course does not result in much resulting strain on your part, then your "learning modulus" for physics is quite large. Of course, the value of this learning modulus varies with the specific material too (be it you or the topic in physics)!

QUESTION 1

Would you expect Young's modulus for rubber to be greater than or less than that for copper? Make a few crude measurements to determine the order of magnitude of Young's modulus for the elastic material used to make rubber bands.

EXAMPLE 11.1

A wire of diameter 0.617 mm and length 0.984 m is suspended vertically from a rigid support. The wire is subjected to a tensile force by hanging an 8.00 kg mass from its end. Careful measurement* indicates that the wire stretches 1.30 mm. Determine Young's modulus of the material.

---

*Such small distances can be measured precisely using an optical lever. You might ask your laboratory instructor about using such a neat device.

Solution

Young's modulus is the ratio of the tensile stress to the tensile strain. The tensile stress is the magnitude of the force per unit cross-sectional area of the wire:

$$\text{tensile stress} = \frac{F}{A}$$

The magnitude of the applied tensile force is the magnitude of the weight of an 8.00 kg mass:

$$w = mg$$
$$= (8.00 \text{ kg})(9.81 \text{ m/s}^2)$$
$$= 78.5 \text{ N}$$

The cross-sectional area of the wire of diameter d is

$$A = \pi r^2 = \pi \frac{d^2}{4}$$
$$= \pi \frac{(6.17 \times 10^{-4} \text{ m})^2}{4}$$
$$= 2.99 \times 10^{-7} \text{ m}^2$$

Hence the tensile stress is

$$\text{tensile stress} = \frac{F}{A}$$
$$= \frac{78.5 \text{ N}}{2.99 \times 10^{-7} \text{ m}^2}$$
$$= 2.63 \times 10^8 \text{ N/m}^2$$

The tensile strain is

$$\text{tensile strain} = \frac{\Delta \ell}{\ell}$$
$$= \frac{1.30 \times 10^{-3} \text{ m}}{0.984 \text{ m}}$$
$$= 1.32 \times 10^{-3}$$

Therefore Young's modulus is

$$E = \frac{\text{tensile stress}}{\text{tensile strain}}$$
$$= \frac{2.63 \times 10^8 \text{ N/m}^2}{1.32 \times 10^{-3}}$$
$$= 1.99 \times 10^{11} \text{ N/m}^2$$

The wire likely is made of steel (see Table 11.1).

11.3 FLUID PRESSURE

Scuba divers can only go so deep before the effects of increasing pressure become quite apparent. The concept of pressure is central to an understanding of fluids. We introduced the idea in our discussion of the bulk modulus in the previous section.

The effects of fluid pressure are apparent when scuba diving.

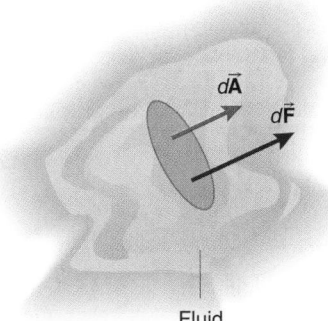

FIGURE 11.6 The fluid on the left of a differential area exerts a differential force on the fluid to the right of the area.

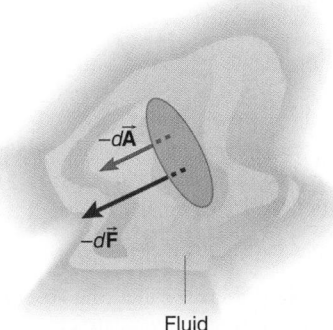

FIGURE 11.7 The fluid on the right of a differential area exerts a differential force on the fluid to the left of the area.

Consider a differential area $d\vec{A}$ at any location and with any orientation in a fluid, as in Figure 11.6. The differential force $d\vec{F}$ exerted by one part of the fluid on another part of the fluid perpendicularly across the area $d\vec{A}$ is proportional to the size of the differential area. We write this as

$$d\vec{F} = P \, d\vec{A}$$

where the quantity P is a *scalar* called the pressure; it has the dimensions of force per unit area. In Figure 11.6 the fluid to the left of the area exerts a force on the fluid to the right of the area. Likewise, in Figure 11.7 the fluid to the right of the area exerts a force on the fluid to the left of the area. These two forces are in opposite directions, are of equal magnitude, and form a Newton's third law force pair. In terms of the magnitudes of the quantities,

$$dF = P \, dA$$

or

$$P = \frac{dF}{dA} \qquad (11.7)$$

Experimental observations indicate that this relationship is true regardless of the orientation of the differential area in the fluid.

> The pressure is a scalar quantity that is the magnitude of the force per unit area acting on *any* differential area, regardless of its orientation within the fluid.

Hence the fluid exerts a force on the walls of its container that is perpendicular to its walls at all points; from Newton's third law, we see that the walls exert a force of equal magnitude in the opposite direction on the fluid.

The SI unit of pressure is N/m^2. Since pressure is such a central concept in studying fluids, the SI unit of pressure is defined to be a **pascal** (Pa):

$$1 \ N/m^2 \equiv 1 \ Pa \qquad (11.8)$$

The unit of pressure is named after the 17th-century French scientist Blaise Pascal (1623–1662) who did important early work investigating fluids.

There are several other common, *convenient* units of pressure. A historically convenient unit for measuring pressure is the *atmosphere* (atm),* but it is not an SI unit. The atmosphere unit of pressure stems from the magnitude of the average weight per horizontal square meter of the atmosphere of the Earth at sea level. An atmosphere of pressure is defined to be

$$1 \ atm \equiv 1.013\,25 \times 10^5 \ Pa \qquad (11.9)$$

The *actual* pressure in our atmosphere at any time may differ from this standard atmosphere because of meteorological factors and elevation. We shall use the pressure unit atm occasionally for convenience, much as we do grams and centimeters.

---

*In banking, an "atm" means automated teller machine!

You may encounter other common pressure units in other technical contexts, but we will not use them in this text (except on rare occasions) because they are not SI units.

1. The *torr** is defined to be the pressure difference between the top and bottom of a column of mercury exactly 1 mm in height when at sea level. Torr, therefore, also are referred to as *millimeters of mercury* (mm Hg)

$$1 \text{ torr} \equiv 1 \text{ mm Hg} \qquad (11.10)$$

Experimentally, a column of mercury 76.0 cm in height corresponds to one atmosphere pressure. Hence the conversion to the SI pressure unit Pa is found from the identity:

$$1.013\,25 \times 10^5 \text{ Pa} \equiv 1 \text{ atm} \equiv 760 \text{ torr} \equiv 760 \text{ mm Hg} \quad (11.11)$$

The pressure unit mm Hg commonly is encountered in meteorological work and is frequently mentioned in weather reports. Your blood pressure also is commonly measured in mm Hg.

2. In many areas of engineering, the **gauge pressure** is used. The gauge pressure is the actual pressure P minus whatever the actual pressure of the atmosphere (≈ 1 atm) is at the location of the gauge, because the other side of the gauge mechanism is exposed to the atmosphere:

$$\text{gauge pressure} \equiv P - 1 \text{ atm} \qquad (11.12)$$

The actual pressure P then is called the **absolute pressure**. Gauge pressure is the pressure indicated by a common tire pressure gauge, which is why it is occasionally a useful unit of convenience in physics and engineering. A totally flat tire has a gauge pressure equal to zero. The absolute pressure P of a thoroughly flat tire is the *actual* value of the atmospheric pressure at the location of the tire, which is approximately 1 atm.

11.4 STATIC FLUIDS

It is the pressure that limits the depth to which submarines can dive. Mountain climbers likewise know that as they ascend high peaks, the pressure of the atmosphere decreases, making it difficult for their bodies to secure adequate oxygen. Commercial airplanes are equipped with systems to increase the cabin pressure above the ambient outside air pressure at high altitudes, so that passengers can breathe normally. Air travelers are reminded before every flight of emergency procedures to follow in the event of a sudden decrease of cabin pressure. Here we discover how pressure varies with altitude in a fluid.

Consider a fluid at rest near the surface of the Earth. We want to figure out how the pressure in the fluid changes either with height or depth from a reference level. We assume that any changes in height or depth are small when compared with the radius of the Earth, so that the magnitude of the acceleration due to gravity g at the surface can be taken as constant.[†] Choose a

---

*The unit named after Evangelista Torricelli (1608–1647), the inventor of the mercury barometer; see Example 11.3.

[†] A change in altitude of approximately 32 km changes the magnitude of the local acceleration due to gravity by only about 1%. You should be able to show this!

High-altitude climbing requires the use of supplemental oxygen because of the low atmospheric pressure.

coordinate system with $\hat{\jmath}$ directed vertically up. Examine the static equilibrium of a differentially thin slice of the fluid of area A and vertical thickness dy; see Figure 11.8. Let $P(y)$ be the pressure at a height y. When y is increased slightly to $y + dy$, let the pressure at the new location be written as $P(y) + dP$.[‡]

There are several forces acting on the thin slice (see Figure 11.9):

1. The weight of the fluid in the thin slice. Let ρ be the density of the fluid at this height. The mass of material in the slice is its density times the volume of the slice: $\rho(A\,dy)$. Thus the weight of the slice is of magnitude

$$(\rho\,A\,dy)g$$

where g is the magnitude of the local acceleration due to gravity (the magnitude of the local gravitational field). With the coordinate system chosen in Figures 11.8 and 11.9, the weight is directed along $-\hat{\jmath}$. Thus the weight is

$$(-\rho\,A\,dy\,g)\hat{\jmath}$$

[‡]The pressure actually decreases as we go up, but this will emerge naturally from our analysis.

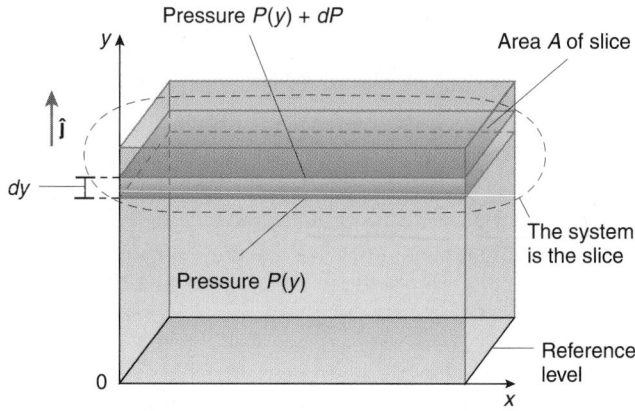

FIGURE 11.8 The pressure on either side of a thin slice of fluid.

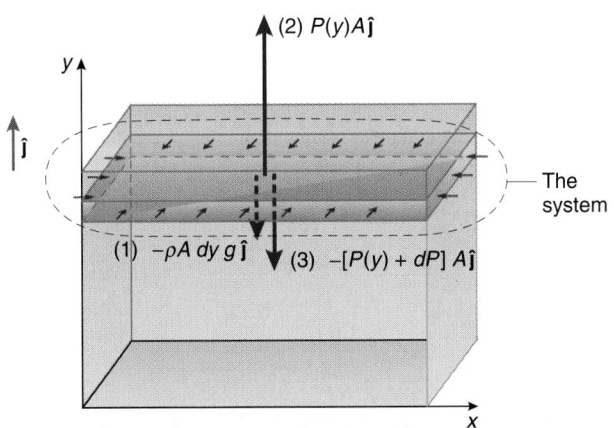

FIGURE 11.9 Forces on a thin slice of fluid.

2. The pressure at the bottom of the slice is $P(y)$. The magnitude of the force exerted by the rest of the fluid on the bottom of the slice is $P(y)A$ and is directed upward ($+\hat{j}$). Thus the force on the bottom of the slice is

$$P(y)A\hat{j}$$

3. The pressure at the top of the slice is $P(y) + dP$, so that the force of the rest of the fluid on the top of the slice is directed downward and is

$$[P(y) + dP]A(-\hat{j})$$

4. Forces also are exerted horizontally by the rest of the fluid on the tiny lateral area of the thin slice, as shown in Figure 11.9, but these horizontal forces pairwise sum to zero and can be omitted from further consideration.

Since the slice is in mechanical equilibrium, the sum of the forces acting in the vertical direction must be zero according to the first condition of equilibrium. Thus

$$F_{y\,total}\hat{j} = 0 \text{ N}$$
$$\{P(y)A - [P(y) + dP]A - \rho A\,dy\,g\}\hat{j} = 0 \text{ N}$$

Rearranging, we get

$$-A\,dP - \rho A\,dy\,g = 0 \text{ N}$$

Thus the pressure changes with height according to the following differential equation:

$$\frac{dP}{dy} = -\rho g \qquad (11.13)$$

The minus sign means that the *pressure decreases with increasing height* in the fluid: a small positive height change dy results in a small negative value for the pressure change dP. For finite changes in height, we have to solve this differential equation to find the pressure.

Two situations arise according to whether the fluid is quite incompressible (most liquids) or quite compressible (gases).

Incompressible Liquids

The bulk modulus of most liquids is quite large (see Table 11.1), and so to a first approximation liquids can be considered to be

incompressible. With this assumption the density of the liquid is constant, regardless of height or depth. Equation 11.13 then can be integrated directly to find the pressure at any elevation y in the liquid:

$$dP = -\rho g\,dy$$

Let the pressure at the origin (i.e., the reference level), $y = 0$ m, be P_0 and the pressure at elevation y be $P(y)$. The definite integrals are

$$\int_{P_0}^{P(y)} dP = -\int_{0\,m}^{y} \rho g\,dy$$

Performing the integrations, we find

$$P(y) - P_0 = -\rho g(y - 0 \text{ m})$$

This can be rearranged into the following form:

$$P(y) = P_0 - \rho gy \qquad (\hat{j} \text{ up}) \qquad (11.14)$$

This equation means that as we increase height in the liquid, the pressure decreases. Or equivalently, pressure increases in direct proportion to depth in an incompressible liquid.

Since Equation 11.14 depends only on the vertical coordinate y, the pressure is the same at all points at the same depth or elevation regardless of their horizontal location, as shown in Figure 11.10.

PROBLEM-SOLVING TACTIC

11.1 You have the freedom to choose where to place the origin of the coordinate system; you simply need to have $\hat{j}$ directed up to use Equation 11.14. The place where you choose the origin is called the *reference level* and the pressure at this level (where $y = 0$ m) is P_0. The pressure at *any* coordinate y within the liquid then is $P(y)$, whether the coordinate y is above the origin (positive values of y) or below the origin (negative values of y).

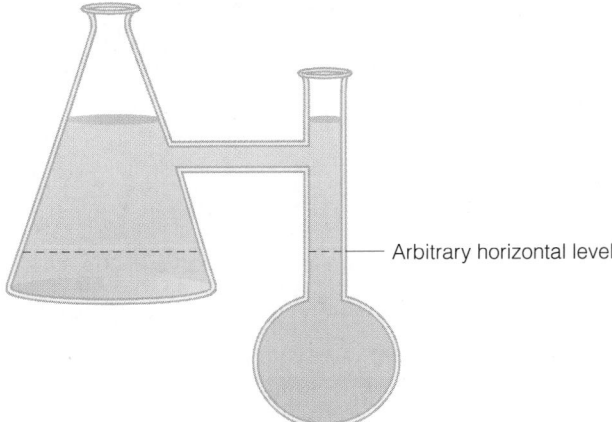

FIGURE 11.10 The pressure is the same at any horizontal level within a connected fluid.

When designing a dam it is *not* the surface area of the lake or its volume of water behind the dam that is the critical design element, but rather the depth of the water at the dam site and the vertical projection of the surface area of the dam face. Explain why.

STRATEGIC EXAMPLE 11.2

Find the fluid pressure in units of atmospheres at the depth of the ill-fated, allegedly unsinkable *Titanic*. The huge ship sank after hitting an iceberg on its maiden voyage from Southampton, England, to New York on 14 April 1912 with great loss of life. The ocean liner is located approximately 3.00×10^3 m below the surface of the northern Atlantic Ocean at latitude 41.8°N and longitude 50.2°W. Sea water has a density of 1.025×10^3 kg/m³.

The *Titanic* embarking on its ill-fated maiden voyage in 1912.

Solution

Use Equation 11.14 for the pressure at any coordinate y in a liquid:

$$P(y) = P_0 - \rho g y \qquad (\hat{\jmath} \text{ up})$$

Method 1

Choose the origin of the coordinate system (the reference level in the fluid) to be at the location of the ship at the bottom of the sea, as shown in Figure 11.11. Then P_0 is the pressure at this level; you are trying to find the value of P_0. Solve Equation 11.14 for P_0:

$$P_0 = P(y) + \rho g y \qquad (1)$$

The pressure at the surface of the sea (at coordinate y) is $P(y)$. The water surface is at coordinate $y = 3.00 \times 10^3$ m and the pressure $P(y)$ at this coordinate location is atmospheric pressure, 1.0 atm. The term $\rho g y$ is

$$\rho g y = (1.025 \times 10^3 \text{ kg/m}^3)(9.81 \text{ m/s}^2)(3.00 \times 10^3 \text{ m})$$
$$= 3.02 \times 10^7 \text{ Pa}$$

This can be converted to atmospheres by using the conversion

$$1 \text{ atm} = 1.013 \times 10^5 \text{ Pa}$$

Hence

$$\rho g y = 298 \text{ atm}$$

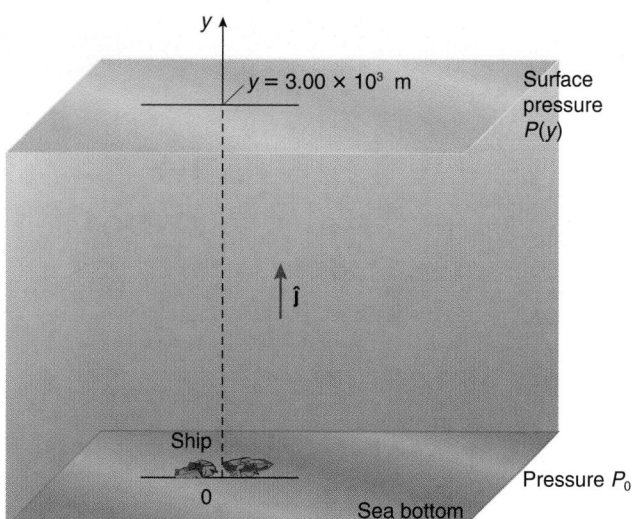

FIGURE 11.11

Since $P(y)$ is 1.0 atm, equation (1) is

$$P_0 = 1.0 \text{ atm} + 298 \text{ atm}$$
$$= 299 \text{ atm}$$

Method 2

Choose the surface of the sea as the origin of the coordinate system, as shown in Figure 11.12. In this case the reference pressure P_0 is 1.0 atm. We want the pressure $P(y)$ at the coordinate y, and so Equation 11.14,

$$P(y) = P_0 - \rho g y$$

is employed with $y = -3.00 \times 10^3$ m, the value of the y-coordinate at the location of the ship.

Thus

$$P = P_0 - (1.025 \times 10^3 \text{ kg/m}^3)(9.81 \text{ m/s}^2)(-3.00 \times 10^3 \text{ m})$$
$$= P_0 + 3.02 \times 10^7 \text{ Pa}$$

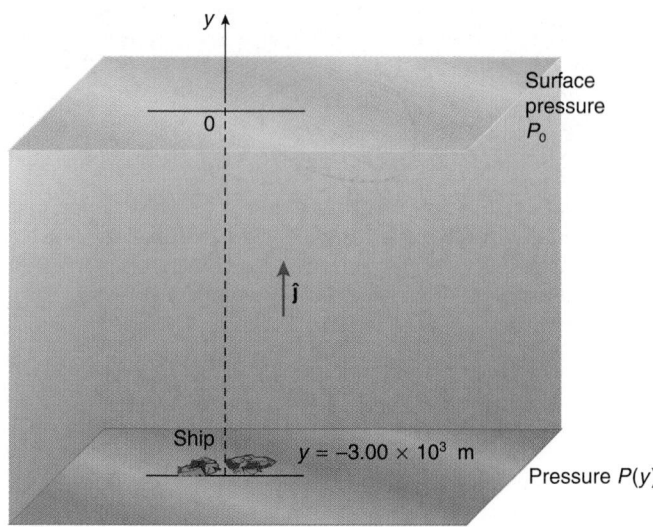

FIGURE 11.12

Converting this pressure to atmospheres, and substituting 1.0 atm for P_0, you get

$$P = 1.0 \text{ atm} + 298 \text{ atm}$$
$$= 299 \text{ atm}$$

EXAMPLE 11.3

A common device for measuring the pressure in the surrounding atmosphere is the mercury barometer. It is constructed by filling a uniform tube and a pan partially with mercury; see Figure 11.13. When the tube is inverted with its open end in the pan, some of the mercury flows out of the tube, as shown in Figure 11.14.

FIGURE 11.13

FIGURE 11.14

   The closed end of the tube essentially is a vacuum.* **The pressure acting on the top surface of the mercury in the inverted column thus is zero to an excellent approximation. The height of the mercury column is measured to be 76.0 cm. Find the atmospheric pressure.**

Solution
Method 1
Using Problem-Solving Tactic 11.1, choose a coordinate system with the origin at the elevation of the pool of mercury, as in Figure 11.15. The origin locates the reference level in the fluid. You must have $\hat{\jmath}$ up, consistent with the choice made in deriving Equation 11.14.

   The pressure at the level of the pool of mercury open to the air is atmospheric pressure. Since this is our reference level, this pressure is P_0 in Equation 11.14:

$$P(y) = P_0 - \rho g y \qquad (\hat{\jmath} \text{ up})$$

The pressure at the top of the column of mercury is 0 Pa since there is essentially a vacuum above the surface at that point. So $P(y) = 0$ Pa. Thus

$$0 \text{ Pa} = P_0 - \rho g y$$

---

*The only pressure in this region is from the small vapor pressure of mercury, which we neglect here.

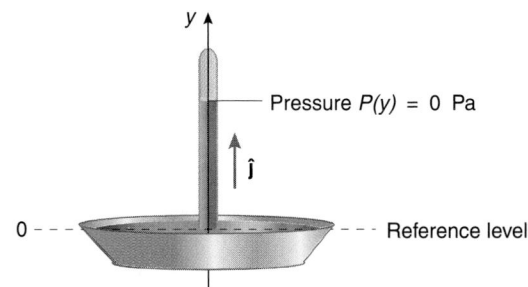

FIGURE 11.15

Hence

$$P_0 = \rho g y$$

You must take care to use consistent SI units. The density of mercury is $\rho = 13.6 \times 10^3$ kg/m³ and of course $g = 9.81$ m/s². The observed height of the mercury column is 76.0 cm or 0.760 m. Substituting these numerical values, you find

$$P_0 = (13.6 \times 10^3 \text{ kg/m}^3)(9.81 \text{ m/s}^2)(0.760 \text{ m})$$
$$= 1.01 \times 10^5 \text{ kg/(m·s}^2)$$

The units kg/(m·s²) are equivalent to Pa since

$$\text{Pa} = \text{N/m}^2 = \frac{\text{kg·m}}{\text{s}^2 \cdot \text{m}^2} = \frac{\text{kg}}{\text{m·s}^2}$$

Thus

$$P_0 = 1.01 \times 10^5 \text{ Pa}$$

Method 2
Choose the origin of the coordinate system to be at the top of the column, as shown in Figure 11.16. The pressure P_0 in Equation 11.14 is the pressure at the level of the origin in the fluid. This is now at the top of the column where the vacuum exists, and so $P_0 = 0$ Pa. Hence Equation 11.14 becomes

$$P(y) = P_0 - \rho g y \qquad (\hat{\jmath} \text{ up})$$
$$P(y) = 0 \text{ Pa} - \rho g y$$
$$= -\rho g y$$

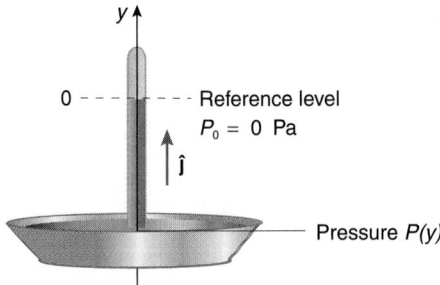

FIGURE 11.16

   The pressure $P(y)$ is the pressure at the coordinate y. But the y-coordinate of the surface of the mercury pool is $y = -0.760$ m. Hence you obtain

$$P = -(13.6 \times 10^3 \text{ kg/m}^3)(9.81 \text{ m/s}^2)(-0.760 \text{ m})$$
$$= 1.01 \times 10^5 \text{ Pa}$$

A Gaseous Atmosphere

Equation 11.13 also may be applied to gases such as the atmosphere of the Earth. Gases are easily compressed. For such situations, the density ρ of the gas is not constant but is a function of height $\rho(y)$, which is why climbers find it more difficult to breathe at high altitudes and may need supplementary oxygen tanks. Make a crude (and unrealistic) assumption that the temperature of the atmosphere is independent of height; the pressure at any height then is found to be proportional to the density. Once again, call the pressure P_0 where $y = 0$ m and the density of the gas there ρ_0. Then

$$P_0 \propto \rho_0$$

Likewise at height y, where the pressure is $P(y)$ and density is $\rho(y)$, we have

$$P(y) \propto \rho(y)$$

Dividing these two proportions, we find

$$\frac{P(y)}{P_0} = \frac{\rho(y)}{\rho_0}$$

or

$$\rho(y) = \rho_0 \frac{P(y)}{P_0}$$

We substitute this expression into Equation 11.13 for ρ and solve for dP; we find

$$dP = -\rho_0 \frac{P(y)}{P_0} g \, dy$$

If we now divide by $P(y)$, the equation can be integrated:

$$\frac{dP}{P} = -\frac{\rho_0}{P_0} g \, dy$$

We integrate the left-hand side from P_0 to $P(y)$ with the corresponding limits for y on the right-hand side:

$$\int_{P_0}^{P(y)} \frac{dP}{P} = -\frac{\rho_0}{P_0} g \int_{0 \text{ m}}^{y} dy$$

$$\ln P(y) - \ln P_0 = -\frac{\rho_0}{P_0} gy$$

$$\ln \frac{P(y)}{P_0} = -\frac{\rho_0}{P_0} gy$$

We put this into exponential form:

$$P(y) = P_0 \exp\left(-\frac{\rho_0}{P_0} gy\right) \qquad (11.15)$$

The pressure, here the magnitude of the weight per unit area of the *remaining* gases overhead, decreases exponentially with height in the atmosphere.

Note that since an exponent must be a pure number with no dimensions, the quantity

$$\frac{P_0}{\rho_0 g}$$

must have the dimensions of a length, or m in SI units. This characteristic length is called the **scale height** in the atmosphere:

$$\text{scale height} \equiv \frac{P_0}{\rho_0 g} \qquad (11.16)$$

When the height y above the origin equals the scale height, the pressure is e^{-1} of the pressure at the position of the origin, where $y = 0$ m. In fact the scale height also depends on the temperature, though this certainly is not apparent since we assumed the temperature of the atmosphere is constant in deriving Equations 11.15 and 11.16.

EXAMPLE 11.4

The surface atmospheric pressure of the Earth is approximately $P_0 = 1.013 \times 10^5$ Pa and the surface air density is about $\rho_0 = 1.29$ kg/m³. Calculate the scale height of the atmosphere measured from the surface of the Earth.

Solution
Use Equation 11.16, choosing the origin to be at the surface of the Earth:

$$\text{scale height} \equiv \frac{P_0}{\rho_0 g}$$

$$= \frac{1.013 \times 10^5 \text{ Pa}}{(1.29 \text{ kg/m}^3)(9.81 \text{ m/s}^2)}$$

$$= 8.00 \text{ km}$$

You might convince yourself that the units indeed check out. This result means that at a height of 8.00 km, the pressure is $e^{-1} = 0.368$ of the surface atmospheric pressure. By way of comparison, the peak of Mt. Everest is approximately 8.85 km above sea level.

EXAMPLE 11.5

The scale height of the atmosphere of the Earth is 8.00 km (see Example 11.4). What is the ratio of the atmospheric pressure in Denver (altitude 1.6 km above sea level) to that in Seattle (at sea level)?

Solution
Choose the origin to be at sea level, so that the pressure at sea level is P_0 in Equation 11.15. Substituting the scale height of the atmosphere, Equation 11.15 becomes

$$P(y) = P_0 \exp\left(\frac{-y}{8.00 \text{ km}}\right)$$

where the altitude y must be in km, and so the exponent is a pure number. Since Denver has an altitude of $y = 1.6$ km above sea level, you get

$$\frac{P(y)}{P_0} = \exp\left(\frac{-1.6 \text{ km}}{8.00 \text{ km}}\right)$$

$$= 0.82$$

The atmospheric pressure in Denver is only about 82% that in Seattle, which is why people (especially visiting athletes) typically have to adjust to the reduced pressure there.

11.5 PASCAL'S PRINCIPLE

Notice in Equation 11.14 that if we increase $P(y)$ by some amount for a liquid *enclosed* in and totally filling a container, say by squeezing on it, then P_0 also must increase by the same amount to maintain the equality; the converse also is true.

> Hence, if we increase the pressure on the surface of an enclosed liquid, the pressure increases by the same amount at all points throughout the liquid and at the walls of its container, independent of the shape of the fluid container.

This result, discovered by Pascal in the 17th century, is called **Pascal's principle**.

A common application of Pascal's principle is in hydraulic lifts and machinery. A force of magnitude F_1, applied to a piston of area A_1 (see Figure 11.17), increases the pressure in the entire liquid by F_1/A_1. Hence, at a second piston of area A_2, the pressure *increase* is the *same*, even if it is at a different elevation in the fluid. The pressure increase at the second piston is F_2/A_2 where F_2 is the force on the second piston and A_2 is its area. Thus, according to Pascal's principle,

pressure increase at piston 1 = pressure increase at piston 2

$$\frac{F_1}{A_1} = \frac{F_2}{A_2}$$

Hence

$$F_2 = \frac{A_2}{A_1} F_1 \qquad (11.17)$$

If $A_2 > A_1$, a small-magnitude force F_1 applied to a small piston results in a large-magnitude force F_2 at the other, larger piston. The second piston may be used to lift a car (the hydraulic lift common in many automotive service shops), move a blade or large bucket on a bulldozer or front loader, or allow you to stop your car by pushing with your foot on the hydraulic brake piston with a force of small magnitude. In each case a small-magnitude force on a small piston results in tremendous forces on a larger piston.

Blaise Pascal discovered the principle underlying hydraulic machinery. He also is noted for his writings on religion, philosophy, and literature.

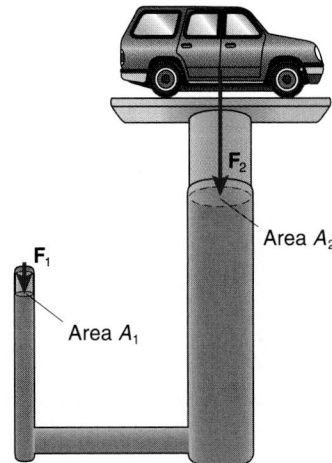

FIGURE 11.17 Apply a force of magnitude F_1 to a piston of area A_1.

EXAMPLE 11.6

A service station hydraulic lift has a large piston of radius 15 cm and a small piston of radius 2.5 cm. What magnitude of force increase must be applied to the small piston to lift a car of mass 2.00×10^3 kg?

Solution
The weight of the car is of magnitude

$$mg = (2.00 \times 10^3 \text{ kg})(9.81 \text{ m/s}^2)$$
$$= 1.96 \times 10^4 \text{ N}$$

Use Equation 11.17:

$$F_2 = \frac{A_2}{A_1} F_1$$

$$1.96 \times 10^4 \text{ N} = \frac{\pi (0.15 \text{ m})^2}{\pi (0.025 \text{ m})^2} F_1$$

Solving for F_1,

$$F_1 = 5.4 \times 10^2 \text{ N}$$

11.6 ARCHIMEDES' PRINCIPLE

Have you ever wondered why some things float while others sink? Throw a penny into the water and it sinks like a rock, but a huge steel aircraft carrier floats. Why is that? What factors determine the answer to this old question? The answer lies with Archimedes' principle, formulated over two millennia ago.

The story goes that Archimedes (c. 287–212 B.C.) was given the problem of determining whether a new crown of King Hieron II was made of pure gold or of a cheaper gold–silver alloy (known as electrum). This was one of the first known problems in consumer testing. Since the crown was irregularly shaped and Archimedes understandably was prohibited from cutting or drilling into it, the problem was not an easy one to solve. Archimedes allegedly thought of a way to solve the problem while entering a communal bath and noting either the

Archimedes formulated one of the oldest principles of classical physics. He died at the hands of an unknown Roman soldier during the capture of Syracuse in 212 B.C.

overflow of the water or the apparent "lightness" of his limbs in the water. Legend has it that he ran streaking stark naked from the bath, shouting *Eureka! Eureka!* ("I have found it!"). Appropriately enough, the word eureka now is the state motto of California, a state whose citizens are known for their uninhibited inventiveness.

Archimedes' principle states that a system submerged or floating in a fluid has a *buoyant force* acting on it whose magnitude is equal to that of the weight of the fluid displaced by the system. As its name implies, the buoyant force acts on the system in the vertically upward direction.

For us in these latter days, Archimedes' principle is not a new fundamental law of physics, nor is the buoyant force a new, fundamental force of nature. The principle is merely a consequence of one of the conditions for static equilibrium that, in turn, stem from Newton's laws of motion. To show this, consider an arbitrarily shaped region of a fluid at rest, as in Figure 11.18. The vector sum of all the forces acting on the fluid within this volume must be zero because the fluid is in

mechanical equilibrium. The forces on this region of the fluid arise from two sources:

1. the *weight* $\vec{\mathbf{w}}$ of the region of fluid (that is, the gravitational force of the Earth on the fluid); and
2. the force $\vec{\mathbf{F}}_{\text{buoy}}$ arising from the *pressure of the rest of the fluid* on this region of fluid.

Since these forces vector sum to zero for a fluid in equilibrium, they must be of equal magnitude and in opposite directions. In other words, the force exerted by the rest of the fluid on this region is equal in magnitude to the weight of the fluid inside the region, but is directed upward.

Now imagine replacing the fluid in the region with an object of exactly the same shape; we consider this new object to be the system, as in Figure 11.19. The force caused by the pressure exerted by the rest of the fluid on this region is the same as before: it could care less what was actually inside the region. The magnitude of the force exerted by the rest of the fluid on this region we found to be equal to that of the weight of the fluid within the region. So the system that replaces (we say *displaces*) the fluid has an upward buoyant force acting on it equal in magnitude to that of the weight of the fluid displaced by the system.

The foregoing argument and conclusion are the same regardless of the shape of the region or the object. Hence any system immersed completely in a fluid has at least two forces acting on it in the vertical direction: the upward buoyant force and the downward weight of the system itself. If the magnitude of the weight of the system is greater than that of the buoyant force on it, as in Figure 11.20, then the total force on the system is in the downward direction and the system sinks. If the magnitude of the buoyant force is greater than that of the weight of the system, as in Figure 11.21, then the total force on the system is in the upward direction and the system rises in the fluid. If the two forces are equal in magnitude, as in Figure 11.22, the total force in the vertical direction is zero. Thus, when the system is released, it stays at the vertical location where it is placed, a situation called **neutral buoyancy**.

To find an expression for the buoyant force on a system, let the fluid density be ρ_{fluid} and the average density of the

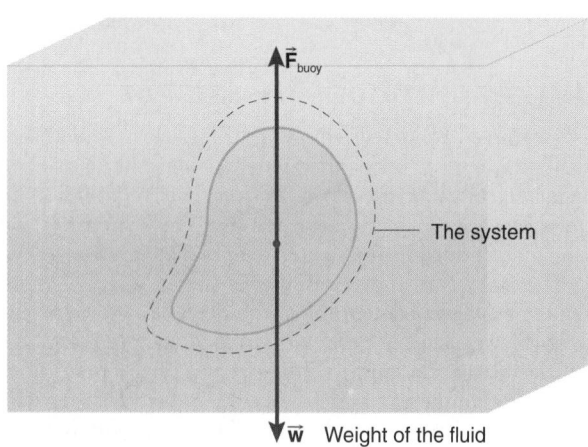

FIGURE 11.18 The forces on a region of fluid at rest.

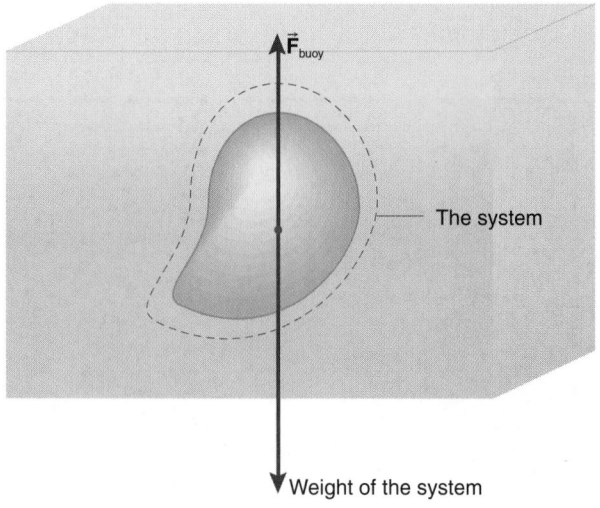

FIGURE 11.19 Forces on a submerged object with the same shape as the region of fluid in Figure 11.18.

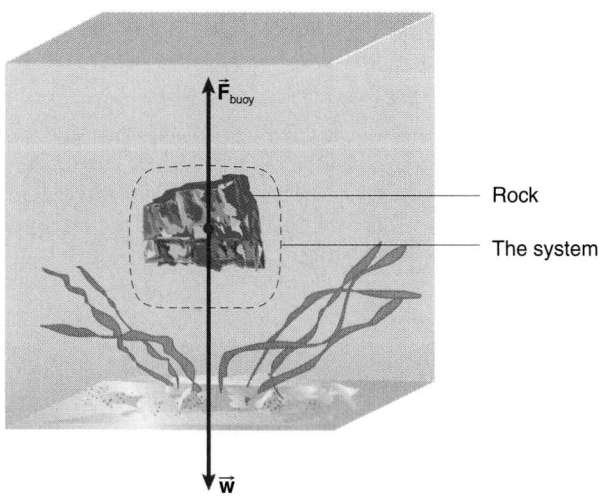

FIGURE 11.20 A system sinks if the magnitude of its weight is greater than the magnitude of the buoyant force.

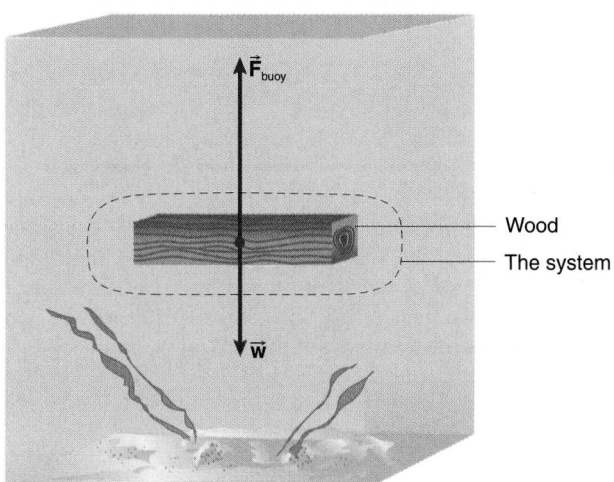

FIGURE 11.21 If the magnitude of the buoyant force is greater than the magnitude of the weight, the system rises in the fluid.

system be ρ_{system}. Let the total volume of the system be V. Then the mass of the system is

$$\rho_{\text{system}} V$$

and its weight is of magnitude

$$\rho_{\text{system}} V g$$

Let V' be the volume of the system *immersed in the fluid*; if the system is totally submerged, $V' = V$. The weight of the fluid displaced by the system is the weight of a region of fluid of volume V'.

Thus the buoyant force on the system is of magnitude

$$F_{\text{buoy}} = \rho_{\text{fluid}} V' g \qquad (11.18)$$

and is directed vertically upward.

When a system floats, the system experiences neutral buoyancy. Of course, the system does not literally float *on the*

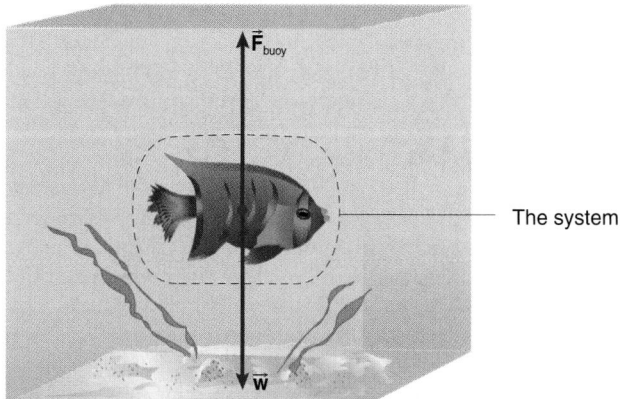

FIGURE 11.22 When the magnitude of the buoyant force is equal to the magnitude of the weight, neutral buoyancy occurs.

surface, although we probably drew childhood pictures of floating boats this way! See Figure 11.23. You know better now. When placed in the fluid, the system sinks into the fluid until the magnitude of the buoyant force on the system (equal to the magnitude of the weight of the liquid displaced by the system) is equal to the magnitude of the weight of the system, as shown in Figure 11.24. That is, the system floats when

$$\rho_{\text{fluid}} V' g = \rho_{\text{system}} V g$$

or

$$\rho_{\text{fluid}} V' = \rho_{\text{system}} V$$

Since V' can at most be equal to V, a system will float in any liquid with a density greater than the average density of the system itself.

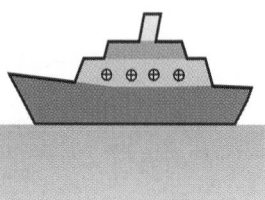

FIGURE 11.23 Typical child's drawing of a floating system.

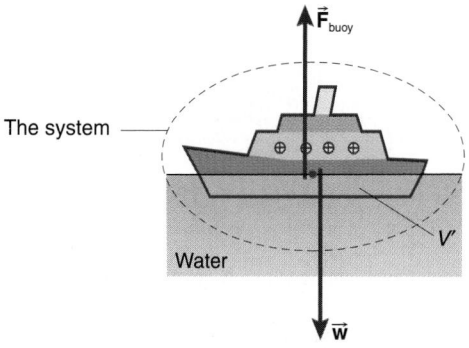

FIGURE 11.24 Forces on a floating system.

For neutral buoyancy or for floating partially submerged, the system thus has an upward buoyant force acting on it (exerted by the rest of the fluid) equal in magnitude to the downward weight of the system itself. The two forces then vector sum to zero and the system floats—that is, it is in vertical equilibrium.

Applications of Archimedes' principle are ubiquitous. Swim bladders in fish, gas-filled chambers in the nautilus shell (*Nautilus pompilius*), and the act of stone swallowing in crocodiles all control the buoyancy of these marine species. Submarines and hot air balloons regulate their altitude in their respective fluids by controlling the magnitude of their weight vis-à-vis that of the buoyant force. Archimedes' principle of displacement also is used to determine the volume of irregular objects. In medicine, the principle is used to assess the amount of swelling in limbs as well as in hydrotherapy for arthritic and other patients.

QUESTION 3

The buoyant force on a system and the weight of the fluid displaced by the system are of equal magnitudes but point in opposite directions. Do these forces form a Newton's third law force pair? Explain why or why not. If they are not a third law force pair, specify the third law force pairs of these forces.

EXAMPLE 11.7

A rock is suspended from a spring scale in air and is found to have a weight of magnitude w; see Figure 11.25a. The rock then is submerged completely in water while attached to the scale, which is found to give a new reading of w_{submrg}; see Figure 11.25b. Find the density of the rock ρ_{rock} in terms of the scale readings and the density of water ρ_{water}.

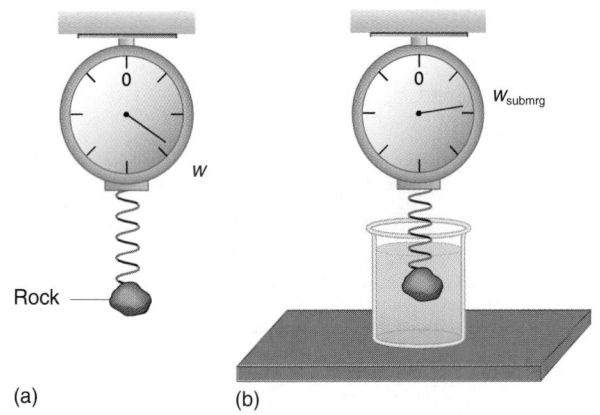

Rock

(a)   (b)

FIGURE 11.25

Solution

Let the system be the rock whose volume is V; its density is given as ρ_{rock}. Then the magnitude of its weight $\vec{w}$ is its mass $(\rho_{rock}V)$ times the magnitude of the local gravitational field (g):

$$w = \rho_{rock}Vg$$

When the rock is submerged in the water while attached to the scale, the rock experiences an additional upward buoyant force with a magnitude equal to that of the weight of the water displaced by the system. Equation 11.18 states that the buoyant force has a magnitude

$$F_{buoy} = \rho_{water}V'g$$

with $V' = V$, since the rock is totally submerged. The magnitude of the total downward force on the system when it is in the water is the new scale reading, w_{submrg}. Thus the new scale reading is the difference between the magnitude of the weight of the rock and the magnitude of the buoyant force on it:

$$w_{submrg} = w - F_{buoy}$$
$$= \rho_{rock}Vg - \rho_{water}Vg$$

Form the ratio

$$\frac{w_{submrg}}{w} = \frac{\rho_{rock}Vg - \rho_{water}Vg}{\rho_{rock}Vg}$$

$$= \frac{\rho_{rock} - \rho_{water}}{\rho_{rock}}$$

Solving for the density of the system (i.e., the rock),

$$\rho_{rock} = \frac{w}{w - w_{submrg}}\rho_{water}$$

EXAMPLE 11.8

A cylindrical object with a weight of magnitude w is placed in two different liquids; in each liquid, it floats with the axis of the cylinder vertical.* When it is placed in an unknown liquid of density ρ, it sinks to a depth h and floats; when placed in water (of known density ρ_{water}), it sinks to a depth h_{water} and floats. What is the density ρ of the unknown liquid in terms of the two depths and the density of water? This example illustrates the operation of a hydrometer, used to measure the relative density of a fluid.

Solution

A floating object sinks until it displaces a volume of liquid sufficient to create a buoyant force of magnitude equal to that of the weight of the object itself. For a cylindrical tube floating with its long axis vertical, the volume of the liquid displaced is proportional to the depth of the tube in the liquid. Indeed, if the cylinder is of cross-sectional area A and sinks to a depth h_{water} when placed vertically in water (see Figure 11.26), the volume of liquid displaced is Ah_{water}. The buoyant force on the floating cylinder thus is of magnitude

$$\rho_{water}(Ah_{water})g$$

and this is equal to the magnitude of the weight of the cylinder since it is floating.

When the cylinder is allowed to float in a liquid of density ρ, it sinks to a depth h, and the buoyant force on the cylinder now has magnitude

$$\rho Ahg$$

---

*The cylinder will need to be supported along its side to avoid capsizing. The long-axis vertical floating position is unstable; the reason for this will be examined in Section 11.7.

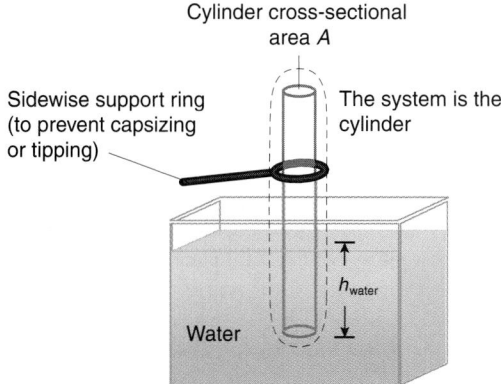

Sidewise support ring
(to prevent capsizing
or tipping)

Cylinder cross-sectional
area A

The system is the
cylinder

h_{water}

Water

FIGURE 11.26

The raising of the remains of the hull of the *Mary Rose* outside Portsmouth, England.

which also is equal to that of the weight of the cylinder. Since the buoyant forces on the floating cylinder in water and in the unknown liquid are both equal in magnitude to that of the weight of the cylinder, these buoyant forces themselves have equal magnitudes:

$$\rho A h g = \rho_{\text{water}} A h_{\text{water}} g$$

Solving for the density of the unknown liquid,

$$\rho = \rho_{\text{water}} \frac{h_{\text{water}}}{h}$$

In this way the density of the unknown liquid can be found in terms of the density of water. Of course another liquid of known density also can be used instead of water.

11.7 THE CENTER OF BUOYANCY*

In July 1545 the great warship of King Henry VIII, the *Mary Rose*, was sailing forth from Portsmouth harbor in southern England in a light breeze on its way to engage a superior French fleet poised for invasion. The *Mary Rose* quite unexpectedly capsized after listing slightly and quickly sank. In horror, the king, his court, and the wife of the captain all witnessed the disaster, which took place only about a mile offshore (oops! make that about 1.6 km). Approximately 700 sailors and soldiers perished; there were only about a dozen survivors. The vessel had been newly outfitted and refitted with massive guns and armament; unfortunately the additions made the vessel unstable and it capsized when slightly rocked.[†] More recently, on 28 September 1994, the ferry *Estonia* lost its bow doors during a fierce storm in the Baltic Sea; sea water rushed in and quickly made the ferry unstable. It capsized and sank in minutes with the loss of over 900 lives; only a few passengers and crew survived.

These tragic sinkings illustrate that for floating objects to be in *stable* equilibrium, any torques about the center of mass arising from a tilting or rocking from the equilibrium position must be directed to restore or rotate the system back to its original upright floating position. If not, the system will capsize.

The stability of a system that is floating (i.e., in neutral buoyancy) depends on the relative positions of the points of application of the weight of the system and of the buoyant force. These points of application are different.

The weight acts at the center of mass point of the system. The position of the center of mass depends on the specific shape and distribution of the mass of the system. On the other hand, the buoyant force acts at the position of the center of mass of the *fluid* imagined to be in the *hole* created by the system in the fluid, a point known as the **center of buoyancy**, as shown in Figure 11.27.

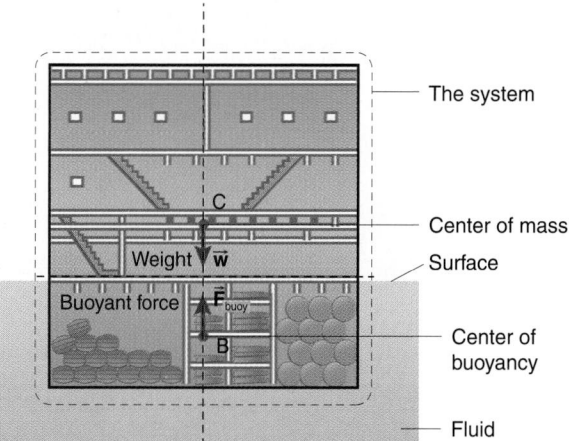

The system

C — Center of mass

Weight $\vec{\mathbf{w}}$ — Surface

Buoyant force — $\vec{\mathbf{F}}_{\text{buoy}}$

B — Center of buoyancy

Fluid

FIGURE 11.27 The point of application of the weight is the center of mass C; the buoyant force acts at the center of buoyancy B.

[†]The remains of the vessel and many nautical and personal artifacts were recovered during the early 1980s and now are displayed in a fascinating museum in Portsmouth.

For a system at rest in equilibrium in neutral buoyancy, the weight and buoyant forces are collinear, having the same line of action; thus the center of mass and the center of buoyancy lie along the same vertical line. This condition ensures that there is zero torque about the center of mass of the system, and so no spin results, fulfilling the second condition for static equilibrium. But what happens if the system is tilted slightly? Under what conditions will the system return to its original, upright, floating position rather than capsize? The answer depends on whether the system is completely submerged or floating, and on the shape of the submerged hull.

Totally Submerged Systems

For simplicity, consider a system, say a submarine, with a circular cross section, as shown in Figure 11.28. The submarine is not symmetrically loaded and we represent this with a large mass at the bottom; the center of mass of the submarine is located below the geometric center of the circular cross section, say at point C. The center of buoyancy B is located at the center of mass of the fluid imagined to be in the *hole* created in the fluid by the submarine. In this case that point is at the center of the circle, point B, since the fluid is of constant density. The center of mass thus lies *below* the center of buoyancy.

If the submarine is tilted slightly, as in Figure 11.29, the buoyant force then produces a torque about the center of mass of the submarine that tends to rotate the submarine back to its original equilibrium position. Thus the original orientation of the submarine is stable.

Suppose, on the other hand, that the center of mass of the submarine is located *above* the elevation of the center of buoyancy, as in Figure 11.30. If the submarine is tilted, then in this case the torque of the buoyant force about the center of mass accentuates the tilt and the system is in unstable equilibrium; the submarine will roll over.

Therefore, for a submerged system to be in stable equilibrium, the center of mass must be *below* the elevation of the center of buoyancy.

Floating Systems

The stability of a floating system is more complicated because the center of buoyancy may be located at different positions relative to the center of mass, depending on the shape of the system *and* the specific position in which it is floating. For example, Figure 11.31 depicts a square cross section of a level ship that has its center of mass C *above* the center of buoyancy B.

According to what we learned about submerged systems, you might think this is an unstable configuration that would result in a capsize. But if the ship is tilted slightly, as in Figure 11.32, the center of mass at C remains in the same location on the ship but the center of buoyancy *shifts* in the direction of the tilt (to the right in Figure 11.32). The triangular shaped region DAE, which was part of the *original* buoyant volume, is transferred to the new buoyant volume GAF. Since the center of buoyancy is at the center of mass of the *hole* created by the system in the fluid, the center of buoyancy must move to the right. Thus, in Figure 11.32, we see that the torque of the buoyant force about the center of mass point will *return* the ship to its original upright position; the ship is in stable equilibrium.

So what condition then determines whether a floating ship is in stable equilibrium or not?

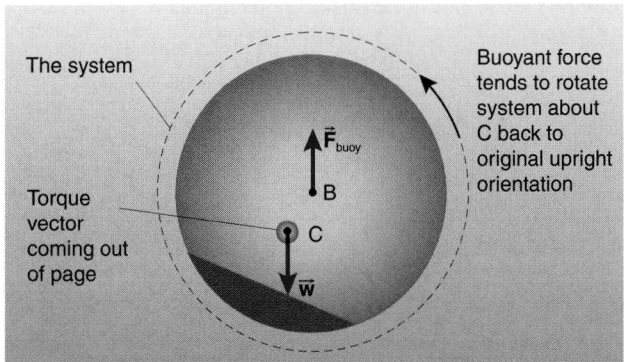

FIGURE 11.29 The torque of this buoyant force about the center of mass C will rotate the system back to the equilibrium position.

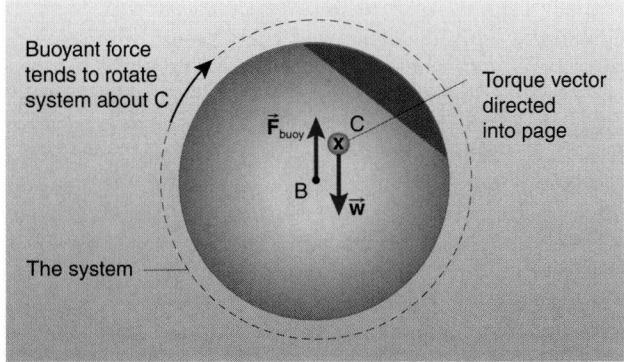

FIGURE 11.30 If the center of mass C of the submerged system is above the center of buoyancy B, the torque of the buoyant force about the center of mass will cause the system to roll directly toward equilibrium.

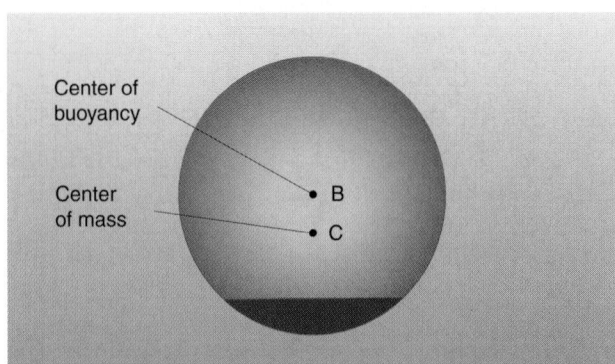

FIGURE 11.28 The center of mass C of this submerged system lies below the center of buoyancy B.

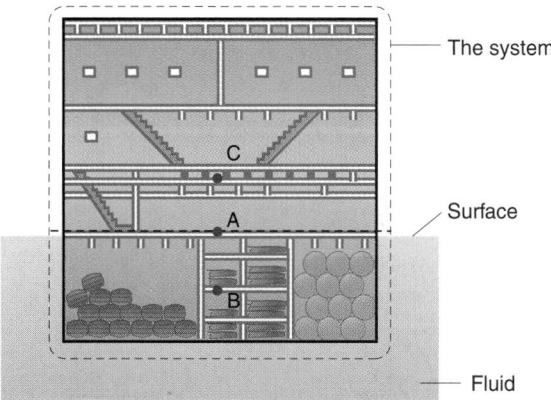

FIGURE 11.31 A cross section of a ship with its center of mass C above the center of buoyancy B.

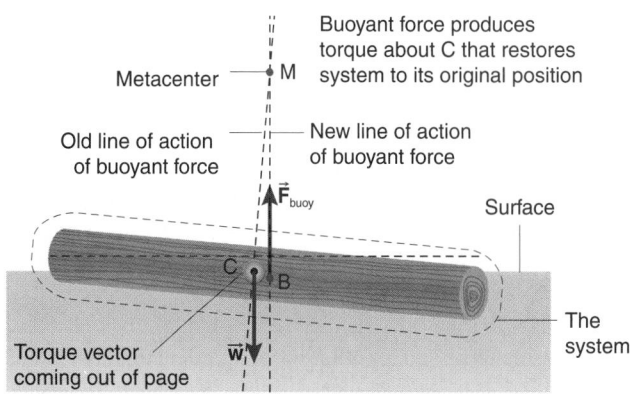

FIGURE 11.33 When a horizontal pole is tilted, the metacenter M is above the center of mass C.

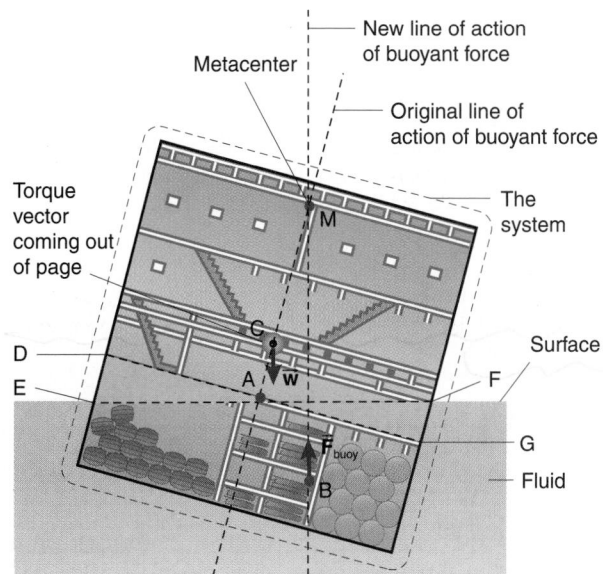

Torque of buoyant force about C rights the ship.

FIGURE 11.32 When tilted the center of buoyancy B shifts in the direction of the tilt.

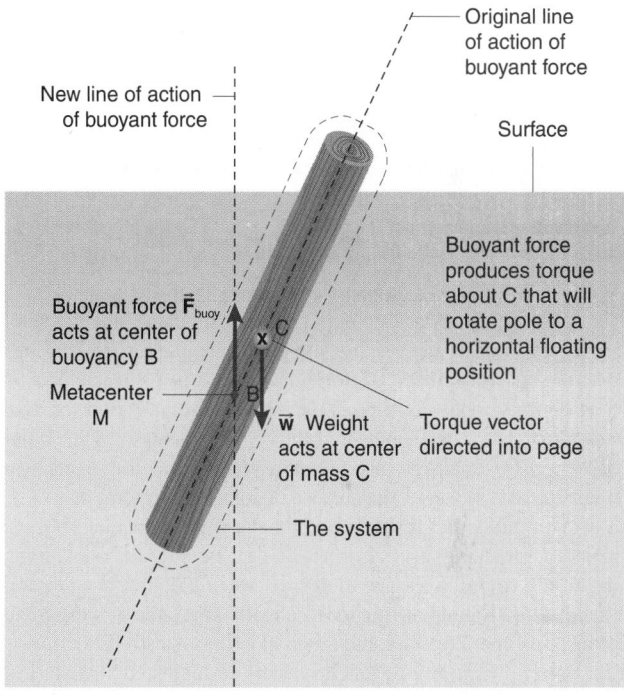

FIGURE 11.34 When a pole is vertical, the metacenter M lies below the center of mass C.

The point of intersection of the *original* line of action of the buoyant force (a line imagined to rotate with the ship when it is rocked) with the *new* line of action of the buoyant force (a vertical direction) when the system is tilted is known as the **metacenter**; see Figure 11.32.

If the elevation of the metacenter is *above* the elevation of the center of mass of the ship, the system will return to its original orientation when tilted and the floating equilibrium position is stable. If the elevation of the metacenter is *below* that of the center of mass, the system is in unstable equilibrium because the torque of the buoyant force then accentuates the tilt; the system will roll over (capsize).

This is seen in the following example.

A long, uniform, cylindrical wooden telephone pole floats with its long axis horizontal rather than vertical. When the long pole is horizontal and is tilted, as shown in Figure 11.33, we see that the location of the metacenter is above the center of mass, indicating the system will return to its original configuration.

Suppose, on the other hand, that the telephone pole is floating with its long axis vertical, as shown in Figure 11.34. If the system is tilted slightly, the location of the center of buoyancy shifts to the left of the center of mass (because the long axis of the pole is vertical). The metacenter now is *below* the center of mass. The torque of the buoyant force about the center of mass point now accentuates the tilt. The system does not return to its original vertical configuration but assumes the stable horizontal floating configuration. Thus the vertical floating position of the pole is unstable.

Hence we see that naval architects must be sure that the following condition is satisfied for stable floating equilibrium:

> A floating system is in stable equilibrium if the metacenter is located above the center of mass of the system—say, a ship— for every possible roll of the ship in heavy seas. The same is true for submerged systems.

For a sailboat and for other boats experiencing high winds, the problem is more complicated, since we must also account for the torque on the boat caused by forces arising from the wind on the sails and/or ship.

Needless to say, these principles were not known in the time of Henry VIII (who lived over a century before Newton), much to the peril of the unfortunate souls aboard the *Mary Rose*.

QUESTION 4

Given enough depth, a sinking ship (such as the *Titanic*) likely will settle on the bottom in an upright position even if it first capsizes while sinking. Explain why.

11.8 SURFACE TENSION*

Water striders (*Hygrotrechus conformis*) literally can walk on water. Many other water insects have the same ability. Such insects are not floating. They are able to execute such feats not because of Archimedes' principle but because of surface tension. Surface tension also is responsible for the spherical shape of small water droplets. More domestically, if you are careful, with a pair of tweezers perhaps you can gently lay a needle from your mending kit on top of the surface of a glass of water.

The molecules inside a liquid have molecular neighbors on all sides; molecules at the surface do not. As a result, if you imagine trying to pull a surface molecule out of a liquid, the molecular bonds with neighboring molecules are stretched and the given molecule experiences a force much like that of a stretched spring. Correspondingly, if a surface molecule is depressed slightly into the liquid, the molecule also experiences a restoring force. The surface of the liquid thus behaves much like an elastic rubber sheet. Small water droplets are spherical in shape because the restoring forces act to minimize the surface area of the drop.

If you gently pull upward on a needle lying on the surface of a glass of water using a calibrated, sensitive spring, you will find that the magnitude of the force needed is proportional to the length ℓ in contact with the liquid. The length of needle edge in contact with the liquid is *twice* the length of the needle because there is liquid on both sides of the needle. When the needle is on the verge of being pulled from the surface, the magnitude of the upward force (indicated by the extension of the spring) is equal to the magnitude of the weight of the needle (mg) plus that of the lifted liquid (which we neglect in comparison to the weight of the needle in a crude first approximation)[†] plus the magnitude $\gamma\ell$ of a *surface tension* force. The quantity γ is the magnitude of the force per unit length in contact with the liquid, known as the **coefficient of surface tension**. Values of γ for representative liquids are given in Table 11.2.

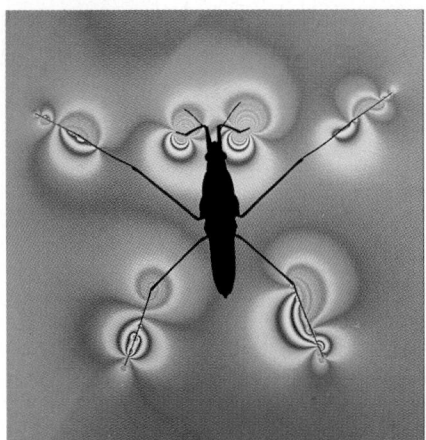

A water strider walking on water.

A more accurate method for measuring the coefficient of surface tension is the Wilhelmy slide method.[‡] A microscope slide (or other thin flat plate) is suspended from a spring or other sensitive balance and the magnitude of the weight w of the slide in air is noted. The slide is lowered slowly until it makes contact with the fluid; the scale reading then increases to w'. The two scale readings are related by

$$w' - w = \gamma p$$

where p is the perimeter of the slide and γ is the coefficient of surface tension.

QUESTION 5

Given the data in Table 11.2, why might water striders have trouble walking on acetone?

EXAMPLE 11.9

A thin, circular wire ring of diameter 4.0 cm and total mass 0.70 g is gently pulled vertically from a water surface by means

---

[†]Typically metal needles or rings are used. Since the density of metals is much greater than that of most liquids, neglecting the weight of the lifted liquid is a reasonable, although crude, first approximation. This extraction method (and that of Example 11.9) typically yields coefficients of surface tension that may have uncertainties of about 25%.

[‡]The method was invented in 1863 by Ludwig Wilhelmy (1812–1864).

TABLE 11.2 Coefficient of Surface Tension at 20 °C

Liquid	γ (N/m)
Acetone	0.024
Ammonia	0.020
Benzene	0.029
Glycol	0.048
Methanol	0.023
Water	0.073

of a sensitive spring with spring constant $k = 0.75$ N/m, as shown in Figure 11.35. When the spring is stretched 3.4 cm from its equilibrium extension in air with *no* ring attached, the ring is on the verge of being pulled free from the water surface. Find the coefficient of surface tension of water. Neglect the mass of the lifted water.

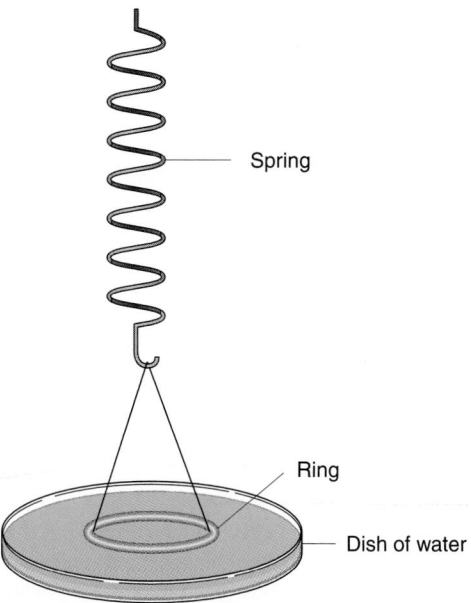

FIGURE 11.35

Solution
Let the wire ring be the system. When the ring is on the verge of breaking free from the surface, the magnitude of the upward force of the spring on the ring (kx) is equal to the sum of the magnitude of the weight of the ring (mg) and the magnitude of the force of surface tension ($\gamma\ell$):

$$kx = mg + \gamma\ell$$

The length in contact with the water surface is *twice* the circumference of the ring since there is fluid on both sides of the ring:

$$\ell = 2(2\pi r)$$

where r is the radius of the ring. Hence the coefficient of surface tension is

$$
\begin{aligned}
\gamma &= \frac{kx - mg}{4\pi r} \\
&= \frac{(0.75 \text{ N/m})(3.4 \times 10^{-2} \text{ m}) - (7.0 \times 10^{-4} \text{ kg})(9.81 \text{ m/s}^2)}{4\pi(2.0 \times 10^{-2} \text{ m})} \\
&= \frac{0.026 \text{ N} - 0.0069 \text{ N}}{0.25 \text{ m}} \\
&= \frac{0.019 \text{ N}}{0.25 \text{ m}} \\
&= 0.076 \text{ N/m}
\end{aligned}
$$

11.9 CAPILLARY ACTION*

The forces between a molecule of a liquid and other molecules of the liquid are called **cohesive forces**. Forces between a molecule of a liquid and other materials such as the side of a container are known as **adhesive forces**.

The **contact angle** θ_{cont} between the vertically upward direction of the wall of a container and the surface of the liquid at the wall (see Figure 11.36) gives a measure of the relative magnitudes of the cohesive and adhesive forces. When the adhesive forces are greater in magnitude than are the cohesive forces, we say the liquid **wets** the surface. In such circumstances the liquid climbs the wall of the container and produces a concave **meniscus**, like that of water around the edges of a measuring cup. The contact angle is less than 90° when wetting occurs. For water and glass, the contact angle is close to 0°.

If the cohesive forces have magnitudes greater than those of the adhesive forces, no wetting occurs and the meniscus is convex. The contact angle is greater than 90° when no wetting occurs. Liquid mercury in glass has this characteristic; the contact angle is about 140°.

When wetting occurs, surface tension will draw a liquid up into a thin tube inserted into the liquid, a phenomenon called **capillary action**; see Figure 11.37. The liquid rises into the tube until the vertical component of the surface tension

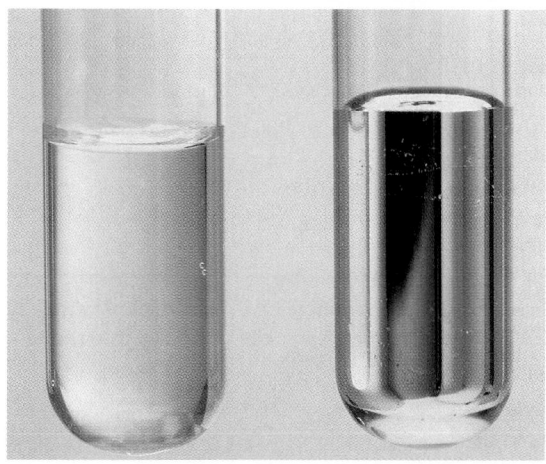

Water in a glass container has a concave meniscus with a contact angle close to 0°. On the other hand, liquid mercury produces a convex meniscus with an obtuse contact angle.

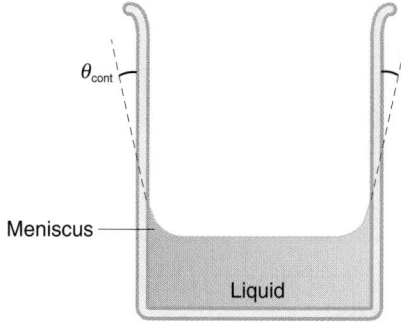

FIGURE 11.36 The contact angle between a liquid and the wall of its container.

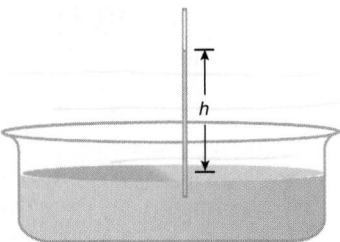

FIGURE 11.37 Capillary action.

force has a magnitude equal to that of the weight of the liquid in the column:

$$\gamma \ell \cos \theta_{\text{cont}} = mg \qquad (11.19)$$

If the column has a circular cross section of radius r, then

$$\ell = 2\pi r$$

Measure the height h of liquid in the column to the bottom of the meniscus. Neglect the small amount of liquid in the concave surface of the liquid. Then the mass of liquid of density ρ in a liquid column of height h is approximately

$$m = \pi r^2 h \rho$$

Substituting for ℓ and m in Equation 11.19 and solving for h, we find

$$h = \frac{2\gamma \cos \theta_{\text{cont}}}{\rho g r} \qquad (11.20)$$

Capillary action is what draws kerosene up into the wick of an emergency lamp or melted wax up to the flame of a candle. Capillary action also is used in the analysis of biological molecules using techniques called chromatography.

EXAMPLE 11.10

A glass tube of inside diameter 1.0 mm is placed vertically into a glass of water. How high will the water rise into the tube due to capillary action?

Solution
Assume the contact angle between the water surface and the glass wall is 0°. Use the value of the coefficient of surface tension for water given in Table 11.2. The density of water is about 1.00×10^3 kg/m³. Using Equation 11.20,

$$h = \frac{2\gamma \cos \theta_{\text{cont}}}{\rho g r}$$

$$= \frac{2\,(0.073\ \text{N/m}) \cos 0°}{(1.00 \times 10^3\ \text{kg/m}^3)(9.81\ \text{m/s}^2)(1.0 \times 10^{-3}\ \text{m})/2}$$

$$= 0.030\ \text{m}$$

$$= 3.0\ \text{cm}$$

11.10 FLUID DYNAMICS: IDEAL FLUIDS

The motions of real fluids are quite complex, as the observation of a fast flowing river indicates. Turbulence, eddies, and viscous

Fuel is drawn up to the flame by capillary action in both candle strings and kerosene lamp wicks.

frictional effects make whitewater enthusiasts happy, but complicate the study of the kinematics and dynamics of real fluids. Many of these complications are best left for advanced courses in fluid mechanics.

Thus we initially confine ourselves to the kinematics and dynamics of **ideal fluids** that have the following characteristics:

1. No viscous, frictional effects are present. Thus the fluid experiences frictionless flow, analogous to the ideal of frictionless surfaces we commonly assumed in our study of mechanics.
2. The flow is steady; that is, the velocity of the fluid at every point in space is time independent. However, the velocity may have a *spatial* dependence. Different regions of the fluid may be moving at different velocities, but wherever we are, the velocity of the fluid at that point in space, whatever its value, does not vary with time.
3. The density of the fluid is constant in space and time; this is called **incompressible flow**.
4. No eddies or rotational flow are present; this condition is termed **irrotational (nonturbulent) flow**. The center of mass of a small pinwheel set loose at any point in the fluid would move, but the pinwheel itself will not spin if the flow is irrotational.

Fluid flow can exhibit complex motions, which make whitewater sports fun and exciting—and occasionally dangerous.

11.11 EQUATION OF FLOW CONTINUITY

A small particle of a flowing, incompressible fluid follows a path known as a **streamline**. For steady flow the flow rate and streamlines are static (not changing with time). Streamlines are like the paths followed by cars on a multilane freeway when the cars maintain fixed separation distances (incompressible traffic!).

Consider the set of streamlines in Figure 11.38, which delineate the motion of a fluid. The set of streamlines form a **flow tube** that may be either part of a larger volume of surrounding fluid or may represent flow inside a physical pipe. At a particular place along the flow tube, consider a section of the fluid of cross-sectional area A and differential length ds (see cross section #1 in Figure 11.38). The amount of mass dm in this small volume is the product of the density ρ and the volume $A \, ds$:

$$dm = \rho A \, ds \qquad (11.21)$$

A differential time interval dt later, the fluid has flowed just past cross section #1, as shown in Figure 11.39. Take Equation 11.21 and divide by dt, obtaining

$$\frac{dm}{dt} = \rho A \frac{ds}{dt}$$
$$= \rho A v$$

where v is the speed of the fluid. For steady flow the rate at which mass passes any given location in the flow tube must be constant. That is, if mass dm passes cross section #1 during a time dt, then the same amount of mass dm also must pass cross section #2 further along the flow tube during dt; see Figure 11.40.

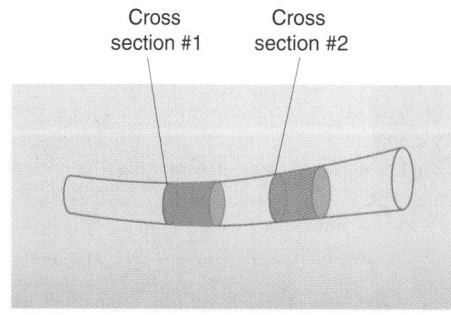

FIGURE 11.40 The same amount of fluid mass must flow past cross sections #1 and #2 along a flow tube during dt.

If this were not the case, then the mass would be either increasing or decreasing, leading to accumulations or losses of mass during an interval of time. Thus

$$\frac{dm}{dt}$$

must be constant for steady flow. Since

$$\frac{dm}{dt} = \rho A v$$

the product $\rho A v$ must be constant at any point along the flow tube.

Therefore, at any two points along the flow tube,

$$\rho_1 A_1 v_1 = \rho_2 A_2 v_2$$

For incompressible fluids the density is constant, and so

$$A_1 v_1 = A_2 v_2 \qquad (11.22)$$

This equation is known as the **equation of flow continuity**.

Equation 11.22 represents conservation of mass: the amount of mass per second flowing past all cross-sectional areas of a flow tube must be the same.

For such incompressible fluids, the equation of flow continuity means that the **flow rate** $Q = Av$, the volume of fluid flowing past any cross-sectional area in the flow tube during one second, is constant:

$$Q_1 = Q_2$$

The volume flow rate is measured in m³/s in SI units, or liters/s using units of convenience. One liter is a cube exactly 10 cm on a side, so that 1 liter = 10^{-3} m³.

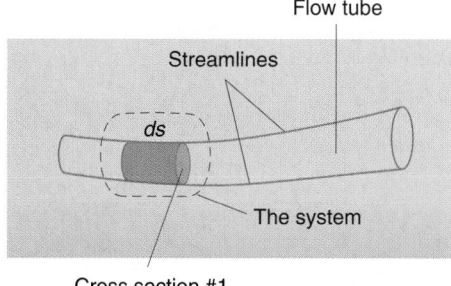

FIGURE 11.38 A small amount of fluid in a flow tube.

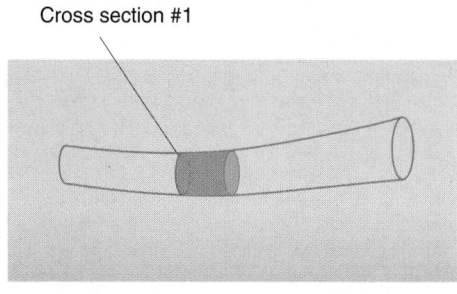

FIGURE 11.39 Fluid flow past a cross section of a flow tube.

11.12 BERNOULLI'S PRINCIPLE FOR INCOMPRESSIBLE IDEAL FLUIDS

Your experience using a garden hose with and without a nozzle (or your thumb) indicates that even an ideal fluid filling and moving through a smooth pipe can race or creep depending on changes in the cross-sectional area of the pipe, the pressure inside the pipe, and changes in elevation of the pipe itself. Here we discover the connecting principle, first formulated by Daniel

Bernoulli (1700–1782), that lets us understand the flow throughout the pipe.

Consider an incompressible ideal fluid (i.e., one with steady flow and no viscous or rotational effects), which permits us to ignore any complications from thermal effects associated with turbulent fluid flow. The work done by "other" (nonconservative) forces acting on the fluid results only in changes in the mechanical energy of the fluid, and so we can use the faithful CWE theorem of Chapter 8. Recall that this theorem states that the work done by the "other" (i.e., nonconservative) forces acting on the system, W_{other}, is equal to the change in the mechanical energy of the system (Equation 8.27):

$$W_{\text{other}} = \Delta KE + \Delta PE$$

Examine a section of fluid in a flow tube, as in Figure 11.41. Let the pressure at cross section #1 in the tube be P_1 and let the cross-sectional area of the flow tube here be A_1. Let y_1 be the average height of the tube (the height of the midpoint of the tube) above some reference level. The speed of the flow here is v_1. At cross section #2 along the flow tube, let the corresponding quantities be P_2, A_2, y_2, and v_2.

The fluid to the left of cross section #1 exerts a force $\vec{F}_1$ on the section of fluid between cross sections #1 and #2. If the fluid undergoes a change in its position vector $\Delta\vec{r}_1$, then the work done by this force is

$$\vec{F}_1 \cdot \Delta\vec{r}_1$$

Since $\vec{F}_1$ is parallel to $\Delta\vec{r}_1$, the scalar product reduces to simply

$$F_1 \, \Delta r_1$$

The fluid to the right of cross section #2 exerts a force $\vec{F}_2$ on the fluid located between the two cross sections #1 and #2. The fluid at cross section #2 experiences a change in position vector $\Delta\vec{r}_2$, which may be different from $\Delta\vec{r}_1$ because of the different cross-sectional areas of the flow tube. The work done by the force $\vec{F}_2$ is

$$\vec{F}_2 \cdot \Delta\vec{r}_2$$

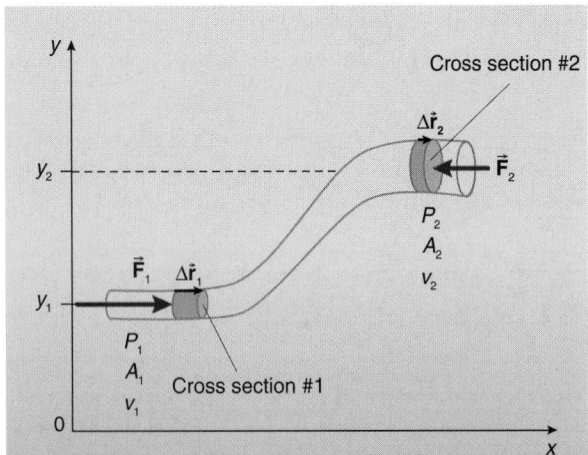

FIGURE 11.41 Fluid flow at two different cross sections along a flow tube.

But here the two vectors are antiparallel, so that the scalar product is

$$F_2 \, \Delta r_2 \cos 180°$$

or

$$-F_2 \, \Delta r_2$$

The work done by other pressure forces on the fluid between these two points is zero because these forces are perpendicular to the changes in the position vectors along the flow tube. These other forces are caused either by the rest of the fluid (surrounding the flow tube) or by the walls of the constraining pipe. So, gathering these results, the CWE theorem here becomes

$$F_1 \, \Delta r_1 + (-F_2 \, \Delta r_2) = \Delta KE + \Delta PE$$

The magnitude of the force $\vec{F}_1$ on the fluid at cross section #1 is the average pressure P_1 at this point times the cross-sectional area A_1;* and likewise at cross section #2: $F_2 = P_2 A_2$. With these substitutions the CWE theorem becomes

$$P_1 A_1 \, \Delta r_1 - P_2 A_2 \, \Delta r_2 = \Delta KE + \Delta PE$$

Notice that $A_1 \, \Delta r_1$ is the volume of the fluid pushed past cross section #1; likewise $A_2 \, \Delta r_2$ is the volume of fluid pushed past cross section #2. Since the fluid is incompressible, these volumes must be the same by the principle of flow continuity. Call this volume of fluid ΔV; so the CWE theorem becomes

$$(P_1 - P_2) \, \Delta V = \Delta KE + \Delta PE \qquad (11.23)$$

Fluid flow means that some fluid enters the region (between cross sections #1 and #2) at cross section #1 and some leaves this region at cross section #2. The same amount of fluid exists between the two cross sections as before, because of flow continuity. Thus the total *change* in the mechanical energy of the entire amount of fluid between the two cross sections takes place at cross sections #1 and #2. The *change* in the kinetic energy of the system is the kinetic energy of a volume ΔV of fluid at cross section #2 minus the kinetic energy of the same volume of fluid at cross section #1. That is,

$$\Delta KE = \frac{1}{2} (\rho \, \Delta V) v_2^2 - \frac{1}{2} (\rho \, \Delta V) v_1^2$$

The change in the gravitational potential energy of the fluid is the potential energy of the fluid in volume ΔV when at cross section #2 minus that at cross section #1:

$$\Delta PE = (\rho \, \Delta V) g y_2 - (\rho \, \Delta V) g y_1$$

With these substitutions, Equation 11.23 becomes

$$(P_1 - P_2) \, \Delta V = \frac{1}{2} \rho \, \Delta V \, v_2^2 - \frac{1}{2} \rho \, \Delta V \, v_1^2 \\ + \rho \, \Delta V \, g y_2 - \rho \, \Delta V \, g y_1$$

---

*Since the pipe or flow tube has some finite size in the vertical dimension, there is a variation in pressure over the cross-sectional area; hence we let P_1 and P_2 be the *average* pressures over the respective cross-sectional areas.

The volume change ΔV divides out of the equation. Shuffling the terms involving cross section #1 to one side of the equation and those involving #2 to the other yields

$$P_1 + \frac{1}{2}\rho v_1^2 + \rho g y_1 = P_2 + \frac{1}{2}\rho v_2^2 + \rho g y_2 \quad (11.24)$$

Equation 11.24 means that the quantity

$$P + \frac{1}{2}\rho v^2 + \rho g y$$

has the same value at every point along the flow tube! This conservation law, essentially a consequence of the CWE theorem applied to the fluid, is known as **Bernoulli's principle** for incompressible flow of an ideal fluid.

Let's look at a few special applications.

Static Incompressible Fluids

If the fluid is not moving, both v_1 and v_2 are zero and Bernoulli's principle reduces to

$$P_1 + \rho g y_1 = P_2 + \rho g y_2$$

Call cross section #1 the reference level where $y_1 = 0$ m, so that the pressure here is P_0. Then let cross section #2 have a pressure P at an elevation y. So the previous equation becomes

$$P_0 + \rho g(0 \text{ m}) = P + \rho g y$$

or

$$P = P_0 - \rho g y$$

which is identical to what we secured before (Equation 11.14) for static incompressible fluids.

Incompressible Fluids Moving in a Horizontal Pipe

For this case (see Figure 11.42), there is no change in the gravitational potential energy of the fluid. Although some fluid in the larger-diameter pipe lowers its gravitational potential energy, an equivalent amount raises its gravitational potential energy, and so the net change is zero. Another way to look at this is that the mean elevation of both pipes is the same. So Bernoulli's principle becomes

$$P_1 + \frac{1}{2}\rho v_1^2 = P_2 + \frac{1}{2}\rho v_2^2 \quad (11.25)$$

The equation of flow continuity for incompressible fluid flow, Equation 11.22, implies that the speed of flow in the small-diameter pipe must be greater than that in the larger-diameter pipe. Since $v_2 > v_1$, Bernoulli's principle (as expressed in Equation 11.25) then indicates that the pressure $P_2 < P_1$.

That is, if the speed of the fluid *increases*, the pressure *decreases*.

Solve the flow continuity equation (Equation 11.22) for v_2:

$$v_2 = \frac{A_1}{A_2}v_1$$

Substitute for v_2 in Equation 11.25:

$$P_1 + \frac{1}{2}\rho v_1^2 = P_2 + \frac{1}{2}\rho\left(\frac{A_1}{A_2}v_1\right)^2$$

This can then be solved for v_1:

$$v_1 = \left[\frac{2(P_1 - P_2)}{\rho\left(\dfrac{A_1^2}{A_2^2} - 1\right)}\right]^{1/2}$$

This means that by measuring the pressure difference between the two regions and knowing the cross-sectional areas of the two pipes, the speed of the fluid v_1 in the pipe can be determined (for a fluid of known density). A device that enables you to make such a measurement is known as a **Venturi meter**.

Lift

Bernoulli's principle also can explain, at least qualitatively, how an airplane wing develops an upward force called **lift** to oppose the downward force of gravitation. To a first approximation, the air flow around a wing is incompressible. Wings are designed so that the flow *over* the wing is faster than the flow *under* the wing, as shown in Figure 11.43.

This means that the pressure in the region above the wing is less than the pressure in the region under the wing. Thus the upward force of the air acting on the underside of the wing is greater in magnitude than that of the downward force of the air on the upper surface. As a result, there is a net upward force on the wing: lift. One aspect of aeronautical engineering is to design wing surfaces and cross sections that maximize the lift for a given

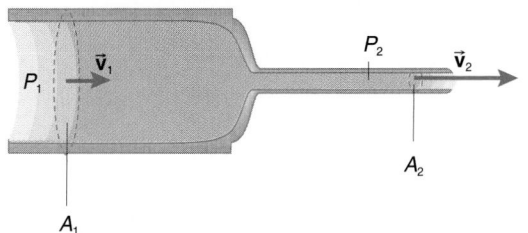

FIGURE 11.42 A horizontal pipe with two different cross-sectional areas.

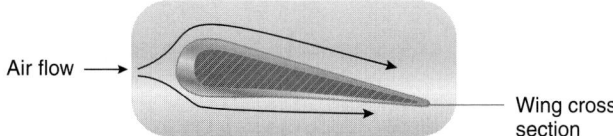

FIGURE 11.43 A wing is designed so the air flow over the wing is faster than the flow under it.

speed. The problem is not as easy as Bernoulli's principle makes it out to be (since we assumed, unrealistically, air flows like an ideal fluid), but at least you can get an inkling of how airplanes manage to get off the ground and stay aloft. Winging it (wing design) is not easy!

Explain the operation of a soda straw. Could such a straw be used to draw liquids on the Moon?

EXAMPLE 11.11

A desperado sneaks into town and at the train depot shoots a small hole in a full, open water tank at a depth h below the water surface, as shown in Figure 11.44. What is the speed of the water as it exits the hole?

FIGURE 11.44

Solution

Choose a coordinate system with the origin at the level of the hole, as shown in Figure 11.45. Use Bernoulli's principle at the top of the tank and the hole:

$$P_{top} + \frac{1}{2}\rho v_{top}^2 + \rho g y_{top} = P_{hole} + \frac{1}{2}\rho v_{hole}^2 + \rho g y_{hole}$$

Since the hole is small, the vertical speed of the fluid at the top of the tank is negligible compared with the speed of the fluid exiting the hole, so take $v_{top} = 0$ m/s. Since both the top of the tank and the hole are open to the atmosphere, $P_{top} = P_{hole} \approx 1.00$ atm $= 1.01 \times 10^5$ Pa. With the coordinate system chosen, $y_{top} = h$ and $y_{hole} = 0$ m. With these substitutions, you have

$$\rho g h = \frac{1}{2}\rho v_{hole}^2$$

Solving for v_{hole}, you find

$$v_{hole} = (2gh)^{1/2}$$

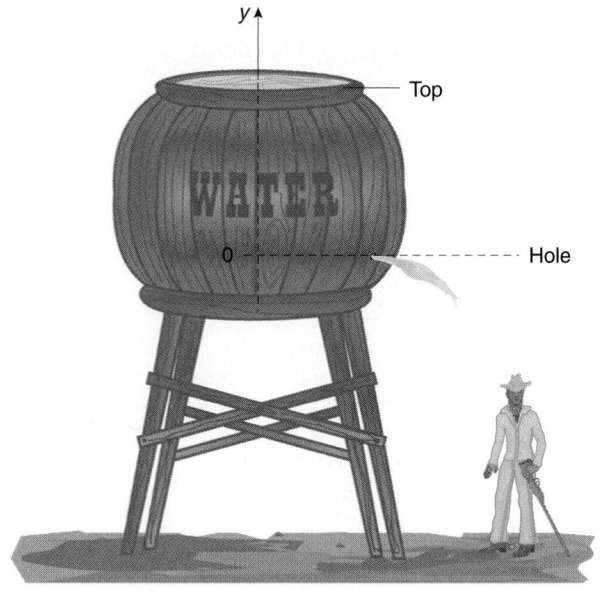

FIGURE 11.45

Notice that the *horizontal* speed of the liquid emerging from the hole is the same as the *vertical* speed it would have as a result of a free-fall from rest through a height h. This observation is known as **Torricelli's law**, after Evangelista Torricelli (1608–1647), who also invented the barometer (see Example 11.3).

11.13 NONIDEAL FLUIDS*

Slow as molasses in January

Southern folk saying

Turn an open bottle of soda pop upside down and the fluid quickly runs out all over you and your textbook in just a few seconds. On the other hand, a comparable bottle filled with molasses will empty itself much more slowly if turned upside down, but still creates a real mess. We say the molasses is more *viscous* than the soda pop. The **coefficient of viscosity** of a fluid is a measure of its resistance to flow. Here we will see how this coefficient of viscosity can be measured.

Consider a thin fluid layer, as in Figure 11.46. A plate of area A, separated by a small distance Δy from the bottom of the

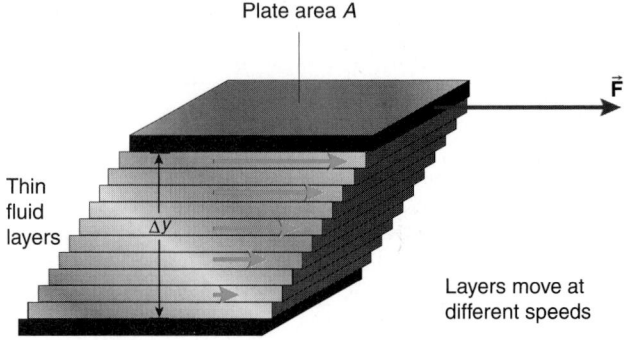

FIGURE 11.46 Fluid flow velocities between a moving and fixed surface.

layer of fluid, is dragged across the fluid at constant speed v. Experiments indicate that the fluid immediately adjacent to the moving plate moves essentially at the speed of the plate, while the fluid immediately adjacent to the bottom of the layer has essentially zero speed. Therefore there exists a speed change Δv across the thin fluid layer of thickness Δy.

Plane layers of fluid parallel to the plate slide over each other and exert friction-like, viscous forces on each other. The existence of the viscous forces means that an applied horizontal force of constant magnitude F is needed to drag the upper moving plate at constant speed, much as a constant applied force is needed to drag an object along a rough surface at constant speed because of friction. Experiments indicate that for many fluids, the magnitude F of the applied force needed to drag the moving plate at constant speed v is proportional to the speed and to the area of the plate but inversely proportional to the separation of the plates as long as the separation distance Δy is small. This empirical dependence is written as

$$F = \eta \frac{A \, \Delta v}{\Delta y}$$

where $\Delta v = v - 0 \text{ m/s} = v$. The proportionality constant η depends on the particular fluid used and is called the coefficient of viscosity of the fluid. Thus the coefficient of viscosity η is

$$\eta = \frac{F}{A} \frac{\Delta y}{\Delta v}$$

or

$$\eta = \frac{F/A}{\Delta v/\Delta y} \qquad (11.26)$$

The numerator of Equation 11.26 is the shear stress applied to the fluid, while the denominator, the variation in the speed of the fluid over a small distance measured perpendicular to the direction of the flow, is called the speed gradient.[†]

Inspecting Equation 11.26, the SI units for the coefficient of viscosity[‡] are $(\text{N/m}^2)/[(\text{m/s})/\text{m}] = \text{N·s/m}^2 = \text{Pa·s}$. Table 11.3 indicates the value of the coefficient of viscosity for various fluids under various conditions.

11.14 VISCOUS FLOW*

When a fluid flows through a long horizontal pipe of constant cross-sectional area, as in Figure 11.47, Bernoulli's equation (Equation 11.24) implies that pressure is constant along the length of the pipe. For real fluids, however, the pressure decreases along the direction of the flow, as shown in Figure 11.48. This has significant ramifications in the firefighting profession if long lengths of hose are used, or in providing water to a house at the end of a long supply pipe.

[†]The term gradient arises by making an analogy with a mountain slope. We say there is a steep gradient if the slope makes a considerable angle with the horizontal.

[‡]An old unit for the viscosity that still appears in some reference manuals is the poise. The conversion is

$$1 \text{ Pa·s} = 10 \text{ poise}$$

The poise unit was named after Jean Louis Marie Poiseuille (1799–1869), a French physician *and* physicist.

TABLE 11.3 Coefficient of Viscosity of Various Liquids

Liquid	Coefficient of viscosity η (Pa·s)
Benzene (20 °C)	0.65×10^{-3}
Ethanol (20 °C)	1.2×10^{-3}
Glycerin (20 °C)	1.41
Mercury (20 °C)	1.6×10^{-3}
Methanol (20 °C)	0.60×10^{-3}
Molasses (20 °C)	~ 100
Water (0 °C)	1.8×10^{-3}
Water (20 °C)	1.0×10^{-3}
Water (100 °C)	0.28×10^{-3}
Glacial ice	12×10^{11}
Blood (whole) (37 °C) [average values]*	
Newborns	5.5×10^{-3}
Men	4.6×10^{-3}
Women	4.0×10^{-3}
Sweat*	0.93×10^{-3}
Tears*	1.3×10^{-3}
SAE 10 motor oil (30 °C)	200×10^{-3}

Geigy Scientific Tables (1981); all other data from *Handbook of Chemistry and Physics* (34th edition).

The pressure drop occurs because of *viscous forces*, exerted by adjacent layers of the fluid on each other as they rub or flow past each other. The flow is not uniform across the cross section of the pipe. For a fluid confined to a pipe, the fluid flows with greatest speed at the center of the pipe and essentially zero speed at the walls. Fluid flow in a straight river is similar: the current is greatest at midstream and decreases as you approach the banks.

Experimental observations indicate that the absolute value of the pressure drop $|\Delta P|$ along the direction of flow is proportional to the volume flow rate Q (in m^3/s), the length ℓ of the pipe, and the coefficient of viscosity η of the fluid and further depends the shape of the cross section of the pipe:

$$|\Delta P| = KQ\eta\ell$$

where K is a geometrical factor depending on the cross-sectional shape of the pipe.

For circular pipes, Jean Louie Marie Poiseuille was able to show that[§]

$$|\Delta P| = \frac{8}{\pi r^4} Q\eta\ell$$

now known as **Poiseuille's law** for simple fluids undergoing steady, nonturbulent flow and whose viscosity does not depend on the speed of the flow. The very strong dependence on the radius of the pipe has implications in medicine. If the radius of an artery gradually decreases with time due to arteriosclerosis, the absolute value of the pressure drop along an artery increases dramatically. To maintain the same flow rate the heart (a biological pump) must do more work per unit time (power), a condition that may result in heart failure.

[§]A derivation of this relationship can be found in Leonard Leyton, *Fluid Behavior in Biological Systems* (Oxford, 1975), pages 198–199.

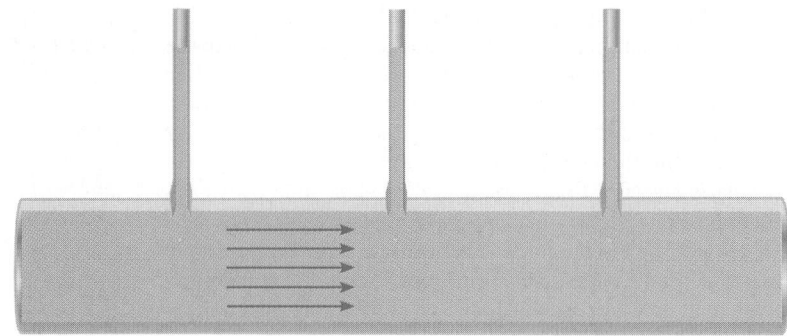

FIGURE 11.47 For an ideal fluid, the pressure is constant along the length of a pipe.

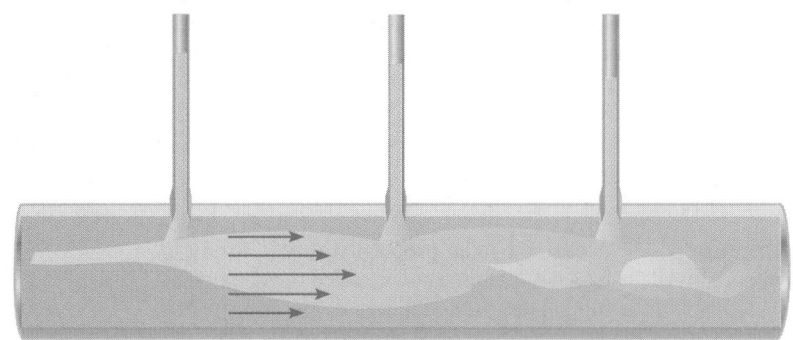

FIGURE 11.48 The pressure of a real fluid decreases gradually along the length of a pipe.

In fact blood is not a simple fluid of constant viscosity. Blood consists of a variety of distinct ingredients, among them biconcave disklike red blood cells (erythrocytes, approximately 8 μm in diameter and 2 μm thick) that make up about 50% of the total volume of the blood. The red blood cells are suspended in a yellowish-clear, cell-free plasma fluid.* The red

_____

*Blood plasma is *not* the same as the state of matter a physicist calls a plasma.

blood cells are randomly oriented at low speeds but the planes of the disks line up at higher flow speeds. As a result the viscosity of blood at higher speeds is smaller than at lower speeds. Flow speeds in the human circulatory system range from about 0.1 m/s in major arteries (1–10 mm in diameter) down to about 10^{-3} m/s in capillaries (about 8 μm in diameter). Other blood constituents, such as the white blood cells and platelets, make up only a small volume fraction and do not significantly affect the flow properties.

CHAPTER SUMMARY

The *tensile stress* applied to a rope or wire of cross-sectional area A is the magnitude of the tensile force divided by the cross-sectional area:

$$\text{tensile stress} = \frac{\text{magnitude of the tensile force}}{\text{cross-sectional area}}$$

The *tensile strain* of a rope or wire of length ℓ, when subjected to a tensile force, is the change in length $\Delta\ell$ divided by ℓ itself:

$$\text{tensile strain} = \frac{\Delta\ell}{\ell}$$

Young's modulus E for a solid material is the tensile stress divided by the tensile strain:

$$E \equiv \frac{\text{tensile stress}}{\text{tensile strain}} \qquad (11.3)$$

The *pressure P* in a fluid is the magnitude of the force per unit area on any surface in any orientation within or surrounding the fluid. The pressure is a scalar quantity whose SI unit is the *pascal* (Pa), equal to one newton per square meter (N/m²).

The pressure P at any vertical coordinate y in a static liquid of constant density ρ near the surface of the Earth is

$$P = P_0 - \rho g y \qquad (\hat{\jmath} \text{ up}) \qquad (11.14)$$

where P_0 is the pressure at the elevation where the coordinate y is chosen to be zero (the origin); g is the magnitude of the local acceleration due to gravity near the surface of the Earth (g = 9.81 m/s²).

Pascal's principle states that if the pressure is increased at some point in an enclosed liquid, the pressure increases by the same amount at all points within the liquid.

Archimedes' principle states that if a system is partially or totally submerged in a fluid, the fluid exerts an upward buoyant force on the system whose magnitude is equal to that of the weight of the fluid displaced by the system. The buoyant force acts at the *center of buoyancy*, located at the center of mass of the fluid imagined to fill the hole created in the fluid by the system.

A submerged system is in stable equilibrium if the elevation of its center of mass is below that of the center of buoyancy.

If a floating system is tilted slightly from its equilibrium position, the intersection of the new line of action of the buoyant force with the original line of action of the buoyant force is known as the *metacenter*. A floating (or submerged) system is in stable equilibrium if the elevation of the metacenter is above that of the center of mass of the system.

Surface tension forces account for capillary action.

An *ideal fluid* is nonviscous and its flow is steady and irrotational (nonturbulent). The flow rate is the product of the cross-sectional area A of a flow tube and the fluid speed v at that location. If the fluid also is incompressible, the flow rate Q of such an ideal fluid is constant:

$$A_1v_1 = A_2v_2 \qquad (11.22)$$

which is known as the *equation of flow continuity*. Such ideal incompressible flow also satisfies *Bernoulli's principle*:

$$P_1 + \frac{1}{2}\rho v_1^2 + \rho gy_1 = P_2 + \frac{1}{2}\rho v_2^2 + \rho gy_2 \quad (11.24)$$

which follows from applying the CWE theorem to the fluid in a flow tube.

The *viscosity* of a real fluid is a measure of its resistance to flow. The *coefficient of viscosity* is the shear stress applied to a fluid divided by the speed gradient (the variation of fluid speed in a direction perpendicular to its flow).

Summary of Problem-Solving Tactics

11.1 (page 497) You have the freedom to choose where to place the origin of the coordinate system when using Equation 11.14:

$$P(y) = P_0 - \rho gy \qquad (\hat{j} \text{ up}) \qquad (11.14)$$

You simply need to have $\hat{j}$ directed up. You can choose any convenient elevation for the origin. The pressure where $y = 0$ m is P_0.

Questions

1. (page 494); 2. (page 498); 3. (page 504); 4. (page 508); 5. (page 508); 6. (page 514)

7. To demonstrate the effect of increasing pressure with depth, Torricelli took a sturdy wine cask, which, when filled with water, held fast as in Figure Q.7a. He then connected a thin tube to it (as in Figure Q.7b) and filled it to a great height, bursting the cask to the astonishment of a surrounding audience, but not to Torricelli—or to you. Explain why. Why did he use water and not wine?

8. Mercury is a very toxic and hazardous material, yet barometers invariably use mercury rather than water as the working substance. What are two reasons that such water barometers are not used in practice?

9. In Example 11.4 we found that the scale height of the atmosphere measured from the surface of the Earth is 8.00 km. If you place an origin on top of Mt. Everest, is the numerical value of the scale height the same? Explain your answer.

10. Using the technique of Example 11.8, how can the density of the unknown liquid be determined if the cylinder *sinks* in the unknown liquid?

11. A rock of volume V is suspended as indicated in Figure Q.11a. The weight of the rock and its support is equal to the weight of the water and the beaker on the opposite pan of the balance. Describe what happens to the balance if the rock is lowered into the water as indicated in Figure Q.11b.

12. A swimmer floating in the Dead Sea (located between Jordan and Israel) floats higher out of the water than in the nearby Mediterranean Sea. What does this imply about the relative densities of the water in the two seas?

(a)　　　　　(b)

FIGURE Q.7

13. Typical large icebergs are made of frozen fresh water. Look up the density of freshwater ice and that of sea water in a reference such as *The Handbook of Chemistry and Physics*. Explain why approximately 90% of the mass of an iceberg lies submerged. This gives rise to the colloquial expression about seeing only the "tip of the iceberg."

14. If you make the observation that very little of the human body floats out of the water when doing a dead-man's float, what is the approximate average density of the human body?

15. Explain the operation of a perfume atomizer.

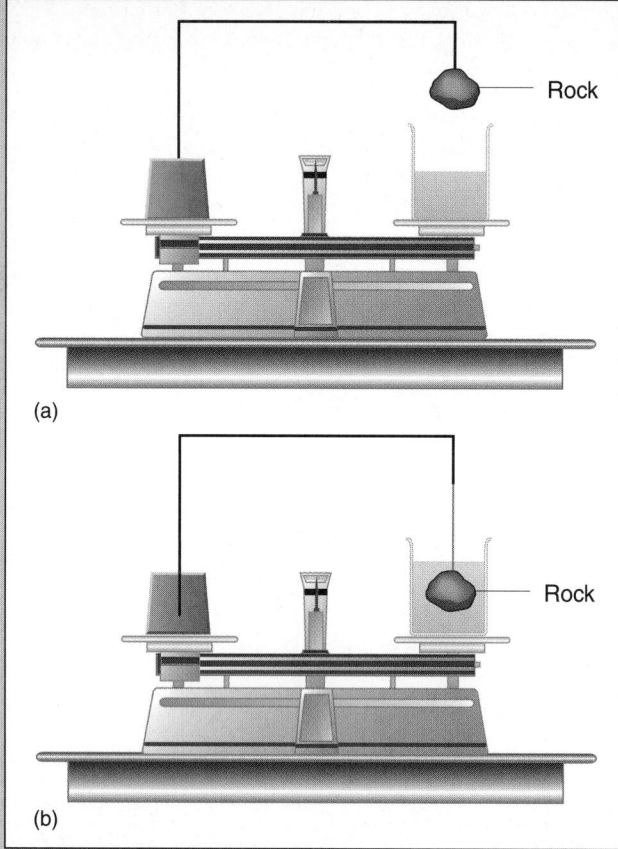

(a)

(b)

FIGURE Q.23

FIGURE Q.24

16. Explain why airplanes take off going *into* the wind rather than *with* the wind. When landing, do planes head into or with the wind?

17. Two cubes are of identical dimensions. One is made of granite and the other of lead. When submerged in water, what is the relationship between the buoyant forces on each cube?

18. To carry large panes of glass, suction cups are used to attach handles to the glass surface. Explain the operation of a suction cup.

19. A woman of mass 50 kg wears spiked high-heel shoes. Making reasonable estimates, calculate the pressure of each heel on the floor and compare it with that exerted by the feet of an elephant.

20. Explain why a ship made of steel (which is much denser than water) can float.

21. You are relaxing peacefully on a foam float or boogie board in a fraternity hot tub and note the level of the water in the tank carefully. Now you slip into the water and float beside the board. What happens to the level of the water in the tank? Explain your answer.

22. In a shower with the water on full blast, the bottom of the shower curtain is drawn slightly into the shower rather than hanging vertical (as it will when the water is off). Explain why.

23. Using a blowdryer, you can suspend a ping-pong ball in midair. Explain the physics. See Figure Q.23.

24. Tall, old farm silos were constructed of vertical timbers with steel bands wrapped around the periphery. Explain why the spacing

of the bands is greatest at the top of the silo and decreases as the bottom of the silo is approached. See Figure Q.24.

25. Is Archimedes' principle applicable in a spacecraft orbiting the Earth? Explain carefully.

26. One pail of water is filled to the brim with water. Another identical pail of the same mass has a wooden toy ship floating in it and it also is filled to the brim with water. Both pails are weighed. What is the relationship between the weights of the two pails?

27. An oil barge floats with its bottom 3.0 m below the surface of a calm ocean. As the barge is towed from the ocean up a large river, will the barge float at the same level, higher in the water, or lower in the water? Explain.

28. After your election to the prestigious ΦBK academic honor society, your mother shows you her cherished ΦBK key. She cannot remember if the key is solid gold or gold-plated steel. With your growing knowledge of physics, describe an experiment to determine from which material the key was made.

29. A smoky fire is burning in a fireplace. Explain why the smoke is drawn up the chimney better if there is a wind blowing horizontally across the top of the chimney.

30. Astronomy books often state (ridiculously) that if there were an ocean big enough (there really isn't), the planet Saturn

could float in it. What does this imply about the density of Saturn?

31. Explain why logs float with their long axis horizontal rather than vertical.

32. Under what circumstances will a damaged and sinking ship capsize before sinking?

33. Why do sailboats and other vessels have massive keels?

34. Kayaks tip over easily; rowboats do not. Explain why.

35. Modern aircraft carriers look decidedly top heavy, yet do not capsize even in heavy seas. Explain how this is accomplished. See Figure Q.35.

FIGURE Q.35

36. Describe a procedure and the equipment needed to determine the volume of an irregularly shaped object by immersing it in a liquid. What assumption must be made about the object for the procedure to be valid?

37. Examine a hydraulic car jack (common at many service stations) and explain the physics associated with its operation.

38. When pitching curve balls in baseball, does the pitcher give the ball a spin about a horizontal or vertical spin axis? Explain your answer.

39. When hitting golf balls, is it desirable to give the ball a spin around a vertical or a horizontal axis? Does this give greater range to the ball?

40. Is Equation 11.14,

$$P = P_0 - \rho g y$$

valid in an orbiting spacecraft with a suitably adjusted value for g? Explain your answer.

41. Describe an experimental procedure that will determine whether the crown of King Hieron II was made of pure gold.

42. A cylindrical steel barrel is of diameter 0.50 m and height 1.0 m. Describe how you could cause the air around the barrel to collapse it.

43. A jumbo jet is cruising at an altitude of 10.7 km. The cabin pressure is that at sea level. Is there a buoyant force on the jet? Explain.

44. Explain the physics involved in the operation of a water pistol.

45. Why is it so easy to get a painful paper cut?

46. What principle is involved in squeezing the toothpaste out of a tube?

47. Take a pin and insert it through the center of an index card. Insert the pin in the end of a spool of thread as shown in Figure Q.47. The purpose of the pin is simply to keep the index card centered on the spool. Now blow hard through the opposite end of the spool. Explain why the index card is *not* blown off the other end; indeed, the harder you blow, the more difficult it is to move the index card from the other end.

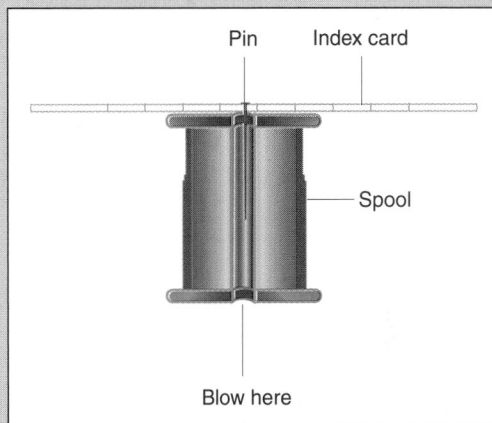

FIGURE Q.47

48. Cork, of course, floats easily on water. A shipment of 10^6 corks is loaded into a freighter for transport to a winery. Does the boat float higher or lower in the water as a result of its highly buoyant cargo? Explain the reasoning behind your answer.

49. If the value of the local acceleration due to gravity g is increased, will the depth to which a ship sinks into the water before floating also change? If so, will the ship sink to a greater or lesser depth? Justify your answer.

50. Explain why the canvas top of a speeding convertible bulges outward.

51. Because of the viscosity of air, spinning baseballs, tennis balls, golf balls, and Ping-Pong balls drag some of the air close to the ball around with it. Left-handed and right-handed pitchers spin the ball in different directions around a vertical axis for a curve ball. What direction is the angular velocity vector of the ball for each pitcher? Which way will the ball curve for each pitcher as the pitcher looks at home plate? If a golf shot hooks to the right, discuss the likely direction of the spin angular velocity vector of the ball.

52. Blood consists of fluid plasma in which various cells are carried. Blood near the walls has a smaller speed than blood near the center of an artery or vein. Explain why this speed profile tends to push large red blood cells toward the center of the arteries and veins.

53. After you wax your car, and it inevitably rains within the hour, do the water droplets on it indicate that cohesive or adhesive forces dominate? What about when there is no wax on your car finish?

PROBLEMS

Sections 11.1 States of Matter
 11.2 Stress, Strain, and Young's Modulus for Solids

1. A vertical steel wire with a diameter of 1.0 mm is 2.00 m long. What mass should be hung from the wire to stretch it 1.0 mm?

2. A stool in a diner is supported by a solid steel column of diameter 4.0 cm and length 65 cm. When a 90 kg trucker sits on the chair with his feet off the floor, determine the compression of the column. Neglect the mass of the seat of the stool.

3. A swing is suspended by means of two nylon ropes from a tree limb as indicated in Figure P.3. The ropes have a diameter of 1.0 cm and are each 8.0 m long. When a passing 25 kg tyke gets on the swing, the ropes stretch 1.0 cm. Use this information to find Young's modulus for the nylon ropes.

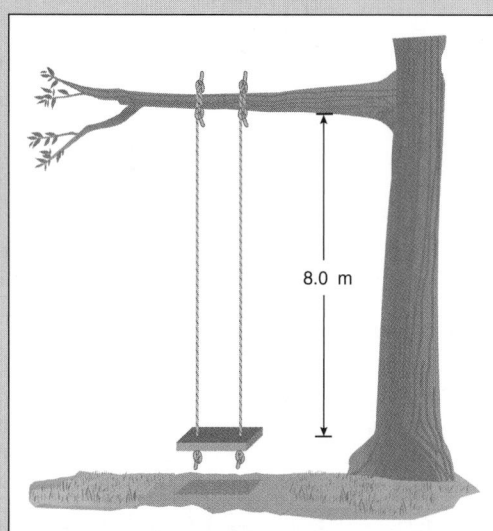

8.0 m

FIGURE P.3

4. A small grouper fish (*Epinephelus striatus*) of mass 2.5 kg hangs from a vertical 1.50 m length of fishing line and stretches it 1.2 cm. The diameter of the line is 0.50 mm. What is Young's modulus for the line?

•5. A 2.00 m long bungee cord of 8.0 mm diameter stretches 250 cm when an 80 kg college-level turkey is suspended from a hook on one end of the cord. Find Young's modulus for the cord.

•6. The well-known civil engineering firm of Rivers, Rhodes, and Waters is designing a walkway that is to be suspended by means of four steel rods of diameter 2.0 cm and length 3.00 m. The stretch of the rods is not to exceed 0.50 cm under any circumstances. What is the maximum additional mass that the 1000 kg walkway can carry to meet the design specifications? Assume that the load is uniformly distributed over the walkway so that each rod carries an equal share of the load.

•7. A 1.00 m³ volume of water is subjected to a pressure of 1.08×10^8 Pa. What is the change in the volume of the water, expressed in cm³? This pressure is approximately the pressure at the deepest part of the ocean in the Mariana Trench (approximately 11 km deep) off the southeastern coast of Japan.

•8. What pressure increase will decrease the volume of a sample of water by 1.0%?

ⵥ9. A mass m is suspended from a fixed support by means of a wire of diameter d and length ℓ made from material whose Young's modulus is E. Show that the frequency v of small vertical oscillations is

$$v = \left(\frac{Ed^2}{16\pi\ell m} \right)^{1/2}$$

Sections 11.3 Fluid Pressure
 11.4 Static Fluids
 11.5 Pascal's Principle

10. (a) What is the pressure at the surface of a swimming pool? (b) What is the pressure at the bottom of a filled swimming pool with a depth of 2.50 m? (c) What is the gauge pressure at the top and at the bottom?

11. At what depth does a diver experience a gauge pressure of 1.00 atm?

12. The Mariana Trench in the Pacific Ocean off the southeastern coast of Japan has water to approximately 11 km deep. Calculate the pressure in atm at the bottom of the trench.

•13. Assume your mass is 60.0 kg and your two bare footprints have a total area of about 4.0×10^{-2} m². How high a water column produces a gauge pressure on the floor equal to the pressure you exert on the floor?

•14. Normal atmospheric pressure is 1.00 atm = 760 mm Hg. A meteorologist finds that a storm decreases the pressure by 25 mm Hg. What is the atmospheric pressure inside the storm in pascals?

•15. A vertical water pipe extends from the bottom to the top of the World Trade Center in the Big Apple (New York City), a distance of about 420 m. Calculate the difference in pressure in atm between the bottom and the top of the water pipe.

•16. Hydraulic equipment makes use of Pascal's principle to create what is called a hydraulic lever, to magnify or increase forces; see Figure P.16. Use the CWE theorem to show that if the place where a force of constant magnitude F_1 is applied moves slowly through a distance d_1, the place where force F_2 (also of constant magnitude) is applied moves through a distance d_2 given by

$$d_2 = \frac{F_1}{F_2} d_1$$

Assume the fluid is incompressible.

•17. Imagine the atmosphere of the Earth to be incompressible and therefore of constant density (similar to a liquid) equal to its value at sea level (about 1.30 kg/m³). (a) What total depth of atmosphere would yield a pressure at sea level equal to the present value of 1.00 atm? (b) Mt. Everest has a height of about 8.85 km. Would Mt. Everest poke through this uniform atmospheric blanket? If so, by how much?

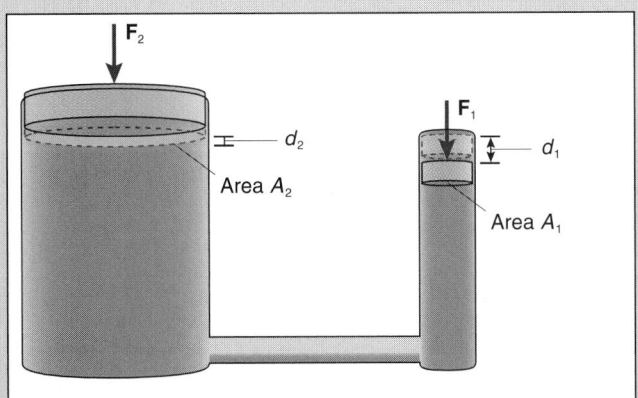

FIGURE P.16

•**18.** With a quantitative calculation, explain why even with the best efforts, it is not possible to use a soda straw longer than about 10.3 m for sipping water. How is this length determined?

•**19.** In 1654 Otto von Guericke, the inventor of the air pump, made a spectacular demonstration of his new device before the Emperor Ferdinand III. Von Guericke took two hollow hemispheres of copper that could be fitted snugly to a gasket between them (see Figure P.19). He then used his air pump to evacuate the air from inside the sphere formed from the two hemispheres so that the atmospheric pressure held them together, preventing them from falling apart. Thirty horses, fifteen pulling on each hemisphere, were unable to tear the hemispheres apart. Assume the hemispheres used were 1.00 m in diameter and that Guericke was able to completely evacuate the interior. (a) Find the magnitude of the force of the atmosphere on each hemisphere. (b) What maximum pull per horse would still be incapable of separating

the hemispheres? Hemispheres that can be so evacuated now are known as *Magdeburg hemispheres* after the town of von Guericke's birth and where he was burgomaster. Von Guericke's invention of the air pump was such a sensation that it is now immortalized as a constellation called *Antlia*, or Air Pump, in the southern celestial hemisphere.

•**20.** You and five classmates are drinking by means of straws from a large tankard of a beverage derived from hops and other assorted spices and ingredients. The beverage has essentially the density of water. The level of liquid in the various straws is as indicated in Figure P.20. The levels are stationary and the liquid is not in motion. The levels of liquid in the straws are lifted or depressed the same distance d relative to the open liquid surface. The distance d corresponds to a pressure difference of 1.00×10^{-2} atm. (a) Calculate the value of d. (b) The pressures in atmospheres at various locations are denoted by the symbols P_1, P_2, P_3, and so forth. Indicate what the following pressures or pressure differences are (in units of atmospheres); the *signs* are important:

$$P_1 =$$
$$P_2 - P_1 =$$
$$P_3 =$$
$$P_5 - P_4 =$$
$$P_5 - P_6 =$$
$$P_7 - P_1 =$$
$$P_8 - P_1 =$$

•**21.** A hydraulic jack is filled with an oil of density 0.80 times that of water ($\rho_{water} = 1.00 \times 10^3$ kg/m^3). The input piston has a cross-sectional area that is 1/10 that of the output piston; see Figure P.21. In its operation: (a) Calculate the ratio of the pressures at the output and input pistons. (b) Calculate the ratio of the magnitudes of the forces at the output and input pistons.

FIGURE P.19 A drawing of the Guericke demonstration. It depicts an incorrect number of horses pulling on each hemisphere.

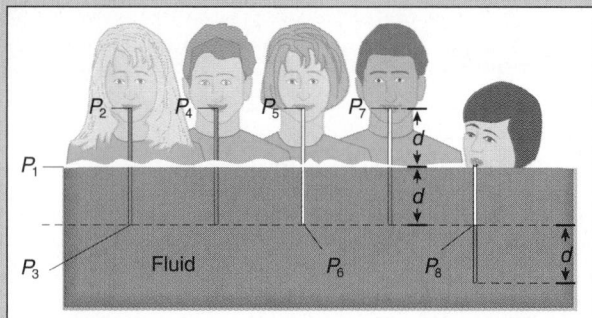

FIGURE P.20

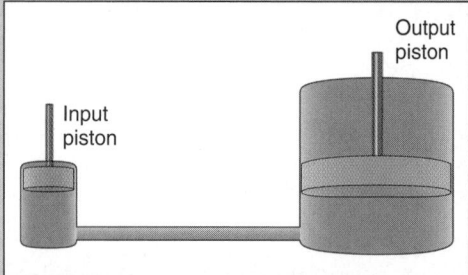

FIGURE P.21

•**22.** The gauge pressure of the air in an automobile tire is 2.08×10^5 Pa.[†] The tire supports one-fourth of the weight of a car of mass 1.50×10^3 kg. What is the area of the tire that is in contact with the road? This area is called the footprint of the tire. As a tire goes flat, explain what must happen to the footprint of the tire.

•**23.** A barber chair is supported by a piston that is 6.0 cm in diameter. The chair and its occupant have a total mass of 100 kg. (a) What pressure is required to raise the chair and its occupant? (b) If the pressure is produced by a plunger whose diameter is 1.5 cm, what force is needed to accomplish the task?

•**24.** A cylindrical steel barrel has a diameter of 0.50 m and a height of 1.00 m. Calculate the magnitude of the maximum force that air pressure can exert on the entire barrel from the outside.

•**25.** The Italian scientist Evangelista Torricelli originally made a barometric device such as examined in Example 11.3 with water instead of mercury. What height of water column corresponds to one atmosphere pressure?

‡**26.** A right-circular cone of radius R and height h (see Figure P.26) is filled with a fluid whose density is ρ. The top of the cone is exposed to the atmosphere. (a) Determine the magnitude of the weight of the fluid in the cone. (b) What is the gauge pressure at the bottom of the fluid? (c) What is the magnitude of the force of the fluid on the container's bottom caused by the gauge pressure? By how much is this greater than the magnitude of the weight of the fluid in the cone? Explain and resolve this paradoxical result, first noted by Simon Stevin (1548–1620). This problem was inspired by the article by Alpha E. Wilson, "The hydrostatic paradox," *The Physics Teacher*, 33, #8, pages 538–539 (November 1995).

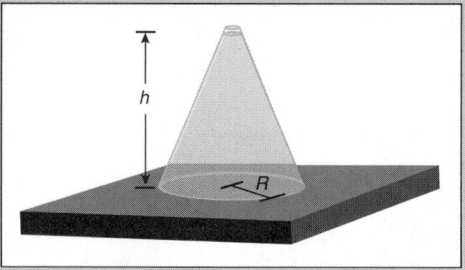

FIGURE P.26

Sections 11.6 Archimedes' Principle
11.7 The Center of Buoyancy*

27. A 5.0 kg stone with a density of 5.3×10^3 kg/m³ is placed on a scale when totally submerged in water. What does the scale indicate (in kg)?

28. A boulder of granite of mass 200 kg is weighed in air and then when submerged in a pool of water. What is the difference in the readings of the scale used to perform the measurements? ($\rho_{\text{granite}} = 2.7 \times 10^3$ kg/m³.)

29. A cube of wood 15.0 cm on an edge floats in water (density 1.00×10^3 kg/m³) with 2.5 cm of a vertical edge above the surface. (a) Draw a second law force diagram indicating all the forces acting on the cube. (b) Determine the density of the wood.

30. A block with a density of 550 kg/m³ and a volume of 0.600 m³ is held completely under water by means of a rope. Determine the magnitude of the force of the cord on the block.

•**31.** What is the magnitude of the buoyant force on a cube of iron 0.100 m on a side when it is totally submerged in a pool of mercury? The density of iron is 7.9×10^3 kg/m³ and that of mercury is 13.6×10^3 kg/m³.

•**32.** A goldsmith alleges that a beautiful bracelet is made of links of pure gold. The mass of the bracelet is easily determined to be 85.00 g. You are a bit suspicious of the pure gold claim and decide to check it out. Quite by chance you happen to have a small graduated cylinder in your briefcase. By submerging the bracelet in water, you determine the bracelet has a volume of 6.1 cm³. Should you patronize the merchant or call the fraud squad? Explain the reasoning that leads to your conclusion. (Gold has a density of 19.3×10^3 kg/m³.)

•33. An olive is perched atop a cubical ice cube 3.0 cm on a side so that the cube barely floats in a martini; see Figure P.33. Consider the martini to be quite dilute (essentially water). What is the mass of the olive?

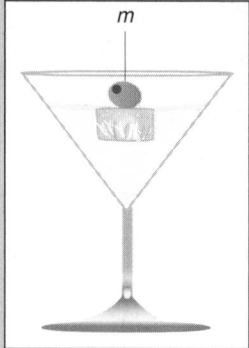

FIGURE P.33

[†]This corresponds to 30 pounds per square inch (psi). Common tire pressure gauges in the United States indicate the gauge pressure in pounds per square inch.

•**34.** An object of density ρ ($> \rho_{water}$) and volume V contains a cavity of air of volume V'. Volume V does *not* include V'; see Figure P.34. How big must V' be (in terms of V, ρ_{water}, and g) so the object barely floats in water? This is the principle underlying the swim bladders of many fish.

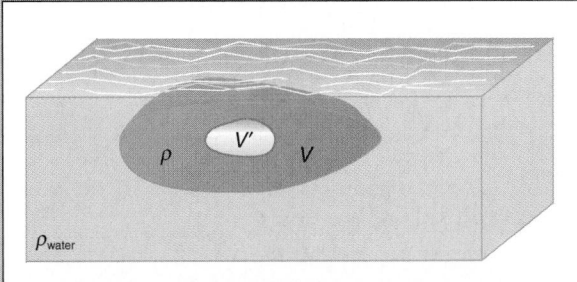

FIGURE P.34

•**35.** The 40.0 kg steel ball of a Foucault pendulum is suspended in air as shown in Figure P.35. (a) Draw a second law force diagram indicating the forces on the ball. (b) Determine the magnitude of the tension in each cord. (c) If the system were submerged in water, draw a second law force diagram indicating the forces on the ball and determine the tension in each cord. Do you need to know the density of steel to do this calculation?

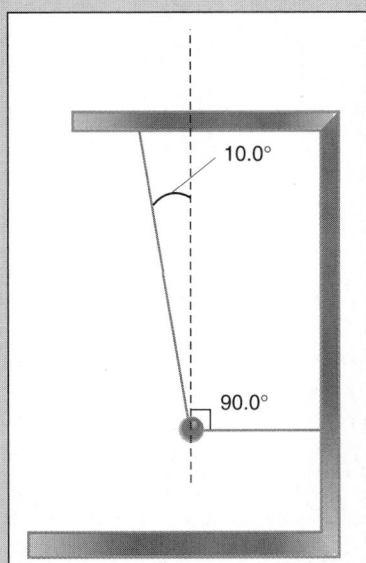

FIGURE P.35

•**36.** Model a prone human being as a rectangular box of dimensions 15 cm × 20 cm × 180 cm with the density of water, as indicated in Figure P.36a. (a) If the model then is placed in a huge pool of mercury, to what depth will the model sink? This may give you an indication of the extent to which you can float on mercury rather than on water. (b) If the model is placed in the mercury pool in the orientation indicated in Figure P.36b, to what depth will the model sink?

•37. A rectangular barge has a depth of 1.00 m, a floor area of 40.0 m², and a mass of 1.00×10^4 kg. (a) Will the barge float on water? Prove it. (b) If the barge will float, what is the maximum mass of cargo that the barge can carry (barely) without sinking?

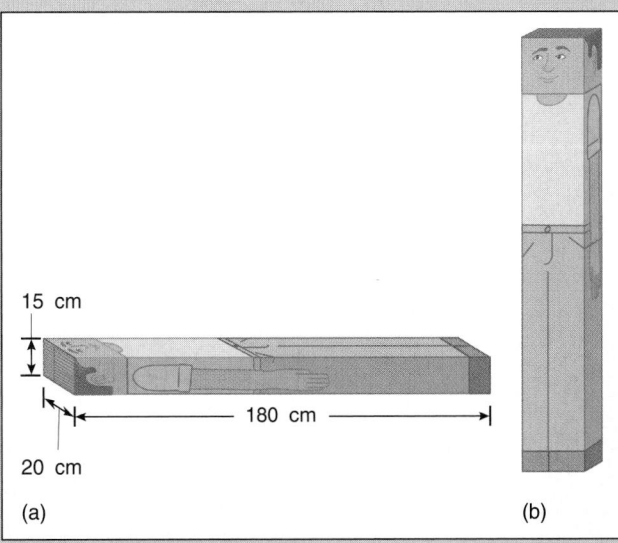

FIGURE P.36

•**38.** One way of moving a large undersea rock or mooring is to fasten a closed, empty oil drum to the rock while at low tide as in Figure P.38a. As the tide comes in, the drum is drawn down into the water (see Figure P.38b), increasing the buoyant force acting on the system of the rock and drum. If the drum is of sufficient size (or if multiple drums are used), they are capable of lifting the rock or mooring, so that it can easily be floated to another location and released. Suppose a cylindrical drum of radius 25 cm, length 100 cm, and mass 20 kg is used. What is the maximum mass of a rock (of density 5.0×10^3 kg/m³) that the drum can lift from the bottom? Assume the density of sea water is the same as that of pure water.

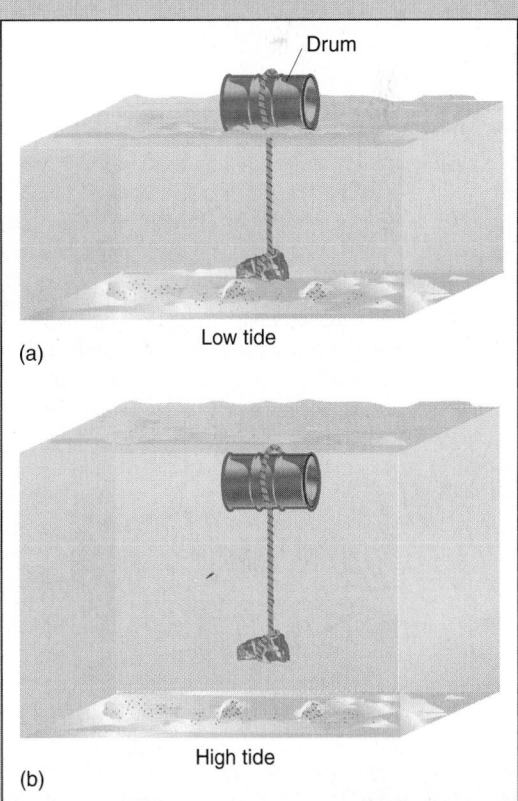

FIGURE P.38

•**39.** A solid student in the shape of a rectangular parallelopiped of dimensions 30.0 cm × 20.0 cm × 2.000 m floats horizontally in a calm hot tub filled with warm water; 3.00 cm of the parallelopiped (along the 20.0 cm side) is above the water surface. Find the average density of the student.

•**40.** A cubical block of a solid material is suspended from a spring balance over a vat of water as indicated in Figure P.40a. The block and scale are gradually lowered into the vat until the block is completely submerged. The scale reading as a function of distance x is graphed in Figure P.40b. (a) What is the magnitude of the weight of the cube in air? (b) How far was the block lowered before it came into contact with the water? (c) What is the length of a side of the cube? (d) What is the density of the cube in kg/m^3? (e) What is the maximum magnitude of the buoyant force on the cube? (f) What is the magnitude of the weight of the water displaced by the object?

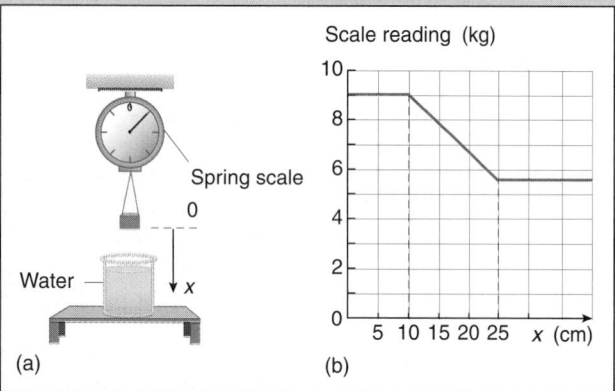

FIGURE P.40

•41. A cork of volume 3.00×10^{-5} m^3 and density 120 kg/m^3 floats on water. No surprises here. (a) Sketch a second law force diagram indicating the forces acting on the cork as it floats. (b) What fraction of the volume of the cork is submerged as it floats? (c) The cork now is tied, using an ideal string, to a suspended piece of aluminum (of density 2.70×10^3 kg/m^3) of such mass that the cork is just completely submerged without sinking. What mass of aluminum is needed? (d) What is the magnitude of the tension in the string connecting the cork and aluminum when the cork is just barely completely submerged?

•**42.** Archimedes, prove your stuff. A royal crown (cola?) of pure gold is suspended in air by a thread and the tension in the thread is of magnitude T. When the crown is totally immersed in water, the magnitude of the tension in the thread is $0.872T$. (a) Sketch a second law force diagram indicating the forces acting on the crown when suspended in air. (b) Sketch a second law force diagram indicating the forces acting on the crown when suspended in water. (c) Is the crown made of pure gold or of another metal with just a gold plating? (The density of pure gold is 19.3×10^3 kg/m^3.) Justify your answer.

•**43.** You are strapped to a large floating cushion, enjoying the pool at a luxury resort in Tahiti. The float is submerged to half its depth. A friend (?) approaches from underneath and playfully upsets the cushion, flipping you and it over. You are still strapped to the cushion and now are in danger of drowning. Is the float still submerged to half, less than half, or more than half its thickness? Indicate your reasoning.

•**44.** A sample of pure gold is found to have a scale weight when submerged in water that is 95% of its weight in air. From this information, calculate the ratio of the density of gold to the density of water (this ratio is called the *specific gravity* of the substance).

•45. Liquid mercury has a density of 13.6×10^3 kg/m^3. A heavy metal band floats on mercury with 80% of its volume above the surface. Find the average density of the heavy metal band.

•**46.** A huge, 8.00 m long, uniform log of wood (density 700 kg/m^3) of mass 1000 kg is forced to float vertically in a clear mountain lake because of a massive iron ball attached to one end. Only 40 cm of the log remains above the surface of the lake. What mass of iron is attached to the end?

•**47.** You are idly fishing while sitting on the bank of a river. An empty soda can floats by and you decide to practice physics. The can has a mass of 20.0 g and a volume of 200 cm^3. You intend to just barely sink the can by adding an appropriate mass of sand while the can has the orientation shown in Figure P.47. (a) Is it necessary to know the density of sand to determine the mass of sand to add? Explain. (b) Find the minimum mass of sand needed to sink the can, without tipping it over. Recover the can so you don't litter!

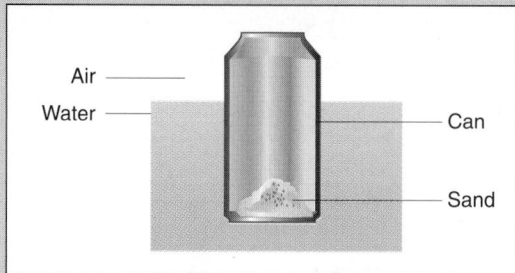

FIGURE P.47

•**48.** You (mass 70.0 kg) decide to take off—literally—with a helium balloon of mass 30.0 kg. The densities of air and helium are $\rho_{air} = 1.321$ kg/m^3 and $\rho_{helium} = 0.179$ kg/m^3. What volume of helium is needed to levitate you and the balloon? Is it reasonable to neglect your volume in making this calculation?

•49. A conical waxed paper drinking cup of negligible mass has a diameter equal to its height (see Figure P.49). The cup holds 0.040 kg of beach sand of density 2.00×10^3 kg/m^3. How deep will the conical cup float in water?

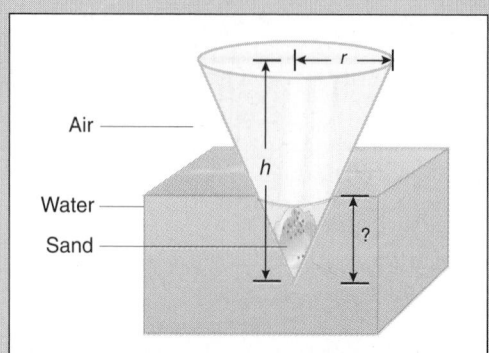

FIGURE P.49

50. A cube of wood (of density ρ) with horizontal and vertical edges of length ℓ is floating on water of density ρ_{water}. (a) Show that the cube sinks to a depth x_0 given by

$$x_0 = \frac{\rho}{\rho_{\text{water}}}\ell$$

(b) An additional downward force of magnitude F now is applied that sinks the cube an *additional* distance x. Show that the magnitude of the force F is

$$F = \rho_{\text{water}}\ell^2 gx$$

Notice that the magnitude of F is proportional to the additional distance x. (c) Show that if the applied force is removed, the cube will begin simple harmonic oscillation with a period T equal to

$$T = 2\pi\left(\frac{\rho}{\rho_{\text{water}}}\frac{\ell}{g}\right)^{1/2}$$

Sections 11.8 Surface Tension*
11.9 Capillary Action*

51. A water strider (*Hygrotrechus conformis*) has a mass of 1.00 g and six legs. Assume the legs are identical. For each leg, what is the minimum length of leg that must be in contact with a water surface so the insect can walk on water? Such water striders actually have hairs on the ends of their legs that significantly increase the length actually in contact with the water surface.

52. Humans have two feet, of course. Measure the approximate perimeter of one foot. If your mass is 60.0 kg, what would the perimeter of your foot have to be if you could just barely stand on water because of surface tension? This is a rather large shoe size!

53. Measure the inside diameter of a thin drinking straw. Place the straw vertically in a glass of water and measure the height to which the water is drawn by capillary action. Use this data to calculate the surface tension of water.

54. A pipette of uniform inside diameter of 0.500 mm is inserted into a beaker of water. To what height is the water drawn by capillary action?

55. A water surface is located 2.0 cm below the rim of a cup. What is the inside diameter of a circular tube that can be inserted into the water so that water is drawn by capillary action to the same height as the rim of the cup?

Sections 11.10 Fluid Dynamics: Ideal Fluids
11.11 Equation of Flow Continuity
11.12 Bernoulli's Principle for Incompressible Ideal Fluids

56. To quench a ravenous thirst, a notorious desperado shoots a hole 2.00 m below the level of the water in a tank as indicated in Figure P.56. What is the speed of the water as it emerges from the tank?

57. A wind is blowing parallel to one face of a low building as indicated in Figure P.57. How large a wind speed is needed to cause a pressure difference of 10% between the inside and outside of the building? Such pressure differences are the

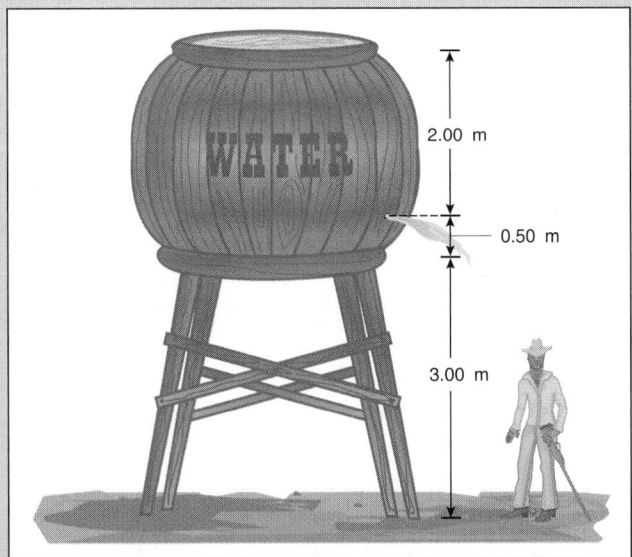

FIGURE P.56

principal cause of the damage to buildings during tornados; the buildings literally explode because of the pressure difference between the inside and outside!

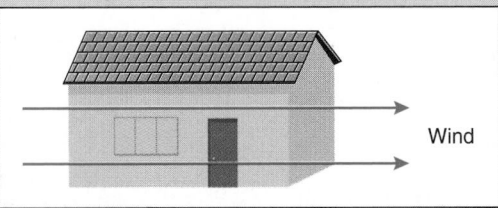

FIGURE P.57

58. Water flows at speed 1.0 m/s through a level pipe with a circular cross-sectional area of 10 cm²; the water continues into a pipe of twice the diameter as shown in Figure P.58. Assume the water is an ideal fluid. (a) What is the volume flow rate expressed in liters/s? (b) What is the pressure difference between the two sections of the pipe? (c) Which section of pipe has the fluid at higher pressure?

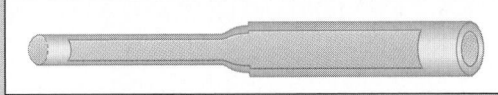

FIGURE P.58

59. Two horizontal pipes of diameters 4.00 cm and 0.500 cm are coupled as indicated in Figure P.59. A pail collects the outflow and you measure 5.0 liters = 5.0×10^{-3} m³ of water flowing out of the pipe every minute. (a) Determine the speed of the flow in each pipe. (b) What is the difference in pressure of the fluid in the two pipes? (c) Which pipe has the fluid at higher pressure?

60. The pressure in a uniform pipe of flowing water is enough to hold up 4.00 m of water in an open, vertical tube as shown in Figure P.60. Assume that water is an ideal fluid. The pipe leads 3.00 m upstairs and empties into the open air through an

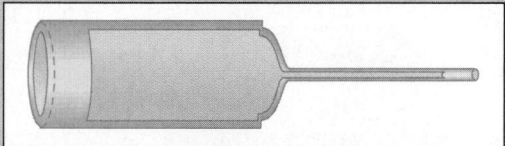

FIGURE P.59

opening with an area half that of the uniform pipe. (a) At the exit opening of the pipe, what is the pressure? (b) What is the pressure at the point where the vertical tube is connected to the pipe? (c) What is the exit speed of the water from the pipe?

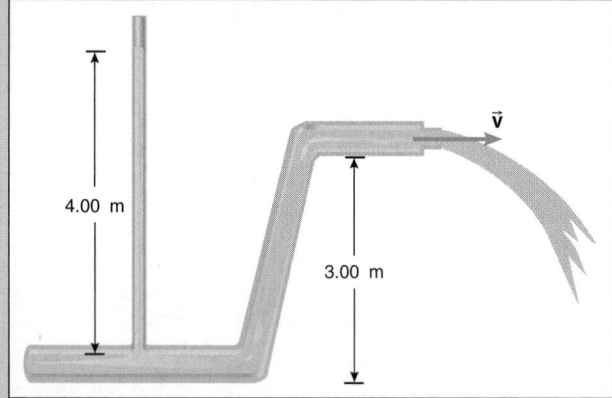

FIGURE P.60

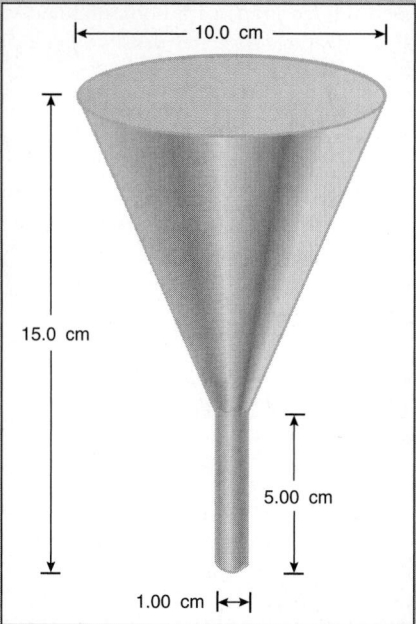

FIGURE P.62

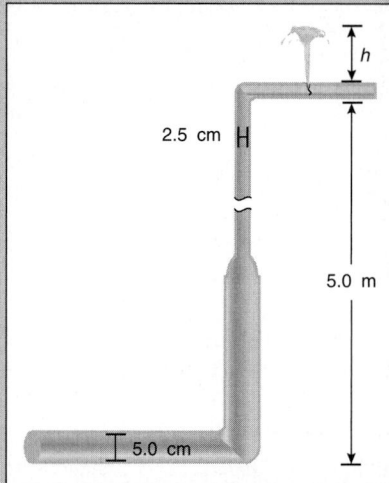

FIGURE P.63

•61. Beer is flowing in a brewery through a horizontal pipe of circular cross section and (inside) diameter 5.0 cm. The pressure in this pipe is 9.60×10^5 Pa. The pipe tapers to (inside) diameter 2.0 cm and pressure 5.50×10^5 Pa. What is the volume flow rate in the pipe in m³/s? Assume the density of beer is that of water (1.00×10^3 kg/m³).

•62. The funnel in Figure P.62 is filled with water. You initially have an index finger covering the outlet of the funnel to prevent the water from draining out. Assume that water is an ideal fluid. (a) What is the gauge pressure at the bottom of the funnel? Now remove your thumb, so water begins to drain out of the funnel. (b) What is the ratio of the speed of the fluid exiting the funnel to the downward speed of the surface of the fluid at the top of the funnel? (c) What is the initial exit speed of the fluid from the funnel? (d) What is the initial rate at which the water exits the funnel when the funnel is full? Express your result in liters/s.

•63. The plumbing contractor Flood, Leeks, and Drain uses a 5.0 cm diameter, high-pressure pipe to supply a building. The pressure inside the pipe is 6.0 atm and the speed of the water flowing past the inlet valve is 3.00 m/s. On a floor 5.0 m above the inlet pipe, the single pipe has only a 2.5 cm diameter; see Figure P.63. Assume that water is an ideal fluid. (a) What is the speed of the water in the smaller pipe? (b) What is the pressure in the smaller pipe at the 5.0 m height? (c) If the smaller pipe ruptures, propelling water vertically upward as in the figure, to what further height *h* will the water rise? Neglect air resistance.

•64. Just after its completion, the glass-walled skyscraper known as the John Hancock tower in Boston (see Figure P.64) was faced with an embarrassing problem: its windows kept falling

out on windy days! The building soon became known as the "plywood palace," since at one point plywood covered a significant percentage of the exterior of the building as a temporary replacement for the lost windows. Needless to say, the 2.00 m × 3.00 m glass panes were quite a safety hazard as they cascaded to the pavement below. Fortunately, no casualties resulted from the flying panes. If a wind with a speed of 60.0 km/h is blowing parallel to the face of one of the windows, what is the magnitude of the net outward force on each window caused by the pressure difference between the inside and outside of the building? Compare the magnitude of this force to the magnitude of the weight of a 70.0 kg insurance salesperson.

•65. The architectural firm of Steele, Woods, and Glass is designing a decorative fountain for the new terminal building of U.S. Snarelines at Kennedy International Airport in New York. The fountain is fed by a water pipe of 10.0 cm inside diameter; water

FIGURE P.64

will be projected vertically as indicated in Figure P.65. Assume that water is an ideal fluid. (a) At what speed must the water be projected from the pipe to reach a height of 20.0 m? (b) How many liters per second of water will emerge from the pipe if it is fed directly into the air with no nozzle? (c) If the fountain is gravity-fed from the indicated elevated, open reservoir, what must be the altitude of the top of the water level in the reservoir measured with respect to the nozzle of the fountain?

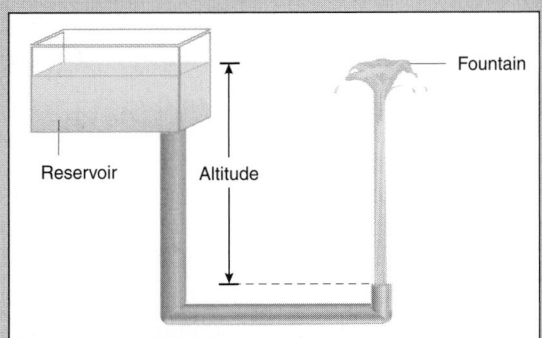

FIGURE P.65

•**66.** The diameter of the aorta emerging from the human heart is approximately 2.0 cm. Assume the aorta is of circular cross section and that blood is an ideal fluid. The heart pumps about 5.0 liters of blood per minute at an average pressure of 100 mm Hg (= 1.33×10^4 Pa). (a) Determine the approximate speed of the blood in the aorta. (b) The blood is distributed to billions of capillaries in the circulatory system. Assume the average diameter of a capillary is about 7.0 μm. If there are, say, 5×10^9 capillaries, what is the average speed of the blood in the capillaries?

•**67.** A Pitot tube can be used to determine the speed of an airplane relative to the surrounding air. The device consists of a U-shaped tube containing a fluid with a density ρ; see Figure P.67. One end of the U-shaped tube (A) opens to the air at the side of the plane; the other end (B) is open to the air in the direction

the plane is flying so the air is stagnant (at zero speed) at this location. Show that the speed of the plane can be expressed in terms of the difference in the heights h of the fluid in the tube as

$$v = \left(2gh \, \frac{\rho}{\rho_{\text{air}}} \right)^{1/2}$$

where ρ_{air} is the density of air. Why will this not serve to measure the speed of a satellite in Earth orbit?

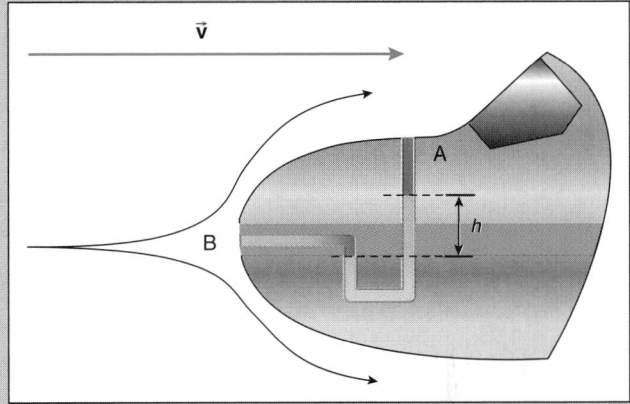

FIGURE P.67

•**68.** An airplane of mass 1.00×10^3 kg has a total wing area of 30 m². (a) What pressure difference between the upper and lower sides of the wing is needed to keep the plane in level flight? (b) If the speed of the air relative to the underside of the wing is 300 km/h, what is the speed of the air over the wing? (The density of air is about 1.30 kg/m³.)

‡**69.** A perfume dispenser consists of a tube that dips into the liquid. Air blown across the top of the tube causes the perfume to rise in the tube and be dispersed out the nozzle (see Figure P.69). The device is called an aspirator. The tube is 6.0 cm long with its top 5.0 cm above the level of the liquid perfume (of density 900 kg/m³). The density of air is 1.30 kg/m³. What is the air speed needed to raise the perfume to the top of the tube? Neglect capillary action. Paint sprayers also employ similar physical principles.

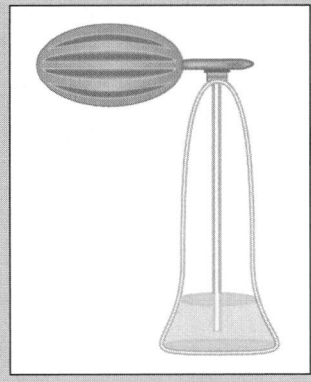

FIGURE P.69

‡70. One of the more remarkable devices involving fluids is the siphon (see Figure P.70). The device is used to transfer fluid out of a container that (for one reason or another) cannot be easily tipped. Let the density of the liquid be ρ and let atmospheric pressure be P_0. The tube (that is called the siphon) is filled with the liquid. One end of the filled tube then is inserted into the tank and the other end held at a level below that of the liquid in the tank as shown. Liquid will flow (seemingly all by itself) out of the tank until the level of the tank reaches the level of the output orifice. (a) Calculate the speed at which the liquid leaves the orifice shown in Figure P.70. (b) Let A be the point at which the siphon has its greatest elevation h_1 above the level of the fluid in the tank. What is the pressure at the point A in the siphon? (c) What is the maximum height h_{max} that A can be above the surface of the liquid? (Hints: When A is at maximum height, the speed of the fluid in the siphon is zero. What is the smallest pressure that can exist at point A?) (d) If the fluid is water, what is the maximum height h_{max} for the point A?

Sections 11.13 Nonideal Fluids*
11.14 Viscous Flow*

•71. A rectangular metal plate has dimensions of 10 cm by 20 cm. A thin film of oil of thickness 0.20 mm separates the plate from a horizontal surface. An ideal string is attached to the plate (see Figure P.71) and passes over an ideal pulley to a mass m. When $m = 125$ g, the metal plate moves at a constant speed of 5.0 cm/s across the horizontal surface. Determine the coefficient of viscosity of the oil.

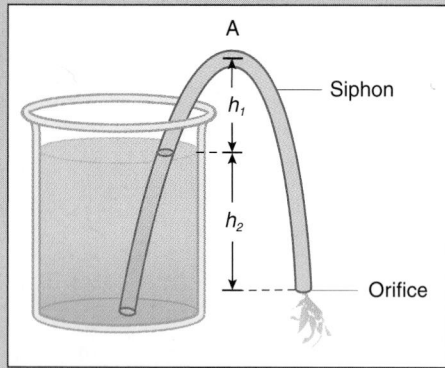

FIGURE P.70

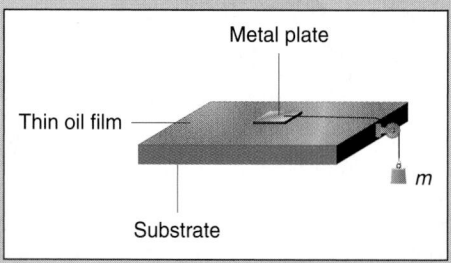

FIGURE P.71

INVESTIGATIVE PROJECTS

A. Expanded Horizons

1. Investigate the aspects of Bernoulli's principle applicable to sailing.
Rachel P. Gray, "Investigation of the airflow around a sail," *Physics Education,* 21, #1, pages 10–13 (January 1986).
Paul Jackson, "What limits the speed of a sailboat," *The Physics Teacher,* 18, #3, pages 224–225 (March 1980).
George C. Goldenbaum, "Equilibrium sailing velocities," *American Journal of Physics,* 56, #3, pages 209–215 (March 1988).
Sabinus H. Christensen, "Sailing into the wind," *The Physics Teacher,* 17, #7, pages 416, 418 (October 1979).

2. Investigate the operation of a hydraulic ram, a simple device to pump water uphill to, say, a storage reservoir.
H. Richard Crane, "The hydraulic ram: how to make water go uphill," *The Physics Teacher,* 25, #4, pages 245–247 (April 1987).

3. Investigate the application of Bernoulli's principle to lifting forces in aerodynamics.
Klaus Weltner, "Bernoulli's law and aerodynamic lifting force," *The Physics Teacher,* 28, #2, pages 84–86 (February 1990).
Klaus Weltner, "A comparison of explanations of the aerodynamic lifting force," *American Journal of Physics,* 55, #1, pages 50–54 (January 1987).

4. Investigate the physics associated with the circulatory system.
Alan Van Heuvelen, "Physics of the circulatory system," *The Physics Teacher,* 27, #8, pages 590–596 (November 1989).
L. J. Bruner, "Cardiovascular simulation for the undergraduate physics laboratory," *American Journal of Physics,* 47, #7, pages 608–611 (July 1979).

5. At a bend in a flowing river, the depth of the river along the outer bank (the outside of the curve) is deeper than along the inner bank (the inside of the curve). Investigate how this arises.
Robert W. Pohl, *Physical Principles of Mechanics and Acoustics* (Blackie and Sons, London, 1951), pages 189–190.
Albert Einstein, *Ideas and Opinions* (Crown, New York, 1954), pages 249–253.

6. By consulting appropriate sources, find out how the SAE (Society of Automotive Engineers) rating of the viscosity of various motor oils is related to the SI unit of viscosity.

B. Lab and Field Work

7. Design and conduct appropriate experiments to determine the elastic limits and breaking points of various types of fishing line.
Graig A. Spolek, "The mechanics of flycasting: the fly line," *American Journal of Physics,* 54, #9, pages 832–836 (September 1986).

8. Silk has extraordinary mechanical properties. Many insects secrete silks, among them, of course, silkworms (*Bombyx mori*) and spiders. The golden orb-weaving spider (*Nephila clavipes*) secretes silk that has attracted much interest in attempts to design and manufacture artificial silks. Design and perform an experiment to determine Young's modulus, the elastic limit, and the breaking point for spiderweb silk.
Richard Lipkin, "Artificial spider silk," *Science News,* 149, pages 152–153 (9 March 1996).
Somdev Tyagi and Arthur E. Lord Jr., "Simple and inexpensive apparatus for Young's modulus measurement," *American Journal of Physics,* 48, #3, pages 205–206 (March 1980).

9. Investigate the strength of bone material to determine its Young's modulus and the forces necessary to fracture various bones in the human body such as a femur, tibia, or fibula. Are bones intrinsically stronger under tension or compression? Do most fractures occur because of tensile forces, compressive forces, or shear forces? Use chicken bones to measure the Young's modulus of a readily available bone material; does it make a difference if the bone is fresh or dry?

John R. Cameron and James G. Skofronick, *Medical Physics* (Wiley, New York, 1978), pages 46–55.

10. Design and perform experiments to measure the terminal speed of a small spherical bead falling through liquids such as motor oils or cooking oils with various coefficients of viscosity. See if you can determine if there is a correlation between the terminal speed and the coefficient of viscosity. You might want to use small air bubbles rising through the fluid instead of falling beads.

Metin Yersel, "A simple demonstration of terminal velocity," *The Physics Teacher*, 29, #6, pages 334–335 (September 1991).

Jim Nelson, "About terminal velocity," *The Physics Teacher*, 22, #4, pages 256–257 (April 1984).

11. Use the Wilhelmy slide method (see Section 11.8) to measure the coefficient of surface tension for several liquids. Compare the results with the simple pulled ring method of Example 11.9.

R. D. Edge, "Surface tension," *The Physics Teacher*, 26, #9, pages 586–587 (December 1988).

S. Y. Mak and K. Y. Wong, "The measurement of surface tension by the method of direct pull," *American Journal of Physics*, 58, #8, pages 791–792 (August 1990).

12. Devise and perform an experiment to measure the coefficient of viscosity of various motor oils at room temperature and near 0 °C.

13. Investigate how capillary action is used in paper chromatography, which has applications in chemistry and biology.

Richard J. Block, *Paper Chromatography* (Academic Press, New York, 1958).

14. The physics of granular materials (like sand and food grains) is quite different from the physics of fluids. Use vertical pipes with convenient diameters to investigate how the pressure at the bottom of a pipe varies with the depth of dry sand in it and with the diameter of the pipe. Present your results in graphical form. In what way is the pressure of such granular materials *not* like that of a fluid? What might account for your results?

Heinrich M. Jaeger, Sidney R. Nagel, and Robert P. Behringer, "The physics of granular materials," *Physics Today*, 49, #4, pages 32–38 (April 1996).

15. Design an experiment to measure the pressure decrease along the length of a copper water pipe of small diameter.

C. Communicating Physics

16. When your blood pressure is measured, two numbers are quoted—for example, 120/80. What do these numbers mean and represent? What is the physics underlying an instrument used to measure your blood pressure? Make your explanation appropriate for the patients of a general practitioner.

17. Investigate and explain the operation of an automotive battery hydrometer; consider your audience to be the employees of an automotive repair facility.

Michael Davis, "A straw hydrometer," *The Physics Teacher*, 25, #3, page 184 (March 1987).

18. Consult the department of civil engineering at your university and/or appropriate engineering materials resources to determine the advantage of using prestressed concrete for bridge decks, floors, and ceilings. Explain why prestressed concrete is not used for walls and columns. Consider the audience for your explanation to be fellow classmates in an introductory physics course.

CHAPTER 12

WAVES

Life is a *wave* which in no two consecutive moments of its existence is composed of the same particles.

John Tyndall (1820–1893)*

We all like to make waves. Nature does too. At the beach, the light we detect when viewing the sea, the characteristic sounds we hear of the surf, and the ocean waves in which we frolic illustrate three very different types of waves. More sobering, we experience with terror the motions of seismic waves produced by earthquakes. The technological marvels of radio waves bring us music, entertainment, and information from around the globe, while their natural counterparts inform radio astronomers about cataclysmic events in the distant cosmos. Microwaves have meant new convenience in our kitchens and dorms. You read this page by detecting waves of light. It is difficult to imagine the practice of medicine without X-rays and gamma rays.[†] Ultrasound waves enable obstetricians to view developing fetuses in the womb and other medical specialists to practice lithotripsy. These examples indicate the wide variety of wave phenomena in our natural and technological environments. We will see in Chapter 27 that the waves in quantum mechanics take on a decidedly unfamiliar, almost mystical cast. So the study of waves is an essential aspect of physics, many other sciences, and even our very culture.

We begin this chapter with an examination of wave phenomena in general. Then we investigate a special class of waves called sinusoidal (harmonic) waves that form the mathematical building blocks for modeling complex, realistic wave phenomena such as acoustic waves from oboes, pianos, and mellophoniums or seismic waves from earthquakes. We also examine several aspects of waves on strings as well as sound waves, including the acoustic Doppler effect. Finally, we see what happens when two or more waves of the same physical type are present simultaneously at the same location; interesting and unusual physical effects result from their superposition.

12.1 WHAT IS A WAVE?

Undulate around the world, serenely arriving, arriving. . .
                          Walt Whitman (1819–1892)[‡]

You all know what a wave is . . . at least until you have to write a paragraph defining one, illustrating once again that physics is as much language as mathematics. In the sciences it is often quite difficult to generalize or explain precisely what is meant by our observations using the deep and common pool of words from the well of our linguistic heritage.

We need a paradigm for a wave that encompasses the many varieties of waves that occur in nature. If you think of graceful ocean swells, or the ripples generated when you drop a pebble into a calm pond, it is apparent that water waves are an oscillating disturbance that moves, we say *propagates*, across the water surface. When such waves make contact with you or a canoe floating in the water, it is evident that they give or transfer energy and momentum to you or the canoe. A wave involves energy and momentum transport at a characteristic speed that has little direct relationship to the speeds of the entrained par-

*(Chapter Opener) *Fragments of Science* (Hooper, Clarke, Chicago, 1900), Chapter XXVI, "Vitality," page 379.

[†]Radio waves, microwaves, infrared light, visible light, ultraviolet light, X-rays, and gamma rays all are electromagnetic waves. We will study electromagnetic waves in Chapter 21.

[‡]*Leaves of Grass, Memories of President Lincoln, When lilacs last in the dooryard bloom'd*, stanza 14, line 29.

Experimenting with the right wave can be exciting!

ticles. In a water wave well away from the beach, the dominant motion of an individual water particle is up and down, perpendicular to the direction the water wave is moving. Such waves are much more complex as they "break" near a beach!

As we know, energy and momentum also can be transferred by the movement of a particle, such as when a pitcher throws a baseball to the catcher. Energy and momentum can be transferred by one, two, or a whole stream of particles; but such particle streams are not what we imagine as waves.

A propagating wave disturbance may involve any manner of physical quantity. A **mechanical wave** is a propagating disturbance in and of a material (such as a solid, liquid, or gas) from its equilibrium configuration. The term mechanical wave indicates its close association with matter. These waves involve oscillatory motions of the atoms and/or molecules that make up the material, without requiring net transport of the material itself over long distances in the direction of motion of the wave. Such mechanical waves represent propagating oscillations from equilibrium, be they departures from equilibrium positions (e.g., water waves and acoustic waves), or pressure or density oscillations (also acoustic waves). An ocean wave predominantly moves water up and down with little or no average motion of the water along the direction of travel of the wave. A huge locomotive engine generates mechanical waves (vibrations and sound waves) in the steel rails that propagate speedily along their length; although the particles in the rail move slightly, the rail itself does not progressively move as do the waves.

Other types of waves, capable of propagating even through a vacuum, involve propagating oscillations of *fields* of various kinds: gravitational fields (gravity waves), or electric and magnetic fields (electromagnetic waves—i.e., light). Such waves are not mechanical waves.

Generalizing from these observations, we characterize a **classical wave** as a propagating disturbance that transfers energy and momentum at its own characteristic speed from one region of space to another with little if any mass transfer.

Mechanical waves and waves of oscillating fields are both classical waves because they satisfy a general **classical wave equation**, which we will develop in Section 12.6. In this chapter we confine our study principally to mechanical waves. We see how electromagnetic waves arise in Chapter 21. Later, when we study quantum mechanics in Chapter 27, we shall see that the peculiar waves associated with quantum mechanics satisfy a different wave equation and are not themselves directly observable. The waves in quantum mechanics are not classical waves. We also shall see in that now distant chapter that there is a subtle, mysterious, almost oxymoronic connection between particles and waves called the wave–particle duality: ultimately (and surprisingly) we cannot have waves without particles, nor particles without waves!

QUESTION 1 _____

The word wave is used in many colloquial contexts. Examples include:

- Don't make waves.
- Wave goodbye.
- Whew! What a heat wave we had this summer!
- Waves of troops invaded.
- Her hair has beautiful waves.
- Flag waving children greeted the President.

Perhaps you can think of others. In what ways does the use of the word wave in these contexts differ from the meaning of the word in physics? Check out the meaning of the word *wave* in the *Oxford English Dictionary*; you may be surprised at the many and varied uses of the word!

12.2 LONGITUDINAL AND TRANSVERSE WAVES

There are two different kinds of "pure" waves. In this section we contrast them.

Longitudinal Waves

Take a slinky and stretch it across your desk, holding one end in your hand with the other end fixed to a wall. Quickly push or compress the end you hold along the axis of the slinky, as in Figure 12.1, without returning your hand back to its original position. The single **compression** or pulse propagates down the axis of the slinky at a speed called the *speed of the wave*.

If you now quickly extend the end you hold, as in Figure 12.2, again without returning your hand to its original position, an anti-compression pulse called a **rarefaction** is transmitted down the axis of the slinky. Multiple successive compressions and extensions by your hand produce multiple compressions and rarefactions that move along the axis of the slinky. For such disturbances, the oscillation (jiggling) of the particles of the slinky is along the line of propagation of the wave.

> Waves whose oscillation or jiggling is along the line the wave propagates are called **longitudinal waves**.

A given particle of the slinky has a maximum speed when the coils are most compressed or rarefied. This is characteristic of longitudinal mechanical waves: particle speeds are a maximum at the instants of maximum compression and rarefaction. This observation is *not* obvious, so do not worry about it for now; we will explore this aspect of longitudinal waves in greater detail in Section 12.13.

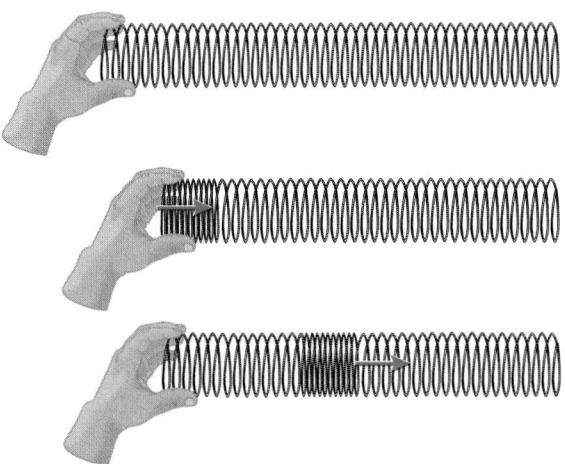

FIGURE 12.1 Compress an end of a slinky and a compression pulse propagates along it.

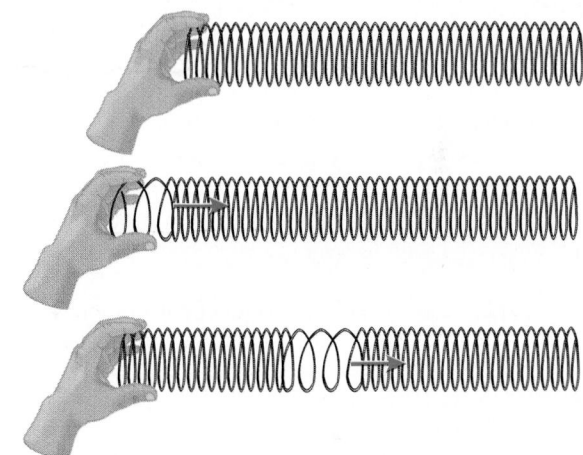

FIGURE 12.2 Extend your end of the slinky quickly and a rarefaction propagates along the slinky.

Common sound waves in air or other gases are longitudinal waves that can be viewed as propagating density or pressure variations that result from longitudinal motions of the particles that make up the medium through which the sound wave travels. Regions of increased density (or pressure) are compressions; regions of reduced density (or pressure) are rarefactions. Such *variations* or oscillations in density (or pressure) about the ambient or mean value are positive for compressions and negative for rarefactions. The density (or pressure) itself is never negative. The density (or pressure) variations are correlated to the particle motions in ways that will be discussed in Section 12.13.

Transverse Waves

As another experiment, take a thin clothesline or climbing rope and stretch it taut across your room. If you quickly move your hand up and down once, perpendicular to the axis of the rope, the disturbance you generate propagates along the rope; see Figure 12.3. But in this case the disturbance (the jiggling—i.e., the motion of the particles that make up the rope) is perpendicular to the direction the wave disturbance propagates.

Quick motion of hand

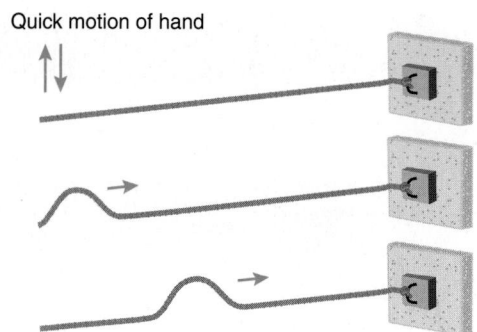

FIGURE 12.3 A propagating pulse on a rope.

Move your hand up and down several times quickly, as in Figure 12.4, and you produce multiple oscillations that propagate along the rope.

> Waves whose oscillation or jiggling is perpendicular to the line along which the wave propagates are called **transverse waves**.

Waves on a rope or string are transverse waves.

> The plane in which the jiggling of a transverse wave disturbance occurs is called the **plane of polarization** of the transverse wave; see Figure 12.4.

Only transverse waves have planes of polarization; we say such waves are **plane polarized**. Light also is a transverse wave, even though it is not a mechanical wave but a wave of oscillating electric and magnetic fields.*

Hence the primary physical distinction between longitudinal and transverse waves is the direction of the jiggling or oscillations of the particles:

Transverse wave: the oscillations are perpendicular to the line along which the wave propagates.

---

*We explore the properties of light waves in greater detail in Chapter 24, including the evidence that suggests that light is indeed a wave (it is *not* obvious that this is the case) and what observations (polarization) enable us to conclude that it is a transverse wave. The polarization of ordinary light is not apparent because typical sources emit many waves with many different planes of polarization present simultaneously.

Longitudinal wave: the oscillations are along the line along which the wave propagates.

Earthquakes produce *both* longitudinal and transverse mechanical waves, called **seismic waves**, that propagate from the **focus**, the source of the quake. The **epicenter** is the place on the surface of the Earth directly above the focus of the quake. The longitudinal wave is called a **P-wave**, where the P means *p*ressure wave. The transverse wave is called an **S-wave**, where the S means *s*hear wave. The S- and P-waves travel at different speeds, on the order of 3 km/s and 5 km/s respectively.[†] Because of this difference, seismometers detect the arrival of the P-wave before the arrival of the S-wave, as in Figure 12.5. The difference in the time of arrival enables us to estimate the distance from the seismometer to the site of the earthquake that generated the waves. The S-waves do not penetrate the outer core of the Earth, since it is fluid and cannot sustain shear or transverse forces.

Do *not* get the impression that all waves are either longitudinal or transverse! Certain types of waves exhibit *both* characteristics simultaneously, and so are neither purely transverse nor purely longitudinal in character. Surface water waves are good examples of this mixed type of wave. If you carefully follow the jiggling of water particles as a water wave passes in deep water well away from the shore,[‡] the particles generally move up and down but also exhibit smaller motions parallel and antiparallel to the direction of propagation of the wave, as shown in Figure 12.6.

[†]The speeds vary depending on the density of the rock through which they travel.
[‡]This can be done by using a water suspension of small particles (dirty water!) and illuminating the dirt particles with a strong light.

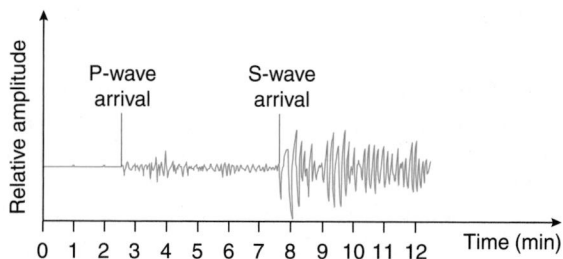

FIGURE 12.5 A record of a moderate earthquake in Chile on 3 September 1993, made by a seismograph at Washington and Lee University in Virginia; note the different arrival times of the P- and S-waves. The time $t = 0$s is not when the earthquake occurred in Chile but here is an arbitrary choice for the time axis.

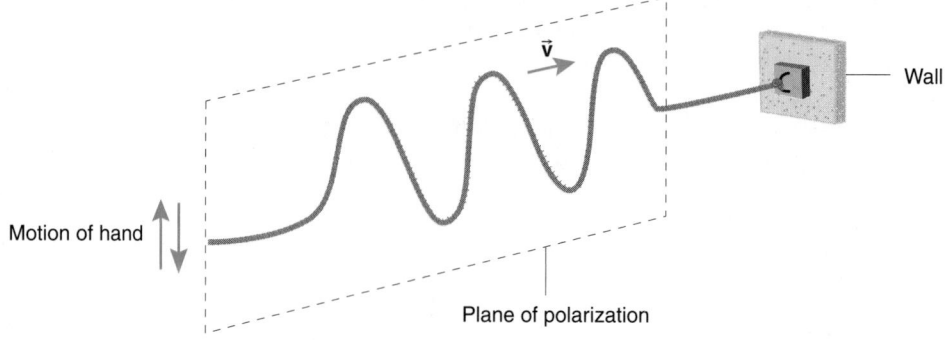

FIGURE 12.4 The plane of polarization of a transverse wave.

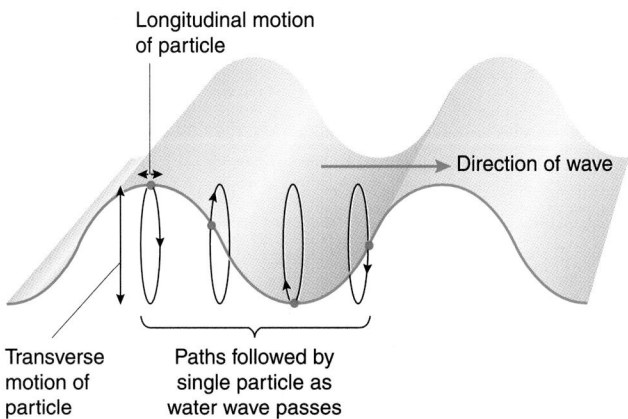

Longitudinal motion of particle

Direction of wave

Transverse motion of particle

Paths followed by single particle as water wave passes

FIGURE 12.6 A surface water wave has both a transverse and longitudinal character.

FIGURE 12.7 A waveform of a water wave shows its shape at an instant of time.

Under certain circumstances, even light exhibits both transverse and longitudinal characteristics (for example, when microwaves propagate in certain ways inside pipes called wave guides).

QUESTION 2

Grab the end of a taut rope whose other end is tied to a wall. Whirl the end you are holding around in a circle in a plane perpendicular to the axis of the rope. Is the wave on the rope transverse, longitudinal, or neither?

12.3 WAVEFUNCTIONS, WAVEFORMS, AND OSCILLATIONS

Waves are closely related to oscillations.* Both words imply undulatory (back-and-forth) motion but there is a distinction. Unlike the oscillatory behavior we studied in Chapter 7, waves are *propagating* oscillatory disturbances.

We denote the propagating wave disturbance by the Greek letter Ψ, called psi (pronounced like the English word "sigh"), whatever the wave may represent physically. This way we can avoid being too parochial and can, without bias, think of Ψ as representing *any* kind of specific wave disturbance. We call Ψ the **wavefunction** since it represents the wave disturbance. Since a wave propagates through space and time, Ψ must be a function of space and time: $\Psi(x, y, z, t)$.

If you "freeze" the wave in time, say by taking an imaginary picture of it at time instant t_0, the resulting spatial pattern of the disturbance $\Psi(x, y, z, t_0)$ is called a **waveform**. A photograph taken after you shake the end of a rope or a photograph of a water wave depicts the waveforms of the respective waves (see Figure 12.7).

On the other hand, if you imagine a specific point in space (x_0, y_0, z_0) and examine the behavior there as a function of time, we have a *dynamical picture* of the *oscillatory motion* at that point (like a video camera viewing one location, without panning the camera). For example, we can take a video of the oscillatory motion of a given point on a rope you shake. Likewise we could take a video of your canoe lazily bobbing up and down as a water wave passes, since the boat responds to the oscillatory behavior of the water particles at its location.

12.4 WAVES PROPAGATING IN ONE, TWO, AND THREE DIMENSIONS

Waves can propagate in one or more spatial dimensions. A longitudinal wave on a slinky is a good example of a one-dimensional wave. Let the x-axis be the line along which the wave propagates. Then the wave disturbance Ψ is a function of the position along the x-axis as well as the time: $\Psi(x, t)$.[†] Look at the waveform at some instant, as shown in Figure 12.8. The **wave train** (also called the **pulse train**) is an indication of the spatial region over which the waveform is nonzero at an instant. The **carrier wavelength** is the distance between successive compressions or successive rarefactions measured on the waveform.

A transverse wave on a taut rope is another example of a one-dimensional wave; see Figure 12.9. The wave propagates along a line, the x-axis, coincident with the undisturbed rope and the disturbance Ψ is a measure of the position, perpendicular to the x-axis,[‡] where a particle of the rope is located. Again Ψ

---

*An oscillation is any kind of back-and-forth motion. Recall that simple harmonic oscillation (Chapter 7) is a very specific type of oscillation. Not all oscillations are simple harmonic oscillations.

[†]Here is a good example of why we call the wave disturbance Ψ. For such a longitudinal compression wave, the disturbance is along the same direction the wave propagates, say x. By using Ψ to represent the disturbance, we avoid likely notational confusions such as $x(x, t)$.

[‡]We can measure this perpendicular distance along the y-axis, so $\Psi(x, t) \equiv y(x, t)$, but we prefer to call all wave disturbances Ψ.

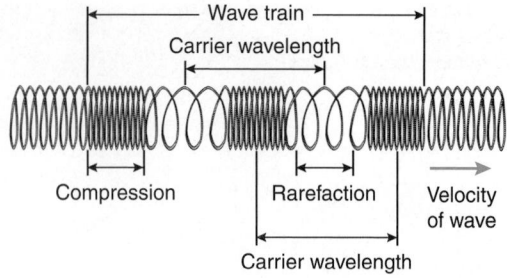

FIGURE 12.8 The waveform of a compressional wave on a slinky.

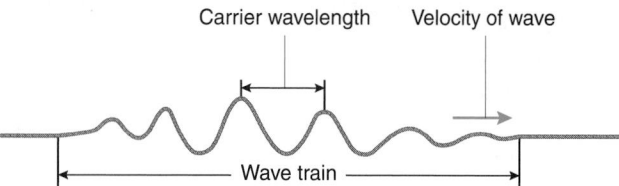

FIGURE 12.9 The waveform of a wave on a rope.

depends on both the position x along the rope and t: $\Psi(x, t)$. The wave train is the distance over which the waveform is nonzero. The carrier wavelength is the distance between successive corresponding points of the wave along the waveform.

Drop a rock into a calm pond and the resulting surface water wave is two-dimensional, since it propagates in all directions across the surface of the water (see Figure 12.10). In this case we have $\Psi = \Psi(x, y, t)$. The wave train is the radial distance between the leading and trailing edges of the waveform—that is, the region over which the disturbance is nonzero. Taking a cue from water waves, we call the peaks of a waveform **crests**, and the minima **troughs**. The **carrier wavelength** is the distance between successive peaks, troughs, or other corresponding successive points on the ripples. The carrier wavelength may vary along the wave train, particularly for complex waveforms.

Scream and shout and the resulting acoustic wave in the air is three-dimensional, since it propagates in all spatial directions;

thus, $\Psi = \Psi(x, y, z, t)$. The wave train is the distance over which the sound is nonzero at some instant, and the carrier wavelength is the distance between successive compressions or rarefactions on the waveform.

With the slinky, if you compress the end once, a single, one-dimensional, longitudinal, compressional pulse propagates down the axis. A one-dimensional transverse pulse is generated if you shake a taut rope side-to-side once. The sound of a rifle shot is approximately a three-dimensional, longitudinal, acoustic wave pulse. A tsunami closely approximates a surface water wave pulse. Waveforms with wave trains of finite extent are called **wavepackets**.

If the waveform is essentially random (see Figure 12.11), then the wave is called **noise**, whether the wave be acoustic or not. Noise has no single, well-defined carrier wavelength. The waveforms of the sound of a waterfall or the crackle of static from your radio are good examples of waveforms with no definitive carrier wavelength.

A crest or trough or any locus of connected points where Ψ has a constant value defines a region of **constant phase**, known as a **wavefront**, shown in Figure 12.12. Directed lines drawn perpendicular to wavefronts and parallel to the velocity of the wave are called **rays**.

QUESTION 3

Make a quick sketch of the waveform of a water swell on an otherwise calm ocean, well away from the shore. Indicate what is meant by a crest, trough, wavefront, and ray.

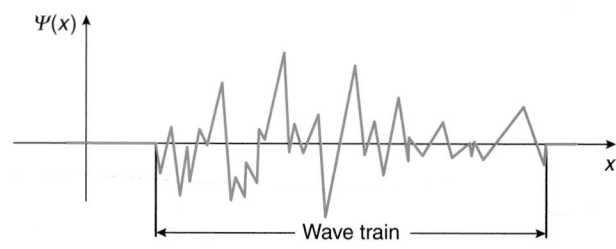

FIGURE 12.11 The waveform of noise.

FIGURE 12.10 A two-dimensional surface water waveform.

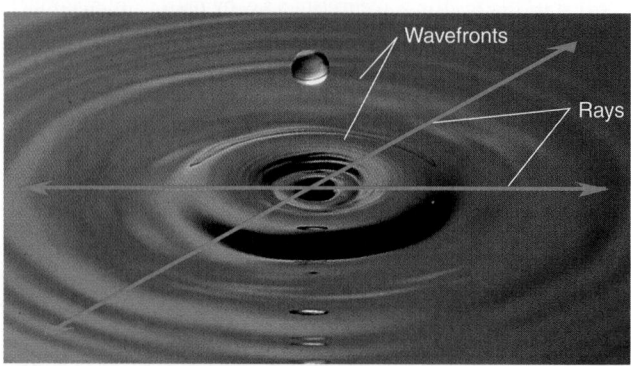

FIGURE 12.12 Wavefronts and rays of a waveform.

12.5 ONE-DIMENSIONAL WAVES MOVING AT CONSTANT VELOCITY

In this section we seek to find the way the wavefunction $\Psi(x, t)$ depends more specifically on x and t for a one-dimensional wave propagating along the x-axis at constant velocity.

Imagine a waveform of Ψ with a wavefront crest of maximum size located at $x = 0$ m when $t = 0$ s, as shown in Figure 12.13. We always can arrange the coordinate system and begin measuring time when this largest peak is at the origin, so that there is no loss of generality in imposing this restriction. This largest wavefront crest of this disturbance Ψ occurs where the argument of Ψ is zero, that is, where the stuff inside the functional parentheses is all zeros. Let's keep track of this wavefront as the wave moves at constant velocity. Tracking a particular wavefront means following a point of constant phase; the speed v of the wave is thus called a **phase speed**.

At a later instant t the wavefront has moved along the x-axis, say in the direction of increasing x, a distance equal to vt, so that the wavefront we are following is now at coordinate $x = vt$, as shown in Figure 12.14. Thus we have

$$x - vt = 0 \text{ m}$$

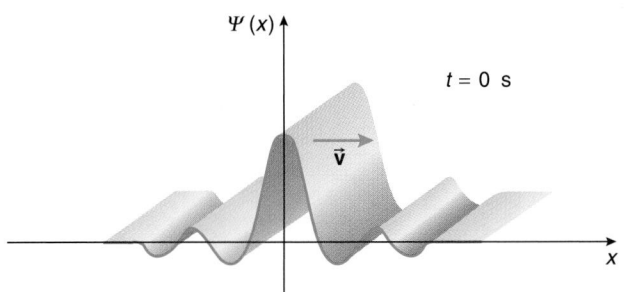

FIGURE 12.13 A large wavefront crest at $x = 0$ m when $t = 0$ s.

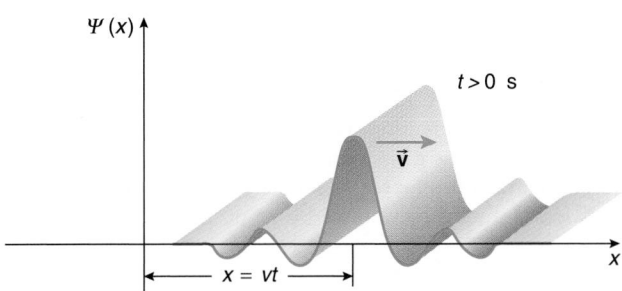

FIGURE 12.14 The wavefront has moved to coordinate $x = vt$ when the time is t.

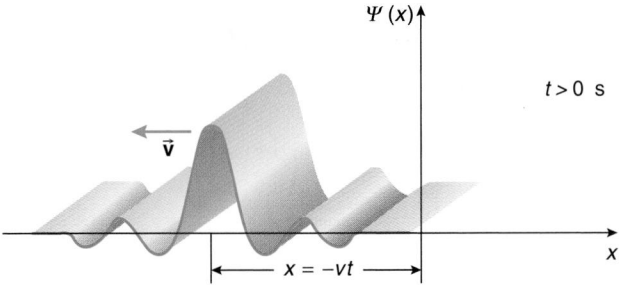

FIGURE 12.15 A wave moving toward decreasing values of x.

From a mathematical viewpoint, if $x - vt$ holds a fixed value, here 0 m, this value specifies a particular wavefront of the disturbance.

The wave is moving at speed v along $\hat{\imath}$; the wavefront is thus at the coordinate $x = vt$ at time t, and $x - vt$ is therefore constant for a given wavefront. Thus we describe the disturbance $\Psi(x, t)$ more specifically as a function of the quantity $x - vt$:

$$\Psi(x - vt)$$

In this way, whenever the argument $x - vt$ of the function is equal to zero, we have $\Psi(0 \text{ m})$ and have located this chosen wavefront of the disturbance.

> Therefore a one-dimensional wave propagating at constant speed v in the direction of *increasing* values of x is described by the wavefunction
> $$\Psi(x - vt) \qquad (12.1)$$

If you want to describe the same disturbance moving in the direction of *decreasing* values of x, as in Figure 12.15, the wavefront will have moved out along the negative x-axis for times $t > 0$ s and the wavefront is at coordinate $x = -vt$. So

$$x + vt = 0 \text{ m}$$

locates the chosen wavefront that originally was at the origin when $t = 0$ s.

> Hence the wavefunction
> $$\Psi(x + vt) \qquad (12.2)$$
> represents a one-dimensional disturbance moving at constant speed v in the direction of *decreasing* values of x.

PROBLEM-SOLVING TACTIC

12.1 The sign between x and vt determines the direction Ψ is moving along the x-axis. When the sign is negative, $\Psi(x - vt)$, the wave is moving toward *increasing* values of x; when the sign is positive, $\Psi(x + vt)$, the wave is moving toward *decreasing* values of x.

12.6 THE CLASSICAL WAVE EQUATION FOR ONE-DIMENSIONAL WAVES*

It is quite remarkable that all one-dimensional classical waves propagating with constant velocity satisfy a single equation known as the **classical wave equation**. In this section we formulate this special equation.

One-dimensional waves propagating at constant phase speed v in the direction of increasing values of x are described by a wavefunction $\Psi(x - vt)$. Notice that Ψ is a function of two variables: x and t. The calculus you have learned to date likely involved just one independent variable (the ubiquitous "x" in calculus courses). When there are two independent variables (such as x and t here), you can take derivatives of Ψ with respect to either x or t. When differentiating with respect to x, the other variable t is treated as a constant, and vice versa. Such derivatives are called **partial derivatives.*** The same familiar rules of differentiation apply, but the derivative symbol is changed from the familiar

$$\frac{d}{dx}$$

to the more stylish

$$\frac{\partial}{\partial x}$$

If Ψ is differentiated with respect to x, treating t as a constant, we call the result the partial derivative of Ψ with respect to x and write it as

$$\frac{\partial \Psi}{\partial x}$$

We could also differentiate Ψ with respect to the other independent variable, t, in which case you treat x as a constant and apply the usual differentiation rules. This is the partial derivative of Ψ with respect to t, written

$$\frac{\partial \Psi}{\partial t}$$

We know that a wave Ψ propagating in the direction of increasing values of x is a function of the quantity $x - vt$:

$$\Psi(x - vt)$$

Simply from a mathematical standpoint, we might suspect that there is a relationship between the partial derivatives of Ψ with respect to x and those of Ψ with respect to t. Let's see if we can discover what that relationship is. For convenience, we define a parameter u to be

$$u \equiv x - vt \qquad (12.3)$$

To differentiate Ψ partially with respect to x, we use the chain rule from calculus: differentiate Ψ with respect to u and then multiply the result by the derivative of u with respect to x:

$$\frac{\partial \Psi}{\partial x} = \frac{\partial \Psi}{\partial u} \frac{\partial u}{\partial x}$$

From Equation 12.3 we see the partial derivative of u with respect to x is just 1:

$$\frac{\partial u}{\partial x} = 1$$

(remember that t is treated as a constant in partial differentiations with respect to x). The wave speed v is a constant for the class of waves with which we are concerned. So we have

$$\frac{\partial \Psi}{\partial x} = \frac{\partial \Psi}{\partial u} 1$$

---

*There is nothing really "partial" about the differentiation process, however, since the same differentiation rules of calculus are used for either choice of the variable.

For good measure we take the second partial derivative of Ψ with respect to x (again using the chain rule), getting

$$\begin{aligned}
\frac{\partial^2 \Psi}{\partial x^2} &= \left[\frac{\partial}{\partial u} \left(\frac{\partial \Psi}{\partial x} \right) \right] \frac{\partial u}{\partial x} \\
&= \left[\frac{\partial}{\partial u} \left(\frac{\partial \Psi}{\partial u} \right) \right] 1 \\
&= \frac{\partial^2 \Psi}{\partial u^2} \qquad (12.4)
\end{aligned}$$

Now we look at the partial derivatives of Ψ with respect to t (in which case we treat x as a constant). Once again, we use the chain rule:

$$\frac{\partial \Psi}{\partial t} = \frac{\partial \Psi}{\partial u} \frac{\partial u}{\partial t} \qquad (12.5)$$

But from Equation 12.3 we have

$$\frac{\partial u}{\partial t} = -v$$

Equation 12.5 thus becomes

$$\frac{\partial \Psi}{\partial t} = -v \frac{\partial \Psi}{\partial u}$$

Since

$$\frac{\partial \Psi}{\partial u} = \frac{\partial \Psi}{\partial x}$$

we could stop here and say that the desired relationship between the partial derivatives is

$$\frac{\partial \Psi}{\partial t} = -v \frac{\partial \Psi}{\partial x} \qquad (12.6)$$

Unfortunately, if we did stop here, we would find a *different* relationship between the partial derivatives for a wave traveling in the opposite direction, toward decreasing values of x! (See Problem 12.) This is mathematically and physically inelegant (and becomes even more of a problem when considering waves in two and three dimensions). So we push on and see if we can find a relationship between the second partial derivatives.

Taking the second partial derivative of Ψ with respect to t, applying the chain rule once again, we get

$$\begin{aligned}
\frac{\partial^2 \Psi}{\partial t^2} &= \frac{\partial}{\partial t} \left(\frac{\partial \Psi}{\partial t} \right) = \frac{\partial}{\partial u} \left(\frac{\partial \Psi}{\partial t} \right) \frac{\partial u}{\partial t} \\
&= \frac{\partial}{\partial u} \left(-v \frac{\partial \Psi}{\partial u} \right)(-v) = v^2 \frac{\partial^2 \Psi}{\partial u^2}
\end{aligned}$$

Equation 12.4 enables us to write this as

$$\frac{\partial^2 \Psi}{\partial t^2} = v^2 \frac{\partial^2 \Psi}{\partial x^2}$$

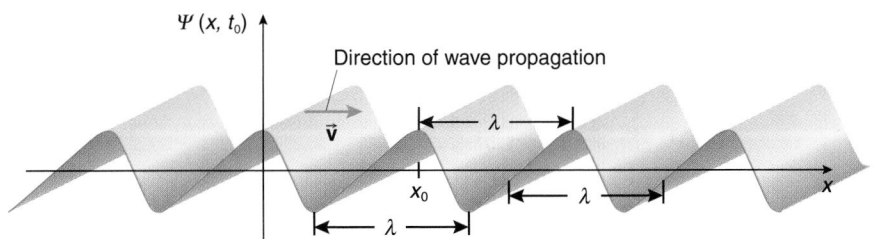

FIGURE 12.16 The wavelength λ of a periodic waveform.

With some slight rearrangement, we find the *classical wave equation* for one-dimensional waves is

$$\frac{\partial^2 \Psi}{\partial x^2} - \frac{1}{v^2} \frac{\partial^2 \Psi}{\partial t^2} = 0 \qquad (12.7)$$

Such an equation is called a partial differential equation for Ψ, since it involves the partial derivatives of the function. This particular partial differential equation is very famous in physics and engineering and is given its own name: the classical wave equation for one-dimensional waves. Any classical real wave Ψ, propagating at constant velocity $\vec{v}$ in one dimension (along the x-axis), must satisfy this relationship among its partial derivatives.*

12.7 PERIODIC WAVES

Many waves, such as ocean swells, periodically repeat themselves. A **periodic wave** is a wave with characteristic, constant length and time scales. The waveform of a periodic wave repeats itself along a characteristic, constant scale of length known as the **wavelength** λ. The wavelength λ is the distance between successive corresponding points along the waveform, as shown in Figure 12.16. In SI units, λ is expressed in meters. Note that Figure 12.16 is a graph of the *waveform* at a particular instant t_0 of time, plotting Ψ as a function of x.

On the other hand, if you fix your attention at any given position x_0, the *oscillatory* behavior of a periodic wave passing that position has a constant **period** T, as shown in Figure 12.17. In SI units, the period is measured in seconds. Note that the graph in Figure 12.17 depicts the *dynamic*, oscillatory behavior of Ψ at a fixed position as a function of time.

---

*The peculiar waves encountered in quantum mechanics are complex waves (complex in the mathematical sense: they involve $i = \sqrt{-1}$). We will see in Chapter 27 that the waves in nonrelativistic quantum mechanics do not satisfy the classical wave equation, but another wave equation known as the Schrödinger equation.

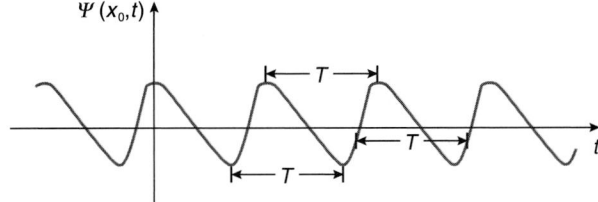

FIGURE 12.17 The period T of a periodic wave.

The period T is the interval of time needed to execute one complete oscillation. The **frequency** v is the number of oscillations made during exactly one second. The SI unit of frequency is the hertz (Hz). The frequency is the reciprocal of the period:

$$v = \frac{1}{T} \qquad (12.8)$$

Let the constant phase speed of the wave be v. The wave moves through a distance equal to the wavelength during the time interval of one period. Since the distance is the product of the speed and time interval, we have

$$\lambda = vT \qquad (12.9)$$

Rewriting Equation 12.9 in terms of the frequency instead of the period (using Equation 12.8), we obtain

$$v = v\lambda \qquad (12.10)$$

Equation 12.10 is valid for any periodic wave moving at constant speed.

A particularly important class of periodic waves is **sinusoidal waves** (also known as *harmonic waves*); their waveform has the shape of the sine or cosine trigonometric functions. The oscillatory disturbance at any position then is simple harmonic oscillation, also described by sinusoidal functions. Such sinusoidal waves are idealizations, since they have infinite wave trains and exist at all times. No real wave, of course, exists over all space for all time. Nonetheless, sinusoidal waves are important because they form the building blocks for the analysis of more complex waves, as we will see in Section 12.21. We look at sinusoidal waves in greater detail in the following section.

STRATEGIC EXAMPLE 12.1

A periodic wave traveling at a constant speed 10.0 m/s toward increasing values of x has the waveform at the instant $t = 0$ s sketched in Figure 12.18.

a. **What is the wavelength λ of the periodic wave?**
b. **Sketch the corresponding oscillatory behavior, Ψ versus t, at a fixed position, say the origin.**
c. **What is the period T of the periodic wave?**
d. **What is the frequency v of the periodic wave?**

Solution
a. The wavelength λ is the distance between successive corresponding points along the waveform. This is 6.0 m.
b. Consider the origin in Figure 12.18. Notice that as the wave moves to the right at constant speed, the time to reach the peak

FIGURE 12.18

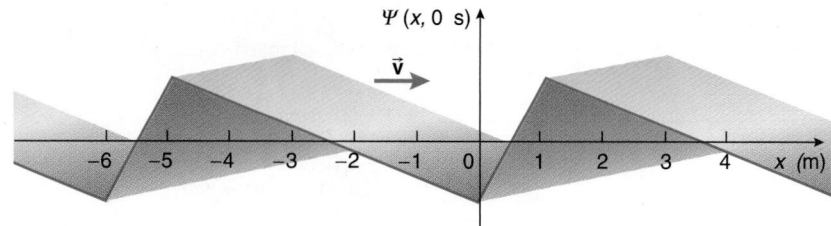

will be five times as long as the time to fall to the trough of the wave. Since the wave is moving at speed 10 m/s to the right in Figure 12.18, it will take 0.50 s for the first peak to the left of the origin in Figure 12.18 to reach the origin; 0.10 s later, the first trough to the left of the origin in Figure 12.18 reaches the origin. Hence the oscillatory behavior at the origin has the appearance shown in Figure 12.19.

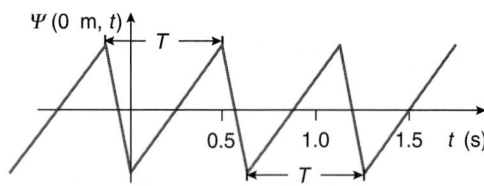

FIGURE 12.19

c. The period T of the wave is the time between successive corresponding points on the graph of Ψ versus t. The period is 0.60 s.
d. There are two ways to find the frequency.

Method 1

From Equation 12.8, the frequency is the inverse of the period:

$$\nu = \frac{1}{T}$$

$$= \frac{1}{0.60 \text{ s}}$$

$$= 1.7 \text{ Hz}$$

Method 2

Since you know the speed v and the wavelength λ, you can use Equation 12.10 to find the frequency ν:

$$v = \nu\lambda$$

Solving for the frequency,

$$\nu = \frac{v}{\lambda}$$

$$= \frac{10 \text{ m/s}}{6.0 \text{ m}}$$

$$= 1.7 \text{ Hz}$$

EXAMPLE 12.2 ━━━━━━━━

A periodic wave, traveling toward *decreasing* values of x at speed 4.0 m/s, produces oscillatory behavior at the origin of the x-axis depicted in Figure 12.20.

a. **What is the period of the wave?**
b. **What is the frequency of the wave?**

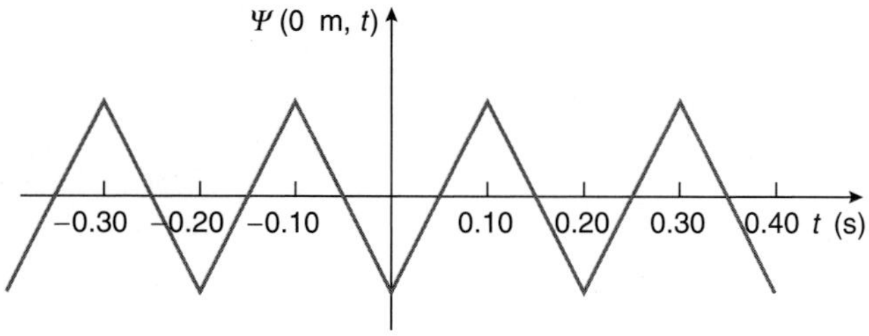

FIGURE 12.20

c. **What is the wavelength of the wave?**
d. **Sketch the waveform, Ψ versus x, when $t = 0$ s.**

Solution
a. The period of the wave is the time between successive corresponding points on the Ψ versus t graph. This time is $T = 0.20$ s.
b. The frequency of the wave is the inverse of the period:

$$\nu = \frac{1}{T}$$

$$= \frac{1}{0.20 \text{ s}}$$

$$= 5.0 \text{ Hz}$$

c. The wavelength can be found in two equivalent ways.

Method 1

The wave travels at speed v through a distance equal to one wavelength during one period. Hence the wavelength is

$$\lambda = vT$$

$$= (4.0 \text{ m/s})(0.20 \text{ s})$$

$$= 0.80 \text{ m}$$

Method 2

Since you know the frequency and the speed, you can find the wavelength using Equation 12.10:

$$v = \nu\lambda$$

Solving for the wavelength,

$$\lambda = \frac{v}{\nu}$$

$$= \frac{4.0 \text{ m/s}}{5.0 \text{ Hz}}$$

$$= 0.80 \text{ m}$$

d. The oscillatory pattern at the origin, given in Figure 12.20, indicates that Ψ increases linearly with time from a trough when $t = 0$ s to a peak when $t = 0.10$ s. (Notice that the waveform is moving toward negative values of x.) The wave then decreases to a trough after another 0.10 s. Since the wave is moving at constant speed, the waveform of Ψ must have linear segments of equal length. Since the wavelength is 0.80 m and, when $t = 0$ s, Ψ is at a trough at the origin, a graph of Ψ versus x has the appearance shown in Figure 12.21.

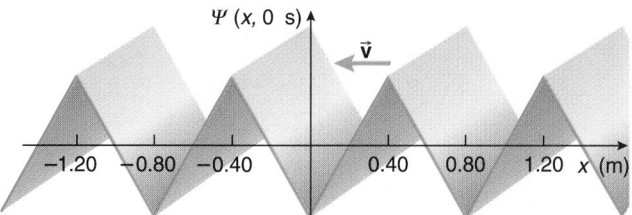

FIGURE 12.21

12.8 SINUSOIDAL (HARMONIC) WAVES

Since any one-dimensional wave traveling at constant speed v toward increasing values of x has a wavefunction depending on the combination of variables $x - vt$, a **sinusoidal wave** traveling in the direction of increasing x will have the form

$$\Psi(x, t) = A \cos[k(x - vt)] \qquad (12.11)$$

where we have to determine the meaning and physical significance of k. Since the cosine can be taken only of angles (which have units but no dimensions), the product of k with x is dimensionless; hence k itself must have the dimensions of inverse length.

> By convention the angular argument of the cosine function—that is, $k(x - vt)$—is expressed in radian units of angular measure, *not* degrees.

Although we use the cosine function in Equation 12.11, the wave still is called a sinusoidal wave because the sine and cosine functions have the same general shape and characteristics; they simply are shifted by $\pi/2$ rad with respect to each other. One could just as well use a sinusoidal wavefunction Ψ of the form $A \sin[k(x - vt)]$. However, by convention, we will use Equation 12.11 for sinusoidal waves.

Let's examine the waveform and oscillatory characteristics of a sinusoidal wave.

Waveform of a Sinusoidal Wave

Take a "snapshot" of the wave when $t = 0$ s and make a graph of the waveform of Ψ as a function of x. That is, when $t = 0$ s, we have

$$\Psi(x, 0 \text{ s}) = A \cos(kx) \qquad (12.12)$$

A graph of this waveform as a function of x is a cosine function, as shown in Figure 12.22.

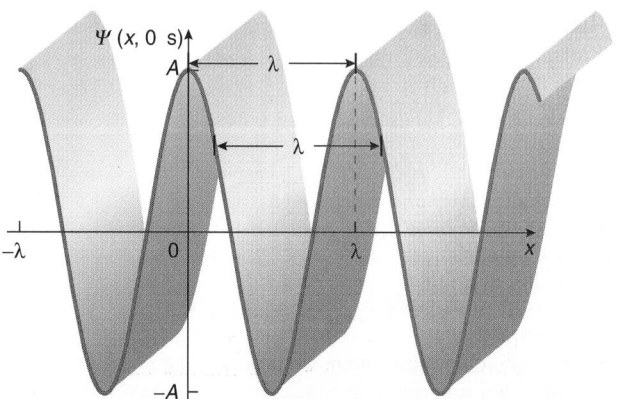

FIGURE 12.22 The waveform of the sinusoidal wave given by Equation 12.11 when $t = 0$ s.

Since the maximum value of a cosine function is 1, the maximum value of Ψ is equal to A, known as the **amplitude** of the sinusoidal wave. The amplitude has the same dimensions as Ψ itself (*not* necessarily a distance, since Ψ represents whatever particular physical wave is under discussion). Notice that the variation of Ψ from trough to peak, the so-called peak-to-peak amplitude, is 2A.

A sinusoidal waveform has a fixed wavelength. Notice that a shift in the value of x by λ in either direction along the x-axis yields the same value again for Ψ. For example, at $x = 0$ m the value of Ψ is A according to Equation 12.12. When $x = \pm\lambda$, the value of Ψ is again A. The argument of the cosine function changes by 2π rad as x changes from 0 m to λ; in other words, when $x = \lambda$ the angle (or argument) of the cosine is 2π rad. That is, from Equation 12.12

$$k\lambda = 2\pi \text{ rad}$$

The constant k therefore is

$$k = \frac{2\pi \text{ rad}}{\lambda} \qquad (12.13)$$

The constant k is known as the **angular wavenumber**.* Physically, the angular wavenumber represents the number of wavelengths in a distance of exactly 2π meters.†

---

*The angular wavenumber also commonly is called the *propagation vector*. The latter term bears some explanation. The propagation vector k for a one-dimensional wave is a scalar; it represents the magnitude of a propagation vector. For waves in two or three dimensions, k takes on vector properties, which is the reason that it is called the propagation vector. In fact, the product kx in Equations 12.11 and 12.12 is the one-dimensional scalar product of a vector $\vec{k}$ with the position vector $\vec{r}$:

$$\vec{k} \cdot \vec{r} = kx \qquad \text{(in one dimension)}$$

So we have a case of trickle-down terminology for the one-dimensional case. To avoid calling k a vector when it is a scalar for a one-dimensional situation, we use the terminology *angular wavenumber*.

†The reciprocal of the wavelength, $1/\lambda$, without the factor of 2π, is called the *wavenumber*. The wavenumber is the number of wavelengths in a distance of exactly one meter.

The Oscillatory Behavior of a Sinusoidal Wave

You can monitor the oscillatory behavior of the wave passing a particular location, say $x = 0$ m. The wavefunction Ψ then is a function of time. Equation 12.11 becomes

$$\Psi(0 \text{ m}, t) = A \cos(-kvt) \qquad (12.14)$$

Since the cosine of a negative angle is the same as the cosine of the positive angle $[\cos(-\theta) = \cos(\theta)]$, Equation 12.14 becomes

$$\Psi(0 \text{ m}, t) = A \cos(kvt)$$

whose graph also is sinusoidal, as shown in Figure 12.23. Notice that the graph is periodic in time. The constant period T of the wave is the time between any two successive corresponding points on the graph shown in Figure 12.23.

As t increases from $t = 0$ s to $t = T$, the argument of the cosine goes through 2π rad. Hence

$$kvT = 2\pi \text{ rad} \qquad (12.15)$$

The period is the time for Ψ to execute a complete oscillation at any given location. The inverse of the period is the frequency ν of the wave. Since the period is the time for one complete oscillation, the frequency is the number of oscillations completed during one second. Substituting $1/\nu$ for T in Equation 12.15,

$$\frac{kv}{\nu} = 2\pi \text{ rad}$$

But from Equation 12.13, $k = (2\pi \text{ rad}/\lambda)$, so that

$$\frac{2\pi \text{ rad}}{\lambda} \frac{v}{\nu} = 2\pi \text{ rad}$$

and we regain Equation 12.10, valid for any periodic wave:

$$v = \nu\lambda$$

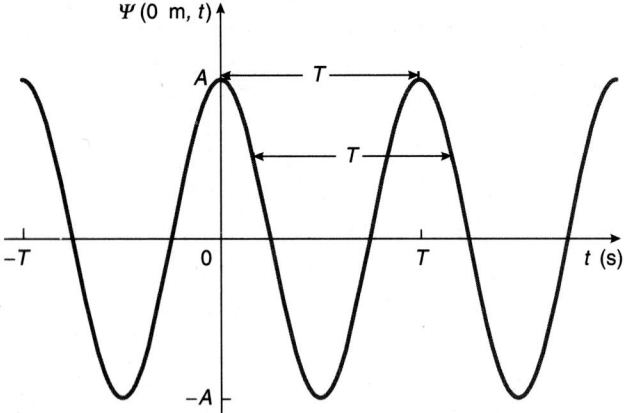

FIGURE 12.23 The oscillatory behavior of a sinusoidal wave at a fixed location, here the origin where $x = 0$ m.

Note that the product

$$kv = \frac{2\pi}{\lambda} v\lambda = 2\pi\nu = \omega$$

The product $2\pi\nu$ is called the angular frequency ω of the wave from the analogous quantity in circular motion.

The sinusoidal wavefunction of Equation 12.11 can be written in many equivalent forms:

$$\Psi(x, t) = A \cos[k(x - vt)]$$
$$= A \cos(kx - kvt)$$

$$\Psi(x, t) = A \cos(kx - \omega t) \qquad (12.16)$$
$$= A \cos\left[2\pi\left(\frac{x}{\lambda} - \nu t\right)\right]$$
$$= A \cos\left[2\pi\left(\frac{x}{\lambda} - \frac{t}{T}\right)\right]$$

To avoid notational overload, we choose Equation 12.16 as the standard form for a sinusoidal wave because of its notational simplicity.

The sinusoidal wavefunction contains an incredible amount of information within its rather humble mathematical form.

PROBLEM-SOLVING TACTIC

12.2 To find the value of $\cos(kx - \omega t)$ in Equation 12.11 or 12.16, be sure your calculator is set for radian measure of angle, not degrees.

 STRATEGIC EXAMPLE 12.3

A sinusoidal wave representing the transverse oscillations on a string has the wavefunction

$$\Psi(x, t) = (0.40 \text{ m}) \cos[(6.00 \text{ rad/m})x - (10.0 \text{ rad/s})t]$$

a. What is the amplitude of the wave?
b. What is the wavelength of the wave?
c. Find the frequency of the wave.
d. Find the speed of the wave.
e. What is the value of Ψ at the position $x = 5.00$ m when $t = 2.00$ s?
f. Plot the waveform of Ψ as a function of x over the domain $x = 0$ m to $x = 3.00$ m when $t = 2.00$ s. Indicate the wavelength λ on your graph.
g. Plot the oscillation at $x = 5.00$ m as a function of t over the domain $t = 0$ s to $t = 3.00$ s. Indicate the period T on your graph.

Solution

a. The amplitude is the coefficient of the cosine term. Thus the amplitude is 0.40 m.

b. The coefficient of x in the argument of the cosine is the angular wavenumber k. Thus

$$k = 6.00 \text{ rad/m}$$

But according to Equation 12.13, k is $(2\pi \text{ rad})/\lambda$; hence, you have

$$\frac{2\pi \text{ rad}}{\lambda} = 6.00 \text{ rad/m}$$

Solving for the wavelength λ,

$$\lambda = \frac{\pi}{3.00} \text{ m}$$
$$= 1.05 \text{ m}$$

c. The coefficient of t in the argument of the cosine is the angular frequency ω. Thus

$$\omega = 10.0 \text{ rad/s}$$

But the angular frequency ω is related to the frequency ν by

$$\omega = 2\pi\nu$$

Therefore

$$10.0 \text{ rad/s} = 2\pi\nu$$

Solving for the frequency,

$$\nu = \frac{5.00}{\pi} \text{ s}^{-1}$$
$$= 1.59 \text{ Hz}$$

d. According to Equation 12.10, the speed of the wave is the product of the frequency and the wavelength. Thus

$$v = (1.59 \text{ Hz})(1.05 \text{ m})$$
$$= 1.67 \text{ m/s}$$

e. Use the given wavefunction and substitute for x and t:

$$\Psi(5.00 \text{ m}, 2.00 \text{ s}) = (0.40 \text{ m}) \cos[(6.00 \text{ rad/m})(5.00 \text{ m})$$
$$- (10.0 \text{ rad/s})(2.00 \text{ s})]$$
$$= (0.40 \text{ m}) \cos(10.0 \text{ rad})$$
$$= (0.40 \text{ m})(-0.839)$$
$$= -0.34 \text{ m}$$

f. Make the substitution $t = 2.00$ s into the given wavefunction, which yields the waveform

$$\Psi(x, 2.00 \text{ s}) = (0.40 \text{ m}) \cos[(6.00 \text{ rad/m})x - 20.0 \text{ rad}]$$

A graph of this appears in Figure 12.24 with the wavelength λ indicated.

g. Making the substitution $x = 5.00$ m into the given wavefunction yields the oscillatory behavior at this location:

$$\Psi(5.00 \text{ m}, t) = (0.40 \text{ m}) \cos[30.0 \text{ rad} - (10.0 \text{ rad/s})t]$$

A graph of this is shown in Figure 12.25 with the period T indicated.

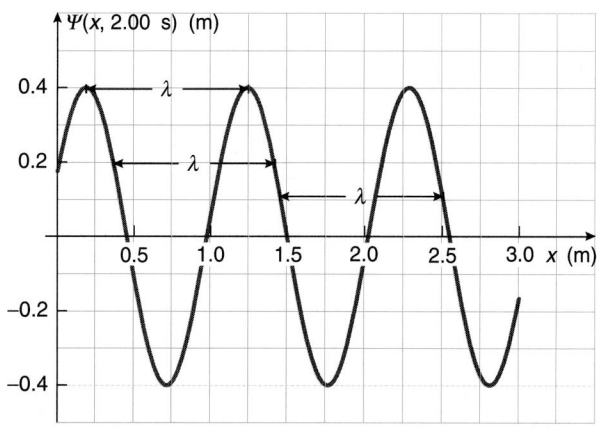

FIGURE 12.24

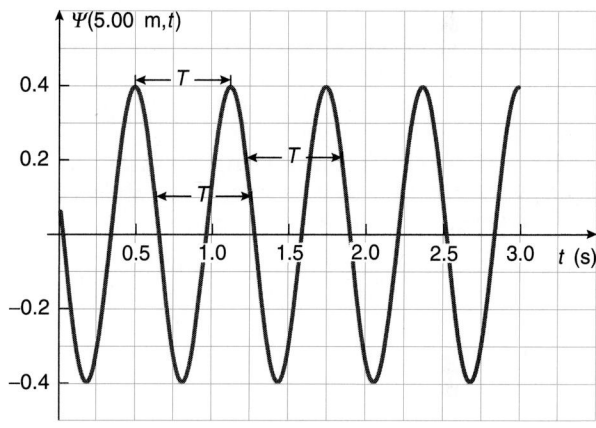

FIGURE 12.25

EXAMPLE 12.4

Show that a sinusoidal wavefunction of the form

$$\Psi(x, t) = A \cos\left(kx - \omega t - \frac{\pi}{2} \text{ rad}\right)$$

is equivalent to

$$\Psi(x, t) = A \sin(kx - \omega t)$$

Solution

Use the double-angle trigonometric formula

$$\cos(\alpha - \beta) = \cos\alpha \cos\beta + \sin\alpha \sin\beta$$

and the given wavefunction with $\alpha = kx - \omega t$ and $\beta = \pi/2$ rad to obtain (omitting the radian units on $\pi/2$)

$$\Psi(x, t) = A\left[\cos\left(kx - \omega t\right)\cos\frac{\pi}{2} + \sin\left(kx - \omega t\right)\sin\frac{\pi}{2}\right]$$
$$= A \sin(kx - \omega t)$$

This means that a sinusoidal wave of the form

$$\Psi(x, t) = A \sin(kx - \omega t)$$

can be written in the cosinusoidal form by subtracting $\pi/2$ rad from the argument of the sine.

12.9 WAVES ON A STRING

Waves on a string or wire under tension are involved in the rich musical texture of all stringed instruments, and so an understanding of the physics associated with such waves can add to your appreciation of the guitar, banjo, violin, or piano that you might play.

When you stretch a string taut and send a transverse wave along it by wagging your hand back and forth, you can easily observe that the speed of the wave increases if you increase the tension in the cord. If you use a cord of greater mass per unit length, the speed decreases. Let's use our knowledge of physics to see how the wave speed depends on the tension and mass per unit length of the cord.

A transverse wave Ψ propagating along a string represents a departure of sections of the string from their equilibrium positions along the straight original position of the string. Such transverse departures from the equilibrium position typically are small compared with the length of the string; we will consider only such small-amplitude waves.

Place a uniform string under a tension of magnitude T. Let μ be the mass per unit length of the string, found by dividing the total mass m of the string by its total length ℓ:

$$\mu = \frac{m}{\ell}$$

If the end of the string suddenly is moved slightly to the side, a kink in the string propagates along the string at speed v. We want to examine the transverse motion of the bit of string caused by the propagating kink. Consider a differential length dx of the string just encompassing the kink during a small time interval dt, so that

$$dx = v \, dt$$

The mass dm of this small section of the string is the product of the length dx and the mass per unit length μ:

$$dm = \mu \, dx$$
$$= \mu v \, dt$$

Let the transverse component of the velocity of the string element be v_y, where we choose the y-axis to be perpendicular to the length of the string in the direction of motion of the mass element of the string. During the interval dt, the change in the y-component of the momentum of the mass element is

$$d\vec{p} = (\mu v \, dt) v_y \hat{\jmath} - (0 \text{ kg·m/s}) \hat{\jmath}$$
$$= (\mu v \, dt) v_y \hat{\jmath}$$

The forces acting on this bit of string are the tension exerted by other parts of the string and the weight of the string segment.

The strings of this Bösendorfer grand piano are under great tension. This piano also is pictured in Question 5 of Chapter 1 (page 12).

Since the segment of string is small, we neglect the weight compared with the tension; we then have the second law force diagram shown in Figure 12.26. Our results, therefore, will not depend on whether the axis of the string is horizontal, vertical, or inclined at some angle to the horizontal, which is fortunate for musicians! Since the tension has a component $T \sin \theta$ in the transverse direction, the impulse of the tension in the y-direction during the interval dt is

$$(T \sin \theta) \, dt \, \hat{\jmath}$$

The impulse–momentum theorem from Chapter 9 states that the total impulse is equal to the change in the momentum. For the y-direction we have

$$(T \sin \theta) \, dt \, \hat{\jmath} = (\mu v \, dt) v_y \hat{\jmath}$$
$$T \sin \theta = \mu v v_y \qquad (12.17)$$

From Figure 12.27 we have

$$\tan \theta = \frac{v_y}{v}$$

$$\frac{\sin \theta}{\cos \theta} = \frac{v_y}{v} \qquad (12.18)$$

Using Equation 12.18 for $\sin \theta$ in Equation 12.17, we find

$$(T \cos \theta) \frac{v_y}{v} = \mu v v_y$$

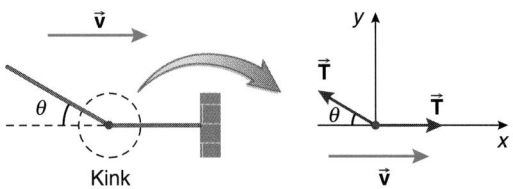

FIGURE 12.26 A propagating kink and the forces on the small mass element at the kink.

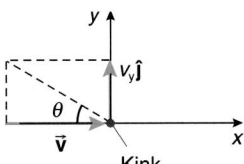

FIGURE 12.27 The velocity components at the kink.

Isolating the wave speed v, we obtain

$$v^2 = \frac{T}{\mu} \cos \theta$$

Since the motion of the string to the side was small, the angle θ also is small, in which case $\cos \theta \approx 1$. Hence, to a good approximation, we have

$$v^2 = \frac{T}{\mu}$$

> The wave speed v is
>
> $$v = \left(\frac{T}{\mu} \right)^{1/2} \qquad (12.19)$$
>
> Notice that the greater the tension, the greater the speed; and the greater the mass per unit length, the smaller the speed of the wave on the string, both in keeping with our qualitative observations for a stretched cord.

In fact, Equation 12.19 is a special case of a general relationship for the speed of mechanical waves in material media. The speed is proportional to the square root of the magnitude of some stiffness quantity related to a force divided by an inertial quantity related to the mass:

$$v \propto \left(\frac{\text{magnitude of a force factor}}{\text{mass factor}} \right)^{1/2}$$

For example, Young's modulus E is the ratio between tensile stress and tensile strain in a material (see Equation 11.3). Young's modulus is a force factor. A corresponding mass factor for a bulk material ought to be the density ρ. A dimensional analysis of the ratio of Young's modulus to the density gives the square of a speed; in SI units,

$$\frac{[E]}{[\rho]} = \frac{\text{N/m}^2}{\text{kg/m}^3} = \frac{(\text{kg} \cdot \text{m/s}^2)/\text{m}^2}{\text{kg/m}^3} = \text{m}^2/\text{s}^2$$

In fact, a rigorous derivation finds that the speed of sound waves in solids is

$$v_{\text{solid}} = \left(\frac{E}{\rho} \right)^{1/2} \qquad (12.20)$$

In a similar way, the speed of sound waves in liquids is related to the bulk modulus B:

$$v_{\text{liquid}} = \left(\frac{B}{\rho} \right)^{1/2} \qquad (12.21)$$

EXAMPLE 12.5

A nylon clothesline of length 5.0 m and mass 0.076 kg is subjected to a tension of magnitude 5.1 N. If the end of the rope is flicked slightly to the side, what is the speed of the pulse that travels along the rope?

Solution
The mass per unit length of the rope is

$$\mu = \frac{m}{\ell}$$

$$= \frac{0.076 \text{ kg}}{5.0 \text{ m}}$$

$$= 0.015 \text{ kg/m}$$

Use Equation 12.19 to find the speed:

$$v = \left(\frac{T}{\mu} \right)^{1/2}$$

$$= \left(\frac{5.1 \text{ N}}{0.015 \text{ kg/m}} \right)^{1/2}$$

$$= 18 \text{ m/s}$$

EXAMPLE 12.6

The bulk modulus of water is 0.20×10^{10} N/m^2 and its density is 1.00×10^3 kg/m^3. What is the speed of sound in water?

Solution
The speed of sound in liquids is related to the bulk modulus and the density through Equation 12.21:

$$v_{\text{liquid}} = \left(\frac{B}{\rho} \right)^{1/2}$$

$$= \left(\frac{0.20 \times 10^{10} \text{ N/m}^2}{1.00 \times 10^3 \text{ kg/m}^3} \right)^{1/2}$$

$$= 1.4 \times 10^3 \text{ m/s}$$

12.10 REFLECTION AND TRANSMISSION OF WAVES

Echoes are reflected sound waves; when you hear traffic noise through the wall of your room, the sound waves were transmitted through the wall. So longitudinal waves can be reflected and transmitted at boundaries between two media.

Likewise, take a rope and tie one end to a rigid wall. If you send a pulse down the rope toward the wall, the pulse will be reflected from the wall and return toward you. Notice that if the pulse incident on the wall is an "up" pulse, the reflected pulse is a "down" pulse, as shown in the series of sketches in Figure 12.28.

We say the pulse experiences a **change of phase** of π rad (=180°) on reflection. The phase change is caused by the force of the wall on the rope. Imagine the particle of the rope at the wall. As the wave pulse impinges on the wall, the wave tries to force the rope upward, and so the rope exerts an upward force on the wall. Thanks to Newton's third law, the wall exerts a force of the same magnitude in the opposite direction; hence the wall exerts a downward force on the rope, which produces the reflected pulse in the downward direction.

On the other hand, if we tie the rope to a ring that is free to slide on a frictionless, vertical post, an incident "up" pulse is reflected as an "up," as shown in the series of sketches in Figure 12.29, quite unlike the reflection from the rigid wall. There is no phase change on reflection.

If we tie the rope to another rope with greater mass per unit length and send a pulse down the rope, as in Figure 12.30(a), we discover that a pulse is transmitted to the second rope *and* a pulse is reflected. Both the transmitted and reflected pulses have smaller amplitudes than the original pulse, indicating that the energy in the original wave is somehow divided between the transmitted and reflected waves. Since the ropes have the same tension, but different masses per unit length, Equation 12.19 implies that the speed of the pulse along the rope with greater mass per unit length is slower. The characteristic length of the pulse also is smaller on this rope. An incident harmonic wave has a smaller wavelength on the rope of greater mass per unit length.

Such simultaneous reflection and transmission of a wave occurs whenever there is a change in the medium in which a wave propagates. Sound waves are both reflected and transmitted at walls and windows, and so you can hear a neighbor's conversation or stereo, much to their consternation or yours.* Light

---

*Intelligence agencies have developed ways to detect (from some distance) the small vibrations of window panes induced by the chatter within.

waves are both reflected and transmitted at boundaries between two transparent media such as air and water.

The simultaneous reflection and transmission of a wave is quite unlike the corresponding case for a particle; an incident particle is *either* totally reflected *or* totally transmitted, not both simultaneously.

QUESTION 4

A surface water wave impinges on a sturdy sea wall. Does the reflected wave experience a phase change at the wall?

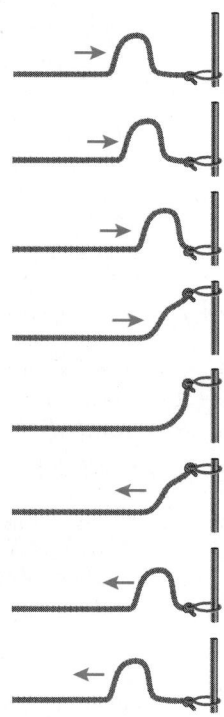

FIGURE 12.29 No phase change on reflection.

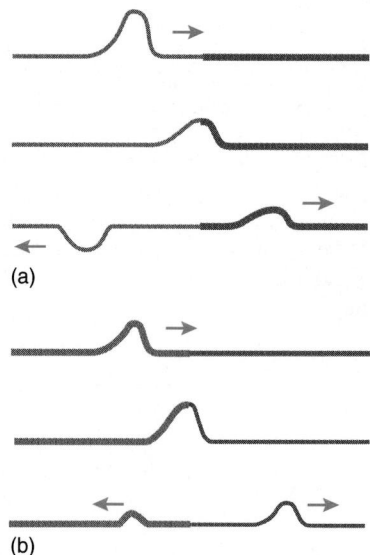

(a)

(b)

FIGURE 12.30 Reflection and transmission of a pulse at the boundary between two ropes with different masses per unit length. (a) A change of phase occurs when reflecting from a thicker rope. (b) No change of phase occurs when reflecting from a thinner rope.

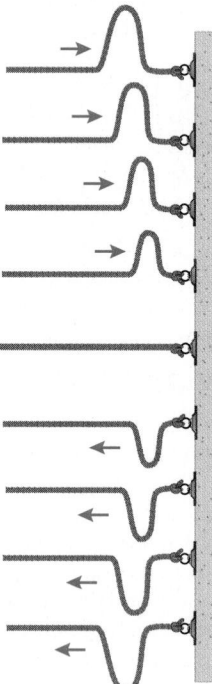

FIGURE 12.28 An incident pulse reflects from a rigid wall.

12.11 ENERGY TRANSPORT VIA MECHANICAL WAVES

Mechanical waves involve masses in motion, and so it should come as no surprise that energy is associated with waves. What is unusual about repetitive mechanical waves is that energy is transported from one point to another along the direction of the wave velocity without the net transport of mass between the two points: there are no particle streams.

Harmonic acoustic waves involve longitudinal simple harmonic oscillations of the particles that compose the medium. Likewise, for transverse harmonic waves on string, the mass particles of the string execute simple harmonic motion. We saw in Chapter 7 that simple harmonic oscillation of a mass m has a total mechanical energy E proportional to the square of the amplitude of the oscillation:

$$E = \frac{1}{2} kA^2 \qquad (12.22)$$

where k is the effective spring constant, equal to $m\omega^2$ via Equation 7.10. For a differential amount of mass dm, there is a differential amount of mechanical energy dE, and we write Equation 12.22 in the form

$$dE = \frac{1}{2} dm\, \omega^2 A^2 \qquad (12.23)$$

Let dm be the mass of a small length dx of the string, so that

$$dm = \mu\, dx$$

where μ is the mass per unit length of the string. Hence the total mechanical energy on the length dx is

$$dE = \frac{1}{2} \mu\, dx\, \omega^2 A^2 \qquad (12.24)$$

Light waves are both reflected and transmitted at the boundary between two transparent materials.

Since the wave is traveling in the x-direction at speed v,

$$v = \frac{dx}{dt}$$

so

$$dx = v\, dt$$

Substituting for dx in Equation 12.24, we find

$$dE = \frac{1}{2} \mu v\, dt\, \omega^2 A^2 \qquad (12.25)$$

The time rate of energy transport,

$$\frac{dE}{dt}$$

is the *power* P transmitted (transferred) by the wave.

Hence the power is

$$P = \frac{1}{2} \mu \omega^2 A^2 v \qquad (12.26)$$

The power transfer is proportional to the square of the angular frequency (equivalently, to the square of the frequency ν, since $\omega = 2\pi\nu$), to the square of the amplitude, and to the propagation speed of the wave.

The power transfer of any classical wave has the same dependence on frequency, amplitude, and speed, although we will not show this explicitly.

QUESTION 5

To double the power of a wave, by what factor must its amplitude increase if other factors are unchanged?

EXAMPLE 12.7

A nylon clothesline of length 5.0 m and mass 76 g has a tension of magnitude 50 N. One end of the string is wiggled back and forth in simple harmonic motion with a frequency of 10 Hz, producing a transverse harmonic wave of amplitude 2.0 cm on the string. What is the power transferred from one end of the string to the other?

Solution

To apply Equation 12.26, you first need to find the speed of the waves on the string from Equation 12.19:

$$v = \left(\frac{T}{\mu}\right)^{1/2}$$

$$= \left[\frac{50 \text{ N}}{(0.076 \text{ kg})/(5.0 \text{ m})}\right]^{1/2}$$

$$= 57 \text{ m/s}$$

The angular frequency ω of the oscillation is

$$\omega = 2\pi\nu$$

$$= 2\pi(10 \text{ Hz})$$

$$= 63 \text{ rad/s}$$

From Equation 12.26, the power is

$$P = \frac{1}{2} \mu \omega^2 A^2 v$$

$$= \frac{1}{2} \frac{0.076 \text{ kg}}{5.0 \text{ m}} (63 \text{ rad/s})^2 (0.020 \text{ m})^2 (57 \text{ m/s})$$

$$= 0.69 \text{ W}$$

12.12 WAVE INTENSITY

The **intensity** of a wave is the average power transmitted by the wave through one square meter oriented perpendicular to the direction the wave is propagating; see Figure 12.31.

The SI units of intensity are, therefore, watts per square meter (W/m^2).

Generalizing from Equation 12.26, the intensity of a wave is proportional to the square of its amplitude and the square of its frequency.

For waves moving out isotropically* in three dimensions, the power is distributed over the *area* of a sphere of radius r centered on the source, as shown in Figure 12.32. Since the area

---

*The word isotropic means "the same in all directions."

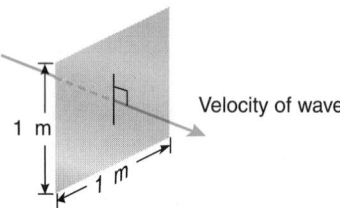

FIGURE 12.31 The intensity is the average power through one square meter perpendicular to the direction of propagation of the wave.

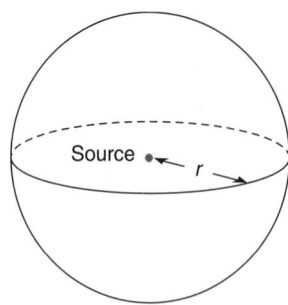

FIGURE 12.32 The power of a pointlike wave source is distributed uniformly over the area of sphere centered on the source.

of sphere is $4\pi r^2$, the intensity of the waves at a distance r from the source is

$$I = \frac{P}{4\pi r^2} \tag{12.27}$$

The intensity decreases with the inverse square of the distance from the source, and so the intensity of a *spherical* wave follows an inverse square law.

EXAMPLE 12.8

A source emitting waves isotropically in three dimensions has an intensity I_0 at a distance r_0 from the source. At what distance from the source is the intensity $0.100 I_0$?

Solution
Since the wave is propagating isotropically in three dimensions, the intensity follows the inverse square law of Equation 12.27:

$$I_0 = \frac{P}{4\pi r_0^2} \tag{1}$$

Likewise at distance r, the intensity is

$$I = \frac{P}{4\pi r^2}$$

where the power P of the source is the same. Since you want $I = 0.100 I_0$, you have

$$0.100 \, I_0 = \frac{P}{4\pi r^2}$$

Using equation (1) to substitute for P, you have

$$0.100 \, I_0 = \frac{1}{4\pi r^2} 4\pi r_0^2 I_0$$

Solving for r, you obtain

$$r = 3.16 r_0$$

Hence if you increase the distance from the source by a factor of 3.16, the intensity decreases by a factor of 10.0.

12.13 WHAT IS A SOUND WAVE?*

We say a **sound** exists in an elastic material if a propagating disturbance is a variation in the ambient density caused by a bulk shift in the positions of the particles of the material away from their nominal equilibrium positions. The disturbance Ψ can represent either the density variation or the departure of the particles from their erstwhile positions. For sound propagating in gases, the density variation also can be interpreted as a variation in pressure from the ambient pressure.

When you think of sound, you likely consider sound propagating through air (a gas), but it equally well can propagate through other media such as liquids and solids. It is difficult to imagine a gas such as air as an elastic material; if you swing your hand through the air, you do not feel much of a countervailing restoring force. However, take a tire pump, raise the plunger, and then hold a finger tightly over the end of the hose. When you press on the plunger, a restoring force is quite apparent: the air has elastic-like properties. Sound waves in air or other gases represent such rapid oscillations in density that the inertial properties of the air make it quite elastic. The propagating variations in density are correlated with the bulk particle motions, much like the longitudinal compressional waves along the axis of a stretched slinky. But, as we will see, the density wave and particle wave disturbances are not in phase (in step) with each other.

The air particles are randomly moving on a microscopic basis but, for clarity, we ignore these random motions here.* We represent the particles of the air as a uniform distribution of fixed dots that represents the ambient density, as shown in Figure 12.33.

Imagine a membrane (such as the diaphragm of a speaker) or wall executing rapid, back-and-forth, simple harmonic motion with period T in the gas. As the wall moves from its equilibrium position to the right, it accelerates the air particles next to it, compressing them because the particles beyond them have not yet shifted their positions. The compression is represented by a crowding of dots near the wall in Figure 12.34, indicating that the density is greater than the ambient density in this region. Dots further from the wall are as yet undisturbed, and the density there remains at its ambient value.

---

*We will examine the microscopic motions of the particles in a gas in greater detail in Chapter 14. Here we are interested in bulk motions of the particles of the gas.

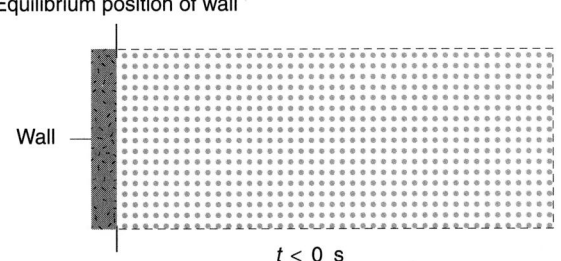

FIGURE 12.33 Undisturbed air is represented as a uniform distribution of fixed dots.

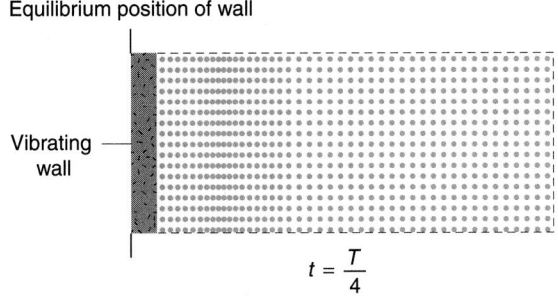

FIGURE 12.34 The shift in the membrane causes a bunching of the dots near it.

In addition to their microscopic, random motion momentum, the particles also now have some momentum to the right, transferred to them by the moving wall. The momentum is transferred to other air particles to the right by collisions, and in this way, the initial compression propagates through the gas.

The motion of the wall in the opposite direction causes a rarefaction in the distribution of particles near it. A rarefaction has a density less than the ambient value, indicated by the increased spacing between the dots shown in Figure 12.35, causing them to accelerate in the opposite direction. As the wall oscillates back and forth, the process is repeated faithfully, and successive compressions and rarefactions propagate out into the gas, as shown in Figure 12.36. The particle motions are longitudinal—that is, along the line the wave is propagating.

Let's see how the variations in the ambient density correlate with the positions of the particles right and left of their original positions, and with the velocity components of the particles. Since the wall is executing simple harmonic oscillation and the gas is

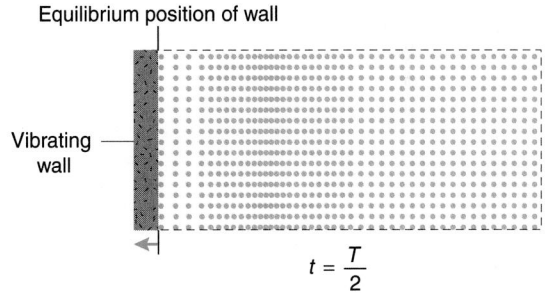

FIGURE 12.35 The formation of a rarefaction.

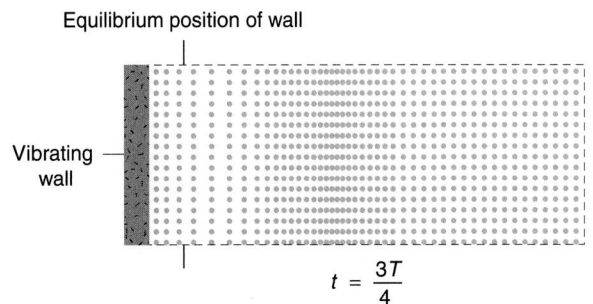

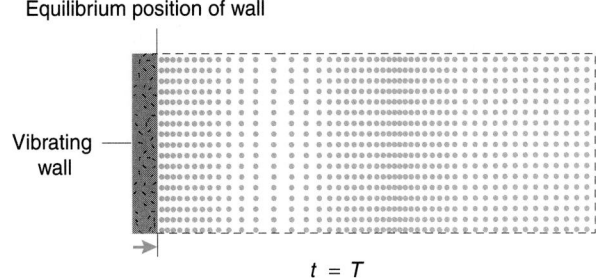

FIGURE 12.36 Successive compressions and rarefactions propagate into the gas.

considered elastic, the particles also execute simple harmonic oscillations about their original nominal equilibrium positions. Recall that in simple harmonic oscillation of a particle, the speed is zero at greatest amplitude; a maximum speed exists when passing through the equilibrium position. The maximum density occurs at the center of the compression, where the particles are at their equilibrium position but are moving with greatest velocity to the right in our pictorial representation of the rightward traveling sound wave. The particle at the center of the compression must be moving faster than the ones on its right and left; we represent this by velocity arrows with different lengths in Figure 12.37. In this way the particle at the center of the compression is catching up to the particle on its right while pulling away from the particle on its left. The result is that the compression moves to the right.

Likewise in a rarefaction, where the variation in density has its greatest value below the ambient value, the particle at the center of the rarefaction must be moving faster to the left than the ones to its left and right, as indicated in Figure 12.38. In this way the given particle is catching up to the particle on its left and pulling away from the particle on its right. As a result the rarefaction also propagates to the right.

Thus we see that the velocity component of the particles is in phase (in step) with the density variation: greatest compression occurs at the same times and places where the velocity of the particles is greatest to the right, while the center of a rarefaction occurs where the velocity of the particles is greatest to the left.

We summarize the information about the relationship among the variations in density above (+) and below (−) the ambient value, the velocity component to the right (+) or left (−), and the variation in position of the particles from their ambient positions, taking + to the right and − to the left in the graphical representation of their respective waveforms shown in Figure 12.39. We reach several conclusions from this analysis. In the regions where the waves exist,

- the density variation and the longitudinal velocity component of the particles have their maxima and minima at those locations where the particles are at their former equilibrium positions; and

- the particles have maximum departure from equilibrium where the density is the ambient value (zero variation in density).

Center of compression is at
equilibrium position of particle's
simple harmonic oscillation with
particle moving to the right

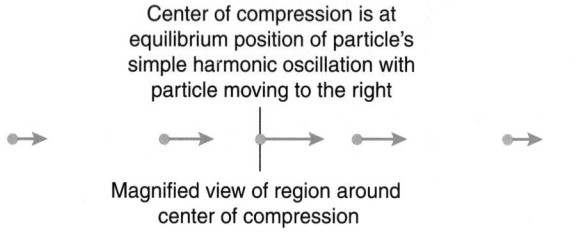

Magnified view of region around
center of compression

FIGURE 12.37 The velocities of particles at and on either side of the maximum compression.

Center of rarefaction is at
equilibrium position of particle's
simple harmonic oscillation with
particle moving to the left

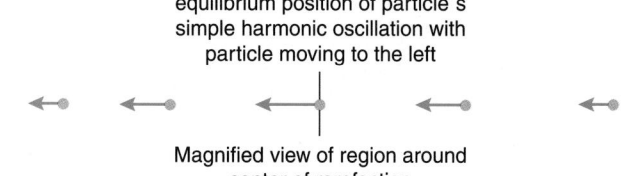

Magnified view of region around
center of rarefaction

FIGURE 12.38 The velocities of particles at and on either side of a rarefaction.

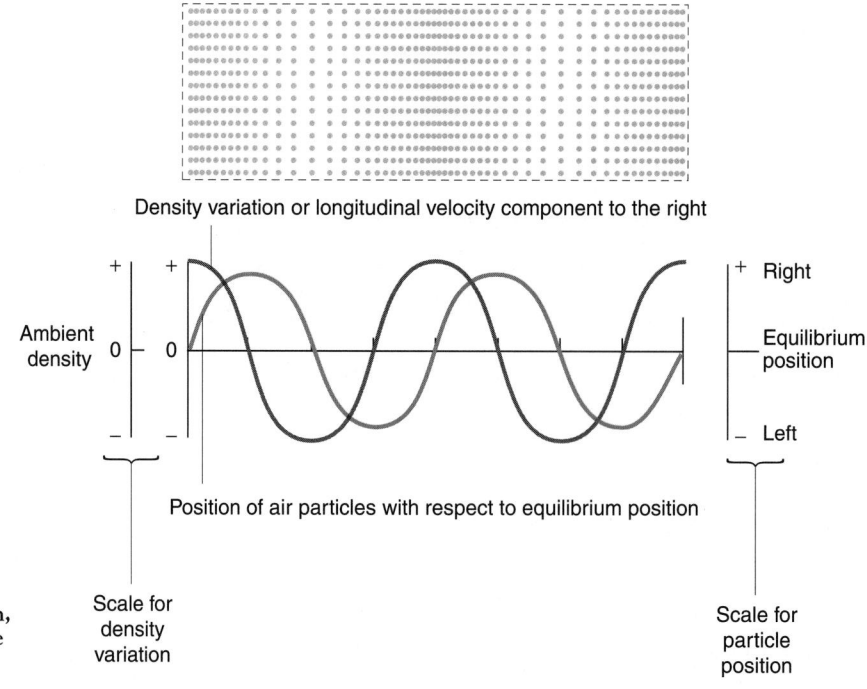

FIGURE 12.39 Graphs of the density variation, velocity component, and position (relative to the equilibrium position) of the particles.

Hence the density wave disturbance Ψ_{density} is 90° (= $\pi/2$ rad) out of phase with the particle position wave disturbance Ψ_{position}.

12.14 SOUND INTENSITY AND SOUND LEVEL*

The Sounds of Silence

Paul Simon and Art Garfunkel[†]

From the sound of gently rustling leaves to the incredible roar of a space shuttle launch, the ear is a very sensitive detector of sound wave intensities. The **threshold for hearing** of the human ear for acoustic waves is an intensity of only about 10^{-12} W/m² at a frequency of about 1 kHz.[‡] At the other extreme, the intensity of sound waves that causes deafening pain (such as encountered frequently at rock concerts or a shuttle launch) is about a factor of 10^{12} greater than the threshold of hearing and is quite independent of frequency.

The enormous range of intensities detectable by the human ear makes the introduction of a more practical measure of sound level a matter of convenience. Logarithms are useful for this purpose. If a quantity α increases to $10^{12}\alpha$, the change in the value of the logarithm is only a factor of 12, rather than 10^{12}:

$$\log 10^{12}\alpha - \log \alpha = 12$$

The threshold of hearing intensity is defined to be the intensity

$$I_0 \equiv 10^{-12} \ \text{W/m}^2$$

and is used as a reference for sound intensity. The **sound level** β of a sound intensity I is defined to be

$$\beta \equiv (10 \ \text{dB}) \log \frac{I}{I_0} \qquad (12.28)$$

The ratio I/I_0 is used since logarithms are exponents and must be dimensionless. The units for the sound level β are **decibels** (dB) one-tenth of the larger, but rarely used unit called a bell (B) named after the famous inventor of the telephone, Alexander Graham Bell (1847–1922).[#] From Equation 12.28 with $I = I_0$, the threshold of hearing thus has a sound level of 0 dB.

Each *factor* of 10 increase in sound intensity I produces an *additive* change in sound level of 10 dB; see Table 12.1.

[†]Title of song on album titled *Sounds of Silence* CS9269 Columbia Records CL2469 (Eclectic Music Company 1965).

[‡]The threshold intensity is a function of frequency; see Figure 12.40.

[#]The decibel unit also can be used for other types of waves if a reference intensity I_0 is chosen for those waves. The unit also is used in electronics for comparing quantities other than wave intensity.

PROBLEM-SOLVING TACTIC

12.3 Logarithms are exponents, so the expression

$$Y = \log X$$

means, after taking the antilogarithms,

$$X = 10^Y$$

Therefore when you take the antilogarithms of Equation 12.28, you have, after first dividing by 10 dB,

$$\frac{I}{I_0} = 10^{\beta/(10 \ \text{dB})} \qquad (12.29)$$

The human ear is not equally sensitive to sounds at all frequencies, as indicated in the graph in Figure 12.40. The ears of other lifeforms have different sensitivities to sounds, as you likely know if you have a dog or cat.

QUESTION 6

Dogs have very sensitive ears, capable of detecting sound intensities less than the threshold for human hearing. What sound levels does this correspond to in dB?

TABLE 12.1 Approximate Sound Levels and Sound Intensities

Sound	Sound level β (dB)	Sound intensity I (W/m²)
Threshold of hearing (definition)	0	10^{-12}
Cat purring	10	10^{-11}
Whisper at 1 m	20	10^{-10}
Normal conversation	60	10^{-6}
Urban traffic	70	10^{-5}
Subway train	90	10^{-3}
Chain saw at 1 m	100	10^{-2}
Loud rock concert	120	1
Threshold of pain	120	1
Jet engine at 10 m	150	10^{3}
Space shuttle launch at 50 m	200	10^{8}

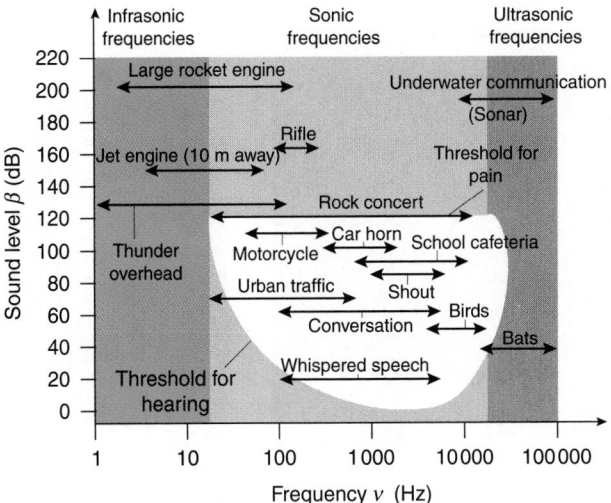

FIGURE 12.40 The frequency dependence of the threshold of hearing and of pain for a typical human ear.

Example 12.9

If a sound level increases by 3.00 dB, by what factor does the sound intensity increase?

Solution

Let β be the sound level with sound intensity I. Then, according to Equation 12.28,

$$\beta = (10 \text{ dB}) \log \frac{I}{I_0}$$

Dividing by 10 dB and then taking antilogarithms, you have

$$\frac{I}{I_0} = 10^{\beta/(10 \text{ dB})} \qquad (1)$$

Let I' be the sound intensity when the sound level is $\beta + 3.00$ dB; hence,

$$\beta + 3.00 \text{ dB} = (10 \text{ dB}) \log \frac{I'}{I_0}$$

Dividing by 10 dB and then taking the antilogarithms, you have

$$\frac{I'}{I_0} = 10^{(\beta + 3.00 \text{ dB})/(10 \text{ dB})} \qquad (2)$$

Dividing equation (2) by equation (1),

$$\begin{aligned}
\frac{I'}{I} &= \frac{10^{(\beta + 3.00 \text{ dB})/(10 \text{ dB})}}{10^{\beta/(10 \text{ dB})}} \\
&= 10^{0.300} \\
&= 2.00
\end{aligned}$$

Therefore an increase in sound level of 3.00 dB increases the intensity by a factor of 2.00.

12.15 THE ACOUSTIC DOPPLER EFFECT*

If not by name, the qualitative aspects of the **acoustic Doppler effect** likely are familiar to you. If a source of sound (siren, train whistle, or screaming Mimi,[†] etc.) is approaching at high speed, you hear a higher frequency than when the source is at rest. Conversely, if the source is receding from you, a lower frequency is heard. The change in frequency is quite apparent as the source whooshes by you. The effect is qualitatively the same if the source is stationary and the observer is approaching the source (higher frequency heard) or receding from it (lower frequency heard).

A quantitative description of the acoustic Doppler effect first was devised in 1842 by Johann Christian Doppler (1803–1853). To verify the theory, Buys Ballot at Utrecht used trumpeters moving on a railroad flatcar and stationary observers who had perfect pitch.[‡] The experiment must have made a curious sight—and sound.

As Doppler himself speculated, the effect is a general one for all types of waves (sound, light, water, and other types of waves), although the quantitative description of the changes in frequency depends on the nature of the particular wave phenomena. The quantitative expressions for the frequency change are different for sound and for light. The distinctions arise because, unlike sound, light waves do not need a medium through which to travel, as we will see in Chapters 22 and 25. Here we examine the acoustic Doppler effect. The Doppler effect for electromagnetic waves (light) is best approached using relativity and will be examined when we tackle that fascinating aspect of physics in Chapter 25.

[†]Heroine of the opera *La Bohème* by Puccini (1896).

[‡]Perfect pitch is the ability to distinguish and name musical notes (frequencies) simply by hearing them.

Johann Christian Doppler first quantified the acoustic effect that now bears his surname.

The Doppler effect first was tested using moving and stationary musicians!

Let v be the speed of sound. The speed of sound in air varies with temperature and is about 343 m/s at 20 °C. At other temperatures $t_{celsius}$, the speed of sound in air is given approximately by the following empirical equation:

$$v = 331 \text{ m/s} + [0.60 \text{ m/(s·°C)}]t_{celsius}$$

The speed of sound in various other materials is tabulated in Table 12.2.

We need to consider motion of the source, the observer, and the medium through which the sound waves travel. For convenience, the speeds will be measured with respect to a fixed reference frame, say the ground. Although this seems to imply the existence of a preferred reference frame, we use this device as a means of keeping the conceptual development clear; it is easier to apply the resulting equations with this approach.

Motion of the Observer with the Source and Medium at Rest

Let a source emit sound of frequency v. The waves propagate toward an observer (see Figure 12.41). The lines represent circles of constant phase (wavefronts) and are spaced one wavelength apart; they propagate toward the observer at a speed v through the medium. If the observer is at rest, v wavefronts (crests) are received each second because the frequency is the number of such crests or wavelengths that are received during each second.

If the observer is moving with a speed v_{obs} toward the stationary source, as in Figure 12.42, the observer moves a distance $v_{obs}t$ during a time interval t. During one second, then,

TABLE 12.2 Approximate Value of the Speed of Sound in Various Materials

Material	Speed (m/s)
Gases	
Air (0 °C)	331
Air (20 °C)	343
Nitrogen (20 °C)	351
Oxygen (20 °C)	328
Liquids	
Ethyl alcohol (20 °C)	1.17×10^3
Mercury (20 °C)	1.45×10^3
Petroleum (15 °C)	1.33×10^3
Water (fresh) (20 °C)	1.48×10^3
Water (sea) (20 °C)	1.52×10^3
Solids	
Aluminum	6.42×10^3
Brass	4.25×10^3
Glass (crown glass)	5.66×10^3
Gold	3.24×10^3
Granite	6.0×10^3
Silver	3.60×10^3
Steel	6.10×10^3
Wood (oak)	4.10×10^3

Myer Kutz (editor), *Mechanical Engineer's Handbook* (John Wiley, New York, 1986), pages 692–693.

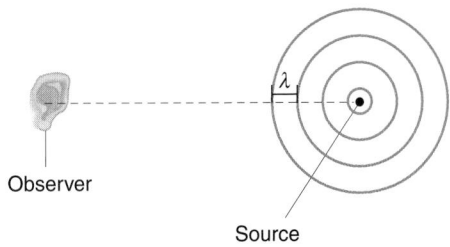

FIGURE 12.41 Wave propagating from a stationary source toward a stationary observer.

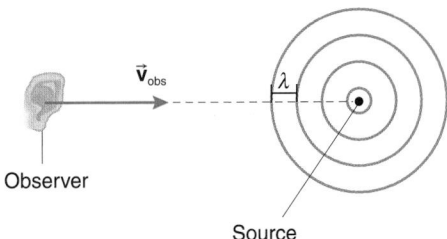

FIGURE 12.42 Observer moving toward a stationary source increases the effective frequency.

the observer moves a distance equal to $v_{obs}(1 \text{ s})$. There are $v_{obs}(1 \text{ s})/\lambda = v_{obs}/\lambda$ crests in this distance. So the observer receives an additional v_{obs}/λ crests each second because of motion toward the source.

Hence the total number of wavefronts the observer collects during the one second interval (which is the frequency v' of the waves detected by the observer) is

$$v' = v + \frac{v_{obs}}{\lambda}$$

Factoring the right-hand side,

$$v' = v\left(1 + \frac{v_{obs}}{\lambda v}\right) \qquad (12.30)$$

But the product of the wavelength and frequency is the speed of the wave (Equation 12.10), here the speed of sound v. Thus Equation 12.30 becomes

$$v' = v\left(1 + \frac{v_{obs}}{v}\right)$$
$$= v\left(\frac{v + v_{obs}}{v}\right) \qquad (12.31)$$

If the observer is moving away from the source, fewer wavefronts are received during each second. Indeed, instead of gaining v_{obs}/λ additional crests each second, this same number is lost; the number of wavefronts received by the observer during one second is thus

$$v' = v - \frac{v_{obs}}{\lambda}$$

Hence the frequency detected by the receding observer is

$$v' = v\left(1 - \frac{v_{obs}}{v}\right)$$

$$= v\left(\frac{v - v_{obs}}{v}\right) \qquad (12.32)$$

We can combine Equations 12.31 and 12.32 into the following form:

$$v' = v\left(\frac{v \pm v_{obs}}{v}\right) \qquad (12.33)$$

where we use the + sign if the observer is approaching the source (higher frequency heard), and the − sign if the observer is receding from the source (lower frequency heard).

PROBLEM-SOLVING TACTIC

12.4 The sign to use in Equation 12.33 can be determined as long as you remember that the frequency v' is higher than v if the observer is approaching the source (so the + sign must be used) and less than v when receding (use the − sign).

Boaters are quite aware of similar Doppler-like effects when riding with or against the waves.

Motion of the Source with the Observer and Medium at Rest

Once again, let v be the frequency emitted by the sound source, λ its wavelength, and v the speed of sound. Equation 12.10 still is valid: $v = v\lambda$. Let the speed of the source toward the observer be v_{source}.

The motion of the source toward an observer results in a decrease in (shortening of) the wavelength. Once a crest (wavefront) is emitted, the wavefront travels at speed v through the medium toward the observer independent of the motion of the source. If T is the period of the sound waves, the next crest is emitted after the source has traveled a distance $v_{source}T$ toward the observer. So instead of being a distance λ apart, the crests are a distance $\lambda - v_{source}T$ apart, as shown in Figure 12.43. Thus the effective wavelength λ' of the sound that reaches the observer is

$$\lambda' = \lambda - v_{source}T$$

The frequency v' of the sound reaching the observer is v/λ'. Hence

$$v' = \frac{v}{\lambda - v_{source}T}$$

$$= \frac{v}{\lambda - \dfrac{v_{source}}{v}}$$

$$= \frac{vv}{\lambda v - v_{source}}$$

$$= v\left(\frac{v}{v - v_{source}}\right) \qquad (12.34)$$

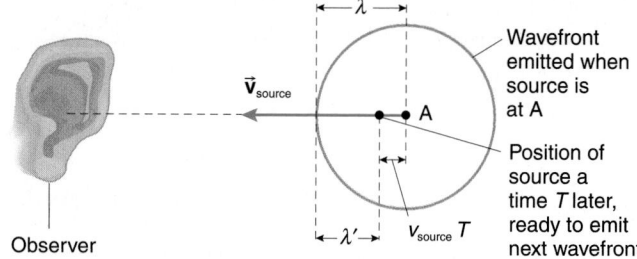

FIGURE 12.43 Motion of the source toward the observer decreases the effective wavelength of the sound.

If the source is receding from the observer, the wavelength is increased because of the motion of the source. In this case the wavelength λ' of the sound reaching the observer is

$$\lambda' = \lambda + v_{source}T$$

In a manner similar to that just shown, we find the frequency v' the observer detects is

$$v' = v\left(\frac{v}{v + v_{source}}\right) \qquad (12.35)$$

For the motion of the source, Equations 12.34 and 12.35 can be combined into the single expression:

$$v' = v\left(\frac{v}{v \mp v_{source}}\right) \qquad (12.36)$$

where we use the − sign for an approaching source (higher frequency heard by the observer), and the + sign for a receding source (lower frequency heard by the observer).

PROBLEM-SOLVING TACTIC

12.5 The sign to choose in Equation 12.36 can be determined by whether the observer hears a higher or lower frequency (as the source approaches or recedes, respectively).

The equations for the motion of the observer, Equation 12.33, and for the motion of the source, Equation 12.36, can be combined into a single equation:

$$v' = v\left(\frac{v \pm v_{obs}}{v \mp v_{source}}\right)$$

If the source is not moving, $v_{source} = 0$ m/s and we have Equation 12.33 for the motion of the observer. If the observer is not moving, $v_{obs} = 0$ m/s and we have Equation 12.36 for the motion of the source. If both are moving, we put in both speeds and choose the signs, as if only one or the other was moving (see Example 12.10).

Motion of the Medium (a Wind) with the Source and Observer at Rest

Let the medium be moving along the line from the source to the observer with a speed v_{med}, as shown in Figure 12.44. The

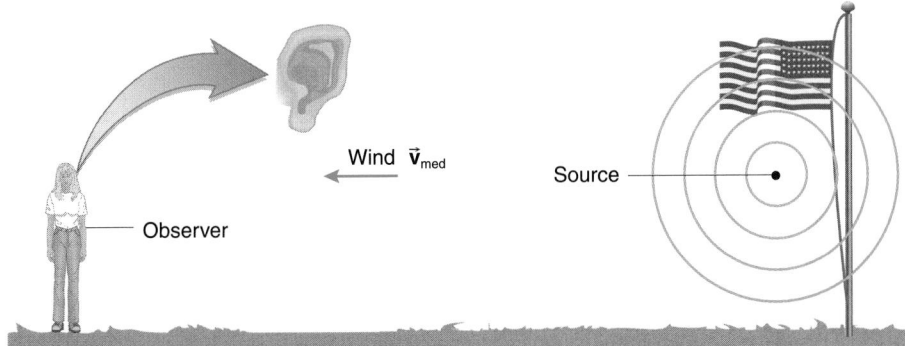

FIGURE 12.44 A wind has no effect for a stationary source and stationary observer. However, the wind *does* have an effect if the source and/or observer is moving.

motion of the medium past the source and observer (at rest relative to each other) is a wind. In this case the wavefront crests are spaced further apart as they leave the source; the wavelength is increased.

This might lead you to conclude that the frequency heard by the observer decreases, but this is *not* the case. Although the distance between the wavefronts crests is longer, the wavefronts are carried to the observer at a faster speed: $v + v_{med}$. The two effects exactly compensate for each other, and so the observer receives exactly the same number of crests per second as with no wind.

Likewise, if the wind is blowing from the observer to the source, the distance between the crests of the waves decreases, which leads you to think you would hear a higher frequency. But the speed the crests move toward the observer also decreases (and is $v - v_{med}$). The two effects again exactly compensate for each other and the observer hears no change in frequency.

> Thus the wind, acting alone (with the source and observer at rest with respect to each other), has *no effect* on the frequency and $v' = v$.

Motion of the Observer and/or Source with a Wind

The effect of the wind is to change the speed of the waves propagating from the source to the observer. Ironically, if either the source or observer is moving, the wind *does* have an effect on the frequency detected! That is, if the wind is blowing in a direction from the source to the observer, we replace v with $v + v_{med}$; on the other hand, if the wind is blowing from the observer and toward the source, we replace v with $v - v_{med}$.

> Therefore the most general equation for the acoustic Doppler effect is
>
> $$v' = v \left(\frac{v \pm v_{med} \pm v_{obs}}{v \pm v_{med} \mp v_{source}} \right) \quad (12.37)$$
>
> Notice that if the source and observer are at rest, $v_{obs} = v_{source} = 0$ m/s, and then the wind speed v_{med} has no effect: $v' = v$. On the other hand, if v_{obs} and/or v_{source} is not zero, then v_{med} *does* have an effect on the frequency detected by the observer.

PROBLEM-SOLVING TACTIC

12.6 To use the general equation for the Doppler effect (Equation 12.37), there are a few things to keep in mind. These conditions may remind you of the infamous IRS 1040 tax form in view of their written complexity:

a. If there is a wind blowing from the source toward the observer, then the + sign is used for the v_{med} term since the effective speed of the waves increases; if the wind is blowing from the observer toward the source, the – sign is used for the v_{med} term since the effective speed decreases. This accounts for the change in the speed of the waves relative to a stationary source and observer.

b. If the source is moving away from the observer (thought of as being fixed), then the frequency heard by the observer is lower and we need to use the + sign in the v_{source} term in the denominator. Likewise, if the source is moving toward the observer (thought of as at rest), then the frequency heard by the observer is higher and we need to use the – sign in the v_{source} term in the denominator.

c. If the observer is moving toward the source (thought of as being at rest), then the observer hears a higher frequency and so the + sign must be used with the v_{obs} term in the numerator. On the other hand, if the observer is moving away from the source (thought of as being at rest), then the frequency heard by the observer is lower and so the – sign must be used with the v_{obs} term in the numerator.

Our treatment of the acoustic Doppler effect involved motion of the source, observer, and medium along the line connecting the source and observer. If the motions are not along this line, then it is the component of the motion along this line that is operative in the acoustic Doppler effect.

QUESTION 7

An observer is moving with respect to a source of sound yet *no* Doppler effect is detected (no change in frequency is heard). Describe the motion of the observer with respect to the source.

EXAMPLE 12.10

A 261.6 Hz source (the famous middle C), an observer, and a wind are moving as indicated in Figure 12.45 with respect to the ground. What frequency is heard by the observer?

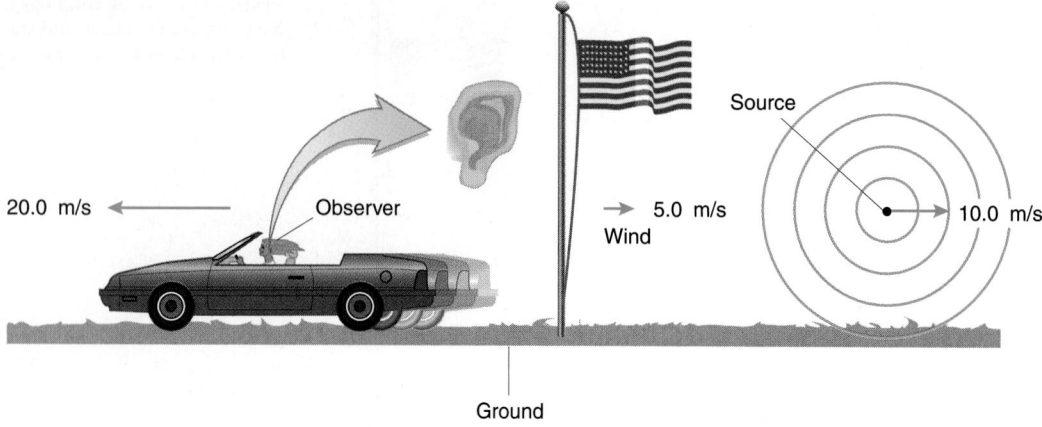

FIGURE 12.45

Solution

Since the wind is blowing in a direction from the observer to the source, use the minus sign attached to the v_{med} term in Equation 12.37 with v_{med} as 5.0 m/s since the effective speed of sound is decreased. The source is moving away from the observer (considered to be fixed); thus you use the plus sign with the v_{source} term in the denominator, since the frequency heard by the observer is lower than normal. The observer is moving away from the source (considered to be fixed); thus you use the minus sign with the v_{obs} term in the numerator of Equation 12.37, since a receding observer results in a decrease in perceived frequency. Now the frequency heard by the observer can be found from Equation 12.37:

$$v' = (261.6 \text{ Hz}) \left(\frac{343 \text{ m/s} - 5.0 \text{ m/s} - 20.0 \text{ m/s}}{343 \text{ m/s} - 5.0 \text{ m/s} + 10.0 \text{ m/s}} \right)$$

$$= 239 \text{ Hz}$$

An observer-musician would say the note is not middle C but between B♭ and B.

12.16 SHOCK WAVES*

As with most equations in physics, Equation 12.37 has limitations. Consider a situation in which

a. the medium is at rest, so that $v_{med} = 0$ m/s;
b. an observer is at rest, so that $v_{obs} = 0$ m/s; and
c. a source is moving toward the observer, so that the minus sign in front of term for v_{source} must be used in Equation 12.37 for the general acoustic Doppler effect.

Under these conditions, the general equation for the Doppler effect becomes

$$v' = v \left(\frac{v}{v - v_{source}} \right) \qquad (12.38)$$

As the speed of the source increases, the denominator decreases and the frequency v' heard by the observer increases. But notice that as v_{source} approaches the speed of sound v, Equation 12.38 diverges (when $v_{source} = v$) and then yields negative frequency values for v when $v_{source} > v$. These results mean that the speed

of the source must be less than the speed of sound in the medium for Equations 12.37 and 12.38 to apply.

As v_{source} approaches v, the distance between the wavefronts ahead of the source approaches zero (the wavelength approaches zero) and the source is able to keep pace with all the wavefronts in the forward direction, as indicated in Figure 12.46.

What happens if the speed of the source exceeds the speed of sound v in the medium? If you blindly use Equation 12.38, it yields a negative answer for the frequency: an unphysical result. The limitation to Equation 12.38 certainly is not apparent mathematically, since there is nothing mathematically wrong with the negative result for the frequency if $v_{source} > v$. Something new must happen physically when the speed of the source exceeds the speed of the wave in the surrounding medium: the source escapes from inside its propagating wavefront before the next one is emitted, as shown in Figure 12.47.

In this case the envelope of the propagating wavefronts forms a cone, known as the **Mach cone,*** with its apex at the

---

*The term is named for Ernst Mach (1836–1916), an Austrian physicist.

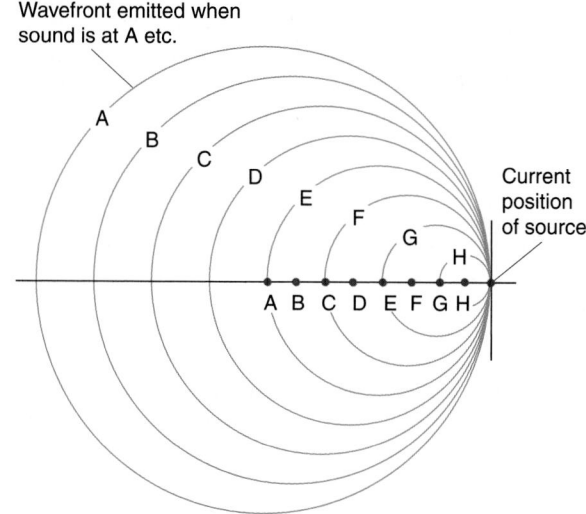

FIGURE 12.46 When the source speed approaches the speed of sound, the wavelength in the forward direction approaches zero.

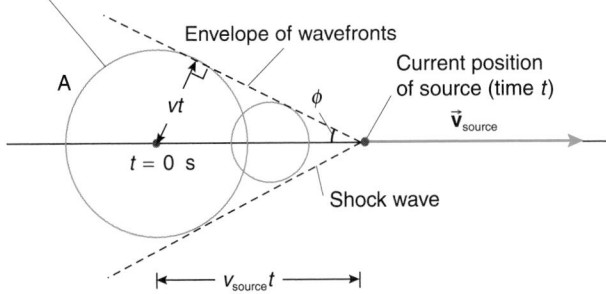

Wavefront emitted when $t = 0$ s when source is at current position at time t

Envelope of wavefronts

Current position of source (time t)

$\vec{v}_{source}$

A

vt

ϕ

$t = 0$ s

Shock wave

$v_{source}t$

Supersonic regime $v_{source} > v$

FIGURE 12.47 A source exceeding the speed of sound escapes from the wavefront before the next one is emitted.

source. If the wavefront labeled A in Figure 12.47 is emitted when $t = 0$ s, then the radius of its circular wavefront at time t is vt, since the wavefront moves at the speed v through the medium. The source traveling at speed v_{source} moves a distance $v_{source}t$ during the same time interval. Thus the half-angle ϕ of the Mach cone (half the apex angle of the cone itself) is found from the trigonometric relation

$$\sin \phi = \frac{v}{v_{source}} \qquad (v_{source} > v) \qquad (12.39)$$

The envelope of the wavefronts forms a **shock wave**.

Such acoustic shock waves commonly are called **sonic booms**, familiar to us from the passage of supersonic aircraft. Cracking a whip is another way to generate a sonic boom. When the whip cracks, its tip is briefly traveling at a speed greater than the speed of sound in air. Notice that since $v_{source} \geq v$ in Equation 12.39, the largest angle ϕ for the shock wave is 90°: when $v_{source} = v$, $\sin \phi = 1$, corresponding to the situation depicted in Figure 12.46. As the speed of the source increases past v, the half-angle ϕ of the Mach cone decreases from 90°.

The **Mach number** is the ratio of the speed of the source to the speed of sound in the medium:

$$\text{Mach number} = \frac{v_{source}}{v} \qquad (12.40)$$

Note that the half-angle ϕ of the Mach cone is the sine of the reciprocal of the Mach number:

$$\sin \phi = \frac{1}{\text{Mach number}} \qquad (\text{Mach numbers} \geq 1) \qquad (12.41)$$

Of course, this equation is legitimate only for Mach numbers greater than or equal to one. (Why? What is $\sin \phi$ for Mach numbers less than 1?)

Shock waves also are apparent in other wave phenomena as well. When electrically charged particles such as electrons or protons travel through a material faster than the speed of light in that medium, an unusual type of shock wave called Cerenkov

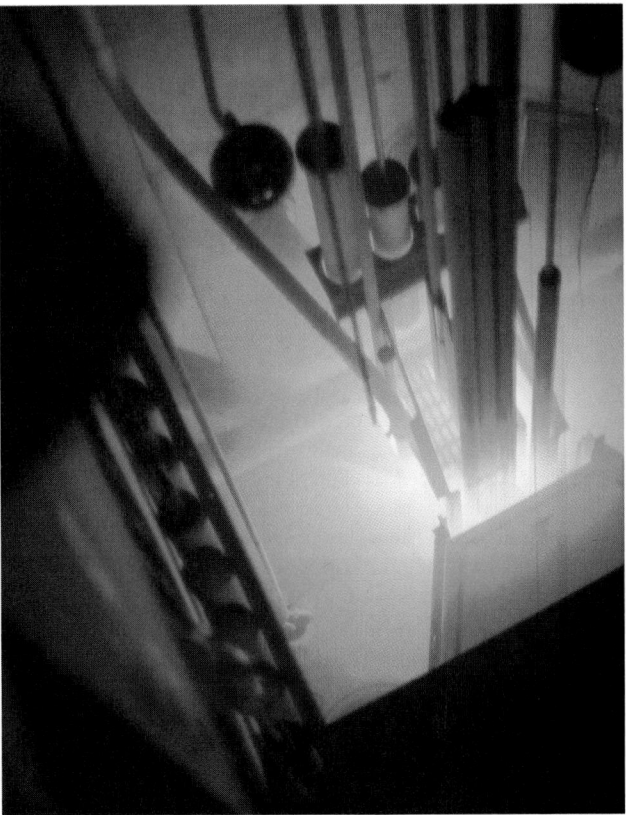

Cerenkov radiation (light) is formed when charged particles travel faster than the speed of light in a material.

Bow waves (wakes) are quite complex waves.

radiation is formed: light. Such Cerenkov radiation is visible in nuclear reactors that use water as a radiation shield.

The angle of the Mach cone has a strong dependence on the speed. Observations of the waves produced by boats, swimmers, or ducks show that their V-shaped patterns have an apex angle of about 40° over a fairly wide range of speeds. Such bow waves are *not* shock waves caused by motion faster through the water than a fixed propagation speed. These bow waves or **wakes** are quite complex physical waves that arise when wave speed varies with wavelength, a phenomenon called **dispersion**. You might recall that small-wavelength ripples on a pond propagate more slowly than large-wavelength ocean waves.

12.17 DIFFRACTION OF WAVES

When waves pass through an opening whose size is comparable to the wavelength, the waves spread out on the other side of the opening, as shown in Figure 12.48.

> The spreading out of waves after they pass through or around such apertures or obstacles is called **diffraction**.

Diffraction is a characteristic feature of any wave. If the opening is made larger, as in Figure 12.49, the diffraction becomes less pronounced, except near the edges of the opening. When the opening is much larger than the wavelength, diffraction is minimal and the waves pass essentially straight through the opening, following the path of the incident rays.

If a person is talking outside your open door but is not visible, you still can easily hear the conversation. Hence audible sound waves must have wavelengths on the order of magnitude

of the width of the door. For middle C with a frequency of 261.6 Hz, the wavelength is found using Equation 12.10:

$$v = \nu\lambda$$
$$\lambda = \frac{v}{\nu}$$
$$= \frac{343 \text{ m/s}}{261.6 \text{ Hz}}$$
$$= 1.31 \text{ m}$$

This is about the order of magnitude of a door opening. The sound waves diffract, which is why you can hear them around corners. Visible light waves, however, travel in straight lines through openings such as doors or keyholes, producing well-defined shadows behind the objects. Hence the wavelengths of visible light waves must be quite small compared with the sizes of common objects. We will study light diffraction effects in greater detail in Chapter 24. Here we are content to note that diffraction effects give a clue to the size of the wavelength of any wave.

QUESTION 8

You may have noticed that behind a small hill AM radio stations can be received more easily than FM radio stations. What does this observation imply about the relative sizes of the wavelengths of these radio waves?

12.18 THE PRINCIPLE OF SUPERPOSITION

And spend her strength with over-matching waves.
William Shakespeare (1564–1616)*

In this section we see what happens when and where two or more waves of the same type[†] meet. For example, if you drop two pebbles into a pond simultaneously at slightly different locations, two wave trains are produced that propagate across the water surface and overlap. Likewise, if you and a friend hold opposite ends of a taut rope and each flick your respective ends in the same plane to generate two wave pulses, the two waves propagate along the rope in opposite directions; see Figure 12.50. Where and when they reach each other, the multiple waves combine to produce a more complicated disturbance. If the amplitudes are not too large, the total wave disturbance Ψ_{total} at any point x and time t is the sum of the individual wave disturbances:

$$\Psi_{\text{total}}(x, t) = \Psi_1(x, t) + \Psi_2(x, t) + \Psi_3(x, t) + \cdots \quad (12.42)$$

Wave disturbances that follow this linear **principle of superposition** are called linear waves and are the only types of waves we will consider.[‡]

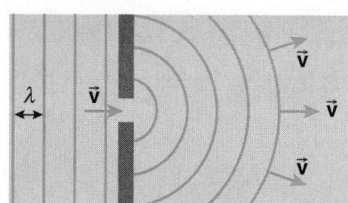

FIGURE 12.48 Diffraction of waves at an aperture; the size of the aperture is comparable to the wavelength.

FIGURE 12.49 Diffraction is minimal when the opening is much larger than the wavelength.

Henry VI, Part III, Act I, Scene 4, line 21.
[†]Two (or more) water waves, waves on a rope, sound waves, light waves, and so on. For transverse waves, the waves must have the same plane of polarization. If not, the waves must be summed like vectors, not scalars.
[‡]If the amplitudes are large, then nonlinear superposition of the waves occurs. Such nonlinear superposition is beyond the scope of an introductory course.

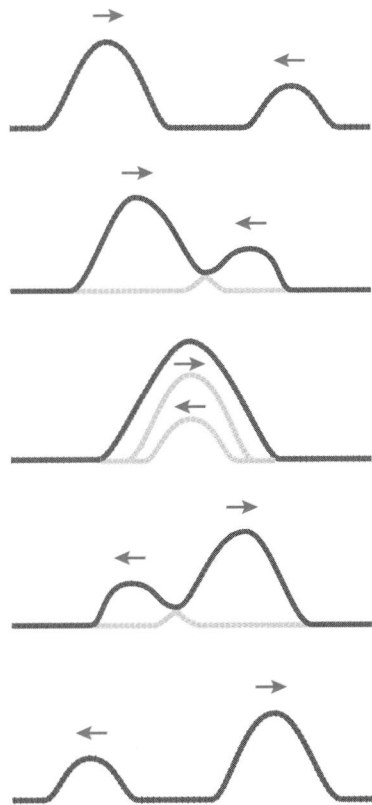

FIGURE 12.50 The interference of two pulses on a rope.

When two or more similar types of waves exist simultaneously at a point in space, we say the waves **interfere** with each other. If the resulting wave disturbance at a point is greater than that produced by any of the waves acting alone, we say the waves exhibit **constructive interference**. If the resulting disturbance is less than that produced by any alone, the waves exhibit **destructive interference**.

Superposition implies that on a string two wave pulses in the same plane moving in opposite directions proceed past each other without being changed in any way except where and when the two waves overlap in space and time; see Figure 12.50. Perhaps you also have observed such constructive and destructive interference of water waves while boating on a busy waterway.

In the next two sections we examine two special kinds of wave interference using the principle of superposition.

12.19 STANDING WAVES

. . . there shall thy proud wave be stayed.

Book of Job*

When you relax by plucking a guitar or banjo, or tickle the ivories on a piano, the waves you generate on the strings have rather remarkable properties. The plucking of a string that is fixed at both ends produces at least two harmonic (sinusoidal)

---
*Chapter 38, verse 11.

waves of the same amplitude that travel in opposite directions. The superposition of the two waves and their reflections from the fixed ends results in a rather astounding total wave disturbance.

Let the x-axis be along the length of the string. A harmonic (sinusoidal) wave of amplitude A moving toward increasing values of x has the form

$$\Psi_1(x, t) = A \cos(kx - \omega t)$$

A similar harmonic wave traveling toward decreasing values of x has the form

$$\Psi_2(x, t) = A \cos(kx + \omega t)$$

The resulting disturbance at any x and t is, according to the principle of superposition,

$$\Psi_{\text{total}}(x, t) = \Psi_1(x, t) + \Psi_2(x, t)$$
$$= A[\cos(kx - \omega t) + \cos(kx + \omega t)] \quad (12.43)$$

That's all the physics. To see what is going on, we dust off a trigonometric identity that indicates how to add two cosine functions together:

$$\cos \alpha + \cos \beta = 2 \cos\left(\frac{\alpha + \beta}{2}\right) \cos\left(\frac{\alpha - \beta}{2}\right) \quad (12.44)$$

We apply this identity to the sum of the cosines in Equation 12.43:

$$\Psi_{\text{total}}(x, t) = 2A \cos\left[\frac{(kx - \omega t) + (kx + \omega t)}{2}\right]$$
$$\times \cos\left[\frac{(kx - \omega t) - (kx + \omega t)}{2}\right]$$
$$= 2A \cos(kx) \cos(-\omega t) \quad (12.45)$$

Since the cosine of a negative angle is the same as the cosine of the positive angle, we rewrite Equation 12.45 in the following form:

$$\Psi_{\text{total}}(x, t) = 2A \cos(kx) \cos(\omega t) \quad (12.46)$$

This is a remarkable type of total wave disturbance. For certain positions, namely where the product kx is an *odd* integral multiple of $\pi/2$ rad, the wave disturbance $\Psi_{\text{total}}(x, t)$ vanishes for all times t! Let n be an odd integer. Hence if

$$kx = n\frac{\pi}{2} \text{ rad} \quad (n \text{ odd integer}) \quad (12.47)$$

the total disturbance is zero at all times, since

$$\cos\left(n\frac{\pi}{2} \text{ rad}\right) = 0 \quad (n \text{ odd integer})$$

The locations where the total wave disturbance is zero at all times are called **nodes**. Since the angular wavenumber k is $(2\pi \text{ rad})/\lambda$, Equation 12.47 implies that the nodes are located at those positions that satisfy the following condition:

$$\frac{2\pi \text{ rad}}{\lambda} x_{\text{node}} = n\frac{\pi}{2} \text{ rad} \quad (n \text{ odd integer})$$

Thus

$$x_{\text{node}} = n\frac{\lambda}{4} \quad (n \text{ odd integer})$$

By substituting n and then $n + 2$ into this expression, we can show the following:

Successive nodes are separated in space by *half* a wavelength.

The maximum amplitude of the resulting superposition of the two waves occurs where the $\cos(kx)$ factor in Equation 12.46 is equal to unity. The amplitude at these locations is then $2A$, or twice what each individual wave could contribute. The positions where the amplitude of the resulting motion has its maximum are called **antinodes**. The antinodes occur where kx is an *even* integral multiple of $\pi/2$ rad, since then the $\cos(kx)$ term becomes

$$\left| \cos\left(n\frac{\pi}{2} \right) \right| = 1 \qquad (n \text{ even integer})$$

That is, the antinodes occur where

$$kx_{\text{antinode}} = n\frac{\pi}{2} \qquad (n \text{ even integer})$$

Since $k = 2\pi/\lambda$, the antinodes are located where

$$x_{\text{antinode}} = n\frac{\lambda}{4} \qquad (n \text{ even integer})$$

Note that successive antinodes along the string also are separated by half a wavelength.

The resulting disturbance has the appearance shown in Figure 12.51, where the double-ended arrows indicate the amplitude of the oscillations at that position. The string oscillates in a stationary envelope pattern and has lost all indication of the underlying traveling waves!

The disturbance resulting from the superposition of two waves of equal amplitude and frequency traveling in opposite directions is called a **standing wave**.

Notice that the amplitude of the resulting standing wave disturbance depends on the location x. At any given position x, the oscillation is simple harmonic oscillation, varying sinusoidally with time [thanks to the $\cos(\omega t)$ term] with amplitude

$$2A \cos(kx)$$

Plucked guitar and banjo strings illustrate standing waves on strings; their dynamical behavior is fun to watch using a strobe light. Since the string is tied down or fixed at both ends, there must be nodes at the extremes of the string. Let ℓ be the distance between the fixed ends of the string.* Since the nodes of a standing wave are separated by half a wavelength, the longest wavelength λ_1 that can be established on the string satisfies the condition

$$\ell = \frac{\lambda_1}{2}$$

or

$$\lambda_1 = 2\ell$$

Since the speed of the wave is the product of the frequency times the wavelength,

$$v = \nu\lambda$$

the lowest frequency ν_1 of the vibrating string is

$$\nu_1 = \frac{v}{\lambda_1}$$

Using Equation 12.19 for the wave speed, the lowest frequency is found to be

$$\nu_1 = \frac{1}{2\ell}\left(\frac{T}{\mu} \right)^{1/2}$$

---

*For stringed instruments, ℓ may be different from the total length of the string, used to determine the mass per unit length μ, because the string may be wrapped around pegs, pass over bridges, or be shortened by a finger fixing one end, such as in instruments like guitars, banjos, and violins.

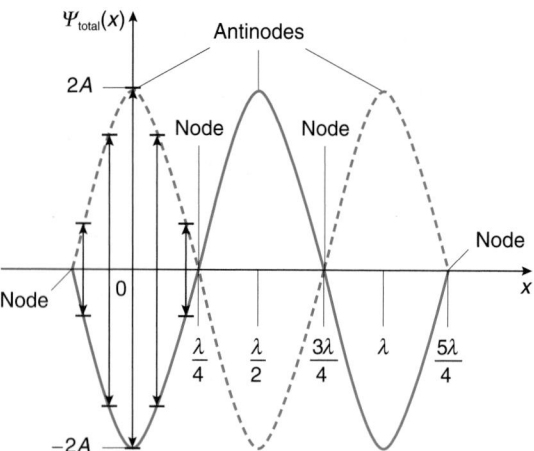

FIGURE 12.51 A graph of a standing wave on a string. The amplitude is a function of position along the string.

The nodes and antinodes of a standing wave on a string fixed at both ends are quite apparent in this photograph.

where T is the magnitude of the tension in the string. The lowest frequency is called the **fundamental frequency** of the string or, equivalently, the first harmonic (whence the subscript 1). For low fundamental frequencies, an instrument must have long strings with considerable mass per unit length like the bass strings on a piano or double bass. *Tuning* a stringed instrument involves adjusting the magnitude T of the tension in each string to obtain precisely the desired note or frequency.

Standing waves of other frequencies, with different amplitudes, also are permitted simultaneously on the string. Since the fixed ends must always be nodes, the only condition that must be met is that the distance ℓ between the fixed ends of the string must be an integral number of half wavelengths:

$$\ell = n \frac{\lambda_n}{2}$$

Thus the permitted wavelengths are given by

$$\lambda_n = \frac{2\ell}{n} \qquad (n \text{ is a positive integer}) \qquad (12.48)$$

The frequencies of these vibrations are then

$$\nu_n = \frac{v}{\lambda_n}$$

Using Equation 12.48 for λ_n and 12.19 for v, we get

$$\nu_n = n \frac{1}{2\ell} \left(\frac{T}{\mu} \right)^{1/2} \qquad (12.49)$$

When $n = 1$, we have the fundamental frequency (i.e., the first harmonic). The second harmonic corresponds to $n = 2$; the third harmonic to $n = 3$; and so forth. See Figure 12.52. So the integer n numbers the various harmonics.

The frequencies of the higher harmonics are integral multiples of the fundamental frequency ν_1.

Not all frequencies are present on the string—only certain special, characteristic frequencies.

The frequencies that may exist on the string are known as the **eigenfrequencies** of the system, from the German word *eigen*, meaning *characteristic*.

The collection of allowed frequencies and their respective amplitudes is called the **frequency spectrum** of the system.

For a vibrating string, the frequency spectrum consists of the fundamental and integral multiples of it. The amplitudes of the various eigenfrequencies may vary, producing the unique sound associated with various instruments. Figure 12.53 indicates the high and low frequency range of many instruments.

As a practical matter, the transverse standing waves on strings generally produce little sound: the coupling between the string and the air is very small because the longitudinal cross section of the string is tiny. The vibrating string alone cannot transfer energy effectively to the air to cause an acoustic wave with an appreciable sound level. Hence the vibration of the

Two-dimensional standing waves are produced on a vibrating drumhead.

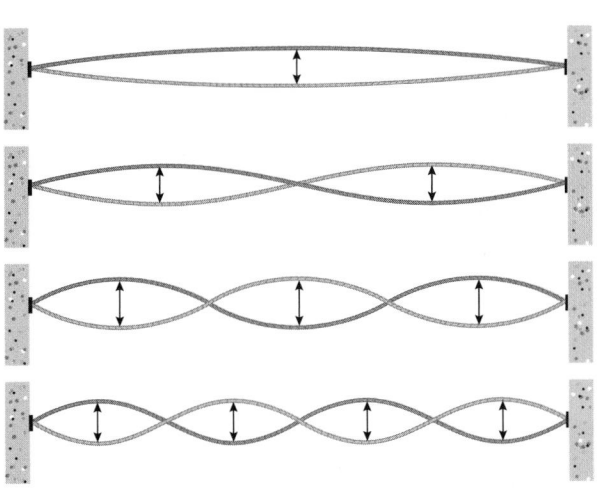

Fundamental (first harmonic) ν_1	$n = 1$	$\lambda_1 = \dfrac{2\ell}{1}$	
Second harmonic $2\nu_1$	$n = 2$	$\lambda_2 = \dfrac{2\ell}{2}$	
Third harmonic $3\nu_1$	$n = 3$	$\lambda_3 = \dfrac{2\ell}{3}$	
Fourth harmonic $4\nu_1$	$n = 4$	$\lambda_4 = \dfrac{2\ell}{4}$	**FIGURE 12.52** Standing wave pattern of the fundamental and several harmonics on a string fixed at both ends.

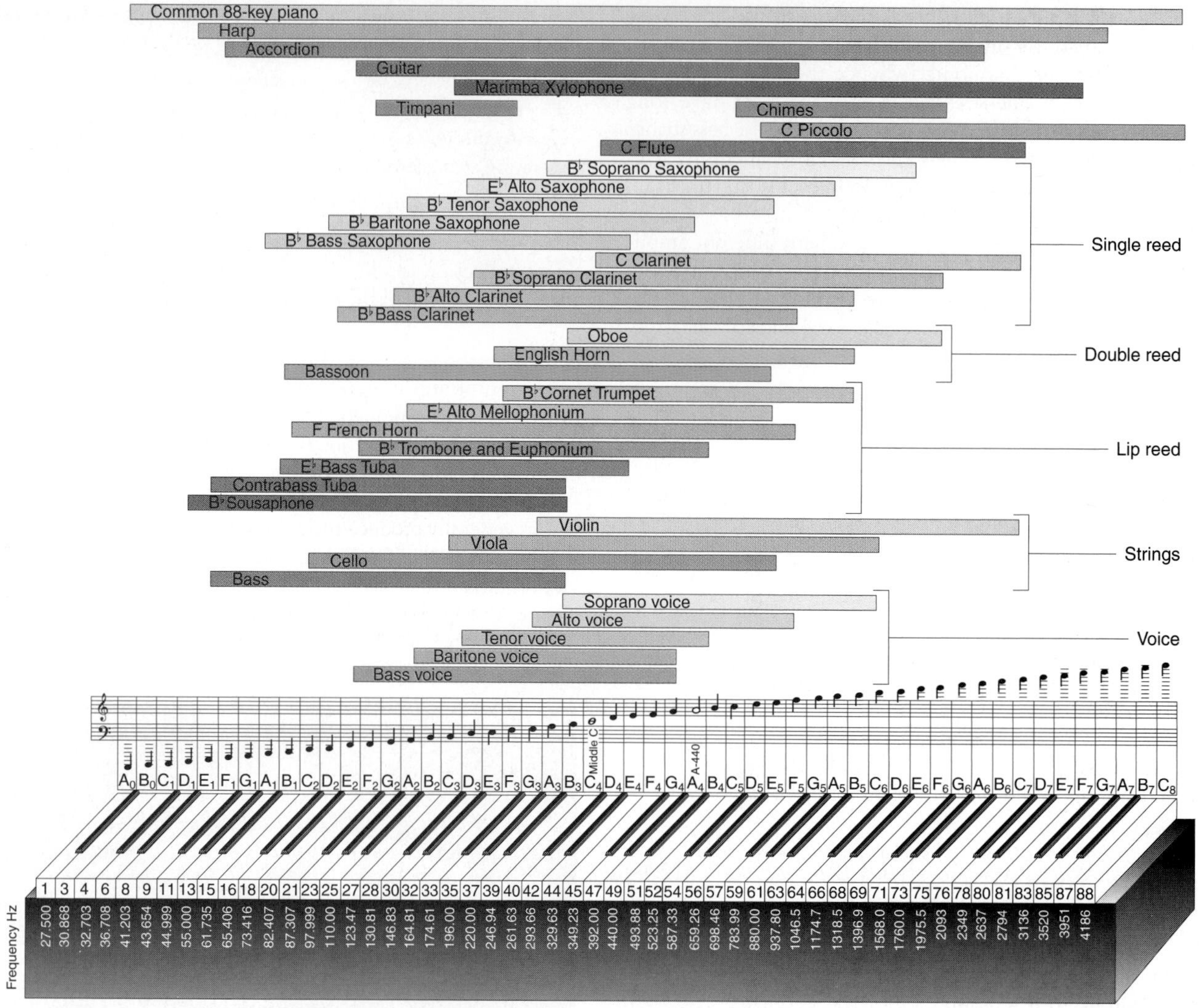

FIGURE 12.53 Frequency ranges of various musical instruments.

string must be made more effective to create louder (more intense) sounds. The amplification can be done mechanically via sounding boards (in pianos and harpsichords), open-backed membranes (in banjos), resonant sound boxes (violins, guitars, etc.), or electronically (electric guitars and synthesizers). The sounding boards and boxes have much greater surface area and couple the vibration to the air with greater efficiency than the naked string, much like the vibrating wall in Section 12.13. Musical synthesizers are capable of closely simulating electronically the frequency spectrum of many musical instruments without the need for the strings, reeds, and air columns associated with acoustic instruments.

Sound waves in organ pipes are examples of resonant, longitudinal acoustic standing waves. Consider a pipe closed at one end by water whose level can be adjusted by means of a reservoir, as shown in Figure 12.54. A tuning fork that vibrates at frequency ν is held over the open end of the pipe. The fork emits sound waves of wavelength λ. The sound waves propagate down the pipe and are reflected from the water surface. The reflected and incident waves superimpose to form a longitudinal standing wave pattern in the pipe.

If the water level in the pipe is adjusted so that maximum sound is heard, a condition of **resonance** is established in the column. At resonance, there is a node of the standing longitu-

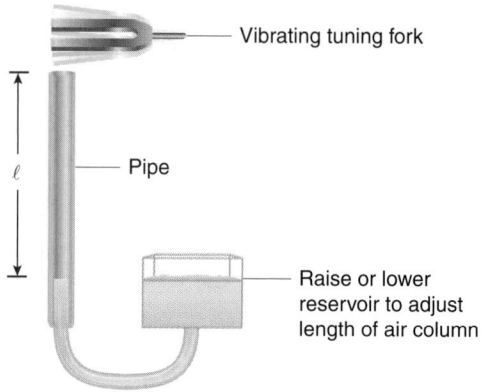

FIGURE 12.54 A way to generate acoustic standing waves in a pipe closed at one end.

dinal vibratory particle position waves at the water surface and an antinode at the open end. The distance between nodes (or between antinodes) is $\lambda/2$, and so the distance between a node and an antinode is $\lambda/4$. Therefore there must be an odd number of quarter wavelengths in the length of the resonant air column:

$$\ell = n\,\frac{\lambda_n}{4} \qquad (n \text{ an odd integer}$$

Thus the permitted wavelengths for a closed pipe are

$$\lambda_n = \frac{4\ell}{n} \qquad (n \text{ an odd integer})$$

The longest resonant wavelength corresponds to $n = 1$ and produces the smallest (lowest) frequency, known as the fundamental frequency of the pipe:

$$\begin{aligned} \nu_1 &= \frac{v}{\lambda_1} \\ &= \frac{v}{4\ell} \end{aligned}$$

Higher odd integers correspond to odd-numbered harmonics of the fundamental; not all harmonics are possible. Closed organ pipes are closed at one end like the pipe in Figure 12.54. They produce the fundamental frequency and the associated odd-numbered harmonics of the fundamental, as shown in Figure 12.55.

A pipe open at both ends, as shown in Figure 12.56, has a fundamental frequency at resonance corresponding to antinodes of the longitudinal vibratory particle position standing waves at each end of the pipe with a single node in the middle. The length of the pipe then is equal to $\lambda/2$. Since $v = \nu\lambda$, the fundamental frequency is

$$\nu_1 = \frac{v}{2\ell}$$

The standing wave pattern with two nodes between antinodes at the ends has a wavelength equal to the length of the pipe. The corresponding frequency is twice the fundamental and so this is the second harmonic. A standing wave pattern with three nodes between the antinodes at the end is the third harmonic. Thus *all* harmonics now are possible.

The wavelengths of the harmonics of an open pipe are given by

$$\lambda_n = \frac{2\ell}{n} \qquad (n \text{ any positive integer})$$

Open organ pipes, open at both ends, are constructed this way.

Since it is possible to have all the harmonics of the fundamental with open pipes rather than with closed pipes (in which only the odd-numbered harmonics are possible), open-piped instruments produce richer musical tones than instruments with closed pipes.

QUESTION 9

Examine Equation 12.46. Are there instants t when the wave disturbance is zero for every x? If so, how many times during one period of the wave disturbance? Sketch such a waveform.

Note: The acoustic waves are *longitudinal* standing waves, not transverse waves (which are indicated for clarity only).

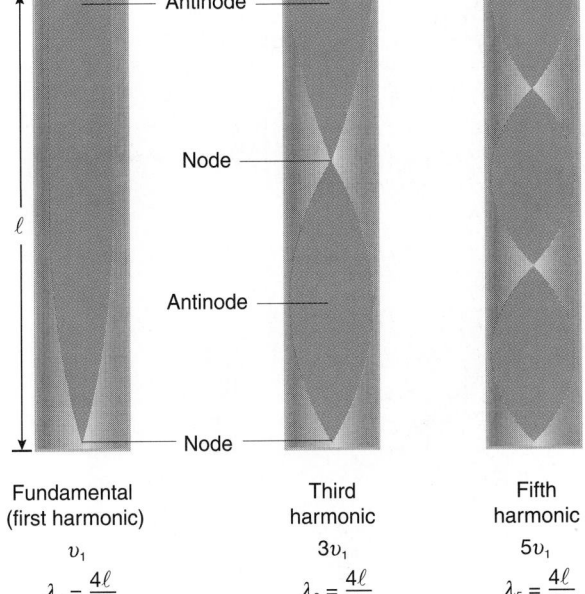

FIGURE 12.55 A pipe closed at one end, its fundamental frequency, and several harmonics. Only odd harmonics are possible.

Note: The acoustic waves are *longitudinal* standing waves, not transverse waves (which are indicated for clarity only).

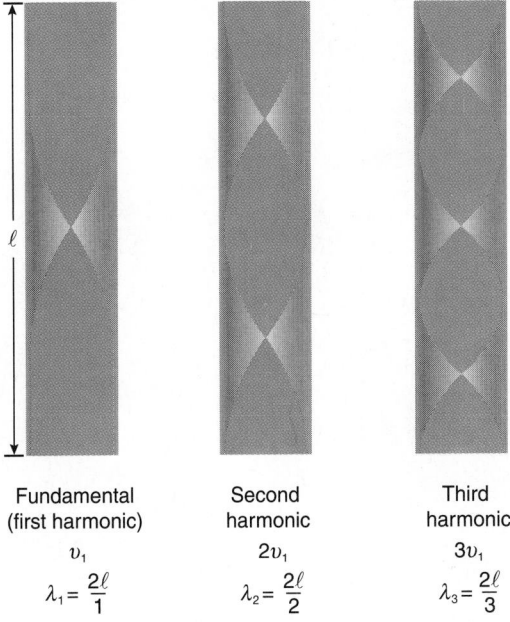

FIGURE 12.56 A pipe open at both ends, its fundamental frequency, and several harmonics. All harmonics are possible.

Why do the strings for the lowest-frequency notes of a piano have wire wrapped tightly around them?

EXAMPLE 12.11

A banjo string with a length 0.585 m between its fixed ends has a mass per unit length of 5.74×10^{-5} kg/m. It is tuned to play the note A_4 (with a frequency of 440 Hz) as the fundamental.

a. What is the tension in the string?
b. What is the wavelength of the fundamental?
c. What is the frequency of the second harmonic?
d. What is the distance between the nodes of the second harmonic?

Solution

a. The frequency, tension, length, and mass per unit length are related by Equation 12.49 with $n = 1$ for the fundamental:

$$v = \frac{1}{2\ell}\left(\frac{T}{\mu}\right)^{1/2}$$

Solving for T, you find

$$T = 4\ell^2 v^2 \mu$$
$$= 4(0.585 \text{ m})^2(4.40 \times 10^2 \text{ Hz})^2(5.74 \times 10^{-5} \text{ kg/m})$$
$$= 15.2 \text{ N}$$

b. The wavelength of the fundamental can be found in two ways.

Method 1

The fundamental has nodes only at each fixed end of the string. The distance between nodes is half a wavelength. Hence the wavelength of the fundamental is twice the distance between the fixed ends, or 1.17 m.

Method 2

First find the speed of the waves on the string via Equation 12.19:

$$v = \left(\frac{T}{\mu}\right)^{1/2}$$
$$= \left(\frac{15.2 \text{ N}}{5.74 \times 10^{-5} \text{ kg/m}}\right)^{1/2}$$
$$= 5.15 \times 10^2 \text{ m/s}$$

Then, since the frequency is known, use Equation 12.10 to find the wavelength:

$$v = \nu\lambda$$
$$\lambda = \frac{v}{\nu}$$
$$= \frac{5.15 \times 10^2 \text{ m/s}}{440 \text{ Hz}}$$
$$= 1.17 \text{ m}$$

c. The frequency of the second harmonic is twice the fundamental frequency, or 880 Hz.

Optical standing waves exist in laser cavities.

Note how the bass strings on a piano are wrapped with wire.

d. The distance between nodes of the second harmonic is half a wavelength. For the second harmonic, the wavelength is equal to the distance between the fixed ends of the string. Therefore the distance between nodes is (1/2)(0.585 m) = 0.293 m; hence, there is a node at the middle of the string.

EXAMPLE 12.12

You are designing a closed organ pipe to play the first A above middle C, with a frequency of 440 Hz.

a. **What is the length of the pipe?**
b. **What are the frequencies of the next two permitted harmonics?**

Solution

a. For a closed organ pipe, the length of the pipe is one-fourth of the wavelength of the fundamental frequency:

$$\ell = \frac{\lambda}{4}$$

The speed of sound is the product of the frequency and the wavelength:

$$v = \nu\lambda$$

Hence the length of the pipe is

$$
\begin{aligned}
\ell &= \frac{v}{4\nu} \\
&= \frac{343 \text{ m/s}}{4\,(440 \text{ Hz})} \\
&= 0.195 \text{ m}
\end{aligned}
$$

b. Since only odd-numbered harmonics can exist in a closed organ pipe, the next higher frequencies are 3 and 5 times the fundamental, or 1.32×10^3 Hz and 2.20×10^3 Hz.

12.20 WAVE GROUPS AND BEATS*

Those of you who play musical instruments likely know that tuning such instruments frequently involves listening for very low frequency **beats** between two acoustic frequencies that are nearly the same. One frequency is adjusted until the beating disappears, indicating that the two notes have the same frequency. Here we see how such beats arise for sound and other types of waves and oscillations.

Consider the superposition of two waves with different frequencies traveling in the *same* direction. For simplicity, we assume the waves have equal amplitudes. Let k_1 and ω_1 be the angular wavenumber and angular frequency of one wave traveling toward increasing values of x:

$$\Psi_1(x, t) = A \cos(k_1 x - \omega_1 t)$$

Since the wave is a function of $x - vt$, the speed can be determined by factoring out k_1 in the argument of the cosine:

$$\Psi_1(x, t) = A \cos\left[k_1\left(x - \frac{\omega_1}{k_1}t\right)\right]$$

Thus the speed of the wave is

$$v_1 = \frac{\omega_1}{k_1}$$

This is just a fancy way of writing Equation 12.10, the familiar $v = \nu\lambda$, since

$$v_1 = \frac{\omega_1}{k_1} = \frac{2\pi\nu_1}{2\pi/\lambda_1} = \nu_1\lambda_1$$

Let k_2 and ω_2 be the angular wavenumber and angular frequency of another wave of the same amplitude A, traveling in the same direction as the first wave (toward increasing values of x):

$$\Psi_2(x, t) = A \cos(k_2 x - \omega_2 t)$$

The speed of this second wave is

$$v_2 = \frac{\omega_2}{k_2}$$

The speeds of the individual waves are known as phase speeds. The phase speeds of the individual waves may or may not be the same. Whether they are the same or not depends on the particular kind of wave being considered and the medium through which the waves are propagating.

> *If* the speeds of waves of different frequency are *not* the same, we say there is **dispersion**; dispersion means that the speeds of individual waves depend on the particular wavelength (or frequency).[†]

It is not necessary for the superimposed waves to have the same amplitudes, but this condition makes the effects of such a superposition easier to discuss both from an analytical and from a physical viewpoint. The principle of superposition implies that the resulting wave disturbance is

$$
\begin{aligned}
\Psi_{\text{total}}(x, t) &= \Psi_1(x, t) + \Psi_2(x, t) \\
&= A[\cos(k_1 x - \omega_1 t) + \cos(k_2 x - \omega_2 t)] \quad (12.50)
\end{aligned}
$$

[†] Light waves traveling in a vacuum exhibit no dispersion; all light wavelengths travel with the same speed, c, the speed of light in vacuum. On the other hand, in material media (such as glass and water), the speed of light is a function of wavelength and dispersion is therefore present. This simply means that the index of refraction of a material (perhaps familiar to you from a high school physics course and a concept we will introduce in Chapter 23) depends on the wavelength; among other things, when light passes through a prism, the component colors (wavelengths) of the light are separated.

We use the trigonometric identity for the sum of two cosines (Equation 12.44):

$$\cos\alpha + \cos\beta = 2\cos\left(\frac{\alpha+\beta}{2}\right)\cos\left(\frac{\alpha-\beta}{2}\right)$$

Applying this identity to the two cosine terms in Equation 12.50, we find

$$\Psi_{\text{total}}(x,t) = 2A\cos\left[\frac{\left(k_1 x - \omega_1 t\right)+\left(k_2 x - \omega_2 t\right)}{2}\right]$$
$$\times\cos\left[\frac{\left(k_1 x - \omega_1 t\right)-\left(k_2 x - \omega_2 t\right)}{2}\right]$$

Collecting similar terms in the arguments of the cosines,

$$\Psi_{\text{total}}(x,t) = 2A\cos\left[\frac{\left(k_1+k_2\right)}{2}x - \frac{\left(\omega_1+\omega_2\right)}{2}t\right]$$
$$\times\cos\left[\frac{\left(k_1-k_2\right)}{2}x - \frac{\left(\omega_1-\omega_2\right)}{2}t\right]$$

This is a rather ungainly expression! We can tidy it up a bit by calling

$$\frac{k_1+k_2}{2}\equiv k$$

which is the average value of the two angular wavenumbers. Likewise we call

$$\frac{\omega_1+\omega_2}{2}\equiv\omega$$

which is the average value of the angular frequencies of the two waves. In a similar fashion, we call

$$|k_1-k_2| = \Delta k$$

and

$$|\omega_1-\omega_2| = \Delta\omega$$

With these cosmetic changes, the superposition of the two waves becomes somewhat prettier:

$$\Psi_{\text{total}}(x,t) = 2A\cos(kx-\omega t)\,\cos\left[\left(\frac{\Delta k}{2}\right)x - \left(\frac{\Delta\omega}{2}\right)t\right]$$
$$(12.51)$$

So what is the meaning of all this? *Both* cosine terms represent waves, since they both have the proper general form of Equation 12.16 for a wave traveling toward increasing values of *x*. Let's look at each harmonic wave individually. The term

$$\cos(kx-\omega t)$$

is a wave much like the two individual waves in the superposition. This wave term has an angular wavenumber *k* equal to the average of the angular wavenumbers of the component waves and an angular frequency equal to the average of the angular frequencies of the two waves superimposed. If these angular frequencies and angular wavenumbers are only slightly different from each other, ω and *k* are essentially equal to the corresponding quantities of either component wave. The speed of this wave is found from

$$v = \frac{\omega}{k}$$

Substituting for ω and *k*,

$$v = \frac{\dfrac{\omega_1+\omega_2}{2}}{\dfrac{k_1+k_2}{2}} = \frac{\omega_1+\omega_2}{k_1+k_2} \qquad (12.52)$$

We factor this expression in the following awkward way:

$$v = \frac{\omega_1\left(1+\dfrac{\omega_2}{\omega_1}\right)}{k_1\left(1+\dfrac{k_2}{k_1}\right)} \qquad (12.53)$$

If the frequencies and wavelengths of the individual waves are only slightly different from each other, then $\omega_1 \approx \omega_2$ and $k_1 \approx k_2$, and Equation 12.53 reduces to

$$v \approx \frac{\omega_1}{k_1} = v_1$$

Here *v* is essentially the phase speed of *either* wave in the superposition.

The other factor resulting from the superposition is

$$2A\cos\left[\left(\frac{\Delta k}{2}\right)x - \left(\frac{\Delta\omega}{2}\right)t\right] \qquad (12.54)$$

and also is in the form of a traveling wave. It has a lower angular frequency than the two component waves because $\Delta\omega/2 < \omega_1$ or ω_2. This wave represents a slowly varying amplitude factor for the superposition, which travels at its own **group speed** v_{group}. We write the factor now as

$$2A\cos\left[\frac{\Delta k}{2}\left(x - \frac{\Delta\omega}{\Delta k}t\right)\right] \qquad (12.55)$$

and can identify this group speed as the coefficient of *t*:

$$v_{\text{group}} = \frac{\Delta\omega}{\Delta k}$$

In the limit of small differential differences between the frequencies and wavelengths of the constituent waves, we get

$$v_{\text{group}} = \frac{d\omega}{dk} \qquad (12.56)$$

where ω is related to k via the phase speed equation:

$$v_{\text{phase}} = \frac{\omega}{k}$$

or

$$\omega = k v_{\text{phase}}$$

We substitute for ω in Equation 12.56 for the group speed:

$$v_{\text{group}} = \frac{d}{dk}(k v_{\text{phase}})$$
$$= v_{\text{phase}} + k \frac{dv_{\text{phase}}}{dk} \qquad (12.57)$$

If the phase speed is independent of wavelength (i.e., independent of k, since $k = 2\pi/\lambda$)—that is, if there is *no* dispersion—then the phase and group speeds are equal since

$$\frac{dv_{\text{phase}}}{dk} = 0 \ \ \text{m}^2/\text{s}$$

On the other hand, if there is dispersion, the phase speed depends on wavelength (or k, since $k = 2\pi/\lambda$), and the phase and group speeds are different.

We write the amplitude term in the standard form of a traveling wave:

$$2A\cos(k_A x - \omega_A t)$$

The angular wavenumber is k_A and ω_A is the angular frequency of the wave. From Equation 12.54 we get

$$k_A = \frac{\Delta k}{2}$$

and

$$\omega_A = \frac{\Delta \omega}{2}$$

Let's see what these mean. Call λ_A the long overall wavelength of this amplitude factor wave. Since the angular wavenumber is related to the wavelength by

$$k_A = \frac{2\pi \ \text{rad}}{\lambda_A}$$

we can relate the wavelength λ_A to the wavelengths of the individual waves involved in the superposition:

$$k_A = \frac{\Delta k}{2} = \frac{|k_1 - k_2|}{2}$$

or

$$\frac{2\pi}{\lambda_A} = \frac{\left| \dfrac{2\pi}{\lambda_1} - \dfrac{2\pi}{\lambda_2} \right|}{2}$$

Solving for λ_A, we find

$$\lambda_A = \frac{2\lambda_1 \lambda_2}{|\lambda_2 - \lambda_1|}$$

If the two wavelengths superimposed are not very different from each other, the wavelength λ_A is much larger than either λ_1 or λ_2.

Also the angular frequency ω_A of the long-wavelength wave is related to the angular frequencies of the two waves superimposed:

$$\omega_A = \frac{\Delta \omega}{2} = \frac{|\omega_1 - \omega_2|}{2}$$

In terms of the frequencies, we have

$$\nu_A = \frac{|\nu_1 - \nu_2|}{2}$$

If the frequencies of the two waves superimposed are not very different from each other, the frequency ν_A is much lower than the frequencies of either of the waves superimposed.

To see what the entire superposition looks like as a function of time, we consider a fixed position and then plot the resulting disturbance as a function of time. For convenience, we choose the position $x = 0$ m. Then the superimposed wave of Equation 12.51 becomes the oscillation

$$\Psi_{\text{total}}(0 \ \text{m}, t) = 2A\cos(-\omega t)\cos\left(-\frac{\Delta \omega}{2}t\right)$$

Since the cosine of a negative angle has the same value as the cosine of the positive angle, we have

$$\Psi_{\text{total}}(0 \ \text{m}, t) = 2A\cos(\omega t)\cos\left(\frac{\Delta \omega}{2}t\right) \qquad (12.58)$$

This expression is plotted in Figure 12.59 in Example 12.13 for two waves with frequencies of 8.000 Hz and 8.250 Hz.

Notice that the high-frequency cosine term

$$\cos(\omega t)$$

is modulated by a low-frequency *envelope* given by the term

$$\cos\left(\frac{\Delta \omega}{2}t\right)$$

The envelope is of maximum amplitude when this cosine term has the value ±1. These maxima are known as **beats**. Since the cosine assumes the values ±1 during each cycle, the beats occur at *twice* the frequency ν_A of the wave itself. The *beat frequency* ν_B thus is

$$\text{beat frequency} \equiv \nu_B \equiv 2\nu_A = 2\left(\frac{|\nu_1 - \nu_2|}{2}\right)$$
$$= |\nu_1 - \nu_2|$$

> The beat frequency is equal to the absolute value of the difference in frequencies of the two waves involved in the superposition.

As mentioned at the beginning of this section, the beating between two waves of slightly different frequency is, among other things, the method used by musicians to tune instruments.

The note in question is played along with a standard reference tone or note. If the frequencies of the two sound waves are not identical, a trained ear can detect the small beat frequency oscillations that are a direct measure of the difference in the frequencies of the two notes. If the two notes differ in frequency by, say, 2 Hz, then the beat frequency is 2 Hz. The frequency of the note on one instrument is adjusted (by changing the tension in a string or other means for other instruments) until no more beats are heard. At that point the difference in the two frequencies is zero and the two notes are in tune with each other.

The beating phenomenon between waves or oscillations of slightly different frequencies is manifested in many aspects of nature, not just acoustics. In other words, it is a general characteristic of wave or other periodic phenomena. For example, Figure 12.57 is a graph of the time interval between successive full moons (called the synodic period of the Moon) versus time (or the calendar date) over about a 30-year period. The graph suggests a beating between two slightly different frequencies associated with the Sun–Earth–Moon system. A detailed analysis of the graph* indicates that the periods associated with the beating frequencies are 366 days (about a year) and 411 days (the time it takes the Sun to return to the same position along the major axis of the lunar orbital ellipse).

EXAMPLE 12.13

Two harmonic waves with frequencies 8.000 Hz and 8.250 Hz are superimposed. The waves each have amplitude A. Consider the position where $x = 0$ m.

a. **On the same graph, plot $\Psi_1(0$ m$, t)$ and $\Psi_2(0$ m$, t)$ as functions of time over the domain $t = 0$ s to 8.00 s.**
b. **On a separate graph, plot $\Psi_{\text{total}}(0$ m$, t)$ versus t over the domain $t = 0$ s to 8.00 s.**
c. **What is the beat frequency?**
d. **What is the beat period?**

Solution
a. Each harmonic wave has the form

$$\Psi(x, t) = A \cos(kx - \omega t)$$

*See Ronald Lane Reese, George Y. Chang, and David L. Dupuy, "The oscillation of the synodic period of the Moon: a 'beating' phenomenon," *American Journal of Physics*, 57, #9, pages 802–807 (September 1989).

At the origin, $x = 0$ m, so that

$$\Psi(x, t) = A \cos(-\omega t)$$
$$= A \cos(\omega t)$$

Since $\omega = 2\pi\nu$, the two oscillations at the origin are

$$\Psi_1(0 \text{ m}, t) = A \cos[2\pi(8.000 \text{ Hz})t]$$
$$= A \cos[(50.27 \text{ rad/s})t]$$

and

$$\Psi_2(0 \text{ m}, t) = A \cos[2\pi(8.250 \text{ Hz})t]$$
$$= A \cos[(51.84 \text{ rad/s})t]$$

These functions are plotted on the same grid in Figure 12.58.
b. At the origin, the superimposed waves have the form of Equation 12.51, with $x = 0$ m:

$$\Psi_{\text{total}}(0 \text{ m}, t) = 2A \cos(\omega t) \cos\left(\frac{\Delta\omega}{2} t\right) \qquad (1)$$

Here ω is the average of two angular frequencies:

$$\omega = \frac{\omega_1 + \omega_2}{2}$$
$$= \frac{2\pi(\nu_1 + \nu_2)}{2}$$
$$= 51.05 \text{ rad/s}$$

and $\Delta\omega$ is the difference in the two angular frequencies:

$$\Delta\omega = \omega_1 - \omega_2$$
$$= 2\pi(\nu_2 - \nu_1)$$
$$= 1.571 \text{ rad/s}$$

Hence

$$\frac{\Delta\omega}{2} = 0.7855 \text{ rad/s}$$

Equation (1) thus takes on the specific form

$$\Psi_{\text{total}}(0 \text{ m}, t) = 2A \cos[(51.05 \text{ rad/s})t] \cos[(0.7855 \text{ rad/s})t]$$

This oscillation is graphed in Figure 12.59. Notice how the resulting oscillation in Figure 12.59 is large when the

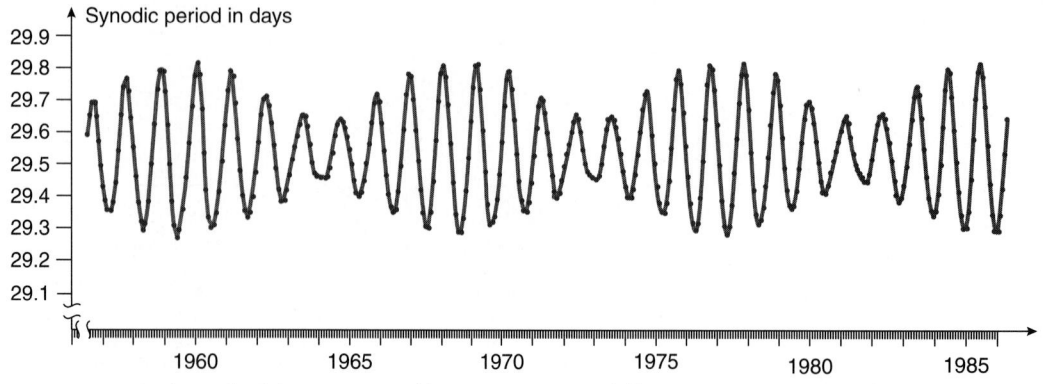

FIGURE 12.57 A graph of the time interval between successive full moons versus time reveals a beating of two frequencies. Other frequencies are present too, but have smaller amplitudes.

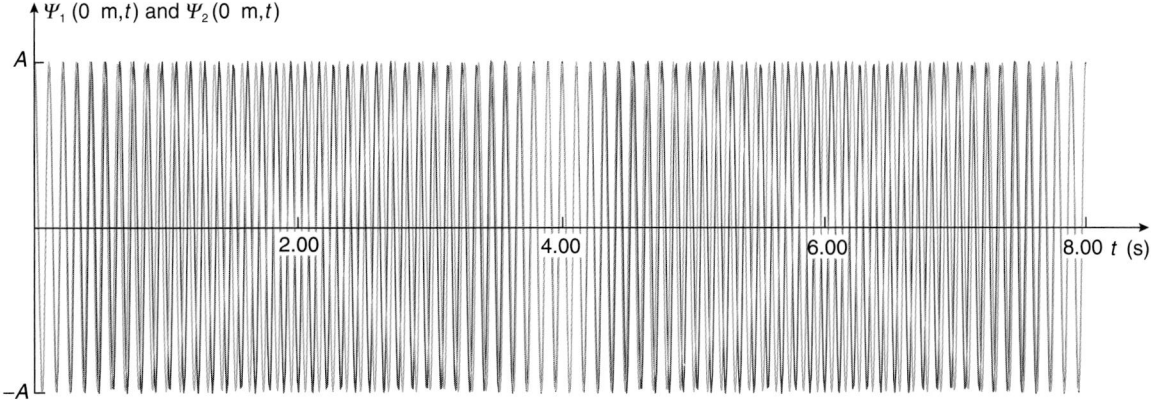

FIGURE 12.58

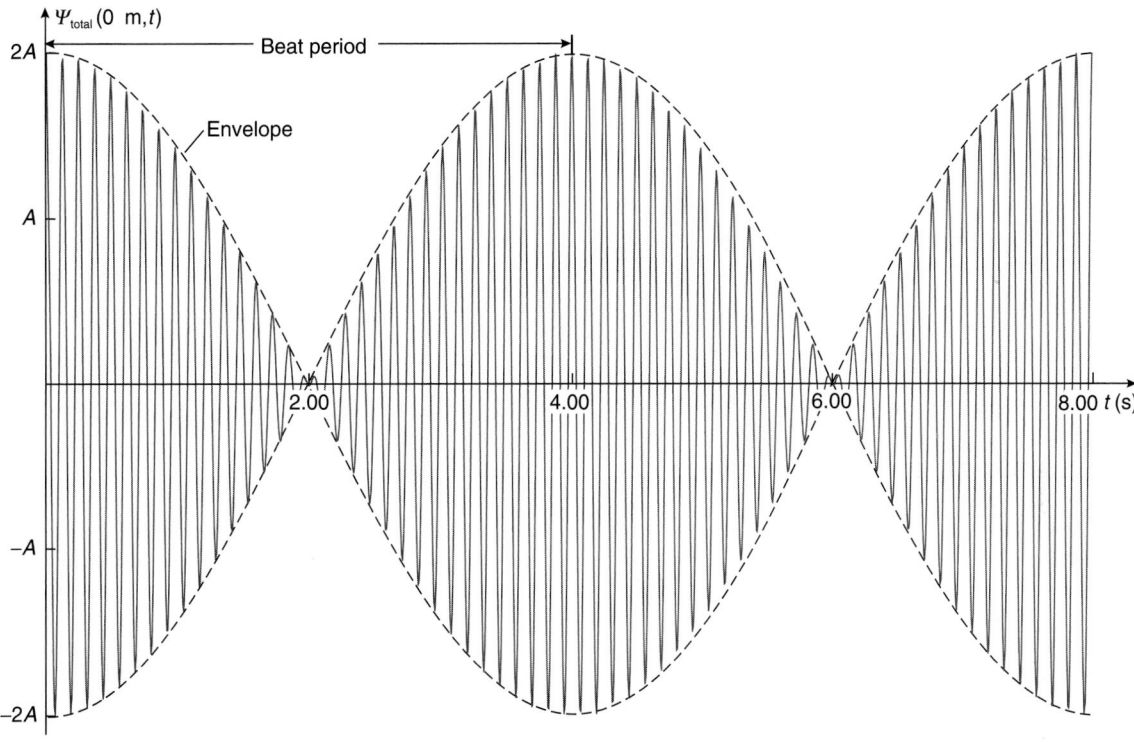

FIGURE 12.59

two oscillations in Figure 12.58 are in phase ("in step") with each other. The resulting oscillation is small when the two oscillations are out of phase ("out of step") with each other.

c. The beat frequency is the difference between the two superimposed frequencies, or 0.250 Hz. In Figure 12.59 the beat frequency is the number of "bumps" of the envelope during an interval of one second.

d. The beat period is the inverse of the beat frequency, or 4.000 s. This is the time interval between the "bumps" of the envelope shown in Figure 12.59.

12.21 FOURIER ANALYSIS AND THE UNCERTAINTY PRINCIPLES*

Ambiguous undulations . . . .

Wallace Stevens (1879–1955)[†]

Sinusoidal waves are like Legos. From sinusoidal waves, any manner of more complex periodic waves can be built. More complicated periodic waves may have waveforms such as those

[†]From the poem *Sunday Morning*, stanza VIII, line 14. In *The Collected Poems of Wallace Stevens* (Alfred A. Knopf, New York, 1954), pages 66–70; quotation is on page 70.

shown in Figure 12.60 (along with their corresponding oscillatory behavior) or many others.

It was Jean Baptiste Joseph Fourier (1768–1830) who realized that *any* periodic oscillation (or waveform) can be represented by a suitable superposition (perhaps with an infinite number of terms) of sinusoidal oscillations (or sinusoidal waveforms) that are harmonics of the fundamental frequency of the oscillation (or harmonics of the fundamental angular wavenumber of the periodic waveform). His work created a branch of mathematics now called **Fourier analysis**.

To see how combinations of sinusoidal functions are used to represent more complicated periodic functions, we examine a particular example of such a superposition. To this end, we look at the oscillation caused by a complicated periodic wave at a particular position. We also could look at the waveform at a particular time. The oscillatory behavior and waveform are complementary aspects of wave motion, as we have seen. The oscillation viewed at a fixed position will be sufficient to present the general ideas underlying Fourier analysis. The analysis is similar in both cases; by examining periodic oscillatory behavior, we can generalize the results so they apply to periodic waveforms as well.

Periodic but nonharmonic oscillatory behavior is imagined as a superposition of harmonic oscillations of various amplitudes.

> The frequencies in a Fourier representation of a periodic oscillation always are harmonics of the fundamental frequency of the periodic oscillation.

The periodic oscillation shown in Figure 12.61 is known, not surprisingly, as a sawtooth. The graph represents the oscillation produced by a sawtooth-shaped traveling wave at a particular place, say $x = 0$ m. The methods used to find the Fourier representation of complicated oscillations such as this one are slightly beyond the level of this course.* Here we will pluck the mathematical result out of thin air, just to show you how such a superposition leads to a representation of more complicated wave shapes.

The results of a Fourier analysis of the sawtooth oscillation indicate that it is represented by the superposition of an infinite number of progressively smaller sinusoidal oscillations that are harmonics of the fundamental frequency (or an-

---

*The methods are standard recipes that involve integrations of the given periodic function with sinusoidal functions.

Fourier was almost executed during the French Revolution. He went to Egypt with Napoleon and contracted a disease (likely myxedema) that eventually proved fatal.

gular frequency). In Figure 12.61 the period of the sawtooth is 2.00 s; its fundamental frequency is the inverse of the period: 0.500 Hz. The fundamental angular frequency is 2π times the fundamental frequency or

$$(2\pi \text{ rad/s})(0.500 \text{ Hz}) = \pi \text{ rad/s}$$

The representation of the sawtooth oscillation involves harmonics of the fundamental angular frequency, integral multiples of π rad/s:

$$\Psi(0 \text{ m}, t) = \frac{8}{\pi^2}\left\{\cos(\pi t) + \frac{1}{9}\cos(3\pi t)\right.$$
$$+ \frac{1}{25}\cos(5\pi t) + \frac{1}{49}\cos(7\pi t) + \cdots$$
$$\left. + \frac{1}{(2n+1)^2}\cos\left[(2n+1)\pi t\right] + \cdots\right\} \quad (12.59)$$

where n is an integer greater than or equal to zero and each π has the units of rad/s. For this particular case, only odd-numbered harmonics contribute. The right-hand side of Equation 12.59 is called the **Fourier series** of the sawtooth periodic oscillation.

To show you that the series begins to approximate the sawtooth oscillation as more and more terms are superimposed, Figures 12.62–12.64 are graphs of successively more terms in the

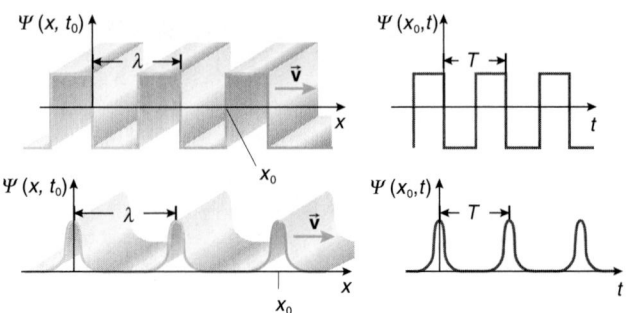

FIGURE 12.60 More complicated periodic waveforms and oscillations.

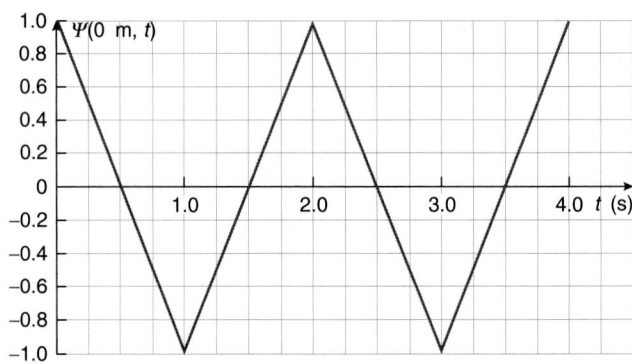

FIGURE 12.61 A sawtooth oscillation.

series on the right-hand side of Equation 12.59. Figure 12.62 is just the first term on the right-hand side of Equation 12.59. Figure 12.63b is the superposition of the first two terms. Figure 12.64b is the superposition of the first three terms. And we could continue in the same manner. As more and more terms are included, the resulting superposition approaches that of the desired sawtooth oscillation. The amplitudes and frequencies of the terms used to create the more complicated oscillation make up the frequency spectrum of the oscillation.[†]

The frequency spectrum of any periodic oscillation is discrete: only integral multiples of the fundamental frequency are present. For the sawtooth, the discrete spectrum contains only the odd-numbered harmonics.

The Fourier representation of a sawtooth waveform, rather than a sawtooth oscillation, has the same form as Equation 12.59, with x replacing t and k replacing ω. In this case we have a spectrum of angular wavenumbers that are integral multiples of the fundamental angular wavenumber of the waveform. The fundamental angular wavenumber is related to the fundamental wavelength of the periodic waveform via $k = 2\pi/\lambda$.

> Although this is a particular example, the important thing to realize is that *any* periodic function can be represented by a Fourier sum of harmonic, sinusoidal functions having a discrete spectrum. This is why the study of sinusoidal waveforms and oscillations is so important: it is the foundation for the quantitative treatment of more complicated waveforms and oscillations.

Even nonperiodic functions, such as a single pulse oscillation or pulse waveform, shown in Figure 12.65, can be represented by

[†]We saw in Section 12.19 how the complex sounds of various musical instruments have a frequency spectrum that consists of integral multiples of the first harmonic. The frequency spectrum indicates the amplitudes and frequencies of the various terms of the Fourier superposition that represents the oscillation.

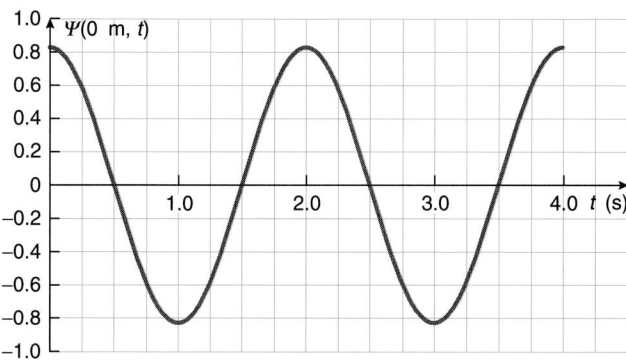

FIGURE 12.62 The first term in the Fourier series of a sawtooth.

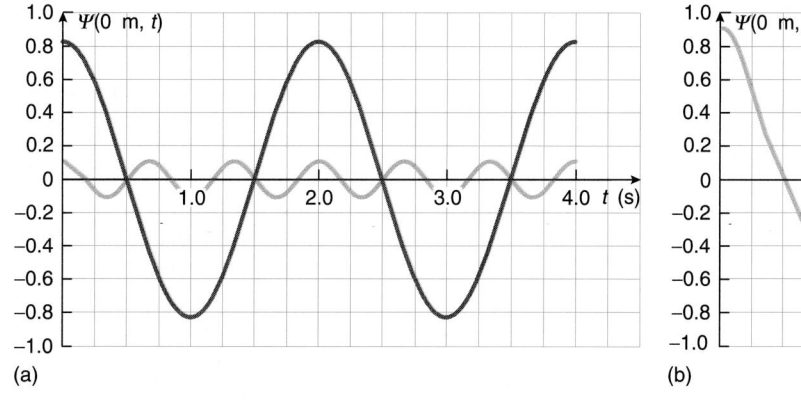

(a)

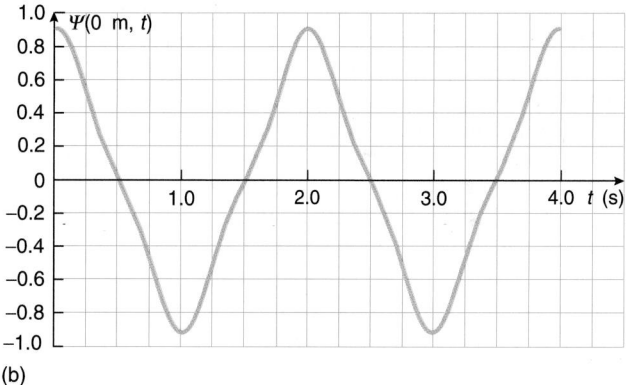

(b)

FIGURE 12.63 (a) The first two terms in the Fourier series of a sawtooth oscillation and (b) their sum.

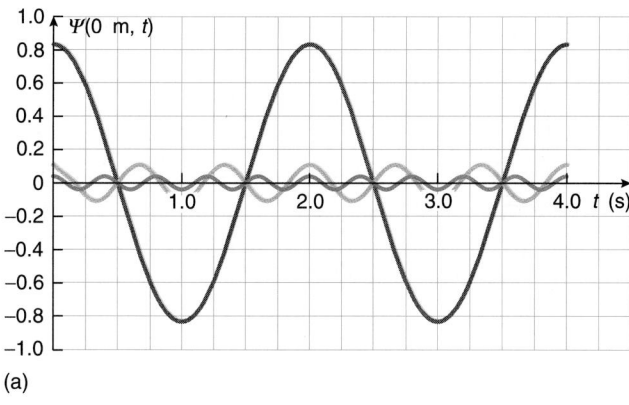

(a)

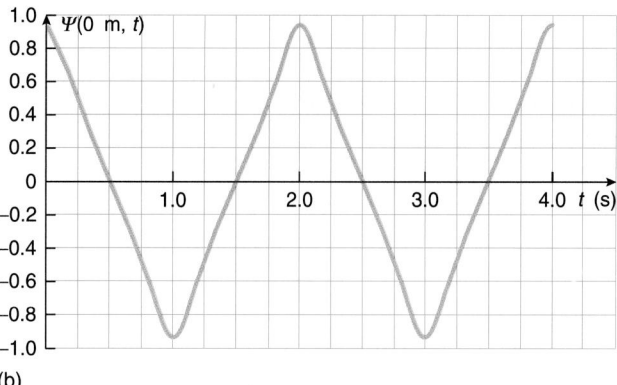

(b)

FIGURE 12.64 (a) The first three terms in the Fourier series of a sawtooth oscillation and (b) their sum.

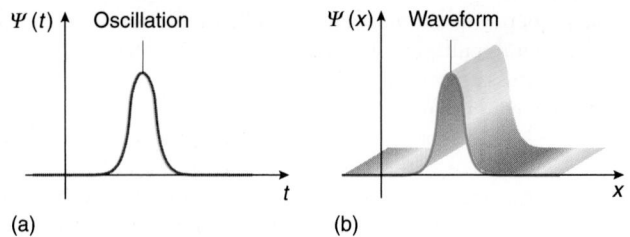

(a)   (b)

FIGURE 12.65 (a) A pulse oscillation and (b) a pulse waveform.

sinusoidal functions. Such nonperiodic oscillations and waveforms are imagined as periodic functions with a very long fundamental period T or fundamental wavelength λ, as shown in Figure 12.66. A long fundamental period T implies a small fundamental frequency ν and small fundamental angular frequency ω since

$$\nu = \frac{1}{T}$$

$$\omega = (2\pi \text{ rad}) \nu$$

Hence the harmonics of the fundamental angular frequency become more closely spaced in angular frequency as the period T increases. Indeed, as $T \to \infty$ s, the fundamental angular frequency $\omega \to 0$ rad/s, and so the Fourier series sum becomes a continuous sum, an integral,* and the frequency and angular frequency spectrum are *continuous* rather than discrete.

One finds in Fourier analysis that if the time interval Δt over which the pulse is nonzero is finite, the angular frequency spectrum also is nonzero over a finite range: $\Delta\omega = \omega_{\max} - \omega_{\min}$. The superposition of terms is completely destructive except during the interval of the pulse. In other words, the frequencies are out of phase at the beginning of the pulse, come into phase at the peak of the pulse, and are out of phase at the end of the pulse. This condition is closely met if the number of periods associated with $\omega_{\max}$ during the interval Δt of the pulse is about one more than number of periods associated with $\omega_{\min}$ during the same interval.† The number of periods T during an interval Δt is

$$\frac{\Delta t}{T} = \Delta t \frac{\omega}{2\pi \text{ rad}}$$

for an oscillation with angular frequency ω. So the condition regarding the relative number of periods can be expressed as

$$\frac{\Delta t}{2\pi \text{ rad}} \left(\omega_{\max} - \omega_{\min}\right) \approx 1$$

---

*The integral is known as the *Fourier transform*.
†This may not be obvious to you, so don't fret it.

Simplifying, we have

$$\Delta t \, \Delta\omega \approx 2\pi \text{ rad} \qquad (12.60)$$

> The shorter the duration of the pulse, the greater the spread of angular frequencies in its Fourier representation, and vice versa.

For a pure harmonic oscillation, the interval Δt over which it is nonzero is very long (strictly speaking, it is infinite), and so the spread of the angular frequency spectrum $\Delta\omega$ is very small: namely, the oscillation is represented by a single angular frequency.

Similar arguments also can be made for a pulse waveform: a long fundamental wavelength λ implies a small fundamental angular wavenumber k since

$$k = \frac{2\pi \text{ rad}}{\lambda}$$

Thus the harmonics of the fundamental angular wavenumber become more closely spaced as λ increases. As $\lambda \to \infty$ m, $k \to 0$ rad/m and the Fourier series sum becomes an integral.‡ The spectrum of angular wavenumbers becomes continuous. Since k and x in the waveform play the roles of ω and t in the oscillation, an argument similar to that made for the oscillatory behavior means that we have a relation analogous to Equation 12.60 for x and k. If the waveform is nonzero over a wavetrain of length Δx, the range of angular wavenumbers Δk in its Fourier representation satisfies

$$\Delta x \, \Delta k \approx 2\pi \text{ rad} \qquad (12.61)$$

Equations 12.60 and 12.61 are called **uncertainty relations** and are inexorably linked with all wave phenomena. In Chapter 27 (an introduction to quantum mechanics) the notions of waves and particles become intertwined, and the uncertainty relations limit the precision with which we can simultaneously measure the position and momentum of a particle, or measure its energy. The word "uncertainty" is used is the sense of "indeterminable," not capable of being fixed or measured precisely. In the context of quantum mechanics, the uncertainty relations are called **Heisenberg uncertainty principles**, after Werner Heisenberg (1901–1976), who first recognized their importance in measurement processes in quantum mechanics.

---

‡This integral also is called a Fourier transform.

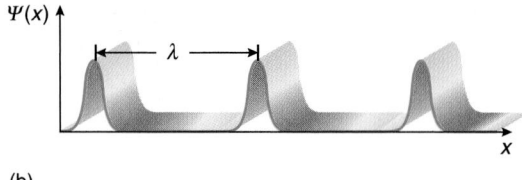

(a)   (b)

FIGURE 12.66 A periodic function with a very long period or wavelength approximates a pulse oscillation or pulse waveform.

CHAPTER SUMMARY

A *classical wave* is a propagating disturbance that transfers energy and momentum at its own characteristic speed from one region of space to another with little if any mass transfer. Particle streams also transfer energy and momentum but are not classical waves. A *mechanical wave* is a propagating disturbance in and of a material. Other types of classical waves, capable of propagating even through a vacuum, involve propagating fields.

Waves in which oscillations are along the line that the wave is propagating are called *longitudinal waves*. Sound waves in gases are longitudinal, as are compressional waves on a slinky or spring. *Transverse waves* have oscillations perpendicular to the velocity of the wave. Waves on strings under tension as well as light waves are transverse waves. Some waves may be both longitudinal and transverse.

A wave is represented by a *wavefunction* Ψ whose value indicates the size of the disturbance, whatever it may involve physically. Any one-dimensional wave propagating at constant speed v toward increasing values of x is represented by a wavefunction that is a function of the quantity $x - vt$:

$$\Psi(x - vt) \tag{12.1}$$

A wave propagating toward decreasing values of x is represented by a wavefunction of the form

$$\Psi(x + vt) \tag{12.2}$$

There are two complementary ways to view a wave. The *waveform* of the wavefunction is found by fixing $t = t_0$ and plotting the wavefunction Ψ as a function of x:

$$\Psi(x, t_0)$$

This graph is a spatial representation of the shape of the wavefunction. The *oscillatory* behavior of the wave at a fixed position x_0 is found by plotting the wavefunction Ψ as a function of time t at that position:

$$\Psi(x_0, t)$$

Any classical one-dimensional wave traveling at constant velocity $\vec{v}$ along the x-axis satisfies the *classical wave equation*:

$$\frac{\partial^2 \Psi}{\partial x^2} - \frac{1}{v^2} \frac{\partial^2 \Psi}{\partial t^2} = 0 \tag{12.7}$$

Periodic waves have a well-defined (constant) *wavelength* λ and *period* T. The wavelength is the distance between corresponding successive points on the waveform. The period is the time for one complete oscillation. The *frequency* ν is the inverse of the period and is the number of oscillations completed during an interval of exactly one second. For a wave traveling at speed v, the wave moves a distance equal to a wavelength during one period. Hence the speed is the product of the frequency and the wavelength:

$$v = \nu \lambda \tag{12.10}$$

Sinusoidal (harmonic) waves, propagating toward increasing values of x, have the form

$$\Psi(x, t) = A \cos(kx - \omega t) \tag{12.16}$$

where k is the *angular wavenumber*:

$$k = \frac{2\pi \text{ rad}}{\lambda} \tag{12.13}$$

and ω is the angular frequency:

$$\omega \equiv (2\pi \text{ rad})\nu$$

The angular wavenumber is the number of wavelengths in a distance of 2π meters.

The speed v of a transverse wave on a string with mass per unit length μ and a tension of magnitude T is

$$v = \left(\frac{T}{\mu} \right)^{1/2} \tag{12.19}$$

The power P transferred by a wave is proportional to the wave speed v, the square of the amplitude A, and the square of the angular frequency. For transverse waves on a stretched string, the power transferred by the wave is

$$P = \frac{1}{2} \mu \omega^2 A^2 v \tag{12.26}$$

The *intensity* I of a wave is the average power transmitted through one square meter oriented perpendicular to the direction the wave is propagating. For a wave propagating in three dimensions, the intensity a distance r from a point source of waves follows an inverse square law:

$$I = \frac{P}{4\pi r^2} \tag{12.27}$$

where P is the power of the source.

The *sound level* β in *decibels* (dB) of an acoustic wave of intensity I is

$$\beta = (10 \text{ dB}) \log \frac{I}{I_0} \tag{12.28}$$

where I_0 is the average human *threshold of hearing*; $I_0 \equiv 10^{-12} \text{ W/m}^2$.

Motion of a source and/or observer changes the observed frequency of acoustic waves, a phenomenon known as the *acoustic Doppler effect*. If the source frequency is ν, the frequency ν' detected is

$$\nu' = \nu \left(\frac{v \pm v_{\text{med}} \pm v_{\text{obs}}}{v \pm v_{\text{med}} \mp v_{\text{source}}} \right) \tag{12.37}$$

where v is the speed of sound, v_{med} is the wind speed (use the positive sign for a wind blowing from the source toward the observer, use the negative sign for the opposite direction), v_{obs} is the speed of the observer, and v_{source} is the speed of the source. If the distance between source and observer is decreasing (increasing), the detected frequency is greater (less) than the source frequency and you should use the top (bottom) signs in front of v_{obs} and v_{source} in Equation 12.37.

If the speed of a source of sound v_{source} is greater than the speed of sound v in the medium, a *shock wave* (sonic boom) is formed. The half-angle ϕ of the *Mach cone* defining the shock wave is found from

$$\sin \phi = \frac{v}{v_{\text{source}}} \qquad (v_{\text{source}} > v) \tag{12.39}$$

The *Mach number* is

$$\text{Mach number} = \frac{v_{\text{source}}}{v} \tag{12.40}$$

Two or more waves together produce a total disturbance that is the sum of the individual wave disturbances:

$$\Psi_{\text{total}}(x, t) = \Psi_1(x, t) + \Psi_2(x, t) + \cdots$$

This is the linear *principle of superposition*. We consider only waves that follow such a linear principle of superposition. You should be aware that not all waves do so.

Standing waves are produced by the superposition of two sinusoidal waves of equal amplitude and frequency propagating in opposite directions; if the waves are transverse, they must have the same plane of polarization. The *nodes* of a standing wave are the locations where there is zero wave disturbance at all times. Nodes are separated in space by half a wavelength. *Antinodes* are places of maximum disturbance and also are separated by half a wavelength.

Standing waves on a string fixed at both ends have frequencies given by

$$\nu_n = n \frac{1}{2\ell}\left(\frac{T}{\mu}\right)^{1/2} \qquad (n \text{ is an integer}) \qquad (12.49)$$

where n is an integer that numbers the harmonics, μ is the mass per unit length of the string, T is the magnitude of the tension of the string, and ℓ is the distance between the fixed ends of the string. The frequency corresponding to $n = 1$ is called the *fundamental frequency* or, equivalently, the *first harmonic*.

Two similar waves of different frequency propagating in the same direction produce *beats*. The beat frequency is equal to the absolute value of the difference between the frequencies of the two waves.

Sinusoidal (harmonic) waves are used to describe more complex periodic waves through *Fourier analysis*. The Fourier analysis leads to two *uncertainty relations*:

$$\Delta t \, \Delta \omega \approx 2\pi \text{ rad} \qquad (12.60)$$

$$\Delta x \, \Delta k \approx 2\pi \text{ rad} \qquad (12.61)$$

SUMMARY OF PROBLEM-SOLVING TACTICS

12.1 (page 537) The sign between x and νt determines the direction $\Psi(x, t)$ is moving along the x-axis. When the sign is negative, $\Psi(x - \nu t)$, the wave is moving toward *increasing* values of x; when the sign is positive, $\Psi(x + \nu t)$, the wave is moving toward *decreasing* values of x.

12.2 (page 542) To find the value of $\cos(kx - \omega t)$ in Equation 12.11,

$$\Psi(x, t) = A \cos(kx - \omega t) \qquad (12.11)$$

be sure your calculator is set for radian measure of angle, not degrees.

12.3 (page 551) Logarithms are exponents, so that the expression $Y = \log X$ means $X = 10^Y$. Therefore the sound level

$$\beta = (10 \text{ dB}) \log \frac{I}{I_0}$$

means

$$\frac{I}{I_0} = 10^{\beta/(10 \text{ dB})}$$

12.4 (page 554) The sign to use in Equation 12.33,

$$\nu' = \nu\left(\frac{\nu \pm \nu_{\text{obs}}}{\nu}\right) \qquad (12.33)$$

can be determined as long as you remember that the frequency ν' is greater than ν if the observer is approaching the source (so the + sign must be used) and less than ν if receding (use the − sign).

12.5 (page 554) The sign to use in Equation 12.36

$$\nu' = \nu\left(\frac{\nu}{\nu \mp \nu_{\text{source}}}\right) \qquad (12.36)$$

can be determined by whether the observer hears a higher or lower frequency (as the source approaches or recedes, respectively).

12.6 (page 555) To use the general equation for the Doppler effect (Equation 12.37), recall the conventions associated with various signs on page 555.

QUESTIONS

1. (page 533); 2. (page 535); 3. (page 536); 4. (page 546); 5. (page 547); 6. (page 551); 7. (page 555); 8. (page 558); 9. (page 563); 10. (page 564)

11. Fans at various sporting events like to make "the wave" as shown in Figure Q.11. Is this a classical wave as we define it in physics? Explain.

12. A line of cars is zooming along a freeway at a constant speed of 100 km/h. One car brakes suddenly and a traffic pulse ripples along the line of cars. Is this traffic pulse a wave? What factors determine its speed?

13. If a radio wave (a light wave) and a sound wave have the same wavelength, which has the greatest frequency?

14. A flash of lightning is observed; t_0 seconds later, a crash of thunder is heard. Given that the speed of sound is about 340 m/s, devise an *approximate* expression for the distance in *kilometers* to the lightning flash. If you wish, devise an approximate expression for the distance in miles.

15. A transverse wave of amplitude A is propagating along a rope. Imagine a small segment of the rope. At what positions in its transverse motion is the magnitude of its acceleration greatest? At what positions is the acceleration of zero magnitude? At what positions is its speed a maximum? At what positions is its speed zero?

16. If you double the tension in a string, by what factor does the speed that a wave propagates on the string increase?

17. When you drop a rock into a calm pond, the circular water waves decrease in amplitude as they radiate outward. Explain why.

And like circles on the water, which, as they grow fainter, expand.
Herman Melville (1819–1891)*

Moby Dick (University of California Press, Berkeley, 1979), Chapter 110, page 486.

FIGURE Q.11

18. In the acoustic Doppler effect, the effect of an approaching and then receding sound source is a transition from a higher to lower frequency as time passes. Explain why a transition *from* a lower *to* a higher pitch never is observed for a source with a constant velocity.

19. Examine a guitar or banjo and consider one of its strings under a constant tension. Measure the lengths of the string corresponding to two frets an octave apart and see if the frequencies found using Equation 12.49 (for $n = 1$) differ by a factor of two (an octave).

20. Pianos come in various sizes, from baby grands to concert grands. As you progress to lower-frequency notes, at what note are the strings wire-wrapped? Is it the same note for all piano sizes? Does your answer depend on whether the piano is made by Steinway, Bösendorfer, Yamaha, or another manufacturer?

21. The high-frequency string on a standard guitar is tuned to the E above middle C and has a frequency of 329.63 Hz.* The lowest-frequency string is tuned to the E two octaves lower. If the strings of a guitar all were made of wire with the same mass per unit length, strung with the same magnitude tension, what would be the ratio of the longest to shortest string?

22. When you tune a guitar or banjo, what parameter in Equation 12.49 are you changing?

23. When you use different frets along a guitar or banjo string, what parameter in Equation 12.49 are you changing?

24. Standing waves are generated on a string fixed at both ends. The distance between the fixed ends is ℓ. You observe a node located

a distance $\ell/3$ from each fixed end as shown in Figure Q.24. What harmonic of the fundamental frequency is oscillating on the string?

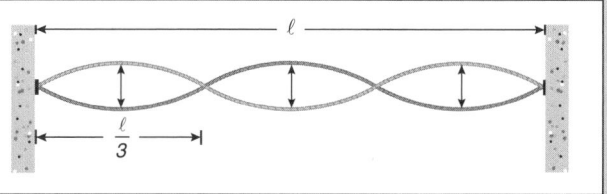

FIGURE Q.24

25. Do the locations of the nodes and antinodes of a standing wave move?

26. Describe an experiment to see if there is dispersion of sound waves in air, that is, to see whether the speed of sound is independent of frequency.

27. Pianos are strung with many strings under considerable tension. A frame is used on which to string the strings. Explain why the transition from wood frames to steel frames, pioneered in the mid-1800s by the now famous Steinway piano factory, revolutionized piano making technology, enabling musicians to have their pianos tuned much less frequently.

28. Consult an appropriate book on acoustics to distinguish between the musical concepts of pitch, loudness, and tone.

29. Two-dimensional standing waves exist on vibrating surfaces such as drum heads. Describe an experimental procedure that can be used to visually locate the nodes and antinodes on the surface if it is horizontal. (See the picture on page 561.)

30. A space shuttle in orbit releases a rocket from its cargo bay and moves 200 m away before the rocket engine is ignited. Should the crew wear hearing protectors? Explain.

31. A xylophone is an instrument with metal bars of various lengths that produce the notes of the musical scale. A vibraphone is similar except that open-ended tubes of appropriate lengths are placed beneath each metal bar to enhance the sound. If you strike a xylophone key, the sound will persist for almost a minute. If you strike the same note on a vibraphone, the sound is much louder than with a xylophone but persists for only a small fraction of the interval for a xylophone. Explain.

32. The strings of a properly tuned guitar play the notes E_2, A_2, D_3, G_3, B_3, and E_4 as fundamentals. Use the frequency chart in Figure 12.53 on page 562 to determine the ratio of the frequencies of the notes to the lowest frequency as a reference.

33. Measure the mass and total length of each string on a guitar to determine its mass per unit length μ. Suppose each string is adjusted to have the same magnitude tension. What is the ratio of the frequencies of the fundamental notes on each string to that of the lowest frequency?

34. In the same way as a boat is set into oscillation by a passing water wave, a sound wave induces oscillations in objects in its path. An opera singer "tweaks" the rim of a wine glass and listens to its ringing. She then sings a note of the same frequency. Explain how the singer can shatter the glass if she sings loud enough near the glass. See Figure Q.34.

---

*In guitar music, this E typically is written for convenience as the E located one octave higher on the treble clef.

In the case of a cock putting its head into an empty utensil of glass where it crowed so that the utensil thereby broke, the payment must be in full. . . .

The Babylonian Talmud*

FIGURE Q.34

35. What observational evidence can you cite that indicates sound is a wave rather than a stream of particles?

36. What observational evidence indicates that radio waves really are waves and not streams of particles?

37. If you inhale helium and then talk, your voice has a falsetto tone that sounds like a chipmunk. Why?

38. Does the speed of sound increase, decrease, or remain the same as a function of altitude in the atmosphere of the Earth? Explain the reasons for your answer.

39. As you fill a thermos with coffee, the frequency of the sound you hear changes. Does the frequency increase or decrease? Explain why.

40. Write a brief answer to the old question that goes like this: A tree falls in a forest. Is there a sound if no one is there to hear it?

41. If the sound intensity doubles, what is the increase in the sound level to the nearest tenth of a decibel? Do you need to know the original sound level to make this calculation?

42. Two sources of sound have the same frequency. Under what circumstances could you hear beats between them?

43. From observations of water wave ripples on a pond and water waves on the open ocean, do surface water waves exhibit dispersion or no dispersion? If so, does the speed increase or decrease with wavelength?

44. For standing acoustic waves, what is the distance between a node of the pressure variation wave and a node of the longitudinal particle position wave? Use wavelength as the unit of measurement.

PROBLEMS

For problems where it is germane, assume the speed of sound in air is 343.0 m/s unless otherwise indicated.

Sections 12.1 What Is a Wave?
       12.2 Longitudinal and Transverse Waves
       12.3 Wavefunctions, Waveforms, and Oscillations
       12.4 Waves Propagating in One, Two, and Three Dimensions
       12.5 One-Dimensional Waves Moving at Constant Velocity
       12.6 The Classical Wave Equation for One-Dimensional Waves*
       12.7 Periodic Waves

1. At a fireworks celebration on the fourth of July, you notice the flash from a cherry bomb and 0.50 s later hear the sound of the acoustic pulse from the explosion. The speed of sound waves in air is about 343 m/s while that of light is about 3.0×10^8 m/s. What was the distance between you and the explosion?

•2. A transverse wave pulse on a string has the wavefunction

$$\Psi(x, t) = \frac{0.250 \text{ m}^3}{2.00 \text{ m}^2 + [x - (5.00 \text{ m/s}) t]^2}$$

(a) What is the speed of the pulse? (b) On the same plot, graph the waveform of the pulse at the instants when $t = 0$ s, 1.00 s, and 2.00 s to show that the pulse is moving toward increasing values of x at the speed you found in part (a). Let the domain of your graph be $x = -10.0$ m to 20.0 m. (c) What is the maximum value of the disturbance? (d) If you change the wavefunction to

$$\Psi(x, t) = \frac{0.250 \text{ m}^3}{2.00 \text{ m}^2 + [x + (5.00 \text{ m/s}) t]^2}$$

show with another series of plots of the waveform when $t = 0$ s, 1.00 s, and 2,00 s that the pulse is moving toward decreasing values of x. Let the domain of this plot be $x = -20.0$ m to 10.0 m.

•3. One-dimensional waves propagating toward increasing values of x are functions of the quantity $x - vt$, so we write $\Psi = \Psi(x - vt)$. Consider the following functions of $x - vt$ where A, α, and v are constants: (i) $A[x - vt]$; (ii) $\log[\alpha(x - vt)]$; (iii) $Ae^{-\alpha(x - vt)^2}$. (a) When $t = 0$ s, sketch the waveform of these functions. (b) At $x = 0$ m, sketch the dynamical behavior of these functions (if any) for $t \geq 0$ s. (c) Two of these functions cannot represent realistic waves; one can. Explain which and why.

•4. An asymmetric transverse pulse on a taut string has the waveform shape indicated in Figure P.4 when $t = 0$ s, where Ψ represents the transverse position of the string in meters, measured from its equilibrium position. The value of Ψ represents the position of a particle of the string along the y-axis, perpendicular to the velocity of the wave. The velocity of the pulse is $(2.0 \text{ m/s})\hat{\imath}$. (a) For the position $x = 6.0$ m in Figure P.4, sketch the corresponding oscillatory behavior on a Ψ versus t graph. (b) At the instant shown in Figure P.4 ($t = 0$ s), make a sketch of the transverse velocity component of the particle as a function of position x along the string. (c) For the position $x = 6.0$ m, sketch the transverse velocity component as a function of time. (d) From your graphs, at what value of the transverse position Ψ does a particle of the string have a velocity component equal to zero? (e) When $t = 0$ s, at what positions along the x-axis is the transverse speed of a particle on the string equal to zero?

This problem was inspired by Arnold Arons, *A Guide to Introductory Physics Teaching* (Wiley, New York, 1990), page 203.

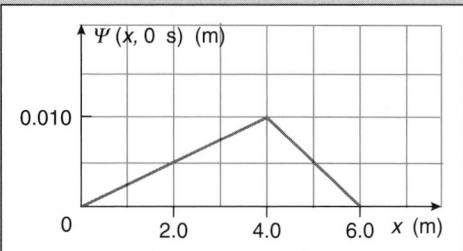

FIGURE P.4

•5. An earthquake generates two principal types of waves known as S-waves (for shear waves) and P-waves (for *p*ressure waves) that propagate away from the focus of the quake. The P-waves travel at a speed of about 5.0 km/s whereas the S-waves poke along at only about 3.0 km/s. A seismometer can detect these waves at sites remote from the focus (or origin) of the quake. The P-waves arrive at a seismometer before the S-waves from the same quake (see Figure 12.5). The time interval between the arrival of the waves is an indication of the distance of the seismometer from the focus of the quake. (a) If the P-waves arrive at the seismometer 4.00 min before the S-waves, what is the distance to the focus of the quake? (b) The previous part determined the distance to the focus of the quake but not the specific location. Why not? (c) By using several seismometers at widely separated locations it is possible to pinpoint the specific location of the quake; indicate how.

•6. A periodic transverse wave moving at 4.00 m/s toward decreasing values of x has the waveform indicated in Figure P.6 when $t = 0$ s. (a) What is the wavelength? Indicate the wavelength on the waveform. (b) Draw a graph of the disturbance at $x = 0$ m as a function of time. Indicate the period of the wave on this graph. What is the period in seconds? (c) Use the results of part (b) to determine the frequency and confirm that the speed is the product of the frequency and wavelength as Equation 12.10 states. (d) What is the maximum transverse speed?

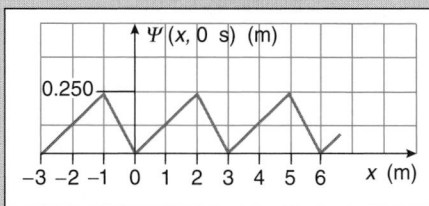

FIGURE P.6

•7. While bottom fishing, you notice that your anchored boat completes 60 complete up-and-down motions during one minute because of the passage of a wave. You estimate the speed of the waves to be 7.0 m/s. Another nearby anchored boat is located in a direction parallel to the velocity of the waves. You notice that the other boat is in a wave trough while yours is at a crest, and vice versa. At what possible distances from you is the other boat?

•8. A perfectly elastic ball bounces on a rigid floor and continually rebounds to a height of 1.20 m. Each time the ball strikes the floor, a sensor emits a brief beep. (a) Are the intervals between successive beeps the same? (b) What is the period of the acoustic beeps? (c) What is the wavelength associated with the periodic acoustic beeps?

•9. A periodic wave moving with a speed of 3.00 m/s toward increasing values of x passes the position $x = 0$ m and the disturbance at that location is graphed as a function of time in Figure P.9. (a) Indicate the period of the wave on the graph. (b) What is the frequency of the wave? (c) At the instant when $t = 0$ s, make a graph of the waveform as a function of x and indicate the wavelength of the wave on your graph. (d) Verify that the speed of the wave is the product of the frequency and wavelength.

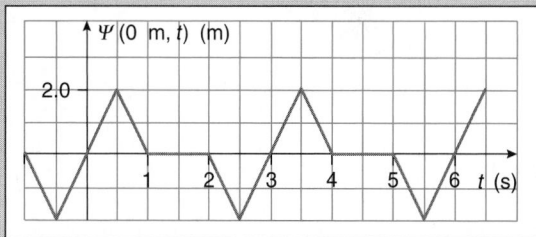

FIGURE P.9

•10. Repeat Problem 9 but with the wave moving toward decreasing values of x.

•11. Show that the sinusoidal wave function

$$\Psi(x, t) = A \cos(kx - \omega t)$$

satisfies the classical wave equation (Equation 12.7).

•12. (a) Show that a wavefunction of the form

$$\Psi(x + vt)$$

representing a wave traveling at constant speed v in a direction toward decreasing values of x has the following relationship between its first partial derivatives:

$$\frac{\partial \Psi}{\partial t} = +v \frac{\partial \Psi}{\partial x}$$

Notice that this result is different from the relationship for a wave traveling in the direction of increasing values of x (Equation 12.6). (b) Show that $\Psi(x + vt)$ satisfies the same classical wave equation (Equation 12.7) as a wave $\Psi(x - vt)$ traveling in the opposite direction, toward increasing values of x.

‡13. An unusual wave is described by the rather formidable wavefunction:

$$\Psi(x, t) = A e^{-[\alpha(x - vt)]^2} + A e^{-[\alpha(x + vt)]^2}$$

Use the following values for the parameters A, α, and v:

$$A = 3.00 \text{ m}$$
$$\alpha = 0.500 \text{ m}^{-1}$$
$$v = 1.00 \text{ m/s}$$

Make a series of graphs of the waveform of this wavefunction over the domain $x = -10.00$ m to $x = +10.00$ m at each of the following times: $t = -6.00$ s; $t = -3.00$ s; $t = 0.00$ s; $t = 3.00$ s; $t = 6.00$ s. (b) What is the meaning of the constant A on the graph? (c) Describe how Ψ changes as time advances during the interval from $t = -6.00$ s to 6.00 s. Note the successive positions of a particular crest of the wave. The series of waveforms illustrates the principle of linear superposition, discussed in greater detail in Section 12.18. (d) Investigate the effect on the graphs of changing the parameter α to 1.00 m^{-1} and v to 2.00 m/s. Make the plots over the domain $x = -15.00$ m to $x = +15.00$ m for the same five instants as in part (a).

Section 12.8 Sinusoidal (Harmonic) Waves

14. What is the approximate frequency of a sinusoidal sound wave that has a wavelength equal to the straight-line distance between your ears?

•15. Alas, after a sybaritic festival, the cheap upright piano in your fraternity house is found upright at the bottom of the house swimming pool. You decide to play Handel's *Water Music* but first test the sound of middle C (261.6 Hz). The speed of sound in water is 1.48×10^3 m/s. What is the wavelength of the sound wave corresponding to middle C in the pool? What assumptions are needed to make the determination?

•16. A transverse wave on a string is described by the wavefunction

$$\Psi(x, t) = (0.300 \text{ m}) \cos[(12.57 \text{ rad/m})x - (251.3 \text{ rad/s})t]$$

Find: (a) the amplitude of the wave; (b) the angular wavenumber of the wave; (c) the wavelength of the wave; (d) the angular frequency of the wave; (e) the frequency of the wave; (f) the period of the wave; (g) the speed of the wave; (h) the wave disturbance at $x = 3.00$ m when $t = 1.50 \times 10^{-3}$ s.

•17. An aging hippie is making good vibes that are described by the wavefunction

$$\Psi(x, t) = (6.00 \times 10^{-3} \text{ Pa}) \cos[(8.06 \text{ rad/m})x - (2.76 \times 10^3 \text{ rad/s})t]$$

(a) What is the amplitude of the wave? (b) What is the angular wavenumber of the wave? (c) How many wavelengths will fit into a distance of 2π meters? (d) What is the wavelength of the waves? (e) What is the angular frequency of the wave? (f) What is the frequency of the wave? (g) What is the period of the oscillation? (h) What is the speed of the wave? (i) Plot the waveform when $t = 1.00$ s. Choose an appropriate domain for x to show several wavelengths. (j) Plot the oscillation produced by the wave at the position $x = 2.00$ m as a function of time. Choose an appropriate domain to show several oscillations. (k) What is the wave disturbance Ψ at $x = 1.00$ m when $t = 0$ s? Explain the physical significance of the negative pressure value of Ψ.

•18. Figure P.18 depicts the waveform of a traveling sinusoidal wave at two instants, when $t = 0$ s and 0.250 s later. The wave is traveling toward increasing values of x. Find $\Psi(x, t)$.

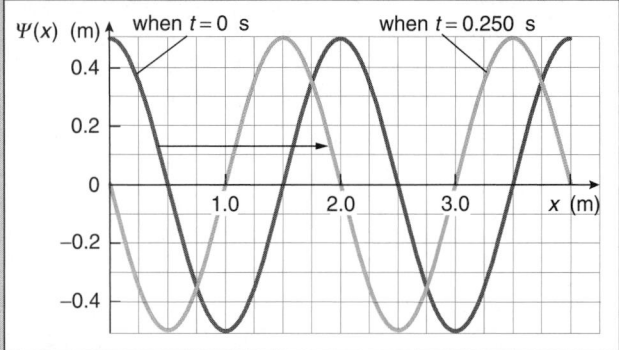

FIGURE P.18

•19. A very slow acoustic sales pitch emitted by a traveling wave salesperson has the wavefunction

$$\Psi(x, t) = (2.00 \times 10^{-3} \text{ Pa}) \cos[(8.00 \text{ rad/m})x - (6.00 \text{ rad/s})t]$$

For the Better Business Bureau, find: (a) the amplitude of the wave; (b) the angular wavenumber of the wave; (c) the angular frequency of the wave; (d) the frequency of the wave; (e) the speed of the wave; and (f) the direction the wave is moving. (g) Just to be ornery, find the value of the disturbance at $x = 2.00$ m when $t = 1.50$ s.

•20. Duck season has arrived. A Harlequin duck (*Histrionicus histrionicus*) is bobbing vertically up and down due to a passing water wave. The duck bobs through a total vertical distance of 10.0 cm and completes 7.00 cycles during 10.0 s according to your Mickey Mouse watch. You also note that a particular crest of the wave travels the 9.00 m from the duck to your canoe during 5.00 s. (a) What is the frequency of the water wave? (b) What is the speed of the wave? (c) What is the wavelength of the waves? (d) Write a mathematical expression for the wavefunction $\Psi(x, t)$ assuming it is a sinusoidal wave with the duck located at the origin for x and at a peak of the wave when $t = 0$ s. Express all distances in meters and times in seconds.

•21. You like to make waves and succeed in generating small, sinusoidal waves that propagate across a liquid surface at a speed of 0.600 m/s. The ripple crests are 0.200 m apart and are 1.00 cm in height (from trough to peak). (a) Find the frequency of the waves. (b) Write an equation for the vertical position Ψ as a function of horizontal position x and time t if the origin in the vertical direction is at the water surface when it is calm. Use SI units. A peak is at the origin when $t = 0$ s.

•22. By careful measurements, you determine that the lecture sound wave propagating from your professor to your ears is described by the following wavefunction:

$$\Psi(x, t) = (1.60 \times 10^{-3} \text{ Pa}) \cos[(11.0 \text{ rad/m})x + (3.77 \times 10^3 \text{ rad/s})t + 10.0 \text{ rad}]$$

Your professor is suitably impressed with your experimental prowess, but now asks you to find: (a) the wavelength; (b) the frequency, (c) the speed; and (d) the direction of propagation of the wave.

•23. During a hairy calculation by the famous classicist I. M. Pompous Mediocritus, an astute physics student detects a "hand wave" propagating across the blackboard. The hand wave is described by the wavefunction

$$\Psi(x, t) = (3.00 \text{ obfuscations}) \cos[(18.0 \text{ rad/m})x - (36.0 \text{ rad/s})t]$$

(a) Determine the wavelength and frequency of the hand wave. (b) Determine the speed of the hand wave (otherwise known as the "sleight of hand").

Section 12.9 Waves on a String

24. Transverse waves with a speed of 25 m/s are to be generated on a 6.50 m length of clothesline of mass 100 g. What magnitude of tension is needed?

25. A string has a tension of magnitude 50 N and transverse waves propagate along its length with a speed of 35 m/s. What magnitude of tension is needed to have the waves propagate with speed 45 m/s?

•26. A 14 gauge copper wire has a diameter of 0.2032 cm. Transverse waves on the wire are to travel at a speed of 125 m/s. What magnitude tension is needed? The density of copper is $8.50 \times 10^3 \text{ kg/m}^3$.

•27. A rope with linear mass density μ_1 is spliced to another rope with linear mass density μ_2; one end of the latter rope is attached rigidly to a wall. You pull on the end of the first rope with a tension of magnitude T, directed along the rope. What is the ratio of the speeds v_1/v_2 of transverse waves on the two strings?

•28. Two separate strings have linear mass densities μ_1 and μ_2. Transverse waves travel at equal speeds along each string. What is the ratio of the magnitudes of the tensions T_1/T_2 in the strings?

•29. You and a buddy hold opposite ends of a rope of length of 10.00 m and mass 156 g. You exert a force of magnitude 40.0 N on your end of the rope, directed along the rope; the rope is stationary. A brief interval Δt after you initiate a pulse down the rope, your buddy does the same. The two pulses meet 1.00 m from the center of the rope. What was the time interval Δt between initiation of the pulses?

•30. When $t = 0$ s, a pulse on a string is described by the equation

$$\Psi(x, 0 \text{ s}) = \frac{1.00 \text{ m}^3}{2.50 \text{ m}^2 + x^2}$$

(a) Sketch $\Psi(x, 0 \text{ s})$ as a function of x over the domain $-10.00 \text{ m} \leq x < 10.00 \text{ m}$. (b) The pulse moves parallel to $-\hat{\imath}$ with a speed of 8.00 m/s. Find $\Psi(x, t)$.

Sections 12.10 Reflection and Transmission of Waves
12.11 Energy Transport via Mechanical Waves
12.12 Wave Intensity

31. You clap your hands once at the top of a deep well shaft and hear the echo 0.40 s later. How far below the top of the well is the water surface?

•32. A river canyon is 150 m across as shown in Figure P.32. You are located on the river surface and clap your hands once to generate a brief acoustic pulse. You detect a time interval of 0.408 s between the twin echoes recorded on tape, analyzed when back home (?) in a physics lab. Where were you relative to the canyon walls? (Neglect multiple echoes.)

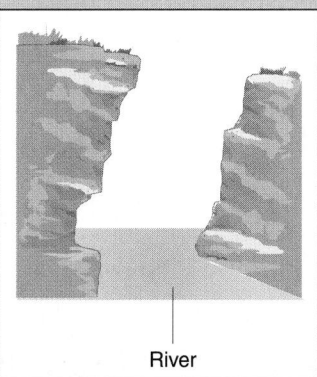

River

FIGURE P.32

•33. The ancient Great Pyramid at Giza, Egypt, built to be the tomb of Khufu, the first pharaoh of the Fourth Dynasty (reigning from 2789–2767 B.C.), had smooth polished stone on each side. Most of the fine stone facing was stripped away by scavengers during the subsequent millennia for other building projects near and in Cairo. The smooth facing remains only near the peak. The tomb also was plundered; so much for eternally resting in peace. The faces of the pyramid now reveal huge blocks of stone that form large steps as shown in Figure P.33. Imagine yourself located on a nearby sand mound on a line that is perpendicular to the side of one face of the pyramid. You generate a very short acoustic pulse by shouting a short "Yip!", setting off a small firecracker, or shooting a starter's gun. The pulse propagates to and is reflected from each step. About how many reflected pulses per second do you hear? You may be able to do a similar experiment on a local scale using stepped bleachers at your university stadium.

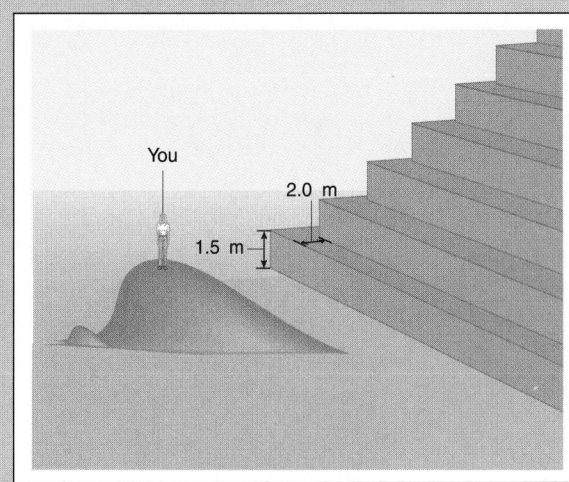

You

2.0 m

1.5 m

FIGURE P.33

•34. You are scuba diving 3.00 m below the surface of the Caribbean Sea. Fun! A support boat is located directly above you. A bell on the boat is located 1.5 m above the water line.

(a) How long does it take the sound of the bell to propagate to you? (b) If the sound reaches the ear of the skipper in the same time, how far is the skipper from the bell? See Table 12.2 (on page 553) for appropriate values of the speed of sound.

•35. The intensity of solar light waves impinging on the Earth is measured by a spacecraft to be 1.36 kW/m². This intensity is called the *solar constant*, and its rather significant value is what motivates much research into tapping such solar energy. Consider the Sun to be a point source of light. The Earth is located 149.6 × 10⁶ km from the Sun. Use this information to calculate the *solar luminosity*, the power output of the Sun.

•36. An acoustic source emits sound waves with a power of 10.0 W in all directions. What is the intensity of the sound waves at distances of 5.00 m and 25.0 m from the source?

•37. The inside of a long straight tunnel has a strip of fluorescent lights along its length; see Figure P.37. The lights are each ℓ meters long and have a power of P watts of light. What is the light intensity at a distance r from the lamps, measured perpendicular to the axis of lights? Neglect reflections from the walls of the tunnel.

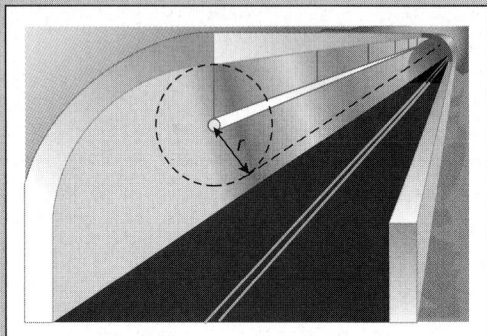

FIGURE P.37

Sections 12.13 What Is a Sound Wave?*
12.14 Sound Intensity and Sound Level*

38. A speaker emits 50 W of power in acoustic waves that propagate away from the source uniformly in all directions. (a) What is the intensity of the sound waves 20 m from the source? (b) What is the sound level in dB at this location?

39. An acoustic source has an output sound level of 96 dB at a distance of 2.00 m from the source. What is the intensity of the sound?

•40. If the sound level increases by 5.0 dB, by what factor does the sound intensity increase?

•41. A sound source that emits acoustic waves in all directions has a sound level of 120 dB at a distance of 5.0 m from the source. (a) What is the power of the source? (b) At what distance is the sound level 90 dB?

•42. A sound level meter indicates that a single trombone player produces a sound level of 75 dB, measured a certain distance away. (a) If two identical trombonists now are used, both the same distance away, what sound level will the meter indicate? (b) If 76 trombones lead the big parade, what sound level will they produce if they all are approximately the same distance from the sound level meter?

•43. As a precaution at a rock concert or in other noisy environments, you wear ear protectors that reduce the sound level by 20 dB. What percentage of the intensity of the sound is blocked?

•44. Professional screamers (a.k.a. small siblings and/or other whiners) can attain sound levels approaching 100 dB, measured 2.0 m away. (a) What is the corresponding sound intensity? (b) Assuming (unrealistically) there is no absorption of sound waves by the air, how far away is the sound level of the scream equal to the threshold of hearing?

•45. A rock concert is given in an auditorium with front row seats 6.0 m from the speakers. The rear row seats are 30.0 m from the speakers. If the sound level is 120 dB in the front row, what is the sound level in the rear row? Assume (unrealistically) that the speakers propagate sound uniformly in all directions.

•46. An experiment called the Acoustic Thermometry of Ocean Climate Project (ATOC) attempted to measure global warming through its effect on long-term changes in ocean water temperatures. The experiment involved measuring the propagation times of sound waves through sea water over very large distances, upward of 2 × 10⁴ km. The speed of sound is a function of temperature, so that the travel times of sound waves to distant receivers vary as ocean temperatures change. The long distances give good average values for ocean temperatures that were monitored over time. Low-frequency sound waves (≈ 57 Hz) were used to minimize interference with marine mammal communication frequencies among other things. The powerful acoustic source was placed deep in the ocean (~ 1 km) to avoid dangerous sound levels for nearby fish and marine mammals and to minimize interference with their communications, but the experiment nonetheless sparked environmental controversy and was abandoned. The source emitted sound waves in all directions. The power of the source was about 1.0 × 10⁵ W. At what distance from the source did the sound level become 100 dB? Assume no absorption of the sound waves by sea water; this assumption is unrealistic.

Sections 12.15 The Acoustic Doppler Effect*
12.16 Shock Waves*

•47. Surface water waves with a wavelength of 3.0 m travel across the surface of Lake Superior with a speed of 7.0 m/s. A ship is moving with a speed of 2.5 m/s. Both speeds are measured with respect to the bottom of the lake. What is the frequency of the waves according to the ship if it is moving: (a) parallel to the velocity of the waves? (b) antiparallel to the velocity of the waves? (c) perpendicular to the velocity of the waves?

•48. A collection of standing passengers is spaced uniformly 1.50 m apart along a moving sidewalk at a busy air terminal. The speed of the sidewalk is 2.0 m/s. An executive is running at speed 5.0 m/s to catch a flight. (a) What is the number of people per second that move past a coffee shop beside the walkway? (b) If the executive on the walkway runs parallel to the velocity of the walkway, how many people per second does she pass?

•49. After a particularly hairy literature exam covering the many works by the famous author Anonymous, you run in a straight line from the examination room at constant speed while screaming. Your scream normally has a frequency of 2.200 × 10³ Hz but your fellow sacrificial lambs, who remain in the room enduring the exasperating examination, somewhat out of tune hear a frequency that differs from this frequency by 20 Hz. (a) What frequency is heard by your classmates? (b) Determine the speed of your departure.

•**50.** A policewoman is chasing a speeder along a straight interstate highway. Both the speeder and trooper are moving at 150.0 km/h. The siren on the police cruiser has a frequency of 2.000×10^3 Hz. What frequency is detected: (a) by the policewoman? (b) by the speeder? (c) by a doomed, dumb possum (*Didelphis marsupialis*) watching the cars approach from a vantage point in the middle of the road?

•**51.** The submarine *U. S. S. Blowfish* lies peacefully at rest just off Tahiti. A fast yellow submarine is moving directly toward the *Blowfish* at a speed of 75.0 km/h using sonar (sound waves in water) of frequency 40.0 kHz to detect the *Blowfish*. The *Blowfish* detects the sonar from the yellow submarine. (a) What frequency does the *Blowfish* detect? Use 1.520×10^3 m/s for the speed of sound in sea water. (b) If the yellow submarine was stationary and the *Blowfish* was moving directly toward it at a speed of 75.0 km/h, what frequency would the *Blowfish* detect?

•**52.** The Doppler effect was first tested experimentally by using moving and stationary musicians. A violinist plays middle C with a frequency of 261.6 Hz. (a) If the violinist is placed on an open railway carriage moving directly away from you (at rest), will the note you hear be sharp or flat? Sharp means a higher frequency; flat means a lower frequency. (b) If you also are playing a violin at the frequency 261.6 Hz, at what speed must the railway carriage travel so you hear a beat frequency (see Section 12.20) of 2.0 Hz from the two violins?

•53. An ambulance siren emits a wail at 1200 Hz. Responding to an accident, the ambulance is rushing to the scene at a speed of 130.0 km/h. For the usual reasons, a lawyer is chasing the ambulance at a speed of 140.0 km/h. (a) What frequency is heard by the paramedic in the ambulance? (b) What frequency does the barrister hear? (c) After the lawyer passes the ambulance, what frequency does she hear?

•**54.** You have a source that emits sound waves with a frequency of 440.0 Hz. (a) At what speed must the source be set in motion either toward or away from you so that you hear a frequency of 439.0 Hz? (b) With the speed calculated in part (a), determine the frequency at which you hear the harmonic of 880.0 Hz.

•**55.** A new jet engine, developed by musically inclined aeronautical engineers, emits sound precisely at a frequency of 440.0 Hz; this thing really hums! A plane equipped with such engines is directly approaching you at 250 km/h on its takeoff roll along a runway, roars past, and recedes into the distance without turning. (a) What is the frequency you hear as the plane approaches? (b) What is the frequency you hear as the plane recedes? (c) Is the frequency *difference* on approach equal to the frequency *difference* as it recedes? Explain your answer.

•**56.** An audio transmitter and receiver are mounted side by side as shown in Figure P.56. The transmitter emits sound of frequency v. A distant flat plate approaches the instrument at speed v_{target} and the receiver detects sound waves of frequency v' reflected from the target. The difference in frequency between the emitted and detected frequencies can be used as a motion detector.

(a) Show that

$$\frac{v'}{v} = \frac{v + v_{target}}{v - v_{target}}$$

(b) In many cases $v_{target} \ll v$. For such cases, show that

$$\frac{v'}{v} \approx 1 + 2\,\frac{v_{target}}{v}$$

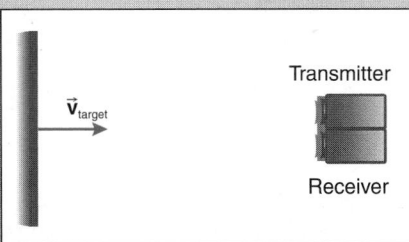

FIGURE P.56

•57. A motion detector (see Problem 56) emits an inaudible sound wave with a frequency of 30.0 kHz. A dog trots toward the stationary detector with a speed of 1.00 m/s. What is the frequency detected by the motion detector?

•**58.** A jet fighter in level flight passes directly overhead at a speed corresponding to Mach number 1.30. The sonic boom is heard by you on the ground 12.0 s later. What is the altitude of the plane? Assume (unrealistically) the speed of sound does not change with altitude.

•**59.** After taking measurements from Figure P.59, which shows the shock wave created by a bullet moving in air, determine the speed and Mach number of the bullet.

FIGURE P.59

•**60.** Some fighter planes can attain speeds of Mach number 2.5 at an altitude of 15.0 km. The speed of sound at that altitude is less than at sea level and is about 2.9×10^2 m/s.* (a) What is the

---

*The speed of sound as a function of altitude depends on the season of the year as well. See Allan D. Pierce, *Acoustics* (McGraw-Hill, New York, 1981), pages 389 and 395 for profiles of the speed of sound as a function altitude and season.

half-angle of the Mach cone formed by the plane? (b) What is the speed of the plane in km/h?

•61. The Concorde supersonic passenger jet attains speeds of about 2150 km/h at a cruising altitude of 19.8 km. The speed of sound at that altitude is about 2.9×10^2 m/s. (a) What is the Mach number of the aircraft at its cruising altitude? (b) What is the half-angle of the Mach cone formed by the plane?

‡62. A source of sound of frequency ν (perhaps your brother singing a note from a favorite aria) is perched on the rim of a merry-go-round of radius R spinning with frequency ν_0. You are located at rest a sufficient distance away that all points on the rim are essentially in the same direction to you; see Figure P.62. Find an expression for the frequency ν' you detect as a function of time if the source is at the position shown in Figure P.62 when $t = 0$ s.

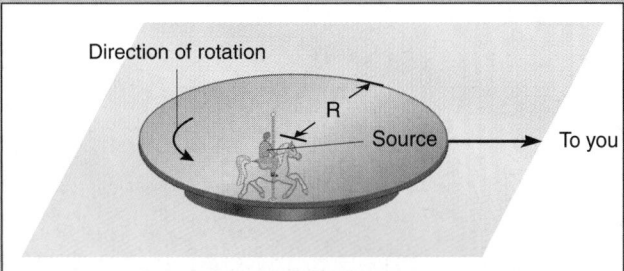

Direction of rotation

R

Source

To you

FIGURE P.62

Section 12.17 Diffraction of Waves

63. The entrance to a marina is approximately 10 m in width, as shown in Figure P.63. What is the order of magnitude of the range of wavelengths of water waves that can pass into the marina with minimal diffraction effects?

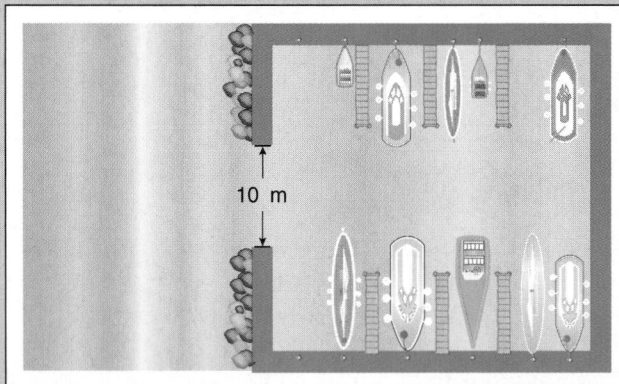

10 m

FIGURE P.63

•64. If the wavelength emitted by a source such as a loudspeaker is much greater than the size of the source (say, at least ten times greater), the waves propagate outward uniformly in all directions and the source is effectively a point source. Conversely, if the wavelength is quite small compared with the size of the source (say, at least ten times smaller), the source is directional and the waves propagate outward in the direction the speaker faces. Two speakers are mounted in a console as shown in Figure P.64. One speaker has a diameter of 25 cm and the other 10 cm. (a) For what range of frequencies is each speaker a point source? (b) For what range of frequencies is each a directed source?

FIGURE P.64

Sections 12.18 The Principle of Superposition
12.19 Standing Waves
12.20 Wave Groups and Beats*

65. A hollow, gas-filled glass cylinder has a piston at each end as shown in Figure P.65. The bottom of the tube has a sprinkling of finely ground cork particles littered along its length. One piston is connected to a oscillator that moves it back and forth at frequency ν, generating sound waves of the same frequency inside the tube. The position of the other piston is adjusted until standing sound waves exist in the tube. The bits of cork particles collect in small piles at the nodes of the standing waves. Let ℓ be the distance between the nodes. Show that the speed of sound v in the gas-filled tube is

$$v = 2\ell\nu$$

This procedure, called *Kundt's method*, was first used by August Adolph Kundt (1839–1894) in 1866 to measure the speed of sound in various gases. Your physics department may have such an apparatus in its attic.

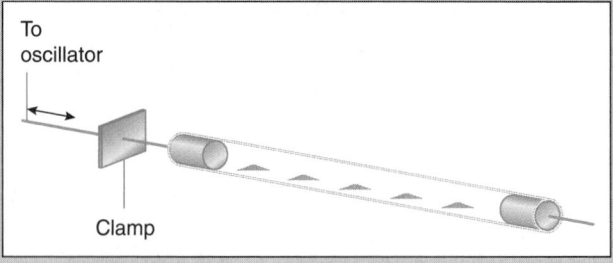

To oscillator

Clamp

FIGURE P.65

66. Two strings, call them A and B, have equal mass. String A has twice the length of string B. (a) If string A has a tension of magnitude twice that of string B, what is the ratio of their fundamental frequencies ν_A/ν_B? (b) What fraction of the tension of string B must string A have to produce a fundamental frequency half that of string B?

67. The base string for the low A (27.5 Hz) on a Model 225 Bösendorfer piano has a total length 2.00 m and a total mass of 0.50 kg; the distance between the ends of the clamped string is 1.90 m. What tension is needed for the properly tuned note?

•**68.** Repeat Problem 13 part (a) but with

$$\Psi(x, t) = Ae^{-[\alpha(x-vt)]^2} - Ae^{-[\alpha(x+vt)]^2}$$

for the wavefunction. Explain the meaning of each term in the wavefunction in terms of your graphs.

•**69.** (a) Describe the physical difference between a wave described by the wavefunction

$$\Psi_1(x, t) = (3.00 \times 10^{-2} \text{ m}) \cos[(9.00 \text{ rad/m})x - (4.00 \text{ rad/s})t]$$

and one whose wavefunction is

$$\Psi_2(x, t) = (3.00 \times 10^{-2} \text{ m}) \cos[(9.00 \text{ rad/m})x + (4.00 \text{ rad/s})t]$$

(b) If these two waves are superimposed, will a standing wave pattern be formed? If so, what is the distance between the nodes of the standing wave? If no standing wave is formed, explain why. (c) If a standing wave is formed, determine the x-coordinates of the first three nodes going from the origin toward increasing values of x.

•**70.** A wave is described by the wavefunction

$$\Psi(x, t) = (2.00 \times 10^{-2} \text{ m}) \cos[(18.0 \text{ rad/m})x - (24.0 \text{ rad/s})t]$$

(a) What wave, when added to the given wave, will form a standing wave pattern? (b) For such a standing wave, what is the distance between the nodes of the wave? (c) For such a standing wave pattern, what is the distance between the antinodes? (d) What is the distance between a node and the nearest antinode?

•**71.** The middle C strings on a piano typically have a length of about 0.60 m. Middle C has a frequency of 261.6 Hz. (a) If the same kind of wire and the same magnitude tension is used to string the C located four octaves higher (the highest note on the piano), what length string is needed? In practice, this string is too short to use for this note. A longer string with less mass per unit length is used in order to accommodate other aspects involved in the construction of a piano. (b) If the same kind of wire and same magnitude of tension as the middle C string are used to produce the lowest note on the common piano (the A with a frequency of about 27.5 Hz), what string length is needed? In practice, this string is much too long to make an affordable (let alone an aesthetic looking) piano even if you had the huge rooms of the Schönbrunn palace in Vienna* to accommodate the instrument.

•**72.** An open organ pipe has a fundamental frequency of 440 Hz. A closed organ pipe has a fundamental frequency that is the second harmonic of the open organ pipe. What are the lengths of the two pipes?

_____
*This palace is where Mozart played for Maria Teresa.

•**73.** An instrument has a string of length ℓ that produces a fundamental frequency ν. A finger shortens the length of the string to produce higher frequencies. If the string is shortened to a length αℓ, where α is a pure number less than or equal to 1, what is the frequency of the fundamental corresponding to the new length?

•**74.** The strobe lights on top of two tall radio towers both have a period of about 1.00 s. When turned on simultaneously, a keen observer notes that during an 800 s interval, the lights gradually fall out of synchronization and then come back into flashing synchronously. What is the interval ΔT between the periods of the two beacons?

•**75.** Two wavefunctions have the following descriptions:

$$\Psi_1(x, t) = (2.00 \times 10^{-3} \text{ Pa}) \cos[(1.00 \text{ rad/m})x - 1.00 \text{ rad/s})t]$$

$$\Psi_2(x, t) = (2.00 \times 10^{-3} \text{ Pa}) \cos[(1.01 \text{ rad/m})x - 1.03 \text{ rad/s})t]$$

(a) Find the speed of each wave. (b) If the waves are superimposed, find the beat frequency and the approximate value of the group speed. (c) Explain what is meant by the phase speed and the group speed.

‡**76.** Two waves with slightly different frequencies combine to form beats. (a) Show that the group speed, given by Equation 12.57, also can be expressed as

$$v_{group} = -\lambda^2 \frac{dv}{d\lambda}$$

(b) If the waves exhibit no dispersion, that is, if all frequencies and wavelengths travel at the same constant speed $v_{phase} = v\lambda$, show that the phase and group speeds are identical. (c) For waves that exhibit no dispersion, make a rough sketch of a graph of the frequency ν versus λ.

Section 12.21 Fourier Analysis and the Uncertainty Principles*

•**77.** The waveform of a periodic traveling wave when t = 0 s has the Fourier series

$$\Psi(x, 0 \text{ s}) = \frac{4.00 \text{ m}}{\pi} \left\{ \frac{\sin[(1.00 \text{ rad/m})x]}{1} + \frac{\sin[(3.00 \text{ rad/m})x]}{3} + \frac{\sin[(5.00 \text{ rad/m})x]}{5} + \cdots + \frac{\sin[((2n+1) \text{ rad/m})x]}{2n+1} + \cdots \right\}$$

where n is a positive integer. (a) Make a graph of the sum of the first six terms of the Fourier series. Choose a domain for x to show several wavelengths of the periodic waveform. (b) As more and more terms are included in such a graph, what periodic waveform does the Fourier series represent? (c) What is the wavelength of the periodic waveform represented by the Fourier series?

•78. A traveling wave produces the following function for the oscillation at $x = 0$ m:

$$\Psi(0 \text{ m}, t) = \frac{6.00 \text{ m}}{\pi} \left\{ \sin\left[(5.00 \text{ rad/s}) t\right] + \frac{\sin\left[(10.00 \text{ rad/s}) t\right]}{2} \right.$$
$$\left. + \cdots + \frac{1}{n} \sin\left[n(5.00 \text{ rad/s}) t\right] + \cdots \right\}$$

where n is a positive integer. (a) Make a graph of the sum of the first ten terms of the series. (b) As more and more terms are included in such a plot, what periodic oscillation does the Fourier series represent? (c) What is the period of the oscillation?

•79. Digital communication systems use short-duration electronic pulses to transmit data. One such system can transmit as many as 10^8 pulses per second. (a) If the pulses do not overlap, what is the maximum duration of a single pulse? (b) According to the uncertainty relations, what range of frequencies is involved in a Fourier representation of such pulses?

INVESTIGATIVE PROJECTS

A. Expanded Horizons

1. Investigate the use of ultrasonic technology for medical applications.

   Peter N. T. Wells, "Medical ultrasonics," *IEEE Spectrum*, 21, #12, pages 44–51 (December 1984).

   Gilbert B. Devy and Peter N. T. Wells, "Ultrasound in medical diagnostics," *Scientific American*, 238, #5, pages 98–112, 172 (May 1978).

   Stephen J. Riederer, "Resource letter MI-1: Medical imaging," *American Journal of Physics*, 60, #8, pages 682–693 (August 1992).

   Russell K. Hobbie, "Resource letter MP-1: Medical physics," *American Journal of Physics*, 53, #9, pages 822–829 (September 1985).

   Hylton B. Meire and Pat Farrant, *Basic Ultrasound* (John Wiley, New York, 1995).

   Steven Webb, *The Physics of Medical Imaging* (Institute of Physics, Philadelphia, 1988).

2. Investigate how bats use sound to detect flying insects.

   Gerhard Neuweiler, "How bats detect flying insects," *Physics Today*, 33, #8, pages 34–40 (August 1980).

   Don E. Wilson and Merlin D. Tuttle, *Bats in Question: The Smithsonian Answerbook* (Aubrey Books, Silver Spring, Maryland, 1997).

3. Noise: it is everywhere, it seems. Investigate the nature of acoustic noise and efforts to combat environmental noise from various sources.

   Thomas D. Rossing, "Resource letter ENC-1: Environmental noise control," *American Journal of Physics*, 46, #5, pages 444–454 (May 1978); this contains many references to these topics.

   John I. Shonle, "Resource letter PE-1: Physics and the environment," *American Journal of Physics*, 42, #4, pages 267–273 (April 1974); this contains references to these topics.

   Sukhbir Mahajan, Michael Shea, D. L. Robinson, and Dave Mathes, "Environmental noise: an undergraduate research project," *American Journal of Physics*, 45, #10, pages 987–990 (October 1977).

   A. E. Lord Jr., "Research projects involving inner city high school students," *American Journal of Physics*, 43, #7, pages 602–605 (July 1975).

   Anthony A. Silvidi and Louis J. Kristopson, "Noise measurements in a university: an open-ended student experiment," *American Journal of Physics*, 41, #7, pages 909–913 (July 1973).

4. Investigate the structure and function of the various parts of the human ear. What are the minimum and maximum motions of the ear membrane at audible frequencies? Are there different forms of ears among animals and fish? Can plants hear or respond to acoustic waves?

   T. S. Littler, *Physics of the Ear* (Pergamon Press, Oxford, 1965).

   Neal P. Rowell, "The response of the ear," *The Physics Teacher*, 18, #7, pages 531–532 (October 1980).

   D. W. Kammer and J. A. Williams, "Some experiments with biological applications for the elementary laboratory," *American Journal of Physics*, 43, #6, pages 544–547 (June 1975).

   C. Daniel Geisler, *From Sound to Synapse: Physiology of the Mammalian Ear* (Oxford University Press, New York, 1998).

5. Investigate an area of the acoustics of music that interests you. There are many aspects to musical acoustics, including the perception of sound, the physics of various musical instruments, and the physical principles underlying the design of concert auditoriums.

   Thomas D. Rossing, "Resource letter MA-2: Musical acoustics," *American Journal of Physics*, 55, #7, pages 589–601 (July 1987); this contains an extensive bibliography to many of these areas.

   John I. Shonle, "Implementing a course on the physics of music," *American Journal of Physics*, 44, #3, pages 240–243 (March 1976).

   Neville H. Fletcher and Thomas D. Rossing, *The Physics of Musical Instruments* (Springer-Verlag, New York, 1998).

   Arthur H. Benade, *Fundamentals of Musical Acoustics* (Dover, Mineola, New York, 1990).

6. Investigate how seismic waves are used to glean information about the structure of the interior of the Earth and earthquakes.

   Paul A. Bender, "Resource letter SEG-1: Solid-earth geophysics," *American Journal of Physics*, 44, #10, pages 903–911 (October 1976); this contains references to these topics.

   Fred Carrington, "The Northridge earthquake—a giant physics laboratory," *The Physics Teacher*, 32, #4, pages 212–215 (April 1994).

   A. J. Berkhout, *Applied Seismic Wave Theory* (Elsevier, New York, 1987).

B. Lab and Field Work

7. Design and perform an experiment to determine the range of frequencies and minimum sound levels to which your ears respond. Is there a difference between your left and right ears? Investigate how the range of frequencies to which the ear responds changes with age from infancy to old age.

   Thomas D. Rossing, *The Science of Sound* (Addison-Wesley, Reading, Massachusetts, 1982), Chapter 6.

8. Using a ripple tank and other necessary equipment, conduct a series of experiments to measure the dispersion of surface water waves, the variation of the speed of the waves with wavelength.

9. Write a proposal for an experiment to determine if audible sound waves exhibit dispersion. Your proposal should include a statement of the objective of the experiment, a detailed discussion of the methods to be used, and the reasons for any equipment needed to perform the experiment. If your proposal is acceptable to your professor and/or laboratory instructor, proceed to perform the experiment. Check to see if such experiments have been made before at the undergraduate laboratory level by checking such journals as the *American Journal*

of Physics, The Physics Teacher, and the "Amateur Scientist" section of *Scientific American*. Do you think it reasonable to assume that ultrasonic sound waves (those with frequencies higher than we can hear) and infrasonic sound waves (those with frequencies lower than what we can hear) also exhibit (or do not exhibit) dispersion?

10. Use a sound level meter and an experienced musician (perhaps yourself) to measure the approximate sound levels for music marked *pp, p, mf, f, ff*, and so on at a standard distance from the instruments. To what extent do your results depend on the instrument?

11. Conduct a series of experiments to determine if the acute angle that the crests of a bow wave make with the line along which a boat travels is a function of the speed of the boat.

12. Contact a piano manufacturer to secure samples of the wire used to string the 88 notes on the standard piano keyboard. Measure the mass per unit length of the various strings. Some notes are strung with multiple strings. Use the frequencies indicated in Figure 12.53 for the various notes to estimate the magnitude of the total tension that the piano frame must withstand due to all the strings.
Anders Askenfelt (editor), *Acoustics of the Piano* (Swedish Academy of Music, Stockholm, 1990).

13. Measure the mass and total length of the E_2 string on a guitar (the lowest-note string). (a) Determine its mass per unit length μ. The frequency of this note is 82.4 Hz. (b) Calculate the tension in the string when it is properly tuned. (c) Calculate the speed of the waves on the string.

14. Carefully measure the positions of the frets along one string of a guitar, banjo, mandolin, or another fretted string instrument. Calculate the various fundamental frequencies that can be produced from the string in terms of the fundamental frequency of the unfretted string. (See Problem 73.)

15. Use a sound level meter to survey the sound levels at various locations on your campus including (a) a class bell while ringing, (b) a car horn, (c) a fume hood in a chemistry lab, (d) cafeteria chatter, (e) a crowded reception hall, (f) a rock concert, and (g) a passing bus, train, or subway. Do any sound levels you survey exceed the threshold for pain?

C. Communicating Physics

16. A group of physicians has commissioned you to write a brief but articulate article for a lay audience (their patients) about the history and physical principles that underlie the common medical stethoscope. Your article will replace the outdated magazines in their waiting room, so you will need to make your writing clear and engaging.
John R. Cameron and James G. Skofronick, *Medical Physics* (Wiley, New York, 1978), pages 263–265.

THE STATIONARY DOPPLER EFFECT

CHAPTER 13

TEMPERATURE, HEAT TRANSFER, AND THE FIRST LAW OF THERMODYNAMICS

Thermodynamics is a funny subject. The first time you go through it, you don't understand it at all. The second time you go through it, you think you understand it, except for one or two points. The third time you go through it, you know you don't understand it, but by that time you are so used to the subject, it doesn't bother you anymore.

Arnold Sommerfeld (1868–1951)*

Among other things, you can now describe the flight of a baseball and understand the mechanics of what happens when it is hit by a bat. In considering the dynamics of a system, the paramount concerns were changes in physical quantities with time. The fundamental principles of mechanics or dynamics are Newton's laws of motion. The CWE theorem, conservation of momentum, and conservation of angular momentum also stem, in one way or another, from Newton's laws of motion.

In our study of dynamics, we ignored or neglected thermal effects. Such an approach is tantamount to assuming that everything in the universe is at the same, constant temperature. But such a universe is a strange one indeed. In it many things we take for granted would not be possible: the stars and the seasons, engines and refrigerators, even life itself could not exist! Thus, to complement our study of mechanics, we now change our perspective and study *thermodynamic*† effects on a system.

In thermodynamics, one of the central parameters is *temperature* rather than time. Effects (such as thermal expansion) and processes (such as heat transfer) that cause or arise from changes or differences in temperature are important aspects of thermodynamics. But what is temperature? Is it the same as heat? In a word: *No!* In this chapter, we will first explore the distinct meanings of these words and then investigate aspects of a system that are affected by them. The fundamental principles of thermodynamics are known as the *laws of thermodynamics*; we encounter two of them in this chapter, reserving others for Chapter 15.

For the study of thermodynamics, we must introduce a new fundamental unit of the SI system, one which is as hallowed as the now familiar kilogram, second, and the meter: the SI unit for quantifying temperature, the *kelvin* (K).‡

13.1 SIMPLE THERMODYNAMIC SYSTEMS

A **thermodynamic system** is defined as a collection of many particles such as atoms and/or molecules.§ A cup of coffee, a can of cola, the gas in a laser discharge tube, and even *you* are examples of thermodynamic systems. In this text, with few exceptions, we consider only **simple thermodynamic systems**.

A simple thermodynamic system is a system that is macroscopic, homogeneous, isotropic, uncharged, chemically inert, and experiences no change in its total mechanical energy. The system is sufficiently large that surface effects can be neglected. No electric or magnetic fields are present, and gravitational fields are irrelevant.

Gravitational effects are irrelevant if the system experiences no variation in the gravitational field over the physical dimensions of the system and experiences no changes in its gravitational po-

tential energy. For simplicity, we omit chemical, electrical, magnetic, and gravitational effects from this introductory treatment of thermodynamics. The thermodynamics of *complex* systems, those that are not simple, is best deferred (gladly, you might feel) to more advanced courses.# While a cup of coffee and a can of cola at rest are *simple* thermodynamic systems, the gas in a laser discharge tube and *you* are not.

QUESTION 1

Explain why *you*, even at rest, are not a simple thermodynamic system. Is a corpse a simple thermodynamic system?

13.2 TEMPERATURE

Some Like It Hot¶

*If you don't like the heat, get out of the kitchen.***

"Whew! It's *hot*!" is a common remark in midsummer; in contrast, a winter day in Maine or Minnesota might elicit the opposite comment: "Brrrr! It's *cold*!" Of course, what you are referring to is the temperature. Just as we all have an intuitive sense of the meaning of time, we all have an intuitive feeling for the meaning of temperature. To determine what is hot and cold, we could use our fingers (carefully!), or we could rely on the numerical readings of an appropriate thermometer, such as the common mercury household or laboratory thermometer. For our purposes early in this chapter, this understanding of the temperature as a quantitative measure of the hotness or coldness of a system will suffice. However, a precise definition of temperature, like a precise definition of time, requires some clear thinking to formulate, which we will do in Section 13.5.

Colloquially the word *heat* is often confused with *temperature*; in physics and the sciences, the two words are *not* synonymous.

The next section investigates the distinction between these terms.

13.3 WORK, HEAT TRANSFER, TEMPERATURE, AND THERMAL EQUILIBRIUM

. . . my bones are burned with heat.

The Book of Job††

When you catch a soda can tossed to you by a buddy, the can loses kinetic energy because of the (negative) work you do on it. When you place the warm can in the refrigerator to cool, we often hear it said that the can "loses heat" as it cools. How can we distinguish between heat and work and between heat and temperature?

---

*(Chapter Opener) A statement (possibly apocryphal) attributed to Arnold Sommerfeld; quoted in Stanley W. Angrist and Loren G. Hepler, *Order and Chaos: Laws of Energy and Entropy* (Basic Books, New York, 1967), page 215.

† The word stems from the Greek θερμός (*thermos*), meaning "hot."

‡ The unit formerly was called the *degree kelvin*, but the word degree and its symbol (°) now are considered superfluous when using the kelvin unit.

§ By many particles, we mean the number of particles in the system is huge.

#We do consider an occasional complex system (in Section 13.14 for example, and in several problems) to illustrate the connection between dynamics and thermodynamics.

¶The title of a 1959 motion picture starring Marilyn Monroe, produced and directed by Billy Wilder.

**This remark is attributed to Harry S. Truman, 33rd president of the United States (the remark evidently was first quoted in *Time* magazine, 28 April 1952 (volume LIX, #17, page 19). Truman attributed the origin of the expression to Major General Harry Vaughan.

††Chapter 30, verse 30.

Work

When we examined the process of forces doing work on a system in Chapter 8, we saw that it was possible to change the kinetic energy of a system thanks to the work done *on it* by mechanical forces. The process of *energy transfer* to a system via work was neatly summarized in the CWE theorem. The work done by the total force is positive if $\Delta KE > 0$ J and negative if $\Delta KE < 0$ J. Work done by forces on a system is one way of transferring energy from external agents *to* a system. We say work is done *on* the system. You can think of work as that (energy) which is transferred from external agents to the system by the total force acting on it over a distance. But when a system is in mechanical equilibrium, we *cannot* say that the system "has" an amount work in the same way that the system *can* be said to have a definite amount of kinetic energy. There is no instantaneous value for the work; there is no mathematical work function associated with a system in mechanical equilibrium by virtue of its position in space or speed. However, there *is* an instantaneous value of the kinetic energy of a system in mechanical equilibrium: the by now well-known kinetic energy functions

$$KE_{trans} \equiv \frac{1}{2} mv^2$$

and

$$KE_{rot} \equiv \frac{1}{2} I\omega^2$$

In the same way, the potential energy of a system is a function of its position at a point in space.

In our study of the CWE theorem in Chapter 8, we noted that the work done on the system by the total force is equal to the change in the kinetic energy of the system. Changes in kinetic energy arise because of the process we call the work of the total force.

Hence work can change the state of mechanical equilibrium of the system. Later in the chapter we shall see that work *also* can change something we call the *internal energy* of system, quite apart from the change in its kinetic energy.

Heat and Heat Transfer

Imagine that you and a friend have different amounts of money in your wallet, say $25 and $15 respectively. To equalize the riches, you generously hand $5 to your friend as a donation, so that you each have the same wealth, $20. Once the money is transferred from you to your friend, the donation no longer exists (except in your mind—and for tax purposes). The donation can be considered to be both the *amount* transferred as well as the act (or *process*) of donating. Now suppose you have two systems with different temperatures in contact with each other.

Heat is defined to be that (energy) which is transferred between the two systems simply because they are at different temperatures.

To emphasize that heat is only involved in a process that occurs between two systems at different temperatures, rather than being a characteristic property of either system, *we will use the term* **heat transfer** *instead of the single word* heat. Heat transfer (like your donation) can be considered to be both the *amount* of energy transferred as well as the *process* itself. Work has this same dichotomous meaning.

The *process* of heat transfer continues until both systems have the same temperature. As with your monetary donation, once the systems have the same temperature, the process of heat transfer no longer exists—it is zero; the energy (the *amount* of heat transfer) manifests itself in other ways, which we will discuss in due course in Section 13.13. Heat transfer is essentially a microscopic atom-to-atom transfer of energy that commonly is not macroscopically apparent to the eye, because no macroscopic motion typically is involved. Thus, when you place a warm soda can in the refrigerator, energy is exchanged between the warm can and the cool air inside the refrigerator since they are initially at different temperatures; the process continues until they have the same temperature. But this energy exchange is not readily visible on a macroscopic basis: the can just sits passively on the shelf.

The energy exchange cools the can and may temporarily warm the inside of the refrigerator. The energy exchanged in this process is what we call a heat transfer. The heat transfer is *from* the warm can *to* the surrounding cool air. The energy exchange continues until a common temperature is reached. Once this common temperature is attained, there is no more energy exchange, no further heat transfer (process) from one system (the can) to the other (the air inside the refrigerator)—in other words, *zero* heat transfer. Nonetheless, both now have the same finite temperature; thus the concept of temperature is quite distinct from heat transfer.

Contrast heat transfer with what happens when a force on the can does work on it and changes its kinetic energy. You can easily see the effects of this energy exchange reflected in the

Using your eyes, you cannot see the heat transfer taking place between a warm soda can and the cool air within a refrigerator.

changing speed of the can. Thus macroscopic *work* is distinctly different from *heat transfer*.

It is not necessary to have direct physical contact between the systems to have heat transfer; for example, there is heat transfer between the Sun and the Earth through the vacuum of space (via a mechanism called *radiation*, as we will see in Section 13.11).

> If it is *possible* for two systems to exchange energy via heat transfer (whether or not they actually do), the systems are said to be in **thermal contact** with each other.

> If two systems at the *same* temperature are placed in thermal contact with each other, there is zero heat transfer between them. Such systems then are said to be in **thermal equilibrium** with each other.

By definition, systems in thermal equilibrium with each other have the *same temperature*. In Chapter 15, we shall see that the common temperature is *never* 0 K, yet the heat transfer rate becomes *zero* when they reach the common temperature. If the combined system is isolated from its surroundings (i.e., if the system is well **insulated**), the common temperature does not change with time. Thermal equilibrium in thermodynamics is analogous to static equilibrium in mechanics.

Heat transfer exists when systems at different temperatures are placed in thermal contact with each other. Heat transfer occurs *from* the hotter system (the one at the higher temperature) *to* the cooler system (the one at the lower temperature). The reason heat transfer occurs from the system at the higher temperature to the system at the lower temperature, and not the other way around, is a subject we will examine in due course (in Chapter 15). For now, take this as an incontrovertible experimental fact; like guano, it just happens that way.

> Just as you *cannot* say that a system in mechanical equilibrium has an amount of work, you *cannot* say a thermodynamic system in thermodynamic equilibrium has an amount of heat.

There is *no* mathematical heat function associated with a system in thermodynamic equilibrium. As stated earlier, heat transfer from one system to another system is nonzero *only* if there is a temperature difference between the systems.

The dimensions of heat transfer are the same as those for work, since both represent energy transferred to (or from) a system. Hence heat transfer is expressed in joules (J) in SI units.

Historically, several other energy units are associated with heat transfer, though rarely with work. We will not use these other units in this text, precisely because they are antiquated and are not SI units. We mention these traditional units, however, because they occur frequently in the literature that you may encounter in your scientific careers.

1. One of these units is the **calorie** (cal), originally defined as the amount of energy needed to be transferred to precisely one gram of pure liquid water to raise its temperature by exactly one kelvin. The precise amount of energy transfer needed to accomplish the one kelvin change in temperature depends on the original temperature of the water. Hence, to

avoid the problem of having a unit tied to a specific substance, the calorie now is defined to be exactly 4.186 J:

$$1 \text{ cal} \equiv 4.186 \text{ J} \quad \text{(an exact conversion)}$$

2. The food **Calorie** (Cal), of concern to sports enthusiasts and anyone on a diet, is the kilocalorie (kcal) or 4186 J:

$$1 \text{ Cal} \equiv 1 \text{ kcal} \equiv 4186 \text{ J} \quad \text{(exact conversion)}$$

Note that the food Calorie (Cal) is spelled and abbreviated with a capital C.

3. Another common energy unit associated with heat transfer is the **British thermal unit** (Btu) in the English system of units. The Btu unit frequently is used in engineering applications, particularly in the United States. The British thermal unit is the energy transfer needed to raise the temperature of one pound* of water one degree fahrenheit (°F). If needed, the appropriate conversion factor is

$$1 \text{ Btu} = 1055 \text{ J}$$

You may have noticed when purchasing an air conditioning unit that its cooling capacity is quantified in Btu.

QUESTION 2

Cookbooks often include baking instructions such as "cook at moderate heat" or "bake at high heat." Is this terminology correct from the standpoint of a physical scientist? Carefully explain your answer.

QUESTION 3

Two systems at different temperatures will eventually assume the same temperature when brought into thermal contact. Specify another physical quantity that becomes the same when two systems are combined or connected together.

13.4 THE ZEROTH LAW OF THERMODYNAMICS

If a can of cola and a bottle of milk each have the same temperature as the inside of your refrigerator, then the milk and cola have the same temperature as well. This common, almost trivial observation is called the **zeroth law of thermodynamics**[†]:

> If systems A and B each are in thermal equilibrium with a third system C, then A and B are in thermal equilibrium with each other.

---

*Actually, the English unit of mass is not the pound but the *slug*; the pound is the English unit for force. What you think of as one pound of sugar has a mass of approximately 1/32 slug, no matter where it may be located. You can see why we try to avoid the English system of units in the sciences!

[†]The importance of this observation for the formal structure of thermodynamics was recognized only long after the first and second laws of thermodynamics were formulated, yet this law must logically precede them. Hence, to avoid a confusing renumbering of the laws, this observation about systems in thermal equilibrium is called the zeroth law of thermodynamics after a suggestion by R. H. Fowler and E. A. Guggenheim [*Statistical Thermodynamics* (Cambridge University Press, Cambridge, England, 1939), page 56].

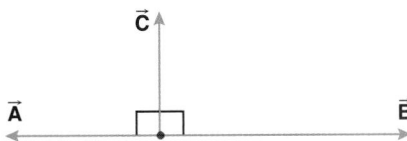

FIGURE 13.1 If vectors $\vec{A}$ and $\vec{B}$ each are perpendicular to $\vec{C}$, this need not mean that $\vec{A}$ and $\vec{B}$ are perpendicular to each other.

The law really stems directly from the definition of thermal equilibrium, but the law is *not* as obvious as it may seem. For example, suppose we return to the now familiar vectors of Chapter 2. If vectors $\vec{A}$ and $\vec{B}$ each are perpendicular to a vector $\vec{C}$, this fact does *not* necessarily mean that $\vec{A}$ is perpendicular to $\vec{B}$: see Figure 13.1.

Likewise, you know sociologically that if both Daniel and Andrew like Anna, and Anna likes them both, this does not necessarily mean that Daniel likes Andrew, particularly if they both are trying to court her! Thus the zeroth law of thermodynamics is not immediately obvious; it is a generalization made from countless experiments. No experimental systems ever have been observed to violate or disagree with the zeroth law of thermodynamics.

13.5 THERMOMETERS AND TEMPERATURE SCALES

Very high and very low temperature extinguishes all human sympathy and relations. It is impossible to feel affection beyond 78° or below 20° of Fahrenheit; human nature is too solid or too liquid beyond these limits.

Sydney Smith (1771–1845)*

The measurement of temperature and the definition of an appropriate temperature scale are not as simple as they might seem initially. When you think of temperature, you think about the reading indicated on a common thermometer. But what exactly *is* a thermometer and what does its reading tell you?

Types of Thermometers

Constructing a reliable and accurate **thermometer** (any device for measuring temperature) relies on using some physical property of matter, known as a **thermometric property**, that changes with temperature. The physical property might be a volume of mercury whose changes with temperature are indicated by changes in the length of a thin, uniform column of mercury connected to the volume, as in common household and laboratory thermometers.

Another useful thermometer of significance, but likely less well known to you, is the **constant volume gas thermometer**, pictured in Figure 13.2. The device keeps a gas at constant volume by means of a column of liquid. The thermometer can measure temperature in the following way. When brought into thermal contact with a system, if the temperature of the gas increases, the pressure increases inside the constant volume. The pressure increase manifests itself as an increase in the height of the liquid column needed to maintain the gas at constant

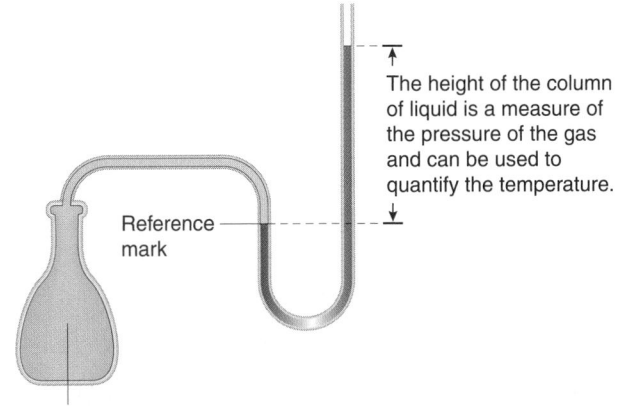

The height of the column of liquid is a measure of the pressure of the gas and can be used to quantify the temperature.

Reference mark

Gas (kept at constant volume)

FIGURE 13.2 A constant volume gas thermometer.

volume. Thus the column height of the liquid provides a way to quantify the temperature.

Another convenient thermometer uses electrical resistance as a thermometric property; *platinum resistance thermometers* and common medical thermometers operate in this fashion. Still other thermometers use an electrical potential difference (known colloquially as a voltage) as the thermometric property; such a device is called a *thermocouple*. An example of a device that uses such electrical thermometers is shown in Figure 13.3.

Calibrating Thermometers

Typically any thermometer, such as the common mercury thermometer, is calibrated by bringing it successively into thermal contact with a large thermodynamic system at two different temperatures, known as **fixed points**. The process proceeds as follows:

1. The thermometer is brought into thermal contact with a massive ice–water slurry at one atmosphere pressure. When the length of the mercury column in the thermometer remains constant, the thermometer is in thermal equilibrium

FIGURE 13.3 Weather balloons use thermometers based on electrical thermometric properties so that data can be easily transmitted to ground-based observers.

with the slurry, and the location of the top of the mercury column can be arbitrarily marked as zero, as in Figure 13.4a.

2. Then the thermometer is brought into thermal contact with a huge mass of boiling water, also at one atmosphere pressure, and the new location of the top of the mercury column is noted and conveniently labeled 100, as in Figure 13.4b.

3. Now divide the change in the length of the uniform mercury column into 100 equally spaced divisions and you have created what we call the celsius scale; see Figure 13.4c. The boiling point of water at one atmosphere pressure is 100 °C and its normal freezing point is 0 °C.* The celsius scale is a *convenient* temperature scale and we shall use it occasionally.

The same procedure can be undertaken to calibrate other thermometers, such as the constant volume gas thermometer discussed earlier. All thermometers calibrated this way will indicate that the freezing point of water is 0 °C and the boiling point of water is 100 °C at a pressure of one atmosphere, because these are the fixed calibration points.

A significant question arises: if a mercury thermometer indicates that the temperature of a system is, say, 37 °C, does a constant volume gas thermometer (or another type of thermometer) *also* indicate 37 °C? Precision experiments indicate that the thermometers *differ slightly* in their temperature readings of a system with a temperature that is not at one of the fixed points. The discrepancy increases for temperatures increasingly removed from the calibration points. This observation raises two new questions: (1) Which thermometer indicates the correct temperature? and (2) How can you be sure that the actual temperature of the system is independent of the particular means or substance used to determine the temperature?

---

*The scale is named after Anders Celsius (1701–1744), a Swedish astronomer who originally proposed having 100 degrees between the boiling and freezing points of water. Curiously, the original proposal of Celsius had the scale inverted from current practice, so that the zero was at the boiling point and 100 was at the freezing point. The celsius scale also was known in the past as the centigrade scale (meaning 100 gradations or divisions between the normal freezing and boiling points of water); the term celsius now is used rather than the word centigrade.

Experiments indicate that the pressure P inside a constant volume gas thermometer, particularly for a dilute gas, is a linear function of the temperature T it indicates (after calibration at the fixed points) as long as the gas is not near its liquefaction temperature. We express this linear relationship between the temperature and pressure in this type of thermometer as

$$T = aP + b \qquad (13.1)$$

The constants a and b are determined by calibrating the thermometer at the two fixed points at a known pressure. Equation 13.1 indicates that a graph of T versus P is linear with a slope equal to a and an intercept b, as shown in Figure 13.5, which presents the results of using three different dilute gases inside the bulb of a constant volume gas thermometer.

Further experiments indicate that the temperature of a system indicated by the constant volume gas thermometer is *independent* of the particular gas used in the calibrated thermometer as long as the gas is dilute and far from its liquefaction temperature. This means that the *intercept b* in Equation 13.1 is the *same* for all gases. The temperature intercept corresponds to $P = 0$ Pa and is found by extrapolating experimental data with a variety of gases to a temperature lower than the liquefaction points of all gases. One finds, on the celsius scale, that the intercept b is equal to −273.15 °C.

Precisely calibrating a thermometer is difficult. The freezing and boiling points of water are experimentally difficult to reproduce because they depend on the pressure on the water surface. A common pressure cooker exploits the fact that the boiling temperature increases with the pressure on the surface of the water, decreasing the cooking time. On the other hand, climbers know that water boils at a lower temperature at high altitude (lower pressure) than at sea level, which increases the cooking time.

To avoid these difficulties, a *single-point* calibration of a thermometer is made. Liquid water, ice, and water vapor all coexist simultaneously in thermal equilibrium at a unique temperature and pressure called the **triple point** of water. The triple point of water has a temperature defined to be 0.01 °C; the pressure of the mixture at the triple point is experimentally found to be 611.73 Pa.

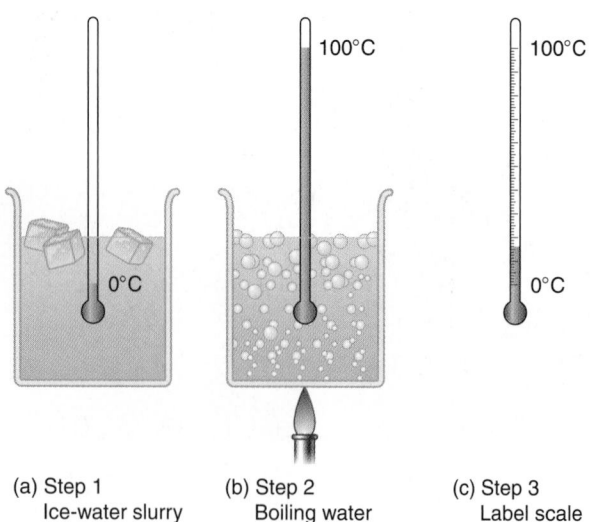

(a) Step 1
Ice-water slurry

(b) Step 2
Boiling water

(c) Step 3
Label scale

FIGURE 13.4 One way to calibrate a thermometer.

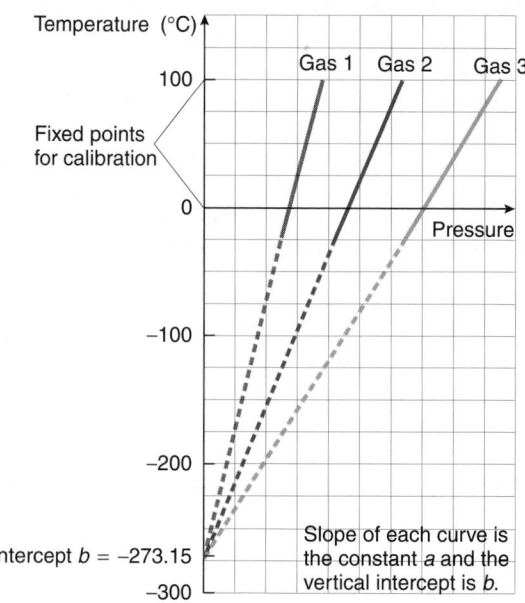

FIGURE 13.5 Graphs of the temperature versus pressure for different dilute gases used in a constant volume gas thermometer.

The boiling temperature of water is lower at a high altitude than at sea level because of the reduced pressure on the water surface; therefore, longer cooking times are necessary.

The **ideal gas temperature scale** is defined by setting the temperature of the triple point of water to be 273.16 K, in which case the intercept in Figure 13.5 is 0 K, called **absolute zero**.

> Thus one **kelvin** is defined to be 1/273.16 of the temperature of the triple point of water.*

> A degree on the celsius scale is the same size as a degree on the kelvin scale, but the zeros of each scale are different.

The zero of the kelvin scale is shifted by 273.15 degrees, and so the normal freezing point of water (at a pressure of exactly one atmosphere) is 273.15 K (= 0 °C) and the triple point is 0.01 °C. That is, a temperature T in kelvin on the ideal gas temperature scale is

$$T = t_{celsius} + 273.15 \text{ K} \qquad (13.2)$$

where $t_{celsius}$ is the temperature on the celsius scale.

> *Temperature changes* on the celsius and kelvin scales are the same because the degrees are the same size, but the two scales *never* indicate precisely the same numerical reading for the temperature of a system because of the different locations of the zero on each scale.

The ideal gas temperature scale means that if the gas in a constant volume gas thermometer has a pressure P when it is in thermal equilibrium with a system at temperature T, the temperature T is found from

$$T = \frac{273.16 \text{ K}}{P_3} P \qquad (13.3)$$

where P_3 is the pressure of the gas in the *same* thermometer when in thermal equilibrium with water at its triple point. The pressure P_3

is *not* the pressure of the triple point of water (611.73 Pa), but the pressure of the gas in the constant volume gas thermometer when it is in thermal equilibrium with a system of water, ice, and water vapor at its triple point. The pressure P_3 in Equation 13.3 depends on the specific amount of gas in the constant volume gas thermometer, but not the specific kind of gas.

The lowest temperatures that can be measured with a constant volume gas thermometer are on the order of a few kelvin, in which case the dilute gas used is helium, since it has the lowest liquefaction temperature of any gas.

When we study the second law of thermodynamics in Chapter 15, we shall see that it is possible to define an **absolute temperature scale** that uses the kelvin and does *not* depend on any thermometric property of *any* substance; the absolute temperature scale also is coincident with the ideal gas temperature scale over the latter's wide range of applicability.

> The absolute temperature scale thus is a truly thermodynamic temperature scale and is used to quantify any temperature, including those close to the absolute zero of temperature (0 K). There is no upper limit on its range.†

Choosing the Right Measuring Device

From a practical standpoint, the particular thermometer chosen to measure the temperature of a system depends on the situation. What would be an appropriate thermometer for one task is completely inappropriate for another. For example, since mercury freezes at −39 °C at 1 atm pressure, a mercury thermometer is useless for measuring temperatures lower than that at the same pressure.

There are other considerations involved in the selection of an appropriate thermometer. For example, the ubiquitous laboratory mercury column thermometer can be used to determine your internal body temperature with no problem. However, it would be ridiculous to use such a large thermometer to measure the internal temperature of, say, a tiny baby mouse. Typically, the mass of the thermometer used to measure the temperature should be *much smaller* than the mass of the thermodynamic system whose temperature is to be measured, because otherwise the thermometer will change the temperature of the system.‡

We want the *rate* of heat transfer to the thermometer to be large, but the actual amount of heat transfer to be small, so that the temperature of the system is essentially unaffected. For instance, suppose you have an isolated thimbleful of hot water and an isolated bucketful of hot water. An ordinary mercury thermometer can measure the temperature of the water in the bucket accurately, as the mass of the thermometer is much less than the mass of the water (the system). But the same thermometer is inappropriate for measuring the temperature of the water in the thimble, since here the mass of the thermometer is comparable to or even greater than the mass of the system. If the amount of heat transfer is large,§ then the thermometer ceases

---

*The kelvin unit is named for Lord Kelvin (William Thomson) (1824–1907), who first suggested (in 1854!) defining a temperature scale with only one fixed point. This procedure was finally sanctioned by the International Committee on Weights and Measures a century later in 1954.

†The temperature of the universe a mere 10^{-43} s after the Big Bang is estimated to have been 10^{32} K. Hot indeed!

‡ The thermometer and system can be of comparable mass if the combined system is in thermal contact with a thermodynamic reservoir (see Section 13.10 for the meaning of the term reservoir).

§ We clarify this relationship further in Section 13.10.

to act as a negligible probe: it significantly affects the thermodynamic system itself.

The question of the effect of a measuring instrument itself on the behavior of the system is one that occurs in many aspects of physics. In classical physics, the effects of a measuring instrument on a system are, in principle at least, able to be calculated, though occasionally with great difficulty. In the quantum mechanics of small systems such as atoms, molecules, and their constituent particles, the act of measuring unavoidably alters the system but, in contrast to classical physics, the alteration in quantum mechanics is in ways that may *not* be able to be calculated, even *in principle*.

QUESTION 4

The temperature inside the core of the Sun is about 1.5×10^7 K. Could you also say, legitimately, that the temperature was 1.5×10^7 °C? Explain your answer.

13.6 TEMPERATURE CONVERSIONS BETWEEN THE FAHRENHEIT AND CELSIUS SCALES*

In the United States another temperature scale is in common use: the **fahrenheit scale**, first introduced by the German scientist Daniel Gabriel Fahrenheit (1686–1736). The fahrenheit scale is defined in terms of the same two fixed points of water used by Celsius: the freezing point of water is 32 °F and its boiling point is 212 °F, both at one atmosphere of pressure. Thus there are $212 - 32 = 180$ fahrenheit degrees between the boiling and freezing points of water.

The basic distinction between the celsius and fahrenheit scales can be thought of in the following way. On a typical thermometer, one can note the column length at which water freezes and where water boils. With the celsius system, the freezing point was labeled 0 °C and boiling point 100 °C, so that the length between the points contained 100 celsius degrees. The divisions on the Celsius thermometer were introduced *after* locating these two fixed points on the thermometer.

On the other hand, Fahrenheit labeled his scale another way. The peculiar numbers for the freezing and boiling points of water came about because Fahrenheit set the zero of his scale at the coldest temperature he could obtain with a freezing mixture of salt and water. Following a suggestion of Newton, he originally took the temperature of the human body to be 12° on his scale, but this resulted in degrees that were too large for practical use. For finer graduations, the 12° interval was divided into eighths and the markings were renumbered, so that the human body temperature became $8 \times 12° = 96°$. Fahrenheit then discovered that pure water froze near the 32° mark and boiled near the 212° mark on his revised scale, with 180° between them (representing the "opposite" properties of water); he then took these as the two fixed points for his scale. With the recalibration of the values associated with the two fixed points, the average human body temperature (no longer a fixed point) became not 96 °F but 98.6 °F.

Since there are 180 fahrenheit degrees and 100 celsius degrees between the boiling and freezing points of water, the celsius degree is larger in size than a fahrenheit degree by the factor 180 °F/100 °C = 9 °F/5 °C. Thus, to convert celsius temperatures to fahrenheit, we do the following: (1) multiply the celsius temperature by the factor 9 °F/5 °C to get a larger number of

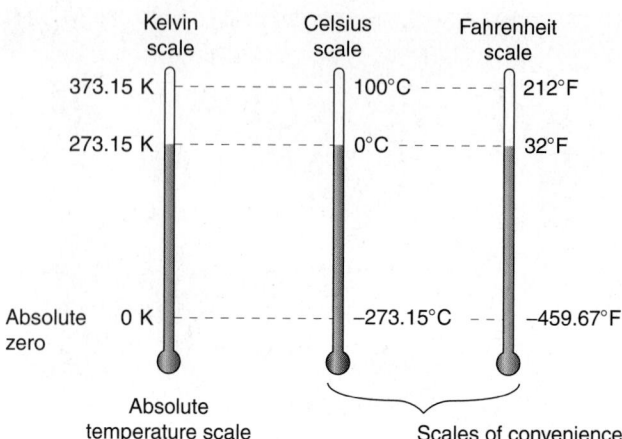

FIGURE 13.6 Schematic comparison of the kelvin (absolute), celsius, and fahrenheit temperature scales.

degrees; (2) then add 32 °F to the result to account for the difference in the location of the zeros of the two scales. That is,

$$t_{\text{farenheit}} = \frac{9 \text{ °F}}{5 \text{ °C}} t_{\text{celsius}} + 32 \text{ °F} \qquad (13.4)$$

To make the conversion from fahrenheit to celsius, we solve Equation 13.4 for t_{celsius}:

$$t_{\text{celsius}} = \frac{5 \text{ °C}}{9 \text{ °F}} \left(t_{\text{farenheit}} - 32 \text{ °F} \right) \qquad (13.5)$$

We think of this latter conversion in the following way: subtract 32 °F from the fahrenheit temperature to readjust the zero location, and multiply the result by 5 °C/9 °F to obtain a smaller number of the larger celsius degrees.

Conversion of degrees celsius to the absolute temperature T in kelvin is accomplished through Equation 13.2:

$$T = t_{\text{celsius}} + 273.15 \text{ K}$$

Corresponding temperatures on the kelvin (absolute), celsius, and fahrenheit scales are shown schematically in Figure 13.6.

PROBLEM-SOLVING TACTIC

13.1 If you recall that the sizes of the degrees on the celsius and fahrenheit scales differ by the factor 9 °F/5 °C or 5 °C/9 °F, depending on which way the comparison is made, you should be able to deduce either Equation 13.4 or 13.5 without memorizing them.

EXAMPLE 13.1

The average normal human body temperature is about 98.6 °F. What is this temperature on the celsius and kelvin scales?

Solution

The human body temperature is 66.6 fahrenheit degrees above the freezing point of water: 98.6 °F − 32.0 °F. Since the celsius degree is a larger size degree, fewer of them quantify this temperature difference. Therefore multiply by the factor 5 °C/9 °F to find the temperature on the celsius scale:

$$\left(\frac{5}{9}\ \text{°C/°F}\right)(66.6\ \text{°F}) = 37.0\ \text{°C}$$

Since the zero of the kelvin scale is 273.15 celsius or kelvin degrees lower than the zero on the celsius scale, the temperature on the kelvin scale is found by adding 273.15 to the celsius temperature. Thus the average normal human body temperature is

$$310.2\ \text{K}$$

13.7 THERMAL EFFECTS IN SOLIDS AND LIQUIDS: SIZE

Most substances expand when their temperature is raised and contract when cooled.* Such thermal effects on the length of materials must be considered in many areas of engineering. Bridge designers must allow for the expansion of steel caused by seasonal changes in temperature; typically, this is done by means of expansion joints, such as the one pictured in Figure 13.7.

The common household thermostat, used to regulate heating and cooling systems, contains a coil of material known as a bimetallic strip; see Figure 13.8. The strip is composed of two different metal strips back-to-back in thermal contact. Different metals expand by different amounts for the same temperature changes, and so the strip will bend or curl as the temperature is raised or lowered. The curling is used to make or

---

*A common exception to the rule is rubber. If a rubber band is chilled while stretched and then placed in a warm environment, you can easily observe that the band contracts as its temperature rises.

FIGURE 13.8 In a thermostat, a coiled bimetallic strip expands and contracts with temperature changes, making or breaking an electrical circuit.

break an electrical contact and thereby act as a temperature-sensitive switch.

Old railroad track rails were relatively short to permit many small expansion joints (see Figure 13.9), giving rise to the familiar clickity-clack of the wheels, romanticized in song and legend. The development of specialty steels with small thermal expansion properties permitted the introduction of continuous weld rails, thus eliminating the necessity for many expansion joints and the rather annoying clickity-clack. All high-speed rail systems use such materials.

Changes in the Length of a Solid

Consider an essentially one-dimensional solid of length ℓ at temperature T, shown in Figure 13.10. If the temperature changes by ΔT, experiments indicate that the change in the length $\Delta \ell$ of the rod is directly proportional to the temperature change as well as to the length ℓ of the rod.

FIGURE 13.9 A thermal expansion joint between two railroad rails.

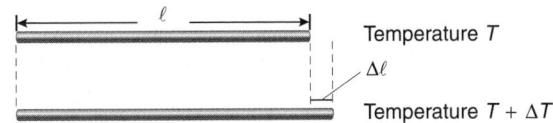

FIGURE 13.10 Thermal expansion of an essentially one-dimensional material.

FIGURE 13.7 A common thermal expansion joint on a New Hampshire bridge (in need of some maintenance).

We express these experimental facts as

$$\Delta \ell = \alpha \ell \, \Delta T \qquad (13.6)$$

The quantity α is known as the **coefficient of linear expansion** of the material.

Those materials that *contract* when their temperature increases have *negative* coefficients of linear expansion (see Table 13.1 for rubber); for these materials if $\Delta T > 0$ K, then $\Delta \ell < 0$ m.

Solving Equation 13.6 for α, we find

$$\alpha = \frac{\Delta \ell / \ell}{\Delta T}$$

We can interpret α as the fractional change in length per kelvin (or equivalently, per celsius degree). The coefficient of linear expansion *at a particular temperature* is found by taking the limit of Equation 13.6 as ΔT approaches zero:

$$\alpha = \lim_{\Delta T \to 0} \frac{1}{\ell} \frac{\Delta \ell}{\Delta T}$$

$$= \frac{1}{\ell} \frac{d\ell}{dT} \qquad (13.7)$$

For most solids and liquids, the value of α is not sensitive to the pressure but may have a temperature dependence. Average values of the coefficient of linear expansion for various materials over a range of temperatures are tabulated in Table 13.1. For all but the most precise work, the average values of α are sufficiently accurate to use when finding the change in length of a substance

TABLE 13.1 Approximate Coefficients of Linear Expansion of Some Elements and Materials at Approximately Room Temperature (293 K)

Element or material	α (K^{-1})
Aluminum	24×10^{-6}
Brass (65% Cu–35% Zn)	21×10^{-6}
Cast iron (ASTM A-47)	12×10^{-6}
Concrete	9.9×10^{-6}
Copper	17×10^{-6}
Ethyl alcohol	3.7×10^{-4}
Glass (ordinary)	9×10^{-6}
Glass (pyrex)	3×10^{-6}
Gold	14×10^{-6}
Invar (Ni–Fe alloy*)	1.3×10^{-6}
Lead	29×10^{-6}
Mercury	61×10^{-6}
Quartz (fused)	0.5×10^{-6}
Rubber (natural)	-620×10^{-6}
Silver	19×10^{-6}
Stainless steel (AISI 302)	17×10^{-6}
Steel (structural ASTM A36)	12×10^{-6}
Titanium	8.6×10^{-6}
Water	
Liquid	69×10^{-6}
Ice	51×10^{-6}

*Invar was designed for a very low thermal expansion coefficient. The name is contracted from the word *invariant*.

using Equation 13.6. Notice that the units attached to the coefficient of linear expansion are K^{-1} or, equivalently, °C^{-1} since the kelvin and celsius degrees are of equal size.

EXAMPLE 13.2

The Golden Gate Bridge, spanning the entrance to San Francisco Bay, has a span of 1.28×10^3 m between its two massive support towers. Calculate the resulting change in the length of the steel span for a seasonal temperature change of perhaps 30 K. Assume the value of α for structural steel in Table 13.1.

Solution

The coefficient of linear expansion for structural steel is, from Table 13.1,

$$\alpha_{\text{steel}} = 12 \times 10^{-6} \text{ K}^{-1}$$

Substitute into Equation 13.6 to find the change in length:

$$\Delta \ell = \alpha \ell \, \Delta T$$
$$= (12 \times 10^{-6} \text{ K}^{-1})(1.28 \times 10^3 \text{ m})(30 \text{ K})$$
$$= 0.46 \text{ m}$$

The expansion joints on the bridge deck are quite large!

Changes in Area of Solids

When pouring concrete for the vast expanses of tarmac and runways at airports, provision must be made for the thermal expansion of the concrete from winter to summer. To accomplish this, the concrete is poured in square or rectangular slabs, with a pliable expansion board around the perimeter (see Figure 13.11).

Temperature-induced changes in area are found by applying linear expansion to each of the two dimensions of the area. For example, consider a rectangular area of dimensions ℓ_1 and ℓ_2, forming an area A, as shown in Figure 13.12. The area is

$$A = \ell_1 \ell_2 \qquad (13.8)$$

When the temperature changes, the change in the area can be found by differentiating Equation 13.8 with respect to the temperature T:

$$\frac{dA}{dT} = \ell_1 \frac{d\ell_2}{dT} + \frac{d\ell_1}{dT} \ell_2$$

FIGURE 13.11 To allow for the thermal expansion in area, large expanses of concrete are poured in smaller slabs surrounded by pliable expansion boards.

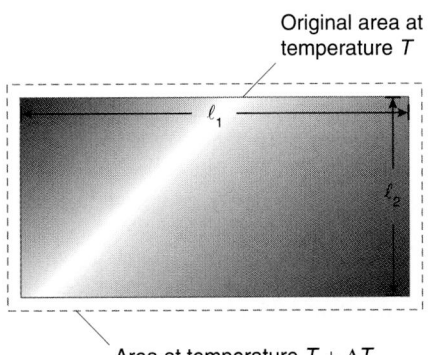

FIGURE 13.12 Thermal expansion of a rectangular area.

Using Equation 13.7 for each of the derivatives, we find

$$\frac{dA}{dT} = \ell_1 \alpha \ell_2 + \alpha \ell_1 \ell_2$$
$$= 2\alpha \ell_1 \ell_2$$

If α is constant over the temperature range of interest, this equation can be integrated:

$$\int_{A_i}^{A_f} dA = 2\alpha \ell_1 \ell_2 \int_{T_i}^{T_f} dT$$
$$\Delta A = (2\alpha)A\,\Delta T \qquad (13.9)$$

The change in the area is proportional to the area and to the change in the temperature.

> The **coefficient of area expansion** is 2α—that is, twice the coefficient of linear expansion.

Equation 13.9 can be used for areas of arbitrary shape and is not restricted to rectangular areas.

One of the curious aspects of area expansion is that for an area with a hole in it, such as a metal washer, the area of the *hole* expands with temperature as if the hole were completely filled with the material. This rather perplexing result can be demonstrated by considering a flat plate of the material with no hole in it, as in Figure 13.13. A circle is drawn on the plate; it has an area A at temperature T.

If the temperature of the system now is increased to $T + \Delta T$, so that the plate expands, the circular area A increases according to Equation 13.9. Thus the *periphery* of the circular area moves further from the center of the circle, as shown in Figure 13.14. Hence, whether the circle is filled with material or not (i.e., is a hole in the larger plate), its periphery expands in the same way.

EXAMPLE 13.3

A structural steel plate of dimensions 2.00 m by 3.00 m is initially at temperature 20 °C. The temperature of the plate is raised to 50 °C. What is the increase in the area of the plate?

Solution
The area of the plate at 20 °C is

$$A = (2.00 \text{ m})(3.00 \text{ m})$$
$$= 6.00 \text{ m}^2$$

The change in the area of the plate is found via Equation 13.9:

$$\Delta A = (2\alpha_{steel})A\,\Delta T$$

The temperature change is $\Delta T = 30$ K, and from Table 13.1,

$$\alpha_{steel} = 12 \times 10^{-6} \text{ K}^{-1}$$

Making these substitutions into Equation 13.9, you find
$$\Delta A = 2(12 \times 10^{-6} \text{ K}^{-1})(6.00 \text{ m}^2)(30 \text{ K})$$
$$= 4.3 \times 10^{-3} \text{ m}^2$$
$$= 43 \text{ cm}^2$$

Changes in Volume of Solids and Liquids

Gasoline is sold by volume. Because of thermal expansion, a given mass of gasoline has a larger volume when warm than when cold. Hence, when possible, a smart consumer will purchase gas at the coolest time of the day.

FIGURE 13.13 A flat area with no hole, but with a circular area marked on it.

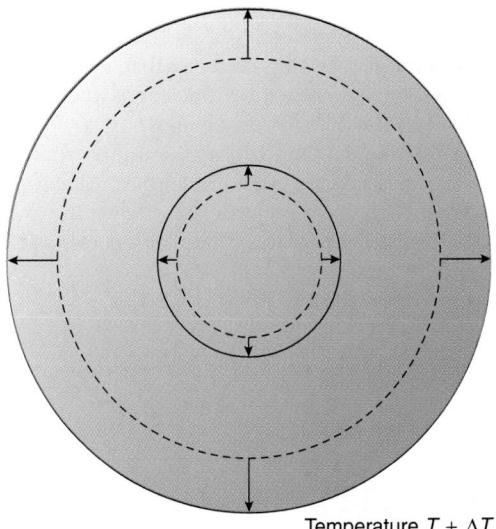

FIGURE 13.14 The periphery of the area expands if the temperature increases.

Think of the volume of a rectangular parallelopiped of sides ℓ_1, ℓ_2, and ℓ_3:

$$V = \ell_1 \ell_2 \ell_3$$

If a temperature change occurs, the change in volume can be found by differentiating this equation with respect to T:

$$\frac{dV}{dT} = \ell_1 \ell_2 \frac{d\ell_3}{dT} + \ell_1 \frac{d\ell_2}{dT} \ell_3 + \frac{d\ell_1}{dT} \ell_2 \ell_3$$

Using Equation 13.7 for the various derivatives, we find

$$\frac{dV}{dT} = \ell_1 \ell_2 (\alpha \ell_3) + \ell_1 (\alpha \ell_2) \ell_3 + (\alpha \ell_1) \ell_2 \ell_3$$

$$= (3\alpha) \ell_1 \ell_2 \ell_3$$

If α is constant over the temperature range of interest, this equation can be integrated to obtain

$$\int_{V_i}^{V_f} dV = 3\alpha \ell_1 \ell_2 \ell_3 \int_{T_i}^{T_f} dT$$

$$\Delta V = (3\alpha) V \, \Delta T \qquad (13.10)$$

Thus the change in volume is proportional to the volume and the change in temperature. Although we derived Equation 13.10 for the specific shape of a rectangular parallelopiped, the result is quite a general one and can be used for volumes of any shape.

The **coefficient of volume expansion** is three times the coefficient of linear expansion.

As for areas, voids or holes in a volume of solid material expand as if the void or hole were filled with the material.

Liquids also change their volume with temperature changes. Indeed, their coefficients of volume expansion (see Table 13.2) are much greater than for solids. Thus the effects of changes in temperature on the volume of liquids are quite substantial. This can be easily observed by filling a soda bottle to the brim with hot water; when the system cools to room temperature, the level of liquid in the bottle is noticeably below the rim. The decrease in the volume of the glass bottle is much less than the decrease in the volume of the liquid, because the coefficient of volume expansion for glass is much less than for most liquids.

Water has unusual thermal expansion properties that may explain why life likely developed first in an aqueous environment. You probably have noticed when swimming and diving that warmer water lies near the surface with cooler water at the

bottom. The temperature of the water decreases with depth. Why, then, does ice form at the top of ponds and not at the bottom?

Most materials contract when cooled, thus decreasing their volume. Since the mass remains constant, a decrease in volume increases the density of these materials as they are cooled. Liquid water does this too. A graph of the *volume* of 1000 kg of water as function of temperature is shown in Figure 13.15.

As the temperature decreases, the volume decreases, causing the density of the water to increase. This is why cooler, more dense water sinks, and warmer, less dense water rises to the top of a pond or lake. The volume is a minimum at about 4 °C.

Below 4 °C, however, the volume *increases* as the freezing point (0 °C) is approached, causing the very cold water to migrate to the surface of the liquid. Thus, when water freezes at 0 °C, ice forms at the top rather than the bottom of ponds and lakes. Once the ice forms, the ice acts to insulate the water below from temperature changes above its surface. In this way, water remains in the liquid state below the ice layer on top. This is fortunate for aquatic life. Consider the alternative scenario: if water continued to contract until it froze, then ice would form at the *bottom* of ponds and lakes and the entire mass of water might freeze during the winter, along with the aquatic life in it. Once frozen solid, it is difficult to melt the entire mass again with solar energy incident only on top. When spring and summer come, the ice at the top might melt, but the ice further down could remain frozen all year. Fortunately this scenario does not happen, because the volume of water actually increases when cooled below 4 °C.

QUESTION 5

A bimetallic strip is made of two adjacent materials with different coefficients of linear expansion. The strip is straight at a temperature T. The temperature now is raised by an amount $\Delta T > 0$ K and the strip curves. Is the material with the greater coefficient of linear expansion on the inside or outside of the bend?

QUESTION 6

To get a greater mass of petrol, should you fill the tank of your car when the temperature is cold or hot?

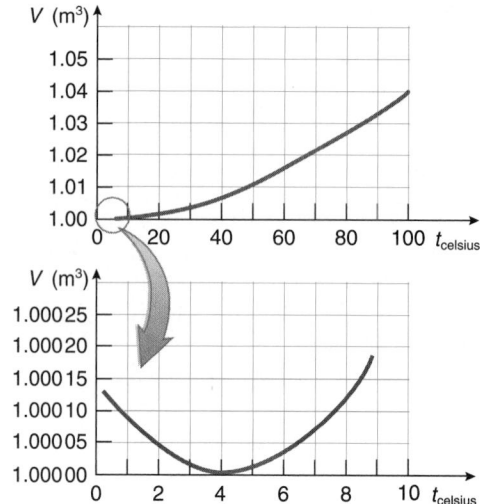

FIGURE 13.15 A graph of the volume of 1000 kg of water versus temperature.

TABLE 13.2 Approximate Coefficients of Volume Expansion for Various Liquids at Approximately Room Temperature

Liquid	3α (K^{-1})
Acetone	14.9×10^{-4}
Benzene	12.0×10^{-4}
Ethyl alcohol	11.2×10^{-4}
Mercury	1.83×10^{-4}
Petroleum (density 846 kg/m^3)	9.6×10^{-4}
Water	2.1×10^{-4}

Liquid water may exist below the ice-covered surface of Europa, a moon of Jupiter.

EXAMPLE 13.4

A petrol station takes delivery of 10 m³ of gasoline (1.0×10^4 liters) at a temperature of 0 °C. If the delivery of the same mass of petroleum were made when its temperature was 30 °C, what would be the increase in the volume of liquid delivered?

Solution

Using Table 13.2, petroleum has a coefficient of volume expansion of about 9.6×10^{-4} K⁻¹. The change in the volume of the petrol is found via Equation 13.10:

$$\Delta V = (3\alpha)V \, \Delta T$$

Making the appropriate substitutions,

$$\Delta V = (9.6 \times 10^{-4} \text{ K}^{-1})(10 \text{ m}^3)(30 \text{ K})$$
$$= 0.29 \text{ m}^3$$
$$= 2.9 \times 10^2 \text{ liters}$$

Since petroleum is sold by volume, not mass, such dramatic temperature changes in the volume have instigated vituperative arguments between petroleum wholesalers and retailers about the timing of deliveries.

EXAMPLE 13.5

The (structural) steel petrol tank of your car has a capacity of 80 liters of petrol when the tank and petrol are at 0 °C.

a. What is the change in the volume of the tank when its temperature is increased to 30 °C?
b. What is the change in the volume of the petrol when its temperature is increased to 30 °C?
c. At which temperature will you get the greater mass of petrol? How much greater?

Solution

a. Note that the petrol tank is a three-dimensional solid with a large hole removed. The change in the volume of the steel tank is

$$\Delta V_{\text{tank}} = (3\alpha_{\text{steel}})V \, \Delta T$$
$$= 3(12 \times 10^{-6} \text{ K}^{-1})(80 \text{ liters})(30 \text{ K})$$
$$= 0.086 \text{ liters}$$

b. When petrol is warmed from 0 °C to 30 °C, the change in the volume of petroleum is, using the data in Table 13.2,

$$\Delta V_{\text{petrol}} = (9.6 \times 10^{-4} \text{ K}^{-1})(80 \text{ liters})(30 \text{ K})$$
$$= 2.3 \text{ liters}$$

c. The mass of petroleum you get when the tank is filled at 30 °C is less than the mass you get at 0 °C. The difference in mass is equal to the mass of about

$$2.3 \text{ liters} - 0.086 \text{ liters} = 2.2 \text{ liters}$$

of 30 °C petrol.

13.8 THERMAL EFFECTS IN IDEAL GASES

In 1660 Robert Boyle (1627–1691) empirically discovered a remarkable property of dilute gases. If the temperature of a fixed amount of a gas is kept constant, its pressure is inversely proportional to the volume: the smaller the volume, the greater the pressure. We usually express this by saying that the product of the pressure and volume is constant:

$$PV = \text{constant} \qquad \text{(at constant temperature)} \qquad (13.11)$$

This approximate empirical relationship is characteristic of all gases at low densities and is now known as **Boyle's law**. If the experiments are performed at another temperature, the numerical value of the constant changes in a way that is directly proportional to the absolute temperature T. In other words,

$$PV = DT \qquad (13.12)$$

where the constant D depends on the amount of gas used in the experiments. Equation 13.12 implies that at constant volume, the pressure of the gas is proportional to the absolute temperature, a result we noted before in formulating Equation 13.3 for the constant volume gas thermometer.

Let's see if we can determine how the constant D in Equation 13.12 depends on the amount of gas present. Take two identical samples of the same gas that have the same volume, temperature, and pressure. Equation 13.12 implies that the constant D must be the same for both individual samples. If the two samples now are brought together, experimentally we find that we have a combined sample with *twice* the volume at the *same* temperature and pressure. Hence, if we apply Equation 13.12 to the combined sample, the constant D must double; this implies that the constant D is directly proportional to the *amount* of gas present. If N is the number of molecules in the sample of gas, we write $D = kN$, and so Equation 13.12 becomes

$$PV = NkT \qquad (13.13)$$

Experiments indicate that k is independent of the particular gas used as well as the amount of the gas. The constant k is known as **Boltzmann's constant** and has the value

$$k = 1.380\,66 \times 10^{-23} \text{ J/K} \qquad (13.14)$$

in SI units. It is named after Ludwig Boltzmann (1844–1906), a pioneer in the theoretical development of thermodynamics during the latter part of the 19th century.

The amount of a gas typically is expressed in moles (mol). The mass of one mole of a substance is called its molar mass M.* The molar mass in *grams per mole* is found by summing the molar masses of the atoms that make up the molecule. Oxygen gas is composed of the oxygen molecule, O_2, with two oxygen atoms. The molar mass of the oxygen atom (O) is 16 g, and so the molar mass of the oxygen molecule is 16 g + 16 g = 32 g. For carbon dioxide (CO_2), we add the molar mass of carbon (12 g) to that of O_2 (32 g), getting 44 g.

For n moles, the mass m of the sample is

$$m = nM \tag{13.15}$$

The molar mass of several gases is given in Table 13.3.

In Chapter 1 we learned that one mole is Avogadro's number N_A of particles (be they atoms or molecules), where

$$N_A = 6.022\ 137 \times 10^{23} \text{ particles/mol} \tag{13.16}$$

Let n be the number of moles of a substance; hence, the number of particles N is equal to nN_A:

$$N = nN_A$$

In this way, we can rewrite Equation 13.13 as

$$PV = nN_A kT \tag{13.17}$$

The product of Avogadro's number and Boltzmann's constant is known as the **universal gas constant** R:

$$R \equiv N_A k \tag{13.18}$$
$$= (6.022\ 137 \times 10^{23} \text{ particles/mol})(1.380\ 66 \times 10^{-23} \text{ J/K})$$
$$= 8.314\ 52 \text{ J/(mol·K)} \tag{13.19}$$

Equation 13.17 commonly is written as

$$PV = nRT \tag{13.20}$$

Equation 13.20 (or its equivalent forms, Equation 13.13 or 13.17) is known as the **ideal gas law**.

We shall have more to say about the characteristic features of an ideal gas in Chapter 14.

*The molar mass also is known as the *molecular weight* (particularly in chemistry); in this context the word weight is *not* the force we call weight in physics. We will not use the term molecular weight.

TABLE 13.3 Molar Masses of Some Common Gases[†]

Gas	Molar mass	
	grams/mole	kilograms/mole
Oxygen (O_2)	32	0.032
Nitrogen (N_2)	28	0.028
Carbon dioxide (CO_2)	44	0.044
Carbon monoxide (CO)	28	0.028
Hydrogen (H_2)	2	0.002
Helium (He)	4	0.004

[†]Molar masses given on the periodic table of the elements reflect the relative natural abundances of the isotopes of each element and so may be given with a precision of fractions of a gram per mole.

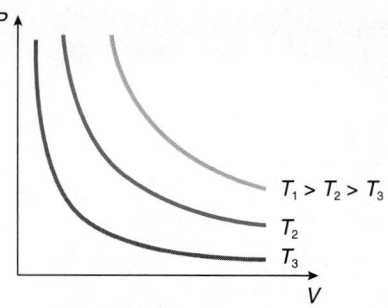

FIGURE 13.16 Isotherms of an ideal gas are hyperbolic in shape.

Notice that at constant temperature, a graph of the pressure versus the volume (called a **P–V diagram**) of an ideal gas results in a hyperbolic curve:[‡] Such curves at constant temperature are called **isotherms**; see Figure 13.16.

According to the ideal gas law, Equation 13.20, the ratio

$$\frac{PV}{nT}$$

should be constant, independent of pressure and equal to the gas constant R. The behavior of *real gases* at pressures up to several atmospheres is closely approximated by the ideal gas law, as shown in Figure 13.17.

Any gas at a temperature of 0 °C (273.15 K) and 1 atm pressure (1.013×10^5 Pa) is said to be at **standard conditions**.

PROBLEM-SOLVING TACTIC

13.2 Be careful with the units when using the ideal gas law. To use the value of R = 8.314 52 J/(mol·K), the pressure P must be in pascals, V in cubic meters, and n in moles.

Different units for P and V (such as expressing P in atmospheres and V in liters, typically done for convenience) mean the gas constant has other numerical values than the one we have quoted; this is frequently the case in chemistry, where the gas

[‡]Recall from analytic geometry that a mathematical expression of the form

$$xy = \text{constant}$$

is a hyperbola.

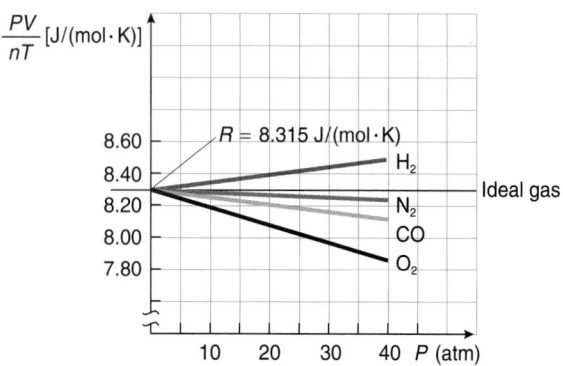

FIGURE 13.17 Real gases approach the characteristics of an ideal gas for pressure up to several atmospheres.

constant has values that change like a chameleon as various units are used. In physics, we typically stick to SI units, in which case the units for pressure and volume should be converted to standard SI units (pascals and cubic meters) when doing most problems.

PROBLEM-SOLVING TACTIC

13.3 When using the ideal gas law, be sure to express the temperature in kelvin (not degrees celsius). If you (incorrectly!) use negative or zero celsius degrees in the ideal gas law, you discover that the product of the pressure and volume (both intrinsically positive quantities) yields a negative or zero result. Such a result is physically absurd and should indicate that something you did is amiss.

EXAMPLE 13.6

Determine the volume of 1.00 mol of an ideal gas at a pressure of 1.000 atm and a temperature of 0 °C.

Solution
To use the ideal gas law, Equation 13.20, the temperature *must* be expressed in kelvin, following Problem-Solving Tactic 13.3, and so $T = 273.15$ K. Problem-Solving Tactic 13.2 indicates you need to convert 1.000 atm pressure to SI units: 1.013×10^5 Pa. Solve the ideal gas law for V:

$$V = \frac{nRT}{P}$$

Substituting the proper SI values, you get

$$V = \frac{(1.00 \text{ mol}) [8.315 \text{ J}/(\text{mol} \cdot \text{K})] (273.15 \text{ K})}{1.013 \times 10^5 \text{ Pa}}$$

$$= 2.24 \times 10^{-2} \text{ m}^3$$

$$= 22.4 \text{ liters}$$

EXAMPLE 13.7

The temperature of a 4.00 liter sample of an ideal gas initially at standard conditions is increased 50.0 °C and its volume reduced to 3.00 liters. Determine the new pressure of the gas.

Solution
The ideal gas law, Equation 13.20, states that

$$PV = nRT$$

Rearrange this into the form

$$\frac{PV}{T} = nR$$

Since the right-hand side is constant for a given sample of gas, the ratio

$$\frac{PV}{T}$$

must be same for the initial and final states of the gas. That is,

$$\frac{P_i V_i}{T_i} = \frac{P_f V_f}{T_f}$$

Solve for P_f:

$$P_f = \frac{V_i}{V_f} \frac{T_f}{T_i} P_i$$

Notice that the answer depends on the volume ratio. Here you do not need to convert to cubic meters to determine this ratio. However, the ratio of the temperatures *must* be calculated using *absolute temperatures* in kelvin. The final pressure emerges in the same units you use for the initial pressure. Since the initial pressure is in atmospheres, the final pressure will be in atmospheres and you do not need to convert to pascals. (You might ask yourself why the temperature ratio must be converted to SI units while the volume and pressure ratios do not).

The ratio of the initial to final volumes is (4.00 liters)/ (3.00 liters). The initial temperature is 273.15 K and the final temperature is 323.2 K. The initial pressure is 1.00 atm. Hence

$$P_f = \frac{4.00 \text{ liters}}{3.00 \text{ liters}} \frac{323.2 \text{ K}}{273.15 \text{ K}} (1.00 \text{ atm})$$

$$= 1.58 \text{ atm}$$

EXAMPLE 13.8

a. How many moles are in a 0.100 kg sample of carbon dioxide?
b. How many molecules are in the sample?

Solution
a. From Table 13.3, the molar mass of carbon dioxide, CO_2, is 44 g/mol or 0.044 kg/mol. Hence the number of moles in the sample of 0.100 kg is

$$n = \frac{0.100 \text{ kg}}{0.044 \text{ kg/mol}}$$

$$= 2.3 \text{ mol}$$

b. Each mole has Avogadro's number of molecules, and so the number of molecules N is

$$N = nN_A$$

$$= (2.3 \text{ mol})(6.02 \times 10^{23} \text{ molecules/mol})$$

$$= 1.4 \times 10^{24} \text{ molecules}$$

13.9 CALORIMETRY

Place a few ice cubes in a warm soft drink and in a few minutes you have a nice cool treat while you study. The soda is cooled and some or all of the ice melts. This observation is so common that we take it for granted, but the observation illustrates an important aspect about heat transfer between two systems initially at different temperatures.

When heat transfer occurs to or from a system, several things typically happen:

1. The temperature of the system may change (the soda cools off).
2. A *change of state* may occur such as from a solid to a liquid (the ice melts) or from a liquid to a gas (such as boiling water into

water vapor). Such changes of state are examples of **phase transitions*** and take place at a constant temperature, characteristic of the material and its environment (e.g., pressure).

Calorimetry experiments probe both of these aspects. In some such experiments (such as when ice cubes are added to soda at room temperature), one thermodynamic system at a cool temperature is placed in thermal contact with another thermodynamic system at a higher temperature; heat transfer then occurs between the two systems until the combined system reaches a common temperature and is in thermodynamic equilibrium.[†]

> The heat transfer *from* the warmer system is *to* the cooler system. By convention, heat transfer *from* a system is considered *negative*; heat transfer *to* a system is considered *positive*.

We typically ensure, by means of insulation, that the heat transfer of the combined system to the environment is negligible (for instance, by using an insulated cup). For such *isolated systems*, the total heat transfer to the combined system is zero.

Given the initial temperatures and masses of, say, the ice and the soda pop, we want to be able to predict the final temperature of the mixture when they are combined (or placed in thermal contact with each other). To do this, we have to see how heat transfer changes the temperature of a system or its phase.

Temperature Change

Put a thimbleful of water on a hot stove burner for 10 s and the water increases in temperature dramatically. Place a huge pot of water on the same hot stove burner for 10 s and the temperature increase of the water is quite small. From such observations, we see that the temperature change of a system in response to heat transfer is inversely proportional to the mass of the system. That is, for a differential heat transfer dQ to a system, the differential temperature change dT that the system experiences is directly proportional to the heat transfer and inversely proportional to the mass of the system. In other words,

$$dT \propto \frac{dQ}{m}$$

This relationship traditionally is written as

$$dQ \propto m \, dT$$

Different materials, all with the same mass, have different dT for a given dQ. Thus the proportionality constant needed to make the relationship above an equality depends on the specific material used in the experiment. We write

$$dQ = mc \, dT$$

where c is the **specific heat** of the substance. The specific heat *at* a specified temperature is defined as

$$c = \frac{1}{m} \frac{dQ}{dT} \qquad (13.21)$$

The dimensions for c must therefore be those of energy divided by the product of a mass and temperature. In SI units, the specific heat is expressed in joules per kilogram-kelvin [J/(kg·K)].

If the specific heat is independent of temperature, then Equation 13.21 can be integrated to obtain

$$Q = mc \, \Delta T \qquad (13.22)$$

If the amount of material is given in moles, rather than in kilograms, then the molar specific heat must be used. Equation 13.22 then becomes

$$Q = nc_{\text{molar}} \, \Delta T \qquad (13.23)$$

where n is the number of moles of material in the system. Analogous to Equation 13.21, the molar specific heat is defined as

$$c_{\text{molar}} \equiv \frac{1}{n} \frac{dQ}{dT} \qquad (13.24)$$

The molar specific heat is given in SI units of joules per mole-kelvin [J/(mol·K)].

> You can think of the specific heat (or molar specific heat) as the heat transfer to one kilogram (or one mole) of the material needed to raise its temperature by one kelvin.

 PROBLEM-SOLVING TACTIC

13.4 Be careful in looking up the specific heat of a substance in tables in technical literature; other common energy and mass units also are used. Check the units carefully.

Table 13.4 lists the specific heat of several common substances and materials.

TABLE 13.4 Approximate Specific Heat of Various Substances at Approximately Room Temperature (20 °C ≈ 293 K)

Substance	J/(kg·K)	J/(mol·K)
Aluminum	896	24.2
Brass	385	
Copper	386	24.5
Gold	132	26.8
Ice (−20 °C)	2050	36.9
Iron	481	24.6
Lead	128	26.4
Mercury	138	27.9
Silver	233	24.9
Water (15 °C)	4186	75.3

The molar specific heat of many metallic elemental solids is approximately $3R = 24.9$ J/(mol·K). The reason for this is explained in Chapter 14.

---

*Phase transitions also are possible without a change of state. For example, when a solid crystalline structure (such as quartz) changes the form of its crystalline structure, the material undergoes a phase transition with no change of state, since both the initial and final structures are solids.

[†] Other types of calorimetric experiments transfer heat at a known rate and measure the corresponding changes in temperature as a function of time. Such devices are called scanning calorimeters.

A change in temperature of a system is always taken as

$$\Delta T = T_f - T_i$$

When $T_f > T_i$, then $\Delta T > 0$ K and, according to Equations 13.22 or 13.23, $Q > 0$ J. If the final temperature of the system is less than the initial temperature, then $Q < 0$ J.

Especially for gases, the value of the specific heat depends on *how* the heat transfer occurs: experimentally, it could be accomplished at constant pressure or at constant volume.* That is, we need to distinguish between *two* distinct specific heats. When the heat transfer is at constant volume, we designate the specific heat as

$$c_V = \frac{1}{m}\left(\frac{dQ}{dT}\right)_V \qquad (13.25)$$

where the subscripts indicate that the process takes place at *constant volume*. When the heat transfer is at constant pressure, we write

$$c_P = \frac{1}{m}\left(\frac{dQ}{dT}\right)_P \qquad (13.26)$$

where the subscripts indicate the process takes place at *constant pressure*. The molar specific heats are similar:

$$c_{\text{molar V}} = \frac{1}{n}\left(\frac{dQ}{dT}\right)_V \qquad (13.27)$$

$$c_{\text{molar P}} = \frac{1}{n}\left(\frac{dQ}{dT}\right)_P \qquad (13.28)$$

The specific heat of solids and liquids typically is determined at constant pressure (one atmosphere), because it is very difficult to keep these phases of matter at constant volume.† The specific heat at constant volume for most solids and liquids typically is within a few percent of the specific heat at constant pressure (but it *is* different). Unless extraordinary precision is needed, one can neglect the difference. This will be done in this text (for solids and liquids, but *not* gases), but you should be aware of the distinction.

On the other hand, with gases it is quite easy experimentally to keep them under either constant pressure or constant volume; the distinction is important for gases, since the two specific heats differ substantially, as we will see in the next chapter.

The specific heat of a substance usually varies with the temperature, so that the value of the specific heat at one temperature is not necessarily the same as at another. If this is the case, Equations 13.22 or 13.23 are not appropriate and we must revert to the more general defining Equations 13.21 and 13.24 and integrate them, accounting for the variation in the specific heat with temperature.

> In this text we will consider the specific heat to be independent of temperature as long as the substance does not change phase; hence, Equations 13.22 and 13.23 may be used.

The temperature dependence of the specific heat—that is, the way the value of the specific heat changes with temperature—is of great interest because it is a *macroscopic* manifestation of changes in the material on the *microscopic* level. Dramatic changes in the value of the specific heat of a material are indications of significant changes on the microscopic level. For example, note in Table 13.4 that the specific heat of ice is about half that of liquid water. The specific heat changes dramatically through the phase transition. This is characteristic of many types of phase transitions.

QUESTION 7
Explain why regions of land near large bodies of water such as oceans have a more moderate climate than inland areas at the same latitude.

EXAMPLE 13.9
The temperature of 4.50 kg of brass is raised from 20.0 °C to 150.0 °C. What heat transfer to the brass is needed to accomplish this, assuming no heat transfer to the surroundings?

Solution
Use Equation 13.22:

$$Q = mc\,\Delta T$$

The specific heat of brass c_{brass} is found from Table 13.4 to be

$$c_{\text{brass}} = 385 \ \text{J/(kg·K)}$$

The change in the temperature of the brass is

$$\Delta T = T_f - T_i$$
$$= 130.0 \ \text{K}$$

Making these substitutions into Equation 13.22, you get

$$Q = (4.50 \ \text{kg})[385 \ \text{J/(kg·K)}](130.0 \ \text{K})$$
$$= 2.25 \times 10^5 \ \text{J}$$

Changes of Phase

Place a pot of water on the stove and bring the water from room temperature to the boiling point. A remarkable thing happens once the boiling point is reached. Additional heat transfer to the boiling water does *not* increase the temperature of the water. Rather, the heat transfer goes into changing the *phase* of the water from a liquid to a gas.‡ Converting water at 100 °C to steam at 100 °C involves considerable heat transfer, but no change in the temperature. Likewise, melting ice at 0 °C to water at 0 °C also involves considerable heat transfer to the ice system, but no change in its temperature.

The heat transfer needed to change the phase of one kilogram of a substance is called the **latent heat**§ L of the phase transition. If the phase change is from a solid to a liquid (or vice versa), the latent heat is called the **latent heat of fusion** L_f. If the phase change is from a liquid to a gas (or vice versa), the latent heat is called the **latent heat of vaporization** L_v. The SI units of

---

*It also could be accomplished without controlling either pressure or volume.
†Extraordinary pressures are needed to keep the system at constant volume because the forces causing thermal expansion are quite large.

---

‡The energy is needed to weaken the atomic or molecular attractions between the particles of the system as it changes from a liquid to a gas (or, in the case of melting, from a solid to a liquid).
§The word comes from the Latin *latere*, meaning "hidden."

TABLE 13.5 Latent Heats of Fusion and Vaporization and Phase Transition Temperatures for Various Materials at 1.00 atm Pressure

Material	Melting point (K)	Latent heat of fusion L_f (J/kg)	Boiling point (K)	Latent heat of vaporization L_v (J/kg)
Aluminum	933	3.96×10^5		
Carbon dioxide*			195	5.73×10^5
Copper	1356	2.05×10^5	2839	47.26×10^5
Ethyl alcohol	159	1.09×10^5	351	8.79×10^5
Gold	1336	0.64×10^5	3081	17.01×10^5
Helium			4.2	0.24×10^5
Lead	601	0.247×10^5	2023	8.58×10^5
Nitrogen	63	0.257×10^5	77	1.99×10^5
Oxygen	55	0.138×10^5	90	2.13×10^5
Silver	1234	1.05×10^5	2436	23.23×10^5
Water	273.15	3.335×10^5	373.15	22.57×10^5

*At 1 atm, carbon dioxide has no liquid state; it sublimates directly from a solid to a gas. The values quoted are for this transition.

Handbook of Chemistry and Physics (73rd edition).

latent heat are J/kg, or else J/mol if the molar quantity of material is used rather than the mass. As always, it is necessary to check the units carefully when using values of the latent heats found in the technical literature. Table 13.5 lists the latent heats of fusion and vaporization associated with several phase transitions.

Thus, to change the phase of a mass m, the heat transfer to the mass is

$$Q = mL \quad (13.29)$$

where L is the appropriate latent heat associated with the particular phase transition in J/kg. If you use moles instead of kilograms, then the appropriate equation is

$$Q = nL_{molar} \quad (13.30)$$

where n is the number of moles of material, and L_{molar} is in J/mol.

PROBLEM-SOLVING TACTIC

13.5 There are no ready signals in Equations 13.29 and 13.30 to indicate whether Q should be positive or negative. Hence we impose the same sign convention as with temperature changes and Equations 13.22 and 13.23: if the heat transfer is *to* the system to change the phase, then Q is positive. If the heat transfer is *from* the system, then Q is negative.

To convert water to steam, heat transfer $Q = mL_v$ must accrue *to* the water (where L_v is the latent heat of vaporization associated with this phase transition), and so Q is *positive* ($Q = mL_v$); see Figure 13.18. As steam condenses into water, the heat transfer is *from* the (steam) system, and so Q is *negative* for the steam system ($Q = -mL_v$); see Figure 13.19.

Likewise, when ice at 0 °C melts to liquid water at 0 °C, the heat transfer Q to the ice system is positive ($Q = mL_f$), where L_f is the latent heat associated with this phase transition. When liquid water at 0 °C freezes into ice at 0 °C, the heat transfer is from the liquid water system, and so Q is negative ($Q = -mL_f$).

Not all phase transitions involve latent heats. Those that do are called **first-order phase transitions**. Phase transitions that have zero latent heat ($L = 0$ J/kg) are called **second-order phase**

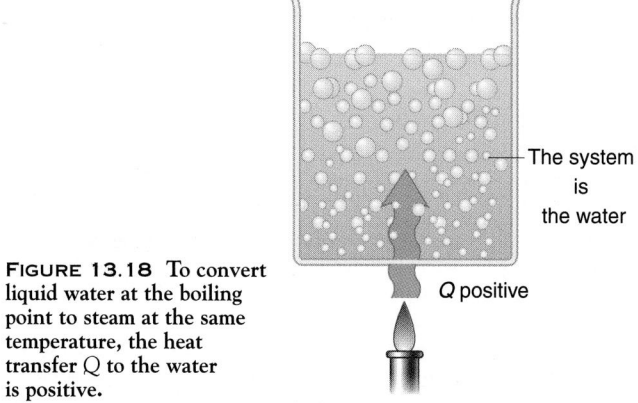

FIGURE 13.18 To convert liquid water at the boiling point to steam at the same temperature, the heat transfer Q to the water is positive.

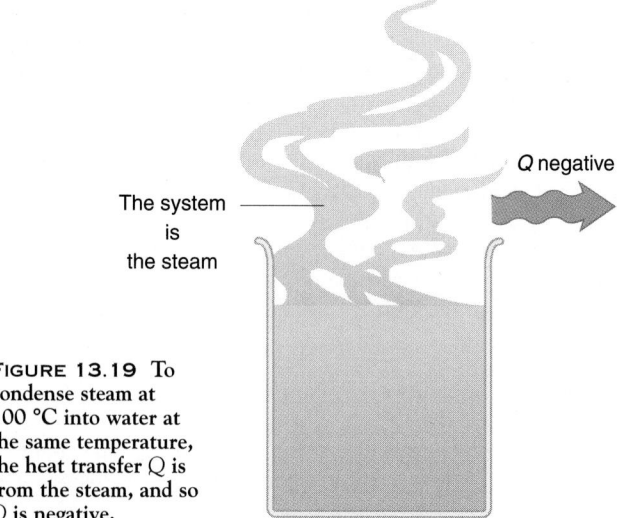

FIGURE 13.19 To condense steam at 100 °C into water at the same temperature, the heat transfer Q is from the steam, and so Q is negative.

transitions. Many structural phase transitions in crystals (changes in the microscopic crystal structure of a material) are second-order phase transitions with no measurable latent heat associated with the change of crystalline phase.

EXAMPLE 13.10

a. A mass of 3.00 kg of water is boiling. What heat transfer to the water is needed to completely vaporize the water?
b. If a 1.00 kW heater is used to supply this heat transfer, how long will the process take?

Solution

a. Use Equation 13.29, with the latent heat of vaporization of water found in Table 13.5:

$$L_{v\,water} = 22.57 \times 10^5 \text{ J/kg}$$

Hence the heat transfer Q to the water is

$$\begin{aligned}
Q &= mL_{v\,water} \\
&= (3.00 \text{ kg})(22.57 \times 10^5 \text{ J/kg}) \\
&= 6.77 \times 10^6 \text{ J}
\end{aligned}$$

b. A 1.00 kW heater supplies energy at a rate of 1.00×10^3 J/s. To supply 6.77×10^6 J will take a time

$$\begin{aligned}
t &= \frac{6.77 \times 10^6 \text{ J}}{1.00 \times 10^3 \text{ J/s}} \\
&= 6.77 \times 10^3 \text{ s} \\
&= 1.88 \text{ h}
\end{aligned}$$

EXAMPLE 13.11

How much ice at -10.0 °C must be added to 4.00 kg of water at 20.0 °C to cause the resulting mixture to reach thermal equilibrium at 5.0 °C? Assume no heat transfer to the surrounding environment, so that heat transfer occurs only between the water and ice.

Solution

The problem is a matter of accounting: the heat transfer to the ice system is from the warm water system; thus, the total heat transfer to (or from) the combined ice and water system is zero.

Let m be the mass of ice needed. The specific heat of ice is found from Table 13.4. Use Equation 13.22. To warm the ice from -10 °C to 0 °C takes a heat transfer to the ice system of

$$\begin{aligned}
Q_1 &= mc_{ice}\,\Delta T \\
&= m[2.050 \times 10^3 \text{ J/(kg·K)}][0 \text{ °C} - (-10.0 \text{ °C})] \\
&= m(2.050 \times 10^4 \text{ J/kg})
\end{aligned}$$

Notice that since only the *change* in temperature is involved, you can find the change in temperature using the celsius temperatures (the temperature *change* is the same on both kelvin and celsius scales).

To melt the ice at 0 °C to water at 0 °C is a phase transition involving a latent heat of fusion $L_f = 3.335 \times 10^5$ J/kg, from Table 13.5. Thus the heat transfer to the ice is

$$\begin{aligned}
Q_2 &= mL_f \\
&= m(3.335 \times 10^5 \text{ J/kg}) \\
&= m(33.35 \times 10^4 \text{ J/kg})
\end{aligned}$$

To warm the newly formed liquid water to the final temperature of 5.0 °C involves an additional heat transfer to the mass m of

$$\begin{aligned}
Q_3 &= mc_{water}\,\Delta T \\
&= m[4186 \text{ J/(kg·K)}](5.0 \text{ °C} - 0.0 \text{ °C}) \\
&= m(2.1 \times 10^4 \text{ J/kg})
\end{aligned}$$

Thus, to warm the ice, melt the ice, and then warm the resulting liquid to 5.0 °C, the total heat transfer to the ice is

$$\begin{aligned}
Q_1 + Q_2 + Q_3 &= m(2.050 \times 10^4 \text{ J/kg}) + m(33.35 \times 10^4 \text{ J/kg}) \\
&\quad + m(2.1 \times 10^4 \text{ J/kg}) \\
&= m(37.6 \times 10^4 \text{ J/kg})
\end{aligned}$$

In cooling the 4.00 kg of water at 20.0 °C to 5.0 °C, the heat transfer from the water is

$$\begin{aligned}
Q_4 &= mc\,\Delta T \\
&= (4.00 \text{ kg})[4186 \text{ J/(kg·K)}](5.0 \text{ °C} - 20.0 \text{ °C}) \\
&= -25.1 \times 10^4 \text{ J}
\end{aligned}$$

where the minus sign indicates that the heat transfer is from the warm water.

Since the combined system of the ice and water is isolated from the environment, the total heat transfer to or from the isolated system is zero. Thus

$$Q_1 + Q_2 + Q_3 + Q_4 = 0 \text{ J}$$
$$m(37.6 \times 10^4 \text{ J/kg}) + (-25.1 \times 10^4 \text{ J}) = 0 \text{ J}$$

Solving for m, you find

$$m = 0.668 \text{ kg}$$

STRATEGIC EXAMPLE 13.12

One (1.00) liter of boiling water at 100.0 °C is poured into an insulated container that contains 20.0 kg of ice at 0.0 °C. Determine the final temperature of the system and, if any ice is left, how much ice remains.

Solution

Assume the density of water does not change with temperature (a reasonable first approximation). One (1.00) liter of water (1.00×10^{-3} m³) has a mass of 1.00 kg, since the density of water is 1.00×10^3 kg/m³.

Let $t_{celsius}$ be the final temperature of the combined water and ice system in celsius degrees.

The heat transfer to the ice to melt it all is

$$\begin{aligned}
Q_1 &= mL_f \\
&= (20.0 \text{ kg})(3.335 \times 10^5 \text{ J/kg}) \\
&= 6.67 \times 10^6 \text{ J}
\end{aligned}$$

To warm the melted ice to the final temperature $t_{celsius}$ takes heat transfer

$$\begin{aligned}
Q_2 &= mc_{water}\,\Delta T \\
&= (20.0 \text{ kg})[4186 \text{ J/(kg·K)}](t_{celsius} - 0 \text{ °C}) \\
&= (8.37 \times 10^4 \text{ J/K})t_{celsius}
\end{aligned}$$

To cool the hot water to temperature $t_{celsius}$ takes

$$\begin{aligned}
Q_3 &= mc_{water}\,\Delta T \\
&= (1.00 \text{ kg})[4186 \text{ J/(kg·K)}](t_{celsius} - 100.0 \text{ °C}) \\
&= (4.19 \times 10^3 \text{ J/K})(t_{celsius} - 100.0 \text{ °C})
\end{aligned}$$

Notice that this heat transfer is negative because the final temperature of the originally hot water is less than the initial temperature.

The total heat transfer to or from the isolated, combined water and ice system is zero. Thus

$$Q_1 + Q_2 + Q_3 = 0 \text{ J}$$

$$6.67 \times 10^6 \text{ J} + (8.37 \times 10^4 \text{ J/K})t_{\text{celsius}}$$

$$+ (4.19 \times 10^3 \text{ J/K})(t_{\text{celsius}} - 100.0 \text{ °C}) = 0 \text{ J}$$

Solving for t_{celsius}, you obtain

$$t_{\text{celsius}} = -71.1 \text{ °C}$$

This is a ridiculous answer! There is no way that the final temperature of the system can be less than the coldest material in the problem. This result means that not all the ice melts. This calculation reminds us to question whether the result obtained is reasonable. This answer is not!

We have to begin anew. Since not all the ice melts, the final temperature of the system must be 0 °C. Let m be the mass of the ice that does melt. Then to melt this amount of ice takes

$$Q_1 = mL_f$$
$$= m(3.335 \times 10^5 \text{ J/kg})$$

The heat transfer from the hot water in cooling from 100 °C to 0 °C is

$$Q_2 = mc_{\text{water}} \Delta T$$
$$= (1.00 \text{ kg})[4186 \text{ J/(kg·K)}](0.0 \text{ °C} - 100.0 \text{ °C})$$
$$= -4.19 \times 10^5 \text{ J}$$

The total heat transfer to or from the combined ice and water must be zero, because the combined system is isolated:

$$Q_1 + Q_2 = 0 \text{ J}$$
$$m(3.335 \times 10^5 \text{ J/kg}) + (-4.19 \times 10^5 \text{ J}) = 0 \text{ J}$$

Solving for m, you get

$$m = 1.26 \text{ kg}$$

So 1.26 kg of ice melts. Thus the amount of ice left is

$$20.0 \text{ kg} - 1.26 \text{ kg} = 18.7 \text{ kg}$$

13.10 RESERVOIRS

If a thimbleful of boiling water is brought into thermal contact with a swimming pool full of cool water, the temperature change of the water in the pool is negligible and the thimble of water cools to the initial temperature of the pool. Likewise, place mom's all-American apple pie to bake in a hot oven and the pie eventually warms up to the temperature of the oven. A warm six-pack placed in a cool refrigerator eventually has the same temperature as the inside of the refrigerator. These examples illustrate what is meant by the term **reservoir** in thermodynamics. One meaning of the word reservoir is "a great supply of something"; we will use this definition for thermodynamic purposes.

A reservoir is a special thermodynamic system. We use a reservoir to bring other systems that are in thermal contact with it to a fixed temperature: the temperature of the reservoir. The large pool of water is a good reservoir for the thimble; the inside of an oven and a refrigerator are other types of reservoirs, maintained at a fixed temperature by electrical means. Ovens and refrigerators may be called *dynamic* reservoirs, whereas large masses of material, such as the pool, may be called *passive* reservoirs. The distinction between active and passive reservoirs is not important.

> A system in thermal contact with a reservoir experiences heat transfer to or from the system until it has the same temperature as the reservoir. The temperature of the reservoir does not change, despite positive or negative heat transfer to it.

A passive reservoir can be designed as follows. Consider a large mass m of material at a temperature T; let c be the specific heat of the material. If heat transfer Q occurs to the mass (from another system at a different temperature placed in thermal contact with the reservoir), the temperature of the reservoir material changes by an amount ΔT, where

$$Q = mc \, \Delta T$$

according to Equation 13.22. The temperature change of the reservoir material thus is

$$\Delta T = \frac{Q}{mc} \qquad (13.31)$$

The product in the denominator of this expression, mc, is called the **thermal mass** of the material. If the thermal mass of the material is sufficiently large, then the temperature change of the reservoir material can be made arbitrarily small. A passive thermodynamic reservoir is a thermodynamic system whose thermal mass is so large that its temperature change is negligible despite heat transfer *to* the reservoir system ($Q > 0$ J) or heat transfer *from* the reservoir system ($Q < 0$ J).

In a strict sense, such ideal passive reservoirs do not exist, since they are never of infinite thermal mass. But from a practical standpoint, there is no problem if the reservoir has a sufficiently large thermal mass (see Example 13.13). Therefore, a passive thermodynamic reservoir has a huge thermal mass. A good passive reservoir must have a thermal mass much larger than the thermal mass of any system placed in thermal contact with it.

> The word reservoir in thermodynamics means either (a) a huge thermal mass or (b) a region (such as the inside of an oven or a refrigerator) that maintains a fixed temperature despite heat transfer to or from it from other thermodynamic systems.

Experimentally, refrigeration and heating units, whose temperatures can be adjusted over a wide range, are used to provide dynamic reservoirs of small total thermal mass.

When the temperature of a system is measured with an appropriate thermometer, the system should act as a thermodynamic reservoir as far as the thermometer is concerned, since we do not want the presence of the thermometer to affect the temperature of the system itself.

EXAMPLE 13.13

A passive reservoir is needed. A heat transfer to the reservoir of 1.50×10^4 J must not change its temperature by more than 2.00×10^{-3} K. How many kilograms of water are needed to ensure a passive reservoir of adequate thermal mass?

Solution

Use Equation 13.22 and solve for m:

$$m = \frac{Q}{c_{\text{water}} \Delta T}$$

The specific heat of water is found in Table 13.4; substituting all values,

$$m = \frac{1.50 \times 10^4 \text{ J}}{[4186 \text{ J/(kg·K)}](2.00 \times 10^{-3} \text{ K})}$$
$$= 1.79 \times 10^3 \text{ kg}$$

This represents about 1.79 m³ of water, which is a cube of water about 1.21 m on a side.

13.11 MECHANISMS FOR HEAT TRANSFER*

Heat transfer between systems at different temperatures occurs by the mechanisms of (1) conduction, (2) convection, and (3) radiation.

Conduction

Hold one end of a silver spoon and put the other end into a hot liquid, such as a delicious soup simmering on a stove. You quickly become aware that the end of the spoon you are holding becomes hot. This illustrates heat transfer by **conduction**. Yet do the same experiment with a wooden spoon of identical dimensions and you know the end you hold does not become hot, even after many minutes of stirring the soup. Hence heat transfer by conduction depends critically on the material bridging or connecting the warmer and cooler regions.

Imagine two reservoirs at slightly different temperatures connected by means of a thin piece of material of uniform cross-sectional area, as pictured in Figure 13.20. The material is in thermal contact with each reservoir.

Heat transfer occurs *through* the material from the hot reservoir to the cold reservoir. The energy is transported from atom to atom through the material by means of atomic vibrations and/or collisions. There is no bulk, macroscopic motion of the material itself. This mechanism of heat transfer is conduction.

Let dT be the differential difference in temperature between the warmer and cooler reservoirs, respectively, and let A be the area of the material in thermal contact with each reservoir. Let the length of material connecting the reservoirs be dx, with the x-axis directed from the warmer to the cooler reservoir. Experiments indicate that the heat transfer per second

$$\frac{dQ}{dt}$$

called the **heat flow (expressed in J/s ≡ W)**, between the reservoirs is directly proportional to the temperature difference dT and the area of thermal contact A, and inversely proportional to the length of material dx. It makes a great deal of difference whether the material is copper or Styrofoam; thus the propor-

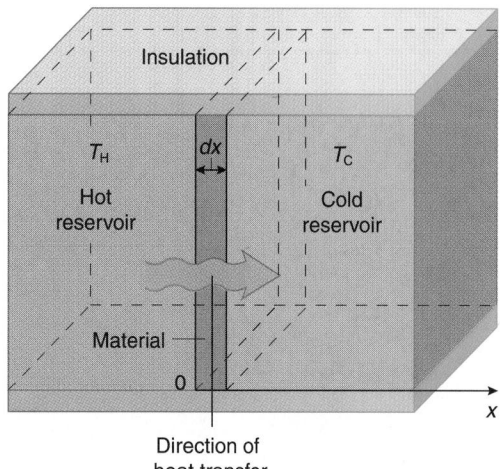

FIGURE 13.20 A material in thermal contact with two reservoirs.

tionality constant k depends on the material; k is called the **thermal conductivity** of the material. We summarize these experimental observations with the following equation:

$$\frac{dQ}{dt} = -kA \frac{dT}{dx} \tag{13.32}$$

Why the minus sign? The heat flow is from the warmer to the cooler reservoirs, so that as *x increases*, the temperature *decreases* through the material. Thus

$$\frac{dT}{dx}$$

is *negative*; we want the heat flow

$$\frac{dQ}{dt}$$

from the hot reservoir to the cold reservoir to be positive, since this is the direction in which energy is transferred.[†] The minus sign in Equation 13.32 thus makes the heat flow positive. Notice also that large values of the thermal conductivity mean a large heat flow (a good thermal conductor). Values of the thermal conductivity k for various materials are tabulated in Table 13.6.

The way the temperature changes with position in the material,

$$\frac{dT}{dx}$$

is known as the **temperature gradient**. Equation 13.32 indicates that the heat flow through a material is proportional to the temperature gradient, the area in contact, and the thermal conductivity of the material.

[†]The reason the heat transfer is in this direction and not the other way (from the cold to the hot region) will be examined in Chapter 15.

TABLE 13.6 Thermal Conductivity k of Various Materials at Approximately Room Temperature and 1 atm

Material	k [W/(m·K)]
Air	0.026
Aluminum	237
Asbestos	0.08
Concrete	1.4
Copper	401
Gold	317
Glass (window)	0.78
Hydrogen	0.18
Ice	2.1
Iron	80.2
Lead	35.3
Mercury	8.18
Nitrogen	0.026
Oxygen	0.027
Sheetrock (dry wall)	0.13
Silver	429
Stainless steel	15
Water	0.613

Myer Kutz (editor), *Mechanical Engineer's Handbook* (John Wiley, New York, 1998).

Let the temperatures of the two reservoirs be T_H and T_C and the length of material through which heat transfer occurs be d, as shown in Figure 13.21.

For a steady-state situation, the temperature at any position x inside the material must remain constant. This means the heat flow

$$\frac{dQ}{dt}$$

past any point in the material must be independent of the position x in the material. If the heat flow were not constant at all points x, then the temperature at any point inside the material would be time dependent and we would not have a steady-state situation. So, for a steady-state situation, in a material of uniform cross section, both

$$\frac{dQ}{dt}$$

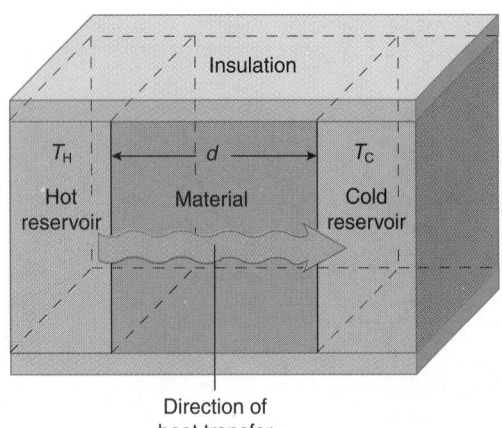

FIGURE 13.21 Two reservoirs connected by a material of length d.

and

$$\frac{dT}{dx}$$

are independent of x. Write the constant temperature gradient as

$$\frac{dT}{dx} \equiv -S \qquad (13.33)$$

where S is a constant. The derivative of T with respect to x is the same everywhere; this means that the temperature decreases linearly with x as we go through the material from the hot to the cold reservoir. We can see this mathematically by integrating Equation 13.33; rearranging it slightly, we have

$$dT = -S\, dx$$

Integrating this between corresponding limits,

$$\int_{T_H}^{T} dT = -S \int_{0\,m}^{x} dx$$

or

$$T - T_H = -S[x - 0\ \text{m}]$$
$$= -Sx$$

Solving for T, we find

$$T = T_H - Sx$$

This shows that the temperature decreases linearly with x through the material from the hot to cold regions. Solving for the temperature gradient S, we obtain

$$S = \frac{T_H - T}{x}$$

When $x = d$, the length of the material connecting the reservoirs, the temperature T is the temperature of the cold reservoir, T_C; hence

$$S = \frac{T_H - T_C}{d} \qquad (13.34)$$

Equation 13.32 for the heat flow becomes, using Equation 13.33:

$$\frac{dQ}{dt} = -kA\frac{dT}{dx}$$
$$= kAS$$

Substituting for the temperature gradient S using Equation 13.34, we get

$$\frac{dQ}{dt} = kA\frac{T_H - T_C}{d} \qquad (13.35)$$

Dividing both the numerator and denominator of the right-hand side of this equation by the thermal conductivity k, we obtain

$$\frac{dQ}{dt} = A\frac{T_H - T_C}{d/k}$$

The quantity d/k is known as the **R-value** R of the material, and represents a thermal resistance to heat flow*:

$$R \equiv \frac{d}{k} \qquad (13.36)$$

---

*It is called the R-value because it is the thermal *Resistance*. Do not confuse the R-value with the universal gas constant.

Note that the smaller the thermal conductivity, the larger the R-value, and the smaller the heat flow between the reservoirs. The R-value incorporates the length (thickness) of the material connecting the two reservoirs; thus if the thickness d of material is doubled, the R-value doubles and the heat flow is halved. Building materials typically are rated for their insulating effect—that is, their resistance to heat flow—by quoting their R-value; the higher the R-value, the greater the insulating effect of the material and the smaller the heat flow between the hot and cold reservoirs. Thus

$$\frac{dQ}{dt} = A\,\frac{T_H - T_C}{R} \qquad (13.37)$$

The approximate R-values of some materials are summarized in Table 13.7.

Materials in Series

In building construction, several layers of different materials typically make up the wall separating the inside and outside of a building. Think of the inside and outside as two reservoirs at different temperatures. We want to determine the effect of such layering on the heat flow between the reservoirs.

Let two layers of material of different thermal conductivities and thicknesses connect the two reservoirs, as shown in Figure 13.22. Such a layered system represents materials in **thermal series**, since the heat flow must unavoidably be through each material in succession (or series) to get from the hotter to the cooler reservoir. Let the R-values of the materials be

$$R_1 = \frac{d_1}{k_1} \quad \text{and} \quad R_2 = \frac{d_2}{k_2}$$

When a steady-state situation is reached, the heat flow through each material is the same. Let T be the temperature of the

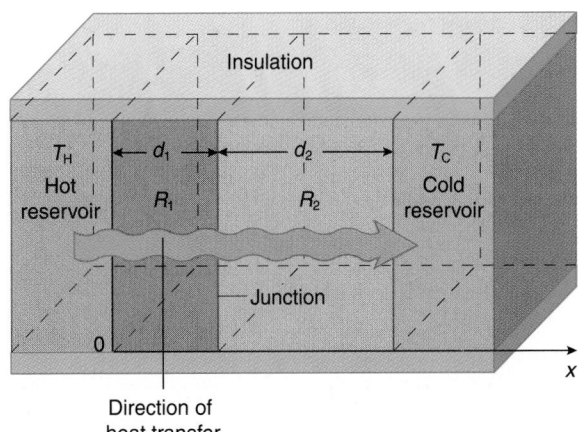

FIGURE 13.22 Two reservoirs connected by two materials in series.

junction between the two materials. The heat flow through the first material is

$$\left(\frac{dQ}{dt}\right)_1 = A\,\frac{T_H - T}{R_1} \qquad (13.38)$$

and that through the second material is

$$\left(\frac{dQ}{dt}\right)_2 = A\,\frac{T - T_C}{R_2} \qquad (13.39)$$

Since the heat flows are equal in the steady state, we have

$$A\,\frac{T_H - T}{R_1} = A\,\frac{T - T_C}{R_2}$$

We can solve for the temperature T at the interface between the materials. After a bit of algebra (famous last words!), we find that

$$T = \frac{1}{R_1 + R_2}\left(R_2 T_H + R_1 T_C\right)$$

Now we use Equation 13.38 and substitute for the temperature T at the interface, since this temperature is not typically of interest in determining the heat flow between the two reservoirs. The resulting expression for the heat flow involves only the temperatures of the two reservoirs, the area of thermal contact, and the R-values of the two materials. After making the substitution for the junction temperature T, the heat flow becomes (after a bit more algebraic simplification)

$$\frac{dQ}{dt} = A\,\frac{T_H - T_C}{R_1 + R_2}$$

Notice that the effect of an additional piece of material between the reservoirs is to increase the total R-value in the denominator of the heat flow equation.

Thus, if other layers of material are added in series between the two reservoirs, the effective (or total) R-value of the composite is the sum of the R-values of the individual layers:

$$R_{\text{total}} = R_1 + R_2 + R_3 + R_4 + \cdots \quad \text{(layers in series)} \quad (13.40)$$

TABLE 13.7 Approximate R-Values of Selected Building Materials

Material	R-value $(\text{m}^2\cdot\text{K}\cdot\text{s/J} = \text{m}^2\cdot\text{K/W})$
Glass (single-pane window glass) (thickness 3.2×10^{-3} m)	0.21
Brick (thickness 1.0×10^{-1} m)	0.93
Fiberglass batting (thickness 8.9×10^{-2} m) (thickness 1.5×10^{-1} m)	2.53 4.37
Air space (thickness 8.9×10^{-2} m)	0.23
Pine board (thickness 1.9×10^{-2} m)	0.22
Plywood (thickness (1.2×10^{-2} m)	0.14
Polyurethane foam (thickness 2.54×10^{-2} m)	2.3
Sheetrock (dry wall) (thickness 1.3×10^{-2} m)	0.10

In the United States, the R-value of a material typically is given in the antiquated English system of units. In the English unit system, the temperature difference is expressed in degrees fahrenheit, the area in square feet, and the R-value is in the singularly awkward units of feet$^2 \cdot {}^\circ$F$\cdot$h/Btu. The heat flow then emerges in Btu/h rather than watts. To convert the R-values in SI units to the English system, multiply each value by

$$5.68\,\frac{\text{feet}^2 \cdot {}^\circ\text{F}\cdot\text{h/Btu}}{\text{m}^2\cdot\text{K/W}}$$

The heat flow is then found from

$$\frac{dQ}{dt} = A\frac{T_H - T_C}{R_{total}} \qquad (13.41)$$

Materials in Parallel

If two different materials *each* connect the hot and cold reservoirs, as in Figure 13.23, then the materials are said to be connected thermally in **parallel**.

For example, both windows and walls connect the inside and outside of a building; the windows and walls are thermally in parallel. We neglect any heat transfer along the interface between the two materials. The heat flow through material 1 is

$$\left(\frac{dQ}{dt}\right)_1 = A_1\frac{T_H - T_C}{R_1} \qquad (13.42)$$

where A_1 is the area of this material in thermal contact with the reservoirs. Likewise, for the second material,

$$\left(\frac{dQ}{dt}\right)_2 = A_2\frac{T_H - T_C}{R_2} \qquad (13.43)$$

where A_2 is the area of this material in thermal contact with the reservoirs.

The total heat flow from the hot to the cold reservoir then is the sum of the individual heat flows:

$$\left(\frac{dQ}{dt}\right)_{total} = \left(\frac{dQ}{dt}\right)_1 + \left(\frac{dQ}{dt}\right)_2 \qquad (13.44)$$

Substituting from Equations 13.42 and 13.43, we find

$$\left(\frac{dQ}{dt}\right)_{total} = A_1\left(\frac{T_H - T_C}{R_1}\right) + A_2\left(\frac{T_H - T_C}{R_2}\right)$$

$$= (T_H - T_C)\left(\frac{A_1}{R_1} + \frac{A_2}{R_2}\right) \quad \text{(materials in parallel)}$$

$$\qquad (13.45)$$

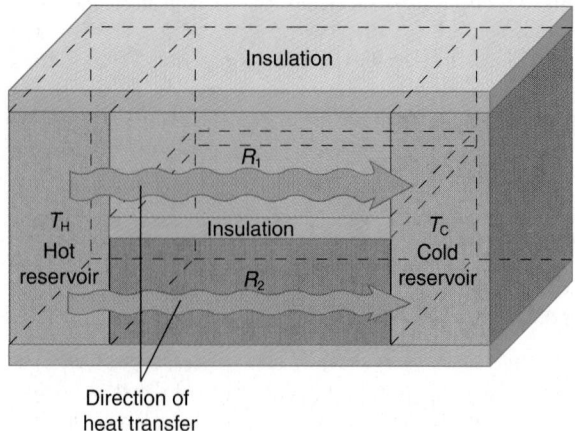

FIGURE 13.23 Two reservoirs connected by two materials in parallel.

Insulation

$\bar{R}_1$

Insulation

T_H
Hot reservoir

R_2

T_C
Cold reservoir

Direction of heat transfer

EXAMPLE 13.14

A single-pane glass window of dimensions 1.00 m × 1.50 m connects a room at temperature 20 °C with outside air on a blistering summer day (temperature 37 °C). Find the heat flow through the window.

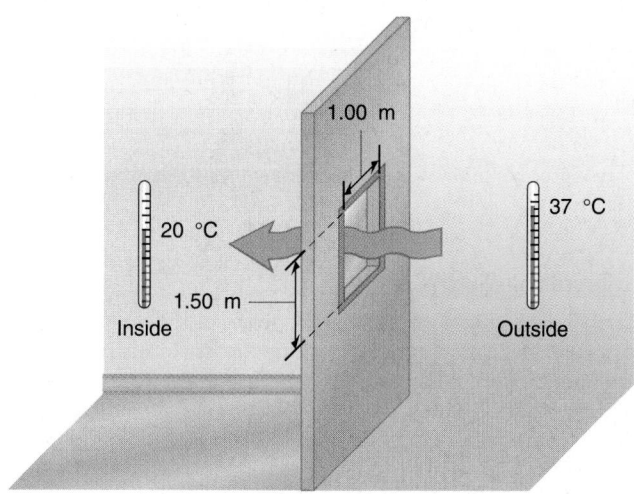

1.00 m

20 °C

1.50 m

Inside

37 °C

Outside

Heat transfer through a window.

Solution

The hot reservoir is the outside air and the cold reservoir is the inside of the room, maintained at its temperature by an air conditioning system. Thus $T_H - T_C = 17$ K. From Table 13.7, we see that the R-value for a single-pane window is

$$R = 0.21 \text{ m}^2 \cdot \text{K/W}$$

Use Equation 13.37 for the heat flow:

$$\frac{dQ}{dt} = A\frac{T_H - T_C}{R}$$

where A is the area of the window, here 1.00 m × 1.50 m = 1.50 m².

Making the substitutions, you have

$$\frac{dQ}{dt} = (1.50 \text{ m}^2)\frac{17 \text{ K}}{0.21 \text{ m}^2 \cdot \text{K/W}}$$

$$= 1.2 \times 10^2 \text{ W}$$

EXAMPLE 13.15

A wall of a house consists of sheetrock (dry wall of thickness 1.3×10^{-2} m), fiberglass insulation (thickness 8.9×10^{-2} m), and plywood (thickness 1.2×10^{-2} m) separating a room from the outdoors. The wall is 2.00 m high and 5.00 m long and is windowless. If the inside temperature is 20 °C and the outside temperature is −25 °C on a cold winter day, what is the heat flow through the wall?

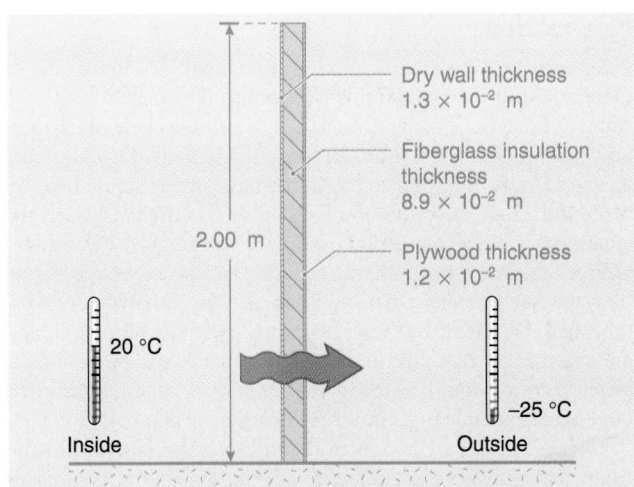

Heat transfer through a composite wall.

Solution

The materials are layered in series, and so the total R-value of the layers is the sum of the individual R-values of the material, each found from Table 13.7:

$$
\begin{aligned}
R_{\text{total}} &= R_{\text{dry wall}} + R_{\text{fiberglass}} + R_{\text{plywood}} \\
&= 0.10 \ \text{m}^2 \cdot \text{K/W} + 2.53 \ \text{m}^2 \cdot \text{K/W} + 0.14 \ \text{m}^2 \cdot \text{K/W} \\
&= 2.77 \ \text{m}^2 \cdot \text{K/W}
\end{aligned}
$$

The temperature difference is

$$
T_{\text{H}} - T_{\text{C}} = 45 \ \text{K}
$$

and the wall area is

$$
\begin{aligned}
A &= (2.00 \ \text{m})(5.00 \ \text{m}) \\
&= 10.0 \ \text{m}^2
\end{aligned}
$$

Using Equation 13.41 for the heat flow, you get

$$
\begin{aligned}
\left(\frac{dQ}{dt}\right)_{\text{total}} &= A \frac{T_{\text{H}} - T_{\text{C}}}{R_{\text{total}}} \\
&= \frac{(10.0 \ \text{m}^2)(45 \ \text{K})}{2.77 \ \text{m}^2 \cdot \text{K/W}} \\
&= 1.6 \times 10^2 \ \text{W}
\end{aligned}
$$

EXAMPLE 13.16

A wall of a dormitory cell has overall dimensions 2.00 m by 3.50 m. A single-pane window of dimensions 0.75 m by 1.20 m is in the wall. The effective R-value of the opaque wall is 2.10 m² · K/W. If the inside temperature is 20 °C and the outside air temperature is −10 °C, determine the heat flow through the entire wall.

Solution

The wall and window connect the two reservoirs in parallel, and so the total heat flow is the sum of the heat flow through the window with the heat flow through the wall. You could use Equation 13.45 directly, but let's look at what is happening with

each parallel conductor separately and add the heat flows. The temperature difference is $T_{\text{H}} - T_{\text{C}} = 30$ K. The area of the window is

$$
\begin{aligned}
A_{\text{window}} &= (0.75 \ \text{m})(1.20 \ \text{m}) \\
&= 0.90 \ \text{m}^2
\end{aligned}
$$

The R-value of the single-pane window is found from Table 13.7 to be

$$
R_{\text{window}} = 0.21 \ \text{m}^2 \cdot \text{K/W}
$$

The heat flow through the window thus is, from Equation 13.37,

$$
\begin{aligned}
\left(\frac{dQ}{dt}\right)_{\text{window}} &= A_{\text{window}} \frac{T_{\text{H}} - T_{\text{C}}}{R_{\text{window}}} \\
&= (0.90 \ \text{m}^2) \frac{30 \ \text{K}}{0.21 \ \text{m}^2 \cdot \text{K/W}} \\
&= 1.3 \times 10^2 \ \text{W}
\end{aligned}
$$

The area of the wall itself (not including the window) is

$$
(2.00 \ \text{m})(3.50 \ \text{m}) - (0.75 \ \text{m})(1.20 \ \text{m}) = 6.10 \ \text{m}^2
$$

The heat flow through the wall is found using Equation 13.37:

$$
\begin{aligned}
\left(\frac{dQ}{dt}\right)_{\text{wall}} &= A_{\text{wall}} \frac{T_{\text{H}} - T_{\text{C}}}{R_{\text{wall}}} \\
&= (6.10 \ \text{m}^2) \frac{30 \ \text{K}}{2.10 \ \text{m}^2 \cdot \text{K/W}} \\
&= 87 \ \text{W}
\end{aligned}
$$

The total heat flow is the sum of the heat flow through the wall and that through the window:

$$
\begin{aligned}
\frac{dQ}{dt} &= 1.3 \times 10^2 \ \text{W} + 87 \ \text{W} \\
&= 2.2 \times 10^2 \ \text{W}
\end{aligned}
$$

EXAMPLE 13.17

The air temperature above a calm pond with water at 0 °C near its surface suddenly plummets from 0 °C to −10 °C.

a. Show that the rate at which the thickness x of ice increases with time is

$$
x = \left(\frac{2k \, \Delta T}{L_{\text{f}} \rho_{\text{ice}}} t\right)^{1/2}
$$

where ΔT is the temperature difference $T_{\text{H}} - T_{\text{C}}$, ρ_{ice} is the density of ice, and L_{f} the latent heat of fusion of ice.

b. How long will it take to form a layer 1.0 cm thick? The density of ice is 920 kg/m³.

Solution

a. Let A be an area on the surface of the pond. In a differential thickness $|dx|$ of ice, there is a differential mass dm, where

$$
dm = \rho_{\text{ice}} A \, |dx|
$$

To solidify this mass of ice, the heat transfer from the water is

$$dQ = -L_f \, dm$$
$$= -L_f \rho_{ice} A \, |dx|$$

Let x be the thickness of the ice. The heat flow is upward through the ice from the water to the air. The additional ice formation occurs at the underside of the ice and proceeds downward. Choose the $+x$-direction downward; so dQ, which flows upward, passes through $-dx$ (i.e., toward negative x). Thus, to eliminate the absolute value signs, we write

$$dQ = -L_f \rho_{ice} A \, (-dx)$$
$$= L_f \rho_{ice} A \, dx$$

The heat flow is given by Equation 13.35:

$$\frac{dQ}{dt} = kA \frac{\Delta T}{x}$$

Substituting for dQ from the preceding, you have

$$L_f \rho_{ice} A \frac{dx}{dt} = \frac{kA \, \Delta T}{x}$$

or

$$x \, dx = \frac{k \Delta T}{L_f \rho_{ice}} dt$$

Integrating this expression,

$$\int_{0\,\text{m}}^{x} x \, dx = \frac{k \, \Delta T}{L_f \rho_{ice}} \int_{0\,\text{s}}^{t} dt$$

Solving for x, you obtain

$$x = \left(\frac{2k \, \Delta T}{L_f \rho_{ice}} t \right)^{1/2}$$

b. The temperature difference ΔT is 10 K. The thermal conductivity k of ice is (see Table 13.6)

$$k = 2.1 \ \text{W/(m·K)}$$

The heat of fusion for water is found in Table 13.5:

$$L_f = 3.335 \times 10^5 \ \text{J/kg}$$

The density of ice is

$$\rho_{ice} = 920 \ \text{kg/m}^3$$

Making these substitutions, you have

$$x = (3.7 \times 10^{-4} \ \text{m/s}^{1/2})\sqrt{t}$$

Substitute $x = 0.010$ m and solve for t:

$$0.010 \ \text{m} = (3.7 \times 10^{-4} \ \text{m/s}^{1/2})\sqrt{t}$$

or

$$t = 7.3 \times 10^2 \ \text{s}$$
$$= 12 \ \text{min}$$

Convection

Heat transfer resulting from the movement or circulation of material is called **convection**. For example, in a hot water tank the water in immediate contact with the hot heating element expands and becomes less dense than the surrounding fluid. Buoyant forces then cause the hot water to rise in the tank; cooler liquid takes its place and a circulation of water is established in the tank. Such circulation is called a **convection current**. Likewise, the air in contact with a flame or a hot radiator expands and becomes less dense; buoyant forces make it rise, and convection currents can be established. During the summer near the beach, hot air over inland areas rises, drawing cooler air inward from across the water surface, producing on-shore breezes in the afternoon. Such winds are a large-scale example of a convection current of air.

Heat transfer by convection can be *passive* (such as in the examples just mentioned) or *forced* with fans or pumps. Quantitative descriptions of convection involve considerations of fluid dynamics that are beyond the scope of an introduction to thermodynamics.

Radiation

Bask in sunlight at poolside and you experience heat transfer by **radiation**. The same radiation mechanism of heat transfer is involved in common heat lamps used to keep food warm in your college cafeteria and fast food restaurants while awaiting voracious student appetites. You likely have experienced the romantic effects of radiative heat transfer curled up in front of a roaring fire with a good physics text for company, or more likely with a boon companion.

> While both conduction and convection require a material medium for heat transfer, radiation does not.

It is well known that we receive energy from the Sun despite the virtually total absence of any material between the Sun and the Earth. The solar energy is transported by (electromagnetic) radiation: light in the most general sense of the term. The particles of light, called **photons**, carry the energy from an emitter (such as the Sun) to an absorber (your body frying beside the pool).

Every object emits energy by radiation.

Radiative heat transfer between the Sun and poolside Sun worshipers.

The rate at which a system emits radiation is found experimentally to be proportional to the *fourth* power of its absolute temperature.

This remarkable result, discovered by Josef Stefan (1835–1893) in 1879, has deep historical significance in the history of physics. It was through a detailed study of the physics of such radiation that Max Planck (1858–1947) discovered (in 1901) a totally new and unanticipated fundamental constant of nature, known formally as the quantum of action, that now bears his name (Planck's constant). Planck's discovery was the opening salvo in the revolution of physics known as *quantum mechanics* (Chapters 26 and 27).

The rate at which energy is *radiated* by a system also depends directly on its surface area A. The heat flow via radiation is written as

$$\frac{dQ}{dt} = -eA\sigma T^4 \tag{13.46}$$

As before, the radiative heat flow

$$\frac{dQ}{dt}$$

is expressed in joules/second (J/s) or watts (W). The constant e in Equation 13.46, called the **emissivity** of the surface,* is a pure number (with no dimensions) between 0 and 1. The emissivity quantifies the relative effectiveness of the surface as a radiator. A perfect radiator has an emissivity $e = 1$; such a system is known as a *blackbody radiator* or, more simply, a **blackbody**.[†] The remaining constant in Equation 13.46, σ, is known as the **Stefan–Boltzmann constant**; experiments indicate it has the numerical value[‡]

$$\sigma = 5.670 \times 10^{-8} \ \text{W/(m}^2 \cdot \text{K}^4)$$

The minus sign in Equation 13.46 indicates that the system is emitting rather than absorbing the radiation, in keeping with our convention that heat transfer from a system is negative. Equation 13.46 is known as **Stefan's law**.

A system also absorbs radiation. The radiation absorbed by a system is proportional to the fourth power of the absolute temperature of its *surroundings*. We write the heat flow to a system as

$$\frac{dQ}{dt} = +aA\sigma T^4$$

where a is a pure number between 0 and 1 indicating the relative ability of the surface to absorb radiation from its surroundings. A surface with $a = 1$ is a perfect absorber.

If you were to place a hot object in a surrounding region at a cooler temperature and insulate the resulting system, as in

---

*Do *not* confuse the emissivity e with the base of natural logarithms (the pure number $e = 2.71828...$) or with the magnitude of the charge on the proton and electron ($e = 1.602 \times 10^{-19}$ coulombs).

[†] The object may not appear black in color at all. An excellent approximation to a blackbody radiator is a closed chamber at temperature T with a small hole drilled into it, through which the radiation is emitted.

[‡] The Stefan–Boltzmann constant can be expressed in terms of other fundamental physical constants (in particular, the Boltzmann constant k, Planck's constant h, and the speed of light c). Thus the Stefan–Boltzmann constant is not truly a fundamental physical constant on a par with h, c, or G.

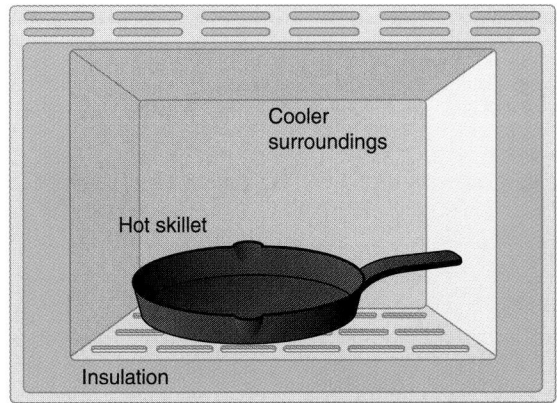

FIGURE 13.24 A hot object in a region of cooler surrounding temperature and insulated from everything else.

Figure 13.24, heat transfer would occur only between the hot object and the surrounding cooler region. The hot object radiates energy and cools, while the surroundings absorb radiant energy and increase in temperature.

Eventually, the object and its surroundings reach the same temperature T and the combined system (object and its surroundings) is in thermal equilibrium. At this point, the object must be radiating an amount of energy equal to the energy it is absorbing. When the combined system is in thermal equilibrium, the total heat flow (via radiation and absorption) is zero. That is, at thermal equilibrium,

$$\text{heat flow radiated} + \text{heat flow absorbed} = 0 \ \text{W}$$
$$-eA\sigma T^4 + aA\sigma T^4 = 0 \ \text{W}$$

Hence

$$e = a$$

In other words, the effectiveness of an object as a radiator (e) is the *same* as its effectiveness as an absorber (a). Good radiators are good absorbers; poor radiators are poor absorbers. We really need only one constant to describe both absorption and radiation; the constant chosen is e, the emissivity.

If an object is at a temperature T_{object} that is different from that of its surroundings, $T_{\text{surroundings}}$, then the object *radiates* energy at the rate

$$-eA\sigma T^4_{\text{object}}$$

and *absorbs* energy at the rate

$$+eA\sigma T^4_{\text{surroundings}}$$

The total heat flow via radiative heat transfer to the object from its surroundings then is

$$\frac{dQ}{dt} = eA\sigma\left(T^4_{\text{surroundings}} - T^4_{\text{object}}\right) \tag{13.47}$$

QUESTION 8

Before the advent of microwave ovens, a common household trick when baking potatoes in a standard oven was to insert a (clean) nail through each potato while baking. Explain how this helps to bake the potatoes faster.

EXAMPLE 13.18 ━━━━━━━━━━━━━━━━━

The surface temperature of the spherical Sun is 5.8×10^3 K and its radius is 6.96×10^8 m. Consider it as a blackbody radiator. What is the heat flow from the Sun via radiation?

Solution

Heat flow via radiation is found using Stefan's law, Equation 13.46:

$$\frac{dQ}{dt} = -eA\sigma T^4$$

For a blackbody radiator, the emissivity $e = 1$. The surface area of the Sun is that of a sphere of radius r:

$$\begin{aligned}
A &= 4\pi r^2 \\
&= 4\pi (6.96 \times 10^8 \text{ m})^2 \\
&= 6.09 \times 10^{18} \text{ m}^2
\end{aligned}$$

Make these substitutions into Stefan's law and use the value of the Stefan–Boltzmann constant $\sigma = 5.670 \times 10^{-8}$ W/(m²·K⁴); you find

$$\begin{aligned}
\frac{dQ}{dt} &= (-6.09 \times 10^{18} \text{ m}^2)[5.670 \times 10^{-8} \text{ W/(m}^2 \cdot \text{K}^4)] \\
&\quad \times (5.8 \times 10^3 \text{ K})^4 \\
&= -3.9 \times 10^{26} \text{ W}
\end{aligned}$$

The Sun is a rather powerful radiator! The minus sign indicates that the heat transfer per second (the heat flow) is *from* the Sun. The absolute value of the heat flow from the Sun via radiation is known as the *solar luminosity*.

EXAMPLE 13.19 ━━━━━━━━━━━━━━━━━

An object at an absolute temperature T_{object} finds itself in a region where the absolute temperature is only slightly greater:

$$T_{\text{surroundings}} = T_{\text{object}} + \Delta T$$

where ΔT is small compared with either T_{object} or $T_{\text{surroundings}}$. Show that the total heat flow to the object from its surroundings via radiative heat transfer is proportional to the *first* power of the temperature difference ΔT.

Solution

Use Equation 13.47. The total heat flow to the object via radiative heat transfer is

$$\frac{dQ}{dt} = eA\sigma \left[\left(T_{\text{object}} + \Delta T \right)^4 - T_{\text{object}}^4 \right]$$

Factor out T_{object} in the first term to obtain

$$\frac{dQ}{dt} = eA\sigma \left[T_{\text{object}}^4 \left(1 + \frac{\Delta T}{T_{\text{object}}} \right)^4 - T_{\text{object}}^4 \right] \qquad (1)$$

Now expand the expression

$$\left(1 + \frac{\Delta T}{T_{\text{object}}} \right)^4$$

using the binomial theorem:

$$(1 + x)^4 = 1 + 4x + 6x^2 + 4x^3 + x^4$$

If x is small, you need keep only the first two terms of the expansion:

$$(1 + x)^4 \approx 1 + 4x$$

Thus you have

$$\left(1 + \frac{\Delta T}{T_{\text{object}}} \right)^4 \approx 1 + 4\frac{\Delta T}{T_{\text{object}}}$$

Substituting this into equation (1) for the total heat flow, you obtain

$$\begin{aligned}
\frac{dQ}{dt} &= eA\sigma \left[T_{\text{object}}^4 \left(1 + 4\frac{\Delta T}{T_{\text{object}}} \right) - T_{\text{object}}^4 \right] \\
&= eA\sigma \left(4 T_{\text{object}}^3 \, \Delta T \right) \qquad (2)
\end{aligned}$$

If the surroundings are at a higher temperature, $\Delta T > 0$ K and the total heat flow is positive and represents heat flow to the object. Correspondingly, if the surroundings are at a cooler temperature, $\Delta T < 0$ K, the total heat flow is negative, indicating the object is an emitter radiating heat flow to the surroundings.

Equation (2) indicates that for small temperature differences between the object and its surroundings, the heat transfer per second (the heat flow via radiation processes) absorbed by the object from its surroundings is directly proportional to the temperature difference ΔT.

EXAMPLE 13.20 ━━━━━━━━━━━━━━━━━

The average human body temperature is 37 °C. If you assume that the human body is a blackbody radiator, estimate the total heat flow of the naked human body into a room at temperature 20 °C.

Solution

For a blackbody, the emissivity is $e = 1$. The temperature difference is

$$\begin{aligned}
\Delta T &= T_{\text{surroundings}} - T_{\text{object}} \\
&= -17 \text{ K}
\end{aligned}$$

The absolute temperature of the human body is about

$$\begin{aligned}
T_{\text{object}} &= 273.15 \text{ K} + 37 \text{ K} \\
&= 310 \text{ K}
\end{aligned}$$

We need to estimate the surface area of the human body. This might make an interesting experiment. Suppose we guesstimate the surface area as about 2.5 m².*

Using equation (2) from Example 13.19, you find the total heat flow is

$$\frac{dQ}{dt} = eA\sigma \left(4 T_{\text{object}}^3 \, \Delta T \right)$$

---

*Average values for the surface area of the human body as a function of height and weight can be found in *Geigy Scientific Tables*, edited by Cornelius Lentner (Medical Education Division, Ciba-Geigy Corp., West Caldwell, New Jersey, 1984), volume 3, page 329. The number used here corresponds to a body mass of 60 kg and a height of 1.80 m.

Making the appropriate substitutions,

$$\frac{dQ}{dt} \approx (2.5 \text{ m}^2)\,[(5.670 \times 10^{-8}(\text{W/m}^2 \cdot \text{K}^4)](4)(310 \text{ K})^3(-17 \text{ K})$$

$$\approx -2.9 \times 10^2 \text{ W}$$

The minus sign indicates that the total heat flow of the human is *from* the human (the system) to the cooler surrounding room.

13.12 THERMODYNAMIC PROCESSES

A thermodynamic process is any way that a system changes from one state of thermal equilibrium to another such state. Heat transfer and work are two processes that *change* the state of thermodynamic equilibrium of a system. For example, a gas may exist in thermal equilibrium at some specified temperature, pressure, and volume. Heat transfer and/or work can change these thermodynamic parameters to other values. To analyze the *way* the change occurs, we distinguish between two very different types of thermodynamic processes: quasi-static (reversible) processes and irreversible processes.

Quasi-static (Reversible) Processes

If a thermodynamic system undergoes a change from one state of thermal equilibrium to another slowly enough so at any instant the entire system essentially is in thermal equilibrium (though at different temperature, pressure, and/or volume), then we say the process is **quasi-static**. Strictly speaking, quasi-static processes do not exist: they are idealizations.

They are convenient fictions, much like the frictionless surfaces and ideal strings used so prolifically in classical mechanics. We use these artificial constructs to make calculations that *approximate* reality.

In order to *change* the temperature, pressure, and/or volume of the gas from some initial values to different final values, the system is going to have to depart from its initial state of thermal equilibrium. The critical factor here is the speed of the change. As long as the changes occur slowly enough so the whole system can adjust continuously to the change and pass through a succession of thermal equilibrium states, the process can be considered quasi-static. The meaning of slow enough depends on the particular thermodynamic system under consideration. For example, if a piston compressing a gas moves too quickly, causing turbulent eddies in the gas, then the process is not quasi-static. The point is to move the piston slowly enough during the compression so the gas can adjust to the change and, at any instant, be in a state of thermal equilibrium. Quasi-static processes are called **reversible processes** if it is assumed that at any instant the small differential changes in the system can be reversed.

With few exceptions, the thermodynamics we study in this text is **equilibrium thermodynamics**, which applies to *processes that are quasi-static and reversible* (we shall use the word quasi-static; it is understood that the processes also are reversible). We assume these conditions apply in our study of thermodynamics. Irreversible processes involve a more complicated thermodynamic formalism known as *nonequilibrium thermodynamics*.

There are four special types of quasi-static processes that are useful in equilibrium thermodynamics. The processes are characterized by the thermodynamic parameter that is kept constant during the process (while other parameters change):

a. **Isothermal process**: the *temperature* of the system is kept constant.

b. **Isobaric process**: the *pressure* of the system is kept constant.

c. **Isochoric process**: the *volume* of the system is kept constant.

d. **Adiabatic process** (also called an **isentropic process**): there is no heat transfer either to or from the system (the system is well insulated from the environment).

We study adiabatic processes in greater detail in Chapters 14 and 15.

Irreversible Processes

Any process that is not quasi-static and reversible is called **irreversible**.

In nature, all thermodynamic processes really are irreversible.

But to paraphrase the pigs in George Orwell's *Animal Farm*, some processes are more irreversible than others. For example, the bursting of a balloon (called a free expansion of a gas) is a classic example of an irreversible process (not quasi-static) and one to which we shall return later; the process does not spontaneously go the other way, nor does it occur slowly enough for the gas to be considered in a state of thermodynamic equilibrium at any instant during the expansion.

13.13 ENERGY CONSERVATION: THE FIRST LAW OF THERMODYNAMICS AND THE CWE THEOREM

As you know, energy is an important concept in physics. Here we formulate a unifying concept about **conservation of energy** and explore its relationship to something old, the CWE theorem from mechanics, and something new, the first law of thermodynamics.

The total energy of a system has two distinct contributions:

1. A macroscopic energy: the total *mechanical energy* E of the system, familiar to you from our study of classical mechanics (dynamics). The total mechanical energy is associated with the *macroscopic position and motion of the system as a whole*. The mechanical energy E is the sum of

   - the kinetic energy of translation of the center of mass point and the rotational kinetic energy about the center of mass point, as well as

   - potential energies associated with the position of the center of mass of the system in space, be they gravitational, elastic, or other potential energies yet to come (such as electrical potential energy).

Recall that the place where the zero of potential energy is located is arbitrary.* Once all the appropriate choices for the zeros of the potential energies are made, the system has a definite value for the total mechanical energy E when in a state of mechanical equilibrium. Because of this freedom of choice, it is only *changes* in the total mechanical energy that are physically significant when the system changes its state of mechanical equilibrium, say, through the action of forces on the system. In other words, the thing that is physically important is ΔE, not E itself, as reflected in the CWE theorem of Chapter 8.

2. The other contribution to the total energy is a vast collection of microscopic energies, known collectively as the **internal energy** U of the system. The internal energy is the sum of the individual kinetic and potential energies associated with the motions and interactions of all the individual particles (atoms and/or molecules) of the system itself. Many of these interactions involve rather complicated potential energy functions on a microscopic distance scale. In principle, after appropriate choices are made for the zeros of these microscopic potential energy functions, we can talk about a definite value for the internal energy of the system when it is in a state of thermodynamic equilibrium, but this is a rather complicated endeavor. When the system changes its state of thermodynamic equilibrium, it is only *changes* in the internal energy ΔU that are physically significant in thermodynamics. This is analogous to the situation in dynamics, where only changes in the total mechanical energy are physically significant. We will see that, with few exceptions (ideal gases, considered in Chapter 14), the *changes* in the internal energy are much easier to calculate than the actual value of the internal energy itself.

The total mechanical energy E of a system occasionally is referred to as *ordered* energy because of its macroscopic character. The internal energy U of a system is called *disordered* energy because of the microscopic nature of the motions of the system's many constituent particles.

Energy can be *transferred* to a system (from other systems) by two mechanisms or processes: (a) via heat transfer, instigated by temperature differences, and (b) via work, instigated by macroscopic forces. Let's look at each of these two processes in more detail.

Heat Transfer to a System

We have seen that heat transfer to a system can occur by three distinct mechanisms: conduction, convection, and radiation. For conduction and convection, the system gains energy by being brought into physical contact with another system at a higher temperature. For radiation, the transfer occurs via photons without the necessity of an intervening medium. Mechanisms for *heat transfer* are fundamentally microscopic (atomic and molecular) in character. The direction of the heat transfer always is from higher to lower temperatures; the reason for this unidirectional aspect of heat transfer will be explained when we investigate the consequences of the second law of thermodynamics (in Chapter 15).

---

*In some cases the zero is chosen at a particular location *by convention* (such as at $x = 0$ m for a mass on a spring, or at $r = \infty$ m for the general form of the gravitational potential energy).

> Do not confuse the concept of heat transfer with the internal energy.

Heat transfer occurs because of a temperature difference. Heat transfer is accomplished by random molecular collisions and other essentially molecular interactions (by conduction, convection, or radiation).

> As we have stated before, you *cannot* think of heat transfer as a fundamental property of the state of the system itself; a system in thermal equilibrium does *not* have an amount of heat, which is why we prefer not to use the word heat as a noun. On the other hand, a system in thermal equilibrium *can* be said to have a specific internal energy (at least in principle), just as the system can be said to have a definite temperature, pressure, and volume. The value of the internal energy thus is a characteristic of the state of the system itself. We distinguish between these concepts by saying that heat transfer Q is *not* a **state variable**; the internal energy U *is* a state variable, as are the temperature, pressure, and volume.

Physical Work Done *on* the System

Physical work W' done *on* the system can change the internal energy U as well as the total mechanical energy E of the system.

As an example of work resulting in a change of just the internal energy, consider a gas enclosed in an insulated container with a movable piston on one side, as shown in Figure 13.25. The insulation ensures that there is no heat transfer either to or from the gas (the system). An external agent (you!) compresses the gas by pushing on the piston, much like using a bicycle pump, thus doing work on the system.

The work done *on* the system (the gas) by the external agent (you) results in an increase in the temperature of the system, indicative of an increase in the internal energy of the gas. Although we might say, colloquially, that the ideal gas "heats up" (since its temperature increases), the increase in the internal energy and the rise in the temperature of the system do *not* come about because of heat transfer, but because of macroscopic work done on the system by the external agent.

It also is possible for the work done *on* the system to change the mechanical energy E of the system: this is just the CWE theorem of mechanics. We will return to explore this relationship again shortly. Thus the work done on the system is a *macroscopic* mechanism for transferring energy to a system; heat transfer to a system is fundamentally a *microscopic* mechanism for transferring energy to a system.

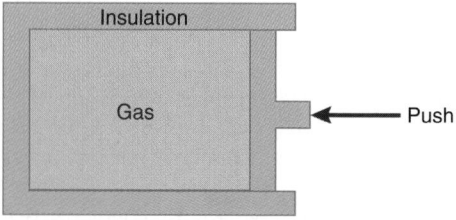

FIGURE 13.25 Compressing a gas by pushing on a piston.

Although heat transfer and work are quite distinct physical ideas, they share several common features: (1) both represent energy transfer to a system; and (2) neither is a state variable. Just as a system in thermal equilibrium cannot be said to have an amount of heat, the system in equilibrium cannot be said to have an amount of work. Once a system has gained the energy (either mechanical energy E or internal energy U) provided by either heat transfer or work processes, there is no way that the system can tell from what process or source the energy came. The energy transferred to the system by heat transfer Q and work W' goes into a common, dichotomous energy pool: the internal energy and the total mechanical energy of the system.

The situation is similar to your bank account. You can deposit money (like joules) into your checking or your saving account (like internal energy and mechanical energy) by two processes: via cash or checks (like heat transfer and work). Once the money is in each account, the account balance cannot distinguish between the means by which the money arrived; the dollars (and likely you!) care not how they got there. But the internal energy *can* be distinguished from the total mechanical energy (your checking account is distinct from your savings account), since the former is microscopic while the latter is macroscopic in character.

A system may gain energy via other mechanisms too—for example, by chemical or nuclear reactions or other, more exotic mechanisms.* Some of these alternative processes for the addition of energy are fundamentally microscopic and some are macroscopic; the former exotic forms of energy transfer to the system are considered as heat transfer, and the latter as work.

Fundamental Energy Conservation Law

We gather these ideas about energy into a fundamental *energy conservation law*. Energy is transferred *to* a system by (1) heat transfer Q and (2) work W' done *on the system by external* macroscopic, *nonconservative* forces. The total energy transfer to the system $(Q + W')$ appears as a *change* in the internal energy, ΔU, and/or as a *change* in the total mechanical energy of the system, $\Delta E \, (= \Delta KE + \Delta PE)$. Any work done by macroscopic *conservative* forces is already present in the total mechanical energy term ΔE, since this term includes changes in the potential energies associated with conservative forces. Thus the W' here is the work done by what we called (in Chapter 8) "other" macroscopic, nonconservative forces acting on the system.

Therefore our fundamental energy conservation law is

$$Q + W' = \Delta U + \Delta E \qquad (13.48)$$

This is the most general statement of energy conservation in classical physics.

Let us see how this statement about energy conservation relates to what we learned in "pure" classical mechanics and how it specializes to the "pure" thermodynamics of *simple* thermodynamic systems.

When we studied pure classical mechanics in Chapters 1–12, we *neglected* so-called thermal effects. That is, we restricted ourselves to situations where (1) there was no heat transfer to the system ($Q = 0$ J) *and* (2) no change in the internal energy ($\Delta U = 0$ J).

Recall that the CWE theorem states that total work done by *all* forces acting *on* the system is equal to the change in the macroscopic kinetic energy of the system:

Work done by all forces acting *on* the system $= \Delta KE$

The forces were separated into those of conservative and nonconservative type, and the work done by each also segregated so that

Work done by conservative forces *on* the system	$+$ Work done by other forces *on* the system (nonconservative forces) (this is W' above)	$= \Delta KE$

The work done by the conservative forces was defined to be the *negative* of the change in the potential energies associated with those conservative forces:

Work done by conservative forces *on* the system $= \, - \Delta PE$

and so

Work done by other forces acting on the system $= \Delta KE + \Delta PE$ (nonconservative forces, i.e., W')

$$W' = \Delta KE + \Delta PE$$

or

$$W' = \Delta E \qquad (13.49)$$

where E is the total mechanical energy of the system.

We now make the connection with the general energy conservation statement of Equation 13.48. With thermal effects neglected ($Q = 0$ J and $\Delta U = 0$ J), the most general statement of energy conservation, Equation 13.48, reduces to the CWE theorem of Equation 13.49. Section 13.14 explores the connection between the general statement (Equation 13.48) of energy conservation and the CWE theorem via a specific example.

In pure thermodynamics (the thermodynamics of *simple* systems), we consider only systems whose total mechanical energy does *not* change: $\Delta E = 0$ J. Therefore macroscopic conservative forces (with their associated potential energy terms) are irrelevant. The forces doing work on the system are then just "other" (nonconservative) forces acting on the system that do work W', such as the work done by the force we use to compress the gas in Figure 13.25. Thus the general statement of energy conservation (Equation 13.48) becomes (since $\Delta E = 0$ J)

$$Q + W' = \Delta U \qquad (13.50)$$

In pure thermodynamics, *it is useful to change the reference regarding the work term.* Instead of considering the work done *on* the system *by* surrounding external agents, as we did in mechanics, in pure thermodynamics we are more interested in the work done *by* the system *on* its surrounding environment.

---

*Recall that we restricted our treatment of thermodynamics in this text to *simple thermodynamic systems*, excluding by definition these more exotic mechanisms for transferring energy to the system.

Forces always appear in pairs (according to Newton's third law of motion). Each member of a force pair has the same magnitude, but the pair are in opposite directions and act on different systems. Hence the work W' done *on* the system by the surroundings is the *negative* of the work W done *by* the system *on* the surroundings:

$$\text{Work done } on \text{ the system} \atop by \text{ the surroundings} = \text{— Work done } by \text{ the system} \atop on \text{ the surroundings}$$

$$W' = -W$$

Whoever thought prepositions could be so important in physics, of all places!*

Substitute $-W$ for W' in Equation 13.50.

Therefore, for pure thermodynamic systems, we have

$$Q - W = \Delta U$$

or

$$\boxed{Q = \Delta U + W} \qquad (13.51)$$

where W *is the work done by the system on its surroundings.* Equation 13.51 is the **first law of thermodynamics** for simple thermodynamic systems.

There are several things to note carefully about the statement of the first law of thermodynamics:

1. Stripped of everything else, the first law of thermodynamics is a statement of conservation of energy for pure (i.e., simple) thermodynamic systems.

2. The equation assumes there are no changes in the total mechanical energy E of the system. That is, there are no changes in the sum of the macroscopic kinetic and potential energies of the system. In this and the next two chapters on thermodynamics, we will only consider simple thermodynamic systems that have $\Delta E = 0$ J, so that we can use $Q = \Delta U + W$. The restriction $\Delta E = 0$ J is what we mean by pure thermodynamics. Essentially, the first law of thermodynamics says that heat transfer Q *to* the system can change the internal energy of the system and/or the system can do work on its surroundings.

3. The heat transfer Q is positive for heat transfer *to* the system; W is the work done *by* the system on its surrounding environment; and ΔU is the change in the internal energy of the system. The work term in the first law is the *negative* of the work done by "other" nonconservative forces, W', that we used in the CWE theorem of mechanics.

4. Since the internal energy U is a state variable and the heat transfer Q and work W are not state variables, it may seem remarkable that the change in the internal energy is equal to the difference between the heat transfer to the system and the work done by the system. As the system changes from one state of thermodynamic equilibrium (an initial state) to another such state (a final state), we shall see that both the heat transfer to the system and the work done by the system depend on the particular way the system interacts with its

environment in executing the change from the initial to the final state (such as via one of the specific types of processes mentioned in Section 13.12). But the change in the internal energy of the system ΔU is *independent of the way the change is made*. Once the change is accomplished, the system has no memory of how the energy was acquired or lost: via heat transfer or via work.

5. You should be aware that the sign convention for the work term in the first law of thermodynamics (Equation 13.51) is not universal; you may encounter other texts[†] that treat the work done *on* the system as positive, rather than the work done *by* the system. The convention we use is clearest for an introduction to thermodynamics. Regrettably, like politics, there is a certain amount of flip-flopping in any field.

QUESTION 9

Energy conservation is an important physical principle in physics. When environmentalists talk about energy conservation, is the term being used in the same sense as it is used in physics? Explain your answer.

13.14 THE CONNECTION BETWEEN THE CWE THEOREM AND THE GENERAL STATEMENT OF ENERGY CONSERVATION

For greater insight into the relationship between the general statement of energy conservation (Equation 13.48) and the specialized CWE theorem of Chapter 8 (Equation 13.49), we need to consider a system that is *not* a simple thermodynamic system, as we defined it in Section 13.1, but one that also has a change in its total mechanical energy. An inelastic collision meets this criterion.

In particular, consider a mass m, initially traveling at speed v on a frictionless surface, making a completely inelastic collision with an initially stationary mass M; see Figure 13.26. Both particles initially are at the same temperature and in thermal equilibrium with their environment. Take the system to be the two masses.

Total momentum is conserved in collisions. Hence the total momentum of the particles before the collision is equal to the total momentum after the collision:

$$\vec{P}_{\text{total bfr}} = \vec{P}_{\text{total aft}}$$
$$mv\hat{i} = (M + m)\vec{v}'$$

---

† Some formal, advanced texts in thermodynamics make the opposite choice for the sign convention of the work term in the first law of thermodynamics.

---

*Such prepositions were critical in studying forces in mechanics, as you know!

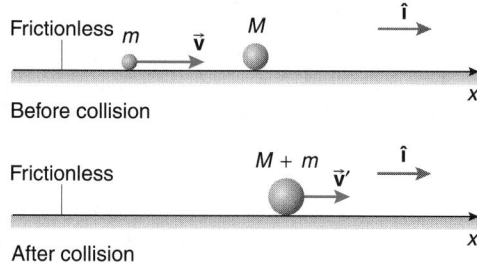

FIGURE 13.26 An inelastic collision of two particles.

where $\vec{\mathbf{v}}'$ is the velocity of the composite particle after the collision. We solve for $\vec{\mathbf{v}}'$:

$$\vec{\mathbf{v}}' = \frac{m}{M + m} v\hat{\mathbf{i}}$$

Kinetic energy is not conserved in such inelastic collisions, as we learned in Chapter 9. The *change* in the kinetic energy of the system is

$$\Delta KE = KE_{aft} - KE_{bfr}$$
$$= \frac{1}{2}(M + m)(v')^2 - \frac{1}{2}mv^2$$

Substituting for v' and simplifying, we find

$$\Delta KE = -\left(\frac{M}{m + M}\right)\frac{1}{2}mv^2 \qquad (13.52)$$

The change in the kinetic energy is negative: the kinetic energy of the system after the collision is less than it was before the collision.

We now can determine what happens to the missing kinetic energy. The CWE theorem of classical mechanics is of no help in solving this riddle of the missing kinetic energy. Indeed, if we apply the CWE theorem to this problem, we obtain a rather puzzling result. Recall that the CWE theorem states that the work done by *all* the nonconservative forces on the system is equal to the change in the total mechanical energy of the system:

$$W'_{all} = \Delta E$$

The forces acting on the system are (see Figure 13.27)

1. the weights of the two masses, and
2. the normal forces of the surface on the masses.

The force of m on M and the force of M on m during the brief time interval of the collision are *internal* to the system. They form a Newton's third law force pair, and together do zero total work.

The work done by the gravitational forces on the masses is zero since these forces are perpendicular to the change in the position vectors of the two masses as they collide; thus there is no change in the gravitational potential energy. The work done by the normal forces is zero since they are zero-work forces (normal forces, when they exist, always are perpendicular to any change in the position vector). Therefore the work done on the system by nonconservative forces is zero; but the change in the kinetic energy is negative. The CWE theorem (Equation 13.49) implies that zero is equal to a negative number! What is going on?

The resolution of the dilemma lies not with the CWE theorem, but with the general statement of energy conservation, Equation 13.48. The CWE theorem is a *special case* of this general statement of energy conservation that ignores thermal effects. Such thermal effects cannot be ignored here, and so the CWE theorem does not apply.

To resolve the dilemma of the missing kinetic energy, let's apply the general statement of energy conservation, Equation 13.48, to this problem:

$$Q + W' = \Delta U + \Delta E$$

Since both particles are initially at the same temperature and in thermal equilibrium with their environment, there is no heat transfer Q to the system: $Q = 0$ J. The quantity W' is the total work done on the system by all the nonconservative forces on the system, here just the normal force; but the normal force is a zero-work force, and so $W' = 0$ J. The work done by conservative forces is built into the change in total mechanical energy term via changes in the potential energy terms (here zero). The quantity ΔU is the change in the internal energy of the system (the change in the collective sum of all the microscopic kinetic and potential energies of the myriad particles making up the system). The change in the total mechanical energy of the system is ΔE and this represents

$$\Delta E = \Delta KE + \Delta PE$$

There was no change in the (macroscopic) potential energy of the system in the collision, indicating that the gravitational forces do no work here; the potential energy here is the gravitational potential energy, which is constant because the two masses do not change their vertical coordinates during the brief collision. Therefore the change in the mechanical energy is just the change in the kinetic energy:

$$\Delta E = \Delta KE + PE$$
$$= \Delta KE + 0 \text{ J}$$

Making these substitutions into Equation 13.48, we have

$$0 \text{ J} + 0 \text{ J} = \Delta U + \Delta KE$$

Hence

$$\Delta U = -\Delta KE$$

Since the change in the kinetic energy is negative for our problem (see Equation 13.52), the change in the internal energy of the system is positive.

In other words, the missing kinetic energy appears as an increase in the internal energy of the system. The increase in the internal energy of the system typically manifests itself as an increase in the temperature of the system, but the increase in the internal energy also could be manifested by a phase change (e.g., melting) with no change in temperature. The increase in temperature (or phase change) occurs *as if* heat transfer to the system occurred, although *no heat transfer* actually does occur to the system of particles.

As the temperature of the system increases above the ambient environment because of the increase in internal energy, heat transfer *then* occurs from the system to the environment until they reach a common temperature.

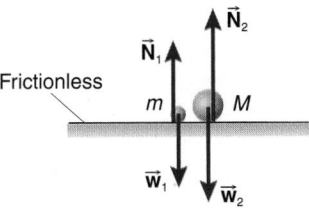

FIGURE 13.27 External forces on the system of colliding particles.

Thus we see that the CWE theorem *cannot* be applied willy-nilly to all systems; the theorem has limitations. It is a special case of the more fundamental statement of energy conservation, Equation 13.48.

13.15 WORK DONE BY A SYSTEM ON ITS SURROUNDINGS

It is always nice when something does work for us! The work done by a system such as a gas in a cylinder confined by a movable piston has enormous ramifications for our personal mobility, since it is the basis of the internal combustion engine. In this section we see how to calculate the work done *by* a system, one of the terms in the first law of thermodynamics, Equation 13.51.

Consider a gas at pressure P confined to a volume V by a movable piston, as shown in Figure 13.28. Let the temperature of the gas increase slightly, so that the gas expands, moving the piston a differential distance dx.

The differential work dW done *by* the gas *on* the piston is, from the definition of work,

$$dW = \vec{F} \cdot d\vec{r}$$

where $\vec{F}$ is the force of the gas on the piston and $d\vec{r}$ is the change in the position vector of the center of mass of the piston. The force $\vec{F}$ is provided by the pressure of the gas. With the coordinate system indicated in Figure 13.28, the force is

$$\vec{F} = PA\,\hat{\imath}$$

where A is the area of the piston. Recall that the change in the position vector $d\vec{r}$ always is written (in one dimension) as

$$d\vec{r} = dx\,\hat{\imath}$$

because the limits of integration account for the direction the piston actually moves along the x-axis. The differential work done by the gas is then

$$dW = PA\,\hat{\imath} \cdot dx\,\hat{\imath}$$
$$= PA\,dx$$

The product $A\,dx$ is the change in volume dV of the gas. Thus the differential work done by the gas is

$$dW = P\,dV \qquad (13.53)$$

The work done by the gas is *positive* for *increases* and *negative* for *decreases* in the volume of the gas. For finite changes in the volume, we integrate Equation 13.53.

The work done by the gas is

$$W = \int_{V_i}^{V_f} P\,dV \qquad (13.54)$$

Although we formulated this relationship for the work done by the gas using a one-dimensional change, Equation 13.54 is a general result. Since the arguments used to derive it could equally well apply to changes in the volume of liquids or solids, Equation 13.54 also can be used to calculate the work done by solids and liquids if they experience a change in volume caused by (1) thermal expansion or (2) changes of phase.

The pressure and volume of the gas are conveniently plotted on a P–V diagram. A gas in thermodynamic equilibrium is represented by a point on such a diagram. Let the gas, initially with volume V_i and pressure P_i, change quasi-statically to another volume V_f and pressure P_f. There are many ways (thermodynamic processes) that the gas can travel between its initial and final states through other states of thermodynamic equilibrium on the P–V diagram. We represent one such way schematically in Figure 13.29.

Whatever quasi-static path the gas follows from the initial state to the final state, the work W done by the gas has a convenient geometrical interpretation on a P–V diagram:

The work done by the gas is the area under the path on the P–V diagram. The area under the path on the P–V diagram has the dimensions of pressure times volume or, using SI units, joules:

$$\text{Pa} \cdot \text{m}^3 = (\text{N/m}^2) \cdot \text{m}^3 = \text{N} \cdot \text{m} = \text{J}$$

PROBLEM-SOLVING TACTIC

13.6 Caution! Be sure the volume of the gas is in cubic meters and the pressure in pascals when calculating work (in joules) using the geometric approach on a P–V diagram. Many P–V graphs have the pressure in atmospheres and the volume in liters; if this is the case, then appropriate unit conversions to pascals and cubic meters must be made, so that the area under the curve emerges as joules.

Let's investigate several special paths connecting an initial and final state on the P–V diagram. These paths represent various special thermodynamic processes. We want to find the work done by the gas in each process.

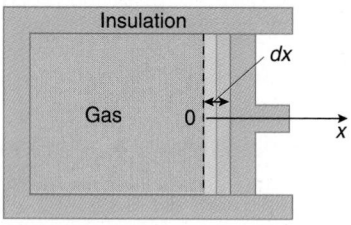

FIGURE 13.28 A gas confined to a cylinder with a movable piston. The gas expands slightly, moving the piston.

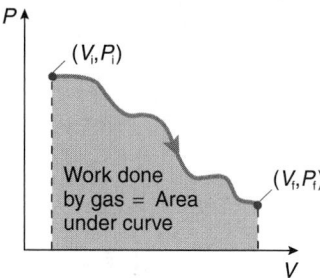

FIGURE 13.29 A gas changes from one state of thermodynamic equilibrium to another.

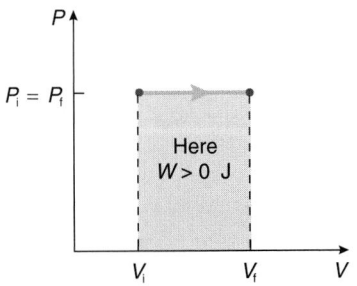

FIGURE 13.30 An isobaric process.

Work Done by the Gas in an Isobaric Process

Let the gas travel along a horizontal path on the *P–V* diagram, as shown in Figure 13.30. A horizontal path on a *P–V* diagram is an isobaric process (meaning constant pressure).

The work done *by* the gas is found from Equation 13.54. Since the pressure is constant (and equal to P_i) along an isobaric path, the integral becomes

$$W = P_i \int_{V_i}^{V_f} dV$$

Now we perform the integration.

The work done by the gas is

$$W = P_i(V_f - V_i)$$
$$= P_i \Delta V \qquad \text{(isobaric process)} \qquad (13.55)$$

Notice that if the gas expands, $V_f > V_i$, so that $\Delta V > 0$ m³ and the work done *by* the gas is *positive*. For a gas that is compressed, $V_f < V_i$, so that $\Delta V < 0$ m³ and the work done *by* the gas is *negative*. The work done *by* the gas is represented on the *P–V* diagram by the rectangular area under the isobaric path on the diagram. The area is positive or negative depending on whether the gas expands or is compressed.

Work Done by a Gas in an Isochoric Process

An isochoric process means constant volume. This is represented as a vertical line on a *P–V* diagram, as shown in Figure 13.31.

Since the volume does not change, Equation 13.54 indicates that the work done *by* the gas in such a process is *zero*. Notice that along an isochoric path segment there is zero area under the path on the *P–V* diagram. It makes no difference whether the

pressure change is positive or negative: the work done by the gas in an isochoric process is zero.

For an isochoric process, there is zero work done by the gas:

$$W = 0 \text{ J} \qquad \text{(isochoric process)} \qquad (13.56)$$

Work Done by a Gas in an Isothermal Process

In an isothermal process (constant temperature), *both* the pressure and volume change along the path. The initial and final pressures and volumes then are located along the same isotherm on the *P–V* diagram of the gas, indicated in Figure 13.32.

To calculate the work done by the gas, we begin again with Equation 13.54. The equation of state of the gas relates its pressure, volume, and temperature. For an *ideal* gas, the equation of state is the ideal gas law, Equation 13.20:

$$PV = nRT$$

We solve for the pressure:

$$P = \frac{nRT}{V}$$

Thus the work done by the gas is

$$W = \int_{V_i}^{V_f} P \, dV$$
$$= \int_{V_i}^{V_f} \frac{nRT}{V} \, dV$$

The number of moles n and the gas constant R are constants, and since the process is isothermal, the absolute temperature T also is constant. Hence we have

$$W = nRT \int_{V_i}^{V_f} \frac{dV}{V}$$
$$= nRT(\ln V) \Big|_{V_i}^{V_f}$$

Hence the work done by an ideal gas in an isothermal process is

$$W = nRT \ln \frac{V_f}{V_i} \qquad \text{(ideal gas isothermal process)} \qquad (13.57)$$

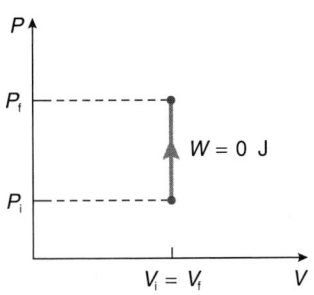

FIGURE 13.31 An isochoric process.

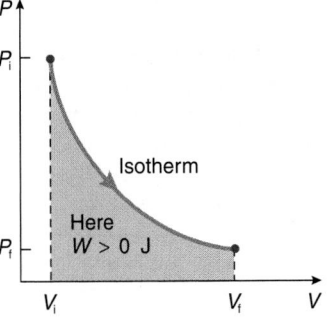

FIGURE 13.32 An isothermal process.

If $V_f > V_i$, the gas expanded and the work done by the gas is positive. Conversely, if the gas is compressed, the work done is negative, since the logarithm then is of a number less than 1.

As before, the work done by the gas is the area under the path on the *P–V* diagram and is positive or negative depending on whether the gas expands or is compressed.

EXAMPLE 13.21

With the exit valve of a tire pump blocked, you compress a gas isothermally at 300 K from an initial volume of 0.50 liter (at 1.00 atm pressure) to a final volume of 0.25 liter. What is the work done by the gas during its compression? Explain the sign of your answer.

Solution

The work done in an isothermal process is given by Equation 13.57:

$$W = nRT \ln \frac{V_f}{V_i}$$

Everything but the quantity of gas present is known. Use the ideal gas law, Equation 13.20, to determine the number of moles of the gas from the initial conditions:

$$PV = nRT$$

Solving for *n*, you have

$$n = \frac{PV}{RT}$$
$$= \frac{(1.00 \ \text{atm})(1.013 \times 10^5 \ \text{Pa/atm})(0.50 \times 10^{-3} \ \text{m}^3)}{[8.315 \ \text{J/(mol·K)}](300 \ \text{K})}$$
$$= 2.0 \times 10^{-2} \ \text{mol}$$

Note that the volume ratio $V_f/V_i = 0.50$. The work done by the gas then is

$$W = (2.0 \times 10^{-2} \ \text{mol})[8.315 \ \text{J/(mol·K)}](300 \ \text{K})(\ln 0.50)$$
$$= -35 \ \text{J}$$

The work done *by* the gas is negative; the negative result indicates that positive work was done *on* the gas by the agent (you) that compressed it.

13.16 WORK DONE BY A GAS TAKEN AROUND A CYCLE

Engines involve taking a gas around a closed path on a *P–V* diagram (see Figure 13.33), so that the process can be repeated many times with useful work done around each closed path. A closed curve on a *P–V* diagram is called a **cycle**, since the gas returns to its initial state (the initial point on the *P–V* diagram, regardless of where the process begins along the cycle).

Let the gas go around the cycle in the clockwise sense in Figure 13.33. As the gas expands from V_1 to V_2 in the figure, the work W_1 done *by* the gas is positive; as we have seen, this work is represented by the area under the curve on the *P–V* diagram. When the gas is compressed from volume V_2 back to V_1, along the lower part of the cycle, the work W_2 done *by* the gas is negative. This work is represented geometrically by the now negative area under the curve on the *P–V* diagram, as in Figure 13.34.

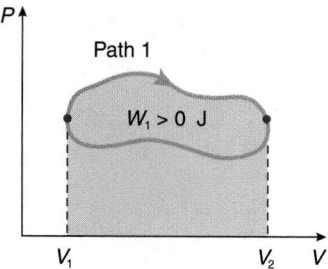

FIGURE 13.33 Positive work is done by the gas in expanding from volume V_1 to V_2.

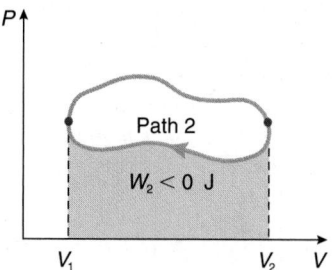

FIGURE 13.34 Negative work is done by the gas when compressed from volume V_2 back to V_1.

Hence the *total work* $W_{\text{total}} = W_1 + W_2$ done *by* the gas in executing the closed curve is represented by *the area enclosed by the cycle curve* on the *P–V* diagram.

If the cycle is completed in a clockwise manner, as in Figure 13.35, $W_{\text{total}} > 0$ J.

Since the gas has returned to the same place on the *P–V* diagram, the internal energy of the gas is *unchanged* by the whole procedure; that is, $\Delta U = 0$ J. The first law of thermodynamics, Equation 13.51, states that the total heat transfer Q_{total} to the system is equal to the sum of the change ΔU in the internal energy of the system and the work W_{total} done by the system:

$$Q_{\text{total}} = \Delta U + W_{\text{total}}$$

Because the change in the internal energy is zero for the complete cycle, the first law becomes

$$Q_{\text{total}} = W_{\text{total}} \qquad \text{(complete cycle)} \qquad (13.58)$$

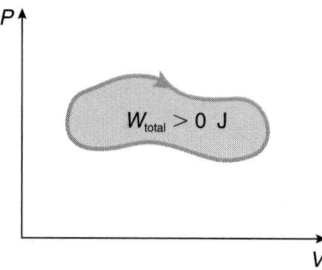

FIGURE 13.35 The total work done during the cycle is the area of the cycle on the *P–V* diagram and is positive if the path around the cycle is clockwise.

This means that the *total heat transfer* to the system is equal to the *total work* done by the system in the cycle. For the clockwise path indicated in Figure 13.35, the total work done *by* the system is positive; thus the total heat transfer around the cycle is positive (i.e., heat transfer was *to* the system).

QUESTION 10 _____

What can be said about the work done *by* a gas (and the heat transfer to a gas) if the gas is taken counterclockwise around a cycle on a P–V diagram?

EXAMPLE 13.22 _____

A gas is taken clockwise around a circular path on its P–V diagram between the volumes and pressures shown in Figure 13.36. What is the total work done by the gas around one cycle?

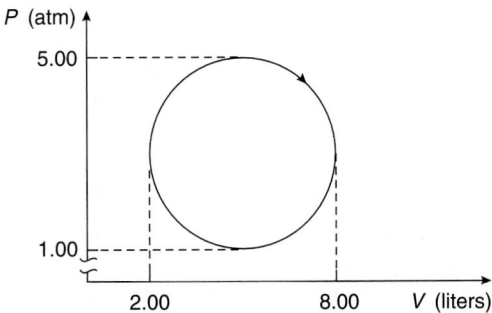

FIGURE 13.36

Solution

The total work done by the gas is the area of the cycle on the P–V diagram. Since the path is clockwise, the total work is positive. The area of a circle is π times the square of the radius. Use a pressure for one radius and a volume for the other, as shown in Figure 13.37.

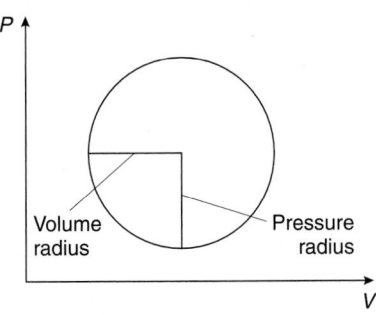

FIGURE 13.37

In order for the work to emerge in joules, the pressure must be expressed in pascals and the volume in cubic meters. Hence the total work done is

$$W_{\text{total}} = \pi(2.00 \text{ atm} \times 1.013 \times 10^5 \text{ Pa/atm})$$
$$\times (3.00 \text{ liters} \times 10^{-3} \text{ m}^3/\text{liter})$$
$$= 1.91 \times 10^3 \text{ J}$$

13.17 APPLYING THE FIRST LAW OF THERMODYNAMICS: CHANGES OF STATE

Here we explore the implications of the first law of thermodynamics for changes of state. When you boil water or melt ice, how much of the heat transfer to the water goes into work by the water (because of its expansion) and how much into changing its internal energy?

Vaporization (Boiling)

To change a system of mass m of liquid to vapor, the heat transfer Q to the system is m times the latent heat of vaporization L_v, as you learned earlier (Equation 13.29):

$$Q = mL_v$$

The process takes place at constant pressure (typically one atmosphere), and so the work done *by* the system is the work in an isobaric process (Equation 13.55):

$$W = P \Delta V$$

where ΔV is change in the volume—that is, the difference between the final volume (the volume of the vapor) V_{vapor} and the initial volume (the volume of the liquid) V_{liquid}:

$$W = P(V_{\text{vapor}} - V_{\text{liquid}})$$

Notice that this work is positive since $V_{\text{vapor}} > V_{\text{liquid}}$.

The first law of thermodynamics, Equation 13.51, states that

$$Q = \Delta U + W$$

Thus the change in the internal energy of the system is

$$\Delta U = Q - W$$

or

$$\Delta U = mL_v - P(V_{\text{vapor}} - V_{\text{liquid}})$$

To get a feeling for the relative numerical values of these quantities, consider the vaporization of 1.00 g = 1.00×10^{-3} kg of water at 100 °C to steam at 100 °C. The heat transfer needed is $Q = mL_v$, where the latent heat of vaporization is found in Table 13.5:

$$Q = (1.00 \times 10^{-3} \text{ kg})(22.57 \times 10^5 \text{ J/kg})$$
$$= 2.26 \times 10^3 \text{ J}$$

To calculate the work done *by* the system in the vaporization, we need to find the volume of 1.00×10^{-3} kg of steam at 100 °C. We use the ideal gas law (Equation 13.20):

$$PV = nRT$$

$$V = \frac{nRT}{P}$$

A mole of water (H_2O) has a mass of 18 g. So 1.00×10^{-3} kg = 1.00 g represents 1/18 mol = 0.0556 mol. The pressure is one atmosphere:

$$P = 1.000 \text{ atm}$$
$$= 1.013 \times 10^5 \text{ Pa}$$

and the absolute temperature at which the vaporization takes place is $T = 373$ K. The gas constant has the value $R = 8.315$ J/(mol·K). We substitute into the ideal gas law to find the volume of the vapor:

$$V_{vapor} = \frac{nRT}{P}$$

$$= \frac{(0.0556 \text{ mol})[8.315 \text{ J/(mol·K)}](373 \text{ K})}{1.013 \times 10^5 \text{ Pa}}$$

$$= 1.70 \times 10^{-3} \text{ m}^3$$

Since the density of water is 1.00×10^3 kg/m^3 = 1.00 g/cm^3, the volume of the liquid (1.00 g of water) is 1.00 cm^3, or

$$V_{liquid} = 1.00 \times 10^{-6} \text{ m}^3$$

Thus the work done *by* the water system in the vaporization is

$$W = P(V_{vapor} - V_{liquid})$$

where P is one atmosphere. That is, the work is

$$W = (1.013 \times 10^5 \text{ Pa})(1.70 \times 10^{-3} \text{ m}^3 - 1.00 \times 10^{-6} \text{ m}^3)$$

$$= 172 \text{ J}$$

The work done *by* the system is positive since the volume of the system increased.

According to the first law of thermodynamics,

$$Q = \Delta U + W$$

or

$$2.26 \times 10^3 \text{ J} = \Delta U + 172 \text{ J}$$

Thus the change in the internal energy of the system is

$$\Delta U = 2.09 \times 10^3 \text{ J}$$

Notice that most of the heat transfer went into the change in the internal energy of the system; indeed, 92% went to change the internal energy and only 8% went into work done by the system on its surroundings. In vaporization, much energy is needed to weaken the intermolecular bonds within the liquid.

Melting

This process is similar to the vaporization process. The heat transfer needed to melt a mass m is the product of the mass and the latent heat of fusion (Equation 13.29):

$$Q = mL_f$$

The work done *by* the system (since the process is accomplished at constant pressure) is

$$W = P(V_{liquid} - V_{solid})$$

The first law of thermodynamics states that

$$Q = \Delta U + W$$

and so the change in the internal energy of the system is

$$\Delta U = Q - W$$

$$= mL_f - P(V_{liquid} - V_{solid})$$

Once again, let us get a feeling for the sizes of the various terms by considering the melting of 1.00 g = 1.00×10^{-3} kg of ice at 0 °C to liquid water at 0 °C. The heat transfer needed to the ice is

$$Q = mL_f$$

where the latent heat of fusion of water is found in Table 13.5:

$$Q = (1.00 \times 10^{-3} \text{ kg})(3.335 \times 10^5 \text{ J/kg})$$

$$= 334 \text{ J}$$

To calculate the work, we need to know the difference in volume between 1.00 g of ice and 1.00 g of water. We can get this from the densities. The density of ice is 920 kg/m^3, and so 1.00×10^{-3} kg of ice has a volume of

$$V_{solid} = \frac{1.00 \times 10^{-3} \text{ kg}}{920 \text{ kg/m}^3}$$

$$= 1.09 \times 10^{-6} \text{ m}^3$$

The density of water at 0 °C is 1.00×10^3 kg/m^3, so that the volume of one gram of water is

$$V_{liquid} = 1.00 \times 10^{-6} \text{ m}^3$$

Thus the work done *by* the system in melting is

$$W = P(V_{liquid} - V_{solid})$$

$$= (1.013 \times 10^5 \text{ Pa})(1.00 \times 10^{-6} \text{ m}^3 - 1.09 \times 10^{-6} \text{ m}^3)$$

$$= -9 \times 10^{-3} \text{ J}$$

Notice that the work done *by* the system here is negative since the system *decreased* its volume in melting (the ice-to-water phase transition is one of just a few that have this characteristic). The work done by the system is quite small compared with the heat transfer to the system. Indeed the absolute value of the work done by the system is only about 0.003% of the heat transfer to melt the ice.

The first law of thermodynamics states that

$$Q = \Delta U + W$$

The change in the internal energy of the system is

$$\Delta U = Q - W$$

$$= 334 \text{ J} - (-9 \times 10^{-3} \text{ J})$$

The work term is very small compared with Q, and so

$$\Delta U \approx 334 \text{ J}$$

Essentially all of the heat transfer to the system in the melting goes into changing the internal energy of the system, rather than work done by it. This is a reasonable result since the change in the volume of the system in melting is much smaller than for vaporization.

CHAPTER SUMMARY

The fundamental SI unit for measuring temperature is the *kelvin* (K). The temperature of the *triple point* of water is defined to be 273.16 K, and so the size of a kelvin (degree) is 1/273.16 of the temperature of the triple point of water.

A temperature scale of convenience is the *celsius scale*. Temperatures t_{celsius} on the celsius scale are related to those on the kelvin scale T by

$$T = t_{\text{celsius}} + 273.15 \text{ K} \tag{13.2}$$

The size of a celsius degree is equal to the size of a kelvin, and so temperature *changes* have the same numerical value using both temperature scales: $\Delta T = \Delta t_{\text{celsius}}$.

Two systems at the same temperature are said to be in *thermal equilibrium* with each other.

Heat is defined as that (energy) which is transferred between two systems simply because they are at *different temperatures*. To emphasize that heat really is a process between two systems rather than a characteristic property of either system, we prefer to use the term *heat transfer* instead of the single word heat as a noun. By convention, heat transfer Q *to* a system is considered positive; heat transfer *from* a system is negative.

The *zeroth law of thermodynamics* states that if system A and system B are each in thermal equilibrium with a third system C, then systems A and B are in thermal equilibrium with each other.

A temperature change causes a change in the size of materials. A change in length $\Delta \ell$ is found using

$$\Delta \ell = \alpha \ell \, \Delta T \tag{13.6}$$

where ℓ is the original length of the material and α is the temperature *coefficient of linear expansion*. The temperature *coefficient of area expansion* is twice the coefficient of linear expansion, so that changes in area are found using

$$\Delta A = (2\alpha)A \, \Delta T \tag{13.9}$$

The *coefficient of volume expansion* is three times the coefficient of linear expansion; thus, changes in volume are found using

$$\Delta V = (3\alpha)V \, \Delta T \tag{13.10}$$

Holes exhibit thermal expansion as if the hole were filled with the material that surrounds the hole.

An *ideal gas* has an *equation of state* that relates the pressure, volume, absolute temperature, and the quantity of gas:

$$PV = nRT \tag{13.20}$$

where n is the number of moles of gas and R is the *universal gas constant*:

$$R = 8.314 \ 52 \text{ J/(mol·K)} \tag{13.19}$$

Real gases are approximately ideal gases at low pressures (up to several atmospheres).

Heat transfer Q to a system (expressed in joules) can change the temperature of the system and/or its phase. If no phase change occurs, the change in temperature is found using

$$Q = mc \, \Delta T \tag{13.22}$$

where c is the *specific heat* of the material:

$$c \equiv \frac{1}{m}\frac{dQ}{dT} \tag{13.21}$$

You also may use the molar specific heat:

$$c_{\text{molar}} = \frac{1}{n}\frac{dQ}{dT} \tag{13.24}$$

in which case the heat transfer to the system is

$$Q = nc_{\text{molar}} \, \Delta T \tag{13.23}$$

Heat transfer to a system at its freezing or boiling temperature changes the *phase* of the material without a change in its temperature. The heat transfer Q needed to change the phase of a mass m of a substance is

$$Q = mL \tag{13.29}$$

where L is the *latent heat* associated with the phase transition.

An ideal thermodynamic *reservoir* is a system that maintains a fixed temperature despite any heat transfer to or from it.

Heat flow is the time rate of heat transfer and is expressed in watts:

$$\frac{dQ}{dt}$$

Heat flow by *conduction* between two reservoirs at different temperatures is proportional to the temperature gradient

$$\frac{dT}{dx}$$

and is

$$\frac{dQ}{dt} = -kA\frac{dT}{dx} \tag{13.32}$$

where A is the area of thermal contact with each reservoir and of the conductor's cross section, and k is the *thermal conductivity* of the material connecting the two reservoirs. The steady-state heat flow is

$$\frac{dQ}{dt} = A\frac{T_{\text{H}} - T_{\text{C}}}{R} \tag{13.37}$$

where R is the *R-value* of the material connecting the warmer reservoir at temperature T_{H} with the cooler reservoir at temperature T_{C}. The total R-value of materials in series is the sum of the R-values. For materials in parallel, the heat flows add.

Heat transfer also may occur via radiation, in which case the heat flow radiated by a system at absolute temperature T is proportional to the fourth power of T, which is *Stefan's law*:

$$\frac{dQ}{dt} = -eA\sigma T^4 \tag{13.46}$$

where e is the *emissivity* of the object (a dimensionless number between 0 and 1), A is its surface area, and σ is the *Stefan–Boltzmann constant*. An object with an emissivity $e = 1$ is known as a *blackbody*.

The *internal energy* U of a system is the sum of the many microscopic kinetic energy and potential energy terms associated with all the individual particles (atoms and/or molecules) of the system. It is only *changes* in the internal energy that are physically significant on a macroscopic basis.

The most general statement of energy conservation in physics is

$$Q + W' = \Delta U + \Delta E \tag{13.48}$$

where Q is the heat transfer to a system, W' is the work done on the system by macroscopic nonconservative forces, ΔU is the change in the internal energy U of the system, and ΔE is the change in the total mechanical energy E of the system ($E = \text{KE} + \text{PE}$).

The CWE theorem of classical mechanics neglects thermal effects ($Q = 0$ J and $\Delta U = 0$ J in Equation 13.48):

$$W' = \Delta E \tag{13.49}$$

For "pure" thermodynamics, there is no change in the total mechanical energy of the system ($\Delta E = 0$ J). The *first law of thermodynamics*, a restatement of the general energy conservation principle for pure thermodynamic systems, then becomes

$$Q = \Delta U + W \qquad (13.51)$$

where W is the work done *by* the system ($= -W'$) on its surroundings.

The work done *by* a system is

$$W = \int_{V_i}^{V_f} P \, dV \qquad (13.54)$$

Geometrically, this is represented by the area under the path that the system follows on a P–V diagram between the initial and final states of thermodynamic equilibrium. The work is positive for expansions in volume and negative for compressions.

For an isobaric process, the work done by a system is

$$W = P_i \, \Delta V \qquad \text{(isobaric process)} \qquad (13.55)$$

For an isochoric process, the work done by the system is zero:

$$W = 0 \text{ J} \qquad \text{(isochoric process)} \qquad (13.56)$$

For an isothermal process, the work done by an ideal gas system is

$$W = nRT \ln \frac{V_f}{V_i} \qquad \text{(ideal gas isothermal process)} \qquad (13.57)$$

The work done by a system in executing a cycle on a P–V diagram is equal to the area of the cycle on the P–V diagram and is positive (negative) if the system executes the cycle in the clockwise (counterclockwise) fashion.

SUMMARY OF PROBLEM-SOLVING TACTICS

13.1 (page 594) If you recall that the sizes of the degrees on the celsius and fahrenheit scales differ by the factor 9 °F/5 °C or 5 °C/9 °F, depending on which way the comparison is made, you should be able to deduce either Equation 13.4 or 13.5 without memorizing them.

13.2 (page 600) Be careful with the units when using the ideal gas law. To use the value of $R = 8.314\ 52$ J/(mol·K), the pressure P must be in pascals, V in cubic meters, and n in moles.

13.3 (page 601) When using the ideal gas law, be sure to express the temperature in kelvin (*not* degrees celsius).

13.4 (page 602) Be careful in looking up the specific heat of a substance in tables in technical literature; other common energy and mass units also are used.

13.5 (page 604) There are no ready signals in Equations 13.29 and 13.30 to indicate whether Q should be positive or negative. The heat transfer Q is positive for heat transfer *to* a system and negative for heat transfer *from* a system.

13.6 (page 620) Caution! Be sure the volume of the gas is in cubic meters and the pressure in pascals when calculating work (in joules) using the geometric approach on a P–V diagram.

QUESTIONS

1. (page 588); 2. (page 590); 3. (page 590); 4. (page 594); 5. (page 598); 6. (page 598); 7. (page 603); 8. (page 613); 9. (page 618); 10. (page 623)

11. If a system is hot, can you say that it "has" a large amount of heat transfer? Explain your answer carefully.

12. The celsius and kelvin temperature scales yield quite different numerical values for the temperature of a system in thermodynamic equilibrium. Why then are temperature *changes* the same on both scales?

13. When two pieces of steel are being riveted together, it is common to use hot rivets. Explain why.

14. Why is it incorrect to use the celsius temperature scale in the ideal gas law rather than the kelvin scale, even though the size of a degree on both scales is the same?

15. Using a good dictionary such as the *Oxford English Dictionary*, trace the roots of the word *degree*.

16. For the common mercury laboratory thermometer, is the thermometric property the length of thin mercury column or the total volume of mercury in the mercury bulb at its base and in the column? Explain.

17. The tile floor in a bathroom is in thermal equilibrium with a fleecy bathroom rug on the floor. When you step on the tile floor, it *feels* much colder than the rug even though both are at the same temperature. Explain why.

18. Stainless steel pots occasionally are manufactured with a copper cladding on the bottom of the outside of the pot. In view of the relative thermal conductivities of these metals (see Table 13.6), why is this done? Is the reason purely aesthetic?

19. A common laboratory mercury thermometer initially at 20 °C suddenly is placed in boiling water to measure the temperature. The mercury column initially *descends* slightly before shooting upward. Explain why. See if you can observe this effect.

20. Explain how a blacksmith can fit an iron hoop as a rim over a wooden wheel so that the iron rim grips the wheel without additional fastenings.

21. If a pendulum clock is adjusted to keep accurate time at 20 °C, will the clock run slow, fast, or be accurate at 30 °C?

22. In northern climates, explain why it is good to tune pianos at least twice a year. At what times of the year should the tuning typically be done?

23. When home canning fruits and vegetables, jars are filled to within about one centimeter of the top rim. The lids are placed loosely on the jars and secured loosely with a screw cap. The jars then are covered completely with water and boiled vigorously for about 15 minutes. They then are removed and the cover is screwed on tight. Explain how this process creates a partial vacuum in the jar and how, when the jars cool, atmospheric pressure helps keep the jars sealed.

24. A tight-fitting, twist-off jar lid occasionally can be loosened by holding the lid under very hot water. Explain why.

25. Explain why a single block of ice melts more slowly than the same mass of ice in the form of ice cubes when both are placed in a warm room.

26. A heat-gun paint stripper can easily bubble old paint applied on wood surfaces but is unable to blister or bubble the same paint applied to metallic surfaces. Explain why.

27. Why is it more dangerous to be burned by steam at 100 °C than by boiling water at the same temperature?

28. Will coffee with cream cool faster than, slower than, or at the same rate as the same amount of black coffee in an identical cup?

29. Explain why the information on this highway sign is accurate during cold rainy or snowy weather:

FIGURE Q.29

30. What physical property of a pot holder prevents you from getting burned when taking a scalding dish out of a hot oven?

31. Does the value of the coefficient of thermal expansion depend on the particular units used to measure the lengths?

32. If the temperature is measured using fahrenheit degrees, by what factor do the coefficients of linear expansion in Table 13.1 change?

33. The water at great depths in the Great Lakes has a temperature of about 4 °C year round. Explain why.

34. In northern climates, explain why it is advisable to drain all water from pipes in unheated summer cottages for the winter.

35. In the northern hemisphere, the Sun is above the local horizon for the greatest interval on the day of the summer solstice (~ 22 June) yet maximum average daily temperatures of lakes and the ocean and on land typically are in late July and early August. Explain what causes the time lag.

36. Explain why the measurement of the change in volume of a liquid when its temperature is raised is not as accurate using a calibrated, steel, graduated cylinder as when a graduated cylinder of pyrex glass is employed.

37. When fruit blossoms are in danger of freezing because of late spring frosts, farmers spray the trees and blossoms with water. Why might this procedure succeed in saving the blossoms from freezing?

38. A thermos bottle containing a cold beverage is shaken vigorously and the temperature of its contents rises slightly. Is there heat transfer to the system in this process? What causes the rise in temperature of the liquid?

39. When a mass m of ice melts at 0 °C to water at 0 °C, there is positive heat transfer to the system (equal to $Q = mL_{f\,water}$). Yet heat transfer was defined as energy transfer that arises solely because of a *difference* in temperature. Resolve this apparent dilemma.

40. Explain how it is possible to increase the temperature of a cup of coffee *without* heat transfer to it.

41. Would a retail gasoline dealer in petroleum be more inclined to want a delivery of petroleum from a wholesaler on a very hot afternoon or on a cool morning?

42. A thermometer placed in direct sunlight indicates a temperature higher than that of the surrounding air. Explain why.

43. Your index finger is a crude thermometer. Describe some circumstances under which your finger can tell approximately if two objects are at the same temperature and circumstances when the finger thermometer is quite inaccurate.

44. When a furnace increases the temperature inside a house, the volume of the gas inside the house does not change, nor does the pressure change. Does this violate the ideal gas law? Explain.

45. Imagine a thermodynamic system divided into smaller regions of various sizes. A thermodynamic parameter is called *intensive* if its value is independent of the mass of the sample in which it is measured. A thermodynamic parameter is *extensive* if its value depends on the mass of the region in which it is measured. Classify the following parameters as intensive, extensive, or *neither*: pressure; volume; internal energy; temperature; mass; work; heat transfer; molar mass.

46. Explain why it is easier to measure c_P rather than c_V for solids and liquids.

47. Write a paragraph to distinguish among temperature, heat transfer, and internal energy.

48. Your instructor asks you to sketch a graph of PV/T versus P for a sample of an ideal gas. Go to it. What is the value of the vertical intercept?

49. Does the length of the neck of a giraffe contract if the ambient temperature falls? What about the length of a snake?

50. A deep snowfall can protect many plants from freezing. Explain how.

51. Explain why a pressure cooker is able to cook beets in fifteen minutes while it might take two hours if they were simply boiled in a simple covered pot.

52. Why does freezing typically kill?

53. Can *you* be considered to be a thermodynamic reservoir?

54. Heat transfer causes so-called "thermal patterns" in snow. Cite several examples of common thermal patterns in snow.

55. What property of a system is measured by a thermometer?

56. Two systems in thermal contact have different values of the property measured by a thermometer. As a consequence, what is transferred between the two systems?

PROBLEMS

Sections 13.1 Simple Thermodynamic Systems
13.2 Temperature
13.3 Work, Heat Transfer, Temperature, and Thermal Equilibrium
13.4 The Zeroth Law of Thermodynamics
13.5 Thermometers and Temperature Scales
13.6 Temperature Conversions Between the Fahrenheit and Celsius Scales*

1. The normal average human body temperature is 98.6 °F. What is the corresponding temperature on the celsius scale?

2. A student walks into the campus infirmary and Dr. Payne quickly assesses that the temperature of the poor soul is 103.5 °F. What is this temperature on the celsius scale?

3. Calculate the fahrenheit temperature corresponding to the absolute zero of temperature 0 K.

4. What kelvin temperature corresponds to 0 °F?

5. You awaken to the infernal clanging of your alarm clock and quickly get ready to trek off to another thrilling 8 A.M. physics class. A glance at the thermometer outside your dorm window indicates that the temperature is a bone-chilling −25 °C. Your roommate wants to know what the temperature is on the fahrenheit scale. Determine it.

•6. At what temperature is the numerical value the same on the celsius and fahrenheit scales?

•7. At what temperature is the numerical value the same on the kelvin and fahrenheit scales?

•8. A physicist devises a customized temperature scale that assigns a temperature reading of zero to the normal melting point of mercury (−39 °C) and a temperature reading of 100 to its boiling point (357 °C). (a) Formulate an equation to convert from degrees celsius to the degrees on this custom scale. (b) What temperature on the custom scale corresponds to 0 °C?

•9. At what temperature celsius is the sum of the celsius temperature, the kelvin temperature, and the fahrenheit temperature equal to 495.15?

•10. Wow are you hungry! You rush off to the Sweet Things ice cream shop and devour a 2000 Calorie ice cream sundae. Yum! (a) How many joules of energy does this food represent? (b) How many 10.0 kg cement blocks must you lift through a vertical distance of 2.00 m to do an equivalent amount of work on the blocks? This calculation indicates that it is much easier to lose

weight simply by not eating highly Caloric foods rather than by exercising to "work it off."

Section 13.7 Thermal Effects in Solids and Liquids: Size

11. A steel beam of length 30 m is used to span a creek. Provisions must be made for changes in the length of the beam in the supporting structure. How much will the length of the beam change if the temperature variation between winter and summer ranges from −20 °C to 40 °C?

12. A copper electrical cable of length 0.500 km is strung over a river between two electrical pylons. If the temperature changes from −10 °C to +40 °C between winter and summer, by how much does the length of the cable change?

•13. A 100.000 m steel measuring tape was marked and calibrated at 20 °C. At a temperature of 30 °C, what is the percentage error in a 100.000 m distance when using this tape?

•14. An optical engineering firm needs to ensure that the separation of two mirrors is unaffected by temperature changes. Such a device is called a Fabry–Perot interferometer. The mirrors are attached to the ends of two bars of different materials that are welded together at one end as shown in Figure P.14. The surfaces of the bars in contact are lubricated. (a) Show that the distance ℓ does not change with temperature if $\alpha_1 \ell_1 = \alpha_2 \ell_2$, where α_1 and α_2 are the respective thermal coefficients of linear expansion. (b) If the materials are brass and steel, which material should be the shorter bar? (c) If the distance ℓ_1 is 2.66 cm at 20 °C, what is ℓ_2 and what is ℓ?

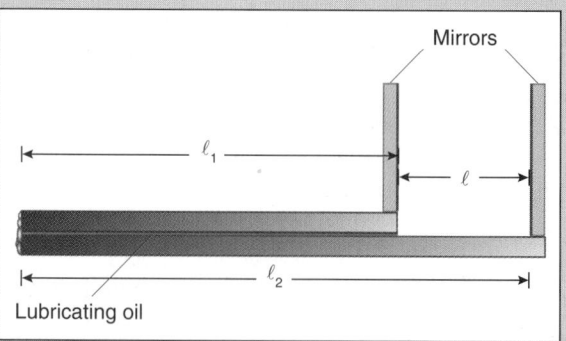

FIGURE P.14

•15. (a) Calculate the length of a pendulum that has a period equal to 1.000000 s. Assume *g* is exactly 9.81 m/s². (b) If the mass on the end of the pendulum is supported by means of a steel wire, what is the period of the pendulum if the temperature increases by 10 °C? (c) Does this result in the clock running slow or fast? (d) After what time interval will the clock differ from the correct time by one minute?

•16. A clock has a pendulum supported by a steel wire. The pendulum has a period of precisely 1.000000 s at 0 °C. (a) As the temperature rises, will the clock run slow or fast? (b) Use the expression for linear expansion to derive an equation for the *change* in the period of the pendulum as a function of the temperature change from 0 °C. (c) The clock runs accurately at 0 °C. How many minutes per day will it gain or lose at 30 °C?

•17. A stretched wire precisely 1.000 m in length is tuned to play middle C (261.6 Hz). (a) If the tension in the wire is kept constant, by how much is it necessary to change the length of the wire so the note is flat by 0.20 Hz? (b) If the wire is made of steel, what temperature change is needed to accomplish this change in length? Temperature-induced changes in the tuning of stringed instruments are more complicated than this example indicates. An increase in the length of the string induced by a rise in temperature changes the tension in the string because there is more slack in the string.

•18. Imagine a hypothetical steel wire girdling the globe around the equator. Assume the Earth is a perfect sphere of radius 6370 km. The wire is at a uniform temperature of 25.0 °C. What temperature change is needed to increase the length of the wire by 1.0 km?

•19. The railroad and highway bridge over the dramatic gorge just downstream from Victoria Falls (also known as *mosi oa tunya*, "the smoke that thunders") in Zimbabwe was constructed at the turn of the 19th to 20th centuries. See Figure P.19. The steel bridge was fabricated in England and shipped in pieces for assembly on site. Working from each bank, the last section to be inserted was the center span. Unfortunately, it was about 10 cm too long to drop into place because of the sag of the halves of the bridge protruding from each bank. To pry apart the two ends of the bridge, long steel cables were attached to each and anchored inland. The cables were tightened during the hottest part of the day, when the temperature was about 25 °C. The following morning the temperature was 10 °C. The thermal contraction of the cables pried the two sections of the bridge apart just enough for the center span to be inserted. What length steel cable was used for each to contract 5 cm? Neglect any thermal contraction of the bridge structure itself.

FIGURE P.19

•20. A solid brass sphere has a radius of 5.00 cm at 20 °C. (a) Calculate the area of the sphere. (b) Calculate the volume of the sphere. (c) The temperature of the sphere is increased to 150 °C. Find the change in the radius, area, and volume of the sphere.

•21. Under a spreading chestnut tree, a farrier wants to put a steel rim on a wagon wheel. The wheel is 1.000 m in diameter but the rim is only 0.995 m in diameter. What temperature change of the rim is necessary for it to just fit over the wheel?

•22. (a) Calculate the surface area of a steel cubical box 0.500 m on a side. (b) If the temperature increases by 20 °C, by what percentage does the surface area increase? (c) By what percentage does the volume increase?

•23. The density of a solid is the amount of mass per cubic meter of the solid. Since a change in temperature changes the volume of a solid, the density also is temperature dependent. If we let ρ be the density of the solid at temperature T, show that the change in the density of the solid when its temperature changes by ΔT can be expressed as

$$\Delta \rho = -3\alpha\rho \, \Delta T$$

where α is the coefficient of linear expansion. Why is the minus sign necessary?

•24. On a chilly 0 °C day, you purchase 50.0 liters of gasoline at a pump that is assumed to measure volume accurately at all temperatures. You casually note that the thermal coefficient of volume expansion for gasoline is 0.0012 K⁻¹, duly posted on the pump to repel tort lawyers. (a) Will you get more or fewer *kilograms* of gasoline when you buy another 50.0 liters on a hot day at 30 °C? (b) What is the difference in the *volumes* of the two purchases in liters after both the purchases cool to 0 °C? Specify which volume is greater. Ignore evaporation effects.

•25. The area of a circle is $A = \pi r^2$. If the radius changes by a small amount Δr, show that the small change ΔA in the area of the circle is approximately

$$\Delta A = 2\pi r \, \Delta r$$

If the change in radius comes about because of a temperature change ΔT, show that the foregoing expression yields

$$\Delta A = (2\alpha)A \, \Delta T$$

where α is the (constant) coefficient of linear expansion. (Hint: Differentiate the expression for the area with respect to r and let the differentials represent the small changes in radius and area.) Thus Equation 13.9 is not restricted to a specific shape for the area.

•26. The volume of a sphere is $V = (4/3)\pi r^3$. Show that if the radius changes by a small amount Δr, the change in the volume ΔV is approximately

$$\Delta V = 4\pi r^2 \, \Delta r$$

If the change in radius occurs because of a change in temperature, show that the foregoing expression becomes

$$\Delta V = (3\alpha)V \, \Delta T$$

where α is the (constant) coefficient of linear expansion. (See the hint in Problem 25.)

•27. Use the graph in Figure 13.15 to determine the range for the average density of an object that will float in water at 10.0 °C yet sink in hot water at 90.0 °C. How might you construct a device of the appropriate average density that will demonstrate this effect?

•28. A thin rod of length ℓ and cross-sectional area A is at temperature T. The rod is clamped at both ends to *prevent* thermal expansion or contraction. If the temperature now is decreased, the tension in the rod must increase by an amount sufficient to produce the *same* fractional change in length as would occur if the rod could contract as a result of the temperature decrease. Show that the increase in the magnitude of the tension in the rod is

$$\alpha A E |\Delta T|$$

where E is Young's modulus (see Chapter 11).

•29. A 2.00 m steel rod is of radius 0.0050 m. The temperature of the rod is cooled from 100 °C to 20 °C. What force is needed to *prevent* the rod from changing its length?

Section 13.8 Thermal Effects in Ideal Gases

30. A 3.00 liter sample of an ideal gas has a pressure of 2.00 atm at 10 °C. If the volume remains constant, what is the pressure of the gas when the temperature is 40 °C?

31. A 2.50 mol sample of an ideal gas at 1.00 atm pressure has a temperature of 20 °C. What is the volume occupied by the gas?

32. A sample of an ideal gas occupies a volume of 4.00 liters at 20.0 °C with a pressure of 3.00 atm. How many moles are in the sample?

•33. For an ideal gas, show that the density of the gas at constant temperature is proportional to the pressure.

•34. An air bubble released from a scuba diver at a depth of 35 m rises to the surface. The water temperature is 5.0 °C in the vicinity of the diver and 20.0 °C at the surface. If you assume that the bubble is in thermal equilibrium with the water at each position, by what factor does the volume of the bubble change as it rises to the surface? Assume the air is an ideal gas.

•35. A spherical bubble of radius r_0 has a volume V_0 and pressure P_0 when just below the surface of a liquid of density ρ; see Figure P.35. Let F_0 be the magnitude of the buoyant force on the bubble when it is at this location. Assume P_0 is atmospheric pressure and that the temperature of the liquid and bubble do not change with depth. Let the y-axis be vertical, with $\hat{j}$ directed up and the origin at the surface. Notice that positions

within the liquid correspond to negative values of y. (a) Show that the magnitude of the buoyant force on the bubble as a function of its vertical position $|y| \gg r_0$ in the liquid is

$$F_0 \frac{P_0}{P_0 - \rho g y}$$

(b) If the liquid is water, show that the buoyant force is half the surface value at a depth of 10.3 m.

•36. (a) For an ideal gas, show that the coefficient of volume expansion at constant pressure, defined by

$$\frac{1}{V}\left(\frac{dV}{dT}\right)_P$$

is the reciprocal of the absolute temperature. Evaluate the numerical value of the coefficient of volume expansion at constant pressure for an ideal gas at 0 °C. (b) If the gas is confined to a tube whose length is much greater than its diameter, what is the coefficient of linear expansion (at constant pressure) for the gas in the tube?

•37. The ambient temperature in outer space is about 3.0 K. Most of the matter in space is hydrogen gas at a density of about one atom per 3.0 m^3. What is the pressure of this dilute gas if it is in thermal equilibrium at the ambient temperature?

•38. A florist has a 5.00×10^{-2} m^3 tank of helium gas at a pressure of 100 atm. Spherical balloons are filled from the tank to a pressure of 1.20 atm and have a diameter of 30 cm. How many balloons can be filled from the tank?

39. A sophisticated laboratory vacuum system creates a partial vacuum with a pressure of 2.00×10^{-6} atm in a 1.50×10^{-3} m^3 chamber. The chamber has a temperature of 20.0 °C. How many gas particles are in the chamber?

•40. A cylinder with a diameter of 5.00 cm has a movable piston attached to a spring as shown in Figure P.40. The cylinder contains 1.50 liters of an ideal gas at 20.0 °C and 1.00 atm pressure. Under these conditions, the spring is unstretched. The temperature of the gas is increased to 100 °C and it is found that the spring compresses 3.50 cm. Determine the spring constant k.

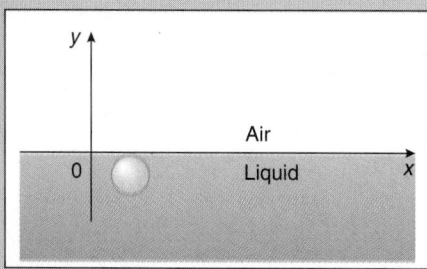

FIGURE P.35

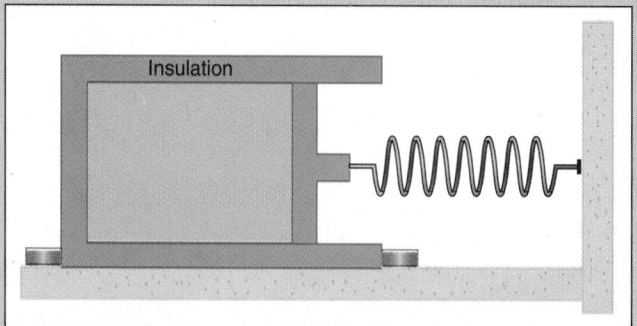

FIGURE P.40

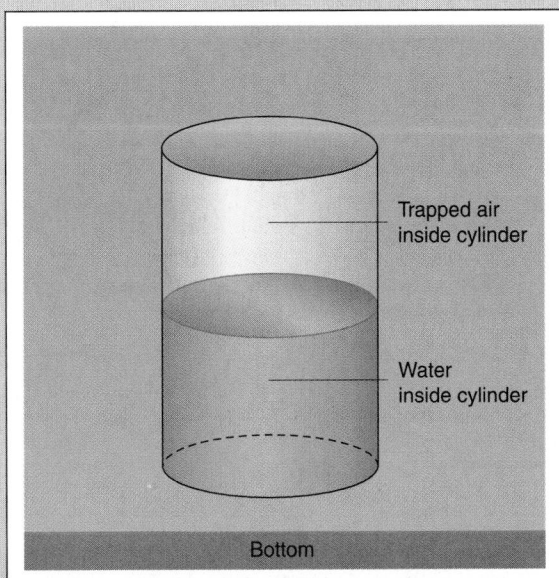

Trapped air
inside cylinder

Water
inside cylinder

Bottom

FIGURE P.41

•41. Early devices for going underwater were called diving bells. They were used for salvage purposes before the advent of modern diving equipment, but were quite limited in the depth they could attain. A cylindrical diving bell has a closed upper end and an open lower end. (See Figure P.41.) At the surface, air inside the cylinder has a temperature of 20.0 °C. The cylinder is lowered into the cold waters of Lake Superior. At what depth has the water risen halfway into the cylinder? Assume the gas in the submerged bell is in thermal equilibrium with the water at 4.0 °C.

Section 13.9 Calorimetry

42. What heat transfer to a 2.0 kg pot of water is necessary to raise its temperature from 20 °C to 100 °C for a coffee break? Neglect the mass of the pot.

43. What heat transfer to 2.0 kg of ice at 0 °C is needed to melt it to water at 0 °C?

•44. You have 0.50 kg of water at 20.0 °C. Assuming no heat transfer to the environment, how much ice at −10.0 °C should be added to chill the water to 0.0 °C with no ice remaining?

•45. A 0.500 kg block of ice at −10.0 °C is placed in a punch bowl with 4.00 kg of water (punch) at 20.0 °C. Assume no heat transfer to the surroundings and neglect the thermal effects of the punch bowl itself. (a) Does all the ice melt? If not, how much is left? (b) What is the final temperature of the punch plus ice system?

•46. A 2.0 kW electric heater is used in a well-insulated hot water tank of capacity 200 liters to raise the temperature of water from 5.0 °C to 60.0 °C. How long will it take to change the temperature of the water in the tank?

•47. An immersion heater supplies a constant 100 W to 0.200 kg of water initially at 10 °C. Assume there is no heat transfer to the surrounding environment. What time elapses until there is 0.190 kg of liquid water left?

•48. You are very particular about the temperature of beverages. You want to cool 2.00 liters of hot tea (consider it to be water) from a temperature of 98.0 °C to 1.0 °C by adding ice from a freezer at −20.0 °C. How much ice should be added to accomplish the task? Neglect any heat transfer to the environment.

49. A typical household water heater has a capacity of about 0.50 m³ of water. The water is initially at a temperature of 15 °C and is warmed to 60 °C for the pleasure of long, hot showers. (a) What heat transfer is necessary to warm the water? (b) If the heating element is rated at 5.0 kW, how long will it take to warm the water?

⁛50. Dense materials often have specific heats expressed in joules per kilogram-kelvin [J/(kg·K)]. Among the materials listed here, test the possibility that equal volumes of different materials might require equal heat transfer to raise the temperature by one kelvin. Calculate the specific heat of each in J/(m³·K). Present your results in an orderly table and state your conclusions. What conclusions can be drawn if a comparison is made among the specific heats of these substances in joules per mole-kelvin [J/(mol·K)]?

Substance	Specific heat [J/(kg·K)]	Density (kg/m³)	Molar mass (kg/mol)	Specific heat [J/(m³·K)]	Specific heat [J/(mol·K)]
Aluminum	896	2.70×10^3	27.0×10^{-3}		
Copper	386	8.9×10^3	63.5×10^{-3}		
Iron	481	7.8×10^3	55.8×10^{-3}		
Lead	128	11.0×10^3	207×10^{-3}		
Mercury	138	13.6×10^3	201×10^{-3}		

⁛51. At very low temperatures, near 0 K, the specific heat [in J/(kg·K)] of solids is strongly temperature dependent and is given by the *Debye equation*:

$$c = \beta T^3$$

where β is a constant and T is the absolute temperature. What heat transfer is necessary to raise the temperature of a solid of mass m from 2.00 K to 5.00 K? Your result will be in terms of m and β.

Sections 13.10 Reservoirs
13.11 Mechanisms for Heat Transfer*

52. If the absolute temperature of a blackbody radiator is doubled, by what factor does the heat flow radiated by the body increase? Can the same be said if the celsius temperature of the blackbody radiator is doubled?

53. A mobile home is 3.0 m wide, 2.5 m high, and 25.0 m long. See Figure P.53. The average *R*-value for the walls, roof, and floor of the dwelling is 3.0 m²·K/W. If the temperature inside the home is kept at 20 °C, find the heat flow to the exterior on a cold winter night with an outside temperature of −20 °C.

FIGURE P.53

54. It is −20 °C outside but nice and toasty inside (20 °C) where you are diligently working on physics homework due tomorrow. The windowless outside wall in your dormitory cell has dimensions 2.40 m by 3.60 m and has an effective *R*-value of 6.0 m²·K/W. Calculate the heat flow through the wall.

55. Two reservoirs, one at 0.0 °C and the other at 100.0 °C, are connected by means of two materials as indicated in Figure P.55. The relative *R*-values of the materials are specified. What is the temperature at the interface between the two materials?

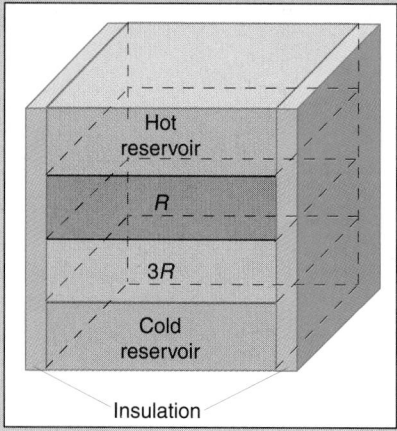

FIGURE P.55

56. The metabolic rate of an average, clothed adult human being is about 100 W (this number corresponds to a diet of about 2000 kcal/day per human). This means that a human being is like a 100 W heater. Twenty-nine students and a professor are in a classroom 10.0 m long, 10.0 m wide, and 5.0 m high, with an air temperature initially at 20.0 °C. Assume the classroom is well insulated. Neglecting any heat transfer between the air and the walls, calculate how much the temperature of the air in the room rises over the course of a 55 min lecture. Assume the process takes place at constant volume in an airtight lecture room The specific heat of air at constant volume is about 20 J/(mol·K). A mole of air represents about 24.1 liters at room temperature and pressure. Neglect professorial histrionics.

57. A wooden box of frozen fish is covered by equal thicknesses of cork and wood (with the cork on the inside of the box). The ratio of the thermal conductivities is

$$\frac{k_{wood}}{k_{cork}} = 3.0$$

The temperature of the inside of the box is −20 °C and the outside is at 25 °C. (a) When a steady state exists, what is the ratio of the heat flow through the wood to the heat flow through the cork? (b) What is the temperature of the surface at which the wood and cork are joined?

58. A window pane has dimensions 1.00 m × 0.50 m × 2.0 mm. The inside of the window is at a comfortable 20 °C and the outside is a numbing −20 °C. (a) If the width and height are doubled, but not the thickness, by what factor is the heat flow changed? (b) If each dimension (height, width, and thickness) of the window is doubled, by what factor is the heat flow through the window changed? (c) If the height and thickness are doubled, leaving the width unchanged, by what factor is the heat flow changed?

59. A solid rectangular block of lead has the dimensions 2.00 cm × 3.00 cm × 4.00 cm. A temperature difference of 100 °C can be applied to any pair of opposite faces that you choose. (a) Which choice of separation between the faces produces the largest heat flow? (b) Calculate the heat transfer in joules during 60.0 s between the opposite faces mentioned in part (a). (c) Which choice of separation between faces produces the least heat flow?

60. Show that to convert the SI *R*-value to the English system *R*-value, the conversion factor is as indicated at the bottom of Table 13.7 on page 609.

61. Warm-blooded animals maintain a constant body temperature, warmer than their surroundings. Ancient and medieval families occasionally built their dwellings above sheltered space for their animals. Everyone lived in the barn. In winter, the animals were a natural furnace, but not because of bovine flatulence. The result of Example 13.19 indicates that the heat flow from the animal to the surrounding environment is proportional to the surface area of the animal for a constant temperature difference. Making crude but reasonable assumptions, show that the heat flow likely is proportional to the two-thirds power of the mass *m* of the animal. That is,

$$\frac{dQ}{dt} \propto m^{2/3}$$

According to these assumptions, what is the ratio of the heat flow from a 1.20×10^3 kg steer compared with that from a 60 kg serf?

•62. Let T_n be the normal temperature in kelvin of the human body. When the body has a fever, let its temperature be $T_n + \Delta T$. Show that the ratio of the heat flow radiated by the body when it has a fever to that when the body temperature is normal is, approximately,

$$1 + \frac{4 \, \Delta T}{T_n}$$

Evaluate this ratio for a sick soul with a fever of 40 °C; normal body temperature is 37 °C.

•63. The *solar constant* is the amount of energy from the Sun we receive on the Earth during each second on a 1.000 m² area oriented perpendicular to the direction of the sunlight. The value of the solar constant is about 1.37 kW/m². Imagine sunlight illuminating an asphalt pavement as indicated in Figure P.63. The ambient temperature is 25 °C. What is the equilibrium temperature of the asphalt? Assume the asphalt is a blackbody.

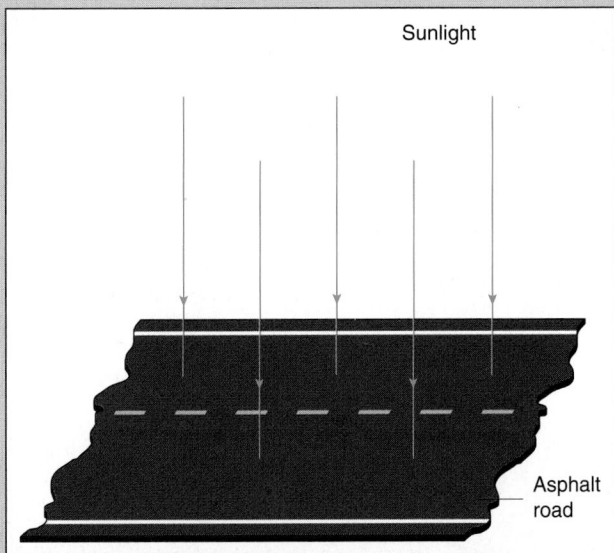

FIGURE P.63

§64. Copper rod A, 2.00 mm in diameter and 30.0 mm long, is joined to copper rod B, which is 3.00 mm in diameter and 20.0 mm long, making a conductor 50.0 mm long. The lateral area of the rods is fully insulated. The rods connect two reservoirs as indicated in Figure P.64. (a) Determine the temperature at the junction between the two rods. (b) Find the heat flow in each rod.

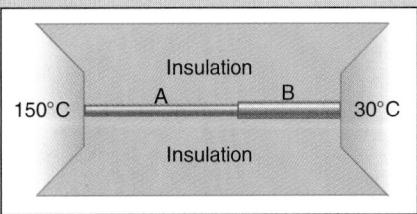

FIGURE P.64

Sections 13.12 Thermodynamic Processes
13.13 Energy Conservation: The First Law of Thermodynamics and the CWE Theorem
13.14 The Connection Between the CWE Theorem and the General Statement of Energy Conservation
13.15 Work Done by a System on Its Surroundings
13.16 Work Done by a Gas Taken Around a Cycle
13.17 Applying the First Law of Thermodynamics: Changes of State

•65. An ideal gas goes from an initial state of thermodynamic equilibrium (1) to a final state (2) via one of the three paths indicated on the *P–V* diagram in Figure P.65. What a gas! (a) Calculate the work done by the gas along path a. (b) Calculate the work done by the gas along path b. (c) Calculate the work done by the gas along path c. The gas now is taken from state (2) back to state (1) via one of the three paths. (d) Calculate the work done by the gas along path a. (e) Calculate the work done by the gas along path b. (f) Calculate the work done by the gas along path c.

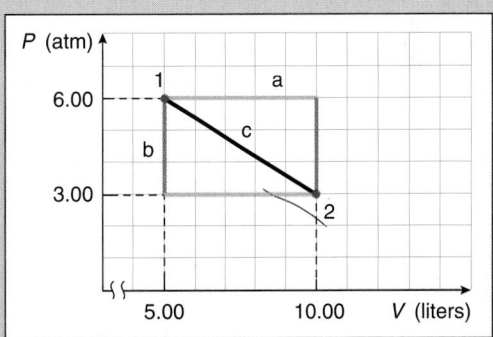

FIGURE P.65

•66. Just for fun, you cause a sample of an ideal gas to move around a closed cycle on the *P–V* diagram beginning and ending at point A indicated in Figure P.66. (a) Find the total work done by the gas. (b) What is the change in the internal energy of the gas? (c) What is the total heat transfer to the gas in executing this cycle? If the gas is taken around the cycle in the opposite direction: (d) find the total work done by the gas; (e) find the change in the internal energy of the gas; and

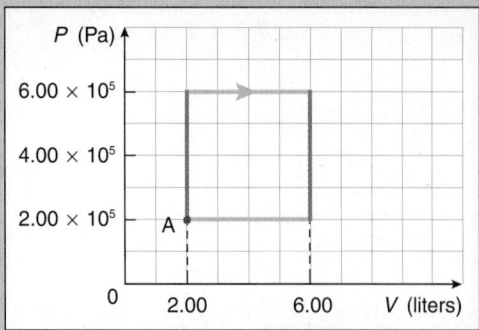

FIGURE P.66

(f) find the total heat transfer to the gas in executing the cycle in this direction.

•**67.** A gas on vacation takes a circle tour of a *P–V* diagram, beginning at point A as indicated in Figure P.67. (a) Find the total work done by the gas after one trip around the cycle. (b) What is the change in the internal energy of the gas in executing one cycle? (c) What is the total heat transfer to the system around the cycle? If the gas takes the circle tour in the opposite direction: (d) Find the total work done by the gas after one trip around the cycle. (e) What is the change in the internal energy of the gas in executing one cycle? (f) What is the total heat transfer to the gas around the cycle?

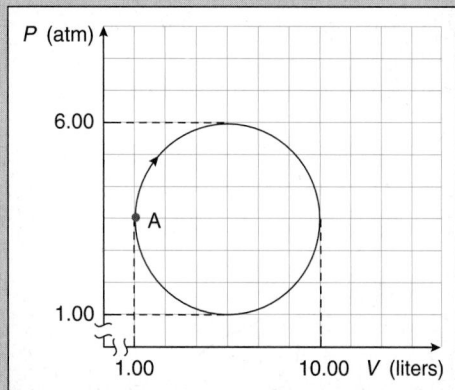

FIGURE P.67

•**68.** A gas is taken from thermal equilibrium state 1 to state 2 on a *P–V* diagram via the semicircular route indicated in Figure P.68. (a) Calculate the work done by the gas in going from state 1 to state 2. (b) What is the work done by the gas if it goes from state 2 back to state 1, retracing the semicircular path?

•69. A gas is taken from thermal equilibrium state 1 to state 2 on a *P–V* diagram via a semicircular route as indicated in Figure P.69. (a) Calculate the work done by the gas in going from state 1 to

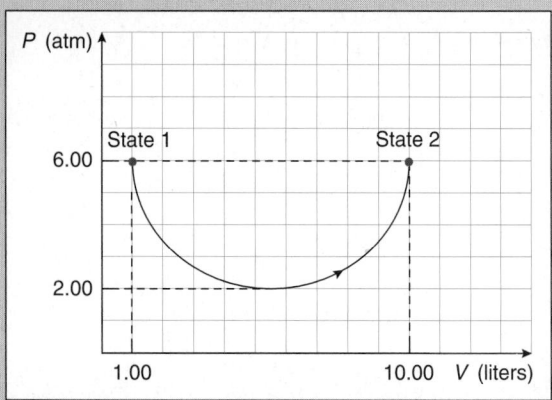

FIGURE P.68

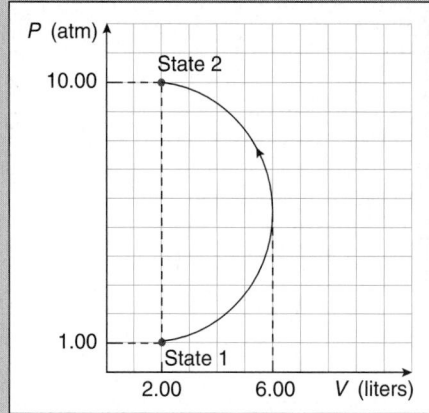

FIGURE P.69

state 2. (b) What is the work done by the gas if it goes from state 2 back to state 1 retracing the semicircular path?

•**70.** A gas is taken around a closed cycle on a *P–V* diagram as indicated in Figure P.70. (a) Find the total work *W* done by the gas. (b) What is the change in the internal energy of the gas? (c) What is the total heat transfer to the gas in executing this cycle? If the gas is taken around the cycle in the opposite direction: (d) find the total work done by the gas; (e) find the change in the internal energy of the gas; and (f) find the total heat transfer to the gas in executing the cycle in this direction.

•**71.** Three (3.00) moles of an ideal gas is taken along the excursion on a *P–V* diagram indicated in Figure P.71. The first part of the path is an isobaric process and the second part of the path is isothermal. (a) What is the initial absolute temperature of the gas? (b) Is the work done by the gas positive or negative over the isobaric segment of the path? (c) Calculate the work done by the gas along the isobaric segment of the path. (d) What is the temperature of the gas along the isotherm? (e) Is the work done by the gas positive or negative along the isothermal segment of the path? (f) Determine the final volume V_f of the

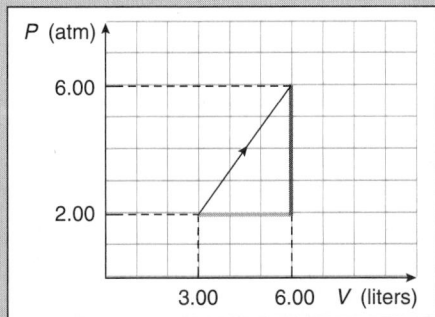

FIGURE P.70

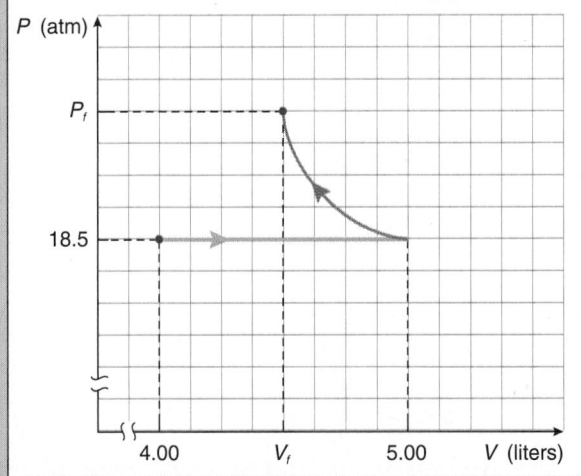

FIGURE P.71

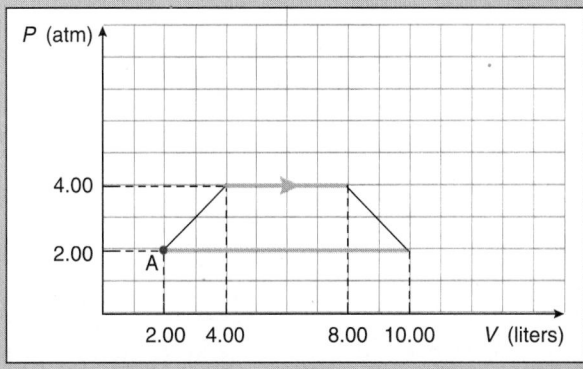

FIGURE P.72

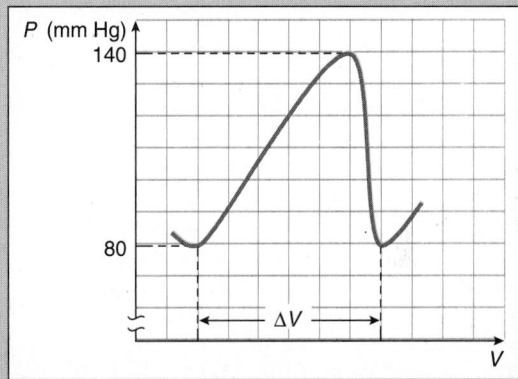

FIGURE P.76

gas so that the total work done by the gas is zero. Express your result in liters. (g) What is the final pressure P_f of the gas? Express your result in atmospheres.

•72. A gas undergoes processes that trace out the path indicated in the P–V diagram of Figure P.72, beginning at point A. (a) What is the total work done by the gas during one cycle? (b) What is the change in the internal energy of the gas during one cycle? (c) What is the total heat transfer to the gas during each cycle? (d) Answer the same questions if the cycle is executed in the opposite sense.

•73. A steel ball of mass 0.100 kg and specific heat 460 J/(kg·K) drops to a hard floor from a height of 10.00 m and rebounds with 0.7071 of its impact speed. (a) Show that the ball attains a height of 5.00 m after the impact. (b) Assume all kinetic energy lost by the ball in the impact is retained by the ball as an increase in its internal energy (as if heat transfer occurred to the ball, although none does). Calculate the rise in temperature of the ball caused by the single bounce. (c) If the bouncing is allowed to continue until the ball is at rest, calculate the total rise in the temperature of the ball. Make the same assumptions as in part (b).

•74. A snowball at 0 °C moving at speed v just barely misses you and smashes into a wall, coming to a complete halt. Assume the snowball retains the lost kinetic energy as a rise in its thermodynamic internal energy (as if heat transfer occurred to the snowball, although none does), with no heat transfer to the wall or the environment. Find the (large) speed v that makes the snowball melt completely into water at 0 °C.

•75. Victoria Falls on the border between Zimbabwe and Zambia in southern Africa has a height of about 130 m. The falls are visible in the background of Figure P.19. Assume that all the kinetic energy gained by the falling water appears (eventually) as an increase in the internal energy of the water (as if heat transfer occurred to the water, although none does). What is the temperature change of the water between the top and bottom of the falls?

•76. When your heart contracts to pump blood through your circulatory system, the blood pressure reaches a peak value about 140 mm Hg (called the systolic blood pressure value); when the heart muscle relaxes, the blood pressure falls to 80 mm Hg (called the diastolic blood pressure value). Your blood pressure is said to be 140/80, using medical parlance. The variation is indicated schematically in the P–V diagram of Figure P.76. The change in the volume of the heart is roughly 90 cm³ per contraction. (a) Convert these medical pressure and volume values to SI units. See Section 11.3. (b) From the P–V diagram, estimate the work done by the heart during each contraction cycle. (c) If a patient has a pulse rate of 80 heart-beats/min, what is the approximate time rate at which the heart is doing work (in watts)?

●77. A well-insulated waterbomb, with a mass of 10 kg and an average specific heat of 4186 J/(kg·K), is dropped from rest and undergoes free fall from the top of the Washington Monument (170 m). After the (intact) waterbomb comes to rest on impact, how much will its temperature change if you assume that its lost kinetic energy manifests itself completely as a change in the internal energy of the water (as if heat transfer occurred to the water, although none actually does)? Assume no heat transfer to the environment.

●78. You toss a lighted firecracker (of negligible mass) into an insulated steel box [of mass 1.8 kg and specific heat 460 J/(kg·K)] and then quickly seal the lid tight. Hang on. The firecracker explodes inside without rupturing the box and you note that the box warms from 20.0 °C to 22.0 °C, assuming no heat transfer to the environment. The air in the box (consisting mostly of N_2 and O_2) has an effective molar mass of 29 g/mol and a volume of 22.4 liters. The molar specific heat of air is about 20 J/(mol·K). (a) How many moles of air are in the

box? (b) Show that the box, rather than the air, has most of the energy released by the firecracker. (c) How much energy was released by the firecracker? (d) Name and describe the processes by which the energy of the exploding firecracker is transferred to the walls of the container.

●79. A steady flow of water vapor from a pot of boiling water passes into and out of an aluminum cup and eventually raises the temperature of the cup from 20 °C to 100 °C. (a) Explain why water vapor condenses in the aluminum cup. (b) When the cup reaches a temperature of 100 °C, explain why no more water vapor will condense in the cup. (c) If the metal cup has mass M and the condensed water in the cup has mass m, calculate the ratio m/M. Can measuring m/M give the latent heat of vaporization?

●80. A silver bullet [mass 0.010 kg with a specific heat of 233 J/(kg·K)] is fired by the Lone Ranger into a fixed block of lead [mass 1.00 kg with a specific heat of 128 J/(kg·K)] and comes to rest within it. Both objects initially are at a temperature of 30.0 °C. Tonto observes that the temperature of the total mass rises to 35.0 °C and also that the block (with the bullet in it) remains at rest. Assume there is no heat transfer to the external world. (a) How much energy is needed to change the temperature of the silver bullet? (b) How much energy is needed to change the temperature of the lead block? (c) What was the kinetic energy of the bullet before impact? (d) What was the speed of the bullet before impact?

INVESTIGATIVE PROJECTS

A. Expanded Horizons

1. Investigate the ways in which very low and very high absolute temperatures are measured experimentally.
   Robert Otani and Peter Siegal, "Determining absolute zero in the kitchen sink," *The Physics Teacher*, 29, #5, pages 316–319 (May 1991).
   Kurt Mendelssohn, *The Quest for Absolute Zero: The Meaning of Low Temperature Physics* (John Wiley, New York, 1977).
   Terry J. Quinn, *Temperature* (2nd edition, Academic Press, London, 1990).
   Thomas D. McGee, *Principles and Methods of Temperature Measurement* (Wiley, New York, 1988).
   James R. Leigh, *Temperature Measurement and Control* (Peregrinus, London, 1988).

2. After reading several of the following references, research and report on the various physical aspects of hyperthermia and hypothermia.
   William W. Forgey, *The Basic Essentials of Hypothermia* (ICS Books, Merrillville, Indiana, 1991).
   Albert A. Bartlett and Thomas J. Brown, "Death in a hot tub: the physics of heat stroke," *American Journal of Physics*, 51, #2, pages 127–132 (February 1983).
   G. M. Hahn, "Hyperthermia for the engineer: a short biological primer," *IEEE Transactions on Biomedical Engineering*, BME-31, #1, pages 3–8 (January 1984).
   Gilbert Nussbaum (editor), "Physical aspects of hyperthermia," AAPM *Monograph Number 8* (American Institute of Physics, New York, 1983).
   Best and Taylor's Physiological Basis of Medical Practice, edited by John B. West (12th edition, Williams and Wilkins, Baltimore, 1990).

3. Investigate aspects of heat transfer in medicine and biology.
   Avraham Shitzer and Robert C. Eberhart, *Heat Transfer in Medicine and Biology: Analysis and Applications* (Plenum, New York, 1985).
   T. H. Benzinger, "Thermodynamics of living matter: physical foundations of biology," *American Journal of Physiology*, 244, #6, pages R743–R750 (June 1983).

4. The average energy per day used by animals simply to act as "couch potatoes" (their basic metabolic rate) is expressed by *Kleiber's law*. The relationship indicates that the metabolic rate is proportional to the 0.75 power of the mass of the animal. Investigate this relationship and its relationship to the prediction of Problem 61.
   Knut Schmidt-Nielsen, *Animal Physiology: Adaptation and Environment* (5th edition, Cambridge University Press, New York, 1997).
   Malcolm S. Gordon, *Animal Physiology: Principles and Applications*, (Macmillan, New York, 1977), pages 70–75.

5. The significance of heat transfer by blackbody radiation was alluded to in the text. The story of the study of such radiation, leading to the discovery of Planck's constant h, is an interesting one and worth looking into in greater detail. (See Chapter 26 of this text.)
   Thomas S. Kuhn, *Black-Body Theory and the Quantum Discontinuity, 1894–1912* (University of Chicago Press, Chicago, 1987).
   Stephen T. Thornton and Andrew Rex, *Modern Physics for Scientists and Engineers* (Saunders, Fort Worth, Texas, 1993).

M. Russell Wehr, James A. Richards Jr., and Thomas W. Adair III, *Physics of the Atom* (4th edition, Addison-Wesley, Reading, Massachusetts, 1984), pages 71–85.

6. Early ideas about the nature of heat transfer included a hypothesis that heat was a substance (called caloric) transferred between hot and cold systems. Investigate the experiments of Benjamin Thompson (Count Rumford) (1753–1814) that disproved this hypothesis. (Count Rumford's legacy lives on with Count Rumford Baking Powder, still available in some supermarkets.)

   Arnold Arons, *Development of Concepts of Physics* (Addison-Wesley, Reading, Massachusetts, 1965), pages 411–415.

   Freeman J. Dyson, "What is heat?" *Scientific American*, 191, #3, pages 58–63, 188 (September 1954).

   Sanborn C. Brown, *Benjamin Thompson, Count Rumford: Count Rumford on the Nature of Heat* (Pergamon, Oxford, New York, 1967).

   Sanborn C. Brown, *Count Rumford: Physicist Extraordinary* (Greenwood Press, Westport, Connecticut, 1979).

   Sanborn C. Brown, *Benjamin Thompson, Count Rumford* (MIT Press, Cambridge, Massachusetts, 1979).

7. Investigate the reason that life is possible only over a fairly narrow range of temperatures.

   Harold Weinstock, "Thermodynamics of cooling a (live) human body," *American Journal of Physics*, 48, #5, pages 339–344 (May 1980).

   Frank H. Johnson, "Heat and life," *Scientific American*, 191, #3, pages 64–68, 188 (September 1954).

B. Lab and Field Work

8. With the assistance of your student health service, measure the human body temperature of a large sample of healthy students to a precision of at least 0.1 °C. You may need special permission from a committee at your university to conduct such an experiment since it involves human subjects. After deciding on an appropriate way to display the data, make three histogram graphs of the results, one for male students, one for female students, and one for all. Are there any distinctions between the histograms?

9. Devise an experiment to precisely measure the volume of 1.00 kg of water as a function of temperature between 0 °C and 100 °C. Note that the graph in Figure 13.15 is not what you would obtain using Equation 13.10. Hence the coefficient of volume expansion of water must be temperature dependent over this range of temperatures.

10. Butter melts at about 31 °C. Design an experiment to determine the melting point temperature more precisely and to determine whether there is a latent heat of fusion associated with the melting. Do your results depend appreciably on whether the raw material is sweet butter or lightly salted butter, or on the brand of butter?

11. Your physics department likely has a precision mercury laboratory thermometer as well as a platinum resistance thermometer (or other equivalent electronic thermometer). Use distilled water to make an ice–water slurry as well as a beaker of boiling water. Use these two fixed points to calibrate the two thermometers. Now measure the temperature of a beaker of warm water (at about 50 °C) with each and see if the two thermometers agree.

12. Using a narrow, long-necked bottle or flask and water, make a crude thermometer. Use an ice–water slurry and boiling water as the two calibration points. What is the precision of your thermometer?

C. Communicating Physics

13. Imagine you are asked by a local fourth-grade science teacher to explain the distinction between heat transfer and temperature, why masses in thermal equilibrium have a temperature but no heat, and the similarities and differences between heat transfer and work. Prepare a 15 minute talk for the class, remembering that the attention span of a fourth-grader is short, so that you will need to spice your talk with many simple demonstrations. As a community service, volunteer to give the talk at a local elementary school. You might be surprised (and aghast) at the number of elementary school science books that talk about heat transfer as if a mass has a "heat energy" when in thermal equilibrium.

"I CAN MAKE IT SIT, BEG, FETCH, AND PLAY DEAD JUST FINE, BUT I STILL NEED TO FINE-TUNE ITS COEFFICIENT OF VOLUME EXPANSION!"

CHAPTER 14
KINETIC THEORY

Ludwig Boltzmann [1844–1906], who spent much of his life
studying statistical mechanics, died in 1906, by his own hand.
Paul Ehrenfest [1880–1933], carrying on the work, died similarly
in 1933. [So did another disciple, Percy Bridgman (1882–1961).]
Now it is our turn to study statistical mechanics.

Perhaps it will be wise to approach the subject cautiously.

*David L. Goodstein**

We have studied the mechanics of single particles and the interaction of several particles in collisions. When we approach the mechanics associated with the many particles in systems such as gases, liquids, and solids, we are faced with analyzing the dynamics of a *huge* number of particles. The dynamics of such many-particle systems is called *statistical mechanics* since, as we shall see, aspects of statistics come into play to facilitate the analysis.

The game involved in studying a system with a large number of particles is similar to what happens after every physics test. Of course, you are interested in your individual grade, but you also want to know the class average to see how you fit into the grand scheme.[†] The average characterizes the group as a whole. The average stems from the large number of individual grades via a suitable arithmetic, statistical procedure. In like manner, with huge numbers of particles, suitable averages of the physical characteristics and motions of individual particles provide information about the macroscopic behavior of the system as a whole. We forgo details about individual particles and concentrate on the collective behavior. So, borrowing a term from the founder of communism, Karl Marx (1818–1883), statistical mechanics may be viewed as a collectivization of physics.

The kinetic theory that we study in this chapter is a special aspect of the statistical mechanics of large numbers of particles. We begin with the simplest model for a monatomic ideal gas, and extend the results to molecular gases, even applying what we can of the theory to solids as well.

14.1 BACKGROUND FOR THE KINETIC THEORY OF GASES

One thing bears mentioning right away. It is *not* at all obvious that a gas consists of very tiny particles (atoms and molecules) moving around at high speeds. We certainly cannot directly see or sense that this is the case. Indeed, during the late 17th century, Newton himself imagined a very different model for a gas, retrospectively called a **static model**. In this model[‡] the corpuscular particles of the gas occupy fixed positions and fill the entire space between them, somewhat like foam squeegee-cubes. The corpuscular squeegees were imagined to expand and contract at their locations as the volume of the entire gas expands or contracts. With this model, Newton even was able to show that if the corpuscles repel each other with a force whose magnitude is inversely proportional to their separation, qualitatively similar to (but not quantitatively identical to) the way a spring resists compression, then the pressure is inversely proportional to the volume. The latter result neatly accounted for the isothermal pressure and volume experiments on gases made by Newton's contemporary, Robert Boyle (known as Boyle's law; see Section 13.8).

Other models of a gas emerged after Newton, but it was a **kinetic model**, first proposed by Daniel Bernoulli (1700–1782) in 1758, that was resurrected with great success during the 19th century by James Clerk Maxwell, Ludwig Boltzmann, and others. The kinetic model envisions a gas to be composed of many tiny particles, freely moving at high speed, whose pressure arises from the innumerable collisions of the particles with each other and with the walls of the container. Because the kinetic model gives a microscopic interpretation not only of pressure but also of absolute temperature and the specific heats of gases, it joined the pantheon of successful theories in physics and convinced many of the reality of the atomic view of matter. Nonetheless, like most physical theories, classical kinetic theory has its limitations and failures too, as we shall see.

Only in the 20th century did the rise of quantum mechanics account for the complex temperature dependence of the specific heats of gases and solids, as well as the structure of atoms and the mechanisms for molecular bonding. Fortunately, such ideas are not needed here for a kinematic understanding of key thermodynamic ideas. We shall find that the underlying microscopic motions of the particles in an ideal gas are mostly (but not entirely) within the reach of classical physics rather than quantum mechanics.

If we accept the atomic nature of matter, why do the particles in a solid resist both compression and tension? A brick is, after all, hard to compress and hard to pull apart. There must exist spring-like forces between the particles in solids; the particles must exist at equilibrium locations when not subjected to tension or compression. Nineteenth-century scientists thought such forces between the particles in solids (and liquids) likely were fundamentally electrical in nature, despite the fact that matter usually is electrically neutral. In view of our current view of the nuclear atom with its surrounding electrons, the ability of a solid to resist tension forces is all the more remarkable because electrons repel each other electrically. You might think that the atoms in solids should naturally fly apart of their own accord, which they clearly do not do because of molecular bonding and their overall electrical neutrality. On the other hand, the ability of solids and liquids to resist compression indicates that the particles likely are quite close together, packed cheek-to-cheek, if you will.

The trait that distinguishes gases from both liquids and solids is the ease with which gases can be compressed. The densities of gases are on the order of 1000 times smaller than those of liquids and solids. The low densities of gases (compared with solids and liquids) is a clue that the spacing of the particles in a gas likely is on the order of a factor of 10, the cube root of 1000, greater than in solids and liquids.[§]

Open a vial of perfume and soon the fragrance is detectable at some distance away from the vial. The same effect is apparent with obnoxious odors such as from cigarettes, cigars, skunks, or (embarrassingly) flatulence. This **diffusion** of smoke and odors through gases indicates that the particles in a gas likely are in motion, rather than occupying fixed positions, as in the static model of Newton. Gases also quickly occupy the entire volume allotted to them rather than settling down to just the bottom of a container in the manner of a liquid.

A consequence of the microscopic motion of the particles in a gas is extraordinary: the motion is *perpetual*, and for that we can be very glad. Yet on the macroscopic scale of our laboratories *no* motion is perpetual, thanks to frictional effects, as you can easily observe. The pressure of a gas in thermal equilibrium inside a container never runs down or decreases with time, with the gas particles gradually settling into the bottom of the container because of the gravitational force. If the pressure is due to the huge number of

---

[*](Chapter Opener) *States of Matter* (Prentice-Hall, Englewood Cliffs, New Jersey, 1975), page 1.

[†] Grading certainly *is* a scheme!

[‡] Isaac Newton, *Principia*, Proposition XXIII, theorem XVIII.

[§] The cube root arises since volumes are the cube of a characteristic spatial dimension.

Daniel Bernoulli proposed a kinetic model of gases. He is also known for Bernoulli's principle of ideal fluids (Chapter 11).

collisions of the particles with each other and with the walls of the container, the collisions must be modeled as completely elastic.

Yet collisions between particles on a macroscopic scale are invariably inelastic. Put a bunch of BBs or tennis balls in motion inside a container in an orbiting spacecraft and the particles eventually decrease their speeds because of inelastic collisions with each other (and the walls of the container). So there is something fundamentally different about the dynamics of the small atomic and molecular particles that are the constituents of a gas, compared with the dynamics of collections of macroscopic masses such as the BBs or tennis balls. The forces involved in the collisions of the particles of the gas with each other and with the walls of the container must be conservative, briefly acting only during the short intervals of the collisions and essentially zero otherwise. Thus the total of all their microscopic mechanical energies—the internal energy—remains constant if the system is in a state of thermodynamic equilibrium.

The constancy of the pressure of a gas in thermal equilibrium also indicates that if the velocity of one of the particles in the gas changes because of a collision, there likely is another particle in the collection that, in another collision, makes a velocity change in the opposite direction, so that the development of winds or currents in the gas does not occur spontaneously. This principle of **detailed balancing** is a subtle concept that underlies much of kinetic theory.

The temperature of the gas must somehow be related to this molecular motion, because an increase in temperature, at constant volume, results in an increase in pressure, thus implying that the collisions of the particles with the walls of the container are more frequent and/or more violent. The heat transfer to the gas at constant volume, raising the temperature of the gas, thus must increase the translational kinetic energy of the particles, *provided* that the particles have no internal degrees of freedom* such as rotation or vibration to soak up the incoming energy in other ways. If such extra motions of the particles can be brought into play (we say these motions become active, excited, or populated), we expect the thermodynamic behavior of the gas, in particular its specific heat, to differ from that in the situation where the particles are treated merely as point particles. As we will see, this is indeed the case.

_____

*We define this term more carefully later in the chapter.

One of the goals of kinetic theory is to discover the connection between temperature and pressure, and even a suitable average speed of the particles. For simplicity, we begin by assuming the gas particles are pointlike and see what results arise from studying the dynamics of such a collection. This is the kinetic theory of a monatomic ideal gas, a dilute gas whose particles are single atoms rather than molecules. The noble gases—helium, neon, argon, krypton, xenon, and radon—are examples of monatomic gases. Later, we will be able to extend the theory to include diatomic and polyatomic gases.

14.2 THE IDEAL GAS APPROXIMATION

In Chapter 13 we defined an ideal gas as one that obeyed the ideal gas equation of state:

$$PV = nRT \qquad (14.1)$$

where P is its pressure, V is its volume, n is the number of moles of the gas, R is the universal gas constant, and T is the absolute temperature of the gas. Real gases behave like an ideal gas if they are sufficiently dilute.

We describe an ideal gas as a suitable collection of pointlike particles and make the following assumptions:

1. The number of particles N in the gas is very large.

Of course any macroscopic mass of gas will satisfy this condition, because the number of particles in a mole of a gas is Avogadro's number ($N_A = 6.02 \times 10^{23}$ particles/mol). Even a trillionth (10^{-12}) of a mole has many particles:

$$10^{-12} \text{ mol} \times (6.02 \times 10^{23} \text{ particles/mol})$$
$$= 6.02 \times 10^{11} \text{ particles}$$

This is about 600 billion particles, three times the number of stars in an average-sized galaxy!

2. The volume V containing the gas is much larger than the total volume actually occupied by the gas particles themselves.

That is, if we let $V_{particle}$ be the volume of a single particle of the gas, then

$$NV_{particle} \ll V$$

This ensures that the separation of the particles in the gas is large compared with their size and that long-range forces between the particles are negligible. In other words, the particles are essentially pointlike and do not feel each other unless they collide.

3. The dynamics of the particles is governed by Newton's laws of motion.

You might wonder if the smallness of the particles requires quantum mechanics, but classical Newtonian physics is sufficient to describe most of their dynamics. The absence of long-range interactions (forces) means that the particles move freely (at constant velocity) between collisions with either the walls of the container or each other.

> 4. The particles are equally likely to be moving in any direction.

That is, the directions of the motions of the particles are random. There is no bulk motion, wind, or turbulence within the gas, which is consistent with our restriction to consider only simple thermodynamic systems.

> 5. The gas particles (be they atoms or molecules) interact with each other and with the walls of the container only via elastic collisions.

From our study of elastic collisions (Chapter 9), this means that both the total momentum and the total kinetic energy of the participants in each collision are conserved.

> 6. The gas is in thermal equilibrium with its surroundings.

This assumption is necessary because the thermodynamics we are considering is equilibrium thermodynamics. This assumption does not prohibit changes in the state of the gas; it simply means that any changes that occur are quasi-static, so that at any instant the gas is in a state of thermodynamic equilibrium.

> 7. The particles of the gas are identical and indistinguishable.

The gas is not a mixture of different constituents. This assumption is made merely to simplify the initial analysis before extending the theory to mixtures.

All pure gases, whether monatomic, diatomic, or polyatomic, that are sufficiently dilute exhibit behavior approaching this ideal gas approximation.

EXAMPLE 14.1 _____

One mole (1.00 mol) of helium gas is at a warm room temperature (300 K) and 1.000 atm pressure. How many helium atoms are in each cubic millimeter of the gas?

A cubic millimeter is not a very large volume!

Solution
Use the ideal gas law, Equation 14.1, to determine the volume of a mole of the gas. The pressure must be in pascals:

$$PV = nRT$$

$$V = \frac{nRT}{P}$$

$$= \frac{(1.00\ \text{mol})[8.315\ \text{J/(mol·K)}](300\ \text{K})}{1.013 \times 10^5\ \text{Pa}}$$

$$= 2.46 \times 10^{-2}\ \text{m}^3$$

One cubic millimeter is a volume $V' = 1.00 \times 10^{-9}\ \text{m}^3$. Since the particles are distributed uniformly throughout the container, the fraction of the particles that are in a cubic millimeter is the same as the volume ratio V'/V. One mole of gas has 6.02×10^{23} particles. Hence the number of helium atoms in one cubic millimeter is

$$N = \frac{V'}{V} N_A$$

$$= \frac{1.00 \times 10^{-9}\ \text{m}^3}{2.46 \times 10^{-2}\ \text{m}^3} (6.02 \times 10^{23}\ \text{atoms})$$

$$= 2.45 \times 10^{16}\ \text{atoms}$$

This is a huge number indeed. For comparison, the age of the universe (about 15 billion years) is only about 10^{17} s.

14.3 THE PRESSURE OF AN IDEAL GAS

From the ideal gas law, Equation 14.1, we already know that a sample of an ideal gas has the pressure

$$P = \frac{nRT}{V} \tag{14.2}$$

In the ideal gas approximation, the gas consists of a vast number of particles moving randomly and colliding occasionally with each other and with the walls of the container; we hope to account for the pressure, that is, derive an independent expression for the pressure, by examining the multitude of tiny impulses delivered to the walls of the container from the many collisions of the particles with the walls.

We consider the gas to be confined to a rectangular parallelopiped of volume *V*, as shown in Figure 14.1. Let *A* be the area of the right wall, perpendicular to the *x*-axis, and let ℓ be the length of the box along the *x*-axis.

We begin by examining the motion of one particle of the gas moving with a speed *v*. Let v_x be the *x*-component of the velocity of this particle. Assume that $v_x > 0$ m/s, as shown in Figure 14.2.

We need to assess the probability that the particle will collide with, say, the right-hand wall during a time interval Δt. If the particle is too far from the wall, the particle cannot reach

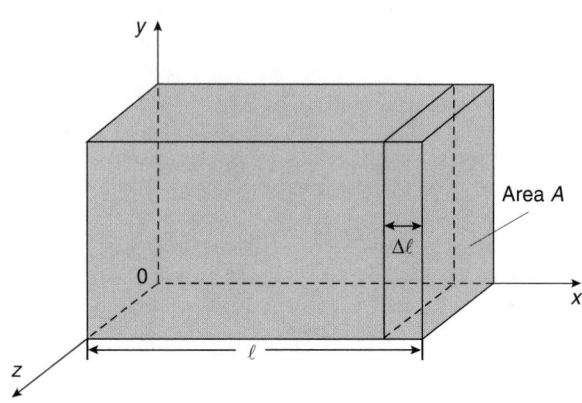

FIGURE 14.1 A rectangular parallelopiped volume of gas.

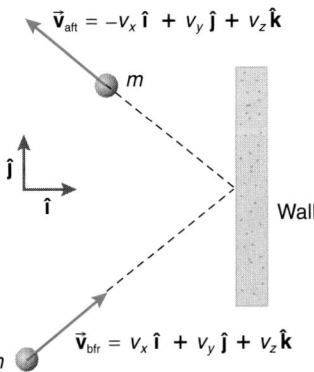

$$\vec{v}_{\text{aft}} = -v_x \hat{\imath} + v_y \hat{\jmath} + v_z \hat{k}$$

m

$\hat{\jmath}$

$\hat{\imath}$

Wall

$$\vec{v}_{\text{bfr}} = v_x \hat{\imath} + v_y \hat{\jmath} + v_z \hat{k}$$

m

FIGURE 14.2 A gas particle collides with the wall of the container.

and collide with the wall during the interval Δt. The particle can travel a distance $\Delta \ell = v_x \Delta t$ along the x-axis during the interval Δt. Thus, if the particle is within a distance $\Delta \ell$ of the wall (see Figure 14.1) and moving toward increasing values of x (that is, $v_x > 0$ m/s), it will collide with the wall during the interval Δt. If the particle is further from the wall than the distance $\Delta \ell$, then it will not collide with the wall during the interval Δt even though the particle may be moving toward the wall.

Since the particle has an equal probability of being found anywhere within the container, the probability that it lies within $\Delta \ell$ of the right-hand wall is given by the ratio of the volume of a slab of thickness $\Delta \ell$ to the volume of the container:

$$\frac{\text{volume of slab of thickness } \Delta \ell}{\text{volume of the container}} = \frac{A \, \Delta \ell}{A \, \ell}$$

$$= \frac{\Delta \ell}{\ell}$$

Of all the particles in the slab, half are expected to have positive x-components for their velocity (in which case they will collide with the wall if they are within the slab) and half will have negative x-components for their velocity [in which case they will not collide with the wall even if they are within the slab, because they are moving away from the (right-hand) wall]. Thus the probability that a particle with velocity component v_x is within $\Delta \ell$ of the right-hand wall and moving toward it is

$$\frac{1}{2} \frac{\Delta \ell}{\ell}$$

or, after substituting $\Delta \ell = v_x \Delta t$,

$$\frac{v_x \Delta t}{2\ell} \qquad (14.3)$$

This also is the probability that a particle with an x-component of velocity v_x will collide with the right-hand wall of the container during an interval Δt.

Now we look at the dynamics of a collision of a particle with this wall. The collisions are elastic according to our assumptions (Section 14.2). Let m be the mass of each individual particle of the gas. The momentum of the particle before the collision with the wall is

$$\vec{p}_{\text{bfr}} = mv_x \hat{\imath} + mv_y \hat{\jmath} + mv_z \hat{k}$$

Since the wall is fixed, its momentum is zero. After an elastic collision with the fixed wall, the x-component of the velocity of the particle is reversed (and becomes $-v_x$) while the y- and z-components of the velocity are unchanged. Thus the momentum of the particle after the collision is

$$\vec{p}_{\text{aft}} = -mv_x \hat{\imath} + mv_y \hat{\jmath} + mv_z \hat{k}$$

The change in the momentum $\Delta \vec{p}$ of the particle is

$$\begin{aligned} \Delta \vec{p} &= \vec{p}_{\text{aft}} - \vec{p}_{\text{bfr}} \\ &= -mv_x \hat{\imath} + mv_y \hat{\jmath} + mv_z \hat{k} - (mv_x \hat{\imath} + mv_y \hat{\jmath} + mv_z \hat{k}) \\ &= -2mv_x \hat{\imath} \end{aligned}$$

The particle's momentum change comes from the impulse it received from the right-hand wall during its collision with this wall. However, because of Newton's third law of motion, the impulse given to the particle by the wall has the same magnitude but is directed opposite to the impulse given to the wall by the particle. Hence, *the impulse given to the wall by the particle* during the collision is

$$\text{Impulse given to wall by particle in a single collision} = +2mv_x \hat{\imath}$$

The probability that a particle with an x-component of velocity v_x actually makes such a collision during an interval Δt is given by Equation 14.3. Thus the impulse we can *expect* to be delivered to the wall during the interval Δt from a particle with an x-component of velocity v_x is

$$\begin{aligned} \begin{array}{l} \text{Expected impulse given to wall} \\ \text{in time interval } \Delta t \text{ from a particle} \\ \text{with } x\text{-component of velocity } v_x \end{array} &= \frac{v_x \Delta t}{2\ell} 2mv_x \hat{\imath} \\ &= \left(\frac{mv_x^2}{\ell} \Delta t \right) \hat{\imath} \end{aligned}$$

Now we have to sum over all N particles of the gas because they have various values for their x-components of velocity. Denote the particles with numerical subscripts: v_{1x} is the x-component of the velocity of particle 1, v_{2x} is the x-component of the velocity of particle 2, and so forth. Thus the *total impulse* delivered to the wall during the interval Δt by all the particles that collide with the wall during the time interval Δt is

$$\begin{array}{l} \text{Total impulse} \\ \text{delivered to wall} \\ \text{during interval } \Delta t \end{array} = \left[\frac{m}{\ell} \left(v_{1x}^2 + v_{2x}^2 + \cdots + v_{Nx}^2 \right) \Delta t \right] \hat{\imath} \qquad (14.4)$$

There are N terms (a *great* number) in the sum of the squares of the x-components of the velocities. If we take the sum of the squares of the velocity components and divide by the total number of particles (N), we obtain an average value for the square of the x-component of the velocity:

$$\langle v_x^2 \rangle = \frac{v_{1x}^2 + v_{2x}^2 + \cdots + v_{Nx}^2}{N}$$

Thus

$$v_{1x}^2 + v_{2x}^2 + \cdots + v_{Nx}^2 = N \langle v_x^2 \rangle$$

Making this substitution into Equation 14.4, we find

$$\text{Total impulse delivered to wall during interval } \Delta t = \left(\frac{m}{\ell} N \langle v_x^2 \rangle \Delta t \right) \hat{\mathbf{i}} \qquad (14.5)$$

The impulse given to the wall by the particles can be written as the *average total force* $\vec{\mathbf{F}}_{\text{ave total}}$ of all the particles on the wall, multiplied by the interval Δt during which the collisions took place:

$$\vec{\mathbf{F}}_{\text{ave total}} \Delta t = \text{Total impulse given to wall during interval } \Delta t$$

Substituting for the total impulse using Equation 14.5, we find

$$\vec{\mathbf{F}}_{\text{ave total}} \Delta t = \left(\frac{m}{\ell} N \langle v_x^2 \rangle \Delta t \right) \hat{\mathbf{i}}$$

Hence the magnitude of the average total force exerted on the wall by all the collisions is

$$F_{\text{ave total}} = \frac{m}{\ell} N \langle v_x^2 \rangle \qquad (14.6)$$

Since the particles are moving randomly throughout the volume without any bulk motion (wind, turbulence, or currents), there should be nothing sacred about the *x*-direction. Symmetry implies that the average value of the square of the *x*-component of the velocity must be equal to the average value of the square of the *y*- or *z*-components of the velocity of the particles; that is,

$$\langle v_x^2 \rangle = \langle v_y^2 \rangle = \langle v_z^2 \rangle$$

But the square of the velocity itself is $\vec{\mathbf{v}} \cdot \vec{\mathbf{v}} = v^2$:

$$v^2 = v_x^2 + v_y^2 + v_z^2$$

Also, the average value of a sum is the sum of the average values:

$$\langle v^2 \rangle = \langle v_x^2 + v_y^2 + v_z^2 \rangle$$
$$= \langle v_x^2 \rangle + \langle v_y^2 \rangle + \langle v_z^2 \rangle$$

The quantity $\langle v^2 \rangle$ is called the **mean square speed**. If we want a typical speed, we take the square root of the mean square speed and call it the **root-mean-square (rms) speed** v_{rms} of the particles of the gas:

$$v_{\text{rms}} \equiv \sqrt{\langle v^2 \rangle} \qquad (14.7)$$

Since the three averages of the velocity components are all equal because of symmetry, we have

$$\langle v^2 \rangle = 3 \langle v_x^2 \rangle$$

$$\langle v_x^2 \rangle = \frac{\langle v^2 \rangle}{3}$$

Hence Equation 14.6 for the magnitude of the average force exerted on the right-hand wall of the container can be written as

$$F_{\text{ave total}} = \frac{m}{\ell} N \frac{1}{3} \langle v^2 \rangle \qquad (14.8)$$

Since A is the area of this wall of the container, the *pressure* P exerted by the gas on this wall is

$$P = \frac{F_{\text{ave}}}{A}$$

We use Equation 14.8 for F_{ave}; the pressure becomes

$$P = \frac{m}{3A\ell} N \langle v^2 \rangle$$

But the product $A\ell$ is the volume V of the container in which the gas resides. Thus

$$P = \frac{m}{3V} N \langle v^2 \rangle \qquad (14.9)$$

Since $mN = mnN_A$, and $mN_A = M$, the molar mass, we can write Equation 14.9 in the form

$$P = \frac{nM}{3V} \langle v^2 \rangle \qquad (14.10)$$

Thus we see that the pressure of the gas arises from the impulses provided by the gas particles colliding with the walls of the container of the gas.

QUESTION 1

In an elastic collision of a particle of an ideal gas with the wall of the container, the momentum of the particle *changes*. Yet we have seen that momentum is conserved in elastic (and inelastic) collisions. Why is the momentum of the particle not conserved?

EXAMPLE 14.2

One mole (1.00 mol) of helium at temperature 300 K has a volume of 24.6 liters = 2.46×10^{-2} m³ and a pressure of 1.00 atm. This volume can be calculated via the ideal gas law (see Example 14.1). Determine the mean square speed of the particles and the root-mean-square speed.

Solution
Use Equation 14.10, and solve for the mean square speed:

$$\langle v^2 \rangle = \frac{3PV}{nM}$$

One mole of helium has a molar mass of 4.00 g/mol = 4.00×10^{-3} kg/mol. One atmosphere pressure is 1.01×10^5 Pa. Making the substitutions, you find

$$\langle v^2 \rangle = \frac{3 \, (1.01 \times 10^5 \text{ Pa}) \, (2.46 \times 10^{-2} \text{ m}^3)}{(1 \text{ mol})(4.00 \times 10^{-3} \text{ kg/mol})}$$
$$= 1.86 \times 10^6 \text{ m}^2/\text{s}^2$$

The root-mean-square speed is the square root of the mean square speed:

$$v_{\text{rms}} = \sqrt{\langle v^2 \rangle}$$
$$= 1.36 \times 10^3 \text{ m/s}$$

The rms speed in m/s of this typical motion is quite large. For comparison, the speed of sound in helium under comparable conditions is about 980 m/s. When sound waves exist in the gas, the bulk motions of the particles caused by the sound waves are superimposed on the random thermal motions of the particles.

14.4 THE MEANING OF THE ABSOLUTE TEMPERATURE

We saw in the previous section that the pressure of a gas arises from the multitude of submicroscopic mechanical impulses resulting from the collisions of the gas particles with the walls of the container. In this section we discover a microscopic interpretation of the absolute temperature.

Rewriting Equation 14.9 slightly,

$$PV = \frac{Nm}{3} \langle v^2 \rangle \tag{14.11}$$

But wait. We also have an experimental relation for the product PV that came from laboratory measurements of gases. The kinetic theory expression for PV must be identical to the ideal gas law expression for PV, the equation of state, Equation 13.13:

$$PV = NkT$$

where k is Boltzmann's constant, 1.38×10^{-23} J/K. Thus the right-hand sides of Equations 14.11 and 13.13 must be identical, even though one came from kinetic theory and the other from the gas law:

$$NkT \equiv \frac{Nm}{3} \langle v^2 \rangle$$

Solving for the absolute temperature of the gas, we find

$$T = \frac{m}{3k} \langle v^2 \rangle \tag{14.12}$$

The *average* translational kinetic energy of a *single* particle of the gas is

$$\text{KE}_{\text{ave}} = \frac{1}{2} m \langle v^2 \rangle$$

Hence Equation 14.12 for the temperature can be written as

$$T = \frac{2}{3k} \text{KE}_{\text{ave}}$$

or

$$\text{KE}_{\text{ave}} = \frac{3}{2} kT \tag{14.13}$$

The temperature of all the gas (a macroscopic quantity) is a manifestation of the average translational kinetic energy of each particle (a submicroscopic quantity).

A measurement of temperature is a measurement of the average translational kinetic energy of any particle of the gas.

One of the triumphs of kinetic theory is this revelation of the meaning of the absolute temperature.

Equation 14.13 shows that the average translational kinetic energy of a particle in an ideal gas is the same for all particles and is equal to

$$\frac{3}{2} kT$$

Two different gases at the same temperature have the same average translational kinetic energy per particle regardless of the masses of the particles of the two gases.

Thus a gas composed of more massive particles will have slower average particle speeds than a gas composed of smaller-mass particles at the same temperature, but the same average kinetic energy per particle.

From Equation 14.12, the mean square speed is

$$\langle v^2 \rangle = \frac{3kT}{m}$$

Hence the rms speed of the particles of the gas is

$$v_{\text{rms}} = \left(\frac{3kT}{m} \right)^{1/2} \tag{14.14}$$

Since $k = R/N_A$ from Equation 13.18, we can also write this as

$$v_{\text{rms}} = \left(\frac{3RT}{mN_A} \right)^{1/2} \tag{14.15}$$

Also, since m is the mass of an individual particle of the gas, the product mN_A is the molar mass M. Thus Equation 14.15 becomes

$$v_{\text{rms}} = \left(\frac{3RT}{M} \right)^{1/2} \tag{14.16}$$

The particles of a gas have a range of speeds. The average speed is found by taking the average of the speeds of all the particles at a given instant (remember that the speed is a positive scalar since it is the magnitude of the velocity):

$$\langle v \rangle = \frac{v_1 + v_2 + v_3 + \cdots + v_N}{N} \tag{14.17}$$

The rms speed is found from

$$v_{\text{rms}} \equiv \sqrt{\langle v^2 \rangle} = \left(\frac{v_1^2 + v_2^2 + v_3^2 + \cdots + v_N^2}{N} \right)^{1/2} \tag{14.18}$$

The average speed is less than the rms speed.

The inequality arises because the squared numbers averaged in forming the rms speed typically are *larger* than the corresponding numbers used in calculating the average speed;* see Example 14.3.

---

*Even if most of the speeds are less than 1 m/s, the average speed is less than the rms speed, in this case because of the square root in v_{rms}. This can be verified by taking an example with all the particle speeds less than 1 m/s; see Problem 12.

> **PROBLEM-SOLVING TACTICS**
>
> **14.1** When using Equation 14.16, if R is expressed as 8.315 J/(mol·K), be sure the molar mass M is expressed in kilograms per mole rather than grams per mole.
>
> **14.2** When using Equations 14.14–14.16, be sure you use the absolute temperature T in kelvin (not celsius).

QUESTION 2

The average speed of all the particles in a gas at a given instant is *not* zero, whereas the average velocity of all the particles *is* zero. Explain why.

QUESTION 3

A gas consists of a mixture of oxygen molecules (O_2) and nitrogen molecules (N_2). Which molecule has, on average, the greater rms speed?

EXAMPLE 14.3

Consider an 11 000-particle gas system with speeds distributed as follows:

1000 particles each with speed 100 m/s,
2000 particles each with speed 200 m/s,
4000 particles each with speed 300 m/s,
3000 particles each with speed 400 m/s, and
1000 particles each with speed 500 m/s.

Find the average speed and the rms speed of the particles.

Solution

The *average speed* is found using Equation 14.17:

$$\langle v \rangle = [1000\,(100 \text{ m/s}) + 2000\,(200 \text{ m/s})$$
$$+ 4000\,(300 \text{ m/s}) + 3000\,(400 \text{ m/s})$$
$$+ 1000\,(500 \text{ m/s})]/11\,000$$
$$= 309 \text{ m/s}$$

The *rms speed* is found from Equation 14.18:

$$v_{rms} = \{[1000\,(100 \text{ m/s})^2 + 2000\,(200 \text{ m/s})^2$$
$$+ 4000\,(300 \text{ m/s})^2 + 3000\,(400 \text{ m/s})^2$$
$$+ 1000\,(500 \text{ m/s})^2]/11\,000\}^{1/2}$$
$$= 328 \text{ m/s}$$

Note that the average speed is less than the rms speed.

EXAMPLE 14.4

A sample of helium gas is at a warm room temperature (300 K) and a pressure of 0.500 atm. What is the average kinetic energy of a particle of the gas?

Solution

The average kinetic energy of a particle in the gas is found from Equation 14.13,

$$KE_{ave} = \frac{3}{2} kT$$

where k is Boltzmann's constant. Notice that the average kinetic energy per particle depends only on the temperature, so that the pressure and identity of the particles are irrelevant. Substituting, you get

$$KE_{ave} = \frac{3}{2}\,(1.38 \times 10^{-23} \text{ J/K})\,(300 \text{ K})$$
$$= 6.21 \times 10^{-21} \text{ J}$$

EXAMPLE 14.5

A sample of helium gas at a warm room temperature (300 K) has a pressure of 1.00 atm. A sample of neon is under the same conditions. The molar mass of helium is 4.00 g/mol and that of neon is 20.2 g/mol.

a. Find the rms speed of the helium atoms and of the neon atoms.
b. What is the average kinetic energy per particle in each sample?

Solution

a. Following Problem-Solving Tactic 14.1, the molar mass of helium in SI units is 4.00 g/mol = 4.00×10^{-3} kg/mol. Use Equation 14.16 for the rms speed:

$$v_{rms} = \left(\frac{3RT}{M}\right)^{1/2}$$
$$= \left(\frac{3\,[8.315 \text{ J/(mol·K)}]\,(300 \text{ K})}{4.00 \times 10^{-3} \text{ kg/mol}}\right)^{1/2}$$
$$= 1.37 \times 10^3 \text{ m/s}$$

Notice that the rms speed depends on the *absolute temperature* but not at all on the pressure.

The molar mass of neon in SI units is 20.2×10^{-3} kg/mol. Use Equation 14.16 for the rms speed once again:

$$v_{rms} = \left(\frac{3RT}{M}\right)^{1/2}$$
$$= \left(\frac{3\,[8.315 \text{ J/(mol·K)}]\,(300 \text{ K})}{20.2 \times 10^{-3} \text{ kg/mol}}\right)^{1/2}$$
$$= 609 \text{ m/s}$$

Notice that the more massive neon atoms have significantly slower speeds than the less massive helium atoms at the same temperature.

b. According to Equation 14.13, the average kinetic energy per particle depends solely on the temperature. Since the two samples are at the same temperature, they have the *same* kinetic energy per particle:

$$KE_{ave} = \frac{3}{2}kT$$
$$= \frac{3}{2}(1.38 \times 10^{-23} \text{ J/K})(300 \text{ K})$$
$$= 6.21 \times 10^{-21} \text{ J}$$

The more massive neon atoms move more slowly than the helium atoms but have the same average kinetic energy.

EXAMPLE 14.6

At what temperature will the particles in a sample of helium gas have an rms speed of precisely 1.00 km/s?

Solution
Use Equation 14.16, and solve for the absolute temperature:

$$T = \frac{Mv_{rms}^2}{3R}$$

The molar mass of helium is 4.00 g/mol = 4.00×10^{-3} kg/mol. Thus the absolute temperature is

$$T = \frac{(4.00 \times 10^{-3} \text{ kg/mol})(1.00 \times 10^{3} \text{ m/s})^2}{3[8.315 \text{ J/(mol·K)}]}$$
$$= 160 \text{ K}$$

Chilly.

14.5 THE INTERNAL ENERGY OF A MONATOMIC IDEAL GAS

The temperature of a monatomic ideal gas is a measure of the average kinetic energy per particle. This energy is associated with the translational motion of the particles, not with any internal motions such as vibrations and/or rotations of the particles, because it is the momentum of the whole particle that contributes to the total gas pressure. We can therefore imagine a monatomic ideal gas simply as a collection of pointlike particles. Thus, if N is the number of particles in the monatomic ideal gas, the internal energy U is N times the average kinetic energy per particle. Using Equation 14.13 for the average kinetic energy per particle, we find that the internal energy is

$$U = N\left(\frac{3}{2}kT\right) \quad \text{(monatomic ideal gas)} \quad (14.19)$$

The number of particles N is equal to the number of moles n times Avogadro's number N_A:

$$N = nN_A$$

Equation 14.19 thus becomes

$$U = nN_A\left(\frac{3}{2}kT\right) \quad \text{(monatomic ideal gas)}$$

But from Equation 13.18, the product $N_A k$ is the universal gas constant R. Thus the internal energy is

$$U = \frac{3}{2}nRT \quad \text{(monatomic ideal gas)} \quad (14.20)$$

The internal energy of a monatomic ideal gas is a function only of the absolute temperature T and the quantity n of gas.

14.6 THE MOLAR SPECIFIC HEATS OF AN IDEAL GAS

With a knowledge of the internal energy of a monatomic ideal gas based on kinetic theory, we can predict the molar specific heats of the gas.

Specific Heat at Constant Volume for a *Monatomic* Ideal Gas

If a differential amount of heat transfer dQ occurs to n moles of a gas at constant volume, there is a differential change dT in the temperature of the gas. The *molar* specific heat at constant volume is defined as (Equation 13.27)

$$c_V = \frac{1}{n}\left(\frac{dQ}{dT}\right)_V$$

We take differentials of the first law of thermodynamics, Equation 13.51:

$$dQ = dU + dW$$
$$= dU + P\,dV$$

However, the heat transfer here is at constant volume, and so $dV = 0$ m^3 and there is no work done by the gas ($dW = P\,dV = 0$ J). Therefore

$$dQ = dU$$

and the molar specific heat at constant volume is

$$c_V = \frac{1}{n}\frac{dU}{dT} \quad (14.21)$$

We use Equation 14.20 for the internal energy of the monatomic ideal gas:

$$c_V = \frac{1}{n}\frac{d}{dT}\left(\frac{3}{2}nRT\right)$$
$$c_V = \frac{3}{2}R \quad \text{(monatomic ideal gas)} \quad (14.22)$$

Substituting the numerical value for R, we find

$$c_V = 12.47 \text{ J/mol}$$

Table 14.1 indicates experimental values of the molar specific heat at constant volume for various monatomic gases near room temperature; the values are satisfyingly close to the value predicted by using kinetic theory.

TABLE 14.1 Measured Molar Specific Heats of Various Gases at Approximately Room Temperature (300 K)

Gas	c_V (J/mol·K)	c_P (J/mol·K)	$c_P - c_V$ (J/mol·K)
Monatomic gases			
Helium (He)	12.52	20.79	8.27
Neon (Ne)	12.68	20.79	8.11
Argon (Ar)	12.45	20.79	8.34
Krypton (Kr)	12.45	20.79	8.34
Xenon (Xe)	12.52	20.79	8.27
Diatomic gases			
Hydrogen (H_2)	20.44	28.82	8.38
Nitrogen (N_2)	20.80	29.12	8.32
Oxygen (O_2)	20.98	29.37	8.39
Carbon monoxide (CO)	20.74	29.04	8.30
Polyatomic gases			
Carbon dioxide (CO_2)	28.17	36.62	8.45
Nitrous oxide (N_2O)	28.39	36.90	8.51
Hydrogen sulfide (H_2S)	27.36	36.12	8.76

Specific Heat at Constant Pressure for *Any* Ideal Gas

Now let's try predicting the specific heat at constant pressure for any ideal gas, not necessarily a monatomic gas. If differential heat transfer dQ occurs to n moles of an ideal gas at constant pressure, the temperature changes by dT. The molar specific heat at constant pressure is, from Equation 13.28,

$$c_P = \frac{1}{n}\left(\frac{dQ}{dT}\right)_P$$

The first law of thermodynamics indicates that

$$dQ = dU + dW$$
$$= dU + P\,dV$$

In this case work *is* done by the gas since its volume changes. Substituting for dQ in the expression for the molar specific heat at constant pressure, we have

$$c_P = \frac{1}{n}\left[\left(\frac{dU}{dT}\right)_P + P\left(\frac{dV}{dT}\right)_P\right]$$
$$= \frac{1}{n}\left(\frac{dU}{dT}\right)_P + \frac{1}{n}P\left(\frac{dV}{dT}\right)_P \qquad (14.23)$$

where the subscripts indicate that the derivatives are taken at constant pressure. For the first term, the subscript is immaterial, since for an ideal gas the internal energy depends only on the absolute temperature T, not the pressure or volume. From Equation 14.21, the first term on the right-hand side of Equation 14.23 is the molar specific heat at constant *volume*. Thus

$$c_P = c_V + \frac{1}{n}P\left(\frac{dV}{dT}\right)_P \qquad (14.24)$$

The ideal gas equation of state is $PV = nRT$, so that at constant pressure

$$\left(\frac{dV}{dT}\right)_P = \frac{nR}{P}$$

Substituting this into Equation 14.24, we find

$$c_P = c_V + R \qquad \text{(any ideal gas)} \qquad (14.25)$$

For any ideal gas, the molar specific heat at constant pressure is *greater* than the molar specific heat at constant volume by an amount equal to the gas constant.

There is an easy way to see (qualitatively) that the molar specific heat at constant pressure must be greater than the molar specific heat at constant volume. When heat transfer occurs to the system at constant volume, $dV = 0 \text{ m}^3$ and no work is done by the gas: the heat transfer goes *solely* into the internal energy (a function of the absolute temperature T alone). However, when heat transfer to the gas occurs at constant pressure, some of the energy goes into the internal energy and some of it is manifested as work done by the gas. Since only a part of the heat transfer goes into internal energy, it is necessary to have more heat transfer at constant pressure than at constant volume in order to get a specified rise in temperature; hence $c_P > c_V$.

If we specialize to a *monatomic* ideal gas, for which we have found that $c_V = 3R/2$, the specific heat at constant pressure is

$$c_P = \frac{3}{2}R + R$$
$$= \frac{5}{2}R \qquad \text{(monatomic ideal gas)} \qquad (14.26)$$
$$= 20.79 \text{ J/(mol·K)}$$

Experimental values of the specific heat at constant pressure for various monatomic gases are also tabulated in Table 14.1, and once again we see that the predictions of kinetic theory are very good.

The central assumption involved in deriving Equation 14.25 was that the internal energy U was a function of only the absolute temperature (and the quantity of the gas). Hence for *any* gas whose internal energy U is a function of T (and, of course, the quantity n of gas), the relationship between the molar specific heats at constant pressure and at constant volume is given by Equation 14.25:

$$c_P = c_V + R \qquad [\text{any gas for which } U = U(T)]$$

For such gases, the *difference* between the molar specific heats is equal to the gas constant R:

$$c_P - c_V = R \qquad [\text{any gas for which } U = U(T)]$$
$$(14.27)$$

Note that the differences between the experimentally determined molar specific heats for each of many gases come close to this prediction of kinetic theory, as indicated in the last column of Table 14.1.

14.7 COMPLICATIONS ARISE FOR DIATOMIC AND POLYATOMIC GASES

In Table 14.1 we saw that kinetic theory predicts the molar specific heats of monatomic gases quite closely, at least for the temperatures at which the experiments were conducted (300 K, near room temperature). On the other hand, the table also indicates that diatomic and polyatomic gases have molar specific heats whose values are quite different from those of monatomic gases; yet the *differences* between the molar specific heat at constant pressure and that at constant volume are quite close to the value of the gas constant, R, as with monatomic gases. In view of the previous section, this indicates that the internal energy of diatomic and polyatomic gases also is a function of the absolute temperature (and the quantity n of gas present), but not the same function as Equation 14.20 for monatomic gases. How can we determine the temperature dependence of the internal energy for such more complex gases?

Table 14.1 indicates that the molar specific heats of diatomic gases and polyatomic gases are *greater* than the corresponding values for monatomic gases. Hence the heat transfer to diatomic and polyatomic gases must be going not only into kinetic energy associated with the motion of the center of mass of the particles (as measured by the temperature of the gas), but also into other types of motions *internal* to the particles of the gas themselves, such as rotations and vibrations about the center of mass of each particle. These types of motions we consider in the next section.

14.8 DEGREES OF FREEDOM AND THE EQUIPARTITION OF ENERGY THEOREM

The idea of **degrees of freedom**, as the name implies, suggests restrictions or limits. There are degrees of freedom associated with our behavior as children, teenagers, and adults. In classical physics, the term degrees of freedom refers to a pure number that quantifies the number of distinct quadratic **generalized coordinates** needed to specify the *microscopic* total mechanical energy of one particle of the system (be it an atom or molecule). The various contributions to the total mechanical energy of a particle are continually shared and exchanged among all the parts of the interacting particle system. The sum of the microscopic total mechanical energies of each of the particles is the total internal energy U of the many-particle system.

The phrase quadratic generalized coordinates might encompass traditional coordinates such as a squared position (as in the potential energy associated with Hooke's force law), and also squared velocity and angular velocity components (such as in the expressions for the translational and rotational kinetic energies). Consider a few macroscopic examples:

1. A pointlike bead confined to slide along a fixed straight wire in the horizontal plane is said to have 1 **active** degree of freedom; see Figure 14.3. The gravitational potential energy of the bead is constant and we can choose it to be zero. A single squared generalized coordinate, say that of the x-component of the velocity of the particle, is needed to specify the total mechanical energy of the particle, which in this case is simply the kinetic energy associated with the translation of the center of mass of the particle.

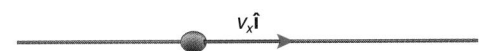

FIGURE 14.3 A pointlike bead on a wire has 1 degree of freedom.

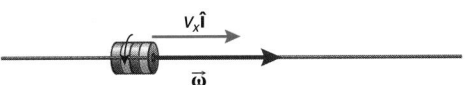

FIGURE 14.4 A large bead that can spin about and slide along a wire has 2 degrees of freedom.

2. A large bead that can slide along the wire *and* also *rotate* about the axis of the wire has 2 active degrees of freedom; see Figure 14.4. One squared velocity component is needed to describe the kinetic energy associated with the translation of the center of mass, and a squared angular velocity component is needed to describe the kinetic energy associated with the spin of the bead about the axis of the wire.

3. A pointlike water droplet on a hot skillet has 2 active degrees of freedom, because it is confined to move only across the two-dimensional area of the skillet. Two squared velocity components are needed to specify the kinetic energy associated with the motion of its center of mass. If the droplet also spins about its center of mass, another degree of freedom comes into play (becomes active): one associated with the rotational kinetic energy of the drop about its center of mass.

In three dimensions, three squared velocity components are needed to specify the total mechanical energy (here purely translational kinetic energy) of a point particle; thus, a point particle in three dimensions has 3 active degrees of freedom.

The idea of a degree of freedom seems simple enough. We will see, however, that there are some subtleties associated with this concept that warrant further comment.

The importance of being able to specify the number of degrees of freedom associated with each particle in a thermodynamic system arises from what is called the **equipartition of energy theorem**. We will not prove the theorem here, since it is a consequence of more sophisticated ideas in statistical mechanics. Nonetheless, we state the theorem:

> For a system of particles in thermal equilibrium, the average energy associated with each active degree of freedom of a particle is the same. For a system with f active degrees of freedom, the average energy per particle is f times as great as for a system with a single degree of freedom.

We can use our model of an ideal gas to find out how much energy is associated with each active degree of freedom. We already have found that the internal energy of n moles of a monatomic ideal gas of point particles is

$$U = \frac{3}{2} nRT$$

Since these particles each have 3 active degrees of freedom, and there are $nN_A = N$ particles in n moles, the energy per particle may be written as

$$\frac{U}{N} = \frac{3nRT}{2N}$$

$$= 3\frac{RT}{2N_A}$$

$$= 3\frac{kT}{2}$$

where k, Boltzmann's constant, is conveniently related to R as

$$k = \frac{R}{N_A}$$

The average kinetic energy per particle of the monatomic ideal gas of point particles is thus

$$\frac{1}{2}kT$$

for each of its three active degrees of freedom.

> Hence the average energy per particle associated with each active degree of freedom of a system is
>
> $$\frac{1}{2}kT$$

We will see later that polyatomic gases have the 3 translational degrees of freedom *and* some for their possible internal motions like vibration and rotation. The expressions for their average internal energies per particle contain more than the foregoing three units of $kT/2$ and are of different form.

14.9 SPECIFIC HEAT OF A SOLID*

It is quite surprising that the equipartition of energy theorem can account for the molar specific heat of a *solid* with all its close neighbors and the restricted motion of its constituent particles.

A crude model of an ideal solid is to imagine each atom having an equilibrium position and connected to its neighbors by springs. The springs permit each atom to execute simple harmonic oscillations about its equilibrium position. The oscillations can be resolved into equivalent oscillations along each of the three Cartesian coordinate axes. These equivalent oscillations along the three directions likely have different spring constants and oscillation frequencies. In contrast to an ideal gas, here we have both potential and kinetic energies to account for. There are 2 degrees of freedom associated with *each* oscillator axis, arising from the squared generalized coordinates in the kinetic and potential energy terms of simple harmonic oscillation:

$$KE + PE = \frac{1}{2}mv_x^2 + \frac{1}{2}k_x x^2$$

Thus each atom in the solid has $3 \times 2 = 6$ active degrees of freedom. The equipartition of energy theorem states that the average energy per particle associated with each degree of freedom

is $kT/2$. The entire solid sample has a huge number N of these atoms. The total internal energy of a solid system of atoms is then

$$U = N\left[6\left(\frac{1}{2}kT\right)\right]$$

$$= 3NkT$$

Since $Nk = nN_A k = nR$, where n is the number of moles and N_A is Avogadro's number, the internal energy is

$$U = 3nRT \qquad \text{(ideal solid)} \qquad (14.28)$$

Now consider the implications of the first law of thermodynamics. The differential heat transfer dQ to the system is equal to the differential change dU in the internal energy plus the differential work dW done by the system on its surroundings:

$$dQ = dU + dW$$

Here $dW = P\,dV$, so that

$$dQ = dU + P\,dV$$

The work done by a solid when there is heat transfer to the system is quite small because the change in the volume of the solid is tiny, thanks to the small numerical value of the coefficient of thermal expansion (volume) for solids. If we ignore the work done by the solid, then

$$dQ \approx dU$$

This is equivalent to saying that there is no difference between the specific heat at constant volume and that at constant pressure for a solid. For differential heat transfer dQ to a system of n moles, the temperature rises by dT, and the *molar* specific heat is (from Equation 13.24):

$$c = \frac{1}{n}\frac{dQ}{dT}$$

But we found that if the work done by the solid is zero (constant volume) or negligible (at constant pressure), then $dQ \approx dU$, and so

$$c \approx \frac{1}{n}\frac{dU}{dT}$$

From Equation 14.28,

$$\frac{dU}{dT} = 3nR$$

and so the molar specific heat of the ideal solid is

$$c \approx \frac{1}{n}3nR$$

$$\approx 3R \qquad \text{(ideal solid)} \qquad (14.29)$$

$$= 24.95 \text{ J/mol·K}$$

This result implies that the molar specific heat of an ideal solid composed of point particles is constant and equal to $3R$. This remarkable result was first discovered by Pierre Dulong (1785–1838) and Alexis Petit (1791–1820) in their experiments on certain elemental metals in the early part of the 19th century, and thus is called the **law of Dulong and Petit**. Table 13.4

(p. 602) indicates that the molar specific heat at room temperature of many metallic elements is close to this prediction. Molecular solids cannot be regarded as point-particle solids, and so we do not expect the prediction of Dulong and Petit to apply to them [notice that the values for water and ice in Table 13.4 are not close to $3R = 24.95$ J/(mol·K)].

14.10 SOME FAILURES OF CLASSICAL KINETIC THEORY

Atoms are not truly point particles. The particles of a monatomic gas (single atoms) have finite size. So the question arises, what about the rotation of the particles? In classical physics it is possible to imagine painting a little dot or planting a flag on the atom to track its rotation. In classical physics it is possible, in principle at least, to follow the rotation of atoms (by watching the dot or flag) about three mutually perpendicular coordinate axes. Three squared angular velocity components are needed to describe the rotational kinetic energies about the three axes. Thus you might think that the monatomic gas particle has a total of 6 active degrees of freedom: 3 associated with the kinetic energy of the center of mass, and 3 with rotational kinetic energy about the center of mass. Maybe you also can imagine the atoms vibrating, twisting, and otherwise contorting.

If f is the number of degrees of freedom, classical kinetic theory and the equipartition of energy theorem predict that the average energy per particle is

$$ f\left(\frac{1}{2}kT\right) \tag{14.30} $$

The internal energy is then

$$ U = Nf\left(\frac{1}{2}kT\right) $$
$$ = f\frac{nRT}{2} \tag{14.31} $$

The molar specific heat at constant volume is

$$ c_V = \frac{1}{n}\frac{dU}{dT} $$
$$ = \frac{1}{n}\frac{d}{dT}\left(f\frac{nRT}{2}\right) $$
$$ = \frac{1}{n}f\frac{nR}{2} $$
$$ = f\frac{R}{2} \tag{14.32} $$

Hence classical physics and the equipartition of energy theorem would attribute an average energy of at least

$$ 6\left(\frac{1}{2}kT\right) = 3kT $$

to each particle of an ideal monatomic gas. Using Equations 14.32 and 14.21, such an energy leads to a specific heat at constant volume of

$$ 6\frac{R}{2} = 3R $$

This is *twice* as large as experiments indicate it is, $3R/2$ (from Equation 14.22 and Table 14.1). This situation is a clear failure of classical theory. You then might try to argue that to the extent we regard the particles as points, the energies associated with such possible rotational motions scarcely compare with the other energies (such as the kinetic energy of translation). Pointlike atoms move only as points. With such an argument, you might think you could ignore the 3 extra (rotational) degrees of freedom and reduce the degrees of freedom to 3 (those associated with the translation of the pointlike particles), which yields the correct molar specific heat for monatomic gases, as we saw in Section 14.6.

The argument, however, is not legitimate. The equipartition of energy theorem associates an energy of $kT/2$ with each active degree of freedom, *regardless* of the contribution, however small, each makes to the total mechanical energy. The problem only can be resolved with quantum mechanics, which indicates that the little dots or flags we imagined to follow the rotation of such monatomic particles cannot, in fact, be so imagined.

> A monatomic gas atom has only 3 degrees of freedom associated with translational motion, each with its average energy $kT/2$; these atoms show no further degrees of freedom, rotation or otherwise, according to quantum mechanics.

What about common *diatomic gases* such as H_2, O_2, or N_2? Here the elementary particles of the gas are molecules. There are 3 translational degrees of freedom associated with the motion of the center of mass of each molecule. But things get more complicated because such molecules really can rotate and vibrate. Let us look at the effects of these separately. Such diatomic molecules can be imagined as dumbbells, as shown in Figure 14.5. The axis of the dumbbell is a symmetry axis of the molecule.

Rotation

In classical physics, three squared angular velocity components are needed to specify the rotational kinetic energies about the various axes of, say, a standard Cartesian coordinate system. If you choose the z-axis to represent the symmetry axis along which the two atoms lie, the angular velocity components are (a) the angular velocity component ω_z about the z-axis; (b) the angular velocity component ω_x associated with rotation of the molecule about the x-axis; and (c) the angular velocity component ω_y associated with rotation about the y-axis.

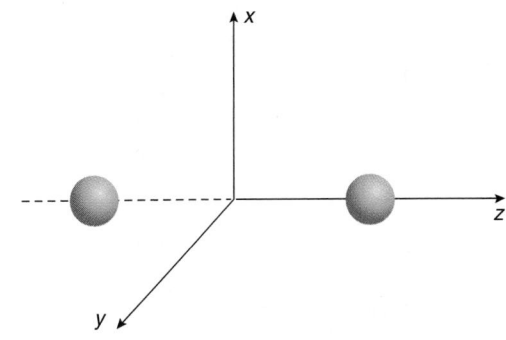

FIGURE 14.5 A diatomic molecule.

Quantum mechanics excludes the rotation about the axis of the molecule from consideration. An old (but incorrect) classical way to exclude this degree of freedom argues that the pointlike atoms have negligible moments of inertia about the symmetry axis because they lie on it; the rotational kinetic energy associated with the rotation about this axis thus is negligible. However, as with the rotation of monatomic gas particles, this argument still conflicts with the equipartition of energy theorem (and for the same reason): just because the moment of inertia and rotational energy are small, it does not dismiss the degree of freedom from the equipartition theorem.

> It is quantum mechanics that indicates that diatomic dumbbell molecules have only 2 (not 3) rotational degrees of freedom, associated with the rotational kinetic energy about the x- and y-axes.

Vibration

If you model the force between the two atoms in a diatomic molecule by means of a Hooke's force law between them, the oscillations of the spring mimic the vibration of the molecule along the symmetry axis, as shown in Figure 14.6. There are potential and kinetic energies associated with this oscillation, and so *2 vibrational* degrees of freedom exist.

Thus, for a diatomic molecule, there appear to be

3 degrees of freedom associated with the motion of the center of mass of each molecule,
2 rotational degrees of freedom associated with the rotational kinetic energies about the two significant rotational axes, and
2 vibrational degrees of freedom associated with the kinetic and potential energies of vibration along the symmetry axis.

This makes a total of 7 degrees of freedom for such molecules.

Now invoking the equipartition of energy theorem, we see that an energy $kT/2$ is associated with each degree of freedom, and the average energy per molecule is therefore

$$7\left(\frac{1}{2}kT\right)$$

The total internal energy of a system of N such molecules then would appear to be

$$U = N\left[7\left(\frac{1}{2}kT\right)\right]$$
$$= \frac{7}{2}nRT$$

FIGURE 14.6 Vibration along the symmetry axis of a diatomic molecule is modeled by a Hooke's law spring.

Using Equation 14.21 for the molar specific heat at constant volume, we predict

$$c_V = \frac{1}{n}\frac{dU}{dT}$$
$$= \frac{7}{2}R$$
$$= 29.1 \text{ J/(mol·K)}$$

Consult the experimental values of c_V in Table 14.1. We see that the molar specific heat *at constant volume* for diatomic gases at room temperature is measured to be about 20.8 J/mol, which is very close to 5R/2, *not* 7R/2 as just predicted.

> Evidently, *only five* degrees of freedom are active for diatomic molecules at room temperature, since $c_V \approx 5R/2$.

Classical kinetic theory again is unable to account for this discrepancy in the number of degrees of freedom for a diatomic gas. Only quantum mechanics correctly explains why 5, not 7, degrees of freedom are available (see the next section): the vibrational degrees of freedom are not active or excited at room temperature. The permitted vibrational energies are quantized and there is insufficient energy available at room temperature to excite the vibrations. Effectively, at room temperatures, diatomic molecules translate and rotate but do not vibrate: they are like two masses connected by a very stiff spring or a massless rigid rod. We say the vibrational degrees of freedom are **frozen out** and are not active.

The temperature dependence of the molar specific heat of diatomic gases bears out this behavior; see Figure 14.7. At high temperatures the vibrational degrees of freedom become excited, resulting in the full 7 active degrees of freedom (translation, rotation, and vibration), and the molar specific heat at constant volume of diatomic gases approaches 7R/2. At very high temperatures, the molecule breaks apart. At room temperature, only 5 degrees of freedom are active (translation and rotation), and so the molar specific heat at constant volume is 5R/2.

The energies associated with the rotation also are quantized, but with a closer energy spacing than for the vibrational energies

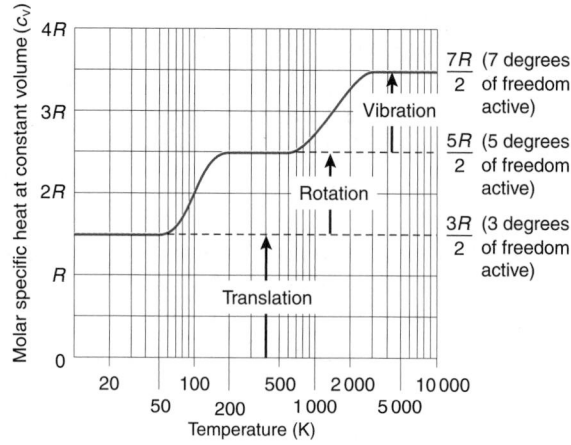

FIGURE 14.7 The molar specific heat at constant volume of a diatomic gas as a function of temperature.

(see Section 14.11) so rotation is activated at a lower temperature than is vibration. At very low temperatures, even the rotational degrees of freedom are frozen out and the diatomic gas behaves like a monatomic gas with only 3 active degrees of freedom, those associated with the translational motion of the center of mass of the particles. The transitions between the various regions in Figure 14.7 are not abrupt, indicating that the average number of active degrees of freedom in these regions is not an integer, a phenomenon that cannot be explained classically.

In this text, we confine the temperatures of the gases to regions where the ratio γ of the molar specific heats of an ideal gas is the following: For a *monatomic* ideal gas, the ratio of the molar specific heats is

$$\gamma \equiv \frac{c_P}{c_V} = \frac{5R/2}{3R/2} = \frac{5}{3} = 1.67 \qquad \text{(monatomic ideal gas)} \tag{14.33}$$

For *diatomic* ideal gases, the ratio of the molar specific heats is

$$\gamma \equiv \frac{c_P}{c_V} = \frac{7R/2}{5R/2} = \frac{7}{5} = 1.40 \qquad \text{(diatomic ideal gas)} \tag{14.34}$$

The situation for triatomic and polyatomic molecular gases is still more complicated. Although we once again have 3 translational degrees of freedom, the situation becomes more involved for counting the various rotational and vibrational degrees of freedom; we will not consider such complex polyatomic molecular gases.

With solids, we saw that the molar specific heat of many metals at room temperature is close to $3R$, as predicted. But the molar specific heat of many molecular solids is not $3R$, indicating that something more complex is going on. Indeed, as with gases, the specific heat is a function of temperature; this is not what the classical arguments predict for an ideal solid. The specific heat varies from solid to solid and is a function of temperature, approaching zero as the absolute temperature approaches zero, as shown in Figure 14.8.

Once again, it is quantum mechanics that accounts for this behavior. As the temperature is lowered, various degrees of freedom evidently are frozen out because of quantum mechanical effects. Dulong and Petit found the specific heats of solids at high temperatures approach the value $3R$. At low temperatures, this is clearly not the case. As the temperature increases, many molecular solids *melt* before their specific heat approaches the value predicted by the law of Dulong and Petit.

We have seen how the temperature dependence of the specific heat of a substance is a macroscopic reflection of changes taking place at the microscopic level. We saw this in our study of gases, in the different specific heats of the same substance in its various phases (e.g., ice, water, steam), and in the latent heat associated with phase transitions. Dramatic temperature variations of the specific heat are clear indications that some new physics (quantum mechanics) is lurking. It is rather surprising that early hints of the need for a new mechanics at the atomic and molecular level (quantum mechanics) were signaled by the failures of classical kinetic theory; however, the problems were not readily understood by the wise minds that formulated classical thermodynamics.

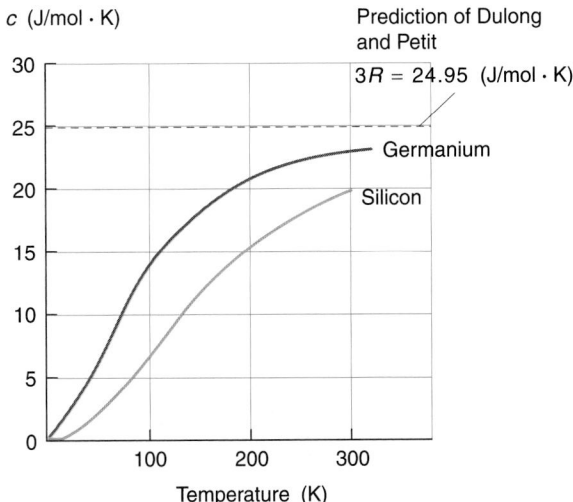

FIGURE 14.8 Molar specific heat of a solid as a function of temperature.

14.11 QUANTUM MECHANICAL EFFECTS*

In this section we present some of the features of quantum mechanics that serve to explain the peculiar behavior of the specific heat.

Think of a particle such as an atom confined to rattling around in a box of very small size (perhaps only a few atomic diameters across). According to quantum mechanics, particles confined to a box are not allowed to have just any amount of energy; in fact, most values of energy are forbidden. We say that the energy is *quantized* and that only certain values of the energy are permitted or allowed, as shown in Figure 14.9. The special energies that are allowed are given many equivalent names: **energy levels**, **energy states**, or **eigenstates of the energy**. There are forbidden values of energy (**energy gaps**) between these allowed values of the energy.

Let the difference between the permitted energies be called E_{gap}(box);[†] energy added to the system must come in lumps of E_{gap}(box). Thus E_{gap}(box) represents a kind of bundle of energy: energy is added (or taken away) from a particle of the system only in lumps of E_{gap}(box).

The larger the dimensions of the box, the more closely spaced the energy levels of a particle become. That is, E_{gap}(box) becomes smaller and smaller as the dimensions of the box increase. For molecules in a box of macroscopic dimensions, the size of the energy gap E_{gap}(box) is vanishingly small compared with the energy $kT/2$ associated with a degree of freedom in the equipartition of energy theorem; this is illustrated schematically in Figure 14.10.

[†] The differences between the permitted energies actually are not all the same for a particle confined to a box, but we make that assumption here for simplicity.

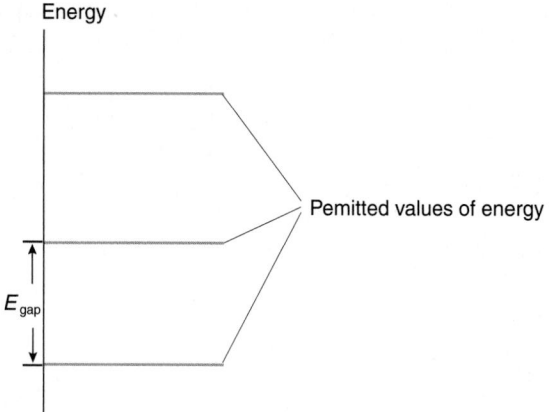

FIGURE 14.9 Allowed energy values, with energy gaps between them.

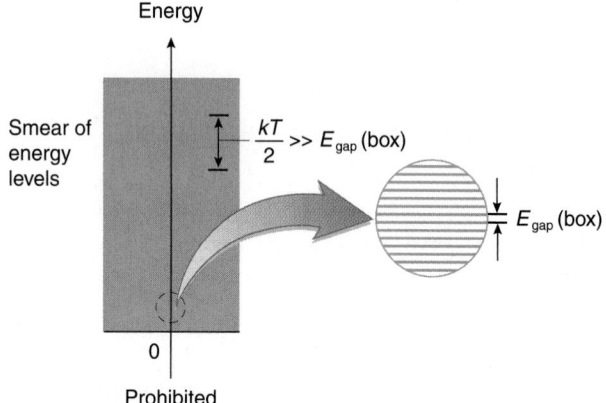

FIGURE 14.10 For a particle in a box of macroscopic size, the size of the energy gap between allowed values of the energy is vanishingly small.

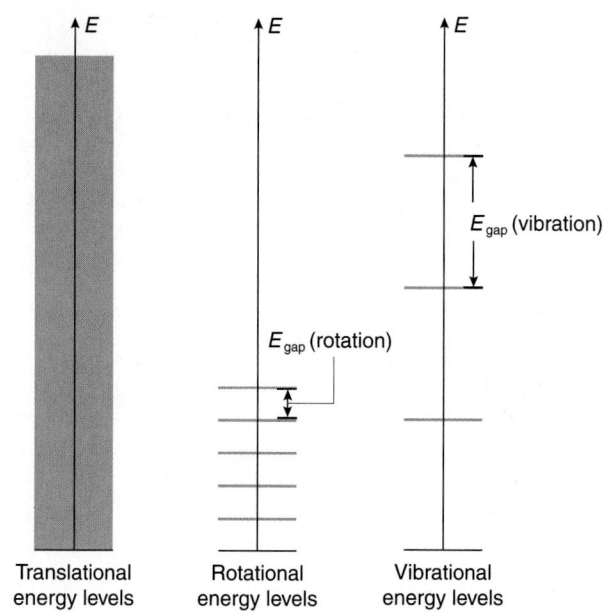

FIGURE 14.11 Energy gaps associated with vibration are much larger than the gaps associated with rotation.

Thus, for particles inside a large box, essentially any translational kinetic energy greater than zero is permitted; negative values for the kinetic energy are not allowed since this implies a speed that is an imaginary number. Speeds must be real numbers because they are measurable quantities.

The rotational and vibrational energies of atoms in molecules also are quantized. For these cases, the magnitudes of the spacing of the permitted energy values, E_{gap}(vibration) and E_{gap}(rotation), are much larger than for the translational motion of the particles confined to a large box. Furthermore, the typical energy gap for vibrational energy levels, E_{gap}(vibration), is larger than for rotational energy levels, E_{gap}(rotation), as shown schematically in Figure 14.11.

The significance of the discrete nature of the energy levels becomes clear when we compare the spacing of the permitted energies with the energy $kT/2$ in the equipartition of energy theorem. If the energy $kT/2 \ll E_{gap}$, then it is not possible for the system to gain energy via this mechanism. We say that the degree(s) of freedom associated with this E_{gap} are frozen out and not active. In other words, the temperature is so low that those degrees of freedom are

unreachable by the system. This means that the number of active degrees of freedom is a function of temperature.

As the temperature increases, $kT/2$ grows in value, and more degrees of freedom become accessible to the system. Thus at very low temperatures, a diatomic gas has only 3 active (translational) degrees of freedom. As the temperature increases and $kT/2$ approaches the size of E_{gap}(rotation), a diatomic gas then can populate 5 degrees of freedom (3 translational and 2 rotational). As the temperature increases further and $kT/2$ approaches the size of E_{gap}(vibrational), the system has 7 active degrees of freedom (3 translational, 2 rotational, and 2 vibrational). Of course, this assumes the gas does not condense at the low temperatures and does not disassociate at the high temperatures.

At very high temperatures (thousands of kelvin), further degrees of freedom come into play as the energy $kT/2$ becomes on the order of the sizes of the quantum mechanical energy gaps in typical atoms. At still higher temperatures (millions of kelvin), $kT/2$ is on the order of the energy gaps associated with particles in the nucleus, and (previously frozen) nuclear degrees of freedom join the others and become active.

It is not easy to describe the gradual way that the specific heat changes with temperature in Figure 14.7. The temperature regions where new degrees of freedom come into play actually are more complex than our treatment indicates.

For simplicity, we will avoid these transition regions and consider the number of degrees of freedom available to a system as an integer. In particular, we will say that the temperatures are such that

1. monatomic gases have 3 degrees of freedom, and

2. diatomic gases have 5 degrees of freedom.

14.12 AN ADIABATIC PROCESS FOR AN IDEAL GAS

An **adiabatic process** involves zero heat transfer between a system and its surroundings.

We mentioned such a process briefly in Section 13.12, but deferred considering them until now. An adiabatic process is, of course, an idealization, since a perfect insulator does not exist. Nonetheless, the process is a useful idealization if, for example, (1) the system is well insulated from its surroundings to minimize heat transfer, or (2) the changes take place so rapidly that there is essentially no time for heat transfer to occur before the change is completed yet slow enough for the process to still be quasi-static. The compressions of gases that occur in engines and refrigerators are examples that may approximate the latter situation.

We investigate engines and refrigerators in the next chapter. Other examples of adiabatic processes include the rapid compressions and expansions that occur when sound waves propagate through a medium. Even clouds involve adiabatic processes: they form when moist rising air cools in an adiabatic expansion, enabling water vapor to condense into tiny fog droplets.

A quasi-static adiabatic process is one that takes place slowly enough for the system to be in thermodynamic equilibrium at any instant, but rapidly enough for heat transfer between the system and its surroundings to be negligible. This is commonly achievable.

The quasi-static adiabatic expansion of an ideal gas is represented on a P–V diagram as shown in Figure 14.12. Note that the pressure, volume, and temperature *all* change during an adiabatic process. We seek to discover how these thermodynamic parameters are related to each other at the beginning and end of such a quasi-static adiabatic process. Since the system is

Cloud formation involves an adiabatic expansion.

in thermodynamic equilibrium at any instant during the process, we assume the ideal gas law is valid at any instant.

We begin with the first law of thermodynamics in differential form:

$$dQ = dU + dW$$
$$= dU + P\,dV$$

In an adiabatic process the heat transfer to the system is zero, and so $dQ = 0$ J. Thus the first law of thermodynamics becomes

$$0\text{ J} = dU + P\,dV \qquad \text{(adiabatic process)} \qquad (14.35)$$

In other words, for an adiabatic process, the differential change in the internal energy of the gas is the negative of the differential work done by the gas:

$$dU = -P\,dV \qquad \text{(adiabatic process)} \qquad (14.36)$$

From Equation 14.21, the differential change in internal energy can be written as

$$dU = nc_V\,dT$$

The quick compression of a gas in an engine cylinder is approximately adiabatic.

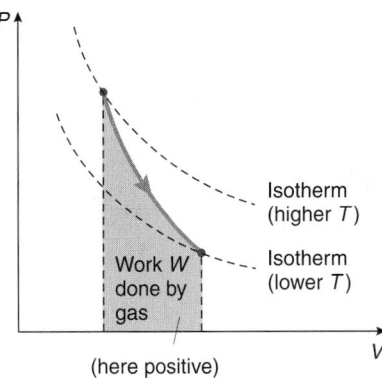

FIGURE 14.12 An adiabatic expansion of an ideal gas on a P–V diagram.

Making this substitution into Equation 14.35, we have

$$0 \text{ J} = nc_V \, dT + P \, dV$$

or

$$dT = -\frac{P \, dV}{nc_V} \qquad (14.37)$$

The equation of state for an ideal gas is Equation 14.1:

$$PV = nRT$$

Since the pressure, volume, *and* temperature all change in an adiabatic process, if we take differentials of the equation of state, we get

$$P \, dV + V \, dP = nR \, dT \qquad (14.38)$$

Using Equation 14.37 for dT, Equation 14.38 becomes

$$P \, dV + V \, dP = nR \left(-\frac{P \, dV}{nc_V} \right)$$

$$= -RP \frac{dV}{c_V} \qquad (14.39)$$

Since the universal gas constant R is equal to the *difference* between the molar specific heats of an ideal gas (Equation 14.27), we can replace R in Equation 14.39 with

$$R = c_p - c_V$$

Thus Equation 14.39 becomes

$$P \, dV + V \, dP = -\left(\frac{c_p - c_V}{c_V} \right) P \, dV$$

$$= -(\gamma - 1) P \, dV \qquad (14.40)$$

where γ is the ratio of the molar specific heats c_p/c_V.

Equation 14.40 simplifies to

$$V \, dP = -\gamma P \, dV$$

Rearranging this slightly, we find

$$\frac{dP}{P} = -\gamma \frac{dV}{V} \qquad (14.41)$$

which is in a form suitable for integration. We integrate it from an initial pressure P_i and volume V_i to some final pressure P_f and volume V_f:

$$\int_{P_i}^{P_f} \frac{dP}{P} = -\gamma \int_{V_i}^{V_f} \frac{dV}{V}$$

$$\ln P_f - \ln P_i = -\gamma(\ln V_f - \ln V_i)$$

Collecting final and initial terms together,

$$\ln P_f + \gamma \ln V_f = \ln P_i + \gamma \ln V_i$$

Taking antilogarithms, we find

$$P_f V_f^\gamma = P_i V_i^\gamma \qquad (14.42)$$

We see that in an adiabatic process, the product PV^γ remains constant:

$$PV^\gamma = \text{constant} \qquad \text{(adiabatic process)} \qquad (14.43)$$

If we couple this result with the equation of state for an ideal gas, $PV = nRT$, we can find similar relationships between the temperature and the volume and between the temperature and pressure:

1. We use Equation 14.43 and eliminate P with the ideal gas law; this yields

$$TV^{\gamma-1} = \text{constant} \qquad \text{(adiabatic process)} \qquad (14.44)$$

2. We use Equation 14.43 and eliminate V using the ideal gas law; this yields

$$P^{1-\gamma}T^\gamma = \text{constant} \qquad \text{(adiabatic process)} \qquad (14.45)$$

The work done by the gas during an adiabatic process can be found in the following way. In an adiabatic process $dQ = 0$ J, so that the first law of thermodynamics becomes

$$0 \text{ J} = dU + dW$$

or

$$W = -\Delta U \qquad (14.46)$$

PROBLEM-SOLVING TACTIC

14.3 The work done by an ideal gas in a quasi-static adiabatic process can be found most directly by calculating the change in the internal energy of the system.

For an ideal gas, the internal energy is a function of temperature only. By finding the initial and final temperatures of the system, the change in the internal energy can be determined. The work done by the system then is just the negative of the change in the internal energy. Indeed, from Equation 14.21, we have $dU = nc_V \, dT$. Since the molar specific heat of an ideal gas is constant, we can integrate this to obtain

$$\Delta U = nc_V \, \Delta T \qquad (14.47)$$

Employing Equation 14.46, we find that the work done by an ideal gas in an adiabatic process is

$$W_{\text{adiabatic}} = -nc_V \, \Delta T \qquad (14.48)$$

QUESTION 4 _____

An ideal gas is compressed adiabatically. What happens to the temperature of the gas? Since there is no heat transfer to the gas in an adiabatic process, what is the source of the energy that changes the temperature of the gas?

STRATEGIC EXAMPLE 14.7

A diatomic gas in an engine is initially at 300 K, 1.000 atm pressure, and occupies a volume 900 cm³. The gas is compressed adiabatically by a piston in a cylinder to 0.100 its initial volume.

a. Find the final pressure of the gas.
b. Find the final temperature of the gas.
c. What is the work done by the gas?

Solution

a. For an adiabatic process, use Equation 14.43 to find the final pressure:

$$PV^\gamma = \text{constant} \qquad \text{(adiabatic process)}$$

Hence

$$P_f V_f^\gamma = P_i V_i^\gamma$$

$$P_f = P_i \left(\frac{V_i}{V_f}\right)^\gamma$$

The volume ratio is $1/0.100 = 10.0$. Since the gas is diatomic, the ratio γ of its molar specific heats is 1.40 (Equation 14.34). Hence you have

$$P_f = P_i (10.0)^{1.40}$$
$$= (1.000 \text{ atm}) \, 25.1$$
$$= 25.1 \text{ atm}$$

b. To find the final temperature, use Equation 14.44,

$$TV^{\gamma-1} = \text{constant} \qquad \text{(adiabatic process)}$$

Hence

$$T_f V_f^{\gamma-1} = T_i V_i^{\gamma-1}$$

$$T_f = T_i \left(\frac{V_i}{V_f}\right)^{\gamma-1}$$

The initial to final volume ratio is 10.0, and so

$$T_f = T_i (10.0)^{1.40-1}$$
$$= (300 \text{ K})(10.0)^{0.40}$$
$$= 7.5 \times 10^2 \text{ K}$$

In diesel engines, the increase in temperature accompanying an adiabatic compression is sufficient to make the fuel spontaneously explode without the necessity of the spark needed in ordinary gasoline engines.

c. To find the work done by the gas, use Equation 14.48, which stemmed from the change in the internal energy of the gas:

$$W_{\text{adiabatic}} = -nc_V \, \Delta T$$

You have to calculate the number of moles of gas present. Use the initial values of temperature, pressure, and volume in the ideal gas law to find n:

$$PV = nRT$$

Solve for n and remember to use pascals for the pressure:

$$n = \frac{PV}{RT}$$
$$= \frac{(1.013 \times 10^5 \text{ Pa}) (9.00 \times 10^{-4} \text{ m}^3)}{[8.315 \text{ J/(mol·K)}] (300 \text{ K})}$$
$$= 3.65 \times 10^{-2} \text{ mol}$$

The molar specific heat at constant volume for a diatomic gas is $5R/2 = 20.1$ J/(mol·K). Making these substitutions into the foregoing expression for the work, you find

$$W_{\text{adiabatic}} = -(3.65 \times 10^{-2} \text{ mol})[20.1 \text{ J/(mol·K)}]$$
$$\times (7.5 \times 10^2 \text{ K} - 300 \text{ K})$$
$$= -3.3 \times 10^2 \text{ J}$$

Negative work is done by the gas (since it is compressed), indicating that positive work was done *on* the gas by the piston.

CHAPTER SUMMARY

The *kinetic theory of an ideal gas* is based on the following assumptions:

1. The number of particles N in the gas is huge.
2. The volume V containing the gas is much larger than the total volume actually occupied by the gas particles themselves.
3. The dynamics of the particles is governed by Newton's laws of motion.
4. The particles are equally likely to be moving in any direction.
5. The gas particles (be they atoms or molecules) interact with each other and with the walls of the container only via elastic collisions.
6. The gas is in thermal equilibrium with its surroundings.
7. The particles of the gas are identical and indistinguishable.

The pressure of a gas arises from the myriad collisions of its constituent particles with the walls of the container. For an ideal gas of N particles each of mass m in a volume V, the pressure is

$$P = \frac{m}{3V} N \langle v^2 \rangle \qquad (14.9)$$

where $\langle v^2 \rangle$ is the average of the square of the speeds of the particles. The average kinetic energy per particle of the gas is

$$KE_{\text{ave}} = \frac{3}{2} kT \qquad (14.13)$$

where k is Boltzmann's constant and T is the absolute temperature of the gas. Hence the temperature is a measure of the average kinetic energy per particle.

The *root-mean-square (rms) speed* v_{rms} of the particles of mass m in the gas is

$$v_{rms} = \left(\frac{3kT}{m} \right)^{1/2} \qquad (14.14)$$

$$= \left(\frac{3RT}{M} \right)^{1/2} \qquad (14.16)$$

where R is the universal gas constant and M is the molar mass (the mass of one mole) of the gas.

The internal energy of an ideal gas is function of the absolute temperature T and the quantity n of gas. For a monatomic ideal gas, the internal energy is

$$U = \frac{3}{2} nRT \qquad \text{(monatomic ideal gas)} \qquad (14.20)$$

The number of *degrees of freedom* populated by (i.e., active in) a system indicates how many quadratic generalized coordinates are needed to specify the microscopic total mechanical energy of a particle.

The *equipartition of energy theorem* states that the average energy per particle is the same for each degree of freedom, and this energy is equal to

$$\frac{1}{2} kT$$

For an ideal gas with f active degrees of freedom, the average energy per particle is

$$f \frac{1}{2} kT \qquad (14.30)$$

The internal energy is

$$U = f \frac{nRT}{2} \qquad (14.31)$$

and the molar specific heat at constant volume is

$$c_V = f \frac{R}{2} \qquad (14.32)$$

An ideal monatomic gas has 3 active degrees of freedom per particle. Thus the molar specific heat at constant volume of a monatomic ideal gas is

$$c_V = \frac{3}{2} R \qquad \text{(monatomic ideal gas)} \qquad (14.22)$$

At room temperature, a diatomic gas has 5 active degrees of freedom; thus, its molar specific heat is

$$c_V = \frac{5}{2} R \qquad \text{(diatomic ideal gas)}$$

The difference between the molar specific heat at constant pressure and that at constant volume is equal to the universal gas constant:

$$c_P - c_V = R \qquad \text{[any gas for which } U = U(T)\text{]} \qquad (14.27)$$

This applies to *any ideal gas*, monatomic, diatomic, or polyatomic, for which the internal energy is a function of the absolute temperature, $U = U(T)$, independent of pressure and volume.

For an ideal gas undergoing a quasi-static adiabatic process,

$$PV^\gamma = \text{constant} \qquad (14.43)$$

where γ is the ratio of the molar specific heats:

$$\gamma \equiv \frac{c_P}{c_V}$$

We restrict the gas to a temperature range for which the value of γ is

$$\gamma = \frac{5}{3} = 1.67 \qquad \text{(monatomic ideal gas)} \qquad (14.33)$$

$$\gamma = \frac{7}{5} = 1.40 \qquad \text{(diatomic ideal gas)} \qquad (14.34)$$

The work done by an ideal gas in an adiabatic process is the negative of the change in the internal energy of the gas:

$$W = -\Delta U \qquad \text{(adiabatic process)} \qquad (14.46)$$

$$= -nc_V \Delta T \qquad \begin{array}{l} \text{(adiabatic process} \\ \text{for an ideal gas)} \end{array} \qquad (14.48)$$

SUMMARY OF PROBLEM-SOLVING TACTICS

14.1 (page 646) When using Equation 14.16, if R is expressed as 8.315 J/(mol·K), be sure the molar mass M is expressed in kilograms per mole rather than grams per mole.

14.2 (page 646) When using Equations 14.14–14.16, be sure you use the absolute temperature T in kelvin (not celsius).

14.3 (page 656) The work done by an ideal gas in a quasi-static adiabatic process can be found most directly by calculating the change in the internal energy of the system.

QUESTIONS

1. (page 644); 2. (page 646); 3. (page 646); 4. (page 656)

5. If the particles of an ideal gas made completely inelastic collisions with the walls of the container, what would happen to the pressure of the gas as a function of time?

6. Before Galileo (1564–1642), it was thought that a continual force was needed just to keep a system moving at constant velocity. Now you know from Newton's first law of motion that a system will continue at constant velocity indefinitely if there is zero total force on it. A similar question arises in kinetic

theory. Does a gas in thermal equilibrium need a continual positive heat transfer just to maintain its temperature? Explain your answer using ideas from the kinetic theory of gases.

7. One of the following statements is true, the other is false. Explain why each is which: (a) The temperature is a measure of the kinetic energy of a particle of a gas. (b) The temperature is a measure of the average kinetic energy of a particle of a gas.

8. At a given temperature, the particles of a gas have many different speeds rather than identical speeds. Give arguments to justify this statement.

9. (a) In what average direction do the gas molecules move in a closed room? (b) Explain why their average kinetic energy is not zero even though their average velocity is zero.

10. Explain why at low altitudes the temperature decreases with increasing altitude in the atmosphere of the Earth. How can this observation be reconciled with the fact that hot air rises by convection?

11. To triple the rms speed of the particles of an ideal gas, by what factor does the temperature need to be increased?

12. The pressure of an ideal gas is tripled while the volume is kept constant. By what factor does the average kinetic energy of a particle in the gas increase?

13. The volume of an ideal gas is tripled while the pressure is kept constant. By what factor does the average kinetic energy of a particle in the gas change?

14. The volume of a gas is decreased in an isothermal process. On the basis of kinetic theory, explain why the pressure of the gas increases.

15. The planet Mercury has a relatively small mass and is the planet closest to the Sun. Give some arguments that might explain why it has no atmosphere.

16. Some of the molecules in the upper atmosphere of the Earth have speeds that exceed the escape speed from the Earth. Does the atmosphere of the Earth gradually dissipate into space?

17. In order to evaporate from a liquid such as water, a molecule must have a certain minimum kinetic energy. (a) Explain why evaporation cools the remaining liquid. (b) How is it possible for a liquid to evaporate completely from an open dish?

18. Is it legitimate to talk about the temperature of a *single* molecule? If not, about how many molecules must be present before the concept of temperature becomes meaningful?

19. Explain why the molar specific heat at constant pressure for gases is greater than the molar specific heat at constant volume.

20. Is the molar specific heat at constant pressure for solids and liquids greater than the molar specific heat at constant volume? Explain.

21. Would you expect the *difference* between the specific heat at constant pressure and the specific heat at constant volume to be greater for liquids than for solids? Explain. (Hint: Think of the relative amounts of thermal expansion between solids and liquids.)

22. One thousand BBs, each of mass 10^{-4} kg, are in an evacuated cubical container of edge 10.0 cm. The container is in an orbiting spacecraft. The system is shaken vigorously and the average speed of a BB is 3.00 m/s. Is it legitimate to talk of a temperature of the BB system? If so, calculate the temperature. Is it likely that the particles will continue to move perpetually as they do in a real gas? Explain.

23. A gas-filled football is thrown by a quarterback to a wide receiver. Is the average velocity of the particles in the gas zero from the point of view of the quarterback? Explain.

24. On the basis of kinetic theory, explain why there is a minimum temperature but no maximum temperature. [This is true even though particles have an upper speed limit, the speed of light. However, this speed limit is not apparent at this stage in your study of physics; classical kinetic theory breaks down at relativistic speeds (speeds that approach the speed of light).]

25. If an ideal gas is simultaneously warmed (by heat transfer) and compressed, how is the molar specific heat for this process qualitatively related to the molar specific heats c_p and c_V? That is, is the molar specific heat for the process greater than both c_p and c_V, less than both, or greater than one and less than the other (which one?)? Explain your reasoning.

26. For an ideal gas, will the difference between the specific heat at constant pressure and that at constant volume be the same for all gases if the specific heats are *not* molar specific heats but the specific heat expressed as joules per kilogram-kelvin [J/(kg·K)]? Explain your answer.

27. A macroscopic mass m slides across a rough surface and comes to rest. On the basis of kinetic theory, where does the kinetic energy of the mass go? Why is it very unlikely that this energy ever will return of its own accord to become macroscopic kinetic energy of the mass m?

28. A thermometer is placed in a region that is a vacuum. Will the thermometer indicate 0 K? Explain why the thermometer will indicate the temperature of the material container enclosing the vacuum. Does this mean that a vacuum has a temperature? Is a vacuum *really* empty? Explain.

29. Explain why *only* the translational kinetic energy of the particles in a gas is proportional to the absolute temperature, and not other contributions to the internal energy such as rotations or vibrations of the gas particles.

30. The kinetic theory of an ideal gas assumes the total volume of the particles themselves is negligible. What effect might the finite volume of the particles have on the ideal gas law?

31. The kinetic theory of an ideal gas assumes that the particles of the gas exert no forces on each other except during the brief time intervals of collisions. What effect might you expect on the ideal gas law from the existence of long-range forces between the particles?

32. The difference between the molar specific heats of an ideal gas is equal to the universal gas constant:

$$c_p - c_V = R$$

Does this also apply for liquids and solids? If so, explain why. If not, explain why not.

33. In what ways are 100 third-graders in bumper cars, confined to a bumper car rink, similar to the particles in a gas? In what ways are they different? Is it possible to develop a kinetic theory of bumper cars?

34. Saturn's most massive moon, Titan, has an atmosphere. Explain why it is more likely for hydrogen gas (H_2) to be found in the Titan atmosphere than in the atmosphere of the much more massive Earth.

35. On the basis of kinetic theory, explain why the temperature of a gas decreases in an adiabatic expansion.

36. In an adiabatic expansion of a system, there is no heat transfer between the system and its environment, yet the temperature and pressure of the gas decrease. Explain how this occurs on the basis of kinetic theory.

PROBLEMS

Sections 14.1 Background for the Kinetic Theory of Gases
14.2 The Ideal Gas Approximation
14.3 The Pressure of an Ideal Gas
14.4 The Meaning of the Absolute Temperature

1. A mole of a gas contains Avogadro's number N_A of particles:

$$N_A = 6.022 \times 10^{23} \text{ particles/mol}$$

If you counted the particles in a mole at the rate of one particle per second, how many years would it take to complete the count? For comparison, the age of the universe is estimated to be about 15 billion years.

2. If 1.00 mol of particles is spread uniformly over the surface of the Earth, how many particles are on each square meter?

3. If the particles in 1.00 mol were dispersed to a density of one particle per cubic meter, what radius of sphere would contain these particles? Express your result in kilometers.

4. An ideal gas is at 1.00 atm pressure and 0 °C. Under these conditions, the number of particles per cubic *centimeter* is known as the *Loschmidt number*. Calculate the Loschmidt number.

5. Hydrogen fusion within the core of the Sun occurs at a temperature of about 15×10^6 K. What is the average kinetic energy of hydrogen atoms at this temperature? [The hydrogen atoms actually are completely ionized at this temperature, so this is the average kinetic energy of the nucleus (a proton).]

6. (a) In order to double the rms speed of the particles in a gas, by what factor must the temperature of the gas be increased? (b) Does this result depend on the temperature scale used? (c) Does this result depend on the units used to measure the speed of the particles?

7. A fast sprinter can attain a speed of 10 m/s. At what temperature do helium atoms have an rms speed of 10 m/s?

8. At what temperature do atoms of helium have an rms speed equal to 1.00% of the speed of light?

9. Calculate the rms speed of the particles in a gas of molecular hydrogen (H_2) at temperature 300 K. By what factor does this differ from the rms speed of oxygen molecules (O_2) at the same temperature?

10. Show that equal volumes of gas at the same temperature and pressure have equal numbers of particles. This statement is called *Avogadro's law*.

11. Show that the number of particles in a cubic millimeter of a gas at temperature 273 K and 1.00 atm pressure is 2.69×10^{16}. To get a feeling for the order of magnitude of this number, calculate the age of the universe in seconds assuming it is 15 billion years old.

12. Six particles have speeds 0.10 m/s, 0.20 m/s, 0.30 m/s, 0.40 m/s, 0.50 m/s, and 0.60 m/s. Show that their average speed is less than their rms speed.

13. A sample of oxygen gas (O_2) is at temperature 300 K and 1.00 atm pressure. One molecule, with a speed equal to the rms speed, makes a head-on elastic collision with your nose. Ouch! What is the magnitude of the impulse imparted to your schnozzle?

14. For an ideal gas at temperature 300 K and 1.00 atm pressure, what are the dimensions of a cube that contains 1000 particles of the gas?

15. To increase the rms speed of a gas by 1.0%, by what percentage must the temperature increase?

16. Assume dust grains have masses about 10^9 times that of a helium atom. At temperature 300 K, what is the rms speed of a dust speck if you treat the dust as an ideal gas?

17. One (1.00) mole of an ideal gas is at temperature 300 K and at 1.00 atm pressure. (a) What is the volume of the gas? (b) Estimate the average spacing between the particles of the gas.

18. Five lonely gas molecules occupy a volume V. Their y- and z-velocity components are all zero and their x-velocity components are −250.0 m/s, −40.0 m/s, 40.0 m/s, 90.0 m/s, and 160.0 m/s. (a) Calculate their average *velocity*. (b) Calculate their average speed. (c) Calculate their rms speed.

19. A census is made of the speeds of a random sample of 1010 oxygen molecules (O_2) inside a box. The results are presented in the following table:

Speed (m/s)	Number of particles with the given speed
100	40
200	125
300	180
400	215
500	180
600	125
700	75
800	40
900	20
1000	10

(a) Find the average speed of the particles. (b) Find the rms speed of the particles. (c) Make a graph of the number of particles versus the speed and indicate on the graph the locations of the average and rms speeds. (d) What is the temperature of the gas?

•20. An ideal gas of molar mass M is at temperature T. Show that if the temperature increases by a small amount ΔT, the increase in the rms speed of the gas particles is

$$\Delta v_{\text{rms}} = \frac{1}{2}\left(\frac{3R}{MT}\right)^{1/2}\Delta T$$

•21. Nitrogen (N_2) gas is at 300.0 K. If the temperature of the gas increases by 1.0 K, what is the increase in the rms speed?

•22. Raindrops of mass 1.00 mg fall vertically at a constant speed of 10.0 m/s, striking a horizontal skylight at the rate of 1000 drops/s and draining off. The window size is 15.0 cm × 25.0 cm. Assume the collisions of the drops with the window are completely inelastic. Calculate (a) the magnitude of the average force of the raindrops on the window, and (b) the resulting pressure developed by the raindrop collisions.

•23. One (1.00) mole of helium at 1.00 atm pressure and at temperature 0 °C is in a rigid container with a volume of 22.4 liters. (a) Calculate the rms speed of the particles in the gas. (b) The sample is warmed to 546 °C. By what factor does its pressure change? (c) By what factor does the rms speed of the particles change? (d) What heat transfer to the system produced the temperature change?

•24. The molar mass of nitrogen gas (N_2) is 28.0 g. (a) Find the mass of one nitrogen molecule. (b) Find the rms speed of a nitrogen molecule at a temperature of −23 °C. (c) Some hydrogen gas (H_2) (molar mass = 2.00 g/mol) also is present in the same container. What is the temperature of the hydrogen gas? (d) What is the rms speed of the hydrogen molecules? (e) What *new* temperature would cause the rms speed in part (b) to be greater by a factor of 2? (f) Does the *ratio* of the hydrogen to nitrogen rms speeds change with temperature?

•25. At what temperature do helium atoms have an rms speed equal to the escape speed from the surface of the Earth ($v_{\text{escape}} = 11.2$ km/s)?

26. An energy unit commonly used in atomic, molecular, and nuclear physics is the *electron-volt*, abbreviated eV. One electron-volt is equivalent to 1.602×10^{-19} J. Show that the average translational kinetic energy of a particle in a gas at temperature 300 K is approximately 1/25 eV.

•27. At what temperature does $kT/2 = 1.00$ eV? (1 eV = 1.602×10^{-19} J) (See Problem 26.)

•28. Classically, at what temperature would the rms speed of helium atoms be equal to the speed of light? (Note: Particles with mass cannot attain speeds equal to the speed of light, as we will see when we study relativity in Chapter 25.)

•29. Calculate the rms speed of hydrogen gas (H_2) at temperature 300 K in the atmosphere and compare it with the escape speed from the Earth (11.2 km/s). Since hydrogen is the least massive gas, hydrogen molecules will have the highest rms speeds at a given temperature. How can this calculation explain why there is essentially no hydrogen gas in the atmosphere of the Earth?

•30. Consider helium gas at temperature 300 K near the surface of the Earth. (a) Calculate the average kinetic energy of one of the helium atoms. (b) Calculate the gravitational potential energy of a single helium atom near the surface of the Earth. Choose the zero of gravitational potential energy to be infinitely far away from the Earth. (c) What is the absolute value of the ratio of the gravitational potential energy of the helium atom to its average kinetic energy? Is it justifiable to neglect the gravitational potential energy in kinetic theory? Why or why not?

•31. A wind is howling at 30.0 m/s at a temperature of 25.0 °C. Consider a 1.00 kg mass of this moving air. Air has an effective molar mass of 29 g/mol; consider this value to be exact. (a) If the pressure is 0.97 atm, what is the density of the air? (b) How many joules of kinetic energy does the 1.00 kg mass of air have because of the wind? (c) How many joules of kinetic energy does the 1.00 kg mass of air have because of its internal energy? (d) Other than the difference in their magnitudes, what is the main distinction between the wind velocity and the thermal velocities?

•32. One (1.00) beautiful mole of helium gas at 300 K finds itself in a cubical box of sides 10.0 cm. (a) What is the rms speed of the particles of the gas? (b) If there were no collisions along the way, how long would it take the average helium atom in the container to travel from one side of the box to the other? (c) What is the pressure inside the container? (d) What is the magnitude of the average force that the particles exert on a side of the box?

•33. The temperature in outer space is about 2.7 K and the matter there consists mainly of isolated hydrogen atoms with a density of about 0.3 atoms per cubic meter. What is the pressure of this gas?

•34. The average kinetic energy of a particle in a gas at temperature T is given by Equation 14.13:

$$KE_{\text{ave}} = \frac{3}{2}kT$$

or

$$\frac{1}{2}m\langle v^2\rangle = \frac{3}{2}kT$$

The special theory of relativity states that there is an upper limit on the speed of any particle: the speed of light $c = 3.00 \times 10^8$ m/s. For a gas of hydrogen *atoms*, the immediately preceding equation implies an upper limit on the temperature. Find the absolute temperature such that $\langle v^2\rangle$ for a gas of hydrogen atoms is equal to the square of the speed of light. In fact, there is *no upper limit* on the temperature; so the classical expression for the kinetic energy cannot be valid for speeds approaching the speed of light.

•35. The isotope of uranium used in some nuclear fission reactors and bombs is uranium 235, with a molar mass of exactly 235 g/mol. The more common uranium isotope is uranium 238, with a molar mass of exactly 238 g/mol. One way of separating the isotopes is to form the gas uranium hexafluoride (UF_6) and exploit the rms speed difference of the two molecules made from the two uranium isotopes. This method of isotope separation is called gaseous diffusion and was first secretly employed in the Manhattan Project during World War II (at a site in Oak Ridge, Tennessee) to secure the material for the nuclear weapon used at Hiroshima near the end of the war with Japan;

the Nagasaki bomb was produced with plutonium made at the Hanford nuclear site in Washington State. Take the molar mass of fluorine (F) to be exactly 19 g/mol. Determine the *ratio* of the faster to slower rms speeds of the two molecules at room temperature (300 K). Does the ratio depend on the temperature?

36. A quantity of oxygen gas has a volume V, a temperature T, and pressure P_1. A quantity of nitrogen gas has the same volume V, identical temperature T, and pressure P_2. The two gases are mixed in a common volume of the same size V and have the same temperature T. Give arguments, based on kinetic theory, to show that the pressure of the mixture is $P_1 + P_2$. This result is known as *Dalton's law of partial pressures*: the total pressure of a mixture of gases is the sum of the partial pressures of the individual gases (occupying the same volume at the same temperature).

37. A helium atom has a mass approximately four times that of a hydrogen atom. Let a helium atom collide head-on with a hydrogen atom in a one-dimensional elastic collision. (a) If the two particles initially have equal kinetic energies, which atom (if either) gains kinetic energy in the collision? (b) If the two atoms initially have momenta of equal magnitudes, which (if either) gains energy in the collision?

Sections **14.5 The Internal Energy of a Monatomic Ideal Gas**
14.6 The Molar Specific Heats of an Ideal Gas
14.7 Complications Arise for Diatomic and Polyatomic Gases
14.8 Degrees of Freedom and the Equipartition of Energy Theorem
14.9 Specific Heat of a Solid*
14.10 Some Failures of Classical Kinetic Theory
14.11 Quantum Mechanical Effects*
14.12 An Adiabatic Process for an Ideal Gas

38. Evaluate the energy $kT/2$ per degree of freedom that appears in the equipartition of energy theorem for each of the following temperatures. Express your results in both joules and in another common energy unit known as the electron-volt (eV), defined as 1 eV = 1.602×10^{-19} J. (a) 3.0 K. This is approximately equal to the present ambient temperature of the radiation left over from the creation of the universe as a whole (the Big Bang). The temperature is just a bit above absolute zero. (b) 300 K. (c) The surface temperature of a cool star: 3000 K.

39. The ionization energy of the hydrogen atom is 13.6 eV = 2.18×10^{-18} J. This is the energy needed to remove the electron from the neutral atom. What temperature corresponds to this energy per degree of freedom according to the equipartition of energy theorem?

40. A nonspinning bead moves around a circular horizontal wire. How many degrees of freedom does the system have?

41. A nonspinning water droplet is free to slide around a frying pan. How many degrees of freedom does it have?

42. One (1.00) mole of a monatomic ideal gas at temperature 300 K occupies a volume of 5.00 liters. The gas now expands adiabatically till its volume is doubled. What is the final pressure of the gas?

43. One (1.00) mole of hydrogen gas (H_2) is warmed isobarically from an initial temperature of 273 K, doubling its volume. (a) What is the final temperature of the gas? (b) What is the heat transfer to the gas in the experiment?

44. A box of volume 1.00×10^{-3} m³ has helium gas at a temperature of 300 K and a pressure of 1.00 atm. (a) What is the average kinetic energy per particle of the gas? (b) What is the total kinetic energy of the entire collection of atoms? This total is the internal energy of the gas.

45. How many moles of helium at temperature 300 K and 1.00 atm pressure are needed to make the internal energy of the gas 100 J?

46. Brrrr . . . it is freezing in the lab today. The temperature is 10 °C and the pressure is 1.00 atm. The lab is 5.0 m by 4.0 m by 3.0 m in size. A 1.0 kW heater is turned on to warm the room. (a) If no air escapes from the room, is the process at constant volume, constant pressure, or neither? (b) Assume the air is a diatomic gas. How long will it take the heater to raise the temperature of the room air from 10 °C to 20 °C?

47. The molar specific heat of a gas at constant pressure is measured to be 29.2 J/(mol·K). The molar mass of the gas is 28 g/mol. (a) What is the molar specific heat of the gas at constant volume? (b) What is the ratio γ of molar specific heats for the gas? (c) Is it possible to tell if the gas is monatomic or diatomic with the given information? If so, how? If not, what additional information is needed to make this determination?

48. An ideal gas has f degrees of freedom. (a) Show that the molar specific heat at constant volume is

$$c_V = \frac{f}{2} R$$

(b) Show that the molar specific heat at constant pressure is

$$c_P = \frac{f+2}{2} R$$

(c) Show that the ratio of the molar specific heats is

$$\gamma = \frac{c_P}{c_V} = \frac{f+2}{f}$$

49. Show that for a monatomic ideal gas

$$PV = \frac{2}{3} U$$

50. The *compressibility* κ of a substance is defined as

$$\kappa \equiv -\frac{1}{V} \frac{dV}{dP}$$

and represents the fractional change in volume for a given change in pressure. (a) Because of the minus sign in the definition, κ is positive. Explain why. (b) Show that if the gas is compressed *isothermally*, the compressibility is

$$\kappa_{\text{isothermal}} = \frac{1}{P}$$

(c) Show that if the gas is compressed *adiabatically*, the compressibility is

$$\kappa_{\text{adiabatic}} = \frac{1}{\gamma P}$$

Therefore

$$\gamma \kappa_{\text{adiabatic}} = \kappa_{\text{isothermal}}$$

•51. A truck is carrying a tank of helium gas of volume 0.250 m³ at a pressure of 100 atm. The temperature is 300 K. You observe the truck traveling at a speed of 100 km/h in the direction of increasing values of *x*. (a) What is the average value of the *x*-component of the velocity of the particles in the gas $\langle v_x \rangle$ according to you? (b) What are the average values of the *y*- and *z*-components of the velocity of the particles in the gas according to you? (c) Can you apply the equipartition of energy theorem to the particles of the gas? Why or why not? (d) *According to the driver of the truck*, what are the average values of the *x*-, *y*-, and *z*-components of the velocity of the particles of the gas? (e) Can the driver of the truck apply the equipartition of energy theorem to the particles of the gas? Why or why not?

•52. You compress an ideal gas isothermally in such a way that 500 J of work is done *by* you *on* the gas. (a) What is the change in the internal energy of the gas? (b) What was the heat transfer to the gas during this heroic feat?

53. A well-insulated 4.00 liter box contains a partition dividing the box into two equal volumes as shown in Figure P.53. Initially, 2.00 g of molecular hydrogen gas (H₂) at 300 K is confined to the left-hand side of the partition, and the other half is a vacuum. (a) What is the rms speed of the particles in the gas? (b) What is the initial pressure of the gas? (c) The partition is removed or broken suddenly, so that the gas now is contained throughout the entire box. Assume that the gas is ideal. Does the temperature of the gas change? What is the change in the internal energy of the system? (d) When the gas reaches equilibrium, what is the final pressure?

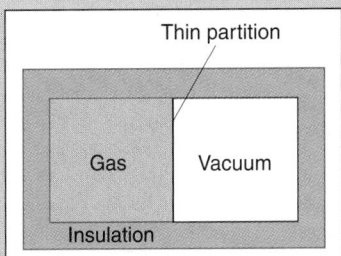

Thin partition

Gas   Vacuum

Insulation

FIGURE P.53

•54. You expand *n* moles of an ideal gas adiabatically from temperature T_i to temperature T_f. Use the first law of thermodynamics to show that the work done by the gas is

$$-nc_V \, \Delta T$$

•55. A typical internal combustion engine compresses the gas by a factor of about 7. Assume that the initial temperature of the gas is 20 °C and it is initially at 1.00 atm pressure. Find the

final temperature and pressure of the gas assuming an adiabatic compression. The gas is diatomic.

•56. A gas initially at temperature T_i, pressure P_i, and volume V_i expands to a final volume V_f via one of the following three types of processes: (1) isothermal; (2) isobaric; (3) adiabatic. (a) Sketch each process schematically on a *P–V* diagram. (b) In which process is the work done by the gas the greatest? Explain your reasoning. (c) In which process is the heat transfer to the gas the greatest? Explain your reasoning. (d) In which process is the change in the internal energy of the gas the greatest?

•57. A gas initially at temperature T_i, pressure P_i, and volume V_i has its pressure reduced to a final value P_f via one of the following types of processes: (1) isochoric; (2) isothermal; (3) adiabatic. (a) Sketch each process schematically on a *P–V* diagram. (b) In which process is the work done by the gas zero? (c) In which process is the work done by the gas the greatest? (d) Show that the ratio of the absolute magnitude of the heat transfer to the gas during the isothermal process to that during the isochoric process is

$$\frac{|Q_{\text{isothermal}}|}{|Q_{\text{isochoric}}|} = \frac{RP_i}{c_V \, |\Delta P|} \left| \ln \frac{P_i}{P_f} \right|$$

(e) In which process is the absolute magnitude of the change in the internal energy of the gas the greatest? Explain your reasoning.

•58. Begin with the definition of the work done by a gas:

$$W = \int_{V_i}^{V_f} P \, dV$$

Use Equation 14.43 for an adiabatic process and show that the work done by the gas in an adiabatic process can be expressed as

$$W_{\text{adiabatic}} = \frac{-1}{\gamma - 1} \left(P_f V_f - P_i V_i \right)$$

where γ is the ratio of the molar specific heats c_P/c_V.

•59. A 2.00 mol sample of a monatomic ideal gas undergoes an adiabatic expansion from a pressure of 2.00 atm at 300 K to a pressure of 1.00 atm. (a) Find the initial volume of the gas. (b) Find the final volume of the gas. (c) What is the final temperature of the gas? (d) What work is done by the gas?

60. An ideal gas at 300 K is compressed adiabatically to half its initial volume. (a) What is the final temperature of the gas if it is monatomic? (b) What is the final temperature of the gas if it is diatomic?

‡61. At any point on a *P–V* diagram, show that the slope of an adiabatic curve through the point is γ times the slope of the isothermal curve through the same point, where γ is the ratio of the molar specific heats ($\gamma = c_P/c_V$). That is,

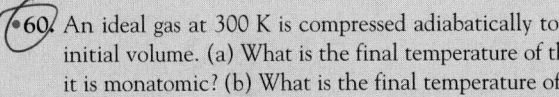

$$\left(\frac{dP}{dV} \right)_{\text{adiabatic}} = \gamma \left(\frac{dP}{dV} \right)_T$$

INVESTIGATIVE PROJECTS

A. Expanded Horizons

1. Investigate the distribution of speeds among the particles in a gas, known as the *Maxwell speed distribution*.

   Frederick Reif, *Statistical Physics*, Berkeley Physics Course, volume 5 (McGraw-Hill, New York, 1967), pages 231–246.

   Frederick Reif, *Fundamentals of Statistical and Thermal Physics* (McGraw-Hill, New York, 1965).

   R. P. Bonomo and F. Riggi, "The evolution of the speed distribution for a two-dimensional ideal gas: a computer simulation," *American Journal of Physics, 52*, #1, pages 54–55 (January 1984).

2. The *principle of detailed balancing* manifests itself in many aspects of physics. In this chapter we found that if the velocity of a particle in a gas changes in an collision, another particle of the gas likely experiences the opposite change, so that the average velocity of the particles in the gas is unchanged in thermal equilibrium (and is zero). Research other aspects of the principle of detailed balancing and formulate a general statement of the principle.

   Earle H. Kennard, *Kinetic Theory of Gases* (McGraw-Hill, New York, 1938), pages 55–58.

   Frank S. Crawford, "Using Einstein's method to derive both the Planck and Fermi–Dirac distributions," *American Journal of Physics, 56*, #10, pages 883–885 (October 1988).

3. The motion of a particle in a gas in thermal equilibrium can be modeled by means of what is called a *random walk*. Random walk problems can be formulated either in one, two, or three dimensions. A one-dimensional random walk problem is analogous to a drunk moving randomly from a lamppost along one dimension such as a narrow alley. The drunk takes steps of equal length d randomly either to the left or right. The problem is to determine the probability of the drunk being at various distances nd from the lamppost. Investigate random walk problems and their relevance to various problems in physics.

   Frederick Reif, *Fundamentals of Statistical and Thermal Physics* (McGraw-Hill, New York, 1965), pages 4–40.

4. *Globular clusters* are roughly spherical collections of hundreds of thousands of stars held together by their mutual gravitational attraction.* The clusters do not collapse because the stars are in motion (just as the solar system does not collapse because the planets are in motion). Pairs of stars in such clusters make "collisions" with each other that change the di-

rections of their motion. They rarely actually collide; rather, the collisions are more like scattering events that send the stars off in different directions (much like a comet passing the Sun). These collisions can be modeled as elastic. Can one regard the collection of stars in such a cluster as an ideal gas? Develop a kinetic theory of a star cluster and thereby define a temperature and pressure of the star particle system. What would happen to the average kinetic energy of a star particle if the cluster as a whole shrank or expanded? Can a star evaporate from the cluster? How could you define the evaporation speed?

Lyman Spitzer, *Dynamical Evolution of Globular Clusters* (Princeton University Press, Princeton, New Jersey, 1987).

Martin Harwit, *Astrophysical Concepts* (3rd edition, Springer-Verlag, New York, 1997).

B. Lab and Field Work

5. In 1827 the English botanist Robert Brown (1773–1858) noticed through a microscope that tiny pollen grains suspended in water exhibit constant, randomly changing motions. Smoke particles in a gas exhibit similar behavior when examined in a small chamber. Such motion now is called *Brownian motion*. Investigate how Avogadro's number is determined experimentally from observations of Brownian motion. Design and perform such an experiment.

   John le P. Webb, "Einstein and Brownian motion—a student project," *Physics Education, 15*, #2, pages 116–121 (March 1980).

   Haym Kruglak, "Brownian motion—a laboratory experiment," *Physics Education, 23*, #5, pages 306–309 (September 1988).

   Robert Stoller, "Viewing Brownian motion with laser light," *American Journal of Physics, 44*, #2, page 188 (February 1976).

   Bill Reid, "Viewer for Brownian motion," *The Physics Teacher, 29*, #1, pages 52–53 (January 1991).

   F. Landis Markley and David Park, "Microscopic and macroscopic views of Brownian motion," *American Journal of Physics, 40*, #12, pages 1859–1860 (December 1972).

   Max Born, *Atomic Physics* (8th edition, Dover, Mineola, New York, 1989).

6. In 1929 Eduard Rüchhardt developed an ingenious way of measuring the ratio γ of molar specific heats by measuring the period of oscillation of a ball bearing sliding in a vertical glass tube inserted through a stopper into a large, gas-filled jar. The diameter of the glass tube is essentially the same as the ball bearing. Investigate this method and, if feasible, construct the apparatus and measure γ for a monatomic gas and for a diatomic gas.

   Mark W. Zemansky and Richard H. Dittman, *Heat and Thermodynamics* (7th edition, McGraw-Hill, New York, 1996).

7. Devise a kinetic theory for an ideal gas with particles confined to move in two dimensions. In particular, derive expressions for the pressure in terms of the rms speed of the particles.

---

*There are about sixty such globular clusters in our Milky Way Galaxy. The most easily visible such cluster in the northern celestial hemisphere is called M-13 in the constellation Hercules (a small telescope is needed to see it, since at most locations it is too dim to see with the naked eye). In the southern celestial hemisphere the globular cluster to see is called the ω Centauri cluster, visible to the naked eye as a small smudgy patch in the constellation Centaurus.

Design an experiment to test aspects of your model using small air pucks on a laboratory airtable. Summarize your theoretical results and experimental proposal. Perhaps your professor can secure the necessary equipment for you to perform your experiment.

Richard L. Liboff, *Kinetic Theory: Classical, Quantum and Relativistic Descriptions* (2nd edition, John Wiley, New York, 1998).

8. Write a proposal to investigate the dynamical behavior of a collection of BBs confined to a box in the orbiting space shuttle and how such a collection of particles differs from those of an ideal gas.

C. Communicating Physics

9. Your roommates play devil's advocates and say, "Atoms do not exist." How could you convince them otherwise? Write an essay describing the experimental evidence for the existence of atoms.

Edwin R. Jones Jr. and Richard L. Childers, "Observational evidence for atoms," *The Physics Teacher*, 22, #6, pages 354–360 (September 1984).

George Gamow, *Mr Tompkins Explores the Atom* (The University Press, Cambridge, England, 1961).

George Gamow, *Mr Tompkins in Paperback* (Cambridge University Press, Cambridge, England, 1993).

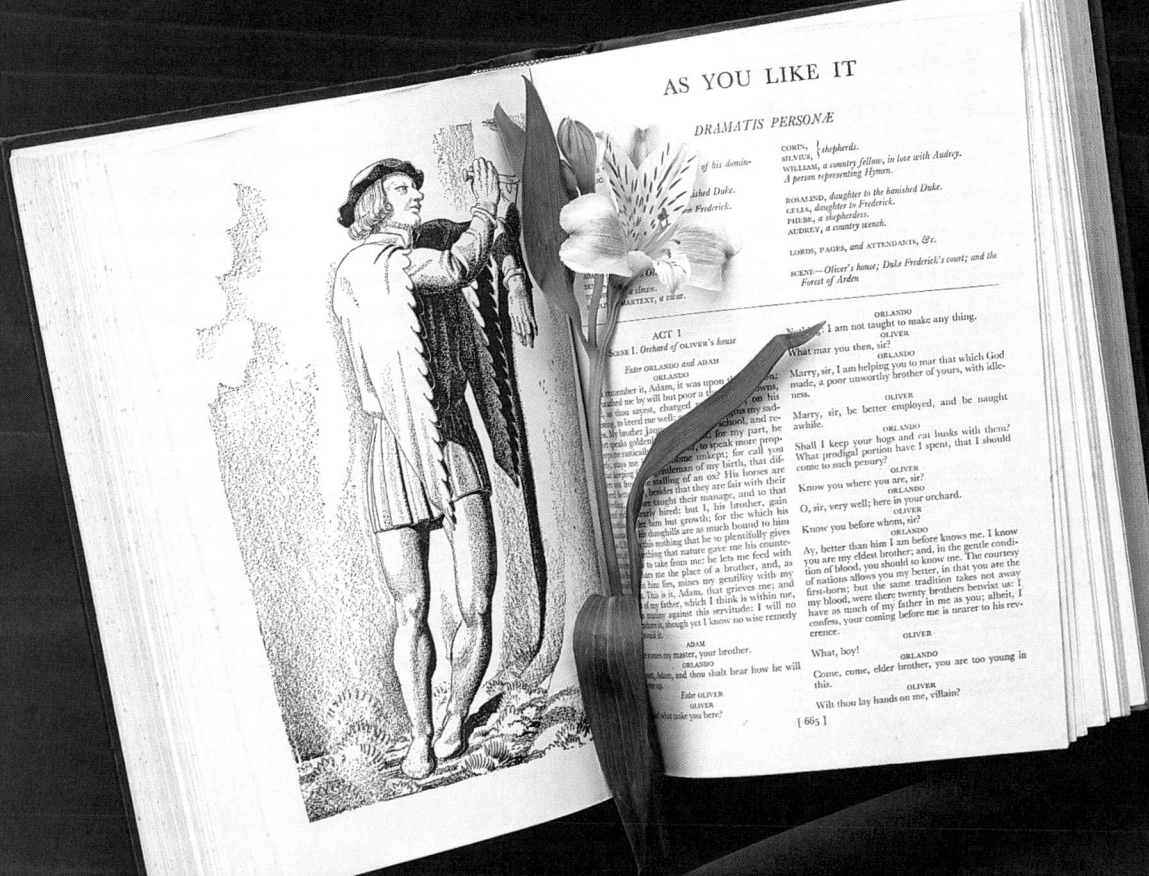

CHAPTER 15

THE SECOND LAW OF THERMODYNAMICS

A good many times I have been present at gatherings of people who, by the standards of traditional culture, are thought highly educated and who have with considerable gusto been expressing their incredulity at the illiteracy of scientists. Once or twice I have been provoked and have asked the company how many of them could describe the Second Law of Thermodynamics. The response was cold: it was also negative. Yet I was asking something which is about the scientific equivalent of: *Have you read a work of Shakespeare's?*

C. P. Snow (1905–1980) *

ir pollution from automotive exhaust, smoke emitted from coal-fired electrical power plants, and thermal emissions from nuclear power plants all have a significant impact on the environment. For decades scientists and politicians have struggled with finding solutions to these important problems. You might ask, why doesn't Congress just outlaw all such discharges and be done with them? Plug up the exhaust pipes, cap the chimneys, zero emissions, no more problem. It *is* a problem, and a big one at that. In this chapter we see why any such legislation, would be physically absurd; Congress might just as well try to repeal the law of universal gravitation or Newton's laws. In the two centuries of its existence, Congress has passed some ridiculous laws, but outlawing all emissions would take the cake. Should Congress or a politician even suggest such a solution, they would be demonstrating their complete ignorance of the second law of thermodynamics.

In this chapter we study this great integrating principle of physics. As C. P. Snow alludes to in the opening quotation, an understanding of the second law is a condition for scientific literacy, much as a knowledge of Shakespeare might be taken as a measure of literacy in the humanities. Hence we will look at the second law of thermodynamics carefully and closely to appreciate both the breadth of its applicability and the depth of its insight into the workings of nature.

15.1 WHY DO SOME THINGS HAPPEN, WHILE OTHERS DO NOT?

Why, sometimes I've believed as many as six impossible things before breakfast.

Lewis Carroll (1832–1898)[†]

Why is it that rocks do not jump up spontaneously and save us the trouble of lifting them? Wouldn't it be nice if pizzas warmed themselves: no need for ovens in the dorm. And of course we all want to win the lottery, but unfortunately it never seems to happen to us. Nuts.

One of the sacred truths of physics is the principle of energy conservation. Some things cannot happen in nature simply because they violate this fundamental principle. We first encountered energy conservation in our study of ("pure") mechanics with the CWE theorem (Chapter 8), which ignored thermal effects. In Chapter 13 we discovered that the CWE theorem in fact is a special case of a more general statement of energy conservation (Equation 13.48). The general statement then was restricted in other ways, so that we could consider simple ("pure") thermodynamic systems; the statement then was called the first law of thermodynamics (Equation 13.51).

Why a Rock Does Not Jump

A rock at the base of a cliff does not, of its own accord, jump spontaneously up to the top of the cliff; such an occurrence would violate the CWE theorem of mechanics. Recall that the CWE theorem states that

Work done by nonconservative forces = ΔKE + ΔPE

*(Chapter Opener) *The Two Cultures and the Scientific Revolution* (Cambridge University Press, New York, 1961), pages 15–16.

[†]*Through the Looking Glass*, Chapter 5, Wool and Water (Boni and Liveright, New York, 1925).

The rock initially resting at the bottom of the cliff has no kinetic energy; likewise, it has no kinetic energy when at rest at the top of the cliff. The change in the kinetic energy is zero: ΔKE = 0 J. We choose the zero of gravitational potential energy to be at the initial location of the rock, as shown in Figure 15.1. With this choice of the origin, the rock has zero local gravitational potential energy at its initial position and its final potential energy is $+mgh$. The change in the local gravitational potential energy thus is ΔPE = PE$_f$ − PE$_i$ = mgh − 0 J = mgh. If the rock spontaneously jumps to the top of the cliff without our effort (or force), the only force acting on the rock is the gravitational force; the work done by this force is accounted for by the change in the gravitational potential energy term in the CWE theorem. There is zero work done by other forces: there are no other forces. Therefore the CWE theorem here becomes

Work done by nonconservative forces = ΔKE + ΔPE

$$0 \text{ J} = 0 \text{ J} + mgh$$

Thus we can only have h = 0 m. In other words, the rock stays where it is. If we want to get the rock to the top of the cliff, we need to *transfer* energy to the system (the rock) via work: we have to lift the rock (doing work on the system), so that the work done by "other" forces (our force on the rock) is not zero. The work we do on the rock equals the change in its potential energy according to the CWE theorem (since here there is no change in the kinetic energy).

Why a Pizza Does Not Warm Itself

A cold pizza placed in an insulated container does not, of its own accord, spontaneously increase its temperature, thereby cooking itself. To do so would violate the principle of energy conservation in the form of the first law of thermodynamics. The first law, Equation 13.51, states that

$$Q = \Delta U + W$$

Since the system is isolated from its environment, there is no heat transfer to the pizza: Q = 0 J. Certainly the pizza does no work on the surroundings, since the pizza is isolated; thus, W = 0 J as well. Hence the first law becomes

$$0 \text{ J} = \Delta U + 0 \text{ J}$$

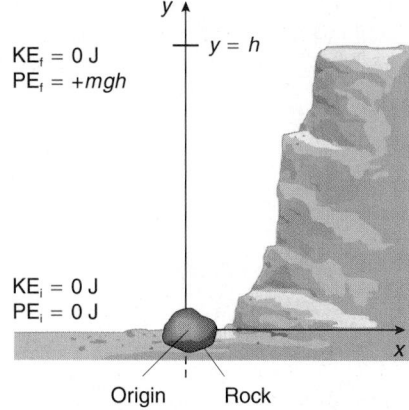

FIGURE 15.1 A rock at the base of a cliff does not jump up of its own accord; to do so would violate the CWE theorem.

and thus $\Delta U = 0$ J. The internal energy of the pizza does not change; its temperature must remain constant. If the pizza were to spontaneously warm itself, then $\Delta U \neq 0$ J and the first law of thermodynamics would be violated. To change the internal energy of the pizza (i.e., to cook it), we must effect heat transfer Q to the pizza system. We place it in a hot oven, so the energy exchange known as heat transfer is nonzero; the heat transfer to the pizza from the hot oven reservoir changes the internal energy of the pizza and thereby raises its temperature.

Why We Never Seem to Win the Lottery

Some things do not happen, not because they violate energy conservation, but because they are simply very unlikely. Such is the case with winning the lottery, at least for most individuals (including the author). *We* never seem to win the lottery; *somebody* wins the lottery but, unfortunately, it is not likely to be us. The reason is that winning the lottery is a very unlikely event for a given individual. It is not that it is impossible: it is just highly improbable. Everyone who purchases a ticket has an equal chance of winning. If we purchase many lottery tickets, the odds of winning improve: there are more ways for us to hit the jackpot because each ticket has an equal chance.

In thermodynamics, something similar occurs, with *vastly* longer odds than the typical lottery. Imagine something that is *permitted* according to energy conservation but still does not happen. Divide an isolated pizza into halves. Why can't one half of the pizza spontaneously warm the other half? Half of the pizza warms up at the (energy) expense of the other half. The energy given up by the cooling half warms the other half. For the pizza as a whole the total heat transfer is zero, since the heat transfer is from one half to the other half. If this were possible, we would end up with half a hot pizza and half a frozen pizza; half a hot pizza is better than none, after all! But, of course, we know from experience that this does not happen naturally. One half of the pizza will not spontaneously cool while warming the other half; it just does not happen.*

Notice that there is no violation of the principle of energy conservation; there is nothing in the first law of thermodynamics that prohibits this spontaneous process from taking place. It just does not happen. Why? A small child might say, "just because I said so," but we seek a more profound explanation! The process does not happen because it is so *exceedingly unlikely* that it effectively never occurs; it is comparable to one ticket in a vast lottery with a humongous number of tickets.

We know (from experiment!) that a hot pizza placed in thermal contact with a cold pizza results in two tepid pizzas: the hot one cools and the cold one warms. Energy is exchanged (heat transfer) until the two reach the same intermediate temperature. Heat transfer occurs from hot to cold objects, not from cold to hot objects. Why is this? We raised this question in Chapter 13 and finally answer it in this chapter.

The key to why these more subtle processes can or cannot take place lies *not* in the first law of thermodynamics (energy conservation), but in another fundamental law of thermodynamics: the **second law of thermodynamics**. The answer

- to our lack of lottery winnings,
- to our half-warmed, half-frozen pizza dilemma, and

- to the never observed spontaneous heat transfer from cold to hot objects

lies in the fact that such occurrences are exceedingly unlikely, so unlikely that these events effectively never happen to a given individual, to a given pizza, or in a given physical system. We shall see later in this chapter that "lotteries" exist in the natural world as well and are closely tied to heat transfer via a statistical interpretation of the second law of thermodynamics.

There are a few equivalent ways to state the second law of thermodynamics; we shall mention several and eventually show that they all are equivalent. We will then examine a statistical interpretation of the law. Several ways of stating the second law have to do with the workings of heat engines and refrigerator engines; it is difficult to see initially how this could relate to our half-hot, half-cold pizza problem or to the direction of heat transfer. But have faith while your appetite grows and Domino's responds (though no longer in thirty minutes, thanks to legal largesse); soon you will get a feast of insight into nature, if not pepperoni. But, as with Domino's, it may take longer than a thirty minute delivery time.

QUESTION 1

A rock cannot spontaneously jump up a flight of stairs, but you can decide to run up them anytime you want. What is the difference between you and the rock? Do you violate energy conservation when you spontaneously run up a flight of stairs?

15.2 HEAT ENGINES AND THE SECOND LAW OF THERMODYNAMICS

What does a heat engine do? Work, work, and more work. Just like students—and professors—but in a different sense! Here we try to examine the general features of all heat engines, not just particular ones like the one inside your mother's new Porsche or her John Deere tractor.

It is easy to change macroscopic physical work into an increase in the internal energy of the environment. For example, when a vehicle slows down, kinetic frictional forces do work on the vehicle, thereby changing (decreasing) its kinetic energy. The frictional work increases the internal energy of the brake lining and tires (they become warmer), resulting in heat transfer to the cooler environment and an eventual increase in the internal energy of the environment. The macroscopic ordered kinetic energy of the car is transformed into disordered internal energy of the environment.

The function of a **heat engine** is to turn this process around: to transform disordered internal energy of, say, a gas system into macroscopic work using the heat transfer between reservoirs at different temperatures.

We use the word **engine** to mean a system such as a gas that performs a closed cycle on a *P–V* diagram. If the cycle is performed clockwise, we call it a **heat engine**; if counterclockwise, we call it a **refrigerator engine** (or just a **refrigerator**).

---

*There also is the problem of how the pizza decides *which* half to warm and which to cool, since the halves can be formed in many ways.

The process involved with a heat engine is diagrammed schematically in Figure 15.2. Heat transfer Q_H occurs to a heat engine from a high-temperature reservoir; part of Q_H is transformed into macroscopic physical work W done *by* the heat engine, with the rest, Q_C, transferred to a cold-temperature reservoir (think of Q_C as exhaust).

If we consider the heat engine to be the thermodynamic system, the heat transfer Q_H to it is positive while Q_C is negative, since the convention is that heat transfer to a system is positive and heat transfer from a system is negative. The heat engine operates in a *cycle* such that it can repeat the process again and again. Whatever the heat engine actually does, since it is operated cyclically it returns to the same initial thermodynamic equilibrium state after completion of each cycle.

> Since the heat engine operates in a cycle, the change in the internal energy of the heat engine system over the complete cycle is zero.

If it were not zero, the engine would be blazing hot (or frozen) after many cycles.

The first law of thermodynamics implies that the total heat transfer to any engine during each cycle is equal to the total change in the internal energy of the engine plus the total work W it does:

$$Q_{\text{total}} = \Delta U + W$$

Since the change in the internal energy of the engine is zero around a cycle ($\Delta U = 0$ J),

$$Q_{\text{total}} = W \qquad (15.1)$$

The total heat transfer to a heat engine is the difference between the heat transfer *to* the heat engine from the hot reservoir ($|Q_H|$, where $Q_H > 0$ J for the heat engine) and the heat transfer *from* the heat engine to the cold reservoir ($|Q_C|$, $Q_C < 0$ J for the heat engine):

$$|Q_{\text{total}}| = |Q_H| - |Q_C| \qquad (15.2)$$

The absolute value signs are used for the following reason. The heat engine is the thermodynamic system under consideration. Heat transfer Q_H to the heat engine is positive according to our convention that heat transfer to a system is positive. The exhaust heat transfer from the heat engine to the cold reservoir, Q_C, is *negative* for the heat engine system according to our sign convention. Thus, for the heat engine as a system, we could write Equation 15.2 without the absolute value signs as $Q_{\text{total}} = Q_H + Q_C$, since Q_C actually is negative for the heat engine system. But by writing $|Q_{\text{total}}|$ explicitly as $|Q_{\text{total}}| = |Q_H| - |Q_C|$, we perhaps avoid some confusion with signs.

> The fraction of the heat transfer to the heat engine from the hot reservoir that is transformed into work done by the heat engine during each cycle is a measure of the **efficiency** of the heat engine. The efficiency ε is defined as
>
> $$\varepsilon \equiv \frac{|W|}{|Q_H|} \qquad (15.3)$$

Since both W and Q_H are positive for the heat engine system, there really is no need for the absolute value signs, but we put them in anyway for reinforcement. The efficiency ε is a positive pure number between 0 and 1. Inspection of Equations 15.1 and 15.2 indicates that a **perfect heat engine** is one for which Q_C is zero, in which case $W = Q_H$. Thus a perfect heat engine has an efficiency of 1. *Nothing* in the *first* law of thermodynamics forbids a perfect heat engine, since the first law is a statement of energy conservation. A perfect heat engine simply converts *all* of the heat transfer $|Q_H|$ into work W with zero exhaust $|Q_C|$ in each cycle.

This is where the **second law of thermodynamics** comes into play.

> The second law of thermodynamics states that *perfect heat engines do not exist.*

This is another way of saying that heat engines with an efficiency of 1 do not exist. All heat engines must have finite, nonzero exhaust heat transfer Q_C from the heat engine to a cooler reservoir. This statement about the nonexistence of perfect heat engines is the first of several equivalent statements of the second law that we will examine.

You likely noted these things during this discussion:

1. The second law of thermodynamics says something *very different* from the first law.
2. The second law means that *the efficiency of real heat engines is always less than 1*. We will address the maximum efficiency that a heat engine can have in Section 15.3.
3. The second law means that no heat engine operating in a *cycle* can transform heat transfer $|Q_H|$ from a hot reservoir completely into work W. That is, there always is some unutilized exhaust, heat transfer $|Q_C|$ from the heat engine to a colder reservoir during the cycle.
4. The word *cycle* is essential in this regard. In a cycle, the heat engine or system returns to its initial state of pressure, volume, and temperature.

If the process is not a cycle, it is quite possible to make a complete conversion of heat transfer into work. For instance,

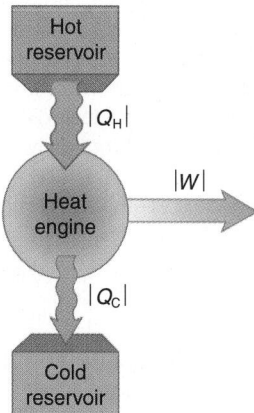

FIGURE 15.2 A schematic representation of the function of a heat engine.

consider the isothermal expansion of an ideal gas from some initial volume V_i to some final volume V_f, as shown in Figure 15.3. The work done by the gas is represented by the shaded area under the isotherm on the P–V diagram. Since the temperature of the ideal gas is unchanged in the isothermal process,* the change in the internal energy of the gas is zero. Thus the first law of thermodynamics, $Q = \Delta U + W$, becomes $Q = W$. The heat transfer to the gas from a reservoir at the temperature of the isotherm is equal to the work done by the gas. A complete (total) conversion of the heat transfer into work has been accomplished, with no exhaust heat transfer to a colder reservoir. This does *not* violate the second law of thermodynamics, because *the gas has not yet been returned to its initial thermodynamic state or condition.* The gas has not yet completed a cycle on the P–V diagram, and so there is no violation of the second law. To return the gas to its initial thermodynamic state, there *always* will be nonzero $|Q_C|$.

For the gas to return to its initial thermodynamic state, the gas must follow a closed path (a cycle) on a P–V diagram, which, according to our definition, makes the system an engine. The work done by the gas around such a closed path on a P–V diagram is represented by the area enclosed by the path on the diagram, as we saw in Chapter 13. Since we want the total work done by the gas to be positive, the gas must follow the closed path on the P–V diagram in the clockwise sense to be a heat engine, as shown in Figure 15.4. If the gas executes the closed path in the counterclockwise sense, we have a refrigerator engine and the total

work done by the gas is negative, meaning that the environment did work on the gas system rather than the other way around.

Equation 15.3 indicates that the efficiency ε of any heat engine is the total work done by the system around one cycle divided by the heat transfer to the heat engine from the hot reservoir during the cycle:

$$\varepsilon = \frac{|W|}{|Q_H|}$$

Since the system (the gas) is taken around a complete cycle (making it an engine) and returns to its initial thermodynamic state, the change in its internal energy over one cycle is zero. The first law of thermodynamics, $Q_{total} = \Delta U + W_{total}$, becomes $|Q_H| - |Q_C| = 0$ J $+ W$. We use this expression for $|W|$ in Equation 15.3.

Hence the efficiency of *any* heat engine can be written as

$$\varepsilon = \frac{|Q_H| - |Q_C|}{|Q_H|}$$

$$= 1 - \frac{|Q_C|}{|Q_H|} \qquad \text{(any heat engine)} \quad (15.4)$$

---

*Recall that the internal energy of a given quantity of an ideal gas depends only on its absolute temperature.

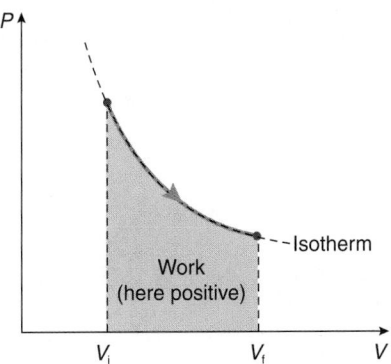

FIGURE 15.3 An isothermal expansion of an ideal gas.

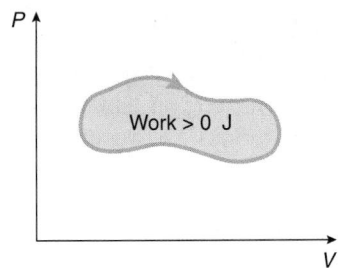

FIGURE 15.4 To do positive work, a heat engine must execute a cycle in the clockwise sense on a P–V diagram.

EXAMPLE 15.1

During each complete cycle of the "Little Engine that Could," 2000 J of heat transfer accrues to the heat engine from a hot reservoir, with 1700 J exhausted by the heat engine to a colder reservoir. Determine the efficiency of the heat engine.

Solution
I think I can. . . I think I can. . . I think I can.

Method 1
Since the heat engine completes one cycle, the change in the internal energy of the heat engine is zero. Apply the first law of thermodynamics to the heat engine:

$$Q_{total} = \Delta U + W$$
$$= 0 \text{ J} + W$$

where Q_{total} is the total heat transfer to the heat engine around the cycle; this is

$$Q_{total} \equiv |Q_H| - |Q_C|$$
$$= 2000 \text{ J} - 1700 \text{ J}$$
$$= 300 \text{ J}$$

The total work done by the heat engine during one cycle is 300 J, since $Q_{total} = W$ as was just shown.

The efficiency of the heat engine is, from Equation 15.3,

$$\varepsilon = \frac{|W|}{|Q_H|}$$
$$= \frac{300 \text{ J}}{2000 \text{ J}}$$
$$= 0.150$$

Method 2

Use Equation 15.4 directly with the given information:

$$\varepsilon = 1 - \frac{|Q_C|}{|Q_H|} \quad \text{(any heat engine)}$$

$$= 1 - \frac{1700 \text{ J}}{2000 \text{ J}}$$

$$= 0.1500$$

The first method is preferable, since it displays greater understanding of the physics. Method 2 is tantamount to "plug-and-chug," which we try to avoid when possible.

15.3 THE CARNOT HEAT ENGINE AND ITS EFFICIENCY

In 1824 the French scientist Nicolas Léonard Sadi Carnot (1796–1832) brilliantly devised an idealized, theoretical thermodynamic cycle that now bears his name: the **Carnot cycle**. We determine the efficiency of this heat engine in this section, deferring the relationship between the Carnot heat engine and the second law of thermodynamics to Section 15.7.

> A Carnot cycle takes an ideal gas around a cycle consisting of alternating isothermal and adiabatic quasi-static processes, as shown on the *P*–*V* diagram in Figure 15.5.

According to Equation 15.4, to calculate the efficiency of the Carnot heat engine we need to find Q_C and Q_H. To do this we need to examine each of the four processes of the Carnot cycle individually with the gas as the system. We begin with the isothermal expansion.

Isothermal Expansion

This expansion is accomplished by placing the gas in thermal contact with a reservoir at temperature T_H. During the expansion there is (positive) heat transfer Q_H to the gas from the hot reservoir; the gas does positive work W_1, as indicated in Figure 15.6.

The work done by an ideal gas in an isothermal expansion was found in Chapter 13 to be (Equation 13.57)

$$W_1 = nRT_H \ln \frac{V_2}{V_1}$$

At the comparatively youthful age of 36, Carnot tragically contracted cholera and died within hours. Almost all of his personal papers and possessions were burned to prevent the spread of the disease, a tragedy for historians of science.

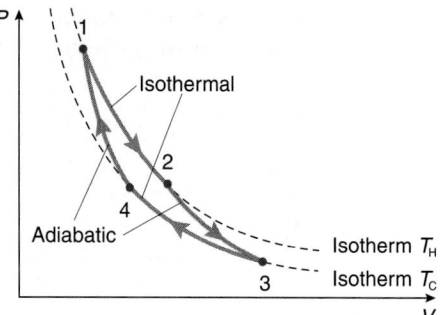

FIGURE 15.5 A Carnot cycle on a *P*–*V* diagram.

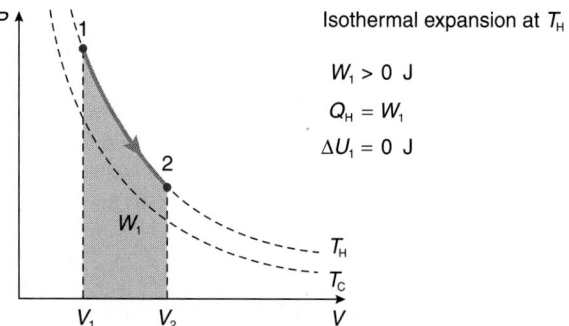

FIGURE 15.6 The work done by the gas in an isothermal expansion is positive.

This work is represented by the area under the isotherm between V_1 and V_2 on the *P*–*V* diagram of Figure 15.6. The internal energy of the ideal gas is purely a function of the temperature. Since the temperature is constant in an isothermal process, the internal energy is constant: $\Delta U_1 = 0$ J. The first law of thermodynamics, $Q = \Delta U + W$, here becomes

$$Q_H = 0 \text{ J} + W_1$$

$$= W_1$$

$$= nRT_H \ln \frac{V_2}{V_1} \quad (15.5)$$

This indicates that *all* of the heat transfer Q_H to the gas is converted into the work W_1. This is *not* a violation of the second law of thermodynamics, because the gas has not yet completed a cycle on the *P*–*V* diagram.

Adiabatic Expansion

In an adiabatic process the system is totally insulated from its environment, so that there is zero heat transfer either to or from the system. Alternatively, the process takes place fast enough so that heat transfer has no time to occur, yet slow enough to be quasistatic and reversible. During an adiabatic process $Q = 0$ J. Positive work W_2 is done since the gas expands; as usual, the work W_2 is represented by the area under the adiabatic curve on the *P*–*V* diagram; see Figure 15.7.

Since $Q = 0$ J, the first law of thermodynamics, $Q = \Delta U + W$, becomes

$$0 \text{ J} = \Delta U_2 + W_2$$

$$\Delta U_2 = -W_2 \quad (15.6)$$

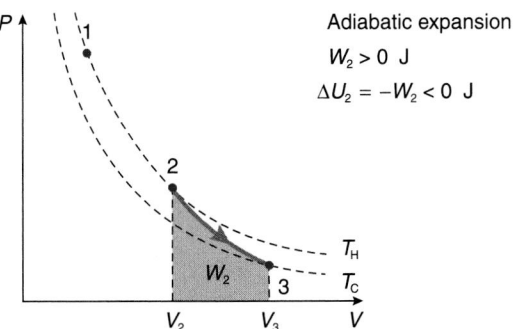

FIGURE 15.7 The work done by the gas during an adiabatic expansion is positive.

Since W_2 is positive (the gas expands), the change in the internal energy of the gas is negative; the internal energy decreases. Since the internal energy of an ideal gas is directly proportional to the temperature, the temperature of the gas decreases to T_C, as indicated on the P–V diagram in Figure 15.7.

Isothermal Compression

Once again, since the process is isothermal the internal energy of the gas does not change: $\Delta U_3 = 0$ J. The gas does negative work W_3, as shown in Figure 15.8.

We use Equation 13.57 again for the work done by an ideal gas in an isothermal process:

$$W_3 = nRT_C \ln \frac{V_4}{V_3} \tag{15.7}$$

The work is negative because V_4 is less than V_3 (the logarithm of a number less than 1 is negative). The first law of thermodynamics, $Q = \Delta U + W$, here becomes

$$Q_C = 0 \text{ J} + W_3$$
$$= nRT_C \ln \frac{V_4}{V_3} \tag{15.8}$$

Since W_3 is negative, Q_C also is negative, indicating that the heat transfer is from the gas (the system) to the reservoir at temperature T_C during this process.

Adiabatic Compression

The final adiabatic compression takes the gas back to its initial state, as shown in Figure 15.9, so the cycle can be repeated. There is zero heat transfer to (or from) the system because this is an adiabatic process. The gas does negative work W_4 since the volume decreases, and this work is the negative area under this adiabatic curve on the P–V diagram.

The first law of thermodynamics, $Q = \Delta U + W$, here becomes

$$0 \text{ J} = \Delta U_4 + W_4$$
$$\Delta U_4 = -W_4 \tag{15.9}$$

Since W_4 is negative, ΔU_4 is positive, reflecting that the temperature of the gas increases in this process from T_C back to T_H.

Now bring the four parts of the cycle together and apply the first law to the *complete cycle*. Since the gas returns to its initial state, the total change in the internal energy of the gas is zero. That is,

$$\Delta U_{total} = \Delta U_1 + \Delta U_2 + \Delta U_3 + \Delta U_4$$
$$= 0 \text{ J} + \Delta U_2 + 0 \text{ J} + \Delta U_4$$
$$= 0 \text{ J}$$

This indicates that the two internal energy changes that occurred during the two adiabatic processes of the cycle are of equal magnitude but opposite sign. This has to be the case, since for an ideal gas the internal energy is a function of temperature alone: in one adiabatic process, the temperature of the gas decreased from T_H to T_C; and in the other adiabatic process, the temperature increased from T_C back to T_H. The first law applied to the whole cycle with the gas as the system therefore is

$$Q_{total} = W_{total} + \Delta U_{total}$$
$$Q_H + Q_C = W_1 + W_2 + W_3 + W_4 + 0 \text{ J}$$

Since Q_C, W_3, and W_4 are all negative for the gas system, we write this as

$$|Q_H| - |Q_C| = |W_1| + |W_2| - |W_3| - |W_4|$$

The left-hand side of the equation is the total heat transfer to the gas in the cycle (positive); the right-hand side is the total work done by the gas (also positive and represented by the area of the cycle on the P–V diagram in Figure 15.5).

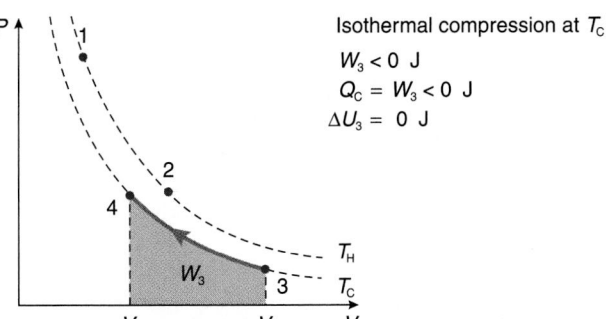

FIGURE 15.8 The work done by the gas in an isothermal compression is negative.

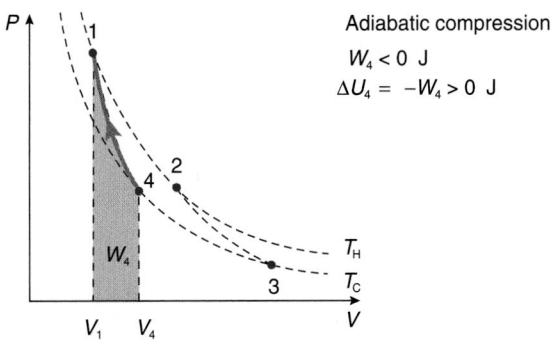

FIGURE 15.9 The work done by the gas during an adiabatic compression is negative.

We wrap things up by calculating the efficiency of the Carnot heat engine using Equation 15.4:

$$\varepsilon = 1 - \frac{|Q_\text{C}|}{|Q_\text{H}|}$$

We use the expressions for both Q_H and Q_C, Equations 15.5 and 15.8:

$$Q_\text{H} = nRT_\text{H} \ln \frac{V_2}{V_1}$$

Since Q_H is positive, $|Q_\text{H}|$ is the same thing.

For Q_C we found

$$Q_\text{C} = nRT_\text{C} \ln \frac{V_4}{V_3}$$

and this is *negative* since $V_4 < V_3$. We can write the absolute value of Q_C by flipping the numerator and denominator in the natural logarithm, since $\ln(a/b) = -\ln(b/a)$:

$$\left| Q_\text{C} \right| = nRT_\text{C} \ln \frac{V_3}{V_4}$$

To calculate the efficiency we need the ratio of $|Q_\text{C}|$ to $|Q_\text{H}|$:

$$\begin{aligned} \frac{|Q_\text{C}|}{|Q_\text{H}|} &= \frac{nRT_\text{C} \ln (V_3 / V_4)}{nRT_\text{H} \ln (V_2 / V_1)} \\ &= \frac{T_\text{C} \ln (V_3 / V_4)}{T_\text{H} \ln (V_2 / V_1)} \end{aligned} \qquad (15.10)$$

This expression can be simplified further by making use of a result from Section 14.12. There we found that during an adiabatic process (Equation 14.44), the product $TV^{\gamma-1}$ remains constant. Hence, if we form the product $TV^{\gamma-1}$ at the beginning of the adiabatic process, the number obtained is the same as the corresponding product at the end of the process. In other words, for the adiabatic process from point 2 to point 3 (in the P–V diagram of Figure 15.7), write Equation 14.44 for point 2; this expression will have the same value at point 3:

$$T_\text{H} V_2^{\gamma-1} = T_\text{C} V_3^{\gamma-1} \qquad (15.11)$$
$$\text{Point 2} \qquad \text{Point 3}$$

We do the same thing for the adiabatic process from point 4 back to point 1 on the P–V diagram of Figure 15.9:

$$T_\text{H} V_1^{\gamma-1} = T_\text{C} V_4^{\gamma-1} \qquad (15.12)$$
$$\text{Point 1} \qquad \text{Point 4}$$

Dividing Equation 15.11 by Equation 15.12, we find the following remarkable result:

$$\frac{V_2}{V_1} = \frac{V_3}{V_4}$$

Thus the two natural logarithms in Equation 15.10 are the natural logarithms of the *same number*. Hence Equation 15.10 reduces to

$$\frac{|Q_\text{C}|}{|Q_\text{H}|} = \frac{T_\text{C}}{T_\text{H}} \qquad \text{(Carnot cycle only)} \qquad (15.13)$$

We use Equation 15.13 in Equation 15.4.

The efficiency of the Carnot heat engine takes on the remarkably simple form

$$\varepsilon_\text{Carnot} = 1 - \frac{T_\text{C}}{T_\text{H}} \qquad \begin{array}{l} \text{(Carnot heat engine only)} \\ \text{(use temperatures in kelvin)} \end{array} \qquad (15.14)$$

Notice that the efficiency of a Carnot heat engine increases and approaches 1 if $T_\text{H} \gg T_\text{C}$.

PROBLEM-SOLVING TACTIC

15.1 In Equation 15.14 for the efficiency of a Carnot heat engine, be sure to use absolute temperatures in kelvin, not celsius temperatures.

QUESTION 2

Equation 15.14 must use absolute temperatures. Is the equation still valid if one uses another temperature scale whose zero is at 0 K but whose degree size is different than a kelvin? For example, the *Rankine* temperature scale has its zero at 0 K but uses degrees equal in size to Fahrenheit degrees. Is Equation 15.14 also legitimate using temperatures on the Rankine scale?

EXAMPLE 15.2

A Carnot heat engine is operating between two reservoirs, one at 0 °C and the other at 100 °C. What is the efficiency of the Carnot heat engine?

Solution

The efficiency of a Carnot heat engine depends only on the *absolute* temperatures of the two reservoirs according to Equation 15.14. Hence

$$\begin{aligned} \varepsilon &= 1 - \frac{T_\text{C}}{T_\text{H}} \\ &= 1 - \frac{273 \text{ K}}{373 \text{ K}} \\ &= 0.268 \end{aligned}$$

15.4 ABSOLUTE ZERO AND THE THIRD LAW OF THERMODYNAMICS

The equation for the efficiency of a Carnot heat engine (Equation 15.14) is used to define the zero point of the absolute temperature scale. The absolute zero of temperature is the temperature of a cold reservoir for which the efficiency of a Carnot heat engine is exactly 1. If $T_\text{C} = 0$ K, then according to Equation 15.13 the Carnot heat engine has zero exhaust heat transfer $|Q_\text{C}|$. Such a heat engine is a perfect heat engine; but such perfect heat engines do not exist according to the second law of thermodynamics.

Hence the absolute zero of temperature is an *unattainable* temperature; reservoirs with $T_\text{C} = 0$ K *cannot exist*. This conclusion is the **third law of thermodynamics**.

The absolute temperature scale (the kelvin scale we use in virtually all the equations of thermodynamics) thus is truly a thermodynamic temperature scale, because its zero of temperature is fundamentally connected with the second law; 0 K is a coldest temperature that exists only as a limit. The size of a kelvin degree does not have a fundamental thermodynamic significance; it simply was chosen to be equal to the size of the common and convenient celsius degree. Recall that the zero of temperature on the celsius scale was chosen for convenience and has no truly fundamental thermodynamic significance.

15.5 REFRIGERATOR ENGINES AND THE SECOND LAW OF THERMODYNAMICS

. . . hell must be isothermal, for otherwise the resident engineers and physical chemists (of which there must be some) could set up a . . . refrigerator to cool off a portion of their surroundings to any desired temperature.

Henry A. Bent*

There is another way to view the second law of thermodynamics that involves refrigerators.[†] The idea of a **refrigerator engine** is to effect heat transfer from a cold reservoir (the inside of what we commonly think of as a refrigerator) to a hot reservoir (the outside of the refrigerator, such as the room in which it is located). The heat transfer from the cold reservoir originally arises from heat transfer to this reservoir from warm objects that we place in the refrigerator compartment. The heat transfer from the cold interior to the warm exterior will not happen spontaneously (by itself), since the natural direction of heat transfer is from hot to cold objects. We have yet to explain this direction of heat transfer, but the explanation is coming soon.

A refrigerator engine (also called simply a refrigerator) accomplishes the unnatural heat transfer from a cold to a hot reservoir. We want the refrigerator engine to operate in a cycle so that the process can be done repetitively. Thus the change in the internal energy of the system (the refrigerator engine) around each complete cycle is zero. The process is represented schematically in Figure 15.10.

Work is done on the device or engine by external agents; in other words, the refrigerator engine system does *negative* work W according to our thermodynamic convention that work done *by* the system is positive. Since W is *negative*, the direction of energy transfer via work is *into* (or on) the engine. Heat transfer $|Q_C|$ occurs from the cold reservoir (the inside of the refrigerator) to the refrigerator engine and heat transfer $|Q_H|$ is exhausted from the refrigerator to the hot reservoir (the outside of the refrigerator). Since the refrigerator engine is the system, the heat transfer Q_C to the refrigerator engine (from the cold reservoir) is positive; the heat transfer Q_H from the refrigerator system (to the hot reservoir) is negative.

The first law of thermodynamics states that the total heat transfer to the refrigerator engine system during one cycle is equal to the change in the internal energy of the system plus the work done by the system. In other words,

$$Q_{total} = \Delta U + W$$

*The Second Law (Oxford University Press, Oxford, 1965), page 313.
[†]In the context of this section, freezers are equivalent to refrigerators.

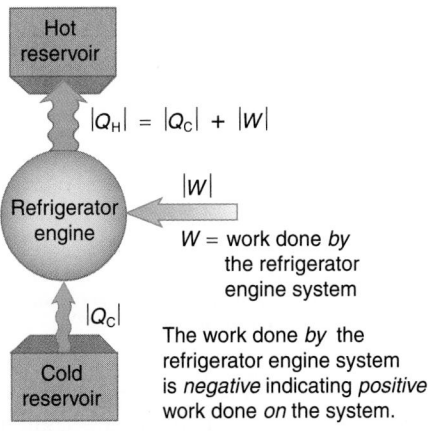

FIGURE 15.10 Schematic representation of a refrigerator engine operating between two reservoirs.

Since the change in the internal energy of the refrigerator engine system around a cycle is zero, $\Delta U = 0$ J, the first law becomes

$$Q_{total} = 0 \text{ J} + W$$
$$= W$$

For refrigerator engines, *both* Q_{total} and W are *negative* according to our sign conventions. To eliminate sign confusion, we use absolute values:

$$|Q_{total}| = |W| \qquad (15.15)$$

The absolute value of the total heat transfer to the system is

$$|Q_{total}| = |Q_H| - |Q_C| \qquad (15.16)$$

since $|Q_H|$ is larger than $|Q_C|$. Substituting $|W|$ for $|Q_{total}|$ from Equation 15.15 into Equation 15.16 and rearranging yields

$$|Q_H| = |Q_C| + |W| \qquad (15.17)$$

in keeping with energy conservation.

The performance of a refrigerator engine is assessed by taking the ratio of the heat transfer to the refrigerator engine from the cold reservoir to the absolute value of the (negative) work that must be done by the system to accomplish the unnatural heat transfer from the colder to the hotter reservoir. The **coefficient of performance** K of a refrigerator engine is defined as

$$K = \left| \frac{Q_C}{W} \right| \qquad (15.18)$$

where absolute values are used to avoid sign confusion.

For the refrigerator engine system Q_C is positive since the heat transfer Q_C is to refrigerator engine system (from the cold reservoir), so that the absolute value signs are superfluous for this quantity. But W itself is negative, because the work done *by* the refrigerator engine system is negative (meaning positive work must be done *on* the refrigerator engine system by external agents). The coefficient of performance K is a positive pure number. A good refrigerator engine is one that maximizes $|Q_C|$ while

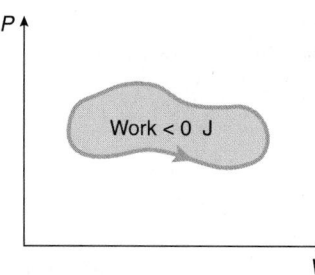

FIGURE 15.11 A refrigerator engine traverses a counterclockwise cycle on a *P–V* diagram and does negative work *W*.

minimizing $|W|$ and therefore has a large coefficient of performance. The examples we study in this chapter indicate that typical values for K are only in single digits. A **perfect refrigerator engine** is one that has $|W| = 0$ J and an *infinite* coefficient of performance.

Nothing in the first law of thermodynamics (energy conservation) prohibits a perfect refrigerator engine. Such a refrigerator engine simply has $|W| = 0$ J, so that $|Q_C| = |Q_H|$.

> *Perfect refrigerator engines do not exist*; this is another way of stating the second law of thermodynamics.

Among the implications of this statement of the second law are the following:

1. It is impossible to construct a system or device (a refrigerator engine) operating in a cycle to effect heat transfer from a cold reservoir to a hot reservoir without negative work being done *by* the device (i.e., positive work done *on* the system).
2. Since the refrigerator engine does *negative* work, its cyclical path on, say, a *P–V* diagram is traversed in the *counterclockwise* sense, as shown in Figure 15.11.

The coefficient of performance of a refrigerator engine is defined by Equation 15.18:

$$K = \frac{|Q_C|}{|W|}$$

But, as we found using the first law of thermodynamics (Equation 15.17), in one cycle $|Q_H| = |Q_C| + |W|$.

> Therefore the coefficient of performance of *any* refrigerator engine also can be written as
>
> $$K = \frac{|Q_C|}{|Q_H| - |Q_C|}$$
>
> $$= \frac{1}{\dfrac{|Q_H|}{|Q_C|} - 1} \qquad \text{(any refrigerator engine)} \quad (15.19)$$

The coefficient of performance of a refrigerator engine must be less than infinity; we address how large it can actually be in Section 15.7.

EXAMPLE 15.3

Two (2.00) kilograms of water at 20.0 °C are to be frozen into an ice cube at 0.0 °C using a refrigerator-freezer with a coefficient of performance equal to 4.00.

a. Determine the work done by the refrigerator-freezer.
b. Determine the heat transfer to the hot reservoir.

Solution

a. Take the water as the system. To cool the water from 20.0 °C to 0.0 °C, the heat transfer Q_1 to it is

$$Q_1 = mc_{\text{water}} \Delta T$$

Recall that to calculate ΔT you can use either the celsius or kelvin scales, since the sizes of the kelvin and celsius degree are the same. From Table 13.4, the specific heat of water is 4186 J/(kg·K). Thus

$$Q_1 = (2.00 \text{ kg})[4186 \text{ J/(kg·K)}](0.0 \text{ °C} - 20.0 \text{ °C})$$
$$= -1.67 \times 10^5 \text{ J}$$

The negative sign indicates that the heat transfer really is from the water, not to it.

To freeze the water involves an additional (negative) heat transfer Q_2 from it:

$$Q_2 = -mL_{f\,\text{water}}$$

Finding the latent heat of fusion of water from Table 13.5, you have

$$Q_2 = -(2.00 \text{ kg})(3.335 \times 10^5 \text{ J/kg})$$
$$= -6.67 \times 10^5 \text{ J}$$

So the total heat transfer from the water is

$$Q = Q_1 + Q_2$$
$$= -8.34 \times 10^5 \text{ J}$$

The heat transfer is from the water *to* the refrigerator engine. Thus positive heat transfer $Q_C = -Q$ accrues *to* the refrigerator engine:

$$Q_C = +8.34 \times 10^5 \text{ J}$$

This corresponds to Q_C in Figure 15.10. The coefficient of performance K is defined by Equation 15.18,

$$K = \frac{|Q_C|}{|W|}$$

$$4.00 = \frac{|8.34 \times 10^5 \text{ J}|}{|W|}$$

Hence

$$|W| = 2.09 \times 10^5 \text{ J}$$

The work W done *by* the refrigerator engine is *negative* because you (or another external agent) have to do work *on* the refrigerator engine for it to do its thing. So the work done by the refrigerator engine is

$$W_{\text{refrig}} = -2.09 \times 10^5 \text{ J}$$

b. The absolute value of the heat transfer from the refrigerator engine to the hot reservoir is found using Equation 15.17:

$$\begin{aligned} |Q_{\text{H}}| &= |Q_{\text{C}}| + |W| \\ &= 8.34 \times 10^5 \text{ J} + 2.09 \times 10^5 \text{ J} \\ &= 10.43 \times 10^5 \text{ J} \end{aligned}$$

The heat transfer to the hot reservoir system is positive, while that from the refrigerator engine system is negative. This energy is exhausted into the room (the hot reservoir), which is why a running refrigerator warms the kitchen.

15.6 THE CARNOT REFRIGERATOR ENGINE

A Carnot cycle executed in counterclockwise fashion on a *P–V* diagram is a **Carnot refrigerator engine**.

We previously showed that the absolute value of the heat transfer to a *Carnot* heat engine from the hot reservoir, $|Q_{\text{H}}|$, divided by that exhausted from the heat engine to the cold reservoir, $|Q_{\text{C}}|$, is equal to the ratio of the absolute temperatures of the reservoirs (Equation 15.13, slightly rearranged):

$$\frac{|Q_{\text{H}}|}{|Q_{\text{C}}|} = \frac{T_{\text{H}}}{T_{\text{C}}} \qquad \text{(Carnot cycle only)}$$

This is *not* true of other heat engine cycles. The result is true even if the Carnot cycle is performed in the counterclockwise sense on the *P–V* diagram. Thus, we substitute this ratio in Equation 15.19.

The coefficient of performance of a Carnot refrigerator engine is

$$K_{\text{Carnot}} = \frac{1}{\dfrac{T_{\text{H}}}{T_{\text{C}}} - 1} \qquad \begin{array}{l}\text{(Carnot refrigerator only)} \\ \text{(use temperatures in kelvin)}\end{array} \quad (15.20)$$

Notice that as the temperature difference between the two reservoirs *increases*, the coefficient of performance of a Carnot refrigerator engine *decreases*.

PROBLEM-SOLVING TACTIC

15.2 In Equation 15.20 for the coefficient of performance of a Carnot refrigerator engine, be sure to use absolute temperatures in kelvin, not celsius temperatures.

EXAMPLE 15.4 ▬▬▬▬▬▬▬

A Carnot refrigerator engine is operating between temperatures −20 °C and 20 °C. What is its coefficient of performance?

Solution

The coefficient of performance of a Carnot refrigerator engine is found using Equation 15.20:

$$K_{\text{Carnot}} = \frac{1}{\dfrac{T_{\text{H}}}{T_{\text{C}}} - 1}$$

where absolute temperatures *must* be used. Making the temperature conversions and substitutions, you find

$$K_{\text{Carnot}} = \frac{1}{\dfrac{293 \text{ K}}{253 \text{ K}} - 1}$$

$$= 6.3$$

15.7 THE EFFICIENCY OF REAL HEAT ENGINES AND REFRIGERATOR ENGINES

What is the most efficient heat engine or refrigerator that can operate between two given reservoirs? This is the problem that motivated Carnot when he began his research into the thermodynamics of engines at the beginning of the Industrial Revolution.

The Carnot heat engine, described in some detail in Section 15.3, is an idealized heat engine. The efficiency of a *real* heat engine operating between two reservoirs at temperatures T_{H} and T_{C} always is *less* than that of a Carnot heat engine operating between the same two reservoirs. In other words:

The efficiency of a Carnot heat engine is the maximum efficiency of a heat engine operating between the two given reservoirs.

We will prove this statement by contradiction. That is, we assume there is a prototype heat engine with an efficiency greater than the Carnot heat engine efficiency and then show that a violation of the second law of thermodynamics results.

Thus imagine we have a prototype heat engine (see Figure 15.12) with an efficiency ε greater than the Carnot heat engine

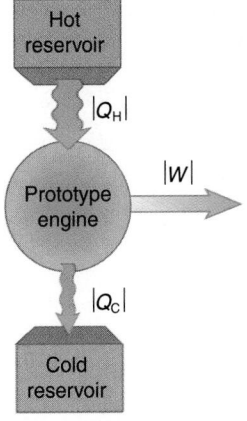

FIGURE 15.12 A prototype heat engine operating between two reservoirs.

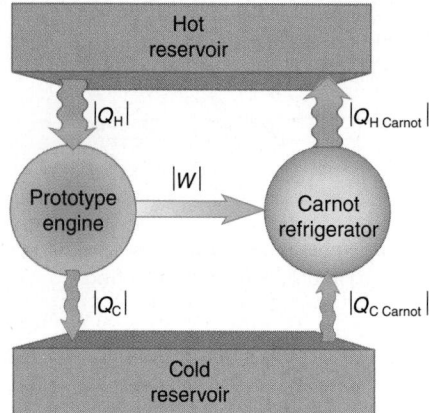

FIGURE 15.13 Use the prototype heat engine to run a Carnot refrigerator engine between the same two reservoirs.

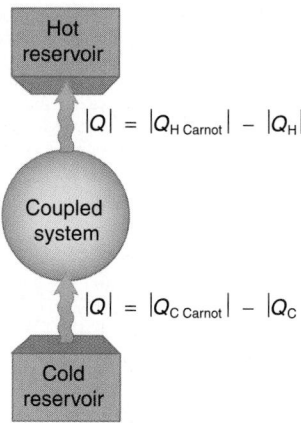

FIGURE 15.14 The total effect of the coupled prototype engine and Carnot refrigerator.

efficiency $\varepsilon_{\text{Carnot}}$ operating between the two given reservoirs at temperatures T_H and T_C. The prototype heat engine accrues heat transfer $|Q_H|$ from the hot reservoir, does work $|W|$, and exhausts heat transfer $|Q_C|$ to the cold reservoir. Now we use the work of the prototype heat engine to run a Carnot refrigerator engine operated between the same two reservoirs, as shown in Figure 15.13.

Heat transfer $|Q_{C\,\text{Carnot}}|$ occurs to the Carnot refrigerator engine from the cold reservoir.* Heat transfer $|Q_{H\,\text{Carnot}}|$ is from the Carnot refrigerator engine to the hot reservoir.† The work needed to run the Carnot refrigerator engine is provided by the prototype heat engine. By hypothesis, the prototype heat engine efficiency is greater than that of the Carnot heat engine operating between the same reservoirs; so from the definition of heat engine efficiency (Equation 15.3), we have

$$\frac{|W|}{|Q_H|} > \frac{|W|}{|Q_{H\,\text{Carnot}}|}$$

Thus

$$|Q_{H\,\text{Carnot}}| > |Q_H| \qquad (15.21)$$

Now we invoke the first law of thermodynamics. The work done by the prototype heat engine during one cycle is, from Equation 15.17,

$$|W| = |Q_H| - |Q_C|$$

(recall that in a cycle $\Delta U = 0$ J for any engine). The work done by the Carnot refrigerator engine is

$$|W| = |Q_{H\,\text{Carnot}}| - |Q_{C\,\text{Carnot}}|$$

The work done by the prototype heat engine is used to run the Carnot refrigerator engine, and so the two expressions for $|W|$ must be the same:

$$|Q_H| - |Q_C| = |Q_{H\,\text{Carnot}}| - |Q_{C\,\text{Carnot}}| \qquad (15.22)$$

---

*$Q_{C\,\text{Carnot}}$ is positive for the Carnot refrigerator engine system and negative for the cold reservoir system.

†$Q_{H\,\text{Carnot}}$ is negative for the Carnot refrigerator engine system and positive for the hot reservoir system.

Rearranging, we obtain

$$|Q_{H\,\text{Carnot}}| - |Q_H| = |Q_{C\,\text{Carnot}}| - |Q_C| \qquad (15.23)$$

The left-hand side of Equation 15.23 is positive according to the hypothesis (Equation 15.21); call it Q:

$$|Q_{H\,\text{Carnot}}| - |Q_H| = |Q_{C\,\text{Carnot}}| - |Q_C| \equiv |Q| \qquad (15.24)$$

The effect of the coupled system of the prototype heat engine and Carnot refrigerator engine is diagrammed schematically in Figure 15.14. The coupled system effects heat transfer $|Q|$ from a cold reservoir to a hot reservoir *with no net work* done by the system, since the work done by the prototype heat engine is used by the Carnot refrigerator. This combination is a perfect refrigerator, but *such perfect refrigerators do not exist* according to the second law of thermodynamics. Thus the quantity $|Q|$ in Equation 15.24 *must* be zero, and

$$|Q_H| = |Q_{H\,\text{Carnot}}|$$

which contradicts the hypothesis formulated in Equation 15.21.

Therefore the efficiency of the prototype heat engine is only at best identical to (but not greater than) the efficiency of the Carnot heat engine operating between the same two reservoirs.

A similar argument can be used to show the following:

> A Carnot refrigerator engine has the maximum coefficient of performance of any refrigerator engine operating between the same two reservoirs.

QUESTION 3

An appliance salesman receives a call from a customer to whom a refrigerator was just sold. The customer is complaining because, even though it is cold inside the refrigerator chamber, there is hot air coming into the kitchen from the vent strip at the bottom of the refrigerator. Is the refrigerator broken? Explain.

EXAMPLE 15.5

Two reservoirs have temperatures 20 °C and 300 °C.

a. What is the maximum possible efficiency for a heat engine operating between the two reservoirs?

b. What is the maximum value of the coefficient of performance of a refrigerator engine operating between the two reservoirs?

Solution

a. The maximum efficiency occurs with a Carnot heat engine operating between the two reservoirs. Equation 15.14 gives the efficiency of a Carnot heat engine:

$$\varepsilon_{\text{Carnot}} = 1 - \frac{T_C}{T_H} \qquad \text{(Carnot heat engine only)}$$

where the temperatures *must* be absolute temperatures. Therefore

$$\varepsilon_{\text{Carnot}} = 1 - \frac{293 \text{ K}}{573 \text{ K}}$$

$$= 0.489$$

b. The maximum value of the coefficient of performance is that of a Carnot refrigerator engine operating between the two reservoirs, given by Equation 15.20:

$$K_{\text{Carnot}} = \frac{1}{\dfrac{T_H}{T_C} - 1} \qquad \text{(Carnot refrigerator engine only)}$$

$$= \frac{1}{\dfrac{573 \text{ K}}{293 \text{ K}} - 1}$$

$$= 1.0$$

15.8 A NEW CONCEPT: ENTROPY

Thus far our study of thermodynamics has considered a number of thermodynamic state variables such as temperature, pressure, volume, and internal energy. A system in thermodynamic equilibrium has definite values for each of these physical quantities, which is why they are called state variables.

The internal energy of a system is a peculiar state variable. From a submicroscopic viewpoint it is possible, in principle, to talk about a specific value for the internal energy of the system in equilibrium. However, from a macroscopic, classical standpoint only *changes* in the internal energy are significant (and only they appear in the first law of thermodynamics).

Other thermodynamic concepts, such as the work W done by a system and the heat transfer Q to a system, are *not* thermodynamic state variables. As we have seen, a system in equilibrium *cannot* be said to have a certain amount of work or heat: there is no mathematical work function or heat function that is characteristic of the system when it is in thermal equilibrium. Heat transfer and work are meaningless concepts for a system in thermodynamic equilibrium. Work and heat transfer are energy transfers to or from a system, thus *changing* its state of thermodynamic equilibrium. We have to distinguish clearly between heat transfer and temperature, between work and internal energy, and between the submicroscopic internal energy and the macroscopic bulk motions of the system as a whole (mechanical energy).

We now come to a different classical physical concept known as the **entropy**.* The entropy is a state variable (as are the inter-

nal energy, volume, temperature, and pressure), which means that a system in equilibrium can be said to have a definite entropy.

From a classical thermodynamic viewpoint, it is *only changes* in entropy that are significant when a system changes its state of thermodynamic equilibrium.

In this respect the entropy is similar to the concept of internal energy, since it is only changes in internal energy that are physically significant when a system changes its state of thermodynamic equilibrium. However, despite these similarities, the entropy is *not* the same thing as the internal energy, as we will see.

The zeroth law of thermodynamics was concerned with thermal equilibrium. The first law of thermodynamics was concerned with energy conservation. The second law of thermodynamics is, as we will see, ultimately concerned with entropy changes.

Once we formulate just what is meant by an entropy change, we can restate the second law using the entropy concept and then show that the previous statements of the second law (the nonexistence of perfect heat engines and perfect refrigerator engines) are manifestations of this new formulation. We have our work[†] cut out for us!

Recall that a Carnot heat engine executes an alternating sequence of isothermal and adiabatic reversible processes that form a closed path on a P–V diagram, traversed in the clockwise sense. (A Carnot refrigerator engine also executes alternating isothermal and adiabatic processes, but the closed path is navigated in the counterclockwise sense on a P–V diagram.) We discovered that for a Carnot heat engine operating between two reservoirs at temperatures T_H and T_C, heat transfer $|Q_H|$ accrues to the engine along the isothermal path at T_H and heat transfer $|Q_C|$ is exhausted by the engine to the colder reservoir along the isothermal path at temperature T_C. No heat transfer occurs along the adiabatic paths. Specifically, we found (Equation 15.13)

$$\frac{|Q_H|}{|Q_C|} = \frac{T_H}{T_C} \qquad \text{(Carnot cycle only)}$$

or

$$\frac{|Q_H|}{T_H} = \frac{|Q_C|}{T_C} \qquad \text{(Carnot cycle only)} \qquad (15.25)$$

The *same* result applies to Carnot refrigerator engines, although in the interests of brevity, we did not show this explicitly. The heat transfers Q_H and Q_C have *opposite* signs for a heat engine system and for a refrigerator engine system. Using our convention that the heat transfer to a system is positive,

- for a Carnot heat engine, Q_H is positive and Q_C is negative;
- for a Carnot refrigerator engine, Q_C is positive and Q_H is negative.

Thus, for *either* a Carnot heat engine or a Carnot refrigerator engine, we can rewrite Equation 15.25 *without* the absolute value signs as

$$\frac{Q_H}{T_H} + \frac{Q_C}{T_C} = 0 \text{ J/K} \qquad \text{(Carnot cycle)} \qquad (15.26)$$

---

*The word stems from the Greek τροπή ("transformation") and was first proposed by Rudolph Clausius (1822–1888)

[†]We do *not* mean W in this sentence!

We derived this equation specifically for a Carnot cycle. To accomplish our objective of a general statement of the second law of thermodynamics, we need to generalize Equation 15.26 to *any* kind of (quasi-static) reversible cycle.

We represent an arbitrary quasi-static reversible cycle (closed path) on a *P–V* diagram as shown in Figure 15.15. We approximate the actual path on the *P–V* diagram by means of a succession of small, alternating isothermal and adiabatic paths.

Let Q_i be the heat transfer to the system along the isothermal at temperature T_i. Notice that the alternating isothermal and adiabatic paths along the upper part of the actual cycle on the *P–V* diagram can be paired with those on the lower part of the cycle to form abutting Carnot cycles on the diagram. The contributions to Q/T arising from the parts of the isotherms *common* to adjacent Carnot cycles sum to zero, since this part of the isotherm is traversed *twice*, once in each direction. So, along the common parts of each of the various isotherms, one Carnot cycle has a positive contribution Q/T and its abutter has a negative contribution of the same magnitude. In this way, the actual cycle can be approximated by the shaded series of Carnot cycles. Since Equation 15.26 can be written for each Carnot cycle, we write the following for the collection of Carnot cycles:

$$\sum_i \frac{Q_i}{T_i} = 0 \text{ J/K} \tag{15.27}$$

The approximation to the actual quasi-static reversible cycle improves by taking tiny *differential* isothermal and adiabatic paths, as indicated in Figure 15.16. Such differential isothermal and adiabatic paths and Carnot cycles then approach the actual cycle path on the *P–V* diagram as a limit. With this limiting procedure, Equation 15.27 becomes

$$\oint_{\substack{\text{reversible} \\ \text{cycle}}} \frac{dQ}{T} = 0 \text{ J/K} \tag{15.28}$$

where the elegant circle through the integral sign indicates that the integral is over a *reversible cycle* on the *P–V* diagram—that is, a closed path on the *P–V* diagram. The quantity

$$\frac{dQ}{T}$$

integrated around a reversible cycle on a *P–V* diagram is zero.

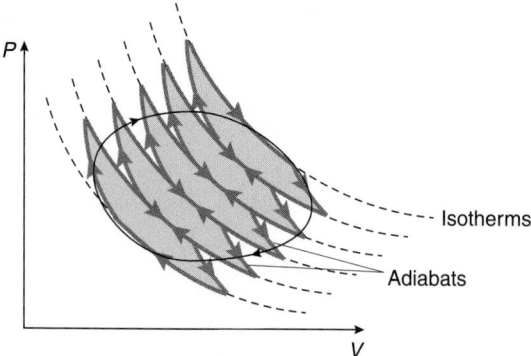

FIGURE 15.15 A cycle on a *P–V* diagram is approximated by abutting Carnot cycles.

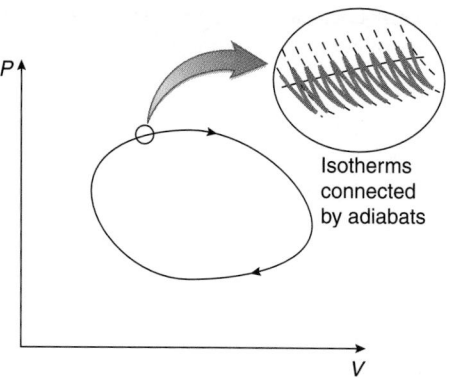

FIGURE 15.16 The actual cycle is approximated by a series of differential isothermal and adiabatic paths.

Now we consider a system undergoing a quasi-static reversible change from some initial equilibrium state to a final equilibrium state, as shown in Figure 15.17. There are many ways to get between these states; two arbitrary paths are indicated.

Suppose we go from the initial state to the final state via path (1) and then return to the initial state via path (2) in Figure 15.17. Since this is a reversible cycle, Equation 15.28 states that

$$\underset{\text{path (1)}}{\int_i^f \frac{dQ}{T}} + \underset{\text{path (2)}}{\int_f^i \frac{dQ}{T}} = 0 \text{ J/K}$$

Now we interchange the limits in the second integration, thus changing the sign of the integral:

$$\underset{\text{path (1)}}{\int_i^f \frac{dQ}{T}} - \underset{\text{path (2)}}{\int_i^f \frac{dQ}{T}} = 0 \text{ J/K}$$

or

$$\underset{\text{path (1)}}{\int_i^f \frac{dQ}{T}} = \underset{\text{path (2)}}{\int_i^f \frac{dQ}{T}} \tag{15.29}$$

Since paths (1) and (2) were *arbitrary* paths, Equation 15.29 means that the *change* in the quantity

$$\frac{dQ}{T}$$

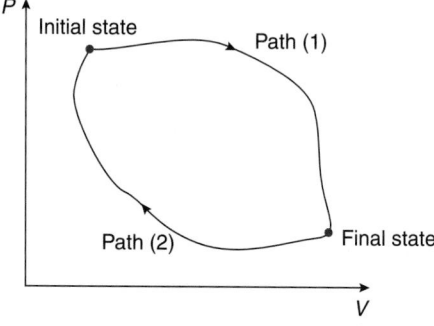

FIGURE 15.17 Proceed from an initial equilibrium state to a final equilibrium state via path (1) and return via path (2).

summed from an initial state to a final state, is *independent of the path* taken between the two states. This quantity

$$\frac{dQ}{T}$$

is defined to be the differential change dS in the entropy of the system when differential heat transfer dQ accrues to the system* at temperature T:

$$dS \equiv \frac{dQ}{T} \tag{15.30}$$

For a reversible cycle, Equation 15.29 means that

$$\oint_{\substack{\text{reversible} \\ \text{cycle}}} dS = \oint_{\substack{\text{reversible} \\ \text{cycle}}} \frac{dQ}{T} = 0 \text{ J/K} \tag{15.31}$$

The entropy change of a system between an initial equilibrium state and a final equilibrium state is found from

$$\int_i^f dS = \int_i^f \frac{dQ}{T}$$

$$\Delta S = \int_i^f \frac{dQ}{T} \tag{15.32}$$

$$\text{reversible path}$$

The dimensions of entropy are those of heat transfer divided by temperature; in SI units, J/K.

Notice that the entropy change of a system as it changes from one thermodynamic equilibrium state to another is

- positive if heat transfer accrues *to* the system ($dQ > 0$ J);

- negative if the heat transfer is *from* the system ($dQ < 0$ J);

- zero for adiabatic processes ($dQ = 0$ J); such processes also are called *isentropic processes*, meaning constant entropy processes;

- zero if the system executes a cycle (since the initial and final states are the same and entropy is a state variable).

Equation 15.32 gives the entropy change of a system undergoing a quasi-static (reversible) process between initial and final equilibrium states. Such quasi-static (reversible) processes are idealizations; no such processes really exist in nature. So how

do we calculate the entropy change of a system undergoing a (realistic) *irreversible* change from an initial state to a final state?

Since the entropy is a state variable, the entropy change between two equilibrium states is the same regardless of how the change is executed.

Therefore, to *calculate* the entropy change between the two states for an irreversible process, we calculate the entropy change for *any* quasi-static (reversible) path connecting the two states. Table 15.1 summarizes the entropy changes that occur in various thermodynamic processes; see Examples 15.6–15.12.

TABLE 15.1 Summary of Entropy Changes in Various Thermodynamic Processes[†]

(a) Ideal gas in a quasi-static reversible process

$$\Delta S = nc_V \ln \frac{T_f}{T_i} + nR \ln \frac{V_f}{V_i}$$

(b) Melting or boiling a mass m:

$$\Delta S = \frac{mL}{T}$$

(where T is the absolute temperature of the phase transition and L is the latent heat associated with the phase transition; $L > 0$ J/kg)

(c) Freezing or condensation of a mass m:

$$\Delta S = -\frac{mL}{T}$$

(where T is the absolute temperature of the phase transition and is the latent heat associated with the phase transition; $L > 0$ J/kg)

(d) Warming or cooling solid or liquid of mass m:

$$\Delta S = mc \ln \frac{T_f}{T_i}$$

(where c is the specific heat of the substance)

(e) Heat transfer $|Q|$ *to* a reservoir at temperature T_C:

$$\Delta S = +\frac{|Q|}{T_C}$$

(f) Heat transfer $|Q|$ *from* a reservoir at temperature T_H:

$$\Delta S = -\frac{|Q|}{T_H}$$

(g) Heat engine or refrigerator engine in one complete cycle:

$$\Delta S = 0 \text{ J/K}$$

(h) Adiabatic process:

$$\Delta S = 0 \text{ J/K}$$

[†] Use absolute temperatures in all equations.

---

*As we have mentioned, since a system in thermal equilibrium *cannot* be said to have a certain amount of heat, there is no heat function associated with a system. The quantity dQ, therefore, represents an infinitesimal amount of heat transfer, and it is called an *inexact* differential dQ. It is quite amazing that the inexact differential dQ, when divided by the absolute temperature T, yields the change dS in the entropy of the system, which *is* a state function. The situation is analogous to transfers of energy via work. Think of the CWE theorem of pure mechanics. A differential amount of work done by all the forces on a system is equal to the differential change in the kinetic energy of the system; a system in equilibrium *cannot* be said to have a certain amount of work, but it *can* be said to have a certain amount of kinetic energy.

PROBLEM-SOLVING TACTIC

15.3 When using the equations in Table 15.1 to calculate entropy changes, be sure to use absolute temperatures in kelvin, not celsius temperatures.

QUESTION 4

Work and heat transfer are the energy transferred to a system by forces or a temperature difference, respectively. The differential entropy change dS is defined as

$$\frac{dQ}{T}$$

Is there value or a reason in dynamics for defining a corresponding concept

$$\frac{dW}{T}$$

or dW divided by some other corresponding parameter? On what basis do you make such decisions?

EXAMPLE 15.6

A 150 g ice cube at 0 °C melts into water at 0 °C. What is the entropy change of the mass?

Solution
The mass is the system. The process takes place at the constant temperature $T = 273.15$ K, and so the entropy change of the system is

$$\Delta S = \int_i^f \frac{dQ}{T}$$
$$= \frac{Q}{T}$$

where Q is the heat transfer to the ice to melt it. You must use temperatures in kelvin; notice how nonsensical it would be to use the 0 °C temperature in this equation. From our study of calorimetry,

$$Q = mL$$

where L is the latent heat of fusion of water: 3.335×10^5 J/kg. Hence

$$Q = (0.150 \text{ kg})(3.335 \times 10^5 \text{ J/kg})$$
$$= 5.00 \times 10^4 \text{ J}$$

The entropy change of the mass is

$$\Delta S = \frac{Q}{T}$$
$$= \frac{5.00 \times 10^4 \text{ J}}{273.15 \text{ K}}$$
$$= 183 \text{ J/K}$$

STRATEGIC EXAMPLE 15.7

a. Find the entropy change for an ideal gas when it changes its state of thermodynamic equilibrium from an initial state at temperature T_i and volume V_i to a final state at temperature T_f and volume V_f.
b. Specialize the result of part (a) for an isochoric process and for an isothermal process.

Solution
a. Imagine taking the ideal gas from its initial state to a final state via a quasi-static, reversible process. Since the entropy is a state variable, the entropy change will be the same regardless of how the gas proceeds between the two states. According to Equation 15.32, the entropy change of the gas is

$$\Delta S = \int_i^f \frac{dQ}{T}$$
$$\text{reversible path}$$

The first law of thermodynamics, Equation 13.51, in differential form states that

$$dQ = dU + dW$$

According to Equation 14.21, the differential change in internal energy dU of an ideal gas is

$$dU = nc_V \, dT$$

and (from Equation 13.53)

$$dW = P \, dV$$

The first law of thermodynamics then can be written in differential form as

$$dQ = nc_V \, dT + P \, dV$$

The entropy change is

$$\Delta S = \int_i^f \frac{dQ}{T}$$

Substituting for dQ, you get

$$\Delta S = nc_V \int_i^f \frac{dT}{T} + \int_i^f \frac{P}{T} \, dV$$

But the ideal gas equation of state is (Equation 13.20)

$$PV = nRT$$

so that

$$\frac{P}{T} = \frac{nR}{V}$$

Substituting this into the second term of the entropy change, you obtain

$$\Delta S = nc_V \int_i^f \frac{dT}{T} + nR \int_i^f \frac{dV}{V}$$

Performing the integrals, you find

$$\Delta S = nc_V \ln \frac{T_f}{T_i} + nR \ln \frac{V_f}{V_i} \qquad (1)$$

b. Notice that equation (1) simplifies to a single term if the two states of thermodynamic equilibrium are reached by various special processes:

1. *Isochoric process* (constant volume), as shown in Figure 15.18. Here $V_f = V_i$, and the second term in equation (1) is zero because $\ln 1 = 0$:

$$\Delta S = nc_V \ln \frac{T_f}{T_i} \qquad (2)$$

(notice that $\Delta S < 0$ J/K if $T_f < T_i$—that is, if the system cools).

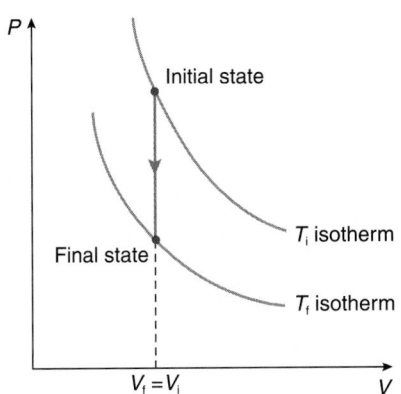

FIGURE 15.18

2. *Isothermal process* (constant temperature), as shown in Figure 15.19. Here $T_f = T_i$, and the first term in equation (1) is zero, again because $\ln 1 = 0$:

$$\Delta S = nR \ln \frac{V_f}{V_i} \qquad (3)$$

(notice that $\Delta S < 0$ if $V_f < V_i$—that is, if the system is compressed).

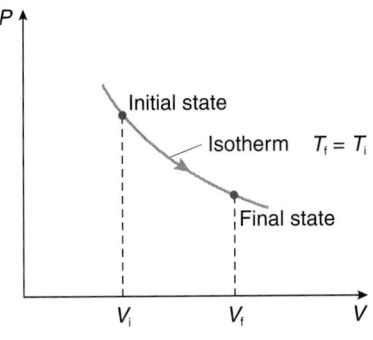

FIGURE 15.19

STRATEGIC EXAMPLE 15.8

An ideal gas is confined by a membrane to a volume V_i in the left-hand side of a well-insulated container of volume V_f, as shown in Figure 15.20a. The right-hand side of the container is initially empty (a vacuum). The entire system is thermally insulated from its environment. The membrane is punctured and the gas rushes to fill the entire container (see Figure 15.20b); this process is called a *free expansion* of a gas, much like the popping of a balloon, though the latter is less scientifically sanitized. Such a free expansion is an irreversible process; it is *not* a quasi-static (reversible) process. Find the entropy change associated with this free expansion of an ideal gas.

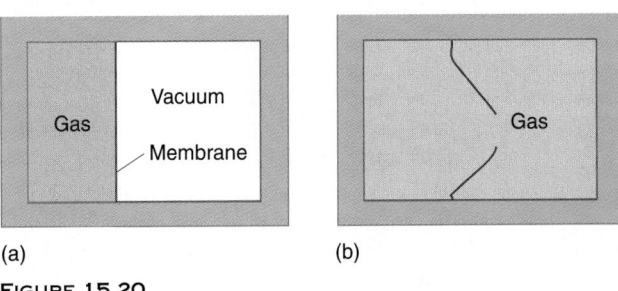

(a)   (b)

FIGURE 15.20

Solution
Begin by looking at the first law of thermodynamics; apply it to the gas as the system:

$$Q = \Delta U + W$$

That is, the total heat transfer to the system is the sum of the change in the internal energy of the system and the work done *by* the system on its surroundings. Since the system is thermally isolated by the insulation, there is no heat transfer to or from the system, and so $Q = 0$ J. No work is done by the gas on its surroundings either, because the gas is isolated; thus, $W = 0$ J. The first law then indicates that the change in the internal energy of the gas is zero. Since the internal energy of an ideal gas is purely a function of temperature (see Chapter 14), the temperature of the ideal gas is unchanged in the free expansion.

You want to calculate the entropy change of the system between the initial state (the gas confined to the volume V_i) and the final state (the gas occupying the entire volume V_f). Since the system (the gas) is thermally insulated from its environment, there is zero heat transfer to the system. That is, $dQ = 0$ J. You saw that the temperature T also does not change. It might appear, at first glance, that the entropy change of the system is zero since (from Equation 15.32)

$$\Delta S = \int_i^f \frac{dQ}{T}$$
_{reversible path}

Since the temperature is constant, the integral becomes

$$\Delta S = \frac{1}{T} \int_i^f dQ$$
_{reversible path}

With $dQ = 0$ J, this seems to imply that $\Delta S = 0$ J/K; *but this is not the case!* Why is that? *To calculate the entropy change using Equation 15.32, you must use a quasi-static reversible process* connecting the initial and final states of thermodynamic equilibrium. A free expansion is *not* such a process. Thus, to calculate the entropy change of the ideal gas in the free expansion, you use a quasi-static (reversible) process connecting the initial and final states of the gas and use *that* process to calculate the entropy change. That is, use equation (1) of Example 15.7, since it was derived for a quasi-static reversible process for an ideal gas:

$$\Delta S = nc_V \ln \frac{T_f}{T_i} + nR \ln \frac{V_f}{V_i}$$

Since the temperature of the ideal gas is unchanged in the free expansion, the first term is zero ($\ln 1 = 0$) and the entropy change of the system is

$$\Delta S = nR \ln \frac{V_f}{V_i}$$

So here is an example where $dQ = 0$ J, but $dS \neq 0$ J/K. Strange but true.

Note that the entropy change for the irreversible free expansion is positive since $V_f > V_i$ (the natural logarithm of a number greater than 1 is positive; the natural logarithm of numbers between 0 and 1 is negative).

EXAMPLE 15.9

Find the entropy change of a system of two reservoirs when heat transfer $|Q|$ occurs from the warmer reservoir at temperature T_H to the cooler reservoir at temperature T_C.

Solution
The process is irreversible. However, the entropy change is calculated by imagining slow quasi-static (reversible) heat transfer.

Recall that a reservoir is capable of losing or accruing any heat transfer without changing its temperature. Since the temperature of the reservoir is constant, the warmer reservoir has an entropy change

$$\Delta S = \int_i^f \frac{dQ}{T}$$
$$= -\frac{|Q|}{T_H}$$

The minus sign arises because the heat transfer is *from* the hotter reservoir. The entropy change of the hotter reservoir is *negative*.

The cooler reservoir also does not change its temperature when accruing heat transfer $|Q|$; the entropy change of the cold reservoir is

$$\Delta S = \int_i^f \frac{dQ}{T}$$
$$= \frac{|Q|}{T_C}$$

where the positive result arises since the heat transfer is *to* the cold reservoir. The entropy change of the cold reservoir is *positive*.

The *total* entropy change of both reservoirs is the sum of the two entropy changes:

$$\Delta S_{total} = \frac{|Q|}{T_C} - \frac{|Q|}{T_H} \tag{1}$$

Note that the total entropy change of the system of two reservoirs is positive, since the first term always is greater than the second term because $T_C < T_H$.

STRATEGIC EXAMPLE 15.10

A mass m_1 at temperature T_H is placed in thermal contact with another mass m_2 at temperature T_C. The entire system is isolated, so that heat transfer occurs only between the two masses. The final (common) equilibrium temperature of the system is T, where $T_C < T < T_H$. Find the total entropy change associated with the heat transfer between the hot and cold masses. The specific heats [in J/(kg·K)] of the masses are c_1 and c_2 respectively.

Solution
Once again, the process is irreversible. However, you can imagine each mass brought to the equilibrium temperature T by placing it in thermal contact with a succession of reservoirs with temperatures differing differentially between the initial temperature of the mass and the equilibrium temperature. In this way you construct a quasi-static (reversible) series of processes connecting the initial and final states of each mass.

The entropy change of the initially hot mass is found from

$$\Delta S_1 = \int_i^f \frac{dQ}{T}$$
$$= m_1 c_1 \int_{T_H}^T \frac{dT}{T}$$
$$= m_1 c_1 \ln \frac{T}{T_H}$$

where it is assumed that the specific heat of the substance is not a function of temperature. Since $T < T_H$, this entropy change is *negative* (the natural logarithm of a number between 0 and 1 is negative).

The entropy change of the initially cold mass is found in a similar way:

$$\Delta S_2 = \int_i^f \frac{dQ}{T}$$
$$= m_2 c_2 \int_{T_C}^T \frac{dT}{T}$$
$$= m_2 c_2 \ln \frac{T}{T_C}$$

where again it is assumed that the specific heat is not a function of temperature. This entropy change is *positive*, since $T > T_C$.

The *total* entropy change is the sum of the entropy changes:

$$\Delta S_{total} = \Delta S_1 + \Delta S_2$$
$$= m_1 c_1 \ln \frac{T}{T_H} + m_2 c_2 \ln \frac{T}{T_C} \tag{1}$$

The first term is negative and the second is positive. *In every case one finds that the total entropy change is positive*; in other words, the positive term is always larger than the negative term. This is yet another consequence of the second law of thermodynamics, as we shall see in the next section.

EXAMPLE 15.11

Find the entropy change associated with the mixing of identical liquids (or identical gases) at different temperatures.

Solution

Equation (1) of Example 15.10 also can be used to calculate the entropy change when two *identical liquids* (or identical gases) at different temperatures are mixed together and reach an equilibrium temperature. The word *identical* means mixing water with water, orange juice with orange juice, vodka with vodka, oxygen with oxygen, and so on (but not water with orange juice, orange juice with vodka, milk with coffee, or similar nonidentical liquids or gases). It is *not* obvious, but the mixing of *nonidentical* substances introduces an *additional term* into the total entropy change (a term known as the entropy of mixing); but to calculate this term is beyond the scope of an introductory course. You have earned the right to be grateful for that.

EXAMPLE 15.12

A 3.00 kg sample of water is warmed from 0 °C to 100 °C. What is the entropy change of the water?

Solution

Since $dQ = mc_{water} \, dT$, the entropy change is

$$\Delta S = mc_{water} \int_{T_i}^{T_f} \frac{dT}{T}$$
$$= mc_{water} \ln \frac{T_f}{T_i}$$

Make the appropriate substitutions and remember that absolute temperatures *must* be used:

$$\Delta S = (3.00 \text{ kg})(4186 \text{ J/kg·K}) \left(\ln \frac{373 \text{ K}}{273 \text{ K}} \right)$$
$$= 3.92 \times 10^3 \text{ J/K}$$

15.9 ENTROPY AND THE SECOND LAW OF THERMODYNAMICS

A theory is the more impressive the greater the simplicity of its premises is, the more different kinds of things it relates, and the more extended is its area of applicability. Therefore the deep impression which classical thermodynamics made upon me. It is the only physical theory of universal content concerning which I am convinced that, within the framework of the applicability of its basic concepts, it will never be overthrown (for the special attention of those who are skeptics on principle).

Albert Einstein (1879–1954)*

---

*Albert Einstein: Philosopher-Scientist, edited by Paul Arthur Schilpp (The Library of Living Philosophers, Evanston, Illinois, 1949), Volume VII, page 33.

We stated two versions of the second law of thermodynamics earlier in our study of heat engines and refrigerator engines:

1. *Perfect heat engines do not exist.*
2. *Perfect refrigerator engines do not exist.*

Now that we have a grasp on the peculiar and less intuitive thermodynamic state variable known as the entropy, we make a more general statement of the second law and show that it is equivalent to the two previous formulations.

The most general statement of the second law of thermodynamics is as follows:

3. *The total entropy change of an **isolated** system is always greater than or equal to zero:*

$$\Delta S_{total \ isolated} \geq 0 \text{ J/K} \qquad (15.33)$$

The total entropy change is equal to zero only for reversible processes.

A few things need to be emphasized regarding this statement of the second law:

1. Notice that the statement concerns an *isolated* system; the entropy change of a system that is not isolated can very well be negative. If the system is not isolated, it can be incorporated into a new and larger isolated system by including appropriate surroundings. When the original nonisolated system is expanded to become an isolated system, the *total* entropy change of the now isolated system, as a whole, is *always* greater than or equal to zero.
2. The total entropy change is the algebraic sum of the individual entropy changes of the various components of the isolated system. The entropy changes of the individual components of the isolated system may be positive, negative, or zero. But the *total* entropy change (the algebraic sum of the individual entropy changes) of the *isolated* system is greater than or equal to zero.
3. The total entropy change of an isolated system is equal to zero only when the system undergoes a quasi-static reversible process.
4. The total entropy change of an isolated system in an *irreversible* process is always *greater than zero*.

Now let us show that our third version of the second law (Equation 15.33) incorporates our earlier versions.

Perfect Heat Engines Do Not Exist

A perfect heat engine is schematically represented in Figure 15.21. Such a perfect heat engine system would accrue positive heat transfer Q_H (> 0 J) from a reservoir at temperature T_H and do (positive) work W with no exhaust heat transfer Q_C (= 0 J) from the engine to a cooler reservoir at temperature T_C. The *isolated* system consists of the perfect heat engine *and* the two reservoirs.

We look at the entropy change of the isolated system for one cycle of the perfect engine. The entropy is a state variable, and so when the heat engine system returns to its initial thermodynamic

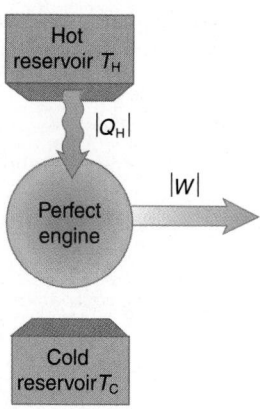

FIGURE 15.21 A perfect heat engine.

state, the entropy of the heat engine returns to its initial value. The entropy change of the heat engine itself is zero:

$$\Delta S_{\text{engine}} = 0 \text{ J/K}$$

The entropy change of the reservoir system at temperature T_H is

$$\Delta S_{\text{hot reservoir}} = -\frac{|Q_H|}{T_H}$$

which is negative because the heat transfer $|Q_H|$ is from this reservoir system.

For a perfect engine, no heat transfer occurs to the lower-temperature reservoir, and so the entropy change of the lower-temperature reservoir system is zero:

$$\Delta S_{\text{cold reservoir}} = 0 \text{ J/K}$$

The total entropy change of the *isolated* system is the sum of the individual entropy changes:

$$\Delta S_{\text{total isolated}} = \Delta S_{\text{engine}} + \Delta S_{\text{hot reservoir}} + \Delta S_{\text{cold reservoir}}$$

$$= 0 \text{ J/K} - \frac{|Q_H|}{T_H} + 0 \text{ J/K}$$

The total entropy change of the isolated system is *negative*. This violates the second law: the entropy change of an isolated system must be greater than or equal to zero. Thus perfect heat engines cannot exist.

Perfect Refrigerator Engines Do Not Exist

A perfect refrigerator engine is represented schematically in Figure 15.22. The perfect refrigerator engine effects heat transfer $|Q|$ from a reservoir at temperature T_C to a hotter reservoir at temperature T_H, doing no work in the process of each refrigerator cycle. The *isolated* system consists of the perfect refrigerator engine and the two reservoirs.

Let us calculate the entropy change for one cycle of the perfect refrigerator engine system. Since the entropy is a state variable and the refrigerator engine operates in a cycle, returning

to its initial state, the entropy change of the refrigerator engine in each cycle is zero:

$$\Delta S_{\text{refrig}} = 0 \text{ J/K}$$

The cold reservoir system suffers an entropy change

$$\Delta S_{\text{cold reservoir}} = -\frac{|Q|}{T_C}$$

which is negative because the heat transfer $|Q|$ is *from* this reservoir. The warmer reservoir system experiences an entropy change

$$\Delta S_{\text{hot reservoir}} = +\frac{|Q|}{T_H}$$

since the heat transfer $|Q|$ is *to* this reservoir. Thus the total entropy change of the *isolated* system is

$$\Delta S_{\text{total isolated}} = \Delta S_{\text{refrig}} + \Delta S_{\text{cold reservoir}} + \Delta S_{\text{hot reservoir}}$$

$$= 0 \text{ J/K} - \frac{|Q|}{T_C} + \frac{|Q|}{T_H}$$

This expression is negative, since $T_C < T_H$. The negative result violates the second law: the total entropy change of an isolated system must be greater than or equal to zero. Thus perfect refrigerator engines cannot exist.

FIGURE 15.22 A perfect refrigerator engine.

QUESTION 5

Explain why the freezing of water into ice does not violate the second law of thermodynamics. What is the isolated system?

EXAMPLE 15.13

A Carnot heat engine, in executing one cycle, uses 2000 J of heat transfer from a reservoir at 500 K, performs work, and effects heat transfer to a low-temperature reservoir at 300 K.

a. Find the efficiency of the Carnot heat engine.
b. Determine the work done by the heat engine in one cycle.
c. Calculate the entropy change of the heat engine and each reservoir during one cycle of the engine.
d. Find the total entropy change of the isolated system of the heat engine and two reservoirs and comment on the answer.

Solution

a. The efficiency of a Carnot heat engine is found using Equation 15.14:

$$\varepsilon_{\text{Carnot}} = 1 - \frac{T_C}{T_H} \quad \text{(Carnot cycle only)}$$

$$= 1 - \frac{300 \text{ K}}{500 \text{ K}}$$

$$= 0.400$$

b. The work done by the heat engine is found from the definition of the efficiency, Equation 15.3:

$$\varepsilon = \frac{|W|}{|Q_H|}$$

where Q_H is the heat transfer to the engine from the hot reservoir. Since the efficiency is known from part (a), you have

$$0.400 = \frac{|W|}{2000 \text{ J}}$$

or

$$|W| = 800 \text{ J}$$

The work done by a heat engine is positive, so that

$$W = 800 \text{ J}$$

c. The entropy change of the heat engine system during one cycle is 0 J/K, because the heat engine returns to its initial thermodynamic state and entropy is a state variable.

The heat transfer from the hot reservoir system is −2000 J but, as a reservoir, no change in its temperature occurred. Hence the entropy change of the hot reservoir is negative:

$$\Delta S_H = -\frac{|Q|}{T_H}$$

$$= -\frac{2000 \text{ J}}{500 \text{ K}}$$

$$= -4.00 \text{ J/K}$$

Of the 2000 J of heat transfer to the engine from the hot reservoir, 800 J of work was done by the engine during each cycle. Hence the heat transfer to the cool reservoir is, according to the first law of thermodynamics (energy conservation),

$$2000 \text{ J} - 800 \text{ J} = 1200 \text{ J}$$

The cool reservoir—indeed, any reservoir—does not change its temperature. The entropy change of the cool reservoir system is positive, because the heat transfer to it is 1200 J. Hence the entropy change of the cool reservoir is

$$\Delta S_C = +\frac{|Q|}{T_C}$$

$$= \frac{1200 \text{ J}}{300 \text{ K}}$$

$$= 4.00 \text{ J/K}$$

d. The total entropy change of the isolated system consisting of the heat engine and two reservoirs is the sum of the individual entropy changes (with proper signs):

$$\Delta S_{\text{total isolated}} = 0 \text{ J/K} + 4.00 \text{ J/K} + (-4.00 \text{ J/K})$$

$$= 0 \text{ J/K}$$

The result is not surprising, since the cycle is a Carnot cycle (reversible) and the total entropy change of an isolated system is zero *only* for such reversible cycles.

STRATEGIC EXAMPLE 15.14

A 2.00 kg block of ice is taken out of a freezer at −20.0 °C, melts, and warms to the temperature of a large room at 25.0 °C. Calculate the entropy change of the ice and the entropy change of the room. Verify that the total entropy change of the (isolated) ice and room system is consistent with the second law of thermodynamics.

Solution

The entropy change of the ice must be calculated in three stages:

1. warming the ice from −20 °C to 0 °C;
2. melting of the ice at 0 °C to water at 0 °C; and
3. warming of the cold water at 0 °C to the temperature of the room.

Consider each stage:

1. Warming the Ice from −20 °C to 0 °C

The entropy change is

$$\Delta S = \int_i^f \frac{dQ}{T}$$

where the differential heat transfer dQ to warm the ice is

$$dQ = mc_{\text{ice}} \, dT$$

and c_{ice} is the specific heat of ice. Thus

$$\Delta S = mc_{\text{ice}} \int_{T_i}^{T_f} \frac{dT}{T}$$

where T_i is the initial absolute temperature of the ice and T_f is the final absolute temperature of the ice before melting. Integrating,

$$\Delta S = mc_{\text{ice}} \ln \frac{T_f}{T_i}$$

From Table 13.4, you have

$$c_{\text{ice}} = 2050 \text{ J/(kg·K)}$$

Making the appropriate substitutions, you find

$$\Delta S = (2.00 \text{ kg})(2050 \text{ J/kg·K}) \left(\ln \frac{273 \text{ K}}{253 \text{ K}} \right)$$

Thus to warm the ice to its melting point results in an entropy change of the ice of

$$\Delta S = 312 \text{ J/K}$$

2. Melting the Ice at 0 °C to Water at 0 °C

This process takes place at constant temperature. Therefore the entropy change is

$$\Delta S = \int_{i}^{f} \frac{dQ}{T}$$
$$= \frac{Q}{T}$$

where Q is the heat transfer to the ice to melt it. The latent heat of fusion of water is found from Table 13.5 to be

$$L_{f\ water} = 3.335 \times 10^5 \text{ J/kg}$$

Therefore

$$Q = mL_{f\ water}$$
$$= (2.00 \text{ kg})(3.335 \times 10^5 \text{ J/kg})$$
$$= 6.67 \times 10^5 \text{ J}$$

The entropy change in melting the ice then is

$$\Delta S = \frac{6.67 \times 10^5 \text{ J}}{273 \text{ K}}$$
$$= 2.44 \times 10^3 \text{ J/K}$$

3. Warming the Cold Water to Room Temperature

This process is similar to the warming of the ice, and so you can use the same equation as in part (1), replacing the specific heat of ice with the specific heat of water (see Table 13.4) and accounting for the different initial and final temperatures:

$$\Delta S = mc_{water} \ln \frac{T_f}{T_i}$$
$$= (2.00 \text{ kg})(4186 \text{ J/kg·K})\left(\ln \frac{298 \text{ K}}{273 \text{ K}} \right)$$
$$= 734 \text{ J/K}$$

The ice and room are an isolated system. The heat transfer to the ice is from the room. Let Q be the heat transfer to the room; Q is negative since the heat transfer is really *from* the room. The total heat transfer to the isolated ice–room system is zero:

$$Q + (mc_{ice}\ \Delta T_1 + mL_{f\ water} + mc_{water}\ \Delta T_2) = 0 \text{ J}$$

where ΔT_1 is the temperature change of the ice as it warms to its melting point, and ΔT_2 is the temperature change of the melted ice water as it warms to room temperature from the melting point. Thus

$$Q = -(mc_{ice}\ \Delta T_1 + mL_{f\ water} + mc_{water}\ \Delta T_2)$$

Inserting appropriate numerical values of the quantities, you get

$$Q = -\{(2.00 \text{ kg})[2050 \text{ J/(kg·K)}](20.0 \text{ K})$$
$$+ (2.00 \text{ kg})(3.335 \times 10^5 \text{ J/kg})$$
$$+ (2.00 \text{ kg})[4186 \text{ J/(kg·K)}](25.0 \text{ K})\}$$
$$= -9.58 \times 10^5 \text{ J}$$

The entropy change of the room, since its temperature is essentially unaffected and it acts as a reservoir, is

$$\Delta S = -\frac{|Q|}{T}$$
$$= -\frac{9.58 \times 10^5 \text{ J}}{298 \text{ K}}$$
$$= -3.21 \times 10^3 \text{ J/K}$$

The *total* entropy change of the ice-room system is the sum of all the entropy changes:

$$S_{total} = 312 \text{ J/K} + 2.44 \times 10^3 \text{ J/K} + 734 \text{ J/K}$$
$$- 3.21 \times 10^3 \text{ J/K}$$
$$= 2.7 \times 10^2 \text{ J/K}$$

The entropy change of the entire (isolated) ice–room system is greater than zero, consistent with the second law of thermodynamics.

15.10 THE DIRECTION OF HEAT TRANSFER: A CONSEQUENCE OF THE SECOND LAW

The entropy statement of the second law (Equation 15.33) finally enables us to explain why heat transfer is *from* hotter objects *to* cooler objects and not the other way around. If heat transfer $|Q|$ occurred spontaneously from a cold to a hot reservoir, represented schematically as shown in Figure 15.23, the total entropy change of the isolated system of reservoirs would be

$$\Delta S_{total\ isolated} = \Delta S_{cold\ reservoir} + \Delta S_{hot\ reservoir}$$
$$= -\frac{|Q|}{T_C} + \frac{|Q|}{T_H}$$

This expression is negative because $T_C < T_H$, and so the process *violates* the second law of thermodynamics; the process cannot occur.

FIGURE 15.23 Spontaneous heat transfer from colder to warmer systems violates the second law of thermodynamics.

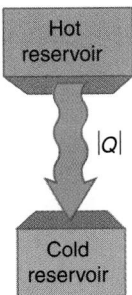

FIGURE 15.24 Spontaneous heat transfer from warmer to cooler systems conforms to the second law of thermodynamics.

Such unobserved heat transfer is just like a perfect refrigerator, which we know cannot exist.

For heat transfer $|Q|$ *from* a hot reservoir *to* a cold reservoir, as shown schematically in Figure 15.24, the total entropy change of the isolated system of two reservoirs is

$$\Delta S_{\text{total isolated}} = \Delta S_{\text{cold reservoir}} + \Delta S_{\text{hot reservoir}}$$
$$= +\frac{|Q|}{T_C} - \frac{|Q|}{T_H}$$

Since $T_H > T_C$, the total entropy change of the isolated system is indeed greater than zero, in conformity with the second law of thermodynamics.

Thus it is the second law of thermodynamics that finally explains why heat transfer occurs spontaneously from hotter to cooler systems and not the other way around.

15.11 A STATISTICAL INTERPRETATION OF THE ENTROPY*

In a real sense maximum disorder was our equilibrium.
T. E. Lawrence (1888–1935)[†]

In introducing the classical concept of entropy in Section 15.8, we mentioned that a physical system in equilibrium has a definite entropy, since it is a state variable (a property of the system). On a macroscopic basis, however, when a system *changes* its state of thermodynamic equilibrium, it is only changes in the entropy that are significant. The second law of thermodynamics is a statement about *changes* in the entropy of an *isolated* system.

The entropy a system has when in thermal equilibrium has a statistical interpretation. For simplicity, we use an analogy to explore such a statistical interpretation.[‡] We employ a macroscopic rather than a microscopic system, because the former is more familiar to you, and experiments using such a system are easier to visualize. Also, by using a macroscopic system we avoid quantum mechanical complications that might arise with microscopic

systems. The essential features of the macroscopic system we consider also have analogies at the microscopic level.

The system we consider is a collection of pennies in a box. Consider each penny as a particle with two sides: heads and tails. If the pennies are placed in a covered box and the container is shaken vigorously for a few moments, there are various possibilities for the number of heads and tails that will emerge (we exclude edge-on pennies). We begin with a system of just a few pennies; then we increase the number of pennies to see what happens for a system with many particles.

With a collection of four pennies, there are five possible states for the system after shaking the box:

1. 4 heads and 0 tails
2. 3 heads and 1 tail
3. 2 heads and 2 tails
4. 1 head and 3 tails
5. 0 heads and 4 tails

We call this set of possibilities a delineation of the various **macrostates** of the 4-penny system. There are five macrostates for the 4-penny system. Associated with *each* macrostate is a collection of **microstates**. A microstate specifies *which individual penny* has a head or a tail. Thus macrostate #1 (4 heads and 0 tails) has only one microstate: every penny must be heads. Macrostate #2 has four microstates: each of the four pennies could be the one that is tails, with the others heads. The various microstates corresponding to each 4-penny macrostate are delineated in Table 15.2. We also can make a histogram depicting the number of microstates associated with each macrostate of the 4-penny system; see Figure 15.25.

> Each of the microstates is equally likely to occur; but the macrostates are *not* equally likely to occur.

Indeed the number of *microstates* associated with a *given macrostate*, divided by the *total* number of *microstates*, is a measure of the *probability* of that macrostate occurring in a toss. Thus the macrostate with 2 heads and 2 tails is the most likely to occur, because it has the greatest number of associated microstates: 6 out of a total of 16 microstates; the probability of the 2-heads, 2-tails macrostate occurring is $6/16 = 0.375$ ($= 37.5\%$). The macrostate with all heads has just 1 microstate out of the 16 total possible microstates, so that the probability of all heads occurring is only $1/16 = 0.0625$ ($= 6.25\%$). Thus the probabilities of the various *macrostates* occurring are not all the same.

> After the box is shaken, the pennies are most likely to be found in the macrostate with the greatest number of associated microstates.

The microstates are like lottery tickets: each is equally likely to win. The number of lottery tickets you have is a macrostate; the more tickets you have, the greater the likelihood of your winning.

The statistical interpretation of entropy, developed by Ludwig Boltzmann during the late 19th century, defines the entropy of the system in a given macrostate to be Boltzmann's constant k times the natural logarithm (whose base is the pure number $e \equiv 2.718\,281\,828\ldots$) of the number of microstates Ω associated with a given macrostate.

[†]*Seven Pillars of Wisdom* (Doubleday, Doran, Garden City, New York, 1936), Chapter LIX, page 338.
[‡]Problem 65 explores a real example.

TABLE 15.2 Four-Penny Macrostates and Microstates

Macrostate (1): 4 heads (H) and 0 tails (T)			
Penny 1	Penny 2	Penny 3	Penny 4
H	H	H	H
(1 microstate)			

Macrostate (2): 3 heads and 1 tail			
Penny 1	Penny 2	Penny 3	Penny 4
H	H	H	T
H	H	T	H
H	T	H	H
T	H	H	H
(4 microstates)			

Macrostate (3): 2 heads and 2 tails			
Penny 1	Penny 2	Penny 3	Penny 4
H	H	T	T
H	T	T	H
T	T	H	H
T	H	H	T
T	H	T	H
H	T	H	T
(6 microstates)			

Macrostate (4): 1 head and 3 tails			
Penny 1	Penny 2	Penny 3	Penny 4
H	T	T	T
T	H	T	T
T	T	H	T
T	T	T	H
(4 microstates)			

Macrostate (5): 0 heads and 4 tails			
Penny 1	Penny 2	Penny 3	Penny 4
T	T	T	T
(1 microstate)			

We define the entropy of a given macrostate to be

$$S \equiv k \ln \Omega \qquad (15.34)$$

where Ω is the number of microstates associated with the given macrostate.

The number of microstates for a system with many particles is staggeringly large. We shall see that for a 100-penny system, hardly a large number of particles, the total number of microstates is $2^{100} \approx 1.27 \times 10^{30}$, a fantastically huge number. This is one reason that the entropy is defined as the natural logarithm of the number of microstates.[*][†]

Figure 15.25 is a histogram of the number of microstates associated with each macrostate of the 4-penny system. The entropy of each macrostate is Boltzmann's constant k times the natural logarithm of the number of microstates associated with that macrostate. The entropy of the system has a maximum at the macrostate that is most likely, the macrostate with the greatest number of associated microstates—namely, the macrostate with half heads and half tails.

Now we perform another experiment. We set the four pennies in a box, all with heads up initially, as shown in Figure 15.26.

---

[*]The *natural logarithm* (ln) is used, rather than logarithms to base 10 (log), because the natural logarithm emerges naturally from the detailed statistical mechanical analysis of a system with a large number of particles. This emergence is difficult to show here.

[†]The logarithm also ensures that the entropy of two systems, when combined into a single system, is the sum of the individual entropies. To see this, suppose we have two individual systems with entropies

$$S_1 = k \ln \Omega_1$$

and

$$S_2 = k \ln \Omega_2$$

where Ω_1 and Ω_2 are the respective numbers of microstates associated with the particular macrostates that the systems are in. If the two systems are brought together, the number of associated microstates is the product $\Omega_1 \Omega_2$. The entropy of the new system thus is

$$\begin{aligned} S &= k \ln(\Omega_1 \Omega_2) \\ &= k \ln \Omega_1 + k \ln \Omega_2 \\ &= S_1 + S_2 \end{aligned}$$

which is the sum of the individual entropies of the two systems.

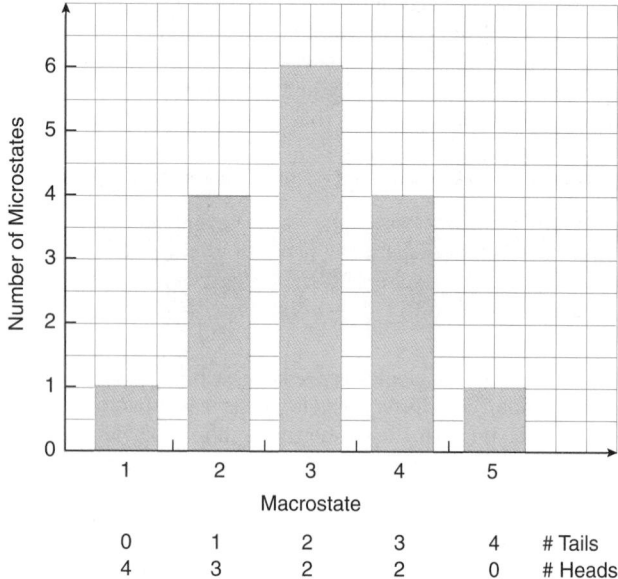

FIGURE 15.25 Number of microstates associated with each macrostate of a 4-penny system.

0	1	2	3	4	# Tails
4	3	2	2	0	# Heads

The epitaph for Ludwig Boltzmann in the Zentralfriedhof in Vienna, Austria, includes his famous entropy equation. The equation engraved on the stone, $S = k \log W$, is what we now write as $S = k \ln \Omega$.

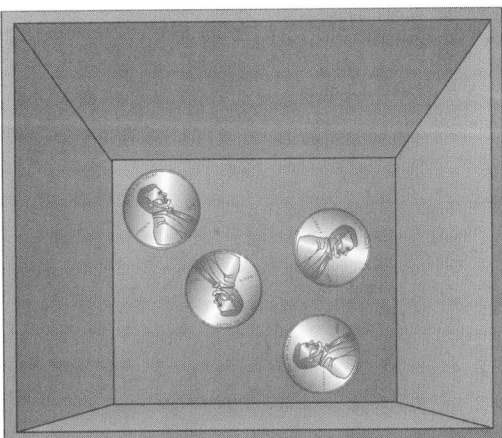

FIGURE 15.26 One of the two 4-penny macrostates with smallest entropy; the other is the macrostate with four tails.

This macrostate has the smallest entropy (since there is only one microstate associated with this macrostate).

We shake the box and then examine the pennies to determine which macrostate the pennies are in. It is more likely that the system is in a macrostate with a higher entropy. In other words, in this physical process the system likely changed to a macrostate with greater entropy. Indeed, the most likely outcome is the macrostate of maximum entropy, because it corresponds to the macrostate with the greatest number of associated microstates.

Now we begin the experiment with the pennies arranged in the macrostate with maximum entropy (2 heads and 2 tails), as shown in Figure 15.27. In this case it is *unlikely* that the system will end up, after shaking, in a macrostate with significantly lower entropy; it is certainly *possible* for the system to end up in a state of lower entropy, but it is less likely to do so than to remain in the macrostate of maximum entropy.* So once a system

is in a state of maximum entropy, it most likely will remain there through natural processes (represented here by shaking the box).

The macrostates with the smallest entropy have only one associated microstate (4 heads and 0 tails, or 0 heads and 4 tails); thus, the exact behavior of each particle (penny) is known. These low-entropy macrostates are the most strictly organized; they have the greatest order. In contrast the macrostate with the greatest entropy (2 heads, 2 tails) has the largest number of associated microstates; there are six different ways the four particles (pennies) can be arranged consistent with this macrostate. Thus this macrostate is the least rigidly determined; it has the least order or, as it is generally phrased, the greatest disorder. This is why the entropy of the system is occasionally referred to as a measure of the disorder associated with the system. The greater the entropy, the greater the disorder.

These ideas can be extended to greater numbers of pennies. For example, for a 10-penny system there are 11 macrostates and $2^{10} = 1024$ microstates (see Table 15.3; Example 15.15 indicates how to determine the number of microstates for each macrostate). Each of the microstates is equally likely to occur, but the macrostates do not have equal probabilities of occurring. A histogram of the number of microstates associated with each macrostate for a 10-penny system is shown in Figure 15.28. The entropy of a macrostate is $S = k \ln \Omega$, where Ω is the number of microstates associated with the particular macrostate.

If the 10-penny system is prepared in a state of low entropy (say, the macrostate with all heads) and the box is shaken, the system is unlikely to remain in that macrostate. You would certainly never make a bet about it remaining in that state unless you really like long odds! In fact, the probability of remaining in that state is 1/1024, since each of the 1024 microstates is equally likely to occur. It is much more likely that the entropy of the system will increase and the system will be found in one of the macrostates of greater entropy, one of the macrostates near the half-heads, half-tails macrostate.

Correspondingly, if the system is in one of the macrostates of large entropy [for example, the macrostate with the greatest entropy (5 heads and 5 tails)], it is unlikely that the system will (after shaking the box) be found in a macrostate of low entropy (say all heads). It is *possible* to end up with 10 heads, but it is *quite unlikely*.

---

*Actually, for this system with just 4 pennies, the odds are 10/16 = 62.5% that the system will end up in a macrostate that is not the 2-heads, 2-tails macrostate, but these odds decrease dramatically as the number of pennies in the system increases (see the discussion that follows).

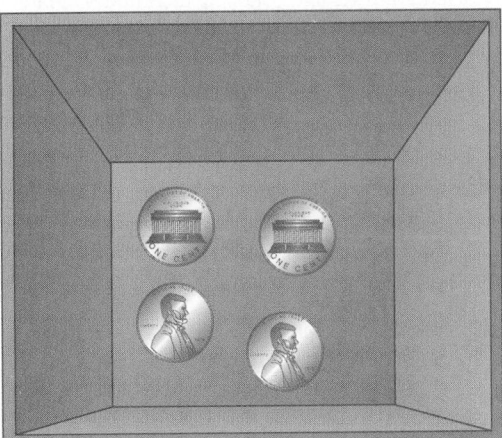

FIGURE 15.27 The macrostate with greatest entropy; half heads, half tails.

TABLE 15.3 Macrostates and Number of Microstates for a 10-Penny System

Macrostate	H	T	# microstates Ω
1	10	0	1
2	9	1	10
3	8	2	45
4	7	3	120
5	6	4	210
6	5	5	252
7	4	6	210
8	3	7	120
9	2	8	45
10	1	9	10
11	0	10	1
		Total	$1024 = 2^{10}$

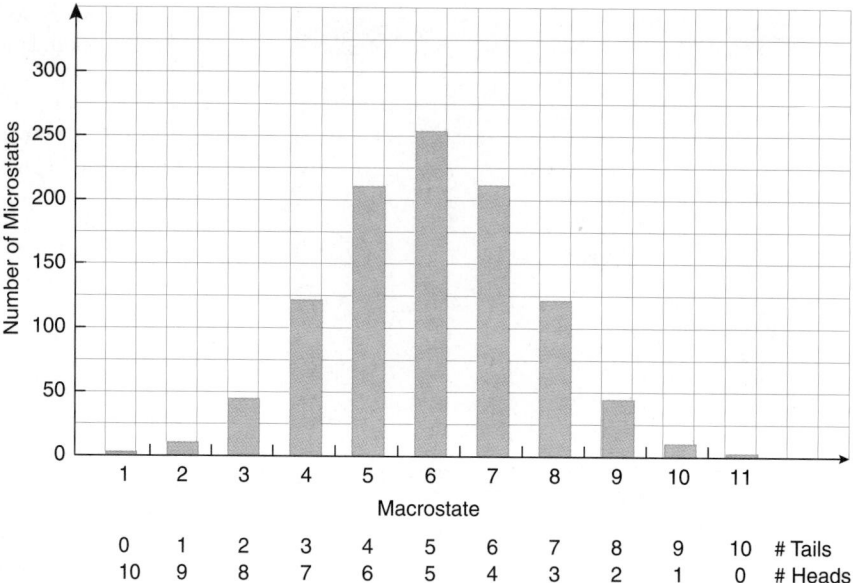

FIGURE 15.28 The number of microstates associated with each macrostate for a 10-penny system.

0	1	2	3	4	5	6	7	8	9	10	# Tails
10	9	8	7	6	5	4	3	2	1	0	# Heads

If we increase the number of pennies further (to say 50, 100, 1000, or many billions), several things happen to the shape of the histogram depicting the number of microstates associated with each macrostate; similar things happen to the shape of the corresponding entropy graph. Figure 15.29 presents the graph of the number of microstates associated with each macrostate for a 50-penny system, a 100-penny system, and a 1000-penny system. We note the following from these graphs:

1. The peak of the curve remains located at the macrostate with half the coins heads and half tails, since this is the macrostate with the greatest number of associated microstates and the greatest entropy.
2. The peak of the distribution becomes sharper and narrower as the number of pennies in the system increases. The distribution falls precipitously as we move to macrostates differing appreciably from the half-heads, half-tails macrostate.

> The implication of the sharp peak is that if a system with a large number of coins is prepared in a state of low entropy (say all heads) and then shaken, it becomes *exceedingly unlikely* that the system will remain in such a state.

Once the shaking occurs, the system of pennies emerges with a distribution that is essentially half heads and half tails. The system went from a macrostate of low entropy to a macrostate of high entropy: the entropy of the system increased.

> Likewise, once the system is in one of the macrostates near the peak, it is *exceedingly unlikely* that it will emerge from the shaking in a macrostate of low entropy (say all heads, or just a few heads or tails), for this entails a significant decrease in the entropy. Thus the system naturally maximizes its entropy.

The entropy form of the second law of thermodynamics is a statement about probabilities. An isolated system naturally will progress to a macrostate of higher entropy because such macrostates are much more probable. Thus, in the free expansion of an ideal gas, once the membrane is punctured it is *exceedingly unlikely* for the particles of the gas to remain confined to one side of the container. In principle, it is *possible* for the particles to remain on one side of the container, but it is so fantastically unlikely that we can say with absolute confidence that it *never* occurs.* Once the membrane is punctured, the macrostate corresponding to all the particles on one side of the box becomes an exceedingly unlikely macrostate. The macrostate with the particles distributed essentially uniformly throughout the container has the greatest number of associated microstates; it is the macrostate with the maximum entropy, and the system quickly and naturally evolves to that macrostate.

The game played in **statistical thermodynamics** is to find ingenious ways of counting the number of microstates associated with a given macrostate. A macrostate is, for example, a specified number of moles of a gas with a specific volume, pressure, and temperature. The entropy of the system then is found analogously to the way we found the entropy for our system of pennies. We leave such sophisticated counting techniques to later courses in statistical physics.

The relationship between the *statistical definition of the entropy* of a system in thermal equilibrium, Equation 15.34,

$$S = k \ln \Omega$$

and the *classical definition of differential entropy changes*, Equation 15.30,

$$dS = \frac{dQ}{T}$$

between different equilibrium states is difficult to fathom with the tools we have at our disposal. Alas, for that reason we must defer

---

*Using our penny system, it is similar to preparing a box with Avogadro's number of pennies initially all heads, shaking the box, and finding that all the pennies *still* are heads. It is possible but *exceedingly* unlikely.

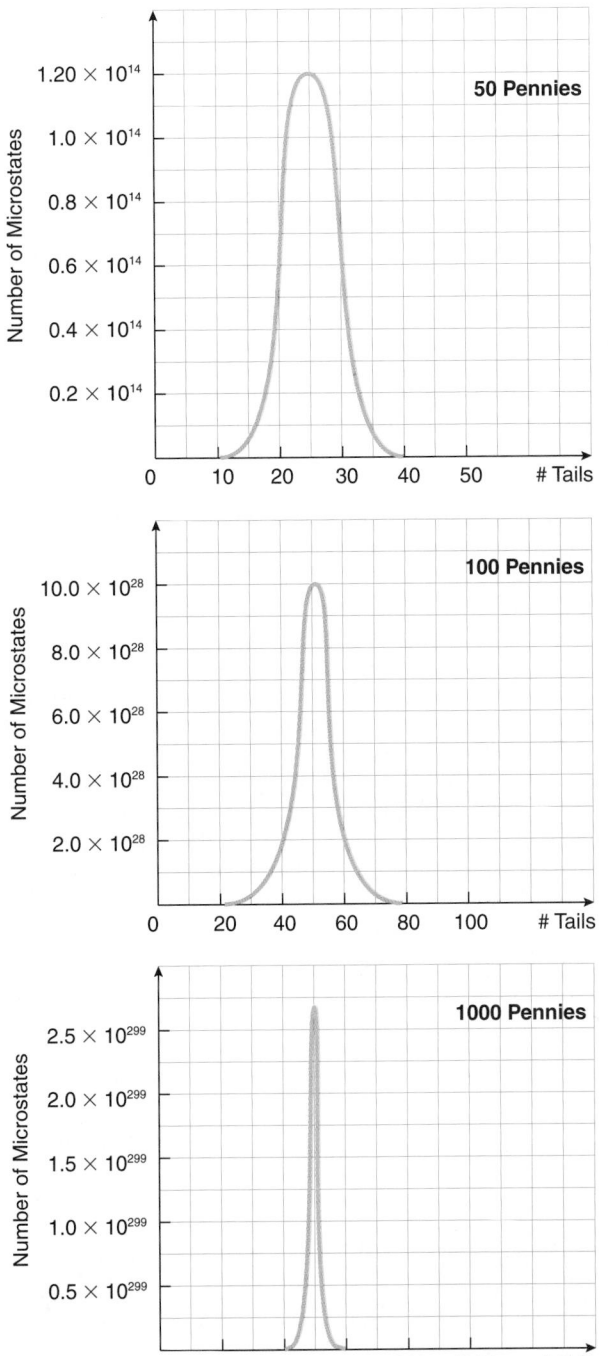

FIGURE 15.29 The number of microstates associated with each macrostate for a 50-penny, a 100-penny, and a 1000-penny system.

the connection between these apparently disparate ideas to your first course in statistical physics.

EXAMPLE 15.15 _____

Consider a system of N pennies.

a. Determine the total number of macrostates and microstates that exist for the system in terms of N.

b. **Find an expression for the number of microstates associated with each macrostate.**

Solution

a. The macrostates are delineated by specifying the number of pennies that are heads and tails. Thus the macrostates are as follows:

Number of heads	Number of tails
N	0
$N - 1$	1
$N - 2$	2
⋮	⋮
2	$N - 2$
1	$N - 1$
0	N

There are $N + 1$ macrostates. Notice that for the 4-penny system, there are $4 + 1 = 5$ macrostates; for the 10-penny system, there are $10 + 1 = 11$ macrostates; and so on. To count the total number of microstates, we observe that each penny has two possible orientations: heads or tails. For N pennies, there are then a total of 2^N possible *micro*states. For the 4-penny system, there are $2^4 = 16$ microstates; for the 10-penny system, there are $2^{10} = 1024$ microstates. For a 100-penny system, the number of microstates is $2^{100} \approx 1.27 \times 10^{30}$. The number of microstates increases dramatically as the number of particles increases.

b. For a macrostate, it does not matter which pennies are heads or tails, only that there are a specified number of heads (say n) and tails ($N - n$). The microstates are penny-specific and delineate how many different ways we can have, say, n heads and $N - n$ tails, keeping track of the individual pennies (as we saw in our example with four pennies). The number of microstates associated with each macrostate involves calculating the number of *different* ways we can have, say, n heads (and $N - n$ tails). This is a problem in combinatorics. The number is the so-called binomial coefficient:

$$\binom{N}{n} \equiv \frac{N!}{n!\,(N - n)!}$$

where N is the total number of pennies, n is the number of heads, and $N - n$ is the number of tails. The binomial coefficient is what determines the number of microstates associated with each macrostate of the penny problem. Thus, for example, in the 10-penny collection, the number of microstates associated with the macrostate of 6 heads and 4 tails is

$$\binom{10}{6} \equiv \frac{10!}{6!\,4!}$$

$$= \frac{3\,628\,800}{(720)\,(24)}$$

$$= 210$$

In other words, there are 210 different microstates associated with the 6-heads, 4-tails macrostate.

15.12 ENTROPY MAXIMIZATION AND THE ARROW OF TIME*

It would be more impressive if it flowed the other way. [When commenting on the grandeur of Niagara Falls]
                                        Oscar Wilde (1854–1900)[†]

It is easy to tell whether a movie is run backward or forward. Watching a video run backward is funny. We chuckle at seeing the process of water flowing uphill, watching the process of a bomb re-forming from its many fragments, noting the process by which an object appears magically out of flame and smoke while burning in reverse. A biographical movie run backward begins with a resurrection from the dead, with the individual gradually growing younger and smaller. Incredible. In the macroscopic world, the temporal direction of a process usually is easily recognized. This direction of time (forward or backward) is known as the **arrow of time**.

Yet, surprisingly, for some processes involving just a *few* particles, it is difficult if not impossible to determine the direction or arrow of time. A collision between two billiard balls can be viewed forward or backward in time with essentially no difference. Either view is quite plausible; there are no laughs either way (and both are, quite finally, pretty boring).

Likewise, a movie sequencing in time the results of a 2-penny or 3-penny shaking experiment is indistinguishable when run forward or backward; there is nothing funny about either direction. No arrow of time is apparent.

Correspondingly, if you could imagine a movie that shows the motion of the particles in a gas in *thermal equilibrium*, it would be difficult to tell whether the movie was running forward or backward. Likewise with a large number of pennies in a box: with the collection initially 50% heads and 50% tails, the direction of a movie sequencing the results of shaking the box is difficult to determine because the collection remains essentially

in a 50–50 distribution either way. No time arrow is apparent. So, for a system in *thermal equilibrium* (its entropy already maximized), the arrow of time likewise is not apparent.

Yet when a system with a large number of particles *changes* from one equilibrium state to another (from one permissible macrostate to another), the direction of time, or of a movie of the process, is easy to determine *if* the system is not already in the macrostate of maximum entropy (or one quite near it). The system naturally evolves *to* a macrostate of maximum entropy (the one with the greatest number of microstates), since that macrostate simply is more probable than any other. The system is exceedingly unlikely to do anything else. Hence in a movie of such a process or change, the direction or arrow of time *is* apparent.

If a many-particle system initially is in the state of maximum entropy (or one close to it), it is difficult to tell whether a movie is running forward or backward; the system is in thermal equilibrium and will not likely change to a macrostate of low entropy. However, if the many-particle system initially is in a macrostate of low entropy (say, all pennies heads) and is given the opportunity to *change* (such as by shaking the penny system), the probability is *overwhelming* that the system will change to a macrostate of maximum entropy. Thus it is easy to tell if a movie sequencing the change is running forward or backward.

A free expansion of gas can be analyzed in the same way. If the many-particle gas is confined initially to the left half of the container, this macrostate becomes *exceedingly unlikely* when the membrane is broken, and the system evolves or changes to the macrostate of highest entropy: the one with the particles equally likely to be found anywhere in the box. The arrow of time is very apparent, since the direction of a movie of the process is easily distinguished; running forward (see Figure 15.30a) the movie is believable but boring, whereas running backward (Figure 15.30b) it is unbelievable. Once in the state of maximum entropy, a movie of the system (in thermal equilibrium) has no discernable arrow of time.

Thus there is an intimate connection between the way isolated many-particle systems *change* macrostates and the arrow

---

[†]Attributed to Oscar Wilde by Daniel Patrick Moynihan, *Science*, **197**, #4305, page 742 (19 August 1977); the quotation possibly is apocryphal, although quite in the spirit of Wilde's notorious visit to Niagara Falls in the late 19th century. Despite extensive searching, the author has not been able to trace this comment directly to Wilde's writings or to news reports of his visit.

It is quite easy to tell if a movie of a waterfall is being shown forward or backward.

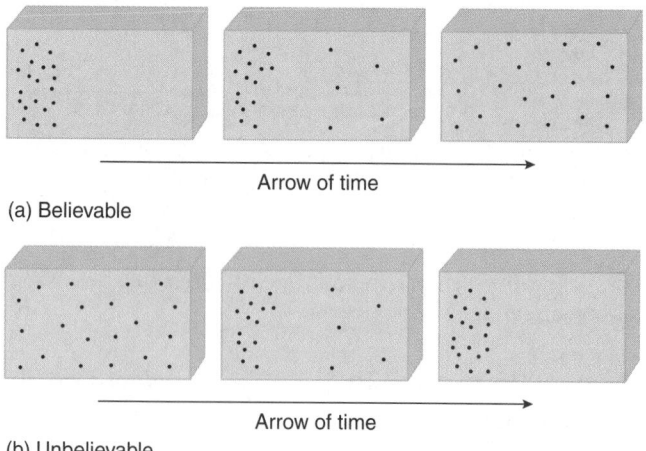

Arrow of time

(a) Believable

Arrow of time

(b) Unbelievable

FIGURE 15.30 Discerning the arrow of time.

of time. Changes to macrostates of *higher* entropy clearly indicate the direction of the arrow of time; the movie of the process is running forward. If the entropy change is *zero*, the arrow of time is not apparent; the system is in thermal equilibrium and is *exceedingly unlikely* to change its macrostate of maximum entropy. If the entropy change of the isolated system is *negative* (violating the second law of thermodynamics!), the direction of the arrow of time is reversed, and the movie becomes amusing and absolutely unbelievable. The second law of thermodynamics says a lot about why some things happen and why others do not.

15.13 EXTENSIVE AND INTENSIVE STATE VARIABLES*

Our study of thermodynamics has considered a number of thermodynamic state variables.

> The state variables whose values are independent of the size of the system are called **intensive**; those whose values are proportional to the size of the system are called **extensive**.

We summarize intensive and extensive state variables in Table 15.4.

TABLE 15.4 Intensive and Extensive State Variables

Intensive variables		*Extensive variables*	
Pressure	P	Volume	V
Temperature	T	Internal energy	U
Specific heat	c	Entropy	S
Latent heat	L	Mass	m

Work W
Heat transfer Q are *not* state variables

You should not come away from thermodynamics thinking that these are the only thermodynamic state variables that exist. Many others also will be encountered if you pursue more advanced work in thermodynamics. Our study of thermodynamics necessarily has been introductory and concerned with *simple* thermodynamic systems (as defined in Chapter 13). The fundamental precepts of thermodynamics are applicable to any *physical system* with a large number of particles. In other words, thermodynamics is not just for gases! Thermodynamic ideas even can be applied to photons (particles of light) or *any* large collection of particles such as those in solids and liquids.

CHAPTER SUMMARY

We use the word *engine* to mean a system such as a gas that performs a closed cycle on a P–V diagram. If the cycle is performed clockwise, we call it a *heat engine*; if counterclockwise, we call it a *refrigerator engine*. The function of a *heat engine*, while executing a cycle, is to do useful work $|W|$ using heat transfer $|Q_H|$ from a high-temperature reservoir with heat transfer $|Q_C|$ exhausted to a colder reservoir. The *efficiency* ε of a heat engine is the absolute magnitude of the work done by the engine divided by the absolute magnitude of the heat transfer to the engine from the hot reservoir during each cycle:

$$\varepsilon \equiv \frac{|W|}{|Q_H|} \tag{15.3}$$

$$= 1 - \frac{|Q_C|}{|Q_H|} \quad \text{(any heat engine)} \tag{15.4}$$

The efficiency is a pure number between 0 and 1.

A *perfect heat engine* has zero heat transfer as exhaust ($|Q_C| = 0$ J) and an efficiency equal to 1. One statement of the *second law of thermodynamics* is as follows:

Perfect heat engines do not exist.

The function of a *refrigerator engine*, while executing a cycle, is to accomplish the unnatural heat transfer from a cold to a hot reservoir. The work done by a refrigerator engine during a cycle is negative. The *coefficient of performance K* of a refrigerator engine is the absolute magnitude of the heat transfer to the refrigerator engine from the cold

reservoir, $|Q_C|$, divided by the absolute magnitude of the work done by the refrigerator engine, $|W|$:

$$K \equiv \frac{|Q_C|}{|W|} \tag{15.18}$$

$$= \frac{1}{\dfrac{|Q_H|}{|Q_C|} - 1} \quad \text{(any refrigerator engine)} \tag{15.19}$$

A *perfect refrigerator engine* has an infinite coefficient of performance. Another statement of the second law of thermodynamics is the following:

Perfect refrigerator engines do not exist.

A *Carnot cycle* is an ideal cycle consisting of an alternating series of isothermal and adiabatic processes on a P–V diagram.

The efficiency of a Carnot heat engine operating between two reservoirs at absolute temperatures T_H and T_C is

$$\varepsilon_{\text{Carnot}} = 1 - \frac{T_C}{T_H} \quad \text{(Carnot heat engine only)} \tag{15.14}$$

Real heat engines have efficiencies *less* than a Carnot heat engine operating between the same two temperatures.

A Carnot refrigerator engine, operating between two reservoirs at absolute temperatures T_C and T_H, has a coefficient of performance

$$K_{\text{Carnot}} = \frac{1}{\dfrac{T_H}{T_C} - 1} \quad \text{(Carnot refrigerator engine only)} \tag{15.20}$$

Real refrigerator engines have coefficients of performance *less* than a Carnot refrigerator engine operating between the same two temperatures.

If a system changes from an initial state of thermodynamic equilibrium to a final equilibrium state, the entropy change ΔS of the system is defined to be

$$\Delta S = \int_i^f \frac{dQ}{T}$$

along a quasi-static reversible path between the two states. The entropy change is *independent of the path* connecting the two states on a *P–V* diagram. The entropy is a new state variable of a thermodynamic system, and (as for internal energy) it is only entropy *changes* that are significant in classical thermodynamics.

The entropy changes associated with various thermodynamic processes are summarized in Table 15.1 on page 681. Such entropy changes may be positive, negative, or zero.

The most general statement of the *second law of thermodynamics* (one that encompasses the two previous statements) is the following: *The total entropy change of an* **isolated** *system is always greater than or equal to zero:*

$$\Delta S_{\text{total isolated}} \geq 0 \text{ J/K} \tag{15.33}$$

The total entropy change is equal to zero only for reversible processes.

The entropy of a system in thermal equilibrium has a statistical interpretation. The entropy is

$$S \equiv k \ln \Omega \tag{15.34}$$

where k is Boltzmann's constant and Ω is the number of *microstates* associated with a given *macrostate* of the system. In any process, it is overwhelmingly probable that an isolated system with a large number of particles will change to a macrostate at or very near that with maximum entropy if the system is not already in that macrostate; any other change is exceedingly unlikely.

SUMMARY OF PROBLEM-SOLVING TACTICS

15.1 (page 674) In Equation 15.14 for the efficiency of a Carnot heat engine, be sure to use absolute temperatures in kelvin, not celsius temperatures.

15.2 (page 677) In Equation 15.20 for the coefficient of performance of a Carnot refrigerator engine, be sure to use absolute temperatures in kelvin, not celsius temperatures.

15.3 (page 682) When using the equations in Table 15.1 to calculate entropy changes, be sure to use absolute temperatures in kelvin, not celsius temperatures.

QUESTIONS

1. (page 669); 2. (page 674); 3. (page 678); 4. (page 682); 5. (page 686)

6. Explain why a rock cannot secure the energy to spontaneously jump to a higher elevation by simply decreasing its temperature.

7. Explain what this statement means: temperature is a state variable. Is time also a state variable?

8. Your un-air-conditioned dorm room is getting very hot on a sultry May afternoon. In desperation, you open the refrigerator door to cool off. Explain why leaving the refrigerator door open will not, in the long run, cool the room but actually will increase the temperature of the room.

9. Can a cool kitchen be heated by turning the oven on and leaving the oven door open? Why is this different from the previous question?

10. To increase the efficiency of power plants, it is desirable to increase the temperature of the hot reservoir. Explain how this increases the efficiency.

11. Two parents return from work on Friday only to discover that their 3-year-old prodigy has maximized chaos in the house, while the lazy baby-sitter stared blankly at a TV all day. Arrg! They subsequently spend the weekend restoring the house to a much-ordered macroscopic state. Explain why this does not violate the second law of thermodynamics.

12. Diesel heat engines compress air and fuel mixtures more than gasoline heat engines. Which heat engine compresses the gases

to a higher temperature? What does this imply about the relative theoretical maximum efficiency of these heat engines?

13. Satellites in low Earth orbit experience kinetic frictional forces because of the atmosphere. Such satellites initially experience an *increase* in kinetic energy as they fall. Why? Does this violate energy conservation? Explain.

14. Explain what could be done to a mass of a material to increase its entropy. Explain what can be done to decrease its entropy. Does the latter violate the second law of thermodynamics? Explain.

15. How many isothermal curves and how many adiabatic curves pass through any given point on a *P–V* diagram? Explain your answer.

16. When you emerge from a swimming pool, your skin and your wet bathing suit spontaneously decrease in temperature. The effect is real, since you occasionally react to the temperature change by shivering. Does this violate the second law of thermodynamics? Why does the temperature decrease? Explain.

17. The presence of kinetic frictional forces increases the temperature of the surfaces in contact. Does the temperature increase occur because of heat transfer, work, and/or because of changes in internal energy that manifest themselves as a rise in temperature? Explain.

18. Can internal energy be converted into mechanical energy? Can mechanical energy be converted into internal energy? If so, give examples of each.

19. In its wisdom (?), Congress decrees that to prevent pollution, *no* exhaust is permissible from *any* engine. Explain why the honorable members desperately need to take a course in introductory physics.

20. In the internal combustion heat engine in most automobiles, what is the high-temperature reservoir and what is the low-temperature reservoir?

21. Winter is colder than summer. Yet automobile heat engines are *less* efficient in winter than in summer. Why is this so?

22. You are a remarkable thermodynamic system. You take in food, water, air, and sunlight, converting such disorganized materials into well-organized brains (well-ordered logic), brawn (well-ordered work), and occasionally braggadocio (well-ordered bragging). Does this violate the second law of thermodynamics? Explain.

23. In an unspecified thermodynamic process, the entropy of a system *decreases* by 1000 J/K. Does this violate the second law of thermodynamics? Explain.

24. A can containing nuts of various sizes, with the nuts initially dispersed randomly throughout the can, is shaken vigorously. The can is opened and the nuts are neatly arranged by size with largest on top and the smallest on the bottom. Explain how this occurs. Does this violate the second law of thermodynamics?

25. What factors make *real* heat engines and refrigerator engines less efficient than Carnot heat engines and refrigerator engines?

26. An ideal gas expands isothermally in a reversible process. Does the entropy of the gas increase, decrease, or remain the same? What about in an adiabatic expansion?

27. Are *we* heat engines? Explain your answer.

28. A sample of gas at volume V expands to volume 2V by either an isothermal or an adiabatic process. Which results in a gas at lower pressure? What happens to the entropy of the gas in each process?

29. Does the cooling that occurs with evaporation violate the second law of thermodynamics? Explain.

30. Use the results of Example 15.8, the concept of entropy changes, and the second law of thermodynamics to explain why the particles in a gas confined to a volume V do not spontaneously collect in, say, the left half of the container. Is this situation *absolutely impossible* in principle or just *exceedingly* unlikely?

31. In the free expansion of a gas there is no heat transfer to or from the gas, yet the entropy change of the isolated gas system is not zero. Explain why.

32. A scientist claims to have *once* observed a process that violated the second law of thermodynamics. Since science is predicated on the ability of scientists to *repeat* observations made by others, explain why the initial claim must be discounted.

33. A student states, "The second law of thermodynamics means that entropy always increases." Your professor marks the answer wrong. What must be added to the statement to make it correct?

34. To distinguish between the internal energy and the entropy, name a process involving an ideal gas for which (a) $\Delta U = 0$ J but $\Delta S \neq 0$ J/K; (b) $\Delta U \neq 0$ J but $\Delta S = 0$ J/K.

35. Qualitatively, which has the greater entropy: one kilogram of ice or one kilogram of liquid water? On what basis do you reach your conclusion?

36. Can it be said that the only long-term physical effect of your existence on the universe is to increase its entropy? Discuss.

37. The work done by a conservative force as a system moves between two locations in space is, by definition, independent of the path taken by the system. The entropy change of a system between two equilibrium states on a *P–V* diagram also is independent of the path taken. Yet work is *not* a state variable while entropy is. Discuss and explain, making reference to the concepts of potential energy and heat transfer.

38. In the free expansion of an ideal gas, which of the following thermodynamic quantities does *not* change?

entropy	pressure	internal energy
volume	density	

PROBLEMS

Sections 15.1 Why Do Some Things Happen, While Others Do Not?
  15.2 Heat Engines and the Second Law of Thermodynamics
  15.3 The Carnot Heat Engine and Its Efficiency
  15.4 Absolute Zero and the Third Law of Thermodynamics
  15.5 Refrigerator Engines and the Second Law of Thermodynamics
  15.6 The Carnot Refrigerator Engine
  15.7 The Efficiency of Real Heat Engines and Refrigerator Engines

1. A heat engine operates between the temperatures of –20 °C and 300 °C. What is the maximum theoretical efficiency for a heat engine operating between these temperatures?

2. Once upon a time there were plans to create a heat engine that would operate between temperature differences that exist between different depths of the ocean. Very deep ocean water has a temperature of about 4 °C while surface water (in tropical areas) has a temperature of about 20 °C. What is the maximum theoretical efficiency for a heat engine operating between these temperatures?

3. What is the maximum efficiency of a heat engine operating between the temperatures of 100 °C and 0 °C?

4. A refrigerator engine operates between the temperatures 1 °C (the inside of the refrigerator) and 25 °C (the outside of the refrigerator). What is the greatest possible value for the coefficient of performance of the refrigerator engine?

5. Consider your body to be a heat engine operating between a reservoir at 37 °C and room temperature at 20 °C. What is the

FIGURE P.6

greatest possible efficiency for a heat engine operating between these two temperatures?

6. Natural deposits of high-temperature steam exist within the Earth. See Figure P.6. Such geothermal sources have been suggested for running a heat engine. If the source of steam is at 200 °C and is exhausted to a reservoir at a temperature of 20 °C, what is the maximum theoretical efficiency of a heat engine operating between these two temperatures?

•7. A Carnot heat engine operates between a reservoir at 950 K and has heat transfer of 400 J exhausted to another reservoir at 300 K during each cycle. What is the amount of work done by the heat engine during each cycle?

•8. A patent application has been received from Professor Mediocratus, who claims to have developed a heat engine that uses 150 J of heat transfer from a reservoir at 150 °C with 50 J of heat transfer to a reservoir at −60 °C while doing 100 J of useful work. As a patent examiner, you have to decide whether to grant the patent or not. (a) Does the heat engine violate the first law of thermodynamics? (b) What is the claimed efficiency of the invented heat engine? (c) What is the efficiency of a Carnot heat engine operating between the same two temperatures? (d) Should the patent application be approved or denied? Explain your decision.

•9. A Carnot heat engine with an efficiency of 0.200 operates between two temperatures that differ from each other by 100 K. What are the temperatures between which the cycle operates?

•10. A refrigerator engine with a coefficient of performance of 3.0 accrues 1000 J of heat transfer from a warm pizza placed in the freezer compartment. What is the heat transfer to the room in which the refrigerator engine is located?

•11. Heat transfer of 200 J occurs to a refrigerator engine from a cold reservoir and heat transfer of 350 J occurs to a higher-temperature reservoir during each cycle. (a) What is the coefficient of performance of the refrigerator engine? (b) If the refrigerator engine performs five complete cycles each second, what power is required from external agents to operate the refrigerator engine?

•12. A Rube Goldberg heat engine uses 3000 J of heat transfer from a hot reservoir at 400 K and does 500 J of useful (?) work during

each cycle. (a) What heat transfer occurs to a cooler reservoir? (b) What is the efficiency of the Rube Goldberg heat engine?

•13. A 100 MW power planet (heat engine) operates with an efficiency of 0.350. The 100 MW refers to the rate at which energy is generated for useful purposes. (a) What is the heat flow to the environment? (b) The exhausted heat flow is to water drawn from a local river. To protect the aquatic environment in the river, the water returned to the river should be no more than 3.0 °C warmer than the water secured to do the cooling. What rate of water use is needed to accomplish this? Express your result in kg/s and m³/s.

•14. One (1.00) mole of an ideal diatomic gas (with $\gamma = 1.40$) initially at 20.0 °C and 1.00 atm pressure is taken around the following cycle (see Figure P.14):

Path (1): an *isochoric* increase in pressure until the temperature of the gas is 150 °C and the pressure is P;
Path (2): an *adiabatic* expansion until the pressure returns to 1.00 atm;
Path (3): an *isobaric* compression until the gas reaches its original volume.

(a) What is the original volume of the gas at the beginning and end of the cycle? (b) What is the pressure of the gas at the completion of path (1)? (c) What is the volume of the gas at the completion of path (2)? (d) What is the temperature of the gas at the completion of path (2)? (e) Calculate the work done by the gas during each path of the cycle and the total work done by the gas. (f) Calculate the heat transfer to the gas during each path of the cycle and the total heat transfer to the gas over the cycle. (g) Find the efficiency of the cycle. (h) Calculate the maximum efficiency that a heat engine could have if it operated between the hottest and coldest temperatures encountered by this gas in this cycle.

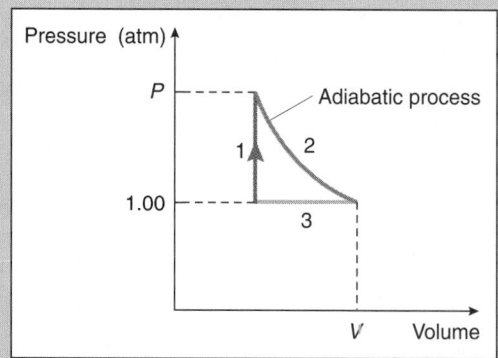

FIGURE P.14

15. One (1.00) mole of an ideal diatomic gas (with $\gamma = 1.40$) initially at 20.0 °C and 1.00 atm pressure is taken around the following cycle (see Figure P.15):

Path (1): an *isochoric* increase in pressure until the temperature of the gas is 150 °C and the pressure is P;
Path (2): an *isothermal* expansion until the pressure returns to 1.00 atm;
Path (3): an *isobaric* compression until the gas reaches its original volume.

(a) What is the original volume of the gas at the beginning and end of the cycle? (b) What is the pressure of the gas at the completion of path (1)? (c) What is the volume of the gas at the completion of path (2)? (d) What is the temperature of the gas at the completion of path (2)? (e) Calculate the work done by the gas during each path of the cycle and the total work done by the gas. (f) Calculate the heat transfer to the gas during each path of the cycle and the total heat transfer to the gas over the cycle. (g) Find the efficiency of the cycle. (h) Calculate the maximum efficiency that a heat engine could have if it operated between the hottest and coldest temperatures encountered by this gas in this cycle.

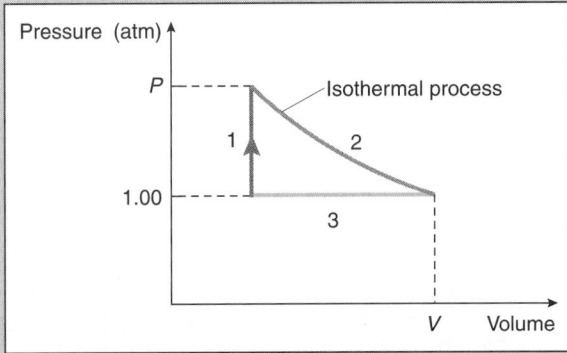

FIGURE P.15

●**16.** To increase the efficiency of a Carnot heat engine, does it make a difference whether the temperature of the hot reservoir is *increased* by 10 K or the temperature of the cold reservoir is *decreased* by 10 K? If there is a difference, which is better to do?

●17. To increase the coefficient of performance of a Carnot refrigerator engine, is it better to increase the temperature of the colder reservoir by, say, 10 K or decrease the temperature of the warmer reservoir by 10 K, or does it make any difference which you do?

●**18.** A Carnot heat engine has an efficiency $\varepsilon_{\text{Carnot}}$ and operates between two reservoirs at temperatures T_H and T_C. When the cycle is run in the reverse direction as a Carnot refrigerator engine, its coefficient of performance is K_{Carnot}. Show that

$$\varepsilon_{\text{Carnot}} K_{\text{Carnot}} = \frac{T_C}{T_H}$$

●**19.** A heat engine operates between two reservoirs at temperatures 700 K and 300 K. (a) What is the efficiency of a Carnot heat engine operating between the same two reservoirs? (b) If the actual efficiency of the heat engine is 1/3 the Carnot efficiency, how many joules of work does the heat engine perform if it accrues 1000 J of heat transfer from the high-temperature reservoir during each cycle?

●**20.** A small refrigerator engine produces 400 W of heat flow to the kitchen when 333 W is transferred to the refrigerator engine from the cooler reservoir. The inside of the refrigerator is at 0 °C and the kitchen in which it lies is at 25 °C. (a) What is the co-efficient of performance of the refrigerator engine? (b) What is the power of the refrigerator engine?

●21. A peculiar "V-8" engine (see Figure P.21) has an efficiency of 0.250 when operating between reservoirs at temperatures 500 K and 300 K. (a) What is the theoretical maximum efficiency for

a heat engine operating between these two reservoirs? (b) If the "V-8" heat engine does 800 J of work per cycle, what heat transfer occurs to the engine from the high-temperature reservoir and what heat transfer is exhausted to the low-temperature reservoir during each cycle?

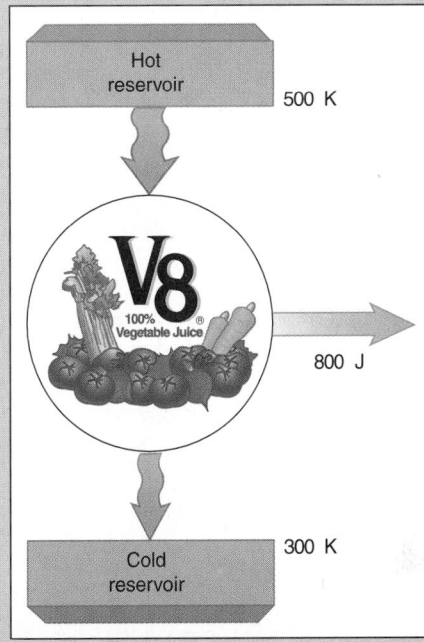

FIGURE P.21

●**22.** The coefficient of performance of a small refrigerator engine is 4.50 and its heat flow into the kitchen is 250 W. At what rate is work done by the refrigerator engine?

●**23.** A Carnot heat engine has an efficiency of 0.300 operating between a high-temperature reservoir at temperature T_H and a low-temperature reservoir at 20 °C. To increase the efficiency of the Carnot heat engine to 0.400, by how many kelvin should the temperature of the hot reservoir be increased?

●**24.** The heat flow to a Carnot heat engine from a reservoir at 100 °C is 5.00×10^3 W. The exhaust heat flow is to a reservoir at temperature 300 K. (a) At what rate is work being done by the heat engine? (b) What is the heat flow to the cooler reservoir?

●25. Nitrogen liquefies at temperature 77 K. To liquefy nitrogen already at 77 K, a refrigerator engine is operating in a room at temperature 300 K. (a) For each joule transferred from the gaseous nitrogen at 77 K, what is the *minimum* number of joules exhausted to the room? Do you need to know the latent heat of vaporization of nitrogen to answer this question? Explain why this is a *minimum* number of joules. (b) Repeat the calculation for a helium liquefier (helium liquefies at 4.2 K).

●**26.** A *heat pump* is essentially a refrigerator engine that uses the inside of a building as a hot-temperature reservoir in the winter, *or* the outside of the building as the hot-temperature reservoir in the summer. Neat! (a) For a Carnot refrigerator engine operating between reservoirs at 25 °C and 40 °C in the summer, what is its coefficient of performance? (b) For each joule of heat transfer from the cooler reservoir per cycle, how many joules of work are done by the Carnot refrigerator engine? Is this work positive or negative according to our conventions? (c) What additional information is needed to "size" or determine an appropriate heat pump for a given building?

•27. Air conditioners are essentially refrigerator engines that exhaust to the outside of a building rather than to the inside. Central air conditioners are rated by saying they have a certain tonnage capacity. To be specific, consider an air conditioning engine rated at 1-ton capacity. This tonnage terminology means that the heat transfer is equivalent to that required to freeze or melt one English ton (~ 909 kg) of ice at 0 °C during one day. (a) What heat transfer is required to melt 909 kg of ice? (b) If this is accomplished during one day, what is the heat flow to the ice in watts?

•28. The oceans contain a vast store of internal energy. (a) Estimate the heat transfer from the water if the temperature of one cubic kilometer of water is decreased by 1.00 K. Assume the oceans are pure water. (b) Explain why such heat transfer is not employed to do useful work via a heat engine.

•29. A thing-a-ma-bob-that-does-the-job has an efficiency $\varepsilon = 0.35$. Five hundred (500) joules of heat transfer occurs to a cool reservoir during each cycle. (a) What is the work done by the heat engine? (b) What is the heat transfer to the heat engine from the hot reservoir?

•30. A heat engine gizmo lifts a 5.00 kg package vertically at constant speed through a distance 1.50 m during each cycle of the heat engine. (a) What work is done by the heat engine during each cycle? (b) What is the change in the internal energy of the heat engine during one cycle? (c) What is the total heat transfer to the heat engine during each cycle? (d) If the engine exhausts 90.0 J of heat transfer per cycle, what is the heat transfer to the heat engine from the high-temperature reservoir? (e) What is the efficiency of the heat engine?

‡31. No two adiabatic curves can intersect on a *P–V* diagram. To show this, construct a hypothetical schematic *P–V* diagram for an ideal gas on which two different adiabatic curves intersect. Connect these adiabatic curves with an isothermal curve to form a cycle. Show that although a heat engine running around such a cycle satisfies the first law of thermodynamics, the engine violates the second law of thermodynamics.

Sections 15.8 A New Concept: Entropy
      15.9 Entropy and the Second Law
           of Thermodynamics
      15.10 The Direction of Heat Transfer:
           A Consequence of the Second Law

32. One (1.00) kilogram of ice melts at 0.00 °C to water at 0.00 °C. Find the entropy change of the mass.

33. One (1.00) kilogram of steam at 100 °C condenses into water at 100 °C. Calculate the entropy change of the mass. Does your result violate the second law of thermodynamics? Explain your answer.

34. Two (2.00) moles of a monatomic ideal gas is warmed slowly from 300 K to 400 K at constant volume. Determine the entropy change of the gas.

35. If 500 J of heat transfer occurs from a reservoir at 450 K to a reservoir at 300 K, what is the change in the entropy of the system of two reservoirs?

•36. A 0.100 kg ice cube at −10.0 °C is placed in a well-insulated container with 1.00 kg of water at 20.0 °C. (a) Find the final temperature of the ice and water system. (b) Calculate the entropy change of the warm water as it cools to the final temperature. (c) Calculate the entropy change of the ice as it warms to the final temperature. (d) What is the total entropy change of the isolated ice and water system?

•37. You throw an ice cube of mass 0.100 kg and temperature −5.00 °C into the ocean, which has a temperature of 10.00 °C. Assume the specific heat of salt water is the same as fresh water. (a) What is the entropy change of the ice? (b) What is the entropy change of the ocean? (c) What is the total entropy change of the ice–ocean system?

•38. Two cars each of mass 1500 kg traveling at 100 km/h in opposite directions collide head-on and come to a disastrous halt; see Figure P.38. Assume that the surrounding air and ground temperature remains fixed at 20 °C. Calculate the total entropy change of the cars and environment system that results from the collision.

FIGURE P.38

•39. A 2.00 kg ripe tomato falls 1.50 m to the ground. The tomato, air, and ground all are at a temperature of 300 K. What is the total entropy change of the tomato, air, and ground system?

•40. You are carefully carrying a 2.00 kg mass of copper at temperature 200 °C along a lakeside dock for no good reason. You stumble clumsily and the copper drops into the lake at temperature 5 °C. Sssssssst! (a) What is the entropy change of the copper? (b) What is the entropy change of the lake? (c) What is the total entropy change of the copper and lake system?

•41. Rather than graph the pressure versus the volume for a gas undergoing a cycle, it occasionally is useful to plot the absolute temperature *T* (on the vertical axis) versus the entropy *S* (on the horizontal axis), forming what is called a *T–S* diagram. (a) Make a schematic sketch of a Carnot cycle on such a *T–S* diagram. (b) What is the physical significance of the area enclosed by the cycle on the *T–S* diagram? (c) How can the efficiency of a Carnot heat engine be obtained *graphically* from a *T–S* diagram of the cycle?

•42. Show that the slope of an *isochoric* process on a *T–S* diagram (see Problem 41) is equal to

$$\frac{T}{nc_V}$$

where c_V is the molar specific heat at constant volume.

•**43.** Show that the slope of an *isobaric* process on a T–S diagram (see Problem 41) is equal to

$$\frac{T}{nc_\mathrm{p}}$$

where c_p is the molar specific heat at constant pressure.

•**44.** An inventor, Charles LeTan, is attempting to convince a loan officer to make a several megadollar loan to Fly-by-Night Enterprises Ltd. to manufacture a heat engine that takes a gas around a cycle on a P–V diagram with the following consequences:

- total (net) heat transfer to the gas during n complete cycles $|Q_\mathrm{H}| - |Q_\mathrm{C}| = 9.21 \times 10^6$ J;
- total work done *by* the gas used in the machine during the same n cycles $= 10.0 \times 10^6$ J.

You have been hired as a technical consultant to the loan officer. Should you advise that the loan application be approved by the loan officer or not? Indicate the reasoning behind your decision.

•**45.** A heat engine with an efficiency of 0.250 operates between two reservoirs at 1000 K and 400 K. The power output of the heat engine is 1.00 kW. (a) What is the heat flow to the heat engine from the high-temperature reservoir? (b) What is the time rate of increase in entropy of the system consisting of the heat engine and the two reservoirs?

•**46.** One end of a large piece of copper is in thermal contact with a reservoir at temperature 100 °C and the other end is in thermal contact with another reservoir at 0 °C. The heat flow between the reservoirs through the copper is 500 W. (a) Determine the rate at which the entropy of each reservoir is changing with time. (b) Does the entropy of the copper rod change with time? Explain your answer.

•**47.** A 1250 kg car speeding at 150 km/h crashes completely inelastically into a concrete bridge pylon when the temperature of the car, ambient air, and concrete is 20.0 °C. The car is totaled. (a) What happens to the kinetic energy of the car? (b) What is the entropy change of the car, bridge, and air system?

•**48.** One (1.00) mole of an ideal monatomic gas is initially at temperature 300 K and has a volume of 0.0100 m³. The volume of the gas increases to 0.0200 m³ by one of the following quasistatic reversible processes:

- isothermal expansion
- adiabatic expansion
- isobaric expansion

(a) Construct an *accurate* graph of the absolute temperature of the gas versus volume (a T–V graph) that represents *each* of these processes. (b) In which (if any) processes is the entropy change of the gas positive? Zero? Negative? Explain your reasoning. (c) For each process, sketch the path of another process that produces an identical entropy change in the gas.

•**49.** Since you are a warm-blooded mammal, your heat flow via radiation into the ambient environment (at temperature 293 K) is about 100 W when clothed. What is the increase in the entropy of the system consisting of you and the environment each day?

•**50.** Heat transfer from a reservoir at 150 °C is used to turn 1.00 kg of ice at 0 °C to steam at 100 °C, all at a pressure of 1.00 atm.

(a) What is the entropy change of the water? (b) What is the entropy change of the reservoir? (c) What is the entropy change of the system consisting of the water and reservoir? (d) Explain why the sign of the answer to part (c) is consistent with the second law of thermodynamics.

•**51.** One (1.00) mole of an ideal monatomic gas is taken around the cycle shown in Figure P.51. (a) What is the work done by the gas in going from point (1) to point (2)? (b) What is the work done by the gas in going from point (2) to point (3)? (c) What is the change in the internal energy of the gas in going from point (1) to point (3) along the path (1) → (2) → (3)? (d) What is the entropy change of the gas in going from point (1) to point (3)? (e) What is the change in the internal energy and entropy of the gas in one complete cycle?

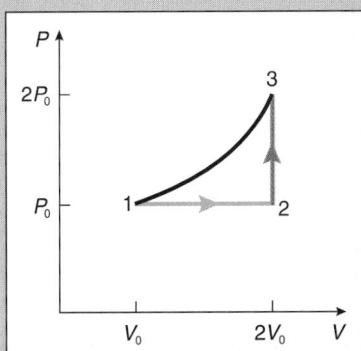

FIGURE P.51

•**52.** Heat transfer of 4500 J occurs from a reservoir at 100 °C to one at 20 °C. (a) What is the entropy change of the hot reservoir? Explain why your result does not violate the second law of thermodynamics. (b) What is the entropy change of the cool reservoir? (c) What is the entropy change of the system consisting of both reservoirs? (d) What is the maximum efficiency of a heat engine operating between the two reservoirs?

•**53.** Two (2.00) moles of a diatomic ideal gas expand adiabatically, doubling its volume while doing 2000 J of work. (a) What is the total heat transfer to the gas during the process? (b) Determine the change in the internal energy of the gas. (c) Determine the entropy change of the gas. (d) Did the temperature rise or fall? (e) Is there sufficient information given to determine the temperature change? If so, determine the temperature change. If not, what additional information is needed?

•**54.** A container with 1.00 liter of water at 20 °C is added to a well-insulated thermos jug holding 2.00 liters of water at 90 °C. Neglect heat transfer to the thermos itself. What is the increase in the entropy of the 3.00 liter water system?

•**55.** Heat transfer of 5000 J occurs to the (heat) engine of progress from a reservoir at 500 °C and 3500 J is exhausted to another reservoir at 20 °C during each engine cycle, while doing some useful work. (a) What work is done by the heat engine during each cycle? (b) What is the efficiency of the heat engine? (c) What is the efficiency of a Carnot heat engine operating between the same two reservoirs? (d) During one cycle of the heat engine, calculate the total entropy change of the system consisting of the two reservoirs and the heat engine.

•**56.** Two (2.00) moles of a diatomic ideal gas (with $\gamma = 1.40$) at 300 K and 3.00 atm pressure expands isothermally until the pressure drops to 2.00 atm. The gas then expands adiabatically until the pressure reaches 1.00 atm. (a) What is the initial volume of the gas? (b) What is the volume of the gas after the isothermal expansion? (c) Find the work done by the gas during the isothermal expansion. (d) Find the change in the internal energy of the gas during the isothermal expansion. (e) What is the final temperature of the gas? (f) Find the change in the internal energy of the gas during the adiabatic expansion. (g) What is the work done by the gas during the adiabatic expansion? (h) What was the total heat transfer to the gas during the entire expansion? (i) What is the total entropy change of the gas?

57. Two containers labeled 1 and 2 each contain n moles of the same monatomic ideal gas and initially are isolated from each other. The two containers have the same pressure but different initial temperatures, T_1 and T_2. The two containers are brought into thermal contact with each other and the pressure is maintained at the same value P. What is the total change in entropy? Show that the total entropy change is not negative. Hint: Consider the following inequality:

$$(T_1 - T_2)^2 \geq 0 \text{ K}^2$$

58. Two containers labeled 1 and 2 each contain n moles of the same monatomic ideal gas and initially are isolated from each other. Each container is at the same temperature T. The volume of container 1 is V_1 and that of container 2 is V_2. The two containers are brought into thermal contact with each other and the partition separating them is removed. What is the change in the entropy of the system of two gases when they are in thermal equilibrium? Show that the change in entropy is positive.

59. A mass m initially at temperature T_1 is placed in thermal contact with an equal mass of the same material initially at temperature T_2. The combination is well insulated from the environment, so that heat transfer occurs only between the two masses. (a) Show that the entropy change of the isolated system of two masses is

$$\Delta S = mc \ln \left[\frac{(T_1 + T_2)^2}{4 T_1 T_2} \right]$$

where c is the specific heat of the material in J/(kg·K). (b) Show that $\Delta S \geq 0$ J/K. Hint: Consider the following inequality:

$$(T_1 - T_2)^2 \geq 0 \text{ K}^2$$

Sections 15.11 A Statistical Interpretation of the Entropy *
15.12 Entropy Maximization and the Arrow of Time*
15.13 Extensive and Intensive State Variables*

60. Consider a 6-penny system in a treasure box. (a) How many macrostates are there? (b) Determine the number of microstates associated with each macrostate of the 6-penny system. (c) Make a histogram of the number of microstates associated with each macrostate. (d) Calculate the quantity S/k for each macrostate. (e) What is the probability that six heads will result when the box is shaken?

61. Suppose one penny in a 4-penny shaking experiment is a counterfeit two-headed penny (with no tails). How does this

affect the number of macrostates and microstates? If such a penny is used unknowingly in a 4-penny experiment, how will the experimental results indicate that the system is not a true 4-penny system?

62. Six pennies are in a box. The box is shaken repeatedly and the macrostate of the system (the number of heads and the number of tails) determined after each shake. Unbeknownst to you, one of the coins actually is glued to the bottom of the box. Estimate the number of times you need to shake the box before this becomes apparent from the survey of each macrostate after each shake, ignoring the locations of the pennies in the box.

63. Take a pair of dice. When the dice are rolled, the possible total sums that emerge on the dice are 2, 3, 4, 5, 6, 7, 8, 9, 10, 11, and 12. These are the macrostates associated with a toss. (a) Construct a table indicating the various possible microstates associated with each macrostate. (b) What is the total number of microstates associated with a toss? (c) What is the probability of rolling a 7? (d) Construct a graph of S/k for each macrostate.

64. Six eggs are to be placed in an egg carton with spaces for 12 eggs. (a) If the eggs are indistinguishable, is there any difference between a macrostate and microstate of the system? If so, indicate the number of microstates associated with each macrostate. (b) If the eggs are distinguishable, is there a difference between a macrostate and a microstate? If so, indicate the number of microstates associated with each macrostate.

65. Six particles of a *very* dilute gas are confined to a box. Divide the box into a left half and a right half by an imaginary partition. (a) Specify the number of ways the six particles can be distributed between the two sides of the box. Let these be the macrostates of the system. Find the number of microstates associated with each macrostate. (b) What is the entropy of the macrostate in which all the particles are on the left side of the box? (c) What is the entropy of the most probable macrostate? Notice that the entropy of the most probable macrostate is greater than the entropy of the macrostate with all particles on one side of the box. The entropy change between these two macrostates is similar to the entropy change associated with a free expansion of gas initially confined to the left half of the box (see Example 15.8). The entropy change of a free expansion (of six particles) and that of this problem are the same order of magnitude but are not quantitatively identical, because this problem of simply distributing the particles between the two halves neglects energy considerations and the fundamental indistinguishability of the particles of a gas. Alas, the quantitative disagreement illustrates the difficulty in exploring the connection between the classical definition of an entropy change and that associated with statistical thermodynamics.

66. Time to play cards with a standard 52 card deck with 13 cards of each suit (clubs, diamonds, hearts, and spades). (a) What is the probability that the first four cards dealt from a shuffled deck are the four aces? (b) What is the probability that the first four cards dealt are spades?

67. Consider a row of ten lockers in a locker room. Define macrostates of the system corresponding to the following situations:
1. two lockers immediately adjacent to each other
2. two lockers with one other locker separating them

3. two lockers with two other lockers separating them
4. two lockers with three other lockers separating them
5. two lockers with four other lockers separating them
6. two lockers with five other lockers separating them
7. two lockers with six other lockers separating them
8. two lockers with seven other lockers separating them
9. two lockers with eight other lockers separating them

Let a microstate be a way each macrostate can be arranged within the row of ten lockers. (a) Delineate the number of microstates associated with each macrostate. (b) What is the total number of microstates? (c) Ten persons are assigned the lockers. Two of these individuals go to the locker room at the same time. What is the probability that they will have immediately adjacent lockers and, therefore, be in each other's way?

•68. A 4-penny system is shaken and examined 250 times; the head and tail distribution of the pennies are shown in following table:

Macrostate	Heads	Tails	Number of times the macrostate occurs
4	0	4	0
3	1	3	31
2	2	2	93
1	3	1	94
0	4	0	32

What can you conclude from the data?

‡69. (a) Specify the macrostates for a 20-penny system, the number of microstates associated with each macrostate, and the value of S/k for each macrostate. (b) Make a graph of the number of microstates versus the macrostates. (c) Make a graph of S/k versus the macrostates.

INVESTIGATIVE PROJECTS

A. Expanded Horizons

1. Investigate the notion of perpetual motion machines and how they necessarily violate the laws of thermodynamics; thus they are impossible machines. Distinguish between so-called perpetual motion machines of the *first kind* and those of the *second kind*.
   Arthur W. J. G. Ord-Hume, *Perpetual Motion: The History of an Obsession* (St. Martin's Press, New York, 1977).
   Stanley W. Angrist, "Perpetual motion machines," *Scientific American*, 218, #1, pages 114–122 (January 1968).
   Henry Dircks, *Perpetuum Mobile* (Rogers and Hall, Chicago, 1916).
   Clifford B. Hicks, "Why won't they work?" *American Heritage*, XII, #3, pages 78–83 (April 1961).
   John Phin, *The Seven Follies of Science* (D. Van Nostrand, New York, 1906), pages 36–78.

2. Investigate and write a report about the famous Maxwell's demon, a hypothetical gremlin imagined by James Clerk Maxwell (1831–1879) as a way of violating the second law of thermodynamics.
   Harvey S. Leff and Andrew F. Rex, "Resource Letter MD-1: Maxwell's demon," *American Journal of Physics*, 58, #3, pages 201–209 (March 1990); this contains an extensive bibliography on the subject.
   Martin J. Klein, "Maxwell, his Demon, and the second law of thermodynamics," *American Scientist*, 58, #1, pages 84–97 (January–February 1970).
   Alfred Brian Pippard, *Elements of Classical Thermodynamics* (Cambridge, England, 1957), pages 99–100.
   George Gamow, *Mr Tompkins Explores the Atom* (Macmillan, New York, 1945), pages 1–19.
   Allen L. King, *Thermophysics* (W. H. Freeman, San Francisco, 1962), pages 250–251.
   Mark W. Zemansky, *Heat and Thermodynamics* (5th edition, McGraw-Hill, New York, 1968), pages 271–272.
   Richard P. Feynman, Robert B. Leighton, and Matthew Sands, *The Feynman Lectures on Physics* (Addison-Wesley, Reading, Massachusetts, 1963), Volume I, Chapter 46.

3. Investigate in further detail the relationship between the second law of thermodynamics and the arrow of time. Write a précis about the following discussion of the topic.
   Joel L. Lebowitz, "Boltzmann's entropy and time's arrow," *Physics Today*, 46, #9, pages 32–38 (September 1993).

   Paul Davies, "The arrow of time," *Sky and Telescope*, 72, #3, pages 239–242 (September 1986).
   David Layzer, "The arrow of time," *Scientific American*, 233, #12, pages 56–69, 148 (December 1975).
   T. Gold, "The arrow of time," *American Journal of Physics*, 30, #6, pages 403–410 (June 1962).
   Stephen Hawking, *A Brief History of Time* (Bantam, New York, 1988).
   Richard Feynman, *The Character of Physical Law* (MIT Press, Cambridge, Massachusetts, 1965).

4. Write a short paper about the physics underlying the idea of *negative* absolute temperatures.
   Ralph Baierlein, "The meaning of temperature," *The Physics Teacher*, 28, #2, pages 94–96 (February 1990).

5. Investigate the energy and *P–V* cyclical action of automobile heat engines. Present an oral report on your findings to your class.
   Gene Waring, "Energy and the automobile," *The Physics Teacher*, 18, #7, pages 494–503 (October 1980).

B. Lab and Field Work

6. Design and perform an experiment to see if the temperature of an ideal gas remains the same in a free expansion.

7. Practical refrigerator engines make use of the *Joule–Kelvin effect* (also known as the porous-plug experiment): a gas seeping through a porous plug under the appropriate circumstances experiences a decrease in temperature. The effect also is used to liquefy various gases. Research the effect and, if appropriate materials are available, demonstrate the effect in the laboratory.
   Mark W. Zemansky, *Heat and Thermodynamics* (5th edition, McGraw-Hill, New York, 1968).

8. What measurements are needed to determine the coefficient of performance of a typical household refrigerator? Make the needed measurements and compare the refrigerator's coefficient of performance with that of a Carnot refrigerator engine operating between the same two temperatures.

9. Take a covered shoe box or some other such container and place 50 pennies in the box, all heads up. Perform an experiment to see how to shake the coins for them to reach their most probable distribution with the fewest shakes.

10. Roll a single die 100 times and keep a record of the number of times each face appears. Make a histogram of the results. How can you be sure the die is not loaded in some way? If the die is legitimate, what is the minimum numerical value of the sum of the values of the 100 throws? What is the maximum numerical value of the sum? Approximately, what is the most likely value for the sum?

C. Communicating Physics

11. Reread C. P. Snow's comment at the beginning of the chapter. Follow C. P. Snow's injunction and take up the challenge to "describe the Second Law of Thermodynamics" in language appropriate for a gathering of well-educated but scientifically illiterate peers.

12. Write a one-page essay describing what is meant by the term entropy in physics. Assume your audience is a class of humanists who know little of the world of physics.

CHAPTER 6

ELECTRIC CHARGES,
ELECTRICAL FORCES,
AND THE ELECTRIC FIELD

Electricity is [the] Soul of the Universe

John Wesley (1703–1791)*

It is hard to imagine life without electricity. Try it sometime. A blackout will do. Backpacking in the wilderness is another way (no flashlights permitted!). No lights, washing machines, dish-washers, microwave ovens, central heat, air conditioners, or snooze alarms either. No stereos, radios, TV, Xerox machines, ink jet printers, no convenient plastic kitchen wrap—essentially no advanced technology. There is one small benefit: no annoying static cling in our clothes. It is a primitive existence without electricity. But it is much worse than that: the electrical force is central to our very being. Without the electrical force, atoms would not exist, no molecules, no chemistry (no chemistry tests!), no life. Worst of all, there would be no physicists.[†] The harnessing of electricity for pro-ductive use is a marvel of our collective technological prowess and keeps many of us gainfully employed and challenged.

What is electricity? Simple, you might say. *Everyone knows what a dragon looks like*[‡]; everyone knows what electricity is. It is positive and negative charge, protons and electrons, and "that kind of stuff." So, *what is charge?* Come to think of it, maybe it is not so simple to explain what electricity is . . . at least if you have to explain it under the watchful eye of your professor! The impor-tance of language, not equations, comes back to haunt us again. The concept of electricity, or what we casually call electric charge, is not quite as tangible or as easy to fathom as mass; and as we have seen, even the concept of mass is quite subtle and mysterious.

How do we know electricity even exists? Unlike gravitation, it is not readily apparent in our natural environment[§]; you have to look for it. For years, you likely have been told that electric charge comes in two varieties or flavors: positive charge and negative charge. Protons have positive charge, electrons have negative charge. It is hard to imagine two more different particles than the electron and the proton (they differ in mass by a factor of 1836), yet they have exactly the same absolute value (magnitude) of electric charge—why is that? We address that question in Section 16.4. As a consequence of the equality, normal atoms (with equal numbers of protons and electrons) are electrically neutral (zero total charge), which is why electricity is not as apparent as gravitation.

In this chapter, we first address what we mean by electric charge. We then explore how charges interact quantitatively, and introduce a useful construct for imagining how the electrical inter-action is transmitted between charges: the *electric field*. Finally, we discover some of the characteristic properties of the electric field.

16.1 THE DISCOVERY OF ELECTRIFICATION

Hence have arisen some new terms among us: we say B (and bodies like circumstanced) is electrized positively; A, negatively. Or rather, B is electrized plus; A minus. And we daily in our experi-ments electrize plus or minus, as we think proper.

Benjamin Franklin (1706–1790)[#]

---

*(Chapter Opener) *The Desideratum: or Electricity Made Plain and Useful* (London, 1759); from the Contents of the 2nd edition (Bailliere, Tindall, and Cox, London, 1871). Wesley was the founder of the Methodist Protes-tant Christian religious denomination.

† This is a line made famous by George Wald in a taped lecture, "Design in the Universe," part of a lecture series *Cosmic Evolution: Are We Alone in the Universe* (American Association for the Advancement of Science, 1974).

‡ The title of a children's book by Jay Williams (Four Winds, New York, 1976).

§ It was only during the 18th century that Benjamin Franklin first demon-strated that lightning had electrical properties.

The wit and wisdom of Benjamin Franklin still are widely admired. He was the first American scientist to gain an international reputation (for his research into electricity).

The discovery of electrical phenomena begins with an unlikely material: amber, the beautiful, yellowish-brown, fossilized resin, valued for ornamental jewelry. A sample of amber is shown in Figure 16.1 encasing a venerable entomological intruder.

The ancient Greeks discovered, likely while polishing the material, that when rubbed vigorously, amber has the ability to at-tract small bits of matter such as straw and grain hulls.[¶] From such chaff was born a revolution in science and technology, though several millennia later. The interaction between the amber and chaff only exists after the amber is rubbed (the chaff is not rubbed) and, even more perplexing, the force gradually dissipates, leaks away, or diminishes as time passes, particularly on humid days. This is quite unlike gravitation; the gravitational interaction

---

Letter to Peter Collinson, 1 September 1747; Benjamin Franklin, *Experi-ments and Observations on Electricity* (E. Cave, London, 1751), page 15. More recently in Duane Roller and Duane H. D. Roller, "The develop-ment of the concept of electric charge," *Harvard Case Studies in Experi-mental Science* (Harvard University Press, Cambridge, Massachusetts, 1954), volume 2, pages 541–693; the quotation is on page 598.

¶ In modern parlance, the chaff is initially *not* charged but is electrically neutral. We address in Section 16.2 how the attraction between electrified material and neutral material arises.

FIGURE 16.1 **A sample of amber with an ancient insect trapped inside.**

The 16th-century age of exploration precipitated much interest in magnetism and it attracted William Gilbert to explore the relationship between it and amber-like effects.

between masses always is present and certainly does not depend on the weather. There is evidently quite a different interaction going on between the amber and chaff.

The concept of electricity and the word itself arose during the 17th century from careful observation of amber-like effects in other materials. In 1600, a century before Newton, William Gilbert (1540–1603), a scientist of Renaissance interests and physician to Queen Elizabeth I, discovered that glass and many other substances, when rubbed with silk, attracted small bits of matter, just as amber did. He described the observations by saying the materials had become *electrified*, meaning they had "become like amber" or "amberized"; *electron* [ἤλεκτρον] is the Greek word for amber.

Could this attraction be what we know as magnetism? Another naturally occurring material known as lodestone [composed of the mineral magnetite (Fe_3O_4)] was known to the ancient Chinese and Greeks (though not as Fe_3O_4!), and has peculiar attractive properties for a restricted list of materials, typically containing iron. Lodestones gave rise to *magnetism* [the word comes from Magnesia, the region of Asia Minor (Anatolia) in present-day Turkey where the mineral was found]. We study

magnetic effects later (Chapter 20). For now, we simply note that *every* magnet *attracts* iron that is not itself a magnet (resembling the attraction of rubbed amber for bits of straw).

So why is electricity different from magnetism? If you play with any *two* magnets as you likely did as a child,* or *two* electrified materials, there is a distinct difference. Any two magnets can attract *or* repel each other depending on the orientation of the ends of the magnets (what we now call the poles of the magnet) that are brought closest together or presented to each other, as shown in Figure 16.2.

This phenomenon is very different from what is observed with electrification forces. By playing with two different pieces of rubbed amber, you find they *always* repel each other; it makes no difference which end of each piece is presented to the other, as indicated in Figure 16.3.

Electric and magnetic effects also were seen to be different phenomena because rubbing seemed intimately connected with electrical effects in amber and other materials, whereas magnetic effects are essentially permanent in lodestones. During the 19th century, electrical and magnetic effects were shown to be closely related. Rubbing is not necessary for some electrical effects, nor are all magnets permanent. We now consider both electrical and magnetic effects to be aspects of **electromagnetism**, to reflect their intimate connection. We shall see how they are related in Chapter 21, after first exploring them separately.

It was the Frenchman Charles-François Cisternay Dufay (1698–1739) who discovered in 1733 that while two glass rods electrified by rubbing with silk *repelled* each other (see Figure 16.4), and two rubber or amber rods electrified by rubbing with fur *also* repelled each other (see Figure 16.5), an electrified glass rod *attracted* an electrified rubber rod or amber (see Figure 16.6). (The latter two figures are on page 709.)

In other words, identical materials electrified by the same procedure always repel each other. Different materials, electrified by different means (rubbing with different materials such as silk or fur), may repel or attract each other depending on the specific materials and means of electrification. If the materials are not electrified by appropriate rubbing, they apparently do not interact (except weakly by gravitation, of course). Thus electrification

---

*No need to stop such playing now. A fascination with magnets was one of the things that led Albert Einstein into physics.

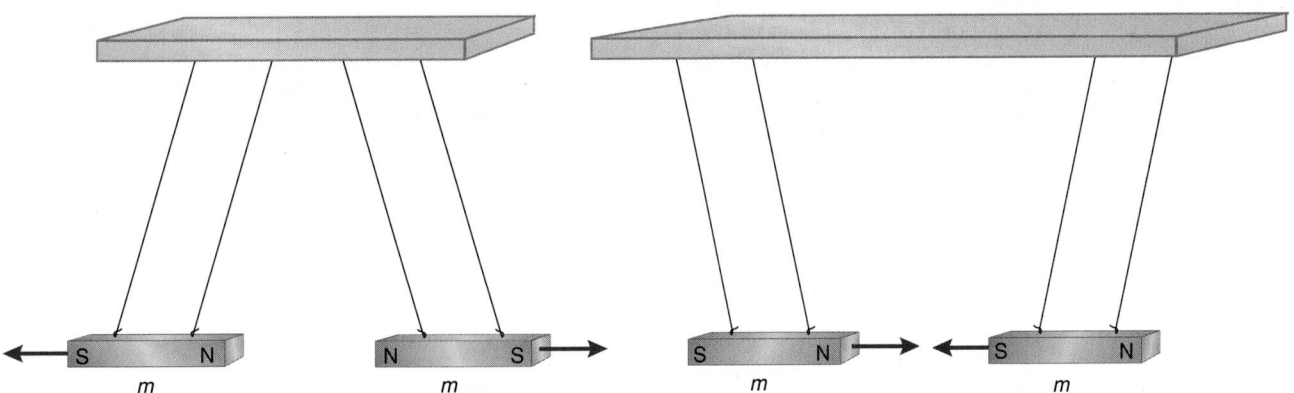

FIGURE 16.2 Two magnets will attract or repel each other depending on which way they face each other.

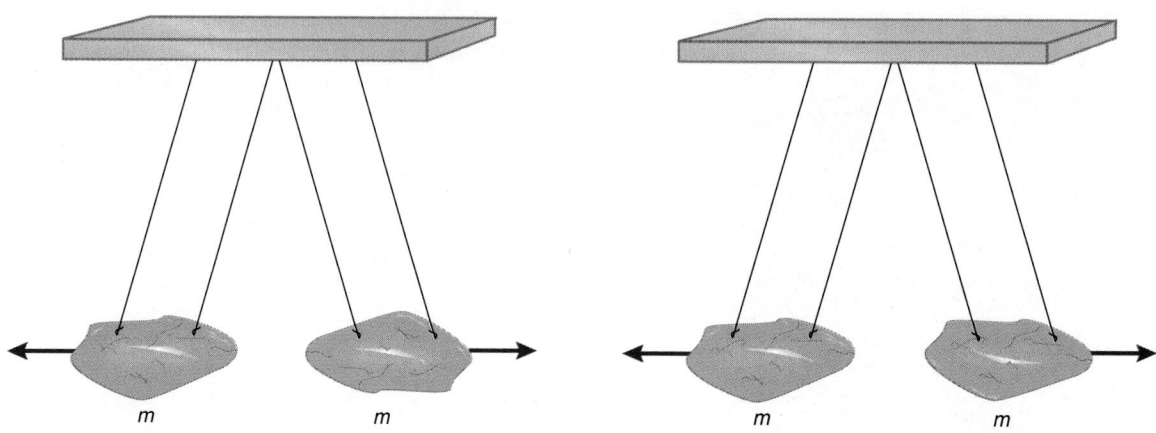

FIGURE 16.3 Two pieces of rubbed amber always repel each other, regardless of which way they face each other.

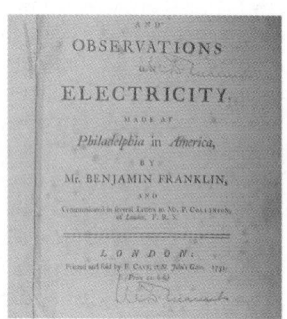

Benjamin Franklin published his researches into electrical phenomena in 1751, well before the American Revolution.

forces are attractive, repulsive or zero.* Confusing? Yes. But Dufay was the first person to realize the following:

> *Two* different kinds of electrification properties are needed to explain *all* of these observations.

It is fortunate for us that all electrical phenomena can be described by assuming only two (and not more) different kinds of electrification properties. Dufay called the two types of electrification properties *vitreous electricity* (that similar to glass) and *resinous electricity* (that similar to amber). One never observes an electrified material that repels (or attracts) both an electrified glass rod and an electrified rubber rod; such an observation would imply a third type of electrified state or electrification property. Since this never has been observed, we can state with confidence that only two types of electrified state or property exist.

Furthermore, electrical forces are quite distinct from gravitational forces in several respects:

1. Electrical forces between electrified materials are quite apparent even with small pieces of electrified matter, whereas the gravitational force between such small masses is almost

FIGURE 16.4 Two glass rods rubbed with silk repel each other.

negligible and detectable only with the most sensitive types of equipment. Thus electrical forces evidently are intrinsically much stronger than gravitational forces (which are appreciable only when one or both of the masses is huge by laboratory standards).

2. Gravitation is always and only observed as an attractive force, and so we have need for only one kind of mass, positive mass.† Electrical forces are observed to be *either* attractive *or* repulsive, hence the need for two types of electrification property, the vitreous and resinous electrifications of Dufay.

---

*Of course, you know that magnetic forces have this dichotomy also, and so it really is a tribute to these experimentalists that they recognized differences between electrical and magnetic effects.

---

† Even antiparticles such as the positron (a positively charged electron) and the antiproton (a negatively charged proton) have positive mass.

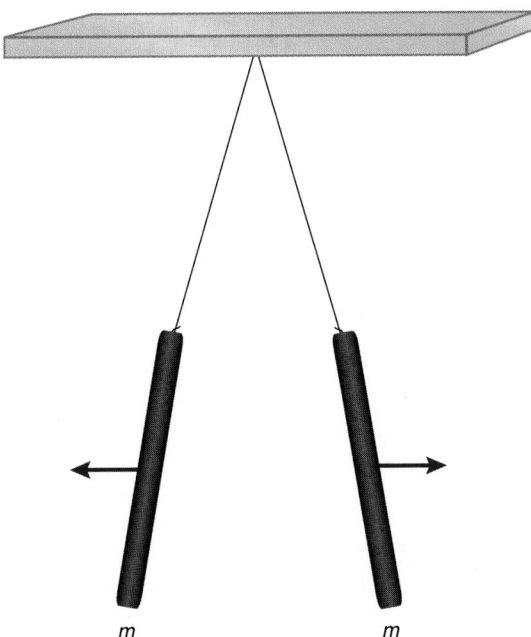

FIGURE 16.5 Two rubber or amber rods rubbed with fur repel each other.

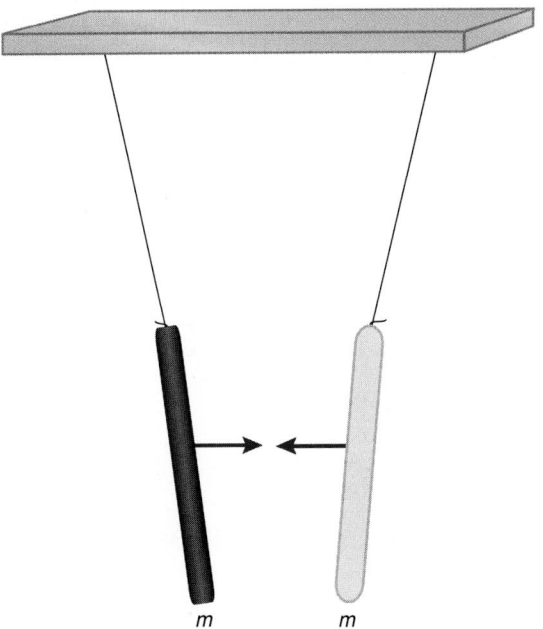

FIGURE 16.6 An electrified glass rod attracts an electrified rubber or amber rod.

The electric and gravitational forces are *similar* in two ways:

1. Both are observed to be *central forces*. They act along the line connecting pointlike materials causing the force.
2. Both are *conservative forces*. The work done by the force around a closed path is zero (equivalently, the work done by the force along a path connecting any two points in space is independent of the path between the two points). We

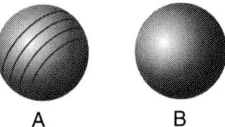

(a) An electrified conducting and an unelectrified conducting sphere.

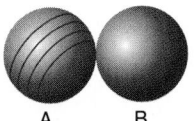

(b) Touch them together.

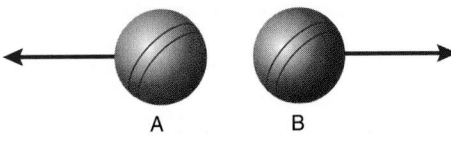

(c) Separate them.

FIGURE 16.7 Transferring charge.

explore the conservative nature of the electrical force in more detail in the next chapter.

It was Benjamin Franklin who introduced the now common definitions associated with the two properties of electrification:

A particle or mass is said to be **positively electrified** if it is repelled by a glass rod that has been freshly rubbed with silk.

All glass rods thus rubbed also have positive electrification, since they individually repel each other.

A particle or mass is said to be **negatively electrified** if it is repelled by rubber or amber that has been freshly rubbed with fur.

Hence the rubber or amber itself has a negative electrification property.

The names for the electrification properties are arbitrary. Dufay called them vitreous and resinous electricity; Franklin called them positive and negative electricity. One could have called the two electrification properties *fat* and *lean* electricity, *hairy* and *bald* electricity, or even *male* and *female* electricity.

The two electrification properties are easily transferred through and shared among materials. Some materials easily let the electrification property move from one place to another; these materials are called **conductors** (from the verb conduct, to transport). With other materials the electrical property lacks mobility (at least over short time intervals); these are called **insulators** (from the verb insulate, to isolate). Insulators also are called **dielectrics**.

If an electrified conducting sphere and an identical but unelectrified conducting sphere (see Figure 16.7a) are brought into contact (Figure 16.7b), and then separated (Figure 16.7c), we find that *both spheres now are electrified* and repel each other. We can measure the repulsive force that exists between the two spheres at a fixed separation. If two more identical conducting but unelectrified spheres are each now brought into contact with one of the identically electrified spheres, as in Figure 16.8, we find that the repulsive force between *any two* of the four electrified spheres, when separated by the same distance, is one-fourth what it was between the original two electrified spheres (see Figure 16.9).

> The condition of electrification thus is quantifiable as measured by the forces.

Combining materials having equal amounts of opposite electrification properties exactly cancels their total effectiveness.

> The two electrification states or properties thus are quantifiable and behave algebraically and arithmetically as scalars.

This makes Franklin's positive and negative terminology much more convenient than the vitreous and resinous characterizations of Dufay and explains why the latter terms quickly were abandoned after Franklin's time. We examine the quantification of the electrification property in Section 16.3. Electrical forces of attraction and repulsion on materials are, of course, vector quantities, as are all forces.

Analogously to gravitation, we could call the electrification properties of matter the positive or negative *electrical masses,** but to avoid confusion with gravitation, the name used for the two electrification properties is **electrical charge.**

> The word **charge** means to endow with electricity (or the electrification property).†

---

*In fact this was the term used by Charles Coulomb in his experiments during the late 18th century (see Section 16.3).

†What we call mass also could be called gravitational charge (endowed with gravitation) or inertial charge (endowed with inertia).

When something *has* charge, it is endowed with one or the other electrification property. We designate the electric charge property of matter symbolically by q (or Q), which may be either a positive or negative *scalar* according to Franklin's convention.

The concept of electric charge is no less nor more abstract than the concept of mass. Mass itself quantifies the property we called inertia or resistance to a change in motion. Both concepts thus are defined operationally by experiment. We just are more familiar with the concept of mass because we think of it as substance, and so regard it as more real and tangible than charge. Mass and charge both are abstractions used to describe the way things in nature behave in certain experiments. We use such technical terms to disguise our fundamental ignorance of what the property really is. The terms mass and charge (and those naming other properties‡) are our ways of describing the response of a simple or complex system to certain types or classes of experiments.

Thus, when we say an object has a **positive charge**, we mean the object has the electrification property that makes it repelled by a glass rod that has been freshly rubbed with silk, as shown in Figure 16.10. When we say an object has a **negative charge**, we mean the object has the electrification property that makes it repelled by a rubber rod that has been freshly rubbed by fur, as shown in Figure 16.11. This is what we mean by the terms positive or negative charge; and, like Humpty Dumpty, *nothing more.* (See quote on page 3.)

From our perspective on fundamental particles at the dawn of the third millennium of the common era, we find that those with nonzero mass§ are the only kinds of fundamental particles that exhibit electric charge (i.e., can have one or the other electrification property), but that not all particles with mass have nonzero total charge (e.g., the neutron has zero total charge).#

---

‡Another example is the *spin* of fundamental particles.

§The particles of light (photons) have zero mass.

#Another distinction between magnetic and electrical phenomena is worth mentioning here: so-called magnetic poles (we define them in Chapter 20) *always* occur in pairs, whereas electric charge is capable of being isolated as either positive or negative.

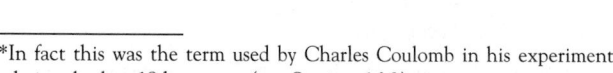

FIGURE 16.8 Touch two more identical conducting spheres to those in Figure 16.7c.

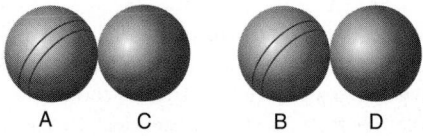

FIGURE 16.9 Any two of the four spheres now repel each other with a force one-fourth as great as that in Figure 16.7c.

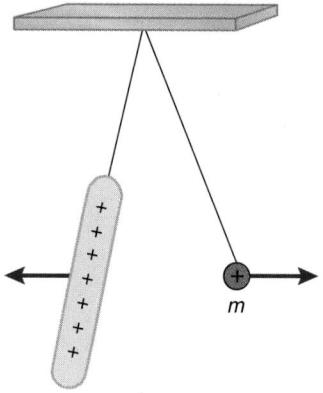

FIGURE 16.10 A positive charge is repelled by a glass rod freshly rubbed with silk.

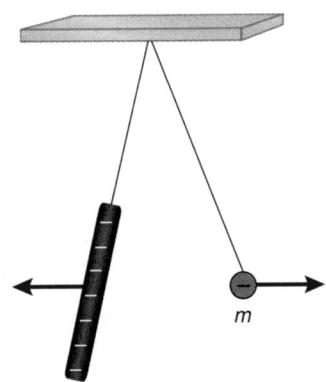

FIGURE 16.11 A negative charge is repelled by a rubber rod freshly rubbed with fur.

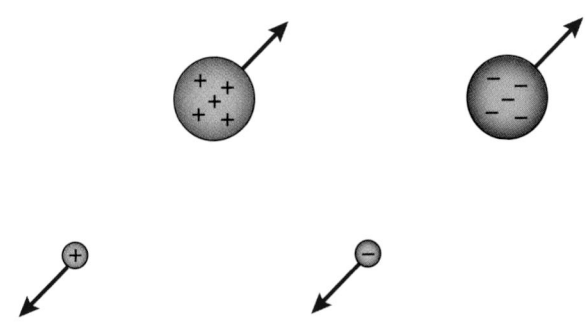

FIGURE 16.12 Like charges repel each other with forces of equal magnitude.

With this terminology, the defining experiments indicate that two particles with the same electrification property (flavor), either both positive or both negative, will feel repulsive electrical forces of equal magnitude on each. If two particles have the same kind of electrification property (flavor) (i.e., both positive or both negative), we say the charges are **like charges**; this does *not* mean that charges are of equal magnitude, only that they have the same type of electrification property. Any like charges produce *repulsive electrical forces* of the same magnitude on each other (see Figure 16.12).

If the electrical charges are of *opposite* flavors (one charge positive, the other negative), the electrical force of the charged particles on each other is attractive (see Figure 16.13). If the two particles have opposite electrification flavors (one positive, one negative), we say the charges are **unlike charges**; these produce attractive electrical forces of equal magnitude on each other.

The electrical forces that two charged masses exert on each other are of equal magnitude and opposite in direction, *regardless* of the quantity of charge each has, in accordance with Newton's third law of motion.

Founding father Franklin also discovered that when glass is rubbed with silk, the glass and silk have opposite charges: the charge of the glass rod is positive (by definition) and the silk negative. Likewise when rubber is rubbed with fur, the rubber and fur have opposite charges: the rubber is negative (by definition) and the fur is positive. The implication of this observation is that the process of electrification by rubbing involves charge transfer, say, from the glass rod to the silk or from the fur to the rubber. We now know that the negatively charged electrons are the charges transferred in the rubbing electrification process. Electrons are transferred from the glass to the silk. The silk acquires a negative charge (because of a surplus of electrons), and the glass acquires an equal magnitude of positive charge (because the positive nuclear charges in the glass slightly outnumber the remaining negative electron charges).

Franklin's many ingenious experiments were the first hint of a profound aspect of the electric charge property, now well demonstrated by experiment:

The total amount of electric charge is *conserved* in any process involving an isolated system; we call this observation **conservation of charge**.

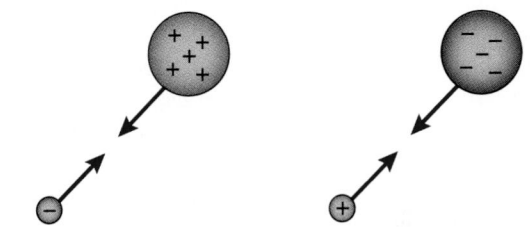

FIGURE 16.13 Unlike charges attract each other with forces of equal magnitude.

Charge can be shuffled around among the constituents of the isolated system, but the total amount of charge at the beginning and end of a process involving an isolated system is the same. These processes may be chemical reactions, particle decays, nuclear transformations, particle creations, collisions, or what not. For example, an isolated, naked neutron (n) is unstable and breaks up or decays into a proton (p), an electron (here designated by β^-), and another exotic particle known as an antineutrino ($\bar{\nu}$). The reaction is written symbolically as

$$n \rightarrow p + \beta^- + \bar{\nu}$$

The total charge initially is zero (the neutron has zero charge); the total charge resulting from the reaction also is zero because the antineutrino has no charge and the charges on the proton and electron are of equal magnitude (absolute value) but opposite sign. We began with zero total charge and ended with zero total charge in this process. No experiment ever has violated the principle of conservation of charge, and so we take it as no more nor less fundamental than energy conservation. Nature apparently just behaves this way.

QUESTION 1

What is the charge of your cat after petting it with rubber gloves? What experiment can you do to see if the cat has the same type of charge when you pet it with your bare hands?

16.2 POLARIZATION AND INDUCTION

Two objects with the same state of electrification, that is, both positive or both negative, repel each other. Two objects with opposite states of electrification (one positive, the other negative) attract each other. But the original experiments involving electricity (the ones with amber and chaff) indicate that any charged object (whether positive or negative) attracts uncharged, electrically neutral matter. This may seem quite surprising; we have to explain how this comes about.

We must be careful to distinguish (via experiments exploring the interaction of the materials with charged glass and rubber rods) whether *attractive* electrical forces (1) arise because the objects have opposite states of electrification, or (2) occur because any charged object will attract an electrically neutral object. An electrically neutral object will be attracted *either* to a positively charged glass rod or to a negatively charged rubber rod. But a charged object will be attracted to one and repelled by the other, in which case the sign of the charge on the unknown object can be determined. If it is repelled by the glass rod, the unknown charge is positive; if it is repelled by the rubber rod, the unknown charge is negative.

> Hence it is *repulsion*, not attraction, that enables us to determine the sign of an unknown charge.

How is it that electrically neutral matter is attracted to a charged object? If, for example, an electrically neutral object is brought close to a positively charged probe, some type of *charge separation* must occur in the neutral object so that negative charges on the neutral object are closer to the positive probe. The total electrical charge on the neutral object is zero; charge separation is the only phenomenon that occurs. The result of the charge separation is (1) an attraction between the positive probe and the closer negative charges on the neutral object, and (2) a repulsion between the more distant positive charges on the neutral object and the positive probe.

The first force is greater in magnitude than the second, and so the total force is attractive; this implies that the electrical force depends inversely on some power of the distance between the charges. We explore the distance dependence in the next section. If a negative probe is used, charge separation occurs in the neutral object so that positive charge is closer to the negative probe, thus causing the attraction.

> Charge separation in electrically neutral materials, caused by the presence of another nearby charged object, is called electrical **polarization**.

It is *not possible* to tell from these experiments which electric charges (positive, negative, or both) in the neutral material are mobile and thus responsible for the charge separation. In fact, one can explain these experiments assuming that *either* positive or negative charges or both are mobile; you might try the explanations as an exercise.

We model electrical polarization phenomena in the following way, assuming the modern atomic and molecular view of matter. In *conducting materials*, we find (using other experiments we discuss in Chapter 20) that the mobile charge carriers are elec-

trons, producing charge separation because of electron migration. When such an uncharged conductor is brought close to a positively charged probe, electrons in the conductor migrate to the region of the conductor closest to the positive probe, leaving behind electrically unbalanced positively charged atoms at the remote end of the conductor. The attractive force on the (closer) negative electrons is greater than the repulsive force on the (more distant) positive charges, resulting in a total force on each that is attractive, as shown in Figure 16.14.

Similarly, if a negative probe is used (see Figure 16.15), some electrons in the conductor are repelled to the remote regions of the conductor. The attractive force on the closer positive charges is stronger than the repulsive force on the more distant negative charges, and the resultant total force is attractive.

So in conducting materials, charge separation occurs over the *macroscopic* distances on the order of the physical size of the conductor itself. Since the conductor as a whole is uncharged, the macroscopic charge separation creates what we call a macroscopic **electric dipole**: charges of equal magnitude, but opposite sign, separated by some distance.

Insulators, however, are by definition materials on which the electrification property (i.e., charge) is not mobile. It is enormously more difficult to transfer charge in them. So how are electrically neutral insulators attracted to charged probes in the vicinity? In this case charge separation—that is, electrical polarization—must occur only on a *local* atomic or molecular scale. But we have great numbers of atoms or molecules to distort slightly. Throughout the interior, this small distortion of charge position for each atom or molecule is balanced out by the same effect on all its nearest neighbors (see Figure 16.16). But at the surface the cancellation is not complete. The excess of positive or negative charge over the near and far surfaces is what gives a resultant total force of attraction toward the probe. Thus an electrically neutral insulator, just as an electrically neutral conductor, is attracted to the charged probe whether the probe is positive or negative.

Let's try yet another experiment.

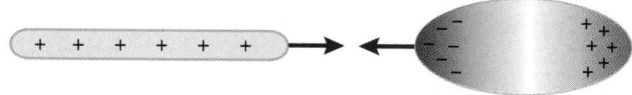

FIGURE 16.14 Attraction of an uncharged conductor to a positively charged object.

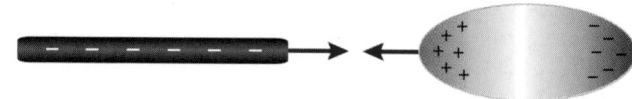

FIGURE 16.15 Attraction of an uncharged conductor to a negatively charged object.

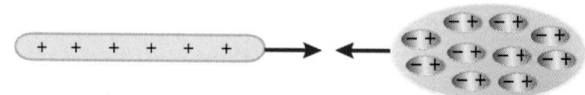

FIGURE 16.16 Attraction of an uncharged insulator to a charged object.

1. We take a negatively charged object and hold it fixed near an uncharged conductor. The uncharged conductor becomes polarized as we explained previously. Electrons migrate in the conductor to the more distant end of the conductor, since they are repelled by the negative probe, as shown in Figure 16.17a.
2. Now we attach another conductor (a wire) to the conductor with the other end of the wire connected to the Earth (a reasonably good conductor itself), as shown in Figure 16.17b. The electrons repelled by the negative probe now can escape to the Earth itself, which is part of the total conducting body. Connecting a conductor to the giant Earth by means of another conductor (typically a wire) is called **grounding** the conductor.
3. If the wire now is removed (see Figure 16.17c), the electrons have no way to get back to the conductor from which they fled, and so the conductor now has a positive charge and is no longer electrically neutral. The negative probe finally can be removed (Figure 16.17d), and the initially neutral conductor remains positively charged and entirely separated. This process is called charging by **induction**.

You might be able to see that the process of electrification by induction operates equally well if begun with a positive probe and a neutral conductor; the result in this case is a negatively charged conductor. Try to explain the corresponding process in your own words.

QUESTION 2

A small object is attracted to an electrified object. Does this imply that the small object is charged opposite to the electrified object? How can you determine whether the small object is charged or not and, if so, whether it is positively or negatively charged?

16.3 COULOMB'S FORCE LAW FOR POINTLIKE CHARGES: THE QUANTIFICATION OF CHARGE

Between 1785 and 1787, Charles Augustin Coulomb (1736–1806) performed a critical and difficult series of experiments on electrified materials using a sensitive torsion balance that he invented for measuring small forces.* He discovered that, as was true with gravitation, the magnitude of the mutual electrical force of attraction or repulsion on each of two small, pointlike, electrified objects varied inversely as the square of the distance of their separation. The magnitude of the electrical force on each pointlike charge is an *inverse square law*, just like gravitation:

$$F_{elec} \propto \frac{1}{r^2}$$

Modern experiments have verified the inverse square nature of the electrical force to better than 16 significant figures for the exponent of the distance.[†] By means of experiments similar to those we described earlier with identical conducting spheres, Coulomb discovered that the electrification property (charge) is quantifiable in terms of the forces that electrified objects exert on each other. Coulomb also found that the magnitude of the electrical force that two pointlike charges q and Q exert on each other is proportional to the product of the magnitudes[‡] of the charges:

$$F_{elec} \propto \frac{|q||Q|}{r^2} \quad (16.1)$$

---

*A similar balance was used by Henry Cavendish during the 18th century to accurately measure gravitational forces between small masses and thus determine the numerical value of the universal gravitational constant G.
[†] See Investigative Projects 1 and 2 at the end of this chapter.
[‡] The word magnitude, when referring to electric charge, means *absolute value*, rather than the magnitude of a vector, though both quantities are nonnegative scalars. Electric charge is a scalar, not a vector.

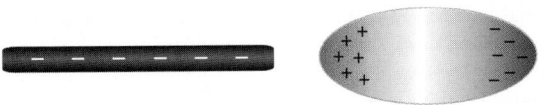

(a) An uncharged conductor placed near a charged object.

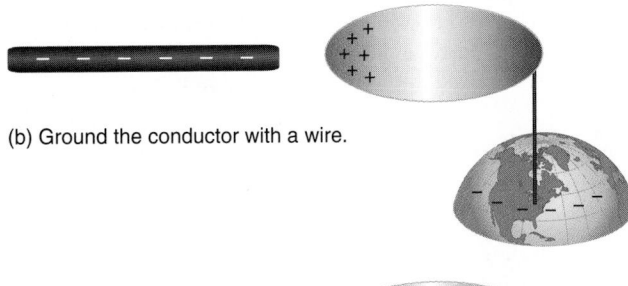

(b) Ground the conductor with a wire.

(c) Remove the grounding wire.

(d) Remove the charged probe and the isolated conductor remains charged.

FIGURE 16.17 Charging by induction.

Aside from his notable researches into electrical forces, Coulomb also did significant work on the mechanical properties of materials and engineering consulting work with water systems, canals, and harbors.

The direction of the force on each charged mass is determined using the observation that like charges repel, unlike charges attract each other.

The forces that *any* two pointlike charges exert on each other are of equal magnitude but in opposite directions, in accordance with Newton's third law. The force on each charge has the same magnitude even if q and Q are of different magnitude, as shown in Figure 16.18.

The units for quantifying the electric charge property of matter depend on the units used for measuring force and distance. In the SI system of units, the unit of charge is defined in terms of *charge flow* or **electric current**. The SI unit for electrical current is the *ampere* (A) and will be defined later more precisely (in Chapter 20). A current of one ampere means that one **coulomb** (C) of charge flows past a specified region during one second. The coulomb turns out to be a very big charge unit; smaller divisions, such as millicoulombs (mC) (= 10^{-3} C), microcoulombs (μC) (= 10^{-6} C), and nanocoulombs (nC) (= 10^{-9} C) also are employed.*

If we measure electric charge using the SI unit, the coulomb, the value of the proportionality constant in Coulomb's force law on each of two pointlike charges, Equation 16.1, is found to be

$$8.987\ 552\ 425 \times 10^9\ \text{N·m}^2/\text{C}^2$$

For reasons of convenience that are not apparent at this juncture,[†] we write this proportionality constant in SI units in a peculiar way as

$$8.987\ 552\ 425 \times 10^9\ \text{N·m}^2/\text{C}^2 \equiv \frac{1}{4\pi\varepsilon_0} \qquad (16.2)$$

whose value usually is taken to be approximately

$$9.00 \times 10^9\ \text{N·m}^2/\text{C}^2$$

unless very precise calculations must be made. The constant ε_0 thus has the value (this value is exact)[‡]

$$\varepsilon_0 = \frac{1}{4\pi\ (8.987\ 552\ 425 \times 10^9\ \text{N·m}^2/\text{C}^2)}$$

$$\equiv 8.854\ 187\ 817\ ... \times 10^{-12}\ \text{C}^2/(\text{N·m}^2) \qquad (16.3)$$

For abstruse reasons we will not pursue, ε_0 is called the **permittivity of free space**.

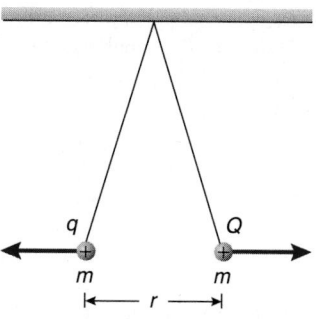

FIGURE 16.18 **The electrical force on each of two charges satisfies Newton's third law.**

Therefore, in the SI unit system, Coulomb's law for the *magnitude* of the force that two pointlike charges exert on each other is

$$F_{\text{elec}} = \frac{1}{4\pi\varepsilon_0}\frac{|q|\,|Q|}{r^2} \qquad (16.4)$$

Coulomb's law is very similar in form to the gravitational force law that gives the magnitude of the force that two pointlike masses m and M exert on each other:

$$F_{\text{grav}} = G\frac{Mm}{r^2}$$

If more than two pointlike charges are present, as in Figure 16.19, experiments indicate that the electrical force on any specific charge is found from the vector sum of the forces that *each* of the *other* charges exerts on the specific charge as if *only* that pair of charges were present; see Figure 16.20. The *total electrical force* on the given charge then is the vector sum of the electrical forces caused by the other charges, calculated as if each acted alone, as indicated in Figure 16.21.

This result is known as the **principle of superposition**. The principle is applicable as long as the charges $Q_1, Q_2, \ldots$ exerting forces on q are fixed in their positions—that is, *static*.

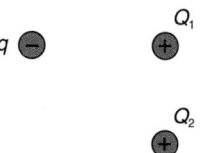

FIGURE 16.19 **Three pointlike charges.**

---

*The coulomb is such a large unit of charge that it is almost as awkward as assessing your tuition in units of the Gross Domestic Product (GDP), about a nanoGDP = 10^{-9} GDP.

†The reason has to do with making one of the fundamental equations of electromagnetism, Gauss's law for the electric field, look prettier. We examine Gauss's law in Section 16.11.

‡The effect of defining the meter to make the speed of light exactly $2.997\ 924\ 58 \times 10^8$ m/s ripples into electromagnetism and makes the value of ε_0 exact as well; this connection may not be apparent to you at this juncture, so do not fret about it needlessly.

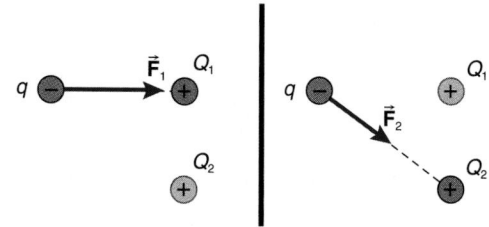

FIGURE 16.20 **To find the force on q, find the force that each of the other charges individually exerts on q.**

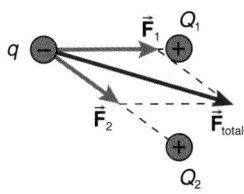

FIGURE 16.21 The total force on q is the vector sum of the individual forces on q.

PROBLEM-SOLVING TACTICS

16.1 When using Coulomb's law for the force between pointlike charges, be sure to express the distance between the charges in meters, not centimeters, and the charge in coulombs. Equation 16.4 is appropriate for SI units, which means the distance must be in the SI unit, meters, and the charge in coulombs.

16.2 To avoid many confusing problems with signs, employ the following procedure when using Coulomb's law. Calculate the *magnitude* of the force (a *positive scalar* quantity) using Coulomb's force law, Equation 16.4. Sketch the situation. Then determine the direction of the force on each charge after examining the types of the charges involved:

- if the charges are *both* positive *or both* negative (so-called like charges), the force on each is repulsive; or

- if the charges are of *opposite* flavors (unlike charges), the force on each is attractive.

Indicate on your sketch the direction of the forces on the charges. Finally, if appropriate, introduce a Cartesian coordinate system and write the forces in standard Cartesian vector form.

QUESTION 3

Two cats named Skimbleshanks and Mr. Mistoffolees are petted vigorously.* Can the electrical force of the cats on each other be calculated using Coulomb's law? Explain why or why not. What assumption must be made to calculate the approximate magnitude of the electrical force of the cats on each other? Why is this an approximation to the actual force?

EXAMPLE 16.1

A pointlike charge, $-2.00 \ \mu C$, is located 15.0 cm from another pointlike charge, $+3.50 \ \mu C$.

a. What is the magnitude of the electrical force that each exerts on the other?
b. Sketch the situation and indicate the direction of the electrical force on each charge.

Solution
a. Since the charges are pointlike, use Coulomb's law (Equation 16.4) for the magnitude of the force that each exerts

on the other. Keep in mind Problem-Solving Tactic 16.1—use meters for the distance and coulombs for the charge:

$$F_{\text{elec}} = \frac{1}{4\pi\varepsilon_0} \frac{|q||Q|}{r^2}$$

$$= (9.00 \times 10^9 \ \text{N·m}^2/\text{C}^2)$$

$$\times \frac{|-2.00 \times 10^{-6} \ \text{C}||3.50 \times 10^{-6} \ \text{C}|}{(0.150 \ \text{m})^2}$$

$$= 2.80 \ \text{N}$$

b. Since the charges are unlike charges, the force that each exerts on the other is an attractive force, as indicated in Figure 16.22. Notice that the force each exerts on the other satisfies Newton's third law.

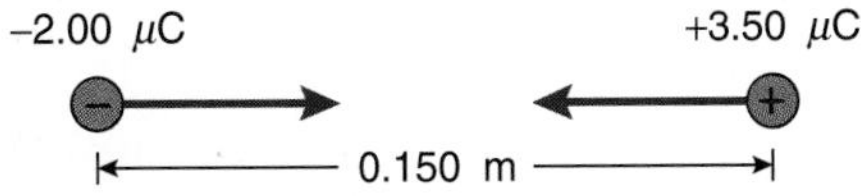

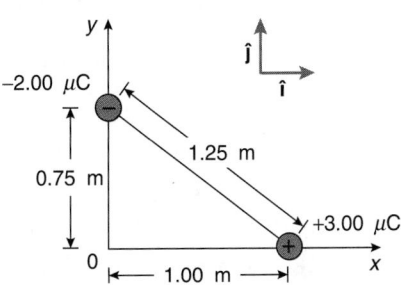

FIGURE 16.22

EXAMPLE 16.2

Two pointlike charges are located as shown in Figure 16.23.

FIGURE 16.23

a. Find the electrical force that each charge exerts on the other and express these forces in terms of the indicated coordinate system.
b. Do these forces represent a Newton's third law force pair?

Solution
a. The magnitude of the electrical force that each charge exerts on the other is found by using the absolute values of the charges and substituting into Coulomb's law, Equation 16.4. Distances must be expressed in meters and the charges in coulombs (Problem-Solving Tactic 16.1):

$$F = \frac{1}{4\pi\varepsilon_0} \frac{|q||Q|}{r^2}$$

$$= (9.00 \times 10^9 \ \text{N·m}^2/\text{C}^2)$$

$$\times \frac{|-2.00 \times 10^{-6} \ \text{C}||3.00 \times 10^{-6} \ \text{C}|}{(1.25 \ \text{m})^2}$$

$$= 3.46 \times 10^{-2} \ \text{N}$$

---
*The names are from T. S. Eliot, *Old Possum's Book of Practical Cats* (Faber and Faber, London, 1939).

Following Problem-Solving Tactic 16.2, the pair of charges are unlike charges (one positive, one negative), and so the forces are attractive and are directed as indicated in Figure 16.24.

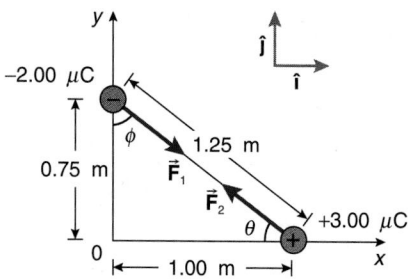

FIGURE 16.24

The forces are expressed in terms of the indicated coordinate axes by writing the force vectors in terms of the Cartesian unit vectors:

$$\vec{F}_1 = (F \sin \phi)\hat{i} - (F \cos \phi)\hat{j}$$
$$= (3.46 \times 10^{-2} \text{ N})(0.800)\hat{i} - (3.46 \times 10^{-2} \text{ N})(0.600)\hat{j}$$
$$= (2.77 \times 10^{-2} \text{ N})\hat{i} - (2.08 \times 10^{-2} \text{ N})\hat{j}$$

Likewise

$$\vec{F}_2 = (-F \cos \theta)\hat{i} + (F \sin \theta)\hat{j}$$
$$= (-3.46 \times 10^{-2} \text{ N})(0.800)\hat{i} + (3.46 \times 10^{-2} \text{ N})(0.600)\hat{j}$$
$$= (-2.77 \times 10^{-2} \text{ N})\hat{i} + (2.08 \times 10^{-2} \text{ N})\hat{j}$$

b. Notice that $\vec{F}_1 = -\vec{F}_2$. The forces form a Newton's third law force pair because they represent the forces that each charge exerts on the other. The forces are of equal magnitude and opposite direction and act on different systems.

STRATEGIC EXAMPLE 16.3

Three charges are located at fixed positions as indicated in Figure 16.25.

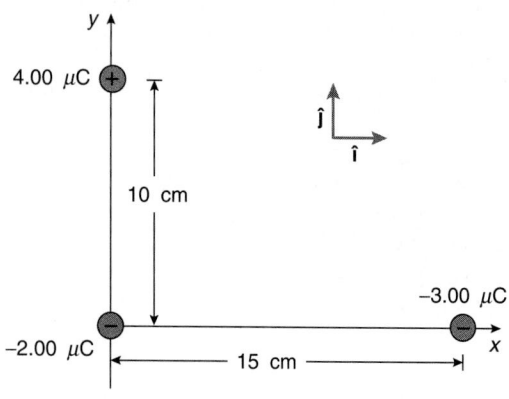

FIGURE 16.25

a. **Find the total electrical force on the charge at the origin.**
b. **Find the magnitude of the total force on this charge.**
c. **Find the angle that the total force makes with the positive x-axis.**

Solution

a. Use Coulomb's law and the principle of superposition. You *must* use SI units to use Coulomb's law as given by Equation 16.4. Hence the *distances must be expressed in meters and the charges in coulombs* (Problem-Solving Tactic 16.1).

1. First, find the magnitude of the force $\vec{F}_1$ that the charge out along the x-axis exerts on the charge at the origin, pretending that the other charge (the one out along the y-axis) is not present; see Figure 16.26 The magnitude of the force is found by using Coulomb's law (Equation 16.4):

$$F_1 = \frac{1}{4\pi\varepsilon_0} \frac{|q||Q|}{r^2}$$

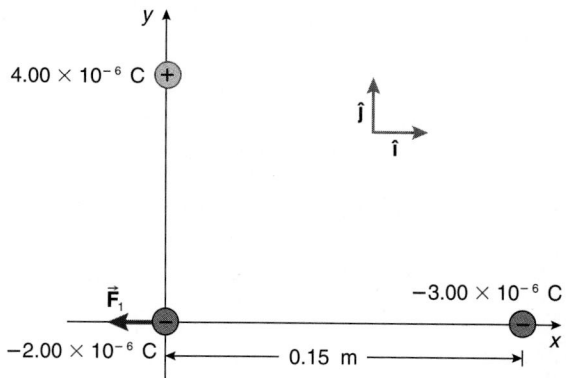

FIGURE 16.26

We take q to be the charge at the origin and Q to be the charge out along the x-axis. Making the substitutions, you get

$$F_1 = (9.00 \times 10^9 \text{ N·m}^2/\text{C}^2)$$
$$\times \frac{|-2.00 \times 10^{-6} \text{ C}||-3.00 \times 10^{-6} \text{ C}|}{(0.15 \text{ m})^2}$$

$$= 2.4 \text{ N}$$

Since the charge out along the x-axis and the charge at the origin are like charges (both are negative), the force on each is repulsive; hence, the force on the charge at the origin is directed toward $-\hat{i}$. Thus the force that the charge out along the x-axis exerts on the charge at the origin is

$$\vec{F}_1 = (-2.4 \text{ N})\hat{i}$$

2. Next, find the magnitude of the force $\vec{F}_2$ exerted by the charge out along the y-axis on the charge at the origin, pretending that the other charge (the charge out along the x-axis) is not present, as shown in Figure 16.27. The magnitude of the force is found using Coulomb's law (Equation 16.4):

$$F_2 = \frac{1}{4\pi\varepsilon_0} \frac{|q||Q|}{r^2}$$

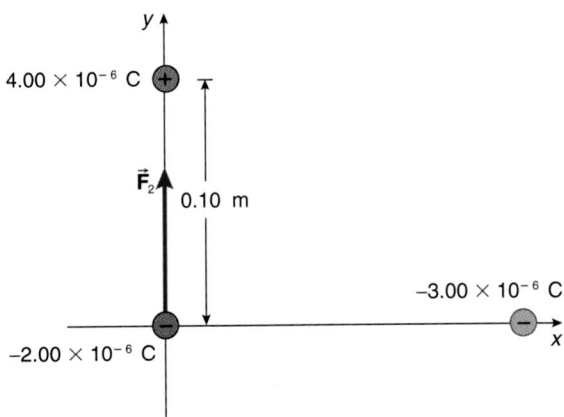

FIGURE 16.27

where we again let q be the charge at the origin and now let Q be the charge out along the y-axis. Making the substitutions,

$$F_2 = (9.00 \times 10^9 \ \text{N} \cdot \text{m}^2/\text{C}^2)$$
$$\times \frac{\left| -2.00 \times 10^{-6} \ \text{C} \right| \left| 4.00 \times 10^{-6} \ \text{C} \right|}{(0.10 \ \text{m})^2}$$
$$= 7.2 \ \text{N}$$

Since the charge out along the y-axis and the charge at the origin are unlike charges (one is positive, the other negative), the force of these two charges on each other is attractive. Thus the force on the charge at the origin is directed toward $+\hat{\jmath}$. Hence

$$\vec{F}_2 = (7.2 \ \text{N})\hat{\jmath}$$

3. The total force on the charge at the origin is the *vector sum* of the forces $\vec{F}_1$ and $\vec{F}_2$ (see Figure 16.28):

$$\vec{F}_{\text{total}} = \vec{F}_1 + \vec{F}_2$$
$$= (-2.4 \ \text{N})\hat{\imath} + (7.2 \ \text{N})\hat{\jmath}$$

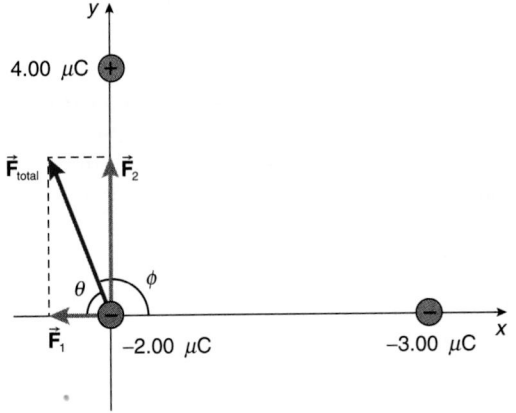

FIGURE 16.28

b. The total force is of magnitude

$$F_{\text{total}} = [(-2.4 \ \text{N})^2 + (7.2 \ \text{N})^2]^{1/2}$$
$$= 7.6 \ \text{N}$$

c. The total force makes an angle θ with the *negative x-axis* that is found from (see Figure 16.28)

$$\tan \theta = \frac{\left| F_{y \ \text{total}} \right|}{\left| F_{x \ \text{total}} \right|}$$
$$= \frac{7.2 \ \text{N}}{2.4 \ \text{N}}$$
$$= 3.0$$

or

$$\theta = 72°$$

The angle ϕ that the total force makes with the *positive x-axis* is

$$\phi = 180° - \theta$$

or

$$\phi = 108°$$

16.4 CHARGE QUANTIZATION

I believe there are
15,747,724,136,275,002,577,605,653,961,181,555,468,
044,717,914,527,116,709,366,231,425,076,185,631,031,296
protons in the universe, and the same number of electrons.
Arthur Eddington (1882–1944)*

The development of chemistry during the 19th century, particularly the concept of valence associated with chemical bonding, the rise of the atomic view of matter with developments in the kinetic theory of gases (Chapter 14), as well as the discovery of charged particle radiation and radioactivity in 1896, led to a search for an elementary, smallest quantity of charge. The search finally bore fruit with the work of Robert A. Millikan (1868–1953) in an epic series of experiments begun in 1909.

By means of tedious, eye-straining microscopic observations of individual tiny oil droplets, charged by squirting (rubbing) them through a perfume bottle atomizer,[†] Millikan discovered that every tiny droplet carried only a positive or negative *integral* multiple of a small, elemental amount of charge. Millikan was awarded the 1923 Nobel prize in physics for this work. We now call the magnitude of this elemental quantity of the electrification property (i.e., charge) the **fundamental unit of electric charge**, designated[‡] e; the unit of elemental charge e is considered intrinsically positive. The value of the fundamental unit of charge, in SI units, is approximately

$$e = 1.602\,177 \times 10^{-19} \ \text{C} \qquad (16.5)$$

The Philosophy of Physical Science (Macmillan, New York, 1939), page 170.
[†]He could change the charge on each drop by using radiation from radium.
[‡]Do not confuse the fundamental unit of charge e with the base of natural logarithms, $e = 2.718\,282\ldots$.

Coming from a classic midwestern farm background, Millikan wrote a widely used physics text and from there took on the difficult challenge of determining if an elemental quantity of charge existed.

This elemental amount of charge e is the magnitude of the charge on both the electron and proton.

> All free electric charges in nature are *integral multiples* of the fundamental unit of electric charge e. We say charge is **quantized**, meaning that only certain values of charge are permitted or allowed to exist.

Charge comes in integral multiples of the discrete and elemental bit of charge e. Any isolated particle in nature has a charge q that is

$$q = ne \qquad (16.6)$$

where n is a positive or negative integer, known as the **charge quantum number**. The electron has a charge $q = -e$, and the proton has a charge $q = +e$, so their charge quantum numbers are -1 and $+1$ respectively. Particles or masses with zero charge (such as the neutron) have a charge quantum number $n = 0$. Masses with zero total electric charge are said to be electrically **neutral**. The proton and electron therefore can be viewed, respectively, as having one charge unit e added to or subtracted from a state of zero charge. The equality of the magnitudes of the charges on the proton and electron thus comes about naturally from this quantum viewpoint.

Contemporary physics views the proton and neutron (and other particles called hadrons) as *composite* particles, made of more elementary particles called quarks. Quarks have *fractional* units of the fundamental unit of charge: $\pm e/3$ and $\pm 2e/3$, thus seeming to violate charge quantization. In this model the proton consists of two quarks with charge $+2e/3$, whimsically called *up* quarks, and one quark with charge $-e/3$, called a *down* quark, for a total charge of

$$\frac{2}{3}e + \frac{2}{3}e - \frac{1}{3}e = e$$

The neutron is composed of one up quark with charge $+2e/3$ and two down quarks (each with charge $-e/3$), for a total charge of

$$\frac{2}{3}e - \frac{1}{3}e - \frac{1}{3}e = 0 \text{ C}$$

Isolated quarks have *never* been observed and there are theoretical reasons to think they never will be observed. Thus the statement that *isolated* particles have integral multiples of the fundamental unit of charge still holds; the quark model does not violate the charge quantization hypothesis.

The amount of charge e is very small. For macroscopic quantities of charge, the charge quantum number may be so huge that the discrete nature of charge is undetectable and the quantity of charge is considered essentially continuous in the mathematical sense; mass has similar properties on the microscopic and macroscopic scales.

 PROBLEM-SOLVING TACTIC

16.3 Remember that we treat the fundamental unit of charge $e = 1.602 \times 10^{-19}$ C as positive, and so the charge on the proton is $+e$, while that on the electron is $-e$.

QUESTION 4
Explain the difference between the terms charge quantification and charge quantization.

EXAMPLE 16.4

What is the charge quantum number of a particle with a charge of $+1.00 \ \mu$C?

Solution
Since charge is quantized, use Equation 16.6:

$$q = ne$$

The charge quantum number n is

$$
\begin{aligned}
n &= \frac{q}{e} \\
&= \frac{1.00 \times 10^{-6} \text{ C}}{1.602 \times 10^{-19} \text{ C}} \\
&= 6.24 \times 10^{12}
\end{aligned}
$$

This is a huge charge quantum number and demonstrates why charge quantization typically is not detectable with charges produced by macroscopic methods such as rubbing or induction.

EXAMPLE 16.5

It is interesting to compare the magnitude of the repulsive electrical force that two electrons exert on each other when they are a distance r apart with the magnitude of the attractive gravitational force of the two for each other at the same separation. Calculate the numerical value of this ratio.

Solution
The magnitude of the electrical force is found using Coulomb's law, Equation 16.4:

$$F_{\text{elec}} = \frac{1}{4\pi\varepsilon_0} \frac{|-e||-e|}{r^2}$$

The magnitude of the gravitational force between the two electrons is

$$F_{grav} = G\,\frac{m_{electron}m_{electron}}{r^2}$$

Thus the ratio of the magnitudes of the forces is

$$\frac{F_{elec}}{F_{grav}} = \frac{1}{4\pi\varepsilon_0}\,\frac{e^2}{Gm_{electron}^2}$$

Notice that the ratio is *independent* of the separation of the two electrons. The ratio thus is a measure of the relative intrinsic strength of the electric and the gravitational forces.

Substituting numerical values for the various quantities, you find the ratio is

$$\frac{F_{elec}}{F_{grav}} = 4.17 \times 10^{42}$$

This incredibly large number means that the electrical force repelling the electrons is about 10^{42} times stronger than their attractive gravitational force. If we represent the gravitational force vector by an arrow a mere 1.00 cm long, as in Figure 16.29, the arrow representing the electrical force is 10^{42} cm long; 10^{42} cm = 10^{40} m, *far, far greater* than the size of the observable universe ($\sim 10^{26}$ m)! Needless to say, we cannot draw this arrow to the appropriate scale in Figure 16.29.

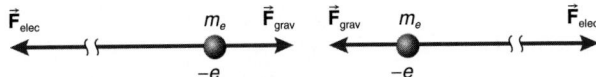

FIGURE 16.29

In the hydrogen atom, if you compute the ratio of the magnitude of the electrical force of attraction that a proton and an electron exert on each other to the magnitude of their mutual gravitational force of attraction, the result is 2.28×10^{39}, still so large that we can comfortably neglect the gravitational force. As a result, *the gravitational force is irrelevant in most atomic and molecular physics.* It is the electrical force that dominates the interactions.

EXAMPLE 16.6

The Bohr model of the electrically neutral hydrogen atom consists of an electron orbiting a single proton. The proton is the only charged constituent of the nucleus of this atom. Actually *both* the electron *and* the nucleus orbit the center of mass of the two-particle system. However, since the mass of the proton is 1836 times that of the electron, the center of mass of the system is essentially coincident with the proton. Thus we say the electron orbits the proton (thinking of the latter as relatively fixed in position). The electron has a number of different possible (or allowed) orbital paths around the central nucleus, but let us consider the electron to be in the orbit closest to the nucleus. This orbit has an average radius of 5.29×10^{-11} m.

a. Find the magnitude and direction of the electrical force on the electron in the hydrogen atom when it is in this orbit.
b. Calculate the magnitude of the acceleration of the electron in this orbit, and compare this magnitude with that of the local acceleration due to gravity ($g = 9.81$ m/s^2).

Solution

a. The *magnitude* of the force on the electron is found using Coulomb's law, Equation 16.4:

$$F = \frac{1}{4\pi\varepsilon_0}\,\frac{|q||Q|}{r^2}$$

Since the electron and the proton each have the same *magnitude* of charge (the fundamental unit of charge e), the magnitude of the force is

$$F = \frac{1}{4\pi\varepsilon_0}\,\frac{e^2}{r^2}$$

Substituting numerical values for the various quantities, you obtain

$$F = \frac{(9.00 \times 10^9 \text{ N·m}^2/\text{C}^2)\,(1.602 \times 10^{-19} \text{ C})^2}{(5.29 \times 10^{-11} \text{ m})^2}$$
$$= 8.25 \times 10^{-8} \text{ N}$$

This is the magnitude of the force *on* the electron *by* the proton (as well as the force of the electron on the proton, thanks to Newton's third law). The force on the electron is directed radially toward the proton because the two are unlike charges.

b. The electrical force is essentially the *only* relevant force acting on the electron since the gravitational force between them is so small; see Example 16.5. Newton's second law enables you to find the magnitude of the acceleration of the electron. In terms of the magnitudes of the total force and the acceleration,

$$F_{elec} = m_e a$$

The acceleration is parallel to the total force and so is directed toward the nucleus. The magnitude of the acceleration thus is

$$a = \frac{F_{elec}}{m_e}$$

Use the result of part (a) for the electrical force on the electron and substitute for the mass of the electron; you obtain

$$a = \frac{8.25 \times 10^{-8} \text{ N}}{9.11 \times 10^{-31} \text{ kg}}$$
$$= 9.06 \times 10^{22} \text{ m/s}^2$$

Dividing by the magnitude of the local acceleration due to gravity ($g = 9.81$ m/s^2), you find the ratio of the accelerations is

$$\frac{a}{g} = \frac{9.06 \times 10^{22} \text{ m/s}^2}{9.81 \text{ m/s}^2}$$
$$= 9.24 \times 10^{21}$$

The magnitude of the acceleration of the electron in the hydrogen atom thus is on the order of 10^{22} so-called "gs." That is *some* acceleration!

16.5 THE ELECTRIC FIELD OF STATIC CHARGES

The Coulomb force law between charges is similar in form to the gravitational force law. We exploit the similarity by recalling several features of the gravitational force that we studied in Chapter 6.

When we discussed gravitational forces in Chapter 6, we saw that a mass m experiences a gravitational force because of the presence of other masses. The effect of these other masses on m is conveyed by a gravitational field $\vec{g}$ created in space by the other masses. The gravitational field $\vec{g}$ at a point in space was defined as the gravitational force per unit mass at the point in question:

$$\vec{F}_{\text{grav on } m} \equiv m\vec{g} \tag{16.7}$$

The gravitational field is the free-fall acceleration due to gravity at the point in question. In Chapter 6 we also found that the specific expression for the gravitational field depends on the shape and distribution of the masses creating the field, *not* the mass m placed *in* the field. The gravitational fields of various mass distributions were summarized in Table 6.1. The only condition subtly attached to the definition of the gravitational field, Equation 16.7, is the assumption that the presence of m *does not alter the shape or distribution of the masses creating the field.* This is typically the case in most gravitational situations,* because in many instances the mass m that we place in the gravitational field is much smaller than the mass that creates the field into which m is placed. For example, the masses m that we place in the gravitational field of the Earth, such as rocks and rockets, humans and heros, beams and beans, even the Moon, have masses that are much less than that of the Earth. These small masses do not affect the distribution of mass on or within the Earth.† Likewise, planetary masses m in the gravitational field of the Sun are very small compared with the mass of the Sun, and so the presence of a planetary mass does not affect the mass distribution within the Sun.

Analogously to gravitation, we introduce an **electric field $\vec{E}$** to convey the electrical force.

The electric field $\vec{E}$ at a point in space is a vector defined via

$$\boxed{\vec{F}_{\text{elec on } q} \equiv q_{\text{test}}\vec{E}} \tag{16.8}$$

where $\vec{F}_{\text{elec on } q}$ is the electrical force on q_{test} at the point where it is located.

The electric field at the point in space where q_{test} is placed is created by other charges, not q_{test} itself, and so the value of the field at a point in space does not depend on the value of the charge placed in the field. We consider the charges that create the field to be fixed in space, and so we call them **static charges** and the field $\vec{E}$ an **electrostatic field**. The field created by these static charges depends on their locations and distributions in space (just as the specific form for the gravitational field depends on the shape and distribution of the mass creating the field, as indicated in Table 6.1).

The reason for introducing the field concept is the same as with the gravitational field: we can talk about the electrical effects of the static charges creating the field, without the presence of another charge (q_{test}) complicating the picture. We think of the field as conveying or transmitting the force to the mass or charge placed in the gravitational or electric field respectively. In the gravitational case, we could talk about the gravitational field of the Earth (the gravitational force per kilogram, or equivalently, the local acceleration due to gravity) without necessarily having to place a mass m in the field.

What is the reason for the subscript "test" on q_{test} in the defining Equation 16.8? Just as in gravitation, we need to ensure that q_{test} is small enough so that its presence does not alter the locations or distribution of the other charges that are producing the field. Since charges on conductors readily move, *this condition is not as easily ensured with electrical phenomena* as it is with gravitation. But we can imagine q_{test} to be a very small magnitude of charge that innocently tests or determines the value of the field at a point in space through the force on it at that location.‡

> From the defining Equation 16.8, we see that the electric field at a point in space is the force per coulomb (force on one coulomb) at that point. Thus the SI units of the electric field $\vec{E}$ are N/C.

However, the coulomb is such a large quantity of charge that, in practice, q_{test} almost never can be considered on the order of a coulomb.

> For our treatment of electrical phenomena, we will always assume that the presence of a test charge q_{test} in an electric field $\vec{E}$ does not alter the distribution or motion of the charges creating the field.

Thus all electric fields we consider in this chapter are electrostatic fields. This assumption ensures that the force on a charge q_{test} placed in an electric field $\vec{E}$ (created in space by other charges, not q_{test} itself) can be found using Equation 16.8. So we will eliminate the subscript on the test charge placed in the electric field and simply call it $q \equiv q_{\text{test}}$.

> It is important to realize that the charge q placed in the field need not be fixed or static but may accelerate in response to the electrical force it experiences. It is all the many *other* charges that create the field that are static, not necessarily the charge placed in the field of these static charges.

> Just as the gravitational field depends on the shape and distribution of the masses producing the field (e.g., the shape and distribution of mass within the Earth), the electric field $\vec{E}$ depends on the specific arrangement of the charges that create the field.

---

*Binary stars with small spatial separations are exceptions.
†The Moon does cause very small tidal bulges on the fluid oceans of the Earth. The heights of these familiar tidal bulges are approximately 2 m.

---

‡It is the same with gravitation. We *measure* the gravitational field (the force per kilogram or the local acceleration due to gravity) only by placing a test mass m in the gravitational field and measuring the force on or the acceleration of m.

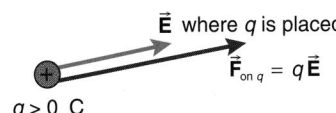

FIGURE 16.30 A positive charge placed in an electric field experiences an electrical force parallel to the field.

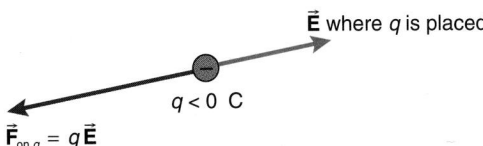

FIGURE 16.31 A negative charge placed in an electric field experiences a force antiparallel to the field.

We will later see how to calculate the electric field for certain distributions of charges. Whatever the field is, the force on *another* charge q, placed *in* the field created by these other charges, is found using Equation 16.8 (as long as the presence of q does not affect the distribution of charge creating the field in which q is placed*).

In the gravitational case (Equation 16.7), the gravitational force on a mass m is *always parallel* to the gravitational field because there is only one type of mass: positive mass. For the electrical case, the direction of the force relative to the field depends on the sign of the charge q placed in the field. If the charge q is *positive*, the electrical force on q is *parallel* to the field $\vec{E}$; as shown in Figure 16.30, the right-hand side of Equation 16.8 then is just a positive scalar times the vector $\vec{E}$. On the other hand, if the charge q is *negative*, the right-hand side of Equation 16.8 is a negative scalar times the vector $\vec{E}$; thus the electrical force on a negative charge q is directed *antiparallel* to $\vec{E}$, as shown in Figure 16.31.

PROBLEM-SOLVING TACTIC

16.4 The algebraic sign of the charge q is very important in Equation 16.8. The sign of the charge q determines whether the force is directed parallel (q positive) or antiparallel (q negative) to the direction of the electric field.

QUESTION 5

The gravitational field has two interpretations: (1) the force per kilogram at a point in space, and (2) the local acceleration of gravity at the point in space. The electric field is defined as the electrical force per coulomb at a point in space. Can the electric field also be interpreted as an acceleration of the charged mass caused by electrical forces? Explain your answer.

EXAMPLE 16.7

a. A proton finds itself in a uniform electric field of magnitude 125 N/C directed as indicated in Figure 16.32. Find the force on the proton.

---

*If the presence of q *does* alter the distribution of charges creating the field, then q experiences the field produced by the *new arrangement* of the original charges. For our purposes, we will never consider such effects.

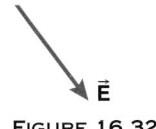

FIGURE 16.32

b. Repeat the problem if the charge placed in the field is an electron.

Solution

Introduce the coordinate system shown in Figure 16.33. With this coordinate system, the electric field vector is

$$\vec{E} = (-125 \text{ N/C})\hat{\jmath}$$

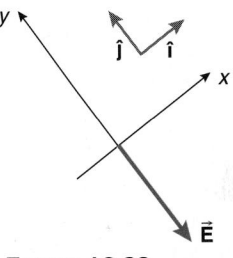

FIGURE 16.33

a. The force on the proton is found using Equation 16.8:

$$\vec{F} = q\vec{E}$$

where $q = +1.602 \times 10^{-19}$ C. Make the appropriate substitutions:

$$\vec{F} = (+1.602 \times 10^{-19} \text{ C})[(-125 \text{ N/C})\hat{\jmath}]$$
$$= (-2.00 \times 10^{-17} \text{ N})\hat{\jmath}$$

Notice that the force on the positively charged proton is *parallel* to the direction of the electric field in which it is placed (see Figure 16.34), in keeping with Equation 16.8 and Problem-Solving Tactic 16.4.

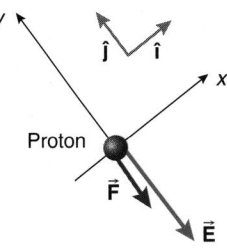

FIGURE 16.34

b. The force on an electron also is found using Equation 16.8:

$$\vec{F} = q\vec{E}$$

But now $q = -1.602 \times 10^{-19}$ C and $\vec{E}$ is unchanged: $(-125 \text{ N/C})\hat{j}$. With these substitutions, you find

$$\vec{F} = (-1.602 \times 10^{-19} \text{ C})[(-125 \text{ N/C})\hat{j}]$$
$$= (2.00 \times 10^{-17} \text{ N})\hat{j}$$

Notice that the force on the negatively charged electron is *antiparallel* to the direction of the electric field in which it is placed (see Figure 16.35), in keeping with Equation 16.8 and Problem-Solving Tactic 16.4.

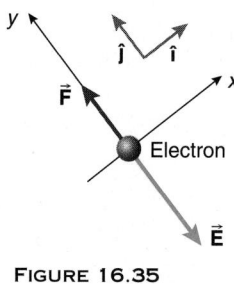

FIGURE 16.35

16.6 THE ELECTRIC FIELD OF POINTLIKE CHARGE DISTRIBUTIONS

Protons and electrons are essentially pointlike charges, and so it is of some import in physics and chemistry to know how to find the electric field of pointlike charges. In this section we determine the electric field of a static pointlike charge Q.

We first assume Q is positive. Another positive pointlike charge q finds itself in the pleasant company of Q, as shown in Figure 16.36. Since the charges are like charges (here both positive), q experiences a repulsive electrical force caused by Q.

The magnitude of the force on q is found using Coulomb's law:

$$F_{\text{elec on } q} = \frac{1}{4\pi\varepsilon_0} \frac{|q||Q|}{r^2} \qquad (16.9)$$

The absolute value signs here are redundant since both charges are positive. The charge Q also experiences a force of the same magnitude (but in the opposite direction, according to Newton's third law of motion). Here we are *not* interested in the force on Q but, rather, the force *of* Q on q.

We want to look at the force on q differently. The fixed charge Q establishes an electric field $\vec{E}$ in the surrounding space.

The field then acts on the charge q that is placed in the field of Q. According to this view, the force on q when it is placed in the field $\vec{E}$ is found from the defining Equation 16.8:

$$\vec{F}_{\text{elec on } q} = q\vec{E}_{\text{of } Q}$$

The magnitude of the force is

$$F_{\text{elec on } q} = |q| E_{\text{of } Q} \qquad (16.10)$$

We equate Equations 16.9 and 16.10 for the magnitude of the force on q since they represent the same force:

$$|q| E_{\text{of } Q} = \frac{1}{4\pi\varepsilon_0} \frac{|q||Q|}{r^2}$$

Hence we find that the magnitude of the electric field created by the positive point charge Q is

$$E_{\text{of } Q} = \frac{1}{4\pi\varepsilon_0} \frac{|Q|}{r^2}$$

Notice that the magnitude of the electric field of Q depends only on Q, and not on the charge q that was placed in the field.

What about the direction of the electric field created by Q at the point where q was placed? Since the charge q placed in the field $\vec{E}$ is positive, the force on q is *parallel* to the direction of the field $\vec{E}$ according to the defining equation for the field (Equation 16.8): $\vec{F}_{\text{elec on } q} = q\vec{E}_{\text{of } Q}$. Since the force on q is repulsive, we see that the electric field created by the positive point charge Q must point *radially away* from positive charge Q, as shown in Figure 16.37.

Thus the electric field of the positive charge Q at a distance r from Q can be written in vector form as

$$\vec{E} = \frac{1}{4\pi\varepsilon_0} \frac{Q}{r^2} \hat{r} \qquad (16.11)$$

where $\hat{r}$ is a unit vector pointing in a radial direction away from Q.

If the positive charge q is placed in the vicinity of a *negative* charge Q (i.e., $Q < 0$ C), the electrical force on q now is an *attractive* force since the two are unlike charges (one is positive, one is negative), as indicated in Figure 16.38. In this case the electrical force on q when it is placed in the field of Q still is given by Equation 16.8: $\vec{F}_{\text{elec on } q} = q\vec{E}_{\text{of } Q}$. Since q is positive, the force on q is *parallel* to the field of Q. The force is attractive. Thus the electric field of a negative charge Q must point *toward* Q, as indicated in Figure 16.39.

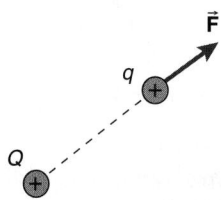

FIGURE 16.36 Positive charge q finds itself near a static, positive charge Q.

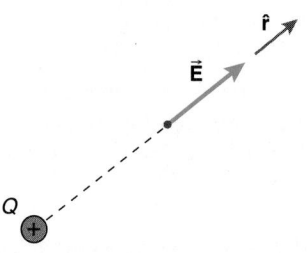

FIGURE 16.37 The electric field of a positive, pointlike charge Q points radially away from Q.

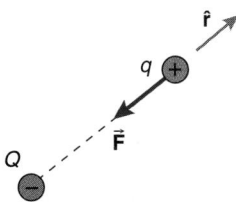

FIGURE 16.38 Positive charge q is attracted to negative charge Q.

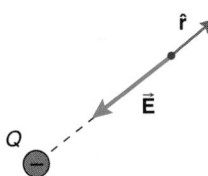

FIGURE 16.39 The electric field produced by a negative, pointlike charge Q is directed toward Q.

Notice that in Equation 16.11, if Q is negative, we have

$$\vec{E} = -\frac{1}{4\pi\varepsilon_0} \frac{|Q|}{r^2} \hat{r}$$

and so the field of Q is directed radially *toward* Q, in the direction of $-\hat{r}$.

> Therefore Equation 16.11 gives the electric field of a point charge Q for both positive and negative charges Q.

This result is very nice! Notice also that since we have already defined $\vec{E}$ in terms of the force on a positive test charge, we can come back with any new test charge q of either sign and $q\vec{E}$ will give the correct direction (and magnitude) of the force on q by the field $\vec{E}$.

PROBLEM-SOLVING TACTIC

16.5 Use the following three-step procedure to find the electric field a distance r from a pointlike charge.

1. First find the *magnitude* of the field at the point in question:

$$E = \frac{1}{4\pi\varepsilon_0} \frac{|Q|}{r^2}$$

This always is a positive quantity, because it is the magnitude of a vector.
2. Determine the direction of the field at the desired point and indicate it in a sketch:
   • radially away from Q if Q is positive
   • radially toward Q if Q is negative
3. Finally, introduce a coordinate system and write the electric field vector $\vec{E}$ in terms of your chosen coordinate system.

If several pointlike charges Q_1, Q_2, ... are distributed in space (see Figure 16.40) and we need the field of the collection at a point P in space, we use the principle of superposition. The

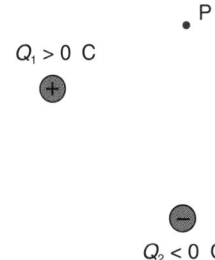

FIGURE 16.40 Several pointlike charges.

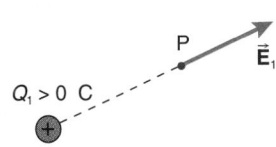

FIGURE 16.41 Charge Q_1 produces electric field $\vec{E}_1$ at the given point P.

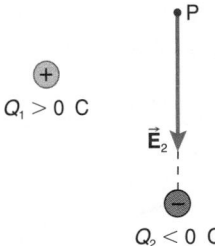

FIGURE 16.42 Charge Q_2 produces an electric field $\vec{E}_2$ at point P.

charge Q_1 *acting alone* produces an electric field $\vec{E}_1$ at point P, as indicated in Figure 16.41. Charge Q_2 *acting alone* produces an electric field $\vec{E}_2$ at the *same* point P; see Figure 16.42.

The total electric field $\vec{E}_{total}$ at the point when *all* the pointlike charges are present simultaneously at their fixed positions is the *vector sum* of the individual fields (see Figure 16.43):

$$\vec{E}_{total} = \vec{E}_1 + \vec{E}_2 + \cdots \quad (16.12)$$

The superposition of fields stems from the underlying principle of superposition associated with the forces: the total force is the vector sum of the individual forces acting on the system.

As always, a charge q, whether positive or negative, placed in this field experiences a force found from Equation 16.8:

$$\vec{F}_{on\ q} = q\vec{E}_{produced\ by\ other\ charges,\ not\ q}$$

EXAMPLE 16.8

a. Determine the magnitude of the electric field of a proton (the nucleus of a hydrogen atom) a distance 5.29×10^{-11} m away from the proton. Write the field in vector form.
b. Sketch the array of electric field arrows at this distance from the proton.

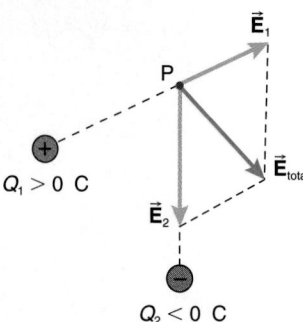

FIGURE 16.43 The total electric field is the vector sum of the individual fields.

c. Place an electron at a point this distance from the proton. Find the force on the electron placed in the field at this location. Sketch the force on a diagram.

Solution

a. The proton is a pointlike charge, and so you can use Equation 16.11. The magnitude of the field is

$$E = \frac{1}{4\pi\varepsilon_0} \frac{|Q|}{r^2}$$

where $Q = +e = 1.602 \times 10^{-19}$ C and $r = 5.29 \times 10^{-11}$ m. Making these substitutions, you find

$$E = (9.00 \times 10^9 \text{ N·m}^2/\text{C}^2) \frac{1.602 \times 10^{-19} \text{ C}}{(5.29 \times 10^{-11} \text{ m})^2}$$

$$= 5.15 \times 10^{11} \text{ N/C}$$

This is an electric field with a very large magnitude!

The electric field of the proton is directed radially away from the proton, since it is a positive charge. Therefore

$$\vec{E} = (5.15 \times 10^{11} \text{ N/C})\hat{r}$$

b. The field arrows are shown in Figure 16.44.

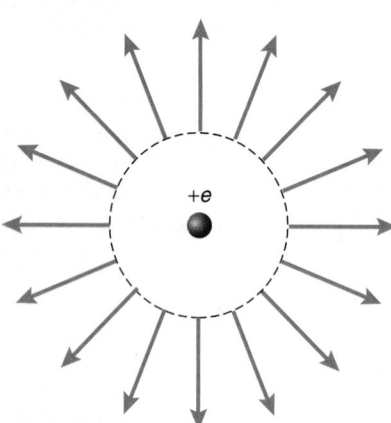

FIGURE 16.44

c. An electron placed in this field experiences a force given by

$$\vec{F} = q\vec{E}$$
$$= (-1.602 \times 10^{-19} \text{ C})[(5.15 \times 10^{11} \text{ N/C})\hat{r}]$$
$$= (-8.25 \times 10^{-8} \text{ N})\hat{r}$$

The force on the electron is directed along $-\hat{r}$, antiparallel to the electric field in which it is placed (which is along $\hat{r}$); the force is toward the proton, as shown in Figure 16.45.

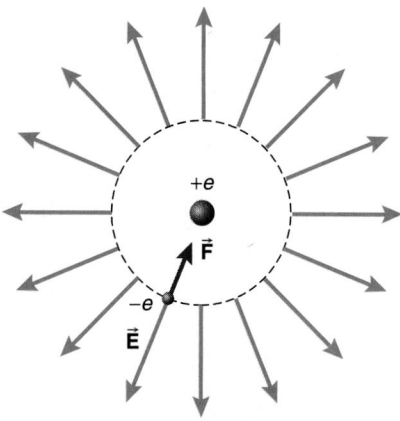

FIGURE 16.45

STRATEGIC EXAMPLE 16.9

a. Find the total electric field at the origin caused by the two pointlike charges shown in Figure 16.46. Determine the magnitude of the total field.

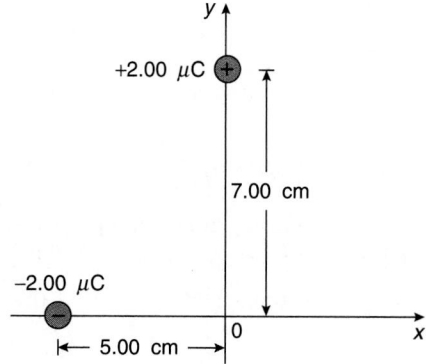

FIGURE 16.46

b. Determine the force on a charge $q = -3.00$ μC placed at the origin. What is the magnitude of the force on this charge?

Solution

a. First find the electric field of each charge at the origin (as if each charge were acting alone) with no charge at the origin. The total electric field is then the vector sum of these fields.

1. **The Electric Field of the -2.00 μC Charge**

Pretend that the other charge (the $+2.00$ μC charge) does not exist; see Figure 16.47. The -2.00 μC charge is a pointlike charge. The field of the charge has a magnitude

$$E_1 = \frac{1}{4\pi\varepsilon_0}\frac{|Q|}{r^2}$$

where $Q = -2.00 \times 10^{-6}$ C and the distance r between the charge and the origin is 5.00 cm or 5.00×10^{-2} m. *You must use coulombs for all charges and meters for all distances.* Making these substitutions,

$$E_1 = \frac{(9.00 \times 10^9 \text{ N·m}^2/\text{C}^2)\left|-2.00 \times 10^{-6} \text{ C}\right|}{(5.00 \times 10^{-2} \text{ m})^2}$$

$$= 7.20 \times 10^6 \text{ N/C}$$

This field is directed toward the negative charge, as shown in Figure 16.48. Using the given coordinate system, write $\vec{\mathbf{E}}_1$ as a vector:

$$\vec{\mathbf{E}}_1 = (-7.20 \times 10^6 \text{ N/C})\hat{\imath}$$

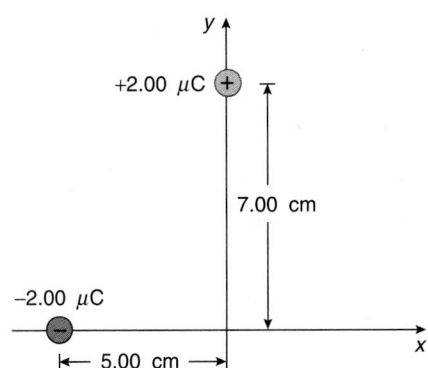

FIGURE 16.47

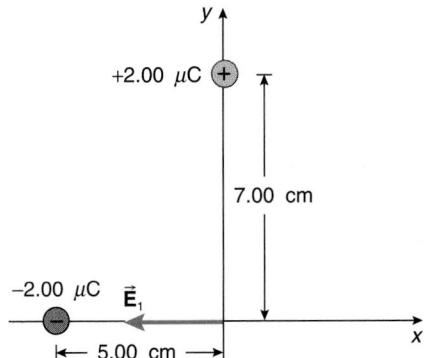

FIGURE 16.48

2. **The Electric Field of the +2.00 μC Charge**

Now pretend that the -2.00 μC charge does not exist, as in Figure 16.49. The $+2.00$ μC charge is a pointlike charge. The field of the charge has magnitude

$$E_2 = \frac{1}{4\pi\varepsilon_0}\frac{|Q|}{r^2}$$

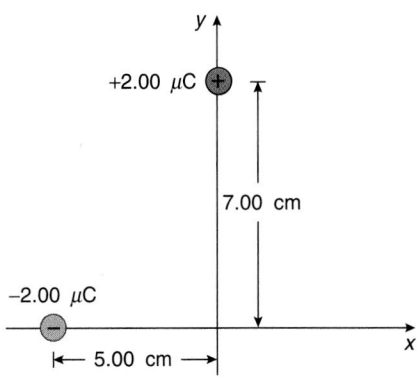

FIGURE 16.49

where $Q = +2.00 \times 10^{-6}$ C and the distance r between the charge and the origin is 7.00×10^{-2} m. You must use coulombs for the charge and meters for all distances. Making these substitutions,

$$E_2 = \frac{(9.00 \times 10^9 \text{ N·m}^2/\text{C}^2)\left|+2.00 \times 10^{-6} \text{ C}\right|}{(7.00 \times 10^{-2} \text{ m})^2}$$

$$= 3.67 \times 10^6 \text{ N/C}$$

This field is directed away from the positive charge, as indicated in Figure 16.50. Using the given coordinate system, write $\vec{\mathbf{E}}_2$ as a vector:

$$\vec{\mathbf{E}}_2 = (-3.67 \times 10^6 \text{ N/C})\hat{\jmath}$$

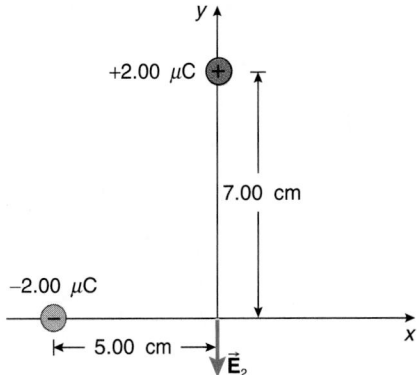

FIGURE 16.50

The total electric field at the origin when both charges are present simultaneously is the *vector sum* of the fields $\vec{\mathbf{E}}_1$ and $\vec{\mathbf{E}}_2$, as indicated in Figure 16.51:

$$\vec{\mathbf{E}}_{\text{total}} = \vec{\mathbf{E}}_1 + \vec{\mathbf{E}}_2$$
$$= (-7.20 \times 10^6 \text{ N/C})\hat{\imath} - (3.67 \times 10^6 \text{ N/C})\hat{\jmath}$$

The magnitude of this field is

$$E_{\text{total}} = [(-7.20 \times 10^6 \text{ N/C})^2 + (-3.67 \times 10^6 \text{ N/C})^2]^{1/2}$$
$$= 8.08 \times 10^6 \text{ N/C}$$

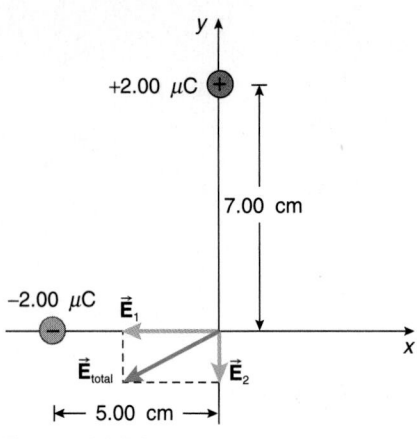

FIGURE 16.51

b. Since you know the electric field at the origin, the force on charge q placed at the origin is found using Equation 16.8:

$$\vec{F} = q\vec{E}$$
$$= (-3.00 \times 10^{-6} \text{ C})[(-7.20 \times 10^{6} \text{ N/C})\hat{\imath}$$
$$- (3.67 \times 10^{6} \text{ N/C})\hat{\jmath}]$$
$$= (21.6 \text{ N})\hat{\imath} + (11.0 \text{ N})\hat{\jmath}$$

This force is shown in Figure 16.52.

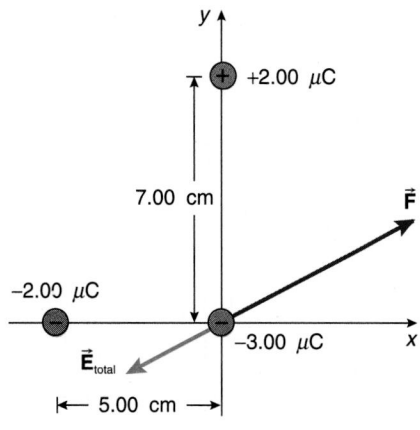

FIGURE 16.52

The magnitude of the force can be found two ways.

Method 1
Find the magnitude of the vector $\vec{F}$:
$$F = [(21.6 \text{ N})^2 + (11.0 \text{ N})^2]^{1/2}$$
$$= 24.2 \text{ N}$$

Method 2
Use the magnitude of the total field and take the magnitude of $\vec{F} = q\vec{E}$:
$$F = |q|E$$
$$= (3.00 \times 10^{-6} \text{ C})(8.08 \times 10^{6} \text{ N/C})$$
$$= 24.2 \text{ N}$$

16.7 A WAY TO VISUALIZE THE ELECTRIC FIELD: ELECTRIC FIELD LINES

The electric field is a vector. We represent vectors geometrically by arrows with the arrow pointing in the direction of the vector and with the length of the arrow drawn proportional to the magnitude of the vector quantity. The field vectors at several different locations surrounding a positive and negative pointlike charge are indicated in Figures 16.53 and 16.54. Nothing new about this; we did the same for the gravitational fields in Chapter 6.

There is a useful alternative geometric representation of the electric field. In this scheme, we draw continuing straight or curved *lines* that at any point are *tangent* to the direction of the field vector at that point; we then place an arrowhead somewhere along the line to indicate the direction to associate with the field vectors at all points along the line. These lines are called **electric field lines**.* For a positive pointlike charge, the electric field lines are radially symmetric, as shown in Figure 16.55, with arrows pointing away from the positive charge since the electric field of a positive pointlike charge is directed away from the charge. Electric field lines therefore are said to *begin* on positive charges.

A similar pattern of lines is associated with a negative pointlike charge, except that the arrows now are directed *toward* the negative charge (see Figure 16.56), since the electric field of such a charge is directed toward the charge. Therefore we say that electric field lines *end* on negative charges.

The *number* of electric field lines drawn we make proportional to (not necessarily equal to) the magnitude (absolute value) of the charge.

---

*The scheme is frequently used for gravitational fields as well, giving rise to *gravitational field lines*.

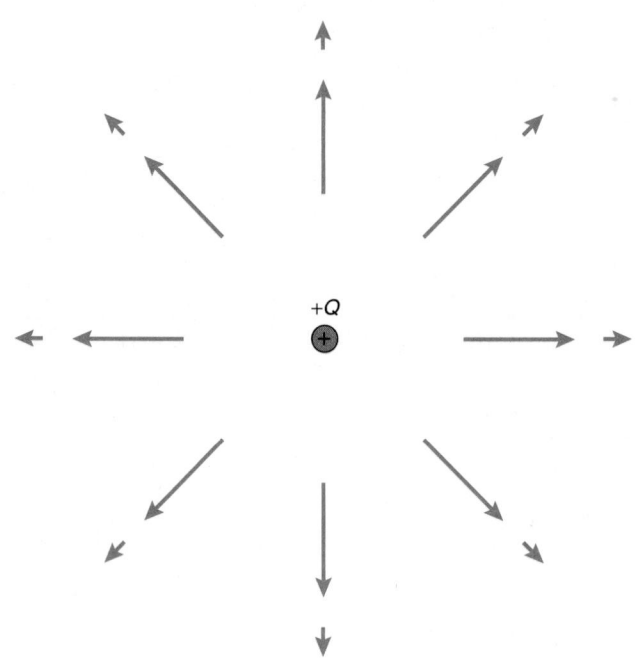

FIGURE 16.53 Electric field vectors at various distances from a positive pointlike charge. (For any arrow, take r as the distance to its tail.)

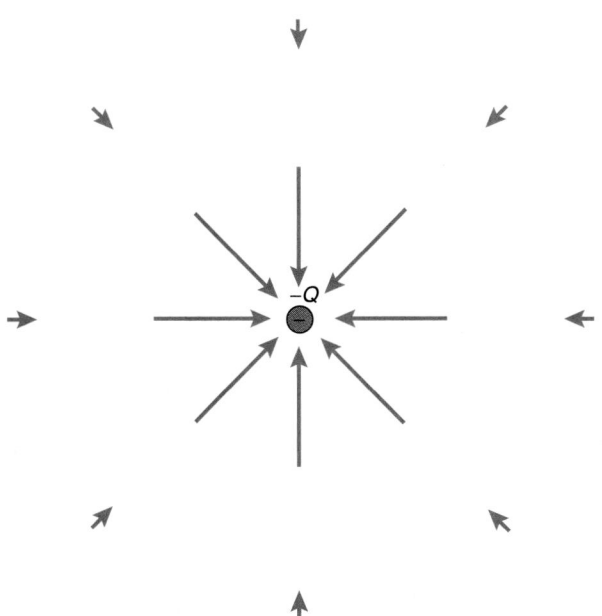

FIGURE 16.54 Electric field vectors at various distances from a negative pointlike charge.

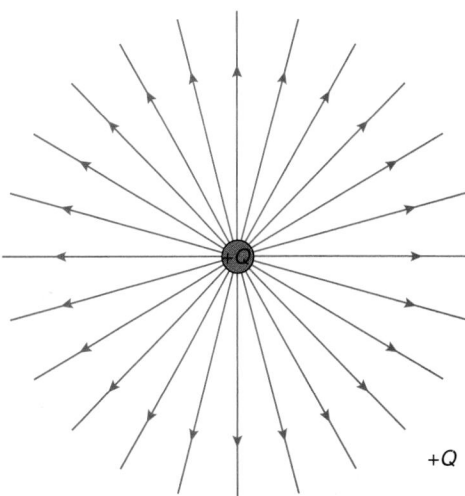

FIGURE 16.55 Electric field lines of a positive pointlike charge.

The area of a sphere of radius r surrounding such pointlike charges is $4\pi r^2$, and so the area *increases* with the square of the radial distance. The electric field of a pointlike charge *decreases* in magnitude as $1/r^2$. Hence the *number of lines* per square meter on the sphere, measured perpendicular to the lines, is a measure of the relative magnitude of the electric field at that location.

There is no reason to restrict the approach to single pointlike charge distributions. With the following rules, you can draw electric field lines for *any* distribution of charges:

1. The electric field lines begin on positive charges and end on negative charges.

2. Very close to pointlike charges, the lines are radially symmetric and their number is a measure of the magnitude of the charge.

3. The number of lines passing through a square meter oriented perpendicular to the lines is proportional to the magnitude of the electric field in that region.

Since the electric field has a *unique* direction at every point in space, no two field lines can intersect. If they were imagined to intersect, there would be two tangents to the field lines at that point, thus implying two different directions for the electric field vector at that point and two different force directions on a charge placed there, which is never the case.*

---
*Electric field line patterns can be shown in the laboratory by using small bits of insulating or conducting threads suspended in oil to prevent their translational motion. The bits of thread become polarized by the field and align themselves along the direction of the electric field at the points where they are suspended in the oil.

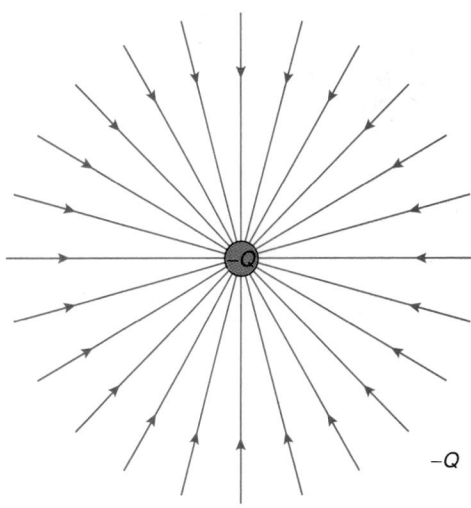

FIGURE 16.56 Electric field lines of a negative pointlike charge.

Using the foregoing conventions, the electric field lines surrounding several pointlike charge distributions can be determined in a qualitative way.

For a positive charge Q and negative charge $-Q$ of equal magnitude, the number of lines leaving the positive charge is equal to the number of lines approaching the negative charge, since the charges have equal magnitude. Very close to each pointlike charge, the lines are radially symmetric. As the distance from both charges increases, we can see that *every* electric field line leaving the positive charge eventually ends on the negative charge, as shown in Figure 16.57.

For two positive charges Q of equal magnitude, the number of lines leaving each charge is the same and the patterns are radially symmetric very close to either charge; see Figure 16.58. Very far from the charges, the distribution is similar to that of a single pointlike charge of magnitude $2Q$, for the pattern becomes essentially radially symmetric, with a total number of lines equal to twice the number emerging from each charge.

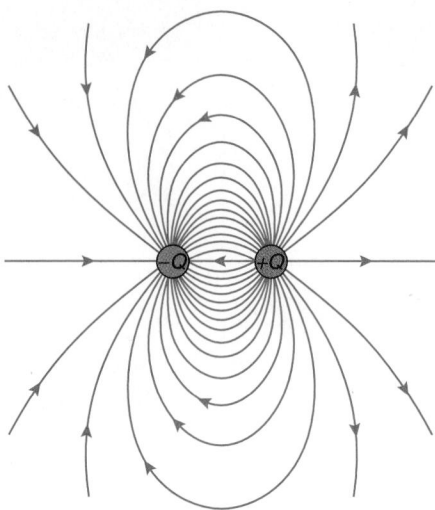

FIGURE 16.57 Electric field lines in the vicinity of a positive and negative charge of equal magnitude.

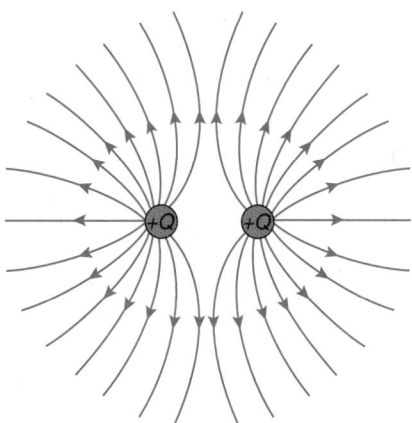

FIGURE 16.58 Electric field lines in the vicinity of two positive charges of equal magnitude.

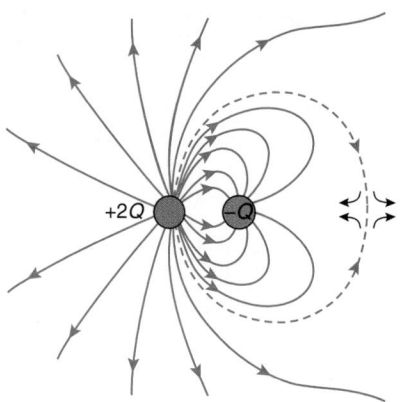

FIGURE 16.59 Electric field lines in the vicinity of a positive and negative charge of unequal magnitude.

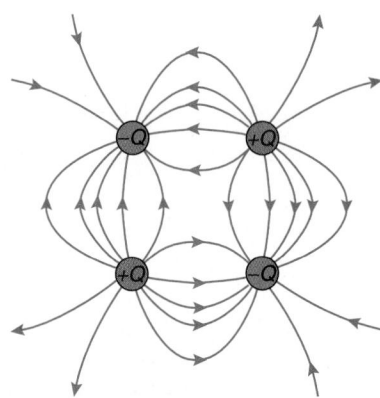

FIGURE 16.60 Electric field lines in the vicinity of four charges of equal magnitude, two of which are positive and two negative.

Figure 16.59 shows the electric field line pattern for a positive charge of magnitude $2Q$ and a negative charge of magnitude $-Q$. In this case the number of lines leaving the positive charge is twice that entering the negative charge, reflecting their difference in magnitude. The remaining half of the lines emerging from charge $2Q$ are essentially radial at great distances from the charge distribution, indicative of the fact that at such distances the charge distribution looks like a single pointlike charge of total charge $2Q - Q = Q$.

The electric field lines of four point charges of equal magnitude, two of which are positive and two negative, arranged on the corners of a square, are shown in Figure 16.60.

QUESTION 6 _____
Explain why electric field lines of static charges (a) never intersect each other and (b) never form closed loops.

16.8 A COMMON MOLECULAR CHARGE DISTRIBUTION: THE ELECTRIC DIPOLE

Many simple molecules like water and more complex molecular structures like DNA, while electrically neutral as a whole, have their positive and negative charges distributed so that they can be modeled as one or more electric dipoles. The prevalence of such molecular structures makes a study of electric dipoles of some importance in chemistry, biology, and physics.

An electric dipole consists of two pointlike charges of equal magnitude $|Q|$, but opposite sign, separated by a distance d, as shown in Figure 16.61. The dipole as a whole is electrically neutral. This implies that the electric field approaches zero at

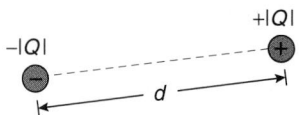

FIGURE 16.61 An electric dipole.

points far removed from the dipole (at distances much greater than the separation d of the charges). However, since the charges are not coincident with each other, the dipole produces a nonzero and distinct electric field in the space immediately surrounding the dipole (at distances comparable to the separation distance d).

In this section we first examine the electric field produced by an electric dipole. Then we examine what happens when an electric dipole is placed in an electric field caused by *other* distributions of charge, not that of the particular dipole itself.

It will be convenient to introduce a new vector called the electric **dipole moment** $\vec{p}$, defined as the product of the magnitude $|Q|$ of either charge times the position vector $\vec{d}$ of the positive charge with respect to the negative charge, as indicated in Figure 16.62:

$$\vec{p} \equiv |Q|\vec{d} \qquad (16.13)$$

Therefore the direction of the dipole moment vector is *from the negative charge to the positive charge*. A mnemonic for remembering this direction is to think of the plus sign + as composed of two short intersecting line segments; the head of the vector arrow also is composed of two short line segments: →.

As an example, Figure 16.63 shows how the water molecule is modeled as an electric dipole. The fractional units of the

fundamental unit of charge e do not violate charge quantization. Rather, the fractions arise because the surrounding electrons spend more time near the oxygen atom than the hydrogen atoms, giving rise to these effective values of the charge distribution. The magnitude of the electric dipole moment of the water molecule is

$$\begin{aligned}
p_{water} &= 0.66ed \\
&= 0.66(1.602 \times 10^{-19} \text{ C})(0.057 \times 10^{-9} \text{ m}) \\
&= 6.0 \times 10^{-30} \text{ C·m}
\end{aligned}$$

The magnitudes of the electric dipole moments of several common molecules and molecular bonds are indicated in Table 16.1. In chemistry and biology, dipole moments frequently are quoted not in the SI unit of C·m, but in another unit: the *debye* (D). We will not use the debye unit but, if needed, the conversion factor is

$$1 \text{ D} = 3.33 \times 10^{-30} \text{ C·m}$$

Electric Field of a Dipole

An expression for the electric field along the axis of the dipole is derived in Example 16.10. The electric field in the plane that is the perpendicular bisector of the line connecting the two charges is derived in Example 16.11. The results of these examples show that the magnitude of the electric field of a dipole along these directions is proportional to the magnitude of the dipole moment and decreases with the *inverse cube* of the distance from the center of the dipole. While the examples demonstrate this distance dependence of the field for directions along the axis or in the bisecting plane, the statement is valid for other directions as well; but in these other directions the field also depends on the angle between the axis of the dipole and the direction from the center of the dipole to the point where the field is to be calculated. The electric field lines associated with an electric dipole have the appearance shown in Figure 16.57.

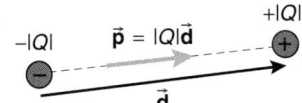

FIGURE 16.62 The electric dipole moment $\vec{p}$.

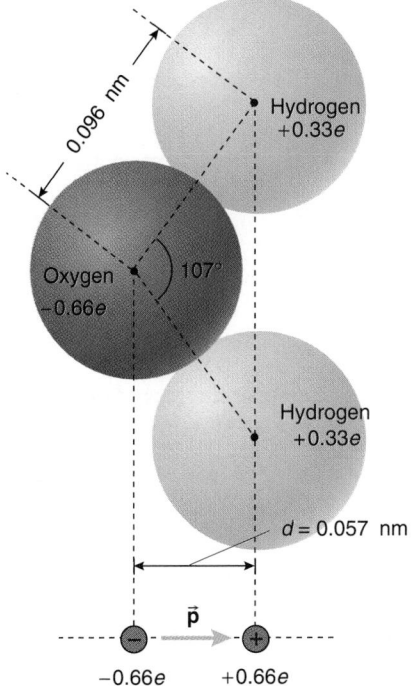

FIGURE 16.63 The water molecule and its dipole moment.

TABLE 16.1 The Magnitude of the Electric Dipole Moment for Selected Molecules and Molecular Bonds

Molecule	p
Acetone (C_3H_6O)	9.7×10^{-30} C·m
Ammonia (NH_3)	5×10^{-30} C·m
Carbon dioxide (CO_2)	None
Carbon monoxide (CO)	0.4×10^{-30} C·m
Ethyl alcohol (C_2H_5OH)	5.7×10^{-30} C·m
Hydrochloric acid (HCl)	3.6×10^{-30} C·m
Oxygen (O_2)	None
Sulfur dioxide (SO_2)	5.5×10^{-30} C·m
Water (H_2O)	6.0×10^{-30} C·m

Molecular bonds	p
H—O	5×10^{-30} C·m
C—Cl	5×10^{-30} C·m
C—O	2.5×10^{-30} C·m
C=O	7.7×10^{-30} C·m
C—Cl	4.9×10^{-30} C·m
C—N	0.73×10^{-30} C·m
H—N	3.0×10^{-30} C·m

A Dipole in a Uniform Electric Field

If an electric dipole is placed in a *uniform* electric field (produced by other charges), the total force on the dipole is *zero*; see Figure 16.64. The forces on each of the charges are of the same magnitude but are oppositely directed and thus vector sum to zero.

> While the total force on the dipole is zero in a uniform field, the total torque on the dipole depends on the orientation of the dipole moment vector with respect to the direction of the electric field.

Example 16.13 shows that the total torque $\vec{\tau}$ on the dipole in a uniform electric field $\vec{E}$ is

$$\vec{\tau} = \vec{p} \times \vec{E} \qquad (16.14)$$

Problem 42 asks you to show that Equation 16.14 is the torque on the dipole *independent of the origin* about which the torque is taken.

Equation 16.14 implies that if the dipole moment $\vec{p}$ is not parallel to $\vec{E}$, and the dipole is free to rotate, it will spin to orient itself until $\vec{p}$ is parallel to $\vec{E}$. When $\vec{p}$ is parallel to $\vec{E}$, the torque is zero.* Equation 16.14 yields the torque on the molecular dipoles in any electric field that is approximately uniform over the small physical size of such dipoles.

The electric field is quite large in the immediate vicinity of molecular electric dipoles (see Example 16.12). The strong nearby field means that *other* electric dipoles or charges in the vicinity experience large electrical forces. Molecular electric dipole–dipole interactions are essential in the study of inter- and intramolecular bonding forces in chemistry and biology.

The unusually high dipole moment of water (for a nonacid and a nonbase) makes it an excellent polar solvent. The relatively strong electric field in the vicinity of a water molecule is able to exert electrical forces on other molecules (particularly ions and molecules with large dipole moments) that disrupt their molecular or intermolecular bonding, causing them to go easily into solution. Hydrocarbons (gasoline and motor oils) have small dipole moments and are not easily dissolved in water. Indeed the ability of water to act as a polar solvent for many materials accounts for its importance in biology for the transport of nutrients and wastes in living organisms. Water is the only solvent from which blood can be made.

---

*If the dipole moment is antiparallel to $\vec{E}$, the total torque also is zero—but this is an unstable equilibrium position. The slightest jitter, and the dipole will flip around so that $\vec{p}$ is parallel to $\vec{E}$. We show this from an energy viewpoint in the next chapter.

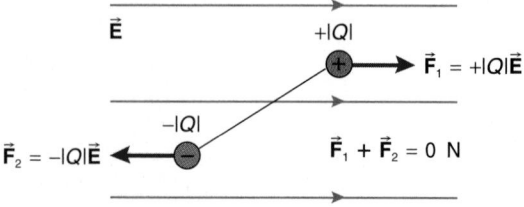

FIGURE 16.64 The total force on a dipole in a uniform electric field is zero.

EXAMPLE 16.10

Calculate the electric field of a dipole at a point P located a distance z from the center of the dipole along its axis, as shown in Figure 16.65. The axis of a dipole customarily is chosen to be the z-axis of a Cartesian coordinate system.

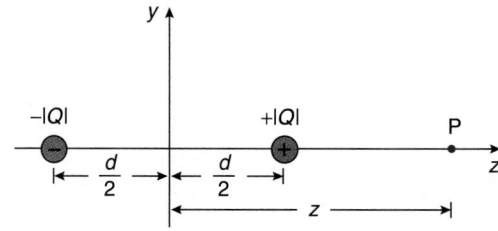

FIGURE 16.65

Solution

The positive charge is a distance

$$z - \frac{d}{2}$$

from point P. The positive charge $+|Q|$ produces a field $\vec{E}_1$ at P that is directed away from $+|Q|$, as shown in Figure 16.66.

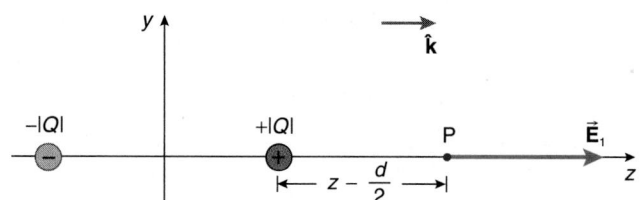

FIGURE 16.66

Expressing Equation 16.11 for the electric field of a pointlike charge in terms of the Cartesian coordinate system here, you get

$$\vec{E}_1 = \frac{1}{4\pi\varepsilon_0} \frac{|Q|}{\left(z - \dfrac{d}{2}\right)^2} \hat{k}$$

The negative charge $-|Q|$ is a distance

$$z + \frac{d}{2}$$

from the point P. The negative charge produces a field $\vec{E}_2$ at P that is directed toward $-|Q|$, as shown in Figure 16.67.

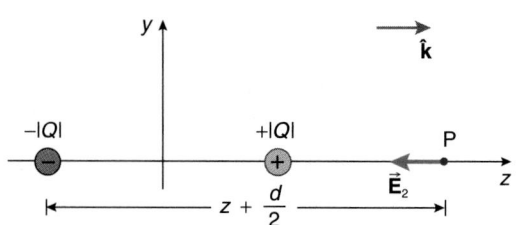

FIGURE 16.67

Using Equation 16.11, again adapted to the coordinate system and direction of the field of the negative charge, you find

$$\vec{E}_2 = \frac{1}{4\pi\varepsilon_0} \frac{|Q|}{\left(z + \dfrac{d}{2}\right)^2} \left(-\hat{k}\right)$$

The total field at P is the vector sum of these fields:

$$\vec{E} = \vec{E}_1 + \vec{E}_2$$

$$= \frac{1}{4\pi\varepsilon_0} \left[\frac{|Q|}{\left(z - \dfrac{d}{2}\right)^2} - \frac{|Q|}{\left(z + \dfrac{d}{2}\right)^2} \right] \hat{k}$$

After some factoring and rearranging, you have

$$\vec{E} = \frac{1}{4\pi\varepsilon_0} \frac{|Q|}{z^2} \left[\left(1 - \frac{d}{2z}\right)^{-2} - \left(1 + \frac{d}{2z}\right)^{-2} \right] \hat{k}$$

If $z \gg d$, you can use the binomial expansion to simplify this unwieldy expression. Specifically

$$(1 + \alpha)^n \approx 1 + n\alpha \qquad \text{for } \alpha \ll 1$$

With this approximation, the field is approximately

$$\vec{E} \approx \frac{1}{4\pi\varepsilon_0} \frac{|Q|}{z^2} \left\{ \left[1 + (-2)\left(\frac{-d}{2z}\right)\right] - \left[1 + (-2)\left(\frac{d}{2z}\right)\right] \right\} \hat{k}$$

$$\approx \frac{1}{4\pi\varepsilon_0} \frac{|Q|}{z^2} \frac{2d}{z} \hat{k}$$

Expressed in terms of the magnitude of the dipole moment ($p = |Q|d$), the field is

$$\vec{E} \approx \frac{1}{4\pi\varepsilon_0} \frac{2p}{z^3} \hat{k} \qquad (1)$$

For $z \gg d$, the electric field of the dipole decreases as the inverse cube of the distance from the dipole.

EXAMPLE 16.11

Calculate the total electric field caused by the two charges of an electric dipole at a point P in the plane that is the perpendicular bisector of the line connecting the two charges; see Figure 16.68. This plane customarily is chosen to be the x–y plane of a Cartesian coordinate system.

Solution

Because the charges are of the same magnitude and the point P in the bisector plane is equidistant from both charges, the electric field of *each* of the two charges (acting alone) at the point P has the same magnitude. From Equation 16.11, the magnitude of the field of each pointlike charge is

$$E = \frac{1}{4\pi\varepsilon_0} \frac{|Q|}{r^2}$$

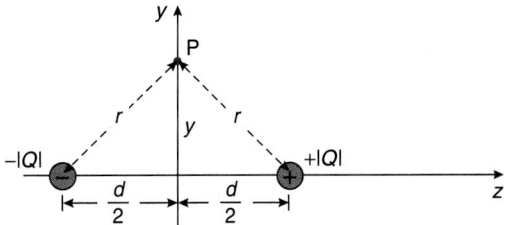

FIGURE 16.68

From the geometry of Figure 16.68 and the Pythagorean theorem,

$$r^2 = y^2 + \frac{d^2}{4}$$

Hence the magnitude of the field each charge produces at P is

$$E = \frac{1}{4\pi\varepsilon_0} \frac{|Q|}{\left(y^2 + \dfrac{d^2}{4}\right)}$$

The field at P of the positive charge points away from the charge, while the field at P of the negative charge points toward the charge, as indicated in Figure 16.69.

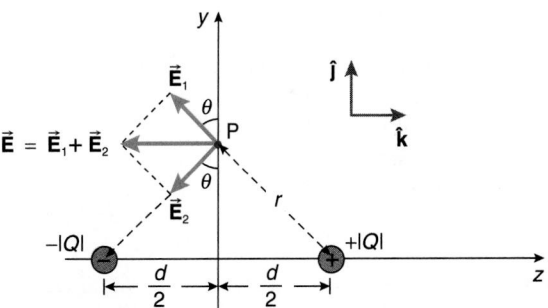

FIGURE 16.69

Use the coordinate system in Figure 16.69. The individual fields in vector form are

$$\vec{E}_1 = \frac{1}{4\pi\varepsilon_0} \frac{|Q|}{\left(y^2 + \dfrac{d^2}{4}\right)} \left[(\cos\theta)\,\hat{j} - (\sin\theta)\,\hat{k}\right]$$

and

$$\vec{E}_2 = \frac{1}{4\pi\varepsilon_0} \frac{|Q|}{\left(y^2 + \dfrac{d^2}{4}\right)} \left[(-\cos\theta)\,\hat{j} - (\sin\theta)\,\hat{k}\right]$$

According to the principle of superposition, the total electric field at the point P when *both charges* are present is the *vector sum* of the fields of the individual charges:

$$\vec{E}_{total} = \vec{E}_1 + \vec{E}_2$$

$$= \left[\frac{-2}{4\pi\varepsilon_0} \frac{|Q|}{\left(y^2 + \frac{d^2}{4}\right)} \sin\theta \right] \hat{k} \qquad (1)$$

But from the geometry of Figure 16.69,

$$\sin\theta = \frac{d/2}{\left(y^2 + \frac{d^2}{4}\right)^{1/2}}$$

Making this substitution for $\sin\theta$ in equation (1), you find the total field at the point P is

$$\vec{E}_{total} = \frac{-1}{4\pi\varepsilon_0} \frac{|Q|d}{\left(y^2 + \frac{d^2}{4}\right)^{3/2}} \hat{k}$$

Since the magnitude of the dipole moment $\vec{p}$ is equal to $|Q|d$, the field at P is

$$\vec{E}_{total} = \frac{-1}{4\pi\varepsilon_0} \frac{p}{\left(y^2 + \frac{d^2}{4}\right)^{3/2}} \hat{k} \qquad (2)$$

For distances y from the dipole that are much greater than the separation of the charges in the dipole—that is, for $y \gg d$—the d^2 term in the denominator of the field can be neglected; the field then becomes

$$\vec{E}_{total} \approx \frac{-1}{4\pi\varepsilon_0} \frac{p}{y^3} \hat{k} \qquad (y \gg d) \qquad (3)$$

For such great distances y, the electric field of the dipole decreases as the inverse *cube* of the distance.

By comparing equation (1) of Example 16.10 and equation (3) here, you can see that at equally great distances from the dipole, the magnitude of the field in the bisector plane is only half that along the axis of the dipole.

EXAMPLE 16.12

The water molecule has a dipole moment of magnitude 6.0×10^{-30} C·m, as indicated in Table 16.1. Find the magnitude of the electric field at a distance of 1.00×10^{-9} m = 1.00 nm from the axis of the dipole in the plane of the perpendicular bisector to the line connecting the two charges representing the dipole.

Solution
The electric field in the plane of the perpendicular bisector was calculated in Example 16.11. Use the result of that example here [equation (2) of Example 16.11] as well as the data

indicated in Figure 16.63 for the water molecule. Specifically for the magnitude of the field, you have

$$E_{total} = \frac{1}{4\pi\varepsilon_0} \frac{p}{\left(y^2 + \frac{d^2}{4}\right)^{3/2}}$$

where $p = 6.0 \times 10^{-30}$ C·m, $y = 1.00 \times 10^{-9}$ m, and $d = 0.057$ nm $= 0.057 \times 10^{-9}$ m. Making these substitutions, you find

$$E_{total} = 5.4 \times 10^7 \text{ N/C}$$

Using the approximation of equation (3) of Example 16.11 yields the same result since $y \gg d$. This is an electric field of substantial magnitude!

EXAMPLE 16.13

Show that an electric dipole in a uniform electric field $\vec{E}$ experiences a torque $\vec{\tau}$ about the midpoint of the dipole that is given by

$$\vec{\tau} = \vec{p} \times \vec{E}$$

where $\vec{p}$ is the dipole moment vector. See Figure 16.70.

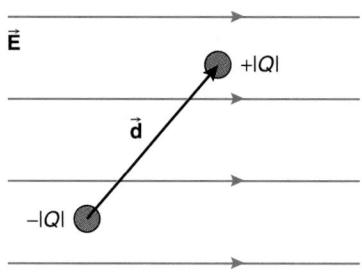

FIGURE 16.70

Solution
Each charge of the dipole experiences an electrical force:

the force on the positive charge is $\vec{F}_1 = +|Q|\vec{E}$; and
the force on the negative charge is $\vec{F}_2 = -|Q|\vec{E}$.

The torque of a force is $\vec{\tau} = \vec{r} \times \vec{F}$, where $\vec{r}$ is the position vector locating the point of application of the force with respect to the point about which the torque is taken. The example asks you to take torques about the midpoint of the dipole; see Figure 16.71.

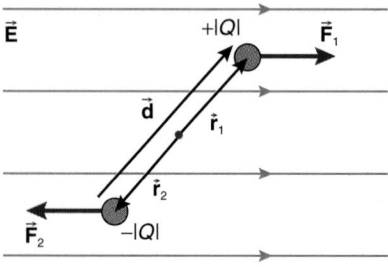

FIGURE 16.71

The position vectors of the points of application of the forces are the locations of the charges themselves:

$$\vec{r}_1 = \frac{\vec{d}}{2}$$

$$\vec{r}_2 = -\frac{\vec{d}}{2}$$

Hence the torque $\vec{\tau}_1$ of $\vec{F}_1$ is

$$\vec{\tau}_1 = \frac{\vec{d}}{2} \times \vec{F}_1$$

$$= \frac{\vec{d}}{2} \times |Q|\vec{E}$$

The torque $\vec{\tau}_2$ of $\vec{F}_2$ is

$$\vec{\tau}_2 = \vec{r}_2 \times \vec{F}_2$$

$$= -\frac{\vec{d}}{2} \times \left(-|Q|\vec{E}\right)$$

$$= \frac{\vec{d}}{2} \times |Q|\vec{E}$$

The total torque $\vec{\tau}$ is the vector sum of the two torques:

$$\vec{\tau} = \vec{\tau}_1 + \vec{\tau}_2$$

$$= \frac{\vec{d}}{2} \times |Q|\vec{E} + \frac{\vec{d}}{2} \times |Q|\vec{E}$$

$$= |Q|\vec{d} \times \vec{E}$$

Since $|Q|\vec{d}$ is the dipole moment $\vec{p}$, you find

$$\vec{\tau} = \vec{p} \times \vec{E}$$

16.9 THE ELECTRIC FIELD OF CONTINUOUS DISTRIBUTIONS OF CHARGE

For charge distributions that are not pointlike in character, we need to invoke the principle of superposition to determine the total electric field at a point in space. The charge distribution is imagined to be composed of a continuous sea of pointlike charges, each of which contributes a bit to the total electric field at the point in space where the field is to be calculated; see Figure 16.72. Each small, differential bit of charge dq produces a differential bit of electric field $d\vec{E}$ at the point P given by a variation of Equation 16.11 for a pointlike charge:

$$d\vec{E} = \frac{1}{4\pi\varepsilon_0} \frac{dq}{r^2} \hat{r} \qquad (16.15)$$

where $\hat{r}$ is a unit vector pointing *from* the bit of charge dq *to* the point P where the field is to be found.

The contribution $d\vec{E}$ points away from dq if it is positive and toward dq if it is negative, just as with all pointlike charges.

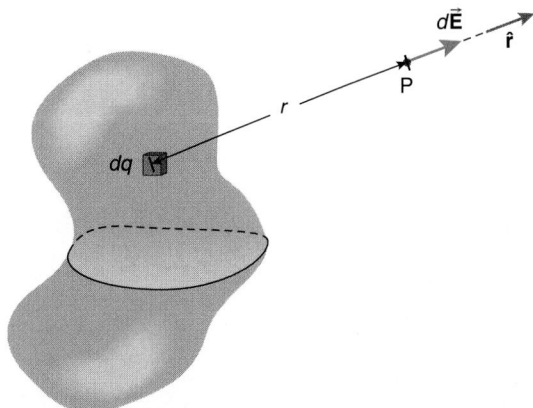

FIGURE 16.72 Each differential bit of charge produces a differential bit of electric field at point P. The situation pictured is for $dq > 0\ \text{C}$

According to the principle of superposition, the total electric field is found by a *vector sum* of all the contributions from all the bits of charge that make up the charge distribution. Here the vector summation is a *vector integration* (a continuous sum) over the physical extent of the charge distribution:

$$\vec{E} = \frac{1}{4\pi\varepsilon_0} \int_{\substack{\text{charge}\\\text{distribution}}} \frac{dq}{r^2} \hat{r} \qquad (16.16)$$

You have to be careful when performing this integration because the direction of the unit vector $\hat{r}$ typically changes as dq sweeps over the charge distribution. The distance r between dq and P also changes in many cases. Such vector integrations are trickier to carry out than ordinary scalar integrations, as we saw in our study of the gravitational field in Chapter 6. The reason for the trickiness is that the contributions $d\vec{E}$ typically point in different directions and we have to account for this fact while doing the vector summation implicit in the integration.

In actually doing such integrations, a certain amount of foresight is involved, foresight that only can be gleaned by carefully studying and profiting from a few examples of the technique. Consequently, in the following examples, we calculate the electric field of several different charge distributions to help you gain the needed insight to do similar calculations.

The examples illustrate that the electric field at a point in space depends on several factors:

• the charge creating the field and its sign;

• the geometric shape of the charge distribution creating the field;

• the distance of the point from the charge distribution; and

• the placement of the point with respect to the charge distribution (in particular, the location of the point with respect to any symmetry axes associated with the charge distribution).

Table 16.2 summarizes the electric fields created by various charge distributions.

TABLE 16.2 The Electric Field of Various Charge Distributions

Pointlike charge:

$$\vec{E} = \frac{1}{4\pi\varepsilon_0} \frac{Q}{r^2} \hat{r}$$

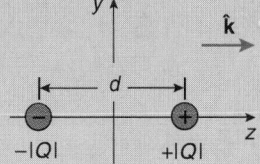

Dipole along the axis ($z \gg d$):

$$\vec{E} \approx \frac{1}{4\pi\varepsilon_0} \frac{2p}{z^3} \hat{k}$$

in the perpendicular bisector plane:

$$\vec{E}_{total} = \frac{-1}{4\pi\varepsilon_0} \frac{p}{\left(y^2 + \dfrac{d^2}{4}\right)^{3/2}} \hat{k}$$

$$\vec{E}_{total} \approx \frac{-1}{4\pi\varepsilon_0} \frac{p}{y^3} \hat{k} \qquad (y \gg d)$$

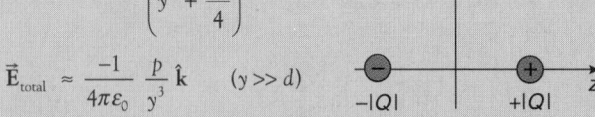

Uniformly charged ring (on the axis):

$$\vec{E} = \frac{1}{4\pi\varepsilon_0} \frac{zQ}{(z^2 + R^2)^{3/2}} \hat{k}$$

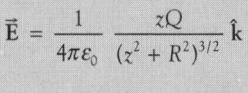

Uniformly charged disk (on the axis):

$$\vec{E} = \frac{1}{4\pi\varepsilon_0} \frac{2Q}{R^2} \left[1 - \frac{z}{(z^2 + R^2)^{1/2}}\right] \hat{k}$$

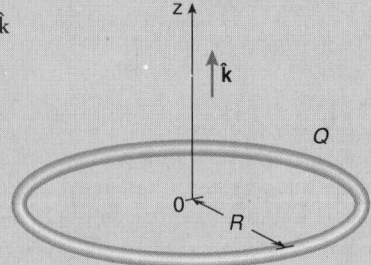

Uniformly charged infinite sheet:

$$E = \frac{\sigma}{2\varepsilon_0}$$

Two infinite uniformly charged sheets with opposite charge:

$$E = \frac{\sigma}{\varepsilon_0}$$

Uniformly charged spherical shell:

$$\vec{E} = 0 \text{ N/C} \qquad (\text{inside, } r < R)$$

$$\vec{E} = \frac{1}{4\pi\varepsilon_0} \frac{Q}{r^2} \hat{r} \qquad (\text{outside, } r > R)$$

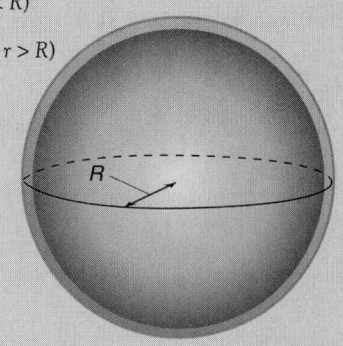

Uniformly charged sphere:

$$\vec{E} = \frac{1}{4\pi\varepsilon_0} \frac{Q}{R^3} r \hat{r} \qquad (r < R)$$

$$\vec{E} = \frac{1}{4\pi\varepsilon_0} \frac{Q}{r^2} \hat{r} \qquad (r > R)$$

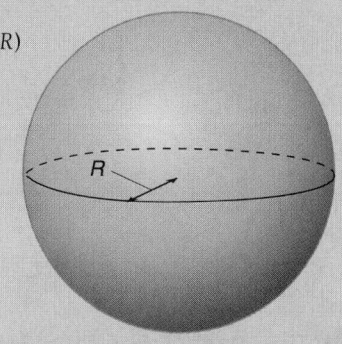

An infinitely long straight-line charge:

$$\vec{E} = \frac{1}{4\pi\varepsilon_0} \frac{2\lambda}{r} \hat{r}$$

(where $\hat{r}$ is a unit vector pointing radially away from the line charge)

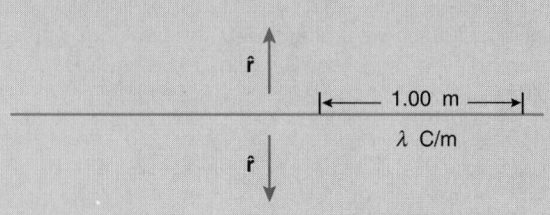

A ring of radius R has a total charge Q smeared out uniformly along its circumference, as shown in Figure 16.73. Calculate the electric field of the ring at a point P along the axis of the ring, a distance z from the plane of the ring.

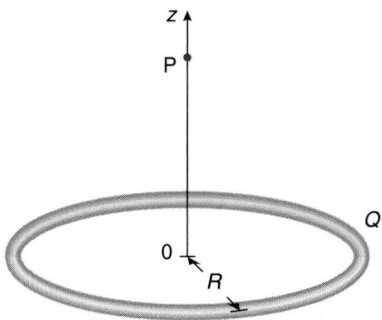

FIGURE 16.73

Solution

The charge Q is smeared out uniformly along the circumference of the ring. If the total charge Q is divided by the circumference of the ring ($2\pi R$), you obtain the number of coulombs of charge along each meter of the circumference of the ring; this quantity is known as a *linear charge density* λ:

$$\lambda = \frac{Q}{2\pi R}$$

Look at the contribution to the field at P from a small differential length ds of the circumference, as shown in Figure 16.74. The length ds has a charge dq on it equal to the product of the linear charge density λ and ds:

$$dq = \lambda \, ds$$

Assume the charge Q on the ring is positive; then this pointlike charge dq produces a bit of electric field $d\vec{E}$ at the axial point P that is directed away from dq. If dq is negative, the field points in the opposite direction.

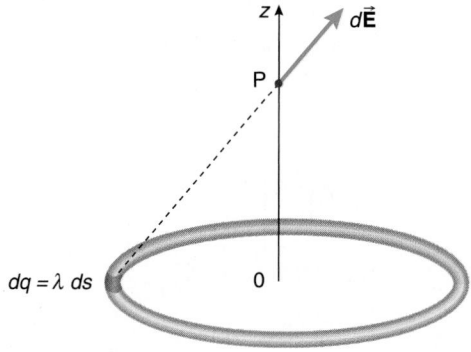

FIGURE 16.74

The magnitude of $d\vec{E}$ is found using Equation 16.15:

$$dE = \frac{1}{4\pi\varepsilon_0} \frac{dq}{r^2}$$
$$= \frac{1}{4\pi\varepsilon_0} \frac{\lambda \, ds}{r^2}$$

Notice that a corresponding piece of the ring on the opposite side *also* produces a bit of field of the same magnitude at the point P, as shown in Figure 16.75. The components of the two differential fields perpendicular to the axis will vector sum to zero (see Figure 16.76), but the components of the fields along the axis add. Hence, of the piece of field $d\vec{E}$ produced at P by the charge dq on ds, *only* the component of the field along the axis will survive the vector summation as you integrate around the ring. All the differential components of the field perpendicular to the axis pairwise vector sum to zero.

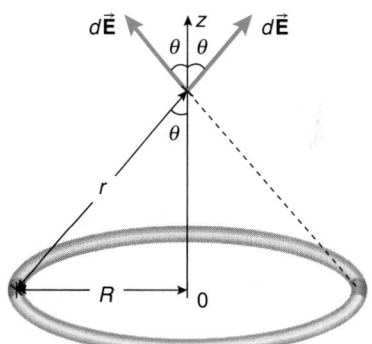

FIGURE 16.75

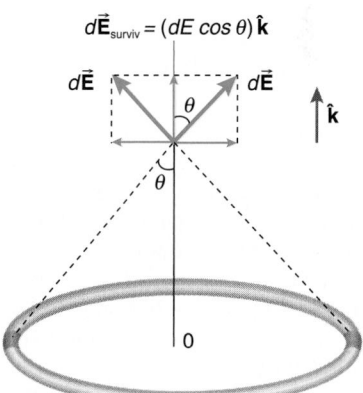

FIGURE 16.76

Hence the *surviving* part of the field caused by the charge on ds is directed along $\hat{\mathbf{k}}$ and is

$$d\vec{E}_{surviv} = \left(\frac{1}{4\pi\varepsilon_0} \frac{\lambda \, ds}{r^2} \cos \theta \right) \hat{\mathbf{k}}$$

Since all the surviving pieces of the field are parallel to each other, the original vector integral over the ring is now a scalar integral for the surviving part of the field:

$$\vec{E}_{surviv} = \frac{1}{4\pi\varepsilon_0} \int_{ring} \left(\lambda \frac{ds}{r^2} \cos\theta \right) \hat{k}$$

As you integrate around the ring, r is a constant because all the differential charges on the ring are equidistant from the field point P. Likewise, $\cos\theta$ also is constant as you integrate around the ring. Thus the integral for the surviving part of the field reduces to

$$\vec{E}_{surviv} = \frac{1}{4\pi\varepsilon_0} \lambda \frac{\cos\theta}{r^2} \int_{ring} ds \, \hat{k}$$

The integral of ds (the arc length) around the ring is the circumference of the ring: $2\pi R$. Thus the surviving field is

$$\vec{E}_{surviv} = \left(\frac{1}{4\pi\varepsilon_0} \lambda \frac{\cos\theta}{r^2} 2\pi R \right) \hat{k}$$

But $2\pi R\lambda$ is the total charge Q on the ring. The Pythagorean theorem gives us r^2:

$$r^2 = z^2 + R^2$$

The geometry in Figure 16.75 implies that

$$\cos\theta = \frac{z}{\left(z^2 + R^2 \right)^{1/2}}$$

Making these substitutions into the expression for $\vec{E}_{surviv}$, you get

$$\vec{E}_{surviv} = \frac{1}{4\pi\varepsilon_0} \frac{Qz}{(z^2 + R^2)^{3/2}} \hat{k} \qquad (1)$$

For positive Q, and for $z > 0$ m, the field points along $+\hat{k}$; for $z < 0$ m, the field is parallel to $-\hat{k}$. In both cases the field is directed along the axis and away from the positively charged ring. You should convince yourself that equation (1) gives the appropriate direction for $\vec{E}$ for a negatively charged ring ($Q < 0$ C) as well: toward the ring.

Notice that when $z = 0$ m in equation (1), the point is at the center of the ring and the surviving field there is zero. Symmetry considerations yield the same conclusion. As one moves out along the axis of the ring, the field first increases and then eventually decreases.

Let's check to see that equation (1) for $\vec{E}$ makes sense if $z \gg R$. When we are very far from the ring, the field of the ring is essentially that of a pointlike charge, because the ring then looks so small. If $z \gg R$, you can neglect R^2 compared with z^2 in the denominator of equation (1). In this case $\vec{E}_{surviv}$ reduces to

$$\vec{E}_{surviv} \approx \frac{1}{4\pi\varepsilon_0} \frac{Qz}{(z^2)^{3/2}} \hat{k}$$

$$\approx \frac{1}{4\pi\varepsilon_0} \frac{Q}{z^2} \hat{k}$$

which is the field of a pointlike charge at a distance z along the z-axis.

To derive an equation for the field of the ring at a point P that is not on the axis is quite difficult and beyond the scope of a course in introductory physics. You may see such an analysis in more advanced courses or use appropriate software to approximate such a field.

EXAMPLE 16.15

Find the electric field at a distance z along the axis of a uniformly charged circular disk of radius R and charge Q (see Figure 16.77).

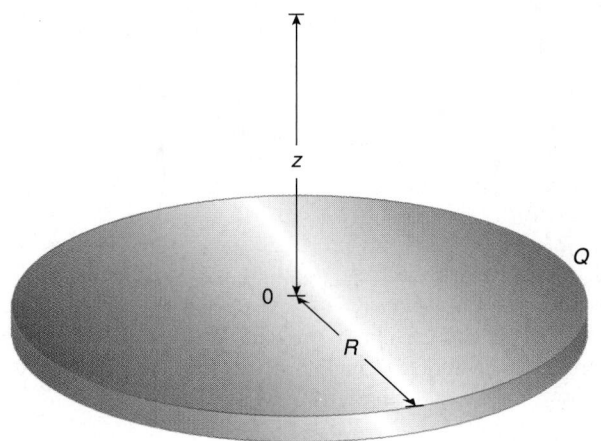

FIGURE 16.77

Solution
The charge is smeared out uniformly over the disk so that the *surface charge density* σ (the number of coulombs on each square meter of the disk) is constant and equal to

$$\sigma = \frac{Q}{\pi R^2}$$

Use the result for the electric field along the axis of a uniformly charged circular ring (Example 16.14). Look at a circular ring of radius r and of differential width dr on the disk; see Figure 16.78.

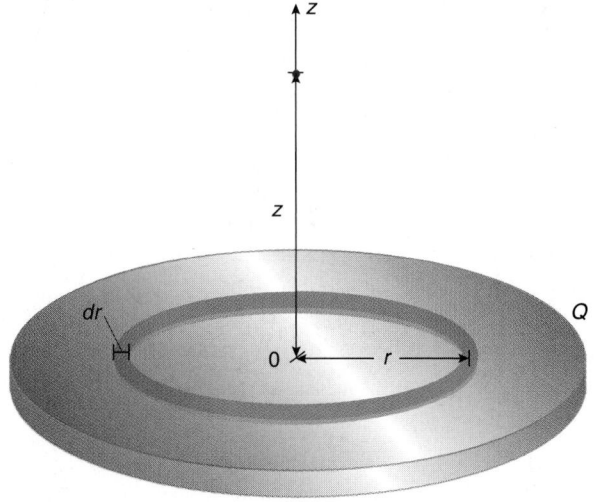

FIGURE 16.78

Imagine unwrapping the ring into a straight-line strip, as in Figure 16.79. The strip has length $2\pi r$ and width dr, and so the differential ring has area

$$2\pi r \, dr$$

FIGURE 16.79

The charge dq on the differential ring is equal to its area times the surface charge density:

$$dq = (2\pi r \, dr)\sigma$$

The charge on the differential ring produces an electric field directed along the axis of the disk. Adapt equation (1) of Example 16.14 for the field of a ring of charge, but let r be the radius of the ring; also, the charge Q on the ring is now $dq = (2\pi r \, dr)\sigma$. Hence the differential ring of charge produces a differential electric field along the axis that is given by

$$d\vec{E} = \frac{1}{4\pi\varepsilon_0} \frac{z \, (2\pi r \, dr \, \sigma)}{(z^2 + r^2)^{3/2}} \hat{k}$$

Now integrate over the disk, since each ring produces an electric field along the axis as well:

$$\vec{E} = \frac{1}{4\pi\varepsilon_0} \int_{0 \text{ m}}^{R} \frac{z \, (2\pi r \, dr \, \sigma)}{(z^2 + r^2)^{3/2}} \hat{k}$$

After removing constant terms from the integral, you obtain

$$\vec{E} = \frac{1}{4\pi\varepsilon_0} z \, (2\pi\sigma) \int_{0 \text{ m}}^{R} \frac{r \, dr}{(z^2 + r^2)^{3/2}} \hat{k}$$

The integral is a standard form of the type

$$\int u^n \, du = \frac{u^{n+1}}{n + 1}$$

Performing the integration, you find

$$\vec{E} = \left[\frac{1}{4\pi\varepsilon_0} z\left(2\pi\sigma\right) \frac{-1}{(z^2 + r^2)^{1/2}} \right]\Bigg|_{0 \text{ m}}^{R} \hat{k}$$

After evaluating the limits and simplifying slightly, you have

$$\vec{E} = \frac{1}{4\pi\varepsilon_0} 2\pi\sigma \left[1 - \frac{z}{(z^2 + R^2)^{1/2}} \right] \hat{k} \qquad (1)$$

Expressing this in terms of the total charge Q on the disk,

$$\vec{E} = \frac{1}{4\pi\varepsilon_0} \frac{2Q}{R^2} \left[1 - \frac{z}{(z^2 + R^2)^{1/2}} \right] \hat{k} \qquad (2)$$

The expression for the field is a bit complicated, but there is nothing that guarantees that all results or equations will be either simple or pretty!

Note that for $Q > 0$ C, the field along the axis is directed away from the disk; for $Q < 0$ C, it is directed along the axis toward the disk.

EXAMPLE 16.16

Find the electric field of a uniformly charged, infinite sheet of charge a distance z from the plane of the sheet (see Figure 16.80).

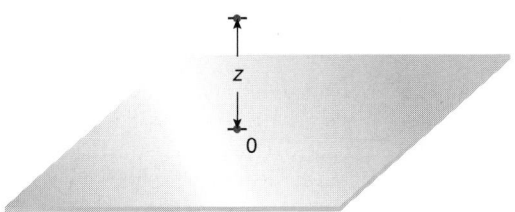

FIGURE 16.80

Solution

Equation (1) of Example 16.15 for the electric field of a uniformly charged circular disk can be used to find the electric field of the uniformly charged infinite sheet by letting the radius R of the disk approach infinity:

$$\vec{E} = \lim_{R \to \infty \text{ m}} \frac{1}{4\pi\varepsilon_0} 2\pi\sigma \left[1 - \frac{z}{(z^2 + R^2)^{1/2}} \right] \hat{k}$$

As the radius R increases without bound with z fixed at any chosen value, the second term in the brackets becomes vanishingly small and you are left with a field of magnitude

$$E = \frac{\sigma}{2\varepsilon_0} \qquad (1)$$

If the charge on the sheet is positive, the field is directed perpendicularly *away* from the sheet, as shown in Figure 16.81. If the sheet has negative charge, the field is directed toward the sheet, as indicated in Figure 16.82.

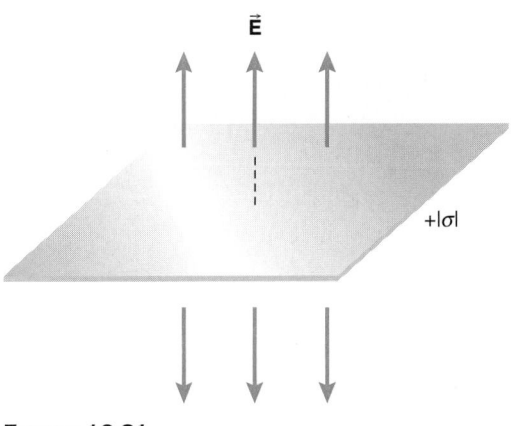

FIGURE 16.81

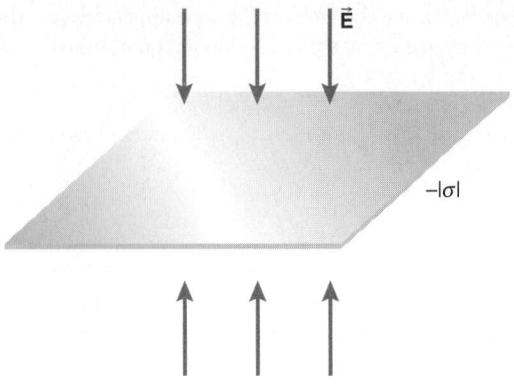

FIGURE 16.82

Such an infinite charged sheet produces a constant electric field. Since the sheet is of infinite extent and no finite distance is significant when compared with infinity, the field is independent of the *x–y* location *and* the distance *z* from the sheet.

EXAMPLE 16.17 ━━━━━━━━━━

Two oppositely charged infinite sheets are separated by a distance *d*; see Figure 16.83. Find the electric field both in the region between the plates and the regions on either side of the plates.

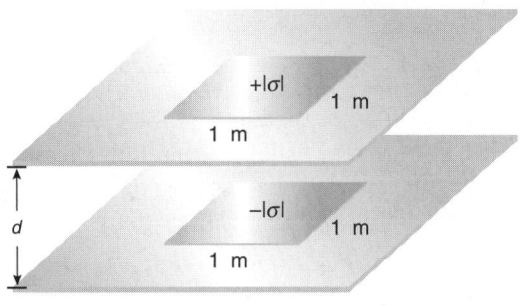

FIGURE 16.83

Solution
We find the electric field by using the principle of superposition. The fields produced by each sheet are indicated separately in Figure 16.84 for three regions.

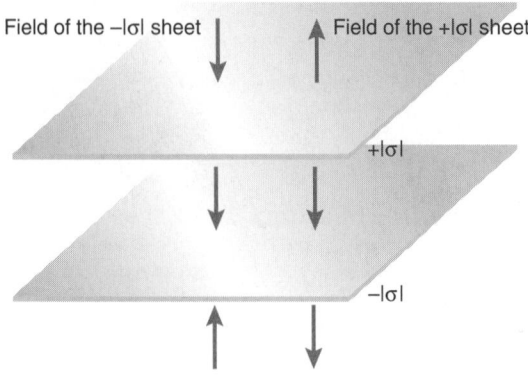

FIGURE 16.84

Notice that below the negative sheet as well as above the positive sheet, the fields of the individual sheets vector sum to zero: the two fields of equal magnitude are antiparallel. In the region between the two sheets, the two fields are parallel. The resulting field between the sheets is then of magnitude

$$E = \frac{\sigma}{2\varepsilon_0} + \frac{\sigma}{2\varepsilon_0}$$

$$= \frac{\sigma}{\varepsilon_0} \qquad (1)$$

where σ is the magnitude of the surface charge density on either sheet.

Such an arrangement of twin, parallel, oppositely charged, infinite sheets is known as an infinite, *parallel plate capacitor* and is the way in which constant electric fields typically are produced in the laboratory. Of course, truly infinite sheets of charge are a bit unwieldy (even if they could be constructed!), and so in practice two finite sheets are placed close together so that the sheets effectively look infinite as long as the location of interest between them is not too near the edges. In the central region between such finite sheets, the electric field is essentially uniform and of magnitude σ/ε_0.

EXAMPLE 16.18 ━━━━━━━━━━

Find the electric field at a point that is a distance *r* from the center of a uniformly charged, hollow spherical shell of radius *R* with charge *Q*; see Figure 16.85. Consider both of the situations *r* < *R* and *r* > *R*.

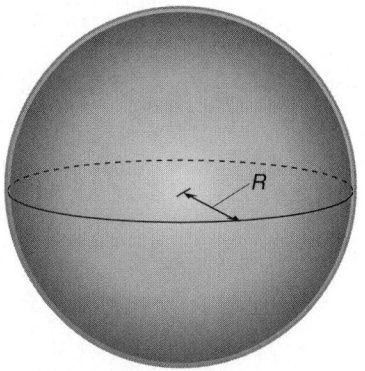

FIGURE 16.85

Solution
Both the gravitational force between pointlike masses and the electrical force between pointlike charges are inverse-square-law forces. The formal similarity between the mathematical expressions for the gravitational force and the electrical force means that the calculations we did in Chapter 6 in deriving the form for the gravitational fields of certain distributions of mass can be applied equally well to the electric fields of the same shape charge distribution.

For example, we discovered (Sections 6.3 and 6.4 and Appendix A) that the gravitational force on a mass *m* located inside a uniform spherical mass shell was zero:

$$\vec{F}_{m \text{ inside uniform shell}} = 0 \text{ N}$$

Since the gravitational force on m is related to the gravitational field $\vec{\mathbf{g}}$ where m is placed by $\vec{\mathbf{F}}_{\text{grav on } m} = m\vec{\mathbf{g}}$, the gravitational field inside the shell must be zero:

$$0 \text{ N} = m\vec{\mathbf{g}}$$

so

$$\vec{\mathbf{g}} = 0 \text{ N/kg}$$

Ultimately, this striking result came about because of the inverse-square-law nature of the gravitational interaction.

Since the electrical interaction also is inverse-square-law, we can say without doing the entire proof again that the electrical field inside a uniform spherical shell of charge also is zero:

$$\vec{\mathbf{E}}_{\text{inside uniform shell}} = 0 \text{ N/C} \qquad (\text{any point } r < R)$$

Hence a charge q placed inside the shell (that does not disturb the uniform distribution of charge on the shell) experiences zero total electrical force:

$$\begin{aligned}
\vec{\mathbf{F}}_{\text{elec on } q} &= q\vec{\mathbf{E}} \\
&= q(0 \text{ N/C}) \\
&= 0 \text{ N}
\end{aligned}$$

Correspondingly the other shell theorem in Chapter 6 means that here the electric field of a uniformly charged spherical shell is the same as that of an equal point charge located at the center of the sphere:

$$\vec{\mathbf{E}} = \frac{1}{4\pi\varepsilon_0} \frac{Q}{r^2} \hat{\mathbf{r}} \qquad (\text{any point } r > R)$$

EXAMPLE 16.19

Find the electric field at a distance r from the center of a sphere with charge Q distributed uniformly throughout the volume of the sphere of radius R; see Figure 16.86. Consider both cases: $r < R$ and $r > R$.

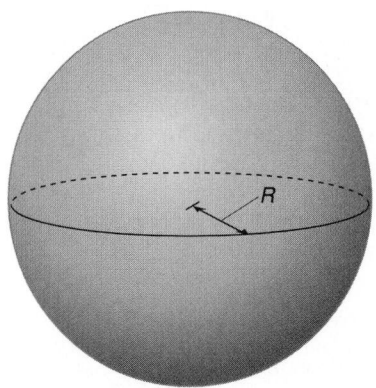

FIGURE 16.86

Solution

Analogously to the gravitational situation, the electric field at a point r *within* a uniformly charged sphere is caused only by the charge located closer to the center of the sphere than the point itself. We modify the case from Table 6.1:

$$\vec{\mathbf{E}} = \frac{1}{4\pi\varepsilon_0} \frac{Q}{R^3} r\hat{\mathbf{r}} \qquad (r < R)$$

Likewise, the electric field at a point *outside* the spherical charge distribution, a distance r from its center, is the same as the field produced by an equal point charge located at the center of the sphere:

$$\vec{\mathbf{E}} = \frac{1}{4\pi\varepsilon_0} \frac{Q}{r^2} \hat{\mathbf{r}} \qquad (r > R)$$

16.10 MOTION OF A CHARGED PARTICLE IN A UNIFORM ELECTRIC FIELD: AN ELECTRICAL PROJECTILE

The motion of a charged particle in a uniform electric field has important technological applications in low-energy particle accelerators, television tubes, and CRT displays, among other things. As we have seen (Example 16.17), a uniform electric field exists in the region between two large, oppositely charged parallel plates. There is a strong similarity between the motion of a charged particle in a uniform electric field and the more familiar motion of a projectile in a uniform gravitational field. Since the electric field is uniform, the force is constant and the acceleration of the charge is constant. You can use the kinematic equations for motion under a constant acceleration (Chapters 3 and 4).

QUESTION 7

A charge q is moving with velocity $\vec{\mathbf{v}}$ perpendicular to an electric field. Is the electrical force on the charge zero? Describe the analogous gravitational situation.

STRATEGIC EXAMPLE 16.20

A particle of mass m and charge q is released at rest close to one of two parallel plates in a region where there is a uniform electric field of magnitude E_0. The particle accelerates toward the other plate, a distance d away. Determine the speed at which it makes impact with the opposite plate.

Solution

Use the coordinate system indicated in Figure 16.87. The electric field then is

$$\vec{\mathbf{E}} = -E_0\hat{\mathbf{j}}$$

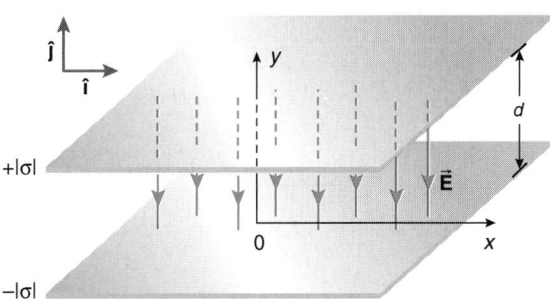

FIGURE 16.87

The charge q experiences an electrical force

$$\vec{F} = q\vec{E}$$
$$= q(-E_0\hat{j})$$
$$= -qE_0\hat{j}$$

We want to determine the speed at which it makes impact with the other plate. The particle is a one-dimensional electrical projectile.

Method 1
If q is positive, the force on it is parallel to the field direction, and so the particle should be released near the upper plate in Figure 16.87. If negative, the force is antiparallel to the field, and the particle should be released near the lower plate in Figure 16.87. The acceleration $\vec{a}$ of the charge is found using Newton's second law:

$$\vec{F} = m\vec{a}$$

The force is constant, and so the acceleration is constant as well:

$$\vec{a} = \frac{\vec{F}}{m}$$
$$= -\frac{qE_0}{m}\hat{j}$$

The acceleration is purely along the y-direction, and the y-component of the acceleration (a_y) is constant and equal to $-qE_0/m$. Since the acceleration is constant, you can apply the one-dimensional kinematic equations of Chapter 3 to determine the impact speed of the particle as it hits the other plate. The problem is analogous to releasing a mass at rest in a uniform gravitational field. Specifically, for motion along the y-direction,

$$v_y = v_{y0} + a_yt \qquad (1)$$

and

$$y = y_0 + v_{y0}t + \frac{1}{2}a_yt^2 \qquad (2)$$

Since the particle was released at rest, $v_{y0} = 0$ m/s. If you assume q is positive, it is released at the coordinate $y_0 = d$. Thus equation (1) simplifies to

$$v_y = a_yt$$
$$= -\frac{qE_0}{m}t \qquad (3)$$

and equation (2) becomes

$$y = d + \frac{1}{2}a_yt^2$$
$$= d - \frac{1}{2}\frac{qE_0}{m}t^2 \qquad (4)$$

Impact occurs at coordinate $y = 0$ m. Thus equation (4) becomes

$$0\text{ m} = d - \frac{1}{2}\frac{qE_0}{m}t^2$$

or

$$d = \frac{1}{2}\frac{qE_0}{m}t^2$$

Solving for t,

$$t = \left(\frac{2dm}{qE_0}\right)^{1/2}$$

The velocity component on impact, using equation (3), is

$$v_y = -\frac{qE_0}{m}\left(\frac{2dm}{qE_0}\right)^{1/2}$$
$$= -\left(2d\frac{qE_0}{m}\right)^{1/2}$$

Method 2
You also can use the CWE theorem to determine the speed. The work done by all the forces acting on the particle is equal to the change in the kinetic energy of the particle:

$$W_{\text{all}} = \Delta KE$$

The only force acting is the electrical force

$$\vec{F} = q\vec{E}$$
$$= -qE_0\hat{j}$$

which is constant. The work W done by this constant force as the particle zips between the plates, changing its position vector by $\Delta\vec{r}$, is

$$W = \vec{F} \cdot \Delta\vec{r} \qquad \text{(constant force } \vec{F}\text{)}$$

If the particle has a positive charge, it is released near the upper plate, and so the change in the position vector of the charge is

$$\Delta\vec{r} = (0\text{ m})\hat{j} - d\hat{j}$$
$$= -d\hat{j}$$

The work is then

$$W = (-qE_0\hat{j}) \cdot (-d\hat{j})$$
$$= qE_0d$$

The particle has no initial kinetic energy, and its kinetic energy just before impact is

$$\frac{mv_y^2}{2}$$

Thus the CWE theorem

$$W_{\text{all}} = \Delta KE$$

becomes

$$qE_0d = \frac{mv_y^2}{2} - 0\text{ J}$$

Solving for v_y, you find

$$v_y = -\left(2d\frac{qE_0}{m}\right)^{1/2}$$

The negative root is chosen because the positive charge is moving toward decreasing values of y.

If a small hole is bored through the plate toward which the particle is traveling, the particle emerges and traverses the region outside the plates at constant velocity, since there is no electric field in that region. We have created a charged particle pea-shooter, an electrical rifle, known more elegantly as a charged particle accelerator. As long as the particles are released from rest, they will emerge from the hole at the constant speed v given by

$$v = \left(2d \, \frac{qE_0}{m} \right)^{1/2} \tag{5}$$

EXAMPLE 16.21

A particle of mass m with charge q, initially moving at constant velocity, encounters a region with a uniform electric field at right angles to its initial velocity; see Figure 16.88. Determine the deflection of the particle when it emerges from the region of uniform field. Assume it does not strike either charged sheet.

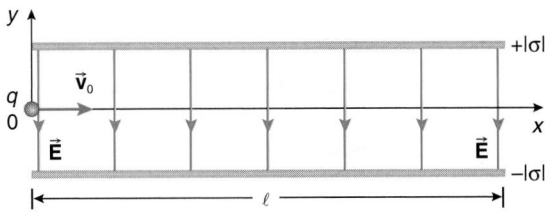

FIGURE 16.88

Solution

Use the given coordinate system. The initial velocity is

$$\vec{\mathbf{v}}_0 = v_{x0}\hat{\mathbf{i}}$$

The uniform electric field is

$$\vec{\mathbf{E}} = -E_0\hat{\mathbf{j}}$$

The electrical force on the particle is found from the field:

$$\vec{\mathbf{F}} = q\vec{\mathbf{E}}$$
$$= q(-E_0\hat{\mathbf{j}})$$

The force is perpendicular to the initial velocity of the particle. Therefore the particle continues with a constant x-velocity component v_{x0}, but the force in the y-direction causes it to accelerate in the y-direction. You have a two-dimensional electrical projectile. If q is positive, the particle is deflected in the direction of the field; if q is negative, the particle is deflected antiparallel to the field.

The particle is within the region of field during a time interval t found by dividing the length ℓ by the constant x-velocity component v_{x0}:

$$t = \frac{\ell}{v_{x0}}$$

During this interval, the particle is deflected along the y-direction. The deflection of the particle as it leaves the field is found using the one-dimensional kinematic equation for constant acceleration:

$$y = y_0 + v_{y0}t + \frac{1}{2}a_y t^2$$

With the coordinates in Figure 16.88, you have $y_0 = 0$ m and $v_{y0} = 0$ m/s. Hence the equation for y reduces to

$$y = \frac{1}{2}a_y t^2$$

The acceleration component a_y is the force component F_y divided by the mass:

$$a_y = -\frac{qE_0}{m}$$

The time of flight you have already found to be ℓ/v_{x0}. Thus the deflection is

$$y = \frac{1}{2}\left(-\frac{qE_0}{m}\right)\left(\frac{\ell}{v_{x0}}\right)^2 \tag{1}$$

The direction of the deflection is different for positive and for negative charges.

By controlling the magnitude of the field, the deflection can be varied at will. This is the essential principle governing the operation of all CRT (cathode ray tube) displays. TV, some computer screens, oscilloscopes, and the like are based on the motion of charged particles in electric fields. In such devices the particles used are negative charges: electrons. In equation (1), note that the deflection depends inversely on the mass of the charged particle. The small mass of electrons means that they respond to the action of electrical forces more dramatically than would more massive protons in the same electric field.

16.11 GAUSS'S LAW FOR ELECTRIC FIELDS*

Here we extend the idea of Gauss's law to electric fields. We want to know what result is obtained for the flux of the electric field through a closed surface. Fortunately, most of the calculations have been done already. We can build on what we did in formulating Gauss's law for the gravitational field, though we have to account for the fact that there are two different kinds of electric charge but only one kind of mass.

We first encountered the idea of the flux of a vector in our study of the gravitational field in Chapter 6. The differential flux $d\Phi$ of any vector $\vec{\mathbf{A}}$ through a differential area $d\vec{\mathbf{S}}$ is a measure of the extent to which the vector $\vec{\mathbf{A}}$ passes through the area $d\vec{\mathbf{S}}$. Specifically we defined the differential flux of a vector $\vec{\mathbf{A}}$ via the scalar product:

$$d\Phi = \vec{\mathbf{A}} \cdot d\vec{\mathbf{S}}$$

The flux of the vector through a finite surface S then is found by integrating $d\Phi$ over the area in question:

$$\Phi = \int_{\text{area } S} \vec{\mathbf{A}} \cdot d\vec{\mathbf{S}}$$

Mathematically, a flux can be associated with any vector; but only certain vectors have a flux that is significant from a physical viewpoint. One of those vectors is the gravitational field $\vec{g}$; another is the electric field $\vec{E}$.

In our study of the gravitational field in Chapter 6, we discovered (in Section 6.15) a curious general relationship between the flux of the total gravitational field through *any closed surface* and the amount of mass trapped within the surface. Specifically this relationship is *Gauss's law for the gravitational field* (Equation 6.55):

$$\int_{\substack{\text{clsd} \\ \text{surface } S}} \vec{g} \cdot d\vec{S} = -4\pi G M_{\text{within } S}$$

Recall that for a closed surface, the direction of the differential area vector $d\vec{S}$ is perpendicular to the surface and directed outward. In the derivation of Gauss's law in Chapter 6, the minus sign arose from the scalar product of $\vec{g}$ with $d\vec{S}$: with $\vec{g}$ directed into the surface (toward the mass M) and $d\vec{S}$ directed perpendicularly outward from the surface, the angle between the two vectors is greater than 90° (see Figure 16.89) and thus the scalar product is negative (the cosine of angles greater than 90° is negative).

The negative right-hand side of Gauss's law for the gravitational field means that the gravitational field lines have a net influx through the surface S; they thread in but not out. If the surface S has a total mass M within it, then the flux of the gravitational field through the closed surface S is $-4\pi G M$.

If the closed surface S has no mass within it, the flux of the gravitational field through S is zero. Every line of $\vec{g}$ enters the surface but also leaves it (see Figure 16.90); there is no net threading of the surface.

The shape of the closed surface S is arbitrary; what matters is whether there is mass trapped inside it or not. For a weirdly shaped surface (such as in Figures 16.89 and 16.90), the surface integral on the left-hand side of Gauss's law is difficult to evaluate. But the *result* of the complicated integration is just $-4\pi G M_{\text{within } S}$. That is, the right-hand side of Gauss's law is easy to evaluate, because it is simple to account for the mass within the surface S.

The parallels between gravitational and electrical phenomena arise because of the similarity of the underlying force laws between masses and between charges. There are differences between the interactions, of course, notably (1) in the intrinsic strength of the interactions (cf. Example 16.5) and (2) in the direction of the forces: gravitation *always* is an attractive force (there is only one kind of mass, positive mass); whereas electrical forces can be *either* attractive or repulsive, as there are two types of charges.

When a (positive) mass M is *anywhere* inside the closed surface, the flux of the gravitational field through the surface is

$$\int_{\substack{\text{clsd} \\ \text{surface } S}} \vec{g} \cdot d\vec{S} = -4\pi G M$$

The gravitational field $\vec{g}$ is directed toward the mass M as shown in Figure 16.89. The electric field of a *negative charge* Q also points toward the charge, as shown in Figure 16.91.

So to convert Equation 6.55 to its electrical analog, replace

1. the field $\vec{g}$ with the field $\vec{E}$;
2. the mass M with the charge $|Q|$; and
3. the universal gravitational constant G with the corresponding electrical constant $1/4\pi\varepsilon_0$.

Thus we have

$$\int_{\substack{\text{clsd surface} \\ \text{enclosing } Q < 0\,\text{C}}} \vec{E} \cdot d\vec{S} = -4\pi \frac{1}{4\pi\varepsilon_0} |Q|$$

$$= -\frac{|Q|}{\varepsilon_0}$$

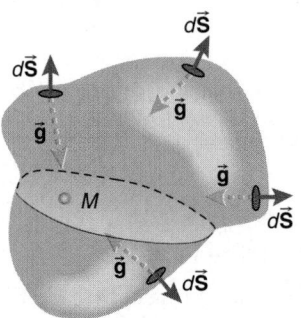

FIGURE 16.89 For the gravitational case with M inside S, the field $\vec{g}$ is inward, while $d\vec{S}$ is outward.

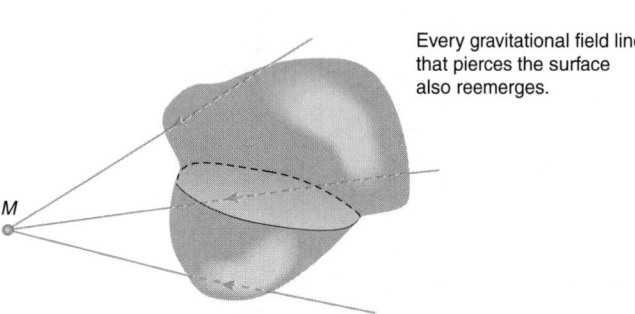

Every gravitational field line that pierces the surface also reemerges.

FIGURE 16.90 If M is outside the closed surface, there is no net threading of the gravitational field through the surface.

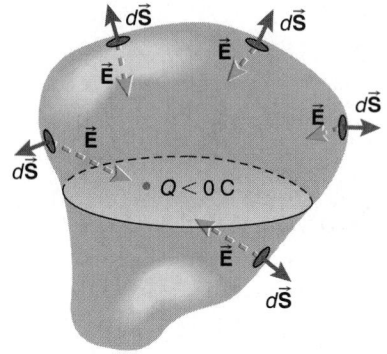

FIGURE 16.91 The electric field of a negative charge points toward the charge.

However, since Q itself is negative, we eliminate the absolute value signs by simply writing this as

$$\int_{\substack{\text{clsd surface} \\ \text{enclosing } Q < 0\,\text{C}}} \vec{\mathbf{E}} \cdot d\vec{\mathbf{S}} = \frac{Q}{\varepsilon_0}$$

What happens if a *positive charge* is inside the closed surface? Now the electric field of the charge points radially away from the charge (see Figure 16.92), opposite to the direction of the gravitational field $\vec{\mathbf{g}}$.

Thus, for a positive charge within closed surface S, we can convert Equation 6.55 to the electrical case by

1. changing the field from $\vec{\mathbf{g}}$ to $-\vec{\mathbf{E}}$;
2. changing $|M|$ to $|Q|$; and
3. changing the universal gravitational constant G to the electrical constant $1/4\pi\varepsilon_0$.

Thus the flux of the electric field through a closed surface surrounding a positive charge is

$$-\int_{\substack{\text{clsd surface} \\ \text{enclosing } Q > 0\,\text{C}}} \vec{\mathbf{E}} \cdot d\vec{\mathbf{S}} = -4\pi \frac{1}{4\pi\varepsilon_0} \left| Q \right|$$

or

$$\int_{\substack{\text{clsd surface} \\ \text{enclosing } Q > 0\,\text{C}}} \vec{\mathbf{E}} \cdot d\vec{\mathbf{S}} = 4\pi \frac{1}{4\pi\varepsilon_0} \left| Q \right|$$

Since Q is itself positive, we can remove the absolute value signs and write

$$\int_{\substack{\text{clsd surface} \\ \text{enclosing } Q > 0\,\text{C}}} \vec{\mathbf{E}} \cdot d\vec{\mathbf{S}} = \frac{Q}{\varepsilon_0}$$

In the gravitational case, if there is zero mass within the closed surface, there is no net flux of the gravitational field through the closed surface. The same is true for the electrical case: if there is zero total charge within the closed surface S, there is no net flux of the electric field vector through S.

The results for both a positive or negative charge or zero charge within the closed surface can be summarized with the same equation:

$$\int_{\substack{\text{clsd} \\ \text{surface } S}} \vec{\mathbf{E}} \cdot d\vec{\mathbf{S}} = \frac{Q_{\text{within } S}}{\varepsilon_0} \tag{16.17}$$

where the appropriate sign is used for Q.

What happens if several different charges are enclosed by the surface S, as in Figure 16.93? Since the electric fields of the charges follow a linear principle of superposition, the fluxes of each field through the surface add algebraically. That is, the flux of the total field $\vec{\mathbf{E}}$ through the closed surface is

$$\int_{\substack{\text{clsd} \\ \text{surface}}} \vec{\mathbf{E}} \cdot d\vec{\mathbf{S}} = \int_{\substack{\text{clsd} \\ \text{surface}}} \vec{\mathbf{E}}_1 \cdot d\vec{\mathbf{S}} + \int_{\substack{\text{clsd} \\ \text{surface}}} \vec{\mathbf{E}}_2 \cdot d\vec{\mathbf{S}}$$
$$+ \int_{\substack{\text{clsd} \\ \text{surface}}} \vec{\mathbf{E}}_3 \cdot d\vec{\mathbf{S}} + \cdots$$
$$= \frac{Q_1}{\varepsilon_0} + \frac{Q_1}{\varepsilon_0} + \frac{Q_1}{\varepsilon_0} + \cdots$$

where it is important to use the appropriate sign for the individual charges.

Any charge *outside* the closed surface S (such as charge Q' in Figure 16.93) contributes zero flux through the surface; we already showed this for mass in Chapter 6.

> Thus Gauss's law for the electric field can be summarized in the following way:
>
> $$\int_{\substack{\text{clsd} \\ \text{surface } S}} \vec{\mathbf{E}} \cdot d\vec{\mathbf{S}} = \frac{\text{total (net) charge enclosed by } S}{\varepsilon_0} \tag{16.18}$$
>
> Gauss's law is a very general relationship between the flux of the *total* electric field through *any closed surface* and the *total (net) charge enclosed* by the surface.

PROBLEM-SOLVING TACTIC

16.6 To evaluate the flux of the electric field through a closed surface, you need only tally the total (i.e., net) charge enclosed within the surface and divide by ε_0. You need not perform the typically complicated integration on the left-hand side of Gauss's law.

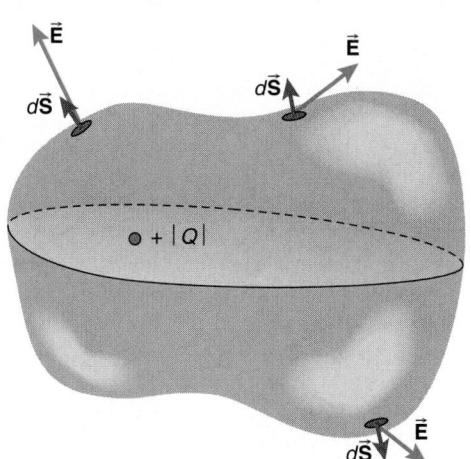

FIGURE 16.92 The electric field of a positive charge points away from the charge.

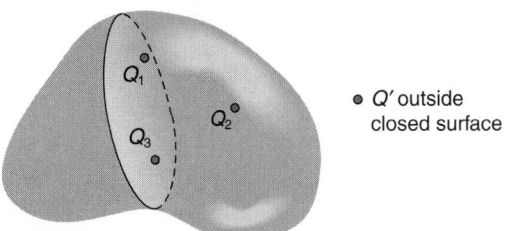

FIGURE 16.93 Several charges within a closed surface.

It is important to realize that Φ is the flux of the field caused by all charges anywhere, even those not enclosed by the surface. The imaginary closed surface involved in Gauss's law is known as a **Gaussian surface**, and we can choose its shape any way that seems useful.

Gauss's law (Equation 16.18) is one of four fundamental equations of electromagnetism called the **Maxwell equations**, named after James Clerk Maxwell (1831–1879), the great 19th-century Scottish physicist who first merged electric and magnetic phenomena into a single encompassing theory. Gauss's law stems from the Coulomb force law between charges; but physicists regard Gauss's law (rather than Coulomb's law) as the fundamental equation, since it expresses the relationship between electric fields and electric charge.

We also finally can see why the peculiar numerical factor of 4π was introduced into Coulomb's law (via $1/4\pi\varepsilon_0$). With the 4π in Coulomb's law (Equation 16.4), there is *no* factor of 4π in Gauss's law for the electric field (Equation 16.18). On the other hand, in the gravitational context, the gravitational force law *lacks* the factor of 4π because the force law has the constant written simply as G, and so a factor of 4π appears in the gravitational version of Gauss's law. The reason a similar juggling of factors of 4π was not done with the gravitational force law was simply because of the difficulty of changing notation. By the time Gauss's law was formulated for electromagnetism during the 19th century, the gravitational theory of Newton had been around for two centuries in the form we still know it. It was just too difficult to change notation at that point in time. This gives you a feeling for the collective inertia associated with the notation of any established theory or practice!

QUESTION 8

If Benjamin Franklin had defined electric charge the other way around, so that protons had negative charge and electrons had positive charge, how would this have affected the form of Gauss's law?

EXAMPLE 16.22

What is the flux of the total electric field through each of the closed Gaussian surfaces in Figures 16.94a–d?

Solution

According to Gauss's law (Equation 16.18), in each case you tally the total (net) charge enclosed within each closed Gaussian surface and divide by ε_0.

a. The total charge enclosed by the surface is zero, and so the flux of the electric field through the closed surface is zero.
b. The total charge enclosed by the surface is $+2.00 \ \mu$C. Thus the total flux is

$$\Phi = \frac{+2.00 \times 10^{-6} \ \text{C}}{\varepsilon_0}$$

$$= \frac{2.00 \times 10^{-6} \ \text{C}}{8.85 \times 10^{-12} \ \text{C}^2/(\text{N} \cdot \text{m}^2)}$$

$$= 2.26 \times 10^5 \ \text{N} \cdot \text{m}^2/\text{C}$$

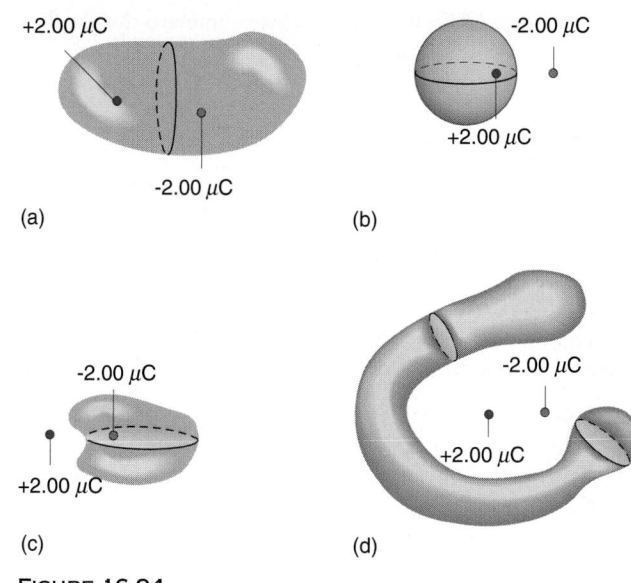

FIGURE 16.94

c. The total charge enclosed by the surface is $-2.00 \ \mu$C, and so the flux of the electric field through the surface is

$$\Phi = \frac{-2.00 \times 10^{-6} \ \text{C}}{8.85 \times 10^{-12} \ \text{C}^2/(\text{N} \cdot \text{m}^2)}$$

$$= -2.26 \times 10^5 \ \text{N} \cdot \text{m}^2/\text{C}$$

d. The total charge enclosed by the surface is zero, and so the flux of the electric field through the surface is zero.

16.12 CALCULATING THE MAGNITUDE OF THE ELECTRIC FIELD USING GAUSS'S LAW*

Now we discover how Gauss's law gives us amazingly rapid answers for the magnitude of the electric field for several highly symmetric distributions of charge. This method avoids the use of the complicated vector integration involved with Equation 16.16.

> The idea of using Gauss's law to calculate the magnitude of an electric field stems from the very generality of the law: the shape of the closed surface involved is arbitrary. This means that in some situations we may be able to judiciously *choose* a surface that will assist us in finding the magnitude of the field.

The idea behind the method is to concoct an imaginary closed Gaussian surface that enables us to explicitly calculate the flux integral on the left-hand side of Gauss's law. We want the integral for the flux to be as easy as possible. Thus we devise a closed surface of such a shape that a constant-magnitude electric field exists over all or at least part of the surface. The flux integral on the left-hand side of the law then is calculated and set equal to the net charge enclosed by the surface, divided by ε_0.

The symmetry of the charge distribution is a clue to choosing the shape of the Gaussian surface over which to evaluate the flux integral. But it is only for situations involving high symmetry that we can use Gauss's law to calculate the magnitude of the

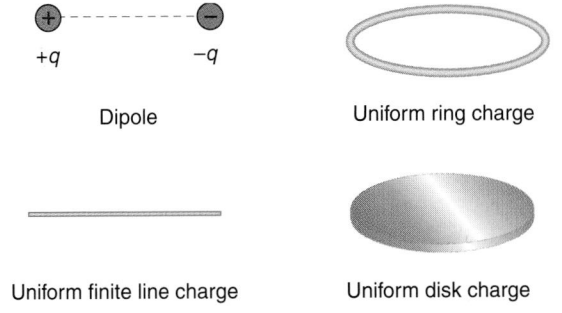

FIGURE 16.95 Charge distributions that *lack* the necessary high symmetry needed to use Gauss's law to find their electric fields.

field. What we mean by the term *high symmetry* is best seen by studying a number of specific examples. There is only a small collection of charge distributions with sufficient symmetry to permit Gauss's law to be usefully employed to find the magnitude of the electric field. In most situations, the charge distribution does not possess enough symmetry to enable us to find a convenient Gaussian surface over which the field has a constant magnitude. Several common charge distributions that *lack* sufficient symmetry are shown in Figure 16.95.

Gauss's law *still applies* to any closed surface; we just cannot easily use the law to find the field because the flux integral on the left-hand side of the law is too complicated to evaluate with ease or in closed form. However, we still can use Gauss's law to find the value of the total electric flux. For these geometries, we must resort to the vector integration techniques of Equation 16.16 to calculate the electric field at any point. This was just what we did in Section 16.9.

The Magnitude of the Electric Field of a Single, Pointlike Charge

We already know what this field is, but the point here is to see how the field can be found using Gauss's law in a neat way. Consider Q to be a positive charge. The field will point radially away from Q, as shown in Figure 16.96.

The symmetry of the situation is the key. Symmetry implies that as long as we remain at a fixed distance r away from the charge, the field $\vec{E}$ has a constant magnitude. Thus the magnitude of the field is constant over the entire surface of a sphere of radius r centered on the charge. So the spherical symmetry indicates that we should choose a sphere of radius r, centered on Q, to be the mathematical Gaussian surface over which to calculate explicitly the flux of the electric field—that is, over which to evaluate the integral on the left-hand side of Gauss's law.

The field is constant in magnitude over the entire surface and directed radially outward. The various differential surface area vectors $d\vec{S}$ also are directed radially outward and so are parallel everywhere to the local $\vec{E}$ at their locations. Thus the scalar product on the left-hand side of Gauss's law becomes

$$\int_{\text{sphere}} \vec{E} \cdot d\vec{S} = \frac{Q_{\text{net enclosed}}}{\varepsilon_0}$$

$$\int_{\text{sphere}} E \, dS \cos 0° = \frac{Q_{\text{net enclosed}}}{\varepsilon_0}$$

$$\int_{\text{sphere}} E \, dS = \frac{Q_{\text{net enclosed}}}{\varepsilon_0}$$

Since the magnitude of the electric field is constant over the entire surface, E can be brought out from the integral and we can integrate the *left*-hand side of Gauss's law, obtaining

$$E\left(4\pi r^2\right) = \frac{Q_{\text{net enclosed}}}{\varepsilon_0}$$

because $4\pi r^2$ is the surface area of the spherical Gaussian surface.

The charge enclosed by the Gaussian surface is just Q. Thus Gauss's law becomes

$$E\left(4\pi r^2\right) = \frac{Q}{\varepsilon_0}$$

When we solve for the magnitude of the electric field E, a familiar result emerges:

$$E = \frac{1}{4\pi\varepsilon_0} \frac{Q}{r^2}$$

If the pointlike charge is *negative*, finding the field using Gauss's law proceeds along similar lines but with a few variations. The field of a *negative* charge points radially toward the charge, as shown in Figure 16.97.

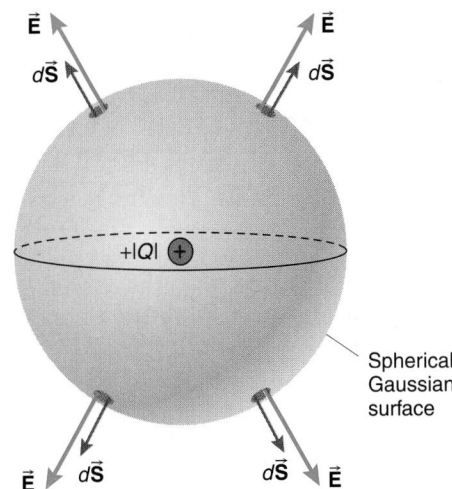

FIGURE 16.96 The electric field a distance r from a pointlike charge.

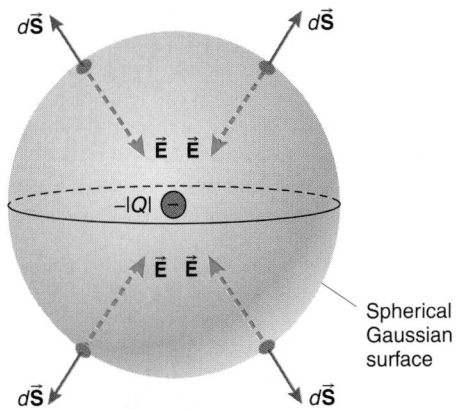

FIGURE 16.97 The electric field of a negative pointlike charge points radially toward the charge.

The symmetry indicates that at a fixed distance r away from the charge, the magnitude of the field is constant. Thus the magnitude of the field is constant over the surface area of a sphere centered on the charge. As before, we choose this sphere as the Gaussian surface over which to evaluate the flux of the electric field. Let E be the magnitude of the field. The differential surface area vector $d\vec{S}$ is *antiparallel* to $\vec{E}$ over the entire Gaussian surface. Thus the left-hand side of Gauss's law becomes

$$\int_{\text{sphere}} \vec{E} \cdot d\vec{S} = \frac{Q_{\text{net enclosed}}}{\varepsilon_0}$$

$$\int_{\text{sphere}} E\, dS \cos 180° = \frac{Q_{\text{net enclosed}}}{\varepsilon_0}$$

$$-\int_{\text{sphere}} E\, dS = \frac{Q_{\text{net enclosed}}}{\varepsilon_0}$$

Since E is constant over the entire surface, the left-hand side of the law becomes

$$-E \int_{\text{sphere}} dS = \frac{Q_{\text{net enclosed}}}{\varepsilon_0}$$

$$-E\left(4\pi r^2\right) = \frac{Q_{\text{net enclosed}}}{\varepsilon_0}$$

The charge enclosed by the surface is $-|Q|$. Gauss's law becomes

$$-E\left(4\pi r^2\right) = -\frac{|Q|}{\varepsilon_0}$$

and so the magnitude of the field (a positive number!) is

$$E = \frac{1}{4\pi\varepsilon_0} \frac{|Q|}{r^2}$$

Once again, a familiar result.

The Magnitude of the Electric Field of an Infinite, Uniform Line Charge Distribution

Let there be $+|\lambda|$ coulombs of charge on each meter of the line, so that the linear charge density is $+|\lambda|$, as shown in Figure 16.98.

The symmetry of the charge distribution indicates that the field will everywhere point radially away from the line and that if we maintain a fixed distance r from the line, the magnitude of the field will be constant. The cylindrical symmetry indicates that an appropriate Gaussian surface for this problem is a cylinder of radius r, concentric with the line charge, as shown in Figure 16.99.

The magnitude of the electric field is constant over the lateral area of the Gaussian surface but is not constant over the ends. This latter fact may seem like a problem, but we will see, in fact, it is not.

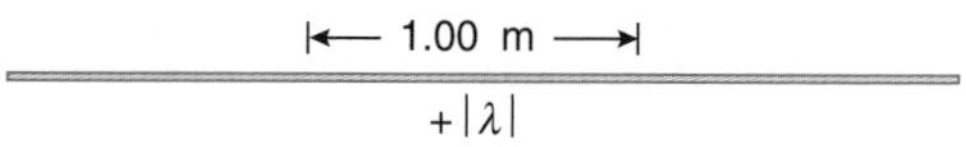

The flux of the electric field through this closed, cylindrical Gaussian surface consists of the flux through the cylindrical lateral area plus the flux through the two ends:

$$\int_{\text{lateral}} \vec{E} \cdot d\vec{S} + \int_{\text{end 1}} \vec{E} \cdot d\vec{S} + \int_{\text{end 2}} \vec{E} \cdot d\vec{S}$$

Over the lateral area, the differential surface area vectors $d\vec{S}$ are everywhere parallel to $\vec{E}$ at their locations. The field $\vec{E}$ also is constant in magnitude over the lateral area, and so the flux through the lateral area is

$$\int_{\text{lateral}} \vec{E} \cdot d\vec{S} = \int_{\text{lateral}} E\, dS = E \int_{\text{lateral}} dS = E\left(2\pi r \ell\right)$$

where ℓ is the length of the Gaussian cylinder. Over each end surface, the differential area vectors $d\vec{S}$ always are perpendicular to the direction of $\vec{E}$ at every location. Thus over the ends, the scalar product everywhere is zero:

$$\vec{E} \cdot d\vec{S} = 0 \text{ N} \cdot \text{m}^2/\text{C}$$

Hence there is zero flux through the end surfaces. (Notice that the lines of $\vec{E}$ do not thread the end surfaces of the Gaussian cylinder.) Hence the flux of the electric field vector through the entire closed surface is just that through the lateral area. The charge enclosed by the Gaussian surface is the charge on the length ℓ of the line within the Gaussian cylinder: $\lambda\ell$. Thus Gauss's law becomes

$$E\left(2\pi r \ell\right) = \frac{\lambda\ell}{\varepsilon_0}$$

Solving for the magnitude of the field, we find

$$E = \frac{\lambda}{2\pi r\varepsilon_0} \qquad (16.19)$$

$$E = \frac{1}{4\pi\varepsilon_0} \frac{2\lambda}{r}$$

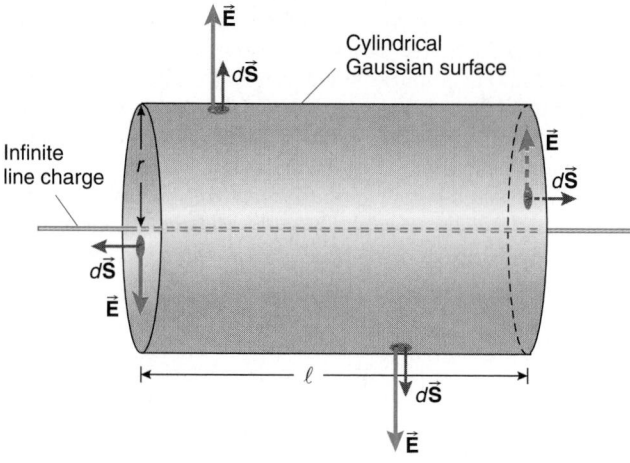

FIGURE 16.99 A cylindrical Gaussian surface concentric with the line charge.

|←—— 1.00 m ——→|

$+|\lambda|$

FIGURE 16.98 An infinite line charge with λ coulombs of charge per meter of length.

The Magnitude of the Electric Field Inside a Uniformly Charged Solid Sphere

Let charge $+|Q|$ be distributed uniformly throughout a spherical volume of radius R. We want to find the magnitude of the electric field inside the charge distribution at a distance $r < R$ from the center, as shown in Figure 16.100.

The spherical symmetry indicates that the electric field will be in the radial direction. The magnitude of the field will be same for a constant value of r. Hence we choose a spherical Gaussian surface of radius r concentric with the spherical charge distribution. The field has a constant magnitude over this surface and is in the same direction as the differential surface area vectors $d\vec{S}$ at all points on the surface. Thus the left-hand side of Gauss's law is

$$\int \vec{E} \cdot d\vec{S} = \frac{Q_{\text{net enclosed}}}{\varepsilon_0}$$

$$E \int dS = \frac{Q_{\text{net enclosed}}}{\varepsilon_0}$$

$$E\left(4\pi r^2\right) = \frac{Q_{\text{net enclosed}}}{\varepsilon_0}$$

The amount of charge within the Gaussian surface is just that fraction of the total charge that lies closer to the center than r. Let ρ be the volume charge density of the charge distribution:

$$\rho \equiv \frac{|Q|}{\frac{4}{3}\pi R^3}$$

Then the charge within the Gaussian sphere of radius r is

$$\frac{4}{3}\pi r^3 \rho$$

or

$$\frac{r^3}{R^3}|Q|$$

Hence Gauss's law becomes

$$E\left(4\pi r^2\right) = \frac{\frac{r^3}{R^3}|Q|}{\varepsilon_0}$$

Solving for the magnitude of the field, we find

$$E = \frac{1}{4\pi\varepsilon_0}\frac{|Q|}{R^3}r \qquad (16.20)$$

Notice that the field increases in magnitude linearly with r until $r = R$. This is analogous to the gravitational field inside a uniform sphere of mass, which also increases linearly with r inside the sphere, as indicated in Table 6.1. If the charged sphere is positive, the field is directed radially outward; if negative, the field is directed radially inward.

The Magnitude of the Electric Field of a Uniformly Charged Infinite Sheet

Let the uniform surface charge density be $+|\sigma|$. The symmetry of the problem indicates that the electric field will point perpendicularly away from the sheet. As long as we are the same distance z from the sheet, the field should be constant. For this symmetry we can choose a Gaussian surface in the shape of a cylindrical hockey puck, as shown in Figure 16.101.*

The flux of the electric field vector through this surface is the sum of the flux through each end of the puck plus the flux through the lateral area. The flux through the lateral area, however, is zero because the electric field everywhere is

---

*This shape sometimes is called a pillbox, since it looks like the small boxes persons occasionally use to store pills.

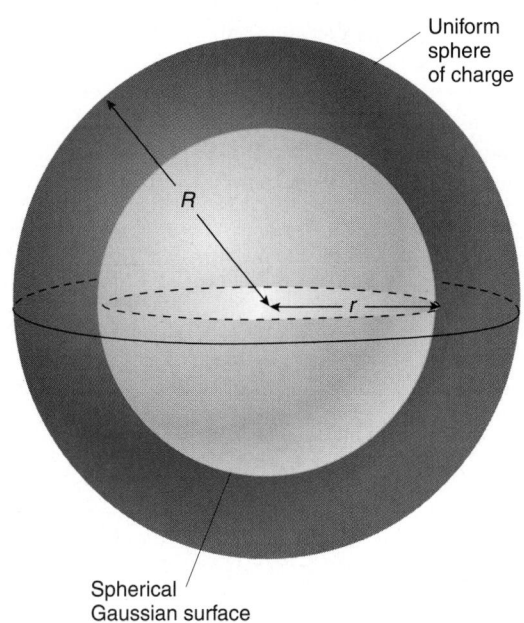

FIGURE 16.100 A uniform spherical distribution of charge.

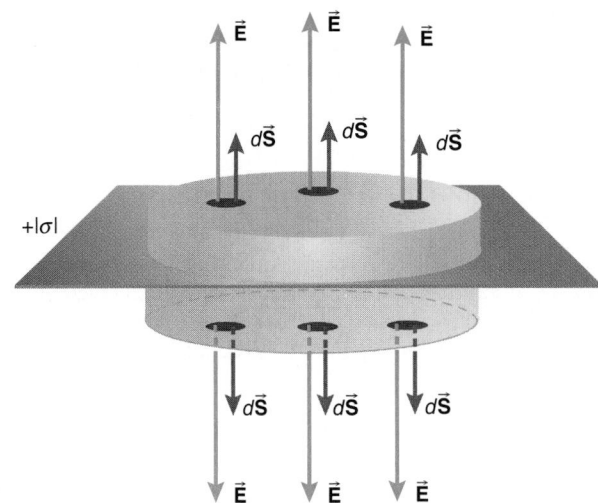

FIGURE 16.101 A cylindrical Gaussian surface intersecting the sheet of charge.

perpendicular to the differential surface area vectors $d\vec{S}$ over the lateral area. Thus

$$\int_{\text{lateral area}} \vec{E} \cdot d\vec{S} = 0 \ \text{N·m}^2/\text{C}$$

Over the entire top of the puck, the electric field vector is constant and parallel to every $d\vec{S}$. Thus the flux of $\vec{E}$ through the top of the puck is

$$\int_{\text{top}} \vec{E} \cdot d\vec{S} = \int_{\text{top}} E \, dS = E \int_{\text{top}} dS = EA$$

where A is the area of the top of the puck. The electric field vector is constant over the entire bottom of the puck too; it also is parallel to $d\vec{S}$ over the entire area of the bottom of the puck. Thus the flux through the bottom of the puck is

$$\int_{\text{bottom}} \vec{E} \cdot d\vec{S} = \int_{\text{bottom}} E \, dS = E \int_{\text{bottom}} dS = EA$$

The total flux through the hockey puck surface thus is

$$EA + EA = 2EA$$

and this is the left-hand side of Gauss's law.

The charge enclosed by the puck is the area A times the surface charge density $+|\sigma|$. Thus Gauss's law becomes

$$2EA = \frac{|\sigma| A}{\varepsilon_0}$$

The magnitude of the electric field is, therefore,

$$E = \frac{|\sigma|}{2\varepsilon_0}$$

as we found before (Example 16.16) for an infinite charged sheet.

QUESTION 9

To calculate the magnitude of the electric field of a uniformly charged infinite sheet using Gauss's law, is it permissible to use a cubical Gaussian surface with the charged sheet intersecting the cube parallel to two of its faces? Is it necessary to have the sheet bisect the cube?

16.13 CONDUCTORS*

A conductor is a material in which the electrons at the outer periphery of an atom have no great affinity for any particular individual atom; they are not bound or tied to individual atoms. These so-called **conduction electrons** are essentially free to move readily and quickly in response to electric fields.

The Electrostatic Field Inside a Conductor

Let a conducting plate be placed in an electric field, as shown in Figure 16.102. The conduction electrons in the plate respond to the field and quickly move to the left, opposite to the field direction; the electrons leave behind positive ions. The positive ions in turn steal conduction electrons from atoms further to the

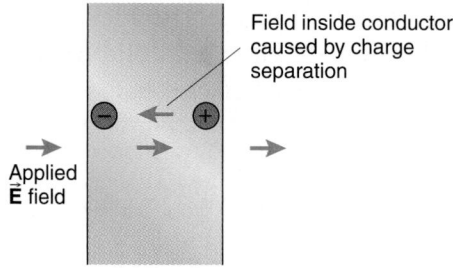

FIGURE 16.102 A conducting plate placed in an electric field.

right in the material. When charges are in motion we have an **electrodynamic** situation. In short order, what results is a negative charge accumulation on the left surface of the conductor in Figure 16.102 and a positive charge accumulation on the right surface. The effect of this charge separation within the conductor is to create another electric field within the conductor, directed opposite to the external field.

Charge separation continues until such time that the total electric field within the conductor is zero.

When the total electric field inside the conductor is zero, we have an **electrostatic** situation with no further charge flow.

If electric charge is placed on a conductor (so that the net charge on the conductor is not zero), the charge distributes itself on the conductor to ensure that the electric field within the conductor is zero. Any such so-called free charge on a conductor must reside on the surface of the conductor. To see why this is the case, we consider a Gaussian surface located entirely *just inside* the conductor but infinitesimally close to the surface, as shown in Figure 16.103.

Since the electric field is zero *inside* the conductor in electrostatics, $\vec{E} = 0$ N/C at *every* $d\vec{S}$ over the entire Gaussian surface. The left-hand side of Gauss's law vanishes, because the field itself is zero over the whole Gaussian surface. Therefore the right-hand side of Gauss's law also must be zero. In other words, there is no net electric charge within this Gaussian surface. Other smaller Gaussian surfaces can be considered any and everywhere within the conductor, as in Figure 16.104, and the same argument shows that no net charge exists within them either.

Hence no free charge can exist anywhere within the conductor.

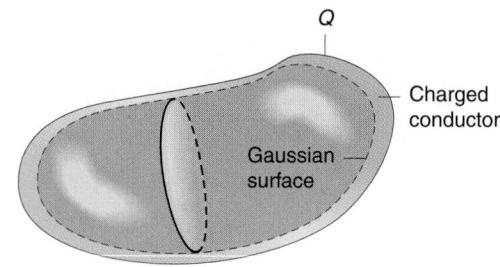

FIGURE 16.103 A Gaussian surface located just inside the surface of a conductor.

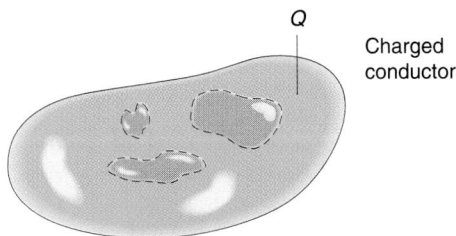

FIGURE 16.104 Additional Gaussian surfaces within the conductor.

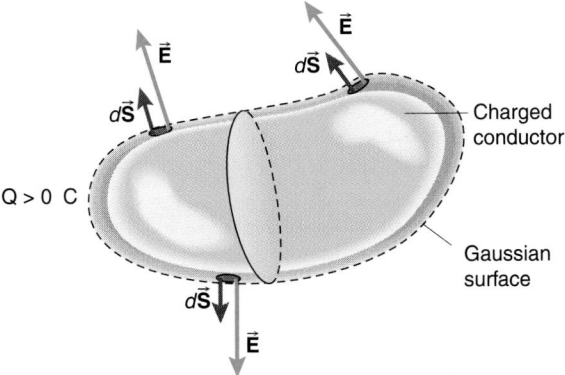

FIGURE 16.105 A Gaussian surface just outside the surface of a conductor.

A Gaussian surface placed differentially just *outside* the surface of the charged conductor, as in Figure 16.105, certainly will have a flux through it, since there is an electric field *outside* the charged conductor (due to the free charge Q on the conductor). The flux through this Gaussian surface must be Q/ε_0 according to Gauss's law.

Thus the free charge Q on a conductor does not lie within it but must reside on its surface.

The Orientation of the Electric Field at the Surface of a Conductor

We also can show that the angle that an electrostatic electric field makes with the actual differential surface area vector at any point on the surface of a conductor is 0°. If the angle were anything other than 0°, as in Figure 16.106, the electric field at the actual surface of the conductor would have a component parallel to the surface. This tangential field then would cause the conduction electrons in the conductor to move. The situation then is not electrostatic, since charges are in motion.

So in electrostatics, the electric field vector at the surface of a conductor must be perpendicular to the surface.

The Magnitude of the Electric Field at the Surface of a Conductor

Examine a small section of the surface of a charged conductor, as in Figure 16.107. Let σ be the value of the surface charge density

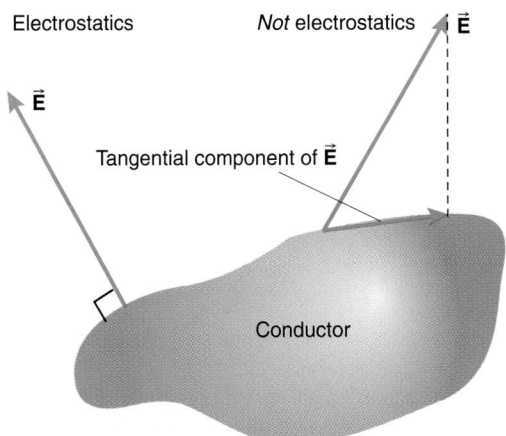

FIGURE 16.106 In electrostatics, the electric field must be perpendicular to the surface of a conductor.

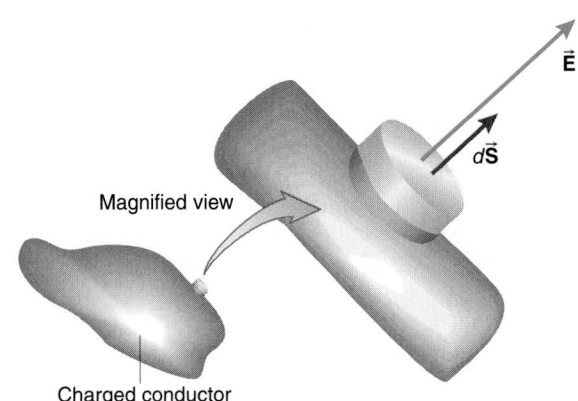

FIGURE 16.107 A tiny Gaussian surface that intercepts the surface of the conductor.

at this location on the surface of the conductor. Assume for simplicity that $\sigma > 0$ C/m^2. Imagine a very small Gaussian hockey puck surface, one end of which is *inside* the conductor and the other end just *outside* the conductor. Let the differential area $d\vec{S}$ of the ends of the puck be small enough so that the field is essentially constant across the end of the closed Gaussian surface.

The flux through the end of the puck lying just outside the conductor then is just $E\,dS$, since $\vec{E}$ is parallel to the differential area vector on this part of the Gaussian surface (since we assumed $\sigma > 0$ C/m^2). Since the field is zero inside the conductor, there is no flux through the end of the puck inside the conductor.

There also is no flux through the tiny portion of the lateral area just outside the conductor, because the electric field vector is perpendicular to the differential area elements over this lateral area. The flux through the portion of the lateral area within the conductor is zero since the field is zero inside this region.

Thus the total flux through the Gaussian surface is $E\,dS$. The area dS is small enough so that the surface charge density is constant over the differential area of the end of the puck. The total charge enclosed by the puck is $\sigma\,dS$. Hence Gauss's law becomes

$$E\,dS = \frac{\sigma\,dS}{\varepsilon_0}$$

The magnitude of the field is

$$E = \frac{\sigma}{\varepsilon_0}$$

> Therefore the magnitude of the field at any location on the surface of any conductor is equal to the magnitude of the local surface charge density σ divided by ε_0.

You might try to reach the same conclusion if the surface charge density is assumed to be negative.

The surface charge density on a conductor is independent of position (i.e., is constant or uniform) *only* for isolated spherically shaped conductors, infinite cylindrical conductors, or infinite conducting planes. For differently shaped conductors, or for all shapes of conductors in the presence of other conductors or point charges, the surface charge density varies with position on the conductor.

16.14 OTHER ELECTRICAL MATERIALS*

In contrast to conductors, in insulators (dielectrics) electrical charges remain fixed in position (except for polarization effects).

Semiconductors (such as germanium and silicon) are neither good conductors nor good insulators, thus making their name appropriate. They could equally well be called semi-insulators, but this terminology is not used. The exploitation of semiconducting materials through the judicious addition of small amounts of other elements (a process called **doping**) led to the invention of the transistor (in 1949) and the electronic revolution of the last half of the 20th century.

Superconductors are quite unusual. These materials lose all resistance[†] to the flow of electric charges below characteristic temperatures that vary with the material. These materials are literally perfect conductors. Superconductivity was discovered in 1911 by Kamerlingh Onnes (1853–1926) when he found that mercury lost all electrical resistance below the rather frigid temperature of 4.2 K. Such extremely cold temperatures are difficult and expensive to achieve (and are typically achieved through the use of liquid helium that, coincidentally, has a boiling point also at about 4.2 K). This low temperature has restricted the use of such materials to applications that can justify the expense involved. The discovery in 1987 of materials that exhibit superconductivity at warmer temperatures [above 100 K, well above the boiling point of liquid nitrogen (77 K), a much cheaper refrigerant than liquid helium] has led to a resurgence of interest in superconductivity and its potential technological applications in a wide variety of fields, including electrical power transmission, medical imaging technologies, and high-speed computing. Explanations of superconductivity involve aspects of quantum mechanics.

[†]We introduce this concept formally in Chapter 19.

CHAPTER SUMMARY

An *electric charge* is said to be *positive* if it is repelled by a glass rod that has been freshly rubbed with silk; the charge is *negative* if it is repelled by a rubber rod (or amber) that has been freshly rubbed with fur. *Like charges* have the same sign (but may have different amounts of charge); *unlike charges* have opposite signs (and also may have different amounts of charge).

Conductors are materials in which charges are free to move; silver, copper, gold, and iron are examples of conducting materials. *Insulators* (*dielectrics*) are materials in which charges are not free to move; examples of insulators include glass, rubber, and wood.

The total electric charge is conserved in any process in an isolated system; this is called *conservation of charge*.

The magnitude of the electrical forces that two pointlike charges q and Q exert on each other is found using *Coulomb's law*:

$$F_{\text{elec}} = \frac{1}{4\pi\varepsilon_0} \frac{|q||Q|}{r^2} \qquad (16.4)$$

where, in SI units,

$$\frac{1}{4\pi\varepsilon_0} = 8.987\,552 \times 10^9 \ \text{N·m}^2/\text{C}^2$$

$$\approx 9.00 \times 10^9 \ \text{N·m}^2/\text{C}^2$$

The SI unit for the quantity of charge is the *coulomb* (C). The electrical force is repulsive for like charges and attractive for unlike charges.

Charge is *quantized*. All free charges q in nature are integral multiples of the *fundamental unit of charge* $e = 1.602 \times 10^{-19}$ C:

$$q = ne \qquad (16.6)$$

where n is an integer known as the *charge quantum number*. The proton has a charge quantum number of $+1$, while the electron has a charge quantum number of -1; the neutron has a charge quantum number of zero. Macroscopic amounts of charge have huge charge quantum numbers.

The electrical force is conveyed by an *electric field* $\vec{E}$ created in space by a *static* (i.e., fixed) arrangement of charges. The electric field is the force per unit charge on a small test charge placed at the point where the field exists. The electric field thus is measured in newtons per coulomb (N/C) when using SI units. We make the assumption that the test charge is small enough so that its presence does not affect the distribution of charges creating the field at the point in question. In this way, a charge q, placed at a point in space where the electric field is $\vec{E}$, experiences a force that is

$$\vec{F}_{\text{elec on } q} = q\vec{E} \qquad (16.8)$$

The electric field in space depends on the geometric arrangement of the charges producing the field. In particular, the electric field at a point in space a distance r from a pointlike charge Q is

$$\vec{E} = \frac{1}{4\pi\varepsilon_0} \frac{Q}{r^2} \hat{\mathbf{r}} \qquad (16.11)$$

The field points radially away from an isolated positive charge ($Q > 0$ C), and radially toward a negative charge ($Q < 0$ C).

The electric field vector of a static arrangement of charges obeys a principle of linear superposition: the total field at any point is the vector sum of the individual fields produced by each charge acting alone:

$$\vec{E}_{total} = \vec{E}_1 + \vec{E}_2 + \cdots \qquad (16.12)$$

The electric field can be pictured by drawing *electric field lines*. Such field lines begin on positive charges and end on negative charges. The electric field at any point is tangent to the electric field line at the location in question and has a magnitude proportional to the number of field lines per square meter oriented perpendicular to the lines.

A special arrangement of charges called an *electric dipole* consists of two charges $\pm |Q|$ of equal magnitude, separated by a distance d. The electric *dipole moment* $\vec{p}$ is defined to be

$$\vec{p} \equiv |Q|\vec{d} \qquad (16.13)$$

where the vectors $\vec{d}$ and $\vec{p}$ are directed from the negative to the positive charge. A dipole placed in an electric field $\vec{E}$ experiences a torque $\vec{\tau}$ given by

$$\vec{\tau} = \vec{p} \times \vec{E} \qquad (16.14)$$

The principle of superposition implies that the electric field created by a continuous distribution of charge is found by a vector integration (summation) of the pointlike contributions to the field, where the integration is taken over the extent of the charge distribution:

$$\vec{E} = \frac{1}{4\pi\varepsilon_0} \int_{\substack{charge \\ distribution}} \frac{dq}{r^2} \hat{r} \qquad (16.16)$$

The electric fields created in space by various continuous distributions of charge are summarized in Table 16.2 on page 734.

Gauss's law for the electric field of a static arrangement of charges states that the flux of the total electric field vector through a closed Gaussian surface is equal to the total (net) charge enclosed by the surface, divided by ε_0:

$$\int_{\substack{clsd\ Gaussian \\ surface}} \vec{E} \cdot d\vec{S} = \frac{Q_{net\ enclosed}}{\varepsilon_0} \qquad (16.18)$$

For charge distributions with sufficiently high symmetry, Gauss's law can be used to find the magnitude of the electric field of the charge distribution.

Static charges on conductors reside on their surface. The static electric field inside a conductor is zero. The static electric field at any point on the surface of a conductor is perpendicular to the surface and of magnitude $|\sigma|/\varepsilon_0$, where $|\sigma|$ is the magnitude of the local surface charge density at the location in question.

SUMMARY OF PROBLEM-SOLVING TACTICS

16.1 (page 715) When using Coulomb's law for the force between pointlike charges, be sure to express the distance between the charges in meters, not centimeters, and the charge in coulombs.

16.2 (page 715) To avoid many confusing problems with signs, employ the following procedure when using Coulomb's law. Calculate the *magnitude* of the force (a positive scalar quantity) using Coulomb's force law, Equation 16.4. Sketch the situation. Then determine the direction of the force on each charge after examining the types of the charges involved:
- if the charges are *both* positive *or* both negative (so-called like charges), the force on each is repulsive; or
- if the charges are of *opposite* flavors (unlike charges), the force on each is attractive.

Indicate on your sketch the direction of the forces on the charges. Finally, if appropriate, introduce a Cartesian coordinate system and write the forces in standard Cartesian vector form.

16.3 (page 718) Remember that we treat the fundamental unit of charge $e = 1.602 \times 10^{-19}$ C as positive, and so the charge on the proton is $+e$, while that on the electron is $-e$.

16.4 (page 721) The algebraic sign of the charge q is very important in Equation 16.8.

16.5 (page 723) Use the following three-step procedure to find the electric field a distance r from a pointlike charge.

1. First find the *magnitude* of the field at the point in question:

$$E = \frac{1}{4\pi\varepsilon_0} \frac{|Q|}{r^2}$$

This always is a positive quantity, because it is the magnitude of a vector.

2. Determine the direction of the field at the desired point and indicate it in a sketch:
- radially away from Q if Q is positive
- radially toward Q if Q is negative

3. Finally, introduce a coordinate system and write the electric field vector $\vec{E}$ in terms of your chosen coordinate system.

16.6 (page 743) To evaluate the flux of the electric field through a closed surface, you need only tally the total (i.e., net) charge enclosed within the surface and divide by ε_0.

QUESTIONS

1. (page 711); 2. (page 713); 3. (page 715); 4. (page 718); 5. (page 721); 6. (page 728); 7. (page 739); 8. (page 744); 9. (page 748)

10. What experimental evidence exists to suggest that there are two, and only two, types of charge?

11. Is the vitreous electricity of Dufay positive or negative charge? What is his resinous electricity?

12. If a bit of material is attracted to a glass rod freshly rubbed with silk, can you be certain that the material is negatively charged?

Explain. If the material is repelled by the glass rod, can you be certain it is positively charged? Explain.

13. The gravitational force is intrinsically much weaker than the electrical force. Electricity was discovered well before gravitation, yet we are much more aware of gravitational forces than electrical forces in our common experience. Explain why.

14. Two balloons are of identical size. One is now charged by rubbing. Are the balloons still exactly the same size?

15. Explain why it can be said correctly that like charges dislike each other, and that unlike charges like each other.

16. How is the process of electrification similar to heat transfer? How is it different? What is transferred in each process?

17. A thin stream of water falling from a pipette is attracted to a negatively charged rubber rod as indicated in Figure Q.17. Explain why. If a positively charged glass rod is used instead, is the stream repelled? Explain.

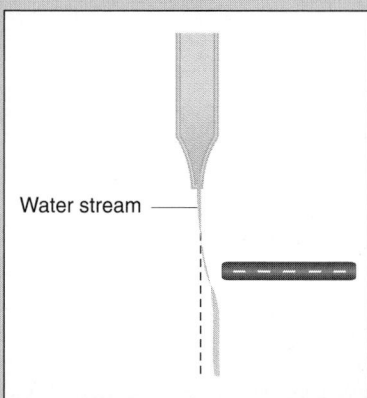

Water stream

FIGURE Q.17

18. An electroscope (see Figure Q.18a) is a device consisting of twin thin gold foils connected to a metal sphere via a conducting wire. The system typically is enclosed in a glass case to avoid drafts and to prevent us from touching the delicate gold foils with our grubby paws. The gold foils normally hang vertically. (a) A charged rubber rod is brought near the metal sphere. The two gold foils are found to separate and remain so as long as the rod is near. If the rod is removed, the foils return to their initial vertical position (see Figure Q.18b). Explain why. (b) The charged rubber rod now is brought in contact with the metal sphere and the foils separate once again. While maintaining contact with the metal sphere, a grounding wire is connected between the sphere and a nearby water pipe. The foils return to the vertical position as soon as this is done. Explain why. The grounding wire now is removed, and then the charged rubber rod is removed. What happens to the two gold foils? Is the electroscope charged? If so, with what type of charge? Explain.

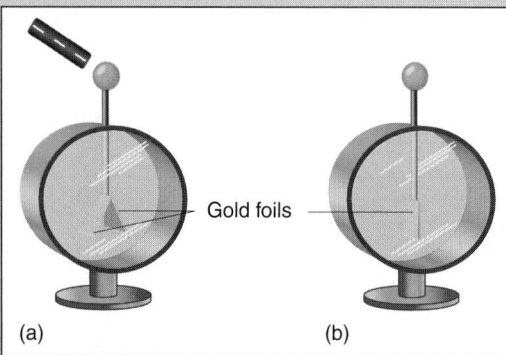

Gold foils

(a)   (b)

FIGURE Q.18

19. A piano factory polishes each piano when its assembly is complete. Why will the piano be likely to attract dust after the buffing? Can you prevent the same phenomenon from occurring after you polish your car? If so, how?

20. Does the term *total charge* on a conductor mean the same thing as the *free charge* on the conductor? Explain.

21. A pencil lies near this text. Explain why we do not normally notice the gravitational and electrical forces that each exerts on the other.

22. The fundamental unit of charge *e* occasionally is called the *quantum of charge*, meaning it is the smallest bit of isolated charge that exists in nature. Such quantum ideas are useful in other fields as well. (a) What is the quantum of currency in your country? (b) Could a biologist talk about a quantum of life? (c) Could you define a quantum of the written word or are many quanta necessary? (d) Could a computer scientist define a quantum of information or is more than one necessary?

23. Carefully distinguish between the terms *no charges present* and *zero charge present*.

24. In the electric field line diagram of a positive pointlike charge (Figure 16.55 on page 727), where are the negative charges on which the field lines end?

25. If a pointlike mass *m* with charge *q* is released at rest in an electric field, will the path followed by the particle always be *necessarily* coincident with the entire electric field line through the point of release? If so, explain why. If not, specify a situation where the particle does follow the field line and another situation where it does not.

26. Describe an experiment which can demonstrate that the electrical force between two charges is independent of their mass.

27. It is more difficult to electrify (charge) an object on humid days than on days with low humidity. What does this imply about the type of electrical material air is on these days?

28. Notice in Equation 16.11 (on page 722) that the magnitude of the electric field of a pointlike charge increases dramatically and approaches an infinite magnitude as we approach the charge in Figure 16.37; that is, as $r \rightarrow 0$ m, $|E| \rightarrow \infty$ N/C. Why does the charge not experience an infinite force? Explain carefully.

29. An electron is placed at a point in an electric field, and the electrical force on it and its acceleration are measured. The electron is removed and a proton is placed at the same location, and the electrical force on it and its acceleration also are measured. How are the forces on the electron and proton related to each other? How are the accelerations related to each other?

30. An electron is moving in the electric field of another pointlike charge *Q*. Is the orbital angular momentum of the electron about the charge *Q* conserved? Explain why or why not. Does your answer depend on the sign of the charge *Q*?

31. You likely have noticed that when clothes are removed from a dryer, they tend to cling together and to you. Explain why. Is the effect changed if a fabric softener is used in the drying process? How might you explain this phenomenon?

32. In Gauss's law, is the electric field in the flux integral the field of only the charges inside the closed Gaussian surface, or is it the total field of all the charges present whether inside or outside the Gaussian surface? Explain your answer.

33. Examine Table 16.2 on page 734 summarizing the electric fields of several charge distributions. A pointlike charge q placed in any of these fields experiences a force $\vec{F} = q\vec{E}$. Many of these forces are *not* inverse-square-law forces. Is this a violation of Coulomb's law? Explain.

34. The model of the water molecule shown in Figure 16.63 (on page 729) uses *fractional* values of the fundamental unit of charges e. The fractional units in the model are about the same magnitude as the charges on the up and down quarks, but this is purely fortuitous (coincidental). Do the fractional values of the fundamental unit of charge in the water molecule violate charge quantization?

35. A pseudo-scientist claims that the force keeping us on the Earth is an electrical force between unlike charges rather than gravity. What experiments can you cite or perform to show that this hypothesis is incorrect?

36. The electric field at a point P located a distance r from the center of a uniform, spherically symmetric charge distribution of total charge Q is given in Table 16.2. If the charge distribution shrinks symmetrically to one-third its former diameter, what happens to the value of the electric field at points $r > R$?

37. At one point along the line connecting two pointlike charges, the total electric field of both charges is found to be zero. Are the charges like or unlike charges? With this information alone, is it possible to tell if the charges are of equal magnitude or what the signs of the individual charges are?

38. The protons in an atom are all in the nucleus. The surrounding electrons are bound to the atom by the electrical force. The protons, of course, repel each other with a substantial electrical force, yet they do not fly out of the nucleus. Explain why these observations indicate the need for another fundamental force that is intrinsically even stronger than the electrical force, yet which must have a very limited range of effectiveness. This short-range, but very strong force, is (uninspiringly) called the *strong force*.

39. A rectangular strip of paper of length ℓ and width d is placed in a uniform electric field of magnitude E_0 as indicated in Figure Q.39. What is the flux of the electric field through the surface of the paper? Does this violate Gauss's law? Why or why not?

40. Take a rectangular strip of paper of length ℓ and width d, twist one end 180° about the long axis of the paper, and tape it to the other end. Such a surface is called a *Möbius strip* (see Figure Q.40). If you run a magic marker along the strip, you will find that the marker comes back to where you began; no side of the paper lacks a marker line, and so the strip can be considered one-sided. If such a Möbius strip surface is imagined to be in a uniform electric field of magnitude E_0, what is the total flux of the electric field through the Möbius strip surface?

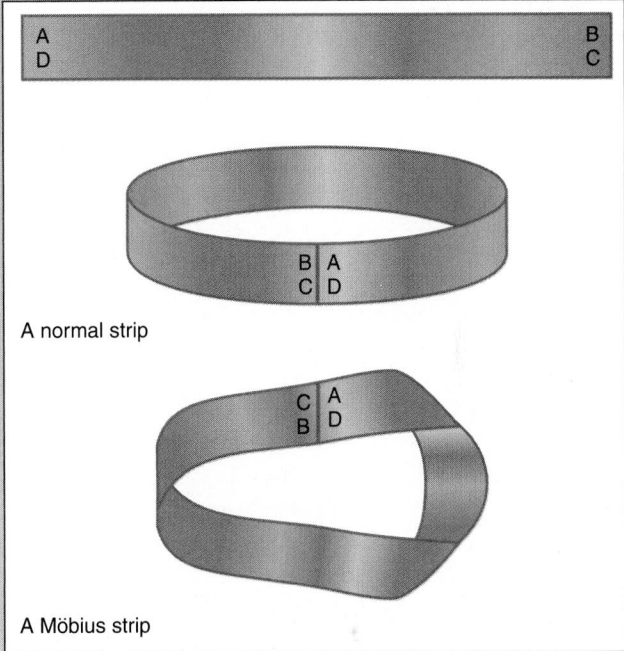

A normal strip

A Möbius strip

FIGURE Q.40

41. A *Klein bottle* is an extension of the Möbius strip to three dimensions; it is a surface with no inside or outside. A point charge Q is located near a Klein bottle as indicated in Figure Q.41. Can a Klein bottle be used as a Gaussian surface in Gauss's law?

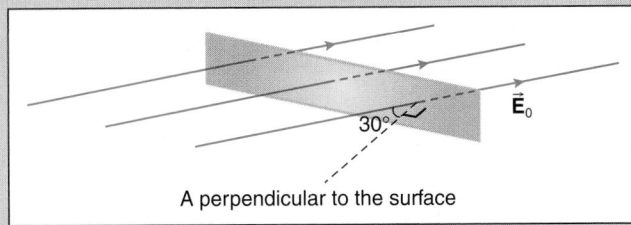

30°

$\vec{E}_0$

A perpendicular to the surface

FIGURE Q.39

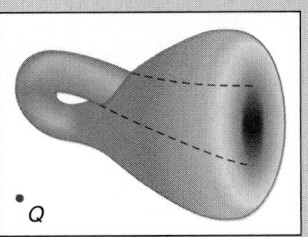

Q

FIGURE Q.41

42. A point charge lies at the center of a spherical Gaussian surface. If the charge is moved to another location inside the Gaussian surface, is the flux of the electric field through the Gaussian surface changed when the charge is in its new location? Explain your answer.

43. If the flux of the electric field through a closed surface is zero, does this imply that the electric field is zero on the surface? Could it be zero on parts of the surface? Explain your answer.

44. A cube with a charge of +Q has a uniform volume charge density. To calculate the magnitude of the electric field of the cubical charge distribution, a student suggests using Gauss's law with a cubical Gaussian surface with surfaces equidistant from each face of the charged cube. Will this enable the student to calculate the magnitude of the electric field of the cube? Why or why not? What is the flux of the electric field through such a Gaussian surface?

45. Two charges of opposite sign and unequal magnitude are the only two charges inside a closed and cozy Gaussian surface. (a) Is the flux of the electric field through the Gaussian surface changed if the positions of the two charges are interchanged? (b) Is the value of the electric field at various points on the Gaussian surface changed by such a procedure? Explain why or why not.

46. The gravitational force, the electrical force, and the Hooke's law spring force all are conservative forces. When studying gravitation, we introduced a gravitational field as a mechanism for the transmission of the force between masses; in this chapter, we introduced an electric field for similar purposes. Discuss whether it is appropriate and/or useful to invent a "spring field" for the study of the Hooke's law force.

47. If you accept the Eddington quote at the beginning of Section 16.4 on page 717, what is the largest charge quantum number?

PROBLEMS

Sections 16.1 The Discovery of Electrification
16.2 Polarization and Induction
16.3 Coulomb's Force Law for Pointlike Charges:
The Quantification of Charge
16.4 Charge Quantization

1. A mole of electrons has what total charge? The magnitude of this charge is known as a *faraday* of charge.

2. Find the charge quantum number associated with a macroscopic charge of 1.000 nC.

3. What is the total positive charge contained in 1.00 kg of hydrogen gas (H_2)?

4. What is the total positive charge contained in 1.00 kg of water (H_2O)?

5. How much positive charge (in coulombs) exists in Avogadro's number of hydrogen atoms?

6. How many protons are in 1.00 C of positive charge? How many electrons are in a negative charge of -1.00 C?

•7. Electrical liftoff. A charge q is glued on top of your head and another like charge of equal magnitude is glued to the floor beneath your shoes. If you are 1.90 m tall and have a mass of 70.0 kg, what magnitude charge q will make the normal force of the floor on you equal to zero?

•8. Two M&M candies (yummy!), each of mass m, have equal charges $q > 0$ C bestowed on them and are placed on a frictionless surface a distance d apart as indicated in Figure P.8. (a) What is the magnitude of the electrical force on each M&M? What are the directions of the forces on each? (b) What is the magnitude of the acceleration of each M&M? (c) If the mass of each M&M is 8.6×10^{-4} kg and d is 0.100 m, how large must q be so that the acceleration of each M&M is 0.50 m/s²?

•9. The nucleus of a gold atom has a charge quantum number of +79. An α-particle (a helium nucleus) has a charge quantum number of +2. For this problem *neglect* the electrons surrounding the nucleus of each atom. (a) What is the magnitude of the electrical force of the gold nucleus on the α-particle when it is

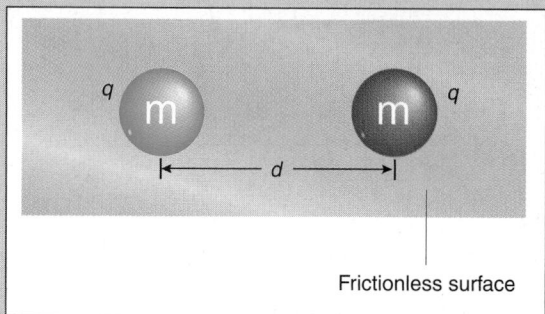

FIGURE P.8

1.00×10^{-14} m from the nucleus? (b) What is the magnitude of the electrical force of the α-particle on the gold nucleus when the α-particle is at the same position as in part (a)?

•10. After tedious measurements, you find that five oil droplets in a Millikan-type experiment have the following charges on them:

Drop number	Charge
1	1.282×10^{-18} C
2	1.762×10^{-18} C
3	2.083×10^{-18} C
4	3.364×10^{-18} C
5	4.006×10^{-18} C

Without knowing or using the value of the fundamental unit of charge, describe and use a procedure for determining the numerical value of e in coulombs *from these data alone*, making only the assumption that each charge is an integral multiple of the (unknown) fundamental unit of charge.

•11. Two small, identical, pointlike conducting spheres have charges $q_1 = -2.00$ nC and $q_2 = +6.00$ nC. (a) Calculate the magnitude of the electrical force on each when they are separated by 3.00 cm. Are the forces attractive or repulsive? (b) The two spheres are brought momentarily into contact and then separated by a distance of 3.00 cm. Now what is the magnitude of the electrical force on each? Are the forces attractive or repulsive?

•12. A pointlike charge $q_1 = 2.00 \times 10^{-10}$ C is 0.300 m from a second pointlike charge $q_2 = -3.00 \times 10^{-10}$ C. The attractive force on q_1 is found to be of magnitude 6.00×10^{-9} N. (a) What is the ratio of the magnitude of the electrical force on q_1 to the magnitude of the electrical force on q_2? (b) Calculate the ratio

$$\frac{Fr^2}{|q_1|\,|q_2|}$$

(c) Is the ratio in part (b) the same in another experiment with different values for the charges and distances? Explain.

•13. Pretend gravity did not exist. Consider the Earth and Sun as pointlike masses and assume the Sun is fixed in position with the Earth in a circular orbit. If the Sun and the Earth had opposite charges of the same magnitude $|q|$, the Earth could orbit the Sun in much the same manner that it does under the influence of the gravitational force, as shown in Figure P.13. What magnitude of charge on each is necessary to accomplish this?

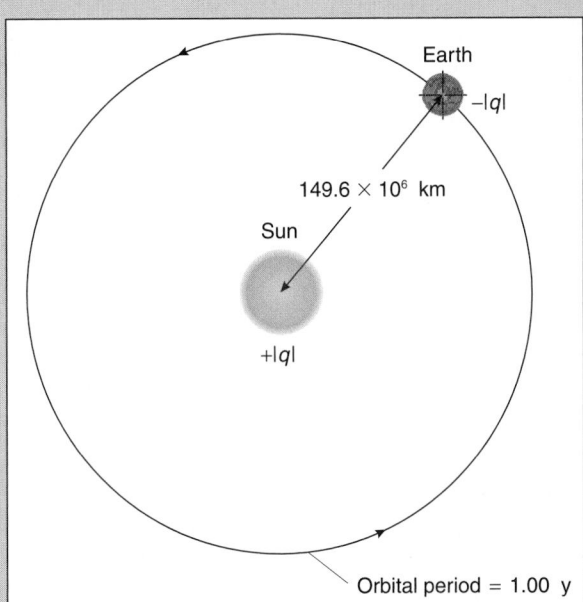

Earth
$-|q|$

149.6×10^6 km

Sun

$+|q|$

Orbital period = 1.00 y

FIGURE P.13

•14. You and a friend each have mass 60 kg and each are holding 1.00 C of charge. Watch out! At what distance from each other is the magnitude of the electrical force on the charges equal to the magnitude of your weight on the surface of the Earth?

•15. Two spheres of identical mass m and radius have charges $+2.00 \times 10^{-9}$ C and -8.00×10^{-9} C. The spheres are attached to insulating, uncharged pucks of mass M that are free to glide on a frictionless airtable in a physics laboratory as indicated in Figure P.15. (a) Draw four second-law force diagrams and indicate all the forces acting on each sphere and each puck.

If forces are of equal magnitude, draw the arrows of equal length. Use words such as "this is the force of _____ on _____" to describe each force. (b) In the four second-law force diagrams in (a), indicate which forces form Newton's third law force pairs. (c) For the horizontal direction, write Newton's second law for one sphere and for the puck on which it rides, letting a be the magnitude of their common horizontal acceleration.

This problem is modeled after and is an extension of a problem suggested by Arnold Arons, *A Guide to Introductory Physics Teaching* (Wiley, New York, 1990), pages 156–157; the figure is from page 157.

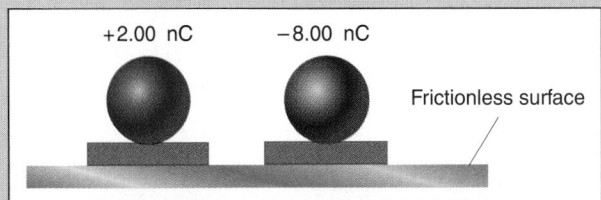

+2.00 nC   −8.00 nC

Frictionless surface

FIGURE P.15

•16. Two positive pointlike charges are a fixed distance r apart. The *total* charge of the pair is Q. (a) What is the charge on each such that the electrical force they exert on each other is of *maximum* magnitude? (b) What charge is on each such that the electrical force they exert on each other is of *minimum* magnitude?

•17. Two equal pointlike masses m, each with positive charge q, are suspended from a common point by threads of equal length ℓ and negligible mass. Because of their mutual repulsion, the threads make an angle θ with the vertical direction as indicated in Figure P.17. Do *not* neglect the weight of the masses. (a) Draw a second law force diagram indicating the forces on each mass. (b) Show that the charge on each mass can be expressed as

$$q = (2\ell \sin \theta)(4\pi\varepsilon_0 mg \tan \theta)^{1/2}$$

(c) Evaluate the charge if $m = 10$ g, $\ell = 0.500$ m, and $\theta = 15.0°$. (d) With the data in part (c), what is the approximate charge quantum number of each charge?

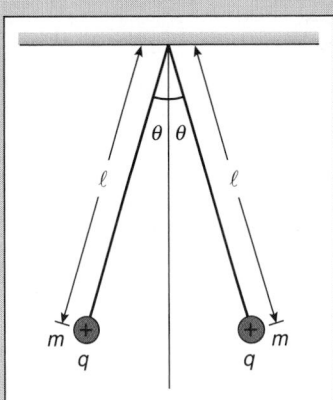

θ θ

ℓ ℓ

m $+$ $+$ m
q q

FIGURE P.17

•18. Three pointlike charges Q are located on three successive vertices of a regular hexagon with sides ℓ as indicated in Figure P.18. Find the electrical force on another charge q located at the center of the hexagon. Assume all the charges are like charges and express your result in terms of the coordinate system indicated.

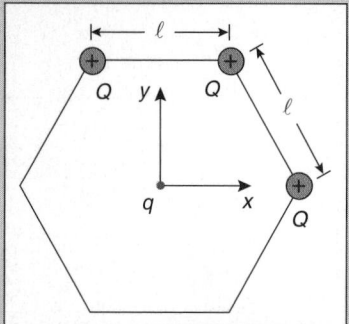

FIGURE P.18

•19. The $+3.00 \, \mu C$ charge of the light brigade is situated relative to two other charges as shown in Figure P.19. (a) Find the total force acting on the charge of the light brigade. (b) Determine the magnitude of the total force. (c) Determine the acute angle that the total force makes with the x-axis.

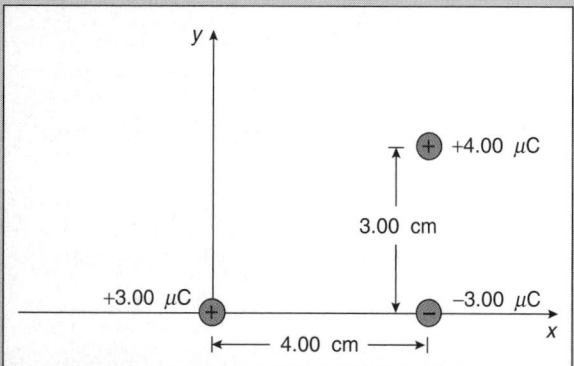

FIGURE P.19

‡20. A positron is a particle with the same mass as an electron but with a *positive* charge. Positrons are created in many radioactive particle decays and are present in cosmic rays (where they were first discovered). A positron and an electron can briefly form an unusual atom known as *positronium*. Imagine a situation where the two particles are in a circular orbit about their center of mass as indicated in Figure P.20. Since the particles have equal mass, the center of mass is midway between them. Let r be the *separation* of the particles (so the orbits are each of radius $r/2$). (a) Show that the orbital period T is related to the separation distance r by

$$T^2 = \frac{16\pi^3 \varepsilon_0}{e^2} \frac{m_e m_p}{m_e + m_p} r^3$$

This is a consequence of Kepler's third law for electrical orbits. This expression simplifies slightly because the mass of the

electron m_e is equal to the mass of the positron m_p:

$$T^2 = \frac{8\pi^3 \varepsilon_0 m_e}{e^2} r^3$$

(b) Show that if an electron and a *proton* are in circular orbits about *their* center of mass (which is *not* at the midway point between them but much closer to the proton), then the same expression results:

$$T^2 = \frac{16\pi^3 \varepsilon_0}{e^2} \frac{m_e m_p}{m_e + m_p} r^3$$

where m_p is now the mass of the proton: In this case $m_p \gg m_e$, so that the expression simplifies to

$$T^2 = \frac{16\pi^3 \varepsilon_0 m_e}{e^2} r^3$$

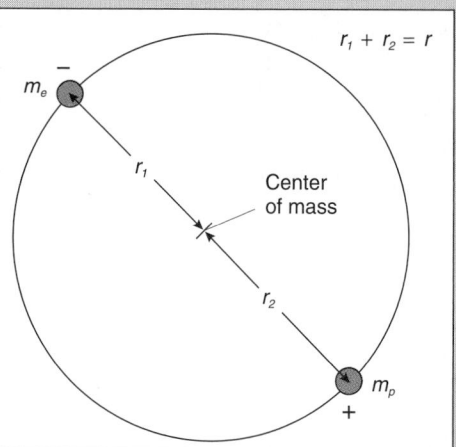

FIGURE P.20

‡21. When considering collections of masses and extended distributions of mass in Chapter 9, we introduced the concept of the center of mass. Since mass and charge play similar roles in their respective force laws, the idea of a *center of charge* may be useful. Discuss how you could extend the definition of the location of the center of mass to define an analogous position vector for the center of charge of a charge distribution. Under what circumstances might the concept of the center of charge be undefined?

Sections 16.5 The Electric Field of Static Charges
 16.6 The Electric Field of Pointlike
 Charge Distributions

22. A uniform electric field of magnitude 500 N/C exists in a region of space, directed as indicated in Figure P.22. (a) An electron is placed in the field. Indicate the direction of the force on the electron relative to the field direction. Find the magnitude of the force and determine the magnitude of the acceleration of the electron. (b) A proton is placed in the field. Indicate the direction of the force on the proton relative to the field direction. Calculate the magnitude of the force on the proton as well as the magnitude of its acceleration. (c) Calculate the ratio of the magnitude of the acceleration of the electron to

the magnitude of the acceleration of the proton when they are placed in the same field. Neglect any electrical interaction between the proton and electron. Compare your result with the ratio of the proton to the electron masses. Explain.

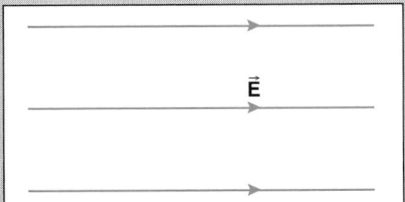

FIGURE P.22

23. At what distance from a proton is the magnitude of its electric field equal to 3.0×10^6 N/C?

24. (a) What is the magnitude of the weight of a proton on the surface of the Earth? (b) What magnitude electric field is sufficient to create an electrical force on the proton of magnitude equal to the magnitude of its weight? (c) How far away should another proton be placed to create an electric field of this magnitude?

25. A botany professor is in an electric field. The uniform electric field has a magnitude of 500 N/C. An electrical force of magnitude 6.0 N is found to act on the professor of mass 75 kg in the direction opposite to that of the field. What is the charge of the professor?

•26. Assume the electron in the ground state of the hydrogen atom is located 5.29×10^{-11} m from the nuclear proton. (a) What is the magnitude of the electric field of the proton at the position of the electron? Make a sketch to indicate the direction of this electric field at the position of the electron. (b) What is the magnitude of the electrical force on the electron? Make a sketch to indicate the direction of this force on the electron. (c) What is the magnitude of the electric field of the electron at the position of the proton? Make a sketch to indicate the direction of this electric field. (d) What is the magnitude of the electrical force on the proton? Make a sketch to indicate the direction of this force on the proton.

•27. A +2.00 μC charge is located at the origin and a −3.00 μC charge is located as indicated in Figure P.27. (a) Find the total electric field at the point P. (b) Suppose Pickett's charge of −4.00 μC now is placed at point P. What is the total force on this charge? (c) What is the magnitude of the force on this charge?

•28. Imagine the two electrons in the helium atom are in a circular orbit at the same radius ($r = 2.64 \times 10^{-11}$ m) about the central nuclear pair of protons but on opposite sides of the nucleus, as shown in Figure P.28. The nucleus is *quite* small compared with the size of the electron orbit; thus the nuclear charge can be considered as a point charge at the center of the orbit. (a) What is the magnitude of the total electric field at the position of one of the electrons? Make a sketch to indicate the direction of this field. (b) What is the magnitude of the electrical force on the electron at this location? Make a sketch to indicate the direction of the force on the electron. (c) What is the magnitude and direction of the electric field caused by each of

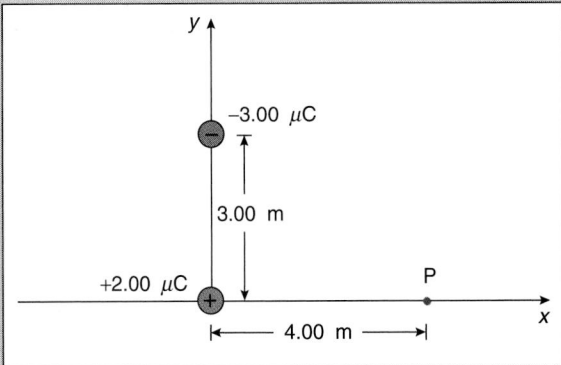

FIGURE P.27

the electrons at the position of the nuclear protons? What is the total electric field of the two electrons at the position of the nucleus? (d) What is the magnitude of the total electrical force that the two electrons exert on the nuclear protons?

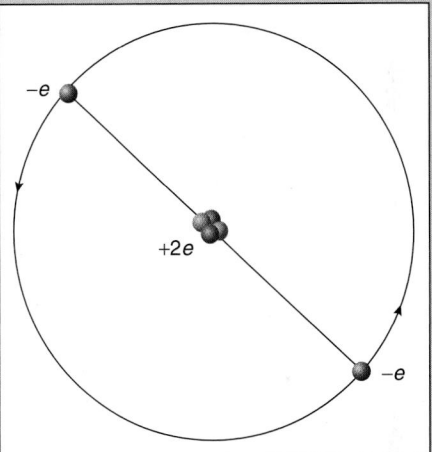

FIGURE P.28

•29. A small charged sphere of mass 20.0 g is suspended at rest by means of a thin (massless) thread of length 50.0 cm in a uniform electric field of magnitude 100 N/C directed as indicated in Figure P.29. (a) What is the sign of the charge on the sphere? (b) Draw a second law force diagram indicating all the forces acting on the little sphere. (c) Find charge on the sphere.

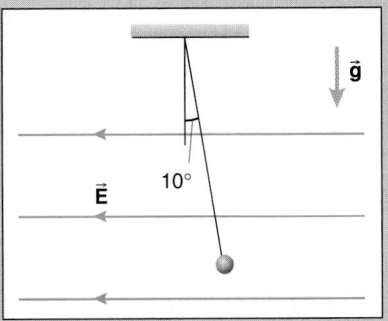

FIGURE P.29

•**30.** To produce an electric field, you arrange four charges of equal magnitude at the corners of a square corral of side ℓ as shown in Figure P.30. (a) Introduce an appropriate Cartesian coordinate system to analyze the problem, clearly indicating the origin and the direction of the positive axes. (b) Find the total electric field at the position of the charge in the upper right corner of the corral caused by the other three charges, expressing your result in Cartesian vector form. Schematically indicate the direction of the electric field on a sketch. (c) Calculate the force on the charge in the upper right corner of the corral and sketch the direction of the force in relation to the electric field calculated in part (b).

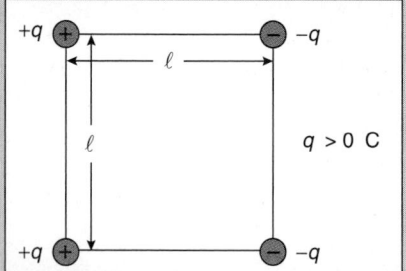

FIGURE P.30

•**31.** You have a pointlike charge of *magnitude* 3.00×10^{-3} C that is attracted to Milly Coulomb yet repelled by Billy Coulomb as shown in Figure P.31. (a) What is the sign of your charge? (b) Find the electric field of Milly and Billy at the position of your charge. Express your result as a Cartesian vector. (c) Find the total electrical force on your charge. Express your result as a Cartesian vector. (d) What is the magnitude of the electrical force on your charge?

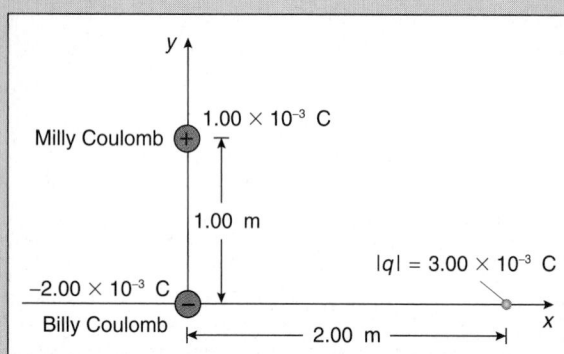

FIGURE P.31

•**32.** You are out cultivating a few charges and find two of them located as shown in Figure P.32. (a) Calculate the total electric field at the origin, expressing it in appropriate Cartesian vector form. (b) What is the magnitude of the electric field at the origin? (c) If a proton now is placed at the origin, find the force on the proton, expressing it in appropriate Cartesian vector form. (d) Find the *magnitude* of the acceleration of the proton.

Section 16.7 A Way to Visualize the Electric Field:
Electric Field Lines

•**33.** Figure P.33 depicts the electric field lines associated with a distribution of three pointlike charges. Determine the signs of the

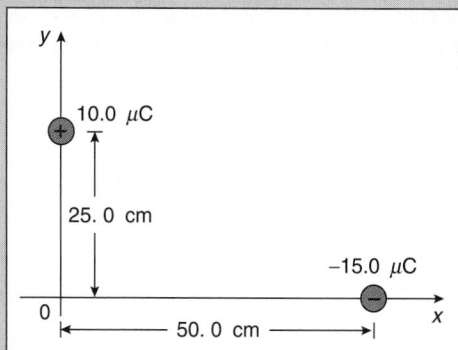

FIGURE P.32

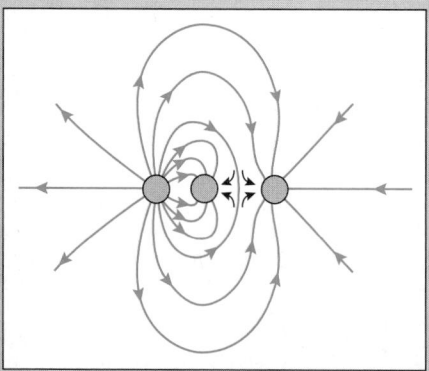

FIGURE P.33

various charges and their relative magnitudes.
Diagram from William Taussig Scott, *The Physics of Electricity and Magnetism* (Wiley, New York, 1959), page 27.

•**34.** Using the appropriate rules for drawing electric field lines, sketch the electric field lines in the vicinity of two negative charges of equal magnitude placed a distance d apart.

•**35.** Three charges of equal magnitude lie on the corners of an equilateral triangle. Two of the charges are positive and one is negative. Make a schematic diagram of the electric field lines of this charge distribution.

Section 16.8 A Common Molecular Charge Distribution:
The Electric Dipole

•**36.** Show that the total electrical force on an electric dipole in a *uniform* electric field is zero, regardless of the orientation of the dipole.

•**37.** Notice in Table 16.1 on page 729 that the carbon dioxide molecule (CO_2) has no dipole moment but each carbon–oxygen bond does. What does this information imply about the structure of the molecule?

•**38.** During a physics lab, you and your paramour construct the beautiful electric dipole shown in Figure P.38. (a) Calculate the electric field at the point P, expressing it in Cartesian vector form. (b) A -5.00 μC charge is now placed at the point P. What is the electrical force on this new charge? Express your result in Cartesian vector form. (c) What is the magnitude of the force on the -5.00 μC charge?

•**39.** Determine the torque on the electric dipole shown in Figure P.39.

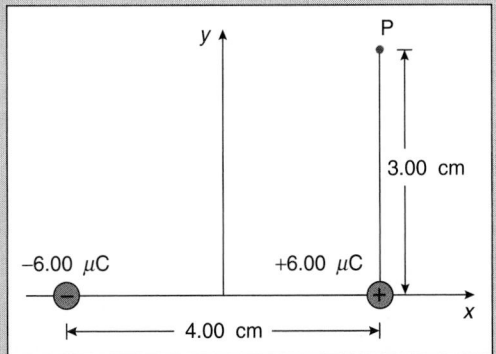

FIGURE P.38

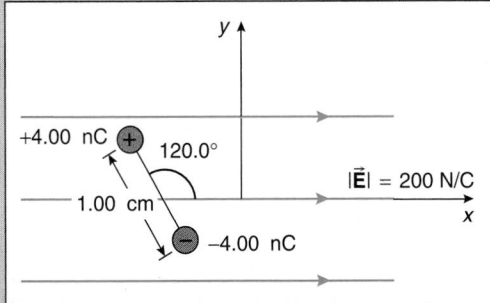

FIGURE P.39

•**40.** The water molecule has an electric dipole moment with a magnitude of about 6.0×10^{-30} C·m. The molecule is placed in a uniform electric field of magnitude 250 N/C with the orientation shown in Figure P.40. What is the torque on the water molecule in its present orientation?

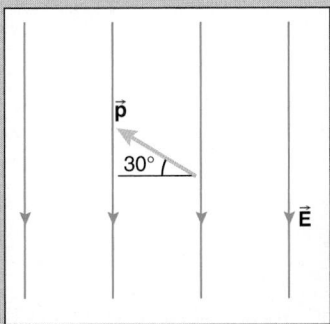

FIGURE P.40

•**41.** A pointlike charge $Q = -2e$ exists at the origin. A water molecule with a dipole moment of magnitude 6.0×10^{-30} C·m is oriented with its dipole moment vector directed toward Q as indicated in Figure P.41. The charge Q is 10.0 nm from the effective positive charge of the dipole. (Figure 16.63 on page 729 may provide pertinent information for this problem.) (a) What is the approximate magnitude of the total force on the water molecule? (b) What is the magnitude of the torque on the water molecule when it is in this orientation?

•**42.** Show that the torque on an electrical dipole placed in a uniform electric field is

$$\vec{\tau} = \vec{p} \times \vec{E}$$

independent of the origin about which the torque is calculated.

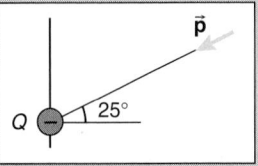

FIGURE P.41

•**43.** The general definition of the dipole moment of two charges Q_1 and Q_2 is

$$\vec{p} \equiv Q_1\vec{r}_1 + Q_2\vec{r}_2$$

where $\vec{r}_1$ and $\vec{r}_2$ are the position vectors of the two charges with respect to some origin. Show that the dipole moment of two charges $+|Q|$ and $-|Q|$, separated by a distance d, is independent of the choice of the origin of the coordinate system and is equal to $\vec{p} = |Q|\vec{d}$, as in Equation 16.13 where $\vec{d}$ is the position vector of the positive charge with respect to the negative charge.

•**44.** One model of the hydrogen atom imagines the electron in a circular orbit about a fixed proton. The arrangement consists of two charges of equal magnitudes but opposite sign, separated by a fixed distance. Explain why the arrangement has a time-averaged dipole moment of zero.

‡**45.** Use $\vec{\tau} = I\vec{\alpha}$ to show that if an electric dipole with dipole moment of magnitude p and moment of inertia I is oriented with its dipole moment making a small angle θ with the direction of an external electric field of magnitude E, the dipole will execute simple harmonic oscillations about the field direction with a frequency ν given by

$$\nu = \frac{1}{2\pi}\left(\frac{pE}{I}\right)^{1/2}$$

Section 16.9 The Electric Field of Continuous Distributions of Charge

•**46.** At what points along the symmetry axis perpendicular to the plane of a uniformly charged circular ring is the electric field of maximum magnitude?

•**47.** At what points along the symmetry axis perpendicular to the plane of a uniformly charged circular disk of radius R is the magnitude of the electric field a maximum?

•**48.** Show that the electric field along the symmetry axis perpendicular to the plane of a uniformly charged circular disk approaches the field of an infinite sheet as the distance z approaches zero.

•**49.** A 2.00 g cork with a charge +3.00 μC floats motionless 1.50 cm above a large, uniformly charged, horizontal pane of glass near the surface of the Earth. What is the surface charge density of the glass pane, assuming it to be an infinite sheet?

•**50.** Air becomes a good conductor if the magnitude of the electric field in air exceeds about 3.0×10^6 N/C; the precise value depends on many factors such as the relative humidity and pressure. (a) For oppositely charged parallel plates, what surface charge density is sufficient to create a field of this magnitude? Thunderstorms can be modeled crudely in this way. (b) How many fundamental units of charge e per square millimeter does this surface charge density represent?

•**51.** Imagine a +1.00 nC charge spread uniformly over the surface of a conducting sphere of radius R. If R is sufficiently large, the magnitude of the electric field on the surface of the conductor is less than 3.0×10^6 N/C, at which value air becomes a good conductor. For what value of R is the magnitude of the electric field on the surface of the conductor equal to 3.0×10^6 N/C? This problem indicates that the electric field at the surface of severely curved (sharply pointed) conductors is greater than in more rounded conductors with the same charge. Explain how this principle is used to advantage in lightning rods.

•**52.** A uniformly charged semicircular hoop of radius R has a total charge Q. Show that the magnitude of the electric field at the center of the semicircle is

$$E = \frac{1}{4\pi\varepsilon_0} \frac{2Q}{\pi R^2}$$

•**53.** A uniformly charged semicircular ring of length 50.0 cm has a total charge of −8.00 nC. Find the magnitude of the electric field at the center of the semicircle. Indicate its direction in a sketch.

•**54.** A rod with λ coulombs of charge per meter of its length has the shape of a circular arc of radius R. The rod subtends an angle θ as indicated in Figure P.54. Show that the magnitude of the electric field at the center of the circular arc is

$$E = \frac{1}{4\pi\varepsilon_0} \frac{2\lambda}{R} \sin\left(\frac{\theta}{2}\right)$$

If $\lambda > 0$ C/m, indicate the direction of the field in a sketch.

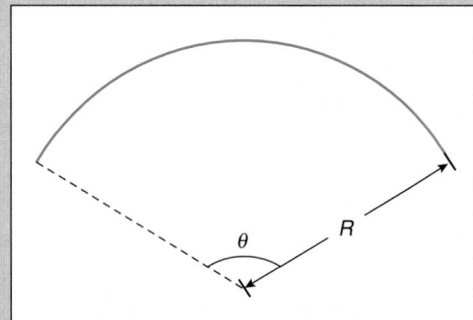

FIGURE P.54

•**55.** A long straight line of uniform positive linear charge density λ coulombs per meter of length begins at the origin and extends out to infinity along the +x-axis as shown in Figure P.55. Find the electric field at the point $x = -a$.

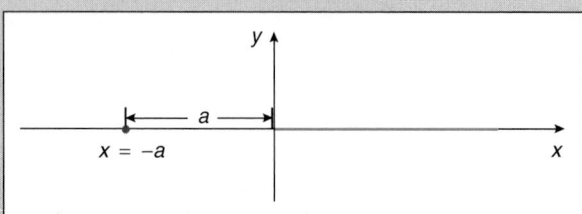

FIGURE P.55

•**56.** Show by direct integration of Equation 16.15 on page 733 that the magnitude of the electric field a distance a from an infinitely long, uniformly charged line charge (with λ coulombs per meter) is given by

$$E = \frac{\lambda}{2\pi a\varepsilon_0}$$

‡**57.** A tiny mass m with charge q is attached to an infinite sheet of charge with surface charge density σ by means of an insulating, massless cord of length ℓ. Neglect gravitational effects. (a) Specify the signs of the charge q and σ such that the cord is taut. (b) Show that if the mass is pulled slightly in a direction parallel to the plane and then released, the mass executes simple harmonic oscillation with frequency ν, where

$$\nu = \frac{1}{2\pi} \left(\frac{q\sigma}{2\varepsilon_0 m\ell} \right)^{1/2}$$

Section 16.10 Motion of a Charged Particle in a Uniform Electric Field: An Electrical Projectile

•**58.** A proton is initially at rest in a uniform electric field of magnitude 300 N/C. (a) How long will it take the proton to travel 1.00 m? (b) How long will it take an electron to do the same thing? (c) What is the ratio of the proton time to the electron time? How is this ratio related to the mass ratio of the proton to the electron?

•**59.** When the magnitude of the electric field in air exceeds about 3×10^6 N/C, it becomes a good conductor. (a) What magnitude acceleration would an electron experience in a field of this magnitude? (b) How long will it take such an electron, initially at rest, to reach a speed equal to 1.00% of the speed of light? Through what distance will such an electron travel to reach this speed?

•**60.** A medical x-ray tube accelerates electrons from rest to a speed of 8.00×10^6 m/s using a uniform electric field of magnitude 5.0×10^3 N/C. Through what distance do the electrons move to acquire this speed?

•**61.** An electron traveling at a speed of 5.00×10^6 m/s is shot into a uniform electric field of magnitude 100 N/C directed along the initial path of the electron to slow the electron from its blistering pace. (a) Is the electric field directed parallel to or antiparallel to the velocity of the electron? (b) Introduce a Cartesian coordinate system to analyze the motion of the electron after it enters the electric field. (c) Calculate the acceleration component of the electron while it is in the field. (d) How much time does it take for the electron to come to rest? (e) What distance will the electron travel into the field region to reach zero speed? (f) Will the electron remain at rest once stopped? If not, what happens to it?

•**62.** An electron is projected at speed 2.00×10^7 m/s into a region with a constant electric field directed perpendicular to the initial velocity of the electron as indicated in Figure P.62. (a) What is the direction of the electrical force on the electron? (b) What is the maximum magnitude for the electric field so that the electron does not strike the sides of the apparatus generating the electric field?

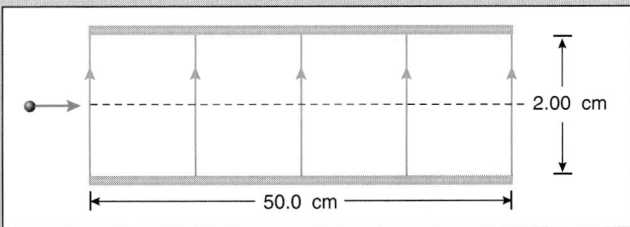

FIGURE P.62

•**63.** An electron finds itself about to begin an exciting trip in a uniform electric field of magnitude 500 N/C, directed as shown in Figure P.63. (a) What is the magnitude of the electrical force on the electron? (b) What is the magnitude of the acceleration of the electron? (c) If the electron begins its journey at rest, how long will it take the electron to travel 1.00 m in the field? (d) Indicate at which end of the field the electron began its journey.

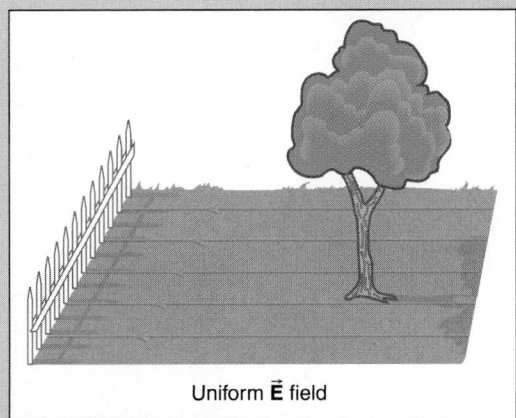

Uniform $\vec{E}$ field

FIGURE P.63

•**64.** A *positron* has the same mass as an electron (9.11×10^{-31} kg) but a *positive* charge ($+1.602 \times 10^{-19}$ C). (a) In the arrangement shown in Figure P.64, near which plate should a positron be released at rest so it will accelerate toward the other plate? (b) What is the speed of the positron just before impact? (c) If the positron is released from rest, what is its kinetic energy the instant before striking the other plate? Express your result in electron-volts (1 eV $\equiv 1.602 \times 10^{-19}$ J).

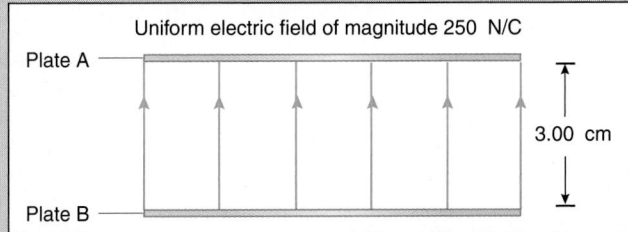

Uniform electric field of magnitude 250 N/C

Plate A

3.00 cm

Plate B

FIGURE P.64

•**65.** An electron is projected with an initial speed of 5.00×10^6 m/s into a uniform electric field of magnitude 500 N/C as shown in Figure P.65. How far into the field will the electron travel before reversing direction?

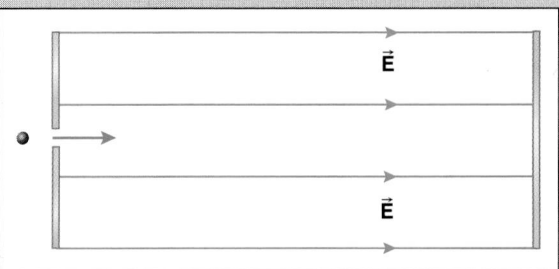

FIGURE P.65

•**66.** An electron is projected with an initial speed 5.00×10^6 m/s into a uniform electric field of magnitude 250 N/C. The angle between the initial velocity vector and the electric field is 30.0° as indicated in Figure P.66. Assume the electron does not strike the upper plate. What horizontal distance does the electron travel before striking the lower conducting plate?

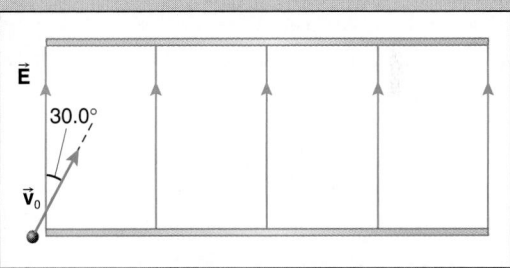

FIGURE P.66

Sections 16.11 Gauss's Law for Electric Fields*
   16.12 Calculating the Magnitude of the Electric Field Using Gauss's Law*
   16.13 Conductors*
   16.14 Other Electrical Materials*

67. A uniform electric field of magnitude 100 N/C threads a map of the state of Colorado of dimensions 0.030 m by 0.040 m. See Figure P.67. What is the flux of the vector through the area? This is a state of flux (groan!).

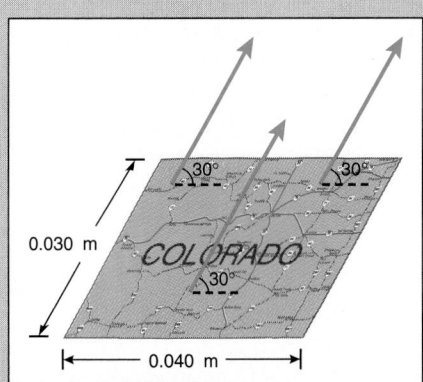

FIGURE P.67

68. For the sheer pleasure of racking up brownie points, calculate the flux of the indicated electric field vector through the surface indicated in Figure P.68.

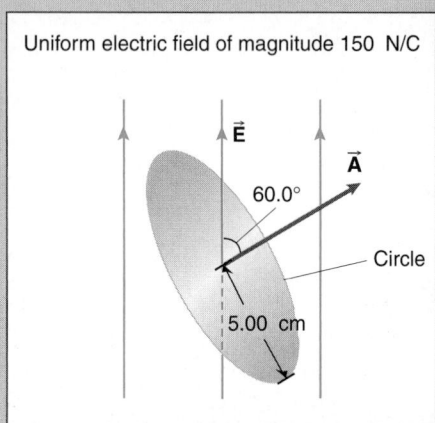

FIGURE P.68

69. What is the total flux of the electric field through a surface that completely encloses an electric dipole?

70. What is the total flux of the electric field vector through the closed surface *S* in Figure P.70?

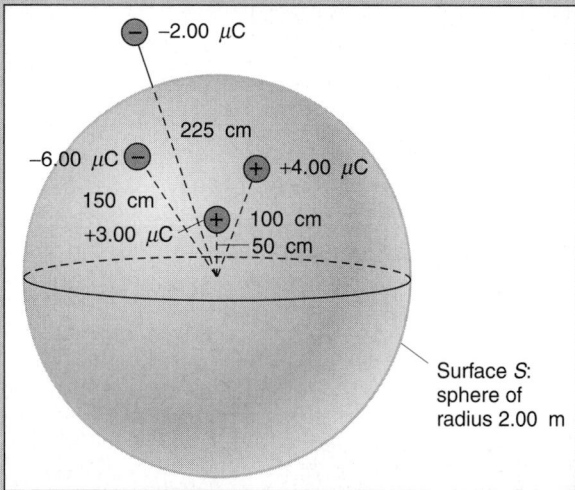

FIGURE P.70

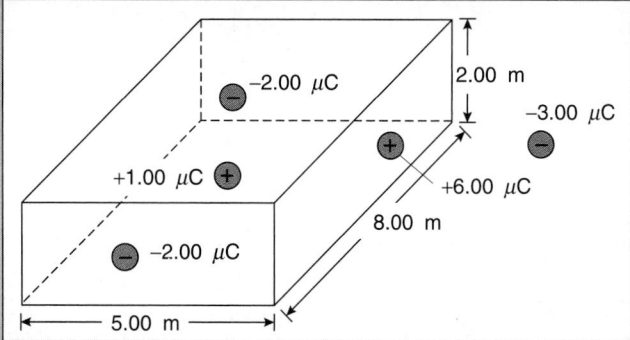

FIGURE P.71

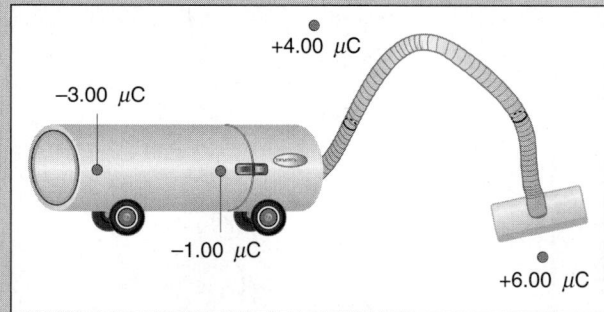

FIGURE P.72

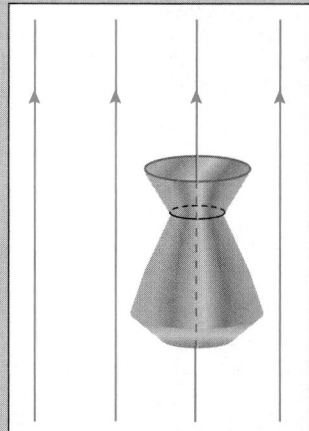

FIGURE P.73

•**71.** The great physicist Dr. E. Fields has discovered that one of her four charges has escaped from the lecture room as indicated in Figure P.71. Free at last! Calculate the total flux of the electric field from all the charges through the walls, floor, and ceiling of the classroom.

•**72.** Calculate the total electrolux . . ., rather, the electric flux through the closed surface indicated in Figure P.72.

•**73.** Consider a surface shaped like a spittoon. The circular open top has a radius of 6.0 cm and a narrower radius of 4.0 cm along its neck. The symmetry axis of the surface is aligned with a uniform electric field of magnitude 200 N/C as shown in Figure P.73. What is the flux of the electric field through the surface?

•**74.** A small copper spherical BB of radius *a* is located at the center of a larger hollow copper spherical shell of inner radius *b* and outer radius *R* as shown in Figure P.74. A charge of +*q* is on the small BB. The hollow copper shell has zero charge on it. (a) What is the electric field within the BB (for radii *r* < *a*)? (b) What is the electric field inside the copper shell (that is, for radii *r* that satisfy *b* < *r* < *R*)? (c) Draw a closed Gaussian surface within the copper of the shell as indicated in Figure P.74. What is the total flux of the electric vector through this Gaussian surface? This result implies that charge must lie on the inside surface of the spherical shell. What charge must reside on the inside surface of the copper shell? Since the copper shell has a total charge of zero, what charge must reside on the outer surface of the copper shell?

•**75.** Two concentric, fixed, hollow, thin uniform spherical shells have charges and radii as indicated in Figure P.75. (a) What is the total force on a pointlike charge *q* = 10.0 nC if it is located

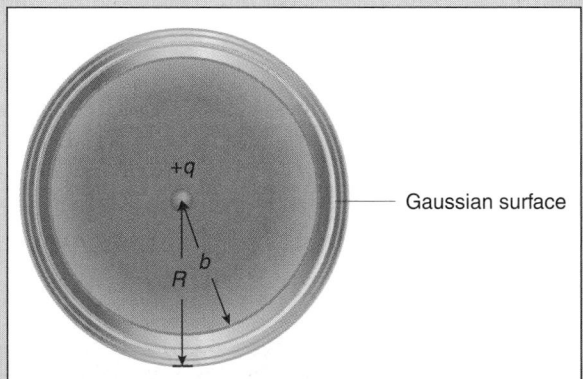

FIGURE P.74

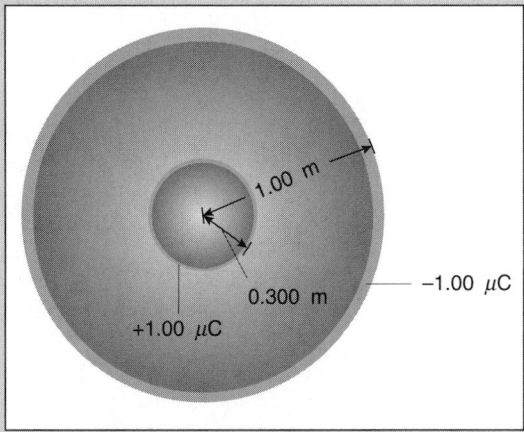

FIGURE P.75

at $r = 0.100$ m from the center? (b) What is the total force on the same charge q if it is located at $r = 0.800$ m from the center? (c) What is the total force on the same charge q if located $r = 2.00$ m from the center? (d) At what value of r does q have the maximum magnitude force on it? Calculate this maximum force.

• **76.** Early in the 20th century, there were experimental indications that atoms had internal structure. A crude model of a nuclear atom consists of a central pointlike nucleus of charge $+Ze$ surrounded by a thin uniform spherical shell of radius R and charge $-Ze$, where Z is what we call the atomic number (the number of protons in the nucleus). (a) Sketch the model. (b) Use Gauss's law to determine the magnitude of the electric field caused by this charge distribution for distances $r > R$. (c) Use Gauss's law to determine the magnitude of the electric field caused by this charge distribution for distances $r < R$.

• **77.** Instead of the model atom proposed in Problem 76, imagine a model consisting of a central pointlike nucleus of charge $+Ze$ surrounded by a sphere of radius R with charge $-Ze$ uniformly distributed throughout its volume. (a) Sketch the model. (b) Use Gauss's law to determine the magnitude of the electric field caused by this charge distribution for distances $r > R$. (c) Use Gauss's law to determine the magnitude of the electric field caused by this charge distribution for distances $r < R$.

‡ **78.** An insulating, thin, hollow sphere has a uniform surface charge density σ. (a) Show that the magnitude of the electric

field at the surface of the sphere is $|\sigma|/\varepsilon_0$ (b) A tiny hole is drilled through the shell, thus removing a negligible bit of the charge. Show that the magnitude of the electric field in the hole is $|\sigma|/2\varepsilon_0$.

‡ **79.** An early model of the hydrogen atom, known picturesquely as the plum pudding model, imagined a sphere of radius R with positive charge $+e$ distributed uniformly throughout its volume with an electron embedded, say, at the center of the sphere. (a) Using Gauss's law, calculate the magnitude of the electric field of the uniform spherical distribution of positive charge at a distance r from the center of the sphere, where $r < R$. (b) Using Newton's second law, show that an electron of mass m at a distance r from the center of the sphere (with $r < R$) executes simple harmonic motion about the center. That is, show that the motion of the electron of mass m is described by an equation of the form

$$\frac{d^2 r}{dt^2} + \omega^2 r = 0 \ \text{m/s}^2$$

where the angular frequency ω of the motion is

$$\omega = \left(\frac{1}{4\pi\varepsilon_0} \frac{e^2}{R^3 m} \right)^{1/2}$$

(c) Calculate the frequency of the simple harmonic oscillation of the electron using $R \approx 1.0 \times 10^{-10}$ m as an estimate of the size of the atom. It was originally thought that the oscillating electron in this model might account for light emission from the hydrogen atom. The model predicts a single oscillation frequency for the electron and for the emitted light. The model had to be abandoned because it was unable to account for the observation that hydrogen emits light at many different specific frequencies.

‡ **80.** Some time ago, R. A. Lyttleton and H. Bondi investigated the consequences of assuming the charges on the electron and proton were of slightly different magnitude.* Such an asymmetry could account for the observed expansion of the universe using purely Newtonian physics, as we demonstrate in this problem. The idea is intriguing, but the *actual* expansion of the universe we now know is *not* caused by this mechanism. Nonetheless, an exploration of Lyttleton and Bondi's hypothesis is an interesting application of electricity and dynamics. For simplicity, assume that the entire universe is composed of hydrogen.† Assume also that the charges on the electron and proton are of slightly different magnitude. A hydrogen atom then has a slight charge of magnitude δe, where δ is a purely numerical factor that likely is very small, since matter appears to be quite closely electrically neutral. Imagine such a slightly charged hydrogen atom of mass m on the surface of a large, uniform spherical mass M of radius r of similar hydrogen atoms.

---

*R. A. Lyttleton and H. Bondi, "On the physical consequences of a general excess of charge," *Proceedings of the Royal Society of London*, A252, pages 313–333, (1959).

† The universe is composed of about 70% hydrogen and 30% helium with just a smattering of heavier elements. For Lyttleton and Bondi's hypothesis, it is not necessary to assume the entire universe is composed of hydrogen but it is easier to do so.

(a) Sketch the situation. (b) Show the giant spherical mass has a charge of magnitude

$$\frac{M}{m}\,\delta e$$

(c) Show that the hydrogen atom (with charge δe) on the surface of M experiences an electrical *repulsive* force of magnitude

$$\frac{1}{4\pi\varepsilon_0}\,\frac{M}{m}\,\frac{\delta^2 e^2}{r^2}$$

Charged matter lying *outside* the large mass M, if also spherical in shape, has no effect on the atom in question because of the shell theorems associated with inverse-square-law forces that we examined in our study of gravitation in Chapter 6. (d) Show that the *ratio* of the magnitude of the repulsive electrical force to the magnitude of the attractive gravitational force on m is

$$\frac{\text{repulsion}}{\text{attraction}} = \delta^2\,\frac{1}{4\pi\varepsilon_0}\,\frac{e^2}{Gm^2}$$

$$\approx 1.23 \times 10^{36}\,\delta^2$$

$$\equiv \mu\delta^2$$

where μ is unitless and has the numerical value of about 1.23×10^{36}. (e) The forces of repulsion and attraction are of *equal* magnitude if the ratio in part (d) is equal to 1. Show that the ratio is equal to 1 if δ has the value

$$\delta \approx 9.02 \times 10^{-19}$$

$$\approx 10^{-18}$$

In other words, if the magnitudes of the charges on the electron and proton differ by about one part in 10^{18}, then the repulsive electrical force on the atom on the surface of the sphere is equal in magnitude to the attractive force of gravitation. (f) Assume that δ is bigger than the value calculated in part (e). Let ρ be the mass density of M. Show that the magnitude of the total repulsive force on the atom by the sphere is

$$F_{\text{repulsive}} = \frac{4}{3}\pi\rho Gm(\mu - 1)\,r$$

The important point to note here is that the magnitude of the net force of repulsion on the atom is proportional to the radius of the sphere; the magnitude of the force is proportional to the distance.* Write this relation as

$$F_{\text{repulsive}} = Kr$$

(g) Apply Newton's second law to m to show that

$$Kr = m\,\frac{d^2 r}{dt^2}$$

The second derivative of r with respect to t is proportional to r itself. Show that one solution to this equation is[†]

$$r(t) = Ae^{Ht}$$

where

$$H \equiv \frac{K}{m}$$

(h) Show that the expansion (repulsive) speed v of the atom in its motion is

$$v = \frac{dr}{dt}$$

$$= Hr$$

Thus the speed of recession of the atom is proportional to its distance. It has been known since the late 1920s that the universe is expanding and that the recessional speed of distant galaxies is proportional to their distance from us; the proportionality constant H is known as the *Hubble constant*. Although the mechanism of charge asymmetry leads to an expanding universe, the actual expansion of the universe is not caused by this mechanism, but by effects associated with the creation of the universe in the so-called Big Bang event.

---

*This is not the same as the Hooke's law spring force. The spring force is attractive; the force here is repulsive.

[†] In this solution e represents the base of natural logarithms, *not* the fundamental unit of charge.

INVESTIGATIVE PROJECTS

A. Expanded Horizons

1. The procedures used by Charles Coulomb in the latter part of the 18th century to determine the force law between static charges are well worth a look on your part. Write a précis about them.
   Peter Heering, "On Coulomb's inverse square law," *American Journal of Physics*, 60, #11, pages 988–994 (November 1992).
   Steven Dickman, "Could Coulomb's experiment result in Coulomb's law?" *Science, 262*, #5133, pages 500–501 (22 October 1993).

2. Experiments with increasing precision test the accuracy of the inverse-square nature of Coulomb's law. Trace the development of these experiments and report on the current precision.
   Peter Heering, "On Coulomb's inverse square law," *American Journal of Physics*, 60, #11, pages 988–994 (November 1992).

John David Jackson, *Classical Electrodynamics* (2nd edition, John Wiley and Sons, New York, 1975), pages 5–9.
E. R. Williams, J. E. Faller, and H. A. Hill, "New experimental test of Coulomb's law: a laboratory limit on the photon rest mass," *Physical Review Letters, 26*, #12, pages 721–724 (22 March 1971).
R. E. Crandall, "Photon Mass Experiment," *American Journal of Physics, 51*, #8, pages 698–702 (August 1983).
Jibayo Akinrimisi, "Note on the experimental determination of Coulomb's law," *American Journal of Physics, 50*, #5, pages 459–460 (May 1982).
P. H. Wiley and W. L. Stutzman, "A simple experiment to demonstrate Coulomb's law," *American Journal of Physics, 46*, #11, pages 1131–1132 (November 1978).

3. Investigate how electrostatics is applied to xerography. Describe the principles in a cogent report.

   H. Richard Crane, "Physics in the copy machine," *The Physics Teacher*, 22, #7, pages 454–461 (October 1984).

   Charles D. Hendricks, "Electrostatic imaging," *Electrostatics and Its Applications*, edited by A. D. Moore (Wiley, New York, 1973), pages 281–306.

   John H. Dessauer and Harold E. Clark (editors), *Xerography and Related Processes* (Focal Press, New York, 1965).

   J. Mort, *The Anatomy of Xerography: Its Invention and Evolution* (McFarland, Jefferson, North Carolina, 1989).

   Edgar M. Williams, *The Physics and Technology of Xerographic Processes* (Krieger, Malabar, Florida, 1992).

4. Investigate and report on the physics underlying the use of electrostatic precipitators in reducing smokestack emissions.

   Myron Robinson, "Electrostatic precipitation," *Electrostatics and Its Applications*, edited by A. D. Moore (Wiley, New York, 1973), pages 180–220.

   Harry J. White, *Industrial Electrostatic Precipitation* (Addison-Wesley, Reading, Massachusetts, 1963).

   David A. Lloyd, *Electrostatic Precipitator Handbook* (Adam Hilger, Philadelphia, 1988).

5. Investigate the electrostatics associated with lightning as well as the classic kite experiment of Benjamin Franklin that demonstrated that lightning was electricity. Only by luck did Franklin escape electrocution by lightning. Describe the experiment.

   Richard E. Orville, "The lightning discharge," *The Physics Teacher*, 14, #1, pages 7–13 (January 1976).

   Martin A. Uman, *All About Lightning* (Dover, New York, 1986).

   Martin A. Uman, *The Lightning Discharge* (Academic Press, Orlando, 1987).

   Martin A. Uman, *Lightning* (Academic Press, Orlando, 1987).

   Leon E. Salanave, *Lightning and Its Spectrum* (University of Arizona Press, Tucson, 1980).

   Rudolph Heinrich Golde, *The Physics of Lightning* (Academic Press, New York, 1977).

   Peter E. Viemeister, *The Lightning Book* (MIT Press, Cambridge, Massachusetts, 1972).

6. Electrostatics is used to sort seeds by size, to remove dirt, hulls, and rodent excrement from cereal grains, to concentrate various minerals in mining, and in recycling reusable wastes. Investigate some of these applications in greater detail.

   A. D. Moore, "Electrostatics," *Scientific American*, 226, #3, pages 46–58 (March 1972); additional references on page 126.

   James E. Lawver and W. P. Dyrenforth, "Electrostatic separation," *Electrostatics and Its Applications*, edited by A. D. Moore (Wiley, New York, 1973), pages 221–249.

7. Tiny charged ink droplets are used in common computer ink jet printers. Investigate the elements involved in the design of such printers.

   Larry Kuhn and Robert A. Myers, "Ink-jet printing," *Scientific American*, 240, #4, pages 162–178 (April 1979); additional references on page 190.

   Hewlett-Packard Journal, 45, #1 (February 1994); the entire issue is devoted to ink jet technology.

   A. J. Rogers, "Ink jet takes off," *Byte*, 16, #10, pages 163–168 (October 1991).

8. Some fish, among them sharks, eels, catfish, and torpedo fish, use electric fields to detect and/or stun and kill prey. Investigate the electric activity of such fish.

   Joseph Bastien, "Electrosensory organisms," *Physics Today*, 47, #2, pages 30–37 (February 1994).

   Chau H. Wu, "Electric fish and the discovery of animal electricity," *American Scientist*, 72, #6, pages 598–607 (November–December 1984).

   Harry Grundfest, "Electric fishes," *Scientific American*, 203, #4, pages 115–125 (October 1960); additional references on page 220.

   Louis Roule, *Fishes and Their Ways of Life* (Norton, New York, 1935) pages 156–171.

   William N. McFarland, F. Harvey Pough, Tom J. Cade, and John B. Heiser, *Vertebrate Life* (Macmillan, New York, 1979), pages 240–248.

   David Hafemeister, "Resource Letter BELFEF-1: Biological effects of low-frequency electromagnetic fields," *American Journal of Physics*, 64, #8, pages 974–981 (August 1996); this contains many references.

   Peter Moller, *Electric Fishes: History and Behavior* (Chapman & Hall, London, 1995).

   Victor Percy Wittaker, *The Cholinergic Neuron and Its Target: The Electromotor Innervation of the Electric Ray "Torpedo" as a Model* (Birkhäuser, Boston, 1992).

B. Lab and Field Work

9. Next time you dry your clothes in a dryer and find your socks sticking together, perform some simple electrical experiments to determine the charge on the socks. Explain your procedure and your results. Investigate the dependence of the effect on the type of fabric.

10. Investigate the physics underlying the Millikan oil drop experiment, the method first used to determine the magnitude of the fundamental unit of charge e. Most physics departments have an apparatus to perform an experiment similar to that of Millikan. Design and perform an experiment to determine e with such an apparatus.

    Robert Andrews Millikan, *The Electron* (University of Chicago Press, Chicago, 1917).

    R. A. Millikan, "On the elementary electrical charge and Avogadro's constant," *Physical Review*, II, #2, pages 109–143 (August 1913).

    Ray C. Jones, "The Millikan oil-drop experiment: making it worthwhile," *American Journal of Physics*, 63, #11, pages 970–977 (November 1995).

    William M. Fairbank Jr. and Allan Franklin, "Did Millikan observe fractional charges on oil drops?," *American Journal of Physics*, 50, #5, pages 394–397 (May 1982).

    Mark A. Heald, "Millikan oil-drop experiment in the introductory laboratory, *American Journal of Physics*, 42, #3, pages 244–246 (March 1974).

11. The electric analog of a magnet is called an *electret* or *ferroelectric* material. Ferroelectrics are not as readily available as common permanent magnets (ferromagnets) because only certain materials (among them barium titanate [$BaTiO_3$]) can sustain a permanent electric polarization near room temperature. Other materials have ferroelectric properties at cryogenic (very low) temperatures (122 K for potassium dihydrogen phosphate [KH_2PO_4]). Secure a sample crystal of barium titanate and explore the electric field in its vicinity.

    Oleg D. Jefimenko and David K. Walker, "Electrets," *The Physics Teacher*, 18, #9, pages 651–659 (December 1980).

    R. N. Varrey and H. T. Hahn, "Electrets and electrostatic measurement," *American Journal of Physics*, 43, #6, pages 509–513 (June 1975).

    Deborah Schurr and Tim Usher, "Demonstrating hysteresis in ferroelectric materials," *The Physics Teacher*, 33, #1, pages 30–31 (January 1995).

    Thomas Kallard, *Electret Devices for Air Pollution Control* (Optosonic Press, New York, 1972).

C. Communicating Physics

12. Electrostatic motors and generators date back to Benjamin Franklin in the 18th century. The most well-known such device was invented about 1878 by James Wimshurst (1832–1903) and is called a Wimshurst machine. Investigate how this historic device operates. Your physics department may have such a Wimshurst electrostatic generator in its attic or museum. As an exercise in technical writing, write an extensive explanation describing how the machine separates and accumulates charge. Design a poster explaining its operation for use in a hallway display of a Wimshurst machine in your science building or a local science museum.

    Encyclopedia Britannica (11th edition, 1910), volume 9, pages 178–179.
    A. D. Moore, *Electrostatics* (Anchor Doubleday, Garden City, New York, 1968).
    Oleg D. Jefimenko, "Electrostatic motors," *Electrostatics and Its Applications*, edited by A. D. Moore, editor (Wiley, New York, 1973), pages 131–147.
    M. J. Mulcahy and W. R. Bell, "Electrostatic generators," *Electrostatics and Its Applications*, edited by A. D. Moore, editor (Wiley, New York, 1973), pages 148–179.

13. Write a paragraph comparing and contrasting the gravitational and electrical *forces* and another paragraph comparing and contrasting the gravitational and electrical *fields*.

CHAPTER 17

ELECTRIC POTENTIAL ENERGY AND THE ELECTRIC POTENTIAL

Vive la différence!

In this chapter we bring the concepts of work, potential energy, the CWE theorem, and also a new and convenient idea (that of the electric potential) to bear on electrical phenomena. Such notions are used in the study of electricity and its technological applications as much as or more than they are in gravitation and mechanics. Electrical phenomena are ubiquitous in physics, engineering, and even biology, chemistry, and medicine, particularly at the cellular and molecular level. It is therefore absolutely essential to grasp the ideas of work, energy, and the electric potential, because they manifest themselves in all the sciences—there is no avoiding them.

Fortunately, as we saw in Chapter 16, there are strong parallels between the electrical force on charges and the gravitational force on masses (though, as we have seen, the forces also have distinctly different features). In this chapter we continue to exploit these parallels in the realms of work and energy.

We found in Chapter 8 that the concepts of work and energy were useful and convenient ways of reformulating mechanics that permitted us to solve many problems without a detailed, point-by-point knowledge of the forces on a particle as it moved along a path. In some cases, such knowledge of the forces is unavailable, difficult to determine, or analytically complex. Neglecting thermal effects, the CWE theorem often is a useful tool for the study of the mechanics of particles. The energy approach is particularly useful for the cases where only conservative and/or zero-work forces act on the system, for then the total mechanical energy of the system, $E \equiv KE + PE$, is conserved.

17.1 ELECTRICAL POTENTIAL ENERGY AND THE ELECTRIC POTENTIAL

In Chapter 8 we defined a conservative force in the following equivalent ways:

1. It is a force that does work whose quantity is independent of the path followed by a particle as it moves from an initial to a final position.
2. It is a force that does zero work around any closed path.

We here can make quick use of what we already learned in Chapter 8, where we found that the gravitational force (an inverse-square-law force) is a conservative force. The gravitational and static electrical forces have the same mathematical form; it is merely the letter symbols that formally distinguish them. Since the gravitational force is a conservative force, identical reasoning shows the following:

> The electrical force caused by static charges is a conservative force.

Therefore (1) the work done by the static electrical force is independent of the path that a charge follows from an initial to a final position; equivalently, (2) the work done by the static electrical force around any closed path is zero.

The work done by a conservative force on a particle is the negative of the change in the potential energy associated with the conservative force:

$$W_{consrv} \equiv -\Delta PE$$
$$= -(PE_f - PE_i)$$
(17.1)

The motivation behind the introduction of a *scalar* potential energy associated with a conservative force was so that the work done by the force could be accounted for via the negative of the change in the associated potential energy. The change in the potential energy then can join the change in the kinetic energy on the right-hand side of the by-now-famous CWE theorem. Only the work done by nonconservative forces then remains on the left-hand side of the theorem. That is, the left-hand side of the CWE theorem is divided into the work done by nonconservative forces and the work done by conservative forces:

$$W_{noncon} + W_{consrv} = \Delta KE$$

Invoking Equation 17.1, relating the work done by conservative forces to changes in their respective potential energy, implies that

$$W_{noncon} + (-\Delta PE) = \Delta KE$$

and so

$$W_{noncon} = \Delta KE + \Delta PE$$
(17.2)

Following Equation 17.1, we define the work done by the static electrical force in moving a charge q from an initial position to a final position as the negative of the change in the electrical potential energy of the charge q:

$$W_{elec} \equiv -\Delta PE$$
(17.3)

Once again, notice that it is only a *change* in the potential energy of the charge that is physically significant. The place where the electrical potential energy is set equal to zero is arbitrary. We can choose it to be zero anywhere we like; however, in some situations, the choice is made for us by convention.* The electrical potential energy of a charge (just like the gravitational potential energy of a mass) is a function of the location or position of the charge in space relative to *other* charges.

Using the general definition of the work done by a force, Equation 8.2, we see that the work done by the electrical force on a charge q as it moves from an initial to a final location is the integral over the path of the scalar product of the electrical force with the differential change in the position vector $d\vec{\mathbf{r}}$ of the charge. Equation 17.3 thus becomes

$$\int_i^f \vec{\mathbf{F}}_{elec} \cdot d\vec{\mathbf{r}} = -\Delta PE$$

$$= -\left(PE_f - PE_i\right)$$
(17.4)

The integral on the left-hand side of Equation 17.4 is performed over the path connecting the initial and final locations of the charge. Since the electrical force is conservative, the result of the integration is independent of the path used to get from the initial to final locations.

The electrical force on a charge q, when it finds itself in an electric field $\vec{\mathbf{E}}$ created by other static (fixed) charges, is given by Equation 16.8[†]:

---

*We encountered these choices and conventions in the gravitational context as well.

[†] It is good to recall, as we mentioned in Chapter 16, that we have made the assumption that the presence of the charge q does not affect the distributions of the charges that create the electric field $\vec{\mathbf{E}}$ in which q is placed. The charge q, therefore, occasionally is referred to as a small "test" charge.

$$\vec{F}_{elec} = q\vec{E}$$

Substituting this into Equation 17.4, we find

$$q \int_i^f \vec{E} \cdot d\vec{r} = -(PE_f - PE_i) \quad (17.5)$$

Equation 17.5 involves both the charge q and the electric field $\vec{E}$ it experiences, caused by other fixed (static) charges. We would like to eliminate the reference to q and focus exclusively on the electric field created by the static charges. We do this by dividing Equation 17.5 by the charge q:

$$\int_i^f \vec{E} \cdot d\vec{r} = -\left(\frac{PE_f}{q} - \frac{PE_i}{q}\right) \quad (17.6)$$

The right-hand side of Equation 17.6 represents the difference in the *potential energy per unit charge*, or the potential energy per coulomb. Now we introduce new terminology.

Define the **electric potential** V at a point in space to be the potential energy per unit charge at that point*:

$$V \equiv \frac{PE_{of\,q}}{q} \quad (17.7)$$

The SI unit for the electric potential thus is joules/coulomb (J/C). The unit is used so frequently that it is given its own special name: the **volt** (V) = J/C.

Perhaps the volt is a familiar unit, but now you know what it really means and where it comes from in physics. The unit honors the Italian physicist Alessandro Volta (1745–1827) who performed important experiments on electrical phenomena, particularly concerned with the development of a crude, early battery called the voltaic pile.

To find the electric potential energy of a charge q, when q is placed where the electric potential has the value V, we rearrange Equation 17.7 to yield

$$\boxed{PE_{of\,q} = qV} \quad (17.8)$$

Equation 17.8 is a very important definition of the relationship between the potential energy of the charge q and the potential V at the location where the charge is placed. Equation 17.6 then can be rewritten in terms of the difference in electrical potential between the two points:

$$\int_i^f \vec{E} \cdot d\vec{r} = -\left(V_f - V_i\right)$$

The change in the electric potential between two points thus is

$$\boxed{V_f - V_i = -\int_i^f \vec{E} \cdot d\vec{r}} \quad (17.9)$$

The change in the value of the electric potential between two points also is called the **potential difference** between the two points. Since the static electrical force is conservative, the line integral of the static electric field on the right-hand side of Equation 17.9 is independent of the path used to get from the initial to final locations in space.

Notice that Equation 17.9 implies that the SI unit for quantifying the electric field, newtons per coulomb (N/C), also can be expressed equivalently as volts per meter (V/m): N/C = V/m.

In the gravitational situation near the surface of the Earth, we often choose the zero for the gravitational potential energy to be at ground level. We did not always do this, nor did we have to.

In the electrical case, any place where the electric potential energy and, from Equation 17.7, the electric potential are zero is defined to be an **electrical ground**.[†]

We typically have a choice about what point to take as the zero of the electric potential energy and electric potential, but once the choice is made, that place is an electrical ground. In other words, an electrical ground means that $V = 0$ V at that point, and if a charged particle is placed at that point, the electric potential energy of that charge is 0 J there as well.

There are a number of important points that need to be stressed about the relationship between the electrical *potential* and the electric *potential energy*:

PROBLEM-SOLVING TACTICS

17.1 The electric potential is not the same thing as the electric potential energy. Do not confuse them. The electric potential is the potential energy per unit charge and is expressed in joules per coulomb (J/C), or volts (V) in the SI unit system. Electric potential is a property of a *point in space*, whether or not a charge is placed at that point in space. The numerical value of the potential can be positive, negative, or zero. Electrical potential energy is expressed in joules (J) and is something that a charged particle has by virtue of its location in space relative to other charges.[‡]

17.2 It is important to use the appropriate sign for the charge q in Equation 17.8:

$$PE = qV \quad (17.8)$$

The numerical value of the potential energy of the charge q depends on *both* the sign of q and the sign of V at the point where q is placed.

17.3 The electric potential and the electric potential energy both are scalar quantities, not vectors. It is the electric field and electrical force that are vectors.

---

*We also could have defined a gravitational potential energy per unit mass, a gravitational potential, but such a potential is used infrequently (except in celestial mechanics) and we omitted mentioning it in Chapters 6 and 8.

[†] Some countries, such as Great Britain, use the word *earth* for an electrical ground.

[‡] The same is true of gravitational potential energy. The gravitational potential energy of a mass m depends on its location in space relative to other masses (say, the Earth).

Alessandro Volta resisted familial pressure to become an attorney and followed his interests and talents into electrical research. Although active in politics, it is unlikely he wore a toga during the 18th and 19th centuries!

PROBLEM-SOLVING TACTIC

17.4 The direction of the electric field always is from regions with higher values of the electric potential to regions with lower values of the electric potential. This fact is demonstrated in Example 17.1.

QUESTION 1

Your professor asks the class why the electrical force of static charges is conservative. Three students respond as follows:

- because the force is proportional to the product of the charges;
- because the force is a central force;
- because the force is an inverse-square-law force.

Which student(s), if any, provided a correct response?

STRATEGIC EXAMPLE 17.1

To a first and crude approximation, the bottom of a thundercloud and the Earth can be modeled as a pair of large, parallel, charged sheets (plates) with a large air gap between them.

a. Calculate the electric potential at a point x between two infinite, uniformly charged plates, separated by a distance d, that create a uniform electric field of magnitude E between them. For convenience, orient the charged sheets vertically with the x-axis horizontal, as shown in Figure 17.1.
b. What is the potential difference between the two charged plates?
c. Choose the place for the electrical ground to be at one or the other plate, and graph $V(x)$ versus x.

Solution

The electric field is directed from the positively charged plate to the negatively charged plate. With the coordinate system

The bottom of a charged thundercloud and the surface of the Earth can be modeled as a pair of large, parallel, charged sheets.

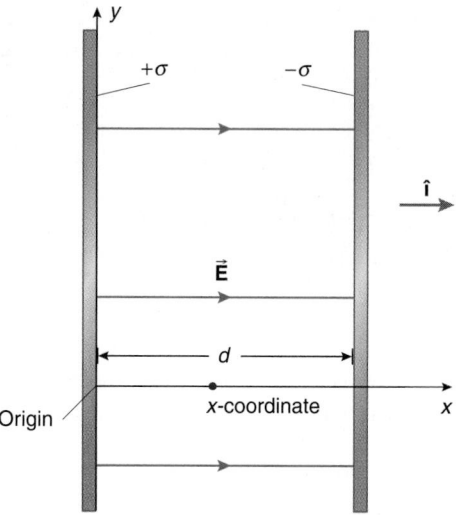

FIGURE 17.1

indicated in Figure 17.1, the electric field between the two plates is, in vector form,

$$\vec{E} = E\hat{\imath}$$

where E is positive

a. To calculate the electric potential at point x, begin with Equation 17.9. Take the origin as the initial position and co-ordinate x as the final position:

$$V(x) - V(0 \text{ m}) = -\int_{0 \text{ m}}^{x} \vec{E} \cdot d\vec{r} \qquad (1)$$

As always, take the one-dimensional differential change in the position vector $d\vec{r}$ to be

$$d\vec{r} = dx\,\hat{\imath}$$

and let the limits of the integration indicate which way you move along the x-axis from the initial to final positions of the charge. Equation (1) becomes

$$V(x) - V(0 \text{ m}) = -\int_{0 \text{ m}}^{x} E\hat{\imath} \cdot dx\,\hat{\imath}$$

Since E is constant, you can bring it outside the integral:

$$V(x) - V(0 \text{ m}) = -E \int_{0 \text{ m}}^{x} dx$$

The integral is the kind we all like to see: quite easy! Thus

$$V(x) - V(0 \text{ m}) = -Ex \qquad (2)$$

The quantity $-Ex$ is negative, since E is the magnitude of the field and the coordinate x is positive according to Figure 17.1. You can conclude that location x has a lower value of the electric potential than that at the origin where $x = 0$ m.

Equation (2) indicates that the electric potential decreases linearly with x in the same direction as the uniform field. Therefore the direction of the electric field is *from* regions with higher values of the electric potential *to* regions of lower values of the electric potential. This is *always* the case, no matter what the charge configuration.

b. The position $x = d$ is that of the negatively charged plate and, from equation (2), you have

$$V(d) - V(0 \text{ m}) = -Ed \qquad (3)$$

which is the potential difference between the locations $x = d$ and $x = 0$ m.

c. Let's consider the ways you might choose the ground.

Choice 1

Since the position $x = d$ is at a lower electric potential than the origin, you might choose the point $x = d$ to be the electrical ground. This means $V(d) = 0$ V. With this choice, equation (3) becomes

$$V(0 \text{ m}) = Ed$$

and equation (2) becomes

$$V(x) - Ed = -Ex$$

or

$$V(x) = Ed - Ex \qquad (4)$$

The potential $V(x)$ versus x is graphed in Figure 17.2.

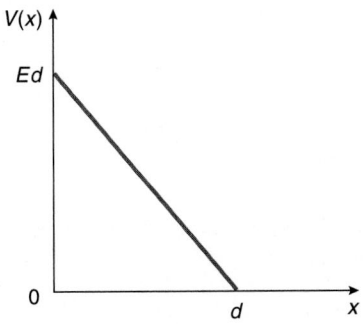

FIGURE 17.2

Choice 2

You can choose the electric ground to be anywhere that is handy! Suppose you choose $x = 0$ m to be the electrical ground. Then $V(0 \text{ m}) = 0$ V, and equation (3) becomes

$$V(d) - 0 \text{ V} = -Ed$$

or

$$V(d) = -Ed$$

With this choice for the ground, equation (2) becomes

$$V(x) - 0 \text{ V} = -Ex$$

or

$$V(x) = -Ex \qquad (5)$$

This function is graphed in Figure 17.3. Notice that location $x = d$ still is at a lower electric potential than is $x = 0$ m. The difference in the electric potential, $V(d) - V(0 \text{ m}) = -Ed$, is the same regardless of where you choose the ground. It is only the *difference* in the electric potential between two points that is physically significant; the change in electric potential between the points $x = d$ and $x = 0$ m is the same in both Figure 17.2 and Figure 17.3.

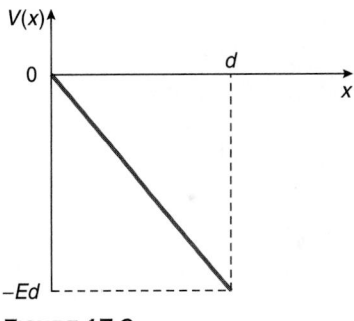

FIGURE 17.3

17.2 THE ELECTRIC POTENTIAL OF A POINTLIKE CHARGE

Electrons and protons are essentially pointlike charges, and so it is of some import to know the electric potential in the vicinity of a pointlike charge. The electric field a distance r away from a point charge Q is given by Equation 16.11:

$$\vec{E} = \frac{1}{4\pi\varepsilon_0} \frac{Q}{r^2} \hat{r}$$

The field is directed radially away from the charge if $Q > 0$ C and toward the charge if $Q < 0$ C. We want to find the potential difference between the points at $\vec{r}_f$ and $\vec{r}_i$, shown schematically in Figure 17.4.

We begin with the defining equation for the electrical potential difference between two points, Equation 17.9:

$$V(r_f) - V(r_i) = -\int_{i}^{f} \vec{E} \cdot d\vec{r}$$

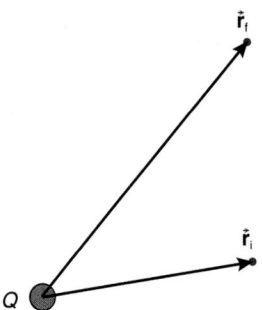

FIGURE 17.4 The vicinity of a pointlike charge Q.

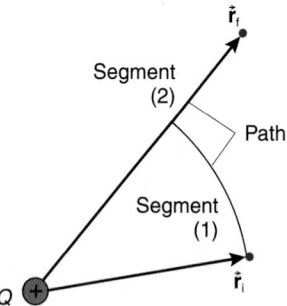

FIGURE 17.5 A path between $\vec{r}_i$ and $\vec{r}_f$ consisting of a circular arc and a radial segment.

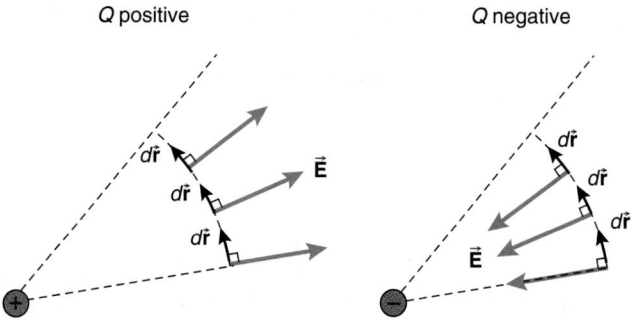

FIGURE 17.6 The scalar product $\vec{E} \cdot d\vec{r}$ is zero at each point along the circular arc.

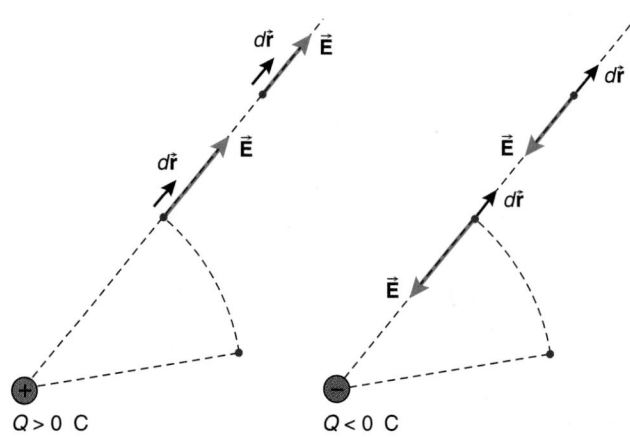

FIGURE 17.7 The field $\vec{E}$ is parallel or antiparallel to $d\vec{r}$ along the radial segment of the path, as Q is positive or negative respectively.

The potential difference between the two points is *independent of the particular path used* to get from $\vec{r}_i$ to $\vec{r}_f$, so we can choose a path to ease the evaluation of the integral. We use a path consisting of an arc of a circle of radius r_i centered on the point charge and subsequently a radial segment out to the point at the tip of $\vec{r}_f$, as indicated in Figure 17.5.

At every point along the circular segment of the path, the scalar product $\vec{E} \cdot d\vec{r}$ is zero because the radially directed electric field $\vec{E}$ is perpendicular to the change in the position vector $d\vec{r}$ along this segment (see Figure 17.6). Thus we need only consider the second segment of the path. The change in the position vector along this segment of the path is in the radial direction, and so we write $d\vec{r}$ as

$$d\vec{r} = dr\,\hat{r}$$

where as always the limits of integration take care of the direction we are moving in the radial direction. Along this second segment, the electric field is along the path, parallel to $d\vec{r}$ if $Q > 0$ C and antiparallel to $d\vec{r}$ if $Q < 0$ C, as shown in Figure 17.7. Equation 17.9 thus becomes

$$
\begin{aligned}
V(r_f) - V(r_i) &= -\frac{1}{4\pi\varepsilon_0} Q \int_i^f \frac{\hat{r}}{r^2} \cdot dr\,\hat{r} \\
&= -\frac{1}{4\pi\varepsilon_0} Q \int_{r_i}^{r_f} \frac{dr}{r^2} \\
&= -\frac{1}{4\pi\varepsilon_0} Q \left(-\frac{1}{r} \right) \Big|_{r_i}^{r_f} \\
&= \frac{1}{4\pi\varepsilon_0} Q \left(\frac{1}{r_f} - \frac{1}{r_i} \right)
\end{aligned}
\tag{17.10}
$$

For the electric potential of a point charge, it is convenient to choose the location of the electrical ground to be at infinity. In other words, for a point charge, the location of the zero of the electric potential is chosen for us by convention.

That is, when $r_f = \infty$ m, $V(r_f) = 0$ V. This is the *only* choice made for the ground of a point charge. With this choice for the location of the zero of the potential, Equation 17.10 becomes

$$0 \text{ V} - V(r_i) = 0 \text{ V} - \frac{1}{4\pi\varepsilon_0} \frac{Q}{r_i}$$

or

$$V(r_i) = \frac{1}{4\pi\varepsilon_0} \frac{Q}{r_i}$$

Since the point r_i could be anywhere, we can drop the subscripts and say the following:

The electric potential of a point charge Q a distance r from the charge is

$$V(r) = \frac{1}{4\pi\varepsilon_0} \frac{Q}{r} \tag{17.11}$$

Notice that as $r \to \infty$ m, the electric potential of the point charge Q approaches 0 V, as we chose.

The electric potential at all points surrounding a positive point charge ($Q > 0$ C) is positive. If the point charge Q is negative, the potential is negative at all points.

In the x-y plane, where $r = (x^2 + y^2)^{1/2}$, the electric potentials of positive and negative pointlike charges have rather spectacular graphs (see Figures 17.8 and 17.9). These potentials approach zero for large r and diverge as $r \to 0$ m.

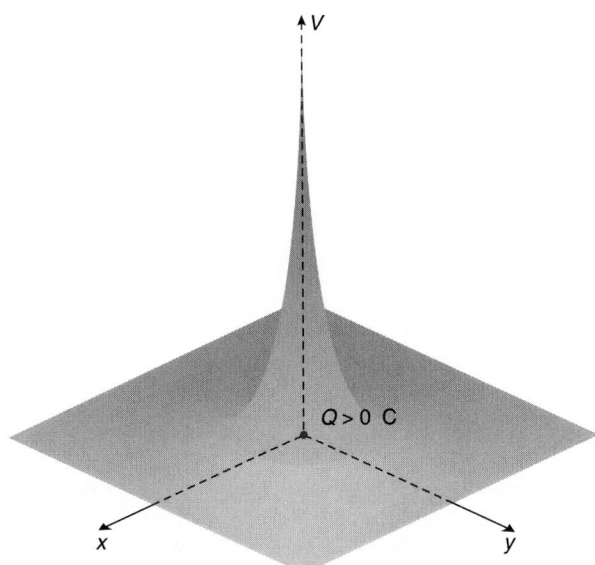

FIGURE 17.8 A graph of the electric potential V of a positive pointlike charge at the origin in the x-y plane.

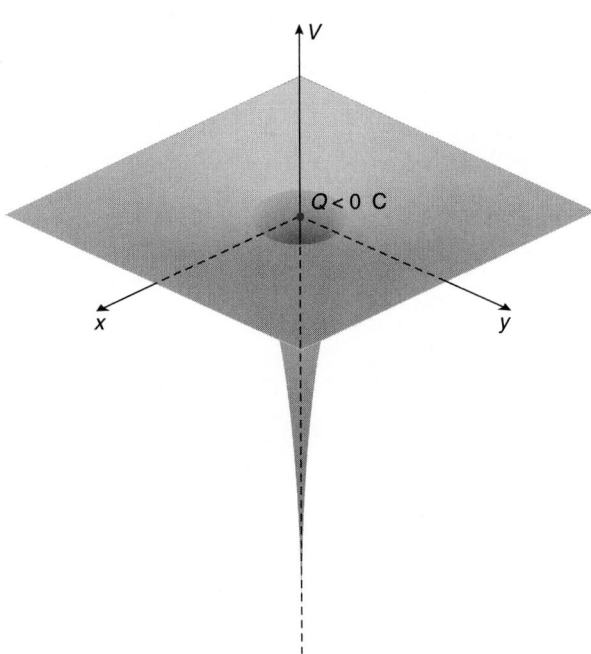

FIGURE 17.9 A graph of the electric potential V of a negative pointlike charge at the origin in the x-y plane.

PROBLEM-SOLVING TACTICS

17.5 Note that in Equation 17.11 for the electric potential of a point charge, the distance r in the denominator is to the first power. The expression for the electric *field* of a point charge (Equation 16.11) has r to the *second* power in the denominator. Remember also that the electric potential is a scalar whereas the electric field is a vector. Keep these distinctions straight and you will avoid many pitfalls in problem solving.

17.6 In Equation 17.11 for the electric potential of a point charge, as well as for other charge distributions in Table 17.1 on page 779, remember to use the appropriate sign for Q.

EXAMPLE 17.2

a. Calculate the electric potential caused by a proton (the nucleus of the hydrogen atom) at a point located 5.29×10^{-11} m away.
b. What is the electric potential energy of an electron placed at this distance from the proton? The electron in hydrogen "orbits" the nucleus at approximately this radius when closest to the nucleus.

Solution

a. The electric potential of a point charge is given by Equation 17.11:

$$V(r) = \frac{1}{4\pi\varepsilon_0} \frac{Q}{r}$$

Since the charge on the proton is $+e$, the electric potential is

$$V = \left(9.00 \times 10^9 \text{ N}\cdot\text{m}^2/\text{C}^2\right) \frac{1.602 \times 10^{-19} \text{ C}}{5.29 \times 10^{-11} \text{ m}}$$

$$= 27.3 \text{ J/C}$$

$$= 27.3 \text{ V}$$

b. The potential energy of any charge placed at this location is found from Equation 17.8. Remember that an electron is a negative charge, $q = -e$:

$$PE = qV$$

$$= (-1.602 \times 10^{-19} \text{ C})(27.3 \text{ V})$$

$$= -4.37 \times 10^{-18} \text{ J}$$

EXAMPLE 17.3

A pointlike small rubber ball is rubbed vigorously with fur and secures a -3.00 nC charge. Find the electric potential at a point 5.00 cm away from the ball.

Solution

All units *must* be SI units, so the distance r is 5.00×10^{-2} m and the charge is -3.00×10^{-9} C. Use Equation 17.11 for the potential of a pointlike charge:

$$V(r) = \frac{1}{4\pi\varepsilon_0} \frac{Q}{r}$$

and make the appropriate substitutions, remembering that the charge Q is negative:

$$V = \left(9.00 \times 10^9 \text{ N} \cdot \text{m}^2/\text{C}^2\right) \frac{(-3.00 \times 10^{-9} \text{ C})}{5.00 \times 10^{-2} \text{ m}}$$

$$= -540 \text{ V}$$

17.3 THE ELECTRIC POTENTIAL OF A COLLECTION OF POINTLIKE CHARGES

For a collection of point charges, such as that in Figure 17.10, the total electric potential at a point P is the algebraic scalar sum of the potentials of each charge, taken individually as if each were the only charge present. That is,

$$V = V_1 + V_2 + V_3 + \cdots \qquad (17.12)$$

$$V = \frac{1}{4\pi\varepsilon_0} \frac{Q_1}{r_1} + \frac{1}{4\pi\varepsilon_0} \frac{Q_2}{r_2} + \frac{1}{4\pi\varepsilon_0} \frac{Q_3}{r_3} + \cdots$$

$$= \frac{1}{4\pi\varepsilon_0} \left(\frac{Q_1}{r_1} + \frac{Q_2}{r_2} + \frac{Q_3}{r_3} + \cdots \right)$$

Some terms in the sum may be positive (those for positive charges) and some negative (those for negative charges). Equation 17.12 is a scalar sum because the electric potential is a scalar quantity. Electric fields and forces are vectors, but electric potential and potential energy are scalars.

17.4 THE ELECTRIC POTENTIAL OF CONTINUOUS CHARGE DISTRIBUTIONS OF FINITE SIZE

The beauty and utility of the electric potential is that it is a scalar quantity. You do not have the complication of worrying about associating any direction with it. Huzza! We saw that the electric potential at a distance r from a point charge Q was given by Equation 17.11:

$$V(r) = \frac{1}{4\pi\varepsilon_0} \frac{Q}{r}$$

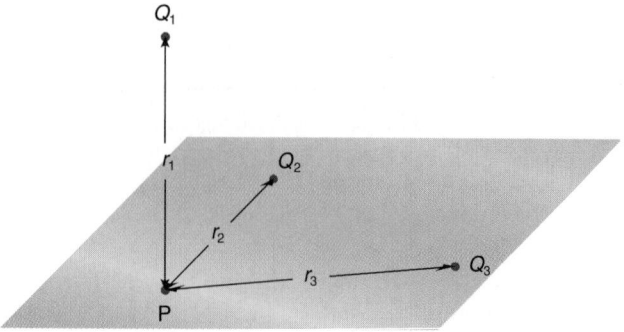

FIGURE 17.10 The electric potential at point P produced by a collection of pointlike charges is the scalar sum of the individual electric potentials at that location.

Electrons and protons are well approximated as pointlike charges. However, macroscopic charges are hardly pointlike: certainly glass and rubber rods are not points, let alone your cat, unless you are very far away from them. If the charge Q is smeared out over some region or object of finite size, as in Figure 17.11, we break up the extended charge distribution into a large collection of differential pointlike charge elements dQ. Each differential charge element dQ produces a differential amount of electric potential dV at point P:

$$dV = \frac{1}{4\pi\varepsilon_0} \frac{dQ}{r}$$

To find the total electric potential at the point P, we then sum these *scalar* contributions from all the differential charge elements. The sum is a continuous one, however: an integration over the charge distribution, be it a line, surface, or volume:

$$V = \frac{1}{4\pi\varepsilon_0} \int_{\substack{\text{finite charge} \\ \text{distribution}}} \frac{dQ}{r} \qquad (17.13)$$

This is a scalar integration because V is a scalar quantity. It is unlike the situation for calculating the electric field of an extended charge distribution in Chapter 16, which involved a trickier vector integration. We need not worry about any directions for V: there are none, a cause for celebration.

There is only one string attached to Equation 17.13, but it is an important one: *the charge distribution must be of finite size.*

If the charge distribution extends off to infinity, you cannot use Equation 17.13. Why this restriction? Equation 17.13 was formulated by considering a superposition of pointlike differential charge elements. For pointlike charges, the electric potential was chosen conveniently to be zero at infinity. If there is charge at infinity (as there would be for a charge distribution of infinite extent), this choice does not make sense.

If the charge distribution does extend to infinity [e.g., infinite sheets of charge, infinite linear (line) charge distributions], we must revert to the defining Equation 17.9 to calculate the electric potential difference between two points. Notice that in Example 17.1 for the twin infinite sheets of charge (producing a uniform electric field between them) you did this: you calculated the potential at any point x between the infinite plates using Equation 17.9.

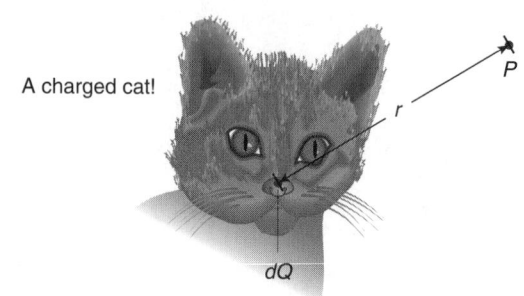

A charged cat!

FIGURE 17.11 Each differential bit of charge dQ produces a differential amount of potential dV at point P.

Table 17.1 (on page 779) summarizes the electric potential in the vicinity of variously shaped charge distributions.

EXAMPLE 17.4

a. Find the electric potential at a point P located a distance z along the axis of a uniformly charged circular ring of radius R with total charge Q (see Figure 17.12).

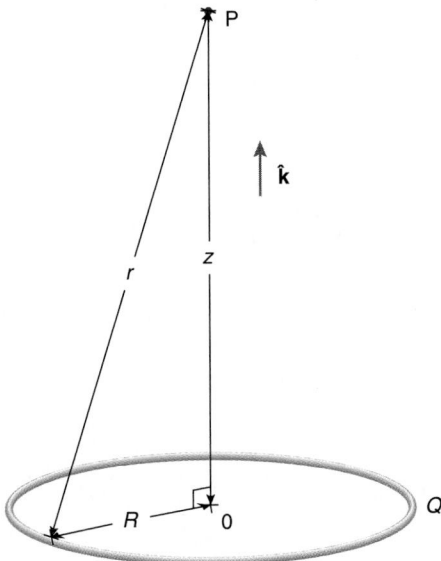

FIGURE 17.12

b. Make two schematic graphs of V as a function of z, one for positive Q and one for negative Q.

Solution

a. All of the charge Q is at the same distance r from the point at coordinate z along the axis. Since the potential is a scalar, you can sum it over the ring of charge. From the geometry, you have

$$r = (z^2 + R^2)^{1/2}$$

The potential at the point z thus is

$$V = \frac{1}{4\pi\varepsilon_0} \int \frac{dQ}{r}$$

$$= \frac{1}{4\pi\varepsilon_0} \frac{Q}{r}$$

$$= \frac{1}{4\pi\varepsilon_0} \frac{Q}{\left(z^2 + R^2\right)^{1/2}}$$

Notice that if $Q > 0$ C, $V > 0$ V for all z; if $Q < 0$ C, then $V < 0$ V for all z.

b. Graphs of V as a function of z for positive and negative Q are shown in Figure 17.13.

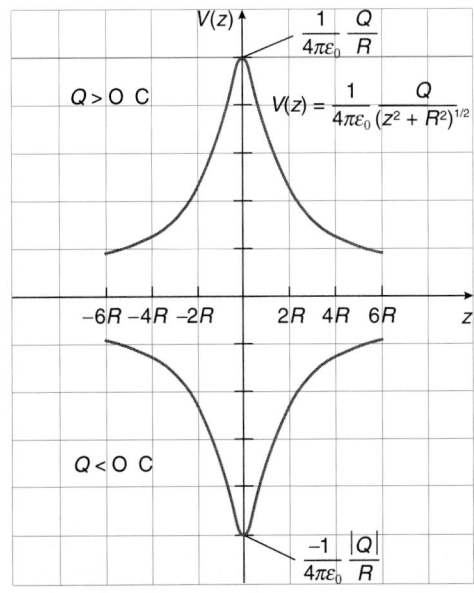

FIGURE 17.13

STRATEGIC EXAMPLE 17.5

a. Find the electric potential at a point P located on the axis of a uniformly charged circular disk of radius R and charge Q a distance z from the plane of the disk; see Figure 17.14.
b. Make two schematic graphs of V as a function of z, one for positive Q and one for negative Q.

Solution

a. The disk of radius R with total charge Q has a surface charge density of

$$\sigma = \frac{Q}{\pi R^2}$$

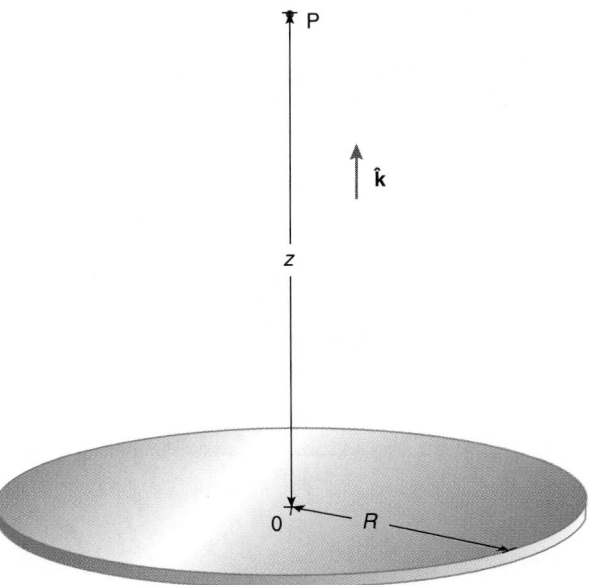

FIGURE 17.14

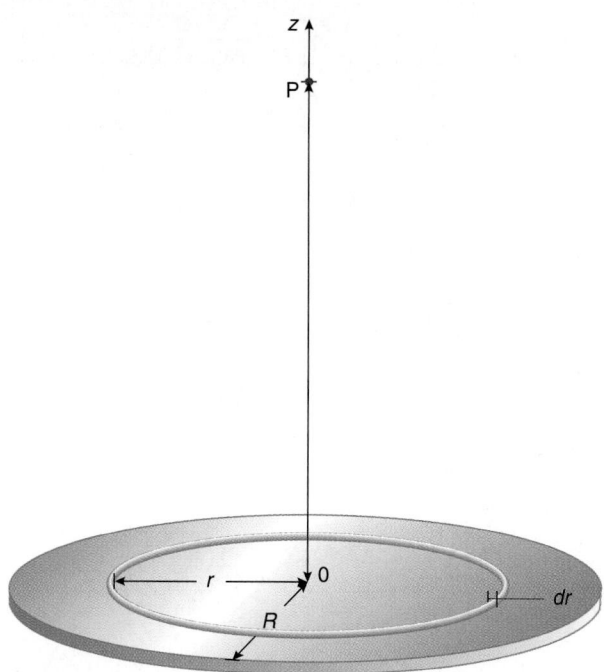

FIGURE 17.15

A ring of radius r and differential width dr (see Figure 17.15) has an area of

$$(\text{circumference})(\text{width}) = 2\pi r\, dr$$

The charge dQ on this differential ring is given by the product of the differential area and the surface charge density:

$$dQ = \sigma(2\pi r)\, dr \qquad (1)$$

Adapt the result of Example 17.4 to find the contribution to the electric potential at point P from this ring of charge:

$$dV = \frac{1}{4\pi\varepsilon_0} \frac{dQ}{(z^2 + r^2)^{1/2}}$$

Now substitute equation (1) for dQ and then integrate over the disk:

$$V = \frac{1}{4\pi\varepsilon_0} \int_{0\text{ m}}^{R} \frac{\sigma\,(2\pi r)\, dr}{(z^2 + r^2)^{1/2}}$$

The integral is of the form $u^n\, du$, where $u = (z^2 + r^2)$ and $n = -1/2$. Performing the integration, you obtain

$$V = \frac{1}{4\pi\varepsilon_0}\, 2\pi\sigma \left[(z^2 + R^2)^{1/2} - z\right] \qquad (2)$$

To express this result in terms of the total charge Q on the disk, substitute for the surface charge density $\sigma = Q/\pi R^2$:

$$V = \frac{1}{4\pi\varepsilon_0} \frac{2Q}{R^2}\left[(z^2 + R^2)^{1/2} - z\right] \qquad (3)$$

How V on the axis of a charged disk changes with the radius R of the disk may be seen from the previous two equations. Two different scenarios of V variation with R are possible: (1) either hold the surface charge density constant and include more total charge as R increases; or (2) hold the total charge constant and spread it across a greater area as R increases. The first choice of constant σ is more informative: if z is constant, V will increase slowly for small R, but eventually grows linearly with R. On the other hand, if Q is constant, it spreads more thinly over a disk as the radius grows, and then V will eventually decrease as $1/R$ for large values of R.

b. Schematic graphs of equation (3) for V as a function of z for fixed Q and R are shown in Figure 17.16 for both positive and negative Q.

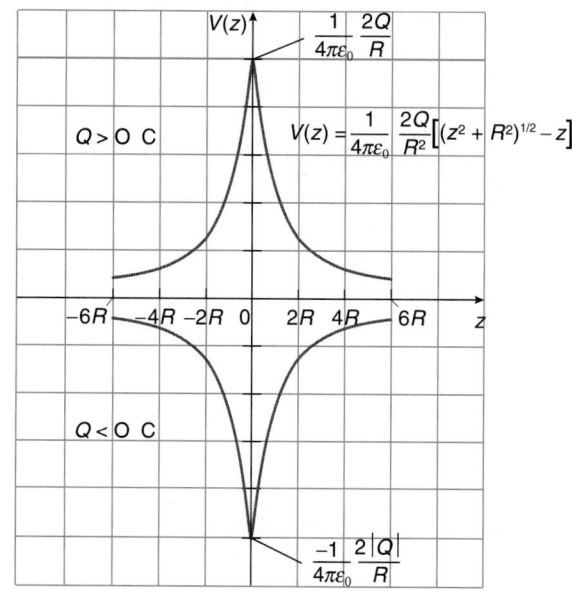

FIGURE 17.16

EXAMPLE 17.6

Find the electric potential at a point P located a distance r from a sphere of radius R that has a total charge Q distributed uniformly throughout its volume, where $r > R$; see Figure 17.17. Find the value of the electric potential on the surface where $r = R$.

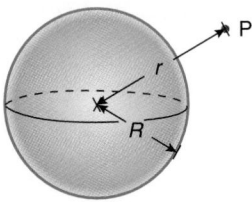

FIGURE 17.17

Solution

You can use Equation 17.13 to find the electric potential since the charge distribution is of finite extent. However, the integration over the sphere is complicated. To avoid this complexity, it is easier to use Equation 17.9 to find the electric potential, since you know the electric field of such a charge distribution from Chapter 16 (Example 16.19):

$$\vec{E} = \frac{1}{4\pi\varepsilon_0} \frac{Q}{r^2} \hat{r}$$

Use Equation 17.9 to find the electric potential at r by integrating the field from an infinite distance away to the radial distance r:

$$V(r_f) - V(r_i) = -\int_i^f \vec{E} \cdot d\vec{r}$$

where the initial position is at infinity and the final position is a point a distance r from the center of the spherical charge distribution. As usual, let the differential change in the position vector $d\vec{r}$ be $dr\,\hat{r}$ and let the limits of integration account for the direction you are moving (here, toward decreasing values of r). Remember that the result is independent of the path (since the electrical force is conservative), so you can come in along a radial direction to the final position. Thus

$$V(r) - V(\infty\ \text{m}) = -\frac{1}{4\pi\varepsilon_0} Q \int_{\infty\ \text{m}}^r \frac{\hat{r}}{r^2} \cdot dr\,\hat{r}$$

$$= -\frac{1}{4\pi\varepsilon_0} Q \int_{\infty\ \text{m}}^r \frac{dr}{r^2}$$

$$= -\frac{1}{4\pi\varepsilon_0} Q \left(-\frac{1}{r} \right) \Bigg|_{\infty\ \text{m}}^r$$

$$= \frac{1}{4\pi\varepsilon_0} \frac{Q}{r}$$

By convention, zero potential is chosen to be where $r = \infty$ m, so $V(\infty\ \text{m}) = 0$ V and you have

$$V(r) = \frac{1}{4\pi\varepsilon_0} \frac{Q}{r} \qquad (1)$$

Notice that this expression is the same as the potential of a pointlike charge Q located at the center of the sphere. So for distances $r > R$, the potential of a spherically symmetric charge distribution of radius R is the same as that of a point charge Q located at the center of the sphere. The sphere can be either a uniformly charged conducting sphere (with the charge on the outer surface of the conductor) or an insulating sphere with a uniform charge distribution either on its surface or distributed throughout its volume, as long as the charge distribution is spherically symmetric.

The potential on the surface of the charge distribution is found by setting $r = R$:

$$V(R) = \frac{1}{4\pi\varepsilon_0} \frac{Q}{R} \qquad (2)$$

EXAMPLE 17.7

a. If the spherical charge distribution of Example 17.6 is on a *conducting* sphere, what is the potential within the sphere at distances $r < R$?
b. Make a graph of V as a function of r, including regions where $r < R$ and $r > R$.

Solution

a. The static charge on a conductor resides on its surface; there is no charge within the conductor for $r < R$, even if the conductor is hollow (see Section 16.13). The interior of a conductor has the same potential everywhere. The potential is constant within the sphere and equal to the value of the potential on its surface:

$$V(r) = \frac{1}{4\pi\varepsilon_0} \frac{Q}{R} \qquad (r < R) \qquad \text{(conductor)}$$

b. The electric potential of the uniformly charged conducting sphere is graphed as a function of r in Figure 17.18 for both the positively and negatively charged cases.

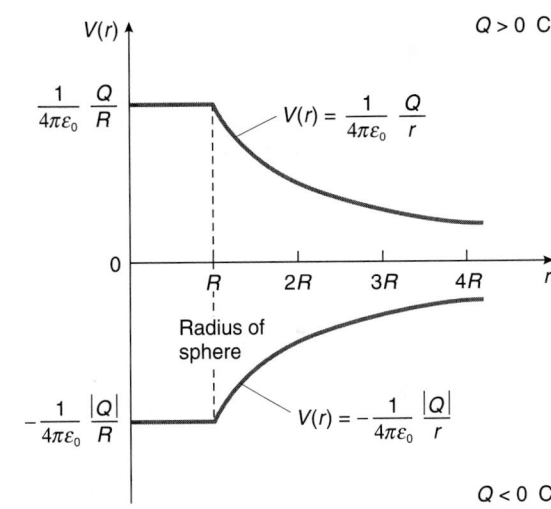

FIGURE 17.18

EXAMPLE 17.8

a. Find the electric potential at a distance r from the center of a *nonconducting* (i.e., insulating) sphere of radius R where the charge Q is distributed uniformly throughout the volume of the sphere and $r < R$; see Figure 17.19.

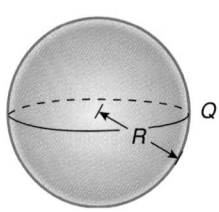

FIGURE 17.19

b. Make a graph of V as function of r, including the regions where $r < R$ and $r > R$.

Solution

a. To avoid a complicated integral over the sphere, begin with Equation 17.9 for the definition of the electric potential difference between two points:

$$V(r_f) - V(r_i) = -\int_i^f \vec{E} \cdot d\vec{r}$$

We found in Section 16.12 that the electric field within such a charge distribution a distance r from the center is given by

$$\vec{E} = \frac{1}{4\pi\varepsilon_0} \frac{Q}{R^3} r \, \hat{r}$$

Integrate from the surface of the sphere R to the interior point r. Once again, let $d\vec{r} = dr \, \hat{r}$ and the limits of integration will account for the radial direction you are moving between the initial and final positions. Hence

$$V(r) - V(R) = -\frac{1}{4\pi\varepsilon_0} \frac{Q}{R^3} \int_R^r r \, \hat{r} \cdot dr \, \hat{r}$$

$$= -\frac{1}{4\pi\varepsilon_0} \frac{Q}{R^3} \int_R^r r \, dr$$

$$= -\frac{1}{4\pi\varepsilon_0} \frac{Q}{R^3} \frac{r^2}{2}\bigg|_R^r$$

$$= \frac{1}{4\pi\varepsilon_0} \frac{Q}{R^3} \left(\frac{R^2}{2} - \frac{r^2}{2} \right)$$

The potential $V(R)$ at the surface of such a spherical charge distribution is given by equation (2) of Example 17.6; substituting this expression, you get

$$V(r) - \frac{1}{4\pi\varepsilon_0} \frac{Q}{R} = \frac{1}{4\pi\varepsilon_0} \frac{Q}{R^3} \left(\frac{R^2}{2} - \frac{r^2}{2} \right)$$

After some algebraic rearrangement, you find

$$V(r) = \frac{1}{4\pi\varepsilon_0} \frac{Q}{2R} \left(3 - \frac{r^2}{R^2} \right) \quad \begin{matrix} (r < R) \end{matrix} \quad \begin{matrix} \text{(uniformly charged} \\ \text{insulating sphere)} \end{matrix}$$

Notice that when $r = R$, you get the potential on the surface of such a spherical charge distribution. The potential at the center of the spherical insulator distribution, where $r = 0$ m, is

$$V(0 \text{ m}) = \frac{1}{4\pi\varepsilon_0} \frac{3Q}{2R}$$

b. Graphs of the potential as a function of r for $r < R$ and for $r > R$ are shown in Figure 17.20.

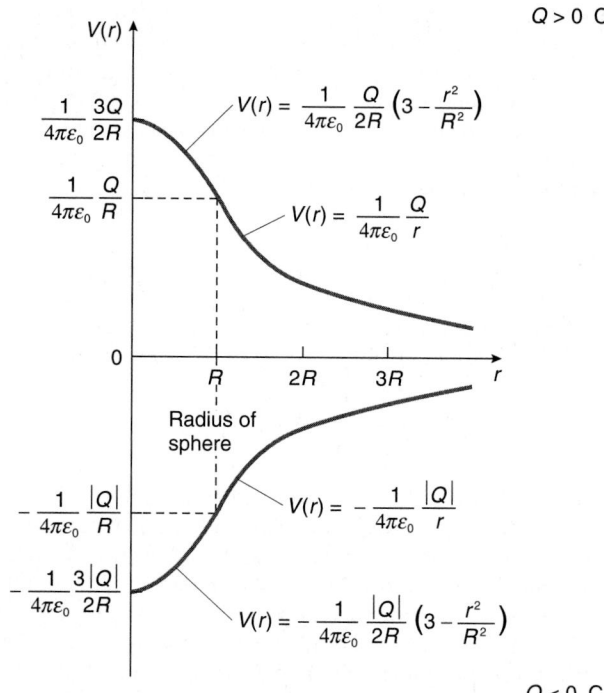

FIGURE 17.20

17.5 EQUIPOTENTIAL VOLUMES AND SURFACES

The electric potential difference between two points is defined by Equation 17.9:

$$V(r_f) - V(r_i) = -\int_i^f \vec{E} \cdot d\vec{r}$$

Let the initial and final positions be very closely (differentially) spaced; indeed, if the points are separated by the differential change in the position vector $d\vec{r}$ (see Figure 17.21), then the differential change in the electric potential is

$$dV = -\vec{E} \cdot d\vec{r} \qquad (17.14)$$

Equation 17.9 is the integral of Equation 17.14.

The locus of points or regions for which the electric potential has a constant value are called **equipotential regions** or, more simply, **equipotentials**. Such equipotentials can be volumes, surfaces, and even lines.

FIGURE 17.21 Two points separated by differential distance $d\vec{r}$ have a potential difference $dV = -\vec{E} \cdot d\vec{r}$.

TABLE 17.1 Summary of the Electric Potential of Various Charge Distributions

Two infinite sheets, oppositely charged (uniform electric field):

$$V(x) - V(0) = -Ex$$

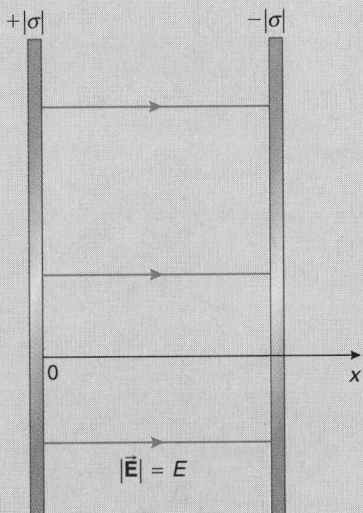

Pointlike charge:

$$V(r) = \frac{1}{4\pi\varepsilon_0} \frac{Q}{r}$$

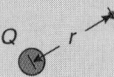

Uniformly charged ring, a distance z along the axis:

$$V = \frac{1}{4\pi\varepsilon_0} \frac{Q}{\left(z^2 + R^2\right)^{1/2}}$$

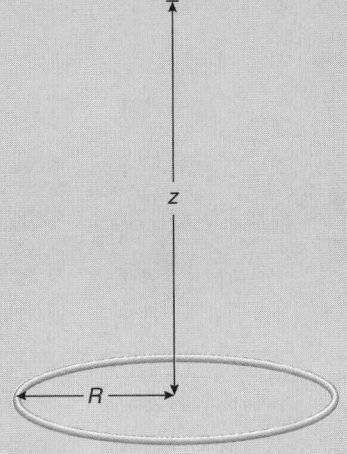

Uniformly charged disk, a distance z along the axis:

$$V = \frac{1}{4\pi\varepsilon_0} \frac{2Q}{R^2} \left[\left(z^2 + R^2\right)^{1/2} - z\right]$$

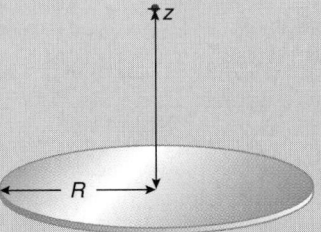

Uniformly charged sphere, $r > R$:

$$V(r) = \frac{1}{4\pi\varepsilon_0} \frac{Q}{r}$$

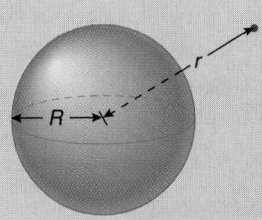

Uniformly charged conducting sphere, $r < R$:

$$V(r) = \frac{1}{4\pi\varepsilon_0} \frac{Q}{R} \quad \text{(constant)}$$

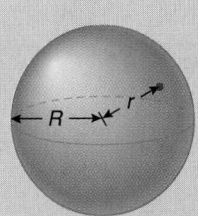

Uniformly charged insulating sphere, $r < R$:

$$V(r) = \frac{1}{4\pi\varepsilon_0} \frac{Q}{2R} \left(3 - \frac{r^2}{R^2}\right)$$

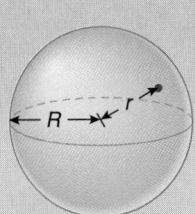

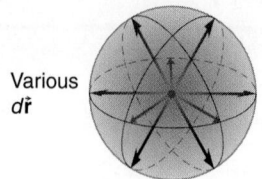

FIGURE 17.22 Within any equipotential volume, the change in the electric potential in any direction is zero.

Within an equipotential volume, the value of V is the same everywhere. If you move in any direction away from any point within the volume (see Figure 17.22), the change in the potential is zero. Hence

$$dV = -\vec{E} \cdot d\vec{r}$$
$$0 \text{ V} = -\vec{E} \cdot d\vec{r}$$

regardless of the direction of the differential change in the position vector $d\vec{r}$.

The only way $\vec{E} \cdot d\vec{r}$ can be zero for every $d\vec{r}$ radiating from a given point is for $\vec{E}$ itself to be zero at the point. This argument is applicable to any point within an equipotential volume.

> Hence the electric field must be zero within an equipotential volume.

We saw in Chapter 16 that in an electrostatic situation, the electric field inside a conductor is zero. This means that the interior of a conductor is an example of an equipotential volume. You discovered this fact in Example 17.7.

The same is true if the conductor is *hollow* and there are no charges inside the hollow.* To show this, imagine a Gaussian surface within the conductor surrounding the cavity, as shown in Figure 17.23.

Since the static electric field is zero within the conductor, the left-hand side of Gauss's law is zero, implying there is *zero total*

---

*No charges are present at all within the hollow; this may be distinct from a situation of zero charge, where the total charge sums to zero.

This stereo amplifier is encased in a metal box to shield its circuitry from outside electrical influences.

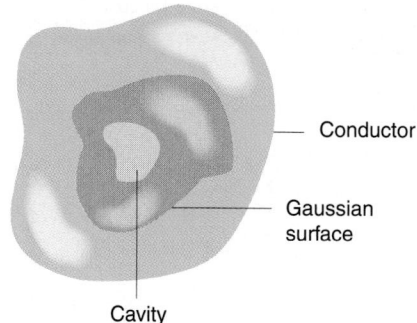

FIGURE 17.23 A Gaussian surface surrounding a hollow in a conductor.

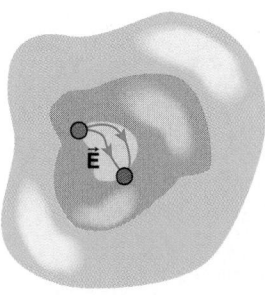

FIGURE 17.24 Electric field lines from a positive to a negative charge.

charge within the Gaussian surface. We imagine some positive charge to be on one side of the inside surface of the hollow conductor and some negative charge on the other, as shown in Figure 17.24, so the total charge is zero, consistent with the prediction of Gauss's law. If such charge separation exists on the inside surface of the cavity, there will be electric field lines within the hollow from the positive to the negative charges.

If we use Equation 17.9 and integrate the electric field along one of these field lines from a positive charge to a negative charge, the result will *not be zero* since $d\vec{r}$ is parallel to $\vec{E}$ over the whole path. This implies that there is a potential difference

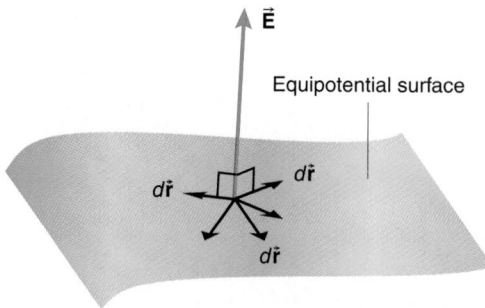

FIGURE 17.25 If you move in any direction on an equipotential surface, the potential is unchanged. The electric field must be perpendicular to an equipotential surface.

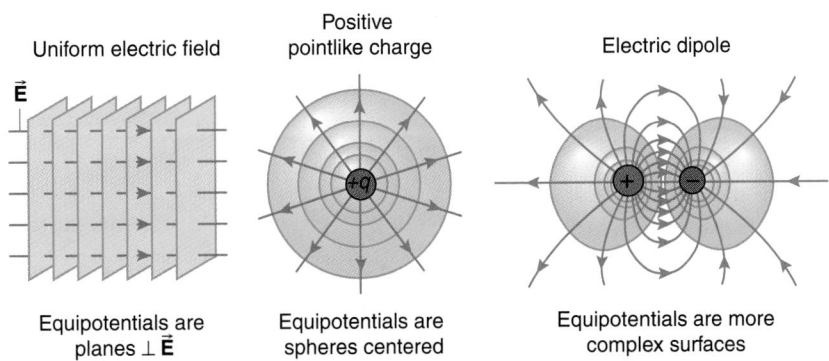

Uniform electric field

Equipotentials are planes ⊥ $\vec{\mathbf{E}}$

Positive pointlike charge

Equipotentials are spheres centered on charge

Electric dipole

Equipotentials are more complex surfaces

FIGURE 17.26 Equipotential surfaces.

between the two points at opposite ends of the field line on the inside surface of the cavity. But the conductor is an equipotential, so there can be no such potential difference. Hence there can be no electric field lines anywhere within the hollow: the electric field inside the cavity is zero.

Therefore, electric field–free regions of space can be created by surrounding the region with a conductor, as long as there are *no charges* within the region. Nowhere inside such metal cavities or boxes is there any electric field. This procedure is called **electrostatic shielding**. Sensitive electronic instruments are manufactured with their circuitry inside metal boxes specifically to shield various components from electrical influences (interference) from charges outside the box.

On an equipotential surface, if we move in any direction on the surface (see Figure 17.25), the value of the potential is unchanged. That is,

$$dV = -\vec{\mathbf{E}} \cdot d\vec{\mathbf{r}}$$
$$0 \ V = -\vec{\mathbf{E}} \cdot d\vec{\mathbf{r}}$$

for any $d\vec{\mathbf{r}}$ confined to the surface. Therefore any electric field $\vec{\mathbf{E}}$ must have a perpendicular equipotential surface associated with it at every point. The surface of a conductor in electrostatics (i.e., with electric charges at fixed locations on its surface) exemplifies an equipotential surface.

Equipotential surfaces need not be physical surfaces (such as that of a conductor). Any imaginary surface over which the electric field is everywhere perpendicular to it is called an equipotential surface. Various equipotential surfaces along with the associated electric fields are shown in Figure 17.26 for a number of different charge distributions. The surfaces are chosen and drawn so that the potential difference between successive surfaces is a constant value. Two-dimensional diagrams of equipotential surfaces are called lines of equipotentials; they also are typically drawn with a fixed value for the potential difference between the lines, much like the contour interval between contour lines on topographic maps.

Any equipotential volume is enclosed within a surface that itself must be an equipotential surface. The converse, however, is *not* true: an equipotential surface need not surround an equipotential volume.

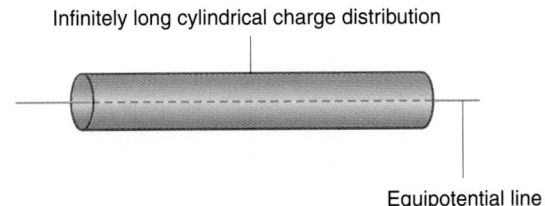

Infinitely long cylindrical charge distribution

Equipotential line

FIGURE 17.27 An equipotential line exists along the axis of a cylindrical charge distribution.

Since a conductor is an equipotential volume, the surface of a conductor is an equipotential surface and, therefore, any electric field at the surface of the conductor must be perpendicular to the surface for any static distributions of charge.

The only charge distribution that can produce a true equipotential line in space is a uniformly charged infinite cylindrical shell or cylindrical volume, as shown in Figure 17.27. An equipotential line exists along the axis of the cylinder; in this rather exotic situation, the electric field is zero along the axis as well.

QUESTION 2
Explain why two equipotential surfaces cannot intersect.

17.6 THE RELATIONSHIP BETWEEN THE ELECTRIC POTENTIAL AND THE ELECTRIC FIELD

In the previous section, we saw that the differential change dV in the electrical potential between two points that are separated by the differential change $d\vec{\mathbf{r}}$ in the position vector is given by Equation 17.14:

$$dV = -\vec{\mathbf{E}} \cdot d\vec{\mathbf{r}}$$

It frequently happens that the electric field is a function of only one coordinate, say s, and has a single vector s-component; this can be seen by examining Table 16.2, which lists the electric field for charge distributions with various geometries. Because of the scalar product in Equation 17.14, the greatest decrease in the value of the electric potential is in the direction parallel to the field direction. If we choose $d\vec{\mathbf{r}}$ to be in this s-direction, then

$$dV = -E_s \, ds$$

where E_s is the component of the field in the direction of the coordinate s. The electric field has *only* this s-component, since we took the direction of the coordinate s to be parallel to the field. Hence we have

$$E_s = -\frac{dV}{ds} \tag{17.15}$$

> Therefore the single s-component of the electric field is the negative of the derivative of the electric potential with respect to the single s-coordinate.

Geometrically, Equation 17.15 implies that the component of the field at a given point is the negative of the slope of a graph of V versus s at that point.

For example, we saw that with the geometry of Figure 17.1, the electric potential associated with the uniform field between two plates is, from Example 17.1,

$$V(x) - V(0 \text{ m}) = -Ex$$

where $V(0 \text{ m})$ is the electric potential where $x = 0$ m. Hence

$$V(x) = V(0 \text{ m}) - Ex$$

Here the coordinate x plays the role of the coordinate s just discussed.

If we use Equation 17.15, adapted to the coordinate x, we find the (single) component of the electric field is

$$\begin{aligned}
E_x &= -\frac{dV}{dx} \\
&= -\left(-E\right) \\
&= E
\end{aligned}$$

The electric field, therefore, is $\vec{\mathbf{E}} = E_x \hat{\mathbf{i}} = E\hat{\mathbf{i}}$, as in Figure 17.1. Note also in Figures 17.2 and 17.3, that the field component E is the negative of the slope of the graph of V versus x.

If the potential function is a function of the single radial coordinate r—that is, $V(r)$—then the radial component of the electric field is

$$E_r = -\frac{dV}{dr} \tag{17.16}$$

For the case of a pointlike charge Q, we found the electric potential is a function of the radial coordinate r only (Equation 17.11):

$$V = \frac{1}{4\pi\varepsilon_0} \frac{Q}{r}$$

We find the (single) electric field component E_r using Equation 17.16:

$$\begin{aligned}
E_r &= -\frac{dV}{dr} \\
&= -\frac{1}{4\pi\varepsilon_0}\left(-\frac{Q}{r^2}\right) \\
&= \frac{1}{4\pi\varepsilon_0} \frac{Q}{r^2}
\end{aligned}$$

So the electric field vector is

$$\begin{aligned}
\vec{\mathbf{E}} &= E_r \hat{\mathbf{r}} \\
&= \frac{1}{4\pi\varepsilon_0} \frac{Q}{r^2} \hat{\mathbf{r}}
\end{aligned}$$

as we know from Equation 16.11. The component of the field, E_r, is the negative of the slope of the graph of $V(r)$ versus r.

We can just as easily generalize to three dimensions. If the electric potential is a function of the three Cartesian coordinates x, y, and z, the various components of the electric field are

$$E_x = -\frac{\partial V}{\partial x}$$

$$E_y = -\frac{\partial V}{\partial y} \tag{17.17}$$

$$E_z = -\frac{\partial V}{\partial z}$$

where the derivatives are partial derivatives. Each Cartesian component of the field is the negative of the slope of the graph of V versus that coordinate.

A special differentiation operator, known as the **del operator** ∇ is defined as

$$\nabla \equiv \frac{\partial}{\partial x}\hat{\mathbf{i}} + \frac{\partial}{\partial y}\hat{\mathbf{j}} + \frac{\partial}{\partial z}\hat{\mathbf{k}} \tag{17.18}$$

Then Equations 17.17 can be summarized by writing

$$\vec{\mathbf{E}} = -\nabla V \tag{17.19}$$

The operation ∇V produces what is called the **gradient** of the scalar potential function V. The gradient is a vector.

Thus we find there is an easier way to find the electric field than via the typically complicated vector integration techniques of Chapter 16. First find the electric potential V, which typically is easier than finding $\vec{E}$ by vector integration, because V is a *scalar* function. Then get $\vec{E}$ from V by taking the negative gradient of V using the ∇ operator.

But remember, we can only find V using Equation 17.13 for charge distributions that are of finite extent:

$$V = \frac{1}{4\pi\varepsilon_0} \int_{\substack{\text{finite charge} \\ \text{distribution}}} \frac{dQ}{r}$$

If the charge distribution extends to infinity, then we can only find V from the defining Equation 17.9:

$$V(r_f) - V(r_i) = -\int_i^f \vec{E} \cdot d\vec{r}$$

which means we have to know the field to begin with; the vector integration techniques to obtain the field then cannot be avoided.

EXAMPLE 17.9

For a uniformly charged circular ring at a point z along the axis of the ring, you found the potential to be (see Example 17.4)

$$V = \frac{1}{4\pi\varepsilon_0} \frac{Q}{\left(z^2 + R^2\right)^{1/2}}$$

Use the potential to find the electric field along the z-axis of the ring.

Solution

The potential is a function of only the coordinate z; hence the partial derivative of V with respect to z is the same as the ordinary derivative of V with respect to z. The electric field along the axis thus has only a z-component, and it is found using Equation 17.15 where the general coordinate s is here z

$$
\begin{aligned}
E_z &= -\frac{\partial V}{\partial z} \\
&= -\frac{dV}{dz} \\
&= -\frac{1}{4\pi\varepsilon_0} Q\left(-\frac{1}{2}\right) \frac{2z}{\left(z^2 + R^2\right)^{3/2}} \\
&= \frac{1}{4\pi\varepsilon_0} \frac{Qz}{\left(z^2 + R^2\right)^{3/2}}
\end{aligned}
$$

The electric field is $\vec{E} = E_z\hat{k}$. This result just what you found in Example 16.14.

17.7 ACCELERATION OF CHARGED PARTICLES UNDER THE INFLUENCE OF ELECTRICAL FORCES

The motion of charged particles under the influence of the electrical force has important technological applications in devices as diverse as x-ray tubes, particle accelerators, and TV tubes. The CWE theorem provides a simple way to analyze the dynamics of such motions.

If the charged particle moves *only* under the influence of the conservative electrical force, the work done by nonconservative forces is zero. Thus the left-hand side of the CWE theorem

$$W_{\text{noncon}} = \Delta KE + \Delta PE \qquad (17.20)$$

is zero, and the theorem simplifies to

$$
\begin{aligned}
0 \text{ J} &= \Delta KE + \Delta PE \\
0 \text{ J} &= \Delta(KE + PE) \qquad (17.21)
\end{aligned}
$$

> In other words, the total mechanical energy of the charged particle is conserved throughout its motion under the influence of the static electrical force.

Consider a charged particle q in space, initially at a point where it has kinetic energy KE_i and electric potential energy $PE_i = qV_i$. When the particle has moved under the influence of the electrical force to another position where the electric potential is V_f, the potential energy of the charge is $PE_f = qV_f$ and its kinetic energy is KE_f. Conservation of mechanical energy, Equation 17.21, implies that the change in the kinetic energy ΔKE is

$$
\begin{aligned}
0 \text{ J} &= \Delta KE + (qV_f - qV_i) \\
0 \text{ J} &= \Delta KE + q \Delta V
\end{aligned}
$$

or

$$\Delta KE = -q \Delta V \qquad (17.22)$$

Equation 17.22 indicates that a *positive* charge in an electric field, moving under the influence of only the electrical force, will *increase* its kinetic energy (while decreasing its electric potential energy) if it moves to regions of *lower* electric potential (since then $V_f < V_i$ and $\Delta V < 0$ V, and so $\Delta KE > 0$ J). The total mechanical energy is constant according to the CWE theorem. Correspondingly, a *negative* charge moves to regions of *higher* electric potential to increase its kinetic energy (and decrease its electric potential energy) (since then $V_f > V_i$ and $\Delta V > 0$ V but $q < 0$ C, giving $\Delta KE > 0$ J).

QUESTION 3

A charge, initially at rest, moves under the action of only the electrical force in a direction toward increasing values of the electric potential. What can be said, if anything, about the sign of the charge? What happens to the potential energy of the charge?

STRATEGIC EXAMPLE 17.10

An electron enters a region with a uniform electric field created by two parallel, uniformly charged plates at potentials $+100$ V and -150 V, as shown in Figure 17.28. The initial speed of the electron is 5.00×10^6 m/s. Determine the speed of the electron when it emerges from the field. This type of apparatus is used to increase the kinetic energy of electrons in TVs, CRT (cathode ray tube) computer displays, and pieces of laboratory equipment you may encounter even in this course, such as an apparatus to measure the charge to mass ratio (e/m) of the electron.

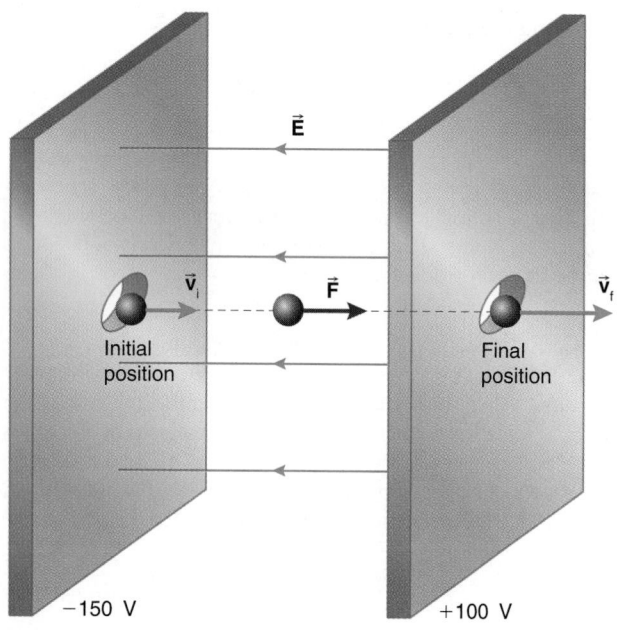

FIGURE 17.28

Solution

The electric field between the charged plates is directed from regions of higher electric potential to regions of lower electric potential. The electric field thus is antiparallel to the initial velocity of the electron. The electrical force on a charged particle in an electric field $\vec{E}$ is given by Equation 16.8:

$$\vec{F} = q\vec{E}$$

For the negatively charged electron ($q = -e$), the electrical force is antiparallel to the field. Hence in this problem the electrical force on the negatively charged electron is parallel to its initial velocity; you can anticipate that the electron will emerge from the device with a higher speed. You find the new speed by applying the CWE theorem to the initial and final positions of the electron.

Method 1

Use the CWE theorem in the guise of Equation 17.21, slightly rearranged:

$$KE_f + PE_f = KE_i + PE_i$$

$$\frac{mv_f^2}{2} + qV_f = \frac{mv_i^2}{2} + qV_i$$

$$\frac{mv_f^2}{2} = \frac{mv_i^2}{2} + q\left(V_i - V_f\right)$$

$$= \frac{mv_i^2}{2} + (-e)\left(V_i - V_f\right)$$

$$= \frac{mv_i^2}{2} + e\left(V_f - V_i\right)$$

Method 2

You also could use Equation 17.22 directly:

$$\Delta KE = -q\,\Delta V$$

$$\frac{mv_f^2}{2} - \frac{mv_i^2}{2} = -(-e)\left(V_f - V_i\right)$$

$$\frac{mv_f^2}{2} = \frac{mv_i^2}{2} + e\left(V_f - V_i\right)$$

Substituting numerical values, you find

$$\frac{mv_f^2}{2} = \frac{(9.11 \times 10^{-31}\ \text{kg})(5.00 \times 10^6\ \text{m/s})^2}{2}$$
$$+ (1.602 \times 10^{-19}\ \text{C})\left[100\ \text{V} - (-150\ \text{V})\right]$$
$$= 5.15 \times 10^{-17}\ \text{J}$$

Solving for the square of the final speed of the electron,

$$v_f^2 = \frac{2 \times 5.15 \times 10^{-17}\ \text{J}}{9.11 \times 10^{-31}\ \text{kg}}$$
$$= 1.13 \times 10^{14}\ \text{m}^2/\text{s}^2$$

The final speed then is

$$v_f = 1.06 \times 10^7\ \text{m/s}$$

17.8 A NEW ENERGY UNIT: THE ELECTRON-VOLT

When specifying the energy of electrons, protons, and other fundamental particles with small charge quantum numbers, it is convenient to use another energy unit called the **electron-volt** (eV). An electron-volt is defined to be the *change* in the kinetic energy of a negative fundamental unit of charge $-e$ when it accelerates through a potential difference of exactly one volt. From Equation 17.22, we see that if an electron accelerates through a potential difference of exactly one volt, the change in the kinetic energy in joules is

$$\Delta KE = -q\, \Delta V$$
$$= -(-1.602 \times 10^{-19}\ \text{C})(1.000\ \text{V})$$
$$= 1.602 \times 10^{-19}\ \text{J}$$

By definition this change in the kinetic energy is one electron-volt. Therefore the conversion factor relating the SI energy unit J to the *convenient* energy unit called the electron-volt (eV) is approximately

$$1\ \text{eV} = 1.602 \times 10^{-19}\ \text{J} \qquad (17.23)$$

Note that the conversion factor is, by definition, *numerically* equal to the number representing the fundamental unit of charge in coulombs.

Although the electron-volt energy unit is defined using the negative fundamental unit of charge, the energy unit can and is equally well applied to protons, helium nuclei, or other charged particles.

The utility of the new energy unit manifests itself by examining Equation 17.22,

$$\Delta KE = -q\, \Delta V$$

If the particle has a charge $q = ne$, where n is the charge quantum number, then

$$\Delta KE = -ne\, \Delta V \qquad (17.24)$$

To convert the kinetic energy change in joules to the new energy unit of electron-volts, use the conversion given by Equation 17.23. The effect of this conversion is to divide e out of Equation 17.24, leaving us with

$$\Delta KE_{\text{in eV}} = -n\, \Delta V \qquad (17.25)$$

where n is the charge quantum number of the particle.

It is interesting to note the following special case of Equation 17.25.

> Protons and electrons have charge quantum numbers ± 1 respectively. If they are accelerated by electrical forces alone between two points, the *absolute value* of the change in their kinetic energy between the two points, expressed in electron-volts, is numerically equal to the *absolute value* of the difference in electric potential between the two points.

The same is true for any particle with charge quantum number ± 1. That is,

$$|\Delta KE_{\text{in eV}}| = |\Delta V| \quad \text{(for any particle with charge}$$
$$\text{quantum number } n = \pm 1\text{)}$$

As noted previously, positive particles such as protons increase their kinetic energy (and decrease their electric potential energy) by moving to regions of lower electric potential; negative charges increase their kinetic energy (and decrease their electric potential energy) by moving to regions of higher electric potential.

PROBLEM-SOLVING TACTICS

17.7 Kinetic energies in electron-volts (eV) must be converted to joules (J) to calculate speeds in meters per second (m/s). Not to do so results in inconsistent units, which should become apparent as you substitute numerical values.

QUESTION 4

A proton and an electron are accelerated (in opposite directions) through a potential difference of the same absolute magnitude. What is the ratio of the change in their kinetic energies?

EXAMPLE 17.11

Calculate the change in the kinetic energy of the electron in Example 17.10 in both electron-volts and joules.

Solution

Since an electron has a charge of $q = ne = -e$, the charge quantum number of an electron is $n = -1$. The change in the electric potential is

$$\Delta V = V_f - V_i$$
$$= 100\ \text{V} - (-150\ \text{V})$$
$$= 250\ \text{V}$$

Use Equation 17.25 for the change in kinetic energy in electron-volts, being careful with the signs:

$$\Delta KE_{\text{in eV}} = -n\, \Delta V$$
$$= -(-1)(250\ \text{V})$$
$$= 250\ \text{eV}$$
$$= (250\ \text{eV})(1.602 \times 10^{-19}\ \text{J/eV})$$
$$= 4.01 \times 10^{-17}\ \text{J}$$

EXAMPLE 17.12

A proton is released at rest in the apparatus shown in Figure 17.29, which consists of two flat, conducting plates at the indicated potentials. The proton accelerates under the action of the electrical force to the other plate.

a. Near which plate should it be released to gain the most speed?
b. Calculate the kinetic energy of the proton as it strikes the other plate, expressing your result in electron-volts.
c. Determine the impact speed in meters per second.
d. Did we forget to give the plate separation?

Solution

a. The electric field is directed from higher to lower electric potential regions, so that it is directed as shown in Figure 17.30. The proton has a positive charge, and so the electrical force on a proton is parallel to the field according to $\vec{F} = q\vec{E}$. Hence the proton should be released near the plate with potential +250 V.

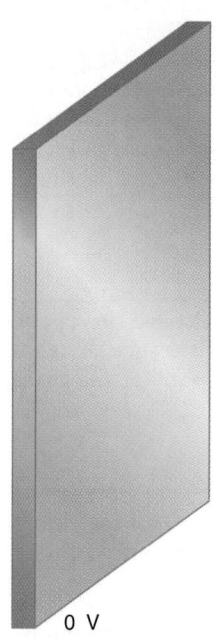

FIGURE 17.29

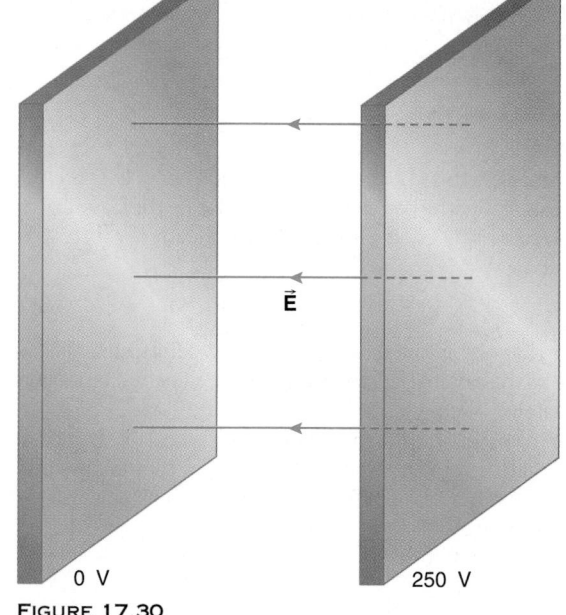

FIGURE 17.30

b. Since the charge on the proton is $q = +e$, the charge quantum number of the proton is $n = +1$. Use Equation 17.25 to find the change in the kinetic energy in electron-volts:

$$\Delta KE_{in\ eV} = -n\ \Delta V$$
$$= -n(V_f - V_i)$$
$$= -1(0\ V - 250\ V)$$
$$= 250\ eV$$

The proton is released at rest and so has zero initial kinetic energy. The final kinetic energy thus is equal to the change in the kinetic energy:

$$\Delta KE = KE_f - KE_i$$
$$= KE_f - 0\ J$$

The final kinetic energy in electron-volts thus is

$$KE_{f\ in\ eV} = 250\ eV$$

c. To find the speed, the kinetic energy must be expressed in joules as mentioned in Problem-Solving Tactic 17.7. Using the conversion indicated in Equation 17.23, you obtain

$$KE = \frac{mv^2}{2}$$

$$(250\ eV)(1.602 \times 10^{-19}\ J/eV) = \frac{mv^2}{2}$$

$$4.01 \times 10^{-17}\ J = \frac{mv^2}{2}$$

Substitute for the mass of the proton, and solve for the speed:

$$v = \left(\frac{2 \times 4.01 \times 10^{-17}\ J}{1.67 \times 10^{-27}\ kg} \right)^{1/2}$$
$$= 2.19 \times 10^5\ m/s$$

d. There is no need to know the separation of the plates!

EXAMPLE 17.13

An α-particle (with charge $+2e$) is accelerated from rest through a potential difference of -1.00 kV. What is the final kinetic energy of the α-particle?

Solution

The charge quantum number n of the α-particle is $+2$ since its charge is $q = +2e$. The change in the kinetic energy of the α-particle is found from Equation 17.25:

$$\Delta KE_{in\ eV} = -n\ \Delta V$$
$$= -(+2)(-1.00 \times 10^3\ V)$$
$$= 2.00\ keV$$

Since the initial kinetic energy of the α-particle is zero, the final kinetic energy is 2.00 keV, or

$$(2.00 \times 10^3\ eV)(1.602 \times 10^{-19}\ J/eV) = 3.20 \times 10^{-16}\ J$$

17.9 AN ELECTRIC DIPOLE IN AN EXTERNAL ELECTRIC FIELD REVISITED

Recall from Section 16.8 that an electric dipole consists of two charges of equal magnitude but opposite sign separated by a distance d, as pictured in Figure 17.31. The electric dipole moment $\vec{p}$ was defined by Equation 16.13:

$$\vec{p} = |Q|\vec{d}$$

where $\vec{d}$ is the vector that locates the positive charge with respect to the negative charge (taken as an origin).

Many molecules such as water and DNA have electric dipole moments; hence this charge distribution is important in physics, chemistry, and molecular biology.

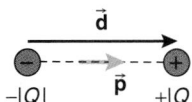

FIGURE 17.31 An electric dipole.

We saw (Example 16.13) that if an electric dipole is in an electric field $\vec{E}$, the field exerts forces on the charges (see Figure 16.71) in opposite directions, creating a torque $\vec{\tau}$ on the dipole given by

$$\vec{\tau} = \vec{p} \times \vec{E}$$

The torque is zero when $\vec{p}$ is parallel to $\vec{E}$ and when $\vec{p}$ is antiparallel to $\vec{E}$, as shown in Figure 17.32.

However, if the dipole moment $\vec{p}$ is antiparallel to $\vec{E}$, any slight change in the orientation allows the forces on the charges to flip the dipole into the parallel position. The antiparallel position thus is a position of unstable equilibrium. We now show this is the case from energy considerations.

Consider a dipole in a uniform electric field, as shown in Figure 17.33. (Uniformity is not really a restriction, because molecular dipoles are so small that most laboratory fields can be considered uniform over the physical size of the dipole.) For a

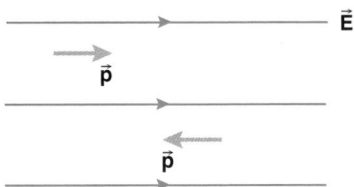

FIGURE 17.32 The torque on a dipole is zero when $\vec{p}$ is either parallel or antiparallel to $\vec{E}$.

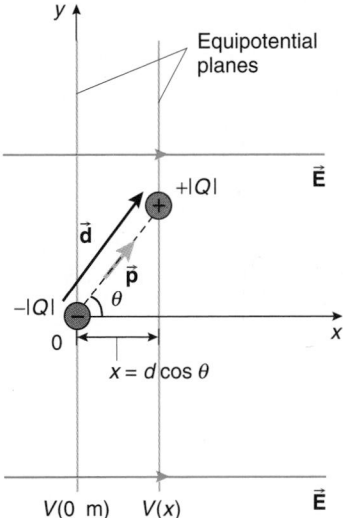

FIGURE 17.33 A dipole in a uniform electric field.

uniform field of magnitude E, the electric potential $V(x)$ at any position x (with the coordinate choice of Figure 17.33) is given (from Example 17.1) by

$$V(x) - V(0 \text{ m}) = -Ex$$

or

$$V(x) = V(0 \text{ m}) - Ex$$

The potential energy of the charge $-|Q|$ located at the origin is

$$PE_{-|Q|} = -|Q|V(0 \text{ m})$$

The potential energy of the positive charge $+|Q|$ at position x is

$$PE_{+|Q|} = +|Q|V(x)$$
$$= |Q|[V(0 \text{ m}) - Ex]$$

The total potential energy of the dipole in the field is the sum of the individual potential energies of the two charges:

$$PE_{dipole} = PE_{-|Q|} + PE_{+|Q|}$$
$$= -|Q|V(0 \text{ m}) + |Q|V(0 \text{ m}) - |Q|Ex$$
$$= -|Q|Ex$$

But geometrically (see Figure 17.33), $x = d\cos\theta$, and so

$$PE_{dipole} = -|Q|Ed\cos\theta$$

Note that $|Q|d$ is the magnitude of the dipole moment p, and so

$$PE_{dipole} = -pE\cos\theta$$

We can write this relationship conveniently using the scalar product because θ is the angle between the vectors $\vec{p}$ and $\vec{E}$.

Thus the potential energy of the dipole in an electric field is written succinctly as

$$PE_{dipole} = -\vec{p} \cdot \vec{E} \qquad (17.26)$$

Notice that when $\vec{p}$ is parallel to $\vec{E}$, the potential energy is

$$PE_{dipole\ parallel} = -pE\cos 0°$$
$$= -pE$$

When $\vec{p}$ is antiparallel to $\vec{E}$, the potential energy is

$$PE_{dipole\ antiparallel} = -pE\cos 180°$$
$$= +pE$$

So the potential energy is a maximum when $\vec{p}$ is antiparallel to $\vec{E}$ and is a minimum when $\vec{p}$ is parallel to $\vec{E}$. This confirms what we stated about the antiparallel orientation: if the dipole is free to rotate, the dipole will lower its electrical potential energy by rotating to the parallel orientation. The antiparallel orientation is unstable because it has the maximum potential energy.

Thus when a dipole is placed in an electric field, it will tend to orient itself so that $\vec{p}$ is parallel to $\vec{E}$. Molecular dipoles behave in a similar fashion.

17.10 THE ELECTRIC POTENTIAL AND ELECTRIC FIELD OF A DIPOLE*

In Chapter 16, you calculated the electric field caused by a dipole for two special geometries: (1) along the axis of the dipole (see Example 16.10), and (2) in the equatorial perpendicular bisecting plane of the dipole (see Example 16.11). Using the concept of the potential, it is possible to obtain a concise expression for the electric field of a dipole at *any* location r as long as $r \gg d$, the separation of the charges in the dipole. Here we see how that is accomplished by taking advantage of the fact that the potential is a scalar.

The dipole produces an electric potential at a point in space that is the algebraic sum of the electric potentials at the point in question caused by each of the two charges (see Figure 17.34):

$$V_{dipole} = \frac{1}{4\pi\varepsilon_0} \frac{|Q|}{r_+} + \frac{1}{4\pi\varepsilon_0} \frac{(-|Q|)}{r_-} \qquad (17.27)$$

where r_+ is the distance of the point P from the positive charge and r_- is the distance of P from the negative charge of the dipole.

Equation 17.27 can be rearranged slightly into the form

$$V_{dipole} = \frac{|Q|}{4\pi\varepsilon_0} \frac{r_- - r_+}{r_+ r_-} \qquad (17.28)$$

If the distance of point P from the dipole is much greater than the separation of the charges d, the dipole is called a pointlike dipole; see Figure 17.35.

Far from a pointlike dipole, we can make the good approximation that

$$r_+ \approx r_- \approx r$$

where r is the distance of the point P from the center of the pointlike dipole. The geometry of Figure 17.35 indicates that when $r \gg d$, the difference in the distances of the charges to the point P is approximately

$$r_- - r_+ \approx d \cos \theta$$

where the angle θ is measured from the axis of the dipole, parallel to the dipole moment $\vec{p}$. Molecular dipoles are very small, so that these approximations are excellent for such dipoles.

We use these approximations in Equation 17.28. The potential of the dipole at P then is

$$V_{dipole} = \frac{|Q|}{4\pi\varepsilon_0} \frac{d \cos \theta}{r^2}$$

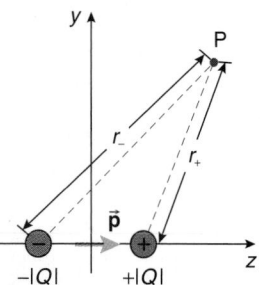

FIGURE 17.34 The electric potential at P is the sum of the potentials caused by $+|Q|$ and $-|Q|$.

But $|Q|d$ is the magnitude of the dipole moment vector $\vec{p}$. Therefore we have

$$V_{dipole} = \frac{1}{4\pi\varepsilon_0} \frac{p \cos \theta}{r^2} \qquad (17.29)$$

The potential is zero in the equatorial plane of the dipole, where $\theta = 90°$. At a fixed distance r, the potential has a maximum positive value when $\theta = 0°$, that is, along the axis of the dipole closer to the positive charge; the potential has its maximum negative value when $\theta = 180°$, along the axis of the dipole closer to the negative charge. The potential is symmetric about the axis of the dipole.

To calculate the electric field of the dipole at a point that is a great distance from the dipole itself, that is, for values of $r \gg d$, we use Equation 17.29 for the potential. To use the del operator in Cartesian coordinates, the potential also must be expressed in Cartesian coordinates. We introduce the Cartesian coordinate system indicated in Figure 17.35, with the dipole moment $\vec{p}$ oriented parallel to $\hat{k}$ along the z-axis and with the x–y plane the equatorial plane of the dipole. Then, from the geometry,

$$r = (x^2 + y^2 + z^2)^{1/2}$$

and

$$\cos \theta = \frac{z}{r}$$

$$= \frac{z}{(x^2 + y^2 + z^2)^{1/2}}$$

Substituting for r and $\cos \theta$ in Equation 17.29, we have

$$V = \frac{p}{4\pi\varepsilon_0} \frac{z}{(x^2 + y^2 + z^2)^{3/2}} \qquad (17.30)$$

Using Equations 17.30 and 17.17 to find the Cartesian components of the electric field (after appropriate differentiations), we obtain

$$E_x = -\frac{\partial V}{\partial x} = \frac{p}{4\pi\varepsilon_0} \frac{3zx}{(x^2 + y^2 + z^2)^{5/2}} \qquad (17.31)$$

$$E_y = -\frac{\partial V}{\partial y} = \frac{p}{4\pi\varepsilon_0} \frac{3zy}{(x^2 + y^2 + z^2)^{5/2}} \qquad (17.32)$$

$$E_z = -\frac{\partial V}{\partial z}$$

$$= -\frac{p}{4\pi\varepsilon_0} \left[\frac{1}{(x^2 + y^2 + z^2)^{3/2}} - \frac{3z^2}{(x^2 + y^2 + z^2)^{5/2}} \right] \qquad (17.33)$$

Along the axis of the dipole, where $x = y = 0$ m, the field has purely a z-component; Equations 17.31–17.33 reduce to the result in Example 16.10, which was calculated using vector methods in Chapter 16. In the equatorial plane, with $z = 0$ m, the field reduces to the result in Example 16.11 (with $x = 0$ m for the y-axis).

One also could find the field of an electric dipole using the potential in polar form as in Equation 17.29.* However, there is

---

*In physics, when working in three dimensions, the polar angle θ is measured with respect to the z-axis. This differs from the convention typically used in mathematics for three dimensions. In only two dimensions (say x and y), the polar angle is measured with respect to the x-axis both in physics and mathematics.

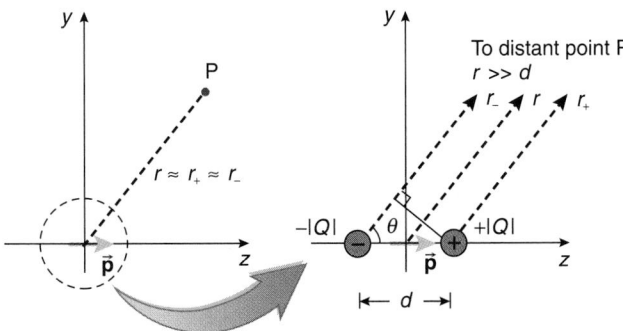

FIGURE 17.35 A pointlike dipole.

one complication. Equation 17.29 for the potential is expressed in polar coordinates r and θ. To use Equation 17.19 to calculate the electric field,

$$\vec{E} = -\nabla V$$

the del operator ∇ also must be expressed in polar coordinates. The expression for ∇ in polar coordinates is more complicated than in Cartesian coordinates (Equation 17.18), since the polar unit vectors $\hat{r}$ and $\hat{\theta}$ change their orientation in space depending on the location of the point in question, as indicated in Figure 17.36.

The del operator in polar coordinates has the form[†]

$$\nabla = \frac{\partial}{\partial r} \hat{r} + \frac{1}{r} \frac{\partial}{\partial \theta} \hat{\theta}$$

where at any point $\hat{r}$ is the unit vector in the direction of increasing r and $\hat{\theta}$ is a unit vector in the direction of increasing θ.

We use the expression for the del operator in polar coordinates; the dipole electric field then is

$$\vec{E} = -\nabla V$$
$$= -\left(\frac{\partial}{\partial r} \hat{r} + \frac{1}{r} \frac{\partial}{\partial \theta} \hat{\theta} \right) \frac{1}{4\pi\varepsilon_0} \frac{p \cos \theta}{r^2}$$

Factoring out the constant terms, we have

$$\vec{E} = -\frac{p}{4\pi\varepsilon_0} \left(\frac{\partial}{\partial r} \hat{r} + \frac{1}{r} \frac{\partial}{\partial \theta} \hat{\theta} \right) \frac{\cos \theta}{r^2}$$

[†] This identity is proved in many texts on vector analysis. See, for example, Mary Boas, *Mathematical Methods in the Physical Sciences* (2nd edition, Wiley, New York, 1983), page 252.

Now we perform the indicated partial differentiations, remembering that when differentiating with respect to r, we treat θ as a constant, and when differentiating with respect to θ, we treat r as a constant. The field is

$$\vec{E} = -\frac{p}{4\pi\varepsilon_0} \left[(\cos \theta)\left(-\frac{2}{r^3} \right) \hat{r} - \frac{\sin \theta}{r^3} \hat{\theta} \right]$$
$$= \frac{p}{4\pi\varepsilon_0} \left(\frac{2 \cos \theta}{r^3} \hat{r} + \frac{\sin \theta}{r^3} \hat{\theta} \right)$$

Notice that when $\theta = 0°$, the point is along the axis of the dipole (see Figure 17.36), and the field is purely along $\hat{r}$ since $\sin 0° = 0$:

$$\vec{E}_{\text{axis}} = \frac{1}{4\pi\varepsilon_0} \frac{2p}{r^3} \hat{r}$$

This corresponds to the result calculated using vector methods in Example 16.10.

Likewise, in the equatorial plane of the dipole, $\theta = 90°$, and the field has no radial component but is purely in the direction of $\hat{\theta}$ (see Figure 17.36):

$$\vec{E}_{\text{equatorial plane}} = \frac{1}{4\pi\varepsilon_0} \frac{p}{r^3} \hat{\theta}$$

This corresponds to a result derived by different vector methods in Example 16.11.

17.11 THE POTENTIAL ENERGY OF A DISTRIBUTION OF POINTLIKE CHARGES

Work is required to assemble a collection of point charges distributed in space, as in Figure 17.37.

Initially all the charges are infinitely far away from each other and so do not interact. We take the potential energy of this initial (widely dispersed) distribution to be zero. Since there is no electric field present initially, no work is done by any electrical force to bring the first charge Q_1 to its new but isolated location in the distribution of Figure 17.38.

The same cannot be said for the second charge. The electric potential created in space by Q_1 at the closer location where Q_2 is next to be placed is

$$V = \frac{1}{4\pi\varepsilon_0} \frac{Q_1}{r_{12}}$$

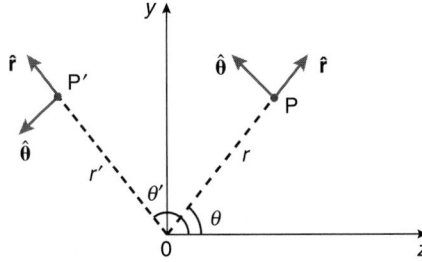

FIGURE 17.36 The orientation of the polar coordinate unit vectors $\hat{r}$ and $\hat{\theta}$ depend on where the point P is located.

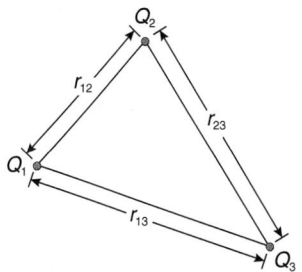

FIGURE 17.37 A collection of point charges.

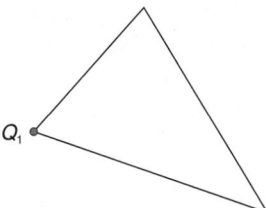

FIGURE 17.38 No work is required to bring the first charge to its location.

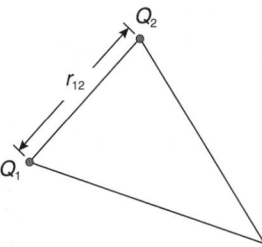

FIGURE 17.39 The second charge is placed at its location.

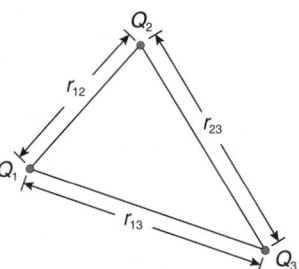

FIGURE 17.40 The third charge is placed at its location.

where r_{12} is the distance between Q_1 and the location where Q_2 is to be placed; see Figure 17.39. When Q_2 is placed at this location, Q_2 has the potential energy $Q_2 V$ or

$$PE = \frac{1}{4\pi\varepsilon_0} \frac{Q_1 Q_2}{r_{12}} \qquad (17.34)$$

The change in the potential energy of the second charge is

$$\Delta PE = PE_f - PE_i$$

The initial potential energy of Q_2 is zero (since it was infinitely far away from the other charge), and its final potential energy is given by Equation 17.34. The change in the potential energy is the negative of the work done by electrical forces as Q_2 is brought from infinity to its location at a distance r_{12} from Q_1. This work is independent of the path taken by Q_2 because the static electrical force is a conservative force. The potential energy change also can be imagined as the work *we* would need to do to drag Q_2 back from its location near Q_1 to infinity with no change in its kinetic energy; you should try to convince yourself of this statement from the CWE theorem. Equation 17.34 also results if we imagine bringing in Q_2 first (no work done by electrical forces to do this), followed by Q_1.*

The electric potential caused by Q_1 and Q_2 at the location where the third charge Q_3 is to be placed is

$$V' = \frac{1}{4\pi\varepsilon_0} \frac{Q_1}{r_{13}} + \frac{1}{4\pi\varepsilon_0} \frac{Q_2}{r_{23}}$$

where r_{13} is the distance between Q_1 and the place where Q_3 is to be placed, and r_{23} is the distance between Q_2 and the place where Q_3 is to be located; see Figure 17.40. If Q_3 now is brought in from infinity in the presence of the other two charges, it has a potential energy of $Q_3 V'$.

The change in the potential energy of Q_3 is

$$PE_f - PE_i = \left(\frac{1}{4\pi\varepsilon_0} \frac{Q_1 Q_3}{r_{13}} + \frac{1}{4\pi\varepsilon_0} \frac{Q_2 Q_3}{r_{23}} \right) - 0 \text{ J}$$

and so the *total* potential energy of the three charges in place is[†]

$$PE_{total} = \frac{1}{4\pi\varepsilon_0} \frac{Q_1 Q_2}{r_{12}} + \frac{1}{4\pi\varepsilon_0} \frac{Q_1 Q_3}{r_{13}} + \frac{1}{4\pi\varepsilon_0} \frac{Q_2 Q_3}{r_{23}}$$

$$= \frac{1}{4\pi\varepsilon_0} \left(\frac{Q_1 Q_2}{r_{12}} + \frac{Q_1 Q_3}{r_{13}} + \frac{Q_2 Q_3}{r_{23}} \right) \qquad (17.35)$$

The addition of a fourth charge leads to an expression similar to Equation 17.35 but with six terms, one for each distinct pair of charges present. Appropriate signs for the respective charges must be used in Equation 17.35 and similar expressions.

The work done by the electrical force is, by definition, the negative of the change in the potential energy:

$$W_{elec} = -\Delta PE$$
$$= -(PE_{total\,f} - PE_{total\,i})$$

Since the initial potential energy was zero with the charges infinitely separated, $PE_{total\,i} = 0$ J and

$$W_{elec} = -PE_{total\,f}$$

If the total final potential energy of such a collection of charges is negative, electrical forces did positive work in assembling the charges and, according to the CWE theorem, *we* will need to do positive work to separate the charges back to infinite distances apart with zero change in their kinetic energies.

QUESTION 5

Explain the distinction between the electric potential of two pointlike charges and the electric potential energy of two pointlike charges.

EXAMPLE 17.14

The charges that model the water molecule have a magnitude of $0.66e$ and an effective separation of 0.057 nm as was shown in Figure 16.63.

a. What is the electric potential energy of this dipolar charge configuration? Express your result in both J and eV.
b. Did electrical forces do positive, negative, or zero work when the dipole formed from charges initially separated by infinite distances?

---

*This result is reflected in the symmetry of Q_1 and Q_2 in Equation 17.34.
[†] Equation 17.35 also does not depend on the order the particles are moved to their positions, which is manifested by the symmetry of Q_1, Q_2, and Q_3 in the equation.

Solution

a. When the charges are part of the dipole, their potential energy is found from Equation 17.34:

$$PE = \frac{1}{4\pi\varepsilon_0} \frac{Q_1 Q_2}{r_{12}}$$

The two charges are $+0.66e$ and $-0.66e$, with a separation of 0.057 nm. Making these substitutions, you find

$$PE = \left(9.00 \times 10^9 \ \text{N}\cdot\text{m}^2/\text{C}^2\right) \frac{(-0.66e)(+0.66e)}{0.057 \times 10^{-9} \ \text{m}}$$

$$= -1.8 \times 10^{-18} \ \text{J}$$

$$= \frac{-1.8 \times 10^{-18} \ \text{J}}{1.602 \times 10^{-19} \ \text{J/eV}}$$

$$= -11 \ \text{eV}$$

b. When the charges initially are infinitely far apart, they have zero potential energy. Since the final potential energy of the dipolar molecule is negative and its initial potential energy was zero, the change in the potential energy is negative. This means that electrical forces did positive work when the dipole formed (since $W_{\text{elec}} = -\Delta PE$). *You* would need to do positive work to move the charges back to an infinite separation. Energy must be supplied to separate the charges to infinite distances. This is why the water molecule does not fall apart of its own accord (fortunately for us!).

EXAMPLE 17.15

The helically shaped strands of the important genetic macromolecule DNA (deoxyribonucleic *acid*) are held together by electrical forces. Along the double helix are many pairs of adenine and thymine molecules. Each adenine molecule in one DNA strand has a dipole moment of magnitude 3.0×10^{-30} C·m, caused by a hydrogen–nitrogen bond, while each thymine molecule in the other DNA strand has a dipole moment of 7.7×10^{-30} C·m, caused by a double carbon–oxygen bond. The pair of dipoles is arranged with their dipole moments parallel, as shown in Figure 17.41.

The absolute magnitude of the effective charge on the adenine dipole is $0.19e$, while that on the thymine dipole is $0.40e$. Note that the magnitude of the two dipole moments can be computed with the information in Figure 17.41. Calculate the electric potential energy of one dipole in the presence of the other when they are configured as in Figure 17.41. Consider each dipole to be already formed and that the potential energy is that of one dipole in the presence of the other.*

Solution

You might anticipate that the potential energy of this charge formation is negative because the two dipoles, in the orientations shown, attract each other. The positive charge of the adenine

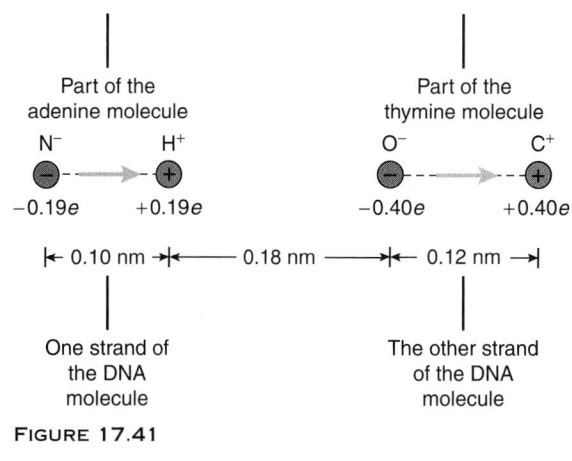

FIGURE 17.41

dipole is closer to the negative rather than to the positive charge of the thymine dipole. Electrical forces therefore did positive work to bring them together; the total potential energy (change) of the system is negative since $W_{\text{elec}} = -\Delta PE$.

The dipoles are quite close to each other in space, and so you *cannot* use Equation 17.29 for the electric potential of a dipole. That equation was derived assuming the distance r is much greater than the separation of the charges in the dipole.

Although four charges are present, since you consider each dipole to be already formed, you are interested in the potential energy of each charge in one dipole in the presence of the two charges in the other dipole.

In the adenine (H—N) dipole, let Q_H and Q_N represent the effective charges on the hydrogen and nitrogen atoms respectively, so that

$$Q_H = 0.19e$$
$$Q_N = -0.19e$$

In the thymine (C=O) dipole, let Q_C and Q_O represent the effective charges of the carbon and oxygen atoms respectively, so that

$$Q_C = 0.40e$$
$$Q_O = -0.40e$$

The potential energy of one dipole in the presence of the other is found using a variation of Equation 17.35:

$$PE = \frac{1}{4\pi\varepsilon_0} \left(\frac{Q_H Q_C}{r_{HC}} + \frac{Q_H Q_O}{r_{HO}} + \frac{Q_N Q_C}{r_{NC}} + \frac{Q_N Q_O}{r_{NO}} \right) \quad (1)$$

where the distances between the various atoms are indicated with subscripts. In Figure 17.41, these distances are

$$r_{HC} = 0.30 \ \text{nm}$$
$$r_{HO} = 0.18 \ \text{nm}$$
$$r_{NC} = 0.40 \ \text{nm}$$
$$r_{NO} = 0.28 \ \text{nm}$$

---

*This problem is modeled after one in Douglas C. Giancoli, *General Physics* (Prentice-Hall, Englewood Cliffs, New Jersey, 1984), page 474.

Making these substitutions using proper SI units into equation (1) for the potential energy, you have

$$PE = \left(9.00 \times 10^9 \ \text{N} \cdot \text{m}^2/\text{C}^2\right) \left[\frac{(0.19e)(0.40e)}{0.30 \times 10^{-9} \ \text{m}} \right.$$

$$+ \frac{(0.19e)(-0.40e)}{0.18 \times 10^{-9} \ \text{m}} + \frac{(-0.19e)(0.40e)}{0.40 \times 10^{-9} \ \text{m}} + \left. \frac{(-0.19e)(-0.40e)}{0.28 \times 10^{-9} \ \text{m}} \right]$$

$$= \left(9.00 \times 10^9 \ \text{N} \cdot \text{m}^2/\text{C}^2\right) \frac{(0.19e)(0.40e)}{10^{-9} \ \text{m}}$$

$$\times \left(\frac{1}{0.30} - \frac{1}{0.18} - \frac{1}{0.40} + \frac{1}{0.28} \right)$$

$$= -2.0 \times 10^{-20} \ \text{J}$$

$$= -0.13 \ \text{eV}$$

Since the potential energy is negative, electrical forces did positive work as the dipoles were brought together; equivalently, *you* need to do positive work to separate the two dipoles. Therefore, the dipole–dipole interaction between such pairs of adenine and thymine molecules along the double helix helps keep the strands of DNA together.

17.12 LIGHTNING RODS

One of the first practical inventions involving electrical phenomena was the lightning rod by Benjamin Franklin in the 18th century; see Figure 17.42. Not wishing to profiteer from such a practical device, Franklin altruistically placed his patent in the public domain to expedite its immediate use by humanity.

Such sharp-pointed conducting rods are used atop structures and even large trees, such as at George Washington's Mount Vernon estate on the banks of the Potomac River in

Virginia, to protect them from the destructive effects of lightning. The conducting lightning rod is connected to the Earth by means of a grounding wire. Without lightning rods, portions of trees or buildings literally can explode and/or catch fire when struck by lightning, because of the vaporization and rapid expansion of water in the wood. The purpose of the rod is to discharge nearby charged clouds harmlessly by providing an alternative conducting path to the Earth that is not through the tree or structure itself. Though it may sound like an oxymoron, such rods "attract" lightning (by literally becoming "the lightning rod") and make the effects of lightning quite harmless (other than the noise!). The lightning discharge in air rapidly heats and expands the air along its path, creating dramatically loud acoustic wave pulses: a complicated way to say thunder!

To see how lightning rods remarkably and safely initiate an electrical discharge, first consider two spherical conductors of different radii r and R connected by a long conducting wire, as shown in Figure 17.43. Thanks to the connecting wire, the system represents a single conductor. Since the two spheres and the connecting wire are electrically a single conductor, the two spheres must be at the same electric potential V because the surface of a conductor is an equipotential surface. Let Q be the charge on the conductor of radius R, and q the charge on the conductor of radius r. Since the two spheres are well separated in space, the charge distribution on them is essentially uniform and spherically symmetric, and we can treat them as spherically symmetric charges for purposes of calculating their common electric potential. Therefore for the small sphere we have

$$V = \frac{1}{4\pi\varepsilon_0} \frac{q}{r}$$

while for the larger sphere we have

$$V = \frac{1}{4\pi\varepsilon_0} \frac{Q}{R}$$

FIGURE 17.42 A fancy lightning rod with a nearby elaborate weather vane.

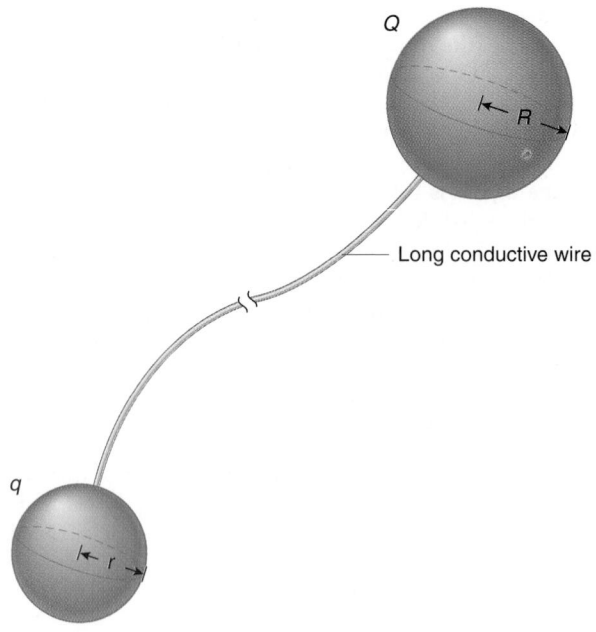

FIGURE 17.43 Two conducting spheres of different radii connected by a long conducting wire.

The potentials V are the same; hence, by dividing the preceding two equations, we find that the ratio of the charges on the two spherical conductors is

$$\frac{q}{Q} = \frac{r}{R} \qquad (17.36)$$

The electric field magnitudes at the surface of each spherical conductor also can be approximated by that of spherically symmetric charges. Call them $E_{\text{at surface of } R}$ and $E_{\text{at surface of } r}$. Then

$$E_{\text{at surface of } r} = \frac{1}{4\pi\varepsilon_0} \frac{q}{r^2}$$

and

$$E_{\text{at surface of } R} = \frac{1}{4\pi\varepsilon_0} \frac{Q}{R^2}$$

Dividing these equations, we find the ratio of the field magnitudes:

$$\frac{E_{\text{at surface of } r}}{E_{\text{at surface of } R}} = \frac{q/r^2}{Q/R^2}$$

But we know $q/Q = r/R$ from Equation 17.36. Therefore we finally get

$$\frac{E_{\text{at surface of } r}}{E_{\text{at surface of } R}} = \frac{R}{r} \qquad (17.37)$$

> This means that the magnitude of the electric field is greater at the surface of the sphere of smaller radius.

This result is quite significant. In a qualitative way, it shows that for a conductor of arbitrary shape, the electric field at its surface has the greatest magnitude near those portions

FIGURE 17.44 The magnitude of the electric field is greater near parts of a conductor with a small radius of curvature (sharper points).

with the smallest radius of curvature; see Figure 17.44.* In other words, if a conductor has a sharp point, the electric field near the point is of much greater magnitude than near other regions of the conductor, as reflected in the number of field lines. This is the purpose of the sharp points on lightning rods. Charged clouds in the vicinity attract or repel electrons in the conducting rod. Since the rod has a sharp tip, the magnitude of the electric field in the vicinity of the tip easily will reach the critical magnitude of about 3×10^6 N/C at which air ceases to be an insulator and becomes an conductor. Thus the electrical discharge is initiated near the tip of the lightning rod rather than near other objects such as a building structure or tall tree.

QUESTION 6

Explain why a lightning rod is connected to the Earth with a grounding wire for the rod to safely do its thing.

*There are exotic exceptions to this rule; see the article by Richard H. Price and Ronald J. Crowley, "The lightning-rod fallacy," *American Journal of Physics*, 53, #9, pages 843–848 (September 1985).

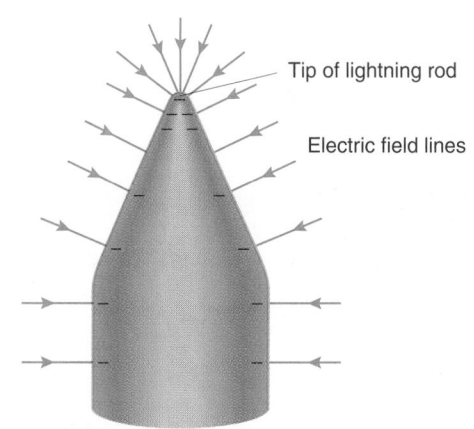

Tip of lightning rod

Electric field lines

CHAPTER SUMMARY

The electrical force caused by static charges is a conservative force. The work done by the electrical force on a charge q is the negative of the change in the electric potential energy of q:

$$W_{\text{elec}} = -\Delta\text{PE} \qquad (17.3)$$

The electrical force on q is $\vec{F} = q\vec{E}$, where $\vec{E}$ is the electric field at the point where q is located, and so

$$q\int_i^f \vec{E} \cdot d\vec{r} = -(\text{PE}_f - \text{PE}_i) \qquad (17.5)$$

The potential energy per unit charge at a point in space is defined to be the *electric potential* V at that point:

$$V \equiv \frac{\text{PE}_{\text{of } q}}{q} \qquad (17.7)$$

or

$$\text{PE}_{\text{of } q} = qV \qquad (17.8)$$

Hence the difference in the electric potential between two points in space is

$$V_f - V_i = -\int_i^f \vec{E} \cdot d\vec{r} \qquad (17.9)$$

The electric potential of a pointlike charge Q is

$$V(r) = \frac{1}{4\pi\varepsilon_0} \frac{Q}{r} \qquad (17.11)$$

The electric potential at a point caused by a collection of pointlike charges Q_i is found from the principle of superposition:

$$V = V_1 + V_2 + V_3 + \cdots \qquad (17.12)$$

where V_i is the potential caused at the point by charge Q_i as if it were the only charge present.

The electric potential of a continuous distribution of charge of finite extent is found by extending the principle of superposition from a sum to an integral over the charge distribution:

$$V = \frac{1}{4\pi\varepsilon_0} \int_{\substack{\text{finite charge} \\ \text{distribution}}} \frac{dQ}{r} \qquad (17.13)$$

The electric potential at a point in space arising from various charge distributions is summarized in Table 17.1 on page 779.

A volume of space in which the electric potential is constant is called an *equipotential volume*. The electric field is zero within an equipotential volume. The interior of a conductor is an equipotential volume. A hollow conductor, with no charges within the cavity, also is an equipotential volume, and thus is shielded from the electric fields of charges outside or on the outer surface of the conductor.

A surface over which the electric potential has a constant value is called an *equipotential surface*. Electric field lines always are perpendicular to an equipotential surface. The surface of a conductor is an example of an equipotential surface. Imaginary equipotential surfaces can be constructed around any charge distribution.

If the electric potential depends on a single coordinate s, the electric field has a single component along s that is

$$E_s = -\frac{dV}{ds} \qquad (17.15)$$

More generally, the electric field is the negative gradient of the potential:

$$\vec{E} = -\nabla V \qquad (17.19)$$

The potential energy associated with a pair of two pointlike charges Q_1 and Q_2 is

$$PE = \frac{1}{4\pi\varepsilon_0} \frac{Q_1 Q_2}{r_{12}} \qquad (17.34)$$

where r_{12} is the separation of the two charges. This potential energy represents the negative of the work done by the electrical force on Q_2 in bringing Q_2 from an infinite distance to its location a distance r_{12} from Q_1. The equation can be extended to larger collections of pointlike charges by writing similar terms for each distinct pair of charges in the charge distribution.

A new energy unit of convenience, used particularly for atomic and molecular physics, is the *electron-volt* (eV). One electron-volt is the change in the kinetic energy of an electron when it freely accelerates through a potential difference of exactly one volt. One electron-volt is approximately 1.602×10^{-19} J:

$$1 \text{ eV} = 1.602 \times 10^{-19} \text{ J} \qquad (17.23)$$

SUMMARY OF PROBLEM-SOLVING TACTICS

17.1 **(page 769)** The electric potential is not the same thing as the electric potential energy.

17.2 **(page 769)** It is important to use the appropriate sign for the charge q in Equation 17.8:

$$PE = qV \qquad (17.8)$$

17.3 **(page 769)** The electric potential and the electric potential energy both are scalar quantities, not vectors.

17.4 **(page 770)** The direction of the electric field always is from regions with higher values of the electric potential to regions with lower values of the electric potential.

17.5 **(page 773)** Note that in Equation 17.11 for the electric potential of a point charge, the distance r in the denominator is to the first power.

17.6 **(page 773)** In Equation 17.11 for the electric potential of a point charge, as well as for other charge distributions in Table 17.1, remember to use the appropriate sign for Q.

17.7 **(page 785)** Kinetic energies in electron-volts (eV) must be converted to joules (J) to calculate speeds in meters per second (m/s).

QUESTIONS

1. (page 770); 2. (page 781); 3. (page 783); 4. (page 785); 5. (page 790); 6. (page 793)

7. Show that the SI units for the electric field N/C and V/m are equivalent.

8. In what direction, relative to the direction of an electric field line, is it possible to move a charge so that the electrical force does zero work on it?

9. Given an electric field line of a charge distribution, in what directions (relative to the field line) is it possible for another charge to move without changing its potential energy?

10. The electric potential varies linearly with a coordinate z through a region. What is the nature of the electric field in this region?

11. Shuffle your feet across a carpet with a deep pile. Now bring your head slowly very near a water pipe or faucet and describe what happens to your hair. Move your head away and then bring your finger close to the faucet. A spark may now jump between your finger and the faucet. Explain why.

12. Two positive charges are separated by a distance ℓ. Is there a point on the line connecting the charges where the electric field is zero? Is there a point on the same line where the electric potential is zero? If your answer to both questions is yes, are the two points at the same location?

13. A particular charged particle lowers its electric potential energy while moving to regions of higher electric potential. Is this possible under the action of electrical forces alone? If so, what is the sign of the charge?

14. Consider the following technical terms: electric potential; electric potential difference; electric potential energy; change in electric potential energy. Carefully explain what is meant by each term.

15. How would you define a *proton*-volt unit of energy so that it has the same value as the electron-volt?

16. A hollow, closed conductor has no charges within the hollow. Is it possible to create a static electric field within the hollow using charges exterior to the hollow conductor? If so, explain how. If not, explain why not.

17. A hollow metal conductor can shield the interior from electrical forces caused by charges exterior to the conductor. A corresponding hypothetical gravitational shield would insulate masses from gravitational forces of exterior masses. Since a mass anywhere inside a hollow spherical shell experiences zero gravitational force due to the shell, is such a shell a gravitational shield in the same sense as the electrical shield? Explain.

18. For the electric field line distributions of Figures 16.55–16.60 in Chapter 16, sketch some equipotential lines representing end-on views of equipotential surfaces.

19. Let the bottom of a 10 m high aluminum flagpole be an electrical ground. What is the potential at the top of the pole?

20. What is the sign of the potential energy of a charge distribution of: (a) Two like charges? (b) Two unlike charges? (c) What is the significance of the signs of the two potential energies in (a) and (b)?

21. (a) Is the sign of the electric potential caused by two negative charges the same at every point in space? If so, what is it? If not, explain why the sign may vary depending on the point in space. (b) Is the sign of the electric potential caused by two positive charges the same at every point in space? If so, what is it? If not, explain why the sign may vary depending on the point in space. (c) If two charges are unlike charges, is the sign of the electric potential caused by the two charges the same at every point in space? If so, what is it? If not, explain why the sign may vary depending on the location of the point in space.

22. The electric field is zero at a particular point. Is the electric potential necessarily zero at the same point? If so, explain why. If not, cite an example that illustrates the contrary situation.

23. The electric potential is zero at a particular point. Is the electric field necessarily zero at the same point? If so, explain why. If not, cite an example that illustrates the contrary situation.

24. From the graphs of the potentials in Figures 17.13 and 17.16, explain how you can tell that the magnitude of the electric field at the center of the circular ring is zero where $z = 0$ m while that of the charged disk is not zero where $z = 0$ m.

25. The static electric field is always zero within a conductor. Does it necessarily follow that the potential within the conductor is zero? If so, explain why. If not, explain why not.

26. The electric potential energy of two like charges is positive whereas that of two unlike charges is negative. What does this imply about the work done by electrical forces in forming the respective charge distributions in the two situations?

27. Is it possible to have two pointlike charges, separated by a finite distance, with zero potential energy? If so, explain how. If not, explain why not.

28. Is it possible to have three pointlike charges, separated by finite distances, with zero potential energy? If so, explain how. If not, explain why not.

29. Few people survive being struck by lightning (zap!). What steps or precautions might you take while hiking if it is not possible to reach shelter before a thunderstorm overtakes you? Such precautions might save your life and are well worth discussing. See "Don't get shocked," *Safety & Health*, *149*, #4, page 89 (April 1994), for some tips.

30. If two points in space are at the same electric potential, what is the work done by electrical forces in moving a charge from one point to the other? Does it necessarily follow that the two points are on the same equipotential surface?

31. Let's make an analogy between the diagrams indicating electric field lines and equipotentials and the topographic maps used frequently by hikers, campers, and geologists. On a topographic map, what corresponds to equipotentials? What corresponds to the electric field? In view of this analogy, explain why the word *gradient* is appropriate when expressing the relationship between the electric field and the electric potential, as in $\vec{E} = -\nabla V$.

32. The proper operation of a sensitive instrument necessitates an environment free from stray electric fields. How can you ensure that the instrument is located in such a region of space?

33. Why do you suppose many electrical instruments are housed in metal boxes?

34. Printed circuit boards typically are stored and shipped in metal foil packages. Why?

35. A word used in early times (and occasionally in some countries today) for the electric potential is the *electrical tension*. You likely have heard high-voltage power lines called high-tension lines. Discuss the ways in which electric potential is similar to and different from the mechanical concept of tension.

36. Is it proper to say that a conductor in electrostatics has a given fixed value for the potential everywhere on its surface? Is it meaningful to say that an insulator has a given fixed value of the potential everywhere on its surface? Explain.

PROBLEMS

Sections 17.1 Electric Potential Energy and the Electric Potential
17.2 The Electric Potential of a Pointlike Charge
17.3 The Electric Potential of a Collection of Pointlike Charges

1. What work is done by the electrical force in moving a $-5.00~\mu C$ charge from an electrical ground to a place where the electric potential is 150 V?

2. Two large parallel plates are separated by 1.00 cm. The potential difference between the plates is 120 V. What is the magnitude of the electric field between the plates?

3. In the Bohr model of the hydrogen atom, the circular electron orbit closest to the nucleus has a radius of 5.29×10^{-11} m. (a) Find the electric potential of the proton at the position of this orbit of the electron. (b) Calculate the electric potential energy of the electron at this location.

•4. A potential difference of 3.00 kV exists between two parallel conducting plates. What plate separation produces an electric field of magnitude 3.0×10^6 N/C in the region between the plates? This magnitude of field causes air to conduct electricity, and a lightning discharge between the plates will neutralize them.

•5. A thunderstorm cloud is 1.00 km overhead. Air will conduct electricity if the field exceeds a magnitude of about 3.0×10^6 V/m. Model the cloud–ground system as a pair of parallel plates. (a) What is the potential difference between the cloud and ground when the field has this magnitude? (b) If the bottom of the cloud has an area of 2.0 km², what static charge resides on the cloud?

•6. Consider an atom with a single electron in a circular orbit of radius r about a nuclear charge of $+Ze$, where Z is the atomic number (identical to the charge quantum number of the nucleus). Such single-electron atoms are known as hydrogenic atoms. (a) What is the electric potential of the nucleus at the position of the electron? (b) What is the potential energy of the electron at this location? (c) Use the Coulomb force law for the interaction between the orbiting electron and the nucleus and write Newton's second law of motion for the orbiting electron, remembering that the acceleration in circular motion is a centripetal acceleration of magnitude v^2/r, where v is the speed of the orbiting electron. Show that the kinetic energy of the electron and its electric potential energy are related by

$$KE = -\frac{1}{2}PE$$

This result is a consequence of a general theorem in theoretical mechanics, known as the virial theorem, applied to inverse-square-law forces. (cf. Chapter 8, Problem 41.)

•7. (a) Find the electric potential at the origin in Figure P.7. (b) Find the electric field at the origin.

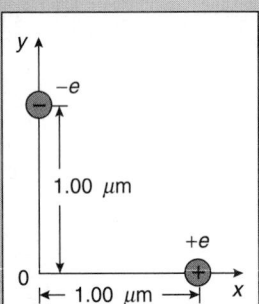

FIGURE P.7

•8. Two charges are nailed by a carpenter to the coordinate grid in Figure P.8 at the indicated locations. (a) Find the total electric potential at the origin. (b) Find the total electric potential an infinite distance away from the origin. (c) What is the potential energy of a proton placed at the origin? (d) How much work is done by electrical forces if the proton of part (c) is moved from the origin to infinity? (e) Does your answer to part (d) depend on the particular path used to take the charge to infinity? (f) If released at the origin, will the proton wander off to infinity of its own accord (under only the influence of the electrical force) or will *we* have to drag it there kicking and screaming all the way?

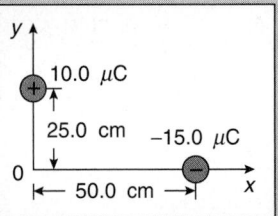

FIGURE P.8

•9. (a) Use the definition of the electric potential as a guide to define carefully the gravitational potential at a point in space. (b) An initial point is located 3.00 m above the ground. A final point is located 8.00 m above the ground and 2.50 m horizontally east with respect to the initial point. What is the gravitational potential difference between the two points?

•10. A constant electric field $\vec{E} = (2.0 \times 10^3 \text{ N/C})\hat{i}$ exists in a region of space. What is the potential difference between an initial point at $x = 0.00$ cm and a final point $x = 3.00$ cm in the field?

•11. To pacify your professor, consider the electric dipole in Figure P.11. (a) Calculate the total electric potential at the point P. (b) What is the potential energy of a proton placed at the point P? (c) How much work is done by the electrical force on the proton if it is dragged from point P to infinity?

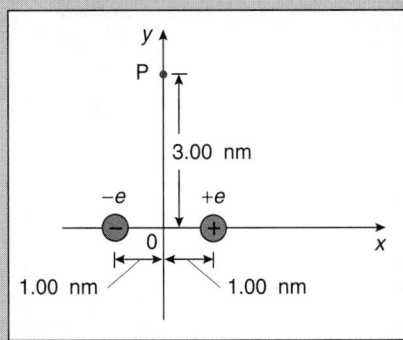

FIGURE P.11

•12. A charge $+2q$ is at the origin while another charge $-q$ is 2.00 m away, out along the x-axis as indicated in Figure P.12. Assume $q > 0$ C. (a) At what point along the x-axis will the electric field vanish? Is this point unique or are there other such points along the x-axis? (b) At what point along the x-axis will the electric potential vanish? Is this point unique or are there other such points along the x-axis?

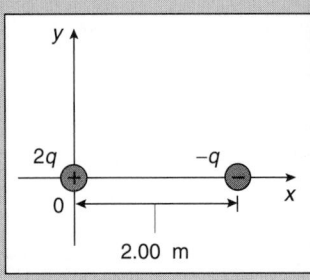

FIGURE P.12

•13. Two large (effectively infinite) horizontal metal plates are separated by 10.0 cm and the electric field between them has a magnitude 1.00×10^4 N/C, directed as shown in Figure P.13. (a) What charge is on each square meter of the upper surface? (b) What charge is on each square meter of the lower surface? (c) What is the electric potential of the upper plate if the lower plate is grounded? (d) A small particle of mass 1.00×10^{-5} kg, placed in the vacuum between the plates, has a charge q and is in equilibrium under the influence of gravitational and electrical forces. What is the charge q?

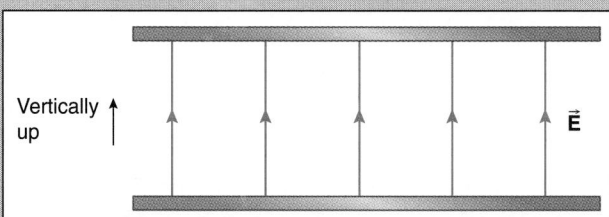

FIGURE P.13

•14. Two charges are located as indicated in Figure P.14. (a) Find the total electric field at point P. (b) Find the magnitude of the electric field at P. (c) Find the angle that the total field at P makes with $\hat{\jmath}$. (d) Find the total electric potential at P. (e) Another charge, $-3.00~\mu C$, now is placed at P. Find the force on this charge and its potential energy.

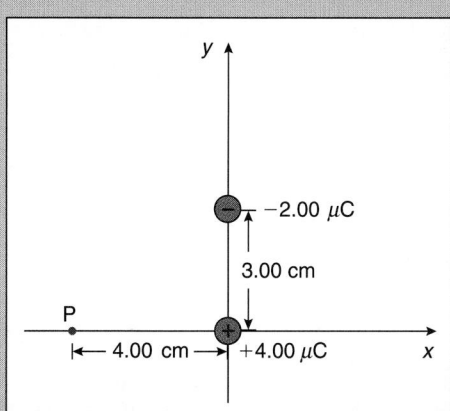

FIGURE P.14

•15. Charge separation (polarization) typically exists across cell membranes such as nerve cells. The inside of the membrane has a low concentration of potassium ions (K^+) while the outside has a high concentration of sodium ions (Na^+). Thus the inside of the membrane is at a lower electric potential than the outside; the potential difference typically is about 90 mV across an insulating membrane roughly of thickness 5 nm. (a) What is the magnitude of the electric field in the membrane if you model the system as a pair of parallel plates? (b) Is the answer to part (a) greater or less than the magnitude of electric field ($\sim 3 \times 10^6$ V/m) that causes air to break down and become a good conductor? When the nerve is stimulated by a mechanical, thermal, or electrical stimulus, the potential difference across the membrane changes (biologists say *depolarizes*) over a time interval of milliseconds in the way schematically illustrated in Figure P.15, and this depolarization propagates along

the nerve cell. Sodium ions migrate (biologists say are *pumped*) to restore the original potential difference, so that the cell is ready for further stimulation.

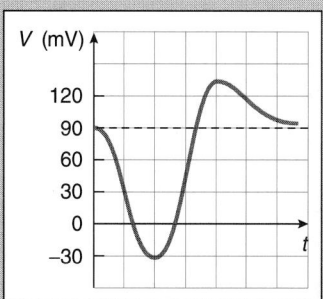

FIGURE P.15

•16. The fundamental unit of charge ($e = 1.602 \times 10^{-19}$ C) was discovered in a classic series of experiments by Robert Andrews Millikan early in the 20th century. In the experiments, tiny oil droplets are sprayed into a region in which there is a constant electric field as shown in Figure P.16. The individual oil droplets are large enough to be seen with a microscope using strong illumination, but small enough to have little weight and charge. The droplets obtain a static electrical charge from frictional effects (rubbing) as they emerge from the tiny orifice into the chamber. Surface tension causes the droplets to assume a spherical shape. To a first approximation, the droplets are subjected to two forces in the region between the plates: (1) the electrical force and (2) the gravitational force. Neglect the buoyant force of the air on the droplet. Let a droplet with mass m and charge $+q$ be in equilibrium within the chamber under the action of the two forces. (a) Show that the charge q is

$$q = \frac{mgd}{V_0}$$

(b) Let ρ be the density of the oil. Express the mass m of the drop in terms of its density and radius. (c) If (i) the oil has a density of 800 kg/m³; (ii) the potential difference between the plates is 2000 V when the drop is in equilibrium; (iii) the separation of the plates creating the field is 2.00 cm; and (iv) the radius of the oil drop is measured (with a microscope) to be 3.629 μm; how many fundamental units of charge are on the drop?

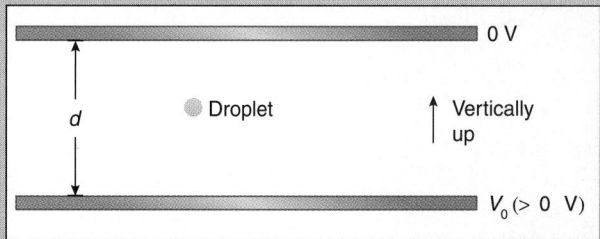

FIGURE P.16

•17. The electric potential is the electric potential energy per coulomb at a point in space. Analogously, the gravitational potential is the gravitational potential energy per kilogram at a point in space. Follow Example 17.1 to formulate the gravitational potential for a uniform gravitational field, such as that near the surface of the Earth.

Sections 17.4 The Electric Potential of Continuous Charge Distributions of Finite Size
17.5 Equipotential Volumes and Surfaces

18. An isolated conducting sphere of radius 3.00 cm is found to have a potential of 300 V on its surface. What charge is on the sphere?

19. Two parallel plates a distance 5.00 cm apart have potentials of −100 V and +100 V. Equipotential surfaces are drawn corresponding to differences in potential of 10 V. How far apart are the equipotential surfaces?

•20. An isolated conducting sphere of radius 3.00 cm has a potential of 300 V on its surface. What are the radii of the equipotential surfaces corresponding to 200 V, 100 V, and 1.00 V respectively?

•21. A spherical conductor is to have an electric potential of 25.0 kV without exceeding the maximum electric field magnitude of 3.0×10^6 N/C that causes air to become a conductor. What is the minimum radius for the sphere? Explain why this is a minimum radius and not a maximum radius.

•22. An insulating rod of length ℓ is bent into a circular arc of radius R that subtends an angle θ from the center of the circle; see Figure P.22. The rod has a charge Q distributed uniformly along its length. (a) Find the electric potential at the center of the circular arc. (b) The rod now is stretched along its length so that it forms a complete circle of the same radius. The total charge on the rod remains unchanged. What is the potential at the center of the circle?

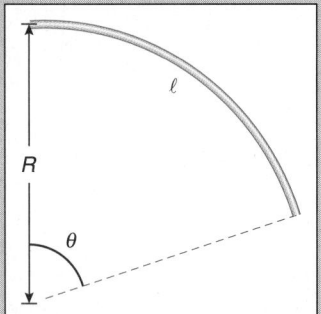

FIGURE P.22

•23. A metal sphere of radius R with a charge q is surrounded by a concentric metal spherical shell as shown in Figure P.23. The outer surface of the spherical shell has a charge of $+30.0 \times 10^{-6}$ C and the inner surface of the shell has a charge of $+25.0 \times 10^{-6}$ C. (a) Find q. (b) Sketch qualitative graphs of (i) the radial electric field component E_r and (ii) the electric potential V as functions of r.

•24. An early crude nuclear model of the atom imagined a very small central nucleus of charge $+Ze$ surrounded by a thin uniform spherical shell of radius R with charge $−Ze$. Show that the electric potential of this charge distribution is

$$\text{(i)} \quad V = 0 \text{ V} \quad \text{for distances } r > R$$

and

$$\text{(ii)} \quad V = \frac{Ze}{4\pi\varepsilon_0}\left(\frac{1}{r} - \frac{1}{R}\right) \quad \text{for distances } r < R$$

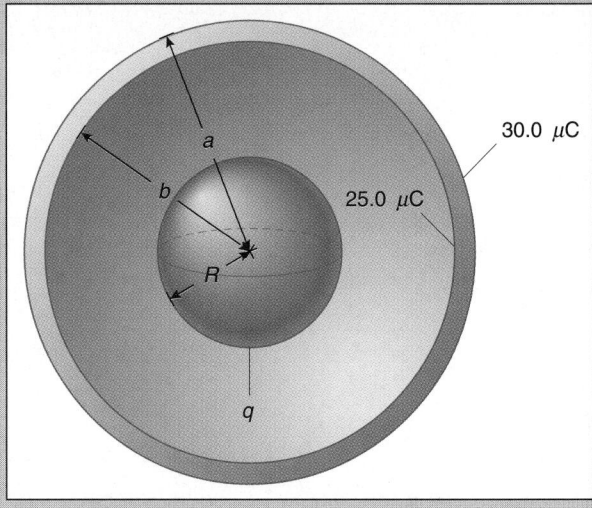

FIGURE P.23

•25. Instead of the model atom proposed in Problem 24, imagine a model consisting of a very small central nucleus of charge $+Ze$ surrounded by a sphere of radius R with charge $−Ze$ uniformly distributed throughout its volume. Find the electric potential of this charge distribution at positions: (a) $r > R$; (b) $r < R$.

•26. At a distance 10.0 cm from the center of a uniformly charged sphere, the electric potential is found to be −100 V. On the surface of the sphere, the potential is −250 V. (a) What is the charge on the sphere? (b) What is the radius of the sphere?

•27. What surface charge density σ on a conductor is needed to create an electric field with a magnitude sufficient to cause air to become a conductor (3.0×10^6 N/C)?

•28. A hollow spherical conductor of radius 5.00 cm has 6.00 nC of charge distributed uniformly over its surface. (a) What is the electric potential at the center of the sphere? (b) If the conductor is solid rather than hollow, what is the potential at the center of the sphere?

•29. A solid, insulating sphere of radius 5.00 cm has a charge of 6.00 nC distributed uniformly throughout its volume. (a) What is the electric potential at the surface of the sphere? (b) What is the electric potential at the center of the sphere?

•30. A hollow, spherical conductor of radius R has a charge of Q. Inside the cavity and concentric with the shell is another conductor of radius $r_0 < R$ with a charge of q as shown in Figure P.30. (a) What is the electric field between the inner conductor and the outer shell? (b) Use the result of (a) and Equation 17.9 to show that the potential difference between the two spheres is

$$V(r_0) - V(R) = \frac{1}{4\pi\varepsilon_0}\, q\left(\frac{1}{r_0} - \frac{1}{R}\right)$$

Explain why this potential difference is *independent* of the charge Q on the outer spherical shell.

•31. (a) At what distance from a +3.00 nC charge is the equipotential surface corresponding to 100 V located? (b) Where is the equipotential surface corresponding to 200 V? (c) Is the distance between the 100 V and 200 V equipotential surfaces the same as the distance between the 200 V and 300 V equipotential surfaces? Explain why it is or is not.

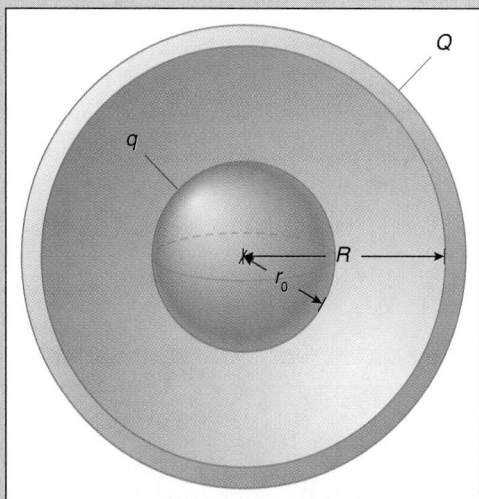

FIGURE P.30

•32. A long thin conducting wire connects two conducting spheres of radii 1.00 m and 0.10 m; see Figure P.32. The total charge on the connected pair is -2.00×10^{-7} C. (a) What is the electric field *inside* the wire? (b) What is the difference in potential between the ends of the wire? (c) What is the ratio of the charge on the larger sphere to the charge on the smaller sphere?

FIGURE P.32

•33. Two conducting spheres of radii 2.00 cm and 4.00 cm are far apart and each has charge +3.00 nC. (a) Calculate the approximate value of the electric potential on the surface of each conductor. (b) The two spheres now are connected by a conducting wire. What is the approximate potential on each surface now? What quantity of charge was exchanged by the two spheres?

•34. The gap between the electrodes in a spark plug is approximately 1 mm. Estimate the potential difference needed to produce an electric field of magnitude 3×10^6 N/C between the electrodes, sufficient to cause the air to become a conductor (and create a spark).

‡35. A charge Q is uniformly distributed along an insulating straight wire of length ℓ as shown in Figure P.35. Find an expression for the electric potential at a point located a distance d from the distribution along its perpendicular bisector.

Section 17.6 The Relationship Between the Electric Potential and the Electric Field

•36. A graph of the electric potential as a function of x is given in Figure P.36. Make an accurate plot of the electric field component E_x as a function of x over the same domain for x. Ignore the points on the graph of V where the slope of the graph changes discontinuously.

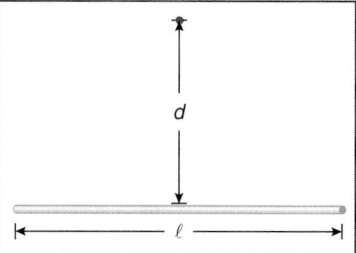

FIGURE P.35

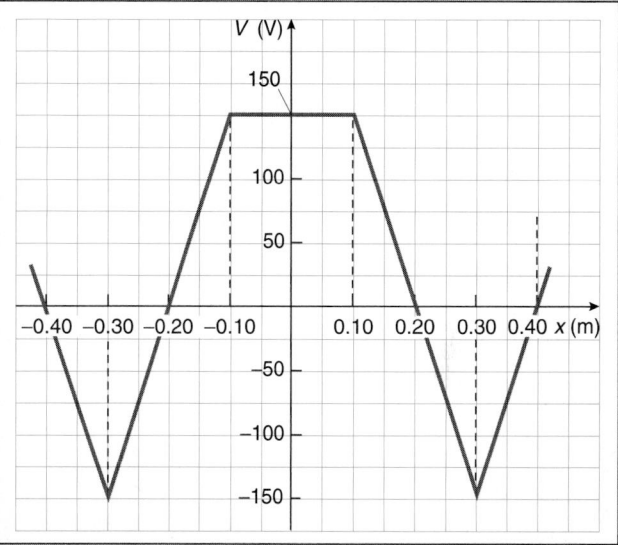

FIGURE P.36

•37. (a) Using the expression for the electric field of a uniformly charged, infinite line charge (see Table 16.2), and choosing the electrical ground to be at a distance a from the line, find the electric potential at a distance r from the line (see Figure P.37). (b) For what values of r is $V(r) > 0$ V? (c) For what values of r is $V(r) < 0$ V? (d) Explain why we cannot choose the electrical ground to be at $r = 0$ m or at $r = \infty$ m. (e) Verify that

$$E_r = -\frac{dV}{dr}$$

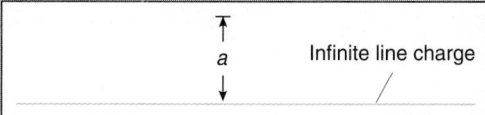

FIGURE P.37

•38. Explain why the graph of the electric potential as a function of x in Figure P.38 is not possible.

•39. The electric potential within a specific spherical charge distribution varies with the radial coordinate r as

$$V(r) = V_0 \frac{r^2}{2R^2}$$

where V_0 and R are constants and $r \le R$. (a) Find $E_r(r)$. (b) Is the charge on the sphere positive or negative? Explain how you make this determination. (c) Make a schematic graph of E_r versus r for 0 m $\le r \le R$.

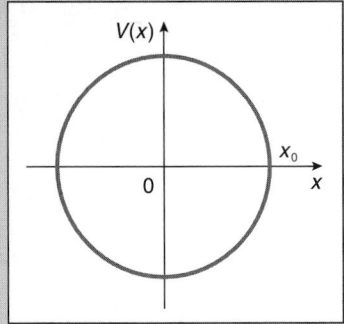

FIGURE P.38

40. An insulating washer (see Figure P.40) with an inner radius a and an outer radius b has a charge Q uniformly distributed on its surface. (a) Calculate the electric potential at a point P located a distance z from the washer along the symmetry axis perpendicular to the washer. (b) What is electric field component E_z as a function of z? (c) What is the electric field at $z = 0$ m?

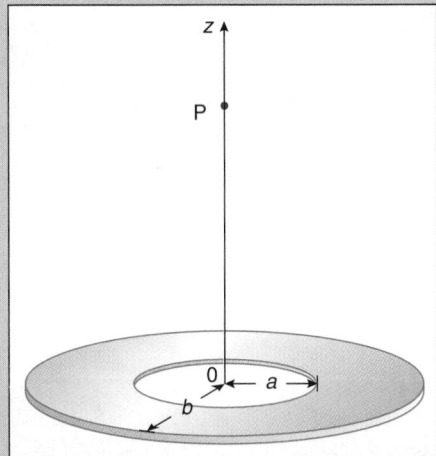

FIGURE P.40

Sections 17.7 Acceleration of Charged Particles Under the Influence of Electrical Forces
17.8 A New Energy Unit: The Electron-Volt

41. An electron is accelerated through a potential difference of 1.00 kV. (a) What is the increase in the kinetic energy of the electron in eV? (b) If the electron began the process at rest, what is its final speed?

42. Express the average kinetic energy of a hydrogen molecule (H_2) at temperature 300 K in electron-volts. Will oxygen molecules (O_2) have the same average kinetic energy per molecule at the same temperature?

43. An electron at rest is surprised to find itself suddenly released very close to one of the conducting infinite sheets shown in Figure P.43 (which are at the indicated potentials). (a) What is the direction of the electric field between the sheets? (b) What is the potential energy of the electron near the top sheet? (c) What is the potential energy of the electron near the bottom sheet? (d) Near which sheet should the electron be released so that it accelerates toward the other sheet? (e) What is the kinetic energy of the electron the instant before it strikes the sheet opposite the one from which it was released? Express your answer in both joules and electron-volts. (f) Calculate the speed of the electron just before impact.

FIGURE P.43

44. An electron is projected at speed 4.00×10^6 m/s into a region with a uniform electric field of magnitude 300 N/C directed parallel to the initial velocity of the electron; see Figure P.44. How far into the region will the electron travel before reversing its direction?

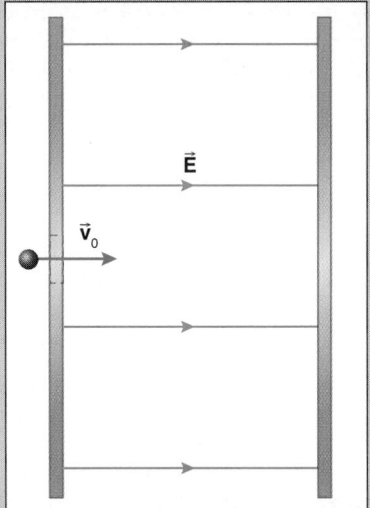

FIGURE P.44

45. A positron has the same mass as an electron but a positive charge of $+e$. (a) In the arrangement shown in Figure P.45, near which conducting sheet should the positron be released so that it accelerates toward the other sheet? (b) If the positron is released at rest, what is its kinetic energy the instant before striking the other charged sheet? Express your result in both electron-volts and joules. (c) What is the speed of the positron at the instant before impact?

FIGURE P.45

•**46.** What kinetic energy in eV must an α-particle (a helium nucleus with charge quantum number +2) have so that it is capable of coming to within a distance of 4.7×10^{-15} m of a gold nucleus (with charge quantum number +79) before coming momentarily to rest? This small distance, much smaller than the size of the atom ($\sim 10^{-10}$ m), is about the radius of the gold nucleus. Experiments with such α-particles (produced by radioactive decay) led Ernest Rutherford (1871–1937) to propose the nuclear model of the atom in 1910.

•**47.** The two infinite conducting sheets shown in Figure P.47 have the indicated potentials. (a) Indicate on a sketch the direction of the electric field between the two sheets. (b) Calculate the magnitude of the electric field between the sheets. (c) What is the electric potential midway between the sheets? (d) At what distance from the lower sheet is the electric potential equal to zero? (e) If a proton is placed at rest midway between the sheets, what is the potential energy of the proton in eV? (f) If the proton in part (e) is released, toward which of the sheets does it accelerate? (g) What is the kinetic energy of the proton just before it strikes the appropriate sheet? (h) What is the speed of the proton just before impact?

FIGURE P.47

•**48.** The Stanford Linear Accelerator (SLAC), located near the San Andreas Fault near Stanford University in California, can accelerate electrons from rest to kinetic energies of about 2.0×10^{10} eV over a straight-line distance of approximately 1.6 km. Assume the acceleration is caused by a uniform electric field directed along the length of the accelerator (in fact, not so). Determine the magnitude of this electric field.

•49. An electron is propelled at a speed of 5.0×10^7 m/s into a region where there is a uniform electric field as indicated in Figure P.49. (a) Sketch the direction of the electric field in the region between the two plates. (b) Indicate the direction of the electrical force on the electron in the region of the field. (c) What is the total mechanical energy of the electron at point A? (d) What is the total mechanical energy of the electron at point B? (e) Determine the speed at which the electron exits the device.

•**50.** A tiny snowflake in air has a charge quantum number of -1.00×10^6 and floats at rest in a vertical electric field of magnitude 1.00×10^5 N/C. (a) Is the electric field directed up or down? (b) What is the mass of the snowflake? (c) In a like manner, is it possible to suspend a charged droplet of oil mist at rest in the same electric field if the mass of the droplet is 2.50×10^{-15} kg? Give quantitative reasoning to justify your answer. (d) Describe how a field like this could be produced with only a potential difference of 2.00 kV available. Provide

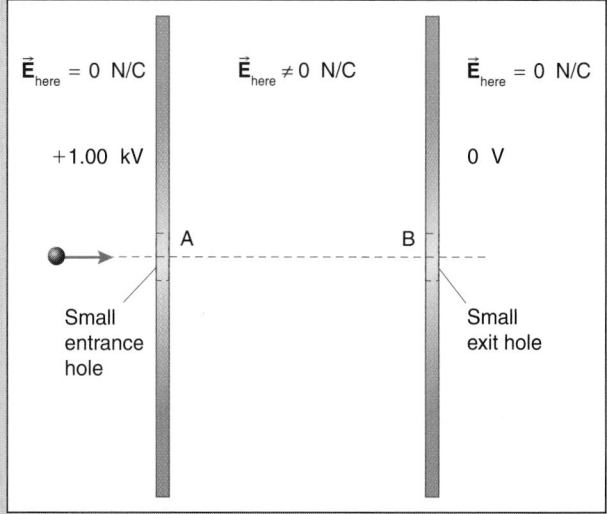

FIGURE P.49

a labeled diagram, appropriate dimensions, and numerical information.

•**51.** Two positive charges Q are held a distance d apart. (a) One charge now is released at rest and escapes to an infinite distance away. What is the kinetic energy of the escaped charge when it is a great distance away from the fixed charge? (b) The second charge now is released at rest. What happens to it?

•**52.** Three equal charges Q are fixed initially at the corners of an equilateral triangle with sides ℓ. (a) One charge is released at rest from its position and is allowed to escape to an infinite distance away. What is the final kinetic energy of this charge? (b) The second charge now is released at rest from its position. What is the final kinetic energy of this charge after it has escaped to an infinite distance? (c) The third charge now is released at rest from its position. What happens to it?

•53. (a) An α-particle (a helium nucleus with charge quantum number +2) moves through a potential difference of +20 V. What is the change in the kinetic energy of the α-particle in electron-volts? (b) If the potential difference was −30 V, what is the change in the kinetic energy of the α-particle, expressed in electron-volts?

•**54.** (a) What potential difference is needed to change the kinetic energy of an α-particle by +500 eV? (b) Is the α-particle moving to regions of higher or lower electric potential as its kinetic energy changes? Explain your reasoning.

•**55.** The nuclear structure of the atom was deduced by sending high-energy α-particles (helium nuclei each with a charge quantum number of +2) into a very thin gold foil. Gold was chosen not because such a foil looks pretty and is expensive, but rather because gold is very malleable and thus can be made into exceedingly thin foils. Most of the α-particles impinging on the foil pass through it with only small deflections. An occasional α-particle, however, makes a head-on approach to a gold nucleus (with a charge quantum number of +79). In such a head-on encounter, the α-particle is brought momentarily to rest before retreating along the path it came; such occurrences are called *back scattering*, to use the lingo of the trade. It is with these privileged α-particles that we can estimate the size of the gold nucleus. (a) Let an incident α-particle initially have a kinetic energy of 5.0×10^6 eV. What is its kinetic energy in joules? The initial electrical potential energy of the α-particle is zero, since it is essentially infinitely far away from the nucleus (and in any case the distant α-particle sees an atom that is electrically neutral, so that the electrical potential is essentially zero outside the atom itself). Thus the initial total mechanical energy of the incident α-particle is purely kinetic energy. (b) What is the electric potential of the gold nucleus at a distance r from the nucleus? (Consider the nucleus to be a spherical distribution of charge of radius $R < r$. Neglect any effects due to surrounding electrons.) (c) Use the CWE theorem to find the distance of closest approach for an incident head-on α-particle. This distance is a measure of the *upper limit* on the size of the gold nucleus since $R < r$. When such α-particle scattering experiments were first performed by Geiger and Marsden between 1909 and 1913, the results yielded a distance of closest approach that was so much smaller than the known size of an atom (about 10^{-10} m) that Rutherford proposed the nuclear model of the atom.

•56. A collection of electrons has a speed 6.00×10^7 m/s; this is too fast for an experiment you wish to perform. Using a uniform electric field provided by two parallel charged plates, you wish to slow them to 4.00×10^7 m/s. (a) Sketch the arrangement, indicating the direction of the electric field and the direction of the velocity of the electrons. (b) Indicate which charged plate is at the higher electric potential. (c) What potential difference $|\Delta V|$ is necessary to slow the electrons to the desired speed?

‡57. A charge $+Q$ is distributed uniformly throughout a sphere of radius R. (a) What is the potential on the surface of the sphere? (b) A small negative charge $-|q|$ of mass m is placed on the surface of the sphere. What is the potential energy of the charge $-|q|$ at this location? (c) Now give the negative charge a speed v_{escape} in the radial direction sufficient to (barely) escape to infinity from the sphere of charge $+Q$. Use the CWE theorem to derive an expression for the escape speed v_{escape} of the small charge in terms of $|q|$, R, Q, m, and appropriate electrical constants (such as ε_0). (d) In Chapter 8 we derived an expression for the analogous escape speed for the gravitational situation. The *gravitational* escape speed is *independent* of the mass m of the escaping object. The electrical escape speed [in part (c)] depends on the mass of the escaping particle. Why is there a difference between the two situations? (e) If the small negative charge is an electron and the positively charged sphere has the dimensions of a large nucleus such as uranium, where $R \approx 10^{-14}$ m, what charge Q makes the escape speed equal to the speed of light? About how many protons does this charge represent? This is the electrical analog of a gravitational black hole. (f) The answer to part (e) is a surprisingly small number of protons. Why, then, are most atoms not black holes? To approach this question, recalculate part (e) if the electron is placed initially at a distance from the nucleus equal to a typical *atomic dimension* ($\approx 10^{-10}$ m).

Sections 17.9 An Electric Dipole in an External Electric Field Revisited
17.10 The Electric Potential and Electric Field of a Dipole*

•58. Calculate the work done by the electrical force to bring a charge $+2e$ from an infinite distance to the midpoint of the line connecting the charges of an electric dipole composed of charges $\pm e$ separated by 2.00 nm.

•59. How far from a dipole along its axis must a point be located so that the approximate expression for the electric potential along the axis, Equation 17.29 (with $\theta = 0°$), is within 1.0% of the exact expression for the potential, Equation 17.27 (applied to a point on the axis). Express your result in terms of d, the separation of the charges in the dipole.

•60. The water molecule has an electric dipole moment of about 6.0×10^{-30} C·m. The molecule is placed with the orientation shown in Figure P.60 between two infinite conducting sheets with the indicated potentials. (a) Indicate the direction of the electric field between the sheets. (b) Find the magnitude of the electric field between the sheets. (c) What is the torque on the water molecule in its present orientation? (d) What is the potential energy of the water molecule in its present orientation? Express your result in electron-volts. (e) What is the minimum value for the potential energy of the dipole in the field?

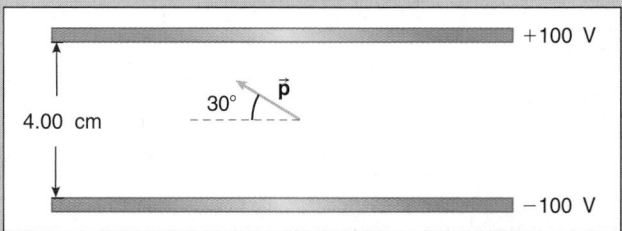

FIGURE P.60

•61. (a) Find the total electric field at the point P in Figure P.61. (b) Find the electric potential at the same point. (c) A dipole with dipole moment $\vec{p} = (6.0 \times 10^{-30} \text{ C·m})\hat{j}$ is placed at this point. Find the potential energy of the dipole in its initial orientation. (d) Calculate the torque on the dipole.

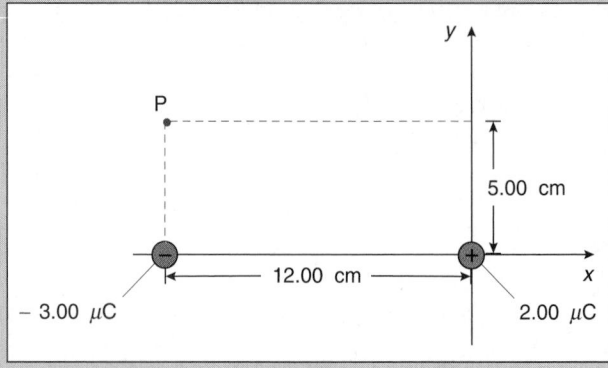

FIGURE P.61

‡62. A *linear quadrupole* is an arrangement of two oppositely directed dipole moments as shown in Figure P.62. (a) At a distance $|z| > d$ along the axis of the quadrupole, calculate the electric potential. (b) Show that if $|z| >> d$, the expression for V becomes

$$V(z) \approx \frac{|Q|}{4\pi\varepsilon_0} \frac{2d^2}{z^3}$$

The expression $2|Q|d^2$ is known as the quadrupole moment. (c) Use the potential found in (b) to find the electric field component E_z when $z >> d$.

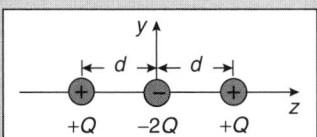

FIGURE P.62

‡63. Show that the potential of the linear quadrupole at point P in Figure P.63, when $r >> d$, is approximately

$$V \approx \frac{1}{4\pi\varepsilon_0} Qd^2 \frac{3\cos^2\theta - 1}{r^3}$$

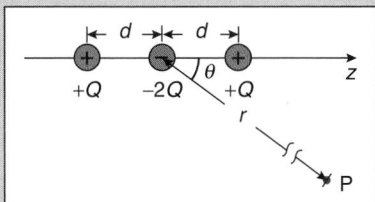

FIGURE P.63

Section 17.11 The Potential Energy of a Distribution of Pointlike Charges

64. What is the electric potential energy of the proton and electron charge distribution in the hydrogen atom when they are normally separated by 5.29×10^{-11} m? Express your result in both joules and electron-volts.

•65. A proton is thought to be a composite particle composed of two so-called up quarks with charges $2e/3$ and one down quark with charge $-e/3$. Imagine the three quarks to be at the vertices of a small equilateral triangle with sides 1.30×10^{-15} m. Determine the electric potential energy of the charge distribution.

•66. Singly ionized helium has a single electron orbiting the nucleus at a distance of about 2.65×10^{-11} m. The nuclear charge is $+2e$. What is the electric potential energy of this charge configuration? Express your result in electron-volts.

•67. When the nucleus of the uranium isotope $^{235}_{92}$U captures a slow neutron, the nucleus splits (we say *fissions*), into two so-called fission fragments. Assume that just after fission, the fragments are of equal mass and charge ($+46e$) and find themselves initially at rest separated by a distance of about 7.0×10^{-15} m. (a) Calculate the initial electric potential energy of the fragments. Express this energy in both joules and electron-volts. (b) What will the electrical force do to the fragments? (c) What is the kinetic energy of each fragment when the two are effectively at an infinite distance apart? (d) The kinetic energy of the fission fragments is used for destructive purposes in bombs or for constructive purposes in nuclear power plants. Assume the fission fragments have most of the energy released in the fission process (this assumption is in fact reasonable). How many fissions per second are needed to provide 10 MW of power in a reactor? How many fissions are needed to create a bomb that releases 9.8×10^{13} J of energy?* What mass of $^{235}_{92}$U corresponds to this number of fission reactions?

•68. Two identical 1.00 kg masses have opposite charges q and $-q$ separated by distance r in space, well away from other masses (such as the Earth or Sun). What magnitude of charge q makes their electric potential energy equal to their gravitational potential energy?

‡69. Two pointlike dipoles are oriented with their dipole moments p_1 and p_2 parallel to each other and separated by a distance r that is much greater than the separation of the two charges in each dipole (see Figure P.69). Show that the potential energy of one dipole in the presence of the other is approximately

$$PE \approx -\frac{1}{4\pi\varepsilon_0} \frac{2p_1 p_2}{r^3}$$

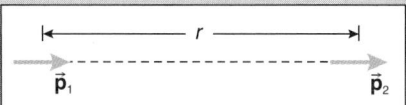

FIGURE P.69

‡70. Nuclear fusion is difficult to achieve because the nuclei of atoms are all positively charged and repel each other with the electrical force. In order to fuse, nuclei must be brought to within about 10^{-15} m of each other, at which distance the strong nuclear force becomes effective and fuses them. A deuteron is the nucleus of a hydrogen atom that consists of a proton and a neutron (so-called heavy hydrogen). (a) If two deuterons are 1.0×10^{-15} m apart, what is the potential energy of the pair? (b) If the deuterons are momentarily at rest at this location, what will be the kinetic energy of each deuteron when they are infinitely far away from each other? This is the kinetic energy each deuteron needs in order to approach to 1.0×10^{-15} m from one another so that the strong nuclear force can fuse the nuclei. (c) If the deuterons are part of a hot gas and have this average kinetic energy, what is the temperature of the gas? The actual fusion process is more complicated than this calculation, but your result indicates why fusion reactions of a controlled nature are difficult to produce on the Earth. Uncontrolled fusion produces a so-called hydrogen bomb.

---

*This corresponds to about 20 000 tons of the explosive TNT and is about the size of the bomb dropped on the unfortunate city of Hiroshima, Japan, in August 1945. The energy released in a nuclear bomb typically is expressed in kilotons or megatons of TNT. There are approximately 4.9×10^{12} joules per kiloton of TNT.

INVESTIGATIVE PROJECTS

A. Expanded Horizons

1. Large electric potentials can be produced with a Van de Graaff generator, invented by Robert Jemison Van de Graaff (1901–1967). Investigate and report on how such generators are able to produce high potentials.

   Richard E. Berg, "Van de Graaff generators: theory, maintenance, and belt fabrication," *The Physics Teacher, 28, #5*, pages 281–285 (May 1990).

   Peter H. Rose and Andrew B. Wittkower, "Tandem Van de Graaff accelerators," *Scientific American, 223, #2*, pages 24–33 (August 1970); additional references are on page 128.

   A. D. Moore, *Electrostatics* (Anchor Doubleday, Garden City, New York, 1968).

   Oleg D. Jefimenko, "Electrostatic motors," *Electrostatics and Its Applications*, edited by A. D. Moore (Wiley, New York, 1973), pages 131–147.

   M. J. Mulcahy and W. R. Bell, "Electrostatic generators," *Electrostatics and Its Applications*, edited by A. D. Moore (Wiley, New York, 1973), pages 148–179.

   A. W. Simon, "Theory of the frictional Van de Graaff electrostatic generator," *American Journal of Physics, 43, #12*, pages 1108–1110 (December 1975).

2. The electrical analog of a permanent magnet (a ferromagnet) is known as a ferroelectric (also called an electret); it is a material in which the molecular electric dipole moments all are aligned permanently in the same direction. Investigate the nature of ferroelectricity and its technological importance.

   Charles Kittel, *Introduction to Solid State Physics* (7th edition, Wiley, New York, 1996).

   Werner Känzig, *Ferroelectrics and Antiferroelectrics* (Academic Press, New York, 1957).

3. Stresses applied to certain materials (such as quartz and tourmaline) produce an electric polarization in the material, an effect known as *piezoelectricity*. Such materials have significant technological importance in, among other things, strain gauges. Investigate the piezoelectric effect and some of its applications.

   Walter Guyton Cady, *Piezoelectricity* (Dover, New York, 1964), Volumes I and II.

   Charles Kittel, *Introduction to Solid State Physics* (7th edition, Wiley, New York, 1996).

4. Static electricity is a major hazard in many industrial and chemical settings, causing fires and explosions of chemicals, aerosols, and airborne dust particles such as in grain elevators. Explore the ways such electrically induced fires and explosions can be prevented.

   Paul Cartwright, "Electrical hazards!" *Chemtech, 21, #1*, pages 682–685 (November 1991); other references are in this article.

5. The charge distributions in the heart muscle that give rise to an electrocardiogram can be modeled electrostatically. Those of you interested in medical careers might enjoy investigating and reporting on the model developed in the first reference below.

   Russell K. Hobbie, "The electrocardiogram as an example of electrostatics," *American Journal of Physics, 41, #6*, pages 824–831 (June 1973).

   D. W. Kammer and J. A. Williams, "Some experiments with biological applications for the elementary laboratory," *American Journal of Physics, 43, #6*, pages 544–547 (June 1975).

   Robert Paine, *Generation and Interpretation of the Electrocardiogram* (Lea & Felsriger, Philadelphia, 1988).

B. Lab and Field Work

6. One way of separating proteins, amino acids, and DNA fragments is via a process called *gel electrophoresis*, which uses electric fields. It is a common technique in the biological sciences. Investigate the technique and its application of electrostatic principles. Visit a member of the biology department who uses the technique. Discuss, design, and perform an experiment using the technique, paying particular attention to the physics of the process in your report of the experiment.

   Shyamsunder Erramilli, Fredrik Österberg, and Bruce Vogelaar, "Undergraduate laboratory: principles of gel electrophoresis," *American Journal of Physics, 63, #7*, pages 639–643 (July 1995).

   John M. Clark Jr. and Robert L. Switzer, *Experimental Biochemistry* (2nd edition, Freeman, New York, 1977), pages 43–55.

   Frederick A. Ausubel, *Current Protocols in Molecular Biology* (Current Protocols, New York, 1994), pages 2.5.1–2.5.15.

   Henry M. Zeidan and William V. Dashek, *Experimental Approaches in Biochemistry and Molecular Biology* (W. C. Brown, Dubuque, Iowa, 1996).

   Rodney F. Boyer, *Modern Experimental Biochemistry* (2nd edition, Benjamin Cummings, Redwood City, California, 1993).

C. Communicating Physics

7. Most nonscientists know of the unit of electric potential, the volt, but have no idea what it represents. Ask a few nonscientist friends or relatives to explain what they think the term means. Take notes or tapes of their explanations, without passing judgment on them. Use your survey as a guide and, by making appropriate analogies to gravitation, explain the meaning of the unit of electric potential in language appropriate to a nonscientist. Have your nonscientist friends critique your explanation to improve its clarity, and then make any necessary revisions.

Reese University Physics

CHAPTER 18
CIRCUIT ELEMENTS, INDEPENDENT VOLTAGE SOURCES, AND CAPACITORS

I sing the body electric
Walt Whitman (1819–1892)*

Storing energy for future use is an important aspect of wise energy management, even survival. Rural residents in northern areas who heat their homes with wood spend considerable time gathering and storing fuel for use during the winter, lest they freeze. Primitive cultures even in tropical climates need to accumulate and store wood for cooking. Squirrels and other animals store acornzs and other foods (sources of energy) for future use. Our bodies store surplus food as fat.

Mechanical energy also can be stored. A rapidly spinning flywheel stores kinetic energy. Lift and hold a set of barbells over your head and you increase and store the gravitational potential energy of the barbells, energy that can be suddenly tapped by simply dropping them. If you either compress or stretch a spring, you increase and store the potential energy of a mass attached to it. In this chapter we examine devices that increase and store the electrical potential energy of charges.

Devices that increase the electrical potential energy of charges are known by several names: **sources of emf**[†] or, equivalently, **independent voltage sources**. The terms mean the same thing. The common battery is an example of such a device. Calculators, automobiles, flashlights, portable CD players, notebook PCs—all use batteries in abundance. Batteries provide a convenient and portable source of electrical energy that originates from electrochemical reactions within the battery. In the next chapter, we will see how we can model real batteries that, much to our occasional consternation, age and die and need to be replaced (usually when we most need them, according to Murphy's law). Other examples of independent voltage sources (or sources of emf) include, among other things:

- electric generators used to provide the potential differences at the electrical outlets common in our homes, classroom, labs, and offices;
- solar cells used on spacecraft and in remote geographic locations; and
- fuel cells used on board the space shuttle and other spacecraft.

Electrical potential energy is accumulated and stored in devices called **capacitors**. Though they may be less familiar to you than batteries, capacitors are used widely in electronic applications. Capacitors are used in flash units on cameras to accumulate electrical potential energy to be released suddenly in the flash lamp when the shutter is pressed (analogous to dropping a set of barbells). In laser power supplies, as well as in medical defibrillation units used by medical trauma and emergency personnel, capacitors also are used to accumulate large amounts of electrical potential energy for sudden use.

Capacitors have other electronic uses too. When you change a radio station or TV channel by turning the dial (or using the remote), you are changing the value of a capacitor in

Solar panels on spacecraft are independent voltage sources. This is the famous orbiting Hubble Space Telescope.

a special electronic circuit known as an electrical oscillator or tuning circuit. Electric utility companies use capacitors to maintain the desired relationship between various electrical parameters in the electric grid around town. Many PC keyboards use capacitors to sense which key is depressed, so that capacitors are literally at your fingertips every time you sit at your PC. Capacitors also are used to count traffic and in smart traffic light systems to sense when a car approaches an intersection. Capacitors are indeed very useful electrical devices!

In this chapter we also study how independent voltage sources (sources of emf) and capacitors, as examples of two-terminal circuit elements, are connected together in various useful ways. This serves as a precursor to a systematic approach to the study of electric circuits in the next chapter and Chapter 22.

18.1 TERMINOLOGY, NOTATION, AND CONVENTIONS

We have seen many times now that language is an important component of physics because it explains what we mean when we discuss the laws of physics; mathematics is employed as quantitative shorthand. It is important to be sure we have a good common understanding of the meaning of various technical terms and phrases.

Imagine two points A and B that are at electric potentials V_A and V_B respectively, as shown in Figure 18.1. The *difference* in the electric potential between the two points is called the **potential difference**[‡] between the two points. Since the unit of electric potential in the SI system of units is the volt, the potential difference between two points also is expressed in volts. Of course, differences can be taken two ways. We need

V_A
•

V_B
•

FIGURE 18.1 The potential at two points in space.

---

*(Chapter Opener) Title of a poem in his collection *Leaves of Grass* (David McKay, Philadelphia, 1900) page 98.

†The term emf arises from the antiquated phrase *electromotive force*, which arose early in the history of explorations into electromagnetism. Emfs are not forces and the term electromotive force is misleading. The term independent voltage source, while also not ideal, is used in more advanced courses in circuits and electronics, but this term is more indicative of what the device actually does: raise the electric potential energy of charges passing through it.

‡In engineering, the term *voltage* is used frequently to mean potential difference. We prefer to use the term potential difference to emphasize that the values of the electric potential at *two different points in space* are involved.

some convention to indicate which way the difference is to be taken: $V_A - V_B$ or $V_B - V_A$.

> When you want the potential difference between A and B, take the potential at A and subtract from it the potential at B:
>
> potential difference between A and B $\equiv V_A - V_B$

When you want the potential difference between B and A, take the potential at B minus the potential at A:

$$\text{potential difference between B and A} \equiv V_B - V_A$$
$$= -(V_A - V_B)$$

The potential difference between two points can have a value that is positive, negative, or zero. For example, if the electric potential at the point A is 100 V and that at B is 50 V, the potential difference between A and B is

$$100 \text{ V} - 50 \text{ V} = 50 \text{ V}$$

On the other hand, the potential difference between B and A is

$$50 \text{ V} - 100 \text{ V} = -50 \text{ V}$$

Thus you must be careful in the way you express the potential difference between two points. Two points at the same electric potential have a potential difference of 0 V.

> A pair of **polarity markings** (+) and (−) is used to designate which of two given points is at the higher electric potential (+) and which is at the lower potential (−).

You likely have noticed such paired polarity markings on common batteries.

Since the location of the zero of electric potential is arbitrary, the (+) symbol at a point does *not* necessarily mean that the potential at the point has a positive value; the symbol means that the point is at a *higher* electric potential than the point marked with the (−). Likewise the (−) symbol does *not* necessarily mean that the value of the electric potential at that point is negative; it means that the point is at a lower electric potential than the point marked (+). This convention regarding polarity markings is illustrated in Example 18.1.

Polarity markings are typically indicated on most batteries.

In prior chapters, we used the letter V to designate the value of the electric potential at a *single point* in space. Thus the potential at point A is designated V_A. Alas, we need to shift notational gears, not because it is particularly desirable to do so, but because the tide of convention dictates we have no choice but to go with the flow.

> In this chapter and the next, the symbol V typically represents the *potential difference* between two points.

We really should call the potential difference ΔV, but the idea of the potential difference between two points occurs so frequently in the study of practical electric devices that it becomes a chore to continually write the delta, meaning difference or change; hence the delta is dropped for notational simplicity. Thus we have to now think of V as typically the *difference* in the electric potential between *two* points. Keep in mind, however, that we occasionally still will need to designate the potential at a *single* point in space and we continue to use, for example, V_A to designate the potential at a point labeled A. The context typically will indicate whether V is the potential at a *single* point in space, or the potential difference between *two* points in space. Of course, in cases where we choose to call $V_B = 0$ V, then $\Delta V \equiv V = V_A - V_B = V_A - 0$ V all have the same value.

QUESTION 1

Point A has an electric potential of −50 V. Point B has a potential 10 V lower. (a) What is the potential at point B? (b) Indicate which of the two points has the positive (+) and the negative (−) polarity markings. (c) What is the potential difference between A and B? (d) What is the potential difference between B and A?

EXAMPLE 18.1

Two points A and B have the potentials indicated in Table 18.1.

TABLE 18.1

B	A
• 50 V	• 100 V
• 0 V	• 50 V
• −25 V	• 25 V
• −50 V	• 0 V
• −75 V	• −25 V

a. For each case, indicate which point is at the higher electric potential with a (+) polarity marking and the one with the lower electric potential with a (−) polarity marking.
b. For each case, determine the potential difference V between A and B.

Solution

Point B (polarity markings in parentheses)	Point A (polarity markings in parentheses)	Potential difference between points A and B
(−) • 50 V	(+) • 100 V	100 V − 50 V = 50 V
(−) • 0 V	(+) • 50 V	50 V − 0 V = 50 V
(−) • −25 V	(+) • 25 V	25 V − (−25 V) = 50 V
(−) • −50 V	(+) • 0 V	0 V − (−50 V) = 50 V
(−) • −75 V	(+) • −25 V	−25 V − (−75 V) = 50 V

In this particular example, note that the potential difference V between the given points A and B is the same regardless of the various specific values of the potentials at A or B. This is *not* always the case.

18.2 CIRCUIT ELEMENTS

With Lego building blocks, you can build simple or amazingly complex structures. Circuit elements are the electronic building blocks used in the many simple and complex applications of electricity that surround us. You likely know of several circuit elements: batteries and light bulbs* are circuit elements. There are others as well. As is often the case in physics, we need to distinguish between ideal and real circuit elements: the difference is analogous to the difference between the ideal frictionless surfaces and real surfaces in mechanics. Real circuit elements are modeled by using combinations of ideal circuit elements.

An **ideal circuit element** has several characteristics:

1. Think of a generic ideal circuit element schematically as a box with two **ideal wires**, made from perfectly conducting material,† coming out of it; see Figure 18.2. The two lines coming out of the schematic symbol represent the two ideal wires and are called the **terminals** of the circuit element or, equivalently, the **leads**‡ of the circuit element.

---

*Light bulbs are a special class of resistors; we study general resistor circuit elements in the next chapter.

†We shall see in the next chapter that a perfectly conducting material has zero electrical resistance. Materials with zero resistance are called superconductors. Such materials in fact exist but are rarely used for the actual leads of real circuit elements.

‡Leads is pronounced "leeds" like the verb, not like the chemical element Pb.

2. An ideal circuit element cannot be subdivided into other ideal circuit elements.
3. Specific symbols are used to represent each distinct kind of ideal circuit element. The symbol used in Figure 18.2 simply means *any* ideal circuit element.

> By convention, we consider every symbol for a circuit element to be an ideal circuit element.

The two leads of a circuit element may be at (or have) different electric potentials; when this situation arises, the circuit element has a nonzero potential difference between its two terminals or leads. When this is the case, we say there exists a potential difference *across* the circuit element, conventionally taken as the higher minus the lower value of the electric potential.

18.3 AN INDEPENDENT VOLTAGE SOURCE: A SOURCE OF EMF

> An independent voltage source (equivalently, a source of emf) is a circuit element that has the following electrical characteristic: come hell or high water, it always maintains a constant potential difference V_0, also known as its emf, between its two terminals.§

Think of an independent voltage source as an ideal battery, one like the Energizer battery; as the pink bunny says, "It just keeps going and going." It never "dies."

The circuit symbol for an independent voltage source is shown in Figure 18.3. The terminal of the independent voltage source at the higher electrical potential is marked with a (+) polarity sign while that at the lower electrical potential is marked with a (−) polarity sign. The emf V_0 of the independent voltage source is the potential difference between the terminal marked (+) and that marked (−); the value V_0, expressed in volts, is positive.

We use the word independent voltage source in a general way. The term encompasses not only the common batteries (assuming they are ideal) you are most familiar with for your personal electronic gadgets, but also any device that maintains a constant potential difference between its two terminals. Electric generators, solar cells, fuel cells, and other devices

---

§Some physics books use the symbol $\mathscr{E}$ for the emf or, equivalently, the potential difference between the terminals of an ideal independent voltage source. Since the emf is measured in volts, for notational simplicity we prefer to use the symbol V_0 for the potential difference between the terminals of an independent voltage source. This is the notational convention used in engineering; there is no reason for us to use a different notation in physics.

FIGURE 18.2 A generic ideal circuit element.

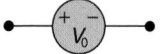

FIGURE 18.3 The circuit symbol for an independent voltage source.

An independent voltage source never "dies."

can be thought of as independent voltage sources in an ideal or first approximation.

An independent voltage source is an unusual device: it raises the electrical potential energy of positive charges moving from the lower-potential terminal, marked (−), to the higher-potential terminal, marked (+).* The potential difference V_0 is the work done by the independent voltage source on one coulomb of positive charge as it moves from the negative to the positive terminal inside or through the source. The potential difference in volts thus is measured in joules per coulomb.

In gravitational physics, when you lift one kilogram of mass through a vertical distance d to increase its gravitational potential energy by an amount $(1 \text{ kg})gd$, you are acting as a gravitational battery. The increase in gravitational potential energy per kilogram is gd, and so gd in that circumstance is analogous to what we mean by V_0 for an electrical independent voltage source. For common batteries, the increase in electrical potential energy of a positive charge moving from the minus to plus terminal through the battery arises from chemical reactions within the battery itself; in other independent voltage sources, the energy originates from other sources such as sunlight (for solar cells) or mechanical energy (for electric generators).

An independent voltage source maintains the specified potential difference V_0 between the (+) terminal and the (−) terminal, no matter what else is connected to its terminals, or for how long. An independent voltage source thus has an infinite lifetime. This surely is unrealistic in actual situations, but is just as useful an idealization as frictionless surfaces are in mechanics.

*Equivalently, it raises the electrical potential energy of negative charges moving from its (+) terminal to its (−) terminal.

Real voltage sources such as real batteries do not maintain a constant potential difference V_0 between their terminals as they are used over very long periods of time, sometimes to our frustration. Real batteries age and die and must be replaced. Real batteries also have a potential difference between their terminals that is smaller than V_0 under most circumstances. Later, in the next chapter, we see how you can model the behavior of *real* batteries with an ideal independent voltage source and another ideal circuit element (a resistor). We use ideal circuit elements to *model* real circuit elements.

QUESTION 2

Examine the packages of batteries for sale in your campus bookstore, supermarket, or discount outlet. What is the potential difference between the terminals of a common D-cell battery? Is it the same for C-cell, AA, AAA batteries, and common calculator and camera batteries? Make a list of the types of batteries and the labeled potential difference between the terminals of each type.

18.4 CONNECTIONS OF CIRCUIT ELEMENTS

Circuit elements are not of much use in isolation; they need to be connected together for various useful purposes. In this section we will learn how to recognize whether circuit elements are connected in several common ways. It will be important to be able to recognize these connections when we study electric circuits.

The place where the leads (terminals) of two or more circuit elements are connected together is called a **node**; see Figure 18.4. A node always is at a specific electric potential whose value depends on, among other things, the reference location where the electric potential is chosen to be zero (the electrical *ground*).

For our purposes, circuit elements may be connected together in several different ways.[†]

Series Connection

If the collection of circuit elements is connected together as shown in Figure 18.5, the circuit elements are said to be connected in **series**. As we march along the chain of circuit

[†] More advanced courses in electric circuits consider other types of specialized connections known as Y and Δ connections.

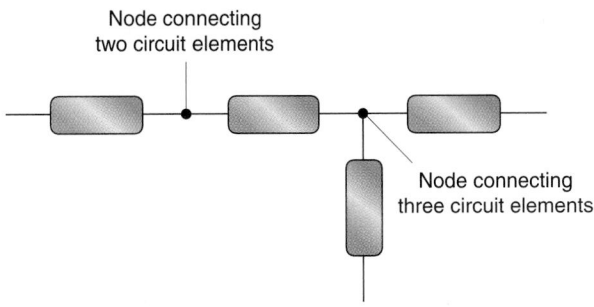

FIGURE 18.4 **Nodes connecting two or three circuit elements.**

FIGURE 18.5 Circuit elements connected in series.

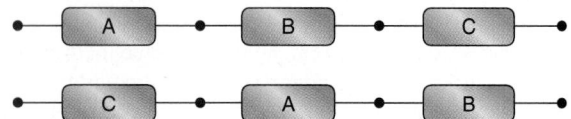

FIGURE 18.6 The order of circuit elements in series is not important.

elements, we encounter each circuit element in turn, or in a serial fashion, one after the other.

The specific order of circuit elements in series is not important (see Figure 18.6), since it has no effect on the potential difference across each one.*

Parallel Connection

If the terminals of various circuit elements are connected to the *same* two distinct nodes (see Figure 18.7), the circuit elements are said to be connected in **parallel**. The parallel connection shown in Figure 18.7a is equivalent to that shown in Figure 18.7b. We prefer schematic drawings like Figure 18.7b because of an aesthetic preference for straight lines and sharp corners in our diagrams.

Since all the leads from one side of each circuit element are connected together, they are all at the same electric potential. The same is true for the leads from the other side of each circuit element in parallel.

> Thus the potential difference *V* across each circuit element in a parallel connection is the same.

Connections Neither in Series nor in Parallel

The circuit elements in Figure 18.8 are connected neither in series nor in parallel. In other words, do not get the impression that everything must be in series or parallel!

We also use another notational convention when we picture on a flat page any three-dimensional arrangement of wires. If two wires or leads are connected together, a dot symbolizes the connection, as indicated to the left of Figure 18.9. If two wires are not connected but merely cross over each

---

*The electric current through each, a concept we introduce in the next chapter, also is unaffected by permuting the order of circuit elements in series.

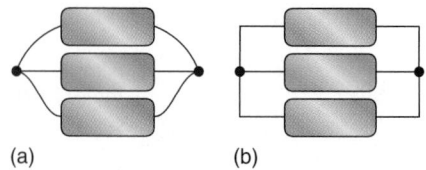

(a)	(b)

FIGURE 18.7 Two equivalent ways of drawing three circuit elements in parallel; that shown in (b) is preferred.

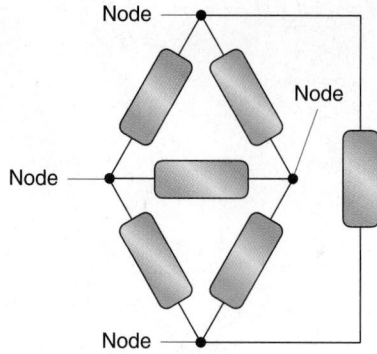

FIGURE 18.8 Six circuit elements connected neither in series nor in parallel.

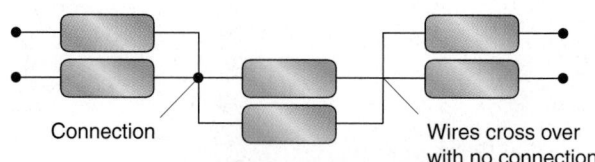

FIGURE 18.9 A dot symbolizes a connection of two or more wires. Two wires that cross over each other but are not connected have no dot drawn where they meet on the diagram. The dots on the far left and right indicate places where other circuit elements may be connected to the given collection.

FIGURE 18.10 Dots are omitted for a series connection.

other (like a freeway overpass), no dot is drawn, as indicated to the right of Figure 18.9. In a series connection of circuit elements, the dots typically are omitted since it is apparent that the leads are unambiguously connected together; see Figure 18.10.

The number of dots is *not* an indication of the number of nodes. For example, the two dots on each side of Figure 18.11a, symbolizing electrical connections, really are single nodes, as shown in Figure 18.11b. Both dots on the left (and right) side of Figure 18.11a are at the same electric potential. The extra ideal wire segment connecting the two dots ensures that they are at the same electric potential. Thus, be careful in counting the number of nodes; it is *not* necessarily the same as the number of dots in the diagram.

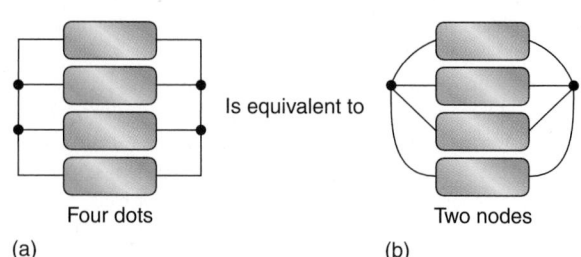

(a)	(b)

FIGURE 18.11 The number of dots may not be the same as the number of nodes.

Shorting Circuit Elements

Imagine a connection of circuit elements like that shown in Figure 18.12. Later, we shall see that if a wire is connected between two nodes that are at different electric potentials, such as in Figure 18.13, the wire very quickly makes the potential difference between the two nodes equal to zero. The act of doing this is called **shorting** the nodes, or shorting the circuit elements connected to those nodes.

When nodes are shorted, the previously distinct nodes reach the same electric potential and become a single node. Any circuit elements connected to two shorted nodes then must have zero potential difference between their terminals; such circuit elements are called **shorted out**. The two leads of a shorted circuit element now are connected to the *same* node.

The effect of shorting is to make any circuit element connected to the shorted nodes irrelevant because the leads of such a circuit element now are connected to the same node. There is 0 V potential difference between the terminals of a shorted circuit element. Shorted circuit elements can be removed with no effect on the remaining circuit elements, as indicated in Figure 18.14.

In Figure 18.14 the dots at (1) and (2) are obviously equivalent. Thus the extra wire connecting the dots at (1) and (2) in Figure 18.13 effectively obliterates any circuit elements connected to nodes (1) and (2) and renders them the same node. The connection in Figure 18.13 between what were distinct nodes (1) and (2) in Figure 18.12 is what is known as a **short circuit**; it means the new arrangement of circuit elements is simpler (shorter).

Notice that an ideal independent voltage source (a source of emf), which always maintains the potential difference V_0

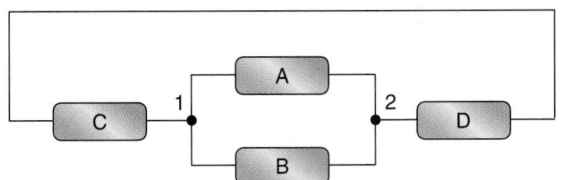

FIGURE 18.12 A connection of circuit elements.

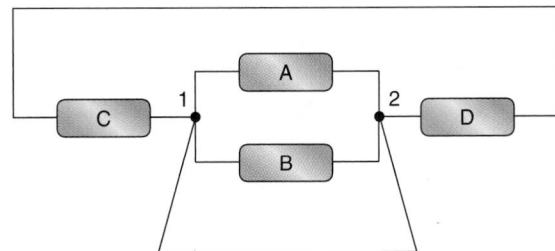

FIGURE 18.13 Shorting circuit elements A and B.

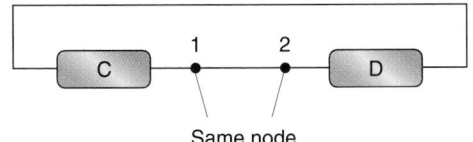

FIGURE 18.14 The circuit of Figure 18.13 with the shorted circuit elements removed.

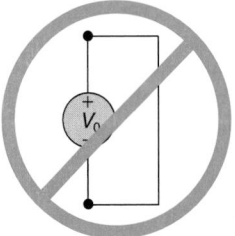

FIGURE 18.15 Never short-circuit an independent voltage source.

between its terminals, should *never* be shorted (see Figure 18.15) since it leads to a logical contradiction.*

QUESTION 3

In a carefully worded sentence or two, explain what is meant by the term node. How does the use of the word in this context differ from its use when associated with standing waves?

QUESTION 4

Carefully explain what is meant by the term shorting out a circuit element.

18.5 INDEPENDENT VOLTAGE SOURCES IN SERIES AND PARALLEL

Many times you likely have replaced the multiple batteries in a flashlight, radio, or portable CD player without thinking about it much, except for the bother. The multiple batteries in these electronic gadgets are connected in series. On the other hand, if you ever have had the misfortune to discover that your car battery is dead, and used jumper cables with the battery in a friend's car or service truck to start your car, the two batteries were connected

---

*For *real* independent voltage sources such as real batteries, shorting the terminals results in very large electric currents that may destroy the battery, even explosively. We shall see why when we examine real batteries in Chapter 19.

The batteries in this device are connected in series.

To jump start a car, the batteries are connected in parallel.

in parallel. Here we discover what happens when you connect independent voltage sources (sources of emf) in series or parallel.

If several independent voltage sources are placed in series, as shown in Figure 18.16, they can be replaced with a single equivalent independent voltage source (see Figure 18.17). The potential difference across the terminals of the equivalent independent voltage source is equal to the algebraic sum of the potential differences of the individual independent voltage sources.

For example, consider the series of independent voltage sources in Figure 18.16. The potential difference of the equivalent independent voltage source is found by starting at terminal A and going to terminal B: if the positive terminal of an independent voltage source is encountered first, the emf of that source is accounted for in the equation for the emf of the equivalent source with a plus sign; if the negative terminal is encountered first, that source of emf is entered into the equation for the emf of the equivalent source with a minus sign. So, for the series combination of Figure 18.16, the potential difference (or emf) of the equivalent independent voltage source is

$$V_{eq} = V_1 + V_2 + V_3$$

For the series combination shown in Figure 18.18, the equivalent independent voltage source has a potential difference (or emf) of

$$V_{eq} = V_1 + V_2 + (-V_3)$$

From a practical viewpoint, independent voltage sources connected in series are arranged as in Figure 18.16, with their polarity markings all oriented the *same* way.

You probably already know this. When multiple batteries are placed in flashlights, radios, or other electronic equipment, the batteries are inserted in series with the positive terminal of each one connected to the negative terminal of the next one in the series. This arrangement results in the greatest combined potential difference.

Independent voltage sources also occasionally are connected in parallel, as shown in Figure 18.19.

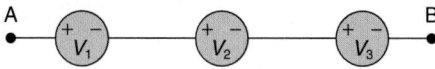

FIGURE 18.16 Independent voltage sources connected in series.

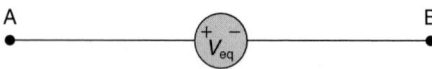

FIGURE 18.17 The equivalent independent voltage source of a series connection of such sources.

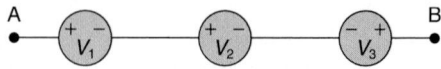

FIGURE 18.18 A possible (but impractical) series connection of independent voltage sources.

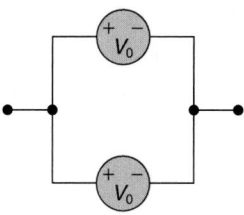

FIGURE 18.19 Independent voltage sources in parallel must have the same emf and must be oriented with similar polarities connected to the same node.

When in parallel, the individual sources *must* have the same emf (or potential difference) between their terminals and must be connected with negative terminals together at one node and positive terminals together at the other node.

The equivalent single independent voltage source has the *same* potential difference (or emf) as either of the sources in parallel.

Never connect two independent voltage sources the way shown in Figure 18.20, since this produces a logical impasse, with each source implying the other node is at the higher potential.

If ideal independent voltage sources with *different* potential differences (or emfs) are connected in parallel, we encounter another logical contradiction (see Figure 18.21). Each source implies a different potential difference between the same two terminals, and yet the potential difference across any two elements in parallel must be the same.

From a practical viewpoint, real batteries with identical emfs are connected in parallel to increase the charge flow (electric current) that the equivalent battery can provide.

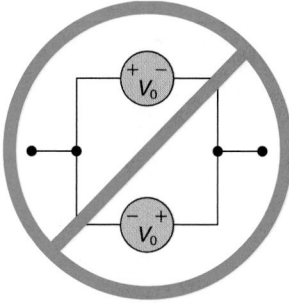

FIGURE 18.20 Never connect two independent voltage sources this way.

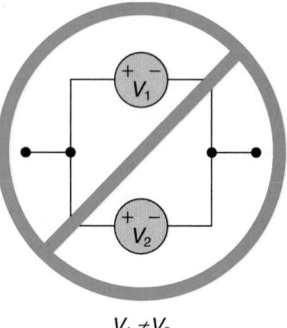

$$V_1 \neq V_2$$

FIGURE 18.21 Never connect in parallel two independent voltage sources with different emfs.

Thus, when your car is jump started, a good real battery is connected in parallel with the dead real battery. The good battery effectively compensates for, or replaces, the dead battery by providing the charge flow needed by the starter motor of the car. Both the good and dead batteries should have the same emf; if this advice is not heeded with real batteries, they may explode. We will see why when we model a real battery in the next chapter.

Real independent voltage sources (such as these batteries) come in a wide variety of sizes and emfs.

EXAMPLE 18.2

Four batteries, modeled as independent voltage sources, each with a potential difference (or emf) of 1.50 V between their terminals, are connected in series for a flashlight, as shown in Figure 18.22.

FIGURE 18.22

a. **What is the potential difference (or emf) of the equivalent battery?**
b. **If one of the batteries is accidentally connected backward (as in Figure 18.23), what is the potential difference (or emf) of the equivalent battery?**

FIGURE 18.23

Solution

a. In Figure 18.22, the batteries are connected in series with the positive terminals connected to the corresponding negative terminals of the next battery in the series. Moving from terminal A to terminal B, the potential difference

between A and B is the potential difference (or emf) of the equivalent battery V_{eq}:

$$V_{eq} = 1.50 \text{ V} + 1.50 \text{ V} + 1.50 \text{ V} + 1.50 \text{ V}$$
$$= 6.00 \text{ V}$$

b. In Figure 18.23, the potential difference between terminal A and B is the equivalent potential difference (or emf) V_{eq}:

$$V_{eq} = 1.50 \text{ V} + 1.50 \text{ V} + (-1.50 \text{ V}) + 1.50 \text{ V}$$
$$= 3.00 \text{ V}$$

This potential difference could be obtained with simply two of the batteries properly connected in series with their polarities correspondingly oriented, as shown in Figure 18.24.

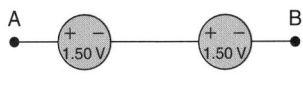

FIGURE 18.24

18.6 CAPACITORS

A **capacitor** is a circuit element that stores charge and electrical potential energy. It (typically) consists of two isolated (separated) conductors with equal and opposite charges, as shown in Figure 18.25. The two conductors of a capacitor are known as **plates** (thin, flat, circular disks) even though they often do not possess that geometry.

The circuit symbol for an ideal capacitor is shown in Figure 18.26. The straight line of the capacitor symbol represents the conductor that is, or will be, at a higher electrical potential than the second conductor, represented by the slightly curved line.

When the two conductors have equal and opposite charges $\pm|Q|$, the capacitor is said to be **charged**, in which case polarity markings are used to indicate which conductor is at the higher electric potential. If both conductors of the capacitor have zero charge, the capacitor is said to be **uncharged**, in which case the polarity markings are superfluous.

Some capacitors, typically electrolytic capacitors, have their polarity markings inscribed on them. Others, typically ceramic capacitors, do not have polarity markings and can be used with either terminal at the higher potential.

Notice that, whether charged or uncharged, the total electric charge on a capacitor as a whole is zero. A charged capacitor keeps opposite charges of equal magnitudes separated in space. For an isolated, ideal, charged capacitor, the charge separation can last indefinitely.

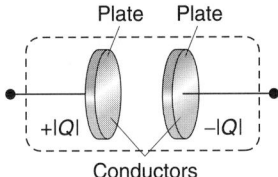

FIGURE 18.25 A capacitor consists of two separated conductors.

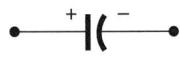

FIGURE 18.26 The circuit symbol for a capacitor.

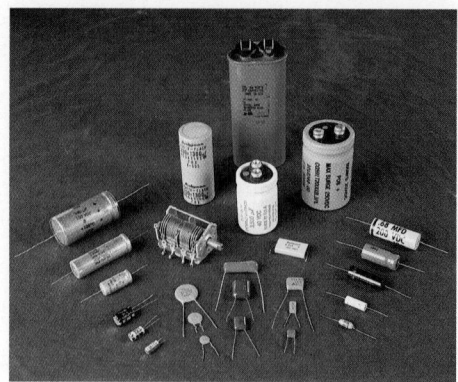

Some of these capacitors have polarity markings on them. When such capacitors are used, the (+) polarity should be at the higher electric potential.

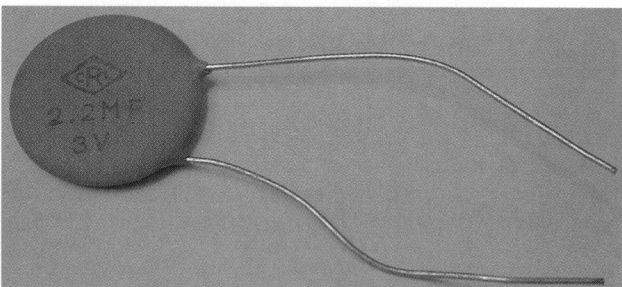

Some capacitors lack polarity markings, and either side may be at the higher electric potential when used.

There is a difference in electric potential between the conductors of a charged capacitor. Let the potential difference between the plate at the higher potential and the plate at the lower potential be V (≥ 0 V).

> The **capacitance** C of a capacitor is defined to be the ratio of the absolute value of the charge on either conductor to the absolute value of the potential difference between them:
>
> $$C \equiv \frac{|Q|}{|V|} \qquad (18.1)$$
>
> The capacitance always has a positive value.* The SI unit of capacitance is the coulomb/volt (C/V) and is renamed a **farad** (F).

The farad unit is named after Michael Faraday (1791–1867), an early 19th-century experimental genius who studied electrical and magnetic phenomena extensively. The farad (F) is a very large unit; most common capacitances are measured in microfarads (μF $= 10^{-6}$ F) or picofarads (pF $= 10^{-12}$ F), although it now is technologically possible to fabricate small-sized capacitors with capacitances exceeding a farad.

The capacitance of a capacitor is a measure of its capacity for holding (storing) charge. A large capacitance means the capacitor has the capacity to hold a significant amount of charge.

---

*The absolute value sign on the potential difference V is superfluous since we said V was the potential difference between the plate at the higher potential and the plate at the lower potential. We include the absolute value signs on V simply to emphasize that a positive value of the potential difference between the two plates is to be used to find the capacitance.

Ironically, despite the fact that Equation 18.1 defining the capacitance involves both Q and V, the value of the ratio is *independent* of either one! The potential difference between the conductors always is proportional to the charge on them (a manifestation of the principle of superposition), and so neither Q nor V independently affects C. The capacitance depends on the geometric arrangement of the conductors, the physical size and shape of the conductors, and the material medium separating them.

This is seen in the examples that follow.

QUESTION 5

Your roommate says that a charged capacitor carries a charge $|Q|$. You say the charged capacitor has zero charge. You both are right. Explain.

EXAMPLE 18.3

a. Find the capacitance of a large parallel plate capacitor whose plate areas each are A, separated by a distance d; see Figure 18.27. Similar capacitors frequently are used in electronic devices such as oscilloscopes, TVs, and other CRT displays.

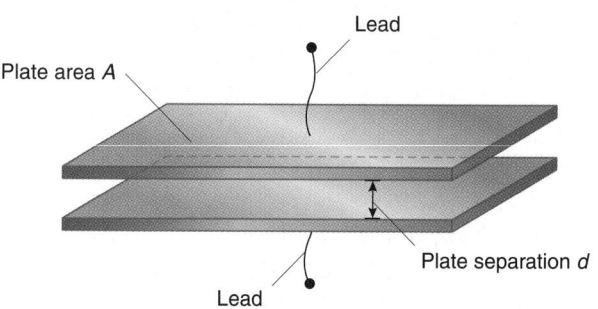

FIGURE 18.27

b. If the capacitor consists of two square plates, each 10.0 cm on an edge, separated by 0.500 cm, find the capacitance.

Solution

a. Imagine charge $+|Q|$ to be on one plate of the capacitor and charge $-|Q|$ on the other. Ignoring the small edge effects, the surface charge density σ on each plate has a magnitude $|Q|/A$. If the spacing d is much less than $\sqrt{A}$, the arrangement is tantamount to a set of parallel infinite plates. The electric field between two such plates is uniform and of magnitude $E = \sigma/\varepsilon_0$ (see Table 16.2). The potential difference between such an arrangement of plates has a magnitude $|V| = |Ed|$, from Example 17.1. The capacitance of the arrangement is, by definition (from Equation 18.1),

$$C = \frac{|Q|}{|V|}$$

Substituting $V = Ed$, you get

$$C = \frac{|Q|}{Ed}$$

Then, using $E = \sigma/\varepsilon_0$,

$$C = \frac{|Q|}{\dfrac{\sigma d}{\varepsilon_0}}$$

Finally, substituting for σ using $\sigma = |Q|/A$, you find

$$C = \frac{|Q|}{\dfrac{|Q|d}{A\varepsilon_0}}$$

$$= \varepsilon_0 \frac{A}{d} \qquad (1)$$

Equation (1) shows that the capacitance is a function of geometric specifications, such as the area of the plates and their separation. The dependence on the medium between the plates is represented through ε_0, the permittivity of free space, since the medium between the capacitor plates was assumed here to be a vacuum.

b. Use the given dimensions and substitute into equation (1), remembering to use SI units (m) rather than units of convenience (cm):

$$C = \frac{\left[8.85 \times 10^{-12} \ \text{C}^2/(\text{N} \cdot \text{m}^2)\right](0.100 \ \text{m})(0.100 \ \text{m})}{5.00 \times 10^{-3} \ \text{m}}$$

$$= 1.77 \times 10^{-11} \ \text{C/V}$$

$$= 17.7 \ \text{pF}$$

The SI units $\text{C}^2/(\text{N} \cdot \text{m}) = \text{C}^2/\text{J} = \text{C/V}$ are equivalent to farads (F).

EXAMPLE 18.4

A conducting sphere of radius R with a charge Q is a charged capacitor even though only one conductor is evident; see Figure 18.28. The second conductor is considered to be at infinity (with charge $-Q$). To a first approximation, the Earth itself can be considered to be such a capacitor (see Problem 18).

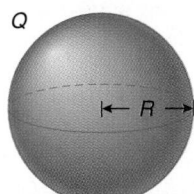

FIGURE 18.28

a. **Find the capacitance of the sphere.**
b. **Evaluate the capacitance for a sphere of radius 10.0 cm.**

Solution

a. The electric potential at infinity is 0 V. That on the surface of the conducting sphere is, from Table 17.1,

$$V = \frac{Q}{4\pi\varepsilon_0 R}$$

Thus the potential difference between the conductors has absolute value

$$|V| = \frac{|Q|}{4\pi\varepsilon_0 R}$$

Using Equation 18.1, you find the capacitance of the system is

$$C = \frac{|Q|}{|V|}$$

$$= \frac{|Q|}{\dfrac{|Q|}{4\pi\varepsilon_0 R}}$$

$$= 4\pi\varepsilon_0 R \qquad (1)$$

Once again you see that the capacitance depends on the geometric specifications of the capacitor (here the radius of the sphere) and the medium (a vacuum), the latter through the permittivity of free space ε_0.

b. For a sphere of radius 10.0 cm, $R = 0.100$ m, since you must use SI units. Making this substitution in equation (1) for the capacitance, you have

$$C = 4\pi\varepsilon_0 R$$

$$= [4\pi \times 8.85 \times 10^{-12} \ \text{C}^2/(\text{N} \cdot \text{m}^2)](0.100 \ \text{m})$$

$$= 1.11 \times 10^{-11} \ \text{F}$$

$$= 11.1 \ \text{pF}$$

From equation (1), note that the permittivity of free space, ε_0, has the equivalent SI units of farads per meter (F/m).

EXAMPLE 18.5

A cylindrical capacitor consists of two concentric conducting cylinders of radii a and b, as shown in Figure 18.29. A Geiger counter, used frequently in nuclear physics to detect subatomic particles, typically has this geometry. Find the capacitance assuming the length ℓ of the cylinders is much greater than either radius.

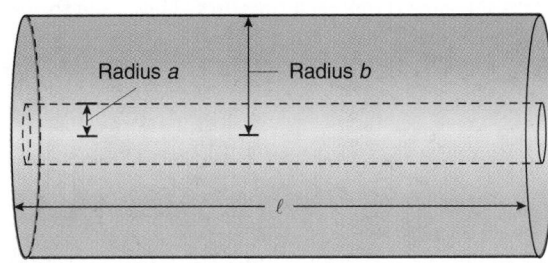

FIGURE 18.29

Solution

Model the system as a pair of infinite concentric cylinders. Imagine a charge of $+|Q|$ on the inner conductor and of $-|Q|$ on the outer conductor. The field between the two is approximately that of an infinitely long wire, and so you can use Equation 16.19 in vector form:

$$\vec{E} = \frac{1}{4\pi\varepsilon_0} \frac{2\lambda}{r} \hat{r}$$

where λ is the charge per unit length $\lambda = |Q|/\ell$.

Since the model charge distribution is not of finite extent, you must use Equation 17.9 to find the potential difference between the two cylinders:

$$V_{outer} - V_{inner} = -\int_a^b \vec{E} \cdot d\vec{r}$$

$$= -\frac{1}{4\pi\varepsilon_0} \int_a^b \frac{2\lambda}{r} \hat{r} \cdot dr\, \hat{r}$$

$$= -\frac{2}{4\pi\varepsilon_0} \lambda \ln\left(\frac{b}{a}\right)$$

The magnitude of the potential difference between the conductors is

$$|V| = \frac{2}{4\pi\varepsilon_0} \lambda \ln\left(\frac{b}{a}\right)$$

$$= \frac{2}{4\pi\varepsilon_0} \frac{|Q|}{\ell} \ln\left(\frac{b}{a}\right)$$

The capacitance is found using the definition, Equation 18.1:

$$C = \frac{|Q|}{|V|}$$

$$= \frac{4\pi\varepsilon_0 \ell}{2 \ln\left(\frac{b}{a}\right)}$$

18.7 SERIES AND PARALLEL COMBINATIONS OF CAPACITORS

It is occasionally useful to connect capacitors in parallel or series to secure larger or smaller capacitances.

Parallel Combinations of Capacitors

Consider a collection of, say, four capacitors in parallel, as shown in Figure 18.30. Since the leads of circuit elements in parallel are connected to the same two nodes, the potential difference is the same across each capacitor. Thus, from the definition of the capacitance (Equation 18.1), we can write the following for each capacitor:

$$C_1 = \frac{|Q_1|}{|V|} \qquad C_2 = \frac{|Q_2|}{|V|} \qquad C_3 = \frac{|Q_3|}{|V|} \qquad C_4 = \frac{|Q_4|}{|V|}$$

(18.2)

To simplify the arrangement, we want to find a single equivalent capacitor of capacitance C_{eq} that can be placed between the same two terminals (A and B) and maintain the same potential difference between them as the four given capacitors. That is, we want to replace the four capacitors with a single capacitor C_{eq} such that

$$C_{eq} = \frac{|Q|}{|V|}$$

where $|V|$ is the same potential difference that is across each of the individual capacitors; see Figure 18.31.

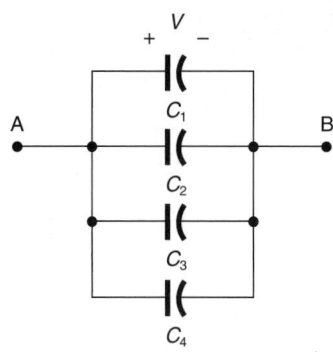

FIGURE 18.30 Four capacitors connected in parallel.

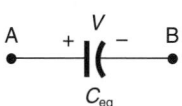

FIGURE 18.31 The equivalent capacitance.

Since all the conducting plates on one side of the capacitors in Figure 18.30 are connected by conducting wires, the total charge on one side of the equivalent capacitor must be the sum of the charges on the individual capacitor plates; the same is true for the other side of the capacitors. That is, the magnitude of the charge $|Q|$ on each plate of the equivalent capacitor must be

$$|Q| = |Q_1| + |Q_2| + |Q_3| + |Q_4|$$

Substituting for the charges using Equation 18.2, we obtain

$$C_{eq}|V| = C_1|V| + C_2|V| + C_3|V| + C_4|V|$$

and so

$$C_{eq} = C_1 + C_2 + C_3 + C_4 \qquad \text{(capacitors in parallel)} \quad (18.3)$$

> The equivalent capacitance is the simple sum of the individual capacitances.

Since the capacitance always is positive, we have no need to worry about signs when using Equation 18.3. The form of Equation 18.3 applies to any number of capacitors connected in parallel; we simply add the capacitances:

$$C_{eq} = C_1 + C_2 + C_3 + \cdots \qquad \text{(capacitors in parallel)} \quad (18.4)$$

Series Connection of Capacitors

Four capacitors in series are shown in Figure 18.32. Let their capacitances be $C_1, C_2, C_3,$ and C_4. Let V_1 be the potential difference between the points A and D in Figure 18.32; this is the potential

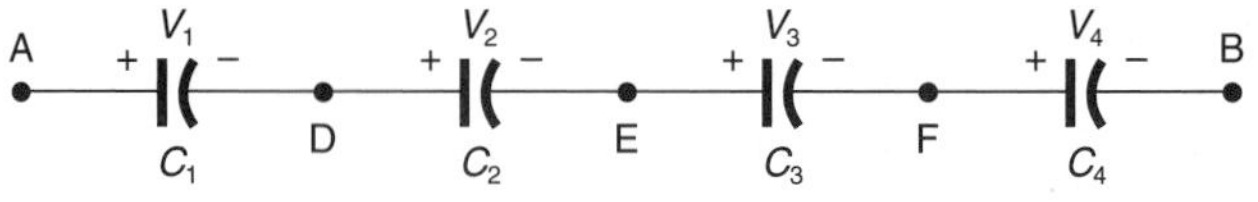

![Four capacitors in series with labels A, V_1, C_1, D, V_2, C_2, E, V_3, C_3, F, V_4, C_4, B]

FIGURE 18.32 Four capacitors in series.

difference across capacitor C_1. Likewise, let V_2 be the potential difference between points D and E in Figure 18.32; and so on for the other two capacitors. The total potential difference V between the points A and B is the sum of the potential differences across each capacitor:

$$V = V_1 + V_2 + V_3 + V_4 \qquad (18.5)$$

Once again we want to find a single equivalent capacitor C_{eq} that can be placed between points A and B to maintain the same potential difference V between the two points; see Figure 18.33.

A charge $+|Q|$ placed on the left plate of capacitor C_1 in Figure 18.32 causes charge separation to take place on the isolated and neutral extended conductor that comprises the right-side plate of C_1 and the left-side plate of capacitor C_2. The charge $+|Q|$ on the left side of C_1 attracts a charge $-|Q|$ to the right side of C_1 and repels $+|Q|$ to the left side of C_2. This charge separation ripples through the series of capacitors. In this way we see that each capacitor in the series collection has the *same* magnitude charge on each plate.

The left and right plates of the equivalent capacitor have the same charge as the left plate of C_1 and the right plate of C_4 because they are connected to the points A and B, respectively, in Figures 18.32 and 18.33. Thus the equivalent capacitor and the four individual capacitors have the same magnitude of charge $|Q|$ on each of their plates.

From the definition of the capacitance, Equation 18.1, we have

$$C_{eq} = \frac{|Q|}{|V|}$$

and

$$C_1 = \frac{|Q|}{|V_1|} \qquad C_2 = \frac{|Q|}{|V_2|} \qquad C_3 = \frac{|Q|}{|V_3|} \qquad C_4 = \frac{|Q|}{|V_4|}$$

Using these equations, substitute into Equation 18.5; we find

$$\frac{|Q|}{|C_{eq}|} = \frac{|Q|}{C_1} + \frac{|Q|}{C_2} + \frac{|Q|}{C_3} + \frac{|Q|}{C_4}$$

Dividing by $|Q|$, we have

$$\frac{1}{C_{eq}} = \frac{1}{C_1} + \frac{1}{C_2} + \frac{1}{C_3} + \frac{1}{C_4} \qquad \text{(capacitors in series)}$$

$$(18.6)$$

The reciprocal of the equivalent capacitance is the sum of the reciprocals of the individual capacitances.

This rule applies to any number of capacitors connected in series:

$$\frac{1}{C_{eq}} = \frac{1}{C_1} + \frac{1}{C_2} + \frac{1}{C_3} + \cdots \qquad \text{(capacitors in series)} \quad (18.7)$$

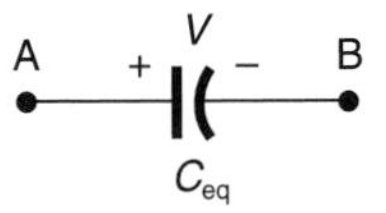

FIGURE 18.33 The equivalent capacitance.

One implication of Equation 18.7 for capacitors in series is that the equivalent capacitance always is less than the smallest capacitor in the series combination.

PROBLEM-SOLVING TACTICS

18.1 When using Equation 18.7, be sure you remember how to add fractions. In particular, for two capacitors in series, note that

$$\frac{1}{C_1} + \frac{1}{C_2} \neq \frac{1}{C_1 + C_2}$$

In other words, don't do crazy things like saying

$$\frac{1}{2} + \frac{1}{3} = \frac{1}{5} \qquad \text{(no!)}$$

This may seem like a trivial point and an insult to your intelligence, but you would be surprised at what is done under the pressure of an examination!

18.2 Be sure to realize that after summing the reciprocals of the individual capacitances in Equation 18.7 that the equivalent capacitance is the reciprocal of the sum. You might correctly add the reciprocals of the individual capacitances but then forget to use the $1/x$ key on your calculator to find the final result for C_{eq}.

18.3 The equivalent capacitance of two capacitors in series is the product of their capacitances divided by the sum. This convenient rule follows from Equation 18.7 applied to two capacitors. The equivalent capacitance is (using Equation 18.7)

$$\frac{1}{C_{eq}} = \frac{1}{C_1} + \frac{1}{C_2}$$

Put the right-hand side over a common denominator and simplify:

$$\frac{1}{C_{eq}} = \frac{C_2 + C_1}{C_1 C_2}$$

or

$$C_{eq} = \frac{C_1 C_2}{C_1 + C_2} \qquad \text{(only two capacitors in series)} \quad (18.8)$$

This is an easier way to calculate the equivalent capacitance because the product divided by the sum usually can be performed mentally. But be aware that this rule (Equation 18.8) is appropriate only for finding the equivalent of *two* capacitors at a time in series. For three capacitors in series, you can either revert to Equation 18.7 or combine them pairwise (see Example 18.6).

QUESTION 6

Three capacitors are to be connected together to produce various equivalent capacitances for a medical defibrillation device. (a) How should they be connected together to produce as large an equivalent capacitance as possible? (b) How should the three be connected to produce the minimum equivalent capacitance?

STRATEGIC EXAMPLE 18.6

You find three capacitors in your electronics shop inventory, 3.0 μF, 6.0 μF, and 8.0 μF, but they all are too large for your purposes. You combine them in series to make a smaller capacitance. Find the equivalent capacitance of the three capacitors in series (see Figure 18.34).

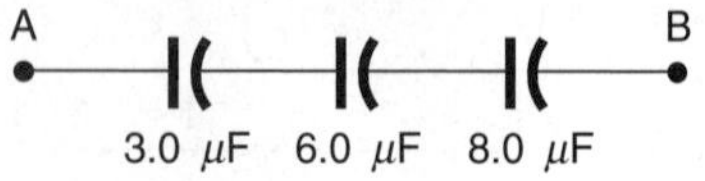

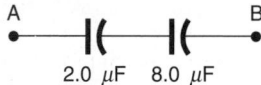

FIGURE 18.34

Solution
Method 1

You can use the shortcut mentioned in Problem-Solving Tactic 18.3 to combine the two capacitors in series on the left side of Figure 18.34. They have an equivalent capacitance given by the product of their capacitances divided by their sum:

$$\frac{(3.0 \ \mu\text{F})(6.0 \ \mu\text{F})}{3.0 \ \mu\text{F} + 6.0 \ \mu\text{F}} = 2.0 \ \mu\text{F}$$

Note that the equivalent capacitance of the two in series is less than the smallest one that was combined. You now have a collection of two capacitors in series, as shown in Figure 18.35.

FIGURE 18.35

This pair can then be combined using the same rule (product divided by the sum):

$$\frac{(2.0 \ \mu\text{F})(8.0 \ \mu\text{F})}{2.0 \ \mu\text{F} + 8.0 \ \mu\text{F}} = 1.6 \ \mu\text{F}$$

You also first could combine the 6.0 μF and 8.0 μF capacitors as a pair, to get 3.4 μF. Then combining the resulting 3.4 μF capacitor in series with the remaining 3.0 μF capacitor, you get 1.6 μF. In other words, for capacitors in series, you can combine them by pairs beginning with *any two* of them; they need not be taken in the order in which they appear in the series connection.

Method 2

Use Equation 18.7 for the original arrangement of three capacitors in series (Figure 18.34):

$$\frac{1}{C_{eq}} = \frac{1}{C_1} + \frac{1}{C_2} + \frac{1}{C_3}$$

$$= \frac{1}{3.0 \ \mu\text{F}} + \frac{1}{6.0 \ \mu\text{F}} + \frac{1}{8.0 \ \mu\text{F}}$$

$$= \frac{15}{24 \ \mu\text{F}}$$

Don't forget to take the reciprocal (Problem-Solving Tactic 18.2):

$$C_{eq} = \frac{24}{15} \ \mu\text{F}$$

$$= 1.6 \ \mu\text{F}$$

This is the same result as that obtained by Method 1.

EXAMPLE 18.7

You need a 250 μF capacitor for manufacturing each of a collection of laser power supplies, but no 250 μF capacitors are in stock or available except by special order at significant cost. An electronics supply firm is having a special on 50 μF capacitors and you realize it will be cheaper to use several of them rather than a single 250 μF capacitor even after accounting for the additional time for assembly. How many will you need for each power supply and how should the capacitors be connected to secure the desired 250 μF capacitance? Your supervisor is well pleased with your cost-consciousness.

Solution

Connect the capacitors in parallel because then the equivalent capacitance is the simple sum of the individual capacitances. You will need five of the 50 μF capacitors:

$$C_{eq} = 50 \ \mu\text{F} + 50 \ \mu\text{F} + 50 \ \mu\text{F} + 50 \ \mu\text{F} + 50 \ \mu\text{F}$$
$$= 250 \ \mu\text{F}$$

EXAMPLE 18.8

One part of a complicated circuit diagram has the combination of capacitors indicated in Figure 18.36. Find the equivalent capacitance of the three capacitors.

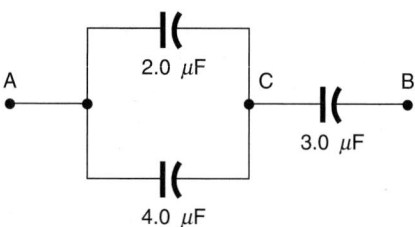

FIGURE 18.36

Solution

The 2.0 μF and 4.0 μF capacitors are in parallel and thus have an equivalent capacitance equal to their sum:

$$2.0 \ \mu\text{F} + 4.0 \ \mu\text{F} = 6.0 \ \mu\text{F}$$

The 6.0 μF equivalent is in series with the 3.0 μF capacitor, as shown in Figure 18.37. The final equivalent capacitance of this pair in series is their product divided by their sum:

$$C_{eq} = \frac{(6.0 \ \mu\text{F})(3.0 \ \mu\text{F})}{6.0 \ \mu\text{F} + 3.0 \ \mu\text{F}}$$
$$= 2.0 \ \mu\text{F}$$

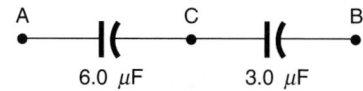

FIGURE 18.37

18.8 ENERGY STORED IN A CAPACITOR

One widespread use of capacitors is to store electrical potential energy for sudden use in devices such as flash attachments to cameras, laser power supplies, or defibrillators. Here we see how the energy storage arises.

Capacitors do not charge themselves. To take a capacitor from an uncharged to a charged state, work must be done to accomplish the charge separation. We want to figure out how much work must be done to charge a capacitor.

Represent a capacitor schematically as shown in Figure 18.38. Since the location of the zero of electric potential (and potential energy) is arbitrary (it is only changes that are physically significant), we choose plate B to be at zero potential and let plate A have a potential V. In this way V represents the potential difference between plates A and B. If we assume $V > 0$ V, the electric field is directed from plate A to plate B because the electric field always points toward regions of lower electric potential.

Plate A has a positive charge and plate B has a negative charge. Imagine slowly carrying a small, differential positive charge dq from plate B to plate A. The work done by the electrical forces on dq will be independent of the path, since the force is conservative. Obviously, though, we are going to have to do work on the charge to get it over to plate A, or something else is going to have to do the work for us. The charge dq will not go of its own accord since the electrical force on it is directed toward plate B, not toward plate A.

Imagine pulling the charge over to plate A so the kinetic energy of the charge is unchanged. The force we have to exert on the charge then is equal in magnitude to the electrical force on dq but, of course, must point in the opposite direction.* The familiar CWE theorem states that

$$W_{\text{total}} = \Delta KE$$

---

*You might ask yourself if the equality of the force magnitudes stems from Newton's third law or from the second law.

or for differential changes,

$$dW_{\text{total}} = d(KE)$$

There is no change in the kinetic energy because we move the charge slowly at constant speed. We do differential work dW_{us} and the electrical force on dq does differential work dW_{elec} on the charge. Thus the CWE theorem has the form

$$dW_{\text{us}} + dW_{\text{elec}} = 0 \text{ J} \qquad (18.9)$$

But the work done by the electrical force is the negative of the change in the electrical potential energy of the charge. We take differentials of Equation 17.3, obtaining

$$dW_{\text{elec}} = -d(PE)$$

Thus Equation 18.9 becomes

$$dW_{\text{us}} - d(PE) = 0 \text{ J}$$

or

$$dW_{\text{us}} = d(PE)$$

The differential work we do is equal to the differential change in the potential energy of the charge. But the potential energy of any charge is the product of the charge times the electric potential at the point in question (Equation 17.8). The initial potential energy of dq at plate B is

$$dq(0 \text{ V}) = 0 \text{ J}$$

and the final potential energy of dq at plate A is

$$dq\, V$$

So the change in the potential energy of the charge (final minus initial values, as usual) is

$$d(PE) = V\, dq - 0 \text{ J}$$
$$= V\, dq \qquad (18.10)$$

From the definition of the capacitance, Equation 18.1, the potential difference across the capacitor is related to the charge on the plates by

$$C = \frac{|q|}{|V|}$$

Since both q and V are positive quantities here, we can dispense with the absolute value signs. Solving for V, we get

$$V = \frac{q}{C}$$

Thus Equation 18.10 becomes

$$d(PE) = \frac{q}{C}\, dq$$

Now we integrate this expression from an uncharged state to a final charge Q on the capacitor. The capacitance is a constant for a given capacitor and depends neither on q nor on V, so that we have

$$\int_{0 \text{ J}}^{PE} d(PE) = \int_{0 \text{ C}}^{Q} \frac{q}{C}\, dq$$

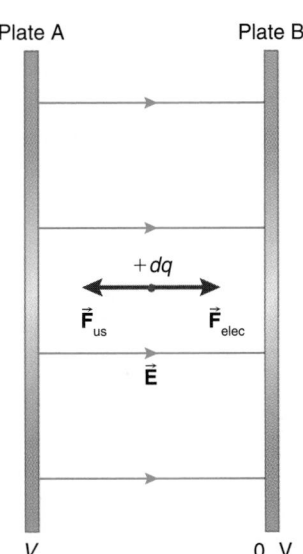

FIGURE 18.38 A schematic representation of a capacitor.

Take the initial electrical potential energy of the uncharged capacitor to be 0 J. After integration, the accumulated potential energy of the charged capacitor is

$$PE = \frac{1}{2}\frac{Q^2}{C} \qquad (18.11)$$

Hence in charging a capacitor the electrical potential energy of the capacitor system increases.

With the definition of the capacitance, Equation 18.11 can be written in a number of equivalent ways:

$$PE = \frac{1}{2}\frac{Q^2}{C} = \frac{1}{2}|Q||V| = \frac{1}{2}CV^2 \qquad (18.12)$$

Only one version of Equation 18.12 need be remembered, since the others can be found from it using the definition of the capacitance.

Consider the special case of a parallel plate capacitor. The capacitance is (from Example 18.3)

$$C = \varepsilon_0 \frac{A}{d}$$

where A is the area of the plates and d is their separation. The magnitude of the potential difference between the plates is related to the magnitude of the electric field between the plates via

$$V = Ed$$

Substituting into the expression for the electrical potential energy stored in the charged capacitor, Equation 18.12, we obtain

$$PE = \frac{1}{2}CV^2$$
$$= \frac{1}{2}\varepsilon_0\frac{A}{d}(Ed)^2$$
$$= \frac{1}{2}\varepsilon_0 E^2 Ad$$

Where is the potential energy stored? We consider it to be distributed throughout the volume of the space wherever there is some electrical field. The product Ad is the volume of the capacitor. If we divide the potential energy by the volume of the capacitor, we have a potential energy per unit volume or an **energy density**. The energy density is then

$$\frac{1}{2}\varepsilon_0 E^2$$

The electrical potential energy density stored in the capacitor is considered to be stored in the electric field.

This result turns out to be quite general:

Any electric field $\vec{E}$ is said to have associated with it a potential energy density given by

$$\frac{1}{2}\varepsilon_0 E^2 \qquad (18.13)$$

The energy stored in the field of a charged capacitor is quite apparent if the capacitor is discharged suddenly by connecting a wire from one plate (terminal) to the other, thus shorting the

capacitor.* The wire provides a path for the charges to decrease their electrical potential energy. Indeed, large capacitors used in common electronic instruments and devices are quite dangerous; if your fingers accidentally become the "wire" that shorts the terminals of a charged capacitor, you can receive quite an electrical shock, even a lethal one if sufficient energy is stored in the charged capacitor. A good capacitor can hold its charge for hours after a piece of equipment is shut off, and so the danger of electrocution is not diminished just because the device is turned off. In brief, *extreme caution* is needed to ensure that all large capacitors are first discharged by connecting a wire to bridge the two terminals (shorting the terminals momentarily) before servicing any piece of electronic equipment.

EXAMPLE 18.9

One of several 100 μF capacitors in a defibrillator has a potential difference of 220 V between its plates.

a. What is the magnitude of the charge on each plate of the capacitor?
b. What is the energy stored in the capacitor?
c. If this energy were fully utilized to lift a 100 g doughnut with no change in its kinetic energy, how high could the doughnut be raised?

Solution
a. From the definition of the capacitance (Equation 18.1), you have

$$C = \frac{|Q|}{|V|}$$

and so

$$|Q| = C|V|$$
$$= (100 \times 10^{-6}\ \text{F})(220\ \text{V})$$
$$= 2.20 \times 10^{-2}\ \text{C}$$

b. The energy stored is found from Equation 18.12:

$$PE = \frac{1}{2}CV^2$$
$$= \frac{1}{2}(100 \times 10^{-6}\ \text{F})(220\ \text{V})^2$$
$$= 2.42\ \text{J}$$

c. To raise a mass through a vertical distance Δy takes work $mg\,\Delta y$, where $\hat{j}$ is vertically up. Hence

$$mg\,\Delta y = 2.42\ \text{J}$$

Substituting values, you get

$$(0.100\ \text{kg})(9.81\ \text{m/s}^2)\,\Delta y = 2.42\ \text{J}$$

which gives

$$\Delta y = 2.47\ \text{m}$$

EXAMPLE 18.10

An isolated, charged, parallel plate capacitor of plate area A and plate separation d has opposite charges of magnitude

---

*We investigate this discharging phenomenon in Section 19.16.

$|Q|$ on its plates. Now increase the separation of the plates to $2d$.

a. By what factor does the electrical potential energy stored in the capacitor change?
b. Where does the increase in potential energy come from, or the decrease go to?

Solution

a. The capacitance of a parallel plate capacitor is, from Example 18.3,

$$C_{\text{parallel plate}} = \varepsilon_0 \frac{A}{d}$$

When the separation between the plates is increased to $2d$, the capacitance falls to half its previous value.

The energy stored in the capacitor is found from Equation 18.12:

$$PE = \frac{1}{2} \frac{Q^2}{C}$$

When the separation between the plates of the isolated capacitor is increased, the charge on the plates is unaffected since the capacitor is isolated. Thus, with $|Q|$ fixed and the capacitance decreasing to half its former value, the potential energy stored increases by a factor of two.

b. The plates of the capacitor have equal and opposite charges on them. They therefore attract each other under the influence of the electrical force. Thus, in order to increase the plate separation, *you* are going to have to pull the plates apart. The work you do to separate the plates is the source of the increase in the electrical potential energy of the system.

The problem is analogous to increasing the gravitational potential energy of a mass. To increase the gravitational potential energy of your physics book, you have to lift the book to a higher elevation. The work you do is equal to the increase in the gravitational potential energy of the text, provided the work is done so that the kinetic energy of the text does not change.

18.9 ELECTROSTATICS IN INSULATING MATERIAL MEDIA*

In our study of electrostatics we have assumed that the charges were located in a vacuum. Indeed, the constant ε_0 that appears in so many of our equations is known as the permittivity of free space; the term free space is the electromagnetic buzzword for a vacuum. The equations for Coulomb's law for point charges, the electric field and electric potential of various charge distributions, Gauss's law, and the explicit expressions we calculated for the capacitance early in this chapter—all involve the constant ε_0, the permittivity of free space.

What happens to all these equations if the electrostatic charges are located not in a vacuum but in an insulating material medium?

Empirically, in many insulating materials we find that all the changes can be accounted for by replacing the permittivity of free space ε_0 with a multiple of it[†]:

$$\varepsilon = \kappa\varepsilon_0 \qquad (18.14)$$

where ε is the **permittivity of the material**.

So all the equations look identical except that ε appears instead of ε_0. The pure number κ (not necessarily an integer) is called the **dielectric constant** of the material. Recall from Chapter 16 that a dielectric[‡] is the technical term for a material that is commonly called an insulator. The dielectric constant of a vacuum is exactly 1. Table 18.2 gives the value of the dielectric constant for various materials. The dielectric constant for air (under standard conditions) is very close to 1, indicating that the fundamental electrical equations in air and in a vacuum do not differ appreciably.

The specific value of the dielectric constant κ depends not only on the material but also on a host of environmental factors such as temperature and pressure, and in electrodynamic situations other factors such as the frequency of the oscillating fields. In this respect the dielectric constant is analogous to the specific heat in thermodynamics: changes in the dielectric constant are macroscopic manifestations of submicroscopic (atomic or molecular) changes in the material. Thus studies of the dielectric constants of materials and their variation with temperature, pressure, and other parameters are subjects of ongoing interest to physicists, engineers, chemists, and biologists.

[†] In anisotropic materials such as certain crystals, the relationship is more complicated; we will not consider such materials here.
[‡] The term dialectic is a philosophical term, unrelated to the term dielectric.

TABLE 18.2 Approximate Values of the Dielectric Constant and Dielectric Strength of Various Materials

Material	Dielectric constant κ (dimensionless) at $\approx 20\,°C$	Dielectric strength (N/C = V/m)
Air	1.0006	3×10^6
Aluminum oxide	8.5	670×10^6
Bakelite	4.9	24×10^6
Barium titanate ($BaTiO_3$)	$\sim 10^4$	
Epoxy resin	~ 4	
Ethyl alcohol	26	
Lucite	~ 3	
Mica	~ 5	$\sim 10–50 \times 10^6$
Paper (kraft)	3.7	16×10^6
Paraffin	~ 2.5	
Polyethylene	~ 2.3	50×10^6
Pyrex glass	5.6	14×10^6
Quartz	~ 3.8	8×10^6
Rubber	~ 3	
Silicone oil	2.5	15×10^6
Tantalum oxide	26	500×10^6
Teflon	2.1	60×10^6
Transformer oil	~ 5	$\sim 10 \times 10^6$
Vacuum	1 (exact)	∞
Water	80	

18.10 CAPACITORS AND DIELECTRICS*

Most capacitors in electronic applications have dielectric materials between the two conductors. In this section we see that the effect of completely filling the space between the plates of a capacitor with an insulating material of dielectric constant κ is to increase the capacitance by the factor κ over that with a vacuum between the plates. We saw previously that the capacitance depends on the geometry of the capacitor (in Examples 18.3–18.5) and the material between the conductors; we explore the latter topic in this section. From a practical viewpoint, the insulating material also conveniently provides some structural rigidity to the capacitor.

How does the increase in the capacitance come about with a dielectric between the plates? Insulating materials are classified into two broad families depending on the nature of the atoms or molecules of the material: polar and nonpolar materials.

Polar Materials

Polar materials have molecules with a permanent electric dipole moment. Water is such a molecule. The center of charge (analogous to the center of mass) of the positive charges in the molecule is at a different location than the center of charge for the negative charges in the molecule. This separation of the centers of the two types of charges means that the molecule acts like a small electric dipole (see Figure 16.63).

Recall that the potential energy of a dipole in an electric field is $PE = -\vec{p} \cdot \vec{E}$. When such molecules are placed in an electric field, they tend to orient themselves with the dipole moment parallel to the field in order to lower their electrical potential energy. The alignment is not complete or perfect because of competing thermal motions of the molecules, but increasing the electric field increases the degree of alignment of the molecules with the field.

Nonpolar Atoms or Molecules

Nonpolar materials have no permanent electric dipole moment; the centers of charge for the positive and negative charges in the molecule normally coincide. However, if the material is placed in an electric field, the (much less massive) negatively charged electrons and their center of charge are displaced slightly with respect to the center of the positive charge distribution (formed from the much more massive, less mobile, positively charged nuclei). The result is an induced electric dipole moment in the molecules of the material.

Now we perform two different experiments.

Experiment 1: Adding a Dielectric to a Capacitor

We begin with an isolated and charged parallel plate capacitor with air or a vacuum between its plates, as shown in Figure 18.39.

Let $+Q_0$ be the charge on one plate, $-Q_0$ the charge on the other plate (we consider Q_0 itself to be positive). The surface charge density on the plates then is of magnitude $\sigma = Q_0/A$. The surface charge density on the plates is the **free surface charge density** because the charges on the conducting plates are free to move.

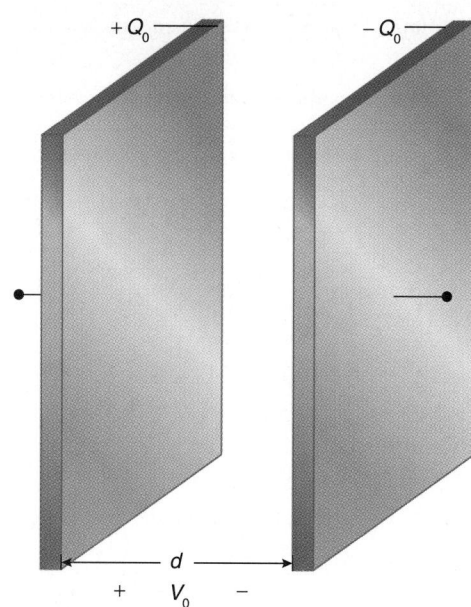

FIGURE 18.39 **An isolated, charged parallel plate capacitor with air or a vacuum between its plates.**

Let V_0 be the potential difference between the plates, considered so $V_0 > 0$ V. Since the capacitor is isolated, the charge on each plate is fixed. The capacitance of the system is

$$C_0 = \varepsilon_0 \frac{A}{d}$$

as we found before (in Example 18.3). The magnitude E_0 of the electric field between the plates is related to the potential difference by

$$V_0 = E_0 d$$

as we have used on a number of occasions.

Now insert a dielectric slab (with dielectric constant κ), completely filling the space between the plates of the capacitor, as shown in Figure 18.40.

The electric field orients the dipoles of a polar dielectric or induces dipoles in the nonpolar dielectric, resulting in the situation depicted schematically in Figure 18.41. On the surfaces of the dielectric facing the plates of the capacitor lie some unbalanced charges: negative charges on the dielectric surface facing the capacitor plate with charge $+Q_0$, and positive charges on the dielectric surface facing the plate with charge $-Q_0$. These unbalanced charges induced on the surfaces of the dielectric are not free to move around, since they are tied to the molecules (if the charges were free to move, we would have a conductor, not an insulator). These charges on the surface of the dielectric are called **bound charges**; there are **bound surface charge densities** $\pm\sigma_{bound}$ on the surfaces of the dielectric facing the capacitor plates.

The bound surface charges produce an electric field of their own in the direction opposite to the original field of the charges on the conducting plates of the capacitor. The electric field of the bound surface charges is of magnitude $\sigma_{bound}/\varepsilon_0$ while that due to the free charges on the capacitor plates is of magnitude σ/ε_0.

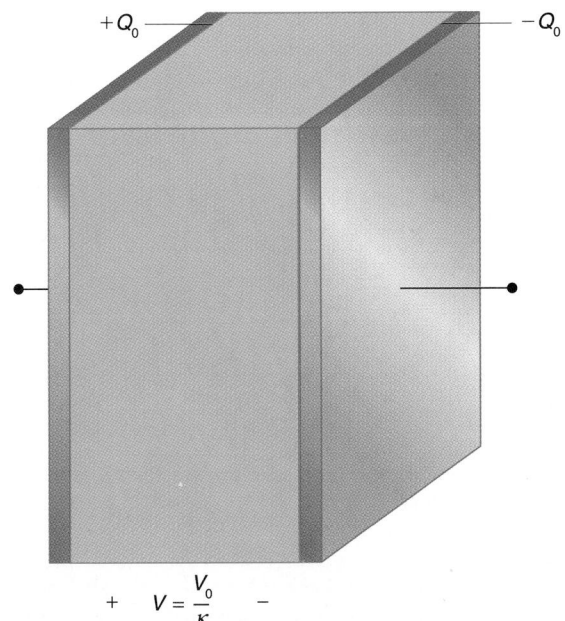

FIGURE 18.40 A slab of dielectric constant κ completely fills the space between the plates of the capacitor.

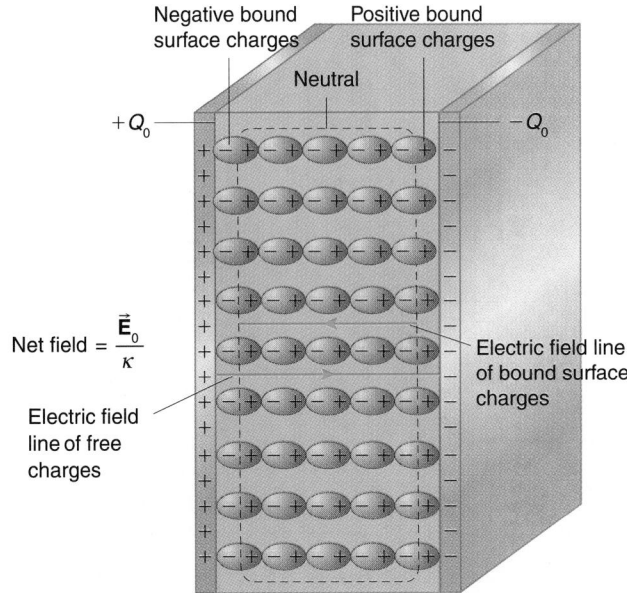

FIGURE 18.41 The free and bound charges each produce an electric field within the dielectric.

Experimentally we find that the potential difference between the capacitor plates is decreased by the factor κ by the presence of the dielectric material:

$$V = \frac{V_0}{\kappa}$$

Since the plate separation is unchanged, the field between the plates must decrease by the same factor κ since $V_0 = E_0 d$:

$$V = \frac{E_0}{\kappa} d \qquad (18.15)$$

The total electric field between the plates of the capacitor is evidently reduced by the factor of the dielectric constant:

$$E = \frac{E_0}{\kappa}$$

The charge Q_0 on the conducting plates of the capacitor is unaffected by all this because the capacitor was isolated. The capacitance of the arrangement is

$$C = \frac{Q_0}{V}$$

Substituting for V using Equation 18.15, we find

$$C = \frac{Q_0}{\frac{E_0 d}{\kappa}}$$

$$= \kappa \frac{Q_0}{E_0 d}$$

$$= \kappa \frac{Q_0}{V_0}$$

$$= \kappa C_0$$

This is a remarkable result: the capacitance increases by the factor of the dielectric constant κ.

We also can find the relationship between the free and bound surface charge densities to see which one is greater. The total electric field between the plates is of magnitude E_0/κ, and this is the difference between the magnitude of the field of the free charges and that of the bound charges because these fields are in opposite directions:

$$E_0 - E_{\text{bound}} = \frac{E_0}{\kappa}$$

Writing the field magnitudes in terms of the surface charge densities, we have

$$\frac{\sigma}{\varepsilon_0} - \frac{\sigma_{\text{bound}}}{\varepsilon_0} = \frac{\sigma/\varepsilon_0}{\kappa}$$

Solving for σ_{bound}, we find

$$\sigma_{\text{bound}} = \frac{\kappa - 1}{\kappa} \sigma \qquad (18.16)$$

The bound surface charge density always is less than the free surface charge density on the plates of the capacitor. If $\kappa = 1$, a vacuum, we have no dielectric and zero bound surface charge.

Experiment 2: Adding an Independent Voltage Source

Now we take a parallel plate capacitor with a vacuum or air between its plates and connect it to an independent voltage source (a source of emf), as shown in Figure 18.42.

The independent voltage source maintains a constant potential difference V_0 between its terminals. Since the capacitor is connected to the same two terminals, the two are in parallel

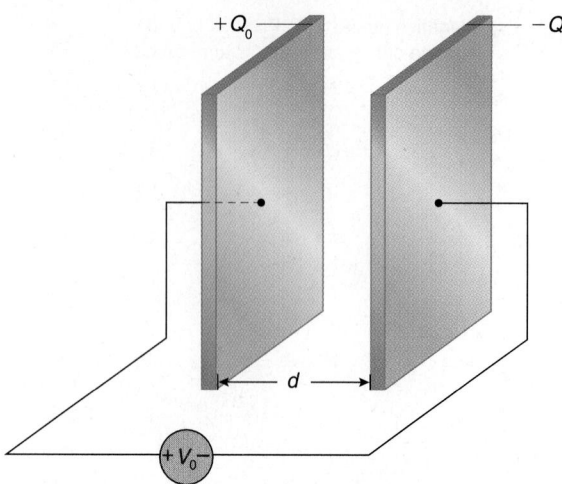

FIGURE 18.42 A parallel plate capacitor with a vacuum or air between its plates connected to an independent voltage source.

and the potential difference across the capacitor also must be V_0. Let $\pm Q_0$ be the charges on the capacitor plates. Thus the capacitance is

$$C_0 = \frac{|Q_0|}{|V_0|}$$

If a dielectric now is inserted into the space between the plates of the capacitor (see Figure 18.43), completely filling them as in Experiment 1, the independent voltage source maintains the same potential difference V_0 between the plates as before. Thus the magnitude of total electric field between the plates must remain equal to E_0, since $V_0 = E_0 d$.

Once again, the dielectric becomes polarized in the field and produces bound surface charge on the surfaces facing the capacitor plates. The bound surface charge produces an electric field in the direction opposite to the original field between the plates. To maintain the same total field as before, additional charge must be supplied to the conducting plates of the capacitor by the battery. The superposition of the field from the charges on

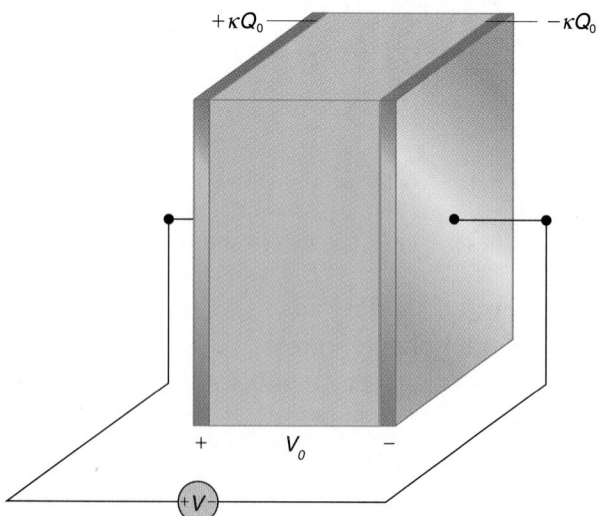

FIGURE 18.43 A dielectric is inserted between the plates.

the conducting plates and the field caused by the bound charges on the surfaces of the dielectric produces a total field equal to the field that existed between the plates when no dielectric was between the plates. The magnitude of charge on the capacitor plates with the dielectric in place is found to be $|Q| = \kappa |Q_0|$. The potential difference between the plates still is V_0. So the new capacitance of the capacitor is

$$C = \frac{|Q|}{V_0}$$
$$= \frac{|\kappa Q_0|}{V_0}$$
$$= \kappa C_0$$

Again, the capacitance increases by the factor κ with the insertion of the dielectric.

QUESTION 7

On the conducting plates of a capacitor with a dielectric filling the space between the plates, what dielectric constant is needed in order to make the bound surface charge density of the dielectric surface half the free surface charge density?

EXAMPLE 18.11

Coaxial cables consist of a central cylindrical wire surrounded by an insulating layer, another cylindrical conductor, and finally another layer of insulating material, as shown in Figure 18.44. Such cables are common in electronics and in communication systems, such as the cable bringing TV signals to your set. A coaxial cable can be approximated as a long cylindrical capacitor. The capacitance per unit length of the cable is important in determining the electronic properties of the cable and the character of the TV signal received.

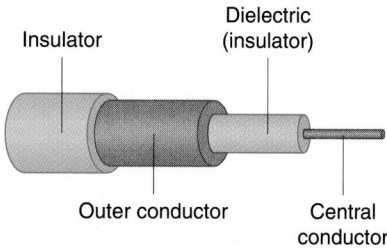

FIGURE 18.44

Use the result of Example 18.5 to determine the capacitance per meter of length of RG59 coaxial cable (used for common household cable TV) that has an inner conductor of diameter 0.812 mm and an outer conductor of diameter 3.66 mm. The insulating material between the two conductors has a dielectric constant of 1.35.

Solution

By the result of Example 18.5, the capacitance of a cylindrical capacitor with a vacuum (or air) between the plates is

$$C = \frac{4\pi\varepsilon_0 \ell}{2\ln\left(\dfrac{b}{a}\right)}$$

In a coaxial cable with a material of dielectric constant κ between the conductors, the capacitance increases by the factor κ. Hence the capacitance of the cable is

$$C = \kappa \frac{4\pi\varepsilon_0 \ell}{2 \ln\left(\dfrac{b}{a}\right)}$$

The capacitance per unit length thus is

$$\frac{C}{\ell} = \kappa \frac{4\pi\varepsilon_0}{2 \ln\left(\dfrac{b}{a}\right)}$$

Substituting the numerical values given, you have

$$\frac{C}{\ell} = (1.35) \frac{4\pi [8.85 \times 10^{-12} \quad C^2/(N\cdot m^2)]}{2 \ln\left[\dfrac{(3.66 \times 10^{-3} \ m)/2}{(0.812 \times 10^{-3} \ m)/2}\right]}$$

$$= 4.99 \times 10^{-11} \ F/m$$

$$= 49.9 \ pF/m$$

or about 50 pF/m.

Lightning is caused by the dielectric breakdown of air in an extremely strong electric field.

18.11 DIELECTRIC BREAKDOWN*

If a dielectric material is subjected to an electric field of sufficient magnitude, the outer electrons of the atoms and molecules in the field are ripped free by the electrical force and then travel through the material (in the direction opposite to the electric field, of course). These liberated electrons in turn collide with other atoms, freeing additional electrons. This multiplication, or avalanche effect, is **dielectric breakdown**. The dielectric then no longer is an insulator but a conductor, because the charges freely move through the material. The maximum electric field that a dielectric can sustain before dielectric breakdown occurs is known as the **dielectric strength** of the material. Table 18.2 indicates the dielectric strength of various dielectric materials. The most spectacular natural example of dielectric breakdown is lightning.

Dielectric breakdown also creates the spark in a spark plug that explodes the gasoline in the pistons of the engine in your car, boat, or lawn mower.[†]

[†] Diesel engines do not use spark plugs; they ignite the fuel by compressing it to the point where it spontaneously explodes from the increase in temperature.

The spark in a spark plug creates radio wave static that can be detected by nearby radio telescopes. Hence, engines with spark plugs are prohibited near such telescopes.

QUESTION 8

In a cylindrical capacitor such as a coaxial cable, if the potential difference between the conductors is too great, dielectric breakdown will occur. Will such breakdown begin near the inner or the outer conductor? Explain your reasoning.

CHAPTER SUMMARY

Polarity markings indicate which of two points is at the higher (+) or lower (−) electric potential. A polarity marking (+) does *not* necessarily imply the potential at the indicated point is positive; the marking merely indicates that the point is at a higher electric potential than the corresponding point with the (−) polarity marking.

An *ideal circuit element* is imagined as a device with two *ideal*

wires, called *leads* or *terminals*, emerging from it. Real circuit elements are modeled with ideal circuit elements.

An *independent voltage source*, also known as a *source of emf*, is an ideal circuit element that supplies and maintains a potential difference, called its *emf*, between its two terminals. An ideal battery is an example of an independent voltage source.

The places where the leads or terminals of two or more circuit elements are connected together are called *nodes*. Circuit elements connected in *parallel* are connected to the same two distinct nodes and have the same potential difference across them. Circuit elements connected in *series* are strung along sequentially like beads on a string.

When independent voltage sources are connected in series, the effective equivalent independent voltage source has an emf equal to the algebraic sum of the emfs of the collection in series:

$$V_{eq} = V_1 + V_2 + V_3 + \cdots$$

Independent voltage sources are most commonly connected with their polarities all directed in the same sense (plus to minus, plus to minus, etc.).

When two or more ideal independent voltage sources are connected in parallel, they all *must* have the same emf and be connected with their positive polarity terminals at the same node.

A *capacitor* is a circuit element that consists of two conductors, called *plates*, separated in space. If the two plates have equal and opposite charges on them, the capacitor is said to be *charged*; if each plate has zero charge, the capacitor is said to be *uncharged*.

The *capacitance* of a capacitor is the ratio of the absolute magnitude of the charge $|Q|$ on either plate to the absolute magnitude of the potential difference $|V|$ between the plates:

$$C \equiv \frac{|Q|}{|V|} \tag{18.1}$$

The SI unit of capacitance is the farad (F), which is equivalent to a coulomb per volt (C/V). The capacitance of a capacitor depends on the physical size and shape of the conducting plates, their geometric arrangement in space, and the medium separating the plates.

The equivalent capacitance of a parallel connection of capacitors is the sum of the individual capacitances:

$$C_{eq} = C_1 + C_2 + C_3 + \cdots \quad \text{(capacitors in parallel)} \tag{18.4}$$

A series connection of capacitors has an equivalent capacitance of

$$\frac{1}{C_{eq}} = \frac{1}{C_1} + \frac{1}{C_2} + \frac{1}{C_3} + \cdots \quad \text{(capacitors in series)} \tag{18.7}$$

For *two* capacitors in series, the equivalent capacitance is the product of their capacitances divided by the sum.

Work is required to charge a capacitor. The electrical potential energy stored in a charged capacitor is

$$PE = \frac{1}{2}\frac{Q^2}{C} = \frac{1}{2}|Q||V| = \frac{1}{2}CV^2 \tag{18.12}$$

If an insulating material, called a dielectric, is placed between the plates of a capacitor, the capacitance increases by a factor κ, known as the *dielectric constant* of the material.

When *dielectric breakdown* occurs, a dielectric material ceases to be an insulator and becomes a conductor. The *dielectric strength* is the maximum magnitude of electric field a dielectric material can sustain before breakdown occurs.

SUMMARY OF PROBLEM-SOLVING TACTICS

18.1 **(page 817)** When using Equation 18.7, be sure you remember how to add fractions.

18.2 **(page 817)** Be sure to realize that after summing the reciprocals of the individual capacitances in Equation 18.7 that the

equivalent capacitance is the *reciprocal* of the sum.

18.3 **(page 817)** The equivalent capacitance of two capacitors in series is the product of their capacitances divided by the sum.

QUESTIONS

1. (page 807); 2. (page 809); 3. (page 811); 4. (page 811); 5. (page 814); 6. (page 817); 7. (page 824); 8. (page 825)

9. For each of the two points A and B in Figure Q.9, label the appropriate polarity signs (+) and (−) taking into account the indicated potentials at the two points. Also indicate the potential difference V between A and B, with its correct sign.

10. In a carefully worded sentence or two, explain what is meant by the terms series connection and parallel connection.

11. Draw an arrangement of two circuit elements that are simultaneously in series and parallel with each other.

12. What does it mean when we say a circuit element is shorted out? What is the potential difference between the leads of the circuit element when it is shorted?

13. Why is it meaningless to short out an independent voltage source? Shorting a voltage source also can be dangerous, as we will see in Chapter 19.

14. Two independent voltage sources have potential differences $V_1 \neq V_2$. They can be connected in series but should never be connected in parallel. Why one way and not the other?

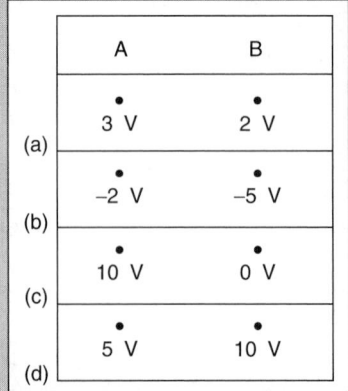

	A	B
(a)	• 3 V	• 2 V
(b)	• −2 V	• −5 V
(c)	• 10 V	• 0 V
(d)	• 5 V	• 10 V

FIGURE Q.9

15. Is Figure 18.8 the only way six circuit elements can be connected together so that they all are neither in series nor in parallel?

16. How many nodes are in the arrangement of circuit elements shown in Figure Q.16?

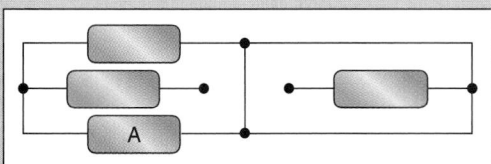

FIGURE Q.16

17. What must be done in Figure Q.16 to short out the circuit element labeled A?

18. Several independent voltage sources are connected as in Figure Q.18. What must be the potential differences V_1, V_2, and V_3?

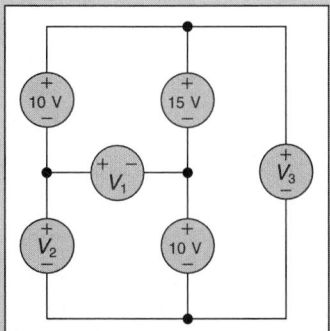

FIGURE Q.18

19. Explain why the capacitance of a capacitor is independent of both the magnitude of the charge on its plates and the potential difference between the plates even though the capacitance is defined as their ratio (Equation 18.1).

20. (a) Sketch a graph of the magnitude of the charge on either plate of a capacitor versus the magnitude of the potential difference between the plates. (b) What is the slope of the curve,

$$\frac{d|Q|}{d|V|}$$

and what are alternative units for the slope besides C/V?

21. Sketch a graph of the magnitude of the potential difference between the plates of a capacitor and the magnitude of the charge on each plate. This is the reverse of the plot in the previous question. Suppose a capacitor has charges $\pm Q_0$ on its plates and a potential difference of magnitude $|V_0|$ between them. Schematically indicate this point on the graph. What is the physical significance of the area under the curve between the origin and the point ($|Q_0|$, $|V_0|$)?

22. The electrical potential energy of a charge q placed at a point where the electrical potential is V is qV (see Equation 17.8). Why, then, is the potential energy stored on a capacitor not $|Q||V|$, rather than the correct expression

$$\frac{1}{2}|Q||V|$$

(Equation 18.12)?

23. If two capacitors, with capacitances $C_1 = 2C_2$, store equal amounts of potential energy, what is the relationship between the potential differences across them? Can you exclude the possibility that these capacitors are connected in series? In parallel? Explain.

24. A thin conducting sheet is carefully inserted between but not touching the plates of a parallel plate capacitor as shown in Figure Q.24. Will the capacitance increase, decrease, or be unaffected? Justify your answer.

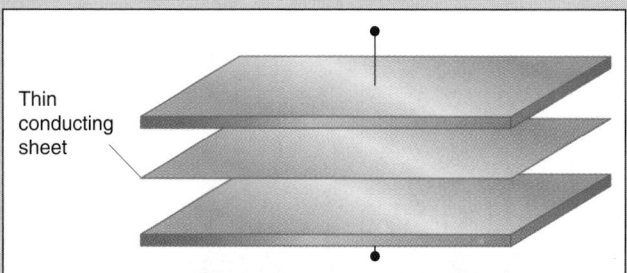

Thin conducting sheet

FIGURE Q.24

25. Your instructor likely can provide you with a variable capacitor used for tuning certain circuits in radios. An example of such a capacitor is shown in Figure Q.25. Examine the device carefully. Are the various capacitors that form the variable capacitor connected in series or parallel?

FIGURE Q.25

26. Let C be the capacitance of a parallel plate capacitor of plate area A and plate separation d. If four plates of area A are arranged as in Figure Q.26, separated by distance d, what is the equivalent capacitance?

27. Let C be the capacitance of a parallel plate capacitor of plate area A and plate separation d. If four plates of area A are arranged as in Figure Q.27, separated by distance d, what is the equivalent capacitance?

28. For a polar material, would you expect the dielectric constant to increase or decrease if the temperature is raised? Give arguments to support your answer.

29. Water has a dielectric constant that is quite large (see Table 18.2) but almost never is used as a dielectric between the plates of a capacitor. Why?

30. A charged capacitor has a large potential difference across its plates and a large capacitance. Explain why it is dangerous to touch the terminals of such a capacitor, even well after the battery used to charge it has been disconnected.

31. An independent voltage source maintains a constant potential difference across the plates of a parallel plate capacitor. Does the potential energy increase, decrease, or remain the same if

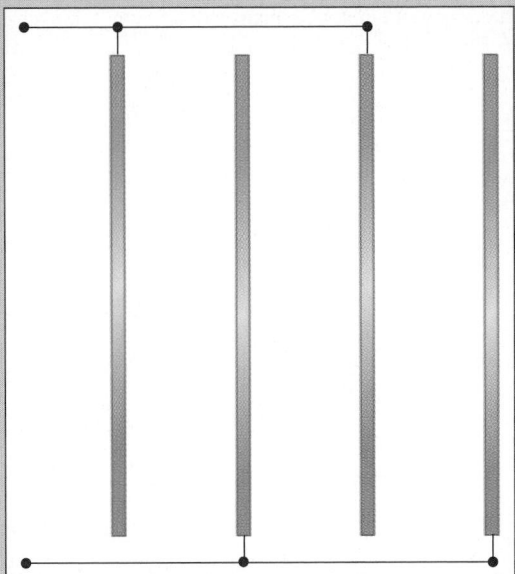

FIGURE Q.26

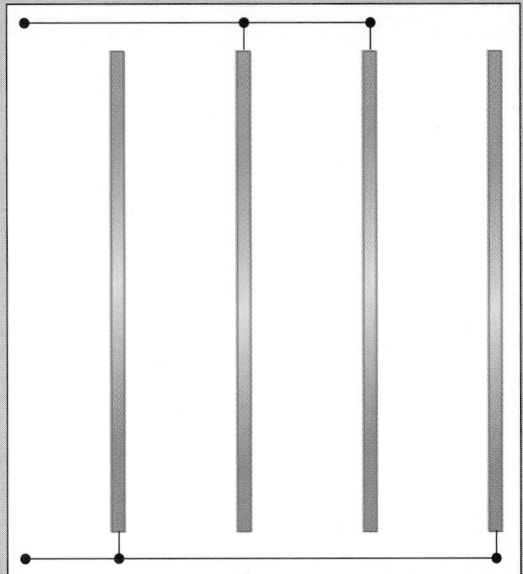

FIGURE Q.27

a sheet of glass is inserted between the plates? If the energy increases, where does the increase in potential energy come from? If it decreases, where does the potential energy go?

32. How is a spark in a spark plug similar to lightning?

33. Notice that, from Example 18.3, the capacitance of a parallel plate capacitor can be increased by decreasing the plate separation d. For a given potential difference V between its plates, what happens to the magnitude of the electric field between the plates as d decreases? If the distance between the plates

becomes too small for an air-filled capacitor (or for any capacitor filled with a dielectric), what will happen?

34. Shuffle you feet across a fluffy rug on a nice dry day and then slowly bring your finger near a metal doorknob. Estimate the distance between your finger and the doorknob when the spark jumps between them. From this distance, *estimate* the charge you accumulated. State what assumptions you make to arrive at your estimate.

PROBLEMS

Sections 18.1 Terminology, Notation, and Conventions
18.2 Circuit Elements
18.3 An Independent Voltage Source: A Source of Emf
18.4 Connections of Circuit Elements

1. How many nodes are in the connection of circuit elements shown in Figure P.1?

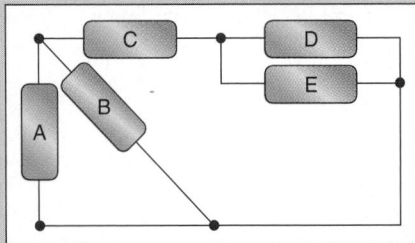

FIGURE P.1

2. How many nodes are in the connection of circuit elements shown in Figure P.2?

3. How many nodes are in the connection of circuit elements shown in Figure P.3?

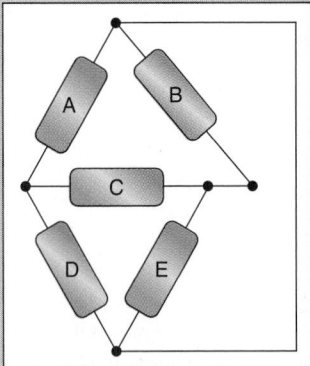

FIGURE P.2

4. Which circuit elements (if any) in Figure P.1 are in series with each other? Which (if any) are in parallel?

5. Which circuit elements (if any) in Figure P.2 are in series with each other? Which (if any) are in parallel?

6. Which circuit elements (if any) in Figure P.3 are in series with each other? Which (if any) are in parallel?

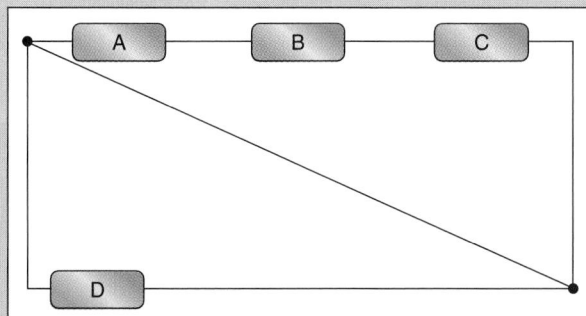

FIGURE P.3

● **7.** In Figures P.1, P.2, and P.3, are any circuit elements shorted out? If so, specify which one(s) and redraw the circuit with extraneous circuit elements removed.

● **8.** (a) In the collection of circuit elements shown in Figure P.8, what is the potential difference between points A and B? (b) What is the potential difference between points B and C? (c) What is the potential difference between points C and A? (d) What is the potential difference between points A and C?

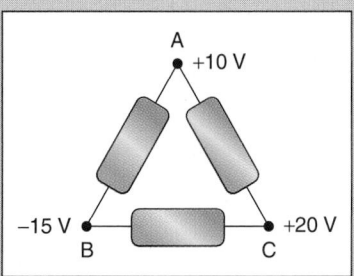

FIGURE P.8

Section 18.5 Independent Voltage Sources in Series and Parallel

9. The lead–acid electrochemical cells used to create the emf in automotive batteries (an independent voltage source, to a first approximation) have an emf of about 2.0 V. How many such cells are needed to create an independent voltage source with an emf of 12.0 V? Sketch how the cells should be connected together to produce the 12.0 V emf.

10. Four D-cell batteries, each with an emf of 1.5 V, are connected in series in a flashlight, with their positive terminals all facing the same way. What is the equivalent emf of the four batteries?

● **11.** The instructions with a set of battery jumper cables say to connect the red wire to the positive polarity terminals and the black wire to the negative polarity terminals of each battery. (a) Are the batteries connected in series or parallel? (b) Will there be any problem if you use the black wire in place of the red wire and the red wire in place of the black wire?

● **12.** A truck driver asks you to help jump start his engine with the help of the 12.0 V battery in your car. You notice, however, that the truck engine has several 12.0 V batteries connected in series as indicated in Figure P.12. (a) Indicate on a sketch a per-

missible way to connect your battery to the collection. (b) Indicate on another sketch several ways you should *not* connect your battery to the collection and explain why each connection is impermissible.

Section 18.6 Capacitors

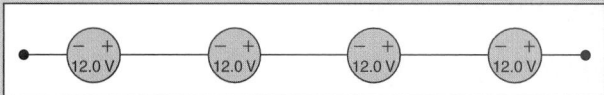

FIGURE P.12

13. A 15 pF capacitor has a potential difference of 1.50 V between its plates. What is the absolute magnitude of the charge on each plate?

14. The plates of a capacitor have charges of ±8.0 nC. A potential difference of −120 V exists between the plates. Find the capacitance.

15. What is the charge on each plate of a 10 μF capacitor with a potential difference of 120 V between its plates?

● **16.** To show you how difficult it is to make capacitors with large capacitances, consider a hypothetical parallel plate capacitor with a capacitance of 1.0 F with plates separated by 1.0 cm. (a) If the plates are square, what is the length of the sides of the square? (b) If the plates are circular, what is the radius of the circles?

● **17.** A parallel plate capacitor consists of two circular disks of radius R and separation $R/1000$. By what factor does the capacitance change if R is doubled, affecting both the area and separation?

● **18.** Calculate the capacitance of a sphere with a radius equal to the average radius of the Earth.

● **19.** What is the radius of a spherical capacitor with a capacitance of 1.00 F? What is the ratio of this radius to the average separation of the Earth and the Moon?

● **20.** The temperature of a brass spherical capacitor of radius 5.00 cm increases by 100 °C. By how much does the capacitance increase? (Consult Table 13.1 on page 596.)

● **21.** A conducting sphere of radius a is surrounded by a concentric, thin conducting spherical shell of radius b. Show that the capacitance of the system is

$$C = 4\pi\varepsilon_0 \, \frac{ab}{b - a}$$

● **22.** Show that the result of Problem 21 reduces to the capacitance of a parallel plate capacitor if the separation distance $d \equiv b - a$ is much smaller than the radius a of the inner conductor.

● **23.** Use the result of Problem 21 to show that as $b \to \infty$ m, the capacitance approaches that of a sphere of radius a.

Section 18.7 Series and Parallel Combinations of Capacitors

For Problems 24–32, find the single equivalent capacitance that can be placed between the terminals A and B.

24.

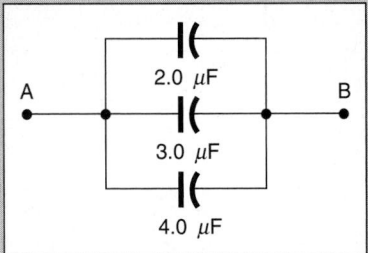

FIGURE P.24

25.

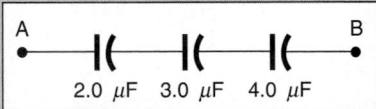

FIGURE P.25

•26.

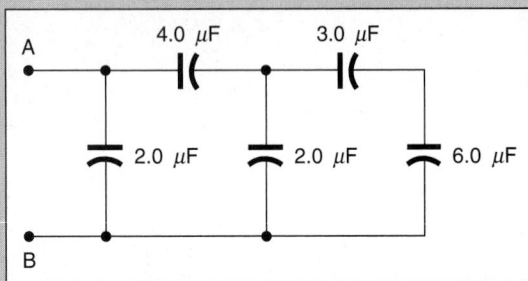

FIGURE P.26

•27.

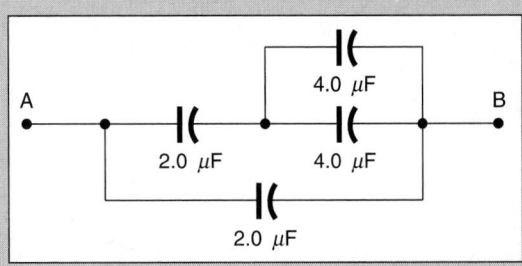

FIGURE P.27

•28.

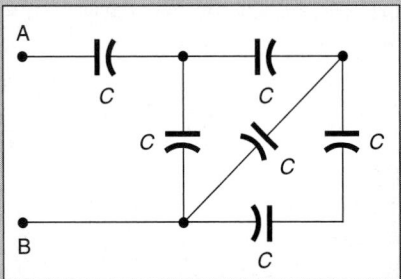

FIGURE P.28

•29.

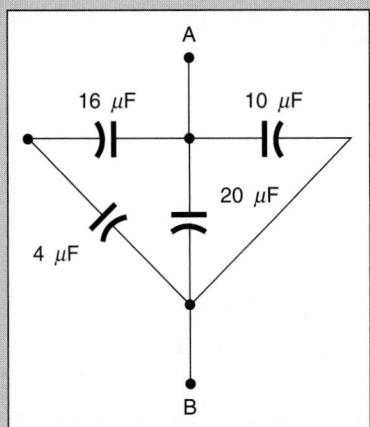

FIGURE P.29

•30.

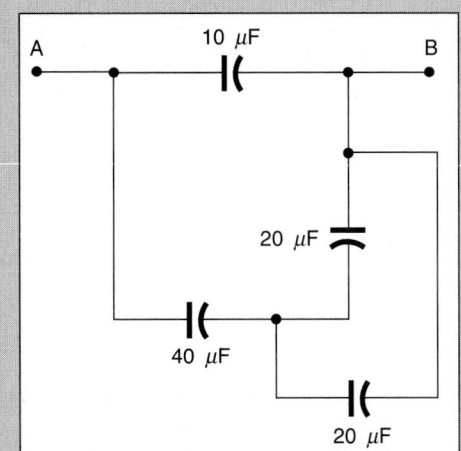

FIGURE P.30

•31.

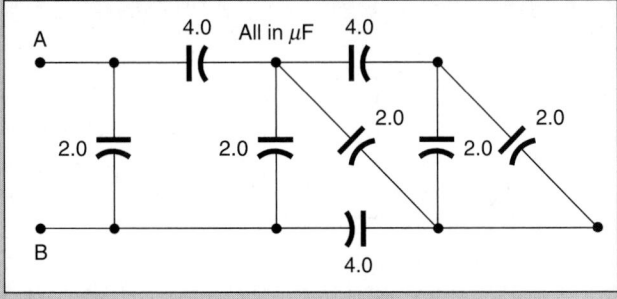

FIGURE P.31

•32.

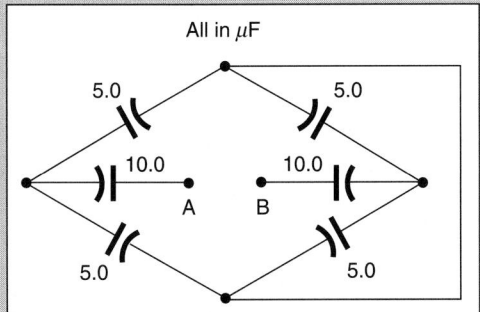

FIGURE P.32

•33. (a) If a single 10 μF capacitor has a potential difference between its plates of 120 V, what magnitude of charge is on its plates? (b) How can you connect other 10 μF capacitors to this one to store additional charge? (c) How many 10 μF capacitors connected to a 120 V independent voltage source are needed to store a total charge of magnitude 1.0 C?

•34. (a) In Figure P.34, find the magnitude of the potential difference across each capacitor. Indicate the polarities of the potential differences. (b) Find the magnitude of the charge on the plates of each capacitor.

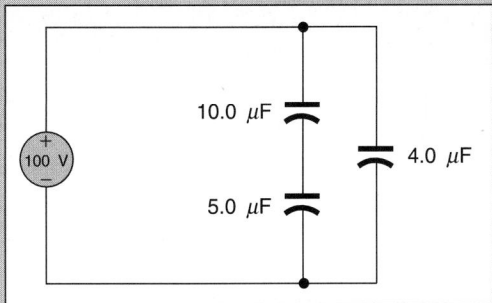

FIGURE P.34

•35. (a) In Figure P.35, what is the potential difference across each capacitor? (b) Determine the magnitude of the charge on each of the plates of the capacitors. (c) What is the equivalent capacitance of the two capacitors?

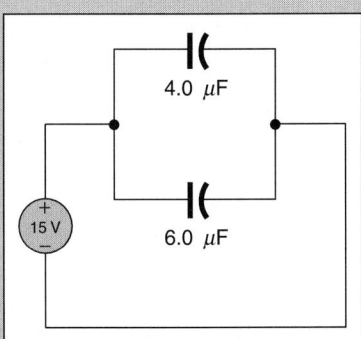

FIGURE P.35

•36. Two capacitors connected in parallel have an equivalent capacitance of 12 μF. When connected in series, the equivalent capacitance is one-third the capacitance of one of the pair. What are the values of the two capacitances?

•37. Nuts. You need a 10 μF capacitor but all you have is a bin of 15 μF capacitors. How can you use these 15 μF capacitors to make one with an equivalent capacitance of 10 μF? Sketch the way the capacitors should be connected together.

•38. Two capacitors are connected in series with an independent voltage source as indicated in the Figure P.38. Show that the potential differences across the capacitors are

$$V_1 = \frac{C_2}{C_1 + C_2} V_0$$

and

$$V_2 = \frac{C_1}{C_1 + C_2} V_0$$

In *real* capacitors (as opposed to the ideal capacitors we have treated), this result is rarely the case because real capacitors also have leakage resistance.* For a discussion of the situation with real capacitors, see L. Kowalski, "A myth about capacitors in series," *The Physics Teacher, 26,* #5, pages 286–287 (May 1988).

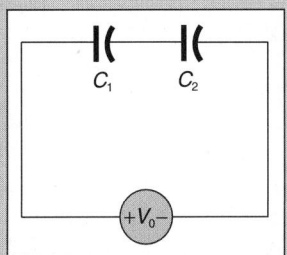

FIGURE P.38

‡39. Show that the equivalent capacitance for a collection of capacitors connected in series is smaller than the smallest capacitance in the series.

Section 18.8 Energy Stored in a Capacitor

40. A capacitor has a potential difference V between its plates. To double the potential energy stored by the capacitor, by what factor must the potential difference be changed?

41. For a laser power supply, you want to store 1.0 J of energy in a capacitor with a potential difference of 120 V between its plates. What capacitance is needed?

42. A 12 V automotive battery stores 2.0 MJ of energy. If a large parallel plate capacitor, with a plate separation of 0.50 cm and potential difference between its plates of 120 V, stores the same amount of energy, what is the length ℓ of a side of its plates if they are square? The size of this parallel plate capacitor is quite unrealistic.

•43. To store about 10 J of electrical potential energy for a cardiac defibrillator, a capacitor bank is used. The bank consists of a parallel connection of 100 μF capacitors with a potential

_____

*We consider this in more detail in the next chapter.

difference of 100 V across each of them. How many capacitors are in the bank?

●**44.** Professor Milly Coulomb finds it takes 2.0 J of work to drag, at constant speed, a −2.0 mC charge between the plates of a parallel plate capacitor. (a) What is the work done by the electrical force in taking the charge between the plates? (b) What is the potential difference between the plates of the capacitor?

●**45.** You are given an independent voltage source and two identical capacitors. Sketch how you would arrange these circuit elements so that the capacitors store the most total electrical potential energy. Prove that this arrangement is the best choice.

●**46.** Near the surface of the Earth there is an electric field with a magnitude of about 100 N/C directed vertically down. Assume this field is caused by a net free charge on the Earth. (a) What are the magnitude and sign of the charge on the total Earth that produces this field? (b) What electrostatic energy density is stored in the field of the Earth near its surface?

●**47.** Three capacitors in series, with capacitances 3.0 μF, 5.0 μF, and 6.0 μF, are connected in series to a 21.0 V independent voltage source. (a) Sketch the arrangement. (b) What is the equivalent capacitance? (c) What charges are on the plates of each capacitor? Indicate which charge is on what plate of your sketch. (d) To the nearest volt, what is the potential difference across each capacitor? (d) Which capacitor is storing the greatest amount of electrical potential energy? What is this energy?

●**48.** Let n identical capacitors be connected (i) in series and (ii) in parallel. (a) If the equivalent capacitor has a potential difference V between its plates, what is the total electrical potential energy stored by each collection? (b) If the equivalent capacitor has a charge of magnitude $|Q|$ on its plates, what is the total electrical potential energy stored by each collection?

Sections 18.9 Electrostatics in Insulating Material Media*
       18.10 Capacitors and Dielectrics*
       18.11 Dielectric Breakdown*

49. An air-filled capacitor is connected to the terminals of an independent voltage source. If an insulator of dielectric constant κ is next inserted to completely fill the air gap between the plates, by what factor does the potential energy stored in the capacitor increase?

50. A small ceramic capacitor has plates 1.20 cm in diameter. A dielectric 0.25 mm thick, with a dielectric constant $\kappa = 5.0$, completely fills the space between the plates. Find the capacitance.

●**51.** Real capacitors have a maximum potential difference V_{max} that can be placed across their plates before dielectric breakdown ensues. You have a collection of identical capacitors each with a V_{max} of 50 V and a capacitance of 10 μF. You need an equivalent capacitor with a capacitance of 10 μF, but with a potential difference of 100 V across it. Devise an appropriate connection of the 50 V, 10 μF capacitors to do the job.

●**52.** You are faced with a problem: a number of 0.50 μF capacitors are available but the potential difference across each cannot exceed 200 V without causing dielectric breakdown. A

capacitor of 0.50 μF capacitance is needed but must be connected across a potential difference of 400 V for who knows what purpose. Show via a diagram how an equivalent capacitor having the desired capacitance can be obtained using the given collection of capacitors, so that the maximum potential difference across any one of them is 200 V.

●**53.** A parallel plate capacitor with plate area A and plate separation d is filled with two dielectrics as shown in Figure P.53. Show that the capacitance of the arrangement is

$$C = \frac{\varepsilon_0 A}{d} \frac{\kappa_1 + \kappa_2}{2}$$

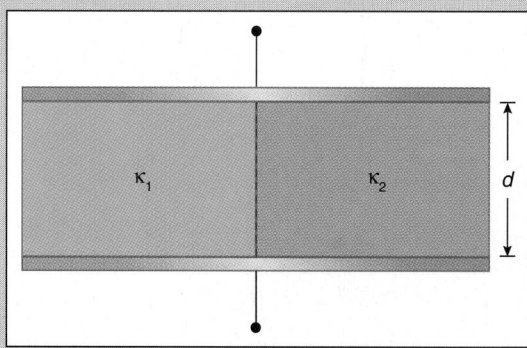

FIGURE P.53

●**54.** A parallel plate capacitor with plate area A and plate separation d is filled with two dielectrics as shown in Figure P.54. Show that the capacitance of the arrangement is

$$C = \frac{2\varepsilon_0 A}{d} \frac{\kappa_1 \kappa_2}{\kappa_1 + \kappa_2}$$

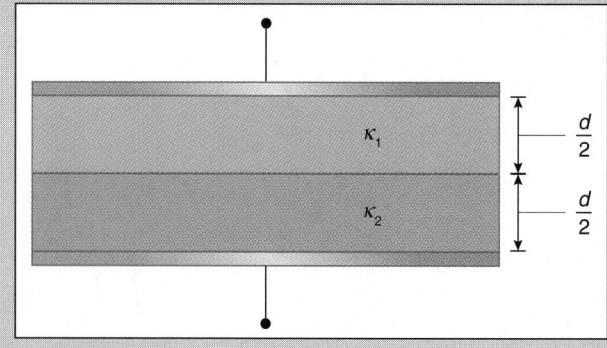

FIGURE P.54

●**55.** A block of paraffin is placed in an electric field of magnitude 1.20×10^4 N/C. What is the magnitude of the electric field within the paraffin? Refer to Table 18.2 on page 821.

●**56.** A charged 50 μF capacitor with a potential difference between its plates of 100 V is about to be connected to an initially uncharged 50 μF capacitor as shown in Figure P.56. (a) Determine the energy stored in the initially charged capacitor. The switch at S now is closed. (b) Are the two capacitors connected in parallel? (c) Determine the magnitude of the potential difference V across each capacitor. (d) Determine the total potential energy stored in the new arrangement. The two energies in (a) and (d) are not equal to each other. This might seem to contradict conservation of energy. With idealized capacitors and wires, it is not possible to explain where the missing energy went. For *real* capacitors and wires, the difference in the two energies appears as an increase in the thermodynamic internal energy of the wires connecting them (because of their finite electrical resistance*), with a small fraction also radiated away as electromagnetic radiation. See R. A. Powell, "Two-capacitor problem: a more realistic view," *American Journal of Physics, 47, #5*, pages 460–462 (May 1979).

•57. Thunderstorms are quite spectacular. Given the dielectric strength of air in Table 18.2, and that the flat bottom of a

---

*We consider the concept of electric resistance in the next chapter.

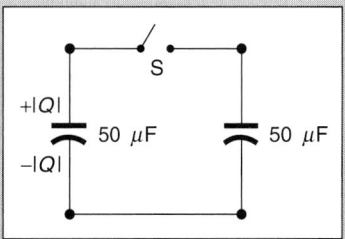

FIGURE P.56

thundercloud of area 25 km^2 is on the order of 2 km from the ground, estimate the capacitance of the cloud–Earth system and the electrical potential energy stored in the electric field between the cloud and the Earth as breakdown ensues.

•58. A 100 μF capacitor has tantalum oxide (dielectric constant $\kappa = 26$) completely filling the space between its plates. The capacitor has a potential difference of 50.0 V between its plates. If the tantalum oxide is removed from the capacitor, what is its capacitance?

INVESTIGATIVE PROJECTS

A. Expanded Horizons

1. Investigate and write a report on the uses of capacitors in modern integrated circuit chip technology.
   Donald M. Trotter Jr., "Capacitors," *Scientific American, 259, #1*, pages 86–90B (July 1988).

2. A *Leyden jar* was the first type of capacitor. Investigate and prepare a report on the history of its discovery.
   Donald M. Trotter Jr., "Capacitors," *Scientific American, 259, #1*, pages 86–90B (July 1988).
   Encyclopaedia Britannica (11th edition, 1910), volume 16, pages 528–529.

3. Investigate the ways capacitors are fabricated and the progress toward manufacturing capacitors of small dimensions with increased capacitance.
   Ray Marston, "Capacitors," *Electronics Now, 64, #3*, pages 47–64 (March 1993).
   Josef Bernard, "All about capacitors," *Radio-Electronics, 60, #5*, pages 49–53 (May 1989); and #8, pages 56–59 (August 1989).
   William Hyland, "Tantalum capacitors keep getting better," *Electronics, 61, #11*, pages 93–95 (May 1988).

4. Lightning is and example of dielectric breakdown. Investigate and prepare a report on the nature of this spectacular natural phenomenon, including lightning sprites.
   Earle Williams, "The electrification of thunderstorms," *Scientific American, 259, #5*, pages 88–99 (November 1988).
   Martin A. Uman, *The Lightning Discharge* (Academic Press, Orlando, 1987), International Geophysics Series, volume 39.
   Martin A. Uman, *All About Lightning* (Dover, New York, 1986).
   Tom Koppel, "Lightning lure," *Scientific American, 268, #2*, page 105 (February 1993).

William R. Newcott, "Lightning: nature's high-voltage spectacle," *National Geographic, 184, #1*, pages 80–103 (July 1993).
Jeff Rosenfeld, "Lightning: among the sprites and jets," *Weatherwise, 48, #6*, pages 9–10 (December 1995).
Dennis J. Boccippio, "Sprites, ELF transients, and positive ground strokes," *Science, 269, #5227*, pages 1088–1091 (25 August 1995).
Richard A. Kerr, "Lofty flashes come down to Earth," *Science, 270, #5234*, page 235 (13 October 1995).

B. Lab and Field Work

5. Design an experiment to measure the dielectric constant of an insulator such as glass or plastic. Specify the equipment needed for the experiment. Make a few back-of-the-envelope calculations to determine the approximate magnitudes of the quantities to be measured to determine the dielectric constant. Secure the equipment from your professor, perform the experiment, and write a report of your results, detailing the design, procedure, and results. The report should be complete enough to guide a peer in reproducing the experiment.

6. Your local electric company uses physically large capacitors in the electric grid around your town. The function of these capacitors is not energy storage, but rather is associated with keeping the oscillatory potential differences and currents in step with each other. Call or visit the engineering department of the utility to determine the capacitances of their capacitors and the potential difference at which each typically is used. Discover what is used as the dielectric material and the value of its dielectric constant. If connected to an independent voltage source with this emf, what maximum energy is stored by the capacitors?

7. Your physics lab may have a meter that measures capacitance. If so, measure the capacitance of several capacitors provided by your instructor to see how close their measured value is to the nominal value typically imprinted somewhere on the capacitor itself.

8. Design an experiment to test the premise and results of Problem 38.

C. Communicating Physics

9. Ask a few nonscience student compatriots what is meant by the term *shorting* in an electrical context. Given this information (or misinformation), write a short paragraph, appropriate for a lay audience, explaining what is meant by the term, specifically addressing any misconceptions you elicit from your fellow students.

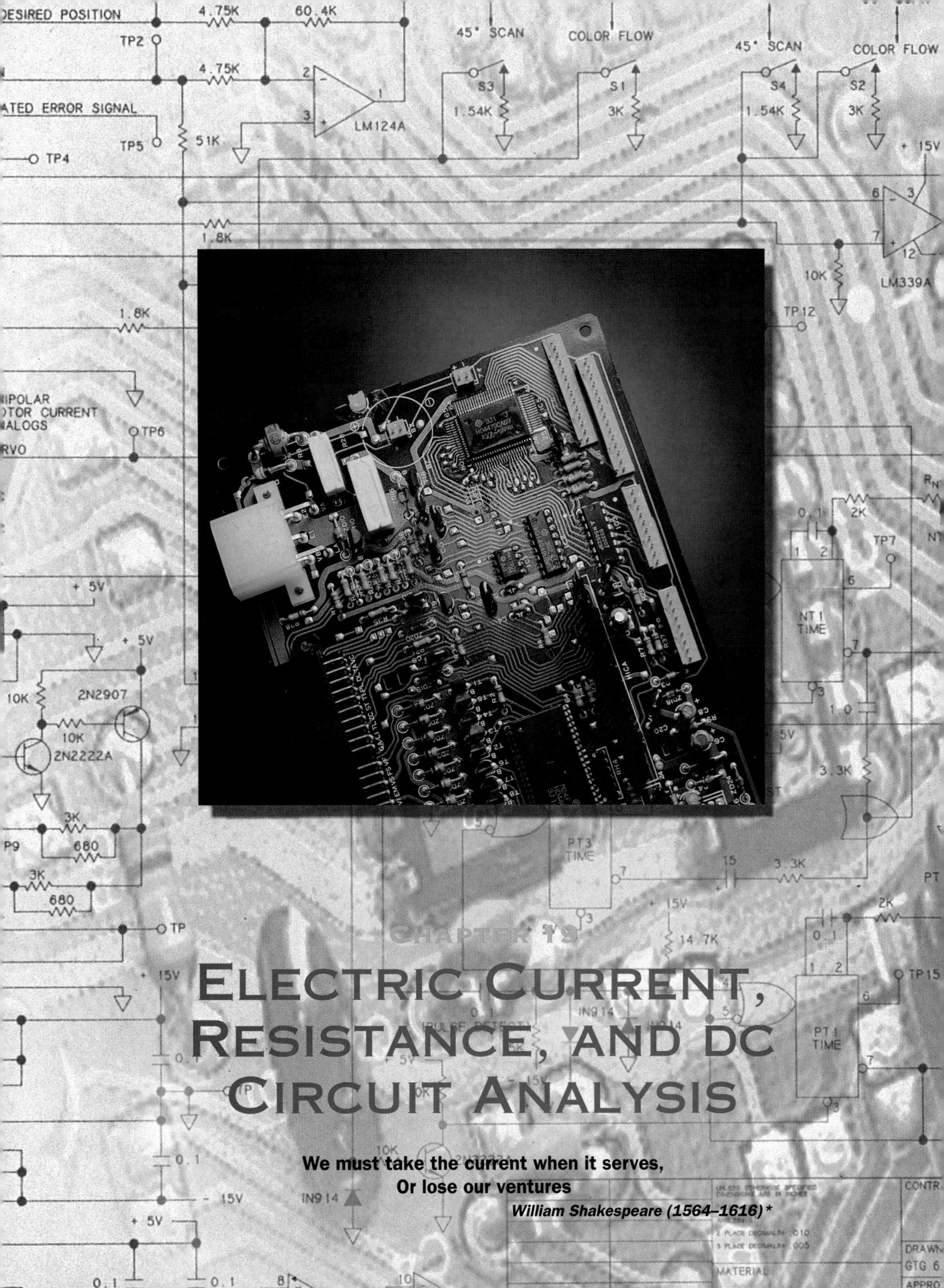

ELECTRIC CURRENT, RESISTANCE, AND DC CIRCUIT ANALYSIS

We must take the current when it serves,
Or lose our ventures

William Shakespeare (1564–1616) *

In the previous three chapters, we were concerned principally with electric charges at rest: electrostatic situations. Here we begin to see what happens as electrical charge moves or flows from one place to another, creating an electric current. Electric currents are double-edged swords: beneficial, such as when they are used to revive persons suffering from cardiac arrest, or deadly when they cause electrocution.

Electric currents are associated with many physical phenomena, from lightning to our central nervous system, from toasters to steel furnaces, from TV tubes to the Internet, from the motion of electrons in atoms to the vast whorls of swirling, charged matter in stars and galactic nebulae. We shall see in the next chapter that electric currents also are responsible for magnetism.

Charged particles may flow unimpeded through a vacuum, for example, as particle beams of electrons inside a TV or computer monitor. When charged particles flow in materials such as common wires, friction-like effects, called resistance, decrease the electrical potential energy of the charges. In superconducting materials at very low temperatures, charge flows without such resistance.

The concepts of current and resistance lead us to a study of electric circuits and a panoply of electronic devices. In this chapter we study steady "direct" currents of electrical charge, used widely in many devices such as computer circuits. The study and use of such *direct* (steady) currents form what is called *dc electronics*. Electric currents that reverse direction regularly with time, so-called *alternating* currents (*ac electronics*), we reserve for Chapter 22.

19.1 THE CONCEPT OF ELECTRIC CURRENT

The concept of electric current, or the flow of the electric charge property of matter, has been with our technological culture so long that it is difficult to appreciate how the concept arose before the advent of the atomic view of matter and the discovery of the electron and proton. A few experiments can show us the origins of the idea of an electric current.[†] We first examine a charged capacitor and see how it can be discharged or neutralized by transferring charge from one plate to the other. We then examine a few experiments with a battery and some light bulbs that also elucidate aspects of a current of electric charge.

Experiments with a Charged Capacitor

Experiment 1: A Capacitor and a Pendulum

We begin with an uncharged, isolated, parallel plate capacitor having its plates maintained at a fixed distance apart and with an isolated independent voltage source.[‡] Experimentally, by connecting the two plates of the uncharged capacitor momentarily to the independent voltage source and then disconnecting the source, we are left with a charged and isolated capacitor. The two plates have equal magnitudes of charge but with

*(Chapter Opener) *Julius Caesar*, Act IV, Scene 3, lines 223–224.
†The author is indebted to Arnold Arons, *A Guide to Introductory Physics Teaching* (John Wiley, New York, 1990) for these ideas.
‡A common battery will do, since, to a first approximation, it is an independent voltage source.

The motion of charged particles is an electric current even if there is no wire, as in this nebula in the constellation Orion.

opposite signs: $\pm |Q|$. Now imagine one plate fixed with the other plate connected to a spring that measures the attractive force that each plate exerts on the other due to their opposite charges (see Figure 19.1). The stretch of the spring needed to keep the plates a fixed distance apart can tell us if the capacitor is charged (nonzero force) or uncharged (zero force). A small ball of cork, covered with a conducting foil, is suspended by an insulating thread between the two plates of the capacitor as a simple pendulum.

If the ball initially is closer to the positive plate, it will be slightly attracted to that plate because of induction. On contact with the positive plate, some of the plate's positive charge is transferred to the ball by charge sharing, since they then briefly form a common conductor. The positively charged ball then is repelled by the positive plate and attracted to the negative plate. Upon reaching the negative plate, the kinetic energy of the ball is mostly converted into thermodynamic internal energy of the negative plate. The positive charge on the ball neutralizes some of the negative charge on the negative plate. The ball also then becomes negatively charged by charge sharing and subsequently is repelled by the negative plate and attracted back to the positive plate.

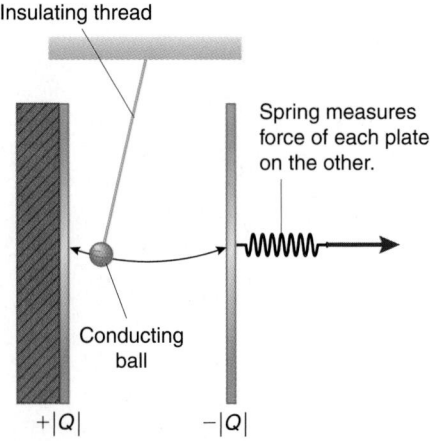

FIGURE 19.1 A spring measures the force needed to keep the plates of a charged capacitor a fixed distance apart.

The process continues with the electric pendulum swinging back and forth between the two plates until essentially all of the charge on the capacitor is neutralized and the capacitor is discharged. We imagine positive charge transferred one way, negative charge the other way until the two plates are discharged. The spring enables us to monitor the force that each plate exerts on the other as a function of time. We observe that the force decreases with each swing of the pendulum, confirming our account of the neutralization or discharge of the two plates. Once discharged, the plates of the capacitor are at the same electric potential, the field between them is zero, they exert zero electric force on each other, and the ball hangs motionless.

If we repeat the experiment with an initially charged capacitor but, instead of using the pendulum to transfer charge, connect a conducting wire between the two charged plates and short the plates, the force of each plate on the other decreases to zero almost instantaneously. The charge transfer occurs much faster via the wire than with the oscillating pendulum. We also might wonder about the fate of the potential energy that used to be stored on the charged capacitor; more about this problem toward the end of this chapter.*

Experiment 2: A Capacitor, a Pendulum, and an Independent Voltage Source

Now we repeat the experiment with the independent voltage source continuously connected to the charged capacitor; see Figure 19.2.

The pendulum once again is set swinging, and we discover that the spring now indicates there is *no* decrease in the force of each capacitor plate on the other. The pendulum swings indefinitely. Even though the pendulum still transfers positive charge to the negative plate and negative charge to the positive plate, just as before, there is no decrease in the total charge on each plate, because the force of each plate on the other remains constant, as

*To give the punch line away, the energy appears as an increase in the thermodynamic internal energy of the wire because of its resistance: the wire becomes warm.

confirmed by the spring. Hence we must conclude that the independent voltage source replenishes the charges on the plates as fast as they are transferred. We have a primitive electric **circuit**, a closed conducting pathway that permits the continuous transfer or flow of charge. In these experiments, the pendulum provides a way for charge to move or flow across the insulating air gap between the capacitor plates.

Some Experiments with Light Bulbs

Now we perform a few other experiments, this time using an independent voltage source, some thick wire (imagined to be ideal wire[†]), and several small, identical light bulbs. A light bulb consists of a thin filament of wire, one end of which is connected to the outer metal sheath and the other connected to the metal contact at the bottom of the bulb, as shown in Figure 19.3.

The contact at the bottom of the bulb is electrically insulated from the outer metal sheath. A light bulb, therefore, is a circuit element with two leads: one is the outer metal sheath and the other is the contact at the bottom of the bulb. We represent the light bulb schematically with the symbol shown in Figure 19.4.

Experiment 1: A Light Bulb and an Independent Voltage Source

Construct the arrangement shown in Figure 19.5. If a string or wood pencil connects points A and B, we find that the light bulb does not light. Materials such as string or wood are insulators.

On the other hand, if a metal wire connects points A and B, as shown in Figure 19.6, the bulb lights. We say an electric circuit exists: a continuous closed conducting path or loop connecting

[†]An ideal wire has no resistance; we introduce the concept of resistance later in the chapter.

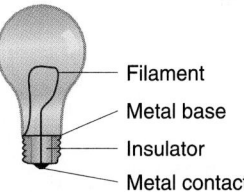

FIGURE 19.3 A light bulb.

FIGURE 19.4 Schematic representation of a light bulb.

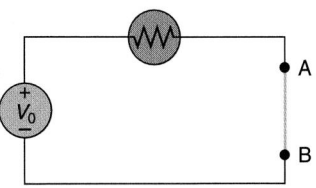

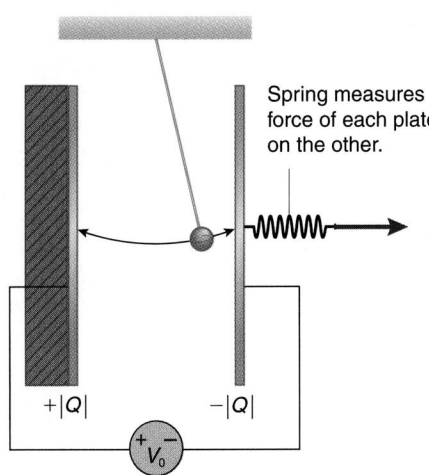

FIGURE 19.2 An independent voltage source connected to the capacitor.

FIGURE 19.5 If an insulator connects points A and B, the bulb does not light.

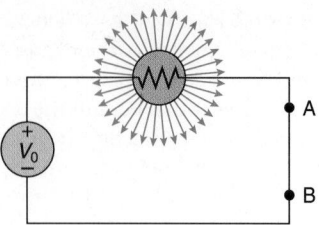

FIGURE 19.6 If a conductor connects points A and B, the bulb lights.

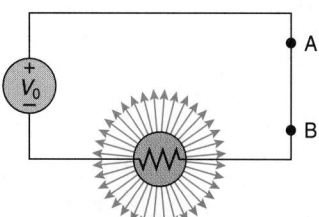

FIGURE 19.7 The bulb lights up with the same brightness, no matter where the bulb is placed.

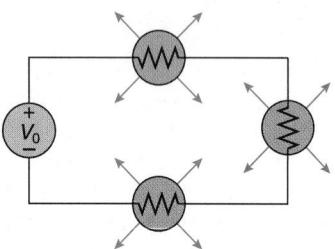

FIGURE 19.8 Several bulbs all light up equally bright, but not as bright as with just one bulb in the conducting loop.

the circuit elements. Note also that the light bulb becomes quite hot, indicating that energy is transferred continuously to the bulb and then goes to the environment or to our fingers via heat transfer. These observations indicate that whatever is happening is not a static electrical effect but is fundamentally dynamic, since energy transfer is occurring continuously. It makes no difference where the bulb is placed along the loop; the effect is the same, as shown in Figure 19.7. The bulb lights up equally bright regardless of where it is placed along the loop.

Experiment 2: Several Light Bulbs

Place several identical bulbs along the closed loop path, as indicated in Figure 19.8. All the bulbs light up equally bright, but not as bright as with only a single bulb.

This observation indicates that whatever is taking place is happening at all points equally along the closed conducting path, not just at one point. The more bulbs used, the dimmer each is, but all the dim bulbs in a series are equally bright.

Experiment 3: Real Wire and an Independent Voltage Source

Similar effects can be observed with no bulb and some real copper or aluminum wires of various lengths connected to a small battery. In Figure 19.9, a very long such real wire is

FIGURE 19.9 A long piece of real wire connected to an independent voltage source becomes equally warm along its entire length.

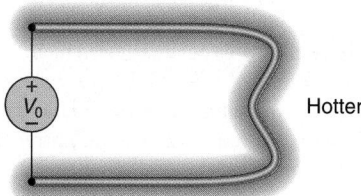

FIGURE 19.10 A shorter wire gets even hotter than a longer wire.

connected to the two terminals of the battery. The real wire is found to become equally warm at all points along its length.

If a shorter length of real wire made of the same material and with the same cross section is used, as in Figure 19.10, the wire becomes even hotter than the very long wire. Hence the length of the real wire somehow affects the amount of energy transfer per unit length to the real wire.

Such experiments and many others indicated to early researchers that whatever was happening was rather like the flow of a fluid through a closed loop of pipe. The battery acts like a pump, increasing the potential energy of the charges transferred through it. The ideal wires act as pipes, and the light bulb or the real wire lengths act like frictional impediments or resistances in the pipe that decrease the potential energy of the charges passing through them. The loss of electrical potential energy increases the thermodynamic internal energy of the material (it becomes warmer), and the energy eventually is passed along to the environment via heat transfer.

From such experiments arose the concepts of an electric **current** and electric **resistance**. From other experiments, such as those we performed with a charged capacitor, we know that the flow involved is a flow of electric charge. Typically it is of negative electrons, though this is not apparent from the experiments mentioned here.* We now explore the concept of electric current, or the flow of electric charge, in quantitative detail.

19.2 ELECTRIC CURRENT

The flow of air is called a wind; we could just as well call the flow of the charge property an electric wind, but such terminology is not used. Moving charges constitute an electric current, just as

---

*In Chapter 20 we will investigate an experiment (the Hall effect) that is used to determine the sign of the charge carriers involved in electric currents. In conducting materials, the moving charge carriers are electrons. In semiconducting materials, positive and/or negative charge carriers are present.

the flow of water is called a current. A quantitative measure of the current is the amount of the charge property that passes a given location per unit time. You can imagine a current meter as something like a traffic counter that tallies the coulombs of charge that move by some point during one second.

> The time rate of charge flow is defined to be the electric current I:
>
> $$I \equiv \frac{dQ}{dt} \qquad (19.1)$$
>
> The SI unit for current is the **ampere** (A), one coulomb per second (C/s).

The unit is named for the Frenchman André-Marie Ampère (1775–1836), an early researcher into electrical phenomena.

The particles that carry the charge property are called *charge carriers*. Charge carriers in motion constitute an electric current. The carriers may have positive charge or negative charge. In some materials the charge carriers are exclusively negative charge carriers (ordinary conductors); in other materials the charge carriers are predominantly positive charge carriers, predominantly negative charge carriers, or both positive and negative charge carriers. A current exists anytime there is a charge in motion. The charges need not be in conducting wires. For example:

1. The electron orbiting the proton in the hydrogen atom represents a charge in motion and therefore constitutes an electric current.
2. The moving collection of charged particles in electron beams, proton beams, positron beams, antiproton beams, or ion beams—each constitutes an electric current. Charged particles moving in a vacuum or a material are a current whether the velocity of the particles is constant or not.

Ampere was fascinated by numbers even as a toddler. The deeply religious Ampere made contributions to mathematics, philosophy, botany, taxonomy, chemistry, and the physics of electricity and magnetism.

3. Electrons not attached to any particular atoms are abundantly available in metals, and currents result from their motion in response to an applied electric field.
4. In a pure semiconductor (called an intrinsic semiconductor), currents result from the motion of equal numbers of negative and positive charge carriers in response to an applied electric field; the negative charges are electrons, the propagating vacancies left by electron migration (which act like they have positive charge) are called holes. Small amounts of impurity elements are added to pure semiconducting materials, a process called **doping**, to make two other types of semiconducting materials: (a) n-type semiconducting materials, in which most of the moving charges are electrons (the n stands for *negative* charge carrier); (b) p-type semiconducting materials, in which most of the moving charges are the positive holes (p stands for *positive* charge carrier).

When considering the atomic view of matter, we must be careful with what we mean by the motion or flow of charge. If we take an isolated uncharged material, whether a conductor or not, and set it in motion, there are charges in motion (the positive nuclei and the surrounding electrons), but the current is zero because there is no motion associated with the net or total charge. On the other hand, a charged particle in motion does represent a current.

Likewise, for an uncharged conductor at rest, we find that some (but not all) electrons are continually moving randomly through the conductor, because they are very loosely attached to the nuclei. The thermodynamic internal energy of the material is sufficient to liberate the outer electrons from individual atoms, enabling the electrons to travel about in the material. But just as many electrons are moving one way as in the opposite direction, and the net flow of charge past any location is zero; there is zero current. This is the situation in electrostatics where the total electric field inside the conductor (whether charged or not) is zero. On the other hand, if an independent voltage source maintains even a slight potential difference between two sides (or ends) of a conductor, the electrons preferentially move in the direction opposite to that of the electric field established in the conductor by the independent voltage source; we have an electrodynamic situation where there is a net motion of charge, and so a current exists.

An electric field causes charges to accelerate. An electric field inside a material can be maintained by connecting the material to the terminals of an independent voltage source (say, an ideal battery), as in Figure 19.11, where we pretend the charges in motion are positive for reasons that will become apparent shortly.

The independent voltage source maintains a fixed potential difference between the ends of the material. The mobile charges

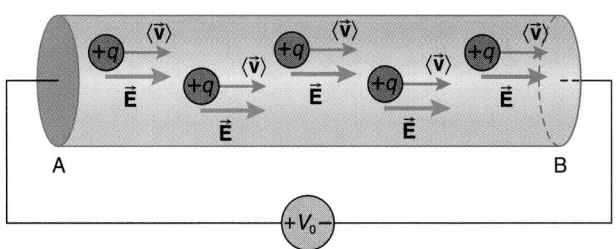

FIGURE 19.11 An independent voltage source maintains a current through a material.

in the material accelerate under the influence of the field (parallel to the field for positive charge carriers, antiparallel to the field for negative charge carriers). However, the accelerating charges collide very frequently with other particles inside the material. Such collisions mean that the velocities of the mobile charge carriers continually change their magnitude and direction almost randomly, though the overall motion or drift is more or less parallel or antiparallel to the field (depending on the sign of the charge carriers).

This rather messy situation is made tractable by introducing an **average drift speed** $\langle v \rangle$ for the mobile charges. As its name implies, the average drift speed represents the average speed at which the mobile charge carriers migrate through the material, parallel or antiparallel to the field. Although $\langle v \rangle$ is called the drift speed, $\langle v \rangle$ is more properly thought of as the (single) *component* of the average velocity vector of the charge carriers.

We wish to relate the current to this velocity component. Let the density of mobile charges (the charge carriers) be n; this represents the number of charge carriers per unit volume of the material. Let q be the charge on each charge carrier. Imagine the material to be cylindrically shaped, as in Figure 19.12. The electric current is the time rate of charge flow:

$$I = \frac{dQ}{dt}$$

where dQ is the number of coulombs of charge that pass a reference location of cross-sectional area A during a time interval dt. Charge conservation implies that the amount of charge per second arriving at and leaving any point along a wire is the same, thus preventing an otherwise embarrassing growing accumulation of charge (a pileup!) at any location within the wire. All the charge carriers within a distance $\langle v \rangle \, dt$ to the left of the reference location in Figure 19.12 will pass the location during the time interval dt. The number of charge carriers within this volume is

(number of charge carriers per unit volume)(volume) = $n(A\langle v \rangle \, dt)$

Each charge carrier has a charge q, and so the charge dQ within this volume is

$$dQ = qnA\langle v \rangle \, dt \qquad (19.2)$$

The current I is the time rate of charge flow past the location, or

$$I = \frac{dQ}{dt}$$

So, from Equation 19.2, the current is

$$I = qnA\langle v \rangle \qquad (19.3)$$

Note that in Equation 19.3, if we reverse the sign of *both* q and its velocity component $\langle v \rangle$, the *same* current I results. This is an important point; it means that positive charge flowing in one direction is *equivalent* to negative charge flowing in the opposite direction: the same current I results.

But if the flow of negative charge in one direction is equivalent to the flow of positive charge in the opposite direction, why, then, is there any current at all? The reason is because equivalent is not the same thing as actual. The two viewpoints can substitute for each other but are not in addition to each other.

Experiments now indicate the charge carriers in metal conductors are electrons, though this is not at all obvious. There is no way we can tell by examining, say, the experiments we performed in the last section whether the electric current in the wires is due to positive charge flowing one way, negative charge flowing the other way, or both.

By a convention dating back to Benjamin Franklin, the direction of a current is taken to be the direction of positive charge flow, whatever the actual sign of the charge carriers.

It is only through other experiments that we know the charge flow or current in conductors is caused by the motion of the negative electrons. In other materials, such as semiconductors, plasmas, or ionic solutions, the charge carriers may be positive, negative, or both.

In Example 19.1 you will calculate the average drift speed of the electrons in a common copper wire. The drift speed is surprisingly small; most snails easily outrun the electrons, and a terrapin is a fleet-footed demon by comparison. It takes on the order of *hours* for the electrons to drift through a distance of several meters in a common wire conductor! This should be surprising to you. Why does a distant light bulb come on almost instantaneously when you turn on the switch, if it takes the electrons so long to travel even one meter? Why does it not take hours for the bulb to light once the switch is turned on?

That is a good question and it warrants more than a lousy answer. An analogy may help. Hook up a long hose to a water faucet and turn the faucet on; it takes some time for the hose to fill with water and for a stream to emerge from the nozzle at the other end; the longer the hose, the longer the wait. On the other hand, if the hose is already full of water and the faucet is turned on, the water emerges from the other end as soon as the pressure increase propagates through the water in the hose to the other end. Since water is quite incompressible, the pressure pulse travels at great speed down the length of the water column in the hose and the water emerges from the nozzle almost immediately after the faucet is turned on.

Within a wire, the charge carriers (electrons) are distributed all along the wire and, like the water in a filled hose, this means that the electrical pipe (the wire) is already filled and ready for the faucet (the switch) to be turned on. The electric field that instigates the charge drift motion, analogous to the pressure pulse in the hose, propagates down the wire at a speed

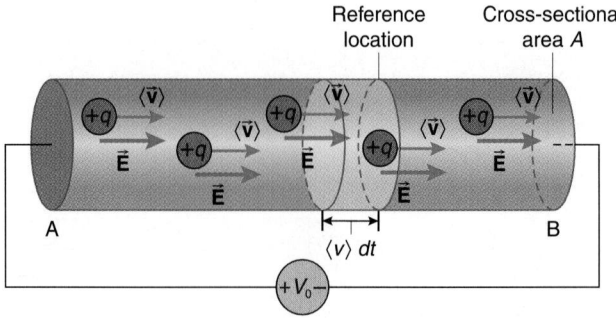

Reference location   Cross-sectional area *A*

$\langle v \rangle \, dt$

FIGURE 19.12 An electric current.

nearly equal to the speed of light. Thus the electrons all along the wire feel the field almost instantaneously and pick up their drift speed very fast. The current appears all along the wire almost instantaneously.

It is occasionally convenient to eliminate the dependence on the cross-sectional area in Equation 19.3, by defining a **current density** J across it to be

$$J = \frac{I}{A} = qn\langle v \rangle \qquad (19.4)$$

The current density J has the SI units of amperes per square meter (A/m^2). We make the current density a full-fledged vector by giving it the direction of the average drift velocity vector $\langle \vec{v} \rangle$:

$$\vec{J} = qn\langle \vec{v} \rangle \qquad (19.5)$$

In general, the current I is the *flux* of the current density through an area:

$$I = \int_{\text{area}} \vec{J} \cdot d\vec{S} \qquad (19.6)$$

The flux of $\vec{J}$ through the cross-sectional area A, shown in Figure 19.13, is easy to calculate, since $\vec{J}$ is parallel to the differential area vector everywhere across the cross-sectional area.

We also assumed the current density was constant over the entire area A, and so we have

$$I = \int_{\text{area}} \vec{J} \cdot d\vec{S} = J \int_{\text{area}} dS = JA$$

Thanks to the scalar product, we obtain the same current, or flux of the current density $\vec{J}$, regardless of the orientation of the area slice through the wire through which the flux is taken (see Figure 19.14). We have to, because it is the same total current everywhere.

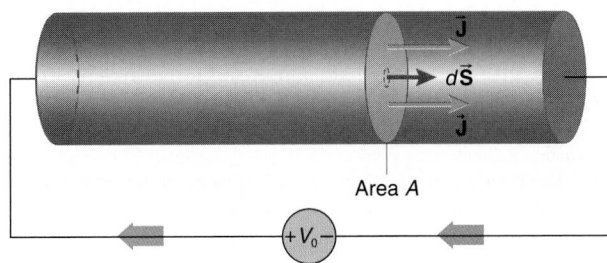

FIGURE 19.13 The flux of the current density through a cross-sectional area of the wire is the current.

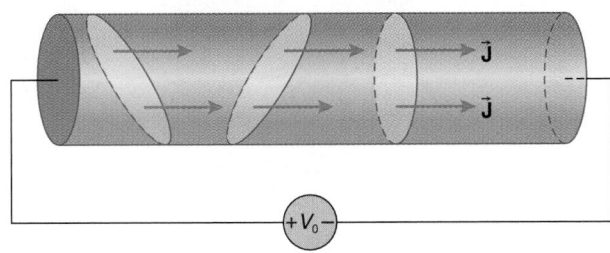

FIGURE 19.14 The current is the same through any cross section of the wire.

QUESTION 1

Geometrically, what represents the electric current at an instant t on a graph of charge Q versus time?

EXAMPLE 19.1

A 12 gauge copper wire, used for some household wiring, has a diameter of 2.05 mm and carries a current of 15.0 A.

a. Assume one electron per copper atom is responsible for the current. Find the drift speed of the electrons in the wire. The density of copper is 8.93×10^3 kg/m^3 and its molar mass is 63.5 g/mol.

b. How long does it take each charge carrier in this wire to travel 1.00 m?

Solution

a. The number of charge carriers per unit volume, n, is the same as the number of copper atoms per unit volume, since you are assuming one charge carrier per atom. One cubic meter of copper has a mass of 8.93×10^3 kg and one mole of copper has a mass of 63.5 g. Hence one cubic meter of copper represents

$$\frac{8.93 \times 10^3 \text{ kg/m}^3}{63.5 \times 10^{-3} \text{ kg/mol}} = 1.41 \times 10^5 \text{ mol/m}^3$$

Each mole has Avogadro's number of copper atoms. Hence the number of copper atoms per cubic meter is

$$(1.41 \times 10^5 \text{ mol/m}^3)(6.02 \times 10^{23} \text{ atoms/mol})$$
$$= 8.49 \times 10^{28} \text{ atoms/m}^3$$

With one charge carrier per atom, the number of charge carriers per cubic meter is this same number:

$$n = 8.49 \times 10^{28} \text{ electrons/m}^3$$

The circular cross-sectional area of the wire is

$$A = \pi r^2$$
$$= \pi[(2.05 \times 10^{-3} \text{ m})/2]^2$$
$$= 3.30 \times 10^{-6} \text{ m}^2$$

The charge carriers are electrons, which have a charge of magnitude $e = 1.602 \times 10^{-19}$ C. From Equation 19.3, the *average drift speed* is

$$\langle v \rangle = \frac{I}{qnA}$$

$$= \frac{15.0 \text{ A}}{(1.602 \times 10^{-19} \text{ C})(8.49 \times 10^{28} \text{ m}^{-3})(3.30 \times 10^{-6} \text{ m}^2)}$$
$$= 3.34 \times 10^{-4} \text{ m/s}$$

This speed is only a few tenths of a millimeter per second! This is a very slow speed. You can see that the current in the wire results not from the rapid flow of a huge charge, but rather from the slow flow of a humongous number of very small charges.

b. The time it takes the electrons to travel one meter in this wire is the distance divided by the average drift speed:

$$t = \frac{1.00 \text{ m}}{3.34 \times 10^{-4} \text{ m/s}}$$
$$= 2.99 \times 10^3 \text{ s}$$
$$= 49.8 \text{ min}$$

or slightly less than the leisurely duration of a typical physics lecture.

19.3 THE PIÈCE DE RÉSISTANCE: RESISTANCE AND OHM'S LAW

Now we come to grips with the resistive effects to charge flow in conducting materials, phenomena analogous to the resistive effect of obstructions or friction experienced by water when flowing in pipes. This introduces a very important new circuit element: a **resistor**.

The potential difference V between the points A and B along a length of material (see Figure 19.15) is found from the defining equation for the potential difference between two points, Equation 17.9:

$$V_B - V_A = -\int_A^B \vec{E} \cdot d\vec{r}$$

If the electric field $\vec{E}$ is constant over the entire length ℓ of the material, Equation 17.9 becomes

$$V_B - V_A = -E\ell$$

where V_A is greater than V_B (note the direction of the electric field in Figure 19.15). By convention, we take the potential difference V between the ends of the material to be the higher value minus the lower value; we say the potential difference is *across* the material:

$$V \equiv V_A - V_B = E\ell \qquad (19.7)$$

For many materials, it is found experimentally that the magnitude of the current density in the material is proportional to the magnitude of the applied electric field:

$$J = \sigma E \qquad (19.8)$$

where σ is a proportionality constant called the **conductivity** of the material.* Such materials are called **ohmic materials**; Equation 19.8 is one version of **Ohm's law**, though we will soon recast the law into a more useful form. It is important to realize that Ohm's law is an empirical law and does *not* characterize all materials. Using the ohmic approximation in Equation 19.7 for the potential difference, we find

$$V = E\ell$$
$$= \frac{J}{\sigma} \ell \qquad (19.9)$$

We assume the current density is uniform across the cross-sectional area of the material.[†] Using Equation 19.4 to substitute for J in Equation 19.9, we find

$$V = I \frac{\ell}{\sigma A} \qquad (19.10)$$

The reciprocal of the conductivity is called the **resistivity** (ρ) of the material:

$$\rho \equiv \frac{1}{\sigma} \qquad (19.11)$$

The resistivity of various materials is listed in Table 19.1.

In terms of the resistivity, Equation 19.10 becomes

$$V = I \frac{\rho \ell}{A} \qquad (19.12)$$

We define the **resistance** R of the material to be

$$R \equiv \frac{\rho \ell}{A} \qquad (19.13)$$

The SI units of resistance are V/A or **ohms** (Ω), after Georg Simon Ohm (1787–1854), a German physicist.

Equation 19.13 indicates that the resistance is proportional to the length ℓ of an ohmic material of constant cross-sectional area A. Empirically, the resistance is found to be inversely proportional to the cross-sectional *area* A of the

---

*Do not confuse the symbol for conductivity with same symbol for the surface charge density in electrostatics.

[†]This assumption is confirmed for dc currents by experiments we mention shortly; the assumption is not necessarily true for ac currents.

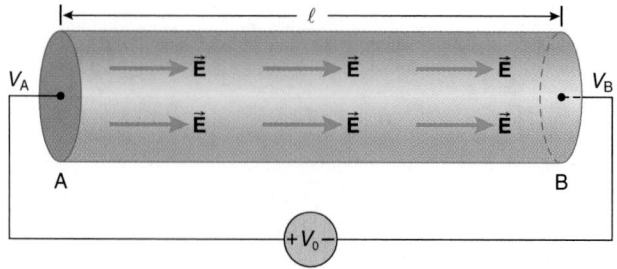

FIGURE 19.15 A material connected to an independent voltage source.

Ohm delved into electrical research, discovering his now famous law, to escape a pedestrian life as a tutor and to secure a university position. You better do your homework for him!

TABLE 19.1 The Resistivity and Temperature Coefficient of Resistivity for Various Materials at 20 °C

Material	Resistivity ρ ($\Omega \cdot m$)	Temperature coefficient of resistivity α (K^{-1})
Aluminum	2.82×10^{-8}	3.9×10^{-3}
Carbon	3500×10^{-8}	-0.5×10^{-3}
Copper*	1.77×10^{-8}	3.8×10^{-3}
Germanium	0.46	-48×10^{-3}
Gold	2.44×10^{-8}	3.4×10^{-3}
Glass	$\sim 10^{12}$	
Iron	10×10^{-8}	5.0×10^{-3}
Lead	22×10^{-8}	4.3×10^{-3}
Mercury	96×10^{-8}	0.9×10^{-3}
Nichrome†	100×10^{-8}	0.4×10^{-3}
Platinum	10×10^{-8}	3.92×10^{-3}
Silicon	640	-75×10^{-3}
Silver	1.59×10^{-8}	3.8×10^{-3}
Tungsten‡	5.6×10^{-8}	4.5×10^{-3}

*Most common electrical wire is made of copper.

†This is a nickel–chromium–iron alloy used widely in electrical heaters such as toasters.

‡Tungsten is used widely for light bulb filaments.

material. If the resistance were found to be inversely proportional to the *radius* of the conductor, one could infer that the current was confined to a certain outer layer of atoms at the perimeter of the circular cross section; instead, it is inversely proportional to the *square* of the radius. Therefore, we can infer that conduction in ohmic materials takes places in the bulk material and not on only its surface, which partially justifies the assumption we made that the current density in the material is uniform across the area A.

For convenience, Table 19.2 lists a few common wire sizes by their gauge numbers, their corresponding diameters in millimeters, and cross-sectional areas in square meters. You may find this table useful in computing the resistance of various real

TABLE 19.2 Wire Gauge Number, Wire Diameter, and Cross-Sectional Area at 20 °C for Common Wire Sizes

Gauge number*	Diameter	Area
0	8.25 mm = 8.25×10^{-3} m	53.5×10^{-6} m²
2	6.54 mm = 6.54×10^{-3} m	33.6×10^{-6} m²
4	5.19 mm = 5.19×10^{-3} m	21.1×10^{-6} m²
6	4.12 mm = 4.12×10^{-3} m	13.3×10^{-6} m²
8	3.26 mm = 3.26×10^{-3} m	8.37×10^{-6} m²
10	2.59 mm = 2.59×10^{-3} m	5.26×10^{-6} m²
12	2.05 mm = 2.05×10^{-3} m	3.31×10^{-6} m²
14	1.63 mm = 1.63×10^{-3} m	2.08×10^{-6} m²
16	1.29 mm = 1.29×10^{-3} m	1.31×10^{-6} m²
18	1.02 mm = 1.02×10^{-3} m	0.823×10^{-6} m²
20	0.812 mm = 0.812×10^{-3} m	0.517×10^{-6} m²
22	0.644 mm = 0.644×10^{-3} m	0.326×10^{-6} m²
24	0.511 mm = 0.511×10^{-3} m	0.205×10^{-6} m²
26	0.405 mm = 0.405×10^{-3} m	0.129×10^{-6} m²
28	0.321 mm = 0.321×10^{-3} m	0.081×10^{-6} m²

*For wire diameters of odd number gauge sizes, consult a handbook on electrical engineering.

wires. Household wiring typically is 14 gauge, but 12 gauge is used for kitchens and baths, which utilize large currents; 8 gauge wire is used for range ovens and clothes dryers. Common extension cords range from 18 gauge to 10 gauge, depending on the amount of current they are expected to carry.

We now combine Equations 19.12 and 19.13.

We find the potential difference across an ohmic material (high value − low value) is proportional to the current through it:

$$V = IR \qquad (19.14)$$

This is the most useful form for Ohm's law.

The standard circuit symbol for an **ideal resistor** is shown in Figure 19.16.* As with every ideal circuit element, we consider the leads to be *ideal wires* with zero resistance—that is, infinitesimal compared with real values for circuit resistors. Charges flow through an ideal wire with no loss of potential energy. We use ideal wires in circuit diagrams to connect discrete circuit elements. We have used the notion of ideal wires already as a way to connect other idealized circuit elements: independent voltage sources (ideal batteries) and ideal capacitors.

The direction of current is taken, by convention, to be that of positive charge flow. If an electric current exists in a resistor, the current enters it at the high-potential end and leaves at the low-potential end.

Thus we make polarity markings on a resistor consistent with the direction of the current through it, as shown in Figure 19.17. The current enters the resistor circuit element at the (+) polarity lead and exits at the (−) polarity lead of the circuit symbol.

---

*You can think of the resistor symbol as representing the filament of a light bulb, but keep in mind that not all resistors are light bulbs!

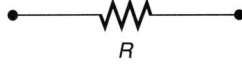

FIGURE 19.16 Circuit symbol for a resistor.

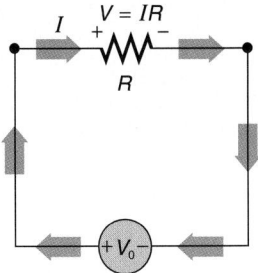

FIGURE 19.17 The polarity marking of a resistor is written so that the current enters it at the (+) side.

The potential difference V between the resistor lead marked (+) and that marked (−) is given by Ohm's law, Equation 19.14.

According to Ohm's law, an ideal resistor has a linear relationship between the potential difference V across it and the current through it. If you plot the current versus the potential difference across an ideal resistor, a straight line is obtained whose slope is the inverse of the resistance, as shown in Figure 19.18.

This linear relationship is an identifying characteristic of all ohmic materials. A negative current means it is directed opposite to that indicated in Figure 19.17; the potential difference $V = IR$ also then is negative, which means that the terminal marked (+) actually is at a lower potential than the one marked (−). This does *not* mean a mistake was made; the significance of these negative values will be noted later in the chapter (Problem-Solving Tactic 19.4 and Example 19.8). Materials or devices with *nonlinear* current–voltage relationships are called **nonohmic** materials or devices.

Off-the-shelf resistors, such as those available at electronic suppliers, have pretty little colored bands encircling them to indicate the approximate numerical value of the resistance, as shown in Figure 19.19. The colored bands are a code to determine the *nominal*

value of the resistance in ohms. The *actual* resistance value is within a certain percentage of the nominal value; the percentage is specified by an additional color band indicating the tolerance.

The resistor color code is as follows. Begin at the lead closest to the color bands. The first two colored bands signify the first two numbers of the value of the resistance according to the scheme in Table 19.3. The third colored band, called the multiplier band, indicates the exponent of the factor 10 by which to multiply the first two numbers to find the nominal value of the resistance in ohms.

The value of the resistance is determined according to the scheme

$$[\text{number number}] \times 10^{(\text{multiplier})}$$

The accuracy associated with the nominal value of the resistance is indicated by a fourth colored band according to the following scheme:

Color	Tolerance
Gold	5%
Silver	10%
(No fourth band)	20%

For examples, the resistors in Figure 19.19 have the colored bands brown–black–red–gold. The first two numbers of the value of the resistance are

$$\text{brown} = 1 \quad \text{and} \quad \text{black} = 0$$

or

$$10$$

and the multiplier band is red (2) so the multiplying factor is 10^2. The nominal resistance thus is

$$[\text{number number}] \times 10^{(\text{multiplier})} = [10] \times 10^2 \ \Omega$$
$$= 1.0 \ \text{k}\Omega$$

The gold band indicates that the actual numerical value of the resistance is within 5% of 1000 Ω. Since 5% of 1000 is 50 Ω, the actual resistance could be anywhere from 950 Ω to 1050 Ω. As an another example, a resistor with bands colored yellow (4), violet (7), and orange (3) with *no* fourth band has a nominal value of its resistance of

$$47 \times 10^3 \ \Omega = 47 \ \text{k}\Omega$$

with a tolerance of 20% (±9.4 kΩ).

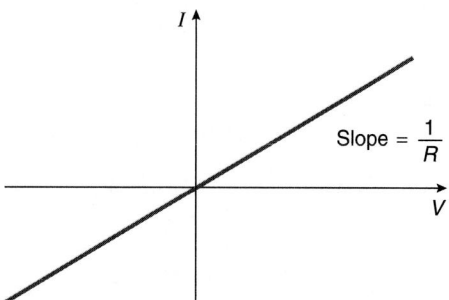

FIGURE 19.18 The current through a resistor plotted versus the potential difference across it; the inverse of the slope is the resistance.

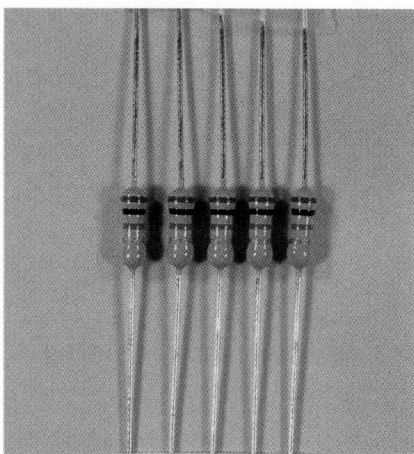

FIGURE 19.19 Resistors with a nominal value of 1 kΩ have the colored bands brown, black, and red with the fourth color band (here, gold) indicating the tolerance.

TABLE 19.3 Resistor Color Code

Color	Number	or Multiplying factor
Black	0	$10^0 = 1$
Brown	1	10^1
Red	2	10^2
Orange	3	10^3
Yellow	4	10^4
Green	5	10^5
Blue	6	10^6
Violet	7	10^7
Gray	8	10^8
White	9	10^9

A resistance measuring device known as an **ohmmeter** is used to measure the actual resistance of individual resistors; we will see how to use an ohmmeter later in the chapter (Section 19.15).

Your physics class might run a contest to create a tasteful and unprofane mnemonic to remember the order of the colors in the resistor color code. For example (and you likely can do better):

"Black Bears Roar, Orangutans Yell, Goats Bleat Violently, Go Weep!"

QUESTION 2

A resistor has the colored bands orange, black, yellow, and gold, in that order. Between what two values is the actual resistance of this resistor supposed to be according to the color code?

EXAMPLE 19.2

Calculate the resistance of 10.0 m of 12 gauge copper wire. Such wire is used frequently in home wiring.

Solution

The resistance is given by Equation 19.13:

$$R = \frac{\rho \ell}{A}$$

The resistivity of copper is found in Table 19.1 on page 843 and the cross-sectional area of a 12 gauge wire is found in Table 19.2. Making the appropriate substitutions, you have

$$R = \frac{(1.77 \times 10^{-8} \ \Omega \cdot m)(10.0 \ m)}{3.31 \times 10^{-6} \ m^2}$$
$$= 5.35 \times 10^{-2} \ \Omega$$

This indicates that the resistance of even long pieces of common copper wire is quite small, so such wires can be treated as ideal wires to an excellent approximation.

EXAMPLE 19.3

Calculate the magnitude of the electric field inside a 10.0 m length of 12 gauge copper wire that carries a current of 15.0 A. Assume the field is uniform along the entire length of the wire.

Solution

In Example 19.2, you found the resistance of such a piece of wire was $5.35 \times 10^{-2} \ \Omega$. The potential difference between the ends of the wire is found using Ohm's law, Equation 19.14:

$$V = IR$$
$$= (15.0 \ A)(5.35 \times 10^{-2} \ \Omega)$$
$$= 0.803 \ V$$

If you assume the field is uniform along its length, the magnitude of the potential difference is related to the magnitude of the electric field by Equation 19.7:

$$V = E\ell$$

Solving for E, you have

$$E = \frac{V}{\ell}$$
$$= \frac{0.803 \ V}{10.0 \ m}$$
$$= 8.03 \times 10^{-2} \ V/m$$

Recall that volts/meter (V/m) are equivalent to newtons/coulomb (N/C). This example shows that the magnitude of the electric field which causes the electric current inside a conductor is quite small. Notice that since both R and V are proportional to the length ℓ, the value of E is the same for any length of wire of a fixed gauge carrying a given current.

19.4 RESISTANCE THERMOMETERS

For many ohmic materials, the resistivity ρ and the resistance R are linear functions of temperature within certain temperature extremes. That is, we have the empirical relations of the following form:

$$\rho = \rho_0[1 + \alpha(T - T_0)] \quad (19.15)$$

and

$$R = R_0[1 + \alpha(T - T_0)] \quad (19.16)$$

where α is a constant known as the **temperature coefficient of resistivity**. Values of α for various materials are tabulated in Table 19.1 on page 843. The quantity R_0 is the resistance of the material at a reference temperature T_0 while ρ_0 is the corresponding resistivity. The temperature dependence of the resistance is exploited to make resistance thermometers, commonly used for medical applications and in physical laboratories.

In a resistance thermometer, a measurement of the resistance is used to find the temperature of the resistor. If the resistor is in thermal equilibrium with its environment, the temperature found via Equation 19.16 also gives the temperature of the environment. Resistance thermometers have several advantages over traditional mercury bulb thermometers:

1. Resistance thermometers can be fabricated with small physical size (dimensions) and mass, so that it is possible to measure the temperature of small systems easily and reliably. Recall that a thermometer typically must be small in mass and size compared with the system whose temperature is to be measured, unless the system is in thermal contact with a thermodynamic reservoir.
2. Resistance is an electrical quantity, so it is easy to use the electrical measurement of temperature as the first stage in the electronic manipulation of data.

If the resistance is not a linear function of temperature, then of course Equation 19.16 cannot be used. Manufacturers of resistance thermometers supply calibration curves that plot empirically determined temperature versus resistance data.

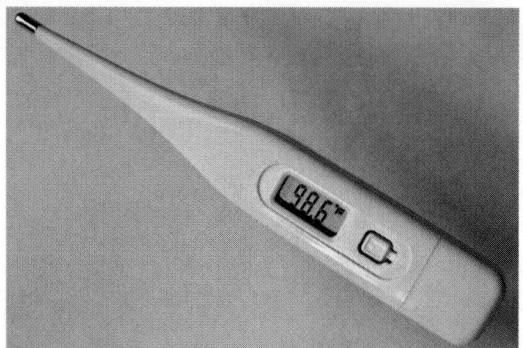

An electric thermometer with a digital readout is quite useful and convenient for both research and medical applications.

A resistance thermometer is an example of a **transducer**, a device for converting a physical quantity into (usually) an electrical quantity or vice versa.

QUESTION 3

Notice in Table 19.1 on page 843 that carbon, germanium, and silicon all have negative temperature coefficients of resistance; the resistance decreases as the temperature increases. Does this imply that their resistance goes to zero at sufficiently high temperatures? What might prevent this?

EXAMPLE 19.4

What temperature change produces a 10% increase in the resistance of a copper wire?

Solution

The resistance of the wire depends on temperature via Equation 19.16:

$$R = R_0[1 + \alpha(T - T_0)]$$

where R is the resistance at temperature T and R_0 is the resistance at temperature T_0. For a 10% increase, you want R to be equal to $1.10R_0$; hence you have

$$1.10R_0 = R_0(1 + \alpha\,\Delta T)$$

Solve for ΔT:

$$\Delta T = \frac{0.10}{\alpha}$$

The temperature coefficient of resistance α of copper is found from Table 19.1 on page 843, and so you obtain

$$\Delta T = \frac{0.10}{3.8 \times 10^{-3}\ \text{K}^{-1}}$$

$$= 26\ \text{K}$$

Hence a temperature increase of 26 K increases the resistance of a copper wire by 10%.

EXAMPLE 19.5

A tungsten filament of a light bulb has a resistance of 18 Ω at a room temperature of 20 °C. The bulb is connected to an independent voltage source, as shown in Figure 19.20, and when the potential difference across the bulb has a magnitude of 30.0 V, the current through it is 0.185 A. What is the temperature of the filament of the bulb?

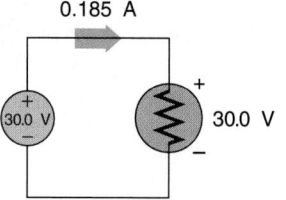

FIGURE 19.20

Solution

The temperature coefficient of resistivity for tungsten is found in Table 19.1 (on page 843) to be

$$\alpha = 4.5 \times 10^{-3}\ \text{K}^{-1}$$

Use Equation 19.16 and let R_0 be the resistance of the bulb at room temperature T_0; that is, $R_0 = 18\ \Omega$ when $T_0 = 293$ K. The resistance of the bulb when lit is found using the given data and Ohm's law, Equation 19.14:

$$V = IR$$

Solve for R:

$$R = \frac{V}{I}$$

$$= \frac{30.0\ \text{V}}{0.185\ \text{A}}$$

$$= 162\ \Omega$$

Using this information in Equation 19.16, you get

$$162\ \Omega = (18\ \Omega)[1 + (4.5 \times 10^{-3}\ \text{K}^{-1})(T - 293\ \text{K})]$$

Remember the sizes of a kelvin and a celsius degree are the same. Solving for T, you find

$$T = 2.1 \times 10^3\ \text{K}$$

Hot stuff! For comparison, this temperature is about 30% of the surface temperature of the Sun (about 6×10^3 K).

19.5 CHARACTERISTIC CURVES

No, this section is not about the grading curve of your professor.

> A graph of the current through a circuit element versus the potential difference across the element is called the **characteristic curve** of the circuit element.

The potential difference is usually the most easily controlled electrical quantity, and so the potential difference always is graphed as the abscissa (the horizontal axis) with the current as the ordinate (the vertical axis).

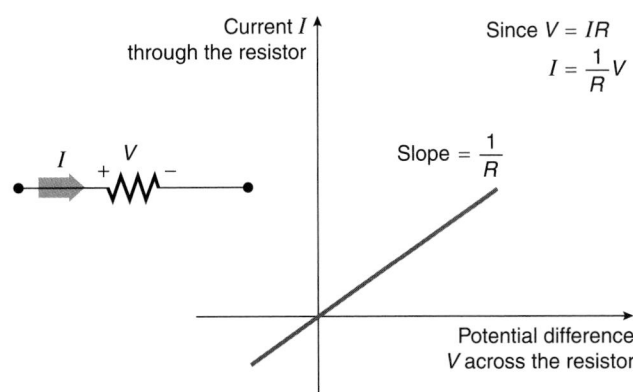

FIGURE 19.21 The characteristic curve of a resistor.

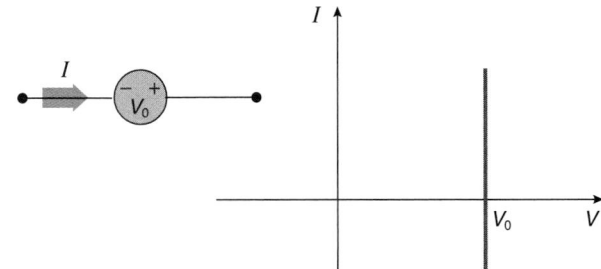

FIGURE 19.22 The characteristic curve of an independent voltage source.

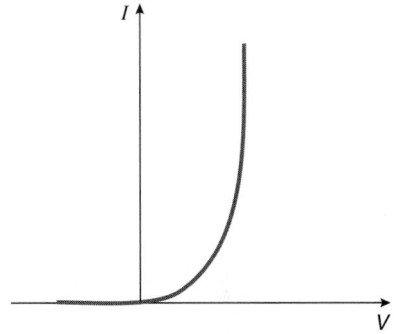

FIGURE 19.23 The characteristic curve of a semiconducting pn junction, a nonlinear device.

From Ohm's law, Equation 19.14, the characteristic curve of an ideal resistor, a plot of I versus V,

$$I = \frac{1}{R} V$$

is a straight line inclined to the axes, as shown in Figure 19.21. An ideal resistor is called a linear device because its characteristic curve is linear with a positive slope. The slope of the characteristic curve of a resistor is the *reciprocal* of the resistance. For negative values of the potential difference, the current is negative and indicates that the current is in the direction opposite to that implied by the polarity markings.

The characteristic curve of an independent voltage source (such as an ideal battery) is a straight vertical line at the specified emf of the battery, as shown in Figure 19.22. The resistance of an ideal battery is zero, since the slope of its characteristic curve is infinite.

A distinctly nonlinear device is shown in Figure 19.23. It is the characteristic curve of a semiconducting device known as a pn junction. Such pn junctions form one of the fundamental building blocks of semiconducting technology. We define the (varying) resistance of the device as the inverse of the slope of the characteristic curve at any point. That is, more generally, the resistance at any point along *any* characteristic curve is defined as the inverse of the slope at that point:

$$R \equiv \frac{1}{\frac{dI}{dV}} = \frac{dV}{dI} \qquad (19.17)$$

For a pn junction, the resistance depends on the value of the potential difference because the (inverse) slope of the curve changes. An ideal resistor has a resistance that is independent of the potential difference across it, since the relationship between the current and the potential difference is linear.

QUESTION 4 _____

For the characteristic curve of the pn junction shown in Figure 19.23, what is the resistance of the device for negative values of V? How much current can pass through a resistance of such magnitude? What does the resistance ap-

proach for increasingly positive values of V? A resistance of zero is equivalent to a short circuit. A pn junction is called a *diode*, because it conducts current easily for positive values of V and has essentially zero current for negative values of V. A diode has the circuit symbol shown in Figure Q.4, where the arrow indicates the direction that charge flows easily through the device (the direction that significant current can pass through the device).

FIGURE Q.4

19.6 SERIES AND PARALLEL CONNECTIONS REVISITED

We now examine a series connection from the standpoint of electric current. Since charge is conserved, the electric current through every element in a series connection must be the same. What goes in one terminal comes out the other, just like water in a single pipe.

> The current is the same in all circuit elements in series with each other.

> Recall that the potential difference is the same across all circuit elements in parallel with each other.

19.7 RESISTORS IN SERIES AND IN PARALLEL

The light bulb resistors in your home are connected in parallel; cheap strings of Christmas tree lights are connected in series. In this section we see how to combine resistors in series and parallel into equivalent single resistors.

Resistors in Series

Figure 19.24 shows a series connection of three resistors between terminals A and B, connected to an independent voltage source. Current I passes through each of the resistors, instigated by the independent voltage source. Because of the direction of the current, we must mark the resistors with the indicated polarities: the current enters at the high-potential end of each resistor [polarity marking (+)] and leaves at the low-potential end [polarity marking (−)].

The potential difference across each resistor is found from Ohm's law:

$$V_1 = IR_1 \qquad V_2 = IR_2 \qquad V_3 = IR_3 \qquad (19.18)$$

We want to replace the series of resistors with a single equivalent resistor R_{eq} connected between the same two terminals A and B, as shown in Figure 19.25.

The current in the equivalent resistor also is I because that is the current between terminals A and B. The total potential difference between A and B is

$$V = V_1 + V_2 + V_3 \qquad (19.19)$$

Since the potential difference between A and B across R_{eq} also must be V, we have $V = IR_{eq}$. Using Ohm's law for each potential difference in Equation 19.19, we obtain

$$IR_{eq} = IR_1 + IR_2 + IR_3$$
$$R_{eq} = R_1 + R_2 + R_3$$

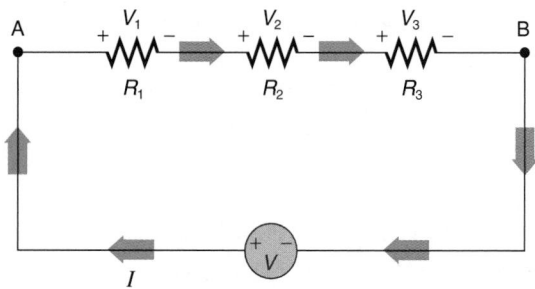

FIGURE 19.24 A series connection of three resistors.

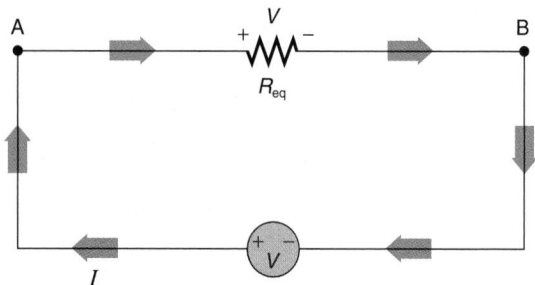

FIGURE 19.25 The equivalent resistor between the same two terminals A and B.

This argument can be extended to any number of resistors in series:

$$R_{eq} = R_1 + R_2 + R_3 + \cdots \qquad \text{(resistors in series)} \qquad (19.20)$$

The equivalent resistance is the sum of the individual resistances. Resistors in series combine like capacitors in parallel.

Resistors in Parallel

If several resistors are connected in parallel, they all are connected to the same two distinct nodes, as shown in Figure 19.26. Because the collection is connected to the same two nodes, the potential difference V across each resistor is the same. We want to replace the set with an equivalent resistor with resistance R_{eq} (see Figure 19.27) so that the potential difference across R_{eq} is the same as across any of the resistors in parallel.

The current I leaving terminal A in Figure 19.26 divides (much like water dividing up among multiple pipes), with only a fraction going through each of the individual resistors. But conservation of charge implies that the currents through the individual resistors must sum to the current entering terminal A:

$$I = I_1 + I_2 + I_3 \qquad (19.21)$$

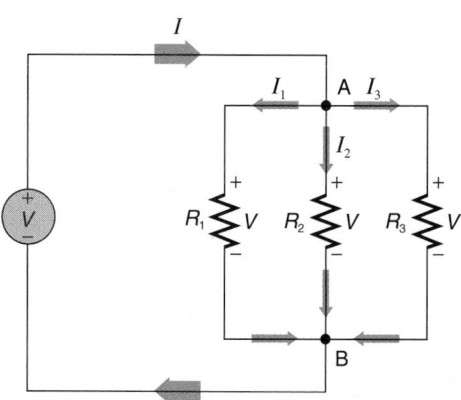

FIGURE 19.26 Three resistors connected in parallel.

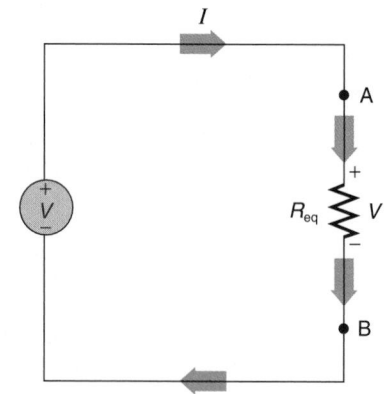

FIGURE 19.27 The equivalent resistor.

For each resistor we write Ohm's law, remembering that the potential difference across each is the same (since they are in parallel):

$$V = I_1R_1 \quad V = I_2R_2 \quad V = I_3R_3 \quad \text{and} \quad V = IR_{eq}$$
$$(19.22)$$

We solve these Ohm's law equations for the individual currents, and substitute into Equation 19.21. We obtain

$$\frac{V}{R_{eq}} = \frac{V}{R_1} + \frac{V}{R_2} + \frac{V}{R_3}$$

$$\frac{1}{R_{eq}} = \frac{1}{R_1} + \frac{1}{R_2} + \frac{1}{R_3} \qquad (19.23)$$

Equation 19.23 can be generalized for any number of resistors in parallel:

$$\frac{1}{R_{eq}} = \frac{1}{R_1} + \frac{1}{R_2} + \frac{1}{R_3} + \cdots \quad \text{(resistors in parallel)} \quad (19.24)$$

So resistors in parallel combine like capacitors in series.

As with the latter, the equivalent resistance R_{eq} always is less than the smallest resistance that was combined.

PROBLEM-SOLVING TACTICS

19.1 As with the capacitors in series, be careful using Equation 19.24 in (1) adding the fractions correctly and (2) remembering that the sum of the reciprocals is the *reciprocal* of the equivalent resistance.

19.2 The equivalent resistance of two resistors in parallel is the product of their resistances divided by the sum. For *two* resistors in parallel we can use the same shortcut we developed for capacitors in series. That is, for two resistors in parallel the equivalent resistance is

$$\frac{1}{R_{eq}} = \frac{1}{R_1} + \frac{1}{R_2}$$
$$= \frac{R_2 + R_1}{R_1R_2}$$
$$R_{eq} = \frac{R_1R_2}{R_1 + R_2} \qquad (19.25)$$

This shortcut only is appropriate for two resistors at a time; parallel combinations of more than two resistors must be tackled pairwise to use this shortcut.

QUESTION 5

Time for the holidays. You and a friend each have bought a set of tree lights. When one bulb on the set you bought blows out, the entire string of lights goes out; just your luck! With your friend's set, if one light blows out, the oth-

ers stay lit. What is the difference between the way the two sets of lights are wired? Are the bulbs in the two sets likely to be interchangeable?

EXAMPLE 19.6

What is the equivalent resistance between the terminals A and B of the resistors shown in Figure 19.28?

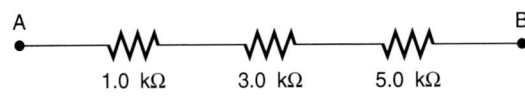

FIGURE 19.28

Solution
The resistors are in series, and so the equivalent resistance between the terminals A and B is the sum of the resistances (Equation 19.20), as shown in Figure 19.29:

$$R_{eq} = 1.0 \text{ k}\Omega + 3.0 \text{ k}\Omega + 5.0 \text{ k}\Omega$$
$$= 9.0 \text{ k}\Omega$$

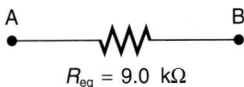

FIGURE 19.29

EXAMPLE 19.7

Find the equivalent resistance of the resistors shown in Figure 19.30.

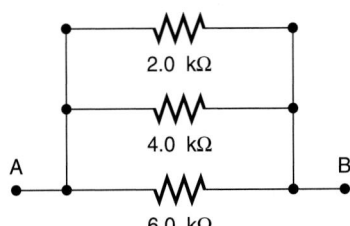

FIGURE 19.30

Solution
All three resistors are connected to the same two distinct nodes and so the three are connected in parallel.

Method 1
Use Equation 19.24:

$$\frac{1}{R_{eq}} = \frac{1}{2.0 \text{ k}\Omega} + \frac{1}{4.0 \text{ k}\Omega} + \frac{1}{6.0 \text{ k}\Omega}$$
$$= \frac{11}{12.0 \text{ k}\Omega}$$

so

$$R_{eq} = \frac{12.0}{11} \text{ k}\Omega$$
$$= 1.1 \text{ k}\Omega$$

Note that the equivalent resistance is smaller than the smallest of the resistances that were combined.

Method 2

Combine the resistances pairwise using the product over the sum rule, Equation 19.25 in Problem-Solving Tactic 19.2. Combining the 2.0 kΩ and 4.0 kΩ resistances, you get an equivalent resistance of

$$\frac{(2.0 \ k\Omega)(4.0 \ k\Omega)}{2.0 \ k\Omega + 4.0 \ k\Omega} = \frac{8.0}{6.0} \ k\Omega = 1.3 \ k\Omega$$

Then combine this pairwise with the 6.0 kΩ resistance using the same product divided by sum rule:

$$\frac{(1.3 \ k\Omega) \ (6.0 \ k\Omega)}{1.3 \ k\Omega + 6.0 \ k\Omega} = 1.1 \ k\Omega$$

The resistors also could be combined pairwise in another order, say, by combining the 2.0 kΩ resistance with the 6.0 kΩ resistance, yielding 1.5 kΩ; then this result with the 4.0 kΩ resistance, again yielding 1.1 kΩ.

19.8 ELECTRIC POWER

The light bulb in your study lamp burning brightly on your desk, long into the night while pouring over this and other texts, likely is a 100 W or 200 W bulb. The wattage rating of a light bulb resistor is an indication of the electric power which is absorbed by (transferred to) the bulb when it is on and then released to the environment (transferred to it) as light and heat transfer. Stove burners and wires that glow inside your toaster also are resistors. Here we see how to calculate the electrical power absorbed by (transferred to) *any* circuit element.

To calculate the power absorbed, we need to account for the changes in the electrical potential energy of charges flowing through a circuit element as a function of time. To do this, we consider a circuit element with a potential difference $V = V_A - V_B$ across it, as shown in Figure 19.31. Point A is at the higher electric potential, so the polarities are marked as indicated.

A small, differential charge dQ at position A has a potential energy $dQ \ V_A$, and at position B it has potential energy $dQ \ V_B$. The differential change in the potential energy of the charge is (final value − initial value):

$$d(PE) = dQ \ V_B - dQ \ V_A$$
$$= dQ \ (V_B - V_A)$$
$$= dQ \ (-V)$$

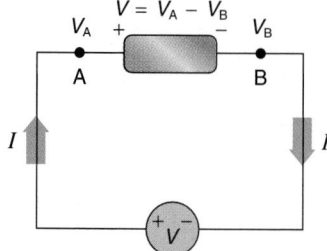

FIGURE 19.31 The potential difference V across a circuit element.

For positive dQ, the change in the electrical potential energy of the charge is negative since $V_B < V_A$. The potential energy of the positive charge decreases. You should convince yourself that the potential energy of a negative charge going the other way (from B to A) *also* decreases. What happens to this loss of electrical potential energy? We say the circuit element *absorbs* the energy; the energy is transferred to it. The circuit element continually converts the electrical potential energy into other forms of energy (typically thermodynamic internal energy and perhaps light) and shares it with the world (via heat transfer).

For a resistor circuit element, the energy increase appears as an increase in the thermodynamic internal energy of the device: it gets warmer. In other words, the energy lost by the charge is gained by the circuit element. The energy gained (absorbed) by the circuit element (transferred to it) is therefore

$$+dQ \ V$$

so that the total energy of the system (charge + circuit element) is conserved. Let the process occur during a differential time dt. The time rate at which the circuit element absorbs energy from the change in the electrical potential energy of the charges flowing through it is the **electric power absorbed** by (transferred to) the circuit element. Hence the electric power absorbed by a circuit element is

$$P = \frac{dQ}{dt} V$$

But

$$\frac{dQ}{dt} = I$$

the current through the circuit element.

> Thus the electrical power absorbed by (transferred to) a circuit element is the product of the current through the element and the potential difference V across the element:
>
> $$P = IV \qquad (19.26)$$

Notice the direction of the current indicated in Figure 19.31 and the polarity of the potential difference across the element.

> To use Equation 19.26 the current should be directed *into* the positive polarity terminal of the circuit element, as shown in Figure 19.32.

If the current is directed *out* of the positive polarity terminal, the current $-I$ is directed into it, as indicated in Figure 19.33. This result follows since a positive charge moving in one direction is equivalent to a negative charge moving in the opposite direction, as we showed in Section 19.2.

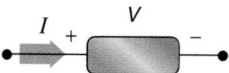

FIGURE 19.32 For Equation 19.26, the current I should be directed into the positive polarity terminal.

FIGURE 19.33 Equivalent representations of the current.

If the result of using Equation 19.26 for the power absorbed is positive, charges passing through the circuit element are losing potential energy (per unit time), the circuit element gains (absorbs) this energy (per unit time), and the energy manifests itself in various other forms. For a resistor, the energy absorbed appears as an increase in its thermodynamic internal energy: the resistor becomes warmer. If an independent voltage source absorbs positive power, it is storing the energy in other forms (such as chemical energy). An independent voltage source absorbing positive electric power is said to be **charging**. When a capacitor absorbs positive power, it stores the energy in the electric field between its plates.

On the other hand, if the result of using Equation 19.26 for the power is negative, we have negative power absorbed: energy (per unit time) is generated (or provided) by the circuit element, transferred from it to electrical charges flowing through it. Negative power absorbed does *not* mean the circuit element gets cooler! It simply means that the circuit element increases the potential energy of positive charges that we imagine constituting the current (the direction of positive charge flow); the energy for this may arise from electrochemical reactions or more exotic sources. Since independent voltage sources typically increase the electrical potential energy of positive charges flowing through them, independent voltage sources absorb negative power under these circumstances.* It is for this reason that independent voltage sources also are called *power supplies*, that is, suppliers of electrical power (electric energy per unit time). When a capacitor absorbs negative power, it is discharging; the capacitor is decreasing the energy stored in the electric field between its plates and increasing the electrical potential energy of the positive charges that we imagine constituting the current.

So electric power absorbed can be either positive or negative. Positive power absorbed is real power absorption by the circuit element (transferred to it) from the positive charges that we imagine constitute the current. Negative power absorbed by a circuit element is power generated by the circuit element (transferred from it) to increase the potential energy of the positive charges that create the current.

Conservation of energy (per unit time) means that the total power absorbed by all the circuit elements in a circuit must be zero:

$$\text{total power absorbed by a circuit} = 0 \text{ W} \quad (19.27)$$

That is, the sum of the individual powers absorbed by all the circuit elements must be zero.

*If a battery is charging, it absorbs positive electrical power.

The common unit known as the **kilowatt-hour** (kW·h) appears inexorably on the bill that you or your parents receive from the electric company every month. A kilowatt-hour is equivalent to a power of exactly 1000 W ≡ 1 kW used for exactly one hour.

Thus a kilowatt-hour is an *energy* unit, *not* a power unit.

The amount of energy in one kilowatt-hour is exactly

$$
\begin{aligned}
1 \text{ kW·h} &= (1000 \text{ W})(1 \text{ h}) \\
&= (1000 \text{ J/s})(3600 \text{ s}) \\
&= 3.600 \times 10^{6} \text{ J} \quad \text{(exact conversion)}
\end{aligned}
$$

A bill from the electric company, therefore, is not a bill for power; it is a bill for energy. Your electric company is an *energy* company.

PROBLEM-SOLVING TACTICS

19.3 Since resistors always become warm or hot when they have a nonzero current, resistors always absorb positive power (or zero power if the current through the resistor is zero). This follows from Equation 19.26. Since Ohm's law states that $V = IR$, with the current I going into the positive polarity terminal of the resistor, the power absorbed by a resistor is

$$
\begin{aligned}
P_{\text{res}} &= IV \\
&= I(IR) \\
&= I^{2}R \quad \text{(for a resistor } \textit{only}) \quad (19.28)
\end{aligned}
$$

which is never negative. When solving a problem, if you ever find that the *power* absorbed by a resistor comes out negative, you can be certain that an error was made somewhere in the calculation, not that you have a new patentable invention that will make you a millionaire overnight!

19.4 If you find the current through a resistor is negative, $-|I|$, Ohm's law leads to a negative potential difference, $-|V|$. The power absorbed by the resistor still is positive. This result follows since

$$
\begin{aligned}
P &= IV \\
&= (-|I|)(-|V|) \\
&= |I||V|
\end{aligned}
$$

See Example 19.8 for an illustration of this.

19.5 Independent voltage sources absorb either positive or negative power depending on the direction of the current through them. Positive power is absorbed when the current *enters* the positive polarity terminal; negative power when the current *emerges from* the positive polarity terminal.

19.6 If a current I is emerging from the positive polarity terminal of a circuit element, current $-I$ is entering the positive polarity terminal. See Figure 19.33.

QUESTION 6

Equation 19.28 states that the power absorbed by a resistor is proportional to R: $P = I^2R$. Using Equation 19.26 for the power, $P = IV$, and using I from Ohm's law,

$$I = \frac{V}{R}$$

the power absorbed by a resistor is

$$\frac{V^2}{R}$$

which shows that the power absorbed by the resistor is inversely proportional to R. How do you explain the apparent contradiction?

 STRATEGIC EXAMPLE 19.8

A 12.0 V independent voltage source is connected to a 10.0 Ω resistor, as shown in Figure 19.34.

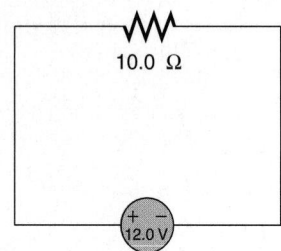

FIGURE 19.34

a. Determine the current through the resistor.
b. Calculate the power absorbed by the resistor.
c. Calculate the power absorbed by the independent voltage source.
d. Verify that the total power absorbed by all the circuit elements is 0 W.

Solution

a. Since the resistor is connected to the same two terminals as the independent voltage source, the potential difference across the resistor is 12.0 V. The two circuit elements are simultaneously in parallel and in series! Use Ohm's law to find the current through the resistor:

$$V = IR$$
$$12.0 \text{ V} = I(10.0 \text{ Ω})$$

Solving for I, you get

$$I = 1.20 \text{ A}$$

The current through a resistor is from the high-potential end to the low-potential end, so the direction of the current and the polarity marking of the resistor are as indicated in Figure 19.35.

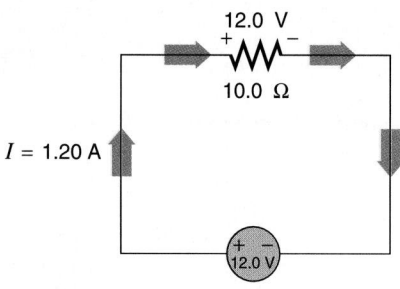

FIGURE 19.35

b. The power absorbed by the resistor can be found in two ways.

Method 1

The power absorbed is the product of the current into the positive polarity terminal of the resistor times the potential difference across the resistor:

$$P = IV$$
$$= (1.20 \text{ A})(12.0 \text{ V})$$
$$= 14.4 \text{ W}$$

Method 2

The power absorbed by a resistor is the product of the square of the current and the resistance (Equation 19.28):

$$P = I^2R$$
$$= (1.20 \text{ A})^2(10.0 \text{ Ω})$$
$$= 14.4 \text{ W}$$

c. The power absorbed by the independent voltage source is the product of the current into its positive polarity terminal times the potential difference across the circuit element. The current 1.20 A is coming out of the positive terminal, hence −1.20 A is going into the positive terminal (Problem-Solving Tactic 19.6), as shown in Figure 19.36. Hence the power absorbed by the independent voltage source is

$$P = (-1.20 \text{ A})(12.0 \text{ V})$$
$$= -14.4 \text{ W}$$

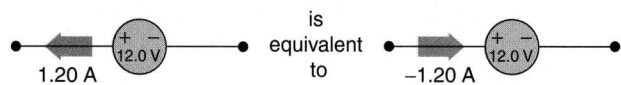

FIGURE 19.36

The power absorbed is negative, indicating that the circuit element actually is generating power for (transferring power to) the rest of the circuit.

d. The total power absorbed by all the circuit elements is

$$P_{total} = P_{res} + P_{independent \text{ voltage source}}$$
$$= 14.4 \text{ W} + (-14.4 \text{ W})$$
$$= 0 \text{ W}$$

19.9 ELECTRICAL NETWORKS AND CIRCUITS

Any collection of circuit elements that are connected together is an electrical **network**, as in Figure 19.37.

> An electrical **circuit** is a network with at least one closed conducting path or loop, as shown in Figure 19.38.

All circuits are networks, but not all networks are circuits. It is possible for a part of a network to be a circuit; in Figure 19.38, part of the network is a circuit. Circuit elements that are not part of a circuit can be disregarded (see Figure 19.39), since there will be no currents through these circuit elements; they are extraneous circuit elements. So the significant part of a network is just the part that is a circuit. We are specifically interested in electrical circuits since it is around such closed paths that electrical currents can exist.

The simplest circuit has one elementary loop of two circuit elements, as indicated in Figure 19.40. The two circuit elements are connected to the same two distinct nodes, are in parallel, and have the same potential difference between their terminals. But

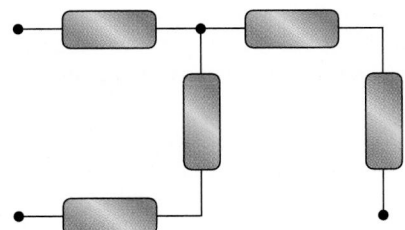

FIGURE 19.37 An electrical network.

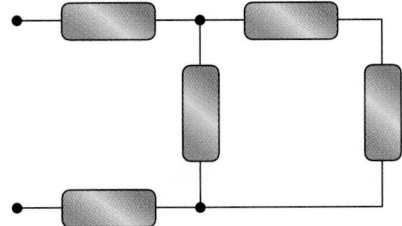

FIGURE 19.38 An electrical circuit.

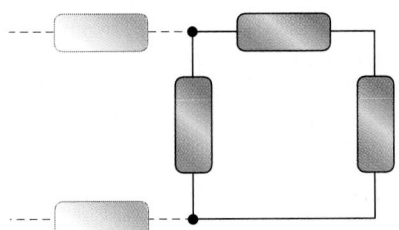

FIGURE 19.39 Circuit elements that are not part of the circuit can be disregarded.

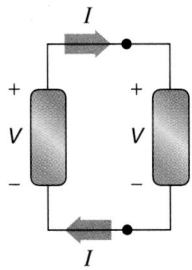

FIGURE 19.40 The simplest circuit: two circuit elements simultaneously in series and parallel.

the current in the loop also has no choice but to flow through one element and then the other. The current thus is the same in both circuit elements. The two circuit elements are *simultaneously* connected in series and parallel.

19.10 ELECTRONICS

Electronics is the application of physics to electric charge, its motion, and circuits. The subject is divided into several broad areas to indicate the principal focus of inquiry.

1. In **analog electronics**, the specific values of the currents through and potential differences across the various circuit elements are important to know and are the focus of concern. Common household circuits, flashlights, and most household appliances are examples of applications of analog electronics.

2. In **digital electronics**, such intimate knowledge of specific potential differences and current values is, by and large, of secondary importance. In digital electronics the objective typically is to determine whether various circuit elements are *on* (meaning either conducting current or with some standard potential difference across them) or *off* (not conducting current or with another standard potential difference across them). Much of the electronics associated with computers and other digital equipment are applications of digital electronics.

We will be concerned in this text with analog electronics because the ideas of digital electronics rest on analog principles. It is important to grasp the fundamental analog ideas first; digital electronics then comes quite easily.

Analog electronics is further subdivided into three broad areas:

1. **dc circuit analysis:** In this area, the various currents through the circuit elements and potential differences across the circuit elements are independent of time, as shown in Figure 19.41. In other words, the various potential differences and currents associated with the various circuit elements have steady, constant values. The term dc comes from the words *direct current*, meaning currents that are not functions of time.

2. **ac circuit analysis:** In this area, the various currents and potential differences are periodic functions of time; see Figure 19.42. That is, they fluctuate in time but periodically repeat the pattern of the fluctuations. The fluctuations need not be sinusoidal; they are merely periodic in time. We study sinusoidally varying ac circuits in Chapter 22.

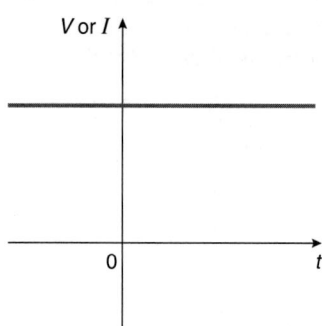

FIGURE 19.41 Currents and potential differences are independent of time in dc circuit analysis.

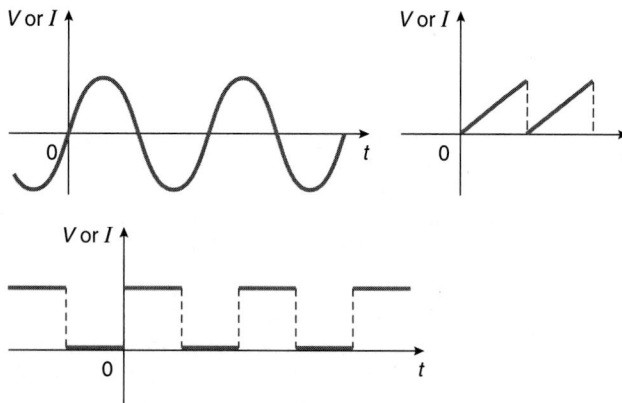

FIGURE 19.42 Periodic currents and potential differences characterize ac circuit analysis.

3. **Transient circuit analysis**: This area investigates what happens when, for example, circuits are turned on or off. Transient analysis investigates the changes that happen when a circuit goes from one steady state (in a dc or ac situation) to another steady state. We examined one aspect of transient circuit analysis in Chapter 18, the charging and discharging of a capacitor.

We begin our study of electronics with an investigation of dc circuits. The fundamental ideas of dc circuit analysis also are used to study ac circuits. We touch only lightly on transient analysis, since the study of transients is perhaps the most difficult of the three fields.

19.11 KIRCHHOFF'S LAWS FOR CIRCUIT ANALYSIS

O, two silver currents, when they join

William Shakespeare*

The analysis of both dc and ac circuits relies on just two pillars: the **Kirchhoff laws** of circuit analysis, named after Gustav Robert Kirchhoff (1824–1887), a German pioneer in the study of electrical phenomena. Kirchhoff's laws are restatements of physical ideas that already are quite familiar to you.

---

*King John, Act 2, Scene 1, line 441.

Kirchhoff was a gifted German researcher in thermodynamics, spectra, and electricity. He was known for his cheerful disposition. An accident left him disabled and he needed crutches or a wheelchair for mobility.

The Kirchhoff Current Law

The Kirchhoff current law (KCL) states that the algebraic sum of the currents leaving any node of a circuit is zero.

One could state, equivalently, that the algebraic sum of the currents entering any node of a circuit is zero. But in courses in engineering, the statement conventionally is expressed in terms of currents leaving a node, and there is no reason to depart from that convention here.

What does the KCL mean physically?

From the standpoint of physics, the KCL is a statement of *charge conservation.*

The node neither can accumulate more and more charge nor go into complete charge bankruptcy and beyond.

The application of the KCL can be seen with a specific example. Figure 19.43 depicts an arbitrary node, extracted from a circuit. A "bird's nest" of wire leads from various circuit elements are tied together at the node. Notice that some currents are leaving the node and some currents are entering the node. The KCL states that the algebraic sum of all the currents leaving a node is zero. Currents actually leaving the node are placed into the sum of the currents with a plus sign. If a current actually is entering a node (such as current I_2), the current $-I_2$ is leaving the node; so the negative current is placed in the sum of the currents leaving the node. The KCL applied to the node depicted in Figure 19.43 results in the following equation:

$$I_1 + (-I_2) + I_3 + (-I_4) + I_5 = 0 \text{ A}$$
$$I_1 - I_2 + I_3 - I_4 + I_5 = 0 \text{ A}$$

The KCL can be applied to *any* node of a circuit. In practice, one uses the KCL only at nodes connecting three or more circuit elements. Why? Consider a node connecting just two circuit

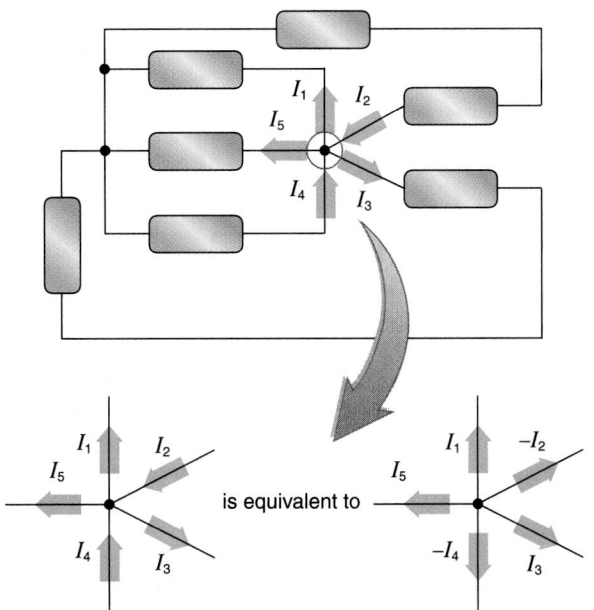

FIGURE 19.43 One node of a circuit.

elements in series, as in Figure 19.44. Let I_1 and I_2 be the currents in the two wires at the node. The KCL applied to this node yields

$$-I_1 + I_2 = 0 \text{ A}$$

or $I_1 = I_2$. That's nice but not too interesting. The two elements are connected in series and we know the current is the same for each element connected in series. So the application of the KCL to such a node did not yield any information that we did not already know.

PROBLEM-SOLVING TACTIC

19.7 The KCL really becomes useful only at nodes connecting three or more circuit elements.

The Kirchhoff Voltage Law

The Kirchhoff voltage law (KVL) states that the algebraic sum of the potential differences around any closed loop of an electrical circuit is zero.

The word voltage in the name of the law really means potential difference. We use the word potential difference consistently in our discussions, but will call the KVL the Kirchhoff voltage law in conformity with common practice.

What does the KVL mean physically? The law is a reflection of the conservative nature of the electrical force: the work done by a conservative force on a charge taken around a closed path is zero.

The KVL is a statement of *conservation of energy*.

The KVL can be applied to *any* loop of a circuit. An elementary loop contains no other closed loops within itself. While a KVL equation can be written for any closed loop in the circuit, in practice we fortunately do not have to write KVL equations for every possible loop in the circuit. The reason for this will be explained shortly. Consider the elementary circuit loop in Figure 19.45. The various circuit elements around the loop have potential differences across them, say, with the indicated values and polarity markings.

The KVL involves walking around the loop and algebraically summing the various potential differences encountered. Three things to note about the stroll around the loop:

1. We can go around the loop in either direction: clockwise (CW) or counterclockwise (CCW); it makes no difference because the overall sum of the potential differences is zero.
2. It also makes no difference where we begin the journey around the loop; we merely have to go all the way around the loop, returning to wherever we began.
3. As we go around the loop, if we encounter the (+) polarity marking of a circuit element first, we place the potential difference across that circuit element into the sum with a plus sign; if we encounter the (−) polarity marking first, the potential difference for that circuit element is placed into the equation for the loop with a minus sign. So in Figure 19.45, if we go around the loop clockwise and begin in the lower left corner, the KVL takes the following form:

$$V_1 + (-V_2) + (-V_3) + V_4 = 0 \text{ V}$$
$$V_1 - V_2 - V_3 + V_4 = 0 \text{ V} \quad (19.29)$$

If we begin in the lower left corner and go around the loop counterclockwise, the KVL takes the form

$$(-V_4) + V_3 + V_2 + (-V_1) = 0 \text{ V}$$
$$-V_4 + V_3 + V_2 - V_1 = 0 \text{ V} \quad (19.30)$$

which differs from Equation 19.29, obtained by executing the loop clockwise, by only a minus sign. Equations 19.29 and 19.30 are not independent equations. So it makes no difference which way we go around the loop. Starting the walk around the loop at a different place in the loop merely

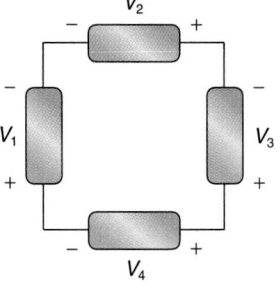

FIGURE 19.45 An elementary loop of a circuit.

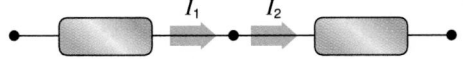

FIGURE 19.44 A node connecting two circuit elements in series.

rearranges the order of the terms in the sum; we do not get a new equation.

PROBLEM-SOLVING TACTIC

19.8 Write KVL equations only for each elementary loop in a circuit.

The parts of an elementary loop that contain either one circuit element or two or more in series with each other are known as *branches*.

The branches of the circuit in Figure 19.46 are indicated. There are three elementary loops to the circuit in Figure 19.46, each highlighted in Figure 19.47. There are four more complex loops, shown in Figure 19.48.

The KVL applies to all the loops. But we write the KVL *only* for the elementary loops. Why? From a mathematical viewpoint, of the seven possible loop equations for this circuit, only three are independent equations; the others are surplus because they are linear combinations of the three independent equations. In other words, the number of *independent* loop equations is equal

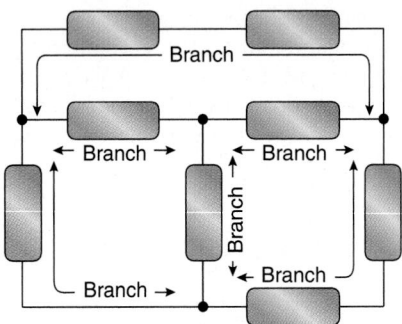

FIGURE 19.46 The branches of a circuit.

to the number of elementary loops in the circuit. It is simply easier to write the KVL equation just for the elementary loops; then we know we have all the independent equations that can be secured using the KVL.

The specific steps to follow for solving any dc circuit problem are now outlined. After reading the procedure, we will apply it to several specific examples to see what is really involved.

1. See if the circuit can be simplified by combining resistors in series or parallel, or independent voltage sources in series or parallel. Not all circuits can be simplified. Sketch the circuit in its simplest form.
2. Locate and identify the significant nodes of the simplified circuit, those nodes connecting three or more circuit elements. Nodes connecting only two circuit elements are not really significant (see our earlier discussion of the KCL).
3. Choose current directions for each distinct branch of the simplified circuit. Introduce appropriate current variables $I_1, I_2, \ldots$ for each branch; we are trying to find these currents. Recognize that circuit elements in series have the same current in every element: thus the current through all circuit elements in a given branch is the same. The directions chosen for the currents in a branch are arbitrary. In practice, for a branch with an independent voltage source, we usually choose the direction of the current to *emerge* from the (+) terminal of the independent voltage source. This choice for current direction is not required, but it usually makes things easier, as we will see. Any circuit element in series with an independent voltage source will have the same current and current direction through it, as it is part of the same branch.
4. Taking account of the current directions, label the polarity for each resistor consistent with the directions chosen for the current through it. That is, for every resistor, the lead along which the current is going into a resistor indicates the (+) terminal of that resistor and the lead along which

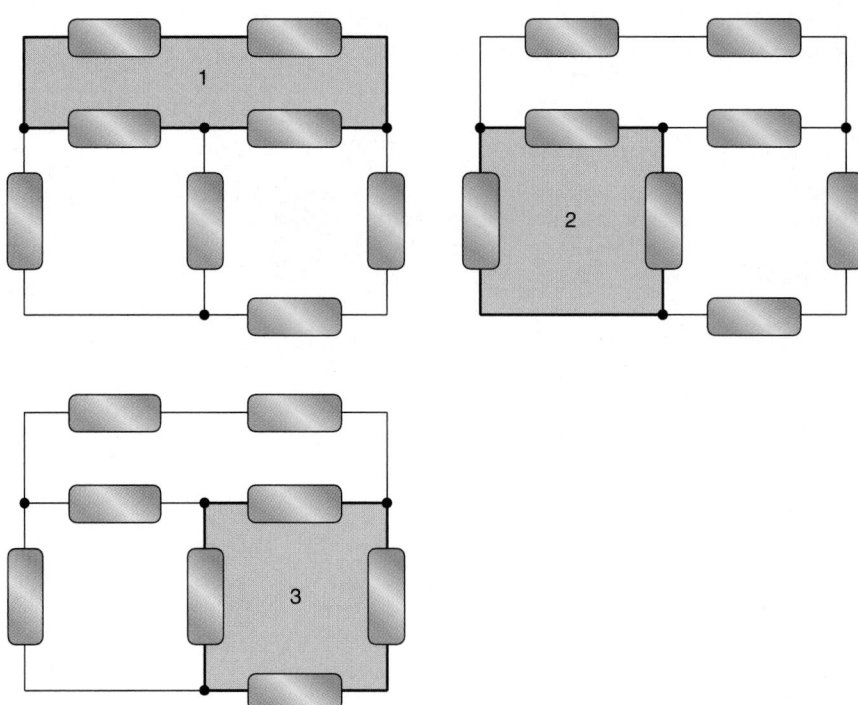

FIGURE 19.47 Elementary loops of the circuit in Figure 19.46.

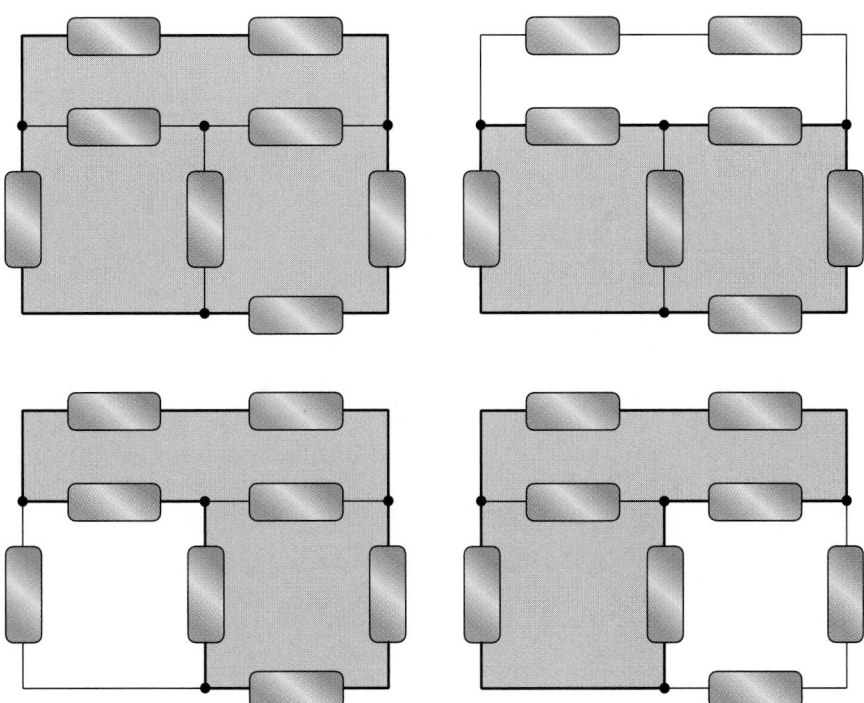

FIGURE 19.48 Four loops that are not elementary loops.

the current is exiting the resistor indicates the (−) polarity terminal of the resistor.

5. Apply the KVL to each elementary loop in the circuit.
6. Assess whether there are sufficient equations to solve for the number of unknown currents. For *n* unknown currents, *n* equations are needed. Determine the number of additional equations needed, if any. Then apply the KCL to the significant nodes (connecting three or more circuit elements) to secure any additional equations needed.
7. Solve the equations for the unknown currents. Once the currents are known, everything about the circuit can be determined from them. If one or more of your currents emerge as negative, there is *no need* to go back and redo anything; a negative current simply means the current actually is directed opposite to the way you chose in your diagram. You can use the negative current to determine any needed potential difference and power absorbed.

STRATEGIC EXAMPLE 19.9

Consider the circuit shown in Figure 19.49, consisting of an independent voltage source (an ideal battery) and a single resistor.

a. **Find the current in the circuit.**
b. **Find the power absorbed by (transferred to) each circuit element.**

Solution

The circuit cannot be simplified. Notice that the resistor is in series with the independent voltage source: the current is the same in both elements since there is no choice of path for the current as it goes around the loop. But this is more interesting than it seems: the resistor also is in parallel with the ideal battery, because the two elements are connected to the same two distinct nodes or terminals. Hence the potential difference across each circuit element is the same. This is an example of the simplest circuit (see Section 19.9), one with only two circuit elements.

Method 1

If an independent voltage source (an ideal battery) is present in a branch, it is convenient to choose the direction of the current to emerge from its positive polarity terminal. Thus choose the current direction indicated in Figure 19.50. This choice for the current direction forces you to label the resistor polarity as indicated in

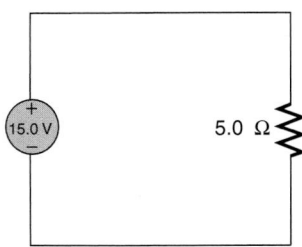

FIGURE 19.49

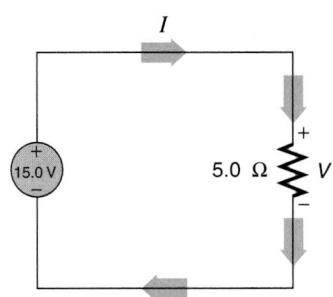

FIGURE 19.50

Figure 19.50, since the current through a resistor always goes from the high- to the low-potential terminal—that is, into the (+) polarity terminal of the resistor and out of its (−) polarity terminal.

a. The potential difference across the ideal battery (an independent voltage source) is 15.0 V; the resistor is in parallel with the ideal battery, and so the potential difference across the resistor also is 15.0 V. Ohm's law, Equation 19.14, states that the potential difference across a resistor is the product of the current through it times the resistance:

$$V = IR$$
$$15.0 \text{ V} = I(5.0 \ \Omega)$$

Thus the current through the resistor and the current in the single loop is

$$I = \frac{15.0 \text{ V}}{5.0 \ \Omega}$$
$$= 3.0 \text{ A}$$

b. The power absorbed by a circuit element is the product of the current into its positive terminal and potential difference across the circuit element (Equation 19.26). For the resistor, you have

$$P = IV$$
$$= (3.0 \text{ A})(15.0 \text{ V})$$
$$= 45 \text{ W}$$

You also could calculate the power absorbed by the resistor using Equation 19.28:

$$P_{res} = I^2 R$$
$$= (3.0 \text{ A})^2(5.0 \ \Omega)$$
$$= 45 \text{ W}$$

The power absorbed by the ideal battery also is the product of the current into the positive terminal and the potential difference across it (its emf), Equation 19.26. The current *out of* the positive polarity terminal of the ideal battery is +3.0 A; therefore, the current *into* the positive polarity terminal is −3.0 A. Hence the power absorbed by the ideal battery is

$$P_{bat} = IV$$
$$= (-3.0 \text{ A})(15.0 \text{ V})$$
$$= -45 \text{ W}$$

The power absorbed by the ideal battery is negative, indicating that the battery actually is generating (or providing) power to the circuit.

Notice that the total power absorbed by the circuit, the sum of the power absorbed by the battery and by the resistor, is zero:

$$P_{total} = -45 \text{ W} + 45 \text{ W}$$
$$= 0 \text{ W}$$

Method 2

a. Choose the current direction indicated in Figure 19.51. Then the polarity of the potential difference across the resistor must

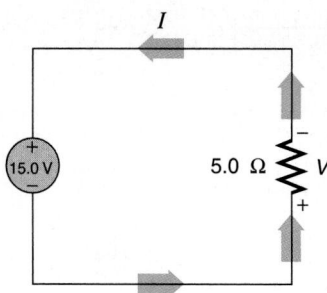

FIGURE 19.51

be labeled as shown in Figure 19.51, because the current goes into the (+) polarity lead of the resistor and out of the (−) polarity lead. Using the KVL around the elementary loop, going clockwise and beginning in the lower left corner, you obtain

$$-15.0 \text{ V} - I(5.0 \ \Omega) = 0 \text{ V}$$

Solving for *I*,

$$I = -3.0 \text{ A}$$

The minus sign indicates that the current actually is in the direction opposite to the one chosen. No matter. You do not have to go back and redo everything! The potential difference across the resistor is, according to Ohm's law,

$$V = IR$$
$$= (-3.0 \text{ A})(5.0 \ \Omega)$$
$$= -15 \text{ V}$$

using the polarity indicated in Figure 19.51. The minus sign here indicates that the terminal marked (+) in Figure 19.51 actually is at a lower potential than the terminal marked (−).

b. To calculate the power absorbed by the resistor, use the power Equation 19.26:

$$P = IV$$
$$= (-3.0 \text{ A})(-15 \text{ V})$$
$$= 45 \text{ W}$$

You also could use Equation 19.28, specifically for the power absorbed by a resistor:

$$P_{res} = I^2 R$$
$$= (-3.0 \text{ A})^2(5.0 \ \Omega)$$
$$= 45 \text{ W}$$

Either way, the power absorbed by the resistor is positive.

The power absorbed by the ideal battery also is found using Equation 19.26 for any circuit element:

$$P_{bat} = IV$$

where the current *I* is directed into the (+) polarity terminal of the ideal battery. Notice that the current direction chosen in Figure 19.51 is going into the positive polarity terminal of the ideal battery; you do not have to reverse its direction to use the power equation, and so you have

$$P_{bat} = (-3.0 \text{ A})(15.0 \text{ V})$$
$$= -45 \text{ W}$$

The total power absorbed by all the circuit elements is zero.

STRATEGIC EXAMPLE 19.10

a. Simplify the circuit in Figure 19.52 as much as possible.
b. Find the potential difference and the current through each circuit element.
c. Find the power absorbed by each circuit element and verify that the total power absorbed by all the circuit elements is zero.

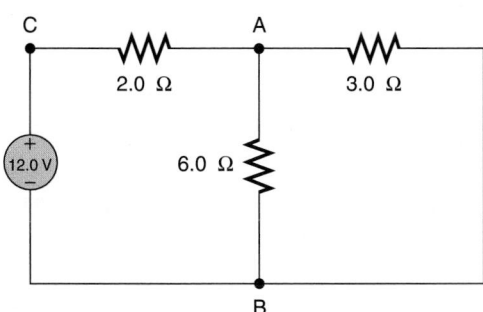

FIGURE 19.52

Solution

a. The 3.0 Ω and 6.0 Ω resistors are in parallel since they are connected to the same two distinct nodes of the circuit (labeled A and B in Figure 19.52). The equivalent resistance of this parallel pair of resistors is (using the product/sum rule for two resistors in parallel, Equation 19.25)

$$\frac{(3.0 \text{ Ω})(6.0 \text{ Ω})}{3.0 \text{ Ω} + 6.0 \text{ Ω}} = 2.0 \text{ Ω}$$

The circuit now has the appearance shown in Figure 19.53. The nodes labeled A and B are the same nodes as in Figure 19.52. The pair of 2.0 Ω resistors in Figure 19.53 are in series and so have an equivalent resistance equal to their sum:

$$R_{eq} = R_1 + R_2$$
$$= 2.0 \text{ Ω} + 2.0 \text{ Ω}$$
$$= 4.0 \text{ Ω}$$

The circuit now has the appearance shown in Figure 19.54, where the nodes labeled C and B are the same as in the original circuit (Figure 19.52). This is the simplest form for the given circuit.

b. The circuit of Figure 19.54 is an example of the simplest circuit, a single loop with two circuit elements. The 4.0 Ω resistor is connected to the same two nodes (C and B) as the independent voltage source, so that the 4.0 Ω resistor has a

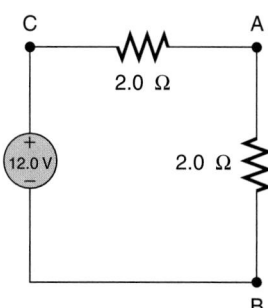

FIGURE 19.53

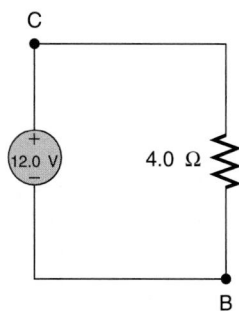

FIGURE 19.54

potential difference of 12.0 V across it with the polarity indicated in Figure 19.55. The current in the single loop circuit is assigned the direction indicated in Figure 19.55, consistent with the resistor polarity. The value of the current is found in one of two ways.

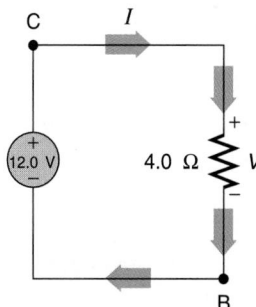

FIGURE 19.55

Method 1
Use Ohm's law applied to the 4.0 Ω resistor:

$$V = IR$$
$$12.0 \text{ V} = I(4.0 \text{ Ω})$$

Solving for *I*, you find

$$I = 3.0 \text{ A}$$

Method 2
You also could first choose the current direction as indicated in Figure 19.55, then label the resistor polarity consistent with this choice, and apply the KVL to the elementary loop.

Going around the loop clockwise beginning in the lower left corner, you get

$$-12.0 \text{ V} + I(4.0 \text{ }\Omega) = 0 \text{ V}$$

which yields

$$I = 3.0 \text{ A}$$

with the direction indicated in Figure 19.55.

Having found the current, now trace your way back to the original circuit (Figure 19.52) by considering each equivalent simplified circuit. Using the simplified circuit of Figure 19.53, each of the 2.0 Ω resistors has the current 3.0 A since they are both in series with the independent voltage source. The potential difference across each 2.0 Ω resistor is the same (since they have the same resistance and current) and is found using Ohm's law:

$$V = IR$$
$$= (3.0 \text{ A})(2.0 \text{ }\Omega)$$
$$= 6.0 \text{ V}$$

The lower 2.0 Ω resistor was the parallel equivalent of the 3.0 Ω and 6.0 Ω resistors in the original circuit. Thus the potential difference across both the 3.0 Ω and the 6.0 Ω resistors of the original circuit is 6.0 V. The potential difference across these resistors is the same since they are connected in parallel with each other. Apply Ohm's law to each of these resistors to find the current through them:

3.0 Ω resistor	6.0 Ω resistor
$6.0 \text{ V} = I_3 (3.0 \text{ }\Omega)$	$6.0 \text{ V} = I_6 (6.0 \text{ }\Omega)$
$I_3 = 2.0 \text{ A}$	$I_6 = 1.0 \text{ A}$

You thus have found the current through each circuit element and the potential difference across each element. These results are summarized in Figure 19.56.

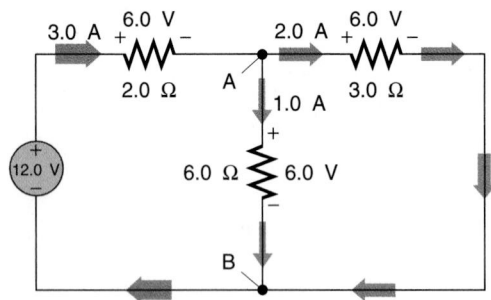

FIGURE 19.56

Note, parenthetically, that the KCL is satisfied at the node at A connecting the three resistors; since currents leaving the node are considered positive, the KCL implies

$$-3.0 \text{ A} + 1.0 \text{ A} + 2.0 \text{ A} = 0 \text{ A}$$

You also can see that the KVL is satisfied around each elementary loop. Going clockwise around each loop beginning in the lower left corner of each, you find the following:

$$\text{Left loop:} \quad -12.0 \text{ V} + 6.0 \text{ V} + 6.0 \text{ V} = 0 \text{ V}$$
$$\text{Right loop:} \quad -6.0 \text{ V} + 6.0 \text{ V} = 0 \text{ V}$$

c. To find the power absorbed by each circuit element, apply Equation 19.26:

$$P = IV$$

where V is the potential difference across the circuit element and I is the current into its positive polarity terminal. Use the directions of the currents to assign polarities to the resistors as indicated in Figure 19.56. Applying the power equation (Equation 19.26) to each resistor, you obtain

$$2.0 \text{ }\Omega \text{ resistor:} \quad P_{2 \text{ }\Omega} = (3.0 \text{ A})(6.0 \text{ V}) = 18 \text{ W}$$
$$3.0 \text{ }\Omega \text{ resistor:} \quad P_{3 \text{ }\Omega} = (2.0 \text{ A})(6.0 \text{ V}) = 12 \text{ W}$$
$$6.0 \text{ }\Omega \text{ resistor:} \quad P_{6 \text{ }\Omega} = (1.0 \text{ A})(6.0 \text{ V}) = 6.0 \text{ W}$$

(Note that each resistor absorbs positive power, as every resistor must.) Equivalently, you can find the power absorbed by each resistor by using Equation 19.28:

$$2.0 \text{ }\Omega \text{ resistor:} \quad P_{2 \text{ }\Omega} = (3.0 \text{ A})^2(2.0 \text{ }\Omega) = 18 \text{ W}$$
$$3.0 \text{ }\Omega \text{ resistor:} \quad P_{3 \text{ }\Omega} = (2.0 \text{ A})^2(3.0 \text{ }\Omega) = 12 \text{ W}$$
$$6.0 \text{ }\Omega \text{ resistor:} \quad P_{6 \text{ }\Omega} = (1.0 \text{ A})^2(6.0 \text{ }\Omega) = 6.0 \text{ W}$$

The power absorbed by the independent voltage source is found from the general power equation $P = IV$, remembering that I must be the current into the positive polarity terminal. The current into the positive terminal of the ideal battery is –3.0 A. Thus the power absorbed by the independent voltage source is

$$P_{\text{bat}} = (-3.0 \text{ A})(12 \text{ V})$$
$$= -36 \text{ W}$$

The total power absorbed by the circuit elements must be zero (cross your fingers!):

$$P_{\text{bat}} + P_{2 \text{ }\Omega} + P_{3 \text{ }\Omega} + P_{6 \text{ }\Omega}$$
$$= -36 \text{ W} + 18 \text{ W} + 12 \text{ W} + 6.0 \text{ W}$$
$$= 0 \text{ W}$$

Ta da! If the sum of the powers were not zero, you could be certain that a mistake was made somewhere in the calculations. On the other hand, if the sum of the powers is zero you can be quite confident (but not certain*) that all is well and you are headed for an A+ on the next test.

---

*The powers might accidentally sum to zero, so while it is necessary they sum to zero, that fact is not sufficient for you to conclude your solution is correct with certainty. Drat.

STRATEGIC EXAMPLE 19.11

Consider the circuit shown in Figure 19.57. Find

- the current through each circuit element;
- the potential difference across each resistor; and
- the power absorbed by each circuit element.

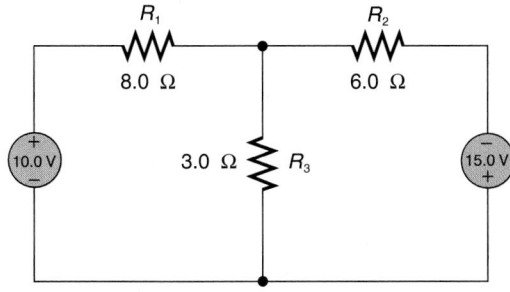

FIGURE 19.57

Solution

Step 1: See if the circuit can be simplified by combining resistors in series or parallel. The resistors in this circuit are neither in series nor parallel with any of the other resistors. The independent voltage sources also are not in series or parallel with each other. This circuit cannot be simplified. Note that the 10.0 V source is in series with the 8.0 Ω resistor, and so they have the same current; the 15.0 V source is in series with the 6.0 Ω resistor, and so they also have a common current.

Step 2: Locate and identify the nodes of the circuit. The nodes of this circuit are indicated in Figure 19.58. You are only interested in the significant nodes, those with three or more leads leaving them. There are two such significant nodes indicated in Figure 19.58.

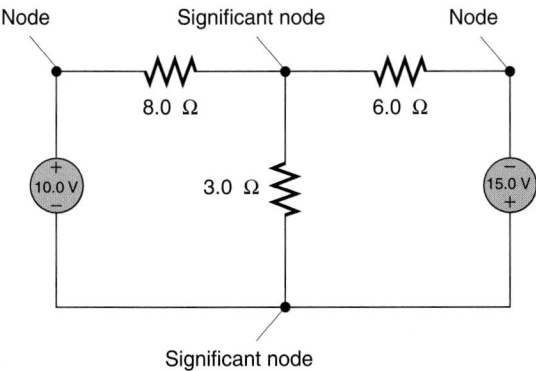

FIGURE 19.58

Step 3: Assign current directions to the individual branches of the circuit. The directions chosen for the currents are arbitrary but, if an independent voltage source is in a particular branch, you should usually choose the current to emerge from its positive polarity terminal. Thus take the directions of I_1 and I_2 to be as indicated in Figure 19.59. The current through the 3.0 Ω resistor can be taken to be in either direction; choose the current I_3 to be in the direction indicated in Figure 19.59.

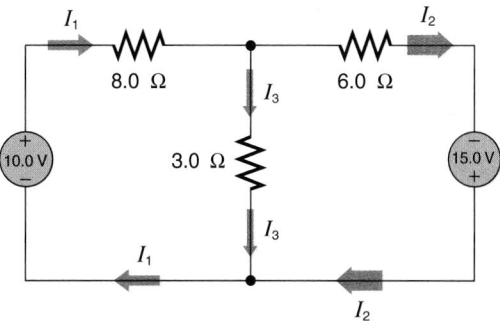

FIGURE 19.59

Step 4: Assign polarities to the resistors consistent with the current direction choices: current enters a resistor at the (+) polarity terminal of the resistor and exits at its (−) polarity terminal. Thus you emerge with the resistor polarities indicated in Figure 19.60.

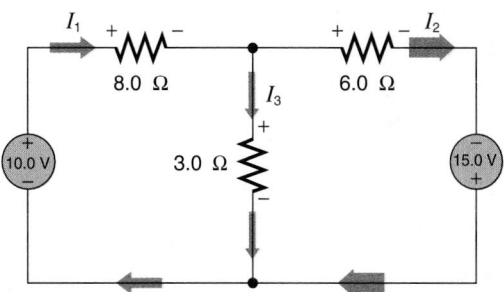

FIGURE 19.60

Step 5: Apply the KVL around each elementary loop. This circuit has two elementary loops; you write a KVL equation for each. The KVL states that the sum of the potential differences around any closed loop is zero. Remember that the direction you travel around the loop is arbitrary and you can begin anywhere along the path of the loop; you simply have to traverse the whole loop. Go around each loop in a clockwise direction beginning in the lower left corner of each loop, as shown in Figure 19.61.

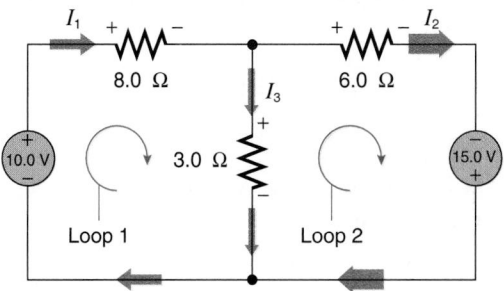

FIGURE 19.61

If a (−) terminal of a circuit element is the terminal first encountered as you go around the loop, the potential difference across that circuit element is put into the KVL equation with a minus sign; if a (+) terminal is the first encountered, that potential difference is put into the KVL equation with a plus sign. The potential difference across each resistor is given by Ohm's law, Equation 19.14. Thus for each loop the following equations emerge:

Loop 1

$$-10.0 \text{ V} + I_1(8.0 \text{ }\Omega) + I_3(3.0 \text{ }\Omega) = 0 \text{ V}$$
$$(8.0 \text{ }\Omega)I_1 + (3.0 \text{ }\Omega)I_3 = 10.0 \text{ V}$$

Loop 2

$$-I_3(3.0 \text{ }\Omega) + I_2(6.0 \text{ }\Omega) - 15.0 \text{ V} = 0 \text{ V}$$
$$(6.0 \text{ }\Omega)I_2 - (3.0 \text{ }\Omega)I_3 = 15.0 \text{ V}$$

Step 6: Assess whether there is sufficient information to solve for the unknown currents. The KVL yielded two equations but you have three unknown currents. Another equation is needed. So apply the KCL to the significant nodes to secure the additional equations needed. Here you need only one additional equation, and so you can apply the KCL to either of the significant nodes (here the same equation will result, except for a minus sign, regardless of which of the two significant nodes is used). Use the upper node, as shown in Figure 19.62.

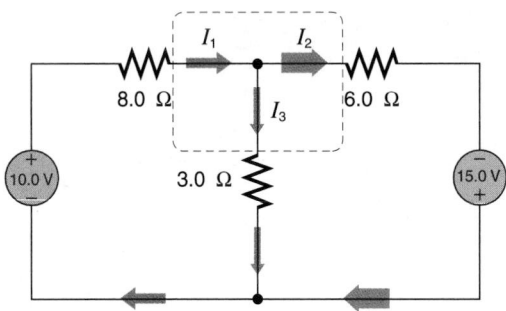

FIGURE 19.62

The KCL states that the sum of the currents leaving the node is zero. Thus the KCL applied to the upper node yields the equation

$$-I_1 + I_2 + I_3 = 0 \text{ A}$$

You now have the three equations needed for the three unknown currents.

Step 7: Solve for the unknown currents; this is a mathematical problem. There are several different ways this can be handled:

a. hire a mathematician;
b. use algebraic brute force;
c. use Kramer's rule methods involving suitable determinants; or
d. use appropriate software to do the grunt work.

Whichever method is chosen, the results for the three currents are

$$I_1 = 1.5 \text{ A}$$
$$I_2 = 2.2 \text{ A}$$
$$I_3 = -0.67 \text{ A}$$

The significance of the minus sign for the current I_3 means that the current through the 3.0 Ω resistor actually is directed opposite to the way indicated in Figure 19.59. No matter; you do *not* need to go back and change this current direction.

With the currents in the circuit known, you can find other quantities of interest.

The potential difference across each resistor is found by using Ohm's law, Equation 19.14:

$$V_1 = I_1R_1 = (1.5 \text{ A})(8.0 \text{ }\Omega) = 12 \text{ V}$$
$$V_2 = I_2R_2 = (2.2 \text{ A})(6.0 \text{ }\Omega) = 13 \text{ V}$$
$$V_3 = I_3R_3 = (-0.67 \text{ A})(3.0 \text{ }\Omega) = -2.0 \text{ V}$$

If a voltmeter* is connected across the various resistors with the (+) lead of the voltmeter on the (+) side of each resistor indicated in Figure 19.60, and the (−) lead on the corresponding (−) side of the resistor, the voltmeter will indicate the values we calculated for V_1, V_2, and V_3—*including the signs!* Here a voltmeter so placed across R_3 will indicate a negative value for V_3. The negative result for V_3 means that the higher-potential end of the 3.0 Ω resistor actually is not the one marked (+); it is the other end. Notice how your results for the currents and potential differences tell you whether the choices made for the current directions are the actual directions or not. A negative current in one direction is identical to a positive current in the opposite direction.

Now calculate the power absorbed by each circuit element. The electrical power absorbed by any circuit element is the product of the current into the (+) polarity terminal of that circuit element and the potential difference across the element (Equation 19.25).

For the 8.0 Ω resistor:

$$P_{8 \text{ }\Omega} = I_1V_1$$
$$= (1.5 \text{ A})(12 \text{ V})$$
$$= 18 \text{ W}$$

For the 6.0 Ω resistor:

$$P_{6 \text{ }\Omega} = I_2V_2$$
$$= (2.2 \text{ A})(13 \text{ V})$$
$$= 29 \text{ W}$$

For the 3.0 Ω resistor (note the signs carefully!):

$$P_{3 \text{ }\Omega} = I_3V_3$$
$$= (-0.67 \text{ A})(-2.0 \text{ V})$$
$$= 1.3 \text{ W}$$

Every resistor absorbs positive power. They get hot because the decrease in the electrical potential energy of the charges passing through them increases the thermodynamic internal energy of the resistors. The power absorbed by the resistors also

---

*We examine the characteristics of voltmeters in Section 19.15. A voltmeter is used to measure the potential difference between two points.

can be calculated using Equation 19.28, $P = I^2R$, with the same results.

Now for the independent voltage sources. Notice that the currents I_1 and I_2 originally were chosen to emerge from the positive terminals of the two independent voltage sources; the power equation needs to have the current *into* the positive terminal. So you must substitute $-I_1$ and $-I_2$ into the power equation for these circuit elements:

For the 15.0 V independent voltage source:

$$P_{15\ V} = (-2.2\ A)(15.0\ V)$$
$$= -33\ W$$

For the 10.0 V independent voltage source:

$$P_{10\ V} = (-1.5\ A)(10.0\ V)$$
$$= -15\ W$$

Both absorb negative power; this fact indicates that they are actually generating or providing (transferring) power to the rest of the circuit, derived from chemical or mechanical energy within the sources.

Now for the crucial test: the sum of the electrical power absorbed by all the circuit elements must be zero to conserve energy (per unit time). You can see if this is the case:

$$P_{8\ \Omega} + P_{6\ \Omega} + P_{3\ \Omega} + P_{15\ V} + P_{10\ V}$$
$$= 18\ W + 29\ W + 1.3\ W - 33\ W - 15\ W$$
$$= 48\ W - 48\ W$$
$$= 0\ W$$

Voilà! Things check.

EXAMPLE 19.12

You have an independent voltage source (an ideal battery) with emf V_0 and two resistors with resistances R_1 and R_2. To have the power absorbed by the resistors be greatest, should they be connected in series to the battery, as shown in Figure 19.63, or in parallel, as in Figure 19.64?

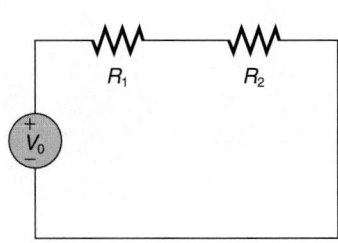

(a) Series connection

(b) Equivalent

FIGURE 19.63

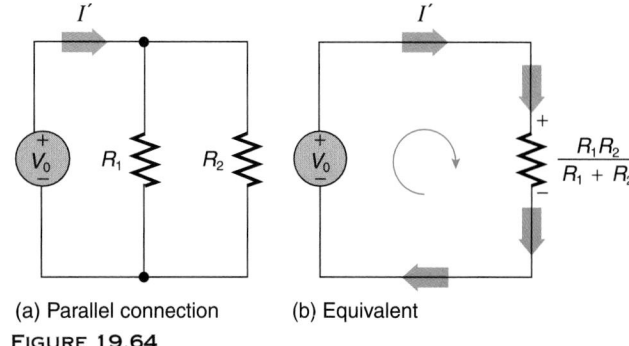

(a) Parallel connection

(b) Equivalent

FIGURE 19.64

Solution

You solve this problem in three steps: (1) compute the power absorbed when the resistances are connected in series; (2) compute the power absorbed for the parallel connection; and (3) compare the computed values of power absorbed.

1. For the resistors in series, the equivalent resistor is the sum of the pair: $R_1 + R_2$. Apply the KVL to the equivalent circuit, with the current direction as chosen in Figure 19.63b; go around the circuit clockwise:

$$-V_0 + I(R_1 + R_2) = 0\ V$$

yielding

$$I = \frac{V_0}{R_1 + R_2}$$

Note that the equivalent resistor is in parallel (and series) with the voltage source. The power absorbed by the equivalent resistor is, from Equation 19.26,

$$P_{res\ series} = IV$$
$$= \left(\frac{V_0}{R_1 + R_2}\right)V_0$$
$$= \frac{V_0^2}{R_1 + R_2}$$

2. For a parallel connection of two resistors, the equivalent resistor is their product divided by their sum:

$$\frac{R_1 R_2}{R_1 + R_2}$$

Apply the KVL around the single equivalent loop shown in Figure 19.64b; go clockwise around the equivalent loop:

$$-V_0 + I'\frac{R_1 R_2}{R_1 + R_2} = 0\ V$$

yielding

$$I' = \frac{V_0(R_1 + R_2)}{R_1 R_2}$$

Note again that the equivalent resistor is in parallel (and series) with the voltage source. The power absorbed by the equivalent resistor then is found from Equation 19.28:

$$P_{\text{res parallel}} = I'V$$

Substituting for I' and the $V = V_0$, you get

$$P_{\text{res parallel}} = \left[\frac{V_0 (R_1 + R_2)}{R_1 R_2} \right] V_0$$

$$= V_0^2 \frac{R_1 + R_2}{R_1 R_2}$$

3. The ratio of the power absorbed by the resistors when they are in parallel to that when they are in series is

$$\frac{P_{\text{res parallel}}}{P_{\text{res series}}} = \frac{V_0^2 \dfrac{R_1 + R_2}{R_1 R_2}}{\dfrac{V_0^2}{R_1 + R_2}}$$

$$= \frac{(R_1 + R_2)^2}{R_1 R_2}$$

$$= \frac{R_1^2 + R_2^2 + 2R_1 R_2}{R_1 R_2}$$

The numerator is larger than the denominator, and so the power ratio is greater than unity. Thus a parallel connection of the two resistors absorbs more power from a given voltage source than a series connection of the same resistors.

19.12 ELECTRIC SHOCK HAZARDS*

Awaiting the sensation of a short, sharp shock
William S. Gilbert (1836–1911)
and Arthur Sullivan (1842–1900)[†]

If you ever have been shocked by an automotive spark plug wire, you have been jolted with a potential difference of over a kilovolt. Potential difference is not what is inherently dangerous to you; it is electric *current* that is dangerous.

You may not think of yourself in these terms, but you are a resistor. If you connect yourself to an independent voltage source (accidentally or intentionally), the current in the circuit depends on both the resistance of your body and the potential difference across it (see Figure 19.65).

Your resistance depends on the nature of the electrical contact the skin makes with the wires that form the rest of the circuit. Dry skin contact with the wires has a much higher resistance than wet skin contact. The condition of your skin and the area of contact determine your effective resistance. Your body resistance can be measured easily with a common laboratory ohmmeter (see Section 19.15) simply by holding each ohmmeter lead with the fingers of each hand. The resistance of the body from one hand to the other is on the order of hundreds of thousands of ohms to several million ohms if the fingers are dry. If the fingers are

[†]*The Mikado* (1885), Act I, #10, "Trio."

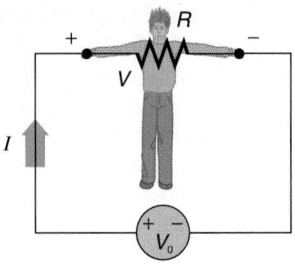

FIGURE 19.65 You can complete a circuit, but it may be dangerous to your health.

wet, the resistance is reduced drastically; the smaller body resistance allows a correspondingly larger current from a fixed voltage source. This means that wet contact with an independent voltage source is much more dangerous than dry contact.

Depending on how you are connected to the independent voltage source, the current through your body may flow across your chest and can trigger dangerous consequences if the current is sufficiently large. A current of only about 1 mA is painful; 10–25 mA is sufficient to prevent you from controlling your muscles in order to release the wires; 50 mA to 3 A can be fatal, since such currents interfere with biological electrical signals regulating your heartbeat. **Ventricular fibrillation** results, a condition in which the heart loses its normal rhythmic pumping action. The ventricle wall ripples and the heart is unable to pump any blood; death ensues within just a few minutes.

Ironically, currents above about 3 A are less dangerous. With such large currents, breathing can stop (which is not too good for the health!); but if breathing can be induced once again quickly via artificial respiration, the odds of surviving are better than with the smaller currents that induce ventricular fibrillation. It takes longer to suffocate than to die from a stopped heart. Indeed, large current pulses are used by hospital trauma physicians and rescue squads to *stop* ventricular fibrillation. A large current pulse causes the heart muscles to contract violently. With the cessation of the current pulse, the heart muscles relax and often resume normal activity. During the procedure, bellows are used to pump air into the lungs of such patients until normal breathing functions can be restored. Most people hit by lightning die because their heart or breathing stops rather than from burns.

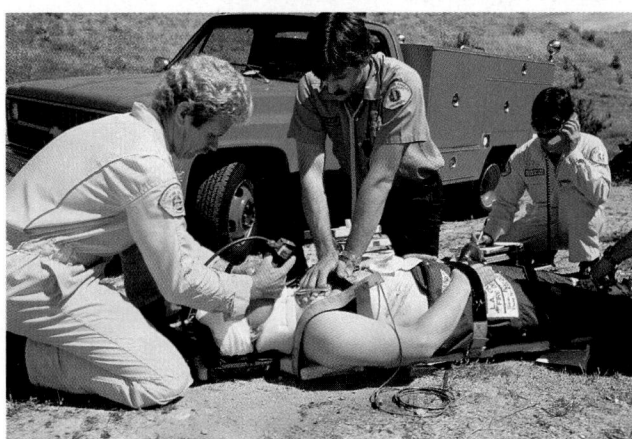

Large currents are used by emergency personnel to stop ventricular fibrillation.

Why are birds not electrocuted when they perch on bare wires at a high electrical potential?

FIGURE Q.7

EXAMPLE 19.13

If your body resistance is $5.0 \times 10^4 \, \Omega$ with wet hands, what potential difference will cause a painful and dangerous current of 1.0 mA through your body?

Solution

Use Ohm's law, Equation 19.14. A current of 1.0 mA through a resistance of $5.0 \times 10^4 \, \Omega$ represents a potential difference across the resistance of

$$V = IR$$
$$= (1.0 \times 10^{-3} \text{ A})(5.0 \times 10^4 \, \Omega)$$
$$= 50 \text{ V}$$

Common electrical outlets have effective potential differences of about 120 V and so are quite dangerous under these circumstances.

19.13 A MODEL FOR A REAL BATTERY

We now can formulate a model of a real battery to explain why they "die."

An independent voltage source such as an ideal battery has a vertical line for its current–voltage characteristic curve, as shown in Figure 19.66. An independent voltage source maintains the same potential difference V_0 between its terminals for all time. The potential difference across its terminals is independent of the current.

Real batteries do not have these ideal characteristics. Real batteries in circuits get warm, indicating they absorb some of the potential energy of the charges that constitute the current.

> To a good approximation, a real battery is modeled by means of an independent voltage source in series with a resistance r, as shown in Figure 19.67.

The series resistor r, known as the **internal resistance** of the real battery, always absorbs power when there is a current through it, thus accounting for the observation that real batteries get warm when used.

The points A and B in Figure 19.67 represent the terminals of the real battery. The potential difference between A and B is called the **terminal potential difference** of the real battery. Notice that if the real battery is not connected to a circuit, the terminal potential difference is the same as the emf of the independent voltage source in the model. No current flows through r, so there is no potential difference across it.

For a fresh real battery, the internal resistance r is very small. When $I > 0$ A, the finite internal resistance makes the terminal potential difference $V_{term} < V_0$, so that the characteristic curve of a fresh real battery is nearly a vertical line with negative slope, as shown in Figure 19.68.

Connect a real battery to an external resistor R (such as a light bulb in a flashlight), making a single loop circuit, as in

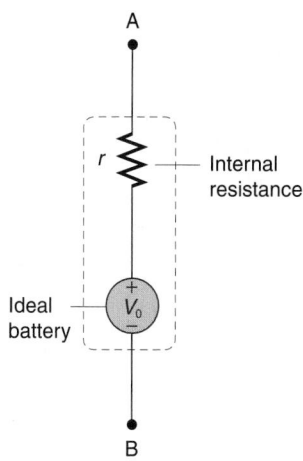

FIGURE 19.67 A model of a real battery.

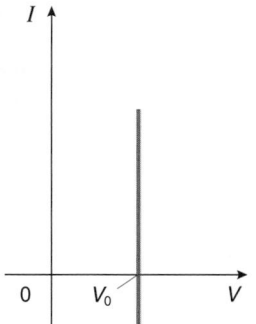

FIGURE 19.66 The characteristic curve of an ideal battery.

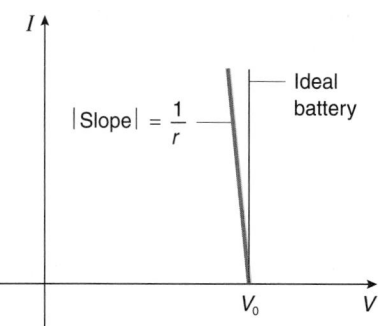

FIGURE 19.68 The characteristic curve of a real battery.

Figure 19.69. The external resistor R sometimes is called a **load resistor**. Since the circuit consists of a single loop, we let I be the current in the loop with the direction chosen as indicated in Figure 19.69. As a result of the current direction choice, the resistors have the polarities indicated in Figure 19.69. Applying the KVL around the single loop in a clockwise direction beginning at point B, we get

$$-V_0 + Ir + IR = 0 \text{ V}$$

and so

$$I = \frac{V_0}{R + r} \tag{19.31}$$

The terminal potential difference of the real battery is the potential difference between A and B. Going from A to B in the circuit and algebraically summing the potential differences, we find

$$-Ir + V_0$$

So the terminal potential difference is

$$V_{\text{term}} = V_0 - Ir \tag{19.32}$$

Using Equation 19.31 for I, the terminal potential difference becomes

$$V_{\text{term}} = V_0 - \frac{V_0}{R + r} r$$

$$= V_0 \frac{R}{R + r} \tag{19.33}$$

For a fresh real battery, the internal resistance r is very small and the terminal potential difference approaches the value of the emf V_0 of the ideal battery. But as a real battery is used, the internal resistance gradually increases with time because of irreversible chemical reactions within the battery.[†]

[†]In a rechargeable battery the chemical reactions can be reversed by charging the battery; this involves forcing a current into the positive terminal of the battery with another independent voltage source, so that the battery absorbs positive power.

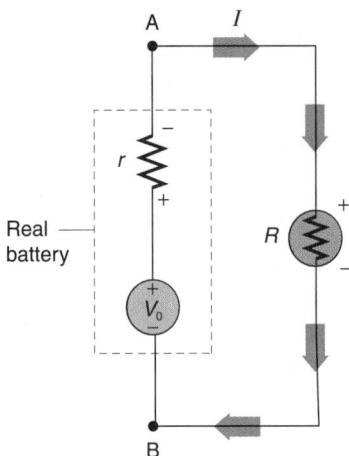

FIGURE 19.69 A real battery connected to an external resistor, such as a light bulb.

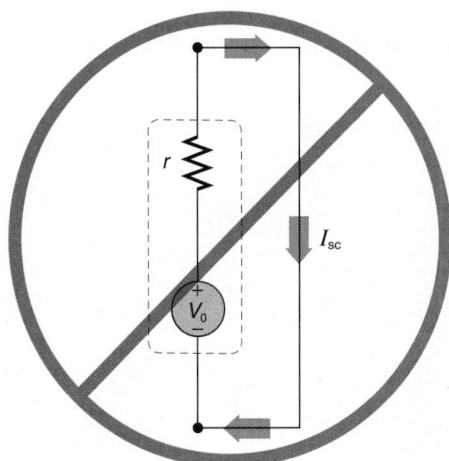

FIGURE 19.70 Shorting a real battery. *Do not do this!*

As a real battery is used, r eventually becomes quite large. As $r \to \infty\ \Omega$, both the current I (see Equation 19.31) and the terminal potential difference (see Equation 19.33) approach zero. The real battery then is called a **dead battery**: there is no current in the circuit to the external load resistor and R absorbs zero power. If R represents the light bulb of your flashlight, the bulb will not light because the current in the circuit is zero.

It is dangerous to short-circuit a real battery. If a fresh real battery with a small internal resistance r is shorted, as shown in Figure 19.70, the external load resistance R is zero and the short circuit current becomes, from Equation 19.31,

$$I_{\text{sc}} = \frac{V_0}{r}$$

Using Equation 19.28, the power absorbed by the internal resistance is

$$P_r = I_{\text{sc}}^2 r$$

$$= \frac{V_0^2}{r}$$

Since r is small, the power absorbed can be quite large. The power absorbed rapidly increases the thermodynamic internal energy of the real battery, which results in a rapid temperature increase, sufficient to explode the battery as chemicals are vaporized within it.

QUESTION 8

Under what circumstances (if any) can the terminal potential difference of a real battery exceed the potential difference of the ideal battery used to model it?

EXAMPLE 19.14

A real battery is connected in series to a 10.0 Ω resistor, as shown in Figure 19.71.

a. Determine the current in the circuit.
b. Find the terminal potential difference of the real battery.
c. Determine the potential difference across the load resistor.
d. What power is absorbed by the internal resistance of the battery?

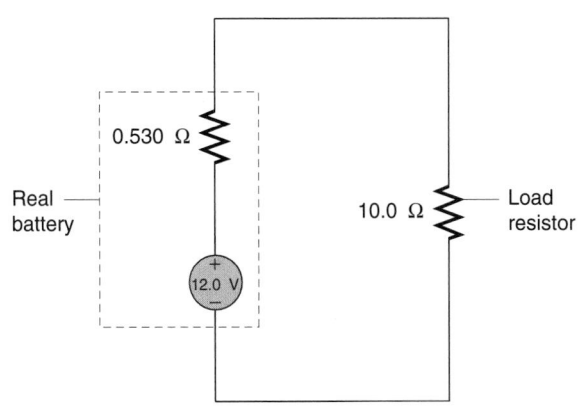

FIGURE 19.71

Solution

a. With the current direction chosen as in Figure 19.72, the resistors have the polarity shown for the potential difference. Use the KVL clockwise around the elementary loop:

$$-12.0 \text{ V} + I(0.530 \ \Omega) + I(10.0 \ \Omega) = 0 \text{ V}$$

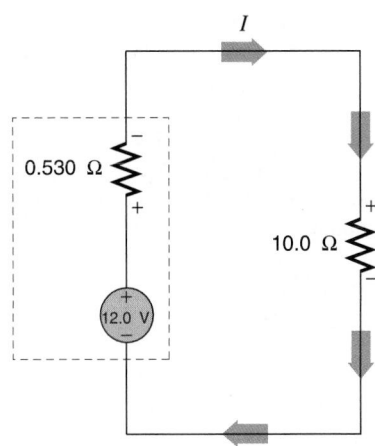

FIGURE 19.72

Solving for I, you get

$$I = 1.14 \text{ A}$$

b. The potential difference across the internal resistance of the real battery is

$$V_r = Ir$$
$$= (1.14 \text{ A})(0.530 \ \Omega)$$
$$= 0.604 \text{ V}$$

From Equation 19.32, the potential difference between the terminals of the real battery then is

$$V_{term} = 12.0 \text{ V} - 0.604 \text{ V}$$
$$= 11.4 \text{ V}$$

c. The potential difference across the load resistor is found using Ohm's law:

$$V = IR$$
$$= (1.14 \text{ A})(10.0 \ \Omega)$$
$$= 11.4 \text{ V}$$

d. The power absorbed by the internal resistance of the real battery is found using Equation 19.28 for the power absorbed by any resistor:

$$P = I^2 r$$
$$= (1.14 \text{ A})^2(0.530 \ \Omega)$$
$$= 0.689 \text{ W}$$

The power absorbed by the internal resistance of the real battery causes it to become warm.

EXAMPLE 19.15

Find current and the power absorbed by the internal resistance of the real battery shown if its terminals are accidentally short-circuited, as in Figure 19.73.

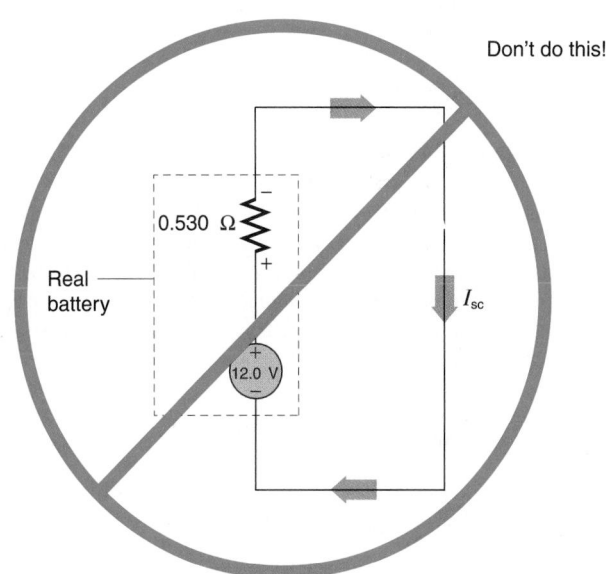

FIGURE 19.73

Solution

The short circuit current I_{sc} in the circuit is found by applying the KVL to the loop:

$$-12.0 \text{ V} + I_{sc}(0.530 \ \Omega) = 0 \text{ V}$$

Solving for I_{sc}, you find

$$I_{sc} = 22.6 \text{ A}$$

The power absorbed by the internal resistance then is found from Equation 19.28 to be

$$P_r = (I_{sc})^2 r$$
$$= (22.6 \text{ A})^2(0.530 \ \Omega)$$
$$= 271 \text{ W}$$

This is a considerable amount of power and may cause the battery to explode.

19.14 MAXIMUM POWER TRANSFER THEOREM

If you are an audiophile, you know that your stereo speakers must be matched to the power supply (the real voltage source) of your audio system for optimal performance. The criterion used is an application of the **maximum power transfer theorem**.

It is important to know what load resistor R maximizes the power absorbed by it from a real battery (see Figure 19.74). The power absorbed by any resistor is found from Equation 19.28:

$$P = I^2R$$

The current through the load resistor we found (Equation 19.31) to be

$$I = \frac{V_0}{R + r}$$

and so the power absorbed by the load resistor R is

$$P(R) = \left(\frac{V_0}{R + r}\right)^2 R \qquad (19.34)$$

Equation 19.34 indicates that $P = 0$ W when $R = 0$ Ω and when $R = \infty$ Ω, and so the power must have a maximum for a finite value of R, as shown schematically in Figure 19.75. Keeping V_0 and r fixed, we want to discover what load resistance R maximizes the power P delivered by the real battery to the load. This is an maximum problem in calculus. So we take the derivative of P with respect to R, set it equal to zero, and solve the resulting equation for R:

$$\frac{dP}{dR} = 0 \ \text{W}/\Omega$$

We use Equation 19.34 for P; its derivative with respect to R is

$$\frac{dP}{dR} = V_0^2 \frac{(R + r)^2 - R\,[2(R + r)]}{(R + r)^4} = 0 \ \text{W}/\Omega$$

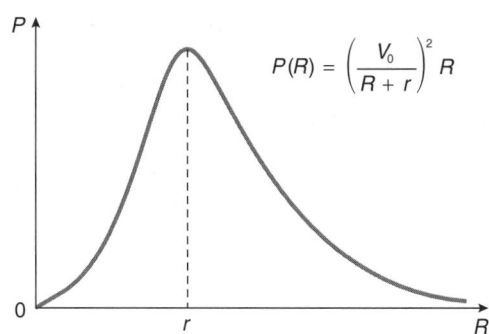

FIGURE 19.75 The power has a maximum at a particular value of R.

Solving for R, we have

$$(R + r)^2 - R(2)(R + r) = 0 \ \Omega^2$$
$$(R + r) - 2R = 0 \ \Omega$$
$$R = r \qquad (19.35)$$

In other words, for a given real battery, the load resistance R that maximizes the power absorbed by R has a value equal to the internal resistance r of the real battery. This result is the maximum power transfer theorem. If R is chosen to match (be equal to) r, the resulting condition is called an **impedance match**. It should be understood that the load resistance R may be a single resistor (as it is in Figure 19.74) or R may be the single *equivalent* resistance of a collection of resistors connected to the terminals of the real battery (see Example 19.16).

The maximum power transfer theorem is quite general and holds for any real voltage source connected to a load, not just the typical real battery. When you match stereo speakers to a power supply you are impedance matching for maximum power transfer to the speakers.

The model for a real battery consisting of an ideal battery (or ideal voltage source) in series with a resistance r is a model used for many complex electronic devices. Indeed, if you pursue electronics by taking courses in circuit analysis, you will learn both analytical and experimental techniques to reduce many complex circuits, such as the "bird's nest" in Figure 19.76, to just an ideal independent voltage source in series with an internal resistance, with the model connected to a load resistor R, as shown in Figure 19.77. The simple model for the complex circuit is known as the **Thévenin equivalent circuit** of the complex circuit.

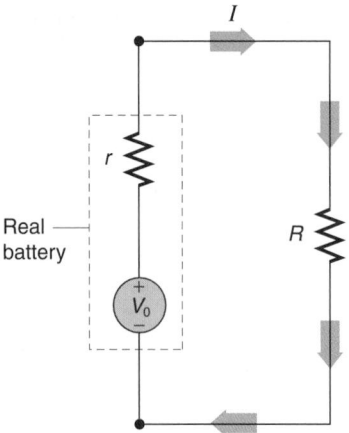

FIGURE 19.74 A load resistor R connected to a real battery.

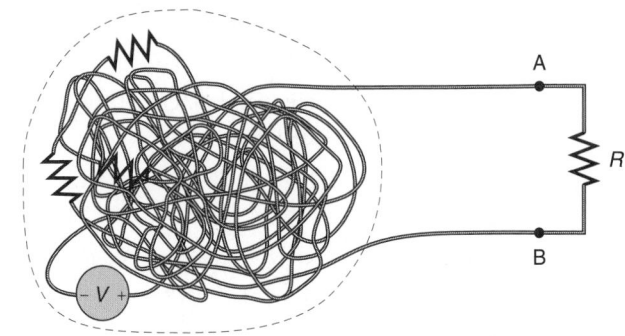

FIGURE 19.76 A complicated circuit.

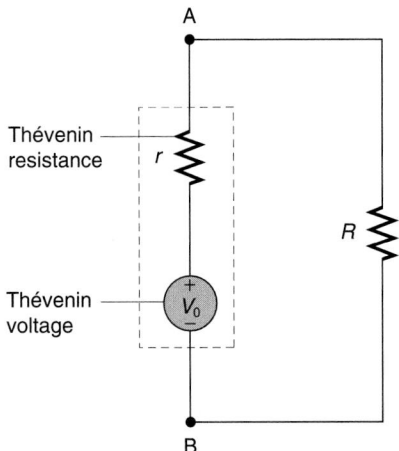

FIGURE 19.77 The Thévenin equivalent of a complicated circuit.

The potential difference of the ideal independent voltage source in the model (or equivalent circuit) is called the **Thévenin voltage** and the internal resistance r of the model is called the **Thévenin resistance**. The resulting single loop circuit depicted in Figure 19.77 is used frequently, and so this simple circuit is one you will encounter in many contexts. (See Problem 60.)

EXAMPLE 19.16

For the circuit indicated in Figure 19.78, what value of R maximizes the power absorbed by all the resistors?

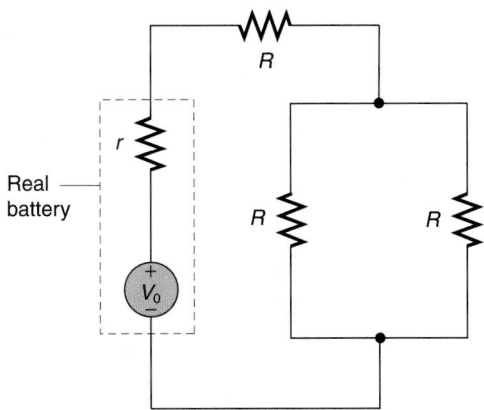

FIGURE 19.78

Solution

To maximize the power absorbed, the total equivalent resistance connected to the real battery must be equal to the internal resistance of the real battery. The parallel pair of resistors has an equivalent resistance equal to their product divided by their sum:

$$R_{eq} = \frac{RR}{R + R} = \frac{R}{2}$$

The $R/2$ equivalent resistance is in series with resistance R. The

total resistance attached to the real battery therefore is

$$R + \frac{R}{2} = \frac{3R}{2}$$

According to the maximum power transfer theorem, the total equivalent resistance external to the real battery must be equal to the internal resistance r of the real battery, and so

$$\frac{3R}{2} = r$$

or

$$R = \frac{2}{3}r$$

19.15 BASIC ELECTRONIC INSTRUMENTS: VOLTMETERS, AMMETERS, AND OHMMETERS

Potential differences are measured with a **voltmeter**, currents are measured with an **ammeter**, and resistances are measured with an **ohmmeter**. The three instruments occasionally are packaged together as a single **multimeter** that can be used for measuring potential differences, currents, or resistances by selecting the appropriate function with a switch.

Such instruments are common in most laboratories, and it is important to have an understanding of their characteristics and how they are used. Ultimately, all three instruments use a sensitive current-detecting device. Historically, a galvanometer was used, consisting of a coil of wire suspended in a magnetic field, but other specialized devices now are employed to measure very small currents. We designate any such sensitive current-detecting device by the symbol shown in Figure 19.79, which was the historic symbol for a galvanometer.

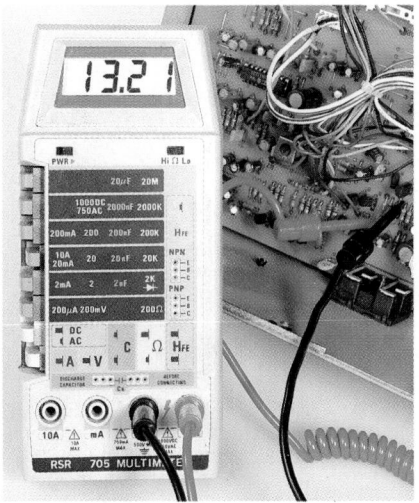

A common laboratory multimeter.

FIGURE 19.79 Symbol for a sensitive current-detecting device.

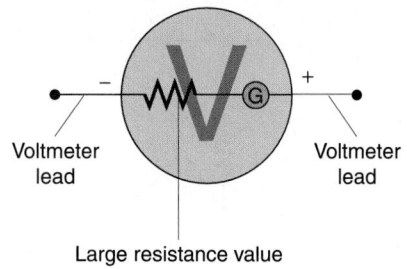

FIGURE 19.82 A voltmeter has a high effective resistance.

Voltmeters

A voltmeter is an instrument used to measure the potential difference between two nodes of a circuit. Such meters are symbolized as shown in Figure 19.80. The positive polarity lead of the meter typically is marked with a (+), is a red-colored terminal, or is red-colored wire. The negative polarity lead of the meter is either marked with a (−), is a black-colored terminal, or is a black-colored wire.

To measure the potential difference across a circuit element or between two nodes, the positive lead (the red wire) of the voltmeter is touched or connected to the high-potential side of the element [the (+) side of the circuit element], while the negative lead (the black wire) of the meter is touched or connected to the low-potential side of the element (−); see Figure 19.81. The meter indicates the value of the potential difference V between the two points. If the voltmeter leads are reversed, the meter will read −V rather than V.

Notice in Figure 19.81 that a voltmeter is connected in *parallel* with the circuit element. This ensures that the potential difference across the voltmeter is the same as the potential difference across the circuit element.

Connecting a voltmeter to an electrical circuit means, of course, that the voltmeter becomes part of the circuit. The circuit is not the same circuit any more! To ensure that the effect of connecting the voltmeter to the circuit is minimal, we ideally want no current in the leads connecting the voltmeter to the circuit. This is accomplished by manufacturing the voltmeter to have a very large effective resistance. Inside the voltmeter, this may be accomplished by placing a high resistance in series with the sensitive current-measuring device, as shown in Figure 19.82. In this way, the current through the

actual circuit is negligibly affected by the presence of the voltmeter in the circuit.

> So the two things to remember about a voltmeter are the following:
>
> 1. It is connected in parallel with the circuit element in question and therefore has the same potential difference across it.
>
> 2. It has a very large effective resistance so as to affect the circuit minimally.

Ammeters

An ammeter is a device for measuring the current through some branch of a circuit. The symbol for an ammeter is shown in Figure 19.83. The positive polarity lead of the meter typically is marked with a (+) sign, is colored red, or is a red-colored wire. The negative polarity lead of the meter is either marked with a (−), is colored black, or is a black-colored wire.

To measure the current, the ammeter must be inserted in series so that all the current in the wire passes through the meter. The (+) side of the ammeter is placed at the higher-potential side of the branch, as shown in Figure 19.84.

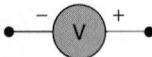

FIGURE 19.80 Symbol for a voltmeter.

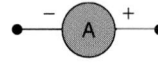

FIGURE 19.83 Symbol for an ammeter.

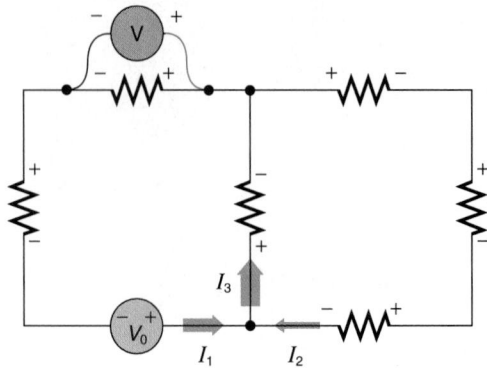

FIGURE 19.81 How to measure the potential difference across a circuit element with the voltmeter.

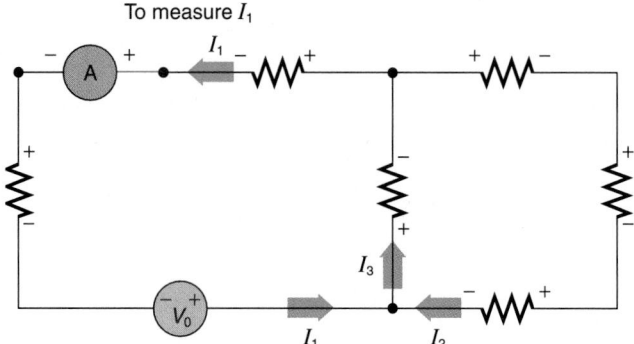

FIGURE 19.84 An ammeter is connected in series in the branch whose current is to be measured.

The original circuit is changed by inserting the ammeter into the circuit. To ensure that the ammeter does not significantly affect the value of the current in the branch of the circuit into which it will be placed, the ammeter must have a very small effective resistance. Then the potential difference across the ammeter itself will be negligible compared with that across the other circuit elements. Within the ammeter itself, a very small resistance is placed in parallel with the sensitive current-detecting device, as shown schematically in Figure 19.85. The small resistance ensures that the overall resistance of the ammeter itself is very small.

Hence an ammeter

1. is connected in series; and

2. must have very small effective resistance.

With the advent of microelectronics or circuits of very small size, it may be physically impossible to place an ammeter in the circuit path. The circuit may be hard-wired, which means that to place the ammeter in the circuit, wires must be cut and the meter inserted. In such situations, placing an ammeter in a circuit may be out of the question, if not just a pain in the neck. So what can be done to determine the current? All hope is not lost. If there is a resistor in the branch along which the current is to be measured, we can measure the potential difference across the resistor with a voltmeter. If the resistance is known (or can be determined from its color code), then the current can be found by using Ohm's law for the resistor, Equation 19.14:

$$V = IR$$

or

$$I = \frac{V}{R}$$

In this way, the current is determined without resorting to an ammeter.

Ohmmeters

An ohmmeter is depicted with the symbol shown in Figure 19.86. To measure the resistance of a resistor, the leads of the ohmmeter are attached to the leads of the resistor, as indicated in Figure 19.87.

The resistor must *not* be part of the circuit in which the resistor is used when such a measurement is made.

If an ohmmeter is used when the resistor is part of the rest of a circuit, as in Figure 19.88, the ohmmeter will indicate the equivalent resistance between the two terminals to which it is attached. The equivalent resistance is the parallel combination of the given resistance with that of the rest of the circuit. The ohmmeter then does not indicate the resistance of the desired resistor.

If the resistor cannot be easily removed from the circuit, its approximate value may be determined from its color code if it is so labeled. The value of a resistance also should be listed on the schematic circuit diagram if one is available

An ohmmeter measures resistance by using a battery to create a small current through the resistor whose resistance is to be measured; see Figure 19.89. The current in the sensitive current detector is inversely proportional to the value of the resistance.

FIGURE 19.86 The symbol for an ohmmeter.

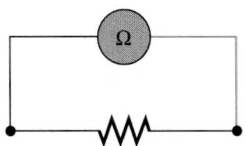

FIGURE 19.87 How to measure the resistance of a resistor with an ohmmeter.

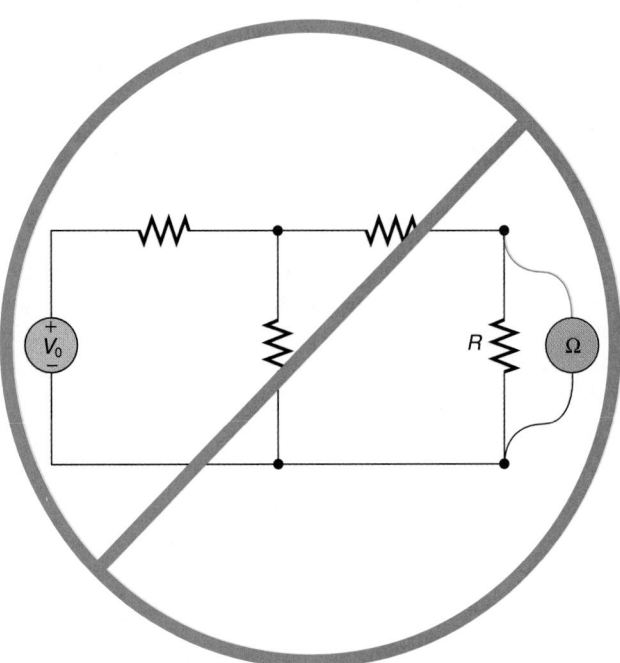

FIGURE 19.88 *Incorrect* way to measure the resistance of a resistor.

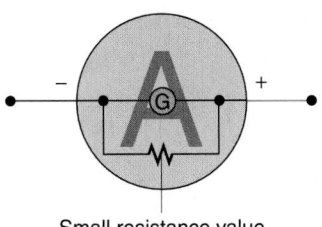

Small resistance value

FIGURE 19.85 An ammeter has a low effective resistance.

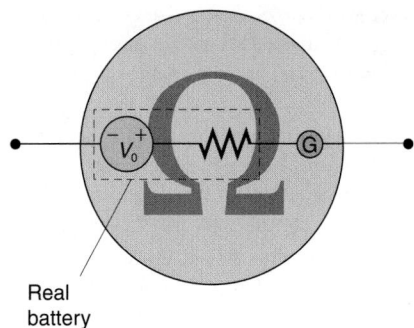

Real
battery

FIGURE 19.89 A model of an ohmmeter.

19.16 AN INTRODUCTION TO TRANSIENTS IN CIRCUITS: A SERIES RC CIRCUIT*

When you turn on the flash unit for your camera, there typically is a delay before the flash unit is ready to fire. Sometimes this delay can be quite annoying if you want to take a candid picture immediately. The delay is caused by the time it takes the battery to fully charge a capacitor used to accumulate the electrical potential energy needed for the flash lamp. This is an example of a transient effect in a circuit.

Transients in circuits are one-shot effects that occur, for example, when a circuit is first completed (turned on) or disconnected (turned off). As an example of such transients, we examine a single loop circuit consisting of an independent voltage source (an ideal battery), a resistor, and a capacitor, as shown in Figure 19.90. The arrangement is called a **series RC circuit**. A switch is included so that we can investigate several situations depending on which way the switch is positioned.

A Charging Capacitor

Let the capacitor in the RC circuit initially be uncharged and the switch open (as shown in Figure 19.90). The system is an electrical network, but not a circuit because there is no closed path. We want to see what happens when, at time $t = 0$ s, the switch is closed to position (1), thus creating the single loop circuit shown in Figure 19.91. In particular, we want to find the current in the circuit as a function of time.

At any instant t after the switch is closed, let $I(t)$ be the current in the circuit, directed as shown in Figure 19.91. According to our conventions, once the current direction is chosen, the resistor polarity is determined as indicated. At time t, let the capacitor have the charges $\pm q$ on its plates as shown (we take $q > 0$ C).

From Equation 18.1, the capacitance is (by definition)

$$C = \frac{|q|}{|V|}$$

so that the potential difference V across the capacitor is

$$V = \frac{q}{C}$$

with the polarity indicated in Figure 19.91.

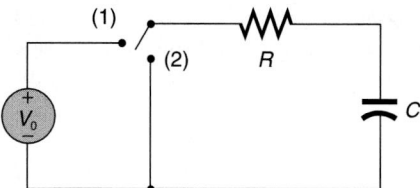

FIGURE 19.90 A series RC network (there is no closed path, so that it is not yet a circuit).

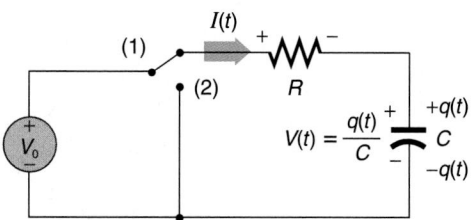

FIGURE 19.91 The switch is closed to position (1) when $t = 0$ s.

We apply the KVL to the single loop, beginning in the lower left corner of the circuit and traversing the loop in the clockwise sense:

$$-V_0 + IR + \frac{q}{C} = 0 \text{ V} \qquad (19.36)$$

But the current I is related to q since

$$I = \frac{dq}{dt}$$

We differentiate Equation 19.36 with respect to t, recognizing that the emf V_0 is a constant; we find

$$0 \text{ V/s} + R\frac{dI}{dt} + \frac{1}{C}\frac{dq}{dt} = 0 \text{ V/s}$$

Substituting I for

$$\frac{dq}{dt}$$

yields

$$R\frac{dI}{dt} + \frac{1}{C}I = 0 \text{ V/s}$$

or

$$\frac{dI}{I} = -\frac{1}{RC}dt$$

This is in a form that can be integrated. We let the initial current in the circuit be I_0 and the current at instant t be I. Then

$$\int_{I_0}^{I} \frac{dI}{I} = -\frac{1}{RC}\int_{0\text{ s}}^{t} dt$$

$$\ln I - \ln I_0 = -\frac{1}{RC}t$$

Taking the antilogarithms, we find

$$\frac{I}{I_0} = e^{-t/(RC)}$$

$$I(t) = I_0 e^{-t/(RC)} \tag{19.37}$$

We still have to figure out the value of the initial current I_0. At the instant when the switch is closed, the capacitor has no charge on it and thus no potential difference across it. Thus, applying the KVL to the loop at the initial instant when current I_0 is in the circuit (see Figure 19.92), we have, going around clockwise,

$$-V_0 + I_0 R + 0 \text{ V} = 0 \text{ V}$$

Solving for I_0,

$$I_0 = \frac{V_0}{R}$$

Hence Equation 19.37 becomes

$$I(t) = \frac{V_0}{R} e^{-t/(RC)} \tag{19.38}$$

The current in the circuit $I(t)$ is plotted as a function of time in Figure 19.93.

When $t = RC$, the current is e^{-1} of its initial value. This time interval is called the **time constant** τ of the circuit. The time constant is used to characterize the decay of the current

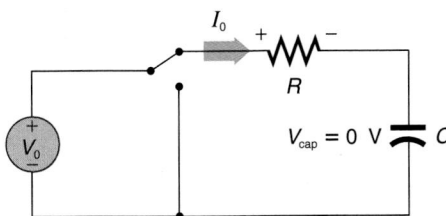

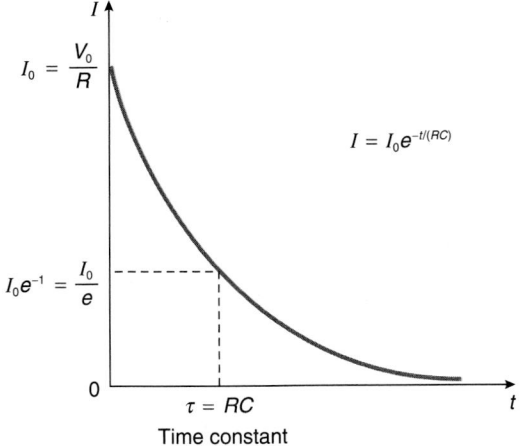

FIGURE 19.92 The situation when $t = 0$ s.

because, in this model, the current never actually reaches zero: notice that as t increases, the current I decreases *asymptotically* to zero. After a time equal to, say, five time constants (that is, when $t = 5\tau$), the current in the circuit is only e^{-5} of its initial value. Since $e^{-5} = 0.006\ 738$, the current effectively is negligible after an interval of about 5τ; it is less than 1% of its initial value. We thus reach the important conclusion that the dc current (the steady, time-independent current) through a capacitor is *zero*.

> Once charged, a capacitor in a dc circuit acts like an open switch in the branch in which it is placed.

That makes sense, because a perfect capacitor has an infinitely good insulator between its two conducting plates. Capacitors are used to prevent dc currents from reaching certain parts of circuits; this property of a capacitor is exploited in many circuits that use transistors, which makes capacitors essential ingredients in modern electronics.

The initial slope of the graph of Equation 19.38—that is, the derivative of I with respect to t, evaluated when $t = 0$ s—has the value

$$-\frac{I_0}{\tau}$$

Hence in Figure 19.93 the line representing the initial slope of the curve intersects the time axis at the time τ, as shown in Figure 19.94. In other words, if the current decreased at this constant rate, the current would be zero at the time τ. But the current actually decreases exponentially, and by the time τ it is $e^{-1} = 0.368$ of its initial value.

The growth of the magnitude of the charge accumulating on each capacitor plate as a function of time in the charging process is found by integrating the current. That is, since

$$I = \frac{dq}{dt}$$

we have

$$dq = I\ dt$$

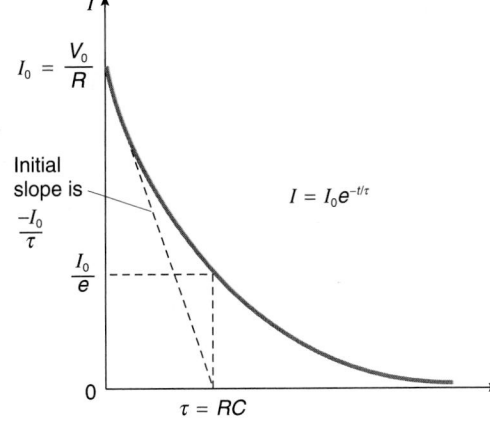

FIGURE 19.93 A graph of the current versus time.

FIGURE 19.94 The initial slope of the current versus time graph is $-I_0/\tau$.

We use Equation 19.38 for I:

$$dq = \frac{V_0}{R} e^{-t/(RC)} dt$$

When $t = 0$ s, the initial charge on the capacitor is zero, and so the integral for the charge takes the form

$$\int_{0\,C}^{q} dq = \frac{V_0}{R} \int_{0\,s}^{t} e^{-t/(RC)} dt$$

Performing the integration, we obtain

$$q(t) = \frac{V_0}{R} \left[e^{-t/(RC)}(-RC) \right]\Big|_{0\,s}^{t}$$

or

$$q(t) = V_0 C[1 - e^{-t/(RC)}] \qquad (19.39)$$

As $t \to \infty$ s, the magnitude of the charge on each plate of the capacitor approaches the value $Q_0 \equiv V_0 C$, and so we can rewrite Equation 19.39 as

$$q(t) = Q_0[1 - e^{-t/(RC)}]$$

This is graphed schematically in Figure 19.95.

The initial slope of this curve—that is, the derivative

$$\frac{dq}{dt}$$

evaluated when $t = 0$ s—is

$$\frac{Q_0}{\tau}$$

where $\tau = RC$, the time constant of the series RC circuit. Hence in Figure 19.95, the line tangent to the curve at $t = 0$ s would reach the value Q_0 at the time τ. The actual value of the charge on the capacitor at the time τ is

$$Q_0(1 - e^{-1}) = 0.632 Q_0$$

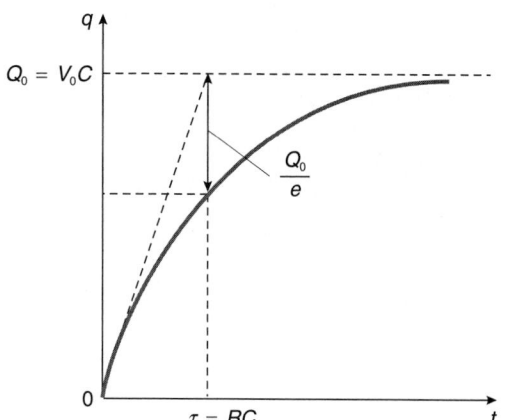

FIGURE 19.95 The magnitude of the charge on the capacitor plates as a function of time.

EXAMPLE 19.17

In Figure 19.93, how long does it take for current to decay to half its value? This time can be considered to be the half-life of the current in a charging series RC circuit. Express your result in terms of the time constant of the circuit.

Solution

The current in a charging series RC circuit is given by Equation 19.37:

$$I(t) = I_0 e^{-t/(RC)}$$

When the current is half its initial value, the equation becomes

$$\frac{I_0}{2} = I_0 e^{-t/(RC)}$$

$$\frac{1}{2} = e^{-t/(RC)}$$

Taking the natural logarithms of both sides, you get

$$\ln 1 - \ln 2 = -\frac{t}{RC}$$

Solving for t, you find

$$t = RC \ln 2$$

Since the time constant $\tau = RC$, you have

$$t = 0.693\tau$$

A Discharging Capacitor

Suppose we have waited long enough for the current in the circuit in Figure 19.96 to vanish. The capacitor in the circuit is fully charged and, since the capacitor acts like an open circuit for dc currents, there is zero current in the circuit.

Since no current is present, according to Ohm's law there is no potential difference across the resistor:

$$\begin{aligned} V_{res} &= IR \\ &= (0\ A)R \\ &= 0\ V \end{aligned}$$

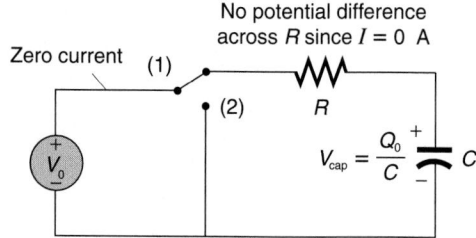

FIGURE 19.96 The capacitor is fully charged and the current is zero.

We apply the KVL to the single loop, going clockwise around the loop beginning in the lower left corner:

$$-V_0 + V_{cap} = 0 \text{ V}$$

so that

$$V_{cap} = V_0$$

The potential difference across the capacitor is equal to the potential difference across the independent voltage source.

Now we open the switch, as in Figure 19.97. We now have an electrical network; there is no circuit. Thus absolutely nothing changes! The capacitor is isolated and will not discharge since the charge is locked onto the capacitor; there is no electrical path for the charges to move from one plate to get to the other plate.

To discharge a capacitor, we need to move the switch to position (2), as shown in Figure 19.98, thus creating a (different!) circuit. We now use the same direction employed before for the current and apply the KVL to the single closed loop (the battery now is not part of the circuit); going clockwise around the new loop, we obtain

$$IR + \frac{q}{C} = 0 \text{ V} \tag{19.40}$$

For variety, let us see what happens to the charge q on the capacitor as a function of time. Since

$$I = \frac{dq}{dt}$$

Equation 19.40 becomes

$$R\frac{dq}{dt} + \frac{q}{C} = 0 \text{ V}$$

giving

$$\frac{dq}{dt} = -\frac{1}{RC}q$$

We rearrange the equation slightly:

$$\frac{dq}{q} = -\frac{1}{RC}dt$$

Now we integrate from $t = 0$ s (when q is Q_0, the charge on the fully charged capacitor) to time t (when the charge on the capacitor is q):

$$\int_{Q_0}^{q} \frac{dq}{q} = -\frac{1}{RC}\int_{0 \text{ s}}^{t} dt$$

$$\ln q - \ln Q_0 = -\frac{t}{RC}$$

Simplifying and taking the antilogarithms, we get

$$q(t) = Q_0 e^{-t/(RC)} \tag{19.41}$$

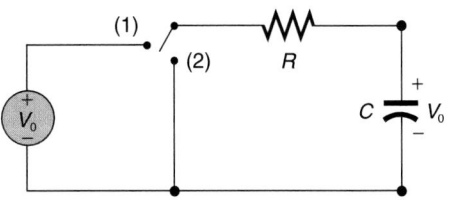

FIGURE 19.97 The switch is opened; nothing changes.

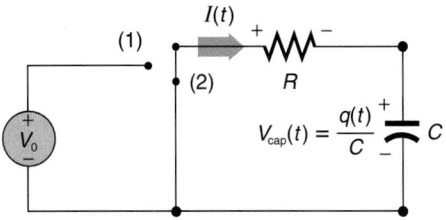

FIGURE 19.98 To discharge the capacitor, move the switch to position (2).

The initial magnitude of the charge on the capacitor is found from

$$C = \frac{Q_0}{V_0}$$

and so

$$q(t) = CV_0 e^{-t/(RC)} \tag{19.42}$$

The charge on the capacitor decreases exponentially with time. The current in the circuit is

$$I = \frac{dq}{dt}$$

$$= CV_0 e^{-t/(RC)}\left(-\frac{1}{RC}\right)$$

or

$$I(t) = -\frac{V_0}{R}e^{-t/(RC)} \tag{19.43}$$

The minus sign indicates that the current in the circuit actually is in the direction opposite to that shown in Figure 19.98. The current decreases in magnitude exponentially with time. The time constant still is $\tau = RC$. After a time of about 5τ, the current once again is negligible.

Note that if the resistance R is very small, as it would be if a simple wire were connected to the terminals of the charged capacitor, the time constant of the circuit is very small and the discharge of the capacitor takes place very quickly. This is what happens when you short-circuit the terminals of a charged capacitor, as we did in Section 19.1 when we first investigated the concept of electric current.

A capacitor is charged by connecting it to a real battery with internal resistance r. Do the final charge on the capacitor and the potential difference between its plates depend on the value of the internal resistance of the real battery? Explain your answer.

EXAMPLE 19.18

The capacitor in the series RC circuit of Figure 19.98 is discharging. At what time (in terms of the time constant τ) is the charge equal to one-half its initial value?

Solution
Use Equation 19.41,

$$q(t) = Q_0 e^{-t/(RC)}$$

Since $\tau = RC$, you have

$$q(t) = Q_0 e^{-t/\tau}$$

You want to know t when $q = Q_0/2$. Hence

$$\frac{Q_0}{2} = Q_0 e^{-t/\tau}$$

or

$$\frac{1}{2} = e^{-t/\tau}$$

Taking the natural logarithms of both sides, you find

$$-0.693 = -t/\tau$$

and so

$$t = 0.693\tau$$

EXAMPLE 19.19

For a discharging capacitor in a series RC circuit, at what time is the potential energy equal to one-half of its initial value? Express your answer in terms of the time constant τ.

Solution
The potential energy stored in the capacitor is found using Equation 18.12,

$$PE = \frac{1}{2}\frac{Q^2}{C}$$

Equation 19.41 describes how the charge on the discharging capacitor decreases with time, where $\tau = RC$. Substituting for the charge in the expression for the potential energy, you have

$$PE = \frac{1}{2C} Q_0^2 e^{-2t/\tau}$$

But

$$\frac{1}{2C} Q_0^2$$

is the initial potential energy PE_0 when $t = 0$ s, when the capacitor starts to discharge. So

$$PE = PE_0\, e^{-2t/\tau}$$

When the potential energy is half its initial value, this becomes

$$\frac{1}{2} = e^{-2t/\tau}$$

Taking the natural logarithms of both sides, you get

$$-0.693 = -2\frac{t}{\tau}$$

$$t = 0.347\tau$$

The potential energy decreases to half its value in half the time it takes the charge to decrease to half its value (see Example 19.18). The reason for this is that the potential energy depends on the square of the charge.

EXAMPLE 19.20

Find the time constant associated with the arrangement of resistors and capacitors shown in Figure 19.99.

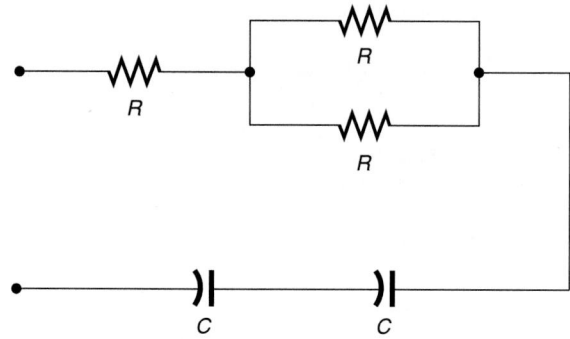

FIGURE 19.99

Solution
Combine the resistors and the capacitors into an equivalent resistor and an equivalent capacitor. The equivalent resistor has a value of $3R/2$ for its resistance and the equivalent capacitor has a value of $C/2$ for its capacitance. The result is a single resistor in series with a single capacitor, as shown in Figure 19.100.

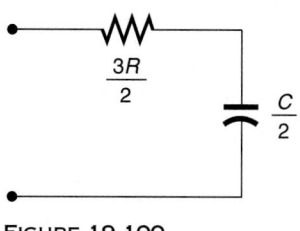

FIGURE 19.100

The time constant of the series combination is

$$\tau = R_{eq}C_{eq}$$

$$= \frac{3R}{2}\frac{C}{2}$$

$$= \frac{3}{4}RC$$

CHAPTER SUMMARY

An *electric current* represents charge in motion. The electric current I is the time rate of positive charge flow past a given location:

$$I \equiv \frac{dQ}{dt} \tag{19.1}$$

The current also can be expressed as

$$I = qnA\langle v \rangle \tag{19.3}$$

where n is the number of charge carriers per cubic meter, q is the charge on each charge carrier, $\langle v \rangle$ is the *average drift speed* (really the average velocity component), and A is the cross-sectional area through which the charges are moving (measured perpendicular to the average drift velocity).

In *ohmic materials*, the potential difference between the ends of a length ℓ of the material is proportional to the current through the material (which is *Ohm's law*):

$$V = IR \tag{19.14}$$

where R is the *resistance*

$$R \equiv \frac{\rho \ell}{A} \tag{19.13}$$

and A is the cross-sectional area of the material. The SI unit of resistance is the *ohm* (Ω). The constant ρ is the *resistivity* of the material and is measured in $\Omega \cdot m$ using SI units.

Resistors connected in series have an equivalent resistance of

$$R_{eq} = R_1 + R_2 + R_3 + \cdots \quad \text{(resistors in series)} \tag{19.20}$$

Resistors connected in parallel have an equivalent resistance obtained from

$$\frac{1}{R_{eq}} = \frac{1}{R_1} + \frac{1}{R_2} + \frac{1}{R_3} + \cdots \quad \text{(resistors in parallel)} \tag{19.24}$$

Two resistors in parallel have an equivalent resistance that is the product of the resistances divided by their sum.

A graph of the current through a circuit element versus the potential difference across it is called the *characteristic curve* of the circuit element. The characteristic curve of an independent voltage source is a straight vertical line; that of a resistor is a line of positive slope whose value is the reciprocal of the resistance.

The electric power P absorbed by (transferred to) a circuit element is the product of the current I into its positive polarity terminal and the potential difference V across it:

$$P = IV \tag{19.26}$$

Resistors always absorb positive power since

$$P_{res} = I^2 R \quad \text{(for a resistor } only\text{)} \tag{19.28}$$

Resistors convert the electric potential energy lost by charges passing through them into thermodynamic internal energy, increasing the temperature of the resistor. Heat transfer to the environment occurs because of this increase in temperature.

If the power absorbed by a circuit element is negative, the circuit element is increasing the electric potential energy of charges passing through it. Negative power absorbed does *not* imply the circuit element is cooling off by decreasing its thermodynamic internal energy.

An electrical *circuit* is a *network* with at least one closed path. Circuits are analyzed using the two Kirchhoff's laws. The *Kirchhoff current law* (KCL) states that the algebraic sum of the currents leaving any node of a circuit is zero. This is a statement of conservation of charge (per unit time). The *Kirchhoff voltage law* (KVL) states that the algebraic sum of the potential differences around any closed loop in a circuit is zero, where appropriate accounting must be made of the polarities of the various potential differences as they are encountered around the path. If a positive polarity terminal of a circuit element is encountered first, the potential difference is entered into the sum with a plus sign; if the negative polarity terminal is encountered first, the potential difference is entered into the sum with a minus sign. The KVL is a statement of conservation of energy (per coulomb of charge). We typically apply the KVL only to the *elementary loops* of a circuit.

The sum of the power absorbed by all circuit elements in a circuit is zero; this statement represents conservation of energy per unit time in the circuit as a whole.

A *real battery* is modeled as an independent voltage source (an ideal battery) in series with a resistance r known as the *internal resistance* of the real battery. Real batteries die because their internal resistance r gradually increases to huge values over the lifetime of the real battery.

The *maximum power transfer theorem* states that the maximum power absorbed by a single (or equivalent) load resistor R connected to the terminals of a real battery occurs when the resistance R has a value equal to the internal resistance r of the real battery.

A *voltmeter* is an instrument for measuring potential differences. It is a device of high effective resistance. A voltmeter is connected in parallel with the circuit element whose potential difference is to be measured. An *ammeter* is a device for measuring current; it is a low-resistance device that is connected in series in the branch of the circuit where the current is to be determined. An *ohmmeter* is an instrument for measuring resistance. To measure the resistance, the resistor must not be part of the circuit in which it is used.

When a capacitor is in series with a resistor, the charging or discharging of the capacitor are not accomplished instantaneously. The charge on the capacitor and the current follow exponential relationships that involve a characteristic time $\tau = RC$, known as the *time constant* of the series RC circuit.

A capacitor acts as an open switch in a dc circuit; the dc current is zero in any branch of a dc circuit that has a capacitor.

SUMMARY OF PROBLEM-SOLVING TACTICS

19.1 (page 849) As with the capacitors in series, be careful using Equation 19.24 in (1) adding the fractions correctly and (2) remembering that the sum of the reciprocals is the *reciprocal* of the equivalent resistance.

19.2 (page 849) The equivalent resistance of two resistors in parallel is the product of their resistances divided by the sum.

19.3 (page 851) Since resistors always become warm or hot when they have a nonzero current, resistors always absorb positive power (or zero power if the current through the resistor is zero).

19.4 (page 851) If you find the current through a resistor is negative, $-|I|$, Ohm's law leads to a negative potential difference, $-|V|$. The power absorbed by the resistor still is positive.

19.5 (page 851) Independent voltage sources absorb either positive or negative power depending on the direction of the current through them. Positive power is absorbed when the current *enters* the positive polarity terminal; negative power when the current *emerges* from the positive polarity terminal.

19.6 (page 851) If a current I is emerging from the positive polarity terminal of a circuit element, current $-I$ is entering the positive polarity terminal.

19.7 (page 855) The KCL really becomes useful only at nodes connecting three or more circuit elements.

19.8 (page 856) Write KVL equations only for each *elementary loop* in the circuit.

QUESTIONS

1. (page 841); 2. (page 845); 3. (page 846); 4. (page 847); 5. (page 849); 6. (page 852); 7. (page 865); 8. (page 866); 9. (page 876)

10. In a conductor, an electric field is necessary to maintain a constant drift speed for the electrons. The electric field provides the force on the charges constituting the current. Does this violate Newton's first law of motion? Prior to Newton, people thought a constant force was needed to maintain a constant speed. Are we harking back to pre-Newtonian physics with our model of currents in conductors? Explain.

11. Why does the gauge number of common wires increase for decreasing wire diameters as in Table 19.2? (Consult an electrical engineers handbook to see on what basis the gauge number is assigned. See also Problem 19.)

12. Construct an analogy between charge, current, and resistance on the one hand and cars and traffic on the other. What plays the role of charge? What plays the role of current? What plays the role of resistance?

13. Which has the greater resistance between opposite faces: a large cube of copper or a small cube of copper? Justify your answer.

14. If a copper wire and an aluminum wire of circular cross section and of the same length have the same resistance, what (if anything) can be said about the ratio of the diameters of the wires?

15. Is it correct to talk about the resistance of copper or the resistivity of copper? What is the distinction?

16. A contractor has a large electric drill for drilling through concrete walls that requires considerable current. The drill has a cord that is too short and you are sent to a hardware store to get an extension cord. You notice that the extension cords available have wire gauges from 10 to 18. To be safe, explain which gauge should you purchase.

17. The cockney accent in the East End district of London is noted for not pronouncing the "h's" at the beginning of a word. Thus "here" is pronounced as "ear," "he" is pronounced as a long-e sound, and so on. This gives rise to a number of interesting expressions since the word "home" then sounds like "ohm." Perhaps you can have fun deciphering Figure Q.17.*

---

*The phrases are from a short note by Marshall Ellenstein in *The Physics Teacher*, 29, #6, September 1991, page 347.

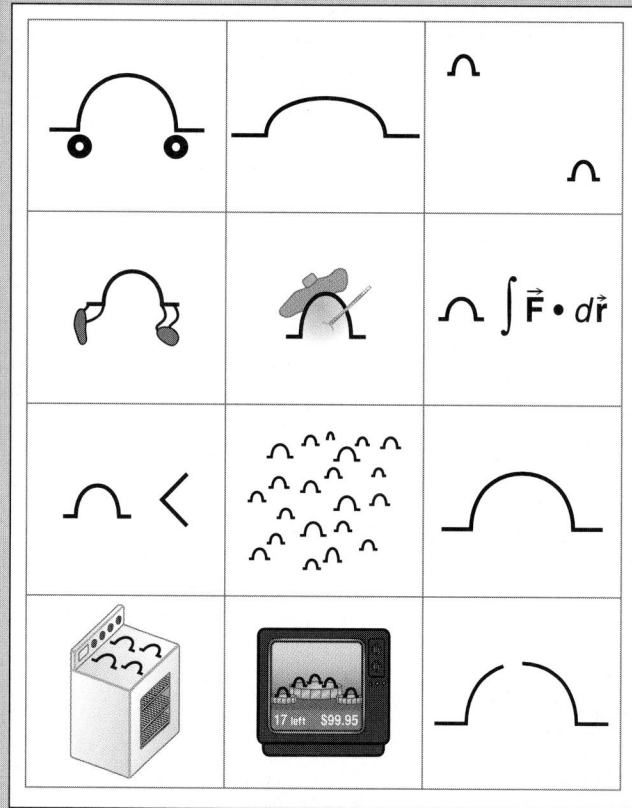

FIGURE Q.17

18. The term brownout means that the electric company reduces the potential difference available at the outlets in our homes. Why and under what circumstances is this done?

19. Fuses are pieces of metal wire in a circuit that melt if the current exceeds a certain value. They are used to prevent excessive currents that may damage other circuit elements. Should a fuse be placed in series or parallel with the circuit element to be protected? Explain the reasoning behind your answer.

20. Two resistors, with $R_1 < R_2$, are connected in parallel with the combination connected to the terminals of an independent voltage source; see Figure Q.20. Which resistor absorbs the most power? Justify your answer.

21. Two resistors, with $R_1 < R_2$, are connected in series with the combination connected to the terminals of an ideal voltage

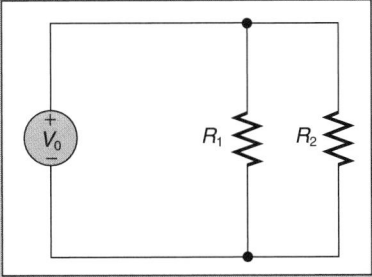

FIGURE Q.20

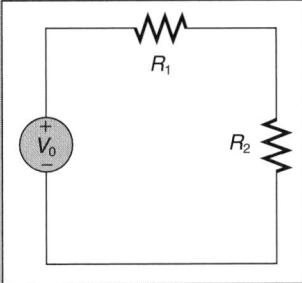

FIGURE Q.21

FIGURE Q.27

source; see Figure Q.21. Which resistor absorbs more power? Justify your answer.

22. A news report of an electrocution states that death occurred "when 14 000 V of electricity passed through the person." What is wrong with this statement?

23. *Future Shock* [title of a book by Alvin Toffler (Random House, New York, 1970)]. A live power line comes down during a violent storm and rests on your parked car while you are entertaining a friend inside the car. Are you both electrocuted? If not, should you get out of your car immediately? What should you do (other than scream and shout)? Explain.

24. If your car is hit by lightning, are you in any danger of electrocution? Is it better to have the windows up or down?

25. A small resistance and a large resistance each is connected in turn to a given independent voltage source. Which resistor absorbs more power? Justify your answer.

26. An electric company constructs a long transmission line from its generators to its customers. The resistance of the lines is symbolized by *r*. To minimize power absorption by the transmission lines themselves, should a large current and a small potential difference be used, or a large potential difference with a small current? Explain your reasoning.

27. Electrical appliances such as hot plates, portable electric ovens, and window-unit air conditioners that require large currents usually come with a warning not to operate them using a common extension cord. (See Figure Q.27.) Why is this admonition good advice? Many fires are caused by ignoring this warning, some with fatal consequences.

28. Carefully explain what is meant by each of the following electrical terms: (a) ground; (b) short circuit; (c) circuit; (d) characteristic curve.

29. Carefully state the Kirchhoff circuit laws and indicate what fundamental law of physics underlies each circuit law.

30. Sketch a schematic characteristic curve for each of the following circuit elements. Clearly label appropriate axes. (a) An ideal battery; (b) a real battery; (c) an ohmic material; (d) a nonohmic material; (e) (optional) the characteristic curve of your professor.

31. A copper-clad pot is set on top of an electric stove burner that is glowing brightly from the electrical power absorbed by the heating element. Why does the copper bottom of the pot not short-circuit the heating element?

32. Incandescent lights frequently burn out when turned on rather than while they are on. See if you can suggest plausible reasons for why this is generally the case.

33. Call your local electric utility to determine the rate structure for using, say 2000 kW·h of energy per month. Determine the cost, excluding and then including any local taxes.

34. Given a battery and three different resistors, how many different circuits can be made using all four circuit elements in each? Sketch the circuits.

35. Three different resistors are to be connected together to produce various equivalent resistances. (a) How many different ways can they be connected together? (b) How should they be connected together to produce as large an equivalent resistance as possible? (c) How should the three be connected to produce the minimum equivalent resistance?

36. Two resistors have resistances such that $R_1 > R_2$. (a) If the two are connected in series to an independent voltage source, which resistor has the greater potential difference across it? (b) If the two are connected in parallel with the ideal battery, which has the greater current?

37. In many homes, it is possible to turn a room light on or off using either of two switches located at different points in the room. Switches used for this purpose are constructed as shown schematically in Figure Q.37. Design a circuit with a battery, two such switches, and a light bulb that does the job. If possible, test your circuit in a laboratory.

38. Several identical light bulbs are arranged in a circuit as shown in Figure Q.38. The switch labeled S is closed. (a) What happens to bulb C? (b) What happens to bulbs A and B?

39. James Thurber (1894–1961) once wrote: "She came naturally by her confused and groundless fears, for her own mother lived

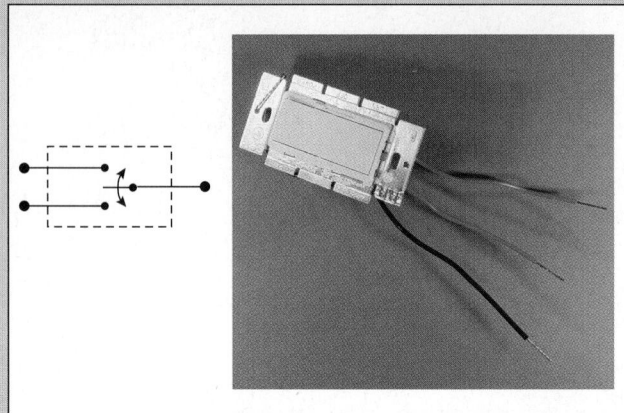

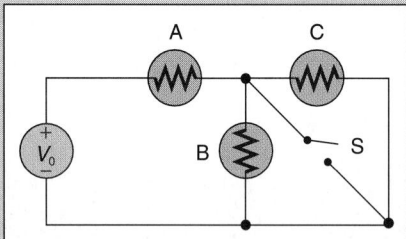

FIGURE Q.38

the latter years of her life in the horrible suspicion that electricity was dripping invisibly all over the house. It leaked, she contended, out of empty sockets as if the wall switch had been left on." [*My Life and Hard Times* (Harper and Brothers, New York, 1933), Chapter 2, "The Car We Had to Push"]. Why is this a groundless fear?

40. Perhaps you have noted that the brightness of a light bulb slowly decreases over its lifetime even though it is connected to an independent voltage source with a constant potential difference. What might account for this observation?

41. In choosing a new car (real) battery, you are offered a choice of various kinds, all of which have a potential difference between their terminals of 12 V. The saleswoman states the various choices differ in ampere-hour (A·h) rating. What is the meaning of the term ampere-hour?

42. You likely are familiar with the various common real battery sizes, called D-cell, C-cell, AA, AAA, and so on. Each has a potential difference between its terminals of about 1.5 V. Other than their physical dimensions and mass, what is the essential difference between them?

43. If you start your car with its headlights on, the lights dim until the motor starts. Suggest reasons for this observation.

44. Explain the distinction between an ammeter and a voltmeter: (a) when comparing the intrinsic electrical characteristics of the instruments; (b) when using the instruments to make a measurement in an electrical circuit.

45. You receive a bill from the electric company for 1500 kW·h. Are you being billed for *power* usage or *energy* usage? Explain.

46. (a) A multimeter M is attached to the resistor as indicated in Figure Q.46. (a) If the multimeter is set so it is a voltmeter, what does it indicate for the potential difference across the resistor? (b) The multimeter is switched to the ohmmeter setting. What will it indicate for the resistance?

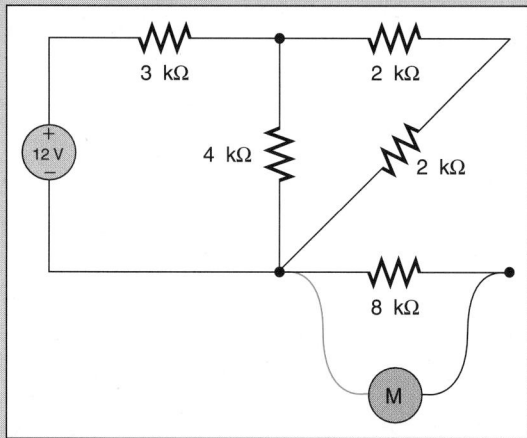

FIGURE Q.46

47. Look at the schematic diagram of an ohmmeter shown in Figure 19.89. Why is it *not* a good idea to leave a resistor connected to the terminals of an ohmmeter for long periods of time?

48. The time constant of a series RC circuit is $\tau = RC$. Show that the product of ohms and farads is seconds: $\Omega \cdot F = s$.

49. What is the power absorbed by a capacitor that is already fully charged?

50. Sketch the characteristic curve (with the dc current on the vertical axis) for a fully charged capacitor.

51. The charges on an isolated, real, charged capacitor gradually decrease with time because of the finite conductivity of most insulating materials separating the plates. The charges gradually leak across the gap separating the two conductors of the capacitor and the capacitor becomes neutralized. This is known as the *leakage current*. (If the capacitor is connected to an independent voltage source, the leaking charges are replenished by the source.) Construct a model of a real capacitor using an ideal capacitor and a resistance. Should the resistance be placed in series or parallel with the ideal capacitor to model the leakage current? Should the resistor have a high resistance or a low resistance? Explain the reasons for your choices.

52. Which graph in Figure Q.52 best represents the qualitative behavior of the charge on a capacitor as a function of time when the initially charged capacitor is connected in parallel with a resistor?

53. In most household and laboratory circuits, the duplex outlet plugs (think of them as independent voltage sources when wired up to the power grid) have *three* wires, one of which is called a ground wire. See Figure Q.53. What is the reason for such a third wire?

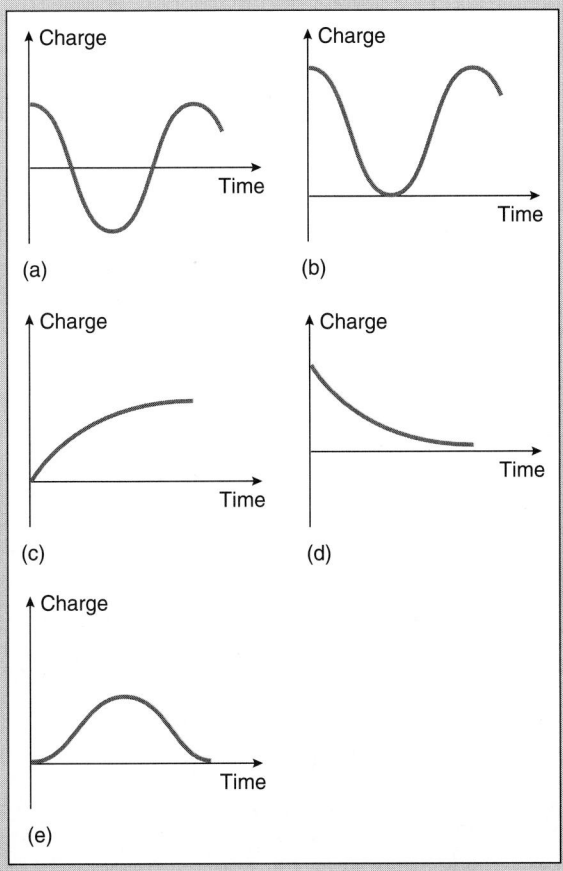

FIGURE Q.52

FIGURE Q.53

PROBLEMS

Sections 19.1 The Concept of Electric Current
19.2 Electric Current

1. The transport of electrons in large molecules has many significant implications. For example, light stimulates the transport of electrons through chlorophyll, the molecule intimately connected to photosynthesis. The conduction of up to 10^6 electrons per second along the axis of segments of the DNA double helix molecule is stimulated by the presence of ruthenium, which binds to ribose, one of the components of the helical DNA strands. What current does this electron transport represent? It is hoped that the development of ways to stimulate electron transport in DNA can lead to the detection of genetic defects and ways of detecting killer viruses, such as the HIV and Ebola viruses.

2. A charge q is in a circular orbit of radius r, moving at speed v. (a) What is the period of the motion? (b) What is the frequency of the motion? (c) What is the current?

•3. A cylindrically shaped beam of protons has a diameter of 2.00 mm and has 1.20×10^6 protons per cubic centimeter. The kinetic energy of each proton is 1.00 keV. (a) What is the speed of each proton in the beam? (b) What is the beam current in amperes?

•4. Protons in a 1.00 μA beam are traveling at a speed nearly equal to that of light. (a) How many protons are along each meter of the beam? (b) If the beam is 1.00 mm in diameter, what is the density of charge carriers (the number of protons per cubic meter)?

•5. (a) Calculate the drift speed in a 12 gauge shiny silver wire (see Table 19.2 on page 843) carrying a current of 1.50 A. Assume there is one charge carrier (an electron) per atom. The molar mass of silver is 108 g and its density is 10.5×10^3 kg/m^3. (b) How many hours does it take the charge carriers to travel one meter?

•6. A charge q is uniformly distributed along a straight line of length ℓ. The line rotates at angular speed ω about an axis through one end as shown in Figure P.6. What is the current past a fixed radial line?

•7. A closed, square insulating ring has sides of length ℓ and charge Q uniformly distributed along its perimeter. The square rotates about one edge at angular speed ω. What is the current past the fixed plane indicated in Figure P.7?

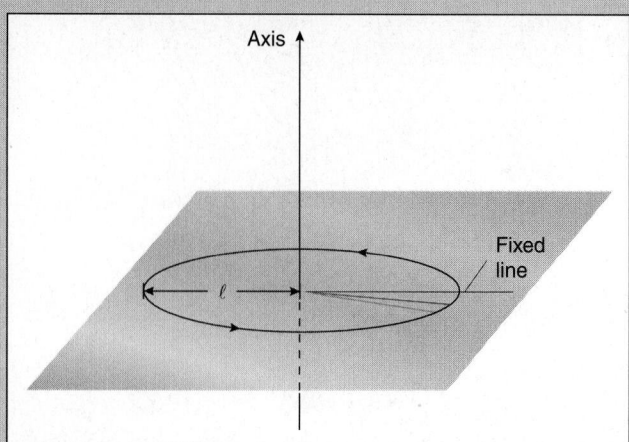

FIGURE P.6

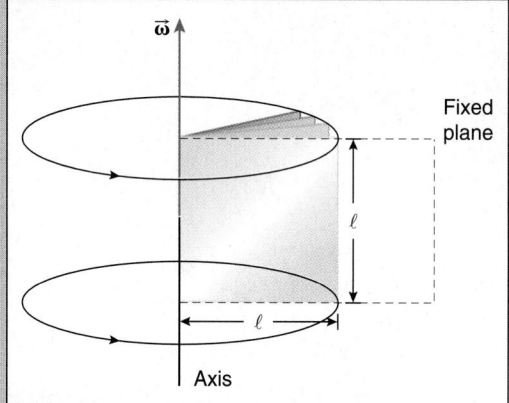

FIGURE P.7

•8. An insulating ring of radius *a* has a charge *Q* spread uniformly around its circumference. (a) The ring spins at constant angular speed ω about a diameter. Find the current past the line momentarily and occasionally coincident with half the circumference of the ring indicated in Figure P.8a. (b) If the ring spins at the same angular speed about an axis through the center of the ring and perpendicular to its plane as shown in Figure P.8b, what is the current past the indicated line?

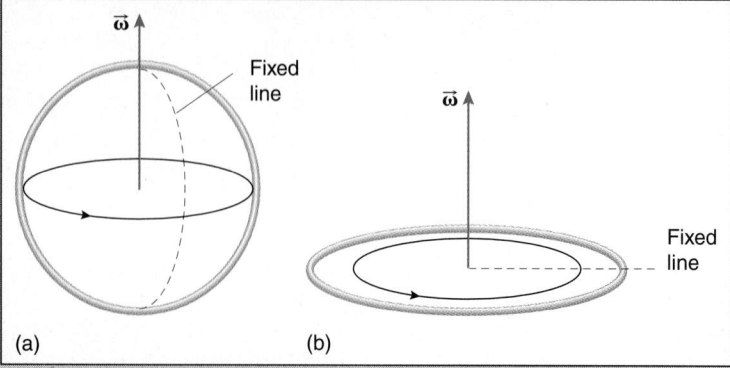

FIGURE P.8

Sections 19.3 The Pièce de Résistance: Resistance and Ohm's Law
19.4 Resistance Thermometers
19.5 Characteristic Curves

9. Resistors are fabricated on integrated circuit chips by suitably tailoring the dimensions of a conducting layer such as gold. Let the resistor to be fabricated be a gold wire of rectangular cross section. If the resistor is 1.0 μm by 10.0 μm in cross section and 400 μm long, what is the resistance of the gold wire?

10. A wire of length ℓ and cross-sectional area A is melted and recast into a wire of length 3ℓ using the same mass of material. (a) What is the new cross-sectional area of the wire? (b) How is the resistivity of the wire affected by the change? (c) How is the resistance of the wire affected by the change?

•11. A copper cable consisting of ten strands of copper wire, each 1.00 mm in diameter, is one of several cables used to supply electrical service to a cow barn. The length of each cable is 200 m. What is the resistance of each cable?

•12. Consider two solid cubes of a material with resistivity ρ. One cube has sides of length ℓ, the other has sides of length 3ℓ. Which cube has the greater resistance between opposite faces, and by what factor is the resistance of this cube greater than the resistance of the other cube?

•13. The Alaska oil pipeline is approximately 1.27×10^3 km long, 120 cm in diameter, and made of steel with a thickness of about 1.0 cm. See Figure P.13. Steel has a resistivity of 1.80×10^{-7} $\Omega \cdot$m. What is the total resistance of the pipeline over its entire length?

•14. A copper wire is stretched uniformly (maintaining constant density and resistivity) so that it is ten times its original length (but of the same volume). By what factor is its resistance changed?

•15. A metal cube has a resistivity of 2.0×10^{-8} $\Omega \cdot$m and a resistance of 1.0 $\mu\Omega$ between opposite faces. (a) What is the length of a side of the cube? The cube now is compressed to form a square plate of the same density as before but of only one-tenth the original thickness. (b) What is the length of a side of the square plate? (c) What is the resistance between the two opposite square faces?

•16. Consult Table 19.2 on page 843. For wires of a given length, if you increase the wire gauge number by 2, by about what

factor does the resistance of a wire increase? Does your result depend on the initial wire gauge?

•17. Two wires are made from the same material, are the same length, and carry equal currents. One wire is half the diameter of the other. Which has the larger average drift speed and by what factor?

•18. Assume the temperature coefficient of resistance of carbon (diamond) is independent of temperature and has the value $-0.5 \times 10^{-3}\,\text{K}^{-1}$ given in Table 19.1 for 20 °C. (a) At what temperature is its resistivity equal to zero? How does this compare with the melting temperature of diamond ($\sim 3.5 \times 10^3$ °C)? (b) What does the result of part (a) likely imply about the assumption of a constant temperature coefficient of resistivity?

•19. Consult Table 19.2 on page 843. For wires of a given length, show that if the wire gauge number is increased by 10, the resistance of the wire increases by about a factor of 10.

•20. A mass of 1.00 kg of copper (density 8.93×10^3 kg/m³) is drawn into 12 gauge wire (see Table 19.2). (a) What is the length of the wire? (b) What is the resistance of this wire?

•21. For a wire of fixed length ℓ and made from a given material of density ρ_{mass}, show that the resistance of the wire is inversely proportional to its mass. What is the proportionality constant?

•22. As a physics student, you consider a cool amber liquid made from hops in ways different from the normal population. You measure the temperature of the liquid using a platinum resistance thermometer. The resistance thermometer has a resistance of 220.25 Ω at 25.0 °C. When immersed in the stein, the resistance of the thermometer is 199.99 Ω. (a) Determine the temperature of the liquid.

(b) If the same thermometer is used to measure one's body temperature (assuming it is normal: 37.0 °C), what resistance will the thermometer indicate?

•23. On the same graph, accurately draw the characteristic curves of a 1.00 kΩ resistor and a 2.00 kΩ resistor when the potential difference across them varies from −10.0 V to 10.0 V.

Sections 19.6 Series and Parallel Connections Revisited
19.7 Resistors in Series and in Parallel

•24. An unusual copper wire is 4.00 m long. For 3.00 m of its length, the diameter is 3.00 mm and for the remaining 1.00 m of its length the diameter of the wire is only 1.00 mm. The wire is covered with a pretty red plastic insulation that is 0.90 mm thick. (a) The wire is connected to the terminals of an independent voltage source that provides 20.0 A current for the thick end of the wire. What current is in the thin end? (b) What is the resistance of the 3.00 m section of the wire? (c) What is the resistance of the 1.00 m section of the wire? (d) What is the total resistance of the wire? (e) What is the potential difference between the ends of the wire?

•25. For each of the resistor combinations shown in Figure P.25, find the resistance of the single equivalent resistor that can be placed between the terminals A and B.

•26. Find the equivalent resistance between the terminals A and B shown in Figure P.26.

•27. Find the single resistance that is equivalent to that of the resistor assortment of Figure P.27 between the terminals A and B.

Section 19.8 Electric Power

28. A resistor has a potential difference of 5.00 V across it and absorbs 15.0 W of power. Determine the current through the resistor and the value of its resistance.

29. Two light bulbs absorb 100 W and 200 W of power when used with a potential difference of 120 V across each of the filaments. Which filament has the greater resistance?

•30. Two resistors with resistances 5.0 Ω and 10.0 Ω are to be connected to an independent voltage source with an emf of 5.0 V. The resistors are first connected in series with the independent voltage source and then in parallel with it. (a) Determine the power absorbed by the two resistors for each arrangement. (b) Which arrangement results in the most total power absorbed by the two resistors?

•31. The 12.0 V battery (an independent voltage source) of your car is connected in series with the starter motor. The battery provides 90.0 A for 2.00 s to the starter motor. What power is absorbed by the motor?

•32. Show that the power absorbed by a resistor R can be written in each of the following equivalent ways:

$$P_{\text{res}} = \frac{V^2}{R}$$

or

$$P_{\text{res}} = I^2 R$$

where V is the potential difference across the resistor and I is the

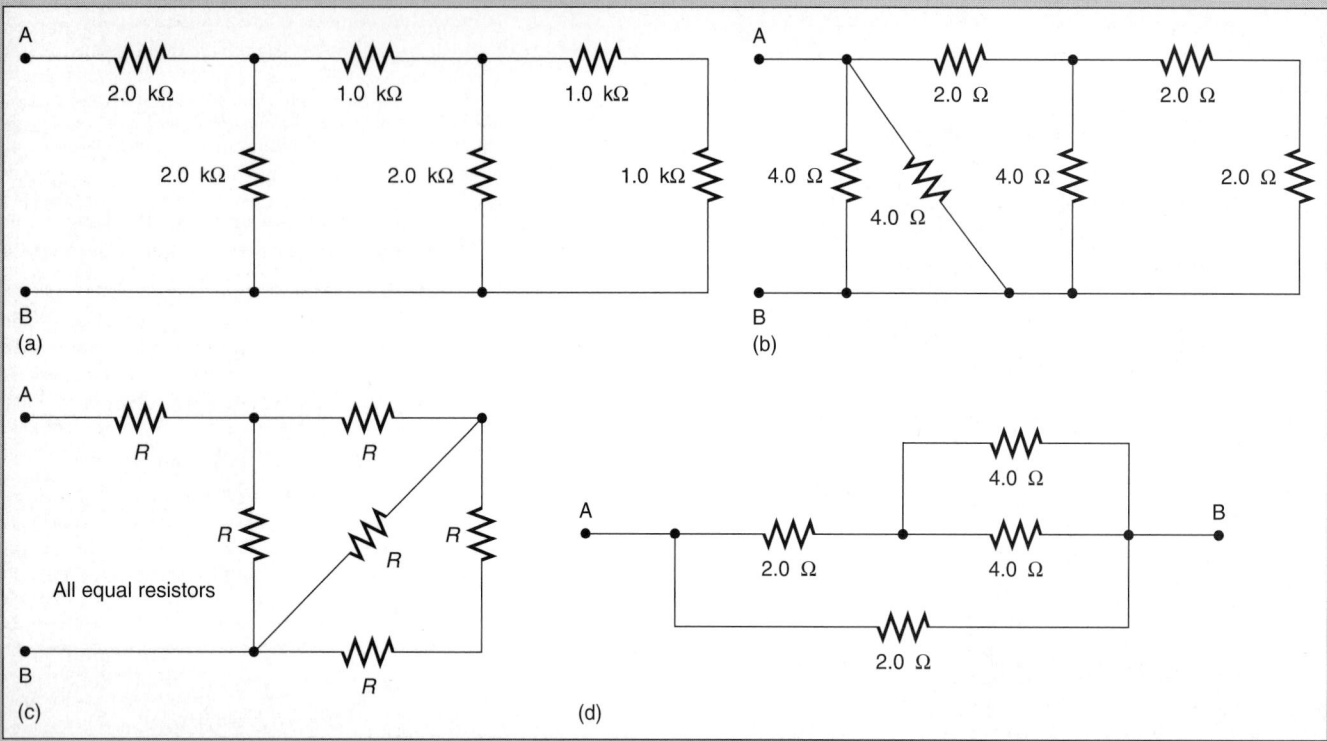

FIGURE P.25

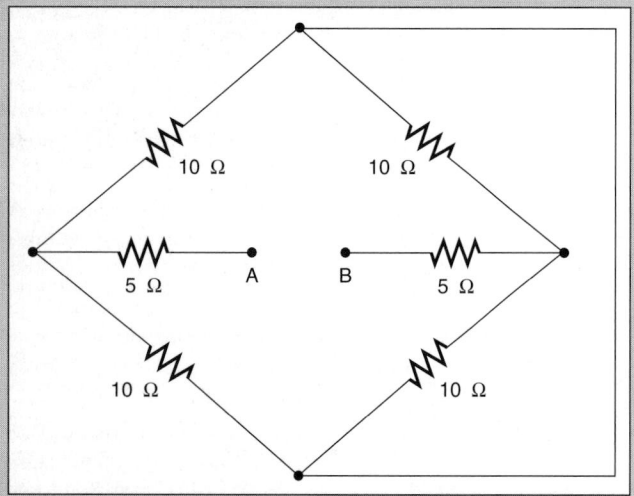

FIGURE P.26

current through it. The implication of the first equation is that, for a given potential difference V, the power absorbed by a resistor is inversely proportional to the resistance. The implication of the second result is that, for a given current I, the power absorbed by a resistor is directly proportional to R. Is anything wrong?

• 33. If house wiring gets too hot, a fire may result. A wire is to carry 15 A and the power absorbed per meter of length is to be no greater than 1.0 W/m. (a) What diameter of copper wire would absorb 1.0 W/m? (b) What is the maximum gauge number of copper wire that should be used (see Table 19.2 on page 842)?

• 34. Your electric toothbrush absorbs 5.0 W of power when used. If you use the brush for 5.0 minutes each day and a kilowatt-hour of energy costs $0.13, how much does it cost to use the cavity fighter for a year?

• 35. If the price of a kilowatt-hour of energy is $0.13, what is the cost to use your 200 W study lamp for 6.0 hours every night for 9 months studying physics?

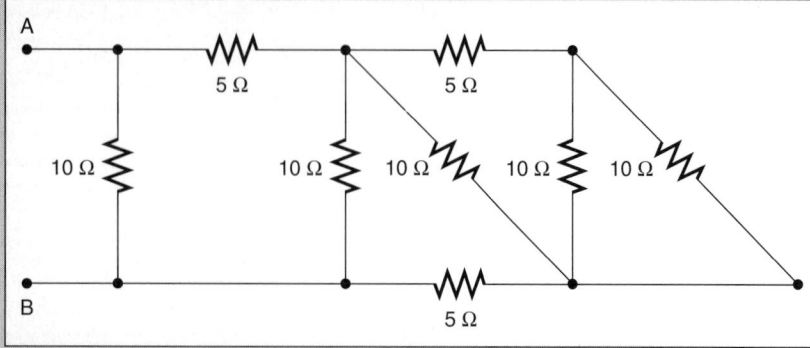

FIGURE P.27

•36. A yellow bellied sapsucker (*Sphyrapicus varius*) perches on a power line that carries a 1.00 kA current. The feet of the bird are 5.0 cm apart. The resistance of the wire per meter of length is 5.0×10^{-5} Ω/m. Assume the wire is copper with a resistivity of 1.77×10^{-8} Ω·m. (a) What is the diameter of the wire? (b) What is the potential difference between the bird's feet? (c) What power is absorbed by the wire section between the feet of the bird?

Sections 19.9 Electrical Networks and Circuits
     19.10 Electronics
     19.11 Kirchhoff's Laws for Circuit Analysis
     19.12 Electric Shock Hazards*
     19.13 A Model for a Real Battery
     19.14 Maximum Power Transfer Theorem

•37. A 100 W light bulb and a 60 W light bulb normally are connected in parallel, so if there is a potential difference of 120 V across each of them, they absorb 100 W and 60 W of power respectively. Suppose, instead, that you wire the two bulbs in series with a 120 V source placed across the combination. Determine the power absorbed by each and indicate which bulb is brightest.

•38. Typical house circuits at 120 V have fuses or circuit breakers rated for a maximum of 15 A. How many 1200 W toaster ovens can be used at full power on the same circuit before the fuse blows? First consider how the toaster ovens are connected to the 120 V source.

•39. Consider the circuit shown in Figure P.39. (a) What is the potential difference across the 15.0 Ω resistor? (Hint: Look at the relationship between the battery and the 15.0 Ω resistor. How are they connected together?) (b) What is the power absorbed by the 15.0 Ω resistor? (c) What is the current through the 10.0 Ω resistors? (d) What is the power absorbed by the battery? (e) What power is absorbed by each of the 10.0 Ω resistors?

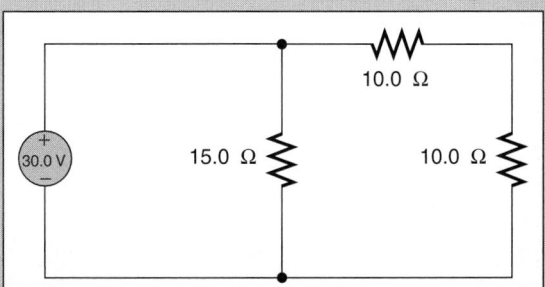

FIGURE P.39

•40. The following questions make reference to the circuit of Figure P.40. The resistors represent identical light bulbs, each with the same resistance R. Refer to the bulbs by number with your answers and explanations. (a) Which bulb(s) has the greatest current? Which bulb(s) absorbs the greatest power? Which is (are) the brightest? (b) Suppose bulb 3 is removed from the circuit. What happens to the brightness of bulbs 1 and 2? (c) Reinsert bulb 3. Now connect a wire from point A to point B. What happens to the brightness of bulbs 2 and 3? What happens to the brightness of bulb 1? (d) Remove the wire inserted in part (c). Connect a wire from point A to point C. What happens to the brightness of bulb 1? What happens to the brightness of bulbs 2 and 3?

This problem was inspired by similar problems in Arnold Arons, *A Guide to Introductory Physics Teaching* (John Wiley, New York, 1990), pages 184–187.

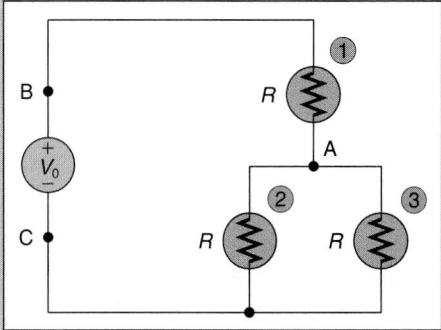

FIGURE P.40

•41. Examine the circuit of ideal batteries and resistors shown in Figure P.41. (a) Indicate clearly in a sketch the current directions you choose for each branch of the circuit to solve the problem. (b) Indicate appropriate polarities associated with each resistor. (c) Use Kirchhoff's laws to find the current through each circuit element. (d) Calculate the electric power absorbed by each circuit element. (e) Check to be sure that the sum of the power absorbed by all the circuit elements is zero.

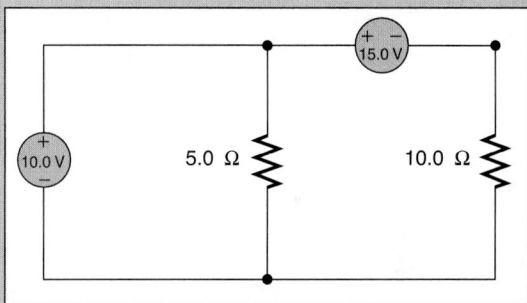

FIGURE P.41

•42. You decide to leisurely toast marshmallows by placing them over the 6.00 Ω resistor in the circuit shown in Figure P.42. (a) Indicate appropriate current directions for each branch of the circuit. (b) Mark the resistor polarities consistent with your choice of the current directions. (c) Find the current through each circuit element. (d) Find the power absorbed by each circuit element. (e) Show that the total power absorbed by all the circuit elements is zero.

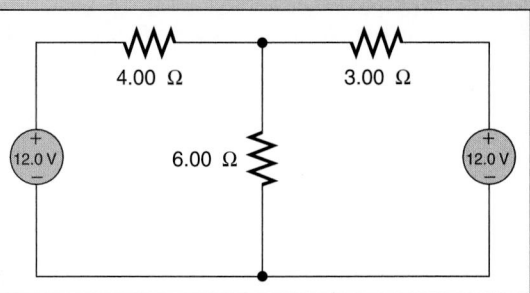

FIGURE P.42

•**43.** The circuit shown in Figure P.43 is known as a bridge circuit.* It is used extensively in applications such as temperature measurement. The purpose of the circuit is to measure the resistance of an unknown resistor R_x (such as the resistance of a resistance thermometer) in terms of known precision variable resistances R_2, R_3, and R_4. The values of R_2, R_3, and R_4 are adjusted until the ammeter indicates zero current; the bridge circuit then is said to be *balanced*. (a) When the ammeter indicates zero current, what must be the relationship between the current in R_x and the current in R_2? What about the relationship between the current in R_4 and the current in R_3? (b) When the ammeter indicates zero current, what is the potential difference between points A and B? (c) Use the result of part (a) to determine the relationship between the potential differences across R_x and R_4. What about the relationship between the potential differences across R_2 and R_3? (d) Using the preceding results, derive an expression for the unknown resistance R_x in terms of R_2, R_3, and R_4.

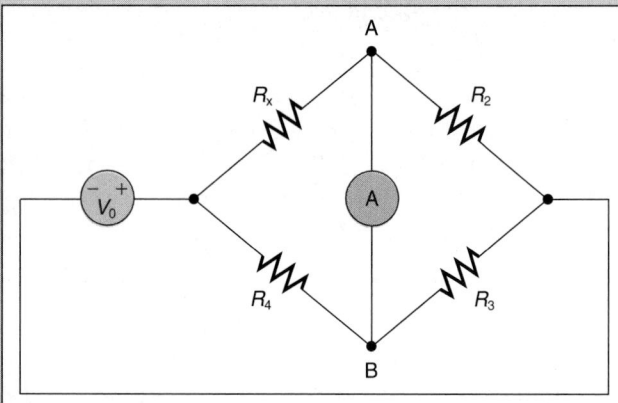

FIGURE P.43

•**44.** Yet another beautiful circuit (see Figure P.44). (a) Indicate clearly on a sketch the direction you choose for the currents in each distinct branch of the circuit. Label resistor polarities in accordance with your choice. (b) Write the KVL for each elementary loop of the circuit. Does this provide a sufficient number of equations to solve for the unknown currents? (c) Use the KCL at a significant node to secure yet another equation relating the unknown currents. (d) Solve the equations for the unknown currents. (e) Calculate the potential difference across each of the resistors. (f) Calculate the power absorbed by each circuit element and check that the total power absorbed by the circuit is zero.

•45. The circuit shown in Figure P.45 is constructed from identical resistances R for your greater enjoyment. The equivalent resistance is found by Zeus to be $13.0 \ \Omega$. (a) Determine the value of the resistance R used in the circuit. (b) Which of the resistors R will have the least current through it? (Specify the resistors by

*This is not the same kind of bridge circuit you or your parents might enjoy with a deck of cards. This bridge circuit was popularized by Charles Wheatstone (1802–1875), although the circuit actually was invented in 1833 by Samuel Christie, only son of James Christie, founder of the well-known auction gallery. Wheatstone, a musician as well, patented the concertina in 1829.

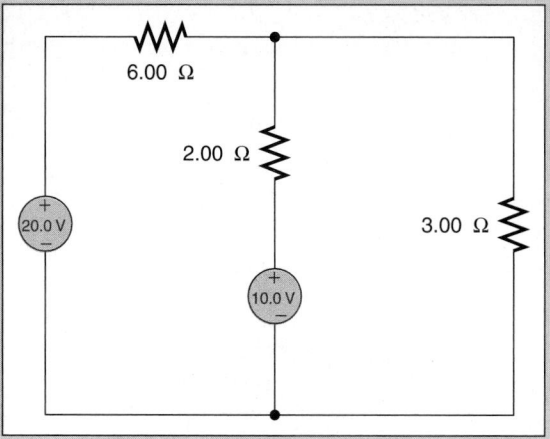

FIGURE P.44

number.) (c) Find the least current through the resistors indicated in part (b). (d) Indicate how you would connect another single resistance R so as to make the total current through the battery the greatest.

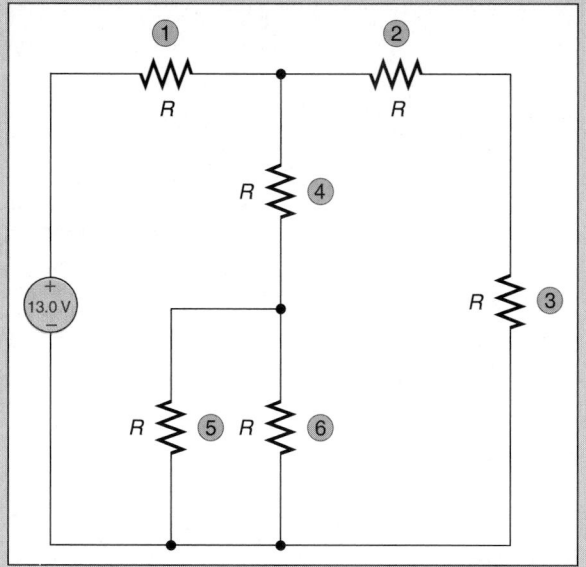

FIGURE P.45

•**46.** Analyze the circuit shown in Figure P.46 in the following manner: (a) Clearly indicate appropriate currents with labels and directions, indicating the resistor polarity markings accordingly. (b) Set up a number of equations sufficient to solve for the unknown currents delineated in part (a). (c) Solve these equations for the unknown currents using appropriate techniques. (d) Calculate the power absorbed by each circuit element and show that these powers sum to zero.

•**47.** The circuit shown in Figure P.47 is called a *voltage divider*. Show that the potential difference across resistance R_1 is

$$\frac{R_1}{R_1 + R_2} V_0$$

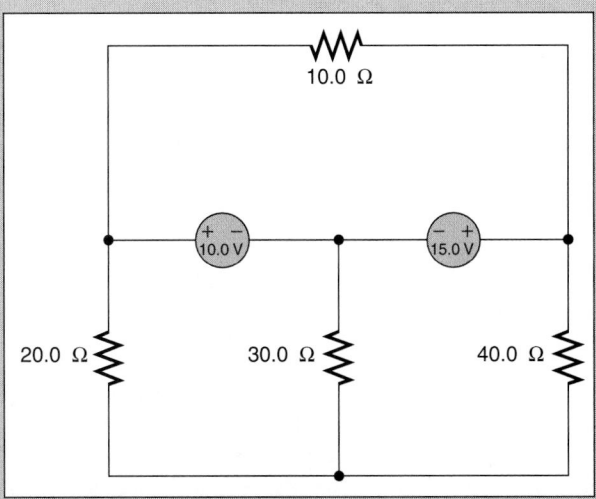

FIGURE P.46

and the potential difference across resistance R_2 is

$$\frac{R_2}{R_1 + R_2} V_0$$

Notice that these results can be summarized in the following way: the fraction of the potential difference V_0 that appears across a given resistance is the given resistance divided by the sum of the two resistances.

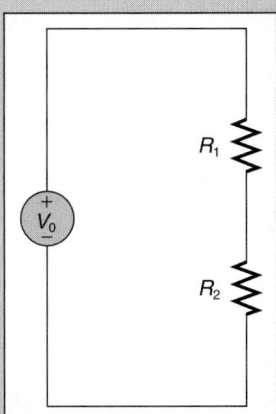

FIGURE P.47

•**48.** In the circuit depicted in Figure P.48, the current I_0 is split (or divided) between the two resistances R_1 and R_2. Show that the current in R_1 is

$$\frac{R_2}{R_1 + R_2} I_0$$

and that the current in R_2 is

$$\frac{R_1}{R_1 + R_2} I_0$$

Hence this circuit is called a *current divider*. Notice that the results can be summarized in the following way: if a current I_0

is split between two resistances, the fraction of the current in a given resistor is the *other* resistance divided by the sum of the two resistors.

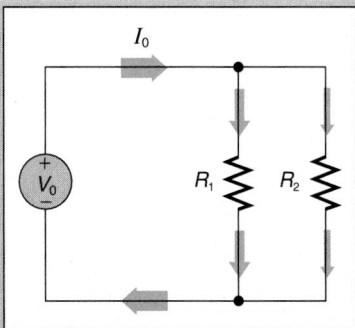

FIGURE P.48

•**49.** Consider the circuit shown in Figure P.49 (a) Determine the current in the 4.0 Ω resistor. (b) Determine the potential difference between points A and B. (c) Since there is a nonzero potential difference between points A and B, why is there no current in the 4.0 Ω resistor? (d) Calculate the power absorbed by the 8.0 V, 12.0 V, and 16.0 V independent voltage sources. (e) Calculate the total power absorbed by all the resistors. This problem was inspired by Robert E. Viens, "A Kirchhoff's rules puzzler," *The Physics Teacher*, 19, #1, page 45 (January 1981).

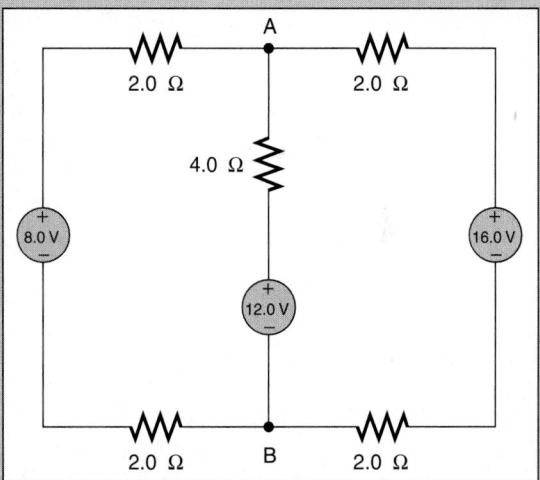

FIGURE P.49

•**50.** You are to design an automobile light circuit that has a 12.0 V battery, a fuse, a switch, two 50.0 W headlights, and two 10.0 W taillights. Each light needs 12.0 V for proper operation. The switch is to be placed so that all four lights are lit when the switch is closed and must ensure that if one light blows out (becomes an infinite resistance or open circuit), the remaining lights are unaffected. (a) Sketch a circuit that will accomplish this objective. (b) What current must the fuse be able to handle without melting?

•**51.** Two long house wires of uniform diameter have resistances $R_2 = 1.0$ Ω and $R_3 = 1.0$ Ω spread uniformly along their lengths. In Figure P.51, the resistance of the wires is lumped into one location for convenience on the circuit diagram. A heater with a resistance of 13.0 Ω is connected between the wires as indicated. An independent voltage source with an emf of 120 V is connected in series with a 0.30 Ω resistor that is a fuse. The fuse will melt (and break, or open, the circuit) if the power absorbed by the fuse exceeds 30 W. The power absorbed along the long house wires safely radiates away via heat transfer unless the currents become too large; this is the purpose of the fuse, among other things. (a) How much current will barely melt the fuse? (b) What current is at point A? (c) What current is at point B? (d) What current passes through the heater? (e) What is the potential difference across R_4? (f) If another 13.0 Ω heater were connected in parallel with the one shown, would the fuse melt? Justify your answer.

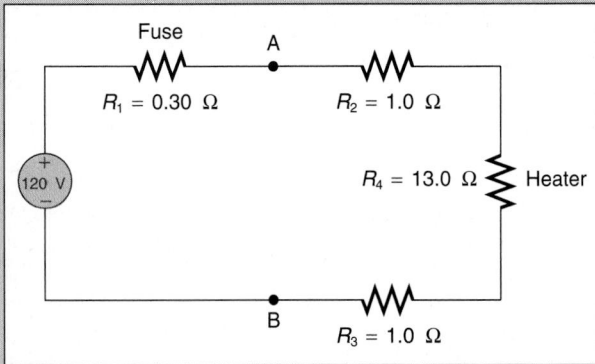

FIGURE P.51

•**52.** As an electrical engineer employed by the firm of Coppers, Gold, Silver, & Shorts you are asked to use a 12.0 V independent voltage source and three 8.0 Ω resistors to provide (in some part of the circuit) a current of 1.0 A and a potential difference of 8.0 V. Sketch a circuit that has these characteristics and indicate where the current is 1.0 A and between what two points the potential difference is 8.0 V.

•**53.** Consider the circuit shown in Figure P.53. (a) Simplify the circuit as much as possible (b) Choose a current direction emerging from the (+) terminal of the independent voltage

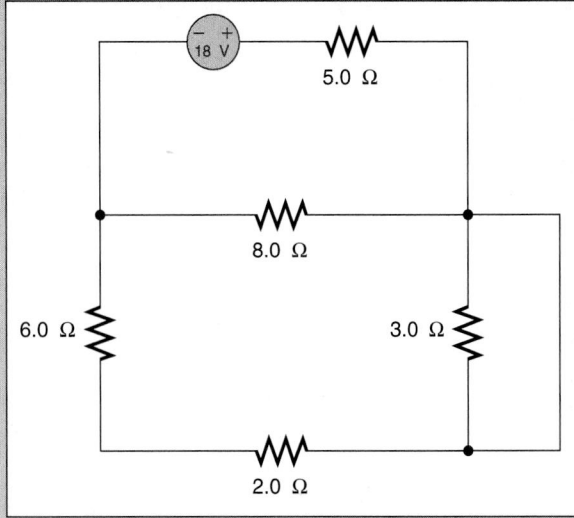

FIGURE P.53

source. Find this current. (c) What power is absorbed by the 8.0 Ω resistor? (d) What power is absorbed by the independent voltage source?

•**54.** A 5.0 W flashlight bulb is powered by two fresh D-cell (real) batteries. The batteries die after 2.0 h of continuous use. (a) How many kilowatt-hours of energy were absorbed by the bulb? (b) If the batteries cost $5.00, what is the cost of one kilowatt-hour of energy supplied by such batteries?

•**55.** An ideal 6.00 V independent voltage source is used to model a real battery. The real battery gets warm when it delivers current to an external load resistance. When the current provided by the battery is 2.00 A, the power absorbed by the battery is found to be 1.16 W. (a) What is the internal resistance of the real battery? (b) When the real battery has a current of 2.00 A, what is its terminal potential difference? (c) Sketch a circuit that will deliver the maximum possible current from this battery. What is this current?

•**56.** Conditions are such that your effective body resistance is 50 kΩ. Assume a 50 mA current is sufficient to induce ventricular fibrillation (see Section 19.12). (a) What potential difference will induce this unfortunate consequence? (b) If your body resistance is lowered to 1.0 kΩ by foolishly having wet contact with wires from a voltage source, what potential difference will induce ventricular fibrillation? (c) Why does a venerable utility wirewoman attempt to do all work with rubber boots, leather climbing belt, and rubber gloves and with one hand at a time?

•**57.** Consider the circuit shown in Figure P.57. (a) If the resistance R is infinite, find: (i) the potential difference between the points A and B; (ii) the power absorbed by the battery. Express your answers in terms of V_0, R_0, and r. (b) If the resistance R is zero, find: (i) the potential difference between the points A and B; (ii) the current emerging from the battery terminal of positive polarity. Express your answers in terms of V_0, R_0, and r.

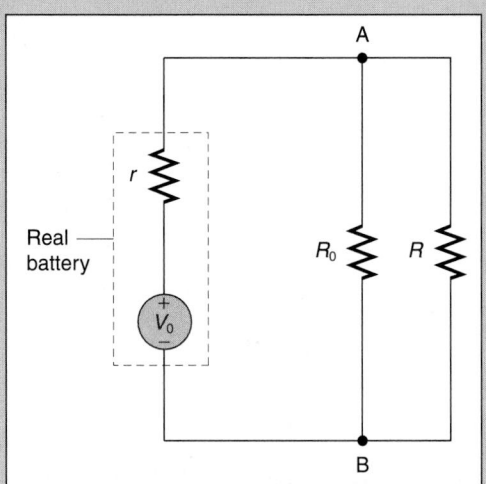

FIGURE P.57

•**58.** In the circuit shown in Figure P.58, what resistance R is needed to maximize the power absorbed by the combination of R and R_0? Express your result in terms of R_0 and r. What resistance R is needed to maximize the power absorbed by R itself?

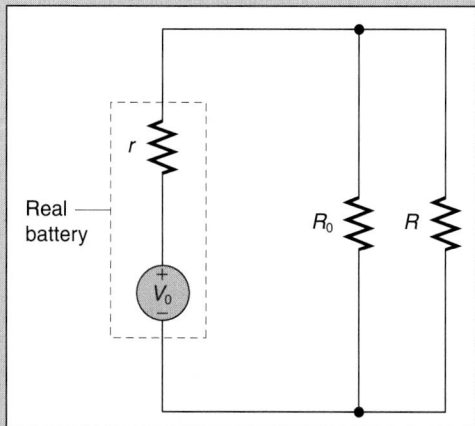

FIGURE P.58

•59. The condition

$$\frac{dP}{dR} = 0 \ \text{W}/\Omega$$

is the condition for a maximum or a minimum for power absorbed by load resistor R in the circuit shown in Figure P.59. To show that the condition indeed is a maximum, what condition must the second derivative

$$\frac{d^2P}{dR^2}$$

satisfy? Show that this condition is met when $R = r$.

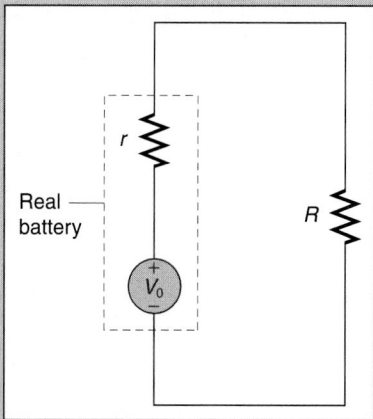

FIGURE P.59

60. An independent voltage source maintains the same potential difference between its terminals no matter what the current is through the device. An *independent current source* is a circuit element that is the current analog of the independent voltage source (an ideal battery). An independent current source maintains a constant current I_0 through it no matter what the potential difference across its terminals. The characteristic curves of the two sources are shown in Figure P.60a, where I_0 is the fixed current supplied by the current source. An ideal current source is indicated with the symbol shown in Figure P.60b, where the arrow indicates the direction of the current. Consider an independent voltage source V_0 in series with a resistance r connected to a load resistor R as in Figure P.60c. The independent voltage

source in series with the resistance r together make up what is called a Thévenin equivalent source. (a) Find the current I through the load resistance R as well as the potential difference V across the load resistor in terms of V_0 and r. (b) When the load resistor R has a resistance equal to zero, the circuit is said to be short-circuited between the terminals A and B. The current then is called the short circuit current I_{sc} between the terminals A and B. Find the short circuit current in terms of V_0 and r. (c) When the load resistance is infinite, the circuit is said to be open. When the load resistance is infinite, what is the current in the circuit? What is the potential difference between the points A and B when $R = \infty \ \Omega$? This potential difference is called the open circuit voltage (or the Thévenin voltage) between the terminals A and B. Now replace the Thévenin equivalent source with an independent current source whose current I_0 is equal to the short circuit current I_{sc} calculated in part (b). The current source is in parallel with the same resistance r used in Figure P.60c. The new arrangement is connected to the load resistance R. The new circuit is known as the *Norton equivalent circuit*. (d) Show that with the circuit of Figure P.60d, the current through the load resistance R and the potential difference across the load resistance R are the same as in part (a). Thus, as far as the load resistance is concerned, it makes no difference whether it is connected to (1) the voltage source in series with r or (2) the current source in parallel with r at terminals A and B (when the current source has a current equal to the short circuit current I_{sc}). Both arrangements result in the same current through R and the same potential difference across R. The two source arrangements (the Thévenin and Norton equivalent sources) are known as *source transformations* of each other. Such source transformations are used occasionally in more advanced courses in circuits to simplify the analysis of more complicated circuits.

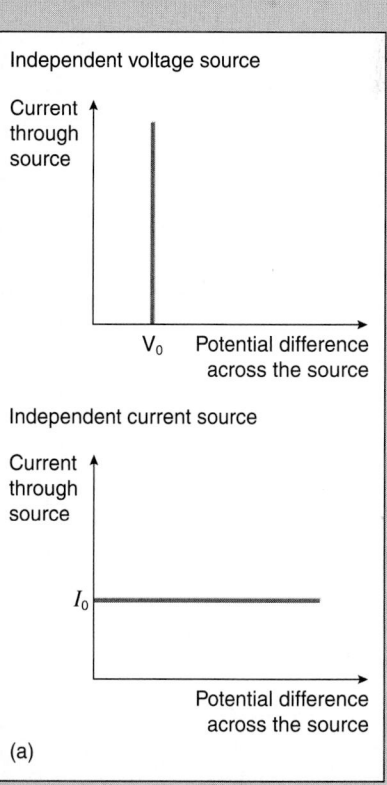

FIGURE P.60   *(Continued on next page.)*

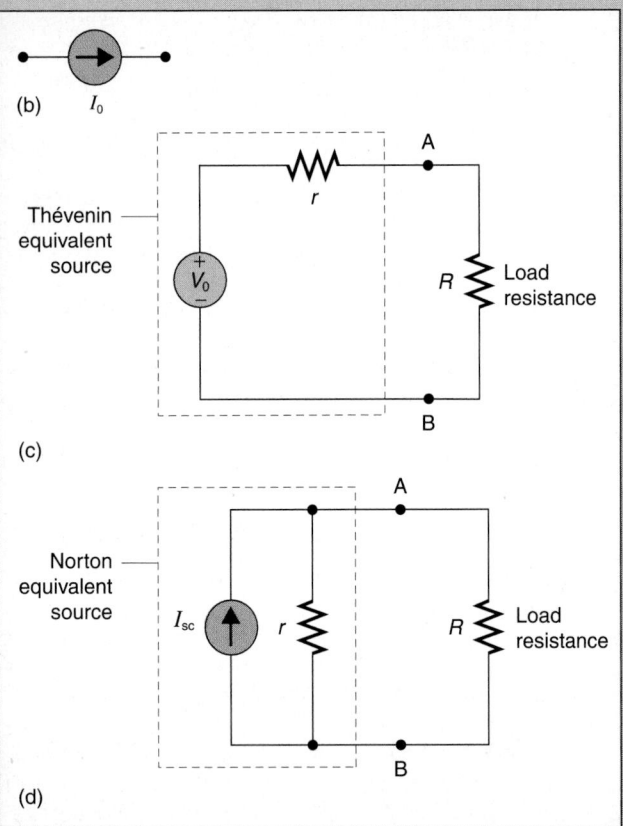

(b) I_0

Thévenin equivalent source

r

A

R Load resistance

V_0

B

(c)

Norton equivalent source

I_{sc} r

A

R Load resistance

B

(d)

FIGURE P.60   *(continued)*

Section 19.15 Basic Electronic Instruments: Voltmeters, Ammeters, and Ohmmeters

61. You have set up the circuit shown in Figure P.61. The voltmeter indicates 5.00 V and the ammeter indicates 0.250 A. (a) Indicate the direction of the current through the resistor. (b) Determine the value of R. (c) Determine the power absorbed by the resistor. (d) What is the emf of the independent voltage source?

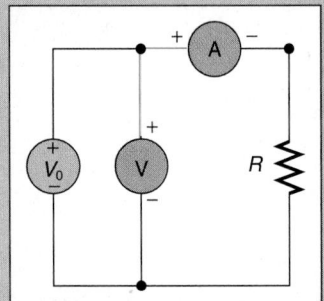

FIGURE P.61

•62. A real battery (Figure P.62) is modeled by an independent voltage source V_0 in series with a resistor r. The real battery is connected to a load resistor R, and a voltmeter is used to measure the potential difference across the load resistor as indicated. The following data are taken meticulously by engineer Milly Watt: (i) When the load resistance is 6.000 Ω,

the voltmeter indicates 1.452 V; (b) When the load resistance is 4.000 Ω, the voltmeter indicates 1.429 V. Use this information to determine the internal resistance r in the model as well as the emf V_0 of the independent voltage source.

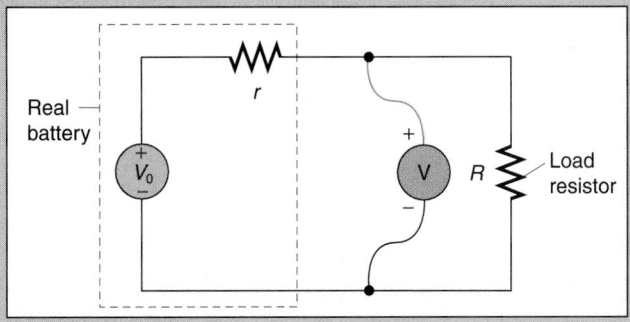

FIGURE P.62

•63. The potential difference V_0 and internal resistance r used to model a real battery can be found by connecting an adjustable resistance (called a *pot*) across the terminals of the real battery as shown in Figure P.63. The potential difference across R is measured for two different but known values of R (as determined with an ohmmeter when R is disconnected from the circuit). An experiment produces the following data: (i) When $R = 1.00$ kΩ, the potential difference across R is 6.00 V; (ii) When $R = 10.0$ Ω, the potential difference across R is 3.00 V. Use these data to determine V_0 and r.

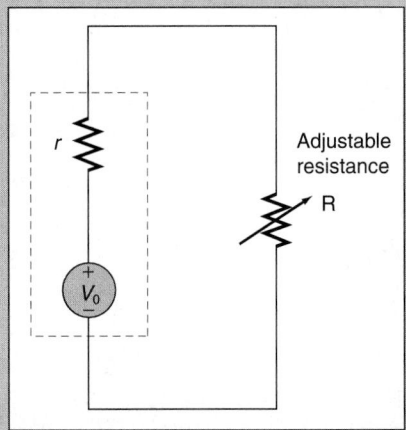

r

Adjustable resistance

R

V_0

FIGURE P.63

•64. Alas. A student has connected the circuit shown in Figure P.64 and his lab instructor is not pleased. Indicate three problems associated with the way the circuit is wired.

•65. In the circuit shown in Figure P.65, voltmeter V_1 indicates 10.0 V when the switch is open. (a) What does voltmeter V_2 indicate? What do the ammeters A_1 and A_2 indicate? What is the emf V_0 of the independent voltage source? (b) The switch is closed and voltmeter V_1 indicates 8.0 V, V_2 indicates 6.0 V, A_1 indicates 0.50 A, and A_2 indicates 0.75 A. Use this information to find the numerical values of R_3, R_2, R_1, and r.

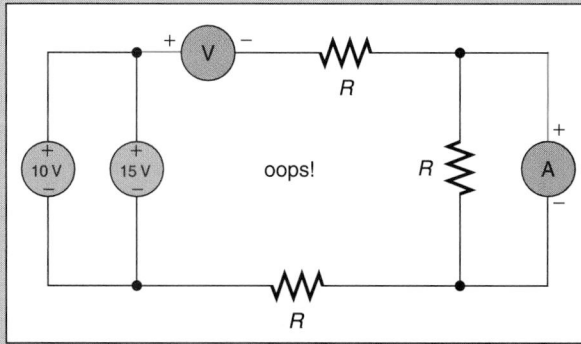

FIGURE P.64

Section 19.16 An Introduction to Transients in Circuits: A Series *RC* Circuit*

66. If you want a series *RC* circuit with a time constant of 60 s and have a capacitor with a capacitance of 100 μF on hand, what resistance is necessary? If you only have 100 kΩ resistors on hand, how do you secure the needed resistance?

•**67.** Show that when a resistor is connected across an isolated charged capacitor to form a single loop, series *RC* circuit, the electrical potential energy stored in the capacitor decays exponentially at a rate twice as fast as the current.

•**68.** Capacitors are used for energy storage, among other things. When collections of capacitors are used for such purposes, they are colloquially referred to as *capacitor banks*. Let there be 1.00 J of electrical potential energy stored in a charged capacitor that has a potential difference of 500 V between its plates. (a) Determine the capacitance of the capacitor. (b) The energy is released by discharging the capacitor via a series *RC* circuit. After about five time constants, the capacitor is effectively discharged. If you want to discharge the capacitor in part (a) so that the capacitor is effectively discharged in 0.100 ms, through what resistance should the capacitor be discharged? (c) What is the average power provided by the capacitor during the 0.100 ms interval?

•**69.** In a series *RC* circuit, at what time (in terms of the time constant τ) is the charge on a charging capacitor equal to one-half its final value? Does it take the same time for the capacitor to lose half its charge when discharging?

•**70.** The capacitors in Figure P.70 are initially charged. When the switch is closed, find the time constant of the circuit.

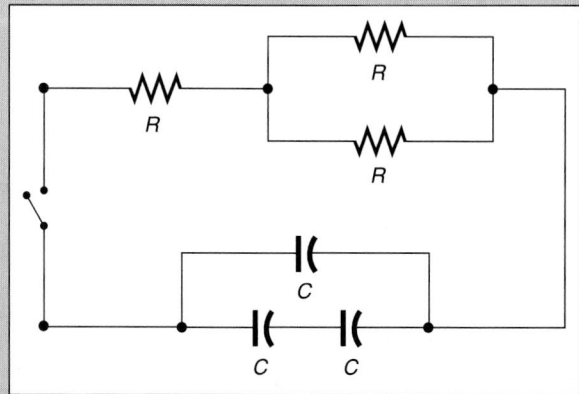

FIGURE P.70

•**71.** For a discharging capacitor in a series *RC* circuit, show that line representing the slope of a graph of the charge *q* versus *t*, evaluated when $t = \tau$, will intersect the time axis when $t = 2\tau$.

•**72.** The circuit of Figure P.72 has been turned on for a long time, so that you can ignore the transient behavior of the capacitor.

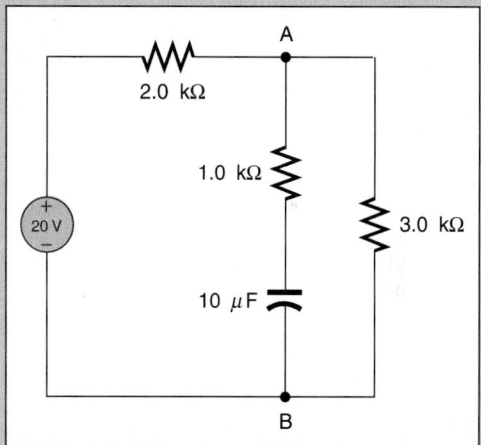

FIGURE P.72

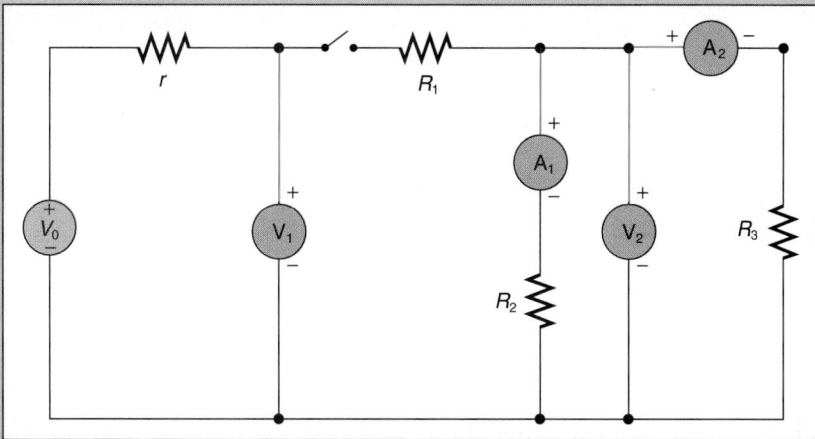

FIGURE P.65

(a) What is the dc current in the branch between nodes A and B that contains the capacitor? (b) What is the potential difference across the 1.0 kΩ resistor? (c) Determine the current through the 2.0 kΩ and 3.0 kΩ resistors. (d) What is the potential difference across the 3.0 kΩ resistor? (e) What is the potential difference across the capacitor? (f) Compute the power absorbed by each circuit element and verify that the sum of the power absorbed by the circuit is zero.

•73. A model of a *real capacitor* consists of a high resistance R, known as a *leakage resistor*, in parallel with an ideal capacitor (see Figure P.73a). The electrical path through R provides a way to account for the observation that all capacitors gradually discharge; the opposite charges on each conductor gradually are neutralized as electrons migrate through the material separating the two conductors of the capacitor. You have two *real* capacitors with leakage resistances R_1 and R_2 in series with an independent voltage source of emf V_0 as shown in Figure P.73b. Determine the potential difference across each capacitor in terms of V_0, R_1, and R_2.

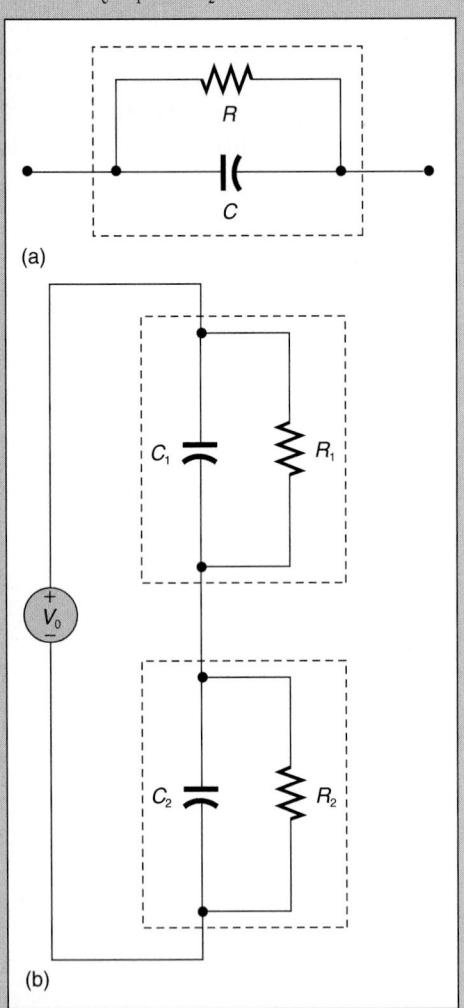

FIGURE P.73

‡74. (a) Calculate the power $P(t)$ absorbed by a charging capacitor in the series RC circuit (Figure P.74) as a function of time.

Express your result in terms of R, V_0, and τ. (b) Integrate $P(t)$ from $t = 0$ s to $t = \infty$ s to show that the energy stored in the capacitor is

$$\frac{1}{2} CV_0^2$$

when it is fully charged.

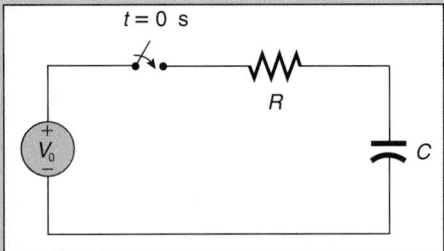

FIGURE P.74

‡75. Having earned a B.S. and an M.S., you are intent on getting the third degree in physics. Here it comes. A capacitor with capacitance C_1 is initially charged so that it has a potential difference V_0 between its plates. The capacitor then is disconnected from the independent voltage source and connected across the terminals of an uncharged capacitor with capacitance C_2 as indicated in Figure P.75. The resistance R models the actual resistance of the connecting wires. (a) Use charge conservation to show that the final potential difference V_f across each capacitor is

$$V_f = \frac{C_1}{C_1 + C_2} V_0$$

(b) Show that the final potential energy PE_f stored in the two-capacitor system is

$$PE_f = \frac{1}{2} \frac{C_1^2 V_0^2}{C_1 + C_2}$$

(c) Show that the energy absorbed by the resistor in this process is

$$\frac{1}{2} \frac{C_1 C_2}{C_1 + C_2} V_0^2$$

This problem is based on a paper by C. J. Macdonald, "Conservation and capacitance," *Physics Education*, *23*, #4, page 202 (July 1988).

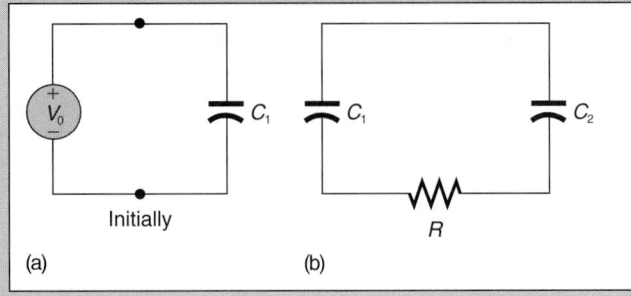

FIGURE P.75

INVESTIGATIVE PROJECTS

A. Expanded Horizons

1. Investigate and report on the physical principles behind polygraphs (lie detectors).
   James Allan Matte, *The Art and Science of the Polygraph Technique* (Thomas, Springfield, Illinois, 1980).
   Eugene B. Black, *Lie Detectors: Their History and Use* (D. McKay, New York, 1977).

2. Investigate the underlying physics behind electrocardiography. You might even want to build your own electrocardiograph!
   W. Bruce Fye, "A history of the origin, evolution, and impact of electrocardiography," *American Journal of Cardiology*, 73, #13, pages 937–949 (15 May 1994).
   H. Edward Roberts, "Electrocardiography," *Radio-Electronics*, 62, #7, pages 31–40 (July 1991); #8, pages 44–49 (August 1991).

3. Electrical trauma still is an active area of research. The following reference is an interesting introduction to the subject; write a summary of it.
   David L. Wheeler, "Understanding electrical trauma," *Chronicle of Higher Education*, 41, #4, pages A10, A14 (21 September 1994).

4. Investigate the electrical characteristics of a defibrillator, used in attempts to restart a heart that has undergone ventricular fibrillation.
   John R. Cameron and James G. Skofronick, *Medical Physics* (Wiley, New York, 1978), page 229.
   Colin Brown, "Electric shock and the human body," *Physics Education*, 21, #6, pages 350–353 (November 1986).

5. Investigate the physics and chemistry of real batteries and the progress being made toward the development of long-lasting batteries with small mass for applications in notebook PCs.
   Karl V. Kordesch and Klaus Tomantschger, "Primary batteries," *The Physics Teacher*, 19, #1, pages 12–21 (January 1981).
   Mark Dewey, "Battery technology," *Radio-Electronics*, 62, #1, pages 45–49 (January 1991).

6. Resistors can be connected in series or parallel, as we have seen. Other types of connections occur as well. Investigate the type of circuit connections known as Y–Δ connections and how they can be transformed into each other.
   James W. Nilsson, *Electric Circuits* (3rd edition, Addison-Wesley, Reading, Massachusetts, 1990), pages 53–57.

7. Investigate how semiconducting materials are doped to become n-type or p-type materials.
   James J. Brophy, *Semiconducting Devices* (McGraw-Hill, New York, 1964).

8. Discover how a semiconducting *pn junction* acts as a *diode*.
   Edward Boyes, "Understanding the p-n junction," *Physics Education*, 25, #1, pages 53–59 (January 1990).
   Albert Paul Malvino, *Electronic Principles* (3rd edition, McGraw-Hill, New York, 1984), Chapters 1, 2.

9. Investigate how semiconducting materials are used to create a bipolar *transistor*. Indicate what is meant by the bipolar transistor terminals known as the *emitter*, the *collector*, and the *base*.
   Albert Paul Malvino, *Electronic Principles* (3rd edition, McGraw-Hill, New York, 1984), Chapter 5.

10. Investigate what is meant by the term bipolar *transistor biasing* and how such biasing is accomplished in electronics. In particular, investigate the type of biasing known as voltage-divider (or ladder) biasing.
    Albert Paul Malvino, *Electronic Principles* (3rd edition, McGraw-Hill, New York, 1984), Chapter 6.

11. Circuit breakers, like fuses, are used to protect electronic devices from excessive currents. The advantage of a circuit breaker is that after experiencing an excessive current it can simply be reset rather than replaced. Investigate and prepare an oral report on the distinction between the ways that a fuse and circuit breaker interrupt or open-switch a circuit. In desperation, some people insert pennies to replace fuses; explain why this practice is *quite* dangerous and can cause electrical fires.

12. Electric fences are common on many farms and ranches as well as for many backyard gardeners. Visit a manufacturer and/or retailer of such fences and determine what factors are involved in the design of such fencing, the potential differences used, and the currents that typically flow when an animal (which may be you!) inadvertently encounters the fence.

B. Lab and Field Work

13. Devise an experiment to measure the resistivity of the conducting jells used when taking electrocardiograms. Secure samples from a local physician or hospital. Newer devices do not use separate jells.

14. Design and perform an experiment to measure the characteristic curve of a light bulb. Determine if the light bulb is an ohmic device. Do your results vary greatly with the wattage rating of the bulb?

15. Devise and perform an experiment to measure the emf V_0 of the ideal battery and internal resistance r used to model a real battery.

16. Devise and test a procedure to measure the capacitance of a capacitor using a series RC circuit with a large known resistance R, an independent voltage source, a voltmeter, and a stopwatch.

17. Write or visit a state or federal prison that enforces the death penalty by electrocution. Determine what potential differences and currents are used to minimize the trauma of the condemned person during their electrocution.

18. Devise and perform an experiment to measure the characteristic curve of a semiconducting pn junction diode.

19. Design and test an apparatus similar to that shown in Figure 19.1 to demonstrate the concept of an electrical current.

C. Communicating Physics

20. Frequently you read about electricity in the newspaper. Journalistic accounts, however, occasionally are tinged with inaccurate science that manifests the scientific ignorance of the reporter. Statements such as "10 000 volts of electrical power," "a 12 000 volt current," "100 amperes of electrical energy," and other such nonsense are, unfortunately, not infrequent; you might even collect a few for the amusement of your class. The editor of your local or campus newspaper has hired you as a technical consultant to provide a 30-minute short-course for the edification the reporting staff on the proper use of electrical terminology. Prepare

such a short-course, designing appropriate visual aids, with a focus on the words power, current, and potential.

21. The electric eel, *Electrophorus electricus*, is able to stun or even kill nearby prey without electrocuting itself, a neat trick. Consult a biology professor at your university and see if the two of you can develop a model for explaining how this fishy trick is accomplished. Present the results of your model in the form of a 15-minute presentation to your physics class.

22. Prepare a brief oral or written report addressing the purpose of a surge protector and discussing a circuit diagram of such a device.

"BETTER REDO THAT POWER CALCULATION"

CHAPTER 20

MAGNETIC FORCES
AND THE MAGNETIC FIELD

We know that the lodestone has a wonderful power. . . . When I
first saw it I was thunderstruck. . . . Who would not be amazed at
the virtue of this stone. . .?

St. Augustine (A.D. 354–430)*

Magnets and magnetic effects have immense natural, technological, and cultural importance. Spectacular auroras arise when charged particles from the Sun collide with the atmosphere; the particles are funneled by the magnetic field of the Earth toward its magnetic poles. Magnetic fields in planetary, stellar, and galactic astronomy lead to almost unbelievably violent cosmic events. Without the simple magnetic compass, navigating across uncharted seas would have been vastly more dangerous and difficult; the Age of Discovery would have taken place with more trepidation, and over a longer time period than it did.

Birds use the magnetic field of the Earth (as well as clues from the Sun and stars) for navigation during long migrations. More immediately, the tapes you play for relaxation and amusement in your Walkman or stereo, the diskettes and devices you use and need to store large quantities of information, magnetic imaging technologies used for life-saving diagnostics in medicine, even the very process used to generate much of the electricity on which so much of our modern society depends— all are based on magnetism and magnetic effects. Thus magnetism is closely associated with a wide variety of technological and natural phenomena.

We first encountered the idea of a field in our study of the gravitational force in Chapter 6. The gravitational force is conveyed by a gravitational field $\vec{g}$. Likewise, in our study of electrical interactions in Chapter 16, we considered the electrical force to be conveyed by an electric field $\vec{E}$. The electric situation is very similar to the gravitational case, except that we have two types of charges (positive and negative) but only one type of mass (positive mass).

The study of magnetism introduces a third important field concept[†]: the magnetic field. It has to do with the *motion* of electric charges (currents). We shall see that while the magnetic field has similarities to the other fields, it also has some unique

features that clearly distinguish it from the others and provide its *raison d'être*. In this chapter we see how the need for the magnetic field concept arose, how charges are affected by its presence, what causes this type of field, and what are the salient characteristics of this field itself.

20.1 THE MAGNETIC FIELD

A wonder of this kind I experienced as a child of 4 or 5 years when my father showed me a compass. That this needle behaved in such a determined way did not at all fit into the kind of occurrences that could find a place in the unconscious world of concepts. . . . I can still remember—or at least believe I can remember—that this experience made a deep and lasting impression on me. Something deeply hidden had to be behind things.

Albert Einstein (1879–1955)[‡]

Magnets are always fun to play with. They fascinated both St. Augustine and a young boy named Albert Einstein.

Undoubtedly you know that magnets have two **poles**, designated north (N) and south (S). The designation arose from the navigational compass. The needle of such a compass is a small, thin magnet.

> The end of a compass needle that points generally in a northerly direction at most places on the Earth is defined to be the needle's **north magnetic pole** N (usually shortened to the **north pole** of the compass needle magnet).[§] The opposite end of the compass needle is defined to be its **south magnetic pole** S (or **south pole**).

In playing with two such permanent magnets, it is easy to determine that opposite poles, called *unlike* poles, attract each other (see Figure 20.1), while poles of the same kind, called *like* poles, repel each other (see Figure 20.2). These observations, coupled with the definition of a north magnetic pole, imply that the magnetic pole of the earth-magnet that is located in the

---

*(Chapter Opener) *The City of God*, translated by Marcus Dods (The Modern Library, New York, 1950), Book XXI, Chapter 4, page 768. Lodestone literally means a stone that leads.

[†]Other fields are encountered in more advanced courses in physics.

[‡]*Albert Einstein: Philosopher-Scientist*, edited by Paul Arthur Schilpp (Library of Living Philosophers, Evanston, Illinois, 1949), page 9.

[§]The *geographic* north and south poles of the Earth define the two points where the spin (rotational) axis of the Earth intersects its surface. The locations of the geographic poles are determined from astronomical observations and are distinct from the magnetic north and south poles of the Earth. The geographic poles exist because the Earth spins. Even if the Earth had no magnetic field, geographic north and south poles would exist.

The compass aided the Age of Discovery.

Magnets are used on many toys.

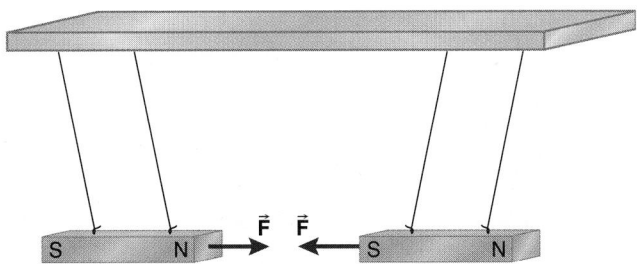

FIGURE 20.1 Unlike poles attract each other.

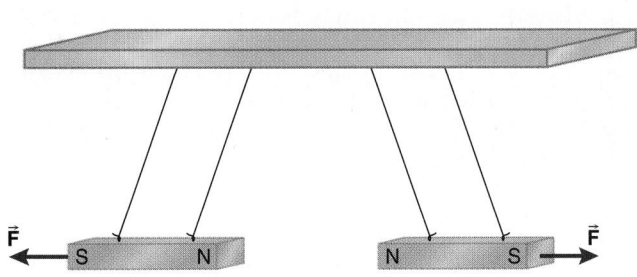

FIGURE 20.2 Like poles repel each other.

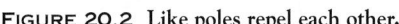

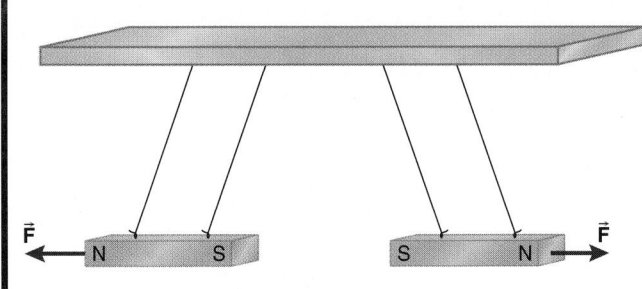

northern geographic hemisphere of the Earth must, therefore, technically be a *south* magnetic pole; see Figure 20.3.*

The magnetic forces of attraction and repulsion of magnetic poles on each other are very similar to the electrical force interactions between electric charges (with the north and south magnetic poles playing the roles of positive and negative charge). We shall see, however, that charges and magnetic poles are *not at all* the same thing.

If we take any bar magnet and sprinkle iron filings around it, a pattern emerges, as shown in Figure 20.4. The pattern indicates that the magnet affects the space in its vicinity. The pattern of lines traces out the magnetic field $\vec{B}$ of the magnet, which we imagine to be present in the space around the magnet, even without the presence of the iron filings. The direction of the magnetic field at any point is tangent to the field line at that point.

Such **magnetic field lines** are similar to the electric field lines we used to trace out the pattern of the electric field surrounding charges (see Section 16.7). By convention, when in the space outside the magnet material, the direction of the magnetic field is tangent to the field line directed *from* the north magnetic end of the magnet *to* its south magnetic end.[†] Just as with electric field, the number density of the magnetic field lines in a region is a measure of the magnitude of the magnetic field there. Since the field lines bunch closer together as you approach either pole end of the magnet, the magnitude of the magnetic field increases accordingly.

---

*Confoundingly, almost everyone (including geoscientists) calls this magnetic pole of the earth-magnet the north magnetic pole anyway. There is no confusion about the science, however.

[†]Within the magnet itself, the field lines run from the south polar end toward the north polar end, preserving their continuity from inside to their curving return pattern outside.

The pattern of magnetic field lines surrounding a magnet is similar to the pattern of electric field lines surrounding an electric dipole (see Figure 16.57). However, if we try to isolate the north and south poles of the magnet by, say, cutting the magnet in two (see Figure 20.5), we find that each piece has a dipolar pattern for the magnetic field surrounding it, albeit a weaker one.

> Indeed, if we keep breaking the magnets into smaller and smaller pieces, we find that it is impossible to isolate either a north or a south magnetic pole; they *always* appear inseparably paired. Hence we say that magnets and the magnetic field always are *dipolar*.

This is quite unlike electric charge. Isolated positive and negative charges (electric **monopoles**) are quite common. Indeed the fundamental unit of charge e exists in positive and negative denominations on particles such as the proton and the electron. With magnets, *no magnetic monopoles ever have been discovered* despite fiendishly valiant and painstaking searches for them in all manner of materials from fossils and ancient rocks to cosmic rays. Magnetic poles always occur in north–south pairs (**magnetic dipoles**) that produce a dipolar magnetic field in the surrounding space. A monopolar magnetic field has never been observed; monopolar electric fields are common.

It is the effect of a magnetic field on charged particles that clearly indicates that magnetic poles and the magnetic field are not the same sort of beast as electric charge and the electric field. Experiments indicate the following:

1. A charge q placed at rest in a magnetic field experiences *zero* force. This single observation means the magnetic field is quite different from the electric field: a charge at rest in an electric field experiences a nonzero force $\vec{F}_{\text{elec}} = q\vec{E}$ (Equation 16.8).
2. If we move the charge along a magnetic field line, either parallel or antiparallel to the direction of the magnetic field, the moving charge again experiences zero force.
3. If the charge is moved at speed v at an angle θ ($\neq 0°$) with the direction of a uniform magnetic field, a nonzero magnetic force exists on the charge. The magnitude of the force is found to be proportional to v (for fixed θ) and proportional to $\sin\theta$ (for fixed v). The magnetic force on q is a *velocity-dependent force*, since it depends on both the speed *and* direction of the velocity vector.
4. If we vary the field magnitude, say, by performing experiments in regions with half or double the number density of magnetic field lines, the force is found to vary with the field

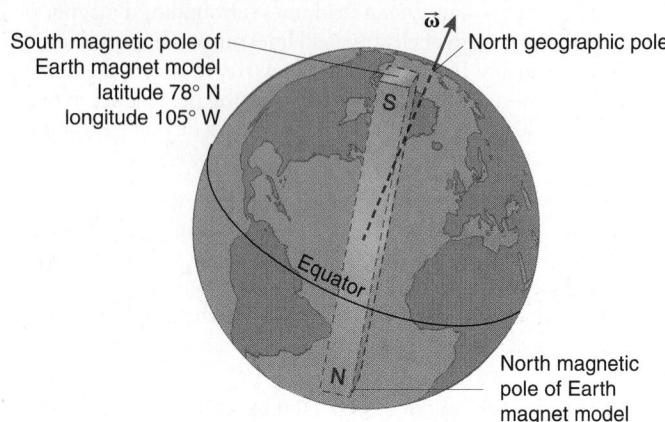

FIGURE 20.3 The magnetic pole of the Earth located in the northern hemisphere is a south magnetic pole because it attracts the north polar end of a compass needle.

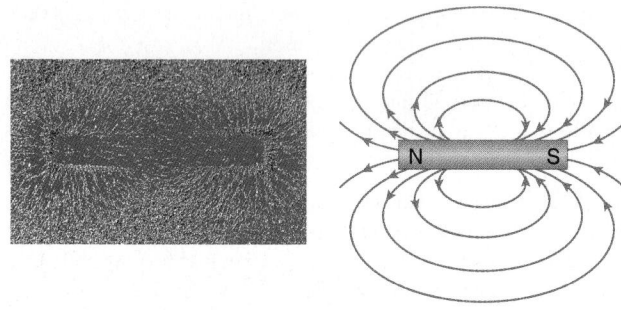

FIGURE 20.4 Iron filings sprinkled around a magnet reveal the presence of its magnetic field.

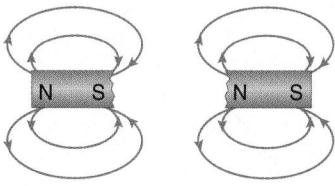

FIGURE 20.5 If we break a magnet in two, we get two dipolar magnets.

magnitude B (as determined from the number of magnetic field lines).

5. The direction of the force on q (determined from its acceleration) depends on the *sign* of the moving charge; charges of opposite sign experience forces in opposite directions. In every case the force (if nonzero) is perpendicular to *both* the velocity vector $\vec{v}$ of the charge and the field direction $\vec{B}$.

The idea of the polyphase induction electric motor occurred to Tesla as a classic "aha!" experience. He emigrated to the United States to develop the idea, arriving with a mere $0.04 in his pockets.

All five observations are summarized eloquently in the following expression for the magnetic force on a moving charge q:

$$\boxed{\vec{F}_{\text{magnet on } q} = q\vec{v} \times \vec{B}} \qquad (20.1)$$

This illustrates yet again the usefulness and cogency of the vector product to express a wealth of information in physics. Equation 20.1 is used to define the magnitude of the magnetic field in SI units, since the units for force, charge, and velocity are known. The dimensions of the magnetic field must be those of force divided by the units for charge and velocity, or

$$\frac{\text{N}}{\text{C} \cdot \text{m/s}} = \text{N} \cdot \text{s}/(\text{C} \cdot \text{m})$$

in SI units. This combination of SI units is defined to be a **tesla** (T), named in honor of Nikola Tesla (1856–1943), a Serbian engineer of Croatian birth who emigrated to the United States to develop his invention of alternating current electric motors.

Another common and convenient unit of magnetic field is the **gauss**, although it is not the SI unit. The gauss is a smaller unit than the tesla; the conversion factor between gauss and teslas is

$$10^4 \text{ gauss} \equiv 1 \text{ T}$$

We shall not use gauss in this text, but the unit is frequently used in geology and geophysics. The magnitude of the magnetic field of the Earth near its surface is quite weak: only about 10^{-4} T, or one gauss.

PROBLEM-SOLVING TACTIC

20.1 When using Equation 20.1, be sure to use the appropriate sign of the charge q. For positive charges, the force is in the direction of $\vec{v} \times \vec{B}$; for negative charges, the force is directed opposite to $\vec{v} \times \vec{B}$.

QUESTION 1

Are the forces shown in Figures 20.1 and 20.2 Newton's third law force pairs?

20.2 APPLICATIONS

Magnetic fields have many practical applications in the sciences; here we examine but a few.

A Velocity Selector

A combination of magnetic and electric fields can be used to sort charged particles according to their speeds. We consider a particle with charge q moving with constant velocity $\vec{v}$. The particle enters a region of space where there exists a uniform magnetic field $\vec{B}$ and a uniform electric field $\vec{E}$ perpendicular to it (see Figure 20.6). The magnetic field is directed perpendicularly into the page, symbolized with the ×'s, representing the crossed tail feathers of a retreating arrow. The electric field is directed as shown in Figure 20.6. The velocity of the particles here is initially perpendicular to *both* the magnetic and electric fields.

With the coordinate system indicated in Figure 20.6, the magnetic field is

$$\vec{B} = -B\hat{k}$$

and the velocity is initially

$$\vec{v} = v_0\hat{i}$$

Thus the initial magnetic force on the charge is, from Equation 20.1,

$$\begin{aligned}\vec{F}_{magnet} &= q\vec{v} \times \vec{B} \\ &= qv_0\hat{i} \times (-B\hat{k}) \\ &= qv_0B\hat{j}\end{aligned}$$

If q is positive, the charged particle experiences a magnetic force parallel to $\hat{j}$ for this choice of the coordinate system; if q is negative, the force on it is toward $-\hat{j}$.

The uniform electric field of magnitude E is directed so that the electrical force on the charge initially is opposite to the direction of the magnetic force on it. With an electric field in the downward direction, the electrical force on q is

$$\vec{F}_{elec} = qE(-\hat{j})$$

The total force initially acting on the charge then is

$$\vec{F}_{magnet} + \vec{F}_{elec} = qv_0B\hat{j} - qE\hat{j}$$

Now we adjust the magnitude of the electric field (by varying the potential difference between the parallel plates causing the uniform electric field) until the total force acting on the charge is zero:

$$\begin{aligned}\vec{F}_{magnet} + \vec{F}_{elec} &= qv_0B\hat{j} - qE\hat{j} \\ &= 0 \text{ N}\end{aligned}$$

Hence, when the total force is zero, we have

$$qv_0B = qE \qquad (20.2)$$

When the total force on the charge is zero, the charged particle continues to move merrily through the fields at a constant velocity and is not deviated by either the magnetic or electrical forces acting on it (since they vector sum to zero). Solving Equation 20.2 for v_0, we find

$$v_0 = \frac{E}{B} \qquad (20.3)$$

This beautifully simple result can be put to practical use. Imagine a beam of charged particles, each with the same charge q, but having a variety or distribution of speeds, all moving in the same direction, as shown in Figure 20.7. For those particles with speeds *smaller* than the speed v_0 given by Equation 20.3, the electrical force is greater in magnitude than the magnetic force. Hence these particles are accelerated in the direction of the electrical force, away from the incident direction. For those particles with speeds *greater* than the speed v_0 given by Equation 20.3, the magnetic force is greater in magnitude than the electrical force. These particles also are accelerated away from the incident direction, in the direction of the magnetic force. *Only* those particles with the speed v_0 given by Equation 20.3 continue to move undeviated through the apparatus, since only particles with this speed have zero total force acting on them. By placing a baffle with a slit or hole at the end of the device, as in Figure 20.7, only the particles with the speed v_0 emerge from it.

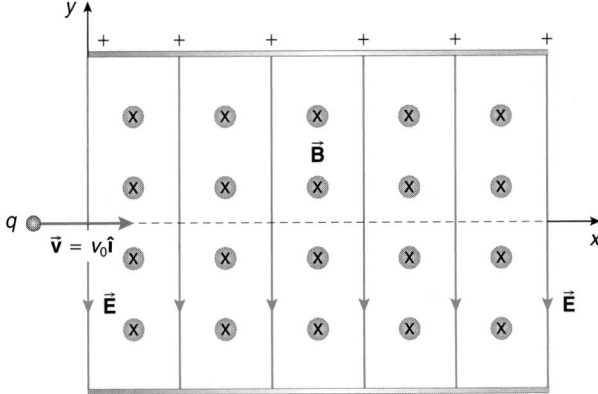

FIGURE 20.6 A charged particle with its velocity initially perpendicular to mutually perpendicular magnetic and electric fields.

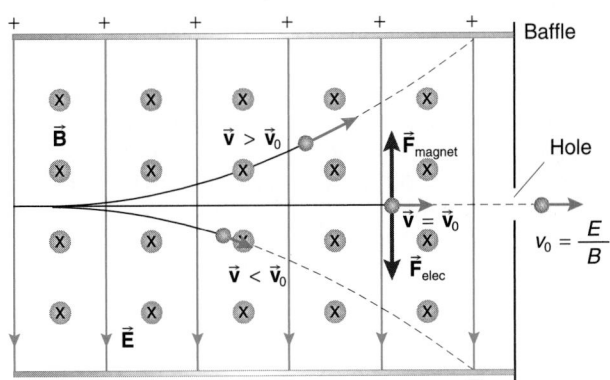

FIGURE 20.7 A velocity selector.

Since we can control the magnitude of the electric field easily, we can select the speed of the particles we want to emerge from the device. In other words, from the incident beam of particles with a dispersion of speeds, we have a way of selecting only those particles with a particular speed. This is a **velocity selector**. Such devices commonly are used to produce mono-energetic particles in experiments and devices that use charged particle beams.

> Notice that the selected speed v_0 given by Equation 20.3
>
> 1. depends only on the magnitude of the fields;
>
> 2. is independent of the identical charge of the particles (as long as the charge is not zero); and
>
> 3. is independent of the mass of the particles.

QUESTION 2

The simultaneous presence of a magnetic and an electric field in a velocity selector is *not* an example of the principle of superposition of *fields*. Explain why. The velocity selector, however, *is* an example of the principle of superposition of *forces*. Explain why.

EXAMPLE 20.1

Ions, each with a charge $+e$, are injected appropriately into a velocity selector containing a magnetic field of magnitude 0.200 T and an electric field of magnitude 2.50×10^5 N/C.

a. What is the speed of the ions that emerge undeflected?
b. If the charged plates creating the electric field are separated by 2.00 cm, what is the potential difference between them?

Solution

a. The undeflected particles in the velocity selector have electric and magnetic forces on them of equal magnitude in opposite directions. Use Equation 20.3 to find the speed of the undeflected particles:

$$v_0 = \frac{E}{B}$$

$$= \frac{2.50 \times 10^5 \text{ N/C}}{0.200 \text{ T}}$$

$$= 1.25 \times 10^6 \text{ m/s}$$

b. Resurrecting ancient material from Chapter 17, the absolute value of the potential difference V between two oppositely charged plates is related to the magnitude of the field E and the plate separation d by

$$V = Ed$$

With the given field and plate separation, the potential difference is

$$V = (2.50 \times 10^5 \text{ N/C})(2.00 \times 10^{-2} \text{ m})$$

$$= 5.00 \times 10^3 \text{ V}$$

A Mass Spectrometer

The different isotopes of an element have different masses, but all behave indistinguishably in chemical reactions. Hence the discovery and separation of isotopes by purely chemical means is not possible. Here we see how it is possible to detect the presence of the various isotopes of a given element.

Send a particle of mass m and charge q into a region that has only a uniform magnetic field. The velocity vector is perpendicular to the field, as indicated in Figure 20.8. We want to determine the trajectory of the particle.

Upon entering the region of the field, the particle experiences a magnetic force given by Equation 20.1:

$$\vec{F}_{\text{magnet}} = q\vec{v} \times \vec{B}$$

The vector product indicates that the direction of the magnetic force on the particle is perpendicular to the velocity vector *and* the magnetic field vector. Thus, if the particle is positively charged, the force initially is directed as shown in Figure 20.9. If the particle is negatively charged, the force is in the opposite direction.

The direction of the velocity vector begins to change in response to the force; but the magnitude of the velocity vector is not affected, since the force is perpendicular to it. However, the magnetic force *always* is perpendicular to the velocity (and the field), and so the direction of the magnetic force also changes as the direction of the velocity vector changes. Recall that a force of constant magnitude that is always perpendicular to the velocity vector produces circular motion; hence the particle here is forced into a circular path, as shown in Figure 20.10.

Now we apply Newton's second law of motion to the particle. The magnetic force is the only force on the particle and produces an acceleration; here, it is a centripetal acceleration.

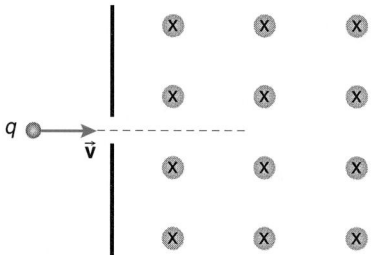

FIGURE 20.8 A charged particle is sent into a region with a uniform magnetic field perpendicular to the velocity.

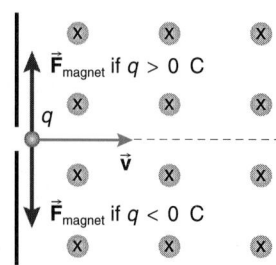

FIGURE 20.9 The initial direction of the magnetic force on the charged particle.

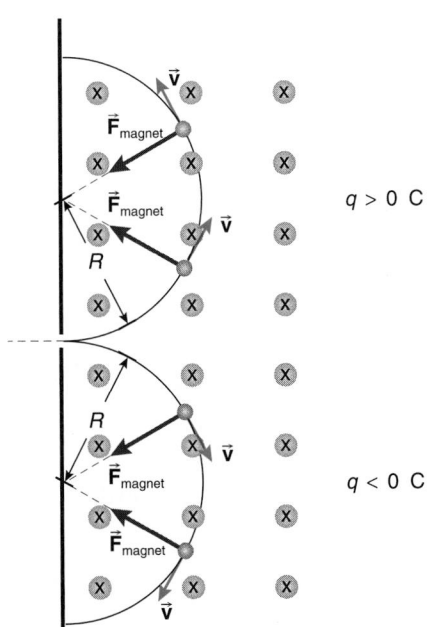

$q > 0$ C

$q < 0$ C

FIGURE 20.10 The trajectory of the charged particle is a circle.

The relationship between the magnitude of the total force and the magnitude of the acceleration is, as we know,

$$F = ma$$

Since the velocity vector is perpendicular to the field, the magnitude of the magnetic force is $F_{magnet} = |q|vB$. The magnitude of the centripetal acceleration is v^2/R, where R is the radius of the resulting circular path of the particle. Substituting for the magnitude of the force into the left-hand side of Newton's second law and for the magnitude of the acceleration on the right-hand side, we obtain

$$\left|q\right|vB = m\frac{v^2}{R}$$

The radius of the circular path of the particle thus is

$$R = \frac{mv}{\left|q\right|B} \qquad (20.4)$$

and is called the **cyclotron radius**, but that is just technical lingo.

This result is of practical import. Imagine a beam of identically charged masses (atoms, molecules, or perhaps other charged particles such as electrons or protons) all moving with the same velocity incident upon the apparatus. The charged particles may represent an ion beam, that is, a collection of atoms of a particular element ionized so that the particles (ions) all have the same positive charge [say, one electron missing from each atom, giving each resulting atom (ion) a charge of $+e$]. By running the ions first through a velocity selector (described earlier), a beam of identically charged ions with identical velocities is formed. Feed this beam into the device pictured in Figure 20.8. According to our analysis, once inside the device the particles move in circular paths whose radii are given by Equation 20.4.

Now here is the punch line:

If the beam is composed of ions of different isotopes of the same element, all with the same charge, each isotope moves in a circle with a different radius.

Equation 20.4 indicates that more massive isotopes have paths of larger radii (since the charge $|q|$ and speed v of all the ion particles are the same; they all are in the same magnetic field $\vec{B}$ as well). The radius of the circular path thus depends only on the mass of the particle. So the device separates the isotopes according to their mass. Indeed the device also is capable of determining the relative abundance of the isotopes by comparing the relative depositions of material at the various terminations of the semicircular paths. Such a mass separator is called a **mass spectrometer** and was instrumental in the discovery of many isotopes of the elements. Such devices also can be used to separate uranium isotopes and enrich nuclear fuel, although this method is quite inefficient and expensive. Mass spectrometers also are used to detect small leaks in high-vacuum systems.

EXAMPLE 20.2

A beam of singly ionized atoms of carbon (each with charge $+e$) all have the same speed and enter a mass spectrometer. The ions accumulate in two different locations (see Figure 20.11), spaced 5.00 cm apart. The more abundant $^{12}_{6}$C isotope traces a path of smaller radius, 15.0 cm. What is the atomic mass number of the other isotope in the beam?

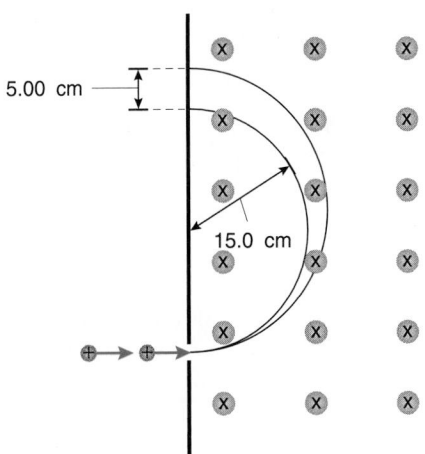

5.00 cm

15.0 cm

FIGURE 20.11

Solution
Let R_1 be the radius of the trajectory of the more abundant isotope $^{12}_{6}$C and R_2 that of the less abundant, unknown isotope. The trajectory of the unknown isotope has a greater radius, and so the mass of the unknown isotope is greater than that of the $^{12}_{6}$C isotope. Since the accumulations are separated by 5.00 cm, the geometry implies that

$$2R_2 - 2R_1 = 5.00 \times 10^{-2} \text{ m}$$

Hence you have

$$R_2 - R_1 = 2.50 \times 10^{-2} \text{ m}$$

Since $R_1 = 15.0 \times 10^{-2}$ m, the larger radius is

$$R_2 = 15.0 \times 10^{-2} \text{ m} + 2.50 \times 10^{-2} \text{ m}$$
$$= 17.5 \times 10^{-2} \text{ m}$$

Write Equation 20.4 for each isotope:

$$R_1 = \frac{m_1 v}{|q|B} \qquad R_2 = \frac{m_2 v}{|q|B}$$

Taking the ratio of these expressions eliminates several unknown quantities:

$$\frac{R_2}{R_1} = \frac{m_2}{m_1}$$

Let A be the atomic mass number of the unknown isotope. The atomic mass number of the known isotope is 12. The masses of the isotopes are proportional to the mass number:

$$m_1 = \alpha(12) \qquad m_2 = \alpha A$$

where α is a constant that converts atomic mass numbers to kilograms. Hence the mass ratio is

$$\frac{m_2}{m_1} = \frac{A}{12}$$

and the ratio of the radii becomes

$$\frac{R_2}{R_1} = \frac{A}{12}$$

Substituting for the radii, you get

$$\frac{17.5 \times 10^{-2} \text{ m}}{15.0 \times 10^{-2} \text{ m}} = \frac{A}{12}$$

Solving for the unknown atomic mass number A, you find

$$A = 14$$

Thus the unknown isotope is $^{14}_{6}\text{C}$. This isotope of carbon is radioactive and is used frequently for dating organic archaeological artifacts, as we shall see in Chapter 26.

The Hall Effect

In 1879 Edwin Hall, working in Cambridge, Massachusetts, discovered an interesting effect that now bears his name. Its importance arises because it provides a way for determining the sign of the charge carriers in a current. The effect also can be used to measure precisely the magnitude of a magnetic field. There is no new physics involved in the so-called classical **Hall effect**: it is another, although sophisticated, application of the magnetic and electrical forces on charged particles. Yet ironically, the Hall effect had some surprises in store for physics almost a century after its discovery.

Consider a conducting rectangular bar of material carrying a current or stream of charged particles moving along the axis of the bar. The same current could be caused by positive carriers moving one way or by negative charges moving oppositely. Which is it? The bar is placed in a uniform magnetic field that is perpendicular to the direction of motion of the charged particles creating the current, as shown in Figure 20.12.

We consider two scenarios.

Scenario 1: Positive Charge Carriers

Say the current in the bar arises from positive charge carriers, each with charge $q > 0$ C. They each experience a magnetic force given by Equation 20.1. With the coordinate system in Figure 20.13, this is

$$\vec{F}_{magnet} = q\vec{v} \times \vec{B}$$
$$= q\langle v \rangle \hat{i} \times B\hat{k}$$
$$= -q\langle v \rangle B\hat{j}$$

The quantity $\langle v \rangle$ is the (single) component of the average drift velocity of the charge carriers. Under the influence of the magnetic force, the positive charges move downward and accumulate near the lower edge of the material, leaving behind an equivalent amount of negative charge near the upper edge of the rectangular bar. The charge separation produces an electric field transverse to the current, directed as in Figure 20.13.

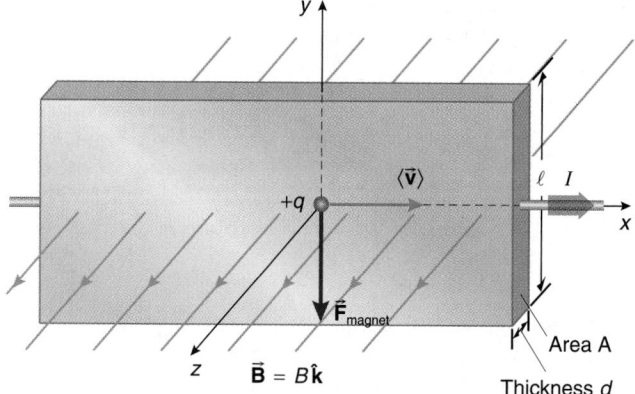

FIGURE 20.12 Positive charge carriers moving in a material in a magnetic field.

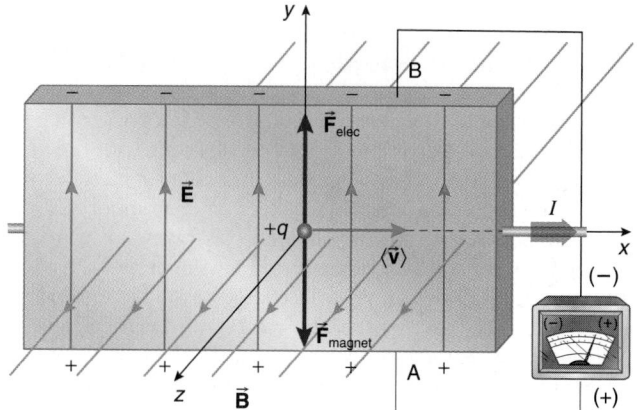

FIGURE 20.13 Charge separation produces an electric field transverse to the current.

As the charge accumulates, this electric field increases in magnitude and produces an electrical force on the positive charge carriers as well:

$$\vec{F}_{elec} = qE\hat{j}$$

The charge accumulation ceases when the magnitude of the opposing electrical force has the same magnitude as that of the magnetic force on each charge carrier. The direction of the electric field (toward $+\hat{j}$) indicates that the lower edge of the bar is at a higher electric potential than the upper edge. If the potential difference between the edges of the bar is measured with a voltmeter (as in Figure 20.13) with the (+) lead of the voltmeter on the lower edge and the (−) lead on the upper edge, the voltmeter will indicate a positive potential difference across the bar.

Scenario 2: Negative Charge Carriers

If the current is caused by negative charge carriers $-|q|$, moving in the opposite direction, then the magnetic force on the negative charge carriers is (see Figure 20.14 for the coordinate system)

$$\vec{F}_{magnet} = (-|q|)(-\langle v \rangle \hat{i}) \times B\hat{k}$$
$$= -|q|\langle v \rangle B\hat{j}$$

The magnetic force on the charge carriers *again* is in the downward direction, as in Scenario 1.

Now it is the negative charge carriers that move downward to the lower edge, leaving behind equivalent positive charges near the top edge. The charge separation produces an electric field directed toward $-\hat{j}$, as indicated in Figure 20.15.

The electric field produces an electrical force on the negative charge carriers:

$$\vec{F}_{elec} = (-|q|)(-E\hat{j})$$
$$= +|q|E\hat{j}$$

The charge separation continues and the electric field (caused by the separated charges) grows in magnitude until the electrical force is equal in magnitude to the oppositely directed magnetic

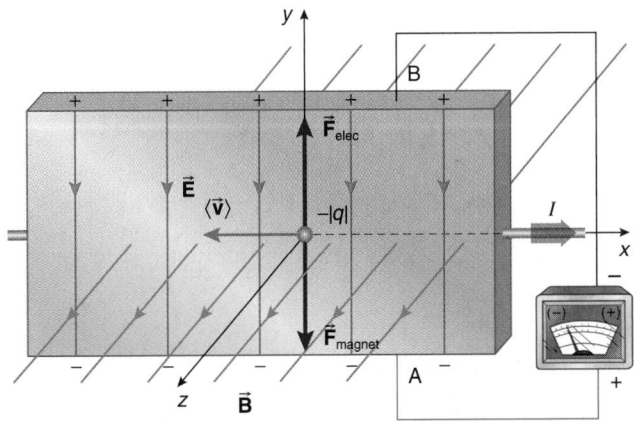

FIGURE 20.15 The electric field produced by charge separation if the charge carriers are negative.

force on each charge carrier. The direction of the electric field indicates that the upper edge of the material is at a higher electric potential than the lower edge; this is the *reverse* of Scenario 1. If a voltmeter is connected (as shown in Figure 20.15) to the edges of the bar the same way as in Scenario 1, the voltmeter now will indicate a *negative* potential difference.

Therefore, if the voltmeter when connected as in Figure 20.13 or 20.15 indicates a positive potential difference between the edges, the charge carriers are positive; if the voltmeter indicates a negative potential difference, the charge carriers are negative.* This is a neat, crucial experiment that says yes or no to an idea (the sign of the charge carriers) even before numbers are obtained.

The results of Hall effect experiments using ordinary metallic conductors indicate the charge carriers in them are negative (they are electrons). Nowadays, we also use semiconducting materials that are contaminated with impurity elements (a process called doping the pure semiconducting material) to create semiconductors in which the dominant charge carriers are either negative (n-type semiconducting material) or positive (p-type semiconducting materials). The Hall effect in doped semiconductors is used to distinguish experimentally between n-type and p-type materials.

An expression for the potential difference between the edges of the bar in the Hall effect can be obtained by equating the magnitudes of the electric and magnetic forces on a charge carrier:

$$|q|E = |q|\langle v \rangle B$$
$$E = \langle v \rangle B \tag{20.5}$$

The potential difference between the edges is found from Equation 17.9. Using Scenario 1,

$$V_B - V_A = -\int_A^B \vec{E} \cdot d\vec{r}$$

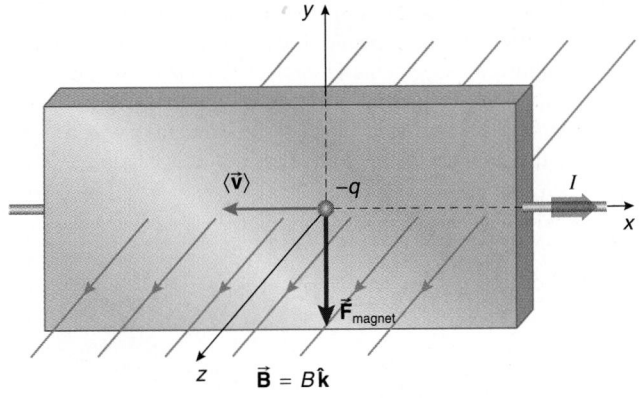

FIGURE 20.14 Negative charge carriers moving in the opposite direction also can account for the same current.

---

*Of course, if the voltmeter is connected to the edges the other way around, a negative potential difference indicates positive carriers and a positive potential difference indicates negative charge carriers. You have to be careful to connect the voltmeter appropriately with respect to the geometry!

where we integrate between the edges (see Figure 20.13). Performing the integration,

$$V_B - V_A = -\int_A^B E\hat{\jmath} \cdot dy\,\hat{\jmath}$$
$$= -E\ell$$

where ℓ is the separation of the edges of the bar. Let $V_{Hall} \equiv V_A - V_B$ be the **Hall voltage**. In this way V_{Hall} is the value the voltmeter reads in Figures 20.13 and 20.15. Thus

$$V_{Hall} = E\ell$$

and Equation 20.5 becomes

$$\frac{V_{Hall}}{\ell} = \langle v \rangle B$$

$$V_{Hall} = \langle v \rangle \ell B \qquad (20.6)$$

This can be put in a more useful form by eliminating the drift speed via its relation to the current (Equation 19.3):

$$I = nqA\langle v \rangle$$

where n is the number of charge carriers per unit volume and A is the cross-sectional area of the wire (here the bar). Solving Equation 19.3 for $\langle v \rangle$,

$$\langle v \rangle = \frac{I}{nqA}$$

Substituting for $\langle v \rangle$ into Equation 20.6, we find the Hall voltage is

$$V_{Hall} = \frac{I\ell B}{nqA} \qquad (20.7)$$

Since the cross-sectional area A of the bar is the product of ℓ and the thickness d, the Hall voltage can be written as

$$V_{Hall} = \frac{IB}{nqd} \qquad (20.8)$$

Note that the Hall voltage is measured across the width ℓ; d is the *other* cross-sectional dimension of the bar.

Semiconducting materials can be fabricated with precisely controlled values for the charge carrier density n. Indeed Equation 20.8 is used to measure n using a known magnetic field. Conversely, once n is determined we can turn the problem around. Let a known current pass through an arrangement like Figure 20.13, now called a **Hall probe**. Currents, of course, are quite easily measured with a sensitive ammeter. Knowing $n, q, d,$ and I means that a measurement of the Hall voltage V_{Hall} permits a determination of the magnitude of the magnetic field. Such Hall probes are one way in which the magnitude of a magnetic field is measured with precision in the laboratory.

The Hall effect still is a topic of current interest and research in physics. Unexpected surprises lurk in many corners of the sciences. In 1980 it was discovered that in very thin sheets of material, the Hall voltage exhibits a steplike behavior as the magnetic field is increased. But Equation 20.7 predicts a smooth, linear relationship between V_{Hall} and B_0. The steplike behavior was quite surprising and unanticipated. The steps in the Hall

voltage are a rare example of a macroscopic manifestation of quantum mechanical effects. The so-called *quantum Hall effect* won its discoverer (Klaus von Klitzing) the 1985 Nobel prize in physics. The quantized Hall effect is used to measure very accurately an important constant of physics known as the fine structure constant; but this is getting a bit over our heads for an introductory course. The point is that the Hall effect certainly is not a dead subject, even though it was discovered well over a century ago.

EXAMPLE 20.3

A copper Hall probe of thickness 125 μm is placed appropriately in a magnetic field (as in Figure 20.13). A 25.0 A current is in the strip. A Hall voltage of -11 μV is measured across the 2.0 cm width of the strip. What is the magnitude of the magnetic field?

Solution
In Example 19.1 on page 841, you found the number of charge carriers per cubic meter for copper is

$$n = 8.49 \times 10^{28} \text{ electrons/m}^3$$

The charge on each carrier is $q = -e$, since they are electrons. Solving Equation 20.8 for the magnitude of the magnetic field, you find

$$B = \frac{nqdV_{Hall}}{I}$$
$$= [(8.49 \times 10^{28} \text{ electrons/m}^3)(-1.602 \times 10^{-19} \text{ C/electron})$$
$$\times (125 \times 10^{-6} \text{ m})(-11 \times 10^{-6} \text{ V})] / 25.0 \text{ A}$$
$$= 0.75 \text{ T}$$

This is a strong magnetic field and illustrates how large the tesla unit for the magnetic field really is. For comparison, the magnetic field of the Earth is only about 10^{-4} T.

20.3 MAGNETIC FORCES ON CURRENTS

Electric currents are charges in motion. If the current is in a wire located in a region where there is a magnetic field, the wire will experience a force that is merely a manifestation of the basic magnetic interaction on the moving charged particles within it.

Imagine a wire segment of length $d\ell$ and cross-sectional area A carrying a current I in the direction indicated in Figure 20.16. The magnetic field at the location of the segment of wire $d\ell$ is $\vec{B}$. Let n be the number of charge carriers per unit volume, $\langle \vec{v} \rangle$ the

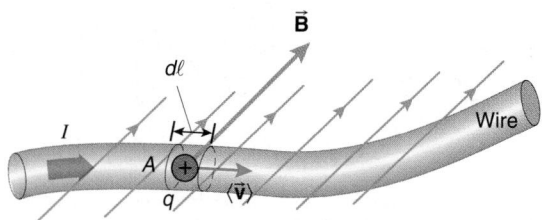

FIGURE 20.16 A current-carrying wire in a magnetic field.

average drift velocity of the charge carriers, and q the charge on each; the current is (from Equation 19.3)

$$I = nqA\langle v \rangle$$

The average magnetic force on a charge carrier in the wire segment is given by

$$q\langle \vec{v} \rangle \times \vec{B}$$

where $\vec{B}$ is the value of the magnetic field at the position of the charge q in the segment $d\ell$.

The number of charge carriers in the length $d\ell$ is the product of the number of charge carriers per unit volume (n) and the volume of the little segment ($A\, d\ell$). Thus the total differential magnetic force $d\vec{F}_{magnet}$ on the segment of wire is

$$d\vec{F}_{magnet} = nA\, d\ell\, q\langle \vec{v} \rangle \times \vec{B} \qquad (20.9)$$

Now we switch vector horses: instead of using the vector drift velocity $\langle \vec{v} \rangle$, we use the magnitude $\langle v \rangle$ of the drift velocity and account for the direction via the length segment $d\ell$. That is, let a vector $d\vec{\ell}$ point in the direction of the drift velocity; this is the same direction as the conventional current (positive charge flow). With this change, we rewrite Equation 20.9 in the equivalent form

$$d\vec{F}_{magnet} = nqA\langle v \rangle\, d\vec{\ell} \times \vec{B}$$

In this way we can identify $nqA\langle v \rangle$ with the current I (Equation 19.3). Hence the differential force on this segment of the wire is

$$d\vec{F}_{magnet} = I\, d\vec{\ell} \times \vec{B}$$

To find the total force on the whole wire, we simply integrate over the length of the wire:

$$\vec{F}_{magnet} = \int_{wire} I\, d\vec{\ell} \times \vec{B}$$

(well, maybe not so simply since this is a vector integration!). Since the current in a wire is constant over its length (charge conservation), I can be brought outside the integral.

The magnetic force on a current-carrying wire is

$$\vec{F}_{magnet} = I \int_{wire} d\vec{\ell} \times \vec{B} \qquad (20.10)$$

If the wire is straight and the field is constant along the length of the wire segment, as in Figure 20.17, then the constant field can be factored out from the integral:

$$\vec{F}_{magnet} = \left(I \int_{wire} d\vec{\ell} \right) \times \vec{B}$$

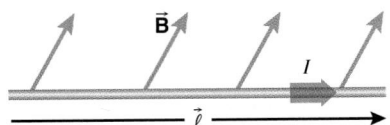

FIGURE 20.17 **A straight wire in a magnetic field that is constant along its length.**

The integration over the straight line segment is the directed length of that segment. Hence the magnetic force on the wire is

$$\vec{F}_{magnet} = I\vec{\ell} \times \vec{B} \qquad \begin{array}{l}\text{(straight wire segment of length } \ell, \\ \vec{B} \text{ constant along its length)} \qquad (20.11)\end{array}$$

EXAMPLE 20.4 _____

A wire of total length ℓ_0 carrying a current I passes through a region permeated by a uniform magnetic field of magnitude B, as shown in Figure 20.18. Only a straight portion of wire, of length ℓ, finds itself in the field.

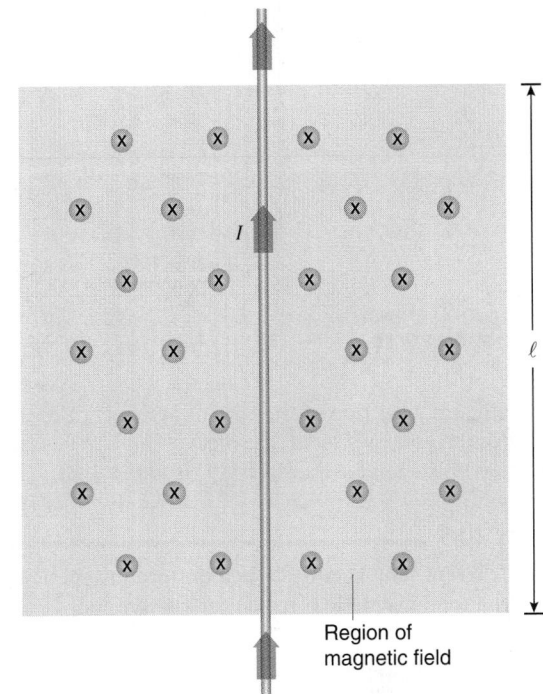

FIGURE 20.18

a. **Calculate the force on the wire.**
b. **What happens if the direction of the current is reversed?**

Solution

a. You need to consider only the portion of the wire in the region of the field. The portion of the wire not in the field has zero magnetic force acting on it. Since the wire is straight and the field is uniform over the portion of the wire in the field, you can use Equation 20.11,

$$\vec{F}_{magnet} = I\vec{\ell} \times \vec{B}$$

The angle between $\vec{\ell}$ and $\vec{B}$ is 90°; hence the magnitude of the magnetic force on the wire is

$$F_{magnet} = I\ell B \sin 90°$$
$$= I\ell B$$

The direction of the vector product is the direction of the magnetic force, found by the vector product right-hand rule. The force is shown in Figure 20.19.

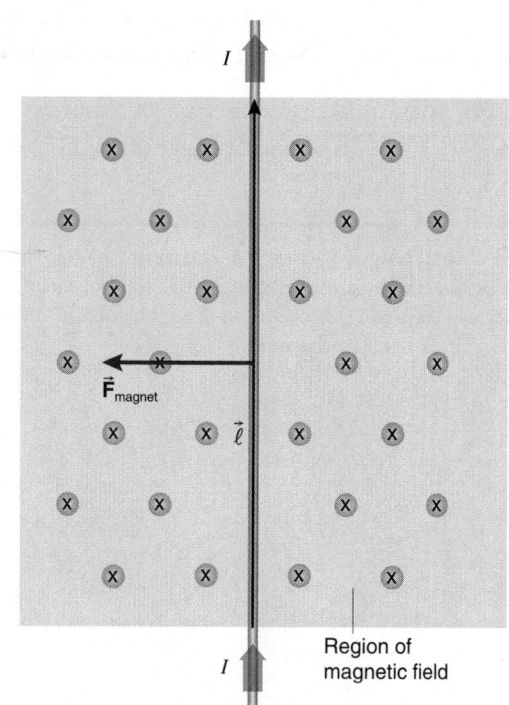

FIGURE 20.19

b. Reversing the direction of the current amounts to reversing the direction of the vector $d\vec{\ell}$; the effect of this is to reverse the direction of the force—that is, $\vec{F}_{magnet}$ is to the right.

EXAMPLE 20.5

A closed current loop is entirely in a uniform magnetic field. What is the total magnetic force on the current loop?

Solution

The force on the loop is found from Equation 20.10:

$$\vec{F}_{magnet} = I \int_{wire} d\vec{\ell} \times \vec{B}$$

Since the magnetic field is constant over the entire loop, you can factor it out from the integral:

$$\vec{F}_{magnet} = I \left(\int_{wire} d\vec{\ell} \right) \times \vec{B}$$

For a closed loop, the integral of $d\vec{\ell}$ around the whole loop is zero since for every $d\vec{\ell}$ pointing in one direction, there is another section of the loop with $d\vec{\ell}$ pointing in the opposite direction and each pair vector sums to zero. Thus the total force on a closed current loop in a uniform field is zero.

20.4 WORK DONE BY MAGNETIC FORCES

A magnetic field that does not vary with time is called a *static magnetic field*. The force it exerts on moving charges is a *static magnetic force*.

The magnetic force on a charge q we found to be (Equation 20.1)

$$\vec{F}_{magnet} = q\vec{v} \times \vec{B}$$

The differential work dW_{magnet} done by this force on the charge when the particle undergoes a differential change $d\vec{r}$ in its position vector is, from the definition of work, Equation 8.1;

$$dW_{magnet} = \vec{F}_{magnet} \cdot d\vec{r}$$
$$= (q\vec{v} \times \vec{B}) \cdot d\vec{r}$$

Recall that the velocity $\vec{v}$ is the time rate of change of the position vector:

$$\vec{v} = \frac{d\vec{r}}{dt}$$

or

$$d\vec{r} = \vec{v}\, dt$$

This means the differential change in the position vector is parallel to the velocity vector. The differential work done by the magnetic force is, therefore,

$$dW_{magnet} = (q\vec{v} \times \vec{B}) \cdot \vec{v}\, dt$$

Since the magnetic force is $q\vec{v} \times \vec{B}$, the force always is perpendicular to the velocity vector of the charge. Hence the *scalar product* of the magnetic force with $\vec{v}$ is always *zero*.

Therefore the differential work done by the static magnetic force always is zero:

$$dW_{magnet} = 0 \text{ J}$$

and, therefore, W_{magnet} also is zero.

PROBLEM-SOLVING TACTIC

20.2 The static magnetic force does no work on a moving charge and, by the CWE theorem, cannot by itself change the kinetic energy of the charge.

EXAMPLE 20.6

An electron moving in and perpendicular to a uniform, static magnetic field executes a circular orbit with a radius equal to the cyclotron radius R (Equation 20.4). What is the work done by the magnetic force on the electron as it executes one orbital circumference?

Solution

The electron has a force acting on it and moves through a distance equal to the circumference. However, the magnetic force is at all times perpendicular to the velocity of the electron. The work done is zero. You should *not* be fooled into thinking the work done by the magnetic force is $F_{magnet}(2\pi R)$! Furthermore, even if the static magnetic field is not uniform and the path is not circular, no work is done on the moving charge by the field.

20.5 TORQUE ON A CURRENT LOOP IN A MAGNETIC FIELD

Electric motors are used in everything from the toy cars and trains we enjoyed as kids (and may still enjoy!) to essential appliances such as refrigerators, washing machines, elevators, and subways. Such motors all are based on a remarkable observation: a current loop in a magnetic field can experience a torque. Couple the current loop appropriately to an axle and we have the beginnings of an elemental motor. Here we see how the torque on a loop of current arises.

Imagine a rectangular current loop of area A and dimensions ℓ_1 and ℓ_2 in a uniform magnetic field $\vec{B}$ that makes an angle θ with a line perpendicular to the plane of the loop, as shown in Figure 20.20. For convenience in describing the orientation of the loop with respect to the field, we introduce an area vector associated with the current loop. The magnitude of the area vector is equal to the area of the loop $A = \ell_1 \ell_2$. The direction of the area vector is perpendicular to the plane of the loop in the sense of the following right-hand rule:

Curl the fingers of your right hand around the perimeter of the current loop in the same sense as the current; then the extended thumb of your right hand indicates the direction of the area vector of the loop. The area vector $\vec{A}$ makes an angle θ with the direction of the magnetic field, as indicated in Figure 20.20.

Example 20.5 showed that the *total* magnetic force on *any* current loop in a *uniform* magnetic field is zero, regardless of the orientation of the loop. While the total force on the loop is zero, there may nonetheless be a nonzero torque on the loop. This torque has significant practical applications in electric motor technology. To find the torque, we first need to find the force on each segment, and then the torque of each force about the center.

Since the field is uniform over each segment of the loop, we can use Equation 20.11 to find the force on each of the four straight segments of the loop:

$$\vec{F}_{\text{magnet}} = I\vec{\ell} \times \vec{B}$$

We use the coordinate system in Figure 20.20 (and Figures 20.21–20.25). The magnetic field everywhere is $\vec{B} = B\hat{k}$.

We start with segment (1) in Figure 20.21. The directed current segment $\vec{\ell}_1$ here is

$$\vec{\ell}_1 = \ell_1\hat{j}$$

and is perpendicular to the magnetic field, as shown in Figure 20.21. The magnetic force on this segment is

$$\begin{aligned}
\vec{F}_1 &= I\vec{\ell}_1 \times \vec{B} \\
&= I\ell_1\hat{j} \times (B\hat{k}) \\
&= I\ell_1 B\hat{i}
\end{aligned}$$

which is also shown in Figure 20.21.

For the corresponding segment (3) on the opposite side of the loop (see Figure 20.22), the directed current segment is $\vec{\ell}_3 = -\ell_1\hat{j}$, and so the magnetic force is

$$\begin{aligned}
\vec{F}_3 &= I\vec{\ell}_3 \times \vec{B} \\
&= I(-\ell_1\hat{j}) \times (B\hat{k}) \\
&= -I\ell_1 B\hat{i}
\end{aligned}$$

The directed wire segment (2) in Figure 20.23 is a bit more complicated. From the geometry of Figure 20.23, we can write

$$\vec{\ell}_2 = \ell_2[(-\cos\theta)\hat{i} - (\sin\theta)\hat{k}]$$

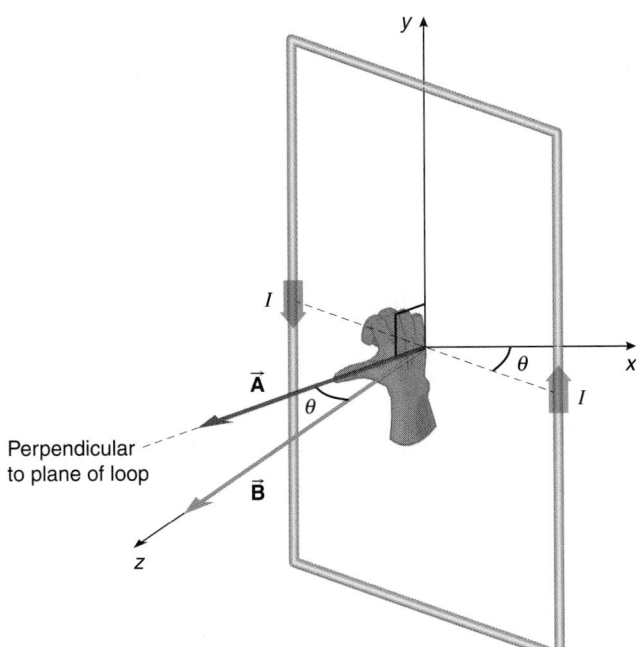

FIGURE 20.20 A current loop with area vector $\vec{A}$ in a magnetic field $\vec{B}$.

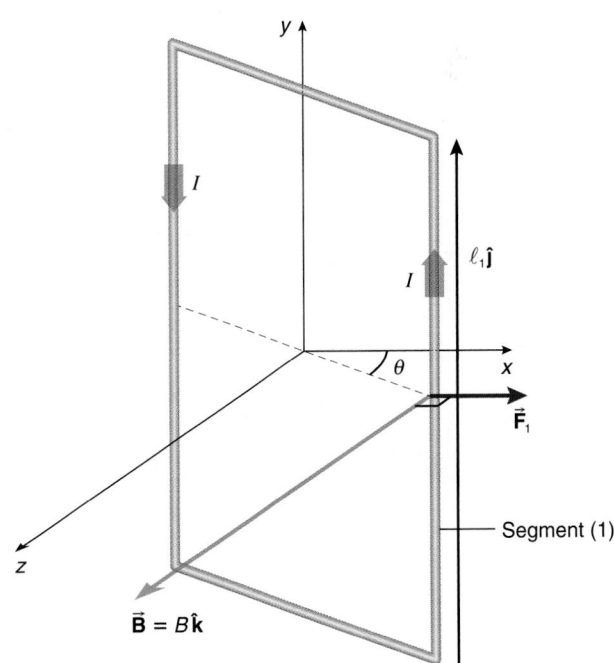

FIGURE 20.21 The force on segment (1).

Thus the magnetic force on this segment is

$$\vec{F}_2 = I\vec{\ell}_2 \times \vec{B}$$
$$= I\ell_2[(-\cos\theta)\hat{\i} - (\sin\theta)\hat{k}] \times B\hat{k}$$
$$= (I\ell_2 B \cos\theta)\hat{\j}$$

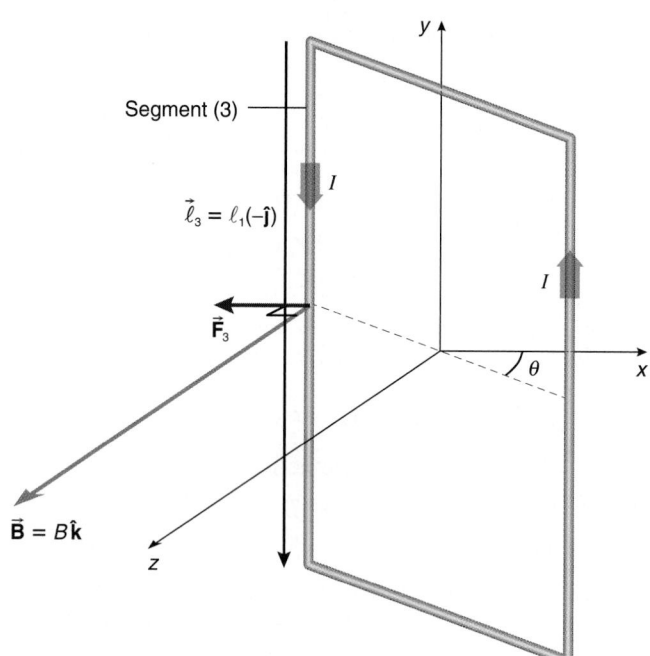

FIGURE 20.22 The force on segment (3).

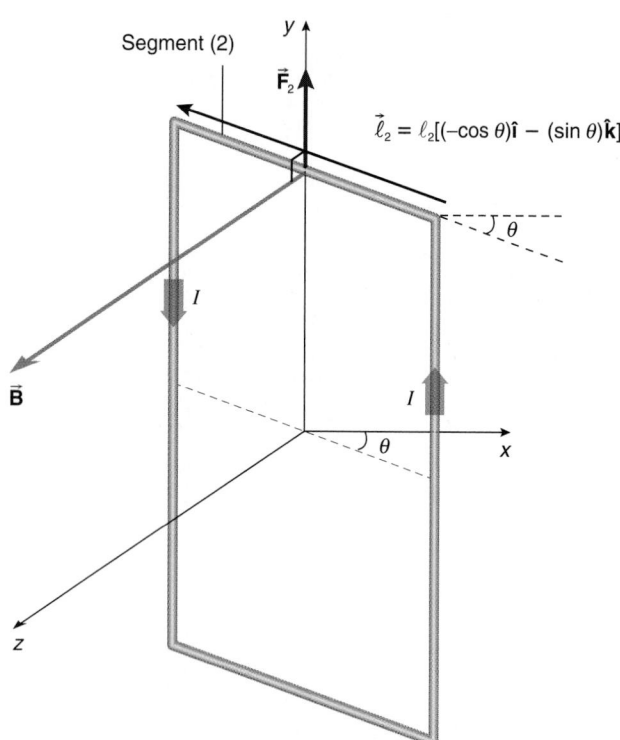

FIGURE 20.23 The force on segment (2).

The current in segment (4) in Figure 20.24 is directed opposite to the current in segment (2). Thus the directed wire segment is (see Figure 20.24)

$$\vec{\ell}_4 = \ell_2[(\cos\theta)\hat{\i} + (\sin\theta)\hat{k}]$$

The force on this segment is

$$\vec{F}_4 = I\vec{\ell}_4 \times \vec{B}$$
$$= I\ell_2[(\cos\theta)\hat{\i} + (\sin\theta)\hat{k}] \times B\hat{k}$$
$$= (-I\ell_2 B \cos\theta)\hat{\j}$$

We note incidentally that the total force on the current loop is the vector sum of the forces on each segment. We find

$$\vec{F}_1 + \vec{F}_2 + \vec{F}_3 + \vec{F}_4 = 0 \text{ N}$$

as we knew from Example 20.5.

Now we can find the torque about the center of the loop. The points of applications of the forces we can take to be the centers of each segment since the field is uniform along each length. To find the torque of each force, we write the position vectors from the center of the loop to the points of application of the forces (see Figure 20.25).

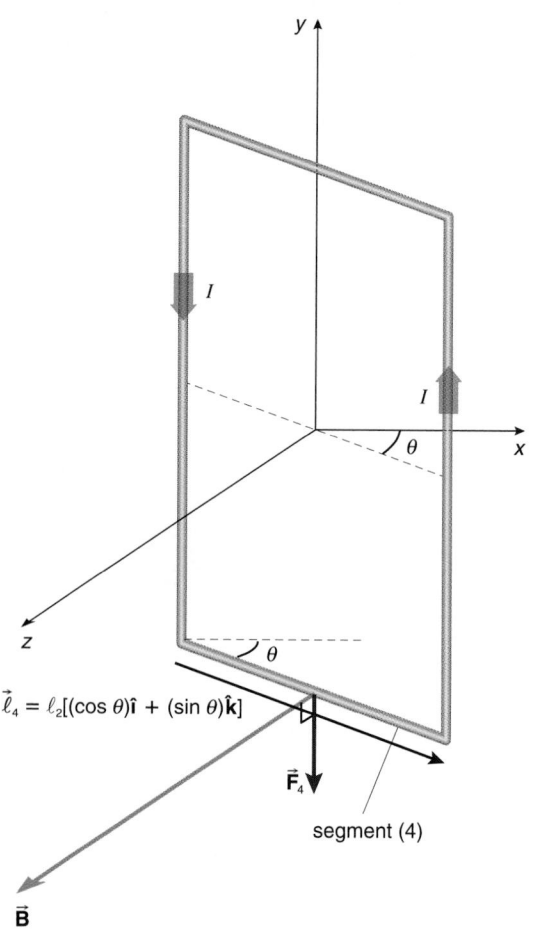

FIGURE 20.24 The force on segment (4).

Note in Figure 20.25 that the lines of action of the forces $\vec{F}_2$ and $\vec{F}_4$ pass through the point about which we are taking the torque, the center of the loop. These forces thus contribute zero torque. Only $\vec{F}_1$ and $\vec{F}_3$ produce a nonzero torque, and so we only need expressions for $\vec{r}_1$ and $\vec{r}_3$. These position vectors are

$$\vec{r}_1 = \frac{\ell_2}{2}\left[(\cos\theta)\,\hat{\imath} + (\sin\theta)\,\hat{k}\right]$$

$$\vec{r}_3 = \frac{\ell_2}{2}\left[(-\cos\theta)\,\hat{\imath} - (\sin\theta)\,\hat{k}\right]$$

The torques of the respective forces are found from the general equation for the torque of a force $\vec{\tau} = \vec{r} \times \vec{F}$:

$$\vec{r}_1 \times \vec{F}_1 = \frac{\ell_2}{2}\left[(\cos\theta)\,\hat{\imath} + (\sin\theta)\,\hat{k}\right] \times I\ell_1 B\,\hat{\imath}$$

$$= \left(\frac{I}{2}\,\ell_1\ell_2 B \sin\theta\right)\hat{\jmath}$$

$$\vec{r}_3 \times \vec{F}_3 = \frac{\ell_2}{2}\left[(-\cos\theta)\,\hat{\imath} - (\sin\theta)\,\hat{k}\right] \times (-I\ell_1 B)\,\hat{\imath}$$

$$= \left(\frac{I}{2}\,\ell_1\ell_2 B \sin\theta\right)\hat{\jmath}$$

The total torque on the entire current loop is the vector sum of the torques of $\vec{F}_1$ and $\vec{F}_3$; we find

$$\vec{\tau} = (I\ell_1\ell_2 B \sin\theta)\hat{\jmath}$$

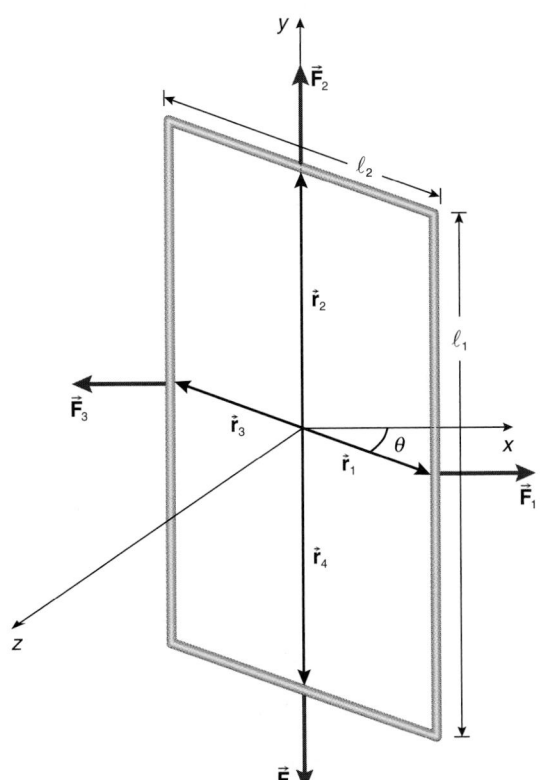

FIGURE 20.25 The position vectors of the points of application of the forces on the four segments.

Since $\ell_1\ell_2$ is the area A of the loop, we can express the torque as

$$\vec{\tau} = (IAB \sin\theta)\hat{\jmath} \qquad (20.12)$$

Whew! A lot of work, but we finally did it. The total torque is along $\hat{\jmath}$; so, if the loop is free to turn, it will begin to spin about the $\hat{\jmath}$ direction. If you align the thumb of your right hand in the direction of the torque, the right-hand fingers indicate the sense that the loop will spin.

Notice from Figure 20.20 that the area vector $\vec{A}$ can be written as

$$\vec{A} = (-A \sin\theta)\hat{\imath} + (A \cos\theta)\hat{k}$$

The magnetic field is $B\hat{k}$. Hence the vector product of $\vec{A}$ with $\vec{B}$ is

$$\vec{A} \times \vec{B} = [(-A \sin\theta)\hat{\imath} + (A \cos\theta)\hat{k}] \times B\hat{k}$$

$$= (AB \sin\theta)\hat{\jmath}$$

Compare this result with Equation 20.12 for the torque. We see that the torque on the current loop can be written elegantly as

$$\vec{\tau} = I\vec{A} \times \vec{B} \qquad (20.13)$$

It is customary to lump the current I with the area vector $\vec{A}$ to form a new vector associated with the current loop. The new vector is defined to be

$$\vec{\mu} \equiv I\vec{A} \qquad (20.14)$$

and is called the **magnetic dipole moment** of the current loop.

This may seem like strange terminology right now, but this is only because we have not yet explored how magnetic fields arise in the first place. We get to that in due course.

Hence we write the expression for the torque on the current loop in its final form:

$$\vec{\tau} = \vec{\mu} \times \vec{B} \qquad (20.15)$$

When the magnetic dipole moment is parallel or antiparallel to $\vec{B}$, the vector product is zero and we have zero torque. When the magnetic dipole moment is perpendicular to $\vec{B}$, the torque has maximum magnitude.

Equation 20.15 looks surprisingly like the expression we found (Equation 16.14 on page 730) for the torque of an electric dipole moment $\vec{p}$ in an electric field $\vec{E}$:

$$\vec{\tau} = \vec{p} \times \vec{E}$$

We also discovered (Equation 17.26 on page 787) that an electric dipole moment in an electric field has a potential energy given by

$$\text{PE} = -\vec{p} \cdot \vec{E}$$

Analogously, the potential energy of a magnetic dipole moment in a magnetic field is*

$$\text{PE} = -\vec{\mu} \cdot \vec{B} \qquad (20.16)$$

*We make the analogy without proof for the sake of brevity. A formal derivation is not easy to find in recent books on electromagnetism; but one can be found in William Taussig Scott, *The Physics of Electricity and Magnetism* (Wiley, New York, 1959), pages 297 and 335.

A current loop (magnetic dipole moment) turns under the action of the torque in order to lower its potential energy. The potential energy is minimized when the magnetic dipole moment vector is parallel to the direction of the magnetic field; the potential energy is maximized with the magnetic dipole moment antiparallel to the field.

Although we derived Equations 20.15 and 20.16 using a rectangular loop, they are valid for planar current loops of *any* shape. The magnetic dipole moment of the loop is the product of the current in the loop and its area vector, directed in the sense of the right-hand rule described at the beginning of this section.

If the idea of a current loop in a magnetic field seems rather esoteric, or far-fetched, recall that the torque on such a current loop is the basic physical principle governing the operation of electric motors, as we mentioned at the beginning of this section.* The effect has enormous technological applications. There also are applications at the atomic level. An electron orbiting the nucleus of an atom can be viewed as a tiny current loop (with the direction of the current opposite to the direction of motion of the electron, because it is a negative charge that is moving). Thus an electron in an atom creates an orbital magnetic dipole moment. The electron and other fundamental particles such as the proton and neutron also have intrinsic magnetic dipole moments associated with their spin. When a collection of such magnetic dipole moments is placed in an external magnetic field, the magnetic dipole moments align with the direction of the field to minimize their potential energies. In this way submicroscopic systems with magnetic dipole moments can be aligned. Many medical imaging technologies exploit this behavior of atomic and nuclear magnetic dipole moments.

PROBLEM-SOLVING TACTIC

20.3 If a wire carrying a current is wrapped into a coil with n turns, each with the same area A, the magnetic dipole moment of the coil is n times that of a single turn. That is, the magnetic moment of a coil is

$$\vec{\boldsymbol{\mu}} = nI\vec{\mathbf{A}} \qquad (20.17)$$

The area A can be circular, rectangular, or whatever. The common direction of $\vec{\boldsymbol{\mu}}$ and $\vec{\mathbf{A}}$ is determined from the same right-hand rule: wrap the fingers of your right hand around the coil in the sense of the (positive) current, and the extended right-hand thumb indicates the direction of both $\vec{\boldsymbol{\mu}}$ and $\vec{\mathbf{A}}$. Motors have coils with many turns of wire to increase the magnetic moment and therefore the torque on the coil if it carries a current in a magnetic field.

EXAMPLE 20.7

A current of 15.0 A is in a circular coil of radius 5.00 cm with 50 turns of wire (see Figure 20.26). The plane of the coil makes an angle of 30.0° with a uniform magnetic field of magnitude 0.150 T.

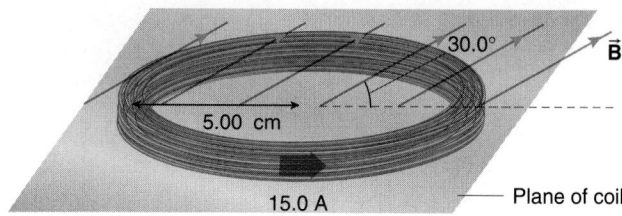

FIGURE 20.26

a. Find the magnitude of the magnetic dipole moment of the coil.
b. Determine the magnitude of the torque on the coil. Indicate in a sketch the direction that the coil will spin if it is free to move.
c. Determine the potential energy of the coil in its given orientation.

Solution

a. According to Problem-Solving Tactic 20.3, the magnitude of the magnetic dipole moment of the coil is

$$\begin{aligned}
\mu &= nIA \\
&= (50)(15.0 \text{ A})[\pi(5.00 \times 10^{-2} \text{ m})^2] \\
&= 5.89 \text{ A} \cdot \text{m}^2
\end{aligned}$$

b. The magnitude of the torque is found from Equation 20.15:

$$\tau = \mu B \sin \theta$$

where θ is the angle between the magnetic dipole moment and the magnetic field. The direction of $\vec{\boldsymbol{\mu}}$ is determined from the right-hand rule; it is shown in Figure 20.27. The angle is between $\vec{\boldsymbol{\mu}}$ and $\vec{\mathbf{B}}$ is seen to be 60.0°. Hence the torque has magnitude

$$\begin{aligned}
\tau &= (5.89 \text{ A} \cdot \text{m}^2)(0.150 \text{ T}) \sin 60.0° \\
&= 0.765 \text{ N} \cdot \text{m}
\end{aligned}$$

The direction of $\vec{\boldsymbol{\tau}}$ and the sense the coil will spin are indicated in Figure 20.27.

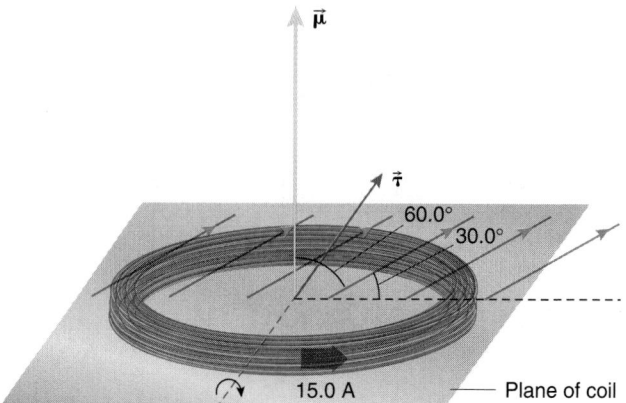

FIGURE 20.27

---

*One also needs to reverse the current direction in the loop periodically to keep the loop turning; the reversals are accomplished by means of a split ring commutator. See Investigative Project 5 in Chapter 22.

c. The potential energy of the coil in its given configuration is found from Equation 20.16:

$$PE = -\vec{\mu} \cdot \vec{B}$$
$$= -\mu B \cos \theta$$
$$= -(5.89 \text{ A·m}^2)(0.150 \text{ T}) \cos 60.0°$$
$$= -0.442 \text{ J}$$

STRATEGIC EXAMPLE 20.8

An electron is in a circular orbit about the nucleus of an atom, as indicated in Figure 20.28 (not to scale!). Find a relationship between the orbital magnetic dipole moment $\vec{\mu}$ and the orbital angular momentum $\vec{L}$ of the electron about the center of its orbit.

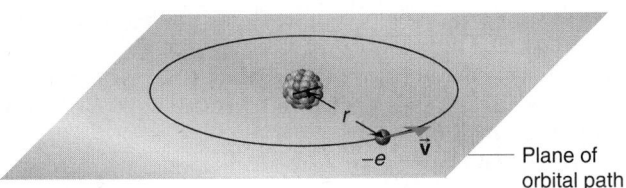

FIGURE 20.28

Solution
Let v be the speed of the electron and r the radius of its orbit. The orbiting electron creates a current opposite to its velocity, because the convention for current is the direction of positive charge motion; see Figure 20.29.

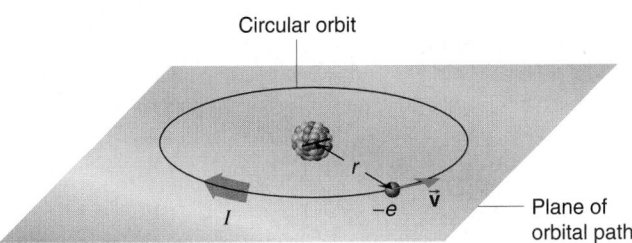

FIGURE 20.29

The current is the number of coulombs that pass any given point along the circumferential path during one second. Hence the current is the fundamental unit of charge e times the frequency ν of the orbital motion (the number of times the electron completes an orbit during one second). That is,

$$I = e\nu$$

The period T of the motion is the inverse of the frequency, so that

$$I = \frac{e}{T} \quad (1)$$

The period is the time to complete one revolution. The electron travels the distance of the circumference $2\pi r$ at the speed v; so the time it takes (the period) is

$$T = \frac{2\pi r}{v}$$

Substituting for T into equation (1) for the current, you find

$$I = \frac{ev}{2\pi r}$$

The magnitude of the magnetic dipole moment of the orbiting electron (a current loop!) is, by definition, the product of the current and the area of the loop (Equation 20.14):

$$\mu = IA$$
$$= \frac{ev}{2\pi r} \pi r^2$$
$$= \frac{evr}{2} \quad (2)$$

The direction of the magnetic dipole moment is perpendicular to the loop in the sense of the right-hand rule; see Figure 20.30.

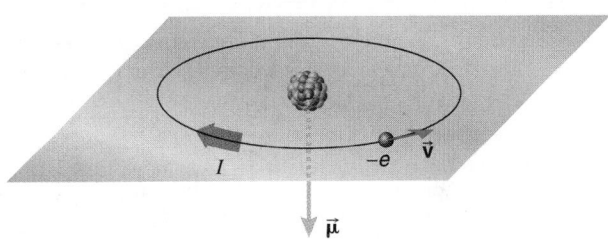

FIGURE 20.30

The orbital angular momentum of the electron is (see Chapter 10)

$$\vec{L}_{orbit} = \vec{r} \times \vec{p}$$

where $\vec{r}$ is the position vector of the electron and $\vec{p}$ is its momentum; see Figure 20.31. In circular motion $\vec{r}$ is perpendicular to $\vec{p}$, and so the magnitude of the orbital angular momentum is

$$L_{orbit} = rp \sin 90°$$
$$= rp$$
$$= mvr \quad (3)$$

where m is the mass of the electron. The direction of the orbital angular momentum is determined from the vector product right-hand rule.

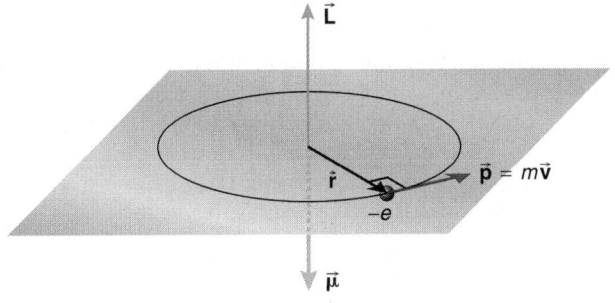

FIGURE 20.31

The orbital magnetic dipole moment and the orbital angular momentum vectors point in opposite directions! Comparing equation (2) with equation (3) shows the two vectors are related by

$$\vec{\mu} = -\frac{e}{2m}\vec{L} \qquad (4)$$

20.6 THE BIOT–SAVART LAW

From our study of the gravitation, we saw that a mass m experiences a gravitational force when in a gravitational field $\vec{g}$. The gravitational field itself is caused, in turn, by other masses M. Think of this relationship schematically as

| masses M | produce | gravitational field $\vec{g}$ | affects | mass m via $\vec{F} = m\vec{g}$ |

Likewise a charge q in an electric field $\vec{E}$ experiences an electrical force. The electric field itself is caused by other charges Q:

| charges Q | produce | electric field $\vec{E}$ | affects | charge q via $\vec{F} = q\vec{E}$ |

We have seen that magnetic fields exert magnetic forces on charges in motion (currents):

| magnetic field $\vec{B}$ | affects | charges in motion (currents) $\vec{F} = q\vec{v} \times \vec{B}$ $\vec{F} = I\int d\vec{\ell} \times \vec{B}$ |

What produces a magnetic field? A glance at the gravitational and electrical cases provides a strong clue. It should come as no surprise that since magnetic fields affect charges in motion (currents), the magnetic field itself is caused by other charges in motion (other currents):

| charges in motion: current | produces | magnetic field $\vec{B}$ | affects | other charges in motion: (currents) |

Here we want to find the relationship between a current (charge in motion) and the magnetic field it produces. Historically, the relationship was not a trivial thing to deduce; it was figured out for the first time during the 19th century (and without the benefit of vector notation!). Indeed much experimentation and head scratching went on before the relationship became apparent.

Part of the reason for this was the nature of the problem. With the gravitational and electrical cases, it was relatively easy to find or calculate the appropriate field of pointlike masses or charges (as well as spherically symmetric mass or charge distributions). If the mass or charge distribution was not one of these geometries, it was more difficult to calculate the gravitational or electrical fields: vector integrations are needed, as we saw in Chapters 6 and 16.

Unfortunately, for the magnetic field there are no macroscopic pointlike currents or spherically symmetric currents; current distributions almost always have messier geometries. To calculate the magnetic field of a current virtually always involves a somewhat intimidating vector integration over the path of the current. Physics, like life, is not always easy.

On a microscopic level, certain particles such as the electron, proton, and neutron (and others) have permanent intrinsic magnetic dipole moments. Just as an electric dipole produces an electric field, a magnetic dipole produces a magnetic field. A magnetic dipole moment has a magnitude that is equal to the current in the current loop times an area, but it is by no means clear what the current or the area is for these particles that create their permanent intrinsic magnetic dipole moment. We simply know the particles have an intrinsic magnetic dipole moment because we can measure it. To the extent that these particles are pointlike particles, the magnetic field their magnetic dipole moments produce can be considered to be caused by very small or pointlike sources of a magnetic field. The fields they produce, however, always are *dipolar* fields: isolated magnetic monopoles never have been detected. These microscopic dipolar elements were, however, unknown during the early 19th century.

By careful observations of the oscillations of a suspended magnet near a long, current-carrying wire, the French scientists Jean-Baptiste Biot (1774–1862) and Félix Savart (1791–1841) experimentally discovered the relationship between a current and the magnetic field it produces in the surrounding space. The relationship now is known as the **Biot–Savart law**.

Consider a small, differential length $d\vec{\ell}$ of a current-carrying wire, as shown in Figure 20.32. The vector $d\vec{\ell}$ is directed in the same sense as the current I. The current element $I\,d\vec{\ell}$ produces a differential bit of magnetic field $d\vec{B}$ at a point P located a distance r away. Let $\hat{r}$ be a unit vector pointing *from* the current element *to* the point P.

Biot and Savart found that the magnetic field at P produced by the current element is expressed (using our modern notation) by

$$d\vec{B} = \frac{\mu_0}{4\pi} I \frac{d\vec{\ell} \times \hat{r}}{r^2} \qquad (20.18)$$

where μ_0 is a scalar constant we discuss in more detail later.* Its role is to specify how successful (i.e., effective) the current element is in producing the magnetic field.

> To find the magnetic field caused by the entire wire, we integrate over the wire:
>
> $$\vec{B} = \frac{\mu_0}{4\pi} I \int_{\text{wire}} \frac{d\vec{\ell} \times \hat{r}}{r^2} \qquad (20.19)$$

This expression may look formidable to you. Actually the law is quite similar to the vector expression for the gravitational field of an extended mass (Equation 6.45 on page 258),

$$\vec{g} = G \int_{\substack{\text{mass} \\ \text{distribution}}} \frac{dM}{r^2} \hat{r}$$

---

*Do not confuse the magnetic dipole moment $\vec{\mu}$, a vector quantity, with the constant μ_0, which is a scalar. The use of the Greek letter μ in two different contexts in magnetism is a regrettable but common notation in each instance. Some texts use $\vec{m}$ for the magnetic dipole moment vector, but the situation is no better since m also is used for mass (a scalar).

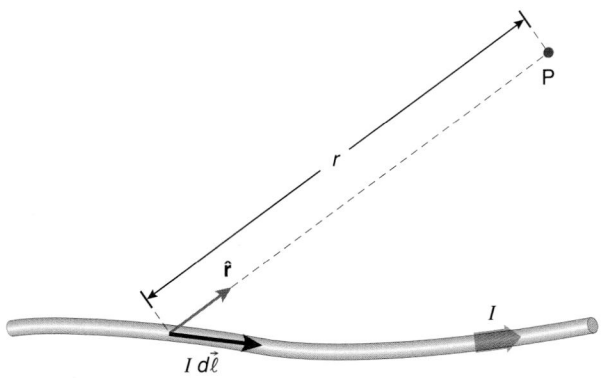

FIGURE 20.32 A current element $I\,d\vec{\ell}$ produces a differential amount of magnetic field $d\vec{B}$ at P.

and to the electric field of an extended charge distribution (Equation 16.16 on page 733),

$$\vec{E} = \frac{1}{4\pi\varepsilon_0} \int_{\substack{\text{charge} \\ \text{distribution}}} \frac{dQ}{r^2}\,\hat{r}$$

Specifically the similarities are as follows:

1. Each has a pointlike source term that causes the field:

$$\text{gravitational case (mass):} \quad dM$$
$$\text{electrical case (charge):} \quad dQ$$
$$\text{magnetic case (current element):} \quad I\,d\vec{\ell}$$

2. Each expression for the field has a dimensional constant that reflects how effective the source is in producing the field:

$$\text{gravitational case:} \quad G$$

$$\text{electrical case:} \quad \frac{1}{4\pi\varepsilon_0}$$

$$\text{magnetic case:} \quad \frac{\mu_0}{4\pi}$$

3. Each is an inverse square law, where r is the distance between the source term and the point where the field is to be calculated.
4. In each case the unit vector $\hat{r}$ is directed *from* the pointlike source of the field *to* the point where the field is to be calculated.
5. Each involves a vector integration over the extended source.

The only additional complication for the magnetic field is the vector product in the numerator of the Biot–Savart law.

The constant in the Biot–Savart law,

$$\frac{\mu_0}{4\pi}$$

reflects what was done with the electrical constant. The factor of 4π is inserted into the constant here so that a factor of 4π will *not* appear in more useful equations later on. The quantity μ_0 has a

fancy name: the **permeability of free space**. (Recall that ε_0 was the *permittivity of free space*; this terminology may be a bit confusing to you at first.) The quantity

$$\frac{\mu_0}{4\pi}$$

has the (exact) numerical value

$$\frac{\mu_0}{4\pi} \equiv 10^{-7} \text{ T} \cdot \text{m/A}$$

with the indicated SI units. These units of μ_0 can be deduced from the Biot–Savart law itself.

The magnetic fields produced by some common current distributions are given in Table 20.1.

EXAMPLE 20.9

A circular wire loop of radius R carries a current I.

a. **Use the Biot–Savart law to find the magnetic field at the center of the loop.**
b. **For a loop of radius 5.0 cm, what current yields a magnetic field of magnitude 1.0×10^{-4} T at the center of the loop (which is about the magnitude of the magnetic field of the Earth).**

Solution
a. The Biot–Savart law (Equation 20.19) states that

$$\vec{B} = \frac{\mu_0}{4\pi} I \int_{\text{wire}} \frac{d\vec{\ell} \times \hat{r}}{r^2}$$

For a current element $I\,d\vec{\ell}$ on the loop, the vector $\hat{r}$ points from the current element to the point where you want to calculate the field: the center of the loop; see Figure 20.33. The vector product $d\vec{\ell} \times \hat{r}$ points upward along the axis of the loop. Choose an origin at the center of the loop and let the z-axis be perpendicular to the plane of the loop.

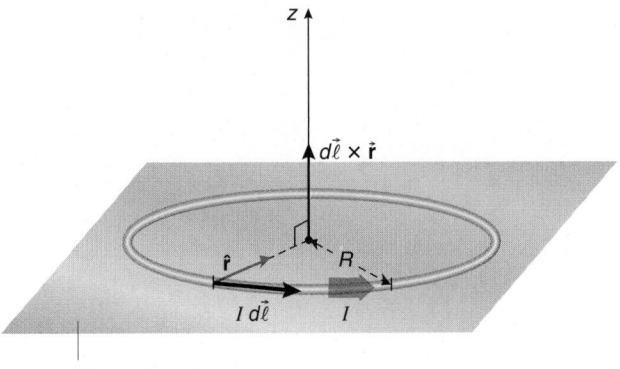

FIGURE 20.33

The angle between $d\vec{\ell}$ and $\hat{r}$ is 90° and is the same for every current element as you go around the loop. Indeed the direction of the vector product also is unchanged as you run

TABLE 20.1 **Some Commonly Used Magnetic Fields**

(1) At the center of circular loop of radius R (see Example 20.9):

$$\vec{B}_{\text{center}} = \frac{\mu_0}{4\pi} \frac{2\pi I}{R} \hat{k}$$

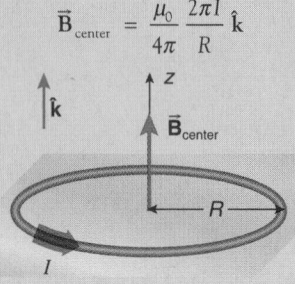

(2) On the axis of a circular current loop (see Example 20.10):

$$\vec{B}_{\text{axis}} = \frac{\mu_0}{4\pi} \frac{I(2\pi R^2)}{(R^2 + z^2)^{3/2}} \hat{k}$$

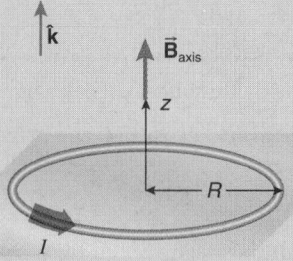

(3) For a circular coil of n loops, all of the same radius, multiply the preceding results by n.

(4) A distance d from an infinite wire (see Example 20.11):

$$B = \frac{\mu_0}{4\pi} \frac{2I}{d}$$

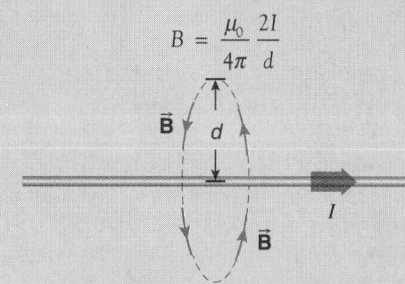

(5) Inside a long solenoid having n turns per meter of its length, each carrying current I, far from its ends (see Example 20.13):

$$B = \mu_0 n I$$

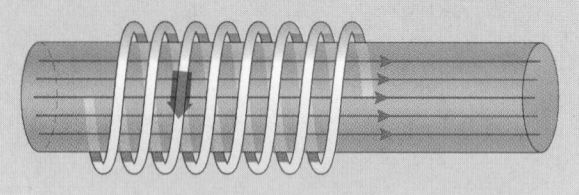

around the loop. The magnetic field, therefore, points along $\hat{k}$. Hence the Biot–Savart law reduces to

$$\vec{B}_{\text{center}} = \frac{\mu_0}{4\pi} I \int_{\text{wire}} \frac{d\ell \ (1) \sin 90°}{r^2} \hat{k}$$

Since the distance r between each current element and the place where you want to find the field is constant and equal

to the radius R of the circle, the integration reduces to

$$\vec{B}_{\text{center}} = \frac{\mu_0}{4\pi} \frac{I}{R^2} \int_{\text{wire}} d\ell \ \hat{k}$$

The integral of $d\ell$ around the loop is the circumference of the loop: $2\pi R$. Thus the field at the center of the loop is

$$\vec{B}_{\text{center}} = \frac{\mu_0}{4\pi} \frac{I}{R^2} 2\pi R \ \hat{k}$$

$$= \frac{\mu_0}{4\pi} \frac{2\pi I}{R} \hat{k} \tag{1}$$

Notice that the direction of the field can be obtained by grasping the wire with your right hand with the thumb in the direction of the current; the curled fingers of your right hand then indicate the way the field surrounds the wire and threads the loop.

The magnetic field at other points in the plane of the loop (whether inside or outside the loop) is much more difficult to calculate.

b. The magnitude of the magnetic field at the center of the loop is

$$B_{\text{center}} = \frac{\mu_0}{4\pi} \frac{2\pi I}{R} \tag{2}$$

Solve this expression for I:

$$I = \frac{4\pi}{\mu_0} \frac{R B_{\text{center}}}{2\pi}$$

Remember to use SI units. Substituting $R = 5.0 \times 10^{-2}$ m and $B_{\text{center}} = 1.0 \times 10^{-4}$ T, you find

$$I = \frac{(5.0 \times 10^{-2} \ \text{m})(1.0 \times 10^{-4} \ \text{T})}{(10^{-7} \ \text{T·m/A}) 2\pi}$$

$$= 8.0 \ \text{A}$$

STRATEGIC EXAMPLE 20.10

Find the magnetic field at a point P located a distance z from the center along the axis of a circular current loop of radius R and current I; see Figure 20.34.

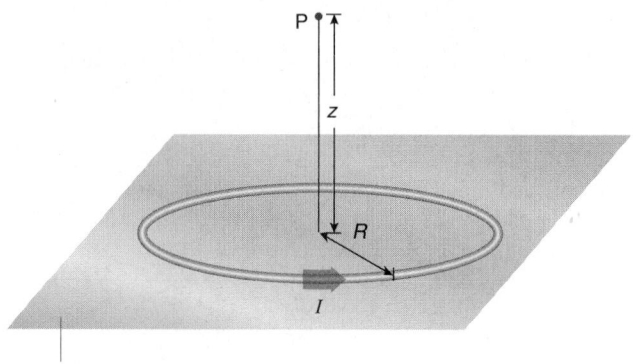

Plane of circular loop

FIGURE 20.34

Solution
Use the Biot–Savart law, Equation 20.19:

$$\vec{\mathbf{B}} = \frac{\mu_0}{4\pi} I \int_{\text{wire}} \frac{d\vec{\ell} \times \hat{\mathbf{r}}}{r^2}$$

The unit vector $\hat{\mathbf{r}}$ points from the current element $I\, d\vec{\ell}$ to the point where you want to compute the field, and r is the distance between the current element and the field point; see Figure 20.35. The angle between $I\, d\vec{\ell}$ and $\hat{\mathbf{r}}$ is 90° for every current element along the wire ring. The distance r also is the same for every current element around the ring. This is beginning to seem like Example 20.9 all over again. But one thing is different: the direction of the vector product $d\vec{\ell} \times \hat{\mathbf{r}}$ is not along the axis, as shown in Figure 20.35.

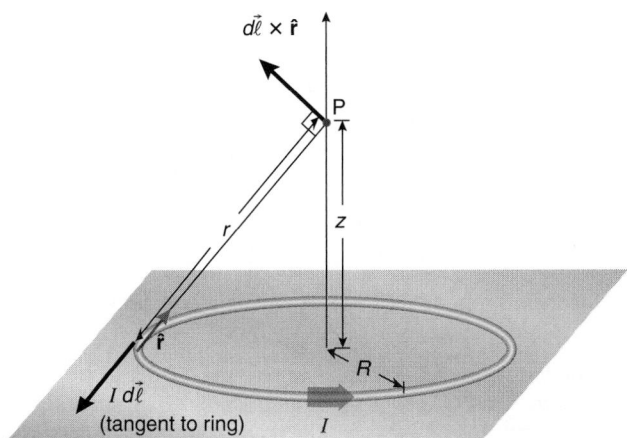

FIGURE 20.35

As you integrate around the loop, the various differential contributions to the magnetic field sweep out a cone-shaped megaphone, pictured in Figure 20.36. The components perpendicular to the axis vector sum pairwise to zero. Only the components along the z-axis survive the integration; the z-component of $d\vec{\ell} \times \hat{\mathbf{r}}$ is $|d\vec{\ell} \times \hat{\mathbf{r}}| \cos\theta$. Hence the total surviving magnetic field at the point P is

$$\vec{\mathbf{B}}_{\text{axis}} = \frac{\mu_0}{4\pi} I \int_{\text{wire}} \frac{d\ell\,(1)\sin 90°}{r^2} \cos\theta\, \hat{\mathbf{k}}$$

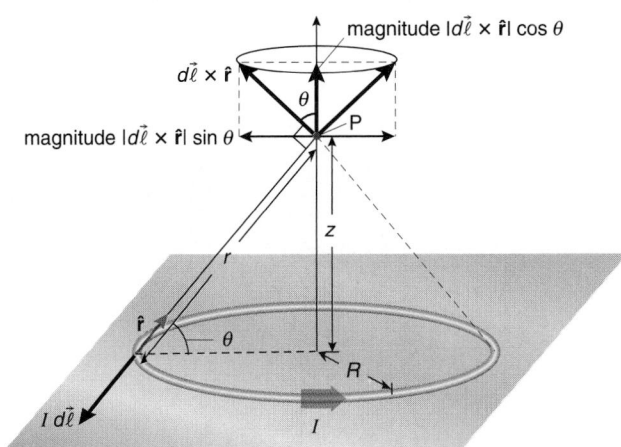

FIGURE 20.36

Since θ and r are the same for every current element around the ring,

$$\vec{\mathbf{B}}_{\text{axis}} = \frac{\mu_0}{4\pi} \frac{I}{r^2} \cos\theta \int_{\text{wire}} d\ell\, \hat{\mathbf{k}}$$

The integration around the wire loop is the circumference $2\pi R$. From the geometry of Figure 20.36, you see

$$\cos\theta = \frac{R}{r}$$

Hence the magnetic field is

$$\vec{\mathbf{B}}_{\text{axis}} = \frac{\mu_0}{4\pi} \frac{I}{r^2} \frac{R}{r}\, 2\pi R\, \hat{\mathbf{k}}$$

The Pythagorean theorem gives

$$r = (R^2 + z^2)^{1/2}$$

and so the field is

$$\vec{\mathbf{B}}_{\text{axis}} = \frac{\mu_0}{4\pi} \frac{I\,(2\pi R^2)}{(R^2 + z^2)^{3/2}}\, \hat{\mathbf{k}} \qquad (1)$$

Notice that when $z = 0$ m you are at the center of the current loop and the result is the same as equation (1) of Example 20.9. It had better be! Note also that the field is in the same direction for both positive and negative values of z, as shown in Figure 20.37. Grasp the wire with your right hand with the thumb in the direction of the current; your fingers thread the loop in the direction of the field along the axis of the loop.

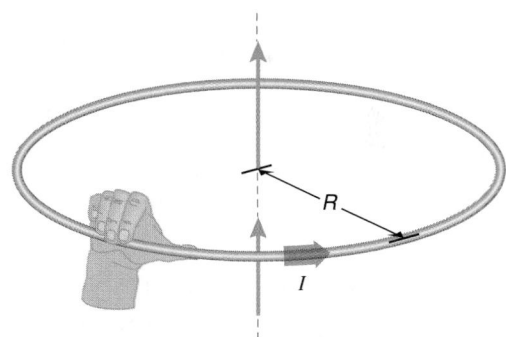

FIGURE 20.37

STRATEGIC EXAMPLE 20.11

a. Find the magnetic field at a point P located a distance d from a long (infinite) wire carrying a current I.
b. If the wire carries a current of 15.0 A, at what distance from the wire is the magnitude of the magnetic field 1.0×10^{-4} T? This is approximately the magnitude of the magnetic field of the Earth.

Solution
a. Begin with the Biot–Savart law, Equation 20.19:

$$\vec{\mathbf{B}} = \frac{\mu_0}{4\pi} I \int_{\text{wire}} \frac{d\vec{\ell} \times \hat{\mathbf{r}}}{r^2}$$

Choose the *x*-axis to be along the wire, as indicated in Figure 20.38. By the right-hand rule, the vector product $I \, d\vec{\ell} \times \hat{r}$ is directed along the *z*-axis for every current element along the wire. You can see this explicitly by noting that the current element $I \, d\vec{\ell}$ is $I \, dx \, \hat{i}$ and, geometrically, the vector $\hat{r}$ can be written as

$$\hat{r} = (\cos \theta)\hat{i} + (\sin \theta)\hat{j}$$

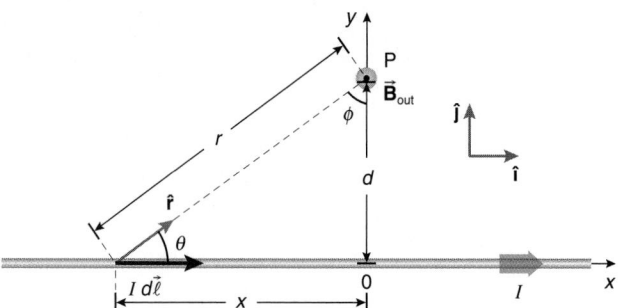

FIGURE 20.38

Hence the vector product is

$$I \, d\vec{\ell} \times \hat{r} = I \, dx \, \hat{i} \times [(\cos \theta)\hat{i} + (\sin \theta)\hat{j}]$$
$$= (I \, dx \sin \theta)\hat{k}$$

The Biot–Savart law becomes

$$\vec{B} = \frac{\mu_0}{4\pi} I \int_{\text{wire}} \frac{dx \sin \theta}{r^2} \hat{k}$$

Geometrically, $r^2 = d^2 + x^2$ and

$$\sin \theta = \frac{d}{r} = \frac{d}{(d^2 + x^2)^{1/2}}$$

With these substitutions in the expression for $\vec{B}$, the field is

$$\vec{B} = \frac{\mu_0}{4\pi} I \int_{\text{wire}} \frac{dx \, d}{(d^2 + x^2)^{3/2}} \hat{k}$$
$$= \frac{\mu_0}{4\pi} Id \int_{\text{wire}} \frac{dx}{(d^2 + x^2)^{3/2}} \hat{k}$$

You have to integrate over the entire wire from $x = -\infty$ m to $x = \infty$ m. Since *x* appears in the integrand to an even power, the integral is symmetric in *x*, and so you can take twice the integral from 0 m to ∞ m. The *magnitude* of the field thus is

$$B = \left(\frac{\mu_0}{4\pi} Id \right) 2 \int_{0 \text{ m}}^{\infty \text{ m}} \frac{dx}{(d^2 + x^2)^{3/2}} \tag{1}$$

The integral has a form that is ripe for trigonometric substitution methods. In particular, substitute

$$x = d \tan \phi \tag{2}$$

(see Figure 20.38 for the geometric interpretation of *ϕ*). For the limits of integration, when $x = 0$ m, $\phi = 0°$ and when $x = +\infty$ m, $\phi = 90°$. To find *dx*, take differentials of equation (2):

$$dx = d \sec^2 \phi \, d\phi$$

With the substitution $x = d \tan \phi$, the denominator of the integrand becomes

$$\begin{aligned} (d^2 + x^2)^{3/2} &= (d^2 + d^2 \tan^2 \phi)^{3/2} \\ &= [d^2(1 + \tan^2 \phi)]^{3/2} \\ &= (d^2 \sec^2 \phi)^{3/2} \\ &= d^3 \sec^3 \phi \end{aligned}$$

Making these substitutions into equation (1) for the magnitude of the magnetic field, you find

$$\begin{aligned} B &= \frac{\mu_0}{4\pi} 2Id \int_{0°}^{90°} \frac{d \sec^2 \phi \, d\phi}{d^3 \sec^3 \phi} \\ &= \frac{\mu_0}{4\pi} \frac{2I}{d} \int_{0°}^{90°} \cos \phi \, d\phi \end{aligned}$$

Perhaps setting up the problem was difficult, but at least the integration is quite easy to do! You finally get

$$\begin{aligned} B &= \frac{\mu_0}{4\pi} \frac{2I}{d} \sin \phi \, \Big|_{0°}^{90°} \\ &= \frac{\mu_0}{4\pi} \frac{2I}{d} \end{aligned} \tag{3}$$

for the magnitude of the magnetic field at a distance *d* from an infinite straight wire.

The direction of the field at P is perpendicularly out of the page in Figure 20.38. Notice that the problem has cylindrical symmetry: no matter where the point P is chosen around the wire (as long as it is a distance *d* from the wire), the field is perpendicular to the wire. Thus the magnetic field lines around the wire are circles, centered on the wire, as shown in Figure 20.39. If you grasp the wire with your right hand with the thumb in the direction of the current, your curled fingers indicate the directional sense in which the magnetic field points around the wire.

b. To find the distance from the wire at which the magnitude of the magnetic field is 1.0×10^{-4} T, solve equation (3) for *d* and substitute appropriate numerical values:

$$\begin{aligned} d &= \frac{\mu_0}{4\pi} \frac{2I}{B} \\ &= \frac{(10^{-7} \text{ T·m/A}) [2(15.0 \text{ A})]}{1.0 \times 10^{-4} \text{ T}} \\ &= 3.0 \times 10^{-2} \text{ m} \\ &= 3.0 \text{ cm} \end{aligned}$$

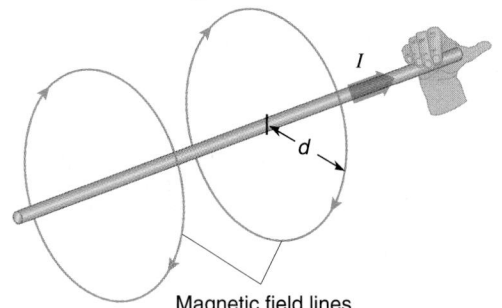

Magnetic field lines

FIGURE 20.39

20.7 FORCES OF PARALLEL CURRENTS ON EACH OTHER AND THE DEFINITION OF THE AMPERE

Two or more wires exert magnetic forces on each other just as two or more charges exert electrical forces on each other. Each wire finds itself in the magnetic field produced by the other wire(s). Here we consider the special situation of two parallel, infinitely long, current-carrying wires separated by a distance d, as shown in Figure 20.40.

Wire (1) has current I_1 and this current produces a magnetic field in the space surrounding the wire. With the coordinate system indicated in Figure 20.40, the magnetic field of wire (1) at the position of wire (2) is (using the results of Example 20.11 and the right-hand rule to determine the direction of the field)

$$\vec{B}_{1 \text{ of wire (1) at the position of wire (2)}} = \frac{\mu_0}{4\pi} \frac{2I_1}{d} \hat{k} \qquad (20.20)$$

Wire (2) has current I_2 and finds itself in the magnetic field of wire (1). The magnetic field of wire (1) is constant along the straight length of wire (2); therefore, we can use Equation 20.11 for the force on a length ℓ of wire (2):

$$\vec{F}_{2 \text{ on length } \ell \text{ of wire (2) due to wire (1)}} = I_2 \vec{\ell} \times \vec{B}_1$$

With the coordinate system in Figure 20.40, this becomes

$$\vec{F}_{2 \text{ on length } \ell \text{ of wire (2) due to wire (1)}} = I_2 \ell \hat{i} \times B_1 \hat{k}$$
$$= -I_2 \ell B_1 \hat{j}$$

Substituting for the magnitude of $\vec{B}_1$ using Equation 20.20, we have

$$\vec{F}_{2 \text{ on length } \ell \text{ of wire (2) due to wire (1)}} = -I_2 \ell \frac{\mu_0}{4\pi} \frac{2I_1}{d} \hat{j}$$
$$= -\frac{\mu_0}{4\pi} \frac{2I_1 I_2}{d} \ell \hat{j} \qquad (20.21)$$

In a similar fashion, wire (2) produces a magnetic field at the position of wire (1). Using the results of Example 20.11 and the right-hand rule to determine the direction of the field, the magnetic field of wire (2) at the position of wire (1) is

$$\vec{B}_{2 \text{ of wire (2) at the position of wire (1)}} = \frac{\mu_0}{4\pi} \frac{2I_2}{d} (-\hat{k}) \qquad (20.22)$$

Wire (1), carrying current I_1, finds itself in the magnetic field of wire (2). The field is constant along the length of wire (1), and

so we can use Equation 20.11 to find the force on a length ℓ of wire (1):

$$\vec{F}_{1 \text{ on length } \ell \text{ of wire (1) due to wire (2)}} = I_1 \vec{\ell} \times \vec{B}_2$$

With the coordinate system in Figure 20.40, this becomes

$$\vec{F}_{1 \text{ on length } \ell \text{ of wire (1) due to wire (2)}} = I_1 \ell \hat{i} \times (-B_2 \hat{k})$$
$$= I_1 \ell B_2 \hat{j}$$

Substituting for the magnitude of $\vec{B}_2$ using Equation 20.22, we have

$$\vec{F}_{1 \text{ on length } \ell \text{ of wire (1) due to wire (2)}} = I_1 \ell \frac{\mu_0}{4\pi} \frac{2I_2}{d} \hat{j}$$
$$= \frac{\mu_0}{4\pi} \frac{2I_1 I_2}{d} \ell \hat{j} \qquad (20.23)$$

Notice that in accordance with Newton's third law, the force of wire (1) on a length ℓ of wire (2) is equal in magnitude but opposite in direction to the force of wire (2) on a length ℓ of wire (1). Indeed, once we found the force of wire (1) on wire (2), there was no need to go back to the beginning to calculate the force of wire (2) on wire (1); we simply could have used Newton's third law to find the other force! We took the scenic route to gain a little more practice in working with magnetic fields and forces.

Notice that the forces that parallel wires exert on each other are attractive if the currents are parallel (see Figure 20.41).

If the currents are antiparallel, the forces on the wires are repulsive, as shown in Figure 20.42.

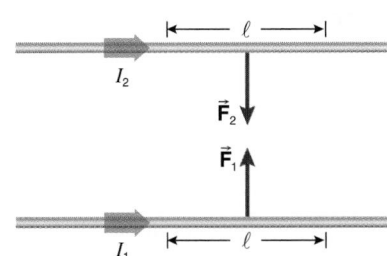

FIGURE 20.41 Parallel currents attract each other.

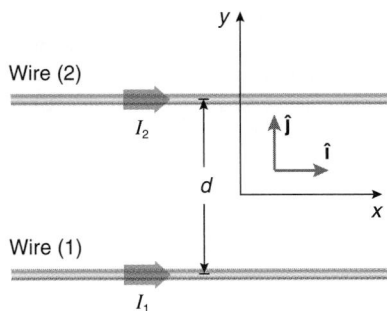

FIGURE 20.40 Two infinite and parallel current-carrying wires.

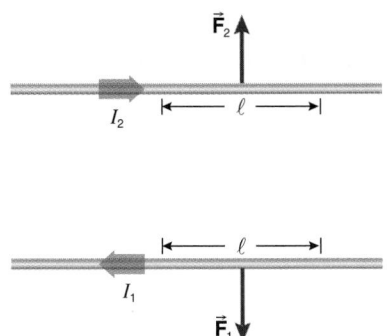

FIGURE 20.42 Antiparallel currents repel each other.

Two infinitely long, parallel, straight current-carrying wires exert forces on a length ℓ of either wire of magnitude

$$F = \frac{\mu_0}{4\pi} \frac{2I_1I_2}{d} \ell \qquad (20.24)$$

from either Equation 20.21 or 20.23. The SI unit of current (the ampere) is defined from Equation 20.24. Let the wires be exactly one meter apart, so that $d \equiv 1$ m. Consider the force on exactly one meter of length of the wires, so that $\ell \equiv 1$ m. Let the currents be of equal magnitude: $I_1 = I_2 = I$. Adjust the size of the equal currents I until the magnitude of the force on each wire is exactly 2×10^{-7} N. The magnitude of the current that yields a force of this magnitude per meter of length is defined to be one ampere. The equation for the magnitude of the force then becomes

$$2 \times 10^{-7} \text{ N} \equiv (10^{-7} \text{ T·m/A}) \frac{2I^2}{1 \text{ m}} (1 \text{ m})$$

Solving for I, we obtain, exactly,

$$I = 1 \text{ A}$$

Since a coulomb is an ampere-second, this procedure also defines the SI unit of charge.

QUESTION 3

Convince yourself that the antiparallel currents in Figure 20.42 produce repulsive forces on each other.

20.8 GAUSS'S LAW FOR THE MAGNETIC FIELD*

We developed Gauss's law for the two other fields we have studied, the gravitational field $\vec{g}$ and the electric field $\vec{E}$. The law involves the flux of the field vector through a *closed surface*.

For the gravitational case, we discovered in Equation 6.55 that the flux of the gravitational field through a closed surface S is equal to -4π times the mass enclosed within the surface:

$$\int_{\text{clsd surface}} \vec{g} \cdot d\vec{S} = -4\pi M_{\text{within } S}$$

The specific location of the mass within the closed surface was irrelevant; what mattered was simply whether the mass was within the surface or not. If there was no mass within the Gaussian surface, the flux of the gravitational field through that closed surface was zero.

For the electric field, we found that the flux of the electric field through a closed surface was equal to the total (net) charge enclosed by the surface divided by the constant ε_0 (Equation 16.18):

$$\int_{\text{clsd surface}} \vec{E} \cdot d\vec{S} = \frac{\text{net charge enclosed by the surface}}{\varepsilon_0}$$

If there is zero net (total) charge within the Gaussian surface, then the flux of the electric field through that surface is zero. It makes no difference where the charges are within the surface, only that they *are* within it. In particular, if the charges within the surface all are electric dipoles, as shown in Figure 20.43,

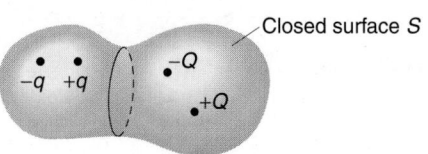

FIGURE 20.43 A collection of electric dipoles produces zero flux of the electric field through a surrounding closed surface.

then the flux of the total electric field vector (resulting from the vector superposition of the individual fields of the individual charges) through the surface is zero.

Gauss's law for the magnetic field likewise involves the flux of the magnetic field vector through a closed surface:

$$\int_{\text{clsd surface}} \vec{B} \cdot d\vec{S}$$

The flux again must fit our observations. We mentioned as we began our study of magnetism that isolated magnetic poles (magnetic monopoles) evidently do not exist: the magnetic field is a *dipolar* field. The poles of the field always are paired. If there are magnets within the surface, the magnets are always dipolar. Thus, in strict analogy with the electric dipole case, we can say the following.

The flux of the magnetic field through *any* closed surface must always be zero:

$$\int_{\text{clsd surface}} \vec{B} \cdot d\vec{S} = 0 \text{ T·m}^2 \qquad (20.25)$$

This is **Gauss's law for the magnetic field**.

Gauss's law for the magnetic field has several implications.

Gauss's law for the magnetic field is, in essence, a statement about the nonexistence of magnetic monopoles (isolated, unpaired north or south magnetic poles).

Mass monopoles exist; electric monopoles exist; but magnetic monopoles evidently do *not* exist. The existence of monopoles is necessary to have a nonzero term on the right-hand side of Gauss's law.

Gauss's law for the magnetic field implies that the magnetic field lines form topologically closed loops.

Magnetic field lines do not begin or end anywhere: they always form closed loops. In gravitation, the field lines representing the field begin at infinity and end at (point toward) the (positive) mass monopoles; there are no negative mass monopoles. Electric field lines begin on positive charge (point away from positive charge) and end on negative charge (point toward negative charge).

Gauss's law for the magnetic field is the second of the four fundamental equations of electromagnetism known collectively as the *Maxwell equations*. Recall that the first of the Maxwell equations was Gauss's law for the electric field (Section 16.11).

20.9 MAGNETIC POLES AND CURRENT LOOPS

We mentioned as we began our study of magnetism that magnets have poles designated N (north) and S (south). Yet we found that it is moving charge (a current) that is the source of the magnetic field. In this section we reconcile these views and see how to determine the north and south poles of a current loop.

The magnetic field produced by a traditional bar magnet is shown in Figure 20.44. The magnetic field produced by a current loop is shown in Figure 20.45.

The magnetic field lines in both cases form closed loops in compliance with Gauss's law for the magnetic field. Outside the bar magnet, the field direction is from the north to the south pole; within the magnet, the field is from the south to the north. The right-hand rule is used to determine the direction of the field in the vicinity of the current loop. The current loop has a north pole side and a south pole side.

You might argue that a current loop is one thing and a permanent magnet is quite another. But the distinction is not apparent, even at the microscopic level. If we keep breaking the bar magnet into smaller and smaller pieces, creating smaller and smaller magnets, eventually we get down to the atomic level where we find that the magnet is really nothing more than a current loop consisting of an electron orbiting a nucleus. A macroscopic permanent magnet is a superposition of many small, generally aligned, submicroscopic current loops represented by the electronic orbital motions inside atoms (the current directions are opposite to the motion of the electrons).

So the idea of magnetic poles is nothing more than an extension of the idea of the two sides of a current loop. The magnetic dipole moment of the current loop is parallel to the axis of a bar magnet we imagine it to represent: the head of the magnetic dipole moment vector represents the north end and the tail of the magnetic dipole moment vector represents the south end.

20.10 AMPERE'S LAW*

The electrical force on a charge is related to the electric field (caused by other charges) by the now well-worn equation

$$\vec{F}_{\text{elec}} = q\vec{E}$$

Just like the gravitational force, the static electrical force is a conservative force. This means that the work done by the static electrical force around any *closed* path is zero:

$$q \int_{\text{clsd path}} \vec{E} \cdot d\vec{r} = 0 \text{ J}$$

Hence we have

$$\int_{\text{clsd path}} \vec{E} \cdot d\vec{r} = 0 \text{ V} \qquad (20.26)$$

In other words, the integral of the static (time-independent) electric field around a closed path is zero.

What about the integral of the magnetic field around a closed path?[†] That is, we want to determine the value of

$$\int_{\text{clsd path}} \vec{B} \cdot d\vec{r}$$

Here we have to be careful. The quantity $\vec{B} \cdot d\vec{r}$ does not represent some physical quantity, and certainly not work. Although the static magnetic *force* does no work on a moving charge, we *cannot* conclude that the path integral of the magnetic *field* around a closed path is zero, since the magnetic field and the magnetic force point in different directions: they are perpendicular to each other. We are just curious about what this analogous line integral amounts to. To see what the path integral of the magnetic field is around a closed contour, we look at a specific example and then generalize from what we learn.

For this purpose, then, consider the relatively simple magnetic field of an infinitely long current-carrying wire (Example 20.11). The magnetic field of the wire at a distance d from the wire is of magnitude

$$B = \frac{\mu_0}{4\pi} \frac{2I}{d}$$

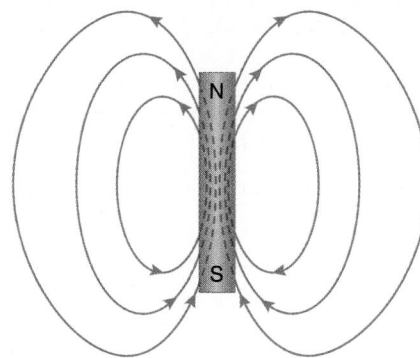

FIGURE 20.44 The magnetic field near a bar magnet.

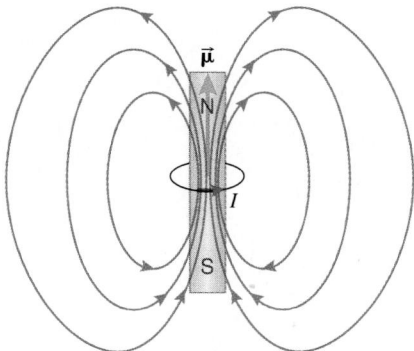

FIGURE 20.45 The magnetic field of a current loop can be imagined as like that of a bar magnet.

[†]We use the term closed path or, equivalently, closed contour rather than the term closed loop here because the path or contour can be a mathematical path rather than a physical entity such as a current loop. We reserve the term loop for current loops.

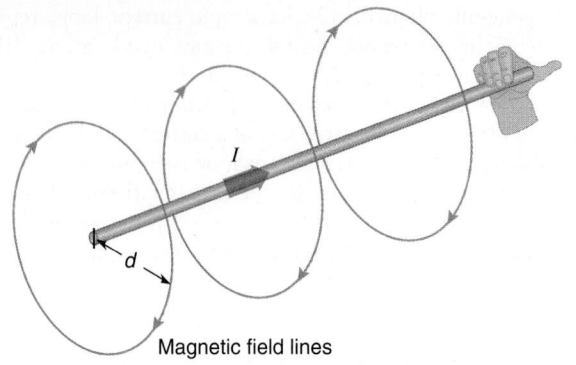

FIGURE 20.46 Integrate the magnetic field around a circle centered on the wire.

The direction of the field is found from the right-hand rule: grab the wire with your right hand with the thumb in the direction of the current. Your curled fingers then circle the wire in the sense of the field; see Figure 20.46.

Case 1: A Circular Path

We integrate the field clockwise around a closed circular path of radius d when viewed along the direction of the current, as shown in Figure 20.46. The magnetic field $\vec{\mathbf{B}}$ is parallel to $d\vec{\mathbf{r}}$ at all points along the path, and so the scalar product in the integrand of the path integral becomes

$$\int_{\text{clsd path}} \vec{\mathbf{B}} \cdot d\vec{\mathbf{r}} = \int_{\text{clsd path}} B \, dr \cos 0° = \int_{\text{clsd path}} B \, dr$$

The magnitude of the field also is constant all along the chosen path, so that

$$\int_{\text{clsd path}} \vec{\mathbf{B}} \cdot d\vec{\mathbf{r}} = B \int_{\text{clsd path}} dr$$

The integral of dr around the entire path is the circumference of the path: $2\pi d$. Thus

$$\int_{\text{clsd path}} \vec{\mathbf{B}} \cdot d\vec{\mathbf{r}} = B(2\pi d)$$

Substituting for the magnitude of the field, we find

$$\int_{\text{clsd path}} \vec{\mathbf{B}} \cdot d\vec{\mathbf{r}} = \frac{\mu_0}{4\pi} \frac{2I}{d} 2\pi d$$
$$= \mu_0 I$$

Clearly the result for the static magnetic field is not like the static electric field result (Equation 20.26). Something new is happening.

If we integrate the field around the circular path in a direction opposite to the field, the scalar product in the integrand then is

$$\int_{\text{clsd path}} \vec{\mathbf{B}} \cdot d\vec{\mathbf{r}} = \int_{\text{clsd path}} B \, dr \cos 180°$$
$$= -\int_{\text{clsd path}} B \, dr$$

and the result is negative:

$$\int_{\text{clsd path}} \vec{\mathbf{B}} \cdot d\vec{\mathbf{r}} = -\mu_0 I$$

Are these results peculiar to this example or to this path? We check out another path.

Case 2: A Path with Circular and Radial Segments

We choose the path indicated in Figure 20.47, which consists of four segments: (1) a circular segment of radius d; (2) a radial segment extending from radius d to radius r; (3) a circular segment at radius r; and (4) a radial segment from radius r back to radius d, thus completing this closed path.

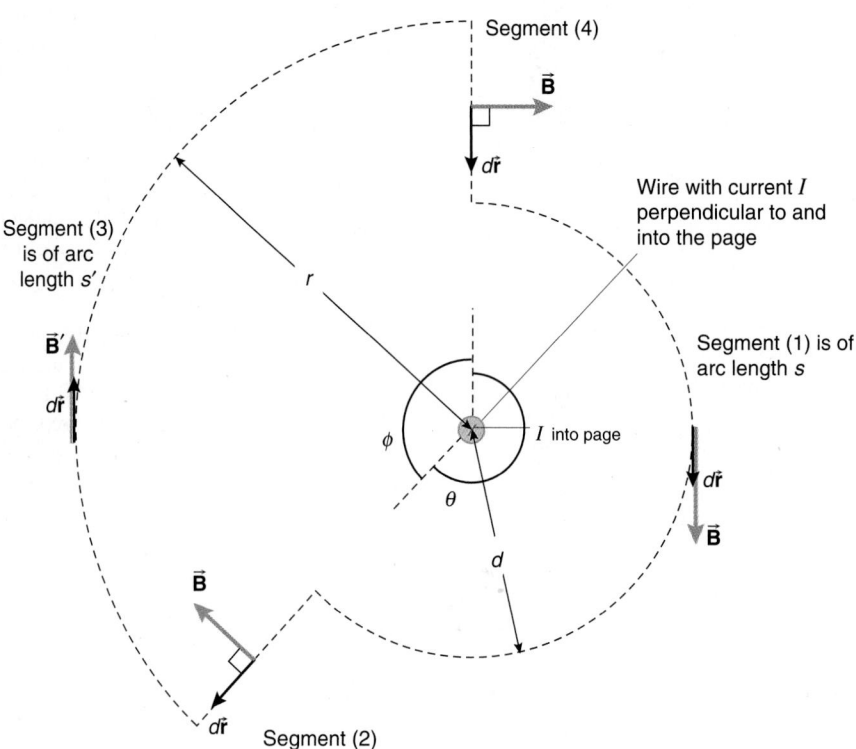

FIGURE 20.47 A more complicated contour around the wire.

The path integral of $\vec{B} \cdot d\vec{r}$ over the whole path breaks down into the integral over each of the four segments:

$$\int_{\text{whole clsd path}} \vec{B} \cdot d\vec{r} = \int_{\text{segment (1)}} \vec{B} \cdot d\vec{r} + \int_{\text{segment (2)}} \vec{B} \cdot d\vec{r}$$
$$+ \int_{\text{segment (3)}} \vec{B} \cdot d\vec{r} + \int_{\text{segment (4)}} \vec{B} \cdot d\vec{r}$$

(20.27)

Along the radial segments (2) and (4), the magnetic field $\vec{B}$ is perpendicular to $d\vec{r}$, making their scalar product zero in the integrands along these segments. No contribution comes from the radial segments. Along segment (1), the magnetic field is parallel to $d\vec{r}$, and so

$$\int_{\text{segment (1)}} \vec{B} \cdot d\vec{r} = \int B \, dr$$

Since the circular segment is centered on the wire, B is constant along this path, giving

$$\int_{\text{segment (1)}} \vec{B} \cdot d\vec{r} = B \int dr$$
$$= Bs$$

where s is the arc length of segment (1). This arc length s is related to the angle θ; indeed, from the definition of a radian, the angle θ in radians is

$$\theta = \frac{s}{d}$$

so that

$$s = \theta d$$

Thus the line integral along this segment is

$$\int_{\text{segment (1)}} \vec{B} \cdot d\vec{r} = Bs$$
$$= B\theta d$$

The magnitude of $\vec{B}$ along this segment is

$$B = \frac{\mu_0}{4\pi} \frac{2I}{d}$$

Thus the line integral along this segment is

$$\int_{\text{segment (1)}} \vec{B} \cdot d\vec{r} = B\theta d$$
$$= \frac{\mu_0}{4\pi} \frac{2I}{d} \theta d$$
$$= \frac{\mu_0}{4\pi} 2I\theta$$

Now for segment (3). Along this segment the magnetic field is different in magnitude than along segment (1). Call the magnetic field $\vec{B}'$ along segment (3). We use the same reasoning as for segment (1) to evaluate the line integral. The field is parallel to $d\vec{r}$; the magnitude B' of the field also is constant along this segment. So the path integral along this segment is

$$\int_{\text{segment (3)}} \vec{B}' \cdot d\vec{r} = \int_{\text{segment (3)}} B' \, dr$$
$$= B' \int_{\text{segment (3)}} dr$$

The integral of dr along this segment is the arc length s'. But the angle ϕ is

$$\phi = \frac{s'}{r}$$

so that

$$\int_{\text{segment (3)}} \vec{B}' \cdot d\vec{r} = B's'$$
$$= B'\phi r$$

The magnitude of $\vec{B}'$ is

$$B' = \frac{\mu_0}{4\pi} \frac{2I}{r}$$

and so

$$\int_{\text{segment (3)}} \vec{B}' \cdot d\vec{r} = B'\phi r$$
$$= \frac{\mu_0}{4\pi} \frac{2I}{r} \phi r$$
$$= \frac{\mu_0}{4\pi} 2I\phi$$

Substituting the results for each segment into Equation 20.27 for the complete path, we find

$$\int_{\text{clsd path}} \vec{B} \cdot d\vec{r} = \frac{\mu_0}{4\pi} 2I\theta + 0 \text{ T·m} + \frac{\mu_0}{4\pi} 2I\phi + 0 \text{ T·m}$$
$$= \frac{\mu_0}{4\pi} 2I(\theta + \phi)$$

But geometrically $\theta + \phi$ is 2π. Thus we obtain

$$\int_{\text{clsd path}} \vec{B} \cdot d\vec{r} = \mu_0 I$$

which is the same result we obtained for the circular contour in Case 1!

Case 3: Paths with Many Circular and Radial Segments

The arguments of Case 2 can be extended to more complex paths encircling the wire such as a closed path consisting of alternating circular and radial segments, as in Figure 20.48.

No contribution to the path integral is obtained from the radial segments (along which the field is perpendicular to $d\vec{r}$), and the sum of the contributions from the circular segments once again will sum to just $\mu_0 I$. Amazing. It is also apparent that if the plane of the path is not perpendicular to the wire, path segments extending parallel to the wire axis would always be perpendicular to $\vec{B}$ and therefore make no contribution to the line integral.

Case 4: Any Closed Path Around the Wire

An arbitrary closed path encircling the wire can be approximated by a series of alternating radial and circular segments, as shown in Figure 20.49.

As the size of these segments becomes smaller and smaller and their number increases indefinitely, the segmented path approaches the actual path as a limit. Thus, for any arbitrary closed path encircling the wire, we still get

$$\int_{\text{clsd path}} \vec{B} \cdot d\vec{r} = \mu_0 I$$

Notice that it makes no difference where the closed path lies relative to the current going through it or, from the opposite

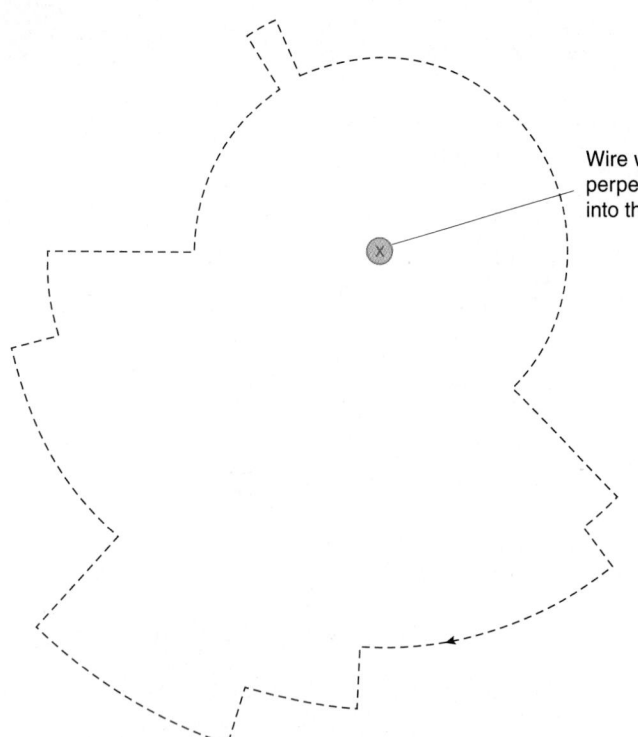

FIGURE 20.48 A complicated path consisting of radial and circular segments.

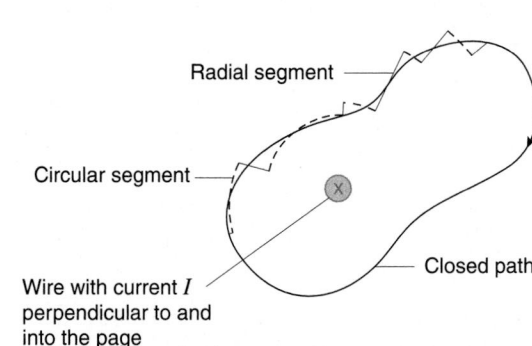

FIGURE 20.49 An arbitrary closed path encircling the wire.

viewpoint, where the current is located within a given path. The essential thing is that the current *threads through* the area bounded by the closed path chosen.

We need to explain more precisely just what is meant by the term: "current threading the path (or contour)." The term means that the current pierces *any surface* that has the contour of the path as a boundary (see Figure 20.50), like open soap bubble surfaces before they leave the rim of the ringlike contour used to blow them. The path (contour) need not be planar. Another way to think of the path is as the *perimeter* of various hat-shaped surfaces; there is a wide variety of hat surfaces for a given hat size (perimeter), as a visit to a hat department in a department store easily will confirm! To thread the path, the current must pierce the bubble surface or hat. The same path can bound many different bubble-shaped or hat-shaped surfaces.

Strictly speaking, the current must pierce the hat-shaped surface an *odd* number of times (see Figure 20.51).*

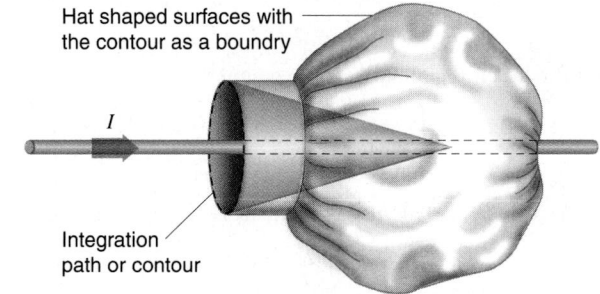

FIGURE 20.50 Hat-shaped surfaces with the contour as boundary.

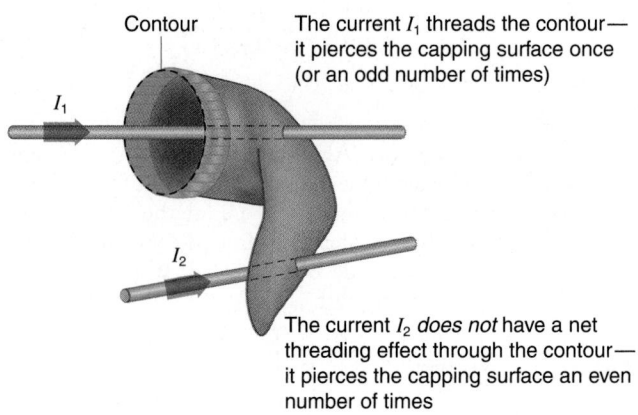

FIGURE 20.51 The current must pierce the surface an odd number of times.

---

*When the same current pierces the surface an even number of times, the net threading effect is zero.

Case 5: A Path Not Threaded by the Current

What happens if the current does *not* thread the path around which we integrate the field, as in Figure 20.52?

There are four segments to the path depicted. No contribution is obtained along the radial segments because the field is perpendicular to $d\vec{r}$ at each point along these segments. Along segment (1) we will get

$$\frac{\mu_0}{4\pi} \frac{2Is}{d}$$

or, since $s = \theta d$,

$$\frac{\mu_0}{4\pi} 2I\theta$$

as we did with case (2). But along segment (3), the field and $d\vec{r}$ are directed opposite to each other. The scalar product in the integrand is

$$\vec{B} \cdot d\vec{r} = B \, dr \cos 180°$$
$$= -B \, dr$$

The result we get will be negative. That is, along segment (3) the result of the integration is

$$\int_{\text{segment (3)}} \vec{B} \cdot d\vec{r} = -\frac{\mu_0}{4\pi} 2I \frac{s'}{r}$$

or, since $s' = \theta r$,

$$= -\frac{\mu_0}{4\pi} 2I\theta$$

The sum of the contributions from the four segments [only segments (1) and (3) contribute nonzero terms] is

$$\int_{\text{clsd path}} \vec{B} \cdot d\vec{r} = \frac{\mu_0}{4\pi} 2I\theta - \frac{\mu_0}{4\pi} 2I\theta$$
$$= 0 \ \text{T·m}$$

Any closed path not encircling the current can be broken up into alternating radial and circular segments as we did for paths circling the current. But if the path does not encircle the current, the grand sum of all the contributions from the segments inevitably will be zero.

Now we summarize all these results into a single statement. The path integral of the static magnetic field around any closed contour,

$$\int_{\text{clsd path}} \vec{B} \cdot d\vec{r}$$

we place on the left-hand side of an equation. The integration is performed by going around the path in either direction.

1. If the current threads the path in such a direction that the magnetic field produced by the current is directed around the wire in the *same* sense as the direction of the path integration, then the path integral gives a contribution of $+\mu_0 I$.
2. If the current threads the path so that the magnetic field produced by the current is directed around the wire in the *opposite* sense to the direction of the path integration, the path integral gives a contribution of $-\mu_0 I$.
3. If the current does not thread the path of the integration, nothing (zero) is contributed to the path integral.

All the verbiage can be elegantly condensed into the following mathematical statement, another example of how we use mathematics to cogently codify much information:

$$\int_{\text{clsd path}} \vec{B} \cdot d\vec{r} = \mu_0 I_{\text{net current threading the path}} \qquad (20.28)$$

This is known as **Ampere's law**.

Although we derived this result using the magnetic field of an infinitely long wire, the result is a very general one and applies to *any static magnetic field*.

If multiple (steady) currents are present simultaneously, as in Figure 20.53, the path integral of the total magnetic field around a closed path is equal to μ_0 times the algebraic sum of the currents threading the path.

If you wrap the fingers of your right hand around the closed path in the direction of the path integration, those currents directed through the path in the direction of your thumb appear on the right-hand side of Ampere's law with a plus sign; those in the opposite direction appear with a minus sign.

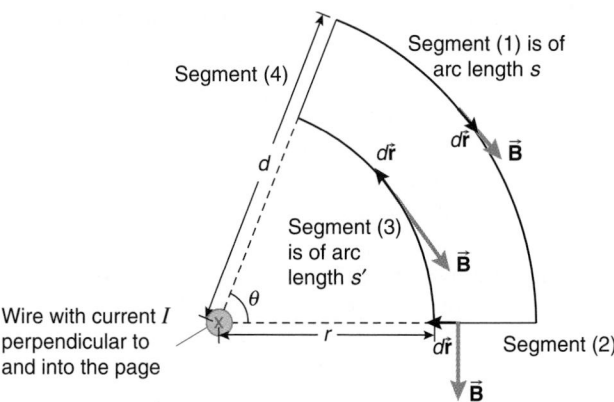

FIGURE 20.52 Zero current threads this contour.

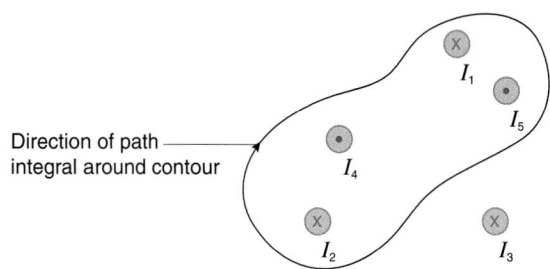

FIGURE 20.53 Multiple currents threading the path.

Thus, in Figure 20.53, currents I_1 and I_2 appear with a plus sign; currents I_4 and I_5 appear with a minus sign. Current I_3 does not thread the path and so does not contribute to the right-hand side of Ampere's law. Ampere's law for the path in Figure 20.53 thus reads

$$\int_{\text{clsd path}} \vec{\mathbf{B}} \cdot d\vec{\mathbf{r}} = \mu_0\left(I_1 + I_2 - I_4 - I_5\right)$$

To perform the actual path integration on the left-hand side of this equation typically is very difficult; but the result of the integration is easy to find because the right-hand side of Ampere's law is the simple matter of an algebraic sum of the currents threading the path. In a few cases, where the magnetic field is constant along the sections of a chosen closed path, Ampere's law often can provide a breathtaking way to calculate the field quickly (see Examples 20.12 and 20.13).

QUESTION 4

Compare and contrast Gauss's law for the static electric field with Ampere's law for the static magnetic field.

EXAMPLE 20.12

Use Ampere's law to find the magnetic field at a distance d from an infinite straight wire carrying current I.

Solution

This was effectively done with Case 1, but let's summarize the argument here. By symmetry, the magnitude of the magnetic field at a fixed distance d from the wire is constant. Hence choose a circular contour of radius d, centered on the wire coincident with a magnetic field line. Integrate around the contour in the same sense as the direction of the magnetic field, as in Figure 20.54.

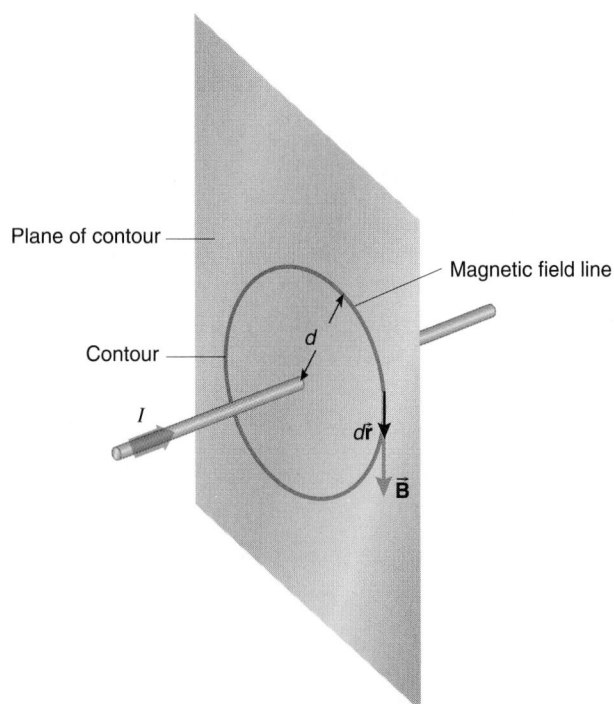

Plane of contour

Magnetic field line

Contour

d

I

$d\vec{\mathbf{r}}$

$\vec{\mathbf{B}}$

FIGURE 20.54

Note that the current threading the contour is the current I in the wire. Ampere's law here is

$$\int_{\substack{\text{circular} \\ \text{contour}}} \vec{\mathbf{B}} \cdot d\vec{\mathbf{r}} = \mu_0 I$$

The magnetic field is parallel to each differential $d\vec{\mathbf{r}}$ along the contour; hence $\vec{\mathbf{B}} \cdot d\vec{\mathbf{r}} = B\,dr$, and you have

$$\int_{\substack{\text{circular} \\ \text{contour}}} B\,dr = \mu_0 I$$

Since the magnetic field has a constant magnitude along the chosen contour, the integral becomes

$$B\int_{\substack{\text{circular} \\ \text{contour}}} dr = \mu_0 I$$

The integral of dr around the circular contour is the circumference of the contour: $2\pi d$. Hence Ampere's law yields

$$B(2\pi d) = \mu_0 I$$

Solving for B, you find the magnetic field has a magnitude given by

$$B = \frac{\mu_0}{2\pi}\frac{I}{d}$$

In order to keep

$$\frac{\mu_0}{4\pi}$$

as a unit (since its value is 10^{-7} T·m/A), rewrite the field magnitude as

$$B = \frac{\mu_0}{4\pi}\frac{2I}{d}$$

You obtained the same result in Example 20.11 using a more complicated vector integration over the current elements along the wire.

STRATEGIC EXAMPLE 20.13

A long wire wound into a helical shape whose length is typically much greater than its diameter is known as a *solenoid*; see Figure 20.55.

FIGURE 20.55

When the wire carries a current I, a magnetic field is created within the solenoid. Use Ampere's law to calculate the field inside the solenoid, far from either end.

Solution

First consider a solenoid with the individual loops of the wire somewhat separated from each other, as in Figure 20.56. For points very close to each turn, the magnetic field will be approximately that of a long wire and will circle the wire as shown.

Notice that in the space between the loops, neighboring loops have fields between them that are in opposite directions.

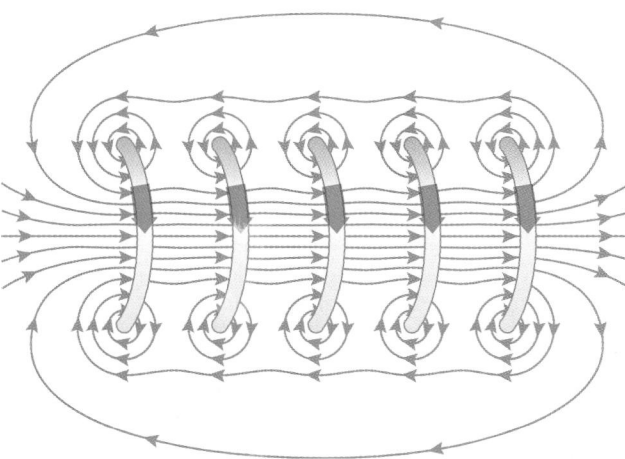

FIGURE 20.56

If the solenoid is tightly wound, as in Figure 20.57, the close-in fields between the loops essentially vector sum to zero. The resulting magnetic field inside the solenoid then is quite uniform and is directed along the axis of the solenoid.

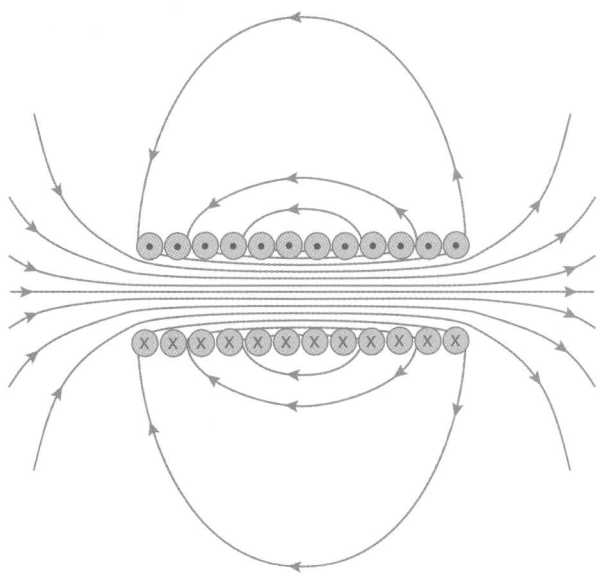

FIGURE 20.57

The magnetic field at points outside the solenoid is quite small compared with the field inside the solenoid. For points near the center of a long solenoid, you can consider the field outside the solenoid to be essentially zero and the field inside the solenoid to be directed along the axis and uniform across the cross-sectional area of the solenoid. These approximations become exact for a close-wound solenoid of infinite length. In practice, as long as the length of the solenoid is much greater than its diameter and you are far from the ends, these approximations are quite good.

Use Ampere's law to calculate the magnetic field inside the solenoid. Ampere's law states that

$$\int_{\text{clsd path}} \vec{\mathbf{B}} \cdot d\vec{\mathbf{r}} = \mu_0 I_{\text{net current threading the path}}$$

To find the magnetic field, you need to choose a path or contour around which to integrate the magnetic field. How do you go about choosing an appropriate path? To explicitly evaluate the left-hand side of the law, you need to pick a path that makes the integration easy. Such paths are those along which the magnetic field is constant over part or all of the path. The shape of the path sometimes reflects the symmetry of the particular situation. A couple of choices might seem appropriate.

Choice 1: A circular path centered on the axis of the solenoid, as shown in Figure 20.58. The magnetic field certainly is constant over the entire contour. The only problem with this choice is that $d\vec{\mathbf{r}}$ along the contour is everywhere perpendicular to the magnetic field. Thus the scalar product $\vec{\mathbf{B}} \cdot d\vec{\mathbf{r}}$ in the integrand on the left-hand side of Ampere's law is zero everywhere. The left-hand side of Ampere's law thus is zero for this contour. Notice that the right-hand side also is zero because no current threads the contour. This shows $0\ \text{T} \cdot \text{m} = 0\ \text{T} \cdot \text{m}$, which is nice and true, but not too informative. So this choice of a contour is not appropriate.

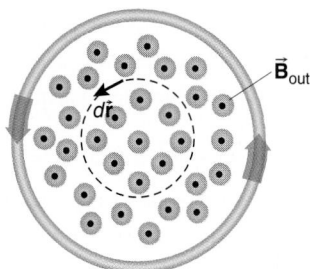

End-on view of solenoid

FIGURE 20.58

Choice 2: A rectangular path within the solenoid, as indicated in Figure 20.59. There are four segments of the rectangular path.

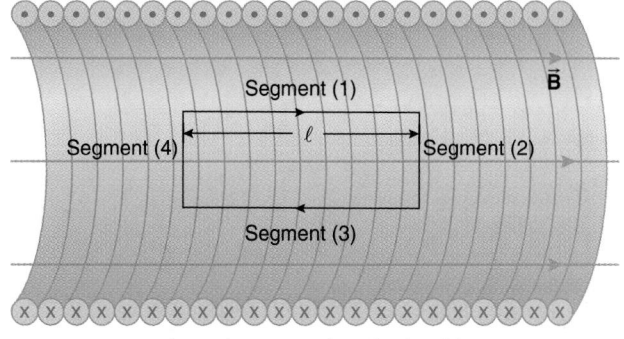

Lateral cross section of solenoid

FIGURE 20.59

Along segments (2) and (4), the magnetic field is perpendicular to $d\vec{r}$, so that the scalar product $\vec{B} \cdot d\vec{r}$ is zero along these segments. Along segment (1), $\vec{B}$ is parallel to $d\vec{r}$, so that the integral of $\vec{B} \cdot d\vec{r}$ along this segment is

$$\int_{\text{segment (1)}} \vec{B} \cdot d\vec{r} = \int_{\text{segment (1)}} B \, dr \cos 0°$$
$$= \int_{\text{segment (1)}} B \, dr$$

The magnitude of the field is constant along this segment, and so you have

$$\int_{\text{segment (1)}} \vec{B} \cdot d\vec{r} = B \int_{\text{segment (1)}} dr$$
$$= B\ell$$

Along segment (3), $\vec{B}$ is antiparallel to $d\vec{r}$, and so along this segment you have

$$\int_{\text{segment (3)}} \vec{B} \cdot d\vec{r} = \int_{\text{segment (3)}} B \, dr \cos 180°$$
$$= -\int_{\text{segment (3)}} B \, dr$$

Once again the field is constant over the length of the segment, so that

$$\int_{\text{segment (3)}} \vec{B} \cdot d\vec{r} = -B \int_{\text{segment (3)}} dr$$
$$= -B\ell$$

Hence the left-hand side of Ampere's law is

$$B\ell + 0 \text{ T·m} - B\ell + 0 \text{ T·m} = 0 \text{ T·m}$$

The right-hand side of Ampere's law also is zero since the contour has no current threading the path. Once again this is an inappropriate path. So ... these choices illustrate that to get anything out of Ampere's law you need to have some current threading the closed path chosen!

Choice 3: Choose a rectangular path that cuts through the side of the solenoid, as indicated in Figure 20.60.

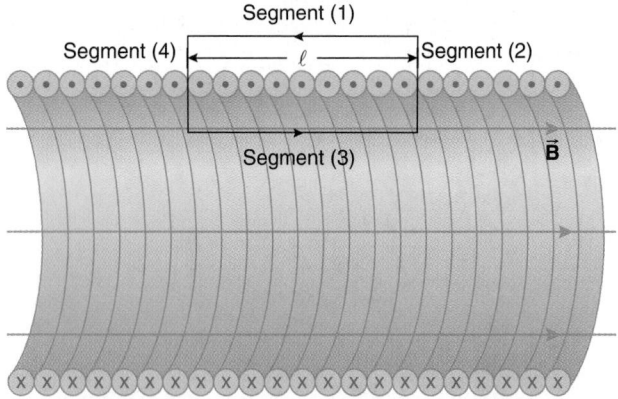

Segment (1)
Segment (4) ℓ Segment (2)

Segment (3) $\vec{B}$

FIGURE 20.60

Now at least you have some current threading the path, so that the right-hand side of Ampere's law will not be zero. To evaluate the right-hand side of Ampere's law, you need to know

how many wires thread the path; each wire carries a current *I*. Let *n* be the number of turns of wire that are wound on each meter of the solenoid. The number of turns of wire in ℓ meters then is just $n\ell$. If you integrate around the contour in the counterclockwise sense in Figure 20.60, each of the currents is a positive current on the right-hand side according to the convention established with Ampere's law. Thus the right-hand side of Ampere's law is

$$\mu_0 n\ell I$$

To evaluate the left-hand side of the law, integrate the magnetic field around all four segments of the path. Along segments (2) and (4), the magnetic field is either perpendicular to $d\vec{r}$ (for points on these segments inside the solenoid) or zero (for points outside the solenoid). Thus there is no contribution to the path integral from these segments. Likewise, the magnetic field is zero along segment (1) because you are outside the solenoid and the field is approximately zero there.

Along segment (3), the magnetic field is parallel to $d\vec{r}$ and constant over the length ℓ of the path. The integration along this segment then is

$$\int_{\text{segment (3)}} \vec{B} \cdot d\vec{r} = \int_{\text{segment (3)}} B \, dr = B \int_{\text{segment (3)}} dr = B\ell$$

So the entire left-hand side of Ampere's law is just $B\ell$. Equate the left-hand side of the law with the right-hand side:

$$\int_{\text{clsd path}} \vec{B} \cdot d\vec{r} = \mu_0 I_{\text{net current threading the path}}$$
$$B\ell = \mu_0 n\ell I$$

Solving for the magnitude of the magnetic field, you find
$$B = \mu_0 nI \qquad (1)$$

EXAMPLE 20.14 ━━━━━━━━━━

Using Ampere's law, find the magnetic field produced by the current in a long wire wrapped around a torus (or doughnut), as shown in the cross-sectional view of Figure 20.61.

Solution
Like the solenoid (Example 20.13), the magnetic field is directed along (around) the toroid. An appropriate closed contour along which to evaluate the left-hand side of Ampere's law is a circle of radius *r*, as indicated in Figure 20.62.

The symmetry implies that the magnitude of the magnetic field is constant along the length of the path. Perform the path integration in the same directional sense as the field. Thus the left-hand side of Ampere's law becomes

$$\int_{\substack{\text{circular} \\ \text{contour}}} \vec{B} \cdot d\vec{r} = \int_{\substack{\text{circular} \\ \text{contour}}} B \, dr \cos 0° = B(2\pi r)$$

The right-hand side of Ampere's law involves summing the currents threading the contour. The number of current-carrying wires threading the contour is equal to the total number of turns of wire on the entire torus; call this number *N*. Then the total current threading the contour is *NI*. Equating the left- and right-hand sides of Ampere's law, you get

$$B(2\pi r) = \mu_0 NI$$

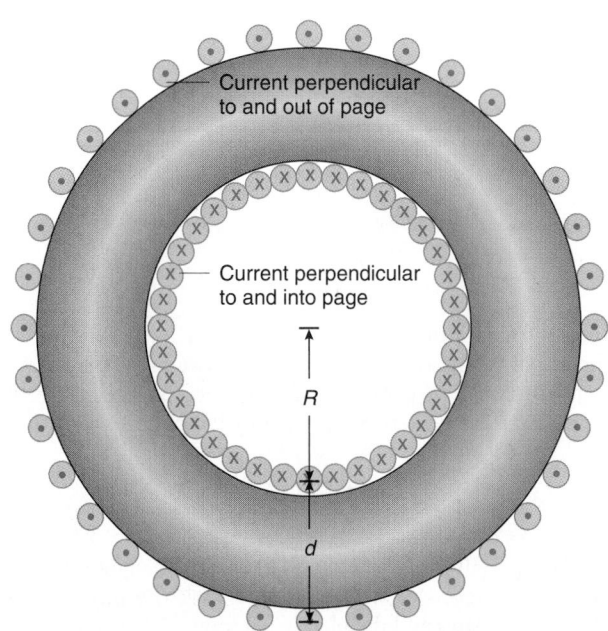

Current perpendicular
to and out of page

Current perpendicular
to and into page

R

d

FIGURE 20.61

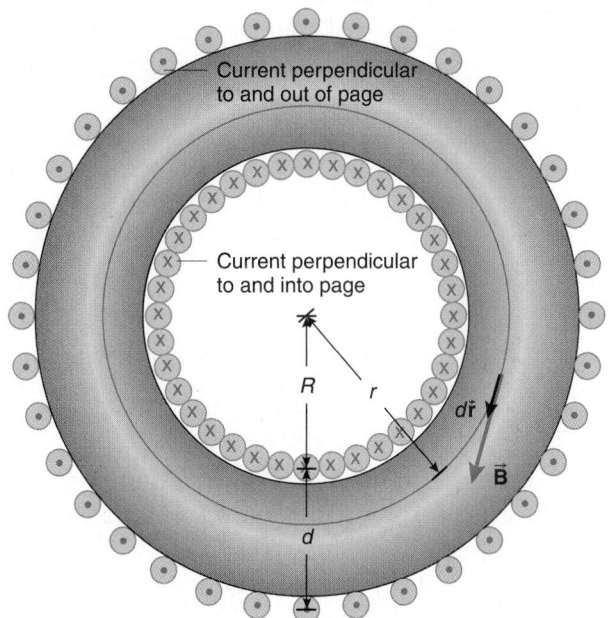

Current perpendicular
to and out of page

Current perpendicular
to and into page

R r

$d\vec{r}$

$\vec{B}$

d

FIGURE 20.62

Solving for the magnitude of the magnetic field, you find

$$B = \frac{\mu_0 NI}{2\pi r} \qquad (1)$$

The magnitude of the magnetic field is not constant over the cross-sectional area of the torus because of the dependence on the radius r. For a torus with a very large radius R when compared with the diameter d of the ring, the magnitude of the field approaches that of a long solenoid (and is essentially

constant over the diameter of the cross section). When R is large, then,

$$\frac{N}{2\pi r} \approx \frac{N}{2\pi R} \approx n = \text{number of turns per meter}$$

and equation (1) for the toroidal geometry becomes

$$B = \mu_0 nI \qquad (2)$$

the same as that for a long solenoid (Example 20.13).

20.11 THE DISPLACEMENT CURRENT AND THE AMPERE–MAXWELL LAW*

The magnetic energy, as developed in the mariner's needle, is, as all know, essentially one with the electricity beheld in heaven. . . .
Herman Melville (1819–1891)[†]

In Section 19.16 we considered what happens when a source of emf (an independent voltage source) is connected to a series combination of a resistor and a capacitor, as in Figure 20.63. With an initially uncharged capacitor and the switch open, there is no current and so no magnetic field is present. When the switch is closed, a current exists for a brief interval in order to charge the capacitor. Such a current produces a magnetic field. Eventually, when the capacitor becomes fully charged, there is no more current and the magnetic field disappears. This brief magnetic field thus is time dependent because the current is a function of time.

Let's look more closely at the charging process, the temporary current, the magnetic field produced by this time-dependent current, and the effect of the time-dependent current and magnetic field on Ampere's law (Section 20.10). This sounds like quite a task! Specifically, consider the charging of the parallel plate capacitor shown in Figure 20.64, extracted from the circuit of Figure 20.63 for clarity. Ampere's law states that the path integral of the magnetic field around a closed contour is equal to μ_0 times the current threading the path. By threading the path we mean piercing any hat-shaped surface that has the path as its boundary or perimeter. We choose the path indicated in Figure 20.64.

During the charging process, a time-dependent conduction current exists in the wire, and this current is equal to the time

[†]*Moby Dick*, Chapter 124, "The Needle." First published in 1851 (University of California Press, Berkeley, 1979).

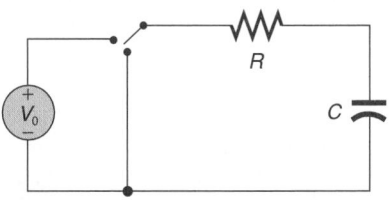

V_0

R

C

FIGURE 20.63 A series RC circuit.

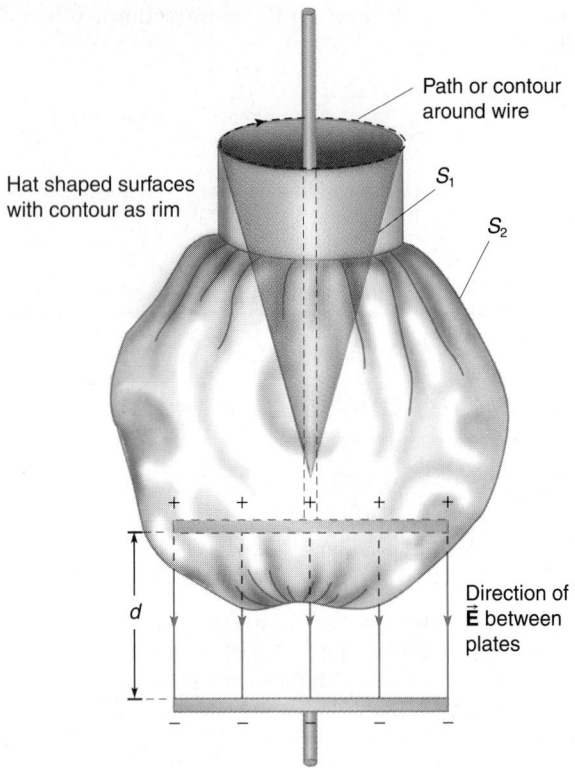

Path or contour around wire

Hat shaped surfaces with contour as rim

S_1

S_2

Direction of $\vec{E}$ between plates

d

FIGURE 20.64 A path around the wire for Ampere's law.

rate at which the charge $Q(t)$ on the capacitor is changing. From the general definition of the current, we have

$$I = \frac{dQ(t)}{dt}$$

This current pierces the dunce cap–shaped surface S_1 indicated in Figure 20.64, and so the right-hand side of Ampere's law is not zero. The left-hand side of Ampere's law therefore is not zero, indicating the presence of a magnetic field surrounding the wire. Nothing new here.

But now we examine the chef's hat–shaped surface S_2 in Figure 20.64, which passes between the plates of the capacitor and has the same path (contour) for its perimeter as the surface S_1. Ampere's law states that the threading current is the current through *any* surface with the contour or path as a boundary or perimeter. The plates of a capacitor are electrically insulated from one another, so that no conduction current can pass between the plates of the capacitor.

We *know* the left-hand side of Ampere's law is not zero (from the previous argument using the dunce cap surface S_1), but the conduction current does not pierce the chef's hat surface S_2. This fact means that the right-hand side of Ampere's law apparently is zero while the left-hand side is not! There is something funny going on here. Something needs to be done (i.e., cooked up) to the right-hand side of Ampere's law to eliminate the paradoxical result with the chef's hat surface S_2. It was James Clerk Maxwell (1831–1879) who first realized what had to be done to eliminate this apparent inconsistency on the right-hand side of Ampere's law. Here is Maxwell's idea.

When the capacitor has a charge Q on its plates, the potential difference V between the plates is found from the definition of the capacitance:

$$C = \frac{Q}{V}$$

We rearrange this slightly:

$$Q = CV$$

Since the capacitance of the capacitor is fixed (and only a function of geometry and the material between the plates, here assumed to be a vacuum), the time-varying charge means the potential difference between the plates also changes with time:

$$\frac{dQ}{dt} = C\frac{dV}{dt} \qquad (20.29)$$

The potential difference V between the plates can be expressed in terms of the magnitude E of the electric field between the plates and the plate separation d (Example 17.1):

$$V = Ed$$

We substitute this into Equation 20.29, obtaining (for fixed plate separation d)

$$\frac{dQ}{dt} = C\frac{dE}{dt}d \qquad (20.30)$$

The capacitance of a parallel plate capacitor is found from Example 18.3:

$$C = \varepsilon_0\frac{A}{d}$$

where A is the area of a plate of the capacitor. We make this substitution for C into Equation 20.30:

$$\frac{dQ}{dt} = \varepsilon_0\frac{A}{d}\frac{dE}{dt}d$$
$$= \varepsilon_0 A\frac{dE}{dt}$$

Since the area of the plates is fixed, we can write this as

$$\frac{dQ}{dt} = \varepsilon_0\frac{d}{dt}(EA) \qquad (20.31)$$

Maxwell realized that the quantity EA was the *flux of the electric field* through the chef's hat surface S_2. Although S_2 extends beyond the plates of the capacitor, the electric field (ideally) is nonzero only between the plates of the capacitor (neglecting edge effects), so that the flux of $\vec{E}$ through the surface S_2 arises only from that portion of S_2 that lies between the capacitor plates. Let Φ_{elec} be the flux of the electric field vector through this surface. Equation 20.31 then is written as

$$\frac{dQ}{dt} = \varepsilon_0\frac{d\Phi_{elec}}{dt} \qquad (20.32)$$

Maxwell regarded the right-hand side of this equation as a new type of current I_D through S_2. Its dimensions and units were right. Namely, he defined the right-hand side of Equation 20.32 to be a **displacement current** I_D:

$$I_D \equiv \varepsilon_0 \frac{d\Phi_{elec}}{dt} \quad (20.33)$$

The name is a bit unfortunate, for it has nothing to do with displacements of any kind. Historically, Maxwell called it the displacement current under the erroneous assumption that it had to do with a real displacement of charges rather than fields; the term is another instance of historical inertia in the notation and terminology of physics. Equation 20.32 indicates that the displacement current through S_2, in quantity and directional sense, is equal to the ordinary conduction current

$$\frac{dQ}{dt}$$

that is present through S_1. The conduction current through S_1 arises from the physical transport of charge; the displacement current through S_2 arises from the time dependence of the flux of the electric field between the plates of the capacitor.

> The right-hand side of Ampere's law should include *both* conduction and displacement currents (if appropriate):
>
> $$\int_{clsd\ path} \vec{B} \cdot d\vec{r} = \mu_0(I + I_D)_{threading\ the\ path} \quad (20.34)$$
>
> This generalization is called the **Ampere–Maxwell law**.

For our example, through surface S_1 there is a conduction current I but no displacement current I_D. On the other hand, through surface S_2 there is a displacement current I_D, equal in quantity and direction to the conduction current, but no conduction current I.

> The crux of Maxwell's argument was the realization that magnetic fields are produced via two distinct mechanisms:
>
> 1. by electric charges in motion (conduction current); and
>
> 2. by time-varying electric fields (via the displacement current).

The Ampere–Maxwell law was a surprising breakthrough in physics and is the third of the four fundamental equations of electromagnetism known as the Maxwell equations.*

Returning to the specific example of the charging parallel plate capacitor, we apply the Ampere–Maxwell law to a circular contour located between the capacitor plates (with the hat-shaped area A' taken to be the plane of the contour, as shown in Figure 20.65. There is no conduction current in this region, and so the Ampere–Maxwell law involves only the displacement current:

$$\int_{clsd\ path} \vec{B} \cdot d\vec{r} = \mu_0 I_D$$

---

*The other two Maxwell equations that we have studied so far are Gauss's law for the electric field and Gauss's law for the magnetic field. We have one more Maxwell equation to go to complete the set; we study this additional equation in Chapter 21.

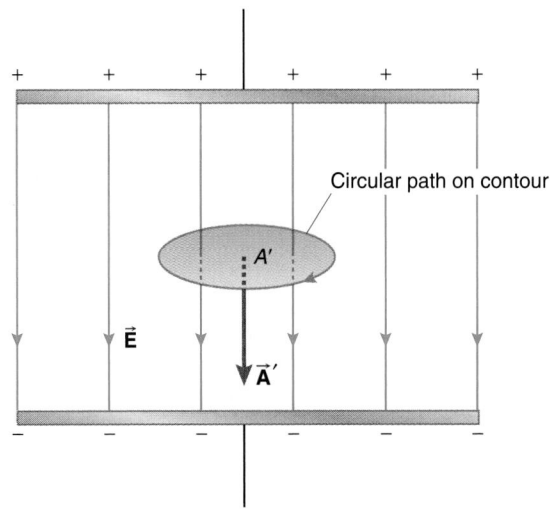

FIGURE 20.65 Apply the Ampere–Maxwell law to the indicated contour.

Using Equation 20.33 for the displacement current, this becomes

$$\int_{clsd\ path} \vec{B} \cdot d\vec{r} = \mu_0 \varepsilon_0 \frac{d\Phi_{elec}}{dt} \quad (20.35)$$

where the electric flux is taken through the area A' bounded by the contour. Since the electric field between the plates of a parallel plate capacitor is the same everywhere between the plates, the flux of the electric vector through the area of the contour is

$$\Phi_{elec} = \vec{E} \cdot \vec{A}'$$
$$= EA'$$

The direction of the area vector $\vec{A}'$ is determined by wrapping the fingers of your right hand around the contour in the direction taken around the path; thus, $\vec{A}'$ is parallel to the field direction. Substituting this expression for the electric flux into Equation 20.35, we have

$$\int_{clsd\ path} \vec{B} \cdot d\vec{r} = \mu_0 \varepsilon_0 \frac{d}{dt}(EA')$$

Since A' does not depend on the time, it comes through unscathed from the differentiation:

$$\int_{clsd\ path} \vec{B} \cdot d\vec{r} = \mu_0 \varepsilon_0 A' \frac{dE}{dt}$$

The displacement current is directed across the entire region between the capacitor plates; if the capacitor is charging, the electric field between the capacitor plates is increasing with time and the derivative

$$\frac{dE}{dt}$$

is positive. The displacement current (in Figure 20.65) is directed down (from the top plate to the bottom plate), but recognize that the displacement current occurs throughout the region between the plates. Place the thumb of your right hand along the direction of the displacement current; your right-hand fingers then indicate the sense of the magnetic field caused by the displacement current (see Figure 20.66).

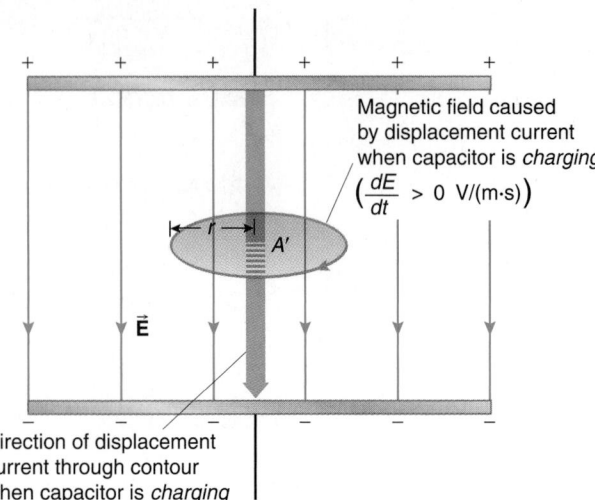

FIGURE 20.66 The direction of the displacement current and the magnetic field it creates for a charging capacitor.

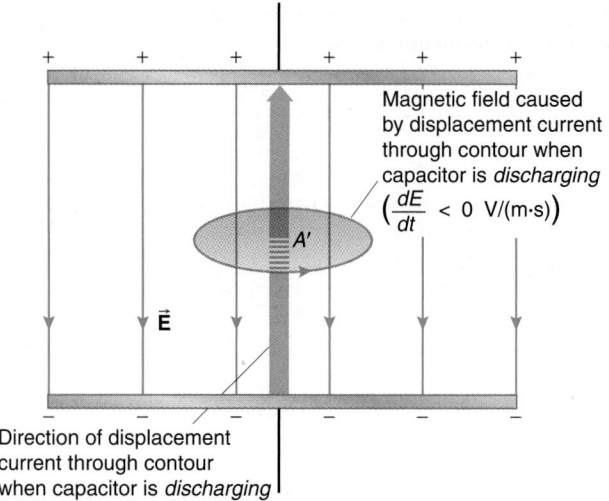

FIGURE 20.67 The direction of the displacement current and the magnetic field it creates for a discharging capacitor.

The cylindrical symmetry implies that the magnetic field has a constant magnitude along the circular contour. If we integrate around the contour in the same direction as the magnetic field, the Ampere–Maxwell law becomes

$$\int_{\text{clsd path}} \vec{B} \cdot d\vec{r} = \mu_0 I_D$$

$$B(2\pi r) = \mu_0 \varepsilon_0 A' \frac{dE}{dt}$$

Since $A' = \pi r^2$, we have

$$B(2\pi r) = \mu_0 \varepsilon_0 \pi r^2 \frac{dE}{dt}$$

$$B = \frac{\mu_0}{2} \varepsilon_0 r \frac{dE}{dt} \qquad (20.36)$$

This explicitly indicates how a changing electric field produces a magnetic field. Note, incidentally, that this magnetic field increases linearly with r, reflecting the fact that as we choose greater r, more displacement current is within the contour because the electric flux through the area of the contour increases.

Equation 20.36 also could be obtained by saying the displacement current is I_D and then using the expression for the magnetic field of a long wire, equation (3) of Example 20.11:

$$B = \frac{\mu_0}{4\pi} \frac{2I_D}{r} \qquad (20.37)$$

The displacement current I_D through the contour is

$$I_D = \varepsilon_0 \frac{d\Phi_{\text{elec}}}{dt}$$

$$= \varepsilon_0 \frac{d}{dt}(EA')$$

$$= \varepsilon_0 A' \frac{dE}{dt}$$

$$= \varepsilon_0 \pi r^2 \frac{dE}{dt}$$

Substituting for I_D into Equation 20.37 gives the same result as Equation 20.36.

This example indicates that a displacement current causes a magnetic field just as effectively as does a normal conduction current of moving charges.

If the capacitor is discharging, then the electric field between the capacitor plates is decreasing with time, and so the derivative

$$\frac{dE}{dt}$$

is negative. The displacement current then is directed upward through the contour (from the bottom plate to the top plate), as shown in Figure 20.67. The magnetic field generated by the displacement current then is in the direction opposite to that when the capacitor was charging.

Thus there is a magnetic field generated by the displacement current between the plates of the capacitor; this field is in one direction when the capacitor is charging and in the reverse direction when the capacitor is discharging, reflecting the different direction of the displacement current.

> Also note that the magnetic field induced by the displacement current is perpendicular to the (changing) electric field that causes it.

When the capacitor is fully charged (or fully discharged), the conduction current in the wires is zero, the displacement current between the plates of the capacitor is zero, and there is no magnetic field.

QUESTION 5

For the parallel plate capacitor, the displacement current I_D between the capacitor plates is equal in quantity and direction to the conduction current I in the wire feeding the plates. Explain why the right-hand side of the Ampere-Maxwell law should *not* be either $2\mu_0 I$ or $2\mu_0 I_D$.

20.12 MAGNETIC MATERIALS*

When we studied electric phenomena and electric fields, we began with electrical effects in vacuum and described the electrical properties of materials by changing the permittivity of free space (a vacuum) ε_0 to the permittivity of the material ε. The ratio of the permittivities was the dielectric constant κ of the material:

$$\kappa = \frac{\varepsilon}{\varepsilon_0}$$

A similar thing is done for the magnetic properties of most materials. The permeability of free space μ_0 changes to the **permeability of the material** μ. The ratio of the permeabilities is known as the **relative permeability**:

$$\kappa_m = \frac{\mu}{\mu_0}$$

(Do not confuse the permeability μ with the magnitude of the magnetic dipole moment vector $\vec{\boldsymbol{\mu}}$.)

Materials are classified magnetically according to the approximate size of the relative permeability (see Table 20.2):

1. **Diamagnetic materials** have κ_m slightly less than 1.
2. **Paramagnetic materials** have κ_m slightly greater than 1.
3. **Ferromagnetic materials** have κ_m significantly greater than 1.

In Example 20.8, we saw that an orbiting electron in an atom is a tiny current loop. The circulating electron thus has an orbital magnetic dipole moment. The electron also has an intrinsic magnetic dipole moment associated with its *spin* (if the spin is imagined literally as a rotating electron, the spin of the charged electron also produces a very tiny current loop). The total magnetic moment of an atom consists of an appropriate (quantum mechanical) vector sum of the orbital and spin magnetic moments of the electrons in the atom.

The protons and neutrons in the nucleus also have magnetic moments, but these are much smaller than the magnetic moments of the electrons because of the much greater mass of those nuclear particles. The (quantum mechanical) vector sum of the magnetic moments of the protons and neutrons in a nucleus produces a (small) nuclear magnetic dipole moment whose value depends on the specific element and isotope. The physics of the interaction of these nuclear magnetic moments

with external magnetic fields is the basis for important chemical and medical technologies such as nuclear magnetic resonance (NMR) and imaging techniques for noninvasive (i.e., nonsurgical) examination of soft tissues. Traditional x-ray imaging typically is inappropriate for soft tissues unless the tissue is doped with a strong x-ray absorber.

We will not go into how these various magnetic moments are combined, but such a summation can lead to two results:

1. The atom as a whole has a nonzero, permanent magnetic dipole moment. Paramagnetic and ferromagnetic materials are of this type.
2. The atom does not have a permanent magnetic dipole moment. Diamagnetic materials are of this type.

The presence of a nonzero magnetic moment in paramagnetic and ferromagnetic materials means that if the material is placed in an external magnetic field, the magnetic moments will tend to align *parallel* to the direction of the field to minimize their potential energies. The alignment is not complete because of random motions due to thermal energy (internal energy). The important factor is the size of the thermal energy per particle (proportional to kT) relative to the magnitude of the potential energy of its magnetic dipole moment in the external field (PE $= -\vec{\boldsymbol{\mu}} \cdot \vec{\mathbf{B}}$). The specific degree of the alignment depends on the particular material and the temperature.

The distinction between paramagnetic and ferromagnetic materials arises when the external field is removed. For paramagnetic materials, the individual magnetic moments of the atoms and molecules become randomly oriented after the external field is removed, and so the total magnetic moment of the material (the vector sum of the magnetic dipole moments of the particles) is zero.

For ferromagnetic materials, the magnetic dipole moments do not become completely randomized with the removal of the external field and a residual total magnetic moment remains: a permanent magnet is produced. A permanent magnet can have its magnetic dipoles completely randomized by increasing the temperature of the material, by sufficiently jarring the material (for instance, by repetitively dropping it on the floor or striking the material several times with a hammer), or by degaussing (using a coil with an opposing magnetic field to neutralize a given field),[†] thus making the total magnetic moment zero once again.

Diamagnetic materials have zero magnetic moment. When placed in a magnetic field, a nonzero magnetic moment is *induced* in the material (much the way electric polarization is induced via electric fields). The induced magnetic moment is directed *antiparallel* to the applied field, because opposite magnetic poles attract each other.

A detailed examination of the magnetism of materials is left for more advanced courses in electromagnetism.

20.13 THE MAGNETIC FIELD OF THE EARTH*

The Earth has a permanent magnetic field. A crude model of the field imagines a huge bar magnet embedded within the Earth, as

TABLE 20.2 Relative Permeability of Selected Materials

Material	κ_m
Diamagnetic materials	
Copper (Cu)	$1 - 9.4 \times 10^{-6}$
Gold (Au)	$1 - 2 \times 10^{-5}$
Mercury	$1 - 3.2 \times 10^{-5}$
Lead	$1 - 1.7 \times 10^{-5}$
Paramagnetic materials	
Aluminum (Al)	$1 + 2.1 \times 10^{-5}$
Magnesium (Mg)	$1 + 5 \times 10^{-5}$
Platinum	$1 + 2.9 \times 10^{-4}$
Sodium (Na)	$1 + 2 \times 10^{-5}$
Ferromagnetic materials	
Iron (Fe)	2×10^2 to 6×10^3
78 Permalloy (78% Ni, 22% Fe)	4×10^3 to 10^5
Supermalloy (5% Mo, 79% Ni, 16% Fe)	10^5 to 10^6

Source: Dale R. Corson and Paul Lorrain, *Introduction to Electromagnetic Fields and Waves* (W. H. Freeman, San Francisco, 1962), page 284. *CRC Handbook of Chemistry and Physics* 78th edition (CRC Press, Boca Raton, Florida, 1997), pages 12-116 and 12-119.

[†]If you have a tape player, frequent degaussing of the recording and playback heads is recommended.

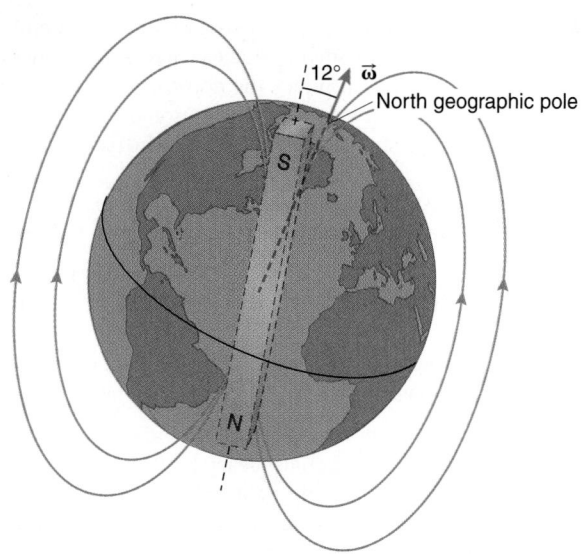

FIGURE 20.68 A crude model for the magnetic field of the Earth.

shown in Figure 20.68. The magnetic axis of the Earth (a line between its south and north magnetic poles) is inclined to its spin axis by about 12°.

A compass needle is a small magnetic dipole that orients itself parallel to the horizontal component of the magnetic field of the Earth. The direction indicated by a line from the center of a compass to the north pole of the compass needle is called **magnetic north**. A line indicating the direction to the north geographic pole is **true north** (or geographic north). The angle between the magnetic and true north directions at a given location is the **magnetic declination**. A compass mounted in the vertical plane aligned along the magnetic north–south direction is called a **dip needle**; it indicates the angle that the magnetic field of the Earth makes with the horizontal direction.

The origin of the magnetic field of the Earth still is a matter of much interest and research. There are only a few aspects of the magnetic field that are known:

1. The field certainly is *not* really caused by a huge, embedded permanent magnet such as in Figure 20.68. Although the core of the Earth consists of much iron and other ferromagnetic material capable of sustaining a permanent magnetic dipole moment on a bulk scale under typical temperatures at the surface of the Earth, the high temperatures in the interior of the Earth prohibit the formation of a permanent magnetic field; the material is not ferromagnetic at these high temperatures.

2. The magnetic field likely is caused by convection currents of ionic material within the outer parts of the fluid core of the Earth. It is thought that these convection currents are closely tied to the rotation of the Earth. The discovery that the fluid outer core of the Earth spins slightly faster than the mantle has added considerable excitement to this research. Slowly rotating planets like Venus have weaker magnetic fields than the Earth; faster rotating planets like Jupiter have

stronger magnetic fields than the Earth. The detailed mechanisms that generate the magnetic field of the Earth still are quite elusive and complex; in other words, we do not really have the answer to the question! The problem is difficult to probe. The *rule of difficulty* is at work once again: if you know the answer, the problem is easy; if you do not know the answer, the problem is hard.

3. The magnetic field of the Earth has undergone *reversals* of magnetic polarity every few million years. What instigates these polarity reversals also is not known. The reversal occurs quite rapidly on a geologic time scale. When reversing direction, the magnetic poles wander relatively quickly (within about 10 000 years) from one hemisphere to the other.*

The discovery of the field reversals was made during the 20th century while exploring the interesting geology near the center of the mid-ocean ridge in the Atlantic. Such ridges exist on the seabed of other oceans as well. The mid-Atlantic ridge feature is a region where hot magma (lava) slowly wells up from the interior of the Earth and cools as it approaches the surface of the Earth (at the bottom of the ocean). Within hot magma, minute particles with a magnetic dipole moment orient themselves parallel to the magnetic field of the Earth at the time. The magnetic orientation of the particles thus is locked into the rocks when they solidify and provides a record of the magnetic field direction of the Earth.

Since the magnetic field of the Earth reverses itself, the present magnetic field at a given location may be greater or smaller than the average magnetic field due to the parallel or antiparallel alignment of the magnetic material in the underlying rocks. A sensitive instrument (a magnetometer) can measure the difference between the actual magnetic field at a given location and the average field. This difference is known as the **magnetic anomaly**. A distinct pattern of alternating magnetic anomalies exists around the regions of the mid-ocean ridges, as shown in Figure 20.69. In particular, if the magnetic anomaly is measured along a line perpendicular to the ridge, a pattern of bands or strips of alternating anomaly is apparent that is quite symmetric about the ridge. The age of the rocks increases with distance from the ridge.

The pattern is consistent with a slow, conveyor belt type of transport of material away from the ridge, indicative of sea-floor spreading, as shown in Figure 20.70. The recently cooled surface rocks record the recent direction of the magnetic field vector and, as they are slowly pushed away from their source ridge, they display a pattern of the past. The existence of these bands is firm proof of the ideas of plate tectonics in geology.

---

*There also is recent evidence (from analyzing ancient volcanic lava flows at Steens Mountain in southern Oregon, latitude 42.5° N, longitude 118.5°W) of very rapid changes in the direction of the magnetic field, with the poles moving as much as 6° in a single day. See R. S. Coe, M. Prévot, and P. Camps, "New evidence for extraordinarily rapid change of the geomagnetic field during a reversal," *Nature*, *374*, #6524, pages 687–692 (20 April 1995).

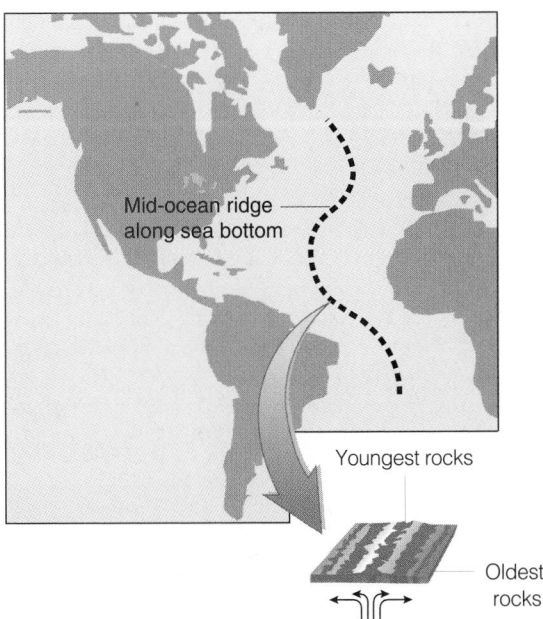

FIGURE 20.69 Strips of alternating magnetic anomaly exist near the ridge.

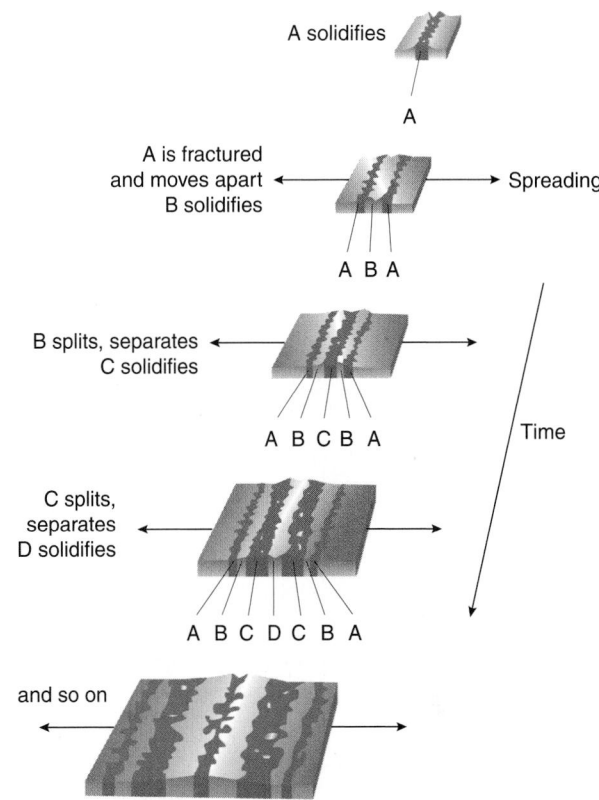

FIGURE 20.70 Formation of the strips of alternating magnetic anomaly.

CHAPTER SUMMARY

The end of a compass needle that points generally in a northerly direction at most places on the surface of the Earth is defined to be a *north* magnetic pole; the opposite end of the needle is a *south* magnetic pole. Like magnetic poles repel each other, unlike poles attract. Magnetic poles always are paired; isolated magnetic poles (magnetic monopoles) evidently do not exist.

Magnets produce a magnetic field $\vec{\mathbf{B}}$ in the surrounding space. The SI unit of magnetic field is the *tesla* (T).

A particle with charge q, moving at velocity $\vec{\mathbf{v}}$ in a magnetic field $\vec{\mathbf{B}}$, experiences a magnetic force $\vec{\mathbf{F}}$ given by

$$\vec{\mathbf{F}} = q\vec{\mathbf{v}} \times \vec{\mathbf{B}} \qquad (20.1)$$

Static magnetic forces do no work on the charge because the force always is perpendicular to the velocity (and thus to the differential change in the position vector of the particle).

A current I in a magnetic field $\vec{\mathbf{B}}$ experiences a force given by

$$\vec{\mathbf{F}} = I \int_{\text{wire}} d\vec{\boldsymbol{\ell}} \times \vec{\mathbf{B}} \qquad (20.10)$$

where the current element $I\,d\vec{\boldsymbol{\ell}}$ is directed in the same sense as the current. If the wire is straight and the magnetic field is constant along its entire length, then Equation 20.10 reduces to

$$\vec{\mathbf{F}} = I\vec{\boldsymbol{\ell}} \times \vec{\mathbf{B}} \qquad \begin{array}{l}\text{(straight wire of length } \ell, \\ \vec{\mathbf{B}} \text{ constant along its length)}\end{array} \quad (20.11)$$

A loop of current I with area A has a magnetic dipole moment $\vec{\boldsymbol{\mu}}$ given by

$$\vec{\boldsymbol{\mu}} \equiv I\vec{\mathbf{A}} \qquad (20.14)$$

where the common direction of $\vec{\boldsymbol{\mu}}$ and $\vec{\mathbf{A}}$ is given by the following right-hand rule: wrap the fingers of your right hand around the loop in the directional sense of the current. Then your extended right-hand thumb indicates the direction of $\vec{\boldsymbol{\mu}}$ and $\vec{\mathbf{A}}$.

A magnetic dipole moment in a magnetic field $\vec{\mathbf{B}}$ experiences a torque $\vec{\boldsymbol{\tau}}$ given by

$$\vec{\boldsymbol{\tau}} = \vec{\boldsymbol{\mu}} \times \vec{\mathbf{B}} \qquad (20.15)$$

and has a potential energy

$$\text{PE} = -\vec{\boldsymbol{\mu}} \cdot \vec{\mathbf{B}} \qquad (20.16)$$

The magnetic field produced by a current I is found from the *Biot–Savart law*:

$$\vec{B} = \frac{\mu_0}{4\pi} I \int_{wire} \frac{d\vec{\ell} \times \hat{r}}{r^2} \qquad (20.19)$$

where $\hat{r}$ is a unit vector pointing from the current element $I\,d\vec{\ell}$ to the point where the field is to be found, and r is the distance between the current element and the field point. The constant μ_0 is the permeability of free space; it satisfies

$$\frac{\mu_0}{4\pi} = 10^{-7} \text{ T·m/A}$$

The ampere, the SI unit of current, is defined in the following way: two infinite parallel wires each carrying one ampere, when spaced one meter apart, exert a force on each other of magnitude exactly 2×10^{-7} N per meter of length.

Gauss's law for the magnetic field states that the flux of the magnetic field $\vec{B}$ through any closed surface always is zero:

$$\int_{clsd \; surface \; S} \vec{B} \cdot d\vec{S} = 0 \text{ T·m}^2 \qquad (20.25)$$

Gauss's law for the magnetic field is a statement that magnetic monopoles do not exist; the magnetic field is dipolar.

Ampere's law states that the path integral of the magnetic field around a closed path is equal to μ_0 times the net current threading the contour:

$$\int_{clsd \; path} \vec{B} \cdot d\vec{r} = \mu_0 I_{net \; current \; threading \; the \; path} \qquad (20.28)$$

The currents should include displacement currents when and where they exist, in which case the law is called the *Ampere–Maxwell law* (see Equation 20.34). Currents are considered positive if they thread the path according to the following right-hand rule: wrap the fingers of your hand around the contour in the same sense as $d\vec{r}$; your extended right-hand thumb indicates the positive sense for currents threading the path.

A time-varying flux of the electric field produces a displacement current I_D given by

$$I_D \equiv \varepsilon_0 \frac{d\Phi_{elec}}{dt} \qquad (20.33)$$

which also creates a magnetic field (via the Biot–Savart law).

To calculate magnetic effects in materials, use the permeability of the material μ rather than the permeability of free space μ_0.

SUMMARY OF PROBLEM-SOLVING TACTICS

20.1 **(page 898)** When using Equation 20.1, $\vec{F} = q\vec{v} \times \vec{B}$, be sure to use the appropriate sign of the charge q. For positive charges, the force is in the direction of $\vec{v} \times \vec{B}$; for negative charges, the force is directed opposite to $\vec{v} \times \vec{B}$.

20.2 **(page 906)** The static magnetic force does no work on a moving charge and, by the CWE theorem, cannot by itself change the kinetic energy of the charge.

20.3 **(page 910)** If a wire carrying a current is wrapped into a coil with n turns, each with the same area A, the magnetic dipole moment of the coil is n times that of a single turn. The common direction of $\vec{\mu}$ and $\vec{A}$ is determined from the same right-hand rule: wrap the fingers of your right-hand around the coil in the sense of the (positive) current, and the extended right-hand thumb indicates the direction of both $\vec{\mu}$ and $\vec{A}$.

QUESTIONS

1. (page 898); 2. (page 900); 3. (page 918); 4. (page 924); 5. (page 930)

6. How do you think the ancient Greeks and Chinese first discovered magnets over two millennia ago? Were they capable of distinguishing between magnetism and electricity?

7. Suggest some experiments that indicate that the interaction between two magnets is not an electrostatic interaction.

8. Indicate in a sketch several magnetic field lines that illustrate: (a) a situation where the magnetic field $\vec{B}$ is constant; (b) a situation where the magnetic field is not constant.

9. The diagrams in Figure Q.9 trace lines of a static electric field and/or a static magnetic field. Indicate which represent electric field lines, which represent magnetic field lines, and which could be either.

10. In electrostatics, the electric field is either parallel or antiparallel to the electrical force on a charged particle. Explain why it is *not* a good idea to define the direction of the magnetic field to be either parallel or antiparallel to the direction of the magnetic force on a charged particle moving in the field.

11. Consider the equation $\vec{F} = q\vec{v} \times \vec{B}$. (a) Which vector quantities always are perpendicular to each other? (b) Which are not necessarily perpendicular to each other?

12. Charged particles from the Sun strike the Earth more frequently near its magnetic poles than elsewhere. Explain why. Collisions of such particles with the atmosphere produce auroras.

13. The electric and magnetic forces on the undeflected ions in a velocity selector have equal magnitudes and opposite directions. Are these forces a Newton's third law force pair? Explain.

14. According to Newton's third law, all forces occur in pairs. What is the force that is the third law counterpart to the magnetic force on a charged particle moving in a magnetic field?

15. A charged particle q is moving in a circle in a uniform magnetic field. An electric field is turned on in the same direction as the magnetic field. Describe and sketch the subsequent trajectory of the particle. Consider both cases: $q > 0$ C and $q < 0$ C.

16. An electron begins initially traveling horizontally west in the United States (to enjoy the scenery) in a region where the magnetic field of the Earth points true north. Describe the initial direction of the force on the electron.

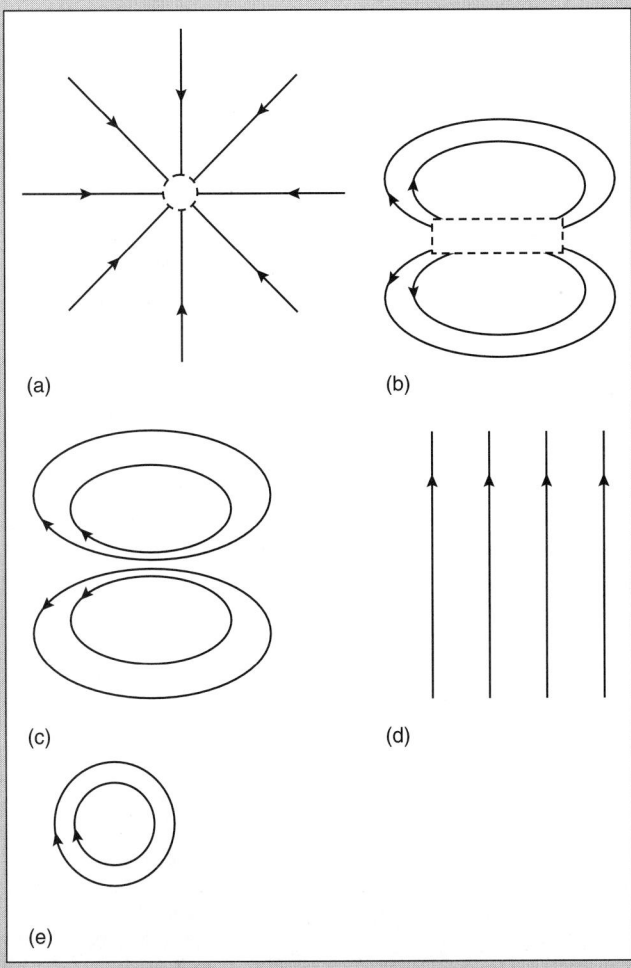

(a)

(b)

(c)

(d)

(e)

17. A charge q is moving at velocity $\vec{v}$ in a magnetic field, yet experiences zero magnetic force. Indicate in a sketch and with a short explanation how can this happen.

18. A charged particle moving with velocity $\vec{v}$ enters a region where there exists either an electric or a magnetic field. The force on the particle is perpendicular to $\vec{v}$. Can you conclude definitively that the field is magnetic rather than electric? Explain.

19. If a charge is at rest in a magnetic field, can it be set in motion with a static magnetic field?

20. Describe an experiment to determine whether a given end of a bar magnet is a north magnetic pole or a south magnetic pole.

21. You suspect the existence of a magnetic field in some region of space. Describe an experiment to verify the existence (or nonexistence) of the magnetic field in the region.

22. Describe the similarities and the differences between static electric and magnetic fields and the effects they have on charged particles.

23. A field exists in a region of space but you are uncertain about whether it is a gravitational, an electric, or a magnetic field. Given a particle of mass m with positive charge q, describe a series of experiments that will enable you to tell if the field is gravitational, electric, or magnetic as well as the direction of the field.

24. A wire carrying a current has zero total charge. How, then, can a magnetic field exert a force on the wire?

25. A wire carries a current. The same current results if you assume that positive charge flows one way or negative charge in the opposite direction. Do the magnitude and direction of the magnetic force on the wire depend on the sign of the charge carrier assumed to be responsible for the current? Explain.

26. A charged particle moves in a magnetic field but the magnetic force on the particle is zero. What can you conclude about the orientation of the velocity of the particle with respect to the magnetic field direction?

27. The magnetic field lines of two parallel, long, current-carrying wires are shown in Figure Q.27, but without directional arrows for the currents. (a) Are the current directions in the wires parallel or antiparallel to each other? (b) Indicate the current directions in the wires.

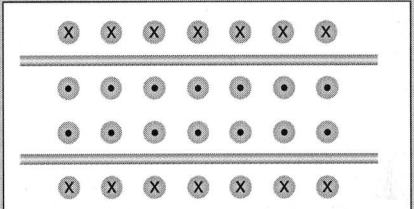

FIGURE Q.27

28. Bring a small, weak bar magnet near a TV, PC monitor, or oscilloscope (while they are turned on!). Explain why the display on the screen is distorted.

29. A moving electron and a proton enter a region with their velocities perpendicular to a uniform magnetic field. For each of the following cases, determine which particle (if either) is deflected into a circle of greater radius: (a) the speeds of the particles are the same; (b) the kinetic energies of the particles are the same; (c) the momenta of the particles are the same.

30. Explain why the following statement is true: A static magnetic field cannot change the *speed* of a charged particle.

31. Explain why the following statement usually is true: A static magnetic field can change the *velocity* of a charged particle. Under what circumstances is the statement false?

32. Describe a situation in which a magnetic dipole moment experiences: (a) zero torque in a magnetic field; (b) nonzero torque in a magnetic field.

33. A jumbo jet is flying horizontally over the Atlantic Ocean near Newfoundland in the direction east as indicated by a magnetic compass. The magnetic field of the Earth makes an angle θ with the horizontal. (a) What is the direction of the horizontal component of the magnetic field of the Earth? (b) What is the direction (up or down?) of the vertical component of the magnetic field? (c) Will the magnetic force cause an accumulation of electrons in the nose, tail, or left or right wing tip (left and right determined facing forward)?

34. A *bubble chamber* is a device in which the paths of tiny charged particles in a fluid can be photographed. A magnetic field is used to deflect the particles from their incident directions. The paths of an electron and positron (a particle with the same mass as an electron but with charge $+e$) are shown in Figure Q.34. (a) If the magnetic field points toward you perpendicular to the page, which is the path for $-e$ and which for $+e$? (b) Notice that the paths are spiral and not circular. What does this imply about the speed of the particles as a function of time? Develop a hypothesis that may account for the spiral paths.

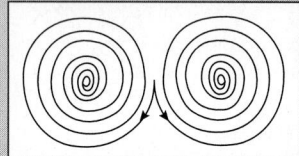

FIGURE Q.34

35. Two long wires carry oppositely directed currents of equal magnitude. To make the magnetic field of the wires a minimum at large distances from the wires, is it better to arrange the wires as in Figure Q.35a or Q.35b, or does it make any difference at all?

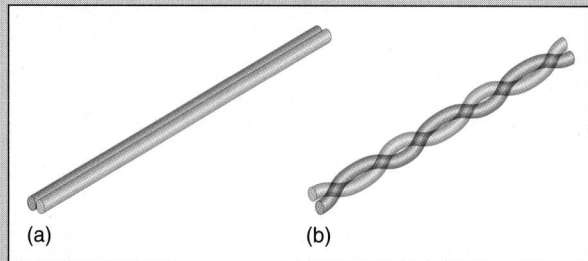

(a)   (b)

FIGURE Q.35

36. Your instructor gives you a piece of wire of length ℓ to conduct a given current. What shape of loop will give you the maximum magnitude of magnetic dipole moment? Is it better to use a single loop or a large number of small coils?

37. Take a highly flexible wire of small mass and lay it on a frictionless surface as in Figure Q.37. Close the switch. Describe and explain what happens to the flexible wire.

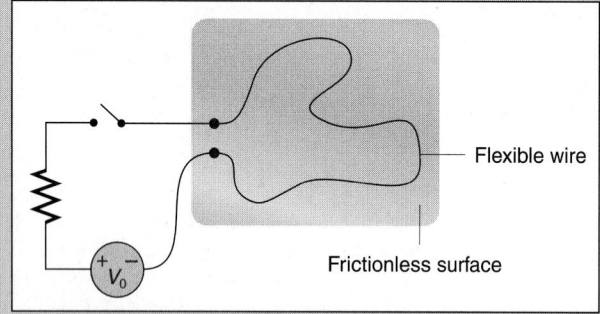

Flexible wire

Frictionless surface

FIGURE Q.37

38. A hollow copper water pipe carries an electrical current I along its length that is uniformly distributed around the circumference of the pipe. What is the magnetic field within the hollow tube of the pipe? What is the magnetic field exterior to the pipe?

39. Why can either pole of a magnet attract an unmagnetized paper clip?

40. Pictured in Figure Q.40 are the magnetic field lines going into the page ● and out of the page ✕ produced by various current distributions. In each case, sketch the placement of the wire and the direction of the current that produces the indicated field. A few words of explanation also may be appropriate.

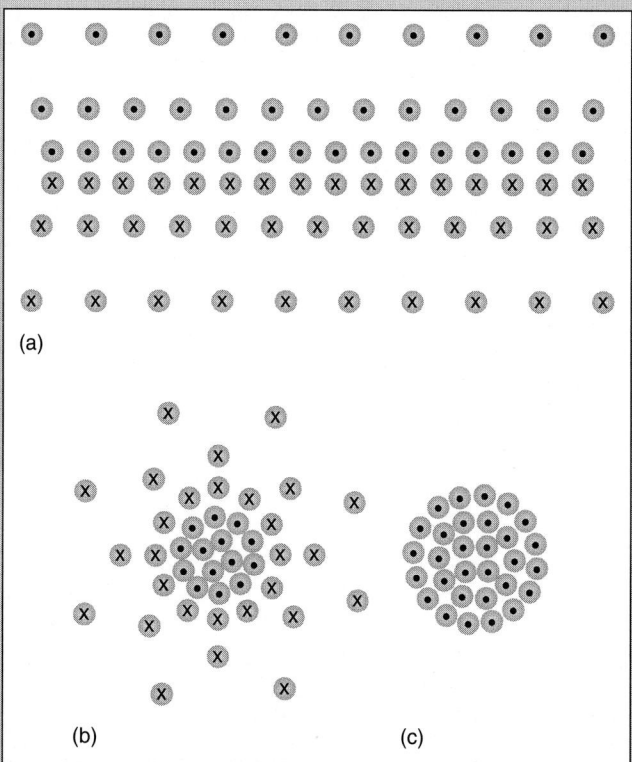

(a)

(b)   (c)

FIGURE Q.40

41. A permanent magnet with the polarities indicated in Figure Q.41 was discovered by Michael Davis ["A magnetic tripole," *The Physics Teacher*, 14, #1, page 34 (January 1976)]. Sketch how a current-carrying wire could be wound into a coil to produce such a tripolar arrangement.

FIGURE Q.41

42. The magnetic axis of the Earth is inclined to its rotational axis by about 12°. Is the magnetic declination angle 12° at all locations on the surface of the Earth? Explain.

PROBLEMS

Sections 20.1 The Magnetic Field
20.2 Applications

1.–6. The wallpaper-like patterns in Figures P.1–P.6 represent various directions for a uniform magnetic field. Charged particles are incident into each region of the field with velocity vectors directed as indicated by the arrows on each charge. Sketch the *initial* direction of the magnetic force that acts on each charged particle.

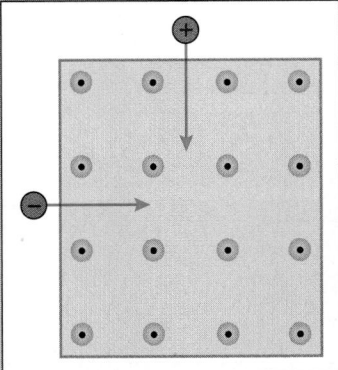

FIGURE P.1

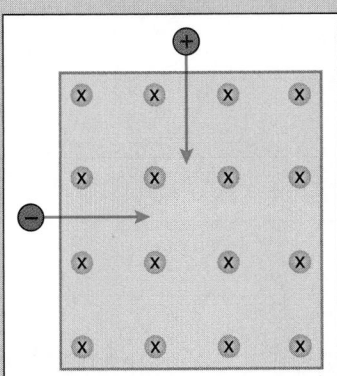

FIGURE P.2

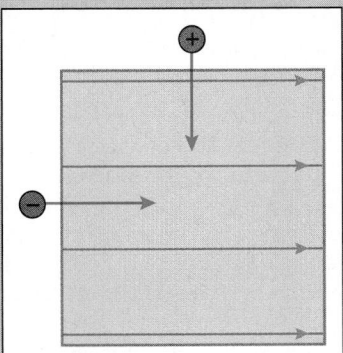

FIGURE P.3

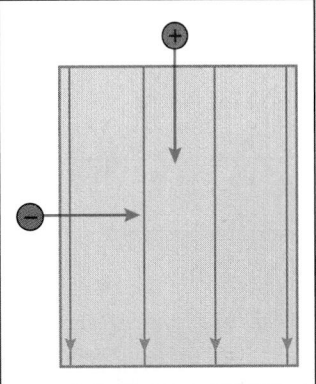

FIGURE P.4

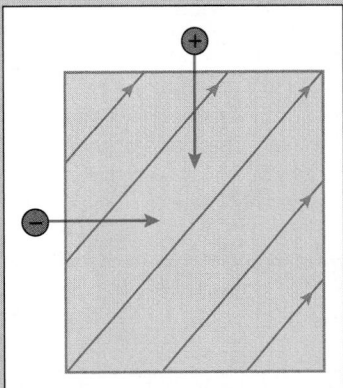

FIGURE P.5

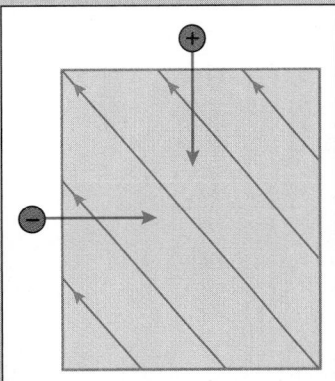

FIGURE P.6

7. An electron is moving at speed 5.0×10^6 m/s in the magnetic field of magnitude 6.0×10^{-2} T indicated in Figure P.7. (a) Find the magnitude and direction of the initial magnetic

force on the electron. (b) What is the magnitude of the acceleration of the electron?

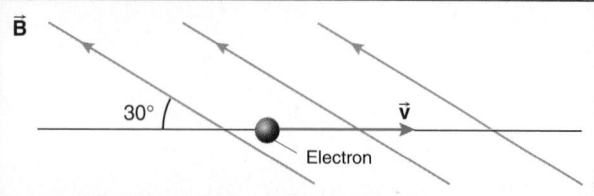

FIGURE P.7

•8. In the beginning, there was a field; and the field was magnetic and uniform; and it was good. Yea, and the field was directed perpendicularly out of the page as indicated in Figure P.8. And charged particles were sent forth into the field with various velocities (all perpendicular to the field direction); and forces and accelerations on these particles resulted. Indeed, the paths of the particles in the field were noted and recorded and none went astray but each to its own place. And each particle was numbered accordingly. (a) At some point along each of the paths, sketch the direction of the magnetic force (if any) acting on the particle. (b) Indicate clearly which particles (if any) were possessed with negative charge, which were possessed with positive charge, and which were possessed with no charge. (c) If all the particles have the same absolute magnitude of charge (if they have nonzero charge) and the same speed, which of the particles has the smallest mass? If this cannot be determined from the sketch, indicate that this is the case and why.

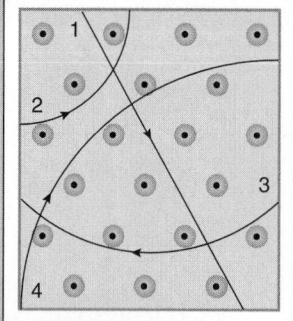

FIGURE P.8

• 9. An electron has velocity $(5.0 \text{ km/s})\hat{\imath} - (6.0 \text{ km/s})\hat{\jmath}$ in a magnetic field. The force on the electron is found to be $(3.84 \times 10^{-19} \text{ N})\hat{\imath} + (3.20 \times 10^{-19} \text{ N})\hat{\jmath} + (2.40 \times 10^{-19} \text{ N})\hat{k}$. The magnetic field lacks an x-component. Find the magnetic field.

•10. An electron, a proton, and a neutron each has a kinetic energy of 1.00 keV. They are projected into a semi-infinite region with a uniform magnetic field of magnitude 50.0 mT directed as indicated in Figure P.10. (a) Determine the speed of each particle. (b) Will the kinetic energy of each particle change in the region of the magnetic field? Explain your answer. (c) Schematically indicate the path that each particle

follows (be sure to indicate which path is appropriate for each particle!). (d) Determine the radii of the path followed by each particle.

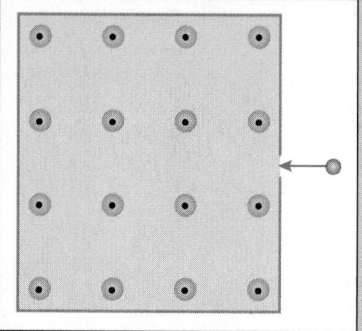

FIGURE P.10

•11. $\vec{\mathbf{B}}$ careful. A charged bee ($q = -2.50 \times 10^{-3}$ C) is streaking at a speed of 22.0 m/s across a uniform magnetic field of magnitude 25 mT as indicated in Figure P.11. What are the magnitude and direction of the magnetic force on the bee?

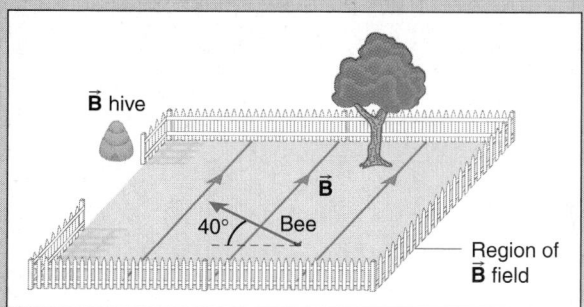

FIGURE P.11

•12. An electron moves at constant velocity in a region where there is an electric field of magnitude 500 N/C and a magnetic field of magnitude 0.150 T. (a) What is the angle between the electric and magnetic fields? (b) What is the minimum value for the speed of the electron? (c) Will a proton moving with the same minimum velocity also move with a constant velocity through the field region?

•13. A happy particle of mass 2.00 mg has charge −10.0 μC and finds itself in a region where the only force on it is a magnetic force. The magnitude of the magnetic field at the location of the particle is 15.0 mT and the speed of the particle is 2.00 km/s. The angle between the directions of the magnetic field and the velocity vector of the particle is 75.0°. (a) Sketch the situation. (b) Determine the magnitude of the magnetic force acting on the particle. Specify the direction of the force. (c) Determine the magnitude of the acceleration of the particle.

•14. Magnetic fields are used to change the direction of charged particle beams. (a) A beam of electrons has a speed of 9.50 × 10^6 m/s. A uniform magnetic field, oriented perpendicular to the velocity of the electrons, is to be used to turn the electron beam through a 90° angle along a circular arc of radius 5.00×10^{-2} m. What is the magnitude of the magnetic field used? (b) If a beam of protons with the same speed is used in the same apparatus, what is the radius of the circular arc along which they move?

•15. An electron is moving nonchalantly with a velocity of $(3.00 \text{ km/s})\hat{\imath} + (4.00 \text{ km/s})\hat{\jmath}$ in a uniform magnetic field directed along the z-axis. The force on the electron is measured to be of magnitude 2.40×10^{-16} N, making an angle of 143.13° with $\hat{\imath}$. (a) Make a schematic diagram of the situation. (b) Determine the magnitude of the magnetic field and express the field in Cartesian vector form.

•16. A tiny particle of mass m with a charge q traveling at speed v enters a semi-infinite region with a uniform magnetic field of magnitude B directed as indicated in Figure P.16. Show that the time t the particle is in the region with the field is independent of the speed v and is given by

$$t = \frac{\pi m}{|q|B}$$

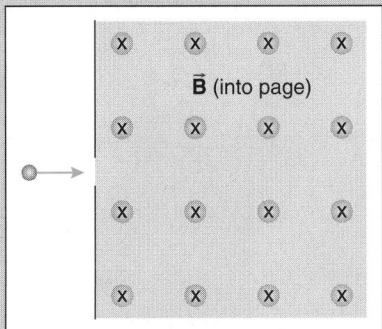

FIGURE P.16

•17. A velocity selector has a magnetic field of magnitude 0.15 T and an electric field of magnitude 2.00×10^4 N/C. A beam of negatively charged morons (each with charge $q = -3e$) enters the field as shown in Figure P.17. (a) If the electric field in the velocity selector is directed as indicated in Figure P.17, sketch the directions of the magnetic and electrical forces acting on each moron. (b) Indicate on the same sketch the orientation

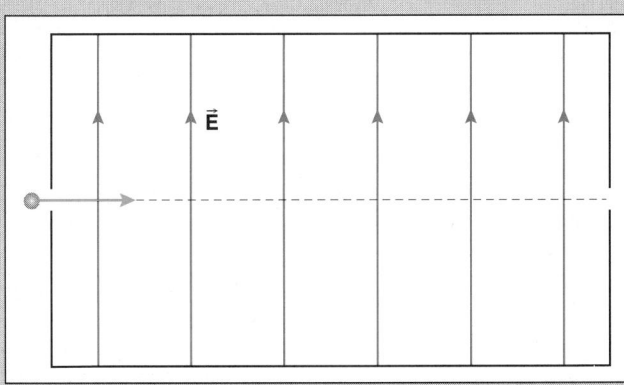

FIGURE P.17

of the magnetic field inside the velocity selector. (c) What is the speed of the morons that emerge undeflected from the velocity selector?

•18. In a mass spectrograph, positive ions of mass m and charge q are accelerated from rest through a potential difference V and then encounter a uniform magnetic field perpendicular to their motion as indicated in Figure P.18. (a) In a sketch, indicate the trajectories of the ions in the region of the magnetic field. (b) Use the CWE theorem to find an expression (in terms of m, q, and V) for the speed of the ions as they exit the region in which they are accelerated. (c) Beginning with Newton's second law, show that the charge-to-mass ratio q/m of the ions is

$$\frac{q}{m} = \frac{8V}{B^2 d^2}$$

where d is the distance from the entrance slit of the magnetic field region to the place where the ions strike the wall of the device.

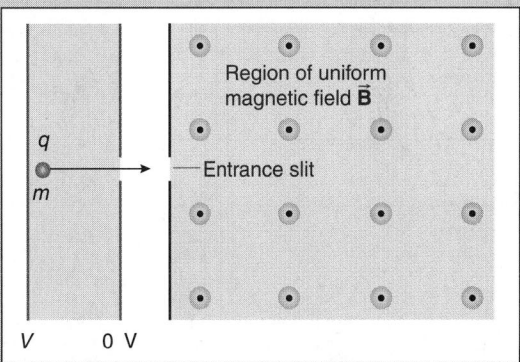

FIGURE P.18

•19. A semiconductor with a rectangular cross section carries a current I along the direction $\hat{k}$ as shown in Figure P.19. A magnetic field is parallel to $\hat{\imath}$. (a) If the charge carriers are predominantly negative (an n-type semiconductor), across which pair of faces of the conductor should the Hall voltage be measured? Which face is at the higher potential? (b) Repeat part (a) if the charge carriers are predominantly positive (a p-type semiconductor).

§20. A particle with mass m and charge q traveling at speed v enters a region with a uniform magnetic field. The velocity vector makes an angle α with the field direction as indicated in Figure P.20. The trajectory of the particle is a helix whose axis is the direction of the magnetic field vector. (a) Show that the radius r of the helix is

$$r = \frac{mv \sin \alpha}{|q|B}$$

(b) Show that the time t for one revolution of the particle around a single turn of the helix is

$$t = \frac{2\pi m}{|q|B}$$

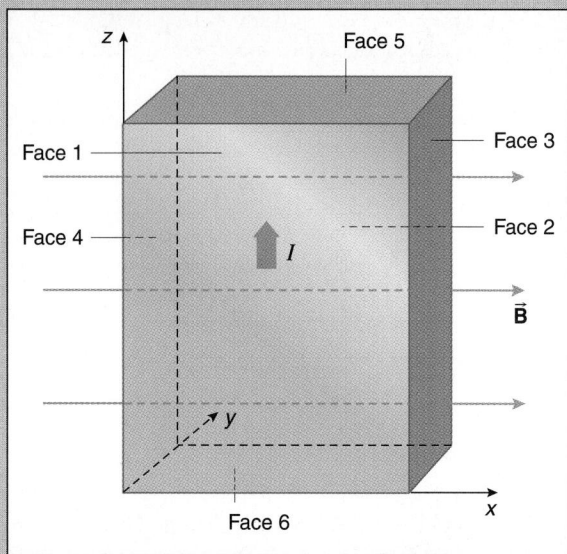

FIGURE P.19

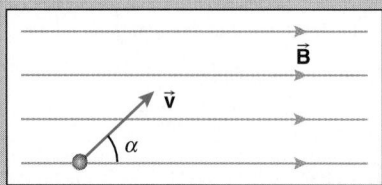

FIGURE P.20

(c) Show that the distance d between successive turns of the helix measured parallel to the axis, known as the pitch of the helix, is

$$d = \frac{2\pi mv \cos \alpha}{|q|B}$$

(d) What effect does the *sign of the charge* have on the spiral motion? Charged particles in space (electrons, ions, cosmic rays) encountering magnetic fields in space (such as that due to the Earth) are confined to such spiral motion. When such particles collide with the atmosphere of the Earth, *auroras* result.

‡21. Electrons of mass m and speed v pass through a small slit and enter a region of uniform magnetic field as shown in Figure P.21. The velocity vectors of the electrons all lie within a small angle $\theta \ll 1$ rad of the direction of the field, so the electrons

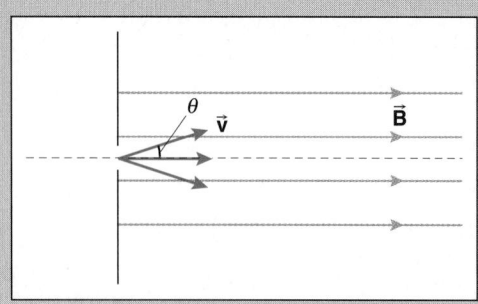

FIGURE P.21

enter the field region initially moving away from each other. Show that the magnetic forces on the electrons will confine the beam to a diameter d given by

$$d \approx \frac{2mv\theta}{eB}$$

where θ is in radians. This illustrates how a magnetic field can be used to keep a charged particle beam confined. The same principle is applicable to beams of other charged particles.

Sections 20.3 Magnetic Forces on Currents
20.4 Work Done by Magnetic Forces
20.5 Torque on a Current Loop in a Magnetic Field

22. A wire of infinite length carries a current of 15.0 A in a uniform magnetic field of magnitude 0.55 T as indicated in Figure P.22. (a) Find the force on a one meter length of the wire. (b) If the angle that the wire makes with the magnetic field is tripled, what happens to the direction of the force? Is the magnitude of the force tripled? Explain your answer.

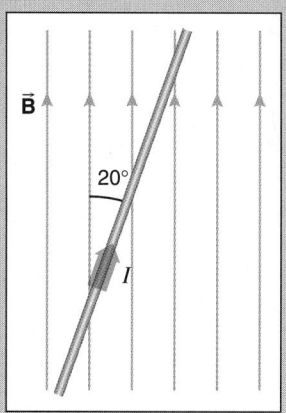

FIGURE P.22

•23. The wire shown in Figure P.23 carries a current I in a uniform magnetic field $\vec{B}$ directed as indicated. Show that the total magnetic force on the wire is zero.

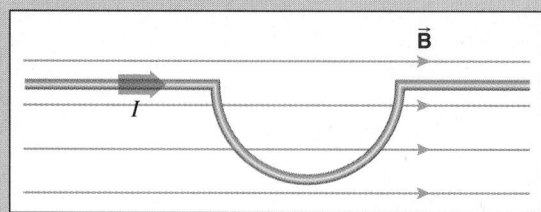

FIGURE P.23

•24. A semicircular wire carries a current I in a uniform magnetic field $\vec{B}$ as shown in Figure P.24. Show that the total magnetic force on the wire is $-2IBR\hat{j}$.

•25. A circular current loop of radius 5.0 cm carries a current of 1.50 A. The loop is situated in a uniform magnetic field of magnitude 0.60 T directed as indicated in Figure P.25. (a) What is the magnitude of the magnetic dipole moment of the current loop?

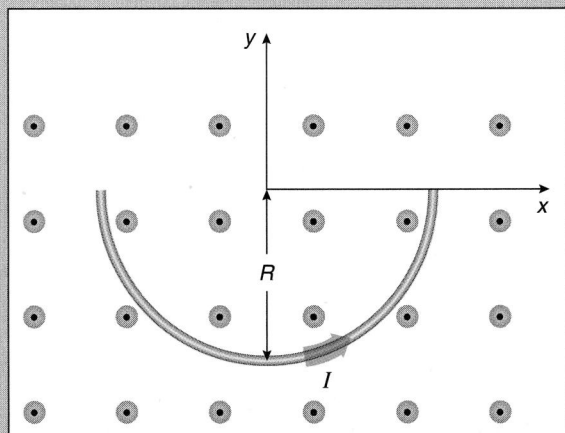

FIGURE P.24

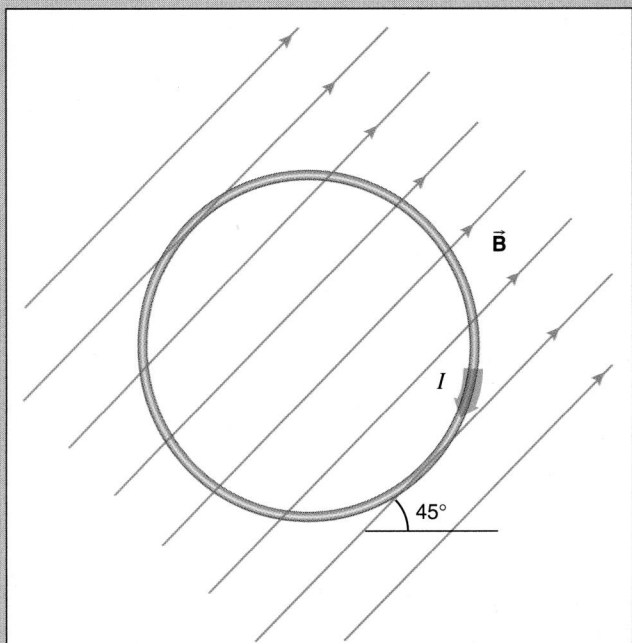

FIGURE P.25

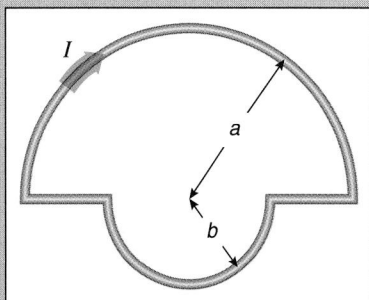

FIGURE P.26

•**27.** A constant magnetic field of magnitude 1.0 mT is directed horizontally. Three independent, flat current loops are suspended to pivot about a vertical symmetry axis that touches the loop at two points. Each current loop carries a current of 15.0 A and each loop is oriented so the plane of the loop makes an angle of 30° with the direction of the magnetic field. The torques on all the loops are identical. One of the loops is a square with sides of 20.0 cm. (a) Find the magnitude of the torque on the loops. (b) Another of the loops is an equilateral triangle with sides ℓ. Find ℓ. (c) The third loop is a circle of radius R. Find R.

•**28.** Calculate the magnitude of the magnetic dipole moment of the current loop indicated in Figure P.28.

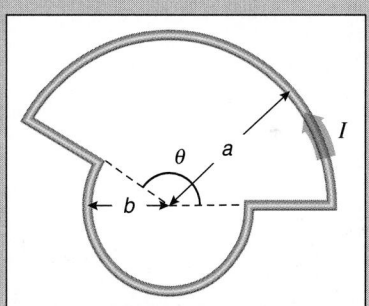

FIGURE P.28

(b) Calculate the magnitude of the torque on the current loop. (c) Indicate clearly in a sketch the axis about which the loop will rotate, as well as the rotational sense of the motion under the influence of the magnetic torque. (d) What is the potential energy of the current loop in the orientation indicated?

•**26.** The current loop in Figure P.26 consists of two semicircles of different radii. (a) Calculate the magnitude of the magnetic dipole moment of the current loop. (b) Specify two directions for a uniform magnetic field that produce a total torque on the loop of 0 N·m. (c) Specify a direction of a magnetic field that produces the maximum magnitude of torque on the current loop. (d) Calculate the magnitude of the maximum torque on the loop.

•29. Two parallel, frictionless rails, separated by distance ℓ and inclined at angle θ to the horizontal, are bridged by a conducting bar of mass m, initially at rest, that is free to move; see Figure P.29. The system lies in a uniform magnetic field $\vec{B}$ directed vertically upward on the surface of the Earth. Neglect the magnetic field of the Earth. When $t = 0$ s, an appropriate source establishes the indicated current I in the system. (a) Determine the magnitude of the magnetic force on m. (b) Draw a second law force diagram indicating all the forces acting on m. (c) Find an expression for the velocity of m as a function of time. (d) What value of B keeps the conducting bar in equilibrium?

•**30.** Two circular current loops have radii r_1 and r_2 as shown in Figure P.30. Each loop carries the same current I. (a) If the

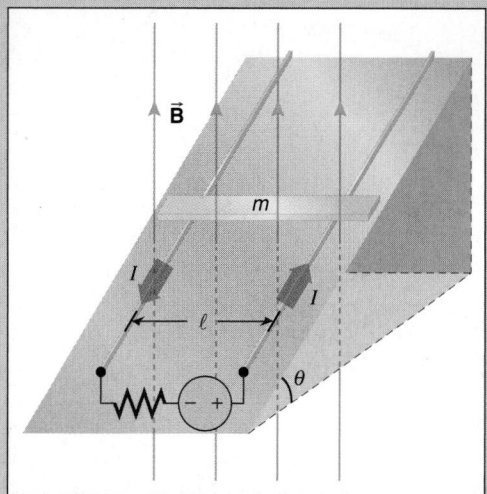

FIGURE P.29

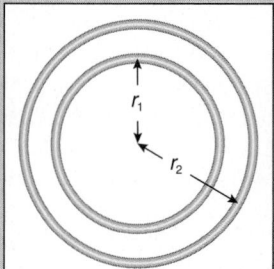

FIGURE P.30

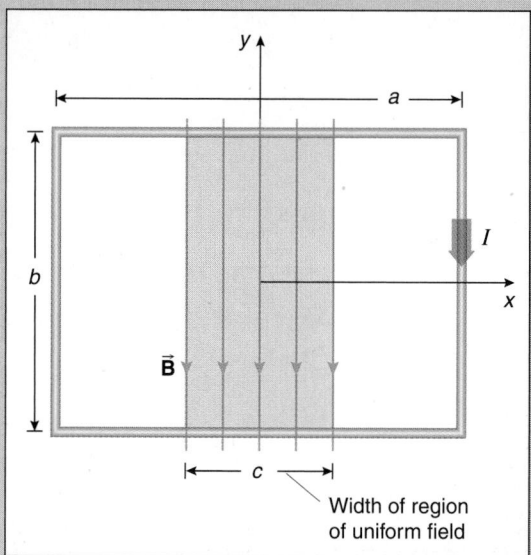

FIGURE P.31

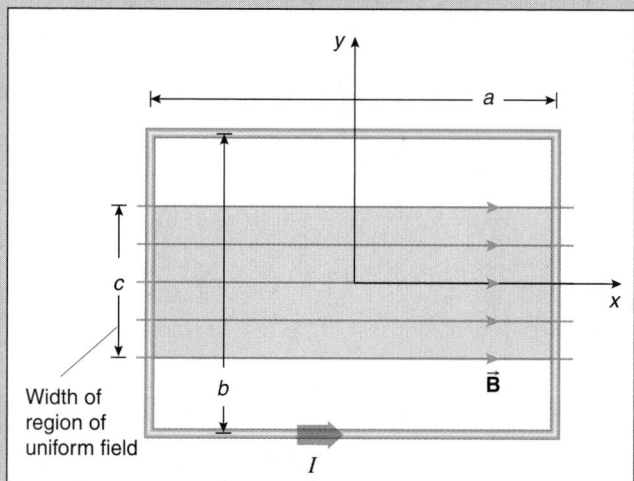

FIGURE P.32

current in each loop is clockwise, determine the magnitude of the magnetic dipole moment of the system of two loops. (b) If the current in the outer loop is clockwise while that of the inner loop is counterclockwise, determine the magnitude of the magnetic dipole moment of the system of two loops.

•**31.** A rectangular current loop carrying a current I is placed in a uniform magnetic field as shown in Figure P.31. (a) Determine the total force on the current loop. (b) Determine the magnetic dipole moment of the current loop. (c) Determine the total torque on the current loop.

•**32.** A rectangular current loop carrying a current I is placed in a uniform magnetic field as shown in Figure P.32. (a) Determine the total force on the current loop. (b) Determine the magnetic dipole moment of the current loop. (c) Determine the total torque on the current loop.

•**33.** Specify the area of a planar coil with 100 turns of wire carrying 25.0 A that can have a maximum torque of magnitude 20.0 N·m when placed in a magnetic field of magnitude 0.200 T.

•**34.** A flat, square piece of wood 20.0 cm on a side lies on a flat table. An external, horizontal field of magnitude 1.2 T is present (see Figure P.34). Around the perimeter of the wood square are wound 100 turns of wire carrying a current I. The total mass of the wood and loop system is 2.00 kg. If the current I is large enough, the square will rise up about one of its edges when the torque of the magnetic force on the loop exceeds the torque of the weight of the system taken about the same edge.

(a) Indicate in a diagram which of the four edges of the square remains in contact with the table when the current I becomes large enough to make the square begin to rotate. (b) Calculate the minimum value for the current I needed to make the square begin to rise about an edge. (c) By what factor would the answer in (b) change if the 20.0 cm square changed to 40.0 cm, with the mass and field staying the same?

•**35.** Electromagnetic rail guns can accelerate small masses to very high speeds extraordinarily quickly. They have the potential to be used for antimissile defense systems as well as for launching materials from the lunar surface to spacecraft in lunar orbit. A rail gun consists of two parallel conducting rails, oriented perpendicular to a strong magnetic field as shown in Figure P.35. Consider the rails to be horizontal. Immediately to the rear of the projectile is a conducting fuse that vaporizes and forms a conducting gas when a large current is initiated briefly in the parallel conductors. The current in the ionized gas experiences a magnetic force that accelerates the gas and the projectile along the rails. Let the constant current pulse be 15 MA and the distance between the

FIGURE P.34

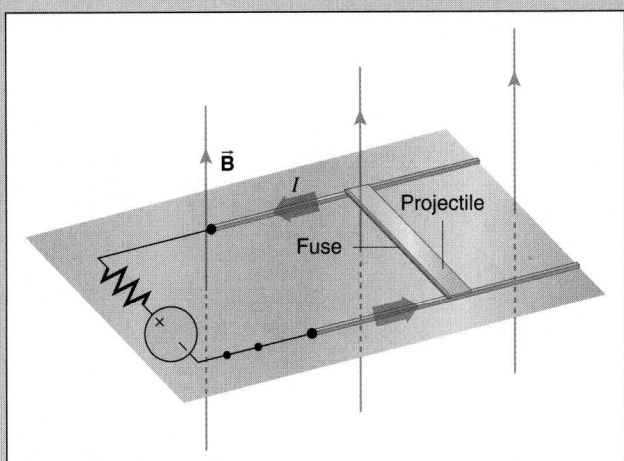

FIGURE P.35

conducting rails be 2.0 cm; the rails are 4.0 m in length. The magnetic field has a magnitude of 2.6 T. (a) Determine the magnitude of the magnetic force on the conducting gas. (b) Assume the force on the gas is transferred without diminution to a projectile of mass 0.10 kg. Neglect any friction between the projectile and rails. Determine the magnitude of the acceleration of the projectile and the launch speed.

‡36. A charge q is distributed uniformly on the surface of a sphere of radius R rotating about a diameter with angular velocity $\vec{\omega}$. Show that the magnetic dipole moment $\vec{\mu}$ of the sphere is

$$\vec{\mu} = \frac{1}{3}\, qR^2\vec{\omega}$$

Section 20.6 The Biot–Savart Law

37. Show that the SI units $A \cdot m^2 \cdot T$ are equivalent to J.

38. A long, straight, 12 gauge copper wire (diameter 2.05 mm) carries a current of 15.0 A. Use the results of Example 20.11 to find the magnitude of the magnetic field at the surface of the wire.

•39. A very long wire carrying 25.0 A has a semicircular deformation as indicated in Figure P.39. Poor thing. (a) What contribution does the straight segment of wire to the left of the semicircle make to the magnetic field at point P? (b) What is the contribution to the magnetic field at P by the straight segment of wire to the right of the semicircle? (c) What is the total magnetic field at point P? (d) *Describe* the difficulties encountered in using the Biot–Savart law to calculate the magnetic field at a point that is *not* at the center of the semicircular loop. (Do not make the calculation.)

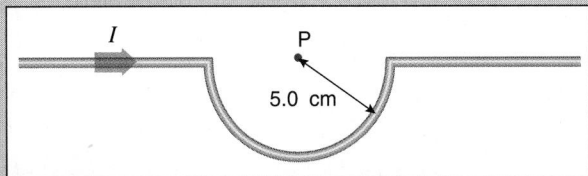

FIGURE P.39

•40. A long wire makes a semicircular U-turn of radius R about point 0 in Figure P.40. If the wire carries a current I, use the Biot–Savart law to find the magnetic field at point 0.

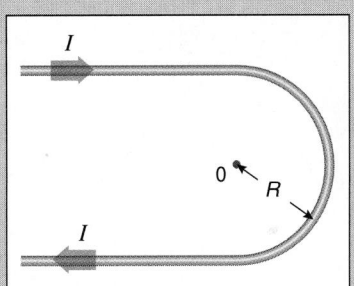

FIGURE P.40

•41. A long wire splits into two identical semicircular segments as shown in Figure P.41. Find the magnetic field at the center of the circle.

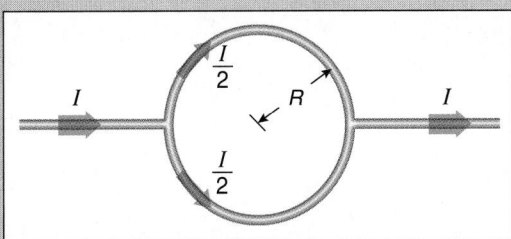

FIGURE P.41

•42. A long wire splits into two identical triangular segments as shown in Figure P.42. Find the magnetic field at point P.

•43. The electron in the hydrogen atom orbits the nuclear proton in a circular orbit of radius 0.529×10^{-10} m with a frequency of 6.58×10^{15} Hz. (a) Sketch the situation; include the orbital direction of the electron and the direction of the current its

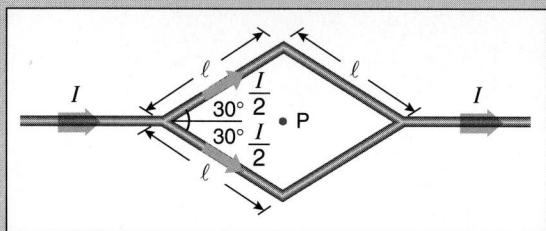

FIGURE P.42

motion represents. (b) What is the magnitude of the magnetic field produced by the electron at the position of the nuclear proton? Indicate the direction of the field in the sketch in part (a). (c) The intrinsic magnetic dipole moment of the proton has a magnitude of

$$\mu_{\text{proton}} = 1.41 \times 10^{-26} \ \text{A} \cdot \text{m}^2$$

The magnetic dipole moment of the proton aligns itself parallel to the magnetic field produced by the circulating electron. What work needs to be done (with no change in kinetic energy) to turn the magnetic dipole moment of the proton so it is antiparallel to the magnetic field? Express your result in eV.

•44. A straight wire segment of length ℓ carries a current I (see Figure P.44). (a) Use the Biot–Savart law to show that the magnitude of the magnetic field at a distance z from the wire along its perpendicular bisector is

$$B = \frac{\mu_0}{4\pi} \frac{2I}{z} \frac{\ell}{(\ell^2 + 4z^2)^{1/2}}$$

(b) Indicate the direction of the field in a sketch. (c) Show that as $\ell \to \infty$ m, this expression for the magnitude of the field approaches that of an infinite wire (the result of Example 20.11).

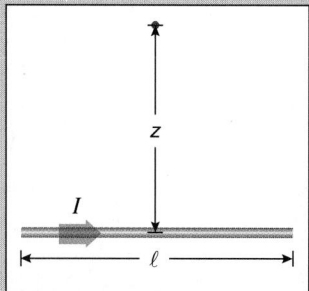

FIGURE P.44

•45. (a) Use the result of Problem 44 to show that magnitude of the magnetic field at the center of a square current loop with sides ℓ, carrying current I (see Figure P.45), is

$$B = \frac{\mu_0}{4\pi} 8\sqrt{2} \frac{I}{\ell}$$

(b) Specify the direction of the field.

•46. (a) Use the Biot–Savart law to show that the magnitude of the magnetic field at a distance z along a perpendicular from the end of a straight wire segment of length ℓ (see Figure P.46), carrying current I, is

$$B = \frac{\mu_0}{4\pi} \frac{I}{z} \frac{\ell}{(\ell^2 + z^2)^{1/2}}$$

(b) Specify the direction of the field.

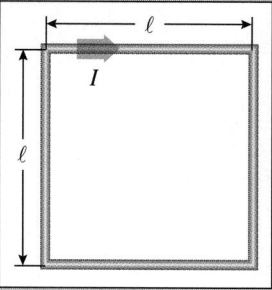

FIGURE P.45

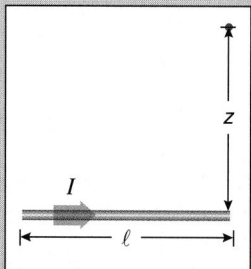

FIGURE P.46

•47. Use superposition, the results of Problem 46, and suitable changes of notation to redo Problem 44.

•48. Two infinitely long straight wires lie in the same plane and carry the currents depicted in Figure P.48. (a) Find the magnetic field at the point P located 25 cm from the intersection of the wires along the bisector of the acute angle between them. (b) Find the magnetic field at the point S, located 25 cm from the intersection of the wires along the bisector of the obtuse angle between the wires.

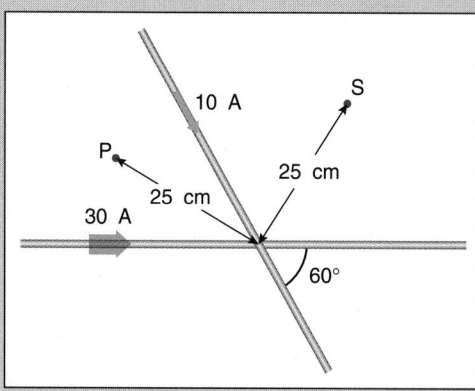

FIGURE P.48

•49. A proton is moving with speed 5.0×10^7 m/s near a long, straight wire carrying a current of 15 A as shown in Figure P.49. (a) What is the magnitude of the magnetic field of the wire at the position of the proton? (b) Determine the magnitude of the force on the proton. Show the direction of the force in a sketch. (c) What is the magnitude and direction of the force of the proton on the wire? (d) What is the speed of the proton after it has moved 5.0 cm from the position shown in Figure P.49?

•50. An otherwise infinite, straight, wire has two concentric loops of radii a and b carrying equal currents in opposite directions as

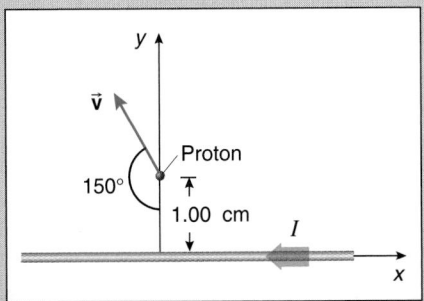

FIGURE P.49

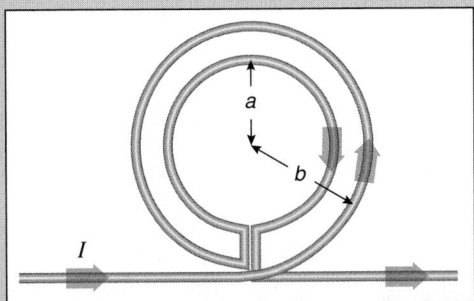

FIGURE P.50

shown in Figure P.50. Show that the magnitude of the magnetic field at the common center of the loops is zero if the radii have the ratio

$$\frac{a}{b} = \frac{\pi}{\pi + 1} \approx 0.7585$$

The idea for this problem comes from Kenneth W. Ford, *Classical and Modern Physics* (Xerox College Publishing, Lexington, Massachusetts, 1973), page 812.

•**51.** A circular wire ring of diameter 20 cm carries a current of 5.0 A directed as indicated in Figure P.51. A moving electron just happens to be passing through the neighborhood. When the electron is at the center of the circular ring and moving at speed 2.50×10^6 m/s in the direction indicated in the sketch, find the acceleration of the electron.

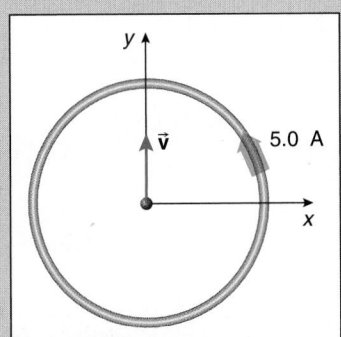

FIGURE P.51

•**52.** (a) For the two infinitely long wires shown in Figure P.52, determine the magnitude and direction of the magnetic field of the top wire at the position of the bottom wire. (b) What is the magnitude and direction of the magnetic field of the bottom wire at the position of the top wire? (c) What is the force of the top wire on 2.00 m of the length of the bottom wire? (d) What is the force of the bottom wire on 2.00 m of the top wire?

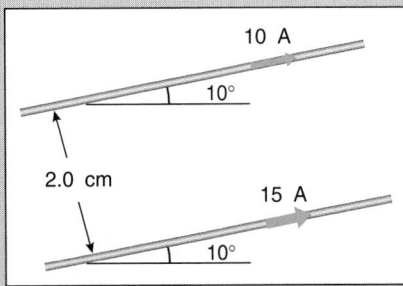

FIGURE P.52

•**53.** The rectangular current loop in Figure P.53 finds itself in the pleasant company of an infinite, straight wire also carrying current. (a) Determine the magnetic field of the infinite wire at the position of the side of the rectangular loop most distant from the infinite wire. (b) Determine the magnetic force of the infinite wire on the side of the rectangle most distant from the infinite wire. (c) Determine the magnetic field of the infinite wire at the position of the side of the rectangular loop closest to the infinite wire. (d) Determine the magnetic force of the infinite wire on the side of the rectangle closest to the infinite wire. (e) By considering symmetrically placed segments of the rectangular wire perpendicular to the infinite wire, show that the total magnetic force on these perpendicular segments is zero. (f) Calculate the total magnetic force on the rectangular current loop due to the magnetic field of the infinite wire. (g) What is the total force exerted on the infinite wire by the (complicated!) magnetic field of the current loop? (Hint: It is not necessary to do an intricate calculation; think of Newton's laws.)

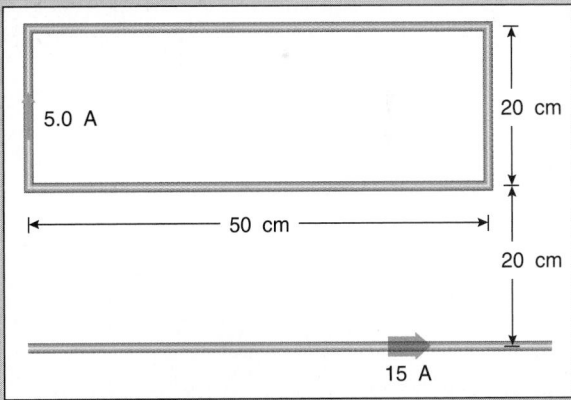

FIGURE P.53

•**54.** An infinitely long, straight wire lies along the z-axis of a Cartesian coordinate system and carries a current of 10.00 A toward increasing values of z. (a) Determine the magnetic field of the wire at the point P described by the coordinates: $x = 3.000$ m, $y = 4.000$ m, $z = 5.000$ m. Express $\vec{B}$ in Cartesian form. (b) An electron is at point P, cavorting along at a speed of 2.000×10^6 m/s parallel to the current in the wire. Determine the magnetic force acting on the electron when it is at this location; express the force in Cartesian form. (c) Show explicitly that $\vec{F} \cdot \vec{v} = 0$ N·m/s and $\vec{F} \cdot \vec{B} = 0$ N·T. What is the meaning of these vanishing scalar products? (d) Will the force in (b) continue to act in the same direction at later times? Explain why or why not.

•**55.** A wire carrying current *I* has the shape indicated in Figure P.55. Beginning with the Biot–Savart law, find the magnetic field at the point P at the center of the circular arcs.

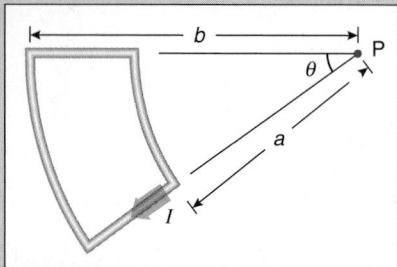

FIGURE P.55

•**56.** The two infinite, insulated wires indicated in Figure P.56 lie in the same plane and carry equal currents *I*, directed as shown. The four positions P_1, P_2, P_3, and P_4 are located symmetrically and are at a distance *d* from the point where the wires cross over each other. (a) Which points have the greatest value for the magnitude of the magnetic field? What is the magnitude of the field at those points? (b) Which points have the smallest value of the magnitude of the magnetic field? What is the magnitude of the field at those points?

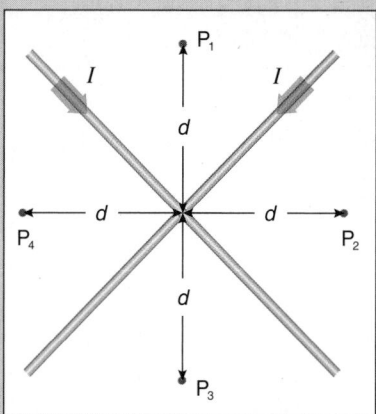

FIGURE P.56

•**57.** A coaxial cable (see Figure P.57) consists of a long straight inner wire of diameter 1.0 mm surrounded by a thin, cylindrical conducting shell of radius 8.0 mm symmetrically placed around the central wire. The cable carries a 3.0 A current via the inner conductor to a resistor some distance away; the current returns through the outer sheath. (a) Find the magnitude of the magnetic field close to the surface of the inner wire. Indicate its direction in a sketch. (b) Find the magnitude of the magnetic field at a distance of 20 mm radially away from the axis of the system.

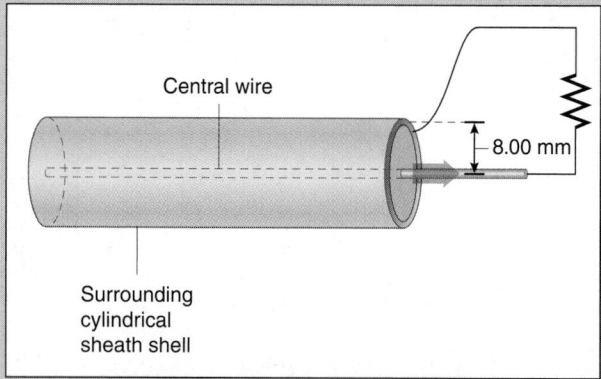

FIGURE P.57

•**58.** (a) Use the results of Example 20.10 to find the magnitude of the magnetic field along the axis of a circular current loop (with current *I*) of radius *R* at a point P a distance $z = R/2$ from the plane of the loop. (See Figure P.58a). (b) If instead of a single current loop, there are *N* loops essentially at the same location, what is the magnitude of the magnetic field at the location in part (a)? (c) If a second identical coil of *N* loops is arranged as in Figure P.58b, show that the magnitude of the magnetic field at point P is

$$B = \frac{\mu_0}{4\pi} \frac{32\pi}{5\sqrt{5}} \frac{NI}{R}$$

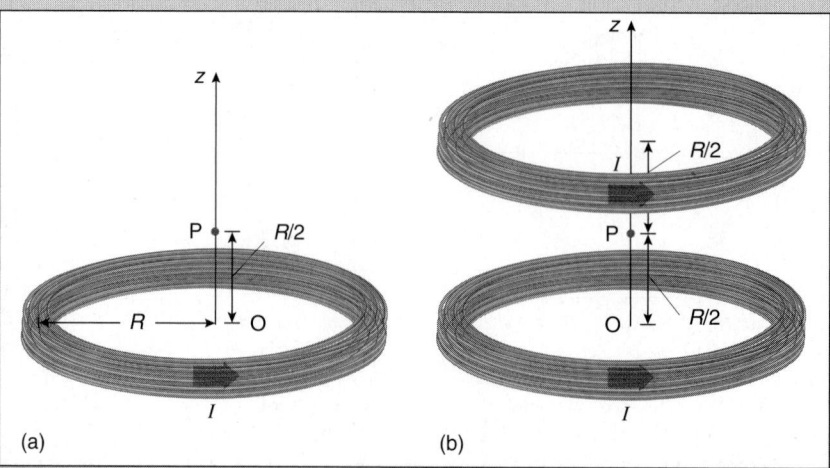

FIGURE P.58

Such an arrangement of twin coils, separated by a distance equal to the radius of the coils, is known as a *Helmholtz coil*. The arrangement commonly is used in the laboratory to produce a magnetic field in the region near the centrally located point P that is quite uniform. For distances along the axis that are up to 10% of the radius of the coil, the magnetic field varies from the value calculated above by less than 1 part in 5000 (0.02%).

•**59.** Two parallel wires, separated by a distance d, are oriented coming out of the page in Figure P.59. Each wire is of infinite length and carries current I. The point P lies equidistant from each wire. Find the magnitude and direction of the magnetic field at point P: (a) if the current in both wires is along $\hat{\mathbf{k}}$; (b) if the current in the left wire is along $\hat{\mathbf{k}}$ and that in the right wire is along $-\hat{\mathbf{k}}$.

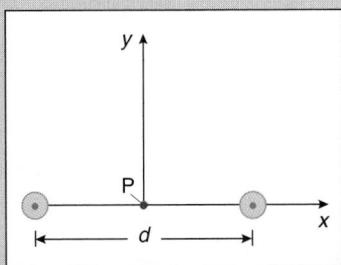

FIGURE P.59

•**60.** Two parallel, infinitely long wires, separated by a distance d, carry parallel currents I and βI, where β is a purely numerical multiple (see Figure P.60). (a) Find the location(s) of the lines along which the total magnetic field is 0 T. (b) Use the result of part (a) to show that if $\beta = 1$, the location of the line is midway between the two wires.

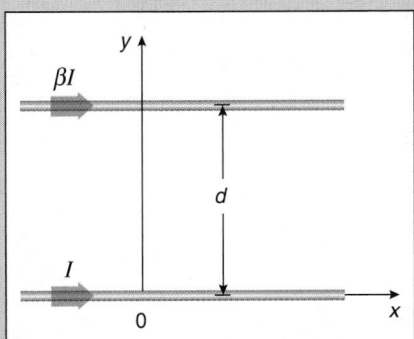

FIGURE P.60

❄**61.** A wire snakes along an arbitrary curved path in the *x*–*y* plane between the origin and point P; see Figure P.61. A uniform magnetic field is parallel to $\hat{\mathbf{k}}$. An appropriate source (not shown) establishes a current I in the wire. (a) By considering a series of infinitesimal steps parallel to $\hat{\mathbf{i}}$ and $\hat{\mathbf{j}}$, show that the total force on the wire is the same as that on a straight wire between the origin and point P. (b) Show that if the wire is a complete loop, so that P is coincident with the origin, the total force on the wire vanishes.

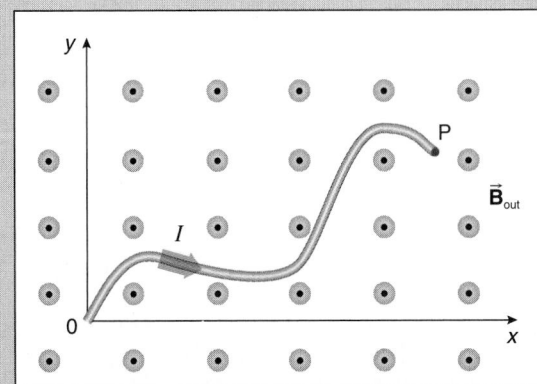

FIGURE P.61

❄**62.** A wire with current I follows the y-axis from $y = +\infty$ m to the origin, then the x-axis out to $x = +\infty$ m, as shown in Figure P.62. Show that the magnetic field in the first quadrant (where $a > 0$ m and $b > 0$ m) is

$$\vec{\mathbf{B}} = \frac{\mu_0}{4\pi} I \left[\frac{1}{a} + \frac{1}{b} + \frac{a}{b\left(a^2 + b^2\right)^{1/2}} + \frac{b}{a\left(a^2 + b^2\right)^{1/2}} \right] \hat{\mathbf{k}}$$

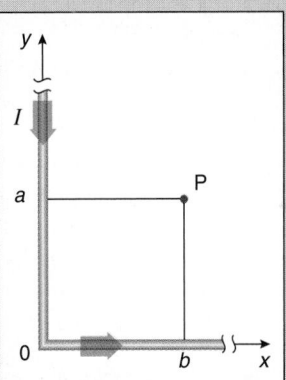

FIGURE P.62

❄**63.** The magnetic field of the Earth has a south magnetic pole in the northern geographic hemisphere and a north magnetic pole in the southern geographic hemisphere. The Earth also rotates in an eastward sense, taking one day to complete a rotation. For simplicity, assume that the line between the magnetic poles is coincident with the rotational axis of the Earth. (a) Imagine that the magnetic field of the Earth is produced by an excess charge distributed uniformly over the surface of the Earth. The spin of the Earth means that the excess charges form current loops. In order to produce the observed polarity of the magnetic field of the Earth, must the excess charge of the Earth be positive or negative? Indicate your reasoning. (b) Imagine (unrealistically) the excess charge Q to be localized in an equatorial ring of radius essentially equal to the radius of the Earth (6370 km). Given that the rotational period of the Earth is $T = 23$ h 56 min, what is the effective current represented by this equatorial distribution of charge? (c) The order of magnitude of the magnetic field of the Earth is about 10^{-4} T. Make the assumptions and use the results of part (b). What must be the magnitude of the charge Q to produce a magnetic field of magnitude $\approx 10^{-4}$ T at the center of the equatorial current loop?

‡64. A charge q moves at constant velocity $v_0\hat{i}$ along the x-axis as shown in Figure P.64. The charge passes the origin when $t = 0$ s. The moving charge is, of course, a current and thus produces a magnetic field. At any given point in space, this magnetic field varies with time. Take a point P to be located at coordinate y along the y-axis. (a) Give arguments to demonstrate that the magnetic field at P is directed along $\hat{k}$ at all times. (b) Show that the magnetic field at P varies with time as

$$\vec{B}(t) = \frac{\mu_0}{4\pi} \frac{qv_0y}{(y^2 + v_0^2t^2)^{3/2}} \hat{k}$$

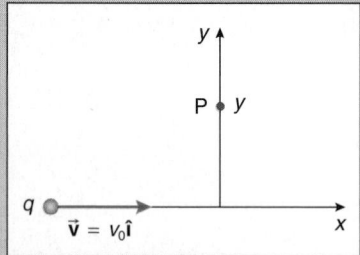

FIGURE P.64

Sections 20.7 Forces of Parallel Currents on Each Other and the Definition of the Ampere
20.8 Gauss's Law for the Magnetic Field*
20.9 Magnetic Poles and Current Loops

•65. You discover an electron promenading around in a circle (see Figure P.65) under the action of a magnetic force. (a) In a sketch, indicate the direction of the magnetic force on the electron. (b) On the same sketch, indicate the direction of the magnetic field causing the circular motion of the electron. (c) Beginning with Newton's second law, find the relationship between the radius r of the circle, the magnetic field magnitude B, and the intrinsic properties of the electron (charge, mass, etc.). (d) Calculate the work done by the magnetic force as the electron traverses one orbital path. (e) A great distance away (so you can effectively ignore the circulating electron), a proton is found at rest in the same field. Describe what happens to the proton.

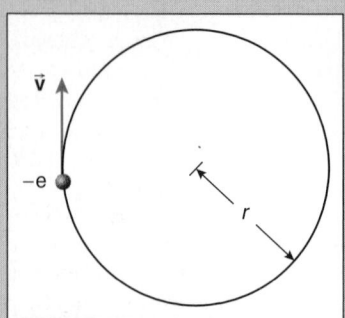

FIGURE P.65

•66. A square current loop 15.0 cm on a side and carrying 20.0 A clockwise in the x–y plane (see Figure P.66) finds itself in a uniform magnetic field $\vec{B} = (4.00 \text{ T})\hat{i} + (5.00 \text{ T})\hat{k}$. (a) What is the magnetic dipole moment of the current loop? (b) What is the torque on the current loop? (c) Will the current loop rotate? (d) What is the potential energy of the current loop in its initial orientation? Is this the minimum value of the potential energy of the current loop? (e) Calculate the flux of the magnetic field through the current loop when it is in its initial orientation.

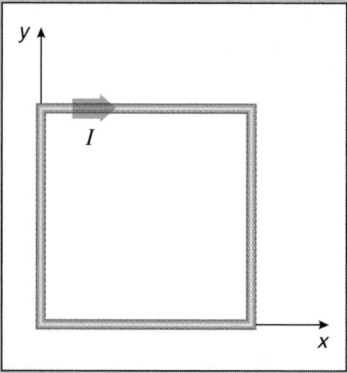

FIGURE P.66

•67. An infinite straight wire carrying a current I produces a magnetic field surrounding the wire. (a) Is the magnetic field uniform over the entire rectangular area depicted in Figure P.67a? (b) Consider the area vector of the rectangle to be in the same direction as the magnetic field at its location. Show that the magnetic flux through the rectangular area is

$$\Phi = \frac{\mu_0}{4\pi} 2I\ell \ln\left(\frac{b}{a}\right)$$

(c) What is the magnetic flux through a circular area of radius R centered on and perpendicular to the wire as in Figure P.67b?

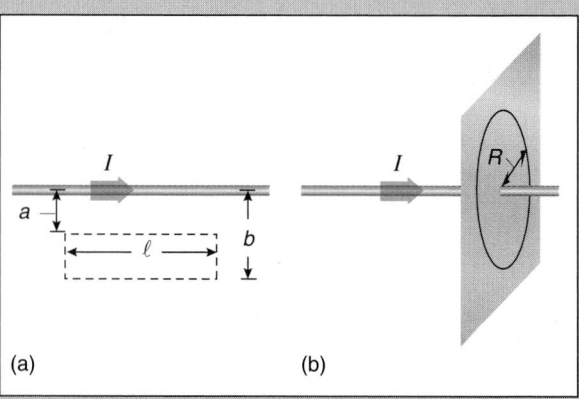

FIGURE P.67

•68. (a) A closed Gaussian surface surrounds the positive charge of an electric dipole as shown in Figure P.68a. What is the flux of the electric field through the Gaussian surface? Sketch the electric field lines in the vicinity of the dipole. (b) A closed Gaussian surface surrounds the north pole end of a bar magnet as shown in Figure P.68b. What is the flux of the magnetic field through the closed surface? Sketch the magnetic field lines around and through the magnet. Why is this result so different from the electrical dipole case in (a)?

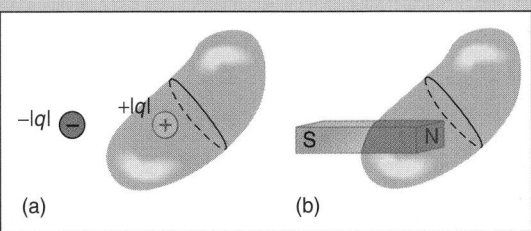

FIGURE P.68

•69. An electron in a TV tube moves with speed 7.50×10^6 m/s into a region with a uniform magnetic field of magnitude 50.0 mT. (a) What is the angle between the velocity of the electron and the magnetic field that produces the maximum magnitude of force on the electron? What is this force magnitude? (b) If the magnitude of the force is only 0.25 the maximum, what is the angle between the velocity of the electron and the magnetic field direction? (c) What is the initial kinetic energy of the electron in eV? (d) What is the kinetic energy of the electron as it leaves the region with the magnetic field?

Sections 20.10 Ampere's Law*
20.11 The Displacement Current
and the Ampere–Maxwell Law*
20.12 Magnetic Materials*
20.13 The Magnetic Field of the Earth*

70. Many parallel wires pierce the page as indicated in Figure P.70. Each wire carries a current of 1.5 A either into or out of the page (as indicated). (a) What is the path integral of the magnetic field taken clockwise around the indicated closed contours? (b) What is the path integral if it is taken counterclockwise around each path?

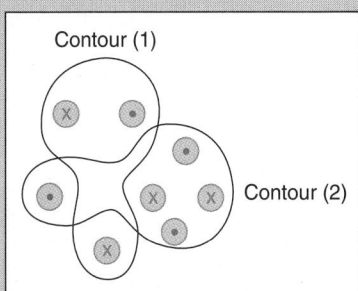

FIGURE P.70

•71. Design a long solenoid to produce a magnetic field of magnitude 50.0 mT along its axis with a current no greater than 15.0 A.

•72. A long, hollow, cylindrical conductor with inner radius a and outer radius b carries a current I uniformly distributed over the cross-sectional area of the conductor (see Figure P.72). (a) Use Ampere's law to show that the magnitude of the magnetic field at a radius r from the axis of the conductor: (a) is zero if $r < a$; (b) is

$$\frac{\mu_0}{4\pi} \frac{2I}{r} \frac{r^2 - a^2}{b^2 - a^2}$$

if $b < r < a$; (c) is equal to the magnitude of the field of an infinite wire,

$$\frac{\mu_0}{4\pi} \frac{2I}{r}$$

if $r > b$. (d) When $a = 1.00$ cm, $b = 1.20$ cm, and $I = 25.0$ A, make an accurate graph of the magnitude of B versus r over the domain 0 cm $< r <$ 2.00 cm.

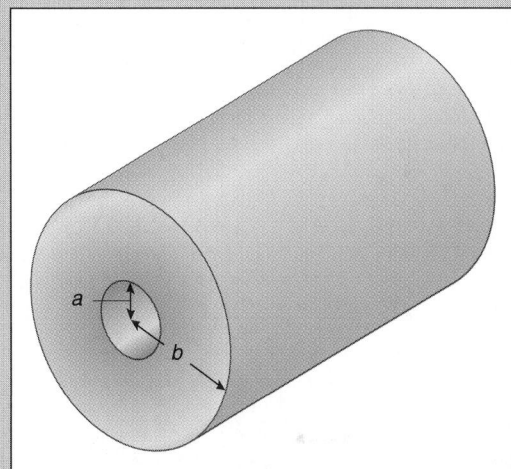

FIGURE P.72

•73. A uniform magnetic field in a region is shown in Figure P.73. The field line diagram implies there is zero field to the right and left of the indicated field lines on the far right and left of the figure, respectively. Using the indicated closed path and

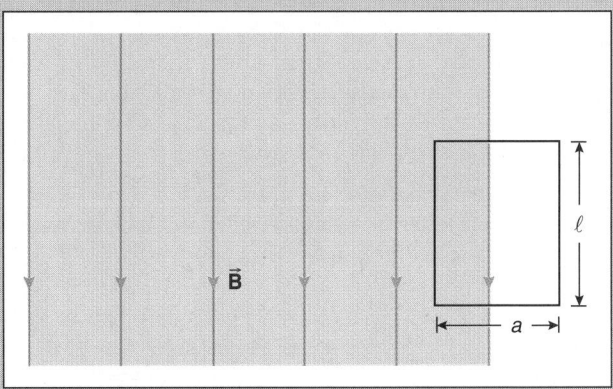

FIGURE P.73

Ampere's law, show that the magnetic field *cannot* drop abruptly to zero as implied by the field line diagram. Real magnets have so-called *fringing fields* to the right and left of the indicated field lines, so the magnetic field gradually approaches zero, rather than abruptly.

•74. Three infinite wires, each carrying a current of 5.0 A, are indicated in Figure P.74. What are the magnitude and direction of the magnetic field at the point $x = 2.0$ cm, $y = 6.0$ cm produced by these currents?

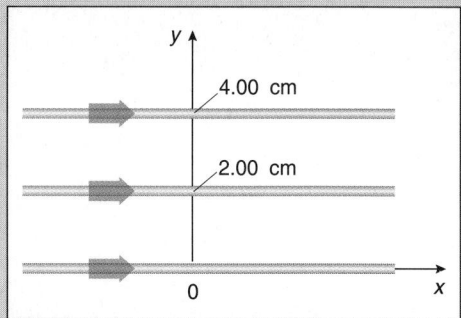

FIGURE P.74

INVESTIGATIVE PROJECTS

A. Expanded Horizons

1. Birds use a number of factors to sense direction when migrating over long distances, among them the magnetic field of the Earth. Investigate and report on the status of research on magnetic effects affecting navigation in migratory birds.
   Thomas Alerstam, *Bird Migration* (Cambridge University Press, New York, 1990).
   Chris Mead, *Bird Migration* (Facts on File, New York, 1983).
   Kenneth P. Able and Mary A. Able, "Daytime calibration of magnetic orientation in a migratory bird requires a view of skylight polarization," *Nature*, 364, #6437, pages 523–525 (5 August 1993).
   Kenneth P. Able and Mary A. Able, "Calibration of the magnetic compass of a migratory bird by celestial rotation," *Nature*, 347, #6291, pages 378–380 (27 September 1990).

2. You might enjoy reading about and summarizing the life of Nikola Tesla for it is an interesting case study of ambition, invention, and failure. He died in poverty as a complete recluse, living in the anonymity of a single-occupancy, resident hotel in New York City.
   John J. O'Neill, *Prodigal Genius: The Life of Nikola Tesla* (Washburn, New York, 1944).
   Stephen S. Hall, "Tesla: a scientific saint, wizard, or carnival salesman?" *Smithsonian*, 17, #3, pages 120–134 (June 1986).
   Bill Lawren, "Rediscovering Tesla," *Omni*, 10, #6, pages 64–68 (March 1988).
   Margaret Cheney, *Tesla: Man Out of Time* (Barnes & Noble, New York, 1993).
   Ben Johnston, editor, *My Inventions: The Autobiography of Nikola Tesla* (Barnes & Noble, New York, 1995)

3. As you know, electricity is used by various biological species for defense and to stun prey (see Investigative Project 8 in Chapter 16). Consult an appropriate faculty member in the biology department at your university to see if there are ways in which biological species exploit magnetism in ways other than for navigation on long migrations (see Investigative Project 1).

4. The characteristics, origin, and evolution of the magnetic field of the Earth still are active areas of research. Survey the problem from both a historical and contemporary viewpoint.
   David W. Strangway, *History of the Earth's Magnetic Field* (McGraw-Hill, New York, 1970).
   Jeremy Bloxham and David Gubbins, "The evolution of the Earth's magnetic field," *Scientific American*, 261, #6, pages 68–75 (December 1989).
   J. A. Jacobs, "The Earth's magnetic field and reversals," *Endeavor*, *New Series Volume 19*, #4, pages 166–171 (December 1995).

J. A. Jacobs, "The Earth's magnetic field," *Contemporary Physics*, 36, #4, pages 267–277 (July 1995).
Ray Ladbury, "Geodynamo turns toward a stable magnetic field," *Physics Today*, 49, #1, pages 17–18 (January 1996).

5. The navigational compass has a long and uncertain history that encompasses Chinese, Arabic, and European sources. Investigate its history and write a brief report about its origins and development.
   Encyclopaedia Britannica (11th edition, 1910), volume 6, pages 806–809.
   H. L. Hitchins and W. E. May, *From Lodestone to Gyro-compass* (Hutchinson, London, 1955).
   A. Crichton Mitchell, "Chapters in the history of terrestrial magnetism," *Terrestrial Magnetism and Atmospheric Electricity*, 37, pages 105–146 (1932); 42, pages 241–280 (1937); 44, pages 77–80 (1939); and 51, pages 323–351 (1946).

6. Detail the search for hypothetical magnetic monopoles.
   Alfred S. Goldhaber and W. Peter Trower, "Resource Letter MM-1: Magnetic Monopoles," *American Journal of Physics*, 58, #5, pages 429–439 (May 1990).
   Henry J. Frisch, "Quest for magnetic monopoles," *Nature*, 344, #6268, pages 706–707 (19 April 1990).

7. Electricity and magnetism have inspired several attempts at constructing perpetual motion machines, none of which can elude the consequences of the laws of thermodynamics. A look at these attempts may be interesting:
   Arthur W. J. G. Ord-Hume, *Perpetual Motion: The History of an Obsession* (St. Martin's Press, New York, 1977), Chapter 5, pages 83–93; also see the bibliography on pages 224–227.

8. Consult appropriate sources in your science library to make a table of the approximate value of the surface magnetic fields on the planets of the solar system, the Sun, and exotic stars such as white dwarfs and neutron stars.

9. For the appropriate orientations, a current experiences a magnetic force in a magnetic field. Imagine a liquid conductor such as mercury or sodium in a pipe. Design a system with a current through the conducting liquid and a magnetic field that produces a magnetic force of maximum magnitude that serves to pump the liquid along the pipe. Such *electromagnetic pumps* are used in some nuclear reactors to facilitate heat transfer. Mercury and sodium are very dangerous substances, and so such pumps must be very carefully constructed to avoid dangerous leaks and adverse environmental consequences.

B. Lab and Field Work

10. Your physics department likely has a small permanent bar magnet. Tape the magnet with its axis along an east–west direction on top of a large sheet of paper on a wooden table, well away from ferromagnetic materials. With a dime-store magnetic compass of small diameter, devise and execute a procedure for tracing out the magnetic field lines surrounding the magnet. The field lines represent the vector superposition of the magnetic field of the bar magnet with the local horizontal component of the magnetic field of the Earth. Determine the locations of the two places where the total magnetic field is zero.

11. Carefully devise and perform a series of experiments on a small charge (such as the charge that can be placed on a small, tethered, conducting ball) in a magnetic field to illustrate the facts on which Equation 20.1 is based.

12. For historical perspective, investigate the methods and experiments used by Biot and Savart to formulate the Biot– Savart law.
Herman Erlichson, "The experiments of Biot and Savart concerning the force exerted by a current on a magnetic needle," *American Journal of Physics*, 66, #5, pages 385–391 (May 1998).

13. For whom the bell tolls. Your life as a student likely is regulated by the periodic ringing of class bells. Your professor likely can provide you with an example of such an electrically activated bell. Explore the construction of such a bell. Sketch a circuit representation of the bell system and explain how a current provided to the bell makes it ring repetitively.

14. Magnetic stirrers are used frequently in chemistry and biology laboratories. Investigate and describe the principles underlying their operation.

15. If your physics department has a Hall probe for measuring the magnitude of magnetic fields, explore the magnitude of the magnetic field between the pole faces of a permanent magnet or within a pair of Helmholtz coils (see Problem 58).

16. Devise and perform an experiment to measure the magnitude and direction of the magnetic field of the Earth at your location.

17. Discover how a degausser is able to eliminate the permanent magnetic field of a material. (See Section 20.12.)

C. Communicating Physics

18. Your former high school physics teacher has invited you back to give a talk about fields to a group of high school physics students. For this audience, create a 30-minute talk about the field concept. Illustrate the talk with a clear discussion of the distinctions among electric, magnetic, and gravitational fields. A few hands-on demonstrations to accompany the talk would enhance its interest for the audience.

19. Write a short paragraph about the origin of the word *lodestone*. Consult the *Oxford English Dictionary*.

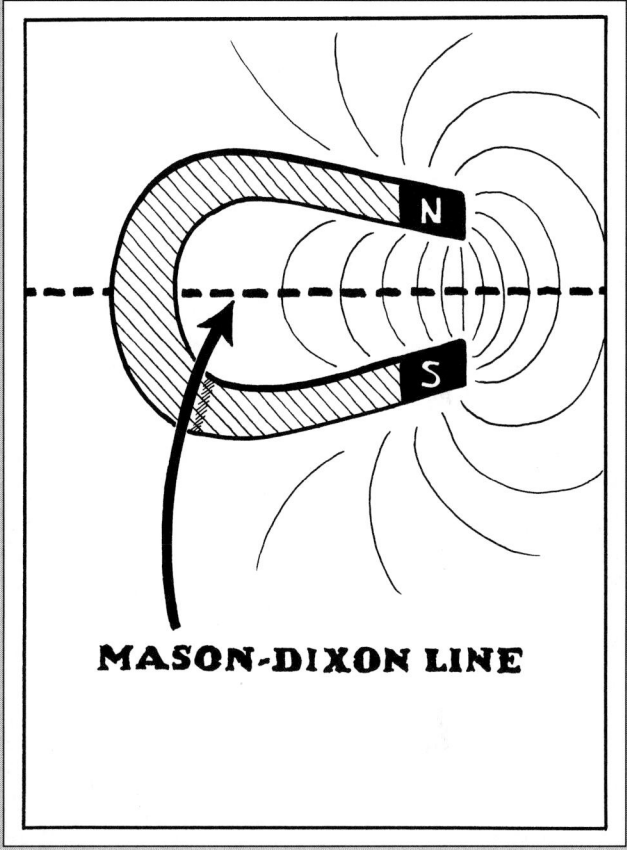

MASON-DIXON LINE

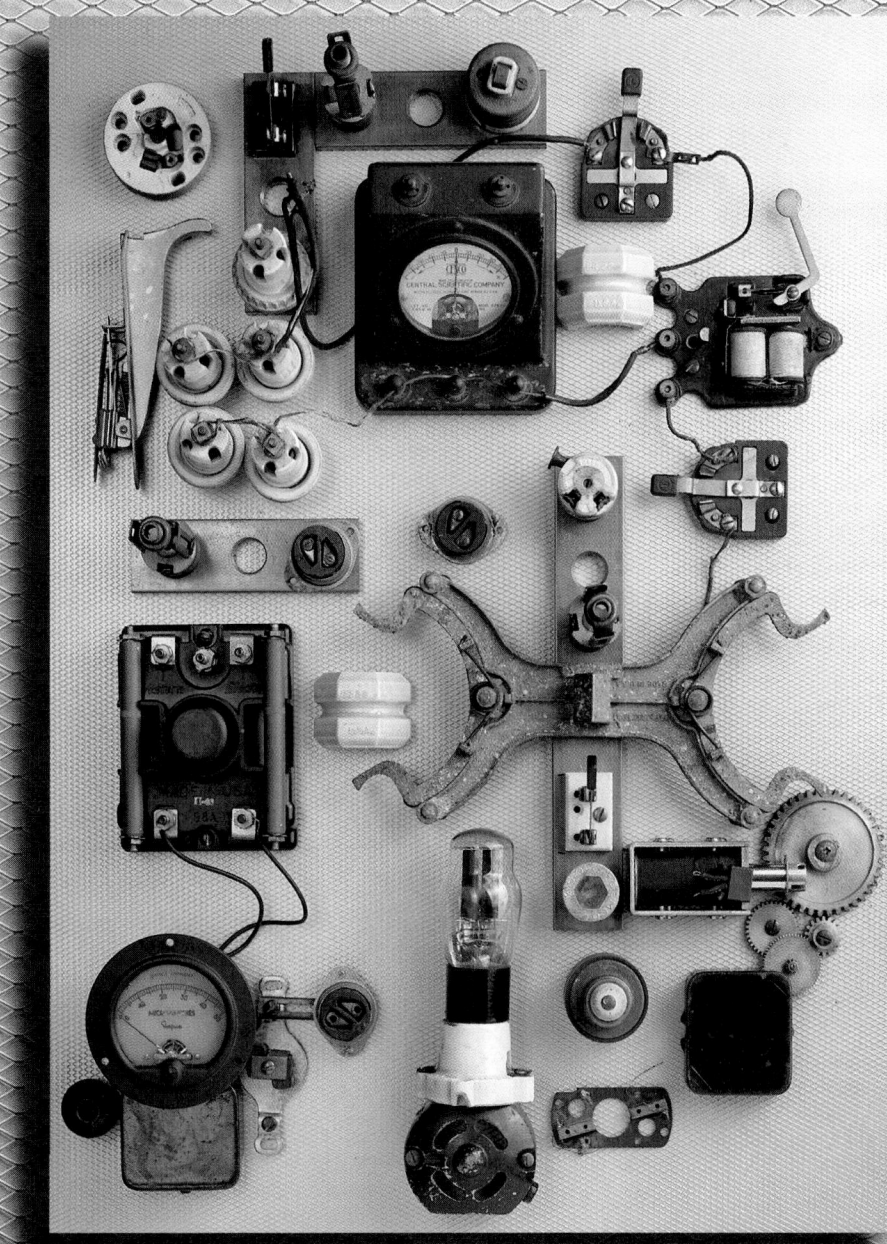

CHAPTER 21

FARADAY'S LAW OF ELECTROMAGNETIC INDUCTION

*And yet it is Faraday's spark which now shines upon our coasts,
and promises to illuminate our streets, halls, quays, squares,
warehouses, and perhaps at no distant day, our homes.*

John Tyndall (1820–1893)*

Without Faraday's law, you likely would be using beeswax and kerosene to study by candle and lamp light. Chemical batteries or Wimshurst machines would be needed to operate every electrical appliance. Electricity would be prohibitively expensive.[†] Generating electricity cheaply for common and geographically widespread use relies on Michael Faraday's remarkable discovery of the law of electromagnetic induction.

There is an interesting story about Faraday (1791–1867), likely apocryphal,[‡] that goes like this. When the British Chancellor of the Exchequer William Gladstone (1809–1898) asked Faraday what earthly good ever could come of his discovery of electromagnetic induction, Faraday is said to have replied: "Why sir, there is every probability that you will soon be able to tax it!" And we thought tax-and-spend was just a recent phenomenon!

21.1 FARADAY'S LAW OF ELECTROMAGNETIC INDUCTION

In the last chapter (Section 20.11) we discovered that the time-varying flux of the electric field produces a magnetic field via the displacement current. The fields are mutually perpendicular

---

[*](Chapter Opener) John Tyndall, *Fragments of Science* (6th edition, D. Appleton, New York, 1897), Volume 2, Chapter XVI, "The Electric Light," page 452. First edition published in 1871.

[†]It costs about a kilodollar ($1000) to provide a kilowatt-hour of energy using batteries!

[‡]See the letters by I. Bernard Cohen, *Nature*, *157*, #3981, pages 196–197 (16 February 1946) and by R. A. Gregory, *Nature*, *157*, #3984, page 305 (9 March 1946), which cast doubt on the authenticity of this story. The quotation is taken from the letter by Cohen.

Faraday came from an impoverished family and his formal education was essentially nonexistent. A deeply religious man, Faraday had a strong sense of community and a deep affection for children (though he had none of his own). Apprenticed as a bookbinder, his interest in electricity was sparked when he chanced upon an article about it while rebinding a volume of the *Encyclopaedia Britannica*.

(orthogonal). What about the converse: does a time-varying flux of the magnetic field produce an electric field?

One of the most technologically significant discoveries of the 19th century occurred when Michael Faraday realized in 1831 that this is indeed the case. Since the displacement current was invented by Maxwell about 30 years after Faraday's work, the originality of Faraday's discovery is all the more remarkable. It also is humbling to know that the most sophisticated mathematics ever used by Faraday in his technical papers and discoveries never went beyond simple algebra and arithmetic. Faraday was a true experimental genius, who insisted on using and exploiting the richness of the English language to the fullest. He was a master wordsmith. It was James Clerk Maxwell (1831–1879) who first used sophisticated mathematics to cogently describe the wealth of experimental information gleaned by Faraday from nature.

We easily can perform two experiments similar to those undertaken by Faraday.

Experiment 1: A Magnet and a Loop of Wire

Hold a permanent magnet near a closed loop or coil of wire, as in Figure 21.1. A very sensitive galvanometer (or current detector) is connected in series with the wire to detect the presence and direction of any, even a very small, electric current in the wire loop. The wire loop represents an electric circuit although there are no batteries or other sources of energy in the circuit. The magnetic field of the magnet produces a magnetic flux through the coil of wire, since the lines of the field thread the coil.[§] The following qualitative observations then are made:

1. If the magnet is held in a fixed position, so that the magnetic flux through the loop is constant (independent of time), the galvanometer indicates there is zero current in the wire loop.
2. If the magnet is moved toward the loop, as shown in Figure 21.1, thus increasing the magnetic flux through the loop of wire, a current is detected in the wire as long as the magnetic flux through the loop is changing with time. The faster the magnet is moved, the greater the time rate of change of the magnetic flux, and the greater the current observed in the loop.
3. If the magnet is moved away from the loop, as in Figure 21.2, thus decreasing the magnetic flux through the loop, the current observed in the wire is in the opposite direction. Once again: the faster the magnet is moved, the greater the change in the magnetic flux through the loop, and the greater the current observed in the wire.

We can summarize these results by saying that a current is observed in the loop as long as the magnetic flux through the loop is changing with time. Indeed, the greater the time rate of change of the magnetic flux, the greater the current in the loop: they are directly proportional. If the magnetic flux through the loop is constant (not changing with time), no current is observed in the loop. The current in the loop is in one direction when the magnetic flux through the loop is increasing with time, and in the opposite direction when the magnetic flux decreases with time. In this experiment the changing magnetic flux is associated with the motion of the magnet.

[§]Remember that the flux is a quantitative measure of the extent to which the field lines thread the loop.

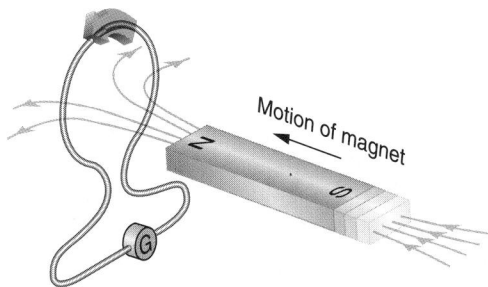

FIGURE 21.1 If the magnet is moved toward the coil, a current is detected in the coil. The curvature of the magnetic field lines through the coil here (and in similar drawings that follow below and on pages 961–962) is exaggerated to clarify the perspective.

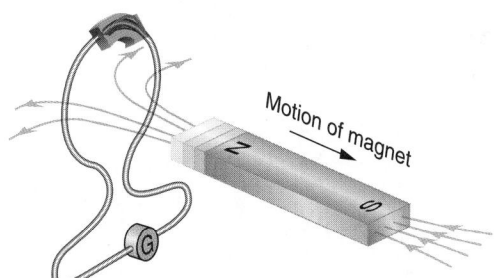

FIGURE 21.2 If the magnet is moved away from the coil, a current is detected in the opposite direction.

Experiment 2: Two Loops of Wire, a Battery, and a Switch

In this experiment no permanent magnet is used. Rather, the coil and galvanometer system used in Experiment 1 is placed near another loop or coil of wire connected to a battery and a switch, as shown in Figure 21.3.

The following observations then are made:

1. With the switch open in the circuit with the battery, there is no current in the wire loop on the right.* No magnetic fields are present. The magnetic flux through the wire loop connected to the galvanometer is zero. No current is detected by the galvanometer in the wire loop.
2. Now we close the switch in the circuit on the right, thus establishing a current in it (see Figure 21.4). The loop on the right can be arranged so that the direction of the magnetic field produced by the current in the wire is the same as that of the magnet in Experiment 1. During the very brief interval during which the current increases in the circuit on the right, the magnetic field of this current also increases and briefly creates an increasing magnetic flux through the loop connected to the galvanometer. The galvanometer detects a current in the wire loop on the left *only* during this brief interval during which the magnetic field of the current in the circuit on the right is changing with time.

   The direction of the current detected by the galvanometer is in the same sense as when the magnetic flux through the loop in Experiment 1 was increasing with time. When

---

*The network on the right actually is not a circuit when the switch is open!

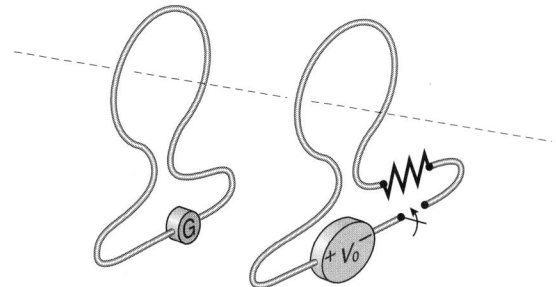

FIGURE 21.3 Two loops or coils, one with a battery and switch.

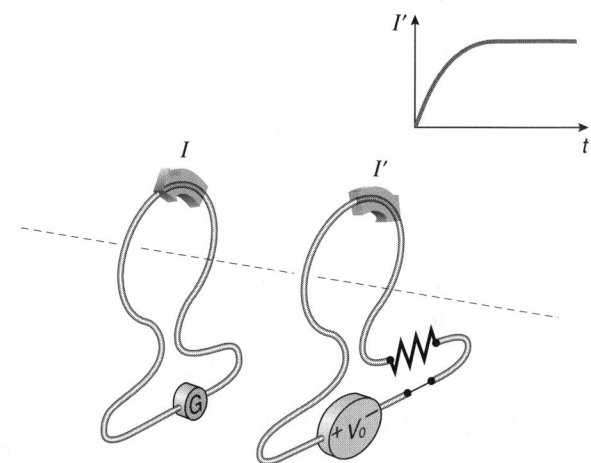

FIGURE 21.4 A current I is detected by the galvanometer only while the current I' in the other coil is increasing with time.

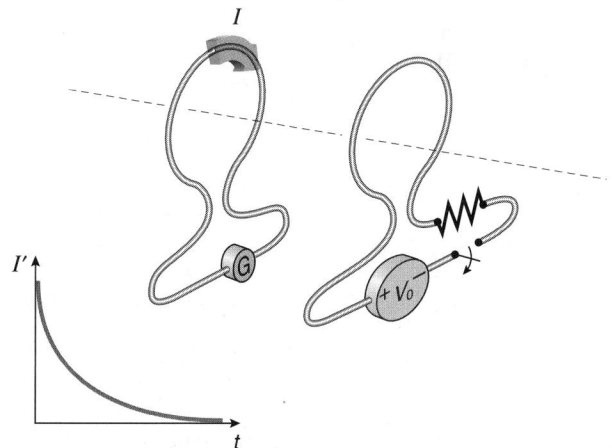

FIGURE 21.5 A current I is detected in the opposite direction when the current I' in the other coil decreases with time.

the current in the circuit on the right reaches a constant value, making the magnetic flux through the loop on the left constant, the galvanometer detects no current in the loop.

3. The switch in the circuit on the right now is opened (see Figure 21.5), and the current in it falls to zero during a very brief time interval as the contact is broken. The magnetic flux through the loop on the left now decreases quickly with

time and the galvanometer detects a current in the opposite direction to the current detected in #2. The current detected by the galvanometer is in the same direction as the current in Experiment 1 when the magnetic flux through the loop was decreasing with time. When the current in the circuit on the right reaches zero (and becomes constant), no current is detected by the galvanometer in the loop on the left.

We summarize these results as follows. During those intervals in which the magnetic flux through the loop with the galvanometer changes with time, a current is detected in the loop. The current is in one direction when the magnetic flux is increasing with time, and in the opposite direction when the magnetic flux through the loop is decreasing with time. In this experiment, the change in the magnetic flux is not caused by the motion of a magnet (as in Experiment 1) but by a change in the magnetic field itself.

A current exists in the wire loop with the galvanometer as long as the magnetic flux through the loop is changing with time. Such a current is called an **induced current**, since it is induced or produced not by conventional batteries but by a changing magnetic flux through the loop.

Electric currents, of course, are charges in motion. Electric charges move in response to electrical forces caused by electric fields* via Equation 16.8:

$$\vec{F}_{elec} = q\vec{E}$$

Hence we can say that a changing magnetic flux through a loop induces (i.e., produces) an electric field—called an **induced electric field**—that causes the charges to move and produce the electric current detected by the galvanometer.

We shall have more to say about this induced electric field shortly; it is an electric field of a very different character than the electric fields we have encountered previously.

The experimental genius of Faraday was complemented by equal theoretical insight into the real significance of his discoveries. Faraday realized that the presence of the wire loop with the galvanometer served only to reveal or manifest the induced electric field. That is, charges present in the conducting wire loop detect the presence of the induced electric field; remove the wire loop and we are left with the mere *mathematical* outline of the closed path. Nonetheless, even if the conductor is absent, the induced electric field (caused by the changing magnetic flux) *still* is present in space.[†]

To investigate the nature of this induced electric field more thoroughly, we consider a more symmetrical situation. A circular wire loop of planar area A is located in a region of space where there exists a uniform magnetic field $\vec{B}$; see Figure 21.6.

We let the area vector associated with the area of the wire loop be directed parallel to the field, so that we have a reference direction associated with the loop. With the thumb of your right hand along the direction of the area vector, the fingers of your right hand define the positive sense around the loop; this sense is counterclockwise in Figure 21.6.

Now, as Faraday did, we perform three experiments: (1) If the magnetic field does not change, there is no induced current observed in the loop. (2) If the magnetic field increases in magnitude, thus increasing the magnetic flux through the loop, the direction of the induced current in the loop is observed to be in the negative sense (here, clockwise around the loop in Figure 21.7). (3) If the magnitude of the magnetic field is decreasing, thus decreasing the magnetic flux through the loop, the direction of the induced current in the loop is in the positive sense (here, counterclockwise around the loop in Figure 21.8).

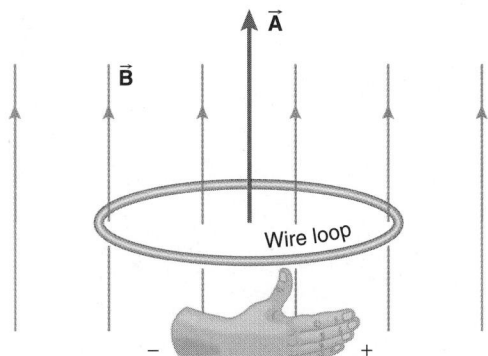

FIGURE 21.6 A circular loop wire loop in a uniform magnetic field.

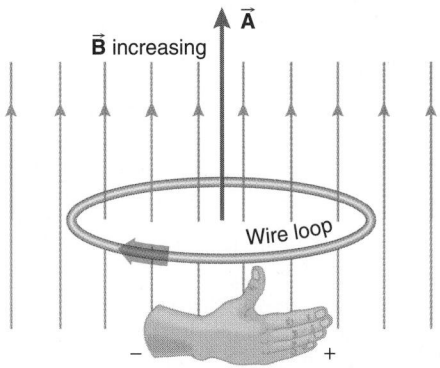

FIGURE 21.7 Increase the magnetic field and the induced current is in the negative sense as the change in flux takes place.

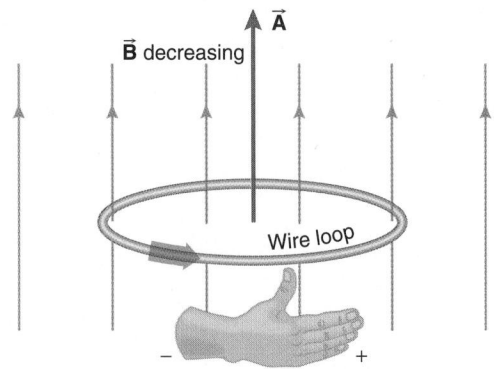

FIGURE 21.8 Decrease the magnetic field and the induced current is in the positive sense as the change in flux takes place.

---

*Recall that a magnetic field exerts zero force on a charge at rest. Only an electrical force can begin to move a charge at rest.

[†]This is just like what happens when a charge Q establishes a static electric field in the surrounding space (that can act on another charge q *if it is present* in the field).

What is the direction of the induced electric field? Since the charges are moving around the loop, we surmise that the induced electric field also is directed around the loop in the same sense as the current (the direction of positive charge motion). You might wonder if the field might be directed in a radial sense; this is not the case because a radial electric field would not cause charge to move around the loop. So the direction of the electric field induced by the changing magnetic flux through the loop is around the circumference of the loop in the direction of the current induced. This is a new type of electric field, quite unlike the static electric field encountered in our study of electrostatic phenomena in Chapter 16. The electric field lines representing the induced electric field form *closed contours*!

The fact that the induced electric field lines form closed loops (just as all magnetic field lines do) means that the flux of the induced electric field through any closed surface is zero (just as the flux of any magnetic field through a closed surface is zero). Gauss's law for the electric field still holds and need not be modified: this new induced electric field has no flux through a closed surface, and so the only electric field that is significant in Gauss's law for the electric field is the *static* electric field arising from stationary electric charges. *Static* electric field lines do *not* form closed loops. The static electric field lines begin on positive charge and end on negative charge.

The induced electric field produces an electrical force on a charge q. Let's calculate the work done by the electrical force in taking a charge q around a circular closed path when the magnetic field through it is increasing in magnitude. This work will be

$$W_{\text{elec}} = \int_{\text{clsd path}} \vec{\mathbf{F}}_{\text{elec}} \cdot d\vec{\ell}$$

Since $\vec{\mathbf{F}}_{\text{elec}} = q\vec{\mathbf{E}}_{\text{induced}}$, we have

$$W_{\text{elec}} = q \int_{\text{clsd path}} \vec{\mathbf{E}}_{\text{induced}} \cdot d\vec{\ell}$$

The induced electric field is parallel to $d\vec{\ell}$ around the whole path, and so the scalar product simplifies:

$$W_{\text{elec}} = q \int_{\text{clsd path}} E_{\text{induced}} \, d\ell$$

The circular symmetry indicates that the magnitude of the induced electric field is constant in magnitude along the path, yielding

$$W_{\text{elec}} = q E_{\text{induced}} \int_{\text{clsd path}} d\ell$$
$$= q E_{\text{induced}} (2\pi r)$$

where r is the radius of the circular loop. The work done by the electrical force due to the induced electric field around a closed path is *not zero*.

The electrical force produced by the induced electric field is *not* a conservative force! The induced electric field is not like an electrostatic electric field caused by charges at rest (static electric fields *are* conservative).

The electric work done per unit charge around the closed path is

$$\frac{W_{\text{elec}}}{q} = E_{\text{induced}} (2\pi r) \qquad (21.1)$$

Faraday discovered that the work done per unit charge by the induced electric field around a closed path is equal to the negative of the time rate of change of the magnetic flux through the same path:

$$\int_{\text{clsd path}} \vec{\mathbf{E}} \cdot d\vec{\ell} = -\frac{d\Phi_{\text{magnetic through the enclosed area}}}{dt} \qquad (21.2)$$

This is **Faraday's law of electromagnetic induction**.

We can immediately generalize the circular path to one of any shape.

The dimensions of the left-hand side of Equation 21.2 are those of the electric field times a distance; using SI units, we have:

$$(\text{N/C})(\text{m}) = \text{J/C} = \text{V}$$

or the same as the electric potential. Notice that the left-hand side of Faraday's law is a path integral of the electric field (around a *closed path*) and thus is similar to the path integral (*not* necessarily around a closed path), Equation 17.9, used to define the (static) electric potential difference between two points. The similarity is the reason that the left-hand side of Faraday's law is called an **induced emf**.

The induced emf is, by definition, the path integral

$$\text{induced emf} \equiv \int_{\text{clsd path}} \vec{\mathbf{E}} \cdot d\vec{\ell} \qquad (21.3)$$

Faraday imagined the motion of charges arising from the induced emf to be motivated or caused by what he picturesquely called an "electromotive force." We use the term *emf* rather than electromotive force because the emf is *not* a force as we now define the word in physics. The work done by the electric field arising from Faraday's law also is *not* to be associated with a potential energy (as it was for electrostatic electric fields) since potential energies *only* can be associated with conservative forces. Unlike the electrostatic force, the electrical force stemming from the induced electric field in Faraday's law is *not* a conservative force: it does finite (nonzero) work around a closed path.

Faraday's law (Equation 21.2) also is commonly written as

$$\text{induced emf} = -\frac{d\Phi_{\text{magnet}}}{dt} \qquad (21.4)$$

Faraday's law of electromagnetic induction is an important, very general law of electromagnetism and is the last of the four fundamental equations of electromagnetism known collectively as the Maxwell equations.

Returning to our specific circular example, the magnetic flux of the uniform magnetic field through the area bounded by the path here is

$$\Phi_{\text{magnet}} = \int \vec{\mathbf{B}} \cdot d\vec{\mathbf{A}}$$
$$= BA$$

Since the flux is changing because the field is changing, we have

$$\frac{d\Phi_{\text{magnet}}}{dt} = A \frac{dB}{dt}$$

The right-hand side of Faraday's law (Equation 21.2) then is

$$-A\frac{dB}{dt}.$$

The left-hand side of Faraday's law is the induced emf, which we found to be given by Equation 21.1:

$$E_{induced}(2\pi r)$$

Thus Faraday's law for this specific case becomes

$$E_{induced}(2\pi r) = -A\frac{dB}{dt}$$

Since $A = \pi r^2$ for the circle, we have, for the induced field,

$$E_{induced} = -\frac{r}{2}\frac{dB}{dt}$$

which shows explicitly how a changing magnetic field can give rise to an induced electric field. You also can note from Figures 21.7 and 21.8 that the induced electric field (directed around the circumference of the circular path) is perpendicular to the changing magnetic field that causes it.

Nature, along with our inventiveness, is neatly symmetric:

• changing magnetic fields give rise to induced electric fields via Faraday's law;

• changing electric fields give rise to magnetic fields via the displacement current and the Ampere–Maxwell law.

These electric and magnetic fields are mutually perpendicular to each other.

PROBLEM-SOLVING TACTIC

21.1 Notice that Faraday's law, as expressed by Equation 21.4, indicates that the induced emf is equal to the negative of the slope of a graph of Φ_{magnet} versus time.

A small, planar wire loop is placed in a magnetic field directed along $\hat{k}$. Sketch an orientation for the loop that produces zero magnetic flux through it. Sketch an orientation for the loop that maximizes the magnetic flux through it.

STRATEGIC EXAMPLE 21.1

The magnetic flux through a single current loop varies with time as indicated in Figure 21.9. Make a graph of the corresponding induced emf in the current loop as a function of time.

Solution

Problem-Solving Tactic 21.1 indicates that the induced emf is equal to the *negative* of the slope of the graph of Φ_{magnet} versus time. The induced emf thus is the negative of the slope of the

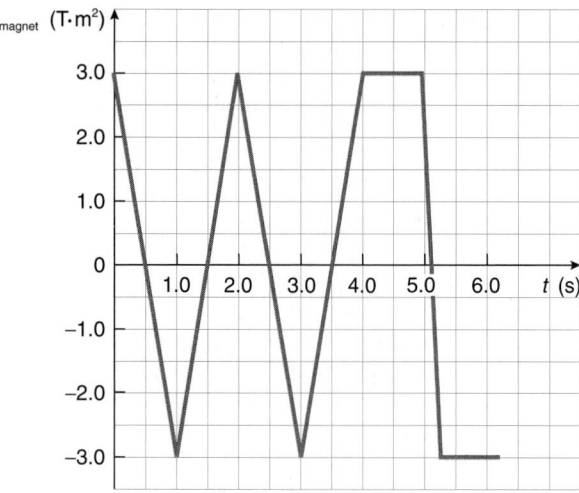

FIGURE 21.9

graph shown in Figure 21.9. The slopes of the various segments are found in the following way.

a. During the interval from $t = 0$ s to $t = 1.0$ s, the slope of the graph is constant and equal to

$$\text{slope} = \frac{-3.0\ \text{T}\cdot\text{m}^2 - (3.0\ \text{T}\cdot\text{m}^2)}{1.0\ \text{s}}$$

$$= -6.0\ \text{T}\cdot\text{m}^2/\text{s}$$

The induced emf is the negative of the slope:

$$\text{induced emf} = -(\text{slope})$$
$$= -(-6.0\ \text{T}\cdot\text{m}^2/\text{s})$$
$$= 6.0\ \text{T}\cdot\text{m}^2/\text{s}$$
$$= 6.0\ \text{V}$$

b. During the time interval from $t = 1.0$ s to $t = 2.0$ s, the slope of the graph is

$$+6.0\ \text{T}\cdot\text{m}^2/\text{s}$$

and the induced emf is

$$\text{induced emf} = -(+6.0\ \text{V})$$
$$= -6.0\ \text{V}$$

c. The slope of the graph during the interval from $t = 2.0$ s to $t = 3.0$ s is the same as from $t = 0$ s to $t = 1.0$ s, and so the induced emf is the same as what you calculated in part (a):

$$+6.0\ \text{V}$$

d. The slope during the interval from $t = 3.0$ s to $t = 4.0$ s is the same as in part (b), and so the induced emf is the same as in (b):

$$-6.0\ \text{V}$$

e. During the interval from $t = 4.0$ s to $t = 5.0$ s, the slope of the graph is zero; the magnetic flux is not changing with time during this interval, and so the induced emf is 0 V.

f. From $t = 5.0$ s to $t = 5.25$ s the slope of the graph is

$$\text{slope} = \frac{-3.0 \ \text{T} \cdot \text{m}^2 - (3.0 \ \text{T} \cdot \text{m}^2)}{0.25 \ \text{s}}$$

$$= -24 \ \text{T} \cdot \text{m}^2/\text{s}$$

and so the induced emf is

$$\text{induced emf} = -(\text{slope})$$

$$= +24 \ \text{V}$$

g. For times $t > 5.25$ s, the magnetic flux is constant; the slope is zero, and the induced emf is 0 V during this interval.

These results are summarized in Figure 21.10.

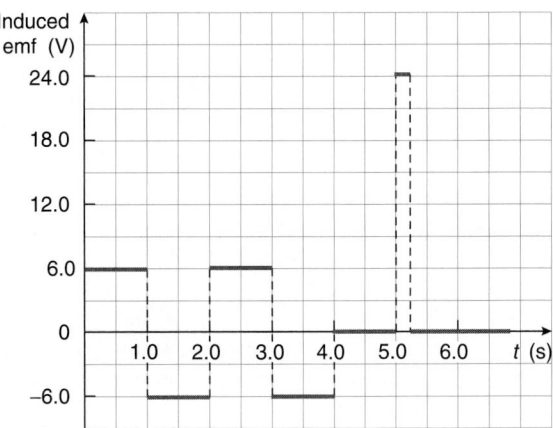

FIGURE 21.10

STRATEGIC EXAMPLE 21.2

A metal rod of length ℓ is rotated at angular velocity $\vec{\omega}$ about an end in a uniform magnetic field $\vec{B}$ that is parallel to $\vec{\omega}$, as shown in Figure 21.11. Find the absolute value of the induced emf and indicate which end of the rod is at the higher electric potential.

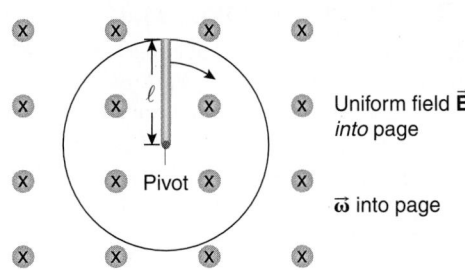

Uniform field $\vec{B}$
into page

$\vec{\omega}$ into page

FIGURE 21.11

Solution
Electrons in the conductor are free to move. Since the electrons find themselves moving in a magnetic field, the electrons experience a magnetic force given by Equation 20.1:

$$\vec{F}_{\text{magnet}} = q\vec{v} \times \vec{B}$$

$$= (-e)\vec{v} \times \vec{B}$$

This force is directed *toward* the pivot. Since $\vec{v}$ is perpendicular to $\vec{B}$, the force has a magnitude of

$$F_{\text{magnet}} = evB$$

The magnetic force produces charge separation in the conducting rod, with the tip of the rod becoming *positive* and the pivot end becoming *negative*.

Method 1
The separation of charge will not occur indefinitely, however, because the separated charges produce an *electric field* directed from the tip of the rod toward the pivot. This electric field produces an electrical force on the electrons that is in the direction opposite to the magnetic force. When the two forces are of equal magnitude, no further charge separation occurs. The result is a potential difference (an induced emf) between the pivot and the tip of the rod, with the outer tip at a higher potential. This effect is similar to the Hall effect discussed in Section 20.2. At equilibrium the electric and magnetic forces are equal in magnitude (but opposite in direction, so the total force is zero):

$$eE = evB$$

so that

$$E = vB$$

For circular motion $v = r\omega$; hence

$$E = r\omega B$$

This electric field is directed toward the pivot, so that

$$\vec{E} = -\omega B r \hat{r}$$

The potential difference between the ends of the rod is found from the definition of the potential difference, Equation 17.9:

$$V_f - V_i = -\int_i^f \vec{E} \cdot d\vec{r}$$

Integrate from the pivot to the tip:

$$V_{\text{tip}} - V_{\text{pivot}} = -\int_{0 \ \text{m}}^{\ell} (-\omega B r \hat{r}) \cdot dr \, \hat{r}$$

$$= \omega B \int_{0 \ \text{m}}^{\ell} r \, dr$$

$$= \frac{\omega B \ell^2}{2}$$

This is the emf induced between the ends of the rod.

Method 2
To calculate the magnitude of the induced emf, determine the flux of the magnetic field through a pie-shaped segment of the circle of angle θ, as shown in Figure 21.12. We express θ in radians.

The area S of the pie-shaped segment is $\theta/2\pi$ times the area of the circle itself:

$$S = \frac{\theta}{2\pi} \pi \ell^2$$

$$= \frac{\theta \ell^2}{2}$$

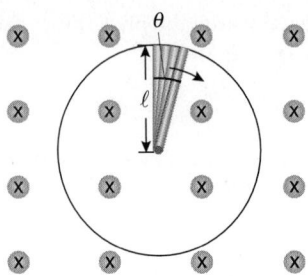

FIGURE 21.12

The flux Φ of the uniform magnetic field through the pie-shaped segment is

$$\Phi = \int \vec{B} \cdot d\vec{S}$$

The differential area vector $d\vec{S}$ can be taken in the same direction as $\vec{B}$ (since the area does not enclose a volume, you have a choice about the direction for the area vector). Thus the scalar product reduces to just $B\, dS$:

$$\Phi = \int B\, dS$$

Since the magnetic field is uniform, the flux becomes

$$\Phi = BS$$
$$= B\,\frac{\theta \ell^2}{2}$$

The absolute value of the induced emf is

$$\left| \text{induced emf} \right| = \left| \frac{d\Phi}{dt} \right|$$

The only time-dependent quantity is the angle θ. Thus

$$\left| \text{induced emf} \right| = \frac{B\ell^2}{2} \left| \frac{d\theta}{dt} \right|$$

But

$$\left| \frac{d\theta}{dt} \right|$$

is the angular speed ω. The absolute value of the induced emf then is

$$\left| \text{induced emf} \right| = \frac{\omega B \ell^2}{2}$$

The outer tip of the rod has the higher potential.

21.2 LENZ'S LAW

We need to find an easy way to determine the direction of the induced electric field caused by a changing magnetic flux. If a conducting loop is coincident with the closed path in Faraday's law, positive charges will flow around the conducting loop in

the same sense as the induced electric field (or negative charges will flow in the opposite direction), producing an induced current. So the way to find the direction of the field is to find a rule for determining the direction of the induced current* if a conducting path is provided. We emphasize again that the induced (nonconservative) electric field is present with or without the presence of a conducting path; a conducting path merely serves to detect the presence of the induced field by providing charges a chance to move in response to it, thereby producing an induced current.

Not long after Faraday discovered the law of electromagnetic induction, Heinrich Emil Lenz (1804–1865) formulated a convenient rule for determining the directional sense of an induced current around a closed conducting path coincident with the closed path in Faraday's law. The rule now is called **Lenz's law**. It is important to realize that Lenz's law contains nothing that is not already implicit in Faraday's law. Lenz's law is merely a rule to help us easily figure out the directional sense of the induced electric field and current (since positive charge flows in the same sense as the induced field) that arise from a changing magnetic flux. Regrettably, the word *law* occasionally is invoked too casually in physics.

> Lenz's law states that the induced current always will be directed so as to *oppose the change* in the magnetic flux that is taking place.

The induction "tries" to preserve the status quo and resists *any* change, much like a two-year-old child; in its own way, the induced current is quite contrarian.

PROBLEM-SOLVING TACTIC

21.2 To see how Lenz's law actually is used is best done by carefully examining a number of specific examples. See the examples that follow. Study them thoughtfully and with good cheer. Once you get a feeling for the examples, you will be able to figure out the directions of induced currents and induced electric fields with confidence.

STRATEGIC EXAMPLE 21.3

Return to the first experiment performed when we introduced Faraday's law. Part of this experiment involved moving a permanent magnet *toward* a loop or coil of wire connected to a galvanometer (used to detect the presence of the induced current and its direction); see Figure 21.13. Use Lenz's law to determine the direction of the current induced in the loop.

Solution
When the magnet is moved toward the loop, the magnetic flux through the loop increases. According to Lenz's law, the induced current is directed so as to oppose the change in the flux. You can determine the direction of the induced current in two ways.

---

*Remember that the conventional direction taken for current is that associated with the motion of positive charge.

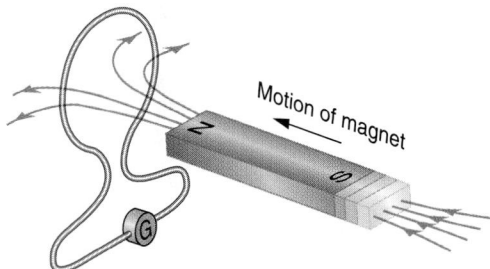

FIGURE 21.13

Method 1

Since the flux is increasing because the field is increasing through the loop, the direction of the induced current will be so that the *magnetic field it creates* is directed *opposite* to the field of the approaching magnet. Since the magnetic field of the magnet is directed through the loop from right to left and is increasing in magnitude (thus increasing the flux through the loop), the magnetic field of the induced current will be directed left to right through the loop to oppose the increasing field (thus decreasing the magnetic flux through the loop).

Grab the loop with your right hand so that your fingers thread the loop from left to right, as in Figure 21.14. The direction of the induced current will be in the direction of your extended right-hand thumb.

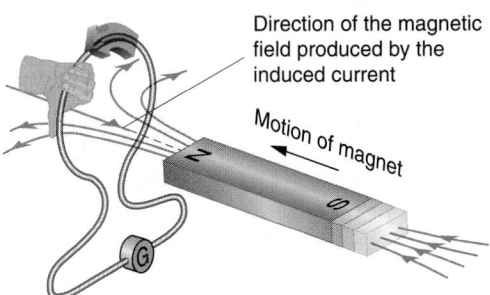

Direction of the magnetic field produced by the induced current

FIGURE 21.14

Method 2

Since the north end of the magnet is approaching the loop, the *magnetic field of the induced current* will be directed to oppose this motion. That is, the direction of the induced current will be such that the magnetic field it creates has a north magnetic pole facing the approaching north pole of the moving magnet; see Figure 21.15. The two like poles then will tend to repel each other, thus opposing the change (the approach of the north pole of the permanent magnet). In other words, the induced current opposes the change taking place. The direction of the induced current must be as indicated in Figure 21.15 to make the side of the loop facing the approaching magnet a north pole.

EXAMPLE 21.4

If the permanent magnet in Example 21.3 now is drawn away from the loop, as shown in Figure 21.16, use Lenz's law to determine the direction of the current induced in the wire loop.

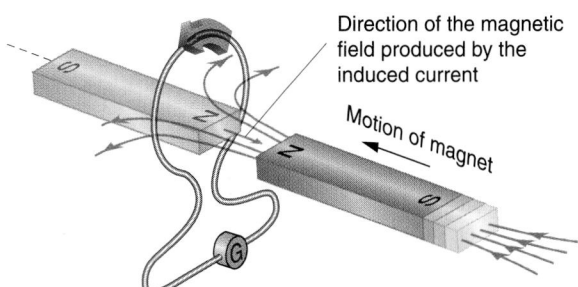

Direction of the magnetic field produced by the induced current

FIGURE 21.15

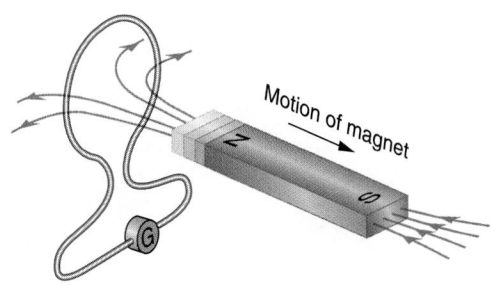

FIGURE 21.16

Solution

The motion of the magnet away from the loop means the magnetic flux through the loop is decreasing with time. According to Lenz's law, the induced current flows in the loop in a direction to oppose this change. You can determine this direction in two equivalent ways.

Method 1

The flux is decreasing through the loop because the magnetic field through the loop is decreasing in magnitude. The *magnetic field caused by the induced current* will be directed so as to reinforce the waning field of the permanent magnet. That is, the field of the induced current will be directed in the *same* direction as the field of the permanent magnet.

This means that you grasp the loop with your right hand so that the fingers thread the loop from right to left, parallel to the field of the permanent magnet through the loop, as shown in Figure 21.17. The extended thumb of your right hand then indicates the direction of the current.

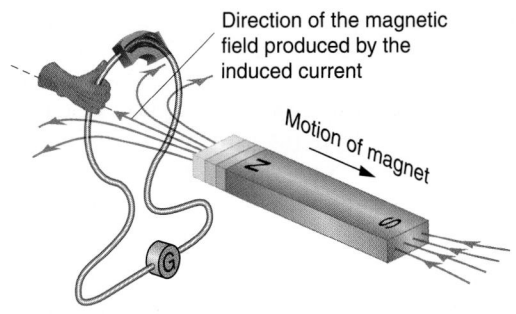

Direction of the magnetic field produced by the induced current

FIGURE 21.17

Method 2

Since the north pole of the permanent magnet is moving away from the loop, the magnetic field of the induced current will oppose this change. So the induced current develops a south magnetic pole on the side facing the departing north pole of the permanent magnet; see Figure 21.18. In this way, the two unlike poles attract each other and the induced current fights or opposes the change (the departure of the north pole of the permanent magnet).

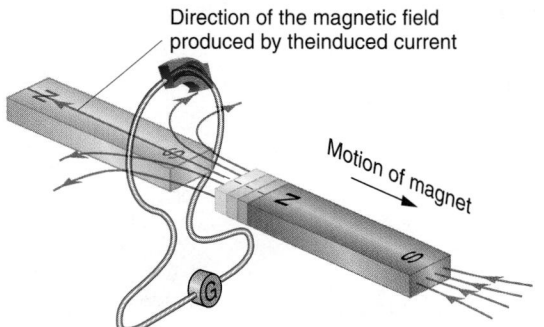

Direction of the magnetic field produced by the induced current

Motion of magnet

FIGURE 21.18

EXAMPLE 21.5

Close the switch in the circuit shown in Figure 21.19. This was the second experiment in Section 21.1. Determine the direction of the induced current in the other wire loop using Lenz's law.

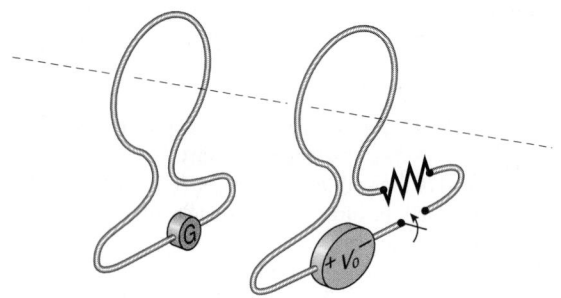

FIGURE 21.19

Solution

During a brief interval, the current in the right-hand circuit increases with time. The magnetic field created by this current, therefore, also increases with time, so that the flux of the magnetic field of this current through the left loop increases. Lenz's law states that the current induced in the loop is directed so as to oppose this change. Since the magnetic field is increasing through the left loop, the magnetic field of the induced current will be directed to fight or oppose the change. The induced current does this by moving in a direction that creates a magnetic field directed opposite to that of the increasing field.

Grasp the loop so that the fingers of your right hand thread the loop from left to right; your extended thumb is directed as indicated in Figure 21.20. This is the direction of the induced current.

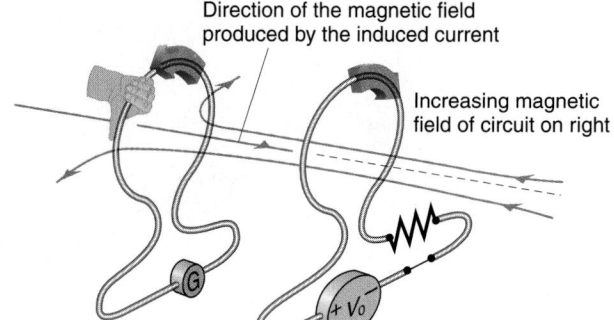

Direction of the magnetic field produced by the induced current

Increasing magnetic field of circuit on right

FIGURE 21.20

EXAMPLE 21.6

When the switch is opened in the circuit on the right of Figure 21.21, determine the direction of the current induced in the wire loop.

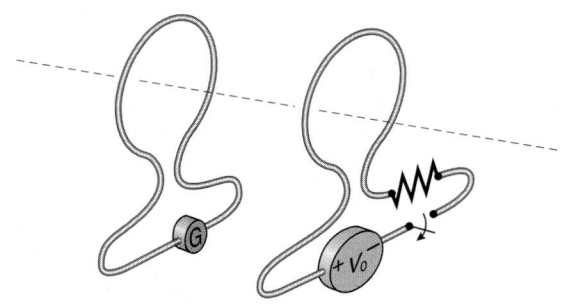

FIGURE 21.21

Solution

As the current falls precipitously to zero, so does the magnetic field created by this current. The magnetic flux of this field through the left loop thus decreases. Lenz's law states that the induced current is directed to oppose the change. Hence the field of the induced current through the left loop must be directed parallel to the field of the falling current in the circuit on the right.

Grasp the loop with the fingers of your right hand threading the loop from right to left (parallel to the existing decreasing field), as in Figure 21.22. The extended thumb of your right hand is directed as indicated. This is the direction of the induced current.

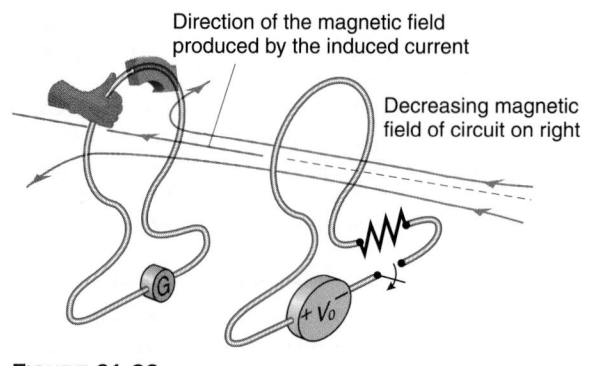

Direction of the magnetic field produced by the induced current

Decreasing magnetic field of circuit on right

FIGURE 21.22

EXAMPLE 21.7

A flat, closed, continuous coil has 50 turns of wire with a total resistance of 0.20 Ω. The area of the coil is 60 cm². The coil is positioned in a uniform magnetic field $\vec{B}$ as shown in Figure 21.23. The magnitude of the magnetic field increases at a constant rate from 0 T to 2.0 T during 1.50 s. Find

a. the emf induced in the coil;
b. the size of the current induced in the coil; and
c. the direction of the induced current.

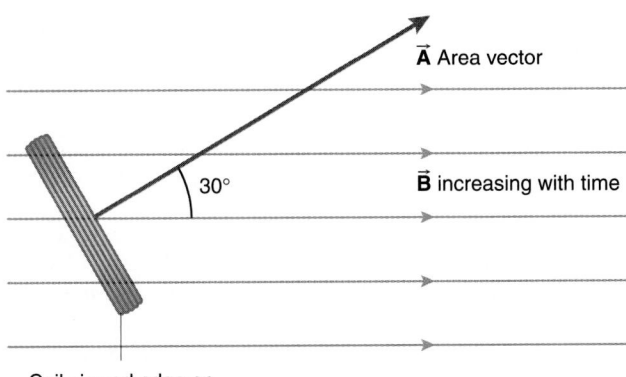

FIGURE 21.23

Solution

a. To calculate the induced emf you need to find the rate at which the magnetic flux through the coil is changing with time. Since the magnetic field is uniform over the area of the coil, the magnetic flux through *each turn* of the coil is

$$\Phi_{\text{single turn}} = \vec{B} \cdot \vec{A}$$
$$= BA \cos \theta$$

where θ is the angle between the magnetic field and the area vector of a loop of the wire; here, $\theta = 30°$. The total magnetic flux through the coil then is

$$\Phi_{\text{total}} = NBA \cos \theta$$

where N is the number of turns in the coil; here, $N = 50$. The induced emf is given by Equation 21.4:

$$\text{induced emf} = -\frac{d\Phi_{\text{total}}}{dt}$$
$$= -\frac{d}{dt}(NBA \cos \theta)$$

Here the only thing that is changing with time is the magnitude of the magnetic field; the induced emf thus is

$$\text{induced emf} = (-NA \cos \theta)\frac{dB}{dt}$$

The rate at which the magnetic field is changing with time is

$$\frac{dB}{dt} = \frac{2.0 \text{ T} - 0 \text{ T}}{1.50 \text{ s}}$$
$$= 1.3 \text{ T/s}$$

With the information provided—$N = 50$, $A = 60$ cm² = 60×10^{-4} m², and $\theta = 30°$—the induced emf is found to be

$$\text{induced emf} = -50(60 \times 10^{-4} \text{ m}^2)(\cos 30°)(1.3 \text{ T/s})$$
$$= -0.34 \text{ V}$$

The minus sign is a manifestation of Lenz's law.

b. The size of the current induced in the coil can be found using Ohm's law:

$$|\text{ induced emf }| = IR$$

or

$$I = \frac{0.34 \text{ V}}{0.20 \text{ Ω}}$$
$$= 1.7 \text{ A}$$

c. The direction of the induced current is determined using Lenz's law. The direction of the induced current will oppose the change in flux. Since the magnetic flux through the coil is increasing with time, the induced current produces a magnetic field opposite to the direction of the applied magnetic field. Use your right hand, and grasp the loops of the coil so that your fingers are directed through the coil in the direction opposite to the direction of the applied field, as shown in Figure 21.24. The thumb of the right hand then indicates the direction of the induced current around the coil loops.

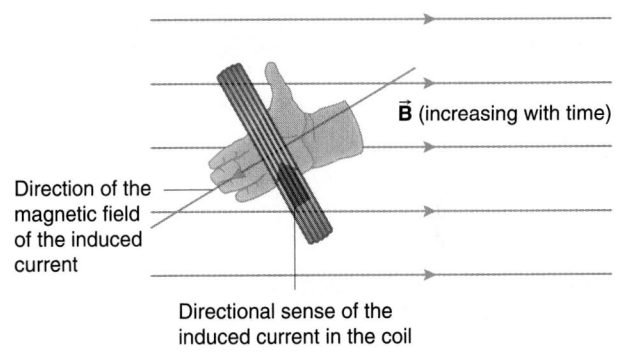

FIGURE 21.24

21.3 AN AC GENERATOR

The phenomenon described by Faraday's law is used for the very practical purpose of generating an alternating (oscillating) emf (an ac independent voltage source). This is the way most electric utilities as well as the alternator in your car generate a source of emf for many useful and convenient electrical applications.

We consider a planar loop of wire with an area vector $\vec{A}$ in a uniform magnetic field $\vec{B}$, as shown in Figure 21.25. The magnetic flux through the loop is

$$\Phi_{\text{magnet}} = \int \vec{B} \cdot d\vec{S}$$
$$= BA \cos \theta \qquad (21.5)$$

where θ is the angle between the area vector $\vec{A}$ and the magnetic field $\vec{B}$. Now let some external agent rotate the loop about its symmetry axis at a constant angular velocity $\vec{\omega}$.

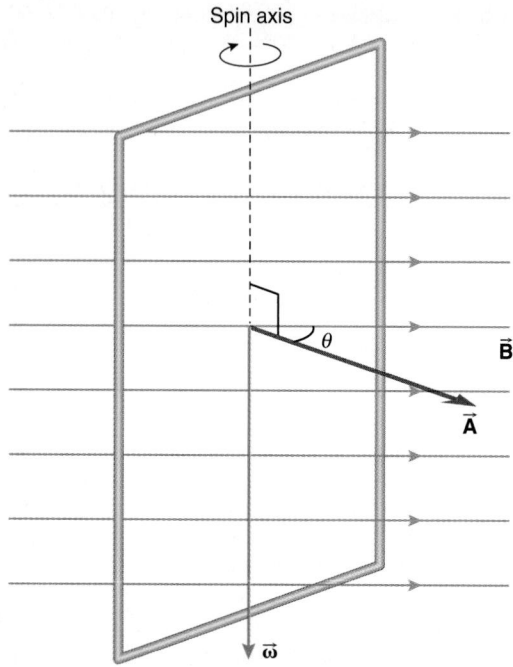

FIGURE 21.25 A wire loop is rotated in a uniform magnetic field.

From our study of uniform circular motion (Chapter 4), if $\theta = 0$ rad when $t = 0$ s, the angle θ at instant t is

$$\theta = \omega t$$

This means that the magnetic flux through the loop (Equation 21.5) is

$$\Phi_{\text{magnet}} = BA \cos(\omega t) \qquad (21.6)$$

Since the magnetic flux through the loop is time dependent, Faraday's law implies that an emf is induced in the loop:

$$\text{induced emf} = -\frac{d\Phi_{\text{magnet}}}{dt}$$

$$= -\frac{d}{dt}\Big[BA \cos(\omega t)\Big]$$

Since the magnitude of the magnetic field and the area of the loop are constant, the differentiation yields

$$\text{induced emf} = -BA[-\sin(\omega t)]\omega$$
$$= BA\omega \sin(\omega t) \qquad (21.7)$$

This induced emf oscillates sinusoidally with time, as shown in Figure 21.26.

If instead of using a single loop, a coil of wire of N loops (of identical area A) is used, the flux through the coil is N times that through a single loop. Therefore the induced emf in the coil then is also N times that of the single loop:

$$\text{induced emf} = NBA\omega \sin(\omega t) \qquad (21.8)$$

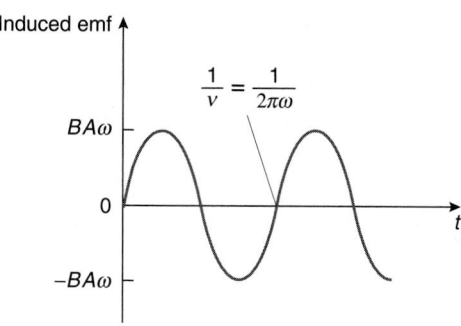

FIGURE 21.26 The emf induced in the wire loop is a sinusoidal function of time.

This is the way that the commercial electrical industry "generates electricity"; Faraday's law is at the very heart of the enterprise!

The rotating coil can be packaged into a box and the ends of the wire run to the outside world. The circuit symbol for a sinusoidal ac independent voltage source (an **ac generator** or an ideal ac battery) is shown in Figure 21.27. The term ac literally means *a*lternating *c*urrent, but we use ac generically to mean oscillating with time.

When associated with a generator, we call the induced emf the **source voltage** $V_{\text{source}}(t)$, and treat it as the ac potential difference associated with the source:

$$V_{\text{source}}(t) \equiv \text{induced emf}$$

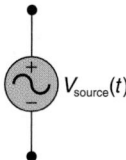

FIGURE 21.27 Circuit symbol for a sinusoidal independent voltage source.

Faraday's law is applied to create an ac independent voltage source in these large commercial generators.

An induced current exists only if the ends of the coil are connected together to an external electrical network to form an electrical circuit.

The electrical network in the United States and the rest of North and South America is based on a standard frequency $\nu = 60$ Hz; the frequency is controlled quite precisely. The coil rotates in the magnetic field 60 times during each second. Since each revolution of the coil represents 2π rad, the angular speed ω (and angular frequency) is related to the frequency ν by

$$\omega = 2\pi\nu$$
$$= (2\pi \text{ rad})(60 \text{ Hz})$$
$$\approx 377 \text{ rad/s}$$

In Europe, Japan, Australia, Asia, and Africa, the standard frequency is 50 Hz, in which case $\omega \approx 314$ rad/s.

Of course, it requires work to rotate the current-generating coil; Lenz's law sees to that! The external agent used to rotate the coil in the magnetic field characterizes the electrical generating plant. Typically these mechanical sources are of three types:

1. Hydroelectric: Falling water is directed against the blades of a turbine that is coupled to the generator coils.
2. Steam: High-pressure steam is directed against the blades of a turbine coupled to the generator coils. The steam is produced by heating water using various types of fuels or energy sources: coal, oil, nuclear energy, geothermal energy, or even wood and garbage.
3. Wind: Wind is used to turn a propeller (turbine) that is coupled to the rotating coils of the generator.

EXAMPLE 21.8

You are commissioned to design an ac generator whose coil is to be turned at 60.0 Hz in a strong magnetic field of 0.150 T. The peak emf of the generator must be at least 170 V. A square coil with sides 10.0 cm is used. What is the number of turns of wire needed in the coil?

Solution

The emf induced in the rotating coil is given by Equation 21.8:

$$V_{source}(t) = NBA\omega \sin(\omega t)$$

The peak emf V_0 is the coefficient of $\sin(\omega t)$. So

$$V_0 = NBA\omega$$

Hence the number of turns in the coil is

$$N = \frac{V_0}{BA\omega}$$

The angular frequency ω is

$$\omega = 2\pi\nu = (2\pi \text{ rad})(60.0 \text{ Hz})$$
$$= 377 \text{ rad/s}$$

Making the substitutions, you find

$$N = \frac{170 \text{ V}}{(0.150 \text{ T})(0.100 \text{ m})(0.100 \text{ m})(377 \text{ rad/s})}$$
$$= 301$$

The coil should have 301 turns of wire.

21.4 SUMMARY OF THE MAXWELL EQUATIONS OF ELECTROMAGNETISM

The four *Maxwell equations* are the pillars of electromagnetism. We have studied them one by one as we progressed through the various electric and magnetic phenomena of the previous chapters. It is useful now to bring the equations together here for convenience, for reflection, and indeed, for celebration. The Maxwell equations are the following:

1. Gauss's law for the electric field:

$$\int_{\text{clsd surface } S} \vec{E} \cdot d\vec{S} = \frac{Q_{\text{net enclosed by } S}}{\varepsilon_0} \qquad (21.9)$$

Gauss's law is a consequence of the inverse-square nature of Coulomb's law for the electrical force interaction between pointlike charges. Recall that there are *two distinct types* of electric fields: (a) the static electric field produced by electric charges; this electric field produces a conservative force on charges; and (b) the induced electric field created by a changing magnetic flux; electrical forces produced by this field are *not* conservative. Only the static electric field need be considered in Gauss's law: the flux of the induced electric field through a closed surface is *zero* because the induced electric field lines are closed loops!

2. Gauss's law for the magnetic field:

$$\int_{\text{clsd surface } S} \vec{B} \cdot d\vec{S} = 0 \text{ T} \cdot \text{m}^2 \qquad (21.10)$$

Unlike Faraday, Maxwell came from a privileged family and published his first paper at age 14! He was the first professor of experimental physics at Cambridge University and established the famous Cavendish Laboratory there.

This is a statement about the nonexistence of magnetic monopoles; magnets are dipolar. Magnetic field lines form closed contours.

3. The Ampere–Maxwell law:

$$\int_{\text{clsd path}} \vec{\mathbf{B}} \cdot d\vec{\ell} = \mu_0(I + I_{\text{D}})$$
$$= \mu_0 I + \mu_0 \varepsilon_0 \frac{d\Phi_{\text{elec}}}{dt} \qquad (21.11)$$

where I and I_{D} are the conduction and displacement currents, respectively, enclosed by the path (or equivalently, piercing any hat-shaped surface that has the path as a perimeter). This law is a statement that magnetic fields are caused by electric conduction currents and/or by a changing electric flux (via the displacement current).

4. Faraday's law of electromagnetic induction:

$$\int_{\text{clsd path}} \vec{\mathbf{E}} \cdot d\vec{\ell} = -\frac{d\Phi_{\text{magnet}}}{dt} \qquad (21.12)$$

This is a statement about how changes in magnetic flux produce (nonconservative) electric fields.

The Maxwell equations as we have written them are in what is called their *integral form*. There also is an equivalent *differential form* for the equations that we will neither write nor pursue, since it involves aspects of differential vector calculus we have not used in this text. A subsequent physics course in electromagnetic theory will develop the useful differential forms for the Maxwell equations in all their rightful glory.

Here we will only point ahead to two of the more remarkable aspects of the Maxwell equations:

1. Maxwell was able to show (as we will in Section 21.5) that electric and magnetic fields can *propagate* themselves through space according to the classical wave equation. The speed c of these traveling *electromagnetic fields* in a vacuum was found by Maxwell to be related to the familiar electric and magnetic constants simply by

$$c = \frac{1}{(\mu_0 \varepsilon_0)^{1/2}} \qquad (21.13)$$

where ε_0 is the permittivity of free space and μ_0 is the permeability of free space. When Maxwell evaluated the numerical value for the speed of these waves by substituting the numerical values for ε_0 and μ_0, a remarkable result was secured:

$$\frac{1}{(\mu_0 \varepsilon_0)^{1/2}}$$

$$= \frac{1}{\left[(4\pi \times 10^{-7} \text{ T·m/A}) \dfrac{1}{4\pi \times 9.00 \times 10^9 \text{ N·m}^2/\text{C}^2} \right]^{1/2}}$$

$$= 3.00 \times 10^8 \text{ m/s}$$

Maxwell recognized this result as very close to the measured speed of light. He then made a great conceptual leap: the waves indeed *were* light.

Light is an **electromagnetic wave** that travels in a vacuum with a speed c that is given by

$$c = \frac{1}{(\mu_0 \varepsilon_0)^{1/2}}$$

This equation is one of the great equations of physics, stemming as it does from the electromagnetic theory of Maxwell; the equation unifies three seemingly disparate fields of physics: electricity, magnetism, and optics.

2. The other remarkable feature of the Maxwell equations is that the later revolution created by relativity (see Chapter 25) had *no effect* on the equations. The equations are, as we say, relativistically correct (a term less disparaging than "politically correct"!). The equations do not need to be modified because of the revolutionary discoveries made by Einstein in 1905 regarding our notions of space, time, and energy. Newtonian mechanics, on the other hand, is fundamentally altered in the relativistic domain (as we have alluded to on occasion and will explore in detail in Chapter 25). In fact, what Einstein did was to take the incompatibility of the Newtonian laws with the Maxwellian equations of electromagnetism and resolve it in favor of Maxwell.

21.5 ELECTROMAGNETIC WAVES*

We all know what light is; but it is not as easy to tell what it is.
Samuel Johnson (1709–1784)[†]

In a vacuum, there is no matter, no charge, and no conduction current; displacement currents exist when the electric flux changes with time. In a vacuum, the four Maxwell's equations (Equations 21.9–21.12 of Section 21.4) take on a symmetric form:

Gauss's law for the electric field:

$$\int_{\text{clsd surface } S} \vec{\mathbf{E}} \cdot d\vec{\mathbf{S}} = 0 \text{ V·m} \qquad (21.14)$$

Gauss's law for the magnetic field:

$$\int_{\text{clsd surface } S} \vec{\mathbf{B}} \cdot d\vec{\mathbf{S}} = 0 \text{ T·m}^2 \qquad (21.15)$$

Ampere–Maxwell law for the magnetic field:

$$\int_{\text{clsd path}} \vec{\mathbf{B}} \cdot d\vec{\ell} = \mu_0 \varepsilon_0 \frac{d\Phi_{\text{elec}}}{dt} \qquad (21.16)$$

Faraday's law for the induced electric field:

$$\int_{\text{clsd path}} \vec{\mathbf{E}} \cdot d\vec{\ell} = -\frac{d\Phi_{\text{magnet}}}{dt} \qquad (21.17)$$

A time-varying electric flux gives rise to a magnetic field and a time-varying magnetic flux gives rise to an electric field. The circularity of this argument suggests the possibility of *self-sustaining,*

[†]*Boswell's Life of Johnson*, edited by George Birkbeck Hill (Oxford at the Clarendon Press, Oxford, 1934), Volume III, page 38. Entry for 12 April 1776.

perhaps propagating, time-dependent, electric and magnetic fields; we call them electromagnetic waves.

The plethora of electromagnetic waves of various kinds likely is familiar to you: radio waves, microwaves, visible light, even x-rays are common examples. With them arise some significant questions:

1. How do we know that these various kinds of light really *are* electromagnetic waves?
2. How are such waves produced?
3. Are the waves longitudinal or transverse?
4. What is oscillating in an electromagnetic wave?

We finally have all the tools at our disposal to answer these important questions. Here we see how the Maxwell equations in a vacuum lead to a wave equation for propagating electric and magnetic fields, as well as how Maxwell made the connection between these propagating electromagnetic disturbances and light, alluded to in the previous section.

Recall (from Chapter 12) that a classical wave disturbance $\Psi(x, t)$, propagating at speed v in either direction parallel to the x-axis, satisfies the classical wave equation (Equation 12.7):

$$\frac{\partial^2 \Psi}{\partial x^2} - \frac{1}{v^2}\frac{\partial^2 \Psi}{\partial t^2} = 0$$

Here we show that Maxwell's equations in a vacuum lead to wave equations for the space- and time-varying electric and magnetic fields. We also show how the fields manage to propagate themselves through a vacuum without the necessity of nearby charges or currents: each field is generated by a time variation in the other field, as implied by the Ampere–Maxwell law and Faraday's law in a vacuum.

To accomplish these objectives, we consider a simple and familiar system: a parallel plate capacitor with plates of *finite* area (see Figure 21.28). If the charges on the capacitor vary with time, perhaps periodically reversing themselves, the electric field in the vicinity of the capacitor also must change its magnitude and reverse its direction accordingly. A component of the electric field in the vicinity of the capacitor, say, in the plane midway between the capacitor plates, must be a function of spatial position and time.* We call it $E_y(x, t)$, as in Figure 21.28.

The time-varying electric field gives rise to a displacement current

$$I_D = \varepsilon_0 \frac{d\Phi_{elec}}{dt}$$

The displacement current in turn gives rise to a magnetic field in the vicinity, whose direction is determined from the now familiar right-hand rule: right-hand thumb in the direction of the current means the right-hand fingers indicate the directional sense of the magnetic field caused by the current. The displacement current is directed from one capacitor plate to the other, and so the magnetic field in the midplane has a component $B_z(x, t)$, as shown in Figure 21.29.

Notice that the electric and magnetic fields are perpendicular to each other: the electric field has a component $E_y(x, t)$ while the magnetic field has a component $B_z(x, t)$. As the charges vary on

---

*Other components also vary with space and time, but we focus on this component for simplicity.

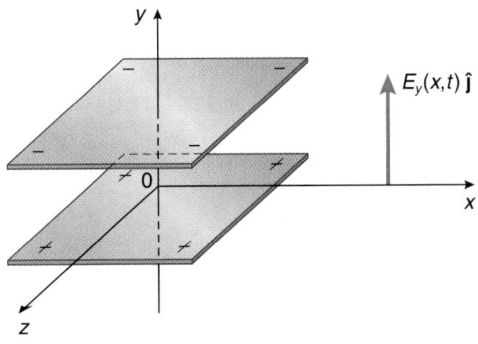

FIGURE 21.28 If the charges on the capacitor vary with time, the electric field in the vicinity is a function of position x and time t.

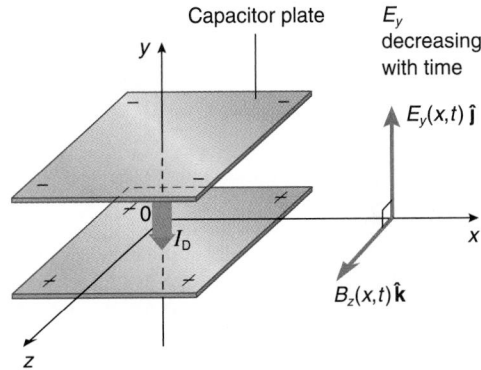

FIGURE 21.29 The displacement current produces a magnetic field.

the capacitor, the fields in the vicinity also must change accordingly. You might ask, quite reasonably, if the changes in the fields some distance away from the capacitor occur instantaneously, or whether there is a finite speed associated with the propagation of the changes. We seek to answer that query. The mutual orthogonality of the electric field, the magnetic field, and the direction of the possible propagation of their changes has important implications that we are now quite able to explore.

Do changes in the fields propagate parallel to the x-axis? Imagine a small, differential rectangle in the x–y plane of width dx and small length ℓ, as shown in Figure 21.30. We apply Faraday's law (Equation 21.17) to this rectangle:

$$\int_{clsd\ path} \vec{E} \cdot d\vec{\ell} = -\frac{d\Phi_{magnet}}{dt}$$

We integrate the electric field around the perimeter of the differential rectangle, say in the counterclockwise sense; Φ_{magnet} is the flux of the magnetic field through the differential area. Along both the narrow widths dx of the differential rectangle, the electric field is perpendicular to $d\vec{\ell}$ (see Figure 21.31), and so the scalar product in the integrand vanishes on the short sides.

Along the left side of the rectangle where the component of the electric field is E_y, $d\vec{\ell}$ is antiparallel to the field (see Figure 21.32), and so the contribution to the integral is $-E_y\ell$. Along the other long side of the differential rectangle, the component of the electric

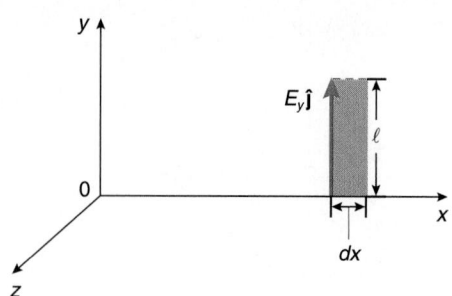

FIGURE 21.30 Apply Faraday's law to the indicated differential rectangle.

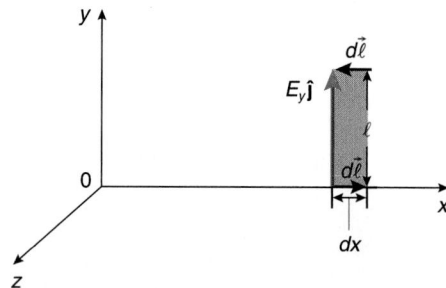

FIGURE 21.31 Along the short sides of the rectangle, $d\vec{\ell}$ is perpendicular to the electric field.

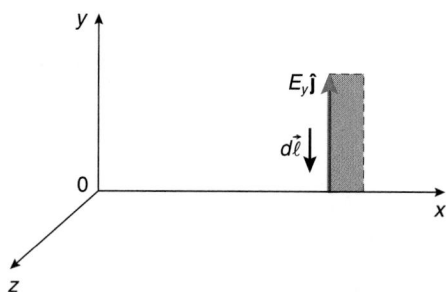

FIGURE 21.32 Along the left side of the rectangle, $d\vec{\ell}$ is antiparallel to the electric field.

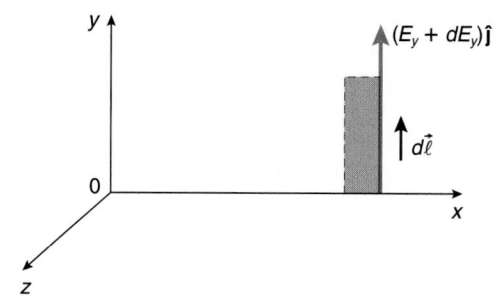

FIGURE 21.33 Along the right side of the rectangle, $d\vec{\ell}$ is parallel to the electric field.

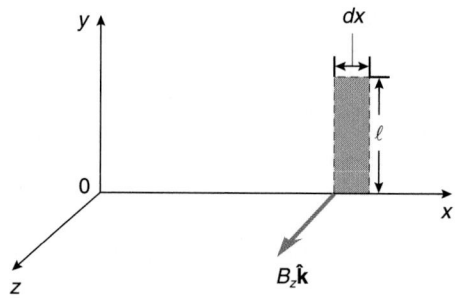

FIGURE 21.34 The magnetic field is perpendicular to the differential rectangle.

field is $E_y + dE_y$ and the field is parallel to $d\vec{\ell}$ (see Figure 21.33); thus, the scalar product yields a contribution of $(E_y + dE_y)\ell$.

Hence the path integral of the electric field around the differential rectangle (the left-hand side of Faraday's law) is

$$-E_y\ell + (E_y + dE_y)\ell + 0 \text{ J/C} + 0 \text{ J/C}$$

or $\ell\, dE_y$.

The magnetic field is perpendicular to the differential area. Since we integrated counterclockwise around the path, the right-hand rule indicates the differential area vector is parallel to the magnetic field (see Figure 21.34). The flux of the magnetic field through the area thus is $B_z\ell\, dx$.

Hence Faraday's law (Equation 21.17) becomes

$$\ell\, dE_y = -\frac{d}{dt}(B_z\ell\, dx)$$

or

$$\frac{dE_y}{dx} = -\frac{dB_z}{dt}$$

Since both fields are functions of x and t, the derivatives really should be written properly as partial derivatives:

$$\frac{\partial E_y}{\partial x} = -\frac{\partial B_z}{\partial t} \qquad (21.18)$$

The rate at which the electric field changes with position is the negative of the rate at which the magnetic field changes with time.

We now perform a similar calculation, applying the Ampere–Maxwell law in a vacuum (Equation 21.16) to a differential rectangle in the x–z plane (see Figure 21.35):

$$\int_{\text{clsd path}} \vec{\mathbf{B}} \cdot d\vec{\ell} = \mu_0\varepsilon_0 \frac{d\Phi_{\text{elec}}}{dt}$$

We integrate counterclockwise around the differential rectangle. Along the short sides of the rectangle, the magnetic field is perpendicular to $d\vec{\ell}$ (see Figure 21.36), and so no contribution to the integral arises there.

Along the left side where the magnetic field component is B_z, $d\vec{\ell}$ is parallel to the field (see Figure 21.37), and so we get a contribution $B_z\ell$ to the path integral. Along the other long side, the value of the magnetic field component is $B_z + dB_z$ and the field is antiparallel to $d\vec{\ell}$ (see Figure 21.38); thus we get a contribution $-(B_z + dB_z)\ell$ to the integral.

The path integral of the magnetic field around the differential rectangle, the left-hand side of the Ampere–Maxwell law in a vacuum, yields $-dB_z\,\ell$.

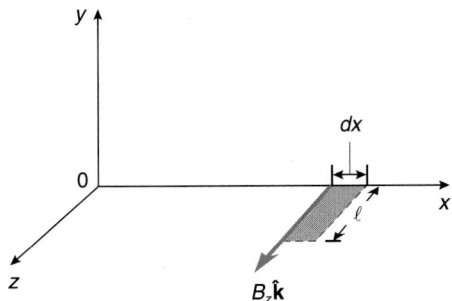

FIGURE 21.35 Apply the Ampere–Maxwell law to a differential rectangle in the x–z plane.

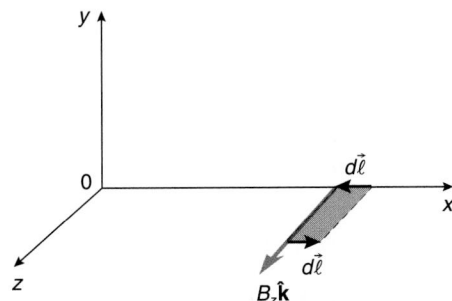

FIGURE 21.36 Along the short sides of the rectangle, $d\vec{\ell}$ is perpendicular to the magnetic field.

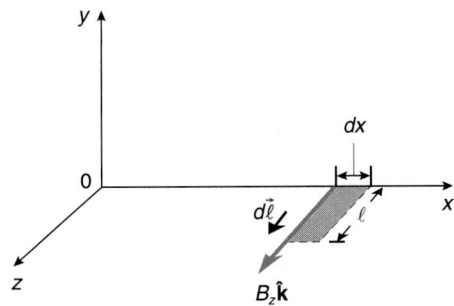

FIGURE 21.37 Along the left side of the rectangle, $d\vec{\ell}$ is parallel to the magnetic field.

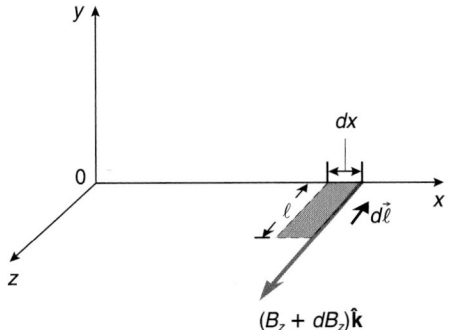

FIGURE 21.38 Along the other long side, $d\vec{\ell}$ is antiparallel to the magnetic field.

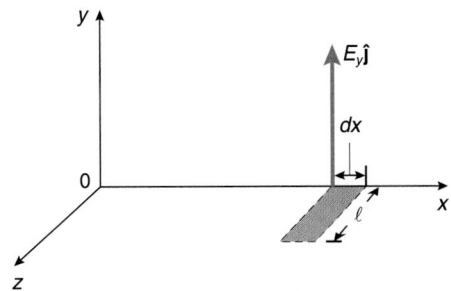

FIGURE 21.39 The electric field is perpendicular to the rectangle in the x–z plane.

Once again, since the fields are functions of both space and time, the derivatives properly are partial derivatives; thus, we write Equation 21.19 as

$$-\frac{\partial B_z}{\partial x} = \mu_0 \varepsilon_0 \frac{\partial E_y}{\partial t} \qquad (21.20)$$

The spatial variation in the magnetic field is proportional to the time rate of change of the electric field.

Mathematically, when taking a partial derivative with respect to t, we keep x constant; likewise, when taking a partial derivative with respect to x, we keep t constant. Therefore it makes no difference whether you first differentiate a function with respect to x and then with respect to t, or first with respect to t and then with respect to x:

$$\frac{\partial^2 E_y}{\partial t\, \partial x} = \frac{\partial^2 E_y}{\partial x\, \partial t} \qquad \frac{\partial^2 B_z}{\partial t\, \partial x} = \frac{\partial^2 B_z}{\partial x\, \partial t} \qquad (21.21)$$

We exploit this fact in the following way. Take Equation 21.18 and differentiate it with respect to x:

$$\frac{\partial^2 E_y}{\partial x^2} = -\frac{\partial^2 B_z}{\partial x\, \partial t} \qquad (21.22)$$

Likewise differentiate Equation 21.20 with respect to t:

$$-\frac{\partial^2 B_z}{\partial t\, \partial x} = \mu_0 \varepsilon_0 \frac{\partial^2 E_y}{\partial t^2} \qquad (21.23)$$

The electric field is perpendicular to the differential rectangle in the x–z plane (see Figure 21.39). Since we integrated counterclockwise around the path, the differential area vector is parallel to the electric field. Hence the flux of the electric field through the area is $E_y \ell\, dx$.

The Ampere–Maxwell law in a vacuum (Equation 21.16) then becomes

$$-dB_z\, \ell = \mu_0 \varepsilon_0 \frac{d}{dt}(E_y \ell\, dx)$$

or

$$-\frac{dB_z}{dx} = \mu_0 \varepsilon_0 \frac{dE_y}{dt} \qquad (21.19)$$

Combining Equations 21.22 and 21.23 using the partial derivative property in Equation 21.21, we find

$$\frac{\partial^2 E_y}{\partial x^2} = \mu_0 \varepsilon_0 \frac{\partial^2 E_y}{\partial t^2}$$

After transposing, we discover that the electric field component satisfies a partial differential equation that has the form of the classical wave equation (Equation 12.7):

$$\frac{\partial^2 E_y}{\partial x^2} - \mu_0 \varepsilon_0 \frac{\partial^2 E_y}{\partial t^2} = 0 \ \text{N/(C·m}^2) \qquad (21.24)$$

In like manner, if we differentiate Equation 21.18 with respect to t and Equation 21.20 with respect to x, and combine them using Equation 21.21, we discover the following:

The magnetic field component also satisfies a partial differential equation of the form of the classical wave equation:

$$\frac{\partial^2 B_z}{\partial x^2} - \mu_0 \varepsilon_0 \frac{\partial^2 B_z}{\partial t^2} = 0 \ \text{T/m}^2 \qquad (21.25)$$

A comparison of Equations 21.24 and 21.25 with the general form for the classical wave equation (Equation 12.7) indicates that the speed of propagation of such both waves of electric and magnetic fields is

$$\frac{1}{(\mu_0 \varepsilon_0)^{1/2}} \qquad (21.26)$$

Substituting the numerical values of μ_0 and ε_0, we find that the speed of these propagating electric and magnetic disturbances is numerically equal to the experimentally determined speed of light. It was from this equality that Maxwell made the conceptual leap and concluded that these propagating electromagnetic disturbances indeed *are* what we know as light.

If the charges move sinusoidally with time, the electromagnetic waves also are sinusoidal (far from the source) and we represent the electric and magnetic fields of the electromagnetic wave with sinusoidal wave functions like those we studied in Chapter 12:

$$E_y(x, t) = E_0 \cos(kx - \omega t) \qquad (21.27)$$
$$B_z(x, t) = B_0 \cos(kx - \omega t) \qquad (21.28)$$

As we noted before, the electric field, the magnetic field, and the direction of propagation all are mutually orthogonal (perpendicular); hence, light is a *transverse* electromagnetic wave.

Indeed, the vector product $\vec{E} \times \vec{B}$ indicates the direction of propagation.*

---

*In a subsequent course on electromagnetism, you will discover that this vector product, called the *Poynting vector*, has a magnitude proportional to the *intensity* of the wave: the power per square meter (oriented perpendicular to the direction of propagation) transmitted by the electromagnetic wave.

Also note that since the magnetic field arises because of *changes* in the electric field, and vice versa, oscillating changes in the fields arise fundamentally from the *acceleration of charges*.[†]

From this we see that the classical electromagnetic theory of Maxwell predicts that *accelerations of charges produce electromagnetic waves* that travel at the speed of light; they *are* light.[‡]

It is this principle that is exploited for generating radio waves, for example. Then, when such electromagnetic waves impinge on a conducting wire (or antenna), the electric field of the wave there causes electrons in the conductor to move back and forth, forming a small alternating current. The current is subsequently amplified and changed, using various electronic tricks, to produce a visual display (TV) or sound (radio).

EXAMPLE 21.9

The antenna for the public radio station WVTF in Roanoke, Virginia, emits electromagnetic waves with a frequency of 89.1 MHz. What is the corresponding wavelength of the electric and magnetic field waves in a vacuum?

Solution

In a vacuum, electromagnetic waves travel with a speed $c = 3.00 \times 10^8$ m/s. The speed c, wavelength λ, and frequency ν of a wave are related by Equation 12.10:

$$c = \nu\lambda$$

The wavelength thus is

$$\lambda = \frac{c}{\nu}$$
$$= \frac{3.00 \times 10^8 \ \text{m/s}}{89.1 \times 10^6 \ \text{Hz}}$$
$$= 3.37 \ \text{m}$$

The frequency of this radio station is near the lower end of the FM band of radio frequencies. Wavelengths of FM radio waves are on the order of meters.

[†]For example, if the charges undergo simple harmonic oscillation, the charges have nonzero acceleration (except at the instant when the charges pass through their equilibrium position).

[‡]We shall see in Chapter 26 that the emission of light by atoms is fundamentally different because of quantum mechanical considerations that have no classical analog.

Maxwell is celebrated on this stamp from San Marino. Now where in the world is San Marino?

21.6 SELF-INDUCTANCE*

We leave the travel of electromagnetic waves for now (though we consider light again in Chapters 23 and 24) to introduce a new thought: the magnetic analog of a capacitor. Actually, the introduction hardly is necessary, for we have been learning about it for some time: a coil of wire. The coil may consist of a single loop or, more likely, multiple turns of wire such as in a solenoid, a toroidal coil, or another such arrangement of wire loops. Such coils of wire are called **inductors** and are represented schematically by the ideal circuit symbol shown in Figure 21.40.

The **self-inductance** L of an inductor (a coil) is defined as the ratio of the total magnetic flux Φ_{magnet} through the coil to the current I in the coil:

$$L \equiv \frac{\Phi_{\text{magnet}}}{I} \qquad (21.29)$$

The current I in the coil is the current that produces the magnetic field threading the coil, which is why L is called the *self*-inductance. The dimensions of inductance are those of magnetic flux (T·m² in SI units) divided by current (A in SI units). The SI unit combination T·m²/A is called a **henry** (H), named for Joseph Henry (1797–1878):

$$\text{T·m}^2/\text{A} \equiv \text{H} \qquad (21.30)$$

Henry apparently discovered electromagnetic induction before Faraday but did not publish his findings.[†]

Analogously to the capacitance, the inductance is a function of the geometric arrangement of the coils of wire and the material (if any) within the coil itself. In other words, although the defining equation for the inductance (Equation 21.29) involves the ratio of the total magnetic flux through the coil to the current producing the flux (via the magnetic field), the ratio is independent of the specific values of Φ_{magnet} and I. We can see the parallels with the capacitance, since C is defined as the ratio Q/V, where the value of the ratio is independent of both Q and V. The capacitance depends only on the geometry (shape) of the capacitor and the material between its conductors (as we saw in Chapter 18).

We rearrange Equation 21.29 to get

$$\Phi_{\text{magnet}} = LI \qquad (21.31)$$

If the current I in the coil changes with time, then the magnetic flux through the coil also changes with time. Faraday's law then implies that there will be an emf induced in the coil:

$$\text{induced emf} = -\frac{d\Phi_{\text{magnet}}}{dt}$$

[†]It is customary in the sciences to give credit to the persons who first publish a finding. The merits of this custom have been vigorously debated over the years, causing many feuds over patents and the priority of discovery. The priority issue for Faraday and Henry is by no means settled among scientific historians. See Investigative Project 15.

FIGURE 21.40 Circuit symbol for an ideal inductor.

Substituting for the magnetic flux using Equation 21.31, we obtain

$$\text{induced emf} = -L\frac{dI}{dt} \qquad (21.32)$$

Imagine a coil such as that shown in Figure 21.41. The current I in the coil creates a magnetic field directed along the axis of the coil. Let the current I be increasing with time; hence, the magnetic flux through the coil also increases with time. Lenz's law states that a current is induced in the coil in a direction that opposes the change taking place. Since the flux is increasing with time, the direction of the induced current (created in response to the induced emf) tends to decrease the flux through the loop by diminishing the current I. Thus the induced current direction opposes that of I, and the polarity of the induced emf also does, as shown in Figure 21.42. If the coil is wound the other way (see Figure 21.43) and I increases with time, we still have the same directions of opposition for the induced current and also for the polarity for the induced emf in the coil.

We now do something that may seem peculiar at first. We define a potential difference V to be

$$V \equiv -\text{induced emf} \qquad (21.33)$$

Then, according to Equation 21.32,

$$V = L\frac{dI}{dt} \qquad (21.34)$$

When

$$\frac{dI}{dt}$$

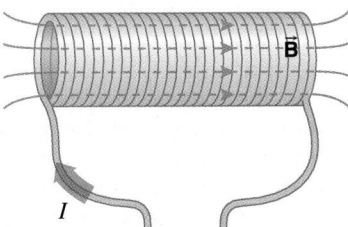

FIGURE 21.41 A current in a coil produces a magnetic field through the coil.

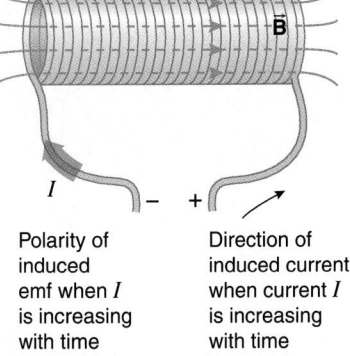

Polarity of induced emf when I is increasing with time

Direction of induced current when current I is increasing with time

FIGURE 21.42 Direction of the induced current and polarity of the induced emf in the coil when the current in it increases with time.

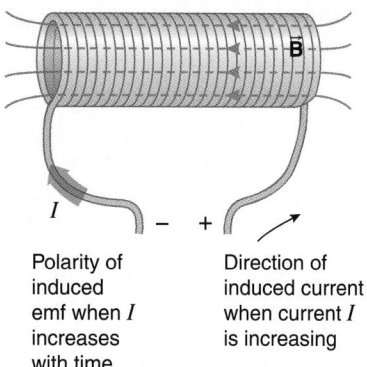

Polarity of
induced
emf when I
increases
with time

Direction of
induced current
when current I
is increasing

FIGURE 21.43 **For a coil wound the other way, the direction of the induced current and polarity of the induced emf are the same as for the coil in Figure 21.42.**

is positive (meaning that I is increasing with time), V has the polarity indicated in Figure 21.44.

The introduction of V to denote the opposition is done for consistency with the sign conventions on other circuit elements such as resistors and capacitors. In other words, we treat V as if it were a potential difference between two points (the two ends of the coil) rather than the negative of an induced emf. An inductor actually behaves like a peculiar battery or voltage source with the current going *into* its positive polarity terminal (unlike a normal battery or voltage source, where we usually think of the current as emerging *from* the positive polarity terminal). The emf is actually distributed continuously along the length of the wire of the coil; with V we just look at the two ends of the wire of the coil.

In this way we can treat the inductor as a common circuit element (an electrical device of some kind with two wires coming out of it). Why are we playing a bit of a semantic and physical shell game? From the viewpoint of physics, one should think of an inductor as an example of Faraday's and Lenz's laws and an induced emf. But from the more practical engineering viewpoint when using inductors in circuits, it is useful and convenient to treat V as if it were just an ordinary potential difference between the terminals of the inductor. It really is not, but doing so simply recognizes the convenient and conventional way inductors are treated in circuits and electronics. By using V instead of the induced emf, we can treat an inductor as a common circuit element: once we choose a current direction, an inductor circuit element is labeled with (+) and (−) polarity

markings in accordance with the direction we choose for the current, with the current going into the terminal (lead) with the (+) polarity sign.

Recall that for a resistor, once we chose the current direction, we labeled the polarity of the potential difference across the resistor so the current enters R at the (+) polarity marking, as indicated in Figure 21.45. For a capacitor, if the current charging the capacitor is directed as shown in Figure 21.46, the potential difference across the capacitor has the indicated polarity. Notice that if the *potential difference* V is not changing with time, the current through the capacitor is zero; the capacitor then acts like an *open circuit* for dc (time-independent voltages and currents).

Now similarly for an inductor, if the current is directed through the inductor as shown in Figure 21.47, then there is a potential difference V across the inductor with the indicated polarity. In this way, the three circuit elements are treated similarly in circuits when applying the Kirchhoff voltage and current laws (see Sections 21.8 and 21.10).

PROBLEM-SOLVING TACTICS

21.3 **Be careful to keep clear the distinction between the induced emf in an inductor (equal to the negative time rate of change of the magnetic flux) and the potential difference**

$$V = - \text{ induced emf} = L\frac{dI}{dt}$$

introduced so that an inductor can be treated with the same sign conventions as resistors and capacitors. The induced emf and the potential difference V are the negatives of each other.

21.4 **For dc situations, an inductor acts as a short circuit.** Since the potential difference across an inductor is proportional to the time rate of change of the current, if the current does not change with time (i.e., the current is constant: a dc current), there then is no potential difference across the inductor and it acts like a short circuit. The inductor then can be mentally replaced by an ideal wire.

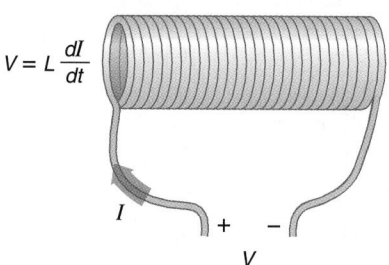

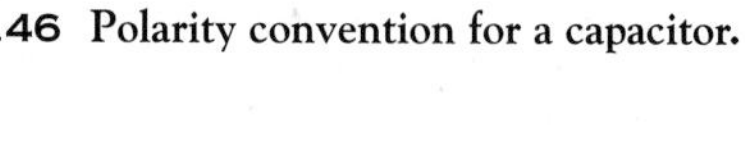

$V = L\dfrac{dI}{dt}$

I

V

FIGURE 21.44 **Polarity of the potential difference V, defined by Equations 21.33 and 21.34.**

$V = IR$ I V

FIGURE 21.45 **Polarity convention for a resistor.**

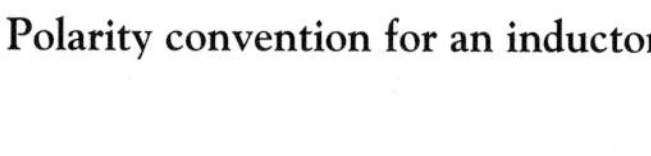

$I = C\dfrac{dV}{dt}$ I V

FIGURE 21.46 **Polarity convention for a capacitor.**

$V = L\dfrac{dI}{dt}$ I V

FIGURE 21.47 **Polarity convention for an inductor.**

QUESTION 2

You have a specified length of wire. (a) Describe a geometric configuration for the wire that makes its self-inductance large. (b) Describe a geometric configuration that minimizes its self-inductance.

EXAMPLE 21.10

Calculate the self-inductance of a long solenoid of length ℓ, with n coils per meter of its length, each with cross-sectional area A. Assume the magnetic field lines are confined to the region within the coil over its entire length.

Solution

The self-inductance is the ratio of the magnetic flux through the coil to the current in the coil (Equation 21.29):

$$L = \frac{\Phi_{magnet}}{I}$$

The magnitude of the magnetic field inside a long solenoid, far from its ends, is (from Example 20.13)

$$B = \mu_0 n I$$

where n is the number of turns of wire per meter of its length. Since the magnetic field is uniform across the cross-sectional area of the solenoid, the flux through a single coil of the solenoid is BA, where A is the cross-sectional area.

The number of loops of wire in a length ℓ of the solenoid is $n\ell$. Thus the total magnetic flux through a length ℓ of the solenoid is $n\ell$ times the flux though a single coil:

$$\begin{aligned}\Phi_{magnet} &= n\ell BA \\ &= n\ell\mu_0 nIA \\ &= n^2\ell A\mu_0 I\end{aligned}$$

The self-inductance of a length ℓ of the long solenoid then is

$$\begin{aligned}L &= \frac{\Phi_{magnet}}{I} \\ &= \mu_0 n^2 A\ell\end{aligned}$$

From this result we see that the self-inductance depends on geometric factors associated with the coil; L is independent of the magnetic flux through the coil and the current in it. The self-inductance of a right-handed coil is identical to that of its mirror image with left-handed windings, because Lenz's law of opposition to a changing current must apply equally to both.

From a practical viewpoint, the self-inductance of a coil is rarely calculated. It is measured experimentally (by various techniques; see Problem 42). The same also could be said of capacitances and resistances: they are rarely calculated from their defining equations except when designing or fabricating them.

21.7 SERIES AND PARALLEL COMBINATIONS OF INDUCTORS*

Just as for other circuit elements, it occasionally is useful to combine inductors into series or parallel combinations to secure different equivalent inductances. For reasons that will become apparent in Section 21.11, we assume here that the magnetic field of each inductor produces zero flux through any other inductor present.

Inductors in Series

A collection of inductors in series is shown in Figure 21.48. Circuit elements in series have the same current through them. Hence for a series connection of inductors, we have, for each inductor,

$$\begin{aligned}V_1 &= L_1\frac{dI}{dt} \\ V_2 &= L_2\frac{dI}{dt} \quad\quad (21.35) \\ V_3 &= L_3\frac{dI}{dt}\end{aligned}$$

We replace the series with a single equivalent inductance L_{eq} with the same potential difference V between the terminals A and B, as indicated in Figure 21.49. Thus we must have

$$V = L_{eq}\frac{dI}{dt} \quad\quad (21.36)$$

The potential difference V is the sum of those in Figure 21.48:

$$V = V_1 + V_2 + V_3$$

We substitute for V on the left-hand side of this equation using Equation 21.36 and use Equation 21.35 for the terms on the right-hand side. The same current is in each inductor in series. The result is

$$L_{eq}\frac{dI}{dt} = L_1\frac{dI}{dt} + L_2\frac{dI}{dt} + L_3\frac{dI}{dt}$$

or

$$L_{eq} = L_1 + L_2 + L_3 \quad \text{(inductors in series)}$$

The preceding argument applies for any number of inductors in series.

Inductors in series combine like resistors in series or capacitors in parallel:

$$L_{eq} = L_1 + L_2 + L_3 + \cdots \quad\quad (21.37)$$

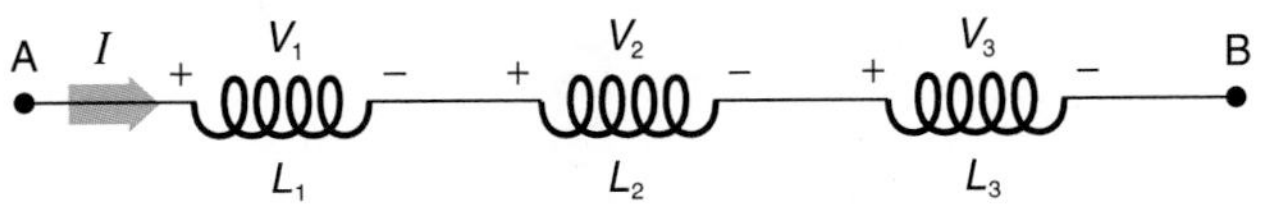

FIGURE 21.48 Inductors in series.

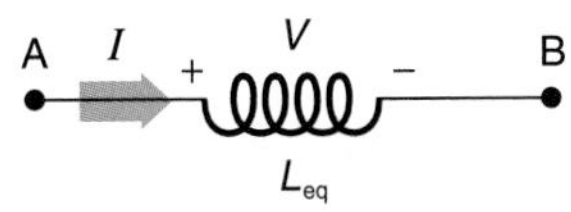

FIGURE 21.49 The equivalent inductor with inductance L_{eq}.

Inductors in Parallel

A collection of inductors in parallel is shown in Figure 21.50. Since circuit elements in parallel are connected to the same two nodes, the same potential difference exists across each circuit element. Thus, for the parallel combination of inductors pictured, we have

$$V = L_1 \frac{dI_1}{dt}$$

$$V = L_2 \frac{dI_2}{dt} \qquad (21.38)$$

$$V = L_3 \frac{dI_3}{dt}$$

We simplify the situation by replacing the collection with a single equivalent inductor L_{eq} having the same potential difference V between its terminals, as shown in Figure 21.51, where

$$V = L_{eq} \frac{dI}{dt} \qquad (21.39)$$

The Kirchhoff current law implies that

$$I = I_1 + I_2 + I_3$$

Differentiating this equation with respect to t, we get

$$\frac{dI}{dt} = \frac{dI_1}{dt} + \frac{dI_2}{dt} + \frac{dI_3}{dt}$$

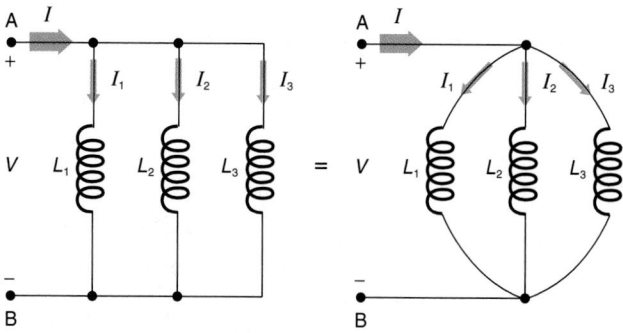

FIGURE 21.50 A parallel collection of inductors.

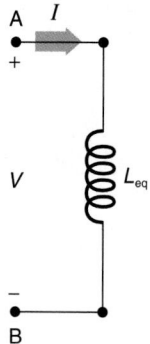

FIGURE 21.51 The equivalent inductor has inductance L_{eq}.

We use Equation 21.39 to substitute for the derivative on the left-hand side of this equation, and Equation 21.38 for the terms on the right-hand side; this yields

$$\frac{V}{L_{eq}} = \frac{V}{L_1} + \frac{V}{L_2} + \frac{V}{L_3}$$

or

$$\frac{1}{L_{eq}} = \frac{1}{L_1} + \frac{1}{L_2} + \frac{1}{L_3} \qquad \text{(inductors in parallel)}$$

The preceding argument applies for any number of inductors in parallel.

> Inductors in parallel combine like resistors in parallel or capacitors in series:
>
> $$\frac{1}{L_{eq}} = \frac{1}{L_1} + \frac{1}{L_2} + \frac{1}{L_3} + \cdots \qquad (21.40)$$

21.8 A SERIES LR CIRCUIT*

Faraday's law ensures that changes in currents cannot be accomplished instantaneously. In this section, we examine the transient behavior of a **series LR circuit** consisting of an independent voltage source V_0, an inductor L, and a resistor R connected in series, as shown in Figure 21.52. The series LR circuit is analogous to the series RC circuit we treated in Chapter 19.

Case 1: Current upon Closing a Switch

With the switches open, there is zero current. We want to determine how the current varies with time after switch (1) is closed when $t = 0$ s (see Figure 21.53). Switch (2) is left open. We choose the direction for the current I shown in Figure 21.53. The choice of a current direction forces us to label the resistor and inductor polarities accordingly, with the current going into the

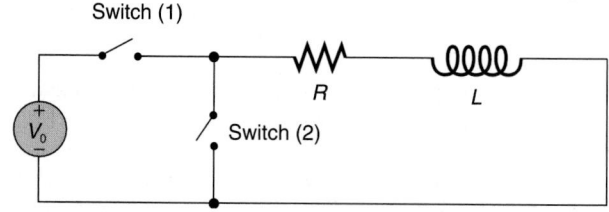

FIGURE 21.52 A series LR network (note there is no closed path, so it is not yet a circuit).

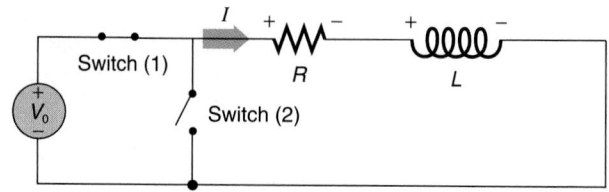

FIGURE 21.53 Switch (1) is closed when $t = 0$ s.

positive polarity terminal of both, as indicated. From Ohm's law, the potential difference across the resistor is IR. The potential difference across the inductor is

$$L \frac{dI}{dt}$$

according to the scheme we instituted in Section 21.6.

We apply the Kirchhoff voltage law (KVL) the same way we have on many occasions in the past (Chapter 19). We traverse the circuit in the clockwise sense, beginning in the lower left corner:

$$-V_0 + IR + L \frac{dI}{dt} = 0 \text{ V} \qquad (21.41)$$

Rearranging the terms slightly, we have

$$\frac{dI}{dt} + \frac{R}{L} I = \frac{V_0}{L} \qquad (21.42)$$

This is another example of a differential equation, an equation involving an unknown variable (I), and its derivatives, here just the first derivative

$$\frac{dI}{dt}$$

We have encountered differential equations before (for example, when studying simple harmonic motion in Chapter 7), but not one of this type. Since you likely have not yet had a formal course in how to solve these equations, we will have to solve this equation using the seat of our pants, your growing knowledge of calculus, and a bit of (mathematically legitimate) sleight of hand. What we do to solve this differential equation may not occur to you, at first, as reasonable. In other words, you may not be able to solve this differential equation on your own just yet, but you should be able to follow what happens here without feeling that any mathematical wool is being pulled over your eyes.

To begin, we change variables. This is the step that may not appear obvious to you at all. In particular, let a parameter x be defined as*

$$x \equiv I - \frac{V_0}{R} \qquad (21.43)$$

Since both V_0 and R are fixed, differentiation with respect to time yields

$$\frac{dx}{dt} = \frac{dI}{dt}$$

We substitute for I and its derivative in Equation 21.42:

$$\frac{dx}{dt} + \frac{R}{L} \left(x + \frac{V_0}{R} \right) = \frac{V_0}{L}$$

Simplifying slightly, we have

$$\frac{dx}{dt} + \frac{R}{L} x = 0 \text{ V/H}$$

---

*Here we use x as an unknown quantity, *not* the Cartesian coordinate x.

Rearranging the terms yields

$$\frac{dx}{x} = -\frac{R}{L} dt$$

which now is in a form that can be integrated. Indeed, the purpose of the change in variables was to achieve just this objective.

We integrate from when the switch is closed at $t = 0$ s to instant (and elapsed time) t, with corresponding limits on x (i.e., from x_0 to x):

$$\ln x - \ln x_0 = -\frac{R}{L} t$$

or

$$\ln \left(\frac{x}{x_0} \right) = -\frac{R}{L} t$$

Taking the antilogarithms, we get

$$\frac{x}{x_0} = e^{-(R/L)t}$$

or

$$x = x_0 e^{-(R/L)t} \qquad (21.44)$$

In Equation 21.43, when $t = 0$ s the current $I = 0$ A, and so

$$x_0 = -\frac{V_0}{R}$$

Substituting for x and x_0 in Equation 21.44, we find

$$I - \frac{V_0}{R} = -\frac{V_0}{R} e^{-(R/L)t}$$

Solving for I yields

$$I(t) = \frac{V_0}{R} \left[1 - e^{-(R/L)t} \right] \qquad (21.45)$$

A few reassuring things to note about this solution for the current:

1. Notice that when $t = 0$ s the current is $I = 0$ A, which of course was the initial condition of the circuit.
2. After a long time, as $t \to \infty$ s, the exponential term approaches 0, and so the current in the circuit eventually becomes equal to V_0/R. In other words, after a long time the current becomes constant, and the inductor *has no effect* (it acts as a short circuit).

When I becomes constant, there is no potential difference across the inductor. An inductor has an effect *only* while the current is *changing* with time. For the steady-state situation, there is no potential difference across an inductor since the current no longer is changing with time.

Indeed, this can be seen from Equation 21.34:

$$V = L \frac{dI}{dt}$$

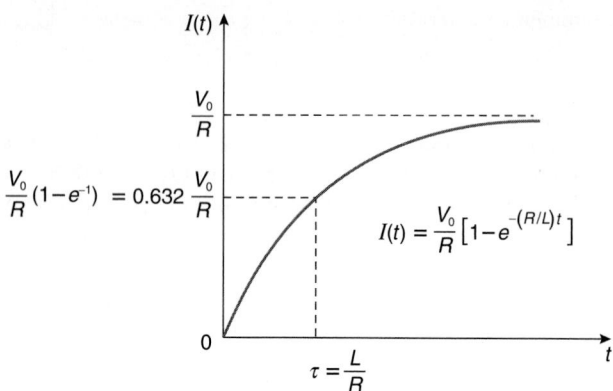

FIGURE 21.54 The current versus time in a series LR circuit.

When I is constant ($= V_0/R$), its derivative is zero, making $V = 0$ V.

3. The exponent for e in Equation 21.45 must always be dimensionless. Thus the quantity L/R must have the SI units of seconds.

This characteristic time is called the **time constant τ** of the circuit:

$$\tau \equiv \frac{L}{R} \qquad (21.46)$$

After a time t equal to five time constants ($t = 5\tau$), the exponential term is then $e^{-5} = 0.006\,74$, and the current I is within 1% of its final value of V_0/R.

A graph of the current versus time is shown in Figure 21.54.

Since *every* circuit consists of at least one loop of wire, there is some (small) self-inductance associated with every circuit, which justifies the behavior of the current we presumed in Experiment 2 in Section 21.1 (see Figure 21.4).

Case 2: Disconnecting the Independent Voltage Source

Wait a sufficient time so the current in Case 1 has reached its steady-state value V_0/R. What happens when the switch (2) is closed just when switch (1) is opened (see Figure 21.55)?

The result of this operation is to disconnect the independent voltage source V_0 from the circuit. The voltage source V_0 is part of a network but the *circuit* now consists of only the resistor and the inductor. Initially (i.e., when $t = 0$ s here) the current in the resistor and inductor is V_0/R. What happens to the current as a function of time? It cannot instantaneously vanish: Lenz's law prohibits it. Once again we label the resistor and inductor polarities in accordance with the direction chosen for the current, as in Figure 21.56.

We apply the KVL to the circuit, going around clockwise beginning in the lower left corner of Figure 21.56:

$$IR + L\frac{dI}{dt} = 0 \text{ V}$$

Fortunately, this differential equation is easier to solve than the one in Case 1! We rearrange the terms to collect those involving I:

$$\frac{dI}{I} = -\frac{R}{L}\,dt$$

This can be integrated immediately. Integrating on the left-hand side from I_0 ($= V_0/R$) to I and on the right-hand side from $t = 0$ s to t yields

$$\int_{I_0}^{I} \frac{dI}{I} = -\frac{R}{L}\int_{0\,\text{s}}^{t} dt$$

$$\ln I - \ln I_0 = -\frac{R}{L}t$$

We combine the logarithms:

$$\ln\left(\frac{I}{I_0}\right) = -\frac{R}{L}t$$

Now we take antilogarithms:

$$\frac{I}{I_0} = e^{-(R/L)t}$$

or

$$I(t) = I_0 e^{-(R/L)t}$$

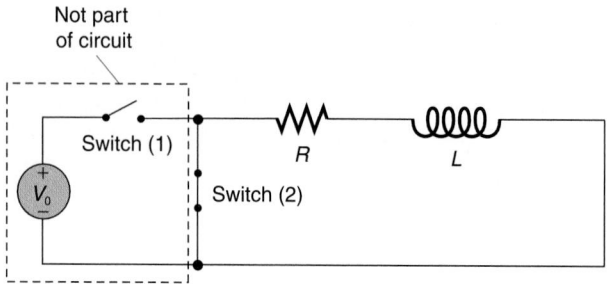

FIGURE 21.55 After a long time, switch (2) is closed while switch (1) is opened.

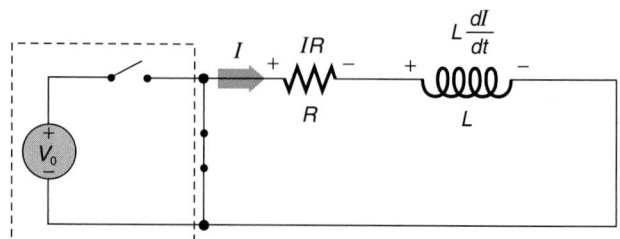

FIGURE 21.56 The polarities are marked consistent with direction chosen for I.

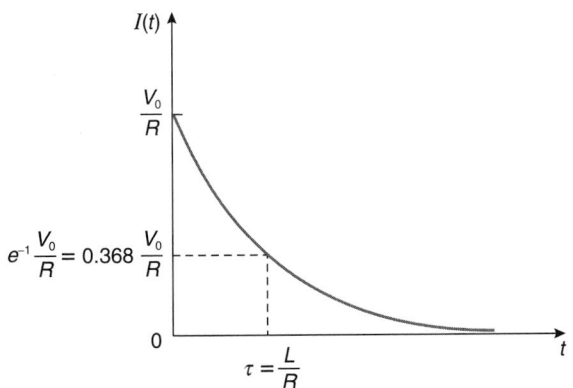

FIGURE 21.57 The current decreases exponentially with time.

Substituting for I_0, we obtain

$$I(t) = \frac{V_0}{R} e^{-(R/L)t} \qquad (21.47)$$

This indicates that the current in the circuit decreases exponentially with time, as shown in Figure 21.57.

Since the exponent must be dimensionless, L/R must have the dimensions of time. Indeed, we have the same time constant that we had for Case 1:

$$\tau = \frac{L}{R}$$

After a time t equal to five time constants, the exponential term is $e^{-5} = 0.006\,74$ and the current I is less than 1% of its initial value.

STRATEGIC EXAMPLE 21.11

Consider the series *LR* network shown in Figure 21.58.

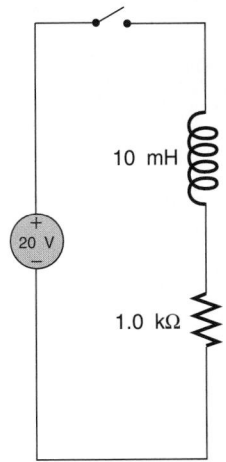

FIGURE 21.58

a. At the instant the switch is closed, determine the following quantities: (i) the current in the resistor; (ii) the current in

the inductor; (iii) the potential difference across the resistor; (iv) the potential difference across the inductor; and (v) the rate at which the current through the inductor is changing with time.

b. After many time constants (i.e., for $t \to \infty$ s), determine the same quantities.

c. Show explicitly that the KVL is satisfied in the circuit at any instant t.

d. Calculate the time constant of the circuit.

e. Make detailed plots of the potential difference across the resistor and inductor as functions of time.

Solution

a. The inductor and resistor are in series and so must have the same current through them at any instant. The switch initially was open and the current was zero. At the instant the switch is closed, the initial value of the current must also be zero. This follows from Equation 21.45 (case 1) for the current through an inductor in a series *LR* circuit:

$$I(t) = \frac{V_0}{R}\left[1 - e^{-(R/L)t}\right]$$

which when $t = 0$ s reduces to

$$I(0 \text{ s}) = \frac{V_0}{R}\left(1 - 1\right)$$
$$= 0 \text{ A}$$

Since R and L are in series, *both* the resistor and inductor have zero current when $t = 0$ s.

Since the resistor has no current when $t = 0$ s, the potential difference across the resistor at this instant is zero (from Ohm's law: $V = IR$). With no potential difference across the resistor when $t = 0$ s, the KVL, taken around the loop (say in the clockwise sense) implies that the potential difference across the inductor must be 20 V at this instant, with the polarity indicated in Figure 21.59. The KVL must be satisfied *at any instant*, since it represents a statement about energy conservation (you will soon see this again explicitly).

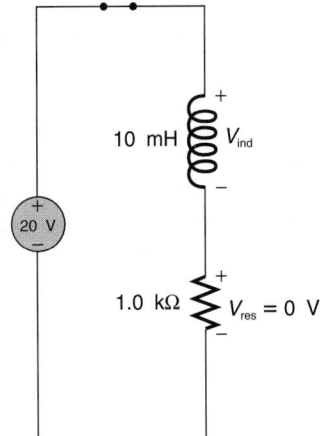

FIGURE 21.59 The situation when $t = 0$ s.

The potential difference V across the inductor depends on how fast the current in the inductor is changing with time. That is, from Equation 21.34,

$$V = L \frac{dI}{dt}$$

$$20 \text{ V} = (10 \times 10^{-3} \text{ H}) \frac{dI}{dt}$$

Solving for

$$\frac{dI}{dt}$$

you find

$$\frac{dI}{dt} = \frac{20 \text{ V}}{10 \times 10^{-3} \text{ H}}$$

$$= 2.0 \times 10^{3} \text{ A/s}$$

This is the initial slope when $t = 0$ s of a graph of I versus t through the inductor. Since the current in the inductor and the resistor are the same (since these circuit elements are in series here), this also represents the rate at which the current through the resistor is changing with time when $t = 0$ s.

You also could obtain this result using Equation 21.45 for the current through the inductor:

$$I(t) = \frac{V_0}{R}\left[1 - e^{-(R/L)t}\right]$$

Differentiating with respect to t, you find

$$\frac{dI}{dt} = \frac{V_0}{R}(-1)\, e^{-(R/L)t}\left(-\frac{R}{L}\right)$$

$$= \frac{V_0}{L}\, e^{-(R/L)t} \qquad (1)$$

Notice that when $t = 0$ s you get the previous result:

$$\frac{dI}{dt} = \frac{V_0}{L}$$

$$= \frac{20 \text{ V}}{10 \times 10^{-3} \text{ H}}$$

$$= 2.0 \times 10^{3} \text{ A/s}$$

b. After many time constants have elapsed, the potential difference across the inductor is zero because the current no longer is changing with time. That is, using equation (1) and letting $t = \infty$ s means that the exponential vanishes, and so

$$\frac{dI}{dt} = 0 \text{ A/s}$$

The current through the inductor then is constant and from Equation 21.45 equal to

$$\frac{V_0}{R} = \frac{20 \text{ V}}{1.0 \times 10^{3} \text{ } \Omega}$$

$$= 20 \text{ mA}$$

This is equal to the current through the resistor since the two circuit elements are in series. Using Ohm's law, the potential difference across the resistor is

$$V_{res} = IR$$

$$= (20 \times 10^{-3} \text{ A})(1.0 \times 10^{3} \text{ } \Omega)$$

$$= 20 \text{ V}$$

The inductor behaves as if it was shorted out.

c. To show that the KVL is satisfied at any time t, we need to find the potential differences across the resistor and inductor at any time t. The potential difference $V_{res}(t)$ across the resistor is

$$V_{res}(t) = I(t)R$$

Using Equation 21.45 for $I(t)$, you get

$$V_{res}(t) = V_0[1 - e^{-(R/L)t}] \qquad (2)$$

The potential difference across the inductor at any time is

$$V_{ind}(t) = L \frac{dI}{dt}$$

which, using equation (1) for $\dfrac{dI}{dt}$, is

$$V_{ind}(t) = L \frac{V_0}{L}\, e^{-(R/L)t}$$

$$= V_0 e^{-(R/L)t} \qquad (3)$$

The polarities are indicated in Figure 21.60.

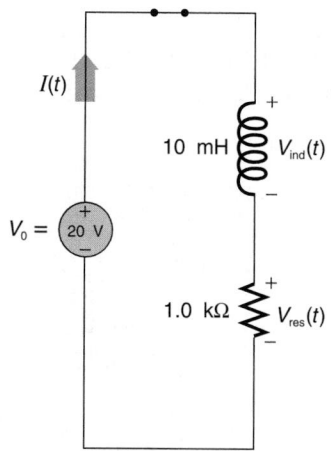

FIGURE 21.60

Now invoking the KVL and marching around the loop (in the CW sense beginning in the lower left corner), you have

$$-V_0 + V_{ind}(t) + V_{res}(t) = 0 \text{ V}$$

Using equations (3) and (2), this becomes

$$-V_0 + V_0 e^{-(R/L)t} + V_0[1 - e^{-(R/L)t}] = 0 \text{ V}$$

$$0 \text{ V} = 0 \text{ V}$$

The KVL is indeed satisfied at any instant.

d. The time constant of the circuit is

$$\tau = \frac{L}{R}$$
$$= \frac{10 \times 10^{-3} \text{ H}}{1.0 \times 10^{3} \text{ } \Omega}$$
$$= 1.0 \times 10^{-5} \text{ s}$$

e. The potential difference across the resistor is

$$V_{res}(t) = V_0[1 - e^{-(R/L)t}]$$
$$= (20 \text{ V})[1 - e^{-(1.0 \times 10^5 \text{ s}^{-1})t}]$$

The potential difference across the inductor is

$$V_{ind} = V_0 e^{-(R/L)t}$$
$$= (20 \text{ V})e^{-(1.0 \times 10^5 \text{ s}^{-1})t}$$

Both are plotted in Figure 21.61.

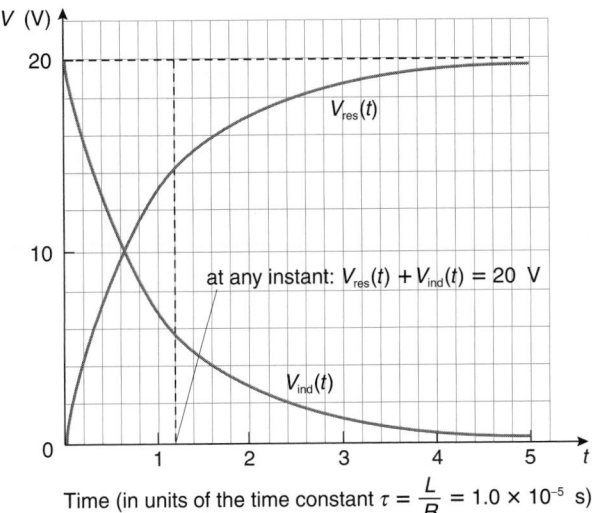

FIGURE 21.61 **The potential difference across the resistor and inductor as a function of time.**

21.9 ENERGY STORED IN A MAGNETIC FIELD*

When we investigated capacitors in Chapter 18, we discovered that electrical potential energy is stored by the capacitor in the electric field between its plates. Here we discover that inductors store energy in their magnetic field, illustrating once again that an inductor is the magnetic analog of a capacitor.

Recall the situation investigated in Case 1 of the previous section. This involved a description of how the current in a series *LR* circuit increases when a switch is closed to complete the circuit. By applying the KVL (which, in closet form, is a consequence of conservation of energy), we found that (from Figure 21.53 and Equation 21.41)

$$-V_0 + IR + L\frac{dI}{dt} = 0 \text{ V}$$

If we multiply this equation by the current *I*, each term has an interesting physical interpretation:

$$-IV_0 + I^2R + IL\frac{dI}{dt} = 0 \text{ W} \qquad (21.48)$$

Recall the sign convention we introduced for the power in Chapter 19: the power absorbed by (transferred to) a circuit element is the potential difference across the circuit element times the current into its positive polarity terminal.

1. The potential difference across the battery is V_0 and the current into its positive polarity terminal is $-I$. Hence the power absorbed by the battery is $-IV_0$, the first term of Equation 21.48. This power is negative because the battery actually is providing power to the circuit.
2. The second term is the power absorbed by the resistor. The power absorbed by the resistor is positive (as it should be), as resistances will always warm up.
3. The third term must also represent power: the power absorbed by the inductor. Once again, the current I is going into the positive terminal of the inductor. Since power is the time rate of change of the energy, the inductor is absorbing (securing) energy U at a rate

$$P = IV$$
$$\frac{dU}{dt} = IL\frac{dI}{dt} \qquad (21.49)$$

or

$$dU = IL \, dI$$

This expression can be integrated to find the energy U stored by the inductor when a current I is passing through the inductor.

The energy stored by the inductor is

$$U = \frac{1}{2}LI^2 \qquad (21.50)$$

when the inductor has a current I passing through it. Notice that the energy does not depend on

$$\frac{dI}{dt}$$

but just on I itself. In other words, the inductor stores energy (as long as $I \neq 0$ A) even when the current is *not* changing with time.

A few important things to glean from Equations 21.49 and 21.50:

The power absorbed by an inductor,

$$P = \frac{dU}{dt} = IL\frac{dI}{dt}$$

depends on the rate at which the current is changing with time.

If the current is increasing with time,

$$\frac{dI}{dt} > 0 \text{ A/s}$$

the power absorbed by the inductor is positive. If the current is decreasing with time,

$$\frac{dI}{dt} < 0 \text{ A/s}$$

the power absorbed is negative, indicating that the inductor actually is providing power (actually returning energy from the field to the electrons). So the power absorbed by an inductor can be positive or negative depending on the sign of

$$\frac{dI}{dt}$$

the time rate of change of the current. If the current is not changing with time,

$$\frac{dI}{dt} = 0 \text{ A/s}$$

and the inductor absorbs zero power.

Table 21.1 illustrates the parallels between a capacitor and an inductor. Recall that a capacitor stores its energy in the electric field between its plates. For an inductor, the energy is stored in its magnetic field. We illustrate this with a long solenoid. A long solenoid has a self-inductance L given by (see Example 21.10)

$$L = \mu_0 n^2 A \ell$$

where A is the cross-sectional area, n is the number of turns per meter, and ℓ is the length of the solenoid. The magnitude of the magnetic field inside a long solenoid, far from its ends, is (from Example 20.13)

$$B = \mu_0 n I$$

where I is the current in the solenoid. The energy stored in the magnetic field of the solenoid is (Equation 21.50):

$$U = \frac{1}{2} L I^2$$

Substituting for L and for I from the two previous equations, we find

$$U = \frac{1}{2} \mu_0 n^2 A \ell \left(\frac{B}{\mu_0 n} \right)^2$$

$$= \frac{1}{2} \frac{A \ell B^2}{\mu_0}$$

But $A\ell$ is the volume of the solenoid. Thus the energy stored per unit volume (assuming the magnetic field is zero outside the

solenoid, an assumption valid for an infinite solenoid) is

$$\frac{U}{A\ell} = \frac{1}{2} \frac{B^2}{\mu_0}$$

The quantity

$$\frac{1}{2} \frac{B^2}{\mu_0} \qquad (21.51)$$

is known as the **magnetic energy density**.

Notice that the magnetic energy density depends on the square of the magnetic field. The result here is analogous to the result we obtained for the energy density of an electric field, Equation 18.13:

$$\frac{1}{2} \varepsilon_0 E^2$$

Although we derived the magnetic energy density using a long solenoid and the electric energy density using a parallel plate capacitor, the results for the energy densities are very general.

Whatever the electric and/or magnetic field is at a particular location, the energy densities associated with them are proportional to the square of the magnitude of the field.

EXAMPLE 21.12

What magnitude magnetic field has an energy density of 1.0 J/m³?

Solution

The energy density of the magnetic field is given by Equation 21.51. Use the given energy density:

$$1.0 \text{ J/m}^3 = \frac{B^2}{2\mu_0}$$

Solve for B^2:

$$B^2 = 2\mu_0 (1.0 \text{ J/m}^3)$$

Substituting for μ_0 and solving for B, you obtain

$$B = 1.6 \text{ mT}$$

21.10 A PARALLEL LC CIRCUIT*

We have investigated the transient behavior of series RC and LR circuits. In those circuits we found that the potential difference across the capacitor in a series RC circuit, or the current in a series LR circuit, increased or decreased exponentially with time with a characteristic time constant ($\tau = RC$ or $\tau = L/R$ as the case may be). We now turn to another possible paired combination of these circuit elements: a **parallel LC circuit**. A few surprises await us!

We begin with the circuit shown in Figure 21.62. Initially, switch (1) is closed and switch (2) is open. This means that initially the inductor is not part of the circuit. The capacitor is in parallel

TABLE 21.1 **Similarities Between a Capacitor and an Inductor**

	Power absorbed	Energy stored
Capacitor	$VC \dfrac{dV}{dt}$	$\dfrac{1}{2} CV^2$
Inductor	$IL \dfrac{dI}{dt}$	$\dfrac{1}{2} LI^2$

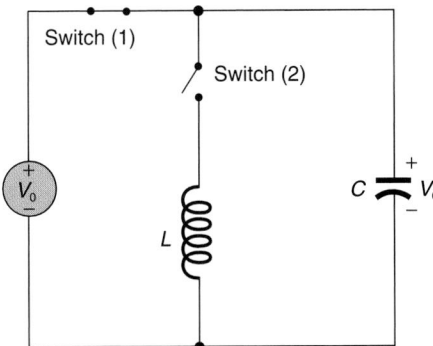

FIGURE 21.62 An initial circuit.

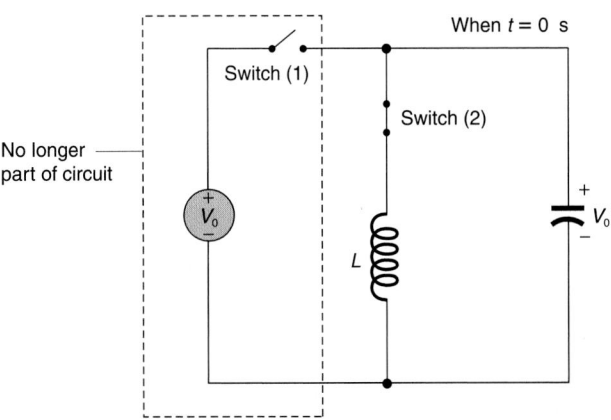

FIGURE 21.63 When $t = 0$ s, the circuit involves just the capacitor and inductor.

with the independent voltage source V_0, since the two circuit elements are connected to the same two nodes (here they also are in series!). Since the capacitor and independent voltage source are in parallel, the two circuit elements have the same potential difference across them; thus the capacitor has a potential difference V_0 between its plates with the polarity indicated in Figure 21.62.

The charge initially on the capacitor is found from the definition of the capacitance:

$$C = \frac{Q_0}{V_0}$$

so that

$$Q_0 = CV_0$$

The initial energy stored in the capacitor is found from Equation 18.12:

$$\frac{1}{2}\frac{Q_0^2}{C}$$

When $t = 0$ s, we close switch (2) and simultaneously open switch (1), securing the circuit shown in Figure 21.63. This takes the voltage source V_0 out of the circuit and places the inductor and capacitor in parallel (also in series, since again we have a one-loop, two-element circuit). We want to discover how the charge Q on the capacitor varies with time as it discharges through the inductor.

Since the capacitor initially has the polarity indicated in Figure 21.64, we choose the current I to be in the direction indicated, consistent with the polarity markings on the capacitor. With this choice for the current direction, we are forced to mark the polarity of the inductor as indicated.*

A differential equation for Q can be found by applying the KVL to the LC loop in Figure 21.64. We go clockwise around the loop, beginning in the lower left corner of the loop:

$$L\frac{dI}{dt} + \frac{Q}{C} = 0 \text{ V}$$

---

*The actual potential difference across the inductor has the reverse polarity since the actual current is opposite to the direction chosen; we choose this polarity to be consistent with our convention for assigning polarity based on the direction of the current.

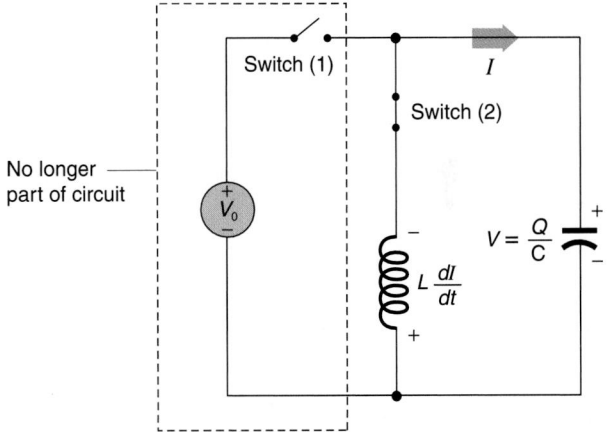

FIGURE 21.64 Polarity markings are assigned consistent with the direction chosen for I.

Then we use

$$I = \frac{dQ}{dt}$$

in the first term to obtain

$$L\frac{d^2Q}{dt^2} + \frac{1}{C}Q = 0 \text{ V}$$

or

$$\frac{d^2Q}{dt^2} + \frac{1}{LC}Q = 0 \text{ A/s} \qquad (21.52)$$

The differential equation for Q (Equation 21.52) is an old friend; it is the differential equation for *simple harmonic oscillation!*

The equation for Q has the same form as the differential equation that describes the oscillation of a mass on a Hooke's law spring (Equation 7.11):

$$\frac{d^2x}{dt^2} + \omega^2 x = 0 \text{ m/s}^2$$

The only difference here is that the differential equation is for the charge Q, not for the position x of a mass on the end of a spring. For the harmonic oscillator in Chapter 7, we found the solution to Equation 7.11 to be

$$x(t) = A \cos(\omega t + \phi) \qquad (21.53)$$

By analogy, we write the solution to Equation 21.52 as

$$Q(t) = Q_0 \cos(\omega t + \phi) \qquad (21.54)$$

The coefficient of Q in the differential equation (Equation 21.52) once again is ω^2.

Thus the angular frequency of the charge oscillations is

$$\omega = \frac{1}{(LC)^{1/2}} \qquad (21.55)$$

We know that when $t = 0$ s the charge on the capacitor was Q_0. Setting $t = 0$ s in Equation 21.54 means that the left-hand side must be Q_0, so that

$$Q_0 = Q_0 \cos(0 \text{ rad} + \phi)$$

This implies that $\cos \phi = 1$, and $\phi = 0$ rad. Thus our solution for Q (Equation 21.54) reduces to

$$Q(t) = Q_0 \cos(\omega t)$$

Since

$$I = \frac{dQ}{dt}$$

a differentiation shows that the current in the circuit also varies sinusoidally:

$$I(t) = -Q_0 \omega \sin(\omega t) \qquad (21.56)$$

The initial value of the current, when $t = 0$ s, thus is 0 A.

Graphs of the charge Q and the current I as functions of time are shown in Figure 21.65. These graphs correspond to the position and velocity component of a mass undergoing one-dimensional simple harmonic oscillation, released from position x_0 when $t = 0$ s.

The energy in the circuit oscillates between the capacitor and the inductor. The energy in the capacitor at any time is

$$\frac{1}{2} \frac{Q^2}{C} = \frac{1}{2} \frac{Q_0^2}{C} \cos^2 \omega t$$

while that in the inductor is

$$\frac{1}{2} LI^2 = \frac{1}{2} LQ_0^2 \omega^2 \sin^2 \omega t$$

At any instant t the sum of these energies is

$$\frac{1}{2} \frac{Q_0^2}{C} \cos^2 \omega t + \frac{1}{2} LQ_0^2 \omega^2 \sin^2 \omega t$$

But since

$$\omega^2 = \frac{1}{LC}$$

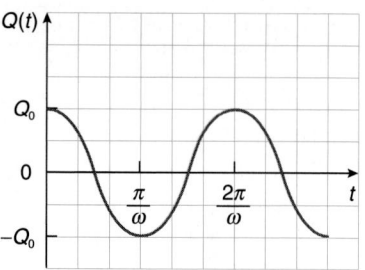

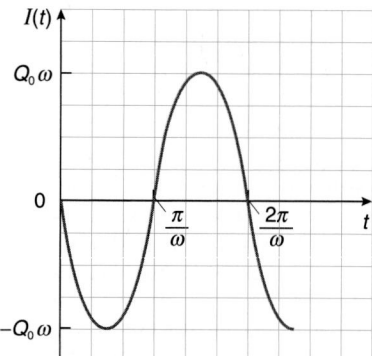

FIGURE 21.65 Graphs of the charge and current oscillations as functions of time.

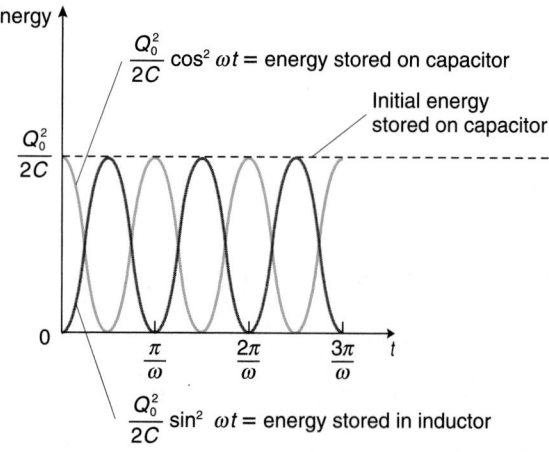

FIGURE 21.66 The energy stored by the capacitor and inductor as functions of time.

the energy sum becomes

$$\frac{1}{2} \frac{Q_0^2}{C} (\cos^2 \omega t + \sin^2 \omega t) = \frac{1}{2} \frac{Q_0^2}{C}$$

which is just the initial energy initially stored on the capacitor. Energy is conserved at every instant, as indicated in Figure 21.66.

In terms of our mechanical example, the energy stored in the capacitor is analogous to the potential energy stored in the spring, and the energy present in the inductor when charge is moving through it is analogous to the kinetic energy of the moving mass.

An *ideal* simple harmonic oscillator oscillates indefinitely. In a *real*, macroscopic harmonic oscillator, frictional effects eventually damp out the motion with the initial mechanical energy eventually transferred to the internal energy of its environment. Similar effects occur in an *LC* circuit. For an *ideal LC* circuit, the charge oscillations continue indefinitely. However, a *real LC* circuit inevitably has some resistance in the circuit, if only from the wire that forms the inductor. The resistor in an *LC* circuit plays the role of the friction in a damped harmonic oscillator. The charge oscillations eventually die out. The resistor absorbs positive power no matter which way the current is directed through it. Thus the initial electrical energy in the circuit eventually appears as an energy transfer to the resistor, increasing its internal energy and passing the energy to the internal energy of the environment via heat transfer.

EXAMPLE 21.13 ⎯⎯⎯⎯⎯⎯⎯⎯

You are asked to design a parallel *LC* circuit that can oscillate with a frequency of 540 kHz, which is in the AM radio band of frequencies. A 15 mH inductor is available. Find the capacitance needed.

Solution
The angular frequency of oscillation ω is given by Equation 21.55, which after squaring is

$$\omega^2 = \frac{1}{LC}$$

The angular frequency ω and the frequency ν are related by

$$\omega = 2\pi\nu$$

Hence you have

$$4\pi^2\nu^2 = \frac{1}{LC}$$

Solving for C, you obtain

$$C = \frac{1}{4\pi^2\nu^2 L}$$
$$= \frac{1}{4\pi^2 (540 \times 10^3 \text{ Hz})^2 (15 \times 10^{-3} \text{ H})}$$
$$= 5.8 \times 10^{-12} \text{ F}$$
$$= 5.8 \text{ pF}$$

21.11 MUTUAL INDUCTANCE*

Here we see how remote circuits can magnetically induce emfs in another circuit via Faraday's law through **mutual inductance**. Since every circuit can be imagined geometrically as at least one complete conducting path or coil, the effects of mutual inductance, however small or large, are present in every circuit unless it is a truly isolated, one-loop circuit. The effects of mutual inductance are hazardous to some sensitive electronic devices (such as heart pacemakers) but provide the *raison d'être* for others such as the transformers we consider in the next section.

In Section 21.6 we investigated what happens when the magnetic flux through a coil changes with time: an emf is induced in the coil. In particular, in Section 21.6 the flux through the coil

was due to the magnetic field caused by the current in the same coil. We described the effect by means of an essentially geometric factor known as the self-inductance *L* of the coil.

Here we want to extend these ideas to two coils in close proximity. We find, perhaps remarkably as Faraday discovered, that changes in one coil affect the other, even though they have no common conducting link. We consider a number of situations.

Case 1: A Closed Circuit near an Open Circuit

Imagine coil (1) with current I_1 with a nearby second coil that is an open circuit (see Figure 21.67); the second coil is a network but not a circuit.

The magnetic field of current I_1 in the first coil produces some magnetic flux Φ_{21} through the second coil. We define the **mutual inductance** M_{21} to be the ratio of the magnetic flux through the second coil (caused by the magnetic field of the current in the first coil) to the current in the first coil:

$$M_{21} \equiv \frac{\Phi_{21}}{I_1} \qquad (21.57)$$

This definition somewhat resembles the definition of the self-inductance (Equation 21.29). Hence, in SI units, the mutual inductance is expressed in henries (like the self-inductance) and also turns out to be a function of the geometric size and arrangement of the coils and the materials (i.e., the medium) between them.

If current I_1 changes with time, the magnetic flux Φ_{21} also changes with time. Faraday's law (Equation 21.4) then implies that an induced emf is produced in the second coil:

$$\text{induced emf}_2 = -\frac{d\Phi_{21}}{dt}$$

We use Equation 21.57 for the magnetic flux:

$$\text{induced emf}_2 = -M_{21}\frac{dI_1}{dt}$$

Just as we did for the emfs induced by self-inductance, we define V_2 to be the *negative* of this induced emf, so that

$$V_2 \equiv M_{21}\frac{dI_1}{dt} \qquad (21.58)$$

The mutual inductance M_{21} relates the changes in the current in the first coil to the resulting effects in the second coil.

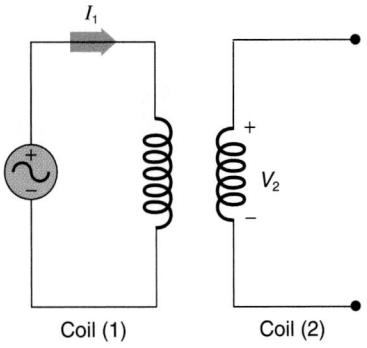

FIGURE 21.67 **A coil in proximity to another coil with a current.**

Case 2: A Reciprocal Arrangement

Now let the first coil be an open circuit and let a current I_2 be in the second coil, as shown in Figure 21.68.

In this case, current I_2 creates a magnetic field and this magnetic field produces a magnetic flux Φ_{12} through the first coil. Now define the mutual inductance M_{12} to be the ratio of the magnetic flux Φ_{12} to the current I_2:

$$M_{12} \equiv \frac{\Phi_{12}}{I_2} \qquad (21.59)$$

Once again, the mutual inductance M_{12} depends only on geometric factors and the material medium between the coils.

If the current I_2 changes with time, the magnetic flux Φ_{12} changes with time, and an induced emf is produced in the first coil according to Faraday's law:

$$\text{induced emf}_1 = -\frac{d\Phi_{12}}{dt}$$

Using Equation 21.59, we have

$$\text{induced emf}_1 = -M_{12}\frac{dI_2}{dt}$$

Call V_1 the negative of this induced emf, so that

$$V_1 \equiv M_{12}\frac{dI_2}{dt} \qquad (21.60)$$

The mutual inductance M_{12} relates the changes in the second coil to the resulting effects in the first coil.

How are the two coefficients of mutual inductance, M_{21} and M_{12}, related to each other? They both depend on geometric factors such as the number of turns and the area of each coil, and on the material medium between the coils. Since the geometry and the medium between the coils are the same in both situations, it is certainly plausible that

$$M_{21} = M_{12}$$

In fact, this is exactly the case, although our argument here is not a rigorous proof. In view of this equality, we write

$$M_{21} = M_{12} \equiv M \qquad (21.61)$$

and call M mutual inductance between the coils.

Case 3: Both Circuits Closed

The situation is more complicated when currents exist in both coils, as shown in Figure 21.69.

Now we have to consider not only the mutual effects, via the mutual inductance, but also the self effects, via the self-inductance. The total induced emf in each coil therefore has two contributions: one from the self-inductance (of the particular coil) and one from the mutual inductance (from the other coil). The total potential difference therefore must also have two contributions. That is,

$$V_1 = L_1\frac{dI_1}{dt} + M\frac{dI_2}{dt} \qquad (21.62)$$

where L_1 is the self-inductance of the first coil. Likewise,

$$V_2 = L_2\frac{dI_2}{dt} + M\frac{dI_1}{dt} \qquad (21.63)$$

where L_2 is the self-inductance of the second coil. Things get a bit involved! The detailed implications of these effects between the coils we leave for a more advanced course in electromagnetic theory and electrical circuits.

> The implication of Equations 21.58 and 21.60 is that it is possible to remotely induce an emf in a circuit by means of a changing magnetic flux through a coil instigated by changes in another circuit.

Unfortunately, mutual inductance is the cause of electronic interference in circuits by time-varying magnetic fields such as those caused by motion of a circuit through a nonuniform magnetic field, or the time-varying magnetic fields of distant lightning, auroras, and even electric machinery.

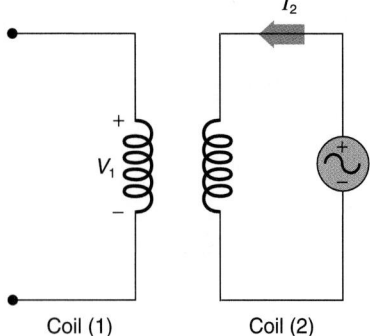

FIGURE 21.68 A reciprocal arrangement of the coils.

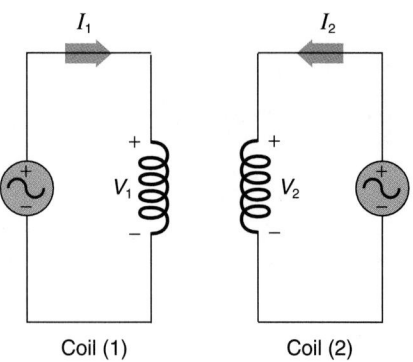

FIGURE 21.69 Currents in both coils.

21.12 AN IDEAL TRANSFORMER*

Transformers are used for many practical purposes; you likely had a small one for your electric train as a child. If you have a low-voltage lighting system in your home, you also will find a transformer in each fixture or a single transformer for the whole collection of lighting fixtures. Open up your stereo amplifier or PC and you also will find transformers. The large, typically cylindrical devices that you see hanging from utility poles all around town are transformers, as are many of the large devices in utility substations. What is the function of a transformer and on what principle of physics are they based?

An ideal transformer has two coils of wire, electrically isolated from each other, and arranged so that all the magnetic field lines produced by the currents in each coil completely thread the other coil, with no leakage of magnetic field lines. In practice, this idealization can be approximated by either (1) tightly wrapping the coils around each other, as in Figure 21.70, or (2) linking the coils with a highly ferromagnetic material such as iron, as in Figure 21.71.

An ideal transformer thus has the following characteristic: the magnetic flux through each *individual* turn (or loop) of wire in both coils is the same. Let Φ be the magnetic flux through a *single* turn of wire; Φ is the same through *every* loop of wire in the entire arrangement, regardless of which coil contains the individual loop. If the two coils are of different cross-sectional area, no matter; then the magnetic field compensates accordingly, so that the flux through the individual loop is the same as that through every other loop. For example, if the second coil has a larger cross-sectional area, then the number of lines of magnetic field per square meter is fewer, and so the total flux through the loop is the same as that through any other loop. The important thing to realize is that Φ is the magnetic flux caused by the total magnetic field arising from whatever currents are in both coils.

The total magnetic flux through the first coil is the number of turns of wire (or loops) N_1 in this coil times the magnetic flux Φ through a single coil:

$$\text{total magnetic flux through coil (1)} = N_1\Phi$$

Likewise the total magnetic flux through the second coil, with N_2 turns, is

$$\text{total magnetic flux through coil (2)} = N_2\Phi$$

If the currents are changing with time, the magnetic flux through the coils changes with time. Faraday's law implies there then is an emf induced in each coil equal to the negative time derivative of the total magnetic flux through the coil. Since the number of turns of wire in each coil is fixed, we have

$$\text{induced emf in coil (1)} = -\frac{d}{dt}\left(N_1\Phi\right)$$
$$= -N_1\frac{d\Phi}{dt}$$
$$\text{induced emf in coil (2)} = -\frac{d}{dt}\left(N_2\Phi\right)$$
$$= -N_2\frac{d\Phi}{dt}$$

Once again, we introduce potential differences V_1 and V_2 equal to the negative of the respective emfs:

$$V_1 = N_1\frac{d\Phi}{dt} \tag{21.64}$$

$$V_2 = N_2\frac{d\Phi}{dt} \tag{21.65}$$

However, since

$$\frac{d\Phi}{dt}$$

Several transformers on a utility pole. The devices on the lower wooden crossbar are fuses.

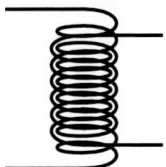

FIGURE 21.70 One coil wrapped around another coil.

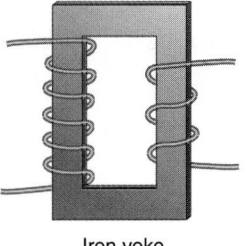

Iron yoke

FIGURE 21.71 Two coils linked with an iron yoke.

is the *same* for every loop of wire in the entire system, we can divide Equations 21.64 and 21.65 to obtain

$$\frac{V_1}{V_2} = \frac{N_1}{N_2} \qquad (21.66)$$

The ratio of the potential differences has the same value as the ratio of the number of coils.

The beauty of the ideal transformer is that we can avoid the more complicated Equations 21.62 and 21.63 we investigated in Section 21.9 involving the self and mutual inductances of the coils themselves (Case 3). The ratio of the potential differences is simply set by the number of turns (coils). The treatment of *real* transformers, however, is more complex than this ideal scenario; we defer their consideration to engineering courses in electrical circuits. Fortunately, with almost all real transformers, the ideal transformer is a reasonable first approximation.

The input coil of a transformer customarily is called the **primary** coil; the output coil of the arrangement is known as the **secondary** coil.

> If the output potential difference V_2 is greater than the input V_1, the transformer is called a **step-up transformer**. From Equation 21.66, we can deduce that $N_2 > N_1$ for a step-up transformer.

> If the output potential difference V_2 is less than the input V_1, then the transformer is called a **step-down transformer**. Equation 21.66 implies $N_2 < N_1$ for a step-down transformer.

By convention, the polarity of V_1 and V_2 is indicated by means of a *dot convention* next to a symbolic representation of the transformer (see Figure 21.72). Dots indicate the terminals with corresponding polarity so the details of the winding need not be examined.

It might seem as if we get something for nothing in a transformer. We can change ac potential differences up or down, but energy conservation prevails, as it must. In an ideal transformer, no electrical energy is gained or lost to the system. The sum of the electrical power absorbed by the primary and secondary coils in an ideal transformer must be zero. Note in Figure 21.73 that the directions chosen for the currents I_1 and I_2 are *into* the terminals marked (+) for both V_1 and V_2 respectively. Thus the power absorbed by the primary is

$$I_1 V_1$$

with our usual sign convention for the power: the current must be into the (+) polarity terminal. The power absorbed by the secondary is

$$I_2 V_2$$

with the same sign convention. The sum of the power absorbed is zero by energy conservation:

$$I_1 V_1 + I_2 V_2 = 0 \text{ W}$$

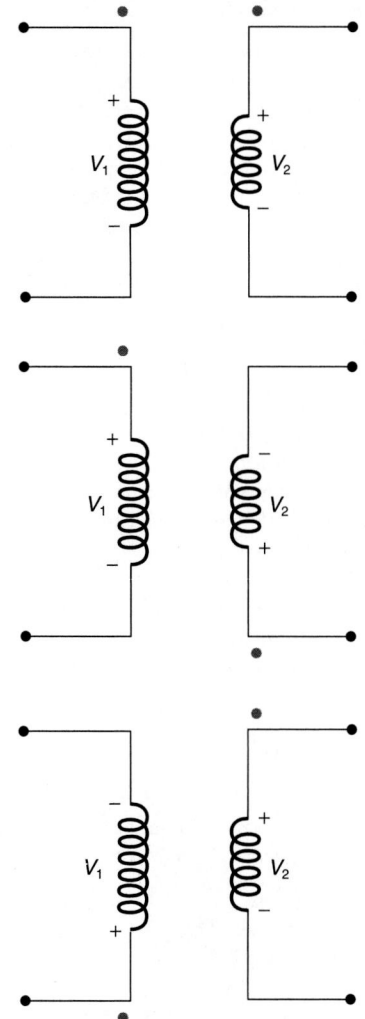

FIGURE 21.72 Dots indicate terminals with the same polarity.

We solve for I_2:

$$I_2 = -\frac{V_1}{V_2} I_1$$

Using Equation 21.66 for the ratio V_1/V_2, we have

$$I_2 = -\frac{N_1}{N_2} I_1$$

or equivalently,

$$N_2 I_2 = -N_1 I_1 \qquad (21.67)$$

The minus sign indicates that if I_1 has the direction shown in Figure 21.73, the current I_2 actually is in the direction opposite to that shown in Figure 21.73. We use these directional and polarity conventions for the currents and potentials indicated in Figure 21.73 because they are the ones used throughout electrical engineering.

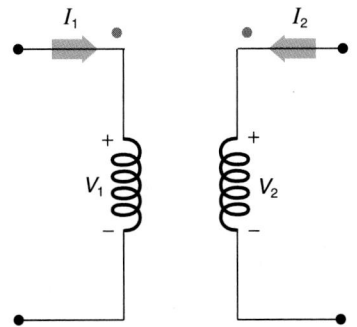

FIGURE 21.73 Conventional directions for the currents.

In a *step-up* transformer, the output potential difference V_2 is greater than the input V_1, but the output current I_2 is less than the input I_1 (in absolute value) in accordance with energy conservation.

Conversely, in a *step-down* transformer, the output potential difference V_2 is less than the input V_1, and the output current I_2 is greater (in absolute value) than the input I_1.

Transformers are used ubiquitously in electronics and electrical power transmission for changing (i.e., transforming) ac potential differences from one value to another.

A transformer will not transform dc (steady) potential differences.

You should be able to explain why!

Transformers are used in the electrical utility industry to facilitate transmission of electrical power over long distances with small resistive power losses. Such transmission involves lengths of wire many kilometers long. Such lengths of wire, of course, have a significant electrical resistance, say R. The power absorbed by a resistance R is proportional to the square of the current: I^2R (Equation 19.28). So, to minimize the power loss caused by the resistance of the transmission wires, the current should be as small as possible.

With a step-up transformer at the generating plant, the output potential difference is large but the corresponding current is small. Thus power transmission over long distances is done at a high potential difference (upward of a megavolt) and small current. Once the wires reach your town, the potential difference is lowered (typically to about 15 kV) with a large step-down transformer, and the current correspondingly increases. The final stage before entering your home is another step-down transformer to secure common household potential differences of 220 V and 120 V.* Transformers have long service lifetimes since they are completely passive devices; there are no moving parts.

Electric arc welding machines use step-down transformers to decrease the potential difference while increasing the current to maximize resistive power absorption (and heating) of steel.

EXAMPLE 21.14

A common step-down transformer has an input potential difference of 14.5 kV and an output of 120 V.

a. Determine the ratio of number of turns of wire in the primary coil to that in the secondary.

b. If the maximum current in the secondary is 300 A, what is the maximum current in the primary?

Solution

a. The input and output potential differences are related to the number of turns in the coils by Equation 21.66:

$$\frac{V_1}{V_2} = \frac{N_1}{N_2}$$

$$\frac{14.5 \times 10^3 \text{ V}}{120 \text{ V}} = \frac{N_1}{N_2}$$

$$\frac{N_1}{N_2} = 121$$

Notice that the secondary coil has many fewer coils since this is a step-down transformer.

b. The currents in the coils are related to the number of turns by Equation 21.67:

$$N_2 I_2 = -N_1 I_1$$

Solving for I_1, you find

$$I_1 = -\frac{N_2}{N_1} I_2$$

$$= -\frac{1}{121} (300 \text{ A})$$

$$= -2.48 \text{ A}$$

The maximum current in the primary is 2.48 A. What is the significance of the minus sign in view of Figure 21.73?

---

*A neat trick is used to provide 220 V. You might ask your professor about it.

CHAPTER SUMMARY

Faraday's law of electromagnetic induction states that an *induced emf* is produced by and is directly proportional to the time rate of change of magnetic flux:

$$\text{induced emf} = -\frac{d\Phi_{\text{magnet}}}{dt} \qquad (21.4)$$

The induced emf itself is the path integral of a (nonconservative) electric field $\vec{E}$ around the closed contour through which the magnetic flux is taken:

$$\text{induced emf} \equiv \int_{\text{clsd path}} \vec{E} \cdot d\vec{\ell} \qquad (21.3)$$

Lenz's law is a rule for determining the direction of the induced electric field and any induced current caused by it. The induced field and current always are directed so as to oppose the change in the magnetic flux that is taking place.

A coil with N turns of wire, each of area A, rotating at angular velocity $\vec{\omega}$ perpendicular to the area vector $\vec{A}$ in a uniform magnetic field $\vec{B}$ (also perpendicular to $\vec{\omega}$) produces an emf given by

$$V_{\text{induced}}(t) = NBA\omega \sin(\omega t) \qquad (21.8)$$

The four Maxwell equations of electromagnetism, together with the equations for the magnetic and electrical forces, describe all electromagnetic phenomena. The force equations (from Chapters 16 and 20) are

$$\vec{F}_{\text{elec}} = q\vec{E} \qquad (16.8)$$

$$\vec{F}_{\text{magnet}} = q\vec{v} \times \vec{B} \qquad (20.1)$$

The Maxwell equations are

1. Gauss's law for the electric field:

$$\int_{\text{clsd surface } S} \vec{E} \cdot d\vec{S} = \frac{Q_{\text{net enclosed by } S}}{\varepsilon_0} \qquad (21.9)$$

2. Gauss's law for the magnetic field:

$$\int_{\text{clsd surface } S} \vec{B} \cdot d\vec{S} = 0 \text{ T} \cdot \text{m}^2 \qquad (21.10)$$

3. The Ampere–Maxwell law:

$$\int_{\text{clsd path}} \vec{B} \cdot d\vec{\ell} = \mu_0(I + I_{\text{D}}) \qquad (21.11)$$

where I is the conduction current threading the closed path and I_{D} is the displacement current. The displacement current is defined as

$$I_{\text{D}} \equiv \varepsilon_0 \frac{d\Phi_{\text{elec}}}{dt}$$

where Φ_{elec} is the flux of the electric field through any hatlike surface that has the closed path as a boundary.

4. Faraday's law of electromagnetic induction:

$$\int_{\text{clsd path}} \vec{E} \cdot d\vec{\ell} = -\frac{d\Phi_{\text{magnet}}}{dt} \qquad (21.12)$$

where Φ_{magnet} is the flux of the magnetic field through any surface that has the closed path as a boundary.

Among the consequences of Maxwell's equations in a vacuum are the existence of electromagnetic waves that propagate with a speed equal to

$$\frac{1}{(\mu_0\varepsilon_0)^{1/2}} \qquad (21.26)$$

This speed is identical to the measured speed of light c. These electromagnetic waves *are* light in the broadest sense of the term, encompassing γ-rays, x-rays, ultraviolet light, visible light, infrared light, microwaves, and radio waves.

A coil of wire, otherwise called an inductor, is the magnetic analog of a capacitor. An inductor has a self-inductance L defined as the ratio

$$L \equiv \frac{\Phi_{\text{magnet}}}{I} \qquad (21.29)$$

where Φ_{magnet} is the flux of the magnetic field through the coil caused by the current I in the coil. If a time-varying current exists in the coil, there is a potential difference V across the coil given by

$$V = L\frac{dI}{dt} \qquad (21.34)$$

The potential difference V is the negative of the emf induced in the coil by the time-varying magnetic flux through the coil. The potential difference V is introduced so that an inductor can be treated as another circuit element with the same sign conventions previously introduced for resistors and capacitors. (See Figure 21.74 for the polarity convention for an inductor.)

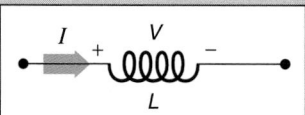

FIGURE 21.74 Polarity convention for an inductor.

Inductors in series can be combined into a single equivalent inductor:

$$L_{\text{eq}} = L_1 + L_2 + L_3 + \cdots \quad \text{(series connection of inductors)} \qquad (21.37)$$

Inductors in parallel also can be combined into a single equivalent inductor:

$$\frac{1}{L_{\text{eq}}} = \frac{1}{L_1} + \frac{1}{L_2} + \frac{1}{L_3} + \cdots \quad \text{(parallel connection of inductors)} \qquad (21.40)$$

A magnetic field stores energy. The energy density associated with a magnetic field is

$$\frac{1}{2}\frac{B^2}{\mu_0} \qquad (21.51)$$

The mutual inductance M of a pair of coils is the ratio of the flux of the magnetic field Φ_{21} through a coil (2) to the current in the *other* coil (1) which created the field:

$$M = \frac{\Phi_{21}}{I_1} \qquad (21.57)$$

If the current in the other coil (1) varies with time, the time-varying flux of its magnetic field through coil (2) induces an emf in the second coil. The potential difference V_2 (the negative of the induced emf) across the given coil (2) is

$$V_2 \equiv M \frac{dI_1}{dt} \qquad (21.58)$$

A similar relationship exists for the other coil.

A transformer has a primary coil with N_1 coils and a secondary with N_2 coils. It is ideal if it does not gain or lose electrical energy per unit time. The potential differences across its primary and secondary coils are related to the numbers of turns:

$$\frac{V_1}{V_2} = \frac{N_1}{N_2} \qquad (21.66)$$

(See Figure 21.75 for the polarity and current conventions for a transformer.) The respective currents in the primary and secondary also are related to their numbers of turns:

$$N_2 I_2 = -N_1 I_1 \qquad (21.67)$$

An ideal transformer absorbs zero total electrical power.

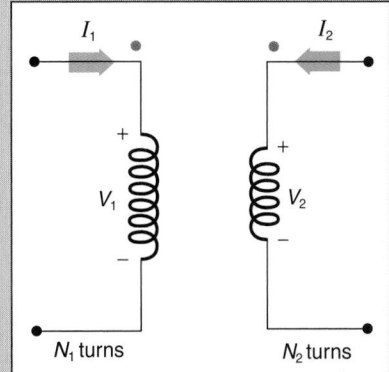

FIGURE 21.75 Polarity and current conventions for an ideal transformer.

SUMMARY OF PROBLEM-SOLVING TACTICS

21.1 (page 958) Notice that Faraday's law, as expressed by Equation 21.4, indicates that the induced emf is equal to the *negative of the slope* of a graph of Φ_{magnet} versus time.

21.2 (page 960) To see how Lenz's law actually is used is best done by carefully examining a number of specific examples.

21.3 (page 972) Be careful to keep clear the distinction between the induced emf in an inductor (equal to the *negative* time rate of change of the magnetic flux) and the potential difference

$$V = -\text{induced emf} = L \frac{dI}{dt}$$

introduced so that an inductor can be treated with the same sign conventions as resistors and capacitors. The induced emf and the potential difference V are the negatives of each other.

21.4 (page 972) For dc situations, an inductor acts as a short circuit.

QUESTIONS

1. **(page 958); 2. (page 973)**

3. The Earth has a magnetic field. What is the total flux of the magnetic field of the Earth through the surface of the Earth?

4. Your friends are confused. As their tutor, explain the distinction between the terms magnetic flux and magnetic field. Is either or both a scalar quantity? Can you say a magnetic field exists at a *point* in space? Can the same be said for the magnetic flux?

5. Two conducting loops have a magnet located along their common axis as indicated in Figure Q.5. The magnet is moved toward loop (1). Indicate the directions of the currents induced in each loop.

6. Figure Q.6 shows the magnetic field inside a very long solenoid (an end-on view). The magnetic field is decreasing with time. For each of the indicated wire loops, indicate the direction of any current induced in the loop.

7. In Figure Q.6 (previous question) the magnetic field does not vary with time. The largest wire loop is moved in such a way

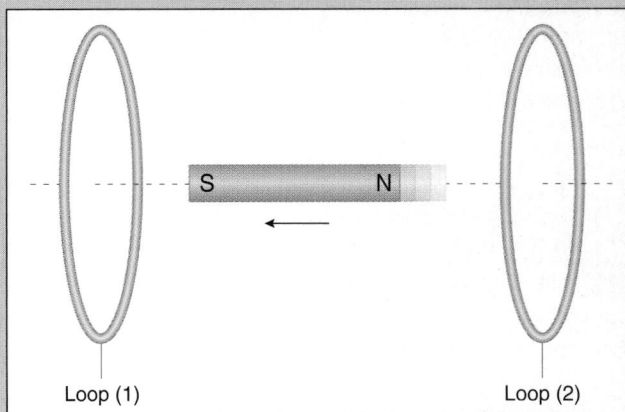

FIGURE Q.5

that every point of the loop perimeter always remains outside the region with the nonzero magnetic field. Explain why there is no current induced in the largest loop.

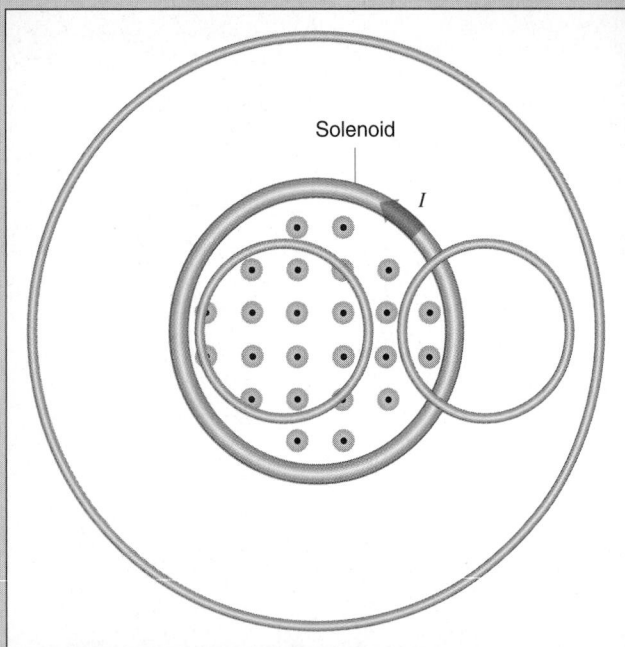

FIGURE Q.6

8. Small gasoline engines, such those used for chain saws, lawn mowers, weed whackers, and small outboard motors, generate a spark for combustion by means of a *magneto*, a permanent magnet attached to the flywheel. A stationary coil of wire is connected to the spark plug. When the magnet passes the coil, a spark is produced across the spark plug gap. Explain how this happens.

9. A bicycle wheel spins about its axis in a constant magnetic field $\vec{B}$ that is in the same direction as the angular velocity vector. Indicate the polarity of the emf induced along a spoke.

10. A single metal ring of diameter 10 cm spins steadily about a diameter with a frequency of 5.0 Hz. A uniform magnetic field of magnitude 0.100 T is perpendicular to the spin axis. What is the orientation of the ring when the induced emf is (instantaneously) zero?

11. In nuclear magnetic resonance (NMR) imaging technology, carefully controlled, oscillating, magnetic fields are used in complex ways to form visible images of soft tissues such as the brain. In the past neurosurgeons occasionally used metal clamps to block cerebral aneurysms. Subsequently, it was found that NMR images in the vicinity of the metal clamps were not as clear as in soft tissues further from the clamps. Explain how Faraday's law gives rise to currents in the clamps that produce magnetic fields that can interfere with the NMR fields, thus spoiling the image. Plastic clamps now are used.

12. AM radio waves have frequencies on the order of 1000 kHz. What is the corresponding order of magnitude of their electromagnetic wavelengths?

13. A scientific charlatan claims sound is an electromagnetic wave. What experimental evidence can you cite to prove that sound is *not* an electromagnetic wave?

14. Briefly state the four Maxwell equations in your own words.

15. Seismographs are used to detect and record vibrations from earthquakes and explosions (deliberetely set to probe for underground reservoirs of oil or to analyze subsurface rock features). One way of detecting ground vibrations is to use a permanent magnet and a coil of wire. Explain how such a detector might function.

16. A planar loop of wire is in a uniform magnetic field oriented perpendicular to the plane of the loop. The loop then is pulled at constant velocity in a direction in the plane of the loop. Is an emf induced in the loop? If the magnetic field is oriented at an angle θ to the plane of the loop and the experiment is repeated, can an emf be induced in the loop? If no emf is induced under these circumstances, explain what (if anything) can be done in the uniform field so an emf *is* induced in the coil.

17. A wire segment AC is moved near the magnet shown in Figure Q.17 (a) Is an emf induced in the wire? If so, proceed with the following questions. If not, you are home free. (b) Which end of the wire segment has the positive polarity? (c) Given the same magnet, describe two things you could do to increase the magnitude of the induced emf. (d) What should you do to reverse the polarity of the induced emf?

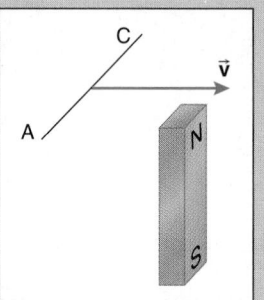

FIGURE Q.17

18. A conducting bar moves along conducting rails as shown in Figure Q.18. Is the direction of the induced current clockwise or counterclockwise? Experimentally, it is found that a constant force is necessary to move the bar at constant velocity; explain why. Does this violate Newton's second law of motion? Explain.

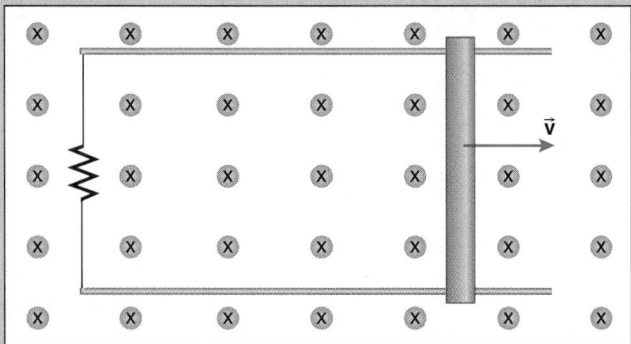

FIGURE Q.18

19. A short bar magnet is dropped through a coil of wire whose plane is horizontal as shown in Figure Q.19. Make a qualitative graph of the emf induced in the coil as a function of time.

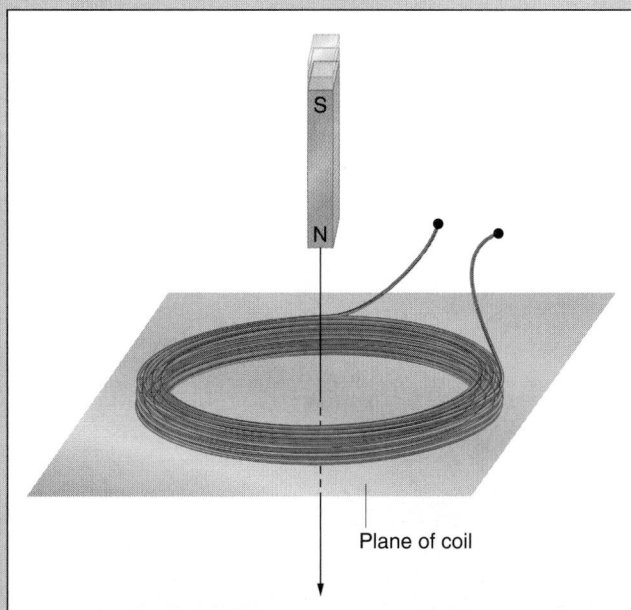

FIGURE Q.19

20. The potential difference V across an inductor is

$$V = L \frac{dI}{dt}$$

Imagine a graph of $V(t)$ versus t, beginning when $t = 0$ s. At an instant t on the graph, what is the geometric interpretation of the current $I(t)$?

21. A metal washer rests on top of a vertical coil as shown in Figure Q.21. The coil has a large self-inductance. When the switch is closed the current increases with time. If the time rate of change of the current is large enough, the metal washer flies vertically off the coil. Explain why.

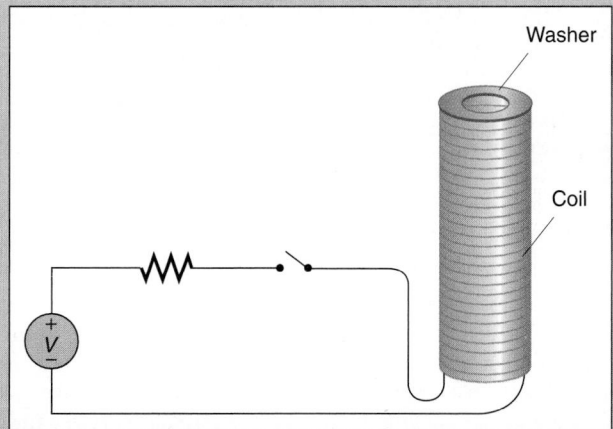

FIGURE Q.21

22. The current through the inductor indicated in Figure Q.22 is increasing with time. Indicate the polarity of the potential difference

$$V = L \frac{dI}{dt}$$

Repeat if the current is decreasing with time.

FIGURE Q.22

23. An LR circuit with a long time constant τ has a constant current. A switch suddenly opens the circuit. Explain why a spark or arc likely is seen as the switch opens. Is the spark likely to be seen if the time constant is very small?

24. An induction stove has a conducting coil imbedded just below a ceramic cooktop surface. An oscillating current exists in the coil. A metal pot with finite resistivity is placed on the stove. Explain how Faraday's law leads to an increase in the temperature of the pot, thus cooking the food within it. The ceramic surface itself may not get very hot in the process.

25. A newspaper account about the safety of induction stoves states that "induction elements work by creating magnetic friction (instant heat) in the pot or pan."* Critique this journalistic account.

26. An independent voltage source V_0 is in a series LR network. The switch is suddenly closed and the current eventually reaches a constant value. The source suddenly is doubled from V_0 to $2V_0$. What is the effect of this latter change on the value of the time constant of the circuit?

27. A coil of wire is rotated at constant angular velocity about a symmetry axis parallel to a uniform magnetic field, as indicated in Figure Q.27. Is there an emf induced in the coil?

Roanoke Times & World News (Roanoke, Virginia), 4 June 1995, page C-1.

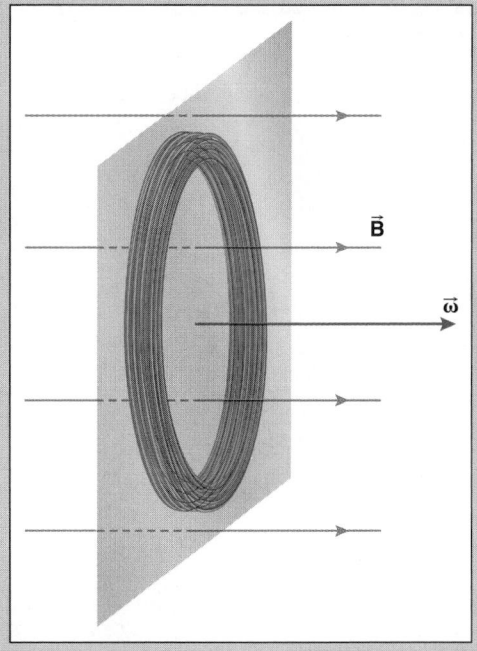

FIGURE Q.27

28. A coil of wire is rotated at constant angular velocity about a symmetry axis perpendicular to a uniform magnetic field, as indicated in Figure Q.28. Is there an emf induced in the coil?

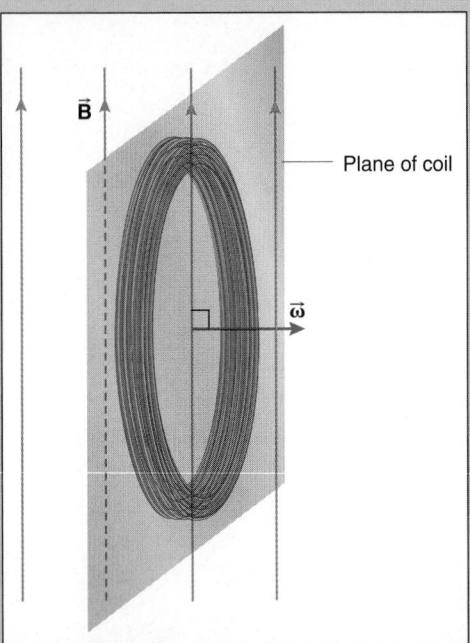

FIGURE Q.28

29. A coil of wire is rotated at constant angular velocity about a symmetry axis in the plane of the coil and perpendicular to a uniform magnetic field, as indicated in Figure Q.29. (a) Is the absolute value of the emf induced in the coil constant throughout one complete rotation of the coil? (b) If the coil forms a circuit, does the induced emf produce an induced current in the same direction in the coil throughout one complete rotation of the coil?

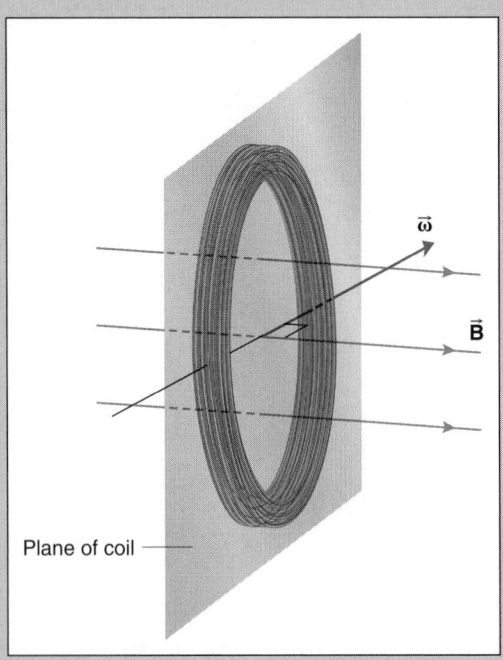

FIGURE Q.29

30. In your physics lab, you drop a magnet through a solid ring of copper, as in Figure Q.30. Neglecting air resistance, is the magnitude of the acceleration of the magnet constant? Does it make a difference whether the magnet falls north pole first or south pole first?

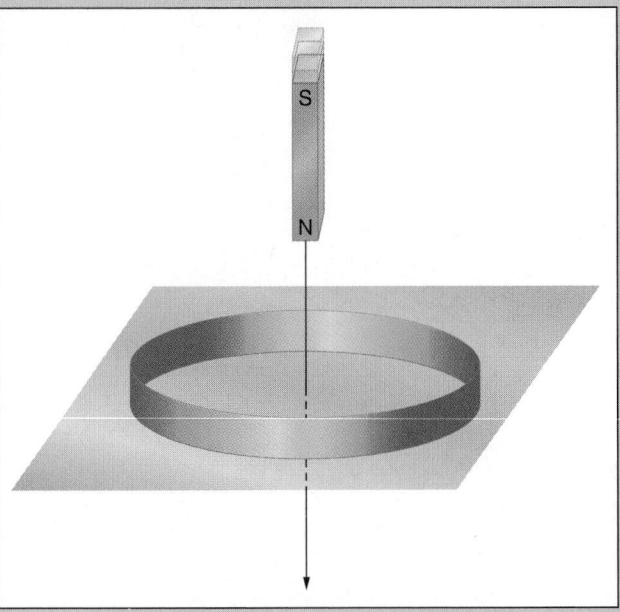

FIGURE Q.30

31. Explain why a small induced emf typically develops between the wing tips of an airplane in flight. Under what circumstances will the induced emf be zero?

32. Explain why a transformer cannot step up or step down dc independent voltage sources such as common batteries.

33. You have two planar coils of wire separated by a small, fixed distance. (a) Describe an orientation for the coils that maximizes their mutual inductance. (b) What orientation minimizes their mutual inductance?

34. Many battery-operated devices such as video cameras and portable PCs typically operate with low-voltage dc independent voltage sources (9 V or other values). An adaptor permits you to use a nearby standard household 120 V ac source, thereby enabling you to save the battery for when true portability is needed. Among other electronic components, what might be inside the adapter?

35. The utility company in your town has distribution lines at about 14.5 kV AC that are stepped down to the much smaller common household potential differences by a step-down transformer near each collection of several homes. The company has a suggestion box and makes cash rewards to employees who invent ways to save money. The suggestion box receives two suggestions for cost cutting: (a) Eliminate the step-down transformers and have each house use the 14.5 kV source directly. (b) Eliminate all the step-down transformers and have the distribution lines throughout town at the smaller value potential differences. As a member of the R&D division, diplomatically explain why each suggestion is not worthy of a monetary award.

36. A transformer is enclosed in a case with four wire leads emerging form it. (a) What can you do to discover whether

two leads represent the opposite ends of one coil or the ends of two different coils? (b) Describe an experiment you can perform to determine the ratio of the number of turns in the two coils and which coil has the larger number of turns.

37. A single coil of wire is wound with two layers of wire. When a current passes through the coil, the current in the two layers is in opposite directional senses, the inner wrapping clockwise,

the outer counterclockwise (or vice versa). Explain why this arrangement minimizes the self-inductance of the coil.

38. A home is served by an ac underground electric cable, but the location of the cable under the property is not known. Suggest a device that can be used to detect the nearby presence of such a cable, thus locating the approximate path of the cable through the property.

PROBLEMS

Sections 21.1 Faraday's Law of Electromagnetic Induction
21.2 Lenz's Law
21.3 An ac Generator

1. Show that the SI units for the time rate of change of the magnetic flux, $T \cdot m^2/s$, are equivalent to volts (V).

2. When the switch in the circuit on the left in Figure P.2 is opened, indicate the direction of the momentary induced current in the resistor in the circuit on the right.

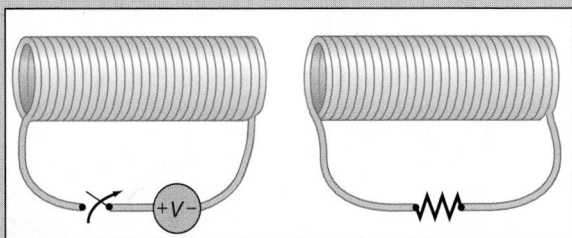

FIGURE P.2

3. While the resistance R in the circuit on the left in Figure P.3 is increased, indicate the direction of the current induced in the circuit on the right.

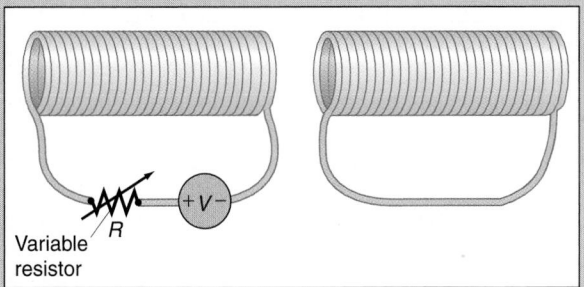

Variable resistor

FIGURE P.3

4. If the metal rod in Example 21.2 has a length of 1.00 m and is spun in a magnetic field whose magnitude is 50 mT, what angular speed ω produces an emf of 1.00 V between the ends of the rod? Express your result in both rad/s and rev/s.

5. An ac generator has a source voltage given by

$$(311 \text{ V}) \sin[(314 \text{ rad/s})t]$$

(a) What is the maximum value of the output? (b) What is the frequency of the generator? This is the common, ac sinusoidal independent voltage source available at electrical outlets in Europe, Africa, Asia, Japan, and Australia.

6. An ac generator has a source voltage given by

$$(170 \text{ V}) \sin[(377 \text{ rad/s})t]$$

(a) What is the maximum value of the output? (b) What is the frequency? This is the common, ac sinusoidal independent voltage source available at electrical outlets in North and South America.

•7. A circular wire loop is held in the horizontal plane as indicated Figure P.7. A magnet is dropped through the loop so that the N pole of the magnet initially faces the loop. You observe the event from the vantage point of a perch above the loop. (a) As the magnet approaches the loop, is the current induced in the loop in the clockwise or counterclockwise sense? (b) As the magnet recedes from the loop (after passing through it), is the induced current in the loop in the clockwise or counterclockwise sense?

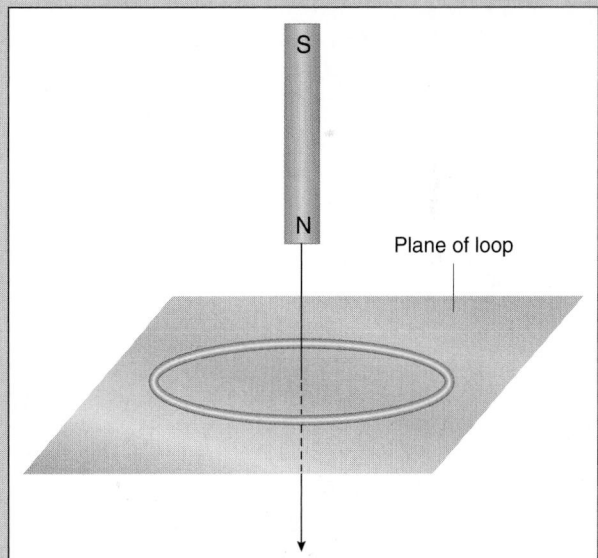

Plane of loop

FIGURE P.7

•8. The magnetic flux through a circular current loop of radius 10 cm changes with time according to the graph in Figure P.8. To gain a few brownie points, make an accurate graph of the emf induced around the loop as a function of time.

•9. A straight conducting rod of length ℓ is moving at constant velocity $\vec{v}$ in a region where there is a uniform magnetic field $\vec{B}$ as indicated in Figure P.9. (a) Electrons in the moving rod experience a magnetic force and are free to move in response to it. In this way, electric charge accumulates on the ends of

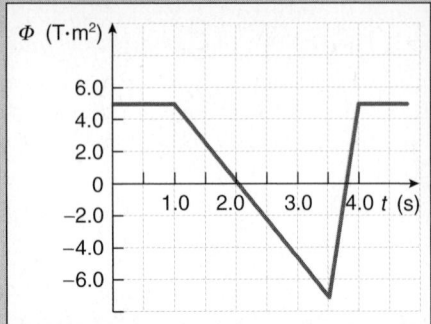

FIGURE P.8

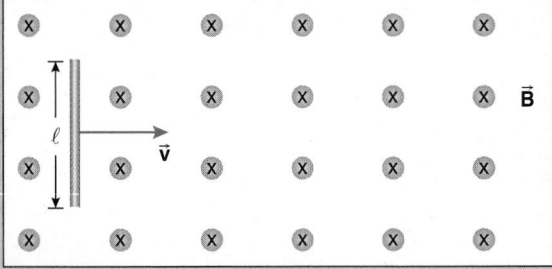

FIGURE P.9

the rod. This accumulation will not continue indefinitely, however. What limits the further accumulation of charge? (b) Which end of the rod accumulates negative charge and which end positive charge? (c) Determine the absolute value of the potential difference between the ends of the rod. (Assume a constant electric field along the length of the rod.) (d) During a time interval Δt, what area is swept out by the rod? (e) Find

$$\frac{|\Delta\Phi|}{|\Delta t|}$$

where $|\Delta\Phi|$ is the absolute value of the magnetic flux swept out by the rod during the interval Δt. Compare this result with the answer you found in part (c).

•**10.** You are not one to miss out on a good physics lecture, and so are racing to class after an early round of golf. You are streaking horizontally in your sports car at 120 km/h. The magnetic field of the Earth has a magnitude of 0.80×10^{-4} T, inclined at 60° to the surface in the vicinity. Your car is 1.5 m wide and 4.0 m long. (a) Will both the horizontal and vertical components of the magnetic field of the Earth contribute to an induced emf between the door handles of the car? Indicate your reasoning. (b) Determine the sign and magnitude of the induced emf between the opposite side door handles of your car if they are separated by 1.5 m. (c) If you connect a sensitive voltmeter between the door handles, which handle is at the higher potential?

•**11.** A U-shaped wire is bridged by a small metal rod AB of length ℓ as indicated in Figure P.11. For your enjoyment, a uniform magnetic field $\vec{\mathbf{B}}$ is directed out of the page. (a) Calculate the magnetic flux through the loop formed by the wire and rod. (b) Now move the bar at constant velocity $\vec{\mathbf{v}}$ to the right as indicated. Any way you can, find the emf induced in the loop

as the rod is moving. (c) Indicate the direction of the current induced in the loop. (d) The induced current in the movable rod is a current-carrying wire moving in a magnetic field. Calculate the direction of the magnetic force on the induced current in the movable bar. Note that this force opposes the force you need to exert on the rod to move it; this means that *you* have to do work on the rod to keep it moving, even at a constant velocity. Does this violate Newton's second law?

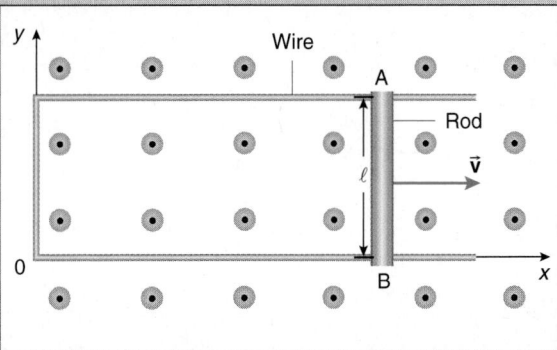

FIGURE P.11

•**12.** A movable metal rod can slide freely on two parallel friction-less conducting rails as in Figure P.12. An independent voltage source produces a current in the circuit, so that the movable rod is a current-carrying wire in a magnetic field. (a) Determine the direction of the magnetic force on the movable rod. (b) The magnetic flux through the circuit is changing with time because of the motion of the rod. Calculate the induced emf and indicate the direction of the induced current in the circuit. Note that this induced current opposes that provided by the voltage source. What are the implications of this phenomenon? Consult the following reference if needed: Mario Iona, "Why Johnny can't learn physics from textbooks I have known," *American Journal of Physics*, 55, #4, pages 299–307 (April 1987); see page 305 in particular.

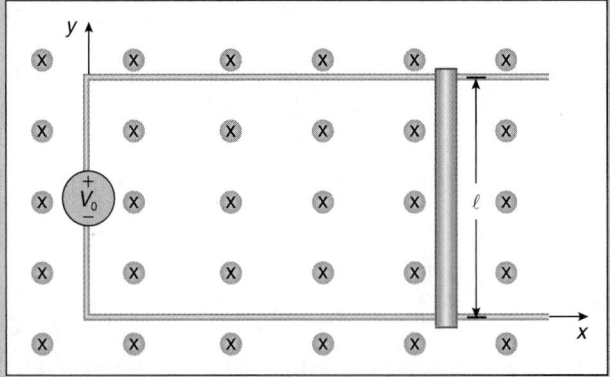

FIGURE P.12

•**13.** A magnetic field $\vec{\mathbf{B}}(t) = B_0 e^{-\alpha t}\hat{\mathbf{k}}$ is perpendicular to a circular path of radius r, which makes an angle θ with the x–y plane, as indicated in Figure P.13. (a) Calculate the flux of the magnetic field through the path as a function of time. (b) Find the emf induced around the path. If a wire is placed coincident with the path, indicate the direction of the current induced in the wire.

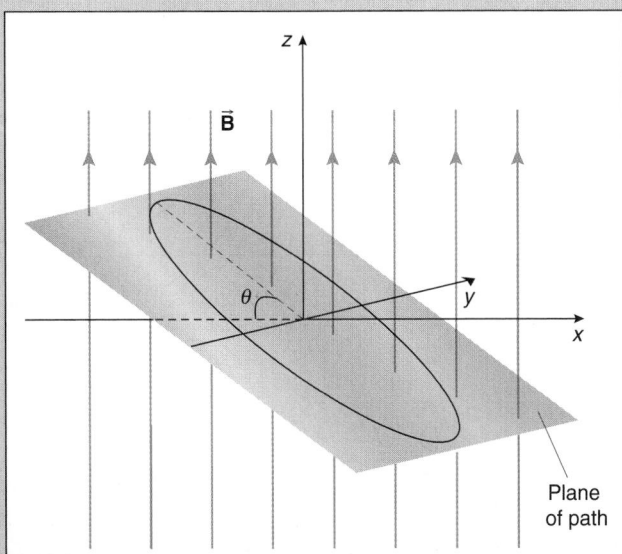

FIGURE P.13

• **14.** A solenoid has 2000 turns of wire and is 125 cm long and 10.0 cm in diameter; see Figure P.14. A circular wire loop of diameter 5.0 cm lies along the axis of the solenoid near the middle of its length as shown. (a) If the current in the solenoid initially is 4.0 A, find the magnetic flux through the smaller loop. (b) If the current in the solenoid is switched off and falls to zero in 2.0 s, calculate the average value of the emf induced in the smaller loop. (c) Indicate the direction of the current induced in the smaller loop as the magnetic field in the solenoid decreases to zero.

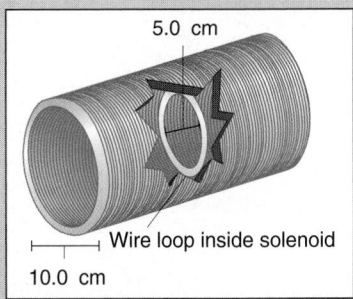

FIGURE P.14

• **15.** A jumbo jet is flying horizontally over the United States at 900 km/h. The vertical component of the magnetic field of the Earth in the vicinity is on the order of 5×10^{-5} T (its precise value depends on the geographic location; the vertical component of the field is greatest near the magnetic poles and zero at the magnetic equator). (a) The wingspan of the jet is 64 m. Estimate the magnitude of the emf induced between the wing tips of the jet as it flies. (b) Recall that the south magnetic pole of the magnetic field of the Earth is in the northern geographic hemisphere. (Why?) In the vicinity of the south magnetic pole of the Earth, which wing tip [port (left) or starboard (right)] of the jet is at the higher electric potential? Justify your answer. (c) How would your answer to part (b) be different over Antarctica? (d) The length of the jet is 71 m. Will there also be an induced emf between the nose and tail? Justify your answer.

• **16.** A helicopter has blades 4.5 m in length, rotating in the horizontal plane at 8.0 Hz. The magnitude of the vertical component of the magnetic field of the Earth in the vicinity is 0.5×10^{-4} T. What is the magnitude of the emf induced between the tip of each blade and the rotor hub?

• **17.** The high-speed TGV train travels horizontally at a speed of 180 km/h in a region where the vertical component of the magnetic field of the Earth is about 0.5×10^{-4} T. A steel axle connects the wheels, separated by the standard western rail gauge of 1.36 m. (a) What emf is induced between the wheels as a result of the motion of the train? (b) Explain why you do not need to know the value of the horizontal component of the magnetic field. (c) Explain why there can be a steady emf present along a highly conducting steel axle.

• **18.** The rectangular wire loop shown in Figure P.18 is moving at constant velocity $\vec{v}$ and enters a semi-infinite region with a uniform magnetic field of magnitude B_0 directed as indicated. The loop begins to enter the field when $t = 0$ s. (a) Make a graph of the magnetic flux through the loop as a function of time for $t \geq 0$ s. Indicate appropriate times on the graph. (b) Make another graph depicting the emf induced in the wire loop for times $t \geq 0$ s.

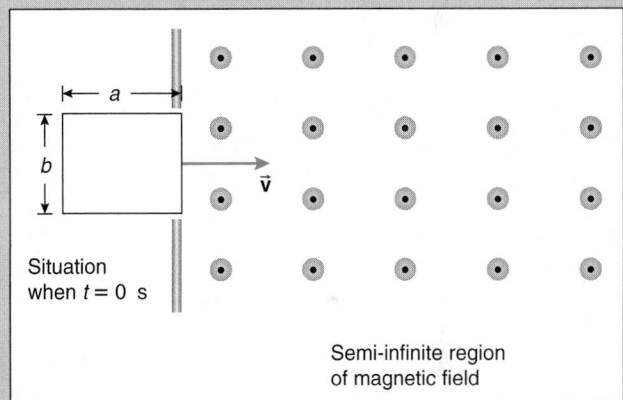

FIGURE P.18

• **19.** You are home on the range in Wyoming. A continuous barbed wire fence encloses a square 10 km on a side with one gate open on the north side of the enclosure. The downward vertical component of the magnetic field of the Earth at the location of the huge ranch is 0.50×10^{-4} T. Suppose the magnetic field of the Earth increases temporarily by 1.0% during a 20 s interval, as it might do on an unusual occasion. (a) If the gate is always open, is the east or west side of the gate at the higher electric potential during the increase? (b) Find the induced emf. (c) During the (different) interval over which the magnetic field later decreases to its original value, must the induced emf be as large as it was in part (b)? Explain.

• **20.** A circular coil of radius 5.0 cm with N turns of wire (all of equal area) is rotated about a diameter at a frequency of 60.0 Hz in a uniform magnetic field of magnitude 0.150 T. The axis of rotation is perpendicular to the direction of the magnetic field. How many turns of wire are needed so that the peak value of the induced emf in the coil is 30 V?

•21. You need to make an ac generator with a sinusoidal output of 12.0 V amplitude and frequency 50.0 Hz. A 10 cm by 20 cm rectangular coil of wire with 300 turns of wire is available. (a) At what angular frequency should the coil be rotated? (b) What magnitude magnetic field is needed?

•22. A rectangular coil 5.0 cm wide and 10.0 cm long has 150 turns of wire. The coil is turned vigorously at constant frequency ν about a symmetry axis (see Figure P.22) perpendicular to the magnetic field of the Earth (of magnitude ~0.60 × 10⁻⁴ T). In order to induce an emf of peak absolute value 1.0 V, what frequency is needed?

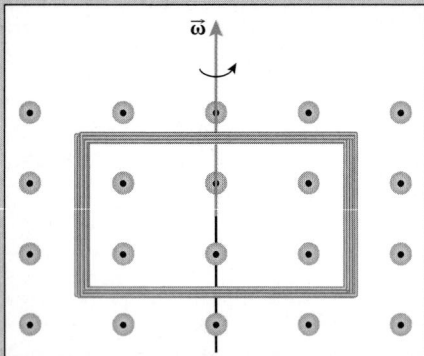

FIGURE P.22

•23. An electrical generator consists of 160 turns of wire wrapped around a frame of area 0.020 m², rotating at a frequency of 10.0 Hz about a symmetry axis perpendicular to a magnetic field of magnitude 0.150 T as shown in Figure P.23. The ends of the coil are connected via slip rings to an external 200 Ω

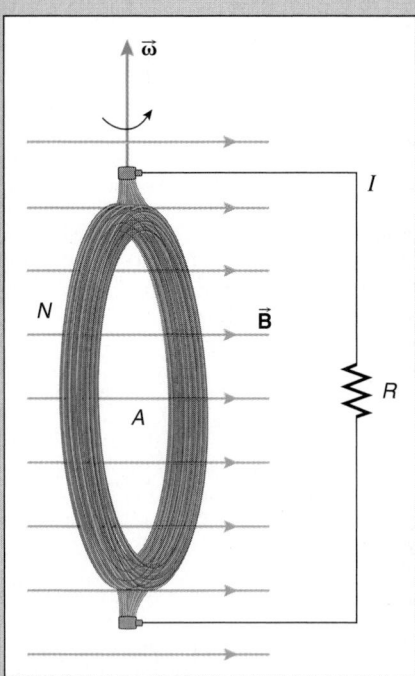

FIGURE P.23

resistor. (a) Calculate the maximum current through the resistor. (b) How many times each second is there zero instantaneous power absorbed by the resistor?

‡24. A thin wire of mass m and length ℓ has a resistance R. The wire can slide freely, with no friction, along twin vertical rails (of negligible resistance) as shown in Figure P.24. A horizontal, uniform magnetic field of magnitude B is perpendicular to the rails and wire. Neglect the magnetic field of the Earth. The wire falls under the action of the gravitational force near the surface of the Earth. Show that the falling wire attains a terminal speed equal to

$$\frac{mgR}{B^2\ell^2}$$

Note that the power and the work done on the wire by the gravitational force are positive. The wire also absorbs positive electrical power because of its finite resistance. Discuss the problem from the standpoint of energy considerations.

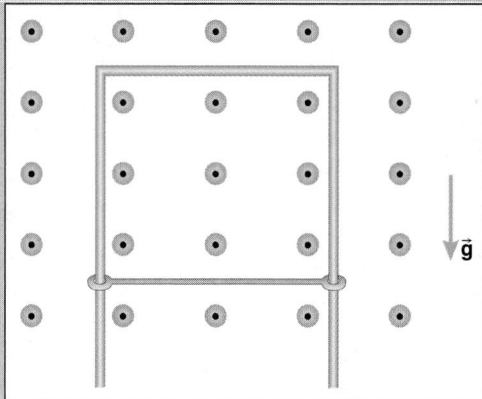

FIGURE P.24

Sections 21.4 Summary of the Maxwell Equations of Electromagnetism*
21.5 Electromagnetic Waves*

25. Calculate the wavelength of the radio waves associated with your favorite radio station.

26. Compare the wavelengths of the radio waves associated with the frequencies 890 kHz in the AM radio band and 89.0 MHz on the FM band.

27. What are the frequencies corresponding to electromagnetic waves with the following wavelengths: (a) 1.00 km; (b) 1.00 m; (c) 1.00 cm; (d) 1.00 mm; (e) 1.00 μm; (f) 1.00 nm.

Sections 21.6 Self-Inductance*
21.7 Series and Parallel Combinations of Inductors*

28. Show that the SI units for μ_0, usually written as T·m/A, also can be expressed as H/m.

29. An inductor has a self-inductance of 150 mH. Determine the potential difference V across the terminals of the inductor (as defined in Section 21.6) when the current in the inductor: (a) is 0.200 A and is increasing at a rate of 60 A/s; (b) is 0.200 A and is decreasing at a rate of 50 A/s; (c) is zero and increasing at a rate of 40 A/s.

30. You have an inductor with an inductance of 75 mH. At what rate must the current change in the inductor to secure a potential difference V across the inductor equal to 3.0 V? Should you have the current increasing or decreasing with time?

•**31.** An inductor has an inductance of 125 mH. A graph of the current in the inductor as a function of time is shown in Figure P.31. Plot the corresponding emf induced in the inductor *and* the potential difference V across the inductor (as defined in Section 21.6) as functions of time.

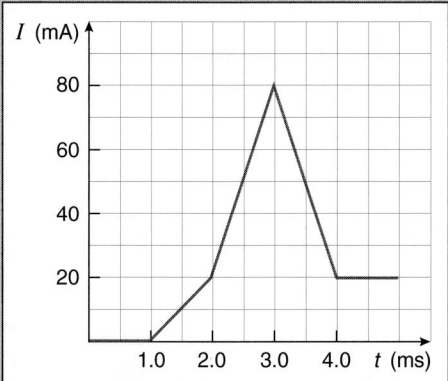

FIGURE P.31

•**32.** When $t = 3.0$ s, the current in a 60 mH inductor is 120 mA and is increasing at a rate of 25 mA/ms. (a) What is the potential difference V across the inductor at this instant? (b) What is the instantaneous power absorbed by the inductor at this time?

•**33.** A coaxial cable consists of an inner conductor of radius R_1 and an outer conductor of radius R_2, as shown in Figure P.33. The currents in the conductors are of equal magnitude but in opposite directions. (a) Calculate the magnetic flux through a rectangular section of length ℓ whose plane contains the axis of the cable (see Figure P.33). (b) Use the result of part (a) to show that the self-inductance per meter of length is

$$\frac{\mu_0}{4\pi} 2 \ln\left(\frac{R_2}{R_1}\right)$$

(c) Calculate the self-inductance per meter of RG58/U coaxial cable, which has an inner conductor of radius 0.41 mm and an outer conductor of radius 1.7 mm. Assume the material separating the conductors has the magnetic properties of a vacuum. This type of coaxial cable frequently is used (with so-called BNC connectors) in scientific laboratories.

•**34.** Two parallel wires, each of radius R, are separated by distance $d \gg R$, as shown in Figure P.34. The wires carry equal currents in opposite directions. Wire of this type is used frequently to connect your TV to its aerial. Assume the material separating the conductors has the magnetic properties of a vacuum. (a) Show that the magnetic flux of the currents through a rectangular area of length ℓ, indicated in Figure P.34, is

$$\Phi = \frac{\mu_0}{4\pi} 4I\ell \ln\left(\frac{d - R}{R}\right)$$

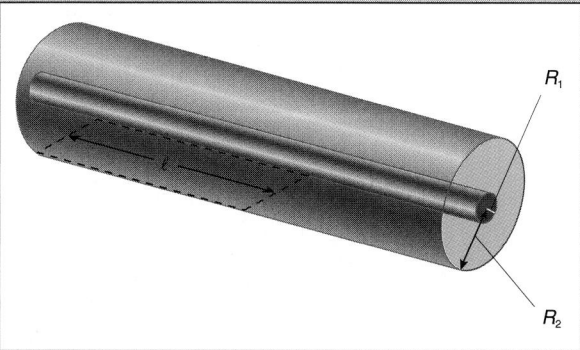

FIGURE P.33

(b) Show that the self-inductance per meter length of the wire is

$$\frac{\mu_0}{4\pi} 4 \ln\left(\frac{d - R}{R}\right)$$

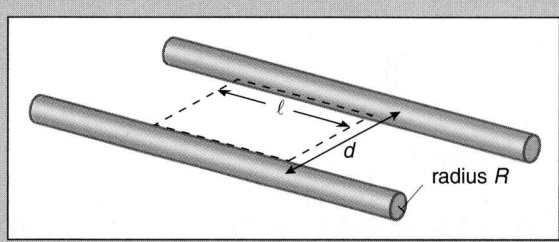

FIGURE P.34

•**35.** A toroidal solenoid (see Example 20.14) of average radius R contains N loops of wire. Each circular loop is of radius r as indicated in Figure P.35. Show that the self-inductance is approximately

$$L \approx \frac{\mu_0}{4\pi} 2\pi \frac{N^2 r^2}{R}$$

if $R \gg r$. (With $R \gg r$, the field within the toroid is essentially uniform).

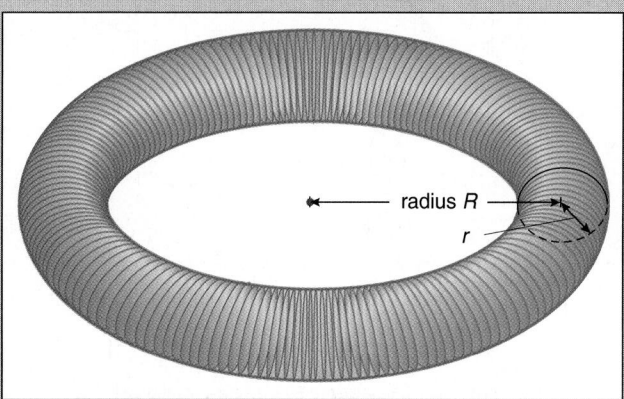

FIGURE P.35

•36. The potential difference V across a 150 mH inductor as a function of time is shown in Figure P.36. Assume the initial value for the current in the inductor is 0 A. (a) Find and graph an expression for the current $I(t)$ as a function of time. (b) What is the current when $t = 2.0$ ms? When $t = 4.0$ ms? (c) How can you check your answers to part (b) by using the *graph* of $V(t)$ versus t? (Hint: Think of the geometric interpretation of I on the graph of $V(t)$ versus t.)

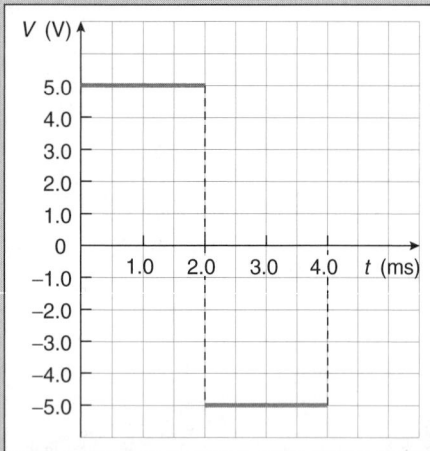

FIGURE P.36

•37. The potential difference across a 150 mH inductor as a function of time is shown in Figure P.37. Assume the initial value for the current in the indictor is 0 A. (a) Find and graph an expression for the current $I(t)$ as a function of time. (b) What is the current when $t = 2.0$ ms? When $t = 4.0$ ms? (c) How can you check your answers to part (b) by using the *graph* of $V(t)$ versus t? (Hint: Think of the geometric interpretation of I on the graph of $V(t)$ versus t.)

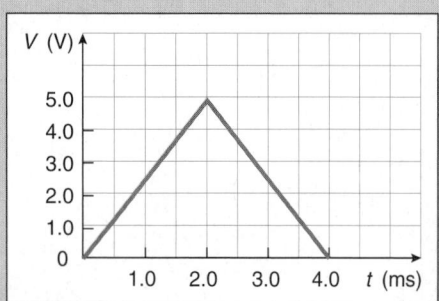

FIGURE P.37

•38. Find the equivalent inductance for each collection of inductors in Figure P.38. Assume each inductor produces negligible magnetic flux through any of the other inductors.

•39. Given three inductors each with self-inductance L: (a) What is the maximum inductance that can be made using all three? (b) What is the minimum inductance that can be made using all three? (c) What other inductances can be made using all three? Sketch the connections. Assume there is no magnetic flux through any inductor from the other inductors.

Sections 21.8 A Series *LR* Circuit*
21.9 Energy Stored in a Magnetic Field*
21.10 A Parallel *LC* Circuit*

•40. When a series *LR* circuit is turned on, the current reaches 25% of its final value in 1.10 s. Find the time constant of the circuit.

•41. You turn on a series *LR* circuit. How many time constants must you wait until the current is 99.44% of its final value?

•42. To measure the self-inductance of an inductor, you connect it in series with a 100 Ω resistor and a constant 15.0 V independent voltage source as shown in Figure P.42. (a) What is the steady-state potential difference across the resistor and across the inductor? (b) What is the steady-state current in the circuit? (c) When $t = 0$ ms, switch (1) is opened while switch (2) is closed. You measure the potential difference across the resistor as a function of time using an oscilloscope. From the oscilloscope trace, you secure the following data:

Time (ms)	Potential difference across the resistor (V)
0.00	15.00
1.00	7.71
2.00	3.96
3.00	2.03
4.00	1.05

Plot the data as a function of time. (d) From the graph, determine the time constant of the circuit. (e) What is the inductance L?

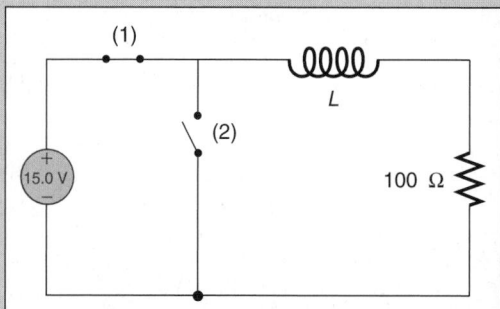

FIGURE P.42

•43. A series *LR* circuit has an inductance of 50 mH and a resistance of 0.25 Ω connected to a 15.0 V dc battery by a switch. The switch is closed when $t = 0$ s. (a) What is the current in the resistor when $t = 0.20$ s? (b) What is the potential difference V across the inductor when $t = 0.20$ s? (c) At what time does the current reach half its long-term value?

•44. When $t = 0$ s, a series *LR* circuit is connected to the independent voltage source shown in Figure P.44. (a) What is the time constant of the circuit? (b) What is the initial time rate of change of the current? (c) What is the time rate of change of the current when $t = \tau$? (d) What is the steady-state current?

•45. A 12.0 V independent voltage source is connected when $t = 0$ s to a series *LR* circuit with $R = 150$ Ω and $L = 250.0$ mH. (a) What is the time constant of the circuit? (b) What is the steady-state current in the circuit? (c) How much time does it take for the current to reach 50% of its steady-state value?

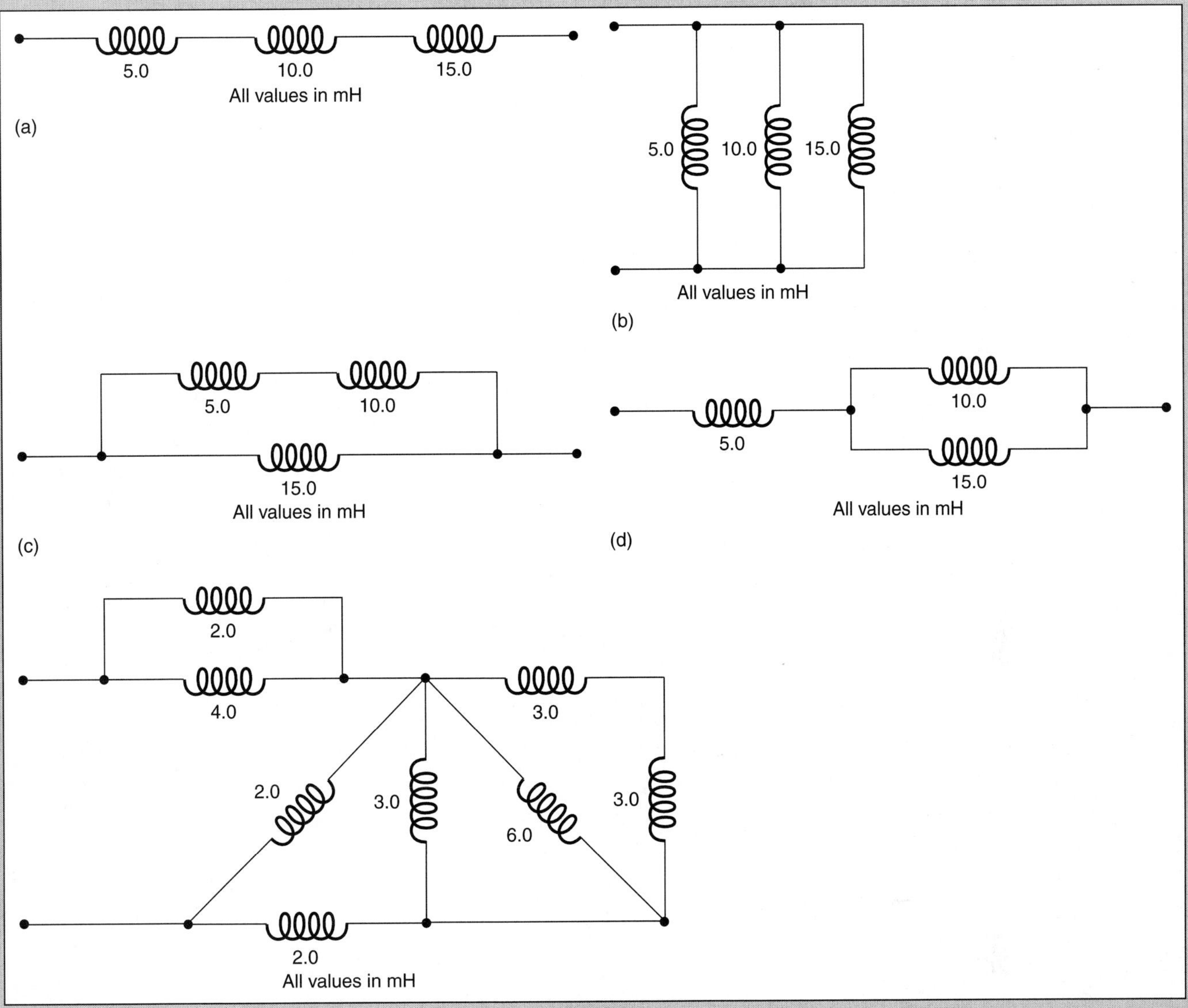

(a)

5.0   10.0   15.0

All values in mH

(b)

5.0   10.0   15.0

All values in mH

(c)

5.0   10.0

15.0

All values in mH

(d)

10.0

5.0

15.0

All values in mH

2.0

4.0

2.0   3.0

6.0

3.0

3.0

2.0

All values in mH

FIGURE P.38

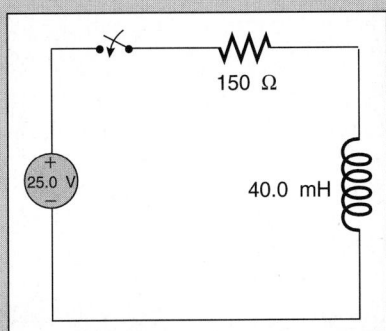

FIGURE P.44

150 Ω

25.0 V

40.0 mH

amount of energy, what potential difference must exist across the capacitor?

•**48.** In the series *LR* circuit indicated in Figure P.48, the switch is closed when $t = 0$ s. (a) Find an expression for the energy $U(t)$ stored in the inductor as a function of time in terms of L, R, and V_0. (b) Let $V_0 = 10.0$ V, $R = 100$ Ω, and $L = 150$ mH. Make accurate graphs of $I(t)$ and $U(t)$ for 0 s $\leq t \leq 0.010$ s. In what ways are the graphs similar? In what ways are they different?

•**46.** If it takes 2.0 ms for the current in an *LR* circuit to reach half its maximum value, what is the time constant of the circuit?

•**47.** (a) Calculate the energy stored in an 800 mH inductor that carries a constant current of 1.50 A. (b) To double the energy stored, by what factor must the current increase? (c) For a 100 μF capacitor to store the same

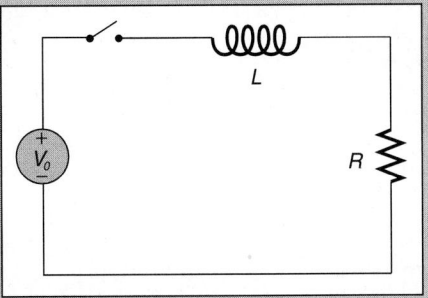

L

V_0

R

FIGURE P.48

•49. A device (not important to this question) provides a current to the two terminals of a circuit element in a sealed box as shown in Figure P.49a. The current varies with time as indicated in Figure P.49b. The potential difference V measured across the terminals of the box varies with time as shown in Figure P.49c. Negative values for the current and potential difference mean the current and polarity markings are opposite to the illustrated direction and/or polarity shown in Figure P.49a. (a) Explain why these graphs indicate that the circuit element is *not* a resistor. (b) Explain why these graphs indicate that the circuit element is *not* a capacitor. (c) Are the graphs consistent with the device being an inductor? Why or why not? (d) Calculate the inductance. (e) Complete the graph of the potential difference versus time for times greater than 2.0 s, indicating proper polarity and magnitude.

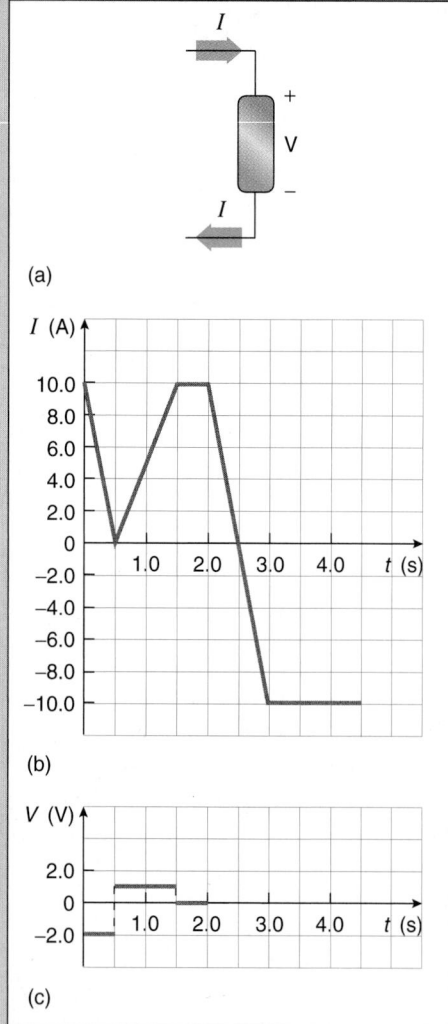

(a)

(b)

(c)

FIGURE P.49

•50. Use the KVL in the *LC* circuit depicted in Figure 21.63 to show that the current *I* also satisfies the differential equation for simple harmonic oscillation:

$$\frac{d^2I}{dt^2} + \frac{1}{LC} I = 0 \text{ A/s}^2$$

•51. The magnitude of the magnetic field of the Earth near its surface is on the order of 0.6×10^{-4} T. What is the energy density of this magnetic field?

•52. What is the oscillation frequency of a parallel *LC* circuit with a capacitance of $1.00 \ \mu$F and an inductance of 100 mH?

•53. AM radio frequencies range from about 500 kHz to about 1600 kHz, a range called the AM band. The tuning circuit of an AM receiver typically is a parallel *LC* circuit with a fixed inductor and a capacitor whose capacitance can be varied to permit the circuit to oscillate at the frequency corresponding to a given radio station. If a receiver has a fixed inductor with $L = 0.33$ mH, over what range of capacitances must the capacitor be able to vary?

‡54. Consider the circuit shown in Figure P.54. (a) When $t = 0$ s the switch is closed. At that instant determine: (i) the current in the inductor; (ii) the current in the 100 Ω resistor; (iii) the current in the 200 Ω resistor; (iv) the current in the switch; (v) the potential difference across the 100 Ω resistor; (vi) the potential difference across the 200 Ω resistor; (vii) the potential difference across the inductor; and (viii) the time rate at which the current is changing through the inductor. (b) After many time constants (effectively for $t \rightarrow \infty$ s), determine the quantities listed in part (a).

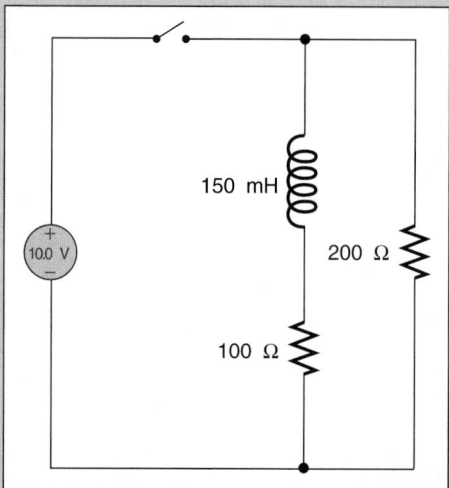

FIGURE P.54

‡55. In the series *LR* circuit of Figure P.55 the current has reached its steady-state value. (a) What energy is stored in the magnetic field of the inductor? (b) When $t = 0$ s, switch (1) is opened while switch (2) is closed and the current decays to zero. Calculate the energy absorbed by the resistor between $t = 0$ s and $t = \infty$ s. Compare this with the answer to part (a).

‡56. Two wires are connected to an unknown circuit element(s) inside a closed box (see Figure P.56). A 6.0 V dc independent voltage source is connected to its leads when $t = 0$ s. The current in the circuit gradually increases from 0 A when $t = 0$ s to 1.0 A when $t = 0.50$ s, and continues to grow until it reaches a steady value of 3.0 A. It is noted that the current does *not* increase linearly with time. The current does not oscillate but grows continuously as described. (a) Does the box contain solely a resistor? Explain your reasoning. (b) Does

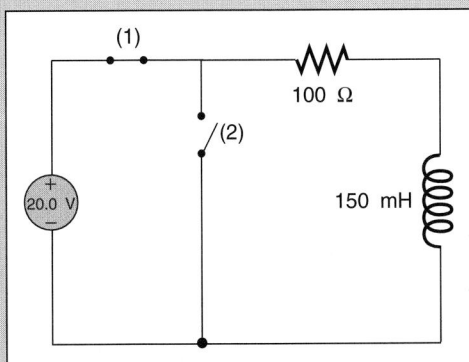

FIGURE P.55

lead of the secondary has the positive polarity of the emf and which lead of the secondary has the positive polarity of the potential difference V.

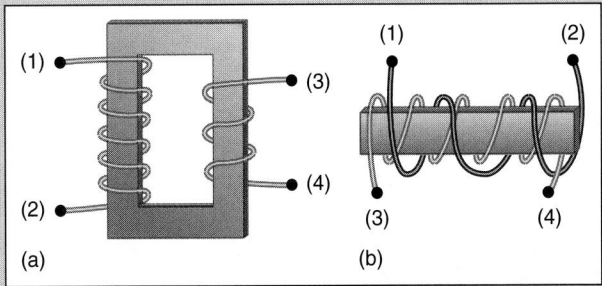

FIGURE P.61

the box contain only a capacitance? Explain your reasoning. (c) Does the box contain only an inductance? Explain your reasoning. (d) What circuit elements are inside the box and how are they connected? Sketch the complete circuit. (e) Determine the expression $I(t)$ for the current as a function of time.

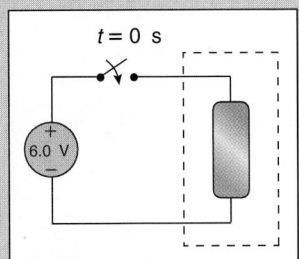

FIGURE P.56

Sections 21.11 Mutual Inductance*
    21.12 An Ideal Transformer*

57. A step-down transformer has an input of 14 kV and an output of 220 V. What is the ratio of the number of turns in the primary to that in the secondary?

58. You have just purchased a low-voltage lighting system that operates at 24 V ac. The independent voltage source available in your home is 120 V ac. Specify the ratio of the number of turns in the primary to those in the secondary for a suitable transformer.

59. Your electric train set operates at 12 V ac, whereas the source in your home is 120 V ac. What is the ratio of the number of turns in the primary to those in the secondary of a suitable transformer?

•60. A solenoid has 250 turns of wire wrapped along its length of 25 cm. The diameter of the solenoid is 1.5 cm. In the middle of the solenoid is a smaller coil of diameter 1.0 cm with 50 turns of wire along its 2.0 cm length. The two coils are coaxial. A current of 15 A is in the larger solenoid with no current in the smaller coil. (a) Determine the magnetic flux through the smaller coil. (b) Determine the mutual inductance of the pair of coils.

•61. For each transformer shown in Figure P.61, imagine the current in the primary coil to be directed from terminal (1) to terminal (2) and increasing with time. Indicate which wire

•62. Transformers can be used to change the effective value of a resistance in an ac circuit. To see how this is done, connect a resistor R across the terminals of the secondary coil of a transformer as shown in Figure P.62. Show that the ratio of V_1 to I_1 in the primary coil is

$$\frac{V_1}{I_1} = \left(\frac{N_1}{N_2}\right)^2 R$$

Thus the effective value of the resistance R in the primary circuit is amplified by the factor

$$\left(\frac{N_1}{N_2}\right)^2$$

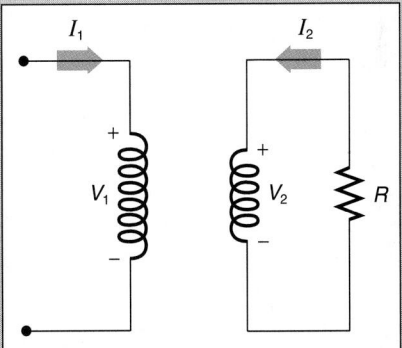

FIGURE P.62

•63. (Refer to the previous problem.) An electrocardiograph (EKG) is used to measure electric potentials associated with currents that traverse the heart muscles. The EKG is modeled as an oscillating independent voltage source in series with a large resistor (40 kΩ). An audio speaker with an effective resistance of 8.0 Ω is to be connected via a transformer to the device (see Figure P.63). To maximize power transfer from the EKG to the speaker, the resistance of the speaker must be amplified by the transformer to equal that of the EKG. Specify the ratio of turns in the primary to that in the secondary for an appropriate transformer.

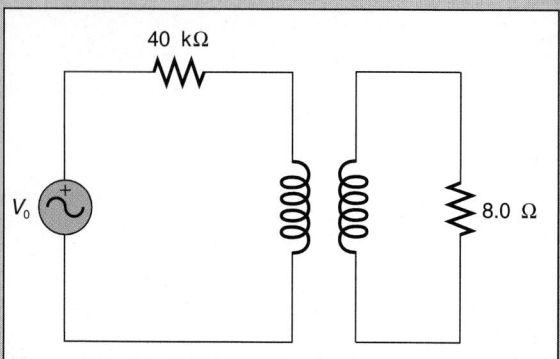

FIGURE P.63

A. Expanded Horizons

1. Hints of a subtle connection between light and electricity were gleaned by Francis Hauksbee (c. 1666–1713) early in the 18th century. A study of his work on electroluminescence and static electricity is a good illustration of the careful observational research made by our scientific forebears. Hauksbee was an artisan (apprenticed as a draper) with little formal education but a keen and observant mind. This was recognized by Isaac Newton, who invited Hauksbee to meetings of the Royal Society to present demonstrations. Hauksbee's work led Dufay and Franklin to the discovery of two types of electric charge (see Chapter 16). Investigate and try to duplicate Hauksbee's experiments.
   Duane Roller and Duane H. D. Roller, "Francis Hauksbee," *Scientific American*, 189, #2, pages 64–69 (August 1953); additional reference on page 100.
   Dictionary of Scientific Biography (Scribner, New York, 1972), volume VI, pages 169–175.

2. If you are interested in learning more about Michael Faraday and his famous and ingenious experiments, see the following references.
   John Meurig Thomas, *Michael Faraday and the Royal Institution* (Adam Hilger, Bristol, 1991).
   Herbert Kondo, "Michael Faraday," *Scientific American*, 189, #4, pages 90–99 (October 1953); additional references on page 122.

3. Heinrich Rudolph Hertz (1857–1894) was a renowned experimental scientist and the first person (at the ripe age of only 30) to create and detect electromagnetic radiofrequency waves in the laboratory (1887). His many ingenious experiments, performed over a very short lifetime, are a rich mine, worthy of study. Investigate and report on his experiments and their significance.
   Joseph P. Mulligan, "Heinrich Hertz and the development of physics," *Physics Today*, 42, #3, pages 50–57 (March 1989).
   Joseph P. Mulligan, "Max Planck and the 'black year' of German physics," *American Journal of Physics*, 62, #12, pages 1089–1097 (December 1994).
   Jed Z. Buchwald, *The Creation of Scientific Effects: Heinrich Hertz and Electric Waves* (University of Chicago Press, Chicago, 1994).

4. The biological effects of electromagnetic fields and waves are of subject of some controversy. You might find it interesting to narrow this topic and investigate it in some detail. For this purpose, the following resource guide will be useful.
   David Hafemeister, "Resource Letter BELFEF-1: Biological effects of low-frequency electromagnetic fields," *American Journal of Physics*, 64, #8, pages 974–981 (August 1996).

5. Although they lack moving parts, real transformers such as those on utility poles hum at audible frequencies when in operation. Investigate why (perhaps by calling your local electric utility company) and report on your findings to your classmates.

B. Lab and Field Work

6. Design an experiment to detect and measure the variation in the acceleration of a magnet falling vertically through a coil of wire. Test to see if it makes any difference which pole of the magnet falls first.

7. Design and perform an experiment to measure the self-inductance of a coil of wire. See Problem 42. Insert a bar of iron into the coil and remeasure its self-inductance.

8. Design and perform an experiment to demonstrate oscillations in an *LC* circuit. Investigate how changing the value of C affects the frequency of the oscillations.

9. Perhaps your professor can arrange for you to use a research magnet in your physics department. After determining the magnitude of the magnetic field available (perhaps with a Hall probe), design and build a hand-cranked ac generator with a peak output of, say, 10 V.

10. Visit and consult with a cardiac surgeon and/or a manufacturer of heart pacemakers to learn more about how electrical energy is supplied to them. On another note, some cordless electric toothbrushes transfer electrical energy using Faraday's law.
    Peter P. Tarjan and Alan D. Bernstein, "An engineering overview of cardiac pacing," *IEEE Engineering in Medicine and Biology Magazine*, 3, #2, pages 10–14 (1984).

11. Visit your local electric utility company to learn more about the variety of common transformers used in the electrical distribution system in your locale. Detail the common potential differences encountered as inputs and outputs and, from these, determine the ratio of primary to secondary turns in the various

transformers. You also might inquire about how switching large ac potential differences is accomplished without initiating large sparks or discharges.

12. Call, write, or e-mail the R&D department of a manufacturer of induction stoves to learn more about their performance with the goal of simulating a *Consumer Reports* type article about such stoves.

13. Visit an automotive shop to investigate the use of induction lights to set the timing of gasoline engines. Demonstrate their use for your classmates.

C. Communicating Physics

14. Most people are little aware of Faraday, let alone of his law of electromagnetic induction, despite the fact that it is "Faraday's spark" that, in large measure, has made a modern, electronic, technological society possible. (See the chapter-opening quotation by John Tyndall on page 953.) Certainly we all complain when the power goes off because of storms! For an audience without technical training (such as most readers of a regional newspaper), write a science feature article and/or essay detailing how Faraday's law maintains and is responsible for a high standard of living.

15. In Section 21.6 we briefly mentioned that Joseph Henry apparently discovered electromagnetic induction before Faraday but did not publish the finding; credit is given to Faraday. Is this fair? Should credit be given to the person who first makes the observation or who first publishes the finding? Write a one page essay detailing your opinion about the controversy. There is risk in publication, because the observations and results are exposed to peer scrutiny and, perhaps, negative criticism and ridicule. Even today, questions about the priority of discovery plague the sciences, mathematics, and engineering. The matter is not purely academic. Huge amounts of money are involved if a discovery has important and patentable implications.

FARADAY'S FLAW

CHAPTER 22

SINUSOIDAL
AC CIRCUIT ANALYSIS

. . . electricity from the square root of minus one.
Vladimir Karapetoff (1876–1948) *

A century ago, one of the great technological debates was whether the electrical distribution system (the world-wide-web of the day) should be ac or dc.[†] The debate was yet to be settled in 1895 when George Vanderbilt constructed a summer retreat called the Biltmore, still the largest private residence in the United States, in Asheville, North Carolina (see chapter opening photo). A technophile with huge financial resources inherited from his grandfather, the shipping and railroad baron Cornelius Vanderbilt, George hedged his bets and had the 220-room mansion built to accommodate both systems.[‡]

Nationally, the choice eventually was settled in favor of ac distribution, in large measure because of the inventive genius of one man: Charles Proteus Steinmetz (1865–1923).[§] A German immigrant to the United States and an electrical engineer of the first rank, Steinmetz realized that the then purely mathematical idea of complex variables could be used to great practical advantage to simplify the analysis of ac circuits. He explained how to do so in a nine-volume, practical treatise. Among other things, this application of pure mathematics to the sciences and engineering illustrates the thesis that the pure math of today is often the physics of tomorrow.

Steinmetz spent a 30-year career working for the relatively new General Electric Company in Schenectady, New York.[#] He came to be known as the Wizard of Schenectady, able to perform amazing calculations in his head. His use of complex variable theory for the practical and quick solution of problems in ac circuits was a precursor of the importance complex variables also were found to play in the development of quantum mechanics during the early decades of the 20th century. For the same reasons, then, we use the algebra of complex variables here to analyze ac circuits; the algebra makes ac circuits much easier to solve than other methods and is essential for the study of quantum mechanics.

We first begin with a short review of the algebra of complex variables. If you already are conversant with this algebra, proceed to Section 22.3. There we see how the algebra is applied to ac circuits with independent voltage sources that vary sinusoidally with time. Sinusoidal, independent voltage sources at the frequency of 60 Hz ($\omega \approx 377$ rad/s) are available from common electrical outlets in North and South America. Sinusoidal voltage sources with a frequency of 50 Hz ($\omega \approx 314$ rad/s) are the European, African, Asian, and Japanese standard.

Sinusoidally varying sources form the basis for the analysis of more complicated periodic (and nonperiodic) time variations, thanks to a branch of mathematics known as Fourier analysis. Fourier analysis enables us to consider *any* time variation as a

During the summer Steinmetz loved to work in a T-shirt, smoking cigars while floating in a canoe, hunched over a thwart. He had little need for large books of mathematical tables (well before the advent of electronic, scientific calculators). Among other mathematical feats, he could find logarithms of numbers to five significant figures in his head.

suitable infinite series or integral of sinusoidal time variations at different frequencies (see Section 12.21). Thus, if we know how to treat a single sinusoidal source in an ac circuit, we have the basic building block for more advanced ac circuit analysis.

22.1 REPRESENTATIONS OF A COMPLEX VARIABLE

Complex variables make use of the imaginary number $i \equiv \sqrt{-1}$.[**]

A **complex variable** z consists of two parts: its **real part** x and its **imaginary part** y. Here, z, x, and y represent *algebraic variables; they do not necessarily represent the Cartesian coordinates.*[††] The variables z, x, and y may represent any physical quantity but must all be dimensionally consistent (i.e., dimensionally the same). In other words, in this introduction to complex variables, think of z, x, and y as algebraic variables, familiar to you from the halcyon days of your algebra and calculus courses. When complex variables are used in physics and engineering, symbols more suggestive of the physical quantity to be represented are employed (as we will see later in this chapter). In particular, complex variables will be used to represent ac independent voltage sources, potential differences, currents, and other electrical parameters. For now, we think of z, x, and y in typically generic, mathematical terms.

Complex variables are written in a number of equivalent representations, each with its own particular usefulness. It is

---

*(Chapter Opener) *Cornell Daily Sun*, 29 October 1923; in an obituary for Charles Proteus Steinmetz.

[†]One only has to see how difficult it is to settle on a unique and universal operating system for computers to realize that such great technological debates continue even today.

[‡]Some computers today are built to accommodate multiple operating systems; history repeats itself.

[§]Others contributed to the development of ac circuit theory as well, but it was Steinmetz who, with passion, made the treatment understandable to a wide audience of engineers.

[#]He also was a great teacher, serving without pay at the nearby, well-respected, Union College. Steinmetz, a dedicated socialist, lived his creed and never pursued wealth for its own sake.

**The letter comes from the first letter of *imaginary*. In electrical engineering, the letter $j \equiv \sqrt{-1}$ is used, which was the notation of Steinmetz, while i is reserved for ac current. We use I for *any* current, and so will use the standard mathematical notation for the imaginary number as is done in quantum mechanics: $i \equiv \sqrt{-1}$. We apologize to any electrical engineers among you, and pray for your understanding of the difficulties associated with choosing appropriate notation.

[††]The letter z is the traditional notation for a general complex variable in mathematics. Likewise its real part traditionally is called x and its imaginary part y.

important for you to be able to change from one representation to another with facility since, as we shall see, various arithmetic and algebraic operations are performed more easily in one representation than another.

Rectangular Form

The complex variable z in **rectangular form** is written as
$$z = x + iy \qquad (22.1)$$

We designate the real part of the complex variable $z = x + iy$ as
$$\text{Re } z \equiv x \qquad (22.2)$$
and its imaginary part, the coefficient of i, as
$$\text{Im } z \equiv y \qquad (22.3)$$

The imaginary part of a complex variable is itself a *real variable*.

Both the real and imaginary parts of a complex variable are *real variables*. The complex nature of z is explicitly and totally contained in the imaginary number i.

From a physical standpoint, the real and imaginary parts of a complex variable must have the same dimensions, since i is a pure number.

EXAMPLE 22.1

Specify the real and imaginary parts of the complex number
$$z = 4.5 - i(2.2)$$

Solution
The real part is equal to 4.5 and the imaginary part is the coefficient of i, or −2.2:
$$\text{Re } z = 4.5 \qquad \text{Im } z = -2.2$$

Graphical Representation of a Complex Number: The Complex Plane

The **complex plane** is a two-dimensional plane that plots the real part of a complex number along the horizontal axis and the imaginary part along the vertical axis.

A complex number thus is a point *on* the complex plane; equivalently, we also say *in* the complex plane. The **graphical representation** of a complex number pictures its location in the complex plane; see Figure 22.1.

If a line is drawn from the origin of the complex plane to the point representing the complex number as in Figure 22.1, the length of this line is the **magnitude** r of the complex number z. Note that the magnitude can be found from the real and imaginary parts using the Pythagorean theorem:
$$r = (x^2 + y^2)^{1/2} \qquad (22.4)$$

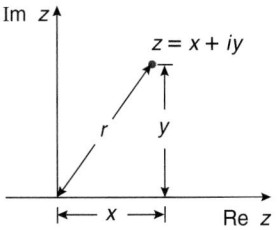

FIGURE 22.1 Graphical representation of a complex number in the complex plane.

The magnitude of the complex number z is the positive square root of the sum of the squares of its real and imaginary parts. The magnitude of a complex number is always a real number greater than or equal to zero. Physically, r is not necessarily a distance (it is only a distance if x and y are distances).

You may note that a complex number is similar to a two-dimensional vector. The magnitude of the complex number is found using the same procedure as that used to find the magnitude of a two-dimensional vector.

EXAMPLE 22.2

Plot the complex number $z = -2.0 + i(3.0)$ on the complex plane and determine its magnitude.

Solution
The real part of the complex number is −2.0 and its imaginary part is 3.0. Its location in the complex plane is shown in Figure 22.2.

The magnitude of the complex number is found from Equation 22.4:
$$\begin{aligned} r &= (x^2 + y^2)^{1/2} \\ &= [(-2.0)^2 + (3.0)^2]^{1/2} \\ &= 3.6 \end{aligned}$$

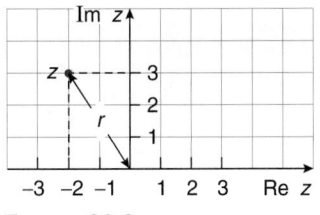

FIGURE 22.2

Polar Form

A complex number also can be represented in **polar form** by specifying its magnitude r and the angle θ that locates the complex number in the complex plane, as shown in Figure 22.3. The polar form is written symbolically as
$$z = r \angle \theta \qquad (22.5)$$

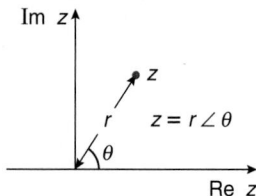

FIGURE 22.3 Polar representation of a complex number in the complex plane.

The angle θ can be expressed either in degrees or radians; for consistency, *we will always use radians*.

From the geometry, the real part of the complex number is

$$x = r \cos \theta \qquad (22.6)$$

and its imaginary part is

$$y = r \sin \theta \qquad (22.7)$$

The angle θ in the polar expression for the complex number is the angle whose tangent is the imaginary part divided by the real part of the complex number:

$$\tan \theta = \frac{\text{Im } z}{\text{Re } z} = \frac{y}{x} \qquad (22.8)$$

Using Equations 22.6, 22.7, and 22.8, we have a way to convert from polar to rectangular form:

$$z = r \angle \theta$$
$$= r \cos \theta + ir \sin \theta \qquad (22.9)$$

If all this seems to have a vaguely familiar ring, you are not mistaken. The algebra of complex numbers is very similar to two-dimensional vector analysis. We never need to extend this on into three dimensions, because there are only two parts to a complex number.

PROBLEM-SOLVING TACTIC

22.1 When converting between rectangular and polar forms, be sure your calculator is in radian mode for performing trigonometric operations. The reason for this is because we choose to express the angles in radians, not degrees.

EXAMPLE 22.3

Convert the complex number $z = 4.00 + i(3.00)$ to polar form.

Solution
From Equation 22.4, the magnitude of the complex number is the (positive) square root of the sum of the squares of its real and imaginary parts:

$$r = (x^2 + y^2)^{1/2}$$
$$= [(4.00)^2 + (3.00)^2]^{1/2}$$
$$= 5.00$$

The polar angle is found from Equation 22.8

$$\tan \theta = \frac{y}{x}$$
$$= \frac{3.00}{4.00}$$
$$= 0.750$$
$$\theta = \tan^{-1}(0.750)$$
$$= 0.644 \text{ rad}$$

Thus the polar form of $4.00 + i(3.00)$ is

$$5.00 \angle (0.644 \text{ rad})$$

Exponential Form

Yet another way to represent a complex number makes use of the amazing Euler identity.* The identity states that

$$e^{i\theta} = \cos \theta + i \sin \theta \qquad (22.10)$$

The proof of this identity stems from comparing the series expansions of the exponential, the sine, and the cosine; the series expansions of the left- and right-hand sides of the Euler relation are seen to be identical.[†]

We multiply both sides of the Euler identity by the magnitude r of the complex number:

$$re^{i\theta} = r \cos \theta + ir \sin \theta \qquad (22.11)$$

The right-hand side of Equation 22.11 we recognize as the rectangular form of the complex number (from Equation 22.9). The left-hand side of Equation 22.11, $re^{i\theta}$, is the **exponential form** of the complex number z.

The angle in the exponent of the exponential form *will always be expressed in radians* because we express the angle in the polar form in like fashion.

In summary, a complex variable z has three distinct representations:

rectangular form:	$z = x + iy$
polar form:	$z = r \angle \theta$
exponential form:	$z = re^{i\theta}$

---

*Leonhard Euler (1707–1783), a great mathematician and scientist, gave us many of our modern mathematical notations, among them i for $\sqrt{-1}$, e for the base of natural logarithms, and $f(x)$ for a function of x.
[†]These series expansions are

$$e^{i\theta} = 1 + (i\theta) + \frac{(i\theta)^2}{2!} + \frac{(i\theta)^3}{3!} + \frac{(i\theta)^4}{4!} + \cdots$$

$$\cos \theta = 1 - \frac{\theta^2}{2!} + \frac{\theta^4}{4!} - \frac{\theta^6}{6!} + \cdots$$

$$\sin \theta = \theta - \frac{\theta^3}{3!} + \frac{\theta^5}{5!} - \frac{\theta^7}{7!} + \cdots$$

QUESTION 1

Discuss the ways the use of the word *imaginary* for $i = \sqrt{-1}$ is simultaneously appropriate and inappropriate.

EXAMPLE 22.4

Find the exponential form for the complex number $4.00 + i(3.00)$.

Solution

In Example 22.3 you found the polar form for this complex number; so you know

$$r = 5.00 \quad \text{and} \quad \theta = 0.644 \text{ rad}$$

The exponential form is thus

$$re^{i\theta} = 5.00 \, e^{i(0.644 \text{ rad})}$$

STRATEGIC EXAMPLE 22.5

a. Graph the complex number $z = i$ in the complex plane.
b. Find the polar form for this complex number.
c. Find the exponential form for this complex number.

Solution

a. In the complex plane, the complex number $z = i$ is located one unit along the imaginary axis, as shown in Figure 22.4

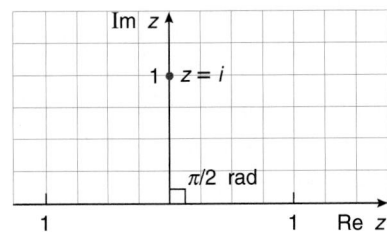

FIGURE 22.4

b. The complex number i has a real part equal to 0 and imaginary part equal to 1. Its magnitude is thus

$$r = [(0)^2 + (1)^2]^{1/2}$$
$$= 1$$

From Equation 22.8, the polar angle has an infinite tangent. Hence

$$\theta = \frac{\pi}{2} \text{ rad}$$

Hence the complex number $z = i$ has the polar form

$$1 \angle \left(\frac{\pi}{2} \text{ rad} \right)$$

c. The exponential form of the complex number i thus is

$$e^{i(\pi/2 \text{ rad})}$$

This is an important example that will be used frequently!

EXAMPLE 22.6

a. Determine the exponential form of the complex number

$$5.00 \angle \left(\frac{\pi}{6} \text{ rad} \right)$$

b. Determine the real and imaginary parts of this complex number.

Solution

a. The given complex number is in polar form $r \angle \theta$. The exponential form is

$$re^{i\theta} = 5.00 \, e^{i(\pi/6 \text{ rad})}$$

b. From the Euler identity, Equation 22.10, we have

$$re^{i\theta} = r \cos \theta + ir \sin \theta$$

$$= 5.00 \cos \left(\frac{\pi}{6} \text{ rad} \right) + i \left[5.00 \sin \left(\frac{\pi}{6} \text{ rad} \right) \right]$$

$$= 4.33 + i(2.50)$$

The real part is 4.33 and the imaginary part is 2.50.

The techniques of this example will be used frequently later in the chapter.

22.2 ARITHMETIC OPERATIONS WITH COMPLEX VARIABLES

The importance of the ability to add, subtract, multiply, and divide complex numbers will become apparent when we use them for ac circuit analysis.

Addition and Subtraction of Complex Variables

The addition of two complex variables,

$$z_1 = x_1 + iy_1$$

and

$$z_2 = x_2 + iy_2$$

produces another complex variable. Operationally,

$$z_1 + z_2 = (x_1 + iy_1) + (x_2 + iy_2)$$

or rearranging,

$$z_1 + z_2 = (x_1 + x_2) + i(y_1 + y_2) \qquad (22.12)$$

The real parts are added together, and likewise for the imaginary parts.

Subtraction proceeds along similar lines:

$$z_1 - z_2 = (x_1 + iy_1) - (x_2 + iy_2)$$

or rearranging,

$$z_1 - z_2 = (x_1 - x_2) + i(y_1 - y_2) \qquad (22.13)$$

The respective real parts are subtracted, as are the imaginary parts.

PROBLEM-SOLVING TACTIC

22.2 Addition and subtraction of complex variables can only be performed when they are in rectangular form. If the complex variables are in polar or exponential form, they *must* be converted to rectangular form in order to perform the addition or subtraction. The sum or difference then can be converted back to polar or exponential form if desired. Addition and subtraction cannot be performed when the complex numbers are in polar or exponential forms.

EXAMPLE 22.7

Find the sum of the complex numbers

$$3 + i(4)$$

and

$$2 - i(2)$$

Solution
The real and imaginary parts are separately added to obtain

$$(3 + 2) + i[4 + (-2)] = 5 + i(2)$$

EXAMPLE 22.8

Find the sum of the complex numbers

$$4.00 \angle \left(\frac{\pi}{6} \text{ rad} \right)$$

and

$$3.00 \angle \left(\frac{\pi}{4} \text{ rad} \right)$$

and convert it into polar and exponential forms.

Solution
You first need to convert the complex numbers into rectangular form:

$$4.00 \angle \left(\frac{\pi}{6} \text{ rad} \right) = 4.00 \cos \left(\frac{\pi}{6} \text{ rad} \right) + i \left[4.00 \sin \left(\frac{\pi}{6} \text{ rad} \right) \right]$$
$$= 3.46 + i(2.00)$$

Likewise,

$$3.00 \angle \left(\frac{\pi}{4} \text{ rad} \right) = 3.00 \cos \left(\frac{\pi}{4} \text{ rad} \right) + i \left[3.00 \sin \left(\frac{\pi}{4} \text{ rad} \right) \right]$$
$$= 2.12 + i(2.12)$$

The sum is then written in rectangular form:

$$[3.46 + i(2.00)] + [2.12 + i(2.12)] = 5.58 + i(4.12)$$

Now convert the sum $5.58 + i(4.12)$ back to polar form. From Equation 22.4 you have

$$r = [(5.58)^2 + (4.12)^2]^{1/2}$$
$$= 6.94$$

The angle is found from Equation 22.8:

$$\tan \theta = \frac{y}{x}$$
$$= \frac{4.12}{5.58}$$
$$= 0.738$$
$$\theta = 0.636 \text{ rad}$$

Thus the polar form of the sum is

$$6.94 \angle (0.636 \text{ rad})$$

The exponential form for the sum is

$$6.94 \, e^{i(0.636 \text{ rad})}$$

Multiplication of Complex Variables

Multiplication of two complex variables is accomplished most easily when they are in exponential or in polar form. For example, let

$$z_1 = r_1 e^{i\theta_1}$$

and

$$z_2 = r_2 e^{i\theta_2}$$

The product is

$$z_1 z_2 = r_1 e^{i\theta_1} r_2 e^{i\theta_2}$$

Rearranging and exploiting the properties of exponentials, we find

$$\boxed{z_1 z_2 = r_1 r_2 e^{i(\theta_1 + \theta_2)}} \qquad (22.14)$$

The product has a magnitude equal to the product of the magnitudes and the angle in the exponential form is simply the sum of the polar angles.

This means that multiplication in polar form also is operationally easy. We multiply the magnitudes and add the polar angles:

$$\boxed{\begin{aligned} z_1 z_2 &= (r_1 \angle \theta_1)(r_2 \angle \theta_2) \\ &= r_1 r_2 \angle (\theta_1 + \theta_2) \end{aligned}} \qquad (22.15)$$

A product also can be performed when the complex numbers are in rectangular form, but the result is algebraically more involved:

$$z_1 z_2 = (x_1 + iy_1)(x_2 + iy_2)$$
$$= (x_1 x_2 - y_1 y_2) + i(x_1 y_2 + x_2 y_1) \qquad (22.16)$$

> ## PROBLEM-SOLVING TACTIC
>
> **22.3 For the multiplication of two complex numbers in rectangular form, it occasionally is easier first to convert them to polar or exponential form, then do the multiplication.** The result can be converted back into rectangular form if desired.

EXAMPLE 22.9

Find the product of the two complex numbers

$$3.00 \angle \left(\frac{\pi}{4} \text{ rad} \right)$$

and

$$4.00 \angle \left(\frac{\pi}{6} \text{ rad} \right)$$

Express the result in both polar and exponential forms.

Solution
In polar form you have

$$(3.00)(4.00) \angle \left(\frac{\pi}{4} \text{ rad} + \frac{\pi}{6} \text{ rad} \right) = 12.0 \angle \left(\frac{5\pi}{12} \text{ rad} \right)$$

The exponential form is
$$12.0 \, e^{i(5\pi/12 \text{ rad})}$$

EXAMPLE 22.10

Find the product of the complex numbers

$$3.00 + i(4.00) \quad \textbf{and} \quad 2.00 - i(3.00)$$

Express the result in rectangular form.

Solution

Method 1
Perform the product directly:

$$[3.00 + i(4.00)][2.00 - i(3.00)]$$
$$= (3.00)(2.00) + (3.00)[-i(3.00)]$$
$$+ [i(4.00)](2.00) + [i(4.00)][-i(3.00)]$$
$$= 18.0 - i(1.00)$$

Method 2
First convert the complex numbers to polar form:

$$3.00 + i(4.00) = 5.00 \angle (0.927 \text{ rad})$$
$$2.00 - i(3.00) = 3.61 \angle (-0.983 \text{ rad})$$

Then perform the product:

$$(5.00)(3.61) \angle [0.927 \text{ rad} + (-0.983 \text{ rad})]$$
$$= 18.1 \angle (-0.056 \text{ rad})$$

Now convert this to rectangular form:

$$18.1 \angle (-0.056 \text{ rad}) = 18.1 \cos(-0.056 \text{ rad})$$
$$+ i[18.1 \sin(-0.056 \text{ rad})]$$
$$= 18 - i(1.0)$$

which is identical (except with two significant figures) to the product found from their rectangular forms.

Division of Complex Numbers

Division of complex variables also is most easily performed when they are in exponential or polar form.

In exponential form, the division is

$$\frac{z_1}{z_2} = \frac{r_1 e^{i\theta_1}}{r_2 e^{i\theta_2}}$$
$$= \frac{r_1}{r_2} e^{i(\theta_1 - \theta_2)} \qquad (22.17)$$

We divide the magnitudes and subtract the polar angles to find the quotient. This result also means that division in polar form is quite straightforward:

$$\frac{z_1}{z_2} = \frac{r_1 \angle \theta_1}{r_2 \angle \theta_2}$$
$$= \frac{r_1}{r_2} \angle (\theta_1 - \theta_2) \qquad (22.18)$$

Division of complex numbers in rectangular form also can be accomplished by making use of complex conjugates (see the next subsection).

> ## PROBLEM-SOLVING TACTIC
>
> **22.4 To divide two complex numbers in rectangular form, it occasionally is easier first to convert them to polar or exponential form, then do the division.** The result can then be put back into rectangular form again if desired.

EXAMPLE 22.11

Divide the two complex numbers

$$5.00 \angle \left(\frac{\pi}{4} \text{ rad} \right)$$

and

$$2.00 \angle \left(\frac{\pi}{6} \text{ rad} \right)$$

Solution
You divide the magnitudes and subtract the polar angles to obtain

$$\frac{5.00}{2.00} \angle \left(\frac{\pi}{4} \text{ rad} - \frac{\pi}{6} \text{ rad} \right) = 2.50 \angle \left(\frac{\pi}{12} \text{ rad} \right)$$

EXAMPLE 22.12

Divide $3.00 + i(4.00)$ **by** $2.00 - i(3.00)$.

Solution
First, convert each complex number to polar form:

$$3.00 + i(4.00) = 5.00 \angle (0.927 \text{ rad})$$
$$2.00 - i(3.00) = 3.61 \angle (-0.983 \text{ rad})$$

Perform the division with their polar forms:

$$\frac{5.00}{3.61} \angle \left[0.927 \text{ rad} - (-0.983 \text{ rad}) \right] = 1.39 \angle (1.910 \text{ rad})$$

This can be put back into rectangular form if needed:

$$1.39 \angle (1.910 \text{ rad}) = 1.39 \cos(1.910 \text{ rad})$$
$$+ i[1.39 \sin(1.910 \text{ rad})]$$
$$= -0.463 + i(1.31)$$

Complex Conjugation

The **complex conjugate** of a complex variable z is often useful. The complex conjugate of z is designated as z^*. The complex conjugate is found by replacing i with $-i$ wherever it is found in z. Thus the complex conjugate of

$$z = x + iy$$

is

$$z^* = x - iy$$

Notice that the complex conjugate of the complex conjugate is the original complex number:

$$(z^*)^* = (x - iy)^*$$
$$= x - (-i)y$$
$$= x + iy$$
$$= z$$

In exponential form, if $z = re^{i\theta}$, then its complex conjugate is

$$z^* = re^{-i\theta}$$

This implies that in polar form if

$$z = r \angle \theta$$

then the complex conjugate is

$$z^* = r \angle (-\theta)$$

Notice that the product of a complex variable and its complex conjugate gives the square of the magnitude of the complex variable:

$$zz^* = re^{i\theta}re^{-i\theta}$$
$$= r^2 e^0$$
$$= r^2$$

QUESTION 2

For a complex number z plotted on the complex plane, where is its complex conjugate z^* located?

EXAMPLE 22.13

Find the complex conjugate of $z = 5 + i(6)$.

Solution
Replace i with $-i$:

$$z^* = 5 - i(6)$$

EXAMPLE 22.14

Find the complex conjugate of $z = 2 - i(3)$.

Solution
Replace i with $-i$ to obtain

$$z^* = 2 - (-i)3$$
$$= 2 + i(3)$$

EXAMPLE 22.15

Find the complex conjugate of $z = 5e^{i(\pi/6 \text{ rad})}$.

Solution
Replace i with $-i$ to obtain

$$z^* = 5e^{-i(\pi/6 \text{ rad})}$$

EXAMPLE 22.16

Find the complex conjugate of $z = 2e^{-i(\pi/4 \text{ rad})}$.

Solution
Replace i with $-i$ to obtain
$$z^* = 2e^{i(\pi/4 \text{ rad})}$$

EXAMPLE 22.17

Find the complex conjugate of

$$z = 4 \angle \left(\frac{\pi}{3} \text{ rad} \right)$$

Solution
Change the sign of the polar angle to obtain

$$z^* = 4 \angle \left(-\frac{\pi}{3} \text{ rad} \right)$$

Rationalization

We all love to rationalize. But rationalization means something special in the mathematics of complex variables, illustrating yet again how common words often are used differently in technical contexts. **Rationalization** is a procedure to collect in the numerator all the i's in a complex-valued expression. Rationalization also is a way that the division of two complex numbers expressed in rectangular form can be performed directly.

The process is best examined with an example. We consider the following quotient of two complex numbers in rectangular form:

$$\frac{2.00 + i(3.00)}{3.00 - i(4.00)}$$

We want to get all the i's into the numerator. This is accomplished by multiplying *both* the numerator and denominator by the *complex conjugate of the denominator*; this procedure multiplies the original expression by unity (in a very special form) and leaves the value of the original quotient unchanged.

The complex conjugate of the denominator here is $3.00 + i(4.00)$. So we multiply the expression by

$$\frac{3.00 + i(4.00)}{3.00 + i(4.00)}$$

In this way, we obtain a real denominator and a different complex numerator:

$$\frac{2.00 + i(3.00)}{3.00 - i(4.00)}\left[\frac{3.00 + i(4.00)}{3.00 + i(4.00)}\right]$$

Performing the multiplications (in rectangular form), we get

$$\frac{6.00 - 12.00 + i(8.00 + 9.00)}{9.00 + i(12.0) - i(12.0) + 16.0} = \frac{-6.00 + i(17.00)}{25.0}$$
$$= \frac{-6.00}{25.0} + i\left(\frac{17.00}{25.0}\right)$$
$$= -0.240 + i(0.680)$$

This puts the original expression into the standard rectangular form for a complex number. The real part of

$$\frac{2.00 + i(3.00)}{3.00 - i(4.00)}$$

thus is -0.240 and the imaginary part is 0.680.

Another way to rationalize is to put both the numerator and denominator of the complex fraction into polar form and perform the division. So, for example, to rationalize

$$\frac{2.00 + i(3.00)}{3.00 - i(4.00)}$$

we express the numerator in polar form:

$$2.00 + i(3.00) = 3.61 \angle\tan^{-1}\left(\frac{3.00}{2.00}\right)$$
$$= 3.61 \angle(0.983 \text{ rad})$$

and the denominator in polar form:

$$3.00 - i(4.00) = 5.00 \angle\tan^{-1}\left(\frac{-4.00}{3.00}\right)$$
$$= 5.00 \angle(-0.927 \text{ rad})$$

Thus

$$\frac{2.00 + i(3.00)}{3.00 - i(4.00)} = \frac{3.61 \angle(0.983 \text{ rad})}{5.00 \angle(-0.927 \text{ rad})}$$

Performing the division in polar form, we obtain

$$\frac{3.61}{5.00} \angle[0.983 \text{ rad} - (-0.927 \text{ rad})] = 0.722 \angle(1.910 \text{ rad})$$

The result can then be expressed in rectangular form:

$$0.722 \angle(1.910 \text{ rad}) = 0.722 \cos(1.910 \text{ rad})$$
$$+ i[0.722 \sin(1.910 \text{ rad})]$$
$$= -0.240 + i(0.681)$$

This differs slightly from the previous result because of rounding errors.

22.3 COMPLEX POTENTIAL DIFFERENCES AND CURRENTS: PHASORS

With a background in complex variables now under your belt, we now see how useful they are for analyzing sinusoidal ac circuits. We first encountered oscillatory behavior in Chapter 7 when we examined the dynamics of simple harmonic oscillation. As we chose to there, here we also represent all sinusoidal oscillations using cosine functions. Thus sinusoidally oscillating potential differences and currents will be written as

$$V(t) = V_0 \cos(\omega t + \theta) \quad \text{and} \quad I(t) = I_0 \cos(\omega t + \phi)$$

Since the product ωt is in *radians*, the **phase angles** θ and ϕ also should be expressed in radians in the arguments of the cosines.* Recall that any sine function can be represented as a cosine function by subtracting $\pi/2$ rad from the argument of the sine function, an identity Example 22.18 will use.

You likely recall that working with trigonometric functions necessitates intimate knowledge of a raft of trigonometric identities that you had the pleasure (?) of proving during your trig course in high school. Working with trigonometric functions is cumbersome from an algebraic viewpoint. Not impossible, mind you, but awkward.

---

*Electrical engineers sometimes express the phase angle in degrees, while keeping the product ωt in radians. To avoid potential confusion, we choose not to follow that convention and always express both ωt *and* the phase angle in radians.

We express all oscillating potential differences or currents in the form of cosine functions to exploit the Euler identity, Equation 22.10, to *invent* complex potential differences and currents whose *real parts* are the actual potential differences and currents.

That is, for a real oscillating potential difference

$$V(t) = V_0 \cos(\omega t + \theta)$$

we construct a *complex potential difference*

$$\mathbf{V}(t) \equiv V_0 e^{i(\omega t + \theta)}$$
$$= V_0 \cos(\omega t + \theta) + iV_0 \sin(\omega t + \theta) \quad (22.19)$$

Notice that only the *real part* of the complex potential difference $\mathbf{V}(t)$ is the actual (i.e., real) oscillating potential difference. The imaginary part of the complex potential difference just comes along for a mathematical ride. But by bringing the imaginary part along, we will see that trigonometric identities can be avoided in favor of exponentials. We will shortly see that the calculations in ac circuits then are no more difficult conceptually that those for dc circuits. This was the great contribution of Steinmetz to the simplification of ac circuit analysis. The beauty and utility of suitable notation is quite remarkable.

For a real oscillating current

$$I(t) = I_0 \cos(\omega t + \phi)$$

we likewise invent a *complex current*

$$\mathbf{I}(t) \equiv I_0 e^{i(\omega t + \phi)}$$
$$= I_0 \cos(\omega t + \phi) + iI_0 \sin(\omega t + \phi) \quad (22.20)$$

The real part of the complex current is the actual oscillating current. The imaginary part of the complex current comes along for the ride.

We do this because the complex potential differences and currents are much easier to work with than trigonometric functions. Exponential functions are easier to multiply, divide, and even differentiate than trigonometric functions. More significantly, we will see that ac circuit equations deduced from the Kirchhoff voltage law (KVL) and the Kirchhoff current law (KCL) turn out to be *algebraic* equations for the complex potential differences and currents. On the other hand, if only the real sinusoidal voltages and currents are used, the circuit equations turn out to involve both integrals and derivatives: integro-differential equations—a horrible mathematical quagmire. Algebraic equations are much easier to solve!

The complex potential difference

$$\mathbf{V}(t) = V_0 e^{i(\omega t + \theta)}$$

has the polar representation

$$\mathbf{V}(t) = V_0 \angle(\omega t + \theta)$$

The complex potential difference $\mathbf{V}(t)$ is called a **potential difference phasor**.* In the complex plane, the complex potential

difference phasor is represented by a length V_0 rotating counterclockwise at the angular frequency ω, as shown in Figure 22.5. At any instant, the real part of the complex potential difference is the horizontal projection of V_0:

$$V_0 \cos(\omega t + \theta)$$

which is the real oscillating potential difference.

Likewise a complex current

$$\mathbf{I}(t) = I_0 e^{i(\omega t + \phi)}$$

has a polar representation

$$\mathbf{I}(t) = I_0 \angle(\omega t + \phi)$$

The complex current $\mathbf{I}(t)$ is called a **current phasor**. Just like the complex potential difference, the complex current can be thought of as a length I_0 rotating counterclockwise in the complex plane at the angular frequency ω, as shown in Figure 22.6. The projection of I_0 along the horizontal axis is the real part of $\mathbf{I}(t)$ and is the real current

$$I_0 \cos(\omega t + \phi)$$

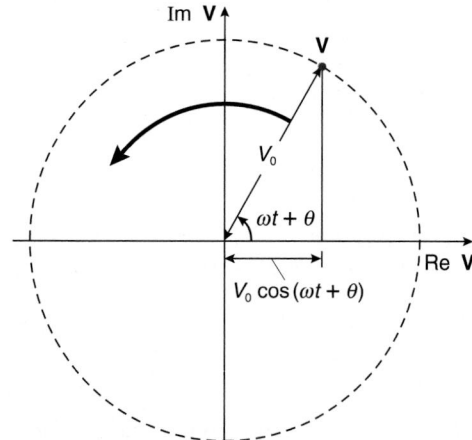

FIGURE 22.5 The potential difference phasor spins counterclockwise at angular frequency ω in the complex plane.

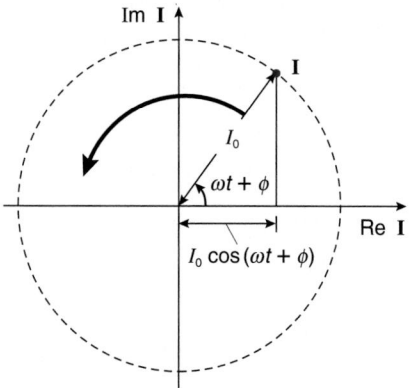

FIGURE 22.6 The current phasor spins counterclockwise at angular frequency ω in the complex plane.

---

*Sorry, Star Trekkers, electrical engineers long ago invented the term phasor. The Star Trek phasors are figments of Hollywood imaginations (pun intended). Groan.

EXAMPLE 22.18

A potential difference varies sinusoidally as given by the equation $V(t) = V_0 \sin(\omega t + 0.157 \text{ rad})$. **Express this in the form**

$$V(t) = V_0 \cos(\omega t + \theta)$$

Solution

To convert sines to cosines, take the cosine of the argument of the sine, minus $\pi/2$ rad. That is, take

$$V(t) = V_0 \cos\left[(\omega t + 0.157 \text{ rad}) - \left(\frac{\pi}{2} \text{ rad}\right)\right] \qquad (1)$$

$$= V_0 \cos(\omega t - 1.414 \text{ rad})$$

The double angle cosine formula from trigonometry verifies that equation (1) is equivalent to the original expression:

$$V(t) = V_0\left[\cos(\omega t + 0.157 \text{ rad})\cos\left(\frac{\pi}{2} \text{ rad}\right)\right.$$

$$\left. + \sin(\omega t + 0.157 \text{ rad})\sin\left(\frac{\pi}{2} \text{ rad}\right)\right]$$

$$= V_0 \sin(\omega t + 0.157 \text{ rad})$$

22.4 THE POTENTIAL DIFFERENCE AND CURRENT PHASORS FOR RESISTORS, INDUCTORS, AND CAPACITORS

In dc circuits and in the transient analysis we investigated in Chapters 19 and 21, the relationship between the (real!) current I through a circuit element and the (real) potential difference V across it were as indicated in Table 22.1.

We next want to determine the relationship between the complex current phasor through these circuit elements and the corresponding complex potential difference phasor across each of them.

Resistor

Take the complex potential difference phasor

$$\mathbf{V}(t) = Ve^{i(\omega t + \theta)} \qquad (22.21)$$

and a complex current phasor

$$\mathbf{I}(t) = Ie^{i(\omega t + \phi)} \qquad (22.22)$$

and substitute them into Ohm's law:

$$\mathbf{V}(t) = R\mathbf{I}(t) \qquad (22.23)$$

In polar form this is

$$V \angle(\omega t + \theta) = RI \angle(\omega t + \phi)$$

Since we have $V = IR$ in step at all times, the phase angle θ of the potential difference must be the same as that of the current, ϕ. We say the current and the potential difference are **in phase** with each other since $\phi = \theta$. In the complex plane, the current and potential difference phasors rotate *together* (the two phasors are on top of each other at all times, as shown in Figure 22.10).

TABLE 22.1 The Real Current and Real Potential Difference Relationships

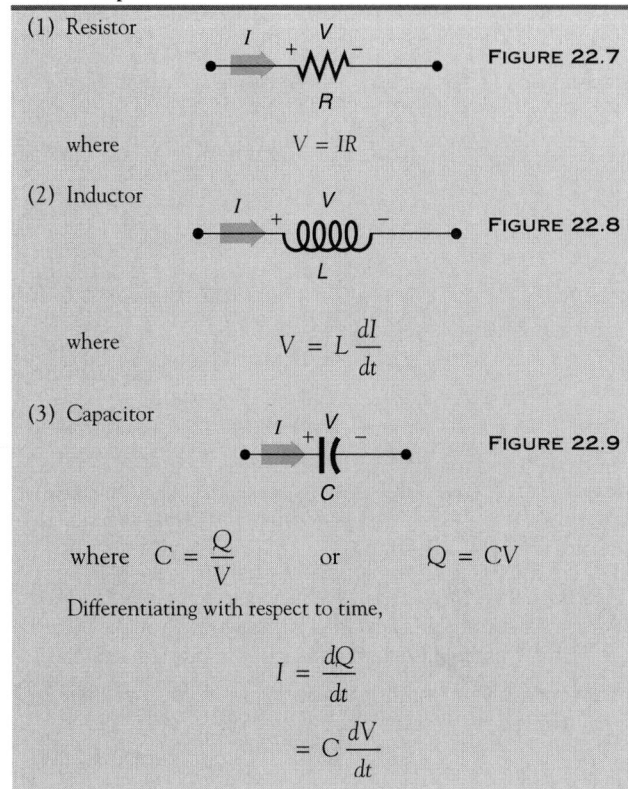

(1) Resistor

where $\qquad V = IR$

(2) Inductor

where $\qquad V = L\dfrac{dI}{dt}$

(3) Capacitor

where $\quad C = \dfrac{Q}{V} \qquad$ or $\qquad Q = CV$

Differentiating with respect to time,

$$I = \frac{dQ}{dt}$$

$$= C\frac{dV}{dt}$$

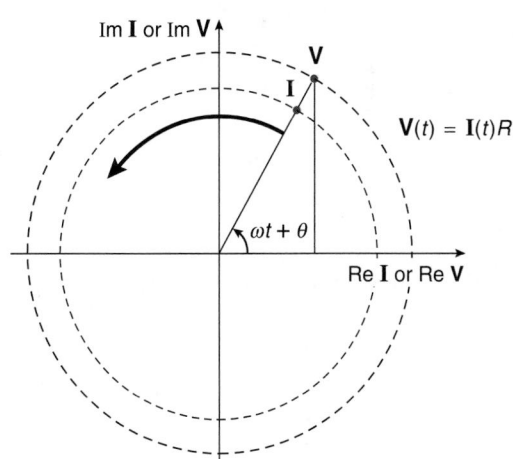

FIGURE 22.10 For a resistor, the potential difference and current phasors are in phase and rotate together in the complex plane.

The phasor magnitudes are related by

$$V = IR$$

In other words, Ohm's law expresses for a resistor both the relationship between the current and potential difference phasors and also their magnitudes. Think of the relationship between the current and potential difference phasors through a resistor in the way shown in Figure 22.11, where

$$\mathbf{V}(t) = \mathbf{I}(t)R \qquad (22.24)$$

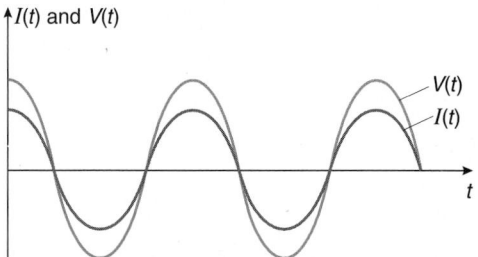

FIGURE 22.11 **The potential difference and current phasors for a resistor.**

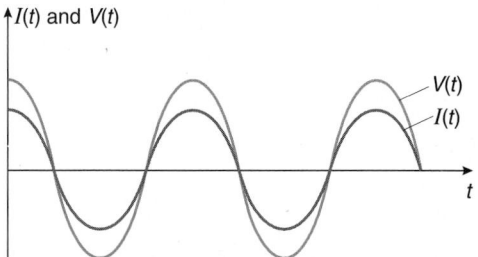

FIGURE 22.12 **The real current in a resistor and the real potential difference across it reach their peak values simultaneously.**

The real parts (and the imaginary parts) of the potential difference and current phasors have the same phase angle since the real (actual) potential difference across the resistor and the real (actual) current through it are in phase, reaching their respective peak values at the same instant, as indicated in Figure 22.12.

Inductor

Let the complex current

$$\mathbf{I}(t) = Ie^{i(\omega t + \phi)}$$

pass through an inductor. The potential difference across the inductor is related to the current through it (see Table 22.1) by:

$$V(t) = L \frac{dI(t)}{dt} \qquad (22.25)$$

Now we can begin to see the advantage of using the complex quantities. We substitute the complex current and voltage phasors and something remarkable happens. Thus

$$\begin{aligned}
\mathbf{V}(t) &= L \frac{d\mathbf{I}(t)}{dt} \\
&= L \frac{dIe^{i(\omega t + \phi)}}{dt} \\
&= Li\omega Ie^{i(\omega t + \phi)} \\
&= i\omega L \mathbf{I}(t) \qquad (22.26)
\end{aligned}$$

What is remarkable here is that this is an *algebraic* relationship between the current and potential difference phasors, thanks to the unique properties of the exponential. The algebraic relationship between the phasors is precisely the motivation for introducing the complex currents and potential differences. If we use the real current and potential difference, the relationship between them in an inductor involves a derivative (Equation 22.25).

Equation 22.26 has some important ramifications. We write the current phasor in polar form:

$$\mathbf{I}(t) = I \angle (\omega t + \phi)$$

and the potential difference phasor in polar form:

$$\mathbf{V}(t) = V \angle (\omega t + \theta)$$

Also we convert the complex number i to polar form (see Example 22.5):

$$i = 1 \angle \left(\frac{\pi}{2} \text{ rad} \right)$$

Substituting these forms into Equation 22.26, we find

$$V \angle (\omega t + \theta) = \left[1 \angle \left(\frac{\pi}{2} \text{ rad} \right) \right] \left[\omega L I \angle (\omega t + \phi) \right]$$

The product of the two complex numbers on the right-hand side means their phase angles add; hence

$$V \angle (\omega t + \theta) = \omega L I \angle \left(\omega t + \phi + \frac{\pi}{2} \text{ rad} \right)$$

From this we can conclude that the phase angles for the current and potential difference are *not* the same in an inductor but differ by $\pi/2$ rad (= 90°):

$$\theta = \phi + \frac{\pi}{2} \text{ rad}$$

Therefore the location of the potential difference phasor in the complex plane is $\pi/2$ rad (= 90°) *ahead* of the current phasor; we say ahead, since the phase angles increase in a counterclockwise sense in the complex plane. Another way of expressing the same thing is to say the current phasor *lags* the potential difference phasor by $\pi/2$ rad in an inductor. As the two phasors rotate in the complex plane, they maintain this $\pi/2$ rad separation; see Figure 22.13.

The magnitudes of the potential difference and current phasors are related by

$$V = \omega L I$$

but the phasors always are at right angles to each other in the complex plane.

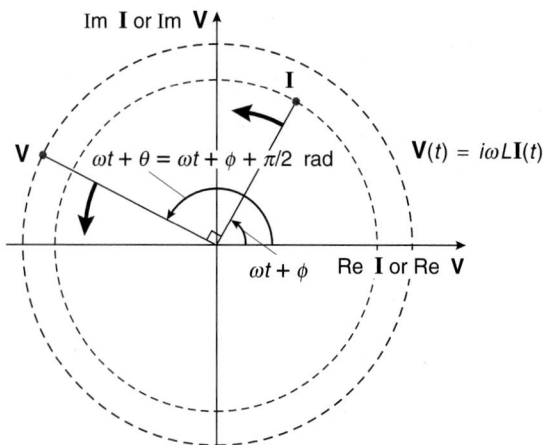

FIGURE 22.13 **The potential difference phasor always is $\pi/2$ rad ahead of the current phasor for an inductor as they both rotate in the complex plane.**

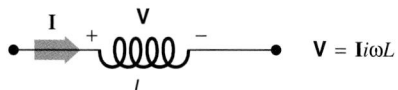

FIGURE 22.14 The potential difference and current phasors for an inductor.

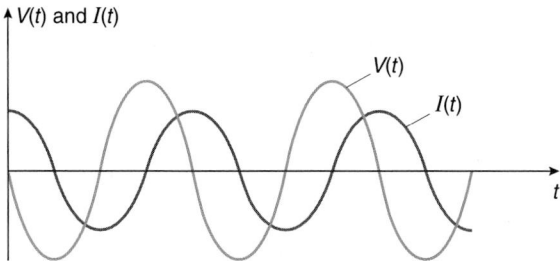

FIGURE 22.15 In an inductor, the real potential difference reaches its peak value ahead of (before) the real current.

Hence for an inductor in an ac circuit, the relationship between the current and potential difference phasors is as indicated in Figure 22.14 and Equation 22.26:

$$\mathbf{V} = \mathbf{I}i\omega L$$

The real parts of the potential difference and current phasors, which represent the real (actual) potential difference across the inductor and the real (actual) current through it, also have a phase difference of $\pi/2$ rad. The potential difference reaches its peak value before (i.e., ahead of) the current, as shown in Figure 22.15.

Capacitor

Let the complex potential difference across a capacitor be

$$\mathbf{V}(t) = Ve^{i(\omega t + \theta)}$$

According to Table 22.1, the current through the capacitor is

$$\mathbf{I}(t) = C\frac{d\mathbf{V}(t)}{dt}$$

We use the complex potential difference and current phasors and perform the differentiation; we find

$$
\begin{aligned}
\mathbf{I}(t) &= C\frac{d\mathbf{V}}{dt} \\
&= C\frac{dVe^{i(\omega t+\theta)}}{dt} \\
&= CVi\omega e^{i(\omega t+\theta)} \\
&= i\omega C\mathbf{V}(t) \qquad (22.27)
\end{aligned}
$$

Once again, notice that the relationship between the current and potential difference phasors is algebraic rather than the differential relationship between the real current and potential difference across the capacitor.

We write the potential difference phasor in polar form:

$$\mathbf{V}(t) = V\angle(\omega t + \theta)$$

the current phasor in polar form:

$$\mathbf{I}(t) = I\angle(\omega t + \phi)$$

and the complex number i in polar form:

$$i = 1\angle\left(\frac{\pi}{2}\ \text{rad}\right)$$

Substituting these forms into Equation 22.27, we find

$$I\angle(\omega t + \phi) = \left[1\angle\left(\frac{\pi}{2}\ \text{rad}\right)\right]\left[\omega CV\angle(\omega t + \theta)\right]$$

The product of the complex numbers in polar form on the right-hand side means the polar angles are added, yielding

$$I\angle(\omega t + \phi) = \omega CV\angle\left(\omega t + \theta + \frac{\pi}{2}\ \text{rad}\right)$$

We see the phase angles are connected by the following relation:

$$\phi = \theta + \frac{\pi}{2}\ \text{rad}$$

The current phasor in a capacitor is $\pi/2$ rad ahead of the potential difference phasor in the complex plane. Equivalently, the potential difference phasor lags the current phasor by $\pi/2$ rad, as shown in Figure 22.16.

The phase relationship between the potential difference and current phasors in a capacitor is just the reverse of that in an inductor. As the phasors rotate in the complex plane, they maintain the $\pi/2$ rad separation in angle.

We solve Equation 22.27 for $\mathbf{V}(t)$ so it has the same form as for the other circuit elements:

$$\mathbf{V}(t) = \frac{\mathbf{I}(t)}{i\omega C} \qquad (22.28)$$

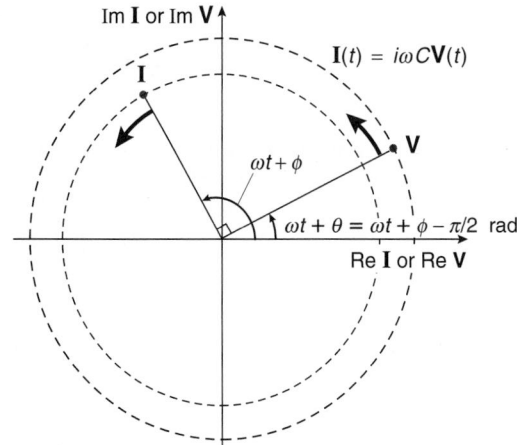

FIGURE 22.16 The current phasor always is $\pi/2$ rad ahead of the potential difference phasor in a capacitor as they both rotate in the complex plane.

Thus, for a capacitor, the relationship between the current and potential difference phasors is as indicated in Figure 22.17, where

$$\mathbf{V} = \mathbf{I}\,\frac{1}{i\omega C}$$

The real parts of the potential difference and current phasors also have a $\pi/2$ rad difference in phase. Hence the real (actual) current through the capacitor and the real (actual) potential difference across it differ in phase by $\pi/2$ rad, with the current reaching its peak value ahead of (i.e., before) the potential difference, as shown in Figure 22.18.

For each circuit element the relationship between the phasors is algebraic and has the general form:

$$\boxed{\mathbf{V} = \mathbf{I}Z} \qquad (22.29)$$

where for a

$$\text{resistor} \quad Z_R = R \qquad (22.30)$$

$$\text{inductor} \quad Z_L = i\omega L \qquad (22.31)$$

$$\text{capacitor} \quad Z_C = \frac{1}{i\omega C} \qquad (22.32)$$

Equation 22.29 is called the **ac version of Ohm's law**.

Equations 22.29–22.32 make the relationship between the phasors $\mathbf{V}$ and $\mathbf{I}$ look like Ohm's law for each of the three circuit elements. This is wonderful!

The quantity Z is called the **impedance** of the circuit element. Using SI units, the impedance is expressed in ohms (Ω). For a resistor, the impedance is just its resistance. For capacitors and inductors, the impedance is a (purely imaginary) complex number that depends on the angular frequency ω.

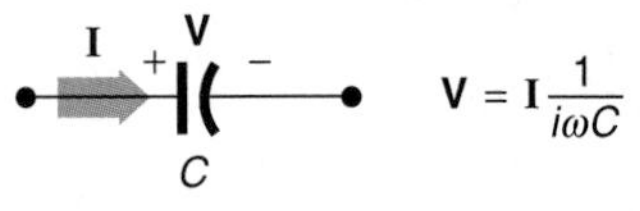

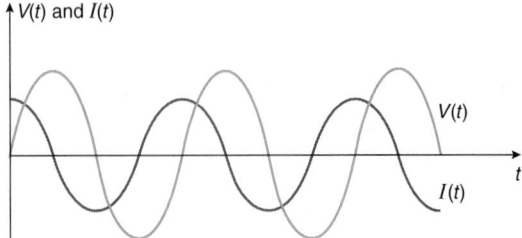

$$\mathbf{V} = \mathbf{I}\,\frac{1}{i\omega C}$$

FIGURE 22.17 The potential difference and current phasors for a capacitor.

FIGURE 22.18 In a capacitor, the current reaches its peak value ahead of (before) the real potential difference.

The impedance of an inductor

$$Z_L = i\omega L$$

occasionally is written as

$$Z_L = iX_L \qquad (22.33)$$

where $X_L \equiv \omega L$ and is called the **inductive reactance**. The inductive reactance is a *real* number, measured in ohms.

The impedance of a capacitor

$$Z_C = \frac{1}{i\omega C}$$

is rationalized by multiplying the numerator and denominator by $-i$ to obtain

$$Z_C = \frac{1}{i\omega C}\left(\frac{-i}{-i}\right)$$

$$= -\frac{i}{\omega C}$$

This is rewritten as

$$Z_C = -iX_C \qquad (22.34)$$

where

$$X_C \equiv \frac{1}{\omega C}$$

X_C is known as the **capacitive reactance**.

A resistor has an impedance that is a *real* number; for capacitors and inductors, the impedance is purely *imaginary*. For combinations of these circuit elements, the equivalent impedance, in general, is a *complex number* (with real and imaginary parts).

The impedance is *not* a phasor and does not rotate in the complex plane, since there is no ωt term associated with it; there is an ωt term for the current and potential difference phasors.

Notice that the impedance of a capacitor is *infinite* when $\omega = 0$ rad/s, which is a dc (time-independent) situation. Capacitors, therefore, act like open switches for dc currents, since no dc current can pass through them.

Any branch of a circuit that contains a capacitor has a (steady-state) dc current equal to zero. Capacitors therefore can be used to isolate or confine dc currents to only certain sections of a complicated circuit. This property of a capacitor is used extensively in transistor circuits to bias transistors properly. To **bias** a transistor is to provide a dc potential difference to ensure its proper operation.

For sufficiently large angular frequencies, the impedance of a capacitor *approaches zero*; at these high angular frequencies, then, the capacitor acts like a *short circuit* for ac currents; this characteristic also is used extensively in transistor circuits.

As a function of angular frequency, the impedance of an inductor behaves in just the reverse manner as that of a capacitor. For $\omega = 0$ rad/s (a dc situation), the impedance of an inductor is *zero*, and so the inductor acts like a *short circuit* for dc currents. If the angular frequency is very high, however, the impedance of an inductor becomes very large and it increasingly behaves like an *open switch* for ac currents (i.e., very small ac current exists in the inductor).

EXAMPLE 22.19

Find the impedance and the capacitive reactance of a 10.0 μF capacitor at a frequency of 60.0 Hz.

Solution

The 60.0 Hz frequency corresponds to an angular frequency of $\omega = 2\pi\nu = 377$ rad/s. From Equation 22.32, the impedance of the capacitor is

$$Z_C = \frac{1}{i\omega C}$$

$$= \frac{1}{i(377 \text{ rad/s})(10.0 \times 10^{-6} \text{ F})}$$

$$= \frac{265 \ \Omega}{i}$$

Rationalizing, you find

$$Z_C = \frac{265 \ \Omega}{i}\left(\frac{-i}{-i}\right)$$

$$= -i(265 \ \Omega)$$

Notice that the impedance is a *complex number*.

The capacitive reactance X_C is a *real number* defined from Equation 22.34 to be

$$Z_C \equiv -iX_C$$

Comparing this with the impedance just calculated shows that the capacitive reactance is

$$X_C = 265 \ \Omega$$

EXAMPLE 22.20

Calculate the impedance and inductive reactance of a 25.0 mH inductor at a frequency of 60.0 Hz.

Solution

From Equation 22.31 the impedance of an inductor is

$$Z_L = i\omega L$$

The angular frequency corresponding to 60.0 Hz is

$$\omega = 2\pi\nu$$

$$= 377 \text{ rad/s}$$

You are given $L = 25.0$ mH; thus the impedance is

$$Z_L = i(377 \text{ rad/s})(25.0 \times 10^{-3} \text{ H})$$

$$= i(9.43 \ \Omega)$$

The inductive reactance X_L is a *real number* defined from Equation 22.33:

$$Z_L = iX_L$$

which means that the inductive reactance here is

$$X_L = 9.43 \ \Omega$$

22.5 SERIES AND PARALLEL COMBINATIONS OF IMPEDANCES

The beauty of the complex variable approach to ac circuits is becoming apparent. The relationship between the current phasor through each of the primary circuit elements (resistors, capacitors, and inductors) and the potential difference phasor across them has the form of the ac version of Ohm's law given by Equation 22.29:

$$\mathbf{V} = \mathbf{I}Z$$

where Z is the impedance of the circuit element. We now exploit this to combine impedances in series and in parallel.

Impedances in Series

Impedances in series combine like resistors did in series.

That is, for the series of impedances shown in Figure 22.19, we can replace the series combination with a single equivalent impedance Z_{eq} :

$$Z_{eq} = Z_1 + Z_2 + Z_3 + \cdots \qquad (22.35)$$

Since the impedances are complex numbers, a complex number addition must be performed to find Z_{eq}.

Impedances in Parallel

Impedances in parallel combine like resistors did in parallel.

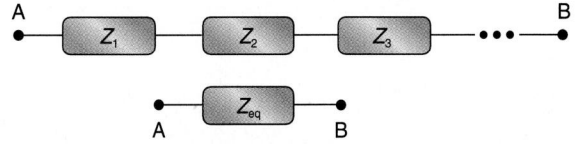

FIGURE 22.19 Impedances in series and their equivalent.

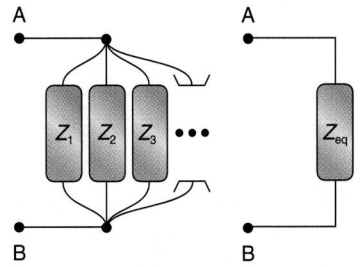

FIGURE 22.20 Impedances in parallel and their equivalent.

For the parallel combination of impedances shown in Figure 22.20, we can replace the collection with an equivalent impedance Z_{eq} found from

$$\frac{1}{Z_{eq}} = \frac{1}{Z_1} + \frac{1}{Z_2} + \frac{1}{Z_3} + \cdots \qquad (22.36)$$

Once again, these are complex number arithmetic operations.

PROBLEM-SOLVING TACTIC

22.5 For two impedances in parallel, you can make use of the shortcut we found for resistors in parallel: the equivalent impedance is the product of the two impedances divided by their sum:

$$Z_{eq} = \frac{Z_1 Z_2}{Z_1 + Z_2} \qquad (22.37)$$

Remember this rule is valid *only for two* parallel impedances at a time.

EXAMPLE 22.21

A circuit is a series combination of a 1.000 kΩ resistor, a 10.0 μF capacitor, a 0.100 H inductor, and a 60.0 Hz ac independent voltage source. Find the equivalent impedance of the series combination of the resistor, capacitor, and inductor.

Solution
First find the impedances of the individual circuit elements. The angular frequency is $\omega = 2\pi \nu = 377$ rad/s. The impedance of the resistor is

$$Z_R = R = 1.000 \text{ kΩ}$$

The impedance of the capacitor is

$$Z_C = \frac{1}{i\omega C}$$
$$= \frac{1}{i(377 \text{ rad/s})(10.0 \times 10^{-6} \text{ F})}$$
$$= \frac{265 \text{ Ω}}{i}$$

Rationalizing, you find

$$Z_C = -i(265 \text{ Ω})$$

The impedance of an inductor is

$$Z_L = i\omega L$$
$$= i(377 \text{ rad/s})(0.100 \text{ H})$$
$$= i(37.7 \text{ Ω})$$

The equivalent impedance of the series combination is the sum

$$Z_{eq} = Z_R + Z_C + Z_L$$
$$= 1000 \text{ Ω} - i(265 \text{ Ω}) + i(37.7 \text{ Ω})$$
$$= 1000 \text{ Ω} - i(227 \text{ Ω})$$

EXAMPLE 22.22

In a circuit operating at 60.0 Hz, a 10.0 μF capacitor is in parallel with a 0.100 H inductor. Find the equivalent impedance of the parallel combination.

Solution
From Example 22.21, the impedance of these circuit elements at the frequency of 60.0 Hz is

$$Z_C = -i(265 \text{ Ω})$$

and

$$Z_L = i(37.7 \text{ Ω})$$

Since the two circuit elements are in parallel, the equivalent impedance is their product divided by their sum (Equation 22.37):

$$Z_{eq} = \frac{Z_C Z_L}{Z_C + Z_L}$$
$$= \frac{\left[-i(265 \text{ Ω})\right]\left[i(37.7 \text{ Ω})\right]}{-i(265 \text{ Ω}) + i(37.7 \text{ Ω})}$$
$$= -\frac{44.0 \text{ Ω}}{i}$$

Rationalizing, you obtain

$$Z_{eq} = -\frac{44.0 \text{ Ω}}{i}\left(\frac{-i}{-i}\right)$$
$$= i(44.0 \text{ Ω})$$

22.6 COMPLEX INDEPENDENT AC VOLTAGE SOURCES

An ideal, ac, sinusoidal, independent voltage source produces a potential difference between its two terminals expressed as

$$V_{source}(t) = V_0 \cos(\omega t) \qquad (22.38)$$

where ω is the angular frequency of the source and V_0 is its *amplitude*. We take the phase angle of the voltage source to be zero.

A *complex ac voltage source*, known as a **voltage source phasor**, has the actual ac voltage source as its real part:

$$\mathbf{V}_{source}(t) = V_0 \cos(\omega t) + iV_0 \sin(\omega t) \qquad (22.39)$$

In exponential form, the complex source is

$$\mathbf{V}_{source}(t) = V_0 e^{i(\omega t)} \qquad (22.40)$$

and in polar form, the source is

$$\mathbf{V}_{source}(t) = V_0 \angle(\omega t) \qquad (22.41)$$

An ac source is symbolized in circuit diagrams as in Figure 22.21.

The terminals nominally have polarity markings (+) and (−) and indicate the polarity of the potential difference between the two terminals when $t = 0$ s.

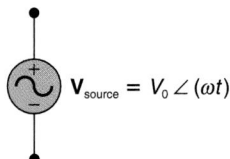

FIGURE 22.21 An ac voltage source phasor symbol.

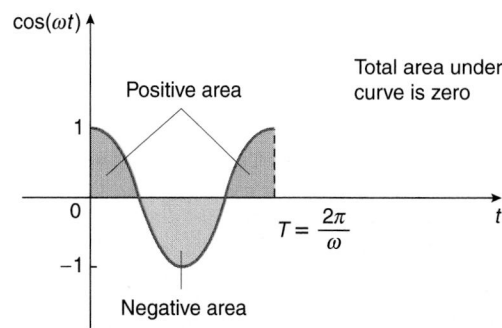

FIGURE 22.22 The time-averaged value of an oscillating cosine (or sine) function over a period (or multiple periods) is zero.

22.7 POWER ABSORBED BY CIRCUIT ELEMENTS IN AC CIRCUITS

As we have seen, power is an important electrical concept. The instantaneous power $P(t)$ absorbed by (transferred to) a circuit element is the product of the potential difference $V(t)$ across it and the current $I(t)$ into its (nominally) positive terminal:

$$P(t) = V(t)I(t) \tag{22.42}$$

If the potential difference and current are oscillating sinusoidally with time, we have

$$V(t) = V_0 \cos(\omega t + \theta) \tag{22.43}$$
$$I(t) = I_0 \cos(\omega t + \phi) \tag{22.44}$$

where V_0 and I_0 are the peak values of the potential difference across and the current through the circuit element. Substituting these quantities into Equation 22.42, we find the instantaneous power is

$$P(t) = V_0 I_0 \cos(\omega t + \theta) \cos(\omega t + \phi) \tag{22.45}$$

In many cases, we are not interested so much in the instantaneous power absorbed as in the *average* power absorbed over a complete cycle (or many complete cycles) of the oscillation. To find the average power absorbed, we resurrect a trigonometric identity for the product of the cosines of two angles:

$$\cos \alpha \cos \beta = \frac{1}{2}\left[\cos(\alpha + \beta) + \cos(\alpha - \beta)\right] \tag{22.46}$$

Applying this identity to the product of the two cosines in Equation 22.45, we obtain

$$\begin{aligned}
P(t) &= V_0 I_0 \frac{1}{2}\Big[\cos(\omega t + \theta + \omega t + \phi) \\
&\quad + \cos(\omega t + \theta - \omega t - \phi)\Big] \\
&= V_0 I_0 \frac{1}{2}\Big[\cos(2\omega t + \theta + \phi) + \cos(\theta - \phi)\Big]
\end{aligned} \tag{22.47}$$

The first term in the brackets in Equation 22.47 is time dependent; the second term is independent of time. Since the first term oscillates with time as a cosine, its time-averaged value over a period (or multiple periods) is zero, as shown in Figure 22.22.

The second term in Equation 22.47, while also a cosine, is independent of time and does not oscillate. Thus the average power $\langle P \rangle$ absorbed by the circuit element is

$$\langle P \rangle = \frac{1}{2} V_0 I_0 \cos(\theta - \phi) \tag{22.48}$$

The cosine term here involves the phase difference $\beta \equiv \theta - \phi$ between the potential difference across the circuit element and the current through it. The cosine of this phase difference is called the **power factor**:

$$\cos(\theta - \phi) \equiv \cos \beta \equiv \text{power factor} \tag{22.49}$$

Hence the average power absorbed by a circuit element is

$$\boxed{\langle P \rangle = \frac{1}{2} V_0 I_0 \cos \beta} \tag{22.50}$$

The peak values are the magnitudes of the potential difference and current phasors.

The average power absorbed by a circuit element also can be expressed in terms of the potential difference and current phasors associated with the circuit element in the complex plane. Consider the expression

$$\frac{1}{2} \text{Re}(\mathbf{V}\mathbf{I}^*) \tag{22.51}$$

The potential difference phasor in exponential form is

$$\mathbf{V} = V_0 e^{i(\omega t + \theta)}$$

and the current phasor is

$$\mathbf{I} = I_0 e^{i(\omega t + \phi)}$$

The complex conjugate of the current phasor then is

$$\mathbf{I}^* = I_0 e^{-i(\omega t + \phi)}$$

Substituting into Equation 22.51, we find

$$\frac{1}{2} \text{Re}\left[V_0 e^{i(\omega t + \theta)} I_0 e^{-i(\omega t + \phi)}\right] = \frac{1}{2} \text{Re}\left[V_0 I_0 e^{i(\theta - \phi)}\right]$$

We take the real part of the complex exponential:

$$\frac{1}{2} V_0 I_0 \cos(\theta - \phi) \equiv \frac{1}{2} V_0 I_0 \cos \beta$$

which is identical to Equation 22.50 for the average power absorbed by the circuit element. Hence an alternative expression for the average power absorbed by a circuit element is

$$\langle P \rangle = \frac{1}{2} \text{Re}(\mathbf{V}\mathbf{I}^*) \tag{22.52}$$

Frequently, we rearrange Equation 22.50 for the average power:

$$\langle P \rangle = \frac{V_0}{\sqrt{2}} \frac{I_0}{\sqrt{2}} \cos \beta \qquad (22.53)$$

The peak values divided by $\sqrt{2}$ are known as the **effective values** of the potential difference and current. They also are known as **rms values** (for *r*oot *m*ean *s*quare values). That is,

$$V_{rms} = \frac{V_0}{\sqrt{2}} \qquad (22.54)$$

$$I_{rms} = \frac{I_0}{\sqrt{2}} \qquad (22.55)$$

The average power then is

$$\langle P \rangle = V_{rms} I_{rms} \cos \beta \qquad (22.56)$$

The oscillatory independent voltage source at most common electrical outlets in the United States is 120 V; this is the rms value V_{rms}. The peak value V_0 is greater by the factor of $\sqrt{2}$.*

On a practical note, multimeters that measure ac potential differences and/or ac currents read the rms values of these quantities (V_{rms} and/or I_{rms}), *not* the peak values (V_0 and/or I_0).

For resistors, the phase difference β between the potential difference and the current is 0 rad. Thus the power factor is

$$\cos \beta = 1$$

and the average power absorbed by a resistor is

$$\langle P \rangle_{res} = \frac{1}{2} V_0 I_0 = V_{rms} I_{rms} \qquad (22.57)$$

For capacitors, the phase difference β between the potential difference and the current is always $-\pi/2$ rad. The power factor is

$$\cos \left(-\frac{\pi}{2} \text{ rad} \right) = 0$$

and the average power absorbed by the ideal capacitor is zero:

$$\langle P \rangle_{cap} = 0 \text{ W}$$

The *instantaneous* power is *not* zero at every instant but the *average* power absorbed *is* zero. This simply means that the capacitor absorbs positive power over half a cycle and negative power over the other half, and the average over each period is zero.

For inductors, the phase difference between the potential difference and current in an inductor is $\pi/2$ rad, so that the power factor is

$$\cos \left(\frac{\pi}{2} \text{ rad} \right) = 0$$

and the average power absorbed by an inductor is zero:

$$\langle P \rangle_{ind} = 0 \text{ W}$$

As in capacitors, the instantaneous power in inductors is not zero at every instant but the average power over any period is zero.

These are characteristics of *ideal* circuit elements. *Real* capacitors and inductors are modeled using ideal circuit elements (see Chapter 19 and 21). Real capacitors and inductors always have some resistance, and it is this resistance that absorbs positive electrical power and accounts for the observation that real capacitors and inductors become warm in use.

We now have all the mathematical and conceptual hardware to analyze a plethora of ac circuits. We will examine just a few to become familiar with the techniques. The important thing to realize is that the methods used closely resemble those employed for dc circuits. The complex variable approach to ac circuits makes this possible.

22.8 A FILTER CIRCUIT

To illustrate the concepts we have learned about ac circuits, we apply them to a special class of circuits of some technological import. Most stereo receivers permit you to selectively eliminate high or low frequencies and create the musical mix most enjoyable to your ears. There is a wide variety of circuits, called **filter circuits**, that let certain frequencies pass relatively unimpeded and filter out or eliminate one or another range of frequencies.

Here we examine the simplest kind of filter circuit: an ac independent voltage source whose single frequency can be varied, a resistor with resistance R, and a capacitor with capacitance C, all connected in series, as shown in Figure 22.23.

The sinusoidal ac independent voltage source varies as

$$V_0(t) = V_0 \cos(\omega t) \qquad (22.58)$$

We want to find the ac current in the circuit and the ac potential differences across both the resistor and the capacitor.

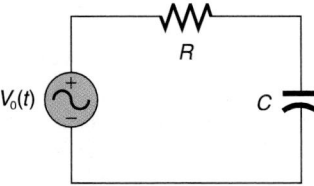

FIGURE 22.23 A series *RC* filter circuit.

---

*In Europe and other countries the rms voltage source is 220 V, so the peak value is $\sqrt{2}$ (220 V) = 311 V. This is the reason that voltage adapters are needed when using U.S. made appliances in Europe to protect the appliances from too high a potential difference. The adapters are really transformers to step down the potential difference to 120 V rms (or 170 V peak). In Japan the rms voltage is 100 V, less than in the United States. No volt-age adapters are needed for U.S. appliances (although the appliances will not operate as effectively because of the reduced voltages and power). The frequency used in Europe and Japan is 50 Hz. In both Europe and Japan, one also needs plug adapters since the topological configuration of electrical plugs varies from country to country.

Finding the Current Phasor

To solve the circuit, we proceed in the following way.

Step 1: We convert the circuit to the complex domain. To accomplish this:

a. We change the independent ac voltage source to a complex voltage source phasor. For a real source voltage $V_0 \cos(\omega t)$, the complex source voltage phasor is

$$\mathbf{V}_{source} = V_0 \angle(\omega t) \qquad (22.59)$$

whose real part is the real voltage source.

b. We indicate the impedances of the resistor and capacitor next to their circuit symbols, as shown in Figure 22.24. The impedance of the resistor is just R; the impedance of the capacitor is $-iX_C$ where X_C is the capacitive reactance

$$X_C = \frac{1}{\omega C}$$

Step 2: We indicate a direction for the current phasor $\mathbf{I}$. The choice of this direction is yours to make but we customarily choose it to be directed out of the nominally positive terminal of the ac voltage source, just as for dc circuits. Once the direction for the current phasor is chosen, the nominal polarities of the impedances are determined and can be marked accordingly: the current phasor goes into the (+) polarity terminal of each impedance and leaves at the (−) polarity terminal, again just as it did for dc circuits. The circuit diagram then has the appearance shown in Figure 22.25.

The potential difference phasor across each of the impedances is found from the ac version of Ohm's law, Equation 22.29:

$$\mathbf{V} = \mathbf{I}Z$$

Hence the potential difference phasor across the resistor is

$$\mathbf{V}_R = \mathbf{I}R \qquad (22.60)$$

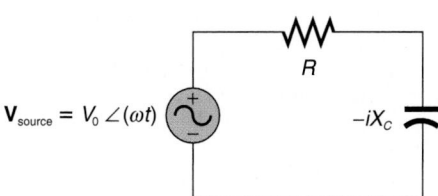

FIGURE 22.24 The circuit in the complex domain.

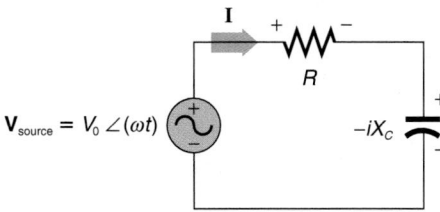

FIGURE 22.25 The phasor current direction is indicated and the polarities of the impedances are marked accordingly.

Across the capacitor, the potential difference phasor is

$$\mathbf{V}_C = \mathbf{I}(-iX_C) \qquad (22.61)$$

with the polarities indicated, consistent with the direction chosen for the current phasor.

Step 3: We use the KVL to go around the loop, using the same conventions we established for dc circuits:

- If, in going around the loop, we first encounter a (+) polarity sign for a circuit element, the potential difference phasor for that circuit element is entered into the KVL equation with a plus (+) sign.
- If, in going around the loop, we first encounter a (−) polarity sign for the circuit element, the potential difference phasor for that circuit element is entered into the KVL equation with a minus (−) sign.

As for dc circuits, we can go around the loop in either direction. We go around the loop of Figure 22.25 clockwise and begin in the lower left corner of the loop. Then the KVL here becomes

$$-\mathbf{V}_{source} + \mathbf{I}R + \mathbf{I}(-iX_C) = 0 \text{ V} \qquad (22.62)$$

Step 4: We solve the equation for the current phasor; we use Equation 22.62:

$$\mathbf{I}(R - iX_C) = \mathbf{V}_{source}$$

The current phasor thus is

$$\mathbf{I} = \frac{\mathbf{V}_{source}}{R - iX_C} \qquad (22.63)$$

We need to put Equation 22.63 into polar form. The denominator is a complex number in rectangular form; we convert the denominator to polar form:

$$R - iX_C = \left[R^2 + (-X_C)^2\right]^{1/2} \angle \tan^{-1}\left(\frac{-X_C}{R}\right) \qquad (22.64)$$

Since $\tan(-\alpha) = -\tan \alpha$,

$$\tan^{-1}\left(-\frac{X_C}{R}\right) = -\tan^{-1}\left(\frac{X_C}{R}\right)$$

We define ϕ to be

$$\phi \equiv \tan^{-1}\left(\frac{X_C}{R}\right) \qquad (22.65)$$

Thus Equation 22.64 becomes

$$R - iX_C = (R^2 + X_C^2)^{1/2} \angle(-\phi)$$

Substituting this into the expression for the current phasor (Equation 22.63), we obtain

$$\mathbf{I} = \frac{\mathbf{V}_{source}}{(R^2 + X_C^2)^{1/2} \angle(-\phi)} \qquad (22.66)$$

Division by complex numbers in polar form means we can move the $\angle(-\phi)$ in the denominator of Equation 22.66 into the numerator by changing the sign of the angle. Thus

$$\mathbf{I} = \frac{\mathbf{V}_{source} \angle \phi}{(R^2 + X_C^2)^{1/2}} \qquad (22.67)$$

Substituting the source voltage phasor in polar form (Equation 22.59) yields

$$\mathbf{I} = \frac{V_0 \angle(\omega t) \angle \phi}{(R^2 + X_C^2)^{1/2}}$$

When multiplying complex numbers in polar form, the angles add, which means that the current phasor becomes

$$\mathbf{I} = \frac{V_0}{(R^2 + X_C^2)^{1/2}} \angle(\omega t + \phi) \qquad (22.68)$$

To find the real current $I(t)$ as a function of time, we take the real part of the current phasor:

$$I(t) = \text{Re } \mathbf{I}$$

$$= \frac{V_0}{(R^2 + X_C^2)^{1/2}} \cos(\omega t + \phi) \qquad (22.69)$$

The angle ϕ is given by Equation 22.65.*

Now that we know the current, we can find the potential differences across the capacitor and the resistor by using the ac Ohm's law for each impedance.

The Potential Difference Across the Capacitor

The phasor representing the potential difference across the capacitor $\mathbf{V}_C$ is found by taking the product of the impedance of the capacitor with the current phasor through it:

$$\mathbf{V}_C = \mathbf{I} \mathbf{Z}_C$$

Using $\mathbf{Z}_C = -iX_C$ and Equation 22.68 for the current phasor, we have

$$\mathbf{V}_C = \frac{V_0 \angle(\omega t + \phi)}{(R^2 + X_C^2)^{1/2}} (-iX_C) \qquad (22.70)$$

---

*If the angular frequency of the source is zero, we have a dc situation. When $\omega = 0$ rad, we say we are at the *dc limit*. Since a capacitor acts as an open circuit for dc current, when $\omega = 0$ rad/s, we expect I to be zero. Does Equation 22.69 for $I(t)$ yield 0 A when $\omega = 0$ rad/s? Let's check. First, we have to evaluate the angle ϕ using Equation 22.65:

$$\phi = \tan^{-1}\left(\frac{X_C}{R}\right)$$

The capacitive reactance is

$$X_C = \frac{1}{\omega C}$$

and when $\omega = 0$ rad/s, $X_C = \infty$ Ω. Thus ϕ is the angle whose tangent is infinite:

$$\phi = \tan^{-1}(\infty)$$

$$= \frac{\pi}{2} \text{ rad}$$

Equation 22.69 for the real current then is

$$I(t) = \frac{V_0}{(R^2 + X_C^2)^{1/2}} \cos\left(0 \text{ rad} + \frac{\pi}{2} \text{ rad}\right)$$

The cosine term is zero *and* the coefficient of the cosine is zero (since $X_C = \infty$ Ω). Thus the current is indeed zero, as it should be for the dc situation.

Since the complex number $-i$ in polar form is

$$1 \angle \left(-\frac{\pi}{2} \text{ rad}\right)$$

Equation 22.70 becomes

$$\mathbf{V}_C = \frac{V_0 \angle(\omega t + \phi)}{(R^2 + X_C^2)^{1/2}} \left[1 \angle \left(-\frac{\pi}{2} \text{ rad}\right)\right] X_C$$

Once again, the product rule for complex numbers in polar form means the angles in the numerator are added together, yielding

$$\mathbf{V}_C = X_C \frac{V_0 \angle(\omega t + \phi - \pi/2 \text{ rad})}{(R^2 + X_C^2)^{1/2}} \qquad (22.71)$$

Notice incidentally that the phase difference between the potential difference phasor and the current phasor is $-\pi/2$ rad (as it always will be for a capacitor).

The real potential difference across the capacitor $V_C(t)$ is the real part of its potential difference phasor $\mathbf{V}_C$:

$$V_C(t) = \text{Re } \mathbf{V}_C$$

$$= X_C \frac{V_0}{(R^2 + X_C^2)^{1/2}} \cos\left(\omega t + \phi - \frac{\pi}{2} \text{ rad}\right) \qquad (22.72)$$

The actual potential difference across the capacitor (Equation 22.72) and actual current (Equation 22.69) also have a phase difference of $-\pi/2$ rad.

We look at the magnitude of the peak potential difference across the capacitor as a function of angular frequency. We call this $V_{C\,\text{peak}}(\omega)$; this is the coefficient of the cosine in Equation 22.72:

$$V_{C\,\text{peak}}(\omega) = \frac{X_C V_0}{(R^2 + X_C^2)^{1/2}}$$

We can simplify this algebraically by taking the X_C in the numerator and putting it in the denominator as $1/X_C$:

$$V_{C\,\text{peak}}(\omega) = \frac{V_0}{\dfrac{1}{X_C}(R^2 + X_C^2)^{1/2}}$$

Putting X_C into the square root, we obtain

$$V_{C\,\text{peak}}(\omega) = \frac{V_0}{\left(\dfrac{R^2}{X_C^2} + 1\right)^{1/2}}$$

Since

$$X_C = \frac{1}{\omega C}$$

the peak potential difference across the capacitor is

$$V_{C\,\text{peak}}(\omega) = \frac{V_0}{\left[1 + (\omega RC)^2\right]^{1/2}} \qquad (22.73)$$

This expression shows how the peak value of the potential difference across the capacitor varies with the angular frequency of the source. In particular, we look at two limiting cases.

Case 1: If the angular frequency is zero, then

$$V_{C\,peak} = V_0$$

which is the amplitude of the voltage source. This is just what we would expect, because when $\omega = 0$ rad/s we have a dc situation. Zero current is present in the circuit since a capacitor acts as an open circuit for dc. There then is zero potential difference across the resistor and so the capacitor effectively is connected in parallel to the dc battery (an ac voltage source at zero frequency).

Case 2: For very large values of the angular frequency of the source, Equation 22.73 indicates that the peak value of the potential difference across the capacitor approaches zero.

> As the angular frequency increases, the impedance of the capacitor decreases. At very large ω, the capacitor essentially has zero impedance and acts more and more like just a piece of wire (with zero potential difference across it). We say the capacitor shorts out for large ω.

A schematic sketch of the peak potential difference across the capacitor as a function of ω is shown in Figure 22.26.

We consider the voltage source to be an input and the potential difference across the capacitor to be an output, as indicated in Figure 22.27.

> Since the peak potential difference across the capacitor (the output voltage) is appreciable only for low (i.e., small) angular frequencies, this output is known as a **low pass filter.**

The term arises because only for *low* angular frequencies is there a significant potential difference passed from the input (the voltage source) to the output (the potential difference across the capacitor).

The ratio of the peak output potential difference to the peak input potential difference is the **voltage gain** of the circuit. Here, the voltage gain is (using Equation 22.73)

$$\text{voltage gain} \equiv \frac{V_{C\,peak}}{V_0} = \frac{1}{\left[1 + (\omega RC)^2\right]^{1/2}} \quad \text{(low pass filter)}$$

$$(22.74)$$

At zero angular frequency, the voltage gain is equal to 1; as the frequency increases, the voltage gain falls, approaching 0 at very high frequencies. At very high frequencies, the impedance of the capacitor approaches zero, so that the capacitor acts like a short circuit (a wire) with essentially no potential difference across it.

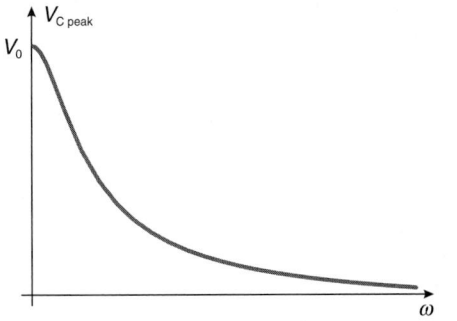

FIGURE 22.26 The peak potential difference across the capacitor as a function of ω.

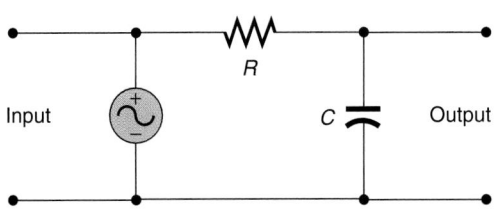

FIGURE 22.27 A low pass filter.

The Potential Difference Across the Resistor

The phasor representing the potential difference across the resistor is the product of the current phasor **I** times the impedance of the resistor:

$$\mathbf{V}_R = \mathbf{I}R$$

The current phasor is the same as that through the capacitor since the two circuit elements are in series. That is, the current phasor is that given by Equation 22.68. Thus

$$\mathbf{V}_R = \frac{V_0 \angle(\omega t + \phi)}{(R^2 + X_C^2)^{1/2}} R$$

$$= \frac{RV_0 \angle(\omega t + \phi)}{(R^2 + X_C^2)^{1/2}}$$

The real potential difference $V_R(t)$ across the resistor is the real part of this potential difference phasor:

$$V_R(t) = \text{Re } \mathbf{V}_R$$

$$= \frac{RV_0}{(R^2 + X_C^2)^{1/2}} \cos(\omega t + \phi) \quad (22.75)$$

Notice that the real current through the resistor (Equation 22.69) and the real potential difference across the resistor (Equation 22.75) have the same phase angle: they are in phase. This is *always* the case in a resistor, as we discovered in Section 22.4.

Now we look at the magnitude of the peak potential difference across the resistor $V_{R\,peak}$ as a function of the angular frequency ω of the source. The peak potential difference is the coefficient of the cosine term in Equation 22.75:

$$V_{R\,peak}(\omega) = \frac{RV_0}{(R^2 + X_C^2)^{1/2}} \quad (22.76)$$

We substitute for the capacitive reactance to see how $V_{R\,peak}$ depends on the angular frequency. Since

$$X_C = \frac{1}{\omega C}$$

the peak potential difference across the resistor is

$$V_{R\,peak}(\omega) = \frac{RV_0}{\left[R^2 + \left(\dfrac{1}{\omega C}\right)^2\right]^{1/2}}$$

After some algebra (that infamous expression), we find

$$V_{R\,peak}(\omega) = \frac{\omega R C V_0}{\left[1 + (\omega R C)^2\right]^{1/2}} \quad (22.77)$$

We look at the low and high angular frequency limits.

Case 1: For angular frequencies approaching zero, the peak potential difference across the resistor approaches zero. When $\omega = 0$ rad/s, the current in the circuit is zero, and there should then be no potential difference across the resistor. This is the dc limit.

Case 2: For very high angular frequencies, the denominator of Equation 22.77 approaches

$$[1 + (\omega R C)^2]^{1/2} \approx [(\omega R C)^2]^{1/2} = \omega R C$$

Hence the peak potential difference across the resistor approaches

$$
\begin{aligned}
V_{R\,peak} &\approx \frac{\omega R C V_0}{\omega R C} \\
&= V_0
\end{aligned}
$$

For high angular frequencies, the impedance of the *capacitor* becomes negligible; we say the capacitor shorts out and acts more and more like just a common wire. In this high angular frequency limit, most of the source potential difference appears across the resistor since the Kirchhoff voltage law must be satisfied at all times in the circuit.

Next let's look at the complement of the circuit by trading the positions of R and C (they are in series so their order is immaterial). Consider the output voltage to be the potential difference across the resistor, as in Figure 22.28. The input is the ac voltage source.

> Since the potential difference across the resistor (the output) is appreciable only for high values of the angular frequency, this arrangement is called a **high pass filter**.

It is only for high frequencies that a significant potential difference appears across the output resistor; see Figure 22.29.

For the high pass filter, the voltage gain is, using Equation 22.77,

$$
\begin{aligned}
\text{voltage gain} &\equiv \frac{V_{R\,peak}(\omega)}{V_0} \\
&= \frac{\omega R C}{\left[1 + (\omega R C)^2\right]^{1/2}} \quad \text{(high pass filter)} \\
&\qquad\qquad\qquad\qquad (22.78)
\end{aligned}
$$

Two things to note about our solution of this ac circuit:

1. The nominal polarities of the potential differences across the capacitor and the resistor are indicated in Figure 22.25.

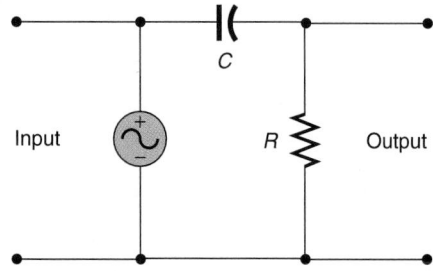

FIGURE 22.28 A high pass filter.

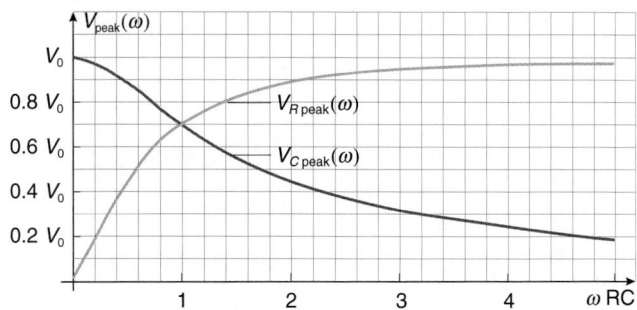

FIGURE 22.29 Peak output potential differences in an *RC* circuit as a function of ωRC.

At any instant t, the KVL around the loop must be satisfied (energy conservation!). That is,

$$-V_0(t) + V_R(t) + V_C(t) = 0 \text{ V} \quad (22.79)$$

This is not at all obvious from the expressions for the actual potential differences (Equations 22.58, 22.72, and 22.75); but it is indeed the case (as you are asked to show in Problem 50). The phasor approach guarantees that Equation 22.79 is the case (if we do the mathematics correctly) since we used the KVL to find the voltage phasors.

2. At any *nonzero* angular frequency ω, the sum of the *amplitudes* (peak values) of the oscillating potential differences across the capacitor and the resistor (the peak values) does *not* equal the amplitude of the source. That is,

$$V_0 \neq \frac{\omega R C V_0}{\left[1 + (\omega R C)^2\right]^{1/2}} + \frac{V_0}{\left[1 + (\omega R C)^2\right]^{1/2}}$$

The reason for inequality is that the respective potential differences are *not* all in phase: they do not reach their peaks simultaneously.

> **PROBLEM-SOLVING TACTIC**
>
> **22.6 To solve an ac circuit, convert the circuit to the complex domain and employ the same techniques and conventions as in dc circuit analysis.** Use impedances, source voltage phasors, current phasors, and potential difference phasors. The real currents and potential differences are the real parts of the complex current and potential difference phasors.

22.9 A SERIES RLC CIRCUIT

One day looking at the formulas in some book or other, I discovered a formula for the frequency of a resonant circuit which was $f = 1/[2\pi(LC)^{1/2}]$, where L is the inductance and C was the capacitance of the circuit. And there was pi, and where was the circle? You laugh, but I was serious then. Pi was a thing with circles, and here is pi coming out of an electric circuit. Where was the circle?

Richard Feynman (1918–1988)[*]

---

[*]Richard P. Feynman, "What is science?" *The Physics Teacher*, 7, #6, pages 313–320 (September 1969); the quote is on page 315.

When you tune your radio or select a TV channel, a **series RLC circuit** may be employed to select the frequency at which a resistor has maximum potential difference—that is, absorbs maximum average power. In order to tune in only one station at a time, the circuit also must respond only to a narrow range (or band) of frequencies around the desired one. A series *RLC* circuit accomplishes these objectives. As its name implies, a series *RLC* circuit consists of a resistor *R*, an inductor *L*, and a capacitor *C* in series with an ac voltage source, as shown in Figure 22.30.

Our focus will be to obtain an expression first for the potential difference across the resistor, as this is amplified and detected by other electronic components (that we will not investigate). The average power absorbed by the resistor is a maximum when the potential difference across it is a maximum (since for a resistor $P = V^2/R$). We know from Section 22.7 that inductors and capacitors each absorb zero average power over each cycle or multiple of it. Hence the average power absorbed by the circuit is the sum of the power absorbed by the resistor and by the source. In keeping with energy conservation and our sign convention for the power, the sum of the average power absorbed by these two circuit elements over a cycle (or multiple of cycles) must be zero:

$$\langle P \rangle_{\text{res}} + \langle P \rangle_{\text{source}} = 0 \text{ W}$$

A resistor always absorbs positive power, so the source absorbs negative power, indicating that it is generating the power for the circuit.

Eventually we calculate the expression for the average power absorbed by the resistor, which is proportional to the square of the potential difference across it. To do this, we need to find expressions for the current through the resistor and the potential difference across it. To accomplish this objective, we proceed as we did with the filter circuit of the previous section.

Step 1: We convert the ac independent voltage source to the complex domain by using a voltage source phasor. Since the source voltage is

$$V(t) = V_0 \cos(\omega t) \tag{22.80}$$

the voltage source phasor is

$$\mathbf{V} = V_0 e^{i\omega t} \quad \text{(exponential form)} \tag{22.81}$$
$$= V_0 \angle(\omega t) \quad \text{(polar form)} \tag{22.82}$$

Also we use the impedance of the resistor, capacitor, and inductor. Converted to the complex domain, the circuit now appears as in Figure 22.31.

Step 2: We simplify the circuit by combining circuit elements that are in series or parallel. In the present case the impedances of the resistor, inductor, and capacitor are in series with each other. The total impedance of the series combination is the sum of the three impedances:

$$Z_{\text{total}} = R + i\omega L + \frac{1}{i\omega C}$$

After rationalizing the impedance of the capacitor, the total impedance becomes

$$Z_{\text{total}} = R + i\omega L - i\frac{1}{\omega C}$$

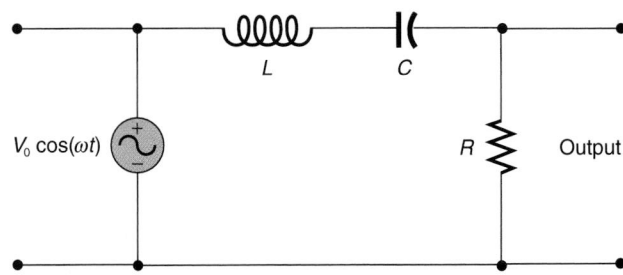

FIGURE 22.30 A series *RLC* circuit.

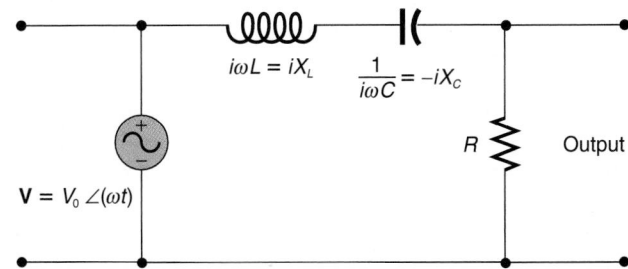

FIGURE 22.31 The series *RLC* circuit in the complex domain.

We use the inductive reactance $X_L = \omega L$ and the capacitive reactance

$$X_C = \frac{1}{\omega C}$$

to write the total impedance more cleanly as

$$Z_{\text{total}} = R + i(X_L - X_C) \tag{22.83}$$

The circuit now appears as shown in Figure 22.32.

Step 3: We choose a direction for the current phasor **I** and mark the polarity of each impedance in accordance with this choice. The choice for the direction of the phasor current is out of the nominally positive terminal of the ac source. The impedance Z_{total} then must be labeled with the polarity indicated in Figure 22.33.

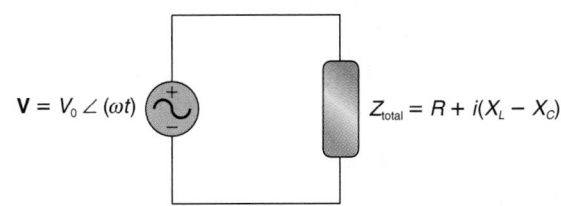

FIGURE 22.32 The simplified circuit in the complex domain.

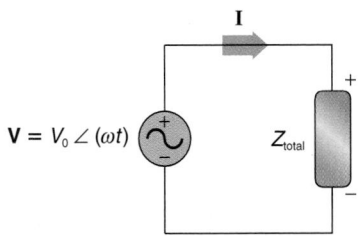

FIGURE 22.33 The direction for the current phasor and the corresponding polarity of the impedance.

Step 4: We use the KVL to form the loop equations. Going around the loop in the clockwise sense and beginning in the lower left corner of the loop in Figure 22.33, the KVL results in the following equation:

$$-\mathbf{V} + \mathbf{I}Z_{total} = 0 \ \text{V} \qquad (22.84)$$

There is only one loop in the circuit, so the KVL yields only one equation. But we have only one unknown current phasor, and so this single equation is sufficient to solve the problem.

Just as with dc circuit techniques, if the application of the KVL to each elementary loop of the circuit does not result in a sufficient number of equations to solve for the unknown currents, the KCL is then employed at one or more significant nodes until the total number of equations (from the KVL and KCL) is equal to the number of unknown current phasors in the problem.

Step 5: We solve the equation(s) for the unknown current(s). (In the present case, there is only one equation and one unknown current.) We solve Equation 22.84 for **I**, yielding

$$\mathbf{I} = \frac{\mathbf{V}}{Z_{total}}$$

Substituting for Z_{total} using Equation 22.83, we find

$$\mathbf{I} = \frac{\mathbf{V}}{R + i(X_L - X_C)} \qquad (22.85)$$

That is the end of the physics, but we have to express **I** in proper polar form to extract its real part, which is the real current. The denominator is a complex number in rectangular form. We express the denominator in polar form:

$$R + i(X_L - X_C) = [R^2 + (X_L - X_C)^2]^{1/2} \ \angle\theta$$

where

$$\theta = \tan^{-1}\left(\frac{X_L - X_C}{R}\right)$$

We make this substitution into the denominator of Equation 22.85 and write the voltage source phasor in polar form; the current phasor then becomes

$$\mathbf{I} = \frac{V_0 \ \angle(\omega t)}{\left[R^2 + (X_L - X_C)^2\right]^{1/2} \ \angle\theta}$$

Division of complex numbers in polar form means that we put the angle in the denominator into the numerator by changing its sign:

$$\mathbf{I} = \frac{V_0 \ \angle(\omega t) \ \angle(-\theta)}{\left[R^2 + (X_L - X_C)^2\right]^{1/2}}$$

The angles in the numerator then are combined according to the product rule for complex numbers in polar form: the angles just add. Thus the current phasor is

$$\mathbf{I} = \frac{V_0 \ \angle(\omega t - \theta)}{\left[R^2 + (X_L - X_C)^2\right]^{1/2}} \qquad (22.86)$$

To find the phasor representing the potential difference across the resistance, we multiply the current phasor **I** by the impedance of the resistor, which is just R:

$$\begin{aligned}\mathbf{V}_{res} &= \mathbf{I}R \\ &= \frac{RV_0 \ \angle(\omega t - \theta)}{\left[R^2 + (X_L - X_C)^2\right]^{1/2}}\end{aligned} \qquad (22.87)$$

Notice that the current and potential difference phasors for the resistor (Equations 22.86 and 22.87) are in phase, as they must be for any resistor.

To find the average power absorbed by the resistor, we use Equation 22.50, modified for the notation here:

$$\langle P \rangle = \frac{1}{2} V_{res} I_{res} \cos \beta$$

where V_{res} is the amplitude of the oscillating potential difference across the resistor, I_{res} is the amplitude of the oscillating current through the resistor, and $\cos \beta$ is the power factor (the cosine of the phase angle difference between the potential difference and the current). The phase angle difference is zero for a resistor, and so the power factor is unity:

$$\cos \beta = \cos(0 \ \text{rad}) = 1 \qquad \text{(for any resistor)}$$

The amplitude of the potential difference across the resistor is the magnitude of the potential difference phasor of Equation 22.87:

$$V_{res} = \frac{RV_0}{\left[R^2 + (X_L - X_C)^2\right]^{1/2}}$$

Likewise the amplitude of the current through the resistor is the magnitude of the current phasor of Equation 22.86:

$$I_{res} = \frac{V_0}{\left[R^2 + (X_L - X_C)^2\right]^{1/2}}$$

Thus the average power absorbed by the resistor is

$$\langle P \rangle = \frac{1}{2} \frac{RV_0^2}{R^2 + (X_L - X_C)^2} \qquad (22.88)$$

This also is the average power absorbed by the series *RLC* combination, because the capacitor and inductor absorb no average power.

We want to determine the angular frequency ω that makes $\langle P \rangle$ as large as possible. From Equation 22.88, $\langle P \rangle$ is greatest when $X_L = X_C$. The dependence of the average power on the angular frequency can be made explicit by substituting for the inductive and capacitive reactances. In particular, the term $X_L - X_C$ in the denominator of Equation 22.88 can be rewritten as

$$\begin{aligned}X_L - X_C &= \omega L - \frac{1}{\omega C} \\ &= \frac{\omega^2 LC - 1}{\omega C}\end{aligned} \qquad (22.89)$$

Recall that when we examined the transient behavior of an LC circuit (in Section 21.10), we discovered that the quantity

$$\frac{1}{(LC)^{1/2}}$$

was an angular frequency. We define this quantity to be the **natural resonant angular frequency** ω_0 of the circuit:

$$\omega_0 \equiv \frac{1}{(LC)^{1/2}} \tag{22.90}$$

The reason for this terminology will become apparent shortly. With this definition, Equation 22.89 can be rewritten, after some algebraic manipulation, as

$$X_L - X_C = \frac{L}{\omega}(\omega^2 - \omega_0^2) \tag{22.91}$$

Substituting this expression into Equation 22.88 for the average power results in

$$\langle P \rangle = \frac{1}{2} \frac{RV_0^2}{R^2 + \dfrac{L^2}{\omega^2}(\omega^2 - \omega_0^2)^2}$$

With a bit more algebraic manipulation, this can be reexpressed as

$$\langle P \rangle = \frac{1}{2} \frac{\omega^2 RV_0^2}{R^2\omega^2 + L^2(\omega^2 - \omega_0^2)^2} \tag{22.92}$$

which is about as tidy (!) as the expression for the average power can be made.

We seek the angular frequency that maximizes the average power in, the current through, and potential difference across R because the latter is amplified by subsequent electronic circuitry. The condition $X_L = X_C$ maximizes the power, which implies the desired angular frequency ω is equal to the natural resonant angular frequency of the circuit ω_0.

There are several aspects of Equation 22.92 that merit comment.

> If the average power is plotted as a function of the angular frequency of the source, the peak of the curve occurs when $\omega = \omega_0$.

This can be seen from Equation 22.92 directly: when $\omega = \omega_0$, the denominator is smallest, so the average power then is greatest. Equivalently, we could take the derivative of $\langle P \rangle$ with respect to ω and then set the result equal to zero to locate the extremum (see Problem 46). A numerical value for ω_0 is determined from the specific values of L and C using Equation 22.90. Since ω_0 is the natural resonant angular frequency, the power is a maximum when the angular frequency of the source is equal to the natural resonant angular frequency. If the angular frequency of the source is too large or too small, the resistor does not absorb as much power.

When $\omega = \omega_0$ in a series RLC circuit, we have an electronic example of **resonance**.* The circuit is analogous to the forced, simple harmonic oscillator that we studied in Chapter 7. If the frequency of the driving force on a child's swing is too fast or too slow, the swing absorbs little energy; if the driving frequency is equal to the natural oscillation frequency of the swing, the energy (and amplitude) of the swing is greatest.

The average power absorbed by the resistor when the angular frequency of the source is equal to ω_0 can be found by substituting $\omega = \omega_0$ into Equation 22.92. The result is

$$\langle P \rangle_{\text{max}} = \frac{1}{2}\frac{V_0^2}{R} \tag{22.93}$$

Since

$$V_{\text{rms}} = \frac{V_0}{\sqrt{2}}$$

the peak power also can be written as

$$\langle P \rangle_{\text{max}} = \frac{V_{\text{rms}}^2}{R} \tag{22.94}$$

> For fixed L and C (and therefore for fixed ω_0), the numerical value of the resistance affects the shape of the graph of $\langle P \rangle$ versus ω. The smaller the resistance R, the more sharply peaked the curve.

This effect is illustrated in Figure 22.34 for two values of R. To make the graphs, the following values were used for the parameters of the circuit:

$$\text{amplitude of the source voltage } V_0 = 10.0 \text{ V}$$
$$\text{inductance } L = 100 \text{ mH}$$
$$\text{capacitance } C = 1000 \text{ pF}$$

With these values for L and C the resonant angular frequency of the circuit is

$$\omega_0 = \frac{1}{(LC)^{1/2}} = 1.00 \times 10^5 \text{ rad/s}$$

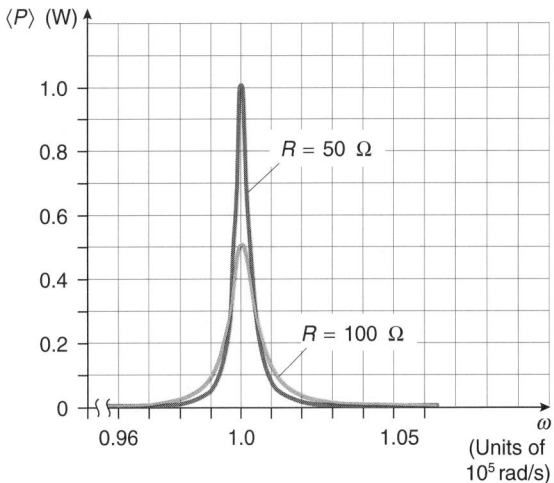

FIGURE 22.34 The value of R affects the shape of the average power versus ω curve.

---

*Operationally, for a fixed source angular frequency (say a radio station), we tune the circuit to the desired angular frequency by adjusting the value of the capacitor in the circuit (which is the tuning knob on our radios and TVs).

The graphs are for $R = 50.0\ \Omega$ and $100\ \Omega$.

The effect of R on the shape of the average power curve also can be explored analytically.

Consider the two angular frequencies ω_1 and ω_2 on either side of ω_0 at which the average power falls to *half* the peak average power, as shown in Figure 22.35. The values of these angular frequencies are found in the following way. At these angular frequencies, the average power is half the peak power or

$$\frac{1}{2}\langle P\rangle_{\text{max}} = \frac{1}{2}\frac{1}{2}\frac{V_0^2}{R}$$

Using this on the left-hand side of Equation 22.92, we obtain

$$\frac{1}{2}\frac{1}{2}\frac{V_0^2}{R} = \frac{1}{2}\frac{\omega^2 R V_0^2}{R^2\omega^2 + L^2(\omega^2 - \omega_0^2)^2} \quad (22.95)$$

Now we solve this equation for ω. This solution for the angular frequencies at half power is a bit tricky algebraically, but eventually we find that Equation 22.95 reduces to

$$L^2(\omega^2 - \omega_0^2)^2 = R^2\omega^2$$

There are both positive and negative choices when we take the square root:

$$L(\omega^2 - \omega_0^2) = \pm R\omega$$

Taking the positive term on the right-hand side yields

$$L(\omega^2 - \omega_0^2) = +R\omega$$

or

$$\omega^2 - \frac{R}{L}\omega - \omega_0^2 = 0\ \text{rad}^2/\text{s}^2$$

Taking the negative term on the right-hand side yields

$$L(\omega^2 - \omega_0^2) = -R\omega$$

or

$$\omega^2 + \frac{R}{L}\omega - \omega_0^2 = 0\ \text{rad}^2/\text{s}^2$$

Each of these equations has two roots:

$$\omega_2 = \frac{\dfrac{R}{L} \pm \sqrt{\left(-\dfrac{R}{L}\right)^2 + 4\omega_0^2}}{2}$$

$$\omega_1 = \frac{-\dfrac{R}{L} \pm \sqrt{\left(\dfrac{R}{L}\right)^2 + 4\omega_0^2}}{2}$$

To avoid negative angular frequency solutions, we must reject the $-$ sign choice for both cases. The positive solutions are

$$\omega_2 = \frac{R}{2L} + \frac{1}{2}\sqrt{\left(-\frac{R}{L}\right)^2 + 4\omega_0^2} \quad (22.96)$$

$$\omega_1 = -\frac{R}{2L} + \frac{1}{2}\sqrt{\left(\frac{R}{L}\right)^2 + 4\omega_0^2} \quad (22.97)$$

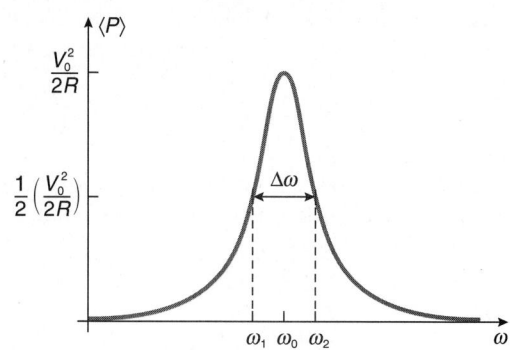

FIGURE 22.35 The frequencies ω_1 and ω_2 at which the average power absorbed falls to half the peak.

The difference between these half-power angular frequencies is called the **angular frequency bandwidth** $\Delta\omega$ of the circuit:

$$\Delta\omega \equiv \omega_2 - \omega_1 \quad (22.98)$$

$$= \frac{R}{L} \quad (22.99)$$

Equation 22.99 shows that for fixed ω_0 (i.e., fixed L and C), the half-power width is directly proportional to R; the smaller R is, the smaller the half-power angular frequency bandwidth, as confirmed in Figure 22.34.

The ratio between the resonant angular frequency ω_0 and the angular frequency bandwidth $\Delta\omega$ is called the **quality factor** Q of the circuit:

$$Q \equiv \frac{\omega_0}{\Delta\omega} \quad (22.100)$$

Substituting for ω_0 using Equation 22.90 and for $\Delta\omega$ using Equation 22.99, the quality factor can be expressed in terms of the circuit components:

$$Q = \frac{\dfrac{1}{(LC)^{1/2}}}{\dfrac{R}{L}}$$

$$= \left(\frac{L}{CR^2}\right)^{1/2} \quad (22.101)$$

A circuit with a large value of Q is one with a small (narrow) angular frequency bandwidth; such a condition is necessary to avoid cross talk, the simultaneous reception of two different radio or TV frequencies.

The quality factors of the circuits whose power curves are plotted in Figure 22.34 are found using Equation 22.101. Using those values of L and C, the quality factor of the circuit is

$$Q = \frac{1.00 \times 10^4\ \Omega}{R}$$

For the $100\ \Omega$ resistor (using the given values for L and C),

$$Q_{100} = 100$$

and for the $50.0\ \Omega$ resistor,

$$Q_{50} = 200$$

indicating that as the resistance is *decreased* the quality factor of the circuit *increases* (for fixed L and C).

Notice in Figure 22.34 that the average power absorbed is more sharply peaked at the resonant angular frequency ω_0 for the 50.0 Ω resistor than for the larger, 100 Ω resistor, reflecting the difference in the quality factors of the circuit with the two different resistors.

Since the resonant angular frequency ω_0 depends on C (see Equation 22.90), changing the value of the capacitance shifts the angular frequency at which the peak in the power curve occurs.

If the capacitance is increased, Equation 22.90 shows that the resonant angular frequency decreases. Note that changing C also affects the quality factor Q of the circuit (see Equation 22.101); increasing C decreases the quality factor Q. The angular frequency bandwidth $\Delta\omega$ is unaffected by changing the value of the capacitance (see Equation 22.99).

After our leisurely tour through the complex algebraic thicket of the series *RLC* circuit, you should realize that it is the basic element behind tuned communication systems, so it has a wide range of applications!

CHAPTER SUMMARY

A *complex variable* z can be represented in *rectangular form* as

$$z = x + iy \tag{22.1}$$

where $i \equiv \sqrt{-1}$. The *real part* of z is x and its *imaginary part* is y:

$$\text{Re } z \equiv x \tag{22.2}$$

$$\text{Im } z \equiv y \tag{22.3}$$

Both the real and imaginary parts are real variables. Here z, x, and y represent general algebraic variables, *not necessarily Cartesian coordinates*.

The *polar form* of a complex variable z has the form

$$z = r \angle \theta \tag{22.5}$$

where r is the *magnitude* of the complex number, also known occasionally as its *amplitude*:

$$r = [(\text{Re } z)^2 + (\text{Im } z)^2]^{1/2}$$
$$= (x^2 + y^2)^{1/2} \tag{22.4}$$

and

$$\tan\theta = \frac{\text{Im } z}{\text{Re } z} = \frac{y}{x} \tag{22.8}$$

The *exponential form* of the complex variable z is

$$z = re^{i\theta}$$

The *Euler identity* is useful for changing from the exponential form to the rectangular form of a complex variable:

$$re^{i\theta} = r\cos\theta + ir\sin\theta \tag{22.11}$$

Two complex variables $z_1 = x_1 + iy_1$ and $z_2 = x_2 + iy_2$ can be added and subtracted in rectangular form:

$$z_1 \pm z_2 = (x_1 \pm x_2) + i(y_1 \pm y_2) \tag{22.12, 22.13}$$

The product of two complex variables in polar form, $z_1 = r_1 \angle \theta_1$ and $z_2 = r_2 \angle \theta_2$, is

$$z_1 z_2 = r_1 r_2 \angle (\theta_1 + \theta_2) \tag{22.15}$$

while their quotient is

$$\frac{z_1}{z_2} = \frac{r_1}{r_2} \angle (\theta_1 - \theta_2) \tag{22.18}$$

The *complex conjugate* z^* of a complex variable z is found by replacing i with $-i$ everywhere. Thus, if

$$z = x + iy = re^{i\theta} = r \angle \theta$$

then its complex conjugate is

$$z^* = x - iy = re^{-i\theta} = r \angle (-\theta)$$

The magnitude r of a complex variable z can be found from

$$r = (zz^*)^{1/2}$$

Sinusoidal, ac potential differences $V(t) = V_0 \cos(\omega t + \theta)$, currents $I(t) = I_0 \cos(\omega t + \phi)$, and ac independent voltage sources are represented by complex *phasors* whose real parts are the actual (i.e., real) potential differences, currents, and sources, respectively:

$$\mathbf{V}(t) \equiv V_0 e^{i(\omega t + \theta)} = V_0 \angle (\omega t + \theta) \tag{22.19}$$

$$\mathbf{I}(t) \equiv I_0 e^{i(\omega t + \phi)} = I_0 \angle (\omega t + \phi) \tag{22.20}$$

$$\mathbf{V}_{\text{source}}(t) \equiv V_0 \angle (\omega t) \tag{22.41}$$

In resistors, inductors, and capacitors, the potential difference phasor and the current phasor are related by the *ac version of Ohm's law*:

$$\mathbf{V} = \mathbf{I}Z \tag{22.29}$$

where Z is the *impedance* of the circuit element:

$$Z_R = R \quad \text{for a resistor} \tag{22.30}$$

$$Z_L = i\omega L \equiv iX_L \quad \text{for an inductor} \tag{22.31}$$

$$Z_C = \frac{1}{i\omega C} \equiv -iX_C \quad \text{for a capacitor} \tag{22.32}$$

The quantity $X_L \equiv \omega L$ is the *inductive reactance* and

$$X_C \equiv \frac{1}{\omega C}$$

is the *capacitive reactance*.

The potential difference phasor across a resistor and the current phasor through it are in phase. In an inductor, the phase difference between the potential difference phasor and the current phasor is $\pi/2$ rad. For a capacitor, the phase difference between the potential difference phasor and the current phasor is $-\pi/2$ rad. The same phase relationships exist between the real potential differences and the real currents in the various circuit elements.

Impedances in series are added to find an equivalent impedance:

$$Z_{\text{eq}} = Z_1 + Z_2 + Z_3 + \cdots \tag{22.35}$$

The equivalent impedance for impedances in parallel is

$$\frac{1}{Z_{\text{eq}}} = \frac{1}{Z_1} + \frac{1}{Z_2} + \frac{1}{Z_3} + \cdots \tag{22.36}$$

If V_0 is the amplitude of the sinusoidal potential difference across a circuit element and I_0 is the amplitude of the sinusoidal current into its nominally positive terminal, the average power ⟨P⟩ absorbed by (transferred to) the circuit element is

$$\langle P \rangle = \frac{1}{2} V_0 I_0 \cos \beta \tag{22.50}$$

$$= V_{rms} I_{rms} \cos \beta \tag{22.56}$$

where $\cos \beta$ is the *power factor*, the cosine of the phase difference between the potential difference and the current. The *rms values* (also called *effective values*) are related to the peak values by

$$V_{rms} = \frac{V_0}{\sqrt{2}} \tag{22.54}$$

$$I_{rms} = \frac{I_0}{\sqrt{2}} \tag{22.55}$$

For a resistor, the power factor is equal to 1. For inductors and capacitors, the power factor is 0; therefore, both inductors and capacitors absorb *zero* average power.* The average power absorbed also can be expressed as

$$\langle P \rangle = \frac{1}{2} \operatorname{Re}(\mathbf{V I}^*) \tag{22.52}$$

In the complex domain, sinusoidal ac circuits are solved with the KVL and KCL using conventions and techniques like those used for dc circuit analysis.

---

*Remember these are *ideal* circuit elements.

SUMMARY OF PROBLEM-SOLVING TACTICS

22.1 (page 1008) When converting between rectangular and polar forms, be sure your calculator is in radian mode for performing trigonometric operations.

22.2 (page 1010) Addition and subtraction of complex variables can only be performed when they are in rectangular form.

22.3 (page 1011) For the multiplication of two complex numbers in rectangular form, it occasionally is easier first to convert them to polar or exponential form, then do the multiplication.

22.4 (page 1011) To divide two complex numbers in rectangular

form, it occasionally is easier first to convert them to polar or exponential form, then do the division.

22.5 (page 1020) For two impedances in parallel, you can make use of the shortcut we found for resistors in parallel: the equivalent impedance is the product of the two impedances divided by their sum.

22.6 (page 1026) To solve an ac circuit, convert the circuit to the complex domain and employ the same techniques and conventions as in dc circuit analysis.

QUESTIONS

1. (page 1009); 2. (page 1012)

3. Considering Lenz's law from Chapter 21, explain why the inductive reactance should increase with increasing angular frequency.

4. Explain why the capacitive reactance should decrease with increasing angular frequency.

5. Explain why a capacitor acts as a short circuit at high angular frequencies and as an open circuit at low angular frequencies.

6. Explain why an instructor, er. . . no, an inductor acts as a short circuit at low angular frequencies and an open circuit at high angular frequencies.

7. A light bulb (a resistor) is connected in series with a capacitor and a sinusoidal ac independent voltage source with a variable angular frequency ω as in Figure Q.7. For small values of ω, will the light bulb be brightly lit or dim? For large values of

ω, will the bulb be brightly lit or dim? Explain your reasoning in each case.

8. A light bulb (a resistor) is connected in series with an inductor and a sinusoidal ac independent voltage source with a variable angular frequency ω as in Figure Q.8. For small values of ω, will the light bulb be brightly lit or dim? For large values of ω, will the bulb be brightly lit or dim? Explain your reasoning in each case.

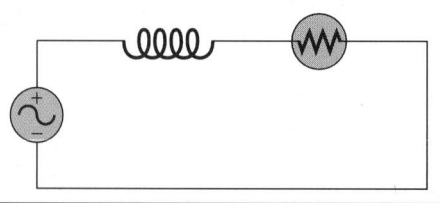

FIGURE Q.8

9. Can an ohmmeter be used to measure the impedance of a resistor? Of a capacitor? Of an inductor?

10. Two sinusoidal ac independent voltage sources have the same amplitude but different angular frequencies:

$$V_0 \cos(\omega_1 t) \qquad V_0 \cos(\omega_2 t)$$

Do they have equal rms amplitudes?

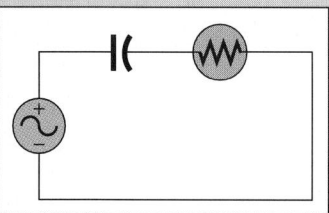

FIGURE Q.7

11. Fluorescent lights use an inductor to limit the glow discharge current through the bulb. Why is this usage more energy efficient than using a resistor to limit the current?

12. The angular frequency in a series *RLC* circuit is doubled. By what factor is the impedance of the resistor changed? Answer the same question for the inductor, the capacitor, and the total impedance.

13. At resonance, what is the total impedance of a series *RLC* circuit?

14. One of the *RL* circuits in Figure Q.14 is a high pass filter and the other is a low pass filter. By considering the qualitative behavior of the impedance of an inductor as a function of the angular frequency ω, deduce which circuit is the high pass filter and which is the low pass filter. Explain your reasoning.

15. Compared with acoustic frequencies, is radio static high frequency or low frequency? To eliminate such static should a low pass or high pass filter be used?

16. In the schematic diagram in Figure Q.16, explain why the capacitor prevents dc currents in each circuit from passing into the other circuit but permits the passage of ac currents. Such circuits are said to be dc isolated and ac capacitively coupled. Capacitors are used frequently in transistor circuits in just this way.

17. In the schematic diagram of Figure Q.17, explain why node A is effectively an electrical ground for sufficiently large angular frequencies. Capacitors used in this way are called bypass capacitors and are used frequently in transistor circuits.

18. The sum of the peak values of the potential differences across the resistor, inductor, and capacitor in an *RLC* circuit is usually greater than the peak value of the source voltage. Does this contradict the KVL? Explain why or why not.

19. Since the ac current and ac potential difference across a light bulb (a resistor) vary rapidly with time, does the temperature of the filament and the consequent light output of the bulb

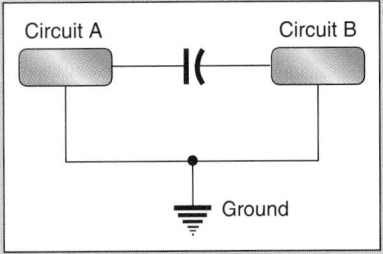

FIGURE Q.16

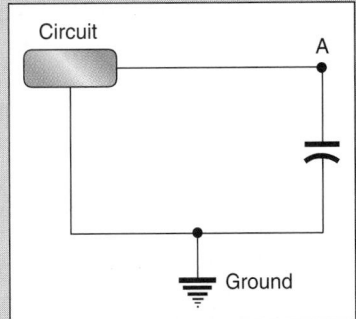

FIGURE Q.17

vary accordingly? What might limit the size of any temperature variation?

20. What is the phase angle θ of a series *RLC* circuit (on page 1028) when the inductive reactance is equal to the capacitive reactance? Is this the condition for resonance?

21. Why are capacitors used more frequently than inductors in integrated circuits?

22. If a source consists of *both* a dc emf and an ac oscillatory emf as well, is it permissible to consider the effects of each separately in determining currents and potential differences in a circuit and then add the results? Explain your answer. This procedure is used frequently in analyzing transistor circuits.

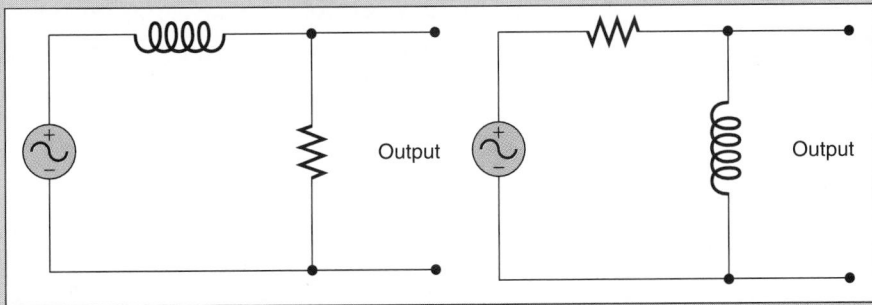

FIGURE Q.14

PROBLEMS

Sections 22.1 Representations of a Complex Variable
22.2 Arithmetic Operations with Complex Variables

•1. Two complex numbers are given by

$$z_1 = 3 + i(4) \quad \text{and} \quad z_2 = 2 - i(5)$$

Find: (a) the magnitude of each complex number; (b) the polar form of each complex number; and (c) the exponential form of each complex number. (d) Locate each complex number on the complex plane.

•2. Perform the following arithmetic operations on

$$z_1 = 3 + i(4) \quad \text{and} \quad z_2 = 2 - i(5)$$

Express each result in rectangular, polar, and exponential form. (a) $z_1 + z_2$; (b) $z_1 - z_2$; (c) $z_1 z_2$; (d) z_1/z_2. (e) Locate the results of (a)–(d) on the complex plane.

•3. For the two complex numbers

$$z_1 = -2 + i(3) \quad \text{and} \quad z_2 = 3 - i(4)$$

find: (a) the magnitude of each complex number; (b) the polar form of each complex number; (c) the exponential form of each complex number; and (d) the location of each complex number on the complex plane.

•4. Perform the following arithmetic operations on the complex numbers

$$z_1 = -2 + i(3) \quad \text{and} \quad z_2 = 3 - i(4)$$

Express each result in rectangular, polar, and exponential form. (a) $z_1 + z_2$; (b) $z_1 - z_2$; (c) $z_1 z_2$; (d) z_1/z_2. (e) Locate the results of (a)–(d) on the complex plane.

•5. Find the simplest form for the following complex expressions: (a) $(1 - i)^4$; (b) $(\sqrt{2} - i) - i(1 - i\sqrt{2})$;

$$\text{(c)} \quad \frac{5}{(1 - i)(2 - i)(3 - i)}.$$

•6. Rationalize the following expressions:

$$\text{(a)} \quad \frac{2}{2 + i}; \quad \text{(b)} \quad \frac{i(5)}{1 - i(2)}.$$

•7. Evaluate the following: (a) $[2 - i(3)]^*$; (b) $[4e^{-i(\pi/4 \text{ rad})}]^*$; (c) $[3.0 \angle(\pi/6 \text{ rad})]^*$.

•8. By using the polar form of the complex number $z = -1 + i$, show that

$$(-1 + i)^7 = -8(1 + i)$$

•9. Show that the addition and subtraction of two complex variables is equivalent to the addition and subtraction of a pair of two-dimensional vectors whose components are the real and imaginary parts of the complex variables.

•10. For two complex variables

$$A = A_x + iA_y \quad \text{and} \quad B = B_x + iB_y$$

show that if A_x and A_y are the two Cartesian components of a two-dimensional vector $\vec{A} = A_x\hat{i} + A_y\hat{j}$, and B_x and B_y are the two Cartesian components of another two-dimensional

vector $\vec{B} = B_x\hat{i} + B_y\hat{j}$: (a) then $\text{Re}(AB^*)$ is the scalar product $\vec{A} \cdot \vec{B}$; and (b) $\text{Im}(AB^*)$ is the single nonzero component of the vector product $\vec{B} \times \vec{A}$.

•11. (a) Show that

$$\frac{1}{2}(1 + i)^2 = i$$

Hence $\sqrt{i}$ can be represented by the complex number

$$\frac{1}{\sqrt{2}}(1 + i)$$

This ensures that there are no "supercomplex" numbers given by multiple roots of i itself; multiple roots of i can be expressed in terms of i itself. (b) By writing i in exponential form, show that the exponential form of $\sqrt{i}$ is

$$\sqrt{i} = e^{i(\pi/4 \text{ rad})}$$

•12. (a) After studying Problem 11, show that

$$(-i)^{1/2} = \frac{1}{\sqrt{2}}(1 - i)$$

(b) Show that

$$(-i)^{1/2} = e^{-i(\pi/4 \text{ rad})}$$

Sections 22.3 Complex Potential Differences and Currents: Phasors
22.4 The Potential Difference and Current Phasors for Resistors, Inductors, and Capacitors
22.5 Series and Parallel Combinations of Impedances
22.6 Complex Independent ac Voltage Sources
22.7 Power Absorbed by Circuit Elements in ac Circuits

13. Show that the inductive reactance has the SI units of ohms.

14. Show that the capacitive reactance has the SI units of ohms.

15. Make an accurate graph of the resistance of a 10.0 kΩ resistor as a function of frequency from 0 Hz to 1.0 kHz.

•16. Make an accurate graph of the capacitive reactance of a 0.010 μF capacitor as a function of frequency ν from 0 Hz to 1.0 kHz.

•17. Make an accurate graph of the inductive reactance of a 0.500 H inductor as a function of frequency ν from 0 Hz to 1.0 kHz.

•18. At what frequency will a 50.0 μF capacitor have a capacitive reactance of 53.05 Ω?

•19. At what frequency will a 150 mH inductor have an inductive reactance of 47.12 Ω?

•20. Given an inductor with inductance L and a capacitor with capacitance C, at what angular frequency is the inductive reactance equal to the capacitive reactance?

•21. For what range of frequencies will a 50 μF capacitor have a capacitive reactance less than 10.0 Ω?

•22. For what range of frequencies will a 50 mH inductor have an inductive reactance less than 10.0 Ω?

•23. Find the impedance of the following circuit elements at a frequency of 1.00 kHz: (a) a resistance with $R = 10.0$ kΩ;

(b) a capacitor with $C = 0.0100\,\mu\text{F}$; and (c) an inductor with $L = 0.500$ H. (d) What is the capacitive reactance of the capacitor at the given frequency? (e) What is the inductive reactance of the inductor at the indicated frequency?

•24. (a) The three circuit elements in Problem 23 are connected in series. Find the equivalent impedance. Express your result in both rationalized rectangular form and polar form. (b) If the three circuit elements are connected in parallel, find the equivalent impedance. Express your result in both rationalized rectangular form and polar form.

•25. (a) An inductor L and a capacitor C are connected in series. At what frequency (in Hz) will the equivalent impedance of the pair be zero? (b) Is there a (finite) nonzero frequency at which the *parallel* combination of the pair will produce an equivalent impedance equal to zero? Justify your answer.

•26. A 150 Ω resistor is in series with a 0.250 H inductor and a capacitor, with the collection connected to an ac independent voltage source at 60.0 Hz. What capacitance results in a total impedance of 150 Ω?

•27. A parallel combination of a 1.00 kΩ resistor and a 250 mH inductor is in series with a capacitor C, as shown in Figure P.27. The combination is connected to a source with a frequency of 1.00 kHz. What value of C produces a total impedance whose real part is ten times greater than its imaginary part?

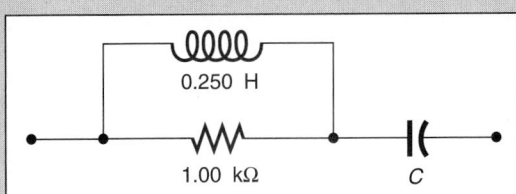

0.250 H

1.00 kΩ C

FIGURE P.27

•28. Show that the time-averaged value of $\cos(\omega t)$, averaged over a period T of the oscillation, is zero. That is, show that

$$\langle \cos(\omega t) \rangle = \frac{1}{T} \int_{0\,\text{s}}^{T} \cos(\omega t)\,dt = 0$$

•29. Three circuit elements are at your disposal:

$$R = 1.00 \text{ k}\Omega$$
$$C = 100 \ \mu\text{F}$$
$$L = 500 \text{ mH}$$

(a) Using a single log–log graph grid,* plot
   (i) the logarithm (base 10) of the capacitive reactance X_C;
   (ii) the logarithm (base 10) of the inductive reactance X_L; and
   (iii) the logarithm (base 10) of the resistance

*Such graph paper *automatically* takes the logarithms of the numbers plotted. If you are unfamiliar with the use of log–log graph paper, consult your instructor.

as functions of the logarithm (base 10) of the angular frequency ω from 1.00 rad/s to 1.00×10^4 rad/s. The logarithm of the angular frequency should be the abscissa. (b) Explain why each graph is a straight line. (c) At what angular frequency is the inductive reactance equal to 1.00 kΩ? What frequency is this in Hz? (d) At what angular frequency is the capacitive reactance equal to 1.00 kΩ? What frequency is this in Hz? (e) At what angular frequency is the inductive reactance equal to the capacitive reactance? What frequency is this in Hz?

Sections 22.8 A Filter Circuit
22.9 A Series *RLC* Circuit

•30. For the sake of your professor, consider the circuit indicted in Figure P.30 (a) What is the frequency v of the source? (b) What is the impedance of the resistor? (c) What is the independent voltage source phasor? Express your result in both rationalized rectangular and polar form. (d) Choose a direction for the current phasor and label the resistor polarity accordingly. (e) Use the KVL to find the current phasor. Express your result in both rationalized rectangular and polar form. (f) What is the potential difference phasor across the resistor? Express your result in both rationalized rectangular and polar form. (g) What is the real current $I(t)$ through the resistor? (h) What is the real potential difference $V(t)$ across the resistor? (i) What is the phase difference between the current through the resistor and the potential difference across it? (j) What is the peak value of the current through the resistor? (k) What is the rms value of the current in the resistor? (l) What is the peak value of the potential difference across the resistor? (m) What is the rms value of the potential difference across the resistor? (n) What average power is absorbed by the resistor?

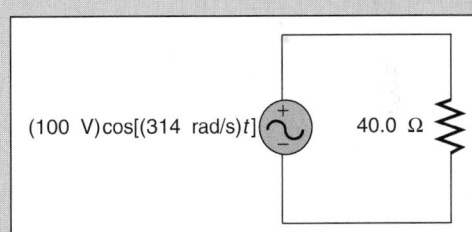

(100 V)cos[(314 rad/s)t]   40.0 Ω

FIGURE P.30

•31. An ac ammeter and an ac voltmeter are used in Problem 30 to measure the current through the resistor and the potential difference across it. (a) Indicate how they should be connected to the circuit. (b) What will be the reading indicated by each meter?

•32. A resistor is connected to an ac independent voltage source as shown in Figure P.32. The peak value of the ac current through the resistor is I_0 when the source frequency is v. If the frequency of the source is changed to $2v$, what is the value of the peak current through the resistor?

•33. Find the dc current I_{dc} through a resistor R that will produce the same power absorbed as an ac current with a peak value of I_0.

•34. Consider the circuit indicated in Figure P.34. (a) What is the frequency v of the source? (b) What is the impedance of the capacitor? (c) What is the independent voltage source phasor? Express your result in both exponential and polar form. (d) Choose a direction for the current phasor and label the capacitor polarity accordingly. (e) Use the KVL to find the

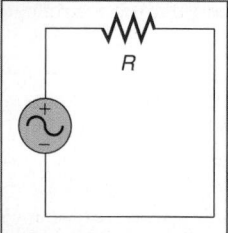

FIGURE P.32

current phasor. Express your result in both polar and exponential form. (f) What is the potential difference phasor across the capacitor? Express your result in both polar and exponential form. (g) What is the real current $I(t)$ through the capacitor? (h) What is the real potential difference $V(t)$ across the capacitor? (i) What is the phase difference between the potential difference across the capacitor and the current through it? (j) What is the peak value of the current through the capacitor? (k) What is the rms value of the current in the capacitor? (l) What is the peak value of the potential difference across the capacitor? (m) What is the rms value of the potential difference across the capacitor? (n) What is the power factor associated with the capacitor? (o) What average power is absorbed by the capacitor?

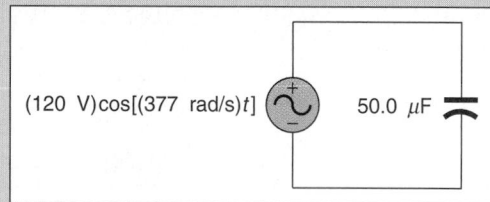

FIGURE P.34

•**35.** An ac independent voltage source described by

$$(20.0 \text{ V}) \cos[(377 \text{ rad/s})t]$$

is connected to a 100 μF capacitor as shown in Figure P.35. (a) What is the impedance of the capacitor? (b) What is the capacitive reactance X_C? (c) What is the *peak* value of the current through the capacitor? (d) At what time $t \geq 0$ s does the current in the capacitor first reach its peak value?

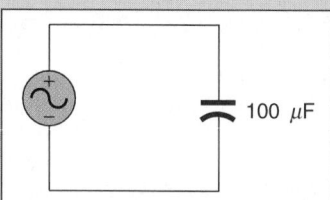

FIGURE P.35

•**36.** A capacitor is connected to an ac independent voltage source as shown in Figure P.36. The peak value of the ac current through the capacitor is I_0 when the source frequency is ν. If the frequency of the source is changed to 2ν, what is the new value of the peak current through the capacitor?

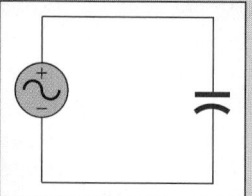

FIGURE P.36

•**37.** Consider the circuit indicated in Figure P.37 (a) What is the frequency ν of the source? (b) What is the impedance of the inductor? (c) What is the independent voltage source phasor? Express your result in both exponential and polar form. (d) Choose a direction for the current phasor and label the inductor polarity accordingly. (e) Use the KVL to find the current phasor. Express your result in both polar and exponential form. (f) What is the potential difference phasor across the inductor? Express your result in both polar and exponential form. (g) What is the real current $I(t)$ through the inductor? (h) What is the real potential difference $V(t)$ across the inductor? (i) What is the phase difference between the potential difference across the inductor and the current through it? (j) What is the peak value of the current through the inductor? (k) What is the rms value of the current in the inductor? (l) What is the peak value of the potential difference across the inductor? (m) What is the rms value of the potential difference across the inductor? (n) What is the power factor associated with the inductor? (o) What average power is absorbed by the inductor?

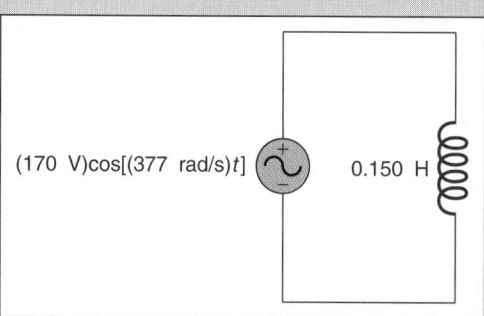

FIGURE P.37

•**38.** An inductor is connected to an ac independent voltage source as shown in Figure P.38. The peak value of the ac current through the inductor is I_0 when the source frequency is ν. If the frequency of the source is changed to 2ν, what is the new value of the peak current through the inductor?

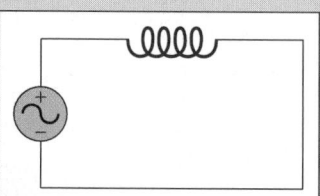

FIGURE P.38

•**39.** An ac independent voltage source described by

$$(20.0 \ V) \cos[(377 \ rad/s)t]$$

is connected to a 50 mH inductor as shown in Figure P.39. (a) What is the impedance Z_L of the inductor? (b) What is the inductive reactance X_L? (c) What is the peak value of the current through the inductor? (d) At what time $t \geq 0$ s does the current in the inductor first reach its peak value?

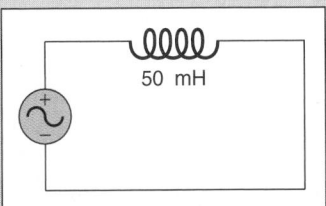

FIGURE P.39

•**40.** A series LR circuit is driven by an ac independent voltage source as shown in Figure P.40. (a) Redraw the circuit in the complex domain indicating the impedances of the circuit elements as well as the polar form of the complex source voltage phasor $\mathbf{V}_{source}$. (b) Choose the direction for the complex current phasor $\mathbf{I}$ so it emerges from the nominally positive terminal of the ac voltage source. Use the KVL to show that the current phasor has the form

$$\mathbf{I} = \frac{\mathbf{V}_{source}}{R + i\omega L}$$

(c) Show that the current phasor in polar form is

$$\mathbf{I} = \frac{V_0}{\left[R^2 + (\omega L)^2\right]^{1/2}} \angle(\omega t - \theta)$$

where

$$\tan \theta = \frac{\omega L}{R}$$

(d) Show that the real current $I(t)$ in the circuit is

$$I(t) = \frac{V_0}{\left[R^2 + (\omega L)^2\right]^{1/2}} \cos(\omega t - \theta)$$

(e) Find the potential difference phasors across both the resistor and the inductor and use them to show that the real potential differences across the resistor and the inductor are

$$V_{res}(t) = \frac{V_0 R}{\left[R^2 + (\omega L)^2\right]^{1/2}} \cos(\omega t - \theta)$$

and

$$V_{ind}(t) = \frac{V_0 \omega L}{\left[R^2 + (\omega L)^2\right]^{1/2}} \cos\left(\omega t - \theta + \frac{\pi}{2} \ rad\right)$$

Notice that the potential difference across the resistor is in phase with the current through it, while the potential difference across the inductor and the current through it are $\pi/2$ rad out of phase. (f) What is the power factor of the inductor? (g) What is the average power absorbed by the inductor? (Think! A detailed calculation is not needed!) (h) What is the power factor of the resistor? (i) What is the average power absorbed by the resistor? (j) What is the power factor of the source? (k) What is the average power absorbed by the independent voltage source?

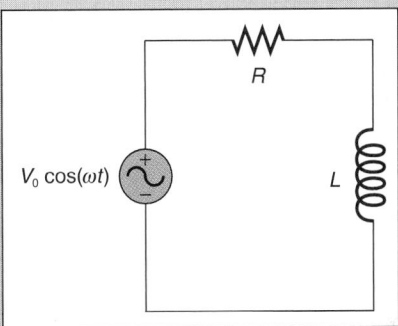

FIGURE P.40

•**41.** For the low pass filter of Figure P.41, what value of the angular frequency produces a peak potential difference across the capacitor equal to half the amplitude of the source voltage? That is, find the value of ω for which

$$V_{C \ peak}(\omega) = \frac{V_0}{2}$$

Under these circumstances, what is the voltage gain?

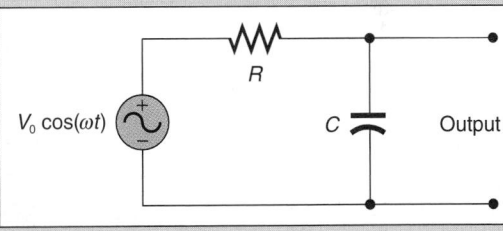

FIGURE P.41

•**42.** For what value of the angular frequency is the peak potential difference across the resistor in a high pass RC filter equal to half the source voltage? That is, find the value of ω for which

$$V_{R \ peak}(\omega) = \frac{V_0}{2}$$

What is the voltage gain under these circumstances?

•**43.** (a) In Equation 22.65 on page 1023, for what value of ω is the phase angle $\pi/2$ rad? For what value of ω is the phase angle $\pi/4$ rad? (b) Sketch a graph of the phase angle ϕ in Equation 22.65 as a function of the angular frequency ω. The angle ϕ represents the phase angle difference between the current in the circuit and the source voltage. Notice that as the angular frequency increases and the impedance of the capacitor decreases, the current and source voltage are increasingly in phase. For very large ω, the capacitor shorts out and the circuit is purely resistive with the current in phase with the source voltage.

•44. A high pass filter has $R = 1.00$ kΩ and $C = 0.0500$ μF as shown in Figure P.44. A graph of the logarithm (base 10) of the voltage gain (Equation 22.78 on page 1026) versus the logarithm (base 10) of the frequency ν ($= 2\pi\omega$) is called a *Bode plot*. (The logarithm of the voltage gain is the ordinate and the logarithm of the frequency is the abscissa.) (a) Construct a Bode plot for this high pass filter over the frequency domain from 10 Hz to 100 kHz. (b) Estimate the frequency at which the voltage gain is 0.50.

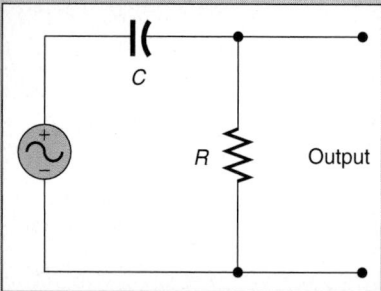

FIGURE P.44

•45. A low pass filter circuit has $R = 1.00$ kΩ and $C = 0.0500$ μF as shown in Figure P.45. A graph of the logarithm (base 10) of the voltage gain (Equation 22.74 on page 1025) versus the logarithm (base 10) of the frequency ν ($= 2\pi\omega$) is called a *Bode plot*. (The logarithm of the voltage gain is the ordinate and the logarithm of the frequency is the abscissa.) (a) Construct a Bode plot for this low pass filter over the frequency domain from 10 Hz to 100 kHz. (b) Estimate the frequency at which the voltage gain is 0.50.

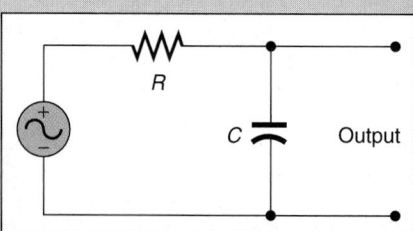

FIGURE P.45

•46. Show that the peak of the average power absorbed by the resistor in a series *RLC* circuit, Equation 22.92 on page 1029, occurs where $\omega = \omega_0$ by taking the derivative

$$\frac{d\langle P \rangle}{d\omega}$$

and setting it equal to zero.

•47. In a series *RLC* circuit, a variable capacitor is used to tune the circuit over the frequencies of the AM radio band, from about 500 kHz to 1600 kHz. (a) If the inductor has an inductance $L = 650$ μH, what must be the range of the variable capacitor to be able to tune in the spectrum of frequencies in the AM radio band? The radio stations of the AM radio band are

separated from each other by 10 kHz in frequency. The bandwidth of the tuning circuit must be smaller than this to ensure that when you tune in one station, you do not pick up neighboring stations (which would result in cross talk, listening to two stations simultaneously!). (b) If the bandwidth of the circuit is to be 2.0 kHz, what is the Q of the circuit when tuned to a station at 500 kHz? What does this Q imply about the value of the resistance R in the circuit? (c) For a bandwidth of 2.0 kHz, what is the Q of the circuit when tuned to a station at the other end of the AM band at 1600 kHz? What does this Q imply about the size of the resistance in the circuit?

‡48. The potential difference across a resistor with resistance R has the time dependence shown in Figure P.48. (a) Determine the power absorbed by the resistor during the interval from when $t = 0$ s to $t = T/2$. (b) Determine the power absorbed by the resistor from when $t = T/2$ to $t = T$. (c) Calculate the average power absorbed by the resistor during the interval from when $t = 0$ s to $t = T$. (d) Determine the average potential difference across the resistor over a complete cycle from when $t = 0$ s to $t = T$. (e) If the average potential difference across the resistor over one cycle now is applied continuously (in a dc way) to the resistor, what is the power absorbed by the resistor?

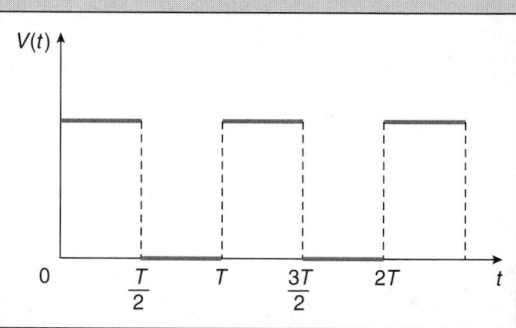

FIGURE P.48

‡49. Figure P.49 depicts a *parallel RLC* circuit. Notice that since the resistor, capacitor, and inductor are in parallel with the ac independent voltage source, all the circuit elements have the same potential difference across them, equal to the potential difference across the source. (a) Convert the circuit to the complex domain by indicating the polar form of the source voltage phasor and the impedances of the resistor, the capacitor, and the inductor. Use X_C for the capacitive reactance of the capacitor and X_L for the inductive reactance of the inductor. (b) Show that the polar form for the current phasors in the resistor, the capacitor, and the inductor are (using the current directions indicated in Figure P.49)

$$\mathbf{I}_{res} = \frac{V_0}{R} \angle (\omega t)$$

$$\mathbf{I}_{ind} = \frac{V_0}{X_L} \angle \left(\omega t - \frac{\pi}{2} \text{ rad} \right)$$

$$\mathbf{I}_{cap} = \frac{V_0}{X_C} \angle \left(\omega t + \frac{\pi}{2} \text{ rad} \right)$$

(c) Show that the equivalent impedance Z_{eq} of the parallel RLC combination satisfies

$$\frac{1}{Z_{eq}} = \frac{1}{R} + i\left(\frac{1}{X_C} - \frac{1}{X_L}\right)$$

(d) Use the expression for the equivalent impedance found in part (c) to show that the peak value of the current in the source is

$$V_0\left[\left(\frac{1}{R}\right)^2 + \left(\frac{1}{X_C} - \frac{1}{X_L}\right)^2\right]^{1/2}$$

(e) What angular frequency ω maximizes the peak value of the total current from the source? This is the resonant angular frequency for the parallel RLC circuit. How does the resonant angular frequency for the *parallel RLC* circuit compare with the resonant angular frequency for the *series RLC* circuit for the same values of R, L, and C?

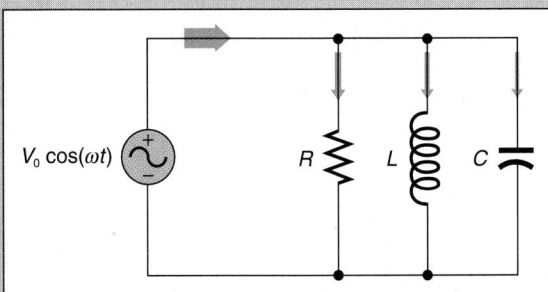

FIGURE P.49

‡50. For the RC filter circuit discussed in Section 22.8, show explicitly that the real potential differences across the voltage source, the resistor, and the capacitor satisfy the KVL at any instant t. See Figure P.50. That is, show that

$$-V(t) + V_R(t) + V_C(t) = 0 \ V$$

where

$$V(t) = V_0 \cos(\omega t)$$

$$V_R(t) = \frac{\omega RC V_0}{\left[1 + (\omega RC)^2\right]^{1/2}} \cos\left(\omega t + \phi\right)$$

and

$$V_C(t) = \frac{V_0}{\left[1 + (\omega RC)^2\right]^{1/2}} \cos\left(\omega t + \phi - \frac{\pi}{2} \ \text{rad}\right)$$

Recall from Equation 22.65 that

$$\tan \phi = \frac{1}{\omega RC}$$

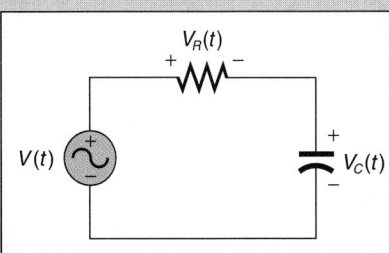

FIGURE P.50

‡51. The electrical grid in the United States is a *three-phase* grid because the ac independent voltage sources provided consist of three sources that can be represented by the following expressions:

$$V_1(t) = V_0 \cos(\omega t)$$

$$V_2(t) = V_0 \cos\left(\omega t - \frac{2\pi}{3} \ \text{rad}\right)$$

$$V_3(t) = V_0 \cos\left(\omega t - \frac{4\pi}{3} \ \text{rad}\right)$$

Show that the potential difference between *any two* of these sources has a peak value of $\sqrt{3}\,V_0$ and oscillates sinusoidally at angular frequency ω. Hint:

$$\cos \alpha - \cos \beta = -2 \sin\left(\frac{\alpha + \beta}{2}\right) \sin\left(\frac{\alpha - \beta}{2}\right)$$

INVESTIGATIVE PROJECTS

A. Expanded Horizons

1. The invention of new math is part of the creative process enjoyed by mathematicians; such math occasionally finds use in physics and engineering. Although it is somewhat removed from physics, investigate the properties of the so-called *hallucinatory numbers*, also called the *perplex numbers*, based on a number h whose absolute value is minus one: $|h| = -1$.
Paul Fjelstad, "Extending special relativity via the perplex numbers," *American Journal of Physics, 54, #5,* pages 416–422 (May 1986).

V. Majernik, "The perplex numbers are in fact the binary numbers," *American Journal of Physics, 56, #8,* page 763 (August 1988).

2. Charles Proteus Steinmetz and Albert Einstein were equally fascinating to the public during the early decades of the 20th century. Steinmetz made significant contributions to lightning research and even wrote a book about special relativity. Someone (perhaps in jest) long ago suggested the frequency unit not be named after Heinrich Hertz, but after C. P. Steinmetz, for then the unit could be abbreviated cps, a cycle per second, from his initials! If you are interested in learning more about Steinmetz and perhaps discovering the reasons that he subsequently has become all but unknown to the general public (unlike Einstein), you might enjoy reading and comparing each of the following biographies, one written just after his death, the other after the passage of some time.

Ronald R. Kline, *Steinmetz: Engineer and Socialist* (The Johns Hopkins University Press, Baltimore, 1992).

John Winthrop Hammond, *Charles Proteus Steinmetz* (The Century Co., New York, 1924).

3. Investigate the electrical effects generated in body tissue with a view toward understanding the electricity and electronics of an electrocardiogram.

John R. Cameron and James G. Skofronick, *Medical Physics* (Wiley, New York, 1978), Chapter 9.

Robert O. Becker and Andrew A. Marino, *Electromagnetism and Life* (State University of New York Press, Albany, 1982).

4. Investigate the applications of high- and low-frequency currents and potential differences in medicine.

John R. Cameron and James G. Skofronick, *Medical Physics* (Wiley, New York, 1978), Chapter 11.

5. Investigate the differences between ac and dc electrical motors.

Richard A. Honeycutt, *Electromechanical Devices* (Prentice-Hall, Englewood Cliffs, New Jersey, 1986), Chapters 5 and 6, pages 111–163.

6. Investigate the differences between ac generators and alternators.

Richard A. Honeycutt, *Electromechanical Devices* (Prentice-Hall, Englewood Cliffs, New Jersey, 1986), Chapter 4, pages 83–110.

B. Lab and Field Work

7. Your physics department likely has an independent voltage source of variable frequency (occasionally called a signal or function generator) and an oscilloscope for measuring time-dependent potential differences. Use a 1.0 kΩ resistor and a 0.50 μF capacitor to design and investigate the frequency dependence of the voltage gain of both a high pass and a low pass filter. See Problems 44 and 45.

8. Design and perform an experiment to investigate the frequency dependence of the phase difference between a sinusoidal ac source and the resulting current in a series *RC* circuit. Your instructor may be able to assist you in devising a method for displaying and measuring the phase difference on an oscilloscope.

9. Given an inductor with an unknown inductance and a capacitor with an unknown capacitance, devise and perform an experimental procedure to measure their reactances and thereby determine the values of the inductance and capacitance.

C. Communicating Physics

10. Write a paper investigating, comparing, and contrasting the advantages and disadvantages of transmitting electrical energy over long distances by dc and by ac methods. In the early part of the 20th century, why did ac methods win out over dc

methods? Will the practical development of relatively low cost, high temperature superconducting materials (which have zero resistance for conduction currents) change the terms of the debate? Write the paper for a nontechnically educated audience.

James P. Rybak, "AC or DC?" *Popular Electronics*, *11*, #9, pages 42–48 (September 1994).

George Westinghouse Commemoration (The American Society of Mechanical Engineers, New York, 1937).

11. Amazingly, the invention of the electric chair was associated with the competition between ac and dc means of transmitting electricity. Thomas Edison (1847–1931) favored dc transmission while George Westinghouse (1846–1914), among others, favored ac. Edison invented an ac electric chair and had the method of execution adopted by New York State, primarily to illustrate the danger associated with ac electricity and to discredit the technology. Ironically, the ploy failed. This amazing story is a fascinating blend of physics, engineering, politics, and the worst elements of human nature. You might find it an interesting topic to investigate and report to your class.

"Edison's Miracle of Light" (WGBH Educational Foundation show, *American Enterprise*, 23 October 1995, Transcript 802).

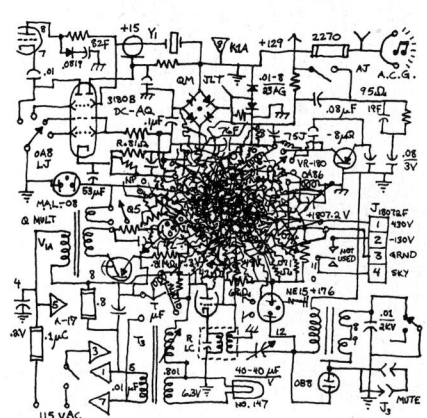

52. Calculate the potential difference across and the power absorbed by each circuit element.

CHAPTER 23

GEOMETRIC OPTICS

But soft! what light through yonder window breaks?
William Shakespeare (1564–1616) *

The nature of light and the way it passes through and interacts with matter are rich veins of physics that have led to many significant discoveries, revolutionary ideas, and technological advances. Newton thought light was a stream of particles and his theory of the mechanics of particles was able to account for both reflection and refraction. In Chapter 21 (Sections 21.4 and 21.5), we saw how the four fundamental equations of electromagnetism led James Clerk Maxwell to conclude light was electromagnetic waves.[†] Later, in Chapter 26, we shall see that light has aspects of *both* particles and waves, a seemingly contradictory situation. In the 20th century the speed of light also assumed an important role in relativity theory (Chapter 25).

But investigations into light and optical instruments had begun millennia before Newton and Maxwell, who published their work in the 17th and 19th centuries. Indeed, reflections from water surfaces likely were noted curiously by prehistoric peoples. Mirrors of polished metal were manufactured by hand in ancient Egypt as well as other ancient middle eastern cultures. The hieroglyph for *life*, called the *ankh*, ☥, is thought to be a stylized rendition of the shape of ancient Egyptian hand mirrors. Today, infants and even some of our pets seem intrigued by their reflections in mirrors. Birds have been known to attack their reflections, with typically frustrating results!

Transparency of certain materials (for example, water and crystals) is a fact of nature. The discovery of glass and glass making arose in many of the early middle eastern civilizations of the Fertile Crescent. The great Greek-Alexandrian mathematician, astronomer, and scientist Ptolemy[‡] (~A.D. 150) wrote a long treatise on the refraction or bending of light at the interface between two transparent materials, even deducing a relationship between the directions of the incident and refracted rays that was correct for small angles.

The invention of lenses arose from unknown sources in the early Middle Ages, likely by North African Islamic Arabs, and resulted in the invention of spectacles to correct vision problems. The word lens come from the Latin *lens*, meaning "lentil bean," because of their similar shapes. Ibn-al-Haitham (965–1038) discovered the relationship between object and image distances in lenses.

The likely accidental invention of the telescope[§] about 1608 and the microscope[#] about the same time began the application of the optical principles of reflection and refraction to more sophisticated instrumentation that continues to this day.

In this chapter we first explore the domains of optics and the empirical laws of reflection and refraction. We will see how these laws are used to understand the image-forming prop-

The ankh hieroglyph ☥, meaning "life," appears on the base of the statue at the Ramasseum near Luxor, Egypt, that inspired Percy Bysshe Shelley's famous poem *Ozymandias*.

Little known outside the Middle East, Ibn-al-Haithan (spellings vary) made significant contributions to optics at the turn of the last millennium.

erties of useful optical devices such as mirrors and lenses, as well as more complicated optical systems consisting of several optical elements.

Manufactured optical instruments such as cameras, microscopes, and telescopes have been and continue to be of immense practical use.[**] Our eye (we have two!) is one of the most remarkable optical instruments known. It is humbling to realize that eyes developed naturally through biological evolution during aeons in which the scientific and inventive prowess of our intellect were nonexistent. Nature is more creative and inventive than we are.

23.1 THE DOMAINS OF OPTICS

To begin our investigations into light, we assume that the directions of motion of light are represented by directed straight lines called **rays**. Rays do not travel, although quite frankly the language we associate with them might imply that they do! Rays really indicate *paths* of light. Rays are perpendicular to the wave crests in a wave model.

---

*(Chapter Opener) *Romeo and Juliet*, Act II, Scene ii.

[†]The Maxwell equations further are able to account for the laws of reflection and refraction. There also were much earlier experiments (by Young in 1800) which implied that light had wavelike properties. We discuss Young's experiments in Chapter 24.

[‡]Ptolemy also codified the then dominant geocentric theory of the solar system, known as the Ptolemaic theory.

[§]The discovery was made independently by three Dutch spectacle makers: Hans Lippershey, Zacharias Jansen, and James Metius.

[#]By Robert Hooke (1635–1703) (and others), whom we also know for Hooke's law; see Chapter 7.

[**]The development of sophisticated CCD cameras, electron microscopes, and new-technology telescopes still are based, fundamentally, on the basic principles we study in this chapter.

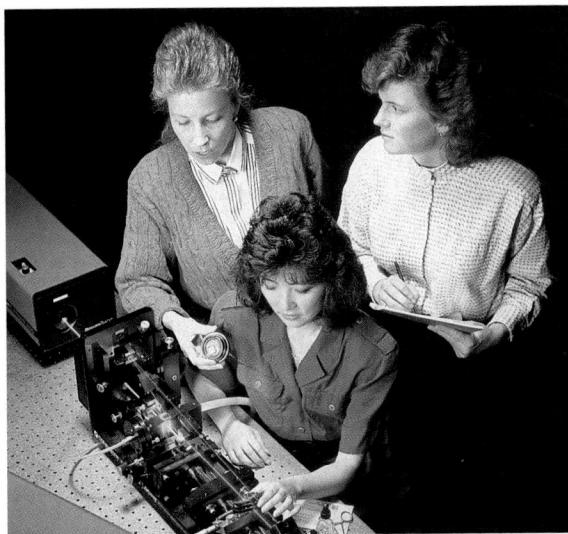

Light rays indicate the paths that light follows. A thin laser beam can be considered as a ray such as in this laboratory.

The study of light can be organized into three broad domains: geometric optics, physical optics, and quantum optics. These domains are not strictly disjoint. The complicated transitions between them are continuous, not sharp; but, for simplicity, we consider them as distinct. We distinguish the domains as follows:

- **geometric optics**: employs only rays

- **physical optics**: employs rays and waves

- **quantum (photon) optics**: employs rays, waves, and energy bundles called photons

The distinction between geometric and physical optics is delineated by comparing the wavelength of the light under study with the characteristic dimensions of the devices or systems interacting with the light.* The distinction between physical and

---

*We briefly touched on these ideas in Section 12.17.

quantum optics will be based on whether the wave or the (yet to be studied) particle aspects of light are manifest.

To see how the dimensions of devices interacting with light are used to distinguish between the geometric and wave domains, consider the following analogy. Imagine a series of ocean wave crests or swells methodically moving across the surface of the sea. Let the wave have a characteristic wavelength λ and speed v. The wave enters a large bay as shown in Figure 23.1.

If we neglect the complicated edge effects near the shore, the wave crests go directly into the bay; they eventually crash into the beach giving much pleasure to local surfers. However, suppose you just invested in a very expensive sailboat and want to moor the boat in the bay. This could be quite hazardous to the boat during severe storms because the boat is unprotected from the ravages of large wave crests entering the wide bay. Your congresswoman obligingly uses her influence to get the Army Corps of Engineers to construct a breakwater across the mouth of the bay as shown in Figure 23.2. The breakwater narrows the entrance of the bay to a fraction of its former size. The incident ocean wave crests now are presented with a different geometry: a much smaller entrance to the harbor. The wave crests now fail to go straight into the harbor in a single direction, but rather spread out as they come through the opening, traveling in *many* different directions. Your boat is safe, though the surfing is ruined. The amplitude of the wave crests in the harbor is greatest in the straight-through direction, but there are wavelets of smaller amplitude propagating in directions quite different from the incident direction of the ocean swell.

This situation can be summarized in the following way. When the aperture of the bay is much greater than the wavelength of the waves, the wave crests go straight into the bay, propagating in a single direction (neglecting the effects at the edges of the wide bay aperture). This is the geometric approximation or limit. On the other hand, when the size of the aperture is on the order of the wavelength or less, the wave crests spread out in the region behind the opening, moving in many different directions. This spreading out of a wave (diffraction) as it passes through an opening or around an obstacle occurs when the size of the opening (or obstacle) is comparable to or smaller than the wavelength of the waves incident on it.

The same ideas apply to light:

Lines along the "crests"
of the waves: Wavefronts

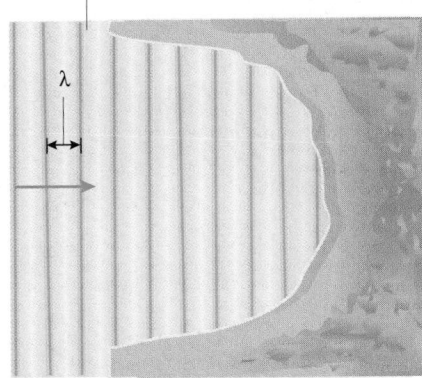

λ

FIGURE 23.1 Waves enter a large bay.

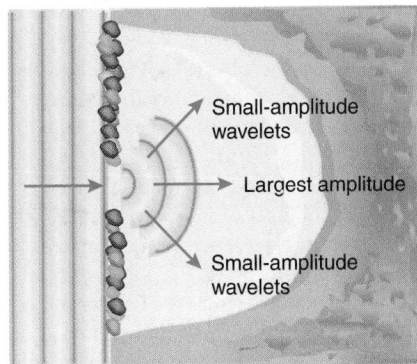

Small-amplitude wavelets

Largest amplitude

Small-amplitude wavelets

FIGURE 23.2 A breakwater across the mouth of the bay causes diffraction.

If its wavelength is much less than the size of the opening or obstacle presented to the light, then the geometric limit is appropriate and each incident light ray subsequently travels in a unique direction or single ray. Well-defined shadows exist in regions where there are no rays. This is the domain of geometric optics.

If its wavelength is on the order of the size of the opening through which the light passes, then the wave limit is appropriate and we must account for diffraction effects. This is the domain of physical optics, which we consider in Chapter 24.

The geometric limit of light is appropriate when the size of its wavelength is negligible, such as when visible light passes through the openings in these clouds.

Although we shall be concerned principally with *visible* light, these ideas also apply equally to other portions of the electromagnetic spectrum. For example, visible light passes straight through large apertures such as door and window openings, even small keyholes, with well-defined shadow regions; the geometric limit is appropriate and we can conclude that the wavelength of visible light is much smaller than the dimensions of these apertures. Thus visible light travels in straight lines by or through these openings, and the ray approximation is an adequate description of the behavior.

On the other hand, if you carry a radio through a door opening or around to the back of your house, the antenna does not lose the station; hence, radio waves must diffract around such openings or obstacles,* indicating that radio waves have wavelengths comparable to or exceeding the dimensions of common objects in our environment. The geometric limit, therefore, is not appropriate for radio waves interacting with such obstacles or apertures. The diffraction of radio waves around objects of macroscopic size such as small hills is why one does not have to be along a clear line of sight to a radio transmitter in order to receive its signal.

In this chapter we study geometric optics, and so we need be concerned here only with the interaction of rays of light with reflecting surfaces (mirrors) and refracting surfaces (abrupt boundaries between transparent media) that are large compared with

---

*This assumes the materials are not transparent to radio waves. In fact, many nonconducting materials are transparent to radio waves, but you likely are able to see the point we are trying to make here.

the wavelength. Wavelengths are occasionally mentioned in this chapter for their familiar use in describing different colors of light, but the observational evidence that light is a wave can be postponed to serve as the real core of the following chapter on physical optics.

23.2   THE INVERSE SQUARE LAW FOR LIGHT

There is one glory of the sun, and another of the moon, and another glory of the stars; for one star differeth from another star in glory.
The New Testament, 1 Corinthians 15:41

Go outside on a clear night and there are few of us so crass that we are not impressed with or humbled by the magnificence of the starry sky. Some stars are quite bright, others are barely detectable to the eye. Stars differ in brightness not only because of their various intrinsic luminosities (their power output) but also because they are at many different distances from us. These common and well-known conclusions (which are suggested by, but are not at all obvious from simply looking at the night sky) raise questions about what we mean by dim or bright and how the light collected by, say, our eyes or telescopes varies with their distance from the light source.

The more distant you are from a pointlike light source, the dimmer it is. A pointlike star, after all, is a distant sun, but it provides precious little nighttime illumination because of its vast distance from us. Consider such a pointlike or spherical source of light, as in Figure 23.3, radiating equally in all directions, much like a distant star or the relatively nearby Sun. Let L be a measure of the total power of the light sent out in all directions by the source, which is the amount of energy emitted per second (in watts), called the **luminosity** of the source.[†]

The wattage of a light bulb is a measure of its luminosity (at all wavelengths, not necessarily just in the visible portion of the electromagnetic spectrum). The light bulb source is only pointlike when viewed from distances r at which its physical size is negligible. Certainly all the power must pass through any imagined sphere of radius r centered on the pointlike source (as in Figure 23.3). Since $4\pi r^2$ is the area of the sphere, the power per unit area on the sphere, called the **intensity** I, thus is

$$I \equiv \frac{L}{4\pi r^2} \tag{23.1}$$

[†]The luminosity of the Sun is 3.83×10^{26} W, a rather bright light bulb!

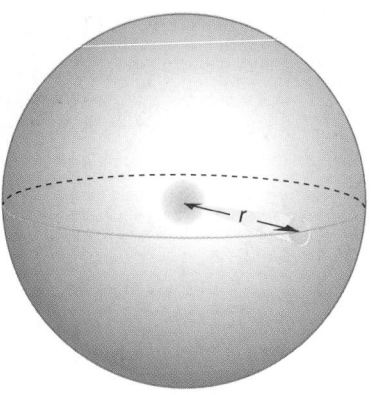

FIGURE 23.3   A pointlike source of light radiates equally in all directions.

The intensity received from the pointlike or spherical source decreases as the inverse square of the distance r, an **inverse square law**.

The inverse square law for light intensity makes no distinction as to whether the light energy is associated with waves or particles. The law applies equally well to *both* viewpoints and cannot be used to distinguish between them; other experiments must do that.

Recall that in Chapter 12 (page 548), we defined the intensity of *any* wave in a similar way: the average power transmitted through one square meter oriented perpendicular to the direction the wave is travelling. The intensity is proportional to the square of the amplitude of the wave.

The **brightness** B is a measure of the power collected by, say, your eye or a telescope of aperture area A oriented perpendicular to the rays from the source:

$$B \equiv IA \qquad (23.2)$$

Hence using a telescope with a larger aperture than your eye increases the brightness of a star, although the intensity of the light incident on both is the same, since the telescope is at the same distance from the source as your eye. Likewise, you see light sources as brighter with a dilated pupil than with it contracted, something you may have experienced uncomfortably during and after an opthamological exam.

From the meaning of the intensity we can deduce that the amplitude of the *spherical wave* diverging from a pointlike or spherical source must decrease as $1/r$. That is, the amplitude of a spherical wave is proportional to the inverse first power of the distance r from the source.

EXAMPLE 23.1

The Sun has a luminosity of 3.83×10^{26} W. What is the intensity of solar light on the Earth, located 1.496×10^8 km from the Sun? This intensity is called the *solar constant*.

Solution
Since the light energy from the Sun is spread equally over a sphere centered on the Sun, the intensity is the luminosity divided by the area of a sphere with a radius equal to the distance of the Earth from the Sun. Use Equation 23.1:

$$I = \frac{L}{4\pi r^2}$$

$$= \frac{3.83 \times 10^{26} \text{ W}}{4\pi (1.496 \times 10^{11} \text{ m})^2}$$

$$= 1.36 \times 10^3 \text{ W/m}^2$$

It is this considerable intensity that motivates the use of solar energy; it is free and nonpolluting as well.

23.3 THE LAW OF REFLECTION

The phenomenon of reflection commonly is observed by us all. Indeed, for better or worse, a reflection from a mirror is one of the first apparitions that greets us each and every morning.

The path of a narrow flashlight beam or laser beam can be represented as a single ray. If the ray encounters a smooth reflecting surface as in Figure 23.4, the direction of the ray is changed abruptly at the surface. A line perpendicular to the surface at the point where the incident ray strikes the surface is called a normal line, or more simply, the **normal** to the surface. The angle θ that the **incident ray** makes with the normal line is the **angle of incidence**. The angle θ' that the **reflected ray** makes with the normal line is the **angle of reflection**.

Simple observations and measurements readily indicate two things:

1. The incident ray, the normal line, and the reflected ray all lie in the same plane.

2. The angle of incidence is equal to the angle of reflection:

$$\theta = \theta' \qquad (23.3)$$

These two observations constitute the **law of reflection**.

Note that a graph of θ' versus θ is linear and makes a 45° angle with each axis, as shown in Figure 23.5.

The first observation is important because it is conceivable (but wrong) that the reflected ray might not lie in the plane of the incident ray and the normal line, even if the two angles were equal ($\theta = \theta'$), as indicated in Figure 23.6.

The place where the incident ray strikes the reflecting surface must be **locally smooth**. This means the differential area where the ray contacts the surface must be locally flat over distances much larger than the wavelength of the incident light,

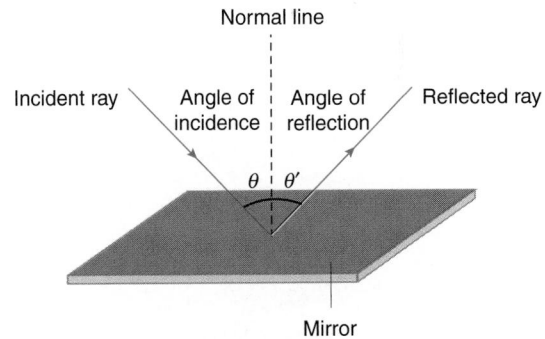

FIGURE 23.4 Reflection of a ray from a surface.

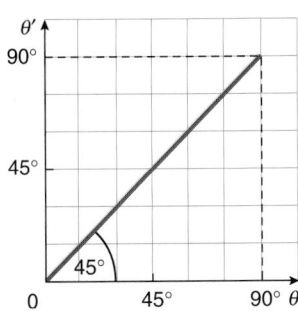

FIGURE 23.5 A graph of the angle of reflection versus the angle of incidence is a straight line with a 45° slope.

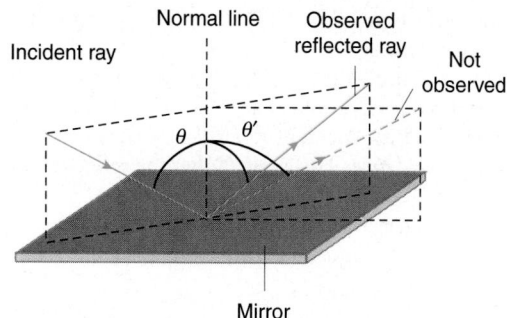

FIGURE 23.6 The reflected ray, normal line, and incident ray all lie in the same plane.

as shown in Figure 23.7. Any deviations from being locally flat are much smaller than the wavelength of the light under consideration. This type of reflection is known as **specular reflection*** and is the type of reflection that principally concerns us.

The local area is **locally rough** if the deviations from a plane in the topography of the local differential surface area are on the order of or larger than the wavelength of the incident light. A ray reflected from locally rough surfaces produces many rays going off at many different angles and directions, producing **diffuse reflection**; see Figure 23.8.

The appearance of the reflection of approaching headlights from a dry pavement surface illustrates diffuse reflection; if the pavement is wet, the reflection is more specular.

A surface that is locally rough for visible light can be locally smooth for longer wavelengths such as microwaves and radio waves. For example, a wire mesh screen is an effective locally

---

*The word comes from the Latin *specularis*, meaning "of or pertaining to mirrors."

Specular reflection

FIGURE 23.7 The differential area from which the incident ray reflects must be locally smooth for the reflected ray to follow the law of reflection.

smooth surface and a good specular reflector for microwaves and radio waves. The holes in the metal screen windows in microwave ovens permit the passage of visible light, but the screen acts as a reflecting surface for microwaves, since their wavelength is on the order of centimeters in size.

An elementary experiment indicates that light rays are reversible: if you reverse the direction of a ray, the light retraces its path, as shown in Figure 23.9.

Sophisticated experiments indicate that if the reflecting surface is moving with a significant nonzero velocity component along the normal line, the angle of incidence no longer is equal to the angle of reflection. If the law of reflection is to hold, the velocity component v of the reflecting surface along the normal line must be negligible compared with the speed c of light: $v \ll c$.[†] Similar experiments also indicate that if the reflecting surface has a velocity *parallel* to the surface itself, the law of reflection still applies.

An unusual mirror called a **corner cube reflector** consists of three plane mirrors that meet at right angles with each other. The terminology comes about because the system of three mirrors would form the corner of a cube. A ray of light reflected from a corner cube mirror system is antiparallel to the incident ray regardless of the angle that the incident ray makes with the first normal line. A corner cube reflector is a generalization to three mirrors of the situation examined in Example 23.3. The device turns the ray around and sends it back from where it came, along a parallel, though slightly displaced, ray.

These devices have many applications:

1. A satellite known as LAGEOS (an acronym for *LAser GEOdynamic Satellite*) orbits the Earth. The satellite is a passive device that bristles with corner cube reflectors over its spherical surface. By measuring the time it takes a laser pulse to go from a point on the Earth to the satellite and back, it is possible to calculate the distance to the satellite very accurately. If this is done simultaneously from two different locations on the surface of the Earth, it is possible to triangulate and obtain an accurate measurement of the distance between the two widely separated points on the Earth.[‡] Such distance determinations are so accurate that the data confirm the theory of plate tectonics: the continents are slowly shifting their positions relative to one another.

[†]A treatment of reflection from high-speed mirrors can be found in R. W. Ditchburn, *Light* (Interscience, New York, 1964), pages 437–438.

[‡]For other ways in which precise locations are determined on the Earth, see the article by Thomas A. Herring, "The global positioning system," *Scientific American*, *274*, #2, pages 44–50 (February 1996).

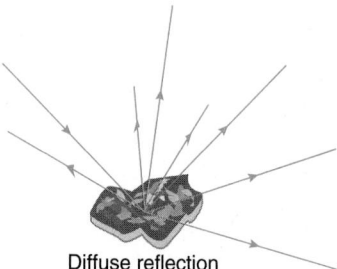

Diffuse reflection

FIGURE 23.8 Diffuse reflection from locally rough surfaces.

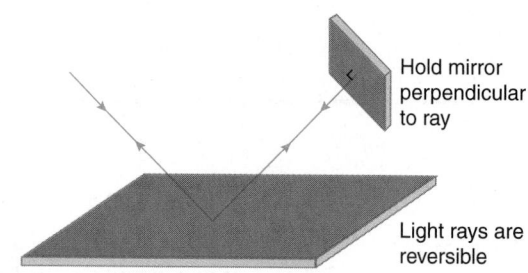

Hold mirror perpendicular to ray

Light rays are reversible

FIGURE 23.9 Reverse the direction of the light and it retraces its path.

The LAGEOS satellite is covered with corner cube reflectors.

A corner cube reflector array was placed on the lunar surface during the Apollo Moon exploration voyages.

2. The first Apollo flights to the Moon set up and left behind small arrays of corner cube reflectors pointed in the general direction of the Earth. The round-trip travel times of laser pulses from the Earth to these corner cube arrays are used to measure the distance between a point on the Earth and the array on the Moon with extraordinary precision (an uncertainty of less than a centimeter). These experiments have been under way for many years now and confirm that the distance between the Moon and the Earth is increasing at the rate of 3.8 centimeters per year. The cause for this slow increase in the Earth–Moon distance is the friction between our ocean tides and the continental land masses (see Problem 53 in Chapter 10).
3. Most modern surveying and leveling equipment routinely uses corner cube reflectors.

QUESTION 1

Describe several surfaces that produce specular reflection and several that produce diffuse reflection for visible light; repeat the question for radio waves (which have much longer wavelengths than visible light).

EXAMPLE 23.2

A light ray is incident at angle θ on the reflecting surface shown in Figure 23.10. The reflecting surface now is turned through an angle ϕ as shown in Figure 23.11. Show that the *change* in the angle between the incident and reflected ray is 2ϕ, independent of the original incidence angle θ.

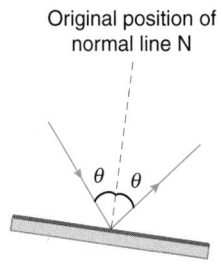

FIGURE 23.10

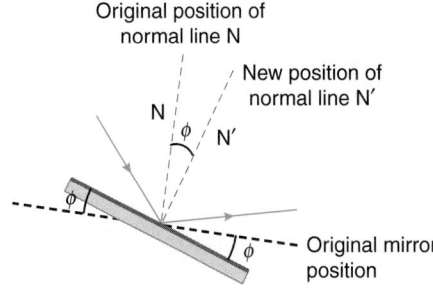

FIGURE 23.11

Solution

For the initial position of the mirror (Figure 23.10), the angle between the reflected ray and the incident ray is 2θ, according to the law of reflection.

When the mirror is rotated through the angle ϕ, the normal line also rotates through the same angle ϕ. The angle of incidence now is $\theta + \phi$, as shown in Figure 23.12.

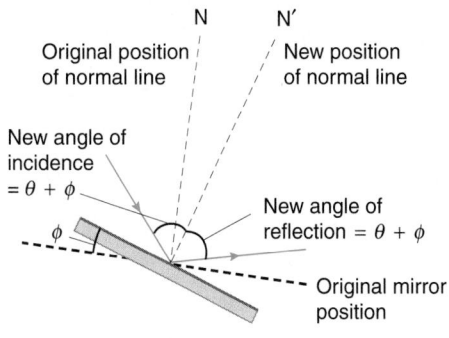

FIGURE 23.12

By the law of reflection, the angle of reflection also now is $\theta + \phi$. The angle between the new reflected ray and the incident ray then is $2(\theta + \phi)$. Hence the *change* in the angle that the reflected ray makes with the incident ray is

$$2(\theta + \phi) - 2\theta = 2\phi$$

This device is called an *optical lever*. The classic Cavendish experiment, used to determine the numerical value of the universal gravitational constant G (see Figure 6.3), uses an optical lever to double the small angular deflection of the mirror.

QUESTION 2

Is the change in the direction of the reflected ray in Example 23.2 independent of the axis about which the mirror is rotated through the angle ϕ?

EXAMPLE 23.3

A ray is incident at angle θ on one of two mirrors that make a 90° angle with each other as shown in Figure 23.13. Determine the direction of the final reflected ray with respect to the incident direction. Such an arrangement of mirrors is called a *corner mirror*. You see corner mirrors frequently in the clothing sections of department stores as well as in mirrored rooms such as the Hall of Mirrors in the palace at Versailles, near Paris.

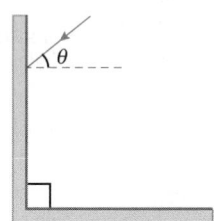

FIGURE 23.13

Solution

The law of reflection is applied at each reflection. Let β be the angle the final reflected ray makes with the horizontal direction, as indicated in Figure 23.14.

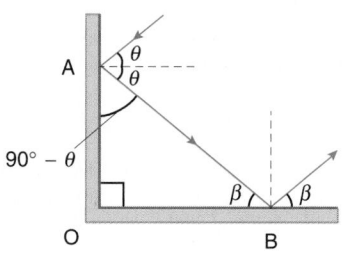

FIGURE 23.14

The angles of triangle AOB in Figure 23.14 are indicated. Since the sum of the angles of any triangle must be 180°, you have

$$\beta + 90° + (90° - \theta) = 180°$$

Solving for β, you obtain

$$\beta = \theta$$

Therefore the reflected ray is *antiparallel* to the incident ray whatever the angle of incidence happens to be. The corner mirror simply reverses the direction of the incident ray; the reflected ray is not coincident with the incident ray but is always antiparallel to it.

23.4 THE LAW OF REFRACTION

Countless wonders are performed by nature in accordance with the laws of these refractions;

Roger Bacon (c. 1214–1294)*

You have no doubt noticed that when a light ray strikes an abrupt boundary between two homogeneous, isotropic, and transparent media, such as that between air and water, the direction of the light ray changes permanently at the interface (unless the light ray approaches along a perpendicular to the surface). Such bending of light rays is called **refraction**.[†]

We must be careful to distinguish between the words

* **homogeneous**, meaning uniform composition and structure throughout the material; and
* **isotropic**, meaning identical or invariant in all directions.

The distinction is illustrated in Figure 23.15 for various patterns. Water, glass, and gases in equilibrium are examples of homogeneous and isotropic materials. Many crystals (such as quartz and calcite) are homogeneous but not isotropic (**anisotropic**).

Opus Majus, translated by Robert Belle Burke (University of Pennsylvania Press, Philadelphia, 1928), Volume I, page 131.
[†]The word comes from the Latin *refringere*, meaning "to break."

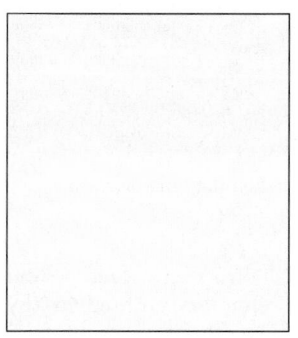
(a)

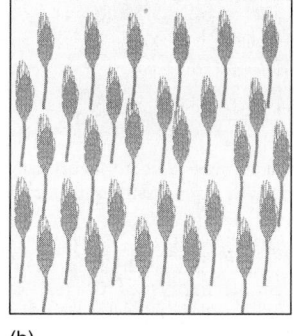
(b)

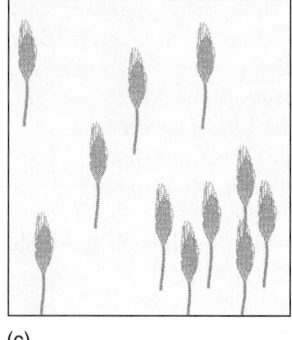
(c)

FIGURE 23.15 Patterns that are (a) homogeneous and isotropic, (b) homogeneous and anisotropic, and (c) inhomogeneous and anisotropic.

Refraction is the reason that a leaning, straight soda straw in a glass of water looks bent when viewed through the level surface (the second and lower image of the straw in water is formed by refraction at the curved surface of the container).

Refraction was noted in ancient Greece as well as in Islamic countries and Europe during the early Middle Ages. Experiments performed by Willebrord Snel* (1580–1626) led to the discovery of the quantitative relationship between the angle of incidence and the angle of refraction, occasionally called **Snel's law.**[†]

As with reflection, the angle of incidence θ_1 is taken in reference to a normal line to the interface, as in Figure 23.16. The ray in the second medium is the **refracted ray**. The **angle of refraction** θ_2 is the angle between the refracted ray and the normal line. We also can notice that some light *reflects* (usually weakly) from these abrupt boundaries between transparent materials, but here we are concerned with the transmitted (refracted) ray and its change of direction.

_____

*Many references spell his name with two l's: Snell; but Snel used only one letter l in the spelling of his last name. The author feels we should use Snel's name the way he spelled it.

[†]In France, Snel's law is called Huygens's law, after Christian Huygens (1629–1695). The priority of the discovery still is in dispute.

In Figure 23.16, clearly θ_2 is not equal to θ_1. To determine the relationship between the angles, we can measure and plot θ_2 for various incident angles θ_1 on a given transparent interface (see Figure 23.17). While θ_2 is proportional to θ_1 for small angles (but not equal to it), a graph of θ_2 versus θ_1 is clearly not simply linear, especially for larger angles of incidence, as shown in Figure 23.17 for an air-to-water interface.

Geometrically, if we draw a circle of any radius r centered on the point where the ray refracts, as in Figure 23.18, experiments indicate that, although θ_1/θ_2 is a changing ratio, the ratio of the ray components *parallel* to the interface, x_1/x_2, shown in Figure 23.18, stays constant at all angles. Hence a graph of $\sin \theta_2 = x_2/r$ versus $\sin \theta_1 = x_1/r$ *is* linear, as shown in Figure 23.19.

Hence the ratio of the sine of the angle of incidence to the sine of the angle of refraction is constant for the given pair of media. Change one of the media and the ratio still is a constant, but with a different numerical value.

If the incidence medium is a vacuum, the constant determined from the ratio of the sine of the angle of incidence θ_{vacuum}

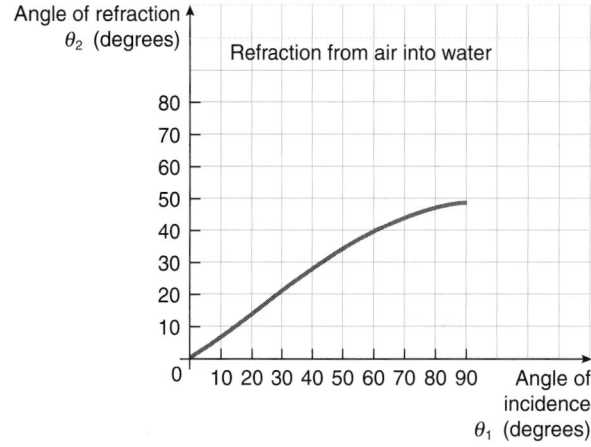

FIGURE 23.17 A graph of the angle of refraction versus the angle of incidence for air-to-water refraction.

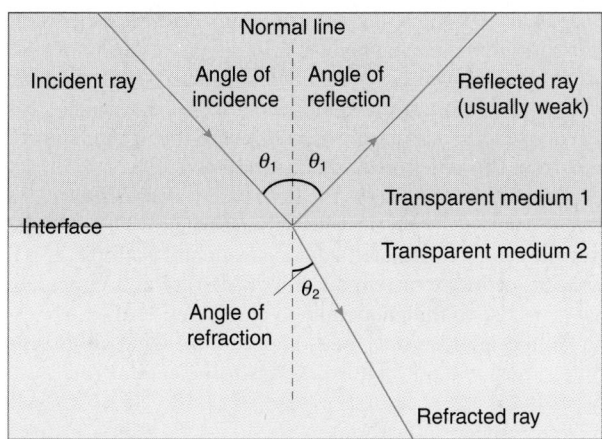

FIGURE 23.16 Refraction (with some reflection) at a transparent interface.

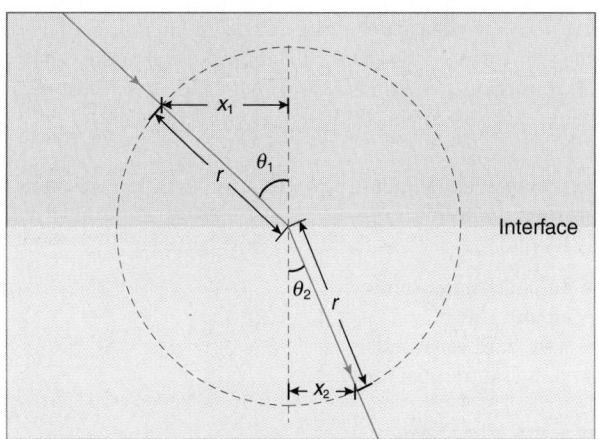

FIGURE 23.18 The ratio of the distances x_1/x_2 is constant for all angles of incidence.

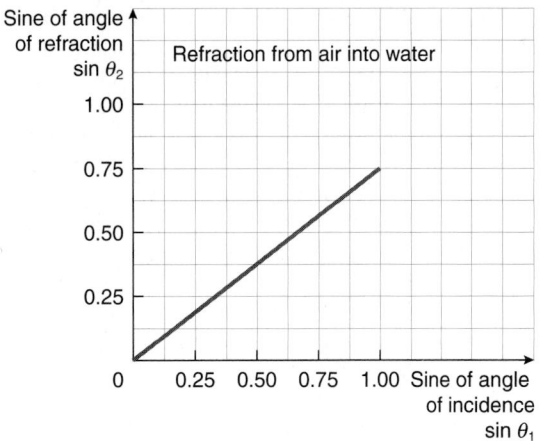

FIGURE 23.19 A graph of the sine of the angle of refraction versus the sine of the angle of incidence is linear.

TABLE 23.1 Approximate Values of the Index of Refraction*

Substance	Index of refraction
Solids	
Amber	1.55
Diamond	2.42
Glass (depends on the type)	~1.5 (about 3/2)
Plastic (depends on the type)	~1.3–1.4
Water (ice)	1.30
Liquids	
Benzene	1.50
Carbon tetrachloride	1.46
Ethanol	1.36
Methanol	1.33
Water (liquid)	1.33 (about 4/3)
Gases	
Air (atmospheric pressure)	1.000 29 (usually taken to be 1.000)
Many minerals have relatively high indices of refraction:	
Hutchinsonite	3.1
Proustite	3.1

*Precise values depend on the temperature and the wavelength of light used.

Source: *Handbook of Chemistry and Physics* (Chemical Rubber Publishing Company, Cleveland, Ohio)

to the sine of the angle of refraction θ_{refract} is a characteristic of the material and defined to be its **index of refraction**[†]:

$$n \equiv \frac{\sin \theta_{\text{vacuum}}}{\sin \theta_{\text{refract}}} \qquad (23.4)$$

The index of refraction of a vacuum thus is equal to exactly 1. Table 23.1 lists the index of refraction of some common transparent materials. When refraction occurs at the interface between two materials with indices of refraction n_1 and n_2, the ratio of the sine of the angle of incidence θ_1 to the sine of the angle of refraction θ_2 is equal to the ratio of n_2 to n_1:

$$\frac{\sin \theta_1}{\sin \theta_2} = \frac{n_2}{n_1} \qquad (23.5)$$

Rearranging this slightly, we obtain

$$\boxed{n_1 \sin \theta_1 = n_2 \sin \theta_2} \qquad (23.6)$$

Equation 23.6, coupled with the observation that the incident ray, the refracted ray, and the normal line all lie in the same plane, as shown in Figure 23.20, constitute the **law of refraction** for homogeneous, isotropic, and transparent materials.

Furthermore, experiments prove that the light retraces its path if its direction is reversed.

Boundaries such as those that exist between

- air and a water surface;
- air and glass;
- water and glass; and
- glass and transparent plastics

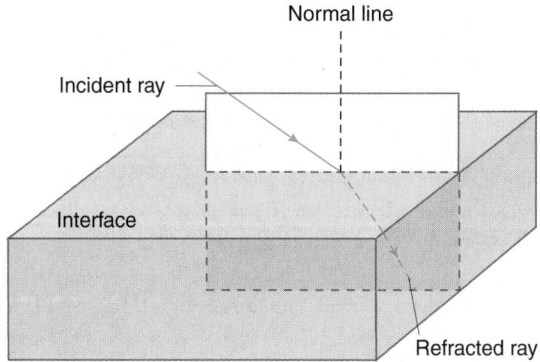

FIGURE 23.20 The incident ray, normal line, and refracted ray all lie in the same plane.

typify abrupt boundaries. The boundary between the atmosphere of a planet and outer space is not abrupt and the refraction of light entering or leaving such an atmosphere results in ray paths that change gradually (continuously), as indicated in Figure 23.21.[‡] Such atmospheres are inhomogeneous (the density decreases with height) and anisotropic (differences in the vertical direction differ from those in horizontal directions).

Nonetheless, as long as the light path is confined to a region small if compared with the atmospheric height, the atmosphere typically can be considered homogeneous and isotropic. It is the variation of index of refraction with density and temperature that gives rise to the phenomenon of mirages.

While most crystals are homogeneous, some are isotropic while others are not. Thus one has to be careful to consider

---

[†]In Chapter 24 we shall see that the index of refraction also is equal to the ratio of the speed of light in vacuum to its speed in the material.

[‡]If a ray is incident tangentially on the atmosphere of the Earth from the vacuum of space, the total change in the direction of the ray is only about 0.5°.

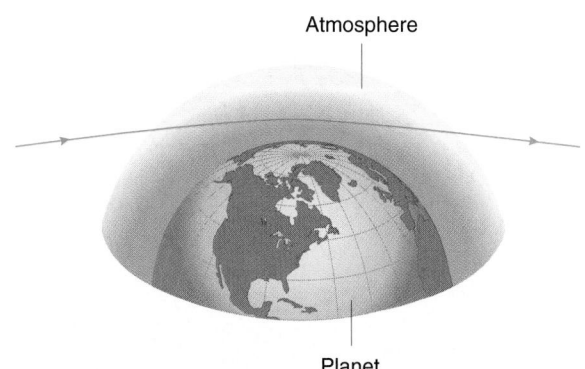

FIGURE 23.21 Refraction of light entering a planetary atmosphere is not abrupt but gradual because of the variation of the density of the atmosphere with height. The thickness of the atmosphere is exaggerated here for clarity.

whether to apply the same law of refraction to everything that is transparent to light. Certain crystals follow the law of refraction; in other crystals, the law needs to be supplemented to account for the anisotropic nature of the material. We consider the description of the refraction of light in these more complex crystals in the next chapter.

PROBLEM-SOLVING TACTIC

23.1 When using the law of refraction,

$$n_1 \sin \theta_1 = n_2 \sin \theta_2 \qquad (23.6)$$

you will find it convenient and less confusing to let the medium in which the incident ray lies always be medium 1 (the first medium) and let the medium in which the refracted ray lies be medium 2 (the second medium).

QUESTION 3

What is the meaning of the slope of a graph of $\sin \theta_1$ (as ordinate) versus $\sin \theta_2$ (as abscissa)?

EXAMPLE 23.4

A light ray in air is incident at an angle of 40.0° to the normal to a surface of water, as shown in Figure 23.22. Determine the *deviation angle* ϕ of the ray.

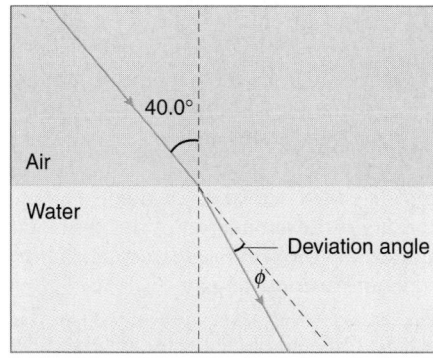

FIGURE 23.22

Solution

Apply the law of refraction to the ray, Equation 23.6:

$$n_1 \sin \theta_1 = n_2 \sin \theta_2$$

The indices of refraction of the materials are found in Table 23.1. The incident ray is in air, so that $n_1 = 1.00$ and $\theta_1 = 40.0°$. The refracted ray is in water, so that $n_2 = 1.33$. Making these substitutions into the law of refraction, you find

$$1.00 \sin 40.0° = 1.33 \sin \theta_2$$
$$\sin \theta_2 = 0.483$$
$$\theta_2 = 28.9°$$

The deviation angle is the difference between the angles of incidence and refraction:

$$\theta_1 - \theta_2 = 40.0° - 28.9°$$
$$= 11.1°$$

QUESTION 4

If, as pictured at the beginning of this section, a straw is inclined to the normal by the angle 40.0° in Example 23.4, is the kink in the straw the same as the 11.1° deviation angle?

23.5 TOTAL INTERNAL REFLECTION

If you have an aquarium, you may have noticed that when you view the tank from certain angles, with your eye well away from a normal to the surface, a fish is not visible through the top surface when it is swimming in certain regions of the tank. Here we see how this disappearing act comes about.

A material with a higher index of refraction than another is known as the more **optically dense** of the two, quite independent of the real relative density of the materials in kg/m³.* The medium with the lower index of refraction is called the less optically dense of the pair. An interesting effect occurs when light passes from a more optically dense medium into a less optically dense material, say from water into air, as in Figure 23.23.

We let n_1 be the index of refraction of the optically more dense medium (here, the water), since it contains the incident ray. We have

$$n_1 > n_2$$

The law of refraction at the interface is given by Equation 23.6:

$$n_1 \sin \theta_1 = n_2 \sin \theta_2$$

Since the light is traveling into the less optically dense medium, the refracted ray is bent farther away from the normal line; that is, since $n_1 > n_2$, we must have $\theta_2 > \theta_1$.

Now we increase the angle of incidence θ_1, as in Figure 23.24. The angle of refraction θ_2 also increases. If we keep increasing θ_1, eventually a situation is reached where the angle of refraction θ_2

---

*Nonetheless, there is a strong correlation between mass density in kg/m³ and optical density.

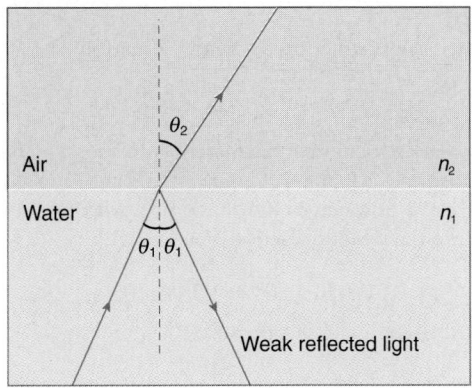

FIGURE 23.23 Light incident from water into air.

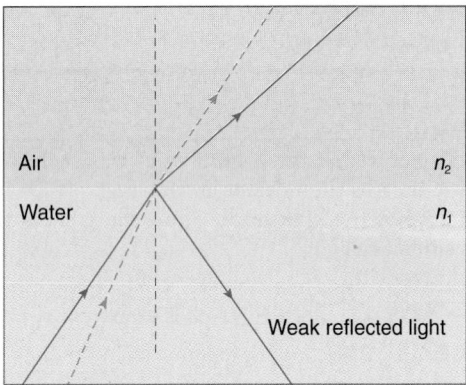

FIGURE 23.24 Increase the angle of incidence and the angle of refraction increases as well.

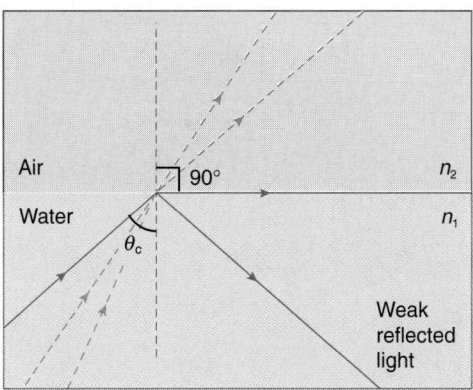

FIGURE 23.25 When the light is incident at the critical angle, an angle of refraction of 90° results.

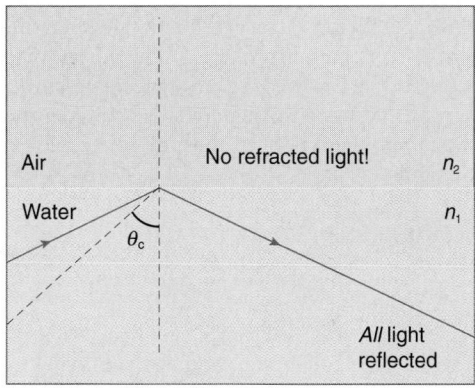

FIGURE 23.26 Total internal reflection.

becomes 90°, as shown in Figure 23.25. The angle of incidence that produces an angle of refraction of 90° is called the **critical angle** θ_c; the refracted ray is along the interface itself as shown in Figure 23.25.

When the angle of incidence is equal to the critical angle, the law of refraction, Equation 23.6, becomes

$$n_1 \sin \theta_c = n_2 \sin 90° \qquad (23.7)$$

Hence the critical angle is found from

$$\sin \theta_c = \frac{n_2}{n_1} \qquad (n_1 > n_2) \qquad (23.8)$$

What happens if the angle of incidence is greater than the critical angle in the more optically dense medium? We have "run out of" angle of refraction, since it cannot be greater than 90°. When the angle of incidence is greater than the critical angle, experimentally we find that the transmission ceases and the weak reflected light at the interface jumps to a strong value that persists at all greater angles of incidence. The light is totally internally reflected at the interface back into the more optically dense medium, as shown in Figure 23.26.

> For incidence angles greater than the critical angle, the transparent interface acts like a good *mirror*; reflection is all that occurs, and so the effect is called **total internal reflection**.

In fact, there is a small "evanescent" disturbance that trickles into the second medium, decreasing exponentially over just a few wavelengths, but there is no net energy transfer into the second medium.*

The phenomenon of total internal reflection is one that has many technological applications, only two of which are mentioned here:

1. Total internal reflection is the basic physical principle that underlies the transmission of light in fiber optic communication systems. Light pulses propagate along the fiber and follow the path of the fiber optic cable, because each time the light is incident on the walls of the fiber the angle of incidence is greater than the critical angle of the fiber. Thus the light is totally internally reflected back into the fiber. Successive total internal reflections along the fiber ensure that the light follows the fiber no matter what its slowly curving shape may be.
2. Light pipes are used in medical instruments for noninvasive (i.e., nonsurgical) examination of the upper and lower intestinal tract and the respiratory tract. Such light pipes also are used in arthroscopic surgery.

---

*This observation is confirmed by a detailed application of the Maxwell equations of electromagnetism at the interface. The evanescent disturbance is an excellent classical example of an analogous phenomenon called *tunneling* in quantum mechanics.

The small optical fiber cables on the left can replace the huge array of conventional copper cables on the right for voice and data transmission because of the high frequency associated with visible light. Total internal reflection keeps light confined to an optical fiber in optical communication systems.

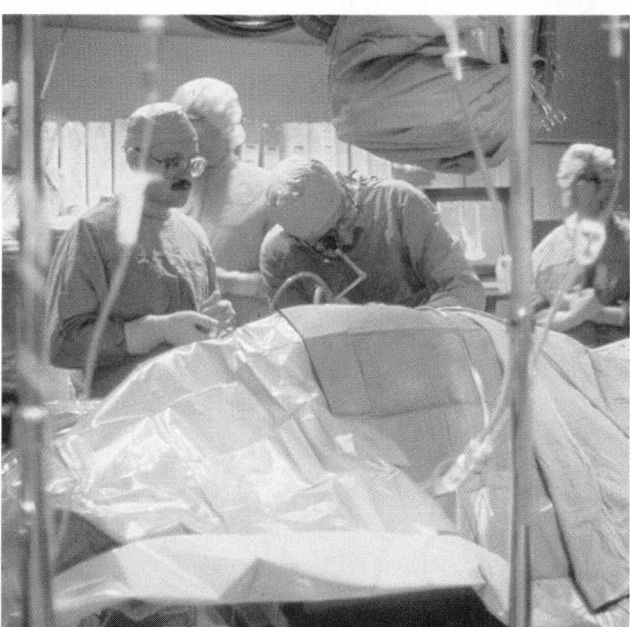

Light pipes that use total internal reflection enable surgeons to view the interior of the body with small incisions.

PROBLEM-SOLVING TACTIC

23.2 In using Equation 23.8 for the critical angle, the material with the greater index of refraction (in which the incident ray lies) is in the denominator. If you confuse the indices of refraction of the materials and place the larger index of refraction in the numerator, the resulting sine of the critical angle is greater than 1, a mathematically impermissible result. This should indicate to you that something is awry. Most calculators even will flag an error on their display if you try to find the angle whose sine is greater than 1.

EXAMPLE 23.5

Calculate the critical angle for a water–air interface.

Solution

From Table 23.1, the index of refraction of water is 1.33 while that for air is 1.00. When light is incident at the critical angle, the angle of refraction is 90°. Hence the law of refraction becomes

$$1.33 \sin \theta_c = 1.00 \sin 90°$$
$$\sin \theta_c = 0.752$$
$$\theta_c = 48.8°$$

23.6 DISPERSION

Rainbows are beautiful. They are caused by the passage of sunlight through water droplets. When sunlight also passes through the prismatic crystals of a chandelier, we may also observe small rainbow-like segments of color on the walls or table. Here we see that such beautiful effects arise because of a variation of the index of refraction of materials with wavelength.

The visible region of the electromagnetic spectrum consists of light with a rather small range of wavelengths: from about 400 nm to about 700 nm in a vacuum (or air). This small range of wavelengths is very important to us since it is the region to which our eyes are sensitive.* There is a correspondence between color and wavelength. White light is composite, consisting of all wavelengths of the visible spectrum (and typically other wavelengths in the near infrared and near ultraviolet regions as well).

If a parallel beam of white light is incident with a nonzero angle of incidence on a transparent interface, a spectrum of colors is produced as shown in Figure 23.27. We say the incident light mixture of wavelengths has been *dispersed* into its component colors or wavelengths.

If the white light is incident on a prism, two refractions take place in the passage through the prism: one at the first air–glass boundary, the second at the glass–air interface on another side of the prism, as indicated in Figure 23.28. We assume that the index of refraction of the air is the same as in a vacuum—that is, $n_1 = 1.00$ (Table 23.1 indicates that this is a good approximation). At the first boundary, all the wavelengths have the same angle of incidence θ_1. If we let n be the index of refraction of the prism, then the law of refraction, Equation 23.6, becomes

$$1.00 \sin \theta_1 = n \sin \theta_2 \qquad (23.9)$$

---

*Not coincidentally, this is the region where the Sun produces most of its output of electromagnetic waves. Our eyes evolved to take advantage of the region of the spectrum where the most light is produced.

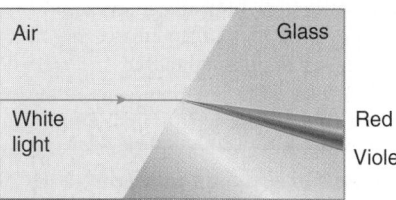

FIGURE 23.27 The refraction of white light produces a spectrum of colors.

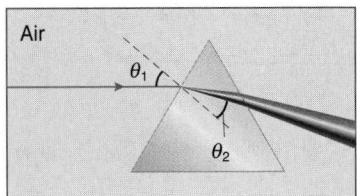

FIGURE 23.28 A prism can be used to produce a spectrum of white light.

Figure 23.28 indicates that the various colors (wavelengths) have *different* angles of refraction θ_2. Since the left-hand side of Equation 23.9 is the same for all the wavelengths, but θ_2 is different for each wavelength, the index of refraction n of the prism must be a function of the wavelength: $n(\lambda)$.

> The variation of the index of refraction with wavelength[†] is known as **dispersion**.

Figure 23.29 is a graph of the variation of n with wavelength for various types of glass and quartz; the variations in these and other materials typically are slight numerically, but significant physically. The wavelengths indicated along the abscissa of Figure 23.29 are the vacuum wavelengths λ of the light.[‡]

Dispersion is the underlying physical principle behind an important instrument known as a prism spectrometer. Each chemical element has a characteristic **emission spectrum** of

---

[†]It is more proper to say that dispersion is variation of the index of refraction with the frequency, because that does not change when light passes from one medium to another. But the wavelength (in air) is the favored variable for colors, probably because it is much easier to measure than the (very high) frequency of visible light.

[‡]In Chapter 24, we show that the wavelength λ_n of the light in a medium with an index of refraction n is given by

$$\lambda_n = \frac{\lambda}{n}$$

This equation is *not* what is meant by dispersion, since it indicates how a particular wavelength λ of light in a vacuum changes to λ_n in a medium with an index of refraction n. Dispersion means that the index of refraction itself varies with the particular wavelength, $n(\lambda)$, usually in a small but complicated way.

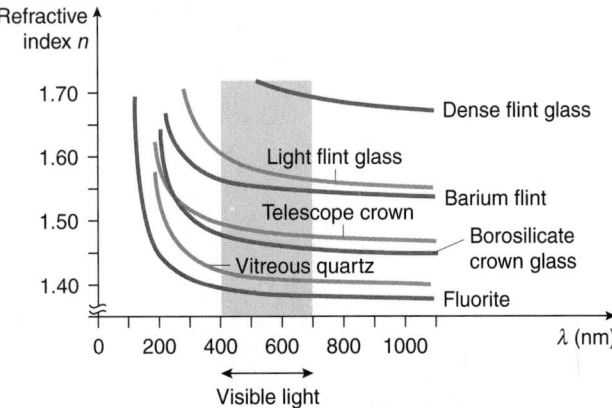

FIGURE 23.29 The variation of the index of refraction with wavelength in various kinds of glasses and quartz.

specific wavelengths corresponding to various transitions of the electrons in the atom; the spectrum is like a natural bar-coding for the element. If the light from a collection of various atoms then is incident on a prism such as in Figure 23.28, the prism separates the light into its constituent wavelengths. By measuring the angles through which the light is deviated by the prism, it is possible to determine the specific wavelengths emitted by the light source. From a table of the known wavelengths emitted by various atoms, the particular element responsible for the emission of the light then can be identified.

QUESTION 5

Since violet light is deviated through a greater angle than red light when both are incident at the same angle on a prism (see Figure 23.28), is the index of refraction of glass for red light greater than or less than the index of refraction for violet light?

EXAMPLE 23.6

A certain sample of flint glass has an index of refraction equal to 1.571 for red light (656 nm) and 1.594 for violet light (434 nm). If white light is incident from air at an incidence angle of 35.0°, what is the angular separation of the red and violet rays in the refracted beam?

Solution

Apply the law of refraction to each ray to determine the angle of refraction of each.

For the red ray:

$$n_1 \sin \theta_1 = n_2 \sin \theta_2$$
$$1.000 \sin 35.0° = 1.571 \sin \theta_2$$
$$\theta_2 = 21.4°$$

For the violet ray:

$$n_1 \sin \theta_1 = n_2 \sin \theta_2$$
$$1.000 \sin 35.0° = 1.594 \sin \theta_2$$
$$\theta_2 = 21.1°$$

The angular separation of the red and violet rays is the difference between the two angles of refraction, only 0.3°.

23.7 RAINBOWS*

> *Rain, rain, and sun! a rainbow in the sky!*
> Alfred, Lord Tennyson (1809–1892)[§]

Among the most beautiful of natural atmospheric effects are rainbows. As we alluded to at the beginning of the previous section, they are caused by dispersion in appropriately placed water drops. Sunlight is essentially white light: a mixture of all colors (wavelengths) in the visible spectrum. The particular white light ray involved in the formation of a rainbow enters a droplet and is refracted and dispersed into its component colors, as indicated at point (1) in Figure 23.30; some light also is reflected at the air–water interface, but this light is not involved in the formation of rainbows.

[§]*Idylls of the King*, The Coming of Arthur, line 402. *The Poetic and Dramatic Works of Alfred Lord Tennyson*, Cambridge edition (Houghton Mifflin, Boston, 1898), page 309.

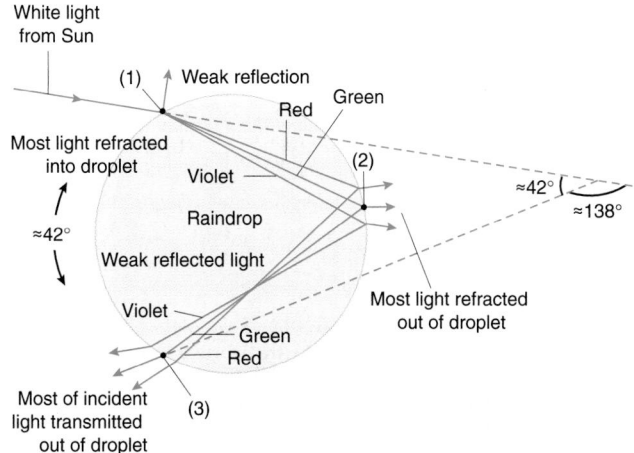

FIGURE 23.30 Path of a ray through a raindrop involved in the primary rainbow.

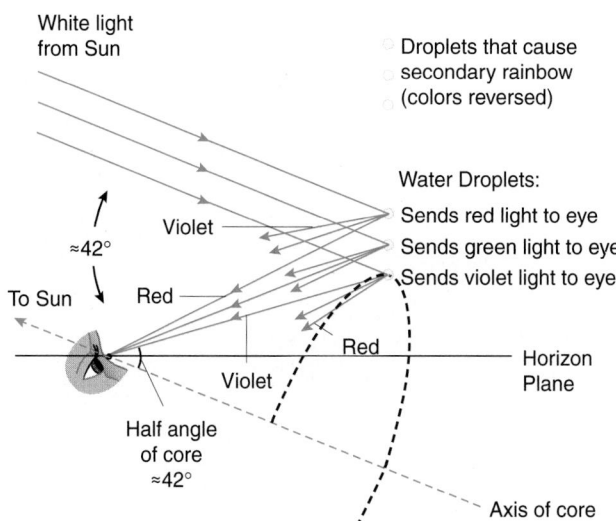

FIGURE 23.31 The formation of the rainbow from a collection of water droplets.

When the dispersed beam of light rays encounters the rear surface of the raindrop at point (2) in Figure 23.30, most of the light is refracted out of the droplet; but some light is reflected from the water–air surface. It is this small amount of light reflected from the back of the droplet that produces the rainbow; rainbows are dim because most of the light is transmitted at the back of the raindrop, not reflected.*

The reflected light is refracted and dispersed additionally at its third encounter with the water–air interface, at point (3) in Figure 23.30. The deviation angle between the incident and emergent beams is about 138°, so the acute angle is about 42° (see Problem 32).

An observer situated to receive red light from the drop also will receive red light from raindrops if they are located on a cone whose vertex is the eye of the observer, whose axis is the extension of the line from the Sun to the eye, and whose half angle is about 42° (see Figure 23.31). A circular arc of light results, centered on the antisolar point opposite the Sun. Droplets located lower than those sending the red light to the eye send the other colors of the **primary rainbow** to the eye. The red arc (at an angular radius of about 42.5°) appears outside the arcs of the other colors, with violet on the inside edge (at an angular radius of about 40°). There is some overlapping of the colors because of the finite angular size of the Sun (about 0.5°).

The rainbow continues to exist as long as new drops replenish those falling through the imaginary cones on the observer's line of sight. An observer on a level plain (whether in Spain or Kansas) sees only a segment of the circular arc (and only along portions of the cone that contain raindrops). Partial rainbows are seen if raindrops exist along only portions of the cone. Observers viewing waterfalls from high cliffs can see more

---

*Some references state (erroneously) that the light is totally internally reflected from the back of the water droplet. This is not the case. Note that each individual dispersed ray inside the water droplet makes equal angles with the two normal lines to the surfaces at (1) and (2) in Figure 23.30. If we reverse the direction of the ray, it is apparent that the angle of incidence is less than the critical angle, so that no total internal reflection occurs at the back surface. The author thanks Professor Richard Ditteon (at Rose-Hulman Institute of Technology) for bringing this common misconception to his attention.

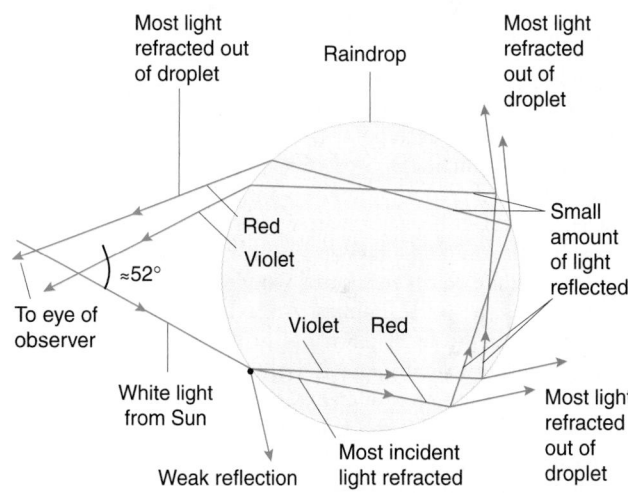

FIGURE 23.32 Light ray paths that form the secondary rainbow.

of a circular arc from a rainbow formed by water droplets from the spray of the waterfall.

A **secondary rainbow** also can be seen occasionally. The secondary rainbow is caused by light that undergoes *two* reflections before emerging from the droplet. In this case the appropriate white light ray enters the droplet on the bottom and leaves on the top, as shown in Figure 23.32. Since only a small amount of light is reflected with each encounter with the water–air interface, the secondary rainbow is *much* dimmer than the primary rainbow and, therefore, even more difficult to see.

The angle between the incident and emergent beams is about 52°. Because of the extra reflection, the colors of the secondary rainbow are the reverse of those in the primary rainbow: red is now on the inside of the bow (at an angular radius of about 50°) and violet on the outside (at about 54°). The width of the secondary rainbow is about twice that of the primary rainbow, another reason secondary rainbows appear fainter and are more difficult to see than primary rainbows.

The primary (brighter) and secondary (dimmer) rainbows. Note the reversal of the color sequence in the two rainbows.

23.8 OBJECTS AND IMAGES

When you look into a mirror, you see an image of yourself. Camera lenses also form images of scenes on film. The lens of a video camera forms an image on an electronic detector. We want to investigate the image-forming properties of optical devices such as mirrors and lenses. To do this, we need to know what is meant by the terms object and image.

When a given scene or **object** is placed in front of an optical device, we think of each point on the object as an **object point**. Each object point is a source of many diverging light rays in many different directions from the point, as shown in the left side of Figure 23.33. It does not matter whether the object point is diffusely reflecting these rays or generating them.

A pointlike object consists of a single object point source of light rays, as the name implies. An **extended object** consists of many different object points, as in the right side of Figure 23.33. If we know what happens to the rays from a single object point, the generalization to an extended object is not difficult.

Some of the light rays from an object point may enter a somewhat wide optical device, as in Figure 23.34. We are not interested in those rays from the object point that miss or do not enter the optical device. The entering rays are processed by the

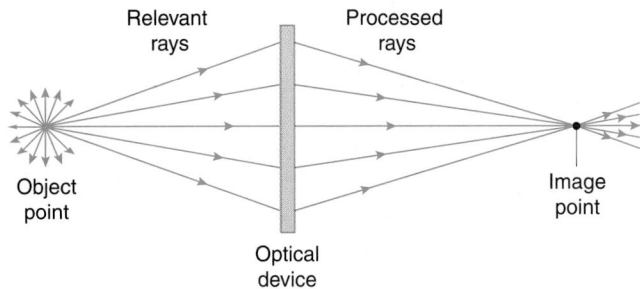

FIGURE 23.34 The processed rays leave the optical device and reveal the location of the image; here the image is real.

device in some fashion, which may involve reflections, refractions, or some combinations of these processes. The processed rays then leave the optical device (see Figure 23.34) and reveal where the image is located.

> If the exiting light rays intersect, or even appear to intersect, at some point, that point is the **image point** of the object point. If these rays physically converge at the image point, the image is a **real image** (the rays *really* intersect there). If the exiting rays only *appear* to intersect at the image point, the image point is a **virtual image.****

For various optical devices such as mirrors, refracting surfaces, and lenses, our goals are the following:

1. to find a relationship between the location of an object and the location of its corresponding real or virtual image;
2. for extended objects, to determine if the resulting extended image is larger, smaller, or the same size as the object—that is, to determine the magnification; and
3. to determine if an extended image is upright or inverted.

To accomplish these goals, it is convenient to introduce a sign convention associated with distances in the problem.

23.9 THE CARTESIAN SIGN CONVENTION

A sign convention is used to facilitate computations of the positions and sizes of images and the assessment of the nature of the image (real or virtual) and its magnification and orientation. Distances and sizes take on positive or negative values depending on certain assumed rules or conventions.

The sign convention we adopt is the **Cartesian sign convention**, known for its simplicity and named for its association with the standard Cartesian coordinate system.† A standard geometry is used.

---

*The word virtual means *not in actual fact*. Unfortunately, the current term virtual reality makes mincemeat out of this distinction.

†This sign convention was recommended by the International Commission on Optics many years ago; see "International Commission of Optics," edited by S. S. Ballard, *Journal of the Optical Society of America, 41,* #2, pages 140–141 (February 1941). Despite its simplicity, the Cartesian sign convention has yet to be universally adopted in the literature. You need to be aware of this confusion in the literature. We will do our part to further the cause of the Cartesian sign convention.

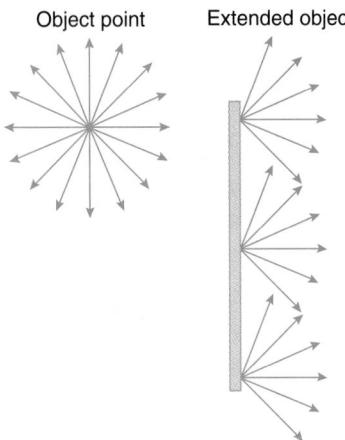

FIGURE 23.33 An object point and an extended object consisting of many object points.

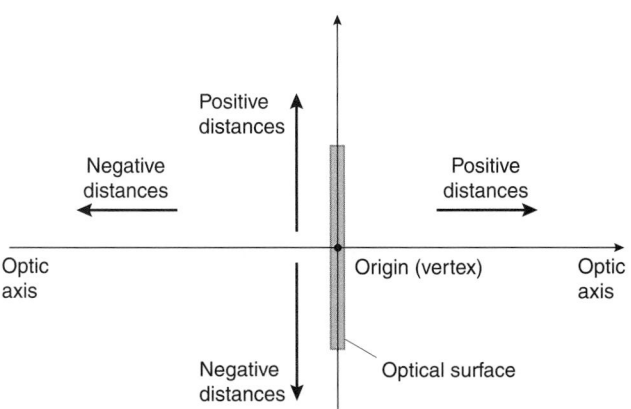

FIGURE 23.35 The Cartesian sign convention.

The object is placed to the left of the optical surface (a mirror or refracting surface). The light incident on the optical device from the object initially is traveling from left to right.

The center of the optical surface, called its **vertex**, is taken to be the origin of a Cartesian coordinate system. The horizontal Cartesian coordinate axis is called the **optic axis**. Distances measured to the right of the origin along the optic axis (i.e., measured to the right of the optical surface) are *positive* distances, since they are along the positive axis of the standard Cartesian coordinate system; see Figure 23.35. Distances measured to the left of the origin along the optic axis (i.e., to the left of the optical surface) are *negative* distances, since they are along the negative Cartesian coordinate axis. Likewise, distances measured upward from the horizontal Cartesian axis also are positive; distances measured downward from the horizontal axis are negative.

That is all there is to the sign convention for distances. Just imagine the vertex of the optical surface to be the origin of a standard Cartesian coordinate system and everything else follows quite readily from there. The same Cartesian sign convention for distances and magnification will be used throughout our discussion of geometric optics.

There also is a sign convention for the magnification.

The **magnification** m is defined as the ratio of the size of the image to the size of the object. If the magnification m is positive, the image of the object is **upright** (right-side-up), meaning that the image has the same vertical orientation as the object. If the magnification is negative, the image is **inverted** (up-side-down) with respect to the vertical orientation of the object.

Assessing whether left and right are preserved or reversed depends on how the image is viewed by the observer (see Investigative Project 9).

23.10 IMAGE FORMATION BY SPHERICAL AND PLANE MIRRORS

Mirror, mirror, on the wall, who is fairest of them all?
From the well-known fairy tale, *Snow White*

When you look	*kool uoy nehW*
into a mirror	*rorrim a otni*
it is not	*ton si ti*
yourself you see,	*,ees uoy flesruoy*
but a kind	*dnik a tub*
of apish error	*rorre hsipa fo*
posed in fearful	*lufraef ni desop*
symmetry.	*.yrtemmys*
	—*ekidpU nhoJ*

John Updike (1932–)*

Here we see how simple polished or coated surfaces, such as a bathroom mirror, a curved makeup mirror, or the curved rearview mirror on the passenger side of your car, use the law of reflection to produce an image and how that image differs from the object.

A **spherical mirror** is a reflecting surface whose shape is part of (a section of) a spherical surface. The radius of the spherical surface is the **radius of curvature** R of the mirror, as shown in Figure 23.36. The **center of curvature** C of the mirror is the center of the sphere of which the mirror surface is part.

Rather than treat a **plane mirror** as something different, we can and do consider it as a special case of a spherical mirror: one with an infinite radius of curvature.

There are two types of spherical mirrors, concave and convex mirrors, shown in Figure 23.37. According to the Cartesian sign convention, with its origin at the mirror, a concave mirror has a negative radius of curvature and a convex mirror has a positive radius of curvature.

Consider an object point in front of a convex spherical mirror, as in Figure 23.38. A line from the object point through the center of curvature C of the mirror defines the optic axis, as indicated in Figure 23.38. The point O on the mirror surface is the vertex of the mirror, and is the origin of the Cartesian coordinate system used to specify the signs of the appropriate distances. The distance of the object from the mirror is the **object distance** s. According to the

---

*The New Yorker, 30 November 1957, page 200.

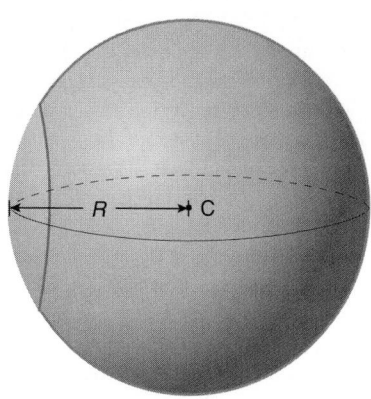

FIGURE 23.36 A spherical mirror is a section of a spherical surface.

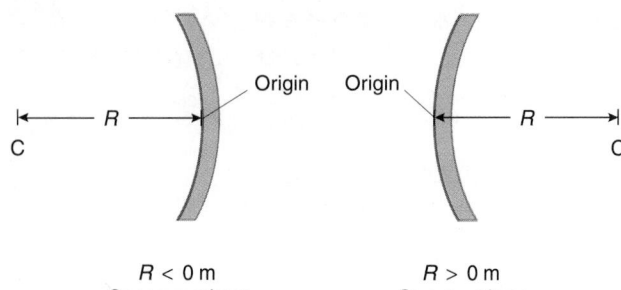

$R < 0$ m
Concave mirror

$R > 0$ m
Convex mirror

FIGURE 23.37 A concave mirror and a convex mirror.

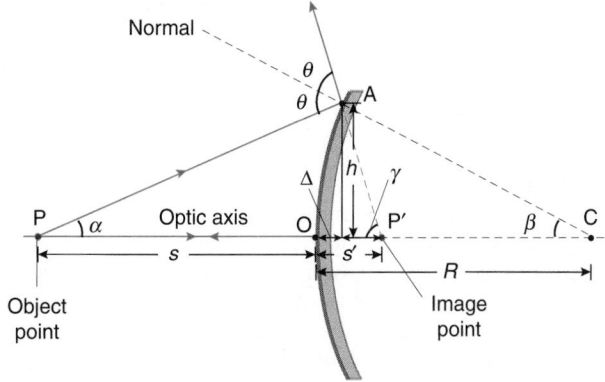

FIGURE 23.38 The formation of the image of point P at P' by a convex mirror.

Cartesian sign convention, the object distance s here is *negative*. That is, here s itself has a negative value. The radius of curvature R of the convex mirror is positive, since the center of curvature of this mirror and the length R are to the right of the origin.

To locate the image point, we examine a few rays diverging from the object point to see what happens to them after they are processed (here, reflected) by the mirror surface. We look at two rays in particular:

1. A ray from the object point at P directed along the optic axis; see Figure 23.38. This incident ray hits the mirror at point O, the mirror vertex. The normal to the mirror surface at this point is the optic axis itself. The angle of incidence is zero and so, applying the law of reflection, the angle of reflection also is zero. The ray merely retraces its path, as shown in Figure 23.38.

2. Another ray from the object point P strikes the mirror at point A in Figure 23.38. The normal to the mirror surface at point A is a line from the center of curvature to point A. The angle of incidence is θ and, applying the law of reflection to this ray, the angle of reflection also is θ as shown. The location of the image point is found by examining the two rays after they have been processed (reflected) by the mirror. The two extended reflected rays appear to have an intersection at point P'; this, therefore, is the image point. The image here is virtual because the two rays do not physically meet at P'; the rays only *appear* to have come from P'. Let s' be the distance of the image from the origin, called the **image distance**. Here, since the image is located to the right of the origin, s' is a positive distance according to the Cartesian sign convention.

Derivation of the Mirror Equation

We want to find a relationship between the object distance s, the image distance s', and the radius of curvature R of the mirror. We

define the angles α, β, and γ as in Figure 23.38. The angle θ is an exterior angle of triangle PAC. The exterior angle of a triangle is equal to the sum of the two remote interior angles.* Thus

$$\theta = \alpha + \beta \qquad (23.10)$$

Likewise, the angle 2θ is an exterior angle of triangle PAP'. Applying the same geometric theorem, we find

$$2\theta = \alpha + \gamma \qquad (23.11)$$

Substituting for θ from Equation 23.10, we find

$$2(\alpha + \beta) = \alpha + \gamma \qquad (23.12)$$

Rearranging this slightly, we get

$$\gamma - \alpha = 2\beta \qquad (23.13)$$

Now we make the assumption that the incident rays from the object point that are processed by the mirror make small angles with the optic axis, and so never are located far from the optic axis. This is called the **paraxial ray approximation** (meaning the rays are almost *parallel* to the optic *axis*). This assumption is a reasonably good first approximation in most cases.

If the angles α, β, and γ are small, we can use the **small angle approximation** to relate them to the distances. For small angles, the angle (in radians) is approximately equal to the tangent of the angle (it is also approximately equal to the sine of the angle). For angles up to about 10° (= 0.17 rad), the difference between the angle (in radians!), the tangent of the angle, and the sine of the angle is less than 1%. Notice in Figure 23.39 that graphs of θ in radians, $\tan \theta$, and $\sin \theta$ all are close together for small θ.

With the small angle approximation and the geometry of Figure 23.38, we have

$$\beta \approx \tan \beta = \frac{h}{R - \Delta} \qquad (23.14)$$

$$\gamma \approx \tan \gamma = \frac{h}{s' - \Delta} \qquad (23.15)$$

$$\alpha \approx \tan \alpha = \frac{h}{-s + \Delta} \qquad (23.16)$$

In Equation 23.16 for the tangent of α, the quantity $-s$ is used in the denominator because s itself is negative according to the Cartesian sign convention; thus, $-s$ is positive.

If the mirror is thin† and the angles are small, the distance Δ in the denominator is negligible compared with

---

*This theorem from plane geometry follows because the sum of the angles of a plane triangle must be 180°. Hence, in the triangle pictured, we must have

$$(180° - \theta) + \alpha + \beta = 180°$$

and so

$$\theta = \alpha + \beta$$

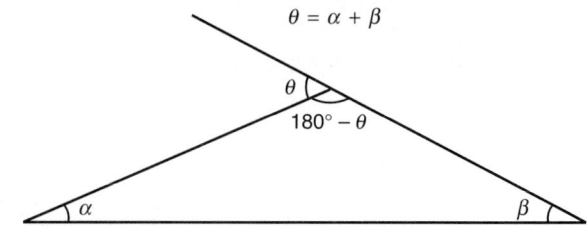

†A thin mirror is only a small section of the sphere from which its surface was cut.

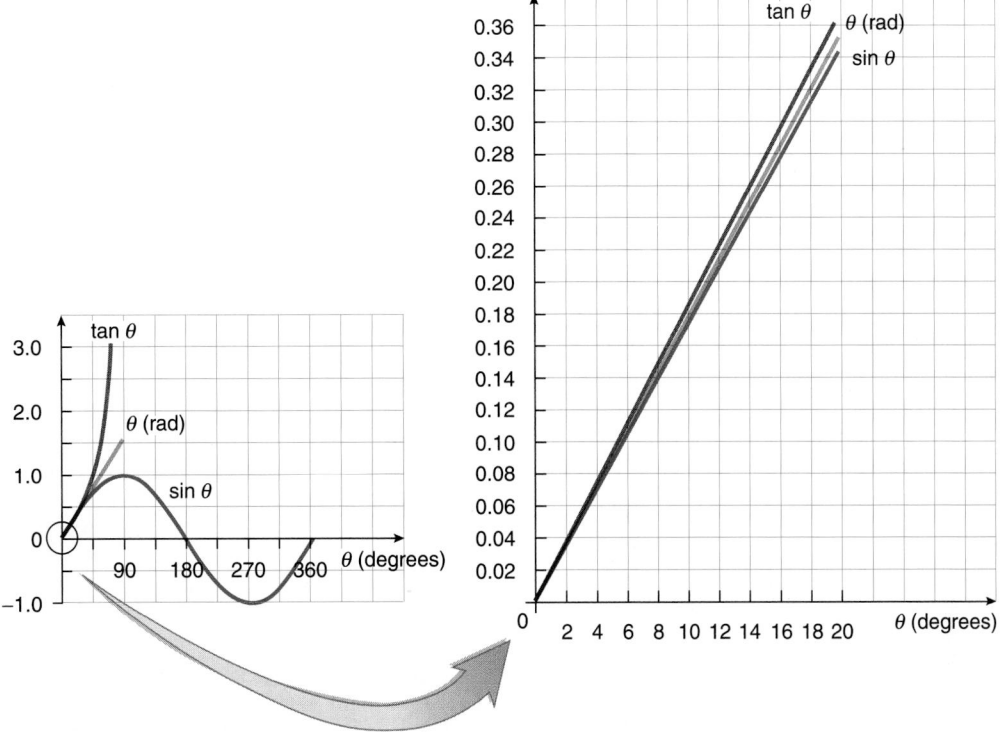

FIGURE 23.39 The graphs of $\sin\theta$, $\tan\theta$, and θ itself (in radians) are almost identical for small angles.

other distances such as s, s', and R. Hence Equations 23.14–23.16 become

$$\beta \approx \frac{h}{R} \qquad (23.17)$$

$$\gamma \approx \frac{h}{s'} \qquad (23.18)$$

$$\alpha \approx \frac{h}{-s} \qquad (23.19)$$

We use the expressions in Equations 23.17–23.19 for the angles in Equation 23.13. The result is

$$\frac{h}{s'} - \frac{h}{-s} = 2\frac{h}{R} \qquad (23.20)$$

We divide by h and perform some minor algebraic housekeeping. The result is the **mirror equation** for spherical mirrors:

$$\frac{1}{s} + \frac{1}{s'} = \frac{2}{R} \qquad (23.21)$$

The Focal Point and Focal Length

One convenient refinement is made to the mirror equation.

> If the object point P is located infinitely far away from the mirror, then $s = -\infty$ m and the position of the image is called the **focal point** F of the mirror. The distance of the focal point from the mirror is the **focal length** f of the mirror; the focal length is given a sign in accordance with the Cartesian sign convention.

If we substitute $s = -\infty$ m into the mirror equation (Equation 23.21), we find

$$\frac{1}{-\infty \text{ m}} + \frac{1}{s'} = \frac{2}{R}$$

Since the image distance for an object infinitely far away is defined to be the focal length f, we put $s' = f$ and obtain

$$0 \text{ m}^{-1} + \frac{1}{f} = \frac{2}{R}$$

$$f = \frac{R}{2} \qquad (23.22)$$

> The focal length of a spherical mirror is half its radius of curvature.

The focal point of the mirror is located halfway between the center of curvature and the mirror itself. Light rays from an object point at increasing distances from the mirror become increasingly close to parallel as they reach the mirror, as shown in Figure 23.40. Incident parallel rays after reflection intersect (or appear to intersect) at the focal point (the image point for an object point infinitely far away), as in Figure 23.41.

> An important point: the image of an object located a *finite* object distance from the mirror is *not* at the focal point! An image is at the focal point *only* when the object is infinitely far away from the mirror.

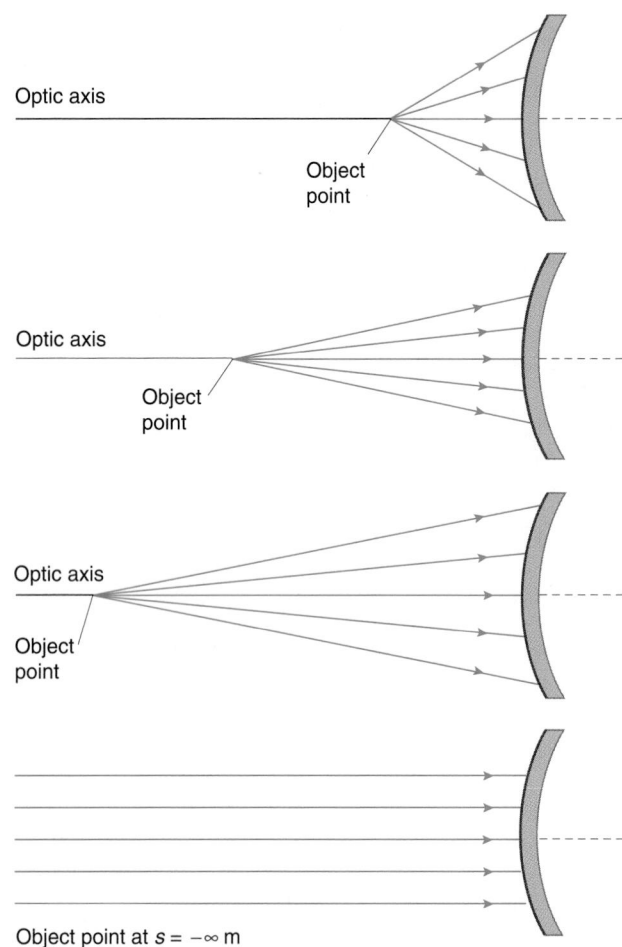

Optic axis

Object point

Optic axis

Object point

Optic axis

Object point

Object point at $s = -\infty$ m

FIGURE 23.40 Light rays from an increasingly distant object point approach the optical surface increasingly close to parallel.

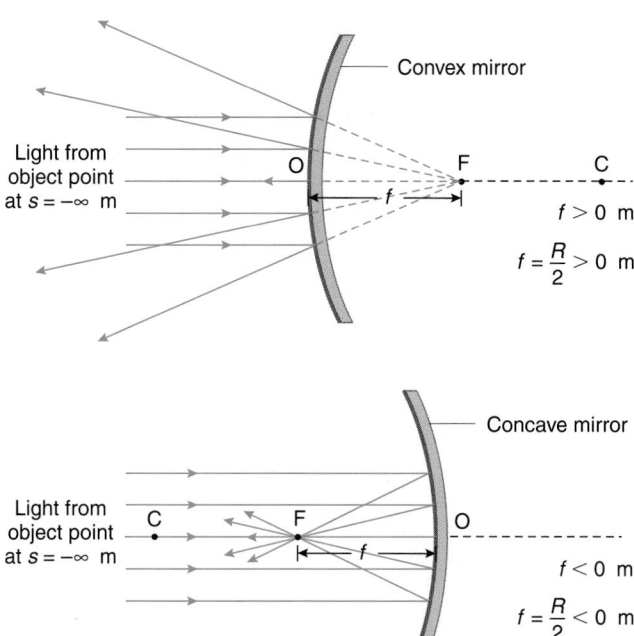

Convex mirror

Light from object point at $s = -\infty$ m

$f > 0$ m

$f = \dfrac{R}{2} > 0$ m

Concave mirror

Light from object point at $s = -\infty$ m

$f < 0$ m

$f = \dfrac{R}{2} < 0$ m

FIGURE 23.41 Parallel incident rays intersect (or appear to intersect) at the focal point after reflection from the spherical mirror.

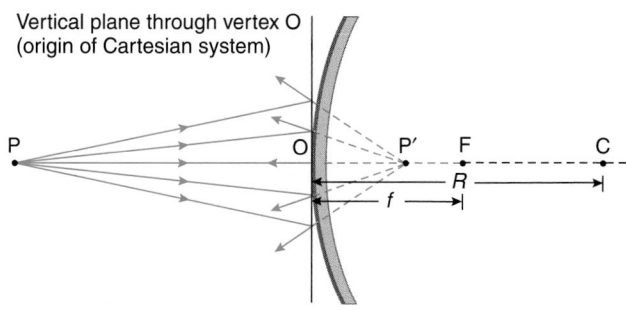

Vertical plane through vertex O (origin of Cartesian system)

FIGURE 23.42 All paraxial rays from an object point intersect (or appear to intersect) after reflection at its image point. Here they appear to intersect at P′.

All paraxial rays from an object point that are processed (reflected) by the mirror either intersect or appear to intersect at the corresponding image point.

Because we consider only paraxial rays, we henceforth draw the reflections of the rays as if they occur at the vertical plane through the vertex of the mirror (see Figure 23.42) rather than at the indicated mirror surface itself. The curvature of the mirror is exaggerated in the figures for clarity.

Using Equation 23.22, we rewrite the mirror equation (Equation 23.21) in its most versatile form:

$$\frac{1}{s} + \frac{1}{s'} = \frac{2}{R} = \frac{1}{f} \quad \text{(spherical mirror equation)} \quad (23.23)$$

Although we derived the mirror equation (Equation 23.23) using a convex mirror, a similar derivation for a concave mirror leads to the same result. Hence Equation 23.23 can be applied to *any* spherical mirror, be it convex or concave. We just must be careful to apply the appropriate signs to the distances according to the Cartesian sign convention.

You can use units of convenience in Equation 23.23 as long as you use the *same* units for all the distances.

If the image is located behind the mirror, then the image must be virtual, since light never really gets in back of the mirror. On the other hand, if the image is located in front of the mirror, the image is real, since the reflected rays really will pass through the image in that case. Thus for any mirror, if the image distance $s' > 0$ m, the image is virtual; if $s' < 0$ m, the image is real. These relationships are not worth memorizing, since they can be determined by thinking about where the convergent light that forms the image really is located.

PROBLEM-SOLVING TACTIC

23.3 When using the general mirror equation (Equation 23.23), recognize that it involves the reciprocals of the various distances. Do not forget to take a final reciprocal with your calculator to find s, s', f, or R.

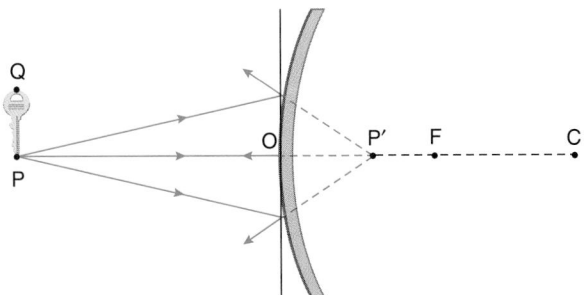

FIGURE 23.43 An object point P on the optic axis has its image point P′ on the optic axis.

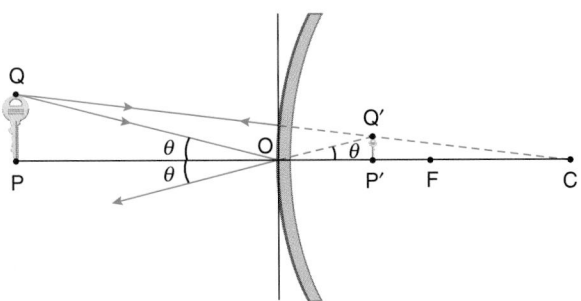

FIGURE 23.44 An object point Q off the optic axis has its image point Q′ off the optic axis.

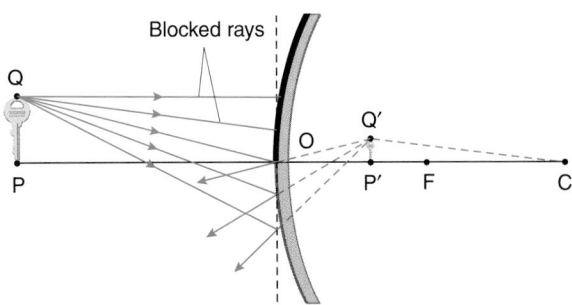

FIGURE 23.45 If we block a portion of the mirror, the entire image still is formed, but is dimmer.

An Extended Object

If an *extended object* is placed in front of the mirror, the location of the image is found by applying the law of reflection to the rays from each point on the object. In particular, the object point P on the optic axis has an image point P′ on the optic axis located by using the general mirror equation (Equation 23.23), as in Figure 23.43.

The point Q at the top of the object (see Figure 23.44) has an image point Q′ that can be located geometrically in the following way. A ray from the top of the object directed toward the center of curvature merely reflects back on itself since it strikes the mirror along a line normal to the mirror at this point. Another ray from the object point Q to the origin O makes an angle θ with the optic axis. By the law of reflection at this point, the reflected ray also makes an angle θ with the optic axis, as indicated in Figure 23.44. The image point Q′ is at the point of intersection from which the reflected rays appear to have come from. The extended image of the extended object is located between the points P′ and Q′.

The image in Figure 23.44 is virtual, since the light never is really in or behind the space in back of the mirror. Since the points on the extended object all are essentially the same distance s from the mirror (if the mirror is thin), the locations of all the image points also are essentially the same distance s' from the mirror as is the image point on the optic axis.

If we block or cover a portion of the mirror, as in Figure 23.45, the image is *dimmer*, since fewer rays from each object point now are processed by the mirror; but *the entire extent of the image still is formed*. Every part of the mirror is involved in forming every part of the image.

Thus, as long as some light from each object point reaches the mirror, each image point is formed; the entire extent of the image is formed. Covering part of the mirror simply reduces the amount of light processed by the mirror; some light from each object point still is processed, and so the entire extent of the image nevertheless is formed.

If half of the *object* is blocked, the image is halved as well. In this instance, no light from the blocked part of the object reaches the mirror, and so the blocked part of the object is irrelevant and does not appear as part of the image.

Magnification

To determine the magnification, consider Figure 23.44. The magnification indicates the size of the image relative to the size of the object. Notice that triangles PQO and P′Q′O are similar triangles. This means, of course, that their corresponding sides are proportional. The distances PQ and P′Q′ in Figure 23.44 both are positive according to the Cartesian sign convention. Since s is negative, $-s$ is positive.

Hence the magnification is

$$m = \frac{\text{image size}}{\text{object size}} = -\frac{s'}{s} \quad \text{(mirror magnification)} \quad (23.24)$$

So for Figure 23.44, Equation 23.24 for the magnification yields a positive number, indicating that the image is upright according to the sign convention for the magnification.

Equation 23.24 for the magnification can be applied to *any* spherical mirror. Moreover, if the appropriate signs are used for the object and image distances s and s', the sign of m automatically will indicate whether the image is upright (if $m > 0$) or inverted ($m < 0$). Wonderful!

The law of reflection is independent of the medium in which the light is traveling. Thus the mirror and magnification equations (Equations 23.23 and 23.24) can be applied whatever the transparent medium surrounding the mirror happens to be. The mirror and magnification equations are the same regardless of whether the mirror is immersed in air, water, gin, or cranberry juice.

QUESTION 6

Why is a plane mirror considered to have an infinite radius of curvature rather than a radius of curvature of zero?

EXAMPLE 23.7

You stand some distance | *s* | in front of a plane mirror. Determine the location of your image, its magnification, and whether it is real or virtual.

Solution

A plane mirror is a spherical mirror with an infinite radius of curvature. Thus the general mirror equation (Equation 23.23) becomes

$$\frac{1}{s} + \frac{1}{s'} = \frac{2}{\infty \text{ m}}$$

$$= 0 \text{ m}^{-1}$$

So

$$s' = -s \tag{1}$$

The object distance *s* is negative according to the Cartesian sign convention (see Figure 23.46). Since the object distance *s*

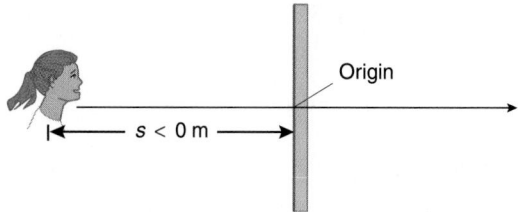

FIGURE 23.46

is negative, equation (1) indicates that the image distance *s'* is positive; the image therefore is located to the right of the mirror and is as far from the mirror as the object. The image location is behind the mirror; no light from the object really reaches this region. Therefore the image is virtual. But you can surely see it!

The magnification of a plane mirror is found using the general magnification equation (Equation 23.24):

$$m = -\frac{s'}{s}$$

Since you found *s'* = −*s*, the magnification is

$$m = -\frac{(-s)}{s}$$

$$= +1$$

The image is upright (since *m* > 0) and the same size as the object. It certainly would make life more difficult if a plane mirror gave an inverted image!

STRATEGIC EXAMPLE 23.8

A lipstick container 6.0 cm in height is placed 100 cm from a concave spherical cosmetic mirror cut from a sphere of radius 50 cm. Determine

a. **the location of the image;**
b. **the nature of the image (real or virtual);**
c. **the magnification and size of the image; and**
d. **the focal length of the mirror.**

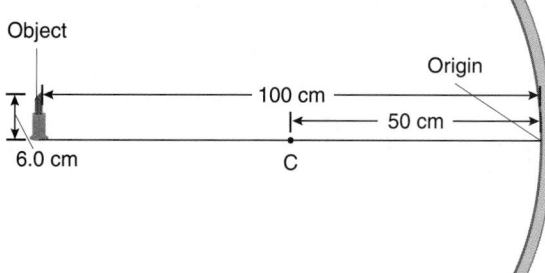

FIGURE 23.47

Solution

a. The situation is depicted with the standard geometry, shown in Figure 23.47. The origin for the Cartesian sign convention is at the vertex of the mirror. Since the object is located to the left of the origin, the object distance *s* is negative:

$$s = -100 \text{ cm}$$

The mirror is concave, so that the center of curvature of the mirror surface also is to the left of the origin. Thus the radius of curvature of the mirror *R* is negative:

$$R = -50 \text{ cm}$$

Use the general mirror equation, Equation 23.23:

$$\frac{1}{s} + \frac{1}{s'} = \frac{2}{R}$$

$$\frac{1}{-100 \text{ cm}} + \frac{1}{s'} = \frac{2}{-50 \text{ cm}}$$

Solving for *s'*, you find

$$\frac{1}{s'} = -\frac{2}{50 \text{ cm}} + \frac{1}{100 \text{ cm}}$$

$$= -\frac{3}{100 \text{ cm}}$$

Problem-Solving Tactic 23.3 cautions you to remember to take the final reciprocal:

$$s' = -\frac{100 \text{ cm}}{3}$$

$$= -33 \text{ cm}$$

b. Since the image distance is negative, the image is located to the *left* of the origin (to the left of the mirror). Since the reflected light actually traverses this region, the image is *real*.

c. The magnification is found from the general mirror magnification equation (Equation 23.24):

$$m = -\frac{s'}{s}$$

$$= -\frac{-33 \text{ cm}}{-100 \text{ cm}}$$

$$= -0.33$$

Since the magnification is negative, the image is inverted. The image size is about 1/3 that of the object, or

$$-0.33(6.0 \text{ cm}) = -2.0 \text{ cm}$$

in height, extending below the optic axis according to the Cartesian sign convention.

d. The focal length of the mirror is half the radius of curvature of the mirror (Equation 23.22):

$$f = \frac{R}{2}$$
$$= \frac{-50 \text{ cm}}{2}$$
$$= -25 \text{ cm}$$

The focal point thus is 25 cm to the *left* of the origin (i.e., to the left of the mirror).

EXAMPLE 23.9

A pretty but deadly 75 cm long coral snake (*Micrurus fulvius*) is stretched out along the optic axis of a convex mirror of radius of curvature 150 cm. Its head is 50 cm from the mirror vertex with its tail further away.

a. Sketch the situation using the standard geometry.
b. Determine the locations of the images of the head and tail of the snake.
c. Is the length of the image of the snake longer or shorter than its actual length?

Solution

a. The situation appears as in Figure 23.48.
b. Use the general mirror equation (Equation 23.23) to locate the images:

To locate the image of the head	To locate the image of the tail
$s = -50$ cm	$s = -125$ cm
$R = 150$ cm	$R = 150$ cm
$\dfrac{1}{s} + \dfrac{1}{s'} = \dfrac{2}{R}$	$\dfrac{1}{s} + \dfrac{1}{s'} = \dfrac{2}{R}$
$\dfrac{1}{-50 \text{ cm}} + \dfrac{1}{s'} = \dfrac{2}{150 \text{ cm}}$	$\dfrac{1}{-125 \text{ cm}} + \dfrac{1}{s'} = \dfrac{2}{150 \text{ cm}}$
$s' = 30$ cm	$s' = 47$ cm

c. The length of the virtual image of the snake is

$$47 \text{ cm} - 30 \text{ cm} = 17 \text{ cm}$$

and so is shorter than the actual snake.

23.11 RAY DIAGRAMS FOR MIRRORS

Ray diagrams are useful geometric constructions that serve as a check on calculations made with the mirror and magnification equations. The diagrams also serve as a useful check on whether the image is real or virtual. In addition, ray diagrams provide an alternative method for solving mirror (and lens) problems.

Ray diagrams begin with an extended object placed in front of a mirror, as in Figure 23.49. The image of the object point P on the optic axis also will lie on the optic axis. Hence we just need to determine where the image of the point Q at the top of the object lies. A perpendicular line from the image of point Q to the optic axis will indicate the extent of the image of the extended object.

Many rays diverge from point Q and are processed by the mirror. *All* the processed rays intersect at the image point Q'. To locate the image point Q', then, we need consider only a few rays. Two rays will suffice, but ray diagrams typically use three or four, just as a check on the construction. Of the bevy of rays diverging from the object point Q, we choose four special rays, called **principal rays**, whose paths are easy to trace geometrically. Ray diagrams also are called **principal ray diagrams** as a result. Since we consider only paraxial rays, the plane of effective reflection is coincident with a plane perpendicular to the optic axis passing through the vertex point of the mirror, the origin. We draw rays as if they reflect from this plane.

The privileged rays that warrant specific attention (because their paths are easy to trace) are the following:

1. A ray from the object point directed toward the center of curvature of the mirror, shown as ray (1) in Figure 23.49. After reflection, this ray simply retraces its path because it strikes the mirror along the normal line to the mirror.
2. A ray from the object point parallel to the optic axis, shown as ray (2) in Figure 23.49. Any ray parallel to the optic axis will pass through (or *appear* to pass through) the focal point; here, this ray after reflection *appears* to pass through the

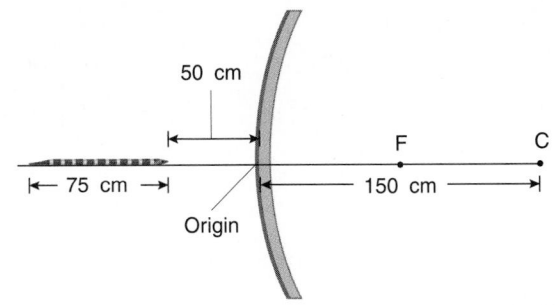

FIGURE 23.48

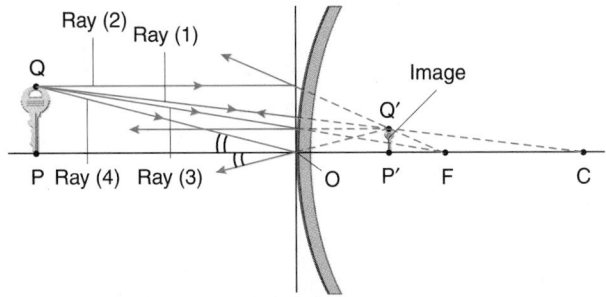

FIGURE 23.49 The principal rays for a mirror ray diagram.

focal point, located halfway between the center of curvature and the mirror itself.

3. A ray from the object point directed toward the focal point F, indicated as ray (3) in Figure 23.49. Since rays are reversible, this ray will emerge after reflection traveling parallel to the optic axis.

4. A ray from the object point to the vertex of the mirror, shown as ray (4) in Figure 23.49. This ray will have equal angles of incidence and reflection above and below the optic axis. This reflected ray also will appear to come from Q′.

Any two of these rays are sufficient to locate the position of the image. Once the position of the image is found, two other things can be assessed easily:

a. The nature of the image (real or virtual) is determined from where the image is relative to the mirror itself. If the image is in front of the mirror, the image is real; if the image is in back of the mirror, the image is virtual.

b. We also are able to see if the magnification is positive or negative and greater or less than unity.

EXAMPLE 23.10 _____

A Baltimore oriole (*Icterus galbula*), a bird, not a baseball player, is perched in front of a concave spherical mirror such that $|s| > |R|$.

a. **Draw the corresponding ray diagram to locate the image of the bird.**

b. **From your ray diagram, indicate whether the magnification is positive or negative, and whether the image is smaller or larger than the object.**

c. **Determine whether the image is real or virtual.**

Solution

a. The situation is depicted schematically in Figure 23.50. Principal ray (1) to the center of curvature reflects back on itself. Principal ray (2), incident parallel to the optic axis, passes through the focal point F. Principal ray (3), incident through the focal point, emerges parallel to the optic axis. Principal ray (4) makes equal angles of incidence and reflection with the optic axis at the mirror vertex.

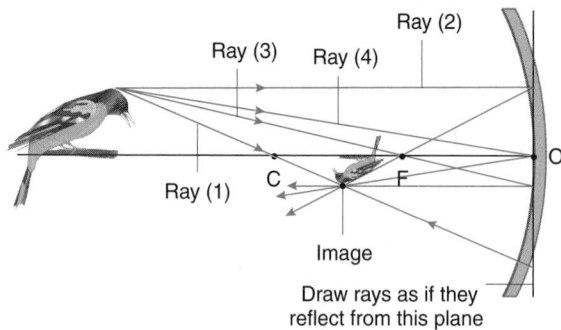

FIGURE 23.50

b. The image is inverted, so the magnification m is negative. The image also is smaller than the object, so $|m| < 1$.

c. The four rays actually pass through the image point, so the image is real.

23.12 REFRACTION AT A SINGLE SPHERICAL SURFACE

One side of either a water-filled, thin-walled spherical fish bowl or a solid crystal ball is an example of a single spherical refracting surface. The law of refraction relates the directions of the incident and refracted rays through the interface. We want to investigate the imaging properties of such single, spherically shaped, refracting surfaces. As with mirrors, we want to locate the image, determine whether it is real or virtual, and find the magnification. Since simple lenses are composed of two refracting surfaces, the work we do here soon will enable us to analyze lenses (in Section 23.13).

Derivation of the Single Surface Refraction Equation

An object point P is located at object distance s from a spherical refracting surface with a radius of curvature R. We formulate the problem with the standard geometry, as in Figure 23.51. Let the medium in which the object lies have an index of refraction n_1 and the second medium have index n_2. The Cartesian sign convention is used for all distances, of course. The sign of R depends on where the center of curvature is located with respect to the surface. For the system pictured in Figure 23.51, the radius of curvature R is positive and the object distance s is negative.

To locate the image of the object point, we need to see where at least two rays, originally diverging from the object point, intersect (or appear to intersect) after they have been processed (here, refracted) by the surface. In particular, we consider the following two rays:

1. A ray from the object point directed along the optic axis, ray (1) in Figure 23.51. This ray strikes the refracting surface at the origin O. We apply the law of refraction to the ray. Since the ray is directed along the normal to the surface at O, the angle of incidence of this ray is zero. Thus the angle of refraction also is zero and the ray proceeds undeviated into the second medium.

2. For a second ray, consider the ray from the object point that strikes the refracting surface at point A; this is ray (2) in Figure 23.51. Once again, we apply the law of refraction to the ray. The normal to the surface where the ray is refracted is a line from the center of curvature to the point A. The angle of incidence is θ_1 and the angle of refraction is θ_2. The two refracted rays intersect at point P′ and, thus, this is the image

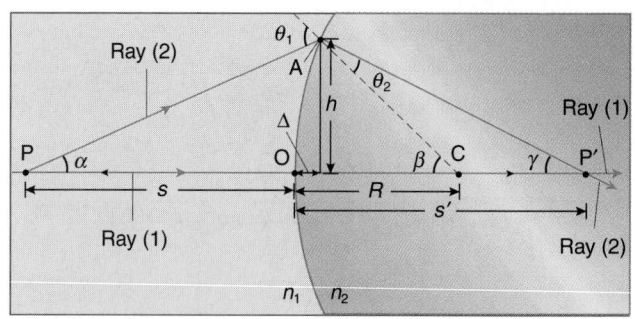

FIGURE 23.51 Geometry for deriving the single surface refraction equation.

point. The law of refraction (Equation 23.6) applied to ray (2) refracted at point A states that

$$n_1 \sin \theta_1 = n_2 \sin \theta_2$$

We restrict ourselves to paraxial rays from the object point, so that the angles θ_1 and θ_2 are small. The small angle approximation means that we can replace the sines in the law of refraction with the angles themselves (in radians). In other words, Equation 23.6 becomes*

$$n_1 \theta_1 = n_2 \theta_2 \qquad (23.25)$$

We introduce the angles α, β, and γ, as shown in Figure 23.51. The angle θ_1 is an exterior angle of triangle PAC. Thus θ_1 is equal to the sum of the two remote interior angles:

$$\theta_1 = \alpha + \beta \qquad (23.26)$$

Likewise β is an exterior angle of triangle P'AC, and so

$$\beta = \theta_2 + \gamma \qquad (23.27)$$

using the same geometric theorem. Substituting for θ_1 and θ_2 in Equation 23.25 using Equations 23.26 and 23.27, we obtain

$$n_1(\alpha + \beta) = n_2(\beta - \gamma)$$

We rearrange this slightly:

$$n_1 \alpha + n_2 \gamma = (n_2 - n_1)\beta \qquad (23.28)$$

Since the angles α, β, and γ are small, we approximate the angles with their tangents (the small angle approximation once again):

$$\gamma \approx \tan \gamma = \frac{h}{s' - \Delta} \qquad (23.29)$$

$$\beta \approx \tan \beta = \frac{h}{R - \Delta} \qquad (23.30)$$

$$\alpha \approx \tan \alpha = \frac{h}{-s + \Delta} \qquad (23.31)$$

For the tangent of α, we use $-s$ in the denominator because s itself is negative; thus, $-s$ is positive.

As we did for mirrors, we assume the refracting surface is a small section of the sphere defining its curvature. Thus we neglect the small distance Δ in comparison with s, s', and R, and Equations 23.29–23.31 become

$$\gamma \approx \frac{h}{s'} \qquad (23.32)$$

$$\beta \approx \frac{h}{R} \qquad (23.33)$$

$$\alpha \approx \frac{h}{-s} \qquad (23.34)$$

---

*It was this small angle form for the law of refraction that was discovered by Ptolemy during the second century A.D., as we mentioned in the introduction to this chapter. For small angles, the angle of refraction is directly proportional to the angle of incidence.

Using Equations 23.32–23.34 for the angles in Equation 23.28, we find

$$n_1 \frac{h}{-s} + n_2 \frac{h}{s'} = (n_2 - n_1)\frac{h}{R}$$

Hence the single surface refraction equation is

$$\boxed{-\frac{n_1}{s} + \frac{n_2}{s'} = \frac{n_2 - n_1}{R}} \qquad (23.35)$$
(single surface refraction equation)

Although we derived Equation 23.35 for the case of a convex surface, a similar derivation shows that it is also applicable to concave spherical refracting surfaces, including a flat surface (in which case $R = \infty$ m). We must use the Cartesian sign convention for all the distances in the single surface refraction equation (Equation 23.35), including R.

Whether the image is real or virtual is assessed using common sense (a dangerous term!).

If the image is located to the right of the origin (positive image distances), then the image must be real since the refracted rays really exist in this region and will converge to the image point. On the other hand, if the image distance is negative, the image is to the left of the origin (the refracting surface). Since the refracted light rays really are on the other side of the surface, the image is virtual; the refracted rays never really intersect at a virtual image point, they only appear to intersect there. Thus a negative image distance here means a virtual image.

Magnification

To determine the magnification caused by a single surface refraction, we need to examine what happens to an extended object. The object point P in Figure 23.52 is imaged at P' on the optic axis; the location of P' is determined using the single surface refraction equation (Equation 23.35). The object point Q is imaged at point Q'. This means that all the rays diverging from object point Q converge on the image point Q'. Once again, because of the small angle approximation, we draw the refractions as if they occur at the plane perpendicular to the optic axis through the vertex of the surface (the origin of the Cartesian system).

We look at just one particular ray from Q to Q', the ray directed from Q to the center of curvature C, shown in Figure 23.53. This ray is undeviated upon refraction because it is directed along the normal to the surface at the point where it strikes the surface.

Notice that triangles PQC and P'Q'C are similar. The corresponding sides of these triangles are proportional. In particular, the magnification is the ratio between the image and object heights:

$$m = \frac{\text{image height}}{\text{object height}} \qquad (23.36)$$

The object height is PQ, and this is a positive distance according to the Cartesian sign convention since it is above the

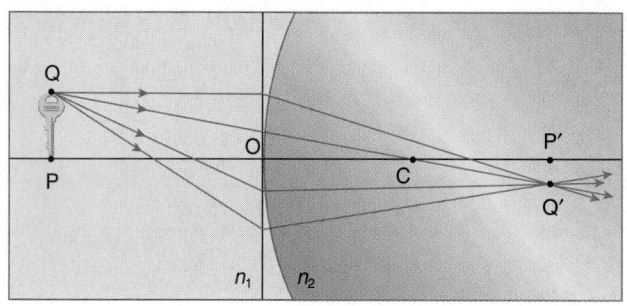

FIGURE 23.52 Formation of the image of an extended object.

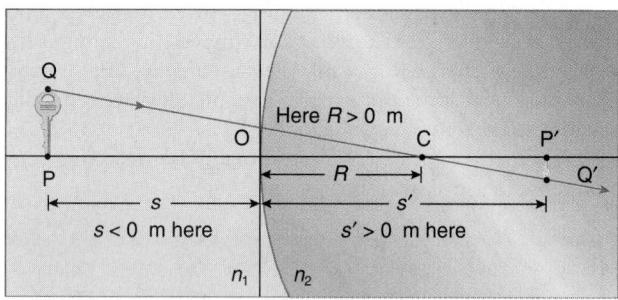

FIGURE 23.53 Consider one particular ray along the normal line; it is not deviated at the surface.

optic axis. The image height is P′Q′ and, since it is below the optic axis, it is negative according to the Cartesian sign convention. So *m* itself will be negative according to Equation 23.36. This is consistent with the sign convention for *m*: a negative magnification means the image is inverted with respect to the object.

It is more convenient to put the magnification in terms of the object and image distances. To do this, we use the proportionality of the sides of similar triangles PQC and P′Q′C:

$$m = \frac{\text{image height}}{\text{object height}}$$
$$= -\frac{s' - R}{-s + R} \qquad (23.37)$$

Some explanation of the signs is needed here. The height of the image in Figure 23.53 is negative according to the Cartesian sign convention. In Figure 23.53, the distances *s*′ and *R* are both positive and $s' > R$, so that $s' - R$ is positive; thus, we substitute $-(s' - R)$ for the image distance. The height of the object in Figure 23.53 is positive. Since the object is to the left of the origin, the object distance *s* is negative; hence, $-s$ is positive and $-s + R$ also is positive and is the appropriate side of the similar triangle. The magnification comes out negative, indicating that the image is inverted in Figure 23.53.

A more convenient equation for the magnification emerges if we eliminate *R* from Equation 23.37 by using the single surface refraction equation (Equation 23.35). To accomplish this goal,

we solve the single surface refraction equation for *R*. This yields

$$R = \frac{ss'(n_2 - n_1)}{-n_1 s' + n_2 s}$$

Substituting this expression for *R* into the magnification equation (Equation 23.37), we obtain

$$m = -\frac{s' - \dfrac{ss'(n_2 - n_1)}{-n_1 s' + n_2 s}}{-s + \dfrac{ss'(n_2 - n_1)}{-n_1 s' + n_2 s}}$$

And this is supposed to be simpler! Not yet. But if we grind through a simplification of this equation (putting things over common denominators and so on), a rather remarkable series of cancellations takes place (check it out!) and this rather horrid looking magnification equation reduces to the following elegant form:

The magnification for single surface refraction is

$$m = \frac{n_1 s'}{n_2 s} \qquad (23.38)$$

(magnification for single surface refraction)

Equation 23.35 for the location of the image and Equation 23.38 for the magnification are applicable to refraction at any spherical refracting surface (and for a flat or plane refracting surface, in which case $R = \infty$ m). Remember that the Cartesian sign convention must be used for all the distances in the equations.

Just as was the case for mirrors, if part of the spherical surface is blocked in some way, as in Figure 23.54, the entire image of an extended object still is formed; the image is dimmer because fewer light rays reach the image.

If more than one refracting surface is present, we take each refracting surface in the order in which it is encountered by the light. A succession of single surface refraction problems then must be solved individually. The location of the image formed by each surface is calculated using the single surface refraction equation (Equation 23.35); the magnification of each surface is found using Equation 23.38.

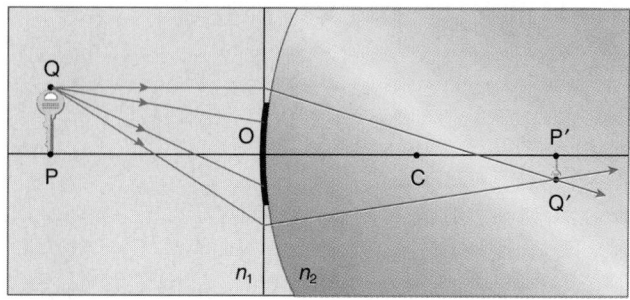

FIGURE 23.54 The entire image is formed, even if part of the refracting surface is blocked.

PROBLEM-SOLVING TACTICS

23.4 When considering refraction (or, for that matter, reflection) of the light from an object by two (or more) surfaces in succession, the image formed by the first surface becomes the object for the second surface. This fact does *not* necessarily mean the image distance of the first surface becomes the object distance of the second surface. For the purpose of assigning appropriate signs and numerical values for the *distances*, the origin of the Cartesian system is always taken to be at the particular refracting surface then under consideration. *Distances are measured from the particular surface under consideration, not from any previous surface.* Numerical values of distances (and their signs) must be adjusted to account for the shift in origin as you move from surface to surface in the problem (see Example 23.12).

23.5 The indices of refraction n_1 and n_2 in Equations 23.35 and 23.38 are the indices of the first medium and the second medium respectively; these change as you progress from surface to surface. That is, if we have two refractions:

Air to glass for the first refraction	and then	Glass to water for the second refraction
$n_1 = 1.00$ $n_2 = 1.50$		$n_1 = 1.50$ $n_2 = 1.33$

23.6 The total magnification of the entire system of surfaces is the product of the magnifications produced by the individual surfaces.

$$m_{\text{total}} = m_1 m_2 m_3 \cdots$$

where $m_1, m_2, m_3, \ldots$ are the magnifications produced by the first, second, third, and successive refracting (or reflecting) surfaces. The sign of the total magnification indicates whether the final image is upright or inverted with respect to the original object.

QUESTION 7

How could you define a focal length f for a single refracting surface? How is the focal length related to the indices of refraction n_1 and n_2 and the radius of curvature R? What conditions on the indices of refraction and the radius of curvature lead to $f > 0$ m? To $f < 0$ m? If an object is at a finite distance from the refracting surface, is the image located at the focal point of the surface?

EXAMPLE 23.11

A heads-up Sacagawea dollar coin lies at the bottom of the shallow end of a swimming pool at a depth of 100 cm, as shown in Figure 23.55.

a. To an observer looking straight down at the coin through the calm surface at a height of 200 cm above the surface, what is the apparent depth of the pool?

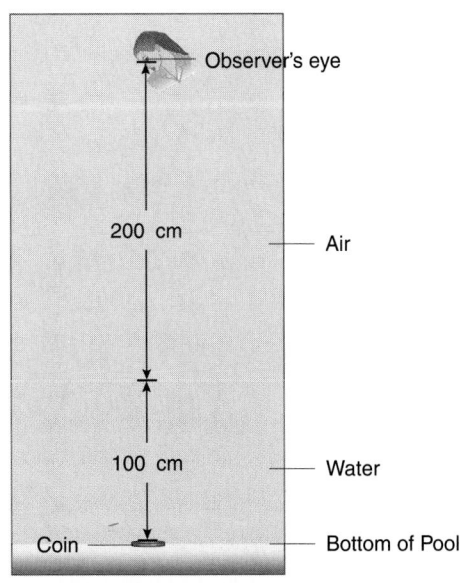

FIGURE 23.55

b. Is the image real or virtual?
c. What is the magnification of the coin?
d. Does the image appear right-side-up or inverted?

Solution

a. A flat surface is treated as a spherical surface of infinite radius. The single surface refraction Equation 23.35 becomes, with $R = \infty$ m,

$$-\frac{n_1}{s} + \frac{n_2}{s'} = \frac{n_2 - n_1}{R}$$
$$= 0 \text{ m}^{-1} \qquad (1)$$

Change your perspective of the pool geometry to put the problem into the standard geometry, shown in Figure 23.56. Since the object (the coin) is in the water and the light is refracted into the air, you must take

$$n_1 = 1.33$$
$$n_2 = 1.00$$

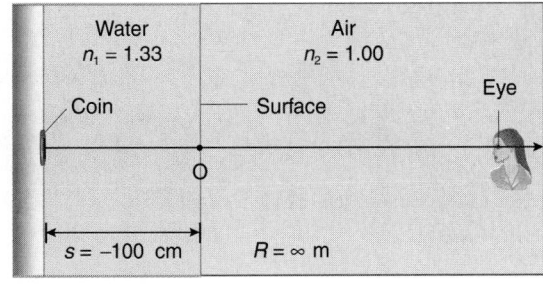

FIGURE 23.56

The object distance is $s = -100$ cm. The height of the observer's eye above the surface of the pool is irrelevant; you want the location of the image of the *coin*; it is this *image* that is seen by the eye (the image becomes the object for the eye). Substituting these numerical values into equation (1), you obtain

$$-\frac{1.33}{-100 \text{ cm}} + \frac{1.00}{s'} = 0 \text{ m}^{-1}$$

Solving for s', you find

$$s' = -75.2 \text{ cm}$$

b. The image is virtual because it is on the same side of the interface as the object and is 75.2 cm below the surface. The water thus appears to be 75.2 cm deep, shallower than it actually is.

c. The magnification of the coin is found from Equation 23.38:

$$m = \frac{n_1 s'}{n_2 s}$$
$$= \frac{1.33 \, (-75.2 \text{ cm})}{1.00 \, (-100 \text{ cm})}$$
$$= 1.00$$

d. Since the magnification is 1.00, the virtual image of the coin is upright and the same size as the object.

QUESTION 8

Does the upright image in Example 23.11 mean merely heads-up in this example (with the writing backward), or that the portrait of Sacagawea has the same orientation as on the object, with the writing in proper direction?

STRATEGIC EXAMPLE 23.12

The 2.0 cm diameter gaudy ring of a charlatan is placed 100 cm in front of a huge, spherical, glass crystal ball of radius 25 cm with an index of refraction equal to 1.50, shown in Figure 23.57. Determine the position of the final image of the ring and its magnification and size.

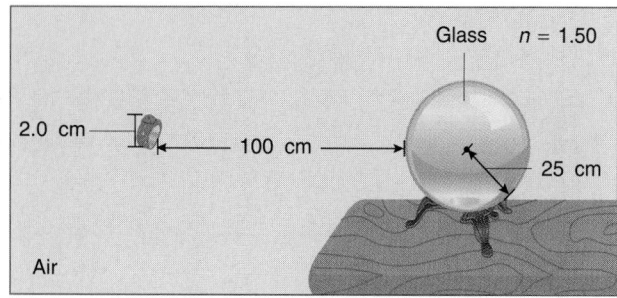

FIGURE 23.57

Solution

Light rays from the object (the ring) are refracted by the crystal ball once as they enter the ball and then on the other side when leaving it. You consider only paraxial rays close to the optic axis, so the single surface refraction equation can be used. There are *two* successive single surface refraction problems to solve.

Refraction at the First Surface

Since the object is in air and the ball is glass, the indices of refraction are

$$n_1 = 1.00 \qquad n_2 = 1.50$$

The object is located 100 cm to the left of the first refracting surface. Since the refracting surface is the origin, the Cartesian sign convention indicates that the object distance is negative and the radius of curvature of the first refracting surface is positive (see Figure 23.58):

$$s = -100 \text{ cm} \qquad R = +25 \text{ cm}$$

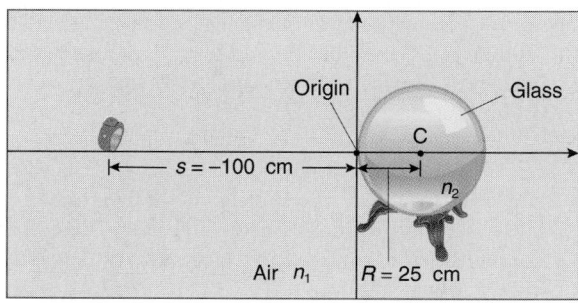

FIGURE 23.58

Substitute these values into the single surface refraction equation (Equation 23.35):

$$-\frac{n_1}{s} + \frac{n_2}{s'} = \frac{n_2 - n_1}{R}$$

$$-\frac{1.00}{-100 \text{ cm}} + \frac{1.50}{s'} = \frac{1.50 - 1.00}{+25 \text{ cm}}$$

Solving for s', you find

$$s' = 150 \text{ cm}$$

The image is located 150 cm to the right of the first refracting surface. The magnification m_1 caused by this first refraction is found using Equation 23.38:

$$m_1 = \frac{n_1 s'}{n_2 s}$$
$$= \frac{1.00 \, (150 \text{ cm})}{1.50 \, (-100 \text{ cm})}$$
$$= -1.00$$

Refraction at the Second Surface

Following Problem-Solving Tactic 23.4, for the refraction at the second surface you must shift the origin of the Cartesian system to the vertex of the second refracting surface. Following Problem-Solving Tactic 23.5, in the single surface refraction equation, the first medium now is the glass, so $n_1 = 1.50$; the second medium (into which the light is going) is the air, so $n_2 = 1.00$. The problem appears as in Figure 23.59.

The image formed by the first surface becomes the object for the second surface. The image formed by the first surface is located

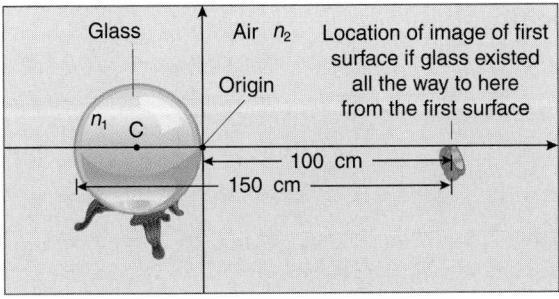

FIGURE 23.59

150 cm to its right or 100 cm to the right of the second refracting surface. Hence the object distance for the refraction at the second surface is

$$s = +100 \text{ cm}$$

This is a *virtual object*! The light forming the image of the first surface is refracted by the second surface before it can actually converge to form the image created by the first surface.

The radius of curvature for the second surface is negative since its center of curvature lies to the left of the origin at the second surface:

$$R = -25 \text{ cm}$$

Making these substitutions into the single surface refraction equation, Equation 23.35, you get

$$-\frac{n_1}{s} + \frac{n_2}{s'} = \frac{n_2 - n_1}{R}$$

$$-\frac{1.50}{100 \text{ cm}} + \frac{1.00}{s'} = \frac{1.00 - 1.50}{-25 \text{ cm}}$$

Solving for s', you find

$$s' = +29 \text{ cm}$$

The image is located 29 cm to the right of the second refracting surface. The final image is real because the light refracted by the second surface actually reaches this point. The magnification m_2 caused by the second surface is

$$m_2 = \frac{n_1 s'}{n_2 s}$$

$$= \frac{1.50 (29 \text{ cm})}{1.00 (100 \text{ cm})}$$

$$= 0.44$$

Following Problem-Solving Tactic 23.6, the total magnification of the original object is the product of the magnifications of each refraction:

$$m_{\text{total}} = m_1 m_2$$
$$= (-1.00)(0.44)$$
$$= -0.44$$

The minus sign indicates that the final image is inverted with respect to the original object. The height of the final image is

$$\text{image size} = m_{\text{total}}(\text{object size})$$
$$= -0.44(2.0 \text{ cm})$$
$$= -0.88 \text{ cm}$$

The minus sign also indicates that the image is located below the optic axis, in keeping with the Cartesian sign convention.

23.13 THIN LENSES

Those of us, including the author, who need eyeglasses or contact lenses to see clearly, certainly appreciate the gift that refracting surfaces give us every waking moment! Without our eyeglasses, the world we see would appear as an impressionistic blur of light and color. While artistically beautiful, that view is annoying—and downright dangerous when driving. Eyeglasses are thin lenses, and here we see how they form an image.

A **thin lens** consists of two closely spaced refracting surfaces. The distance between the surfaces (the thickness of the lens) is considered negligible compared with object distances, image distances, and the radii of curvature of the two refracting surfaces.*

To find the location of the image formed by such a lens, we apply the single surface refraction equation successively to each surface. In particular, in Figure 23.60, let

n_1 = the index of refraction of the first medium (where the object lies)

n_2 = the index of refraction of the material out of which the lens is constructed

n_3 = the index of refraction of the medium on the other side of the lens from the object

Derivation of the Thin Lens Equation

We examine the refraction at each of the two surfaces.

Refraction at the First Surface

Applying the single surface refraction equation (Equation 23.35) here, we find

$$-\frac{n_1}{s} + \frac{n_2}{s_1'} = \frac{n_2 - n_1}{R_1} \qquad (23.39)$$

---

*Our artistic rendition of thin lenses makes them look overly plump on the scale of our sketches.

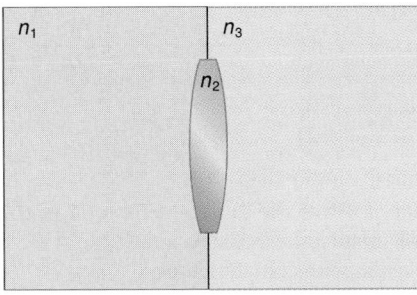

FIGURE 23.60 A thin lens with different media on both sides.

where s is the object distance, s_1' is the image distance of the first surface refraction, and R_1 is the radius of curvature of the first refracting surface. The magnification produced by the first surface refraction is given by Equation 23.38; here, it has the form

$$m_1 = \frac{n_1 s_1'}{n_2 s} \qquad (23.40)$$

Refraction at the Second Surface

The single surface refraction equation here becomes

$$-\frac{n_2}{s_2} + \frac{n_3}{s'} = \frac{n_3 - n_2}{R_2} \qquad (23.41)$$

where s_2 is the object distance of the second surface, s' is the image distance of the second surface, and R_2 is the radius of curvature of this surface. The magnification produced by the second surface refraction is

$$m_2 = \frac{n_2 s'}{n_3 s_2} \qquad (23.42)$$

The position s' of the image formed by the second surface is the location of the final image.

The image of the first surface becomes the object for the second surface. Since the lens is thin, the two surfaces are essentially at the same location and have the *same origin* for the Cartesian sign convention. Thus, because the lens is thin,

$$s_1' = s_2 \qquad (23.43)$$

Two terms in Equations 23.39 and 23.41, rewritten here, are identical (the circled terms below are the same):

$$-\frac{n_1}{s} + \frac{n_2}{s_1'} = \frac{n_2 - n_1}{R_1} \qquad (23.44)$$

$$-\frac{n_2}{s_2} + \frac{n_3}{s'} = \frac{n_3 - n_2}{R_2} \qquad (23.45)$$

If we add Equations 23.44 and 23.45 together, we are left with an equation that involves only the original object distance s and the final image distance s':

$$-\frac{n_1}{s} + \frac{n_3}{s'} = \frac{n_2 - n_1}{R_1} + \frac{n_3 - n_2}{R_2} \qquad (23.46)$$

This eliminates the intermediate step, since we really do not care where the image formed by the first surface lies. We just want to locate the final image.

Equation 23.46 is the most general form of the thin lens equation; it is useful if the media on either side of the lens are not the same, such as with contact lenses.

Magnification

The total magnification resulting from the two refractions is

$$\begin{aligned} m &= m_1 m_2 \\ &= \frac{n_1 s_1'}{n_2 s} \frac{n_2 s'}{n_3 s_2} \end{aligned} \qquad (23.47)$$

Since the lens is thin, the image distance of the first surface s_1' is the same as the object distance for the second surface s_2. Thus Equation 23.47 becomes

$$\begin{aligned} m &= \frac{n_1 s_2}{n_2 s} \frac{n_2 s'}{n_3 s_2} \\ &= \frac{n_1}{n_3} \frac{s'}{s} \end{aligned} \qquad (23.48)$$

A Widely Encountered Special Case

In many instances the media are the same on both sides of the lens. Then $n_3 = n_1$ and Equation 23.46 simplifies to

$$-\frac{n_1}{s} + \frac{n_1}{s'} = \frac{n_2 - n_1}{R_1} + \frac{n_1 - n_2}{R_2}$$

We rearrange this slightly:

$$-\frac{1}{s} + \frac{1}{s'} = \frac{n_2 - n_1}{n_1}\left(\frac{1}{R_1} - \frac{1}{R_2}\right) \qquad (23.49)$$

When $n_3 = n_1$, the magnification Equation 23.48 also simplifies to

$$m = \frac{s'}{s} \qquad (23.50)$$

Focal Point and Focal Length

As we saw in Figure 23.40, if an object point is placed infinitely far away from the lens, the incident rays are parallel. The image then is at the *focal point* of the lens; the distance of the focal point from the lens is called the *focal length*. We use Equation 23.49 with an infinite object distance, letting the image distance be the focal length f:

$$-\frac{1}{-\infty\ m} + \frac{1}{f} = \frac{n_2 - n_1}{n_1}\left(\frac{1}{R_1} - \frac{1}{R_2}\right)$$

$$\frac{1}{f} = \frac{n_2 - n_1}{n_1}\left(\frac{1}{R_1} - \frac{1}{R_2}\right) \qquad (23.51)$$

Equation 23.51 is known as the **lens maker's equation** because it indicates what possible radii to use when manufacturing a lens with a given focal length and material for use in a given medium.

The focal length depends on three things:

a. the curvature of the two refracting surfaces of the thin lens (R_1 and R_2);
b. the index of refraction n_2 of the material out of which the lens is made; and
c. the medium in which the lens is used (n_1).*

Lenses with positive focal lengths are called **converging lenses**, as in Figure 23.61; lenses with negative focal lengths are called **diverging lenses**.

With the lens maker's equation (Equation 23.51) the thin lens equation (Equation 23.49) reduces to the more elegant form:

$$-\frac{1}{s} + \frac{1}{s'} = \frac{1}{f} \qquad \text{(thin lens equation; medium is the same on both sides of lens)} \qquad (23.52)$$

The magnification is given by Equation 23.50:

$$m = \frac{s'}{s}$$

Just as with mirrors and single refracting surfaces, if part of a thin lens is blocked, the entire image still is present; the image is simply dimmer, since less light reaches it than before. This is the principle behind an iris diaphragm. Such a device controls the amount of light through a lens, but does not affect the size or position of the image. It is good that this is the case: otherwise, in

---

*The dependence of the focal length on the medium in which the lens is used is why, among other reasons, your eyes and eyeglasses do not perform properly underwater without a face mask.

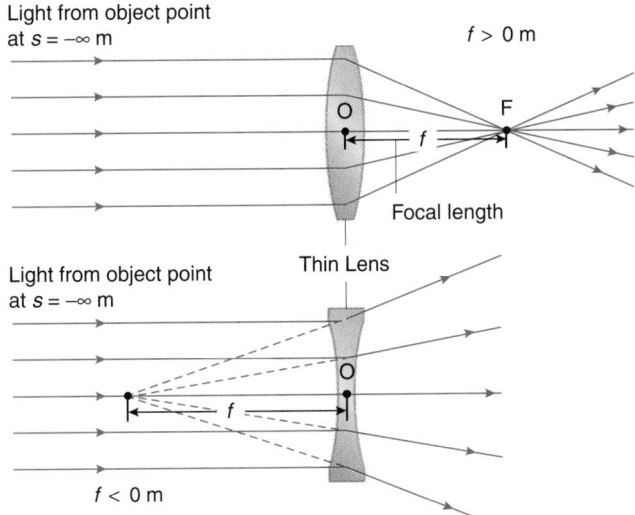

FIGURE 23.61 A converging lens has a positive focal length, while a diverging lens has a negative focal length.

bright sunlight we would have tunnel vision! The size of the iris merely controls the amount of light reaching the retina of the eye or the film of a camera.

If the image distance is positive, the image is to the right of the lens in the standard geometry; this image is real. On the other hand, if the image distance is negative ($s' < 0$ m), then the image is to the left of the lens, even though the refracted rays are on the right side of the lens. Thus the refracted rays do not really pass through the image point if $s' < 0$ m and the image is virtual.

PROBLEM-SOLVING TACTIC

23.7 Be sure to distinguish between the equations for lenses (Equations 23.52 and 23.50) and those for mirrors (Equation 23.23 and 23.24); they are not the same. However, the Cartesian sign convention is used for *both* mirrors and lenses, as well as for single refracting surfaces. The mirror reflects light, but not the Cartesian sign convention, and that makes the equations for lenses and mirrors different.

EXAMPLE 23.13

Determine whether the lenses shown in Figure 23.62 are converging or diverging.

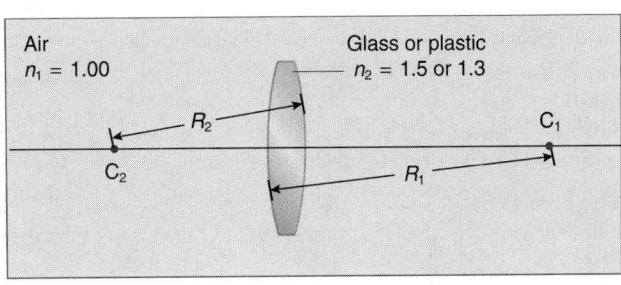

(a)

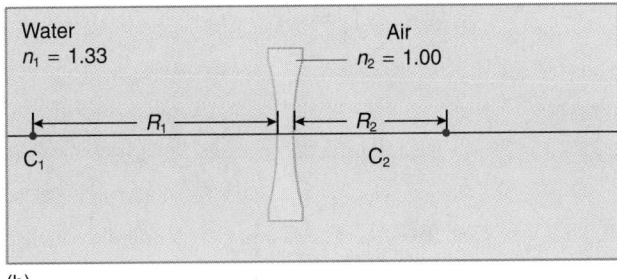

(b)

FIGURE 23.62

Solution

a. Use the lens maker's equation (Equation 23.51):

$$\frac{1}{f} = \frac{n_2 - n_1}{n_1}\left(\frac{1}{R_1} - \frac{1}{R_2}\right)$$

Here $n_2 > n_1$, so that the term $n_2 - n_1$ is positive. The radii of curvature $R_1 > 0$ m and $R_2 < 0$ m; thus, the term with the radii also is positive. The focal length f is therefore greater than zero, and so the lens is a converging lens.

b. Use the lens maker's equation (Equation 23.51). In this case $n_2 < n_1$, so $n_2 - n_1 < 0$. The radii term in the lens maker's equation also is negative, since $R_1 < 0$ m and $R_2 > 0$ m. So the focal length is positive and the air lens in water is converging.

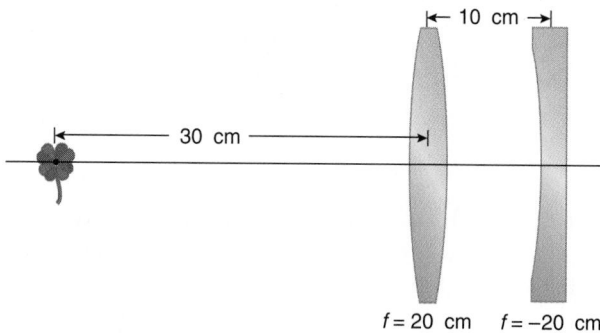

STRATEGIC EXAMPLE 23.14

A lucky four-leaf clover is placed 30 cm in front of the two-lens system shown in Figure 23.63. Determine the location of the final image, its magnification, and whether it is upright or inverted and real or virtual.

FIGURE 23.63

Solution
Take the lenses one at a time, beginning with the first lens encountered by rays from the object, the converging lens.

The First Lens
Place the origin of the Cartesian system at the vertex of the first lens. Then, for this lens, you have $s = -30$ cm and $f = +20$ cm. Use the thin lens equation (Equation 23.52) to determine the position of the image of the first lens:

$$-\frac{1}{s} + \frac{1}{s'} = \frac{1}{f}$$

$$-\frac{1}{-30 \text{ cm}} + \frac{1}{s'} = \frac{1}{20 \text{ cm}}$$

Solving for s', you find
$$s' = 60 \text{ cm}$$

The image of the first lens is 60 cm to its right.
   The magnification produced by the first lens is found using Equation 23.50:

$$m_1 = \frac{s'}{s}$$

$$= \frac{60 \text{ cm}}{-30 \text{ cm}}$$

$$= -2.0$$

The Second Lens
The image of the first lens becomes the object for the second lens. You must transfer the origin for the Cartesian sign convention to

the position of the second lens and adjust for this change in determining how far the image of the first lens is from the second lens (Problem-Solving Tactic 23.4). Hence the object distance for the second lens is $s = +50$ cm, *not* 60 cm. This is an example of a *virtual object*.
   Use the thin lens equation to determine the position of the image of the second lens:

$$-\frac{1}{s} + \frac{1}{s'} = \frac{1}{f}$$

$$-\frac{1}{+50 \text{ cm}} + \frac{1}{s'} = \frac{1}{-20 \text{ cm}}$$

Solving for s', you find

$$s' = -33 \text{ cm}$$

The image formed by the second lens is the final image. This image is 33 cm to the left of the second lens because s' is negative. The light refracted by the second lens is to its right; this light does not physically pass through the image location. Therefore the final image is virtual.
   The magnification produced by the second lens is found using Equation 23.50:

$$m_2 = \frac{s'}{s}$$

$$= \frac{-33 \text{ cm}}{50 \text{ cm}}$$

$$= -0.66$$

The total magnification is the product of the magnifications produced by each lens:

$$m_{\text{total}} = m_1 m_2$$
$$= (-2.0)(-0.66)$$
$$= 1.3$$

The final image thus is upright (since $m_{\text{total}} > 0$) and enlarged (since $|m_{\text{total}}| > 1$).

23.14 RAY DIAGRAMS FOR THIN LENSES

We introduced ray diagrams with spherical mirrors as a useful check on the calculations made to locate the position of the image and determine its nature (real or virtual) and relative size. Ray diagrams also can be drawn for thin lenses to accomplish the same purposes. Since the lens is thin and because of the paraxial ray approximation, we draw ray diagrams as if all the refraction occurs at a plane perpendicular to the optic axis passing through the center (vertex) of the lens.
   The focal point located with the lens maker's equation (Equation 23.51),

$$\frac{1}{f} = \frac{n_2 - n_1}{n_1}\left(\frac{1}{R_1} - \frac{1}{R_2}\right)$$

is the **primary focal point** of a lens. For a converging lens, $f > 0$ m and the primary focal point is to the right of the lens in the standard geometry; see Figure 23.64a. For a diverging lens, $f < 0$ m and the primary focal point is to the left of the lens in the standard geometry; see Figure 23.64b.

The significance of the primary focal point is as follows. *Any* ray incident on the lens parallel to the optic axis (say, from an object point at $s = -\infty$ m), when refracted by the lens, passes through (or appears to pass through) the primary focal point, as shown in Figure 23.64a and 23.64b.

Each lens also has a **secondary focal point** F_2; this is the location for an object that leads to an infinitely distant image. If the medium is the same on both sides of the lens, the secondary focal point is located the same distance (in absolute value) from the lens as the primary focal point, but is on the *opposite* side of the lens as the primary focal point.

The significance of the secondary focal point is as follows. *Any* ray passing through (or appearing to pass through) the secondary focal point emerges from the lens traveling parallel to the optic axis, as shown in Figure 23.65. If we reverse the direction of the light (by using a plane mirror facing and perpendicular to the rays), the light will retrace its path. The secondary focal point is useful when sketching ray diagrams to locate the position of an image, as we will see.

As with mirrors, we use an extended object and locate the position of the image of an object point not lying on the optic axis, say point Q in Figure 23.66. Of the many rays diverging from point Q that are refracted by the lens, we consider just three principal rays whose paths through the lens are easy to trace. Two rays are sufficient to locate the image, but the third ray serves as a useful check of the ray diagram itself.

The three rays used in a ray diagram for thin lenses are the following:

1. A ray from object point Q through the center of the thin lens (the origin for the Cartesian sign convention), as

indicated by ray (1) in Figure 23.66. The central region of a thin lens is approximately a very thin parallel plate. Problem 29 shows that a ray passing through a parallel plate emerges parallel to its incident direction, though translated. Since the lens is thin, the thickness of the essentially

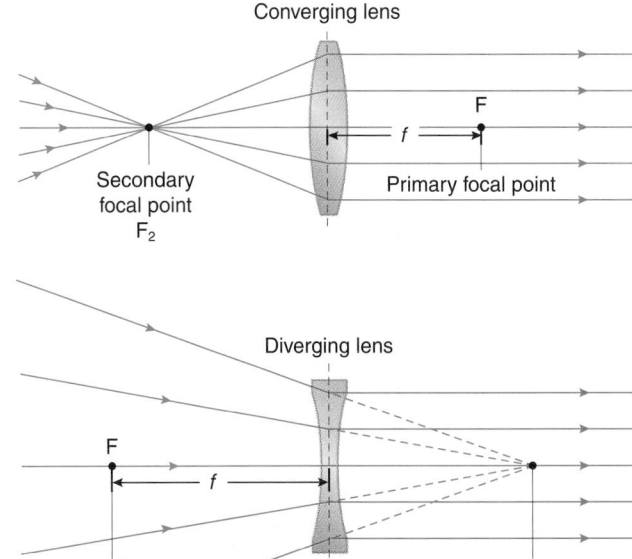

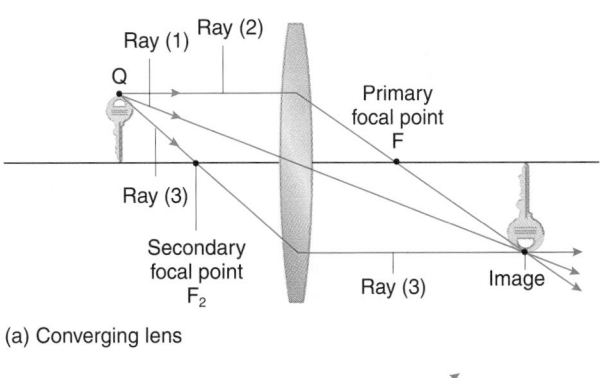

FIGURE 23.65 Secondary focal points F_2 of converging and diverging lenses.

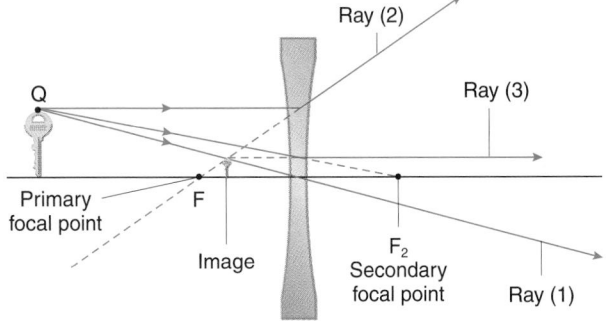

(a) Converging lens

(b) Diverging lens

FIGURE 23.66 Ray diagram for a converging (a) and for a diverging lens (b).

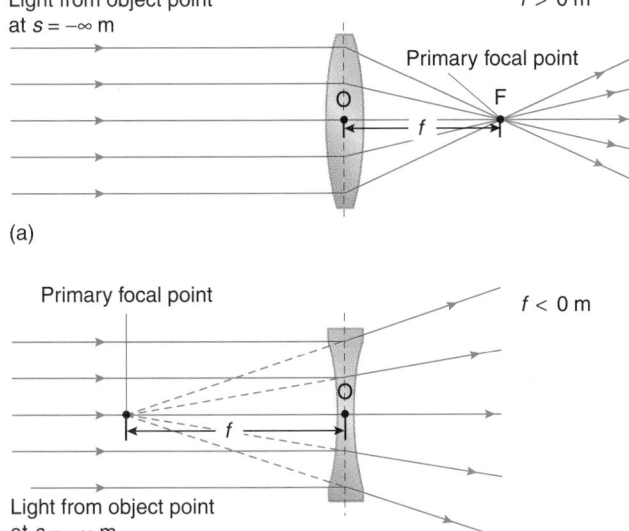

(a)

(b)

FIGURE 23.64 The primary focal point F. (a) Converging lens. (b) Diverging lens.

parallel plate at the center of the lens is thin and the translation of the ray is negligible. Thus this ray passes essentially straight through the lens.

2. A ray from object point Q parallel to the optic axis, ray (2) in Figure 23.66. This ray passes through (or appears to pass through) the primary focal point of the lens. Any ray parallel to the optic axis passes through (or appears to pass through) the primary focal point when refracted by the lens; the single ray from object point Q that approaches the lens parallel to the optic axis is no exception.

3. A ray from object point Q through (or toward) the secondary focal point emerges from the lens parallel to the optic axis; this is ray (3) in Figure 23.66.

The place where the three rays intersect (or appear to intersect) after they are refracted by the lens locates the position of the image. Once the image is located geometrically using the three principal rays, the nature of the image can be distinguished:

- whether it is real or virtual;
- whether it is upright or inverted;
- whether it is larger or smaller than the object.

EXAMPLE 23.15

A small bracket fungus (*Polyporus applanatus*) is placed 10 cm in front of a converging lens of primary focal length 20 cm. Draw a thin lens ray diagram to locate the image and determine whether it is upright or inverted, and real or virtual.

Solution

Draw a diagram approximately to scale. Principal ray (1) passes through the vertex of the lens undeviated, as in Figure 23.67. Principal ray (2), incident parallel to the axis, passes through the primary focal point. Principal ray (3) acts *as if* it came through the secondary focal point, and emerges from the lens parallel to the optic axis.

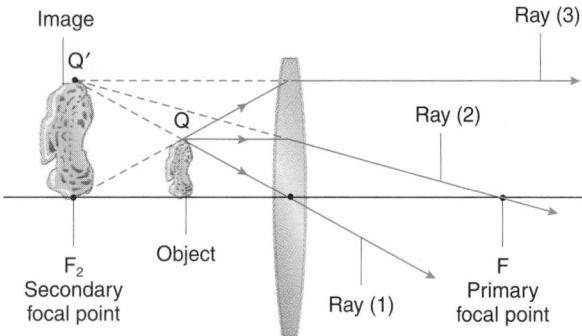

FIGURE 23.67

It is apparent that the three principal rays do not intersect on the right-hand side of the lens; rather they appear to intersect at point Q′ on the left-hand side of the lens. The image, therefore, is virtual. The diagram also shows that the image is upright and enlarged.

23.15 OPTICAL INSTRUMENTS

Where the telescope ends, the microscope begins. Which of the two has the grander view? Choose.

Victor Hugo (1802–1885)*

Here we examine a few optical instruments of particular interest and import in the sciences. These instruments also form the basis of more complex optical systems. A wide variety of more sophisticated optical instruments continually are being invented, illustrating that applications of geometric optics are limited only by our imaginations.

A Simple Magnifier

A magnifying glass may be the first seemingly magical optical instrument you remember from your childhood. A simple magnifier is a converging lens used in a particular way. If we are given a magnifying glass and are asked to demonstrate its use, we bring an object close to the glass and look through it to see a magnified image of the object. How close? To use the lens as a magnifier, the object is placed between the converging lens and its secondary focal point. Then an enlarged virtual image is produced, as shown in the ray diagram of Figure 23.68.

The magnification is calculated using the magnification equation for a thin lens (Equation 23.50):

$$m = \frac{s'}{s}$$

Since both the object and image distances are negative here, the magnification is positive, indicating that the virtual image is upright. This confirms what you already knew: when using a magnifying glass, the object does not have to be turned up-side-down to see the image right-side-up!

There is another way that the magnification of a simple magnifier is specified that involves comparing the *angular size* of the object and image: the **angular magnification** $m_{\angle}$.

The angular size of the object θ_o is defined and measured when the object is exactly 25 cm away from the eye, an arbitrarily chosen standard distance called the **standard eye**

Les Misérables, St. Denis, Book III, The House in the Rue Plumet, Chapter 3, "Foliic ac Frondibus" (A. L. Burt, New York, 1862) page 163.

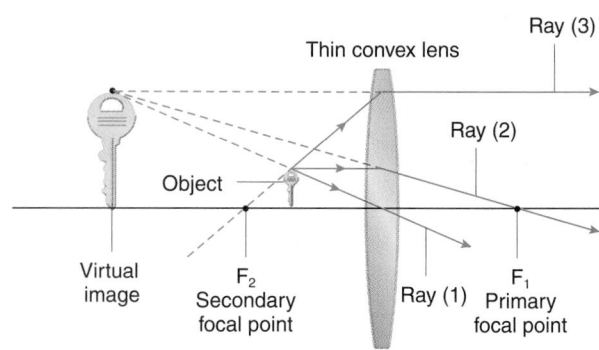

FIGURE 23.68 A simple magnifier.

near point,* as shown in Figure 23.69. From the geometry of Figure 23.69, we have (using the small angle approximation)

$$\theta_o = \frac{h}{25 \text{ cm}}$$

The angular size of the image θ_i when it is at the standard eye near point is found using triangle QOP in Figure 23.70 (using the small angle approximation again):

$$\theta_i = \frac{h}{-s} \qquad (23.53)$$

where the term $-s$ is used in the denominator since s itself is negative.

The thin lens equation (Equation 23.52) is solved for s with the image distance $s' = -25$ cm:

$$-\frac{1}{s} + \frac{1}{-25 \text{ cm}} = \frac{1}{f} \qquad (23.54)$$

All the distances must be in centimeters since the image distance is expressed in centimeters. Solving Equation 23.54 for $-s$, we find

$$-s = \frac{(25 \text{ cm})f}{25 \text{ cm} + f} \qquad (s \text{ and } f \text{ in centimeters})$$

Using this expression for $-s$ in Equation 23.53 for the angular size of the image, we obtain

$$\theta_i = \frac{h(25 \text{ cm} + f)}{(25 \text{ cm}) f}$$

The angular magnification is the ratio of the angular size of this image at the standard eye near point to that of the object when it is at the standard eye near point:

$$m_\angle \equiv \frac{\theta_i}{\theta_o} \qquad (23.55)$$

$$= \frac{\dfrac{h(25 \text{ cm} + f)}{(25 \text{ cm}) f}}{\dfrac{h}{25 \text{ cm}}}$$

$$= \frac{25 \text{ cm}}{f} + 1 \qquad (f \text{ in centimeters}) \qquad (23.56)$$

Another way to use the lens as a magnifying glass is to place the object at the secondary focal point of the lens; the virtual

---

*The actual near point of the eye depends on the individual.

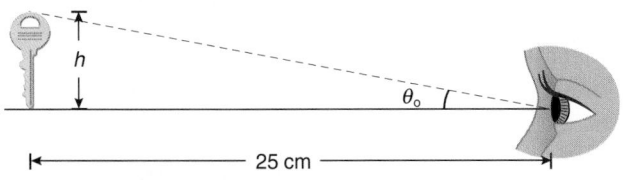

FIGURE 23.69 The angular size θ_o of the object when placed at the standard eye near point.

image then is at $-\infty$ m, as shown in Figure 23.71. With the final image at $-\infty$ m, the ciliary muscles that control the shape of the eye lens are relaxed when we view the image through the lens. In this case the angular size of the image, determined from Figure 23.71, is (using the small angle approximation)

$$\theta_i = \frac{h}{f} \qquad (23.57)$$

The angular magnification, Equation 23.55, then becomes

$$m_\angle = \frac{\dfrac{h}{f}}{\dfrac{h}{25 \text{ cm}}}$$

$$= \frac{25 \text{ cm}}{f} \qquad (f \text{ in centimeters}) \qquad (23.58)$$

This magnification is slightly less than the magnification with the *image* at the standard eye near point (Equation 23.56); however, Equation 23.58 is chosen to typify the magnification because the typical eye is then most relaxed. In adjusting

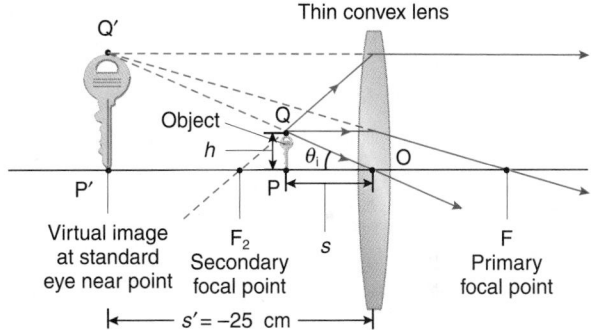

FIGURE 23.70 The angular size θ_i of the image when it is at the standard eye near point.

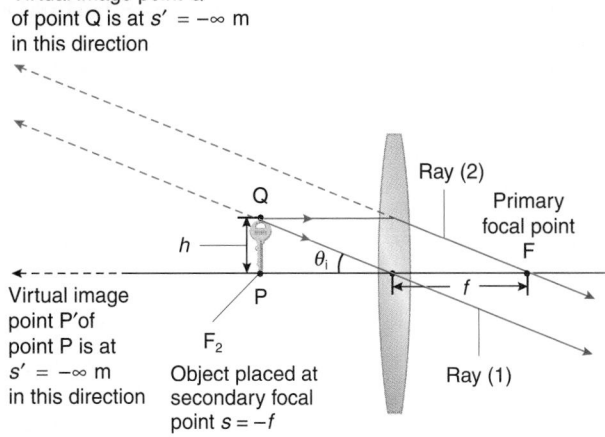

FIGURE 23.71 With the object at the secondary focal point, the image is at $-\infty$ m.

an object distance for clear focus, you should move the object in from too great an object distance and stop at the first clearness of image in order to avoid eye muscle strain during use of the magnifier.

QUESTION 9
Why can you *not* draw principal ray (3) in Figure 23.71?

A Pinhole Camera

One of the more remarkable cameras has an optical element that consists of nothing more than a pinhole. The aperture of a pinhole is very small, of course. Only a small amount of light passes into the lighttight box containing the film. This means that long exposure times may be necessary. The object must be stationary during the exposure, so it is difficult to obtain pictures of moving objects with a pinhole camera.

The ray diagram for an ideal pinhole camera consists of a *single* ray from each object point to each image point, as shown in Figure 23.72. Other than this single ray through the pinhole, no other rays reach a given image point. Thus the image of the object is found at *any* and *every* image distance on the positive distance side of the pinhole.

This means that the image is fairly sharp wherever the film is placed. The position of the film or screen determines the magnification using the magnification Equation 23.50:

$$m = \frac{s'}{s}$$

As the image distance and magnification increase, however, the image becomes dimmer because the same small amount of light through the pinhole is spread over a larger area.

Since a given object can be registered on the film at many image distances, we can turn the problem around and say that for a fixed image distance (corresponding to where the film is placed), objects at many different object distances are equally well imaged on the film simultaneously. This quality is called good **depth of field**. Objects close to and far from the pinhole camera are equally well recorded on the film, as when taking a picture of your friends against a distant mountain background.

We can easily illustrate the principle of pinhole image formation by performing a simple experiment. We form an image of the Sun on a piece of paper with a pinhole in an index card, as

FIGURE 23.73 Imaging the Sun with a pinhole is a safe way to view a solar eclipse.

in Figure 23.73. As the screen is moved closer to the pinhole, the image size decreases and the brightness of the image increases. This technique is a common method for safely observing the Sun to detect sunspots or the partial phases of solar eclipses.

The principle of pinhole projection also is the reason you squint if you are near- or farsighted and do not use your eyeglasses. By squinting, you are trying to make the aperture of the eye small. The small aperture or pinhole over the eye lens reduces the cone of ray convergence that would tend to blur the poorly focused images near and far. The depth of field is thus improved. Try it. If you are highly nearsighted, just look at the (bright) world through a pinhole and you will be surprised at how much more clearly you can see. Just don't drive at night with pinholes!

Standard Cameras

The purpose of the lens of a camera is to form a real image of an object on the film, as indicated in Figure 23.74.

The lens must be a converging lens; a single diverging lens cannot form a real image of real objects, no matter where they are placed (see Problem 61). Of course, a shutter is provided to determine when light is allowed onto the film, and various other bells and whistles are added to increase the flexibility of use (and

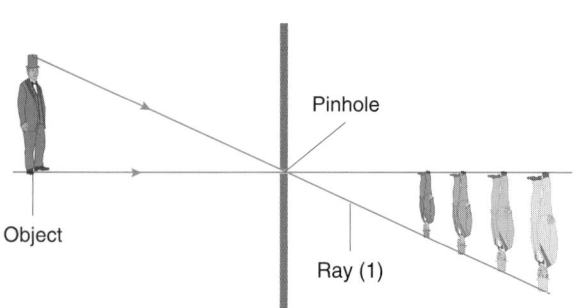

FIGURE 23.72 Ray diagram for a pinhole aperture.

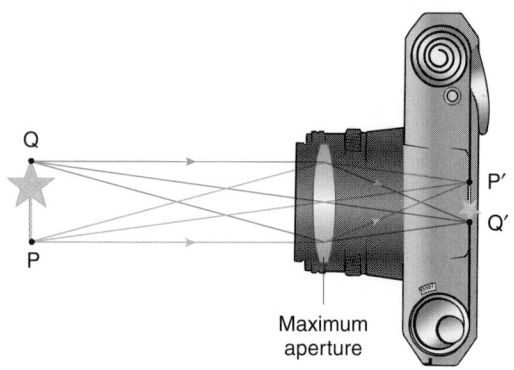

FIGURE 23.74 A camera lens forms a real image on the film.

cost). The essential imaging element is the converging lens. It can be much wider than a pinhole because the refracting lens gathers a wide bundle of rays from each object point and converges them to a single point on the image plane.

An iris diaphragm is used on some camera lenses to regulate the amount of light reaching the film as well as to control the depth of field. The smaller the aperture size, the more the effective part of the lens is like a pinhole, and the greater the depth of field. In other words, to take a picture of your compatriots against a lovely but distant background scene, use a small aperture.

Two numbers are used to specify a camera lens: its focal length (typically quoted in millimeters) and the **f/number** of the lens.* The f/number of the lens is defined to be the focal length divided by the maximum aperture (diameter) of the lens:

$$f/\text{number} \equiv \frac{\text{focal length}}{\text{maximum aperture}} \quad (23.59)$$

The f/number is dimensionless. The slash / can be a bit confusing at first. When a lens is said to be f/1.4, it means that the f/number is 1.4. In other words:

$$1.4 = \frac{\text{focal length}}{\text{maximum aperture}}$$

If the focal length of the lens is 50 mm, the maximum aperture of this lens thus is

$$\text{maximum aperture} = \frac{50 \text{ mm}}{1.4}$$
$$= 36 \text{ mm}$$

Smaller apertures (meaning larger f/numbers) can be obtained with an iris diaphragm. These larger f/numbers are called effective f/numbers since they determine effectively how much of the entire lens is being used. For a lens *specified* as f/1.4, an iris diaphragm can be used to secure *larger* (effective) f/numbers of 2.8, 5.6, 11, and so forth, but not f/numbers smaller than the one in the lens specification. That is, the iris cannot be used to make the f/number of this lens 1.2. The f/number on the spec sheet of the lens is the *smallest* f/number the lens can attain (corresponding to the maximum aperture). Thus an iris diaphragm can be used to give an f/1.4 lens an effective f/number of f/5.6; that is, the effective aperture of the lens has been reduced by the diaphragm, so that

$$5.6 = \frac{\text{focal length}}{\text{effective aperture}}$$

The aperture determines the amount of light entering the camera. The amount of light reaching the film is proportional to the area of the lens opening; the area, in turn, is proportional to the square of the diameter or the square of the aperture (apart from numerical factors). Thus the amount of light reaching the lens is inversely proportional to the square of the f/number.

To have good depth of field, one wants to shoot with a high effective f/number, since then the aperture of the camera is small and approaches the imaging properties of a pinhole. In-

expensive, even disposable cameras have lenses of small, fixed aperture and, consequently, a relatively high f/number to obtain reasonably good depth of field and eliminate the need for focusing hardware.

The Eyes of Vertebrates

The eyes of vertebrates (see Figure 23.75) are certainly some of the more remarkable optical instruments that exist.[†] The eye produces an inverted (!) real image of objects on the retina. The brain then performs another (essentially electronic) inversion of the image to produce the image we perceive in its proper orientation. The retina has two types of cells: **rods**, which specialize in detecting low levels of light; and **cones**, which specialize in color vision and detecting bright light conditions.

Most of the refraction of light rays from an object occurs at the first surface encountered by the rays from the object: the **cornea**. However, for simplicity, we draw the refraction in Figures 23.76 and 23.77 as if the eye lens was the principal culprit. With the eye, *the image distance is fixed* because of anatomical considerations (the eye is essentially of fixed size). (It would be quite amazing if our eye lens poked in and out of our heads to focus, like a camera lens!) For objects at various distances from the eye, then, the focal length of the eye lens must change to ensure that the image lies on the retina. The eye lens tinkers with the refraction to produce a sharp image on the retina. The eye lens is a smart lens: **ciliary muscles** adjust the curvature of the lens, thus changing its focal length (see Equation 23.46) to ensure that the image of the object lies on the retina.

An **iris** regulates the amount of light entering the eye, automatically expanding the aperture (called the **pupil**) of the eye under dim lighting conditions or decreasing the aperture under bright lighting conditions. We do not usually talk about the f/number of the eye because both the focal length of the eye lens and its aperture can change. When you look at something, the eye turns so that the image of the object being examined falls on a central region of the retina known as the **fovea**, containing almost exclusively cones.

[†]Various life forms have different kinds of eyes. For example, insects typically have an eye known as a *compound eye* (see Investigative Project 10).

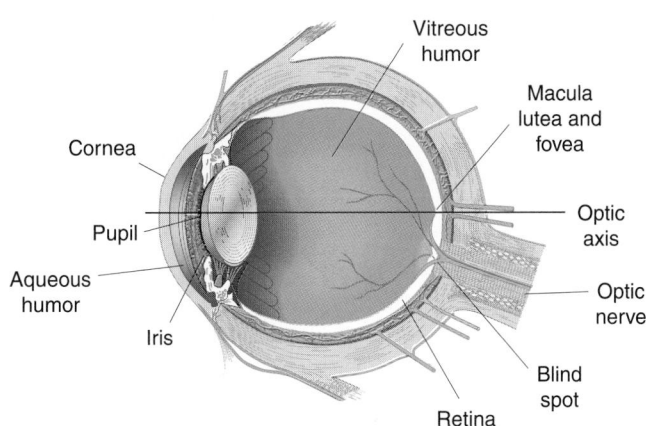

FIGURE 23.75 The vertebrate eye anatomy.

---

*The term also is known as the *focal ratio*; we will not use this terminology.

As many of you likely know from personal experience, not all eyes perform well. Various types of vision defects exist and can be corrected with supplementary optical hardware: a fancy word for eyeglasses or contact lenses. We consider some of the more common vision problems and how they are corrected.

Myopia (Nearsightedness)

The operational problem with **myopia** is an inability to see distant objects clearly. The focal length of the eye lens is too small; even when the ciliary muscles around the lens are relaxed, they are unable to let the focal length of the eye lens be long enough to image distant objects on the retina. For a myopic eye, the image of distant objects is always located between the eye lens and the retina, on the *near* side of the retina (see Figure 23.76). A corrective lens is added to make the final image of the lens and eye optical system appear on the retina. We can see what type of lens is needed from the following analysis.

The most distant (farthest) object that can be seen clearly by a myopic person is called the **far point** d_{far} (consider d_{far} always to be a positive number). A normal eye would have a far point of infinity. The term far point is a bit ironic, for in many cases it actually is not very far from the eye, especially for highly myopic eyes. Indeed, far points of only 15 cm are not uncommon.

The purpose of using a lens in front of the eye is to place the image of distant objects at the far point of the eye. The image provided by the lens becomes the object for the eye; if the object for the eye appears to be within its far point, the eye can see the image clearly (meaning that the final image is on the retina). The lens thus creates a close *virtual image* of a distant object. The virtual image of the lens then becomes a *virtual object* for the eye.

What lens can place the image of a distant object at the far point of the eye? We use the thin lens equation (Equation 23.52) to determine this focal length:

$$-\frac{1}{s} + \frac{1}{s'} = \frac{1}{f}$$

Let the distant object be at $s = -\infty$ m; we want s' to be equal to

the far point—that is, $s' = -d_{far}$, where the minus sign indicates that the image is virtual (since d_{far} was taken to be intrinsically positive). We make these substitutions into Equation 23.52:

$$-\frac{1}{-\infty \text{ m}} + \frac{1}{-d_{far}} = \frac{1}{f}$$

Solving for the focal length, we find

$$f = -d_{far}$$

A diverging lens ($f < 0$ m) whose focal length is equal in absolute value to the far point is needed to correct myopia.

Optometrists and ophthalmologists quote the focal length not in meters or centimeters but in **diopters** (dp). The diopter value of a lens is the reciprocal of the focal length, when the focal length is expressed in *meters*:

$$\text{diopter} \equiv \frac{1}{f_{\text{in meters}}} \qquad (23.60)$$

The diopter value is indicated on your eyeglass prescription (if you need it!). Thus a myopic person with a far point of 15 cm needs a corrective lens of focal length $f = -15$ cm or

$$\frac{1}{-0.15 \text{ m}} = -6.7 \text{ dp}$$

Hyperopia (Farsightedness)

Hyperopia is an inability to see objects clearly that are as close to the eye as perhaps 25 cm. Distant objects are seen clearly, but the simple task of reading is difficult because the material must be placed far from the eye in order to be seen clearly. In this case the eye produces an image of a nearby object that is behind the retina, on its *far* side. The image is not on the retina itself, as shown in Figure 23.77.

The closest that a person with hyperopia can bring an object and still see it clearly is called the **near point**. That is, the eye can

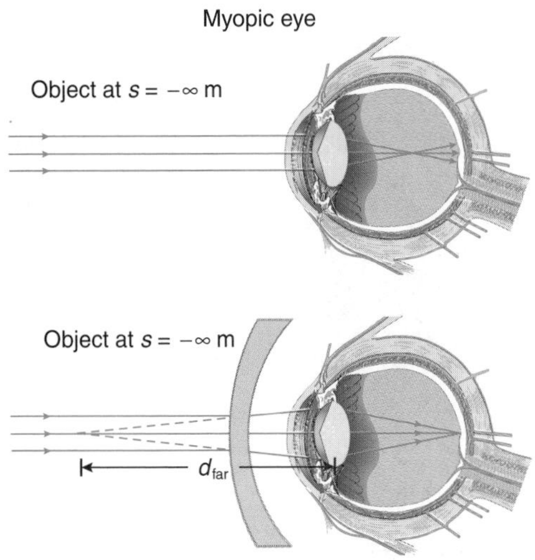

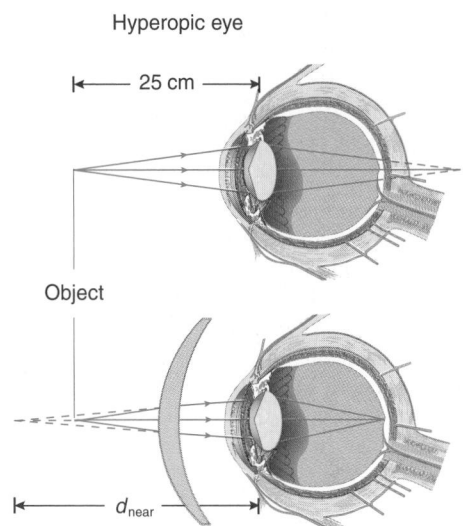

FIGURE 23.76 A myopic eye and a lens to correct for myopia.

FIGURE 23.77 The hyperopic eye and a lens to correct for hyperopia.

see clearly only if the object it is viewing is located *further* from the eye than its near point.

We can see how corrective eyeglasses can be prescribed. Let d_{near} be the closest distance from the unaided eye that reading material can be held and still seen clearly. The function of corrective lenses is to create a virtual image of the reading material essentially at $-d_{near}$; this virtual image then becomes a virtual object for the eye. The eye thus views a virtual object that is far away, and the image produced by the eye is sharp on the retina. We use the lens equation (Equation 23.52) to see what type of lens is needed:

$$-\frac{1}{s} + \frac{1}{s'} = \frac{1}{f}$$

The object for the lens is the reading material, placed a convenient distance d away (consider d to be always a positive distance); therefore, according to the Cartesian sign convention, $s = -d$ and the image distance is to be $s' = -d_{near}$. Making these substitutions in Equation 23.52, we find

$$-\frac{1}{-d} + \frac{1}{-d_{near}} = \frac{1}{f}$$

Since $d < d_{near}$, the focal length is positive. Thus a converging lens is needed to correct for hyperopia.

Astigmatism

Astigmatism is a more complex vision problem caused by variations in the curvature of the corneal surface. We will not study this in greater detail except to mention that astigmatism is corrected with a lens that has different curvatures along two perpendicular axes in the plane of the lens.

Other Vision Problems

A **cataract** is a clouding of the eye lens, rendering it translucent rather than transparent. It is corrected only by surgically removing the natural eye lens and implanting an artificial substitute.

Glaucoma is an abnormally high pressure in the fluid within the eye, which if uncorrected (via drugs or surgery) can irreversibly rupture blood vessels within the eye, causing the death of retinal cells (the rods and cones) and permanent blindness.

A Simple Microscope

A simple microscope is a two-lens system designed to make small objects appear larger. Both lenses are converging lenses of short focal length, as we will see. The lens closest to the object is called the **objective lens**. The object is placed at a distance from the objective lens that is slightly greater than the secondary focal distance, as shown in Figure 23.78.

The objective lens then forms a real, enlarged image with a magnification m_o given by

$$m_o = \frac{s'}{s} \qquad (23.61)$$

The object distance s (<0 m) is essentially the negative of the focal length of the objective lens:

$$s \approx -f_o$$

where $f_o > 0$ m for a converging lens. Since the eyepiece focal length also is short, the image distance s' is essentially the length of the microscope tube ℓ:

$$s' \approx \ell$$

The magnification of the objective, Equation 23.61, thus is essentially

$$m_o \approx -\frac{\ell}{f_o} \qquad (23.62)$$

Since ℓ and f_o are both positive, the magnification of the objective is negative, indicating (as Figure 23.78 shows) that the image of the objective is inverted.

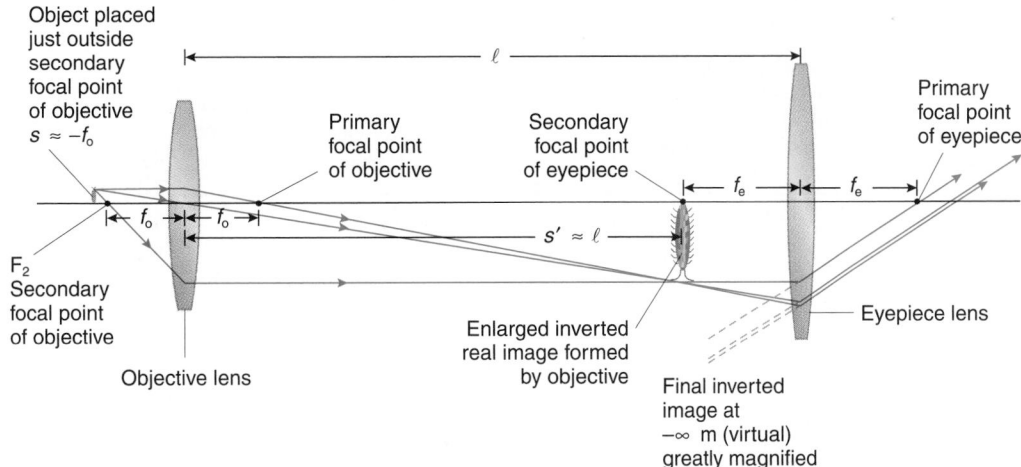

FIGURE 23.78 A microscope.

The image of the objective lens becomes the object for the **eyepiece lens**, which acts as a simple magnifier. Thus the eyepiece produces an angular magnification $m_{\angle e}$ given by Equation 23.58:

$$m_{\angle e} = \frac{25 \text{ cm}}{f_e}$$

where f_e is the eyepiece focal length ($f_e > 0$ cm, since it is a converging lens).

The total magnification of the microscope is the product of the magnifications:

$$m_{total} = m_o m_{\angle e}$$
$$\approx -\frac{\ell}{f_o} \frac{25 \text{ cm}}{f_e} \qquad (23.63)$$

(all distances and focal lengths *must* be in centimeters)

This expression for the magnification of a microscope indicates why the objective and eyepiece lenses are chosen to have small focal lengths.

> The magnification of a microscope is inversely proportional to the product of the focal lengths.

Since ℓ, f_o, and f_e are all positive, the minus sign indicates that the final image is inverted with respect to the original object. The inversion typically is not a problem, since an inverted amoeba (*Amoeba proteus*) looks to all the world like another one that is upright.

Most microscopes have the magnification of the objective element and eyepiece elements inscribed on the barrels of the lenses. A product of the numbers yields the total magnification.

An Astronomical Telescope

A refracting telescope consists of two converging lenses. In this case the objective lens (the one closest to the object) has a long focal length (for reasons that will become apparent) and a large aperture (to collect as much light as possible). A telescope, of course, is designed to look at distant objects and make them appear closer. Hence the object distance is effectively infinite: $s = -\infty$ m.

The thin lens equation (Equation 23.52) indicates that the image distance of the objective lens is the primary focal length. Since the object distance is effectively infinite, the usual equation for the magnification of a lens (Equation 23.50),

$$m = \frac{s'}{s}$$

is inappropriate. With $s = -\infty$ m, the magnification is zero. But the image still has finite size! So we revert to defining an angular magnification. The angular size of the object θ_o is the angle subtended by the object without optical aid (i.e., without the telescope).

Rays from the object point Q approach the objective as parallel rays making the angle θ_o with the optic axis as indicated in Figure 23.79. One of these rays [ray (1)] passes undeviated through the center of the objective. Another ray [ray (3)] passes through the secondary focal point of the objective and emerges parallel to the optic axis.

The extended image P'Q' of the extended object PQ is in the **focal plane**, a plane perpendicular to the optic axis at the primary focal point of the objective. The rays then enter the eyepiece lens, placed so that the final image is a virtual image at $-\infty$ m. In other words, the eyepiece is placed so that the image formed by the objective lens is at the secondary focal point of the eyepiece lens.

The ray incident on the eyepiece parallel to the optic axis passes through the primary focal point of the eyepiece as shown in Figure 23.79. Using the small angle approximation, the angle subtended by the image θ_i is found using triangle FAO':

$$\theta_i = \frac{-h}{f_e} \qquad (23.64)$$

The distance h is negative (according to the Cartesian sign convention), so $-h$ is positive.

Using the small angle approximation again, the angle subtended by the object itself, θ_o, is found from

$$\theta_o = \frac{-h}{f_o} \qquad (23.65)$$

where $-h$ is positive since h itself is negative.

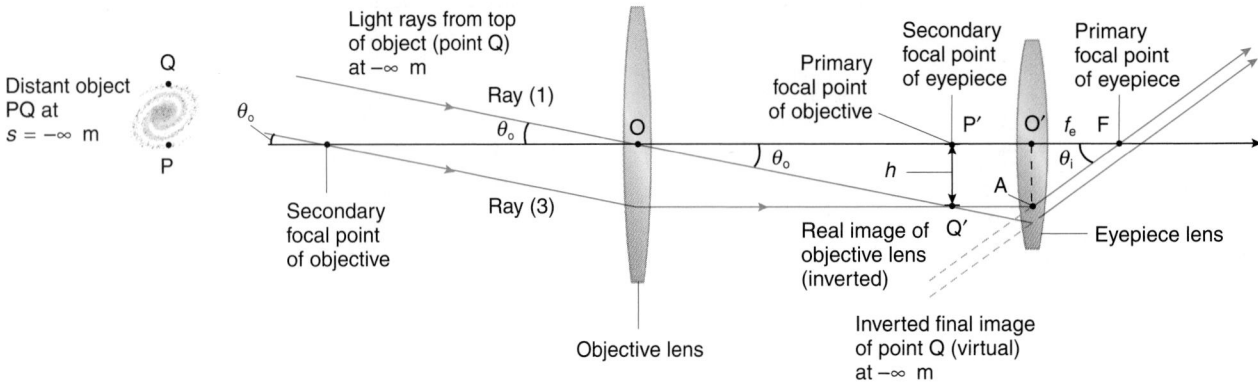

FIGURE 23.79 An astronomical telescope.

The absolute magnitude of the angular magnification of the telescope is the ratio of these angles; a minus sign is introduced since the final image is inverted:

$$m_{\angle} = -\frac{\theta_i}{\theta_o}$$

Substituting for the angles using Equations 23.64 and 23.65, we find

$$m_{\angle} = -\frac{f_o}{f_e} \qquad (23.66)$$

> The magnification of a telescope is the negative of the ratio of the focal length of the objective to that of the eyepiece.

It is difficult, not to mention expensive, to change the magnification of a telescope by changing its objective element because of its large aperture. So, to change the magnification of a telescope, we switch to an eyepiece of different focal length. To increase the magnification, a shorter focal length eyepiece is used.

Despite the Madison Avenue hype, never purchase a telescope on the basis of its magnification. You can get any useful magnification with any telescope with a suitable choice for an eyepiece. In practice, astronomers rarely work with magnifications greater than several hundred; increasing the magnification beyond that only increases blur because of refractive turbulence within the atmosphere along the line of sight. The critical factor to consider (other things aside, such as cost, mechanical performance, and beauty) for an astronomical telescope is aperture, for the area of the objective determines the amount of light collected by the telescope.

Like biologists, astronomers are not bothered by a final image that is inverted: a star, planet, or galaxy looks essentially the same either way. For terrestrial use, however, an inverted final image is awkward, if not inconvenient. Various methods are used to produce a final upright image in a terrestrial telescope:

a. A *field lens* is inserted between the eyepiece and the objective. The purpose of this lens is to invert the inverted image of the objective, so that the simple magnifier eyepiece looks at an upright image.

b. A diverging eyepiece lens is used in conjunction with the objective lens, in place of the traditional converging eyepiece lens, as in Figure 23.80. Such a two-lens system produces a right-side-up final image. This is the design employed in most opera glasses and terrestrial spyglasses.

c. Traditional binocular telescopes (binoculars) use a prism to invert the image of the objective before the eyepiece magnifies the result. Binoculars have two numbers associated with them: for example, 7×35, 7×50, or 8×35. The first number (7, 7, or 8 here) is the angular magnification; the second (35, 50, or 35 here) is the diameter of the objective element in millimeters.

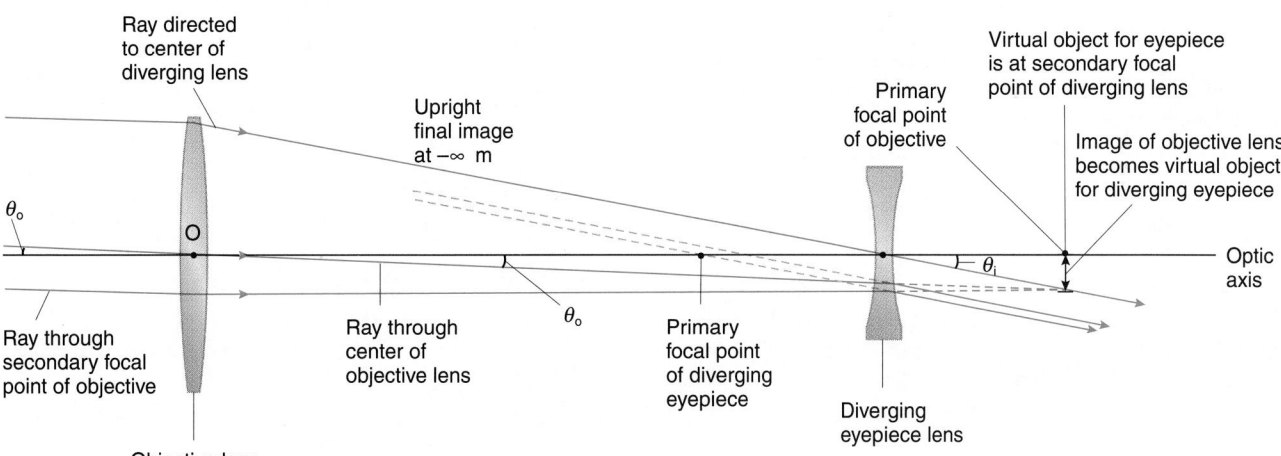

FIGURE 23.80 **A terrestrial spyglass telescope produces an upright final image.**

CHAPTER SUMMARY

Light *rays* are perpendicular to light wavefronts and indicate the path of light. *Geometric optics* is appropriate if light waves interact with apertures or obstacles that are much larger than the wavelength of the light.

The *intensity* of light is the average power transmitted through one square meter oriented perpendicular to the direction the light is travelling. The light intensity I from a pointlike source of *luminosity* L decreases as the inverse square of the distance r from the source:

$$I = \frac{L}{4\pi r^2} \tag{23.1}$$

In SI units, the luminosity L is expressed in watts, the distance r in meters, and the intensity I in watts per square meter.

The *law of reflection* states that the angle of incidence of a light ray is equal to the angle of reflection from a mirror surface; the incident ray, the reflected ray, and the normal line are in a common plane. The angles are measured with respect to a normal to the surface at the point of reflection.

The *law of refraction* at an interface between two transparent media states that the angle of incidence θ_1 of a light ray and its angle of refraction θ_2 are related by

$$n_1 \sin \theta_1 = n_2 \sin \theta_2 \tag{23.6}$$

where n_1 and n_2 are the indices of refraction of the two respective media; the incident ray, refracted ray, and normal line lie in one plane.

When a light ray passes from a more *optically dense* medium with index of refraction n_1 to a less optically dense medium with index of refraction n_2, *total internal reflection* occurs for angles of incidence greater than the critical angle θ_c. There then is no refracted ray. The critical angle is found from

$$\sin \theta_c = \frac{n_2}{n_1} \qquad (n_1 > n_2) \tag{23.8}$$

Dispersion occurs because the index of refraction n of a material is a function of the wavelength of light: $n = n(\lambda)$.

Optical surfaces may cause rays diverging from an object point to reconverge. The point of intersection is called the image point of the object point. If the light rays really pass through the image point, the image is *real*; if the light rays only appear to pass through the image point, the image is *virtual*.

The *Cartesian sign convention* is used for assigning positive and negative signs in geometric optics. The standard geometry assumes light from an object initially travels toward an optical device from left to right. An origin is placed at the vertex of the optical device, with Cartesian coordinates increasing horizontally to the right and vertically upward. Hence distances measured from the origin horizontally to the right or vertically upward are positive; those measured from the origin horizontally to the left or vertically down are negative. A positive value for the magnification indicates that the image is *upright* (the same vertical orientation as the object); a negative magnification shows the image is *inverted* (up-side-down).

When an object is placed at object distance s in front of a spherical mirror of radius R, the image is located at distance s' from the mirror, according to

$$\frac{1}{s} + \frac{1}{s'} = \frac{2}{R} = \frac{1}{f} \quad \text{(spherical mirror equation)} \tag{23.23}$$

where f is the focal length of the mirror, equal to half its radius of curvature. The magnification m is found from

$$m = -\frac{s'}{s} \quad \text{(mirror magnification)} \tag{23.24}$$

The Cartesian sign convention is used to assign positive or negative values to all distances. A ray diagram that traces the path of principal rays is useful for checking calculations.

Refraction at a single spherical surface of radius R is governed by the single surface refraction equation:

$$-\frac{n_1}{s} + \frac{n_2}{s'} = \frac{n_2 - n_1}{R} \quad \text{(single surface refraction)} \tag{23.35}$$

The magnification is found from

$$m = \frac{n_1 s'}{n_2 s} \quad \text{(single surface refraction magnification)} \tag{23.38}$$

The Cartesian sign convention is used to assign positive or negative values to all distances.

The image formed by the first refracting surface encountered by light rays becomes the object for the next surface encountered. The Cartesian origin is transferred to each successive surface in turn, with distances adjusted in value and sign to account for the transposition of the origin. The total magnification of the final image is the product of the magnifications produced by each optical surface.

A *thin lens* is one whose thickness is negligible compared with the radii of curvature of its surfaces and the object and image distances. A thin lens with the same medium on both sides has the object and image distances related by

$$-\frac{1}{s} + \frac{1}{s'} = \frac{1}{f} \quad \text{(thin lens equation)} \tag{23.52}$$

where f is the focal length of the lens. The focal length f is a function of the radii of curvature of the lens surfaces, the index of refraction n_2 of the material from which the lens is made, and the index of refraction n_1 of the medium in which the lens is used:

$$\frac{1}{f} = \frac{n_2 - n_1}{n_1} \left(\frac{1}{R_1} - \frac{1}{R_2} \right) \tag{23.51}$$

This relationship is called the *lens maker's equation*. We use the Cartesian sign convention for thin lenses as well.

Focal lengths occasionally are quoted in *diopters* (dp). The diopter value of the focal length is the *reciprocal* of the focal length in *meters*.

The magnification of a thin lens is

$$m = \frac{s'}{s} \tag{23.50}$$

A thin lens ray diagram is useful for checking calculations.

Note that the mirror equations for distances and magnification (Equations 23.23 and 23.24) are *different* from the thin lens equations (Equations 23.52 and 23.50). This is the price paid for using the same Cartesian sign convention for both mirrors and lenses.

When a simple magnifier of focal length $f > 0$ cm produces a virtual image at $-\infty$ m, the *angular magnification* is

$$m_\angle = \frac{25 \text{ cm}}{f} \qquad (f \text{ in centimeters}) \qquad (23.58)$$

The magnification of a microscope that has an objective lens of focal length $f_o > 0$ cm and an eyepiece lens of focal length

$f_e > 0$ cm is inversely proportional to the product of the two focal lengths:

$$m \approx -\frac{\ell}{f_o} \frac{25 \text{ cm}}{f_e} \qquad \text{(all distances in centimeters)} \qquad (23.63)$$

where ℓ is the length of the tube of the microscope.

The magnification of a telescope with an objective element of focal length $f_o > 0$ cm and an eyepiece lens of focal length f_e is

$$m_\angle = -\frac{f_o}{f_e} \qquad (23.66)$$

SUMMARY OF PROBLEM-SOLVING TACTICS

23.1 (page 1051) When using the law of refraction,

$$n_1 \sin \theta_1 = n_2 \sin \theta_2 \qquad (23.6)$$

you will find it convenient and less confusing to let the medium in which the incident ray lies always be medium 1 (the first medium) and let the medium in which the refracted ray lies be medium 2 (the second medium).

23.2 (page 1053) In using Equation 23.8 for the critical angle, the material with the greater index of refraction (in which the incident ray lies) is in the denominator.

23.3 (page 1060) When using the general mirror and lens equations (Equations 23.23 and 23.52), recognize that they involve the reciprocals of the various distances. Do not forget to take a final reciprocal with your calculator.

23.4 (page 1067) When considering refraction (or, for that matter, reflection) of the light from an object by two (or more) surfaces in succession, the image formed by the first surface becomes the object for the second surface. This fact

does *not* necessarily mean the image distance of the first surface becomes the object distance of the second surface. For the purpose of assigning appropriate signs and numerical values for the *distances*, the origin of the Cartesian system is always taken to be at the particular surface then under consideration. *Distances are measured from the particular surface under consideration, not from any previous surface.*

23.5 (page 1067) The indices of refraction n_1 and n_2 in Equations 23.35 and 23.38 are the indices of the first medium and the second medium respectively; these change as you progress from surface to surface.

23.6 (page 1067) The total magnification of the entire system is the product of the magnifications produced by the individual surfaces or devices.

23.7 (page 1071) Be sure to distinguish between the equations for lenses (Equations 23.52 and 23.50) and those for mirrors (Equation 23.23 and 23.24); they are not the same.

QUESTIONS

1. (page 1047); 2. (page 1048); 3. (page 1051); 4. (page 1051); 5. (page 1054); 6. (page 1061); 7. (page 1067); 8. (page 1068); 9. (page (1076)

10. Is a blind person incapable of detecting *any* electromagnetic waves, or only portions of the electromagnetic spectrum? Explain.

11. Historians study the past. On a clear night, when you look out into space, you literally see history in real time. Explain why.

12. The oscillating electric field of an electromagnetic wave emanating from a pointlike source decreases as the inverse of the radial distance r from the source. In what significant way does this radial dependence differ from the static electric field of a pointlike charge?

13. Starry, starry night .... Clearly explain the distinction between a star that is too faint to see and a star that is invisible because of its color.

14. When you drive your car through an old steel-truss bridge (see Figure Q.14), still common in some areas, radio reception is noticeably poorer than just before entering the bridge or just after leaving it. Does this happen because of diffraction, reflection, or what?

FIGURE Q.14

15. Explain the difference between the appearance of the Moon as seen reflected from the dead-calm surface of a pond and that seen reflected from a surface with small ripples.

16. When driving your car at night, is it more difficult to see the road if the pavement is wet or dry? Explain your answer.

17. Look into a corner cube reflector. If one is unavailable, make one with three small cosmetic mirrors. Is your eyelid on the bottom or top of the image? Explain the observation.

18. Does the law of reflection describe diffuse reflection?

19. Look at the Moon. How can you tell that its surface is locally rough and not locally smooth like a spherical mirror? Why are some of the illuminated areas noticeably darker than others?

20. In Example 23.2, explain why the term *optical lever* is appropriate.

21. When you brush your hair in front of a mirror, is the image of your hair real or virtual?

22. Since acoustic waves cannot exist in a vacuum, how could you define an index of refraction for acoustic waves in various material media such as gases, liquids, and solids?

23. Explain why the refracted ray bends toward the normal line when entering a more optically dense material and away from the normal when entering a less optically dense material.

24. An optical system produces a magnification of −1 for a heads-up penny. Is the image tails-up? Explain.

25. If you place a screen (or camera film) at the location of a virtual image, does the image deliver energy to the screen?

26. When you watch a movie, is the image on the screen real or virtual?

27. Will an air prism in glass (see Figure Q.27) disperse white light into its component colors? If so, sketch the path of red and violet rays through the prism.

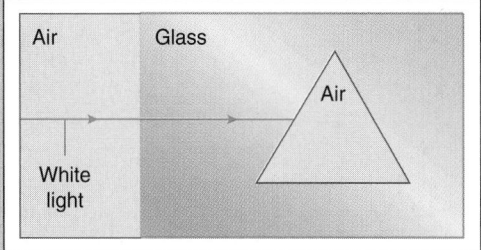

FIGURE Q.27

28. Do other types of waves, such as surface water waves and sound waves, reflect and refract?

29. In Example 23.6, will the angular separation of the red and violet rays increase, decrease, or remain the same if the angle of incidence of the white light increases?

30. Explain the operation of a two-way mirror.

31. Diamond, with its relatively large index of refraction ($n = 2.42$), is cut so rays of light that are refracted into the top face of the gem reemerge through the top face, thus giving diamonds their optical brilliance. Does the light reemerge because the diamond is cut into a corner cube configuration and silvered, or because of total internal reflection of the light within the diamond?

32. What is the focal length of a plane mirror?

33. If a clear solid crystal is placed in a clear liquid with the same index of refraction, is the crystal easily visible? Explain. If strong materials could be made with the same index of refraction as that for most radar wavelengths in air, they would have the ultimate properties for stealth technology in military aircraft.

34. Explain why the total magnification of a series of optical surfaces is given by the product of the individual magnifications, *not* by their sum.

35. An observer looking vertically down into a pool sees a coin lying on the bottom of the pool. The image appears closer to the observer than the coin really is, but the magnification is 1. (See Example 23.11.) Should not the image appear larger than the coin at the bottom of the pool? (Does a diamond grow in value when you view it at close range?)

36. Pure water is colorless and transparent, yet you can easily see water drops if you spill several onto a glass table. Explain.

37. Determine whether the lenses pictured in Figure Q.37 are converging or diverging.

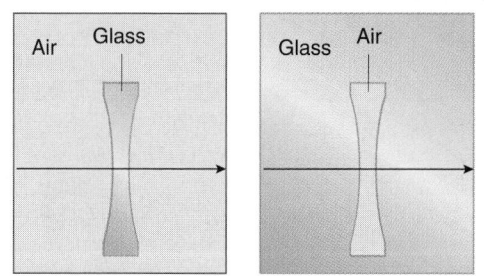

FIGURE Q.37

38. When used in air, a particular concave mirror and a particular converging lens have the same absolute value for their focal lengths. Each is now submerged in water. Which, if any, of their focal lengths change in absolute value?

39. If you look at the eyes of a myopic person when they have their eyeglasses on, do their eyes appear smaller or larger than when the eyeglasses are off? Explain your answer with the assistance of a ray diagram.

40. If you look at the eyes of a hyperopic person when they have their eyeglasses on, do their eyes appear smaller or larger than when the eyeglasses are off? Explain your answer with the assistance of a ray diagram.

41. Many telescopes use a mirror for an objective element to form a real image at its focal point. The eyepiece then magnifies the real image formed by the objective. Explain why it is difficult in practice (though not in theory) to use a mirror for the objective element of a microscope.

42. As an engineer working for the laser development company of Light, Powers, and Watt, you are asked to design a beam expander that will take the thin pencil of rays from a laser and expand it into a wide beam of parallel light as indicated in Figure Q.42. Specify what to place inside the beam expander to accomplish the job, indicating any pertinent spacing of optical components. (The solution is not unique.)

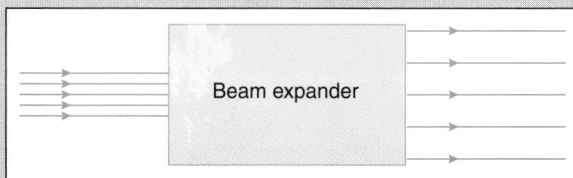

FIGURE Q.42

43. In William Golding's famous novel *Lord of the Flies*,* the boy mockingly called Piggy was quite myopic and his glasses were used "as burning glasses" to start a fire by focusing the rays of the Sun onto dry tinder. If only one lens was used, what is wrong with this description? Devise a way both lenses could be combined to form a "burning glass."

The inconsistency of this literary passage first was pointed out by Sam Prytulak, "Interdisciplinary application," *The Physics Teacher*, 29, #3, page 135 (March 1991).

44. An inventor has submitted a patent application for an unusual device: a brightly polished, solid steel sphere that concentrates the rays of the Sun on a small target and sets it afire. The application has been sent to you for review. Should a patent be granted? Use diagrams or calculations to support your decision.

45. An object is placed in front of a mirror or a lens and a real image is formed of magnification *m*. If the object now is placed where the image was located, where is the new image of the object located? What is the value of the new magnification?

46. Shaving and makeup mirrors are concave. With your face close to such a mirror, what happens to your image as you gradually back away? When sufficiently far from the mirror, your image appears in a different orientation. How far is your face from the mirror relative to its radius of curvature when the change takes place?

47. You see a fish some distance from a river bank in a calm pool. Is the actual location of the fish above or below the image you see? What implications does this have for spear fishing?

48. On the front of an ambulance, the word is spelled backward

with the letters also reversed:

AMBULANCE

Explain why.

49. How should the poem by John Updike at the beginning of Section 23.10 on page 1057 really be printed to indicate accurately the process of reflection?

50. Telephoto lenses foreshorten images. Explain why. (See Figure Q.50.)

51. The passenger side rear view mirror on your car is a convex mirror while that on the driver's side is a plane mirror. What are the advantages and disadvantages of each as a rearview mirror?

52. Examine a good camera lens and tabulate its *f*/numbers. Beginning with the lowest *f*/number, how are the *f*/numbers related to the aperture diameter of the lens? To its area?

53. When you change the *f*/number setting on your camera lens by one full "stop" (or setting), by how much do you change the amount of light reaching the film?

54. How does the depth of field depend on the effective *f*/number of a lens?

55. Perhaps you or your parents wear bifocal eyeglasses. Does this term mean a single lens with two focal points, or what? What vision problems do bifocal eyeglasses attempt to correct?

56. Most astronomical telescopes of large aperture use concave mirrors for the objective. What advantage in cost and optical performance do mirrors have compared with lenses as the objective element?

57. The specifications of most binoculars quote a number known as the field of view. Find out what is meant by this term.

58. If you stand under a leafy tree during a partial solar eclipse, you may observe hundreds of images of the partially eclipsed Sun on the ground. What forms these images? Why do you not normally notice similar images of the uneclipsed Sun?

59. A slide projector has a converging lens that projects a real, inverted image of the slide onto a screen. Imagine you have a slide of the words

THIS END UP   ↑

There are four different ways you can put the slide into the projector: the "proper way," turning the slide left for right, turning the slide top for bottom, and again reversing it left for right. Sketch the four possible ways the image can appear on the screen.

*William Golding, *Lord of the Flies* (Coward-McCann, New York, 1955), pages 45 and 202.

FIGURE Q.50

PROBLEMS

Sections 23.1 The Domains of Optics
23.2 The Inverse Square Law for Light

1. The solar constant, 1.36 kW/m², represents the intensity of solar light, the amount of energy per second received from the Sun on a one square meter area perpendicular to the direction to the Sun, at the distance of the Earth from the Sun (1.496×10^8 km). What is the intensity of sunlight at the position of a space probe in the vicinity of Saturn, located 9.54 times as far from the Sun as the Earth?

•2. The intensity of sunlight is about 1.36 kW/m² at the distance of the Earth from the Sun (1.496×10^8 km). (a) Use only this data to calculate the power output of the Sun in watts, known as the solar luminosity. (b) Determine the value of the solar constant at the spatial average distance of the planet Pluto from the Sun (about 40 times the distance of the Earth from the Sun).

•3. Light spreads out in all directions from a small pointlike source in air and passes points A, B, C, and D located distances 1.00 m, 2.00 m, 8.00 m, and 9.00 m from the source respectively. (a) Find the ratio of the intensity of the light at point A to that at point B. (b) Find the ratio of the intensity of the light at point C to that at point D. (c) Why are the ratios calculated in parts (a) and (b) the same or different? (d) How would the answers to parts (a)–(c) be affected if the source of the light and the points A, B, C, and D were in water?

•4. Over the course of a year, the distance of the Earth from the Sun varies from 1.47×10^8 km on about 3 January to 1.52×10^8 km on about 6 July. (a) By about what percentage does the energy received from the Sun vary between January and July? (b) Is this variation the cause of the seasons?

•5. The luminosity of the Sun is 3.83×10^{26} W. How much solar energy is contained in the light in 1.00 cubic kilometer of space at a distance from the Sun equal to that of the Earth (1.496×10^8 km)?

‡6. Consider a universe uniformly sprinkled with stars throughout its volume.* Let the number density of stars be $\langle n \rangle$, the number of stars per unit volume. Let $\langle L \rangle$ be average luminosity of a star. (a) Imagine a thin spherical shell of such stars centered on the Earth. The radius of the shell is r and its thickness is Δr (where $\Delta r \ll r$) (see Figure P.6). Each star in the shell is essentially the same distance (r) away from the Earth. Show that the amount of light from each star in the shell collected by an aperture of area A on the Earth is

$$\frac{\langle L \rangle}{4\pi r^2} A$$

(b) Show that the total light collected by the aperture from the stars in the shell is

$$\frac{\langle n \rangle \langle L \rangle A}{4\pi} \Delta r \, \Omega$$

where Ω is the solid angle of the shell intercepted by the area A (see Figure P.6). Note about this result is *independent* of the radius of the shell. This means that we get the *same* amount of light from *every* shell surrounding the Earth. If the universe were of infinite extent, the amount of light collected from the infinite number of shells surrounding the Earth also would be infinite and the night sky would be brighter than the surface of the Sun! This calculation and conclusion is the *dark night sky* paradox. Clearly, the calculation produces an amazing result, contradicted by the obvious observation that the night sky is dark. For a complete treatment of the dark night sky paradox, as well as its resolution, see the wonderful book by Edward R. Harrison, *Cosmology* (Cambridge University Press, Cambridge, England, 1981), Chapter 12, pages 249–265. Also see Chapter 1, Problem 74 of this book.

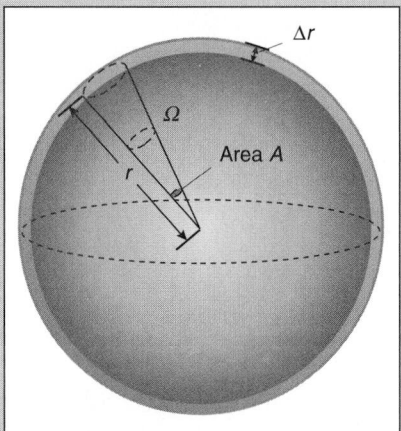

FIGURE P.6

Section 23.3 The Law of Reflection

7. Newton thought light was a stream of particles. Imagine such particles of light approaching a rigid wall (reflecting surface) at angle θ (see Figure P.7). Each particle makes an elastic collision with the wall and rebounds at angle θ'. By conserving the component of the momentum of the particle parallel to the surface, show that $\theta = \theta'$, thus proving a part of the law of reflection. Why is the component of the particle momentum perpendicular to the wall *not* conserved?

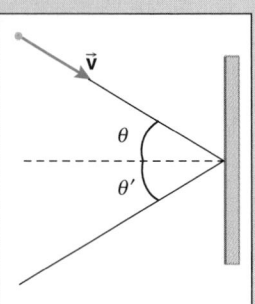

FIGURE P.7

---

*Stars actually are clumped into vast arrays called galaxies that each contain several hundred billion stars. There are estimated to be about 10^{12} such galaxies.

•8. Two plane mirrors are inclined at an angle ϕ to each other as indicated in Figure P.8. An incident ray is reflected from each of the mirrors as shown. (a) Show that $\phi = \theta + \beta$. (b) Show that the exiting ray makes an angle of $180° - 2\phi$ with the incident ray and that this is the case regardless of the angle of incidence (as long as the ray is reflected from both mirrors).

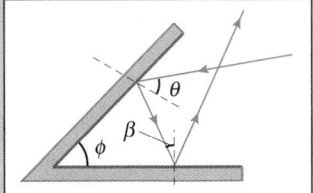

FIGURE P.8

•9. Two parallel rays are incident on two plane mirrors that make an angle ϕ with each other, as shown in Figure P.9. Show that the two reflected rays make an angle 2ϕ with each other, independent of the angles of incidence that the two parallel rays make with the two mirrors.

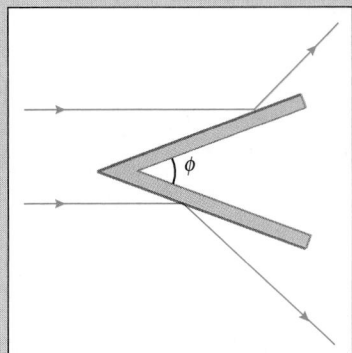

FIGURE P.9

‡10. A paraboloid of revolution is formed by rotating a parabola about its axis of symmetry. Prove that any ray of light parallel to the axis of such a parabolic reflector is reflected to the focus, as shown in Figure P.10. Such parabolic reflectors are used in astronomical telescopes to collect light. Car headlamps have a light source placed at the focus of a parabolic reflector, so that the rays emerge parallel to the axis of the parabola.

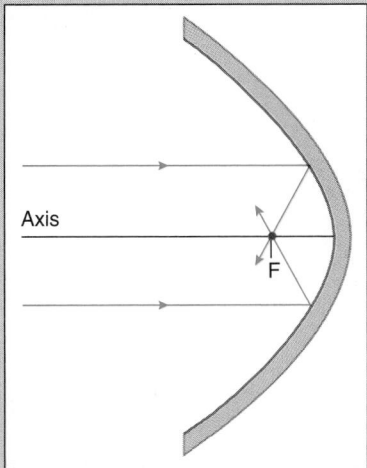

FIGURE P.10

‡11. A long reflecting cavity has a cross section that is an ellipse. Prove that a ray in the plane of the ellipse, emitted from one focus, is reflected to the other focus, as shown in Figure P.11. Such elliptical reflecting cavities are used in some lasers.

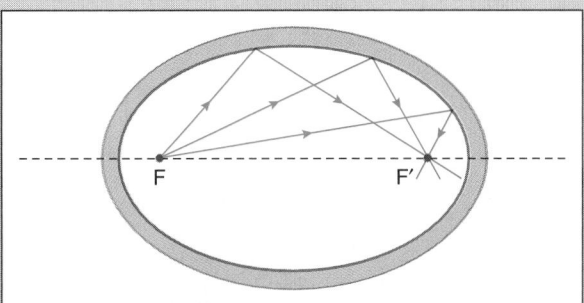

FIGURE P.11

‡12. A hyperbola of revolution is formed by rotating a hyperbola about its axis of symmetry, as in Figure P.12. Show that a ray emitted from one focus of such a reflector is reflected so that it appears to come from the focus of its opposite twin.

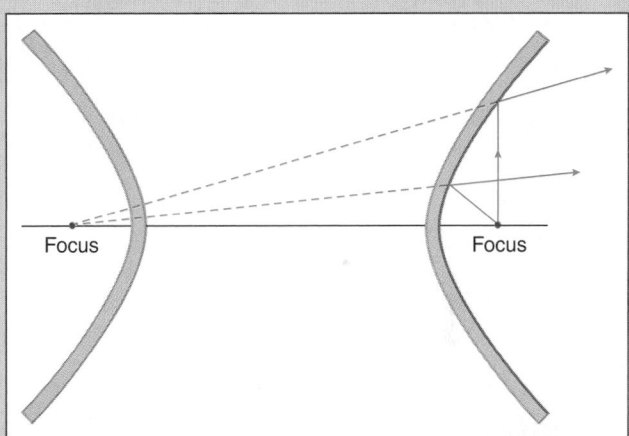

FIGURE P.12

‡13. A hollow cylindrical reflecting surface has diameter D and length ℓ as in Figure P.13. One end of the reflector has a cap with a small hole through which light from a light source can enter the reflecting pipe. An eye at point P views concentric circular rings of light caused by multiple reflections of light from the sides of the tube. (a) Show that the ring observed closest to the axis of the tube is formed by light from the hole making a reflection at a distance $\ell/2$ from the end of the tube with the hole. (b) Show that the next ring out from the axis

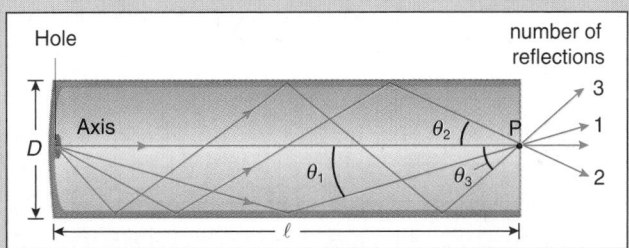

FIGURE P.13

of the tube is formed by a ray making *two* reflections from the sides of the tube, the first one at a distance $\ell/4$ from the end of the tube with the hole. (c) Show that the third ring from the axis of the tube is formed by three reflections from the sides of the tube, the first at a distance of $\ell/6$ from the end of the tube with the hole. (d) Show that the angular radius θ_m of the mth ring of light is found from

$$\tan \theta_m = \frac{Dm}{\ell}$$

where m is the number of reflections that the ray makes with the sides of the tube in propagating to point P.

This problem was inspired by Laurence A. Marschall and Emma Beth Marschall, "Reflections in a polished tube," *The Physics Teacher, 21, #2,* page 105 (February 1983).

Sections 23.4 The Law of Refraction
23.5 Total Internal Reflection
23.6 Dispersion
23.7 Rainbows*

14. (a) On the same piece of graph paper, make an accurate graph of the angle of refraction versus the angle of incidence for light incident from air onto a water surface and from air onto a glass surface. (b) Make similar graphs of the sine of the angle of refraction versus the sine of the angle of incidence. (c) What is meaning of the slope of the graphs in part (b)?

15. Determine the critical angle for a diamond ($n = 2.42$) immersed in water ($n = 1.33$).

• 16. A ray of light is incident in air on a transparent material with an index of refraction n, as shown in Figure P.16. Determine the angle of incidence that makes the angle of refraction *half* the angle of incidence.

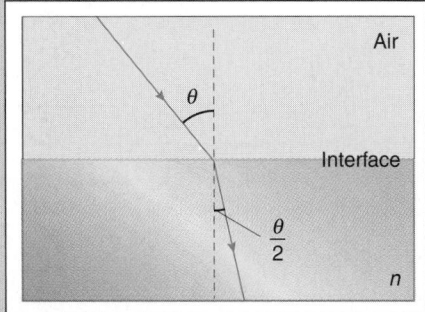

FIGURE P.16

• 17. A laser beam of cross-sectional diameter d is incident at angle θ on a glass plate with index of refraction n, as shown in Figure P.17. What is the width of the beam in the glass?

• 18. A ray of light is incident as shown in Figure P.18 on a 30°–60°–90° glass prism ($n = 1.50$), so that when *inside* the prism the ray is traveling parallel to the hypotenuse. (a) Determine the angle of incidence at which the incident ray struck the short face of the prism. (b) Find the direction of the ray after it leaves the prism.

• 19. An expensive spotlight is located at the bottom of a gold-plated swimming pool of depth 2.00 m (see Figure P.19).

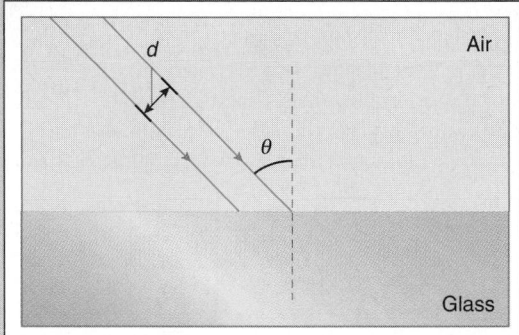

FIGURE P.17

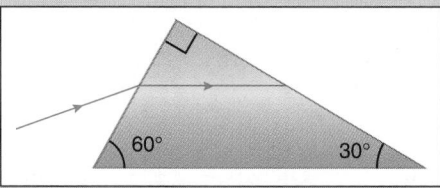

FIGURE P.18

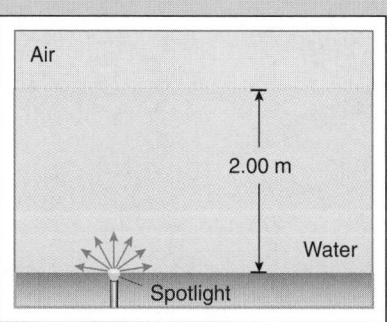

FIGURE P.19

Determine the *diameter* of the circle from which light emerges from the tranquil surface of the pool.

• 20. A ray of white light is incident at an angle of 60° on a glass plate of thickness 5.0 cm as shown in Figure P.20. The index of refraction for red light is 1.500 and that for violet is 1.510. Determine the width of the emerging beam of colors.

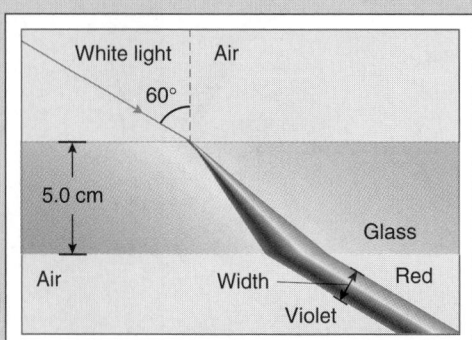

FIGURE P.20

•21. The red light from a helium–neon laser has a wavelength of 632.8 nm. If the light is incident on a plate of glass as indicated in Figure P.21, what is the *deviation angle* ϕ?

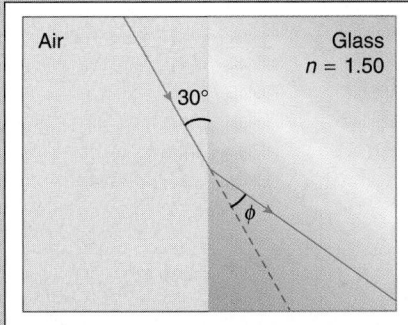

Air — Glass $n = 1.50$ — 30° — ϕ

FIGURE P.21

•22. A ray of light is incident on the transparent prism in Figure P.22. (a) Determine what index of refraction of the prism makes the light incident on the hypotenuse just barely totally reflect from the prism–air interface. (b) If the prism is made of material with an index *greater* than that calculated in part (a), what will happen to the path of the light? (c) If the prism is made of material with an index *less* than that calculated in part (a), what will happen to the path of the light?

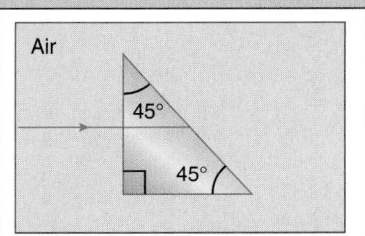

Air — 45° — 45°

FIGURE P.22

•23. Your cat is attentively watching a tasty ichthyological specimen from a vantage point level with the surface of the water and to the right of the opaque side of a water garden pond, as indicated in Figure P.23. The cagey fish, however, remarkably knows some physics. If the fish is at a depth of 5.0 cm, where relative to the edge of the pond can the fish swim so your cat is unable to see it?

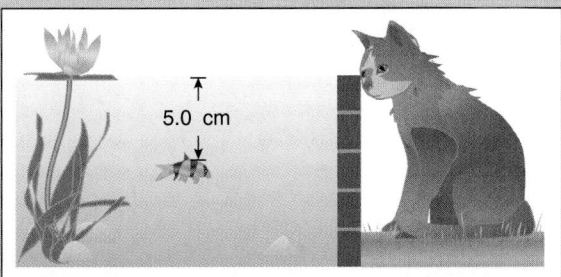

5.0 cm

FIGURE P.23

•24. In Figure P.24, when viewed from S, point P appears to be at point P′ along the perpendicular from P to the interface. The original version of the law of refraction formulated by Willebrord Snel (c. 1621) stated that the ratio of the distance OP to OP′ was a constant for all rays incident on the interface between two given transparent materials. Show that this statement leads to the current version of the law of refraction:

$$\frac{\sin \theta_1}{\sin \theta_2} = \text{constant}$$

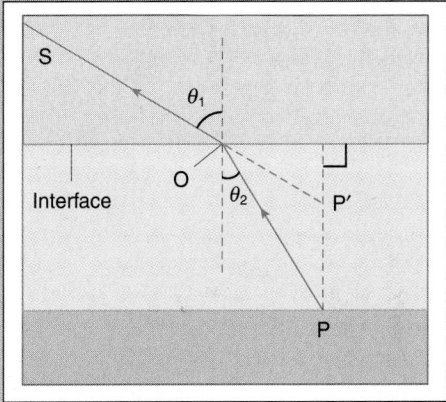

S — θ_1 — O — Interface — θ_2 — P′ — P

FIGURE P.24

•25. Precise measurements of the index of refraction are made using a prism with an apex angle A as shown in Figure P.25. A light ray is directed through the prism. The incident light is refracted twice and undergoes a deviation angle δ, which is a minimum when the ray inside the prism travels parallel to the base of the prism. When this is the case, show that the index of refraction of the prism (for the wavelength of the light used in the experiment) is given by

$$n = \frac{\sin \left[\frac{1}{2} (A + \delta) \right]}{\sin \left(\frac{A}{2} \right)}$$

where δ is the angle of minimum deviation.

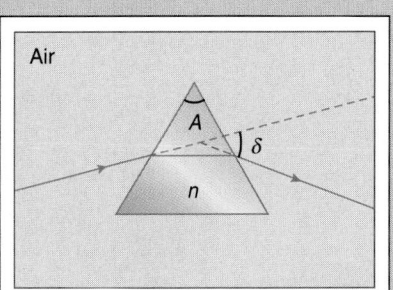

Air — A — δ — n

FIGURE P.25

•26. To calm troubled waters, a sheet of glass 2.0 cm thick is placed in full contact over water as indicated in Figure P.26. (a) What is the critical angle for the glass–air interface? (b) At what angle of incidence θ must light approach the water–glass interface so that the light is incident on the glass–air interface at the critical angle?

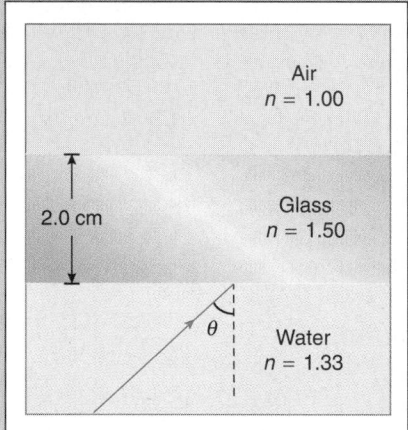

FIGURE P.26

•27. An optical fiber of index of refraction n is surrounded by air. A light ray enters the end of the fiber as shown in Figure P.27. (a) Show that the largest value of ϕ permitted, if the ray in the fiber is incident on the wall at the critical angle for the fiber–air interface, is

$$\sin\left(\frac{\phi}{2}\right) = (n^2 - 1)^{1/2}$$

The angle ϕ defines what is called the *acceptance angle* of an optical fiber. (b) To make the acceptance angle small, would you use a fiber with a high or low index of refraction? Explain your reasoning.

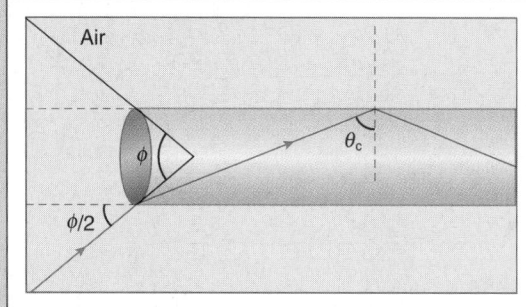

FIGURE P.27

•28. An old abandoned quarry has a vertical cylindrical shaft 20.0 m in diameter and 20.0 m deep. The hole is filled with water completely to ground level. To prevent children from falling into the water, a 1.5 m high fence surrounds the pool 3.0 m from the edge of the water. The water is perfectly calm. (a) Where under water can you lurk and not be seen by any person 2.0 m tall standing at the fence? (b) Where in the pool can you be located in order to see the entire fence height in all directions?

‡29. A parallel plate of thickness d and index of refraction n_2 is surrounded by a medium with index of refraction n_1. A ray of light

is incident on the parallel plate with an angle of incidence θ_1, as indicated in Figure P.29. (a) Show that the emerging ray is parallel to the incident ray. (b) Show that the lateral shift ℓ of the ray is

$$\ell = (d \sin\theta_1)\left[1 - \frac{n_1 \cos\theta_1}{(n_2^2 - n_1^2 \sin^2\theta_1)^{1/2}}\right]$$

Note when $\theta_1 = 0°$, there is no lateral shift. Why? Also, when $n_2 = n_1$, there is no lateral shift, regardless of the angle of incidence. Why?

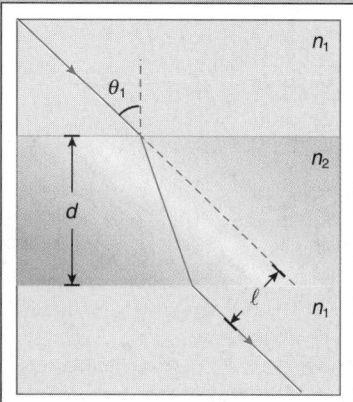

FIGURE P.29

‡30. Light passes from point A (coordinates $x = 0$ m, $y = \ell$) to point B (coordinates $x = \ell$, $y = -\ell$) by refracting somewhere at the surface between two transparent media at the point x as indicated in Figure P.30. Forget the law of refraction and let x vary. (a) Using the coordinate system provided, find an expression for the total time t it takes for light to travel from A to B via point x. (b) From the relation involving x, find the special value of x that makes the elapsed time from A to B a *minimum*. (c) Relate the result of part (b) to trigonometric functions of the angle of incidence θ_1 and the angle of refraction θ_2 to show that the law of refraction emerges from this minimization of the time. This is known as *Fermat's principle*.

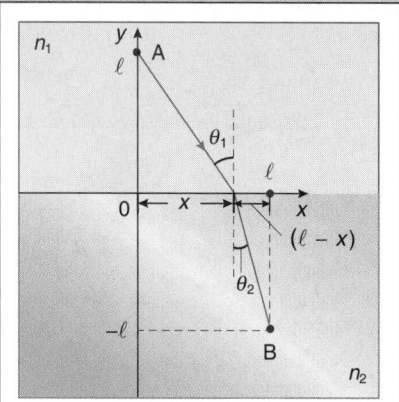

FIGURE P.30

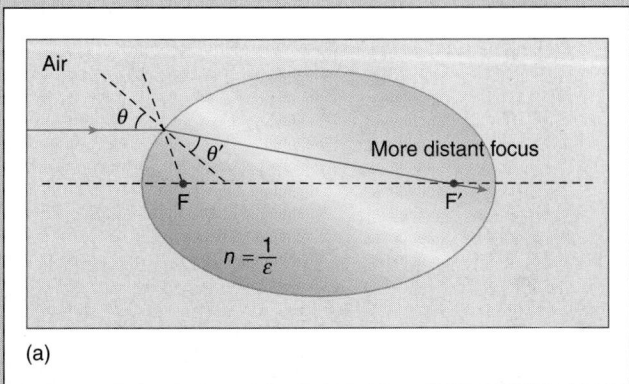

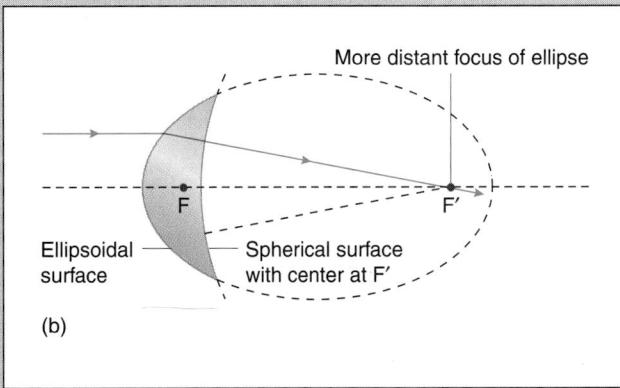

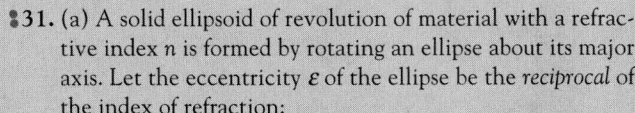

FIGURE P.31

31. (a) A solid ellipsoid of revolution of material with a refractive index n is formed by rotating an ellipse about its major axis. Let the eccentricity ε of the ellipse be the *reciprocal* of the index of refraction:

$$\varepsilon = \frac{1}{n}$$

Show that for such an ellipse a ray of light incident parallel to the major axis is refracted so that it passes through the more distant focus (see Figure P.31a). (b) If such an ellipsoid is truncated to form a lens by the surface of a sphere whose center is the "more distant" focus (see Figure P.31b), the light refracted at the ellipsoidal surface will proceed undeviated through the spherical surface and continue to the focus. (Why?) Such an arrangement has been suggested as a novel solar collector.

Part (a) was inspired by David Mountford, "Refraction properties of conics," *The Mathematical Gazette*, 68, pages 134–137 (1984). For the application in part (b), see A. Tan, "A novel refracting solar collector," *Physics Education*, 22, #3, page 141 (May 1987).

32. The path of a light ray involved in the formation of the primary rainbow is illustrated in Figure P.32. The figure depicts the path of a single wavelength with a refractive index n. The incident light is incident with an *impact parameter* b on a spherical water drop of radius R. The angle of incidence at point A is θ_1 and the angle of refraction is θ_2. At point A, the ray is deviated by an angle α. Some of the light is reflected from the back surface of the drop at point B, and deviated through an angle β at this point. Finally, the ray is refracted out of the drop at point C and deviated through the angle γ at this point. (a) The total angle of deviation of the light ray is the sum of the deviation angles $\alpha + \beta + \gamma \equiv \phi$. Show that

$$\phi = 180° + 2\theta_1 - 4\theta_2$$

(b) Examine the geometry at the point A and show that

$$\sin \theta_1 = \frac{b}{R}$$

(c) Use the law of refraction at point A to show that

$$\sin \theta_2 = \frac{b}{nR}$$

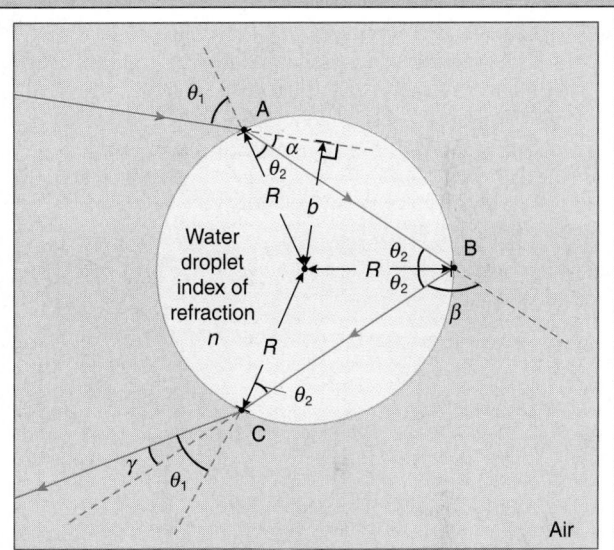

FIGURE P.32

(d) Let the ratio of the impact parameter b to the radius of the drop R be defined as

$$\psi \equiv \frac{b}{R}$$

a dimensionless parameter. Then the results of parts (a), (b), and (c) can be expressed as

$$\phi = 180° + 2 \sin^{-1}(\psi) - 4 \sin^{-1}\left(\frac{\psi}{n}\right)$$

Show that the deviation angle ϕ has a *minimum* when

$$\psi = \left(\frac{4 - n^2}{3}\right)^{1/2}$$

(e) Use the result of part (d) to show that if the spherical drop is water (with $n = 1.33$), then the minimum deviation angle occurs for

$$\psi = \frac{b}{R} = 0.862$$

and the minimum deviation angle then is

$$\phi = 138°$$

The acute angle between the incident and exiting rays then is

$$180° - 138° = 42°$$

Because of dispersion (the variation of *n* with wavelength), the various colors emerge with slightly different angles from the drop and give rise to the rainbow.

This problem was inspired by Colin M. Cartwright, "Rainbows," *Physics Education*, 27, #3, pages 155–158 (May 1992).

Sections 23.8 Objects and Images
23.9 The Cartesian Sign Convention
23.10 Image Formation by Spherical and Plane Mirrors
23.11 Ray Diagrams for Mirrors

•33. A hummingbird (*Calypte anna*) is hovering 50 cm in front of a spherical garden mirror of diameter 20.0 cm. (See Figure P.33.) Hmmm. (a) Determine the location of the image of our feathered friend. (b) Determine the magnification. (c) Is the image real or virtual? (d) Is the image upright or inverted?

FIGURE P.33

•34. A puffin (*Fratercula arctica*) is in front of a concave spherical mirror of radius *R*. Its image is real and the same size. Where is the puffin relative to the mirror?

•35. For a fixed focal length mirror, plot the image distance *s'* versus the object distance *s* < 0 m assuming: (a) *f* = 25 cm; (b) *f* = −25 cm. Let −50 cm ≤ *s* < 0 cm.

•36. A basketball player of height 2.2 m is standing 3.0 m in front of a convex spherical mirror of radius of curvature 4.0 m. (a) What is the focal length of the mirror? (b) Locate the image. (c) Determine the size of the image. (d) Is the image real or virtual? (e) Is the image upright or inverted? (f) Draw a neat ray diagram to confirm the calculations and conclusion drawn in parts (a)–(e).

•37. Dr. Ruth Canal is a dentist who is designing a small dental mirror. When placed 2.0 cm from a molar, the mirror will provide an image that is the same size as the tooth and (of course) upright. (a) Is the image real or virtual? (b) Determine the radius of curvature for the mirror.

•38. A horseshoe crab (*Limulus polyphemus*) in sea water finds itself 3.0 m in front of a convex, spherical mirror of radius of curvature 1.0 m. (a) Locate the position of the crabby image. (b) Is the image real or virtual? (c) What is the magnification? (d) Is the image upright or inverted? Does the crab care? (e) Draw a neat ray diagram to confirm your calculations.

•39. You are preening yourself for the annual Fancy Dress Ball in front of a large plane mirror as indicated in Figure P.39. Show geometrically that the vertical height of the mirror needed to see yourself is completely half your height. This means that a full-length mirror (properly mounted on a wall) need only be half your height. You can never view yourself in entirety in a cosmetic, pocket size mirror.

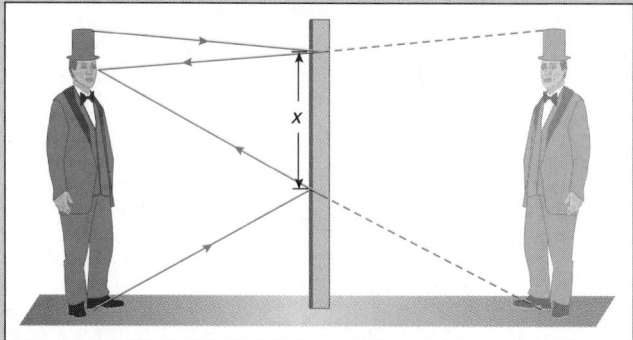

FIGURE P.39

•40. The Wicked Witch of the West is rushing along a normal line toward a plane mirror at speed *v*. At what speed is her image approaching her?

•41. If the witch in the previous problem approaches the mirror at speed *v* at an angle of incidence *θ*, at what speed is her image approaching her?

•42. A convex mirror of focal length +2.00 m is used to form an image of an impala (*Aepyceros melampus*) located 6.0 m from the mirror. (a) Find the location of the image. (b) Determine the magnification. (c) Is the image upright or inverted? (d) Draw a ray diagram to confirm your calculations.

•43. A tasty lollipop is located 200 cm in front of a concave spherical mirror with a focal length of absolute value 50 cm. (a) Locate the image of the sucker. (b) Determine the magnification. (c) Is the image real or virtual? (d) Is the image upright or inverted? (e) Construct a *neat* ray diagram to confirm your calculations and conclusions in parts (a)–(d).

•44. A large, circular mirror has a focal length of +2.0 m and a diameter of 0.50 m. (a) Is the mirror concave or convex? (b) What is the radius of curvature of the mirror? (c) A chowder clam (*Venus mercenaria*) is placed 1.0 m in front of the mirror. Where is the image, what is its magnification, and is the image real or virtual? (d) Draw a ray diagram to confirm the aspects of the image found in part (c). (e) If the entire system is submerged in crystal clear water, how are the answers to (b), (c), and (d) affected? Explain your reasoning.

•45. The "Optic Mirage" toy (Edmund Scientific Co.) produces very realistic real images of objects placed at the vertex of the mirror at the bottom of the two-mirror array. (See Figure P.45.) The two concave mirrors have radii of curvature of identical

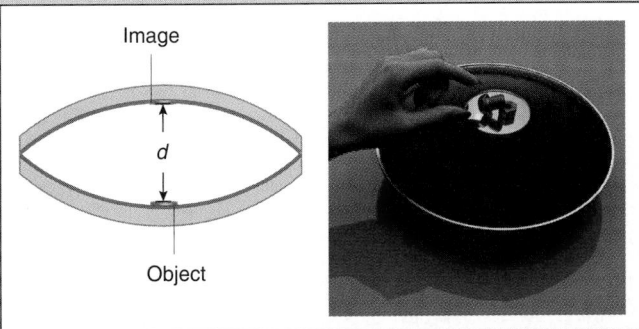

FIGURE P.45

absolute magnitude $|R|$. Show that if the separation d of the mirror vertices is *either* $|R|/2$ or $3|R|/2$, a real image of a heads-up coin is formed at the mirror vertex at which the coin is *not* located. Will the image show heads-up or tails-up? Determine the magnification of the object for each mirror separation.

The idea for this problem originated with the following article: Andrzej Sieradzan, "Teaching geometrical optics with the 'Optic Mirage,'" *The Physics Teacher*, 28, #8, pages 534–536 (November 1990).

•**46.** Two plane mirrors face each other a distance d apart as shown in Figure P.46. You stand a distance x in front of the mirror on the right, admiring a new addition to your wardrobe. (a) How many images are there? (b) Are the images real or virtual? (c) What is the separation of successive images in each mirror? (d) If you have a patch over your left eye, which images have the patch over the left eye of the image and which have the patch over the right eye of the image, or do they all have the patch over the same eye (in which case, which eye?)?

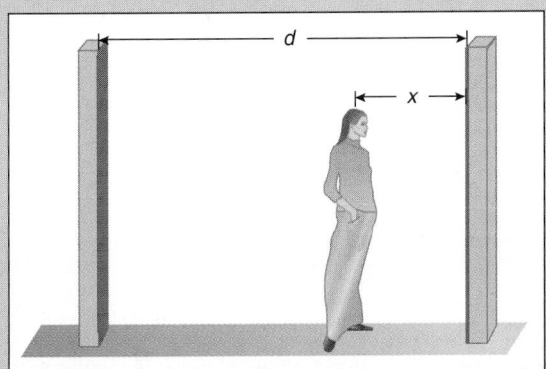

FIGURE P.46

•**47.** Two plane mirrors are in contact with each other along one edge and form an angle of 40° between their faces as shown in Figure P.47. A small American firefly (*Photuris pennsylvanica*) light source is hovering midway between the two mirrors. In a carefully drawn figure, geometrically locate all the images of this marvelous insect. How many are there?

•**48.** A magician uses a clear plate of glass and a large concave mirror in a black enclosure (see Figure P.48) to project a real image of a disembodied head. The plate glass reflects and also transmits light. The center of curvature of the mirror is at C in Figure P.48. (a) The flat glass forms a reflection image. Where is this image and how is it oriented? (b) The curved mirror uses the image of the flat glass plate as its object to form a final real image. Where is the final image, what is its size, and what is its orientation?

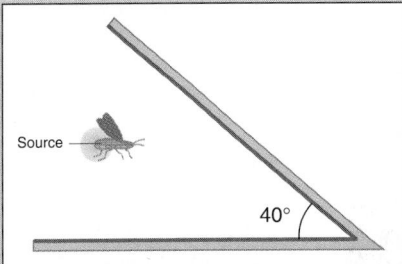

FIGURE P.47

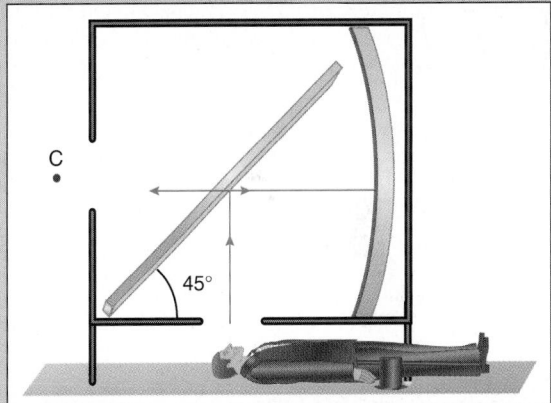

FIGURE P.48

•**49.** Four lollipops of different shapes are in a stand in front of a plane mirror as shown in Figure P.49. Sketch the total image.

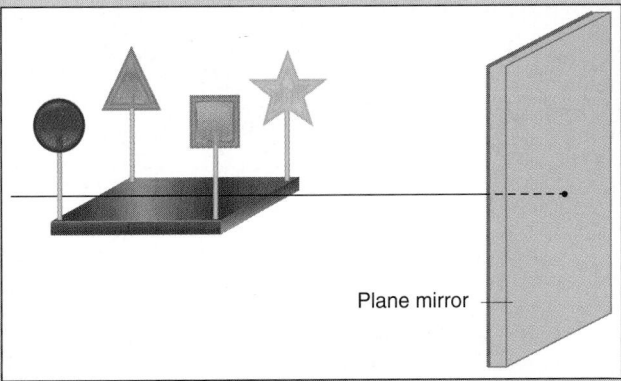

FIGURE P.49

•**50.** A flat semicircle with one quadrant shaded is placed in front of two plane mirrors oriented as shown in Figure P.50. Sketch the orientation and shading of the three images.

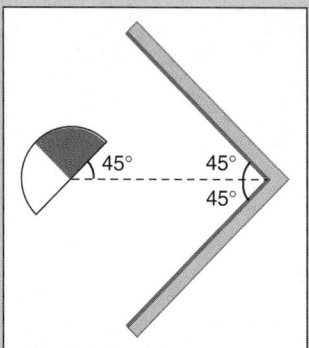

FIGURE P.50

‡**51.** Telescopic mirrors typically are shaped not as sections of a spherical surface but as paraboloids of revolution. Several large new-technology telescope mirrors were made by a new process of spinning a mass of molten glass, a process known as spin casting.* The spinning molten glass, as we will see in this problem, assumes a surface that is a paraboloid of revolution. The melted glass then is cooled slowly while spinning so that it solidifies into the desired paraboloid shape. To see that a spinning surface assumes the shape of a parabola of revolution and to see how the focal length of a mirror surface of this shape is set by the angular speed of the rotation, consider the following: (a) Imagine a small mass element m of the mirror as it is spinning at constant angular speed ω at a radius x from the axis as in Figure P.51a. Two forces act on the mass element: its weight and the normal force of the surface acting perpendicular to the surface arising from the rest of the glass in the melt. Use Newton's second law to show that

$$\tan \theta = \frac{x\omega^2}{g}$$

where θ is the angle that the tangent to the surface makes with the horizontal direction at the location of the mass. (b) Notice that with the coordinate system in Figure P.51a,

$$\tan \theta = \frac{dy}{dx}$$

Use this relation from analytic geometry and the result of part (a) to show that

$$y = \frac{\omega^2}{2g} x^2$$

*See the following articles: Mark Dragovan and Don Alvarez, "Making a mirror by spinning a liquid," *Scientific American*, *270*, #2, pages 116–117 (February 1994); Ben Iannotta, "Spinning images from mercury mirrors," *New Scientist*, *147*, #1986, pages 38–41 (15 July 1995); and Ermanno F. Borra, "Liquid mirrors," *Scientific American*, *270*, #2, pages 76–81 (February 1994).

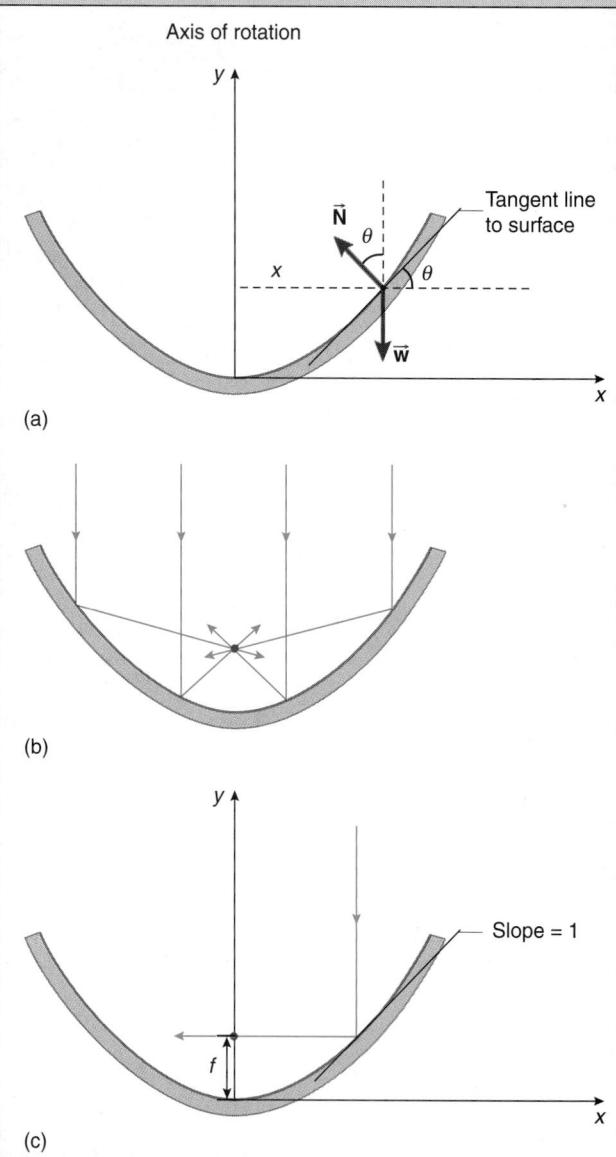

FIGURE P.51

This shows that the shape of the surface is parabolic. (c) Any light ray incident parallel to the axis of the parabola will go through the focus (the focal point!) of the parabola (see Figure P.51b). Consider the special parallel ray incident on the surface where the slope of the surface is equal to unity (Figure P.51c). Show that for this ray

$$x = \frac{g}{\omega^2}$$

(d) Use the result of part (c) and the equation of the parabola from part (b) to show that the focal length f of the paraboloid surface is

$$f = \frac{g}{2\omega^2}$$

(e) Calculate the spin rate in revolutions per minute needed to make a mirror with a focal length of 5.00 m.

Section 23.12 Refraction at a Single Spherical Surface

•**52.** Your pet angelfish (*Pterophyllum scalare*) is located 10.0 cm from the edge of a thin-walled fish tank as shown in Figure P.52. Neglect the effect of the thin glass wall. (a) Determine the location of the image of the fish. (b) Determine the magnification. (c) Is the image real or virtual? (d) Is the image upright or inverted? (e) Why is the thin wall optically unimportant?

FIGURE P.52

•**53.** A small olive is suspended 3.0 cm from the bottom of a large glass filled to the brim with peculiarly flavored water, as shown in Figure P.53. (a) What is the apparent depth of the olive as seen from a point directly above the glass? (b) What is the magnification of the olive? (c) Is the image real or virtual? Upright or inverted?

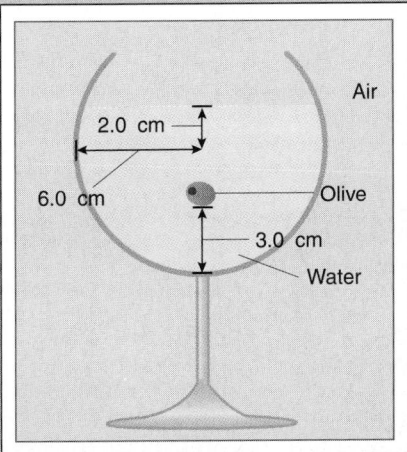

FIGURE P.53

•**54.** A glass crystal ball of radius 10 cm is used by a local wizard to divine your grade for the course. (See Figure P.54.) How far is the focal point of the sphere from its center?

•**55.** The warm and inviting waters of Tahiti beckon you from the rigors of winter. While boating in a calm cove of depth 4.0 m, you spy a pirate treasure poking up slightly from the bottom (Figure P.55). What is the apparent depth of the treasure when viewed along the normal line?

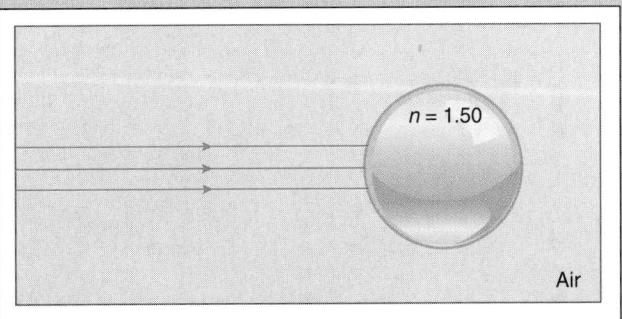

FIGURE P.54

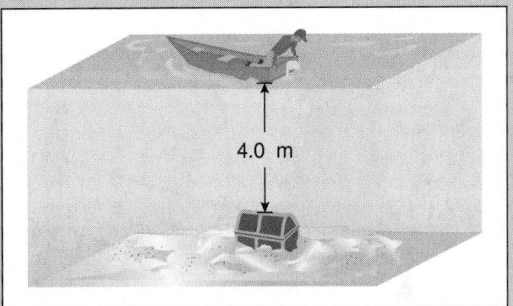

FIGURE P.55

•**56.** While scuba diving (using a face mask with a very thin glass plate) in the tranquil and warm waters off the French Riviera, you stretch your arms straight out in front of you. (a) Will your arms appear to you to be shorter, longer, or the same size than they really are? (b) If your arms really are 70 cm in length, how long do they appear to be? (c) You notice a small fish touching your fingertips. If the fish swims at a speed v toward the face mask, will its observed *image* move at speed v? Faster? Slower?

•**57.** A layer of crystal clear ice 1.00 m thick ($n = 1.30$) floats on the surface of a pond. A brown trout (*Salmo trutta*) lies motionless at the boundary between the ice and water. (a) Sketch the path of the light ray from the trout that spends the shortest time in the ice before passing into the air. (b) Sketch the path of the light ray from the trout that spends the greatest time in the ice before passing into the air. (c) How thick does the ice appear when viewed from above? (d) How thick does the ice appear when viewed from the water underneath?

•**58.** A penny is located 15.0 cm from the surface of a glass crystal ball of radius 5.00 cm with $n = 1.500$. (a) Determine the location of the final image of the penny. (b) Is the image real or virtual? (c) Determine the magnification and orientation of the penny.

•**59.** A parallel beam of laser light is incident on a sphere (surrounded by air). The light is brought to a focus at the rear surface of the sphere as indicated in Figure P.59. What is the index of refraction of the sphere? Does your result depend on the radius of the sphere?

‡**60.** A spinning vertical cylinder of water forms a curved surface that is approximately spherical (actually it is a paraboloid; see

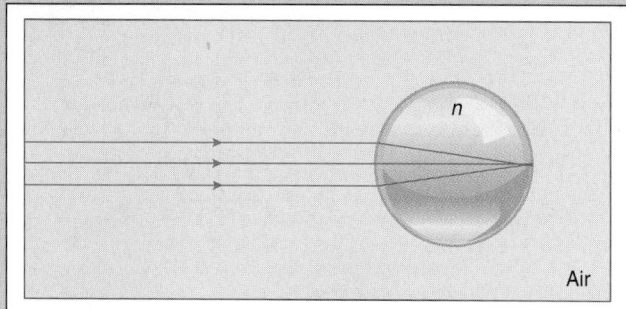

FIGURE P.59

Problem 51) and of radius of curvature 0.40 m. A bright electric spark occurs in the water at a depth of 0.80 m. (a) Where does the spark appear to be located as seen by a person above, looking along the vertical axis of the cylinder? (b) We shall see in Chapter 24 that the index of refraction of light is the ratio of the speed of light in vacuum to that in the material. The speed of the *sound* of the spark in water is approximately four times the speed of sound in air. The sound wave follows the standard refraction laws. Using the speed of sound in air as a reference speed, calculate the *sonic* index of refraction for water relative to air. (c) The sound from the spark is transmitted from the water into the air. Where is the acoustic image of the sound source?

Sections 23.13 Thin Lenses
23.14 Ray Diagrams for Thin Lenses

Assume all lenses are thin unless otherwise stated.

•61. Use the thin lens equation (Equation 23.52) to show that a diverging lens can never form a real image of a real object no matter where the object is placed.

•62. For a fixed focal length lens, sketch a plot of s' versus s assuming: (a) $f = 25$ cm; (b) $f = -25$ cm. Assume a domain for s of -100 cm $< s < 0$ cm.

•63. A lens has radii of curvature of its first and second surfaces of 20 cm and 25 cm, respectively, as indicated in Figure P.63. The lens is made from recycled physic elixir bottles with an index of refraction of 1.50. The lens is surrounded by air. (a) Find the focal length of the lens. (b) Express the focal length in diopters. (c) Is the lens converging or diverging? (d) An uninteresting object is placed 40 cm from the lens with the standard geometry. Find the location of its equally uninteresting image. Is the image real or virtual? Determine the magnification. Is the image upright or inverted? Verify your calculations with a ray diagram.

FIGURE P.63

•64. A small snipe (*Capella gallinago*) finds itself 150 cm in front of a diverging lens. Time for a snipe hunt. (a) If the absolute value of the image distance is 50 cm, find the focal length of the lens. (b) Determine the magnification. (c) Is the image real or virtual? (d) Is the image upright or inverted? (e) Draw a neat ray diagram to locate the position of the image. (f) If the upper half of the lens is now painted black so that no light can pass through this portion of the lens, describe what happens (if anything) to the image.

•65. A small gremlin from Academia finds itself 150 cm in front of a two-lens optical system as indicated in Figure P.65. For light entertainment, determine: (a) the location of the final image; and (b) the total magnification. (c) Is the final image real or virtual? Upright or inverted?

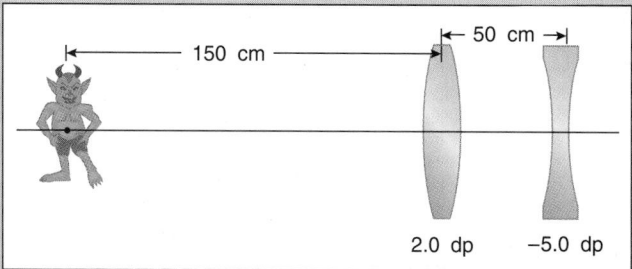

FIGURE P.65

•66. Two thin lenses of focal lengths f_1 and f_2 are placed in contact with each other; the combination still is thin. Show that the equivalent focal length f of the combination is

$$f = \frac{f_1 f_2}{f_1 + f_2}$$

•67. A glass convex lens ($n = 1.50$) has radii of curvature for both of its surfaces of absolute magnitude $|R|$. (See Figure P.67.) (a) Determine the focal length of the lens when used in air. (b) In terms of the focal length in air, determine the focal length of the lens when it is immersed in a medium with an index of refraction n.

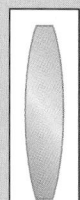

FIGURE P.67

•68. Santa Claus has given you a converging lens with a 10.0 cm focal length. The instructions ask you to place a candy cane 50.0 cm from a screen. (a) Determine where to place the lens so that a real, enlarged image of the candy cane appears on the screen. What is the magnification? Is the image upright or inverted? (b) Determine where to place the lens such that a real, reduced image of the candy cane appears on the screen. What is the magnification? Is the image upright or inverted?

•69. A lens of focal length f_{lens} is placed directly in front of and in contact with a mirror of focal length f_{mirror}. The combination is considered to be thin. What is the effective focal length of the combination?

•70. Most single-lens-reflex (SLR) cameras come with a standard lens of focal length 50.0 mm. The lens is used to form an image of an Oxford gargoyle, located 10.00 m from the lens. (a) Determine the image distance and the magnification. (b) The lens is replaced with another lens of focal length 200 mm. Determine the image distance and the magnification of the gargoyle using this lens. (c) What is the *ratio* of the magnification with the 200 mm lens to that with the standard 50.0 mm lens? This is the reason that camera lenses with focal length longer than the standard 50.0 mm lens are known as *telephoto lenses*. Compare this ratio with the ratio of the focal lengths. (d) If the lens is switched to one with a focal length of 28.0 mm, determine the image distance and the magnification of the gargoyle. Compare the magnification of the 28.0 mm lens with that of the 50.0 mm lens. This is the reason that lenses with focal lengths shorter than the standard 50.0 mm focal length lens are known as *wide angle lenses*.

•71. A candelabra is located 5.00 m from a screen. Determine the focal length and location of a lens needed to form a real, inverted image that is three times the size of the candelabra.

•72. A converging lens is placed at the center of curvature of a concave mirror as indicated in Figure P.72. The absolute magnitude of the focal lengths of both the lens and mirror is 25.0 cm. A ladybug (*Anatis quindecimpunctatum*) is placed 50.0 cm in front of the lens. Determine the location, magnification, and nature of the final image of the bug (real or virtual; upright or inverted).

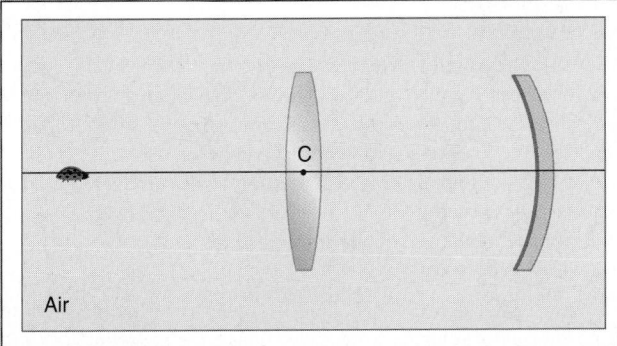

FIGURE P.72

•73. A projector with a lens of 10.0 cm focal length is to produce an image of slide (22.0 mm × 34.0 mm) on a screen (150 cm × 150 cm). The image fills as much of the screen as possible. (a) What is the magnification? (b) Is the image inverted? Explain how to place the slide to yield an image that is properly oriented. (c) How far is the slide from the lens? (d) What is the distance from the lens to the screen?

•74. A diverging lens with a focal length of −20.0 cm and a mirror lie 10.0 cm apart as shown in Figure P.74. The radius of curvature of the mirror is not known. Parallel light rays, from an object infinitely afar, enter the lens as indicated. You note that the final image is located 20.0 cm to the left of the lens. (a) Determine the radius of curvature of the mirror. (b) Is the mirror concave or convex? (c) Is the final image real or virtual?

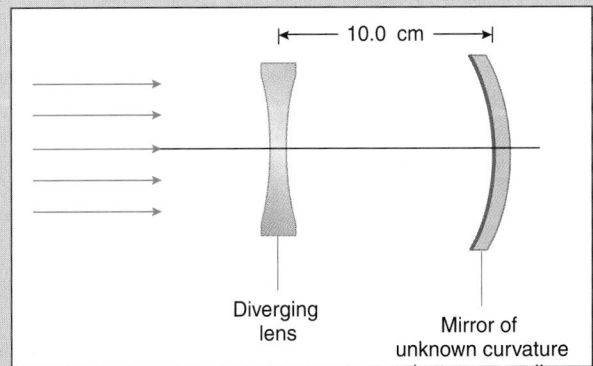

FIGURE P.74

•75. A beam of parallel light is incident on a thin lens of focal length $f = +20.0$ cm. Some distance from the lens is a glass crystal ball ($n = 1.50$) of radius 10.0 cm. The light is brought to a focus at the center of the ball. What is the distance between the center of the ball and the lens?

•76. A black widow spider (*Latrodectus mactans*) is 60.0 cm in front of the first lens of a two-lens system as indicated in Figure P.76. (a) Find the position of the final image relative to the second lens. (b) Determine the total magnification. (c) Is the final image real or virtual? Upright or inverted with respect to the orientation of the original spider object?

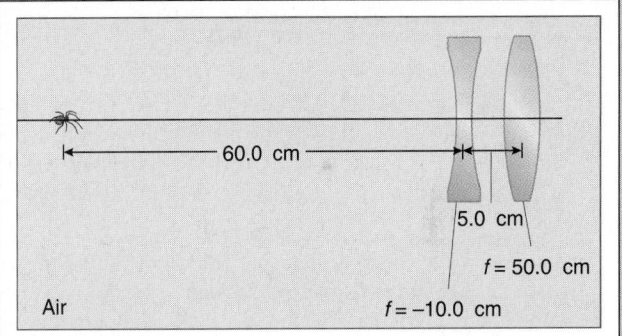

FIGURE P.76

•77. Four lollipops of different shapes are in a small stand 50.0 cm in front of a convex lens of focal length 25.0 cm. See Figure P.77. Consider all the lollipops to be essentially the same distance from the lens. Sketch the image of the suckers.

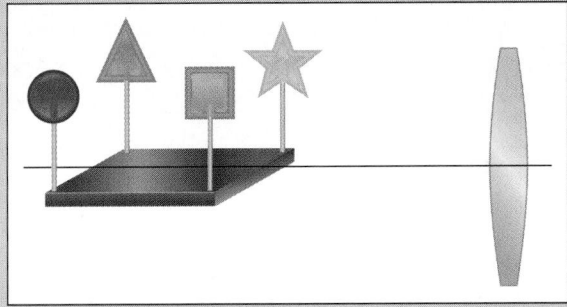

FIGURE P.77

•78. An object is placed at the secondary focal point of a thin converging lens, and the lens is followed (at any distance) by a plane mirror as indicated in Figure P.78. Show that the final image is located at the same position as the object but is inverted. Describe how you can use this result to experimentally find the focal length of a converging lens.

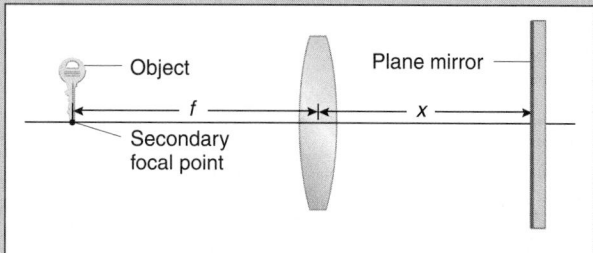

FIGURE P.78

•79. A romantic, lighted candle is placed 40 cm in front of a diverging lens. The light passes through the diverging lens and on to a converging lens of focal length 10 cm that is 5 cm from the diverging lens. The final image is real, inverted, and 20 cm beyond the converging lens. Find the focal length of the diverging lens.

•80. The diameter of the Sun subtends an angle of 0.50° when viewed from the Earth. The intensity of direct sunlight is 1.36 kW/m² on the surface of the Earth. A 5.0 cm diameter lens forms a hot image of the Sun that is 1.0 cm in diameter on a screen. (a) How do you know that this is a real image? (b) What light power goes through the lens? (c) What light power reaches the hot image? (d) What is the intensity of the light in the image W/m²? Your result is much greater than the intensity of direct solar light. Does this result violate energy conservation? Explain. (e) What is the focal length of the lens? Is the lens converging or diverging?

•81. Three thin converging lenses each of focal length 60 cm are placed close together so that the combination still can be considered a thin lens. An object is placed 30 cm in front of the system. Find the location of the image.

‡82. (a) For a converging lens of focal length f (> 0 m), show that the distance d (> 0 m) between a real object and its real image ($d = -s + s'$) is equal to

$$d = -\frac{s^2}{s + f}$$

(b) Show that the distance d is a *minimum* when $s = -2f$.

‡83. (a) Beginning with the thin lens equation, show that if the object distance s changes with time, the image distance s' changes with time according to

$$\frac{ds'}{dt} = m^2 \frac{ds}{dt}$$

(b) Beginning with the mirror equation, show that the corresponding relationship is

$$\frac{ds'}{dt} = -m^2 \frac{ds}{dt}$$

Section 23.15 Optical Instruments

84. A myopic lawyer has a far point of 25 cm. What is the diopter value of suitable corrective eyeglasses?

85. You have misplaced your eyeglasses and are bumbling around the physics building. Your far point is 15 cm. Specify the diopter value of suitable corrective eyeglasses.

86. A telescope with an objective lens of focal length 200.0 cm is provided with an assortment of eyepiece lenses of focal lengths 6.0 mm, 25 mm, and 40 mm. (a) What is the greatest absolute value of magnification that can be achieved? (b) What is the smallest absolute value of magnification that can be achieved?

•87. The film in a pinhole camera is moved from a distance d (from the pinhole) to a distance $2d$. (a) What happens to the magnification of the image? (b) By how much is it necessary to change the exposure time if you need the same total amount of light per unit area on the film during the exposure?

•88. A microscope with an eyepiece lens of focal length 2.00 cm and an objective lens of focal length 1.00 cm is used to examine a section of the brain of a sacrificed rat. The distance between the lenses of the instrument is 22.0 cm and the final image is virtual and at infinity. (a) How far from the objective lens is the object? (b) What is the magnification produced by the objective lens? (c) What is the total magnification produced by the microscope?

•89. You are admiring the views of Mt. Phobos and Mt. Deimos. The twin peaks are 1.0 km apart and 5.0 km distant along the perpendicular bisector of the line joining the peaks. You whip out your 35 mm SLR camera with its 50.0 mm focal length lens to record the panorama. What is the distance separating the twin peaks on the film?

•90. A camera lens with an aperture of 20 mm and a focal length of 50.0 mm is used to photograph your class for an upcoming issue of the alumni magazine. The exposure time is 10 ms. (a) With the 20 mm aperture, what is the f/number used for this exposure? (b) What diameter of lens opening would produce the same exposure on the film in a time of 40 ms? What is the corresponding effective f/number for this exposure?

•91. Two 1.0 cm focal length converging lenses are used to form a microscope with a magnification of −350 with a final virtual image infinitely far from the eye. Find the approximate separation of the lenses.

•92. A hyperopic patient can see objects clearly only if they are farther than 2.00 m away. Distant vision is good. The patient wishes to read a book when it is 25 cm away. (a) Calculate the focal length of the eyeglasses necessary to provide this correction. (b) Where is the image formed by the eyeglasses? Is the image real or virtual? (c) What is the magnification caused by the eyeglasses? (d) Why is the eye now able to see the image in (b) clearly? (Explain with no numbers!) (e) Is the final image on the retina real or virtual? Upright or inverted? With a magnification $|m| > 1$ or < 1?

•93. (a) Take appropriate measurements to determine the following: when viewed at arm's length, about what angle does the four-finger width of your hand fill? Express your result in radians and degrees. (b) The Moon has a diameter of 3.48×10^6 m and is 3.84×10^8 m away. What is the angular size of the Moon? Express your result in radians and in degrees.

A giant telescope has an objective element that is a concave mirror 2.00 m in diameter with a radius of curvature of −24.00 m. (c) What is the focal length of the mirror? (d) What angle in radians (and degrees) does the image of the Moon subtend as measured from the mirror? (e) Is the image of the Moon real or virtual? (f) What is the linear width of the image of the Moon as photographed on film that is placed at the focal plane of the mirror? (g) If the mirror aperture were *square* in shape instead of circular, what would be the shape of the image of the Moon? Explain your reasoning.

•**94.** A microscope consists of two lenses of 15 mm focal length that are 20.0 cm apart. The device is used to examine a cootie that is of diameter 0.10 mm. (a) How far from the objective lens must the cootie be placed for most comfortable viewing through the eyepiece (i.e., for a final image that is virtual and infinitely far away)? (b) What is the angular extent of the final image?

•**95.** A tree 15.0 m high is 1.000 km away from a two-lens system consisting of a converging lens of focal length 12.0 cm followed by a diverging lens of focal length −2.0 cm. The separation of the lenses is 10.3 cm. (a) Find the location of the image of the first lens. (b) Find the angular size of the image of the first lens. (c) Find the object distance for the second lens. (d) Find the image distance of the second lens. (e) Find the magnification provided by the second lens. (f) Find the height of the final image. This optical system is a compact telephoto lens.

•**96.** A bug-a-boo is located 25.0 cm in front of a −10.0 dp lens. Another +5.0 dp lens is located 25.0 cm from the first lens as indicated in Figure P.96. (a) Locate the position of the final image of the system. (b) Determine the total magnification. (c) Is the final image real or virtual? (d) Is the final image upright or inverted?

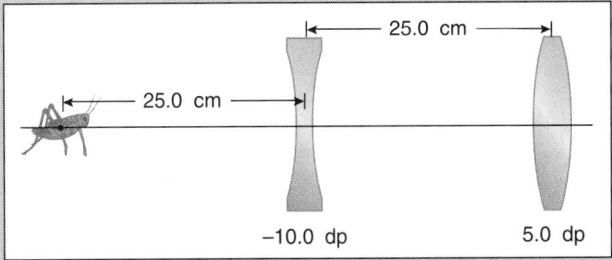

FIGURE P.96

•97. It is fun to look through a telescope backward—that is, with the light from a distant source first entering, say, a 2.0 cm focal length converging lens, followed by a 20.0 cm focal length converging lens. (a) Calculate the magnification of this arrangement and describe the image. (b) How could this backward arrangement be used to look at an object close to you on the table so that the final image has a magnification of absolute magnitude greater than one?

INVESTIGATIVE PROJECTS

A. Expanded Horizons

1. You might enjoy reading and reporting about some of the revolutionary applications of optical fibers in both communications and medicine.

   W. S. Boyle, "Light-wave communications," *Scientific American*, 237, #2, pages 40–48, 140 (August 1977).

   Harry Rheam, "Lightwave communications," *Science Teacher*, 60, #5, pages 26–29 (May 1993).

   Emmanuel Desurvire, "Lightwave communication: the fifth generation," *Scientific American*, 266, #1, pages 114–121 (January 1992).

   Alastair M. Glass, "Fiber optics," *Physics Today*, 46, #10, pages 34–38 (October 1993).

   Gary Stix, "The last frontier," *Scientific American*, 270, #3, pages 105–106 (March 1994).

   Abraham Katzir, "Optical fibers in medicine," *Scientific American*, 260, #5, pages 120–125 (May 1989).

   Abraham Katzir, *Lasers and Optical Fibers in Medicine* (Academic Press, San Diego, 1993).

2. The magnification of many microscopes can be increased by placing a small drop of oil on the microscope slide and bringing the objective lens into contact with the oil drop. Investigate these so-called *oil immersion* techniques in microscopy and determine how the droplet is able to increase the magnification.

   Francis A. Jenkins and Harvey E. White, *Fundamentals of Optics* (4th edition, McGraw-Hill, New York, 1976).

   Max Born and Emil Wolf, *Principles of Optics* (6th edition, Cambridge University Press, Cambridge, England, 1997), pages 253–254.

3. Investigate the cause of the *green flash* in atmospheric optics that occurs occasionally as the Sun sets or rises.

   Roger W. Sinnott, "The green flash," *Sky and Telescope*, 83, #2, pages 200–203 (February 1992).

   Robert Gannon, "Stalking the elusive green flash," *Focus*, 38, #3, pages 10–11 (Fall 1988).

   Sky and Telescope, 87, #2, pages 110–111 (February 1994).

   D. J. K. O'Connell, "The green flash," *Scientific American*, 202, #1, pages 112–122, 189–190 (January 1960).

   Robert Greenler, *Rainbows, Halos and Glories* (Cambridge University Press, Cambridge, England, 1989).

4. What causes the ring seen occasionally around the Moon? What is the angular size of the ring? Does it have colors like the rainbow?

   Walter Tape, *Atmospheric Halos* (American Geophysical Union, Washington, D.C., 1994).

   Bartley L. Cardon, "An unusual lunar halo," *American Journal of Physics*, 45, #4, pages 331–335 (April 1977).

   Robert Greenler, *Rainbows, Halos and Glories* (Cambridge University Press, Cambridge, England, 1989).

   David K. Lynch and William Livingston, *Color and Light in Nature* (Cambridge University Press, Cambridge, England, 1995).

5. Investigate how mirrors occasionally are used for deception by magicians.

   Michael J. Ruiz and Terry L. Robinson, "Mirrors in magic," *The Physics Teacher*, 25, #4, pages 206–212 (April 1987).

   Derek B. Swinson, "Magic mirrors—front and back," *The Physics Teacher*, 32, #6, page 329 (September 1994).

   Marshall Ellenstein, "Magic and physics," *The Physics Teacher*, 20, #2, pages 104-106 (February 1982).

6. Investigate the optical path of light through a spherical rain drop in the formation of a *secondary rainbow*. The region

between the primary and secondary rainbows is noticeably darker than that within the primary or outside the secondary rainbows; the darker region between the rainbows is known as the *Alexander dark band*. Investigate its cause.

Colin M. Cartwright, "Rainbows," *Physics Education*, 27, #3, pages 155–158 (May 1992).

7. Investigate the use of materials with a spatially varying index of refraction (so-called *gradient index materials*) in the fabrication of lenses.

Neil Morton, "Gradient refractive index lenses," *Physics Education*, 19, #2, pages 86–90 (March 1984).

Erich W. Marchand, *Gradient Index Optics* (Academic Press, New York, 1978).

James Evans and Mark Rosenquist, "'F = ma' optics," *American Journal of Physics*, 54, #10, pages 876–883 (October 1986).

James Evans, "The ray form of Newton's law of motion," *American Journal of Physics*, 61, #4, pages 347–350 (April 1993).

K. C. Mamola, Wilhelm F. Mueller, and Bruce J. Regittko, "Light rays in gradient index media: a laboratory exercise," *American Journal of Physics*, 60, #6, pages 527–529 (June 1992).

8. Investigate the geometric optics associated with the appearance of mirages.

Alistair B. Fraser and William H. Mach, "Mirages," *Scientific American*, 234, #1, pages 102–111 (January 1976).

Walter Tape, "The topology of mirages," *Scientific American*, 252, #6, pages 120–129 (June 1985).

David S. Falk, Dieter R. Brill, and David G. Stork, *Seeing the Light* (Harper and Row, New York, 1986), pages 58–62.

Alistair B. Fraser, "Theological optics," *Applied Optics*, 14, #4, pages A92–A93 (April 1975).

E. Khular, K. Thyagarajan, and A. K. Ghatak, "A note on mirage formation," *American Journal of Physics*, 45, #1, pages 90–92 (January 1977).

G. P. Sastry, "Teaching mirages," *American Journal of Physics*, 46, #7, page 765 (July 1978).

9. You look in a plane mirror and see that it reverses right for left but does not reverse up for down. Should a flat mirror know up–down from right–left? Investigate what is meant by this description and whether in a strict sense it is true or not.

J. Ken Gee, "The myth of lateral inversion," *Physics Education*, 23, #5, pages 300–301 (September 1988).

10. The human eye is one of the more remarkable optical instruments. Investigate the physiology of the eye. Compare the invertebrate eye with the compound eyes of many insects such as the common (and annoying) house fly (*Musca domestica*).

Paul L. Pease, "Resource letter CCV-1: Color and color vision," *American Journal of Physics*, 48, #11, pages 907–917 (November 1980); this contains many references to the subject.

11. Stars are classified according to their brightness with a *stellar magnitude scale*. For example, the Sun has an apparent visual magnitude of −26, while the bright star Sirius (α Canis Majoris) is −1.5. Polaris (α Ursae Minoris) has a magnitude of about 2, while the dimmest stars visible to the naked eye have magnitudes of about 6. Consult an astronomy text and discover the basis for the magnitude scale. If two stars differ in magnitude by 1, what is the ratio of their apparent brightness? What is the distinction between the apparent visual magnitude of a star and its absolute magnitude? Prepare a short report on your findings.

B. Lab and Field Work

12. Design and perform an experiment to measure the angle of reflection for various angles of incidence of a light ray on a mirror.

13. Design and perform an experiment to measure the angle of refraction for various angles of incidence for a light ray incident from air into a water-filled, thin-glass-walled fish tank.

14. To prevent fraud, an association of jewelry stores has asked you to design and test a simple experimental procedure to easily distinguish between a real diamond and an imitation diamond (zirconium). Design such a test. Several instruments for this purpose actually exist. Discover the principles underlying their operation.

15. The variation of the index of refraction with wavelength (dispersion) for many materials is approximately given by an empirical equation developed by Augustin Louis Cauchy (1789–1857):

$$n(\lambda) = A + \frac{B}{\lambda^2}$$

Your physics department likely has a prism spectrometer and a light source (such as a mercury lamp) that emits several well-known precise wavelengths. Design and perform an experiment to measure accurately the index of refraction of the glass prism for each wavelength; see Problem 25. Fit your data to Cauchy's equation by plotting n versus $1/\lambda^2$ to determine the value of the constants A and B (and their units, if any).

16. Measure the focal length of a converging lens. A diverging lens does not form a real image of a real object. Devise an experimental technique to measure the focal length of a diverging lens using a converging lens of known focal length.

17. If a transparent material such as glass is immersed in a liquid with the same index of refraction (an experimental technique called *index matching*), the glass appears to disappear. Take a pyrex beaker and place it in a large glass beaker. Under a fume hood, add a solution consisting of equal amounts of benzene and carbon tetrachloride and watch the pyrex apparently disappear as the fluid level rises. Some vegetable oils (such as Wesson oil) can be used instead of the chemical solution.

William R. Gregg, "An old physics demonstration—redone more safely," *The Physics Teacher*, 31, #1, page 40 (January 1993).

18. Visit a ophthalmologist who specializes in the surgical procedure called *radial keratotomy*, which corrects myopia and hyperopia. Determine how the procedure corrects for these vision defects. Learn the risks and long-term prognoses for it.

Raymond Munna, *As I See It: Radial Keratotomy Before, During and After Surgery* (Granite, Metairie, Louisiana, 1985).

19. Investigate the optical characteristics and operation of a *zoom lens*. In particular, determine the *minimum* number and types of thin lenses needed to construct a zoom lens. Construct a simple zoom lens on an optical bench to demonstrate its operation.

Michael J. Ruiz, "Camera optics," *The Physics Teacher*, 20, #6, pages 372–380 (September 1982).

20. The index of refraction has a slight temperature dependence. Scavenge your physics department for an appropriate experimental arrangement to perform an experiment to measure the variation of the index of refraction of glass or water as a function of temperature.

C. Communicating Physics

21. Light has been used as an image by writers and poets in many varied literal and allegorical contexts from biblical times to the present. Choose one of your favorite writers and/or works and explore in an essay how the word *light* is used in the context of the work. Does the author have an understanding of light in a physical context?

22. Rainbows are common in artistic works, and even on clothing. In many cases the order of the colors in such artistic rainbows is not correct. Search for and collect several examples of such incorrect rainbows.

23. Investigate the physics and chemistry of color mixing in the paint industry and prepare a short report of your findings. Kurt Nassau, *The Physics and Chemistry of Color: The Fifteen Causes of Color* (John Wiley, New York, 1983).

THE SECRETS OF REFRACTION REVEALED

CHAPTER 24
PHYSICAL OPTICS

**We see how we may determine their forms [referring to the planets
and stars], their distances, their bulk, and their motions, but we
can never know anything of their chemical . . . [composition].**

Auguste Compte (1798–1857)*

As any politician knows, never say never. Auguste Compte was proved wrong shortly after his now infamous comment on what is impossible for us to know in astronomy. (see chapter opening quotation on page 1103). By inventing techniques for precisely measuring the wavelengths of light, physicists shortly realized that each chemical element emits characteristic wavelengths, thereby enabling astronomers to probe the composition of light sources anywhere in the visible universe.

Behold, I will send my messenger, and he shall prepare the way…
    The Old Testament, Malachi 3:1

Light is the messenger. What we know about the stars and the universe as a whole depends almost entirely on our ability to squeeze as much information as possible from the thin thread of light available to us from such distant sources.[†] In the previous chapter, we saw how light can be manipulated by mirrors and lenses to form images. Here we see how we can exploit the wave nature of light to invent new and useful devices and instruments to glean much additional information about stars, galaxies, and more local materials and phenomena of technological and practical import.

Heretofore only alluded to, we now first present convincing experimental evidence of the wave nature of light that was discovered long before Maxwell developed his theory of electromagnetic waves. We investigate several aspects of physical optics, the study of phenomena that manifest the wave aspects of light: interference, diffraction, and polarization.

24.1 EXISTENCE OF LIGHT WAVES

Look at a distant, bright, and small light source such as a star at night through two closely spaced, narrow, and parallel slits. It is not a hallucination when you see *many* bright spots, not one. From such a simple observation we can deduce that light must spread out considerably from these narrow double slits and, in overlapping, add to give bright and dark regions. The multiple spots are a sure sign of overlapping waves, like overlapping ripples on a pond. Small apertures elucidate the wave nature of visible light, as we alluded to in Section 23.1.

24.2 INTERFERENCE

. . . and light was against light . . .
    The Old Testament, 1 Kings 7:4

A superposition of waves may give rise to variations in the resulting amplitude of the total wave disturbance, known as **interference**; the interference of surface water waves is common. We first encountered interference effects in Chapter 12 when we superimposed

- two waves of the same frequency traveling in opposite directions, producing standing waves; and

- two waves of slightly different frequency traveling in the same direction, producing beats.

Similar effects also are seen with light waves, but only with some experimental and technical finesse.

Here we examine other aspects of interference, restricting ourselves to *sinusoidal* waves that meet the following conditions:

1. The waves must be of the same physical type; in this chapter the sinusoidal waves are light waves, but one also can observe the same effects using sound, water, or other waves.

2. The waves must have the same frequency, and so are described by similar sinusoidal wavefunctions; from an equivalent viewpoint, we say the waves, traveling similarly, have identical sinusoidal wavelengths. A wave with a definite sinusoidal wavelength is called a **monochromatic wave**.[‡]

3. The sources of the waves must be **coherent**, meaning the sinusoidal waves have *a phase difference that is independent of time*. Recall that the phase of a sinusoidal wave is the argument of the sinusoid describing the wave. That is, for two sinusoidal waves of the same frequency, traveling toward increasing values of x, the wavefunctions are

$$\Psi_1(x, t) = A \cos(kx - \omega t)$$
$$\Psi_2(x, t) = A \cos(kx - \omega t + \delta)$$

The phase of each wave is the angle of its respective cosine: $kx - \omega t$ for Ψ_1 and $kx - \omega t + \delta$ for Ψ_2. The **phase difference** δ is the difference between the individual phases of the two waves. Notice that the phase of each wave *is* time dependent, but the phase difference δ between the two waves may or may not be. For the two waves to be coherent, the phase difference δ must be independent of time. By convention, we express the phase of a wave (and any phase difference δ) in radians (rad), *not* degrees.

As we considered in Chapters 12, a line or surface connecting points of constant phase on a wave is called a **wavefront**; for a water wave, a wavefront is imagined as the line along each crest of the wave (or equivalently, each trough), as indicated in Figure 24.1.

Coherent wave sources typically are obtained from a single wave by one of two methods.

1. Each wavefront of the wave is divided by a series of slits or holes into multiple coherent wave sources, a process called **wavefront division**. We see examples of how this is done in Sections 24.3 and 24.8.

2. Each wavefront of the wave is divided into a collection of smaller-amplitude wavefronts by reflection and/or transmission, a process called **amplitude division**. We see an example of how this is accomplished in Section 24.11.

What happens when the waves from the multiple coherent sources arrive at the same place at the same time (i.e., are

---

[†]There are other messengers, too, from the distant recesses of the universe, but they are more difficult to decipher: cosmic rays and neutrinos. Neutrinos are extremely difficult to detect because they interact *extraordinarily* weakly with matter.

*(Chapter Opener) *Cours de philosophie positive*, Book II, Astronomy (1835), from *Auguste Compte and Positivism*, edited by Gertrud Lenzer (Harper and Row, New York, 1975), page 130.

[‡]Note the careful wording here. A *nonsinusoidal*, periodic wave with a definite wavelength, such as

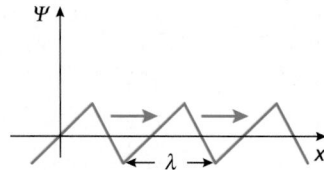

actually contains many sinusoidal frequencies even though it periodically repeats itself. Such nonsinusoidal waves can be treated as collections of sinusoidal waves through *Fourier analysis* but, for simplicity, we do not consider them in this text, except as we did briefly in Section 12.21.

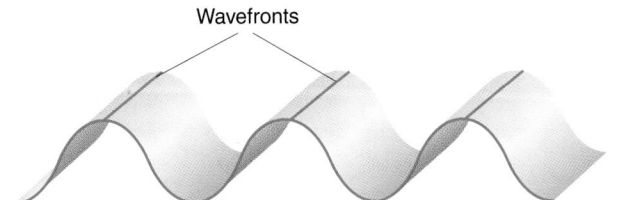

FIGURE 24.1 Wavefronts are lines or surfaces of constant phase.

superimposed)? For light and many other waves, experiments indicate that the algebraic sum of the individual wave disturbances at the same point accurately describes the resulting disturbance; this is the principle of linear superposition. That is, for two disturbances $\Psi_1(x, t)$ and $\Psi_2(x, t)$, the resulting disturbance is

$$\Psi(x, t) = \Psi_1(x, t) + \Psi_2(x, t)$$

When two coherent, monochromatic waves at the same location superimpose and interfere, the resulting wave disturbance depends on their phase difference δ.

If the phase difference between the waves is 0 rad, or an *integral multiple* of 2π rad, the waves are said to be **in phase**. Graphs of the two waves at a fixed x or a fixed t will show the waves to be in step with each other, the maxima occurring at the same times or places, as shown in Figure 24.2.

When the waves are in phase, the interference is called **constructive interference** and the amplitude of the resulting disturbance is the sum of the amplitudes of the individual disturbances. The intensity of the resulting wave is proportional to the square of the resulting amplitude. The resulting disturbance has a maximum intensity when the waves are in phase with each other. If the two interfering waves are of equal amplitude A, the resulting amplitude with constructive interference is $2A$, and the resulting intensity is proportional to $(2A)^2 = 4A^2$, which is *four times* the intensity of each of the individual interfering waves.

If the phase difference δ between the waves is π rad, or any *odd integral multiple* of π rad, the waves are said to be completely **out of phase** with each other. Graphs of the two waves as functions of either space or time will show that the maximum of one wave occurs at the place or time of the minimum of the other wave, as indicated in Figure 24.3.

When the waves are completely out of phase, the interference is called completely **destructive interference**, and the amplitude of the resulting disturbance is the difference in the amplitudes of the individual disturbances. If the two interfering waves are of equal amplitude, the amplitude of the resulting disturbance is zero for completely destructive interference.

One way to produce a phase difference between two coherent waves is to let them travel different distances from their sources to where they are superimposed. The **path difference** causes a phase difference.

A path difference of one wavelength corresponds to a phase difference of 2π rad, because a shift in position of one wavelength along the wave changes its phase by a complete cycle of the cosine—that is, by 2π rad.

Hence, for a path difference Δx, the resulting phase difference δ_{path} in radians is

$$\delta_{\text{path}} = \frac{\Delta x}{\lambda}(2\pi \ \text{rad}) \qquad (24.1)$$

The factor

$$\frac{\Delta x}{\lambda}$$

is the number of wavelengths corresponding to the path difference; each wavelength of path difference represents a phase difference of 2π rad.

QUESTION 1

Does constructive interference violate energy conservation? Does completely destructive interference?

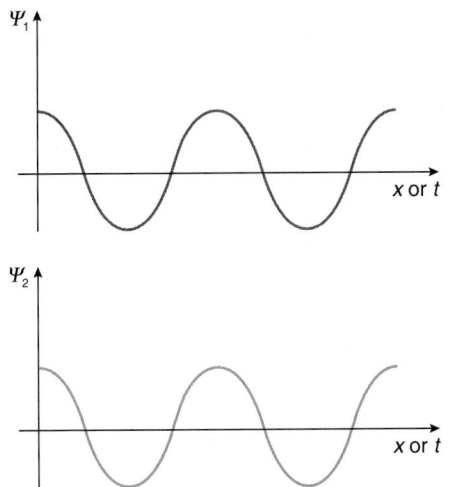

FIGURE 24.2 Two waves in phase: constructive interference results.

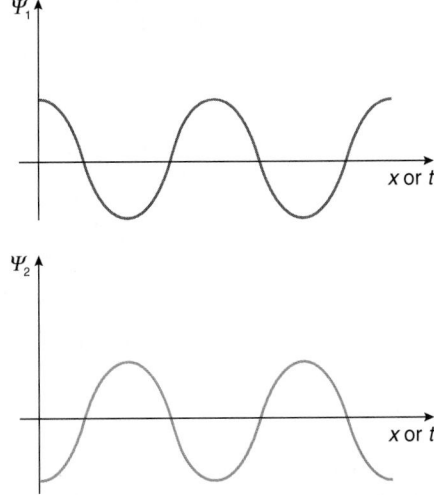

FIGURE 24.3 Two waves out of phase: completely destructive interference results.

24.3 Young's Double Slit Experiment

. . . whenever two portions of the same light arrive at the eye by different routes, either exactly or very nearly in the same direction, the light becomes most intense when the difference of their routes is any multiple of a certain length, and least intense in the intermediate state of the interfering portions; and this length is different for light of different colours.

Thomas Young (1773–1829)*

In 1801 Thomas Young, a medical doctor, scientist, and one of the first to successfully translate Egyptian hieroglyphs (talk about a Renaissance man!), described the results of a now famous experiment that bears his name. **Young's double slit experiment** is one of the classic experiments of physics; it was the first convincing demonstration that light has wavelike characteristics, though the electromagnetic nature of the wave still awaited Maxwell's work some six decades later. Despite its apparent simplicity, the double slit experiment has implications that reverberate into the modern philosophy of quantum mechanics, in which the complementary particle-like (photon) aspects of light must be taken into account. The full significance and importance of Young's experiment is difficult to appreciate in this first encounter; we return to it again briefly in Chapter 27.

The geometry of the experiment is simple. Parallel wavefronts of a monochromatic wave (from a distant pointlike source) are incident on two identical, narrow slits, each of width a, separated by center-to-center distance d, as shown in Figure 24.4.† The slit width a and their separation d are on the order of the wavelength of the incident monochromatic light.

Since each wavefront arrives at the two slits at the same instant, the two slits form coherent sources, secured by wavefront division. There is a time-independent phase difference equal to 0 rad between the two slit sources. The coherent slit sources also have equal amplitudes. The diffracted wavelets (recall Section 23.1) emanating from the two slits interfere with each other where the wavelets overlap. We want to examine the interference effects on a screen parallel to the plane of the slits. We return to other effects of the diffraction in Section 24.7.

How the diffracted waves from the two slits interfere at any point depends only on the path difference from the slits to the point in question. In particular, the point O in Figure 24.5 is equidistant from each slit; the path difference between the waves from each slit thus is 0 m and the two waves from the slits are in phase with each other at O. Constructive interference results and the point O appears bright. We say there is a **bright fringe** at point O.

We consider arrival points on the screen on one or the other side of point O, as indicated in Figure 24.6. From a point P of superposition on the screen, we draw an arc of constant radius through the nearer slit to show the extra path Δx from the more distant slit.

As P moves away from O on the screen, the path difference Δx between the waves from the two sources increases from zero. Eventually the path difference will be half a wavelength. A path

difference of $\lambda/2$ corresponds to a phase difference of (from Equation 24.1)

$$\delta = \frac{\Delta x}{\lambda}(2\pi \text{ rad})$$
$$= \frac{\lambda/2}{\lambda}(2\pi \text{ rad})$$
$$= \pi \text{ rad}$$

The two wavelets reaching the screen at this location then are completely out of phase and therefore interfere completely destructively. We say there is a **dark fringe** at that location.

If we select points farther away from point O, the path difference keeps increasing and eventually becomes equal to one wavelength. A path difference of λ corresponds to a phase difference of 2π (according to Equation 24.1):

$$\delta = \frac{\Delta x}{\lambda}(2\pi \text{ rad})$$
$$= \frac{\lambda}{\lambda}(2\pi \text{ rad})$$
$$= 2\pi \text{ rad}$$

The waves are now back in phase with each other again: the constructive interference results in the formation of a bright fringe. Every time the path difference between the two waves is an integral number of wavelengths, the interference is constructive and a bright fringe appears on the screen.

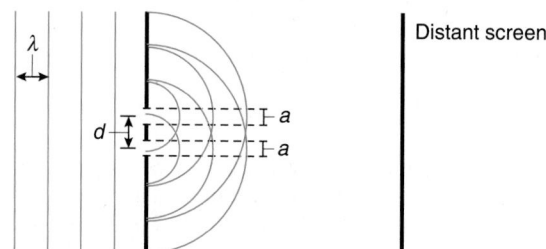

FIGURE 24.4 Geometry for Young's double slit experiment.

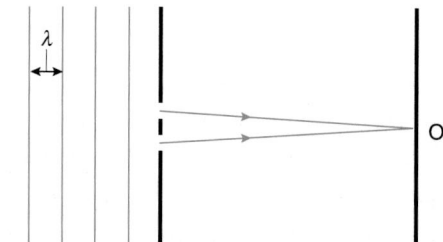

FIGURE 24.5 Zero path difference for waves interfering at point O.

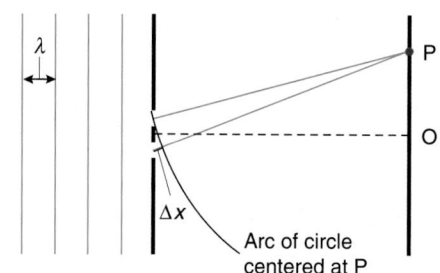

FIGURE 24.6 Nonzero path difference Δx for waves interfering at point P.

---

*Miscellaneous Works of the Late Thomas Young, M.D., F.R.S. . . ., edited by George Peacock (J. Murray, London, 1855), Volume I, page 170.

†More conveniently, we also can put the pointlike source at the secondary focal point of a converging lens, in which case the incident wavefronts (remember rays are perpendicular to wavefronts) on the double slit are parallel.

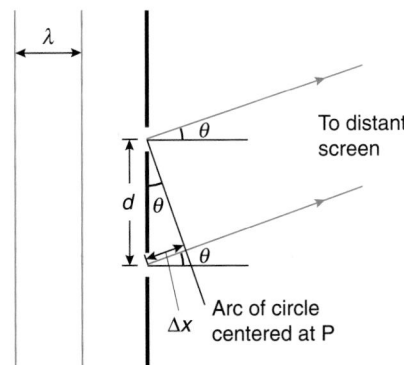

FIGURE 24.7 Magnified view of Young's double slit arrangement for a distant screen geometry.

For a *distant* screen, the small arc of the circle from P to the slits approaches a straight line, as indicated in Figure 24.7.* The path difference then can be expressed trigonometrically in terms of the separation of the slits d and the angle θ from the straight-through direction to any point P on the screen. The path difference is

$$\Delta x = d \sin \theta$$

When the path difference is an integral number of wavelengths, we have constructive interference and a bright fringe on the screen.

Thus the condition for constructive interference on the distant screen is

$$d \sin \theta = m\lambda \qquad (24.2)$$
(condition for constructive interference: bright fringe)

where m is an integer that can take on the values $0, \pm 1, \pm 2, \ldots$.

Each value of m corresponds to a particular bright fringe, as shown in Figure 24.8. Positive values of m correspond to the bright fringes for positive θ (measured counterclockwise from the straight-through direction) using the point O′ halfway between the slits as an origin. The value $m = 0$ is the central bright fringe in the straight-through direction. Negative values of m correspond to negative values for θ.

The absolute value of m is known as the **order of interference**, or equivalently, the **order number**. Thus $|m|$ indicates how many bright fringes we are away in either direction from the central bright fringe at the straight-through direction. Since $\sin \theta \approx \theta$ rad for small values of θ, we also have described the observation that the multiple fringes appear uniformly spaced on a distant screen near the center of the pattern.

*Experimentally, we can place a converging lens beyond the slits and place the screen at the focal point of the lens. Optically, the screen then is effectively an infinite distance from the slits.

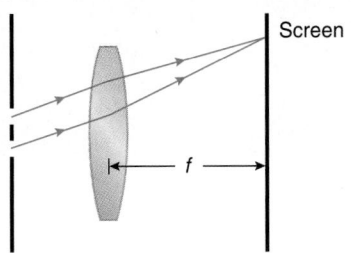

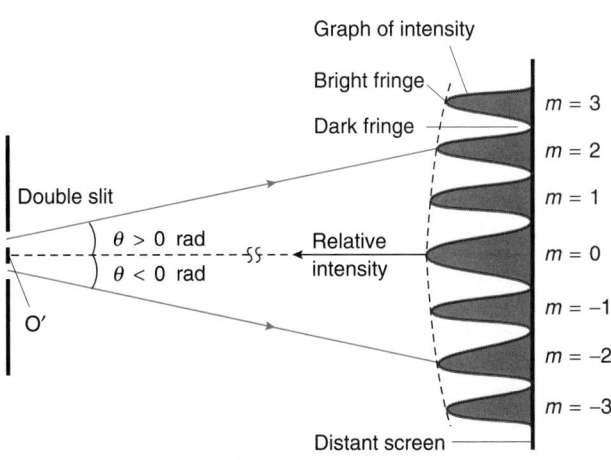

FIGURE 24.8 Pattern of fringes in Young's double slit experiment.

Notice from Equation 24.2 that for a fixed wavelength λ and fixed order number m, if the separation between the slits d now is decreased, the angle θ locating the interference maximum corresponding to any given order number m must increase. So the smaller the slit separation, the greater the angular separation of the interference peaks.

QUESTION 2

The same note from a single sound source is fed to two loudspeakers separated by a distance d. (a) Are the two acoustic sources coherent? (b) Would you expect to hear loud and quiet acoustic interference fringes as you scan your ear across a line parallel to the line between the speakers?

EXAMPLE 24.1

Two parallel slits, each 0.10 mm wide, are separated by 0.50 mm and are illuminated by a parallel beam of laser light with a wavelength of 633 nm. A screen is placed 4.00 m from the slits. What is the separation of the bright interference fringes on the screen?

Solution

Since the distance to the screen is much greater than the distance d between the slits, use Equation 24.2 to find the locations of the interference maxima. In particular, the first-order ($|m| = 1$) fringe is located at an angle θ, where

$$d \sin \theta = 1\lambda$$
$$\sin \theta = \frac{\lambda}{d}$$
$$= \frac{633 \times 10^{-9} \text{ m}}{0.50 \times 10^{-3} \text{ m}}$$
$$= 1.3 \times 10^{-3}$$

Since the sine of θ is so small, the sine is also virtually equal to θ in radians, according to the small angle approximation. Hence the angle is

$$\theta = 1.3 \times 10^{-3} \text{ rad}$$

Imagine a circle centered on the slits with a radius equal to the distance to the screen, as in Figure 24.9. For small angles θ, the arc length is approximately equal to y, and so you have

$$\theta = \frac{y}{\ell}$$

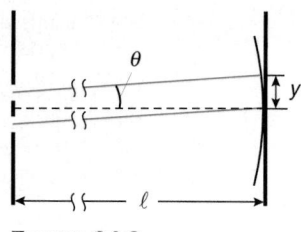

FIGURE 24.9

where y is the distance on the screen between the fringes for order numbers $m = 0$ and $m = 1$. Hence you have

$$y = \theta \ell$$
$$= (1.3 \times 10^{-3} \text{ rad})(4.00 \text{ m})$$
$$= 5.2 \times 10^{-3} \text{ m}$$

Since the fringes are equally spaced for small angles, the separation between the fringes observed on the screen is 5.2 mm.

24.4 SINGLE SLIT DIFFRACTION

When monochromatic light passes through a single slit whose aperture is on the order of the size of the wavelength, the light diffracts, or spreads out, on the other side of the slit. As we have seen, diffraction is characteristic of wave phenomena in general. Here we examine the diffraction of light as it passes through a single slitlike aperture of width a; the results, though, are equally applicable to other types of wave diffraction.

The intensity distribution of the diffracted light on a distant screen is indicated in Figure 24.10. The observed distribution of the light intensity on the screen indicates that the light is not uniformly spread out or diffracted: certain regions on the screen have no light, while others have some. Thus the amplitude of the diffracted light depends on the angle θ from the straight-through direction in Figure 24.10.

To account for this we resort to an old and what might seem to be odd geometric construction, called **Huygens's principle**. It was first used in the 17th century by Christian

Huygens (1629– 1695) to explain how waves propagate from one place to another. Huygens's idea was the following. Imagine a wavefront of any shape at some position in space at a particular time, as in Figure 24.11. According to Huygens's principle, each point on a wavefront acts as a source of a wavelet that propagates outward from the point; the points of the wavefront are coherent emitters of the wavelets. During a time interval Δt, each wavelet of light travels a distance $c \, \Delta t$ from its source point, as shown in Figure 24.11. The new position of the wavefront is determined by the envelope* of the wavelets from the individual points on the original position of the wavefront.

We apply this to the single slit. Using Huygens's principle, we may usefully regard the points on the wavefront filling the slit as a collection of coherent sources of wavelets propagating into the region beyond the slit, as shown in Figure 24.12. Their envelope creates the wavefront anywhere beyond the slit.

Arriving and superimposing at any given point on the screen are the contributions from the multitude of coherent pointlike sources arrayed across the small width of the slit. To visualize the path difference to a point P on the screen, we draw an arc of constant radius centered on the screen point of superposition P through the nearer slit edge, as in Figure 24.13.

If the distance ℓ to the screen is much larger than the tiny slit width a, the extra path Δx from the farther slit edge is part of a small right triangle,[†] blown up for clarity in Figure 24.14. From the right triangle in Figure 24.14 and definition of the sine, the path difference Δx is found to be

$$\Delta x = a \sin \theta$$

If the screen point is at the center (straight-through direction), the contributions from paired point-like sources placed symmetrically on either side of the straight-through direction arrive in phase and we have a maximum amplitude and intensity at this position.

---

*A geometrical *envelope* is a curve (or surface) that is tangent to a set of curves (or surfaces).

[†]To avoid having to put the screen far away to make $\ell \gg a$, we alternatively can put a converging lens near the slit aperture and place the screen at the focal plane of the lens. Then those rays that are superimposed on reaching each point on the screen are indeed exactly parallel on leaving the slit aperture, as if $\ell = \infty$ m.

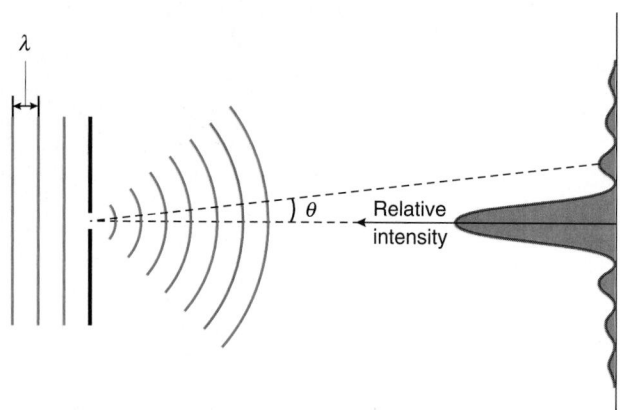

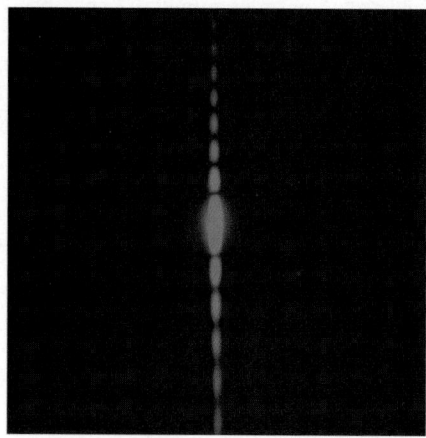

FIGURE 24.10 Diffraction pattern formed by a single slit.

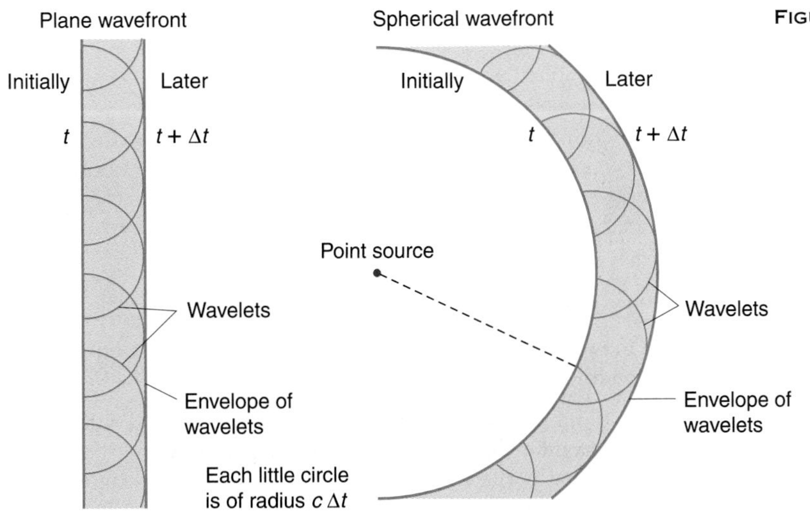

Plane wavefront

Initially | Later

t | $t + \Delta t$

Wavelets

Envelope of
wavelets

Each little circle
is of radius $c\,\Delta t$

Spherical wavefront

Initially | Later

t | $t + \Delta t$

Point source

Wavelets

Envelope of
wavelets

FIGURE 24.11 Huygens's principle.

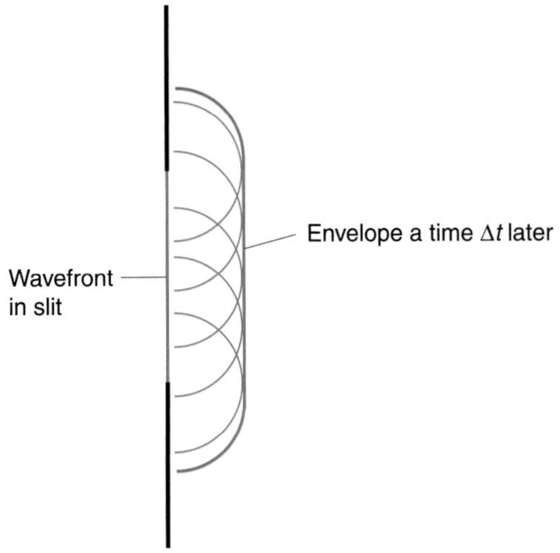

Wavefront
in slit

Envelope a time Δt later

FIGURE 24.12 Huygens's principle applied to a single slit
(magnified view).

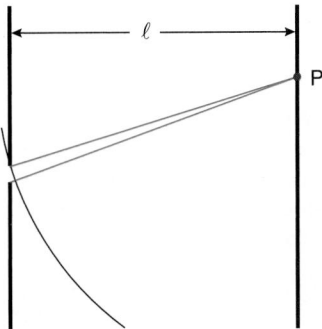

FIGURE 24.13 A circle centered on P helps to delineate the
path difference.

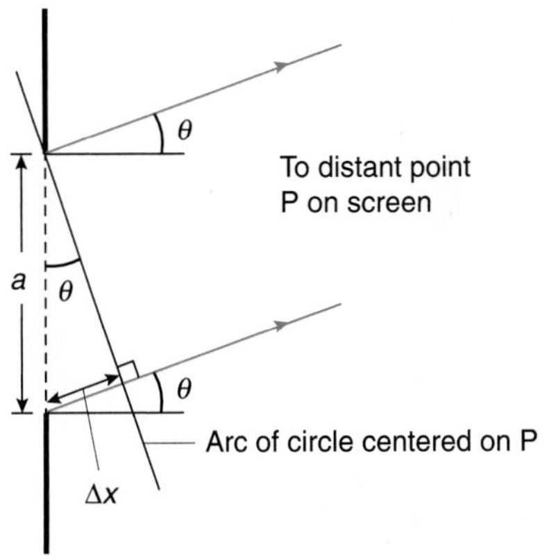

The minima at the sides of the central diffraction maximum
come about in the following way. Take, for example, the first
minimum to the side. The wavelets from the coherent point
sources distributed across the aperture of the slit now travel
different distances to the screen. As we move away from the
central maximum on the screen, eventually we reach a position

θ

To distant point
P on screen

a | θ

θ

Arc of circle centered on P

Δx

FIGURE 24.14 Geometry in the vicinity of the single slit for a
distant screen.

where the path difference to the point P on the screen between
the point source at the near edge (top) of the aperture and the
point source at the *middle* of the slit is half a wavelength, as shown
in Figure 24.15. The half wavelength path difference for the pair
corresponds to a phase difference of π rad (from Equation 24.1).
The two wavelets from these two points then interfere destructively on the screen and contribute nothing.

Now let's consider the pair of point sources in the aperture
just below the first pair. The wavelets from these sources also
have a difference in path to point P of half a wavelength (and a
phase difference of π rad), and so these two wavelets *also* interfere
destructively. As we consider similarly paired points across the
full aperture, *every* pair adds to give zero amplitude. The sum of
any number of zeros is still zero. Thus we have pairwise destructive interference from all the sources in pairs half the aperture
apart, distributed across the entire aperture.

The $\lambda/2$ path difference between a point at the top and a
point at the *middle* of the slit also means that the path difference
between the points at the top and at the *bottom* of the slit is a full
wavelength, as shown in Figure 24.15. Geometrically, this path difference between the extremes of the slit is related to the width a
of the slit and the angle θ_1 locating the first minimum. The path
difference in Figure 24.15 is $a \sin \theta_1$.

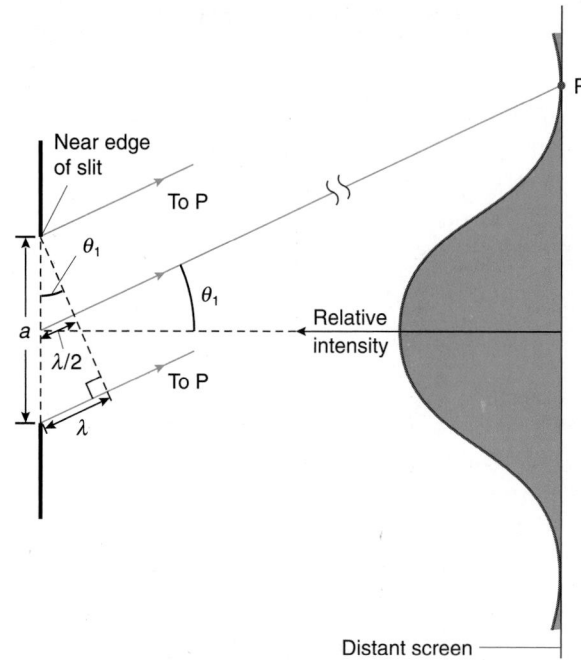

FIGURE 24.15 Path difference to the screen of $\lambda/2$ between the near edge and middle of the slit.

So the *first minimum* of the diffraction pattern occurs where

$$a \sin \theta_1 = \lambda \qquad (24.3)$$
(first minimum, single slit diffraction)

Other diffraction minima exist when the path difference $a \sin \theta$ between opposite edges of the slit is any integral multiple m of the wavelength:

$$a \sin \theta = m\lambda \qquad (m = \text{integer}) \qquad (24.4)$$
(single slit diffraction minima)

For small angles, $\sin \theta \approx \theta$ in radians, so the minima are equally spaced. In between the minima are maxima, but their locations are harder to specify than the minima, except for the central diffraction maximum, and we will not consider them further. These other maxima are progressively weaker than the central maximum.

Notice in Equation 24.3 that for a fixed wavelength, if the slit width a is made smaller, the angle θ_1 locating the first minimum of the diffraction pattern must increase, and so the broad central maximum increases in width.

The *narrower* the slit, the *greater* the total angular width $2\theta_1$ of the complete central diffraction peak.

You can see this by looking through a crack between your fingers at a distant pointlike street light.

Conversely, the greater the width of the slit, the smaller the angle θ_1 to the first minimum of the diffraction pattern. As the slit width a becomes much greater than the wavelength ($a \gg \lambda$), the angle θ_1 approaches zero and we enter the geometric limit of Chapter 23, where we can essentially ignore the wave nature of the light and treat the light as propagating in straight lines (rays).

PROBLEM-SOLVING TACTIC

24.1 Do not confuse the order number m locating the interference *maxima* for Young's double slit experiment in Equation 24.2 with the integer m locating the single slit diffraction *minima* in Equation 24.4. While both m's are integers, each represents a different effect!

EXAMPLE 24.2

Parallel light of wavelength 633 nm is incident on a slit of width 0.12 mm.

a. Find the total angular width of the central diffraction maximum on a screen located 3.50 m from the slit.
b. What is the width of the central diffraction maximum on the screen in centimeters?

Solution

a. Since the distance to the screen is much larger than the width of the slit, you can use Equation 24.3 to find the angle θ_1, which is half the total angular width of the central diffraction maximum:

$$a \sin \theta_1 = \lambda$$

$$\sin \theta_1 = \frac{\lambda}{a}$$

$$= \frac{633 \times 10^{-9} \text{ m}}{0.12 \times 10^{-3} \text{ m}}$$

$$= 5.3 \times 10^{-3}$$

Since the sine of the angle is so small, the small angle approximation implies the angle is

$$\theta_1 = 5.3 \times 10^{-3} \text{ rad}$$

The total angular width of the pattern is

$$2\theta_1 = 1.1 \times 10^{-2} \text{ rad}$$
$$= 0.63°$$

b. The extent y of the central diffraction maximum on the screen is determined by making use of the definition of the radian measure of an angle, as in Figure 24.16. For small angles, the distance y is approximately equal to the arc length of a circle of radius ℓ centered on the slit. Hence

$$2\theta_1 = \frac{y}{\ell}$$

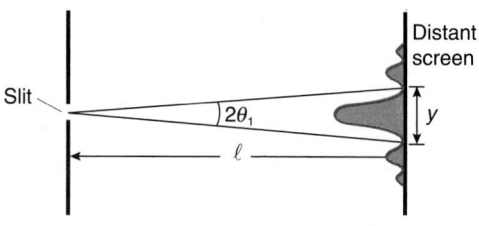

FIGURE 24.16

Therefore you have

$$y = 2\theta_1 \ell$$
$$= (1.1 \times 10^{-2} \text{ rad})(3.50 \text{ m})$$
$$= 3.9 \times 10^{-2} \text{ m}$$
$$= 3.9 \text{ cm}$$

24.5 DIFFRACTION BY A CIRCULAR APERTURE

Circular apertures are common to lenses and our eyes, and so it is of more than passing interest to know something about diffraction by such a geometry.

If monochromatic light is incident on a circular aperture of *diameter a*, rather than on a slit, diffraction of the light through the aperture produces a circularly symmetric pattern on a distant screen, indicated in Figure 24.17. The pattern consists of a central circular bright spot surrounded by a series of increasingly dimmer concentric circles of light.

The concentric dark rings are not equally spaced (as the fringes are for diffraction by a single slit), because the wavelet sources across the diameter have varying lengths (circumferences) and importance (as opposed to the equal segments across a slit), and the effective circle width is somewhat less than the diameter. A detailed mathematical analysis of the locations of the secondary peaks and the minima of the diffraction pattern of a circular aperture is beyond the scope of an introductory course in physics. However, it is of practical import to know the location of the first minimum of the diffraction pattern of such a circular aperture, as we will see in Section 24.6.

> The first minimum of the circular diffraction pattern is located at an angle θ_1 found from the following relation*:
>
> $$a \sin \theta_1 = 1.220\lambda \qquad (24.5)$$
> (first diffraction minimum for a circular aperture)
>
> where a is the *diameter* of the circular aperture.

This is very similar to the equation that locates the first minimum of the diffraction pattern of a single slit of width a (Equation 24.3):

$$a \sin \theta_1 = \lambda \qquad \text{(first diffraction minimum for a single slit)}$$

The distinction between the two equations is the effective value $a/1.220$ for the circular aperture. The number 1.220 arises from a special class of functions, called Bessel functions, that invariably are associated with circular geometries in physics; Bessel functions are analogous to the familiar special functions you know as the sines and cosines.

Notice, as for a single slit, that for a fixed wavelength, the *smaller* the diameter a of the circular aperture, the *greater* the angle θ_1 to the first minimum of the diffraction pattern. In other words, the smaller the aperture, the greater the angular width of the central diffraction peak. Conversely, the greater the size

---

*Note that we did not derive this. Alas, we apologize; to derive the factor 1.220 is not simple.

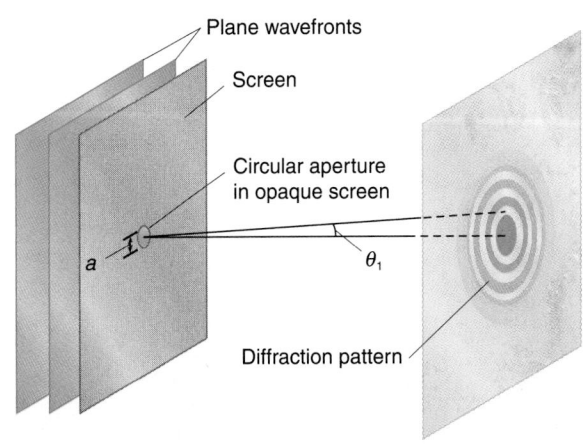

FIGURE 24.17 Diffraction by a circular aperture produces a pattern of concentric rings.

of the aperture, the smaller the angular width of the central diffraction peak and the less apparent the effects of diffraction.

PROBLEM-SOLVING TACTIC

24.2 For circular apertures, be sure to use Equation 24.5 to locate the angle to the first minimum of the diffraction pattern, not Equation 24.3 for slit apertures.

EXAMPLE 24.3

Light with a wavelength of 633 nm is incident on a pinhole of diameter 0.30 mm.

a. What is the angular width of the central diffraction peak on a screen 4.0 m from the pinhole?
b. What is the linear extent of the central diffraction peak on the screen?

Solution

a. The angular width of the central diffraction peak is $2\theta_1$, where θ_1 is the angle from the straight-through direction to the first minimum on either side of the peak. The angle θ_1 is found from Equation 24.5:

$$a \sin \theta_1 = 1.220\lambda$$

$$\sin \theta_1 = \frac{1.220\lambda}{a}$$

The diameter of the pinhole is $a = 0.30 \text{ mm} = 3.0 \times 10^{-4} \text{ m}$. Substituting for a and λ, you obtain

$$\sin \theta_1 = \frac{1.220 (633 \times 10^{-9} \text{ m})}{3.0 \times 10^{-4} \text{ m}}$$

$$= 2.6 \times 10^{-3}$$

Using the small angle approximation, we have

$$\theta_1 = 2.6 \times 10^{-3} \text{ rad}$$

The angular width of the central peak is

$$2\theta_1 = 5.2 \times 10^{-3} \text{ rad}$$
$$= 0.30°$$

b. The linear extent y of the central peak on the distant screen is found from the definition of an angle in radians, as shown in Figure 24.18. Since y is approximately equal to the arc length of a circle of radius ℓ centered on the hole, you have

$$2\theta_1 = \frac{y}{\ell} \quad (\theta_1 \text{ in radians})$$

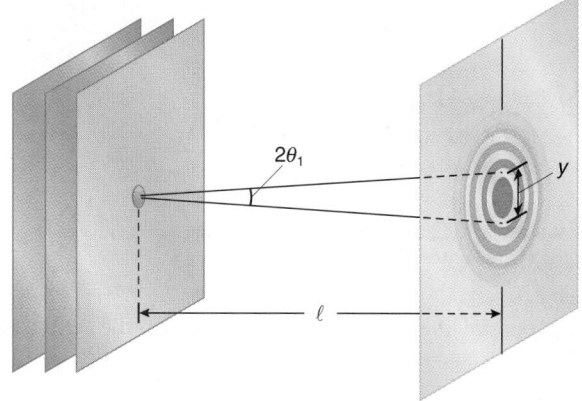

FIGURE 24.18

Solving for y, you find

$$
\begin{aligned}
y &= 2\theta_1\ell \\
&= 2(2.6 \times 10^{-3} \text{ rad})(4.00 \text{ m}) \\
&= 2.1 \times 10^{-2} \text{ m} \\
&= 2.1 \text{ cm}
\end{aligned}
$$

24.6 RESOLUTION

When you see a very distant car approaching at night, it is hard to distinguish the individual headlights; they appear as one source. We say they are **unresolved**. When the car is closer, it is easy to tell there are two headlights; we say they then are **resolved**.* Now look at the starry sky. Almost half the stars visible to the naked eye at night really are not single stars like the Sun but binary stars, two stars gravitationally bound to and orbiting each other. A telescope easily reveals the binary character of many stars that the naked eye perceives as one.[†] What determines whether we can distinguish whether a source is single or binary? For a telescope, does this ability depend on the magnification? At the other extreme, what determines the smallest details you can discern in a microscope? Here we answer these important questions about **resolution**, the ability to separate and distinguish fine details.

Independent binary sources, such as two headlights, two stars, or two amoebas, are **incoherent sources**, unlike a pair of double slits illuminated by monochromatic light. The phase of the light from each source has no correlation with that of the light from the other source; the light from each is emitted by independent atoms and so their phase difference is not constant in time.

We saw in the previous section how, when monochromatic light traverses a circular aperture of diameter a, a diffraction

pattern of concentric rings is produced. This pattern appears in the focal plane of the eye, or the focal plane of the objective element of telescope, where the Huygens's wavelets are superimposed. The angular position of the first minimum of the diffraction pattern of a circular aperture is located using Equation 24.5:

$$a \sin \theta_1 = 1.220\lambda$$

> Two incoherent, monochromatic, pointlike, distant sources each produce their own diffraction pattern.

Since the sources are incoherent, the waves from one source do not consistently interfere with those from the other. Let the angle between the sources be ϕ. The angular separation of the two diffraction patterns is equal to the angle ϕ between the sources, as shown in Figure 24.19.

If the sources are well separated in angle, the diffraction patterns on the distant screen or focal plane are quite distinct from each other and it is easy to tell that two sources of light are present. Under these circumstances, we say the two sources are *well resolved*.

Certainly if the angle between the two sources is zero, the two diffraction patterns are superimposed on top of one another and we cannot tell that there are really two sources producing the pattern on the screen. The two sources are *unresolved*.

The question then arises: what is the minimum separation angle between the sources that allows us to barely resolve or distinguish that there are really two diffraction patterns present, indicating the presence of the two sources? As the angle between the sources increases, the diffraction patterns caused by each become further separated as well. When the angle ϕ is large enough, a dip appears in the combined diffraction pattern that indicates two patterns are present, not one, as in Figure 24.20. In the 19th century. Lord Rayleigh [John William Strutt (1842–1919)] proposed a convenient criterion, now called the **Rayleigh criterion**, for determining the minimum angular separation angle for resolution.

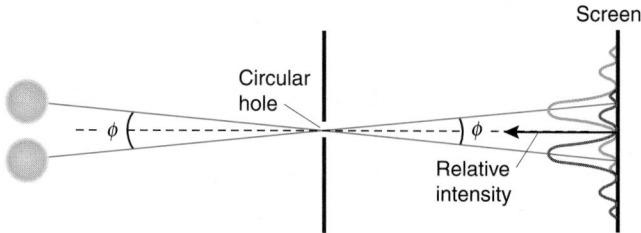

FIGURE 24.19 Two incoherent sources produce two independent diffraction patterns; here the two are well resolved.

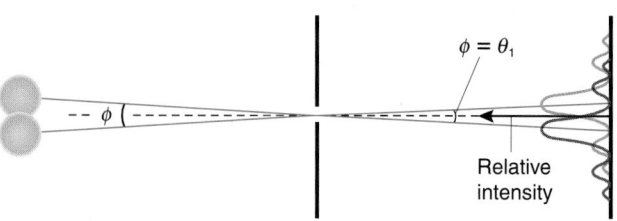

FIGURE 24.20 Two sources barely resolved according to the Rayleigh criterion.

---

*The word comes from one of the meanings of the word resolve: to decide.
[†] The orbital periods for such visual binaries are quite long, even centuries, so the orbital motion is not readily apparent in a telescope.

When the central peak of one diffraction pattern is located at the position of the first minimum of the *other* diffraction pattern, we say the two sources are *barely resolved*.

The angle ϕ between the sources then is equal to the angle θ_1, the angle that locates the first minimum of either diffraction pattern.

Mathematically, for a circular aperture, the sources are barely resolved if their angular separation ϕ is equal to the angle θ_1 in Equation 24.5:

$$a \sin \theta_1 = 1.220\lambda$$

Since the angle θ_1 is almost always very small, the small angle approximation can be used, and we have

$$a\theta_1 = 1.220\lambda \qquad (\theta_1 \text{ in radians}) \text{ (circular aperture)} \quad (24.6)$$

For stars and headlights, which emit many different wavelengths, to determine the angle in the Rayleigh resolution criterion, use an *effective wavelength*, typically in the yellow region of the spectrum ($\lambda \approx 550$ nm) where our eyes are most sensitive.

The eyes of cats and other felines, some other mammals, and a few reptiles like boa constrictors have vertical slit apertures. For these life forms, it is apparently more important to resolve vertical details than horizontal. If the aperture is a slit, the minimum resolution angle is found from Equation 24.3, $a \sin \theta_1 = \lambda$, which for small angles is

$$a\theta_1 = \lambda \qquad (\theta_1 \text{ in radians}) \text{ (slit aperture)} \quad (24.7)$$

Consider a telescope with a circular aperture. The aperture diameter of the objective lens or mirror determines the resolution of the instrument. If the angular separation of two sources is less than θ_1, where θ_1 is determined from the Rayleigh criterion and λ is the effective wavelength of the incident light, then the two sources cannot be distinguished or resolved. Problems such as this occur quite frequently in astronomy. In order to see binary stars as indeed separate sources of light, a telescope of sufficient aperture is needed: one whose Rayleigh criterion minimum resolution angle θ_1 is *smaller* than the angular separation ϕ of the stars.

The resolution is *not* determined by the magnification. This can be easily confirmed by looking at a binary star through a telescope, such as Mizar (ζ Ursae Majoris), the second star from the end of the handle of the Big Dipper (Ursa Major), shown in Figure 24.21.* You will see that Mizar is a close binary in the telescope. Now place a piece of cardboard over the aperture of the telescope with a hole about the diameter of the eye pupil (about 1 cm); you will no longer be able to see the two stars resolved through the telescope, regardless of the magnification you are using! Thus resolution depends on the aperture, *not* the magnification.

As another example, the eye has a circular aperture and it is the size of its aperture that determines the ability of the eye to distinguish two sources as being separate.[†] The function of the

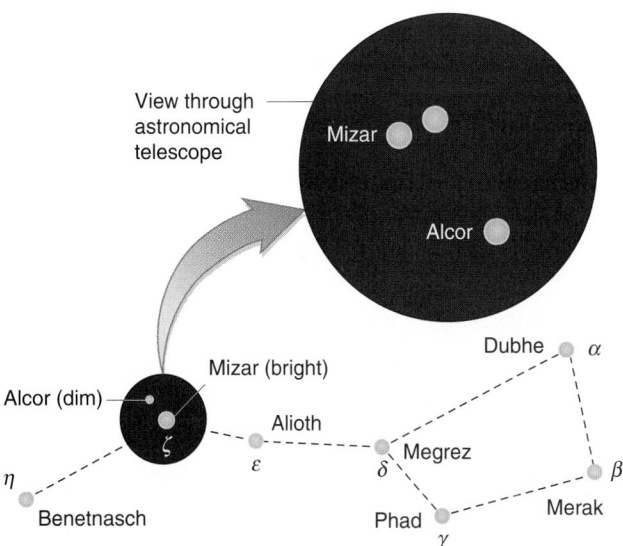

FIGURE 24.21 The location of Mizar in the constellation Ursa Major.

cornea and eye lens is merely to bring the distant screen to a more reasonable finite distance: the screen is the retina. If the separation of the two sources is less than the angle determined from the Rayleigh criterion, then the eye cannot distinguish the presence of the two sources and they are unresolved.

The same considerations apply to microscopes. The detail that can be distinguished or resolved is essentially the ability to tell if two object points are distinct or not. Such resolution ultimately depends on the aperture of the objective lens of the instrument and the wavelength of the light used to view the object. Using smaller-wavelength light enhances the resolution, since for a fixed aperture, decreasing λ decreases the minimum resolution angle θ_1. Electron microscopes use electron wavelengths that are about 1000 times smaller than visible light wavelengths, so such microscopes have extraordinary resolution.

It is diffraction that sets a theoretical limit on the resolution of optical instruments. The resolution is determined by the aperture of the instrument and the wavelength, *not the magnification.*

The *smaller* the angle θ_1 found from the Rayleigh criterion, the *better* the resolution. That is, the easier it is to distinguish the presence of the two point sources. Good resolution means that the angle θ_1 to the first minimum of the diffraction pattern is small, which happens when the circular aperture (or the slit width) is large. So for good resolution, the angle θ_1 between the center and first minimum of each diffraction pattern produced by two incoherent sources is smaller than the separation angle between the two source centers.

---

*Mizar has a dim, nearby, naked eye companion, called Alcor. Fix your attention only on Mizar.

[†]For the eye, there is a biological factor to consider as well: the angular size of the cone cells (measured from the pupil) on the fovial region of the retina. See Example 24.4.

QUESTION 3

A slit has a width equal to the diameter of a circular aperture. Which has the better resolution?

EXAMPLE 24.4

What is the minimum angle of resolution for the human eye as determined from the Rayleigh criterion if the effective wavelength of the light is 550 nm? Consider the maximum aperture of the eye when fully dilated to be about 7.0 mm.

Solution
Use Equation 24.5:

$$a \sin \theta_1 = 1.220\lambda$$

The resolution angle θ_1 is small, so the small angle approximation is appropriate:

$$a\theta_1 = 1.220\lambda \quad (\theta_1 \text{ in radians})$$

$$\theta_1 = \frac{1.220\lambda}{a}$$

$$= \frac{1.220\,(550 \times 10^{-9}\ \text{m})}{7.0 \times 10^{-3}\ \text{m}}$$

$$= 9.6 \times 10^{-5}\ \text{rad} \quad (\text{about 20 arc seconds})$$

The actual resolution limit of the eye is about ten times this result, or about three arc minutes (60 arc seconds ≡ 1 arc minute), because of the size of the cone cells on the fovea region of the retina. The cone cells subtend an angle of about one arc minute from the pupil. Each resolved diffraction pattern must form on a separate cone cell with an unaffected cone cell in between them, and so the resolution limit of the eye actually is about three arc minutes (but varies greatly with the individual).

24.7 THE DOUBLE SLIT REVISITED

This is not a Hollywood sequel.* We must revisit the double slit to account for the simultaneous interference effects between the slits and the diffraction effects of each slit.

Each slit, acting alone, produces a diffraction pattern on a distant screen. If one slit is covered, the incident monochromatic, coherent light only passes through the other slit, and the illuminated slit produces a single slit diffraction pattern on the screen. When each slit is illuminated separately, the individual diffraction patterns of each slit are very slightly separated from each other, as shown in Figure 24.22.

The individual diffraction patterns cannot be observed on the screen simultaneously because they appear only when one or the other slit is illuminated, but not both. Since the slit separation is small, the separation of the individual diffraction patterns is negligible. But when both slits are illuminated by monochromatic light, the slit sources are coherent and the light from one slit now can also interfere with the light from the other slit. A double slit interference pattern, with equally spaced dark and bright fringes (for small angles), is seen on the distant screen.

*Perhaps fortunately, no one has been able to create a credible screenplay for a potential TV show called *L.A. Physics* analogous to *L.A. Law*. Unfortunately, scientists and engineers invariably are depicted by Hollywood as mad or deranged. See Investigative Project 16.

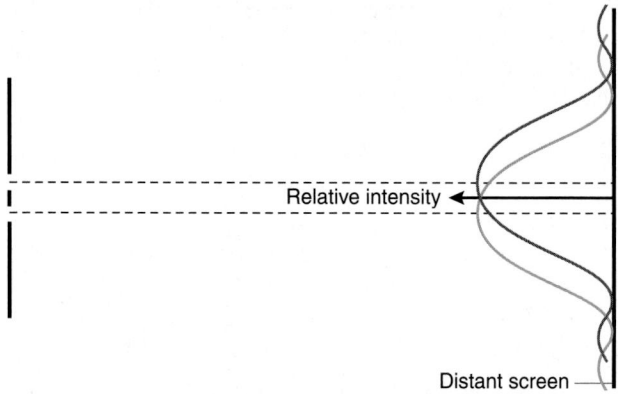

FIGURE 24.22 Two slits each produce a diffraction pattern on a distant screen when illuminated one at a time; the two patterns are very slightly separated from one another.

The intensity of the interference peaks, however, is modulated by the **envelope** of the single slit diffraction pattern, as shown in Figure 24.23.

In other words, the light coming from the slits, which forms the bright fringes of the double slit interference pattern, is reduced because each slit sends out less light to interfere constructively at off-center directions of its own broad diffraction pattern. Diffraction effects are apparent from the modulation of the intensity of the interference peaks. The diffraction effects of each single slit are manifested like the Cheshire Cat: the diffraction disappears with only its "smile" remaining.[†]

We now can play with a few variations.

1. If the slit separation d is fixed and the slit width a is decreased, the interference peaks remain in the same locations, since they are determined by the fixed slit separation d. But since the slit width a is decreased, the location of the first minimum of the diffraction pattern appears at a larger angle θ_1 according to Equation 24.3:

$$a \sin \theta_1 = \lambda$$

With the right-hand side of this equation constant (for a fixed wavelength), decreasing a means that $\sin \theta_1$ must increase, and so θ_1 itself must increase. The result is that more interference peaks now appear under the expanded size of the envelope of the central diffraction peak, as indicated in Figure 24.24.

2. Now we fix the slit width a, but decrease the slit separation d. The envelope of the diffraction pattern now stays fixed in size, since the slit width is kept constant. However, as the slit separation d decreases, the angular position of a fixed order number m of the interference peaks must increase according to Equation 24.2:

$$d \sin \theta = m\lambda$$

With the right-hand side of this equation fixed (for a fixed order number and fixed wavelength), if d decreases, then $\sin \theta$ must increase—and so θ itself must increase—for that

[†]With apologies to Lewis Carroll and his *Alice's Adventures in Wonderland*.

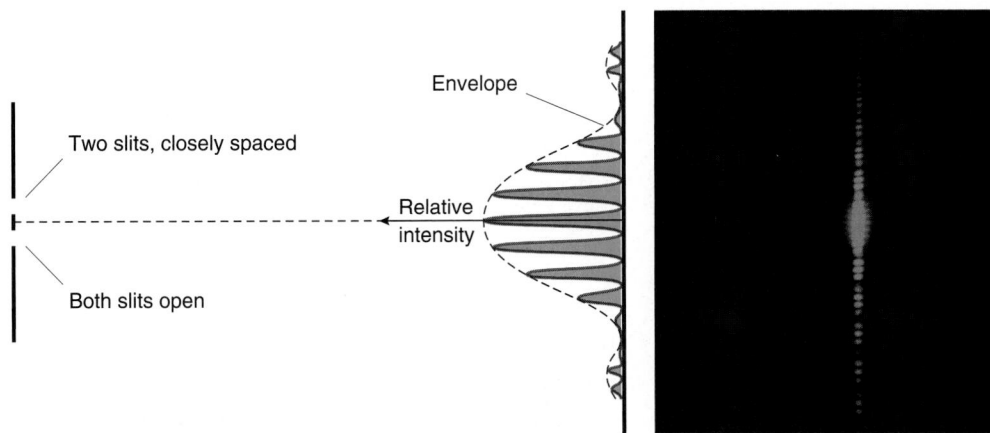

FIGURE 24.23 Double slit interference and diffraction. (Photo is not to the same scale as graph.)

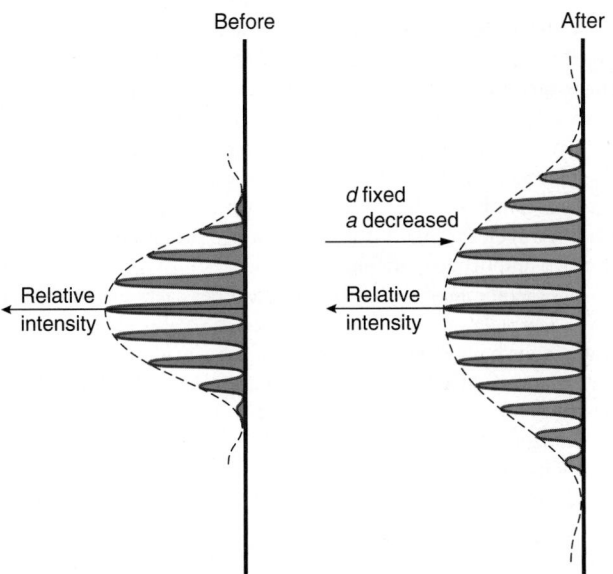

FIGURE 24.24 Decrease the slit width for fixed slit separation; more interference peaks appear under the expanded angular size of the central diffraction envelope.

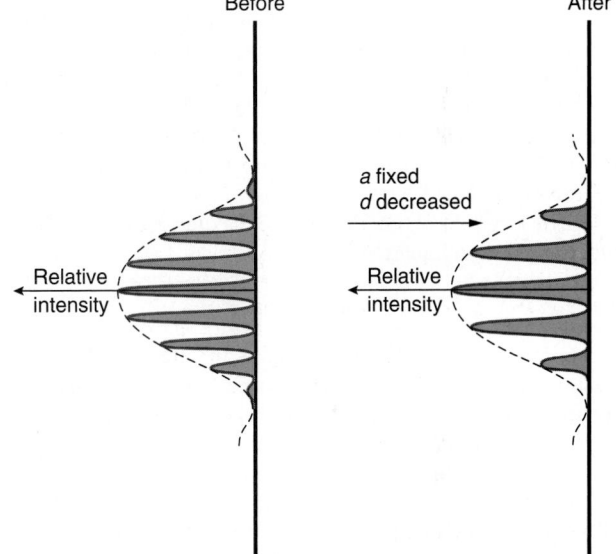

FIGURE 24.25 Decrease the slit separation with fixed slit width; the interference peaks get further apart and the diffraction envelope remains the same.

order. This results in fewer interference peaks under the same size central diffraction envelope, as shown in Figure 24.25.

EXAMPLE 24.5

Two slits, each of width 0.20 mm, are separated by 0.70 mm and illuminated with light of wavelength 633 nm in a Young's double slit experiment. How many bright interference fringes appear under the central diffraction envelope?

Solution

The interference fringes are located using Equation 24.2:

$$d \sin \theta = m\lambda$$

where d is the slit separation. The first minimum of the central diffraction peak is found from Equation 24.3:

$$a \sin \theta_1 = \lambda$$

where a is the slit width. You can find the order number that appears at $\theta = \theta_1$ by substituting θ_1 into Equation 24.2:

$$d \sin \theta_1 = m\lambda$$

Now dividing by Equation 24.3, you find

$$\frac{d}{a} = m$$

With the given values for d and a, this is

$$m = \frac{0.70 \text{ mm}}{0.20 \text{ mm}}$$

$$= 3.5$$

Since the order number must be an integer, this result means that the seven bright fringes corresponding to $m = 0$,

±1, ±2, and ±3 are within the angle ±θ_1, while those with higher absolute values of m are not. Note that the result is independent of λ. Why?

24.8 MULTIPLE SLITS: THE DIFFRACTION GRATING

A single slit produces the single slit diffraction pattern of Figure 24.10. The double slit produces the double slit interference pattern of Figure 24.23, in which the diffraction pattern of a single slit manifests itself as the modulation of the intensity of the interference fringes. What happens to the interference and diffraction patterns if many more additional identical, equally spaced slits are presented to the incident light? Such a multiple arrangement of N slits is a **diffraction grating**.

Since the N slits each are of width a, the diffraction patterns of each slit, taken individually, are identical in shape (but very slightly separated in space by the slit separation). If the slit width a is very small, the resulting diffraction patterns of the individual slits have large angular widths. The envelopes of all the individual single slit diffraction patterns are essentially coincident, since the total width Nd of the entire array of slits typically is not appreciable compared with the width of any single diffraction central maximum.

The locations of the interference maxima of a double slit are determined from Equation 24.2:

$$d \sin \theta = m\lambda$$

where the order of interference m is an integer (or 0). This equation means that for constructive interference the path difference $d \sin \theta$ from the two slits to a point on a distant screen is an integral number of wavelengths (= $m\lambda$). The addition of more identical slits with the same slit separation d does not change this argument: the path difference between adjacent slits still is $d \sin \theta$ as indicated in Figure 24.26, and the resulting intensity is increased.

When the path difference between adjacent slits is an integral number of wavelengths, $m\lambda$, constructive interference again occurs at the same locations as with a double slit with the same slit separation.

The locations of the interference maxima for a multiple slit diffraction grating thus are found using

$$d \sin \theta = m\lambda \qquad (24.8)$$
(diffraction grating maxima)

which is known as the **grating equation**. The slit separation d is called the **grating spacing**.

Does *anything* different happen in going from a double slit to a multiple slit array of identical slits with the same slit separation? Indeed.

As the number of slits increases, the angular widths of the interference maxima decrease. In other words, the places where the interference is constructive become more sharply defined, as shown in Figure 24.27.

The maxima occur for values of θ determined from the grating equation, Equation 24.8. Even small variations from these specific values of θ cause the waves from the various slits to interfere in essentially a completely destructive manner. Why is this? With a large number of slits, as θ varies even slightly from the places where the path difference between adjacent slits is an integral number of wavelengths, the wavelet from any given slit finds a wavelet from another slit that is out of phase with it and the two amplitudes cancel.

For example, if the path difference between *adjacent* slits is, say, 1.1 wavelengths, then two slits spaced 5 slits apart will have waves that have a path difference of 0.5 wavelength and so are out of phase with each other. So slits that are $5d$ apart interfere destructively; we get a pairwise cancellation of waves on the screen from slits separated by five slit separations. Only the few slits left over from all this pairwise cancellation contribute a small amount of light to the region between the interference

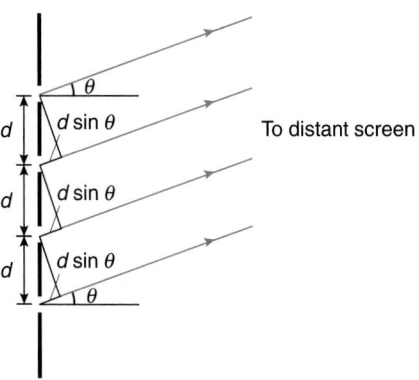

FIGURE 24.26 The path difference $d \sin \theta$ between adjacent slits of a diffraction grating is the same.

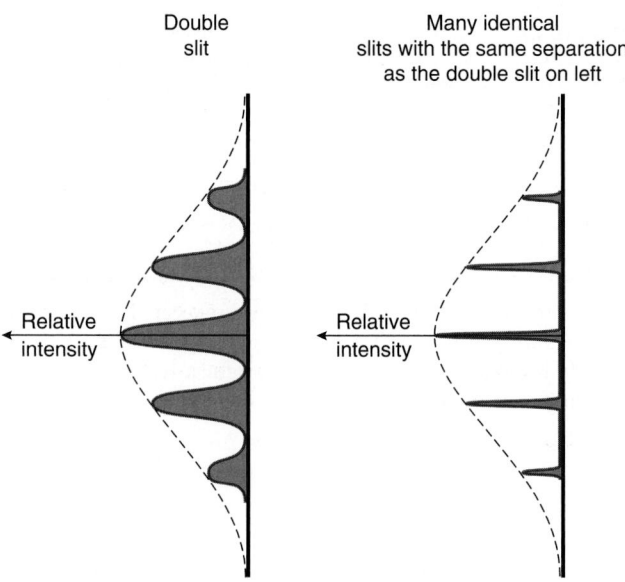

FIGURE 24.27 Increasing the number of slits with the same slit separation sharpens the interference maxima.

maxima. Very narrow and bright fringes remain for the orders of any given wavelength.

We can use the technique we first employed for locating the first minimum of a single slit diffraction pattern to determine the angular width of the central *interference maximum* for a diffraction grating (the maximum corresponding to order number $m = 0$). To find the minima of the single slit diffraction pattern, we broke the slit into many coherent wave sources (using Huygens's principle). We then examined the interference of the myriad of coherent sources. We discovered that the location of the first minimum is found from Equation 24.3:

$$a \sin \theta_1 = \lambda$$

where a is the width of the collection of Huygens's sources (the width of the slit).

A diffraction grating with N slits illuminated also consists of a myriad (actually N) of coherent wave sources as well. Thus the location of the first minimum of the central *interference maximum* of these N coherent sources is found by replacing the slit width a in Equation 24.3 with the width of the *illuminated* grating: Nd. Thus the location θ_{min} of the first minimum of the central interference fringe is

$$Nd \sin \theta_{min} = \lambda$$

Since the angle θ_{min} is small, we use the small angle approximation to write this as

$$Nd\theta_{min} = \lambda$$

or

$$\theta_{min} = \frac{\lambda}{Nd} \qquad (24.9)$$

The angular width of the interference maximum is $2\theta_{min}$. Equation 24.9 shows explicitly that as the number of slits increases greatly, the angular width of the interference maximum decreases drastically and the maxima become very narrow ("sharp"). Note that Equation 24.9 for a single slit ($N = 1$) gives the expected result (where d then is equal to a).

Although we derived Equation 24.9 for the central interference fringe, the result is applicable to *any* of the m interference maxima of the diffraction grating. We just use the projected width of the illuminated grating, $Nd \cos \theta$ (shown in Figure 24.28), as the effective width of the aperture and obtain the line width about the mth maximum:

$$\theta_{min} = \frac{\lambda}{Nd \cos \theta} \qquad (24.10)$$

The quantity N is the number of slits of a grating that are *illuminated* by the incident light. N is different from the total number of slits in the grating if only a portion of the total number of slits actually are illuminated.

EXAMPLE 24.6 ━━━━━━━━

Light waves from a sodium vapor lamp (like the common yellow street light) are incident along the normal to a diffraction grating with 12.0×10^3 slits per centimeter. The first-order

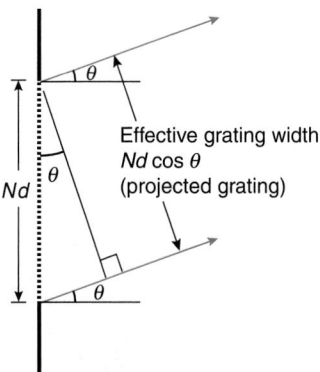

FIGURE 24.28 The projected width of a grating.

interference fringe is found to be located at an angle of 45.0° from the incident, straight-through direction. Determine the wavelength of the light.

Solution

You need to know the grating spacing d. There are 12.0×10^3 slits per centimeter, or 1.20×10^6 slits per meter. The grating spacing is the distance d between slits, which is the *inverse* of the number per meter. Hence

$$d = 8.33 \times 10^{-7} \text{ m}$$

Use the grating equation (Equation 24.8):

$$d \sin \theta = m\lambda$$

Solving for λ, you obtain

$$\lambda = \frac{d \sin \theta}{m}$$

The order number m is $m = 1$. Making the appropriate substitutions, you find

$$\lambda = \frac{(8.33 \times 10^{-7} \text{ m}) \sin(45.0°)}{1}$$
$$= 5.89 \times 10^{-7} \text{ m}$$
$$= 589 \text{ nm}$$

24.9 RESOLUTION AND ANGULAR DISPERSION OF A DIFFRACTION GRATING

Illuminate a diffraction grating with light from a source of wavelength λ_1. The interference maxima are located at angles determined by the grating equation, Equation 24.8:

$$d \sin \theta_1 = m\lambda_1 \qquad (24.11)$$

(Here the angle θ_1 is the angle at which wavelength λ_1 has order number m, *not* the angular position of the first minimum of the diffraction peak of a single slit.) If light of *another* wavelength λ_2

is incident simultaneously on the diffraction grating, its interference maxima are found from

$$d \sin \theta_2 = m\lambda_2 \qquad (24.12)$$

Thus an interference pattern of bright fringes is established for *each* wavelength present in the incident light, as shown in Figure 24.29.

Examples of such multiple-wavelength light sources include gas discharge lamps such as sodium vapor lamps (common yellowish street lights), mercury vapor lamps (common bluish street lights), hydrogen vapor lamps, and neon signs. The atoms in each source act independently (incoherently), but each light wave from an individual atom forms multiple coherent sources at the grating by wavefront division.

We want to determine how close the wavelengths λ_1 and λ_2 can be and still reveal the two interference patterns formed by the grating as separate. In other words, if the difference between the wavelengths $\Delta\lambda = \lambda_2 - \lambda_1$ is too small, the two interference patterns are essentially superimposed on each other and we cannot infer the presence of two different wavelengths. We are interested in determining the *resolution* of the diffraction grating: its ability to distinguish between two incident wavelengths.

Both wavelengths produce an interference fringe at $\theta = 0$ rad corresponding to order number $m = 0$. So we need to consider order numbers m whose absolute values are greater than zero to have any hope of distinguishing between the interference patterns. For the same order number m, the interference fringes for each wavelength are found at

$$d \sin \theta_1 = m\lambda_1 \qquad \text{and} \qquad d \sin \theta_2 = m\lambda_2$$

To see how changing λ changes θ, we use the general grating equation (Equation 24.8)

$$d \sin \theta = m\lambda$$

and take its derivative with respect to λ (remembering that the grating spacing d is constant for a given grating; the order number

m also is constant, since we are considering the *same* order interference maxima for both wavelengths):

$$(d \cos \theta)\frac{d\theta}{d\lambda} = m$$

$$\frac{d\theta}{d\lambda} = \frac{m}{d \cos \theta} \qquad (24.13)$$

This expression enables us to find the small angular separation $d\theta$ of two closely spaced wavelengths (the difference between the wavelengths is $d\lambda$). Equation 24.13 is the way in which θ changes with wavelength λ, and is known as the **angular dispersion** of the grating. Notice that the angular dispersion

- is directly proportional to the order number; the angular separation of two wavelengths increases as the interference order number increases; and

- is inversely proportional to the grating spacing d; the smaller the slit separation d, the greater the angular separation of the wavelengths.

If the angles are small, the cosine is near unity, so the angular dispersion then is approximately

$$\frac{d\theta}{d\lambda} \approx \frac{m}{d} \qquad (24.14)$$

We say the two interference patterns of two different but nearly equal wavelengths are barely resolved if the interference maximum for one wavelength falls on the first interference minimum of the other wavelength: this is analogous to the Rayleigh criterion invoked in considering resolution for single aperture diffraction patterns. This resolution criterion means that the small angular difference $d\theta$ in the locations of the interference maxima is the angle $\theta_{\min}$ in Equation 24.9:

$$\theta_{\min} = \frac{\lambda}{Nd}$$

Making this substitution for $d\theta$ in Equation 24.14 and writing $\Delta\lambda$ for the small difference in wavelength, we find

$$\frac{\lambda}{Nd} \approx \frac{m}{d}\Delta\lambda \qquad (24.15)$$

We can use either λ_1 or λ_2 for the wavelength in this equation since the two wavelengths are virtually the same. Rearranging Equation 24.15 slightly, we find

$$\frac{\lambda}{\Delta\lambda} = Nm \qquad (24.16)$$

The expression $\lambda/\Delta\lambda$ is known as the **resolving power** of the diffraction grating. A grating with a large resolving power can discern between two wavelengths that are not very different from each other.

Although we derived Equation 24.16 for the resolving power of a grating making use of the small angle approximation, Equation 24.16 is legitimate even for large angles (although we have not proved this explicitly).

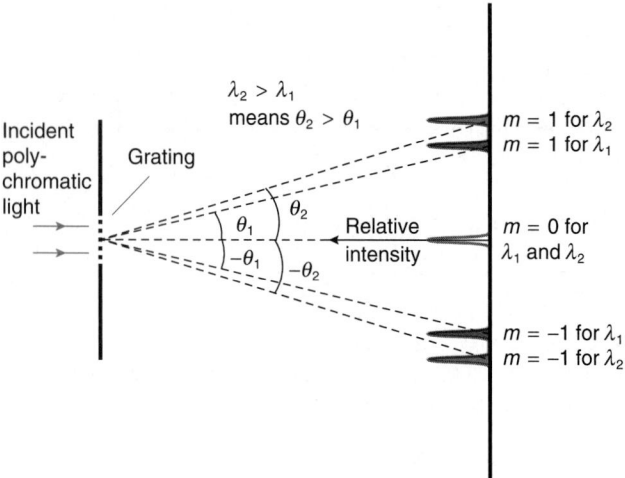

FIGURE 24.29 Each incident wavelength establishes its own interference pattern.

Notice that the resolving power of a grating depends on two factors:

1. The order number m. The resolution increases for larger order numbers.

2. The number of slits N illuminated in the grating. The greater N, the greater the resolution. This is the principal justification for increasing the number of illuminated slits in a grating.

Diffraction gratings are extraordinarily useful instruments. If light with a collection of many different wavelengths is incident on a diffraction grating, it separates the light into its component wavelengths according to the grating equation, Equation 24.8. Each wavelength appears at a different angle θ for a given order number m. The collection of wavelengths present in a source is called its spectrum.* Each chemical element has its own unique spectrum. Thus the determination of the wavelengths present in a source (known as spectroscopy) provides information about the chemical nature of the source, a technique as useful in astronomy as in forensics. A prism also can be used to separate or distinguish the various wavelengths present in a light source (thanks to dispersion—the variation of index of refraction with wavelength), but diffraction gratings are preferred because they are much easier to use, have greater angular dispersion, and have better resolution.

Spectroscopy is an important tool of many fields of physics and astronomy. Spectroscopy is the primary tool of astronomy, for very little comes to us from distant stars and galaxies other than light. Indeed it was spectroscopy that penetrated one of the "impossible problems" of the 19th century: the chemical composition of the stars (cf. the quote by Auguste Compte opening this chapter).

The first precision diffraction gratings were manufactured in the latter part of the 19th century by Henry Rowland at The Johns Hopkins University. In the best spirit and interest of science, he provided them at cost to many research laboratories around the world, likely forgoing a fortune in royalties for the invention. His gratings are as valuable today as then because they have no electronic or moving parts—quite a legacy!

STRATEGIC EXAMPLE 24.7

A grating with 2000 slits spaced equally over a 1.000 cm distance is used to analyze the spectrum of the element mercury. Among other wavelengths, mercury emits light at 576.959 nm and 579.065 nm. What is the angular separation of the two wavelengths in the second-order spectrum?

Solution
Since there are 2000 slits in each centimeter of the grating, the distance d between the slits of the grating (the grating spacing) is

$$d = \frac{1.000 \text{ cm}}{2000}$$
$$= 5.000 \times 10^{-4} \text{ cm}$$
$$= 5.000 \times 10^{-6} \text{ m}$$

---

*The word comes from the Latin *spectrum*, meaning "an appearance."

The second-order spectrum corresponds to order number $m = 2$. If $m = -2$ we get similar results.

Method 1
Use the grating equation (Equation 24.8)

$$d \sin \theta = m\lambda$$

to find the angles at which each wavelength appears in the second order:

$$\sin \theta = \frac{m\lambda}{d}$$

A subtraction of the resulting angles then yields the angular separation of the two wavelengths.

Thus the angle at which the 576.959 nm wavelength appears is

$$\sin \theta_{576.959 \text{ nm}} = \frac{2(576.959 \times 10^{-9} \text{ m})}{5.000 \times 10^{-6} \text{ m}}$$
$$= 0.2308$$

$$\theta_{576.959 \text{ nm}} = 13.34°$$

The 579.065 nm wavelength appears at

$$\sin \theta_{579.065 \text{ nm}} = \frac{2(579.065 \times 10^{-9} \text{ m})}{5.000 \times 10^{-6} \text{ m}}$$
$$= 0.2316$$

$$\theta_{579.065 \text{ nm}} = 13.39°$$

The angular separation of the two wavelengths in the second order thus is

$$\theta_{579.065 \text{ nm}} - \theta_{576.959 \text{ nm}} = 13.39° - 13.34° = 0.05°$$

Note that to use this method, you need to know many significant figures for the wavelengths and grating spacing to be able to perform the subtraction. Method 2 is a better way.

Method 2
Since the two wavelengths are very closely spaced you can *differentiate* the grating equation (Equation 24.8) with respect to λ to see how the angle θ varies with wavelength. This is the angular dispersion, Equation 24.13; slightly rearranged, you have

$$d\theta = \frac{m}{d \cos \theta} d\lambda \qquad (1)$$

Let the small angle and wavelength differences be $\Delta\theta$ and $\Delta\lambda$. You found the slit separation $d = 5.000 \times 10^{-6}$ m. The order number is $m = 2$. The difference between the wavelengths $\Delta\lambda$ is

$$\Delta\lambda = 579.065 \text{ nm} - 576.959 \text{ nm}$$
$$= 2.106 \text{ nm}$$
$$= 2.106 \times 10^{-9} \text{ m}$$

For the angle θ you can use the angle at which *either* wavelength appears:

Using $\theta_{576.959 \text{ nm}} = 13.34°$ in equation (1), you have

$$\Delta\theta = \frac{2(2.106 \times 10^{-9} \text{ m})}{(5.000 \times 10^{-6} \text{ m}) \cos(13.34°)}$$
$$= 8.658 \times 10^{-4} \text{ rad}$$
$$= 0.049\,61°$$

Using $\theta_{579.065 \text{ nm}} = 13.39°$ in equation (1), you have

$$\Delta\theta = \frac{2(2.106 \times 10^{-9} \text{ m})}{(5.000 \times 10^{-6} \text{ m}) \cos(13.39°)}$$
$$= 8.659 \times 10^{-4} \text{ rad}$$
$$= 0.049\,61°$$

The angular separation is 0.049 61°. Notice that you secure a result to four significant figures using the angular dispersion equation, whereas with the grating equation you only secured a result to one significant figure. Notice also that it makes little difference which wavelength is used in the equation for the angular dispersion (as long as the wavelengths are nearly the same).

EXAMPLE 24.8 _____

Among other wavelengths, a sodium vapor lamp emits two slightly different wavelengths: 588.995 nm and 589.592 nm. It is these wavelengths that give sodium vapor lamps their predominantly yellowish pall. How many slits are needed to barely resolve these wavelengths in the first-order spectrum?

Solution
Equation 24.16 determines the resolution:

$$\frac{\lambda}{\Delta\lambda} = Nm$$

The wavelength difference is

$$\Delta\lambda = 589.592 \text{ nm} - 588.995 \text{ nm}$$
$$= 0.597 \text{ nm}$$

You can use either wavelength for λ in Equation 24.16 since their difference is small. The wavelengths are to be resolved in the first order, so that $m = 1$. Substituting into Equation 24.16, you find

$$\frac{589 \text{ nm}}{0.597 \text{ nm}} = N \times 1$$
$$N = 987$$

A grating with a minimum of about 1000 illuminated slits is needed.

24.10 THE INDEX OF REFRACTION AND THE SPEED OF LIGHT

The law of refraction (and the law of reflection), dear to us from geometric optics in Chapter 23, can be derived from the Maxwell equations of electromagnetism. From this derivation (which we leave for a more advanced course on electromagnetism) comes another meaning for the index of refraction of a medium.

> The index of refraction is the ratio of the speed of light in vacuum to its speed in the transparent medium:
>
> $$n = \frac{\text{speed of light in vacuum}}{\text{speed of light in the medium}} = \frac{c}{v} \qquad (24.17)$$

We discussed in Section 21.5 how the speed of light in vacuum is related to the permittivity of free space ε_0 and permeability of free space μ_0, the electrical and magnetic properties of free space (a vacuum). The speed of light in a material also is related to the electrical and magnetic properties of the medium: its permittivity ε and its permeability μ. The reason the electrical and magnetic properties of the materials come into play is that the electric and magnetic fields of the light wave in the two media must each be properly matched on both sides of the interface between the media. This means that the oscillations of the fields of the electromagnetic wave must be at the same frequency in each material. If this were not true, we would find intolerable numbers of oscillations piling up on one side, waiting to be transmitted.

> Therefore the frequency of the light is the same in both media.

It is from this matching of the fields that the law of refraction arises.

Since the index of refraction of visible light in transparent materials is a number greater than 1 (see Table 23.1), Equation 24.17 implies that the speed of light waves in a material medium is *less* than that in a vacuum. The reason has to do with the process of light transmission through such materials. That process involves the absorption and subsequent emission of light photons (energy bundles or particle-waves of light) by atoms. The photons of light travel between the atoms at the vacuum speed of light, but the absorption and emission processes each take a bit of time. The result is that the effective speed of the light through the material is less than if the material were not present. The process is similar to having annoying toll booths spaced every 10 km along a super-highway; they slow down the effective speed of the traffic.

The product of the frequency and the wavelength is equal to the speed of the wave. Thus for light in a vacuum we have

$$c = \nu\lambda \qquad (24.18)$$

where v is the frequency of the light and λ is its wavelength in vacuum. In a medium where the speed of light is v_1, we have a similar relation:

$$v_1 = v\lambda_1 \qquad (24.19)$$

where λ_1 is the wavelength of the light in the medium. The frequency of the light in the medium is the *same* as in the vacuum. Dividing Equation 24.18 by 24.19, we obtain

$$\frac{c}{v_1} = \frac{\lambda}{\lambda_1}$$

The left-hand side of this equation is the index of refraction n_1 of the medium (from Equation 24.17). Thus

$$n_1 = \frac{\lambda}{\lambda_1}$$

After slight rearrangement, this becomes

$$\lambda_1 = \frac{\lambda}{n_1} \qquad (24.20)$$

Experiments indicate that the index of refraction of visible light in transparent materials is greater than 1 (see Table 23.1).

> Hence Equation 24.20 indicates that the wavelength of such light in a material medium is *shorter* (smaller) than the wavelength in vacuum.

EXAMPLE 24.9

The index of refraction of water is about 1.33 (from Table 23.1). Determine the speed of light in water.

Solution
Equation 24.17 relates the speed of light in a material to the index of refraction:

$$n = \frac{c}{v}$$

So

$$
\begin{aligned}
v &= \frac{c}{n} \\
&= \frac{3.00 \times 10^8 \text{ m/s}}{1.33} \\
&= 2.26 \times 10^8 \text{ m/s}
\end{aligned}
$$

This speed still is *quite* fast!

EXAMPLE 24.10

Light with a vacuum wavelength of 632.8 nm enters glass with an index of refraction of 1.50. What is the wavelength of the light in glass?

Solution
The wavelength of light when in a material medium is shorter than that in vacuum by a factor equal to the index of refraction, according to Equation 24.20:

$$
\begin{aligned}
\lambda_{\text{glass}} &= \frac{\lambda_{\text{vacuum}}}{n_{\text{glass}}} \\
&= \frac{632.8 \text{ nm}}{1.50} \\
&= 422 \text{ nm}
\end{aligned}
$$

24.11 THIN-FILM INTERFERENCE*

No light, but rather darkness visible

John Milton (1608–1674)[†]

You may have noticed that fine lenses for cameras, microscopes, and telescopes have transparent coatings on their optical surfaces. What purpose do the coatings serve, or are they purely decorative and protective? Quite apart from this, perhaps while taking a relaxing bubble bath, you have noted that the bubbles have interesting rainbow-like, swirled, colored bands. Less romantically, when filling up at a gas station on a rainy day, you may have noted the delicate rainbow-like colored bands on puddles, a clear indication of some gasoline polluting the water surface. Here we shall see that these apparently disparate observations all are manifestations of **thin-film interference** of light.

Aside from a path difference, there is another way to secure a phase difference. When a wave is reflected, a *phase change* of π rad may, or may not, occur for the reflected wave, as we saw in Section 12.10. Under what circumstances do electromagnetic waves experience such a phase change of π rad when reflected? Experiments indicate two circumstances lead to such phase changes:

a. Reflection from a conducting surface (a metal). This is analogous to the rope wave reflection off a rigid wall, since the conductor cannot sustain an electric field.
b. Reflection from an optically more dense material ($n_2 > n_1$). This situation is like a rope pulse reflecting from a section of

---

[†]*Paradise Lost*, Book I, line 63.

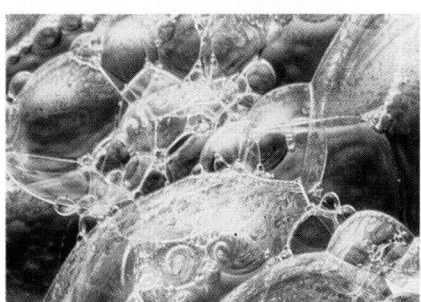

Thin-film interference is apparent in soap bubbles.

rope with a greater mass per unit length than that of the incident pulse. If $n_2 < n_1$, no phase change occurs in the reflected wave. We shall see in the next subsection how these conclusions are sustained by experimental observations.

When light passes the boundary between two transparent media, some light is reflected at the boundary. The amount of light reflected depends on a number of factors, among them the indices of refraction of the two media and the angle of incidence. For light incident along the normal line from air to glass, about 4% of the incident light is reflected back into the air at the boundary; this increases to almost 100% reflection for angles of incidence approaching 90°. At such angles, the glass acts much like a mirror surface.* Here we examine only light incident normally on the boundaries between transparent materials.

We consider, in particular, monochromatic light incident on a thin film of material separating two other transparent media, as shown in Figure 24.30. Some light is reflected at the first interface and some from the second interface.

We are interested in the interference of the two reflected waves when they superimpose in the first material along the normal line. The two reflected waves are monochromatic and also coherent because they arise from the same monochromatic incident light wave via *amplitude division*. Once the wave reflected from the second surface gets back to the wave reflected from the first surface, the two waves interfere, since they are superimposed along the same normal line.

> To determine the nature of the interference of the reflected waves, we need to consider *two* factors that contribute to their phase difference: a path difference and phase changes upon reflection.

We consider each in turn.

---

*Most surfaces reflect most light when the angle of incidence approaches 90°. You can observe this with even a sheet of paper. If you hold the paper up before a light bulb so the light is incident at a large angle to the normal, you will see that much of the light from the bulb is reflected from the paper surface.

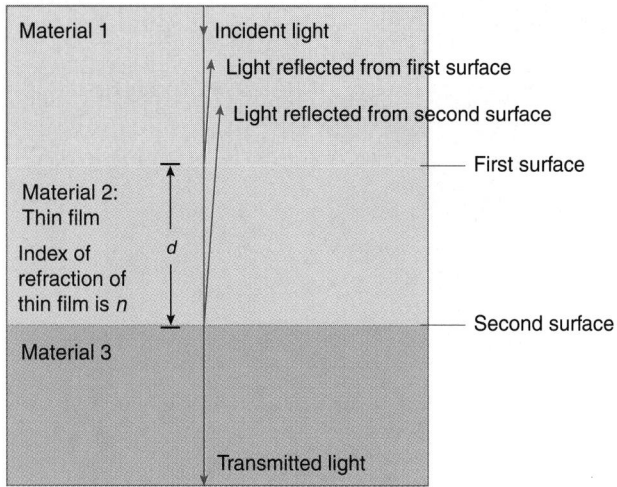

FIGURE 24.30 Light incident along the normal to a thin film separating two other transparent media.

Path Difference

The wave reflected from the lower interface travels an additional distance $2d$ before it is superimposed and interferes with the wave reflected from the top surface, where d is the thickness of the film. Hence the path difference Δx between the two reflected rays is $2d$. From Equation 24.1, this path difference corresponds to a phase difference of

$$\delta_{\text{path}} = \frac{2d}{\lambda_{\text{film}}}(2\pi \text{ rad}) \qquad (24.21)$$

PROBLEM-SOLVING TACTIC

24.3 When investigating thin-film interference, note that the wavelength of the light in the film is *different* from that in vacuum. It is the wavelength of the light in the film that is involved in assessing the phase difference associated with the path difference. So be careful to use the wavelength of the light in the film in Equation 24.21. In particular, we saw from Equation 24.20 that

$$\lambda_{\text{film}} = \frac{\lambda_{\text{vacuum}}}{n}$$

Phase Changes upon Reflection

We also need to assess whether there are any phase changes involved in the two reflections. *This assessment must be done on a case-by-case basis*, taking into consideration the specific indices of refraction of the three media involved. Three situations may arise; the first two are the following:

1. *Neither* reflected wave experiences a phase change upon reflection.
2. *Both* reflected waves experience a phase change upon reflection.

Because the *same* phase shift has occurred to *both* reflected waves in either of these two scenarios, the phase changes upon reflection thus are irrelevant in assessing the status of the interference of the two reflected rays. No *difference* in phase results between the two reflected waves from this cause in these instances. For either of these cases, the interference of the two waves is determined *solely* from the path difference between the two reflected waves.

In particular, if the thickness of the film is such that the phase difference arising from the path difference is a multiple of 2π rad, the interference of the two waves is completely constructive and the reflected light is bright:

$$\frac{2d}{\lambda_{\text{film}}}(2\pi \text{ rad}) = m(2\pi \text{ rad}) \qquad m = \text{integer}$$

$$\text{(constructive interference)} \qquad (24.22)$$

On the other hand, if the thickness of the film is such that the resulting phase difference is an *odd multiple* of π rad, the resulting interference of the two reflected rays is completely destructive. If m is an integer, then $2m + 1$ is an odd integer. Hence the condition for destructive interference is

$$\frac{2d}{\lambda_{\text{film}}}(2\pi \text{ rad}) = (2m + 1)(\pi \text{ rad}) \qquad m = \text{integer}$$

$$\text{(destructive interference)} \qquad (24.23)$$

The third scenario is the following:

3. One of the reflected waves experiences a phase change of π rad upon reflection and the other wave does not.

It makes no difference which one experiences the phase change; the important thing is that one reflected wave does while the other one does not. If this is the case, then the previous conclusions are reversed. That is, the condition for destructive interference is

$$\frac{2d}{\lambda_{film}}(2\pi \text{ rad}) = m(2\pi \text{ rad}) \qquad m = \text{integer}$$

(destructive interference) (24.24)

and the condition for constructive interference is

$$\frac{2d}{\lambda_{film}}(2\pi \text{ rad}) = (2m + 1)(\pi \text{ rad}) \qquad m = \text{integer}$$

(constructive interference) (24.25)

In other words, if the path difference implies a phase difference between the waves of a multiple of 2π rad, initially leading us to think the waves are in phase and yielding constructive interference, the phase change upon reflection of one but not the other wave means the two will be out of phase and the interference actually is destructive. Conversely, if the path difference implies a phase change of an odd multiple of π rad, initially leading us to think the waves are out of phase and interfering destructively, the phase change upon reflection of one wave but not the other means the reflected waves actually are in phase and will interfere constructively.

Combining all the scenarios for phase difference, we write the total phase difference as

$$\delta = \delta_{path} + \delta_{bdy} \qquad (24.26)$$

$$= \begin{cases} \dfrac{\Delta x}{\lambda_{film}}(2\pi \text{ rad}) + 0 \text{ rad} & \text{(for similar reflections)} \\[4mm] \dfrac{\Delta x}{\lambda_{film}}(2\pi \text{ rad}) + \pi \text{ rad} & \text{(for different reflections)} \end{cases}$$

PROBLEM-SOLVING TACTIC

24.4 There is no magic bullet formula to memorize that can treat all situations of thin-film interference. Each situation must be assessed individually to determine (a) the path difference and its resulting phase difference and (b) the phase changes (if any) upon reflection and whether they are relevant or not. The two pieces of information then must be interwoven to determine if the superposition of the reflected waves is constructive or destructive.

Thin-film *destructive* interference is the principle behind coating optical surfaces such as lenses to *minimize* reflections; more light then is transmitted by the lens. Since an uncoated air–glass interface reflects approximately 4% of the light incident along the normal, it does not take many such surfaces to significantly attenuate the incident light, particularly in optical systems consisting of many lenses. However, if each lens surface is coated with an appropriate thickness of a transparent material (such as magnesium fluoride [MgF_2]), the reflected light can be minimized

with a resulting gain in the transmission of the lens system. All fine optical lens systems are so coated.

Thin-film *constructive* interference in the reflected light (at various wavelengths) also explains the colored bands seen when gasoline is on water. Each colored band indicates constructive interference in the light reflected from the air–gasoline and gasoline–water interfaces for that color. Each colored band thus traces out a film of constant thickness of gasoline. It is similar for soap bubbles.

If we place a slightly convex glass surface on a flat glass surface as shown in Figure 24.31, and view the reflected light from the glass–air and the air–glass interfaces, we observe a dark spot at the point of contact. At the point of contact, the rays reflected from the glass–air and air–glass interfaces have zero path difference. The interference nonetheless is seen to be destructive. Since the path difference is zero, the two waves are out of phase because one reflected wave experienced no phase change upon reflection (the wave from the glass-to-air reflection) while the other reflected wave had a phase change of π rad (the one from the air-to-glass reflection). This confirms the conclusions we drew earlier about phase changes upon reflection.

Note that the pattern of the constructive and destructive interference in the reflected light consists of a series of concentric

A thin film coating on a lens minimizes reflections and enhances transmission.

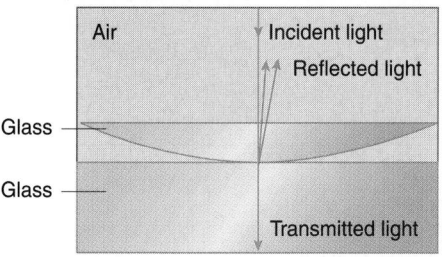

FIGURE 24.31 Newton's rings in reflected light produce a dark spot at the point of contact.

circles, called **Newton's rings**, that trace out contours of constant thickness of the air gap between the surfaces.

If we view the transmitted light, the central spot is bright, as shown in Figure 24.32. There is no path difference between the two transmitted waves, and no phase difference from reflection since the transmitted light is not reflected. The transmitted light also consists of a series of concentric bright and dark rings.

Thin-film interference is used extensively to test the shape of various optical surfaces. The constructive and destructive interference conditions map contours of constant thickness. A convex surface on a flat surface produces concentric rings of constructive and destructive interference fringes centered on the point of contact. If the surface is not flat, or the convex surface is not symmetric about its optic axis, the rings are distorted in shape. See Figure 24.33.

If two optical flats are inclined slightly to each other to form an air wedge between them, the contours of constant thickness should be straight lines; any apparent deviations of the interference fringes from this geometry indicate that one or the other flat (or both) really is not flat.

FIGURE 24.32 Newton's rings in transmitted light produce a bright spot in the center.

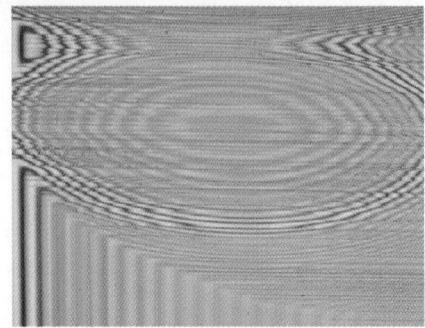

FIGURE 24.33 Distortions in fringe shapes reveal deviations from spherical or flat geometries of the surfaces.

QUESTION 4

If both surfaces of a lens with index of refraction $n > 1$ in air are coated with a material of index n' to minimize reflections, are the coatings of the same thickness? Explain.

EXAMPLE 24.10

Light with a vacuum wavelength 633 nm is incident normally on a thin-film coating with an index of refraction $n = 1.35$ over glass, as in Figure 24.34. What is the minimum thickness d for the film needed to minimize reflected light at normal incidence?

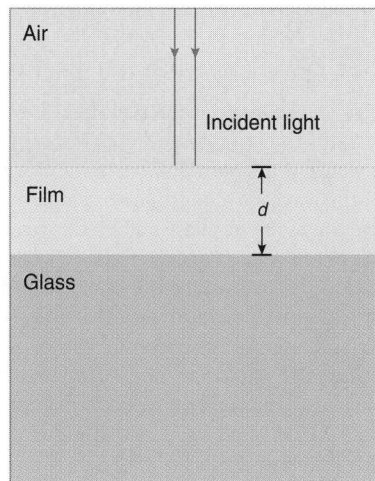

FIGURE 24.34

Solution

Since the index of refraction of the film is greater than that of the air, the wave reflected from the air-to-film interface experiences a change of phase of π rad upon reflection. Likewise, the wave reflected from the second interface (film-to-glass) also experiences a phase change of π rad upon reflection, because the index of refraction of the glass ($n \approx 1.5$) is greater than that of the film. Since *both* reflected rays experience a phase change of π rad upon reflection, the phase changes upon reflection are *irrelevant* for determining the total phase difference between the two reflected waves.

The path difference between the two reflected rays is, therefore, the sole factor that determines the character of the interference. The path difference is $2d$. Since you want to make the two reflected waves interfere destructively, the phase difference δ_{path} corresponding to the path difference $2d$ should be an odd integral multiple of π rad. To keep the thickness of the film to a minimum, the phase difference is made the first odd multiple of π, just π itself. Thus Equation 24.21 becomes

$$\delta_{\text{path}} = \pi \ \text{rad} = \frac{2d}{\lambda_{\text{film}}} (2\pi \ \text{rad})$$

Solving for d, you obtain

$$d = \frac{\lambda_{\text{film}}}{4}$$

The wavelength in the film is related to the vacuum wavelength by Equation 24.20:

$$\lambda_{\text{film}} = \frac{\lambda_{\text{vacuum}}}{n}$$

where n is the index of refraction of the film. Hence the minimum film thickness needed is

$$d = \frac{\lambda_{\text{vacuum}}}{4n}$$
$$= \frac{633 \text{ nm}}{4(1.35)}$$
$$= 117 \text{ nm}$$

EXAMPLE 24.11

The edge of a thin sheet of tissue paper is placed between two optical flats to form an air wedge, as shown in Figure 24.35. Light of vacuum wavelength 589 nm is incident normally on the arrangement and 41 parallel, bright fringes of constructive interference are seen in the reflected light. A bright fringe is seen essentially coincident with the edge of the paper. Find the thickness of the sheet of paper.

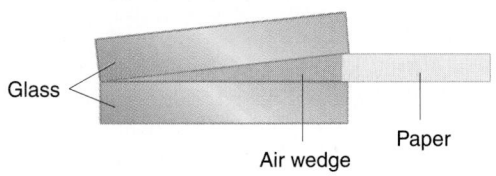

Glass
Paper
Air wedge

FIGURE 24.35

Solution
Here the interference occurs between the wave reflected from the top of the air wedge and the wave reflected from the bottom of the air wedge, as shown in Figure 24.36.

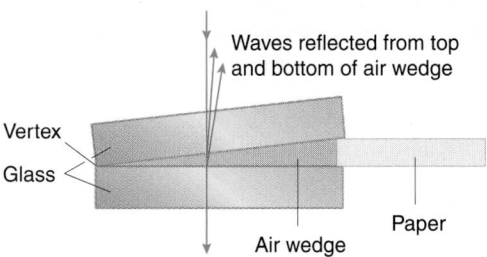

Waves reflected from top and bottom of air wedge
Vertex
Glass
Paper
Air wedge

FIGURE 24.36

The top reflection (at the glass-to-air wedge interface) does not experience a phase change of π rad upon reflection, because the index of refraction of air (the second medium) is less than that of glass (the first medium). The bottom reflection (at the air wedge-to-glass interface) does experience a phase change upon reflection, since the index of refraction of the glass (now the second

medium) is greater than that of the air (now the first medium). Thus you have a situation in which one of the reflected waves experiences a phase change upon reflection while the other does not.

At the vertex of the air wedge, the path difference between the waves is zero. But the two reflected waves here are nevertheless completely out of phase because of the phase change of one of the waves upon reflection. Hence the two reflected waves interfere completely destructively at this location and a dark fringe is seen at the wedge vertex. As you move out from the vertex, when the phase difference corresponding to the path difference $2d$ (twice the thickness d of the wedge at that location) becomes equal to an odd multiple of π rad, the two reflected waves will interfere constructively. Thus, from Equation 24.25, a bright interference fringe occurs whenever

$$\frac{2d}{\lambda_{\text{film}}} (2\pi \text{ rad}) = (2m + 1)(\pi \text{ rad})$$

where m is an integer. The first bright fringe from the vertex corresponds to $m = 0$, the second bright fringe corresponds to $m = 1$, and so forth. Thus the 41st bright fringe (where the paper is located) corresponds to $m = 40$. Thus the thickness d of the air wedge at the location of the paper must satisfy

$$\frac{2d}{\lambda_{\text{film}}} (2\pi \text{ rad}) = [2(40) + 1](\pi \text{ rad})$$

$$= 81\pi \text{ rad}$$

The thin film in this situation is the air wedge. Since the index of refraction of the air is essentially the same as the index of refraction of the vacuum, $\lambda_{\text{film}} = \lambda_{\text{vacuum}} = 589$ nm. Hence the thickness d is

$$d = \frac{81}{4} \lambda_{\text{film}}$$
$$= \frac{81}{4} (589 \text{ nm})$$
$$= 1.19 \times 10^4 \text{ nm}$$
$$= 1.19 \times 10^{-5} \text{ m}$$

24.12 POLARIZED LIGHT*

There are two distinct types of waves: longitudinal and transverse. We first encountered them in Chapter 12. The distinction is made on the basis of the direction of the "jiggling" or oscillatory motion associated with the wave motion. If the jiggling of the waves is parallel and antiparallel to the direction of propagation of the wave, the wave is longitudinal. If the jiggling is perpendicular to the direction of propagation, the wave is transverse. Sound is a longitudinal wave in fluids but can have longitudinal and transverse characteristics in solids. We showed how the Maxwell equations indicate that light is a transverse electromagnetic wave in Section 21.5. Here we see how that can be determined experimentally.

We shall consider many different ways to produce what we shall term "polarized" light in subsequent sections of this chapter. Passing over the many natural processes for the moment, take your polarizing sunglasses and borrow a second pair from a friend in order to make a few observations and conclusions about the nature of light waves. Look at any light source through these polarizing samples held together. Rotate one about the axis of viewing, while keeping the second pair fixed. You see the intensity vary between bright and dark through every 90° of rotation.

That is all the experimenting; now ponder. If light waves were longitudinal (jiggling along the axis of travel, like sound waves in air), there is *no way* they could distinguish the transverse angle of rotation of the transmitter of intensity. But they obviously do. The light waves must have a sense of direction angle *around* the axis, and we now have proof they are *transverse* in character. Light is a transverse (and electromagnetic) wave in vacuum and in most media: the electric and magnetic field oscillations of a light wave are perpendicular to the direction of propagation, as we saw in Chapter 21.*

To describe the **polarization** of the wave is to geometrically describe what is happening to the orientation of the oscillation as a function of time. By convention, it is the oscillations of the electric field vector of a light wave (rather than the associated perpendicular magnetic field vector) that we choose to define the polarization of a light wave. The plane of the magnetic field vector oscillations is perpendicular to the plane of the electric field oscillations.

Only transverse waves can exhibit polarization.

We fix a position in space (i.e., we fix x) and follow the oscillation of the electric field vector associated with the light wave as a function of time. If the oscillations of the transverse wave at a particular location are confined to a line (see Figure 24.37), we say the wave is **plane polarized**.[†]

Other types of polarized light can be created by judicious superposition of plane polarized light. For example, consider two coherent, plane polarized light waves with the following characteristics:

- the waves are of equal amplitude;
- both waves are traveling along the +x-axis;

---

*Light can assume both a transverse *and* longitudinal character under some special circumstances, such as when microwaves propagate down hollow conducting waveguides. We confine ourselves to purely transverse light waves.
[†]Plane polarized light also is known as linearly polarized light.

- each wave is plane polarized, but with the directions of the oscillations at right angles to each other; and
- the waves are $\pi/2$ rad out of phase.

We write the electric vector of one of the waves as

$$\vec{E}_1 = E_0 \cos(kx - \omega t)\hat{\jmath}$$

The other wave is described by

$$\vec{E}_2 = E_0 \cos\left(kx - \omega t + \frac{\pi}{2} \text{ rad}\right)\hat{k} \qquad (24.27)$$

We use the trigonometric formula for the cosine of the sum of two angles to express the second wave as

$$\vec{E}_2 = -E_0 \sin(kx - \omega t)\hat{k}$$

The superposition of the two waves is

$$\vec{E}_1 + \vec{E}_2 = E_0[\cos(kx - \omega t)\hat{\jmath} - \sin(kx - \omega t)\hat{k}]$$

We look at the origin ($x = 0$ m) and see what happens to the electric vector as a function of time:

$$\vec{E}_1 + \vec{E}_2 = E_0[\cos(-\omega t)\hat{\jmath} - \sin(-\omega t)\hat{k}]$$
$$= E_0[\cos(\omega t)\hat{\jmath} + \sin(\omega t)\hat{k}] \qquad (24.28)$$

The resulting electric vector traces out a circle in the y–z plane at the angular frequency ω, schematically shown in Figure 24.38. With the thumb of your right hand pointing along the direction of propagation, the curled fingers of the right hand indicate the direction of the circular motion of the electric vector given by Equation 24.28. Such light is said to be **right circularly polarized**, since the right hand relates the direction of the circular motion to the direction of propagation.

By changing the sign of the $\pi/2$ rad phase difference in Equation 24.27 for $\vec{E}_2$, the electric vector of the resulting superposition of the two waves executes circular motion in the opposite sense. Such light is called *left circularly polarized* light, since the fingers of the left hand indicate the sense of the motion of the electric vector if the thumb of the left hand is pointed along the direction of propagation of the waves. If the amplitudes of the two waves are different but the other conditions (listed

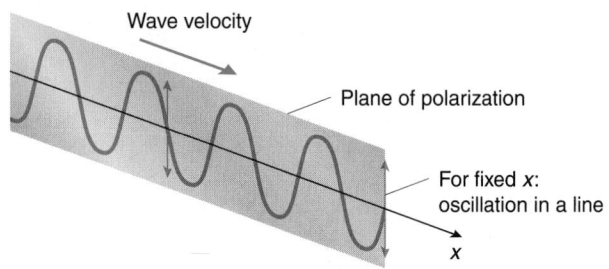

FIGURE 24.37 Plane polarized light.

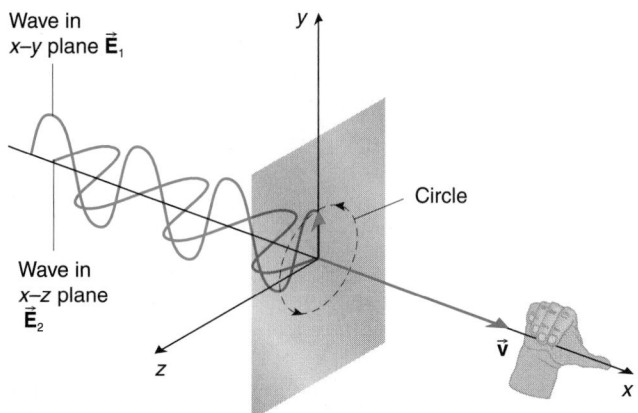

FIGURE 24.38 Right circularly polarized light.

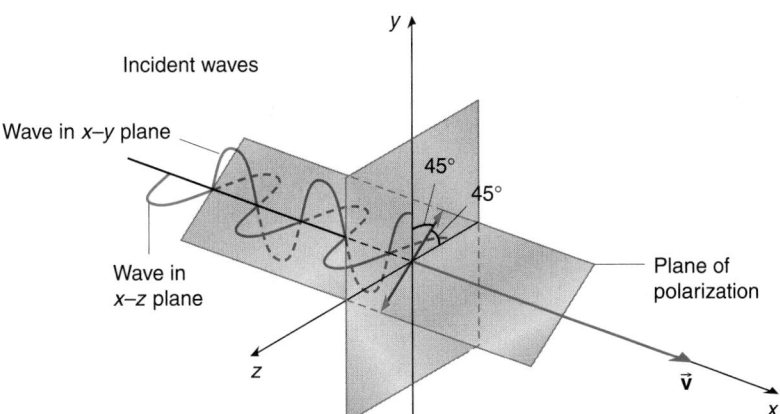

FIGURE 24.39 Plane polarized light.

previously) on the two waves are the same, the motion of the resulting electric vector traces out an ellipse; thus, we can have left and right **elliptically polarized light**.

We now consider two plane polarized light waves, differing in phase by π rad, but oscillating in planes at right angles to each other. The two waves are

$$\vec{E}_1 = E_0 \cos(kx - \omega t)\hat{\jmath}$$

and

$$\vec{E}_2 = E_0 \cos(kx - \omega t - \pi \text{ rad})\hat{k}$$

We use the trigonometric formula for the cosine of the sum of two angles to obtain

$$\vec{E}_2 = E_0[\cos(kx - \omega t) \cos \pi + \sin(kx - \omega t) \sin \pi]\hat{k}$$
$$= -E_0 \cos(kx - \omega t)\hat{k}$$

The superposition of the two waves then is

$$\vec{E}_1 + \vec{E}_2 = E_0 \cos(kx - \omega t)\hat{\jmath} - E_0 \cos(kx - \omega t)\hat{k}$$
$$= E_0 \cos(kx - \omega t)(\hat{\jmath} - \hat{k})$$

We fix our attention at the origin (where $x = 0$ m) and see what happens to the electric vector as a function of time; we have

$$\vec{E}_1 + \vec{E}_2 = E_0 \cos(\omega t)(\hat{\jmath} - \hat{k}) \qquad (24.29)$$

This is *plane polarized light* with the plane of oscillation at 45° to the *y*- or *z*-axis, as shown in Figure 24.39.

The important point to note is that the two waves *do not interfere destructively* despite the π rad (= 180°) phase difference between them. The two waves are coherent because they have a constant phase difference between them. But the two waves do not interfere because the oscillations are perpendicular to each other! If the two oscillations were polarized in the *same* plane, they would interfere destructively: notice that this can be accomplished (mathematically) by simply changing the unit vector $\hat{k}$ to the unit vector $\hat{\jmath}$ in Equation 24.29, in which case the superposition becomes zero at all times (i.e., destructive interference).

Interference between coherent sources can occur only when the sources have the same polarization.

Most light originates from atoms. The light wave emitted during a very short interval from an individual atom is polarized. The next time it emits light, the polarization direction may be different. The atoms in a typical light source also all act independently of each other.[†] The resulting incoherent light from the entire collection of atoms within the source thus consists of countless waves with planes of polarization that are randomly oriented with respect to each other, as schematically illustrated in Figure 24.40. Such a continuum of polarization directions constitutes **unpolarized light**. Light from typical light sources is unpolarized light.

It is convenient to model unpolarized light in another way. We impose two mutually perpendicular directions on the sea of different planes of polarizations associated with unpolarized light. Then we resolve the amplitude of each oscillation into two components along the two mutually perpendicular directions chosen. For a given wave from a particular atom, these mutually perpendicular oscillations are in phase and coherent with each other. However, the *collective* oscillations along each perpendicular direction are incoherent, since they represent oscillations from many different and independently acting atoms in the source. Thus the resultant field oscillations along each mutually perpendicular direction are incoherent with each other. So we model unpolarized light by two mutually perpendicular, collective, incoherent oscillations, as shown in Figure 24.41.

[†]Lasers are *not* typical light sources because the atoms act in concert with each other. Many lasers emit polarized light.

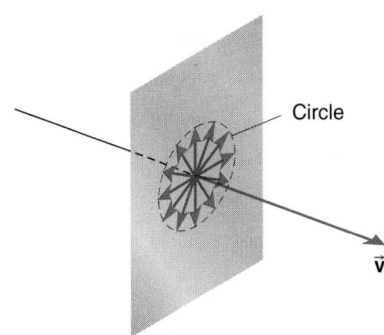

FIGURE 24.40 Unpolarized light consists of incoherent oscillations in every direction perpendicular to the velocity of the light.

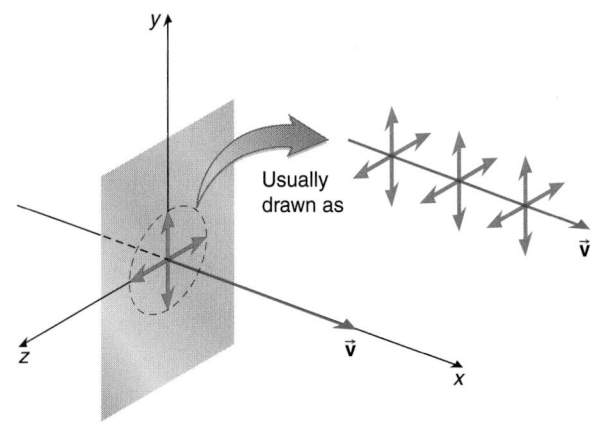

FIGURE 24.41 Unpolarized light is modeled with two mutually perpendicular, incoherent oscillations.

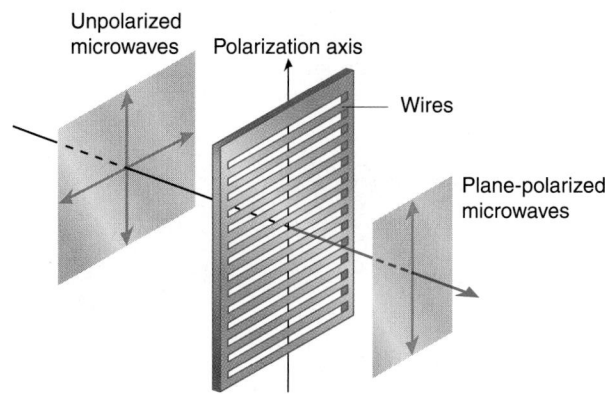

FIGURE 24.42 Unpolarized microwaves are polarized by sending them through a set of parallel conducting wires. The axis of a microwave polarizer is perpendicular to the wires in the plane of the polarizer.

QUESTION 5

Explain why longitudinal waves cannot be said to have their oscillations confined to a single plane and, therefore, why the concept of polarization is appropriate only for transverse waves.

24.13 POLARIZATION BY ABSORPTION*

How can the unpolarized light from a typical light source be polarized? That is, how can one of the two incoherent, mutually perpendicular plane polarizations used to model unpolarized light be selected or separated from the other? Several techniques are used; we explore one here and several others subsequently.

Light in the most general sense consists of the entire electromagnetic spectrum. For the moment, we consider the microwave region of the spectrum. Unpolarized microwaves can be polarized by sending the transverse waves through a grid of parallel conducting wires, as in Figure 24.42. We choose the two perpendicular directions used to represent the unpolarized incident beam to be parallel and perpendicular to the wires. The polarized waves with an electric vector parallel to the conducting wires are absorbed by the wires. The oscillatory field parallel to the wires transfers energy to the electrons that can move along the wires; it is the polarization direction perpendicular to the wires that is transmitted.

So the wire grid acts as a **polarizer**, a device for producing polarized microwaves. The **axis** of a polarizer is the direction parallel and antiparallel to the plane of polarization of the transmitted waves. The axis of a polarizer is not a unique line but simply a direction or a whole collection of lines oriented parallel to each other. Thus, for the wire grid polarizer of microwaves, the axis of the polarizer is a direction in the plane of the polarizer perpendicular to the direction of the wires, as shown in Figure 24.42.*

In 1932 one of the more prolific American inventors of the 20th century, Edwin H. Land (1909–1991), discovered how to polarize visible unpolarized light via a mechanism similar to the wire grid polarizer for microwaves just outlined.[†] He developed a conducting polymeric material (an iodine-impregnated form of polyvinyl alcohol) whose long molecules conduct readily along their length but poorly across their very thin width. The molecules (analogous to the wires in the microwave polarizer) were oriented parallel by stretching the material in nitrocellulose sheets as they solidified; the molecules thereby served as a conducting medium analogous to the wires in the microwave example. Unpolarized incident light with electric vectors oriented along the oriented molecules is absorbed and that perpendicular to the long axis of the molecules is transmitted. Thus the axis of the polarizer once again is perpendicular to the long axis of the molecules, similar to the case for the wires and microwaves. Land called the material he created *Polaroid* and founded the now famous company bearing that name to manufacture it.[‡]

Thus unpolarized light can be polarized by selective absorption in passing it through a polarizing sheet such as a Polaroid®. Half the light intensity is absorbed by the polarizing material and half is transmitted with a plane of polarization parallel to the axis of the material.

Recall that the intensity is related to the energy of the wave, which in turn is proportional to the square of the wave amplitude.

Materials that produce polarized light by selective absorption are known as **dichroic** materials. Certain natural crystals (such as the mineral tourmaline) also have the prop-

---

*This is quite unlike a transverse wave on a rope passing through a picket fence. For the rope, a transverse oscillation parallel to the pickets is transmitted and a transverse wave oscillating perpendicular to the direction of the pickets is absorbed.

[†]For an interesting historical and personal account of the development of polarizing sheets, see Edwin H. Land, *Journal of the Optical Society of America*, 41, #12, pages 957–963 (December 1951).

[‡]Polaroid cameras and Polaroid film are registered trademarks that have nothing to do with polarized light or polaroid material. The cameras and film were other inventions of Edwin H. Land and the Polaroid Corporation, provoked by an impatient question from his young daughter about why camera pictures could not appear "right away."

erty of selectively absorbing one of the two perpendicular oscillations representing unpolarized light. Such crystals are called dichroic crystals.

24.14 MALUS'S LAW*

If unpolarized light is incident on a sheet of polarizing (dichroic) material, the transmitted light is plane polarized in a direction parallel to the transmission axis of the polarizer, as in Figure 24.43.

The polarizer transmits half of the incident light while the other half is absorbed by the sheet. If the linearly polarized light then is incident on a second polarizing sheet, as in Figure 24.44, the second sheet examines or analyzes the incoming light and can tell us something about the nature of the polarized light. The second polarizing sheet thus is called an **analyzer**.

The amount of light transmitted by the analyzer depends on the angle θ between its transmission axis and the direction of the oscillation of the incident plane polarized light. In particular, let E_0 be the amplitude of the incident plane polarized light; see Figures 24.44 and 24.45. The amplitude E_0 is resolved into two components:

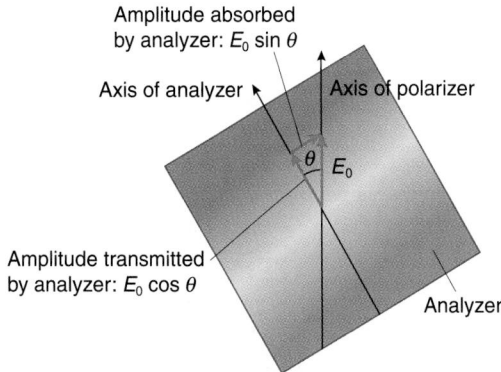

FIGURE 24.45 The action of an analyzer; the view is looking back along the light in Figure 24.44.

1. One component, $E_0 \sin \theta$, perpendicular to the axis of the analyzer. This component is absorbed by the analyzer and is not transmitted.
2. Another component, $E_0 \cos \theta$, parallel to the axis of the analyzer. This component is transmitted by the analyzer.

Thus the amplitude of the transmitted light is $E_0 \cos \theta$. The intensity transmitted is proportional to the square of the amplitude transmitted:

$$\begin{aligned} I_{\text{transmitted}} &= K(E_0 \cos \theta)^2 \\ &= KE_0^2 \cos^2\theta \end{aligned}$$

where K is a proportionality constant. But E_0^2 is proportional to the intensity of the light incident on the analyzer, $I_{\text{incident}} = KE_0^2$.

Thus

$$I_{\text{transmitted}} = I_{\text{incident}} \cos^2\theta \qquad (24.30)$$

This relationship between the incident and transmitted intensities was discovered in the early 19th century by the French physicist Étienne Louis Malus (1775–1812) and is, therefore, called **Malus's law**.

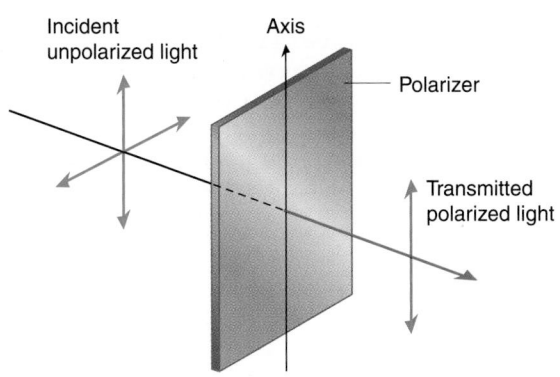

FIGURE 24.43 Production of polarized light by a polarizer.

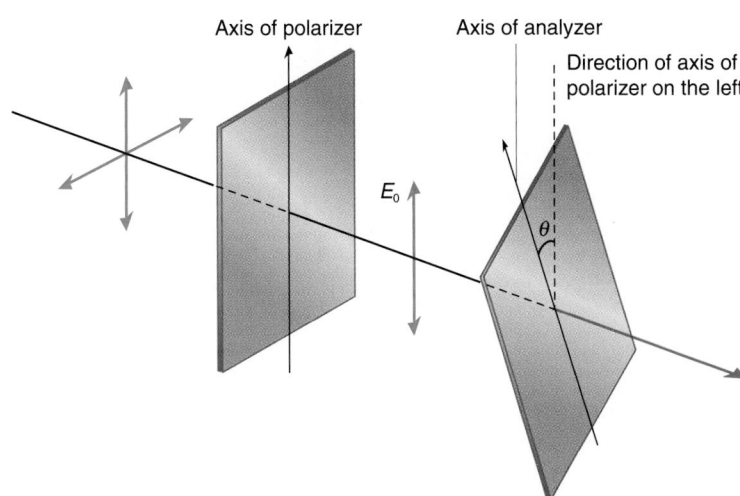

FIGURE 24.44 Polarizer and analyzer.

QUESTION 6

At what angle must the axis of the analyzer be oriented with respect to incident plane polarized light so that *no* light is transmitted by the analyzer?

EXAMPLE 24.12

Plane polarized light is produced by a polarizer. An analyzer is to be oriented at an angle θ to the plane of polarization of the incident polarized light so the intensity transmitted by the analyzer is 25% of the incident intensity. What is the angle θ required?

Solution

The intensity transmitted by the analyzer is determined from Malus's law, Equation 24.30:

$$I_{\text{transmitted}} = I_{\text{incident}} \cos^2\theta$$

Since $I_{\text{transmitted}}$ is to be 25% of I_{incident}, you have

$$0.25 I_{\text{incident}} = I_{\text{incident}} \cos^2\theta$$

$$0.25 = \cos^2\theta$$

Solving for θ, you obtain

$$\theta = 60°$$

24.15 POLARIZATION BY REFLECTION: BREWSTER'S LAW*

Perhaps you have noticed how Polaroid® sunglasses or a polarizing filter on a camera reduce reflected light (glare) from surfaces such as lakes. Here we see how light is polarized by reflection.

When light is incident on a transparent interface, some of the light is refracted into the second medium while some light is reflected back into the first medium. The proportion of the light reflected depends on several factors:

a. the angle of incidence: as it increases, the amount of the light reflected increases;
b. the numerical values of the indices of refraction of the two media: the greater the difference between the indices of refraction, the greater the reflection; and
c. the state of polarization of the incident light.

Let unpolarized light be incident on the interface at a nonzero angle of incidence, as in Figure 24.46. Consider the unpolarized beam to be composed of two mutually perpendicular, incoherent, plane polarized oscillations with

a. oscillations parallel to the interface (perpendicular to the plane of incidence encompassing the incident ray and the normal to the surface); and
b. oscillations in the plane of incidence (parallel to the plane of incidence, the plane containing the incident ray and the normal line).

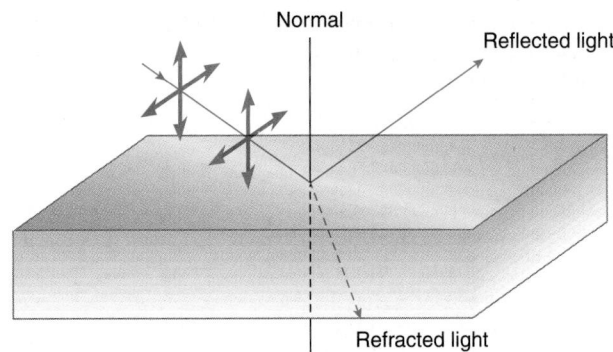

FIGURE 24.46 Unpolarized light incident on a transparent interface.

The following observations can be made about the reflected and refracted light by using a polarizing sheet as an analyzer of the light:

1. If we rotate the analyzer in the refracted beam,[†] the transmitted intensity varies but never to zero intensity, indicating unequal mixtures of the two types of oscillations. Such an unequal mixture of the two mutually perpendicular, incoherent oscillations is called **partially polarized** light. Hence, for angles of incidence greater than 0°, the refracted light is partially polarized.
2. Repeating the same procedure in the reflected beam, we find it, too, is partially polarized. By noting the orientation of the axis of the analyzer (which can be assessed independently if not marked on the analyzer), we find the reflected light has more of the oscillations perpendicular to the plane of incidence; the refracted light has more of the oscillations parallel to the plane of incidence, as in Figure 24.47.

At a particular angle of incidence θ_B, the reflected light is *completely* plane polarized and consists entirely of the oscillation perpendicular to the plane of incidence, as shown in Figure 24.48.

[†]For solid materials, this may be difficult to do! Consider using a liquid material such as water as the medium into which the ray is refracted.

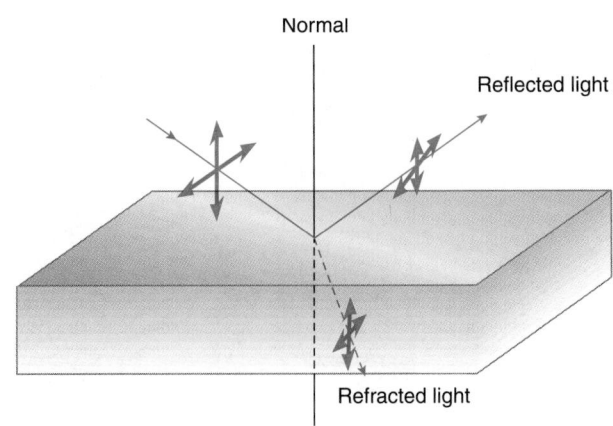

FIGURE 24.47 The reflected and transmitted light are each partially polarized.

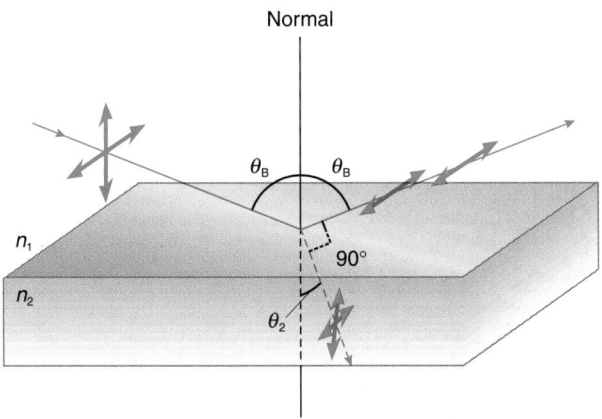

FIGURE 24.48 The Brewster angle of incidence θ_B yields completely plane polarized reflected light with oscillations perpendicular to the plane of incidence (and parallel to the interface).

The refracted light still is partially polarized, because not all of the oscillations perpendicular to the plane of incidence are reflected; even at this special angle of incidence, only *some* of that oscillation is reflected. The refracted beam thus always contains some of both oscillations. At this special angle of incidence θ_B, the angle between the reflected and refracted beam is found to be 90°. These observations were first made by David Brewster (1781–1868) and are known as **Brewster's law.***

> The special angle of incidence that results in a reflected beam that is completely plane polarized is called the **Brewster angle.** The refracted beam makes a 90° angle with this reflected beam.

An expression for the Brewster angle can be found using the law of refraction. When the angle of incidence is equal to the Brewster angle θ_B, the law of refraction (Equation 23.6) becomes

$$n_1 \sin \theta_B = n_2 \sin \theta_2$$

---

*All of these results can be derived from the Maxwell equations of electromagnetism.

The angle of refraction θ_2 can be expressed in terms of the Brewster angle θ_B by noting geometrically in Figure 24.48 that

$$\theta_B + 90° + \theta_2 = 180°$$
$$\theta_2 = 90° - \theta_B$$

Substituting for θ_2 in the law of refraction, we find

$$
\begin{aligned}
n_1 \sin \theta_B &= n_2 \sin(90° - \theta_B) \\
&= n_2(\sin 90° \cos \theta_B - \cos 90° \sin \theta_B) \\
&= n_2 \cos \theta_B
\end{aligned}
$$

We rearrange this slightly to yield

$$\frac{\sin \theta_B}{\cos \theta_B} = \frac{n_2}{n_1}$$

$$\tan \theta_B = \frac{n_2}{n_1} \tag{24.31}$$

Thus the Brewster angle can be found from the indices of refraction.

Perhaps the most widespread use of Brewster's law is in laser technology. The windows of the discharge tubes in gas lasers are inclined to the axis of the tube by the Brewster angle to selectively enhance the polarization of the emitted laser light. Such windows are called **Brewster windows.**

Light reflected from other dielectric surfaces also is partially polarized, with the dominant polarization being the oscillation parallel to the surface or (equivalently) perpendicular to the plane of incidence. Such reflected light often is called *glare*. Most of the light reflected from a surface has an oscillation parallel to the surface. For the case of a horizontal surface, Polaroid® sunglasses and polarizing camera filters reduce glare by selectively blocking (absorbing) mostly the reflected oscillations that are horizontal; the transmission axis of the polarizing material is oriented vertically and thus absorbs most of the reflected oscillations, as in Figure 24.49.

Take a pair of Polaroid® sunglasses and examine strongly reflected light such as that from a water surface or even a well-polished floor; if the sunglasses are rotated, the transmitted light from the reflection is a minimum when the glasses are in a conventional orientation (or up-side-down). The reflected light is

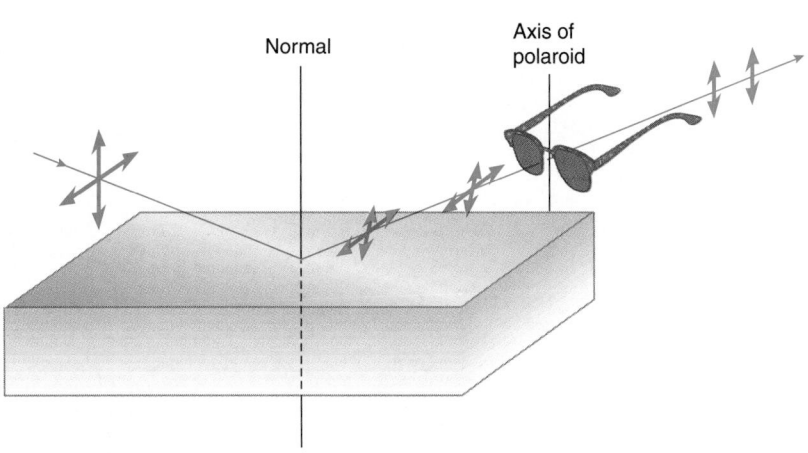

FIGURE 24.49 Polaroid® sunglasses and filters block (i.e., absorb) some reflected light. If the reflected light is at the Brewster angle, all of the light will be blocked.

transmitted strongly when the glasses are rotated 90° from their conventional orientation. Polarizing filters for cameras are essentially sunglasses that eliminate polarized glare from reflections when properly oriented. Such filters also are used by anglers.

QUESTION 7

What specific observations lead you to the conclusion that the refracted beam has more of the polarization oriented in the plane of incidence, while the reflected light has more of the oscillation perpendicular to the plane of incidence?

EXAMPLE 24.13

a. **What is the Brewster angle for an air–water interface?**
b. **What is the Brewster angle for a water–air interface?**

Solution
a. Use Brewster's law in the form of Equation 24.31:

$$\tan \theta_B = \frac{n_2}{n_1}$$
$$= \frac{1.33}{1.00}$$
$$\theta_B = 53.1°$$

b. The role of the indices of refraction is reversed, and so Brewster's law becomes

$$\tan \theta_B = \frac{n_2}{n_1}$$
$$= \frac{1.00}{1.33}$$
$$\theta_B = 36.9°$$

This is less than the critical angle for a water–air refraction, so that there is both a reflected and transmitted beam.

Note that the two angles sum to 90°. Why should they?

24.16 POLARIZATION BY DOUBLE REFRACTION*

Certain anisotropic crystals (such as calcite and many others) exhibit a remarkable type of refraction. When an unpolarized light ray is incident on the crystal in certain directions, *two* refracted rays result! See Figure 24.50. Such **double refraction** also is known, equivalently, as **birefringence**. One of the refracted rays follows the ordinary law of refraction, and so this ray is known as the **ordinary ray** (or O ray). The other refracted ray is called the **extraordinary ray** (or E ray), since it does not follow the law of refraction. Indeed, the extraordinary ray usually goes off in a direction not even in the plane of incidence!

The extraordinary ray experiences an index of refraction that *varies* with the direction the incident ray makes with respect to microscopic symmetry axes of the crystal structure.

> For our purposes, the most important aspect of double refraction is that the two refracted rays are completely plane polarized along two mutually perpendicular directions.

That is, the O ray polarization direction is perpendicular to the E ray polarization direction. Since the beams typically propagate in different directions within the crystal,[†] the two polarized rays can easily be separated in space.

Producing polarized light via double refraction has several advantages over the other methods for producing polarized light:

1. With double refraction, half the incident unpolarized light goes into each of the O and E rays. Polarization by reflection is less efficient, since typically only a small percentage of the light is reflected at the Brewster angle.

[†]If the crystal is cut and illuminated appropriately, the O and E rays will propagate in the same direction within the crystal; for some orientations within the crystal, they propagate in the same direction at the same speed, and in other crystal orientations, they propagate in the same direction at different speeds. This latter property is exploited to make circularly polarized light from linearly polarized light in a device known as a quarter-wave plate (see Investigative Project 12). The analysis of such anisotropic crystals is quite interesting and leads to many practical optical applications.

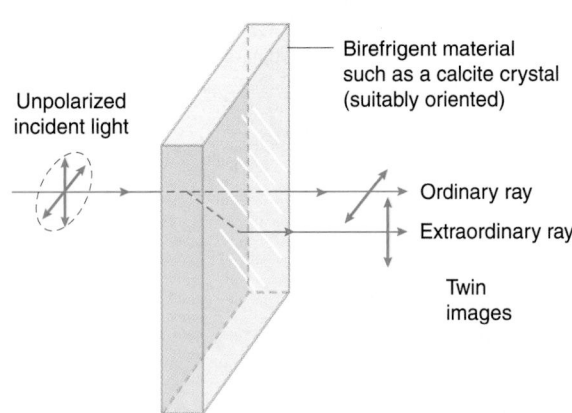

FIGURE 24.50 Double refraction.

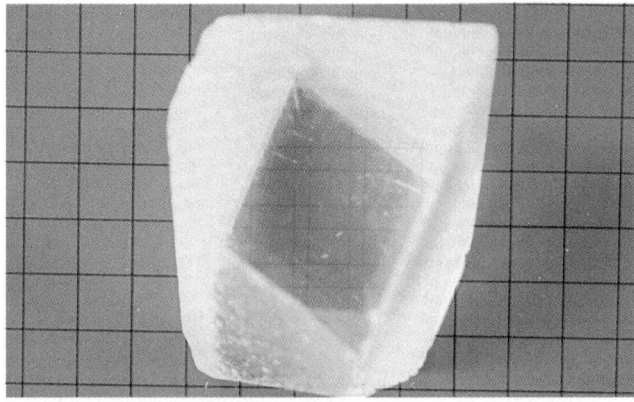

A properly oriented calcite crystal will show twin images. If you rotate the crystal, the image formed by the extraordinary rays rotates about that formed by the ordinary rays.

2. With polarization by double refraction, little of the light is absorbed by the transparent crystal, so it is more suited to high-power applications. Polarization by absorption such as with polarizing sheets only can be used effectively if the amount of light energy to be absorbed does not appreciably warm the polarizer. Excessive energy transfer can melt the material! High-power lasers easily can melt polarizing sheets because of the high energy the film may be called on to absorb.

24.17 POLARIZATION BY SCATTERING*

Take your Polaroid® sunglasses, hold them in front of your eyes, and look at the blue sky, preferably at a region about 90° from the Sun. Rotate the sunglasses and you find the light transmitted by them varies in intensity, but not to zero; hence, the sky light is partially polarized. If the sky light were completely plane polarized, there would be an orientation at which all the skylight was blocked and the view through the sunglasses would look black. The polarization of sky light arises because of **scattering**.

When a beam of light is incident on transparent matter (be it solid, liquid, or gas), some of the light is redirected or scattered from the original direction. The redirected light is called scattered light and the process is called scattering. It is scattered light that makes light rays visible in dusty conditions such as when chalk-filled erasers are clapped together along the path of a laser beam, or when sunbeams are seen streaming through the atmosphere. The scattered light in such circumstances is partially polarized.

We can qualitatively model these scattering processes in the following way. The oscillating electric field vectors of the incident light are always in a plane perpendicular to the ray and cause electrons in atoms or molecules to oscillate and produce electric fields in the same directions. The scattered light thus generated by these particles has electric fields that are the components perpendicular to the direction of viewing. Since sunlight is unpolarized, it has equal amounts of field in all directions normal to its incident ray. Viewed along an angle to the incident ray, its induced accelerations and fields in the scattering medium have larger components normal to the incident ray than parallel to it.

Light scattering by the atmosphere and dust particles makes this laser beam visible from the side.

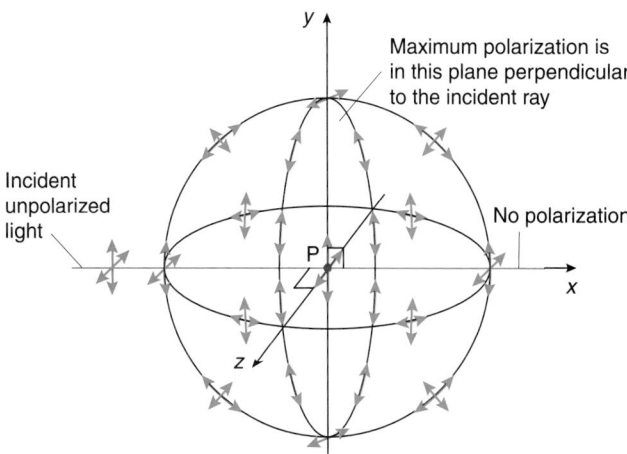

FIGURE 24.51 Polarization of light scattered at various orientations relative to the incident ray.

So the scattered light rays are not polarized if viewed straight ahead or back along the incident ray, but the scattered light at 90° is maximally polarized perpendicular to the rays of the Sun, as shown in Figure 24.51.

Atmospheric light scattering is only partially polarized because of several effects:

1. Multiple scattering takes place. In other words, the light reaching you from the sky has been scattered several times; this complicates the analysis.
2. The scattering molecules are anisotropic, so that the electrons do not necessarily experience oscillations in the same direction as an incident electric field vector of the light wave.
3. The degree of polarization of the scattered light depends on the size of the scattering particle. The smaller the scattering particle, the greater the polarization of the scattered light.

Nevertheless, sunlight scattered from a clear sky shows enough polarization to be well reduced by a properly oriented polarizing filter on your camera or by Polaroid® sunglasses.

24.18 RAYLEIGH AND MIE SCATTERING*

Why is the clear, daytime sky blue? Why are clouds white? What causes the Sun to be red when rising or setting? Here we see that such effects are caused by light scattered by the constituents of the atmosphere.

Light scattering from particles much smaller than the wavelength of the incident light is known as **Rayleigh scattering**, after the 19th-century English physicist Lord Rayleigh, who first quantitatively investigated such scattering.[†] The molecules of our atmosphere satisfy this condition very well for visible light wavelengths. Thus incident light from the Sun is scattered in the

---

[†]Recall from Section 24.6 that Lord Rayleigh also formulated the Rayleigh criterion for resolution.

atmosphere, giving our sky its characteristic blue color for reasons we shall soon see.* On the other hand, the Moon and some other planets (such as Mercury) and other moons lack atmospheres; on these celestial objects the sky between the stars appears black because there are no particles present to scatter the incident solar radiation.

Rayleigh scattering has a dependence on the wavelength of the incident light that accounts for the blue color of the sky. The dependence on wavelength can be deduced from a dimensional analysis of the scattering problem.[†] After pondering the problem, Rayleigh decided that the amplitude of the scattered light likely could depend on several factors:

- the total content or volume V of the scattering particle;
- the relatively large wavelength λ of the incident light;
- the distance r from the scattering particle to the observer; and
- the speed of light.

The ratio of the amplitude of the scattered light to the amplitude of the incident light will, of course, be dimensionless. So for this ratio we seek to formulate a dimensionless combination of the potential contributing factors listed here. The first three factors, V, λ, and r all involve the dimension of length, whereas the speed of light c has the dimensions of length per time. Since c is the *only* factor listed that involves a time dimension, the speed of light must in fact be irrelevant to the amplitude ratio (none of these other factors can cancel the time dimension, since they involve the length dimension only).

Therefore the ratio of the amplitudes must depend on V, λ, and r. Certainly as the particle volume vanishes, the scattering should vanish as well; thus it seems reasonable to suppose (at least to a first approximation) that the scattered wave amplitude is directly proportional to the volume of the scattering particle. The inverse square law governing the intensity of propagating waves (see Section 23.2) implies that the scattered amplitude received should be inversely proportional to the distance r between the scattering particle and the observer: in this way the scattered light intensity decreases as $1/r^2$, in keeping with the inverse square law. So the ratio of the scattered to incident amplitudes should go in part something like

$$\frac{\text{scattered amplitude}}{\text{incident amplitude}} \propto \frac{V}{r}$$

The right-hand side of this relationship is not yet dimensionless. However, if we put the remaining factor, the wavelength

λ, into the denominator as λ^2, we can cause the right-hand-side to be dimensionless:

$$\frac{\text{scattered amplitude}}{\text{incident amplitude}} \propto \frac{V}{\lambda^2 r} \qquad (24.32)$$

Thus the amplitude of the scattered light is proportional to the inverse square of the wavelength.

> The intensity of the scattered light is proportional to the square of the scattered amplitude, and so the scattered light intensity must therefore be proportional to the inverse *fourth* power of the wavelength. That is,
>
> $$\text{scattered intensity} \propto \frac{1}{\lambda^4} \qquad (24.33)$$
>
> omitting the other contributing factors. Therefore short wavelengths (blues) are scattered more efficiently than longer wavelengths (reds).

Both sunlight and our visual sensitivity are maximum around yellow and weaken rapidly toward the violet end of the spectrum. These trends, when multiplied by the scattering ratio which strongly favors short wavelengths, gives us a blue sky as a compromise. Thus the clear sky looks blue and is most blue 90° from the direction to the Sun; see Figure 24.52. When you observe a sunrise or sunset, the incident light traverses a very long atmospheric path (see Figure 24.52) compared with

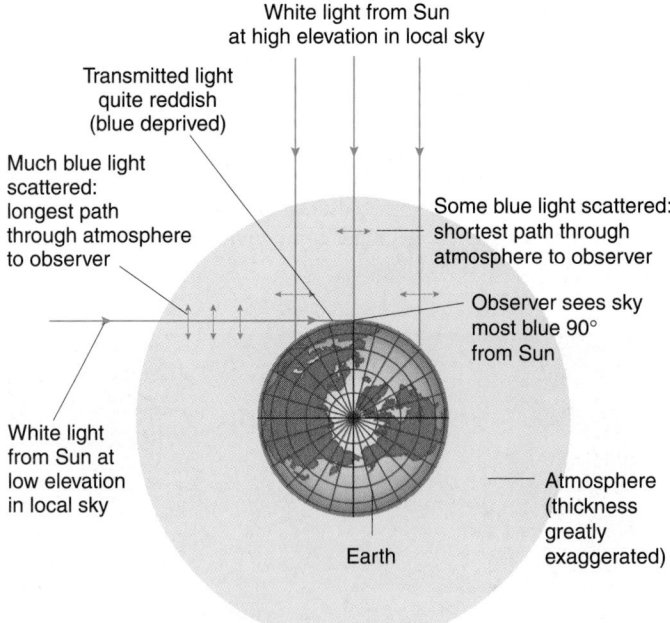

FIGURE 24.52 Light scattering in the atmosphere produces the blue sky and red sunrises and sunsets.

---

*A fine suspension of iron oxide particles gives the atmosphere on Mars a reddish hue.

[†]This argument parallels that first made by Lord Rayleigh in his epochal treatment of the scattering problem: John William Strutt [Lord Rayleigh], "On the light from the sky: its polarization and colour," *Philosophical Magazine, 41,* pages 107–120 (1871).

the path at, say, noontime. More of the blue light is scattered out of the incident light and the resulting light reaching our eyes is blue deficient: sunsets and sunrises are red. Physics is amazing stuff!

The scattering by larger particles, on the order of the wavelength of the incident light, is called **Mie scattering** [after Gustav Mie (1868–1957), who first began investigating this type of scattering quantitatively in 1908[†]]. The wavelength dependence of such scattering is more complex than Rayleigh scattering; but as the particle size increases further, the scattering approaches diffuse reflection. This is the reason that clouds appear white: the scattering from the water particles that make up clouds becomes diffuse reflection, essentially independent of wavelength.

A special circumstance bears mentioning, however. If the scattering particles are between one and two wavelengths in size and are essentially the same uniform size, then Mie scattering predicts that the longer wavelengths are scattered more than the shorter wavelengths, just the opposite of Rayleigh scattering. If clouds of dust in the atmosphere meet these conditions, it is possible to observe the rising or setting Sun (or Moon) as blue or even green in color! Sightings of such colored Moons and Suns were quite common after the cataclysmic volcanic eruption of Krakatoa in the 19th century.[‡] More recent sightings also have

occurred.[§] The rarity of such sightings may be the origin of the colloquial expression "once in a blue moon." *Calendrical* "blue moons" refer to the second full moon of a calendar month, which occur about every three years (although 1999 had two calendrical blue moons). Such calendrical blue moons do not, of course, appear blue in the sky!

24.19 OPTICAL ACTIVITY*

Certain materials and solutions are able to rotate the plane of polarization of the light passing through them as shown in Figure 24.53. This phenomenon is called **optical activity**.

Optical activity occurs in certain crystals (e.g., quartz) when plane polarized incident light is incident along certain directions with respect to symmetry axes associated with the particular crystal structure. Some liquids (e.g., turpentine) and solutions (e.g., sugar solutions) also exhibit optical activity. The extent of the rotation of the plane of polarization depends on the distance ℓ traveled through the crystal or liquid and the concentration c of the solutions.[#] In solutions, the angle of rotation ϕ of the plane of polarization is written empirically as

$$\phi = \alpha \ell c \qquad (24.34)$$

where α is a proportionality constant known as the **specific optical rotating power** and is a function of both wavelength and temperature. The dependence on the wavelength of the light leads to **rotary dispersion**, analogous to the variation of the index of refraction with wavelength. Thus linearly polarized incident white light is dispersed into various planes of polarization by optically active materials, as shown in Figure 24.54.

The directional sense of the rotation of the plane of polarization determines the type of optical activity in the sample. If the rotation of the plane is in a right-handed sense, the material is said to be **dextrorotatory**. Right-handed means that if the thumb of the right hand is pointing in the direction of propagation of the light, the curled fingers indicate the sense of the progressive rotation

[†]Gustav Mie, "Beiträge zur Optik trüber Medien, speziell kolloidaler Metallösungen," *Annalen der Physik*, 25, #3, pages 377–445 (1908).

[‡]The sightings are summarized and discussed in F. A. Rollo Russell and E. Douglas Archibald, "On the unusual optical phenomena of the atmosphere, 1883–1886, including twilight effects, coronal appearances, coloured suns, etc." which appeared as a section of the Report of the Krakatoa Committee of the Royal Society, *The Eruption of Krakatoa and Subsequent Phenomena*, edited by G. J. Symons (Trubner, London, 1888), Section I. (c), "The blue, green, and other coloured appearances of the sun and moon," pages 199–218.

[§]In 1950 a blue-colored Sun was sighted over eastern Canada and the northeastern United States caused by smoke from extensive forest fires in western Canada. Reports of this sighting are found in Rudolf Penndorf, "On the phenomenon of the colored sun, especially the 'blue sun' of September 1950," Geophysical Research Paper #20, Air Force Cambridge Research Center Technical Report 53-7 (April 1953); R. Wilson, "The blue sun of 1950 September," *Monthly Notices of the Royal Astronomical Society of Canada*, 11, pages 447–489 (1951); and William Paul and R. V. Jones, "Blue sun and moon," *Nature*, 168, page 554 (29 September 1951).

[#]Do not confuse the concentration with the speed of light, also designated by c.

Incident plane polarized light (monochromatic)

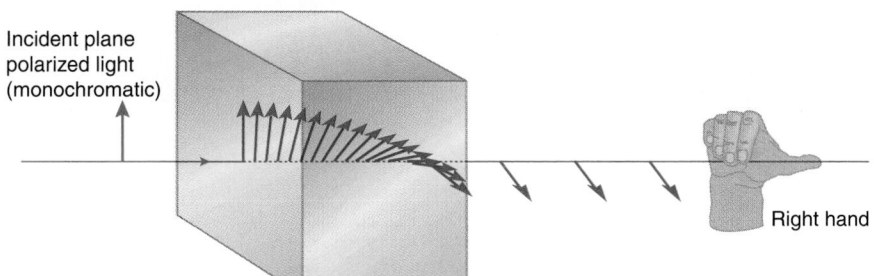

Right hand

FIGURE **24.53** Optical activity. For clarity, only half of the transverse oscillation is shown.

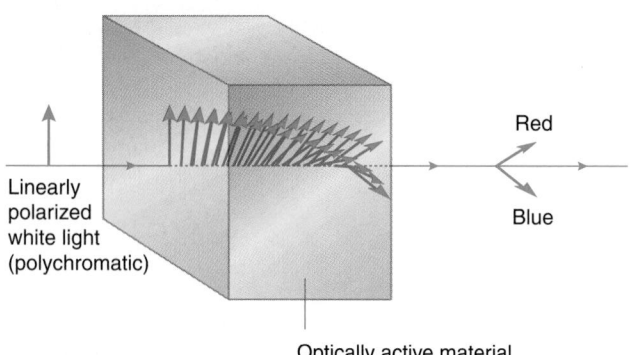

FIGURE 24.54 Rotary dispersion of plane polarized white light. For clarity, only half of the transverse oscillation is shown.

of the plane of polarization; see Figure 24.53.* If the rotation of the plane of polarization is in the opposite (left-handed) sense, the material is **levorotatory**.

Optical activity is used in many of the sciences including geology (crystalline minerals), chemistry (concentrations of solutions, isomeric structures of molecules), biology (biologically important molecules with spiral structures such as DNA and RNA), structural engineering (for optical stress analysis), electrical engineering (liquid crystals), and physics for a wide variety of purposes and applications.

---

*Regrettably, these conventions for the directional senses are not universal, although the one we choose is the most widely used. You need to be sure what convention is being employed when consulting other references. As we have seen before, many conventions in optics have yet to be made standards, used by all, once again illustrating the inherent conservative nature of science as well as the human inertia associated with change.

CHAPTER SUMMARY

Physical optics is the study of phenomena related to the wave nature of light.

A *wavefront* is a line or surface of constant phase in a propagating disturbance and is perpendicular to a ray. Two or more light sources are said to be *coherent* if the phase difference between them is independent of time. Coherent sources of light (or other waves) can be made by *wavefront division* (with two or more slits), or by *amplitude division* (such as in thin-film interference). Light from coherent sources will interfere when superimposed at the same place at the same time. The nature of the superposition depends on the *phase difference* between the waves. *Constructive interference* occurs if the phase difference between the waves is an integral multiple of 2π rad. *Destructive interference* occurs if the phase difference is an odd integral multiple of π rad.

A phase difference can arise from a *path difference* between waves going from coherent sources to the point of their superposition. A path difference of one wavelength corresponds to a phase difference of 2π rad. The phase difference δ_{path} arising from a path difference Δx is

$$\delta_{path} = \frac{\Delta x}{\lambda} (2\pi \text{ rad}) \qquad (24.1)$$

If monochromatic light is incident from a distant source on two parallel slits separated by a center-to-center distance d, constructive interference fringes are seen on a distant screen when the path difference $d \sin \theta$ between the wavelets from the slits is an integral number of wavelengths:

$$d \sin \theta = m\lambda \qquad (m = \text{integer}) \qquad (24.2)$$
$$\text{(double slit, constructive interference)}$$

where the angle θ is measured from the incident direction. The integer $|m|$ is called the *order number* of interference. This is *Young's double slit* experiment.

When monochromatic light from a distant source diffracts through a single slit of width a, the first intensity minimum of the diffraction pattern on a distant screen (measured from the incident direction) is located at an angle θ_1 given by

$$a \sin \theta_1 = \lambda \qquad (24.3)$$
$$\text{(first diffraction minimum, slit aperture)}$$

The quantity $a \sin \theta_1$ is the path difference between opposite sides of the slit to the first minimum on the distant screen.

When monochromatic light from a distant source diffracts through a *circular* aperture of diameter (aperture) a, the first minimum of the diffraction pattern is located at an angle θ_1 found from

$$a \sin \theta_1 = 1.220\lambda \qquad (24.5)$$
$$\text{(first diffraction minimum, circular aperture)}$$

The *Rayleigh criterion* states that two incoherent sources are *barely resolved* if the central diffraction peak from one source is located at the first minimum of the pattern formed by the other source. This angular separation of the sources then is θ_1, which for a circular aperture is found from Equation 24.5, and for a slit aperture is found from Equation 24.3. Good resolution means that the angle θ_1 between the center and first minimum of each diffraction pattern produced by two incoherent sources is smaller than the separation angle between the two source centers.

The intensity of interference fringes in a double (or multiple) slit experiment is modulated by the envelope of the single slit diffraction pattern.

A *diffraction grating* has many parallel, identical, equally spaced slits. Interference maxima occur at those locations where

$$d \sin \theta = m\lambda \qquad \text{(diffraction grating maxima)} \qquad (24.8)$$

where d is the *grating spacing* (the center-to-center distance between adjacent slits) and m is the order number. If the number of slits N is large, the interference peaks are narrow and sharply defined. The *resolution* of a diffraction grating, its ability to discern two nearly equal wavelengths whose difference is $\Delta\lambda$, is directly proportional to the order number m and the number of illuminated slits N of the grating:

$$\frac{\lambda}{\Delta\lambda} = Nm \qquad (24.16)$$

where λ is either of the nearly equal wavelengths.

The *angular dispersion* of a diffraction grating is

$$\frac{d\theta}{d\lambda} = \frac{m}{d\cos\theta} \qquad (24.13)$$

Diffraction gratings are widely used to measure the wavelength of light.

The index of refraction of a material is equal to the ratio of the speed of light c in vacuum to that in the material, v:

$$n = \frac{c}{v} \qquad (24.17)$$

The wavelength λ_1 of light in a medium that has an index of refraction n_1 is shorter than in a vacuum (the frequency of the light is unchanged):

$$\lambda_1 = \frac{\lambda}{n_1} \qquad (24.20)$$

where λ is the wavelength in vacuum.

Light reflected from a metal (a conducting surface), or from an optically more dense material (a medium with an index of refraction greater than that of the medium from which the light is incident), experiences a *phase change* of π rad upon reflection.

In cases of thin-film interference, the phase difference between waves arises from a path difference and from phase changes that may occur from reflections. The relevance of phase changes upon reflection must be assessed on a case-by-case basis.

Light is a transverse wave that can be plane polarized, meaning the oscillations of the electric vector of the light are confined to a plane that includes the ray. Unpolarized light arises from sources whose atoms emit light with planes of polarization in random orientations. For convenience, unpolarized light is imagined to be composed of two mutually perpendicular, incoherent, planes of polarization.

Unpolarized light can be polarized in one of several ways.

A sheet of polarizing material selectively transmits the light polarized parallel to its transmission axis, and absorbs the polarization perpendicular to it. If plane polarized light of amplitude E_0 is incident on a polarizing sheet with an angle θ between the axis of the polarizer and the plane of oscillation of the polarized light, the amplitude of the light transmitted is $E_0\cos\theta$. Hence the light intensity (which is proportional to the square of the amplitude) transmitted by the polarizer follows *Malus's law*:

$$I_{\text{transmitted}} = I_{\text{incident}}\cos^2\theta \qquad (24.30)$$

Reflected light typically is *partially polarized*, with a preponderance of the oscillations perpendicular to the plane of incidence. If the light is incident at the *Brewster angle* θ_B on a refractive medium, the reflected and transmitted rays are 90° apart and the reflected light is completely plane polarized with its (oscillating and propagating) electric field perpendicular to the plane of incidence. The Brewster angle is found from *Brewster's law*:

$$\tan\theta_B = \frac{n_2}{n_1} \qquad (24.31)$$

Certain crystals produce *two* refracted rays, a phenomenon called *double refraction*. The *ordinary ray* follows the law of refraction while the *extraordinary ray* does not. The ordinary and extraordinary rays have plane polarizations perpendicular to each other.

The scattering of light by small particles (such as the atoms and molecules that make up the atmosphere) also produces partially polarized light.

The scattering of light by particles whose size is much smaller than the wavelength is called *Rayleigh scattering*. The intensity of Rayleigh scattering is proportional to the inverse fourth power of the wavelength, which accounts for the blue sky and red sunrises and sunsets. Light scattered by particles whose size is on the order of the wavelength is called *Mie scattering*, and accounts, under special circumstances, for the rare observation of blue- and green-colored Suns and Moons. The water particles that comprise clouds are comparatively large and scatter light independent of wavelength (approaching diffuse reflection), so that clouds appear white.

Certain crystals and solutions can rotate the plane of polarization of incident plane polarized light, a phenomenon called *optical activity*. The effect is slightly dependent on the wavelength of the incident light, leading to *rotary dispersion*.

SUMMARY OF PROBLEM-SOLVING TACTICS

24.1 **(page 1110)** Do not confuse the order number m locating the interference *maxima* for Young's double slit experiment in Equation 24.2 with the integer m locating the single slit diffraction *minima* in Equation 24.4.

24.2 **(page 1111)** For circular apertures, be sure to use Equation 24.5 to locate the angle to the first minimum of the diffraction pattern, not Equation 24.4 for slit apertures.

24.3 **(page 1122)** When investigating thin-film interference, note that the wavelength of the light in the film is *different* from that in vacuum. It is the wavelength of the light in the film that is involved in assessing the phase difference associated with the path difference. So be careful to use the wavelength of the light in the film in Equation 24.21.

24.4 **(page 1123)** There is no magic bullet formula to memorize that can treat all situations of thin-film interference. Each situation must be assessed individually to determine (a) the path difference and its resulting phase difference and (b) the phase changes (if any) upon reflection and whether they are relevant or not.

QUESTIONS

1. (page 1105); 2. (page 1107); 3. (page 1113); 4. (page 1124); 5. (page 1128); 6. (page 1130); 7. (page 1132)

8. When 76 trombones play the same note, is the sound a collection of 76 coherent or incoherent sources of acoustic waves?

9. Describe some simple observations that clearly indicate that the wavelength of visible light is quite short while that of sound is much longer.

10. Glass prisms change the direction of violet light more than that of red light. However, with double slit interference, the least deviated color is violet. Which of these two colors is shown by these facts to have the greater wavelength, and through what reasoning?

11. For a fixed slit separation d and slit width a, what happens to the appearance of the double slit pattern if light of a longer wavelength is first used and next changed to light of shorter wavelength? Draw sketches to illustrate the relative appearance of the two interference patterns.

12. Which color of visible light provides the best resolution in a microscope? In a telescope?

13. In a single slit diffraction pattern, the angular width of the central peak $(2\theta_1)$ depends on several factors: the width of the slit and the wavelength of the light. Sketch rough graphs indicating how the angular width depends on each of these factors (with the other factor held constant).

14. A close pair of binary stars is more easily resolved if their color is predominantly blue rather than red. Explain why.

15. A single slit of width a is a distance $d/2$ above a plane mirror as indicated in Figure Q.15. Monochromatic light of wavelength λ is incident normally on the slit. A screen S is located far from the arrangement. (a) The mirror produces a virtual image of the slit. Is the virtual image *coherent* with the slit itself? (b) Is the pattern on the screen a double slit interference pattern or a pair of single slit diffraction patterns? (c) If the pattern is an interference pattern, is the fringe closest to the mirror surface bright or dark? Explain. This arrangement is known as *Lloyd's mirror*.

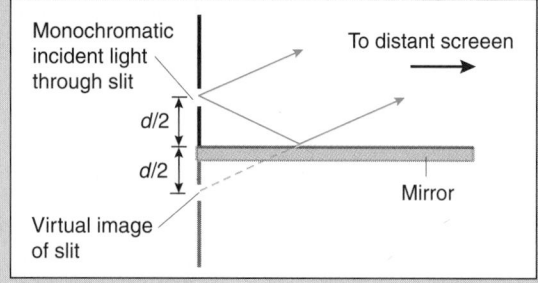

Monochromatic incident light through slit

To distant screeen

$d/2$

$d/2$

Mirror

Virtual image of slit

FIGURE Q.15

16. Take out one of your compact discs and observe white light reflected from the CD at a large angle of incidence. Explain why you see a spectrum of colors.

17. Can a panther with a vertical slit pupil in each eye judge vertical details such as tall grass better than horizontal details such as level tree branches? Explain with a convincing discussion.

18. Explain with the assistance of a diagram the appearance of a single slit diffraction pattern formed by *white light* incident from a distant source.

19. Explain the advantages of using a diffraction grating with: (a) many slits; (b) closely spaced slits.

20. Which visible light color experiences the greatest deflection when refracted through a prism? Which experiences the greatest deflection when using a diffraction grating?

21. To analyze the spectrum of a light source, a diffraction grating typically is preferred over a prism. Discuss several reasons for this.

22. A light source emits several distinct wavelengths of light. The light is incident on a diffraction grating. You note that the second order of one wavelength is coincident in angular position with the third order of another wavelength. What conclusions can you draw from this observation?

23. Your professor gives you a diffraction grating in the laboratory but has no information about its small grating spacing d. Describe a simple experiment that you can perform to measure the grating spacing.

24. Watch a water wave come in and reflect from a rigid sea wall. Does the reflected wave experience a phase change of π rad?

25. A red traffic light is seen by an underwater swimmer. Don't ask how. What color does the light appear to be to the swimmer? Explain.

26. Is the resolution of a circular aperture improved if the aperture is used under water instead of in air? Explain.

27. Is Equation 24.20 what is meant by the term dispersion?

28. Young's double slit experiment is performed in air and then in water with the same pair of slits and light source. Explain the difference (if any) in the interference pattern observed on a distant screen.

29. Discuss the feasibility of creating a thin-film antireflection coating on military aircraft to enhance their "stealth" capability of invisibility to radars (that typically have wavelengths of a few centimeters).

30. Does red or violet light travel more slowly in glass, or do they both travel at the same speed?

31. You have a rather attractive red bathing suit in air (red ≈ 660 nm). Will the color of the suit change when viewed under water?

32. You likely used a soap and glycerin solution and a bubble-blower loop as a child. Resurrect the toy. Notice the light reflected from the soap film on the loop. If the film is very thin at some location, explain why you see reflections from this location or region as black. (See Investigative Project 3.)

33. Describe an observation that clearly demonstrates that light is a transverse wave.

34. By means of a few simple experiments, see if the human eye can distinguish light according to various orientations of its plane of linear polarization. Some flying insects and birds can detect polarization and use such information for navigational purposes, since skylight is partially polarized.

35. A polarizer is placed with its axis parallel to one slit of two parallel slits in a Young's double slit experiment, while another polarizer is placed on the other with its axis perpendicular to the slit. The double slits are simultaneously illuminated with monochromatic, unpolarized light from a slit source. What is seen on a distant screen and why?

36. Describe how a single polarizer can be used to show how the Rayleigh scattered blue sky light on a sunny day is partially polarized.

37. What advantage would there be if the windshield of your car were covered with a polarizing material? What should be the orientation of the axis of the polarizing material to gain greatest advantage? Cite one significant disadvantage of covering the windshield with such a material.

38. You have a sheet of polarizing material but there is no information to indicate its axis. With only an unpolarized light source available, describe an experiment you can perform to determine the orientation of the axis of the polarizer.

39. The difference in the scattering of light by small and large particles (relative to the wavelength of the incident light) is well illustrated by your driving experiences with auto windshields and either (a) headlights from oncoming cars at night or (b) driving "into the Sun" during the day. If the windshield is littered with leaves (particles large compared with the wavelength of visible light), the incident light is more reflected (perhaps diffusely) than scattered. This is called *back scattering* (back to the source). Thus when you sit in the driver's seat your eye is shaded by the large particles. On the other hand, if the particles on the windshield are small, such as with road grit or dust, your eye sees the windshield ablaze with extra light. The scattering then is predominantly in the *forward* direction of the light with little back scattering. The rings of Saturn (located about 9.5 times further from the Sun than the Earth) are clearly visible from the Earth using even a moderately sized telescope. On the other hand, the ring around Jupiter (about 5 times farther from the Sun than the Earth) is invisible from the Earth and only was discovered by a spacecraft that had already *passed* the giant planet (on its way to Saturn), while looking *back* toward Jupiter. What can be said about the likely sizes of the particles in the two very different ring systems?

PROBLEMS

Sections 24.1 Existence of Light Waves
24.2 Interference
24.3 Young's Double Slit Experiment

1. Two narrow slits with 0.14 mm between their centers are illuminated by monochromatic light of wavelength 488 nm. What is the angle between the zeroth- and first-order fringes? Express your result in both radians and degrees.

2. A laser with light of wavelength 632.8 nm illuminates a double slit in a Young's double slit experiment. On a screen 2.00 m from the slits, interference fringes are observed that are separated by 1.0 cm. (a) Determine the angle between the zeroth- and first-order fringes. Express your result in both radians and degrees. (b) Determine the separation d of the two slits.

3. In a Young's double slit experiment using light of wavelength 488 nm, the angular separation of the interference fringes on a distant screen is measured to be 1.0°. What is the separation of the two slits used in the experiment?

4. Monochromatic light of unknown wavelength from a distant source is incident along the normal line to the twin slits of a Young's double slit experiment. The slits are separated by 0.50 mm. A screen located 6.50 m from the slits has interference fringes separated by 7.7 mm. What is the incident wavelength?

•5. The double slit experiment of Young is one of the classic experiments of physics. Light with a wavelength of 632.8 nm is incident on two slits separated by 0.10 mm. A screen 1.50 m by 1.75 m is located 5.00 m from the slits. (a) Calculate the separation of the interference fringes on the screen in centimeters. (b) If the slit separation is doubled, what happens to the separation of the fringes on the screen?

•6. Two light waves begin with identical phases from different places, propagate outward, and arrive at the same location. Each wave has amplitude A, wavelength 500 nm, and speed c, but one wave must travel 600 nm farther to arrive there and superimpose with the other wave. (a) What is the frequency of the resulting disturbance? (b) What is the phase difference (in radians) between the waves when they arrive?

•7. Each of two coherent light waves has an amplitude of 10 V/m and a frequency of 5.0×10^{14} Hz. The waves are identical except for a constant relative phase difference ϕ; the waves now are superimposed. (a) What is the wavelength? Is this within the range of visible light wavelengths? (b) If the phase difference is 0.60 rad, what is the combined amplitude? (c) If this phase difference is caused by a difference in path Δx traveled, find Δx. (d) If this phase difference is caused by a difference Δt in starting time, find Δt.

•8. Two synchronized radio wave transmitters are located 100 m apart along a north–south line and emit radio waves with a wavelength of 50.0 m. (Synchronized simply means that the peaks of the waves emitted occur at the same instant of time.) (a) What is the frequency of the radio waves? (b) Are the sources coherent? Explain why or why not. (c) Consider the full 360° horizontal plane of angles measured from the center point between the transmitters. Find all directions to distant locations for which the radio wave intensity is a maximum and sketch these directions on a map including the transmitters.

•9. The great operatic tenor Luciano Pavarotti is singing a perfect note with a frequency of 262 Hz (middle C). The speed of sound in air is 343 m/s. Some distance away are two slitlike openings in a wall, spaced 2.0 m between their centers, as shown in Figure P.9. (a) What is the wavelength of the sound waves? (b) At what minimum nonzero angle θ from the straight-through direction beyond the openings should you locate yourself to hear his voice loud and clear?

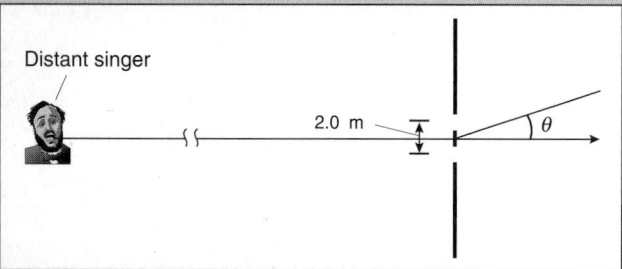

FIGURE P.9

$10. Monochromatic light of wavelength λ is incident at an angle ϕ on a pair of slits separated by a distance d as indicated in Figure P.10. Determine the angles θ at which constructive interference takes place on a distant screen.

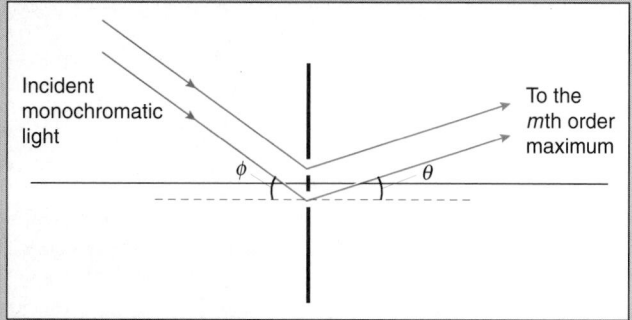

FIGURE P.10

Sections 24.4 Single Slit Diffraction
24.5 Diffraction by a Circular Aperture
24.6 Resolution
24.7 The Double Slit Revisited

11. Find the angle at which the first diffraction minimum is formed when light of wavelength 632.8 nm is sent through: (a) a slit of width 0.10 mm; (b) a circular aperture with a diameter of 0.10 mm.

•12. The separation of the headlights on your car is 110 cm. (a) At what distance are the headlights just barely resolved by a skeptical eye with an aperture of 7.0 mm? Use 550 nm as the effective wavelength of the light emitted by the headlights. Assume the resolution of the eye is diffraction limited, rather than determined by cone cell size on the retina. (b) On the retina of each eye, will there be two diffraction patterns or an interference pattern from two coherent sources?

•13. A distant monochromatic point source of wavelength 600 nm is viewed through a circular hole; the circular central image has an angular radius of 1.0×10^{-3} rad. (a) What is the dia-

meter of the hole? (b) How would you change the diameter of the aperture to produce a smaller image?

•14. Calculate the minimum angular resolution angle for a human eye whose aperture is 7.0 mm for an effective wavelength of 550 nm. Express your result in arc seconds. Assume the resolution of the eye is limited by diffraction rather than by cone cell size. If our eyes had *slit* apertures with a 2.0 mm width instead of circular apertures, would the resolution of the eye be better, worse, or the same? Explain your reasoning.

•15. Estimate the angular separation of two stars that can be barely resolved by a telescope of aperture 10 cm. Express your result in both radians and arc seconds. Use 550 nm as the effective wavelength of the light. Neglect distortion caused by the atmosphere.

•16. The binary star Albireo in the constellation Cygnus (the swan) consists of two stars with an angular separation of 35 arc seconds and is among the most beautiful binary stars to observe. What is the minimum aperture needed for a telescope to resolve the pair? Use 550 nm as the effective wavelength of the light. Neglect distortion caused by the atmosphere.

•17. The Great Wall in China is approximately 5 m wide. Show that it is impossible for an astronaut on the Moon to resolve the thickness of the Great Wall with the unaided eye even when the Moon is at its perigee location in its orbit (its distance of closest approach, 3.63×10^5 km). Use a reasonable estimate of the aperture of the eye of the astronaut.

•18. Georges Seurat (1859–1891) was a French neo-impressionist painter most noted for his pointillist paintings. Aspects of his life were the focal point of a play, *Sunday in the Park with George*, by Stephen Sondheim and James Lapine. Seurat's paintings consist of a myriad of individual colored dots (see Figure P.18 for an example of his work). He must have gone mad painting the millions of dots on his canvases! The dots are on the order of several millimeters in diameter, say, 3.0 mm for the purposes of this problem. If your eye is sufficiently far from the painting, you cannot resolve the dots. Assume an eye with an aperture of 4.0 mm in a moderately lit gallery. For two adjacent red dots (with an effective wavelength of 600 nm), how far from the painting must you stand so you are barely able to resolve the dots if the resolution of the eye is diffraction limited? If two adjacent dots are of blue pigment (with an effective wavelength of 450 nm), what is the distance at which the dots can be barely resolved?

FIGURE P.18

●**19.** Big Brother is watching. A spy satellite at an altitude of 120 km is checking up on your diligence in doing physics homework. The satellite uses a lens with an aperture of 40 cm. Assume light with an effective wavelength of 550 nm. What is the separation of two objects on the ground that can be barely resolved by the lens?

●**20.** The Lilliputians in Jonathan Swift's *Gulliver's Travels* were said to be of a height of "somewhat under six Inches" (say, about 10 cm) and to have eyes that could "see with great Exactness" and "Sharpness."* (a) If we scale the height (say, 2.00 m) and eye diameter (7.0 mm) of a normal human accordingly, what is the aperture of the eye of a Lilliputian? (b) Compare the resolution limit of a Lilliputian eye with that of a normal human, assuming (incorrectly) the resolution of the eye is determined by the Rayleigh criterion. What does this suggest about Swift's knowledge of physics? Of course, Swift published *Gulliver's Travels* in the 18th century, well before Young discovered the wave aspects of light (in 1801), so Swift can be forgiven!

For an amusing article bearing on this question, see David Piggins, "Lilliputian eyes and vision," *Perception*, 7, pages 609–610 (1978).

●**21.** A telescope with an aperture of 10 cm is used to view Mt. Hood from a distance of 150 km. Assume visible light has an effective wavelength of 550 nm, optimal focusing, and no distortions caused by atmospheric effects or turbulence. (a) Will details on the mountain be seen better with the unaided eye or with the telescope? Explain your answer. (b) What is the closest distance apart on the mountain of two bright objects that, in principle, could be distinguished through the telescope? (c) An eyepiece lens with half the focal length is substituted for the original eyepiece in the telescope. What effect does this change have on the magnification? What effect does this change have on your answer to part (b)?

●**22.** (a) Look at the fine detail of a poster on a distant wall, make appropriate length measurements, and determine the smallest angle your eye can resolve at an effective wavelength of 550 nm. (b) Have a friend estimate the diameter of your pupil[†] (or do it by looking in a mirror) and compare your measured angle in part (a) with the theoretical angular resolution limit of your eye.

●**23.** A telescope lens of diameter D is used to form an image of a distant star. The image is a diffraction disk in the focal plane of the objective. The aperture of the telescope is decreased to $D/2$. (a) By what factor does the amount of light collected by the telescope decrease? (b) What happens to the size of the diffraction disk of the star? (c) By what factor does the intensity of the light of the central image of the star change?

●**24.** A reconnaissance aircraft at an elevation of 30 km manages to photograph a railroad yard at a military installation. The railroad ties, spaced 40 cm apart, are just barely resolved. Assume no atmospheric distortion. Calculate the minimum diameter of the (circular) lens used to secure the photograph. Assume the effective wavelength of the light is 550 nm.

*Jonathan Swift, *Gulliver's Travels*, Chapter 6 (Doubleday, Garden City, New York, 1945).

†Be careful to do this safely: no rulers in the eye!

Sections 24.8 Multiple Slits: The Diffraction Grating
24.9 Resolution and Angular Dispersion
of a Diffraction Grating

25. Light of wavelength 488 nm is incident normally on a grating with a grating spacing of 2.0×10^{-6} m. At what angle from the normal line are the first and second orders of interference located?

●**26.** A source of light emits two different wavelengths. One of them is known to be 488 nm but the other wavelength is not known. Careful experimentation by Milly Hertz has determined that the fourth-order interference maximum of the 488 nm wavelength is at the same angular position as the third-order interference maximum of the unknown wavelength. What is the unknown wavelength?

●**27.** You have pirated a sodium vapor lamp to investigate its spectrum. A diffraction grating with 10 000 slits per centimeter is used to separate the light into its component wavelengths. Sodium has two prominent emission wavelengths at 589.0 nm and 589.6 nm that cast an ugly yellow pall over much of our environment at night. (a) Calculate the angular separation of the two wavelengths in the first-order spectrum. (b) What is the minimum number of slits of the grating that must be illuminated by the light to barely resolve the doublet?

●**28.** You have been assigned the task of designing a grating for use in an astronomical research satellite. Lucky you! The grating is to examine the range of visible light from about 400 nm to 680 nm. The grating must be able to resolve (barely) wavelengths differing by only 0.010 nm in the first order. The first-order spectrum of visible light is to occupy an angular range of 30°. Describe the specifications of the grating, showing the calculations used to determine each parameter specified.

●**29.** The eyes of your fellow students in an 8 A.M. class have become glazed over with a thick coating of lecture verbiage in which there occurs complete destructive interference of too many obfuscations. But lo! White light (400 nm to 700 nm) is incident normally on a grating with 10 000 slits per centimeter. Calculate the angular spread of the resulting beautiful first-order spectrum.

●**30.** For cultural enrichment as well as to demonstrate your algebraic prowess, show that the angular dispersion of a grating (Equation 24.13) can be written in the following form:

$$\frac{d\theta}{d\lambda} = \frac{m}{\left[d^2 - (m\lambda)^2\right]^{1/2}}$$

●**31.** Show that the angular dispersion of a diffraction grating, Equation 24.13, can be expressed as

$$\frac{d\theta}{d\lambda} = \frac{\tan \theta}{\lambda}$$

●**32.** A diffraction grating consists of 500 parallel slits per millimeter. The grating is illuminated along the normal to the grating by a parallel beam of monochromatic light. The first bright line fringes are found at 18.44° on either side of the normal. (a) Find the grating spacing d. (b) Find the wavelength of the light.

•33. Two wavelengths of light are incident normally on a diffraction grating. One of the wavelengths is known to be 656 nm but the other is unknown. However, it is found that the third order of the unknown wavelength appears at the same angle as the second order of the 656 nm wavelength. (a) With this information, is it necessary to know the grating spacing d to determine the unknown wavelength? (b) If the answer to part (a) is no, calculate the unknown wavelength. Otherwise you can take the day off.

•34. Two wavelengths are incident normally on an arrangement of parallel slits. The intensity distribution for each of the colors on a distant screen is plotted in Figure P.34 as a function of sin θ. The first maximum for the longer, 600 nm wavelength occurs when sin $\theta = 0.010$. (a) Is it possible to determine the number of slits through which the light passed? If so, what is the number of slits? (b) What is the second wavelength if its interference maxima lie as illustrated relative to the 600 nm pattern? (c) What is the slit spacing d?

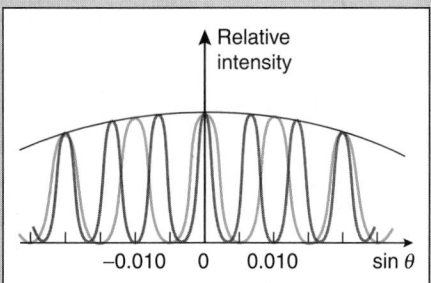

FIGURE P.34

‡35. A diffraction grating is illuminated at normal incidence by light with a frequency of 5.0×10^{14} Hz from a distant source. The first-order interference maximum occurs at 20° on either side of the straight-through direction. (a) How many slits are in one centimeter of the grating? (b) How many interference fringes, including the center and both sides, are formed by this grating? (c) If the overall width of the grating illuminated were doubled, keeping the same slit separation, what effect would this have on the interference pattern? (d) If the whole system were lowered into pure water ($n = 1.33$), at what angle would the first-order interference appear?

‡36. Four identical radio transmitter antennas each consist of a vertical wire that radiates radio waves of wavelength λ that propagate "important messages from our sponsors" horizontally in all directions into the hearts and minds of the population. The antenna array forms a square whose edges are equal to the wavelength (see Figure P.36). If all the antennas radiate in phase, in what horizontal directions will the combined radiation be strongest at great distances from the antennas?

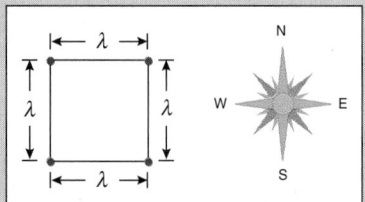

FIGURE P.36

Sections 24.10 The Index of Refraction and the Speed of Light
24.11 Thin-Film Interference*

37. The red light from a helium–neon laser has a wavelength of 632.8 nm. (a) What is the wavelength of this light in glass that has an index of refraction of 1.50? (b) What is the speed of the light in this glass?

38. Light is incident normally on a system of three glass lenses in air. About 4% of the incident light is reflected from each air–glass or glass–air interface. What percentage of the incident light is transmitted by the system?

•39. A long group of runners in single file are spaced 2.0 m apart when each is running at a speed of 5.0 m/s on grass. When they reach a sandy surface, they slow to 2.0 m/s, and when they reach pavement, they run at 6.0 m/s. No runners are added and none is lost as they cruise along. (a) How many runners per second (frequency) pass you on the grass? (b) How many runners per second (frequency) pass you on the sand? (c) How many runners per second (frequency) pass you on the pavement? Do any runners vanish? (d) What is their spacing (wavelength) on sand? (e) What is their spacing (wavelength) on pavement?

•40. Just to pass the time, the steady traffic flow along a highway is watched at many locations. No cars enter or leave the highway, but their speeds over three different sections are 20 m/s, 40 m/s, and 30 m/s. It is found that the cars traveling at 30 m/s are spaced 30 m apart between centers. (a) At what rate (frequency in cars/s) do cars go past three observing points located at the three specified regions of the highway? (b) What is the spacing (wavelength) between car centers in the other two regions? (c) Which quantity—speed, wavelength, or frequency—is the same all along the highway? Does this make physical sense in terms of piling up of cars over time? (d) What does this problem have to do with the passage of light through a succession of transparent media?

•41. Light is incident on a glass sphere of radius 1.00 cm and index of refraction $n = 1.50$. (a) Sketch the path of the light ray that spends the maximum amount of time within the glass sphere. (b) Calculate this maximum time. (c) Considering only paths of transmission, sketch the path of the light ray that spends the least amount of time inside the glass. (d) Calculate the minimum time. (The answer is not zero.)

•42. For the surfaces indicated in Figure P.42, light is incident from the left. What is the least thickness x for a coating that will minimize reflected light with an effective wavelength (in air) of 632.8 nm?

•43. A glass-covered mirror has a thin film (with $n_{film} = 1.60$) to minimize light reflected at normal incidence from the first glass surface, as shown in Figure P.43. The effective wavelength of the light is 550 nm in air. What is the minimum thickness for the coating?

•44. A piece of paper is used to form an air wedge between two optical flats as indicated in Figure P.44. Light of wavelength 633 nm is incident normally. In the reflected light, 25 bright interference fringes are observed from the vertex of the wedge to the paper, with the 25th bright fringe coincident with the edge of the paper. A dark fringe appears at the vertex of the air wedge. (a) Why is the fringe at the vertex dark? (b) Calculate the thickness of the piece of paper. (c) Calculate the small

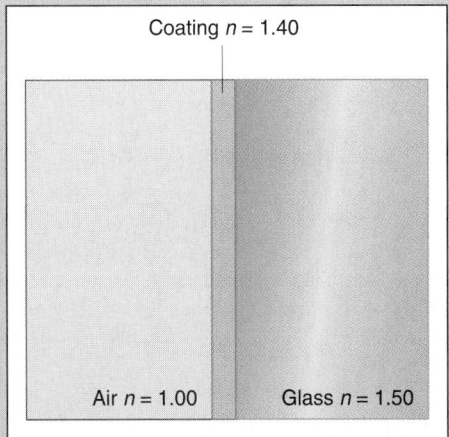

FIGURE P.42

Coating $n = 1.40$

Air $n = 1.00$ Glass $n = 1.50$

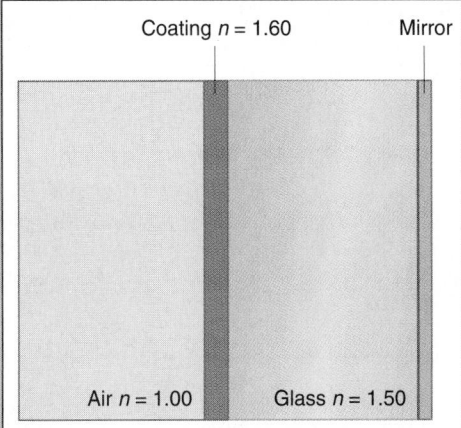

FIGURE P.43

Coating $n = 1.60$ Mirror

Air $n = 1.00$ Glass $n = 1.50$

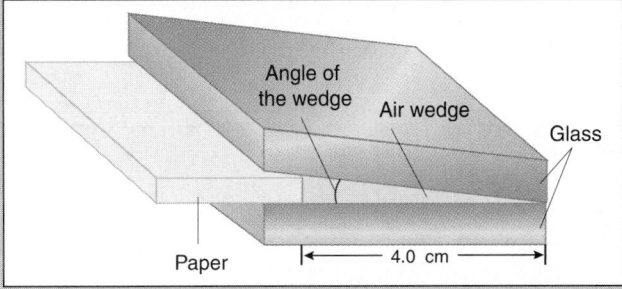

FIGURE P.44

Angle of the wedge Air wedge

Glass

Paper 4.0 cm

apex angle of the wedge. (d) If the air wedge now is filled with brine ($n = 1.31$), what is the order number of the bright fringe located nearest the edge of the paper?

•45. You have just received delivery of two expensive circular optical flats (serial numbers 1 and 2) from the firm Glass, Waters, and Mirage, Inc. Your supervisor asks you to test the flats by placing them each in turn on another piece of glass that is known to be flat. Light of wavelength 632.8 nm illuminates the combination at normal incidence and the patterns indicated in Figure P.45 are seen in the reflected light. (a) Which of the

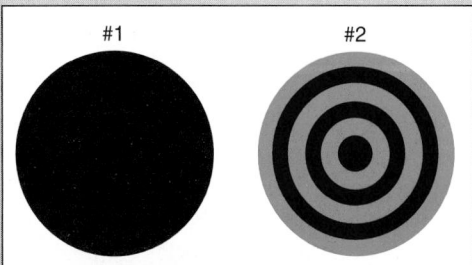

#1 #2

FIGURE P.45

newly received flats really is flat? (b) What is the approximate shape of the other newly received surface?

•46. A thin oil film ($n = 1.30$) lies on level water and is illuminated normally by white light. The reflection at normal incidence in air lacks the wavelengths 400 nm (violet) and 666 nm (red) but has all the rest of the visible range of wavelengths. (a) What color does the reflection appear? (b) What is the thickness of the film?

•47. A thin film of water appears on a glass surface. The water is illuminated by white light at normal incidence. Excruciating measurements find that the thickness of the water film is 8.00×10^{-7} m. What wavelengths are missing in the reflected light in the visible light range of wavelengths (about 400 nm–700 nm)?

•48. A thin transparent layer of magnesium fluoride ($n = 1.3$) is applied to *both* sides of a glass plate (or lens) to reduce reflections and to increase transmission. Assume normal incidence. (a) Sketch the layered arrangement and label the reflections, indicating which reflected beams undergo a phase change upon reflection. (b) If the first coating is 317 nm thick, what (vacuum) wavelength in the visible portion of the spectrum does not reflect from the front side of the lens? (c) If the other coating also is 317 nm thick, does the same color also not reflect from the *back* side of the system after going through the glass? (d) What are the nearest (vacuum) wavelengths less than and greater than the one removed that would have *maximum* reflection?

‡49. A plano-convex lens of index of refraction n is placed on an optical flat as indicated in Figure P.49. The radius of the spherical surface is R. The arrangement is illuminated along the normal with light of wavelength λ. The interference in the light reflected from the two surfaces forming the air gap produces circular fringes known as Newton's rings (although they were apparently first observed by Robert Hooke). (a) Show that the radius r of the mth bright fringe from the center is given approximately by

$$r \approx \left[(2m + 1) \frac{\lambda}{2} R \right]^{1/2}$$

(b) Show that the radius of the mth *dark* fringe from the center is given approximately by

$$r \approx (m\lambda R)^{1/2}$$

‡50. The detector of a radio telescope is located on the Earth's equator 100 m above the sea atop a cliff (see Figure P.50). The telescope is used to detect the radio emissions with a 10.0 cm

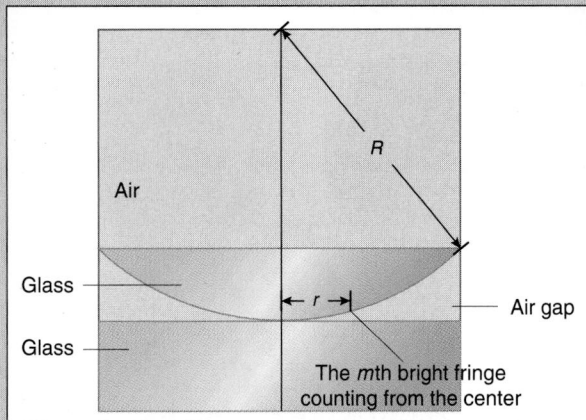

FIGURE P.49

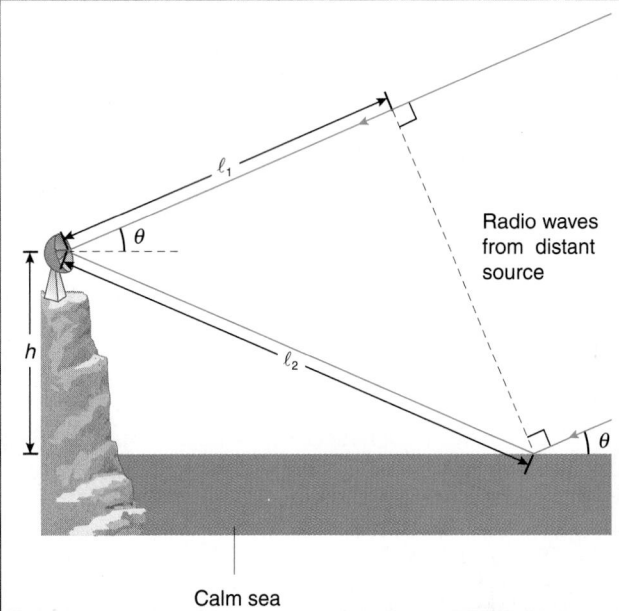

FIGURE P.50

wavelength from a distant star source as it sets in the west. The sea is very calm (i.e., flat). A reflection of the source from the water means that the telescope detects radio waves from *two* sources (the actual source and its image) with a phase difference of π rad between them. The separation of the source and its image decreases with time as the source sets. Neglect atmospheric refraction of the radio waves. (a) Show that the *path difference* between radio waves from the image and those directly from the radio source is (see Figure P.50)

$$\ell_2 - \ell_1 = 2h \sin \theta$$

where h is the height of the detector above the sea. (b) If the path difference is an integral number of wavelengths, will the interference between the double source be constructive or

destructive? (c) At the instant the source sets, will the radio telescope detect a constructive interference peak or a destructive interference valley? (d) Sketch the pattern of interference detected by the radio telescope as a function of time as the source approaches the horizon, indicating clearly the instant the source sets. (e) Through how many degrees per hour does the Earth rotate? (f) Calculate the time interval in seconds between the interference peaks detected by the radio telescope as the source sets. (g) If the telescope were located at, say, latitude 37° N, how would this location qualitatively affect the answer to part (e) assuming all other information provided is the same?

Sections 24.12 Polarized Light*
24.13 Polarization by Absorption*
24.14 Malus's Law*
24.15 Polarization by Reflection: Brewster's Law*
24.16 Polarization by Double Refraction*
24.17 Polarization by Scattering*

51. What is the Brewster angle for glass with an index of refraction of 1.6 when used in air? What is the angle of refraction for the transmitted beam in this case?

52. Determine the Brewster angle for light incident on a glass cube ($n = 1.6$) immersed in water ($n = 1.33$).

•**53.** Unpolarized light is incident on a spherical glass marble (with $n = 1.50$) as indicated in Figure P.53. Indicate the directions (relative to the incident direction) at which the light reflected from the marble is completely plane polarized.

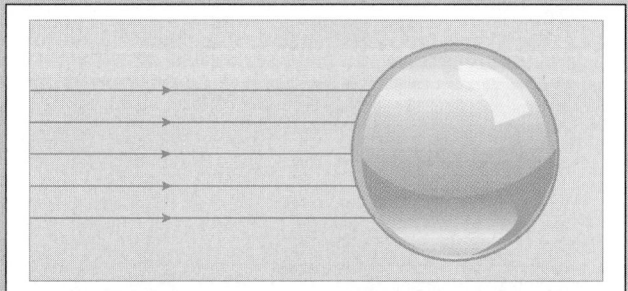

FIGURE P.53

•**54.** A cool loon (*Gavia immer*) floats on a calm lake and surveys the full reflection of the clear bright sky. In what directions down from its eye and around the horizon will the sky image be darkest? Explain.

•**55.** An unpolarized beam of light passes through two polarizers with a 30° angle between their transmission axes. (a) What is the ratio of the amplitude of the light transmitted through both polarizers to the amplitude of the light between the polarizers? (b) What is the ratio of the intensity of the light transmitted through both polarizers to the intensity between the polarizers?

•**56.** Two polarizing sheets are arranged so that initially the light transmitted through the combination is a maximum. (a) Through what angle should the second sheet be rotated so that the transmitted intensity decreases by 50%? (b) Does it make any difference which sheet is rotated through the angle calculated in part (a) if the only goal is to achieve a 50% decrease in the light intensity of the transmitted beam?

•57. Brewster's law states that the angle θ_B at which the reflected beam is completely polarized is related to the indices of refraction of the two media via

$$\tan \theta_B = \frac{n_2}{n_1}$$

where n_1 in the index of refraction of the medium in which the incident ray lies and n_2 is the index of refraction of the medium in which the refracted ray lies. (a) Show that if $n_2 > n_1$, then $\theta_B > 45°$. (b) Show that if $n_2 < n_1$, then $\theta_B < 45°$. (c) If the greatest value of the index of refraction is about 4.50, then find the smallest possible value for the Brewster angle. Describe the experimental situation that yields this minimum value for the Brewster angle. Is the light totally internally reflected?

•58. Given $n_1 > n_2$, show that the critical angle for total internal reflection always is greater than the Brewster angle at the same interface.

•59. Two pieces of polarizing material (labeled #1 and #2) are oriented so that their transmission axes are perpendicular to each other. Unpolarized light of intensity I_0 is incident on the pair. (a) What is the intensity of the light between the two sheets? (b) What is the intensity of the light after the second sheet? (c) A third polarizing sheet (#3) is now placed between the others so that its transmission axis is oriented at an angle θ relative to the transmission axis of the first sheet; see Figure P.59. What is the intensity of the light between the third and second sheets? (d) What is the intensity of the light after sheet #2?

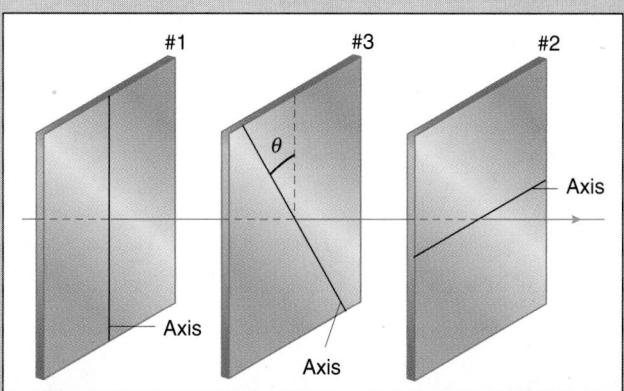

FIGURE P.59

•60. Unpolarized light is incident on three polarizing sheets whose transmission axes are oriented at 30° angles with respect to each other as illustrated in Figure P.60. Let I_0 be the intensity of the incident unpolarized light. What is the intensity of the light emerging from the third sheet?

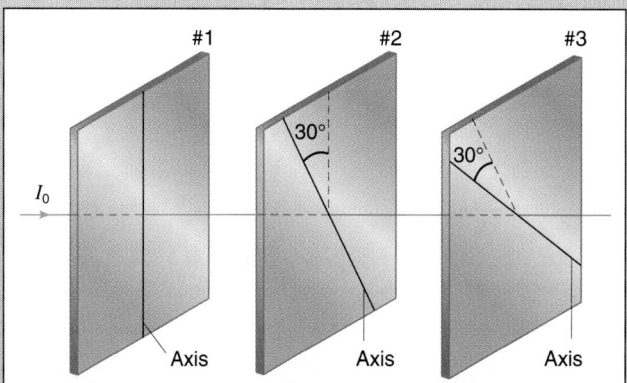

FIGURE P.60

•61. There you are, on the shores of a calm pond, romantically watching the reflection of the full Moon as it silently rises above the eastern horizon. Then you remember a bit of physics: when the elevation of the Moon is just right, the light from the reflected Moon will be completely polarized! To impress your date, and to make manifest the social utility of a knowledge of physics, you explain the situation and take out your Polaroid® sunglasses to demonstrate the effect at the appropriate time. What is the Moon's elevation angle above the horizon when its reflected light is completely polarized?

•62. You are trying to photograph the street scene illustrated in Figure P.62. The vertical store windows, whose normal lines are oriented at about 45° to your line of sight, reflect an annoying bright glare. You wish to reduce this glare relative to the general scene. (a) How should the axis of a polarizing filter be oriented in front of the lens to reduce the glare? (b) By what factor will the presence of the filter reduce the light from the bricks, wood, and other objects in the scene?

FIGURE P.62

•63. On the days of the March and September equinoxes (~20 March and ~22 September), the Sun sets due west. A vertical side window (with $n = 1.5$) of a car is oriented so that the reflected light of the setting Sun is completely plane polarized. In what four possible directions can the car be pointing? (Express your answer in degrees measured clockwise from north.)

INVESTIGATIVE PROJECTS

A. Expanded Horizons

1. Investigate the relationship between light, color, and the nature of color vision by the eye.

   Paul L. Pease, "Resource letter CCV-1: Color and color vision," *American Journal of Physics*, 48, #11, pages 907–917 (November 1980); this contains many references to the subject.

   Marcel G. J. Minnaert, *Light and Color in the Outdoors* (Springer-Verlag, New York, 1993).

2. Investigate the uses of lasers in various aspects of medicine.

   Alan L. McKenzie and John A. S. Carruth, "Lasers in surgery and medicine," *Physics in Medicine and Biology*, 29, #6, pages 619–642 (June 1984).

   Robert A. Weale, "Physics and ophthalmology," *Physics in Medicine and Biology*, 24, #3, pages 489–504 (May 1979).

3. Soap bubbles exhibit many interesting optical and physical effects. Investigate the physics of soap bubbles.

   Cyril Isenberg, *The Science of Soap Films and Soap Bubbles* (Dover, New York, 1992).

   Charles V. Boys, *Soap Bubbles* (Anchor, Garden City, New York, 1959).

   David Lovett, *Demonstrating Science with Soap Films* (Institute of Physics Publishing, Philadelphia, 1994).

   Coran Ramme, "Videotaping the lifespan of a soap bubble," *The Physics Teacher*, 33, #9, pages 558–561 (December 1995).

   Feredoon Behroozi and Dale W. Olson, "Colorful demos with a long-lasting soap bubble," *American Journal of Physics*, 62, #9, pages 856–857 (September 1994).

4. See how polarized light is used for investigating the distribution of strain in materials. Such techniques have been used by engineers to study what elements of Gothic cathedrals carry the most stress, an important consideration when undertaking repairs or remodeling.

   Frank L. Pedrotti and Leno S. Pedrotti, *Introduction to Optics* (Prentice-Hall, Englewood Cliffs, New Jersey, 1987), pages 378–380.

5. Investigate the function of Brewster windows in laser cavities.

   Frank L. Pedrotti and Leno S. Pedrotti, *Introduction to Optics* (Prentice-Hall, Englewood Cliffs, New Jersey, 1987), pages 371–372.

6. Determine why it is generally *not* possible to see interference effects in *thick* films.

   James Trefil, "Thick film interference," *The Physics Teacher*, 21, #2, pages 119–121 (February 1983).

   Joseph C. Amato, Roger E. Williams, and Hugh Helm, "An inexpensive, easy to build Fabry-Perot interferometer and its use in the introductory laboratory," *American Journal of Physics*, 59, #11, pages 992–994 (November 1991).

7. Diffraction gratings come in many varieties; we have investigated the action of a plane, *transmission* grating since the light actually passes through the grating. Other types of gratings include *reflection gratings*, *concave spherical gratings*, and so-called *blazed gratings*. Investigate the characteristics and applications of such gratings.

   Frank L. Pedrotti and Leno S. Pedrotti, *Introduction to Optics* (Prentice-Hall, Englewood Cliffs, New Jersey, 1987), pages 417, 419–423.

8. There are many electrical and magnetic effects associated with light. Effects associated with electric fields are known as *electro-optic effects*; those associated with magnetic fields are called *magneto-optic effects*. A few of these effects are listed here and would make fine topics for further reading and research:

Electro-optic effects	Magneto-optic effects
Stark effect	Zeeman effect
Kerr electro-optic effect	Cotton–Mouton effect
Electric double refraction	Voigt effect
	Faraday effect
	Kerr magneto-optic effect

These specialized effects cause ripples across many of the sciences and fields of engineering. Our survey of physical optics only touches the proverbial tips of many icebergs.

B. Lab and Field Work

9. For some fundamental experiments into the speed of light, which had unforeseen implications for the development of relativity, Albert A. Michelson (1852–1931) invented a new interference device, now called a *Michelson interferometer*. The device subsequently was used for precise measurements of the meter and also for the measurements of the diameters of large but distant stars. The device creates interference fringes from the amplitude division of a light wave. Investigate the construction of a Michelson interferometer and how it creates interference fringes. Your physics department likely has such an interferometer in its optics laboratory. Use it with an appropriate light source to form such interference fringes; also use it to measure precisely small changes in the position of one of the mirrors on the device. Research the historical importance of the interferometer (in the so-called *Michelson–Morley* series of experiments) in showing that light, unlike other waves, needs no medium in which to propagate.

   Francis A. Jenkins and Harvey E. White, *Fundamentals of Optics* (4th edition, McGraw-Hill, New York, 1976).

10. Another interferometer with a wide range of technical applications for the precise measurements of wavelengths in light scattering experiments is the *Fabry–Perot interferometer*, which like the Michelson interferometer (see Investigative Project 9) employs amplitude division of light to create interference fringes. Investigate how a Fabry–Perot interferometer is constructed and how its interference fringes are formed. Determine what is meant by the terms *finesse* and *free spectral range* when associated with this instrument. If your physics department has a Fabry–Perot interferometer available, use it to form interference fringes of an appropriate light source.

    Francis A. Jenkins and Harvey E. White, *Fundamentals of Optics* (4th edition, McGraw-Hill, New York, 1976).

11. Take a sheet of polarizing material and use it as an analyzer. Look through it at the liquid crystal display of your calculator, while rotating the analyzer. What do you observe and what conclusions can you draw from your observation? Investigate the principles and uses of liquid crystals in electronic displays.

    Renate J. Ondris-Crawford, Gregory P. Crawford, and J. William Doane, "Liquid crystals, the phase of the future," *The Physics Teacher*, 30, #6, pages 332–339 (September 1992).

    June E. Ball, "Liquid crystals," *Physics Education*, 15, #2, pages 108–109 (March 1980).

Anthony J. Nicastro, "Demonstrations of some optical properties of liquid crystals," *The Physics Teacher*, *21*, #3, pages 181–182 (March 1983).

Carl H. Hayn, "Liquid crystal displays," *The Physics Teacher*, *19*, #4, pages 256–257 (April 1981).

Renate J. Ondris-Crawford, Gregory P. Crawford, and J. William Doane, "Resource letter LC-1: Liquid crystals: physics and applications," *American Journal of Physics*, *63*, #9, pages 781–788 (September 1995); this paper has a lengthy list of references to the literature on liquid crystals.

Edward F. Carr and James P. McClymer, "A laboratory experiment on interference of polarized light using a liquid crystal," *American Journal of Physics*, *59*, #4, pages 366–367 (April 1991).

12. Doubly refracting (or birefringent) materials are fabricated into special devices known as *quarter-wave plates*. Discover what a quarter-wave plate is and demonstrate several applications of such plates.

Francis A. Jenkins and Harvey E. White, *Fundamentals of Optics* (4th edition, McGraw-Hill, New York, 1976).

Grant R. Fowles, *Introduction to Modern Optics* (2nd edition, Holt, Dover, New York, 1989).

13. Obtain a double refracting crystal (such as calcite) from your instructor. Place the crystal over an X you have made on a piece of paper to form a double image of the X as in Figure 24.50. Rotate the crystal and determine which image is formed from the ordinary ray and which from the extraordinary ray. Now view the images through a polarizer. Rotate the polarizer slowly, describe what happens, and explain your observation.

14. Investigate how chemists use the phenomenon of optical activity to determine the concentration of certain solutions. Design and perform an experiment to determine the concentration of a sugar solution using optical activity.

Thomas M. Lowry, *Optical Rotatory Power* (Dover, New York, 1964).

Francis A. Jenkins and Harvey E. White, *Fundamentals of Optics* (4th edition, McGraw Hill, New York, 1976), pages 572–573, 584–587.

15. Investigate, demonstrate, and explain the optical properties and characteristics of compact discs (CDs).

Christian Nöldeke, "Compact disc diffraction," *The Physics Teacher*, *28*, #7, pages 484–485 (October 1990).

Haym Kruglak, "Diffraction demonstration with a compact disc," *The Physics Teacher*, *31*, #2, page 104 (February 1993).

Thomas D. Rossing, "The compact disc audio system," *The Physics Teacher*, *25*, #9, pages 556–563 (December 1987).

C. Communicating Physics

16. Despite the significant contributions of scientists and engineers to our material well-being, health, and technological prowess, the TV and movie industries generally depict engineers and scientists as mad or deranged nerds, lacking a social conscience and social skills. What might account for such consistently unfavorable depictions of scientific personnel in the entertainment industry? Do you think such attitudes may be a reflection of the unfortunate experience of the tragic poet Sylvia Plath when she took a college physics course? Her comment: *The day I went into physics class it was death.* [*The Bell Jar* (New York, Harper and Row, 1967), Chapter 3.]

CHAPTER 25

THE SPECIAL THEORY OF RELATIVITY

Nothing puzzles me more than time and space; and yet nothing
puzzles me less, for I never think about them.

Charles Lamb (1775–1834) *

L ike Charles Lamb, Einstein also was puzzled deeply by space and time, but unlike Lamb, he *did* think about them, and seriously too. As a result, our ideas about space, time, light, and energy never will be the same. The theories of relativity are beautiful constructions of the human intellect; you will experience much real intellectual satisfaction from an understanding of them.

There is a certain mysterious aura surrounding relativity in our culture. Perhaps it stems from the popular view that equates the very name of Einstein with genius[†]; or perhaps it comes from the perceived difficulty of the theories. In any case, the theories of relativity invented by Einstein certainly are among the greatest of the many revolutions in physics during the 20th century. Relativity also begot enormous social and political ramifications as well, foremost among them fearful and awesome nuclear weapons.

Just what is meant by the word **relativity**? The term refers to the relative motion of two reference frames. Einstein created two theories of relativity: the *special theory* (published in 1905), which we consider in significant detail in this chapter; and the more difficult *general theory* (published during 1914–1916), to which we only briefly allude. There also is classical *Galilean relativity* that stems from the now familiar kinematics and dynamics of Galileo and Newton.

25.1 REFERENCE FRAMES

The world of events forms a four-dimensional continuum.
Albert Einstein (1879–1955) and Leopold Infeld (1898–1968)[‡]

A **reference frame** consists of (1) a coordinate system and (2) a set of synchronized clocks distributed throughout the coordinate grid and at rest with respect to it.

---

*(Chapter Opener) Letter to Thomas Manning, 2 January 1810, in *The Complete Works and Letters of Charles Lamb* (The Modern Library, New York, 1935), letter CLXV, pages 775–777; the quote is on page 776.
[†]We all likely have referred to a gifted individual as "an Einstein."
[‡]*The Evolution of Physics* (Simon & Schuster, New York, 1938), page 219.

One of the most recognizable faces of the 20th century, indeed, of human history!

For example, we can consider a reference frame to be, say, a Cartesian coordinate system and a set of clocks distributed at any and every location throughout the coordinate system, as shown in Figure 25.1.

The clocks all are synchronized, so that they indicate the same time and tick at the same rate. The clocks are like all the timepieces that you and your classmates are wearing; they all indicate the same standard time (or daylight saving time)[§] and progress at the same rate. We think of the clocks as having a stopwatch feature, so that time intervals can be measured conveniently. A reference frame thus has three spatial coordinates, such as x, y, and z, and one time coordinate, t; from a mathematical viewpoint, a reference frame is a four-dimensional space–time affair.

Historical events specify places and dates. The tragic assassination of President John F. Kennedy occurred just outside the Texas School Book Depository in Dallas at 12:30 P.M. on 22 November 1963; the Civil War Battle of Antietam occurred on 17 September 1862 in Maryland near Antietam Creek, a tributary of the Potomac River. Events in physics are no different, but hopefully are less tragic. As we did in Chapter 3 (long ago!), we define an **event** as something that happens at a particular place and instant: where and when. To specify where an event occurs means choosing a coordinate system, say a Cartesian coordinate system. A set of synchronized clocks is needed to specify when. Thus we use a reference frame to specify an event. Different reference frames describe the same event in different ways.

Among other things, relativity is concerned with how an event described in one reference frame is related to its description in another reference frame.

In particular, relativity specifies how the coordinates and times of events (and other physical quantities such as momentum, energy, and electric and magnetic fields) measured or specified in one reference frame are related to the coordinates, time, and corresponding physical quantities in another reference frame.

Given any pair of reference frames:

1. the reference frames may be moving at constant velocity (including zero velocity) with respect to each other; or

---

[§]The clocks in different time zones on the Earth all tick at the same rate but indicate times that differ from each other by an hour (or a half hour, in some instances). Nonetheless, we can set them all to the same time, much like simultaneously started stopwatches. The clocks then fulfill the condition for the clocks in a reference frame: the same time and the same tick-rate.

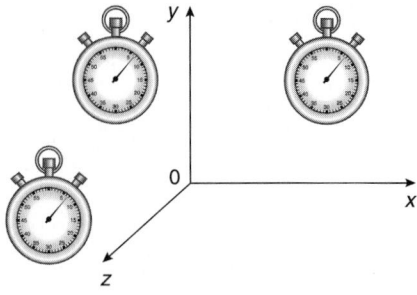

FIGURE 25.1 A reference frame is a coordinate system and a collection of synchronized clocks at rest.

2. one reference frame may be accelerating (which includes rotations) with respect to the other.

Inertial reference frames are the ones in which Newton's first law is true (see Chapter 5). Reference frames that are not inertial are called **noninertial reference frames**; accelerated reference frames are noninertial.

> The **special theory of relativity** is concerned with the relationship between events and physical quantities specified in different inertial reference frames (and only these *special* reference frames). The **general theory of relativity** is concerned with the relationship between events and physical quantities in *any* (i.e., *general*) reference frames.

The equations that indicate how the four space and time coordinates (and physical quantities) specified in one reference frame are related to the corresponding quantities specified in another reference frame are called **transformation equations**. You already are familiar with transformation equations in a mathematical context when changing coordinate systems (say from rectangular to polar coordinates, or vice versa), and so the concept of such transformations should not be intimidating.

> The **standard geometry** we use for the special theory of relativity has two inertial reference frames called S and S', with their x- and x'-coordinate axes collinear (see Figure 25.2). Imagine collections of clocks (stopwatches) distributed at rest throughout each respective frame; the clocks all are set to 0 s when the two origins coincide.

If we imagine ourselves at rest in reference frame S, then frame S' is moving to the right at any chosen constant velocity $v\hat{\imath}$, measured using clocks and rulers in S.

Changing horses, we imagine ourselves at rest in S'; then frame S is moving to the left with a velocity $-v\hat{\imath}'$, as in Figure 25.3, measured using clocks and rulers in S'. Note that the *relative speed v* (a positive scalar!) of the two systems is the same; they simply move in opposite directions with respect to each other.

QUESTION 1

Reference frames attached to (a) a plane flying with a constant velocity in nonturbulent air, (b) a ship traveling with a constant velocity in calm seas, or (c) your car moving at constant velocity across the level plains of Kansas all are examples of inertial reference frames. A reference frame attached to a spinning merry-go-round, an airplane taking off or landing, or your car while braking are examples of noninertial reference frames. Describe a specific experiment to tell whether a given reference frame is an inertial reference frame.

25.2 CLASSICAL GALILEAN RELATIVITY

> Classical **Galilean relativity** relates the space and time coordinates and other physical quantities in two inertial reference frames when the relative speed v of the two frames is much less than the speed of light c—that is, when v << c.

This is the realm of our everyday common experience. The equations that relate the space and time coordinates and other physical quantities, such as velocity components, between the two standard inertial reference frames, when v << c, are the **Galilean transformation equations**.

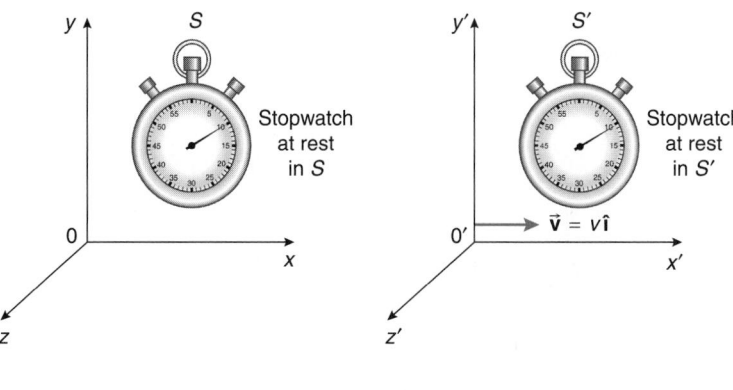

FIGURE 25.2 Frame S' moves to the right as seen from frame S in the standard geometry.

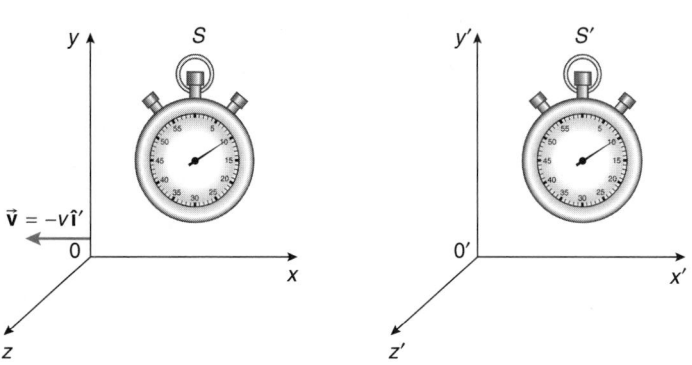

FIGURE 25.3 Frame S moves to the left as seen from frame S' in the standard geometry.

The Galilean Time Transformation Equation

> In classical physics time is a *universal* measure of the chronological ordering of events and the time interval between them.*

Watches in fast sports cars, airplanes, spacecraft, and oxcarts tick at the same rate as those at rest on the ground; this is a common, almost trivial observation, usually unstated and taken for granted. The time interval between two events and the rate at which time passes are independent of the speed of the moving clock; they are the same everywhere. In classical physics all the clocks distributed throughout any two inertial reference frames tick at the same rate and, once synchronized, can be arranged to indicate the same time instants: that is,

$$t' = t \qquad (25.1)$$

The Galilean Spatial Coordinate Transformation Equations

The origins of the two inertial frames having the standard geometry coincide when $t = t' = 0$ s; see Figure 25.4. Therefore an object at position x' in frame S' has the coordinate x in frame S, where

$$x = x' + vt \qquad (25.2)$$

or, solving for x',

$$x' = x - vt \qquad (25.3)$$

Note the slight asymmetry of these equations. The difference in the sign of the vt term arises because frame S' is moving to the *right* as seen by frame S, while frame S is moving to the *left* when viewed from frame S' in the standard geometry, so the velocity components of the frames have opposite signs. The other coordinates are identical; that is,

$$y' = y \qquad (25.4)$$
$$z' = z \qquad (25.5)$$

---

*We shall see in special relativity that this *cannot* be the case (Section 25.4). In special relativity the temporal interval between events depends on the inertial reference frame.

The Galilean space and time transformation equations are Equations 25.1 and 25.3–25.5, summarized here:

$$x' = x - vt \qquad (25.3)$$
$$y' = y \qquad (25.4)$$
$$z' = z \qquad (25.5)$$
$$t' = t \qquad (25.1)$$

The transformation equations for some other physical quantities can be found from Equations 25.1 and 25.3–25.5. For example, we consider the velocity and acceleration component transformations in the next two subsections.

The Galilean Velocity Component Transformation Equations

When you run horizontally at 3 m/s and toss a ball at 5 m/s in the same direction, the velocity component of the ball with respect to the ground is the algebraic sum of the velocity components: 8 m/s. We can see how this comes about from the Galilean transformation equations for position.

Imagine a particle moving to the right in frame S' along the x'-axis with a velocity component

$$u'_x = \frac{dx'}{dt'}$$

in frame S';† see Figure 25.5. The velocity component u_x of the particle from the viewpoint of the frame S is

$$u_x = \frac{dx}{dt}$$

The two velocity components can be related using the transformation Equation 25.3:

$$u_x = \frac{dx}{dt}$$
$$= \frac{d(x' + vt)}{dt}$$

---

†We apologize for the slight change in notation, but in this section (and later in Sections 25.11 and 25.16), it is convenient to use the letter $\mathbf{\bar{u}}$ for velocity of a particle here to distinguish it clearly from the relative speed v of the two inertial reference frames.

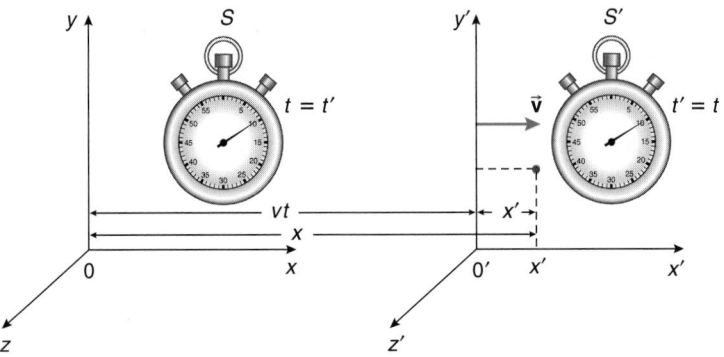

FIGURE 25.4 The view from frame S, when $t = t' > 0$ s.

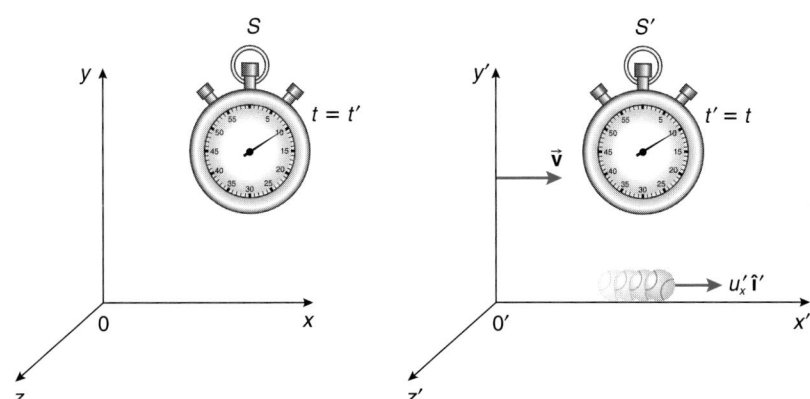

FIGURE 25.5 A particle moving in reference frame S'.

Recall that the relative speed v of the two inertial frames is constant. Hence we have

$$u_x = \frac{dx'}{dt} + v$$

Since $t' = t$ according to Galilean relativity (Equation 25.1), we have

$$u_x = \frac{dx'}{dt'} + v$$

But

$$\frac{dx'}{dt'} = u'_x$$

is the velocity component of the object in the frame S'.

Thus we have

$$u_x = u'_x + v \qquad (25.6)$$

Equation 25.6 agrees with our commonsense ideas about the way the velocity components should add. Indeed, Equation 25.6 is a modified version of the relative velocity addition equation we played with back in Chapter 4 (Equation 4.20).

We imagine a particle moving in the S' frame along the y'-axis, as in Figure 25.6. Its velocity component is

$$u'_y = \frac{dy'}{dt'}$$

as measured in frame S'. In reference frame S, the corresponding velocity component is

$$u_y = \frac{dy}{dt}$$

But since $t = t'$ and $y = y'$, the two velocity components are equal:

$$u_y = \frac{dy}{dt} = \frac{dy'}{dt'} = u'_y \qquad (25.7)$$

The velocity components along a direction *perpendicular* to the motion are the same in the two standard inertial reference frames.

The Galilean Acceleration Component Transformation Equations

We imagine a particle in frame S' that has an acceleration component

$$a'_x = \frac{d^2 x'}{dt'^2}$$

The acceleration component of the particle as measured in frame S is

$$a_x = \frac{d^2 x}{dt^2}$$

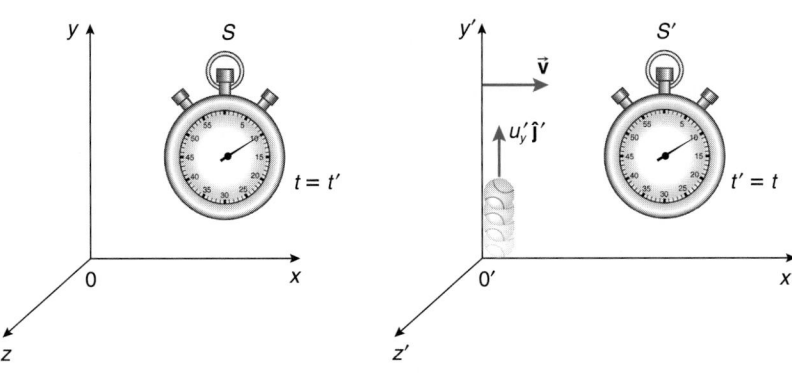

FIGURE 25.6 A particle moving along the y'-axis in frame S'.

We use Equation 25.2 for x and Equation 25.1 for t, remembering that the velocity component v is constant:

$$
\begin{aligned}
a_x &= \frac{d^2 x}{dt^2} \\
&= \frac{d^2 (x' + vt)}{dt^2} \\
&= \frac{d^2 x'}{dt'^2} + 0 \ \text{m/s}^2 \\
&= a_x'
\end{aligned}
\tag{25.8}
$$

In a similar fashion, we can show that

$$
\begin{aligned}
a_y &= a_y' \\
a_z &= a_z'
\end{aligned}
$$

The acceleration components are the same in the two inertial reference frames. (We showed this once before in Section 4.4.)

It is the equality of the accelerations in the two inertial reference frames that restricts the application of Newton's laws of motion to such inertial reference frames.

25.3 THE NEED FOR CHANGE AND THE POSTULATES OF THE SPECIAL THEORY

The relativity theory arose from necessity, from serious and deep contradictions in the old theory from which there seemed no escape. The strength of the new theory lies in the consistency and simplicity with which it solves all these difficulties, using only a very few convincing assumptions.

Albert Einstein and Leopold Infeld*

So what's the deal? What is the need for changing the Galilean ideas of relativity? None of the Galilean results indicates that there is any subtlety about nature or its workings that warrants changing the Galilean transformation equations between the two inertial reference frames. Where $v \ll c$, experiments verify the validity of the equations within the common kinematic world of dynamics; welcome back to the world of 19th-century kinematics and dynamics! The problem is that we have little direct experience with objects or reference frames moving at speeds comparable to that of light. However, electromagnetism provides some subtle clues that all is not right with Galilean relativity.

The Maxwell theory predicts the existence of electromagnetic waves that were shown to travel at a speed equal to the known speed of light; they *are* light. The theory indicates that the speed c of such waves in vacuum stems from the permittivity of free space ε_0 and the permeability of free space μ_0 (see Sections 21.4 and 21.5):

$$
c = \frac{1}{(\mu_0 \varepsilon_0)^{1/2}}
$$

Note that there is no dependence on the speed v of the source of the waves in this equation!

The Evolution of Physics (Simon & Schuster, New York, 1938), page 203.

Now we imagine an experiment that precisely measures the speed of light. Measuring the speed of light presents significant technological challenges, but we ignore them to focus on simply the results of such experiments. The experiment is performed twice, once with light from a stationary source, where the result obtained is c. The reference frame in which we perform this experiment we call S, as shown in Figure 25.7.

The experiment then is performed again when the source (at rest in frame S') is moving at speed v in the same direction as the light is moving, as in Figure 25.8. The measurements again are made using clocks and rulers at rest in reference frame S.

According to Galilean relativity (Equation 25.6), the second experiment should yield $c + v$ for the speed of light, since the velocity components are both positive. In fact, the second experiment indicates that the speed of light is *still* c, the identical value obtained when the source is at rest! Such experiments have been performed many times in many ingenious different ways.[†] The results are always the same.

The measured speed of light in vacuum has the same numerical value $c = 299\ 792\ 458$ m/s, independent of the speed of the source *or* that of the observer, which contradicts the predictions of Galilean relativity.

[†]The most famous of these experiments is known as the Michelson–Morley experiment; see Investigative Project 2. Other experiments also confirm it: atoms moving at many different speeds emit light with the same speed c. A binary star, with one component approaching and the other receding from us, emits light with the same speed c.

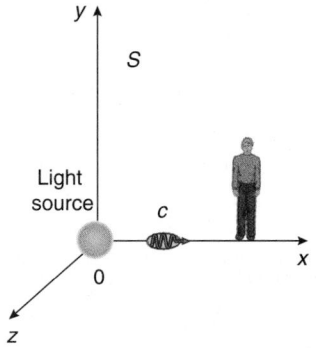

FIGURE 25.7 Measure the speed of light in S with the source at rest in S. The result is c.

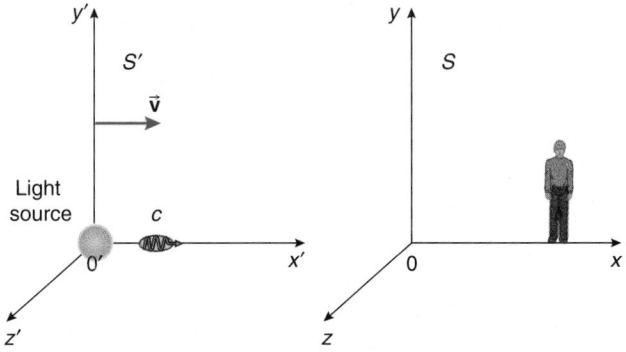

FIGURE 25.8 Measure the speed of light in S from a moving source that is at rest in S'. The result *also* is c.

These results are peculiar, even baffling, for they defy common sense, that is, the prediction of Galilean relativity. Since physics ultimately rests on experiment, the speed-of-light experiments indicate that Maxwell must be right and the speed of light must not depend on the speed of the source or observer.*

> Therefore the Galilean transformation equations between two inertial reference frames need to be modified to accommodate the observed invariance of the speed of light. The modifications needed to make this accommodation truly revolutionized physics. The new transformation equations, called the **Lorentz transformation equations** (which we derive in Section 25.7), together with their implications and consequences, constitute the special theory of relativity.

The entire special theory stems from only two postulates, one rooted in experiment, the other stemming from aesthetics, even natural philosophy as well as experiment. The first bold postulate encompasses all measurements, early and modern, of the speed of light and the prediction of *c* by the Maxwell theory of electromagnetism.

> **Postulate 1:** The speed of light in a vacuum has the same numerical value *c* when measured in any inertial reference frame, independent of the motion of the source and/or observer.

Despite the confounding nature of this statement, it reflects the way nature is; physics must take nature on its own terms.

The second postulate of the theory makes an aesthetic argument about the apparent experimental equivalence of all inertial reference frames.

> **Postulate 2:** The fundamental laws of physics must be the same in all inertial reference frames.

Certainly we can apply the same laws of physics to the gas atoms, electric charges, or even frisbees in an airplane cabin flying at constant velocity as we can when it is at the terminal gate. In so many words, the second postulate states that if the laws of physics are codified into a sacred text (this one!?) in an inertial frame *S*, the *same* sacred text can be used in *any* other inertial reference frame *S'*. Whatever laws of physics are discovered in one inertial frame can be used in any other inertial frame. If this were not the case, new physics texts would have to be written for each inertial reference frame and the task of learning physics would be compounded immensely; that is a recipe for total confusion. So we insist that the laws of physics, whatever they may be, have the same form for all inertial observers.

Descriptions of what happens as a result of the laws of physics may differ from one inertial reference frame to another, but the underlying fundamental physical principles and laws are the same. For example, while running at constant velocity across the room, let your professor toss a ball vertically upward according to her. You see the ball follow a parabolic trajectory. She says it traveled in a straight line vertically up and down. But you both begin your description with the same laws of physics.

The entire edifice of the special theory of relativity rests on the foundation of these two postulates. Without either, the entire theory crumbles.

> *On these two commandments hang all the law and the prophets.*
> The New Testament, Matthew 22:40

QUESTION 2

The effects of special relativity are only apparent for speeds that are significant fractions of the speed of light. Based on your experience, estimate the order of magnitude of the speed of the fastest material thing you have actually seen with your naked eye.

25.4 TIME DILATION

In classical physics it was always assumed that clocks in motion and at rest have the same rhythm, that rods in motion and at rest have the same length. If the velocity of light is the same in all CS [coordinate systems], if the relativity theory is valid, then we must sacrifice this assumption. It is difficult to get rid of deep-rooted prejudices, but there is no other way.

Albert Einstein and Leopold Infeld[†]

Before we tackle the problem of finding the new transformation equations between the inertial reference frames *S* and *S'*, let us explore several interesting and peculiar consequences of the special theory of relativity. If we accept the postulates of the theory, we are hooked and cannot avoid their consequences.

We construct several identical, special clocks to measure time. They all run the same way. Each clock consists of a pulsed laser, a mirror located a distance ℓ_0 from the laser, and, adjacent to the laser, a detector and trigger mechanism to stimulate the next emission of a pulse of light from the laser, as shown in Figure 25.9.

The operation of each light clock is described by three events:

1. emission of the pulse from the laser;
2. reflection of the pulse from the mirror; and
3. detection of the pulse (and emission of the next pulse at the same instant).

The time interval for the laser pulse to travel to the mirror and back to the detector is the total round-trip distance traveled, $2\ell_0$, divided by the speed of light *c*. So the fundamental time interval τ_0, the tick rate of each clock at rest, is

$$\tau_0 = \frac{2\ell_0}{c} \tag{25.9}$$

This time interval is called the **rest time interval** or **proper time interval** of the clock because it is measured in the frame in which the clock is at rest.

To characterize the three events of each clock, place the laser and detector at rest in inertial reference frame *S'* with the laser and detector at its origin, as in Figure 25.10. Let this clock at rest in *S'* move at constant velocity (together with *S'*) along the *x*-axis of another inertial reference frame *S*.

---

*Historically, there is confusion as to what experiments (if any) motivated Einstein, even according to Einstein himself! (See Investigative Project 2.) On the one hand, he says that he was motivated by electromagnetic problems. Later in his life, however, he said that the Michelson–Morley experiments were important in convincing him of the validity of Maxwell's theory. He also wondered what it would be like to ride alongside a light wave.

[†]*The Evolution of Physics* (Simon & Schuster, New York, 1938), page 196.

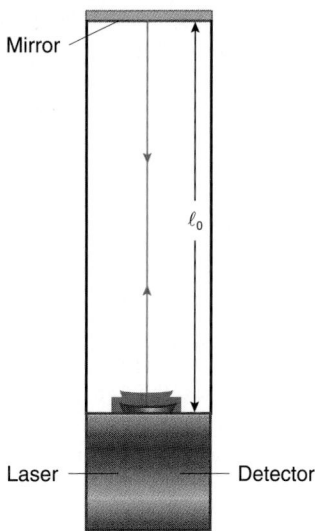

FIGURE 25.9 A special light clock.

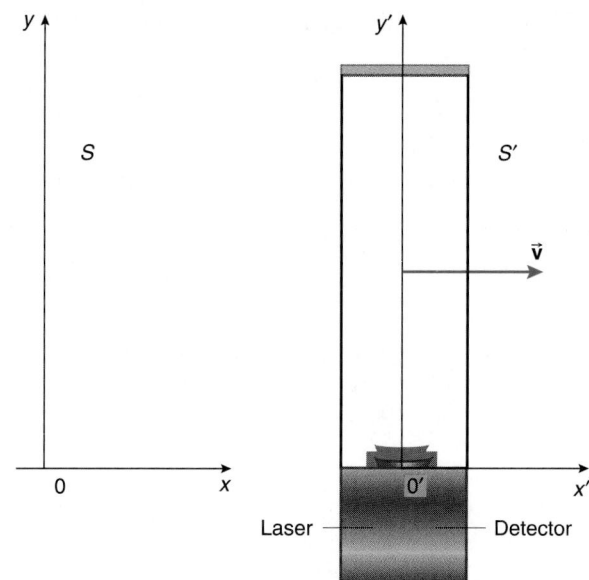

FIGURE 25.10 A clock at rest in S' is moving as seen from S.

The three events are specified by indicating where and when they occur. In the reference frame S', the space and time coordinates of each event are as follows:

1. emission of the pulse from the laser:

$$x'_1 = 0 \text{ m}$$
$$y'_1 = 0 \text{ m}$$
$$z'_1 = 0 \text{ m}$$
$$t'_1 = 0 \text{ s}$$

2. reflection of the pulse from the mirror:

$$x'_2 = 0 \text{ m}$$
$$y'_2 = \ell_0$$
$$z'_2 = 0 \text{ m}$$
$$t'_2 = \frac{\ell_0}{c}$$

3. detection of the pulse and emission of the next pulse:

$$x'_3 = 0 \text{ m}$$
$$y'_3 = 0 \text{ m}$$
$$z'_3 = 0 \text{ m}$$
$$t'_3 = \frac{2\ell_0}{c}$$

Identical clocks are at rest in reference frame S in order to measure time intervals in the S frame. We want to describe the three events associated with the moving clock from the perspective of the S reference frame. In particular, we want to find the time interval τ between Event 3 and Event 1, for this is the tick rate of the *moving* clock as measured by the stationary clocks in S.

We arrange things so that Event 1 (emission of the light pulse of the moving clock) occurs at the instant the S' origin is coincident with the origin in S. We start a stopwatch in S at this instant; see Figure 25.11. The stopwatch keeps time in agreement

with the identical light clock at rest in this frame. Event 1 is described in S by

1. emission of the light pulse:

$$x_1 = 0 \text{ m}$$
$$y_1 = 0 \text{ m}$$
$$z_1 = 0 \text{ m}$$
$$t_1 = 0 \text{ s}$$

While the light pulse is propagating from the laser to the mirror, the clock at rest in reference frame S' is moving to the right according to frame S. When the stopwatch reads $\tau/2$ in frame S, the light is reflected from the mirror; see Figure 25.12.

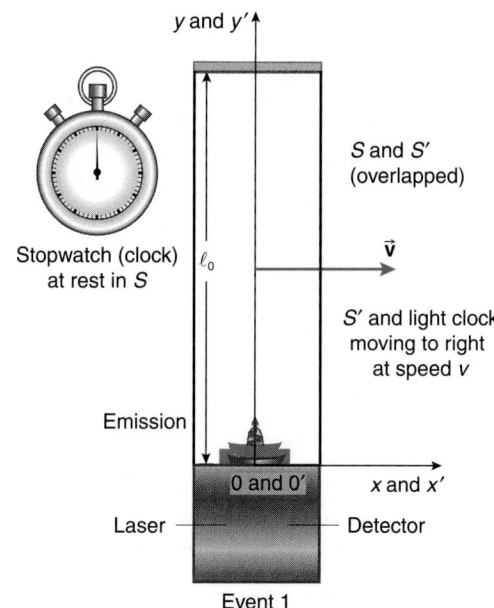

FIGURE 25.11 Emission of the light pulse seen from frame S.

The clock has moved a distance to the right (in S) equal to

$$\frac{v\tau}{2}$$

between Event 1 and Event 2. Event 2 is described in the reference frame S by the following space and time coordinates:

2. reflection of the light by the mirror:

$$x_2 = \frac{v\tau}{2}$$

$$y_2 = \ell_0$$

$$z_2 = 0 \text{ m}$$

$$t_2 = \frac{\tau}{2}$$

Between Events 2 and 3 the clock continues to move to the right in frame S. The light pulse then is detected by the detector, as in Figure 25.13. Event 3 is described in S in the following way:

3. detection of pulse (and emission of the next pulse):

$$x_3 = v\tau$$

$$y_3 = 0 \text{ m}$$

$$z_3 = 0 \text{ m}$$

$$t_3 = \tau$$

Notice in Figure 25.12 that in the S reference frame, the light was emitted at the coordinates $x_1 = 0$ m and $y_1 = 0$ m (Event 1) and reflected at coordinates $x_2 = v\tau/2$ and $y_2 = \ell_0$. From the well-worn Pythagorean theorem, the path following by the light in the reference frame S is of length

$$\sqrt{\ell_0^2 + \left(\frac{v\tau}{2}\right)^2} \qquad (25.10)$$

From the first postulate of special relativity, the light travels this distance at the familiar constant speed c. The time for the light to travel any distance is the distance divided by the speed:

$$\frac{\sqrt{\ell_0^2 + \left(\frac{v\tau}{2}\right)^2}}{c}$$

But this is just the time interval between Events 2 and 1 in S, which is $\tau/2$. Thus

$$\frac{\tau}{2} = \frac{\sqrt{\ell_0^2 + \left(\frac{v\tau}{2}\right)^2}}{c} \qquad (25.11)$$

Now we solve Equation 25.11 for the time interval τ. Recall that τ represents the time between emission and detection of the light pulse as measured by an observer in reference frame S using clocks at rest in that reference frame (the time between Events 3 and 1). We eliminate the square root by squaring Equation 25.11, obtaining

$$\frac{\tau^2}{4} = \frac{\ell_0^2 + \left(\frac{v\tau}{2}\right)^2}{c^2}$$

Now we turn the proverbial mathematical crank to obtain an expression for τ^2. After some algebra we find

$$\tau^2 = \frac{4\ell_0^2}{c^2 - v^2}$$

For convenience we factor out c^2 from the denominator, obtaining

$$\tau^2 = \frac{\dfrac{4\ell_0^2}{c^2}}{1 - \dfrac{v^2}{c^2}}$$

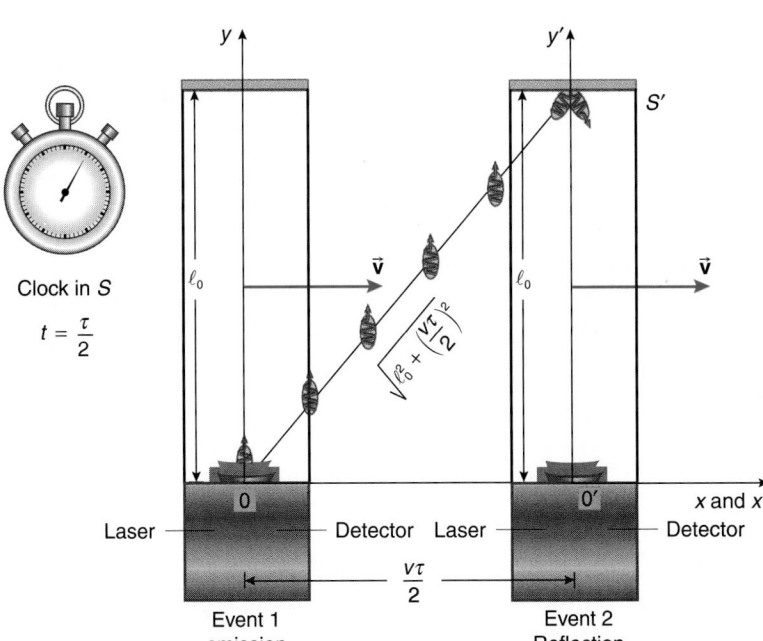

Clock in S

$$t = \frac{\tau}{2}$$

Laser — Detector   Laser — Detector

Event 1
emission

Event 2
Reflection

FIGURE 25.12 Light reflected from mirror, as seen in frame S.

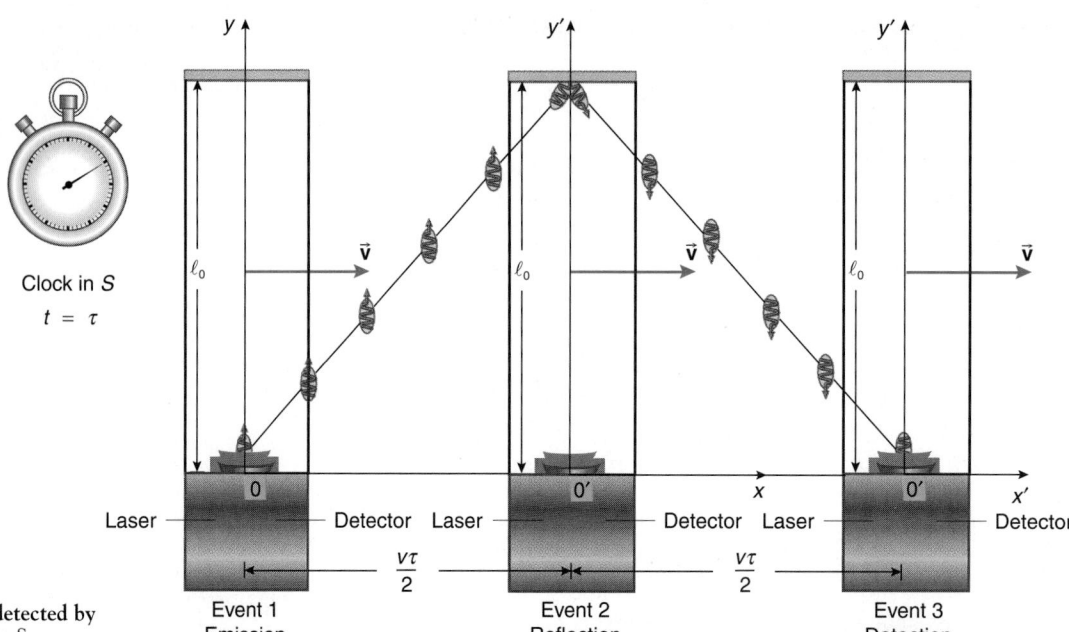

FIGURE 25.13 Light detected by detector, as seen in frame S.

We take the (positive) square root to obtain τ itself:

$$\tau = \frac{\dfrac{2\ell_0}{c}}{\sqrt{1 - \dfrac{v^2}{c^2}}}$$

But the quantity $2\ell_0/c$ is the fundamental tick rate τ_0, the proper time of the clock as measured in the reference frame S' in which the clock is at rest (Equation 25.9). The quantity $2\ell_0/c$ also is the fundamental tick rate of the identical clocks at rest in S.

Thus the relationship between the time intervals is

$$\tau = \frac{\tau_0}{\sqrt{1 - \dfrac{v^2}{c^2}}} \qquad (25.12)$$

The square root is a pure real number less than 1 when $v < c$. Thus the time interval between the emission and detection of the light in the moving clock, as measured by the clock at rest in S, is *longer* than the time interval between the two events as measured in the reference frame S' where the clock stays at rest. The moving clock is seen to tick more slowly than the identical clock at rest. The moving clock has a greater time interval between its ticks than the clock that is at rest; a moving clock runs slow. This effect is called **time dilation**.

Thus time intervals in relativity are not absolute or universal but depend on whether the clock is moving or not: we say time is a relative not an absolute quantity. If the clock is at rest with respect to you, such as the watch on your wrist, it shows no time dilation and nothing is peculiar about its rate of ticking. It is only the tick rate of a clock moving with respect to you that

ticks more slowly than your own clock at rest. The slower tick rate is determined and measured by clocks at rest with respect to you—the clocks in your own reference frame.

Time dilation has been confirmed by many experiments, some using radioactive particles as natural clocks. In another experiment, the U.S. Naval Observatory flew two atomic clocks around the world in opposite directions (first class, mind you, but with no champagne or caviar, even for its government chaperons) and compared them with a clock left behind. The experiment involved both the predictions of special and general relativity; the clocks confirmed the predictions of both theories.

At first glance, you might think you can cheat nature and extend your lifetime using time dilation simply by moving fast. Sorry! Your aging process is determined by natural biological clocks that *always are at rest* with respect to you. The relativity of time is real only for clocks moving with respect to you. Thus your own individual aging always proceeds quite normally, no matter what speed you happen to be traveling.

We get another useful piece of information from this analysis. In particular, notice that if $v > c$ in Equation 25.12, the square root becomes a purely *imaginary number*. So, if the moving clock has a speed $v > c$, τ becomes an imaginary time. Frankly, we do not know what to make physically of this imaginary time. Ordinary time and other physical quantities are represented by real numbers, since the measurements we make from meter sticks, clocks, and other instruments always yield real numbers. Nonetheless, the imaginary prediction when $v > c$ has not prevented theorists from speculating about the possible existence of faster-than-light particles, called **tachyons**. Despite feverish theoretical and experimental activity, no experimental evidence exists for such tachyon particles (see Investigative Project 3).

Later, in Section 25.17, we see that it is impossible for a particle initially traveling at a speed less than c to attain a speed equal to c.* Hence in special relativity the speed of light evidently is the ultimate speed limit.

The inverse of the purely numerical square root in Equation 25.12 occurs so frequently in relativity that it is given its own special symbol: γ.

By definition, we take

$$\gamma \equiv \frac{1}{\sqrt{1 - \dfrac{v^2}{c^2}}} \qquad (25.13)$$

With the definition of γ, the time dilation Equation 25.12 is succinctly written as

$$\tau = \gamma \tau_0 \qquad (25.14)$$

A graph of γ versus the speed ratio v/c is shown in Figure 25.14. Notice that if $v = 0$ m/s, then $\gamma = 1$. As v approaches c, γ increases dramatically in value, diverging to ∞ when $v = c$.

PROBLEM-SOLVING TACTIC

25.1 Be sure to remember that γ is the *inverse* of the square root

$$\sqrt{1 - \frac{v^2}{c^2}}$$

The thing to keep in mind about γ is that it is *always* greater than or equal to 1:

$$\gamma \geq 1$$

When calculating γ with your calculator, do not forget to take the inverse of the square root.

---

*We will see that for a particle initially with speed $v < c$, it takes an *infinite* amount of work to increase its speed to $v = c$. Since such an amount of energy transfer simply is unavailable, any particle with a speed less than c will remain at speeds less than c.

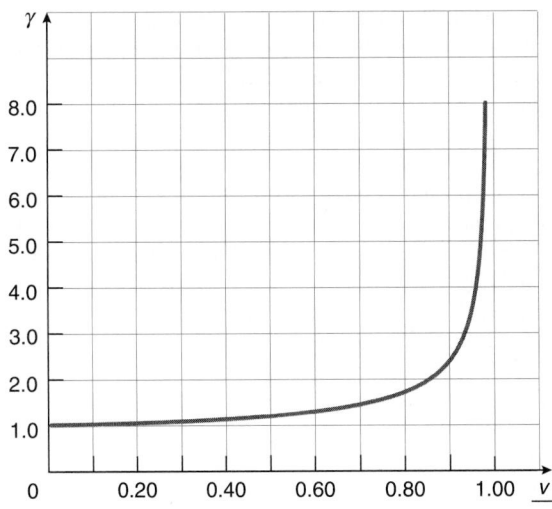

FIGURE 25.14 A graph of γ versus v/c.

QUESTION 3

Moving clocks run slow. Does time dilation depend on the velocity of the clock or simply on its speed?

EXAMPLE 25.1

Clocks in a moving S' reference frame indicate an interval between noon (12 h 0 min 0 s) and 12 h 1 min 0 s. Clocks in reference frame S indicate the same interval as noon till 12 h 2 min 0 s. How fast is frame S' moving with respect to frame S?

Solution

Use Equation 25.14. The one minute on the S' moving clocks is measured to be two minutes according to clocks in S. The proper time τ_0 of the moving clocks is one minute, since this one minute interval is indicated on their dials. In S, clocks indicate the interval τ of the moving clock to be two minutes. Hence Equation 25.14 becomes

$$2.00 \text{ min} = \gamma(1.00 \text{ min})$$

Thus you have

$$\gamma = 2.00$$

Use the definition of γ, Equation 25.13, to find v/c:

$$\gamma = \frac{1}{\sqrt{1 - \dfrac{v^2}{c^2}}}$$

$$2.00 = \frac{1}{\sqrt{1 - \dfrac{v^2}{c^2}}}$$

After squaring, solve for v/c:

$$4.00 = \frac{1}{1 - \dfrac{v^2}{c^2}}$$

$$\frac{v^2}{c^2} = 0.750$$

$$v = 0.866c$$

The S' clocks are moving at 86.6% the speed of light, or 2.60×10^8 m/s.

STRATEGIC EXAMPLE 25.2

A day on the Earth has 24.000 hours. How fast must a spacecraft travel so that the spacecraft clocks tick through 23.000 hours (as indicated by the clock hands on the spacecraft clocks) while the Earth clocks tick through 24.000 hours from the viewpoint of an observer at rest on the Earth?

Solution

Identify the Earth with reference frame S and the spacecraft with reference frame S', as in Figure 25.15. Since the spacecraft (frame S')

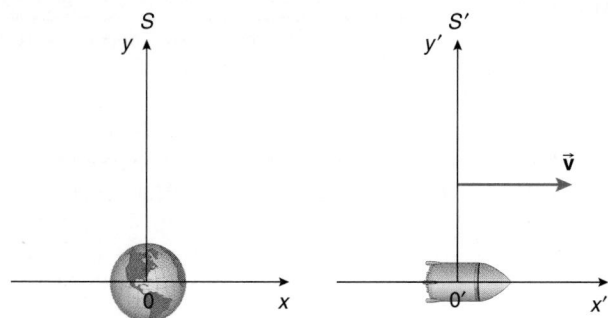

FIGURE 25.15 The standard geometry.

is moving relative to the Earth (frame S), the spacecraft clocks run slow compared with the clocks on the Earth. You can find the value of γ in two ways.

Method 1
The clocks on the spacecraft tick through 23 hours while the Earth clocks tick through 24 hours. Thus the spacecraft hours are dilated or are longer than the Earth hours by the factor

$$\frac{24.000 \text{ h}}{23.000 \text{ h}} = 1.0435$$

In other words, it takes 1.0435 hours according to the Earth clocks for the spacecraft clocks to tick through one hour. In this way, when the Earth clocks have ticked through 24.000 hours, the spacecraft clocks have ticked through only 23.000 hours.

So, summarizing, the moving spacecraft clocks are slow according to an observer at rest with respect to the Earth clocks. Hence the moving spacecraft clock hours are dilated by the factor 1.0435 compared with the Earth clock hours. The dilated time interval τ is related to the proper time interval τ_0 by the time dilation equation (Equation 25.14):

$$\tau = \gamma\tau_0$$

Since $\gamma > 1$, you know $\tau > \tau_0$. Thus you can write

$$1.0435 \text{ h} = \gamma(1.0000 \text{ h})$$

and so

$$\gamma = 1.0435$$

Method 2
Identify the 23.000 h on the spacecraft clock as the proper time interval τ_0, since this clock is at rest at a fixed location on the spacecraft. The Earth observer measures a dilated time interval τ of 24.000 h. The time dilation equation (Equation 25.14) becomes

$$24.000 \text{ h} = \gamma(23.000 \text{ h})$$

and so

$$\gamma = 1.0435$$

Having found γ, you can solve for the speed since

$$\gamma = \frac{1}{\sqrt{1 - \dfrac{v^2}{c^2}}}$$

$$1.0435 = \frac{1}{\sqrt{1 - \dfrac{v^2}{c^2}}}$$

Solving for v/c, you find

$$\frac{v}{c} = 0.2857$$

Hence the speed is

$$v = 0.2857c$$
$$= 0.2857 \times 3.00 \times 10^8 \text{ m/s}$$
$$= 0.857 \times 10^8 \text{ m/s}$$

25.5 LENGTHS PERPENDICULAR TO THE DIRECTION OF MOTION

Wait a minute. There is a residual question that needs to be addressed in connection with our derivation of the relativity of time. If something as basic as time is affected by relative motion, who is to say that lengths are not affected as well? In particular, in analyzing the light clock in the previous section, a tacit assumption was made: we assumed the length ℓ_0 of the clock was unaffected by the motion. That is, the length of the moving light clock attached to frame S′ was assumed (implicitly) to be the same as for the clock at rest in frame S.

If the length of the moving clock is affected by the motion in some way, that could easily account for the difference in the tick rates of the two clocks. In particular, if the length of the moving clock was longer as measured in frame S than its length at rest (in frame S′), then the time interval between emission and detection of the light would, of course, be longer as measured in S. So we have to be sure nothing funny is going on associated with a moving length ℓ_0 oriented perpendicular to the direction of motion. We now show that lengths measured *perpendicular* to the direction of the motion are unaffected by the motion.*

First, we have to clarify what we mean by the term "moving length." Take two identical sticks of length ℓ_0, cut from the same stock at the same time. Glue one of the sticks to the y′-axis in reference frame S′ and the other to the y-axis in reference frame S, as in Figure 25.16.

Now let the S′ reference frame go past the S frame at speed v. An observer in reference frame S compares the length of the stick nailed down to the y-axis in the S frame with the length of the stick nailed down to the y′-axis in the S′ frame. The comparison

---
*In Section 25.6 we shall have something significant to say about lengths *parallel* to (along) the direction of motion.

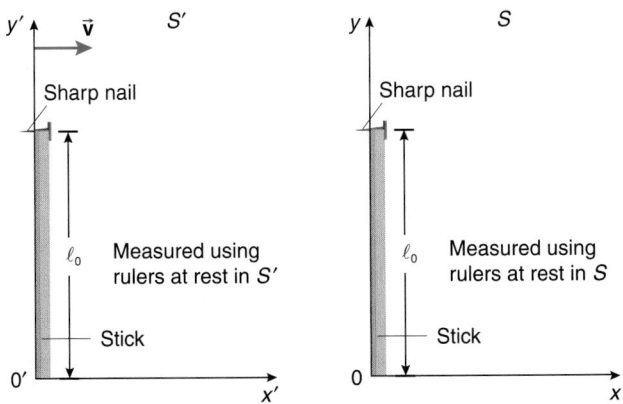

FIGURE 25.16 Two sticks of equal length, each measured in their respective reference frames. Note that S' is here on the left, the origins have not yet coincided, so both t and t' are < 0 s, but we are not concerned with time here, only length.

of the lengths can be made in the following way. A sharp nail is attached to the top of each stick. When the two sticks go by each other, if they are not the same length, the nail on the shorter stick will make a scratch mark on the longer stick.

We prove the sticks are the same length by contradiction. That is, we make the hypothesis that such a length *is* affected by the motion and then show that a contradiction develops, so that the hypothesis must be false.

We make the hypothesis that a moving stick, oriented perpendicular to the direction of motion, is *longer* than an identically placed stick at rest.

An observer in reference frame S then makes the following observations (see Figure 25.17):

1. The stick glued to the y'-axis in S', perpendicular to the direction of motion of the S' frame, is moving.
2. By hypothesis, the moving stick glued to the y'-axis in S' is longer (as measured in S) than the stick at rest in S; this is shown in Figure 25.17.
3. When the moving stick passes by the stick at rest in S, the nail on the stick in S will make a scratch on the stick in S'.
4. The nail on the stick in S' will *miss* the stick in S.

The observer in S therefore concludes

a. the stick in S' has a scratch on it; and
b. the stick in S has no scratch on it.

Now we switch reference frames. An observer stationed in S' sees the S frame moving to the left, as shown in Figure 25.18. The observer in S' makes the following observations:

1. The stick in S, perpendicular to the direction of motion of the S frame, is moving.
2. By hypothesis, the moving length glued to the y-axis in reference frame S is measured in S' to be longer than the stick at rest in S'.
3. When the moving stick passes by the stick in S', the nail on the stick in S' will make a scratch on the stick in the S frame.
4. The nail on the stick in the S frame misses the stick in the S' frame.

Therefore the observer in the S' frame concludes

a. the stick in S has a scratch on it; and
b. the stick in S' has no scratch on it.

The results in each reference frame can be communicated to the other reference frame at their leisure (by radio signals or e-mail), so that a comparison of their conclusions can be made. But notice that the observers in S and S' come to *contradictory* conclusions about which stick has the scratch on it. Thus the hypothesis must be false.

Contradictory conclusions also ensue if we assume that a moving length oriented perpendicular to the direction of its motion is *shorter* than the length at rest; you might want to check the reasoning for this scenario yourself.

Thus the only legitimate conclusion we can draw is the following.

> Lengths measured perpendicular to the direction of motion are *unaffected* by the motion.

The two nails on the tops of the sticks pass by each other at the same height. Thus our assumption about the length of the light clock was legitimate; the length is not affected by the motion

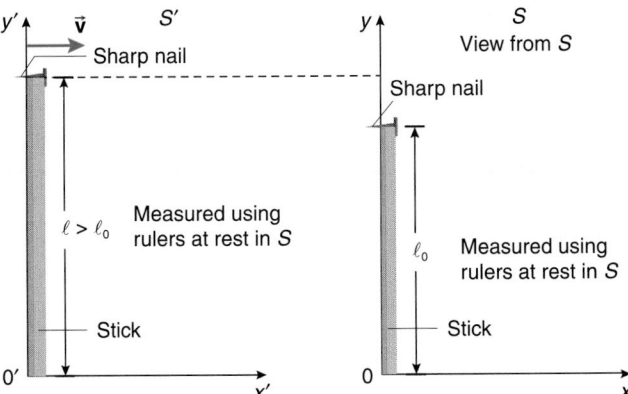

FIGURE 25.17 The view from S: frame S' is moving. Hypothesis: Frame S measures the moving stick (in S') to be longer than the stick at rest in S.

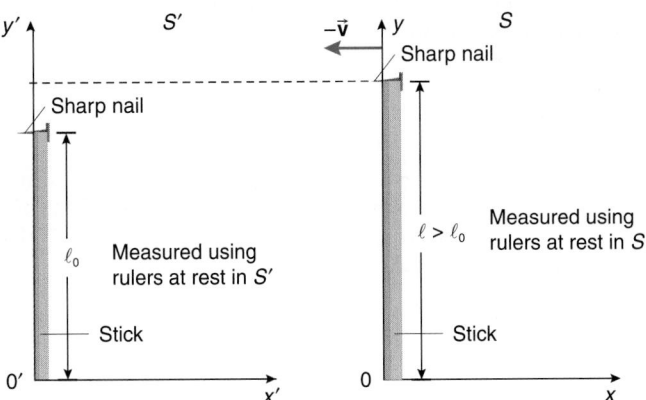

FIGURE 25.18 The view from S': frame S is moving. Hypothesis: Frame S' measures the moving stick (in S) to be longer than the stick at rest in S'.

because the length of the light clocks is oriented perpendicular to the direction of the motion!

25.6 LENGTHS ORIENTED ALONG THE DIRECTION OF MOTION: LENGTH CONTRACTION

In the previous section, we proved that lengths perpendicular to the direction of motion are *unaffected* by the motion. Here we find—surprisingly—that if oriented along the direction of motion, a length *is* affected by the motion!

To show this, we imagine a particle (consider it to be our fairy godmother) that "lives" for only a fixed period of time τ_0 according to her own watch, a clock at rest with respect to her.* The time interval τ_0 thus is a proper time interval or rest time interval.

The fairy godmother particle travels at high speed v across our laboratory during its lifetime and we note that it traverses a distance ℓ_0 (say, from one end of the lab to the other) during its brief existence. The distance ℓ_0 is a **rest length** or **proper length** since the lab is at rest with respect to us when we measure its dimensions with rulers also at rest with respect to us in the lab.

Our clocks in the lab measure the lifetime of the particle to be the dilated time $\tau = \gamma\tau_0$, according to the clocks on the wall of the laboratory, because the particle lifetime (its watch) is a moving clock. We calculate the speed v of the particle by dividing the measured distance ℓ_0 it travels by the time $\tau = \gamma\tau_0$ it exists according to our clocks:

$$v = \frac{\ell_0}{\tau}$$

$$= \frac{\ell_0}{\gamma\tau_0} \tag{25.15}$$

Notice that to determine v, we used a distance and a time measured using rulers and clocks at rest in our reference frame.

Now take the view of our fairy godmother particle. She sees the length of our lab moving past her at the same relative speed v. Using rulers at rest with respect to her, she measures the length of our lab to be ℓ, whose value is to be determined. According to our fairy godmother (using her watch), the lab passes by during her lifetime τ_0. Therefore she determines the speed of the lab by dividing its measured length ℓ by the time τ_0 it takes to pass by her:

$$v = \frac{\ell}{\tau_0} \tag{25.16}$$

Notice that to determine v, she used a distance and a time measured using rulers and clocks at rest in her reference frame.

The relative speeds in Equations 25.15 and 25.16 are the same. Equating them yields

$$\frac{\ell}{\tau_0} = \frac{\ell_0}{\gamma\tau_0}$$

The moving length is

$$\ell = \frac{\ell_0}{\gamma} \tag{25.17}$$

The proper length ℓ_0 is *shorter* when measured in a frame in which it is moving. This is called relativistic **length contraction**.

Length contraction only occurs for those lengths (or components of lengths) oriented along the direction of motion.

Lengths (or components of lengths) perpendicular to the direction of motion are unaffected by it, as we showed in Section 25.5. We measure the contracted moving length using rulers at rest with respect to ourselves.

EXAMPLE 25.3

A javelin with a proper (rest) length of 2.00 m is moving past you at the incredible speed of 0.95c. What do you measure to be the length of the javelin, using rulers at rest with respect to you?

Solution
Identify the frame in which the javelin is at rest as S', where it has the proper length $\ell_0 = 2.00$ m. Your reference frame you identify as S, in which the javelin is moving at speed 0.95c.

Determine the value of the factor γ from its definition, Equation 25.13:

$$\gamma = \frac{1}{\sqrt{1 - \dfrac{v^2}{c^2}}}$$

Since $v/c = 0.95$, this becomes

$$\gamma = \frac{1}{\sqrt{1 - (0.95)^2}}$$

$$= 3.2$$

The javelin is moving in your reference frame and so appears contracted by the factor $1/\gamma$. You measure its length ℓ (using rulers at rest with respect to you) to be

$$\ell = \frac{\ell_0}{\gamma}$$

$$= \frac{2.00 \text{ m}}{3.2}$$

$$= 0.63 \text{ m}$$

25.7 THE LORENTZ TRANSFORMATION EQUATIONS

The old mechanics is valid for small velocities and forms the limiting case of the new one.

Albert Einstein and Leopold Infeld[†]

In Section 25.3 we saw that the Galilean transformation equations relating the space and time coordinates of two inertial reference

---

*Rather than unobservable fairy godmothers, physicists use radioactive particles with a well-defined half-life for this purpose; we discuss radioactivity in Chapter 26.

[†]*The Evolution of Physics* (Simon & Schuster, New York, 1938), page 204.

frames needed to be replaced with another transformation to account for the peculiar constancy of the speed of light, an invariance that violates the Galilean velocity component addition rule. However, since Galilean relativity is quite valid for speeds $v \ll c$, the new transformation equations must approach the Galilean equations in the limit of small speeds. The new transformation equations are known as the Lorentz transformation equations. In this section we develop their form.

The relative motion of the two reference frames is along the x- and x'-coordinate axes in the standard geometry. The y- and y'-axes, as well as the z- and z'-axes all are perpendicular to the direction of motion. Since we showed in Section 25.5 that lengths measured perpendicular to the direction of motion are unaffected by it, the Lorentz transformation must have

$$y' = y \qquad (25.18)$$

and

$$z' = z \qquad (25.19)$$

It remains for us to determine how the spatial coordinates x and x' are related, as well as the time instants t and t'.

The Galilean transformation gives us a clue about these new relationships. Notice in Equation 25.3 that the relationship between x' and x also involves t. Thus it is tempting to write

$$x' = Ax + Bt \qquad (25.20)$$

where A and B are constants (which may depend on the constant relative speed v) whose values we wish to find.

From the viewpoint of reference frame S, the origin of the S' reference frame is located using the equation

$$x = vt$$

because the origins coincide when $t = 0$ s in the standard geometry, as indicated in Figure 25.19. We use this fact in Equation 25.20 in the following way. The origin in S' has coordinate $x' = 0$ m, in which case x has the value vt. Making these substitutions in Equation 25.20, we obtain

$$0 \text{ m} = Avt + Bt$$
$$= t[Av + B] \qquad (25.21)$$

The only way Equation 25.21 can be true for *any* and *every* instant t is if

$$Av + B = 0 \text{ m/s}$$

or

$$B = -Av$$

Thus Equation 25.20 becomes

$$x' = Ax + (-Av)t$$
$$= A[x - vt] \qquad (25.22)$$

For values of v that are small compared with c, the value of A must approach 1, and so the transformation equations reduce to the Galilean transformation, Equation 25.3. Likewise, the inverse equation for x in terms of x' and t' must have the form

$$x = A[x' + vt'] \qquad (25.23)$$

which then also reduces to the Galilean result (Equation 25.2) for small v as A approaches 1.

Now let us use our newfound knowledge about length contraction to determine the value of A. Consider a proper (rest) length ℓ_0 in reference frame S' as shown in Figure 25.20. The x'-coordinates of the ends of the proper length are x_1' and x_2' in S', as indicated in Figure 25.20.

We write Equation 25.22 for each x'-coordinate:

$$x_1' = A[x_1 - vt_1] \qquad (25.24)$$
$$x_2' = A[x_2 - vt_2] \qquad (25.25)$$

Now we subtract Equation 25.24 from 25.25:

$$x_2' - x_1' = A[x_2 - x_1] - Av[t_2 - t_1] \qquad (25.26)$$

But from Figure 25.20, the left-hand side of Equation 25.26 is the proper length ℓ_0. Hence

$$\ell_0 = A[x_2 - x_1] - Av[t_2 - t_1] \qquad (25.27)$$

The length ℓ_0 in S' is moving in frame S. To measure the moving length in S (with rulers at rest in S, of course), it is necessary to determine where the ends of the moving length are *at the same time in S*. It does no good to determine where one end of the moving length is at one instant and where the other end is at another instant; to measure the length of a moving car, we must determine where its bumpers are at the same time. Hence in S, to measure the moving length, we *must* make the measurements simultaneously—that is, when $t_1 = t_2$. Call ℓ the length measured in S (using rulers at rest in S). Then Equation 25.27 becomes, with $x_2 - x_1 = \ell$ and $t_1 = t_2$,

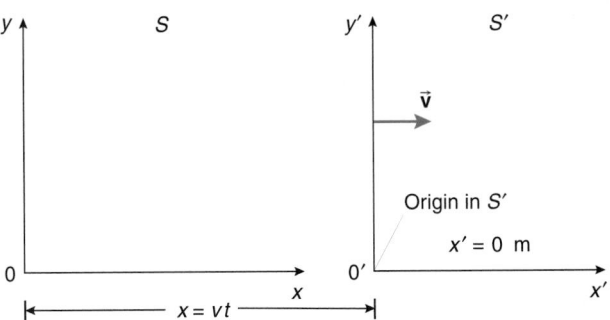

FIGURE 25.19 The view in S of the origin of S'.

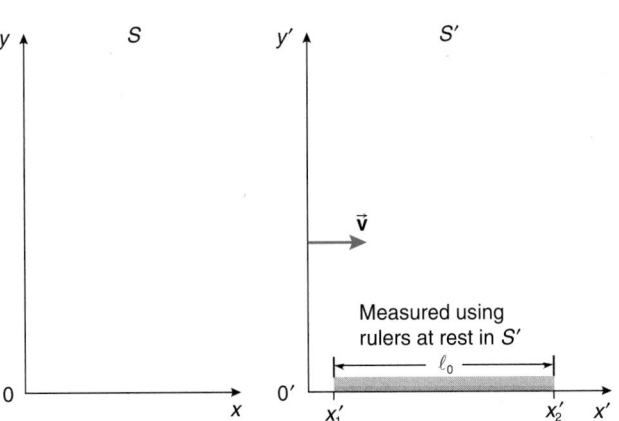

FIGURE 25.20 A proper length in frame S'.

$$\ell_0 = A\ell \qquad (25.28)$$

But we already know that the proper (rest) and moving lengths are related by Equation 25.17:

$$\ell = \frac{\ell_0}{\gamma}$$

Comparing Equations 25.28 and 25.17, we find

$$A = \gamma$$

Thus the Lorentz transformation equations (Equations 25.22 and 25.23) become

$$x' = \gamma(x - vt) \qquad (25.29)$$
$$x = \gamma(x' + vt') \qquad (25.30)$$

To find the relationship between t' and t, we take Equations 25.29 and 25.30 and eliminate x' between them. After some tedious (and somewhat tricky) algebra that should keep you out of trouble for awhile, we eventually find

$$t' = \gamma\left(t - \frac{v}{c^2}x\right) \qquad (25.31)$$

Now we solve Equation 25.31 for t in terms of x' and t', after substituting Equation 25.30 for x. With proper algebraic footwork, we find

$$t = \gamma\left(t' + \frac{v}{c^2}x'\right) \qquad (25.32)$$

We summarize these results.

The set of Lorentz transformation equations is

$$x' = \gamma(x - vt) \qquad (25.33)$$
$$y' = y \qquad (25.34)$$
$$z' = z \qquad (25.35)$$
$$t' = \gamma\left(t - \frac{v}{c^2}x\right) \qquad (25.36)$$

PROBLEM-SOLVING TACTIC

25.2 Equations 25.33–25.36 are useful when the space and time coordinates of an event are known in the reference frame S and you need to know the corresponding coordinates of the same event in S'.

The inverse equations also are useful. Solving Equations 25.33–25.36 for x, y, z, and t in terms of x', y', z', and t', we obtain

$$x = \gamma(x' + vt') \qquad (25.37)$$
$$y = y' \qquad (25.38)$$
$$z = z' \qquad (25.39)$$
$$t = \gamma\left(t' + \frac{v}{c^2}x'\right) \qquad (25.40)$$

PROBLEM-SOLVING TACTIC

25.3 Equations 25.37–25.40 are useful if the space and time coordinates of an event are known in the reference frame S' and the corresponding coordinates of the event in S are needed.

In Equations 25.33–25.36 and Equations 25.37–25.40, the relative speed v is positive. The transformation equations for going from S to S' (Equations 25.33–25.36) and those from S' to S (Equations 25.37–25.40) differ only by the sign of terms involving v, other than switching primes for unprimes. This is because in one case S' is moving at velocity $v\hat{\imath}$ with respect to S, while in the other case S is moving with velocity $-v\hat{\imath}'$ with respect to S'. The only difference is the sign of the velocity component.

For reassurance, notice that when $v \ll c$,

$$\gamma \to 1$$

$$\frac{v}{c^2} \to 0 \text{ s/m}$$

So, for the familiar limit of low speeds, the Lorentz transformation Equations 25.33–25.40 reduce to the familiar Galilean transformation equations (and their inverses):

$$x' = x - vt$$
$$y' = y$$
$$z' = z$$
$$t' = t$$

The Lorentz transformation equations easily give us time dilation and length contraction, which they must since we used these effects to find the equations. For example, we revisit the peculiar relationship between stationary and moving clocks, first addressed in Section 25.4 with a pair of light clocks. We want to show that the results obtained using the light clocks do not depend on our using that special type of clock.

Imagine two events that take place at instants t_1' and t_2' *at the same place* x_0' in reference frame S'. The events might correspond to, say, two successive ticks on any clock in S'; or the events might correspond to the creation of a radioactive particle at rest at instant t_1' and its subsequent decay at t_2', where we identify frame S' as the one with the particle at rest at coordinate x_0'. The time interval τ_0 between the two events in S' is $\tau_0 = t_2' - t_1'$; this is the rest time interval or proper time interval because the two events happen at the same place in S' (the particle is at rest in this frame). We want to calculate the temporal interval between the two events as measured by the clocks in reference frame S.

According to the Lorentz transformation equations (Equation 25.40, in particular), the time instants the events occur in the reference frame S are

$$t_1 = \gamma\left(t_1' + \frac{v}{c^2}x_0'\right)$$

and

$$t_2 = \gamma\left(t_2' + \frac{v}{c^2}x_0'\right)$$

We call τ the temporal separation of the two events as measured by clocks at rest in S; the two events occur at different locations in S because S' is moving, but we are interested here only in the time interval between the events as determined from the clocks distributed throughout the S reference frame. The time interval of interest is

$$\tau = t_2 - t_1$$

$$= \gamma(t_2' - t_1') + \frac{v}{c^2}(x_0' - x_0')$$

$$= \gamma(t_2' - t_1')$$

But $t_2' - t_1' = \tau_0$, the proper time interval between the two events. Hence we have the same relationship (Equation 25.14) between the time intervals that we first derived with the special light clocks of Section 25.4:

$$\tau = \gamma\tau_0$$

Notice that this argument makes no mention of special clocks, and so it applies to *all* clocks. The tick rate of a moving clock is slower than a clock at rest with respect to you.

EXAMPLE 25.4 ─────────────

A firecracker goes off in reference frame S at a point with co-ordinates $x = 20.0$ m, $y = 5.0$ m, and $z = 0.0$ m when clocks (stopwatches) distributed throughout that frame indicate 0 s. Captain Kirk in reference frame S' is traveling by at a speed $v = 0.900c$. What are the space and time coordinates of the event according to rulers and clocks at rest with respect to Captain Kirk?

Solution

First find the factor γ, using its definition, Equation 25.13:

$$\gamma = \frac{1}{\sqrt{1 - \dfrac{v^2}{c^2}}}$$

$$= \frac{1}{\sqrt{1 - (0.900)^2}}$$

$$= 2.29$$

Since the space and time coordinates of the event are known in S and you want those in S', use Problem-Solving Tactic 25.2 and Equations 25.33–25.36. The speed v is $0.900c$ or 2.70×10^8 m/s:

$$x' = \gamma(x - vt)$$

$$= 2.29[20.0 \text{ m} - (2.70 \times 10^8 \text{ m/s})(0 \text{ s})]$$

$$= 45.8 \text{ m}$$

$$y' = y$$

$$= 5.0 \text{ m}$$

$$z' = z$$

$$= 0.0 \text{ m}$$

$$t' = \gamma\left(t - \frac{v}{c^2}x\right)$$

$$= 2.29\left[0 \text{ s} - \frac{(2.70 \times 10^8 \text{ m/s})(20.0 \text{ m})}{(3.00 \times 10^8 \text{ m/s})^2}\right]$$

$$= -1.37 \times 10^{-7} \text{ s}$$

Notice that in reference frame S', the event occurs before $t' = 0$ s on the clocks in S'. The event occurs at the point with coordinates $x' = 45.8$ m, $y' = 5.0$ m, and $z' = 0.0$ m according to rulers at rest in S'.

25.8 THE RELATIVITY OF SIMULTANEITY

The importance of doing things simultaneously is quite apparent when playing in an orchestra. Beethoven's famous Ninth Symphony would be mere cacophony if each musician began to play when they felt like it. Every musician must begin when the conductor commands.

Imagine two musicians in the Cleveland Symphony separated by a distance ℓ_0 along the x-axis in frame S as shown in Figure 25.21. The two musicians begin to play simultaneously and harmoniously when $t = 0$ s, in their frame and that of the conductor. We define two events and write their space and time coordinates in S:

Event 1	Event 2
Musician 1 begins to play	Musician 2 begins to play
$x_1 = 0$ m	$x_2 = \ell_0$
$t_1 = 0$ s	$t_2 = 0$ s

You are rushing past the performance at high speed v carrying reference frame S' along with you; you are the S' reference frame. We use the Lorentz transformation, Equations 25.33 and 25.36, to find the x'- and t'-coordinates of the two events in S'. We find

Event 1	Event 2
Musician 1 begins to play	Musician 2 begins to play
$x_1' = 0$ m	$x_2' = \gamma\ell_0$
$t_1' = 0$ s	$t_2' = -\gamma v\ell_0/c^2$

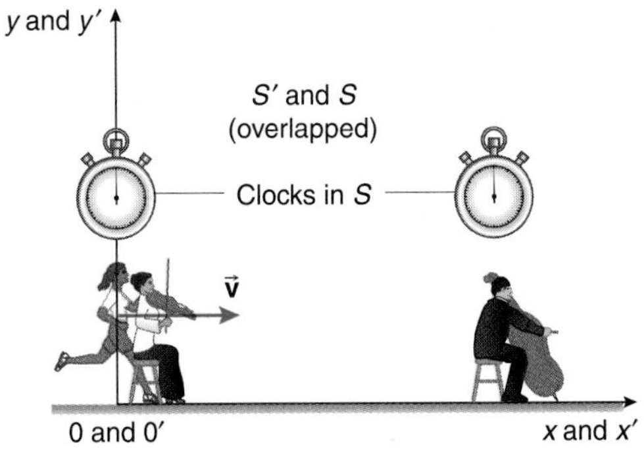

FIGURE 25.21 Two musicians in S, separated by distance ℓ_0 along the x-axis. You are the S' frame and rush by at speed v. Situation shown when $t = 0$ s $= t'$.

Note that the two events are *not simultaneous* in S', unless $\ell_0 = 0$ m, in which case the two events really are the same musician—that is, the same event! This is a characteristic and surprising feature of special relativity: the notion of simultaneity is relative.

> If two events are simultaneous in one inertial reference frame, they will *not* be simultaneous to any other inertial frame moving at nonzero speed v with respect to the first.

25.9 A RELATIVISTIC CENTIPEDE

Here we examine a hypothetical problem that ties together much of the kinematics of relativity. The problem is pedagogically useful and playful at the same time. It involves a high-speed centipede (*Scolopendra cingulata*) and a compassionate butcher (*Homo sapiens*).* The centipede is 10.0 cm long, measured with rulers at rest with respect to the centipede; thus, the proper length of the centipede is 10.0 cm. A butcher holds two meat cleavers 9.0 cm apart. The centipede runs at such a high speed v across a chopping block that the butcher measures the (contracted) length of the centipede to be 8.0 cm, using rulers inlaid on the chopping block. According to length contraction, Equation 25.17, we have

$$\ell = \frac{\ell_0}{\gamma}$$

$$8.0 \text{ cm} = \frac{10.0 \text{ cm}}{\gamma}$$

Note that we can use units of convenience here. Solving for γ, we obtain

$$\gamma = \frac{10.0 \text{ cm}}{8.0 \text{ cm}} = 1.25$$

The definition of γ is Equation 25.13. We use this to solve for v/c:

$$\gamma = \frac{1}{\sqrt{1 - \dfrac{v^2}{c^2}}} = 1.25$$

We find

$$\frac{v}{c} = 0.600$$

Talk about fleet feet! Sports considerations aside (let alone coordinating all the feet), the instant the tail of the centipede is at cleaver A (see Figure 25.22), the butcher immediately swings *both* cleavers instantaneously down on the chopping block simultaneously (kerchunk!) and immediately raises them up again. (Obviously, we need a butcher with quick reflexes to avoid having the centipede collide with cleaver B.) The butcher argues that since the centipede is only 8.0 cm long according to the rulers inlaid on the chopping block in the butcher's reference

---

*The author believes the idea for this problem originates with Professor George Ruff, Bates College, Lewiston, Maine 04240.

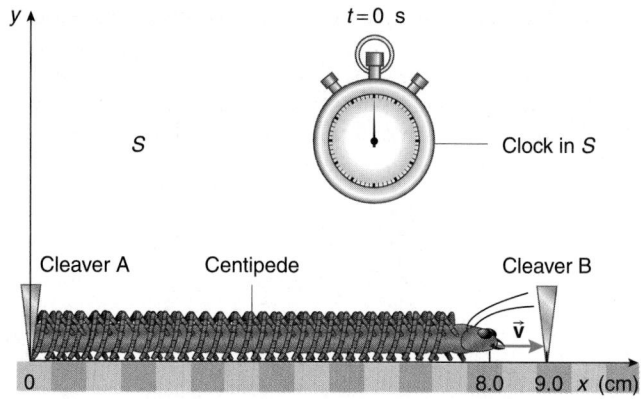

FIGURE 25.22 The view of the butcher, the S frame.

frame, the centipede neatly fits between the cleavers and no bug juice is shed.

On the other hand, the relativistically literate centipede is quite worried about the situation. From the viewpoint of the centipede, the meat cleavers and butcher are approaching at a high speed v. The centipede sees the separation of the cleavers, not as the 9.0 cm according to the butcher, but as a contracted length

$$\ell = \frac{\ell_0}{\gamma}$$

$$= \frac{9.0 \text{ cm}}{1.25}$$

$$= 7.2 \text{ cm}$$

The separation of the cleavers is only 7.2 cm according to the centipede; but the centipede knows she is 10.0 cm long! Uh oh.

Either there *is* bug juice on the chopping block or there is *none*. The butcher says no juice was shed; so how does the centipede manage to wiggle out of this relativistic dual guillotine? We will see in what follows.

We call the frame of the butcher S and that of the centipede S'. We define two events corresponding to the cleavers hitting the chopping block. These events are most easily described in the reference frame S of the butcher. Let the x-coordinates of the cleavers be $x_A = 0$ cm $= 0$ m and $x_B = 9.0$ cm $= 0.090$ m, as in Figure 25.22.

Let the cleavers hit the chopping block when $t = 0$ s according to the butcher. Here we use SI units since we intend to use the Lorentz transformation equations. Then the space–time coordinates of the two events in S are as follows (the y and z coordinates are not relevant so we do not specify them; we can set them both equal to zero):

Event 1. Cleaver A hits the block:

$$x_1 = 0 \text{ m}$$
$$t_1 = 0 \text{ s}$$

Event 2. Cleaver B hits the block:

$$x_2 = 0.090 \text{ m}$$
$$t_2 = 0 \text{ s}$$

The events are *simultaneous* in the reference frame S of the butcher, so that $t_1 = t_2$. The events occur, however, at two different locations in S.*

Now we transform the events into the reference frame S' of the centipede using the Lorentz transformation, Equations 25.33 and 25.36; remember that γ has the value 1.25.

Event 1. Cleaver A hits the block:

$$x_1' = 1.25[0 \text{ m} - v(0 \text{ s})]$$
$$= 0 \text{ m}$$
$$t_1' = 1.25\left[0 \text{ s} - \frac{v}{c^2}(0 \text{ m})\right]$$
$$= 0 \text{ s}$$

Event 2. Cleaver B hits the block:

$$x_2' = 1.25[0.090 \text{ m} - v(0 \text{ s})]$$
$$= 0.11 \text{ m} = 11 \text{ cm}$$

---

*If two events occur at the same time *and* place, then the two events are identical as far as relativity is concerned.

$$t_2' = 1.25\left[0 \text{ s} - \frac{v}{c^2}(0.090 \text{ m})\right]$$

$$t_2' = -2.3 \times 10^{-10} \text{ s}$$

The two events are *not simultaneous* to the centipede, since $t_2' \neq t_1'$. Event 2 (cleaver B hits the block) occurs before Event 1 (cleaver A hits the block). Thus the fact that the separation of the cleavers is only 7.2 cm according to the 10.0 cm centipede is no problem, since the cleavers do not descend simultaneously in her frame. Since the centipede is stretched out in S', the event that occurs first according to the centipede (Event 2) takes place away from her head (her head is located at $x' = 10.0$ cm when cleaver B hits at $x_2' = 11$ cm), as in Figure 25.23.

The subsequent event (Event 1) occurs at the end of her tail (located at $x_2' = 0$ cm). The centipede escapes, as in Figure 25.24.

Note that the separation of the meat cleavers according to the centipede (7.2 cm) is *not* $x_2' - x_1' = 11$ cm, because the events associated with the coordinates x_2' and x_1' *do not happen at the same time in S'.* To measure the separation of the (*moving*) cleavers in S', it is necessary for the centipede to determine the locations of the two cleavers *at the same time in her own reference frame.*

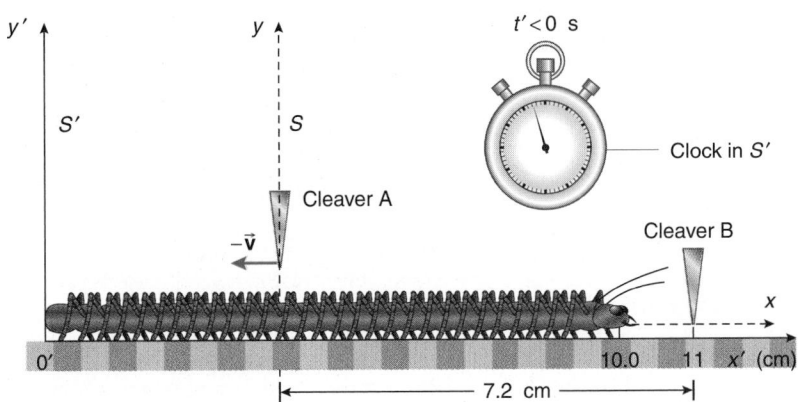

FIGURE 25.23 Event 2 occurs first according to the centipede. All coordinates and lengths are as measured by the centipede in S'.

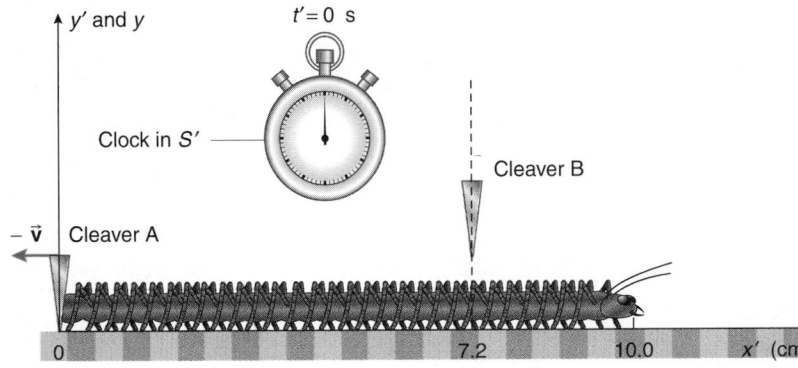

FIGURE 25.24 Event 1 occurs later according to the centipede. Coordinates are according to the centipede in S'.

25.10 A RELATIVISTIC PARADOX AND ITS RESOLUTION*

Speeding through space, speeding through heaven and the stars
Walt Whitman (1819–1892)[†]

Time dilation means that clocks in one inertial reference frame that is moving with respect to another inertial reference frame run more slowly than the clocks at rest in the latter frame. Length contraction means that moving lengths oriented along the direction of motion are shorter than identical lengths at rest. Time dilation and length contraction apply *whenever* clocks or lengths (oriented along the motion) are *moving* with respect to another inertial reference frame.

Now here's the rub.

1. Imagine yourself in reference frame S in the standard geometry, shown in Figure 25.25. Reference frame S' is moving to the right at speed v with respect to you in S. Thus, according to time dilation, you say the clocks in S' tick slow compared with your clocks in S. Likewise, if you compare the length of a meter stick nailed down to the x'-axis in S' with a meter stick nailed down to your x-axis in S, then length contraction means the moving meter stick (in S') is shorter than your own meter stick at rest in S.

2. But now imagine a friend in the S' reference frame. From their viewpoint, you and your reference frame S are moving to the left at the speed v, as in Figure 25.26. Thus, according to time dilation, your friend in S' says *your* clocks in S tick slow compared with the clocks at rest in S'. Likewise, if your friend in S' compares the length of your meter stick nailed down to the x-axis in S with a meter stick nailed down to the x'-axis in S',

---

[†]*Leaves of Grass*, Song of Myself, #33, line 790 (David McKay, Philadelphia, 1900), page 67.

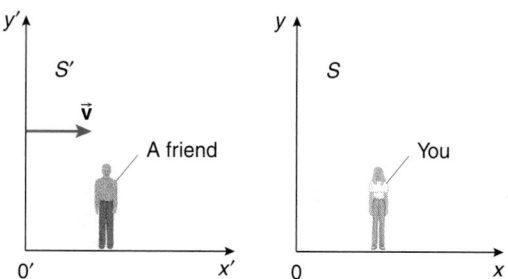

FIGURE 25.25 The standard geometry; the view in the S frame (you).

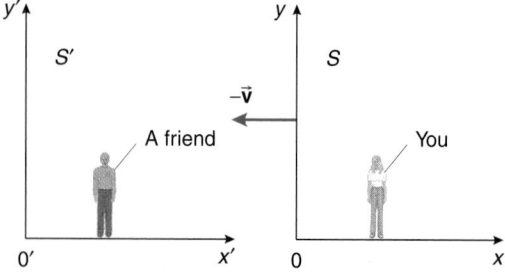

FIGURE 25.26 The standard geometry; the view in the S' frame of a friend.

then length contraction to him means *your* moving meter stick (in S) is shorter than the meter stick at rest in S'.

Each reference frame says

- the clocks in the *other* reference frame run slow compared with the clocks at rest; and

- the lengths in the *other* frame moving along the direction of motion are shorter.

Ridiculous, you might think! How can each of you say the *other* clocks are slow and the *other* lengths are shorter? Yet this apparent paradox is indeed what happens; both views are correct!

To demonstrate there is no contradiction or paradox, we consider a particular example in some detail.

The example involves an interstellar spaceflight, so that it will be convenient to measure time in years (y) and lengths in light-years (LY). A light-year is the distance light travels during an interval of one year. Let t_0 be the number of seconds in one year:

$$t_0 \approx 3.156 \times 10^7 \text{ s}$$

Then one light-year is a distance ct_0:

$$\begin{aligned} 1 \text{ LY} &= ct_0 \\ &= (3.00 \times 10^8 \text{ m/s})(3.156 \times 10^7 \text{ s}) \\ &= 9.47 \times 10^{15} \text{ m} \end{aligned}$$

To convert a distance x in meters to a distance X in light-years, we divide x by ct_0. To convert a time t in seconds to a time T in years, we divide t by t_0. Hence Equation 25.33 of the Lorentz transformation becomes*

$$\frac{x'}{ct_0} = \frac{\gamma(x - vt)}{ct_0}$$

$$X' = \gamma\left[X - \frac{v}{c}T \text{ (LY/y)}\right] \quad (25.41)$$

$$(X \text{ and } X' \text{ in LY, } T \text{ in y})$$

Likewise, Equation 25.36 of the Lorentz transformation becomes[†]

$$\frac{t'}{t_0} = \frac{\gamma\left(t - \frac{v}{c^2}x\right)}{t_0}$$

$$T' = \gamma\left[T - \frac{v}{c}X \text{ (y/LY)}\right] \quad (25.42)$$

$$(T \text{ and } T' \text{ in y, } X \text{ in LY})$$

---

*We explicitly include the units LY/y with the T term so the units for X' emerge properly in LY.

[†]Likewise, we explicitly include the units y/LY in the X term so the units for T' emerge properly in y.

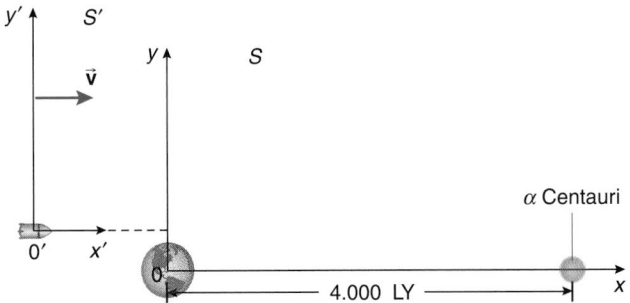

FIGURE 25.27 The standard geometry. We shift the x- and x'-axes slightly here for clarity.

The distance from our solar system to α Centauri (the closest star system to our solar system) is about 4.2 LY, which means light takes 4.2 y to travel this distance, according to clocks at rest on the Earth (or at rest on α Centauri). Just to keep the numbers clean, we consider the distance to be precisely 4.000 LY, so that we can focus on the physics rather than awkward numbers.

Let the Earth and α Centauri define a reference frame S, as in Figure 25.27. The distance between the Earth and α Centauri is 4.000 LY as measured by rulers at rest in the S frame. For convenience, we place the Earth at the origin in S, so that α Centauri is 4.000 LY out along the positive x-axis. Let a spacecraft travel at constant velocity through our region of the galaxy on its way past the Earth and α Centauri. Let S' be a reference frame attached to the spacecraft with it at the S' origin, as in Figure 25.27.

Let the speed of the spacecraft (frame S') be $v = 0.8660c$. After a short calculation, we find that this speed yields $\gamma = 2.000$. We want to resolve the paradox that *each* reference frame says the *other* frame's clocks are slow and the other frame's lengths (oriented parallel to the direction of motion) are contracted.

To compare the results of measurements in each frame, we need to communicate information between the frames. This can be accomplished in the following way: when the spacecraft is flying by the Earth, set both the clocks on the spacecraft and the clocks on the Earth to zero. When the spacecraft reaches α Centauri, a light signal is sent by the spacecraft back to the Earth indicating the time on the spacecraft clock and the results of calculations made in this reference frame, so a comparison can be made with those in the Earth–α Centauri frame S.

We approach the problem by (1) defining several events, (2) writing the space–time coordinates of the events in one of the reference frames (whichever frame in which it is most convenient to do so), and then (3) transforming the space–time coordinates of these events to find the space–time coordinates of the same events in the other reference frame using the customized Lorentz transformation given by Equations 25.41 and 25.42. We then interpret the results.

We define the following events:

Event 1: The spacecraft has the Earth outside its window.
Event 2: The spacecraft has α Centauri outside its window and sends a light signal back to the Earth indicating the time on the spacecraft clocks and the results of its calculations in the S' frame.
Event 3: The Earth receives the light signal sent by the spacecraft when α Centauri was outside its window.

The space–time coordinates of the events are most easily expressed in the reference frame S, the Earth–α Centauri system.

Event 1 is that the spacecraft has the Earth outside its window, shown in Figure 25.28. The space–time coordinates of this event are easy to enumerate:

$$X_1 = 0 \text{ LY}$$
$$T_1 = 0 \text{ y}$$

We use the customized Lorentz transformation, Equations 25.41 and 25.42, to find the space–time coordinates of this event in the S' reference frame (the spacecraft). After substituting for X_1 and T_1, we find

$$X_1' = 0 \text{ LY}$$
$$T_1' = 0 \text{ y}$$

Event 2 is that the spacecraft has α Centauri outside its window and sends a light signal back to Earth, as shown in Figure 25.29. In reference frame S, this event occurs at the location

$$X_2 = 4.000 \text{ LY}$$

The time for the spacecraft to fly to α Centauri, according to Earth clocks, is determined by dividing the 4.000 LY distance that the spacecraft must travel in this frame by the speed of the spacecraft in this frame (0.8660c). Because the distance is expressed in LY, the time involved is (be careful with the units):

$$\frac{\text{distance}}{\text{speed}} = \frac{4.000 \text{ LY}}{0.8660 \text{ LY/y}} = 4.619 \text{ y}$$

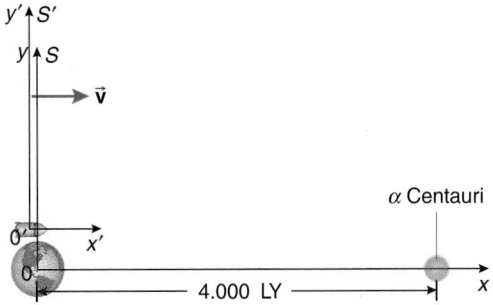

FIGURE 25.28 Event 1 from the viewpoint of the S reference frame

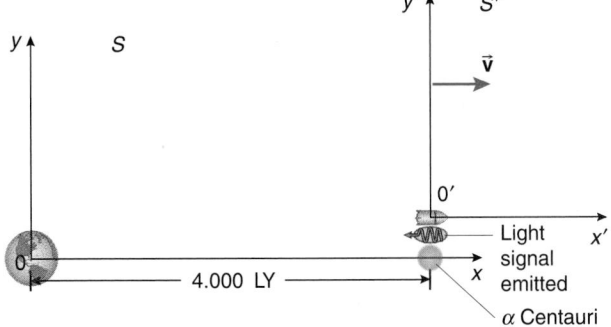

FIGURE 25.29 Event 2 from the viewpoint of the S reference frame.

Thus

$$T_2 = 4.619 \text{ y}$$

We use the customized Lorentz transformation, Equations 25.41 and 25.42, to find the space–time coordinates of this event in the S' frame (the spacecraft):

$$X_2' = 2.000 \left[4.000 \text{ LY} - 0.8660 \ (4.619 \text{ y}) \text{ LY/y} \right]$$
$$= 0 \text{ LY}$$

and

$$T_2' = 2.000 \left[4.619 \text{ y} - 0.8660 \ (4.000 \text{ LY}) \text{ y/LY} \right]$$
$$= 2.310 \text{ y}$$

The result $X_2' = 0$ LY is not surprising since the spacecraft is at the origin in its own S' frame! The result $T_2' = 2.310$ y means that the clocks on the spacecraft indicate that the trip took only 2.310 years. The Earth clocks say the trip took 4.620 y. Notice that the Earth sees the spacecraft moving, and so says the spacecraft clocks tick slower than the Earth clocks; they tick slower by the factor $1/\gamma$, here 1/2.000. According to the Earth, the moving clocks (in the spacecraft and its reference frame) indeed are ticking slower than the clocks on the Earth and in its reference frame.

From the perspective of the spacecraft (S'), the trip is described differently. The spacecraft sees the S frame moving toward it at the speed 0.8660c, as in Figure 25.30. Thus the 4.000 LY proper length measured in S is contracted by the factor $1/\gamma$; the spacecraft measures the moving distance between the Earth and α Centauri (using rulers at rest in S') to be only

$$\ell = \frac{\ell_0}{\gamma} = \frac{4.000 \text{ LY}}{2.000} = 2.000 \text{ LY}$$

When the Earth is outside the spacecraft window, the spacecraft sets its clock to 0 y. The time (according to the spacecraft clocks) for the moving Earth–α Centauri distance (measured to be 2.000 LY in S') to go past the spacecraft thus is*

$$\frac{\text{distance}}{\text{speed}} = \frac{2.000 \text{ LY}}{0.8660 \text{ LY/y}} = 2.309 \text{ y}$$

---

*The slight discrepancy is due to rounding in the calculations.

Hence the S frame (the Earth–α Centauri system) interprets the reading on the spacecraft clock as due to time dilation, because the Earth sees the spacecraft clock as a moving clock; the spacecraft explains the reading on its clock as due to a length contraction of the moving Earth–α Centauri distance.

Now we consider Event 3, in which the light signal from the spacecraft arrives back at the Earth, as in Figure 25.31. The event occurs at the location of the Earth, and so the spatial coordinate in S is

$$X_3 = 0 \text{ LY}$$

According to the Earth clocks, the spacecraft arrived at α Centauri when the time was 4.619 y and then emitted a light signal to travel back to the Earth. The light signal takes 4.000 y to travel the 4.000 LY distance back to the Earth, according to the Earth clocks. Thus the total elapsed time when the light gets back to the Earth (according to the Earth clocks) is

$$4.619 \text{ y} + 4.000 \text{ y} = 8.619 \text{ y}$$

Thus

$$T_3 = 8.619 \text{ y}$$

We use the customized Lorentz transformation, Equations 25.41 and 25.42, to find the space–time coordinates of this event in the S' frame:

$$X_3' = 2.000 \left[0 \text{ LY} - 0.8660 \ (8.619 \text{ y}) \text{ LY/y} \right]$$
$$= -14.93 \text{ LY}$$
$$T_3' = 2.000 \left[8.619 \text{ y} - 0.8660 \ (0 \text{ LY}) \text{ y/LY} \right]$$
$$= 17.24 \text{ y}$$

The light signal sent by the spacecraft returns to the Earth when the Earth is located in S' at coordinate $X_3' = -14.93$ LY and when the spacecraft (S') clock indicates a time of 17.24 y, as shown in Figure 25.32.

Why did it take so long for the light signal to return to the Earth from the viewpoint of the spacecraft? According to it, the light was sent on its way toward an Earth that was receding from the spacecraft at 0.8660c. Thus it took quite some time (17.24 y − 2.310 y = 14.93 y) for the light to travel the 14.93 LY distance to the Earth in the S' frame.

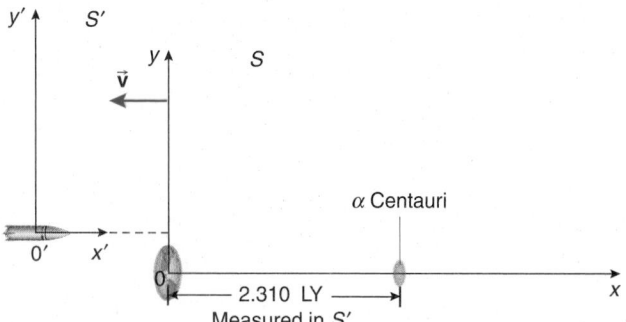

FIGURE 25.30 Perspective of the spacecraft, the S' reference frame.

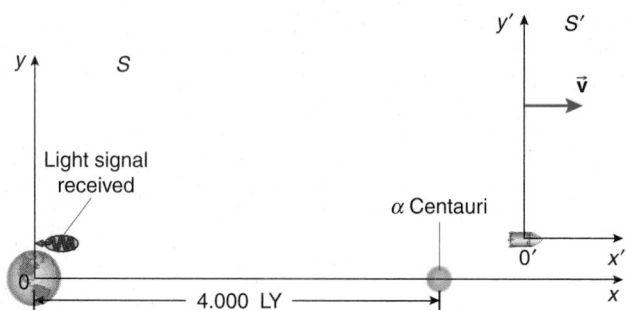

FIGURE 25.31 Event 3: The view from frame S.

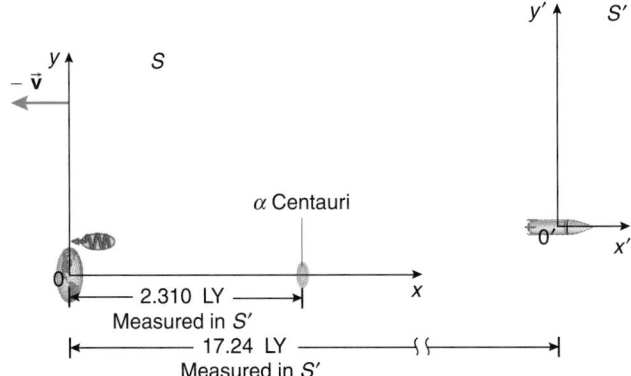

FIGURE 25.32 Event 3: The view from frame S'.

Notice that now the tables are turned. The Earth clock reads 8.619 y when the signal arrives; the spacecraft clock reads 17.24 y. The spacecraft, viewing the moving Earth, says the moving clocks on the Earth tick slower than the spacecraft clocks by the factor $1/\gamma$, here equal to $1/2.000$. The Earth clocks thus read half the time of the spacecraft clocks.*

There is only one residual question to resolve. According to the Earth (S frame), how distant is the spacecraft when the light from the spacecraft returns to the Earth? The signal was sent from the spacecraft when it was 4.000 LY distant (at α Centauri); the light took 4.000 y to reach the Earth (according to the Earth clocks). Meanwhile the spacecraft went on its merry way for an additional 4.000 y at the speed of 0.8660c. During those 4.000 y, the spacecraft thus traveled an additional (0.8660 LY/y)(4.000 y) = 3.464 LY. Thus the distance of the spacecraft in the S frame is 4.000 LY + 3.464 LY = 7.464 LY when the light reaches the Earth. The 14.93 LY proper length between the spacecraft and the Earth as measured in S' is contracted to 14.93 LY/γ = 14.93 LY/2.000 = 7.465 LY[†] according to the Earth, since the Earth sees the 14.93 LY distance as a moving length oriented along the direction of motion.

Thus each frame says that the moving clocks in the other frame run slow compared with clocks at rest. Each frame also says that lengths moving along the direction of motion are contracted. Both are right. There is no paradox.

25.11 RELATIVISTIC VELOCITY ADDITION

The velocity of light forms the upper limit of velocities for all material bodies. . . . The simple mechanical law of adding and subtracting velocities is no longer valid or, more precisely, is only approximately valid for small velocities, but not for those near the velocity of light. The number expressing the velocity of light appears explicitly in the Lorentz transformation, and plays the role of a limiting case, like the infinite velocity in classical mechanics.

Albert Einstein and Leopold Infeld[‡]

---

*There is a slight discrepancy due to rounding in the calculations.
[†]The discrepancy between 7.460 LY and 7.465 LY is due to rounding in the calculations.
[‡]*The Evolution of Physics* (Simon & Schuster, New York, 1938), page 202.

The Achilles' heel of Galilean relativity became apparent in the addition of velocity components. The Galilean rule (Equation 25.6) is not correct except at speed $v \ll c$; we discover the new rule in this section. Just as the Galilean rule stemmed from the Galilean transformation equations, the relativistic rule follows from the Lorentz transformation equations.

Velocity Parallel or Antiparallel to the Direction of Motion of the Two Inertial Reference Frames

We use the standard geometry. Imagine a particle in the frame S' moving along the x'-axis with a velocity component u'_x, as in Figure 25.33. We want to determine the velocity component u_x of the particle as measured in the reference frame S. The Galilean result was (Equation 25.6)

$$u_x = u'_x + v$$

but this only yields the correct result if the velocity components v and u'_x are small compared with the speed of light c.

The velocity component of the object in the S' frame is the time rate at which the position of the object changes in the S' frame:

$$u'_x = \frac{dx'}{dt'} \qquad (25.43)$$

The velocity component of the object in the S frame is the time rate at which the position of the object changes in the S frame:

$$u_x = \frac{dx}{dt} \qquad (25.44)$$

To find u_x we need to find the differentials dx and dt and then take their quotient. The differentials are found from the Lorentz transformation equations (Equations 25.37 and 25.40):

$$x = \gamma(x' + vt') \qquad \text{and} \qquad t = \gamma\left(t' + \frac{v}{c^2}x'\right)$$

We take the differentials; remember that the speed of light c and the relative speed v of the two reference frames are constant:

$$dx = \gamma(dx' + v\,dt') \qquad \text{and} \qquad dt = \gamma\left(dt' + \frac{v}{c^2}dx'\right)$$

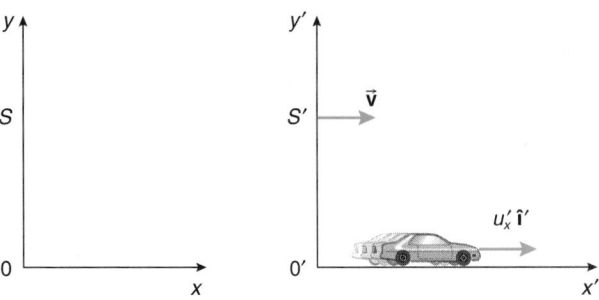

FIGURE 25.33 A particle moving in frame S' with velocity component u'_x.

Taking their quotient, we obtain

$$u_x = \frac{dx}{dt} = \frac{\gamma(dx' + v\,dt')}{\gamma\left(dt' + \dfrac{v}{c^2}\,dx'\right)}$$

Things begin to fall together if we divide the numerator and denominator of the right-hand side by dt':

$$u_x = \frac{\dfrac{dx'}{dt'} + v}{1 + \dfrac{v}{c^2}\dfrac{dx'}{dt'}}$$

But

$$u'_x = \frac{dx'}{dt'}$$

is the velocity component of the object in the S' frame.

> Thus we have for the x-components of the velocities:
>
> $$u_x = \frac{u'_x + v}{1 + \dfrac{vu'_x}{c^2}} \qquad (25.45)$$
>
> Equation 25.45 is the new velocity component addition rule along the common x- and x'- axes.

Notice happily that if the velocity components v and u'_x are small compared with the speed of light c, the second term in the denominator is negligible and Equation 25.45 reduces to the Galilean result (Equation 25.6), as it must for small speeds.

> **PROBLEM-SOLVING TACTIC**
>
> **25.4** If $u'_x > 0$ m/s, then the object is moving to the right in S' toward increasing values of x'. If the object is moving toward decreasing values of x' in S', then substitute a negative velocity component for u'_x in Equation 25.45. A few examples (see the following) will (hopefully!) make the application of Equation 25.45 clear and comfortable for you.

Velocity Perpendicular to the Direction of Motion of the Two Inertial Reference Frames

The Galilean transformation for velocities transverse to the direction of motion v was simple (Equation 25.7):

$$u_y = u'_y$$

> In special relativity, even though the y- and y'-coordinates in the two frames are the same ($y = y'$), the velocity components are *not* the same because of the relativity of time ($t \neq t'$).

In particular, if a particle has a velocity component u'_y in S', then

$$u'_y = \frac{dy'}{dt'}$$

The velocity component of this particle along the y-axis in the S frame is

$$u_y = \frac{dy}{dt}$$

To find the relationship between them, we use the Lorentz transformation equations once again:

$$y = y' \qquad \text{and} \qquad t = \gamma\left(t' + \frac{v}{c^2}\,x'\right)$$

Taking differentials of these equations, we find

$$dy = dy' \qquad \text{and} \qquad dt = \gamma\left(dt' + \frac{v}{c^2}\,dx'\right)$$

From these differentials, we can construct u_y:

$$u_y = \frac{dy}{dt} = \frac{dy'}{\gamma\left(dt' + \dfrac{v}{c^2}\,dx'\right)}$$

Dividing the numerator and denominator of the right-hand side of this result by dt', we obtain

$$u_y = \frac{\dfrac{dy'}{dt'}}{\gamma\left(1 + \dfrac{v}{c^2}\dfrac{dx'}{dt'}\right)} \qquad (25.46)$$

But

$$\frac{dy'}{dt'} = u'_y$$

and

$$\frac{dx'}{dt'} = u'_x$$

and so Equation 25.46 becomes

$$u_y = \frac{u'_y}{\gamma\left(1 + \dfrac{vu'_x}{c^2}\right)} \qquad (25.47)$$

Note that this differs from the transformation for u_x (Equation 25.45).

Thus the transverse velocity components are different in the two reference frames. Notice, in particular, that even if $u'_x = 0$ m/s, so that the particle has only a y'-component for its velocity in S', the y and y' velocity components still are not the same! If $u'_x = 0$ m/s, then

$$u_y = \frac{u'_y}{\gamma} \qquad (25.48)$$

Since the z- and z'-axes also are transverse to the motion, an equation similar to Equation 25.47 exists for the velocity components along z and z':

$$u_z = \frac{u_z'}{\gamma\left(1 + \dfrac{vu_x'}{c^2}\right)} \qquad (25.49)$$

For a particle moving with a velocity

$$\vec{\mathbf{u}}' = u_x'\,\hat{\mathbf{i}}' + u_y'\,\hat{\mathbf{j}}' + u_z'\,\hat{\mathbf{k}}'$$

in frame S', each velocity component must be transformed separately and appropriately to find the velocity of the particle in the frame S:

$$\vec{\mathbf{u}} = u_x\,\hat{\mathbf{i}} + u_y\,\hat{\mathbf{j}} + u_z\,\hat{\mathbf{k}}$$

The u_x and u_x' components are related via Equation 25.45; u_y and u_y' are related via Equation 25.47, and u_z and u_z' via Equation 25.49.

PROBLEM-SOLVING TACTIC

25.5 The relativistic velocity component addition equations always yield velocity components less than or equal to the speed of light. If you ever find that you have calculated a speed greater than c with these equations, you can be sure a mistake has been made, and you can set about finding it and making repairs.

EXAMPLE 25.5

A laser is traveling at a high speed v and emits light in the direction of its motion, as indicated in Figure 25.34. Use Equation 25.45 to show that the speed of light in the reference frame in which the laser is moving still is c.

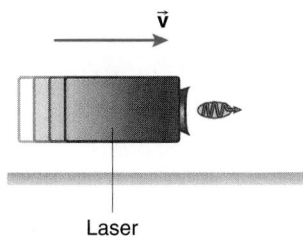

FIGURE 25.34

Solution

The speed of the light with respect to the laser is, of course, c. Let the laser be bolted to frame S', so that you have the standard geometry indicated in Figure 25.35. You can then make the following identification:

$$u_x' = c \qquad \text{(the velocity component of the light moving in } S')$$

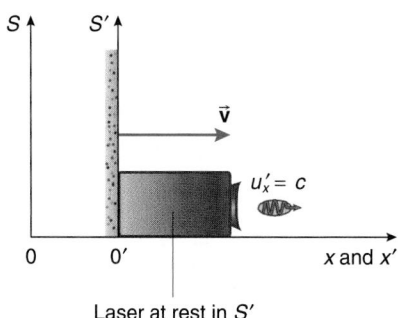

FIGURE 25.35

To find the velocity component of the light with respect to the frame S, use the velocity component addition Equation 25.45:

$$u_x = \frac{u_x' + v}{1 + \dfrac{vu_x'}{c^2}}$$

Making the substitution $u_x' = c$, you find

$$u_x = \frac{c + v}{1 + \dfrac{vc}{c^2}}$$

This simplifies to

$$u_x = \frac{c + v}{\dfrac{c^2 + cv}{c^2}}$$

$$= \frac{c + v}{c(c + v)}\,c^2$$

$$= c$$

The speed of light is c in frame S and also in frame S', in conformity with the first postulate of special relativity.

STRATEGIC EXAMPLE 25.6

Two spacecraft are approaching the Earth from opposite directions with speeds $0.80c$ and $0.50c$, as measured by Earth observers, as shown in Figure 25.36. What is the speed of one spacecraft as seen by the other?

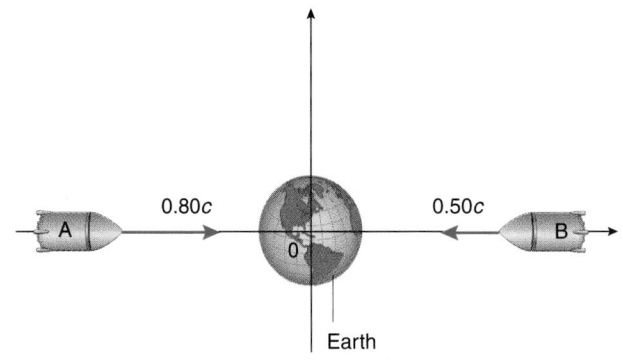

FIGURE 25.36

Solution

Note, incidentally, that the Galilean prediction for the relative speed of approach of the two spacecraft is 1.30c. Their speed of approach must be less than the speed of light according to the special theory of relativity.

Method 1

Assign reference frames and objects so that you have the standard geometry. Imagine yourself inside the spacecraft B on the right in Figure 25.37. Then the Earth and the other spacecraft A are traveling to the right. That is, from the viewpoint of spacecraft B, the situation is as shown in Figure 25.37.

Thus you identify a reference frame attached to spacecraft B as S; the Earth is S′ (and is moving to the right with velocity component v = 0.50c); the spacecraft A is moving to the right in S′ with a velocity component $u_x' = 0.80c$. Thus you can apply Equation 25.45 with all the velocity components positive:

$$u_x = \frac{u_x' + v}{1 + \frac{vu_x'}{c^2}}$$

$$= \frac{0.80c + 0.50c}{1 + \frac{(0.50c)(0.80c)}{c^2}}$$

$$= \frac{1.30c}{1 + 0.40}$$

$$= \frac{1.30}{1.40}c$$

$$= 0.929c$$

Thus spacecraft B sees spacecraft A approaching at a speed of (only!) 0.929c. Spacecraft A sees B approaching at the same speed (but in the opposite direction).

Method 2

For this method you identify the frame S with the Earth and the spacecraft on the left as the frame S′, to have the standard geometry as in Figure 25.38.

Frame S′ thus is moving to the right according to S with a speed of v = 0.80c. The spacecraft on the right is moving in S toward decreasing values of x; thus, the velocity component is negative: $u_x = -0.50c$. Here you want to find u_x', the velocity component of the spacecraft B as seen by spacecraft A. You can

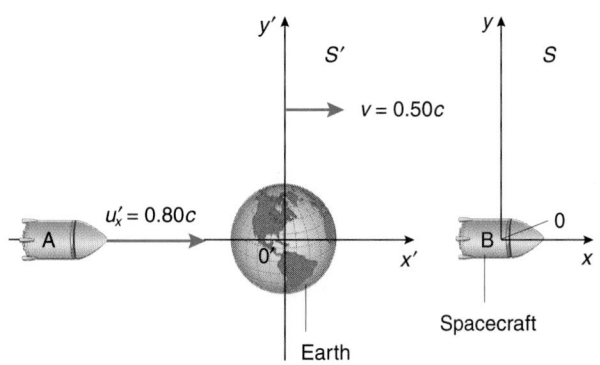

FIGURE 25.37

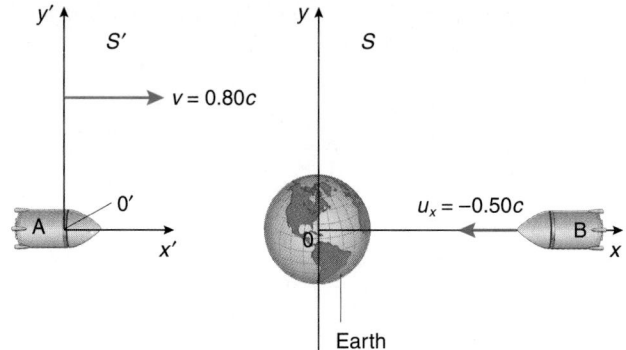

FIGURE 25.38

anticipate that u_x' also will be negative since spacecraft B is moving in the negative x′ direction. Having made these identifications, you can use Equation 25.45:

$$u_x = \frac{u_x' + v}{1 + \frac{vu_x'}{c^2}}$$

$$-0.50c = \frac{u_x' + 0.80c}{1 + \frac{(0.80c)u_x'}{c^2}}$$

Now solve for u_x'. After a bit of algebraic tedium, you will find that

$$u_x' = -\frac{1.30}{1.40}c$$

$$= -0.929c$$

Thus spacecraft A says spacecraft B is approaching at a speed of 0.929c.

25.12 COSMIC JETS AND THE OPTICAL ILLUSION OF SUPERLUMINAL SPEEDS*

Quasars are thought to be young and active galaxies and are among the most distant objects in the universe; they are located billions of light-years from us, far beyond our own Milky Way Galaxy. Some of these quasars emit jets of ionized gas that appear to be moving at speeds faster than the speed of light,[†] an observation that contradicts what we have learned about relative velocity addition in special relativity. In fact, the apparent superluminal speeds of these jets is an *illusion*; here we show how and why.[‡]

Consider a source of light located at point A, a distance d from the Earth located far away at distant point O in Figure 25.39.

[†]See the article by Roger D. Blandford, Mitchell C. Begelman, and Martin J. Rees, "Cosmic jets," *Scientific American*, 246, #5, pages 124–142 (May 1982).
[‡]This argument is based upon a paper by Richard M. Helsdon, "Cosmic jets," *Physics Education*, 18, #4, pages 169–170 (July 1983).

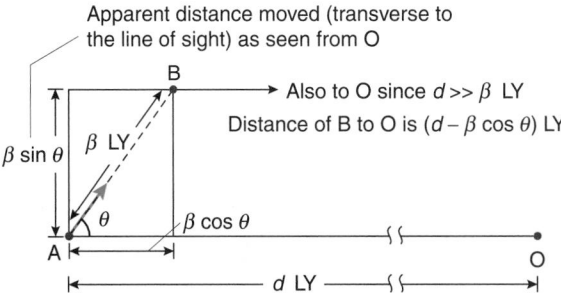

FIGURE 25.39 A very distant light source moves from A to B during one year.

The distance d is very large (billions of light-years). If we express the distance d in light-years, light from the source point A reaches the Earth after a time in years numerically equal to d. Let the source at A move at high speed v, which we express as a fraction of the speed of light: $v/c \equiv \beta$. After *one year*, the source will move a distance in light-years numerically equal to β, to position B. The distance β LY is less than 1 LY because the speed of the source must be less than the speed of light.

Since the distance d is many billions of light-years and the distance β is less than 1 LY, the line from B to the Earth at O is essentially parallel to the line from A to O. Thus, if we examine the geometry of Figure 25.39, the distance (in LY) of point B from the Earth is

$$d - \beta \cos \theta$$

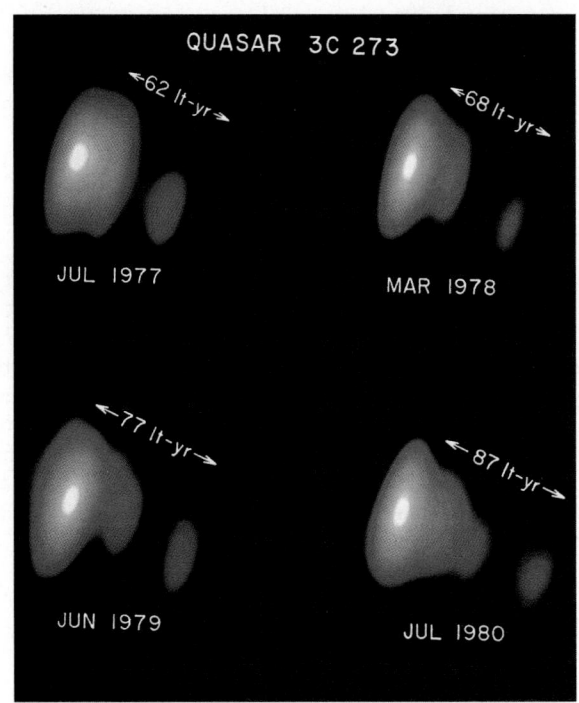

Some celestial objects, such as this quasar, appear to emit material at speeds exceeding the speed of light; the effect is an illusion.

Once light has been emitted at point B, it takes $d - \beta \cos \theta$ years to reach the Earth. Since the source took one year to get from point A to point B, the time of arrival of the light from B at the Earth is $1 + d - \beta \cos \theta$ years after the emission of the light from point A (which takes d years to get to the Earth). Thus the time *interval* ΔT between the arrival at the Earth of the light emitted when the source was at A and the arrival at the Earth of the light emitted when the source was at point B is

$$\Delta T = (1 + d - \beta \cos \theta) - d \quad \text{(in y)}$$
$$= 1 - \beta \cos \theta \quad \text{(in y)}$$

During this time, the source is seen by an observer on the Earth to move perpendicular (transverse) to the line of sight a distance ΔX given by

$$\Delta X = \beta \sin \theta \quad \text{(in LY)}$$

Thus an observer on the Earth says the object moved an apparent distance $(\beta \sin \theta)$ LY during an interval $(1 - \beta \cos \theta)$ y. The *apparent fractional speed* $\beta' = \Delta X / \Delta T$ of the object (expressed as a fraction of the speed of light) is

$$\beta' = \frac{\beta \sin \theta}{1 - \beta \cos \theta} \quad (25.50)$$

Now we fix β and see if the apparent fractional speed β' has a maximum value as a function of θ. We maximize Equation 25.50 as a function of θ by setting its derivative with respect to θ equal to zero:

$$\frac{d\beta'}{d\theta} = 0$$

$$\frac{d\beta'}{d\theta} = \frac{(1 - \beta \cos \theta)(\beta \cos \theta) - (\beta \sin \theta)(\beta \sin \theta)}{(1 - \beta \cos \theta)^2} = 0$$

This yields the following equation:

$$\beta \cos \theta - \beta^2 \cos^2 \theta = \beta^2 \sin^2 \theta \quad (25.51)$$

Since

$$\sin^2 \theta + \cos^2 \theta = 1 \quad (25.52)$$

Equation 25.51 can be reduced to

$$\beta = \cos \theta \quad (25.53)$$

as the condition for maximizing the apparent fractional speed β'.

If $\beta = \cos \theta$, then Equation 25.52 also implies

$$(1 - \beta^2)^{1/2} = \sin \theta \quad (25.54)$$

Using Equation 25.53 and 25.54 in Equation 25.50 for the apparent fractional speed β', we obtain

$$\beta'_{max} = \frac{\beta}{(1 - \beta^2)^{1/2}} \quad (25.55)$$

Now we can see how the apparent fractional speed β'_{max} depends on the real fractional speed β of the object. In particular $\beta'_{max} = 1$, implying an apparent speed of the object equal to the speed of light, when

$$\beta = \frac{1}{\sqrt{2}} = 0.707$$

For this value of β, the angle θ (found from Equation 25.53) is 45°. For values of β greater than 0.707, then, it is possible (for some angle θ) for the apparent fractional speed β'_{max} to be greater than 1, implying an apparent speed greater than the speed of light. For example, if $\beta = 0.90$, then the angle $\theta = 26°$ (since $\beta = \cos\theta$) and the maximum apparent fractional speed is, using Equation 25.55,

$$\beta'_{max} = 2.1$$

The apparent speed of the object to an Earth observer thus is over *twice* the speed of light! But notice that the *actual* speed of the object moving from A to B is 0.90 times the speed of light. Thus the apparent superluminal speed of the real object is not its real speed.

Superluminal speeds do not need galactic astronomy (or nuclear disintegrations) for their occurrence. All you need is the realization that they represent motion, not of real objects, but of ideas. We conceptualize ideas, like points of intersection, which are free to move as fast as fantasy. A garden hose, for example, can easily produce superluminal motion. Hold it as you stand watering the lawn. Water follows a leisurely (parabolic) arc and hits the lawn at a distant spot. Next, aim the hose downward quickly so that the spot of impact is near your feet. With a little skill you can make the stream strike near you while it is still arriving at the previous point of impact. The point of impact can, with practice, be caused to move toward you at any speed, and certainly faster than c. You might call it a superluminal splash! But it is just the intersection point that moves, and that is merely a geometric concept.

25.13 THE LONGITUDINAL DOPPLER EFFECT

We studied the acoustic Doppler effect in Chapter 12. There we discovered that because sound waves travel in a material medium, the change in frequency was different if the source was moving rather than the observer. Light waves need no medium and can travel through a vacuum, since it is a well-known observation that we can see the Sun (on clear days!). We will find that it is only the relative motion of the source and observer that is important in the Doppler effect for light. There is no preferred inertial reference frame, just as Einstein said. In this section we see how the frequency of a light source is affected by the relative motion of the source and observer.

Imagine a well-trained and fleet-winged all-American firefly (*Photuris pennsylvanica*) moving at high speed v relative to an inertial frame S, as in Figure 25.40. Let inertial frame S' be attached to the firefly, with the bright bug at the origin in S'.

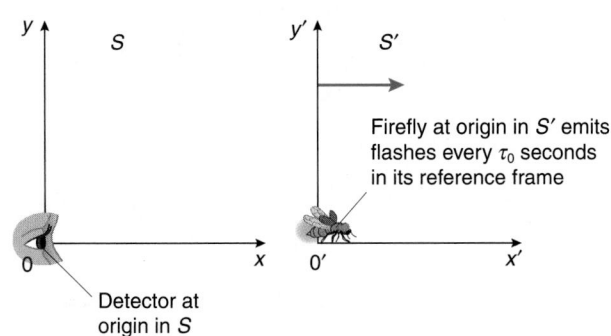

FIGURE 25.40 **The standard geometry.**

The firefly emits a flash of light every τ_0 seconds in its reference frame (S'). Thus the proper period of the firefly (in S') is τ_0 and its proper frequency ν_0 is the reciprocal of τ_0:

$$\nu_0 = \frac{1}{\tau_0} \tag{25.56}$$

We specify a whole series of events in S' corresponding to successive flashes of the firefly. Since the firefly is at the origin of the S' frame, x' and t' space–time coordinates of these events are (if we let the initial flash, the zeroth flash, occur when $t' = 0$ s) as follows:

Event 0. The zeroth flash:

$$x'_0 = 0 \text{ m}$$
$$t'_0 = 0 \text{ s}$$

Event 1. The first flash:

$$x'_1 = 0 \text{ m}$$
$$t'_1 = \tau_0$$

Event 2. The second flash:

$$x'_2 = 0 \text{ m}$$
$$t'_2 = 2\tau_0$$

And so on for additional flashes.

Let your eye (a light detector) be at the origin of inertial reference frame S. You see the firefly receding from your eye at the speed v. We want to determine the frequency ν of the flashes *detected by your eye* (or another suitable light detector) at the origin of frame S.

To find this frequency, we transform the events in S', just enumerated, to frame S using the appropriate Lorentz transformation (Equations 25.37 and 25.40):

$$x = \gamma(x' + vt')$$

and

$$t = \gamma\left(t' + \frac{v}{c^2}x'\right)$$

We substitute the space and time coordinates (x' and t') of each event into these Lorentz transformation equations to determine where and when each event occurs in the frame S. Since the x'-coordinate of every event is zero, this is relatively easy to do (pun intended). The results are as follows.

Event 0. The zeroth flash:

$$x_0 = 0 \text{ m}$$
$$t_0 = 0 \text{ s}$$

Event 1. The first flash:

$$x_1 = \gamma v \tau_0$$
$$t_1 = \gamma \tau_0$$

Event 2. The second flash:

$$x_2 = 2\gamma v \tau_0$$
$$t_2 = 2\gamma \tau_0$$

And so on.

Your eye at the origin in S sees the zeroth flash when $t_0 = 0$ s on your watch, since this flash occurs where your eye is located in S (at $x = 0$ m). But your eye does *not see* the first flash when $t_1 = \gamma\tau_0$. Why not? The first flash occurs at the location $x_1 = \gamma v \tau_0$ when the clocks in the frame S (including your watch) all read time t_1, as shown in Figure 25.41. For your eye at the origin actually to *see* the flash, the light at position x_1 must propagate from x_1 to the origin. The light thus needs to travel the distance x_1 at the speed c, and so takes an additional time $\gamma v \tau_0 / c$ to get to the origin in S. Therefore the total time interval τ between the zeroth flash and the first flash as seen by your eye at the origin in S is the sum of t_1 and the propagation time:

$$\tau = \gamma\tau_0 + \frac{\gamma v \tau_0}{c} \qquad (25.57)$$

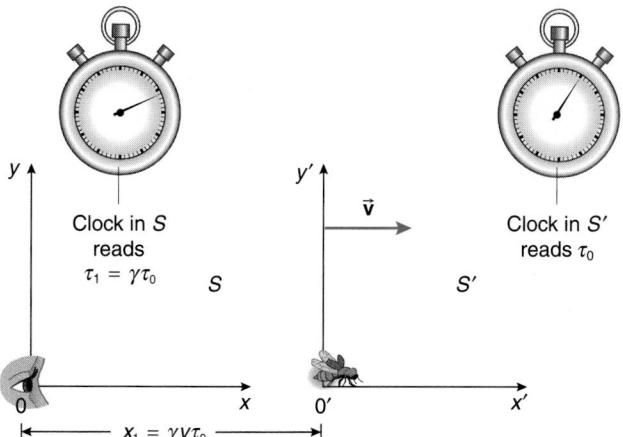

FIGURE 25.41 Event 1 in the S frame.

The second flash (Event 2) occurs in S at the location $x_2 = 2\gamma v \tau_0$ when the clocks in frame S read $t_2 = 2\gamma\tau_0$. This flash also must propagate the distance x_2 to be detected by your eye at the origin in S. Therefore the time on the clocks in S that the second flash arrives at your eye at the origin is

$$2\gamma\tau_0 + 2\frac{\gamma v \tau_0}{c}$$

which is 2τ.

Therefore the time interval between the flashes seen by your eye at the origin in frame S (the period of the flashes observed in S) is τ, given by Equation 25.57. The frequency ν of the flashes seen at the origin is $1/\tau$ or

$$\nu = \frac{1}{\gamma\tau_0 + \frac{\gamma v \tau_0}{c}}$$

$$= \frac{1}{\gamma\tau_0\left(1 + \frac{v}{c}\right)}$$

But $1/\tau_0$ is the proper frequency ν_0 of the firefly. Thus the detected frequency from the receding bug is

$$\nu_{\text{recede}} = \frac{\nu_0}{\gamma\left(1 + \frac{v}{c}\right)}$$

Now we substitute for γ using its definition, Equation 25.13, and simplify:

$$\nu_{\text{recede}} = \frac{\nu_0\sqrt{1 - \frac{v^2}{c^2}}}{1 + \frac{v}{c}}$$

$$= \frac{\nu_0\sqrt{1 - \frac{v}{c}}\sqrt{1 + \frac{v}{c}}}{1 + \frac{v}{c}}$$

Hence the receding source is observed to have a frequency given by

$$\nu_{\text{recede}} = \nu_0\sqrt{\frac{1 - \frac{v}{c}}{1 + \frac{v}{c}}} \qquad \text{(source receding)} \qquad (25.58)$$

The frequency observed by your eye in S is less than the proper frequency of the receding firefly.

If the firefly is *approaching* your eye, you can replace v in Equation 25.58 with $-v$ to obtain

$$v_{appro} = v_0 \sqrt{\frac{1 + \dfrac{v}{c}}{1 - \dfrac{v}{c}}} \qquad \text{(source approaching)} \qquad (25.59)$$

The observed frequency v_{appro} is greater than the proper frequency.

What we have analyzed, as you may have surmised (if only from the title of this section), is the relativistic **longitudinal Doppler effect**. The flashes play the role of the peak values of a light wave.

It makes no difference whether the source is moving away from the observer (detector), or the observer is moving away from the source. In other words, if we redid the problem with a firefly at the origin in S and the detector receding at speed v in S', the frequency of the flashes seen by the detector is less than the proper frequency of the source and is given by Equation 25.58. Likewise it makes no difference whether it is the source approaching the observer, or the observer approaching the source; the frequency seen by the observer is higher than the proper frequency and is given by Equation 25.59.

Of course, since the frequency and wavelength are related by

$$c = v\lambda$$

if the frequency decreases, the wavelength increases.

Any shift to a longer (larger) wavelength (i.e., smaller frequency) is known as a **red shift**,[†] regardless of the actual color or wavelength of the light. Red shifts occur if the distance between the source and observer is increasing with time (i.e., they are *receding* from each other) regardless of whether the source, observer, or both are moving.

Correspondingly, if the frequency increases, the wavelength decreases. Any shift to a shorter (smaller) wavelength (i.e., increased frequency) is known as a **blue shift**, regardless of the actual color or wavelength of the light. Blue shifts occur when the distance between a source and observer is decreasing with time (i.e., they are *approaching* each other) regardless of whether the source, observer, or both are moving.

The relativistic longitudinal Doppler effect for light in a vacuum is much simpler than the corresponding acoustic Doppler effect for sound (see Chapter 12). In the latter case, distinctly different quantitative effects occur even at low velocities (1) when the source is in motion; (2) when the observer is in motion; and (3) if there is a wind or motion of the medium itself in combination with motion of the source and/or observer. No such complications occur in the relativistic Doppler effect for light in a vacuum: there is no medium, and it is only the relative motion of the source and observer that is important. However,

when the source and light travel in a material medium (such as water or glass), the Doppler effect becomes more complicated and takes on several aspects of the acoustical Doppler effect.

EXAMPLE 25.7

A very distant quasar is receding from us. You find that the frequency of the light from hydrogen atoms in the quasar is 75% that from a hydrogen source at rest in your lab. At what speed is the quasar receding? Express your answer as a fraction of the speed of light. All observed Doppler shifts from distant galaxies are red shifts, indicating these sources are receding from us: the universe is "expanding."

Solution
The light is red-shifted. Since $v = 0.75v_0$, Equation 25.58 becomes

$$0.75v_0 = v_0 \sqrt{\frac{1 - \dfrac{v}{c}}{1 + \dfrac{v}{c}}}$$

After squaring, you can solve for v/c. You find that

$$v = 0.28c$$

The quasar source is receding at an incredible 28% the speed of light. Such large speeds for quasars are not uncommon.

25.14 THE TRANSVERSE DOPPLER EFFECT*

In the preceding section, the relative motion of the source or observer was along the line of sight. What happens if our relativistic firefly is moving transverse to the line of sight, as when the firefly is at point P in Figure 25.42?

The proper period of the firefly is τ_0 and its proper frequency (the inverse of the period) is v_0 in its reference frame S', since the firefly is at rest in this frame. We imagine the frequency v_0 is so great that many flashes are emitted while the firefly is in the immediate vicinity of point P in frame S. In this way, the propagation time of the pulses sent from the vicinity of P to your eye at the origin in S is the same for all the flashes because they all travel the same distance.

In S, you can think of the firefly as a moving clock. Thus, because of time dilation, the period τ between the flashes in S is measured to be longer by the factor γ:

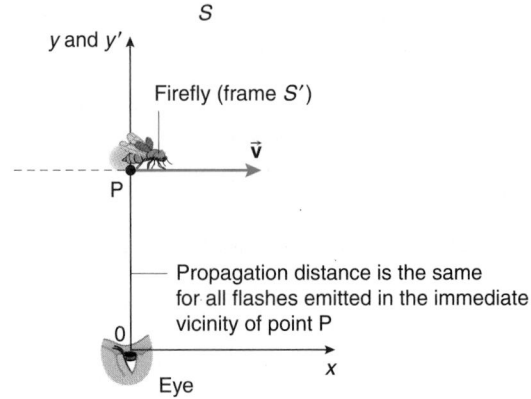

FIGURE 25.42 A source transverse to the line of sight.

[†]The term arose because the red wavelengths are longer than blue wavelengths, defining roughly the extremes of the visible region of the electromagnetic spectrum.

$$\tau = \gamma \tau_0$$

The frequency of the firefly is the inverse of the period, and so the frequency v of the flashes observed at the origin in S when the source is at P is

$$v = \frac{1}{\tau}$$

$$= \frac{1}{\gamma \tau_0}$$

$$= v_0 \sqrt{1 - \frac{v^2}{c^2}} \qquad (25.60)$$

Thus, when the light source is moving transverse to the line of sight, the frequency is *smaller* than the proper frequency; this effect is called the **transverse Doppler effect**. The effect is a consequence of time dilation and thus has no classical analog.

Classically, if a source is moving transverse to the line of sight there is no change in frequency; however, special relativity indicates there is!

According to Equation 25.60, the transverse Doppler effect *always* is a red shift, since the observed frequency is less than the proper frequency (i.e., the observed wavelength is longer than the proper wavelength).

It is interesting to compare the longitudinal and transverse Doppler effect red shifts. Figure 25.43 is a plot of the fractional frequency ratio v/v_0 versus v/c for the two effects.

Note that the transverse Doppler effect red shift always is smaller than the longitudinal Doppler effect red shift for the same v/c (the ratio of the frequencies v/v_0 always is closer to 1 for the transverse effect).

EXAMPLE 25.8

A light source moves at speed v perpendicular to your line of sight. The frequency of the light is only 75% that from a comparable source at rest with respect to you. What is the speed of the source, expressed as a fraction of the speed of light?

Solution

Since the source is moving perpendicular to the line of sight, the frequency changes because of the transverse Doppler effect. Use Equation 25.60 with $v = 0.75v_0$:

$$0.75v_0 = v_0 \sqrt{1 - \frac{v^2}{c^2}}$$

After squaring, you can solve for v/c. You find

$$\frac{v}{c} = 0.66$$

The source is moving at 66% the speed of light across your line of sight. Compare this with Example 25.7. Notice that for a given percentage frequency change, the source must move much faster for the change to occur via the transverse Doppler effect than via the longitudinal Doppler effect. Viewed another way, for a source with constant speed v, the longitudinal Doppler effect produces a greater frequency change (i.e., smaller frequency ratio) than does the transverse Doppler effect, as shown in Figure 25.43.

25.15 A GENERAL EQUATION FOR THE RELATIVISTIC DOPPLER EFFECT*

The red and blue shifts for the longitudinal Doppler effect and the red shift for the transverse relativistic Doppler effect can be incorporated into a single equation. We let the relative motion of the source be as indicated in Figure 25.44.

The angle θ is the angle between the velocity of the source and the line of sight from the source to the observer. Notice that while the source and observer are approaching each other, the angle θ ranges from 0° to 90° (acute); when the two are receding from each other, the angle θ ranges from 90° to 180° (obtuse).

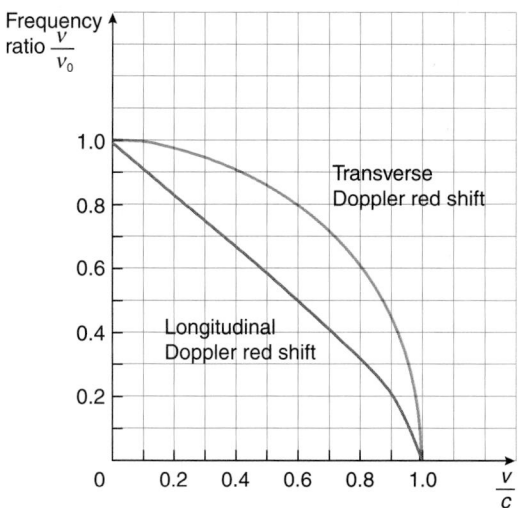

FIGURE 25.43 Longitudinal and transverse Doppler effects.

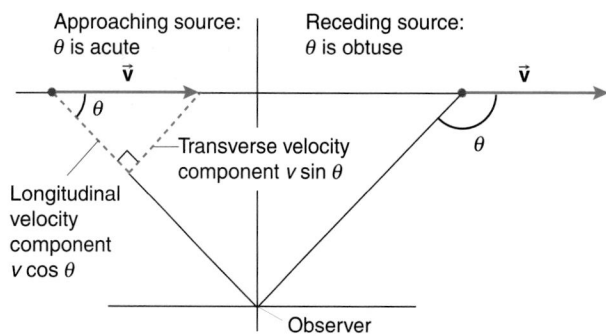

FIGURE 25.44 A source approaching not along the direct line of sight has both transverse and longitudinal velocity components.

With this definition of θ, the general equation for the relativistic Doppler effect is*

$$\nu = \nu_0 \frac{\sqrt{1 - \frac{v^2}{c^2}}}{1 - \frac{v}{c} \cos \theta} \qquad (25.61)$$

When a source is approaching an observer the *longitudinal* Doppler effect predicts a *blue shift* in the spectrum, caused by the component of the velocity along the line of sight. On the other hand, the *transverse* Doppler predicts a *red shift*, caused by the component of the velocity transverse to the line of sight. Thus, somewhere among the possible angles of approach, there must exist an angle θ_0 at which the two effects exactly cancel each other and *no* shift in frequency is observed.

To find this special angle, we take the general equation (Equation 25.61) and set ν equal to ν_0. The resulting expression is

$$1 = \frac{\sqrt{1 - \frac{v^2}{c^2}}}{1 - \frac{v}{c} \cos \theta_0} \qquad (25.62)$$

Solving this expression for $\cos \theta_0$, we find

$$\cos \theta_0 = \frac{1 - \sqrt{1 - \frac{v^2}{c^2}}}{\frac{v}{c}}$$

The solution of this equation for θ_0 for various values of v/c is shown in Figure 25.45.

For small speeds (small v/c), the transition from an approaching blue shift to a receding red shift occurs essentially at an angle of 90°, since the transverse Doppler effect is small compared with the longitudinal effect. As the speed increases, however, the transition from blue to red shift occurs at decreasing values of θ_0.

In other words, for high-speed sources not moving directly along the line of sight, the transition from blue to red shift occurs *before* the source and observer pass each other.

25.16 RELATIVISTIC MOMENTUM

Our analysis of the special theory of relativity thus far has examined kinematical connections between the descriptions of events in space and time in two different inertial reference frames. We now turn to a consideration of the implications of

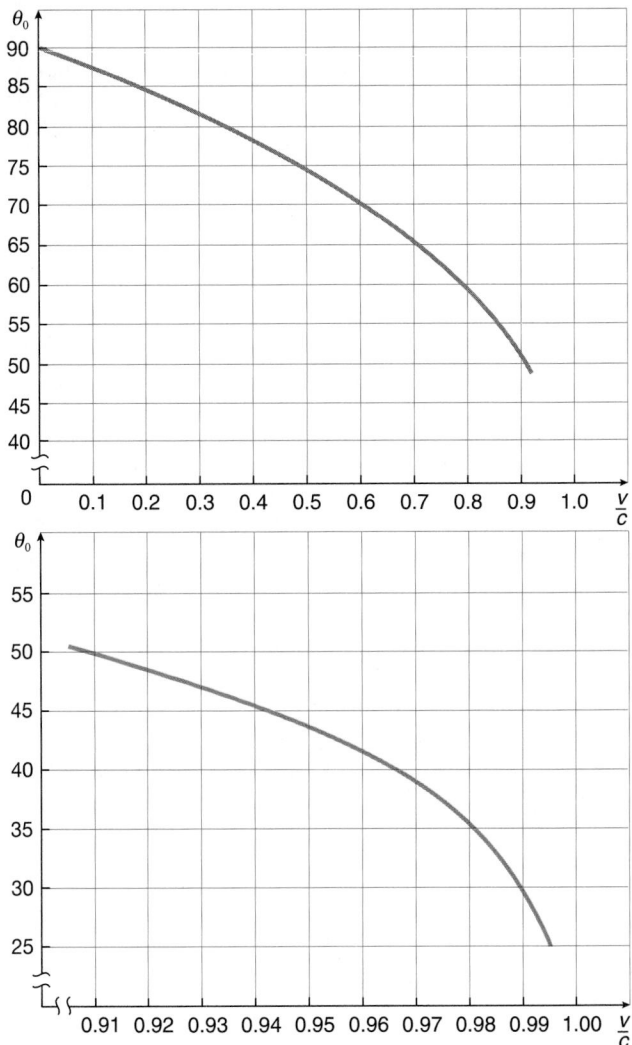

FIGURE 25.45 The angle at which no frequency shift occurs.

relativity for the dynamics of a single particle, the effect of forces, the work done by these forces, and the consequent changes in momentum and kinetic energy of the particle.

Recall the classical momentum $\vec{p}_{class}$ of a particle with a velocity $\vec{u}$ is defined to be

$$\vec{p}_{class} = m\vec{u} \qquad (25.63)$$

We also know from our study of collisions of two or more particles in Chapter 9 that when the total force on the system of colliding particles is zero, the total momentum of system is conserved.

A easily visualized experiment is sufficient to show that the classical definition of the momentum cannot be valid at high speeds if we accept the postulates of the special theory of relativity (in particular, Postulate 2). Imagine you are in reference frame S and a friend is in reference frame S', moving past at speed v with respect to you in the standard geometry (see Figure 25.46).

Each of you throws a particle of the same standard mass m (such as a billiard ball) at speed u_0 along your respective y- or y'-axis to hit the other ball. Each of you determines the magnitude of u_0 using rulers and clocks at rest in your respective reference frames. The particles make an elastic collision and return to each of you

---

*We have not derived this equation, but pluck it from thin air. You can see that when $\theta = 0°$ (source approaching along the line of sight), Equation 25.61 reduces to Equation 25.59. When $\theta = 180°$ (source receding along the line of sight), Equation 25.61 reduces to Equation 25.58. Likewise, when $\theta = 90°$ (source transverse to the line of sight), Equation 25.61 reduces to Equation 25.60. See Problem 34.

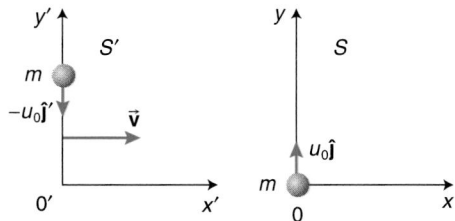

FIGURE 25.46 The standard geometry.

along the y- and y'-axes with their velocities reversed. Because of the symmetry of each of your viewpoints, we would like to think each of you says that the total y-component of momentum of the system of two particles is zero before and after the collision, since each of you says that equal mass particles were thrown with opposite y-velocities in your respective reference frames.

But knowing what we now do about how velocity components transform between the two reference frames (Section 25.11), if you (or your friend) analyze the elastic collision using the classical definition of the momentum, you find that the total momentum is *not* zero before or after the collision in your frame. Why not? The ball you throw has a classical momentum of $mu_0\hat{\mathbf{j}}$ according to rulers and clocks in your reference frame S. (Analogously, your friend says the ball she throws has classical momentum $-mu_0\hat{\mathbf{j}}'$ according to rulers and clocks in her reference frame S'.) But you say your friend's billiard ball has a y-velocity component in your reference frame of

$$-\frac{u_0}{\gamma}\hat{\mathbf{j}}$$

because of the new rule for adding velocity components (Equation 25.48). Hence you say her billiard ball has a classical momentum (when measured using rulers and clocks in your reference frame) of

$$-m\frac{u_0}{\gamma}\hat{\mathbf{j}}$$

and this has a *different* magnitude from that of the momentum of the ball you threw ($mu_0\hat{\mathbf{j}}$) in the opposite direction. As a result, the total classical momentum before the collision is *not* zero, though you both would expect it to be zero because of the planned symmetry of the encounter; nor is the total momentum zero after the collision.

The question then becomes, what expression or definition for the momentum can enable each of you to note that the total momentum before and after the collision is conserved and in fact zero? Measure the velocity of a particle to be $\vec{\mathbf{u}} = u_x\hat{\mathbf{i}} + u_y\hat{\mathbf{j}} + u_z\hat{\mathbf{k}}$ in your reference frame (using, of course, rulers and clocks in your frame).

We define the **relativistic momentum** $\vec{\mathbf{p}}$ of a particle to be

$$\vec{\mathbf{p}} \equiv \gamma m\vec{\mathbf{u}} \qquad (25.64)$$

where

$$\gamma \equiv \frac{1}{\sqrt{1 - \dfrac{u^2}{c^2}}}.$$

Notice that γ involves the *speed* of the particle.* The magnitude of its velocity as measured in your reference frame is $u = (u_x^2 + u_y^2 + u_z^2)^{1/2}$. Observe that if $u \ll c$, in which case $\gamma \to 1$, the relativistic momentum reduces to the familiar classical expression for the momentum: $m\vec{\mathbf{u}}$.

You need to be aware that many texts distinguish between a so-called rest mass m of a particle (what we have called simply the mass, an intrinsic property of a particle) and a so-called relativistic mass $m_{rel} \equiv \gamma m$, so that the relativistic momentum is written as the product of the relativistic mass times the velocity: $\vec{\mathbf{p}} = m_{rel}\vec{\mathbf{u}}$; this makes the equation for the relativistic momentum look identical to the classical form for the momentum. Despite the historical popularity of such an approach, even Einstein suggested simply, "It is better to introduce no other mass concept than the 'rest mass' m."[†] We use this approach: the mass m (the **rest mass**) is the *only* mass we will use for a particle, with the relativistic momentum defined by Equation 25.64.

We now satisfy your curiosity and show that the definition of the relativistic momentum (Equation 25.64) means that the total y-component of the momentum of the two colliding equal mass particles in our collision experiment is zero in each reference frame. The ball you throw along the y-axis has velocity $u_0\hat{\mathbf{j}}$ and speed u_0 as measured by you. Its relativistic momentum then is

$$\vec{\mathbf{p}}_1 = m\gamma u_0\hat{\mathbf{j}} \qquad (25.65)$$

where

$$\gamma = \frac{1}{\sqrt{1 - \dfrac{u_0^2}{c^2}}}$$

since the speed of the particle is u_0 according to you. Since your friend in the S' frame is moving at speed v, the ball she throws at speed u_0 (as measured by her rulers and clocks) along her y'-axis has *two* velocity components in your reference frame (using rulers and clocks in your frame):

$$u_x = v \qquad (25.66)$$

and, according to the rule for transforming y-velocity components, Equation 25.48,[‡]

$$u_y = -u_0\sqrt{1 - \frac{v^2}{c^2}} \qquad (25.67)$$

With Equation 25.64 for the definition of the relativistic momentum, you say that her particle has a y-component of momentum in your reference frame given by

$$p_2 = m\frac{1}{\left[1 - \dfrac{(u_x^2 + u_y^2)}{c^2}\right]^{1/2}}\left(-u_0\sqrt{1 - \frac{v^2}{c^2}}\right) \qquad (25.68)$$

---

*This γ involves the speed of the *particle* and must be distinguished from the γ that involves the relative speed v of the reference frames S and S', unless the particle *is* one of these frames.

[†]The quotation by Einstein appears in an article by Lev B. Oken, "The concept of mass," *Physics Today*, *42*, #6, pages 31–36 (June 1989); the quotation is on page 32.

[‡]Notice that here only the relative speed v of the two reference frames is involved in the square root.

We use Equation 25.66 for u_x and Equation 25.67 for u_y in Equation 25.68. After making these substitutions and simplifying, we find

$$p_2 = m \frac{1}{\left[1 - \dfrac{v^2 + u_0^2 - \dfrac{u_0^2 v^2}{c^2}}{c^2}\right]^{1/2}} \left(-u_0 \sqrt{1 - \frac{v^2}{c^2}}\right)$$

$$= m \frac{1}{\left[1 - \dfrac{v^2}{c^2} - \dfrac{u_0^2}{c^2} + \dfrac{u_0^2 v^2}{c^4}\right]^{1/2}} \left(-u_0 \sqrt{1 - \frac{v^2}{c^2}}\right)$$

$$= m \frac{1}{\sqrt{1 - \dfrac{v^2}{c^2}} \sqrt{1 - \dfrac{u_0^2}{c^2}}} \left(-u_0 \sqrt{1 - \frac{v^2}{c^2}}\right)$$

$$= -m \frac{u_0}{\sqrt{1 - \dfrac{u_0^2}{c^2}}} \qquad (25.69)$$

When you find the total *y*-component of the momentum before the collision in your reference frame, using Equations 25.65 and 25.69, you now pleasingly discover the total *y*-component of the momentum before the collision is zero. A similar argument used by your friend in S′ shows that the total *y*′-component of the momentum of the two particles is zero when measured by her using clocks and rulers in her frame.

Hence the new definition of the momentum, Equation 25.64, is consistent with the postulates of special relativity and we can conserve momentum in collisions within any inertial reference frame.

There is a more heuristic way to see how Equation 25.64 comes about for the relativistic momentum. For simplicity, we consider a particle subject to a constant total force $\vec{\mathbf{F}}_{total}$. In the Galilean and Newtonian theory of motion, a constant total force acting on a particle of constant mass *m* produces a constant acceleration $\vec{\mathbf{a}}$ in the same direction as the total force; this is Newton's second law of motion:

$$\vec{\mathbf{F}}_{total} = m\vec{\mathbf{a}}$$

The constant acceleration changes the velocity of the particle according to the familiar kinematic relation

$$\vec{\mathbf{v}} = \vec{\mathbf{v}}_0 + \vec{\mathbf{a}}t$$

Given enough time, such a force (and resulting acceleration) can increase the speed of the particle without limit, according to this classical view.

Later, in Section 25.17, we shall see that the special theory of relativity indicates there is an *upper limit* to the speed a particle can attain: the speed of light. Thus the question arises, how do we reconcile the Newtonian description of motion and Newton's

second law with the special theory of relativity? The key is through the more general statement of the Newtonian second law of motion: the total force is equal to the time rate of change of the momentum of the particle:

$$\vec{\mathbf{F}}_{total} = \frac{d\vec{\mathbf{p}}}{dt}$$

Given enough time, a constant total force will increase the *momentum* of a particle without limit, but *not* its speed. In classical physics the momentum of a particle was defined to be the product of its mass *m* and its velocity $\vec{\mathbf{u}}$:

$$\vec{\mathbf{p}}_{class} \equiv m\vec{\mathbf{u}}$$

The mass of a particle is an innate characteristic of the particle (much like electric charge). Although under the action of a constant total force, the speed of a particle can increase up to a limit, we want the momentum of the particle to be able to increase without limit under the action of a constant force to follow Newton's law, $\Delta\vec{\mathbf{p}} = \vec{\mathbf{F}}_{total}\,\Delta t$. Thus the classical expression for the momentum of a particle needs to be modified or changed. How do we make the change? Certainly, for small speeds, the classical expression for the momentum is legitimate, and so whatever we take for the relativistic momentum must reduce to the classical expression when $u \ll c$.

We need to multiply the classical expression for the momentum by a dimensionless factor that

- increases without limit as $u \to c$; and
- approaches 1 when $u \ll c$ (or equivalently, as $u \to 0$ m/s).

These constraints are just the characteristics of the ubiquitous factor γ in the special theory of relativity. Hence we try defining the relativistic momentum of a particle with a velocity $\vec{\mathbf{u}}$ to be

$$\vec{\mathbf{p}}_{rel} = \gamma m\vec{\mathbf{u}} = \frac{m\vec{\mathbf{u}}}{\sqrt{1 - \dfrac{u^2}{c^2}}}$$

where *u* is the speed of the particle (the magnitude of its velocity); this is Equation 25.64. With this definition, under the action of a constant force, the momentum of a particle of intrinsic (constant) mass *m* can increase indefinitely as $u \to c$ whereas its speed *u* has an upper bound.

The justification for this definition of the relativistic momentum is, of course, the correct results of all experiments to test it. As we saw earlier, experiments involving the collisions of particles traveling at high speeds indicate that the definition of the relativistic momentum in Equation 25.64 is the only way momentum can be defined so that it is conserved in relativistic collisions.

EXAMPLE 25.9 _____

A cosmic ray proton is traveling with a speed of 0.99900*c*. **Determine the magnitude of its relativistic momentum.**

Solution

First calculate the factor γ associated with its speed:

$$\gamma = \frac{1}{\sqrt{1 - \dfrac{u^2}{c^2}}}$$

$$= \frac{1}{\sqrt{1 - (0.99900)^2}}$$

$$= 22.4$$

The magnitude of the relativistic momentum is found from Equation 25.64:

$$p = \gamma m u$$

$$= 22.4(1.67 \times 10^{-27} \text{ kg})(0.99900 \times 3.00 \times 10^8 \text{ m/s})$$

$$= 1.12 \times 10^{-17} \text{ kg·m/s}$$

25.17 THE CWE THEOREM REVISITED

The CWE theorem (see Chapter 8) states that if we neglect thermal effects, the work done by all the forces acting on a system is equal to the change in the kinetic energy of the system. What implications does this theorem have in relativity?

We consider a single force acting on a single particle system. Let the force have only a positive x-component, $\vec{\mathbf{F}} = F\hat{\mathbf{i}}$, so the problem is one-dimensional. The differential work dW done by the force when the position vector of the particle changes by $d\vec{\mathbf{r}} = dx\,\hat{\mathbf{i}}$ is

$$dW = \vec{\mathbf{F}} \cdot d\vec{\mathbf{r}}$$

$$= F\,dx$$

According to the CWE theorem, the differential work done by the total force on a system is equal to the differential change in the kinetic energy, $d(\text{KE})$, of the system:

$$F\,dx = d(\text{KE}) \qquad (25.70)$$

Keeping Newton's second law, the time rate of change of the relativistic momentum $\vec{\mathbf{p}}$ still is equal to the applied total force:

$$\vec{\mathbf{F}} = \frac{d\vec{\mathbf{p}}}{dt}$$

where the relativistic momentum* is $\vec{\mathbf{p}} = \gamma m \vec{\mathbf{v}}$.

Substituting for F in the CWE theorem, Equation 25.70, and writing the change in the kinetic energy first, we obtain

$$d(\text{KE}) = \frac{dp}{dt}\,dx \qquad (25.71)$$

---

*We switch to the more conventional notation of $\vec{\mathbf{v}}$ for the velocity of the particle, since we can take the particle to be at rest in S'. Remember that the γ in the relativistic momentum involves the speed v of the particle.

We want to integrate this equation over distance to find an expression for the changes in the relativistic kinetic energy caused by the force. We anticipate that since the relativistic momentum of the particle increases without limit as $v \to c$, the relativistic kinetic energy will do likewise. We see already that the classical expression for the kinetic energy,

$$\text{KE}_{\text{class}} = \frac{mv^2}{2}$$

like the classical momentum, *cannot* be the correct expression for the relativistic kinetic energy of a particle, since the speed of a particle cannot exceed c, a fact we justify shortly. The classical expression for the kinetic energy would imply an upper bound, equal to $mc^2/2$, on the kinetic energy of a particle, even though F could do an infinite amount of work on it over unbounded distance. In reality, such an upper bound on the kinetic energy does not exist. The expression for the relativistic kinetic energy must be different from the classical expression. Nonetheless, for speeds $v \ll c$, the relativistic expression must reduce to the classical expression, since the latter expression serves us well in mechanics, as long as the speeds of the particles are such that $v \ll c$.

The problem at hand is to integrate Equation 25.71 to find an expression for the relativistic kinetic energy. It is not immediately apparent how to proceed with this integration. One approach suggests itself: since the relativistic momentum is a function of the velocity of the particle, it is convenient to convert the integration from one involving the change in the position vector component dx to one involving the change in the velocity component dv. To accomplish this change, we use the chain rule from calculus in the following way.

The right-hand side of Equation 25.71 can be expressed as

$$\frac{dp}{dt}\,dx = \frac{dp}{dv}\frac{dv}{dt}\,dx \qquad (25.72)$$

Likewise, the derivative

$$\frac{dv}{dt}$$

can be written with the chain rule as

$$\frac{dv}{dt} = \frac{dv}{dx}\frac{dx}{dt}$$

Hence Equation 25.72 becomes

$$\frac{dp}{dt}\,dx = \frac{dp}{dv}\frac{dv}{dx}\frac{dx}{dt}\,dx$$

But

$$\frac{dx}{dt} = v$$

and so

$$\frac{dp}{dt}\,dx = \frac{dp}{dv}\frac{dv}{dx}\,v\,dx$$

$$= \frac{dp}{dv}\,v\,dv$$

Thus the CWE theorem (Equation 25.71) becomes

$$d(KE) = \frac{dp}{dv} v\, dv \qquad (25.73)$$

Now we just have to find the derivative

$$\frac{dp}{dv}$$

The magnitude p of the relativistic momentum component in one dimension is

$$p = \gamma mv \qquad (25.74)$$

We put in the explicit meaning of the factor γ:

$$p = \frac{mv}{\sqrt{1 - \dfrac{v^2}{c^2}}}$$

Remember that c and m are constants; we take the derivative of p with respect to v, using the usual calculus rules for the derivative of a quotient:

$$\frac{dp}{dv} = \frac{\sqrt{1 - \dfrac{v^2}{c^2}}\; m - mv\left[\dfrac{1}{2}\left(1 - \dfrac{v^2}{c^2}\right)^{-1/2}\left(-2\dfrac{v}{c^2}\right)\right]}{1 - \dfrac{v^2}{c^2}}$$

After a bit of algebra (maybe a big bit!), this ungainly expression reduces to

$$\frac{dp}{dv} = \frac{m}{\left(1 - \dfrac{v^2}{c^2}\right)^{3/2}}$$

Substituting this derivative into Equation 25.73 for the CWE theorem, we obtain

$$d(KE) = \frac{mv\, dv}{\left(1 - \dfrac{v^2}{c^2}\right)^{3/2}}$$

This expression can now be integrated. When $v = 0$ m/s the kinetic energy also is zero, and so the definite integrals are

$$\int_{0\text{ J}}^{KE} d(KE) = m\int_{0\text{ m/s}}^{v} \frac{v\, dv}{\left(1 - \dfrac{v^2}{c^2}\right)^{3/2}}$$

The integral on the right-hand side can be put into the standard form:

$$\int u^n\, du = \frac{u^{n+1}}{n+1}$$

where we let

$$u = 1 - \frac{v^2}{c^2}$$

The result is

$$KE = \frac{m}{\left(-\dfrac{2}{c^2}\right)} \int_{0\text{ m/s}}^{v} \frac{-\dfrac{2}{c^2}\, v\, dv}{\left(1 - \dfrac{v^2}{c^2}\right)^{3/2}}$$

$$= mc^2\left(\frac{1}{\sqrt{1 - \dfrac{v^2}{c^2}}} - 1\right)$$

The first term in the brackets is γ.

Thus the **relativistic kinetic energy** is

$$\boxed{KE = (\gamma - 1)mc^2} \qquad (25.75)$$
$$= \gamma mc^2 - mc^2$$

The second term on the expanded right-hand side of Equation 25.75 is *independent of the speed* of the particle and is intrinsic to the particle itself. Hence mc^2 is called the **rest energy** of the particle.

It is from this term that Einstein made the famous conclusion that energy is the mass times the square of the speed of light. Mass is equivalent to energy. There is a subtlety to this equivalence that we explore subsequently in Section 25.18.

Consider a particle with $v < c$. It has a finite amount of relativistic kinetic energy. If we try to increase its kinetic energy so that $v = c$, at which speed the factor γ becomes infinite, the particle then would have an infinite amount of relativistic kinetic energy. This means we need to do an infinite amount of work on the particle to accomplish this feat! Such an amount of energy clearly is unavailable. Hence a particle initially confined to a speed less than c can *never* attain a speed even equal to c in special relativity.

The speed of light thus is an unreachable upper bound on the speed of a particle (with nonzero mass) in special relativity.

This is much like absolute zero is an unreachable lower bound on the temperature of a system.

Transposing the rest energy term in Equation 25.75, we find

$$\gamma mc^2 = KE + mc^2 \qquad (25.76)$$

The right-hand side of Equation 25.76 contains two terms: the kinetic energy; and the rest energy mc^2, which is independent of the speed of the particle.

The sum of the kinetic energy of the particle and its rest energy is called the **total relativistic energy** E of the particle:

$$E \equiv KE + mc^2 \qquad (25.77)$$

Thus, from Equation 25.76, the total relativistic energy is

$$\boxed{E = \gamma mc^2} \qquad (25.78)$$

The relativistic kinetic energy thus is the difference between the total relativistic energy E and the rest energy mc^2:

$$KE = \gamma mc^2 - mc^2$$
$$= E - mc^2 \qquad (25.79)$$

What is the relationship between the relativistic kinetic energy and the classical expression for the kinetic energy? To elucidate this relationship, we begin with Equation 25.75 and rearrange it as follows:

$$KE = mc^2(\gamma - 1)$$

$$= mc^2 \left(\frac{1}{\sqrt{1 - \dfrac{v^2}{c^2}}} - 1 \right)$$

$$= mc^2 \left[\left(1 - \frac{v^2}{c^2}\right)^{-1/2} - 1 \right] \qquad (25.80)$$

We use the binomial theorem to expand the term

$$\left(1 - \frac{v^2}{c^2}\right)^{-1/2}$$

The binomial theorem states that

$$(1 + x)^n = 1 + nx + \frac{n(n-1)}{2!}x^2 + \frac{n(n-1)(n-2)}{3!}x^3 + \cdots$$

where we assign $x = -v^2/c^2$ and $n = -1/2$. The expansion is

$$\left(1 - \frac{v^2}{c^2}\right)^{-1/2} = 1 + \left(-\frac{1}{2}\right)\left(-\frac{v^2}{c^2}\right) + \frac{\left(-\dfrac{1}{2}\right)\left(-\dfrac{3}{2}\right)\left(-\dfrac{v^2}{c^2}\right)^2}{2!}$$

$$+ \frac{\left(-\dfrac{1}{2}\right)\left(-\dfrac{3}{2}\right)\left(-\dfrac{5}{2}\right)\left(-\dfrac{v^2}{c^2}\right)^3}{3!} + \cdots$$

$$= 1 + \frac{1}{2}\frac{v^2}{c^2} + \frac{3}{8}\frac{v^4}{c^4} + \frac{5}{16}\frac{v^6}{c^6} + \cdots$$

Substituting the expansion into Equation 25.80 for the kinetic energy, we obtain

$$KE = mc^2 \left(1 + \frac{1}{2}\frac{v^2}{c^2} + \frac{3}{8}\frac{v^4}{c^4} + \frac{5}{16}\frac{v^6}{c^6} + \cdots - 1 \right)$$

After removing the parentheses, we find

$$KE = \frac{1}{2}mv^2 + \frac{3}{8}m\frac{v^4}{c^2} + \frac{5}{16}m\frac{v^6}{c^4} + \cdots \qquad (25.81)$$

If the speed v of the particle is small compared with the speed of light c—that is, if $v \ll c$—then only the first term of Equation 25.81 for the kinetic energy is significant. This term we recognize as the classical expression for the kinetic energy:

$$KE \approx \frac{1}{2}mv^2 = KE_{class} \qquad (v \ll c)$$

PROBLEM-SOLVING TACTIC

25.6 Remember that the relativistic kinetic energy is

$$KE = (\gamma - 1)mc^2$$

The relativistic kinetic energy is *not* $(1/2)\,mv^2$.

There are two important and frequently used expressions relating the magnitude of the relativistic momentum $p = \gamma mv$ and the total relativistic energy $E = \gamma mc^2$ of a single particle moving in one dimension. They are exact relations, not just approximations.

1. We solve Equation 25.78 for γ, obtaining

$$\gamma = \frac{E}{mc^2} \qquad (25.82)$$

Likewise, using Equation 25.74 and solving for γ, we find

$$\gamma = \frac{p}{mv} \qquad (25.83)$$

Equating the expressions for γ in Equations 25.82 and 25.83, we find

$$p = \frac{v}{c^2}E \qquad (25.84)$$

2. With Equations 25.78 and 25.74, we form the expression

$$E^2 - c^2p^2 = \gamma^2 m^2 c^4 - c^2\gamma^2 m^2 v^2$$

After substitution for γ^2, and a bit (!) of algebra to remove v, we find that the right-hand side reduces to just m^2c^4. Hence we have

$$E^2 - p^2c^2 = m^2c^4$$

or

$$E^2 = p^2c^2 + m^2c^4 \qquad (25.85)$$

EXAMPLE 25.10 ━━━━━━━━━

Calculate the rest energy of an electron. Express your result in both joules and in millions of electron-volts (MeV).

Solution

The rest energy of an electron is

$$E_{rest} = m_e c^2$$
$$= (9.11 \times 10^{-31} \text{ kg})(3.00 \times 10^8 \text{ m/s})^2$$
$$= 8.20 \times 10^{-14} \text{ J}$$

To convert to electron-volts, use the conversion factor

$$1.602 \times 10^{-19} \text{ J} = 1.000 \text{ eV}$$

Thus the rest energy in electron-volts is

$$E_{rest} = \frac{8.20 \times 10^{-14} \text{ J}}{1.602 \times 10^{-19} \text{ J/eV}}$$
$$= 5.12 \times 10^5 \text{ eV}$$
$$= 0.512 \text{ MeV}$$

STRATEGIC EXAMPLE 25.11

A proton is moving at speed $v = 0.900c$.

a. Find its total relativistic energy E. Express your result in both J and MeV.
b. Find its kinetic energy. Express your result in both J and MeV.
c. Determine the magnitude p of its relativistic momentum.

Solution

First find the factor γ:

$$\gamma = \frac{1}{\sqrt{1 - \dfrac{v^2}{c^2}}}$$
$$= \frac{1}{\sqrt{1 - (0.900)^2}}$$
$$= 2.29$$

a. The total energy is found from Equation 25.78:

$$E = \gamma m c^2$$
$$= 2.29(1.67 \times 10^{-27} \text{ kg})(3.00 \times 10^8 \text{ m/s})^2$$
$$= 3.44 \times 10^{-10} \text{ J}$$

Converting to electron-volts,

$$E = \frac{3.44 \times 10^{-10} \text{ J}}{1.602 \times 10^{-19} \text{ J/eV}}$$
$$= 2.15 \times 10^9 \text{ eV}$$
$$= 2.15 \times 10^3 \text{ MeV}$$

b. The kinetic energy is found using Equation 25.75:

$$KE = (\gamma - 1)m c^2$$
$$= (2.29 - 1)(1.67 \times 10^{-27} \text{ kg})(3.00 \times 10^8 \text{ m/s})^2$$
$$= 1.94 \times 10^{-10} \text{ J}$$
$$= 1.21 \times 10^3 \text{ MeV}$$

c. The magnitude of the relativistic momentum can be found in two ways.

Method 1

Use the definition of the relativistic momentum, Equation 25.64:

$$p = \gamma m v$$
$$= 2.29(1.67 \times 10^{-27} \text{ kg})(0.900 \times 3.00 \times 10^8 \text{ m/s})$$
$$= 1.03 \times 10^{-18} \text{ kg·m/s}$$

Method 2

Use Equation 25.84:

$$p = \frac{v}{c^2} E$$
$$= \frac{v}{c} \frac{E}{c}$$
$$= (0.900) \frac{3.44 \times 10^{-10} \text{ J}}{3.00 \times 10^8 \text{ m/s}}$$
$$= 1.03 \times 10^{-18} \text{ kg·m/s}$$

25.18 IMPLICATIONS OF THE EQUIVALENCE BETWEEN MASS AND ENERGY

The equivalence between mass and energy alluded to in Section 25.17 warrants further discussion.* Mass, like charge, is a property or attribute of a particle. Mass and charge cannot exist independently of a particle. Particles may or may not have the attribute or property called mass; they also may or may not have charge. Likewise, particles may or may not have various kinds of energy.[†] For example:

- Some particles have mass (the electron, proton, neutron, nerf ball, *you*, etc.). If a particle has mass, we imagine the particle as being tangible. If a particle has mass, it has rest energy mc^2 and may or may not have other varieties of energy such as kinetic and potential energy.
- Other particles have energy but zero mass; for example, the particle of light, the photon,[‡] has zero mass but nonetheless has energy (although not rest energy, since $m = 0$ kg means $mc^2 = 0$ J).

In an isolated system the *total* amount of energy possessed by its constituent particles is conserved; this statement is the generalized first law of thermodynamics. The total energy is the sum of the individual varieties of energy of all the particles of the isolated system. This includes the individual rest energies of particles that have mass, kinetic energies of the particles, various potential energies of the particles, and so on.

---

*An extensive discussion can be found in Ralph Baierlein, "Teaching $E = mc^2$," *The Physics Teacher*, 29, #3, pages 170–175 (March 1991).
[†]Waves also can have the attribute of energy.
[‡]We introduce the photon particle in a formal way in Chapter 26.

It is possible to convert an isolated system of particles with mass to a system of particles with less mass, even zero mass, and—remarkably—vice versa. These processes are what is typically meant by statements such as: "mass is equivalent to energy," "mass is converted into energy," or "energy is converted into mass." What is meant is that *rest energy* is converted into other types of energy, or other types of energy are converted into rest energy.

Examples include nuclear fusion and fission processes.[†] Consider the fission process of the $^{235}_{92}\text{U}$ isotope of uranium. With the addition of a slow neutron, the nucleus of this atom spontaneously splits (fissions) into a host of other particles, some with mass, some without mass. The total energy is conserved in the process. The total energy of the two initial particles, the slow neutron and the $^{235}_{92}\text{U}$ atom (including rest energy, kinetic energy, and various potential energies), is equal to the total energy of all the particles resulting from the fission process (once again, including rest energy, kinetic energy, energy associated with photons, and various potential energies).

The point is that the total energy is apportioned among the resulting particles in different ways than it was for the initial particles. One finds that the total *rest energy* of the initial particles is greater than the total rest energy of the resulting particles by some amount $(\Delta m)c^2$. Energy has *not* been created out of mass, since the energy was all there to begin with. The energy merely has been shuffled or distributed among the constituent particles in different ways than it was originally. The idea is no more subtle or profound from a conceptual point of view than the conversion of potential energy into kinetic energy when, say, a rock falls freely under the influence of the gravitational force.

The product particles have less rest energy but a correspondingly greater amount of the other kinds of energy (kinetic energy, energy of photons, etc.). The increase in the other kinds of energy in the product particles is precisely compensated for by the decrease in the rest energies.

The total energy is conserved in the isolated system.

QUESTION 4

When browsing through some artifacts in your parents' attic, you come across the physics book used by your

The conversion of rest energy into other forms of energy is what happens in nuclear explosions.

great-grandfather in 1904. In leafing through its pages, you come across a statement that "matter can neither be created nor destroyed." Is this statement still true in view of the special theory of relativity? Comment.

25.19 SPACE–TIME DIAGRAMS*

A **space–time diagram** is a graphical depiction of the space and time coordinates of a reference frame. Since there are three spatial coordinates and one temporal coordinate, a complete space–time diagram is four-dimensional! Since our powers of visualization in four dimensions are limited, we typically suppress one or more spatial dimensions, and plot the time coordinate vertical with one or two spatial coordinates horizontally, as in Figure 25.47.

An event is represented by a point on a space–time diagram. The path followed by a system on a space–time diagram is called a **world-line**. A space–time diagram is dynamic. Time stops for no man (or woman), and so if you simply stand or lie at a particular point in space as a couch potato, a vertical world-line is traced out on the space–time diagram, as shown in Figure 25.48. If you walk at constant speed along the *x*-axis of the coordinate grid, your world-line is inclined to the vertical, as also shown in Figure 25.48.

[†]Another example is the annihilation of matter with antimatter, as when an electron meets its antiparticle the positron. Both particles are annihilated and form two (or more) particles of light (photons). Since the total energy is conserved, the total energy associated with the electron–positron pair of particles (kinetic energy + rest energy) is transferred to the photon particles that have no rest energy, since they have zero mass. Also, it is possible for particles with no mass to create particles with mass. A sufficiently energetic photon (with zero mass) can create an electron–positron pair in a process known as pair production; this process is seen frequently in nuclear physics. In this case the energy of the photon is transferred to the total relativistic energy associated with the electron and positron. The photon must have an energy at least equal to the sum of the rest energies of the electron and positron for this process to occur (since the minimum value for the total relativistic energy of an electron and positron is the sum of their rest energies with zero kinetic energy).

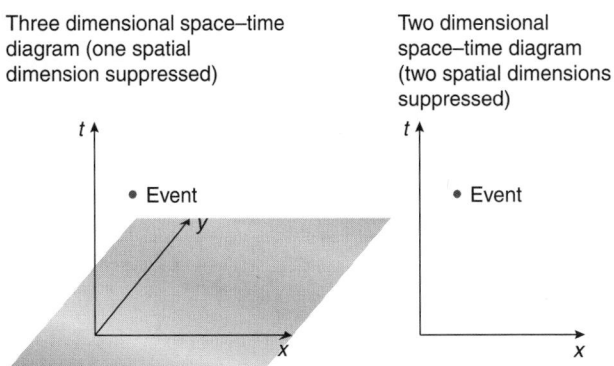

FIGURE 25.47 Space–time diagrams.

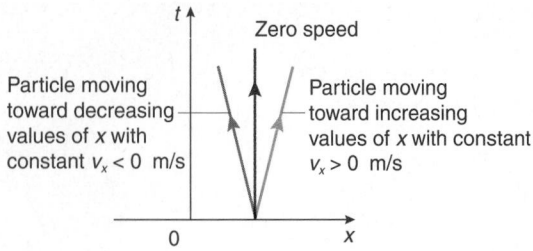

FIGURE 25.48 World-lines of particles moving with zero and nonzero constant speeds.

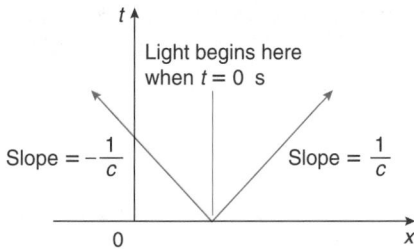

FIGURE 25.49 World-lines of light on a two-dimensional space–time diagram.

The greater the speed, the smaller the slope of the world-line on the space–time diagram. Indeed, the inverse of the slope of the world-line on a space–time diagram is the velocity component of the particle. We saw in Section 25.17 that the ultimate speed is the speed of light. If we let light propagate parallel and antiparallel to $\hat{\mathbf{i}}$, the path of the light on the space–time diagram is represented by straight world-lines whose slopes are $\pm 1/c$, as shown in Figure 25.49. Nothing can have a world-line on a space–time diagram that has a slope with an absolute magnitude less than that of the world-line for light, because it takes an infinite amount of work to change the kinetic energy of a particle with mass so that its speed is c.

A space–time diagram is a simple way to show that travel into your past (a topic of much science fiction) is impossible according to special relativity.[†] The point at the origin on a space–time diagram signifies "here and now." The future is $t > 0$ s and the past is $t < 0$ s. To travel to the past involves a world-line on a space–time diagram similar to that shown in Figure 25.50.

As the slope of the world-line decreases, the speed of the particle increases. Eventually, you reach a point P on the world-line of Figure 25.50 where the slope is equal to that of the light-line. Beyond point P, the slope of the world-line is less than that of the light-line, meaning a speed greater than c has been achieved. But such speeds exceeding c are prohibited according to special relativity because infinite amounts of energy are unavailable. Hence such a loop on a space–time diagram is impossible to complete. Thus travel into the past is *verboten* according to special relativity; the past is gone forever.

[†]There still is considerable theoretical debate about whether time travel into the past is permitted under the *general* theory of relativity, but that is too long a story for us to consider here.

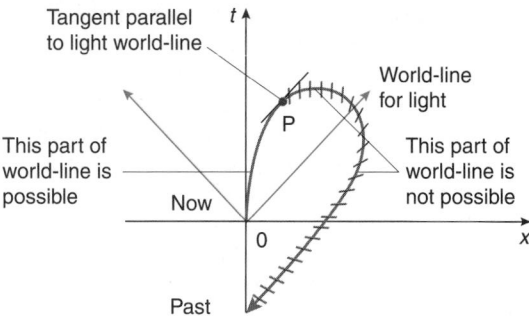

FIGURE 25.50 A hypothetical world-line into the past.

Ironically, nothing in relativity prohibits travel (at speeds $v < c$) into the future region on the space–time diagram. But once there, the future instant becomes the new "now" and travel back to the past is prohibited.

> **PROBLEM-SOLVING TACTIC**
>
> **25.7** Do not confuse a space–time diagram, which is drawn for a particular reference frame, with the standard geometry showing the relative motion of two reference frames. The latter diagrams do not show a time axis.

25.20 ELECTROMAGNETIC IMPLICATIONS OF THE SPECIAL THEORY*

Questions about electromagnetism led Einstein to relativity. One of the problems that perplexed him we examine here.

Imagine a positive charge q located at coordinate y' on the y'-axis some distance from an (infinite) array of closely spaced positive charges Q distributed uniformly along the x'-axis of a reference frame S', as in Figure 25.51. There are λ' coulombs of charge along every meter of the x'-axis in frame S'. The charge q experiences an electrical force $\vec{\mathbf{F}}'_{\text{elec}} = q\vec{\mathbf{E}}'$, where the electric field $\vec{\mathbf{E}}'$ that q experiences is caused by the array of positive charges distributed along the x'-axis. The electrical force $\vec{\mathbf{F}}'_{\text{elec}}$ on q is directed away from the array of like charges along the x'-axis as indicated in Figure 25.51.

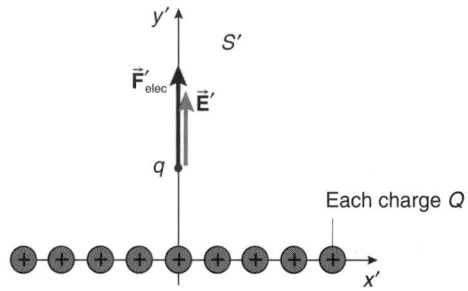

FIGURE 25.51 An array of positive charges in frame S'.

The electric field of an infinite line charge (from Equation 16.19, with $r = y'$) at the position of q has a single component along the y'-axis:

$$E_{y'} = \frac{\lambda'}{2\pi\varepsilon_0 y'}$$

Thus the repulsive force on q, measured in the S' frame, has a y'-component equal to

$$F'_{elec} = \frac{q\lambda'}{2\pi\varepsilon_0 y'} \qquad (25.86)$$

Now we imagine the S' inertial reference frame moving at constant speed v with respect to another inertial reference frame S in the standard geometry of Figure 25.52. From the viewpoint of reference frame S, the array of charges along the x'-axis is moving at speed v; there are *two* forces acting on q in this reference frame.

First, in the frame S, the λ' coulombs of charge on each meter in S' is length contracted into a length $(1\ m)/\gamma$ in frame S. Hence the line charge density in S, the number of coulombs of charge in every meter in S, *increases* to $\lambda = \gamma\lambda'$. The change in the line charge density does not violate charge conservation: the total amount of charge along the entire line is the same in the two systems; it is just distributed differently in the two reference frames. In other words, conservation of charge is unaffected by relativity. The array of charges in S produces an electrical force on q. The electrical force is determined from

$$\vec{F}_{elec} = q\vec{E}$$

where $\vec{E}$ is the electric field produced by the array of moving charges along the x-axis in S. The electric field in S also is that of an infinite line charge and has a single component at the position of q:

$$E_y = \frac{\lambda}{2\pi\varepsilon_0 y}$$

Since $\lambda = \gamma\lambda'$ and $y = y'$ (from the Lorentz transformation equations), the electric field component in S is

$$E_y = \frac{\gamma\lambda'}{2\pi\varepsilon_0 y'}$$

The electrical force on q in frame S thus has only the y-component

$$F_{elec} = \frac{q\gamma\lambda'}{2\pi\varepsilon_0 y'}$$

The electrical force in frame S is repulsive since the charges have the same sign.

Now for the second force on q. Since the charges Q are moving in S, they constitute an electrical current $I = \lambda v = \gamma\lambda'v$. This current produces a magnetic field $\vec{B}$ surrounding the x-axis, directed as shown in Figure 25.53. The magnitude of the magnetic field is that of an infinite straight current (Example 20.11, with $d = y$)

$$B = \frac{\mu_0}{4\pi}\frac{2I}{y}$$

According to frame S, the charge q also is moving with velocity $\vec{v}$ in this magnetic field. Therefore q experiences a magnetic force given by

$$\vec{F}_{magnet} = q\vec{v} \times \vec{B}$$

By using the vector product right-hand rule, the direction of this magnetic force on the positive charge q is *toward* the array of charges (see Figure 25.54). The magnitude of the magnetic force on q is (since $y = y'$)

$$F_{magnet} = qv\frac{\mu_0}{4\pi}\frac{2I}{y}$$
$$= qv\frac{\mu_0}{4\pi}\frac{2\gamma\lambda'v}{y'}$$

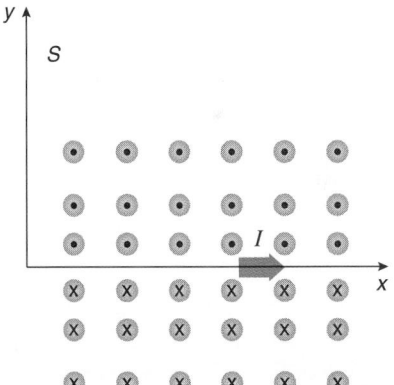

FIGURE 25.53 The magnetic field in S caused by the moving charges.

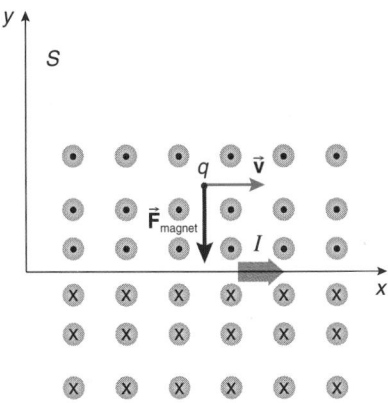

FIGURE 25.54 The magnetic force on q, as viewed in the S frame.

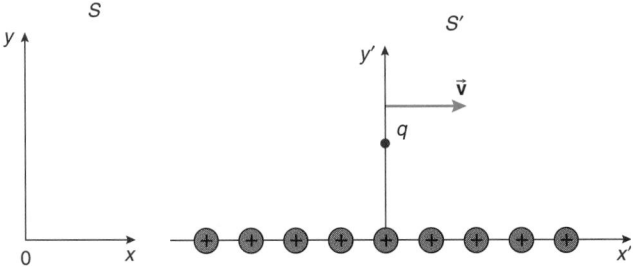

FIGURE 25.52 The standard geometry.

The total force $\vec{F}$ on q in frame S is the vector sum of the magnetic and electrical forces calculated in frame S:

$$\vec{F} = \vec{F}_{elec} + \vec{F}_{magnet}$$
$$= \frac{q\gamma\lambda'}{2\pi\varepsilon_0 y'}\,\hat{j} - qv\,\frac{\mu_0}{4\pi}\,\frac{2\gamma\lambda'v}{y'}\,\hat{j}$$

The total force has a single component F_y directed away from the array of charges along the x-axis:

$$F_y = \frac{q\gamma\lambda'}{2\pi\varepsilon_0 y'} - qv\,\frac{\mu_0}{4\pi}\,\frac{2\gamma\lambda'v}{y'}$$
$$= \frac{q\lambda'}{2\pi\varepsilon_0 y'}\,\gamma(1 - \varepsilon_0\mu_0 v^2)$$

From Equation 21.26,

$$c^2 = \frac{1}{\mu_0\varepsilon_0}$$

We use this to write the component of the total force in S as

$$F_y = \frac{q\lambda'}{2\pi\varepsilon_0 y'}\,\gamma\left(1 - \frac{v^2}{c^2}\right)$$
$$= \left(\frac{q\lambda'}{2\pi\varepsilon_0 y'}\right)\frac{1}{\gamma}$$

The term in brackets is the repulsive electrical force component F'_{elec} on q in the S' frame (Equation 25.86). Hence the repulsive force components in the two reference frames are related by

$$F_S = \frac{F_{S'}}{\gamma} \qquad (25.87)$$

> The point is this. In frame S' the force on q is purely an electrical interaction. On the other hand, in frame S the interaction is electric *and* magnetic. Thus electric fields in one inertial reference frame transform into electric *and* magnetic fields in another inertial reference frame. Likewise, magnetic fields in one inertial frame transform into magnetic *and* electric fields in another inertial frame. Thus magnetic fields can be thought of as manifestations of moving electric fields. There is no reason to take one or the other field as more basic or privileged.

Note that Equation 25.87 implies that as $v \to c$, so $\gamma \to \infty$, the total force on q in S approaches zero. Since the total force in S is the vector sum of the electric and magnetic forces, the two interactions are of equal strength when $v = c$.[†]

Relativity indicates that the way we calculate the electric fields of static charges (essentially as a manifestation of Coulomb's law) and the way in which we calculate the magnetic field of moving charges (currents) (the Biot–Savart law) are really the same fundamental interaction. Thus, instead of talking about electricity and magnetism as if they are distinct, we speak of electromagnetism to emphasize their unity.

The details of the transformation equations associated with electric and magnetic fields are a wonderful story you will learn in a subsequent course in electromagnetic theory or relativity.

25.21 THE GENERAL THEORY OF RELATIVITY*

One of the postulates of the special theory of relativity states that the laws of physics must be the same in all *inertial* reference frames. Einstein eventually was able to broaden this postulate to state that the laws of physics must be the same in *all* reference frames, not only inertial ones. This generalization is called the general theory of relativity. The extension was not easy to accomplish; it took Einstein over a decade to figure it out. Unlike special relativity, the general theory is quite difficult mathematically, and so we consider it only qualitatively and will briefly describe only some of its successful predictions.

The general theory is based on a seemingly common observation about gravity and accelerations. When you drop any mass near the surface of the Earth, the mass accelerates downward with an acceleration of magnitude $g \approx 9.81$ m/s². Recall from Chapter 6 that the local acceleration due to gravity at a point in space is the same thing as the gravitational field at that point. Now imagine yourself in a spaceship, well away from any other masses, so that the gravitational field at the position of the spacecraft is zero. If its rockets are ignited and the spacecraft is given an acceleration $\vec{a} = -\vec{g}$, there is no way, through mechanics experiments strictly confined to the interior of the spacecraft, that you can distinguish the effects of the acceleration on a physical system (such as a free mass within the spacecraft) and the effects we attribute to gravitation near the surface of the Earth; see Figure 25.55.

The general theory of relativity is based on a generalization of this observation, called the **principle of equivalence**[‡]: the effects

[‡]Another (equivalent) way to state the principle of equivalence is to say the gravitational and inertial masses of a system are identical. The principle is restricted to small regions of space over which variations in the magnitude of the gravitational field are negligible.

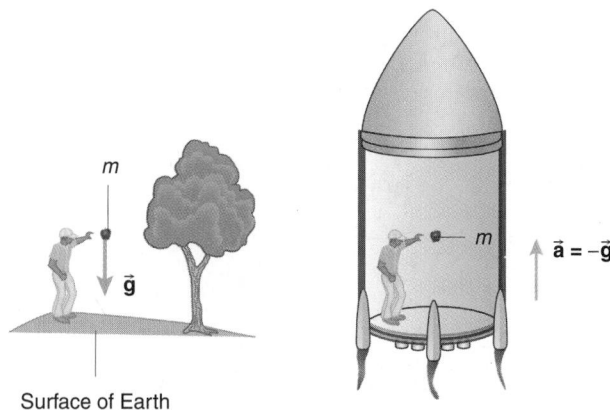

FIGURE 25.55 The effects of gravitation are indistinguishable from those of an acceleration.

of any gravitational field $\vec{g}$ are indistinguishable from the effects of an acceleration $\vec{a} = -\vec{g}$ in the absence of any gravitational field. We explore some of the consequences of the general theory, recognizing that the details involve sophisticated mathematics that we do not confront.

The Gravitational Deflection of Light

One of the predictions of the general theory is that the path of a light ray in a gravitational field will be curved or bent; we call this the **gravitational deflection of light**.

To see why the deflection occurs, imagine yourself in a uniformly *accelerating* spacecraft well away from any other masses, such as the Earth or Sun, as in Figure 25.56. A light pulse traverses the spacecraft in a direction perpendicular to its acceleration. Let the time interval during which the light travels across the spacecraft be divided into a number of equal smaller intervals Δt. Because of its acceleration, the spacecraft moves through increasingly greater distances during each successive interval Δt. If the position of the light pulse is noted after each successive interval Δt, the path of the light pulse across the spacecraft is parabolic, as shown in Figure 25.56.

According to the principle of equivalence, the effects of accelerations are indistinguishable from those of gravitation. Thus Einstein realized that a light beam will be deflected when traveling in a gravitational field (unless it is moving parallel or antiparallel to the field direction). In the gravitational field of the Earth, the deflection of a horizontally directed light ray is very small because the speed of light is so great and the gravitational field is quite weak compared with, say, the gravitational field at the surface of the Sun. Indeed, over a 100 m distance, the vertical deflection of an initially horizontal light beam on the Earth is on the order of the diameter of a typical nucleus of an atom ($\sim 10^{-14}$ m). Thus the deflection on the Earth is negligible and too small to detect.

Einstein suggested that when light from distant stars passes near the Sun (which has a much stronger gravitational field at its surface than the Earth), the light also would be deflected but by an amount that was detectable with the astronomical technology

of that time (c. 1916). For a ray grazing the solar surface tangentially, he predicted the deflection in the path of the light would be about 1.7 arc seconds ($\sim 4.7 \times 10^{-4}$ °). The effect of the deflection is to cause the star to appear at a position slightly removed from its true position, as shown in Figure 25.57.

Einstein also suggested that such a deflection could be detected by comparing star positions in the sky before and during a total solar eclipse. During such an eclipse, the light of the Sun is blocked by the Moon and stars can be seen quite near the position of the Sun in the sky. The successful observation of the deflection in 1919, and its quantitative agreement with the prediction by Einstein,* propelled his name to the fame it has worthily enjoyed ever since. The gravitational deflection of light by the very strong gravitational fields of white dwarf stars, neutron stars, and black holes is much greater than that near the Sun. Indeed, the theory predicts that compact massive objects such as black holes, neutron stars, and galactic cores even can act as *gravitational lenses* for light, producing multiple images of more distant objects along the line of sight. Such gravitationally induced multiple images also have now been observed by astronomers.

---

*One also can predict that light is deflected by the gravity of the Sun using essentially Newtonian physics, but the result obtained is exactly *half* that predicted by Einstein's general theory. The agreement with Einstein's prediction was the cause for celebration.

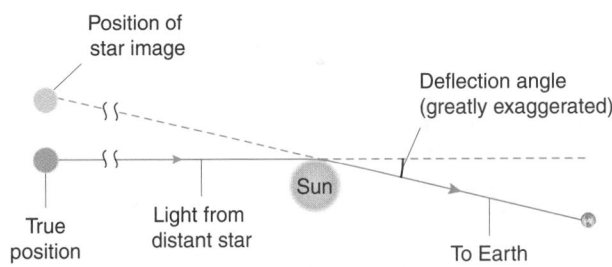

FIGURE 25.57 Gravitational deflection of light grazing the Sun.

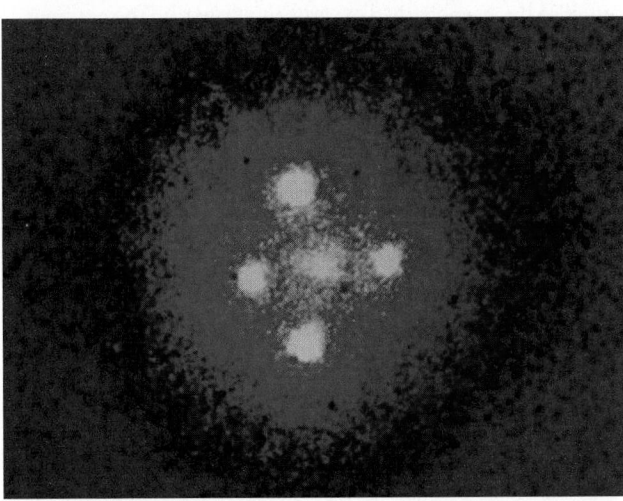

The strong gravitation of a compact object along the line of sight produces multiple images of this very distant quasar.

FIGURE 25.56 Light follows a parabolic path across the accelerating spacecraft.

The Precession of the Perihelion Point of Mercury

The general theory also was able to explain a long-standing discrepancy in the motion of the planet Mercury that was apparent by about 1845. The major axis of the elliptical orbit of Mercury gradually rotates about the Sun (which is at one focus of the ellipse, of course), causing the position of the perihelion and aphelion points of the orbit to change; see Figure 25.58. The angular speed of the rotation is quite slow, about 574 arc seconds per *century*.

> The gravitational influences of the other planets on Mercury cause most of this precession of the perihelion point, about 531 arc seconds per century. The excess precession, amounting to 574 arc seconds − 531 arc seconds = 43 arc seconds per century, is completely explained by the general theory of relativity.

The effect exists for the other planets as well, including the Earth, but is greatest for the planet Mercury because the gravitational field of the Sun that Mercury experiences is greater than for any other planet. Mercury also has the greatest speed of all the planets. The perihelion motion predicted by general relativity also depends on the orbital eccentricity, and Mercury's orbit has an eccentricity second only to Pluto's in magnitude.

Gravitational Red Shifts and Blue Shifts

> General relativity also predicts that clocks run slower in stronger gravitational fields than in weaker ones. Clocks on the basement of your building run slightly slower than those on the roof! The consequences of this for light are a **gravitational red shift** in its wavelength (to longer wavelength) for light escaping from a strong gravitational field, such as from a white dwarf star or a neutron star, and a corresponding **gravitational blue shift** (to shorter wavelength) for light falling on such objects.

The red shift of light escaping from strong gravitational fields was discovered in the 1920s, using the white dwarf companion to the bright star Sirius (α Canis Majoris). The gravitational red shift also has been verified experimentally within the confines of

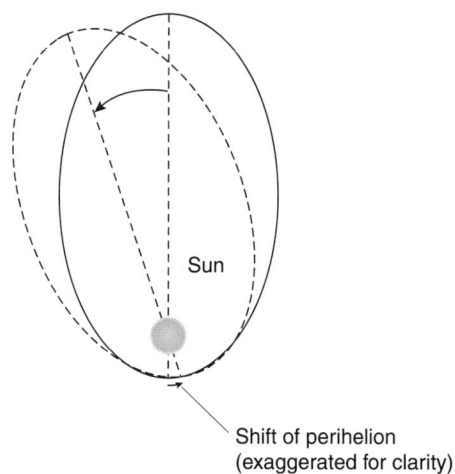

FIGURE 25.58 Long-term motion of the major axis of the orbit of Mercury.

a single laboratory on the Earth in an ingenious series of experiments over vertical distances of only several tens of meters.*

General relativity also predicts bizarre behavior of light and particles near black holes, and other objects such as collapsed star cores and compact galactic cores, whose gravitational fields in their vicinity are exceedingly large.

Despite the technical difficulties of performing experimental tests of general relativity, the theory has met every test to which it has been placed.

QUESTION 5

When the position of a star is changed by the gravitational deflection of its light by the gravitational field of the Sun (see Figure 25.57), is the image of the star we see in the sky a real or virtual image?

---

*These experiments were first made by R. V. Pound and G. A. Rebka Jr., "Apparent weight of photons," *Physical Review Letters*, 4, #7, pages 337–341 (1 April 1960), now called the *Rebka–Pound experiment*.

CHAPTER SUMMARY

A *reference frame* is a coordinate system together with a collection of synchronized clocks, distributed at rest throughout the coordinate system. An *inertial reference frame* is one in which Newton's first law of motion is valid: zero total force on a particle means the particle has zero acceleration.

To specify an *event*, one states where and when it occurs. Once a reference frame is chosen, an event is specified by indicating its three spatial coordinates and its single time coordinate.

The special theory of relativity is based on two postulates:

1. The speed of light in a vacuum has the same numerical value c, equal to exactly 299 792 458 m/s, when measured in any inertial

reference frame, independent of the motion of the source and/or the observer.

2. The fundamental laws of physics must be the same in all inertial reference frames.

Time dilation means that moving clocks run slow when compared with clocks at rest. A *proper (or rest) time interval* τ_0 is registered by a clock at rest. If that clock then has a speed v, the interval is found to be a longer time interval τ when measured using clocks at rest in the frame that sees the moving clock, where

$$\tau = \gamma \tau_0 \qquad (25.14)$$

The quantity γ is a pure number, always greater than or equal to 1:

$$\gamma \equiv \frac{1}{\sqrt{1 - \dfrac{v^2}{c^2}}} \qquad (25.13)$$

Particles with nonzero mass are restricted to speeds less than the speed of light because it takes an infinite amount of work to increase their speed to the speed of light c, at which speed their kinetic energy would be infinite. Thus γ is real and finite, since v must be less than c.

Moving lengths oriented perpendicular to the direction of their motion are unchanged by the motion.

Length contraction means that the length of a moving object (or the component of its length) oriented along the direction of its motion is measured to be shorter than the identical length at rest. A moving *proper length* ℓ_0 (as determined with rulers at rest with respect to it) is measured to be a length ℓ (as determined by rulers at rest in the frame that sees the moving object), where

$$\ell = \frac{\ell_0}{\gamma} \qquad (25.17)$$

We use a *standard geometry* that has an inertial reference frame S' traveling to the right at speed v with respect to another inertial reference frame S, with their x'- and x-axes collinear. Stopwatch clocks in each frame indicate $t = t' = 0$ s the instant the two origins of the two reference frames coincide.

The *Lorentz transformation equations* express the relationship between the space and time coordinates of an event in an inertial frame S and those of the same event in another inertial reference frame S':

$$x' = \gamma(x - vt) \qquad (25.33)$$
$$y' = y \qquad (25.34)$$
$$z' = z \qquad (25.35)$$
$$t' = \gamma\left(t - \frac{v}{c^2}x\right) \qquad (25.36)$$

The corresponding inverse equations are

$$x = \gamma(x' + vt') \qquad (25.37)$$
$$y = y' \qquad (25.38)$$
$$z = z' \qquad (25.39)$$
$$t = \gamma\left(t' + \frac{v}{c^2}x'\right) \qquad (25.40)$$

The Lorentz transformation equations reduce to the *Galilean transformation equations* for speeds $v \ll c$, in which case $\gamma \to 1$:

$$x' = x - vt \qquad (25.3)$$
$$y' = y \qquad (25.4)$$
$$z' = z \qquad (25.5)$$
$$t' = t \qquad (25.1)$$

If a particle is moving with velocity $\vec{u}' = u_x'\,\hat{\imath}' + u_y'\,\hat{\jmath}' + u_z'\,\hat{k}'$ in reference frame S', then the corresponding velocity components of the particle in reference frame S are

$$u_x = \frac{u_x' + v}{1 + \dfrac{vu_x'}{c^2}} \qquad (25.45)$$

$$u_y = \frac{u_y'}{\gamma\left(1 + \dfrac{vu_x'}{c^2}\right)} \qquad (25.47)$$

The expression for the velocity component u_z is similar to that for u_y. The quantity

$$\gamma = \frac{1}{\sqrt{1 - \dfrac{v^2}{c^2}}}$$

involves the relative speed v of the two reference frames.

When $v \ll c$, these velocity component transformation equations reduce to the Galilean transformation equations:

$$u_x = u_x' + v \qquad (25.6)$$
$$u_y = u_y' \qquad (25.7)$$
$$u_z = u_z'$$

Imagine a light source at rest that emits wavelength λ_0, corresponding to frequency ν_0. The *longitudinal Doppler effect* states that if the source is receding from an observer, and/or an observer is receding from the source at relative speed v, the light detected by the observer is *red-shifted* to longer wavelengths λ and smaller frequency ν:

$$\nu = \nu_0 \sqrt{\frac{1 - \dfrac{v}{c}}{1 + \dfrac{v}{c}}} \qquad \text{(source and/or observer receding)} \qquad (25.58)$$

If the source is approaching the observer and/or the observer is approaching the source at relative speed v, the light detected by the observer is *blue-shifted* to shorter wavelength λ and greater frequency ν:

$$\nu = \nu_0 \sqrt{\frac{1 + \dfrac{v}{c}}{1 - \dfrac{v}{c}}} \qquad \text{(source and/or observer approaching)}$$
$$(25.59)$$

The *transverse Doppler effect* indicates that light from a source moving at speed v perpendicular to the line of sight is red-shifted. The red-shifted frequency ν is

$$\nu = \nu_0 \sqrt{1 - \frac{v^2}{c^2}} \qquad (25.60)$$

For a given speed v, the transverse Doppler effect red shift produces a smaller frequency change than a longitudinal Doppler effect red shift.

The *relativistic momentum* $\vec{p}$ of a particle of mass m moving with velocity $\vec{v}$ is

$$\vec{p} = \gamma m\vec{v} \qquad (25.64)$$

where γ involves the speed v of the particle.

The relativistic kinetic energy of such a particle is

$$\text{KE} = (\gamma - 1)mc^2 \qquad (25.75)$$

Only if $v \ll c$ does the relativistic kinetic energy reduce to the familiar classical expression

$$\frac{1}{2}mv^2$$

The *rest energy* of a particle of mass m is mc^2. The *total relativistic energy* E of such a particle is the sum of its kinetic and rest energies:

$$E \equiv \text{KE} + mc^2 \qquad (25.77)$$
$$= \gamma mc^2 \qquad (25.78)$$

The total relativistic energy of a particle and the magnitude of its momentum are related by

$$p = \frac{v}{c^2} E \qquad (25.84)$$

as well as by

$$E^2 = p^2c^2 + m^2c^4 \qquad (25.85)$$

A *space–time diagram* is a graph with the time axis vertical and, typically, one or two spatial coordinate axes horizontal. An event is represented by a point on a space–time diagram. The trajectory of a single particle system on a space–time diagram is called a *world-line*.

The general theory of relativity is based on the *principle of equivalence*, which states that the effects of a gravitational field $\vec{g}$ are equivalent to the effects of an acceleration $\vec{a} = -\vec{g}$ in the absence of a gravitational field. General relativity predicts that the path of light is bent in a gravitational field (if the path is not parallel or antiparallel to the field), called the *gravitational deflection of light*. The general theory also predicts that clocks in stronger gravitational fields run slower than clocks in weaker gravitational fields. This effect results in a *gravitational red shift* of light (to longer wavelengths) when it escapes from massive objects (into regions with a weaker gravitational field) and a corresponding *gravitational blue shift* when light falls onto such objects (into regions with a stronger gravitational field).

SUMMARY OF PROBLEM-SOLVING TACTICS

25.1 (page 1159) Be sure to remember that γ is the *inverse* of the square root

$$\sqrt{1 - \frac{v^2}{c^2}}$$

The thing to keep in mind about γ is that it is *always* greater than or equal to 1:

$$\gamma \geq 1$$

25.2 (page 1164) Equations 25.33–25.36 are useful when the space and time coordinates of an event are known in the reference frame S and you need to know the corresponding coordinates of the same event in S′.

25.3 (page 1164) Equations 25.37–25.40 are useful if the space and time coordinates of an event are known in the reference frame S′ and the corresponding coordinates of the event in S are needed.

25.4 (page 1172) If $u_x' > 0$ m/s, then the object is moving to the right in S′ toward increasing values of x′. If the object is moving toward decreasing values of x′ in S′, then substitute a negative velocity component for u_x' in Equation 25.45.

25.5 (page 1173) The relativistic velocity component addition equations *always* yield velocity components less than or equal to the speed of light.

25.6 (page 1185) Remember that the relativistic kinetic energy is

$$KE = (\gamma - 1)mc^2$$

The relativistic kinetic energy is *not* $(1/2)mv^2$.

25.7 (page 1188) Do not confuse a space–time diagram, which is drawn for a particular reference frame, with the standard geometry showing the relative motion of two reference frames.

QUESTIONS

1. (page 1151); 2. (page 1155); 3. (page 1159); 4. (page 1187); 5. (page 1192)

6. Describe several reference frames that are *not* inertial reference frames. Explain why they are not inertial.

7. Using radio telescopes, we have the technological ability to detect extraterrestrial intelligent civilizations located anywhere in the galaxy, should they wish to make their presence known to us (which in itself is a good topic for discussion). Since our physical theories are creations of the human intellect, can you expect such civilizations to express the laws of physics the same way we do? If so, explain your reasons. If not, does this violate the second postulate of the special theory of relativity?

8. Explain why special relativity implies that a perfectly rigid body cannot exist.

9. If you observe a gold brick moving past you at a very high speed, will the density of the moving brick be measured to be greater than, less than, or equal to the density of a gold brick at rest with respect to you?

10. You are in a spaceship moving at a speed 0.900c away from the Sun. (a) At what speed does sunlight stream past you? (b) Do you measure your pulse rate to be slower than normal? (c) Will an observer on Earth think your pulse rate is slower than normal?

11. Time dilation means that moving clocks run slow, compared with clocks at rest with respect to you. Do observers at rest with respect to the moving clocks perceive time to run slow? Explain.

12. The speed of light was first deduced to be finite by Ole Römer about 1730 by noting a systematic discrepancy between the times the moons of Jupiter were predicted to be eclipsed and the times they actually were eclipsed. Prior to Römer, the speed of light was thought to be essentially infinite. If the speed of light were infinite, what would happen to time dilation, length contraction, and velocity component addition?

13. The equation for the relativistic kinetic energy of a mass m moving at speed v is $KE = (\gamma - 1)mc^2$. A sharp student notes that the equation apparently does not have the speed v

explicitly in it. Does this observation mean the relativistic kinetic energy is independent of the speed v? Explain.

14. The photon is the particle of light (an energy bundle). Why can you never be in a reference frame with the photon at rest?

15. When two events occur simultaneously, must they occur at the same place?

16. Discuss whether it is proper to say that "mass is energy and energy is mass."

17. Explain why a particle of mass m cannot travel at speeds equal to or greater than the speed of light c in special relativity.

18. Explain why it is not possible to travel into the past according to special relativity.

19. A broom handle of length ℓ_0 makes an angle θ' with the x'-axis of a spacecraft moving at $v = 0.99c$ with respect to reference frame S in the standard geometry. In reference frame S, is the angle θ of the broom handle with the x-axis greater than, less than, or the same as θ'? Explain your reasoning.

20. The horizontal beam from a lighthouse sweeps around the horizon at an angular speed of $(\pi/2$ rad$)$/s. The beam sweeps along the hull of a distant ship anchored with its hull perpendicular to the beam. At what distance from the lighthouse is the speed of the beam along the hull equal to the speed of light? If the ship were anchored further away, the speed of the beam along the hull would be *greater* than the speed of light. Does this violate relativity?

21. If you approach a plane mirror along the normal line at a speed of $0.800c$, does your image approach you at $1.60c$ since the image is virtual and not real?

22. Can you say correctly that the proper time interval between two events that occur at the same place is the smallest time interval between the two events?

23. When something will never occur, you may have heard someone say that it "will only occur light-years from now." What is wrong with this all-too-common statement?

24. The total mass of an electron and a proton when they are well separated from each other is slightly greater than when they are bound together in a hydrogen atom. Why?

25. Astronomers can *see* the past, but cannot go into the past. Explain this apparent paradox.

26. When light is red- or blue-shifted in the relativistic Doppler effect or in the gravitational red shift or blue shift, what is shifted and with respect to what?

27. Why can you *not* use length contraction to calculate the shortened wavelength of blue-shifted light in the Doppler effect?

28. The Concorde supersonic airplane is speeding past you at 0.50 km/s. Explain an experimental procedure you can employ to measure its length while it is in motion. If you compared your measurement with its length when at rest, would you expect a detectable difference? Explain.

29. Imagine yourself in a spacecraft orbiting the Earth every 120 minutes flying east. If you readjust your calendar watch every time you cross a time zone, you quite soon are into tomorrow and then the next day. How does the International Date Line [at longitude 180° E (or west)] adjust for such apparent time travel into the future?

30. When you fly west across a time zone, you have to adjust your watch back an hour, thus giving you a longer day than normal. Is this because of time dilation?

31. The speed of light in transparent materials is less than its speed c in vacuum. Can a particle with a nonzero mass travel faster than the speed of light (but less than c) in such materials?

32. For interplanetary spaceflight, NASA must create many complex computer programs based on the dynamics of the spacecraft. The speeds of such spacecraft occasionally are quite large, upward of 25 km/s. Can the computer programmers use mv for the magnitude of the spacecraft momentum, or must they used γmv? Explain your answer.

33. Neutrinos are particles created prolifically in nuclear reactions in stellar interiors as well as during the spectacular death throes of massive stars when they undergo supernova explosions and more nucleosynthesis. Neutrinos may have zero mass, in which case they travel at the speed of light, but there are reasons now to suspect they have a small, nonzero mass. If neutrinos do have a small mass, can they travel at the speed of light? Explain why or why not.

34. (See Question 33 as well.) About 30 000 years ago, a massive star underwent a supernova explosion in the irregular satellite galaxy of the Milky Way called the Large Magellanic Cloud. Light from the explosion first arrived at the Earth on 23 February 1987. Neutrinos are extraordinarily difficult to detect because of their weak interaction with matter. However, slightly over a dozen neutrinos from the explosion were detected on the Earth at essentially the same time as the arrival of the light (within a few hours). Discuss what this observation implies about: (a) the speed of the neutrinos; (b) whether they have nonzero mass.

35. If the chronological ordering of two events in inertial frame S indicates event A occurred before event B, does another inertial reference frame S′ exist that has the chronological ordering reversed? Distinguish between the following situations: two events that are *causally* related (event A *caused* event B) and two events that are not causally related.

36. Will a pendulum clock in an accelerating elevator tick slow or fast if the acceleration is upward? Downward? Is this what we mean by time dilation in special relativity? Is this what is meant by the relativity of time in general relativity?

PROBLEMS

Sections 25.1 Reference Frames
25.2 Classical Galilean Relativity

1. You place a collection of new clocks at rest throughout a co-ordinate system to form a reference frame. Describe an experimental procedure to synchronize the clocks, so that they all indicate the same time.

•2. You are riding your bicycle at speed 5.0 m/s along a straight, level section of the Greenbriar River Trail in West Virginia. You toss a ball at 2.0 m/s horizontally. Determine the speed of the ball with respect to the ground immediately after the ball leaves your hand if you toss the ball: (a) in the forward direction; (b) in the backward direction; (c) to the side, perpendicular to your velocity.

•3. A straight river has a current of speed v_0 that is the same across its width ℓ. You can swim at constant speed v with respect to the water. Assume $v > v_0$. (a) If you swim a distance ℓ downstream, how long an interval does the trip take? (b) How long will it take to swim back to the point you began? (c) What is the round-trip travel time? Assume you reverse direction instantaneously. (d) If you swim in such a direction that your motion with respect to the river bank is directly across the river and back along the same path, how long does the round trip take? Again, assume you can reverse direction instantaneously. (e) Which of the times you calculated in (c) and (d) is the shorter time?

Sections 25.3 The Need for Change and the Postulates
of the Special Theory
25.4 Time Dilation
25.5 Lengths Perpendicular to the Direction
of Motion
25.6 Lengths Oriented Along the Direction
of Motion: Length Contraction

4. Your spaceship is traveling off to another star and moving directly away from the Sun at a speed of $0.90c$. How fast is the light from the Sun passing you?

5. How fast must a clock be moving with respect to you so that it ticks off a 1.00 s interval while your watch indicates 3.00 s have elapsed?

6. An astronaut traveling at a speed of $0.90c$ is holding a meter stick in her hand along the line of motion. What does she measure to be the length of the meter stick?

7. A meter stick whizzes by you at a speed of $0.80c$ with its length oriented parallel to its velocity. (a) What do you measure to be the length of the moving meter stick using rulers at rest relative to you? (b) How long does it take the moving meter stick to pass by you?

•8. A clock at rest in an inertial reference frame S' is moving directly away from you (in another inertial reference frame S) at speed $v = 0.949c$. (a) What is the factor γ? (b) If the moving clock indicates an interval of 24.0 h has elapsed, what interval has elapsed on your watch? (c) Are your answers the same if the moving clock is coming directly toward you?

•9. Show that as $v/c \to 1$, an approximate expression for the relativistic factor γ is

$$\gamma \approx \frac{1}{\left[2\left(1 - \dfrac{v}{c}\right)\right]^{1/2}}$$

Evaluate γ using this expression for $v/c = 0.999\,999$.

•10. Close to the San Andreas Fault south of San Francisco lies the Stanford Linear Accelerator (SLAC for short), used to project subatomic particles to great speed. The accelerator is 3.0 km long and actually passes under a major highway. The accelerator is capable of accelerating electrons to speeds that have a relativistic factor $\gamma = 1.00 \times 10^4$. (a) Calculate the ratio of the speed of the electrons to the speed of light; use the result of Problem 9 to find v/c. Can you easily find v/c on your calculator *without* the result of Problem 9? Comment. (b) In a reference frame in which the electrons are at rest, what is the length of the accelerator? Assume (incorrectly) that the speed of the electrons is constant along the length of the accelerator.

•11. A beam of morons is moving at a speed of 2.90×10^8 m/s. They may be dumb, but they sure are fast. At this speed, the half-life of the morons is measured to be 2.50×10^{-6} s. [The half-life is the time during which half the particles change (we say decay) into other particles.] What is the half-life of a collection of morons that is at rest?

Sections 25.7 The Lorentz Transformation Equations
25.8 The Relativity of Simultaneity
25.9 A Relativistic Centipede
25.10 A Relativistic Paradox and Its Resolution*

•12. In an inertial reference frame S, the following observations are made: (i) your professor passes out a horrible test at the origin and starts a stopwatch simultaneously; (ii) 10.0 s later, you have a fit at $x = 9.0 \times 10^8$ m. The Dean of Deans (located at the origin in another inertial reference frame S' in the standard geometry) is cruising by at a speed of $0.98c$. (a) Calculate the relativistic factor γ. (b) Specify the space and time coordinates of the two events in S. (c) Use the Lorentz transformation equations to find the space and time coordinates of these events in the Dean of Deans' reference frame.

•13. You are rushing off to an 8 A.M. class at a constant speed. Your roommate, at rest, nonchalantly observes that you travel the 100.0 m length of a long corridor in 5.00×10^{-7} s. (a) What is your speed? (b) According to rulers at rest with respect to you, what is the length of the corridor? (c) How long does it take you to cover the distance according to your watch?

•14. A U.S. marshal is riding along a straight trail into town at $0.867c$. Two gunslingers are glaring at each other with only 50.0 m of dirt between them (oriented parallel to the line along which the marshal is riding), according to witnesses cowering behind hitching posts along the street. The two desperadoes draw and fire simultaneously according to them. (a) Determine the value of the relativistic factor γ. (b) How far apart are the bad guys according to the marshal? (c) Carefully sketch and define two appropriate reference frames. Then define two events corresponding to the shots fired by the two gunslingers. Transform the events into the other reference frame. (d) Which gunslinger fired first according to the reference frame of the marshal?

•15. At what speed (expressed as v/c) could an astronaut (at least in principle) travel across the diameter of our Milky Way Galaxy, a distance of about 1.0×10^5 LY, in a mission lifetime of 25 y?

•16. Two plane mirrors are located in the S' reference frame. One mirror M_1 is at $x' = 0$ m while the other (mirror M_2) is at $x' = \ell_0$ as indicated in Figure P.16. The S' reference frame with its mirrors is moving at high speed v relative to a reference frame S in the standard geometry. The origins of the two reference frames coincide when $t = t' = 0$ s. A photon (a particle of light) bounces back and forth between the mirrors and is reflected from mirror M_1 when $t = t' = 0$ s. (a) In reference frame S', when does the photon arrive at mirror M_2? (b) Specify the space–time coordinates of two events in S' corresponding to the photon at mirror M_1 and the photon at mirror M_2. (c) In reference frame S, when does the photon arrive at mirror M_2? (d) In reference frame S, where on the x-axis is mirror M_2 when the photon arrives at that mirror? (e) Find the distance Δx in S that the photon travels while moving from mirror M_1 to mirror M_2. (f) Find the interval Δt in S that it takes the photon to travel from mirror M_1 to mirror M_2. (g) Take the ratio of

$$\frac{\Delta x}{\Delta t}$$

using the results of parts (e) and (f). Is this ratio what you expected to obtain?

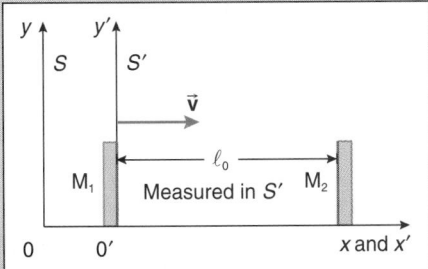

FIGURE P.16

•17. You decide to stop the world and get off by heading for α Centauri, the nearest star to the Sun, located 4.2 LY away. (a) At what constant speed must you travel to get to α Centauri in 3.0 y according to a calendar and watch you carry along in your pocket? (b) How long does your journey take according to your friends left behind on the Earth?

•18. Two unforgettable events occur 1.00 s apart at positions separated by 2.40×10^8 m along the x-axis in the well-known inertial reference frame S. The equally prestigious inertial reference frame S' is moving with the standard geometry. Both events occur at the same position on the x'-axis. (a) Delineate

the space and time coordinates of the two events in the reference frame S. (b) How fast is the S' reference frame moving with respect to S? (c) What is the time interval between the events in S'?

•19. Two events occur in the inertial reference frame S and have space and time coordinates (x_1, t_1) and (x_2, t_2). (a) Find the speed of another inertial reference frame S' in which the two events occur simultaneously. (b) Find the spatial separation $\Delta x'$ of the two events in this reference frame. (c) Discuss the feasibility of finding yet another reference frame in which the two events occur at the same place.

•20. You are in a spaceship (of length 100 m) moving at speed $0.900c$ to visit the bright star Arcturus (α Boötis). Back home, your mother notices that you forgot your socks. She sends a light signal to the spaceship. Let S' be the reference frame of the spaceship and S be the reference frame of home-sweet-home, back on the Earth. The signal arrives at the tail of the spaceship when $t = t' = 0$ s. (a) When does the light signal reach the front of the spaceship according to the spaceship clocks? (b) When does the light signal reach the front of the spaceship according to your mother's clocks? (c) Are the answers to (a) and (b) related by the time dilation equation? Why or why not? (d) The light signal is reflected back to the tail by a mirror in the nose of the spaceship. When does the light signal reach the tail according to the spaceship clocks? When does it arrive at the tail according to the Earth-based clocks? (e) Determine the total distance traveled by the light signal as it travels from the tail to the nose and then back to the tail according to: (i) you, inside the spaceship; (ii) your mother, on the Earth.

•21. A meteor crashes into your lecture room leaving a gaping hole in the ceiling. Precisely 1.10 s later (according to clocks in the classroom), hawk-eyed astronomers at your college notice that a meteor crashes into the Moon (3.84×10^5 km away), creating a fresh crater. A tabloid newspaper hypothesizes that both events were caused by a strafing run from a passing flying saucer dropping meteors when near the Moon and near the Earth. Assume (unrealistically) the fall-time of the rocks from the saucer is negligible. (a) How long does it take light to propagate from the Moon to the Earth according to your classroom clock? (b) What is the time interval between the impacts as measured by clocks in the reference frame of the Earth? Which event occurred first in the Earth–Moon reference frame? (c) Could the impacts have been caused by such a flying saucer? Explain your answer.

•22. When $t = 0$ s, you notice a speeder go by in a new Ferrari at speed $0.200c$. Later, when $t = 10.0$ s, a state trooper goes by at a speed of $0.500c$ with flashing lights atop his Yugo. The lawyers, of course, bring up the rear. (a) At what time (according to your clocks) does the trooper catch up with the speeder? (b) How far from you does the trooper nab the speeder?

•23. In reference frame S', a flashbulb goes off at the origin and the light propagates toward positive x' and negative x' to two detectors, each located 24.0×10^8 m from the S' origin as measured in S' (see Figure P.23). (a) Define two events in S'

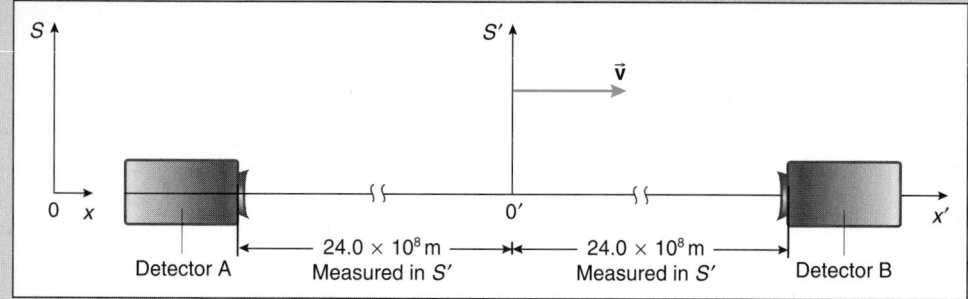

FIGURE P.23

corresponding to the detectors' reception of the light. (b) Reference frame S' is moving at speed $v = 0.995c$ relative to reference frame S in the standard geometry. Find γ. (c) Find the spatial separation Δx between the two events in S. (d) Find the interval Δt between of the two events in S. (e) Which event occurs first in S?

‡24. If the transformation equations between the inertial reference frames S and S' are *not* linear, peculiar things happen. For example, suppose the transformation equation for x has the quadratic form

$$x^2 = Ax' + Bt'$$

where A and B are two constants that depend on the relative speed v of the two frames. Imagine two one-meter sticks nailed down in reference frame S, one meter stick between coordinates $x = 2$ m and $x = 3$ m and the other between $x = 4$ m and $x = 5$ m, as in Figure P.24. (a) Write the given transformation equation for each given value of x. (b) Since the meter sticks are at rest in reference frame S, they are moving in S'. Hence, to measure their length, you must determine where the ends of the sticks are at the same time in S', say $t_1' = t_2' = t_3' = t_4' = 0$ s. Find the lengths of the two sticks in S' and note that they are not the same. This situation is ridiculous: equal lengths in S have *different* lengths in S' depending on where the objects are located! The only way we can avoid this unpleasant result is to have linear transformation equations between the reference frames.

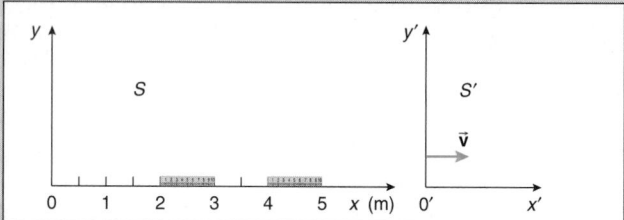

FIGURE P.24

‡25. (a) Use the binomial expansion to show that if $v/c \ll 1$, then the amount $\Delta \ell \equiv \ell - \ell_0$ that a proper length ℓ_0 is contracted is approximately

$$\Delta \ell \approx -\frac{1}{2}\frac{v^2}{c^2}\ell_0$$

(b) The Earth has a diameter of 12.7×10^3 km and an orbital speed of about 30 km/s. By about how much is the diameter of

the planet length contracted because of its orbital motion, when viewed from the Sun? Is every measure of the diameter so contracted?

Section 25.11 Relativistic Velocity Addition
25.12 Cosmic Jets and the Optical Illusion
of Superluminal Speeds*

•26. Two spacecraft are approaching the Earth at equal speeds from opposite directions. Each spacecraft sees that the other spacecraft approaching at a speed $0.90c$. How fast are the spacecraft approaching the Earth?

•27. The chase is on. The Lone Ranger, galloping at $0.950c$, is chasing a desperado, moving at speed $0.900c$, in a direction directly away from Dodge City, Kansas. What is the speed at which the desperado sees the Lone Ranger approaching her?

•28. Two protons in a laboratory particle accelerator are fired toward each other, each with speed $0.990\ 000c$ when measured in the lab. What is the speed at which one proton sees the other approaching?

•29. In inertial reference frame S', a particle is fired along the y'-axis at speed $0.995c$ parallel to $\hat{\jmath}'$, measured using rulers and clocks in the S' frame. The S' reference frame is moving at speed $0.990c$ with respect to inertial reference frame S in the standard geometry. (a) Determine the velocity components u_x and u_y of the particle in S. (b) What is the angle the velocity vector $\vec{u}$ makes with the x-axis in S?

•30. A high-frequency lighthouse beacon sweeps around the horizon with a frequency of 1.43 kHz. At what distance is the beam sweeping perpendicular to the line of sight from the lighthouse with a speed of $1.5c$? Does this result violate the first postulate of the special theory of relativity? Explain.

•31. A particle in the S' reference frame has a velocity $\vec{u}' = 0.779c\hat{\imath}' + 0.450c\hat{\jmath}'$. The S' frame is moving in the standard geometry with speed $0.950c$ with respect to reference frame S. (a) Sketch the situation. (b) What is the angle θ' that $\vec{u}'$ makes with the x'-axis in S'? (c) What is the speed u' of the particle in the S' frame when measured with clocks and rulers at rest in S'? (d) Find the velocity components u_x and u_y of the particle in the S reference frame. (e) What is the speed u of the particle in the S reference frame, measured using rulers and clocks in that frame? (f) What angle θ does the velocity vector $\vec{u}$ make with the x-axis in S?

‡32. A light source at the point P in Figure P.32 is turned on at time $t = 0$ s for a very short interval and then extinguished. (a) At

what time does the light arrive at the origin O? (b) At what time does the light arrive at a point A located a distance x from the origin along the $+x$-axis? (c) The spot of light moves from point O to point A in a time given by the difference between the answer to part (b) and the answer to part (a). Show that the *average* speed $\langle v \rangle$ of the light spot in moving along the x-axis from O to A is

$$\langle v \rangle = c \frac{a + (a^2 + x^2)^{1/2}}{x}$$

(d) Show that when x is small, $\langle v \rangle \to \infty$ m/s, and when $x \to \infty$ m, $\langle v \rangle$ approaches c. The spot of light, therefore, begins to move out along the x-axis at a speed *greater* than the speed of light and *slows down* to the speed of light as $x \to \infty$ m. Does this result contradict the special theory of relativity? Explain.

This problem was inspired by Gilbert W. Kessler, "Shadows," *The Physics Teacher*, 17, #5, pages 315–316 (May 1979).

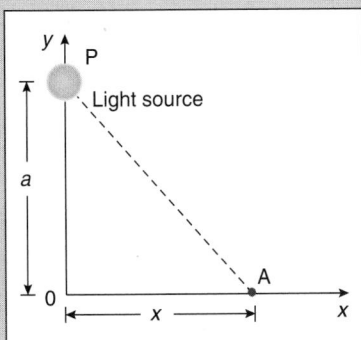

FIGURE P.32

$\boxed{3}$33. In special relativity two reference frames S and S′ are inertial reference frames. A particle with velocity component u_x in inertial reference frame S has an acceleration component

$$a_x = \frac{du_x}{dt}$$

The acceleration component of the same particle in reference S′ is

$$a_x' = \frac{du_x'}{dt'}$$

Corresponding definitions exist for the other components of the acceleration. The relative speed v of the two inertial frames is constant. By differentiating the velocity component transformation equations (Equations 25.45 and 25.47 on page 1172), show that the acceleration components are related by the following ghastly expressions:

$$a_x = \frac{a_x'}{\gamma^3 \left(1 + \dfrac{vu_x'}{c^2}\right)^3}$$

$$a_y = \frac{\left(1 + \dfrac{vu_x'}{c^2}\right) a_y' - \dfrac{vu_y'}{c^2} a_x'}{\gamma^3 \left(1 + \dfrac{vu_x'}{c^2}\right)^3}$$

where

$$\gamma = \frac{1}{\sqrt{1 - \dfrac{v^2}{c^2}}}$$

Sections 25.13 The Longitudinal Doppler Effect
25.14 The Transverse Doppler Effect*
25.15 A General Equation for the Relativistic Doppler Effect*

•**34.** (a) Show that when $\theta = 0°$ in Equation 25.61 on page 1180, a purely longitudinal Doppler blue shift results—that is, Equation 25.61 reduces to Equation 25.59. (b) Show that when $\theta = 180°$ in Equation 25.61, a purely longitudinal Doppler red shift results—that is, Equation 25.61 reduces to Equation 25.58. (c) Show that when $\theta = 90°$ in Equation 25.61, a purely transverse Doppler effect (a red shift) results—that is, Equation 25.61 reduces to Equation 25.60.

•**35.** For a given value of v/c, show that the ratio of the fractional frequency changes v/v_0 produced by the red shifts of the transverse Doppler to the longitudinal Doppler effect is

$$1 + \frac{v}{c}$$

•**36.** On the same graph, plot v/v_0 versus v/c for the red shifts produced by the longitudinal Doppler effect and the transverse Doppler effect for v/c between 0.90 and 1.00. Discuss and interpret the graphs.

•**37.** Using the firefly as in Section 25.13 and the standard geometry, show that a decrease in frequency (a red shift to longer wavelengths) occurs for a detector at the origin in S′, moving away from a stationary firefly at the origin in S. Hint: Define suitable events in S, transform them to S′, and proceed as in Section 25.13.

•**38.** You are vigorously testing a new sports car. Your speed is so great that a red traffic light ($\lambda \approx 620$ nm) appears green ($\lambda \approx 540$ nm). An officer of the law pulls you over and gives you a ticket for running a red light. In court, you contest the charge, indicating with relativistic calculations that the red light certainly looked green as you approached the light. The ticket should have been written for speeding instead; the charge is dismissed on this legal technicality. Now that is practical relativity! Calculate the speed at which you approached the traffic light.

•**39.** Quasars are among the most perplexing objects in astronomy. They are thought to be at the fringes of the observable universe and are being carried away from us by the expansion of the universe itself. Their speeds of recession produce enormous red shifts in their spectra. The red shift z is formally defined as

$$z \equiv \frac{\Delta\lambda}{\lambda_0}$$

where $\Delta\lambda \equiv \lambda - \lambda_0$ and λ_0 is the proper wavelength of the light. Some quasars have red shifts with $z = 4.0$. Calculate the speed at which these objects are receding from us.

•**40.** If $v \ll c$, use the binomial theorem to show that the red shift in wavelength $\Delta\lambda \equiv \lambda - \lambda_0$ caused by the *transverse* Doppler effect is approximately

$$\Delta\lambda \approx \lambda_0 \, \frac{v^2}{2c^2}$$

where λ_0 is the wavelength of the light measured with the source at rest.

•**41.** A star emits light from hydrogen with a wavelength $\lambda_0 = 656.282$ nm. How fast must the star move *transverse* to the line of sight for you to measure its wavelength to be 656.382 nm?

•**42.** If $v \ll c$, use the binomial theorem to show that the red shift in wavelength $\Delta\lambda \equiv \lambda - \lambda_0$ caused by the *longitudinal* Doppler effect is approximately

$$\Delta\lambda \approx \lambda_0 \, \frac{v}{c}$$

where λ_0 is the wavelength of the light measured with the source at rest.

•**43.** A star emits light from hydrogen with a wavelength $\lambda_0 = 656.282$ nm. How fast must the star move *along* your line sight for you to measure its wavelength to be 656.382 nm? Is the distance between the source and receiver increasing or decreasing with time?

Sections 25.16 Relativistic Momentum
25.17 The CWE Theorem Revisited
25.18 Implications of the Equivalence Between Mass and Energy

44. The speed of a particle is increased by a factor of two from 1.25×10^8 m/s to 2.50×10^8 m/s. (a) By what factor does its momentum increase? (b) By what factor does its kinetic energy increase?

45. Calculate your rest energy and the price it would bring at $0.10 per kilowatt-hour.

46. (a) Calculate the rest energy of a penny with a mass of 5.0 g. (b) If a 1.0 kg mass had a kinetic energy equal to the rest energy of the penny, at what speed would the kilogram be moving?

47. The luminosity of the Sun is 3.83×10^{26} W. By how much does the mass of the Sun decrease each second because of its luminosity?

48. How much mass must be completely converted from rest energy into other forms of energy in order to keep a 100 W light bulb burning for one century?

49. Calculate the kinetic energy, total relativistic energy E, and the magnitude of the momentum p of an electron traveling at a speed of $0.600c$.

50. Show that the momentum of a particle with *zero* mass is

$$p = \frac{E}{c}$$

where E is its relativistic energy.

•**51.** A 1.0 g sugar cube travels past you at a speed of $0.90c$. How sweet it is! You measure its density while it is moving. By what numerical factor is the density of the cube changed because of its speed?

•**52.** (a) To double the relativistic kinetic energy, how is the factor gamma at the higher speed (say γ') related to the gamma at the slower speed (say γ)? (b) Show that in the nonrelativistic limit, when $v \ll c$, the expression derived in part (a) implies an increase in speed by a factor of $\sqrt{2}$ doubles the nonrelativistic kinetic energy.

•**53.** (a) At what speed is the total relativistic energy of a proton 10% greater than its rest energy? (b) How much work is necessary to accelerate the proton from rest to this speed?

•**54.** (a) At what speed must an electron travel so that its kinetic energy is equal to its rest energy? (b) At what speed must a proton travel so that its kinetic energy is equal to its rest energy?

•**55.** Show that if E is the total relativistic energy of any particle, then

$$\frac{dE}{dp} = v$$

where p is the magnitude of the relativistic momentum and v is the speed of the particle.

•**56.** One of the most energetic cosmic ray protons detected had a kinetic energy estimated to be about 1.0 J. For such a proton, by what amount is v/c less than exactly 1?

•**57.** An electron at rest is given a present of 1.60×10^{-15} J of kinetic energy. (a) Find the speed of the electron. (b) Express the kinetic energy of the electron in units of keV. (c) What is the magnitude of the momentum of the electron?

•**58.** An electron moves horizontally and parallel to a wall at speed $0.90c$ through your physics lab (of dimensions 10.0 m × 10.0 m × 3.0 m) on its way to its annihilation with a positron next door. (a) What is the volume of your lab as measured by you? (b) What is the volume of your lab in a reference frame at rest with respect to the electron? (c) Find the kinetic energy of the electron as measured in the lab. (d) Through what potential difference did the electron accelerate from rest to attain the speed of $0.90c$?

•**59.** A proton has a total relativistic energy of 1500 MeV. (a) What is its rest energy in MeV? (b) What is its speed? (c) What is the magnitude of its momentum?

•**60.** (a) How much work must be done to bring an electron from rest to a speed of $0.600c$? Express your result in both joules and keV. (b) How much additional work is needed to increase its speed from $0.600c$ to $0.800c$? Express your result in both joules and keV. (c) What is the ratio of the kinetic energy of the electron at the speed $0.800c$ to its kinetic energy at the speed $0.600c$?

•**61.** A politician of mass 70.0 kg is ejected from office and sent at a speed of $0.980c$ to a planet located 20.0 LY from the Earth. (a) What is the kinetic energy of the politician according to us on the Earth? (b) What is the kinetic energy of the politician according to the politician? (c) How many years (measured with clocks at rest on the Earth) will the one-way trip take? (d) According to the politician, how far away will the Earth be when the trip is completed (but the politician is still moving at the same speed)? (e) How many years will the trip take according to clocks at rest with respect to the politician?

•**62.** The energy associated with the eruption of the Mt. St. Helens volcano (in southern Washington state) on 18 May 1980 was estimated to be about 10^{17} J. At about what speed would a 10^3 kg spacecraft have a kinetic energy equal to this?

•**63.** (a) Through what potential difference must an electron be accelerated so that its total relativistic energy is 1.0% greater than its rest energy? (b) Will the result calculated in part (a) be different if the particle is a proton? (c) Repeat the calculation for a total relativistic energy that is 10% greater than the rest energy.

•**64.** Two protons approach each other at equal speeds 0.800c in some reference frame. What is the kinetic energy of one proton as seen from a reference frame with its origin on the other proton?

•**65.** (a) Beginning with the left-hand side of the following relationship, show that it is equal to the right-hand side using the definition of the relativistic factor γ:

$$\frac{\gamma^2 v^2}{\gamma + 1} = c^2(\gamma - 1)$$

(b) Use the result in part (a) to show that the total relativistic energy E can be expressed as

$$E = \gamma mc^2$$
$$= mc^2 + \frac{m\gamma^2 v^2}{\gamma + 1}$$

(c) Use the result of part (b) to show that the relativistic kinetic energy KE is related to the nonrelativistic kinetic energy

$$KE_{class} = \frac{mv^2}{2}$$

in the following way:

$$KE = KE_{class}\frac{2\gamma^2}{\gamma + 1}$$

This problem was inspired by Wendell G. Holladay, "The derivation of relativistic energy from the Lorentz γ," *American Journal of Physics*, 60, #3, page 281 (March 1992).

•**66.** A mass m at rest spontaneously disintegrates into two masses m_1 and m_2 with speeds v_1 and v_2 respectively. Show that conservation of energy implies that $m > m_1 + m_2$.

•**67.** Nuclear weapons convert rest energy into other forms of energy. The nuclear weapon used in the bombing of Hiroshima on 6 August 1945 had an estimated yield equivalent to the detonation of about 10^4 metric tons of TNT,* and so was called a 10 kiloton weapon. Each metric ton of TNT yields about 4.9×10^9 J when detonated. (a) Estimate the amount of mass that vanished when the nuclear weapon exploded. (b) If 10^4 metric tons of TNT exploded, does the same amount of mass vanish?

•**68.** On the same graph, make schematic plots of the total relativistic energy E versus the magnitude p of the momentum for each of the following: (a) A particle with mass m. What is the meaning of the intercept of this graph? For large p, what is the slope of this graph? (b) A particle with zero mass. What is the slope of this graph? (c) A classical particle with zero rest energy. What is the intercept of this curve? What is the shape of this curve?

Section 25.19 Space–Time Diagrams*

69. Draw a space–time diagram with its time axis vertical and one spatial dimension. Schematically illustrate the world-lines that represent you: (a) while sleeping; (b) while walking in a straight line to class; (c) running in a straight line from class; (d) pacing back and forth when nervous about an upcoming exam.

•**70.** Draw a space–time diagram that schematically illustrates the world-line of a ball tossed vertically from the origin near the surface of the Earth. What is the mathematical shape of the world-line?

•**71.** Draw a space–time diagram and schematically sketch the world-line of a particle executing simple harmonic motion. The equilibrium position is at $x = 0$ m. The particle is released at $x = A$ when $t = 0$ s. What is the mathematical shape of the world-line?

•**72.** Draw an appropriate space–time diagram and schematically sketch the world-line of a particle executing uniform circular motion in the x–y plane. Describe the shape of the world-line in a cogent sentence or two.

•73. A flash of light occurs at $x = 0$ m when $t = 0$ s and moves toward increasing values of x. Draw a space–time diagram that accurately plots the world-line of the light.

•**74.** The starship *Enterprise* in *Star Trek* occasionally traveled at "warp speeds" of, say, three times the speed of light (which is not permissible in special relativity). On the same space–time diagram, sketch the world-line of light propagating along the x-axis and the world-line of the starship when traveling at such a fictional speed in the same direction. Explain why such speeds are not possible in special relativity.

*Trinitrotoluene ($C_7H_5N_3O_6$).

A. Expanded Horizons

1. For millennia, the speed of light was thought to be infinite. Ole Römer (1644–1710) was the first to realize that light travels at a finite speed with his investigations of the eclipses of the Galilean satellite Io of the planet Jupiter in 1676. Investigate Römer's discovery and various other historical and contemporary methods for measuring the speed of light.
   Andrzej Wróblewski, "De mora luminis: a spectacle in two acts with a prologue and an epilogue," *American Journal of Physics*, 53, #7, pages 620–630 (July 1985); this article corrects many historical errors in the physics literature regarding the nature of the discovery by Römer.
   Harry E. Bates, "Resource letter RMSL-1: Recent measurements of the speed of light and the redefinition of the meter," *American Journal of Physics*, 56, #8, pages 682–687 (August 1988); this contains an extensive bibliography.

2. Investigate the background, details, and results of the Michelson–Morley experiment (c. 1880s) and the impact (if any) of the experiment on Einstein's development of the special theory of relativity.
   Bernard Jaffe, *Michelson and the Speed of Light* (Anchor Books, Garden City, New York, 1960).
   Yoram Kirsh and Meir Meidav, "The Michelson–Morley experiment and the teaching of special relativity," *Physics Education*, 22, #5, pages 270–273 (September 1987).
   Gerald Holton, "Einstein and the 'crucial' experiment," *American Journal of Physics*, 37, #10, pages 968–982 (October 1969).
   R. S. Shankland, "Michelson–Morley experiment," *American Journal of Physics*, 32, #1, pages 16–35 (January 1964).
   Gerald Holton, "On the origins of the special theory of relativity," *American Journal of Physics*, 28, #7, pages 627–636 (October 1960).
   Isaac Asimov, *How Did We Find Out About the Speed of Light?* (Walker, New York, 1986).
   R. S. Shankland, "Michelson: America's first Nobel Prize winner in science," *The Physics Teacher*, 15, #1, pages 19–25 (January 1977).
   Arthur I. Miller, *Albert Einstein's Special Theory of Relativity: Emergence (1905) and Early Interpretation (1905–1911)* (Springer-Verlag, New York, 1997).

3. Hypothetical particles that travel *faster* than the speed of light (so-called *tachyons*) have intrigued physicists. The characteristics of such hypothetical particles have been investigated in a number of different ways. One way uses some new math involving *perplex numbers* whose absolute value is −1, somewhat like the *complex* numbers stemming from $\sqrt{-1}$. This method is explored in the following:
   Paul Fjelstad, "Extending special relativity via the perplex numbers," *American Journal of Physics*, 54, #5, pages 416–422 (May 1986).
   Other ways of investigating such hypothetical particles can be found in the following references:
   Olexa-Myron Bilaniuk and E. C. George Sudarshan, "Particles beyond the light barrier," *Physics Today*, 22, #5, pages 43–51 (May 1969); 22, #12, pages 47–52 (December 1969).
   Laurence M. Feldman, "Short bibliography on faster-than-light particles (tachyons)," *American Journal of Physics*, 42, #3, pages 179–182 (March 1974).
   Edwin F. Taylor, "Why does nothing move faster than light? Because ahead is ahead!" *American Journal of Physics*, 58, #9, pages 889–890 (September 1990).

4. Investigate the nature of the expansion of the universe. In particular, address the question of whether expansion speeds really are Doppler-type speeds and whether expansion speeds can exceed the speed of light without violating one of the postulates of relativity.
   Edward R. Harrison, *Cosmology* (Cambridge University Press, Cambridge, England, 1981), Chapter 10, pages 206–230.
   H. S. Murdoch, "Recession velocities greater than light," *Quarterly Journal of the Royal Astronomical Society*, 18, #2, pages 242–247 (June 1977).

5. Investigate in greater detail the ideas involved in the general theory of relativity. In particular, discover the postulates of the theory and the investigate the three classical tests of the theory discussed in Section 25.21 in greater detail: (1) the gravitational deflection of light (and how it was first detected during a solar eclipse in 1919); (2) the motion of the perihelion point of the planet Mercury (and why it is that, of all the planets, it is Mercury that is most affected by the theory); and (3) the gravitational red shift of light (and how it was first detected with light from the white dwarf star known as Sirius B in the constellation Canis Major).
   Wolfgang Rindler, *Essential Relativity* (Van Nostrand Reinhold, New York, 1969), pages 130–151.
   Edward R. Harrison, *Cosmology* (Cambridge University Press, Cambridge, England, 1981), Chapter 8, pages 160–184.
   S. Chandrasekhar, "Einstein and general relativity: historical perspectives," *American Journal of Physics*, 47, #3, pages 212–217 (March 1979).
   Clifford M. Will, "Testing general relativity: 20 years of progress," *Sky and Telescope*, 66, #4, pages 294–299 (October 1983).
   P. W. Worden Jr. and C. W. F. Everitt, "Resource letter GI-1: Gravity and inertia," *American Journal of Physics*, 50, #6, pages 494–500 (June 1982); this contains many references to the topic.
   S. Chandrasekhar, "On the derivation of Einstein's field equations," *American Journal of Physics*, 40, #2, pages 224–234 (February 1972).

6. Discover the association between the general theory of relativity, gravity, and geometry. In particular, investigate why the geometry around massive objects (such as the Sun) is not Euclidean but of positive curvature. Also research contemporary ideas about the possible geometry of the universe itself.
   George Gamow, *Gravity* (Anchor, Garden City, New York, 1962), pages 115–146.
   P. K. MacKeown, "Gravity is geometry," *The Physics Teacher*, 22, #9, pages 557–564 (December 1984).
   Edward R. Harrison, *Cosmology* (Cambridge University Press, Cambridge, England, 1981), Chapter 7, pages 147–159.
   Richard H. Price, "General relativity primer," *American Journal of Physics*, 50, #4, pages 300–329 (April 1982); and Edward P. Tryon, "Comment on 'General relativity primer,'" *American Journal of Physics*, 52, #4, pages 366–367 (April 1984).
   P. W. Worden Jr. and C. W. F. Everitt, "Resource letter GI-1: Gravity and inertia," *American Journal of Physics*, 50, #6, pages 494–500 (June 1982); this contains many references to the topic.
   Peter G. Bergmann, *The Riddle of Gravitation* (Dover, New York, 1992).

B. Lab and Field Work

7. There are various ways to measure the speed of light within the confines of a laboratory. Consult your instructor about such methods; then design and perform an experiment to measure c.
   Harry E. Bates, "Resource letter RMSL-1: Recent measurements of the speed of light and the redefinition of the meter," *American Journal of*

Physics, 56, #8, pages 682–687 (August 1988); this contains an extensive bibliography.

8. Visit a professor in the astronomy department at your college or university and secure spectra that illustrate the red and blue shifts of light from stars. Use data from the spectra to calculate the speed of recession and approach of the given stars. Also secure the spectrum of a quasar and determine its recessional speed.

C. Communicating Physics

9. Within the genre of science fiction, it sometimes is difficult for the uninitiated to sort scientifically bogus ideas from those that are scientifically plausible on the basis of the science we now know. Choose a favorite work of science fiction, either in print, film, or video form. Analyze the work, sorting the science into fiction and truth; use your professor as a resource for this if necessary. Can you cite examples of ideas in science fiction that were fiction at the time they were written or produced, but subsequently have been found to be scientifically correct or at least plausible?

CHAPTER 26

AN APERITIF
MODERN PHYSICS

[I] tried immediately to weld the elementary quantum of action *h* somehow into the framework of the classical theory. But in the face of all such attempts, this constant showed itself to be obdurate. . . . My futile attempts to fit the elementary quantum of action into the classical theory continued for a number of years, and they cost me a great deal of effort. Many of my colleagues saw in this something bordering on a tragedy. But I feel differently about it. For the thorough enlightenment I thus received was all the more valuable. I now knew for a fact that the elementary quantum of action played a far more significant part in physics than I had originally been inclined to suspect.

Max Planck*

Physics seemed quite tidy in 1890, quaintly formal and typically Victorian. The approach of the discipline to nature was essentially mechanistic and deterministic. With the successes of Newton's dynamical theory in the 17th century, and the development of thermodynamics, electromagnetism, and more sophisticated ways of solving complicated equations during the 19th century, some physicists seemed almost cocky that the end of physics was almost nigh. What lay ahead could not even be imagined.

Between about 1890 and 1925, a number of startling, perplexing, and seemingly unconnected discoveries were made that eventually converged to provide explanations for a host of divergent phenomena. The time was one of unusual ferment that led us into a totally new and unimagined submicroscopic world and worldview of physics. A study of the scientific history of this period reveals much of how science proceeds: haltingly and with hesitation, carefully and with caution, chaotically and with chance playing a major role. Yet the period was filled with bold new hypotheses and theoretical insight, based fundamentally on the keen and wary eye of scientists at first simply intrigued by unexpected observations coming from seemingly mundane experiments. The many accidental discoveries gradually fueled progress on many unexpected fronts; finally, a clearer picture emerged in the late 1920s of a new and almost incomprehensible submicroscopic world.

It is only with the luxury of hindsight that we can spy and superimpose a somewhat more orderly, but imperfect, pedagogical thread through the maze of new knowledge. The path we choose in presenting this material somewhat distorts the historical record and glosses over the confusion experienced by physicists of the time, but should give you a feeling for how physicists confronted and explored the submicroscopic world of the atom and nucleus. You may get the impression that the development was so logical as to be inevitable, but nothing could be further from the truth. All scientists proceed much like sleepwalkers,[†] simultaneously aware and yet unaware of just what they are doing.[‡] The advance of science is made as much by the "what's that?" and "aha!" experiences as by logic.

26.1 THE DISCOVERY OF THE ELECTRON

Electrons are very small particles, too small to see, too small to feel even if one hit your nose at great speed. How do we know they exist? The discovery of this fundamental particle of nature illustrates how significant discoveries can arise from careful observations made while performing experiments whose purpose is completely different. In this case, the unexpected discovery arose from observations made while investigating the discharge of electricity through low-pressure gases.

The glowing discharge of electricity through gases at low pressure is accomplished by connecting a source of large potential

difference to two terminals sealed into a glass tube with the gas, as in Figure 26.1. The positive terminal of the tube is called the **anode**, and the negative one the **cathode**.[§]

If we gradually evacuate the gas from inside the tube, thus decreasing the pressure, a dark region is observed near the cathode as the tube is emptied. With further evacuation, the region extends farther along the tube length, eventually reaching the opposite end. There a diffuse spot of light can be seen, whose color depends on the kind of glass used to form the tube. If several wire screens are placed inside the tube, the spot is quite well defined. Such observations in the late 19th century indicated that apparently something invisible was being emitted by the cathode, traveling across the tube, and colliding with the glass, causing the emission of light. Such **cathode rays** intrigued a number of investigators.

Early experiments in 1895 by Jean Perrin (1870–1942) found that a transverse electric field deflected the spot in a direction opposite to that of the field, indicating the cathode rays were negatively charged particles. Since charge was always associated with mass, the particles had to have mass.[#]

J. J. Thomson (1856–1940) was able to determine the ratio of the charge to mass of these particles in the following way. By subjecting the particles to mutually perpendicular electric and magnetic fields, as in the speed selector we examined in Section 20.2, he was able to select particles with a known speed $v = E/B$. Then by measuring the deflection of the particles in another magnetic field of known magnitude and extent,[¶] the ratio of the absolute value of the charge to its mass was found to be about $|q|/m \approx 1.8 \times 10^{11}$ C/kg.

The largest charge to mass ratio of any particle then known was about 9.6×10^7 C/kg for hydrogen ions (measured using electrolysis experiments). Thomson conjectured that if the charge of the newly discovered negatively charged particles was typical of those on chemical ions, such as the hydrogen ion, which he had already measured, the large charge to mass ratio of cathode ray particles implied the mass of the individual particles was significantly smaller than that of any atom.

[§] The word anode comes from the Greek ἄνοδος (*anados*), meaning "the way up" (or positive). Cathode comes from the Greek κάθοδος (*kathodos*), meaning "the way down" (or negative).

[#] At the time, no massless particles were known.

[¶] His apparatus was an early mass spectrometer.

---

[*](Chapter Opener) *Scientific Autobiography and Other Papers*, translated by Frank Gaynor (Philosophical Library, New York, 1949), pages 44–45.

[†]This is an analogy wonderfully made by Arthur Koestler in his classic study of the history of astronomy from the ancient Greeks to Galileo. See *The Sleepwalkers* (Penguin, London, 1959).

[‡]A scientist, unknown to the author, once remarked that science is doing what you do when you do not know what you are doing.

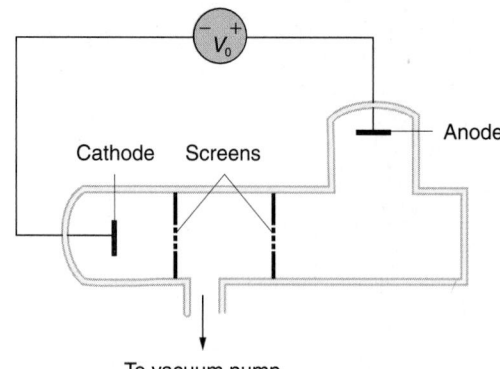

To vacuum pump

FIGURE 26.1   A gas discharge tube.

He called the particles primordial atoms or corpuscles; we now call them **electrons**.

The charge of electrons finally was measured by Robert Millikan (1868–1953) in 1909 with an elegant series of experiments using charged oil mist droplets floating in air and manipulated with known electric fields. The charge of the electron was found to be a constant quantity, and from the known charge to mass ratio their mass also was established. We now know the charge is the negative fundamental unit of charge $q = -e = -1.602 \times 10^{-19}$ C; their mass is 9.11×10^{-31} kg. The first truly fundamental particle of nature had been found.

> Since atoms commonly are electrically neutral, the discovery of the electron, presumed to exist within the neutral atom, was the first clear evidence that the heretofore indivisible atoms of nature likely had structure.

The search for an acceptable atomic model thus began in earnest at the end of the 19th century; we discuss several early such models in Sections 26.6 and 26.7.

26.2 THE DISCOVERY OF X-RAYS

The importance of x-rays in medicine and dentistry is well known, not only for the diagnosis of fractures and caries, but also for the treatment of malignant and inoperable tumors. Certainly the practice of medicine would be much cruder had x-rays not been discovered over a century ago. Their application to clinical medicine occurred within months of their discovery by Wilhelm Röntgen (1845–1923) in 1895; technology transfer was as rapid then as it is now for significant new tools.

The discovery of x-rays was accidental, as are many great discoveries in science. Indeed, if Röntgen had set out to discover a new means to assist physicians in diagnosing fractures, he almost certainly *never* would have been playing with just the right equipment needed for him to produce and, therefore, to discover

x-rays. This shows how pure (basic) research, driven solely by the interest and curiosity of scientists, is in many cases more productive for society in the long run than is directed (applied) research, undertaken with specific uses or goals in mind.

X-rays were discovered while investigating the discharge of electricity through rarefied gases, the same phenomena and equipment through which cathode rays first were noticed by Perrin and Thomson. Indeed, it is likely that x-rays were generated also in their laboratories and others, but Röntgen was the first person to *notice* them as he was investigating the characteristics of cathode rays. This is yet another example of how a researcher might be deflected by a puzzling observation during an experiment, only to realize later that the puzzle actually is more important than the initial investigation. It is, therefore, important for scientists to be constantly on the lookout for puzzling effects, even if they might seem at first to be an annoyance or nuisance.

Röntgen noticed that a paper screen treated with barium platino cyanide [BaPt(CN)$_4$] lit up quite brilliantly when near the discharge tube in which he had generated cathode rays. Intrigued, he found the effect was visible at distances up to two meters from the tube. He attributed the effect to a new, unknown "x" type of radiation, which he appropriately dubbed **x-rays**; the name stuck. He quickly was able to assess that the radiation appeared to originate from the area where the cathode rays (electrons) struck the glass tube. Through a remarkable series of insightful investigations, an experimental tour de force lasting only several weeks, Röntgen discovered many of salient features of x-rays and reported them in his first publication about their discovery. Appropriately, he won the very first Nobel prize in physics in 1901.

X-rays were shown by Röntgen to have the following characteristics:

1. X-rays are generated whenever high-energy cathode rays (electrons) strike solid materials. Generally, the greater the density of the impacted material, the more x-rays produced.
2. Matter is more or less transparent to x-rays. Wood and flesh are very transparent, bone and metals less so, which makes the use of x-rays in medicine so useful.
3. Photographic film is affected by x-rays, so its use as a detector was assured from the beginning.
4. The rays are undeviated by electric and magnetic fields, and so are uncharged.

Subsequent experiments by Hermann Haga and Cornelius H. Wind in 1899 determined that x-rays could be diffracted by extraordinarily narrow slits, on the order of 10^{-6} m wide. X-rays, therefore, are waves of very small wavelengths, much less than visible light, although the electromagnetic nature of x-rays was only demonstrated later.

The wave nature of x-rays makes them useful tools for the study of the structure of crystals and molecules, where the atoms and molecules act as three-dimensional diffraction gratings. Indeed, it was the diffraction of x-rays by crystals [by Max von Laue (1879–1960) in 1912] that proved crystals were regular arrangements of atoms and molecules. X-ray diffraction techniques, which mushroomed during the early decades of the 20th century, pioneered by Walter Bragg (1862–1942), were used to deduce the complex double helical structure of the famous DNA molecule* at

Röntgen's discovery of x-rays precipitated fear that clothing would need to be "x-ray proof" to avoid x-rated images; the fear was groundless. This post card from 1900 depicts "Sun bathing à la Röntgen"!

---

*Deoxyribonucleic *acid*.

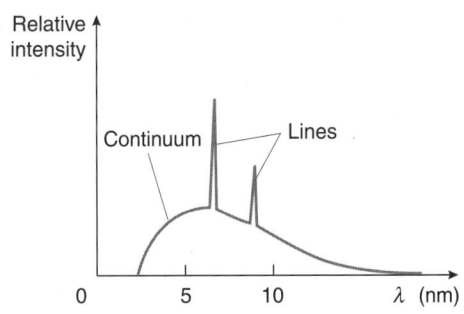

FIGURE 26.2 A typical x-ray spectrum. Note that their wavelengths are about 1/1000 those of visible light.

mid-century. DNA is critical for cellular reproduction, because it carries the molecular blueprints of the genetic code.

The spectrum of x-rays produced when electrons slam into a solid was first determined by Walter Bragg in 1913. The spectrum consists of a small continuous range of wavelengths and several very intense monochromatic (single-wavelength) lines, characteristic of the elements in the solid; an example is shown in Figure 26.2.

The Maxwell theory of electromagnetism predicts that accelerated charges produce electromagnetic radiation. The continuous range of x-ray electromagnetic wavelengths stems from the large and various magnitudes of the accelerations of the electrons as they are slowed by the material. We now know that the monochromatic x-ray wavelengths arise from within and characterize the specific atom itself.

26.3 THE DISCOVERY OF RADIOACTIVITY

The proper disposal of radioactive waste, be it from medical facilities, nuclear power plants, research reactors, or nuclear weapons, is a serious problem of ongoing concern to humanity, if only because of the environmental consequences to us and future generations. However, radioactive materials are *not* solely the result of our inventive, technological prowess; they occur naturally and have been with us on the Earth from the very beginning, albeit in concentrations that typically are not serious environmental hazards.

The discovery of radioactivity, about a century ago, is an interesting tale. In 1896, just a few months after the discovery of x-rays by Röntgen, Henri Becquerel (1852–1908) set out to investigate the way in which the mysterious x-rays caused various substances to emit visible light and expose photographic plates. He took a crystalline sample of a chemical salt of uranium* and placed it near a photographic plate completely wrapped in heavy, opaque black paper, intending to see what happened when he exposed the salts to sunlight. The plates were exposed, as if light had penetrated the heavy wrapping encasing the plates. The effect was the same even when both the wrapped plates and uranium salts were in complete darkness.

Well intrigued by the miraculous exposure of the plates when well shielded from visible light, Becquerel explored the phenomena in some detail. In short order he realized the magnitude of the effect (the exposure) was proportional to the amount of uranium present and that changing the temperature had no effect. The latter observation led him to conclude the

effect was an atomic process (originating *within* the atom) rather than a chemical one, since temperature typically has a dramatic effect on the rate of virtually all chemical reactions.

His publication of the results led others to search for other substances that produced similar effects. In 1898 Marie Curie (1867–1934) noticed that two different uranium ores were significantly more active than pure uranium itself and, as a result, eventually was able to chemically isolate two previously unknown elements, which she called polonium[†] and radium. She called the phenomenon **radioactivity**. Curie discovered the radiations evidently were very energetic, because a sample of radium remarkably was able to maintain itself in thermal equilibrium several degrees above room temperature! Speculation arose that perhaps radioactivity was the source of the Sun's energy (now known to be false) and a source of energy for the hot interior of the Earth (true).

The substances that produce the effect evidently emit several distinct types of radiations. One is a penetrating radiation, dubbed α, that propagates through several centimeters in air and can even penetrate very thin metal foils. Another less penetrating radiation, dubbed β, is easily stopped by even a sheet of paper. Another type, called γ, was discovered in 1900 and is much more penetrating than even the α radiation. Magnetic fields influenced the trajectories of the α and β radiations, and indicated the α radiation consisted of positively charged particles while the β were negatively charged particles. The γ radiation was unaffected by such fields, and so was electrically uncharged.

A series of experiments with an electromagnetic speed selector showed that the β particles from a given substance emerge with a variety of speeds that are significant fractions of the speed of light (over 90% in many cases). Substances that emit α particles produce them with discrete (specific) speeds, also very large. The extraordinary speeds and energies indicated that the processes causing their emission were not chemical reactions of any known type. Charge to mass ratio experiments proved that the β particles were high-speed cathode rays: electrons. The α particles were associated with helium atoms (the nucleus had not yet been discovered); helium always seemed to be present in the radioactive materials that emitted α radiation. In fact, α particles are helium nuclei, not helium atoms.

[†]After Poland, the homeland of Marie Curie. Her maiden name was Marie Sklodowska. She met and married Pierre Curie in Paris.

Becquerel came from a long line of French physicists and occupied the same professorship of physics held by both his father and grandfather!

---

*Potassium uranyl sulfate: $K_2SO_4 \cdot UO_2SO_4 \cdot 2H_2O$.

This symbol indicates the presence of radioactive materials much the way other symbols are used to indicate poisons or biohazards.

By 1908 Ernest Rutherford (1871–1937) was able to count directly the number of α particles emitted per gram of radium* as well as the total charge they represented, concluding correctly that each α possessed twice the magnitude of charge as on a β particle. With a knowledge of the charge and kinetic energy of the α particles emitted by radium, Rutherford had the tools needed to begin probing the structure of the atom using the monoenergetic α particles as high-speed, submicroscopic bullets (see Sections 26.6 and 26.7).

We discuss radioactivity in more detail in Section 26.11 from the viewpoint of the nuclear model of the atom.

26.4 THE APPEARANCE OF PLANCK'S CONSTANT h

Turn on an electric stove heating element and soon it begins to glow a dark red, eventually becoming bright red as its temperature increases to its maximum. Turn on a light bulb and it glows white hot because its temperature is much hotter than the stove burner. The greater the temperature of a solid, the bluer the light emitted by it. The same is true even of very hot gases: bluish-colored stars, such as Rigel (β Orionis) in the constellation

*When α particles collide with a phosphor screen, some light is emitted, so Rutherford and others actually could count the collisions visually.

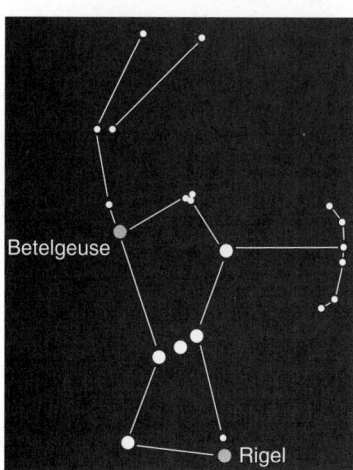

When you next look at the constellation Orion, notice the color difference between bluish Rigel and reddish Betelgeuse, indicative of their different surface temperatures.

Orion, have higher surface temperatures than reddish-colored stars, such as Betelgeuse (α Orionis) also in Orion.

We imagine a closed oven or cavity with a tiny hole through which we can peek, as in Figure 26.3. As the absolute temperature of the cavity gradually is increased, we begin to notice that light emitted from the hole is dark red in color. As we increase the temperature of the cavity still more, the light becomes bright red, then whitish, then progressively bluish. When the entire cavity is all in thermal equilibrium at one absolute temperature T, we can analyze the spectrum (or the domain of wavelengths or frequencies) of the light from the hole, knowing very surely the source temperature. Experiments have shown that the spectrum of the light inside the cavity for any fixed absolute temperature T is independent of the material composing the cavity. The spectrum has a characteristic shape (see Figure 26.4) called a **blackbody spectrum**, even though the cavity, if its temperature is high enough, will hardly look black when we peek into it.

The peak of the spectral curve shifts to shorter wavelengths (higher frequencies) for higher absolute temperatures, thus accounting for the change in the color of the light emerging from the cavity. Let λ_{max} be the wavelength at which the blackbody emission spectrum has its peak for a given absolute temperature T. Wilhelm Wien (1864–1929) thermodynamically analyzed what happened to the light inside the cavity if it behaved adiabatically (thermally isolated). He discovered that the product of the peak

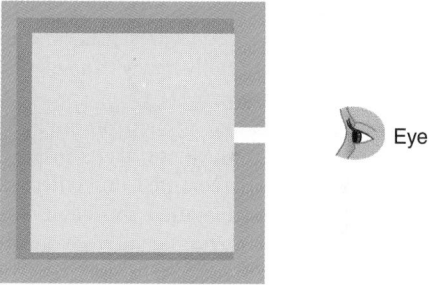

FIGURE 26.3 A cavity.

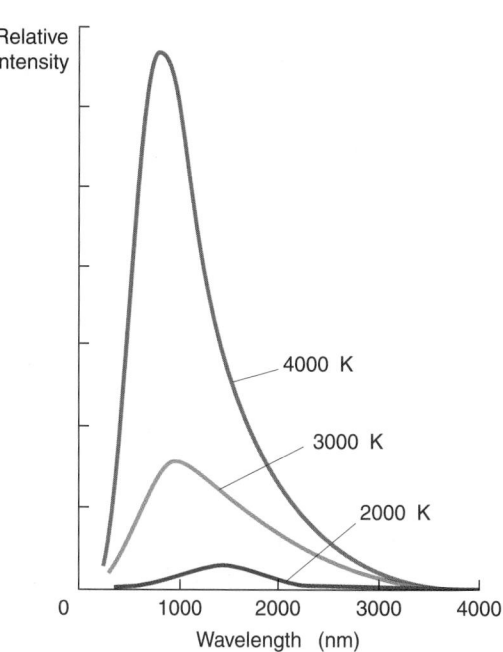

FIGURE 26.4 Blackbody spectra at various absolute temperatures. Visible light has wavelengths from about 400 nm to 700 nm.

wavelength $\lambda_{\max}$ and the absolute temperature T was a constant. By use of experimental spectra, this constant is found to have the approximate value

$$\lambda_{\max} T = 0.289\ 78 \times 10^{-2}\ \text{m}\cdot\text{K} \qquad (26.1)$$

Equation 26.1 is called **Wien's displacement law.***

> Increase the temperature and the peak wavelength shifts to smaller wavelengths.

Arno Penzias and Robert Wilson serendipitously detected a blackbody spectrum from space in the mid-1960s at the Bell Telephone Laboratories in Holmdel, New Jersey, during experiments tracing the source of peculiar static in a large microwave antenna. The spectrum from space has a characteristic absolute temperature of only about 3 K. The detection of this radiation provided experimental evidence to support the Big Bang theory of the creation of the universe.[†] The light left over from the Big Bang (some 10 or 20 billion years ago) has cooled significantly as the universe expanded and now has its peak in the microwave region of the electromagnetic spectrum.

In the late 19th century, understanding the characteristic shape of a blackbody spectrum provoked much theoretical interest and controversy. Since the cavity was in thermal equilibrium, it was hoped that an application of the laws of statistical mechanics and thermodynamics might account for the spectrum of the electromagnetic waves inside and emerging from the cavity. Indeed, Wien initially met with some success with the displacement law (Equation 26.1).

However, other results of such an analysis were very troubling. Since the electromagnetic waves inside a cavity were continually being absorbed, reemitted, and reflected by the walls of the cavity, physicists imagined the cavity as filled with standing electromagnetic waves. Recall from our study of waves (Chapter 12) that standing waves in one dimension (such as those on a guitar string) are formed from the superposition of two identical waves of wavelength λ traveling in opposite directions. When confined to a distance ℓ, say one dimension of a cubical cavity, the wavelengths λ_n that form standing waves are

$$\lambda_n = \frac{2\ell}{n}$$

where n is a positive integer ($n = 1, 2, 3, \ldots$). In particular, note that there is, in principle, no limit on how short the wavelengths can be and, thus, on how high the corresponding frequency can be.

In statistical thermodynamics, when we add a particle to a system, the number of degrees of freedom of the system increases; the Maxwell theory of electromagnetism implies that each standing wave in the cavity would be a degree of freedom. When in thermal equilibrium, the equipartition of energy theorem (Section 14.8) assigns an average energy $kT/2$ to each degree of freedom (provided it is not frozen out).

Recall from Chapter 7 that the energy of a classical oscillator is proportional to the square of its *amplitude* and is *independent* of its frequency. Hence each electromagnetic standing wave oscillator in the cavity should share in the energy inside the cavity. This leads to two equally unpleasant possible conclusions:

a. The energy available in the cavity is finite, but the cavity has an essentially infinite number of standing waves. Thus each standing wave effectively has zero energy and no electromagnetic waves should emerge from the hole! But electromagnetic waves clearly *are* inside the cavity and are emitted from it.

b. Since there are many more standing waves of high frequency than of low frequency, the light emerging from the hole should be at the extreme high frequency end of the spectrum (with blue and violet light, even x-rays). This interpretation, first realized by Lord Rayleigh (John William Strutt) (1842–1919) and James Jeans (1877–1946) in 1900, came to be known as the **ultraviolet catastrophe.**

Every attempt to explain the blackbody spectrum based on electromagnetic theory and thermodynamics failed to predict the shape of the blackbody spectrum. Clearly, something peculiar was happening that was not amenable to the standard physics of the day.

In 1901 Max Planck (1858–1947) finally was able to account for the shape of the blackbody spectrum using a drastic ad hoc hypothesis made, he said, in desperation.

> Quite reluctantly, Planck assumed that the energy E associated with the light inside the cavity was present only in finite packets (bundles) proportional to the frequency ν:
>
> $$E = h\nu \qquad (26.2)$$
>
> where h was an unknown constant that he hoped to be able to set equal to zero after taking appropriate mathematical limits.[‡]

[‡]He was, in essence, trying to approximate a diverging integration by a summation.

Planck had little to do with rocket science, although it would not appear so from this stamp from the Ivory Coast commemorating his Nobel prize. Planck was an accomplished pianist as well as a physicist; many physicists love the arts, and practice them as well!

---

*The term, which arises from the German *Verschiebungsgesetz*, was first used in 1899 by Otto Lummer and Ernst Pringsheim.
[†]They won the 1978 Nobel prize in physics for their discovery.

With this hypothesis, the energy associated with high-frequency light is very large, and so the degrees of freedom of such standing waves must be frozen out. The finite amount of energy in the cavity then can be apportioned among a smaller, finite number of low-frequency oscillators (standing waves). The hypothesis neatly avoided the ultraviolet catastrophe and correctly predicted the shape of the blackbody spectrum.

By fitting his predictions to the experimental blackbody spectrum, Planck had hoped to show that the value of h was zero. Much to his surprise and consternation, he found the constant could not be set to zero but had an approximate numerical value of

$$h \approx 6.6 \times 10^{-34} \text{ J} \cdot \text{s}$$

Tiny for sure, but *not* zero.

> We now know h with more precision to be about
>
> $$h = 6.626 \times 10^{-34} \text{ J} \cdot \text{s} \qquad (26.3)$$
>
> The constant h, now regarded as a fundamental constant of nature, was called by Planck the *quantum of action*, but shortly became simply **Planck's constant**.

The revolutionary and disturbing nature of Planck's hypothesis, Equation 26.2, is apparent in Planck's recollections in the opening quotation of this chapter.

QUESTION 1
Show that the SI units of Planck's constant ($J \cdot s$) are the same as those of angular momentum ($kg \cdot m^2/s$).

EXAMPLE 26.1

Light from the Sun has a blackbody radiation spectrum with its peak at the visible light wavelength 5.0×10^2 nm. What is the surface temperature of the Sun?

Solution
The Wien displacement law, Equation 26.1, relates the wavelength of the peak of a blackbody spectrum to the absolute temperature:

$$\lambda_{max} T = 0.289\,78 \times 10^{-2} \text{ m} \cdot \text{K}$$

Since $\lambda_{max} = 5.0 \times 10^2$ nm, you have

$$T = \frac{0.289\,78 \times 10^{-2} \text{ m} \cdot \text{K}}{5.0 \times 10^2 \times 10^{-9} \text{ m}}$$
$$= 5.8 \times 10^3 \text{ K}$$

A rather toasty temperature!

26.5 THE PHOTOELECTRIC EFFECT

Planck's strange hypothesis about the way energy is carried in bundles by light soon bore fruit in another area, quite unrelated to the blackbody spectrum: the photoelectric effect. The reappearance of Planck's constant in another context solidified its

place as a fundamental constant of physics. Here we investigate this additional thread in the complex tapestry of discoveries at the dawn of the 20th century.

> When light of an appropriate frequency (or correspondingly, of an appropriate wavelength) is incident on a metallic surface, electrons are liberated from the surface. This observation is known as the **photoelectric effect**.*

The electronic detection of light, even incredibly dim light such as that from stars and distant galaxies, is based on the photoelectric effect for some kinds of sensitive photodetectors. Its discovery was contemporaneous with the work with cathode rays.

In 1888 Wilhelm Hallwachs (1859–1922) noticed that a freshly polished metal (he used zinc), when insulated and connected to an electroscope, lost negative charge and became positive when irradiated with ultraviolet light.[†] Initially it was thought the charge might be transferred by the gas particles surrounding the metal sample, but the loss of negative charge persisted even when the sample was illuminated while in a good vacuum, thus eliminating that possibility. Perhaps charged atoms of the metal were being expelled by the light? No, because even after extensive illumination, no traces of metallic material could be found elsewhere inside the chamber in which the metal electrode was placed. By 1899 Philipp Lenard (1862–1947)[‡] had shown that the emitted particles were negative with the same charge to mass ratio as cathode rays, correctly surmising that the light caused the metal to emit electrons.

Experiments with the photoelectric effect typically are performed with an apparatus schematically illustrated in Figure 26.5. When appropriate light is incident on the metal, the ejected electrons are swept to and collected by the positive terminal

---

*The effect really should be called the photo-induced liberation of electrons from metals, but this is shortened for convenience to the photoelectric effect.
[†]The effect evidently was first observed by Heinrich Hertz in the course of his experiments with the production and detection of electromagnetic waves, but he did not follow up on his observations—a missed opportunity!
[‡]Regrettably, Lenard was an anti-Semite, who made vicious personal attacks on many Jewish scientists, even Einstein. An early admirer of Adolph Hitler, Lenard became the bridge between the Third Reich and the German scientific establishment before and during the Second World War.

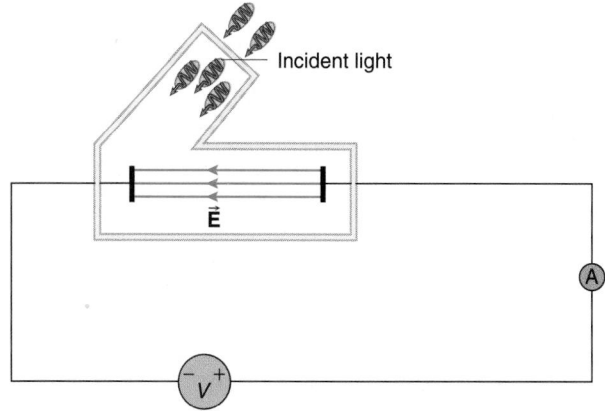

FIGURE 26.5 Apparatus for investigating the photoelectric effect.

(the anode). A small current flows in the circuit and is detected with a sensitive ammeter.

Experiments by Lenard and others discovered a number of significant features about the effect:

1. There exists a critical **cutoff frequency** v_c for the incident light (and corresponding **cutoff wavelength** $\lambda_c = c/v_c$). If the frequency of the incident light is less than v_c, no electrons are liberated from the metal regardless of the intensity of the incident light: there is no current detected by the ammeter.

2. For incident light with frequencies $v > v_c$, the number of electrons liberated per second is directly proportional to the light intensity. That is, the current in the circuit is directly proportional to the incident light intensity, as shown in Figure 26.6.

3. For a given frequency of light illuminating the metal, if we increase the source potential difference V in the circuit of Figure 26.5, no increase in the current occurs (see Figure 26.7 for V > 0 V). The electric field established in the tube by the battery effectively pulls all the liberated electrons to the positive terminal to be collected and registered by the ammeter.

   On the other hand, if we reverse the polarity of the potential difference, the direction of the electric field in Figure 26.5 reverses. As the value of the reversed potential difference increases, the current in the circuit decreases to zero (see Figure 26.7 for V < 0 V). The electric field now prevents some of the electrons from reaching the now negative terminal. When this reversed electric field is large enough, no electrons are collected and the current is zero. This observation indicates that the liberated charge is negative and that these electrons have a smooth variety of kinetic energies, ranging from essentially zero up to a maximum value KE_{max}. Indeed, we will show later in this section that the maximum kinetic energy of the electrons is related to the reversed potential difference V_s at which the current of electrons ceases (see Figure 26.7). The quantity V_s is known as the **stopping potential**, since the electric field it creates in the tube stops *all* the liberated electrons from reaching the opposite terminal, even the most energetic ones.

4. Experiments also show that the value of this stopping potential V_s is directly proportional to the frequency of the incident light for frequencies greater than the critical cutoff frequency, as shown in Figure 26.8.

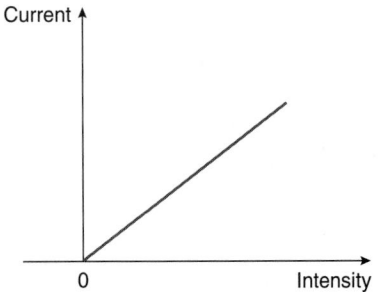

FIGURE 26.6 For $v > v_c$, the current is proportional to the incident light intensity.

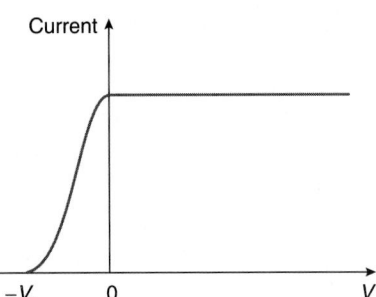

FIGURE 26.7 The current versus the applied potential difference with the arrangement of Figure 26.5 for the photoelectric effect.

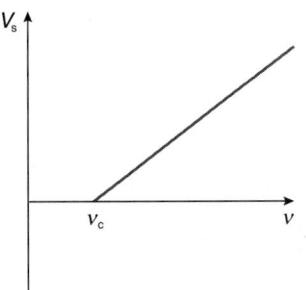

FIGURE 26.8 The stopping potential is proportional to the light frequency for $v > v_c$.

5. The liberated electrons appear promptly when the metal is illuminated, with no time lag. That is, the electrons ejected from the metal appear essentially instantaneously even when the incident light is of very low intensity.

It is difficult to account for the existence of a cutoff frequency (observation #1) and the prompt emergence of the electrons (observation #5) with a wave model for light for two reasons:

a. As we recalled in Section 26.4, the energy of a classical wave is proportional to the square of its amplitude and frequency. Thus, according to this scheme, electrons should be able to absorb energy from incident light of *any* frequency. The photoelectric effect should be independent of frequency; however, the effect is strongly frequency dependent.

b. For low levels of illumination (light waves of very small amplitude or energy), an electron might have to wait for enough energy to arrive before accumulating enough to become liberated from the metallic surface. In other words, for very low levels of light wave illumination, there likely will be a significant and detectable time delay between the onset of the light and the release of the electrons. However, no such delay is observed with low illumination; the electrons always appear promptly.

The problems with explaining the photoelectric effect experiments with a classical wave model for light forced Einstein to abandon it in 1905.* He invented a bold new model that completely accounted for the observations. It was principally for his

---

*You may recall that Einstein also published his special theory of relativity in 1905. It was a very productive year for him—and for physics.

explanation of the photoelectric effect that he won the 1921 Nobel prize in physics. Einstein's explanation is all the more remarkable in retrospect, because the charge on the electron was unknown in 1905. Many of Einstein's ideas about the effect were not confirmed by experiment for another decade.

Einstein built on and extended the ad hoc hypothesis about light energy used by Planck to avoid the ultraviolet catastrophe of the blackbody spectrum (Section 26.4). Einstein imagined light to consist of wavelike particles, packets (bundles) of electromagnetic energy, called **photons**. Planck initially thought that light photons only existed when confined to a cavity, and that once light left the cavity it was an electromagnetic wave. Einstein, on the other hand, extended Planck's idea to say that such photons (light quanta) existed in free space, even outside of any cavity. The particles of light each have an energy E_{photon} proportional to the frequency of the light:

$$E_{photon} = h\nu \qquad (26.4)$$

where h is Planck's constant. A beam of light thus consists of an incredible flood of photons (see Example 26.2).

When light is absorbed by matter (i.e., by particles with mass), Einstein hypothesized that the matter-particle "swallows" the *entire* photon, thus destroying (annihilating) it completely (no leftovers). A matter-particle cannot nibble just a piece of a passing photon: it is all or nothing, like swallowing a pill. Energy is conserved because the matter-particle now has the energy previously possessed by the photon. Of course, the matter-particle also has other forms of energy, such as rest energy; but the change in the energy of the matter-particle is precisely the energy of the photon.

How do these ideas account so neatly for the experimental observations of the photoelectric effect?

When an electron in the metal totally absorbs an incoming photon and acquires its energy, some of it subsequently is lost in collisions as the electron escapes from the metal. This lost energy (which can be any fraction, since it is not associated with a photon but is the kinetic energy of the electron) is transferred to other atoms in the metal.* We sweep the details of these complex interactions under the rug and say that the electron does a certain amount of work to escape the metal, and so loses some of the energy gained from the photon. The leftover energy, after escape, is the kinetic energy of the liberated electron. In other words, we conserve energy for *each* electron that absorbs a photon by writing

$$\begin{pmatrix} \text{energy} \\ \text{of the} \\ \text{photon} \\ \text{absorbed} \end{pmatrix} = \begin{pmatrix} \text{work needed to} \\ \text{get the electron} \\ \text{liberated from} \\ \text{the metal} \end{pmatrix} + \begin{pmatrix} \text{kinetic energy} \\ \text{of the} \\ \text{liberated} \\ \text{electron} \end{pmatrix} \qquad (26.5)$$

Depending on the particular circumstances of an individual electron, it may have to give up part or all of its newly acquired additional energy to get out of the bulk metal. Collectively, the emerging electrons thus appear with a variety of kinetic energies. Also because of energy conservation, the privileged electrons that escape from the metal with the *minimum* amount of energy loss

*Once the photon energy is gained by the electron, the energy is part of a common pool from which any amount can be tapped.

(the first term on the right-hand side of Equation 26.5) must be the ones with the *maximum* amount of kinetic energy after escape, since the energy picked up from the absorbed photon is a fixed amount. For these privileged electrons, Equation 26.5 becomes

$$\begin{pmatrix} \text{energy} \\ \text{of the} \\ \text{photon} \\ \text{absorbed} \end{pmatrix} = \begin{pmatrix} minimum \\ \text{work needed to} \\ \text{get the electron} \\ \text{liberated from} \\ \text{the metal} \end{pmatrix} + \begin{pmatrix} maximum \\ \text{kinetic energy} \\ \text{of the} \\ \text{liberated} \\ \text{electron} \end{pmatrix} \qquad (26.6)$$

The minimum work needed to liberate an electron from a metal is called the **work function** W of the metal. Thus Equation 26.6 is rewritten as

$$\boxed{h\nu = W + KE_{max}} \qquad (26.7)$$

If we now decrease the frequency of the incoming light, thus decreasing the energy of the incoming photons (since $E_{photon} = h\nu$), the maximum kinetic energy of the emergent electrons also decreases. Eventually we reach a frequency ν_c such that the electrons barely escape and (each) have zero kinetic energy:

$$\boxed{h\nu_c = W + 0 \text{ J}} \qquad (26.8)$$

The frequency ν_c is the cutoff frequency for the photoelectric effect. For frequencies $\nu < \nu_c$, the energy of the incoming photons is not sufficient to give the electron even the minimum energy W needed to escape from the metal. As a result, no electrons emerge and the photoelectric effect disappears.

Thus the work function of a metal is equal to the energy of a photon with the cutoff frequency.

Experimentally, this is how the work functions of various metals are determined. Table 26.1 presents the approximate work functions of various metals.

How is the maximum kinetic energy of the emitted electrons measured? We use the apparatus of Figure 26.5, but with polarity of the voltage source reversed, as in Figure 26.9. The voltage source produces an electric field between the terminals inside the tube. The liberated electrons find themselves in this electric field, and thus are subjected to an electrical force

$$\vec{F}_{elec} = q\vec{E} = -e\vec{E}$$

in the direction opposite to the field, since the charge on the electron is negative. This force drives them back to the plate from which they emerged.

As the source potential difference increases from zero toward the value of the stopping potential, the current in the circuit gradually diminishes (see Figure 26.7, to the left of the origin); only the electrons ejected with sufficiently high kinetic energy can reach the opposite plate. The potential difference V_s that stops all current from the ejected electrons is the stopping potential. The liberated electrons are subjected to only the conservative electrical force in the region between the plates. As they leave the metal, the most energetic liberated electrons have

TABLE 26.1 Approximate Photoelectric Work Functions of Various Metals

Metal	Work function (eV)
Aluminum (Al)	4.1
Barium (Ba)	2.5
Cesium (Cs)	2.0
Copper (Cu)	4.7
Gold (Au)	4.8
Iron (Fe)	4.7
Lead (Pb)	4.1
Lithium (Li)	2.4
Mercury (Hg)	4.5
Nickel (Ni)	5.0
Platinum (Pt)	6.4
Potassium (K)	2.2
Rubidium (Rb)	2.1
Silver (Ag)	4.7
Sodium (Na)	2.3
Strontium (Sr)	2.7
Tin (Sn)	3.9
Zinc (Zn)	4.3

Source: CRC Handbook of Chemistry and Physics, 50th edition (CRC Press, Cleveland, Ohio, 1963), pp. 2655–2660.

an initial kinetic energy KE_{max} and an initial electric potential energy $PE_{elec} = qV = (-e)V_s = -eV_s$. When these electrons (barely) reach the opposite plate, they have zero kinetic energy, and an electrical potential energy $PE_{elec} = qV = -e(0 \text{ V}) = 0 \text{ J}$. The CWE theorem becomes

$$W_{other} = \Delta KE + \Delta PE \qquad (26.9)$$
$$0 \text{ J} = \Delta KE + \Delta PE$$
$$= (KE + PE)_f - (KE + PE)_i$$
$$0 \text{ J} = (0 \text{ J} + 0 \text{ J}) - (KE_{max} - eV_s)$$

and so

$$\boxed{KE_{max} = eV_s} \qquad (26.10)$$

The stopping potential therefore is a direct measure of the kinetic energy of the most energetic liberated electrons.

Equation 26.7 for these electrons then can be rewritten as

$$h\nu = W + eV_s \qquad (26.11)$$

The work function W is related to the cutoff frequency by Equation 26.8:

$$W = h\nu_c$$

Now we send light of frequency $\nu_1 > \nu_c$ and measure the stopping potential V_{s1} associated with this incident light. We change the light to another frequency $\nu_2 > \nu_c$ and measure its associated stopping potential V_{s2}. We do this for a number of different frequencies: ν_3, ν_4, ..., all $> \nu_c$. We plot the measured stopping potentials versus the associated frequency. According to Equation 26.11, we have

$$V_s = -\frac{W}{e} + \frac{h}{e}\nu \qquad (26.12)$$

The graph of V_s versus ν will be a straight line (as in Figure 26.8), with a slope equal to h/e and an intercept $-W/e$; see Figure 26.10.

Such an experiment, and its resulting plot of V_s versus ν, is one way to measure Planck's constant, given the electronic charge e. The frequency corresponding to $V_s = 0$ V is the cutoff frequency.

PROBLEM-SOLVING TACTICS

26.1 In using Equations 26.7, 26.8, 26.10, or 26.11 for the photoelectric effect, you must express all the energies either in joules (J) or, if you choose, electron-volts (eV). As a practical matter, the photon energy $h\nu$ in SI units is expressed in joules. The work functions of metals typically are expressed in electron-volts (eV) (see Table 26.1). For photoelectricity, electron-volts are very convenient, since the energy of visible light photons is on the order of just a few electron-volts (see Example 26.2).

26.2 From Equation 26.10, the stopping potential in volts is *numerically* equal to the maximum kinetic energy of the electrons expressed in electron-volts. This is always the case, and is one reason that the electron-volt is a convenient unit of energy. See Example 26.4.

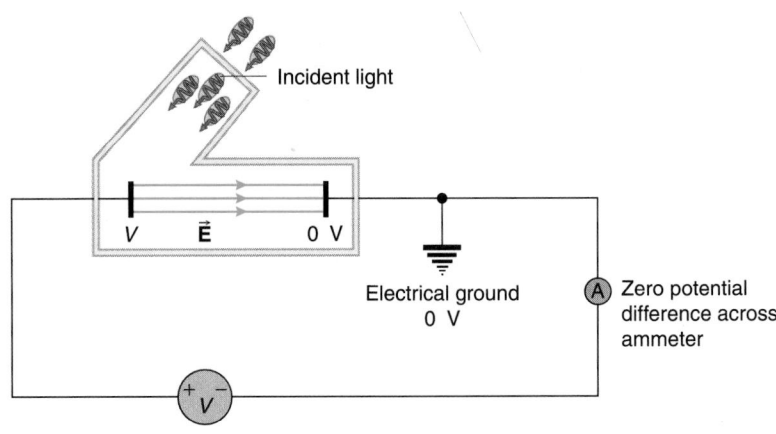

FIGURE 26.9 Apparatus to determine KE_{max}.

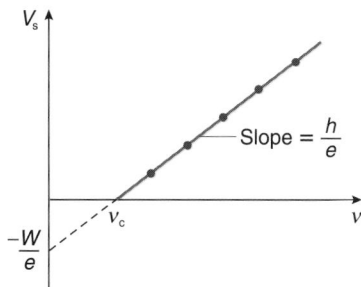

FIGURE 26.10 The stopping potential versus the frequency of the incident light.

Which of the metals in Table 26.1 has the longest cutoff wavelength for the photoelectric effect? What is that wavelength? For what wavelengths is the photoelectric effect observed for that metal?

EXAMPLE 26.2

A low-power helium–neon laser has a power output of 1.00 mW of light of wavelength 632.8 nm.

a. Calculate the energy of each photon, expressing your result in both joules and electron-volts.
b. Determine the number of photons emitted by the laser each second.

Solution

a. The energy E of each photon is proportional to its frequency v, from Equation 26.4:

$$E = hv$$

Since the wavelength and frequency are related to the speed of the light waves by

$$c = v\lambda$$

Equation 26.4 becomes

$$E = \frac{hc}{\lambda}$$

$$= \frac{(6.626 \times 10^{-34} \text{ J·s})(3.00 \times 10^{8} \text{ m/s})}{632.8 \times 10^{-9} \text{ m}}$$

$$= 3.14 \times 10^{-19} \text{ J}$$

To convert to electron-volts, divide by 1.602×10^{-19} J/eV; you obtain

$$E = 1.96 \text{ eV}$$

Photons of visible light have energies on the order of several electron-volts.

b. Since the energy of each photon is quite small, we expect the number N of photons emitted by the laser each second to be quite large. The number is equal to the energy emitted

by the laser each second (its power output in W = J/s) divided by the energy of each photon:

$$N = \frac{1.00 \times 10^{-3} \text{ W}}{3.14 \times 10^{-19} \text{ J/photon}}$$

$$= 3.18 \times 10^{15} \text{ photons/s}$$

This is quite a large number indeed!

EXAMPLE 26.3

Iron has a work function of 4.7 eV (see Table 26.1). Calculate the cutoff frequency and the corresponding cutoff wavelength for the photoelectric effect for this metal.

Solution

The energy of a photon with a frequency equal to the cutoff frequency v_c is equal to the work function W of the metal (see Equation 26.8):

$$hv_c = W$$

The work function is 4.7 eV; converting this to joules, you obtain

$$(4.7 \text{ eV})(1.602 \times 10^{-19} \text{ J/eV}) = 7.5 \times 10^{-19} \text{ J}$$

Therefore Equation 26.8 becomes

$$hv_c = 7.5 \times 10^{-19} \text{ J}$$

Solving for the cutoff frequency, you find

$$v_c = \frac{7.5 \times 10^{-19} \text{ J}}{h}$$

$$= \frac{7.5 \times 10^{-19} \text{ J}}{6.626 \times 10^{-34} \text{ J·s}}$$

$$= 1.1 \times 10^{15} \text{ Hz}$$

The photoelectric effect from iron will occur only with incident light that has a frequency *greater* than 1.1×10^{15} Hz. The wavelength corresponding to this frequency is

$$c = v_c \lambda_c$$

or

$$\lambda_c = \frac{c}{v_c}$$

$$= \frac{3.00 \times 10^{8} \text{ m/s}}{1.1 \times 10^{15} \text{ Hz}}$$

$$= 2.7 \times 10^{-7} \text{ m}$$

$$= 2.7 \times 10^{2} \text{ nm}$$

This is well into the ultraviolet region of the electromagnetic spectrum. The photoelectric effect will occur for iron only if the wavelength of the incident light is *less* than 2.7×10^{2} nm; visible light will not produce a photoelectric effect in iron.

STRATEGIC EXAMPLE 26.4

Ultraviolet light of wavelength 200 nm is incident on a freshly polished iron surface. Find:

a. the stopping potential;
b. the maximum kinetic energy of the liberated electrons; and
c. the speed of these fastest electrons.

Solution

The incident photon has an energy of

$$E = h\nu$$

$$= \frac{hc}{\lambda}$$

$$= \frac{(6.626 \times 10^{-34} \text{ J·s})(3.00 \times 10^8 \text{ m/s})}{200 \times 10^{-9} \text{ m}}$$

$$= 9.94 \times 10^{-19} \text{ J}$$

$$= \frac{9.94 \times 10^{-19} \text{ J}}{1.602 \times 10^{-19} \text{ J/eV}}$$

$$= 6.20 \text{ eV}$$

a. There are two ways to find the stopping potential.

Method 1

Use Equation 26.11 for the photoelectric effect using energy units of joules for each term. The work function W, expressed in joules, is

$$W = (4.7 \text{ eV})(1.602 \times 10^{-19} \text{ J/eV})$$

$$= 7.5 \times 10^{-19} \text{ J}$$

Thus Equation 26.11 for the photoelectric effect equation becomes

$$h\nu = W + eV_s$$

$$9.94 \times 10^{-19} \text{ J} = 7.5 \times 10^{-19} \text{ J} + eV_s$$

Solving for eV_s, you obtain

$$eV_s = 2.4 \times 10^{-19} \text{ J}$$

The stopping potential V_s then is

$$V_s = \frac{2.4 \times 10^{-19} \text{ J}}{1.602 \times 10^{-19} \text{ C}}$$

$$= 1.5 \text{ V}$$

Method 2

The problem also can be solved using electron-volt units for the energy in Equation 26.11 for the photoelectric effect. The photon energy is 6.20 eV; from Equation 26.4, $E_{photon} = h\nu$:

$$h\nu = W + eV_s$$

$$6.20 \text{ eV} = 4.7 \text{ eV} + eV_s$$

The product eV_s, in electron-volts, is

$$eV_s = 1.5 \text{ eV}$$

This can be converted to joules:

$$eV_s = (1.5 \text{ eV})(1.602 \times 10^{-19} \text{ J/eV})$$

$$= 2.4 \times 10^{-19} \text{ J}$$

Then solve for the stopping potential:

$$V_s = \frac{2.4 \times 10^{-19} \text{ J}}{1.602 \times 10^{-19} \text{ C}}$$

$$= 1.5 \text{ V}$$

Note that in accordance with Problem-Solving Tactic 26.2, the maximum kinetic energy, expressed in electron-volts, is *numerically* equal to the stopping potential in volts.

b. The maximum kinetic energy of the photoelectrons is eV_s:

$$eV_s = (1.602 \times 10^{-19} \text{ C})(1.5 \text{ V})$$

$$= 2.4 \times 10^{-19} \text{ J}$$

c. Use the nonrelativistic form of the kinetic energy. Solve for the square of the speed and substitute for the mass of the electron:

$$\frac{mv^2}{2} = 2.4 \times 10^{-19} \text{ J}$$

So

$$v^2 = \frac{2(2.4 \times 10^{-19}) \text{ J}}{9.11 \times 10^{-31} \text{ kg}}$$

$$= 5.3 \times 10^{11} \text{ m}^2/\text{s}^2$$

Hence the speed is

$$v = 7.3 \times 10^5 \text{ m/s}$$

Notice that $v/c = 2.4 \times 10^{-3}$, so that the relativistic factor γ is very close to 1, which justifies using the nonrelativistic expression for the kinetic energy of the liberated electrons.

26.6 THE QUEST FOR AN ATOMIC MODEL: PLUM PUDDING

When diffraction gratings were invented by Henry Rowland (cf. Section 24.9) in the 19th century, they were applied vigorously to the study of light emitted in gas discharges. As we have seen, the study of such discharges led to the discovery of electrons, x-rays, and the photoelectric effect. It was quickly determined that each element emitted a characteristic set of optical wavelengths of light, which could be used to detect the presence of the various elements in light sources such as distant stars.

However, the mechanism for how multiple wavelengths could be emitted simultaneously by a sample of atoms remained totally unknown. Hydrogen, known to be the least massive atom and, therefore, likely the simplest in structure, emits many different characteristic wavelengths. How is light produced in an atom that is supposedly indivisible?* The discovery of the electron and of radioactivity shattered the myth of an indivisible atom, and the search for an acceptable atomic model began in

---

*Recall that the word atom stems from the Greek word ἄτομος, meaning "indivisible."

earnest.* Certainly one test of any such model would be its ability to predict the spectrum of electromagnetic wavelengths emitted by atoms. The α radiation of radium provided the tool for exploring the structure of atoms.

Since negatively charged particles (electrons) came out of atoms, yet atoms were electrically neutral, the first model of the atom [proposed by J. J. Thomson (1856–1940)] imagined the electrons as discrete entities randomly and uniformly distributed throughout a homogeneous spherical glob of positively charged material (see Figure 26.11). The model came to be known picturesquely as the **plum pudding model**, after the English culinary delicacy.

The experiments of Ernest Rutherford (1871–1937), Hans Geiger (1882–1945), and Ernest Marsden (1889–1970) between 1909 and 1913 concerning the scattering of a beam of α particles by thin gold foils[†] were the first convincing evidence that the plum pudding model was wrong. The collision of a high-speed, positively charged α particle with a plum pudding model atom was expected to be undramatic. As the α particle plowed through the pudding, it would just as likely be attracted one way as another, since the positive electrical charge within the atom was assumed to be distributed uniformly throughout the pudding and the negative charges were assumed to be at random locations. The deflection or scattering of the α particles was expected to be slight.

The experiments indicated that most α particles indeed *were* scattered through very small angles; a hasty experimentalist would call this proof that the model is correct. However, Rutherford was not hasty with this pudding. He noticed that a very small number of the α particles (estimated to be about one in ten thousand) were scattered through quite a large angle, even in the backward direction. Rather than dismiss these rare events as "bad data" (always a temptation to be resisted), Rutherford took the problem of these few strays very

seriously. He described the problem of the backward-scattered α particles quaintly:

> It was quite the most incredible event that has happened to me in my life. It was almost as if you fired a 15-inch [diameter artillery] shell at a piece of tissue paper and it came back and hit you. On consideration I realized that this scattering backwards must be the result of a single collision, and when I made calculations I saw that it was impossible to get anything of that order of magnitude unless you took a system in which the greater part of the mass of the atom was concentrated in a minute nucleus. It was then that I had the idea of an atom with a minute massive center carrying a charge.[‡]

The annoying data burned the pudding into oblivion and gave us the nuclear model of the atom.

Since the kinetic energy of the α particles was known, Rutherford could calculate the distance of closest approach of the α particle to a relatively massive and motionless nucleus by using our good friend, the CWE theorem. Far from the pointlike nucleus, the α particle has a known kinetic energy, and its electrical potential energy at great distances is zero since the atom is electrically neutral. Hence the initial total mechanical energy of the α particle is $(KE + PE)_i = KE_i + 0$ J. For a head-on trajectory, the α particle has zero kinetic energy when closest to the nucleus and, at that location, the electrical potential energy of the α particle is

$$PE_f = qV = \frac{1}{4\pi\varepsilon_0}\frac{Qq}{r}$$

where q is the charge on the α particle and V is the electric potential of the nucleus of charge Q at the distance r of closest approach. From the CWE theorem, then, we have

$$KE_i = \frac{1}{4\pi\varepsilon_0}\frac{Qq}{r}$$

From this analysis, Rutherford estimated the diameter of the nucleus to be only about 10^{-15} m. The diameter of an atom, however, was known to be ~10^{-10} m, about 100 000 times larger. Thus most of an atom is empty space.

---

*The nuclear model of the atom is not as obvious as it may seem to you. In fact, the only reason it may seem obvious is that ever since you first encountered the idea back in grade school, you have been *told* that it is the way things are. The nuclear model has been drilled into you since infancy. But how would you convince a nonbeliever?

[†]Gold foils were used because the metal is very malleable, capable of being rolled into extraordinarily thin foils that are quite transparent to visible light!

[‡]Transcript of a lecture by Rutherford, "Forty years of physics," in *Background to Modern Science*, edited by Joseph Needham and Walter Pagel (Macmillan, New York, 1938), pages 68–69.

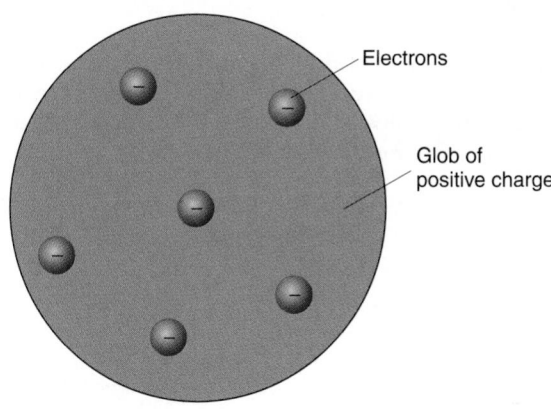

FIGURE 26.11 The plum pudding model of the atom.

The scattering of charged particles by nuclei, now called Rutherford scattering, is celebrated on this Soviet era postage stamp.

The crude pictures typically drawn to represent an atom grossly distort the relative sizes of the nucleus relative to the size of the atom. With a nucleus scaled up and modeled as a sphere a mere 1 mm in diameter, the size of an atom is 100 000 times larger, or 100 m!

QUESTION 3

Explain why it was tempting for Rutherford to dismiss the few wayward, backscattered α particles as bad data. Have *you* ever been tempted to disregard an observation as bad data?

26.7 THE BOHR MODEL OF A HYDROGENIC ATOM

There were serious unresolved questions about how atoms emit light. Most light originates from atoms.* The visible spectra of atoms were studied extensively in the late 19th century with the advent of the diffraction gratings of Henry Rowland. Each element produces its own unique spectrum of light, specific wavelengths characteristic of that element or isotope. The least massive element, hydrogen, has a spectrum that is yet quite complex: many different wavelengths are emitted. Indeed, from our modern perspective, how could an atom as simple as hydrogen, consisting of a single electron and a single proton, produce a complex spectrum of specific wavelengths?

Between 1913 and 1915 Niels Bohr developed a quantitative atomic model for the hydrogen atom that could account for its spectrum. The model incorporated the nuclear model of the atom proposed by Rutherford and his collaborators, as well as the concept of light photons developed by Einstein to explain the photoelectric effect.

> The **Bohr model** was developed specifically for **hydrogenic atoms**: atoms consisting of a nucleus with positive charge $+Ze$ (where Z is the atomic number, the number of protons in the nucleus) and a *single* electron. Thus more complex electron–electron interactions are nonexistent for hydrogenic atoms.

The model is appropriate for hydrogen, singly ionized helium, doubly ionized lithium, and so forth. We shall see that this model was successful in its ability to predict the gross features of the spectra emitted by such hydrogenic atoms.

However, the Bohr model is not a true picture of even these simple atoms. The true picture is a fully quantum mechanical affair that differs from the Bohr model in a number of fundamental ways, which we touch on in Section 26.10. Since the Bohr model incorporates aspects of some classical and some modern physics, it is now called a semiclassical model.† If the Bohr model is not strictly correct, why, then, do we bother to present it here? For several reasons, other than tradition: (1) The model does account for the gross features of the hydrogen spectrum. (2) The

---

*Light also is emitted (a) when charged particles are accelerated (a prediction of the Maxwell theory of electromagnetism), which accounts for the continuous spectrum of x-rays when electrons crash into solids; and (b) when matter and antimatter annihilate, producing γ-rays.

†The model also could be called a semimodern model, but semiclassical has a nicer ring to it.

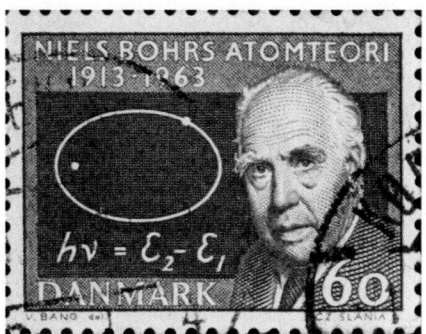

Niels Bohr was smuggled out of Nazi-occupied Denmark during World War II by the Allies. Before leaving, he dissolved his gold Nobel physics prize medal in acid and hid it in an inconspicuous cabinet at his institute in Copenhagen. After the war, the Nobel Institute kindly recast his medal using the very same gold atoms!

model incorporates in a single problem many of the ideas you have learned from a study of introductory physics. (3) The model illustrates how a theoretical physicist occasionally must quite literally ignore certain problems of an approach in hopes of being able to make some predictions. If the predictions of the theory or model agree with experiment, a theoretician then must somehow hope to explain away or rationalize the problems that were ignored along the way. These considerations make the Bohr model still pedagogically fruitful.

The features of the Bohr model are the following.

The Coulomb Force and Newton's Second Law

First, the Bohr model incorporates and analyzes features of the Rutherford nuclear model. In the Rutherford nuclear model of the atom, the single electron in a hydrogenic atom and the nucleus exert electrical forces (the Coulomb electrostatic force) on each other, and both orbit the center of mass of the electron–nucleus system. However, since the mass of the nucleus is so much greater than the mass of the electron, the center of mass of the system is essentially coincident with the nucleus. Thus, to a first approximation, we consider the electron to be in an orbit about a fixed nucleus. The orbit of the electron is assumed to be circular for simplicity; thus, the acceleration of the electron is the centripetal acceleration and has a magnitude of

$$a_c = \frac{v^2}{r} \tag{26.13}$$

where v is the speed of the electron and r is the radius of its circular orbit. The force producing the acceleration of the electron is the Coulomb force exerted by the nuclear charge on the electron. This force has a magnitude of

$$F_{elec} = \frac{1}{4\pi\varepsilon_0} \frac{|q||Q|}{r^2}$$
$$= \frac{1}{4\pi\varepsilon_0} \frac{e(Ze)}{r^2} \tag{26.14}$$

Now we apply Newton's second law to the motion of the electron:

$$\vec{F} = m\vec{a}$$

where m is the mass of the electron. Since the force and acceleration are in the same direction, we take their magnitudes to obtain

$$\frac{1}{4\pi\varepsilon_0}\frac{e(Ze)}{r^2} = m\frac{v^2}{r} \tag{26.15}$$

The Total Energy

In discussing the second ingredient of the Bohr model, we assume that relativistic speeds are not involved, so we can use the classical expressions for the kinetic energy and ignore the constant rest energy of the particles. The total energy E of the electron thus consists of the sum of its kinetic energy and electrical potential energy:

$$E = \frac{mv^2}{2} + PE_{elec} \tag{26.16}$$

The electrical potential energy of the electron is found, as usual, in the following way. The pointlike nuclear charge $+Ze$ produces an electric potential V at each point in the surrounding space:

$$V = \frac{1}{4\pi\varepsilon_0}\frac{+Ze}{r}$$

The electron at distance r from the nuclear charge thus has an electrical potential energy given by

$$
\begin{aligned}
PE_{elec} &= qV \\
&= (-e)V \\
&= (-e)\frac{1}{4\pi\varepsilon_0}\frac{Ze}{r} \\
&= -\frac{1}{4\pi\varepsilon_0}\frac{Ze^2}{r}
\end{aligned}
$$

Thus the total energy E of the electron (Equation 26.16) becomes

$$E = \frac{mv^2}{2} - \frac{1}{4\pi\varepsilon_0}\frac{Ze^2}{r} \tag{26.17}$$

There is nothing fundamentally new in what we have done so far. We used classical electrostatics to determine the force (the Coulomb force), classical dynamics (Newton's second law) to analyze the motion, classical (nonrelativistic) kinetic energy, and once again classical electrostatics to determine the electrical potential energy of the electron.

Here is where Bohr had to turn a blind eye toward one aspect of classical physics and ignore a glaring inconsistency. Classical electrodynamics (the Maxwell theory of electromagnetism) predicts that an accelerated charge will radiate electromagnetic waves and energy as light. Indeed, electromagnetic theory predicts that a charge executing circular motion at frequency ν should radiate light with the same frequency. The electron in the Bohr model is accelerated because of its circular motion; therefore, according to classical electrodynamics, the electron should steadily lose energy by emitting light. From a purely classical viewpoint the electron would have, ages ago, spiraled down into

the nucleus and thus obliterating the model.* Bohr was forced to postulate that the electron in such an orbit was in a *stable* state, and simply would not radiate energy in spite of classical electrodynamics. Bohr called this stable state a *stationary state*, but today we call them simply a **state** of the atom.

Yet Bohr used the Coulomb force law that is part and parcel of the classical theory of electromagnetism! He used one part of the theory of electromagnetism and threw out another part; it is tantamount to keeping one's theoretical cake and eating it too.

The Angular Momentum

Now we reach the third key feature of the Bohr model. Part of the genius of Bohr lay in his masterful understanding of the formal, theoretical aspects of mechanics. Bohr had reason to believe[†] that the magnitude of the orbital angular momentum of the electron was restricted to only certain values; we say the orbital angular momentum of the electron is **quantized**. He therefore took this as a second postulate of the model. In Section 26.15, we will see why his postulate was a reasonable one in retrospect, based on a discovery made almost ten years after he developed his model. For now, we must take the postulate on an ad hoc basis.

Bohr proposed that the magnitude of the orbital angular momentum of the electron L_{orbit} could not have just any value, but instead, only integral multiples of Planck's constant divided by 2π:[‡]

$$L_{orbit} = n\frac{h}{2\pi} \tag{26.18}$$

where n is a positive integer known as a **quantum number**. The quantity $h/2\pi$ occurs so frequently in quantum physics that, for convenience, it is given its own designation $\hbar$, pronounced "h-bar":

$$\hbar \equiv \frac{h}{2\pi} \approx 1.055 \times 10^{-34} \text{ J} \cdot \text{s} \tag{26.19}$$

The Bohr orbital angular momentum postulate then becomes

$$L_{orbit} = n\hbar \tag{26.20}$$

The orbital angular momentum of a particle is defined to be (from Equation 10.1)

$$\vec{L}_{orbit} = \vec{r} \times \vec{p}$$

---

*Of course, we want to be able to explain how light is emitted from the atom. Bohr realized that the light emitted from this spiraling into the nucleus has a *continuous* spectrum (analogous to white light), not the observed *discrete* spectrum with a few specific wavelengths. The lifetime of an atom with a circulating and radiating electron is extremely short ($\sim 10^{-11}$ s).

†The reasons are associated with a peculiar integral in the formal structure of mechanics known as the *action integral*. The details are beyond the level of this course.

‡The factor of 2π arises because there are 2π rad in the circle of its orbit, but this connection in mechanics is not one you would be expected to understand at this point in your studies.

In circular motion, the position vector $\vec{r}$ of a particle is perpendicular to its momentum $\vec{p}$, and so the magnitude of the orbital angular momentum is

$$L_{\text{orbit}} = rp \sin 90°$$
$$= rp$$

The Bohr angular momentum postulate, Equation 26.20, then becomes

$$rp = n\hbar \qquad (26.21)$$

Thus the Bohr model has three ingredients:

1. an application of Newton's second law (Equation 26.15);
2. the expression for the total energy (Equation 26.17); and
3. the angular momentum postulate (Equation 26.21).

If we mix these ingredients appropriately, two results emerge.

First, we take Equation 26.21 for the magnitude of the angular momentum and substitute for the magnitude of the momentum $p = mv$; we obtain

$$rmv = n\hbar \qquad (26.22)$$

Solving for v, we get

$$v = \frac{n\hbar}{mr} \qquad (26.23)$$

We use Equation 26.23 to substitute for v in Equation 26.15 (Newton's second law); this yields

$$\frac{1}{4\pi\varepsilon_0} \frac{e(Ze)}{r^2} = \frac{m}{r} \left(\frac{n\hbar}{mr} \right)^2$$

Now we solve for the radius r:

$$r = \frac{n^2 \hbar^2 (4\pi\varepsilon_0)}{mZe^2} \qquad (26.24)$$

Since n is an integer, Equation 26.24 implies that the electron correspondingly orbits the nucleus only at specific radii (so-called **allowed orbits**).

Now for the second result. Equation 26.15,

$$\frac{1}{4\pi\varepsilon_0} \frac{e(Ze)}{r^2} = m \frac{v^2}{r}$$

can be simplified slightly by multiplying by the radius r; this yields

$$\frac{1}{4\pi\varepsilon_0} \frac{e(Ze)}{r} = mv^2 \qquad (26.25)$$

Using this expression for mv^2 in the energy equation (Equation 26.17), we find

$$E = \frac{mv^2}{2} - \frac{1}{4\pi\varepsilon_0} \frac{Ze^2}{r}$$
$$= \frac{1}{2} \frac{1}{4\pi\varepsilon_0} \frac{e(Ze)}{r} - \frac{1}{4\pi\varepsilon_0} \frac{Ze^2}{r}$$
$$= -\frac{1}{2} \frac{1}{4\pi\varepsilon_0} \frac{Ze^2}{r} \qquad (26.26)$$

Using Equation 26.24 to substitute for the radius of the orbit, we obtain

$$E = -\frac{1}{2} \frac{1}{4\pi\varepsilon_0} \frac{Ze^2}{\dfrac{n^2 \hbar^2 (4\pi\varepsilon_0)}{mZe^2}}$$
$$= -\frac{1}{2} \frac{mZ^2 e^4}{(4\pi\varepsilon_0)^2 n^2 \hbar^2} \qquad (26.27)$$

Beyond all the several constants, we see that since n is an integer, the energy also is quantized.

Only certain values of the energy are permitted.

The rather ungainly expression for the energy (Equation 26.27) can be simplified if we substitute numerical values for the fundamental constants m (the mass of the electron), e (the fundamental unit of charge), ε_0 (the permittivity of free space), and $\hbar$ (Planck's constant divided by 2π). The result of these substitutions is

$$E = -(21.8 \times 10^{-19} \text{ J}) \frac{Z^2}{n^2} \qquad (26.28)$$

This can be converted to electron-volts by dividing by 1.602×10^{-19} J/eV, yielding

$$E = -(13.6 \text{ eV}) \frac{Z^2}{n^2} \qquad (26.29)$$

For hydrogen, $Z = 1$, and we have

$$E_{\text{hydrog}} = -\frac{13.6}{n^2} \text{ eV} \qquad (26.30)$$

Bohr's permitted values of the energy typically are represented graphically on an **energy level diagram**, shown in Figure 26.12. The vertical axis of the graph is an energy scale; there is no horizontal axis, but horizontal lines are drawn to show the appropriate levels (values) of the energy.

The lowest energy state (corresponding to $n = 1$ in Equations 26.29 and 26.30) is the **ground state**. Notice that as the quantum number n increases, the energy of the electron increases and approaches zero as a limit for $n = \infty$. As $n \to \infty$, the radius of the orbit also increases indefinitely according to Equation 26.24. Thus, when $E \geq 0$ J, the electron is no longer associated with the nucleus and is free; the atom has been **ionized**.

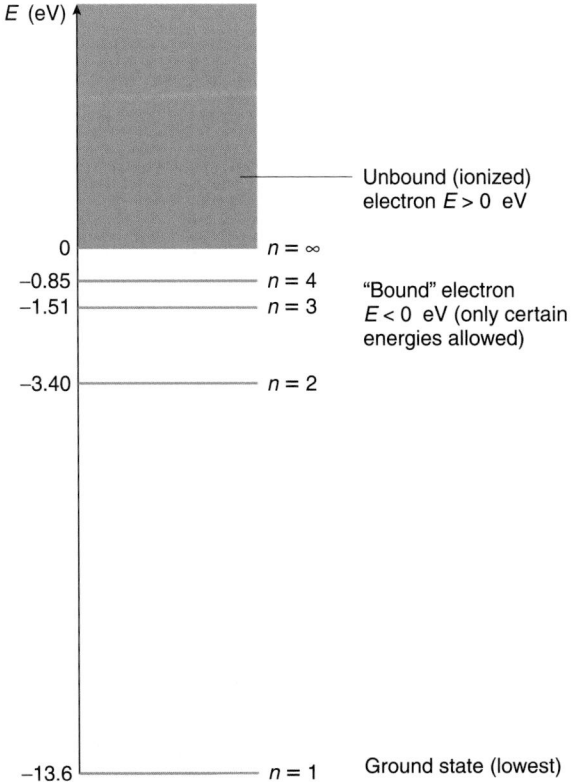

FIGURE 26.12 Energy level diagram for hydrogen.

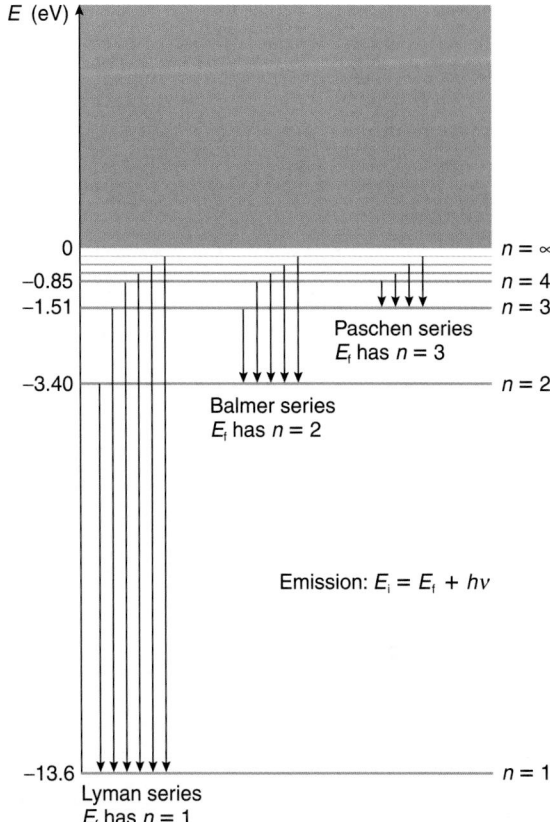

FIGURE 26.13 The discrete emission spectrum of hydrogen. A photon is emitted when the electron makes a transition to an energy state with a lower value of the quantum number n.

The neutral atom has energies $E < 0$ J, with the electron bound to the nucleus.

Recall that Bohr's objective was to explain the discrete spectrum of hydrogen, the fact that only specific wavelengths of light are emitted by a collection of hydrogen atoms. Since Bohr *postulated* that the electron would not radiate electromagnetic waves despite its acceleration, he had to invent a new mechanism for light emission. The energy of the electron in a particular orbit in the atom depends on the quantum number n; in particular, the higher n, the higher the energy.

Thus an electron with an $n > 1$ can lower its energy by changing to an orbit with a lower value of n, an orbit closer to the nucleus according to Equation 26.24. Such a jump is called a **transition**.* In this process, the electron lowers the amount of its energy (loses energy). Bohr postulated that the energy lost by the electron appears as the energy of a photon emitted by the atom.

In this way total energy is conserved. That is, if an electron is in an orbit with $n > 1$ with energy E_i, in making a transition to a lower energy E_f the atom conserves overall energy according to

$$E_i = E_f + E_{\text{emitted photon}} \quad \text{(emission)} \quad (26.31)$$

---

*This is the origin of the pedestrian phrase "quantum jumps."

The discrete emission spectrum of hydrogen therefore corresponds to electrons cascading down to lower energy levels in various atoms of hydrogen (see Figure 26.13).

Each transition of the electron to a lower energy level produces a photon with a specific energy or, equivalently, with a definite frequency or wavelength. All transitions to the deep $n = 1$ state emit light that is in the ultraviolet region of the spectrum; this series of transitions is called the **Lyman series**. The visible light spectrum of hydrogen involves transitions to the $n = 2$ state from $n > 2$ higher states, called the **Balmer series**. There are other series too, those ending on $n = 3$, $n = 4$, and so forth, but it is pointless to name them all since there are an infinite number of them.[†]

Correspondingly, if a photon of exactly the right energy is available, the electron in a lower state can totally absorb the photon (destroying it, but taking its energy) and make a transition to a higher energy level.

Energy again is conserved by writing:

$$E_i + E_{\text{photon absorbed}} = E_f \quad \text{(absorption)} \quad (26.32)$$

[†]Those transitions ending on $n = 3$ form the Paschen series; those on $n = 4$ are the Brackett series; those on $n = 5$ are the Pfund series, after which we gave up naming them.

This neatly explains the experimental observation that the wavelengths (or frequencies) that hydrogen emits are the same as the wavelengths that hydrogen absorbs. That is, if a white light spectrum (a continuous spectrum) is incident on cool hydrogen gas, the hydrogen will absorb just those photon wavelengths that (hot) hydrogen emits. So the transmitted light is missing certain wavelengths; the wavelengths missing are precisely the same ones that hydrogen emits.

The host of energy levels and the multitude of possible transitions thus explain the complexity of the discrete hydrogen spectrum. The crowning achievement of the Bohr model was its ability to predict the wavelengths of the spectrum of single-electron atoms (hydrogenic atoms).

Refinements subsequently were made to the Bohr model, allowing for the possibility of elliptical electron orbits, much like the elliptical orbits of the planets of the Sun. Corrections made for the finite mass of the nucleus, by treating both the electron and nucleus in orbit about their center of mass, led to small corrections in the predicted spectrum and eventually to the discovery of **deuterium**, an isotope of hydrogen with twice the nuclear mass of ordinary hydrogen.

QUESTION 4

Why is the total energy in the Bohr model of a hydrogenic atom negative?

EXAMPLE 26.5

a. Calculate the orbital radius of the electron orbit closest to the nucleus of the hydrogen atom. This distance is called the *Bohr radius* a_0 of the hydrogen atom.
b. Express the other radii of a hydrogenic atom in terms of the Bohr radius. Use this expression to find the radius of the orbit closest to the nucleus of singly ionized helium.

Solution

a. The orbital radius is given by Equation 26.24:

$$r = \frac{n^2 \hbar^2 \ (4\pi\varepsilon_0)}{mZe^2}$$

For hydrogen, $Z = 1$. The orbit closest to the nucleus has the smallest n-value: $n = 1$. Hence Equation 26.24 becomes, for the Bohr radius a_0,

$$a_0 = \frac{\hbar^2 (4\pi\varepsilon_0)}{me^2} \qquad (1)$$

The mass m is that of the electron. Substituting numerical values for the various quantities, you get

$$a_0 = \frac{(1.055 \times 10^{-34} \ \text{J·s})^2}{(9.11 \times 10^{-31} \ \text{kg})(1.602 \times 10^{-19} \ \text{C})^2(9.00 \times 10^9 \ \text{N·m}^2/\text{C}^2)}$$

$$= 5.29 \times 10^{-11} \ \text{m}$$

b. To express the other radii of hydrogenic atoms in terms of the Bohr radius a_0, use equation (1) in Equation 26.24:

$$r = \frac{n^2}{Z} \frac{\hbar^2 4\pi\varepsilon_0}{me^2}$$

$$= \frac{n^2}{Z} a_0 \qquad (2)$$

Singly ionized helium is a hydrogenic atom (a one-electron atom). Helium has $Z = 2$ since there are two protons in its nucleus. The orbit closest to the nucleus has $n = 1$. Hence for helium you have

$$r = \frac{a_0}{2}$$

$$= 2.65 \times 10^{-11} \ \text{m}$$

which is half the Bohr radius of the hydrogen atom.

EXAMPLE 26.6

An electron in the hydrogen atom makes a transition from the $n = 5$ orbit to the $n = 2$ orbit. Calculate the wavelength of the photon emitted by the atom.

Solution

It is easier to use Equation 26.30 for the energy of each state of the hydrogen atom, rather than the more ungainly Equation 26.27. The energy of the $n = 5$ state thus is

$$E_5 = -\frac{13.6}{(5)^2} \ \text{eV}$$

$$= -0.544 \ \text{eV}$$

while that of the $n = 2$ state is

$$E_2 = -\frac{13.6}{(2)^2} \ \text{eV}$$

$$= -3.40 \ \text{eV}$$

Since the total energy is conserved, you have

$$E_{\text{before}} = E_{\text{after}}$$
$$E_5 = E_2 + E_{\text{photon}}$$
$$-0.544 \ \text{eV} = -3.40 \ \text{eV} + h\nu$$
$$h\nu = 2.86 \ \text{eV}$$
$$= (2.86 \ \text{eV})(1.602 \times 10^{-19} \ \text{J/eV})$$
$$= 4.58 \times 10^{-19} \ \text{J}$$

The wavelength λ is found recalling that $c = \nu\lambda$. Hence you have

$$\frac{hc}{\lambda} = 4.58 \times 10^{-19} \ \text{J}$$

Solving for the wavelength, you find

$$\lambda = \frac{hc}{4.58 \times 10^{-19} \ \text{J}}$$

$$= \frac{(6.626 \times 10^{-34} \ \text{J·s})(3.00 \times 10^8 \ \text{m/s})}{4.58 \times 10^{-19} \ \text{J}}$$

$$= 4.34 \times 10^{-7} \ \text{m}$$

$$= 434 \ \text{nm}$$

This is a visible light wavelength; you can see the light as a violet-colored line in the spectrum.

26.8 THE BOHR CORRESPONDENCE PRINCIPLE

Recall that Bohr swept a crucial prediction of classical electromagnetic theory under the rug, namely, that a charge undergoing circular motion (accelerated motion) should radiate light with a frequency equal to that of its orbital motion. Bohr devoted much thought and effort to the problem of how his new ideas about light emission might be compatible with or correspond to the classical predictions in at least some appropriate limiting situation.

To see how he made the correspondence, it is first necessary to find the orbital frequency of the electron in a hydrogenic atom. The electron travels the circumference $2\pi r$ at the speed v, and so the period T of its motion is

$$T = \frac{2\pi r}{v}$$

The frequency of its orbital motion ν_{orbit} is the reciprocal of its period, so that we have

$$\nu_{\text{orbit}} = \frac{v}{2\pi r} \qquad (26.33)$$

The angular momentum postulate,

$$L = pr = mvr = n\hbar$$

indicates that

$$v = \frac{n\hbar}{mr}$$

Making this substitution for v into Equation 26.33 for ν_{orbit}, we find

$$\nu_{\text{orbit}} = \frac{1}{2\pi r} \frac{n\hbar}{mr}$$

$$= \frac{n\hbar}{2\pi mr^2}$$

If we now substitute for the radius using Equation 26.24, we obtain

$$\nu_{\text{orbit}} = \frac{n\hbar}{2\pi m \dfrac{n^4 \hbar^4 (4\pi\varepsilon_0)^2}{m^2 Z^2 e^4}}$$

$$= \frac{1}{2\pi} \frac{mZ^2 e^4}{(4\pi\varepsilon_0)^2} \frac{1}{\hbar^3 n^3} \qquad (26.34)$$

For small quantum numbers n this orbital frequency is *not* the frequency of any of the wavelengths emitted by such a hydrogenic atom. But Bohr wondered what happened to the frequency of light emitted by the hydrogen atom in making a transition from the n to the $n - 1$ state, at large values of the quantum number n. For such a transition from n to $n - 1$ (see Figure 26.14) a photon is emitted with an energy that satisfies Equation 26.31:

$$E_i = E_f + E_{\text{emitted photon}}$$

We use the expression for the energy levels of such an atom (Equation 26.27) in Equation 26.31 to find

$$-\frac{1}{2} \frac{mZ^2 e^4}{(4\pi\varepsilon_0)^2 \hbar^2 n^2} = -\frac{1}{2} \frac{mZ^2 e^4}{(4\pi\varepsilon_0)^2 \hbar^2 (n-1)^2} + E_{\text{emitted photon}}$$

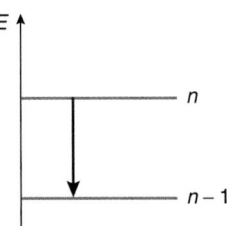

FIGURE 26.14 **The transition from a state with quantum number n to one with quantum number $n - 1$.**

Rearranging, we have

$$E_{\text{emitted photon}} = \frac{1}{2} \frac{mZ^2 e^4}{(4\pi\varepsilon_0)^2 \hbar^2} \left[\frac{1}{(n-1)^2} - \frac{1}{n^2} \right] \qquad (26.35)$$

The expression in the brackets on the right-hand side of Equation 26.35 can be combined into a single term:

$$\frac{1}{(n-1)^2} - \frac{1}{n^2} = \frac{n^2 - (n-1)^2}{n^2(n-1)^2}$$

$$= \frac{2n - 1}{n^4 - 2n^3 + n^2} \qquad (26.36)$$

We take the limit of this expression as n becomes very large. Then the number 1 can be neglected in the numerator of the right-hand side of Equation 26.36, and all but the first term in the denominator also can be neglected,* so that

$$\frac{1}{(n-1)^2} - \frac{1}{n^2} \approx \frac{2n}{n^4}$$

$$\approx \frac{2}{n^3}$$

Thus, for large n, the energy of the light emitted in this transition is, from Equation 26.35,

$$E_{\text{emitted photon}} = \frac{1}{2} \frac{mZ^2 e^4}{(4\pi\varepsilon_0)^2 \hbar^2} \frac{2}{n^3}$$

The frequency of this photon is found from Equation 26.4:

$$E_{\text{photon}} = h\nu$$

Hence the emitted photon has a frequency of

$$\nu_{\text{photon}} = \frac{1}{h} \frac{1}{2} \frac{mZ^2 e^4}{(4\pi\varepsilon_0)^2 \hbar^2} \frac{2}{n^3}$$

---

*If $n = 1000$, then the difference between

$$\frac{2n - 1}{n^4 - 2n^3 + n^2}$$

and

$$\frac{2n}{n^4}$$

is less than 0.15% (which can be verified by direct substitution into the two expressions).

Converting the h to an $\hbar$ by multiplying and dividing by 2π, we obtain

$$\nu_{\text{photon}} = \frac{1}{2\pi} \frac{mZ^2e^4}{(4\pi\varepsilon_0)^2} \frac{1}{\hbar^3 n^3} \qquad (26.37)$$

This frequency is identical to the frequency of the orbital motion of the electron given by Equation 26.34.

Thus Bohr discovered that for large quantum numbers n, the light emitted when the electron jumps to the next lower orbit has a frequency that is the same as the classical prediction of electromagnetic theory. The correspondence between classical physics and what happens with large quantum numbers is called **Bohr's correspondence principle.**

The principle is analogous to the situation encountered in relativity: in the limit of speeds small compared with the speed of light, the Lorentz transformation equations reduce to the classical Galilean relativity equations.

26.9 A BOHR MODEL OF THE SOLAR SYSTEM?*

The motion of the electron around the nucleus of the hydrogen atom in the Bohr model is imagined to be similar to the motion of a planet around the Sun: a planetary atom. What happens if we apply the Bohr model to the two-body gravitational interaction between a planet and the Sun? The only real change is the substitution of the gravitational force for the electrical force. Let's formulate a Bohr model for the Earth in its orbit around the Sun and ask two questions:

1. What is the quantum number for the orbit of the Earth?
2. What is the distance between adjacent permitted orbits for the Earth? Is the next allowed orbit beyond the orbit of the Earth where Mars is located?

We formulate the problem just as we did for the Bohr model of a hydrogenic atom. First, we apply Newton's second law to the motion of the Earth, assuming (to a first approximation) that the orbit of the Earth is a circle of radius r:

$$\vec{F} = m\vec{a}$$

The total force is the gravitational force on the Earth by the Sun. Since the force is in the same direction as the centripetal acceleration, we use the magnitudes of the vectors:

$$\frac{GMm}{r^2} = m\frac{v^2}{r} \qquad (26.38)$$

where M is the mass of the Sun and m is the mass of the Earth.

The second step is to apply Bohr's angular momentum postulate. This postulate states that the magnitude of the angular momentum is quantized in integral multiples of $\hbar$:

$$L = pr = mvr = n\hbar \qquad (26.39)$$

where n is the quantum number of the orbit of the Earth. We solve Equation 26.39 for v:

$$v = \frac{n\hbar}{mr}$$

Substituting for v in the force law expression, Equation 26.38, we find

$$\frac{GMm}{r^2} = \frac{m}{r}\left(\frac{n\hbar}{mr}\right)^2$$

Now we solve for r, obtaining

$$r = \frac{n^2\hbar^2}{GMm^2} \qquad (26.40)$$

To find the quantum number n for the orbit of the Earth, we solve Equation 26.40 for n:

$$n^2 = \frac{GMm^2r}{\hbar^2}$$

We make the following numerical substitutions:

$$G = 6.67 \times 10^{-11} \ \text{N} \cdot \text{m}^2/\text{kg}^2$$
$$M_{\text{Sun}} = 1.99 \times 10^{30} \ \text{kg}$$
$$m_{\text{Earth}} = 5.98 \times 10^{24} \ \text{kg}$$
$$r_{\text{Earth orbit}} = 1.496 \times 10^{11} \ \text{m}$$
$$\hbar = 1.055 \times 10^{-34} \ \text{J} \cdot \text{s}$$

We finally find that

$$n = 2.52 \times 10^{74} \qquad (26.41)$$

The quantum number of the Earth is *quite* large!

Let Δr be the distance between the present orbit of the Earth, with quantum number n, and the next orbit out from the Sun, corresponding to quantum number $n + 1$. From Equation 26.40, we find

$$\Delta r = r_{n+1} - r_n$$
$$= \frac{\hbar^2}{GMm^2}\left[(n + 1)^2 - n^2\right] \qquad (26.42)$$

Since n is so large, we make the following approximation:

$$(n + 1)^2 - n^2 = n^2 + 2n + 1 - n^2$$
$$= 2n + 1$$
$$\approx 2n$$

Equation 26.42 for Δr then becomes

$$\Delta r \approx \frac{\hbar^2}{GMm^2} 2n \qquad (26.43)$$

Now make the appropriate substitutions and use $n = 2.52 \times 10^{74}$. From Equation 26.43 we find that

$$\Delta r \approx 1.2 \times 10^{-63} \ \text{m}$$

This is ~48 *orders of magnitude* smaller than the diameter of a typical nucleus (~ 10^{-15} m)! The distance between the quantized orbits of the Earth is *so* tiny as to be essentially zero; the planet has a continuum of orbits in which to orbit.

The Bohr model thus is irrelevant for the celestial mechanics of the solar system, and the spacing of the planets in the solar system has nothing to do with the quantization of angular momentum.

26.10 PROBLEMS WITH THE BOHR MODEL

Alas, the Bohr model with its planet-like electron must be discarded. While the model correctly predicts the gross features of the spectrum of hydrogenic atoms, in particular, the frequencies of the light emitted or selectively absorbed, the model was just a way station to a completely new mechanics, quantum mechanics, and an improved but stranger model of the atom. The limitations of the Bohr model include the following.

> The Bohr model is applicable only to hydrogenic (single-electron) atoms. It cannot be easily extended, even to mere two-electron atoms such as helium.

Unlike the situation in the solar system, where planet–planet gravitational forces are very small compared with the gravitational force of the Sun on each planet (because the mass of the Sun is so much greater than the mass of any of the planets), the electron–electron electrical force interaction is comparable in magnitude to the electron–nucleus electrical force, because the charges and distances are of the same order of magnitude.

> While the Bohr model correctly predicts the frequencies of the light emitted by hydrogenic atoms, the model says nothing about the relative intensities of the frequencies in the spectrum.

Some of the visible wavelengths have weak intensity, others strong. The question is, of course, why? The experimental fact that the relative intensities of the various frequencies are not the same actually is a relief, since there are infinitely many of them, but it indicates that some transitions are more favored than others. On what basis is this assessment made by the electron? The Bohr model is unable to account for the intensity variations.

The angular momentum postulate of the Bohr model, Equation 26.20, $L = n\hbar$, predicts that the magnitude of the orbital angular momentum of the ground state, corresponding to $n = 1$, is $L_{\text{ground state}} = \hbar$.

> In retrospect, after the advent of a quantum mechanical model of the atom in the late 1920s, it was realized that the ground state has an orbital angular momentum equal to *zero*. This called into question the whole conceptual picture of an orbiting planet-like electron and placed the fundamental premise of the Bohr model in jeopardy.

These problems indicated that the model is not a correct rendition of even the simplest atom, hydrogen. Thus, while the Bohr model has a picturesque simplicity and is a model that can be readily visualized, the model has fundamental flaws and presents an incorrect picture of the atom.

26.11 RADIOACTIVITY REVISITED

The nuclear model of the atom led to an increased understanding of radioactivity, discovered in 1896 by Becquerel, as we saw in Section 26.3. The understanding of radioactivity that emerged was the realization of the age-old quest of ancient and medieval alchemists: the **transmutation** of the elements (the transformation of one chemical element into another). Such transmutations are impossible via chemical techniques, which leave the nucleus quite unaffected.

> Early experiments involving radioactive materials indicated that the particle emission rate (the number of particles per unit time) emitted from a radioactive substance
>
> a. is independent of temperature and pressure;
> b. is not affected by the presence of electric and/or magnetic fields; and
> c. is independent of how the element is chemically bound.
>
> The emission rate depends *solely* on the *concentration* of the radioactive isotope, that is, on the number of atoms of the isotope present. These surprising observations, together with the extraordinary energy of the emitted particles (compared with the energies typically associated with chemical reactions) led to the conclusion that the particles emitted from a radioactive material emerge from the nuclei of the atoms involved.

Atoms are radioactive if their nuclei are unstable and spontaneously (and randomly) emit various particles, the α, β, and/or γ radiations. When naturally occurring nuclei are unstable, we call the phenomena **natural radioactivity**. Other nuclei can be transformed into radioactive nuclei by various means, typically involving irradiation by neutrons; this is called **artificial radioactivity**.

We saw in Section 26.3 that α particles were found to be helium nuclei (two protons and two neutrons), since helium always was present in materials that emitted α particles. Magnetic and electric field experiments with β particles led to the conclusion that they were electrons (called β^- particles). During the 1930s, a particle with the same mass as an electron but with a *positive* charge, $+e$, was discovered in cosmic rays and later in the emission of certain radioactive isotopes. These **positrons** are called β^+ particles. A positron is called the **antiparticle** of the electron.

Gamma particles (γ-rays), unaffected by both electric and magnetic fields, finally were shown to be high-energy photons. Gamma rays have frequencies larger than x-rays, and correspondingly shorter wavelengths; the distinction between γ-rays and x-rays is somewhat arbitrary. X-rays typically result from certain transitions of the electrons surrounding the nucleus in the atom; γ-rays generally are emitted from the nucleus itself.

A radioactive nucleus is called a **parent** nucleus; the nucleus resulting from its decay by particle emission is called the **daughter** nucleus. Daughter nuclei also might be radioactive, producing granddaughter nuclei, and so on. There are no son or grandson nuclei.*

For any nucleus, the atomic number Z is the number of protons in the nucleus, typically written as a numerical subscript together with the chemical symbol for the element. The mass number A is the sum of the number of protons and neutrons in

---

*While the term parent nucleus conforms to contemporary, politically correct trends for nonsexist language, the term daughter nuclei does not. The term "progeny" nuclei just doesn't hack it.

the nucleus, typically written as a numerical superscript along with the chemical symbol for the element. Thus, for example, we write $^{235}_{92}U$ for the isotope of uranium (atomic number = 92) of mass number 235.

α Decay

The emission of an α particle (a helium nucleus, ^4_2He), a process known as α **decay**, means that the mass number of the parent nucleus decreases by 4 and the number of protons by 2. The resulting daughter nucleus thus has a mass number of $A - 4$ and an atomic number of $Z - 2$. An example of this is the decay of radium into radon:

$$^{226}_{88}Ra \quad \rightarrow \quad ^{222}_{86}Rn \quad + \quad ^4_2He$$

$$\begin{array}{ccc} \text{parent} & \rightarrow & \text{daughter} + \alpha \text{ particle} \\ \text{nucleus} & & \text{nucleus} \end{array}$$

The α particles from a given decay are monoenergetic, that is, all emerge with the same energy.

β Decay

The emission of an electron (β^- particle) in radioactive decay is not the emission of an orbiting electron; the β^- particle is ejected from the nucleus. To conserve electrical charge, we imagine a neutron turning into a proton (which remains in the nucleus) with the ejection of an electron. Thus a parent nucleus with atomic number Z and mass number A decays by β^- emission into a daughter with atomic number $Z + 1$ and the same mass number A.

Experiments indicated that the energies of the emitted electrons from a given nuclear species have a continuous spectrum of energies. To conserve energy and momentum, it was necessary to hypothesize that another particle was emitted with each β^- decay: an electron antineutrino, designated $\overline{v}$. The hypothetical particle was not easily detected, but its existence finally was confirmed in the 1950s (the 1995 Nobel prize in physics was awarded to Frederick Reines for the discovery). Such antineutrinos interact extremely weakly with matter and are extraordinarily difficult to detect. They have no charge, and little (if any) mass; the question of their mass still is hotly debated and the subject of much experimental and theoretical inquiry. The consensus now is that the mass is nonzero, but very small. An example of β^- decay, antineutrino and all, is the decay of carbon 14 into nitrogen:

$$^{14}_6C \rightarrow ^{14}_7N + \beta^- + \overline{v}$$

Positron (β^+) emission from a nucleus decreases the atomic number Z by 1 while keeping the same mass number A. The process can be imagined as the conversion of a proton to a neutron within the nucleus, with the positron carrying off the positive charge. Positron emission always is accompanied by the emission of an electron neutrino v that interacts extremely weakly with matter and has properties similar to the antineutrino of β^- decay (the difference between an electron neutrino and an electron antineutrino has to do with another fundamental property of particles known as their spin). An example of positron emission is the decay of another isotope of carbon into boron:

$$^{11}_6C \rightarrow ^{11}_5B + \beta^+ + v$$

Both β^- and β^+ emission are known as β **decay**.

γ Decay

The emission of a γ-ray photon from a nucleus, called γ **decay**, does not transmute the element. Nuclei that emit γ-ray photons typically are the daughter nuclei of another radioactive decay process. The daughter nuclei created in this other decay are in an excited state (analogous to the excited states of the orbiting electrons, except that the energy levels associated with the nucleus have much larger energy differences than those involved with the atomic electrons). The subsequent transition of the nucleus to a lower energy state results in the emission of a γ-ray photon whose energy is the energy lost by the nucleus.

An example of γ decay is that associated with cobalt 60, a radioisotope of cobalt widely used in medicine. The cobalt 60 decays into nickel by β^- decay[†] but the nickel nucleus is formed in an excited state [designated with a star (*) superscript] that subsequently emits two γ-ray photons:

$$^{60}_{27}Co \rightarrow (^{60}_{28}Ni)^* + \beta^- + \overline{v}$$
$$\hookrightarrow \quad ^{60}_{28}Ni + 2\gamma$$

Radioactive Decay Law

Since all the radioactive nuclei of a given radioactive material do not decay instantaneously, but do so over finite, even extended periods of time, the mechanism governing the process has to be fundamentally different from classical physics.

If you drop 1000 marbles all at once, they all fall immediately. Indeed, it would be quite surprising if they did not! With 1000 radioactive nuclei, they do *not* all decay immediately. The mechanism governing radioactivity is *statistical* in nature rather than deterministic. It is impossible to say which specific nuclei will decay during a given time interval; it is only possible to measure how many out of the large population of nuclei do so. The process is much like life. Everyone will die, since life is fatal (we all are unstable particles). But we cannot predict which people will live for how long, although we can deduce statistical averages on the basis of observations of a large number of people.[‡]

Experimentally, the small number of nuclei dN that decay during a short time interval dt is found to be proportional to the number N of nuclei present (that is, to the concentration of the nuclei), as well as to the interval dt. That is, we have

$$dN \propto N \, dt$$

A proportionality constant is introduced, called the **disintegration constant** λ,[§] so that we write

$$dN = -\lambda N \, dt \qquad (26.44)$$

[†]This decay can proceed by two different paths resulting in the emission of β^- particles with two different energies and γ-rays (from the decay of the nickel nucleus) of two different energies. Each individual cobalt nucleus proceeds along one or the other of these two paths.

[‡]The accuracy and dependability of such demographic statistics are what make issuing life insurance a viable economic activity.

[§]Do not confuse the disintegration constant λ with the wavelength of light.

The minus sign indicates that the number of nuclei present (of a particular kind) is decreasing with time. The time rate at which the nuclei decay is

$$\frac{dN}{dt}$$

and is called the **activity** of the sample; it is proportional to the number of nuclei present:

$$\boxed{\frac{dN}{dt} = -\lambda N} \qquad (26.45)$$

The SI unit of activity is the **becquerel** (Bq). One becquerel is one disintegration per second. A more common and convenient unit of activity is the **curie** (Cu), defined to be exactly 36 billion disintegrations per second:

$$1 \ \text{Cu} \equiv 36 \times 10^9 \ \text{Bq} \qquad (26.46)$$

Equation 26.45 can be separated and integrated; we rearrange the equation slightly to find

$$\frac{dN}{N} = -\lambda \ dt$$

If we begin with N_0 nuclei at time $t = 0$ s and have N left at time t, then

$$\int_{N_0}^{N} \frac{dN}{N} = -\lambda \int_{0 \ \text{s}}^{t} dt$$

After integration, we find

$$\ln N - \ln N_0 = -\lambda t$$

or

$$\ln \frac{N}{N_0} = -\lambda t$$

Taking the antilogarithms,* we obtain

$$\frac{N}{N_0} = e^{-\lambda t}$$

or

$$\boxed{N = N_0 e^{-\lambda t}} \qquad (26.47)$$

Equation 26.47 is known as the **radioactive decay law**.

> The number of parent radioactive nuclei decreases exponentially with time, as shown in Figure 26.15.

> The time $\tau_{1/2}$ at which only half the parent nuclei remain is known as the **half-life** of the parent.

We use Equation 26.47 and substitute $N = N_0/2$ when $t = \tau_{1/2}$; this yields

$$\frac{N_0}{2} = N_0 e^{-\lambda \tau_{1/2}}$$

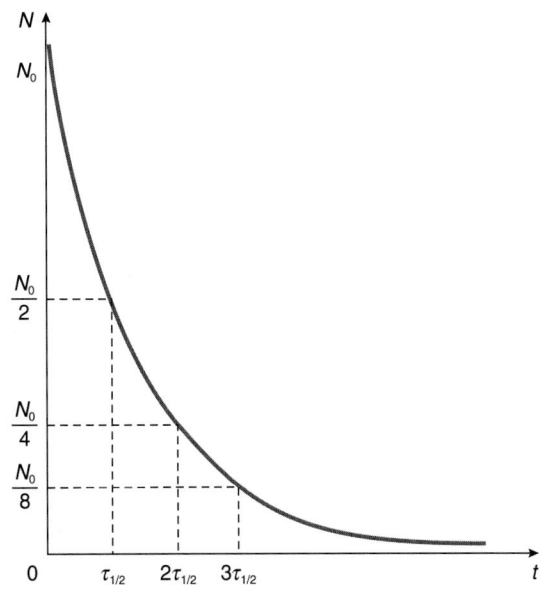

FIGURE 26.15 The number of radioactive nuclei decays exponentially with time.

or

$$\frac{1}{2} = e^{-\lambda \tau_{1/2}}$$

Taking the natural logarithm of this equation, we find

$$\ln 1 - \ln 2 = -\lambda \tau_{1/2}$$

Since $\ln 1 = 0$, we have

$$\boxed{\begin{aligned} \tau_{1/2} &= \frac{\ln 2}{\lambda} \\ &= \frac{0.6931}{\lambda} \end{aligned}} \qquad (26.48)$$

which relates the disintegration constant λ to the half-life. After a time equal to two half-lives, only 1/4 of the original nuclei is present. After three half-lives, only $(1/2)^3 = 1/8$ is present; and so on. After a time equal to ten half-lives, only $(1/2)^{10} = 1/1024$ is present, not quite 0.1%.

Experimentally, it is easier to measure the activity, rather than the number of atoms present. Let the original activity of the sample at $t = 0$ s be

$$\left(\frac{dN}{dt}\right)_0$$

According to Equation 26.45, the original activity is proportional to the original number N_0 of radioactive nuclei present:

$$\left(\frac{dN}{dt}\right)_0 = -\lambda N_0 \qquad (26.49)$$

Later, at time t, the activity is

$$\frac{dN}{dt}$$

---

*Do not confuse the base of natural logarithms e, a pure number approximately equal to 2.718 2818..., with the fundamental unit of charge $e \approx 1.602 \times 10^{-19}$ C.

This activity is proportional to the number of number N present at time t:

$$\frac{dN}{dt} = -\lambda N \qquad (26.50)$$

According to Equation 26.47, $N = N_0 e^{-\lambda t}$. Making this substitution for N in Equation 26.50, we find

$$\frac{dN}{dt} = -\lambda N_0 e^{-\lambda t} \qquad (26.51)$$

But $-\lambda N_0$ is the original activity

$$\left(\frac{dN}{dt}\right)_0$$

according to Equation 26.49. Hence Equation 26.51 becomes

$$\boxed{\frac{dN}{dt} = \left(\frac{dN}{dt}\right)_0 e^{-\lambda t} \qquad (26.52)}$$

In other words, the activity of the sample also decays exponentially with time with the same half-life as the number of nuclei.

The half-life of a sample of radioactive nuclei is the time for half of the sample to disintegrate. The average or **mean life** of the sample is different. Some of the atoms in the sample exist much longer than others before decaying. To determine the mean life, consider an analogy. Imagine a collection of 10 people with the following death statistics: 4 die at age 65 y, 3 at age 75 y, 2 at 85 y, and 1 at a venerable 95 y. The average age is found by multiplying the number dying at each age, summing the results, and dividing the sum by the total sample size:

$$\text{average age} = \frac{4(65 \text{ y}) + 3(75 \text{ y}) + 2(85 \text{ y}) + 1(95 \text{ y})}{10}$$
$$= 75 \text{ y}$$

The average lifetime of an initial sample of N_0 radioactive atoms is found in a similar way. Let N be the number of atoms that still exit at time t. Between t and $t + dt$, we lose a few of these hearty atoms: dN of them decay. Thus the number of atoms that live a time t is dN. The sum in the numerator of the average is really an integration of the quantity $t \, dN$ between N_0 and 0 particles; the denominator is the sum of the particles, or the integration of dN over all the particles:

$$\text{mean or average life} = \frac{\int_{N_0}^0 t \, dN}{\int_{N_0}^0 dN} \qquad (26.53)$$

For the integration in the numerator, we use Equation 26.45 to find dN:

$$\frac{dN}{dt} = -\lambda N$$

$$dN = -\lambda N \, dt$$

We make this substitution for dN in the numerator of Equation 26.53. For the average life, we then have

$$\text{mean or average life} = \frac{-\int_{0 \text{ s}}^{\infty \text{ s}} t \lambda N \, dt}{-N_0} \qquad (26.54)$$

But $N = N_0 e^{-\lambda t}$ from Equation 26.47. Thus Equation 26.54 becomes

$$\text{mean life} = \frac{\int_{0 \text{ s}}^{\infty \text{ s}} t \lambda N_0 e^{-\lambda t} \, dt}{N_0} \qquad (26.55)$$

This integration is done by parts. The result is

$$\text{mean life} = \frac{1}{\lambda} \qquad (26.56)$$

The mean life is the reciprocal of the disintegration constant.

Notice that the mean life is longer than the half-life; from Equation 26.47, the mean life $t = 1/\lambda$ is the time when there are $N_0 e^{-\lambda(1/\lambda)} = N_0 e^{-1} = N_0/e$ nuclei remaining, as shown in Figure 26.16.

PROBLEM-SOLVING TACTIC

26.3 In the radioactive decay law for the **number of nuclei remaining or for the activity, the units you use for** $\tau_{1/2}$ **and** $1/\lambda$ **must be the same.** The dimensions for the half-life are time while those for the disintegration constant λ are inverse time. The units need not be seconds. Example 26.7 at the end of the next section illustrates this point.

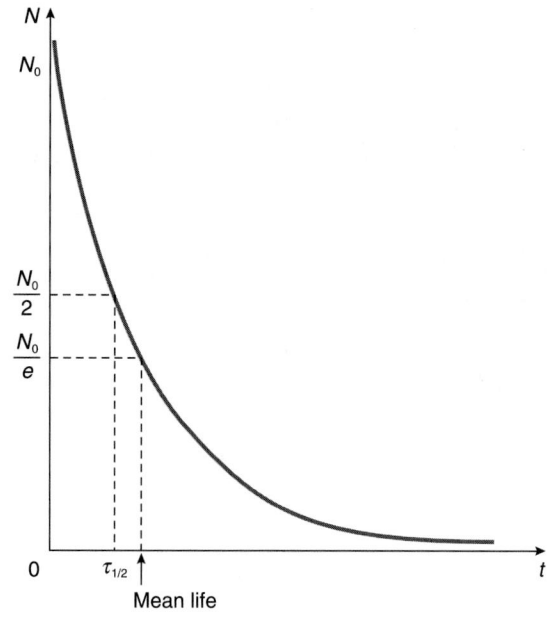

FIGURE 26.16 The mean life is longer than the half-life.

26.12 CARBON DATING

One of the most common methods for dating organic artifacts and remains makes use of a radioactive isotope of carbon, $^{14}_{6}C$, commonly called carbon 14. This radioactive isotope is created in the upper atmosphere continuously via cosmic ray* bombardment from space. Interactions of cosmic ray protons (designated p) with nuclei in the upper atmosphere produce, among many other particles, neutrons. Some of these uncharged neutrons (designated n) are captured easily by plentiful nitrogen nuclei in the atmosphere to form carbon 14 via the nuclear reaction

$$^{14}_{7}N + n \rightarrow \,^{14}_{6}C + p$$

The radioactive carbon 14 decays by β^- emission:

$$^{14}_{6}C \rightarrow \,^{14}_{7}N + \beta^- + \bar{\nu}$$

Since carbon 14 is created by cosmic rays and simultaneously destroyed by radioactive decay, an equilibrium concentration of carbon 14 exists in the atmosphere. The equilibrium concentration is about 1.5×10^{-12} kg of carbon 14 per kilogram of carbon.[†] Atmospheric gases such as carbon dioxide thus contain a small amount of the radioactive isotope. Plant life uses carbon dioxide for photosynthesis; in turn, plants are eaten by other forms of life and the radioactive carbon works its way into and up the food chain. As a result, all living things contain small amounts of carbon 14. The carbon 14 decays, of course, but is replenished continuously with fresh carbon 14 as the plant or organism breathes and eats.[‡] Thus living things have a small equilibrium concentration of carbon 14. Most of the carbon in living things is the much more plentiful and stable isotope of carbon, $^{12}_{6}C$.

It is a sad but well-known fact that, among other things, the simple pleasures of eating and breathing cease when death occurs. At this juncture, the organism no longer can replenish its carbon 14, but the decay of the existing ingested radioactive isotope continues unabated. Thus the concentration of carbon 14 decreases exponentially with time upon the onset of death. Hence old bones have less carbon 14 than new bones; old wood has less carbon 14 than new wood; and so on.

The decay of carbon 14 has a half-life of 5.73×10^3 y. Once the carbon content of the sample is known, we can compare the relative concentration of carbon 14 in the sample (or the activity of the sample) with that of a fresh sample of organically derived carbon. The concentrations (or activities) are related via the radioactive decay law of Equation 26.47 (or Equation 26.52 for the activities). Since little of a radioactive material is left after a time span of ten half-lives (only about 0.1%; see Section 26.11), carbon 14 dating techniques only can be used on samples with ages no greater than about $10 \times 5.73 \times 10^3$ y or about 6×10^4 y B.P. (Before the Present).[§]

---

*Cosmic rays are high-energy charged particles (principally protons— hydrogen nuclei) from space whose origins have mystified astronomers for decades.

[†]*Experimental Nuclear Physics*, edited by Emilio Segrè (John Wiley, New York, 1960), Volume I, page 615.

[‡]It is well known that we cannot exist entirely on purely inorganic materials such as rocks and water.

[§]For some purposes, the time before the present is more important, and using B.P. is preferable to the B.C. and A.D. epochs.

The carbon 14 dating technique has many practical difficulties.

1. The experiments to measure the carbon concentrations and activities involve destroying the sample; in other words, to determine the concentration of carbon present, chemical techniques that destroy the integrity of the sample are necessary. If the sample has artistic, cultural, or intrinsic value of some kind, only a small piece of the object is used. Of course, the smaller the piece used, the smaller the amount of carbon 14 present and the more difficult the measurements become. More recent techniques using particle accelerators count the carbon 14 atoms directly and are capable of determining ages up to about 100 000 y B.P. These techniques also have the advantage of being able to use very small mass samples (measured in milligrams).

2. Another significant problem that must be addressed is assessing if the sample is indeed pure and not contaminated with fresh carbon via leaching from surrounding materials.

3. The technique assumes that the carbon 14 content of the atmosphere is independent of time. Since carbon 14 is created by cosmic rays, the assumption is, then, that the cosmic ray flux on the atmosphere never changes. This is a bold assumption and, unfortunately, is only a first and crude approximation. In fact, the cosmic ray flux does change with time and the ages derived from carbon 14 analysis are only approximate. With the use of samples of carbon from tree rings, however, the carbon 14 dates for the most recent epochs are calibrated against ages determined from tree ring counts. That is, a correction can be applied to the carbon 14 derived dates to determine the age of the sample more accurately. The tree ring corrections, however, only can be used for several millennia B.P. Work continues (with the discovery and analysis of old wood samples) to extend these corrections further back in time.

4. In more recent epochs, the extensive use of fossil fuels has increased the stable carbon isotope content of the atmosphere relative to the carbon 14 content. Such fossil fuels are derived from carbon deposits in which all the carbon 14 effectively has decayed. On the other hand, atmospheric testing of nuclear weapons during the decades immediately after World War II significantly increased the present carbon 14 content of the atmosphere. These changes have affected the background value of the $^{14}C/^{12}C$ ratio.

QUESTION 5
Why can nuclei easily capture neutrons but not protons?

EXAMPLE 26.7

The earthly remains of a long-expired dean have an activity of 13 disintegrations per minute per gram of carbon. A freshly sacrificed dean has an activity of 16 disintegrations per minute per gram of carbon. Determine the approximate date of the administrator's demise.

Solution
Since the activities are known, Equation 26.52 is used:

$$\frac{dN}{dt} = \left(\frac{dN}{dt}\right)_0 e^{-\lambda t}$$

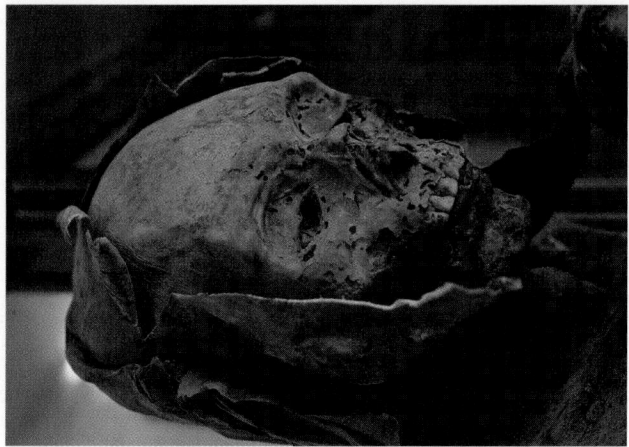

The decay of carbon 14 helps scientists date organic remains, such as this rather well-preserved Incan mummy.

13 disintegrations/min = (16 disintegrations/min)$e^{-\lambda t}$

$$0.81 = e^{-\lambda t}$$

Take the natural logarithm of this equation:

$$\ln 0.81 = -\lambda t$$
$$-0.21 = -\lambda t$$

Solving for t, you get

$$t = \frac{0.21}{\lambda} \tag{1}$$

The disintegration constant λ is related to the half-life $\tau_{1/2}$ via Equation 26.48:

$$\tau_{1/2} = \frac{0.6931}{\lambda}$$

For carbon 14 the half-life is 5.73×10^3 y, so that

$$5.73 \times 10^3 \text{ y} = \frac{0.6931}{\lambda}$$

The disintegration constant thus is

$$\lambda = \frac{0.6931}{5.73 \times 10^3 \text{ y}}$$
$$= 1.21 \times 10^{-4} \text{ y}^{-1}$$

Making this substitution into equation (1) for the age t, you find

$$t = \frac{0.21}{1.21 \times 10^{-4} \text{ y}^{-1}}$$
$$= 1.7 \times 10^3 \text{ y}$$

Notice that even though the activities were given in disintegrations per minute per gram, we can use other units (here years) in the exponentials for both t and $1/\lambda$, in accordance with Problem-Solving Tactic 26.3.

26.13 RADIATION UNITS, DOSE, AND EXPOSURE*

The activity of a sample is measured in becquerels (Bq) [the number of disintegrations per second] or curies (Cu) [one curie is exactly 36×10^9 Bq]. While the curie is a useful unit for activity, it does not give an indication of the amount of energy deposited when a substance absorbs the radiation. For this purpose, a new SI unit, the **gray** (Gy), was introduced about 1975. One gray is equal to one joule of energy absorbed per kilogram of material (see Table 26.2).[†] The unit was named for the English medical physicist Harold Gray who, in 1955, discovered the *oxygen effect*: cells exposed to radiation in the presence of oxygen are more easily killed than the same cells exposed in the absence of oxygen.[‡]

The gray is a physical unit that quantifies the **dose** D of radiation. Unfortunately, the biological effects of radiation are quite difficult to measure and assess. For example, γ radiation deposits energy over relatively long path lengths through material; α particles deposit their energy in quite short path lengths of material. Thus the likelihood of a given sample of biological tissue suffering damage is far greater from a dose of 1 Gy of α particles than from a dose of 1 Gy of γ radiation. To account for these differences, a **quality factor** (QF) is used to measure the amount of energy deposited per unit path length of material.[§] The quality factor depends on (a) the type of radiation, (b) the energy of the radiation, and (c) to some extent the nature of the target material (soft tissue versus bone, etc.). Radiations that deposit relatively little energy per unit path length (β and γ) have quality factors of about 1, while α particles have quality factors of about 20. (See Table 26.3.)

The effect of various radiations in biological tissues depends on the product of the dose D (in grays) and the quality factor (QF), forming the **dose-equivalent** (DE):

$$DE = D \, QF \tag{26.57}$$

For near total confusion, the dose-equivalent is measured in **sieverts** (Sv),[#][¶] named for the 20th-century Swedish radiologist,

---

[†] An older unit used for energy absorption is the rad (for *radiation absorbed dose*); 1 Gy is equivalent to 100 rad. Do not confuse the radiation unit (rad) with the unit abbreviation for radian, also rad.

[‡] The oxygen effect has implications in the treatment of tumors with radiation. Cells at the center of large tumors are oxygen deprived, since the blood flow to them is hogged by tumor cells near the periphery of the tumor. Thus radiation is effective in killing tumor cells on the periphery but not tumor cells at its center, leaving them free to reproduce and make the nasty tumor reoccur.

[§] An older way of measuring the biological effectiveness of various radiations is with a unit called the RBE (relative *biological effectiveness*); the RBE attempts to assess the effects of the radiation in the tissue compared with x-rays that produce the same effects. The RBE is quite difficult to measure and so the quality factor is coming into style.

[#] Alas, another way of measuring the dose-equivalent is to express the dose in rads and use the RBE instead of the quality factor. In this scenario, the dose-equivalent is

$$DE = D \, RBE$$

with the result expressed in a unit known as the rem (for *röntgen equivalent man*). The conversion between sieverts and rems is

$$1 \text{ Sv} = 100 \text{ rem}$$

[¶] One sievert is the γ-ray dose received during one hour at a distance of 1 cm from a point source of 1 mg of radium enclosed in a shell of platinum 0.5 mm thick.

TABLE 26.2 Common Radiation Units and Conversions

	Traditional unit	SI unit
Activity	curie (Cu) = 36 × 10⁹ disintegrations/s	becquerel (Bq) = 1 disintegration/s
Absorbed dose D	rad 100 rad = 1 Gy	gray (Gy)
Dose-equivalent DE	rem 100 rem = 1 Sv	sievert (Sv)

TABLE 26.3 Quality Factors for Various Absorbed Radiation

Radiation	Quality factor
X-rays, β, γ	1
Low-energy protons and neutrons (~keV energies)	2–5
High-energy protons and neutrons (~MeV energies)	5–10
α particles	20

TABLE 26.4 Dose-equivalents of Common Exposures to Radiation

Exposure	Dose-equivalent*
Natural background from cosmic rays and naturally occurring radioactive materials in the environment (soils, concrete, etc.)	0.1–0.2 rem/y 1–2 × 10⁻³ Sv/y
Absorption by bone marrow (very sensitive tissue)	
diagnostic chest x-ray	0.05 rem 5 × 10⁻⁴ Sv
dental x-ray	0.002 rem 2 × 10⁻⁵ Sv

*The International Commission on Radiation Protection recommends a whole-body dose-equivalent of no more than 0.5 rem/y = 5×10^{-3} Sv/ y for the general public and 5 rem/y = 5×10^{-2} Sv/y for people who work with radiation routinely.

Data from Kenneth S. Krane, *Introductory Nuclear Physics* (John Wiley, New York, 1988), page 187.

Rolf Sievert (1896–1966), long active in the problem of quantifying radiation doses in biological materials. Table 26.4 lists the dose-equivalents of some common forms of radiation exposure.

The long-term effects of radiation on biological tissues is a matter of some concern and much research. The problem of assessing safe exposures is quite difficult, if only because the same dose-equivalent can have different effects in different individuals. This is quite apparent in medical treatment; the same radiation dose-equivalent given to two individuals can result in widely different responses and side effects.

26.14 THE MOMENTUM OF A PHOTON

There is no wind in the vacuum of space but, nonetheless, it is possible to sail to the stars (if you have lots of time on your hands). Huh? When light reflects from a surface, a small amount of momentum is transferred to the surface. This idea is the principle behind a concept for effortlessly propelling a spacecraft between the planets or off to the stars by using a large reflecting surface (perhaps several square kilometers in area) to gain the momentum change associated with the reflection of sunlight from the sail.

In this section we see how the idea of momentum associated with a photon of light arose from seemingly totally unrelated experiments concerning the interaction of x-rays with matter.

While investigating the interaction (scattering[†]) of x-rays with graphite in 1923, Arthur H. Compton (1892–1962) noticed that if monochromatic x-rays were incident, the spectrum of the scattered x-rays consisted of the incident wavelength as well as one shifted slightly to a longer wavelength, as shown in Figure 26.17. This is called the **Compton effect**. The reason for the new wavelength intrigued Compton.

[†] Collision experiments involving fundamental particles such as photons and electrons are examples of *scattering experiments* in physics.

For a medical career in radiology, the study of both the physical and biological effects of radiation is necessary. Here a child is having a diagnostic x-ray.

Reflecting photons from sails can propel spacecraft such as in this artist's conception of a solar sail freighter.

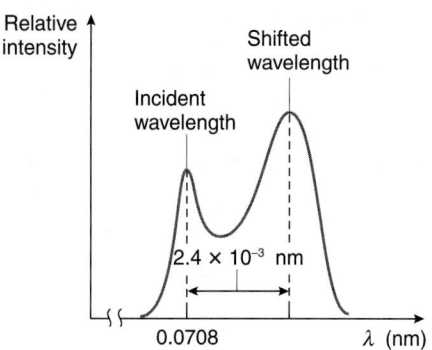

FIGURE 26.17 The scattered x-ray intensity versus wavelength in Compton's original experiment.

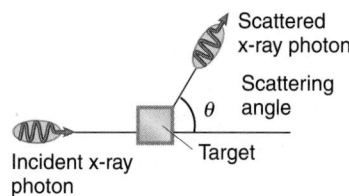

FIGURE 26.18 Geometry for the Compton effect.

The amount of the shift in wavelength depends on the angle θ at which the scattered x-rays were observed, relative to the incident direction, as in Figure 26.18. No shift is observed if the scattering angle θ is 0°, while the maximum shift occurs if scattered x-rays are detected in the backward direction where the scattering angle θ is 180°.

In order to account for these observations, Compton boldly applied the photon concept invented earlier by Einstein for the photoelectric effect, and used by Bohr with great success in his model of the hydrogen atom. Compton's analysis provided further confirmation that the photon concept was real and legitimate.

According to Einstein's treatment of the photoelectric effect, a photon is a particle of light with an energy $E = h\nu$ (Equation 26.4). There is no particle more relativistic (high speed) than a photon, of course, since it travels at the speed of light. We discovered in Chapter 25 that the total relativistic energy E of a particle is related to the magnitude p of its momentum and its mass m via (Equation 25.85)

$$E^2 = p^2c^2 + m^2c^4$$

A photon has zero mass, and so for it this equation yields

$$E = pc \qquad (26.58)$$

Thus the magnitude of the momentum of a photon (or any other particle with zero mass) is its energy divided by the speed of light:

$$p = \frac{E}{c} \qquad (26.59)$$

For a photon, $E = h\nu$, and so the magnitude of the photon momentum is

$$p_{photon} = \frac{h\nu}{c} \qquad (26.60)$$

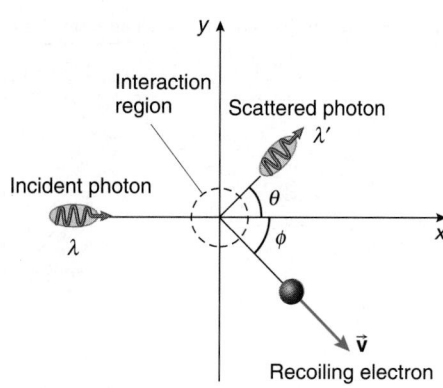

FIGURE 26.19 Geometry for the analysis of the Compton effect.

Since $c = \nu\lambda$, this also can be written in the equivalent form

$$p_{photon} = \frac{h}{\lambda} \qquad (26.61)$$

Compton assumed that an incoming x-ray photon collided elastically with an electron, conserving both energy and momentum. A recoiling electron meant the scattered x-ray photon would have less energy and, therefore, a longer wavelength than the incident x-ray.

What actually happens in the collision is an absorption of the incident photon by the electron and a subsequent (almost instantaneous) reemission of a different photon, conserving energy and momentum in the process; Compton shrewdly avoided the details of the photon absorption and reemission and treated the interaction simply as a collision. The electron binding energy (holding it to the nucleus) is trivial in comparison with the energy of the x-ray photon (see Example 26.8 below). With a collision viewpoint, a photon of definite energy and momentum is sent into a "black box" containing an electron assumed to be essentially at rest,* as in Figure 26.19. After the interaction, a photon with a different energy and momentum is observed leaving the interaction region, as well as a recoiling electron that is not detected in the experiment.†

In an elastic collision between two classical particles we could ignore the details of the interaction between the particles and simply conserve momentum and energy (see Chapter 9). We do likewise here, treating the interaction between the photon and electron particles as an elastic collision to see if what emerges from the calculations agrees with experiment or not. In fact, as we will see, the analysis is confirmed by the observations. To analyze the collision, we set up the coordinate system indicated in Figure 26.19.

---

*This condition is satisfied if the incident photon energy is much greater than any initial kinetic energy of the electron. This is the case for x-ray photons (see Example 26.8). Since the electrons in atoms are moving, however, the scattered x-rays are not strictly monochromatic but have a spread in their wavelengths at a given scattering angle.

†It was the observation of a different frequency emerging from the interaction that led Compton to hypothesize about the momentum of a photon and to treat the interaction as an elastic collision.

First we look at conservation of momentum. The incident photon of wavelength λ has momentum

$$\frac{h}{\lambda}\,\hat{\mathbf{i}}$$

Since the electron is assumed to be initially at rest, it has zero momentum. Thus the total momentum of the system of two particles before the collision is

$$\vec{\mathbf{p}}_i = \frac{h}{\lambda}\,\hat{\mathbf{i}} \qquad (26.62)$$

The photon of wavelength λ' emerging from the experiment at an angle θ to the x-axis has momentum

$$\left(\frac{h}{\lambda'}\cos\theta\right)\hat{\mathbf{i}} + \left(\frac{h}{\lambda'}\sin\theta\right)\hat{\mathbf{j}}$$

The electron, recoiling from the collision at speed v, and making an angle ϕ with the x-axis, has *relativistic* momentum

$$(\gamma m v \cos\phi)\hat{\mathbf{i}} - (\gamma m v \sin\phi)\hat{\mathbf{j}}$$

Thus the total momentum after the collision is

$$\vec{\mathbf{p}}_f = \left(\frac{h}{\lambda'}\cos\theta + \gamma m v \cos\phi\right)\hat{\mathbf{i}} + \left(\frac{h}{\lambda'}\sin\theta - \gamma m v \sin\phi\right)\hat{\mathbf{j}}$$

$$(26.63)$$

Momentum conservation means that the change in the total momentum is zero:

$$\vec{\mathbf{p}}_f - \vec{\mathbf{p}}_i = 0 \ \text{kg}\cdot\text{m/s}$$
$$\vec{\mathbf{p}}_i = \vec{\mathbf{p}}_f \qquad (26.64)$$

Since two vectors are equal if and only if their respective components are equal, Equation 26.64 means the following:

1. The x-components of the total momentum before and after the collision are equal:

$$\frac{h}{\lambda} = \frac{h}{\lambda'}\cos\theta + \gamma m v \cos\phi \qquad (26.65)$$

2. The y-components of the total momentum before and after the collision are equal:

$$0 \ \text{kg}\cdot\text{m/s} = \frac{h}{\lambda'}\sin\theta - \gamma m v \sin\phi \qquad (26.66)$$

Now we examine (relativistic) energy conservation. The incident photon has an energy $h\nu$ or, in terms of its wavelength,

$$\frac{hc}{\lambda}$$

The electron, assumed to be initially at rest, has a total (relativistic) energy equal to its rest energy mc^2, since it has no kinetic energy (actually, a negligible amount when compared with its rest energy). Thus the total initial energy is

$$E_i = \frac{hc}{\lambda} + mc^2 \qquad (26.67)$$

After the collision, the emerging photon has energy

$$\frac{hc}{\lambda'}$$

and the emerging electron has total relativistic energy

$$\gamma mc^2$$

Thus the final total energy is

$$E_f = \frac{hc}{\lambda'} + \gamma mc^2 \qquad (26.68)$$

Energy conservation means that the change in the total energy is zero:

$$E_f - E_i = 0 \ \text{J}$$

or

$$E_i = E_f \qquad (26.69)$$

Making the appropriate substitutions, we find

$$\frac{hc}{\lambda} + mc^2 = \frac{hc}{\lambda'} + \gamma mc^2 \qquad (26.70)$$

Equations 26.65 and 26.66, from momentum conservation, and Equation 26.70, from energy conservation, summarize the physics. Since it is the emerging x-ray photon that is detected, we want to eliminate the speed v and the angle ϕ associated with the recoiling electron in order to find how the emerging photon wavelength λ' is related to the incident photon wavelength λ and the scattering angle θ. The problem is a bit of an algebraic quagmire, but the solution is outlined here in case like an algebraic challenge and have your mathematical boots on. The result is Equation 26.78 if you want to skip the details.

Here is the solution for the emerging photon wavelength; check it out.

1. To eliminate the angle ϕ associated with the recoiling electron:

   a. Solve Equation 26.65 for $\gamma m v \cos\phi$ and square the result. You should get

   $$\gamma^2 m^2 v^2 \cos^2\phi = \left(\frac{h}{\lambda} - \frac{h}{\lambda'}\cos\theta\right)^2 \qquad (26.71)$$

   b. Solve Equation 26.66 for $\gamma m v \sin\phi$ and square the result. You should get

   $$\gamma^2 m^2 v^2 \sin^2\phi = \left(\frac{h}{\lambda'}\sin\theta\right)^2 \qquad (26.72)$$

   c. Add Equations 26.71 and 26.72 to eliminate ϕ. After a bit of algebraic simplification, you should be able to put this into the form

   $$\gamma^2 m^2 v^2 = \frac{h^2}{\lambda'^2} + \frac{h^2}{\lambda^2} - \frac{2h^2\cos\theta}{\lambda\lambda'} \qquad (26.73)$$

2. To eliminate the speed of the recoiling electron v:
   a. Take the energy equation (Equation 26.70) and rearrange

it into the following form (some transposing and canceling is involved):

$$\left(\frac{h}{\lambda} - \frac{h}{\lambda'}\right) + mc = \gamma mc \qquad (26.74)$$

b. Square Equation 26.74, obtaining

$$\left(\frac{h}{\lambda} - \frac{h}{\lambda'}\right)^2 + m^2c^2 + 2mc\left(\frac{h}{\lambda} - \frac{h}{\lambda'}\right) = \gamma^2 m^2 c^2 \qquad (26.75)$$

c. Now subtract Equation 26.73 from Equation 26.75. After some simplification, you should obtain

$$-\frac{2h^2}{\lambda\lambda'} + m^2c^2 + 2mc\left(\frac{h}{\lambda} - \frac{h}{\lambda'}\right) + \frac{2h^2\cos\theta}{\lambda\lambda'}$$
$$= \gamma^2 m^2 c^2 - \gamma^2 m^2 v^2 \qquad (26.76)$$

d. Substitute for γ^2 in the right-hand side of Equation 26.76. The right-hand side of the equation then simplifies to just m^2c^2:

$$-\frac{2h^2}{\lambda\lambda'} + m^2c^2 + 2mc\left(\frac{h}{\lambda} - \frac{h}{\lambda'}\right) + \frac{2h^2\cos\theta}{\lambda\lambda'} = m^2c^2 \quad (26.77)$$

e. Now simplify and rearrange Equation 26.77. You should eventually obtain

$$\lambda' - \lambda = \frac{h}{mc}(1 - \cos\theta) \qquad (26.78)$$

The wavelength of the x-ray photon changes by an amount given by Equation 26.78.

In Equation 26.78, the quantity

$$\frac{h}{mc}$$

where m is the mass of the electron, has the dimensions of a length and is known picturesquely as the **Compton wavelength** of the electron. It is *not* a wavelength of light; the quantity simply has the same dimensions as a wavelength.

The mass of the electron m appears in the denominator of Equation 26.78. This is the reason the change in wavelength is more apparent for scattering from electrons rather than from, say, protons or the nucleus. The proton and nuclear masses are much larger than the electron mass; so any shift in the frequency of the photon in its scattering from protons or nuclei is much less than from electrons and accounts for the unshifted wavelength in Figure 26.17. Another way of saying this is that the Compton wavelength of a proton or a nucleus is much less than the Compton wavelength of an electron because of the drastic difference in the masses of the two particles.

The experiment is not done with a single photon but with a huge number of photons. A beam of photons is incident on a whole collection of electrons. Each photon scatters from an individual electron and scatters to some definite angle θ. The photons emerging from the interaction region at the angle θ have the wavelength λ' given by Equation 26.78. Photons emerging at different angles (from encounters with different electrons) emerge with different wavelengths.

In agreement with experiment, the change in the wavelength is a maximum when $\theta = 180°$, in which case the emerging photon moves antiparallel to the incident photon. In this case the change in wavelength is equal to twice the Compton wavelength of the electron (substitute $\theta = 180°$ into Equation 26.78).

The result of Compton's bold theoretical analysis of the problem in terms of photon energy and momentum, Equation 26.78, beautifully confirmed his experimental results. He received the 1927 Nobel prize in physics for the discovery.

Another way to see that photons of any wavelength have momentum is to observe the force they impart to a reflector. A floodlight can produce measurable movement of a delicate mirror suspension in high vacuum. We reflect a beam of photons off a mirror surface of mass m, as in Figure 26.20. The reflection is essentially a collision between each incoming photon and the mirror. The Compton wavelength of the mirror is extremely small because of its huge macroscopic mass. Thus the reflected photon is effectively unchanged in wavelength. The incident photon momentum is $(h/\lambda)\hat{\imath}$ with the coordinates of Figure 26.20. The exiting photon momentum is $-(h/\lambda)\hat{\imath}$. We call the momentum of the recoiling mirror $\vec{\mathbf{p}}_{\text{mirror}}$. Conserving momentum in the collision means that

$$\vec{\mathbf{p}}_\text{f} - \vec{\mathbf{p}}_\text{i} = 0 \text{ kg}\cdot\text{m/s}$$
$$-\frac{h}{\lambda}\hat{\imath} + \vec{\mathbf{p}}_{\text{mirror}} - \frac{h}{\lambda}\hat{\imath} = 0 \text{ kg}\cdot\text{m/s}$$

Solving for $\vec{\mathbf{p}}_{\text{mirror}}$, we find

$$\vec{\mathbf{p}}_{\text{mirror}} = +2\frac{h}{\lambda}\hat{\imath} \qquad (26.79)$$

The mirror picks up some momentum as a kick from the reflected photon. If instead of a single photon, we reflect a flood

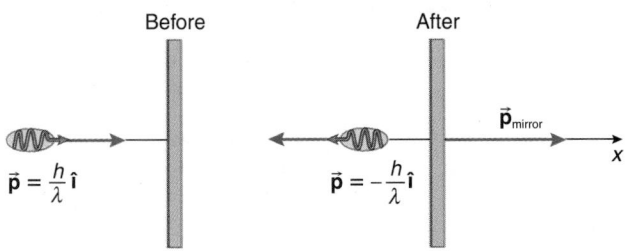

FIGURE 26.20 Reflection of photons from a mirror.

of photons each second from the mirror, the momentum gained by the mirror will increase correspondingly. If we reflect N photons each second from the mirror, the mirror gains momentum each second equal to N times that of Equation 26.79. Since the photons are changing the momentum of the mirror by a certain amount each second, they exert a force on the mirror equal to the time rate of change of the momentum of the mirror (Newton's second law).

The magnitude of the force per unit area of the light on the mirror is known as the **radiation pressure** exerted by light. It is the principle behind sailing between the planets and off to the stars. Radiation pressure also is apparent in some comet tails. When sunlight reflects from small particles, such as those in the tails of comets, it drives the particles in directions away from the Sun. Hence some dusty comet tails are directed away from the Sun, even as the comet recedes from the Sun.*

EXAMPLE 26.8

Compton used photons with an incident wavelength of 0.0708 nm.

a. Determine the energy of each incident photon in joules and in electron-volts. Compare this energy with the sum of the kinetic and potential energies of an electron in the hydrogen atom.
b. Determine the magnitude of the momentum of each incident photon.

Solution

a. The photon energy is

$$E = h\nu$$
$$= \frac{hc}{\lambda}$$
$$= \frac{(6.626 \times 10^{-34} \text{ J} \cdot \text{s})(3.00 \times 10^{8} \text{ m/s})}{7.08 \times 10^{-2} \times 10^{-9} \text{ m}}$$
$$= 2.81 \times 10^{-15} \text{ J}$$
$$= \frac{2.81 \times 10^{-15} \text{ J}}{1.602 \times 10^{-19} \text{ J/eV}}$$
$$= 17.5 \text{ keV}$$

This is much greater than the sum of the kinetic and potential energies of the electron in the hydrogen atom (typically on the order of only a few electron-volts, and negative; see the Bohr model of the hydrogen atom in Section 26.7). Similar arguments apply to electrons in other atoms. This comparison justifies Compton's neglect of the any initial kinetic energy associated with the electron.

b. The magnitude p of the momentum of the photon is found from Equation 26.59:

$$p = \frac{E}{c}$$
$$= \frac{2.81 \times 10^{-15} \text{ J}}{3.00 \times 10^{8} \text{ m/s}}$$
$$= 9.37 \times 10^{-24} \text{ kg} \cdot \text{m/s}$$

EXAMPLE 26.9

Note that in Figure 26.17 the wavelength shift of the x-rays in Compton's original experiment was 2.4×10^{-3} nm. Use Equation 26.78 to determine the approximate angle θ at which Compton measured the scattered x-rays.

Solution

The shift in wavelength of the scattered x-rays is given by Equation 26.78:

$$\lambda' - \lambda = \frac{h}{mc}(1 - \cos\theta)$$

Hence you have

$$1 - \cos\theta$$
$$= \frac{mc}{h}(\lambda' - \lambda)$$
$$= \frac{(9.11 \times 10^{-31} \text{ kg})(3.00 \times 10^{8} \text{ m/s})(2.4 \times 10^{-12} \text{ m})}{6.626 \times 10^{-34} \text{ J} \cdot \text{s}}$$
$$= 0.99$$

Solving for $\cos\theta$, you find

$$\cos\theta = 0.01$$

or

$$\theta \approx 90°$$

Indeed, from Compton's paper in 1923,[†] the scattering angle used was 90°.

26.15 THE DE BROGLIE HYPOTHESIS

Now a purely corpuscular theory [of light] contains nothing that enables us to define a frequency; for this reason alone, therefore, we are compelled in the case of Light, to introduce the idea of a corpuscle and that of a periodicity simultaneously.

On the other hand, determination of the stable motion of electrons in the atom introduces integers; up to this point the only phenomena involving integers in Physics were those of interference and of normal modes of vibration. This fact suggested to me the idea that electrons too could not be regarded simply as corpuscles, but that periodicity must be assigned to them also.

Louis de Broglie (1892–1987)[‡]

You easily can picture a wave if you think of a surface water wave. You also easily can picture a particle, be it a baseball, peanut, or a sand grain, since such particles are quite visible to us. It is vastly more difficult for us to picture a wave (such as light) as a particle (a photon). To do so, we have to extrapolate into a regime that is not directly accessible to our sense experiences. Perhaps then, we should be cautious about extending our common ideas about particles into a realm in which particles are not amenable to our direct senses, the domain of the atom,

---

*Some comets have multiple tails, each caused by different physical effects.

[†] Arthur H. Compton, "The spectrum of scattered x-rays," *Physical Review* (2nd series), *22*, #5, pages 409–413 (November 1923); Figure 26.17 is from page 411.
[‡] Louis de Broglie, *Matter and Light*, translated by W. H. Johnston (Norton, New York, 1939), pages 168–169.

with its constituent electrons, protons, neutrons, and other submicroscopic entities.

The true nature of light is difficult to assess. The double slit experiments by Young in 1801 showed that light exhibited the macroscopic wavelike properties of diffraction and interference, although he was unaware of the nature of the waves themselves. Much later in the century, Maxwell discovered that light was a wave of oscillating electric and magnetic fields: an electromagnetic wave. On the other hand, we have seen in this chapter that the photoelectric and Compton effects indicate that light has the microscopic aspects of a particle, the photon, with both energy and momentum. Thus light exhibits a **wave–particle duality**. This duality profoundly taxes our imagination. Both the photoelectric and Compton effects are not amenable to our direct senses: we cannot literally *see* the processes take place.

The wave–particle duality was extended to particles in the early 1920s with the brilliantly original work of a young French physicist, Louis de Broglie (1892–1987).* His theoretical insight into the nature of particles and waves was the last critical thread in the tapestry that led to the invention of a totally new mechanics of particles: quantum mechanics, which we touch on in Chapter 27.

In a sudden inspiration de Broglie reasoned that if what we think of as waves have particle properties, perhaps what we think of as particles (such as electrons) have wave properties. This symmetry argument is called the **de Broglie hypothesis**.

De Broglie was then able to extend this concept into the analysis of particle motion. Using relativistic arguments, we saw in Section 26.14 with the Compton effect that the particle of light, the photon, has a momentum whose magnitude is given by Equation 26.61:

$$p_{\text{photon}} = \frac{h}{\lambda}$$

Thus the wavelength of light is related to the magnitude of its momentum via

$$\lambda = \frac{h}{p_{\text{photon}}} \qquad (26.80)$$

Taking this fact as a cue, de Broglie proposed that a wavelength is associated with *any* particle and is related to the magnitude of its momentum by an equation of the same form:

$$\lambda = \frac{h}{p_{\text{particle}}} \qquad (26.81)$$

The wavelength associated with a particle is called its **de Broglie wavelength**; it is *not* a wavelength of light, nor is it the so-called Compton wavelength of a particle (of the previous section). He also assumed that the energy E of the particle is proportional to the frequency ν of the associated wave:

$$E = h\nu \qquad (26.82)$$

just as for a photon.

Louis de Broglie began to study history at the Sorbonne in Paris at the age of 17, but he became enthralled with physics. He won the Nobel prize in physics in 1929 for his theoretical ideas about particle-waves.

The true nature of the wave associated with a particle remained a mystery, yet to be determined (much like when Young discovered the wave nature of light). The de Broglie hypothesis implies that the wave–particle duality has a universal and symmetrical character: waves have particle properties, particles have wave properties.

The boldness of de Broglie's hypothesis is even more astounding when you realize that Equation 26.81 implies that the particle speed v (which determines its momentum) is *not* the same as the speed v' of the de Broglie waves associated with the particle! The de Broglie wave speed v' is related to the frequency and wavelength of the de Broglie wave by $v' = \nu\lambda$, so $\lambda = v'/\nu$. From Equation 26.81, we have

$$\frac{v'}{\nu} = \frac{h}{p}$$

or

$$v' = \frac{h\nu}{p}$$

From Equation 26.82, this becomes

$$v' = \frac{E}{p} \qquad (26.83)$$

If we write the energy of the particle as $E = \gamma mc^2$ and the magnitude of its momentum as $p = \gamma mv$, then Equation 26.83 yields

$$v' = \frac{\gamma mc^2}{\gamma mv}$$

or

$$vv' = c^2 \qquad (26.84)$$

Equation 26.84 indicates the wave speed v' thus is *inversely proportional* to the particle speed (through its momentum). Since the particle speed v is restricted to be less than the speed of light c (from relativity), the de Broglie waves associated with the particle apparently travel at superluminal speeds ($> c$)!

These superluminal speeds may seem peculiar until you recall from Chapter 12 that the superposition of a collection of waves with slightly different wavelengths (or frequencies) forms a *wave group*; the speed of the wave group typically is

---

*His last name is pronounced like "broy" (brô y), a close rhyme to "oily." He won the 1929 Nobel prize in physics.

different from that of the waves that form the group.* An individual de Broglie wave thus does not remain long with the particle. We associate the particle speed with the *group speed* of the de Broglie waves, and this always is less than the speed of light.

You might be tempted to think that things really are getting a little far-fetched. What is the wavelength associated with common and easily seen particles such as a baseball, a bowling ball, or even a particle as exotic as a physics student? To get a feeling for the size of the de Broglie wavelength of various particles, look at Examples 26.10 and 26.11. You will see that the de Broglie wavelength of macroscopic particles is completely negligible, but the wavelength associated with electrons and other small particles (that are too small for us to see directly) can be significant on the atomic or nuclear scales of length.

The de Broglie hypothesis casts the Bohr angular momentum hypothesis in a new light, as de Broglie himself realized. Recall that the Bohr angular momentum postulate (Equation 26.20) stated that the magnitude of the angular momentum L of the orbiting electron in a hydrogenic atom was an integral multiple of $\hbar$:

$$L = n\hbar$$

How do the de Broglie waves reconcile with the Bohr model? For an electron in a circular orbit of radius r, the magnitude of its orbital angular momentum $\vec{L} = \vec{r} \times \vec{p}$ is mvr, since the two vectors $\vec{r}$ and $\vec{p}$ are mutually perpendicular; recall that the electron in the Bohr model is nonrelativistic, so the momentum is just mv. Hence the Bohr postulate becomes

$$mvr = n\frac{h}{2\pi} \qquad (26.85)$$

Since the magnitude of the momentum is $p = mv$, Equation 26.85 can be rewritten as

$$2\pi r = n\frac{h}{p} \qquad (26.86)$$

The de Broglie wavelength λ of the electron is given by Equation 26.81, so that Equation 26.86 can be rewritten as

$$2\pi r = n\lambda \qquad (26.87)$$

In other words, the allowed orbits of the electron are precisely those for which an integral number of electron de Broglie wavelengths fit along the circumference. This realization gave credence to the de Broglie hypothesis and to the quantization of angular momentum. Nonetheless, a definitive experiment was needed to confirm the existence of the waves associated with particles such as electrons. De Broglie suggested an experiment with crystals acting as diffraction gratings for the particle-waves, but his idea lay fallow for several years.

Wave phenomena such as diffraction and interference become apparent when the waves interact with (i.e., pass through or around) obstacles or openings on the order of the wavelength in size. Clearly it is impossible for the student in Example 26.10 to pass through or

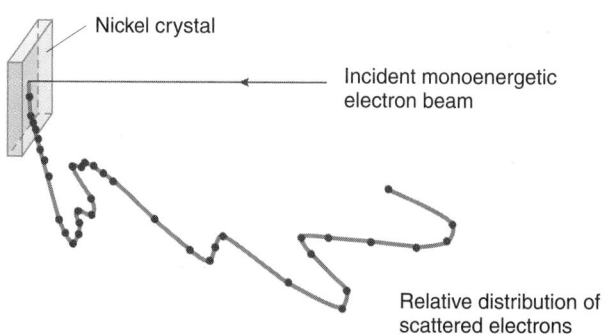

FIGURE 26.21 Distribution (relative number) of electrons scattered at various angles in the Davisson and Germer experiment.

around objects as small as 10^{-36} m (even if objects so small existed!), and so the wavelike aspect of a physics student particle never is manifest. The de Broglie wavelength of any macroscopic particle is *so* small that you always can neglect the wave aspects of these large particles; the wavelike aspects never become apparent.

On the other hand, Example 26.11 indicates that the de Broglie wavelength of electrons can be quite large compared with the sizes of nuclei (10^{-15} m) and on the order of the size of atoms (10^{-10} m = 0.1 nm). The wavelike behavior of electrons thus is very important in atomic physics.

The experimental verification of the wavelike aspects of electrons first was made accidentally by Clinton Davisson and Lester Germer in 1927 in studies of the reflection of electrons from a nickel target, some three years after de Broglie published his conjectures.[†] After some initial experiments, the accidental explosion of a bottle of liquid air oxidized the nickel target. To purge the nickel, the target was subjected to prolonged heating. It then crystallized upon cooling, and the electrons then scattered by the nickel target had an angular distribution totally unlike that detected before the accident. Much to their initial consternation, Davisson and Germer detected maximum and minimum peaks and valleys in the angular distribution of electrons scattered from the crystal array (see Figure 26.21), which they finally and successfully interpreted by treating the crystalline array of nickel atoms as a three-dimensional diffraction grating for the electron de Broglie waves.

Working independently, George Thomson (1892–1975)[‡] also detected the diffraction and interference of electron waves through a thin film of crystallized metal. These discoveries confirmed the de Broglie hypothesis. As we have seen on many occasions, from such innocuous squiggles come Nobel prizes.[§]

An important technological application of the short wavelengths associated with electrons became apparent to Ernst Ruska in 1931: an electron microscope.[#] We saw in Chapter 24 that the limiting resolution angle of a microscope

---

*This is commonly seen with water waves, where the speeds of the waves that form a group are greater than that of the group itself. The waves thus arise at the trailing edge of the group disturbance, propagate through it, and disappear at the leading edge of the group. When at the beach, you might note that every so often, a very large amplitude wave appears; this is the maximum amplitude of the group of waves formed by the superposition of the waves that create the group and travel faster than the group itself. When you are canoeing on a lake and watch waves from a bag distant boat coming toward you, the component waves seem to move faster than the big swell itself.

[†]Clinton Davisson and Lester H. Germer, "Diffraction of electrons by a crystal of nickel," *Physical Review*, 30, #6, pages 705–740 (December 1927).

[‡]Thomson was the son of J. J. Thomson, the inventor of the plum pudding model of the atom (see Section 26.6).

[§]George Thomson shared the Nobel prize with Clinton Davisson in 1937 for the discovery of the electron de Broglie waves. One wonders why Germer did not share in the prize.

[#]He eventually recieved the Nobel prize in physics (in 1986) for the invention along with Gerd Binning and Heinrich Rohrer, who (in 1981) invented an improved and sophisticated version of the instrument known as a scanning tunneling electron microscope.

Electron microscopes exploit the wave nature of electrons to improve resolution.

of circular aperture a depends on the wavelength according to Equation 24.5:

$$\theta = \frac{1.220\,\lambda}{a} \qquad \text{(circular aperture)}$$

Since electron wavelengths can be made so much smaller than visible light wavelengths, the limiting resolution angle is correspondingly that much smaller for electron-waves than for visible light. Electron microscopes thus have the potential for probing smaller details than optical microscopes. The electrons can be manipulated (refracted) much like light waves except that the electromagnetic "lenses" used consist of electric and magnetic fields. The details are involved, as one might imagine, but the essence of the ability of electron microscopes to fathom greater detail lies in the much smaller size of the electron wavelengths used. You might wonder why x-ray or γ-ray microscopes have not been invented or developed to increase the resolution afforded by electromagnetic radiation, since their wavelengths are much smaller than those of visible light. But it is difficult to manipulate or focus such short electromagnetic wavelengths, whereas it is quite easy to manipulate electrons by electric and magnetic fields.*

The de Broglie hypothesis and the wave–particle duality of nature raise several questions:

1. How can a particle be diffracted? What does it mean when we say a *particle* undergoes diffraction because of its wavelike properties?
2. What is it that is jiggling or oscillating in these particle waves? What does the particle-wave represent?
3. Do the particle-waves satisfy the classical wave equation (Equation 12.7) or some other equation?

These are interesting and profound questions about the nature of the wave–particle duality. We shall have more to say about these questions shortly (Chapter 27). We raise them here because you may be wondering about them; you are not alone. Physicists *still* find these questions quite profound and not easily answered.

---

*On the other hand, ingenious x-ray and γ-ray telescopes have been invented over the last 30 years to collect and study these radiations coming from distant sources in the galaxy and beyond.

EXAMPLE 26.10

You are late and rushing off to class at a speed of 3.0 m/s; assume your mass is 60 kg. Determine the de Broglie wavelength of the you-particle.

Solution

The magnitude of your momentum is

$$p = mv$$
$$= (60 \text{ kg})(3.0 \text{ m/s})$$
$$= 1.8 \times 10^2 \text{ kg·m/s}$$

Substituting this into Equation 26.81 for the de Broglie wavelength, you find

$$\lambda = \frac{h}{p}$$
$$= \frac{6.626 \times 10^{-34} \text{ J·s}}{1.8 \times 10^2 \text{ kg·m/s}}$$
$$= 3.7 \times 10^{-36} \text{ m}$$

This is 21 orders of magnitude smaller than the diameter of a typical nucleus ($\sim 10^{-15}$ m)! Your wavelike properties are negligible beyond belief.

EXAMPLE 26.11

Electrons initially at rest are accelerated through a potential difference of 100 V, as in Figure 26.22. Find the de Broglie wavelength of the electrons when they reach their greatest speed.

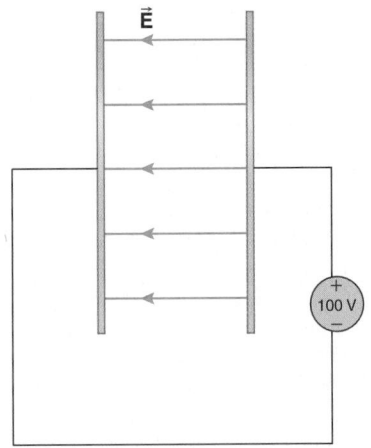

FIGURE 26.22

Solution

To find the de Broglie wavelength of the electrons at their greatest speed, you need to find their momentum; this, in turn, means you have to calculate their maximum speed. The easy way to find the speed is to realize that the kinetic energy of the emerging electrons is changed by 100 eV, because they were accelerated through a potential difference of 100 V; see Equation 17.25. Since their initial kinetic energy is zero, their final kinetic energy is 100 eV. This corresponds to a kinetic energy in joules of

$$(100 \text{ eV})(1.602 \times 10^{-19} \text{ J/eV}) = 1.60 \times 10^{-17} \text{ J}$$

Assume the electrons are traveling at a nonrelativistic speed, so for each electron you can write

$$\frac{mv^2}{2} = 1.60 \times 10^{-17} \text{ J}$$

Substituting for the mass of an electron ($m = 9.11 \times 10^{-31}$ kg) and solving for the speed v, you find

$$v = 5.93 \times 10^6 \text{ m/s}$$

Notice that $v/c = 0.0198$, implying the relativistic factor γ has the value $\gamma = 1.000$. The low speed justifies the use of the non-relativistic expressions for the kinetic energy and momentum.

The electrons each have a momentum of magnitude

$$p = mv$$
$$= (9.11 \times 10^{-31} \text{ kg})(5.93 \times 10^6 \text{ m/s})$$
$$= 5.40 \times 10^{-24} \text{ kg·m/s}$$

The de Broglie wavelength of the electrons then is

$$\lambda = \frac{h}{p}$$
$$= \frac{6.626 \times 10^{-34} \text{ J·s}}{5.40 \times 10^{-24} \text{ kg·m/s}}$$
$$= 1.23 \times 10^{-10} \text{ m}$$
$$= 0.123 \text{ nm}$$

Although the wavelength of the electrons is quite small, it is comparable in size to that of an atom and much larger than the de Broglie wavelength of macroscopic masses (see Example 26.10).

CHAPTER SUMMARY

The discoveries of the *electron* and *radioactivity* were the first clues that atoms had structure.

Planck's constant $h \approx 6.626 \times 10^{-34}$ J·s made its first appearance in physics in the successful attempt by Max Planck to avoid the *ultraviolet catastrophe* associated with blackbody radiation.

When light with a frequency greater than a *cutoff frequency* v_c illuminates a metal, electrons are liberated; this is the *photoelectric effect*. The cutoff frequency depends on the specific metal used in the effect. Einstein explained the features of the effect by assuming light consists of a collection of massless particles, called *photons*, each with an energy proportional to the frequency of the light:

$$E_{\text{photon}} = hv \qquad (26.4)$$

where h is Planck's constant. The emerging electrons, called photoelectrons, have a variety of speeds. For the electrons with the greatest speed, energy conservation implies that

$$hv = W + \text{KE}_{\text{max}} \qquad (26.7)$$

where W is the *work function* of the metal and KE_{max} is the kinetic energy of the fastest emergent electrons. The work function is related to the cutoff frequency v_c for the photoelectric effect, and represents the minimum amount of work needed to liberate an electron from the metal:

$$hv_c = W \qquad (26.8)$$

For incident light of frequency v_c, the (barely) liberated electrons have 0 J of kinetic energy. The maximum kinetic energy of the emergent photoelectrons is related to the stopping potential V_s:

$$\text{KE}_{\text{max}} = eV_s \qquad (26.10)$$

Experiments by Rutherford using α particles scattered from thin gold foils led to the development of the *nuclear model* of the atom.

Bohr developed a nuclear model for *hydrogenic atoms* (single-electron atoms), called the *Bohr model*, that correctly predicted the wavelengths of light emitted and absorbed by such atoms. To do this, Bohr quantized the magnitude of the orbital angular momentum of the electron:

$$L_{\text{orbit}} = n\hbar \qquad (26.20)$$

where n is a positive, nonzero integer called a quantum number, and $\hbar$ is Planck's constant divided by 2π:

$$\hbar \equiv \frac{h}{2\pi} = 1.055 \times 10^{-34} \text{ J·s} \qquad (26.19)$$

As a result of the quantization of angular momentum, the electron orbits the nucleus at only specific radii given by

$$r = \frac{n^2 \hbar^2 (4\pi\varepsilon_0)}{mZe^2} \qquad (26.24)$$

The total energy also is quantized:

$$E_n = -\frac{1}{2} \frac{mZ^2 e^4}{(4\pi\varepsilon_0)^2 n^2 \hbar^2} \qquad (26.27)$$
$$= -(13.6 \text{ eV}) \frac{Z^2}{n^2} \qquad (26.29)$$

The *ground state* has the lowest energy and corresponds to the quantum number $n = 1$. A photon of light is emitted if the electron lowers its energy by changing to an orbit with a lower value of n:

$$E_i = E_f + E_{\text{emitted photon}} \qquad \text{(emission)} \qquad (26.31)$$

An atom absorbs light of the same frequency the atom emits, in which case the electron changes to an orbit with a higher value for n:

$$E_i + E_{\text{absorbed photon}} = E_f \qquad \text{(absorption)} \qquad (26.32)$$

The *Bohr correspondence principle* states that as a quantum number such as n becomes very large, the results of quantum theory approach those secured from classical physics.

The Bohr model is unable to account for the relative intensities of the wavelengths emitted by hydrogenic atoms, and so is an incomplete model of the atom.

Radioactive materials emit three types of radiation:

- α particles (helium nuclei $^4_2$He);
- β particles (electrons β^-; or positrons β^+, which are positively charged electrons); and
- γ rays (high-energy photons).

The emission of α or β particles by a nucleus transmutes the element into a different element, realizing the dream of ancient and medieval alchemy.

The *activity*

$$\frac{dN}{dt}$$

of a radioactive material measures the number of nuclei decaying per unit time. The activity is measured in SI units of *becquerels* (Bq), equal to one disintegration per second. Another common and convenient unit of activity is the curie (Cu), equal to 36×10^9 Bq.

The activity is proportional to the number of nuclei of the radioactive isotope present:

$$\frac{dN}{dt} = -\lambda N \qquad (26.45)$$

where λ is the *disintegration constant*. The number of radioactive nuclei N decreases exponentially with time:

$$N = N_0 e^{-\lambda t} \qquad (26.47)$$

where N_0 is the number of nuclei present when $t = 0$ s. The *half-life* $\tau_{1/2}$ is the time at which only half the original number of nuclei remain:

$$\tau_{1/2} = \frac{\ln 2}{\lambda} \qquad (26.48)$$

The mean lifetime is the average age of the entire sample of nuclei after their decay:

$$\text{mean lifetime} = \frac{1}{\lambda} \qquad (26.56)$$

The mean life is longer than the half-life.

The activity of a sample decays exponentially with time with the same half-life as the number of nuclei:

$$\frac{dN}{dt} = \left(\frac{dN}{dt}\right)_0 e^{-\lambda t} \qquad (26.52)$$

The biological effects of radiation are difficult to assess, predict, and measure. The absorbed dose is measured in *grays* (Gy) while the dose-equivalent is measured in *sieverts* (Sv).

A photon has a momentum of magnitude p, related to its energy E:

$$p = \frac{E}{c} \qquad (26.59)$$

Since the photon energy is $E = h\nu = hc/\lambda$, the momentum and wavelength of a photon are related by

$$p = \frac{h}{\lambda} \qquad (26.61)$$

The *Compton effect* is observed during the scattering of high-energy photons (x-rays) from electrons. The emerging x-ray wavelength λ' is greater than the incident x-ray wavelength λ. The change in the wavelength of the incident photon is related to the scattering angle θ of the emerging x-ray and the *Compton wavelength* h/mc of the electron:

$$\lambda' - \lambda = \frac{h}{mc}(1 - \cos \theta) \qquad (26.78)$$

Compton derived Equation 26.78 by assuming the x-ray photon undergoes an elastic collision with a stationary electron. The Compton wavelength of a particle is *not* a wavelength of light, nor is it the de Broglie wavelength of the particle. The Compton wavelength merely is a way to say that the quantity h/mc has the dimensions of a length. It has no other physical significance.

The *de Broglie hypothesis* associates wavelike properties with particles. The *de Broglie wavelength* λ of a particle with momentum p is

$$\lambda = \frac{h}{p} \qquad (26.81)$$

The de Broglie wavelength is not a wavelength of light nor is it the Compton wavelength of a particle. The Bohr angular momentum postulate implies there are an integral number of de Broglie wavelengths of the electron around the circumference of an allowed orbit.

The de Broglie wavelength of macroscopically massive particles is completely negligible, since it is vanishingly small compared with even the size of a nucleus. On the other hand, the wave nature of particles of small mass, such as electrons, is not negligible on the atomic or nuclear length scales. The wave aspect of material particles was first confirmed in the electron scattering experiments of Davisson and Germer, and of Thomson and led to the development of electron microscopes.

SUMMARY OF PROBLEM-SOLVING TACTICS

26.1 (page 1214) In using Equations 26.7, 26.8, 26.10, or 26.11 for the photoelectric effect, you must express all the energies either in joules (J) or, if you choose, electron-volts (eV).

26.2 (page 1214) From Equation 26.10, the stopping potential in volts is *numerically* equal to the maximum kinetic energy of the electrons expressed in electron-volts.

26.3 (page 1228) In the radioactive decay law for the number of nuclei remaining or for the activity, the units you use for $\tau_{1/2}$ and $1/\lambda$ must be the same.

QUESTIONS

1. (page 1211); 2. (page 1215); 3. (page 1218); 4. (page 1222); 5. (page 1229)

6. Television picture tubes, oscilloscopes, and computer monitors all are called CRTs. Guess or determine what the meaning of CRT is.

7. Why is the term radioactivity somewhat of a misnomer? Invent a new term to describe the phenomenon.

8. All objects emit a blackbody spectrum characteristic of their absolute temperature. Why then can we not see most things in the dark?

9. The stars Antares (α Scorpii in the constellation Scorpius) and Aldebaran (α Tauri in the constellation Taurus) are slightly reddish in color while Vega (α Lyrae in the constellation Lyra) and Arcturus (α Boötis in the constellation Boötes) are slightly bluish in color. What does this observation imply about their relative surface temperatures?

10. What is the smallest unit of currency in your country? Is money quantized? Explain how the smallest unit of money plays the role of Planck's constant in measuring economic activity. What is the role of a quantum number in this case?

11. Is there a smallest unit of a living organism? Is life quantized?

12. Which has greater energy: a microwave photon or a radio wave photon?

13. Ultraviolet light causes sunburn. Visible light does not. Explain this observation in terms of photons.

14. Can you get a sunburn while sitting in your car with the windows up? What does this tell you about the optical properties of glass?

15. Explain why plastic bags do not deteriorate when exposed to incandescent lights inside your home but will degrade when exposed to sunlight for long periods. What photons likely are responsible for the degradation of the plastic?

16. Is the dark night sky empty of all photons?

17. If a light source is approaching you, a frequency higher than the true frequency of the light is observed. This means the photon you observe has a higher energy than the one in a reference frame in which the light source is at rest. Where did the additional energy come from? Does this observation violate energy conservation?

18. You perform the photoelectric effect experiment with light of frequency $v > v_c$ for a metal with a work function W_0. You now use light of the same frequency on another metal with work function W. In order to produce photoelectrons with the replacement metal using light of the same frequency as before, what must be the relationship between W and W_0?

19. What is the difference (if any) between an electron and a photoelectron?

20. Why are many spacecraft wrapped in gold foil rather than much cheaper aluminum or tin foils?

21. Does the stopping potential V_s depend on the frequency of light used in the photoelectric effect? Does V_s depend on the intensity of the light? Explain your answers.

22. Does the current detected by the ammeter in Figure 26.5 depend on the intensity of the light? Does the current depend on the frequency of the light as long as $v > v_c$? Explain your answers.

23. Would you expect the work function of a metal to be a function of the temperature of the metal? If so, in what way?

24. When zinc is illuminated by the light from a mercury vapor lamp, the photoelectric effect is noticed. When a thick piece of glass is placed between the lamp and the zinc, the photoelectric effect stops even though light still reaches the zinc. Develop a hypothesis to explain this effect.

25. The nuclear model of the atom is not as obvious as it may seem. How would you convince a nonbeliever of its reality?

26. A collection of hydrogen atoms is illuminated with high-energy photons with energies greater than 13.6 eV. Can such photons be absorbed by the electrons? What happens if they are?

27. An antihydrogen atom consists of a positron (a positively charged electron) orbiting an antiproton (a negatively charged proton). Will the spectrum of antihydrogen be different from the spectrum of ordinary hydrogen? Can we be certain that the hydrogen in distant stars is ordinary hydrogen and not antihydrogen? One of the great mysteries of the universe is why it is principally composed of ordinary matter and not antimatter. We think we know why; see Edward R. Harrison, *Cosmology* (Cambridge University Press, Cambridge, England, 1981), pages 354–357.

28. The electromagnetic and gravitational forces are long-range forces. Masses exert gravitational forces and charges exert electrical forces on each other whatever their separation. Explain why the existence of multiple protons in a nucleus is evidence for the existence of a very strong nuclear force with a very short effective range.

29. What evidence suggests that radioactivity is not an atomic process but a nuclear process?

30. In the typical carbon dating technique of organic materials, explain why it is typically necessary to destroy (at least part of) the sample in order to determine its age.

31. In β^- decay, what experimental evidence indicates the electron originated from the nucleus rather than from the electrons surrounding the nucleus?

32. Some radioactive materials, like $^{238}_{92}U$, have very long half-lives (for $^{238}_{92}U$ it is 4.47 *billion* years). Obviously, we did not actually wait billions of years to measure such a long half-life, since we have only known about radioactivity for about a century. Describe how it is possible to measure such very long half-lives.

33. What is the difference between the photoelectric effect and the Compton effect?

34. Why is the Compton effect not observed with visible light?

35. If the recoiling electron in the Compton effect has the maximum kinetic energy, in what direction is it moving relative to the direction of the incident x-rays?

36. A proton and electron have the same speed. Which has the longer de Broglie wavelength and by how much?

37. The electron and proton have the same magnitude of electric charge. What factors make the development of a proton microscope less appealing than an electron microscope?

38. A photon in vacuum has a certain momentum. If the same photon is in water, is its momentum changed? This is not an easy question to answer! Why? You might discuss what it means to say a photon is in water.

39. The wavelength of a photon is given by

$$\frac{hc}{\text{total energy}}$$

but the de Broglie wavelength of a neutron is *not* given by

$$\frac{hc}{\text{total energy}}$$

Why not? What property of the neutron makes for the distinction?

Sections 26.1 The Discovery of the Electron
26.2 The Discovery of X-rays
26.3 The Discovery of Radioactivity
26.4 The Appearance of Planck's Constant *h*

1. If you double the temperature of a blackbody radiator, by what factor is the peak wavelength emitted changed?

2. Your normal body temperature is about 37 °C. According to Wien's law, what is the wavelength of the peak of the blackbody spectrum that you emit? In what region of the electromagnetic spectrum is this wavelength?

3. What is the peak wavelength of the blackbody radiation left over from the creation of the universe, whose temperature is 3 K? Is this radio wave, microwave, infrared, visible, ultraviolet, x-ray, or γ-ray electromagnetic radiation?

4. The stars with the highest surface temperatures are called O-type stars and appear bluish in color. The peak of their blackbody spectrum occurs at a wavelength of about 1.0×10^2 nm. What is the approximate surface temperature of such stars?

5. A red giant star such as Antares (α Scorpii in the constellation Scorpius) has the peak of its spectrum at about 6.5×10^2 nm. What is the surface temperature of the star?

6. The tungsten filament of a common light bulb has a temperature of 2.0×10^3 K. What is the wavelength of the peak of its blackbody spectrum? What does this result imply about the relative efficiency of the bulb in producing visible light compared with infrared light?

7. At what wavelength is the peak of the electromagnetic spectrum of a blackbody radiator with a temperature of 20 °C?

Section 26.5 The Photoelectric Effect

8. How many radiofrequency photons per second are emitted by a radio station broadcasting at 89.1 MHz with a power output of 50 kW?

9. The visible light spectrum ranges from about 400 nm to 700 nm. What is the corresponding energy range of visible light photons in electron-volts?

•10. When 280 nm light is incident on a particular metal, the stopping potential is found to be 0.90 V. (a) What is the maximum kinetic energy of the photoelectrons in eV? (b) What is the maximum kinetic energy of the photoelectrons in J? (c) What is the maximum speed of the photoelectrons? (d) What is the work function of the material in eV?

•11. The cutoff wavelength for photoelectric emission from a certain metal is found to be 656 nm. (a) What is the work function of the material in eV? (b) When the material is illuminated with light with a wavelength of 430 nm, what is the maximum kinetic energy of the photoelectrons in eV? (c) What is the corresponding stopping potential? (d) If the number of 430 nm photons illuminating the material is doubled, what happens to the stopping potential?

•12. Photons with an unknown wavelength are used in the photoelectric effect from a sodium surface. Sodium has a work function of 2.3 eV. It is noted that a stopping potential of 5.0 V causes the photocurrent to decrease to zero. (a) Determine the wavelength of the incident photons in nm. (b) Determine the maximum speed of the emitted photoelectrons.

•13. Light with a wavelength of 200 nm falls on an aluminum metallic smile. The work function of aluminum is 4.1 eV. (a) What is the maximum kinetic energy of the photoelectrons in eV? (b) What is the stopping potential? (c) What is the cutoff wavelength in nm?

•14. Photons with an energy of 3.1 eV are incident on a metal surface. Zap. The maximum kinetic energy of the emitted photoelectrons is found to be 1.0 eV. Find the cutoff wavelength for the photoelectric effect on this metal.

•15. After many hours of diligent research, you obtain the following data on the photoelectric effect for a certain material:

Wavelength of light (nm)	Stopping potential (V)
360	1.40
300	2.00
240	3.10

(a) Plot the stopping potential (vertical axis) versus the *frequency* of the light. (b) What is the cutoff frequency for the photoelectric effect in this material? (c) What is the cutoff wavelength? (d) What is the work function (in eV) for the metal used in this experiment? (e) Given the value of the fundamental unit of charge $e = 1.602 \times 10^{-19}$ C, what value for Planck's constant is implied by this data?

•16. You doesn't mess around with low-class metals. When you do the photoelectric effect, you does it with class: on platinum. The work function for platinum is 6.4 eV. (a) What is the cutoff wavelength for the photoelectric effect on platinum? (b) If light of wavelength 488 nm is incident on the platinum surface, what happens?

•17. Light with a frequency of 5.0×10^{14} Hz strikes a metal surface that has a work function of 2.0 eV. (a) Will photoelectrons be ejected or not? Show your reasoning. (b) Sketch a graph of the greatest kinetic energy (in eV) of the photoelectrons versus the frequency of light used in the photoelectric effect on this metal. (c) If the light intensity is doubled, how does this affect the graph of part (b)?

•18. Professor I. M. Shirley Wright has given you the task of designing a photodetector that is sensitive to blue light (~450 nm) but not to red light (~600 nm). (a) In order to choose a suitable photocathode metal, determine the range of possible work functions that would be acceptable for this detector. (b) Sketch a complete simple circuit to detect this light using an ideal voltage source (battery), an ammeter (to detect the photocurrent), wire, and metal parts within the photocell. Show where the incident light hits and which way the standard current flows through the ammeter. (c) Without using selective light-transmitting filters, is it possible to design a photodetector that is sensitive to the red light but not the blue light? Explain your answer.

•19. An aluminum tool is adrift in the vacuum of space, left basking in the unshielded sunlight by a careless butterfingered astronaut. (a) The tool gradually acquires an electrical charge. By what mechanism does it acquire a charge and what is the sign of the charge? (b) Will the amount of charge accumulate indefinitely? Explain. (c) Calculate the range of solar wavelengths that are involved in the process. (d) Solar radiation is most intense at about 550 nm but some radiation exists down to about 300 nm and below. To prevent charge accumulation on a spacecraft, you are asked by your supervisor at NASA to consider coating the craft with a thin layer of gold. The work function for gold is 4.8 eV. What range of wavelengths will *not* produce any photoelectric charging of the gold surface? (e) A gold salesperson insists that a gold layer 1000 times thicker would be 1000 times more effective in preventing charge accumulation by sunlight. Assess this option from the standpoint of the physics involved and make a recommendation to your supervisors about the wisdom of the sales proposal.

Sections 26.6 The Quest for an Atomic Model: Plum Pudding
26.7 The Bohr Model of a Hydrogenic Atom
26.8 The Bohr Correspondence Principle
26.9 A Bohr Model of the Solar System?*
26.10 Problems with the Bohr Model

•**20.** An incident α particle with an initial kinetic energy of 5.0 MeV is in a head-on trajectory with a gold nucleus ($Z = 79$). Use the CWE theorem to calculate the distance of closest approach of the α particle to the gold nucleus.

•**21.** An electron is in the $n = 2$ state of the Bohr model for hydrogen atom. Determine: (a) the orbital radius of the electron; (b) the orbital speed of the electron; (c) the magnitude of the orbital angular momentum of the electron; (d) the magnitude of the centripetal acceleration of the electron; (e) the ratio of the magnitude of the centripetal acceleration to that of the local acceleration of gravity g near the surface of the Earth; (f) the kinetic energy of the electron (in eV); (g) the potential energy of the electron (in eV).

•**22.** The various wavelengths emitted by hydrogen in its spectrum are categorized in the following way. The wavelengths emitted when the final state of electron is the $n = 1$ state are part of the Lyman series; those wavelengths emitted when the final state of the electron is the $n = 2$ state are part of the Balmer series; the wavelengths emitted when the final state of the electron is the $n = 3$ state are part of another series. Strictly speaking, each series of wavelengths contains an infinite number of wavelengths. (a) Explain why the various wavelengths in a given series converge to a short wavelength limit. (b) Calculate the short wavelength limit of the Lyman and Balmer series.

•**23.** In the Bohr model of the hydrogen atom, use Newton's second law and the expressions for the centripetal acceleration and the (nonrelativistic) kinetic energy to find (algebraically) the pure number ratio of the kinetic to the potential energy of the circulating electron.

•**24.** An electron is quite Bohr(ed) tooling around in the lowly $n = 2$ state of the hydrogen atom and decides to make life more exciting by absorbing an appropriate passing photon and making a transition to the $n = 5$ state. Determine the wavelength (in nm) of the photon absorbed.

•**25.** (a) What is the wavelength of a photon that barely can ionize a hydrogen atom with its electron initially in the $n = 2$ state?

(b) Will longer wavelength photons ionize the atom if the electron is initially in the $n = 2$ state?

•**26.** A hydrogen atom becomes excited by a passing photon, absorbs it, and makes a transition from the $n = 1$ state (of Virginia) to the $n = 5$ state (also in Virginia). (a) What energy must be absorbed for the electron to make this transition? (b) What was the wavelength of the photon absorbed?

•**27.** Cool atomic hydrogen just gobbles up (absorbs) light with a wavelength of about 122 nm yet turns its nose up at nearby wavelengths, which pass unimpeded through the gas. Explain why this is the case with the assistance of an appropriate calculation.

•**28.** In the Bohr model of the hydrogen atom: (a) Show that the ratio of the nth orbital radius to the *longest* wavelength light that can be emitted from that state is

$$(5.80 \times 10^{-4}) \left[\frac{n^2}{(n-1)^2} - 1 \right]$$

(b) Show that the ratio of the nth orbital radius to the *shortest* wavelength light that can be emitted from that state is

$$5.80 \times 10^{-4}(n^2 - 1)$$

•**29.** Photons with wavelengths less than about 10 nm typically are classified as x-rays. Prove that none of the light emissions from the transitions of the orbital electron in the hydrogen atom produce x-rays.

•**30.** Hydrogen atoms with very large quantum numbers n are called *Rydberg atoms*. If $n = 1.00 \times 10^3$, find the radial distance of the electron from the nucleus.

‡**31.** An exotic hydrogenic atom consists of an electron and positron (a particle with the same mass as an electron but with a positive charge $+e$; the atom is known as *positronium*. The electron–positron pair orbit their center of mass (located halfway between them (see Figure P.31) at the same speed v. We can develop a Bohr model for positronium, assuming the mutual orbit is circular. Let $2r$ be the distance between the

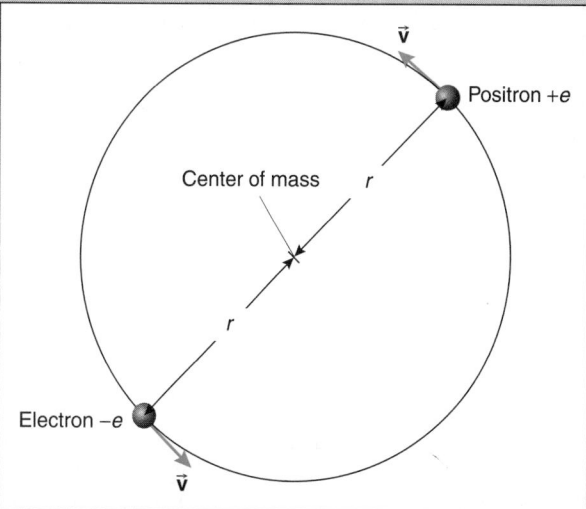

FIGURE P.31

electron and positron, so r is the radius of their mutual circular orbit as indicated in Figure P.31. (a) The force on each particle is the Coulomb force of electrical attraction. Apply Newton's second law to the electron (or the positron) to show that

$$\frac{1}{4\pi\varepsilon_0} \frac{e^2}{4r} = mv^2$$

where v is the common speed of the electron and positron. (b) The total energy E of the system consists of the kinetic energy of the electron and that of the positron (assume the speeds are nonrelativistic), plus the mutual electric potential energy of the pair of charges. Show that the total energy E of the system can be written as

$$E = -\frac{1}{4} \frac{1}{4\pi\varepsilon_0} \frac{e^2}{r}$$

(c) Quantize the magnitude of the total orbital angular momentum of the electron and positron by setting it equal to a positive integer n times $\hbar$. Show that this quantization leads to the expression

$$v = \frac{n\hbar}{2mr}$$

(d) Show that the total energy E then can be written in the form

$$E = -\frac{1}{2}\left[\frac{1}{2} \frac{1}{(4\pi\varepsilon_0)^2} \frac{me^4}{n^2\hbar^2}\right]$$

The expression in the bracket is the total energy for the electron in the normal hydrogen atom. Thus the energy levels for the positronium system are precisely *one-half* those of the normal hydrogen atom. (e) Show that the energy levels of positronium can be expressed in electron-volts as

$$E = -\frac{6.8 \text{ eV}}{n^2}$$

(f) Calculate the wavelength in nm of a photon emitted by positronium in making a transition from the $n = 2$ state to the $n = 1$ state. Is this photon in the visible region of the electromagnetic spectrum? The existence of positronium was experimentally confirmed in 1951 by Martin Deutsch (1917–). After its creation, the positronium atom is quite short-lived before the electron and positron annihilate each other to produce γ-rays.

‡32. A particle called a *muon* has a mass 207 times that of the electron but also with a single negative fundamental unit of charge $(-e)$. The muon can be captured by a proton to form an exotic hydrogenic atom known as *muonium*. (a) Modify the Bohr expression for the energy levels of a hydrogenic atom to find the energy levels of muonium in electron-volts; for simplicity, assume the nucleus is fixed. Explain why this assumption is a crude first approximation. (b) What energy is needed to ionize muonium if the muon is in the ground state? (c) If the muon makes a transition from the $n = 2$ state to the $n = 1$ state, what is the wavelength of the photon emitted?

Sections 26.11 Radioactivity Revisited
26.12 Carbon Dating
26.13 Radiation Units, Dose, and Exposure*

•33. Radon $^{222}_{86}$Rn has a half-life of 3.82 d. It is produced naturally from the decay of uranium, present in small amounts in stone, bricks, concrete, and other building materials. Long-term exposure to even low doses of radon has been implicated as a possible cause of lung cancer. Since radon is a clear, odorless gas, it cannot be detected by its odor in unventilated spaces such as basements. (a) What is the disintegration constant λ for the decay of radon? (b) What fraction of a sample of radon remains after a month (30 days)?

•34. Radium $^{226}_{88}$Ra was first isolated by Marie and Pierre Curie in 1898. Its half-life is 1.60×10^3 y. (a) What is its disintegration constant λ in y^{-1}? (b) Approximately what percentage of their original sample has not yet decayed?

•35. Radium $^{226}_{88}$Ra decays by α particle emission with a half-life of 1.60×10^3 y. (a) What is its half-life in s? (b) Find the disintegration constant λ in s^{-1}. (c) How many atoms of the substance will provide an activity of 1.00 Cu? (d) What is the mass of the collection of atoms found in part (c)? Your answer is an indication that radium in large quantities is an extraordinarily dangerous radioactive material.

•36. For a radioactive material with a half-life $\tau_{1/2}$, how long will it take until the activity of a given sample is only 10% of its initial value? Express your answer in terms of the half-life of the substance.

•37. Provide the details of the integration in Equation 26.55 to show that the mean life is equal to the reciprocal of the disintegration constant.

•38. Show that after a time t equal to the mean life, the number of radioactive particles remaining in the sample is N_0/e, where e is the base of natural logarithms.

•39. Show that the difference between the mean life and the half-life is approximately

$$\frac{0.3069}{\lambda}$$

••40. The element $^{226}_{88}$Ra decays by α emission with a half-life of 1.60×10^3 y. (a) What is its daughter? (b) How long will it take until only 1.0% of an original sample has not decayed?

•41. The isotope $^{238}_{92}$U has a half-life of 5.0×10^9 y. What fraction of the $^{238}_{92}$U present on the Earth 2.5×10^9 years ago has not yet decayed?

•42. Indicate the type of particle emitted in each of the following radioactive decay equations by completing the reaction: (a) $^{56}_{27}$Co $\rightarrow$ $^{56}_{26}$Fe + _____; (b) $^{212}_{84}$Po $\rightarrow$ $^{208}_{82}$Pb + _____; (c) $^{232}_{90}$Th $\rightarrow$ $^{228}_{88}$Ra + _____.

•43. The radioactive isotopes of sodium, $^{22}_{11}$Na and $^{24}_{11}$Na, both decay by b$^-$ emission. Determine the daughter products of the decays of each isotope.

•44. Radioactive materials are used widely in cancer therapy and for biological research. For these purposes, it is useful to know the *biological half-life* of the radioactive substance, defined as the time during which the organism expels half the ingested material. Assume the biological expulsion is similar to the natural physical decay. When ingested, some of the material decays and some is expelled. Assume the activity remaining is proportional to the number of nuclei remaining, via

$$\frac{dN}{dt} = -(\lambda + \lambda_b)N$$

where λ_b is the biological disintegration constant and λ is the physical disintegration constant. Show that the effective half-life τ_{eff} of the sample within the organism is related to the biological half-life by $\tau_b = (\ln 2)/\lambda_b$, and the natural half-life τ by

$$\frac{1}{\tau_{eff}} = \frac{1}{\tau} + \frac{1}{\tau_b}$$

•45. An animal is given a small dose of radioactive iodine $^{131}_{53}I$, which has a physical half-life of 8.1 d. Measurements are made of the activity of the animal and it is determined that the effective half-life of the material in the organism is 6.2 d. What is the biological half-life of the material within the organism? (Refer to Problem 44.)

•46. A 10 mCu pointlike source of $^{60}_{27}Co$ is implanted into a tumor. A detector with a collecting area of 25 cm² is placed 15 cm from the source, external to the patient. Assume the γ-rays emitted by the source are equally likely to be emitted in any direction. What is the approximate activity registered by the detector assuming only a negligible (but biologically significant) fraction of the radiation is absorbed by the tumor and surrounding tissue?

•47. An archeological dig in the basement of a physics laboratory uncovers the bones of a college student who expired some time ago. A careful analysis of the remains indicates a carbon 14 activity of 0.15 Bq per gram of carbon. A freshly sacrificed student has a carbon 14 activity of 0.27 Bq per gram of carbon. The half-life of carbon 14 is 5.73×10^3 y. (a) What is the disintegration constant λ in y^{-1}? (b) What is the age of the old student bones? (c) What will be the activity per gram of carbon of the old student bones when the class of A.D. 3268 graduates?

•48. Because of burnout, a sample of deans decreases its activity from 8000 memos per day to 1000 memos per day in 10 days. Assume their memo output is governed by exponential decay. What is the half-life of the sample?

•49. You uncover a cache of yellowed lecture notes on papyri written by the ancient and apocryphal natural philosopher Claudius Mediocratus. By careful analysis, the notes are discovered to have an activity (all their own) of 750 disintegrations per lecture hour per gram of carbon. Fresh notes on papyrus have an activity of 960 disintegrations per lecture hour per gram of carbon. The half-life of carbon 14 is 5.73×10^3 y. (a) Find the disintegration constant λ. (b) Determine the age of the lecture notes.

•50. An unusual sample population of Methuselahs decays exponentially with half-life of 50 y. (a) How long will it take till only 10% of the original population is left? (b) If the initial collection consisted of 1000 persons, what is the death rate (deaths/year) of the sample when only 10% remain?

•51. Uranium ores on the Earth at the present time typically have a composition consisting of 99.3% of the isotope $^{238}_{92}U$ and 0.7% of the isotope $^{235}_{92}U$. The half-lives of these isotopes are 4.47×10^9 y and 7.04×10^8 y respectively. If these isotopes were equally abundant when the Earth was formed, estimate the age of the Earth.

•52. The radioactive gas radon $^{222}_{86}Rn$ decays by α particle emission with a half-life of 3.82 days. (a) What is the daughter? (b) Compute the activity in curies of 1.00 mg of radon. This activity indicates why radon, even in small concentrations, may be a health hazard.

•53. The half-life of a sample of carbon 14 atoms is 5.73×10^3 y. If you had a collection of only *five* carbon 14 atoms, how many will still be present after 5.73×10^3 y? Explain the meaning of your answer.

Section 26.14 The Momentum of a Photon*

54. Calculate the magnitude of the momentum of a photon with a vacuum wavelength of 500 nm.

55. Photon A has twice the energy of photon B. What is the ratio of their momenta?

56. Massive atoms are struck by x-ray photons each with an energy of 40 keV and electrons are easily knocked away. What is the change in the wavelength of an x-ray photon scattered through an angle of 120° to the direction of the incident beam?

57. An x-ray photon with a wavelength of 0.1000 nm makes an elastic collision with an electron and is scattered through an angle of 90.0°. What is the wavelength of the scattered photon?

•58. Light of wavelength 632.8 nm is incident normally on a black mass and is totally absorbed. The number of photons per second that strike the mass is 1.50×10^{18}. (a) What is the magnitude of the force exerted on the black mass by the photons? (b) What is the power supplied to the black mass?

•59. Light from the Sun has an effective wavelength of 5.5×10^2 nm. The luminosity of the Sun is 3.83×10^{26} W and the distance between the Earth and the Sun is 149.6×10^6 km. (a) About how many joules per second are incident on each square meter of a sphere centered on the Sun with a radius equal to the distance between the Earth and Sun? (b) About how many photons per second are incident on each such square meter? (c) If the photons reflect from a sail oriented perpendicular to the direction to the Sun at the distance of the Earth, how large a square sail area is needed for it to experience a force of magnitude 1.00 N?

•60. A helium–neon laser emitting light at a wavelength of 632.8 nm has a power output of 500 mW. (a) How many photons per second are being emitted by the laser? (b) The laser has a mass of 5.00 kg and is so far away from any large masses that gravitational effects can be neglected. The laser is initially at rest and then is turned on for a year. Calculate the final speed of the laser.

•61. Show that the momentum of a photon can be interpreted as the product of Planck's constant h times the number of wavelengths in a meter. The number of wavelengths per meter is called the *wavenumber*.

•62. If the Earth (of radius $R = 6730$ km) absorbed all photons incident on it from the Sun, what magnitude of force would the solar radiation exert on the Earth? Assume the effective wavelength of solar radiation is 550 nm and the solar luminosity is 3.83×10^{26} W.

‡63. Interplanetary dust in the solar system is simultaneously subjected to at least two forces: the gravitational attraction of the Sun on the particles and a repulsive force caused by the absorption of solar radiation and the consequent transfer of photon momentum to the particle. Here we investigate the conditions under which one or the other of these forces dominates. To do this, first consider the force that solar photons exert on a dust particle. For simplicity, consider a spherical particle of radius R located a distance r from the Sun. Let the solar luminosity be L (equal to the power output of the Sun: 3.83×10^{26} W). Let ν be the average or effective frequency of solar radiation (the frequency of light with a wavelength of about 550 nm). (a) Show that the number of photons per second intercepted by the particle of radius R located a distance r from the Sun is

$$\frac{L}{4\pi r^2}\,\frac{\pi R^2}{h\nu}$$

(b) Each photon has a momentum $h\nu/c$. If all the intercepted photons are absorbed by the particle, show that the magnitude of the force on the particle (the magnitude of the time rate of change of its momentum) due to the solar radiation is

$$F_{\text{radiation}} = \frac{L}{4\pi r^2 c}\,\pi R^2$$

Note that the magnitude of this force of repulsion is proportional to the cross-sectional area of the particle and so increases with the square of its radius. (c) Let ρ be the density of the spherical dust particle. Show that the magnitude of the gravitational force of the Sun (of mass $M = 1.99 \times 10^{30}$ kg) on the particle is

$$F_{\text{grav}} = \frac{GM}{r^2}\,\frac{4}{3}\,\pi R^3 \rho$$

Note that the magnitude of this attractive force is proportional to the volume of the particle and so increases with the cube of its radius. (d) Show that the *ratio* of the magnitude of the repulsive force due to solar radiation to that of the attractive force due to gravitation is

$$\frac{F_{\text{radiation}}}{F_{\text{grav}}} = \frac{3L}{16\pi GMcR\rho}$$

Note that this ratio is *independent of the distance r of the particle from the Sun*. (e) Let the density of each particle be $\rho = 5.0 \times 10^3$ kg/m³, which is on the order of the density of many minerals. Show that the repulsive force of radiation is equal in magnitude to the attractive force of gravitation if the radius R of the particle is on the order of 1.2×10^{-7} m. Particles smaller than this are repelled from the Sun since the solar radiation force dominates; particles larger than this are attracted to the Sun, since the gravitational force then dominates. Radiation forces on small particles were very important factors in the early history of the solar system.

Section 26.15 The De Broglie Hypothesis

64. For a local fund-raiser, you volunteer to be the target at a pie-throwing booth. Calculate the de Broglie wavelength of a 0.50 kg banana cream pie thrown at you with a speed of 10 m/s.

65. For an electron in a Bohr model energy state with quantum number n, determine how many de Broglie wavelengths of the electron fit around the circumference of the orbit.

•66. An electron is accelerated from rest through a potential difference of V to a speed that is nonrelativistic. In terms of V, h, e, and the mass of the electron m_e, find expressions for: (a) the kinetic energy of the electron; (b) the magnitude of the momentum of the electron; and (c) the de Broglie wavelength of the electron.

•67. An electron beam in an electron microscope has electrons with individual kinetic energies of 50 keV. Determine the de Broglie wavelength of such electrons.

•68. A photon and an electron have the same wavelength λ. Assume the electron has a nonrelativistic speed. In terms of λ and fundamental constants: (a) determine the speeds of the photon and the electron; (b) determine the magnitudes of the momenta of the photon and the electron; and (c) determine the energy of the photon and the kinetic energy of the electron.

•69. (a) What is the wavelength of a photon that has the same magnitude of momentum as an electron moving with a speed of 2.0×10^6 m/s? (b) What is the de Broglie wavelength of the electron?

•70. An electron and a proton are accelerated so that they have equal kinetic energies. Find the ratio of their de Broglie wavelengths. Assume that the speeds of the particles are nonrelativistic.

•71. (a) At what speed must a neutron move to have a de Broglie wavelength of 1.5×10^{-11} m? (b) What is the magnitude of the momentum of the neutron of part (a)? (c) At what speed must a photon move to have a wavelength of 1.5×10^{-11} m? (d) What is the magnitude of the momentum of the photon of part (c)?

•72. What is the ratio of the photon wavelength emitted in the $n = 2$ to $n = 1$ transition in the Bohr model of hydrogen to the de Broglie wavelength of the electron in the $n = 1$ state?

INVESTIGATIVE PROJECTS

A. Expanded Horizons

1. Investigate the use of radioactive isotopes in determining the age of the Earth.
   Ludwik Kowalski, "Radioactivity and nuclear clocks," *The Physics Teacher*, 14, #7, pages 409–416 (October 1976).
   Robert J. Packhurst, "Radiometric dating in geology," *Physics Education*, 15, #7, pages 340–343 (November 1980).
   Hans E. Suess, "Radiocarbon geophysics," *Endeavor*, 4, #3, pages 113–117 (1980).

2. Investigate the so-called *natural units* known as the *Planck length*, the *Planck mass*, and the *Planck time*.
   Edward R. Harrison, *Cosmology* (Cambridge University Press, Cambridge, England, 1981), page 33.
   Robert L. Wadlinger and Geoffrey Hunter, "Max Planck's natural units," *The Physics Teacher*, 26, #8, pages 528–529 (November 1988).

3. Write a précis about the following interesting article about the origins of particle waves.
   Bruce R. Wheaton, "Louis de Broglie and the origins of wave mechanics," *The Physics Teacher*, 22, #5, pages 297–301 (May 1984).

4. Investigate the use of particle accelerators in improving the carbon 14 dating technique.
   Harry E. Gove, "A new accelerator-based mass spectrometry," *The Physics Teacher*, 21, #4, pages 237–245 (April 1983).

5. Tree ring counts are used to fine-tune carbon 14 dating techniques. Investigate how this done.
   Hans E. Suess, "Radiocarbon geophysics," *Endeavor*, 4, #3, pages 113–117 (1980).

6. Investigate the use of modern physics in the analysis of art work to decipher painting techniques as well as uncover forgeries.
   Maurice J. Cotter and Kathleen Taylor, "Neutron activation analysis of paintings," *The Physics Teacher*, 16, #5, pages 263–271 (May 1978).

7. Investigate the exposure to radiation from background, diagnostic x-rays, and other sources.
   Stewart C. Bushong, "Radiation exposure in our daily lives," *The Physics Teacher*, 15, #3, pages 135–144 (March 1977).

8. An early model of the universe (from the 1950s) imagined a dual universe, half of which was composed of matter, the neighboring half of antimatter. Investigate this model and report on the evidence that indicates this model is not likely to be correct.
   Hannes Alfvén, *Worlds–Antiworlds: Antimatter Is Cosmology* (W. H. Freeman, San Francisco, 1966).

9. Why is the universe composed of matter and not antimatter?
   Edward R. Harrison, *Cosmology* (Cambridge University Press, Cambridge, England, 1981), pages 356–357, 365.

10. Discover why the curie unit of activity is defined as 36×10^9 Bq rather than some more convenient number of disintegrations per second.

B. Lab and Field Work

11. Design and perform an experiment to measure the photoelectric effect from a metal with a small work function, so that visible light can be used.

12. Estimate the number of photons per second received by your eye from a bright star. Organize your calculation carefully and state what assumptions you make to secure an answer.

13. With an appropriate photometer, measure the average power output of an American firefly (*Photuris pennsylvanica*). Measure the spectrum of its light and determine an approximately effective wavelength of its light. Then estimate the approximate number of photons per second emitted in each flash.

14. Design and perform an experiment to measure the temperature dependence of the work function of a metal in the photoelectric effect.

15. Design and build a scale model of the Bohr model for hydrogen and place it appropriately on your campus. You may need to seek permission from an appropriate campus official!

16. Fairly short-lived radioactive nuclides are readily available for experimental use; consult your professor about them. Design and perform an experiment to measure the half-life of a radioactive material with a reasonably short half-life. Now do the same for a material with a long half-life.

17. Design and perform an experiment to measure the relative penetrating power of α, β, and γ particles.

18. Discover how commercial radon detectors operate. Measure the radon level at various places in the basement of your physics building and perhaps other structures on campus. Report your findings to the director of buildings and grounds on your campus.

19. Small particles can be levitated with the pressure provided by light. Investigate the particle sizes and powers of light sources necessary. Then demonstrate the effect for your class.
    Arthur Ashkin, "The pressure of laser light," *Scientific American*, 226, #2, pages 62–71 (February 1972); page 118 has additional references.

20. By consulting with a local radiologist and/or oncologist, investigate the activities of the sources and dosages associated with radiation therapy for various cancer treatments. How are the dosages measured?
    T. S. Curry, J. E. Dowdey, and R. C. Murry, *Christensen's Introduction to the Physics of Diagnostic Radiology* (Lea and Febiger, Philadelphia, 1984).
    Harold E. Johns and John R. Cunningham, *The Physics of Radiology* (4th edition, C. C. Thomas, Springfield, Illinois, 1983).
    J. W. Boag, "Forty years of development in radiation dosimetry," *Physics in Medicine and Biology*, 29, #2, pages 127–130 (February 1984).
    W. J. Meredith, "Forty years of development in radiotherapy," *Physics in Medicine and Biology*, 29, #2, pages 115–120 (February 1984).

21. The electron microscope is an invaluable tool for research. Investigate the characteristics of an electron microscope likely to be at your university. How are samples prepared? What wavelength electrons are used? What is the resolution of the instrument?

C. Communicating Physics

22. The life of Marie Curie is a case study not only of the trials and tribulations, but also of the successes and triumphs of women in physics. After reading about Curie, interview one or more women physicists at your university (or another university) to see if the climate for women in the field has changed during the last century. On the basis of your interview, write a feature for your campus newspaper.

Dictionary of Scientific Biography (Charles Scribner, New York, 1971), Volume III, pages 497–503; see page 503 for additional references.

23. Poll your colleagues to determine common meanings of the phrase "quantum jump." Is the common usage of the term anything like its use in physics?

24. Soon after the discovery of x-rays, René-Prosper Blondlot (1849–1930) reported the discovery of yet another "unknown" type of radiation, which he dubbed N-rays, after Nancy, France, the city and university where he worked. Many other scientists also claimed to observe the mysterious N-rays. The story of their discovery and the subsequent realization that they simply do not exist is a case study of the danger of scientists seeing what they wanted to see, rather than what was actually there. Investigate the case of the "discovery" of N-rays and the convincing way that American physicist Robert Williams Wood (1868–1955) of The Johns Hopkins University "undiscovered" them by deft sleight of hand. Prepare an interesting oral report about the N-ray fallacy for your departmental seminar.

Robert T. Lagemann, "New light on old rays: N rays," *American Journal of Physics*, 45, #3, pages 281–284 (March 1977).

Spencer Weart, "A little more light on N rays," *American Journal of Physics*, 46, #3, page 306 (March 1978).

Robert Williams Wood, "The n-rays," *Nature*, 70, #1822, pages 530–531 (29 September 1904).

RADIOACTIVE DATING
OF DAUGHTER PRODUCTS

CHAPTER 27

AN INTRODUCTION TO QUANTUM MECHANICS

. . . I think I can safely say that nobody understands quantum mechanics.

The amusing yet sobering statement by Richard Feynman (see previous page), one of the foremost minds in 20th-century physics (and a physics Nobel prize laureate as well), reminds us of Sommerfeld's comments about thermodynamics (cf. the introduction to Chapter 13). We approach quantum mechanics with more than a little awe and certainly respect for the mysteriousness of nature.

In many respects, the world of physics mirrors its surrounding cultural milieu and, to some extent, helps to shape it. The classical and formal art, music, literature, and mathematics created during the Renaissance and Enlightenment periods, until the dawn of the 20th century, were complemented by the classical physics of Newton's dynamics, Maxwell's electromagnetism, and Boltzmann's thermodynamics. Art was realistic, music was harmonious, literature (poetry in particular) followed strict rules of form, meter, and rhyme; mathematics was calculus and number theory. Physics was precise, predictable, and deterministic.

But realistic art gave way to the abstractions of Renoir, Monet, and Picasso. The harmonic and structural formality of Bach, Mozart, and Beethoven evolved into new harmonies and tone poems by Chopin, Liszt, and Mahler as well as the wonderful spontaneity and improvisation of ragtime and jazz. The measured meters of Shakespeare, Shelley, Byron, and Tennyson changed to the free verse of Whitman, Eliot, Frost, and e. e. cummings. Mathematicians developed abstract algebra. And physicists discovered a new abstract formulation of the physical world as Bohr, de Broglie, Schrödinger, Heisenberg, and Dirac elucidated the features of a totally new and unexpected (almost counterintuitive) type of mechanics: quantum mechanics. The new art, music, literature, and mathematics was not always easy to understand or appreciate; it did not make common sense or resonate pleasingly to everyone. The physics of relativity (Chapter 25) and quantum mechanics (which we examine briefly in this chapter) defied common sense as well. Just as creators in the humanities were doing, physicists uncovered and invented a new way of thinking and a new sense of beauty. Strict determinism was replaced by probability, uncertainty, and an unfamiliar new world of nature.

In this chapter we venture a first look into this peculiar natural world. Some of this is quite heady material that will require much reflection, even years to understand; but, as in contemporary art or music, with guidance you certainly are capable of gleaning an appreciation for it. Keep your mind focused on the conceptual forest rather than the mathematical trees.

27.1 THE HEISENBERG UNCERTAINTY PRINCIPLES

I have known what the Greeks did not: uncertainty.
Jorge Luis Borges (1899–1986)[†]

Light exhibits both wave and particle characteristics. The diffraction and interference experiments of Young and the Maxwell theory of electromagnetism clearly indicate that light has wave properties. Yet Einstein's theory of the photoelectric effect and the Compton effect indicate that light, at least on a submicro-

scopic level, exhibits the attributes of a particle as well (a photon), having both energy and momentum. Conversely, the de Broglie hypothesis, the Bohr angular momentum postulate, and the experiments of Davisson and Germer, and Thomson indicate that what we traditionally think of as particles on a submicroscopic level (electrons, protons, neutrons, etc.) also exhibit behavior typically associated with waves (diffraction and interference). Thus nature is bilateral: particles are waves and waves are particles. The particle aspect carries with it the traditional concepts of position and momentum; the wave aspect carries with it the concepts of wavelength and frequency.

Here we shall see that the wave-particle aspects are inextricably interwoven in a way that fundamentally limits our ability to know or measure several of these parameters simultaneously with perfect precision. Nature places natural limits on the precision of our measurements; some knowledge and information forever is shrouded from our prying eyes.

The Position–Momentum Uncertainty Principle

Consider the diffraction of a monochromatic wave through a slitlike aperture of width a. A single slit diffraction pattern is observed on a distant screen, as shown in Figure 27.1. From our analysis of single slit diffraction in Section 24.4 we know that the smaller the aperture of the slit, the greater the angular width of the central diffraction peak. Such diffraction is characteristic of all wave phenomena.

Let's look at the diffraction from a slightly different perspective. To the left of the aperture, the wave has a precise wavelength λ and, according to the de Broglie relation, we have

$$\lambda = \frac{h}{|p_x|} \tag{27.1}$$

The absolute value of the momentum component $|p_x|$ in the incident direction of the wave also is well defined or precisely known. Since the wave is traveling along the x-axis, the wave has zero momentum component along the transverse direction, y. On the other hand, the position of the wave along the x-axis to the left of the aperture can be described by saying that the wave is everywhere. It is not possible to say that the wave is at any particular value of x. The wave simultaneously is at *every* value of x to the left of the aperture. Thus the incident wave has a well-defined wavelength or, equivalently, x-component of momentum, but an ill-defined position.

The passage of the wave through the slitlike aperture results in a single slit diffraction pattern on a distant screen. Most of the wave intensity is confined within the central diffraction peak (though

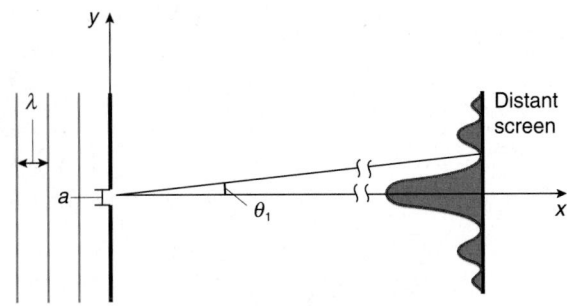

FIGURE 27.1 A single slit diffraction pattern.

several secondary maxima also exist). The location of the first minimum of the diffraction pattern is found from Equation 24.3:

$$a \sin \theta_1 = \lambda$$

If the angle θ_1 is small, the sine is approximately equal to the angle itself (in radians), which is the small angle approximation. Thus Equation 24.3 becomes

$$a\theta_1 = \lambda \qquad (27.2)$$

Using the de Broglie relation (Equation 27.1) to substitute for λ in Equation 27.2, we find

$$a\theta_1 = \frac{h}{|p_x|}$$

Rearranging this slightly, we obtain

$$a\theta_1 \,|p_x| = h \qquad (27.3)$$

At the slit, the slit width a is a measure of the extent to which the position of the wave is confined along the y-axis; certainly as the wave goes through the slit, you can say that the position of the wave was known to within $\Delta y = a$ along the y-axis or transverse direction. Correspondingly, the quantity $\theta_1 \,|p_x|$ is a measure of the degree to which we know the y-component of the momentum of the wave. The *smaller* the aperture, the *larger* θ_1 becomes, which means that the width of the central diffraction peak increases. Hence as θ_1 increases, the greater is the extent to which the wave spreads in the transverse (y) direction and has a nonzero y-component of the momentum.

We call $\theta_1 \,|p_x| = \Delta p_y$ and $a = \Delta y$.* Then Equation 27.3 becomes

$$\Delta y \,\Delta p_y = h \qquad (27.4)$$

At the slit, the more we try to confine the wave along the y-axis (by using a smaller and smaller slit opening $a = \Delta y$), the better we know the position at the slit of the wave along the y-axis. But the result of decreasing the slit width means that the width of the diffraction peak increases, which means we know less about the value of the y-component of the momentum of the wave. If we account for the fact that *secondary* diffraction maxima also exist, the uncertainty in the y-component of the momentum is even greater: Equation 27.4 really is a best-case scenario. To account for the secondary maxima, we really should write

$$\Delta y \,\Delta p_y \geq h \qquad (27.5)$$

Although what we have just discussed certainly is not a rigorous derivation, the treatment is an illustration of a fundamental aspect of the wave–particle duality.

The **Heisenberg uncertainty principle** states that there exists a fundamental limit to the extent to which we can simultaneously determine the position and corresponding momentum component of a wave-particle in any given direction.

The principle is named for Werner Heisenberg (1901–1976), who first elucidated the principle in the late 1920s.

---

*We consider Δy and Δp_y to be intrinsically positive. We really should write $|\Delta y|$ and $|\Delta p_y|$ but prefer to avoid the clutter of the absolute value signs.

There is nothing sacred about the y-axis in Equation 27.5; the principle applies to the position and corresponding momentum components in any direction:

$$\boxed{\Delta x \,\Delta p_x \geq h} \qquad (27.6)$$

Corresponding quantities such as y and p_y, or x and p_x, are known as **complementary variables**.

The Energy–Time Uncertainty Principle

Recall (Equation 12.16) that the mathematical form for a monochromatic wave with wavelength λ and frequency ν is

$$\Psi(x, t) = A \cos(kx - \omega t)$$

where $k = 2\pi/\lambda$ and $\omega = 2\pi\nu$. Writing out these substitutions in Equation 12.16, we get

$$\Psi(x, t) = A \cos\left(\frac{2\pi}{\lambda} x - 2\pi\nu t\right) \qquad (27.7)$$

Now we use the de Broglie relation for particle-waves, Equation 27.1,

$$\lambda = \frac{h}{|p_x|}$$

and substitute for λ in Equation 27.7; we obtain

$$\Psi(x, t) = A \cos\left(\frac{2\pi}{h}\Big|p_x\Big| x - 2\pi\nu t\right) \qquad (27.8)$$

Recall that Einstein hypothesized in the photoelectric effect that the relationship between the energy E of a photon particle and its frequency ν is

$$E = h\nu \qquad (27.9)$$

We take this equation as the relationship between the energy of *any* particle-wave and its associated frequency, as de Broglie did.

Heisenberg received his Ph.D. at the venerable age of 22 and also was a gifted pianist. He was the most prominent active physicist to remain in Nazi Germany during World War II. His role as head of the Nazi efforts to exploit nuclear energy was quite controversial both during and after the war.

Using Equation 27.9 to substitute for the frequency in Equation 27.8, we find

$$\Psi(x, t) = A \cos\left(\frac{2\pi}{h}\left|p_x\right|x - \frac{2\pi}{h}Et\right) \quad (27.10)$$

Notice that the roles of E and t in Equation 27.10 are *mathematically identical* to the roles of x and $|p_x|$. Since there is a Heisenberg uncertainty principle for the complementary pair of variables x and p_x, the mathematical similarity of the quantities E and t in the expression for the wave strongly suggests that an uncertainty principle also exists for the energy and time.

We can find this additional uncertainty principle in the following simplified way. For a free particle with zero potential energy moving at speed v along the x-axis, its (nonrelativistic) energy is

$$E = \frac{1}{2}mv^2$$
$$= \frac{p^2}{2m} \quad (27.11)$$

We take the derivative

$$\frac{dE}{dp}$$

and approximate the differential dE with ΔE and dp with Δp. Then the uncertainty in energy ΔE is related to the uncertainty Δp in the momentum by

$$\frac{\Delta E}{\Delta p} = \frac{2p}{2m}$$

$$\Delta E = \frac{p}{m}\Delta p \quad (27.12)$$

Likewise, the speed v is approximately

$$v = \frac{\Delta x}{\Delta t} \quad (27.13)$$

We use Equation 27.12 for Δp and Equation 27.13 for Δx and substitute into the position–momentum uncertainty principle, Equation 27.6. This yields

$$\Delta x \, \Delta p \geq h$$
$$(v \, \Delta t)\left(\frac{m}{p}\Delta E\right) \geq h$$

Hence we find that the energy–time uncertainty principle is

$$\boxed{\Delta E \, \Delta t \geq h} \quad (27.14)$$

The energy and time have their own uncertainty principle.

In fact Equation 27.14 is a very general equation, not restricted to free particles, although we have not shown this here.

Both the position–momentum and the energy–time uncertainty principles are fundamental manifestations of the underlying mathematics of waves and harmonic functions. We broached

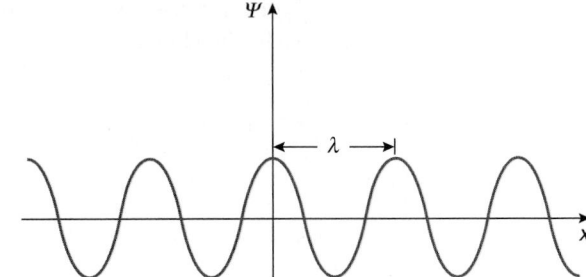

FIGURE 27.2 A sinusoidal waveform.

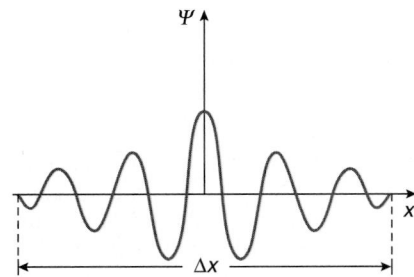

FIGURE 27.3 A wave packet.

this connection in Chapter 12 with a brief introduction to Fourier analysis (Section 12.21). For example, consider a sinusoidal wave with a precisely known wavelength (or momentum since $\lambda = h/p$), as in Figure 27.2. This wave is defined for *all* values of x, and so information about the position of the wave is very uncertain: it is everywhere. On the other hand, if the wave is nonzero over only a very small region of space, and resembles a pulse or wave packet as shown in Figure 27.3, then the wave has a wavelength that is not precisely known (what is called an ill-defined wavelength).

Why is that? Many slightly different sinusoidal wavelengths must be superimposed to create the short waveform. The various wavelengths interfere constructively in the general region where the wave is nonzero and interfere destructively everywhere else. Since many different wavelengths are needed to perform the superposition, the wave pulse has a wavelength that is not precise or, equivalently, a momentum (again since $\lambda = h/p$) that is not precisely known. There is no escape. The Heisenberg uncertainty principles are inextricably tied to the mathematical formalism associated with the description of waves. Equally, they describe the inevitable disturbances caused by all experimental measurements of physical quantities like energy and time, or position and momentum.

27.2 IMPLICATIONS OF THE POSITION–MOMENTUM UNCERTAINTY PRINCIPLE

In classical mechanics it was possible to say that a particle has a well-defined position x and momentum p_x (or equivalently, velocity component v_x) at a time t. In principle, the position and momentum component can be measured with arbitrary precision in classical physics. In quantum mechanics, however, this is not strictly true, because of the wave nature of particles. The position and momentum uncertainty principle was given by Equation 27.6:

$$\Delta x \, \Delta p_x \geq h$$

The more precisely we know the position, the less precisely we can know the corresponding momentum component and vice versa.

The position–momentum Heisenberg uncertainty principle enables us to understand one reason that atoms do not collapse with their electrons spiraling into the nucleus.* Recall from Chapter 26 that the Bohr model of the atom simply *postulated* the stability of the nuclear model of the atom.

Within the context of the Heisenberg uncertainty principle, we say that the electron is confined to a region on the order of size r and abandon the idea of an electron orbit. The position of the electron thus is known with a precision of about $\Delta x \approx r$. The Heisenberg uncertainty principle implies that any time a particle is confined or localized, there is an uncertainty associated with its complementary momentum component. Thus, even if we *imagine* the electron to be at rest with zero momentum, the uncertainty principle states that the momentum *cannot* be exactly zero but must be on the order of

$$p \approx \Delta p \approx \frac{h}{\Delta x} \approx \frac{h}{r}$$

The smaller the value of r, the greater the uncertainty in the momentum. If p increases, so does the kinetic energy of the electron. So what does the electron then do? The electron in the hydrogen atom has an electrical potential energy

$$PE = qV$$
$$= (-e)V$$
$$= (-e)\frac{1}{4\pi\varepsilon_0}\frac{e}{r} \quad (27.15)$$

where V is the electric potential of the nuclear charge ($+e$ for hydrogen) at the position of the electron. The electron has a kinetic energy

$$KE = \frac{1}{2}mv^2$$

which, since $p = mv$, can be rewritten as

$$KE = \frac{1}{2}m\left(\frac{p}{m}\right)^2$$
$$= \frac{p^2}{2m} \quad (27.16)$$

Now we consider the total energy E of the electron,

$$E = KE + PE$$
$$= \frac{p^2}{2m} - \frac{1}{4\pi\varepsilon_0}\frac{e^2}{r}$$
$$= \frac{h^2}{2mr^2} - \frac{1}{4\pi\varepsilon_0}\frac{e^2}{r}$$

and find what value of r minimizes the total energy. To do this, we take the derivative of E with respect to r and set it equal to zero:

$$\frac{dE}{dr} = \frac{h^2}{2m}\left(-\frac{2}{r^3}\right) - \frac{1}{4\pi\varepsilon_0}e^2\left(-\frac{1}{r^2}\right) = 0 \text{ J/m}$$

Solving for r, we find

$$r = 4\pi\varepsilon_0 \frac{h^2}{me^2}$$

This is the same order of magnitude as the fictitious orbits of the Bohr model (Equation 26.24) for low values of the quantum number n. Indeed the first Bohr orbital radius a_0 (with $n = 1$) was found in Example 26.5 to be

$$a_0 = 4\pi\varepsilon_0 \frac{\hbar^2}{me^2}$$

Thus one reason that atoms do not collapse is the uncertainty principle.

If you try to squeeze an atom by forcing its electrons closer to the nucleus, the smaller Δx causes a larger Δp_x due to the uncertainty principle; the larger momentum uncertainty prevails against the confinement.

EXAMPLE 27.1

You measure the diameter of a shotgun pellet to be 1.00 mm ± 0.01 mm with a micrometer. What is the minimum uncertainty of its momentum if the magnitude of its momentum is measured simultaneously?

Solution
According to the Heisenberg uncertainty principle, $\Delta x \, \Delta p_x \geq h$. Here the minimum uncertainty associated with its position is $\Delta x = 0.01$ mm $= 1 \times 10^{-5}$ m. The minimum uncertainty in its momentum thus is

$$\Delta p_x = \frac{h}{\Delta x}$$
$$= \frac{6.626 \times 10^{-34} \text{ J·s}}{1 \times 10^{-5} \text{ m}}$$
$$= 7 \times 10^{-29} \text{ kg·m/s}$$

---

*There is another reason that atoms do not collapse: the *Pauli exclusion principle*, which also arises from quantum mechanics, though we will not show how. The principle prohibits electrons (and other particles belonging to a family of particles known as fermions) from "all doing the same thing," or having identical sets of quantum numbers in an atom. The principle is named for Wolfgang Pauli (1900–1958), who first enunciated it in the late 1920s. It is the Pauli exclusion principle that prevents the electrons in atoms from all ending up in the ground state. The Pauli exclusion principle

creates a pressure that resists atomic collapse. This special pressure (that is independent of temperature and so is very different from a thermal pressure) colloquially is called a packing pressure but is more formally known as the *Fermi pressure* (after the great theoretical and experimental physicist Enrico Fermi [1901–1954]). The Fermi pressure is an important factor in the later stages of the evolution of stars.

This is exceedingly small. As a result, for such a macroscopic particle, the limitations placed on the simultaneous determinations of its position and momentum by the Heisenberg uncertainty principle are irrelevant.

STRATEGIC EXAMPLE 27.2

A proton is known to be in the nucleus of an atom. The size of the nucleus is about 1.0×10^{-14} m.

a. What is the minimum magnitude of the momentum the proton must have?

b. If its momentum is nonrelativistic, what is the minimum speed of the proton?

c. What fraction of the speed of light is the speed calculated in part (b)? Does this result justify the use of the nonrelativistic expression for its momentum?

d. What minimum kinetic energy must the proton have because of its confinement? Express your result in MeV.

Solution

a. The minimum magnitude of momentum is found using the Heisenberg uncertainty principle, Equation 27.6 (with an equality). Since the proton is confined to within 1.0×10^{-14} m, you use this value for Δx, and find Δp accordingly:

$$\Delta p = \frac{h}{\Delta x}$$
$$= \frac{6.626 \times 10^{-34} \text{ J·s}}{1.0 \times 10^{-14} \text{ m}}$$
$$= 6.6 \times 10^{-20} \text{ kg·m/s}$$

b. The magnitude of the nonrelativistic momentum is $p = mv$. Since the minimum uncertainty in the momentum is Δp, this is the minimum possible value for p itself. Hence the minimum proton speed is

$$v = \frac{p}{m}$$
$$= \frac{6.6 \times 10^{-20} \text{ kg·m/s}}{1.67 \times 10^{-27} \text{ kg}}$$
$$= 4.0 \times 10^{7} \text{ m/s}$$

c. The speed is the following fraction of the speed of light:

$$\frac{v}{c} = \frac{4.0 \times 10^{7} \text{ m/s}}{3.00 \times 10^{8} \text{ m/s}}$$
$$= 0.13$$

With this ratio, you can calculate the relativistic factor γ:

$$\gamma = \frac{1}{\sqrt{1 - \dfrac{v^2}{c^2}}}$$
$$= \frac{1}{\sqrt{1 - (0.13)^2}}$$
$$\approx 1.0$$

Since $\gamma \approx 1$, the use of the nonrelativistic momentum is justified.

d. In view of part (c), you can use the nonrelativistic expression for the kinetic energy:

$$\text{KE} = \frac{1}{2}\, mv^2$$
$$= \frac{1}{2}(1.67 \times 10^{-27} \text{ kg})(4.0 \times 10^{7} \text{ m/s})^2$$
$$= 1.3 \times 10^{-12} \text{ J}$$
$$= \frac{1.3 \times 10^{-12} \text{ J}}{1.602 \times 10^{-19} \text{ J/eV}}$$
$$= 8.1 \times 10^{6} \text{ eV}$$
$$= 8.1 \text{ MeV}$$

This is a very substantial amount of energy and indicates that particles in the nucleus have energies much greater than the energies typically associated with the electrons surrounding the nucleus (which are on the order of a few to a few tens of electron-volts). These large energies of nuclear particles also indicate why the comparably high-energy particles emitted in radioactive decay originate from the nucleus.

27.3 IMPLICATIONS OF THE ENERGY–TIME UNCERTAINTY PRINCIPLE

In classical mechanics, it is possible to say that a particle has a definite energy E at time t. The energy of the particle can be precisely determined at a given instant to arbitrary precision. In quantum mechanics, this is not the case. The uncertainty principle for energy and time is (Equation 27.14)

$$\Delta E \, \Delta t \geq h$$

In order to increase the precision with which the energy is determined, it is necessary to increase the time over which the measurement is made. Correspondingly, if we decrease the time over which a measurement of the energy is made, we must put up with less precision in the value of the energy so determined.

There are many implications to the energy–time uncertainty principle; we consider only two here.

The Mass of Fundamental Particles

One implication is the precision with which we can determine the mass of fundamental particles. In relativity we discovered that there is a rest energy associated with a particle with mass:

$$E_{\text{rest}} = mc^2 \qquad (27.17)$$

The uncertainty principle implies that there is a nonzero uncertainty associated with any measurement of an energy, here of the rest energy ΔE_{rest}. The uncertainty Δm associated with the mass of the particle is found by taking differentials of Equation 27.17 and approximating the differential changes with ΔE and Δm:

$$\Delta E_{\text{rest}} = c^2 \, \Delta m \qquad (27.18)$$

The energy–time uncertainty principle (Equation 27.14) becomes

$$c^2 \, \Delta m \, \Delta t \geq h \qquad (27.19)$$

For example, the mean lifetime of a free neutron* is 888 s. We use this for Δt in Equation 27.19. Thus the mass of a neutron can be determined to a precision no better than Δm, where

$$\Delta m = \frac{h}{c^2 \, \Delta t}$$

Evaluating this expression yields

$$\Delta m = \frac{6.626 \times 10^{-34} \text{ J} \cdot \text{s}}{(3.00 \times 10^8 \text{ m/s})^2 (888 \text{ s})}$$
$$= 8.29 \times 10^{-54} \text{ kg}$$

This minimum uncertainty associated with the measurement of the mass of the free neutron is 27 orders of magnitude smaller than the mass of the neutron itself (1.675×10^{-27} kg) and so is of no experimental import, since no experiments currently known can determine the mass of the neutron to such precision. The smallness of the fundamental uncertainty arises because the neutron is quite long-lived as a free particle. What this means from a practical standpoint is that if the masses of a large number of neutrons are determined, all the masses are essentially identical, as shown schematically in Figure 27.4.

On the other hand, there are many exotic, unstable, fundamental particles that have very short lifetimes, with the result that the uncertainty in determining their mass can be a significant fraction of it.

In other words, if the masses of a large number of these very short-lived particles are measured, the values obtained have an intrinsic spread because of the Heisenberg uncertainty principle, as shown schematically in Figure 27.5. The shorter the lifetime of a particle, the greater the uncertainty associated with determining its mass.

The Nature of a Vacuum

Another strange implication of the energy–time uncertainty principle concerns the nature of a vacuum. The idea of a vacuum in quantum physics is quite different from the classical idea of a vacuum as simply a volume with nothing in it.

The quantum mechanical vacuum is a seething sea of particle–antiparticle pairs, called **virtual particles**, since their existence is ephemeral. The pairs of virtual particles well up out of *nothing*, live for a very short time, and then disappear (annihilate each other).

To create a virtual particle–antiparticle pair, each with mass m, we need to borrow or create an amount of energy ΔE at least equal to the sum of the rest energies of the two particles: $2mc^2$. The virtual particles can live only for a time Δt no longer than the minimum time obtained from the Heisenberg energy–time uncertainty principle (Equation 27.14):

$$\Delta E \, \Delta t = h$$

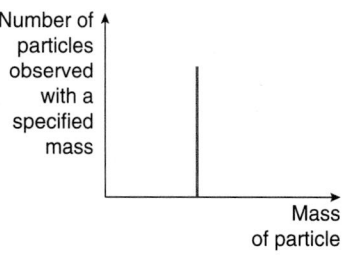

FIGURE 27.4 The mass of the neutron can be determined quite precisely because it is a relatively long-lived particle.

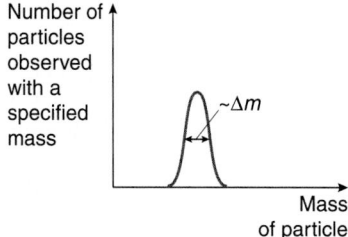

FIGURE 27.5 Short-lived particles have noticeable uncertainty in their masses.

Solving for Δt, we find

$$\Delta t = \frac{h}{\Delta E}$$
$$= \frac{h}{2mc^2} \qquad (27.20)$$

The larger the mass of the virtual particle–antiparticle pair, the shorter the time interval during which the virtual particles exist as elements of the vacuum. For example, a virtual electron–positron pair, the least massive particle–antiparticle pair,[†] can exist only for a time Δt that is at most

$$\Delta t = \frac{6.626 \times 10^{-34} \text{ J} \cdot \text{s}}{2(9.11 \times 10^{-31} \text{ kg})(3.00 \times 10^8 \text{ m/s})^2}$$
$$= 4.04 \times 10^{-21} \text{ s}$$

This is a very short time indeed.

The vacuum process of particle–antiparticle creation and annihilation does not violate the principle of energy conservation. The process is analogous to kiting a check. You can borrow or create money from your empty checking account (a vacuum) by writing a check. As long as you get the money into the account before the check clears (annihilation) there is no problem (although kiting is technically illegal and banks certainly do not like you to do it!). In the quantum vacuum, you can borrow or create an amount of energy ΔE to create a virtual particle–antiparticle pair as long as the energy loan is repaid (by particle annihilation) before a time Δt elapses given by the minimum of the Heisenberg energy–time uncertainty principle: $\Delta E \, \Delta t = h$.

---

*Neutrons not bound inside a nucleus are called free neutrons. Free neutrons are radioactive and decay into a proton, an electron, and an electron antineutrino.

[†]Neutrinos now are thought to have a small nonzero mass, much smaller than that of the electron or positron.

The more massive the virtual particle–antiparticle pair, the shorter the duration of the energy loan. Thus virtual electron–positron pairs can last longer than virtual proton–antiproton pairs.

You might legitimately ask how in heaven such exotic aspects of a vacuum could be verified if the processes occur behind the well-shrouded veil of the uncertainty principle. Suffice it to say that there are experiments that only can be explained via an appeal to such an exotic interpretation of the vacuum. Among these experiments is the **Lamb shift** in the spectrum of hydrogen (for which Willis Lamb Jr. [1913–] won the 1955 Nobel prize in physics.

Near the positively charged nucleus, virtual electron and positron pairs generated from the vacuum ceaselessly are appearing momentarily and disappearing.* Thanks to the positive nucleus, the negative virtual electrons spend more time near the nucleus than the positively charged virtual positrons. This results in a slight or partial shielding (neutralization) of the nuclear charge by a cloak of virtual electrons. Thus the real electron in the ground state feels the electrostatic effect of a nuclear charge that is slightly *less* than $+Ze$. The shielding causes a slight shift in the energy levels and the resulting spectrum of light emitted or absorbed by hydrogen. The shift in the ground state energy is the Lamb shift; it is quite real. The Lamb shift is said to be attributed to a **vacuum polarization**.

EXAMPLE 27.3

For an electron–positron pair created from the vacuum, what is the maximum distance the pair can travel before they annihilate each other?

Solution
You saw in this section that an electron–positron pair can exist for at most 4.04×10^{-21} s. Since the upper limit on their speed is the speed of light, they can travel at most a distance

$$c \, \Delta t \approx (3.00 \times 10^8 \text{ m/s})(4.04 \times 10^{-21} \text{ s})$$
$$\approx 1.21 \times 10^{-12} \text{ m}$$

This distance is about a factor of 10^2 to 10^3 greater than the size of the nucleus, which is what permits vacuum polarization.

EXAMPLE 27.4

The uncertainty in the rest energy of some of the most short-lived elementary particles is estimated to be about 3×10^2 MeV. What is the approximate mean lifetime of such particles?

Solution
The energy–time uncertainty principle, Equation 27.14, states that

$$\Delta E \, \Delta t \geq h$$

*Since virtual electron–positron pairs can last longer than virtual proton–antiproton pairs, the effect of the former is more prevalent.

The uncertainty in the rest energy is 3×10^2 MeV. Hence the mean lifetime is about

$$\Delta t \approx \frac{h}{\Delta E}$$
$$\approx \frac{6.626 \times 10^{-34} \text{ J·s}}{(3 \times 10^8 \text{ eV})(1.602 \times 10^{-19} \text{ J/eV})}$$
$$\approx 1 \times 10^{-23} \text{ s}$$

These are among the shortest-lived particles known.

EXAMPLE 27.5

Light of wavelength 632.8 nm is incident on an extremely fast shutter that chops the beam into pulses. The shutter stays open for only 1.5×10^{-9} s. What is the approximate minimum range of wavelengths $\Delta\lambda$ in the light pulses that pass the shutter?

Solution
The energy–time uncertainty principle states that

$$\Delta E \, \Delta t \geq h$$

For the minimum uncertainty, you have

$$\Delta E \, \Delta t \approx h$$

Since $\Delta t \approx 1.5 \times 10^{-9}$ s, the uncertainty in the energy is

$$\Delta E \approx \frac{h}{\Delta t}$$
$$\approx \frac{6.626 \times 10^{-34} \text{ J·s}}{1.5 \times 10^{-9} \text{ s}}$$
$$\approx 4.4 \times 10^{-25} \text{ J}$$

The energy of light is proportional to its frequency:

$$E = h\nu$$
$$= \frac{hc}{\lambda}$$

To see how E varies with a change in wavelength, take differentials of this expression, approximating the differentials with small changes ΔE and $\Delta\lambda$. You can ignore the minus sign resulting from the differentiation, since only the absolute values are of interest:

$$\Delta E = \frac{hc}{\lambda^2} \Delta\lambda$$

You know ΔE, h, c, and λ, and so you can find $\Delta\lambda$:

$$\Delta\lambda \approx \frac{\Delta E \, \lambda^2}{hc}$$
$$\approx \frac{(4.4 \times 10^{-25} \text{ J})(632.8 \times 10^{-9} \text{ m})^2}{(6.626 \times 10^{-34} \text{ J·s})(3.00 \times 10^8 \text{ m/s})}$$
$$\approx 8.9 \times 10^{-13} \text{ m}$$

The fractional spread in wavelengths, $\Delta\lambda/\lambda$, is quite small but is nonetheless very measurable:

$$\frac{\Delta\lambda}{\lambda} = \frac{8.9 \times 10^{-13} \text{ m}}{632.8 \times 10^{-9} \text{ m}}$$
$$= 1.4 \times 10^{-6}$$

27.4 OBSERVATION AND MEASUREMENT

Physics ultimately must "preserve the phenomena," as the ancient Greek natural philosophers attempted to do with their imaginative models of the night sky and solar system. Physics is based on experiments, which, of course, involve observation and measurement.

In the classical domain, such observations and measurements are conceptually easy, at least in principle. For example, to measure the speed of a moving object we simply look at the object and record its changing position as a function of time. The act of looking does not disturb the object under examination; or, even if the measurement did disturb the system, we could account for the disturbance with classical physical laws and techniques to find out what was happening before our measurement.

In other words, in classical physics, the act of measurement either has a negligible effect on the physical system under observation, or can be accounted for, at least in principle, in predictable ways.

We already have encountered several instances where we had to be careful about the act of measurement in classical physics. Specifically:

1. When we studied thermodynamics and the measurement of temperature with thermometers, we mentioned that caution is needed to be sure that the mass of a thermometer is insignificant with respect to the mass of the system whose temperature is to be determined. If the mass of a thermometer is too large compared with the mass of the system, the act of measurement affects the temperature of the system to be measured because of the energy exchange between the system and the thermometer. (Of course, if the system is in contact with a reservoir, this condition is not necessary.) Classical physics can, at least in principle, account for these effects.

2. When analyzing electrical forces, it is necessary to be careful that the charge q placed in an electrical field is not so large that its presence affects the magnitude or direction of the field in which the charge is placed. Nonetheless, classical physics can, in principle, account for the effect of q on the other charges creating the field in which q is placed.

3. In electrical circuits, when measuring potential differences, it is necessary to use voltmeters with a high internal resistance so as not to affect the potential differences in the circuit under study. The same is true with ammeters: an ammeter with low effective resistance is needed in order to minimize

the effects of the meter on the current to be measured. Again, it is possible in classical physics to account completely for the effect of a measurement or a disturbance on a physical system.

In the quantum physics of very small systems, the act of measurement or observation inescapably plays a significant role. Even simply looking is no longer innocent or trivial. The mere act of inspecting something means that light must be scattered or reflected from it. A reflected or scattered photon is no big deal to a baseball or to a tasty M&M. But to an electron, an atom, or other small particle, the interaction with even a single photon is tantamount to a sledgehammer. Any measurement on a physical system always involves an interaction of some kind between the system and an apparatus.

Furthermore, the act of measurement in the quantum domain inevitably changes the system in ways that cannot be accounted for in a deterministic sense, *even in principle.*

The system changes in a random and unpredictable way. For example, in Section 27.1, we tried to localize the particle-waves by passing them through a slit aperture. But the greater the extent to which we try to localize the particles with a smaller and smaller aperture, the greater the width of the resulting diffraction pattern and the uncertainty associated with its transverse momentum. Thus the uncertainty principle limits our ability to make measurements of physical observables with infinite precision. The role of an observer in quantum physics is vastly more complex and bumbling than in classical physics. It differs fundamentally in that it is impossible, in principle, to account for the changes in a system induced by the act of measurement.

27.5 PARTICLE-WAVES AND THE WAVEFUNCTION

All Nature is but Art unknown to thee;
All chance, direction, which thou canst not see;
All discord, harmony not understood;
All partial evil, universal good;
And spite of Pride, in erring Reason's spite,
One truth is clear, Whatever is, is right.

Alexander Pope (1688–1744)*

We saw in Chapter 26 that particles have aspects of waves (the electron diffraction experiments of Davisson and Germer, and Thomson) and waves have aspects of particles (the Compton effect and Einstein's explanation of the photoelectric effect). When we say that photons, electrons, protons, neutrons, and so forth behave like *particles*, we mean that they have physical properties (which may be zero), such as mass, charge, and momentum, all reminiscent of what we imagine when we think of the very word particle. On the other hand, the behavior of the particles is determined by a wavelike function whose nature we have yet to delineate. That time has come.

---

*An Essay on Man, Epistle I, lines 289–294, The Complete Poetical Works of Alexander Pope, edited by Henry W. Boynton (Houghton Mifflin, Boston, 1903), page 141.

Here we try to address some of the questions we raised at the conclusion of the last chapter. Do the particle-waves represent anything in a physical context? The superluminal speeds of the particle-waves (see Equation 26.84) give us pause for thought. How are the particle-waves to be represented in a mathematical context by a wavefunction $\Psi(x, t)$? What equation governs the propagation of the particle-waves? Is it the classical wave equation? This last aspect is addressed in the next section.

The wave aspects of matter are shown by diffraction and interference experiments. In order for these effects to stand out, the particles must interact with objects whose size is comparable to the de Broglie wavelength of the particles. What happens in such experiments? What is observed in particle diffraction and interference experiments?

Imagine a horde of identical particles, each with the same momentum (that is, with the same de Broglie wavelength), incident on an appropriately sized single slit aperture, as in Figure 27.1. The shape of the pattern that develops on the screen or detector is identical to the single slit diffraction pattern of a classical wave. The pattern indicates the number of particles detected at any given location along the screen. In other words, even if the particles are sent through the slit *one at a time*, each particle goes somewhere on the screen, and their distribution (the diffraction pattern) builds up over time.

The diffraction pattern is a direct measure of the number of particles arriving at various locations on the screen in a given time; it is a number distribution of particles. The pattern, therefore, is an indication of the likelihood or the *probability* that a particle will arrive at each location on the screen. We *cannot predict* where an individual particle will go when it traverses the slit; we can only assess the odds or the probability that it will arrive at a particular location. We *cannot* say that the wave used to predict the pattern on the screen indeed *is* the particle; the wave model determines the shape of the pattern of particles, or equivalently, the pattern displays the relative probability that a single particle will arrive at any particular location on the screen.

The situation becomes even more perplexing if a double slit is used. We might naively anticipate that since an individual particle goes through either one slit or the other, the pattern observed will be simply two slightly separated single slit diffraction patterns as in Figure 27.6. This is not the case.

The experimental results indicate that, just as in the classical optical double slit experiments discussed in Chapter 24, a double slit *interference* pattern results on the distant screen (or detector), as in Figure 27.7. Even if the particles are sent individually (one at a time) through the double slit arrangement, the double slit pattern emerges gradually as the particles accumulate at various positions across the screen, as in Figure 27.8. The same effect is seen with either electrons or photons.

This defies logic and common sense, because it taxes our conceptual notions about the word particle. We get one pattern on a distant screen if the particles go through a single slit, but a completely different pattern if the particles go through two or more closely spaced slits. Thus we cannot think of the particles as going through either one slit or the other in the double slit arrangement. Why not? If each particle went through one slit or the other, we would logically expect two single slit patterns as if from two incoherent sources, not an interference pattern. However, each particle individually somehow senses *both* slits.

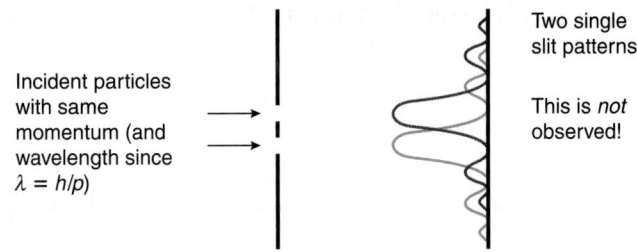

FIGURE 27.6 Two single slit patterns of particles are *not* observed.

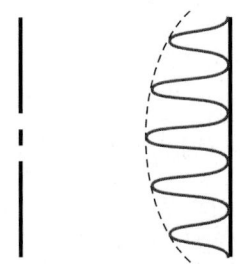

FIGURE 27.7 A double slit pattern of particles *is* observed.

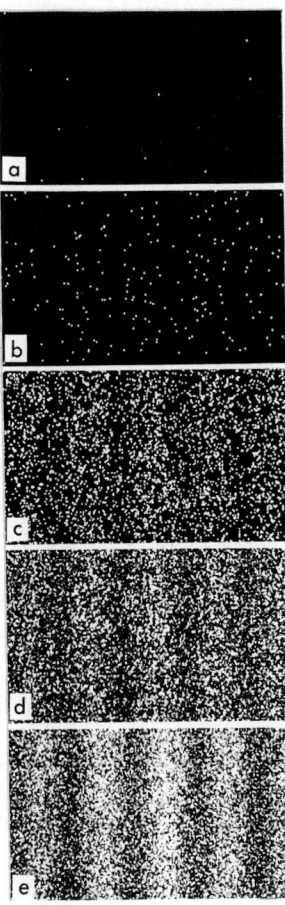

FIGURE 27.8 A double slit pattern gradually emerges as more and more particles are sent through the double slit.

Thus the idea of a particle as a discrete billiard ball entity must be abandoned in favor of some abstract particle-wave which can diffract and interfere with itself.

What characteristics does a wavefunction Ψ representing the particle-wave have? Several come to mind:

1. In order to represent the particle-wave, the amplitude of the wavefunction must be relatively large in regions where the particle is likely to be found and, conversely, of small amplitude in regions where the particle is not likely to be located.

2. The wavefunction must be able to interfere with itself, since a single or individual particle-wave somehow is able to sense the presence of either a single or double slit and know what to do in order to reproduce statistically the single slit diffraction pattern or the double slit interference pattern, as the case may warrant.

3. Since a single particle-wave is capable of sensing the presence of both slits in a double slit experiment, the wavefunction must represent the behavior of single particle-waves and not the statistical distribution of an ensemble of identical particle-waves.

4. For the wavefunction to truly represent all features of the particle-wave, we must be able (somehow) to obtain from the wavefunction Ψ *all* physical characteristics of the particle-wave such as its momentum, energy, and other physical parameters that are observable. In other words, by doing something to the wavefunction, by mathematically operating on it in some way, we ought to be able to extract the ordinary physical parameters or observables that we associate with the particle-wave.

The wavefunction used to predict the pattern on the screen must have something to do with the probability that a particle arrives at any location. The wavefunction itself *cannot* be the probability since, by their very nature, probabilities are intrinsically greater than or equal to zero; while waves are oscillatory in nature and have both positive and negative values.

We consider a special case. What kind of wavefunction can represent an individual particle in free space, traveling in one dimension, say, toward increasing values of x, with a well-defined (i.e., precise) momentum (or equivalently, a well-defined wavelength)?* We ignore any descriptions in the y and z directions. Intuitively, we might think that the monochromatic wavefunction we have used on many occasions in the past could do this:

$$\Psi(x, t) = A \cos(kx - \omega t) \qquad (27.21)$$

Since this function can take on both positive and negative values, perhaps the square of Ψ is the probability of finding a particle at any given place x at time t? Alas, this cannot be the case. Why not? Consider the wavefunction of Equation 27.21 when $t = 0$ s. Then the square of Ψ is

$$\Psi^2 = A^2 \cos^2(kx)$$

which is indeed everywhere positive (see Figure 27.9) but has other unacceptable problems, such as unreal periodic bunchings along the line of motion.

*The Heisenberg uncertainty principle then implies that the wave is everywhere, since $\Delta x = \infty$ m.

For such a particle unconstrained in x, it would be equally likely to find it at *any* position x, since the wave is defined for all values of x. In other words, the probability of finding this particle at *any* position x ought to be independent of x, as in Figure 27.10. But the square of Equation 27.21, depicted in Figure 27.9, certainly is not independent of x. How can we avoid this problem?

Recall that a sinusoidal oscillation also can be represented by a *complex* exponential expression via the Euler identity[†]:

$$\Psi(x, t) = A e^{i(kx - \omega t)} \qquad (27.22)$$
$$= A\,[\cos(kx - \omega t) + i \sin(kx - \omega t)]$$

The squared magnitude of Ψ, found by multiplying Ψ by its complex conjugate Ψ^*, then would give us just what we want: a probability that is independent of x and t, as in Figure 27.10:

$$\Psi^* \Psi = A^* e^{-i(kx - \omega t)} A e^{i(kx - \omega t)}$$
$$= A^* A$$

Although this argument involves a very specific example (a free particle with a well-defined wavelength), there are several important points to note here about the nature of *any* quantum mechanical wavefunction Ψ.

> The wavefunction $\Psi(x, t)$, whatever mathematical form it takes in a particular context, typically is a complex-valued function.

[†]We did something similar when considering ac circuits in Chapter 22, where the real part of the complex exponential represented the real potential difference or current and the imaginary part came along for a mathematical ride. The complex potential differences and currents were introduced simply to make calculations easier, since exponentials are easier to manipulate mathematically than trigonometric functions.

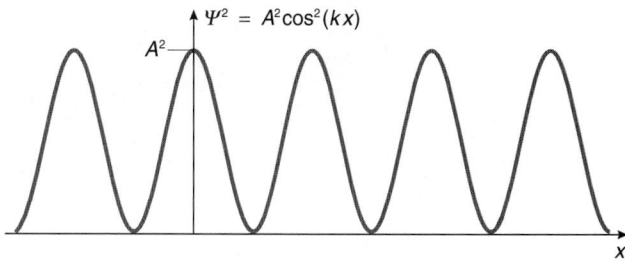

FIGURE 27.9 Graph of the function $A^2 \cos^2(kx)$ versus x.

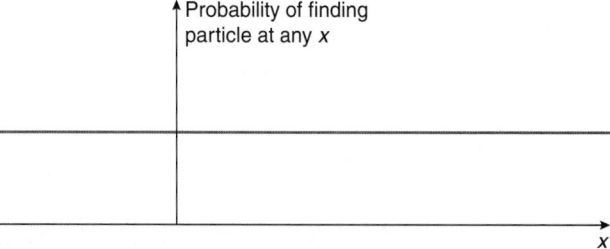

FIGURE 27.10 A free particle is equally likely to be found anywhere.

Since the wavefunction $\Psi(x, t)$ is a complex-valued function, $\Psi(x, t)$ itself is *not* a physical quantity that can be observed directly in any experiment; measurements in experiments always yield *real* numbers.

The quantity $\Psi^*(x, t)\Psi(x, t)\,dx$ is usefully defined as the probability of finding the particle in a region of space between x and $x + dx$ at a particular time t. For this reason, the wavefunction Ψ itself also is known as a **probability amplitude**.

The complex wavefunction for a particle with a well-defined wavelength, Equation 27.22,

$$\Psi(x, t) = Ae^{i(kx - \omega t)}$$

provides clues about the nature of the equation that governs the propagation of the particle-waves. If the particle is traveling in a region where its potential energy is constant, for convenience we can choose the constant potential energy to be zero, as we have many times in the past, because the place where the potential energy is zero is an arbitrary choice. With this choice, the total, nonrelativistic energy E of the particle is just its kinetic energy:

$$E = \frac{1}{2}mv^2$$
$$= \frac{p^2}{2m} \qquad (27.23)$$

The particle energy E is related to its frequency by

$$E = h\nu$$
$$= \hbar\omega \qquad (27.24)$$

The momentum is related to the wavelength via the de Broglie relation

$$p = \frac{h}{\lambda}$$

or in terms of k, since $k = 2\pi/\lambda$,

$$p = \hbar k \qquad (27.25)$$

Thus Equation 27.23 for the energy of the free particle can be put in the form

$$\hbar\omega = \frac{1}{2m}\hbar^2 k^2 \qquad (27.26)$$

To obtain this same relationship by suitably differentiating the wavefunction given by Equation 27.22, taking the *first time derivative* of Ψ will yield a term in ω (like the left-hand side of Equation 27.26), and the *second spatial derivative* of Ψ will yield a term in k^2 (like the right-hand side of Equation 27.26). Thus it is likely that the differential equation governing the propagation of the particle-waves relates the *first time derivative* to the *second spatial derivative* of the wavefunction. The implications of this are profound.

A new wave equation is needed to describe the propagation of the wavefunction representing the particle-wave.

Our old familiar classical wave equation relates the *second time derivative* to the second spatial derivative of the wavefunction:

$$\frac{\partial^2\Psi}{\partial x^2} - \frac{1}{v^2}\frac{\partial^2\Psi}{\partial t^2} = 0 \qquad \text{(classical wave equation)} \qquad (27.27)$$

The new wave equation for the particle-wave function is the called the **Schrödinger equation**, after Erwin Schrödinger (1887–1961), who first deduced the equation in the mid-1920s while on a romantic holiday in the Alps with a paramour. In Section 27.7 we develop the specific form of the Schrödinger equation.

Notice that in the quantum domain we lose the ability to predict with certainty what happens to individual particles such as electrons. We encountered this peculiar aspect of quantum mechanical phenomena previously in our study of radioactive decay: the statistical behavior of a large number of radioactive particles can be ascertained, but we cannot predict when an individual particle will decay.

The determinism of classical physics must be abandoned in the quantum world. Does this imply that everything we learned in classical physics was for naught? Fortunately, this is surely not the case.* The situation is analogous to what we encountered in relativity: relativity is a theory of space and time that, in the limit of speeds small compared with the speed of light, reduces to classical Galilean relativity. No one uses relativity to examine the flight of a football; we use the Newtonian representation because it gives the correct results (and so would a relativistic calculation, but that is much harder!).

In quantum mechanics, large masses usually mean small de Broglie wavelengths (see Example 26.10); as long as the wavelengths of the particles are small compared with other lengths involved in a particular situation, our classical physics is typically quite appropriate, and can be used with confidence and correctness. The de Broglie wavelength of a macroscopic object is many

---

*I can imagine the reaction if you were told, after nearly a full year of hard work, that everything you learned was "wrong"! Arg!

Schrödinger was an Austrian physicist who loved Vienna. His friends at Oxford University helped him to escape the Nazis after the Anschluss. He spent 18 years in Dublin, Ireland, before returning to Vienna in 1956 to a welcoming country and government.

orders of magnitude smaller than even the size of a typical nucleus of an atom; the wavelike behavior of such a particle never is manifest because its wavelength is so small compared with anything interacting with it. No one (in his or her right mind, at any rate) would use quantum mechanics to calculate the motion of a planet in orbit around the Sun.

On the other hand, for an electron in an atom or other atomic, nuclear, or submicroscopic phenomena, the de Broglie wavelength of the particle under consideration typically is comparable to the size of the system itself; then the wavelike aspects are really central to an analysis of the situation, and quantum mechanics is unavoidable for making predictions that agree with experiments.

27.6 OPERATORS*

Perhaps without realizing it, you have been using mathematical operators for some time. A mathematical operator† is a symbolic way of representing a mathematical operation. For example, various mathematical operations can be performed on the function $\Psi(x, t)$. These might include the following:

1. Multiplying the function by a scalar. For example, we could multiply the function $\Psi(x, t)$ by x to get

$$x\Psi(x, t)$$

We symbolically represent the multiplication simply by placing x next to $\Psi(x, t)$. No formal symbol is used.‡

2. Taking derivatives of the function. For example, when we write

$$\frac{\partial \Psi}{\partial x}$$

what we mean is to take the partial derivative§ of $\Psi(x, t)$ with respect to x. The operation is symbolized by

$$\frac{\partial}{\partial x}$$

and so this is called a differential operator. Partial differentiation with respect to time is represented by the differential operator

$$\frac{\partial}{\partial t}$$

There are many other types of mathematical operators as well, but we need not be concerned with them in this brief introduction to quantum mechanics.

It is important to realize that the order in which mathematical operators are written and are performed can be important. For

---

†We do not mean your math or physics professors.

‡We did use formal symbols to represent particular types of multiplication associated with vectors: the familiar scalar product by • and the vector product by ✕.

§Recall that to differentiate Ψ partially with respect to x, we treat t as a constant and use the familiar differentiation rules. Or to differentiate Ψ partially with respect to t, we treat x as a constant and apply the usual differentiation rules.

---

example, the results obtained in the following examples of sequential operations are quite different:

a. First multiply Ψ by x and then take the derivative with respect to x. This is symbolized in the following way:

$$\frac{\partial}{\partial x}(x\Psi) = \Psi + x\frac{\partial \Psi}{\partial x} \tag{27.28}$$

b. Reversing the order, first take the derivative of Ψ and then multiply by x. The result is symbolized by

$$x\frac{\partial \Psi}{\partial x}$$

which is quite different from Equation 27.28.

We say, then, that the operators "multiply by x" and "take the derivative with respect to x" do not commute since the order in which they are executed affects the result.

On the other hand, certain operators do commute: the order in which they are performed is immaterial. For example, the differential operators

$$\frac{\partial}{\partial x} \quad \text{and} \quad \frac{\partial}{\partial t}$$

commute since

$$\frac{\partial^2 \Psi}{\partial x\,\partial t} = \frac{\partial^2 \Psi}{\partial t\,\partial x}$$

Let's play with the wavefunction that represents a monochromatic (fixed wavelength) free particle to see what operators on the wavefunction yield the momentum and energy of the particle. The free particle wavefunction is given by Equation 27.22:

$$\Psi(x, t) = Ae^{i(kx - \omega t)}$$

The Momentum Operator

We operate on Ψ with the differential operator

$$\frac{\partial}{\partial x}$$

We find

$$\begin{aligned}
\frac{\partial}{\partial x}\Psi(x, t) &= \frac{\partial}{\partial x}[Ae^{i(kx-\omega t)}]\\
&= ikAe^{i(kx-\omega t)}\\
&= ik\Psi
\end{aligned} \tag{27.29}$$

Then using the definition of k,

$$k = \frac{2\pi}{\lambda}$$

and the de Broglie relation for the momentum, we get

$$\begin{aligned}
k &= \frac{2\pi}{h}p\\
&= \frac{p}{\hbar}
\end{aligned}$$

Substituting this into Equation 27.29, we obtain

$$\frac{\partial \Psi}{\partial x} = i \frac{p}{\hbar} \Psi$$

Rearranging things slightly, we find

$$\frac{\hbar}{i} \frac{\partial \Psi}{\partial x} = p\Psi$$

We rationalize the complex number on the left-hand side of the equation by multiplying the numerator and denominator by i. This yields

$$-i\hbar \frac{\partial \Psi}{\partial x} = p\Psi \qquad (27.30)$$

Hence the differential operator

$$-i\hbar \frac{\partial}{\partial x}$$

is what extracts the x-component of the momentum from the wavefunction. In quantum mechanics, then, we call the operator

$$-i\hbar \frac{\partial}{\partial x} \qquad (27.31)$$

the (one-dimensional) **momentum operator**.

Although we used a very specific wavefunction to obtain Equation 27.31, the (one-dimensional) momentum operator in quantum mechanics always has this form.

The Energy Operator

We use the differential operator

$$\frac{\partial}{\partial t}$$

on the free particle wavefunction

$$\Psi(x, t) = Ae^{i(kx - \omega t)}$$

We obtain

$$\begin{aligned}\frac{\partial \Psi}{\partial t} &= \frac{\partial}{\partial t}[Ae^{i(kx-\omega t)}] \\ &= -i\omega Ae^{i(kx-\omega t)} \\ &= -i\omega \Psi \end{aligned} \qquad (27.32)$$

Since the particle has an energy $E = h\nu = h(\omega/2\pi) = \hbar\omega$, Equation 27.32 can be rewritten by substituting for ω:

$$\frac{\partial \Psi}{\partial t} = -i \frac{E}{\hbar} \Psi$$

Rearranging this slightly, we obtain

$$-\frac{\hbar}{i} \frac{\partial \Psi}{\partial t} = E\Psi$$

We rationalize the left-hand side of this equation, finding

$$i\hbar \frac{\partial \Psi}{\partial t} = E\Psi \qquad (27.33)$$

Hence the operator

$$i\hbar \frac{\partial}{\partial t} \qquad (27.34)$$

is what extracts the total energy of the particle from the wavefunction, and so is called the **energy operator** in quantum mechanics.

Notice from Equation 27.30 that when the momentum operator acts on the wavefunction, the result is the momentum component times the wavefunction itself. Likewise, when the energy operator acts on the wavefunction, the result is the energy times the wavefunction itself (Equation 27.33). When a mathematical operator acts on a function and simply produces a scalar multiple of the function itself, the relationship is called an **eigenvalue equation**.* An eigenvalue equation is represented symbolically in the following way:

$$\text{operator} \mid \text{function} \rangle = \text{scalar} \mid \text{function} \rangle$$

The fancy brackets $\mid \ \rangle$ around the function serve to set the function apart from the operator and the scalar eigenvalue, and form the basis of a useful and ubiquitous shorthand notation invented for quantum mechanics by Paul Dirac (1902–1984); it is called **Dirac notation**.

---

*The word *eigen* (pronounced eye-gen) is German and has several meanings: *special*, *characteristic*, or *specific*.

Dirac's contributions to quantum mechanics were honored with the 1933 Nobel prize in physics, which he shared with Erwin Schrödinger. Dirac was the first person to foretell the existence of antimatter particles such as the positron and antiproton. Dirac was honored with a plaque in 1995, set next to the tomb of Newton in Westminster Abbey.

The scalar is called an **eigenvalue** and the function is called an **eigenfunction** of the mathematical operator. Eigenvalue equations indicate the way we extract physically meaningful information from the complex-valued wavefunction Ψ of the particle-wave.

> Various mathematical operators correspond to physically observable properties, such as momentum and energy. The operators act on the wavefunction, and the resulting eigenvalues are the appropriate values of that physical observable.[†] For the eigenvalues to represent physically observable properties, the eigenvalues of the operators in quantum mechanics must be real numbers.

From a purely mathematical viewpoint, the real nature of the eigenvalues places restrictions on the type of mathematical operators that are associated with observable quantities in quantum mechanics.[‡][§]

> With the operator formalism, we say that the wavefunction contains *all* the information about the physical system. This explains why such a premium is placed in quantum mechanics on discovering the form for the wavefunction in a particular physical context; the wavefunction Ψ says it all. Discovering what the wavefunction *is*, is the problem we face in quantum mechanics; for many systems, it is not an easy undertaking to find Ψ.

27.7 THE SCHRÖDINGER EQUATION*

If you came this way,
Taking any route, starting from anywhere,
At any time or at any season,
It would always be the same: you would have to put off
Sense and notion.

T. S. Eliot (1888–1965)[#]

When de Broglie first surmised that particles might have wavelike properties, the search for the physical nature of the waves as well as an equation that all such particle-waves

satisfied began in earnest. The procedure for finding such an equation was by no means obvious then, nor is it really clear even in hindsight.

The best that can be done is to surmise an equation for the particle-waves based on what we know of them (which originally was quite little; see Section 27.5) and then test the equation and its predictions in situations that can be compared with experiment. Just as there is little value in pulling Newton's second law out of thin air, there is little value in doing the same with the Schrödinger equation. Perforce, our development of the Schrödinger equation is not rigorous, but it should serve to make it somewhat plausible to you.

Two of the pillars on which quantum mechanics is built are the de Broglie relation between the wavelength and momentum of a particle-wave,

$$\lambda = \frac{h}{p} \tag{27.35}$$

and the relationship between the energy of the particle-wave and its frequency (generalized from Einstein's treatment of the photon particle-wave of light),

$$E = h\nu = \hbar\omega \tag{27.36}$$

Both of these relationships have roots in relativity.

Classical mechanics is dichotomous: there is relativistic mechanics and nonrelativistic mechanics. The latter is a special case of the former, appropriate when the speeds of the particles are much less than the speed of light. Quantum mechanics has a similar dichotomy depending on whether the speeds of the particles are nearly equal to or small compared with the speed of light. The fundamental equation of nonrelativistic quantum mechanics is the Schrödinger equation; the corresponding equation in relativistic quantum mechanics is called the Klein–Gordon equation.[¶] For reasons of simplicity and time, we will examine only the nonrelativistic scenario (the Schrödinger equation).

The distinction between the relativistic and nonrelativistic regimes is needed because the expressions for the total energy and momentum of a particle are quite different in the two domains. For the nonrelativistic limit, the rest energy of the particle is ignored** and the classical nonrelativistic expression for the kinetic energy of the particle is used:

$$\text{KE} = \frac{1}{2}mv^2 = \frac{p^2}{2m} \tag{27.37}$$

Then total energy of a particle is the sum of the kinetic energy and any potential energies appropriate to the problem. Until now, for clarity we have always used PE as the symbolic representation for the potential energy. Alas, this notation now must be abandoned to conform with standard notation in quantum mechanics where V commonly is used for the potential energy of

---

[†]We have only demonstrated this feature with a particular wavefunction, but it is a general feature of quantum mechanics.

[‡]Operators that have real eigenvalues are known as *Hermitian operators* in mathematics.

[§]In an allegorical and sarcastic sense, one of the most useful "operators" known is the *clearly* operator: it has one eigenvalue—the *result*. Thus

clearly | anything⟩ = result | anything⟩

You have seen this "operator" in action many times, in many texts (likely including this one). It typically appears in the following way. At the beginning or in the middle of some complicated calculation, an author will invoke the *clearly* operator, and somehow the *result* magically appears on the very next line of text. Other variations of the *clearly* operator are known as well: "The reader can readily show that ..." or "Obviously, ..." Clearly, it is best *not* to take the contents of this footnote too seriously!

[#]*Four Quartets*, "Little Gidding," lines 39–43 (Harcourt Brace Jovanovich, San Diego, California, 1971) page 50.

[¶]For so-called spin 1/2 particles, the relativistic equation is known as the Dirac equation.

**In the classical limit, the rest energy can be considered constant. It is only changes in energy that are physically significant, so that we can legitimately ignore the rest energy.

a particle.*† Any potential energies involved in a problem are still, of course, related to the presence of conservative forces acting on the particle.

For the nonrelativistic situation, the total energy E of a particle is

$$E = \frac{p^2}{2m} + V(x, t) \qquad (27.38)$$

For simplicity we consider only one-dimensional systems, so the potential energies V and wavefunctions Ψ are functions of only one spatial coordinate (x) and the time (t).

If we take the classical wave equation (Equation 27.27) as a paradigm, we expect the Schrödinger equation to involve the partial derivatives of the wavefunction $\Psi(x, t)$ with respect to x and t. We conjectured in the last section that the equation will involve the first partial time derivative and the second spatial derivative of Ψ.

We assume the Schrödinger equation must be a linear equation. The assumption of linearity ensures that if $\Psi_1(x, t)$ and $\Psi_2(x, t)$ are both solutions to the equation, then the sum $\Psi_1(x, t) + \Psi_2(x, t)$ also is a solution. Thus the principle of superposition for waves is retained and follows from the assumption of linearity. The principle of superposition is necessary to account for the experimentally observed interference of particle-waves, such as in the electron diffraction experiments of Davisson and Germer. To be linear, the Schrödinger equation must involve only the first powers of Ψ, and its partial derivatives. That is, terms such as Ψ^2 or squares of the derivatives of Ψ do not appear in the equation.

The Schrödinger equation for particle-waves can be found from Equation 27.38 for the total energy, here rewritten as

$$\frac{p^2}{2m} + V(x, t) = E \qquad (27.39)$$

We multiply this expression by the wavefunction $\Psi(x, t)$:

$$\left[\frac{p^2}{2m} + V(x, t) \right] \Psi(x, t) = E\Psi(x, t)$$

---

*Do not confuse the potential energy V with the electrical potential, which also is represented by V, as you know. Once again, the problem of suitable notation rears its ugly head. To eliminate the potential (pun!) for this confusion early in your study of physics is the reason we have, until now, used PE for potential energy rather than V.

†A famous professor (said to be Julian Schwinger), having covered the blackboard with ambiguous letters meaning different things, once responded: "Oh, he could use the same symbol for everything, provided we know what it means!" Here, we are trying to be honest with our notational ambiguities!

Now we replace the momentum p and energy E with their differential operators, which we found in Section 27.6 (Equations 27.31 and 27.34):

$$p \rightarrow -i\hbar \frac{\partial}{\partial x} \qquad \text{and} \qquad E \rightarrow i\hbar \frac{\partial}{\partial t}$$

We obtain

$$\left[\frac{1}{2m} \left(-i\hbar \frac{\partial}{\partial x} \right) \left(-i\hbar \frac{\partial}{\partial x} \right) + V(x, t) \right] \Psi(x, t) = i\hbar \frac{\partial \Psi(x, t)}{\partial t}$$

This can be simplified to the following:

$$-\frac{\hbar^2}{2m} \frac{\partial^2 \Psi(x, t)}{\partial x^2} + V(x, t)\Psi(x, t) = i\hbar \frac{\partial \Psi(x, t)}{\partial t} \qquad (27.40)$$

Equation 27.40 is the one-dimensional Schrödinger equation for the wavefunction Ψ.

The mathematical operator formed from the expression for the total energy,

$$\frac{p^2}{2m} + V \qquad \text{which is} \qquad -\frac{\hbar^2}{2m} \frac{\partial^2}{\partial x^2} + V$$

is called the **Hamiltonian** H of the system.‡ Using the Hamiltonian operator, the (one-dimensional) Schrödinger equation is written in a less intimidating and more succinct form as

$$\boxed{H\Psi = i\hbar \frac{\partial \Psi}{\partial t}} \qquad (27.41)$$

Just as Newton's second law encompasses classical nonrelativistic mechanics, the Schrödinger equation is the touchstone of nonrelativistic quantum mechanics. In classical mechanics, a knowledge of the forces acting on a system permitted us to find interesting information about the acceleration, velocity, position, energy, and momentum of the system. The key was our ability to identify and describe the forces acting on the system. In quantum mechanics, the wavefunction is the corresponding key to information about the physical system.

---

‡The term is named for William Hamilton (1805–1865), who during the 19th century developed a formal theory of classical dynamics that showed how to obtain the equations of motion of a system from an expression for its total energy.

CHAPTER SUMMARY

The wavelength λ of a particle-wave is found from the de Broglie relationship:

$$\lambda = \frac{h}{|p_x|} \qquad (27.1)$$

The energy E of a particle-wave, like that of a photon, is proportional to its frequency v, or its angular frequency ω:

$$E = hv = \hbar\omega \qquad (27.9)$$

Particles have wavelike properties and waves have particle-like properties. The *wave–particle duality* naturally leads to two *Heisenberg uncertainty principles*. The uncertainty in the position x and its complementary momentum component p_x satisfy the following inequality:

$$\Delta x \, \Delta p_x \geq h \qquad (27.6)$$

If the wave-particle becomes more localized, Δx decreases and there is an increase in the uncertainty associated with the complementary momentum component. The uncertainty principle fundamentally limits our ability to determine simultaneously the position and corresponding momentum component of a particle-wave. Likewise, if the energy of the particle-wave is determined during a time interval whose uncertainty is Δt, the energy has a corresponding uncertainty ΔE, where

$$\Delta E \, \Delta t \geq h \qquad (27.14)$$

Diffraction and interference of particle-waves are statistical processes. If the particles are sent through a single slit or multiple slits, it is impossible, even in principle, to predict where an individual particle will go. However, the single slit or multiple slit pattern gradually appears as more and more particles are sent through the arrangement. The diffraction or interference pattern that results therefore represents the *relative probability* that a particle is detected at a given position.

Observation and measurement of a system in quantum mechanics is fundamentally different from that in classical physics. In classical physics it is possible, at least in principle, to account for the effects of a measurement on a physical system. However, in quantum mechanics, observation and measurement fundamentally alter the system in ways that cannot be predicted with certainty.

A particle-wave is characterized by a wavefunction $\Psi(x, t)$, known as a *probability amplitude*. Since the wavefunction is complex valued, it is not directly observable.

Observable quantities such as momentum and energy are extracted from the wavefunction by operators in quantum mechanics. The (one-dimensional) *momentum operator* is

$$-i\hbar \frac{\partial}{\partial x} \qquad (27.31)$$

The *energy operator* is

$$i\hbar \frac{\partial}{\partial t} \qquad (27.34)$$

The wavefunction $\Psi(x, t)$ in nonrelativistic quantum mechanics obeys a new wave equation known as the *Schrödinger equation*:

$$-\frac{\hbar^2}{2m} \frac{\partial^2 \Psi(x, t)}{\partial x^2} + V(x, t)\Psi(x, t) = i\hbar \frac{\partial \Psi(x, t)}{\partial t} \qquad (27.40)$$

The Schrödinger equation also can be written as

$$H\Psi = i\hbar \frac{\partial \Psi}{\partial t} \qquad (27.41)$$

where H is an operator called the *Hamiltonian* of the system.

QUESTIONS

1. Particles behave like waves and waves like particles. We have described this situation by calling them both particle-waves. Use your imagination to invent a new, shorter word to associate with the hybrid.

2. An electron, proton, neutron, and α particle all have the same energy. List the particles in the order of increasing de Broglie wavelength.

3. An electron, proton, neutron, and α particle all have the same speed. List the particles in the order of increasing de Broglie wavelength.

4. An electron, proton, neutron, and α particle all have momentum of the same magnitude. What can you say about the de Broglie wavelengths of the particles?

5. Explain why the Bohr model of the atom, with its well-defined, planet-like orbits for the electron, is incompatible with the Heisenberg position–momentum uncertainty principle.

6. Explain why the Heisenberg uncertainty principles are irrelevant for the physics of a macroscopic particle such as a jumping bean.

7. Explain why a radioactive material with a very short half-life corresponds to a large energy associated with the disintegration.

8. The age of the universe is about 15 billion years. What does the uncertainty principle say about the possibility of their being an exact instant when its age was zero?

9. If a single electron is sent through a single slit, explain in what sense the electron forms a single slit diffraction pattern.

10. If a single particle is sent through a double slit, in what sense is a double slit interference pattern formed?

11. If particle-waves are sent one at a time through a double slit arrangement, the double slit interference pattern gradually emerges. Explain why the double slit experiment is absolutely confounding to the notion we have of a particle, which is

what makes quantum mechanics somewhat baffling to our notions of common sense.

12. Discuss the similarities and differences between particle-waves and electromagnetic waves.

13. Is the relationship between geometric optics and physical (wave) optics similar to the relationship between a particle and a particle-wave? Discuss this analogy.

14. Particle-waves have a wavelength. If the wavelength is in the range of visible light wavelengths, do the particle-waves have color? Explain.

15. For a confined particle, its energy never can be zero. What does this imply about the attainability of the absolute zero of temperature?

PROBLEMS

Sections 27.1 The Heisenberg Uncertainty Principles
   27.2 Implications of the Position–Momentum Uncertainty Principle
   27.3 Implications of the Energy–Time Uncertainty Principle
   27.4 Observation and Measurement

1. What are the values of Planck's constant h and $\hbar = h/2\pi$ when expressed in units of $eV \cdot s$ rather than the SI units of $J \cdot s$?

2. Show that the SI units of the product $\Delta x \, \Delta p_x$ and the product $\Delta E \, \Delta t$ are those of angular momentum ($kg \cdot m^2/s$) and that these units also are the same as the SI units of Planck's constant ($J \cdot s$).

3. If the uncertainty Δx in the position of a particle is equal to its de Broglie wavelength, show that the uncertainty in the magnitude of its momentum Δp_x is greater than or equal to the magnitude of its momentum.

•4. Define the *angular wavenumber k* of a de Broglie wave to be

$$k \equiv \frac{2\pi}{\lambda}$$

(a) Show that the magnitude of the momentum of a particle-wave can be expressed as

$$p = \hbar k$$

(b) Show that the position–momentum uncertainty relation can be written as

$$\Delta x \, \Delta k \geq 2\pi$$

(c) Since the energy E of a particle-wave is

$$E = h\nu = \hbar\omega$$

show that the energy–time uncertainty relation can be expressed as

$$\Delta\omega \, \Delta t \geq 2\pi$$

These forms for the uncertainty relationships were discussed in Chapter 12 on waves (Section 12.21).

•5. A particle known unromantically as the $\Delta(1232)$ has a rest energy of 1232 MeV ± 120 MeV. What is the average lifetime of such a particle? These are among the shortest-lived particles known.

•6. An atom represents a region about 2×10^{-10} m wide in which an electron is confined. Use the Heisenberg uncertainty principle to estimate the minimum kinetic energy of the electron, expressing your result in electron-volts (eV).

•7. A passing photon is found to have a length in space (its wavetrain) of about 10 cm. The wavelength of the photon is 633 nm. About how well (to what fractional error $\Delta E/E$) is its energy or frequency determinable?

•8. A 60 kg motorist in a 1940 kg car is accused of speeding at 72 km/h while crossing an intersection 13.2 m wide where the legal speed limit is 36 km/h. A defense lawyer from the firm Chase, Cheatam, and Dunnum contends that because of the Heisenberg uncertainty principle, the speed cannot be exactly determined even with the best of equipment; therefore, the case should be dismissed against her client. Discuss the relevancy of the argument and include a quantitative estimate of the ultimate uncertainty in the speed in this situation.

•9. The lifetime of a certain atom while it exists in an excited state before making a transition to a lower-energy state is about 1.0 ns. The wavelength emitted in the transition is 200 nm. (a) What is the frequency of the light? (b) About how well can the frequency of the light be known? (c) About how well can the wave packet (photon) be localized in space?

•10. (a) The lifetimes of electrons in excited states in atoms are typically on the order of 1.0×10^{-8} s. What is the minimum uncertainty in the energy of states with such lifetimes? (b) What is the corresponding uncertainty in the frequency of the photons emitted in the transition from such a state to the ground state?

•11. Certain atoms have special excited states (called metastable states) in which an electron can exist for relatively long times, on the order of 1.0×10^{-4} s. What is the minimum uncertainty associated with the frequency from such metastable excited states? Such relatively long-lived states make lasers possible; lasers emit light with a very precise frequency.

•12. If the entire universe once were confined to a volume on the order of 1×10^{-90} m across in its early history, what is the corresponding energy uncertainty? Assume this early universe was filled entirely by high-energy photons and no mass.

‡13. (a) About what magnitude of momentum might an electron have if it were confined inside a box about 1.0×10^{-14} m on a side? (This is about the size of the nucleus of an atom.) (b) Estimate the approximate speed of such an electron. (Hint: Use the relativistic expression $p = \gamma m v$ for the magnitude of the momentum.) (c) Estimate the approximate total (relativistic) energy of such an electron; express your result in MeV. Compare this energy with the rest energy of an electron (0.511 MeV).

Sections 27.5 Particle-Waves and the Wavefunction
     27.6 Operators*
     27.7 The Schrödinger Equation*

14. Probabilities are dimensionless. What are the dimensions and SI units of the product $\Psi^*\Psi$ of the one-dimensional wavefunction and its complex conjugate? Use this result to determine the dimensions and SI units of the one-dimensional wavefunction Ψ itself.

•**15.** If $\Psi_1(x, t)$ and $\Psi_2(x, t)$ are both solutions to the Schrödinger equation for a given potential energy function $V(x)$, show that

$$c_1\Psi_1(x, t) + c_2\Psi_2(x, t)$$

also is a solution to the Schrödinger equation, where c_1 and c_2 are constants that may be real or complex numbers. This principle of linear superposition indicates that particle-waves can interfere with each other.

INVESTIGATIVE PROJECTS

A. Expanded Horizons

1. Investigate the statistical behavior of dice with parallelopiped shapes.
Patricia F. Bronson and Robert L. Bronson, "Dice with parallelopiped shapes," *The Physics Teacher*, 28, #5, pages 286–290 (May 1990).

2. Investigate the interesting quantum mechanical problem known as the paradox of *Schrödinger's cat*.
John Gribben, *In Search of Schrödinger's Cat* (Bantam Books, New York, 1984), pages 203–208.
Christopher N. Villars, "The paradox of Schrödinger's cat," *Physics Education*, 21, #4, pages 232–237 (July 1986).

3. Investigate the so-called *cosmic number hypothesis* and clues to several peculiar numerological coincidences among the constants of physics and cosmology.
Edward R. Harrison, *Cosmology* (Cambridge University Press, Cambridge, England, 1981), Chapter 17, pages 329–345.
Herman Bondi, *Cosmology* (Cambridge University Press, Cambridge, England, 1960), Chapter 7.
Edward R. Harrison, "The cosmic numbers," *Physics Today*, 25, #12, pages 30–34 (December 1972).

4. Investigate how the characteristic sizes of many things in nature are related to the magnitudes of fundamental physical constants.
Victor F. Weisskopf, "Of atoms, mountain, and stars: a study in qualitative physics," *Science*, 187, #4177, pages 605–612 (21 February 1975).

5. Investigate speculations about the fundamental constants of nature *varying with time*.
Paul S. Wesson, "Does gravity change with time?" *Physics Today*, 33, #7, pages 32–37 (July 1980).
Thomas C. Van Flandern, "Is gravity getting weaker?" *Scientific American*, 234, #2, pages 44–52, 140 (February 1976).
R. A. Alpher, "Large numbers, cosmology, and Gamow," *American Scientist*, 61, #1, pages 52–58 (January–February 1973).
See also the references under Investigative Project 3.

6. Investigate the effects of gravity on antimatter.
Terry Goldman, Richard J. Hughes, and Michael Martin Nieto, "Gravity and antimatter," *Scientific American*, 258, #3, pages 48–56 (March 1988).

B. Communicating Physics

7. Uncertainty is a perilous concept in life, religion, business, politics, and the legal profession. In an essay, compare and contrast the notion of uncertainty in these contexts with its meaning in physics.

EPILOGUE

**. . . to make an end is to make a beginning.
The end is where we start from.**
T. S. Eliot (1888–1965) *

To paraphrase Matthew 25:21, *well done thou good and faithful student.* We come full circle, like the orobouros, the serpent swallowing its tail.

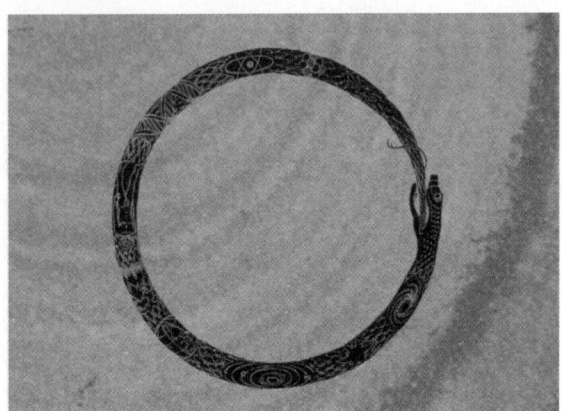

The five great theories of physics with which you now have more than passing acquaintance—classical mechanics, thermodynamics, electromagnetism, relativity, and quantum mechanics—are the beginning, not the end. There really is no end to physics! You may see this either as a horrifying realization or, more optimistically, as a window of opportunity through which to transmit your own creativity and imagination to the field. Whichever, (though I hope it is the latter!), I wish you Godspeed and welcome your correspondence. *Shalom aleichem.*—R.L.R.

*We shall not cease from exploration
And the end of all our exploring
Will be to arrive where we started
And know the place for the first time.*

T.S. Eliot[†]

Four Quartets, "Little Gidding," lines 215–216 (Harcourt Brace Jovanovich, New York, 1971).
[†]*Four Quartets,* "Little Gidding," lines 239–242.

APPENDIX A
PROOFS OF THE GRAVITATIONAL SHELL THEOREMS

A.1 A MASS WITHIN A UNIFORM SPHERICAL SHELL

We seek to prove the following theorem:

> If a mass m is located anywhere within the volume enclosed by a thin uniform spherical shell of mass M, the total gravitational force of the shell on m is equal to zero.

Consider a pointlike mass m positioned anywhere *inside* a thin uniform spherical shell of radius R and mass M, as shown in Figure A.1. Draw two narrow, equal apex-angled, differential cones centered on the mass m, as indicated. These small conical constructions are like two narrow ice cream cones. The cones intercept masses dM and dM' on the spherical shell. The differential gravitational force $d\vec{F}$ of dM on m attracts m toward dM. The differential gravitational force $d\vec{F}'$ of dM' on m attracts m toward dM'.

We need to calculate the ratio of the magnitudes of these two forces on m. The vector addition of the two forces then will be quite easy, since they are in opposite directions because of how we constructed the differential cones.

If the apex angle of the cones is sufficiently small, the masses dM and dM' are essentially pointlike masses. Hence we use the point-mass gravitational force law deduced by Newton, Equation 6.3, to calculate the gravitational effects of dM and dM' on the mass m inside the shell. Thus the magnitude of the differential force of dM on m is

$$dF = \frac{Gm\,dM}{r^2}$$

where r is the distance of dM from the mass m inside the shell as shown in Figure A.1. Likewise, the magnitude of the differential force of dM' on m is

$$dF' = \frac{Gm\,dM'}{r'^2}$$

where r' is the distance between m and dM', also indicated in Figure A.1.

We take the ratio of the magnitudes of the two forces:

$$\frac{dF}{dF'} = \frac{\dfrac{Gm\,dM}{r^2}}{\dfrac{Gm\,dM'}{r'^2}}$$

$$= \frac{dM}{dM'}\frac{r'^2}{r^2} \qquad (A.1)$$

Since the shell is a *uniform* shell, the ratio of the masses intercepted by the differential cones is equal to the ratio of the areas intercepted:

$$\frac{dM}{dM'} = \frac{dS}{dS'}$$

We draw radii R from the center of the sphere to the two areas dS and dS', as shown in Figure A.2. The radii intersect the axis of the double cone, forming a large isosceles triangle with its apex angle at the center of the shell. The angle ϕ that the radii make with the common axis of the cones therefore is the same at area dS and dS'. Let dA be the *projected* area of dS that is perpendicular to the axis of the cone; let dA' be the corresponding projected area at dS'.

The angle between dS and dA is ϕ; likewise, the angle between dS' and dA' is ϕ. The projected areas are related to the areas on the shell itself by

$$dA = dS \cos \phi$$

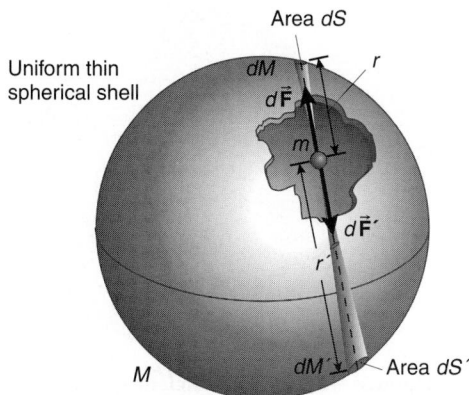

FIGURE A.1 A pointlike mass m inside a thin uniform spherical shell.

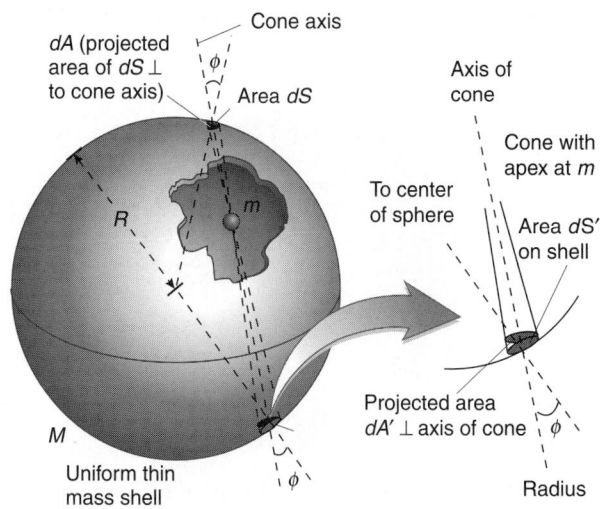

FIGURE A.2 The projected areas are perpendicular to the common axis of the differential cones.

A.1

and

$$dA' = dS' \cos \phi$$

So the ratio of the areas dS/dS' is the same as the ratio of the areas dA/dA'. Since the apex angles of the cones are equal, the sizes of the projected areas are proportional to the square of the distance they are from the apex of the cones; that is,

$$\frac{dA}{dA'} = \frac{r^2}{r'^2}$$

Hence the ratio of the masses intercepted by the cones is the same as

$$\frac{dM}{dM'} = \frac{dA}{dA'}$$
$$= \frac{r^2}{r'^2}$$

The ratio of the magnitudes of the two differential forces is then (from Equation A.1)

$$\frac{dF}{dF'} = \frac{dM}{dM'} \frac{r'^2}{r^2}$$
$$= \frac{r^2}{r'^2} \frac{r'^2}{r^2}$$
$$= 1 \qquad (A.2)$$

Thus the two differential forces on m have exactly the same magnitude, but point in opposite directions. The two differential forces are *not* a Newton's third law pair since they act on the *same* system, the mass m. Hence the vector sum of the two differential forces on m is *zero*. Although m is closer to dA than to dA', the more distant area dA' has more mass in just the right amount to compensate for its greater distance from m.

Now we sweep the differential cones around in all directions to take into account the effect of the entire spherical shell on m. *Every* pair of little differential masses at opposite ends of the differential cones produce zero total gravitational force on the mass m. Thus we reach the startling conclusion that the total gravitational force of the shell on mass m is zero, as indicated in Figure A.3.

Notice that this remarkable result depended *crucially* on the inverse square law nature of the gravitational force law; if the gravi-

tational force had any other power of the distance in the denominator, the distances in Equation A.2 would not cancel and there would be a nonzero total gravitational force on m unless it were at the very center of the shell. Nature was kind in giving us an inverse square law gravitational force.

A.2 A MASS OUTSIDE A UNIFORM SPHERICAL SHELL

We seek to prove the following theorem:

> If a pointlike mass m is located anywhere outside a thin uniform spherical shell of mass M, the total gravitational force of the shell on m is the same as that of a point mass, with a mass equal to the mass M of the shell, located at the center of the shell.

To prove this theorem, we have to calculate the gravitational force of a uniform spherical shell of mass M on a pointlike mass m lying a distance r from the center of the shell, where $r > R$, the radius of the shell (see Figure A.4).

It is necessary to do a vector integration here; we managed to avoid such an integration when the point mass was inside the sphere, but here it is unavoidable. Vector integrations can be difficult because we have to account both for the magnitudes *and* directions of the vectors. So we proceed with some caution; it will be beneficial for you to study this technique with some care, since vector integrations are encountered again later in our course of study. The typical procedure for performing a vector integration involves converting the vector integration into an ordinary scalar integration by subtly looking for ways to exploit or account for the directional effects of the vectors. This only comes with practice and patience.

The amount of mass on any square meter of the shell is known as a *surface mass density* σ. Since the shell is uniform, the surface mass density σ is constant and equal to the total mass of the shell divided by the surface area of the spherical shell:

$$\sigma = \frac{M}{4\pi R^2}$$

We begin by examining a thin differential ring of the shell, as shown in Figure A.5. Every point on the differential ring is the

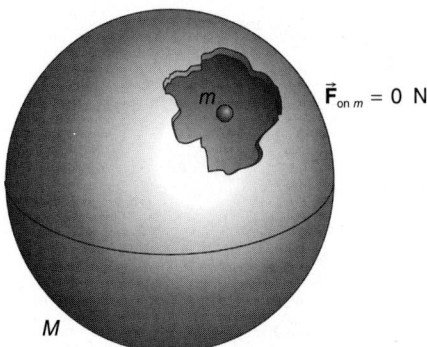

FIGURE A.3 The total gravitational force of the shell on mass m inside the shell is zero.

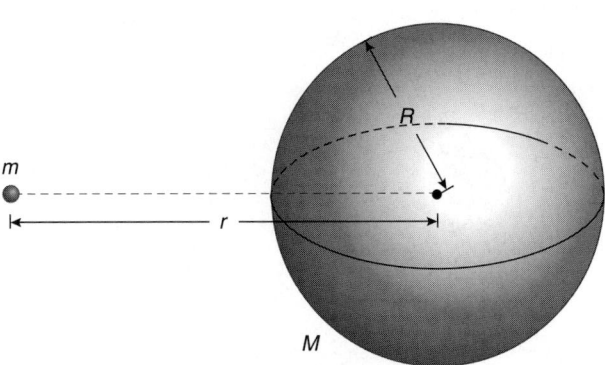

FIGURE A.4 A pointlike mass m outside a thin uniform spherical shell.

same distance s from the point mass m. The radius of the differential ring is $R \sin \theta$. The area of the differential ring is its circumference ($2\pi R \sin \theta$) times its width ($R \, d\theta$), or

$$dA = (2\pi R \sin \theta)(R \, d\theta)$$
$$= 2\pi R^2 \sin \theta \, d\theta$$

The mass dM on this ring of material is the product of the area dA and the surface mass density σ:

$$dM = \sigma \, dA$$
$$= \sigma (2\pi R^2 \sin \theta \, d\theta)$$

Each little segment of the ring produces a gravitational force on mass m that is directed toward that little piece of the ring, as shown in Figure A.6. Across the diameter of the ring is a twin segment of the ring. Each of the two twins produces a differential force on m. Note that the components of the two force vectors that are perpendicular to the line joining m and the center of the shell will vector sum to zero. The components of the force vectors of the twins directed *along* the line between

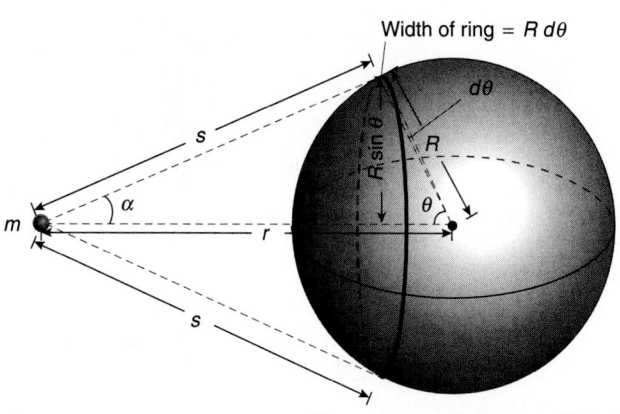

FIGURE A.5 A thin differential ring of the shell.

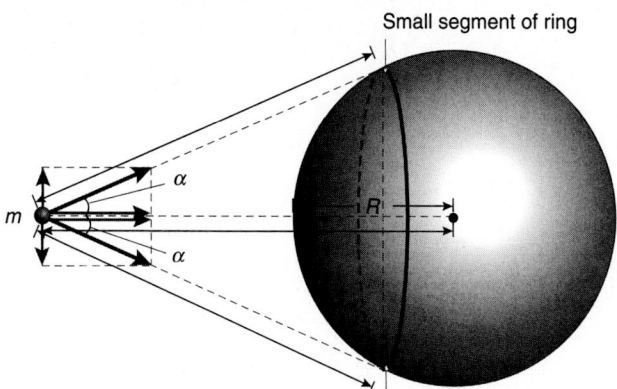

FIGURE A.6 Twin segments produce forces of equal magnitude on m. The force components perpendicular to the line joining m and the center of the shell vector sum to zero; the components along the line add.

m and the center of the shell will add, since they point in the same direction.

Thus, as we vector sum (integrate) around the ring, the components of the differential forces perpendicular to the line will pairwise vector sum to zero, while those along the line will simply add. This is certainly cause for celebration!

The *surviving* component of the differential force on m by the segment of material dM is

$$dF = \frac{Gm \, dM}{s^2} \cos \alpha$$

The $\cos \alpha$ term picks off the surviving part of the differential force as we vector integrate around the ring. Thus we have accounted for the directional effects of the vectors with the $\cos \alpha$ term. The resulting integration is now just an ordinary scalar integration for the *magnitude* of the total gravitational force of the shell on m.

If we substitute for dM, the expression for dF begins to look like a nightmare:

$$dF = Gm \frac{\sigma (2\pi R^2) \sin \theta \, d\theta}{s^2} \cos \alpha \qquad \text{(A.3)}$$

Now we integrate this expression over the shell. For all little rings anywhere on the spherical shell, the differential forces given by Equation A.3 simply add together because they all point in the same direction: toward the center of the shell.

It is convenient to express everything in terms of the distance s. It is not obvious that this is the thing to do, but using the variable s enables us to perform the integration easily and that is why we do it.

The law of cosines from trigonometry will help. In the triangle whose sides are r, s, and R in Figure A.5, the law of cosines implies that

$$s^2 = r^2 + R^2 - 2rR \cos \theta$$

As we integrate over the sphere, s and θ change but r and R are fixed constants. Taking differentials of the law of cosines equation, we find

$$2s \, ds = -2rR(-\sin \theta \, d\theta)$$

or

$$s \, ds = rR \sin \theta \, d\theta$$

Hence

$$\sin \theta \, d\theta = \frac{s \, ds}{rR}$$

We apply the law of cosines in a different way in the same triangle; this enables us to find an expression for $\cos \alpha$. Since

$$R^2 = s^2 + r^2 - 2sr \cos \alpha$$

we can solve for $\cos \alpha$:

$$\cos \alpha = \frac{s^2 + r^2 - R^2}{2sr}$$

Substituting the expressions for $\sin\theta\, d\theta$ and $\cos\alpha$ into the expression for dF in Equation A.3, we really get quite a mouthful:

$$dF = \frac{Gm\sigma\,(2\pi R^2)}{s^2}\,\frac{s\,ds}{rR}\,\frac{s^2 + r^2 - R^2}{2sr}$$

Simplifying this slightly, we obtain

$$dF = \frac{Gm\sigma\pi R}{r^2}\,\frac{s^2 + r^2 - R^2}{s^2}\,ds$$

Now we integrate to find the magnitude of the total force on m due to the entire shell. The limits on s are from $r - R$ to $r + R$ (see Figure A.6). The integral becomes

$$F = \frac{Gm\sigma\pi R}{r^2}\int_{r-R}^{r+R}\left[1 + \frac{(r^2 - R^2)}{s^2}\right]ds$$

The two integrals are not too bad to do, since both r and R are constants:

$$F = \frac{Gm\sigma\pi R}{r^2}\left[\int_{r-R}^{r+R}ds + (r^2 - R^2)\int_{r-R}^{r+R}\frac{ds}{s^2}\right]$$

Performing the integrations, we get

$$F = \frac{Gm\sigma\pi R}{r^2}\left[s - \frac{(r^2 - R^2)}{s}\right]\Bigg|_{r-R}^{r+R}$$

Inserting the limits of integration and simplifying, we find

$$F = \frac{Gm\sigma\pi R}{r^2}\,4R$$

We rearrange this a bit, getting

$$F = \frac{Gm}{r^2}\,4\pi R^2\sigma$$

But the quantity $4\pi R^2\sigma$ is the product of the surface mass density σ and the surface area of the shell. This is the total mass M of the shell. Hence the magnitude of the total force on m is

$$F = \frac{GmM}{r^2}.$$

Voilà! Notice that this is the same expression as Equation 6.3 for the magnitude of the gravitational force between two pointlike masses separated by a distance r.

Thus the magnitude of the gravitational force of the shell on m is found by imagining the mass M of the shell to be located at its center, since the distance r is the distance between m and the center of the shell. Effectively, we can think of the mass of the shell as if it were at the center of the shell, as shown in Figure A.7. What a relief this result must have been to Newton as he wrestled with how to account for the gravitational effects of the shell on a mass m located outside the shell!

To satisfy Newton's third law of motion, the mass m exerts a gravitational force of the same magnitude but opposite direction *on the shell*, so that we have the third law force diagram shown in Figure A.8.

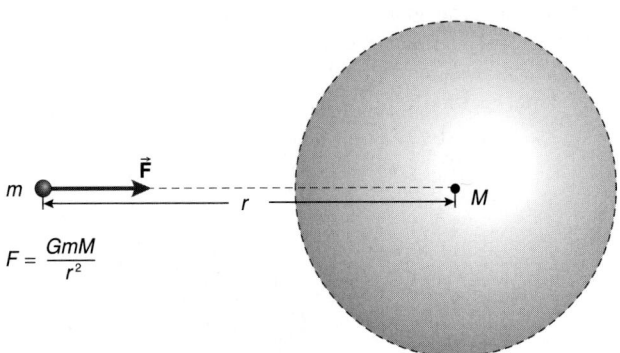

$$F = \frac{GmM}{r^2}$$

FIGURE A.7 **The force of the shell on m can be found by imagining the entire mass of the shell to be at its center.**

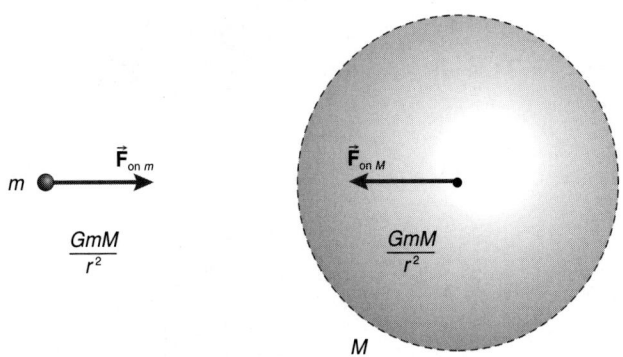

FIGURE A.8 **Third law force diagram.**

ANSWERS TO PROBLEMS

CHAPTER 1

1. 4.6403×10^{-5} m^3

5. peekaboos, nannygoats, military, millionaires, megaphones, microphones, megabucks, centipedes, microfilm, terrapin, megalopolis, deck of cards, microfiche, microwaves, decimate, microeconomics, to kill a mockingbird, terrible, terra firma, megalomaniacs, millinery, microscopes, megabytes

9. (a) 1.83×10^3 (b) 3.33×10^5 (c) 2×10^{11} (d) 3×10^{11}
   Parts (c) and (d) are the same order of magnitude.

13. (a) $\sim 10^6$ m (b) $\sim 10^5$ m

17. (a) A factor of 3 (b) A factor of 9

21. (a) A factor of $\sqrt{3} \approx 1.73$ (b) A factor of $3\sqrt{3} \approx 5.20$

25. 0.447 h·m/(mile·s)

29. 3.4×10^{22} molecules

33. 3.63 m^3

37. (a) 7.77×10^{-4} m^3 (b) 1.6×10^{-4} m^3

41. (a) 6.5×10^8 m^3 (b) 8.7×10^2 m on an edge (c) 3.6×10^8 m^2
   (d) 19 km

45. 1.80×10^{12} furlongs/fortnight

49. (a) 9.74×10^{-4} kg (b) 8.0×10^{-5} m (c) 8.2×10^{-7} m^3
   (d) 9.74×10^5 kg (e) 8.0×10^4 m (f) 1.0×10^7 m^2
   (g) 9.4 m

53. (a) 5×10^{-28} kg/m^3 (b) 3×10^9 m

57. (a) ~ 12 minutes (b) ~ 3 years (c) ~ 10 minutes

61. 1.19×10^{57} proton masses

65. $\sim 3 \times 10^7$ ties, 1×10^6 m^3

69. $\sim 7 \times 10^8$ km^3 assuming an average depth of 2 km

73. $\sim$ several percent

77.

Angle	Sine	Cosine	Tangent
78.0°	0.978	0.208	4.70
78.02°	0.9782	0.2076	4.713
78.024°	0.97823	0.20750	4.7143

CHAPTER 2

1.

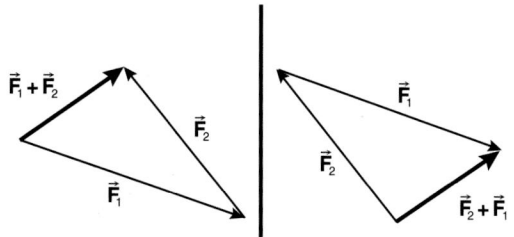

5. 100 m east, 141 m northwest

9. $(\vec{R} + \vec{r}) \cdot (\vec{R} - \vec{r}) = 0$ so $\vec{R} + \vec{r}$ is perpendicular to $\vec{R} - \vec{r}$. The vector $\vec{R} + \vec{r}$ is also perpendicular to $\vec{r} - \vec{R}$.

13. Answers are in Table 2.1.

17. $\dfrac{\hat{i} + \hat{j} + \hat{k}}{\sqrt{3}}$

21.

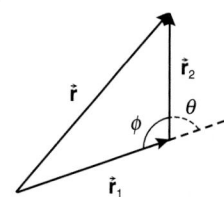

$r^2 = r_1^2 + r_2^2 - 2r_1r_2 \cos \phi$

25. (a) 64.9° (b) 115.1°

29. Answers are given in the statement of the problem.

33. (a) 90° (b) 53° between the vector sum and the vector of magnitude 12. 37° between the vector sum and the vector of magnitude 16

37. $\alpha = -1.00$, $\beta = -10.0$, $\gamma = 9.00$

41. $\hat{k}$ or $\alpha\hat{k}$ where α is any scalar; the solution is not unique.

45. (a) 19.1 (b) $-8.2\,\hat{k}$

49. (a) 0 (b) $6\,\hat{j} - 3\,\hat{k}$; The solution is not unique. (c) 0

53. A proof

57. (a) $-3.0\,\hat{i} + 3.0\,\hat{j} - 4.0\,\hat{k}$ (b) 4.0 (c) 37°

61. Answers are given in the statement of the problem.

65. $\dfrac{5}{2\sqrt{t}}\,\hat{i} - 9\sqrt{t}\,\hat{j}$

69. A proof

CHAPTER 3

1. (a) 0 m (b) 50 m (c) 4.2 m/s (d) 0 m/s

5. 4.0 weeks

9. 6.7 m/s, 2×10^{-2} s

13. Let $\hat{i}$ be in the direction the tanker is moving:
   (a) $\vec{a} = (-9.26 \times 10^{-3}$ m/s$^2)\,\hat{i}$ (b) 3.75 km

17. (a) 2.47 s (b) 24.2 m/s (c) 19.8 m/s (d) 25.1 m

21. (a) $v_x \approx -6.0$ m/s (b) Yes, when $t \approx 1.5$ s. (c) No. $x \approx 5.2$ m.

25. 3×10^{24} m/s^2

29. (a) 10 s (b) No. 10 m

33. (a) $\vec{a}_{ave} = (3.3 \times 10^{-2}$ m/s$^2)\hat{i}$, $\vec{a} = (3.3 \times 10^{-2}$ m/s$^2)\hat{i}$
   (b) $\vec{a}_{ave} = (-6.7 \times 10^{-2}$ m/s$^2)\hat{i}$, $\vec{a} = (-6.7 \times 10^{-2}$ m/s$^2)\hat{i}$
   (c) $\vec{a}_{ave} = (3.3 \times 10^{-2}$ m/s$^2)\hat{i}$, $\vec{a} = (3.3 \times 10^{-2}$ m/s$^2)\hat{i}$
   (d) The barge begins at rest and accelerates to a speed 2.0 m/s when $t = 60$ s. The barge then begins to slow down, stops instantaneously when $t = 90$ s, reverses the direction it is moving and reaches a speed of 2.0 m/s in the opposite

direction when $t = 120$ s. Then the barge again begins to slow down and is brought to a stop when $t = 180$ s.

(e)

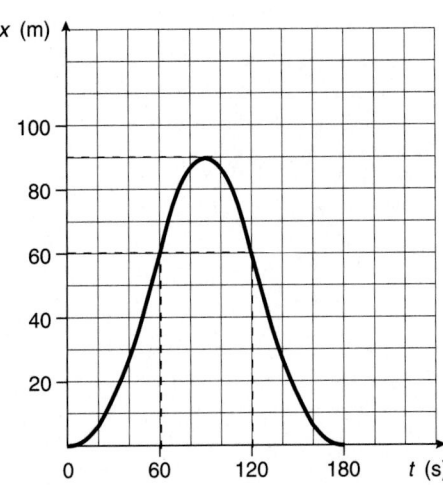

37. (a) When $t = 0$ s and 6.0 s (b) During the interval between $t = 1.0$ s and 4.0 s (c) $\vec{a} = (-2.0 \text{ m/s}^2)\,\hat{\imath}$ (d) 18 m

41. (a) 6.4×10^2 m/s² (b) 48 m, 6.2 s

45. (b) 12.0 s and 20.0 s (c) When $t = 12.0$ s, they first catch up to the back of the train. If they kept running they would pass the back of the train, but when $t = 20.0$ s, the back end of the accelerating train would pass them for the last time.

49. (a) 1.0 ms (b) 6.0×10^9 m/s²

53. (a) Let $\hat{\imath}$ be upward with the origin at the point of release. (b) 31.9 m (c) 5.10 s
(d)

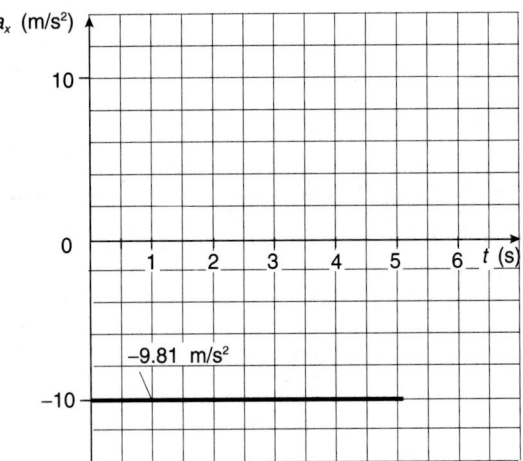

(e)

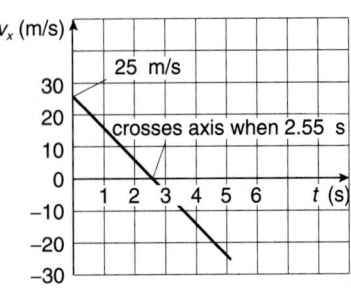

(f)

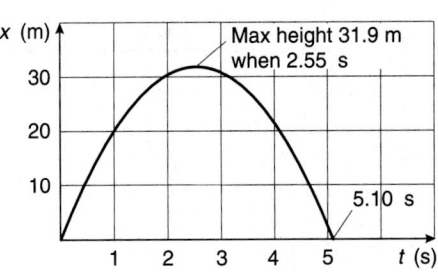

57. 6.5 m

61. 1.7 m

65. 42 m

69. (a) Up (b) Up (c) ≈3.0 s (d) ≈25 m (e) ≈25 m

CHAPTER 4

1. (a) $\vec{r}_i = (-0.15 \text{ m})\hat{\imath}$, $\vec{r}_f = (0.26 \text{ m})\hat{\imath} + (0.15 \text{ m})\hat{\jmath}$
(b) average speed $= 7.5 \times 10^{-3}$ m/s, $\vec{v}_{ave} = (6.8 \times 10^{-3} \text{ m/s})\hat{\imath} + (2.5 \times 10^{-3} \text{ m/s})\hat{\jmath}$. $|\vec{v}_{ave}| = 7.2 \times 10^{-3}$ m/s.

5. Answers are given in the statement of the problem.

9. (a)

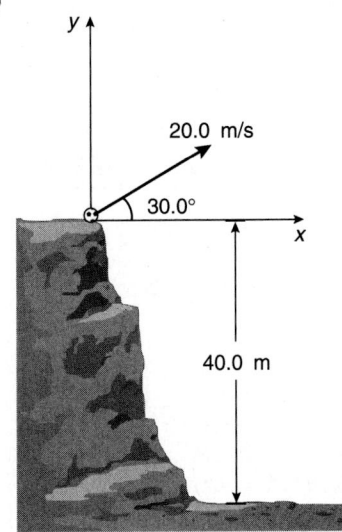

(b) $v_y(t) = 20.0$ m/s $(\sin 30.0°) - gt$, $y(t) = 0$ m $+ 20.0$ m/s $\times (\sin 30.0°)t - (1/2)gt^2$, $v_x(t) = 20.0$ m/s $(\cos 30.0°)$, $x(t) = 20.0$ m/s $(\cos 30.0°)\,t$ (c) 4.05 s (d) Using the coordinate system shown in part (a), the coordinates of the impact point are $x = 70.1$ m and $y = -40.0$ m.

13. 0.56 s and 2.09 s

17. (a)

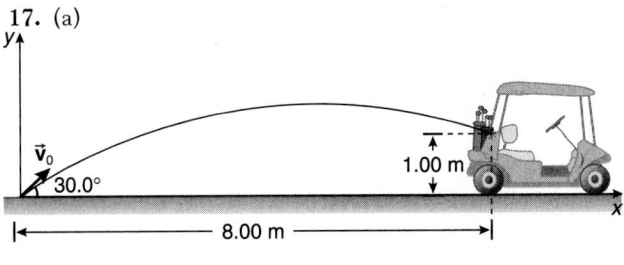

(b) 10.8 m/s (c) 0.856 s

21. (b) 35 m/s   (c) 23 m/s^2

25. Answer is given in the statement of the problem.

29. (a) $(2gh)^{1/2}$ (b) $2h$ (c) $h/2$

33. A proof

37. (a) 36.9° to the current   (b) 120 s

41. (a) Choose î east and ĵ north, $\vec{v}$ = (−1.50 m/s) î + (1.00 m/s) ĵ
   (b) No

45. (a) Let î be east and ĵ be north, (−1095 km/h) î + (495 km/h) ĵ
   (b) 1.20×10^3 km/h

49. 314.2 rad/s

53. 10 m

57. 4.4×10^8 m/s^2

61. 18 km. If the radius decreases, the centripetal acceleration increases, so the curve must have a radius no smaller than 18 km.

65. (a) Second hand: 0.105 rad/s; minute hand: 1.75×10^{-3} rad/s; hour hand: 1.45×10^{-4} rad/s   (b) No   (c) Yes, perpendicular to the clock face directed into the clock   (d) 60   (e) 12

69. (a) 131 rad/s^2 (b) 75.1 rev

73. (a) 25.0 rad/s^2 (b) 3.52×10^5 rev   (c) 8.82×10^6 m/s^2, $a/g = 8.99 \times 10^5$ (d) 43.8 rad/s^2

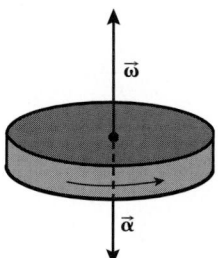

77. (a) A proof   (b) $\pi/4$ rad

81. (a) Let k̂ be parallel to $\vec{\omega}$ (b) 6.28 s   (c) $\alpha_z = 0.159$ rad/s^2
   (d)

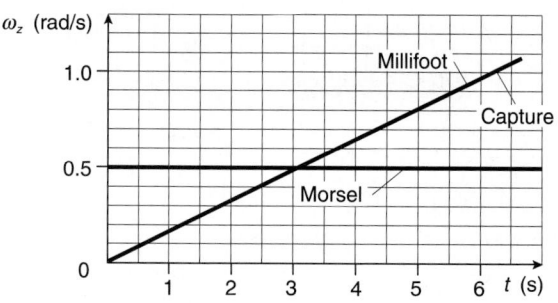

At capture, the areas under each curve are the same size, since both then have traveled through the same angle.

85. Answers are given in the statement of the problem.

CHAPTER 5

1. 188 N

5. (a) 8.4 N, 73° clockwise from the +x-axis   (b) 2.1 m/s^2

9. (a)

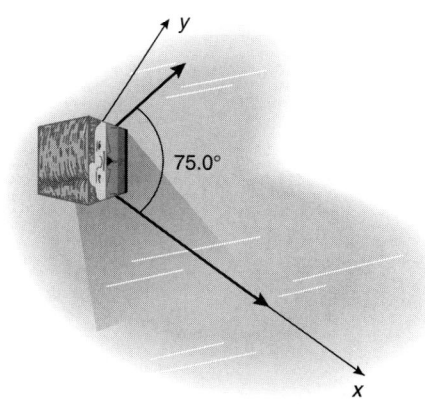

(b) $\vec{F}_{total}$ = (565 N) î + (241 N) ĵ, F_{total} = 614 N
(c) 7.7 m/s^2 (d) 0.920 î + 0.393 ĵ, yes

13. (a)

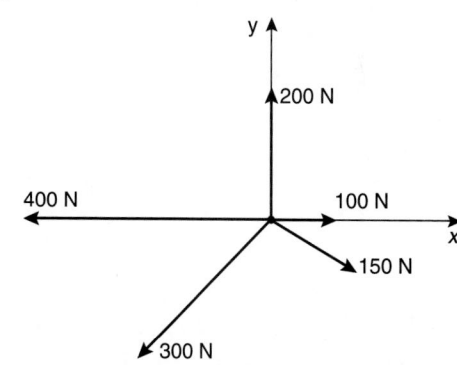

(b) (−4.2 m/s^2) î − (1.0 m/s^2) ĵ   (c) 4.3 m/s^2

17. 0.102 kg

21. (a)

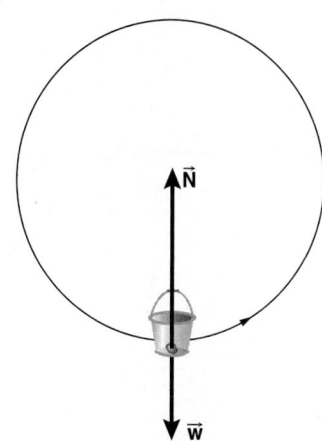

(b) 20.0 N toward the center of the circle   (c) 34.7 N vertically up   (d) It remains in contact with the bottom of the bucket since the normal force of the bucket bottom on the rock is not zero at this location. Its magnitude is 5.3 N, directed vertically down.

25. (a) 5.85°   (b) Yes   (c) 1.00 N

29. 10.2 kg

33. (a) 100 N vertically up

(b)

(c) Force of you on the rope, force of you on the surface, gravitational force of you on the Earth. (d) 687 N (e) 587 N (f) the magnitude of the force you exert on the surface; 59.8 kg.

37. Answers are given in the statement of the problem.

41. (a)

(b) 1.48 m/s², minimum value (c) 4.87 m/s

45. (a) 12.4 kN (b) 1.13 kN. Since the oaf is accelerating upward, the normal force of the floor on the oaf is greater than the oaf's weight.

49. (a)  (b) 785 N (c) 785 N

(d) 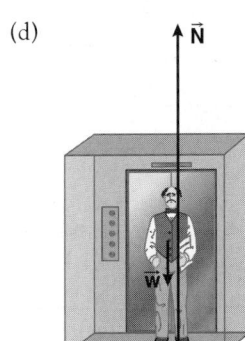 (e) 785 N (f) 865 N (g) 865 N. The scale measures the magnitude of the normal force, not the magnitude of the weight.

(h) 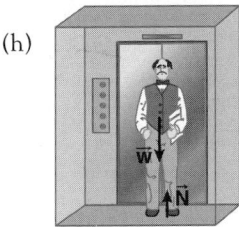 (i) 785 N (j) 705 N (k) 705 N

53. $\mu_s = 0.71$

57. $\mu_k = 0.09$

61. (a)

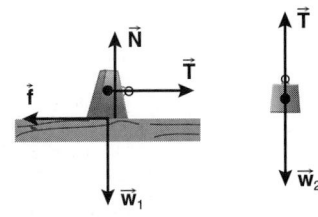

(b) 10 kg (c) 49 N static friction

65. (a)

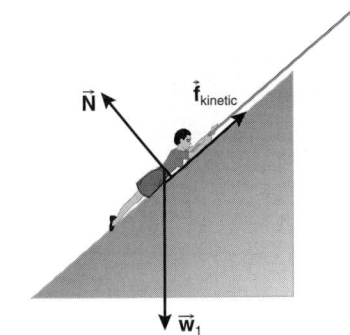

You will slip because the component of your weight down the plane is greater than f_{max}.

(b)

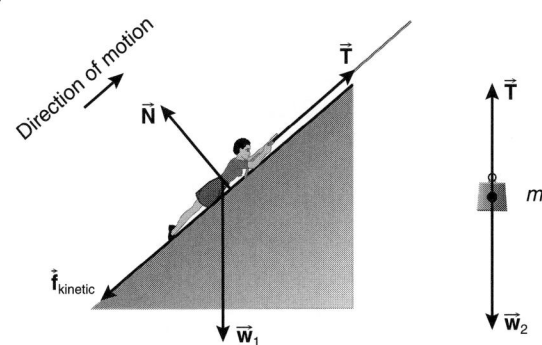

88 kg   (c) 7.3×10^2 N

69. 3.4 m/s^2

73. (a) The total force is directed toward the center of the circle since you have a centripetal acceleration in that direction. (b) 2.0 m/s   (c) 7.8 m   (d) Ground observer: the ball follows a parabolic trajectory. The horizontal component of the velocity of the ball is 2.0 m/s and the vertical component is equal to that given to the ball by the person on the merry-go-round; merry-go-round observer: the ball goes up and hooks to the right as the merry-go-round is spinning counterclockwise when seen from above.

77. (a) and (b)

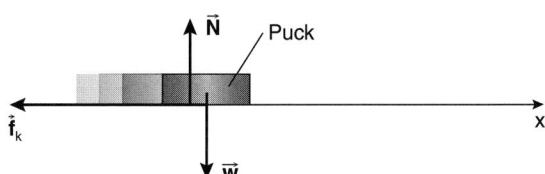

(c) $v_x(t) = v_0 - \mu_k g t$

(d) $t = \dfrac{v_0}{\mu_k g}$

(e) $x(t) = v_0 t - \dfrac{\mu_k g t^2}{2}$

(f) $\dfrac{v_0^2}{2\mu_k g}$

(g) 6.4×10^2 m

81. (a)

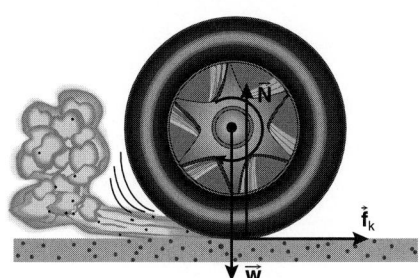

(b) The kinetic frictional force.   (c) Answer is given in the statement of the problem.

85. (a) and (b) Answer is given in the statement of the problem.

(c)

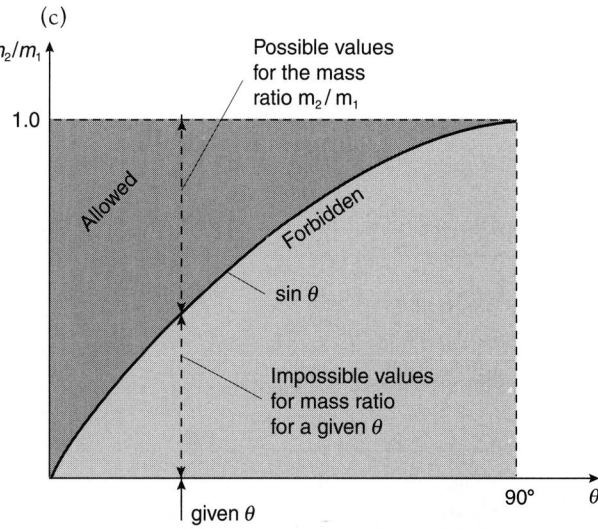

89. 0.0931 — The effects of pseudoforces are more apparent on Jupiter than on the Earth, since this ratio is almost 10%.

CHAPTER 6

1. 1.85×10^{-10} N

5. (a) $v = (GM_{Sun}/r)^{1/2} = 29.8$ km/s   (b) Nothing

9. (a) 1.05×10^{-8} N   (b) 1.05×10^{-8} N   (c) For the 0.400 kg sphere: 2.63×10^{-8} m/s^2; For the 4.00 kg sphere: 2.63×10^{-9} m/s^2

13. (a) Answer is given in the statement of the problem. (b) 0.16 m/s^2 (c) 1.2×10^2 m, 78 s

17. (a) $\rho = \left(\dfrac{9F}{4\pi^2 GR^4} \right)^{1/2}$

(b) No element is dense enough.   (c) 1.61 m, 3.93×10^5 kg

21. 6.43×10^{23} kg, $M_{Mars} = 0.108 \, M_{Earth}$

25. (a) 5.00×10^6 km   (b) Outside the Sun (c) Perihelion: 1.471×10^8 km, aphelion: 1.521×10^8 km (d) 0.334 cm

29. 1.12

33. (a) Answer is given in the statement of the problem. (b) 0.140, 0.417

37. 1.034

41. 4.7 y

45. (a) 17.9 AU   (b) 0.968   (c) 35.2 AU   (d) ≈ 62

49. (a) 4.7×10^3 km from the center   (b) 1.7×10^3 km below the surface

53. (a) $\dfrac{GM^2}{64r^2}$

(b) $\dfrac{GM^2}{4r^2}$

(c) $\dfrac{17}{64} \dfrac{GM^2}{r^2}$

(d) $2\pi r/T$

(e) $T^2 = \dfrac{64\pi^2}{17GM} r^3$

57. 1.86×10^{32} kg, 93.5 solar masses

61. (a) 2.1×10^8 y   (b) About 24 times   (c) 1.9×10^{41} kg, 9.5×10^{10} solar masses

65. (a) 0.900 d, measured from the center of the Earth   (b) No

69. 34 km

73. -6.17×10^{13} m³/s², -5.01×10^{15} m³/s²

CHAPTER 7

1. 50 N/m

5. 2.5×10^2 N/m

9. Answers are given in the statement of the problem.

13. (a) 40.0 rad/s   (b) 6.37 Hz

   (c) $x(t) = (0.050 \ \text{m}) \cos\left[(40.0 \ \text{rad/s}) t + \dfrac{\pi}{2} \ \text{rad}\right]$

   (d) -0.0363 m

17. (a) $\omega = 11.5$ rad/s, A = 0.201 m, $\phi = -1.05$ rad   (b) 0.546 s
   (c) 2.31 m/s, 26.6 m/s²

   (d)

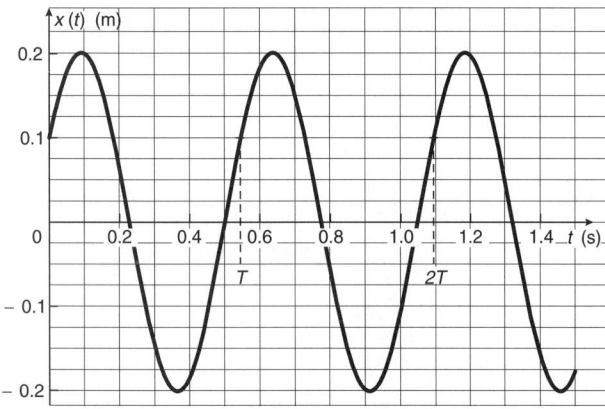

21. $v_x(t) = (-0.500 \ \text{m/s}) \sin\left[(7.07 \ \text{rad/s}) t - \dfrac{\pi}{2} \ \text{rad}\right]$

   $a_x(t) = (-3.54 \ \text{m/s}^2) \cos\left[(7.07 \ \text{rad/s}) t - \dfrac{\pi}{2} \ \text{rad}\right]$

   graphs follow:

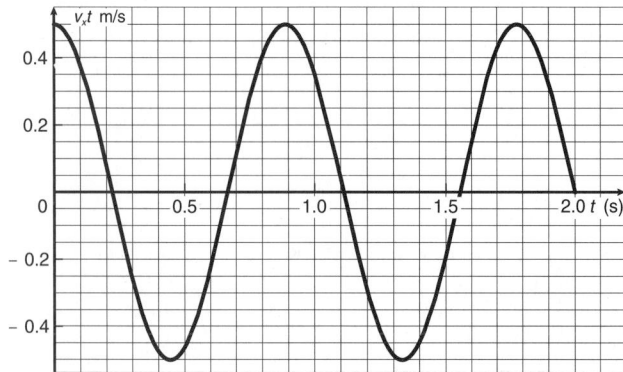

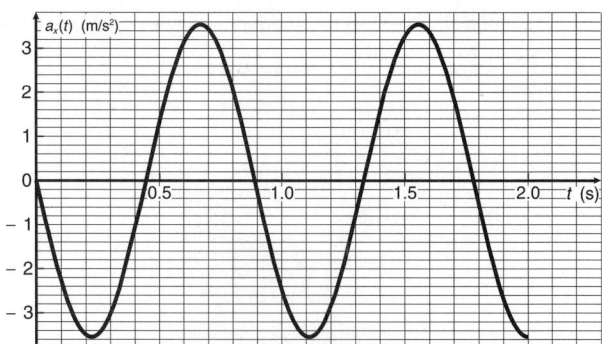

25. (a) 25.1 s, 18.0 s   (b) Approximately 10 s between the instants

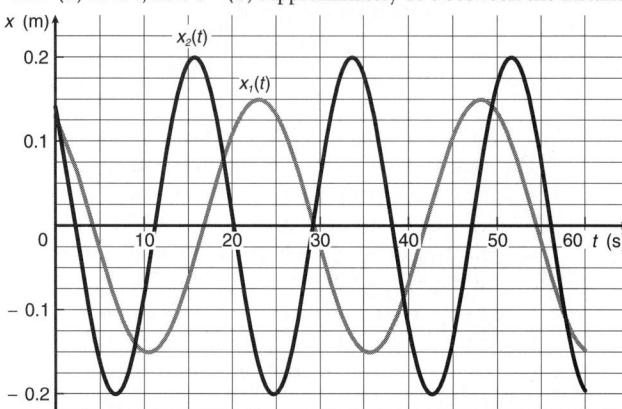

(c) The times change as well as the intervals between them.

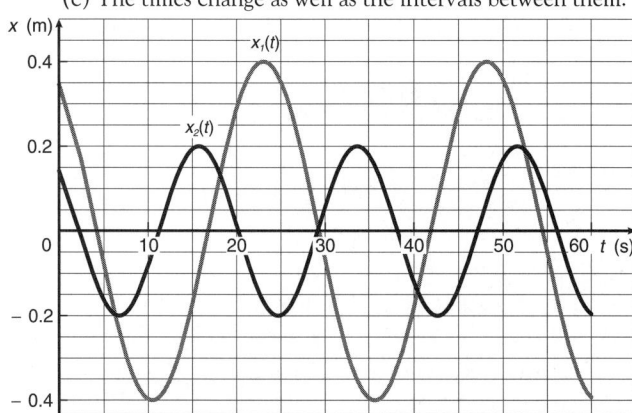

29. 0.496 m from the fixed end

33. (a) 1.30 Hz   (b) At the high point of the oscillation   (c) At the low point of the oscillation   (d) 0.147 m

37. (a) 1.5 m/s   (b) 23 m/s²

41. $R \cos\left(\omega t - \dfrac{\pi}{6} \ \text{rad}\right)$

45. (a) 10 s   (b) 0.10 Hz   (c) no

49. The equation of motion is not like Equation 7.15.

53. 16.5 s

57. $\dfrac{T_1 T_2}{T_1 - T_2}$

61. 9.780 m/s²

65. 526 km

69. An experiment

73. 14 Hz

CHAPTER 8

1. (a) 75.0 J (b) 0 J (c) 65.0 J (d) −53.1 J

5. (a) The graph should be a downward-sloping straight line with a slope of −6.00 N/m, passing through the origin. (b) 0 J

9. 2.4×10^2 J

13. (a) 360 J (b) 0 J (c) −62.5 J (d) 62.5 J

17. Answer is given in the statement of the problem.

21. Answer is given in the statement of the problem.

25. Answer is given in the statement of the problem.

29. 7.91 km/s

33. (a) 7.91 km/s

   (b) $\dfrac{1}{\sqrt{2}} = 0.707$

37. 5.8 km

41. (a) $-\dfrac{GMm}{r}$

   (b) $\left(\dfrac{GM}{r}\right)^{1/2}$

   (c) $\dfrac{1}{2}\dfrac{GMm}{r}$

   (d) $-\dfrac{1}{2}$

45. (a) $(2gh)^{1/2}$ (b) $(gR)^{1/2}$ (c) 5/2 (d) $(2gh)^{1/2}$ (e) mgh (f) Only (e)

49. Answer is given in the statement of the problem.

53. (a) $-\dfrac{GMm}{r_1}$ $-\dfrac{GMm}{r_2}$

   (b) mv_1r_1, mv_2r_2 (c) Answer is given in the statement of the problem.

57. (a) Answer is given in the statement of the problem. (b) 3.33×10^{-3} m/s

61. Answer is given in the statement of the problem.

65. (a) 70.2 N/m (b) 0.585 m

69. (a) 434 W (b) 868 W

73. (a)

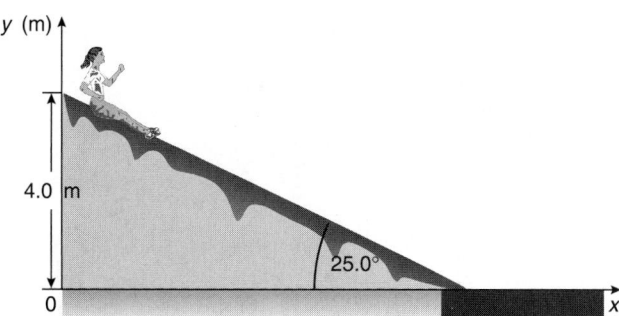

   (b) 3.1×10^3 J (c) 8.9 m/s (d) 20 m (e) 0 J along both paths (f) -3.1×10^3 J (g) -1.4×10^3 W; No, since the velocity is changing.

77. 109 kW

81. (a) J/m (b) −C $\hat{\mathbf{i}}$ for $x > 0$ m, C $\hat{\mathbf{i}}$ for $x < 0$ m (c)

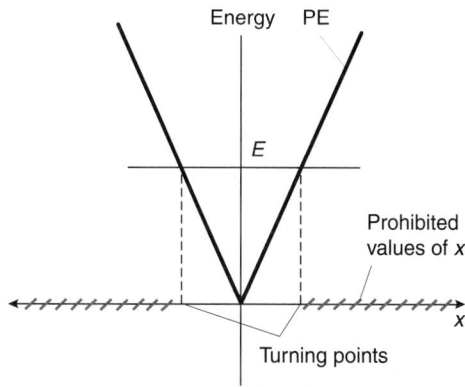

   (d) Two turning points (e) The motion is oscillatory but is not simple harmonic oscillation. This force has a constant magnitude whereas the force in simple harmonic oscillation is a function of position.

CHAPTER 9

1. 3.3×10^2 km/h

5. 4.63 m/s = 16.7 km/h

9. 4.96 kg·m/s

13. (a)

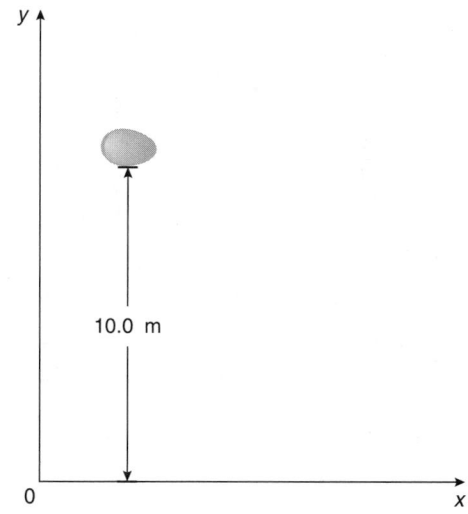

   (b) 98.1 J (c) Yes, since the total force is constant. (d) $(-14.0\ \text{N·s})\,\hat{\mathbf{j}}$

17. (a) $(15.0\ \text{N·s})\,\hat{\mathbf{i}}$ (b) $(7.50\ \text{N})\,\hat{\mathbf{i}}$ (c) $(15.0\ \text{N·s})\,\hat{\mathbf{i}}$ (d) $(3.75\ \text{N})\,\hat{\mathbf{i}}$

21. 87 m/s²

25. 1.718

29. Answer is given in the statement of the problem.

33. (a) $(8.00\ \text{m/s})\,\hat{\mathbf{i}}$ (b) It is an inelastic collision since the kinetic energy is not conserved. $\Delta \text{KE} = -2.40 \times 10^4$ J $\neq 0$ J (c) No. The momentum of each particle in a collision is not conserved; only the total momentum of all the particles is conserved in a collision.

37. (a) A stupid one. Completely inelastic.
(b) $(1.67 \times 10^4 \text{ kg} \cdot \text{m/s}) \, \hat{\imath} + (1.11 \times 10^4 \text{ kg} \cdot \text{m/s}) \, \hat{\jmath}$
(c) The same as (b)   (d) 10.0 m/s   (e) -1.01×10^5 J

41. (a) Take $\hat{\imath}$ east and $\hat{\jmath}$ north, $(19 \text{ m/s}) \, \hat{\imath} - (4 \text{ m/s}) \, \hat{\jmath}$
(b) -5×10^4 J

45. (a) No. The speed of the car was 29.0 m/s.   (b) 0.930 m/s

49. 17

53. A proof

57. $(-0.129 \text{ m}) \, \hat{\imath} + (0.257 \text{ m}) \, \hat{\jmath}$

61.

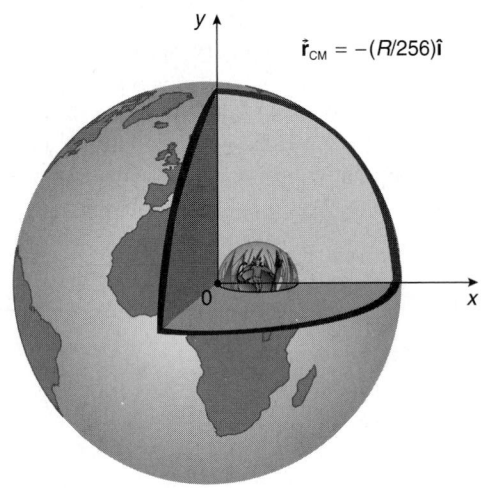

$\bar{r}_{CM} = -(R/256)\hat{\imath}$

65. 0.71 m/s in a direction opposite to that of his daughter

69. (a) 1.8 m/s   (b) $\Delta KE = -3 \times 10^4$ J   (c) 1.8 m/s

CHAPTER 10

1. 0 kg m²/s, orbital

5. Answers are given in the statement of the problem.

9. 4.78×10^4 km

13. $7mR^2/4$

17. Answer is given in the statement of the problem.

21. (a) 1.36×10^6 J   (b) 0.047

25. (a) Answer is given in the statement of the problem.
(b) 2.83×10^{42} J   (c) Answer is given in the statement of the problem.   (d) -7.17×10^{31} W, 1.87×10^5 times the luminosity of the Sun

29. (a) $7mR^2/5$ (b) $3mR^2/2$ (c) $5mR^2/3$

33. (a) 7.293×10^{-5} rad/s   (b) 7.08×10^{33} kg·m²/s   (c) The effect is to lengthen the day (the moment of inertia increases and decreases ω_{spin} because of conservation of angular momentum).   (d) 1.99×10^{-7} rad/s   (e) 2.64×10^{40} kg·m²/s
(f) 3.73×10^6

37. $\dfrac{13}{10}\, \omega_0$

41. (a) Answer is given in the statement of the problem.

(b)

	KE_{rot}	KE_{CM}
highest	cylindrical shell	sphere
↓	spherical shell	disk
	disk	spherical shell
lowest	sphere	cylindrical shell

(c) First to last: sphere, disk, spherical shell, cylindrical shell.

45. (a) 1.96×10^3 N·m   (b) 7.34 rad/s²   (c) No. The torque of the weight of the beam about the hinge varies with the orientation of the beam to the vertical.   (d) 3.84 rad/s

49. $\left(\dfrac{1 - \varepsilon}{1 + \varepsilon}\right)^2$

53. (a) 4.67×10^6 m from the center of the Earth along the Earth–Moon line   (b) 2.81×10^{34} kg·m²/s
(c) 3.47×10^{32} kg·m²/s   (d) 2.37×10^{29} kg·m²/s
(e) 7.08×10^{33} kg·m²/s

57. (a) The spherical shell   (b) The solid sphere reaches the bottom of the incline before the spherical shell if both are released at rest from the same location.

61. $\dfrac{v_0/R}{1 + \dfrac{M}{2m}}$

The lazy-susan rotates in a direction opposite to that in which the mouse is going.

65. (a) 1.0×10^{42} kg·m²/s   (b) 5×10^{43} kg·m²/s   (c) 0.7 d

69. Answer is given in the statement of the problem.

73. 59°

77. 0.71 m

CHAPTER 11

1. 8.0 kg

5. 1.2×10^7 N/m²

9. Answer is given in the statement of the problem.

13. 1.5 m

17. (a) 7.92×10^3 m   (b) 0.93 km

21. (a) 1   (b) 10

25. 10.3 m

29. 0.833×10^3 kg/m³

33. 2×10^{-3} kg

37. (a) It floats.   (b) 3.00×10^4 kg

41. (a)

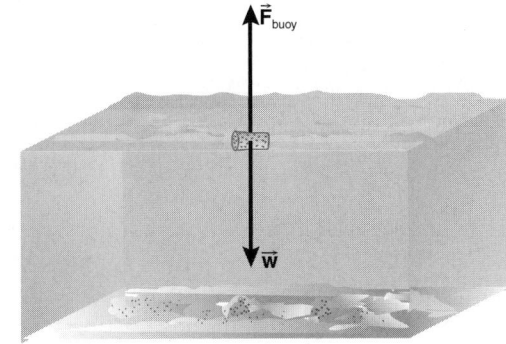

(b) 0.120   (c) 4.2×10^{-2} kg   (d) 0.26 N

45. 2.7×10^3 kg/m³

49. 5.3 cm

53. For 2 mm diameter straw and $h = 1.5$ cm, $\gamma = 0.074$ N/m.

57. 1.2×10^2 m/s

61. 9.0×10^{-3} m³/s

65. (a) 19.8 m/s   (b) 156 liters/s   (c) 20.0 m

69. 26 m/s

CHAPTER 12

1. 1.7×10^2 m

5. (a) 1.8×10^3 km   (b) Only the distance to the focus was found.
(c) Use three seismometers at different locations.

9. (a)

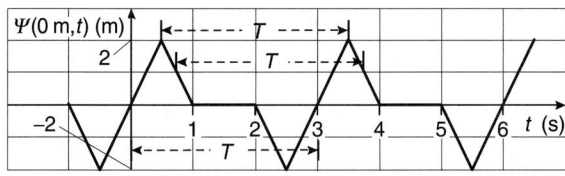

(b) 0.33 Hz

(c)

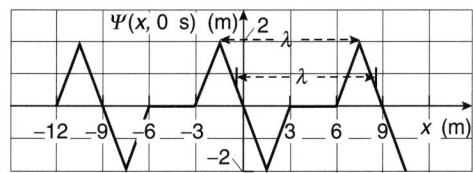

13. (a) and (b)

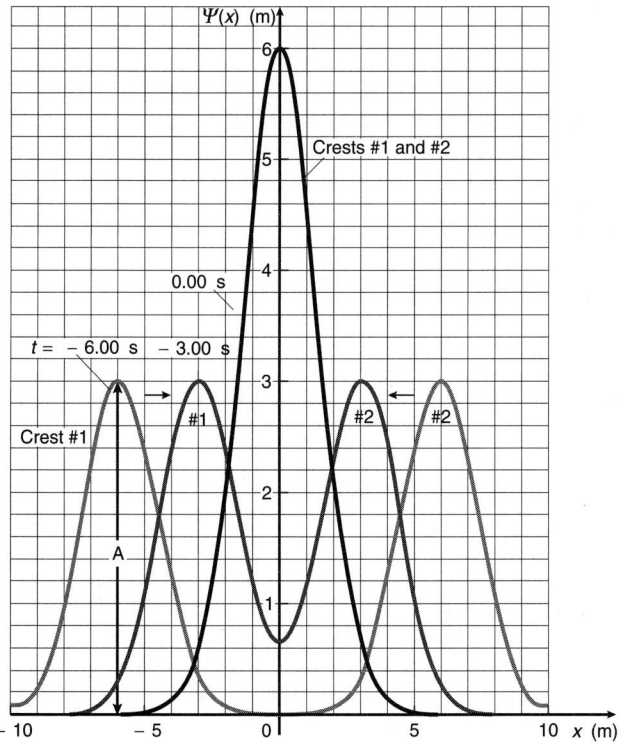

(c)

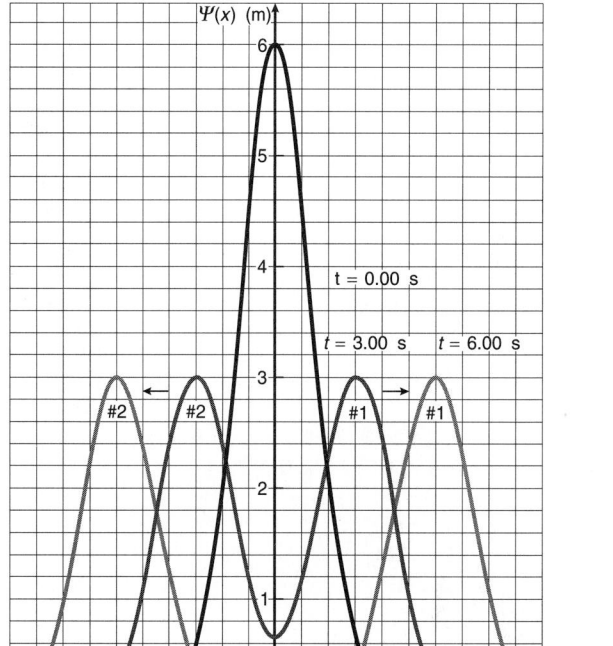

(d)

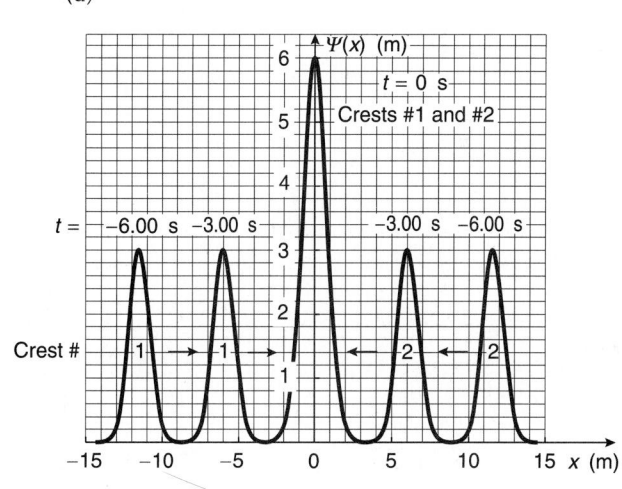

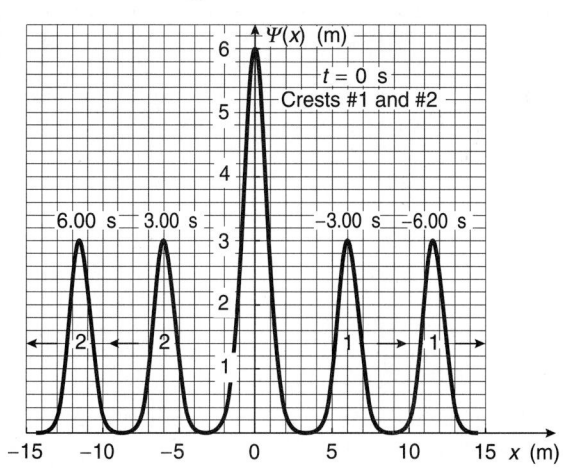

17. (a) 6.00×10^{-3} Pa   (b) 8.06 rad/m   (c) 8.06 wavelengths
(d) 0.780 m   (e) 2.76×10^3 rad/s   (f) 439 Hz
(g) 2.28×10^{-3} s   (h) 342 m/s
(i)

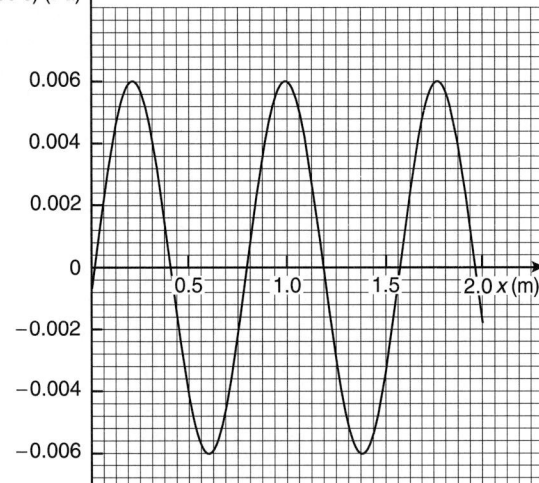

(j)

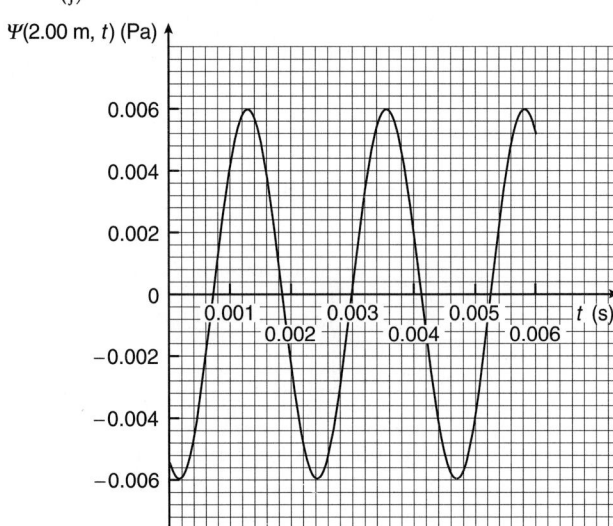

(k) -1.23×10^{-3} Pa. The negative result implies a rarefaction with an absolute pressure below the average value in the medium in which the sound is propagating.

21. (a) 3.00 Hz   (b) $\Psi(x, t) = (0.50 \times 10^{-2}$ m$) \cos [(31.4$ rad/m$) \times x - (18.8$ rad/s$) t])$

25. 83 N

29. 3.95×10^{-2} s

33. 86 Hz

37. $I = \dfrac{P}{2\pi r \ell}$

41. (a) 3.1×10^2 W   (b) 1.6×10^2 m

45. 106 dB

49. (a) 2.180×10^3 Hz   (b) 3 m/s

53. (a) 1200 Hz   (b) 1.21×10^3 Hz   (c) 1.19×10^3 Hz

57. 30.2 kHz

61. (a) Mach number = 2.1   (b) 29°

65. Answer is given in the statement of the problem.

69. (a) The first (second) wavefunction represents a wave moving toward increasing (decreasing) values of x.   (b) The two will form a standing wave; $\lambda/2 = 0.349$ m   (c) 0.175 m, 0.524 m, 0.873 m

73. $\dfrac{v}{\alpha}$

77. (a)

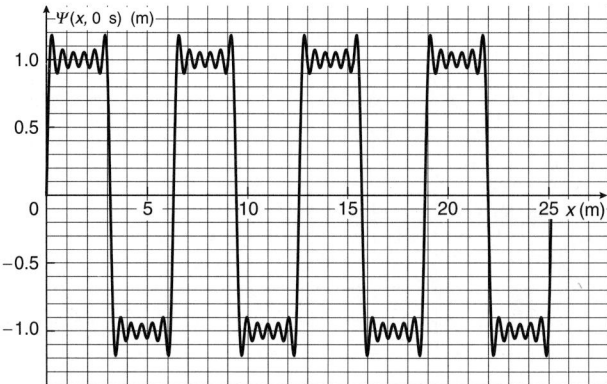

(b)

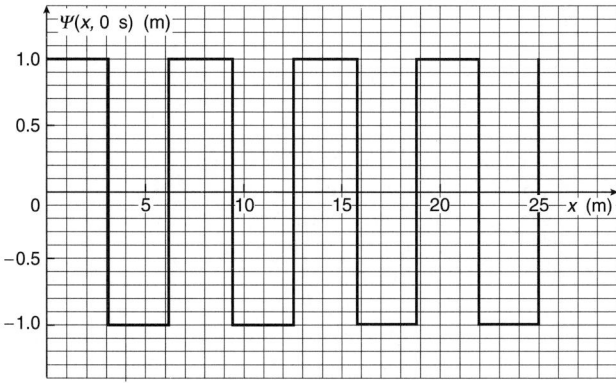

which is called a square wave.

(c) ≈6.3 m

CHAPTER 13

1. 37.0 °C

5. −13 °F

9. 50.00 °C

13. 0.012%

17. (a) 7.6×10^{-4} m   (b) $\Delta T = 63$ K

21. $\Delta T = 4 \times 10^2$ K

25. A proof

29. 1.5×10^4 N

33. A proof

37. 1.4×10^{-23} Pa

41. 9.3 m below the surface

45. (a) It all melts   (b) 8.40 °C

49. (a) 9.4×10^7 J   (b) 5.2 h

53. 3.9×10^3 W

57. (a) 1   (b) 14 °C

61. 7.4

65. (a) 3.04×10^3 J   (b) 1.52×10^3 J   (c) 2.28×10^3 J
(d) -3.04×10^3 J   (e) -1.52×10^3 J   (f) -2.28×10^3 J

69. (a) -2.86×10^3 J   (b) 2.86×10^3 J

73. (a) A proof   (b) 0.11 K   (c) 0.213 K

77. 0.398 K

CHAPTER 14

1. 1.908×10^{16} y

5. 3.1×10^{-16} J

9. 1.93×10^3 m/s; The rms speed of the hydrogen molecules is four times faster than that of the oxygen molecules.

13. 5.14×10^{-23} N·s

17. (a) 2.46×10^{-2} m³   (b) ~3.4×10^{-9} m

21. 0.86 m/s

25. 2.01×10^4 K

29. 1.93 km/s; Although this rms speed is considerably less than the escape speed, some hydrogen molecules will have speeds exceeding the escape speed and can escape, if moving in the right direction, if no collisions occur. Heat transfer from the Earth continually replenishes the high-speed molecules, so they gradually escape. Such a process also explains the phenomenon of evaporation.

33. 1×10^{-23} Pa

37. (a) The helium atom loses KE while the hydrogen atom gains KE.   (b) The KE of each is unchanged.

41. 2

45. 2.67×10^{-2} mol

49. Answer is given in the statement of the problem

53. (a) 1.93×10^3 m/s   (b) 1.25×10^6 Pa   (c) No. $\Delta U = 0$ J.
(d) 6.25×10^5 Pa

57. (a)

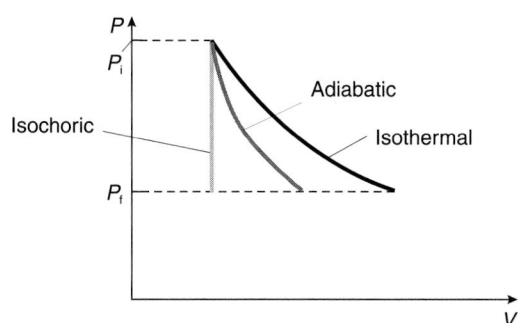

(b) The isochoric process   (c) The isothermal process
(d) Answer is given in the statement of the problem.
(e) The isochoric process

61. A proof

CHAPTER 15

1. 0.558

5. 0.055

9. 400 K and 500 K

13. (a) 186 MW   (b) 1.5×10^4 kg/s, 15 m³/s

17. The coefficient of performance increases more by warming the colder reservoir than by cooling the warmer reservoir.

21. (a) 0.400   (b) $|Q_H| = 3.20 \times 10^3$ J, $|Q_C| = 2.40 \times 10^3$ J

25. (a) 3.9 J; No   (b) 71 J

29. (a) 2.8×10^2 J   (b) 7.8×10^2 J

33. -6.05×10^3 J/K, No

37. (a) 141 J/K   (b) −136 J/K   (c) 5 J/K

41. (a)

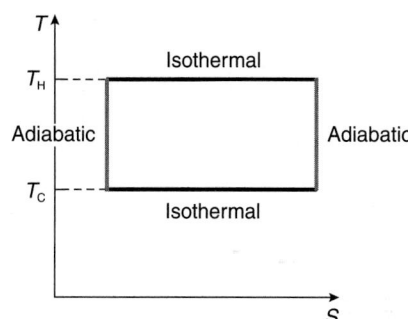

(b) The area enclosed by the cycle is equal to the total heat transfer to the gas system in one cycle.

(c) Take the ratio of $T_H - T_C$ to T_H:

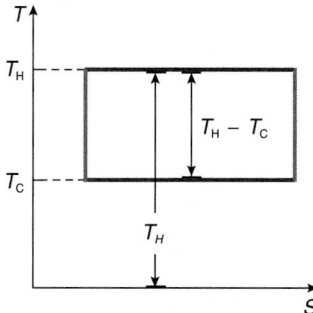

45. (a) 4.00 kW   (b) 3.50 W/K

49. 1.6×10^3 J/K

53. (a) $Q = 0$ J   (b) $\Delta U = -2000$ J   (c) $\Delta S = 0$ J/K   (d) The temperature fell.   (e) $\Delta T = -48.1$ K

57. $\Delta S = nc_P \ln\left[\dfrac{(T_1 + T_2)^2}{4T_1 T_2}\right]$

61. Shake the box many times and assess the experimental probability of securing each macrostate after a shake. If after many shakes, you *never* see a result with 0 heads, you can begin to become suspicious.

65. (a)

Number on left side	Number on right side	Number of microstates
6	0	1
5	1	6
4	2	15
3	3	20
2	4	15
1	5	6
0	6	1

(b) 0 J/K   (c) The macrostate with three particles in each half: 4.137×10^{-23} J/K

69. (a)

Macrostate Number of heads	Number of microstates Ω	$\frac{S}{k}$
0	1	0
1	20	3.00
2	190	5.25
3	1140	7.04
4	4845	8.49
5	15 504	9.65
6	38 760	10.57
7	77 520	11.26
8	125 970	11.74
9	167 960	12.03
10	184 756	12.13
11	167 960	12.03
12	125 970	11.74
13	77 520	11.26
14	38 760	10.57
15	15 504	9.65
16	4845	8.49
17	1140	7.04
18	190	5.25
19	20	3.00
20	1	0

(b) The graph of the number of microstates versus the macrostate is shown below:

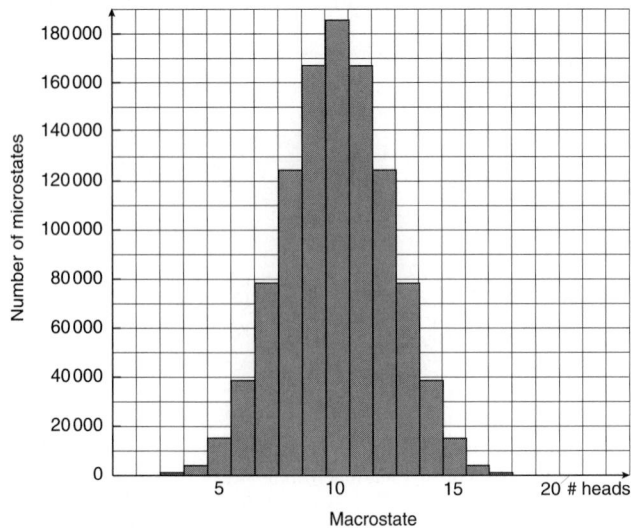

(c) The graph of $\frac{S}{k}$ versus the macrostate is shown below:

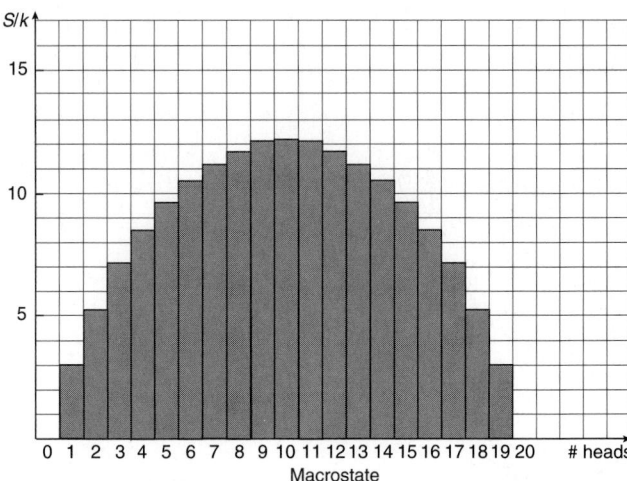

CHAPTER 16

1. 9.6486×10^4 C

5. 9.6486×10^4 C

9. (a) 365 N   (b) 365 N

13. 2.95×10^{17} C

17. (a)

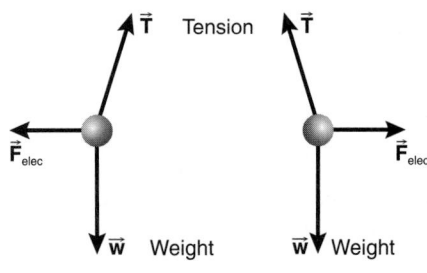

(b) Answer is given in the statement of the problem.
(c) 4.4×10^{-7} C   (d) 2.7×10^{12}

21. (a) $\vec{r}_{\text{center of charge}} = \dfrac{\sum_i q_i \vec{r}_i}{Q_{\text{total}}}$

(b) If the total charge is zero, then $\vec{r}_{\text{center of charge}}$ is undefined.

25. -1.2×10^{-2} C

29. (a) Negative

(b)

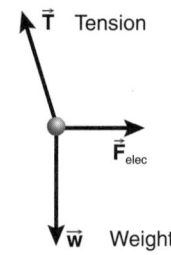

(c) -3.5×10^{-4} C

33. Going left to right: $+2q, -q, -q$

37. The information implies that the three atoms of the molecule are collinear so the two dipole moment of the two carbon–oxygen bonds point in opposite directions and vector sum to zero.

41. (a) Since the distance from Q to the dipole is much greater than the separation of the charges in the dipole, each charge of the dipole feels essentially the same electric field. Hence the magnitude of the force on the dipole is approximately 0 N. (b) $0\ \mathrm{N\cdot m}$

45. Answer is provided in statement of problem.

49. $1.16 \times 10^{-7}\ \mathrm{C/m^2}$

53. $1.81 \times 10^3\ \mathrm{N/C}$

57. (a) Q and σ must be like charges. (b) Answer is given in the statement of the problem.

61. (a) $\vec{\mathbf{F}}$ must be antiparallel to $\vec{\mathbf{v}}$, so $\vec{\mathbf{E}}$ must be parallel to $\vec{\mathbf{v}}$.

(b)

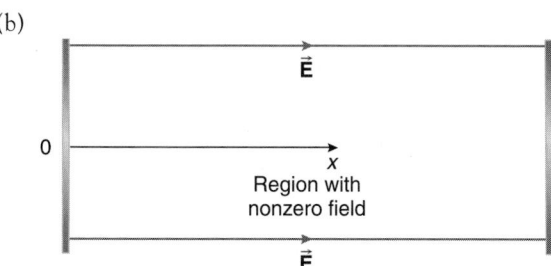

(c) $a_x = -1.76 \times 10^{13}\ \mathrm{m/s^2}$ (d) $2.84 \times 10^{-7}\ \mathrm{s}$ (e) $0.71\ \mathrm{m}$ (f) The electron will not remain at rest. It will retrace its path and emerge from the region of the field with a speed equal to the initial speed of the electron.

65. $0.142\ \mathrm{m}$

69. $0\ \mathrm{N\cdot m^2/C}$

73. $2.3\ \mathrm{N\cdot m^2/C}$

77. (a)

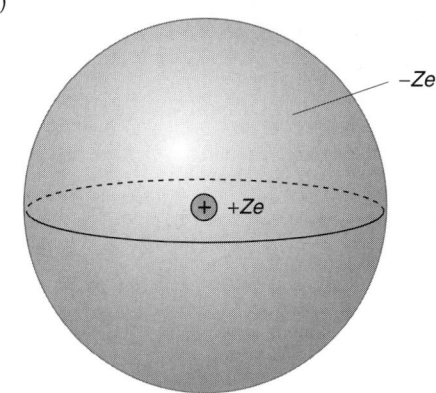

(b) $0\ \mathrm{N/C}$

(c) $E = \dfrac{1}{4\pi\varepsilon_0} \dfrac{Ze}{r^2} \left(1 - \dfrac{r^3}{R^3} \right)$

CHAPTER 17

1. $7.50 \times 10^{-4}\ \mathrm{J}$

5. (a) $3.0 \times 10^9\ \mathrm{V}$ (b) $54\ \mathrm{C}$

9. (a) The gravitational potential energy per kilogram (b) $49.1\ \mathrm{J/kg}$

13. (a) $-8.84 \times 10^{-8}\ \mathrm{C/m^2}$ (b) $8.84 \times 10^{-8}\ \mathrm{C/m^2}$ (c) $-1.00 \times 10^3\ \mathrm{V}$ (d) $9.81 \times 10^{-9}\ \mathrm{C}$

17. Take $\hat{\mathbf{j}}$ to the vertically upward. Then the gravitational potential is gy.

21. $8.3 \times 10^{-3}\ \mathrm{m}$ If the radius were smaller, the magnitude of the field would exceed the maximum magnitude.

25. (a) $0\ \mathrm{V}$

(b) $\dfrac{Ze}{4\pi\varepsilon_0} \left[\dfrac{1}{r} + \dfrac{r^2}{2R^3} - \dfrac{3}{2R} \right]$

29. (a) $1.08 \times 10^3\ \mathrm{V}$ (b) $1.62 \times 10^3\ \mathrm{V}$

33. (a) 2.00 cm sphere: 1.35 kV; 4.00 cm sphere: 675 V (b) 900 V for each, 1.00 nC was exchanged

37. (a) $\dfrac{\lambda}{2\pi\varepsilon_0} \ln\left(\dfrac{a}{r} \right)$

(b) $r < a$ (c) $r > a$ (d) The distribution of charge extends to infinity; likewise, when $r = 0$ m the natural logarithm of ∞ is undefined. (e) A proof.

41. (a) $1.00 \times 10^3\ \mathrm{eV}$ (b) $1.88 \times 10^7\ \mathrm{m/s}$

45. (a) The lower plate (b) $200\ \mathrm{eV} = 3.20 \times 10^{-17}\ \mathrm{J}$ (c) $8.38 \times 10^6\ \mathrm{m/s}$

49. (a) and (b)

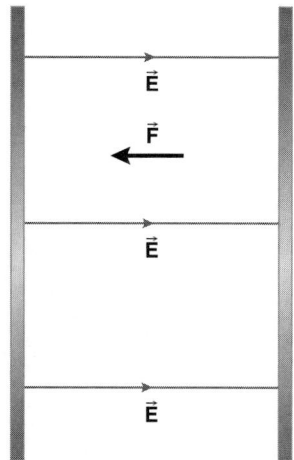

(c) $0.9 \times 10^{-15}\ \mathrm{J}$ (d) $0.9 \times 10^{-15}\ \mathrm{J}$ (e) $4 \times 10^7\ \mathrm{m/s}$

53. (a) $-40\ \mathrm{eV}$ (b) $+60\ \mathrm{eV}$

57. (a) $\dfrac{1}{4\pi\varepsilon_0} \dfrac{Q}{R}$

(b) $-\dfrac{1}{4\pi\varepsilon_0} \dfrac{|q| Q}{R}$

(c) $\left(\dfrac{2|q| Q}{4\pi\varepsilon_0 R m} \right)^{1/2}$

(d) The gravitational potential energy is proportional to the mass, and so appears in both the KE and PE in the CWE theorem and divides out. The electrical potential energy is independent of the mass, and so the mass appears in the KE term, but not the PE term and so does not divide out. (e) 3×10^{-19} C $\approx 2e$ (f) In this case $Q \approx 3 \times 10^{-15}$ C $\approx 2 \times 10^4$ e; Nuclei with such high charge quantum numbers do not exist in nature.

61. (a) $\vec{E} = -(0.01 \times 10^7$ N/C) $\hat{\imath} - (1.04 \times 10^7$ N/C) $\hat{\jmath}$

(b) -4.02×10^5 V   (c) 6.2×10^{-23} J   (d) $(6 \times 10^{-23}$ N·m) $\hat{k}$

65. 0 J

69. Answer is given in the statement of the problem.

CHAPTER 18

1. 3 nodes

5. A and D are in parallel; B and E are in parallel.

9. 6 in series, all connected (+) to (−)

13. 23×10^{-12} C

17. C increases by a factor of 2

21. Answer is given in the statement of the problem.

25. 0.92 μF

29. 33 μF

33. (a) 1.20 mC   (b) In parallel   (c) 8.3×10^2

37.

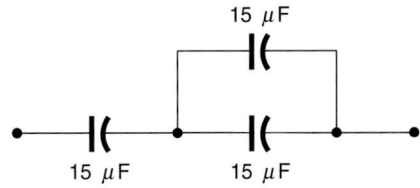

41. 1.4×10^2 μF

45.

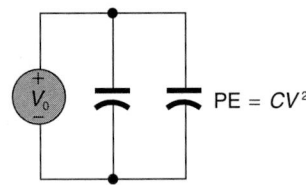

49. κ

53. Answer is given in the statement of the problem.

57. 0.1 μF, 2×10^{12} J

CHAPTER 19

1. About 0.16 Pa

5. (a) 4.84×10^{-5} m/s   (b) 5.74 hours

9. 0.98 Ω

13. 6.0 Ω

17. The smaller-diameter wire has a drift speed four times that of the larger-diameter wire.

21. $R = \dfrac{(\text{resistivity})(\text{density})\,\ell^2}{m}$

25. (a) 3.0 kΩ (b) 1.3 Ω (c) $(13/8)R$ (d) 1.3 Ω

29. The filament of the 100 W bulb has the greater resistance.

33. (a) 2.3×10^{-3} m   (b) Gauge 10

37. The 100 watt bulb absorbs 15 W and the 60 watt bulb absorbs 25 W. The 60 watt bulb is brighter.

41. (a) and (b)

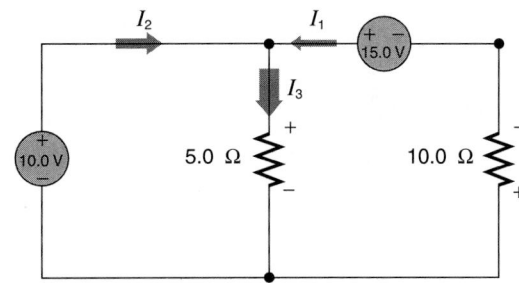

(c) $I_1 = 1.5$ A, $I_2 = 0.50$ A, $I_3 = 2.0$ A
(d) $P_{5\,\Omega} = 20$ W, $P_{10\,\Omega} = 2.5$ W, $P_{10\,V} = -15$ W, $P_{15\,V} = -7.5$ W

45. (a) 7.00 Ω (b) Resistors 5 and 6   (c) 0.29 A

(d)

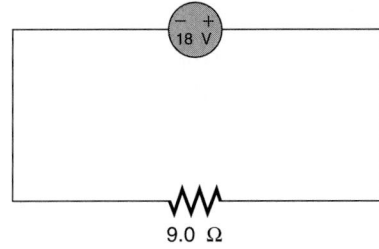

49. (a) 0 A   (b) 12.0 V   (c) The potential difference between A and B is entirely provided by the 12.0 V independent voltage source since there is no current in the 4.0 Ω resistor.   (d) $P_{8\,V} = 8.0$ W, $P_{12\,V} = 0$ W, $P_{16\,V} = -16$ W   (e) $P_{\text{all resistors}} = 8.0$ W

53. (a)

(b) 2.0 A   (c) 8.0 W   (d) -36 W

57. (a) $\dfrac{R_0}{R_0 + r} V_0$

$-\dfrac{V_0^2}{R_0 + r}$

(b) 0 V, $\dfrac{V_0}{r}$

61. (a)

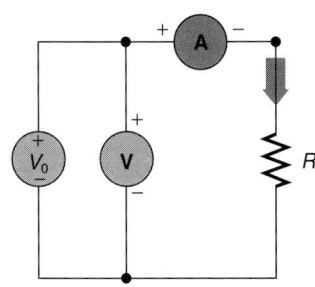

(b) 20.0 Ω (c) 1.25 W (d) 5.00 V

65. (a) 10.0 V, A_1 and A_2 both indicate 0 A. (b) $r = 1.6\ \Omega$, $R_1 = 1.6\ \Omega$, $R_2 = 12\ \Omega$, $R_3 = 8.0\ \Omega$

69. $\tau \ln 2$ ($\approx 0.693\ \tau$); Yes

73. Across C_1: $\dfrac{R_1}{R_1 + R_2} V_0$

Across C_2: $\dfrac{R_2}{R_1 + R_2} V_0$

CHAPTER 20

1.

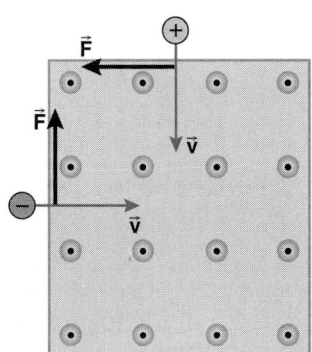

5.

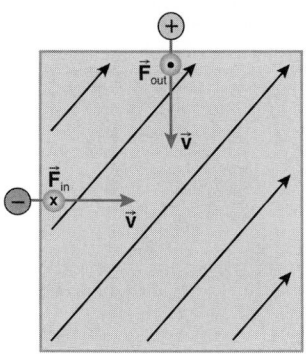

9. $(-3.0 \times 10^{-4}\ \text{T})\,\hat{\jmath} + (4.0 \times 10^{-4}\ \text{T})\,\hat{k}$

13. (a)

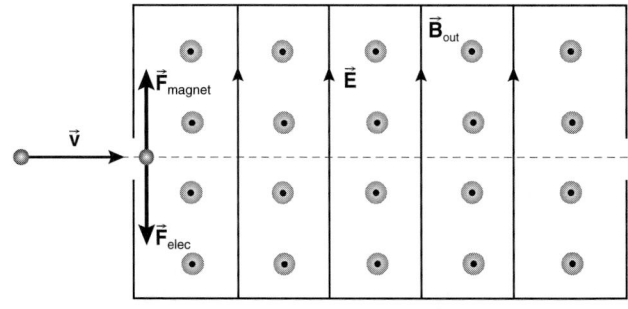

(b) 2.90×10^{-4} N directed perpendicular to and out of the page (c) $0.145\ \text{m/s}^2$

17. (a) and (b)

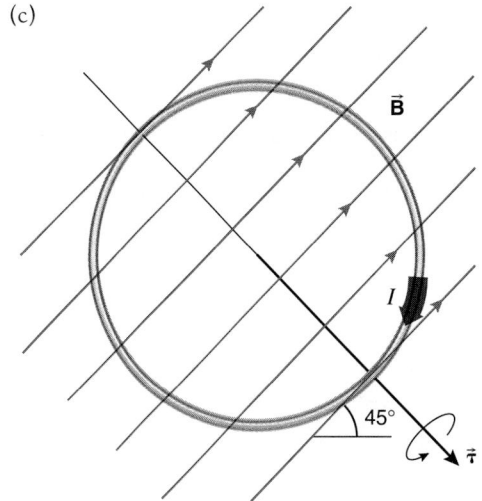

(c) 1.3×10^5 m/s

21. Answers are given in the statement of the problem.

25. (a) $1.2 \times 10^{-2}\ \text{A} \cdot \text{m}^2$ (b) $7.2 \times 10^{-3}\ \text{N} \cdot \text{m}$

(c)

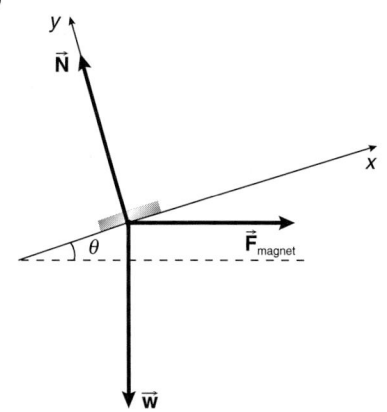

(d) 0 J

29. (a) $I\ell B$

(b)

(c) $\left(\dfrac{I\ell B}{m} \cos \theta - g \sin \theta\right) t\hat{\imath}$

(d) $\dfrac{mg \tan \theta}{I\ell}$

33. The area of the coil must be 4.00×10^{-2} m².

37. Answer is given in the statement of the problem.

41. 0 T

45. (a) Answer is given in the statement of the problem.

(b)

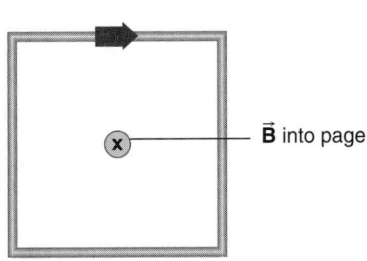

$\vec{B}$ into page

49. (a) 3.0×10^{-4} T   (b) 2.4×10^{-15} N, 30° counterclockwise from $\hat{\imath}$ in sketch   (c) 2.4×10^{-15} N, 30° counterclockwise from $-\hat{\imath}$ in sketch   (d) 5.0×10^{7} m/s

53. (a) 7.5×10^{-6} T perpendicularly out of page   (b) 1.9×10^{-5} N toward the infinite wire   (c) 15×10^{-6} T perpendicularly out of page   (d) 3.8×10^{-5} N away from the infinite wire   (e) 0 N (f) 1.9×10^{-5} N away from the infinite wire   (g) 1.9×10^{-5} N away from the rectangular loop

57. (a) 1.2×10^{-3} T   (b) 0 T

61. Answers are given in the statement of the problem.

65. (a) and (b)

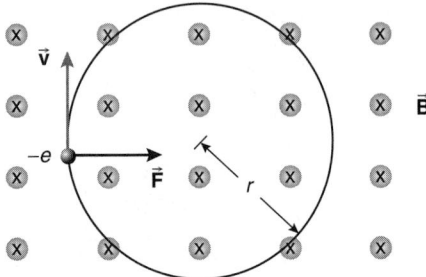

(c) $r = \dfrac{m_e v}{eB}$

(d) 0 J   (e) It remains at rest.

69. (a) 90°, 6.00×10^{-14} N   (b) 14°   (c) 160 eV   (d) The same: 160 eV

73. A proof

CHAPTER 21

1. Answer is given in statement of problem.

5. (a) 311 V   (b) 50.0 Hz

9. (a) The charge separation creates an electric field that produces an electric force on the charges equal in magnitude but opposite in direction to the magnetic force on the charges. These forces are *not* a Newton's third law force pair.

(b)

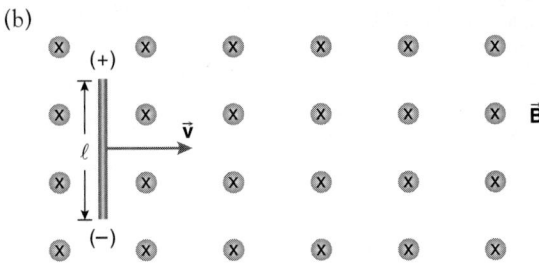

(c) $vB\ell$ (d) $\ell v \, \Delta t$ (e) $vB\ell$, the same as part (c)

13. (a) $\Phi = \pi r^2 (\cos \theta) B_0 e^{-\alpha t}$

(b) Induced emf $= \alpha \pi r^2 (\cos \theta) B_0 e^{-\alpha t}$

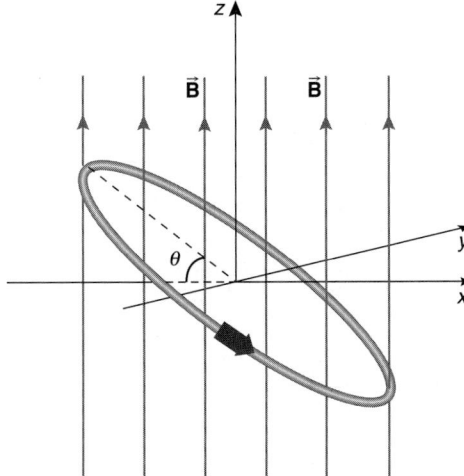

17. (a) -3×10^{-3} V   (b) For a hoizontally moving train, only the vertical component contributes to the flux.

21. (a) 314 rad/s   (b) 6.4×10^{-3} T

25. The answer depends on the frequency of the station. For 89.1 MHz, $\lambda = 3.37$ m.

29. (a) 9.0 V   (b) -7.5 V   (c) 6.0 V

33. (a) $\Phi = \dfrac{\mu_0}{4\pi} 2I\ell \ln\left(\dfrac{R_2}{R_1}\right)$

(b) Answer is given in the statement of the problem.

(c) 2.8×10^{-7} H/m

37. (a)

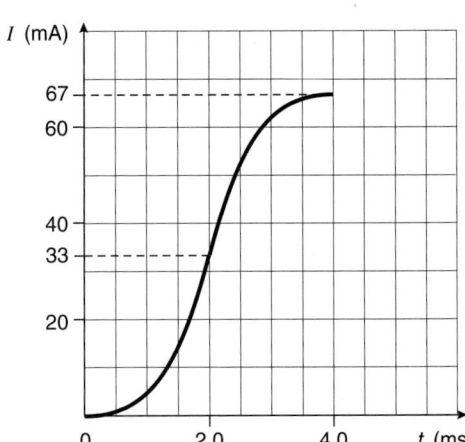

(b) 33 mA, 67 mA

(c) The current is $1/L$ times the area under the $V(t)$ versus t graph.

41. 5.2 time constants

45. (a) 1.67×10^{-3} s (b) 80.0×10^{-3} A (c) $0.69\,\tau = 1.2 \times 10^{-3}$ s

49. (a) For a resistor, V is proportional to I; this is not the case here. (b) For a capacitor, I is proportional to

$$\frac{dV}{dt}$$

this is not the case here. (c) For an inductor V is proportional to

$$\frac{dI}{dt}$$

this *is* the case here. (d) 0.10 H

(e)

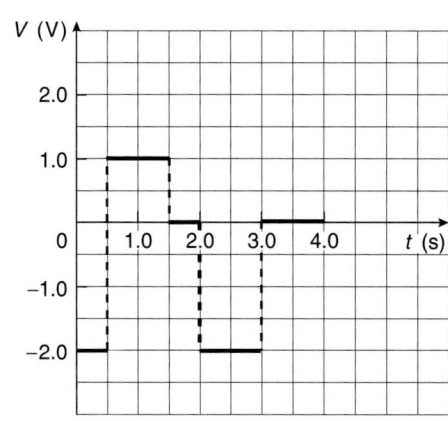

53. 30 pF to 3.1×10^2 pF

57. 64

61.

Sketch 1	Sketch 2
Primary: emf: 2, V: 1	Primary: emf: 2, V: 1
Secondary: emf: 3, V: 4	Secondary: emf: 3, V: 4

CHAPTER 22

1. z_1, z_2: (a) $5, \sqrt{29} \approx 5.4$ (b) $5\,\angle 0.927$ rad, $\sqrt{29}\,\angle(-1.19$ rad) $\approx 5.4\,\angle(-1.19$ rad) (c) $5\,e^{i(0.927\,\text{rad})}, \sqrt{29}\,e^{-i(1.19\,\text{rad})} \approx 5.4\,e^{-i(1.19\,\text{rad})}$

(d)

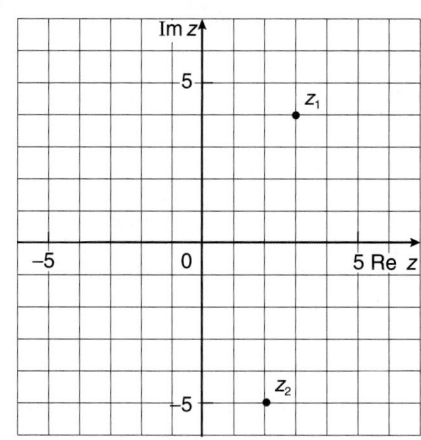

5. (a) -4 (b) $-i(2)$

(c) $\dfrac{i}{2}$

9. A proof

13. A proof

17.

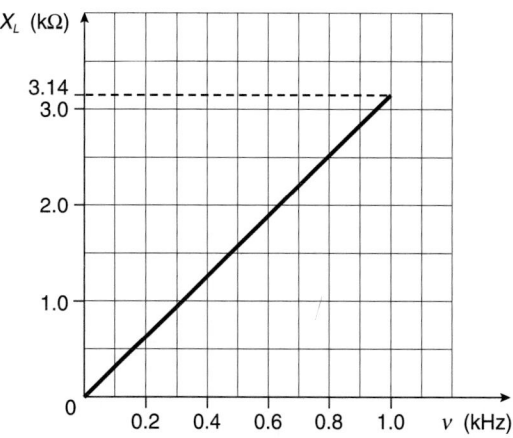

21. $>3.2 \times 10^2$ Hz

25. (a) $v = \dfrac{1}{2\pi\,(LC)^{1/2}}$ (b) no

29. (a)

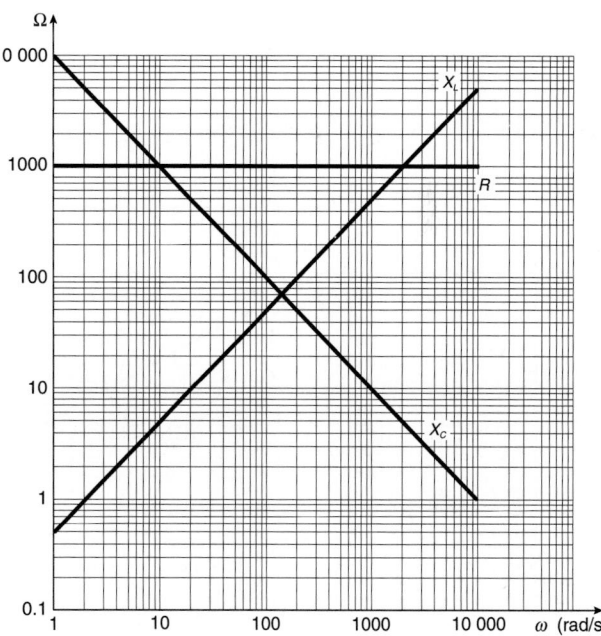

(b) Since $X_L = \omega L$, log X_L will proportional to log ω; Since $X_C = 1/(\omega C) = (\omega C)^{-1}$, log X_C will proportional to $-$log ω; Since R is a constant, the graph of R versus ω will be a horizontal line. (c) 2.00×10^3 rad/s, 318 Hz (d) 10.0 rad/s, 1.59 Hz (e) 141 rad/s, 22.4 kHz

33. $\dfrac{I_0}{\sqrt{2}}$

37. (a) 60.0 Hz (b) $i(56.6\ \Omega)$ (c) $(170\text{ V})\,e^{i\,[(377\,\text{rad/s})\,t]}$, 170 V$\angle[(377$ rad/s$)\,t]$

(d)

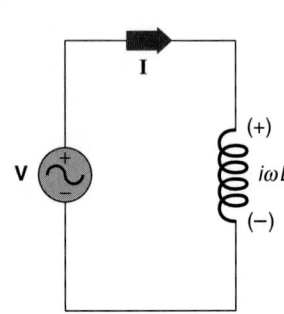

(e) $(3.00 \text{ A}) \, e^{i \, [(377 \text{ rad/s}) \, t \, - \, \pi/2 \text{ rad}]}$

$$(3.00 \text{ A}) \, \angle \left[(377 \text{ rad/s}) \, t \, - \, \frac{\pi}{2} \text{ rad} \right]$$

(f) $(170 \text{ V}) \, e^{i \, [(377 \text{ rad/s}) \, t]}$, $170 \text{ V} \, \angle \, [(377 \text{ rad/s}) \, t]$

(g) $I(t) = (3.00 \text{ A}) \cos \left[(377 \text{ rad/s}) \, t \, - \, \frac{\pi}{2} \text{ rad} \right]$

(h) $V(t) = (120 \text{ V}) \cos [(377 \text{ rad/s}) \, t]$

(i) $\dfrac{\pi}{2}$ rad

(j) 3.00 A   (k) 2.12 A   (l) 170 V   (m) 120 V   (n) 0   (o) 0 W

41. $\omega = \dfrac{\sqrt{3}}{RC}$

voltage gain $= \dfrac{1}{2}$

45. (a)

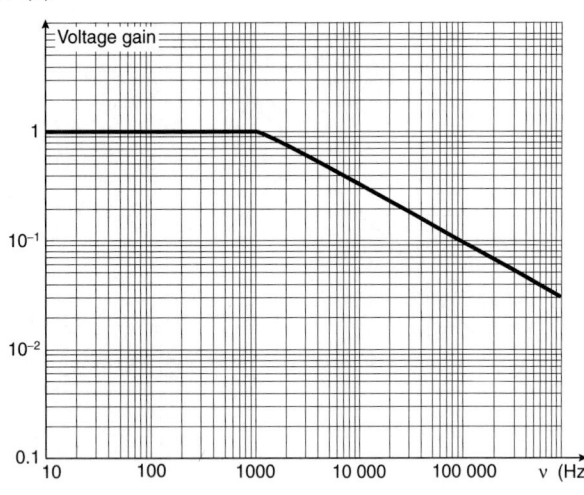

(b) $\nu \approx 5$ kHz; A calculation yields 5.5 kHz.

49. (a)

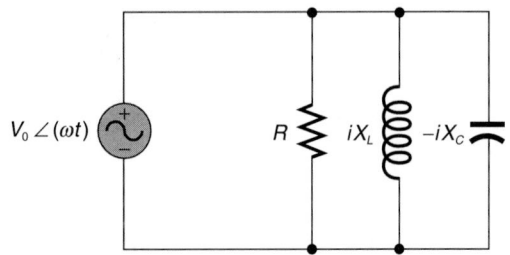

(b) Answers are given in the statement of the problem.
(c) Answer is given in the statement of the problem.
(d) Answer is given in the statement of the problem.

(e) $\omega = \dfrac{1}{(LC)^{1/2}}$

This is the same as the resonant angular frequency of a series *RLC* circuit.

CHAPTER 23

1. 14.9 W/m²

5. 4.54 kJ

9. Answer is given in the statement of the problem.

13. Answer is given in the statement of the problem.

17. $d \left(\dfrac{n^2 - \sin^2 \theta}{n^2 \cos^2 \theta} \right)^{1/2}$

21. 25°

25. Answer is given in the statement of the problem.

29. Answer is given in the statement of the problem.

33. (a) $s' = 4.5$ cm   (b) $m = 0.090$ (c) Virtual   (e) Upright

37. (a) Virtual   (b) $R = \infty$ m

41. $2v \cos \theta$

45. Answers are given in the statement of the problem.

49.

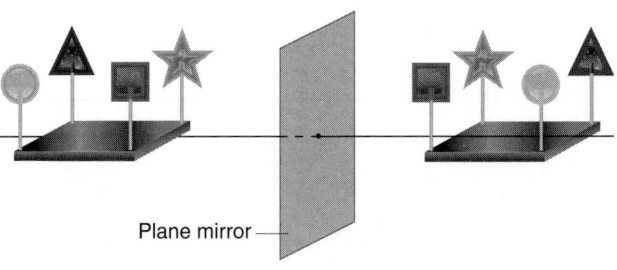

Plane mirror

53. (a) Depth of 3.8 cm   (b) $m = 1.0$ (c) Virtual, upright

57. (a)

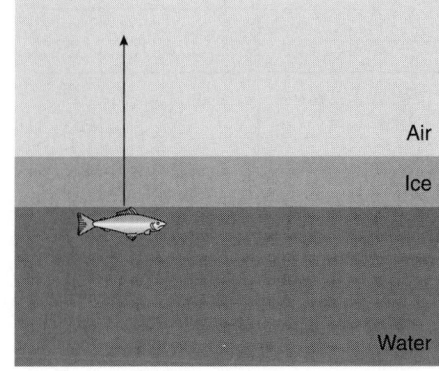

(b)

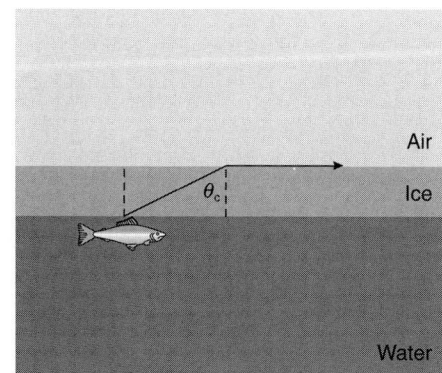

(c) 0.77 m   (d) 1.02 m

61. A proof

65. (a) 1×10^2 cm to the left of the diverging lens
(b) $m_{total} = 2$ (c) Virtual, upright

69. $f_{eff} = \dfrac{f_{lens} f_{mirror}}{f_{lens} - f_{mirror}}$

73. (a) $m = -44.1$ (b) Inverted. Invert the slide.   (c) 10.2 cm
(d) 450 cm

77.

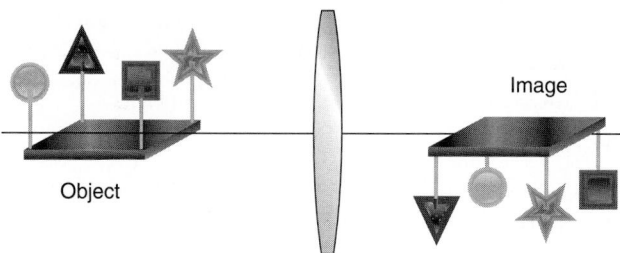

81. The image is 60 cm to the right of the lens system.

85. -6.7 dp

89. 10 mm

93. (a) ≈ 0.1 rad $\approx 6°$ (b) 9.06×10^{-3} rad $= 0.519°$
(c) -12.00 m   (d) 9.06×10^{-3} rad $= 0.519°$ (e) Real
(f) 0.109 m   (g) Circular

97. (a) $m = -0.10$ (b) Place the object near the secondary focal
point of the objective lens; the system then acts as a micro-
scope, with $|m| \approx 14$.

CHAPTER 24

1. 3.5×10^{-3} rad, $0.20°$

5. (a) 3.2 cm   (b) The separation of the fringes is halved to 1.6 cm.

9. (a) 1.31 m   (b) $41°$

13. (a) 0.73 mm   (b) Increase the diameter.

17. A proof

21. (a) The telescope because of its larger aperture   (b) 1.0 m
(c) Magnification is doubled; resolution is unaffected.

25. $14°$ for the first order, $29°$ for the second order

29. $20.8°$

33. (a) no   (b) 437 nm

37. (a) 422 nm   (b) 2.00×10^8 m/s

41. (a) and (c)

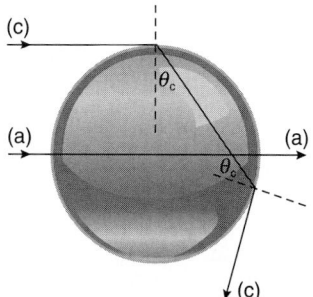

(b) 1.00×10^{-10} s   (d) 7.45×10^{-11} s

45. (a) Flat #1   (b) Slightly spherical

49. Answer is given in the statement of the problem.

53.

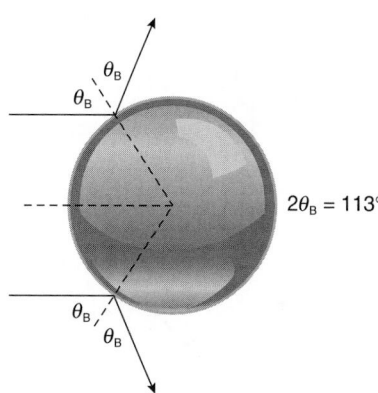

57. (a) and (b) Answers are given in the statement of the problem.
(c) $12.5°$; This is less than the critical angle of $12.8°$.

61. $36.9°$

CHAPTER 25

1. Account for the light travel time between a central clock and
the others.

5. $0.943c = 2.83 \times 10^8$ m/s

9. 7×10^2

13. (a) 2.00×10^8 m/s   (b) 74.6 m   (c) 3.73×10^{-7} s

17. (a) $v = 0.82c = 2.5 \times 10^8$ m/s
(b) 5.1 y

21. (a) 1.28 s   (b) 0.18 s: The Moon is hit first.   (c) No, a single
saucer it would have to travel with $v > c$ to cause them both.

25. (a) A proof   (b) 0.064 m; No, only the diameter measured in
the direction of the motion or the projection of a diameter
along the direction of the orbital motion is length contracted.

29. $u_x = 0.990c$, $u_y = 0.14c$ (b) $8.0°$

33. Answers are given in the statement of the problem.

37. A proof

41. $0.0175c \approx 5.25 \times 10^3$ km/s

45. Assume a mass of 60 kg: $\$1.5 \times 10^{11}$

49. KE $= 2.05 \times 10^{-14}$ J, $E = 1.02 \times 10^{-13}$ J,
$p = 2.05 \times 10^{-22}$ kg·m/s

53. (a) $0.417c$ (b) 1.5×10^{-11} J

57. (a) $0.194\,70c$ (b) 9.99 keV (c) 5.42×10^{-23} kg·m/s

61. (a) 2.5×10^{19} J (b) 0 J (c) 20.4 y (d) 4.0 LY (e) 4.1 y

65. Answers are given in the statement of the problem.

69.

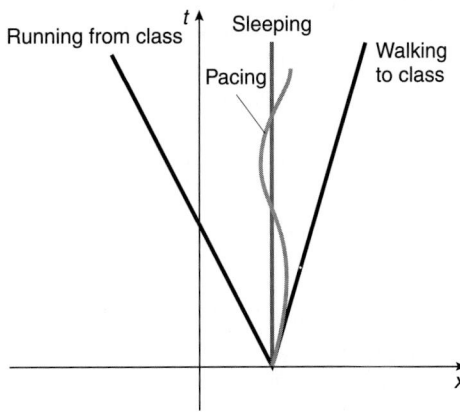

73.

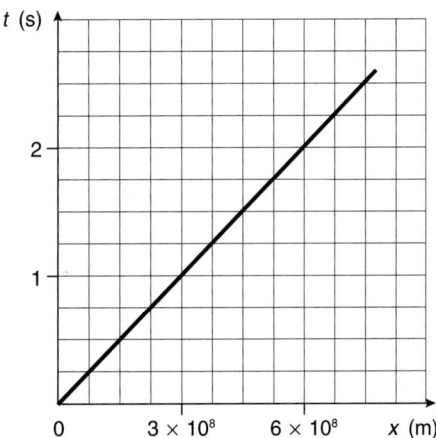

CHAPTER 26

1. λ_{max} is halved.

5. About 4.5×10^3 K

9. For 400 nm: 3.10 eV; For 700 nm: 1.77 eV

13. (a) 2.1 eV (b) 2.1 V (c) 3.0×10^2 nm

17. (a) Yes, since $v > v_c = 4.8 \times 10^{14}$ Hz.

(b)

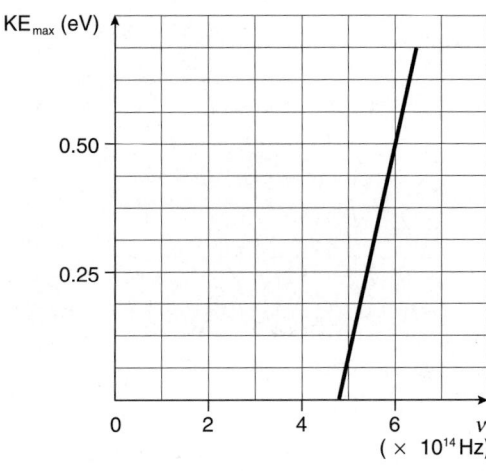

(c) The graph is unaffected.

21. (a) 2.12×10^{-10} m (b) 1.09×10^6 m/s
(c) 2.110×10^{-34} kg·m²/s (d) 5.60×10^{21} m/s²
(e) 5.71×10^{20} (f) 3.38 eV (g) −6.80 eV

25. (a) 365 nm (b) No

29. The shortest wavelength is the series limit of the Lyman series: 91.2 nm. This is too long to be classified as an x-ray wavelength.

33. (a) $0.181\,d^{-1}$ (b) 4.5×10^{-3}

37. Answer is given in the statement of the problem.

41. 0.70

45. 26 d

49. (a) 1.21×10^{-4} y^{-1} (b) 2.04×10^3 y

53. One cannot say with certainty, since the process is statistical. It is possible that all five may have decayed, or none. The most likely number remaining is two or three.

57. 0.1024 nm

61. Answer is given in the statement of the problem.

65. $n = 2\,\pi r/\lambda$

69. (a) 0.37 nm (b) 0.37 nm

CHAPTER 27

1. $h = 4.136 \times 10^{-15}$ eV·s, $\hbar = 6.583 \times 10^{-16}$ eV·s

5. 3.45×10^{-23} s

9. (a) 1.50×10^{15} Hz (b) $\Delta v \geq 1.0 \times 10^9$ Hz (c) 0.30 m

13. (a) 7×10^{-20} kg·m/s (b) $\lesssim c$ (c) 1×10^2 MeV; This energy is approximately 2×10^2 times the rest energy.

QUOTATIONS INDEX

REFERENCE INDEX

GENERAL INDEX

Note: f's, t's, and n's following page numbers refer to figures, tables, and footnotes, respectively.

I.7

implications of, 1186–1187
ordered *vs.* disordered, 616
potential, 334–335
quantization of, and specific heat,
653–654, 654f
as special term in physics, 320
storage of, 806
total mechanical energy, 342
total, of system, 615–616
total relativistic, 1185
Energy conservation, and series LR
circuits, 977, 978
Energy conservation law, fundamental,
617–618
and CWE theorem, 618–620
and first law of thermodynamics, 618
Energy diagrams
of central forces, 361–362
of conservative forces, 359–363f, 360f,
361f
of gravitational force, 360–361, 361f
of simple harmonic oscillation, 359–360,
360f
Energy gaps, 653
Energy level diagram, in Bohr model of
atom, 1220, 1221f
Energy operator, 1262–1263
Energy states, 653
Energy-time uncertainty principle,
1251–1252
implications of, 1254–1257
Energy transport, via mechanical waves,
547–548
Engine. *See also* Heat engine(s);
Refrigerator engine(s)
definition of, 669
function of, 622–623, 622f
internal combustion, 620, 663
Entropy, 679–685
change in, 681
in various thermodynamic processes,
681t
maximization of, in time, 694–695, 694f
and second law of thermodynamics,
685–688
statistical interpretation of, 689–693,
690f–693f
vs. classical definitions of, 692
Envelope
geometrical, definition of, 1108n
of single slit diffraction pattern, 1114,
1115f
Epicenter, of earthquake, definition of,
534
Equality, approximate, symbols for, 16
Equal sign, meanings of, 15–16
Equations, proper use of, 2–3
Equilibrium
mechanical, definition of, 173
static
conditions for, 464–465
definition of, 464
thermodynamic (thermal), 590

Equilibrium position, of spring, 282
Equilibrium thermodynamics, definition
of, 615
Equipartition of energy theorem, 649–650
and solids, 650
Equipotential regions, 779–781, 780f–781f
Equipotentials, 779
Equipotential surfaces, 781, 781f
Equivalence, principle of, 235, 1190–1191
Escape speed, 347–349
Estimation, value of, 17–18
Euler, Leonhard, 1008n
Euler identity, 1014
Event, definition of, 75, 1150
Everything, Theory of, 172, 172f
Exiguus, Dionysius, 30
Expansion, adiabatic, of Carnot heat
engine, 672–673, 673f
Expansion of universe, 155, 1202
Explosions, physics of, 396–397, 396f
Exponential form of complex numbers,
1008–1009
Extended object(s), in optics, definition
of, 1056
Extensive state variables, 695, 695t
External forces, vector sum of, 404–405,
405f
Extraordinary (E) ray, 1132
Eyepiece lens, of microscope, 1079, 1079f
of telescope, 1080, 1080f
Eyes
focussing of, 1077
of vertebrates, functioning of,
1077–1079, 1077f, 1078f
Eyesight, 1077–1079, 1077f, 1078f
defects in, 1078–1079

f number, of camera, 1077
Fabry-Perot interferometer, 628, 1146
Fahrenheit, Daniel Gabriel, 594
Fahrenheit scale, 594
conversion to celsius scale, 594–595,
594f
Farad (unit), definition of, 814
Faraday, Michael, 814, 954f
career of, 4–5, 171, 954, 1003
reference materials on, 1002
Faraday's law of electromagnetic
induction, 954–960, 955f, 956f
AC generators and, 963–965, 964f
definition of, 957–958
and mutual inductance, 983–984
summary of, 966
Far point, in optics, definition of, 1078
Farsightedness, 1078–1079, 1078f
Fermat's principle, 1090
Fermi, Enrico, 6, 17, 32, 1253n
Fermions, 467
Fermi pressure, 1253n
Ferroelectric material, 765, 804
Ferromagnetic materials, 931, 931t
Feynman, Richard, 34, 1250
Fiber optic cables, 1052

Field lens, 1081
Fields. *See* Electric field(s); Gravitational
force; Magnetic field(s)
Fifth fundamental force, 279
Filter circuits, 1022–1026, 1022f
high pass, 1026, 1026f
low pass, 1025, 1025f
First Brillouin zone, 70
First harmonic, definition of, 560–561
First law of motion (Newton), 173–174
First law of planetary motion (Kepler),
244–247
First law of thermodynamics, 618
and changes in state, 623–624
inviolability of, 668–669
First order phase transitions, definition of,
604
Fission, nuclear, 1187
Fixed points (of temperature), definition
of, 591
Flow
incompressible, definition of, 510
irrotational, definition of, 510
nonturbulent, definition of, 510
Flow continuity, equation of, 511
Flow rate, 511
Flow tube, definition of, 511
Fluid dynamics
of ideal fluids, 510–514
of nonideal fluids, 514–516
Fluids
buoyancy in, 501–505, 502f, 503f
definition of, 491
ideal
characteristics of, 510
fluid dynamics of, 510–514
incompressible, Bernoulli's principle
for, 511–514
piped, drop in pressure with distance,
515–516, 516f
pressure of, 494–504, 495f
static, and pressure, 496–499
Flux. *See also* Gauss's law
differential, of vector, definition of, 262
of electric field, 741–744
of gravitational field, meaning of, 266
of magnetic field, 918
of vector, 261–264, 261f–263f
Flywheels, 443–444
Focal length
of spherical mirror, 1059
of thin lens, 1070–1071, 1071f
Focal plane, definition of, 1080
Focal point(s)
of mirror, definition of, 1059
primary and secondary, 1072–1073,
1073f
of thin lens, 1070–1071, 1071f
Foci of ellipse, 245
Focus of earthquake, definition of, 534
Food Calorie (unit), 590
Force(s)
central, potential energy function of, 338

CREDITS

COMMON UNIT ABBREVIATIONS

Unit	Abbreviation	Unit	Abbreviation	Unit	Abbreviation
ampere	A	gram	g	ohm	Ω
atomic mass unit	u	henry	H	pascal	Pa
atmosphere	atm	hertz	Hz	radian	rad
coulomb	C	hour	h	revolution	rev
day	d	joule	J	second	s
degree (angle)	°	kelvin	K	tesla	T
degree celsius	°C	kilogram	kg	volt	V
degree fahrenheit	°F	meter	m	watt	W
electron-volt	eV	minute	min	year	y
farad	F	newton	N		

GREEK ALPHABET

Letter	Lower case	Upper case	Letter	Lower case	Upper case	Letter	Lower case	Upper case
alpha	α	A	iota	ι	I	rho	ρ	P
beta	β	B	kappa	κ	K	sigma	σ	Σ
gamma	γ	Γ	lambda	λ	Λ	tau	τ	T
delta	δ	Δ	mu	μ	M	upsilon	υ	Y
epsilon	ϵ	E	nu	ν	N	phi	ϕ	Φ
zeta	ζ	Z	xi	ξ	Ξ	chi	χ	X
eta	η	H	omicron	o	O	psi	ψ	Ψ
theta	θ	Θ	pi	π	Π	omega	ω	Ω

PLANETARY, LUNAR, AND SOLAR DATA

	Mass	Average radius	Spin period	Semimajor axis of orbit		Eccentricity of orbit	Orbital period
Sun	1.99×10^{30} kg	6.96×10^{8} m	≈ 25 d	—	—	—	—
Mercury	3.34×10^{23} kg	2.45×10^{6} m	58.6 d	5.79×10^{7} km	0.387 AU	0.2056	0.241 y
Venus	4.87×10^{24} kg	6.05×10^{6} m	243 d	1.08×10^{8} km	0.723 AU	0.0068	0.615 y
Earth	5.98×10^{24} kg	6.37×10^{6} m	23.9 h	1.496×10^{8} km	1 AU	0.0167	1.000 y
Moon	7.36×10^{22} kg	1.74×10^{6} m	27.3 d	3.84×10^{5} km	—	0.055	27.3 d
Mars	6.43×10^{23} kg	3.37×10^{6} m	24.6 h	2.28×10^{8} km	1.52 AU	0.0934	1.88 y
Jupiter	1.90×10^{27} kg	6.97×10^{7} m	9.84 h	7.78×10^{8} km	5.20 AU	0.0484	11.9 y
Saturn	5.69×10^{26} kg	5.82×10^{7} m	10.2 h	1.43×10^{9} km	9.54 AU	0.0560	29.5 y
Uranus	8.69×10^{25} kg	2.59×10^{7} m	17.2 h	2.87×10^{9} km	19.2 AU	0.0461	84.0 y
Neptune	1.03×10^{26} kg	2.45×10^{7} m	16.1 h	4.50×10^{9} km	30.1 AU	0.0100	165.8 y
Pluto	1.32×10^{22} kg	1.16×10^{6} m	6.39 d	5.90×10^{9} km	39.4 AU	0.2484	247.7 y